ADVANCED ENGINEERING MATHEMATICS

ADVANCED ENGINEERING MATHEMATICS

SIXTH EDITION

C. Ray Wylie
Professor Emeritus, Furman University

Louis C. Barrett
Professor Emeritus, Montana State University

McGraw-Hill, Inc.

New York St. Louis San Francisco Auckland Bogotá Caracas
Lisbon London Madrid Mexico City Milan Montreal
New Delhi San Juan Singapore Sydney
Tokyo Toronto

ADVANCED ENGINEERING MATHEMATICS

This book is printed on acid-free paper.

3 4 5 6 7 8 9 0 VNH VNH 9 0 9 8 7 6 5

ISBN 0-07-072206-4

This book was set in Times Roman by York Graphic Services, Inc.
The editors were Maggie Lanzillo and Scott Amerman;
the designer was Jo Jones;
the production supervisor was Denise L. Puryear.
New drawings were done by Fine Line Illustrations, Inc.
Von Hoffmann Press, Inc., was printer and binder.

Library of Congress Cataloging-in-Publication Data

Wylie, Clarence Raymond, (date).
Advanced engineering mathematics / C. Ray Wylie, Louis C. Barrett. — 6th ed.
p. cm.
Includes index.
ISBN 0-07-072206-4
1. Mathematics. I. Barrett, Louis C. II. Title.
QA401.W9 1995
515—dc20 94-47667

INTERNATIONAL EDITION

When ordering this title, use ISBN 0-07-113543-X.

ABOUT THE AUTHORS

C. Ray Wylie, professor emeritus of Furman University in Greenville, South Carolina, received a B.A. degree (in mathematics) and a B.S. degree (in chemical engineering) from the College of the City of Detroit, now Wayne State University. He earned an M.S. degree in 1932 and a Ph.D. degree in 1934, both in mathematics, from Cornell University. In 1994 Furman University conferred an honorary D.Sc. degree on Professor Wylie.

From 1934 to 1946 Professor Wylie was a member of the Department of Mathematics of Ohio State University. During World War II, while on leave from Ohio State, he served as a civilian engineer in the Propeller Laboratory at Wright-Patterson Air Force Base. After the war, he served as chairman of the Department of Mathematics and acting dean of the College of Engineering at the base's Air Institute of Technology. He left the Air Institute in 1948 to become chairman of the Department of Mathematics at the University of Utah, in which capacity he served until 1967. Upon his retirement from the University of Utah, he became chairman of the Department of Mathematics at Furman University. From 1971 until his retirement from Furman he was William R. Kenan, Jr., Distinguished Professor of Mathematics.

Ray Wylie has published technical papers in applied mathematics and geometry. In addition to *Advanced Engineering Mathematics,* he is the author or co-author of twelve other books, including the McGraw-Hill texts *Differential Equations, Foundations of Geometry,* and *Introduction to Projective Geometry*. He is listed in *American Men and Women of Science* and *Who's Who in America,* and is a member of Phi Beta Kappa and Sigma Xi.

Louis C. Barrett received a B.A. in mathematics, an M.S. in mathematics and physics in 1951, and a Ph.D. in mathematics and physics from the University of Utah in 1956. He is currently professor emeritus of the Department of Mathematics at Montana State University, where he was professor and department head from 1967 until 1972. Prior to his service at Montana State, Professor Barrett was professor and chairman of the Department of Mathematics at Clarkson College of Technology (1965–1966); associate professor, then professor and department head of the Mathematics Department at South Dakota School of Mines and Technology (1957–1965); associate professor in the mathematics department of Arizona State College (1956–1957); and instructor in the mathematics department of the University of Utah (1953–1956).

In addition to his professorships, Louis Barrett has served as consultant to the Holloman Air Development Center in New Mexico and the Naval Weapons Center in China Lake, California. He has also been a lecturer in eight National Science Foundation Institutes and for the Mathematical Association of America.

Professor Barrett has written numerous technical reports in army and navy contract research. With C. Ray Wylie, he has co-authored the fifth edition of *Advanced Engineering Mathematics.*

CONTENTS

PREFACE

The first edition of this book was written to provide an introduction to those branches of postcalculus mathematics with which average analytical engineers or physicists need to be familiar in order to carry on their own work effectively and keep abreast of current developments in their fields. In the present edition, as in each of the preceding editions, much of the material has been rewritten, but the various additions, deletions, and refinements have been made only because they seem to contribute to the achievement of this goal.

CONTENT OF THE BOOK

Because ordinary differential equations are probably the most immediately useful part of postcalculus mathematics for the student of applied science, and because the techniques for solving simple ordinary differential equations stem naturally from the techniques of calculus, this book begins with a chapter on ordinary differential equations of the first order and their applications. This chapter is followed by one which develops the theory and applications of linear differential equations, especially those with constant coefficients. In the present edition, this chapter has been augmented by two major sections, one dealing with the series solution of linear differential equations, the other covering the method of Frobenius. In earlier editions, this material was postponed until the chapter on Bessel functions. It is now included at this early stage so that the first four chapters can be used as a basis for a self-contained course in ordinary differential equations.

Next, in Chap. 3, to prepare for a discussion of linear differential systems with constant coefficients, there is an introduction to linear algebra. Although this material will be used extensively in Chap. 4, it can be omitted by students who are already familiar with matrices, determinants, and the solution of simultaneous linear algebraic equations. To this chapter there has now been added an introductory section on complex variables. This material will be used in many places throughout the rest of the book, but it can be omitted by students familiar with the properties of complex variables through De Moivre's theorem and the Euler formulas.

Chapter 5 is devoted to numerical methods, and it covers such topics as finite differences, interpolation formulas, numerical differentiation and integration, the numerical solution of ordinary differential equations, featuring the various Runge-Kutta methods and Milne's method, and difference equations. A section on least-squares has been restored in this edition and a new section on the G and Z transformations has been added. It is hoped that the material in this chapter will provide a useful background in classical finite differences on which a more extensive course in computer-oriented numerical analysis can be based. Chapter 6 is a new chapter that has not been included in any of the earlier editions of this text. It deals with the problem of determining such properties of the solutions of a differential equation as periodicity and stability without finding the solutions themselves.

Chapter 7 is devoted to the application to mechanical systems and electric circuits of the ideas developed in the first five chapters. As in the earlier editions, the mathematical identity of these fields is emphasized. The section on systems with more than one degree of freedom has been divided into a section on systems with several degrees of freedom and a section on systems with many degrees of freedom. This new section features the interplay between differential equations and difference equations, with emphasis on wave filters and wave traps. A final section on electromechanical analogies has been restored from the second edition.

Motivated by the work on periodic phenomena in Chap. 7, Fourier series and their applications are discussed in Chap. 8. In particular, in this edition more emphasis has been placed on the use of the jumps of a function and its derivatives to eliminate the need to integrate to determine the Fourier coefficients of a function. Chapter 9 is a new chapter containing in expanded form the material on Fourier integrals that was grouped with Fourier series in earlier editions. The Fourier integral is introduced as the limit of a Fourier series, and then a variety of Fourier transforms, with their basic properties obtained from it. This chapter contains a new section on the Gibbs phenomenon and the convergence of Fourier series and Fourier integrals at the jumps of a function. There is also a new section on singularity functions and their fundamental properties.

In Chapter 10, the Laplace transformation is introduced as a natural outgrowth of the Fourier integral and Fourier transforms. In this edition, the presentation of the requisite theory is a little less abrupt than it was in earlier editions, and examples of particular transforms are given very early. The chapter concludes with a new section in which the nature and properties of Laplace transforms, Fourier transforms, and Z- transforms are compared and contrasted.

Chapters 11 and 12 deal, respectively, with differential equations and boundary-value problems, and Bessel functions and Legendre polynomials. Here, Fourier series play a prominent role in satisfying initial and boundary conditions and provide motivation for the discussion and use of expansions in terms of more general systems of orthogonal functions. In this edition, a new section on the generating functions of J_n and I_n illustrate their use in obtaining many of the identities of these functions. New examples in these chapters include incomplete systems of orthogonal functions, interface Sturm-Liouville systems, and the use of Legendre polynomials in potential problems.

In Chaps. 13 and 14 we return to the subject of linear algebra and discuss vector spaces, linear transformations, the existence of Green's functions for systems of differential equations, and further properties of matrices and their eigenvalues and eigenvectors. An important addition to Chap. 14 is a section on the discrete and fast Fourier transforms, an important topic in the field of signal processing. This work is followed by a chapter on vector analysis developed in the traditional geometric way, much as it was in the fifth edition. New material here includes some interesting topics in differential geometry. Chapter 16 deals with the calculus of variations and its applications to dynamics. New material here includes a section on Hamilton's equations.

The last four chapters provide an introduction to the theory of functions of a complex variable, with applications to fluid mechanics and two-dimensional potential theory, the evaluation of real definite integrals, the complex inversion integral of Laplace transformation theory, stability criteria, conformal mapping, and the Schwarz-Christoffel transformation. The only significant difference between these chapters and the corresponding chapters in the last edition is that the introductory material through De Moivre's equation and the formal content of Euler's formulas has been moved ahead and now appears at the beginning of Chap. 3.

This book falls naturally into three main subdivisions. The first twelve chapters constitute a reasonably self-contained treatment of ordinary and partial differential equations and their applications. The next four chapters, 13 through 16, cover the related areas of linear algebra, vector analysis, and the calculus of variations. The last four chapters, 17 through 20, cover the elementary theory and applications of functions of a complex variable.

FEATURES

This book contains enough material for a two-year course in applied mathematics. However, since we have tried to keep important subjects concentrated in specific chapters rather than diffused throughout the book, selected chapters are well-adapted for use as a text for any of several shorter courses. Following this preface, in the section headed "To the Instructor," there is a detailed Planning Guide showing how this text can be used for a number of courses.

In this edition, as in each of the others, every effort has been made to keep the presentation detailed and clear while at the same time maintaining acceptable standards of precision and accuracy. To achieve this goal, more than the usual number of worked examples and carefully drawn figures have been included, and in every development there has been a conscious attempt to make the transition from step to step so clear that a serious student, working with paper and pencil, should seldom be held up very long. Many new exercises have been added in this edition, and there are now more than 5000. Hints are given in many of the exercises, and answers to the odd-numbered ones are given at the end of the book. A manual containing the answers to the even-numbered ones is available for instructors using this text. More detailed solutions to the odd-numbered exercises are provided in a Student Solutions Manual. As in earlier editions, words and phrases defined informally in the body of the text are set in bold-faced type, and italics are used frequently for emphasis. Illustrative examples are consistently set in type of a different size than that used for the main body of the text.

ACKNOWLEDGMENTS

The indebtedness of the authors to their colleagues, students, and former teachers is too great to catalog, and to all who have given help and encouragement through the years we can offer only a most inadequate acknowledgment of our appreciation. In particular, we are deeply grateful to those users of this book who have been kind enough to write us of their impressions and criticisms of the first five editions and their suggestions for an improved sixth edition and to the following McGraw-Hill reviewers: Barbara Bohannon, Hofstra University; Michael Bryant, University of Texas at Austin; Chung-wu Ho, Southern Illinois University; J. Lubliner, University of California at Berkeley; Gordon Melrose, Old Dominion University; Keith B. Olson, Montana College of Mineral Science and Technology; Mauro Pierucci, San Diego State University; Michael E. Ryan, State University of New York at Buffalo; Duane W. Storti, University of Washington; Mo Tavakoli, Chaffey College; and Arnold Villone, San Diego State University. We also appreciate greatly the invaluable advice and assistance that our editor, Maggie Lanzillo, gave so generously during both the preparation of the manuscript and the subsequent editorial process. Thanks, too, to the production team at McGraw-Hill who did such an admirable job of guiding this project from manuscript to bound book: Scott Amerman, editing supervisor; Denise Puryear, production supervisor; and Jo Jones, designer. Finally, we must express our gratitude to our wives, Ellen and Betty, not only for their assistance and encouragement in this project but for their patience and understanding during our long preoccupation with the manuscript. ◗

C. Ray Wylie

Louis C. Barrett

TO THE INSTRUCTOR

This book contains ample material for a two-year sequence in applied mathematics. It has been written so that important subjects are concentrated in specific chapters and are not covered partially in several different places. By the judicious selection of particular chapters, it is thus readily adaptable as a text for a number of short courses. To assist you in making maximum use of the book, we have prepared the accompanying Course Modules and Planning Guide. It identifies modules suitable for a variety of one-term courses, as well as combinations of modules on which several one-year sequences can be based. It also indicates prerequisite relations for instructors planning their own sequences.

One new feature of this edition is the inclusion in each chapter of an introductory, overview section alerting the student to the material to be covered in the chapter, pointing out portions that may be extensions of topics discussed earlier, and indicating where the new material will be used later in the book. We hope that you will encourage your students to orient themselves to the work in each chapter by reading these introductory statements carefully.

This book contains over 5000 exercises, many of them new. As in each of the other editions, many of these contain extensions of topics in the text or interesting new results that could not be included within the chapter because of space limitations. Since the difficulty of an exercise is often a subjective judgment, we have made no attempt to distinguish ''hard'' problems from ''easy'' ones, nor to arrange the exercises in an assumed order of increasing difficulty. Nonetheless, nearly every set begins with a few routine, practice problems. Answers to the odd-numbered exercises are listed in the back of the book. A manual containing the answers to the even-numbered exercises is available for the instructor, and a manual containing solutions to the odd-numbered exercises is available for the student.

C. Ray Wylie

Louis C. Barrett

PREREQUISITES

COURSE MODULES AND PLANNING GUIDE

Calculus	Ordinary Differential Equations	Complex Numbers	Fourier Analysis	Calculus-Based Physics	COURSE	1. Ordinary Differential Equations of the First Order	2. Linear Differential Equations	3. Complex Numbers and Linear Algebra	4. Simultaneous Linear Differential Equations	5. Numerical Methods	6. Descriptive Theory of Ordinary Differential Equations	7. Mechanical Systems and Electrical Circuits	8. Fourier Series
					1 - Ordinary Differential Equations					Section 5.6			
					2 - Linear Algebra								
					3 - Mathematical Applications and Fourier Analysis	Sections 1.14 & 1.15	Sections 2.11 & 2.12						
					4 - Partial Differential Equations and Boundary Value Problems								
					5 - Numerical Methods								
					6 - Descriptive Theory of Ordinary Differential Equations								
					7 - Vector Analysis			Section 3.2					
					8 - Calculus of Variations								
					9 - Complex Variables			Section 3.1					

ACADEMIC YEAR SEQUENCES

Ordinary Differential Equations: Course 1
Applied Mathematics: Courses 3 & 4
Applied Analysis: Courses 5, 6, 7, & 8
Complex Variables: Course 9

CHAPTER												
9. Fourier Integrals and Fourier Transforms	10. The Laplace Transformation	11. Partial Differential Equations	12. Bessel Functions and Legendre Polynomials	13. Vector Spaces and Linear Transformations	14. Applications and Further Properties of Matrices	15. Vector Analysis	16. Calculus of Variations	17. Analytic Functions of a Complex Variable	18. Infinite Series in the Complex Plane	19. Theory of Residues	20. Conformal Mapping	OPTIONAL TOPICS
												Sections: 1.14, 1.15, 2.11, 2.12, & 5.6
												Sections: 14.2 & 14.4
					Section 14.4							Sections: 11.4 & 11.9
		Sections 11.4 & 11.9			Section 14.2							

TO THE STUDENT

This book has been written to help you in your development as an applied scientist, whether an engineer, physicist, chemist, or mathematician. It contains material that will be of great use to you, not only in the technical courses you have yet to take, but also in your profession after graduation, as long as you deal with the analytical aspects of your field.

We have tried to write a book which you will find not only useful but also relatively easy, at least as easy as a book about advanced mathematics can be. There is a good deal of theory in it, for it is the theoretical portion of a subject which is the basis for the nonroutine applications of tomorrow. But nowhere will you find theory for its own sake, interesting and legitimate as this may be to a pure mathematician. Our theoretical discussions are designed to illuminate principles, to indicate generalizations, to establish limits within which a given technique may or may not be safely used, or to point out pitfalls into which one might otherwise stumble. On the other hand, there are many applications, illustrating with the material at hand the usual steps in the solution of a physical problem: formulation, manipulation, and interpretation. These examples are, without exception, carefully set up and completely worked, with all but the simplest steps included. Study them carefully, with paper and pencil at hand, for they are an integral part of the text. If you do this, you should find the exercises, though challenging, still within your ability to work.

A new feature in this edition is the inclusion in each chapter of an introductory section, giving an overview of the material to be covered, pointing out where we may have encountered some of it before, and indicating where and how it will be used later in the book. Be sure to orient yourself to the work and purpose of each new chapter by reading carefully these introductory sections. Another new feature is the inclusion of subtitles for many of the important examples. These will alert you to the main point of each example and, perhaps more importantly, help you to identify examples to which you may later wish to refer. You will find them listed inside the covers of this book.

Terms defined informally in the body of the text are always indicated by the use of **bold-faced type.** *Italic type* is used for emphasis, much as verbal stress is used when speaking. We suggest that you read each section through for the main ideas before you concentrate on filling in any of the details. You will probably be surprised at how many times a point which seems to hold you up in one paragraph will be explained in the next as the discussion unfolds.

Because this book is long and contains material suitable for various courses, your instructor may begin with any of a number of chapters. However, the overall structure of the book is this: The first twelve chapters are devoted to the general theme of ordinary and partial differential equations and related topics. Here you will find the basic analytical techniques for solving the equations in which physical problems must be formulated when continuously changing quantities are involved. Chapters 13 through 16 deal with the somewhat related topics of linear algebra and matrix theory, vector analysis, and the calculus of variations. Finally, Chaps. 17 through 20 provide an introduction to the theory and applications of functions of a complex variable.

It has been gratifying to receive letters from students who have used this book giving us their reactions to it, pointing out errors and misprints in it, and offering suggestions for its improvement. Should you be inclined to do so, we would be glad to hear from you also.

Finally, we hope that you will find this book in some sense a friendly book. It was written with you in mind, as someone with whom we would like to share not only our knowledge but our enthusiasm. We have written almost entirely in the first person plural. Never are you referred to obliquely and impersonally as ''the student.'' Our use of the word ''we'' indicates that we feel we are exploring something interesting with you. And now good luck and every success.

C. Ray Wylie

Louis C. Barrett

ADVANCED ENGINEERING MATHEMATICS

CHAPTER 1

ORDINARY DIFFERENTIAL EQUATIONS OF THE FIRST ORDER

Differential equations, that is, equations involving rates of change, provide an indispensable tool for anyone studying continuously varying phenomena, such as velocities and accelerations or electric currents. In this chapter, after a review of the concepts of *variable* and *function* (Sec. 1.1), we identify the major types of differential equations (Sec. 1.2). Then, as a necessary foundation, we discuss the general notions of *solution* and *family of solutions* (Secs. 1.3–1.5) and the *existence* and *uniqueness* of solutions (Sec. 1.6).

With these ideas in mind, we begin the study of methods of solving differential equations by learning how to solve all the major forms of ordinary first-order equations. These include *exact equations* (Sec. 1.7), *equations solvable by integrating factors* (Sec. 1.8), *separable equations* (Sec. 1.9), *homogeneous equations* (Sec. 1.10), *linear equations* (Sec. 1.11), and several more special types of equations (Secs. 1.12 and 1.13).

Every section contains a number of examples, but significant practical applications we leave to Secs. 1.14 and 1.15. This organization makes it possible to use the bulk of this chapter for a portion of a course in the theory of differential equations. At the same time, the work on *orthogonal trajectories* in Sec. 1.14 and 8 examples and 120 exercises in Sec. 1.15 provide convincing evidence of the utility of differential equations and ample material for practice in using them to formulate and solve physical problems ranging from heat conduction and fluid flow to orbital motion.

Prerequisite for this chapter: single-variable calculus.

Prerequisite for Sec. 1.15: a calculus-based physics course and general chemistry.

1.1 VARIABLES AND FUNCTIONS

The variety and complexity of the problems which confront today's engineers and scientists have increased remarkably in recent years and, if anything, the increase seems to be accelerating. As a consequence, not only is there a continuing demand for more and more effective computers and better and better experimental facilities, but so too is there a growing need for more, and more thoroughly understood, mathematics to support the whole scientific enterprise. Mathematics demands clarity of thought and clarity of exposition, and so as we begin our study it seems proper that we review briefly the raw material of all our work, *variables* and *functions.*

A **variable** is a symbol identifying elements of a given set. A *function* can be thought of as a rule relating the elements of one set to the elements of a second set, possibly the same as the first. The rule defining the functional values is often a formula of some kind, although other modes of definition are possible. Variables that designate values for which a function is defined are called **independent variables,** and, collectively, these values form the **domain** of the function. Variables which identify values of a function are called **dependent variables** and, collectively, these values form the **range** of the function.

Functions are usually denoted by single letters. For each x in the domain of a function f, the *value* of f at x is denoted by $f(x)$. As is customary, we shall often use the notation $f(x)$ not only to denote a value of f but also to name the function itself, although this is notationally inaccurate. Depending on the domain of a function, which is never empty, the variable x may stand for a number or any other object for which the function is defined.

EXAMPLE 1

Since for all real values of x, $2 + \cos \pi x \geq 1$, the expression $f(x) = \ln (2 + \cos \pi x)$ defines a function f of the variable x having the set of all real numbers as its domain. The value of f at 1 is given by $f(1) = \ln (2 + \cos \pi) = \ln 1 = 0$, and $f(2) = \ln 3$ is the value of f at 2. In this example, the independent variable x may be replaced by any real number. The range of f is made up of all values of the dependent variable $y = f(x)$, namely, all numbers y such that $0 \leq y \leq \ln 3$.

EXAMPLE 2

Let the domain of a function g defined by $g(x) = \int_0^1 x(t)\,dt$ be the set of all functions continuous on the closed interval $[0, 1]$. If $x_1(t) = t$ on $[0, 1]$, the value of g at x_1 is given by $g(x_1) = \int_0^1 t\,dt = \frac{1}{2}$; while if $x_2(t) = \cos \pi t$ on $[0, 1]$, $g(x_2) = \int_0^1 \cos \pi t\,dt = 0$. Here x can be replaced by any function that is continuous on $[0, 1]$. Once x is chosen, the corresponding value of the dependent variable $y = g(x)$ is given by the integral of x from 0 to 1.

Note that in Example 1 the independent variable x stands for a real number, but that in Example 2 it stands for a function. A function whose domain is a set of functions is called a **functional.**

Frequently, values of a given function may be found by making the appropriate substitutions for variables in some analytic expression. Such an expression is commonly called a **representation** of the function on that part of the functional domain over which the representation yields the corresponding functional values. Often, as in Examples 1 and 2, all values of the function are determined by a single representation, but this need not be the case. In a variety of physical problems the very nature of the function requires different representations on different subsets of the domain.

EXAMPLE 3

At time $t = 0$, a unit voltage is suddenly introduced into an electric circuit, as suggested in Fig. 1.1*a*. This voltage is represented for $t < 0$ by the expression $E(t) = 0$ but represented for $0 \le t$ by the expression $E(t) = 1$. Thus on the domain $-\infty < t < \infty$ the voltage E is given by the two representations

$$E(t) = \begin{cases} 0 & t < 0 \\ 1 & 0 \le t \end{cases}$$

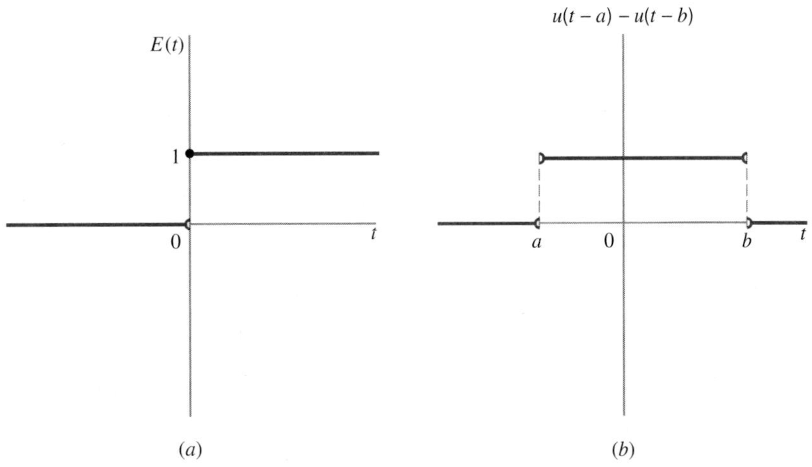

Figure 1.1
Graphs of (*a*) a unit step function voltage; (*b*) a unit pulse of duration $b - a$.

Example 3 illustrates the important fact that whereas the functions we dealt with in calculus were usually either continuous on their domains or had only removable discontinuities, in applied mathematics we must often consider functions with one or more nonremovable discontinuities. Typically, these discontinuities will be what are called *jumps*.

DEFINITION 1 A function f has a **jump** J at $x = s$ if and only if the respective right- and left-hand limits $f(s^+)$ and $f(s^-)$ exist and

$$J = f(s^+) - f(s^-)$$

According as J is positive or negative, the jump is said to be **upward** or **downward.** At a point where f is continuous, the jump J is of course zero. The function E of Example 3 has an upward jump $J = 1$ at $t = 0$ and for all practical purposes is represented by the simple but very important **unit step function** $u(t)$, defined by

$$u(t) = \begin{cases} 0 & t < 0 \\ 1 & 0 < t \end{cases}$$

The unit step function will appear repeatedly in the work ahead of us.

A difference of two unit step functions

$$u(t-a) - u(t-b) = \begin{cases} 0 & t < a \\ 1 & a < t < b \qquad a < b \\ 0 & b < t \end{cases}$$

shown in Fig. 1.1b as a unit pulse of duration $b - a$, is often used as an *analytic filter* to isolate a desired segment of a function. Hence, if $f(t)$ is a function whose domain includes the interval (a, b), the product

$$f(t)[u(t-a) - u(t-b)]$$

represents the "filtered" function defined for all t in the domain of f by

$$\begin{cases} 0 & t < a \\ f(t) & a < t < b \\ 0 & b < t \end{cases}$$

In other words, the **filter function** $u(t-a) - u(t-b)$ reduces f to zero outside the "passband" $a < t < b$ and reproduces f exactly within the passband.

EXAMPLE 4

The absolute value function $|x|$ has the set of all real numbers as its domain. Its values are computed by using the definition

$$|x| = \begin{cases} -x & x \le 0 \\ x & 0 \le x \end{cases}$$

which expresses the function in terms of two alternative representations, one on $(-\infty, 0]$, the other on $[0, \infty)$. In particular, since $-3 < 0$, $|-3| = -(-3) = 3$.

The square root function $\sqrt{x}$ is always nonnegative and has the interval $[0, \infty)$ as its domain. Thus the domain of $\sqrt{x^2}$ is $(-\infty, \infty)$. In fact, $\sqrt{x^2} = |x|$ and of course $\sqrt{(-3)^2} = 3$.

The radical expression $\sqrt{x^4 - 2x^2}$ is undefined at $x = 1$, so it cannot by itself define a function. What is lacking is a suitable domain. By convention, this is to be taken as large as possible when not otherwise specified. The intended domain of $\sqrt{x^4 - 2x^2}$ is therefore the set of all real numbers for which the radicand $x^4 - 2x^2$ is nonnegative. Solving the inequality $x^2(x^2 - 2) \ge 0$, we find the domain to be the union of the intervals $(-\infty, -\sqrt{2}]$ and $[\sqrt{2}, \infty)$ and the isolated point $x = 0$. For every x in this domain.

$$\sqrt{x^4 - 2x^2} = \sqrt{x^2(x^2-2)} = \sqrt{x^2}\sqrt{x^2-2} = |x|\sqrt{x^2-2}$$

hence our function is defined by the three representations

$$\sqrt{x^4 - 2x^2} = \begin{cases} -x\sqrt{x^2-2} & x \le -\sqrt{2} \\ 0 & x = 0 \\ x\sqrt{x^2-2} & \sqrt{2} \le x \end{cases}$$

Its functional values can now be computed by simply using the appropriate representation. For instance, its value at $x = -3$ is $-(-3)\sqrt{(-3)^2 - 2} = 3\sqrt{7}$. This is also its value at $x = 3$ because the value of the function remains unchanged if x is replaced by $-x$.

EXAMPLE 5

A function y has the set of positive integers N as its domain, and for each n contained in N, $y(n) = 2 + \cos n\pi - \sin[(2n-1)\pi/2]$. Among its values are $y(1) = 2 - 1 - 1 = 0$ and $y(2) = 2 + 1 - (-1) = 4$. The value of y corresponding to any other positive integer can be computed using the same formula. However, there is a much easier way. It is to use the identities

$$\cos n\pi = (-1)^n \qquad \text{and} \qquad \sin[(2n-1)\pi/2] = \sin n\pi \cos\frac{\pi}{2} - \cos n\pi \sin\frac{\pi}{2} = (-1)^{n+1}$$

to transform the formula for $y(n)$ into the simpler representation

$$y(n) = 2 + (-1)^n - (-1)^{n+1} = 2[1 + (-1)^n]$$

and thence into the two different representations

$$y(n) = \begin{cases} 4 & n \text{ even} \\ 0 & n \text{ odd} \end{cases}$$

All values of y are now known.

Various properties of a function are often evident from its *graph.* The **graph** of a function f is the set of all points (x, y) such that x is in the domain of f and $y = f(x)$. The voltage function $E(t)$ of Example 3 is partly graphed in Fig. 1.1.

The function of Example 1 and the two functions $|x|$ and $\sqrt{x^4 - 2x^2}$ of Example 4 are *even* functions.

DEFINITION 2 A function f is **even** if and only if, for every x in its domain, $f(-x) = f(x)$.

An important geometric property of every even function f is that its graph is **symmetric in the y axis;** i.e., for every x in the domain of f, the y axis is the perpendicular bisector of the line segment joining the points $(-x, f(-x))$ and $(x, f(x))$, as illustrated in Fig. 1.2*a*.

DEFINITION 3 A function f is **odd** if and only if, for every x in its domain, $f(-x) = -f(x)$.

The graph of an odd function f is **symmetric in the origin;** i.e., for every x in its domain the origin is the midpoint of the line segment joining $(-x, f(-x))$ and $(x, f(x))$, as illustrated in Fig. 1.2*b*.

Figure 1.2
Graphs: (*a*) of an even function; (*b*) of an odd function.

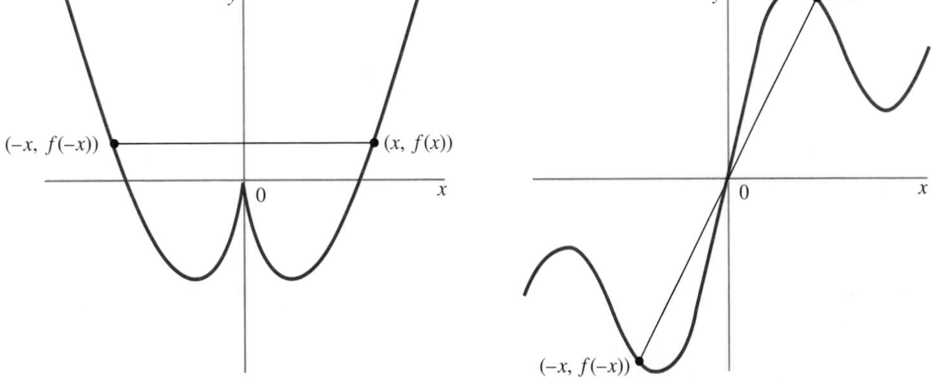

(*a*) Symmetry in the y axis

(*b*) Symmetry in the origin

If $b > 0$ and $\int_{-b}^{b} f(x)\,dx$ exists, then

$$\int_{-b}^{b} f(x)\,dx = \int_{-b}^{0} f(x)\,dx + \int_{0}^{b} f(x)\,dx = \int_{b}^{0} f(-x)\,d(-x) + \int_{0}^{b} f(x)\,dx$$

$$= -\int_{b}^{0} f(-x)\,dx + \int_{0}^{b} f(x)\,dx = \int_{0}^{b} f(-x)\,dx + \int_{0}^{b} f(x)\,dx$$

$$= \int_{0}^{b} [f(-x) + f(x)]\,dx$$

According as f is even or odd, the last *integrand* reduces to $2f(x)$ or 0. Thus we have

THEOREM 1 If f is even, $\int_{-b}^{b} f(x)\,dx = 2\int_{0}^{b} f(x)\,dx$.

THEOREM 2 If f is odd, $\int_{-b}^{b} f(x)\,dx = 0$.

DEFINITION 4 A function f is **trivial** if and only if for every x in its domain $f(x) = 0$.

Our past work in mathematics frequently involved equations containing one or more variables whose solutions we had to find. Familiar examples might be

$$x^2 + 3x + 2 = 0 \qquad \tan\theta = \tfrac{2}{3} \qquad t = e^{-t} \qquad \{v = u^2, 8u = v^2\}$$

We are now going to consider **differential equations,** which are equations containing *derivatives* or *differentials* of one or more variables. Differential equations are of fundamental importance in many areas of pure and applied science and engineering, and much of this text will be devoted to their study.

Here are four examples of differential equations:

(1) $$\frac{dy}{dx} = e^x + \sin x$$

(2) $$y'' - 2y' + y = \cos x$$

(3) $$\frac{\partial^2 u}{\partial x^2} + \frac{\partial^2 u}{\partial y^2} = \frac{\partial u}{\partial t}$$

(4) $$3x^2\,dx + 2y\,dy = 0$$

In (1) and (2), only ordinary derivatives of y with respect to x occur. This signifies that y must be a function of x. Thus y is the *dependent* variable and x is the *independent* variable. In (3), u is the dependent variable and x, y, and t are independent variables, as the partial derivatives of u imply. In (4), either x or y can be thought of as the dependent variable, the other variable then being independent.

EXERCISES

In Exercises 1–8 let f be a function having the given representation.

1. $f(x) = \sqrt{x} + |1 - x| \ln(1 + x)^{1/3}$, $x \geq 0$; find **(a)** $f(0)$, **(b)** $f(1)$, **(c)** $f(4)$
2. $f(x) = \sin x + (\pi - x)\sinh x$, x real; find **(a)** $f(0)$, **(b)** $f(\pi)$, **(c)** $f(-\pi)$
3. $f(x) = |x| + \mathrm{Tan}^{-1} x$, x real; find **(a)** $f(0)$, **(b)** $f(1)$, **(c)** $f(-1)$

4. $f(x) = \pi \cos \pi x + \mathrm{Sin}^{-1} x$, $|x| \le 1$; find **(a)** $f(0)$, **(b)** $f(-1)$, **(c)** $f(\frac{1}{2})$

5. $f(x) = \Sigma_{n=0}^{\infty} [1 + (-1)^n]x^n/n!$, x real; find **(a)** $f(0)$, **(b)** $f(1) - f(-1)$, **(c)** $f(\ln 2)$. *Hint:* Recall the *Maclaurin series*† for e^x.

6. $f(x) = \int_0^2 x^2(t)\, dt$, x continuous on $[0, 2]$; if $x_1(t) = t^{1/2}$, $x_2(t) = t$, $x_3(t) = \sin \pi t$, find **(a)** $f(x_1)$, **(b)** $f(x_2)$, **(c)** $f(x_3)$

7. $f(x, y) = \int_{\pi}^{2\pi} x(t)y(t)\, dt$, x, y continuous on $[\pi, 2\pi]$; if $x_1(t) = \cos t$, $y_1(t) = \sin 2t$, find **(a)** $f(1, y_1)$, **(b)** $f(x_1, 1)$, **(c)** $f(x_1, y_1)$

8. $f(x, y, z) = 2 \int_0^x [3y(t) - z(t)]\, dt$, y, z continuous for all real x; if $y_1(t) = 1/(1 + t^2)$ and $z_1(t) = \mathrm{Tan}^{-1} t$, find **(a)** $f(\infty, y_1, 0)$, **(b)** $f(1, y_1, z_1)$, **(c)** $f(-1, y_1, z_1)$

9. A function f is represented on the set of real numbers by $f(t) = 2(1 - \cos^2 t) + \ln e^t$. Give three other representations of f over the reals which do not contain the logarithmic function.

10. Every 40 min during an 8-h period, a dump truck reloads with roadbase. After t hours, how many times has it been filled if it takes 40 min to get the first load?

11. A high-rise has 15 stories. The number of weeks it took to construct *each* floor is given in the table:

Floors	1–3	4–6	7–9	10–12	13–15
Weeks	3	4	5	6	7

(a) Which floor was under construction during the forty-fifth week?
(b) How many floors were completed at the end of 1 year?
(c) What was the total construction time?

12. Determine whether the following functions are even, odd, or neither.
(a) $x \sin x^2$ **(b)** $e^{|x|} + x \sinh x$
(c) $x \ln x + \mathrm{Tan}^{-1} x$

13. Prove that if a function f is both even and odd, then for every x in the domain of f, $f(x) = 0$.

14. Assuming they have the same domain,
(a) What can be said about the sum or difference of two even functions? Of two odd functions? Of an odd function and an even function?
(b) What can be said about the product of two even functions? Of two odd functions? Of an odd function and an even function?

15. By considering the identity

$$f(x) = \frac{f(x) + f(-x)}{2} + \frac{f(x) - f(-x)}{2}$$

show that any function defined over an interval which is symmetric with respect to the origin can be written as the sum of an even function and an odd function.

16. Express each of the following functions as the sum of an even function and an odd function.
(a) $e^x + \ln |x|$ **(b)** $|x - 1|$

(c) $f(x) = \begin{cases} 0 & x < 0 \\ 2 & 0 \le x \end{cases}$

17. Evaluate

(a) $\displaystyle\int_{-1}^{1} \frac{\cos 2\pi - \cos \pi}{1 + x^2}\, dx$

(b) $\displaystyle\int_{-\pi}^{\pi} \sin 2x \cos 3x \cosh 4x\, dx$

(c) $\displaystyle\int_{-\ln 2}^{\ln 2} (e^{|x|} + \sin^5 \pi x)\, dx$

(d) $\displaystyle\int_{-10}^{10} (6t^{99} - 13t \sin et^2 + 5)\, dt$

Identify the dependent and independent variables in each of the following differential equations.

18. $z'' + zy' + yx = \sec x$ **19.** $3xy'' + \tanh y' = y$

20. $u_{xx} + v_{yy} - xyuv = 0$

1.2 CLASSIFICATION OF DIFFERENTIAL EQUATIONS

Various distinctive features are used to classify differential equations into a number of identifiable types. *Ordinary* and *partial* differential equations are characterized by the number of independent variables and the kind of derivatives they involve.

DEFINITION 1 An **ordinary differential equation** is a differential equation in which all derivatives are ordinary derivatives of one or more dependent variables with respect to a single independent variable.

†Named for the Scottish mathematician Colin Maclaurin (1698–1746).

Clearly, Eqs. (1) and (2) of Sec. 1.1 are ordinary differential equations. The same is true of Eq. (4); for, according as x or y is chosen as independent variable and the equation is divided by the differential dx or the differential dy, it involves one or the other of the ordinary derivatives dy/dx or dx/dy.

DEFINITION 2 A **partial differential equation** is a differential equation containing at least one partial derivative of some dependent variable.

Equation (3) of Sec. 1.1 is a partial differential equation.
Differential equations are also classified according to their *order.*

DEFINITION 3 The **order** of a differential equation is the order of the highest-order derivative which appears in the equation.

Equations (1) and (4) of Sec. 1.1 are first-order differential equations; Eqs. (2) and (3) are second-order equations.

Another broad classification of differential equations is based on the way in which a dependent variable and its indicated derivatives appear in the terms of such an equation.

DEFINITION 4 A differential equation is **linear in a set of one or more of its dependent variables** if and only if each term of the equation which contains a variable of the set or any of their derivatives is of the first degree in those variables and their derivatives.

A differential equation which is not linear in some dependent variable is said to be **nonlinear** in that variable. A differential equation which is not linear in the set of all of its dependent variables is simply said to be **nonlinear.**

EXAMPLE 1

The equation $y'' + 4xy' + 2y = \cos x$ is a linear ordinary differential equation of second order. The presence of the product xy' and the term $\cos x$ does not alter the fact that the equation is linear because, by definition, linearity is determined solely by the way the *dependent* variable y and its derivatives enter into combination among themselves within each term of the equation.

EXAMPLE 2

The equation $y'' + 4yy' + 2y = \cos x$ is a nonlinear equation because of the occurrence of the product of y and one of its derivatives.

EXAMPLE 3

The equation $\partial^2 u/\partial x^2 + \partial v/\partial t + u + v = \sin u$ is linear in the dependent variable v but nonlinear in the dependent variable u because $\sin u$ is a nonlinear function of u. The equation is also nonlinear.

EXAMPLE 4

The equation $d^2x/dt^2 + dy/dt + xy = \sin t$ is linear in each of the dependent variables x and y. However, because of the term xy, it is not linear in the set of dependent variables $\{x, y\}$. As a consequence, the equation is nonlinear.

EXAMPLE 5

As written, the equation $3x^2\, dx + (\sin x)\, dy = 0$ is neither linear nor nonlinear. Division by dx transforms it into the equation $3x^2 + (\sin x)y' = 0$ which is linear in y, but division by dy gives $3x^2\, dx/dy + \sin x = 0$ which is nonlinear in x.

From Definition 4 it follows that the most general ordinary linear differential equation of order n in a single dependent variable is of the form

$$a_0(x)y^{(n)} + a_1(x)y^{(n-1)} + \cdots + a_{n-1}(x)y' + a_n(x)y = f(x) \tag{1}$$

where $a_0(x) \neq 0$ throughout some interval.

EXERCISES

Describe each of the following equations, giving its order and telling whether it is ordinary or partial and linear or nonlinear (a, b constants).

1. $y'' - 5y' + 3y = x^4$
2. $y' + (a + b \sin 3x)y = 0$
3. $y''' - 7y'' + 11y' - 13y = xe^x$
4. $y^{\text{iv}} + x^2y' + y^{1/2} = 0$
5. $x^{3/2}y'' - 9x^{1/2}y' + 5y = \text{Tan}^{-1} x$
6. $d(xy')/dx + xy = 0$
7. $d(axy)/dt + bxy = \ln t$
8. $a^2\, \partial^2u/\partial x^2 = b^2\, \partial^2u/\partial t^2$
9. $\partial^2(x^2\, \partial^2u/\partial x^2)/\partial x^2 = \partial^2u/\partial t^2$
10. $\partial^2u/\partial x^2 = u\, \partial u/\partial t$
11. $\partial^2u/\partial x^2 + \partial^2u/\partial y^2 = \phi(x, y)$
12. $\partial u/\partial t + xu = d^2x/dt^2$
13. $\partial u/\partial x + u\, \partial v/\partial y = v\, \partial^2v/\partial x\, \partial y$
14. $\sec t = t^3x + 2x^{1/2}\, d(x^{1/2})/dt$

For each of the following equations, determine whether or not it becomes linear when divided by dx or dy.

15. $(x + y)\, dy = (x - y)\, dx$
16. $a\, dy + by \sin x\, dx = 0$
17. $3y\, dx + 2x\, dy = 0$
18. $e^x\, dy + xy^{1/3}\, dx = 0$

1.3 SOLUTIONS OF DIFFERENTIAL EQUATIONS

A **solution** of an algebraic or transcendental equation in a single variable x is a *number* which satisfies the equation. On the other hand, solutions of differential equations are *functions,* rather than numbers, which satisfy the equation. Whereas all variables which appear in algebraic or transcendental equations are called ''unknowns,'' only the dependent variables in a differential equation are referred to as ''unknowns.''

EXAMPLE 1

Under certain constraints, the motion of a spring-suspended mass is described by solutions of the differential equation $y'' + k^2y = 0$, where the dependent variable (or unknown) y is the vertical displacement of the mass, the second derivative is taken with respect to the independent variable t (or time) and k is a positive constant. Let f be the function represented by $\sin kt$ on the set of real numbers. Then f'' is represented by $-k^2 \sin kt$. Upon replacing y by $\sin kt$ and y'' by $-k^2 \sin kt$, the given equation is transformed into $-k^2 \sin kt + k^2 \sin kt = 0$, which holds identically for all real values of t. The fact that this replacement process results in an identity over the reals is described by calling $f(t) = \sin kt$ a solution of $y'' + k^2y = 0$ on $(-\infty, \infty)$ or by saying that f is a solution of the differential equation. In other words, the differential equation is satisfied when f is substituted for the unknown y. Another solution is the function g defined on $(-\infty, \infty)$ by $g(t) = \cos kt$.

EXAMPLE 2

The equation $u_x - v_y = 0$ has u and v as unknowns. Functions f and g are defined, for all real values of x and y, by $f(x, y) = \sinh x \sinh y$ and $g(x, y) = \cosh x \cosh y$. Replacement of u_x by $\partial f/\partial x = \cosh x \sinh y$, and of v_y by $\partial g/\partial y = \cosh x \sinh y$, converts the given differential equation into $\cosh x \sinh y - \cosh x \sinh y = 0$. Since this is an identity for all real values of x and y, we say that $f(x, y)$ and $g(x, y)$ are solutions for the respective unknowns u and v and that the set of functions $\{f, g\}$ is a solution of the given equation over the entire xy plane.

The last two examples illustrate the concept of a solution of a differential equation. They also indicate how to substitute a set of functions for the unknowns of such an equation when testing to see if the set is a solution of the equation.

DEFINITION 1 A **solution** of a differential equation **over a region** R is a set of functions which, when they are substituted for the dependent variables in the differential equation, reduce the equation to an identity in the independent variables over R.

At present we shall be concerned primarily with ordinary differential equations in a single unknown. For such equations, the preceding definition may be rephrased as follows.

DEFINITION 2 A **solution** of an ordinary differential equation in one dependent variable **on an interval** I is a function which, when substituted for the dependent variable, reduces the equation to an identity in the independent variable over I.

A solution on I whose values are all equal to 0 is said to be **trivial on** I.

If a differential equation of order n is satisfied by a function f on I, then the nth derivative $f^{(n)}$ of f necessarily exists throughout I. Since a function must be continuous wherever its derivative exists, the existence of $f^{(n)}$ over I implies that f and its derivatives of all orders up to and including $n - 1$ are continuous on I. A function that is not continuously differentiable at least $n - 1$ times on I cannot be a solution over I. In other words

THEOREM 1 Every solution on an interval I of an nth-order differential equation in one dependent variable must be continuously differentiable at least $n - 1$ times on I.

EXAMPLE 3

It is easy to verify that both $y_1 = 2 - x$ and $y_2 = x - 2$ are solutions of $y'' = 0$ on every interval I. Let f be the function defined by $f(x) = |x - 2|$ on $(-\infty, \infty)$. Then f is represented by $y_1(x) = 2 - x$ for $x \leq 2$ and by $y_2(x) = x - 2$ for $x \geq 2$, and f is continuous at $x = 2$. However, f is not a solution of $y'' = 0$ on any open interval containing $x = 2$ because f' is not continuous at $x = 2$. In fact,

$$f'(2) = \lim_{h \to 0} \frac{f(2 + h) - f(2)}{h} = \lim_{h \to 0} \frac{|h|}{h} = \begin{cases} -1 & \text{as } h \to 0^- \\ 1 & \text{as } h \to 0^+ \end{cases}$$

hence $f'(2)$ does not exist.

Some differential equations have no solutions; others have only a trivial solution.

EXAMPLE 4

The equation $|dy/dx| + 1 = 0$ has no solution. The equation $|dy/dx| + |y| = 0$ has only the trivial solution $y = 0$.

Differential equations for which every solution of each equation is a solution of all the others are said to be **equivalent.** If their solutions are the same only on some interval I, we say that the differential equations are **equivalent on** I.

The study of the existence, nature, and determination of solutions of differential equations is of fundamental importance not only to the pure mathematician but also to anyone engaged in the mathematical analysis of natural phenomena. In general, mathematicians consider it a triumph if they are able to prove that a given differential equation possesses a solution and if they can deduce a few of the more important properties of that solution. Engineers and applied scientists, on the other hand, are usually greatly disappointed if a specific expression for the solution cannot be exhibited. The usual compromise is to find some practical procedure by which the required solution can be approximated with satisfactory accuracy.

Not all differential equations are difficult enough to make this necessary, however, and there are several large and very important classes of equations for which solutions can readily be found in terms of functions that are already familiar from calculus.

Even if f is an unfamiliar function, an equation such as

$$\frac{dy}{dx} = f(x)$$

is really a differential equation, and for any constant c,

$$y = \int f(x)\, dx + c$$

is a solution on every interval where f has an antiderivative. This illustrates the fact that the solutions of an ordinary differential equation are usually specified by an expression involving one or more arbitrary constants called **parameters.**

More generally, for suitable functions g, the equation

$$\frac{d^n y}{dx^n} = g(x)$$

is a differential equation for which solutions can be found by n successive integrations. Except in name, the process of integration is actually a process for finding solutions of differential equations.

EXAMPLE 5

All solutions of the equation $y' = \text{Tan}^{-1} x$ must be continuously differentiable on $(-\infty, \infty)$ because $\text{Tan}^{-1} x$ is everywhere continuous. They must also be antiderivatives of $\text{Tan}^{-1} x$. From calculus we know that all antiderivatives of a continuous function differ at most by a constant. Using integration by parts, with $u = \text{Tan}^{-1} x$ and $dv = dx$, we find $x \,\text{Tan}^{-1} x - \frac{1}{2} \ln (1 + x^2)$ to be an antiderivative of $\text{Tan}^{-1} x$. Thus, with c as a parameter, the expression

$$y = x \,\text{Tan}^{-1} x - \tfrac{1}{2} \ln (1 + x^2) + c$$

specifies the set of all solutions of $y' = \text{Tan}^{-1} x$ on $(-\infty, \infty)$ and not just one solution. Nevertheless, this expression for y is customarily referred to as a **solution** of the given differential equation **containing the parameter c.**

EXAMPLE 6

Since $1/x$ is undefined at $x = 0$, there is no solution of the differential equation $y'' = 1/x$ over any interval which contains the origin. To solve this equation, we perform two successive integrations which give

$$y' = \ln |x| + c \qquad \text{and} \qquad y = x \ln |x| - x + cx + c_2$$

where c and c_2 are parameters. Replacing $c - 1$ by the parameter c_1, we find that every solution of the given differential equation on any subinterval of either $(-\infty, 0)$ or $(0, \infty)$ has a representation of the form

$$y = x \ln |x| + c_1 x + c_2$$

Each individual solution of a differential equation is called a **particular solution.** A **general solution** of an ordinary differential equation is a nonempty set of solutions specified by an expression which contains at least one parameter and which becomes a representation of a particular solution when its parameters are all replaced by numbers. A solution is said to be **singular** relative to any given general solution if it does not belong to that solution set. In practice, solution sets and the expressions which specify them are ordinarily spoken of synonymously.

EXAMPLE 7

It is easy to verify that $y = cx + 2c^2$ is a general solution of the nonlinear equation $y = xy' + 2(y')^2$. Another solution is $y_1 = -x^2/8$, and this solution cannot be obtained by particularizing the parameter c. Therefore $y_1 = -x^2/8$ is a singular solution relative to the given general solution.

The set of *all* solutions of a differential equation constitutes its **complete solution.** The set of all solutions of a given differential equation is obviously a unique set, and from this point of view it is correct to speak of *the* complete solution of the equation. On the other hand, various expressions involving parameters may possibly describe the set of all solutions of a given differential equation; and when we speak of one of these expressions as though it were the family of solutions, as we commonly do, we must be careful to speak of it only as *a* complete solution. In many problems, the choice of which complete solution we use, that is, the choice of which expression we use to specify *the* complete solution, often has an important bearing on the ease with which the problem can be solved.

EXAMPLE 8

For every real number c, $y = x^2 + c$ is a solution of the simple equation $y' = 2x$ on $(-\infty, \infty)$. On the other hand, to each specific solution of $y' = 2x$ on $(-\infty, \infty)$ there corresponds a number c such that $y = x^2 + c$ is that solution. The expression $y = x^2 + c$ thus describes the set of all solutions of $y' = 2x$ on every interval I. Hence, $y = x^2 + c$ is not only a general solution but is also a *complete* solution of $y' = 2x$.

For first-order equations, particular solutions often arise from the requirement that a solution $y = y(x)$ satisfy some auxiliary condition, frequently of the form $y(x_0) = k_0$, with the numbers x_0 and k_0 prescribed. For equations of order $n > 1$, particular solutions similarly arise from the requirement that a solution and its first $n - 1$ derivatives satisfy n conditions of the form $y^{(i)}(x_0) = k_i$ for $i = 0, 1, \ldots, n - 1$, where x_0 and k_i are specified numbers and $y^{(0)} = y$. The problem of determining solutions that satisfy conditions of this explicit nature is called an **initial-value problem** since in many applications the independent variable stands for time and the auxiliary conditions are all prescribed at some initial instant. The auxiliary conditions themselves are called **initial conditions.**

As illustrated by the simple equation

$$\frac{dy}{dx} = e^{-x^2}$$

and its solution

$$y = \int e^{-x^2}\,dx + c$$

a solution of a differential equation may depend upon integrals which cannot be expressed in terms of elementary functions.

EXAMPLE 9

Find the solution of the differential equation

$$y'' = \frac{1}{x} + 2x\cos x^2 \tag{1}$$

on $(0, \infty)$ which satisfies the initial conditions $y(1) = 0$ and $y'(1) = 1 + \sin 1$.

Equation (1) can be solved by two successive integrations. The first gives us

$$y' = \ln x + \sin x^2 + c \tag{2}$$

and a second integration yields $y = \int \ln x\, dx + \int \sin x^2\, dx + cx + b$, where c and b are arbitrary constants. Since there is no elementary antiderivative for $\sin x^2$, we take $\int_1^x \sin t^2\, dt$ as an antiderivative and then integrate $\int \ln x\, dx$ by parts, getting

$$y = x \ln x + \int_1^x \sin t^2\, dt + (c - 1)x + b \tag{3}$$

This is a general solution of (1) on $(0, \infty)$. It is also a complete solution in the sense that for suitable values of c and b it will specify any member of the set of all solutions of the given equation on $(0, \infty)$.

Imposing the initial condition $y'(1) = 1 + \sin 1$ on (2), we find that $c = 1$. Subsequently, imposing the condition $y(1) = 0$ on (3), we find that $b = 0$. Thus the required solution is defined on $(0, \infty)$ by

$$y = x \ln x + \int_1^x \sin t^2\, dt$$

EXAMPLE 10

Let S be the set of all functions which have the domain $(-\infty, \infty)$ and are represented by a formula of the type

$$a \cos^2 x + b \sin^2 x + c \cos 2x$$

where a, b, and c are parameters. Since $\cos 2x = \cos^2 x - \sin^2 x$ is an identity, we may substitute for $\cos 2x$, collect terms on $\cos^2 x$ and $\sin^2 x$, and then set $d = a + c$ and $e = b - c$ to obtain the formula

$$d \cos^2 x + e \sin^2 x$$

which not only specifies the functions of S but also contains just the two parameters d and e instead of the original three a, b, and c. This shows that different expressions identifying the same set need not contain the same number of parameters.

Example 10 naturally raises the question of how few, or how many, arbitrary constants may appear in a complete solution of a differential equation. A rule of thumb that is often quoted asserts that *in general, a complete solution of an nth order differential equation contains n arbitrary constants.* As we shall see, there is a measure of truth in this, but as a principle it should be viewed with suspicion and used, if at all, with caution. In fact, we have already seen in Example 4 that there are differential equations which have no solutions containing any arbitrary constants. Moreover, complete solutions containing several parameters can in general be replaced by others containing only a single parameter.† The procedure for doing this is quite complicated, however, and the end result is of little practical value. Thus, for our purposes, we shall rely on familiar identities, rules of algebra, and the renaming of parameters (as in Example 10) to transform general and complete solutions into more convenient forms.

Later on in this chapter we shall learn to recognize and solve various types of differential equations which are next in difficulty after those which can be solved by direct integration. These equations form only a small part of the class of all differential equations, and yet with a knowledge of them a scientist is equipped to handle a great variety of applications. To get so much for so little is indeed remarkable.

---

†See, for instance, R. P. Agnew, *Differential Equations,* 2d ed., pp. 103–105, McGraw-Hill, New York, 1960.

EXERCISES

Verify that each of the following equations has the indicated solution for all values of the constants a and b.

1. $y'' + 9y = 0$ $\quad y = a\cos 3x + b\sin 3x$
2. $y'' - 9y = 0$ $\quad y = a\cosh 3x + b\sinh 3x$
3. $y'' - 6y' + 9y = 0$ $\quad y = ae^{3x} + bxe^{3x}$
4. $y'' + 3y' + 2y = 12e^{2x}$ $\quad y = ae^{-x} + be^{-2x} + e^{2x}$
5. $(\cos 2x)y' + (2\sin 2x)y = 2$ $\quad y = a\cos 2x + \sin 2x$
6. $y'' + (y')^2 + 1 = 0$ $\quad y = \ln\cos(x - a) + b$
7. $y'' - (y')^2 + 1 = 0$ $\quad y = \ln\operatorname{sech}(x - a) + b$
8. $2xy\,dy = (y^2 - x)\,dx$ $\quad y = y_1$ or $y = y_2$, where $y_1 \neq y_2$ and $y_1^2 = y_2^2 = x(a - \ln x)$
9. $\partial^2u/\partial x^2 = \partial u/\partial t$ $\quad u = ae^{-4t}\cos(2x + b)$
10. $9\,\partial^2u/\partial x^2 = \partial^2u/\partial t^2$ $\quad u = af(x + 3t) + bg(x - 3t)$, f and g twice-differentiable on $(-\infty, \infty)$

Find all values of m for which

11. **(a)** $y = e^{mx}$ is a solution of $6y'' - y' - y = 0$ on $(-\infty, \infty)$
    (b) $y = e^{mx}$ is a solution of $y''' - 3y'' - 4y' + 12y = 0$ on $(-\infty, \infty)$
12. $y = m(e^2 - x^2)^{1/2}$ is a solution of $x\,dx + y\,dy = 0$ on $(-e, e)$
13. **(a)** $y = x^m$ is a solution of $x^2y'' - 3xy' + 3y = 0$ on $(-\infty, \infty)$
    (b) $y = x^m$ is a solution of $3x^3y''' - x^2y'' + 2xy' - 2y = 0$ on $(-\infty, \infty)$
14. If n is an integer and $f(x) = \cot x$ for $x \neq n\pi$, find all intervals on which f is a solution of $y' + y^2 + 1 = 0$ and for which $|x - \pi| < 2\pi$.
15. Determine all intervals on which the function f defined by $f(0) = 0$ and $f(x) = e^{-1/x^2}$ for $x \neq 0$ is a solution of the following equations.
    (a) $x^3y' + y = 0$
    (b) $x^3y'' + (3x^2 - 2)y' = 0$
    (c) $x^3y' - 2y = 0$
    (d) $x^3y''' + (6x^2 - 2)y'' + (x^3 + 6x)y' - 2y = 0$
16. Find a solution of $y''' = (3\sec^2 x - 2)\sec^2 x$ such that $y(0) = 0$, $y'(0) = \frac{1}{2}$, and $y''(0) = 0$ and identify all intervals on which the solution is defined.

Find a complete solution of each of the following differential equations.

17. $y' = 1 - 2\sin 2x$
18. $y' = 2/[x(2 - x)]$
19. $y'' = 1 + \tan^2 x$
20. $y'' = e^{-x} + 9\cos 3x$
21. $y' = \ln e^{1+x} - 3xe^{-x^2}$
22. $y^{\text{iv}} = \sinh(x/2)$
23. $y''' = (2/x^3) + (\sec x)(1 + 2\tan^2 x)$
24. $y'' = (3\sin x)/(5 + 3\cos x)^2$

Solve each of the following initial-value problems.

25. $y' = xe^{3x}$; $y(0) = 0$
26. $(\sin x)\,dx - (\cos x)\,dy = 0$; $y(0) = 1$
27. $y' = 2\,\mathrm{Tan}^{-1}x$; $y(0) = 0$
28. $y'' = (1 - x^2)^{-1/2}$; $y(0) = 1$, $y'(0) = 0$
29. $y'' = 9(\cos x)(\sin 2x)$; $y(\pi) = 2$, $y'(\pi) = 6$
30. $y''' = 48\cos^2 x$; $y(0) = 2$, $y'(0) = -6$, $y''(0) = 0$

Find an expression that specifies the same set of functions as the given formula but which contains fewer parameters. Make the number of parameters as small as you can. (A, B, C, a, b, c, d, and k are parameters.)

31. Ae^{x+k}
32. $a + \ln bx$
33. $a\ln x^b$
34. $(ax + b)/(cx + d)$
35. $A\sin(x + b) + C\sin(x + d)$
36. $A[\cos(x + a) + \cos(x - a)]$
37. $a\cosh^2\theta + b\sinh^2\theta + c\cosh 2\theta$
38. $a\sin 3x + b\sin x + c\sin^3 x$
39. $\dfrac{A}{x + 1} + \dfrac{B}{x + 2} + \dfrac{C}{x^2 + 3x + 2}$
40. $a(x - 6y - 7) + b(3x + 4y + 5) + c(5x + 3y + 4)$

1.4 SOLUTION CURVES AND INTEGRAL CURVES

Once a solution of a differential equation has been found, it is possible to exhibit its properties geometrically by plotting its graph for various specific values of any parameters involved. The graph of a *particular* solution of an ordinary differential equation in one dependent variable is called a **solution curve** of that equation.

Most differential equations associated with applied problems are not directly integrable, and special procedures are needed in order to find their solutions. Frequently, these techniques do not yield explicit solutions. Instead, they lead to an *implicit* relation between the dependent and independent variables which is free of derivatives and which defines *one or more* solutions of the relevant differential equation. Such a relation is said to be an **implicit solution** or **integral** of the differential equation. Implicit solutions, as well as explicit solutions, are called **general solutions** if they contain arbitrary constants. A general integral becomes a **particular integral** when its parameters are given appropriate values. The graph of a particular integral is called an **integral curve.**

Although the distinction between a solution curve and an integral curve is seldom crucial, it should be clearly understood. As the next example illustrates, an integral curve of a differential equation is usually not the graph of a single solution but is a graph which includes several solution curves, each the graph of a different solution.

EXAMPLE 1

Implicit differentiation of the relation $x^2 + y^2 = 1$ with respect to x, and subsequent division by 2, give the differential equation $x + yy' = 0$. It is easy to verify that this equation has each of the explicit functions

$$y_1 = \sqrt{1 - x^2} \qquad \text{and} \qquad y_2 = -\sqrt{1 - x^2}$$

as a solution on the interval $-1 < x < 1$. Since both of these solutions are defined implicitly by

$$x^2 + y^2 = 1$$

this relation is a particular integral of $x + yy' = 0$. A circle of radius 1 and center $(0, 0)$ is the corresponding integral curve (Fig. 1.3*a*). The upper semicircular arc of this circle, with $(-1, 0)$ and $(1, 0)$ deleted (Fig. 1.3*b*), is the solution curve determined by the solution $y_1 = \sqrt{1 - x^2}$, and the lower semicircular arc, with the same two points deleted (Fig. 1.3*c*), is the solution curve associated with the solution $y_2 = -\sqrt{1 - x^2}$.

The implicit relation $x^2 + y^2 = c$ is a general integral of the differential equation $x + yy' = 0$ which, for each positive value of the parameter c, becomes a particular integral defining two corresponding solutions of the equation on $-\sqrt{c} < x < \sqrt{c}$. However, if $c \le 0$, $x^2 + y^2 = c$ is not an implicit solution of the differential equation because in this case there is no interval over which $x^2 + y^2 = c$ defines a solution of $yy' + x = 0$.†

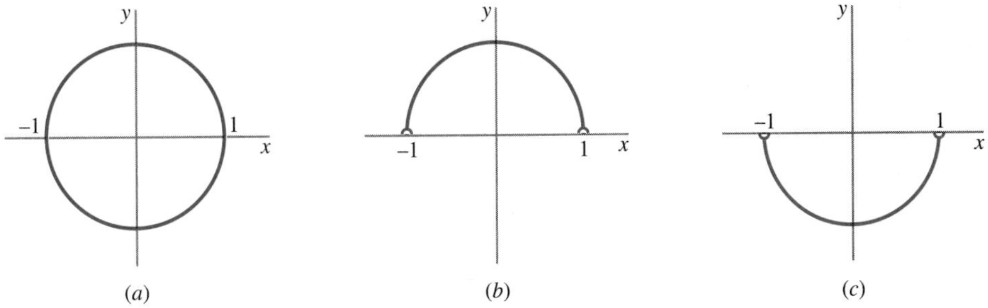

Figure 1.3
The integral and solution curves corresponding to the integral $x^2 + y^2 = 1$ of the equation $yy' + x = 0$, and to the solutions on $-1 < x < 1$ defined by this integral.

In general, the problem of finding a solution of a differential equation is considered to be *solved* if either an implicit solution or a solution over an interval has been found, even though the implicit solution may define several (explicit) solutions. When a set of auxiliary conditions accompanies a differential equation, an implicit solution may have to be analyzed further to determine which of the (explicit) solutions it defines is the one that satisfies the conditions of the problem.

As our experience in solving differential equations grows, we shall occasionally encounter solutions containing more parameters than the order of the equation. A case in point is afforded by the differential equation

$$xy' = 2y \tag{1}$$

On the two separate intervals $(-\infty, 0)$ and $(0, \infty)$ all solutions of (1) except the obvious trivial solution $y = 0$ are solutions of $y'/y - 2/x = 0$, which is equivalent to

$$\frac{d}{dx}(\ln|y| - \ln x^2) = 0 \tag{2}$$

†Theoretically, the question of whether or not an implicit solution defines some solution over an interval is answered by an appropriate implicit function theorem. See, for instance, R. Creighton Buck, *Advanced Calculus,* 3d ed., pp. 362–363, McGraw-Hill, New York, 1978.

One integration of this equation gives

$$\ln|y| - \ln x^2 = a$$

where the parameter a may take on any real value. The logarithms appearing here suggest that we replace a by $\ln|c|$, $c \neq 0$. This substitution is permissible because, given any real number r, if $|c| = e^r$ then $\ln|c| = r$. Thus $\ln|c|$ may stand for any real number just as a did.

The relation $\ln|y| - \ln x^2 = \ln|c|$ may be written successively as

$$\ln\frac{|y|}{|c|} = \ln x^2 \qquad |y| = |c|x^2 \qquad y = \pm cx^2$$

Since c may be any positive or negative number, the two expressions $y = \pm cx^2$ and $y = cx^2$ specify the same set of functions over any given interval. Hence

$$y = cx^2 \tag{3}$$

is a complete solution of (2) on all subintervals of $(-\infty, 0)$, or $(0, \infty)$. However, on any interval containing $x = 0$, (3) is *not* a solution of (2) because ln 0 is undefined. On the other hand, (3) is a *general* solution of (1) on $(-\infty, \infty)$ or on any subinterval thereof.

To obtain a *complete* solution of (1) on $(-\infty, \infty)$ we observe that for suitable values of c_1 and c_2, any solution y on this interval can be represented on $(-\infty, 0)$ by c_1x^2 and on $(0, \infty)$ by c_2x^2. Since (1) has $y = 0$ as a trivial solution on $(-\infty, \infty)$, either c_1 or c_2 may be zero. Now, $y'(0)$ must be finite. At $x = 0$, (1) gives $0 \cdot y'(0) = 2y(0)$ which implies $y(0) = 0$. Thus every solution of (1) on $(-\infty, \infty)$ has 0 as its value at $x = 0$. It follows that

$$y = \begin{cases} c_1x^2 & x \le 0 \\ c_2x^2 & x \ge 0 \end{cases} \tag{4}$$

is a complete solution of (1) since this formula specifies all solutions of (1) on $(-\infty, \infty)$.

Figure 1.4
Arcs of different members of the family $y = cx^2$ pieced together to form solution curves of the differential equation $xy' = 2y$.

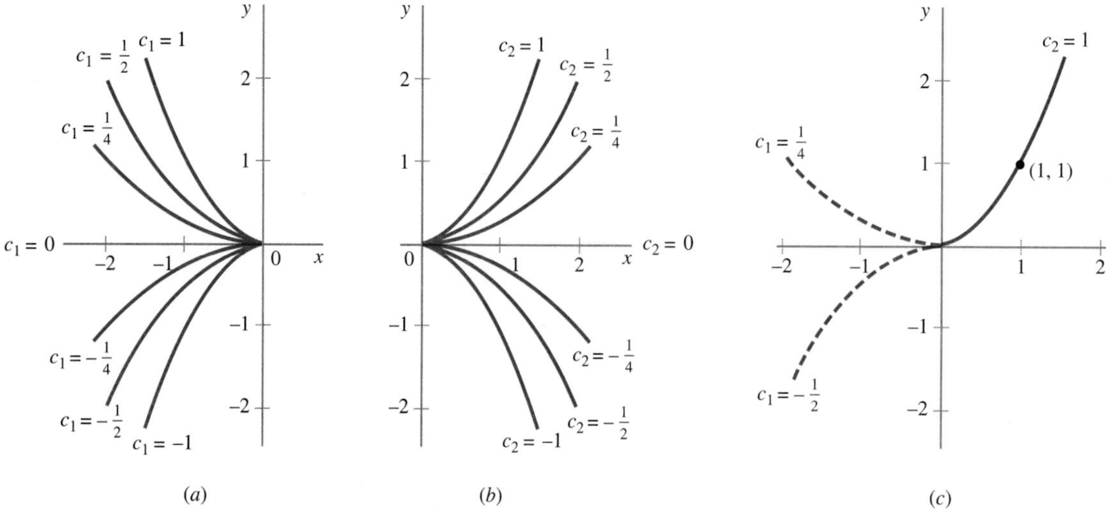

Interpreted geometrically (4) shows that for all nonzero values of c_1 and c_2, the curve obtained by pairing the *left*-hand half of the parabola $y = c_1x^2$ with the *right*-hand half of the parabola $y = c_2x^2$, the origin being included, is a solution curve of (1). If either c_1 or c_2 is zero, the corresponding parabolic arc is to be replaced by the corresponding half of the x axis.

Since $y(0) = 0$ and $y'(0) = \lim_{h\to 0} (c_ih^2 - 0)/h = 0$, whether i is 1 or 2, each solution curve is the graph of a solution which for all x has a continuous derivative. Figure 1.4 shows these observations graphically. A still more striking example of this sort appears in Exercise 12, where a first-order equation with a complete solution containing infinitely many parameters is given.

EXERCISES

1. Find a complete solution of $y' = 4x^{1/3}$ and write an equation of the solution curve through the point $(-1, 3)$.
2. Find a general integral of $yy' = 1$, an equation of the integral curve through $(3, 4)$, and an equation of the solution curve through $(3, -4)$.
3. Find a general integral of $4x + 9yy' = 0$, the particular integral for which $y(0) = 2$, and an equation of the solution curve through $(0, -2)$. *Hint:* $(d/dx)(4x^2 + 9y^2) = 8x + 18yy'$.
4. **(a)** Find a complete solution of the differential equation $y' + xe^{-x^2/2} = 0$.
   (b) Sketch the solution curve through $(1, 1/\sqrt{e})$, showing that portion of the curve where $-2 \le x \le 2$.
5. **(a)** Find a general integral of the equation $yy' - x = 0$. *Hint:* Compare the equation with $(d/dx)(y^2 - x^2) = 0$.
   (b) Identify the family of integral curves.
   (c) Sketch the members of the family which pass through $(1, 0)$, $(-1, 0)$, $(0, 1)$, and $(0, -1)$.
   (d) Sketch all solution curves which pass through $(0, 0)$ and give the corresponding solutions.
6. **(a)** Find a complete solution of the equation $xy' = 3y$ on $(-\infty, \infty)$.
   (b) Sketch three distinct solution curves through $(1, 1)$, showing that portion of the curves for which $-1 \le x \le 1$.
7. **(a)** Determine a general integral for the equation $(\sec y)\, dx - dy = 0$. *Hint:* To solve the equation treat y as the independent variable.
   (b) Find the solution y on $-1 \le x \le 1$ for which $y(0) = 0$.
   (c) Find the solution y on $-1 \le x \le 1$ for which $y(0) = 2\pi$ and sketch the corresponding solution curve.

Find a complete solution on $(-\infty, \infty)$ for each of the following equations by comparing it with $(d/dx)[g(x, y)] = 0$, where $g(x, y)$ is as given.

8. $y' = y \tanh x \qquad g(x, y) = y \operatorname{sech} x$
9. $xy' = 2y + 2 \qquad g(x, y) = (y + 1)/x^2$
10. $xy' = 2y + x \qquad g(x, y) = (y + x)/x^2$
11. $(x^2 - 1)y' = 4xy \qquad g(x, y) = \ln|y| - \ln(x^2 - 1)^2$
12. $(1 - \cos x)y' = (\sin x)y \quad g(x, y) = \ln|y| - \ln|1 - \cos x|$

1.5 DIFFERENTIAL EQUATIONS WITH PRESCRIBED SOLUTIONS

Sometimes it is necessary to derive a differential equation of order n which has a given general solution containing n essential arbitrary constants. This can be done (at least theoretically) by differentiating the given solution n times and then eliminating the parameters by algebraic manipulation of the original and the derived relations. In the process, various familiar identities often prove useful.

EXAMPLE 1

If a and b are arbitrary constants, find a second-order differential equation which has

(1) $$y = ae^x + b\cos x$$

as a general solution on $(-\infty, \infty)$.

By differentiating the given solution twice, we obtain

(2) $$y' = ae^x - b\sin x$$

(3) $$y'' = ae^x - b\cos x$$

Then, by adding and subtracting (1) and (3), we get

$$ae^x = \frac{y + y''}{2} \quad \text{and} \quad b\cos x = \frac{y - y''}{2}$$

Substitution of these into (2), multiplied through by $\cos x$, gives

$$y' \cos x = \frac{y + y''}{2}\cos x - \frac{y - y''}{2}\sin x$$

and finally

(4) $$(\cos x + \sin x)y'' - 2(\cos x)y' + (\cos x - \sin x)y = 0$$

Although direct substitution confirms that (4) has (1) as a solution, (4) is by no means the only equation of which (1) is a general solution. For not only are there different second-order equations, for instance,

(5) $$y'' \cos\left(x - \frac{\pi}{4}\right) - \sqrt{2}\, y' \cos x + y \cos\left(x + \frac{\pi}{4}\right) = 0$$

that are equivalent to (4), but there are also higher-order equations having (1) as a solution. In particular, if (3) is differentiated twice more, we obtain $y^{\text{iv}} = ae^x + b\cos x$ and, comparing this with (1), we see that the given solution (1) is also a general solution of the very simple equation

(6) $$y^{\text{iv}} = y$$

Since Eq. (6) is of fourth order, it presumably possesses general solutions containing more than two arbitrary constants, and it is easy to verify that

$$y = ae^x + b\cos x + ce^{-x} + d\sin x$$

does in fact satisfy (6) on $(-\infty, \infty)$ for all values of the four parameters a, b, c, and d.

Usually, a differential equation is referred to as a *differential equation of a given solution* only if the order of the equation is the same as the number of parameters explicitly appearing in that solution.

A number of geometric problems, especially those having to do with families of curves, also give rise to differential equations.

EXAMPLE 2

A **level line** on a surface is a curve all of whose points are at the same elevation above (or below) some horizontal reference plane. The projections of the level lines onto any horizontal plane are the **contour lines** on the topographic map of the surface. An excavation is planned whose contour lines form a family of concentric ellipses. The major axis of each ellipse is twice as long as the minor axis, and all major axes fall along the same line. The semiminor axis b of the level line s feet below the surface is $(100 - s^2/25)^{1/2}$ feet long (Fig. 1.5). Find an equation for the family of elliptic contours and a differential equation of the family.

Taking the ellipses in standard position with their major axes along the x axis, we obtain

$$\frac{x^2}{[2(100 - s^2/25)^{1/2}]^2} + \frac{y^2}{[(100 - s^2/25)^{1/2}]^2} = 1$$

as an equation of the family of contours. After clearing fractions and simplifying, this equation becomes

(7) $$25x^2 + 100y^2 = 10{,}000 - 4s^2$$

where the parameter s is restricted by the condition $0 \leq s < 50$.

Differentiating (7) implicitly, then dividing by 50, we find

$$x + 4yy' = 0 \tag{8}$$

as a differential equation of the family of contours. Because of the restriction previously noted on s, (7) is a general solution of (8) but *not* a complete solution. A complete solution of (8) is given by

$$x^2 + 4y^2 = k \tag{9}$$

where k may denote any positive number.

As an interesting sidelight, we observe that the total volume V of earth to be excavated is given by

$$V = 2\pi \int_0^{50} \left(100 - \frac{s^2}{25}\right) ds = \frac{20{,}000\pi}{81} \doteq 775.7 \text{ yd}^3$$

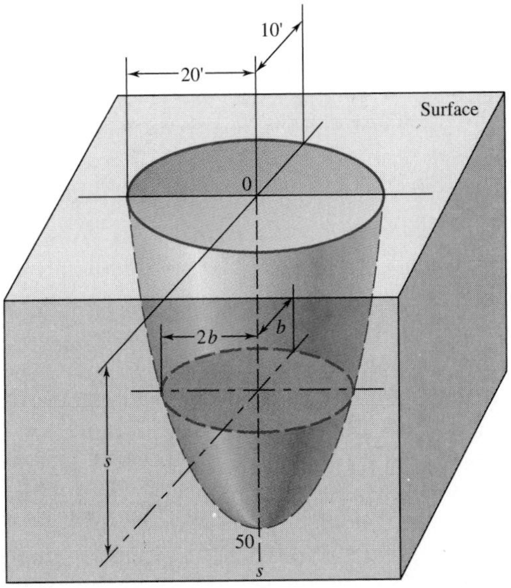

Figure 1.5
A proposed excavation whose contour lines are concentric ellipses.

EXERCISES

1. Find a second-order differential equation which has a general solution given by the implicit relation $a(x + y) + y + b = 0$, where a and b are parameters.
2. Convert Eq. (4) into Eq. (5).
3. (a) The voltage E across a resistor is proportional to the current i through it. Find a differential equation of which the voltage is a general solution involving i as the independent variable and the proportionality constant as a parameter.
   (b) The intensity I of a traveling wave is proportional to the square of its amplitude A. Find a differential equation of which the intensity is a general solution involving A as the independent variable and the proportionality constant as parameter.

Assuming a, b, c, and d are arbitrary constants, find a differential equation of which a general solution is the given expression and whose order is equal to the number of parameters in the expression.

4. $y = ae^{-x} + b\cos x$
5. $y = ae^{-x} + b\sin x$
6. $y = ae^{-2x} + be^{x}$
7. $y = ae^{-2t} + bte^{-2t}$
8. $y = ae^{-t} + be^{t} + ce^{2t}$
9. $y = 2ax + bx^2$
10. $y = a\cosh 2x + b\sinh 2x$
11. $y = \sin(ax + b)$
12. $ax + \ln|y| = y + b$
13. $x^2 + 2bxy + y^2 = d$
14. $y^2 + by + ct = 0$
15. $y = ae^{x} + b\ln|x| + cx + d$
16. $y = ae^{-x} + b/x$
17. $x^2 + axy + bx + c = 0$
18. Find a second-order differential equation of which a general solution containing the five parameters a, b, c, d, and $k \neq 0$ is

$$ky + a\cosh x + be^{x} + ce^{-x} + d\sinh x = 0$$

19. Find a third-order differential equation for which $c_1y + c_2 + c_3x + c_4x^{1/3} = 0$, $c_1 \neq 0$, is a general solution.

20. By solving $x^2 - \text{Tan}^{-1} ay + 2 = 0$ for a and then formally differentiating the resulting equation with respect to x, derive the differential equation $y' \tan(x^2 + 2) - 2xy\sec^2(x^2 + 2) = 0$. Does this equation have $x^2 - \text{Tan}^{-1} ay + 2 = 0$ as a general solution? Explain.

21. A curve within a thin plate along which the temperature is constant is called an **isothermal line,** or **isotherm.** The isotherms of a flat plate so large that it may be assumed to coincide with the xy plane make up the family of all equilateral hyperbolas in standard position. Find a differential equation of the isotherms.

22. Find differential equations of the following families of curves.

(a) All circles centered at the origin.
(b) All circles in the xy plane.
(c) All ellipses in standard position.
(d) All hyperbolas in standard position.
(e) All circles centered on a line of slope m through the origin. Discuss the special cases for which $m = 0$, $m = 1$, and $m = \infty$.

23. Find a second-order differential equation which has as a general solution an expression that defines the family of parabolas tangent to the x axis and having their axes parallel to the y axis.

24. Find a first-order differential equation which has as a general solution an expression that defines the family F of all lines tangent to the parabola defined by $2y = x^2$. Verify that $2y = x^2$ is also a solution of the required equation even though the parabola it defines is not a member of F.

25. Three two-dimensional waves are instantaneously superimposed. One of them has an arbitrary constant displacement $y_1 = a$. Another has a displacement y_2 which varies as $x \sin x$, whereas the displacement y_3 of the remaining wave varies inversely as e^x. Find a third-order differential equation of which the *instantaneous* resultant displacement $y_1 + y_2 + y_3$ is a general solution.

1.6 EXISTENCE AND UNIQUENESS OF SOLUTIONS

Having already encountered differential equations which have no solution, it is natural to ask if it is possible to tell when a solution actually exists. As long as we work exclusively with differential equations for which we can find solutions, the general question of existence of solutions is of course irrelevant. Moreover, if complete solutions can be found, matters concerning uniqueness may be quickly settled. Most differential equations are of such a nature, however, that they can be solved, if at all, only through the use of a computer. But even a computer is of no avail on a problem that has no solution. So before programming a computer or engaging in an analytic investigation for the purpose of solving an unfamiliar equation, it may be wise to first seek answers to certain fundamental questions.

With regard to first-order equations, the fundamental questions are these. Under what conditions does the initial-value problem

$$(1) \qquad y' = f(x, y);\ y(x_0) = y_0$$

have a solution? If a solution exists, for what values of x is it defined? And if there is a solution, is it unique? Answers to these questions are contained in the following theorem,† originally credited to the French mathematician Emile Picard (1856–1941).

THEOREM 1 If both $f(x, y)$ and $f_y(x, y)$ are continuous on a closed rectangle R: $|x - x_0| \le a$, $|y - y_0| \le b$; if $|f(x, y)| \le M$ in R; and if h is the smaller of the two numbers a and b/M; then there exists a unique solution of the initial-value problem $y' = f(x, y)$; $y(x_0) = y_0$, on the interval $|x - x_0| \le h$.

Under the hypotheses of this theorem, the *existence* and *uniqueness* of a solution $y = y(x)$ to the initial-value problem (1) are guaranteed on an interval $x_0 - h \le x \le x_0 + h$ about x_0. In any particular case, the truth of each premise may be examined directly. Of course, we must not read more into the theorem than it actually says. Theorem 1 does not say that the initial-value problem

†For a proof of Theorem 1 see, for instance, C. Ray Wylie, *Differential Equations,* pp. 532–540, McGraw-Hill, New York, 1979.

(1) cannot have a unique solution on an interval around x_0 if the hypotheses of the theorem are not all satisfied. In other words, the hypotheses of Theorem 1 are sufficient but not necessary conditions for the conclusion of the theorem to hold.

EXAMPLE 1

Direct substitution confirms that on $(-\infty, \infty)$, both x and $\ln e^x$ are solutions of the initial-value problem $y' = 1$; $y(1) = 1$.

For all positive real numbers a and b, both $f(x, y) = 1$ and $f_y(x, y) = 0$ are continuous on the closed rectangle R: $|x - 1| \le a$, $|y - 1| \le b$, and $|f(x, y)| \le 1 = M$ in R. Since a and b are unbounded, so too is $h = \min(a, b/M = b)$. It follows from Theorem 1 that on every interval $-h \le x \le h$, the solution of $y' = 1$; $y(1) = 1$, is unique. Consequently, x and $\ln e^x$ represent the same function. In other words, for every real number x, $x = \ln e^x$; that is, $x \equiv \ln e^x$.

EXAMPLE 2

If $f(x, y) = 2xy/(1 + x^2y^2)$ on the rectangle R: $|x| \le \frac{1}{2}$, $|y| \le 1$, verify that the initial-value problem $y' = f(x, y)$; $y(0) = 0$, has a unique solution and determine an interval centered at $x_0 = 0$ over which the solution is defined.

It is evident at once that $(0, 0)$ is the center of R, and it is readily confirmed that both f and f_y are continuous over R. To obtain a bound for $|f(x, y)|$ on R, we observe that

$$|f(x, y)| = \frac{|2xy|}{|1 + x^2y^2|} = \frac{2|x||y|}{1 + x^2y^2} \le \frac{2(\frac{1}{2})(1)}{1} = 1 = M$$

The smaller of the two numbers $a = \frac{1}{2}$ and $b/M = \frac{1}{1} = 1$ is $\frac{1}{2}$. With $h = \frac{1}{2}$, all the hypotheses of Theorem 1 are seen to hold. Therefore, a unique solution of the given initial-value problem $y' = 2xy/(1 + x^2y^2)$; $y(0) = 0$, on $-\frac{1}{2} \le x \le \frac{1}{2}$ is guaranteed by the theorem.

EXAMPLE 3

It is instructive to reconsider Eq. (1) of Sec. 1.4 in the light of Theorem 1. For this equation, we have $f(x, y) = 2y/x$, and neither f nor f_y is defined at $x = 0$. Hence over an interval containing $x = 0$, neither the existence nor the uniqueness of a solution of $xy' = 2y$ is guaranteed by Theorem 1. Actually, as our earlier discussion pointed out, this equation does have solutions which are defined for all values of x. However, as Fig. 1.4*c* illustrates, over any interval which contains $x = 0$, the solution curve which passes through a given point (x_0, y_0), for example $(1, 1)$, is not unique. On the other hand, over any interval which contains x_0 but does not contain $x = 0$, the solution curve passing through the given point (x_0, y_0) is unique.

EXAMPLE 4

After its chutes open, a reentry vehicle, or capsule, of mass m and volume V falls straight down into the ocean and sinks along a vertical path. Impact occurs at time t_0, at which instant the velocity of the capsule is given by $v(t_0) = v_0 > 0$, and a release mechanism frees the chutes. The value of v_0 is so large and the dimensions of the capsule so small that submersion may be assumed to take place instantaneously. While in the atmosphere the vehicle is retarded by a force $f_1 = -k_1 v$ due to friction and the drag of the chutes, whereas in the water the frictional force alone is $f_2 = -k_2 v$, where $k_2 > k_1 > 0$. Taking w as the weight per unit volume of ocean water, find an initial-value problem which is satisfied by v and determine whether or not Theorem 1 guarantees a unique solution of the problem.

First, we recognize that the velocity is positive when the motion is downward. Applying **Newton's second law of motion,**† mass × acceleration = force, we find that the velocity $v = v(t)$ satisfies the initial-value problem

$$(2) \qquad m\frac{dv}{dt} = \begin{cases} mg - k_1 v & t < t_0 \\ mg - k_2 v - wV & t > t_0 \end{cases}; \quad v(t_0) = v_0$$

†Named in honor of Sir Isaac Newton (1642–1727, English), generally acclaimed the greatest mathematician of all time. Differential calculus is primarily the creation of Newton and Gottfried Wilhelm Leibniz (1646–1716, German).

in which g denotes the acceleration of gravity and $-wV$ is the buoyant force of the water upon the capsule, as given by *Archimedes' principle.*† In the notation of Eq. (1) and Theorem 1, (2) becomes

$$v' = f(t, v);\ v(t_0) = v_0 \tag{3}$$

where

$$f(t, v) = \begin{cases} g - k_1 v/m & t < t_0 \\ g - (k_2 v + wV)/m & t > t_0 \end{cases} \quad \text{and} \quad f_v(t, v) = \begin{cases} -k_1/m & t < t_0 \\ -k_2/m & t > t_0 \end{cases}$$

Clearly, f_v is not continuous over any rectangle centered at (t_0, v_0). Therefore the first premise of Theorem 1 fails to hold and, consequently, a unique solution of (3) is not guaranteed by the theorem, though physical intuition suggests that there can be only one solution.

EXERCISES

If possible, find a solution of each of the following initial-value problems and state whether or not it is unique.

1. $y' = e^{\sin x} \cos x;\ y(\pi) = 2$
2. $y' = \pi(\cos \pi x - \pi \sin \pi x);\ y(1) = -\pi$
3. $y' = \cos x \sin \sin x;\ y(-\pi) = 0$
4. $y' = \ln x;\ y(e) = -1$
5. $y' = \mathrm{Sin}^{-1} x;\ y(0) = 5$
6. $y' = \mathrm{Cos}^{-1} x;\ y(0) = \frac{1}{2}$
7. $y' = 1 + y^2;\ y(\pi/4) = 1$ (Solve by inspection.)
8. $(\cos x)\, dx - (\sin x)\, dy = 0;\ y(\pi/2) = 0$ (Solve for y'.)
9. $x\, dx - y\, dy = 0;\ y(0) = 0$ (Differentiate $x^2 - y^2$.)
10. $(x + 1)y' = 2y;\ y(-1) = 0$ [Differentiate $y/(x + 1)^2$.]
11. $(x + 1)y' = 2y;\ y(0) = 1$ [Differentiate $y/(x + 1)^2$.]
12. $x^2y' = xy' + 2x - 1;\ y(-1) = 2 \ln 2$ (Solve for y' to integrate.)
13. $x^2y' = xy' + 2x - 1;\ y(1) = 3$ (Solve for y' to integrate.)
14. $x^2y' = 2(y + 1);\ y(0) = 1$ (Is there a solution?)
15. $x^2y' = 2(y + 1);\ y(2) = -1$ (Seek a constant solution.)
16. $x^3y' = 2y;\ y(0) = 0$ [See Exercise 15(**c**), Sec. 1.3.]

In Exercises 17 through 23, find a complete solution of the differential equation. If possible, also determine a solution which satisfies the given initial conditions.

17. $y'' = \mathrm{sech}^2 x;\ y(0) = y'(0) = 0$
18. $y'' + 2x/(1 + x^2)^2 = 0;\ y(0) = y'(0) = 0$
19. $y'' - x/(1 - x^2)^{3/2} = 0;\ y(2) = y'(2) = 0$
20. $y'' = e^x + 4 \ln x;\ y(1) = -3,\ y'(1) = -4$
21. $y'' = 2 \cos x - x \sin x;\ y(0) = y'(0) = 0$
22. $y'' + 6xe^{-x^2} = 0;\ y(0) = 0,\ y'(0) = 4$
23. $(d/dx)(xy') = 16x^3 + 1;\ y(1) = 2,\ y'(1) = 5$
24. Consider the differential equation $y' + q(x)y = 0$ having $y = y(x)$ as a solution over an interval $x_1 \le x \le x_2$ on which q is continuous. If the graph of the solution intersects the x axis between x_1 and x_2, (**a**) identify the solution, (**b**) describe its solution curve, (**c**) state whether the solution is unique, (**d**) explain your answer to Part (**c**).

For the given values of x_0 and y_0 and for the given function f defined on the rectangle R, determine an interval centered at x_0 over which a unique solution of the initial-value problem $y' = f(x, y);\ y(x_0) = y_0$, is defined.

25. $x_0 = y_0 = 0;\ f(x, y) = (1 + x + y)/(4 + x^2 + y^2)$ on R: $|x| \le 2,\ |y| \le 3$
26. $x_0 = y_0 = 0;\ f(x, y) = (1 + xy)/(1 + \cos xy)$ on R: $|x| \le 2,\ |y| \le 1$
27. $x_0 = y_0 = 1;\ f(x, y) = xy/(1 + x + y^2)$ on R: $|x - 1| \le 1,\ |y - 1| \le 1$
28. $x_0 = -1,\ y_0 = 1;\ f(x, y) = (1 - x - y)/(1 + x^2 + y^2)$ on R: $|x + 1| \le 2,\ |y - 1| \le 2$
29. Use Theorem 1 and the fact that on $(-\infty, \infty)$ both $\sin 2x$ and $2 \sin x \cos x$ are solutions of the initial-value problem $y' = 2 \cos 2x;\ y(0) = 0$, to verify that $\sin 2x \equiv 2 \sin x \cos x$.
30. Use Theorem 1 and the fact that on $(-\infty, \infty)$ both $\sin^2 x$ and $(1 - \cos 2x)/2$ are solutions of the initial-value problem $y' = \sin 2x;\ y(\pi/2) = 1$, to verify that $\sin^2 x \equiv (1 - \cos 2x)/2$.
31. Is Example 1 a proof of the identity $\ln e^x \equiv x$? Explain.

1.7 EXACT FIRST-ORDER EQUATIONS

We shall now begin a systematic study of how to recognize and solve several important kinds of first-order differential equations that cannot be solved at once by direct integration. Most of our

†Archimedes (287?–212 B.C., Greek) was the originator of the integral calculus and the greatest mathematician of antiquity.

equations will be expressible in either the so-called **standard differential form**

$$M(x, y)\,dx + N(x, y)\,dy = 0 \tag{1}$$

or in the **standard derivative form**

$$y' = f(x, y) \tag{2}$$

Clearly, (2) is equivalent to its standard differential form

$$f(x, y)\,dx + (-1)\,dy = 0$$

However, (1) will have the same solutions as its standard derivative form

$$y' = -\frac{M(x, y)}{N(x, y)}$$

only if no solutions are lost as a consequence of dividing (1) through by the function $N(x, y)$. Due caution must therefore be exercised in passing from any standard differential form to a corresponding standard derivative form of a first-order differential equation.

The left-hand member of (1), namely,

$$M(x, y)\,dx + N(x, y)\,dy$$

is called a **first-order differential form** in two variables. It is reminiscent of the **total differential**

$$df = \frac{\partial f}{\partial x}\,dx + \frac{\partial f}{\partial y}\,dy$$

associated with a suitably differentiable function $f(x, y)$. A differential form is said to be an **exact differential** if and only if at each point of some region it is the total differential of a function f. An *exact differential equation* is defined analogously.

DEFINITION 1 An equation $M(x, y)\,dx + N(x, y)\,dy = 0$ is an **exact differential equation** if and only if there is a function f such that $M = \partial f/\partial x$ and $N = \partial f/\partial y$ throughout some region.

Every exact equation can be written in the form

$$\frac{\partial f}{\partial x}\,dx + \frac{\partial f}{\partial y}\,dy = 0$$

hence as

$$d[f(x, y)] = 0$$

from which it follows that

$$f(x, y) = k$$

is a solution containing the arbitrary constant k.

When $M(x, y)$ and $N(x, y)$ are sufficiently simple, it is possible to tell by inspection whether or not there exists a function f with the property that

$$\frac{\partial f}{\partial x} = M(x, y) \qquad \text{and} \qquad \frac{\partial f}{\partial y} = N(x, y) \tag{3}$$

In fact, even when no such function is apparent, it is sometimes possible to disregard the relations (3) and solve an exact equation directly. Our next example illustrates this possibility.

EXAMPLE 1

By an obvious regrouping of terms, the differential equation $(y^3 + 2x)\,dx + (3xy^2 + 1)\,dy = 0$ can be written successively as

$$(y^3\,dx + 3xy^2\,dy) + 2x\,dx + dy = 0$$

$$d(xy^3) + d(x^2) + dy = 0$$

$$d(xy^3 + x^2 + y) = 0$$

Hence a general solution of the equation is $xy^3 + x^2 + y = k$. With $f(x, y) = xy^3 + x^2 + y$, it is easy (though now unnecessary) to verify the relations (3).

Although some exact equations are easy to identify and solve, it is generally impossible to tell by inspection whether a given first-order differential equation is exact or not. What we need is a straightforward test for exactness and a routine procedure for solving an equation when it is exact. Here is a simple criterion for exactness.

THEOREM 1 If $\partial M/\partial y$ and $\partial N/\partial x$ are continuous in a rectangular region R, then the differential equation $M(x, y)\,dx + N(x, y)\,dy = 0$ is exact if and only if

$$\frac{\partial M}{\partial y} = \frac{\partial N}{\partial x} \qquad \text{in } R$$

Once an equation is known to be exact, we can apply Definition 1 to solve it. An example will be helpful before we prove Theorem 1.

EXAMPLE 2

To test the equation $(4x^3 + 6e^y + 2y\cos 2x)\,dx + (3y^2 + 6xe^y + \sin 2x)\,dy = 0$ for exactness, we use Theorem 1 with $M(x, y) = 4x^3 + 6e^y + 2y\cos 2x$ and $N(x, y) = 3y^2 + 6xe^y + \sin 2x$. Since $\partial M/\partial y = 6e^y + 2\cos 2x = \partial N/\partial x$, the given equation is exact. To solve it, we apply Definition 1 which asserts there is a function $f(x, y)$ such that

$$\frac{\partial f}{\partial x} = M(x, y) = 4x^3 + 6e^y + 2y\cos 2x$$

Holding y fixed and integrating with respect to x, we have

$$f(x, y) = x^4 + 6xe^y + y\sin 2x + c(y)$$

where, according to Definition 1, $c(y)$ must be a function such that

$$\frac{\partial f}{\partial y} = 6xe^y + \sin 2x + c'(y) = N(x, y) = 3y^2 + 6xe^y + \sin 2x$$

From this, we conclude that $c'(y) = 3y^2$, $c(y) = y^3 + a$, where a is a parameter, and

$$f(x, y) = x^4 + 6xe^y + y\sin 2x + y^3 + a$$

Equating $f(x, y)$ to a parameter b and then setting $k = b - a$, we obtain as a general solution of the given differential equation

$$x^4 + 6xe^y + y\sin 2x + y^3 = k$$

Differentiation of this relation with respect to x yields the original equation.

The following proof of Theorem 1, based in part on Definition 1, gives two alternative formulas either of which may be used to solve an exact differential equation.

PROOF If the given equation is exact, there is a function f such that $M = \partial f/\partial x$ and $N = \partial f/\partial y$. Hence $\partial M/\partial y = \partial^2 f/\partial y\,\partial x$ and $\partial N/\partial x = \partial^2 f/\partial x\,\partial y$. Since $\partial M/\partial y$ and $\partial N/\partial x$ are continuous by hypothesis, the cross-derivatives $\partial^2 f/\partial y\,\partial x$ and $\partial^2 f/\partial x\,\partial y$ are also continuous and therefore equal. It follows that $\partial M/\partial y = \partial N/\partial x$. This proves the "only if" part of the theorem which says that $M(x, y)\,dx + N(x, y)\,dy = 0$ is exact only if $\partial M/\partial y = \partial N/\partial x$.

To prove the converse, which is the "if" part of the theorem, we utilize Definition 1, just as we did in Example 2, to show equivalently that if $\partial M/\partial y = \partial N/\partial x$, then there is a function f such that $\partial f/\partial x = M$ and $\partial f/\partial y = N$. To do this let us first integrate $M(x, y)$ with respect to x, holding y fixed. Introducing the dummy variable t gives us the expression

$$f(x, y) = \int_{x_0}^{x} M(t, y)\,dt + c(y) \qquad (x_0, y) \text{ in } R \tag{4}$$

in which, since the integration is done with respect to x while y is held constant, the usual integration constant is in this instance actually a function of y to be determined. Clearly, $\partial f/\partial x = M(x, y)$, and our proof will be complete if we can determine $c(y)$ so that $\partial f/\partial y = N(x, y)$.

Now, observing that under the hypothesis that $\partial M/\partial y$ is continuous, the operations of integrating with respect to x and differentiating with respect to y can legitimately be interchanged, and recalling our current assumption that $\partial M/\partial y = \partial N/\partial x$, we have from (4)

$$\begin{aligned}\frac{\partial f}{\partial y} &= \frac{\partial}{\partial y}\int_{x_0}^{x} M(t, y)\,dt + c'(y) = \int_{x_0}^{x}\frac{\partial M(t, y)}{\partial y}\,dt + c'(y) = \int_{x_0}^{x}\frac{\partial N(t, y)}{\partial t}\,dt + c'(y)\\ &= N(x, y) - N(x_0, y) + c'(y)\end{aligned}$$

Thus $\partial f/\partial y$ will equal $N(x, y)$, as required, if $c(y)$ is determined so that $c'(y) = N(x_0, y)$, that is, if

$$c(y) = \int_{y_0}^{y} N(x_0, t)\,dt \qquad (x_0, y_0) \text{ in } R$$

We have thus shown that if $\partial M/\partial y = \partial N/\partial x$, then

$$f(x, y) = \int_{x_0}^{x} M(t, y)\,dt + \int_{y_0}^{y} N(x_0, t)\,dt \tag{5}$$

is a function such that $\partial f/\partial x = M$, $\partial f/\partial y = N$, and even more important,

$$df = \frac{\partial f}{\partial x}\,dx + \frac{\partial f}{\partial y}\,dy = M(x, y)\,dx + N(x, y)\,dy$$

This establishes the "if" assertion of the theorem, and our proof is complete.

The preceding proof tells us that when the equation $M(x, y)\,dx + N(x, y)\,dy = 0$ is exact, its left-hand member is the total differential of the function f defined by (5). Under the same hypotheses, the "if" part of Theorem 1 may be proved by first integrating $N(x, y)$ with respect to y, holding x fixed. In this way we find that the function g represented by

$$g(x, y) = \int_{x_0}^{x} M(t, y_0)\,dt + \int_{y_0}^{y} N(x, t)\,dt \tag{6}$$

on R also has $M(x, y)\,dx + N(x, y)\,dy$ as its total differential.

It follows that an exact differential equation can be solved by integration. The following corollary gives two ways of writing down solutions.

COROLLARY 1 If the differential equation $M(x, y)\,dx + N(x, y)\,dy = 0$ is exact in a rectangular region R, then for (x_0, y_0) an arbitrary point of R, either

(7)
$$f(x, y) = \int_{x_0}^{x} M(t, y)\,dt + \int_{y_0}^{y} N(x_0, t)\,dt = k$$

or

(8)
$$g(x, y) = \int_{x_0}^{x} M(t, y_0)\,dt + \int_{y_0}^{y} N(x, t)\,dt = c$$

is a general solution of the differential equation. For suitable values of the constants c and k, either solution determines the same family of integral curves.

By judiciously choosing the point (x_0, y_0), it is generally possible to decide in any particular case whether (7) or (8) is the more advantageous when it comes to performing the actual integrations involved.

EXAMPLE 3

Work Example 1 using the results of Theorem 1 and its corollary.

To test for exactness, we first make the identifications $M(x, y) = y^3 + 2x$ and $N(x, y) = 3xy^2 + 1$. Since $\partial M/\partial y = 3y^2 = \partial N/\partial x$, the equation is exact and a general solution can be found by using either formula of Corollary 1. Taking $(x_0, y_0) = (0, 0)$, we have $M(t, 0) = 2t$ and $N(0, t) = 1$. Hence, using formula (7), the solution is

$$\int_0^x (y^3 + 2t)\,dt + \int_0^y 1\,dt = (y^3 t + t^2)|_0^x + t|_0^y = y^3 x + x^2 + y = k$$

Similarly, using formula (8), we obtain

$$\int_0^x 2t\,dt + \int_0^y (3xt^2 + 1)\,dt = t^2|_0^x + (xt^3 + t)|_0^y = x^2 + xy^3 + y = c$$

As expected, the two solutions are the same.

EXAMPLE 4

Solve the equation $(x^3 + \cos y + 1/x)\,dy = (y/x^2 - 3yx^2)\,dx$.

By inspection, $y = 0$ is a solution on the two separate intervals $(-\infty, 0)$ and $(0, \infty)$. To find a general solution, we transpose to transform the equation into the standard differential form $(3yx^2 - y/x^2)\,dx + (x^3 + \cos y + 1/x)\,dy = 0$. This equation is exact because

$$\frac{\partial M}{\partial y} = \frac{\partial(3yx^2 - y/x^2)}{\partial y} = 3x^2 - \frac{1}{x^2} = \frac{\partial(x^3 + \cos y + 1/x)}{\partial x} = \frac{\partial N}{\partial x}$$

Since neither M nor N is continuous where $x = 0$, we cannot take $x_0 = 0$, but we can set $(x_0, y_0) = (1, 0)$. This gives $M(t, 0) = 0$ and $N(1, t) = 2 + \cos t$. Formulas (7) and (8) then become

$$\int_1^x \left(3yt^2 - \frac{y}{t^2}\right) dt + \int_0^y (2 + \cos t)\,dt = k \qquad \text{and} \qquad \int_0^y \left(x^3 + \cos t + \frac{1}{x}\right) dt = c$$

respectively. Although each of these gives us

$$x^3y + \frac{y}{x} + \sin y = c = k$$

as a general solution, the integrations in the second formula are somewhat easier to perform.

EXERCISES

Show that the following equations are exact and solve each one.

1. $(3x^2 + y^2)\,dx + 2xy\,dy = 0$
2. $(y^2 - 1)\,dx + (2xy - \sin y)\,dy = 0$
3. $(2xy + x^3)\,dx + (x^2 + y^2)\,dy = 0$
4. $(ye^x - \sin x)\,dx = (y^2 - e^x)\,dy$
5. $(1 + \cos x)\,dx - dy = 0$
6. $(3x^2 - 6xy)\,dx - (3x^2 + 2y)\,dy = 0$
7. $y^2\,dx = (2 + 3y^2 - 2xy)\,dy$
8. $(2xy^4 + \sin y)\,dx + (4x^2y^3 + x\cos y)\,dy = 0$
9. $(2xy + e^y)\,dx + (x^2 + xe^y)\,dy = 0$
10. $3xy^2\,dy = [x^{-1} + (1 - x^2)^{-1/2} - y^3]\,dx$
11. $(x\sqrt{x^2 + y^2} + y)\,dx + (y\sqrt{x^2 + y^2} + x)\,dy = 0$
12. $[\cos(x^2 + y) - 3xy^2]\,dy +$ $[2x\cos(x^2 + y) - y^3]\,dx = 0$
13. $(2x + \mathrm{Tan}^{-1}\,y)\,dx + [x/(y^2 + 1)]\,dy = 0$
14. $(2x + \cosh xy)\,dx + [(xy\cosh xy - \sinh xy)/y^2]\,dy = 0$
15. $(\sin y - y\sin xy)\,dx + (x\cos y - x\sin xy)\,dy = 0$
16. $(1 + \ln xy)\,dx + (1 + x/y)\,dy = 0;\ x > 0,\ y > 0$
17. $(3x^2 + y/x)\,dx + (\ln x + 2y)\,dy = 0;\ x > 0$
18. $(e^x + y/x)\,dx + (\ln x + 1/y)\,dy = 0;\ x > 0$

Find a solution of each of the following equations that satisfies the given initial condition.

19. $(x - y)\,dx + (2y - x)\,dy = 0;\ y(0) = 1$
20. $(2xy - 1)\,dx + x^2\,dy = 0;\ y(1) = 2$
21. $(2xy + e^y)\,dx + (x^2 + xe^y)\,dy = 0;\ y(1) = \ln 2$
22. $(x + y)\,dx + (x + 2y)\,dy = 0;\ y(2) = 3$
23. $(x^2 + y^2)\,y' + (2xy + 1) = 0;\ y(2) = -2$
24. $(x^2 - 2xy)\,dx = (x^2 + 2y)\,dy;\ y(0) = -1$
25. $(x^2 + y/x)\,dx + \ln x\,dy = 0;\ y(e) = -e^3/3$
26. $(\operatorname{sech}\ln x)\,dy - [(y\operatorname{sech}\ln x\tanh\ln x)/x]\,dx = 0;$ $y(1) = 7$
27. $\tan y + y/(1 + x^2) =$ $(2\,\mathrm{Tan}^{-1}\,y - \mathrm{Tan}^{-1}\,x - x\sec^2 y)y';\ y(0) = 1$
28. $(ye^x + 2e^x + y^2)\,dx + (e^x + 2xy)\,dy = 0;\ y(1) = -e$
29. Substitute your solution to Exercise 19 into the differential equation and verify that the equation reduces to an identity.
30. Prove the "if" part of Theorem 1 by first integrating $N(x, y)$ with respect to y.
31. Show that the arbitrary constants x_0 and y_0 appearing in the formulas of Corollary 1, Theorem 1, add no generality to the solution. *Hint:* Consider the partial derivatives with respect to x_0 and y_0 of the left-hand side of the formulas.
32. Show by an example that the integral formulas of Corollary 1, Theorem 1, do not necessarily define a solution of the exact equation $M(x, y)\,dx + N(x, y)\,dy = 0$ for *all* values of c or k.

1.8 INTEGRATING FACTORS FOR FIRST-ORDER EQUATIONS

In addition to being useful in connection with exact equations, the concept of an *exact differential* forms the basis of a fundamental solution technique that extends to various other types of first-order differential equations as well. One important possibility is that an equation which is not exact can be made exact by multiplying it through by a suitable expression.

For example, if the exact equation $2xy^3\,dx + 3x^2y^2\,dy = 0$ is simplified by the natural process of dividing out the common factor xy^2, the resulting equation, namely, $2y\,dx + 3x\,dy = 0$, is *not* exact. However, it is equivalent to the original equation because the only solutions we might have lost when we divided by xy^2 are the two trivial solutions $x = 0$ and $y = 0$ and, by inspection, they are solutions of each equation. Of course, the reduced equation can be restored to its original (exact) form by multiplying it through by xy^2. This illustrates the general result† that every first-order equation which possesses a family of solutions can be made exact by multiplying it by an

†See, for instance, M. Golomb and M. E. Shanks, *Elements of Ordinary Differential Equations,* 2d ed., pp. 52–53, McGraw-Hill, New York, 1965.

appropriate factor called an **integrating factor.** In general, the determination of an integrating factor for a given equation may be difficult or even impossible. However, as the following examples show, in particular cases an integrating factor can be found by inspection.

EXAMPLE 1

Show that $1/(x^2 + y^2)$ is an integrating factor for the equation $(x^2 + y^2 - x)\,dx - y\,dy = 0$ and then solve the equation.

The test provided by Theorem 1, Sec. 1.7, shows that in its present form the given equation is not exact. However, if it is multiplied by the indicated integrating factor, it can be written in the form

$$\left(1 - \frac{x}{x^2 + y^2}\right) dx - \frac{y}{x^2 + y^2}\,dy = 0$$

Testing again, we find that the last equation is exact, and we can now use Corollary 1, Theorem 1, Sec. 1.7, to find a general solution of it. However, it is simpler to observe that it can also be written

$$dx - \frac{x\,dx + y\,dy}{x^2 + y^2} = 0 \qquad \text{or} \qquad dx - \frac{1}{2}\,d[\ln (x^2 + y^2)] = 0$$

Hence, integrating, we obtain the general solution

$$x - \ln \sqrt{x^2 + y^2} = k$$

EXAMPLE 2

Find an integrating factor for the equation $y\,dx + (x^2y^3 + x)\,dy = 0$ and solve the equation.

Since the equation can be rewritten in the form

$$(y\,dx + x\,dy) + x^2y^3\,dy = 0$$

and since $y\,dx + x\,dy = d(xy)$, it is natural to divide the preceding equation by $x^2y^2 = (xy)^2$, getting $d(xy)/(xy)^2 + y\,dy = 0$. This equation can now be integrated by the power rule, and we have

$$-\frac{1}{xy} + \frac{y^2}{2} = k \tag{1}$$

Note that there is no value of k for which this solution implicitly defines the trivial solutions $x = 0$ and $y = 0$ of the original differential equation. These solutions were lost when we divided by x^2y^2.

Upon setting $k = 1/2c$ and clearing of fractions, our general solution becomes

$$c(xy^3 - 2) = xy \tag{2}$$

This is a general solution of the given equation relative to which the solutions $x = 0$ and $y = 0$ are not singular since they now correspond to the value $c = 0$. But now the solutions $y = (2/x)^{1/3}$ and $x = 2/y^3$, which correspond to the value $k = 0$ in (1), are singular relative to (2).

Thus the given equation is solved either by (1) together with the trivial functions $x = 0$ and $y = 0$ or by (2) in conjunction with the functions $y = (2/x)^{1/3}$ and $x = 2/y^3$.

EXAMPLE 3

Find an integrating factor for the equation $x\,dy - y\,dx = (4x^2 + y^2)\,dy$ and solve the equation.

In this equation the terms on the left seem related equally well to

$$d\left(\frac{y}{x}\right) = \frac{x\,dy - y\,dx}{x^2} \qquad \text{or} \qquad d\left(\frac{x}{y}\right) = \frac{y\,dx - x\,dy}{y^2}$$

There is one distinct difference, however. Whereas $x = 0$ is not a solution of the given equation, $y = 0$ is. Hence no solution is lost if we divide by x^2 rather than y^2. Consequently, we shall pursue the first option and multiply by $1/x^2$. This gives

$$d\left(\frac{y}{x}\right) = \left(4 + \frac{y^2}{x^2}\right) dy$$

which is equivalent to the original equation on every interval of the x axis which excludes the origin. Division by $4 + y^2/x^2$ gives us

$$\frac{d\,(y/x)}{4 + (y/x)^2} = dy \qquad \text{or} \qquad \frac{1}{2} \times \frac{d\,(y/2x)}{1 + (y/2x)^2} = dy$$

or finally

$$d\left[\frac{1}{2}\,\mathrm{Tan}^{-1}\left(\frac{y}{2x}\right) - y\right] = 0$$

Integrating, we obtain

$$\frac{1}{2}\,\mathrm{Tan}^{-1}\left(\frac{y}{2x}\right) = y + k$$

as a general solution of the given differential equation.

The results of the last three examples suggest the following observations, which are often helpful.

1. If a first-order differential equation contains the combination $x\,dx + y\,dy = \frac{1}{2}\,d(x^2 + y^2)$, try some function of $x^2 + y^2$ as a multiplier to make the equation integrable.
2. If a first-order differential equation contains the combination $x\,dy + y\,dx = d(xy)$, try some function of xy as a multiplier to make the equation integrable.
3. If a first-order differential equation contains the combination $x\,dy - y\,dx$, try $1/x^2$ or $1/y^2$ as a multiplier to make the equation integrable. If neither of these works, try $1/xy$ or $1/(x^2 + y^2)$, or some function of these expressions, as an integrating factor, remembering that $d \ln (y/x) = (x\,dy - y\,dx)/xy$ and $d\,\mathrm{Tan}^{-1}(y/x) = (x\,dy - y\,dx)/(x^2 + y^2)$.

Two other possibilities for finding an integrating factor for the general first-order differential equation $M(x, y)\,dx + N(x, y)\,dy = 0$ are indicated in the following theorem.

THEOREM 1

a. If $(\partial M/\partial y - \partial N/\partial x)/N$ is a function of x only, say $f(x)$, then $e^{\int f(x)\,dx}$ is an integrating factor of the equation $M\,dx + N\,dy = 0$.

b. If $(\partial N/\partial x - \partial M/\partial y)/M$ is a function of y only, say $g(y)$, then $e^{\int g(y)\,dy}$ is an integrating factor for the equation $M\,dx + N\,dy = 0$.

PROOF An equation $M(x, y)\,dx + N(x, y)\,dy = 0$ will have an exponential function $e^{\int f(x)\,dx}$ of the single variable x as an integrating factor if and only if $(e^{\int f(x)\,dx} M)\,dx + (e^{\int f(x)\,dx} N)\,dy = 0$ is an exact equation. According to Theorem 1, Sec. 1.7, this will be the case if and only if

$$e^{\int f(x)\,dx} M_y = e^{\int f(x)\,dx} N_x + f(x) e^{\int f(x)\,dx} N$$

or, dividing out $e^{\int f(x)\,dx}$, $(M_y - N_x)/N = f(x)$, which proves the first assertion of the theorem. Working similarly with the integrating factor $e^{\int g(y)\,dy}$ establishes the second assertion.

EXAMPLE 4

Find an integrating factor for $e^x \sin y\, dx = 2(\csc y - e^x \cos y)\, dy$ and solve the equation.

As written, the equation is not exact because

$$\frac{\partial M}{\partial y} = \frac{\partial}{\partial y}(e^x \sin y) = e^x \cos y \neq \frac{\partial N}{\partial x} = \frac{\partial}{\partial x}[2(e^x \cos y - \csc y)] = 2e^x \cos y$$

However,

$$\frac{\partial N/\partial x - \partial M/\partial y}{M} = \frac{2e^x \cos y - e^x \cos y}{e^x \sin y} = \frac{\cos y}{\sin y} = g(y)$$

Hence, by Part **b** of the last theorem, an integrating factor is

$$e^{\int (\cos y)/(\sin y)\, dy} = e^{\ln \sin y} = \sin y$$

Except where $\sin y = 0$, the integral curves of the given equation and the exact equation

$$e^x \sin^2 y\, dx + 2(\sin y)(e^x \cos y - \csc y)\, dy = 0$$

are identical.

To solve the latter equation, we take $(x_0, y_0) = (0, \pi/2)$ and utilize Eq. (7) of Corollary 1, Theorem 1, Sec. 1.7, which gives

$$\int_0^x e^t \sin^2 y\, dt + \int_{\pi/2}^y 2(\cos t - \csc t) \sin t\, dt = (e^x - 1)\sin^2 y + \sin^2 y - 1 - 2y + \pi = k$$

Setting $c = k + 1 - \pi$ and simplifying, we obtain

$$e^x \sin^2 y - 2y = c \qquad y \neq 0$$

as a general solution of the original differential equation.

Should none of the preceding observations (or Theorem 1) yield an integrating factor of a given first-order equation, other methods, the most important of which we shall discuss in the following sections, must be tried.

EXERCISES

Solve each of the following equations by first multiplying by a suitable factor u (of either one or two variables) and then integrating.

1. $x\, dx + y\, dy = e^y\sqrt{x^2 + y^2}\, dy$
2. $y(1 + xy)\, dx + (2y - x)\, dy = 0$
3. $(x + x^2y + y^3)\, dx + (y + xy^2 + x^3)\, dy = 0$
4. $(xy^2 + y)\, dx + x(1 - xy)\, dy = 0$
5. $3(y^4 + 1)\, dx + 4xy^3\, dy = 0$
6. $x\, dy + 3y\, dx = xy\, dy$
7. $(x^2 + y^2 + 2x)\, dy = 2y\, dx$
8. $x\, dy - y\, dx + x^2\, dx = 0$
9. $(3x - y)\, dx + (3y + x)\, dy = 0$
10. $y\, dx + y^2\, dx = x\, dy$
11. $2y\, dx + (3y - 2x)\, dy = 0$
12. $(x^2 - y^2)\, dy = 2xy\, dx$

Solve each of the following equations.

13. $x\, dy + y\, dx = (1/y)\, dx - (1/x)\, dy$
14. $2xy^2\, dx + (3x^2y + 2)\, dy = 0$
15. $y(e^{x/y} + \sin xy + xy \cos xy)\, dx = [(x - y)e^{x/y} + x^2y \cos xy]\, dy$
16. $2x \ln |y|\, dx + [(1 + x^2)/y]\, dy = 0$
17. $y(x^2 + \tan y/x)\, dx + x(x^2 - \tan y/x)\, dy = 0$
18. $2xy^3(y\, dx + x\, dy) = (y\, dx - x\, dy)\sin x/y$
19. $e^x \sin y\, dx - (y - 2e^x \cos y)\, dy = 0$
20. $(y \tanh x - \cosh x)\, dx + dy = 0$
21. $(\ln x - x/y)\, dy = (\ln y - y/x)\, dx$
22. $(1 - 3y)\, dx - x\, dy = 0$
23. $y(1 + \ln xy)\, dx + x\, dy = 0$
24. $\sqrt{x^2 + y^2}\, dx = x\, dy - y\, dx$. *Hint:* Multiply by $1/(x\sqrt{x^2 + y^2})$.

25. $(5xy^2 - 4y \sin y)\, dx = (xy \cos y + 3x \sin y - 4x^2y)$.
Hint: Multiply by $x^m y^n$.
26. Solve the equation $(xy^2 - y)\, dx + (x^2y - x)\, dy = 0$ first by integrating it as an exact equation and then by multiplying it by $1/x^2y^2$ before integrating it. Reconcile your results.
27. Show that if the equation of Exercise 26 is multiplied by any differentiable function of the product xy, it is still exact.
28. Show that $\phi(x, y)$ is an integrating factor of the differential equation $M(x, y)\, dx + N(x, y)\, dy = 0$ if and only if ϕ satisfies the partial differential equation

$$M\frac{\partial \phi}{\partial y} - N\frac{\partial \phi}{\partial x} + \left(\frac{\partial M}{\partial y} - \frac{\partial N}{\partial x}\right)\phi = 0$$

29. Show that $f(x, y) = k$ is a general solution of the differential equation $M(x, y)\, dx + N(x, y)\, dy = 0$ if and only if $M(\partial f/\partial y) - N(\partial f/\partial x) = 0$ holds identically.
30. Show that if $\phi(x, y)$ is an integrating factor leading to the solution $f(x, y) = k$ for the differential equation $M(x, y)\, dx + N(x, y)\, dy = 0$, then $\phi F(f)$ is also an integrating factor, where F is an arbitrary differentiable function.
31. Determine ϕ in Exercise 28 if ϕ is a function of **(a)** x only, **(b)** y only.

1.9 SEPARABLE FIRST-ORDER EQUATIONS

In many cases a first-order differential equation can be reduced by algebraic operations to the form

$$f(x)\, dx + g(y)\, dy = 0 \tag{1}$$

The distinguishing feature of this equation is that the variables x and y are *separated* from each other in such a way that x appears in just one of the nonzero terms and y in the other.

DEFINITION 1 A **separable differential equation** is a first-order ordinary equation that is algebraically reducible to a standard differential form in which each of the nonzero terms contains exactly one variable.

An equation of type (1) is exact. In fact, $\int f(x)\, dx + \int g(y)\, dy$ is a function whose partial derivatives with respect to x and y are $f(x)$ and $g(y)$, respectively. An alternative form of (1) is therefore $d[\int f(x)\, dx + \int g(y)\, dy] = 0$. For integrable functions f and g, this gives

$$\int f(x)\, dx + \int g(y)\, dy = c \tag{2}$$

as a general solution of (1) containing the parameter c. From (2) it is evident that (1) can be solved by simply integrating each term. It must be borne in mind, however, that the integrals which appear in (2) may be impossible to evaluate in terms of elementary functions and that numerical or graphical integration may be required before this solution can be put to practical use.

Incidentally, there is no need to express a separable differential equation in the standard differential form before integrating. The integrations may be performed just as soon as the equation has been written in such a way that each nonzero term contains only one variable.

EXAMPLE 1

To solve the differential equation $y' + 4x + (\sin x)/x = 6y^2y' \cos y^3 - 1/(1 + x^2)$, we first write y' as dy/dx and then multiply by dx to obtain

$$dy + 4x\, dx + \frac{\sin x}{x}\, dx = 6y^2 \cos y^3\, dy - \frac{1}{1 + x^2}\, dx$$

In the sense that each nonzero term of this equation contains only one variable, the variables are separated and we may integrate immediately. This gives

$$y + 2x^2 + \int \frac{\sin x}{x}\, dx = 2 \sin y^3 - \mathrm{Tan}^{-1} x + c$$

as a general solution of the original equation. The integral sign has been left in the third term because it is impossible to express $\int (\sin x)/x\, dx$ in terms of elementary functions.

One way of discovering whether or not a given equation is separable is to collect coefficients on the two differentials and see if the result can be put in the form

$$f(x)G(y)\, dx = F(x)g(y)\, dy \tag{3}$$

Another way is to solve for a derivative and compare the result with

$$\frac{dy}{dx} = M(x)N(y) \tag{4}$$

Both of the forms, (3) and (4), should be recognized as being separable.

A general solution of (3) can be found by first dividing by the product $F(x)G(y)$ to separate the variables and then integrating:

$$\int \frac{f(x)}{F(x)}\, dx = \int \frac{g(y)}{G(y)}\, dy + c$$

Similarly, (4) can be solved by first multiplying by dx, then dividing by $N(y)$, and subsequently integrating:

$$\int \frac{dy}{N(y)} = \int M(x)\, dx + c$$

Clearly, the process of solving a separable equation will often involve division by one or more expressions. In such cases the results are valid where the divisors are not equal to zero but may or may not be meaningful for values of the variables for which the division is undefined. Such values require special consideration and, as we shall see in the next example, may lead us to singular solutions.

EXAMPLE 2

Solve the differential equation $dx + xy\, dy = y^2\, dx + y\, dy$.

It is not immediately evident that this equation is separable. In any case, however, the best first step in solving an equation of this sort is to collect terms on dx and dy. This gives

$$(1 - y^2)\, dx = y(1-x)\, dy$$

which is of the form (3). Hence division by the product $(1 - x)(1 - y^2)$ will separate the variables as follows:

$$\frac{dx}{1 - x} = \frac{y\, dy}{1 - y^2}$$

To make certain that no solutions get lost as a consequence of this division, we check to see if $x = 1$, $y = -1$, or $y = 1$ is a solution of the *original* equation. All of them are, because when substituted for the corresponding variable, they satisfy the equation. Their solution curves are of course straight lines (Fig. 1.6). Solutions such as these, which are lost as a consequence of algebraic operations performed on an equation, are sometimes called **suppressed solutions.**

Having noted the suppressed solutions, we proceed to solve the separated equation. Multiplying by -2, then integrating, we obtain

$$2 \ln |1 - x| = \ln |1 - y^2| \pm \ln |\lambda|$$

Use of the plus sign in front of $\ln |\lambda|$ gives $\ln (1 - x)^2 = \ln |\lambda| \, |1 - y^2|$, which leads to the solution

$$(1 - x)^2 = \lambda(1 - y^2) \tag{5}$$

Use of the minus sign, on the other hand, gives

$$\lambda(1 - x)^2 = 1 - y^2 \tag{6}$$

as a general solution. In either of these solutions λ may take on any real value even though $\ln |\lambda|$ was undefined for $\lambda = 0$. For if $\lambda = 0$, (5) yields the suppressed solution $x = 1$, while (6) defines the suppressed solutions $y = -1$ and $y = 1$.

Relative to (6), the solution $x = 1$ is singular because there is no finite value of λ for which (6) particularizes to this solution. Likewise, $y = -1$ and $y = 1$ are singular relative to (5).

If $\lambda \neq 0$, (5) and (6) have the same family of integral curves. Using (5) and rewriting it as

$$\frac{(x - 1)^2}{\lambda} + y^2 = 1 \qquad \lambda \neq 0 \tag{7}$$

we recognize these curves to be a family of conics, typical members of which are shown in Fig. 1.6. If $\lambda > 0$, the curves are all ellipses; if $\lambda < 0$, the curves are all hyperbolas.

In most practical problems a general solution of a differential equation is required to satisfy specific conditions which permit its arbitrary constants to be uniquely determined. For instance, in the present problem we might ask for the integral curve which passes through the point $(-\frac{7}{5}, \frac{13}{5})$. Substituting these values for x and y we have

$$\frac{(-\frac{7}{5} - 1)^2}{\lambda} + \left(\frac{13}{5}\right)^2 = 1$$

from which we find the value $\lambda = -1$ and then the specific solution

$$y^2 = 1 + (x - 1)^2 \tag{8}$$

Equation (8) defines the unique member of the family of curves (7) which passes through the point $(-\frac{7}{5}, \frac{13}{5})$. This curve is a hyperbola, and it is its upper branch that passes through the specified point. However, over any open interval containing both $x = -\frac{7}{5}$ and $x = 1$, there are many solution curves which pass through $(-\frac{7}{5}, \frac{13}{5})$. In fact, the upper branch of *any* curve of the family (7) for $x \geq 1$ can be associated with the upper branch of the hyperbola defined by (8) for $x \leq 1$ to give a solution curve containing $(-\frac{7}{5}, \frac{13}{5})$. This is certainly consistent with Theorem 1, Sec. 1.6, which ensures the uniqueness of a solution satisfying the initial-value problem $y' = (1 - y^2)/y(1 - x)$; $y(-\frac{7}{5}) = \frac{13}{5}$, only over an interval around $x = -\frac{7}{5}$ which excludes $x = 1$ since $(1 - y^2)/y(1 - x)$ is discontinuous where $x = 1$.

As Fig. 1.6 illustrates, infinitely many integral curves pass through $(1, 1)$ and $(1, -1)$. However, none of the conics defined by (7) pass through any point with coordinates $(1, y_0)$, $(x_0, -1)$, or $(x_0, 1)$, where $x_0 \neq 1$ and $y_0 \neq \pm 1$. Therefore had we desired a solution curve containing such a point, it would have been necessary to take recourse to the suppressed solutions. To be specific, $x = 1$, $y = -1$, and $y = 1$ are the solutions which, for any fixed x_0 or y_0 of the kind stipulated, define the solution curves through the respective points $(1, y_0)$, $(x_0, -1)$, and $(x_0, 1)$.

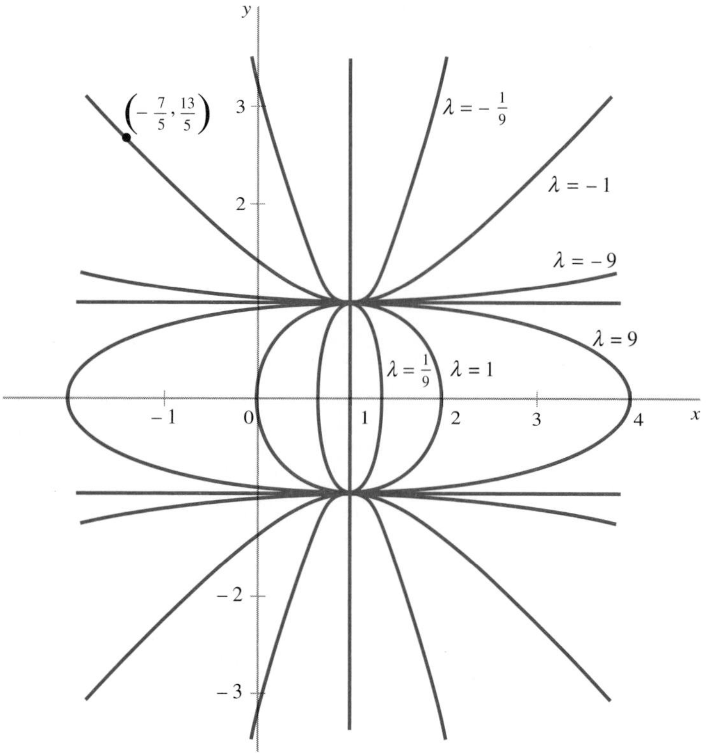

FIGURE 1.6

Typical solution and integral curves associated with the differential equation $(1 - y^2)\,dx = y(1 - x)\,dy$. The straight lines correspond to suppressed solutions; the other curves are conics defined by the equation $[(x - 1)^2/\lambda] + y^2 = 1$.

EXERCISES

Find a general solution of each of the following equations. Test any suppressed solutions to see if they are singular.

1. $(\cos x \cos y)\,dx - (\sin x \sin y)\,dy = 0$
2. $y' = -2xy$
3. $(\sin x)\,dy = 2y(\cos x)\,dx$
4. $y' = 3x^2(1 + y^2)$
5. $x\,dy = 3y\,dx$
6. $yy' = 2(xy + x)$
7. $dx + y\,dy = x^2 y\,dy$
8. $2xy\,dy = (x + 1)(9 - y^2)\,dx$
9. $xe^{x^2+y}\,dx = y\,dy$
10. $e^{x+y}y' = e^{2x-y}$
11. $y' + y^4 \cos x = 0$
12. $2(xy + x)y' = y$
13. $x\,dy = (y^2 - 3y + 2)\,dx$
14. $y' + y^2 \sin x = 0$
15. $(2 \cos y)\,dx + x(\sec y - \sin y)\,dy = 0$
16. $2y' - y^3 \cos x = 0$
17. $y\,dx + (\cot x)\,dy = 0$
18. $(y + x^2 y)\,dy = (xy^2 - x)\,dx$
19. $y\,dx - x\,dy = x(dy - y\,dx)$
20. $y' = (y + 1)^2/(x + 1)^2$
21. $y' = 2(y^2 + y - 2)/(x^2 + 4x + 3)$
22. $(xy^2 + 3xy)\,dx - dy = 0$
23. $(2 \sin^2 3x + 1 + \cos 6x)\,dx - 2\,dy = 0$
24. $3\sqrt{9 - y^2}\,dx + (9 + x^2)\,dy = 0$
25. $(2 \cosh y)\,dx + x(\operatorname{sech} y - \sinh y)\,dy = 0$
26. $2yx^3\,dx + [(x^2 + 1) \ln y^2]\,dy = 0$
27. $(y + y^2)\,dx + (x^2 + 4x + 5)\,dy = 0$
28. $xy'' = y'$. *Hint:* Note that $y'' = dy'/dx$.
29. $y'' + (y')^2 + 1 = 0$
30. $yy'' + (y')^2 = 0$
31. $yy'' = (y')^2$
32. $xy'' = (1 + 2x^2)y'$

Find a solution of each of the following equations which satisfies the indicated conditions.

33. $2xy\,dx + (1 + y)\,dy = 0$; $y(2) = 1$
34. $y' + y^2 \sin x = 0$; $y(0) = 0$
35. $2xy' + y = 0$; $y(4) = 1$
36. $y' + 2y = 0$; $y(0) = 100$
37. $2x\,dx - dy = x(x\,dy - 2y\,dx)$; $y(-3) = 1$
38. $dy = x(2y\,dx - x\,dy)$; $y(1) = 4$
39. $(5x^2 + 1)y' - 20xy = 10x$; $y(0) = \frac{1}{2}$
40. $y\,dx + (\coth x)\,dy = 0$; $y(0) = -1$
41. $3y\,dx = x(1 + 3y)\,dy$; $x(1) = e$
42. $(\tan \theta)\,dr + 2r\,d\theta = 0$; $r(\pi/2) = 13$
43. $r \sin \theta\,dr = (r^2 + 1) \cos \theta\,d\theta$; $r(\pi/2) = 1$
44. $\sin 2\theta\,dr = r \cos 2\theta\,d\theta$; $r(3\pi/4) = -1$
45. Find the solution of $(\sin^2 y)\,dx + (x - x^2)\,dy = 0$ which satisfies the initial condition:
    - **(a)** $x(0) = 0$
    - **(b)** $x(-\pi) = 1$
    - **(c)** $y(e) = -2\pi$
    - **(d)** $y(-2e) = 3\pi$
    - **(e)** Find all suppressed solutions of the differential equation.
    - **(f)** Find a general solution.
    - **(g)** Find the solution for which $x(\pi/2) = 2$.

46. Find a solution of the equation $(1 - x^2)\,dy + 4xy\,dx = 0$ with the property that $y = 9$ when $x = -2$, $y = 2$ when $x = 0$, and $y = 0$ when $x = 2$.

47. Is there a solution of the equation $x\,dy = 3(y - 1)\,dx$ satisfying the two conditions $y(1) = 3$ and $y(2) = 9$? Is there a solution of this equation which satisfies the two conditions $y(-1) = 3$ and $y(2) = 9$? Explain.

48. The initial-value problem $xy' - 3y = 0$, $y(0) = 0$, has $y_1 = 0$ as a solution on $(-\infty, \infty)$. Are there any other solutions of the problem on $(-\infty, \infty)$? If so, find one; if not, explain why.

49. Show that every solution of the equation $y' = ky$ is of the form $y = Ae^{kx}$. *Hint:* Let y be any solution of the given equation and consider the derivative of the fraction y/e^{kx}.

50. We know that there is no loss of generality if the arbitrary constant added when a separable equation is integrated is written in the form $\ln |c|$ rather than just c. Is there any loss of generality if the integration constant is written in the form c^2? $\tan c$? $\sin c$? e^c? $\sinh c$? $\cosh c$?

51. Show that the change of dependent variable from y to v defined by the substitution $v = ax + by + c$ will always transform the equation $y' = f(ax + by + c)$ into a separable equation.

Using the substitution described in Exercise 51, find a general solution of each of the following equations.

52. $y' = (x - y)^2$

53. $y' = e^{2x+y-1} - 2$

54. $y' = (x + y - 3)^2 - 2(x + y - 3)$

55. $y' = (x - y + 1)^2 + x - y$

56. Does the equation $y' = \begin{cases} 1 & x \le 0 \\ y & x > 0 \end{cases}$ have a solution which is continuous on $(-\infty, \infty)$? Does it have a solution with a continuous derivative on $(-\infty, \infty)$?

1.10 HOMOGENEOUS FIRST-ORDER EQUATIONS

If all terms in the coefficient functions $M(x, y)$ and $N(x, y)$ in the standard differential form

$$M(x, y)\,dx + N(x, y)\,dy = 0 \tag{1}$$

of a general first-order differential equation are of the same total degree in the variables x and y, then either of the substitutions $y = ux$ or $x = vy$ will reduce the equation to one which is separable.

More generally, if $M(x, y)$ and $N(x, y)$ have the property that for all *positive* values of λ, the substitution of λx for x and λy for y converts them, respectively, into the expressions

$$\lambda^n M(x, y) \qquad \text{and} \qquad \lambda^n N(x, y)$$

then (1) can always be reduced to a separable equation by either of the substitutions $y = ux$ or $x = vy$.

A function $f(x, y)$ with the property that the substitutions

$$x \to \lambda x \qquad \text{and} \qquad y \to \lambda y \qquad \lambda > 0$$

merely reproduce the representation $f(x, y)$ multiplied by λ^n is called a **homogeneous function of degree n.** While working with first-order equations, only functions of two variables need be considered. Nevertheless, it will be convenient to have a general definition of homogeneity available.

DEFINITION 1 A function f of m variables $x_1, x_2, \ldots, x_m$ is **homogeneous of degree n in these variables** if and only if for all values of the variables and for every positive value of λ for which $f(x_1, \ldots, x_m)$ and $f(\lambda x_1, \ldots, \lambda x_m)$ are defined,

$$f(\lambda x_1, \ldots, \lambda x_m) = \lambda^n f(x_1, \ldots, x_m) \tag{2}$$

In this definition, the degree of homogeneity n need *not* be an integer; it may be any real number.

EXAMPLE 1

The function $f(x, y) = \sqrt{xy}$ is defined throughout the region R in the xy plane which is comprised of the first and third quadrants together with the coordinate axes. For every real number λ and for all values of x and y such that (x, y) is a point of R, $f(\lambda x, \lambda y)$ exists and is given by

$$f(\lambda x, \lambda y) = \sqrt{\lambda^2 xy} = |\lambda|\sqrt{xy} = |\lambda|\, f(x, y)$$

The fundamental relation (2) is not satisfied if λ is negative because in this case $f(\lambda x, \lambda y) = -\lambda f(x, y)$. This has no bearing on the homogeneity of f, however, since Definition 1 requires that λ be positive. For $\lambda > 0$ we have $f(\lambda x, \lambda y) = \lambda f(x, y)$, and so by definition f is homogeneous of degree 1 in the variables x and y.

Of course, if either x or y is held constant, then f becomes a homogeneous function of degree $\frac{1}{2}$ in the remaining variable because $f(\lambda x, y) = f(x, \lambda y) = \lambda^{1/2} f(x, y)$.

EXAMPLE 2

Is the function $F(x, y) = x(\ln \sqrt{x^2 + y^2} - \ln y) + ye^{x/y}$ homogeneous?

To answer this question, we first note that $\ln y$ and $\ln \lambda y$ are both defined only if y and λy are positive. With λ thus restricted, we replace x by λx and y by λy, getting

$$\begin{aligned} F(\lambda x, \lambda y) &= \lambda x(\ln \sqrt{\lambda^2 x^2 + \lambda^2 y^2} - \ln \lambda y) + \lambda y e^{\lambda x/\lambda y} \\ &= \lambda x[(\ln \sqrt{x^2 + y^2} + \ln \lambda) - (\ln y + \ln \lambda)] + \lambda y e^{x/y} \\ &= \lambda[x(\ln \sqrt{x^2 + y^2} - \ln y) + ye^{x/y}] \\ &= \lambda F(x, y) \end{aligned}$$

The given function is therefore homogeneous of degree 1.

As a direct extension of the preceding ideas to first-order differential equations, we have the following definition.

DEFINITION 2 A **homogeneous differential equation** is a first-order ordinary differential equation that is algebraically reducible to a standard differential form in which the coefficients of the differentials are homogeneous functions of the same degree.

Thus (1), for instance, is homogeneous if and only if over some region R both M and N are homogeneous functions of the same degree.

The results asserted at the beginning of this section can now be stated more precisely as follows.

THEOREM 1 If over a region R, $M(x, y)$ and $N(x, y)$ are homogeneous functions of the same degree, then the substitution $y = ux$ will reduce the equation

$$M(x, y)\, dx + N(x, y)\, dy = 0 \tag{1}$$

to a separable equation in the variables x and u. Similarly, the substitution $x = vy$ will reduce Eq. (1) to a separable equation in y and v.

PROOF If (1), assumed now to be homogeneous, is written in the form

$$\frac{dy}{dx} = \frac{-M(x, y)}{N(x, y)}$$

the fraction on the right is a homogeneous function of degree 0 because the same power of λ will multiply both numerator and denominator when the test substitutions $x \to \lambda x$ and $y \to \lambda y$ are made. Then, since

$$\frac{M(\lambda x, \lambda y)}{N(\lambda x, \lambda y)} = \frac{M(x, y)}{N(x, y)}$$

is an identity in x and y and $\lambda > 0$, it follows, by assigning to the arbitrary symbol λ the value $1/x$ if x is positive and the value $-1/x$ if x is negative, that

$$\frac{M(x, y)}{N(x, y)} = \frac{M(\lambda x, \lambda y)}{N(\lambda x, \lambda y)} = \begin{cases} \dfrac{M(1, y/x)}{N(1, y/x)} & x > 0 \\[2ex] \dfrac{M(-1, -y/x)}{N(-1, -y/x)} & x < 0 \end{cases}$$

In either case, it is clear that the result is a function of the fractional argument y/x. Thus an alternative standard form for a homogeneous first-order differential equation is

$$\frac{dy}{dx} = F\left(\frac{y}{x}\right) \tag{3}$$

Now if $y = ux$, then $dy/dx = u + x\,du/dx$ or $dy = u\,dx + x\,du$. Hence under this substitution, (3) becomes $u + x(du/dx) = F(u)$ or

$$x\,du = [F(u) - u]\,dx \tag{4}$$

If $F(u) = u$ holds identically, (3) is simply $dy/dx = y/x$, and this is separable at the outset. If $F(u) \not\equiv u$, we can divide (4) by the product $x[F(u) - u]$, getting

$$\frac{du}{F(u) - u} = \frac{dx}{x}$$

The variables have now been separated, and the equation can be integrated at once. Finally, by replacing u by its value y/x, we obtain an equation defining the original dependent variable y as a function of x.

An almost identical proof for the substitution $x = vy$ (see Exercise 7) establishes the theorem. ◗

In solving homogeneous differential equations, it is not necessary to reduce the equation to the form $dy/dx = F(y/x)$ as we did in the proof of Theorem 1. Instead, it is sufficient to substitute for either y or x as the theorem directs (see Exercise 8), then separate variables and proceed with the solution process.

EXAMPLE 3

Solve the equation $(x^2 + 3y^2)\,dx - 2xy\,dy = 0$.

By inspection, this equation is homogeneous since all terms in the coefficient of each differential are of the second degree. Hence we substitute $y = ux$ and $dy = u\,dx + x\,du$, getting

$$(x^2 + 3u^2x^2)\,dx - 2x^2u(u\,dx + x\,du) = 0$$

or, dividing by x^2 and collecting terms,

$$(1 + u^2)\,dx - 2xu\,du = 0$$

Separating variables, we obtain

$$\frac{dx}{x} - \frac{2u\,du}{1 + u^2} = 0$$

At this point we check to see if $x = 0$ is a suppressed solution of the original equation, and we see that it is. To continue, we integrate the separated equation, getting

$$\ln|x| - \ln(1 + u^2) = \ln|c|$$

$$\ln\frac{|x|}{1 + u^2} = \ln|c|$$

$$\frac{|x|}{1 + u^2} = |c|$$

$$\frac{x}{1 + u^2} = c$$

Finally, replacing u by y/x and clearing fractions, we have

$$x^3 = c(x^2 + y^2)$$

From the preceding steps, it appears that c must be different from zero. However, c is actually unrestricted in the solution we have just found because, with $c = 0$, it particularizes to the suppressed solution $x = 0$.

EXERCISES

Determine which, if any, of the following functions are homogeneous.

1. $\cos\dfrac{x}{x^2 + y^2}$

2. $\dfrac{x - y}{x^2 + y^2}$

3. $\dfrac{x^2 + y^2 + 1}{xy + 3}$

4. $x^{1/3}y^{-2/3}\,\mathrm{Tan}^{-1}\left(\dfrac{x}{y}\right)$

5. $x\left[\ln\dfrac{2x^2 + y^2}{x} - \ln(x + y)\right] + y^2\tan\dfrac{x + 2y}{3x - y}$

6. Determine all sets of variables with respect to which each of the following functions is homogeneous and state the degree of homogeneity in each case.
(a) $x^2 - xy + y^2z$
(b) $zx^2y^2\cosh(y/z) + (x^2y^4/z)\sinh(z/y)$

7. Prove that the substitution $x = vy$ will also transform a homogeneous first-order differential equation into one which is separable.

8. Under what conditions, if any, do you think that the substitution $x = vy$ would be more convenient than the substitution $y = ux$?

9. Show that the product of a homogeneous function of degree m and a homogeneous function of degree n is a homogeneous function of degree $m + n$.

10. Show that the quotient of a homogeneous function of degree m divided by a homogeneous function of degree n is a homogeneous function of degree $m - n$.

11. If $f(x, y, c_1) = 0$ and $f(x, y, c_2) = 0$ define two integral curves of a homogeneous first-order differential equation and P_1 and P_2 are, respectively, the points of intersection of these curves and an arbitrary line, $y = mx$, through the origin, prove that the slopes of these two curves at P_1 and P_2 are equal.

12. If f is nontrivial, are *all* terms of the linear equation (1), Sec. 1.2, of the same degree of homogeneity in y and its first n derivatives? Explain.

Find general solutions of each of the following equations.

13. $2xy' = y - x$

14. $(x^2 + y^2)\,dx = 2xy\,dy$

15. $x^2\,dy = (xy - y^2)\,dx$

16. $(x^2 + 3y^2)\,dx - 2xy\,dy = 0$

17. $y' = (x^2 + y^2)/x^2$

18. $y^2\,dx + (y^2 - xy)\,dy = 0$

19. $(2x^2 - y^2)\,dx + 3xy\,dy = 0$

20. $xy' - y = \sqrt{x^2 - y^2}$

21. $(y^2 - x^2e^{x/y})\,dy + (y^2 + xy)e^{x/y}\,dx = 0$

22. $(x^2 + y^2)\,dx + 3xy\,dy = 0$

23. $y\,dx = [x - y\tanh(x/y)]\,dy$

24. $x^2y' = y^2 + 2xy$

25. $(xy + y^2)\,dx = (x^2 + xy + y^2)\,dy$

26. $[x + y \cot (x/y)]\, dy - y\, dx = 0$
27. $\{y - x[1 + \sin (y/x)]\}\, dx = x\, dy$
28. $[x \cosh^2 (y/x) - y]\, dx + x\, dy = 0$
29. $xe^{x^2/y^2}(x\, dy - y\, dx) = y^2\, dy$
30. $x \cosh (y/x)\, dx + x\, dy - y\, dx = 0$

Solve each of the following equations and describe the family of integral curves.

31. $(x - y)\, dx + (y - x)\, dy = 0$
32. $(x - 2y)\, dy = (2x - y)\, dx$
33. $(x + 3y)\, dy = (x - y)\, dx$
34. $(123x - 25y)\, dx = (25x - 3y)\, dy$
35. $(x - y)\, dy = (x + y)\, dx$
36. $(2x + y)\, dy = (x + 2y)\, dx$

Find a solution of each of the following equations which satisfies the given condition.

37. $x^{1/2}\, dx + (xy)^{1/4}\, dy = 0;\ y(0) = 1$
38. $(x^4 + 4x^2y^2 + y^4)y' = 2xy^3;\ y(0) = 2/e$
39. $xy' = y + \sqrt{x^2 + y^2};\ y(4) = 3$
40. $x^2y\, dx = (x^3 - y^3)\, dy;\ y(1) = 1$
41. $(x^4 + y^4)\, dx = 2x^3y\, dy;\ y(1) = 0$
42. $(3y^3 - x^3)\, dx = 3xy^2\, dy;\ y(1) = 2$
43. $(x^3 + y^3)\, dx = 2xy^2\, dy;\ y(1) = 0$
44. $y' = \sec (y/x) + y/x;\ y(2) = \pi$
45. $\left(x \cos \frac{y}{x} + x - y\right) dx + x\, dy = 0;\ y(2) = \pi$
46. $(3y^2 + 2xy)\, dx - (2xy + x^2)\, dy = 0;\ y(1) = 1$
47. Prove that the differential equation

$$\frac{dy}{dx} = \frac{ax + by + c}{Ax + By + C}$$

can be solved as follows:

(a) If $aB \neq bA$, constants h and k may be determined such that the equation is transformed into a homogeneous equation in the new variables u and v by the substitutions $y = u + h$ and $x = v + k$.

(b) If $aB = bA$, $a^2 + A^2 \neq 0$, $b^2 + B^2 \neq 0$, and $c^2 + C^2 \neq 0$, introduction of a new dependent variable

$$v = ax + by \qquad \text{if } a \neq 0$$

$$v = Ax + By \qquad \text{if } a = 0$$

converts the original equation into one that is separable. *Hint:* Recall Exercise 51, Sec. 1.9.

(c) If $a^2 + A^2 = 0$ or $b^2 + B^2 = 0$, the given equation is separable, and if $c^2 + C^2 = 0$, it is homogeneous.

Using appropriate substitutions, as described in Exercise 47, solve each of the following equations.

48. $y' = \dfrac{x - y + 5}{x + y - 1}$ **49.** $y' = \dfrac{2x + 2y + 1}{3x + y - 2}$

50. $y' = (y - x)/(x - y + 2)$
51. $(2x + y)y' = 1$
52. $y' = (y + 2)/(x + y + 1)$
53. $(2x - y)y' = 4x - 2y - 5$
54. Prove that $b + c = 0$ is a sufficient condition for all solution curves of the equation $y' = (ax + by)/(cx + ey)$ to be conics. Prove further that when this is the case, the conics are all ellipses if $c^2 + ae < 0$ and all hyperbolas if $c^2 + ae > 0$.
55. Show that $b + c = 0$ is not a necessary condition for the solution curves of the equation $y' = (ax + by)/(cx + ey)$ to be conics. *Hint:* Construct a counterexample.
56. If $M(x, y)\, dx = N(x, y)\, dy$ is a homogeneous equation, prove that if it is expressed in terms of the polar coordinates r and θ by means of the substitutions $x = r \cos \theta$ and $y = r \sin \theta$, it becomes separable.

Solve the next two equations using the method described in Exercise 56.

57. $y' = \dfrac{x + y}{x - y}$ **58.** $y' = \dfrac{x + 2y}{2x - y}$

59. If $f(x, y)$ is a homogeneous function of degree n, show that

$$x\frac{\partial f}{\partial x} + y\frac{\partial f}{\partial y} = nf$$

What is the generalization of this result to functions of more than two variables? (This result is commonly referred to as **Euler's theorem for homogeneous functions.**†)

60. Show that if the equation $M(x, y)\, dx + N(x, y)\, dy = 0$ is both homogeneous and exact, its solutions are given by $xM(x, y) + yN(x, y) = k$.
61. If the equation $M(x, y)\, dx + N(x, y)\, dy = 0$ is homogeneous, show that $1/(xM + yN)$ is an integrating factor. *Hint:* Observe that

$$\frac{M\, dx + N\, dy}{xM + yN} = \frac{dx}{x} + \frac{(x\, dy - y\, dx)N}{x(xM + yN)}$$

$$= \frac{dx}{x} + \frac{(x\, dy - y\, dx)/x^2}{M/N + y/x}$$

Use an integrating factor of the kind identified in Exercise 61 to find a general solution for each of the following equations.

62. $3y\, dx + 2x\, dy = 0$ **63.** $(x - y)\, dx + x\, dy = 0$
64. $(2y + 3x)\, dx + x\, dy = 0$
65. $x\, dy - y\, dx = (xy)^{1/2}\, dx$

†Named for the Swiss mathematician Leonard Euler (1707–1783), one of the most prolific and one of the greatest mathematicians of all time.

1.11 LINEAR FIRST-ORDER EQUATIONS

By definition, a linear first-order differential equation in y cannot contain products, powers, or other nonlinear combinations of y or y'. Hence its most general form is

$$F(x)\frac{dy}{dx} + G(x)y = H(x)$$

If we divide this equation by $F(x)$ and rename the coefficients, it appears in the more usual form

$$\frac{dy}{dx} + P(x)y = Q(x) \tag{1}$$

If $P(x) \equiv 0$, this equation can be solved immediately by integration. If $Q(x) \equiv 0$, the equation is separable. And whether P or Q is trivial or not, we have the following theorem.

THEOREM 1 The equation $dy/dx + P(x)y = Q(x)$ has $e^{\int P(x)\,dx}$ as an integrating factor.

PROOF Let us rewrite (1) in the form

$$[P(x)y - Q(x)]\,dx + dy = 0 \tag{2}$$

Comparing this with the standard form $M(x, y)\,dx + N(x, y)\,dy = 0$, we see that $M(x, y) = P(x)y - Q(x)$ and $N(x, y) = 1$. Therefore $(M_y - N_x)/N = P(x)$ and Theorem 1, Sec. 1.8, give $e^{\int P(x)\,dx}$ as an integrating factor of (1).

When (1) is multiplied by $e^{\int P(x)\,dx}$, it can be written in the form

$$\frac{d}{dx}\left(ye^{\int P(x)\,dx}\right) = Q(x)e^{\int P(x)\,dx}$$

The left-hand side is now an exact derivative, hence can be integrated at once. Moreover, the right-hand side is a function of x only and therefore can also be integrated, with at most practical difficulties requiring numerical integration. Thus, on performing these integrations and subsequently dividing by $e^{\int P(x)\,dx}$, we have

$$y = e^{-\int P(x)\,dx}\int Q(x)e^{\int P(x)\,dx}\,dx + ce^{-\int P(x)\,dx} \tag{3}$$

which tells us that *every first-order equation which is linear can be routinely solved by the use of an integrating factor.*

Equation (3) should *not* be remembered as a formula for the solution of (1). Instead, a linear first-order equation should be solved by integrating immediately if $P(x) \equiv 0$. Otherwise, actually carry out the steps we have described.

1. Compute the integrating factor $e^{\int P(x)\,dx}$.
2. Multiply the right-hand side of the given equation by this factor and write the left-hand side as the derivative of [y times the integrating factor].
3. Integrate and then solve the integrated equation for y.

EXAMPLE 1

Find the solution of the equation $(1 + x^2)(dy - dx) = 2xy\,dx$ for which $y = 1$ when $x = 0$.

Dividing the given equation by $(1 + x^2)\,dx$ and transposing, we have

$$\frac{dy}{dx} - \frac{2x}{1 + x^2}y = 1 \tag{4}$$

which is a linear first-order equation. In this case $P(x) = -2x/(1 + x^2)$; hence the integrating factor is

$$\exp\left(\int \frac{-2x}{1 + x^2}\,dx\right) = \exp\,[-\ln\,(1 + x^2)] = e^{\ln\,(1+x^2)^{-1}} = \frac{1}{1 + x^2}$$

Multiplying the right-hand side of (4) by this factor and writing the left-hand side as the derivative of [y times the integrating factor] gives

$$\left[\frac{y}{1 + x^2}\right]' = \frac{1}{1 + x^2}$$

Integrating this and subsequently solving for y, we have

$$\frac{y}{1 + x^2} = \mathrm{Tan}^{-1}\,x + c \qquad \text{or} \qquad y = (1 + x^2)\,\mathrm{Tan}^{-1}\,x + c(1 + x^2)$$

To find the specific solution required, we substitute the given values $y = 1$, $x = 0$ into the general solution, getting $1 = 0 + c$. The required solution is therefore

$$y = (1 + x^2)\,\mathrm{Tan}^{-1}\,x + (1 + x^2)$$

EXERCISES

Find a general solution of each of the following equations.

1. $xy' = x^3 - 2y$
2. $(x - y)\,dx + x\,dy = 0$
3. $y' = e^{2x} + 3y$
4. $x^2\,dy + (2xy - x + 1)\,dx = 0$
5. $2xy' + y = 2x^3$
6. $(2y + x^2)\,dx = x\,dy$
7. $y' + 2xy + x = \exp\,(-x^2)$
8. $y' + y\tan x = \sec x$
9. $y' + y\cot x = \sin 2x$
10. $(1 + e^x)(y' + y) = 1$
11. $(1 - x^2)y' + xy = 2x$
12. $y' + y/(1 - x) = x^2 - x$
13. $y' = 2y/(x + 1) + (x + 1)^3$
14. $xy' + (1 + x)y = e^{-x}$
15. $xy' + 2(1 - x^2)y = 1$
16. $(e^x + 1)y' = y - ye^x$
17. $y' = \sec^3 x + y\tan x$
18. $y' + \sinh xe^x - \cosh xe^x + (x + 1)ye^x = 0$
19. $y^2\,dx + (3xy - 4y^3)\,dy = 0$
20. $(x + e^y)\,dy - dx = 0$
21. $dx = (1 - x)\tanh y\,dy$
22. $y^2\,dx + [(y^2 + 2y)x - 1]\,dy = 0$
23. $y'' - y' = 1$
24. $xy'' + y' = 1 + x^2$
25. $y'' + [y'/(x - 1)] = x - 1$
26. $y'' + (\cot x)y' = 2\cos x$

Find the particular solution of each of the following equations which satisfies the indicated conditions.

27. $y' + y = e^{-x}$; $y(0) = 3$
28. $y' + y = e^x$; $y(0) = 2$
29. $xy' + xy = 1 - y$; $y(1) = 0$
30. $xy' + 2y = (\sin x)/x$; $y(2) = 1$
31. $y' + (1 + 2x)y = \exp\,(-x^2)$; $y(0) = 3$
32. $(1 + x^2)\,dy = (1 + xy)\,dx$; $y(1) = 0$
33. $y' = (1 - 2xy)/x^2$; $y(1) = 2$
34. $(5x^2 + 1)y' - 20xy = 10x$; $y(0) = \frac{1}{2}$
35. $y' = (2x^3 + y)/x$; $y(2) = 0$
36. $y' = y + 4e^x\sin^2 x$; $y(0) = 10$
37. $y' + y\tan x = \sin 2x$; $y(0) = -2$
38. $y'\sin x + y\cos x = \cos 2x$; $y(\pi/2) = \frac{1}{2}$
39. $(x^2 - 1)y' + 2xy - 3x^2 + 1 = 0$; $y(0) = 0$
40. $(x^2 + 1)\,dy = (x^3 - 2xy + x)\,dx$; $y(1) = 1$
41. $y' + y\sec x = \sec x$; $y(0) = 6$
42. $\exp\,(y' + y/x) = \cosh 3x + \sinh 3x$; $y(1) = 4$
43. $3y\,dx = (x + 4y^{5/3})$; $x(1) = 5$
44. $(y\ln y)\,dx = (1 - x)\,dy$; $x(e) = 1 + \ln 2$
45. $y' + (1 + \tanh x)y = 2e^{-x}\sinh x$; $y(\ln 3) = \frac{16}{45}$
46. $y' + (1/x + \tanh x)y = 3x\,\mathrm{sech}\,x$; $y(\ln 2) = \frac{4}{5}\ln^2 2$
47. Find a solution of the equation $x\,dy + (x^2 - 3y)\,dx = 0$ which simultaneously satisfies the conditions $y(-1) = 1$ and $y(1) = -1$.
48. Find a solution of the equation $y\,dx = (3x + y^4)\,dy$ which simultaneously satisfies the conditions $x(-1) = -1$ and $x(2) = 0$.
49. Prove that no extra generality in the final answer results from using

$$\exp\,[\textstyle\int P(x)\,dx + k]$$

instead of just $\exp\,[\int P(x)\,dx]$ as an integrating factor of Eq. (1).
50. Use separation of variables to find a general solution of $y' + P(x)y = 0$.

1.12 SPECIAL FIRST-ORDER EQUATIONS

We are now able to recognize and solve four major types of first-order differential equations, namely, those that are exact, separable, homogeneous, or linear. To recognize the first three types, we look for definitive properties of their standard *differential* forms. To see if an equation is linear, we compare its standard *derivative* form with the **standard first-order linear form**

(1)
$$y' + P(x)y = Q(x)$$

Some nonlinear differential equations can be transformed into linear equations by a change of dependent variable. A case in point is provided by the so-called **Bernoulli equation**†

(2)
$$y' + P(x)y = Q(x)y^n$$

where n may have any real value. If n is either 0 or 1, Eq. (2) is linear; for all other values of n it is nonlinear. To transform a *nonlinear* Bernoulli equation into a linear equation we use the following procedure.

THEOREM 1 The change of dependent variable $z = y^{1-n}$ converts a nonlinear Bernoulli equation $y' + P(x)y = Q(x)y^n$ into the linear equation

(3)
$$\frac{dz}{dx} + (1 - n)P(x)z = (1 - n)Q(x)$$

PROOF We first multiply Eq. (2) through by $(1 - n)y^{-n}$ to get

(4)
$$(1 - n)y^{-n}y' + (1 - n)P(x)y^{1-n} = (1 - n)Q(x)$$

The change of dependent variable $z = y^{1-n}$, together with its derivative $z' = (1 - n)y^{-n}y'$, now converts (4) directly into (3).

After the linear equation (3) has been solved for z, a general solution of (2) can be found by substituting y^{1-n} for z. It is important to note that if $n > 0$, Eq. (2) also has the suppressed solution $y = 0$.

EXAMPLE 1

Solve the differential equation $3xy' + y + x^2y^4 = 0$.

If we divide the equation by $3x$ and then transpose the last term, we obtain

$$y' + \frac{y}{3x} = -\left(\frac{x}{3}\right)y^4$$

which we recognize as a nonlinear Bernoulli equation with $n = 4$ and suppressed solution $y = 0$.

As we have seen, the change of dependent variable described in Theorem 1 is most conveniently made after Eq. (2) has been multiplied through by $(1 - n)y^{-n}$. In the present problem this leads to the equation

$$-3y^{-4}y' - \frac{1}{x}y^{-3} = x$$

†Named for the Swiss mathematician Jakob Bernoulli (1654–1705) although the equation was first solved by Leibniz in 1696.

The required change of variable

$$z = y^{1-n} = y^{-3} \qquad \text{and its derivative} \qquad z' = -3y^{-4}y'$$

now convert the last equation, involving x and y, into the linear equation

$$\frac{dz}{dx} - \frac{z}{x} = x$$

which has the integrating factor

$$e^{\int(-1/x)\,dx} = e^{-\ln x} = e^{\ln(1/x)} = \frac{1}{x}$$

Multiplying the linear equation by this factor gives

$$\left(\frac{z}{x}\right)' = 1$$

and integrating, we obtain

$$\frac{z}{x} = x + c \qquad \text{or} \qquad z = x^2 + cx$$

Finally, since $z = y^{-3}$, we have

$$\frac{1}{y^3} = x^2 + cx \qquad \text{or} \qquad y^3 = \frac{1}{x^2 + cx}$$

The suppressed solution $y = 0$ is not obtained for any value of c.

Another first-order differential equation of some importance is the generalized **Riccati equation†**

(5) $$y' = P(x)y^2 + Q(x)y + R(x)$$

If $R(x) \equiv 0$, Eq. (5) is a Bernoulli equation and can be solved as such by means of the substitution $z = 1/y$ and by noting its suppressed solution $y = 0$. But if R is a nontrivial function, no elementary solution method is available. Yet a general solution of (5) can always be found whenever one specific solution of the equation is known. Here, as with the Bernoulli equation, the strategy is to devise a change of dependent variable that will transform a Ricatti equation into a related first-order equation which is linear, hence solvable. Details of the procedure are outlined in Exercise 35.

An example of a solvable type of first-order equation which in most instances cannot be algebraically reduced to the standard derivative form $y' = f(x, y)$ is provided by the **equation of Clairaut‡**

(6) $$y = xy' + f(y')$$

whose solutions are described in Exercise 43.

†Named for the Italian mathematician J. F. Riccati (1676–1754).

‡Named for the French astronomer and mathematician Alexis Claude Clairaut (1713–1765).

EXERCISES

Solve each of the differential equations.

1. $y' + \frac{2y}{x} = \frac{1}{2yx^4}$

2. $y' - \frac{y}{x} + \frac{y^2}{x} = 0$

3. $y^2y' + x^2y^3 = x^2$

4. $(x - 2y^2)\,dx + 2xy\,dy = 0$

5. $xy^2y' - y^3 = x^2$

6. $dy = (xy^2 + 3xy)\,dx$

7. $y' + y = xy^2$

8. $y\,dy = (x - y^2)\,dx$

9. $y' + y = xy^{2/3}$

10. $x(2x + 1)y' + (8x + 2)y = 2xy^{1/2}$

11. $3xy' + y + x^2y^4 = 0$

12. $x^2y' = y^2 + 2xy$

13. $y' + y\tan x + 2y^2 \sin x = 0$

14. $yx\,dy + (y^2 + 2x^2 + 2)\,dx = 0$

15. $dx = (x^2 - x)\,dy$

16. $(x^2y + y^3)\,dy = x\,dx$

17. $3xy\,dx + (2y^2 - x^2)\,dy = 0$

18. $dx/dy - x\cot y = 2x^2 \csc y$

19. $2yy'' - (y')^2 = 1$

20. $x^2y'' + (y')^2 = 2xy'$

Solve each of the initial-value problems.

21. $y' + y = y^2;\ y(0) = \frac{1}{2}$

22. $yy' + y^2 = x;\ y(0) = \sin(\pi/4)$

23. $yy' + xy^2 - x = 0;\ y(0) = -1$

24. $3x^2\,dx + 2xy\,dy + y^2\,dx = 0;\ y(2) = 1$

25. $xy\,dx + (x^2 + 2y^2 + 2)\,dy = 0;\ y(1) = 1$

26. $xy' + 3y = x^3y^2;\ y(1) = 1$

27. $y' + y = xy^3;\ y(0) = 1$

28. $y' + y/x = y^2 \ln x;\ y(1) = 1$

29. $2xy' + (1 - x)y + 9x^3y^3e^{2x} = 0;\ y(1) = 1/e\sqrt{2}$

30. $\pi y' - (2\,\mathrm{csch}\,2x)y = (y\tanh x)/y^{\pi};\ y(1) = \sqrt[\pi]{\tanh 1}$

31. $2y' + 3y/(x\ln x) = 3y^{1/3};\ y(e) = 0$

32. $(1 + x^2)y' + y/(2\,\mathrm{Tan}^{-1} x) = y^{-1};\ y(1) = \sqrt{\pi}/2$

33. $y' + y\ln x^2 = 2x^{-x}e^xy^{1/2};\ y(2) = e^4/4$

34. $2xy'' + 4y' + x^3(y')^2 = 0;\ y(1) = -1,\ y'(1) = 2$

35. Show that if one solution, say $y = u(x)$, of the Riccati equation $y' = P(x)y^2 + Q(x)y + R(x)$ is known, then the substitution $y = u + (1/z)$ will transform this equation into a linear first-order equation in the new dependent variable z.

Using the substitution described in Exercise 35, find a general solution of each of the following equations.

36. $y' = y^2 + (1 - 2x)y + (x^2 - x + 1);\ u = x$

37. $y' = xy^2 + (1 - 2x)y + x - 1;\ u = 1$

38. $y' = (x + y)(x + y - 2);\ u = 1 - x$

39. $y' = x^3(y - x)^2 + y/x;\ u = x$

40. $y' = (y - 1)(y + 1/x);\ u = 1$

41. $y' = e^{-x}y^2 + y - e^x;\ u = e^x$

42. Show that there are two values of c for which $y = c - x^2$ is a solution of the equation $y' = (x^2 + y - \frac{1}{2})(x^2 + y) - \frac{1}{2} - 2x$. Find these values and solve this equation using each of the particular solutions determined by these values of c. Are the two general solutions equivalent?

43. Verify that for all real values of m Clairaut's equation $y = xy' + f(y')$ has $y = mx + f(m)$ as a general solution. Also show that the parametric equations $x = -f'(t)$ and $y = -tf'(t) + f(t)$ define another solution which is in general singular relative to the preceding general solution. *Hint:* Show that the given parametric expressions for x and y satisfy an equation of the form $ax + by + c = 0$ if and only if $f(t)$ is linear.

44. By showing that the lines defined by the general solution $y = mx + f(m)$ of Clairaut's equation are all tangent to the graph of the singular solution defined by the parametric equations of Exercise 43, prove that the solution curve of the singular solution is the **envelope** of the family of straight lines determined by the general solution.

Find a general solution and a singular solution of each of the following differential equations of the kind described in Exercise 43.

45. $y = xy' - 4(y')^3$

46. $y = xy' + 1/4y'$

47. $y = xy' + 1/(1 + y')$

48. $y = xy' - \exp(y')$

49. $y = xy' - (y')^2/4$

50. $y = xy' - (\cosh 3y' + \sinh 3y')$

1.13 SECOND-ORDER EQUATIONS OF REDUCIBLE ORDER

A variety of applied problems involve differential equations of the second order which can be reduced to first-order equations by an appropriate change of variables. There are two important cases in which the substitution of a new dependent variable for the first derivative of the original dependent variable leads to such a reduction. It turns out that in either case solutions of the second-order equation can be found if two first-order equations can be solved in succession.

The simpler of these two cases is distinguished by the absence of the dependent variable from the given second-order differential equation. By this we mean that the dependent variable itself does not appear in the equation; only derivatives of it appear. This case is briefly described as follows.

CASE 1 The dependent variable is not *explicitly* present in the second-order differential equation.

In this case a second-order equation in y can be regarded as a first-order equation in $y' = v$. If the first-order equation in v can be solved, then a first-order equation in y can be obtained from which, if it can be solved, a solution of the original second-order equation can be found.

EXAMPLE 1

According to our definition of the order of a differential equation, the equation $y'' + (y')^2 + 1 = 0$ is a second-order equation in y. However, observing that y does not appear explicitly in the equation and that y'' is just dy'/dx, we see that this equation is also a first-order equation in $y' = v$, namely,

$$v' + v^2 + 1 = 0$$

This is a separable equation which can easily be solved as follows:

$$\frac{dv}{1+v^2} + dx = 0 \qquad \mathrm{Tan}^{-1}\, v = c_1 - x \qquad v = \tan\,(c_1 - x)$$

Reverting back to y', we have

$$y' = \tan\,(c_1 - x)$$

Integration now gives

$$y = \ln|\cos\,(c_1 - x)| + c_2$$

as a general solution of the original second-order equation.

The second case we shall consider is characterized by the absence of the independent variable from the given second-order equation. That is, except as it might appear as part of the notation symbolizing a derivative of the dependent variable, the independent variable occurs nowhere else in the equation. A concise statement of this case is the following.

CASE 2 The independent variable does not appear *explicitly* in the second-order differential equation.

Every differential equation of the form $y'' = f(y, y')$ is of this type. Such equations frequently arise in problems of dynamics. The mere replacement of y' by v in an equation of this sort is ineffectual since it yields a first-order equation $v' = f(y, v)$ involving not one but *two* dependent variables v and y. However, the following important procedure will always reduce such an equation to a first-order equation in a new dependent variable, with the old dependent variable playing the role of independent variable.

Beginning with the substitution $y' = v$ and using the chain rule, we have

$$y'' = \frac{dy'}{dx} = \frac{dv}{dx} = \frac{dv}{dy}\frac{dy}{dx} = \frac{dv}{dy}y' = v\frac{dv}{dy}$$

Under this substitution, the original equation becomes

$$v\frac{dv}{dy} = f(y, v)$$

and this is a first-order equation in v which it may be possible to solve. In particular, if f is a function of y only, then the last equation is separable and we have

$$v\,dv = f(y)\,dy$$

Integrating and reverting back to y', we obtain

$$\frac{1}{2}(y')^2 = \int f(y)\, dy + c_1$$

From this, by solving for y', separating variables, and integrating again, a general solution for y can always be obtained although it may involve integrals which cannot be evaluated in terms of elementary functions.

EXAMPLE 2

Find the solution of the equation $y'' = -2y + 2y^3$ for which $y(0) = 0$ and $y'(0) = 1$.

None of the obvious constant solutions $y = -1$, $y = 0$, or $y = 1$ satisfy the given initial conditions. Hence we must seek a general solution of the differential equation. Since x does not appear explicitly in the equation, we let $y' = v$ and substitute $v(dv/dy)$ for y'', getting

$$v\frac{dv}{dy} = -2y + 2y^3$$

Separating variables, integrating, and replacing v by y', we find

$$\tfrac{1}{2}(y')^2 = c_1 - y^2 + \tfrac{1}{2}y^4$$

Now we know that $y = 0$ and $y' = 1$ when $x = 0$. Hence by substitution we find

$$c_1 = \tfrac{1}{2} \qquad \text{and} \qquad (y')^2 = 1 - 2y^2 + y^4 = (1 - y^2)^2$$

Therefore $y' = \pm(1 - y^2)$. To be consistent with the data of the problem, namely, $y' = 1$ when $y = 0$, we must select the positive sign. Hence we have

$$y' = \frac{dy}{dx} = 1 - y^2 \qquad \frac{dy}{1 - y^2} = dx \qquad \tanh^{-1} y = x + c_2 \qquad y = \tanh(x + c_2)$$

Finally, since $y = 0$ when $x = 0$, we find that $c_2 = 0$ and

$$y = \tanh x$$

This example worked out more simply than is usually the case because the initial conditions were carefully chosen. Had the initial data not been chosen to make $c_1 = \frac{1}{2}$, the expression for $(y')^2$ would not have been a perfect square and the integral for y could not have been evaluated in terms of elementary functions. In fact, for any value of c_1 except $\frac{1}{2}$, the final integral for y requires what are called **elliptic functions**† for its evaluation. It is important to note, however, that in any case an explicit integral expression for y could have been obtained.

EXERCISES

Find a general solution of each of the following differential equations.

1. $yy'' = (y')^2$
2. $xy'' = y'$
3. $y'' + y' = e^x$
4. $y'' - 2y' = 1$
5. $(\cot x)y'' + y' + 1 = 0$
6. $xy'' + y' = 3x^2 - x$
7. $2yy'' = (2ay + y')y'$
8. $yy'' = (y' + ay^{1+b})y'$
9. $y'' + 2x(1 + y')^2 = 0$
10. $(1 + x^2)y'' - 2xy' = 2x$

†See, for instance, I. N. Bronshtein and K. A. Semendyayev, *A Guide-Book to Mathematics,* p. 407, Verlag Harri Deutsch, Frankfurt/Main, 1971.

Solve each of the following initial-value problems.

11. $y'' = 2yy'$; $y(0) = 1$, $y'(0) = 5$
12. $y'' + 3y^2y' = 0$; $y(1) = 1$, $y'(1) = -1$
13. $2yy'' = 1 + (y')^2$; $y(0) = 2$, $y'(0) = -1$
14. $y'' + (y')^2 + y = 0$; $y(0) = -\frac{1}{2}$, $y'(0) = -1$
15. $y'' = x^2e^{-3y'}$; $y(2) = \ln 4$, $y'(2) = \ln 2$
6. $y'' + y' \tan x = 2\cos^2 x$; $y(\pi/6) = \frac{3}{4}$, $y'(\pi/6) = \sqrt{3}$
17. $xy'' + (2x^2 + 1)y' + 4x = 0$; $y(-1) = 5$, $y'(-1) = 2$
18. $y'' + 2y' \tanh x = 3 \sinh x$; $y(0) = 0$, $y'(0) = 2$
19. $e^x\sqrt{1 - e^{-2x}}\,y'' = \sqrt{1 - (y')^2}$; $y(0) = 2$, $y'(0) = -1$
20. Solve one of the three initial-value problems

$$(1 + y^2)(1 + \mathrm{Tan}^{-1} y)y'' + (y')^2 = 0; \qquad y(0) = 0,\ y'(0) = 1$$

$$(1 + y^2)(1 + y')y'' + (2 + \mathrm{Tan}^{-1} y)(y')^4 = 0; \qquad y(0) = 0,\ y'(0) = 1$$

$$(1 + y^2)y'' + (y')^3 = 0;\ y(0) = 0,\ y'(0) = 1$$

and verify that its solution is a solution of the other two problems.

1.14 ORTHOGONAL TRAJECTORIES

Our experience with first-order differential equations has taught us that such equations often have complete solutions containing a single arbitrary constant. Each such solution defines a corresponding set of integral curves. A nonempty set of plane curves defined by an equation involving just one parameter is commonly called a **one-parameter family** of curves. Of special importance in certain applications are those one-parameter families of curves which are *orthogonal trajectories* of one another.

DEFINITION 1 The curves of a family C are said to be **orthogonal trajectories** of the curves of a family K, and vice versa, if and only if each curve of either family is intersected by at least one curve of the other family and at every intersection of a curve of C with a curve of K the two curves are perpendicular.

For two curves to be perpendicular at a point, their tangent lines must be perpendicular there. The tangent line at any point of a straight line is the line itself. One-parameter families whose curves satisfy the conditions of Definition 1 are sometimes called **orthogonal families** of one another.

In a variety of practical problems involving orthogonal families, each family is defined by a general integral of a first-order differential equation. Moreover, every curve of one family intersects every curve of the other family, and there is at most a finite number of points of some region R where curves from both families fail to intersect.

EXAMPLE 1

ORTHOGONAL TRAJECTORIES OF A FAMILY OF CIRCLES

Find an equation of the family of all orthogonal trajectories of the family of circles which pass through the points $(-1, 0)$ and $(1, 0)$.

Clearly, the center of each circle of the given family is a point $(0, k)$ on the y axis. Hence an equation defining the family is $(x - 0)^2 + (y - k)^2 = 1 + k^2$ or

$$x^2 + y^2 - 1 = 2ky \tag{1}$$

To obtain an equation of the orthogonal trajectories of these circles we must first find an expression for the slope of a general member of the family (1) at a general point. This of course is just the process of finding a differential equation satisfied by a given family of functions, which we discussed in Sec. 1.5. In the present case, implicit differentiation of (1) gives

$$2x + 2yy' = 2ky' \tag{2}$$

If P is either $(-1, 0)$ or $(1, 0)$, every curve of the given family of circles passes through P and the slope at P depends on which curve we consider; that is, the slope depends on k. If P is not a point on the x axis, then a unique circle of the family passes through P, and its slope at P must be independent of k. To find an expression for this slope we divide (1) by y, then differentiate and solve for y', getting

$$y' = \frac{2xy}{x^2 - y^2 - 1} \tag{3}$$

This same result may also be found by eliminating k between (1) and (2).

Now the given circles and their orthogonal trajectories are to be perpendicular at every intersection. Hence at every intersection the slopes of the members of the two families which pass through that point are either negative reciprocals or else one is zero and the other is infinite. The orthogonal trajectory through each point where both slopes are finite is therefore an integral curve of the differential equation

$$y' = -\frac{x^2 - y^2 - 1}{2xy}\dagger \tag{4}$$

This can be written as

$$y' - \frac{1}{2x}y = \frac{1 - x^2}{2x}y^{-1} \tag{5}$$

and solved as a Bernoulli equation with $n = -1$. It is quicker, however, to write it first in the form

$$2xy\,dy - y^2\,dx = (1 - x^2)\,dx \tag{6}$$

and then divide by x^2, getting

$$d\left(\frac{y^2}{x}\right) = \left(\frac{1}{x^2} - 1\right) dx$$

which has $y^2/x = -1/x - x + 2h$ or $x^2 - 2hx + y^2 = -1$ or finally

$$(x - h)^2 + y^2 = h^2 - 1 \qquad h^2 > 1 \tag{7}$$

as a general solution. Every circle of the family defined by (7) is an orthogonal trajectory of the given family of circles (1). Typical members of the two families are shown in Fig. 1.7.

The family (7) does not include all orthogonal trajectories of the family (1), however, for from Fig. 1.7 it is clear that the y axis, defined by $x = 0$, is also perpendicular to each of the original circles but is not included

---

†As we have already observed, the slope of the circle of the given family which passes through a point (x, y), other than $(-1, 0)$ or $(1, 0)$, must be independent of k. Thus it would be incorrect to take the slope determined by (2), namely,

$$y' = \frac{x}{k - y}$$

and use its negative reciprocal

$$y' = -\frac{k - y}{x}$$

as the differential equation of the required orthogonal trajectories. The family of curves obtained by integrating this simple separable equation consists of all straight lines defined by

$$y = cx + k$$

which contains two parameters, c and k. A line of this family and a circle of the original family will intersect at right angles if and only if the line runs through the center of the circle; that is, the line and the circle correspond to the same value of k.

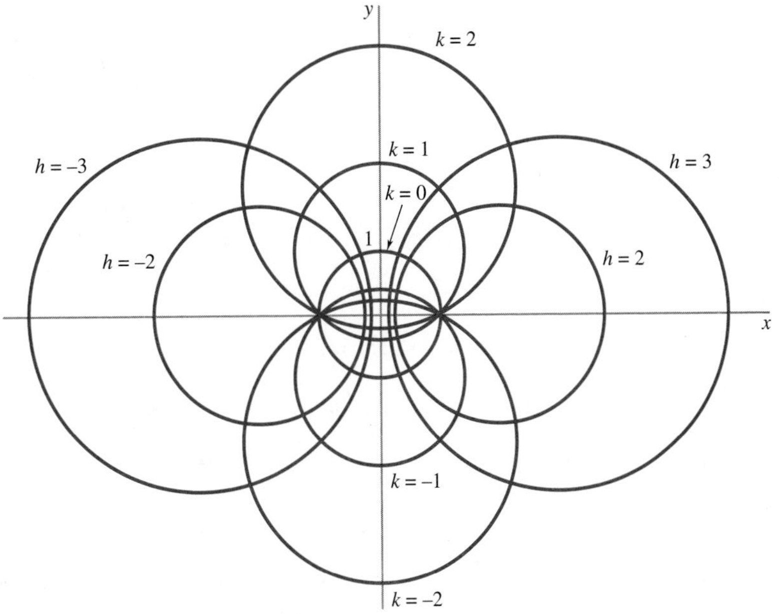

Figure 1.7
The orthogonal trajectories discussed in Example 1.

in (7) for any value of h. Relative to (7), $x = 0$ is a singular solution of the differential equation (6). Thus an equation defining *all* orthogonal trajectories of the given family of circles is

$$x[(x - h)^2 + y^2 - h^2 + 1] = 0 \qquad h^2 > 1$$

None of these trajectories pass through $(-1, 0)$ or $(1, 0)$.

It is also the case that the family of circles (1) does not include all orthogonal trajectories of the circles defined by (7). In fact, it is clear from Fig. 1.7 that the x axis, defined by $y = 0$, is also perpendicular to every circle of the family (7) but $y = 0$ is not included in (1) for any value of k. Relative to (1), $y = 0$ is a singular solution of the differential equation (3). Thus an equation defining *all* orthogonal trajectories of the family (7) is

$$y(x^2 + y^2 - 2ky - 1) = 0$$

Although the concept of orthogonal trajectories appears to be essentially a geometric one, it actually is intimately related to many important physical problems. For instance, in what are known in physics as two-dimensional field problems, the equipotential lines and the lines of flux are orthogonal trajectories. More specifically, if one knows the family of **isothermal curves,** which by definition are the curves joining points at the same temperature, in some problem involving heat flow within a plane region, then their orthogonal trajectories are the curves along which heat flows. In particular, as we shall see in Example 7, Sec. 11.7, the arcs of the circles (1) falling above the x axis are the **isotherms;** i.e., the isothermal curves for heat flow in the upper half-plane when the temperature distribution $T(x)$ maintained along the edge coinciding with the x axis is

$$T(x) = \begin{cases} 0 & |x| > 1 \\ T_0 & |x| < 1 \end{cases}$$

The upper halves of the circles defined by (7), together with the upper half of the y axis, are then the paths along which the heat flows.

Still another interpretation of orthogonal trajectories is this: if the curves of one family are thought of as **contour lines** on a topographic map, which are horizontal projections of lines of constant elevation on the surface being mapped, then their orthogonal trajectories are projections of the lines of steepest descent on the surface, that is, the paths along which rain falling on the hill would run off.

EXERCISES

1. **(a)** Identify the orthogonal trajectories of the family of straight lines $y = c$.
(b) A point (x, y) such that both x and y are integers is called a **lattice point.** How many members of the *two* families pass through lattice points whose distance from the origin is less than 3?
(c) How many members of the *two* families pass through a lattice point on the circumference of the circle of radius 3 centered at the origin?

2. Let C be the one-parameter family of curves defined by $x(y - |cx|) = 0$ and let K be the family of circles given by $x^2 + y^2 = k^2$, c and k real.
(a) Sketch several curves of each family.
(b) Are the curves of C orthogonal trajectories of the curves of K?
(c) Is every orthogonal trajectory of the curves of K a curve of C? Explain.
(d) Do all curves of both families intersect at the origin?
(e) Which curves of the two families have a tangent line at the origin?

3. **(a)** In Eq. (1) of Example 1 set $k = 1/2b$ and clear fractions to obtain another general solution of (3). Relative to this new solution
(b) Is the solution $y = 0$ singular?
(c) Is the solution $x^2 + y^2 = 1$ singular?

4 **(a)** In Example 1, obtain a general solution of (6) which particularizes to $x = 0$. *Hint:* See Exercise 3.
(b) Describe the graph of (7), Example 1, if $h = 1$ or $h = -1$.

Find the orthogonal trajectories of the curves of each of the following families.

5. $x^2 - y^2 = c$
6. $y^2 = cx^3$
7. $y = (x - c)^2$
8. $x^2 + 2y^2 = cy$
9. $y^2 = x^2 + cx$
10. $x^2 + y^2 = cx$
11. $\cos x \cosh y = c$
12. $e^x(x \cos y - y \sin y) = c$

13. A one-parameter family of curves is defined by $y = ce^x$.
(a) Find an equation of all orthogonal trajectories of the given family.
(b) Write equations of the curves of both families which pass through $(0, 1)$.
(c) Sketch the curves through $(0, 1)$.

14. Find an equation of the family of all circles which pass through the origin and have their centers on the x axis. Then find an equation of all orthogonal trajectories of these circles. Finally, sketch the curves of the two orthogonal families that pass through $(-1, -1)$ and give defining equations for them.

15. **(a)** Identify the curves of the family defined by $xy^2 - 4cx^2 = 0$, c real.
(b) Find an algebraic equation defining the family of all orthogonal trajectories of the given family.
(c) Sketch several typical curves of the two orthogonal families.
(d) Does the given family contain all orthogonal trajectories of its orthogonal family?
(e) How many members of each family pass through the origin? Through any other point of the xy plane?

16. Isotherms of a lamina in the xy plane are defined by $x^2 + 3y^2 = c$. **(a)** Find their orthogonal trajectories. Then determine the isotherm and the **flux line** (i.e., the curve along which heat flows) passing through the points **(b)** $(0, 2)$, **(c)** $(-1, 1)$, and **(d)** $(2, 0)$. **(e)** How many isotherms and how many flux lines pass through the origin?

17. A long, narrow mound with a flat, circular top rises from a level plain. On a topographic map, the contour of the zero-level line consists of the straight lines defined by $y^2 = 1$, and the other contours are ellipses defined by $x^2/\cosh^2 c + y^2 = 1$. Figure 1.8 shows some of these contours and indicates that there are two points where the mound drops vertically to the base plain. Clearly, no contours are lost by requiring c to be nonnegative.
(a) Find the orthogonal trajectories of the given family of contours which lie in the same region as the contours themselves.
(b) Write an equation of the contour through $(\sqrt{2}, 1/\sqrt{2})$ and of the orthogonal trajectory through the same point and draw their graphs.

18. A one-parameter family of curves is defined by the equation $y^2 = 4c(x + c)$.
(a) Identify the curves of this family corresponding to finite values of c.
(b) Find a differential equation of the family.
(c) Find a differential equation of all orthogonal trajectories of the given family. *Hint:* After algebraically solving for y' in Part **(b),** take the negative reciprocal and rationalize the denominator or else consider the effect of replacing dy/dx by $-dx/dy$ in the differential equation of Part **(b).**
(d) Is it true that at any point where distinct curves of the given family intersect they intersect at right angles? A family of curves that is its own orthogonal family is said to be **self-orthogonal.**

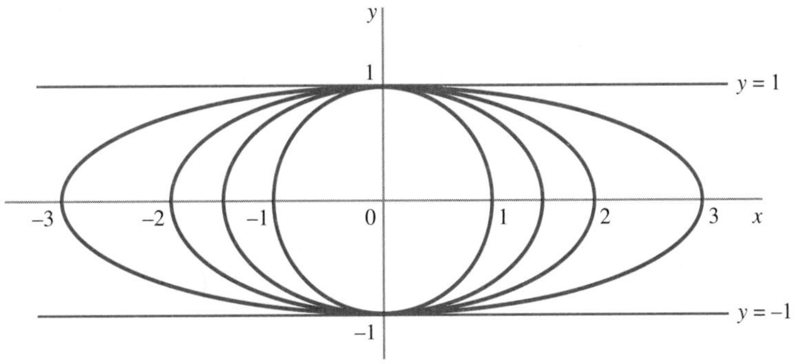

FIGURE 1.8
Contours defined by the equations $y^2 = 1$ and $x^2/\cosh^2 c + y^2 = 1$.

19. Families of orthogonal trajectories are not always found through the use of differential equations. In particular, according to Property 2, Sec. 17.3, if any analytic (i.e., differentiable) function of the complex variable $z = x + iy$ is reduced to the standard form $u(x, y) + iv(x, y)$, where $u(x, y)$ and $v(x, y)$ are real functions, then the curves $v(x, y) = k$ are orthogonal trajectories of the curves $u(x, y) = c$, and vice versa. Verify this fact for the function $1/z$ and compare your results with the results of Exercise 10.

20. Verify the property described in Exercise 19 for the function z^3.

1.15 APPLICATIONS OF FIRST-ORDER DIFFERENTIAL EQUATIONS

Most of the important problems studied in engineering and the physical sciences involve continuously changing quantities such as time, distance, force, temperature, or voltage. On the other hand, many problems in business, economics, and the life sciences deal with aggregates of things such as populations, budgets, accidents, or diseases, which clearly are discrete rather than continuous in nature. Since derivatives, and hence differential equations, are meaningful only for quantities that change continuously, one might expect that differential equations would arise only in the mathematical formulation of physical problems.

This is not the case, however, because sometimes functions that take on only discrete values can be treated as though they actually have derivatives and satisfy differential equations. Whether such an approach is justified depends simply on how well a solution of the mathematical formulation of the problem describes the phenomena being studied. If all sets of observed data check satisfactorily with the mathematical solution and if the solution predicts results which are borne out by further experiments, then the mathematical formulation of the problem constitutes an acceptable **mathematical model** of the real-world problem. Otherwise, the model must be either rejected or refined into one that is more appropriate. It is a further tribute to mathematics that this kind of modeling is becoming increasingly effective in so many areas of inquiry.

The following examples illustrate how the mathematical formulation of problems involving continuously, or discretely, changing quantities leads to differential equations of the types we have now learned to solve.

EXAMPLE 1 **MIXING OF SOLUTIONS**

A tank is initially filled with 100 gal of salt solution containing 1 lb of salt per gallon. Fresh brine containing 2 lb of salt per gallon runs into the tank at the rate of 5 gal/min, and the mixture, assumed to be kept uniform by stirring, runs out at the same rate. Find the amount of salt in the tank at any time t and determine how long it will take for this amount to reach 150 lb.

Let Q pounds be the total amount of salt in the tank at any time t. Then dQ/dt is the rate at which Q is changing. Now dQ/dt is clearly equal to the rate at which salt is entering the tank minus the rate at which salt is leaving the tank. The rate at which salt enters the tank is

$$\frac{5 \text{ gal}}{\text{min}} \times \frac{2 \text{ lb}}{\text{gal}} = \frac{10 \text{ lb}}{\text{min}}$$

At any time t, the amount of salt per gallon of solution is $Q/100$ pounds per gallon. Since the concentration of salt in the mixture running out of the tank is the same as the concentration $Q/100$ in the tank itself, the rate at which salt leaves the tank is

$$\frac{5 \text{ gal}}{\text{min}} \times \frac{Q}{100} \frac{\text{lb}}{\text{gal}} = \frac{Q}{20} \frac{\text{lb}}{\text{min}}$$

Therefore, in pounds of salt per minute,

$$\frac{dQ}{dt} = 10 - \frac{Q}{20}$$

This equation can be written

$$\frac{dQ}{dt} + \frac{Q}{20} = 10 \tag{1}$$

and treated as a linear equation, or it can be solved as a separable equation.

Considering it as a linear equation, we compute the integrating factor

$$e^{\int dt/20} = e^{t/20}$$

which we multiply times both members of (1). The result can be written

$$(Qe^{t/20})' = 10e^{t/20} \tag{2}$$

From this, by integration, we get

$$Qe^{t/20} = 200e^{t/20} + c \qquad \text{or} \qquad Q = 200 + ce^{-t/20}$$

Imposing the initial condition $Q(0) = 100$, we find

$$100 = 200 + c \qquad \text{or} \qquad c = -100$$

Hence

$$Q = 200 - 100e^{-t/20}$$

To find how long it will be before there is 150 lb of salt in the tank, we must find the value of t such that

$$150 = 200 - 100e^{-t/20} \qquad \text{or} \qquad e^{-t/20} = \tfrac{1}{2}$$

From this we have at once

$$-\frac{t}{20} = \ln\frac{1}{2} = -\ln 2 \doteq -0.693 \qquad \text{and} \qquad t \doteq 13.9 \text{ min}$$

EXAMPLE 2

LEAKAGE THROUGH AN ORIFICE

A hemispherical tank of radius R is initially filled with water. At the bottom of the tank, there is a hole of radius r through which the water drains under the influence of gravity. Find the depth of the water in the tank at any time t and determine how long it will take the tank to drain completely.

Let the origin be chosen at the lowest point of the tank and let y be the instantaneous depth, V the instantaneous volume, and x the instantaneous radius of the free surface of the water (Fig. 1.9). Then in an infinitesimal time interval dt, the water level will fall by an amount dy, and the resultant decrease in the volume of the water in the tank will be

$$dV = \pi x^2 \, dy$$

Since y decreases as t increases, both dy and dV are negative quantities. Thus $-\pi x^2 \, dy$ is positive and must equal the volume of water that leaves the orifice during the same interval dt. Now by **Torricelli's law**† the velocity with which a liquid issues from an orifice is

$$v = \sqrt{2gh}$$

where g is the acceleration of gravity and h is the instantaneous height, or **head,** of the liquid above the orifice. In the interval dt, then, a stream of water of length $v \times dt = \sqrt{2gy}\, dt$ and of cross-sectional area πr^2‡ will emerge from the outlet. The volume of this amount of water is

$$\text{Area} \times \text{length} = \pi r^2 \sqrt{2gy}\, dt$$

Upon equating $-\pi x^2 \, dy$ to this volume, we obtain the differential equation

$$-\pi x^2 \, dy = \pi r^2 \sqrt{2gy}\, dt \tag{3}$$

Before this equation can be solved, x must be expressed in terms of y. This is easily done through use of the equation of the circle which describes the vertical cross section of the tank:

$$x^2 + (y - R)^2 = R^2 \qquad \text{or} \qquad x^2 = 2yR - y^2$$

With this, the differential equation (3) can be written

$$\pi(2yR - y^2)\, dy = -\pi r^2 \sqrt{2gy}\, dt$$

This is a simple separable equation which can be solved without difficulty:

$$(2Ry^{1/2} - y^{3/2})\, dy = -r^2\sqrt{2g}\, dt$$

$$\tfrac{4}{3}Ry^{3/2} - \tfrac{2}{5}y^{5/2} = -r^2\sqrt{2g}\, t + c$$

Since $y = R$ when $t = 0$, we find

$$\tfrac{14}{15}R^{5/2} = c$$

and thus

$$\tfrac{4}{3}Ry^{3/2} - \tfrac{2}{5}y^{5/2} = -r^2\sqrt{2g}\, t + \tfrac{14}{15}R^{5/2}$$

To find how long it will take the tank to empty, we must determine the value of t when $y = 0$:

$$0 = -r^2\sqrt{2g}\, t + \tfrac{14}{15}R^{5/2}$$

$$t = \frac{14}{15}\frac{R^{5/2}}{r^2\sqrt{2g}}$$

†Named for the Italian mathematician and physicist Evangelista Torricelli (1608–1647).

‡This neglects the fact that the stream contracts as it leaves the orifice. How much the cross section of the stream decreases depends in a very complicated way upon the size and shape of both the tank and the orifice and also upon the head. However, in most practical problems reasonably accurate answers can be obtained by assuming that the cross section of the stream just after it leaves the orifice is 0.6 times the area of the orifice.

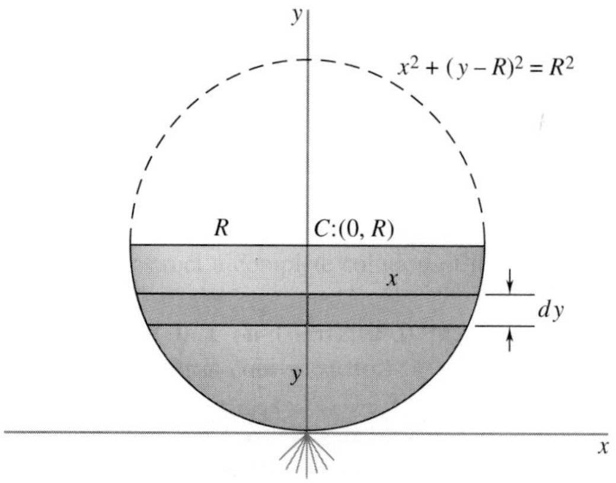

Figure 1.9
A vertical plane section through the center of a hemispherical tank.

EXAMPLE 3

DISSOLVING OF A SOLID IN A LIQUID

Under certain conditions it is observed that the rate at which a solid substance dissolves varies directly as the product of the amount of undissolved solid present in the solvent and the difference between the saturation concentration and the instantaneous concentration of the substance. If 20 lb of solute is dumped into a tank containing 120 lb of solvent and at the end of 12 min the concentration is observed to be 1 part in 30, find the amount of solute in solution at any time t if the saturation concentration is 1 part of solute in 3 parts of solvent.

If Q is the amount of the material in solution at time t, then $20 - Q$ is the amount of undissolved material present at that time and $Q/120$ is the corresponding concentration. Hence, according to the given law,

$$\frac{dQ}{dt} = k(20 - Q)\left(\frac{1}{3} - \frac{Q}{120}\right) = \frac{k}{120}(20 - Q)(40 - Q)$$

This is a simple separable equation, and we have at once

$$\frac{dQ}{(20 - Q)(40 - Q)} = \frac{k}{120}\,dt$$

To integrate the left-hand member it is convenient to use the method of partial fractions and write

$$\frac{1}{(20 - Q)(40 - Q)} = \frac{A}{20 - Q} + \frac{B}{40 - Q} = \frac{A(40 - Q) + B(20 - Q)}{(20 - Q)(40 - Q)}$$

This will be an identity if and only if

$$1 = A(40 - Q) + B(20 - Q)$$

Setting $Q = 20$ and $Q = 40$ in turn, we find that $A = \frac{1}{20}$ and $B = -\frac{1}{20}$. Hence the differential equation can be written

$$\frac{1}{20}\left(\frac{1}{20 - Q} - \frac{1}{40 - Q}\right) dQ = \frac{k}{120}\,dt$$

and, integrating, we have

$$-\ln(20 - Q) + \ln(40 - Q) = \frac{k}{6}t + c \tag{4}$$

To determine the integration constant c we observe that $Q = 0$ when $t = 0$. Hence

$$-\ln 20 + \ln 40 = c \qquad \text{or} \qquad c = \ln 2$$

and (4) can be written

$$\ln \frac{40 - Q}{2(20 - Q)} = \frac{k}{6} t \tag{5}$$

To find the physical constant k we use the fact that when $t = 12$, the concentration $Q/120 = \frac{1}{30}$ or $Q = 4$. Substitution of these values gives

$$\ln \tfrac{36}{32} = 2k \qquad \text{or} \qquad k = \tfrac{1}{2} \ln \tfrac{9}{8} \doteq 0.05889$$

Passing to exponential form from Eq. (5), in order to solve for Q, we have

$$\frac{40 - Q}{40 - 2Q} \doteq e^{0.0098t}$$

and finally

$$Q \doteq \frac{40 - 40e^{0.0098t}}{1 - 2e^{0.0098t}} = \frac{40(1 - e^{-0.0098t})}{2 - e^{-0.0098t}}$$

EXAMPLE 4

HEAT LOSS FROM A PIPE

According to **Fourier's law† of heat conduction,** the amount of heat (in Btu per unit time) flowing through an area is proportional to the area and to the temperature gradient (in degrees per unit length) in the direction of the perpendicular to the area. On the basis of this law, obtain a formula for the steady-state heat loss per unit time from a unit length of pipe of radius r_0 carrying steam at temperature T_0 if the pipe is covered with insulation of thickness w, the outer surface of which remains at the constant temperature T_1. What is the temperature distribution through the insulation; i.e., what is the temperature in the insulation as a function of the radius?

Since the problem tells us that steady-state conditions have been reached, it follows that the heat loss per unit time from a unit length of the pipe is a constant independent of time, say Q. Furthermore, it is reasonable to suppose that heat conduction through the insulation in the direction of the length of the pipe is negligible in comparison with the heat flow in the radial direction; and this we shall assume to be the case. We shall also make the obvious assumption that the heat flow through the insulation has circular symmetry; i.e., we shall assume that the temperature in the insulation depends only on the radial distance r. Let us now consider a typical cross section of the pipe and insulation, as suggested in Fig. 1.10. Clearly, under the assumption that all heat flow through the insulation is radial, it follows that for the unit length of pipe we are considering, all the heat that passes into the insulation through its inner surface will eventually pass into the air through its outer surface. Moreover, on the way, this same amount of heat Q will also pass through every coaxial cylindrical area between r_0 and $r_1 = r_0 + w$. Now if we let T denote the temperature in the insulation at the radius r, it follows that dT/dr is the temperature gradient (or temperature change per unit length) in the direction perpendicular to the cylindrical area of radius r. Hence, by Fourier's law, we have for the (as yet unknown) amount of heat Q flowing through this general area per unit time,

$$\begin{aligned} Q &= \text{thermal conductivity} \times \text{area} \times \text{temperature gradient} \\ &= k(1 \times 2\pi r)\frac{dT}{dr} \end{aligned}$$

†Named for the French mathematician J. B. J. Fourier (1768–1830).

We thus have the exceedingly simple separable equation

$$dT = \frac{Q}{2\pi k}\frac{dr}{r}$$

Hence

$$T = \frac{Q}{2\pi k}\ln r + c$$

To determine the integration constant c, we use the fact that $T = T_0$ when $r = r_0$, from which

$$T_0 = \frac{Q}{2\pi k}\ln r_0 + c \qquad \text{or} \qquad c = T_0 - \frac{Q}{2\pi k}\ln r_0$$

and, substituting and collecting terms,

(6) $$T = T_0 + \frac{Q}{2\pi k}(\ln r - \ln r_0)$$

Furthermore, $T = T_1$ when $r = r_0 + w = r_1$. Hence

$$T_1 = T_0 + \frac{Q}{2\pi k}(\ln r_1 - \ln r_0)$$

from which we find easily that

(7) $$Q = \frac{(T_1 - T_0)2\pi k}{\ln r_1 - \ln r_0}$$

Since k is the (presumably) known thermal conductivity of the insulation, this formula gives the heat loss per unit time, as required.

To find the temperature distribution through the insulation, we merely substitute for $Q/2\pi k$ from (7) into (6), getting

$$T = T_0 + (T_1 - T_0)\frac{\ln r - \ln r_0}{\ln r_1 - \ln r_0}$$

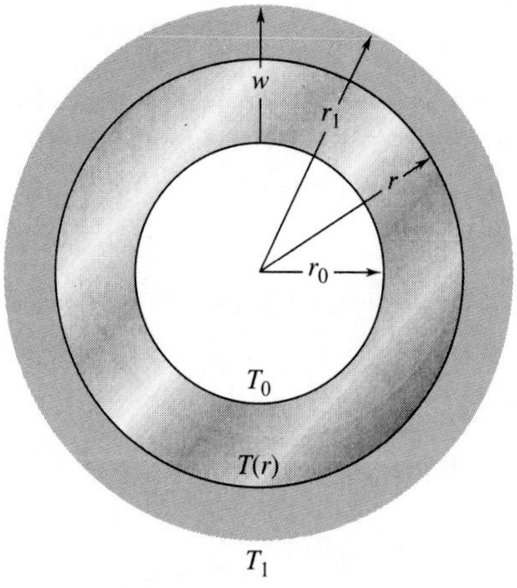

Figure 1.10
A typical cross section of an insulated pipe.

EXAMPLE 5

LAMINAR FLUID FLOW

In fluid mechanics, a flow pattern in which the fluid may be divided into parallel layers which flow past one another is said to be **laminar.** In laminar flow, when adjacent layers move in the same direction with different velocities, each layer exerts a force on the other. The magnitude F of either force is proportional to the area of contact A between the two layers and to the velocity gradient dv/dn perpendicular to the direction of flow; that is,

$$F = \left|\eta A \frac{dv}{dn}\right| \tag{8}$$

The proportionality constant η is called the **coefficient of viscosity.** With these ideas in mind, consider a fluid moving through a uniform cylindrical pipe of inside diameter $2a$ and length l. Assuming the pipe is full of fluid and that the hydrostatic pressures at the intake and outlet ends of the pipe are p_0 and p_1, respectively, determine the velocity of flow as a function of the coordinate r measured radially outward from the axis of the pipe.

Let the axis of the pipe be the x axis with its positive direction in the direction of flow. Since the flow pattern through each cross section has circular symmetry, it is convenient to analyze the forces on the fluid in a cylindrical shell of inner radius r and thickness Δr (Fig. 1.11). Only exterior forces on the shell which act parallel to the x axis need be considered.

Since the curved area of a cylinder of length l and radius r is $2\pi rl$, it follows from (8) that the viscous force distributed over the inner surface of the shell is of magnitude $|\eta 2\pi rlv'(r)|$. This force acts in the positive x direction because the velocity of flow increases toward the center of the pipe, that is, $v'(r) < 0$. Taking forces to be positive or negative, according as they act in the positive or negative x direction, the inner viscous force is given as a function of r by the positive expression

$$-(2\pi\eta l)rv'(r)$$

On the other hand, the viscous force distributed over the outer surface of the shell is a negative force which acts in the negative x direction over a surface of area $2\pi l(r + \Delta r)$ where the velocity gradient is $v'(r + \Delta r)$. This force is given by

$$(2\pi\eta l)(r + \Delta r)v'(r + \Delta r)$$

Hydrostatic forces act at both ends of the shell. The cross-sectional area of either end is $2\pi r\,\Delta r$. A positive hydrostatic force

$$p_0(2\pi r)\,\Delta r$$

acts on the surface of the shell at the intake end, and a negative force

$$-p_1(2\pi r)\,\Delta r$$

acts on the surface at the outlet end.

Under steady-state conditions, the sum of the viscous forces and the hydrostatic forces must be zero. Hence

$$(2\pi\eta l)(r + \Delta r)v'(r + \Delta r) - (2\pi\eta l)rv'(r) + p_0(2\pi r)\,\Delta r - p_1(2\pi r)\,\Delta r = 0$$

Transposing the last two terms and dividing by $2\pi\eta l\,\Delta r$, we obtain

$$\frac{(r + \Delta r)v'(r + \Delta r) - rv'(r)}{\Delta r} = \frac{-r(p_0 - p_1)}{\eta l}$$

Thus in the limit as $\Delta r \to 0$ we have the differential equation

$$\frac{d}{dr}\left(r\frac{dv}{dr}\right) = -\frac{r(p_0 - p_1)}{nl}$$

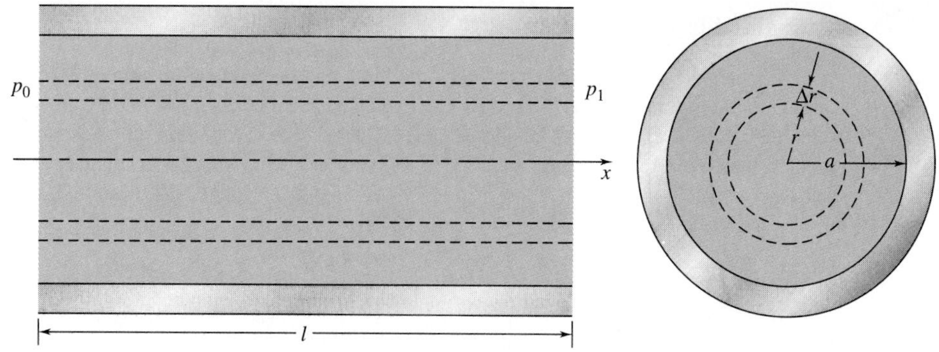

Figure 1.11
Laminar flow of a fluid through a cylindrical pipe.

Integrating this we obtain

$$r\frac{dv}{dr} = -\frac{r^2(p_0 - p_1)}{2\eta l} + c_1$$

Solving for dv/dr and integrating again gives us

$$v = -\frac{r^2(p_0 - p_1)}{4\eta l} + c_1 \ln r + c_2$$

To determine c_1 and c_2, we note first that v must be finite at the center of the pipe, where $r = 0$. Hence $c_1 = 0$. Then we observe that at the inner surface of the pipe, where $r = a$, the velocity is zero. Hence

$$c_2 = \frac{a^2(p_0 - p_1)}{4\eta l}$$

and the final expression for v is

$$v = \frac{a^2(p_0 - p_1)}{4\eta l}\left(1 - \frac{r^2}{a^2}\right)$$

A force whose line of action always passes through a fixed point O and whose magnitude is a function only of the distance r from O is called a **central force.** Such a force is **attractive** or **repulsive** according as it is directed *toward* or *away* from the **center** O. An **inverse square** force is a *central* force whose magnitude is inversely proportional to r^2. Under the sole influence of a central force a body must always traverse a plane curve [Exercise 49(**a**)]. Our next example is a simple version of a classical problem of planetary motion.

EXAMPLE 6

ORBITAL MOTION

Find a polar equation of the orbit traversed by a mass m acted upon by an attractive inverse square force.

Since the motion must take place along a plane curve, we choose the plane of the orbit as an xy plane whose origin is the center O toward which the force is directed and from which the distance r to m is measured. As usual, (x, y) and (r, θ) give the position of m at time t in rectangular and polar coordinates. Thus $x = r\cos\theta$, $y = r\sin\theta$, and $r > 0$.

Now, we are given that the force on m is directed toward O and has magnitude k/r^2, where k is a positive constant. By Newton's second law, the x and y components of the force are

$$m\ddot{x} = -\frac{k}{r^2}\cos\theta \tag{9}$$

$$m\ddot{y} = -\frac{k}{r^2}\sin\theta \tag{10}$$

where each overdot indicates a differentiation with respect to t. Subtracting (9) multiplied through by $r\sin\theta$ from (10) multiplied through by $r\cos\theta$ and simplifying the difference by setting $r\cos\theta = x$, $r\sin\theta = y$ and dividing by m, we get

$$x\ddot{y} - y\ddot{x} = 0 \qquad \text{which can be written} \qquad \frac{d}{dt}(x\dot{y} - y\dot{x}) = 0$$

It follows that $x\dot{y} - y\dot{x} = c$ or, in polar coordinates,

$$r^2\dot{\theta} = c \tag{11}$$

Should $\dot{\theta}$ be zero for some value of t, the parameter c would be zero and both $\dot{\theta} = 0$ and $\theta(t) =$ constant would hold identically. The orbit would then lie along a straight line but not in a *unique* plane. For the orbit to be properly planar we must have $c \neq 0$.

With c thus restricted, we next differentiate $x = r\cos\theta$ twice with respect to t to obtain

$$\ddot{x} = (\ddot{r} - r\dot{\theta}^2)\cos\theta - (r\ddot{\theta} + 2\dot{r}\dot{\theta})\sin\theta \tag{12}$$

We also differentiate (11) which, since $r \neq 0$, gives $r\ddot{\theta} + 2\dot{r}\dot{\theta} = 0$, and this shows that the coefficient of $\sin\theta$ in (12) is zero. Replacing $\dot{\theta}^2$ by c^2/r^4 in what remains of (12), we get $\ddot{x} = (\ddot{r} - c^2/r^3)\cos\theta$ which, when substituted into (9), leads to

$$\ddot{r} - \frac{c^2}{r^3} = \frac{-k}{mr^2} \tag{13}$$

A like reduction of $\ddot{y}$, when substituted into (10), yields this same result. Now, since we are to find a polar equation rather than parametric equations of the orbit, let us transform (13) into an equation having θ instead of t as the independent variable.

To express d^2r/dt^2 in terms of derivatives of r with respect to θ, we use the chain rule twice. Then, simplifying the result by means of (11), we get

$$\frac{d^2r}{dt^2} = \frac{c^2}{r^4}\left[\frac{d^2r}{d\theta^2} - \frac{2}{r}\left(\frac{dr}{d\theta}\right)^2\right]$$

which, when substituted into (13), leads to

$$\frac{d^2r}{d\theta^2} - \frac{2}{r}\left(\frac{dr}{d\theta}\right)^2 = \frac{r}{r_c}(r_c - r) \qquad \text{where } r_c = \frac{mc^2}{k} \tag{14}$$

This equation clearly has $r = r_c$ as a solution, and the corresponding orbit is a circle.

To find solutions of Eq. (14) which are not constant, we first reduce its order by applying the procedure of Case 2, Sec. 1.13. Making the substitutions $dr/d\theta = v$ and $d^2r/d\theta^2 = v\,dv/dr$ in (14), then dividing through by v, we obtain the Bernoulli equation

$$\frac{dv}{dr} - \frac{2}{r}v = \frac{r}{r_c}(r_c - r)v^{-1}$$

which we immediately multiply through by $2v$. The subsequent change of dependent variable $z = v^2$ yields the linear equation

$$\frac{dz}{dr} - \frac{4}{r}z = \frac{2r}{r_c}(r_c - r)$$

A related integrating factor is $1/r^4$, thus

$$(r^{-4}z)' = 2\left(r^{-3} - \frac{r^{-2}}{r_c}\right) \qquad \text{and} \qquad z = r^2\left(ar^2 + 2\frac{r}{r_c} - 1\right)$$

where a is a new parameter.

But $z = v^2 = (dr/d\theta)^2$; hence, upon reverting back to r as dependent variable, the solution of our linear equation becomes

$$\left(\frac{dr}{d\theta}\right)^2 = r^2\left(ar^2 + 2\frac{r}{r_c} - 1\right) = \frac{r^2}{r_c^2}[r^2(1 + ar_c^2) - (r - r_c)^2] \tag{15}$$

from which we see that a can only take on values such that $1 + ar_c^2 > 0$ because $(dr/d\theta)^2 \geq 0$ and $r \not\equiv r_c$. Setting $b = \sqrt{1 + ar_c^2}$ in (15), extracting square roots, separating variables, and rearranging, we have

$$\pm d\theta = \frac{r_c\, dr}{br^2\sqrt{1 - [(r - r_c)/br]^2}} = \frac{d[(r - r_c)/br]}{\sqrt{1 - [(r - r_c)/br]^2}} \tag{16}$$

The final member of (16) has either $\text{Sin}^{-1}[(r - r_c)/br]$ or $-\text{Cos}^{-1}[(r - r_c)/br]$ as an antiderivative. Choosing the latter, we integrate (16) and introduce α as an arbitrary constant, getting

$$\mp(\theta + \alpha) = \text{Cos}^{-1}\frac{r - r_c}{br} \qquad \text{and then} \qquad r - r_c = br\cos(\theta + \alpha)$$

The last equation, when solved for r, gives

$$r = \frac{r_c}{1 - b\cos(\theta + \alpha)} \tag{17}$$

as a polar equation defining the general motion of m. To account for a circular orbit, we let $b = 0$ in (17).

For all parameter values $b \geq 0$, $r_c > 0$, and α real, the orbit defined by Eq. (17) is a conic section. According as $0 < b < 1$, $b = 1$, or $b > 1$, the conic is an ellipse, parabola, or hyperbola of eccentricity b. One focus is at the origin, and r_c/b is the distance from O to the corresponding directrix.

All six of the preceding examples have dealt with continuously changing quantities. In each problem, the essential part of the corresponding mathematical model was a differential equation. The next two examples are concerned with populations which in reality take on only discrete values. Nevertheless, differential equations are used to describe how the populations behave.

EXAMPLE 7

EXPONENTIAL POPULATION GROWTH

An initial colony of N_0 microorganisms reproduces through simple cell division under ideal conditions of unlimited food supply and total absence of predators. If it is observed that the population increases by ρ percent each hour, find an expression for the population at any time t.

According to the statement of the problem, if N is the number of organisms present at any time t, then 1 h later the number of organisms will have increased by the amount $\Delta N = (\rho/100)N$. In other words, for any period of length $\Delta t = 1$ h, the corresponding population change ΔN is proportional to the size of the population at the beginning of the hour. From this it seems plausible that a similar relation should hold for the increase in population in any period of time; that is, the increase in any interval should be proportional to the length of the interval Δt as well as the population size N at the beginning of the period. Under this assumption,

$$\Delta N = \left(\frac{r}{100}N\right)\Delta t \qquad \text{or} \qquad \frac{\Delta N}{\Delta t} = \frac{r}{100}N\dagger$$

We cannot logically consider the limit of the ratio $\Delta N/\Delta t$ as $\Delta t \to 0$ because, for Δt sufficiently small, the number of new organisms appearing in the interval Δt may be either 0 or 1 but nothing in between. Nonetheless, the last expression *suggests* that we explore the equation

$$\frac{dN}{dt} = \frac{r}{100}N \tag{18}$$

as an approximate description of the behavior of the actual system.

Equation (18) is both separable and linear, and it is readily verified that a complete solution is $N = ce^{rt/100}$. When $t = 0$, we know that $N = N_0$. Hence, putting $t = 0$, we find that $c = N_0$ and

$$N = N_0 e^{rt/100} \tag{19}$$

To determine the rate constant r, we must use the given information that when $t = 1$, the population has increased to $N + (\rho/100)N_0$. Thus

$$N_0 + \frac{\rho}{100}N_0 = N_0 e^{r/100} \qquad 1 + \frac{\rho}{100} = e^{r/100} \qquad r = 100 \ln\left(1 + \frac{\rho}{100}\right)$$

With r expressed in terms of the observed growth rate of ρ percent per hour, (19) may be rewritten as

$$N = N_0 e^{t \ln(1+\rho/100)} = N_0 e^{\ln(1+\rho/100)^t}$$

which simplifies to the formula

$$N = N_0\left(1 + \frac{\rho}{100}\right)^t$$

for the population size at any time t.

This example is of course unrealistic since it neglects such important factors as a limited food supply, the presence of other species competing for the food supply, and the presence of predators. However, it does illustrate the highly important **law of exponential growth,** sometimes called the **compound interest law,** which describes the increase of quantities which grow at a rate proportional to their current size. How dramatic this increase may be is shown in Table 1.1 which gives the time it will take a population to double for various rates of increase. Thus a population growing at only 1 percent per year will double in slightly less than 70 years.

---

†The factor r in this equation is not the same as the factor ρ. In fact, if $r = \rho$, then taking $\Delta t = \frac{1}{2}$ as an illustration, the number of organisms present at the end of $\frac{1}{2}$ h would be $N + (\rho/200)N = N[1 + (\rho/200)]$ and the number present at the end of a second half-hour would be

$$N\left(1 + \frac{\rho}{200}\right) + \left[N\left(1 + \frac{\rho}{200}\right)\right]\frac{\rho}{200} = N\left(1 + \frac{\rho}{200}\right)^2 = N\left(1 + \frac{\rho}{100} + \frac{\rho^2}{40{,}000}\right)$$

which is more than the observed number present after 1 h, namely, $N[1 + (\rho/100)]$.

TABLE 1.1

Percentage increase in one time period	Number of periods required for population to double
10	7.27
5	14.21
3	23.45
2	35.00
1	69.66
0.5	138.98

EXAMPLE 8

A FINITE POPULATION MODEL

In a finite world, no population can become infinite; limiting factors of one kind or another must come into play. One such factor is obviously the food supply, because if this is finite, the amount of food available for each individual decreases as the population increases. This in turn, through malnutrition or possibly starvation, should tend to make the death rate increase and the birthrate decrease, thereby reducing the growth rate enough to keep the population finite. One way to incorporate such a factor into a population model is to suppose that in the continuous approximation provided by (18) of Example 7, the growth factor r (arising from the excess of births over deaths) is not constant but decreases as N increases. One possible formula for the variation of r with N, which biologists have found in good agreement with experiments on fruit flies, is

$$\frac{r}{100} = a - bN \qquad a, b > 0$$

Under this assumption, the differential equation (18) describing the growth of the population becomes

$$\frac{dN}{dt} = (a - bN)N \tag{20}$$

At the outset, we note that both $N = 0$ and $N = a/b$ are solutions of (20). These are the unique solutions of (20), guaranteed by Theorem 1, Sec. 1.6, which satisfy the respective initial conditions $N(0) = 0$ and $N(0) = a/b$. The trivial solution $N = 0$ is of no practical interest. The solution $N = a/b$ asserts that if the size of the population is initially $N_0 = a/b$, it will remain stationary at that value.

Other solutions of (20) may be found by separation of variables or by solving it as a Bernoulli equation using the substitution $N^{-1} = z$. Either method gives

$$\frac{1}{N} = \frac{b}{a} + ce^{-at} \tag{21}$$

as a general solution. From this we easily find that the population at time t is

$$N = \frac{aN_0}{bN_0 + (a - bN_0)e^{-at}} \tag{22}$$

assuming N_0 is the number of individuals present when $t = 0$. For the rate of change of N we have

$$\frac{dN}{dt} = \frac{a^2N_0(a - bN_0)e^{-at}}{[bN_0 + (a - bN_0)e^{-at}]^2} \tag{23}$$

When $t = 0$, (22) of course reduces to $N = N_0$. As t becomes infinite, the factor e^{-at} approaches zero and N approaches the limiting value a/b. In fact, if $N_0 < a/b$, (23) implies that $dN/dt > 0$ and the size of the population steadily increases toward a/b as time goes on. Similarly, if $N_0 > a/b$, (23) implies that $dN/dt < 0$

and the size of the population decreases monotonically toward the asymptotic value a/b. Thus our model reflects reality at least to the extent that it predicts a finite limit to the growth of the population. To determine the parameters a and b, we would need to know the size of the population at two times other than $t = 0$. Substituting these data, say (t_1, N_1) and (t_2, N_2), into (22) would give two simultaneous equations in a and b. To solve them would be very difficult, however.

In any event, solution curves corresponding to the three cases of interest, $0 < N_0 < a/b$, $N_0 = a/b$, and $N_0 > a/b$, would have the appearance of the curves depicted in Fig. 1.12.

Although (20) was introduced as an approximate model for a biological population, the same kind of equation might be useful for other purposes, such as the prediction of stock inventories under restricted sources of supply and growing consumer demands or the estimation of economic trends in segments of the economy which are limited by finite resources.

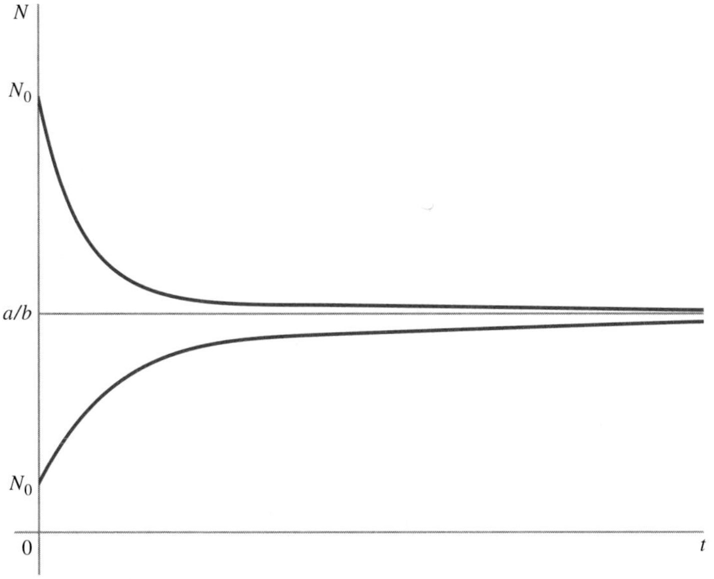

Figure 1.12
Curves showing the effect of initial colony size on the behavior of a population modeled by the equation $dN/dt = (a - bN)N$.

EXERCISES

1. Work Example 1 given that fresh water rather than brine runs into the tank.
2. Work Example 1 given that the tank is initially filled with pure water.
3. A tank contains 100 gal of brine in which 50 lb of salt is dissolved. Brine containing 2 lb/gal of salt runs into the tank at the rate of 3 gal/min, and the mixture, assumed to be kept uniform by stirring, runs out at the rate of 2 gal/min. Assuming that the tank is sufficiently large to avoid overflow, find the amount of salt in the tank as a function of the time t. When will the concentration of salt in the tank reach $\frac{3}{2}$ lb/gal? How much salt is in the tank at the end of 30 min?
4. Work Exercise 3 with the rates of influx and efflux interchanged and find the amount of salt in the tank after 1 h instead of at the end of 30 min. Under these circumstances, when will the tank contain the maximum amount of salt, and what is this amount?
5. A tank contains 200 gal of brine in which 20 lb of salt are dissolved. Brine containing $\frac{1}{4}$ lb of salt per gallon runs into the tank at the rate of 2 gal/min. The mixture, kept uniform by stirring, runs out at the same rate. What is the limiting value approached by the amount of salt in the tank as t increases indefinitely? When, if ever, will the amount of salt in solution be 20 lb?
6. Work Exercise 5 if the tank initially contains 100 gal of brine in which 25 lb of salt are dissolved and if brine containing 1 lb of salt per gallon runs into the tank.
7. A tank contains 1,000 gal of brine in which 500 lb of salt are dissolved. Fresh water runs into the tank at the rate of 10 gal/min, and the mixture, kept uniform by stirring, runs out at the same rate. How long will it be before only 50 lb of salt are left in the tank?
8. A tank contains 500 liters of pure water. Brine containing 1, $\frac{1}{2}$, and $\frac{1}{4}$ g/liter of salt flows into the tank through three pipes at the respective rates of 47, 41, and 37 li-

ters/min. Find **(a)** the amount of salt in the tank Q at time t, **(b)** when Q will equal 194 g, **(c)** $\lim_{t\to\infty} Q(t)$.

9. A lake having a storage capacity of 10,000 acre-ft is lowered until it contains only 1,000 acre-ft of water, at which time it is treated with 2 tons of a toxic chemical to rid it of undesirable aquatic life. New species are to be introduced which can tolerate no more than 10^{-4} lb of chemical per acre-foot of water. Fresh water enters the lake via several streams at the rate of 5 acre-ft/h, and water is released from the lake at the rate of 4 acre-ft/h. Assuming the chemical is quickly and uniformly distributed throughout the lake, when will the amount of chemical reach the tolerance level of the new species so that they may be planted? How much more water will the lake hold at that time?

10. A lake has various streams flowing into it and flowing out of it, the total rates of influx and efflux being equal. For a long time, the streams flowing into the lake were polluted, and pollution in the lake built up to an undesirable level. However, as a result of conservation efforts, the sources of pollution in the streams were eliminated, and now only pure water flows into the lake. If the volume of the lake is V km^3, if the rate of influx and efflux is r km^3/year, and if the pollutants are always uniformly distributed throughout the lake, obtain formulas for the time it will take for the pollution in the lake to be reduced **(a)** to one-half its level at the time of the cleanup and **(b)** to one-tenth its level at the time of the cleanup. Determine numerical values for these times for Lake Erie and Lake Ontario given the following data.

	V	r
Lake Erie	460 km^3	175 km^3/yr
Lake Ontario	1,600 km^3	209 km^3/yr

11. Work Example 2 if the tank has the shape of an inverted right circular cone of radius R and height h.

12. Work Example 2 if the tank is a sphere initially filled with water instead of just a hemisphere. How long will it take for half the water to run out? How long will it take for the tank to drain? (Assume that there is a small hole at the top of the sphere where air can enter the tank.)

13. Work Example 2 if the tank is a right circular cylinder of height h and radius R whose axis is vertical.

14. Work Example 2 if the tank has the shape of a right circular cylinder of radius R and height h and if in addition to a hole of radius r in the bottom of the tank there is also a hole of radius r in the side at a height of $h/2$ above the base.

15. Work Example 2 if the tank is formed by rotating about the y axis the arc of the parabola $y = x^2$ between $x = 0$ and $x = 2$.

16. A cylindrical tank of length l has semicircular end sections of radius R. The tank is placed in an untilted position with its axis horizontal and is initially filled with water. How long will it take the tank to drain through a hole of radius r in the bottom of the tank?

17. What is the shape of a tank which is a surface of revolution if the tank drains so that the water level falls at a constant rate?

18. What is the shape of a perpendicular cross section of a horizontal trough of constant cross section which drains so that the water level falls at a constant rate?

19. A tank having the shape of a right circular cylinder of height h and radius R is filled with water. The tank drains through an orifice whose area is controlled by a float valve in such a way that it is proportional to the instantaneous depth of the water in the tank. Express the depth of the water as a function of time. How long will it take the tank to drain?

20. Work Example 2 if instead of draining through an orifice of constant area the tank drains through an orifice whose area is controlled by a float valve in such a way that it is proportional to the instantaneous depth of the water in the tank.

21. Water flows into a vertical cylindrical tank of cross-sectional area A ft^2 at the rate of Q ft^3/min. At the same time, the water drains out under the influence of gravity through a hole of area a ft^2 in the base of the tank. If the water is initially h ft deep, find the instantaneous depth as a function of time. What is the limiting depth of the water as time increases indefinitely?

22. A vertical cylindrical tank of height h and radius R has a narrow crack of width w running vertically from top to bottom. If the tank is initially filled with water and allowed to drain through the crack under the influence of gravity, find the instantaneous depth of the water as a function of time. How long will it take the tank to empty? *Hint:* First imagine the crack to be a series of adjacent orifices and integrate to find the total efflux from the crack in the infinitesimal time interval dt.

23. Work Example 3 given that the amount of solute dumped into the tank is 40 lb instead of 20 lb.

24. Work Example 3 given that the saturation concentration is 1 part solute to 12 parts solvent.

25. Work Example 3 given that the saturation concentration is 1 part solute to 6 parts solvent.

26. Work Example 3 given that the saturation concentration is $\frac{1}{4}$ instead of $\frac{1}{3}$.

27. Work Example 3 with *concentration* defined as the ratio of solute to solution instead of solute to solvent.

28. Work Exercise 27 if the saturation concentration is 1 part solute to 7 parts solution.

29. Work Example 3 with *concentration* defined as in Exercise 27 given that the saturation concentration is $\frac{1}{4}$ instead of $\frac{1}{3}$.

30. When ethyl acetate in dilute aqueous solution is heated in the presence of a small amount of acid, it decomposes according to the reaction

$$\underset{\text{Ethyl acetate}}{CH_3COOC_2H_5} + \underset{\text{Water}}{H_2O} \longrightarrow \underset{\text{Acetic acid}}{CH_3COOH} + \underset{\text{Ethyl alcohol}}{C_2H_5OH}$$

Since this reaction takes place in dilute solution, the quantity of water present is so great that the loss of the small amount which combines with the ethyl acetate produces no appreciable change in the total amount. Hence of the reacting substances only the ethyl acetate suffers a measurable change in concentration. A chemical reaction of this sort, in which the concentration of only one reacting substance changes, is called a **first-order reaction.** It is a law of physical chemistry that the rate at which a substance is being used up, i.e., transformed, in a first-order reaction is proportional to the amount of that substance instantaneously present. If the initial concentration of ethyl acetate is C_0, find the expression for its concentration at any time t.

31. In some chemical reactions where two substances combine to form a third, the amount of each of the reacting substances changes appreciably. A reaction of this sort is called a **second-order reaction,** and in such cases it is observed that the rate at which the resulting compound is being formed is proportional to the product of the untransformed amounts of the two reacting substances. If two substances combine in the ratio 1:2 (by weight) to form a third substance and if it is observed that 10 min after 10 g of the first substance and 20 g of the second are mixed, the amount of the product that has been formed is 5 g, find an expression for the amount of the product present at any time t. How long will it be before one-half the final amount of the product is formed?

32. Work Exercise 31 given that 20 g of each substance are mixed.

33. Work Exercise 31 given that 10 g of the first substance and 30 g of the second substance are mixed.

34. Most first-order chemical reactions are reversible; that is, not only is substance A being transformed into substance B, but at the same time substance B is being transformed into substance A. If the rate constant for the reaction $A \to B$ is k_1, if the rate constant for the reaction $B \to A$ is k_2, and if initially the amount of substance A is A_0 and the amount of substance B is zero, find the amount of substance B present at any time t. What is the limiting value of the ratio of the amounts of A and B as the reaction approaches equilibrium? *Hint:* Note that the total rate of change of A consists of the rate at which A is being used up by the reaction $A \to B$ and the rate at which A is being produced by the reaction $B \to A$.

35. Some chemical reactions are **autocatalytic;** that is, the product of the reaction catalyzes its own formation. This means that in an autocatalytic reaction in which a substance A is transformed into a substance B, the rate of formation of B is proportional to the product of the instantaneous amounts of both A and B. If the initial amount of substance A is A_0 and the initial amount of B is B_0, find an expression for the amount of B present at any time t.

36. According to **Newton's law of cooling,** the rate at which the temperature of a body changes is proportional to the difference between the instantaneous temperature of the body and the temperature of the surrounding medium. If a body whose temperature is initially 100° C is allowed to cool in air which remains at the constant temperature 20° C and if it is observed that in 10 min the body has cooled to 60° C, find the temperature of the body as a function of time.

37. The temperature of a solid surrounded by air of constant temperature 50° F drops from an initial temperature of 200° F to 150° F in 2 h. Use Newton's law of cooling to express the temperature as a function of time.

38. A body cools in air of constant temperature 20° C according to Newton's law of cooling. Ten minutes after the body began to cool, its temperature was observed to be 75° C, and 10 min later its temperature was 50° C. What was its temperature when it began to cool?

39. Using Fourier's law of heat conduction, obtain a formula for the amount of heat lost under steady-state conditions from 1 ft^2 of furnace wall h feet thick if the temperature in the furnace is T_0, the temperature of the air outside the furnace is T_1, and the thermal conductivity of the material of the furnace wall is k. What is the temperature distribution through the wall?

40. The inner and outer surfaces of a hollow sphere are maintained at the respective temperatures T_0 and T_1. If the inner and outer radii of the spherical shell are r_0 and r_1 and if the thermal conductivity of the material of the shell is k, find the amount of heat lost from the sphere per unit time. What is the temperature distribution through the shell?

41. A tank and its contents weigh 100 lb. The average heat capacity of the system is 0.5 Btu/(lb)(°F). The liquid in the tank is heated by an immersion heater which delivers 100 Btu/min. Heat is lost from the system at a rate proportional to the difference between the temperature of the system, assumed constant throughout at any instant, and the temperature of the surrounding air, the proportionality constant being 2 Btu/(min)(°F). If the air temperature remains constant at 70° F and if the initial temperature of the tank and its contents is 55° F, find the temperature of the tank as a function of time.

42. A perfectly insulated tank of negligible heat capacity contains P lb of brine at T_0° F. Hot brine at T_a° F runs into the tank at the rate of a lb/min, and the brine in the tank, brought instantly to a uniform temperature throughout by vigorous stirring, runs out at the same rate. If the specific heat of the brine is 1 Btu/(lb)(°F), find the temperature of the brine in the tank as a function of time. *Hint:* If h_0 is the amount of heat in 1 lb of brine at the temperature T_0, the total amount of heat in the brine in the tank when its temperature is T is $H =$

$P[(T - T_0) + h_0]$. The change in the heat content of the brine in the tank during the time interval dt is then dH.

43. Work Exercise 42 if the brine, instead of running out at the influx rate of a lb/min runs out at the rate of b lb/min, where $b \neq a$. Verify that the solution to this exercise approaches the solution to Exercise 42 as $b \to a$. *Hint:* With the notation introduced in the hint for Exercise 42, the heat content of the brine in the tank in the present problem is

$$H = [P + (a - b)t][(T - T_0) + h_0]$$

44. A perfectly insulated tank of negligible heat capacity contains 1,000 lb of brine at 60° F. It is necessary that in exactly 20 min the tank contain 2,000 lb of brine at 150° F. To accomplish this, hot brine at 160° F is run into the tank and mixed brine, kept at a uniform temperature by stirring, is run off. What should be the rates of influx and efflux to achieve the desired conditions?

45. A metal ball at temperature 100° C is placed in a tank of negligible heat capacity containing water at 40° C. The tank is perfectly insulated so that no heat is lost from the system, and the only transfer of heat is that from the ball to the water. After 15 min it is observed that the temperature of the water is 50° C and the temperature of the ball is 80° C. Assuming that the temperature of the ball at any instant is the same at all points and that the temperature of the water at any instant is the same at all points, find the temperature of the ball as a function of time. When will the temperature of the ball be 75° C? *Hint:* Contrary to our first impression, numerical values for the mass of the water, the mass of the ball, and the specific heats of the water and the material of the ball are not needed. Introducing convenient symbols for these quantities, compute the amount of heat in the system at $t = 0$, $t = 15$, $t = \infty$, and $t = t$. Then, observing that these are all equal, show that the temperature of the water is a linear function of the temperature of the ball. Finally, use Newton's law of cooling (Exercise 36) to set up the appropriate differential equation.

46. In Example 5, show that the amount of fluid Q issuing from the pipe per unit time is given by the formula $Q = \pi a^4(p_0 - p_1)/8$.

47. In Example 5, show that the average value of the fluid velocity for laminar flow through a straight cylindrical pipe is $\bar{v} = (p_0 - p_1)/8$.

48. In Example 6, suppose $\theta(0) = \dot{\theta}(0) = 0$, $r(0) = R$, $\dot{r}(0) = v_0$, so that $k = mgR^2$, where R is the radius of the earth and $v_0 > 0$ is the speed with which m is launched into space.

(a) Show that if $v_0 < \sqrt{2gR}$, the mass returns to the earth.

(b) Find r as a function of the time if $v_0 = \sqrt{2gR}$.

49. **(a)** The only force acting on a mass m is a central force having O as its center. Initially, the mass is headed along a line l which does not contain O. Show that the orbit of m is always a curve in the plane of l and O.

(b) Verify that (11) holds for a general central force acting on a mass m and verify Kepler's† **law of areas:** *the line joining the center O to m sweeps out area at a constant rate.*

50. Find a polar equation of the planar orbits of a mass m acted upon by a repulsive inverse square force and identify the orbits.

51. In Example 6, show that a simultaneous change of independent variable to θ and dependent variable to $u = 1/r$ converts Eq. (13) into $d^2u/d\theta^2 + u = k/mc^2$ and that this equation becomes $d^2w/d\theta^2 + w = 0$ under the substitution $w = u - k/mc^2$.

52. In Exercise 51, use the procedure of Case 2, Sec. 1.13, to solve the differential equation in w and convert your solution into a polar equation.

53. Verify **Kepler's third law:** *for an elliptical orbit, the square of the period of revolution P varies as the cube of the major axis L. Hint:* An ellipse of eccentricity e has area $\pi\sqrt{1 - e^2}(L/2)^2$, (11) implies $P^2 = \pi^2L^4(1 - e^2)/(2c)^2$, and (17) gives $L = r_{\max} + r_{\min} = 2r_c/(1 - b^2)$, where $r_c = mc^2/k$ and $b = e$.

54. Suppose the population of a bacteria colony is changing at a rate proportional to the population and that after 40 h the population has doubled. When will the population quadruple?

55. For a certain population, both the birthrate and the death rate are constant multiples of the number of individuals instantaneously present. Find the population as a function of time.

56. In a population in which reproduction is bisexual rather than asexual, it is probably more realistic to suppose that the birthrate is proportional to the number of pairs of individuals rather than to the number of individuals. Assuming a birthrate of this nature and a death rate proportional to the number of individuals, set up and solve the differential equation governing the population size N given that the initial size of the population is N_0.

57. Work Exercise 56 if, as in Example 8, the proportionality factor in the death rate, instead of being constant, is assumed to vary as a linear function of the population size N.

58. Discuss the possibility of solving Exercise 56 if the proportionality factors in both the birthrate and death rate vary as a linear function of the population size N.

In Exercises 55 and 56 we assumed birthrates and death rates that were constant multiples of either the population size or the possible number of pairs of individuals in the population.

---

†Named for Johannes Kepler (1571–1630, German), who formulated three basic laws of planetary motion which led Newton to his development of celestial mechanics.

In Example 8 and in Exercises 57 and 58, we assumed that the proportionality factors in these rates varied as linear functions of the population size N. For a population of rational individuals (i.e., humans) there is still another possibility: it may well be that *independent of the size of the population,* changing social values and objectives will act to change the birthrate while at the same time medical progress will act to change the death rate. The natural way to take such influences into account is to assume that the proportionality factors in the birthrate and death rate are functions of the time t, say $k_b(t)$ and $k_d(t)$, so that the fundamental differential equation becomes

$$\frac{dN}{dt} = k_b(t)N - k_d(t)N$$

Solve this equation for the choices of $k_b(t)$ and $k_d(t)$ indicated in the following exercises and determine the limiting behavior of the populations which they predict.

59. $k_b(t) = b_1 - b_2t$; $k_d = d_1 - d_2t$

60. $k_b(t) = b_1e^{-b_2t}$; $k_d(t) = d_1e^{-d_2t}$

61. $k_b(t) = b_1/(b_2{}^2 - t^2)$; $k_d(t) = d_1/(d_2{}^2 - t^2)$

62. A radioactive substance disintegrates at a rate proportional to the amount instantaneously present. During a time interval of 200 years 10 percent of a certain sample disappears. What percentage of the original amount remains after 400 years?

63. If the **half-life** of radium, that is, the time required for one-half of any given amount of radium to disintegrate, is 1,590 years, obtain a formula for the amount of radium present after t years. How long will it be before one-fourth of the original amount has disintegrated? What fraction of the original amount will disintegrate during the first century? During the third century? During the tenth century?

64. Work Exercise 63 for the element plutonium given that the half-life of plutonium is 50 years.

65. Living tissues, both plant and animal, contain carbon derived ultimately from the carbon dioxide in the air. Most of this carbon is the stable isotope ^{12}C, but a small fixed percentage of it is unstable radioactive ^{14}C. It appears that there is little or no segregation of these two forms of carbon in living organisms, and the $^{14}C/^{12}C$ ratio is essentially constant for all types of tissue. When the tissue dies, the vital processes of course end, no more carbon of either form is added to the tissue, and the amount of ^{14}C present at the time of death decreases at a rate proportional to the amount instantaneously present, with a half-life of approximately 5,500 years.

(a) If x_0 is the amount of ^{14}C in a given specimen of tissue at the moment of death (determined as a known percentage of the unchanged amount of ^{12}C calculated by chemical analysis of the specimen) and if x is the amount present in the tissue t years after death, express x as a function of t.

(b) A piece of charcoal found in the Lascaux Cave in France (the cave with the remarkable paintings of prehistoric animals) contained 14.8 percent of the original amount of ^{14}C. Date the occupation of the cave that produced the charcoal.

(c) A charred branch of a tree killed by the eruption that formed Crater Lake, Oregon, contained 44.5 percent of the original amount of ^{14}C. Date the eruption.

66. According to **Lambert's† law of absorption,** when light passes through a transparent medium, the amount absorbed by any thin layer of the material perpendicular to the direction of the light is proportional to the amount incident on that layer and to the thickness of that layer. In his underwater explorations off Bermuda, Beebe observed that at a depth of 50 ft the intensity of illumination, that is, the amount of light incident on a unit area, was 10 cd/ft^2 and that at 250 ft it had fallen to 0.2 cd/ft^2. Find the formula connecting intensity with depth in this case.

67. (a) If p lb/in^2 is the atmospheric pressure and ρ lb/in^3 is the density of air at a height of h in above the surface of the earth, show that $(dp/dh) + \rho = 0$.

(b) If $p = 14.7$ lb/in^2 at sea level and if it has fallen to half this value at 18,000 ft, find the formula for the pressure at any height given that p is proportional to ρ (isothermal conditions). What is the predicted height of the atmosphere under this assumption?

(c) At what altitude under isothermal conditions will the pressure be one-fourth its value at sea level?

(d) Work part **(b)** given that p is proportional to $\rho^{1.4}$ (adiabatic conditions).

68. Although water is often assumed to be incompressible, it actually is not. In fact, using pounds and feet as units, the weight of 1 ft^3 of water under pressure p is approximately $w(1 + kp)$, where $w = 64$, $k = 2 \times 10^{-8}$, and p is measured from standard atmospheric pressure as an origin. Using this information, find the pressure at any depth y below the surface of the ocean. At a depth of 6 mi, by what factor does the actual pressure exceed the pressure computed on the assumption that water is incompressible?

69. Banks often compound interest on savings accounts on a semiannual, quarterly, or even daily basis. If interest at an annual rate of p percent is compounded every Δt years, show that an initial amount P_0 grows to $P_0[1 + (p/100)\,\Delta t]$ in one such (infinitesimal) interest period. Use this result to derive the differential equation satisfied by the amount on deposit P in the limit when interest is compounded continuously. Solve this equation and obtain an expression for P as a function of t.

---

†Named for the German physicist Johann Heinrich Lambert (1728–1777).

70. Five years ago a family was in debt \$9,000. They are now in debt \$10,000. Assuming their indebtedness is changing at a rate proportional to their indebtedness, predict their indebtedness 5 years from now.

71. An inverted conical tank is 12 ft high and has an open top of radius 4 ft. The tank is filled with fluid, and 60 h later the fluid depth is 10.5 ft. Assuming the fluid evaporates at a rate proportional to the surface area exposed to air, find a formula for the fluid depth as a function of time.

72. A mothball loses mass by evaporation at a rate proportional to its instantaneous surface area. If half the mass is lost in 100 days, how long will it be before the radius has decreased to one-half its initial value? How long will it be before the mothball disappears completely?

73. When a volatile substance is placed in a sealed container, molecules leave its surface at a rate proportional to the area of the surface and return at a rate proportional to the amount which has evaporated. If a volatile material is spread evenly to a depth h over the bottom of a closed box, find the depth of the material at any time. Under what conditions, if any, will all the material eventually evaporate?

74. A sphere of volatile material is suddenly introduced into a closed container. Assuming the same laws of evaporation and condensation described in Exercise 73, set up the differential equation whose solution gives the radius of the sphere as a function of time.

75. A unit mass particle moves along the x axis in a resisting medium, and the resistance is proportional to the velocity. If the particle starts from the position $x = x_0$ with velocity v_0, find the limiting position approached by the particle. Take $-kv$, with $k > 0$, as the force of resistance.

76. A 100-lb sled is being pushed along a straight line against the wind by a force of 10 lb. The combined resistance to the motion, due to friction and the wind, is a force whose magnitude in pounds is equal to twice the velocity of the sled in feet per second.

(a) Find the velocity of the sled if its initial velocity in feet per second is v_0.

(b) If the sled starts from rest, find the distance traveled at the end of 1 s.

77. In Exercise 76, suppose the sled weighs 128 lb, that it is being pushed by a 16-lb force, and that the magnitude of the resistance is 4 times the sled velocity.

(a) Find the sled velocity if the initial velocity is 8 ft/s.

(b) Does the sled travel more than 6 ft during the first second?

78. A barge is being towed at 16 ft/s when the towline breaks. It continues thereafter in a straight line but slows down at a rate proportional to the square root of its instantaneous velocity. If 2 min after the towline breaks the velocity of the barge is observed to be 9 ft/s, how far does it move before it comes to rest?

79. It is a fact of common experience that when a rope is wound around a rough cylinder, a small force at one end can resist a much larger force at the other. Quantitatively, it is found that throughout the portion of the rope in contact with the cylinder, the change in tension per unit length is proportional to the tension, the numerical value of the proportionality constant being the coefficient of friction between the rope and the cylinder divided by the radius of the cylinder. Assuming a coefficient of friction of 0.35, how many times must a rope be snubbed around a post 1 ft in diameter for a man holding one end to be able to resist a force 200 times greater than he can exert?

80. A rapidly rotating flywheel, after power is shut off, coasts to rest under the influence of a friction torque which is proportional to the instantaneous angular velocity ω. If the moment of inertia of the flywheel is I and if its initial angular velocity is ω_0, find its instantaneous angular velocity as a function of time. How long will it take the flywheel to come to rest? *Hint:* Use **Newton's law in torsional form**

$$\text{Moment of inertia} \times \text{angular acceleration} = \text{torque}$$

to set up the differential equation describing the motion.

81. The friction torque acting to slow down a flywheel is actually not proportional to the first power of the angular velocity at all speeds. As a more realistic example than Exercise 80, suppose that a flywheel of moment of inertia $I = 7.5$ lb-ft s^2 coasts to rest from an initial speed of 1,000 rad/min under the influence of a retarding torque T estimated to be

$$T = \begin{cases} \dfrac{\sqrt{\omega}}{10} \text{ ft-lb} & 0 < \omega < 100 \text{ rad/min} \\[2ex] \dfrac{1}{10}\left(7.5 + \dfrac{\omega^2}{4{,}000}\right) \text{ ft-lb} & 100 < \omega < 1{,}000 \text{ rad/min} \end{cases}$$

Find ω as a function of time and determine how long it will take the flywheel to come to rest.

82. A stone is dropped from a balloon which is 1,760 ft above the earth and ascending vertically at the rate of 16 ft/s. Neglecting friction, find the equation of motion of the stone referred to an origin fixed in space at the point from which the stone was released. What is the maximum height reached by the stone? When and with what velocity does the stone strike the ground?

83. A body weighing w lb falls from rest under the influence of gravity and a retarding force due to air resistance, assumed to be proportional to the velocity of the body. Find the equations expressing the velocity of fall and the distance fallen as functions of time and verify that these reduce to the ideal laws

$$v = gt \qquad \text{and} \qquad s = \tfrac{1}{2}gt^2$$

when the coefficient of air resistance approaches zero.

84. An object is projected upward from the surface of the earth with velocity 64 ft/s. Air resistance is proportional to the first power of the velocity, the proportionality constant being such that if the body were to fall freely, its velocity would approach the limiting value $v_\infty =$ 256 ft/s. How high does the body rise? When and with what velocity does it strike the ground?

85. Work Exercise 83 given that the retarding force due to air resistance is proportional to the square of the velocity of fall.

86. A body of weight w falls from rest under the influence of gravity and a retarding force proportional to the nth power of the velocity. Show that the velocity of the body approaches the limiting value $v_\infty = \sqrt[n]{w/k}$, where k is the proportionality constant in the law of air resistance. Show also that the time τ that it takes the body to reach one-half its limiting velocity is given by the equation

$$\tau = \alpha \frac{v_\infty}{g}$$

and compute the value of α for

(a) $n = \frac{1}{4}$ **(b)** $n = \frac{1}{3}$ **(c)** $n = \frac{1}{2}$ **(d)** $n = 1$
(e) $n = 2$ **(f)** $n = 3$ **(g)** $n = 4$

87. A particle of mass m moves along the x axis under the influence of a force which is directed toward the origin and proportional to the distance of the particle from the origin. If the body starts from rest at the point where $x = x_0$, find the equations which express its velocity and its distance from the origin as functions of time.

88. Work Exercise 87 **(a)** if the particle starts from the origin with velocity $v = v_0$ and **(b)** if the body starts from the point $x = x_0$ with velocity $v = v_0$.

89. Work Exercise 87 if the force, instead of being directed toward the origin, is directed away from the origin.

90. Work Exercise 87 **(a)** if the particle moves under the influence of a force which is directed toward the origin and inversely proportional to the square of the distance of the particle from the origin and **(b)** if the particle moves under the influence of a force which is directed away from the origin and inversely proportional to the square of the distance of the particle from the origin.

91. A body falls from rest from a height y_0 so great that the fact that the force of gravity varies inversely as the square of the distance from the center of the earth cannot be neglected. Find the equations expressing the velocity of fall and the distance fallen as functions of time in the ideal case in which air resistance is neglected.

92. Under the conditions of Exercise 91, determine the minimum initial velocity with which a body must be projected upward if it is to leave the earth and never return.

93. A cylinder of mass m and radius r rolls, without sliding, down an inclined plane of inclination angle α. Neglecting friction, express the distance the cylinder has rolled down the plane as a function of time. *Hint:* Observe first that since friction is neglected, the principle of the conservation of energy implies that the sum of the kinetic energy and the potential energy of the cylinder remains constant throughout the motion. Then note that the kinetic energy of the cylinder consists of two parts: that due to the translation of the cylinder and that due to its rotation.

94. Work Exercise 93 for a rolling sphere of mass m and radius r.

95. When a capacitor of capacitance C is being charged through a resistance R by a battery which supplies a constant voltage E, the instantaneous charge Q on the capacitor satisfies the differential equation

$$R\frac{dQ}{dt} + \frac{Q}{C} = E\dagger$$

Find Q as a function of time if the capacitor is initially uncharged, i.e., if $Q_0 = 0$. How long will it be before the charge on the capacitor is one-half its final value?

96. In Exercise 95, determine Q if $Q_0 = 0$ and if the battery is replaced by a generator which supplies an alternating voltage equal to $E_0 \sin \omega t$.

97. Determine how an initially charged capacitor will discharge through a resistance; i.e., in Exercise 95, find Q if there is no voltage source in the circuit and the charge on the capacitor is initially some nonzero value Q_0.

98. When a switch is closed in a circuit containing a resistance R, an inductance L, and a battery which supplies a constant voltage E, the current i builds up at a rate defined by the equation

$$L\frac{di}{dt} + Ri = E\ddagger$$

Find the current i as a function of time. How long will it take i to reach one-half its final value? Evaluate $\lim_{t\to\infty} i(t)$. Find i if $i_0 = E/R$.

99. In Exercise 98, determine i if $i_0 = 0$ and if the battery is replaced by a generator which supplies an alternating voltage $E_0 \cos \omega t$.

100. In calculus, the radius of curvature of a curve is defined to be

$$R = \frac{[1 + (y')^2]^{3/2}}{|y''|}$$

Solve this differential equation when R is a constant and show that the family of solution curves is the set of all circles of radius R in the xy plane.

101. Work Exercise 100 if the radius of curvature varies according to the law $R(x) = \sec x$ and if the slope of each curve is to be zero when $x = 0$.

102. What is the equation of the curve in which a perfectly flexible cable, of uniform weight per unit length w, will hang when it is suspended between two points at the same height? *Hint:* Consider a section of the cable, such

---

†For a derivation of this equation see Sec. 7.2.
‡For a derivation of this equation see Sec. 7.2.

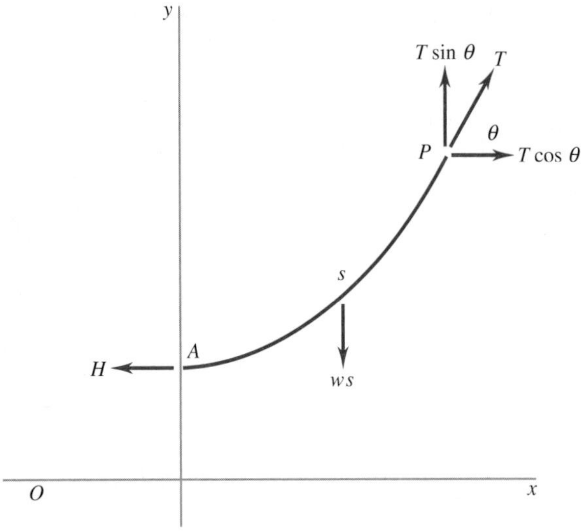

Figure 1.13

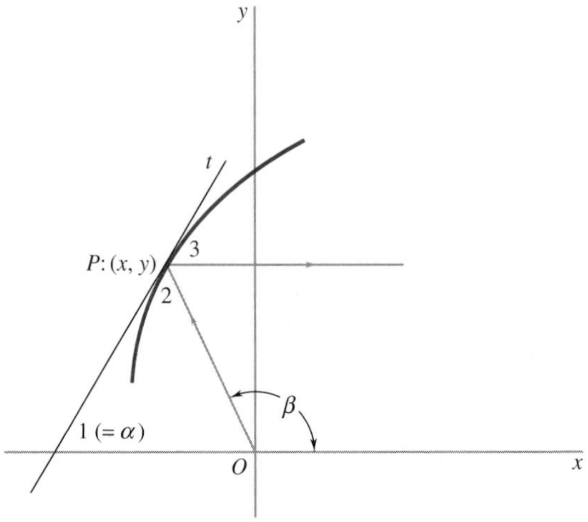

Figure 1.14

as that shown in Fig. 1.13, and note that if H is the horizontal tension in the cable at its lowest point A, if T is the tension in the cable at a general point P, and if s is the length of the cable between A and P, then

$$T \sin \theta = ws \qquad T \cos \theta = H$$

These imply that

$$\tan \theta = y' = \frac{ws}{H} \qquad \text{hence} \qquad \frac{dy'}{dx} = \frac{w}{H}\frac{ds}{dx}$$

After ds/dx is expressed in terms of y' by means of the familiar formula for the differential of arc length, the integrations required to find y' and then y will be simplified if they are carried out in terms of hyperbolic functions.

103. Find the equation of the curve in which a perfectly flexible cable hangs when, instead of bearing a uniform load per unit length of cable, as in Exercise 102, it bears a uniform horizontal load. (This is approximately the case when the cable is part of a suspension bridge carrying a horizontal roadbed whose weight is much greater than the weight of the cable.)

104. Find the equation of a curve with the property that light rays emanating from the origin will all be reflected into rays which are parallel to the x axis. *Hint:* Note in Fig. 1.14 that if t is the tangent to the required curve at a general point P, then the angles 1, 2 and 3 all have the same measure, say α. Then from the exterior-angle theorem $\alpha = \beta - \alpha$, which implies that $\tan \alpha = \tan(\beta - \alpha)$. Expanding $\tan(\beta - \alpha)$ and observing that $\tan \alpha = y'$ and $\tan \beta = y/x$ leads to the differential equation $y = 2xy' + y(y')^2$. This can be solved by first solving for y' and then using an integrating factor, or it can be reduced to a Clairaut equation by multiplying it by y and setting y^2 equal to a new variable u.

105. A vertical cylindrical tank of radius r is filled with liquid to a depth h. The tank is rotated about its axis with constant angular velocity ω. Find the equation of the curve in which the free surface of the liquid is intersected by a plane through the axis of the cylinder, assuming that the tank is sufficiently deep to prevent liquid from spilling over the edge. *Hint:* The normal force maintaining a particle in equilibrium in the surface of the liquid can be resolved into an upward vertical component w, where w is the weight of the particle, and a centripetal force $wx\omega^2/g$ directed radially inward, where x is the distance of the particle from the axis of rotation.

106. In Exercise 105, determine the angular velocity for which the surface of the liquid barely touches the bottom of the container.

107. A weight W is to be supported by a column having the shape of a solid of revolution. If the material of the column weighs ρ lb/ft^3 and if the radius of the upper base of the column is to be r_0 ft, determine how the radius of the column should vary if at all cross sections the load per unit area is to be the same.

108. Work Exercise 107 if the column is hollow, the outer radius of each cross section being h ft more than the inner radius, and the inner radius of the upper base being r_0 ft.

109. A weight is initially located on the y axis at a distance l from the origin. An inextensible chain of length l is attached to the weight, the free end of the chain being initially at the origin. If the free end of the chain is moved slowly along the x axis, find the equation of the path of the weight as it is dragged across the xy plane by the chain. *Hint:* Note that the direction of the chain will always be tangent to the path of the weight. Hence, if P: (x, y) is the instantaneous position of the weight and if s is the length of the path through which the weight has moved, then $dy/ds = -y/l$. (This important curve is known as the **tractrix.**)

110. A Coast Guard vessel is pursuing a smuggler in a dense fog. The fog lifts momentarily, and the smuggler is seen d mi away; then the fog descends, and the smuggler can no longer be seen. It is known, however, that in an attempt to escape, the smuggler will set off on a straight course of unknown direction with constant speed v_1. If the velocity of the Coast Guard boat is v_2 ($>v_1$), what course should it follow to be sure of intercepting and capturing the smuggler? *Hint:* Choose the origin at the point where the smuggler was seen and observe that if the Coast Guard boat spirals the origin in a path such that its distance from the origin is always equal to $v_1 t$, it will necessarily intercept the smuggler. Of course, before the Coast Guard boat can being its spiral, it must first reach a point where its distance from the origin is $v_1 t_0$, where t_0 is the elapsed time since the smuggler was seen. In implementing these suggestions, it will be convenient to use the polar coordinate equations $x = r(t)\cos\theta(t)$ and $y = r(t)\sin\theta(t)$ to describe the path of the Coast Guard vessel.

111. One winter morning it began snowing heavily and continued at a constant rate all day. At noon a snowplow, able to clear c ft^3/h of snow with a blade w ft in width, started plowing. At 1 P.M. it had gone 2 mi, and at 2 P.M. it had gone 1 mi more. When did it start snowing? *Hint:* If t_0 is the number of hours before noon when it began to snow, if t is the time measured in hours after noon, if s ft/h is the constant rate at which the snow is falling, if $y(t)$ ft is the depth of the snow, and if $x(t)$ ft is the distance the snowplow has traveled,

$$y = s(t_0 + t) \quad \text{and} \quad wy\,dx = c\,dt$$

112. In the inversion of cane sugar, raw sugar yields by hydrolysis a mixture of dextrose and levulose at a rate that varies as the amount of raw sugar remaining. If after 10 h, 1,000 kg of raw sugar has been reduced to 800 kg, how much raw sugar will remain at the expiration of 24 h?

113. Let V be the volume of the fluids in the human body and let Q be the concentration of glucose instantaneously present in, and uniformly distributed through, these fluids. Let the tissues of the body absorb glucose at a rate proportional to the instantaneous concentration of glucose in the body fluids. When the glucose level is too low (as it is in many illnesses), glucose solution must be injected into the veins. Suppose that this is done in such a way that A mg of glucose enters the veins per minute but that the accompanying liquid does not appreciably increase the volume of the body fluids. If Q_0 is the initial concentration of the glucose, find an expression for the instantaneous concentration of glucose in the body fluids as a function of time.

114. Consider a fluid-filled cell of volume V and surface area A completely immersed in a fluid in which the concentration of a certain solute is c. The solute diffuses into the cell in accordance with **Fick's law†:**

> *The time rate at which a solute diffuses through a thin membrane in a direction perpendicular to the membrane is proportional to the area of the membrane and to the difference between the concentrations of the solute on the two sides of the membrane.*

Assuming that c remains constant, use Fick's law to show that the diffusion of the solute into the cell is described by the differential equation

$$\frac{dy}{dt} = k\frac{A}{V}(c - y)$$

where y is the concentration of the solute in the cell (assumed uniform throughout the cell) and k is the **permeability coefficient.** Solve this equation given that the initial concentration of the solute in the cell is y_0 ($<c$).

115. In Exercise 114, if the concentration of the solute in the ambient fluid is 0.05, if the initial concentration of the solute in the cell is 0.01, and if after 10 min the concentration is 0.02, how long will it be before the concentration in the cell is 0.03? 0.04?

116. A diffusion model involves two identical compartments each of volume V which share a common boundary surface of area A. The boundary between the two compartments is a permeable membrane, but the rest of the surface of each compartment is impervious to diffusion. Initially the compartments are filled with solutions of a certain solute of differing concentrations. Let x_0 and x denote, respectively, the initial and instantaneous concentrations of the solute in the more dilute solution and let y_0 and y denote the initial and instantaneous concentrations in the more concentrated solution. Use Fick's law to show that the diffusion process between the two compartments is described by the equations

$$x + y = x_0 + y_0 \qquad \frac{dx}{dt} = k\frac{A}{V}(x_0 + y_0 - 2x)$$

Solve these equations and obtain formulas for the concentration of the solute in each compartment as a function of time.

117. If the initial concentrations in the two compartments in Exercise 116 are $x_0 = 0$ and $y_0 = 0.10$ and if after 20 min they are observed to be 0.02 and 0.08, how long will it be before they are 0.04 and 0.06?

118. If the two compartments in Exercise 116 are identical circular cylinders of length 20 cm and radius 4 cm, having one of their circular bases as their common boundary surface, if $x_0 = 0.03$ and $y_0 = 0.12$, and if after

†Named for the German physiologist Adolph Eugen Fick (1829–1901).

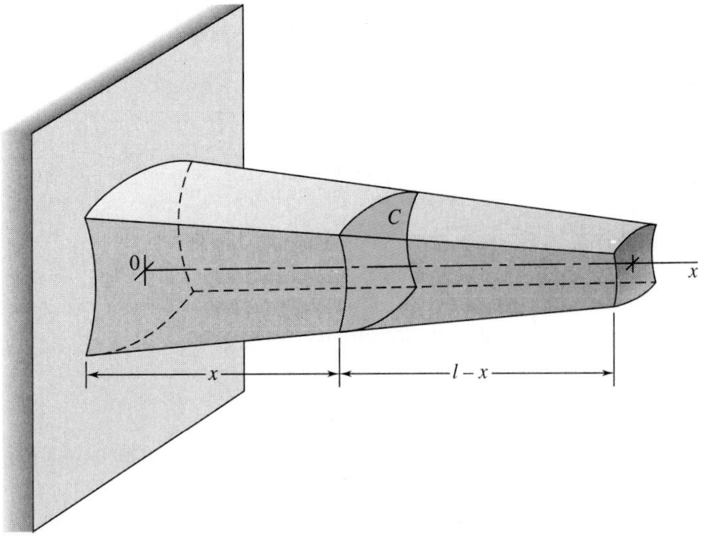

Figure 1.15
A bar with variable cross sections prior to applying a longitudinal force.

30 min it is observed that $x = 0.05$ and $y = 0.10$, what is the value of the permeability coefficient k?

119. Work Exercise 116 if the volume of the compartment containing the more dilute solution is 3 times the volume which contains the more concentrated solution.

120. A bar of homogeneous material, for which E is **Young's modulus of elasticity,**† is of length l. The x axis passes lengthwise through the bar, and the centroid of each transverse cross-sectional area lies on the x axis. The end of the bar at $x = 0$ is firmly secured so that it cannot move. A constant longitudinal force F is uniformly distributed over the end section at $x = l$. Let C be the cross section of the bar which in the absence of F is at a distance x from the secured end (Fig. 1.15). When F is applied, C is displaced a distance $u(x)$ in the x direction. **(a)** Determine $u(x)$ in terms of the area $A(x)$ of C. For $l = 10$, also find $u(10)$ if C is **(b)** a square of side $1 - x/20$ and **(c)** a circle of diameter $1 - x/20$. *Hint:* By definition **stress** is *force per unit area,* **strain** is *stretch per unit length,* and E = stress/strain.

†Named after the English philosopher Thomas Young (1773–1829).

CHAPTER 2

LINEAR DIFFERENTIAL EQUATIONS

In elementary applications, ordinary linear differential equations are undoubtedly the most important. Those with constant coefficients appear immediately in the study of mechanical vibrations and electric circuits (Chap. 7). Those with variable coefficients arise most commonly in the solution of partial differential equations (Chap. 12). The solution of an ordinary linear differential equation with constant coefficients is always a straightforward process, involving only functions familiar to us from calculus (Secs. 2.5–2.7). Linear differential equations with variable coefficients are much harder to solve. For them, the usual method of solution involves the use of infinite series and the method of undetermined coefficients (Secs. 2.9 and 2.10). These series usually represent unfamiliar functions and often become the definitions of important new functions (Chap. 12).

Although the solution procedures differ markedly in form and difficulty, the basic theory of linear differential equations is essentially the same whether their coefficients are constant or variable. Hence we have chosen to discuss the basic theory and such concepts as linear dependence and independence of solutions and families of solutions for the general case, though restricting ourselves to the adequately typical case of second-order equations (Secs. 2.1–2.4). Many of these ideas will reappear when we discuss *systems of linear differential equations* (Secs. 4.3 and 4.4), *difference equations* (Sec. 5.7), and *linear operator equations* (Sec. 13.7).

In Sec. 2.11, we consider from physical principles such as *Hooke's law, Newton's law,* and the *principle of the conservation of energy,* a number of applications such as the motion of a spring-suspended mass, the bending of a beam, and the buckling of a slender column. A detailed discussion of the use of linear differential equations with constant coefficients in studying vibrating mechanical systems and electric circuits we leave to Chap. 7. The applications of several important linear differential equations with variable coefficients (*Bessel's equation* and *Legendre's equation*) will be taken up in Chap. 12.

Finally, in Sec. 2.12, we discuss from a physical point of view what mathematicians call *Green's functions* and engineers call *influence functions.* Physically speaking, these are functions that describe how one component of a system, either mechanical or electrical, is influenced by a disturbance applied to some other component of the system. Green's functions are of considerable theoretical importance, and we shall give a more general matrix treatment of them in Sec. 14.4.

Prerequisite for this chapter: single-variable calculus.

Prerequisite for Secs. 2.11 and 2.12: a calculus-based physics course.

2.1 A FUNDAMENTAL EXISTENCE AND UNIQUENESS THEOREM

In Chap. 1, we learned how to solve various types of first-order differential equations, most of them nonlinear. Many practical applications were also given. When it comes to equations of orders higher than 1, in general, neither linear nor nonlinear differential equations can be solved in terms of known functions. However, there are a number of very important linear differential equations which can be solved by elementary methods or by series. The rest of our work with ordinary differential equations, with the exception of the numerical procedures discussed in Chap. 5 and the qualitative work on nonlinear equations in Chap. 6, will deal almost exclusively with linear equations.

From Eq. (1), Sec. 1.2, we know that an ordinary linear differential equation of order n in the dependent variable y can be put in the form

$$(1) \qquad a_0(x)y^{(n)} + a_1(x)y^{(n-1)} + \cdots + a_{n-1}(x)y' + a_n(x)y = f(x)$$

where $a_0(x) \not\equiv 0$. Because of the presence of $f(x)$, which is unlike the other terms in that it is not homogeneous of degree 1 in the dependent variable y and its derivatives, (1) is said to be a **nonhomogeneous** equation and $f(x)$ is called its **nonhomogeneous term.** This supposes of course that f is a nontrivial function. If f is replaced by a trivial function, (1) becomes

$$(2) \qquad a_0(x)y^{(n)} + a_1(x)y^{(n-1)} + \cdots + a_{n-1}(x)y' + a_n(x)y = 0$$

Every nonzero term of this equation is homogeneous of degree 1, that is, is linear, in y and its derivatives; thus (2) is called a **homogeneous** equation, or the **related homogeneous equation** of (1).

We shall usually assume that the **coefficient functions** $a_0, a_1, \ldots, a_n$, as well as f, are continuous on an interval I over which the value of the **leading coefficient** a_0, which multiplies $y^{(n)}$, is never zero. The interval may be open, closed, or half-open, and either finite or infinite. The following definition of a *normal* differential equation provides a concise way of stating that all of these assumptions are relevant.

DEFINITION 1 A linear differential equation is **normal on an interval I** if and only if its coefficient functions and its nonhomogeneous term, if it has one, are continuous and the value of the leading coefficient is never zero on I.

In Sec. 1.3, it was noted that a solution on an interval I of an nth-order differential equation must be continuously differentiable $n - 1$ times on I. It follows that if (1) is normal on I and $y(x)$ is a solution, then $y^{(n)}(x)$ must also be continuous throughout I.

Neither (1) nor (2) can be solved when $n > 1$, and the theory associated with such special cases as have been studied at length is difficult. However, there are several general theorems about solutions of linear differential equations which do not depend upon how the solutions may be represented or the process by which they are found. Once we have become familiar with these results, we shall use them as a guide in constructing complete solutions of specific linear equations.

As our first and most basic result, we cite a fundamental existence and uniqueness theorem pertaining to (1).†

THEOREM 1 If x_0 is contained in an interval I over which Eq. (1) is normal, and if $k_0, k_1, \ldots, k_{n-1}$ are arbitrary real numbers, then there exists exactly one solution $y(x)$ of (1) on I such that $y(x_0) = k_0, y'(x_0) = k_1, \ldots, y^{(n-1)}(x_0) = k_{n-1}$.

Not only does this theorem specify conditions under which a linear differential equation has solutions, but it also guarantees that over any interval I where the equation is normal there will be precisely one solution corresponding to any assignment of **initial values** $k_0, k_1, \ldots, k_{n-1}$ to a solution and its first $n - 1$ derivatives at some point of I. In other words, no matter how a function has been found, if it satisfies a linear differential equation and a set of initial conditions, in conformity with Theorem 1, then that function is *the* solution of the initial-value problem.

EXAMPLE 1

According to Theorem 1, there is just one solution of the initial-value problem

$$y''' + y'' - 2y' = 4e^{2x} \qquad y(0) = 3,\ y'(0) = 1,\ y''(0) = 5$$

over any interval I containing the origin because the differential equation is normal on $(-\infty, \infty)$. Although we presently have no way of finding the solution, by substituting the appropriate derivatives of

$$y(x) = 1 + e^x + \cosh 2x$$

into the differential equation and evaluating $y(x)$, $y'(x)$, and $y''(x)$ at $x = 0$, it is easy to verify that the unique solution over I is represented by $y(x)$.

EXAMPLE 2

The linear equation $(x + 2)y''' + x^2y'' - xy' + 2y = \cos \ln |x|$ is not normal on any interval which contains -2 or 0 because the leading coefficient $x + 2$ has the value zero at $x = -2$ while the nonhomogeneous term $\cos \ln |x|$ is undefined, hence discontinuous, at $x = 0$. The differential equation is normal over each of the open intervals $(-\infty, -2)$, $(-2, 0)$ and $(0, \infty)$. At an arbitrary point within any of these intervals, say $x_0 = -1$, we may *arbitrarily* prescribe that $y(-1) = \pi$, $y'(-1) = e$, and $y''(-1) = e^{\pi}$, and Theorem 1 asserts that on the interval $(-2, 0)$ there is exactly one solution of the given differential equation which satisfies these three initial conditions. The solution and its first three derivatives are of course continuous on $(-2, 0)$.

The following corollary is an immediate and useful consequence of Theorem 1.

COROLLARY 1 If the homogeneous equation (2) is normal on an interval I, if x_0 is a point of I, and if $y(x)$ is a solution of (2) for which $y(x_0) = y'(x_0) = \cdots = y^{(n-1)}(x_0) = 0$, then $y(x) \equiv 0$ on I.

◖ **PROOF** Clearly, $y = 0$ is a solution of (2) which satisfies the initial conditions, and by Theorem 1 it is unique. ◗

EXERCISES

Find all intervals on which the following differential equations are normal.

1. $(1 + x^2)y''' + xy'' - x^2y' + y = 0$

2. $(1 + x^4)y^{\text{iv}} + x^2y'' + y = e^x$

3. $\sqrt{x}y''' + 13xy' - 11y = \ln(x^2 - 100)$

4. $x^2y''' - 3xy'' + 4y = \cosh x$

5. $y^{\text{iv}} + (1 + |x|)y = \sqrt{x^2}$

6. $\sqrt{x}y'' - 6xy' + 10y = \ln(1 - x)$

†A proof of this theorem may be found in E. L. Ince, *Ordinary Differential Equations,* pp. 73–75, Dover, New York, 1944.

7. $y''' - \sqrt{-x}y'' + |x|y' + y = \ln x$

8. $(\ln\cosh x)y''' + x^2y'' - xy' = \sinh x$

9. $y'' + 7xy' - 11y = \ln\sin\pi x$

10. $(3! - 3\ln x)y''' + (16\cos x)y' + 23y = 5\,\mathrm{Tan}^{-1}x$

11. $(x^2 - e^2)y'' + (\ln|x| - 2\ln e)y' + (\operatorname{csch} x)y = 0$

12. $\sqrt{x^2 - x}y''' - y'e^{-x}\sin x = \ln[(x+1)^y(2-x)]$

13. Show that if every coefficient function of the *homogeneous* equation (2) is constant, then for every solution y and for every integer $m \geq 0$, $y^{(m)}$ is continuous on $(-\infty, \infty)$.

14. Solve the equation $y''\sinh x + y'\cosh x = \cosh x$ and find **(a)** the solutions which have continuous derivatives of all orders on $(-\infty, \infty)$, and **(b)** a solution whose domain does not contain $x = 0$. *Hint:* The equation is a first-order linear equation in y'.

15. Verify that both $y_1 = 1 + \cos 2x$ and $y_2 = 2\cos^2 x$ satisfy the conditions $y(0) = 2$, $y'(0) = 0$, and the equation $y'' + 4y = 4$ on $(-\infty, \infty)$; then deduce that for all real values of x,

$$1 + \cos 2x = 2\cos^2 x$$

16. Verify that both $y_1 = 2\sinh^2 x$ and $y_2 = -1 + \cosh 2x$ satisfy the conditions $y(0) = 0$, $y'(0) = 0$, and the equation $y'' - 4y = 4$ on $(-\infty, \infty)$; then deduce that for all real values of x,

$$\cosh 2x = 1 + 2\sinh^2 x$$

17. Find, by inspection, a solution of $xy'' - xy' + y = 0$ on $(-\infty, \infty)$ such that $y(0) = 0$, $y'(0) = 1$. Does Theorem 1 guarantee the existence and uniqueness of such a solution? Explain.

18. Verify that a *unique* solution of the initial-value problem

$$xy'' - xy' + y = 0 \qquad y(1) = -e,\ y'(1) = 0$$

on $(0, \infty)$ is defined by $y(x) = -e^x + x\int_1^x e^z/z\,dz$.

Find, by inspection, unique solutions of the following initial-value problems.

19. $y''' - 5x^2y'' + xy' - (\cos x)y = 0$
$y(-2) = y'(-2) = y''(-2) = 0$

20. $y'' - 2y' + 2y = e^x$
$y(0) = y'(0) = 1$

21. $y''' - xy'' + 2y' - y = 6 - x^3$
$y(1) = 1,\ y'(1) = 3,\ y''(1) = 6$

22. $y'' + y' = 0$
$y(-\ln 2) = 1,\ y'(-\ln 2) = 0$

23. $x^2y'' + xy' + y = \ln x$
$y(1/e) = -1,\ y'(1/e) = e$

24. Does Theorem 1 guarantee there is a unique solution of the following initial value problem?

$$(12 - x)^2y^{\text{iv}} - 8(12 - x)y''' + 12y'' = 0$$
$$y(12) = y'(12) = y''(12) = y'''(12) = 0$$

25. Let

$$y_1(x) = x^4 - x^3 \qquad -2 \leq x \leq 2$$

and let

$$y_2(x) = \begin{cases} x^3 - x^4 & -2 \leq x \leq 0 \\ x^4 - x^3 & 0 \leq x \leq 2 \end{cases}$$

Show that both y_1 and y_2 are solutions of the equation $x^2y'' - 6xy' + 12y = 0$ which satisfy the conditions $y(1) = 0$, $y'(1) = 1$. Does the fact that y_1 and y_2 are different contradict Theorem 1? Explain.

2.2 FAMILIES OF SOLUTIONS

An expression of the form $c_1y_1 + c_2y_2 + \cdots + c_my_m$, in which $c_1, c_2, \ldots, c_m$ are arbitrary *constants,* is called a **linear combination** of the variables $y_1, y_2, \ldots, y_m$ because each term in the expression is linear, i.e., homogeneous of degree 1, in these m variables. One extremely important property of every homogeneous linear equation is that if y_1 and y_2 are any particular solutions, then for all values of the constants c_1 and c_2, the linear combination $c_1y_1 + c_2y_2$ is also a solution. To verify that

$$a_0(x)y^{(n)} + a_1(x)y^{(n-1)} + \cdots + a_{n-1}(x)y' + a_n(x)y = 0 \tag{1}$$

has this property, we first observe that if $y_1(x)$ or $y_2(x)$ is substituted for the dependent variable y, Eq. (1) becomes an identity in x over every interval I throughout which both y_1 and y_2 are solutions of (1). In other words, the left-hand member of (1) reduces to the *trivial function* on each such interval when y is replaced by y_1 or y_2. We now substitute $c_1y_1 + c_2y_2$ for y, use familiar properties of derivatives, and collect coefficients on c_1 and c_2 to obtain

$$a_0(x)(c_1y_1 + c_2y_2)^{(n)} + a_1(x)(c_1y_1 + c_2y_2)^{(n-1)} + \cdots$$
$$+ a_{n-1}(x)(c_1y_1 + c_2y_2)' + a_n(x)(c_1y_1 + c_2y_2)$$
$$= a_0(x)(c_1y_1^{(n)} + c_2y_2^{(n)}) + a_1(x)(c_1y_1^{(n-1)} + c_2y_2^{(n-1)}) + \cdots$$

$$+ a_{n-1}(x)(c_1y_1' + c_2y_2') + a_n(x)(c_1y_1 + c_2y_2)$$
$$= c_1[a_0(x)y_1^{(n)} + a_1(x)y_1^{(n-1)} + \cdots + a_{n-1}(x)y_1' + a_n(x)y_1]$$
$$+ c_2[a_0(x)y_2^{(n)} + a_1(x)y_2^{(n-1)} + \cdots + a_{n-1}(x)y_2' + a_n(x)y_2]$$
$$= c_1 \cdot 0 + c_2 \cdot 0 = 0$$

where the zeros in the last two steps denote the trivial function over each interval I on which both y_1 and y_2 are solutions of (1). This confirms that for all real values of c_1 and c_2, and on every solution interval common to y_1 and y_2, the linear combination $c_1y_1 + c_2y_2$ is a solution of (1).

Clearly, any constant times a solution of (1) is a solution, as is the sum of any two solutions over the same interval. In fact, by an immediate extension of these ideas, we have the following theorem.

THEOREM 1 If on an interval I, $y_1, y_2, \ldots, y_m$ are m particular solutions of a homogeneous linear differential equation, then for all real values of the parameters $c_1, c_2, \ldots, c_m$,

$$y = c_1y_1 + c_2y_2 + \cdots + c_my_m = \sum_{k=1}^{m} c_ky_k$$

is also a solution of that equation on I.

The significance of Theorem 1 is that it assures us that if we have one or more specific nontrivial solutions of a homogeneous linear differential equation, we can immediately generate an infinite family of solutions by forming an arbitrary linear combination of these solutions. This property does not hold for nonlinear differential equations, and this is one of the reasons why, in general, nonlinear equations are much harder to work with than linear equations.

EXAMPLE 1

Verify that for all values of the constants c_1 and c_2, $y = c_1e^{-x} + c_2e^{2x}$ is a solution of the homogeneous linear equation $y'' - y' - 2y = 0$ on $(-\infty, \infty)$.

By differentiating y, substituting into the differential equation as indicated, and collecting coefficients on c_1 and c_2, we obtain

$$(c_1e^{-x} + 4c_2e^{2x}) - (-c_1e^{-x} + 2c_2e^{2x}) - 2(c_1e^{-x} + c_2e^{2x})$$
$$\equiv c_1(e^{-x} + e^{-x} - 2e^{-x}) + c_2(4e^{2x} - 2e^{2x} - 2e^{2x})$$
$$\equiv c_1 \cdot 0 + c_2 \cdot 0 \equiv 0 \qquad x \text{ real}$$

regardless of the values of c_1 and c_2. In particular, since $y = c_1e^{-x} + c_2e^{2x}$ is a solution for all values of c_1 and c_2, we may first take $c_1 = 1$, $c_2 = 0$, getting $y = e^{-x}$, and then we may take $c_1 = 0$, $c_2 = 1$, getting $y = e^{2x}$. This shows that $y_1 = e^{-x}$ and $y_2 = e^{2x}$ are themselves solutions of the given equation on $(-\infty, \infty)$.

EXAMPLE 2

Show that although $y_1 = e^{-x}$ and $y_2 = e^{2x}$ are solutions on $(-\infty, \infty)$ of the equation $yy'' - (y')^2 = 0$, the linear combination $y = c_1y_1 + c_2y_2$ is a solution there if and only if $c_1 = 0$ or $c_2 = 0$.

To verify that $y_1 = e^{-x}$ is a solution of the given equation, we note that for all real values of x,

$$(e^{-x})(e^{-x}) - (-e^{-x})^2 = e^{-2x} - e^{-2x} = 0$$

Similarly, for $y_2 = e^{2x}$, we have

$$(e^{2x})(4e^{2x}) - (2e^{2x})^2 = 4e^{4x} - 4e^{4x} = 0$$

However, for $y = c_1y_1 + c_2y_2 = c_1e^{-x} + c_2e^{2x}$, we have

$$\begin{aligned}(c_1e^{-x} + c_2e^{2x})(c_1e^{-x} + 4c_2e^{2x}) - (-c_1e^{-x} + 2c_2e^{2x})^2 \\ &= (c_1^2e^{-2x} + 5c_1c_2e^x + 4c_2^2e^{4x}) - (c_1^2e^{-2x} - 4c_1c_2e^x + 4c_2^2e^{4x}) \\ &= 9c_1c_2e^x\end{aligned}$$

and this is not identically zero on $(-\infty, \infty)$ for all values of c_1 and c_2. In fact, it is zero if and only if either c_1 or c_2 is zero.

The reason why an arbitrary linear combination of solutions was always a solution for the equation in Example 1 and is not always a solution for the equation in this example is of course that the equation $y'' - y' - 2y = 0$ of Example 1 is linear, whereas the equation $yy'' - (y')^2 = 0$ of the present example is nonlinear.

EXAMPLE 3

Direct substitution shows that the equation $(x - 1)(2x - 1)y'' + 2xy' - 2y = 0$ has $y_1 = x$ as a solution on $(-\infty, \infty)$, and $y_2 = 1/(x - 1)$ as a solution on $(-\infty, 1)$ and $(1, \infty)$. Therefore, by Theorem 1, $y = c_1y_1 + c_2y_2$ is a general solution on $(-\infty, 1)$ and $(1, \infty)$.

Since the value of its leading coefficient is zero at $x = \frac{1}{2}$ and $x = 1$, the given differential equation is normal on neither $(-\infty, \infty)$ nor $(-\infty, 1)$. Nevertheless, it has nontrivial solutions on these intervals. Thus we must not erroneously suppose that a differential equation has no solutions over intervals on which it is not normal.

In Sec. 1.3, we found that sometimes it is possible to reduce the number of parameters in a general solution of a differential equation by utilizing various identities and renaming constants. This suggests that a linear combination of particular solutions of a linear equation can always be reduced to one containing fewer parameters whenever one or more of the particular solutions is a linear combination of the others. The concepts of *linear dependence* and *linear independence* are fundamental to such reductions in the case of linear combinations. Indeed, these concepts are basic, as we shall see, to a full understanding of how complete solutions of linear equations are usually constructed.

The domain D of a linear combination of functions $c_1f_1 + c_2f_2 + \cdots + c_nf_n$ is, in general, the intersection of the domains of $f_1, f_2, \ldots, f_n$. In many important cases, D is an interval or contains a number of intervals as subsets, and the kind of functions involved is described by one of the following definitions.

DEFINITION 1 Functions $f_1, f_2, \ldots, f_n$ are **linearly dependent on an interval** I if and only if there exist constants $c_1, c_2, \ldots, c_n$, at least one of which is not zero, such that for every x in I,

$$c_1f_1(x) + c_2f_2(x) + \cdots + c_nf_n(x) = 0$$

DEFINITION 2 Functions $f_1, f_2, \ldots, f_n$ are **linearly independent on an interval** I if and only if they are not linearly dependent, i.e., if and only if

$$c_1f_1(x) + c_2f_2(x) + \cdots + c_nf_n(x) \equiv 0 \quad \text{on } I \text{ implies} \quad c_1 = c_2 = \cdots = c_n = 0$$

Definitions 1 and 2 pertain only to functions defined on a common interval. This is to be understood hereafter if no interval is mentioned as we speak of linearly dependent or independent functions.

EXAMPLE 4

Show that the functions $\cos x$ and $\sin x$ are linearly independent.

First we note that the domain of $\cos x$ and $\sin x$ is the set of real numbers. To verify their linear independence we must, according to Definition 2, show that if $c_1 \cos x + c_2 \sin x \equiv 0$ on $(-\infty, \infty)$, then $c_1 = c_2 = 0$. To do this, we note that if the assumed identity holds, it must hold for the particular values $x = 0$ and $x = \pi/2$. Substituting these values in turn we obtain from the assumed relation

$$c_1 1 + c_2 0 = 0 \qquad \text{and} \qquad c_1 0 + c_2 1 = 0$$

Hence $c_1 = c_2 = 0$, which proves that $\cos x$ and $\sin x$ are linearly independent functions as asserted.

It is important to note that the linear independence of $\cos x$ and $\sin x$ does not mean that these functions are unrelated. In fact, for all values of x, they are connected by the *nonlinear* identity $\cos^2 x + \sin^2 x = 1$.

EXAMPLE 5

By convention, the function $f(x) = x$ has domain $(-\infty, \infty)$, and the domain of $g(x) = \sqrt{x}$ is $[0, \infty)$. The intersection of these domains is the interval $[0, \infty)$. If for all $x \geq 0$,

$$c_1 f(x) + c_2 g(x) = c_1 x + c_2 \sqrt{x} = 0$$

then, in particular,

$$c_1(1) + c_2\sqrt{1} = c_1 + c_2 = 0 \qquad \text{and} \qquad c_1(4) + c_2\sqrt{4} = 4c_1 + 2c_2 = 0$$

The equation $c_1 + c_2 = 0$ gives $c_2 = -c_1$, and the equation $4c_1 + 2c_2 = 0$ gives $c_2 = -2c_1$. It follows that $c_1 = c_2 = 0$; hence, by Definition 2, f and g are linearly independent functions on $[0, \infty)$.

EXAMPLE 6

Show that if one of the functions $f_1, f_2, \ldots, f_n$ is trivial, they are linearly dependent.

Suppose $f_k(x) \equiv 0$, k fixed, and $1 \leq k \leq n$, so that f_k is trivial on the relevant interval I. With j as a running index such that $1 \leq j \leq n$, and $j \neq k$, set $c_j = 0$. Take c_k to be nonzero. Then,

$$c_1 f_1(x) + c_2 f_2(x) + \cdots + c_n f_n(x) \equiv c_k f_k(x) \equiv 0$$

and so, according to Definition 1, the functions are linearly dependent.

A nonzero function cannot be proportional to a zero function, but they are linearly dependent. For two nonzero functions f_1 and f_2, linear dependence is equivalent to proportionality. In fact, if f_1 and f_2 are linearly dependent, then, by Definition 1, $c_1 f_1(x) + c_2 f_2(x) \equiv 0$, where, since neither f_1 nor f_2 is trivial, $c_1 \neq 0$ and $c_2 \neq 0$. This gives $f_1(x) \equiv -(c_2/c_1)f_2(x)$ and $f_2(x) \equiv -(c_1/c_2)f_1(x)$, which show that f_1 and f_2 are proportional. Conversely, if f_1 and f_2 are proportional, it is evident that they satisfy an identity of the kind required by Definition 1. Thus linear dependence generalizes the concept of proportionality to include not only a trivial function but also one function or any number of functions.

Informative as Theorem 1 is, it leaves completely unanswered the important question of how many linearly independent solutions a homogeneous linear equation may have; nor does it indicate how all solutions may be determined. These deficiencies are remedied in part by the following theorem.

THEOREM 2 On any interval I over which an nth-order homogeneous linear differential equation is normal, the equation has n linearly independent solutions $y_1, y_2, \ldots, y_n$, and any particular solution of the equation on I can be expressed as a linear combination of these n linearly independent solutions.

Partly for reasons of simplicity and convenience and partly because we shall be primarily concerned with second-order equations, we shall prove this theorem only for $n = 2$, in which case (1) becomes

$$a_0(x)y'' + a_1(x)y' + a_2(x)y = 0 \tag{2}$$

Our method of proof, however, is one that may be extended in an obvious way to higher-order equations.

PROOF Let y_1 and y_2 be the solutions on I of (2), guaranteed by the fundamental existence theorem, which have the properties that at some point x_0 of I

$$y_1(x_0) = 1 \qquad y_2(x_0) = 0$$
$$y_1'(x_0) = 0 \qquad y_2'(x_0) = 1$$

To establish the linear independence of y_1 and y_2 we first observe that if $c_1y_1(x) + c_2y_2(x) = 0$ holds identically on I, then $c_1y_1'(x) + c_2y_2'(x) \equiv 0$ there. Since x_0 is contained in I, we have in particular

$$c_1y_1(x_0) + c_2y_2(x_0) = 0 \qquad \text{and} \qquad c_1y_1'(x_0) + c_2y_2'(x_0) = 0$$

which, in view of the specified properties of y_1 and y_2 at x_0, imply $c_1 = 0$ and $c_2 = 0$. Hence, by Definition 2, the solutions y_1 and y_2 are linearly independent.

To complete the proof we must show that any particular solution y of (2) can be expressed as a linear combination of y_1 and y_2. Since y, y_1, and y_2 are all solutions of (2) on I, so is the function Y given by the linear combination

$$Y = y - y(x_0)y_1 - y'(x_0)y_2$$

where $y(x_0)$ and $y'(x_0)$ are of course the values of the solution y and its derivative y' at x_0. Evaluating Y and Y' at x_0, we have

$$Y(x_0) = y(x_0) - y(x_0)y_1(x_0) - y'(x_0)y_2(x_0) = y(x_0) - y(x_0)1 - y'(x_0)0 = 0$$
$$Y'(x_0) = y'(x_0) - y(x_0)y_1'(x_0) - y'(x_0)y_2'(x_0) = y'(x_0) - y(x_0)0 - y'(x_0)1 = 0$$

It now follows from Corollary 1, Sec. 2.1, that Y is the trivial solution of (2). Hence, for every x in I,

$$0 = y(x) - y(x_0)y_1(x) - y'(x_0)y_2(x)$$

Solving this relation for $y(x)$, we see that y is expressible as the linear combination

$$y = y(x_0)y_1 + y'(x_0)y_2$$

of y_1 and y_2, and the proof for the case $n = 2$ is complete.

In the preceding proof, the convenient numerical values at x_0 of y_1, y_2, and their derivatives made it easy to verify the linear independence of y_1 and y_2. Fortunately, there is a straightforward way of testing any n particular solutions of (1) for linear dependence or independence even though no specific values of the solutions or their various derivatives are known. This test involves a simple application of determinants called *wronskians.*† The notation $W(f_1, f_2, \ldots, f_n)$ will be used to denote an *nth-order wronskian,* i.e., a wronskian of n functions $f_1, f_2, \ldots, f_n$.

†Named in honor of the Polish mathematician Hoene Wronsky (1778–1853).

DEFINITION 3 The **wronskian** of n functions $f_1, f_2, \ldots, f_n$ is the determinant

$$W(f_1, f_2, \ldots, f_n) = \begin{vmatrix} f_1 & f_2 & \cdots & f_n \\ f_1' & f_2' & \cdots & f_n' \\ \cdots & \cdots & \cdots & \cdots \\ f_1^{(n-1)} & f_2^{(n-1)} & \cdots & f_n^{(n-1)} \end{vmatrix} \tag{3}$$

Clearly, an nth-order wronskian will (or will not) exist over an interval according as the functions upon which it depends are (or are not) all differentiable $n-1$ times on that interval.

Our next theorem points out an especially important property of wronskians whose defining functions are solutions of homogeneous differential equations. Since every step which is essential to a proof for wronskians of a general order n is contained in a proof for the case $n = 2$, a proof of the theorem for second-order wronskians will suffice.

THEOREM 3 On any interval over which an nth-order homogeneous linear differential equation is normal, the wronskian of any n particular solutions is either identically zero or its value is never zero.

PROOF Let y_1 and y_2 be particular solutions of (2) on an interval I over which the equation is normal. Then, for every x contained in I,

$$y_i''(x) = -\frac{a_1(x)}{a_0(x)}y_i'(x) - \frac{a_2(x)}{a_0(x)}y_i(x) \qquad i = 1, 2$$

and the wronskian of y_1 and y_2 has the value

$$W(x) = \begin{vmatrix} y_1(x) & y_2(x) \\ y_1'(x) & y_2'(x) \end{vmatrix} = y_1(x)y_2'(x) - y_2(x)y_1'(x)$$

Differentiating $W(x)$ and simplifying, then substituting for $y_2''(x)$ and $y_1''(x)$ and simplifying further, we find

$$\begin{aligned} W'(x) &= y_1(x)y_2''(x) - y_2(x)y_1''(x) \\ &= y_1(x)\left[-\frac{a_1(x)}{a_0(x)}y_2'(x) - \frac{a_2(x)}{a_0(x)}y_2(x)\right] - y_2(x)\left[-\frac{a_1(x)}{a_0(x)}y_1'(x) - \frac{a_2(x)}{a_0(x)}y_1(x)\right] \\ &= -\frac{a_1(x)}{a_0(x)}W(x) \end{aligned}$$

or finally

$$W'(x) + \frac{a_1(x)}{a_0(x)}W(x) = 0$$

Thus W is a solution of a normal homogeneous first-order linear differential equation, and according to Corollary 1, Sec. 2.1, if the value of W is zero at even one point of I, then $W(x) \equiv 0$. In other words, $W(x)$ is either identically zero or never zero, as asserted by the theorem. ◗

By differentiating $W(x)$ and working directly with determinants, it is easy to show that the relation $W'(x) + [a_1(x)/a_0(x)]W(x) = 0$ still holds if $n > 2$ and $W(x)$ is the value of the wronskian $W(y_1, y_2, \ldots, y_n)$ of n particular solutions of (1). Using the integrating factor $\exp\left[\int a_1(x)/a_0(x)\,dx\right]$ to solve for $W(x)$, we obtain **Abel's formula**†

†Named for the Norwegian mathematician Nils Abel (1802–1829).

$$W(x) = ke^{-\int a_1(x)/a_0(x)\, dx} \tag{4}$$

which holds identically over any interval where (1) is normal. Of course the integration constant k depends on the particular solutions involved, and it may have different values on intervals that are separated by zeros of a_0.

EXAMPLE 7

It is readily verified that on every interval of the x axis, $y_1 = e^x$ and $y_2 = x$ are solutions of the equation $(1 - x)y'' + xy' - y = 0$. On the two separate intervals $(-\infty, 1)$ and $(1, \infty)$, neither of which contains the zero $x = 1$ of $a_0(x) = x - 1$, this equation is normal and the value of the wronksian of y_1 and y_2,

$$W(x) = \begin{vmatrix} e^x & x \\ e^x & 1 \end{vmatrix} = (1 - x)e^x$$

is never zero. Now, from Abel's formula,

$$\begin{aligned} W(x) &= k \exp\left(-\int \frac{x}{1-x}\, dx\right) = k \exp\left(\int 1 + \frac{1}{x-1}\, dx\right) = k \exp\,(x + \ln|x - 1|) \\ &= ke^x e^{\ln|x-1|} = k\,|x - 1|\, e^x = \begin{cases} k(x-1)e^x & x > 1 \\ k(1-x)e^x & x < 1 \end{cases} \end{aligned}$$

Clearly, $k = -1$ if x is contained in $(1, \infty)$, or any subinterval thereof, whereas $k = 1$ on any interval to the left of $x = 1$.

We now give a simple criterion for testing n particular solutions of an nth-order homogeneous equation to see if they are linearly independent.

THEOREM 4 Any n particular solutions of (1) on an interval I over which the equation is normal are linearly independent if and only if the value $W(x)$ of their wronskian $W(y_1, y_2, \ldots, y_n)$ is nonzero at each point x of I.

To avoid again the burdensome notation incident to a general proof, a proof is given only for the case $n = 2$.

PROOF Let $a_0(x)y'' + a_1(x)y' + a_2(x)y = 0$ be normal on an interval I over which y_1 and y_2 are any particular solutions. Under the hypothesis that y_1 and y_2 are linearly independent, we are to show that the value of their wronskian is never zero on I. Assume that at some point x_0 of I,

$$W(x_0) = \begin{vmatrix} y_1(x_0) & y_2(x_0) \\ y_1'(x_0) & y_2'(x_0) \end{vmatrix} = 0$$

Then (see Exercise 49) there are numbers k_1 and k_2, not both zero, which when substituted, respectively, for the unknowns c_1 and c_2 satisfy the simultaneous algebraic equations

$$\begin{aligned} y_1(x_0)c_1 + y_2(x_0)c_2 &= 0 \\ y_1'(x_0)c_1 + y_2'(x_0)c_2 &= 0 \end{aligned} \tag{5}$$

that is,

$$k_1 y_1(x_0) + k_2 y_2(x_0) = 0 \qquad \text{and} \qquad k_1 y_1'(x_0) + k_2 y_2'(x_0) = 0$$

The linear combination

$$y = k_1y_1 + k_2y_2$$

formed with these numbers is of course a solution of the given differential equation. Moreover, from the fact that Eqs. (5) hold when $c_1 = k_1$ and $c_2 = k_2$, it follows that

$$y(x_0) = y'(x_0) = 0$$

Thus, by Corollary 1, Sec. 2.1, $y(x) \equiv k_1y_1(x) + k_2y_2(x) \equiv 0$ on I, where either $k_1 \neq 0$ or $k_2 \neq 0$. But this implies that y_1 and y_2 are linearly dependent contrary to hypothesis. Therefore the assumption that $W(x)$ is zero at some point of I must be false. This proves that if y_1 and y_2 are linearly independent, then the value of their wronskian is never zero on I.

To prove the converse, we observe that if $c_1y_1(x) + c_2y_2(x) \equiv 0$ over I, then at any fixed point x_0 on I, Eqs. (5) hold. Since $W(x_0) \neq 0$ by hypothesis, it follows (see Exercise 49) that $c_1 = 0$ and $c_2 = 0$. This proves that if the value of $W(y_1, y_2)$ is never zero on I, then y_1 and y_2 are linearly independent. ◗

An equivalent way of stating Theorem 4 is the following.

THEOREM 5 Any n particular solutions of (1) on an interval I over which the equation is normal are linearly dependent if and only if their wronskian $W(y_1, y_2, \ldots, y_n)$ has the value zero at some point of I.

We now prove, for the case $n = 2$, the following fundamental theorem about families of solutions of homogeneous linear equations.

THEOREM 6 If on an interval I, $y_1, y_2, \ldots, y_n$ are any n linearly independent particular solutions of a normal homogeneous equation

$$a_0(x)y^{(n)} + a_1(x)y^{(n-1)} + \cdots + a_{n-1}(x)y' + a_n(x)y = 0 \tag{1}$$

then every solution of (1) over I is a linear combination of $y_1, y_2, \ldots, y_n$; that is, for every solution y of (1) there exist real numbers $c_1, c_2, \ldots, c_n$ such that for every x contained in I,

$$y(x) = c_1y_1(x) + c_2y_2(x) + \cdots + c_ny_n(x)$$

◖ **PROOF** Suppose y_1 and y_2 are linearly independent solutions and that y is any particular solution of $a_0(x)y'' + a_1(x)y' + a_2(x)y = 0$ on an interval I over which the equation is normal. By Theorem 4, the value of $W(y_1, y_2)$ is never zero on I. Hence for any fixed point x_0 of I, the linear combination

$$Y = y - \frac{\begin{vmatrix} y(x_0) & y_2(x_0) \\ y'(x_0) & y_2'(x_0) \end{vmatrix}}{\begin{vmatrix} y_1(x_0) & y_2(x_0) \\ y_1'(x_0) & y_2'(x_0) \end{vmatrix}} y_1 - \frac{\begin{vmatrix} y_1(x_0) & y(x_0) \\ y_1'(x_0) & y'(x_0) \end{vmatrix}}{\begin{vmatrix} y_1(x_0) & y_2(x_0) \\ y_1'(x_0) & y_2'(x_0) \end{vmatrix}} y_2 \tag{6}$$

of y, y_1, and y_2 is a solution of the differential equation. Evaluating Y and Y' at x_0, we find that $Y(x_0) = Y'(x_0) = 0$ (see Exercise 65), which implies $Y(x) \equiv 0$ throughout I. It follows from (6) that y is a linear combination of y_1 and y_2. ◗

It is important to recognize the difference between Theorems 2 and 6. Theorem 2 guarantees the *existence* of n linearly independent particular solutions of every nth-order homogeneous linear differential equation (1) on any interval over which the equation is normal, while Theorem 6 gives no such assurances. Theorem 2 also says that there are some linearly independent solutions of (1) such that every particular solution of the equation can be expressed as a linear combination of n of them, but it does not say that under conditions of normality every particular solution can be expressed as a linear combination of *any n* linearly independent particular solutions, as Theorem 6 affirms.

Having found that, in general, every solution of (1) can be expressed as a linear combination of any n linearly independent solutions and recalling from Theorem 1 that every linear combination of solutions of (1) is itself a solution, it is now proper to refer to an arbitrary linear combination

$$y = c_1y_1 + c_2y_2 + \cdots + c_ny_n$$

of any n linearly independent particular solutions of (1) as a **complete solution** on any interval over which the equation is normal. This suggests that the maximum number of linearly independent particular solutions of (1) on any interval over which the equation is normal is always equal to the order n of the equation. A proof of this result is asked for in Exercise 66.

We conclude this section with another definition.

DEFINITION 4 Any n linearly independent particular solutions $y_1, y_2, \ldots, y_n$ of an nth-order homogeneous linear differential equation, on an interval over which the equation is normal, are called **fundamental solutions,** or a **basis** for all solutions, of the equation on that interval.

EXERCISES

For each of the following differential equations, specify where the given functions y_1 and y_2 are solutions, where their linear combination $c_1y_1 + c_2y_2$ is a solution, and where the equation is normal.

1. $(x^2 - 9)y'' - 2xy' + 2y = 0$; $y_1 = x$, $y_2 = x^2 + 9$
2. $x(x-1)y'' + 2(2x-1)y' + 2y = 0$; $y_1 = 1/x$, $y_2 = 1/(x-1)$
3. $x^2y'' - xy' + y = 0$; $y_1 = x$, $y_2 = x \ln|x|$
4. $3x^2y'' - 4xy' + 2y = 0$; $y_1 = x^2$, $y_2 = 1/x$
5. $x(x-1)y'' + (x^2-2)y' + (x-2)y = 0$; $y_1 = e^{-x}$, $y_2 = 1/x$
6. $y''(x \ln|x|)^2 - 2xy' \ln|x| + y(2 + \ln|x|) = 0$; $y_1 = \ln|x|$, $y_2 = x\ln|x|$
7. $4x^2y'' + 4xy' - y = 0$; $y_1 = |x|^{1/2}$, $y_2 = |x|^{-1/2}$
8. $y''(e^x - e^{-x})(2 + \sinh 2x) + y'(9\cosh x - \cosh 3x) - 2y(3e^x + e^{-x}) = 0$; $y_1 = e^x$, $y_2 = \coth x$
9. $y'' \cos x \sin^2 x - y' (\sin x)(2 + \cos^2 x) + 3y \cos x = 0$; $y_1 = \sin x$, $y_2 = \tan x$
10. $x^2y'' \sin 2x - 2(x + \sin 2x)(xy' - y) = 0$; $y_1 = x$, $y_2 = x \ln|\sec x|$

In Exercises 11 through 14, verify that on suitable intervals and for all values of the arbitrary constants a and b, the functions y_1 and y_2 satisfy each of the corresponding differential equations but that $y_1 + y_2$ satisfies only the first equation. Explain.

11. $xy'' = y'$, $2yy'' = (y')^2$; $y_1 = a$, $y_2 = bx^2$
12. $2xy'' + y' = 0$, $8x^3(y'')^2 - yy' = 0$; $y_1 = a$, $y_2 = b\sqrt{x}$
13. $(x^2-1)y'' - 2xy' + 2y = 0$, $2yy'' - (y')^2 = 0$; $y_1 = a(x-1)^2$, $y_2 = b(x+1)^2$
14. $y''(\cos 2x + \sin 2x) - 4y' \cos 2x + 4y(\cos 2x - \sin 2x) = 0$, $(y'')^2 - 16y^2 = 0$; $y_1 = a\cos 2x$, $y_2 = be^{2x}$.
15. **(a)** Verify that on $(-\infty, \infty)$ every linear combination of $y_1 = x^2 + 2$ and $y_2 = 3$ is a solution of $xy'' - y' = 0$.
    (b) Are there any linear combinations of y_1 and y_2 that are solutions of
    $$2(y-2)y'' - (y')^2 = 0$$
    on $(-\infty, \infty)$?
16. **(a)** Verify that $\cos x$, $\sin x$, $\cosh x$, and $\sinh x$ are all solutions of $(y'')^2 - y^2 = 0$ on $(-\infty, \infty)$.
    (b) What linear combinations of these solutions are general solutions of the equation?
    (c) Is every linear combination of e^x and e^{-x} also a solution?
17. Prove that $\sin x$ and $\cos x$ are linearly independent by evaluating them for values of x different from those used in Example 4.
18. Prove that $\cosh x$ and $\sinh x$ are linearly independent functions.
19. Show that the polynomials $x + 1$, $x + 2$, and $2x + 1$ are linearly dependent and find a linear differential equation of minimal order which they satisfy.

Determine whether the functions in the following sets, and having the indicated intervals as domains, are linearly dependent or linearly independent. If the functions are linearly dependent, give a relation which characterizes the dependence.

20. $\{e^x, e^{2x}\}$; $(-\infty, \infty)$
21. $\{\sqrt{x}, \ln x\}$; $(0, \infty)$

22. $\{1, x, x^2\}$; $(-\infty, \infty)$ **23.** $\{e^x, x, \cosh x\}$; $(-\infty, \infty)$

24. $\{\cos 2x, \sin^2 x, \cos^2 x\}$; $(-\infty, \infty)$

25. $\{e^x; e^{-x}, \cosh x\}$; $(-\infty, \infty)$

26. $\{x^2, x^2 - 1, x^2 + x + 1)$; $(-\infty, \infty)$

27. $\{x^2 - 1, x^2 + x + 1, x^2 + 3x + 5\}$; $(-\infty, \infty)$

28. $\{\ln (x - 1), 2 \ln (x + 1), 3 \ln (x^2 - 1)\}$; $x > 2$

29. $\{1/(x + 1), 1/x, 1/(x - 2)\}$; $x < -1$

30. $\{x^3, |x|^3\}$; $(-\infty, \infty)$

31. $\{\cos x \sin 2x, 2 \sin x \cos 2x, 3 \sin 3x\}$; $(-\infty, \infty)$

32. Is the function

$$y(x) = \begin{cases} a - x & 0 \le x \le a \\ x - a & a \le x \le b,\ 0 < b \end{cases}$$

a solution of $y'' = 0$ on $0 \le x \le b$? *Hint:* Pay special attention to possible values of a.

33. **(a)** Verify that $\cosh 2x$ and $\sinh 2x$ are linearly independent solutions and that e^{-2x} is also a solution of $y'' - 4y = 0$ on $(-\infty, \infty)$.

(b) Express e^{-2x} as a linear combination of $\cosh 2x$ and $\sinh 2x$.

34. Show that there is no point at which two linearly independent solutions of the equation $y'' + p_1(x)y' + p_2(x)y = 0$ both vanish except possibly where either $p_1(x)$ or $p_2(x)$ is discontinuous.

35. Show that there is no point at which two linearly independent solutions of the equation $y'' + p_1(x)y' + p_2(x)y = 0$ simultaneously take on extreme values except possibly at a point where either $p_1(x)$ or $p_2(x)$ is discontinuous.

36. If the wronskian of two functions is different from zero at every point of an interval, show that there is no point of the interval at which either function has a repeated zero.

37. Given two linearly independent solutions of $y'' + p_1(x)y' + p_2(x)y = 0$, show that between any two consecutive zeros of either solution there is exactly one zero of the other solution. *Hint:* Let y_1 and y_2 be the two solutions, let a and b be two consecutive zeros of y_1, apply Rolle's theorem to the quotient y_1/y_2, and note the contradiction unless $y_2 = 0$ at some point between a and b. This result is known as the **Sturm separation theorem.**†

38. If the quotient of two linearly independent solutions of the equation $y'' + p_1(x)y' + p_2(x)y = 0$ exists at all points of an interval, prove that it either is an increasing function at all points of the interval or is a decreasing function at all points of the interval.

39. For all values of t, the abscissa x and ordinate y of a point P in the xy plane are given by continuously differentiable functions $x(t)$ and $y(t)$, respectively, and $W(t) = x\dot{y} - y\dot{x} \neq 0$.

(a) Is P ever at the origin? Explain.

(b) When, if ever, does the direction of rotation of the line segment joining the origin and P change direction? Explain.

Hint: Express W in terms of polar coordinates $r(t)$ and $\theta(t)$.

40. Show that if y_1, y_2, and y_3 are solutions of $y''' + ky = 0$, k a real parameter, then on $(-\infty, \infty)$ the value $W(x)$ of $W(y_1, y_2, y_3)$ is constant. *Hint:* Find $W'(x)$.

41. If y_1 and y_2 have a nonvanishing wronskian, show that $y_3 = c_1y_1 + c_2y_2$ and $y_4 = k_1y_1 + k_2y_2$ have a nonvanishing wronskian if and only if $c_1k_2 - c_2k_1 \neq 0$.

42. Evaluate the wronskian of the functions e^{mx} and xe^{mx} at a general point x.

43. Evaluate the wronskian of e^x, e^{-x}, and e^{2x} for a general value of x.

44. Show that $y_1 = \sin 2x$ and $y_2 = \cos 2x$ are solutions of $y'' + 4y = 0$ and verify that their wronskian satisfies Abel's formula.

45. Show that $y_1 = e^x$, $y_2 = e^{2x}$, $y_3 = e^{3x}$ are solutions of $y''' - 6y'' + 11y' - 6y = 0$ and verify that their wronskian satisfies Abel's formula.

46. Establish Abel's formula for the case $n = 3$.

47. Explain how Abel's formula can be utilized to find a second solution of a normal equation

$$a_0(x)y'' + a_1(x)y' + a_2(x)y = 0$$

when one solution is known.

48. Using the given solution and the method asked for in Exercise 47, find complete solutions of each of the following equations.

(a) $(1 + x)y'' + xy' - y = 0$; $y = x$

(b) $xy'' + y' = 0$; $y = 1$

(c) $x^2y'' - xy' - 3y = 0$; $y = 1/x$

(d) $y'' + (\tan x)y' - 6(\cot^2 x)y = 0$; $y = \sin^3 x$

(e) $y'' = (2 \cot x - \tan x)y'$; $y = 1$

(f) $x^2y'' + x^3y' - 2(1 + x^2)y = 0$; $y = x^2$

49. Show that two simultaneous homogeneous linear algebraic equations

$$ax + by = 0$$
$$cx + dy = 0$$

in the unknowns x and y have a solution other than $x = 0$, $y = 0$ if and only if the determinant of the coefficients of the system

$$\begin{vmatrix} a & b \\ c & d \end{vmatrix} = ad - bc = 0$$

50. Show that a second-order linear differential equation having y_1 and y_2 as a basis is $W(y_1, y_2, y) = 0$.

51. Show that a second-order linear differential equation having y_1 and y_2 as fundamental solutions is $y''W(y_1, y_2) - y'(d/dx)W(y_1, y_2) + yW(y_1', y_2') = 0$.

Using the equation of Exercise 51, find a second-order linear differential equation having each of the following pairs of functions as a basis on the given interval.

52. $e^{2x} \cos 3x$, $e^{2x} \sin 3x$; $(-\infty, \infty)$

†Named for the Swiss mathematician J. C. F. Sturm (1803–1855).

53. $1, x^3$; $(0, \infty)$
54. x, $1/x$; $(0, \infty)$
55. $x^{7/3}$, e^x; $(0, 7)$
56. x, $\ln|x|$; $(-\infty, 0)$
57. e^{-2x}, $\cosh x$; $(-\infty, \infty)$

58. **(a)** Find solutions y_1 and y_2 of the equation $y'' - y' \tan x - y \sec^2 x = 0$ on $(-\pi/2, \pi/2)$ such that $y_1(0) = 1$, $y_1'(0) = 0$, $y_2(0) = 0$, and $y_2'(0) = 1$.
(b) Evaluate $W(y_1, y_2)$.
(c) Construct a complete solution of the given differential equation on $(-\pi/2, \pi/2)$.
(d) Show that on $(-\pi/2, \pi/2)$ the given differential equation is equivalent to

$$y''W(x) - y'W'(x) - yW^3(x) = 0$$

where $W(x)$ is the value of $W(y_1, y_2)$ at x.

59. On the interval I, either finite or infinite, let $r(x)$ be a positive-valued function with a continuous derivative and let $q_1(x)$ and $q_2(x)$ be continuous nonidentical functions such that $q_2(x) \geq q_1(x)$ on I. Show that if y_1 is any solution of the equation $[r(x)y']' + q_1(x)y = 0$ and y_2 is any solution of the equation $[r(x)y']' + q_2(x)y = 0$, then between any two consecutive zeros of y_1 in I there is at least one zero of y_2. *Hint:* Let a and b be two consecutive zeros of $y_1(x)$ and assume $y_2(x)$ is not zero for any value of x between a and b. Show first that $y_1(x)$ and $y_2(x)$ can both be supposed positive on (a, b). Then, by substituting y_1 and y_2 into the respective differential equations and combining the results, show that

$$r(x)[y_1'(x)y_2(x) - y_2'(x)y_1(x)]\Big|_a^b = \int_a^b [q_2(x) - q_1(x)]y_1(x)y_2(x)\,dx$$

and from this derive a contradiction which will complete the proof. This result is known as the **Sturm comparison theorem.**

60. If $q(x) \leq 0$, show that no nontrivial solution of $y'' + q(x)y = 0$ can have more than one zero. *Hint:* Apply the result of Exercise 59 to the pair of equations $y'' + q(x)y = 0$ and $y'' = 0$.

Verify that each of the following equations has the indicated solutions, and in each case construct two different complete solutions using two different bases.

61. $y'' - y = 0$; $y_1 = e^x$, $y_2 = e^{-x}$
62. $y'' - 3y' + 2y = 0$; $y_1 = e^x$, $y_2 = e^{2x}$
63. $y'' + y = 0$; $y_1 = \sin(x + \pi/4)$, $y_2 = \sin(x - \pi/4)$
64. $x^2y'' + xy' - y = 0$; $y_1 = x$, $y_2 = 1/x$
65. Verify that Eq. (6) can be written in the form

$$Y = \frac{1}{W[y_1(x_0), y_2(x_0)]}\begin{vmatrix} y(x) & y_1(x) & y_2(x) \\ y(x_0) & y_1(x_0) & y_2(x_0) \\ y'(x_0) & y_1'(x_0) & y_2'(x_0) \end{vmatrix}$$

hence verify the assertion that $Y(x_0) = Y'(x_0) = 0$.

66. Theorem 2 ensures that if an nth-order homogeneous equation

$$a_0(x)y^{(n)} + a_1(x)y^{(n-1)} + \cdots + a_{n-1}(x)y' + a_n(x)y = 0$$

is normal on an interval I, then it has at least n linearly independent solutions on I. Prove that such an equation has at most n linearly independent solutions on I.

2.3 SOLUTIONS OF NONHOMOGENEOUS EQUATIONS

So far in this chapter our attention has been centered on homogeneous linear differential equations of order n. If $n > 1$, we have as yet no means of solving such an equation besides inspection and guessing. However, we do know how to form families of solutions from particular solutions and how to determine whether any n particular solutions are adequate to serve as a basis for constructing a complete solution.

In this section we shall learn how to construct a complete solution of a nonhomogeneous equation

$$(1) \qquad a_0(x)y^{(n)} + a_1(x)y^{(n-1)} + \cdots + a_{n-1}(x)y' + a_n(x)y = f(x)$$

when a basis for the related homogeneous equation is known. The fundamental idea behind the procedure is this.

THEOREM 1 If Y is any specific solution of (1) and if $c_1y_1 + c_2y_2 + \cdots + c_ny_n$ is a complete solution of the related homogeneous equation, then a complete solution of the nonhomogeneous equation is $y = c_1y_1 + c_2y_2 + \cdots + c_ny_n + Y$.

If $n > 2$, this theorem may be proved in the same way as is done here for the case $n = 2$.

PROOF Let $y(x)$ represent an arbitrary solution and let $Y(x)$ represent any specific solution of

$$a_0(x)y'' + a_1(x)y' + a_2(x)y = f(x)$$

on any interval I over which the equation is normal. Then, for every x contained in I,

$$a_0(x)y''(x) + a_1(x)y'(x) + a_2(x)y(x) = f(x)$$
$$a_0(x)Y''(x) + a_1(x)Y'(x) + a_2(x)Y(x) = f(x)$$

Subtracting the second equation from the first, we have

$$a_0(x)[y''(x) - Y''(x)] + a_1(x)[y'(x) - Y'(x)] + a_2(x)[y(x) - Y(x)]$$
$$\equiv a_0(x)[y(x) - Y(x)]'' + a_1(x)[y(x) - Y(x)]' + a_2(x)[y(x) - Y(x)] \equiv 0$$

Thus $y - Y$ is a solution of the homogeneous equation

$$a_0(x)y'' + a_1(x)y' + a_2(x)y = 0$$

By Theorem 6, Sec. 2.2, if this equation has $c_1y_1 + c_2y_2$ as a complete solution over I, then $y - Y$ must be expressible in the form $y - Y = c_1y_1 + c_2y_2$. Transposing Y, we obtain

$$y = c_1y_1 + c_2y_2 + Y$$

as the theorem affirms. ◗

The term Y in a complete solution $y = c_1y_1 + c_2y_2 + \cdots + c_ny_n + Y$ of Eq. (1) can be any solution of the equation, no matter how special. It is called a **particular integral** of the nonhomogeneous equation. The linear combination $c_1y_1 + c_2y_2 + \cdots + c_ny_n$, which is a complete solution of the related homogeneous equation, is called a **complementary function** of the nonhomogeneous equation. The steps to be carried out in solving an nth-order nonhomogeneous linear differential equation can now be summarized as follows.

1. Find n particular solutions $y_1, y_2, \ldots, y_n$ of the related homogeneous equation which have a nonvanishing wronksian. Then form a *complementary function* $c_1y_1 + c_2y_2 + \cdots + c_ny_n$ of the given equation.
2. Find one particular solution of the nonhomogeneous equation itself.
3. Add the *particular integral* Y found in Step **2** to the *complementary function* $c_1y_1 + c_2y_2 + \cdots + c_ny_n$ found in Step **1** to obtain a complete solution of the given equation.

Useful as this summary may appear, by itself it is of little practical value. What is needed is a way of performing Steps **1** and **2**, and this often is impossible. There are, however, many important linear differential equations that can be solved as outlined, and much of our subsequent work will deal with these. Quite often the nonhomogeneous term of a linear differential equation will itself be a sum of other terms. When this happens, it may be helpful to make use of the following theorem to find a particular integral of the equation by stages.

THEOREM 2 If for $i = 1, 2, \ldots, m$, Y_i is a particular integral of

$$a_0(x)y^{(n)} + a_1(x)y^{(n-1)} + \cdots + a_{n-1}(x)y' + a_n(x)y = f_i(x)$$

then $Y_1 + Y_2 + \cdots + Y_m$ is a particular integral of

$$a_0(x)y^{(n)} + a_1(x)y^{(n-1)} + \cdots + a_{n-1}(x)y' + a_n(x)y = f_1(x) + f_2(x) + \cdots + f_m(x) \tag{2}$$

Although this theorem does not say so, it is to be understood that $Y_1 + Y_2 + \cdots + Y_m$ is a solution of Eq. (2) on an interval over which the equation is normal. Hereafter, when nothing else is said, we shall always assume that suitable conditions of normality hold for our equations.

Rather than unduly restricting both m and n to prove Theorem 2, let us employ summation notation and give a general proof. We may as well get used to this because the use of series, both finite and infinite, will be indispensable in some of our later work.

PROOF By hypothesis, for $i = 1, 2, \ldots, m$,

$$\sum_{k=0}^{n} a_k(x) Y_i^{(n-k)}(x) = f_i(x)$$

Summing this equation over i, we have

$$\sum_{i=1}^{m} \sum_{k=0}^{n} a_k(x) Y_i^{(n-k)}(x) = \sum_{i=1}^{m} f_i(x)$$

Reversing the order of summation in the double sum, then moving $a_k(x)$ across the inner summation sign (as may be done with any quantity which does not involve the index of summation), and subsequently replacing sums of derivatives by derivatives of sums, we obtain

$$\sum_{k=0}^{n} \sum_{i=1}^{m} a_k(x) Y_i^{(n-k)}(x) = \sum_{k=0}^{n} a_k(x) \left[\sum_{i=1}^{m} Y_i^{(n-k)}(x) \right] = \sum_{k=0}^{n} a_k(x) \left[\sum_{i=1}^{m} Y_i(x) \right]^{(n-k)} = \sum_{i=1}^{m} f_i(x)$$

which states that $Y_1 + Y_2 + \cdots + Y_m$ is a particular integral of (2).

EXAMPLE 1

HARMONIC MOTION

A particle P traverses the circumference of a circle of radius r at a constant angular velocity ω (Fig. 2.1). Since $d\theta/dt = \omega$, $\theta = \omega t + \alpha$, where α is a constant. Thus the rectangular coordinates of P, given by $x = r \cos \theta$ and $y = r \sin \theta$, become

$$x = r \cos (\omega t + \alpha) \qquad \text{and} \qquad y = r \sin (\omega t + \alpha)$$

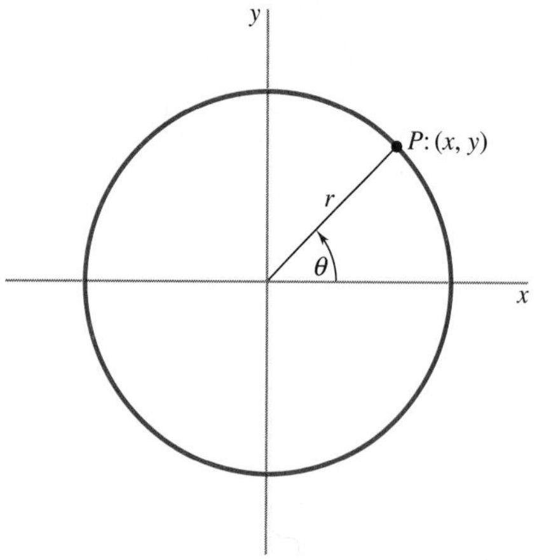

Figure 2.1
A particle traversing a circular path.

Clearly, each of these functions is of amplitude r and period $2\pi/\omega$. Differentiating x and y twice and remembering that r, ω, and α are constants, we find

$$\frac{d^2x}{dt^2} = -\omega^2 r \cos(\omega t + \alpha) = -\omega^2 x \qquad \text{and} \qquad \frac{d^2y}{dt^2} = -\omega^2 r \sin(\omega t + \alpha) = -\omega^2 y$$

This shows that with ω given but r and α arbitrary, both $r \cos(\omega t + \alpha)$ and $r \sin(\omega t + \alpha)$ are general solutions of

$$\text{(3)} \qquad \frac{d^2z}{dt^2} + \omega^2 z = 0 \qquad \omega \neq 0$$

This *type* of equation is called a **harmonic differential equation** or a **differential equation of harmonic motion.** It is the *structure,* and *not* the variables involved, that distinguishes (3) as a harmonic equation. Many problems which have nothing to do with any kind of motion give rise to such an equation.

Using familiar trigonometric identities, the general solutions of (3) just noted can be written as the linear combinations

$$\text{(4)} \qquad \begin{aligned} r \cos(\omega t + \alpha) &= r \cos \alpha \cos \omega t - r \sin \alpha \sin \omega t \\ r \sin(\omega t + \alpha) &= r \cos \alpha \sin \omega t + r \sin \alpha \cos \omega t \end{aligned}$$

of the particular solutions $\cos \omega t$ and $\sin \omega t$ which Eqs. (4) reduce to when $r = 1$ and $\alpha = 0$. Thus both general solutions have been expressed in the form

$$\text{(5)} \qquad z = c_1 \cos \omega t + c_2 \sin \omega t$$

where c_1 and c_2 are arbitrary constants. We find at once that for all values of t the wronskian $W(\cos \omega t, \sin \omega t)$ has the value 1, and so the general solutions (4), and (5) as well, are actually *complete* solutions of (3). Motion described by any nontrivial solution of (3) is called **harmonic motion** or **harmonic oscillation.** As Eqs. (4) indicate, characteristics of the motion are its constant amplitude and periodicity.

EXAMPLE 2

Now that we know how to write a complete solution of any harmonic equation, let us find a complete solution of the nonhomogeneous equation

$$\text{(6)} \qquad y'' + 9y = 3 + 2e^x + \sin x$$

Following the three-step procedure previously outlined, we first observe that a complementary function of (6) is

$$y_c = c_1 \cos 3x + c_2 \sin 3x$$

To obtain a particular integral of the equation, we shall find a particular integral of $y'' + 9y = 3$, then of $y'' + 9y = 2e^x$, and finally of $y'' + 9y = \sin x$. By inspection, we see that the first of these three equations has $Y_1 = \frac{1}{3}$ as a particular integral and that $Y_2 = \frac{1}{5}e^x$ is a particular integral of the second equation. To find a particular integral of the third equation, we observe that the only derivative of y appearing in the related homogeneous equation is of even order. This suggests that we look for a particular integral of the type $Y = A \sin x$, with A constant. Substituting this expression for y and canceling $\sin x$ from each term, we obtain $-A + 9A = 1$, hence $A = \frac{1}{8}$. This gives $Y_3 = \frac{1}{8} \sin x$ as a particular integral of the third equation. To conclude Step **2** we note that, according to Theorem 2, a particular integral of (6) is $Y_1 + Y_2 + Y_3$. Step **3** consists of adding this particular integral and the complementary function given in Step **1** to obtain

$$y = c_1 \cos 3x + c_2 \sin 3x + \tfrac{1}{3} + \tfrac{1}{5}e^x + \tfrac{1}{8} \sin x$$

as a complete solution of (6).

EXERCISES

What are the frequency and amplitude of the harmonic motion defined by the solution of each of the following initial-value problems?

1. $\dfrac{d^2y}{dt^2} + 4y = 0;\ y(0) = -13,\ y'(0) = 0$

2. $\dfrac{d^2y}{dt^2} + 49y = 0;\ y(0) = 0,\ y'(0) = -21$

3. $\dfrac{d^2y}{dt^2} + 64y = 0;\ y(0) = -3,\ y'(0) = 32$

4. $\dfrac{d^2y}{dt^2} + 100\pi^2 y = 0;\ y(0) = 10,\ y'(0) = -100\pi$

5. $25y'' + \pi^2 y = 0;\ y(5) = -12,\ y'(5) = \pi$

6. $9y'' + y = 0;\ y(\pi) = \frac{3}{2} + 2\sqrt{3},\ y'(\pi) = \frac{2}{3} - \sqrt{3}/2$

Find complete solutions of each of the following equations.

7. $y'' + 4y = 16x + 8e^x$

8. $y'' + 25y = 100 + 125 \sin 10x$

9. $4y'' + y = 10e^x + 15 \cos x$

10. $9y'' + 16y = 32 - 50e^x + 40 \cos 2x$

11. $y'' + \pi^2 y = 6\pi^2 \sin \pi x \cos \pi x$

12. $y'' + 36y = 74 + 108x + 36x^2$

13. A curve has parametric equations $x = f(\theta)$, $y = g(\theta)$, which satisfy the respective initial-value problems $x'' + x = 0;\ x(0) = a,\ x'(0) = 0$, and $y'' + y = 0;\ y(0) = 0,\ y'(0) = b$, where $a, b > 0$. Find $f(\theta)$ and $g(\theta)$ and identify the curve.

14. A parametric representation of the hyperbola $x^2 - y^2 = r^2$ is $x = r \cosh \phi$, $y = r \sinh \phi$, where ϕ is $2/r^2$ times the area u of the hyperbolic sector POA (Fig. 2.2). A particle P traverses the positive branch of the curve so that u increases at a constant rate $r^2\omega/2$.

(a) Express x and y as functions of the time t.

(b) Find three different analytic expressions each of which gives a complete solution of the equation $d^2z/dt^2 - \omega^2 z = 0$.

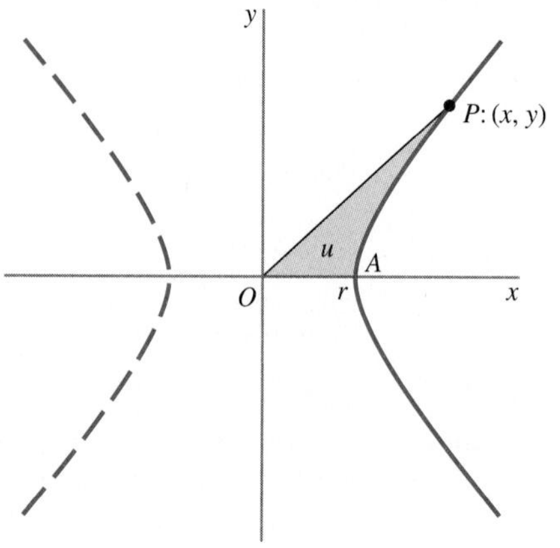

Figure 2.2
A particle traversing a hyperbolic path.

Using Exercise 14, Part **(b)**, find complementary functions for each of the following nonhomogeneous equations.

15. $y'' - 9y = \sin \ln |x|$

16. $y'' - 81y = e^{-x^2}$

17. $11y'' - 396y = \sinh \text{Tan}^{-1} x$

18. $17y'' - 153y = (\cos x)/x$

Solve the following initial-value problems.

19. $y^{\text{iv}} + y'' = 40e^{2x};\quad y(0) = 3,\quad y'(0) = 5,\quad y''(0) = 7,\ y'''(0) = 15$

20. $y^{\text{iv}} - y = 13x^3;\ y(0) = y'(0) = y''(0) = 0,\ y'''(0) = 78$

21. $y^{\text{iv}} - y'' = 2 \sin x;\ y(0) = y'(0) = y''(0) = y'''(0) = 0$

22. $y^{\text{iv}} + y'' = 2 \sinh x;\ y(0) = y'(0) = y''(0) = y'''(0) = 0$

23. $y^{\text{iv}} + y'' = e^{x+\ln 2};\ y(0) = 2,\ y'(0) = 2,\ y''(0) = y'''(0) = 0$

Using the method of Case 2, Sec. 1.13, find complete solutions of the following equations.

24. $\dfrac{d^2y}{dx^2} + \omega^2 y = 0$

25. $\dfrac{d^2y}{dx^2} - \omega^2 y = 0$

2.4 VARIATION OF PARAMETERS AND REDUCTION OF ORDER

Ordinarily it is impossible to find a particular integral of a nonhomogeneous linear differential equation by inspection or by predicting its analytic form. This presents no serious difficulty, however, because there is a general method which can always be used to find a particular integral whenever a complementary function of a normal equation is known. For simplicity, we shall give a detailed discussion of the process only for a second-order equation

$$a_0(x)y'' + a_1(x)y' + a_2(x)y = f(x) \tag{1}$$

It should then be perfectly clear how the process extends to higher-order equations.

The fundamental idea behind the method is very simple. As soon as a complementary function

$$y_c = c_1 y_1 + c_2 y_2 \tag{2}$$

of (1) has been found, we *vary the parameters* c_1 *and* c_2; that is, we replace the arbitrary constants c_1 and c_2 in (2) by functions of x, say u_1 and u_2. Then we require

$$Y = u_1 y_1 + u_2 y_2 \tag{3}$$

to be a solution of (1). In general,

DEFINITION 1 The process of replacing parameters of an analytic expression by functions is called **variation of parameters.**

Having two unknown functions $u_1$and u_2 involved in Y, we need two independent conditions for their determination. One of these will be obtained by substituting Y into (1); the other remains at our disposal. As the analysis proceeds, a convenient way of choosing it will become apparent.

From (3) we have, by differentiation,

$$Y' = (u_1 y_1' + u_1' y_1) + (u_2 y_2' + u_2' y_2) = (u_1 y_1' + u_2 y_2') + (u_1' y_1 + u_2' y_2)$$

Another differentiation will clearly introduce second derivatives of the unknown functions u_1 and u_2, with attendant complications, unless we arrange to eliminate the first-derivative terms u_1' and u_2' from Y'. This can be done if we make

$$u_1' y_1 + u_2' y_2 = 0 \tag{4}$$

which thus becomes the required second condition on u_1 and u_2.

Proceeding now with the simplified expression.

$$Y' = u_1 y_1' + u_2 y_2' \qquad \text{we find} \qquad Y'' = (u_1 y_1'' + u_1' y_1') + (u_2 y_2'' + u_2' y_2')$$

Substituting Y, Y', and Y'' into (1), we obtain

$$a_0(x)(u_1 y_1'' + u_1' y_1' + u_2 y_2'' + u_2' y_2') + a_1(x)(u_1 y_1' + u_2 y_2') + a_2(x)(u_1 y_1 + u_2 y_2) = f(x)$$

or

$$u_1[a_0(x)y_1'' + a_1(x)y_1' + a_2(x)y_1] + u_2[a_0(x)y_2'' + a_1(x)y_2' + a_2(x)y_2] + a_0(x)(u_1' y_1' + u_2' y_2') = f(x)$$

The expressions in brackets vanish because, by hypothesis, both y_1 and y_2 are solutions of the homogeneous equation corresponding to (1). Hence we find for the other condition on u_1 and u_2,

$$u_1' y_1' + u_2' y_2' = \frac{f(x)}{a_0(x)} \tag{5}$$

Solving (4) and (5) for u_1' and u_2', we obtain

$$u_1' = -\frac{y_2}{y_1 y_2' - y_2 y_1'} \frac{f(x)}{a_0(x)} \qquad \text{and} \qquad u_2' = \frac{y_1}{y_1 y_2' - y_2 y_1'} \frac{f(x)}{a_0(x)} \tag{6}$$

The functions y_1, y_2, y_1', y_2', a_0, and f are all known. Hence u_1 and u_2 can be found by a single integration. With u_1 and u_2 known, a particular integral is completely determined by (3).

We should notice of course that if $y_1 y_2' - y_2 y_1' = 0$ or if $a_0(x) = 0$, the solution for u_1' and u_2' may be invalid. However, under our present assumptions, neither of these conditions can hold. For, on any interval where (1) is normal, $a_0(x) \neq 0$. Moreover, since (2) is a complementary function of (1), y_1 and y_2 are linearly independent functions and the value of their wronskian $y_1 y_2' - y_2 y_1'$ cannot be zero.

EXAMPLE 1

Find a complete solution of the equation $y'' + y = \sec x$.

The related homogeneous equation $y'' + y = 0$ is a harmonic differential equation for which $y_1 = \cos x$ and $y_2 = \sin x$ are two linearly independent solutions. Hence a complementary function of the given equation is

$$y_c = c_1 \cos x + c_2 \sin x$$

Using variation of parameters, we set

$$Y = u_1 \cos x + u_2 \sin x$$

and then solve the equations

$$u_1' \cos x + u_2' \sin x = 0$$
$$u_1' (-\sin x) + u_2' \cos x = \sec x$$

for u_1' and u_2' to obtain

$$u_1' = \frac{\begin{vmatrix} 0 & \sin x \\ \sec x & \cos x \end{vmatrix}}{\begin{vmatrix} \cos x & \sin x \\ -\sin x & \cos x \end{vmatrix}} = -\tan x \quad \text{and} \quad u_2' = \frac{\begin{vmatrix} \cos x & 0 \\ -\sin x & \sec x \end{vmatrix}}{\begin{vmatrix} \cos x & \sin x \\ -\sin x & \cos x \end{vmatrix}} = 1$$

Integrating these, we have

$$u_1 = -\int \tan x \, dx = \ln |\cos x| \quad \text{and} \quad u_2 = \int dx = x$$

and thus

$$Y = u_1 y_1 + u_2 y_2 = (\ln |\cos x|) \cos x + x \sin x$$

Note that all integration constants have been omitted here since all we need to construct a complete solution of our problem is any particular integral Y. In fact,

$$\begin{aligned} y &= \text{complementary function} + \text{particular integral} \\ &= c_1 \cos x + c_2 \sin x + (\ln |\cos x|) \cos x + x \sin x \end{aligned}$$

is a complete solution of the original nonhomogeneous equation.

Equations (4) and (5) clearly indicate the following generalization for $n \geq 2$.

THEOREM 1 If $\Sigma_{j=1}^{n} c_j y_j$ is a complementary function of a normal nth-order linear differential equation with leading coefficient $a_0(x)$ and nonhomogeneous term $f(x)$, then for all functions $u_1, u_2, \ldots, u_n$ which simultaneously satisfy the n conditions

$$\sum_{j=1}^{n} u_j' y_j^{(m)} = 0 \qquad m = 0, 1, \ldots, n-2 \qquad \sum_{j=1}^{n} u_j' y_j^{(n-1)} = \frac{f(x)}{a_0(x)}$$

a particular integral of the nonhomogeneous equation is

$$Y = \sum_{j=1}^{n} u_j y_j$$

EXAMPLE 2

Find a complete solution of the equation $y''' + y' = \cot x$.

Writing the related homogeneous equation as a harmonic equation $(y')'' + y' = 0$ in y', we see that every solution of the homogeneous equation has a derivative of the form $y' = -c_1 \sin x + c_2 \cos x$. Hence, integrating, we find that a complementary function of the given equation is

$$y_c = c_1 \cos x + c_2 \sin x + c_3$$

With $n = 3$, $y_1 = \cos x$, $y_2 = \sin x$, and $y_3 = 1$, the conditions of Theorem 1 read

$$\begin{aligned} u_1' \cos x + u_2' \sin x + u_3' &= 0 \\ -u_1' \sin x + u_2' \cos x \quad &= 0 \\ -u_1' \cos x - u_2' \sin x \quad &= \cot x \end{aligned}$$

A direct expansion shows that the determinant of the coefficients of u_1', u_2', and u_3' in these equations has the value 1, and evaluation of the other determinants involved in using Cramer's rule (Theorem 5, Sec. 3.7) to find u_1', u_2', and u_3' gives

$$u_1' = -\cot x \cos x \qquad u_2' = -\cos x \qquad u_3' = \cot x$$

Integrating these and omitting constants of integration, we obtain

$$u_1 = -\cos x - \ln|\csc x - \cot x| \qquad u_2 = -\sin x \qquad u_3 = \ln|\sin x|$$

Theorem 1 now asserts that a particular integral is

$$\begin{aligned} Y &= -(\cos x + \ln|\csc x - \cot x|)\cos x - \sin^2 x + \ln|\sin x| \\ &= -1 - \cos x \ln|\csc x - \cot x| + \ln|\sin x| \end{aligned}$$

Adding Y to y_c and using the identity $\ln|\csc x - \cot x| = \ln|\tan(x/2)|$, we find that a complete solution of the given differential equation is

$$y = c_1 \cos x + c_2 \sin x + c_3 - 1 - \cos x \ln\left|\tan\frac{x}{2}\right| + \ln|\sin x|$$

If somehow one knows a nontrivial particular solution of a homogeneous linear differential equation of order $n > 1$, then, by what is called **reduction of order,** the problem of finding a basis can be reduced to one of solving a homogeneous equation of order $n - 1$. An explanation of the process for a normal second-order equation

$$a_0(x)y'' + a_1(x)y' + a_2(x)y = 0 \tag{7}$$

will make the general procedure clear.

To develop the procedure, let us suppose y_1 is a nontrivial particular solution of (7). Then cy_1 is a general solution containing c as an arbitrary constant. Using variation of parameters, we replace the parameter c by a function u and attempt to determine u such that y_1 and uy_1 will be linearly independent solutions of (7). Substituting $y = uy_1$ into (7) we obtain

$$a_0(x)(uy_1'' + 2u'y_1' + u''y_1) + a_1(x)(uy_1' + u'y_1) + a_2(x)uy_1 = 0$$

or, collecting coefficients of u'', u', and u,

$$a_0(x)y_1u'' + [2a_0(x)y_1' + a_1(x)y_1]u' + [a_0(x)y_1'' + a_1(x)y_1' + a_2(x)y_1]u = 0$$

The last bracketed expression is identically zero since, by hypothesis, y_1 is a solution of (7). Hence uy_1 will be a solution of (7) if $v = u'$ is a solution of the first-order linear equation

$$a_0(x)y_1v' + [2a_0(x)y_1' + a_1(x)y_1]v = 0$$

which has

$$\exp\left[\int\left(2\frac{y_1'}{y_1} + \frac{a_1(x)}{a_0(x)}\right)dx\right] = y_1^2e^{\int a_1(x)/a_0(x)\,dx}$$

as an integrating factor.

Integrating $[y_1^2e^{\int a_1(x)/a_0(x)\,dx}v]' = 0$ and then replacing v by u', we have

$$u' = \frac{c_1}{y_1^2}e^{-\int a_1(x)/a_0(x)\,dx} \tag{8}$$

which holds on any interval over which $y_1(x) \neq 0$. Integrating again, we find

$$u = c_1\int\frac{1}{y_1^2}e^{-\int a_1(x)/a_0(x)\,dx}\,dx + c_2$$

from which we obtain

$$y = uy_1 = c_1y_1\int\frac{1}{y_1^2}e^{-\int a_1(x)/a_0(x)\,dx}\,dx + c_2y_1 \tag{9}$$

as a general solution of (7) involving c_1 and c_2 as arbitrary constants. If $c_1 = 1$ and $c_2 = 0$, this solution particularizes to

$$y_2 = y_1\int\frac{1}{y_1^2}e^{-\int a_1(x)/a_0(x)\,dx}\,dx \equiv y_1u_1 \tag{10}$$

where

$$u_1 = \int\frac{1}{y_1^2}e^{-\int a_1(x)/a_0(x)\,dx}\,dx \qquad \text{and} \qquad u_1' = \frac{1}{y_1^2}e^{-\int a_1(x)/a_0(x)\,dx}$$

For the value of the wronskian of y_1 and y_2, we have

$$\begin{vmatrix} y_1 & y_1u_1 \\ y_1' & y_1u_1' + y_1'u_1 \end{vmatrix} = y_1^2u_1' = e^{-\int a_1(x)/a_0(x)\,dx}$$

Since $a_1(x)/a_0(x)$ is known to be continuous, the last exponential is never zero; hence y_1 and y_2 are fundamental solutions of (7), and (9) is actually a complete solution.

EXAMPLE 3

Find a complete solution of the equation $x^2y'' - 3xy' + 3y = 0$.

By inspection, $y_1 = x$ is a particular solution. From (10), another solution is

$$y_2 = x\int\frac{1}{x^2}e^{-\int(-3x)/x^2\,dx}\,dx = x\int\frac{1}{x^2}e^{3\ln x}\,dx = x\int x\,dx = \frac{x^3}{2}$$

Now y_1 and y_2 are linearly independent, as are y_1 and $2y_2$; thus a complete solution of the given equation is $y = c_1x + c_2x^3$.

EXAMPLE 4

Find a complete solution of $x^3y''' - 3x^2y'' + x(6 - x^2)y' - (6 - x^2)y = 0$.

By inspection, $y_1 = x$ is one solution. Using variation of parameters, we look for another solution of the type $y = ux$. Substituting into the given equation and simplifying, we find that $y = ux$ will be a solution if $(u')'' - (u') = 0$. This second-order equation in u' has the obvious solutions $u_1' = e^x$ and $u_2' = -e^{-x}$. Hence, integrating, we get $u_1 = e^x$ and $u_2 = e^{-x}$. From the substitution $y = ux$, it follows that $y_2 = xe^x$ and $y_3 = xe^{-x}$ are solutions of the given equation.

If on $(-\infty, \infty)$, $c_1y_1(x) + c_2y_2(x) + c_3y_3(x) \equiv x(c_1 + c_2e^x + c_3e^{-x}) \equiv 0$, then, for every $x \neq 0$.

$$c_1 + c_2e^x + c_3e^{-x} = 0$$

Differentiating and multiplying by e^{-x} and then differentiating again, we find

$$c_2 - c_3e^{-2x} = 0 \quad \text{and} \quad 2c_3e^{-2x} = 0$$

The last equation clearly implies $c_3 = 0$, and the two preceding equations imply $c_2 = c_1 = 0$. Thus y_1, y_2, and y_3 are linearly independent and the required solution is

$$y = x(c_1 + c_2e^x + c_3e^{-x})$$

EXERCISES

Using variation of parameters and the indicated solutions of the related homogeneous equation, find a particular integral and then a complete solution of each of the following nonhomogeneous equations.

1. $y'' - 4y' + 3y = e^{-x}$ $\quad y_1 = e^x,\ y_2 = e^{3x}$
2. $y'' - y' - 2y = e^x$ $\quad y_1 = e^{-x},\ y_2 = e^{2x}$
3. $y'' - y' = 2e^{-x}$ $\quad y_1 = 1,\ y_2 = e^x$
4. $y'' + y' = 12e^{3x}$ $\quad y_1 = 1,\ y_2 = e^{-x}$
5. $y'' - y' - 2y = e^{-x}$ $\quad y_1 = e^{-x},\ y_2 = e^{2x}$
6. $y'' - y' - 2y = x$ $\quad y_1 = e^{-x},\ y_2 = e^{2x}$
7. $y'' + y = \sin x$ $\quad y_1 = \sin x,\ y_2 = \cos x$
8. $y'' + 2y' + y = e^{-x}/x$ $\quad y_1 = e^{-x},\ y_2 = xe^{-x}$
9. $y'' + y = \tan x$ $\quad y_1 = \sin x,\ y_2 = \cos x$
10. $y'' + 4y' + 4y = e^{-2x}/x^2$ $\quad y_1 = e^{-2x},\ y_2 = xe^{-2x}$
11. $2xy'' + y' = 15x^{1/4}$ $\quad y_1 = 1,\ y_2 = x^{1/2}$
12. $x^2y'' + xy' - y = x$ $\quad y_1 = x,\ y_2 = 1/x$
13. $x^2y'' + xy' - y = 1/x$ $\quad y_1 = x,\ y_2 = 1/x$
14. $x^2y'' - 2xy' + 2y = x^3e^x$ $\quad y_1 = x,\ y_2 = x^2$
15. $x^2y'' - 6y = x^3 \ln|x|$ $\quad y_1 = x^3,\ y_2 = 1/x^2$
16. $x^2y'' + xy' - y = 1/(1 + x)$ $\quad y_1 = x,\ y_2 = 1/x$
17. $x^2y'' - xy' + y = 1/x$ $\quad y_1 = x,\ y_2 = x\ln|x|$
18. $2x^2y'' - 5xy' + 5y = 30x^{7/2}$ $\quad y_1 = x,\ y_2 = x^{5/2}$
19. $x^2y'' - 2xy' + 2y = x\ln|x|$ $\quad y_1 = x,\ y_2 = x^2$
20. $xy'' + y' = 1 + x$ $\quad y_1 = 1,\ y_2 = \ln|x|$
21. $y'' - 3y' + 2y = -e^{2x}/(1 + e^x)$ $\quad y_1 = e^x,\ y_2 = e^{2x}$
22. $y'' + 2y' + 2y = 2e^{-x}\tan^2 x$
    $y_1 = e^{-x}\cos x,\ y_2 = e^{-x}\sin x$
23. $xy'' - (2x^2 + 1)y' = x^5e^{x^2}$ $\quad y_1 = 1,\ y_2 = e^{x^2}$
24. $(\sin 4x)y'' - 2(1 + \cos 4x)y' = \tan x$
    $y_1 = 1,\ y_2 = \cos 2x$
25. $(2x - 1)y'' - (4x^2 + 1)y' + 2(2x^2 - x + 1)y = 2x(2x - 1)^2e^{x^2+x}$ $\quad y_1 = e^x,\ y_2 = e^{x^2}$
26. Using variation of parameters show that
$$y = c_1\cos kx + c_2\sin kx + \frac{1}{k}\int_0^x \sin k(x - s)f(s)\,ds$$
is a complete solution of the equation $y'' + k^2y = f(x)$, where $k \neq 0$ and f is everywhere continuous. *Hint:* Introduce the dummy variable s in the integrals which define u_1 and u_2. Then move $y_1(x)$ and $y_2(x)$ into the integrands of the respective integrals and combine the two integrals.
27. Use variation of parameters to show that
$$y = c_1\cosh kx + c_2\sinh kx + \frac{1}{k}\int_0^x \sinh k(x - s)f(s)\,ds$$
is a complete solution of the equation $y'' - k^2y = f(x)$, where $k \neq 0$ and f is everywhere continuous. *Hint:* Note the hint given in Exercise 26.
28. Prove Theorem 1 for the case $n = 3$.

Using Theorem 1 and the indicated solutions of the related homogeneous equation, find a particular integral of each of the following equations.

29. $y''' - 3y'' + 3y' - y = e^x/x$
    $y_1 = e^x,\ y_2 = xe^x,\ y_3 = x^2e^x$
30. $y''' - 3y'' + 3y' - y = 2e^x/x^2$
    $y_1 = e^x,\ y_2 = xe^x,\ y_3 = x^2e^x$
31. $y''' + y' = \sec x$ $\quad y_1 = 1,\ y_2 = \cos x,\ y_3 = \sin x$
32. $y''' + y' = \csc x$ $\quad y_1 = 1,\ y_2 = \cos x,\ y_3 = \sin x$
33. $y''' - y' = \operatorname{sech} x$ $\quad y_1 = 1,\ y_2 = \cosh x,\ y_3 = \sinh x$
34. $y''' \cos x + y'' \sin x = \csc x \cot x$
    $y_1 = 1,\ y_2 = x,\ y_3 = \cos x$
35. Solve the initial-value problem $\quad y''' + y'' = 2\cosh x$
    $y(0) = \frac{1}{2},\ y'(0) = \frac{3}{2},\ y''(0) = -\frac{3}{2}$

Using the one solution indicated, find a complete solution of each of the following equations.

36. $y'' - y' - 2y = 0$; $y_1 = e^{-x}$
37. $4x^2y'' - 8xy' + 9y = 0$; $y_1 = x^{3/2}$
38. $(1 - 2x)y'' + 2y' + (2x - 3)y = 0$; $y_1 = e^x$
39. $y''' - y'' - y' + y = 0$; $y_1 = e^x$
40. $x^2y'' + 4xy' - 4y = 0$; $y_1 = x$
41. $y'' - 4xy' + 2(2x^2 - 1)y = 0$; $y_1 = e^{x^2}$
42. $(2x - x^2)y'' + 2(x - 1)y' - 2y = 0$; $y_1 = x - 1$
43. $x^2(1 + x)y'' - 2xy' + 2y = 0$; $y_1 = x$
44. $y'' + 2y' + y = 0$; $y_1 = e^{-x}$
45. $(2x + 1)y'' - (4x + 4)y' + 4y = 0$; $y_1 = e^{2x}$
46. $y''' - 3y'' + 3y' - y = 0$; $y_1 = e^x$
47. $x^2y'' + xy' - 4y = 0$; $y_1 = x^2$

Using the indicated solution of the related homogeneous equation, find a complete solution of each of the next two nonhomogeneous equations.

48. $(x^2 + 1)y'' - 2xy' + 2y = 2(x^2 + 1)^2/x^3$; $y_1 = x$
49. $(2x + 1)y'' - 4(x + 1)y' + 4y = (2x + 1)^2/(x + 1)$; $y_1 = e^{2x}$
50. Show that if y_1 is a nontrivial solution of an nth-order homogeneous linear differential equation, the transformation $y = uy_1$ reduces the equation to a homogeneous linear differential equation of order $n - 1$ in the dependent variable $v = u'$.
51. If y_1 is a solution of the homogeneous equation $a_0(x)y'' + a_1(x)y' + a_2(x)y = 0$, show that the substitution $y = uy_1$ will reduce the nonhomogeneous equation $a_0(x)y'' + a_1(x)y' + a_2(x)y = f(x)$ to a linear first-order equation whose dependent variable is $v = u'$.

Using the procedure described in Exercise 51, find a complete solution of each of the following equations.

52. $y'' - y = e^{2x}$; $y_1 = e^x$
53. $y'' - 3y' + 2y = e^x$; $y_1 = e^x$
54. $xy'' + 2y' = 1/x$; $y_1 = 1$
55. $y'' - 3y' + 2y = e^x$; $y_1 = e^{2x}$
56. $y'' - 2y' + y = xe^x$; $y_1 = e^x$
57. $y'' + y = \cos x$; $y_1 = \sin x$
58. $x^3y'' + xy' - y = 1$; $y_1 = x$
59. $(x^3 + 2x^2)y'' + 2xy' - 2y = (x + 1)^2$; $y_1 = x$
60. Is the procedure described in Exercise 51 effective if y_1 is a solution of the nonhomogeneous equation rather than the homogeneous equation?

2.5 HOMOGENEOUS SECOND-ORDER EQUATIONS WITH CONSTANT COEFFICIENTS

By using variation of parameters, we can now find a particular integral and then construct a complete solution of a nonhomogeneous linear differential equation whenever a complementary function is known. We can also effect a reduction in order of a linear equation if a nontrivial solution is known. For the important second-order case

$$a_0(x)y'' + a_1(x)y' + a_2(x)y = f(x) \tag{1}$$

this means that if we can even find one nontrivial solution of the related homogeneous equation, then the problem of solving (1) can be reduced to that of solving a first-order linear equation, which we learned how to do in Sec. 1.11. We have already said that a linear differential equation of order $n > 1$ usually will be unsolvable in terms of known functions. The problem of solving equations with constant coefficients is not so difficult, however, and we shall encounter many practical applications involving equations of this sort.

The most general linear differential equation with constant coefficients has the form

$$a_0y^{(n)} + a_1y^{(n-1)} + \cdots + a_{n-1}y' + a_ny = f(x) \tag{2}$$

which, in the important second-order case, can be written without subscripts as

$$ay'' + by' + cy = f(x) \tag{3}$$

In our work we shall always suppose that the coefficients are real numbers and that, at present, all nonhomogeneous terms are real-valued functions. Every homogeneous linear differential equation with constant coefficients is normal on $(-\infty, \infty)$. By solving such an equation for the derivative of highest order and then differentiating repeatedly, it is easy to verify that

THEOREM 1 All derivatives of every solution of any homogeneous linear differential equation with constant coefficients are continuous on $(-\infty, \infty)$.

Thus far we have denoted derivatives of a function f by f' and by expressions of the form df/dx or $\dot{f}$. A fourth rather common notation uses the letter D to indicate differentiation. Thus when D is applied to a differentiable function f, it produces f'. More specifically,

DEFINITION 1 The **derivative operator** D is a function such that for every differentiable function f,

$$Df = f'$$

and for each value of x in the domain of f', the value of Df is

$$(Df)(x) = Df(x) = f'(x)$$

Sometimes D is called the **differentiation** or **differential operator.** Of course, Df and f' identify the same function.

In terms of D, the power rule of differentiation reads

$$Dx^r = rx^{r-1} \qquad r \text{ constant} \tag{4}$$

and for functions f and g, where both are differentiable, the basic algebraic properties of derivatives take the form

$$D(fg) = f\,Dg + g\,Df \tag{5}$$

$$D\left(\frac{f}{g}\right) = \frac{g\,Df - f\,Dg}{g^2} \qquad g \neq 0 \tag{6}$$

$$D(af + bg) = a\,Df + b\,Dg \qquad a, b \text{ constants} \tag{7}$$

The last of these formulas implies that D distributes over sums of functions and commutes with parameters. Thus (7) characterizes D as a *linear operator* (see Sec. 13.5).

If f depends on an independent variable other than x, say t, then $Df(t)$ denotes the derivative of $f(t)$ with respect to t. Since $Df = f'$, if $f[g(x)]$ represents a composite function, then $Df[g(x)] \equiv f'[g(x)]$ indicates the derivative of $f[g(x)]$ with respect to $g(x)$, *not* the derivative with respect to x.

EXAMPLE 1

Given the function f defined by $f(x) = \sin^2 3x$, find $Df(x)$, $Df(3x)$, and $Df(\sin 3x)$.

Using the chain and power rules, we find

$$Df(x) = f'(x) = \frac{d}{dx}\sin^2 3x = 6\sin 3x\cos 3x = 3\sin 6x$$

and

$$Df(3x) = f'(3x) = \frac{d}{d\,3x}\sin^2 3x = 2\sin 3x\cos 3x = \sin 6x$$

The power rule alone gives

$$Df(\sin 3x) = f'(\sin 3x) = \frac{d}{d\sin 3x}\sin^2 3x = 2\sin 3x$$

Definition 1 is extended to **derivative operators D^n of order n** for every nonnegative integer n by defining

(8) $$D^0 f = f^{(0)} = f$$

and then imposing the recurrence relation

(9) $$D^n f = D(D^{n-1} f) = f^{(n)} \qquad n \text{ a positive integer}$$

At each point x in the domain of $f^{(n)}$, the value of $D^n f$ is

(10) $$(D^n f)(x) = D^n f(x) = f^{(n)}(x) \qquad n \geq 0$$

Since $D^0 f = f$, D^0 may be treated as a unit multiplier.

EXAMPLE 2

Find a formula for $D^n e^{mx}$ which holds for every nonnegative integer n and parameter m.

Using (10), we set $n = 0$ and substitute e^{mx} for both $f(x)$ and $f^{(0)}(x)$ to obtain

$$D^0 e^{mx} = e^{mx} = m^0 e^{mx} \qquad m \neq 0$$

Differentiating $f(x) = e^{mx}$, we get $f'(x) = me^{mx}$. Again using (10), we set $n = 1$ and substitute e^{mx} for $f(x)$ and me^{mx} for $f'(x)$, getting

$$De^{mx} = me^{mx} = m^1 e^{mx}$$

Anticipating that each differentiation of e^{mx} will give rise to a multiplying factor of m, we let k be any nonnegative integer for which

$$D^k e^{mx} = m^k e^{mx}$$

Then, according to (9), and since D commutes with constants,

$$D^{k+1} e^{mx} = D(D^k e^{mx}) = D(m^k e^{mx}) = m^k D e^{mx} = m^k m e^{mx} = m^{k+1} e^{mx}$$

By induction it follows that for every nonnegative integer n,

(11) $$D^n e^{mx} = m^n e^{mx} \qquad m \text{ constant}$$

It is now possible to use different combinations of derivative operators to form other important operators. For instance, a simple linear combination of D^2, D, and D^0 with D^0 treated as 1, gives us a **second-order polynomial operator**

(12) $$aD^2 + bD + c \qquad a, b, c \text{ constants; } a \neq 0$$

which, when applied to a twice-differentiable function f, transforms it into the corresponding function

(13) $$(aD^2 + bD + c)f = aD^2 f + bDf + cf = af'' + bf' + cf$$

whose value at any point x in the domain of f'' is

(14) $$(aD^2 + bD + c)f(x) = aD^2 f(x) + bDf(x) + cf(x) = af''(x) + bf'(x) + cf(x)$$

Of course, if $a = 0$ and $b \neq 0$, (12) reduces to a **first-order polynomial operator**

(15) $$bD + c \qquad b \neq 0$$

From (13) it is clear that the operator D may be used to write Eq. (3) in the **operator form**

(3a) $$(aD^2 + bD + c)y = f(x)$$

In some respects, derivative operators can be handled as though they were real numbers. For example, under multiplication, superscripts on D add as exponents

(16) $$D^m D^n = D^{m+n} \qquad m, n \text{ nonnegative integers}$$

(see Exercise 26), and if (12) factors algebraically into

(17) $$(hD + k)(rD + s) = (rD + s)(hD + k) \qquad h, k, r, s \text{ real numbers}$$

then each of these products identifies the operator (12) and is therefore referred to as a **factored equivalent** of that operator (see Exercise 27).

EXAMPLE 3

Apply the polynomial operator $3D^2 - 10D - 8$ and its factored equivalents $(3D + 2)(D - 4)$ and $(D - 4)(3D + 2)$ to the function $f(x) = x^2$.

From (14) there follows

$$(3D^2 - 10D - 8)x^2 = 3D^2x^2 - 10Dx^2 - 8x^2 = 3(2) - 10(2x) - 8x^2 = 6 - 20x - 8x^2$$

Applying the factored equivalents to x^2, we obtain

$$\begin{aligned}(3D + 2)(D - 4)x^2 &= (3D + 2)(Dx^2 - 4x^2) = (3D + 2)(2x - 4x^2)\\ &= 3D(2x - 4x^2) + 4x - 8x^2 = 3(2 - 8x) + 4x - 8x^2 = 6 - 20x - 8x^2\\ (D - 4)(3D + 2)x^2 &= (D - 4)(3Dx^2 + 2x^2) = (D - 4)(6x + 2x^2)\\ &= D(6x + 2x^2) - 24x - 8x^2 = 6 + 4x - 24x - 8x^2 = 6 - 20x - 8x^2\end{aligned}$$

This example illustrates how algebraically equivalent forms of a polynomial operator yield identical results when applied to the same function.

It should be noted that the equivalence of the factored forms of polynomial operators, illustrated in Example 3, is not characteristic of all derivative-type operators. Indeed, if any derivative operator in the factors of an operational expression has a *variable* coefficient, then permuting the factors will in general alter the operator originally defined by the expression (see Exercises 30 and 31).

Following the theory of Sec. 2.3, we begin the solution of the second-order equation (3) by searching for a complementary function. Thus we attempt to find a complete solution of the homogeneous equation

(18) $$ay'' + by' + cy = 0$$

or

(18a) $$(aD^2 + bD + c)y = 0$$

In doing so, it is natural to try for a particular solution of the type

$$y = e^{mx}$$

where m is a constant to be determined because, as we know from Example 2, all derivatives of this function are alike except for a numerical coefficient. Substituting into (18) and then factoring e^{mx} from every term, we have

$$e^{mx}(am^2 + bm + c) = 0$$

as the condition to be satisfied if $y = e^{mx}$ is to be a solution. Since e^{mx} can never be zero, it is necessary that

$$am^2 + bm + c = 0 \tag{19}$$

This purely algebraic equation is known as the **characteristic** or **auxiliary equation** of either Eq. (18) or (18*a*) since its roots, called the **characteristic roots,** determine or *characterize* the only possible solutions of the assumed form $y = e^{mx}$. In practice, it is obtained not by substituting $y = e^{mx}$ into the homogeneous equation and then simplifying but rather by substituting m^2 for y'', m for y', and 1 for y in Eq. (18), or by simply equating to zero the operational coefficient of y in (18*a*) and letting the operator D play the role of m:

$$aD^2 + bD + c = 0$$

The characteristic equation (19) is a simple quadratic which will have two distinct real roots, a repeated real root, or conjugate complex roots. To find a complete solution of (18), we now proceed by cases.

CASE 1 The characteristic equation (19) has two distinct real roots m_1 and m_2. In this case, two particular solutions of (18) are

$$y_1 = e^{m_1x} \qquad \text{and} \qquad y_2 = e^{m_2x} \qquad m_1 \neq m_2$$

For every real value of x, the wronskian $W(y_1, y_2)$ of y_1 and y_2 has a value $W(x) = (m_2 - m_1)e^{(m_1+m_2)x}$ which is never zero. Hence, an arbitrary linear combination

$$y = c_1e^{m_1x} + c_2e^{m_2x}$$

of y_1 and y_2 is a complete solution of (18).

EXAMPLE 4

Find a complete solution of the differential equation $y'' + 7y' + 12y = 0$.

The characteristic equation in this case is

$$m^2 + 7m + 12 = 0$$

and its roots are

$$m_1 = -3 \qquad \text{and} \qquad m_2 = -4$$

Since these values are different, a complete solution is

$$y = c_1e^{-3x} + c_2e^{-4x}$$

CASE 2 The characteristic equation (19) has a repeated real root m_1.
In this case, the characteristic equation must be of the form

$$m^2 - 2m_1 m + m_1^2 = 0$$

which implies the differential equation itself may be written as

$$y'' - 2m_1 y' + m_1^2 y = 0$$

One solution of this equation is clearly $y_1 = e^{m_1 x}$. Another solution, as given by Eq. (10), Sec. 2.4, is

$$y_2 = y_1 \int \frac{e^{-\int(-2m_1)\,dx}}{y_1^2}\,dx = e^{m_1 x} \int \frac{e^{2m_1 x}}{(e^{m_1 x})^2}\,dx = xe^{m_1 x}$$

Now, from Sec. 2.4, we know that y_1 and y_2 are linearly independent. Thus a complete solution of (18) is

$$y = c_1 e^{m_1 x} + c_2 x e^{m_1 x}$$

where c_1 and c_2 are arbitrary constants.

EXAMPLE 5

Find a complete solution of the equation $(D^2 + 6D + 9)y = 0$.

The characteristic equation is $m^2 + 6m + 9 = 0$ or $(m + 3)^2 = 0$ and has $m_1 = -3$ as a repeated root. Hence a complete solution of the given equation is

$$y = c_1 e^{-3x} + c_2 x e^{-3x}$$

CASE 3 The characteristic equation (19) has conjugate complex roots, say $m_1 = p + iq$, $m_2 = p - iq$, where p and q are real numbers with $q \neq 0$ and $i^2 = -1$.

As in Case 1, we are inclined to take

$$e^{(p+iq)x} \quad \text{and} \quad e^{(p-iq)x} \tag{20}$$

as solutions of (18). However, these are complex and *not* real-valued functions. To accept them as solutions of (18) would require some complex variable theory and a broadening of our concept of solution. We would also need to reexamine what constitutes a complete solution because the use of complex parameters would enlarge the set of all possible solutions. And even if we were to make these generalizations, the proposed solutions (20) would still be unsatisfactory for many practical purposes because they involve complex exponentials which are awkward to handle and are not tabulated. Thus, for the present, we shall continue to require all solutions to be real-valued functions, and in forming linear combinations, we shall use only real-valued parameters.

Our immediate objective, then, is to find two linearly independent real-valued functions which are particular solutions of (18) when the characteristic equation (19) is of the form

$$(m - p - iq)(m - p + iq) = (m - p)^2 - (iq)^2 = m^2 - 2pm + (p^2 + q^2) = 0$$

which implies that (18) may be written

$$y'' - 2py' + (p^2 + q^2)y = 0 \qquad q \neq 0 \tag{21}$$

To find such functions, let us put $q = 0$ in Eq. (21) and then try using a general solution of the reduced equation

$$y'' - 2py' + p^2y = 0 \tag{22}$$

and variation of parameters to solve Eq. (21) itself.

As we now know from Case 2, Eq. (22) has

$$c_1e^{px} \qquad \text{and} \qquad c_2xe^{px}$$

as general solutions. Choosing the simpler of these, we seek a nontrivial solution

$$y = u(x)e^{px} \tag{23}$$

of Eq. (21). Substituting this and its first two derivatives,

$$y' = u'e^{px} + upe^{px} \qquad \text{and} \qquad y'' = u''e^{px} + 2u'pe^{px} + up^2e^{px}$$

into (21), we obtain the equation

$$(u''e^{px} + 2u'pe^{px} + up^2e^{px}) - 2p(u'e^{px} + upe^{px}) + (p^2 + q^2)ue^{px} = 0$$

which reduces at once to

$$u'' + q^2u = 0$$

the harmonic equation, which has

$$u_1 = \cos qx \qquad \text{and} \qquad u_2 = \sin qx$$

as linearly independent particular solutions (Example 1, Sec. 2.3). Hence, from (23) we have two particular solutions of Eq. (21), namely,

$$y_1 = e^{px}\cos qx \qquad \text{and} \qquad y_2 = e^{px}\sin qx$$

The wronskian test gives $W(y_1, y_2) = qe^{2px} \neq 0$. A complete solution of Eq. (21), that is, of Eq. (18) when $p + iq$ is a complex root of (19), in purely real form is therefore

$$y = c_1e^{px}\cos qx + c_2e^{px}\sin qx = e^{px}(c_1\cos qx + c_2\sin qx) \tag{24}$$

where c_1 and c_2 are real parameters.

EXAMPLE 6

Find a complete solution of the equation $y'' + 2y' + 5y = 0$.

The characteristic equation in this case is

$$m^2 + 2m + 5 = 0$$

and its roots are

$$m_1 = -1 + 2i \qquad \text{and} \qquad m_2 = -1 - 2i$$

Clearly $p = -1$ and $q = 2$. Hence a complete solution as given by (24) is

$$y = e^{-x}(c_1\cos 2x + c_2\sin 2x)$$

TABLE 2.1

Differential equation: $ay'' + by' + cy = 0$ or $(aD^2 + bD + c)y = 0$
Characteristic equation: $am^2 + bm + c = 0$ or $aD^2 + bD + c = 0$

Nature of the roots of the characteristic equation	Condition on the coefficients of the characteristic equation	Complete solution of the differential equation
Real and unequal $m_1 \neq m_2$	$b^2 - 4ac > 0$	$y = c_1 e^{m_1 x} + c_2 e^{m_2 x}$
Real and equal $m_1 = m_2 = m$	$b_2 - 4ac = 0$	$y = c_1 e^{mx} + c_2 x e^{mx}$
Conjugate complex $m_1 = p + iq$ $m_2 = p - iq$	$b_2 - 4ac < 0$	$y = e^{px}(c_1 \cos qx + c_2 \sin qx)$

The complete process for solving the homogeneous equation (3) in all possible cases is summarized in Table 2.1.

In particular applications, the two arbitrary constants in the complete solution must usually be determined to fit given initial conditions on y and y', or their equivalent. The following examples will clarify the procedure.

EXAMPLE 7°

Find the solution of the initial-value problem $y'' - 4y' + 4y = 0$; $y(0) = 3$, $y'(0) = 4$.

The characteristic equation of the differential equation is

$$m^2 - 4m + 4 = 0$$

Its roots are $m_1 = m_2 = 2$; hence a complete solution is

$$y = c_1 e^{2x} + c_2 x e^{2x}$$

By differentiating this, we find

$$y' = 2c_1 e^{2x} + c_2(e^{2x} + 2xe^{2x}) = (2c_1 + c_2)e^{2x} + 2c_2 x e^{2x}$$

Substituting the given data into the equations for y and y', respectively, we have

$$3 = c_1 \qquad \text{and} \qquad 4 = 2c_1 + c_2$$

Hence $c_1 = 3$, $c_2 = -2$, and the required solution is

$$y = 3e^{2x} - 2xe^{2x}$$

EXAMPLE 8

Find the solution of the equation $(4D^2 + 16D + 17)y = 0$ for which $y = 1$ when $t = 0$ and $y = 0$ when $t = \pi$.

In this case, the statement of the problem makes it clear that the independent variable is not x but t. This does not affect the characteristic equation $4m^2 + 16m + 17 = 0$ or its roots $m = -2 \pm \frac{1}{2}i$. However, the solution we construct from these roots must be expressed in terms of t rather than x:

$$y = e^{-2t}\left(c_1 \cos \frac{t}{2} + c_2 \sin \frac{t}{2}\right)$$

Substituting the given conditions into this equation, we find

$$1 = c_1 \qquad \text{and} \qquad 0 = e^{-2\pi}c_2 \qquad \text{or} \qquad c_2 = 0$$

Hence the required solution is $y = e^{-2t}\cos(t/2)$.

Quite often the most efficient way of working a problem involving only real quantities is to use properties of complex quantities. In some applications nothing more is needed than the properties of complex numbers we learned in basic algebra. This material is reviewed in Sec. 3.1. Sometimes ideas and methods of advanced complex analysis are required. Chapters 17–20 are devoted to this subject.

It is natural to call the real function $u(x)$ the **real part** of a complex function $w(x) = u(x) + iv(x)$. Frequently, u is denoted by Re w. The **imaginary part** of $w(x)$ is the *real function* $v(x)$, *not* $iv(x)$. A common notation for v is Im w.

DEFINITION 2 The **conjugate** of a complex quantity $w = u + iv$, u, v real, is $\overline{w} = u - iv$.

This definition says that w and its conjugate $\overline{w}$ have equal real parts but imaginary parts of opposite sign.

Some properties of complex numbers and functions seem to be self-evident. For instance, intuition suggests the law of exponents

$$e^{a+ib} = e^a e^{ib} \qquad a,\ b \text{ real} \tag{25}$$

and the derivative formulas

$$\frac{de^{ix}}{dx} = ie^{ix} \qquad \text{and} \qquad \frac{d^2e^{ix}}{dx^2} = -e^{ix} \qquad x \text{ real}$$

But if these formulas are valid, the harmonic equation $y'' + y = 0$ is satisfied by $y = e^{ix}$. Were this function real, which it does not appear to be, it would have to be a linear combination of $\cos x$ and $\sin x$, giving

$$e^{ix} = c_1\cos x + c_2\sin x \qquad \text{and} \qquad ie^{ix} = -c_1\sin x + c_2\cos x \qquad c_1, c_2 \text{ constants}$$

Setting $x = 0$ in these equations, we find $1 = c_1$ and $i = c_2$. Thus, despite our lack of mathematical rigor, we have arrived at the **Euler identity** or **formula,**

$$e^{ix} = \cos x + i\sin x \tag{26}$$

(see also Exercise 76 and Sec. 17.4).

Using the chain rule to differentiate the expression (20), treating $p + iq$ as though it was real, we find (Exercise 77) that both $e^{(p+iq)x}$ and $e^{(p-iq)x}$ satisfy Eq. (21). So, naturally, these two functions are referred to as **complex solutions** of (21). Adapting (25) and (26) to the first of these, we obtain

$$e^{(p+iq)x} = e^{px}e^{iqx} = e^{px}\cos x + ie^{px}\sin qx \tag{27}$$

The real part of this complex function is $e^{px}\cos qx$, and its imaginary part is $e^{px}\sin qx$. We have already seen that these two functions are fundamental solutions of (21). Consequently, the following simple procedure for solving Eq. (18), when $p + iq$ is a root of (19), is fully justified.

1. As soon as $p + iq$ is known, write the complex solution $e^{(p+iq)x}$.
2. Form a linear combination of the real and imaginary parts of $e^{(p+iq)x}$ as given by (27) to obtain a complete solution of the differential equation.

Note that the conjugate root is not needed. Henceforth, we shall use this procedure and extensions of it whenever it is convenient.

EXAMPLE 9

Solve the initial-value problem $y'' - 10y' + (25 + \pi^2)y = 0$; $y(0) = 0$, $y'(0) = \pi e$.

Using the root $5 + \pi i$ of the auxiliary equation $m^2 - 10m + 25 + \pi^2 = 0$, we write the complex solution

$$e^{(5+\pi i)x} = e^{5x}\cos \pi x + ie^{5x}\sin \pi x$$

Forming a linear combination of the real and imaginary parts of this solution, we get a complete solution which factors as

$$y = e^{5x}(c_1 \cos \pi x + c_2 \sin \pi x)$$

The first initial condition now gives $y(0) = 0 = c_1$, and so

$$y' = c_2 e^{5x}(\pi \cos \pi x + 5 \sin \pi x)$$

The second initial condition gives $y'(0) = \pi e = c_2\pi$; hence $e = c_2$, and the required solution is

$$y = e^{5x+1}\sin \pi x$$

EXERCISES

Use Theorem 1 to prove that each of the following functions is continuous on $(-\infty, \infty)$ by finding a homogeneous linear differential equation with constant coefficients which the function satisfies.

1. e^x **2.** $\cos x$ **3.** $\sin 6x$
4. $x^2 + x$ **5.** $\cosh x$ **6.** $e^{2x}\cos x$
7. $\Sigma_{k=1}^{n} x^k$ **8.** $x + 2\sin^2 x$ **9.** xe^{-x}
10. $x \sin x$ **11.** Prove Theorem 1.

Functions f and g are defined on $(-\infty, \infty)$ by $f(x) = \ln(1 + x^2)$ and $g(x) = 2(1 + x^2)^2$. Find an explicit representation of the function identified by each of the following expressions.

12. $Df(x)$ **13.** $Df(x^2)$ **14.** $Dg(1 + t^2)$
15. $D(g - 4f)(x)$ **16.** $D(fg)(x)$ **17.** $D(f/g)(x)$
18. $g(x)\,D^2f(x)$ **19.** $Dg([1 + x^2]^2)$
20. $D^3f(x)$ **21.** $Dg(t) - 2g(t)\,Df(t)$

22. Find $D^n(x^n)$, n a positive integer.

23. What is the difference between $Df(x)$ and $f(x)D$?

24. The usual rules for adding and multiplying functions are used to define a **second-order linear operator** $L = a_0(x)D^2 + a_1(x)D + a_2(x)$. State conditions under which such an operator is defined and by analogy with (14), specify what $Lf(x)$ is equal to.

25. The usual rules for composition of functions are used to define the **product** L_1L_2 **of two operators** L_1 and L_2 so that $L_1L_2f = L_1(L_2f)$. State conditions under which each of the products L_1L_2 and L_2L_1 exists.

26. Show that $D^mD^n = D^{m+n}$, m, n nonnegative integers.

27. Show that if the polynomial operator $L = aD^2 + bD + c$ is factored algebraically into two first-degree polynomials in D, the product of these first-degree polynomials specifies the same operator as L.

28. Verify that

$$\begin{aligned}(D + 1)(D - 2)\cos x &= (D - 2)(D + 1)\cos x\\ &= (D^2 - D - 2)\cos x\end{aligned}$$

29. Verify that

$$\begin{aligned}(D + 1)(D^2 + 2)\sin 3x &= (D^2 + 2)(D + 1)\sin 3x\\ &= (D^3 + D^2 + 2D + 2)\sin 3x\end{aligned}$$

30. Is $(D + 1)(D + x)e^x = (D + x)(D + 1)e^x$? Explain.

31. Under what conditions, if any, is the following equation valid?

$$[D + r_1(x)][D + r_2(x)]f(x) = [D + r_2(x)][D + r_1(x)]f(x)$$

32. Determine $r(x)$ so that $y = e^x$ will be a solution of the equation $[D + r(x)](D + x)y = 0$.

33. Determine $r(x)$ so that $y = e^x$ will be a solution of the equation $(D + x)[D + r(x)]y = 0$.

34. Find a complete solution of the differential equation $(D - 1/x)(D - 1)y = 0$. *Hint*: Note that any solution of the first-order equation $(D - 1)y = 0$ is also a solution of the given equation since $(D - 1/x)0 = 0$. Solve $(D - 1)y = 0$ and then find a second solution of the given equation by using the method of Sec. 2.4 on reduction of order.

35. Find a complete solution of the equation $(D - 1)(D - 1/x)y = 0$.

Find a complete solution of each of the following equations.

36. $y'' = 0$
37. $y'' - 3y' + 2y = 0$
38. $y'' - 2y' + y = 0$
39. $y'' + y' - 2y = 0$
40. $y'' + 2y' + 10y = 0$
41. $y'' + 5y' + 4y = 0$
42. $y'' - 5y = 0$
43. $y'' + 5y' = 0$
44. $(4D^2 + 4D + 1)y = 0$
45. $(D^2 - 2)y = 0$
46. $y'' - (\pi + 1)y' + \pi y = 0$
47. $y'' - 2\sqrt{3}y' + 3y = 0$
48. $y'' + 4\pi y' + 4\pi^2 y = 0$
49. $(9D^2 - 12D + 4)y = 0$
50. $10y'' + 6y' + y = 0$
51. $y'' + 10y' + 26y = 0$
52. $(225D^2 + 150D + 36)y = 0$
53. $y'' + 6y' + (9 + 2\pi^2)y = 0$

Find the particular solution of each of the following equations which satisfies the given conditions.

54. $y'' + 4y = 0;\ y(0) = 2,\ y'(0) = 6$
55. $y'' - 4y = 0;\ y(0) = 1,\ y'(0) = -1$
56. $y'' + 3y' - 4y = 0;\ y(0) = 4,\ y'(0) = -2$
57. $y'' - y' - y = 0;\ y(0) = 0,\ y'(0) = \sqrt{5}$
58. $16y'' - 8y' + 5y = 0;\ y(0) = 4,\ y'(0) = -1$
59. $5y'' - 3y' = 0;\ y(0) = 10,\ y'(0) = 3$
60. $25y'' + 20y' + 4y = 0;\ y(0) = y'(0) = 0$
61. $y'' + 2y' + 5y = 0;\ y(0) = -4,\ y'(0) = 2$
62. $(D^2 + 6D + 9)y = 0 \quad y = 0,\ y' = 3$ when $x = 0$
63. $(D^2 + 2D + 5)y = 0 \quad y = 1$ when $x = 0$, $y = 0$ when $x = \pi$
64. $(D^2 + 2D + 5)y = 0 \quad y = 1$ when $x = 0$, $y' = 0$ when $x = \pi$
65. $(4D^2 + 16D + 17)y = 0;\ y = 1$ when $x = 0$, $y = 0$ when $x = \pi$
66. Find a basis y_1, y_2 for the set of all solutions of $y'' - 2y' + y = 0$ such that $y_1(0) = 1$, $y_1'(0) = 0$; $y_2(0) = 0$, $y_2'(0) = 1$.
67. The displacement of a particle that moves along the x axis satisfies the differential equation $\ddot{x} + 2\dot{x} + 4x = b\dot{x}$. For what values of b is the motion oscillatory?
68. Show that there is always a unique solution of the equation $y'' + y = 0$ satisfying given conditions of the form $y = y_0$ when $x = x_0$ and $y = y_1$ when $x = x_1$, unless $x_1 = x_0 + n\pi$. What is the situation when $x_1 = x_0 + n\pi$?
69. For what values of λ, if any, are there nontrivial solutions of the equation $y'' + \lambda^2 y = 0$ which satisfy the conditions $y(0) = 0$ and $y(\pi) = 0$? What are these solutions?
70. For what values of λ, if any, are there nontrivial solutions of the equation $y'' + \lambda^2 y = 0$ which satisfy the conditions $y'(0) = 0$ and $y'(\pi) = 0$? What are these solutions?
71. For what values of λ, if any, are there nontrivial solutions of the equation $y'' + \lambda^2 y = 0$ which satisfy the conditions $y(0) = 0$ and $y'(\pi) = 0$? What are these solutions?
72. Show that the only values of λ for which nontrivial solutions of the equation $y'' + \lambda^2 y = 0$ satisfying the conditions $y(0) = 0$ and $y(\pi) = y'(\pi)$ exist are the roots of the equation $\tan \pi\lambda = \lambda$. Show that this equation has infinitely many roots.
73. Work Exercise 69 for the equation $y'' - \lambda^2 y = 0$.
74. Work Exercise 71 for the equation $y'' - \lambda^2 y = 0$.
75. Let $x(t)$ and $y(t)$ be continuously differentiable functions on an interval I over which the value $W(t)$ of their wronskian $W(x, y)$ is never zero.
(a) Using the polar representations $x(t) = r(t)\cos\theta(t)$ and $y(t) = r(t)\sin\theta(t)$, show that $W(t) = r^2\, d\theta/dt$.
(b) Verify that if $r(t) = e^{pt}$ and $\theta(t) = qt$, then $W(t) = qe^{2pt}$.
(c) Prove that between any two consecutive zeros of $x(t)$ there is exactly one zero of $y(t)$, and vice versa. *Hint:* Use the equations of Part **(a)** or observe the geometric behavior of a moving point P having $[x(t), y(t)]$ as rectangular coordinates and $[r(t), \theta(t)]$ as polar coordinates. How does this result differ from the Sturm separation theorem of Exercise 37, Sec. 2.2?
76. Assuming the Maclaurin expansion

$$e^{i\theta} = \sum_{n=0}^{\infty} \frac{(i\theta)^n}{n!}$$

holds for the complex exponential function $e^{i\theta}$ and that the terms of the series may be rearranged, derive the Euler identity.
77. Treating $p + iq$ as though it were a real constant, verify that both

$$e^{(p+iq)x} \quad \text{and} \quad e^{(p-iq)x}$$

satisfy Eq. (21).
78. Show that $y = e^{px}(A\cosh qx + B\sinh qx)$ is a complete solution of the differential equation $ay'' + by' + cy = 0$ when the roots of the characteristic equation are $m = p \pm q$ and $q \neq 0$.
79. If the roots of its characteristic equation are real, show that no nontrivial solution of the differential equation $ay'' + by' + cy = 0$ can have more than one real zero.
80. If the characteristic equation of the differential equation $ay'' + by' + cy = 0$ has distinct roots m_1 and m_2, show that

$$y = \frac{e^{m_1 x} - e^{m_2 x}}{m_1 - m_2}$$

is a particular solution of the equation. Determine the limit of this expression as $m_2 \to m_1$ and discuss its relation to the solution of the differential equation when the characteristic equation has equal roots.
81. **(a)** Show that $(D - a)^2 f(x) = e^{ax}D^2[e^{-ax}f(x)]$.
(b) Use the given formula of Part **(a)** to show that $(D - a)^2[(c_1 + c_2 x)e^{ax}] = 0$.

(c) Explain how the formula of Part (b) can be used to obtain the complete solution of a differential equation whose characteristic equation has equal roots.

82. (a) Show that $D^2(xe^{mx}) = m^2xe^{mx} + 2me^{mx}$.

(b) Using the result of Part (a), show that if $p(D)$ is a quadratic polynomial in D, then

$$p(D)(xe^{mx}) = p(m)xe^{mx} + p'(m)e^{mx}$$

(c) Explain how the formula of Part (b) can be used to obtain a second, nonproportional solution of a differential equation whose characteristic equation has equal roots. *Hint:* Recall that if a polynomial equation $p(x) = 0$ has a double root $x = r$, then $x = r$ is also a root of the equation $p'(x) = 0$.

83. What meaning, if any, do you think can be assigned to D^0? D^{-1}? D^{-2}?

84. Show that the change of dependent variable defined by the substitution $y = -z'/zP$ changes the Riccati equation $y' = P(x)y^2 + Q(x)y + R(x)$ into the linear second-order equation

$$z'' - [Q + (P'/P)]z' + PRz = 0$$

Using the result of Exercise 84, solve each of the following equations.

85. $xy' = x^2y^2 - y + 1$

86. $x^2y' = x^4y^2 + (3x^2 - 2x)y + 2$

87. $y' \sin x = y^2 \sin^2 x + y(4 \sin x - \cos x) + 5$

88. $(\cos x)y' = (\cos^2 x)y^2 + (\sin x - 2 \cos x)y + 5$

89. Can a particular solution of Eq. (21) be obtained by working with the particular solution $y = xe^{px}$ of Eq. (22)?

90. Obtain two linearly independent particular solutions of Eq. (21) by working with the equation obtained from Eq. (21) by setting $p = 0$.

2.6 HOMOGENEOUS EQUATIONS OF HIGHER ORDER

The solution of a homogeneous linear differential equation with constant coefficients and of order greater than 2,

$$a_0y^{(n)} + a_1y^{(n-1)} + \cdots + a_{n-1}y' + a_ny = 0 \qquad n > 2, a_0 \neq 0 \tag{1}$$

parallels the second-order case in all significant details. The substitution $y = e^{mx}$ leads, as before, to the characteristic equation

$$a_0m^n + a_1m^{n-1} + \cdots + a_{n-1}m + a_n = 0 \tag{2}$$

which can be obtained in a specific problem simply by replacing each derivative by the corresponding power of m. The degree of this algebraic equation will be the same as the order of the differential equation (1); hence, counting each distinct root according to its multiplicity, we find that the number of roots will equal the order of the differential equation. Each *real* root r of multiplicity k determines k linearly independent solutions of the differential equation

$$e^{rx}, xe^{rx}, \ldots, x^{k-1}e^{rx} \qquad k \geq 1 \tag{3}$$

(see Exercise 59). Because the coefficients of (2) are real, *complex* roots must occur in conjugate pairs. Each complex root $a + ib$ of multiplicity $k \geq 1$ yields the $2k$ linearly independent solutions

$$\begin{matrix} e^{ax}\cos bx & xe^{ax}\cos bx & \cdots & x^{k-1}e^{ax}\cos bx \\ e^{ax}\sin bx & xe^{ax}\sin bx & \cdots & x^{k-1}e^{ax}\sin bx \end{matrix} \tag{4}$$

(see Exercise 60). Since the conjugate root $a - ib$ yields these same solutions, only one of the conjugate pair $a \pm ib$ is to be used in formulating the solutions (4).

Once all roots of the auxiliary equation have been found and all corresponding solutions of the kinds (3) or (4) taken into account, a complete solution of the homogeneous differential equation can be constructed as follows.

RULE 1 If y_i is a particular solution of (1) normally corresponding to a simple root m_i of the characteristic equation (2), and if this root is of multiplicity k, where $k \geq 1$, then each of the functions $y_i, xy_i, \ldots, x^{k-1}y_i$ is a solution, and an arbitrary linear combination of the solutions associated with the distinct real roots of (2) and the solutions associated with the distinct pairs of conjugate complex roots is a complete solution of (1).

Given any specific homogeneous equation, it is a straightforward matter to verify that the particular solutions formed according to Rule 1, taken together, are fundamental solutions of the equation. This may be done by showing that the wronskian of all the solutions never has the value zero, or it may be more convenient to apply the definition of linear independence given by Definition 2, Sec. 2.2. An example or two should make these ideas clear.

EXAMPLE 1

Find a complete solution of the equation $y''' + 3y'' + 3y' + y = 0$.

In this case, the characteristic equation is

$$m^3 + 3m^2 + 3m + 1 \equiv (m + 1)^3 = 0$$

Hence $m = -1$ is a triple root, and not only are e^{-x} and xe^{-x} solutions of the differential equation but so too is x^2e^{-x}. Since these three solutions have a nonvanishing wronskian (see Exercise 52), it follows that a complete solution of the equation is

$$y = c_1e^{-x} + c_2xe^{-x} + c_3x^2e^{-x}$$

The characteristic equation of a differential equation of order greater than 2 is usually difficult to solve, and in general the characteristic roots must be found by approximate numerical methods. Occasionally, however, as in the last example, close attention to the form of the equation may suggest a method for its exact solution. The next three examples illustrate this possibility in more complicated cases.

EXAMPLE 2

Find a complete solution of the equation $(D^8 + 6D^6 - 32D^2)y = 0$.

The characteristic equation is

$$m^8 + 6m^6 - 32m^2 = m^2(m^6 + 6m^4 - 32) = 0$$

Since the polynomial inside the parentheses vanishes when $m^2 = 2$, it has $m^2 - 2$ as a factor. Dividing $m^6 + 6m^4 - 32$ by $m^2 - 2$ to obtain the other factor, we find that the characteristic equation can be written as

$$m^2(m^2 - 2)(m^4 + 8m^2 + 16) = m^2(m - \sqrt{2})(m + \sqrt{2})(m^2 + 4)^2 = 0$$

which has the unrepeated roots $m_1 = -\sqrt{2}$ and $m_2 = \sqrt{2}$, and the repeated roots $m_3 = 0$, $m_4 = 2i$, and $m_5 = -2i$. Since particular solutions which correspond to one complex root may always be taken the same as those which correspond to the conjugate root, the root $m_5 = -2i$ can be disregarded.

According to Rule 1 and our preceding theory, particular solutions corresponding to the simple roots $m_1 = -\sqrt{2}$ and $m_2 = \sqrt{2}$ are $e^{-\sqrt{2}x}$ and $e^{\sqrt{2}x}$; particular solutions corresponding to the repeated root $m_3 = 0$ are 1 and x; particular solutions corresponding to the double complex root $m_4 = 2i$ are $\cos 2x$, $\sin 2x$, $x \cos 2x$, and $x \sin 2x$; and a complete solution of the given equation is

$$y = c_1e^{-\sqrt{2}x} + c_2e^{\sqrt{2}x} + c_3 + c_4x + (c_5 + c_6x)\cos 2x + (c_7 + c_8x)\sin 2x$$

EXAMPLE 3

Solve the equation $y^{\text{iv}} - 4y''' + 14y'' - 20y' + 25y = 0$.

The characteristic equation is $m^4 - 4m^3 + 14m^2 - 20m + 25 = 0$. To find its roots, we let $f(m)$ denote the left-hand member so as to study the zeros of f. Differentiating $f(m)$ twice, we have

$$f'(m) = 4m^3 - 12m^2 + 28m - 20 = 4(m - 1)[(m - 1)^2 + 4]$$

$$f''(m) = 12m^2 - 24m + 28 = 12(m - 1)^2 + 16$$

Since for all real values of $m, f''(m) > 0$, the entire graph of f is concave upward, and because $f'(1) = 0$ also, the value $f(1) = 16$ is an absolute minimum of f. All roots of the auxiliary equation are therefore complex, and there are at most two conjugate pairs $a \pm ib$ and $c \pm id$. These will satisfy the auxiliary equation if and only if

$$m^4 - 4m^3 + 14m^2 - 20m + 25 = (m^2 - 2am + a^2 + b^2)(m^2 - 2cm + c^2 + d^2)$$

Equating coefficients of like powers of m on the two sides of the equals sign and simplifying, we obtain the four conditions

$$2 = a + c \qquad 14 = (a + c)^2 + 2ac + b^2 + d^2$$
$$10 = c(a^2 + b^2) + a(c^2 + d^2) \qquad 25 = (a^2 + b^2)(c^2 + d^2)$$

for the determination of a, b, c, and d. The first of these conditions gives $a = 2 - c$, which we use to rewrite the next two conditions as

$$10 = 4c - 2c^2 + b^2 + d^2 \qquad \text{and} \qquad 10 = 4c - 2c^2 + cb^2 + 2d^2 - cd^2$$

Equating the right-hand members, canceling terms, and factoring, we find

$$(c - 1)(b^2 - d^2) = 0$$

hence $c = 1$ or $b^2 = d^2$.

If $c = 1$, then $a = 1$; thus b and d must satisfy the two equations

$$8 = b^2 + d^2 \qquad \text{and} \qquad 25(1 + b^2)(1 + d^2)$$

Solving the first of these for d^2, substituting into the second, and then solving for b and d, we find

$$d^2 = 8 - b^2 \qquad 25 = (1 + b^2)(9 - b^2) \qquad (b^2 - 4)^2 = 0 \qquad b, d = \pm 2$$

The alternative hypothesis, $b^2 = d^2$, leads to the equation

$$(c - 1)^2[(c - 1)^2 + 4] = 0$$

which implies that $c = 1$ (see Exercise 58). It follows that $1 \pm 2i$ is a repeated root of the characteristic equation and that a complete solution of the differential equation is

$$y = e^x(c_1 \cos 2x + c_2 \sin 2x + c_3 x \cos 2x + c_4 x \sin 2x)$$

EXAMPLE 4

For what nonzero values of λ, if any, does the equation $y^{\text{iv}} - \lambda^4 y = 0$ have nontrivial solutions which satisfy the four conditions $y(0) = y''(0) = y(l) = y''(l) = 0$? What are these solutions if they exist?

The characteristic equation in this case is $m^4 - \lambda^4 = 0$, and its roots are $m = \pm\lambda, \pm i\lambda$. Hence a complete solution is

$$y = c_1 \cos \lambda x + c_2 \sin \lambda x + c_3 e^{\lambda x} + c_4 e^{-\lambda x}$$

It is more convenient, however, to introduce hyperbolic functions and work with a complete solution of the following form:

$$y = A \cos \lambda x + B \sin \lambda x + C \cosh \lambda x + E \sinh \lambda x$$

Differentiating this twice gives us

$$y'' = \lambda^2(-A \cos \lambda x - B \sin \lambda x + C \cosh \lambda x + E \sinh \lambda x)$$

Hence, substituting the first two of the given conditions, we obtain the relations

$$A + C = 0$$
$$\lambda^2(-A + C) = 0$$

which imply that $A = C = 0$. Using this information and the last two conditions, we have further

$$B \sin \lambda l + E \sinh \lambda l = 0$$

$$\lambda^2(-B \sin \lambda l + E \sinh \lambda l) = 0$$

Dividing out λ^2 and then adding these equations, we find that

$$2E \sinh \lambda l = 0$$

Now the hyperbolic sine is zero if and only if its argument is zero. Moreover, by the statement of the problem we are restricted to nonzero values of λ. Hence $\sinh \lambda l \neq 0$, and therefore $E = 0$, which implies that

$$B \sin \lambda l = 0$$

We have already been forced to the conclusion that $A = C = E = 0$; hence if $B = 0$, the solution would be identically zero, contrary to the requirements of the problem. Thus we must have

$$\sin \lambda l = 0 \qquad \text{or} \qquad \lambda_n = \frac{n\pi}{l} \qquad n \text{ a nonzero integer}$$

For these values of λ, and for these only, there are solutions meeting the requirements of the problem. Clearly, these solutions are all of the form $y = B \sin \lambda x$ or, more specifically,

$$y_n = B_n \sin\left(\frac{n\pi x}{l}\right) \qquad n = 1, 2, 3, \ldots$$

where no solutions have been lost by restricting n to positive integral values because the constants B_n are all arbitrary.

EXERCISES

Find a complete solution of each of the following equations.

1. $y''' - 3y'' + 2y' = 0$
2. $y''' + 3y'' - 4y = 0$
3. $y''' + 2y'' + y' + 2y = 0$
4. $y''' - 3y'' + y' + 5y = 0$
5. $y''' - y'' + y' - y = 0$
6. $y''' + 5y'' + 3y' - 9y = 0$
7. $y''' - 4y'' - 3y' + 18y = 0$
8. $3y''' - y'' - 27y' + 9y = 0$
9. $5y''' - 21y'' - 101y' + 21y = 0$
10. $y''' + y = 0$
11. $y^{\text{iv}} + 4y''' + 7y'' + 6y' + 2y = 0$
12. $(D^4 - 10D^2 + 9)y = 0$
13. $(4D^4 - 4D^3 - 3D^2 + 4D - 1)y = 0$
14. $y^{\text{iv}} + 4y''' + 14y'' + 20y' + 25y = 0$
15. $y^{\text{iv}} - 2y''' + 3y'' - 2y' + y = 0$
16. $y^{\text{iv}} + y = 0$
17. $(D^4 - 5D^3 + 6D^2 + 4D - 8)y = 0$
18. $y^{\text{iv}} - 3y''' - 2y'' + 2y' + 12y = 0$
19. $y^{\text{v}} - 2y^{\text{iv}} - 8y''' + 16y'' + 16y - 32 = 0$
20. $16y^{\text{vi}} + 8y^{\text{iv}} + y'' = 0$
21. $y^{\text{viii}} - 2y^{\text{vi}} + y^{\text{iv}} = 0$
22. $(D - 3)^3(D^2 - 2D + 5)^2y = 0$
23. $D(D + 1)(D - 2)^3(D^2 + 1)^2y = 0$
24. $(D^2 - 25)^2(D^2 - 4D + 13)^3y = 0$

Find a homogeneous linear differential equation with (real) constant coefficients and of minimal order which has the given solution.

25. $e^{3x} - e^{-x} + 2$
26. $\cos 2x - \sin 3x$
27. $xe^{-2x} + 3 \sinh 2x$
28. $x - e^{-x} + e^{2x}$
29. $e^{5x} - x^2e^{2x}$
30. $xe^{-2x} \cos x$
31. $\cosh 9x + 5 \sin 4x$
32. $x^3e^{-x} + \sin 5x$

Determine a solution of each of the following equations which satisfies the given conditions.

33. $(D^3 - 6D^2 + 11D - 6)y = 0$; $y = y' = 0$, $y'' = 2$ when $x = 0$
34. $y''' - 3y' + 2y = 0$; $y = y'' = 0$, $y' = 2$ when $x = 0$
35. $y''' + y'' - y' - y = 0$; $y(0) = y'(1) = 0$, $y'(0) = 1$
36. $y''' + y'' = 0$; $y = 2$, $y' = 1$, $y'' = -1$ when $x = 0$
37. $(D^3 - 2D^2 + D - 2)y = 0$; $y = y' = y'' = 1$ when $x = 0$
38. $y''' - 25y' = 0$; $y(0) = -1$, $y'(0) = 5$, $y''(0) = -25$
39. $y''' + \pi^2y' = 0$; $y(0) = 1$, $y(1) = -1$, $y'(0) = y'(1)$
40. $y^{\text{iv}} + 5y'' + 4y = 0$; $y = y'' = 0$ when $x = 0$; $y = 1$, $y' = 2$ when $x = \pi/2$

41. $8y''' - 4y'' + 6y' + 5y = 0;\ y = 0,\ y' = y'' = 1$ when $x = 0$

42. $y''' - 3y'' + 4y = 0;\ y(0) = 1,\ y'(0) = -8,\ y''(0) = -4$

43. $(D^2 + 10D + 169)^2 y = 0;\ y = y' = y'' = 0,\ y''' = 288$ when $x = 0$

44. $y^{iv} - y''' = 0;\ y(0) = y'(0) = 1,\ y''(1) = 3e,\ y'''(1) = e$

45. $y^{v} - 9y''' + 4y'' + 12y' = 0;\ y = 0,\ y' = 2,\ y'' = -8,\ y''' = 26,\ y^{iv} = -80$ when $x = 0$

46. $(D + 1)^2[(D + 1)^2 + \pi^2]y = 0;\quad y(0) = y(1) = y(-1) = y'(1) = y'(-1) = 0,\ y'(0) = 2$

47. Find the Maclaurin series expansion of the solution of the initial-value problem $y''' - y = 0;\ y(0) = y'(0) = y''(0) = 1$, and identify the solution. *Hint:* Note that $y'''(0) = y(0)$, then differentiate the differential equation to obtain $y^{iv}(0) = y'(0)$ and continue the process.

48. Work Exercise 47 if the initial-value problem is $y''' + y'' - 2y' = 0;\ y(0) = 1,\ y'(0) = -2,\ y''(0) = 4$.

49. For what nonzero values of λ, if any, does the equation $y^{iv} - \lambda^4 y = 0$ have solutions which satisfy the conditions $y(0) = y''(0) = y(1) = y'(1) = 0$ and are not identically zero? What are these solutions if they exist?

50. Work Exercise 49 if the given conditions are $y(0) = y''(0) = y''(1) = y'''(1) = 0$.

51. Prove that the wronskian of the functions e^{m_1x}, e^{m_2x}, and e^{m_3x} is different from zero if and only if m_1, m_2, and m_3 are all different.

52. Prove that the wronskian of the functions e^{mx}, xe^{mx}, and x^2e^{mx} is never equal to zero.

53. Using Definition 2, Sec. 2.2, show that the functions e^{mx}, xe^{mx}, and x^2e^{mx} are linearly independent.

54. Prove that if a and m are real and if a is different from zero, the wronskian of the functions e^{mx}, $\cos ax$, and $\sin ax$ is never equal to zero.

55. Prove that if $\lambda \neq 0$, the wronskian of the functions $\cos \lambda x$, $\sin \lambda x$, $\cosh \lambda x$, and $\sinh \lambda x$ is never equal to zero.

56. Find three linearly independent solutions of the equation $y''' - 2y'' + y' - 2y = 0$ and verify that their wronskian satisfies Abel's formula.

57. If y_1 is a nontrivial solution of $(D - a)y = 0$, and y_2, y_3 form a basis for the solutions of $(D^2 - a^2)y = 0$, are y_1, y_2, y_3 fundamental solutions of $(D - a)(D^2 - a^2)y = 0$? Explain.

58. In Example 3, show that if $b^2 = d^2$, then

$$(c - 1)^2[(c - 1)^2 + 4] = 0$$

so that $c = 1$.

59. Show that for every real number r and positive integer n,

$$(D - r)^n(x^k e^{rx}) = 0 \qquad k = 0, 1, 2, \ldots, n - 1$$

Hint: Show that $(D - r)^{k+1}(x^k e^{rx}) = k!(D - r)e^{rx} = 0$ and set $(D - r)^n = (D - r)^{n-k-1}(D - r)^{k+1}$.

60. Let y_1 and y_2 be the functions

$$y_1 = e^{ax}\cos bx \qquad \text{and} \qquad y_2 = e^{ax}\sin bx \qquad a, b \text{ real}$$

Show that for every positive integer n,

$$(D^2 - 2aD + a^2 + b^2)^n(x^{n-1}y_i) = 0 \qquad i = 1, 2$$

Hint: Use induction and in doing so note that the derivative of either y_1 or y_2 is a linear combination of these two functions.

2.7 NONHOMOGENEOUS EQUATIONS WITH CONSTANT COEFFICIENTS

We have now learned how to solve constant-coefficient homogeneous linear equations, and with this knowledge we can now obtain a complementary function of an nth-order nonhomogeneous equation

$$(1) \qquad a_0 y^{(n)} + a_1 y^{(n-1)} + \cdots + a_{n-1}y' + a_n = f(x)$$

However, we must also have a particular integral of (1) before we can construct a complete solution, namely,

$$y = \text{complementary function} + \text{particular integral}$$

In Sec. 2.4 we developed the method of variation of parameters as a procedure for finding particular integrals of nonhomogeneous equations, and we could of course use it here. However, in most elementary problems another process, called *the method of undetermined coefficients,* is simpler. It does not apply to as large a class of nonhomogeneous terms as does the method of variation of parameters, but it has the advantage of involving differentiation rather than integration and is therefore the method we shall use whenever possible. Initially, the use of undetermined coefficients appears to be based on little more than guesswork, but the procedure can easily be formalized into a well-defined technique applicable to equations whose nonhomogeneous terms belong to a readily identified and very important class of functions.

To illustrate the method, suppose that we wish to find a particular integral of the equation

$$y'' + 4y' + 3y = 5e^{2x} \tag{2}$$

Since differentiating an exponential of the form e^{kx} merely reproduces that function with, at most, a change in its numerical coefficient, it is natural to ''guess'' that it may be possible to determine A so that

$$Y = Ae^{2x}$$

will be a solution of (2). To check this, we substitute $Y = Ae^{2x}$ for y in the given equation, getting

$$4Ae^{2x} + 8Ae^{2x} + 3Ae^{2x} = 5e^{2x} \qquad \text{or} \qquad 15Ae^{2x} = 5e^{2x}$$

which will be an identity if and only if $A = \frac{1}{3}$. Thus the required particular integral is

$$Y = \tfrac{1}{3}e^{2x}$$

Now suppose that instead of $5e^{2x}$ the right-hand side of Eq. (2) is $5 \sin 2x$, so that the equation we have to solve is

$$y'' + 4y' + 3y = 5 \sin 2x \tag{2a}$$

Guided by our previous success in determining a particular solution, we might perhaps be led to try

$$Y = A \sin 2x$$

as a particular integral. Substituting this to check whether or not it can be a solution, we obtain

$$-4A \sin 2x + 8A \cos 2x + 3A \sin 2x = 5 \sin 2x$$
$$-A \sin 2x + 8A \cos 2x = 5 \sin 2x$$

and since $\sin 2x$ and $\cos 2x$ are linearly independent, this cannot be an identity unless, simultaneously, $A = -5$ and $A = 0$, which is absurd. The difficulty here of course is that differentiating $\sin 2x$ introduces the new function $\cos 2x$, which must also be eliminated identically from the equation resulting from the substitution of $Y = A \sin 2x$ for y. Since the one arbitrary constant A cannot satisfy two independent conditions, it is clear that we must arrange to incorporate *two* arbitrary constants in our tentative choice for Y without, at the same time, introducing new terms which will lead to still more conditions. This is easily done by assuming

$$Y = A \sin 2x + B \cos 2x$$

which contains the necessary second parameter yet cannot introduce any further new functions since it already is a linear combination of *all* the independent terms that can be obtained from $\sin 2x$ or from $\cos 2x$ by repeated differentiation. The actual determination of A and B is a simple matter, for substitution into the new equation ($2a$) yields

$$(-4A \sin 2x - 4B \cos 2x) + 4(2A \cos 2x - 2B \sin 2x) + 3(A \sin 2x + B \cos 2x) = 5 \sin 2x$$
$$(-A - 8B) \sin 2x + (8A - B) \cos 2x = 5 \sin 2x$$

and for this to be an identity requires that

$$-A - 8B = 5 \qquad \text{and} \qquad 8A - B = 0$$

from which we find immediately that $A = -\frac{1}{13}$ and $B = -\frac{8}{13}$. Hence, finally,

$$Y = -\frac{\sin 2x + 8\cos 2x}{13}$$

With these illustrations in mind, we now give a rule under which a particular integral can be found using the *method of undetermined coefficients* followed by a more precise *three-step* description of the method itself.

RULE 1 If $f(x)$ is a function for which repeated differentiation yields only a finite number of linearly independent functions, appearing possibly in linear combination, then a particular integral Y for the nonhomogeneous linear equation (1) can be found by the method of undetermined coefficients.

The method, save for one exceptional case which we shall soon identify and learn to handle, proceeds as follows.

1. Assume Y to be an arbitrary linear combination of all the linearly independent functions which arise from $f(x)$ by repeated differentiation.
2. Substitute Y into the given differential equation.
3. Determine the arbitrary constants in Y so that the equation resulting from the substitution is identically satisfied.

The class of functions $f(x)$ possessing only a finite number of linearly independent derivatives consists of the simple functions

k
x^n, n a positive integer
e^{kx}
$\cos kx$
$\sin kx$

and any others obtainable from these by a finite number of additions, subtractions, and multiplications. These functions are precisely those which can occur as solutions of homogeneous linear differential equations with constant coefficients.

If $f(x)$ possesses infinitely many independent derivatives, as is the case, for instance, with the simple function $1/x$, it is occasionally convenient to assume for Y an infinite series whose terms are the respective derivatives of $f(x)$, each multiplied by an arbitrary constant. However, the use of the method of undetermined coefficients in such cases involves questions of convergence which never arise when $f(x)$ has only a finite number of independent derivatives.

When $f(x)$ is a sum of m terms, say $f(x) = f_1(x) + f_2(x) + \cdots + f_m(x)$, we can find a particular integral Y of (1) in either of two ways. If we wish, we can find Y by applying the method of undetermined coefficients to $f(x)$ in its entirety. On the other hand, for $i = 1, 2, \ldots, m$, we can use the method to find a particular integral Y_i when $f_i(x)$ is the nonhomogeneous term, then add these particular integrals to obtain a particular integral $Y = Y_1 + Y_2 + \cdots + Y_m$ of (1) in accordance with Theorem 2, Sec. 2.3.

EXAMPLE 1

Find a particular integral for the equation

(3) $$y'' + 3y' + 2y = 10e^{3x} + 4x^2$$

If we wish, we can find Y by beginning with the expression $Y = Ae^{3x} + Bx^2 + Cx + D$, which means that we are going to handle the various terms in $f(x)$ at the same time. On the other hand, we can also find Y by first finding a particular integral Y_1 for the equation

(3a) $$y'' + 3y' + 2y = 10e^{3x}$$

then finding a particular integral Y_2 for the equation

(3b) $$y'' + 3y' + 2y = 4x^2$$

and finally taking Y to be the sum $Y_1 + Y_2$.

Using the second method (which means that the expressions we have to substitute are not quite so lengthy and the subsequent collection of terms is not quite so involved), we assume $Y_1 = Ae^{3x}$, substitute into Eq. (3a), and determine A so that the resulting equation will be an identity:

$$9Ae^{3x} + 3(3Ae^{3x}) + 2(Ae^{3x}) = 10e^{3x}$$

$$20Ae^{3x} = 10e^{3x}$$

which implies that $A = \frac{1}{2}$ and $Y_1 = \frac{1}{2}e^{3x}$. Then we assume $Y_2 = Bx^2 + Cx + D$, substitute into Eq. (3b), and determine B, C, and D so that again the resulting equation will be an identity:

$$2B + 3(2Bx + C) + 2(Bx^2 + Cx + D) = 4x^2$$

$$2Bx^2 + (6B + 2C)x + (2B + 3C + 2D) = 4x^2$$

$$2B = 4$$

$$6B + 2C = 0$$

$$2B + 3C + 2D = 0$$

Solving these simultaneously, we find at once that

$$B = 2 \qquad C = -6 \qquad D = 7$$

Hence $Y_2 = 2x^2 - 6x + 7$, and finally

$$Y = Y_1 + Y_2 = \frac{e^{3x}}{2} + 2x^2 - 6x + 7$$

As was previously mentioned, there is one important exception to the procedure we have just been outlining, which we must now investigate. Suppose, for example, that we wish to find a particular integral for the equation

(4) $$y'' + 5y' + 6y = e^{-3x}$$

Proceeding in the way we have just described, we would start with

$$Y = Ae^{-3x}$$

and substitute into the left-hand member, getting

$$9Ae^{-3x} + 5(-3Ae^{-3x}) + 6(Ae^{-3x}) = e^{-3x}$$

$$0 = e^{-3x} \qquad (!)$$

Thus $Y = Ae^{-3x}$ satisfies the homogeneous equation corresponding to (4), but not (4) itself. Clearly, it is important that we be able to recognize and handle such cases. The source of the difficulty is easily identified, for the characteristic equation of Eq. (4) is

$$m^2 + 5m + 6 = 0$$

and since its roots are $m_1 = -3$ and $m_2 = -2$, the complementary function of Eq. (4) is

$$y = c_1 e^{-3x} + c_2 e^{-2x}$$

Thus the term on the right-hand side of (4) is proportional to a term in the complementary function; i.e., it is a solution of the related homogeneous equation, hence can yield only zero when it is substituted into the left-hand member.

There are various ways of overcoming this difficulty, several of which are indicated in the next set of exercises. The most natural one for us to try is our powerful general method, variation of parameters. However, despite appearances, it is still possible to use undetermined coefficients to find a particular integral of the problem at hand. In fact, the technique can actually be applied to any nonhomogeneous constant coefficient equation whose nonhomogeneous term itself is a solution of a *homogeneous* linear differential equation with *constant* coefficients. The key to extending the process is to identify a *linear differential operator with constant coefficients* which *annihilates* the nonhomogeneous term.

DEFINITION 1 An expression of the form

$$L = a_0(x)D^n + a_1(x)D^{n-1} + \cdots + a_{n-1}(x)D + a_n(x) \qquad n \geq 1$$

such that on some interval I the coefficient functions $a_0(x), a_1(x), \ldots, a_n(x)$ are defined and $a_0(x) \not\equiv 0$, is called an ***n*th-order linear differential operator on I.**

Such an operator is said to be **normal on** I just in case all of its coefficient functions are continuous, and the leading coefficient $a_0(x)$ is never zero, on I.

DEFINITION 2 A **polynomial operator** is a linear differential operator whose coefficient functions are constants.

Sometimes polynomial operators are referred to as **constant-coefficient linear differential operators.**

When a linear differential operator L is applied to any function f that is differentiable n times on I, it produces a unique function $g = Lf$ such that, for every x in I,

$$g(x) = Lf(x) = a_0(x)f^{(n)}(x) + a_1(x)f^{(n-1)}(x) + \cdots + a_n(x)f(x)$$

Thus, an nth-order linear differential operator (on I) is actually a function whose domain is the set of all functions which have a derivative of order n at each point of I.

DEFINITION 3 A linear differential operator L **annihilates** a function f if and only if $Lf(x) \equiv 0$ on I.

Now suppose

(5) $$Ly = (a_0 D^n + a_1 D^{n-1} + \cdots + a_{n-1}D + a_n)y = f(x)$$

is a nonhomogeneous linear differential equation with constant coefficients and that f satisfies some homogeneous constant-coefficient equation. Then f and all of its derivatives are everywhere continuous and there is a constant-coefficient linear differential operator L_1 which annihilates f. Let $y = y(x)$ be any solution of (5) so that on $(-\infty, \infty)$,

(6) $$Ly(x) \equiv f(x)$$

Operating with L_1 on both members of this identity, we get $L_1Ly(x) \equiv L_1f(x)$ or $L_1Ly(x) \equiv 0$. This shows that every solution of (5) must be a solution of

$$L_1Ly = 0 \tag{7}$$

Hence a particular integral of (5) can be found by particularizing the arbitrary constants appearing in a complete solution

$$y = c_1y_1 + c_2y_2 + \cdots + c_my_m \tag{8}$$

of (7), where m will equal the sum of n and the order of L_1. In fact, since L will annihilate every solution of the homogeneous equation corresponding to (5), a particular integral of (5) can be found by determining the undetermined coefficients in a linear combination of just those linearly independent particular solutions of (7) which are not solutions of $Ly = 0$.

Clearly, this method will lead to a particular integral Y whether or not there exists a term of $f(x)$ one of whose derivatives, if not the term itself, is proportional to some term of the complementary function of (5). Thus we now know what to do when the method of undetermined coefficients as described immediately after Rule 1 fails. We abandon Step **1** and find an annihilator of $f(x)$ because Step **3** cannot be performed if the linear combination assumed for Y is inadequate.

To summarize these observations, we give a more comprehensive *four-step* description of **the method of undetermined coefficients** for finding a particular integral Y of a nonhomogeneous constant-coefficient equation (5) once a complementary function has been found.

1. Find a constant-coefficient linear differential operator that annihilates $f(x)$, say L_1.

2. Find a complete solution of the homogeneous equation $L_1Ly = 0$, delete all terms which L annihilates, and set Y equal to the sum of the remaining terms.

3. Substitute Y into the given equation.

4. Determine the arbitrary constants in Y so that the equation resulting from the substitution is identically satisfied.

Of course, if no term of $f(x)$ or any of its derivatives is proportional to a term of the complementary function, it is usually easier to determine the analytic form of a particular integral by using the method of undetermined coefficients originally outlined in connection with Rule 1.

Returning to Eq. (4), which prompted our discussion of operators, we again take note of the nonhomogeneous term e^{-3x}. Our experience with homogeneous equations has taught us that $y = e^{-3x}$ is a solution of $(D + 3)y = 0$ which has $m + 3 = 0$ as its characteristic equation. Hence the operator $D + 3$ annihilates e^{-3x}. Using this fact, we operate on both members of (4) with $D + 3$, obtaining

$$(D + 3)(D^2 + 5D + 6)y = 0 \tag{9}$$

or

$$(D + 3)^2(D + 2)y = 0 \tag{9a}$$

Every function which satisfies (4) must also satisfy (9) or (9*a*). In other words, every solution of (4) is to be found among the solutions of (9*a*). A complete solution of (9*a*) is

$$y = c_1e^{-2x} + c_2e^{-3x} + c_3xe^{-3x}$$

and, as has already been observed, $c_1e^{-2x} + c_2e^{-3x}$ is a complementary function of (4). Since this function is annihilated by the operator $D^2 + 5D + 6$, Eq. (4) must have a particular integral of the form

$$Y = Axe^{-3x} \qquad A \text{ constant} \tag{10}$$

To determine the undetermined coefficient A, we substitute (10) into (4) for y, getting

$$A(9xe^{-3x} - 6e^{-3x}) + 5A(-3xe^{-3x} + e^{-3x}) + 6Axe^{-3x} = e^{-3x}$$

which will be an identity if and only if $-A = 1$ or $A = -1$. A particular integral of (4) is therefore $Y = -xe^{-3x}$, and a complete solution of the given equation is

$$y = c_1e^{-2x} + c_2e^{-3x} - xe^{-3x} \tag{11}$$

Here are several more examples which further illustrate how nonhomogeneous linear differential equations with constant coefficients may be solved.

EXAMPLE 2

Find a complete solution of the equation $y'' + 5y' + 6y = 3e^{-2x} + e^{3x}$.

The roots of the characteristic equation

$$m^2 + 5m + 6 = (m + 2)(m + 3) = 0$$

are $m_1 = -2$ and $m_2 = -3$. Hence a complementary function is

$$c_1e^{-2x} + c_2e^{-3x}$$

Now e^{-2x} occurs in a term here as well as in the right-hand member of the differential equation; thus we look for an operator that will annihilate $3e^{-2x} + e^{3x}$. Since $D + 2$ annihilates e^{-2x}, while $D - 3$ annihilates e^{3x}, $(D + 2)(D - 3)$ is such an operator. Applying this operator to both members of the original equation we obtain the homogeneous equation

$$(D + 2)(D - 3)(D^2 + 5D + 6)y = (D + 2)^2(D - 3)(D + 3)y = 0$$

which has e^{-2x}, xe^{-2x}, e^{3x}, and e^{-3x} as fundamental solutions. Both e^{-2x} and e^{-3x} appear in the complementary function. Hence, omitting these, the given equation has a particular integral of the form

$$Y = Axe^{-2x} + Be^{3x}$$

Substituting this into the differential equation, we have

$$(4Axe^{-2x} - 4Ae^{-2x} + 9Be^{3x}) + 5(-2Axe^{-2x} + Ae^{-2x} + 3Be^{3x}) + 6(Axe^{-2x} + Be^{3x}) = 3e^{-2x} + e^{3x}$$

or

$$Ae^{-2x} + 30Be^{3x} = 3e^{-2x} + e^{3x}$$

Equating coefficients of like functions, we find $A = 3$ and $B = \frac{1}{30}$. Hence

$$Y = 3xe^{-2x} + \frac{e^{3x}}{30}$$

and a complete solution is

$$y = c_1e^{-2x} + c_2e^{3x} + 3xe^{-2x} + \frac{e^{3x}}{30}$$

EXAMPLE 3

Find a complete solution of the equation $y'' + y = 3 \sin x$.

Since a term of the complementary function $c_1 \cos x + c_2 \sin x$ duplicates the nonhomogeneous term of the given equation, we apply the operator $D^2 + 1$, which annihilates $3 \sin x$, to both members of the given equation. This gives us

$$(D^2 + 1)^2y = 0$$

which has $Ax\cos x + Bx\sin x + C\cos x + E\sin x$ as a complete solution. Deleting the terms $C\cos x$ and $E\sin x$, which satisfy $y'' + y = 0$, we set

$$Y = x(A\cos x + B\sin x)$$

and substitute into the nonhomogeneous equation, getting

$$[x(-A\cos x - B\sin x) + 2(-A\sin x + B\cos x)] + x(A\cos x + B\sin x) = 3\sin x$$

or, collecting terms,

$$-2A\sin x + 2B\cos x = 3\sin x$$

Hence $A = -\frac{3}{2}$, $B = 0$, $Y = -\frac{3}{2}x\cos x$, and a complete solution is

$$y = c_1\cos x + c_2\sin x - \tfrac{3}{2}x\cos x$$

EXAMPLE 4

Find a complete solution of the equation $(D^4 + 8D^2 + 16)y = -\sin x$.

The characteristic equation here is

$$m^4 + 8m^2 + 16 = 0 \qquad \text{or} \qquad (m^2 + 4)^2 = 0$$

This equation has $2i$ and $-2i$ as double roots. Hence a complementary function contains not only $\cos 2x$ and $\sin 2x$ but also these functions multiplied by x and therefore is

$$c_1\cos 2x + c_2\sin 2x + c_3x\cos 2x + c_4x\sin 2x$$

Note carefully that no term of this expression is proportional to $\sin x$. In other words, $\cos 2x$, $\sin 2x$, $x\cos 2x$, $x\sin 2x$, and $\sin x$ are linearly independent functions despite the superficial resemblance of $\sin x$ and $\sin 2x$.

Repeated differentiation of $\sin x$ gives rise to only two functions, namely, $\cos x$ and $\sin x$. Thus, to find a particular integral, we bypass the operator approach and immediately try $Y = A\cos x + B\sin x$, which on substitution into the differential equation gives

$$(A\cos x + B\sin x) + 8(-A\cos x - B\sin x) + 16(A\cos x + B\sin x) = -\sin x$$

or

$$9A\cos x + 9B\sin x = -\sin x$$

This will be an identity if and only if $A = 0$ and $B = -\frac{1}{9}$. Therefore

$$Y = -\frac{\sin x}{9}$$

and a complete solution is

$$y = c_1\cos 2x + c_2\sin 2x + c_3x\cos 2x + c_4x\sin 2x - \frac{\sin x}{9}$$

Since the given differential equation contains only derivatives of even order, we could have foreseen that A would vanish and that Y would need to contain only a sine term in order to match the sine function on the right-hand side and that $Y = B\sin x$ would therefore be a satisfactory trial solution. This simplification should be clearly understood, for it can often be applied to the analysis of vibrating mechanical systems with negligible friction or electric circuits with negligible resistance. Of course, it cannot be applied when the nonhomogeneous term duplicates a term in the complementary function. (See Example 3.)

EXAMPLE 5

Using the method of variation of parameters, derive a formula for a particular integral of the equation

$$y''' - 7y' + 6y = f(x)$$

Since the method of variation of parameters is based on the use of particular solutions of the related homogeneous equation, our first step must be to find three independent solutions of the equation $y''' - 7y' + 6y = 0$. The characteristic equation in this case is $m^3 - 7m + 6 = 0$, and by inspection $m = 1$ is one root. Removing the factor $m - 1$ from the left-hand side, we obtain the quadratic equation

$$m^2 + m - 6 = 0$$

whose roots are $m = 2, -3$. Since these values of m are all distinct, the corresponding solutions

$$y_1 = e^x \qquad y_2 = e^{2x} \qquad y_3 = e^{-3x}$$

are linearly independent.

We must now find three functions of x, say u_1, u_2, u_3, such that

$$Y = u_1y_1 + u_2y_2 + u_3y_3 = e^xu_1 + e^{2x}u_2 + e^{-3x}u_3$$

will be a solution of the given nonhomogeneous equation. The conditions of Theorem 1, Sec. 2.4, particularize in this case to the three equations

$$\begin{aligned} e^xu_1' + e^{2x}u_2' + e^{-3x}u_3' &= 0 \\ e^xu_1' + 2e^{2x}u_2' - 3e^{-3x}u_3' &= 0 \\ e^xu_1' + 4e^{2x}u_2' + 9e^{-3x}u_3' &= f(x) \end{aligned}$$

which may be written down in general by noticing that the determinant of the coefficients of the derivatives of the unknown functions is the wronskian of the basis being used to construct a complete solution of the related homogeneous equation. Of course, the right-hand members are all zero except the last one, which will always be the nonhomogeneous term divided by the leading coefficient of the differential equation.

The wronskian of the basis e^x, e^{2x}, e^{-3x} has the value

$$W(x) = \begin{vmatrix} e^x & e^{2x} & e^{-3x} \\ e^x & 2e^{2x} & -3e^{-3x} \\ e^x & 4e^{2x} & 9e^{-3x} \end{vmatrix} = 20$$

Hence, using Cramer's rule to solve the three simultaneous linear equations in u_1', u_2', and u_3', we have

$$u_1' = \frac{\begin{vmatrix} 0 & e^{2x} & e^{-3x} \\ 0 & 2e^{2x} & -3e^{-3x} \\ f(x) & 4e^{2x} & 9e^{-3x} \end{vmatrix}}{W(x)} = -\frac{5e^{-x}f(x)}{20} \qquad \text{and} \qquad u_1 = -\frac{1}{4}\int e^{-x}f(x)\,dx$$

$$u_2' = \frac{\begin{vmatrix} e^x & 0 & e^{-3x} \\ e^x & 0 & -3e^{-3x} \\ e^x & f(x) & 9e^{-3x} \end{vmatrix}}{W(x)} = \frac{4e^{-2x}f(x)}{20} \qquad \text{and} \qquad u_2 = \frac{1}{5}\int e^{-2x}f(x)\,dx$$

$$u_3' = \frac{\begin{vmatrix} e^x & e^{2x} & 0 \\ e^x & 2e^{2x} & 0 \\ e^x & 4e^{2x} & f(x) \end{vmatrix}}{W(x)} = \frac{e^{3x}f(x)}{20} \qquad \text{and} \qquad u_3 = \frac{1}{20}\int e^{3x}f(x)\,dx$$

Therefore the required particular integral is

$$\begin{aligned} Y &= e^xu_1 + e^{2x}u_2 + e^{-3x}u_3 \\ &= -\frac{e^x}{4}\int e^{-x}f(x)\,dx + \frac{e^{2x}}{5}\int e^{-2x}f(x)\,dx + \frac{e^{-3x}}{20}\int e^{3x}f(x)\,dx \end{aligned}$$

EXERCISES

Find a complete solution of each of the following equations.

1. $y'' + 4y' + 3y = x - 1$
2. $y'' + 4y' + 5y = 2e^x$
3. $2y'' + 3y' - 2y = 10 + 3e^x$
4. $y'' + y' = x + 2$
5. $21y'' - 4y' - y = x^2$
6. $y'' + 5y' + 6y = e^{-3x}$
7. $y'' - 3y' + 2y = 2 + e^x$
8. $y'' + 2y' + y = e^{-x}/x$
9. $y'' - y = e^x + 2e^{2x}$
10. $y'' + 2y' + y = 1 - 12e^{-x}$
11. $y'' + 5y' + 6y = 3e^{-2x} + e^{3x}$
12. $y'' + y = \cos x + 3 \sin 2x$
13. $y'' - 4y' + 5y = e^{2x} + 24 \sin x$
14. $y'' + 3y' = \sin x + 2 \cos x$
15. $y'' + 4y' + 13y = \cos 3x - \sin 3x$
16. $(D^2 + 4D + 4)y = xe^{-x}$
17. $y'' + 2y' + 10y = 25x^2 - 3e^{-x}$
18. $y'' - 2y' + y = xe^x - e^x$
19. $(D^2 + 1)y = e^x \sin x$
20. $y'' + 2y' + 5y = e^{-x} \sec 2x$
21. $y'' - 2y' + 5y = e^x \tan 2x$
22. $y'' - 2y' + 5y = x^2 + \sin x$
23. $y'' + 4y' + 4y = e^{-2x}/x^2$
24. $y'' - 5y' + 6y = \cosh x$
25. $y'' - 5y' + 4y = \cosh x$
26. $25y'' - 15y' + 2y = 4x - 65e^{2x/5}$
27. $4y'' + y = 4{,}225xe^x \sin x$
28. $10y'' - 6y' + y = 30 \sin x \cos x$
29. $y'' + 2y' + y = \cos^2 x$. *Hint:* Recall that $\cos^2 x = (1 + \cos 2x)/2$.
30. **(a)** Show that $Y = -\cosh x$ is a particular integral of the equation

$$y'' + y' - 2y = e^{-x}$$

(b) Determine A so that $Y = A \sinh x$ will be a particular integral of this equation.

Solve each of the following initial-value problems.

31. $y'' + 4y' + 3y = 4e^{-x}$; $y(0) = 0$, $y'(0) = 2$
32. $y'' + 2y' + 5y = 17 \sin 2x$; $y(0) = -4$, $y'(0) = 2$
33. $y'' - 4y' + 4y = e^{2x}$; $y(0) = 0$, $y'(0) = \frac{1}{2}$
34. $y'' + 5y' + 6y = e^{-3x}$; $y(0) = -\frac{1}{2}$, $y'(0) = 0$
35. $y'' + 4y' + 5y = 20e^x$; $y(0) = y'(0) = 0$
36. $y'' + 4y' + 4y = 8x - 10$; $y(0) = 2$, $y'(0) = 0$
37. $y'' + 2y' + y = xe^{-x}$; $y(0) = 1$, $y'(0) = -2$
38. $y'' + 2y' + 5y = 10 \cos x$; $y(0) = 5$, $y'(0) = 6$
39. Show that $Y = (\sin \lambda t - \sin kt)/(k^2 - \lambda^2)$ is a particular integral of the equation $y'' + k^2 y = \sin \lambda t$ and investigate the limiting case when $\lambda \to k$.
40. Construct a solution of the differential equation $y'' - 2ay' + a^2 y = e^{\lambda t}$ which will approach a solution of the equation $y'' - 2ay' + a^2 y = e^{at}$ as $\lambda \to a$.
41. If y_1 and y_2 are two solutions of the nonhomogeneous equation $y'' + P(x)y' + Q(x)y = R(x)$, determine for what values of c_1 and c_2, if any, $y = c_1 y_1 + c_2 y_2$ is a solution of this equation.

Find a complete solution of each of the following equations.

42. $(D^4 - 16)y = e^x$
43. $(D^3 + 6D^2 + 11D + 6)y = 6x - 7$
44. $y''' - 2y'' - 3y' + 10y = 40 \cos x$
45. $y^{iv} + 10y'' + 9y = \cos 2x$
46. $y^{v} - 3y''' - 2y'' = 12x - 2$
47. $y''' + 5y'' + 9y' + 5y = 5 + 26e^{-x}$
48. $y'' + 2y' + 2y = e^{-x} \cos x$
49. $y'' - 2y' + y = xe^x - e^x$
50. $4y'' + 8y' + 5y = e^{-x} \sec x/2$
51. $y'' + 2y' + y = e^{-x} \ln |x|$
52. $y'' - 2y' + y = x^3 e^x$
53. $(D^3 + D^2 + 3D - 5)y = e^x$
54. $(D^3 - 7D + 6)y = e^x - 5e^{2x}$
55. $y'' + y = 1/(1 + \sin x)$
56. $y''' + 5y'' + 9y' + 5y = 3e^{2x}$
57. $y''' + y'' - 8y' - 12y = 36x + 25e^{3x}$
58. $y''' - 3y'' + 2y = x + e^x$
59. $y''' - y' = \operatorname{sech} x$
60. $y''' - y'' - y' + y = x + e^{-x}$
61. $y''' - 3y'' + 2y' = x + e^x$
62. $(D - 1)^3 y = 2e^x/x^2$
63. $y^{iv} + 3y''' + 3y'' + y' = 2x + 8$
64. $y^{iv} + 8y'' + 16y + \sin x = 0$
65. $(D^3 + D)y = 4 \cos x$
66. $(D^3 - 3D^2 + 4D)y = 4e^x - 18e^{-x}$
67. $D^2(D^2 + 1)y = 3x^2 + 4 \sin x - 2 \cos x$
68. $(D^2 + 3D + 2)y = e^{-x}/x$
69. $(D + 1)^3 y = 1 - 3e^{-x}$
70. $(D + 2)(D^2 + 2D + 2)y = x - \sin x$
71. $(D^4 + 8D^2 - 9)y = 9x^2 + 5 \sin 2x$
72. $(D^4 + 2D^3 - 3D^2 - 4D + 4)y = e^x$
73. $y^{iv} - 16y = e^x$
74. $y^{iv} + y'' = 3x^2 + 4 \sin x - 2 \cos x$
75. $(D^8 + 6D^6 - 32D^2)y = 512e^{2x}$
76. $y^{iv} + 4y = \cos x + \sin 2x$. *Hint:* By adding and subtracting the appropriate term on the left-hand side of the characteristic equation, rewrite it as the difference of two squares.
77. Using a method of undetermined coefficients, find a particular integral of $y'' - y = 1/x$. For what values of x, if any, is this solution meaningful?
78. Using a method of undetermined coefficients, find a particular integral of $y'' + y = x^{1/2}$. For what values of x, if any, is this solution meaningful?
79. **(a)** Explain how the result of Part **(b)** of Exercise 82, Sec. 2.5, can be used to obtain a particular integral of the equation $ay'' + by' + cy = e^{rx}$ when $m = r$ is a simple root of the characteristic equation.

(b) Generalize the results of Parts **(a)** and **(b)** of Exercise 82, Sec. 2.5, to show that if $p(D)$ is a quadratic polynomial in D, then $p(D)(x^2e^{mx}) = p(m)x^2e^{mx} + 2p'(m)xe^{mx} + p''(m)e^{mx}$.

(c) Explain how the result of Part (b) can be used to obtain a particular integral of the equation $ay'' + by' + cy = e^{rx}$ when $m = r$ is a double root of the characteristic equation.

80. Using variation of parameters, or by means of operators, verify that Column 2 of Table 2.2 prescribes a correct form for a particular integral Y of $ay'' + by' + cy = f(x)$ for the types of functions $f(x)$ specified and explain how the undetermined coefficients of Y may be determined.

81. Under what conditions, if any, can Table 2.2 be used to find a particular integral Y of a constant-coefficient nth-order nonhomogeneous linear differential equation if $n > 2$?

In Exercises 82 through 93 construct a linear combination of functions suitable for finding a particular integral of the given differential equation using a method of undetermined coefficients. Specific values of the coefficients need not be found.

82. $y''' - 2y'' - y' - 2y = \cosh 3x$
83. $[(D+1)^2 + 9]y = e^x \sin 3x$
84. $D^2(D+5)y = x + e^{-5x}$
85. $y''' - 6y'' + 12y' - 8y = x^2e^{3x} - 7xe^{2x}$
86. $(24D^4 - 14D^3 - 13D^2 + 2D + 1)y = e^{-x} - xe^x$
87. $y^{\text{iv}} + y = 64 \cos 2x$
88. $(D-1)^2(D^2-4)y = 2x \sinh x - \cosh 2x$
89. $(D^2+4)^3(D+1)y = 10e^x + 3x^2 \sin 2x$
90. $y^{\text{iv}} + 3y'' - 4y = \sinh x - \sin^2 x$
91. $(D-1)^3y = (x+1)^2e^x - 3x \cos x$
92. $[(D+2)^2 + \pi^2]^2y = e^{-2x} \sin \pi x - \cos x$
93. $y^{\text{vi}} + 2y^{\text{v}} + 5y^{\text{iv}} = 5x^3 - x^2e^{-x} + e^{-x} \cos 2x$

Find the solution of each of the following equations which satisfies the given conditions.

94. $(D^3 + 2D^2 - D - 2)y = 10 \sin x$
$y = 1,\ y' = -2,\ y'' = -1$ when $x = 0$
95. $y''' - 2y'' + 10y' = 3xe^x$
$y = -\frac{1}{3},\ y' = 0,\ y'' = \frac{1}{3}$ when $x = 0$
96. $(D-1)^3y = 6e^x$
$y(-1) = 0,\ y'(-1) = 3/e,\ y''(-1) = 0$
97. $(D^3 + 5D^2 - 6D)y = 3e^x$
$y(0) = 1,\ y'(0) = \frac{3}{7},\ y''(0) = \frac{6}{7}$
98. $(D^3 + D)y = 2x^2 + 4 \sin x$
$y(0) = 0,\ y'(0) = -2,\ y''(0) = -4$
99. $y^{\text{iv}} - y''' - 2y'' = 12x - 6$
$y(0) = 1,\ y'(0) = -3,\ y''(0) = 6,\ y''(1) = 0$
100. $y''' + 3y'' + 7y' + 5y = 16e^{-x} \cos 2x$
$y(0) = 2,\ y'(0) = -4,\ y''(0) = -2$
101. $y^{\text{iv}} + y'' = e^{x + \ln 2}$
$y(0) = y'(0) = 2,\ y''(0) = y'''(0) = 0$
102. $y''' - y' = 2/(1 + e^x)$
$y(0) = \ln 16,\ y'(0) = -1,\ y''(0) = -1 + \ln 4$
103. $y'' - y' \tan x - y \sec^2 x = 2 \cos x$
$y(0) = y'(0) = 0$. *Hint:* One solution of the related homogeneous equation is $y_1 = \tan x$.

104. Use variation of parameters to find a complete solution of Eq. (4) and compare the result with (11).

105. Using variation of parameters, show that if $r_1 \neq r_2$, the coefficient A can always be determined so that $Y = Axe^{r_1x}$ is a solution of the equation

$$y'' - (r_1 + r_2)y' + r_1r_2y = ae^{r_1x}$$

106. Using variation of parameters, show that A can always be determined so that $Y = Ax^2e^{rx}$ is a solution of the equation $y'' - 2ry' + r^2y = ae^{rx}$.

TABLE 2.2

Differential equation: $ay'' + by' + cy = f(x)$ or $(aD^2 + bD + c)y = f(x)$

$f(x)$†	Normal choice for the trial particular integral Y‡
1. α	A
2. αx^n (n a positive integer)	$A_0x^n + A_1x^{n-1} + \cdots + A_{n-1}x + A_n$
3. αe^{rx} (r either real or complex)	Ae^{rx}
4. $\alpha \cos kx$§ 5. $\alpha \sin kx$	$A \cos kx + B \sin kx$
6. $\alpha x^n e^{rx} \cos kx$ 7. $\alpha x^n e^{rx} \sin kx$	$(A_0x^n + \cdots + A_{n-1}x + A_n)e^{rx} \cos kx + (B_0x^n + \cdots + B_{n-1}x + B_n)e^{rx} \sin kx$

†When $f(x)$ consists of a sum of several terms, the appropriate choice for Y is the sum of the Y expressions corresponding to these terms individually.

‡Whenever a term in any of the Y's listed in this column duplicates a term in the complementary function, all terms in that Y expression must be multiplied by the lowest positive integral power of x sufficient to eliminate all such duplications.

§The hyperbolic functions $\cosh kx$ and $\sinh kx$ can be handled either by expressing them in terms of exponentials or by using formulas entirely analogous to those in lines 4–7.

Using variation of parameters, obtain a formula for a particular integral of each of the following equations.

107. $(D^3 - 6D^2 + 11D - 6)y = f(x)$

108. $y''' - y'' + y' - y = f(x)$

109. $y''' - y'' - y' + y = f(x)$

110. $y^{\text{iv}} - 5y'' + 4y = f(x)$

Find a complete solution of each of the following equations.

111. $y'' + 2ay' + (a^2 - b^2)y = f(x)$

112. $y'' + 2ay' + a^2y = f(x)$

113. $y'' + 2ay' + (a^2 + b^2)y = f(x)$

114. $y'' - (a + b)y' + aby = f(x);\ a \neq b$

115. Solve the equation $y''' + y' = 1 + \csc x$.

116. Using the method of variation of parameters, find a particular integral of the equation $y'' - y = 1/x$. How does this result compare with the result of Exercise 77?

117. Using the method of variation of parameters, find a particular integral of the equation $y'' + y = x^{1/2}$. How does this result compare with the result of Exercise 78?

118. **(a)** Use the result of Exercise 111 to show that if $0 < b < a$ and if $|f(x)|$ is bounded for $x \geq x_0$, then the absolute value of every solution of the equation $y'' + 2ay' + (a^2 - b^2)y = f(x)$ is bounded for $x \geq x_0$.

(b) If $0 < b < a$, if $|f(x)|$ is bounded for $x \geq x_0$, and if $\lim_{x\to\infty} f(x) = L$, show that if y is any solution of the equation $y'' + 2ay' + (a^2 - b^2)y = f(x)$, then

$$\lim_{x\to\infty} y = \frac{L}{a^2 - b^2}$$

2.8 THE EULER-CAUCHY DIFFERENTIAL EQUATION

So far, this chapter has been devoted to a study of linear differential equations with constant coefficients. However, there is one type of linear equation with variable coefficients which it is appropriate to discuss at this point because by a simple change of independent variable it can always be transformed into a linear equation with constant coefficients. This is the so-called **Euler-Cauchy† equation**

$$a_0x^ny^{(n)} + a_1x^{n-1}y^{(n-1)} + \cdots + a_{n-1}xy' + a_ny = f(x) \tag{1}$$

in which the coefficient of each derivative is a constant multiple of the corresponding power of the independent variable. As we shall soon see, the change of independent variable defined by

$$|x| = e^z \qquad \text{or} \qquad z = \ln|x| \qquad x \neq 0$$

will always convert this equation into a linear equation with constant coefficients. This in turn can always be solved by the methods which we have developed in the preceding sections of this chapter. Finally, by replacing z by $\ln|x|$ in the solution of the transformed equation, we obtain the solution of the original differential equation.

Before discussing the solution in the general, nth-order case, let us illustrate the process as applied to a particular second-order differential equation.

EXAMPLE 1

Find a complete solution of the equation $x^2y'' + xy' + 9y = 0$.

Under the transformation $|x| = e^z$ or $z = \ln|x|$, we have by a straightforward application of the chain rule

$$y' = \frac{dy}{dx} = \frac{dy}{dz}\frac{dz}{dx} = \frac{1}{x}\frac{dy}{dz} \tag{2}$$

$$y'' = \frac{d(y')}{dx} = \frac{d}{dx}\left(\frac{1}{x}\frac{dy}{dz}\right) = -\frac{1}{x^2}\frac{dy}{dz} + \frac{1}{x}\frac{d^2y}{dz^2}\frac{dz}{dx} = -\frac{1}{x^2}\frac{dy}{dz} + \frac{1}{x^2}\frac{d^2y}{dz^2} \tag{3}$$

†Augustin Louis Cauchy (1789–1857) was a great French mathematician.

Substituting these into the given differential equation, we have

$$x^2\left(\frac{1}{x^2}\frac{d^2y}{dz^2} - \frac{1}{x^2}\frac{dy}{dz}\right) + x\left(\frac{1}{x}\frac{dy}{dz}\right) + 9y = 0$$

or simplifying and collecting terms,

$$\frac{d^2y}{dz^2} + 9y = 0$$

The characteristic equation of the last equation is $m^2 + 9 = 0$, and from its roots, $m = \pm 3i$, we obtain immediately the complete solution

$$y = c_1 \cos 3z + c_2 \sin 3z$$

Finally, replacing z by $\ln |x|$, we obtain a complete solution of the original equation:

$$y = c_1 \cos (3 \ln |x|) + c_2 \sin (3 \ln |x|) \qquad x \neq 0$$

By repeated use of the chain rule, derivatives of y with respect to x of any order can be expressed in terms of derivatives of y with respect to z, as we illustrated in Example 1 for $y' = dy/dx$ and $y'' = d^2y/dx^2$. However, the process soon becomes very complicated if we carry it out as we did in Example 1. On the other hand, if we use the operational notation which we introduced in Sec. 2.5, we obtain elegant formulas for the various derivatives which make the transformation of the differential equation a very simple matter.

The clue to this approach comes from an inspection of the formulas for y' and y'' which we obtained in Example 1. If we let $D = d/dx$ and $\mathfrak{D} = d/dz$, then formulas (2) and (3) of Example 1 can be written

$$y' = \frac{dy}{dx} = Dy = \frac{1}{x}\mathfrak{D}y \qquad \text{or} \qquad x\,Dy = \mathfrak{D}y$$

and

$$y'' = \frac{d^2y}{dx^2} = D^2y = \frac{1}{x^2}(\mathfrak{D}^2y - \mathfrak{D}y) \qquad \text{or} \qquad x^2\,D^2y = \mathfrak{D}(\mathfrak{D} - 1)y$$

These suggest the generalization

$$x^n\,D^ny = \mathfrak{D}(\mathfrak{D} - 1)\cdots(\mathfrak{D} - n + 1)y \qquad n \geq 1 \tag{4}$$

To prove (4), which we have already verified for $n = 1$ and $n = 2$, we use mathematical induction. Assuming k to be any positive integer such that

$$x^k\,D^ky = \mathfrak{D}(\mathfrak{D} - 1)\cdots(\mathfrak{D} - k + 1)y \tag{5}$$

we differentiate both sides with respect to x, getting

$$\frac{d}{dx}(x^k\,D^ky) = \frac{d}{dz}[\mathfrak{D}(\mathfrak{D} - 1)\cdots(\mathfrak{D} - k + 1)y]\frac{dz}{dx}$$

or

$$x^k\,D^{k+1}y + kx^{k-1}\,D^ky = \mathfrak{D}[\mathfrak{D}(\mathfrak{D} - 1)\cdots(\mathfrak{D} - k + 1)y]\frac{1}{x}$$

Multiplying the last equation through by x, then applying the inductive hypothesis (5) to the second term on the left and subsequently subtracting that term from both sides, we have

$$\begin{aligned} x^{k+1} D^{k+1} y &= \mathfrak{D}[\mathfrak{D}(\mathfrak{D}-1)\cdots(\mathfrak{D}-k+1)y] - k[\mathfrak{D}(\mathfrak{D}-1)\cdots(\mathfrak{D}-k+1)y] \\ &= \mathfrak{D}(\mathfrak{D}-1)\cdots(\mathfrak{D}-k+1)(\mathfrak{D}-k)y \end{aligned}$$

which implies that (5) holds when k is replaced by $k+1$. Thus the induction is complete.

The next example illustrates the application of formula (4) to a nonhomogeneous Euler-Cauchy equation of order greater than 2.

EXAMPLE 2

Find a complete solution of the equation

$$x^3 \frac{d^3y}{dx^3} + 4x^2 \frac{d^2y}{dx^2} - 5x \frac{dy}{dx} - 15y = x^4$$

Using Eq. (4) to transform the various terms on the left-hand side of the given equation, we have

$$\mathfrak{D}(\mathfrak{D}-1)(\mathfrak{D}-2)y + 4\mathfrak{D}(\mathfrak{D}-1)y - 5\mathfrak{D}y - 15y = (e^z)^4 = e^{4z}$$

From this, by expanding the various operational products and then collecting terms, we find

$$(\mathfrak{D}^3 + \mathfrak{D}^2 - 7\mathfrak{D} - 15)y = e^{4z} \tag{6}$$

The characteristic equation of this equation is

$$m^3 + m^2 - 7m - 15 = (m-3)(m^2 + 4m + 5) = 0$$

From its roots, $m_1 = 3$, $m_2, m_3 = -2 \pm i$, we obtain the complementary function

$$y = c_1 e^{3z} + e^{-2z}(c_2 \cos z + c_3 \sin z)$$

For a particular integral we try $Y = Ae^{4z}$:

$$\begin{aligned} 64Ae^{4z} + 16Ae^{4z} - 7(4Ae^{4z}) - 15(Ae^{4z}) &= e^{4z} \\ 37Ae^{4z} &= e^{4z} \\ A &= \tfrac{1}{37} \end{aligned}$$

Therefore $Y = \frac{1}{37}e^{4z}$, and a complete solution of Eq. (6) is

$$y = c_1 e^{3z} + e^{-2z}(c_2 \cos z + c_3 \sin z) + \tfrac{1}{37}e^{4z}$$

Finally, replacing z by $\ln|x|$, we have as a complete solution of the given equation

$$\begin{aligned} y &= c_1 e^{3(\ln|x|)} + e^{-2(\ln|x|)}[c_2 \cos(\ln|x|) + c_3 \sin(\ln|x|)] + \tfrac{1}{37}e^{4(\ln|x|)} \\ &= c_1 x^3 + \frac{1}{x^2}[c_2 \cos(\ln|x|) + c_3 \sin(\ln|x|)] + \frac{x^4}{37} \qquad x \neq 0 \end{aligned}$$

EXERCISES

Find a complete solution of the following equations.

1. $x^2y'' + xy' - y = 0$
2. $x^2y'' - 6y = 1 + \ln|x|$
3. $x^2y'' - xy' + y = x^5$
4. $x^2y'' - xy' + 5y = x + \ln|x|$
5. $x^2y'' - xy' + 2y = 1 + \ln^2|x|$
6. $x^3y'' + x^2y' - xy = 3x^3$
7. $2x^2y'' + 5xy' + y = 3x + 2$
8. $x^3y''' + 2x^2y'' - xy' + y = 0$
9. $3x^3y''' - x^2y'' + 2xy' - 2y = 0$
10. $x^3y''' - 3x^2y'' + 7xy' - 8y = 0$
11. $x^2y'' + xy' + y = \tan\ln|x|$
12. $x^2y'' + xy' + y = \csc\ln|x|$
13. $x^3y''' - 3x^2y'' + 6xy' - 6y = 20x$
14. $x^3y''' + 4x^2y'' - 5xy' - 15y = x^3$
15. $(3x^3D^3 - 8x^2D^2 + 7xD + 9)y = 21x^2$
16. $(x^4D^3 - 3x^3D^2 + 6x^2D - 6x)y = 1$
17. $(x^3D^3 + x^2D^2 - 4xD)y = 2 + 3x^2 + \ln|x|$
18. $(9x^4D^4 + 54x^3D^3 + 143x^2D^2 + 89xD - 9)y = 80\cosh\ln|x|$
19. $8x^4y^{\text{iv}} + 52x^3y''' + 82x^2y'' + 85xy' - 34y = 25 - 34\ln|x|$
20. $x^4y^{\text{iv}} + 6x^3y''' + 15x^2y'' + 9xy' - 9y = 3\ln|x| + 1/x^2$

Find a solution of each of the following equations which satisfies the given auxiliary conditions.

21. $x^3y'' - x^2y' + xy = 1$; $y(1) = \frac{1}{4}$, $y(e) = e + 1/4e$
22. $x^2y'' + xy' + y = 5x^2$; $y(1) = 1$, $y'(1) = 3$
23. $x^3y'' + x^2y' - xy = 3x^3$; $y(1) = 1$, $y'(1) = 2$
24. $(2x^2D^2 - xD + 1)y = 0$; $y(1) = \pi$, $y(4) = 2\pi$
25. $x^2y'' + 2xy' - 2y = 6x$; $y(1) = 1$, $y'(1) = 0$
26. $x^2y'' + 2xy' - 2y = x$; $y(1) = \frac{1}{3}$, $\lim_{x\to 0^+} y(x)$ is finite
27. $x^2y'' - 3xy' + 3y = \ln x$; $y(1) = 1$, $y'(1) = 2$
28. $x^2y'' - 3xy' + 3y = \ln|x|$; $y(-1) = \frac{4}{9}$, $y'(-1) = \frac{1}{3}$
29. $x^2y'' - 2xy' + 2y = x\ln x$; $y(1) = 1$, $y'(1) = 0$
30. $x^2y'' + 2xy' - 6y = 10x^2$; $y(1) = 1$, $y'(1) = -6$
31. $x^3y''' + 2x^2y'' + xy' - y = 15\cos(2\ln x)$; $y(1) = 2$, $y'(1) = -3$, $y''(1) = 0$
32. $(x^3D^3 + 23xD - 75)y = 104x$; $y(-1) = 2$, $y'(-1) = -12$, $y(-e^{\pi/5}) = 2e^{\pi/5}$
33. $x^2y''' + xy'' + 4y' = 4$; $y(1) = \frac{4}{5}$, $y'(1) = y''(1) = 0$
34. $y''' + (3/x)y'' = 6/x$; $2y(1) - y(2) + 1 = 0$, $2y'(1) - y'(2) = 0$, $y''(1) - y''(2) = 0$
35. $4x^3y''' + 4x^2y'' - 5xy' + 2y = 26 + 15x^2$; $y(1) = 13$, $y(e) = 13 + e^2$, $y(0^+)$ is finite
36. Find all values of r for which $y = x^r$ satisfies $5x^2y'' - 79xy' - 17y = 0$ and write a complete solution of the differential equation.
37. Work Exercise 36 for the equation $(x^4D^4 + 6x^3D^3 - 5x^2D^2 - 11xD + 27)y = 0$.
38. Using reduction of order and variation of parameters, find a complete solution of $11x^2y'' - xy' + y = 21x^2$ and verify your result by solving the equation as an Euler-Cauchy equation.
39. Show that the substitution $|Ax + B| = e^z$ or $z = \ln|Ax + B|$ will reduce the equation

$$a(Ax + B)^2\frac{d^2y}{dx^2} + b(Ax + B)\frac{dy}{dx} + cy = 0$$

to a linear equation with constant coefficients. For $n > 2$, can the equation

$$a_0(Ax + B)^n\frac{d^ny}{dx^n} + a_1(Ax + B)^{n-1}\frac{d^{n-1}y}{dx^{n-1}} + \cdots + a_{n-1}(Ax + B)\frac{dy}{dx} + a_ny = 0$$

be solved in a similar fashion?

40. Using the results of Exercise 39, find a complete solution of the equation

$$(x - 2)^2y'' + 2(x - 2)y' - 6y = 0$$

41. The change of variable $|x| = e^z$ transforms the second-order Euler-Cauchy equation

$$(7) \qquad x^2D^2y + axDy + by = 0$$

into the related **constant-coefficient equation**

$$(8) \quad [\mathfrak{D}(\mathfrak{D} - 1) + a\mathfrak{D} + b]y = 0 \qquad \mathfrak{D} = d/dz$$

whose characteristic equation

$$(9) \qquad m(m - 1) + am + b = 0$$

is called the **indicial equation** of (7). Show that each row of Column 2, Table 2.3, provides two linearly independent solutions of (7) for roots of (9) as described in the same row of Column 1.

TABLE 2.3

Nature of the roots	Linearly independent solutions of (7)
Real and unequal $m_1 \neq m_2$	$y_1 = \lvert x\rvert^{m_1}$, $y_2 = \lvert x\rvert^{m_2}$
Real and equal $m_1 = m_2 = m$	$y_1 = \lvert x\rvert^{m}$, $y_2 = \lvert x\rvert^{m}\ln\lvert x\rvert$
Conjugate complex $m_1 = \alpha + i\beta$, $m_2 = \alpha - i\beta$	$y_1 = \lvert x\rvert^{\alpha}\cos(\beta\ln\lvert x\rvert)$, $y_2 = \lvert x\rvert^{\alpha}\sin(\beta\ln\lvert x\rvert)$

2.9 POWER-SERIES SOLUTIONS

Having learned in Sec. 1.11 how to solve general linear first-order differential equations with variable coefficients, and having learned in the last section how to solve Euler-Cauchy-type equations, let us now explore the possibility of solving more general variable-coefficient linear differential equations of order 2 or higher. Since a nonhomogeneous equation can generally be solved whenever the corresponding homogeneous equation is solvable, and because much of the theory of second-order equations can be extended to equations of higher order, we shall (except in a few exercises) consider in detail only the general homogeneous second-order equation

$$a_0(x)y'' + a_1(x)y' + a_2(x)y = 0 \tag{1}$$

In our work, with at most rare exceptions, a_0, a_1, and a_2 will be real-valued functions of a real variable. These restrictions are not necessary, however, because the definitions, theorems, and properties we are about to state can be generalized in terms of complex functions and complex variables.

A clue to the solution process we shall soon develop for solving (1) is found in Theorem 1, Sec. 2.5, which tells us that if the coefficient functions a_0, a_1, and a_2 are constants, then all derivatives of every solution of (1) are continuous on $(-\infty, \infty)$. This being the case, if we are given initial values for $y(0)$ and $y'(0)$, we can evaluate $y''(0)$, $y'''(0)$, . . . , $y^{(n)}(0)$, . . . , by solving the differential equation for y'', then differentiating repeatedly and evaluating successive derivatives at $x = 0$. With these values known, we can construct a Maclaurin series for *the* solution $y(x)$, which has the prescribed initial values, by using the formula we learned in calculus (Exercise 36):

$$y(x) = y(0) + y'(0)x + \cdots + \frac{y^{(n)}(0)}{n!}x^n + \cdots = \sum_{n=0}^{\infty} \frac{y^{(n)}(0)}{n!}x^n \tag{2}$$

An almost identical chain of reasoning leads to the following important conclusion.

THEOREM 1 Every solution of any homogeneous ordinary linear differential equation with constant coefficients is an *entire function.*

DEFINITION 1 A function f is an **entire function** if and only if for all finite values of x it is represented by its Maclaurin series.

In particular, polynomials are entire functions, as are the familiar functions

$$(a)\ e^x = \sum_{n=0}^{\infty} \frac{x^n}{n!} \qquad (b)\ \cos x = \sum_{n=0}^{\infty} \frac{(-1)^n x^{2n}}{(2n)!} \qquad (c)\ \sin x = \sum_{n=0}^{\infty} \frac{(-1)^n x^{2n+1}}{(2n+1)!} \tag{3}$$

whose Maclaurin series we recall from calculus.

The same process of differentiation and evaluation described in the preceding paragraph, when applied to an initial-value problem involving a constant-coefficient equation (1) and specified values of $y(x_0)$ and $y'(x_0)$ at any point $x_0 \neq 0$, leads to a Taylor-series representation of the solution $y(x)$ to the problem:

$$y(x) = y(x_0) + y'(x_0)(x - x_0) + \cdots + \frac{y^{(n)}(x_0)}{n!}(x - x_0)^n + \cdots = \sum_{n=0}^{\infty} \frac{y^{(n)}(x_0)}{n!}(x - x_0)^n \tag{4}$$

EXAMPLE 1

Find the Taylor-series representation of the solution to the equation $y'' - 2y' + y = 0$ determined by the initial conditions $y(1) = 0$, $y'(1) = 1$.

Solving the differential equation for y'', then repeatedly differentiating and evaluating successive derivatives at $x = 1$, we get

$$y''(1) = 2y'(1) - y(1) = 2$$
$$y'''(1) = 2y''(1) - y'(1) = 3$$
$$y^{\text{iv}}(1) = 2y'''(1) - y''(1) = 4$$
$$y^{\text{v}}(1) = 2y^{\text{iv}}(1) - y'''(1) = 5$$
$$\cdots\cdots\cdots$$

and it is easy to prove by induction that $y^{(n)}(1) = n$. Substituting these values into (4), with $x_0 = 1$, we obtain the required Taylor series

$$y(x) = (x-1) + \cdots + \frac{n}{n!}(x-1)^n + \cdots = \sum_{n=0}^{\infty} \frac{n}{n!}(x-1)^n$$

Simplifying coefficients in this series, then changing the index of summation to $m = n - 1$, taking a factor $x - 1$ out of each term, and finally noting the effect of replacing x by $x - 1$ in (3*a*), we have

$$y(x) = \sum_{n=1}^{\infty} \frac{(x-1)^n}{(n-1)!} = \sum_{m=0}^{\infty} \frac{(x-1)^{m+1}}{m!} = (x-1)\sum_{m=0}^{\infty} \frac{(x-1)^m}{m!} = (x-1)e^{x-1}$$

Direct substitution and two simple differentiations show that $(x-1)e^{x-1}$ satisfies $y'' - 2y' + y = 0$ on $(-\infty, \infty)$ and both of the given initial conditions.

Every Maclaurin series is a Taylor series with *center of expansion* $x_0 = 0$, and every Taylor series in powers of $x - x_0$ is a **power series** of the type

$$\sum_{n=0}^{\infty} a_n(x-x_0)^n = a_0 + a_1(x-x_0) + a_2(x-x_0)^2 + a_3(x-x_0)^3 + \cdots \tag{5}$$

with constant coefficients a_n and **center of expansion** x_0. If taken literally, the first term of (5) would read $a_0(x-x_0)^0$; however, we have written it as a_0 because when $(x-x_0)^0$ occurs in a power series, it is always assigned the value 1, even if $x = x_0$.

Dividing the general equation (1) by $a_0(x)$ and renaming coefficients, we obtain its so-called **standard form,** or **normal form**

$$y'' + p(x)y' + q(x)y = 0 \tag{6}$$

which is equivalent to (1) on any interval where $a_0(x) \neq 0$. If $y(x)$ is the solution of (1) on a normal interval I containing x_0, for which the values of $y(x_0)$ and $y'(x_0)$ are prescribed, then from (6),

$$y''(x_0) = -p(x_0)y'(x_0) - q(x_0)y(x_0)$$

Continuing the by now familiar process of solving (6) for y'' and then differentiating, we have formally

$$y''' = -p(x)y'' - [p'(x) + q(x)]y' - q'(x)y$$

and for this to be meaningful, the derivatives of $p(x)$ and $q(x)$ must exist. Up to this point, all we have assumed about the coefficient functions of (1) is that they are continuous and that $a_0(x)$ is not zero on the interval I, and this is not enough to guarantee the existence of even the first derivatives, much less the higher derivatives of $p(x)$ and $q(x)$. Thus if we are to construct a series representation of $y(x)$ by evaluating $y'''(x_0)$ and repeatedly differentiating y''' to eventually find a formula for $y^{(n)}(x_0)$, we must assume that $p(x)$ and $q(x)$ are infinitely differentiable. Fortunately, this assumption is not too severe, for as we shall soon show, it is characteristic of all *analytic* functions.

DEFINITION 2 A function f is **analytic at a point** x_0 if and only if it has a Taylor series in powers of $x - x_0$ which represents it on some open interval containing x_0. A function which is analytic at every point of an interval I is said to be **analytic on** I.

As previously mentioned:

PROPERTY 1 A function which is analytic at a point has derivatives of all orders on an open interval containing that point.

Using properties of power series, we shall also show that:

PROPERTY 2 On an open interval I where they are all analytic, sums, differences, and products of analytic functions are analytic; and the quotient of two analytic functions is analytic provided the divisor is not zero on I.

In the remainder of this section, and in Sec. 2.10, we shall assume that the coefficient functions in (1) are analytic.

Now let I_0 be any open interval, finite or infinite, on which the coefficient functions of Eq. (1) are analytic. Using the leading coefficient $a_0(x)$, we classify each number in I_0 by means of the following criteria.

DEFINITION 3 A number x_0 in I_0 is an **ordinary point** of Eq. (1) if and only if $a_0(x_0) \neq 0$.

DEFINITION 4 A number x_0 in I_0 is a **singular point** of Eq. (1) if and only if $a_0(x_0) = 0$.

Since so much of the work in this and the following section deals with power series, it will be helpful to digress and recall from calculus some of their basic properties. We begin with:

THEOREM 2 For every power series

$$\sum_{n=0}^{\infty} a_n(x - x_0)^n = a_0 + a_1(x - x_0) + a_2(x - x_0)^2 + a_3(x - x_0)^2 + \cdots \tag{5}$$

just one of the following statements is true.

a. The series converges only at $x = x_0$.

b. There is a finite number $R > 0$, called the **radius of convergence,** such that for all values of x, if $|x - x_0| < R$ the series converges absolutely, but if $|x - x_0| > R$ the series diverges.

c. For all values of x, the series converges absolutely, in which case $R = \infty$; that is, the **radius of convergence is infinite.**

The **interval of convergence** of a power series (5) is the open interval $(x_0 - R, x_0 + R)$ if R is finite, and is $(-\infty, \infty)$ if $R = \infty$. The set of all numbers x for which (5) is convergent is the *domain* of a function f represented by

$$f(x) = \sum_{n=0}^{\infty} a_n(x - x_0)^n$$

Thus, for each such x, $f(x)$ stands for the corresponding sum of the series.

Since Theorem 1 tells us that on its interval of convergence a power series converges absolutely, and since the terms of an absolutely convergent series can be rearranged at pleasure, we have:

THEOREM 3 The terms of a power series can be rearranged in any way without affecting the sum of the series anywhere on its interval of convergence.

Power series having the same interval of convergence can be added termwise and combined as linear combinations on that interval.

EXAMPLE 2

Find a power-series representation of $\cosh x$.

By definition, $\cosh x = \frac{1}{2}(e^x + e^{-x})$. Expressing the right-hand member of this identity as a linear combination of e^x and e^{-x}, then using (3a), termwise addition, a rearrangement of terms, and finally simplifying, we have

$$\cosh x = \frac{1}{2}e^x + \frac{1}{2}e^{-x} = \frac{1}{2}\sum_{n=0}^{\infty}\frac{x^n}{n!} + \frac{1}{2}\sum_{n=0}^{\infty}\frac{(-x)^n}{n!} = \sum_{n=0}^{\infty}\frac{x^n}{2(n!)} + \sum_{n=0}^{\infty}\frac{(-x)^n}{2(n!)} = \sum_{n=0}^{\infty}\frac{x^n + (-x)^n}{2(n!)}$$

$$= \sum_{n=0}^{\infty}\frac{1+(-1)^n}{2(n!)}x^n = \sum_{n=0}^{\infty}\frac{1+(-1)^{2n}}{2[(2n)!]}x^{2n} + \sum_{n=0}^{\infty}\frac{1+(-1)^{2n+1}}{2[(2n+1)!]}x^{2n+1}$$

$$= \sum_{n=0}^{\infty}\frac{x^{2n}}{(2n)!} \qquad |x| < \infty$$

The final series is the familiar Maclaurin expansion of the hyperbolic cosine function. A like procedure, starting with $\sinh x = \frac{1}{2}(e^x - e^{-x})$, shows that the hyperbolic sine function is represented by

$$\sinh x = \sum_{n=0}^{\infty}\frac{x^{2n+1}}{(2n+1)!} \qquad |x| < \infty$$

In our work with power series, we will frequently have to multiply them. To make this process as systematic as possible, we first observe that a power series in powers of $x - x_0$ can always be reduced to one in powers of t by the translation $t = x - x_0$. Thus, for simplicity, we shall take $x_0 = 0$ in (5) and restrict our discussion to series in powers of x

$$(7) \qquad f(x) = \sum_{n=0}^{\infty} a_n x^n \qquad |x| < R$$

having $(-R, R)$ as their common interval of convergence, where R may be finite or infinite. Preparatory to defining a product for power series, we next consider two infinite sequences, regarded as functions f and g, whose values for each integer k, namely,

$$(8) \qquad (a)\ f(k) = u_k(x) \qquad \text{and} \qquad (b)\ g(k) = v_k(x) \qquad 0 \le k$$

are functions of x which commute under addition and multiplication on an interval I. Then we introduce the idea of the *convolution* of two such sequences.

DEFINITION 5 The **convolution** of f with g on the range 0 to n is the sum

$$(9) \qquad s_n(x) = \sum_{k=0}^{n} u_k(x)v_{n-k}(x) = u_0(x)v_n(x) + u_1(x)v_{n-1}(x) + \cdots + u_n(x)v_0(x)$$

of all products $u_j(x)v_k(x)$ for which $j + k = n$.

For all $n \geq 0$, the function s_n represented on I by (9) is commonly denoted by $f * g$, with no mention of the range whenever the range is understood. The convolution of g with f, represented by

$$g * f(x) = \sum_{k=0}^{n} v_k(x)u_{n-k}(x) = v_0(x)u_n(x) + v_1(x)u_{n-1}(x) + \cdots + v_n(x)u_0(x)$$

is just $s_n(x)$ with the order of its terms reversed and the factors in each term permuted; hence $f * g = g * f$.

EXAMPLE 3

Find the convolution on the range 0 to n of the sequences $\{a_k x^k\}_0^\infty$ and $\{b_k x^k\}_0^\infty$.

With $f(k) = a_k x^k$ and $g(k) = b_k x^k$, the sum in (9) becomes

$$\sum_{k=0}^{n} a_k x^k b_{n-k} x^{n-k} = \sum_{k=0}^{n} a_k b_{n-k} x^n = x^n \sum_{k=0}^{n} a_k b_{n-k} = x^n a * b$$

The required convolution is thus the product of x^n and the convolution of the sequences of the coefficients $\{a_k\}_0^\infty$ and $\{b_k\}_0^\infty$ on the range 0 to n.

Using Definition 5, we can now define the product of two infinite series as a sum of convolutions.

DEFINITION 6 The **product,** or **Cauchy product,** of two infinite series

$$(a)\ \sum_{n=0}^{\infty} u_n(x) \qquad \text{and} \qquad (b)\ \sum_{n=0}^{\infty} v_n(x)$$

is the sum over n

(10) $$\sum_{n=0}^{\infty} s_n(x) = \sum_{n=0}^{\infty} \sum_{k=0}^{n} u_k(x)v_{n-k}(x)$$

of all convolutions $s_n(x)$ given by (9).

If for each x in I the series (a) and (b) of Definition 6 are absolutely convergent with respective sums $u(x)$ and $v(x)$, it can be shown that (10) is absolutely convergent with sum $u(x)v(x)$†. We then have

(11) $$\left[\sum_{n=0}^{\infty} u_n(x)\right]\left[\sum_{n=0}^{\infty} v_n(x)\right] = \sum_{n=0}^{\infty} \sum_{k=0}^{n} u_k(x)v_{n-k}(x) = u(x)v(x) \qquad x \text{ in } I$$

The **binomial expansion**

(12) $$(a + b)^n = \sum_{k=0}^{n} \binom{n}{k} a^k b^{n-k}$$

containing the so-called **binomial coefficients**

(13) $$\binom{n}{k} = \frac{n!}{k!(n-k)!} \qquad n \geq 0$$

†See, for example, R. Creighton Buck, *Advanced Calculus,* 3d ed., pp. 246–247, McGraw-Hill, New York, 1978.

can be written

$$(a+b)^n = n! \sum_{k=0}^{n} \frac{a^k}{k!} \frac{b^{n-k}}{(n-k)!}$$

wherein the right-hand member is the product of $n!$ and the convolution from 0 to n of

$$f(k) = \frac{a^k}{k!} \qquad \text{and} \qquad g(k) = \frac{b^k}{k!}$$

Setting $a = b = 1$ in (12) and subsequently taking $a = -1 = -b$, we obtain

$$\text{(14)} \qquad (a)\ 2^n = \sum_{k=0}^{n} \binom{n}{k} \quad n \ge 0 \qquad \text{and} \qquad (b)\ 0 = \sum_{k=0}^{n} (-1)^k \binom{n}{k} \quad n \ge 1$$

respectively. These two identities have many practical uses, especially in probabilitity and statistics. With n replaced by $2n + 1$, Eqs. (14) become

$$2^{2n+1} = \binom{2n+1}{0} + \binom{2n+1}{1} + \binom{2n+1}{2} + \cdots + \binom{2n+1}{2n} + \binom{2n+1}{2n+1}$$

$$0 = \binom{2n+1}{0} - \binom{2n+1}{1} + \binom{2n+1}{2} - \cdots + \binom{2n+1}{2n} - \binom{2n+1}{2n+1}$$

Subtracting the second of these equations from the first and dividing the difference through by 2, we obtain

$$\text{(15)} \qquad 2^{2n} = \sum_{k=0}^{n} \binom{2n+1}{2k+1} \qquad n \ge 0$$

This result will be useful to us in our next example.

EXAMPLE 4

By multiplying power-series representations of the sine and cosine functions, show that $2 \sin x \cos x = \sin 2x$.

Substituting the series (3*c*) for $\sin x$, and (3*b*) for $\cos x$, then multiplying the series and simplifying their product by means of (15), we get

$$\begin{aligned} 2 \sin x \cos x &= 2\left[\sum_{n=0}^{\infty} \frac{(-1)^n x^{2n+1}}{(2n+1)!}\right]\left[\sum_{n=0}^{\infty} \frac{(-1)^n x^{2n}}{(2n)!}\right] \\ &= 2 \sum_{n=0}^{\infty} \sum_{k=0}^{n} \frac{(-1)^k x^{2k+1} (-1)^{n-k} x^{2(n-k)}}{(2k+1)![2(n-k)]!} \\ &= 2 \sum_{n=0}^{\infty} (-1)^n x^{2n+1} \sum_{k=0}^{n} \frac{1}{(2k+1)![2(n-k)]!} \\ &= 2 \sum_{n=0}^{\infty} \frac{(-1)^n x^{2n+1}}{(2n+1)!} \sum_{k=0}^{n} \binom{2n+1}{2k+1} = 2 \sum_{n=0}^{\infty} \frac{(-1)^n x^{2n+1}}{(2n+1)!} 2^{2n} \\ &= \sum_{n=0}^{\infty} \frac{(-1)^n (2x)^{2n+1}}{(2n+1)!} = \sin 2x \end{aligned}$$

which completes the required verification.

In addition to the convergence and algebraic properties of power series we have noted, it is important to remember that

THEOREM 4 A power series can be repeatedly differentiated or integrated term by term any number of times on its interval of convergence.

EXAMPLE 5

Use the **geometric series**

(16) $$\frac{1}{1-x} = \sum_{n=0}^{\infty} x^n = 1 + x + x^2 + x^3 + x^4 + x^5 + \cdots \qquad |x| < 1$$

to expand the rational function $4/(2 + x)^3$ as a power series in x.

When written as $4(2 + x)^{-3}$, the rational function to be expanded appears to be related to the second derivative of $(2 + x)^{-1}$; indeed,

$$\frac{4}{(2+x)^3} = 2\frac{d^2}{dx^2}(2+x)^{-1} = \frac{d^2}{dx^2}\left(\frac{1}{1+x/2}\right)$$

Equation (16), with x replaced by $-x/2$, becomes

$$\frac{1}{1+x/2} = \sum_{n=0}^{\infty} \frac{(-1)^n}{2^n} x^n = 1 - \frac{x}{2} + \frac{x^2}{2^2} - \frac{x^3}{2^3} + \frac{x^4}{2^4} - \frac{x^5}{2^5} + \cdots \qquad |x| < 2$$

Differentiating this series termwise twice and then replacing n by $n + 2$ in the resultant series, we find that

$$\frac{4}{(2+x)^3} = \sum_{n=0}^{\infty} \frac{(-1)^n}{2^{n+2}}(n+2)(n+1)x^n = \frac{1}{2} - \frac{3}{4}x + \frac{3}{4}x^2 - \frac{5}{8}x^3 + \cdots \qquad |x| < 2$$

gives the required expansion.

EXAMPLE 6

Expand $\ln(1 - x)$ as a power series in x.

Starting with the integral formula

$$\ln(1-x) = -\int_0^x \frac{dt}{1-t} = \int_x^0 \frac{dt}{1-t} \qquad x < 1$$

we use (16) to convert the latter integrand into a power series in x and then integrate term by term, getting

$$\ln(1-x) = \sum_{n=0}^{\infty} \int_x^0 t^n\, dt = \sum_{n=0}^{\infty} \frac{1}{n+1}[t^{n+1}]_x^0 = -\sum_{n=0}^{\infty} \frac{x^{n+1}}{n+1} \qquad |x| < 1$$

Replacing n by $n - 1$ in the last series, we have

$$\ln(1-x) = -\sum_{n=1}^{\infty} \frac{x^n}{n} = -x - \frac{x^2}{2} - \frac{x^3}{3} - \frac{x^4}{4} - \frac{x^5}{5} - \cdots \qquad |x| < 1$$

as the required expansion. A change of x to $-x$ converts the last expansion into

$$\ln(1+x) = \sum_{n=1}^{\infty} \frac{(-1)^{n-1}}{n} x^n = x - \frac{x^2}{2} + \frac{x^3}{3} - \frac{x^4}{4} + \frac{x^5}{5} - \cdots \qquad |x| < 1$$

From Definition 2, and since every Taylor series is a power series, Properties 1 and 2 now emerge as corollaries of Theorem 4.

Power series having a common interval of convergence are especially nice to work with because within that interval the series behave very much like polynomials. In particular:

THEOREM 5 A power series (5) is identically zero on its interval of convergence if and only if the coefficient of each power of $x - x_0$ is zero; and two such series with the same interval of convergence are identical if and only if coefficients of corresponding powers of $x - x_0$ are equal.

For power series in x whose interval of convergence is I, Theorem 5 gives

(17) $$\sum_{n=0}^{\infty} a_n x^n \equiv 0 \qquad \text{on } I \text{ if and only if } a_n = 0,\ n \geq 0$$

and

(18) $$\sum_{n=0}^{\infty} a_n x^n \equiv \sum_{n=0}^{\infty} b_n x^n \qquad \text{on } I \text{ if and only if } a_n = b_n,\ n \geq 0$$

Various tests are available for finding the radius of convergence of a power series

$$\sum_{n=0}^{\infty} a_n (x - x_0)^n$$

In particular, the ratio test gives

(19) $$R = \lim_{n\to\infty} \left| \frac{a_n}{a_{n+1}} \right|$$

provided the specified limit exists and is not 0. If the limit is $+\infty$, then $R = \infty$. The root test, on the other hand, gives

(20) $$R = \frac{1}{\lim_{n\to\infty} \sqrt[n]{|a_n|}}$$

provided the limit in the denominator exists. If its value is 0, then $R = \infty$.

We are now ready to return to the problem of solving Eq. (1). In this section we shall consider only the problem of finding solutions around ordinary points of (1). In the next section solutions near a singular point will be investigated. Under the assumption that x_0 is an ordinary point of (1), the coefficient functions are analytic at x_0 and $a_0(x_0) \neq 0$. Hence a_0, a_1, and a_2 are continuous, and $a_0(x) \neq 0$ on an open interval I containing x_0. On this interval Eq. (1) is normal and therefore has solutions which, according to Theorem 2, Sec. 2.2, are all linear combinations of a pair of linearly independent solutions. We may choose I such that on this interval both p and q in the standard form

(6) of Eq. (1) will have derivatives of all orders (Properties 1 and 2). On such an interval (1) and (6) are equivalent.

Now let $y(x)$ denote any particular solution of Eq. (6) for which

$$y''(x) \equiv -p(x)y'(x) - q(x)y(x) \qquad x \text{ in } I$$

Since $y''(x)$, $p'(x)$, and $q'(x)$ all exist on I, so does $y'''(x)$. Differentiating $y''(x)$, we have

$$y'''(x) \equiv -p(x)y''(x) - [p'(x) + q(x)]y'(x) - q'(x)y(x) \qquad x \text{ in } I$$

Reasoning with $y'''(x)$, $p''(x)$, and $q''(x)$ in the same way and continuing on inductively, we reach the conclusion that $y(x)$ has derivatives of all orders on I. It follows that $y(x)$ has a *Taylor-series* expansion centered at x_0, which of course is a power series. But we cannot be sure without further proof that this series converges or represents the solution $y(x)$. Nevertheless it does. In fact, *all* power series centered at x_0 which satisfy (6) have the same radius of convergence R, and R is *not less than* the smaller radius of convergence of the Taylor series centered at x_0 which corresponds to p and q.† The equivalence of (1) and (6) then gives us the following fundamental result.

THEOREM 6 At an ordinary point $x = x_0$ of the differential equation

(1) $$a_0(x)y'' + a_1(x)y' + a_2(x)y = 0$$

every solution is analytic and has a power-series representation

$$y(x) = \sum_{n=0}^{\infty} c_n(x - x_0)^n$$

whose radius of convergence R is at least as great as the distance from x_0 to the nearest singular point of (1).

Power series that represent solutions of a differential equation are commonly referred to as **power-series solutions.** More specifically, such a series with center of expansion x_0 is called a **power-series solution around,** or **about,** x_0. Theorem 6 sets a lower bound for the radius of convergence R of each power-series solution of (1) around an ordinary point x_0. The precise value of R may be found by using a suitable convergence test such as the ratio or root test if necessary.

Theorem 6 also guarantees the existence of power-series solutions of (1) about an ordinary point x_0, but it does not tell us how to find them. One familiar way is to compute the successive derivatives of a solution y from the standard form of the given differential equation and then use these derivatives to construct the Taylor series for y. However, generally this process is much too complicated because of the difficult differentiations involved. Thus a better way of solving (1) is needed. We next proceed to outline a method that serves this purpose very nicely. But before doing so, let us first observe that if $x_0 \neq 0$, the change of independent variable $x = t + x_0$ in Eq. (1) leads to a differential equation with analytic coefficient functions having $t = 0$ as an ordinary point and solutions represented by power series in t. These series, with t replaced by $x - x_0$, become solutions of Eq. (1) around x_0. In this way, the problem of solving (1) around a nonzero ordinary point is reduced to that of solving the same kind of equation around the origin. Hence for simplicity we shall suppose $x_0 = 0$.

---

†See, for instance, Earl D. Rainville, *Intermediate Differential Equations,* 2d ed., pp. 67–71, Macmillan, New York, 1964.

To obtain solutions of Eq. (1) around $x = 0$ as an ordinary point, we substitute a power series in x with *undetermined coefficients* c_n

$$\sum_{n=0}^{\infty} c_n x^n$$

into (1) for y, differentiate term by term, equate to zero the coefficients of each power x^n of x, and then solve for the c_n in succession. This procedure is called **the power-series method** or **the method of undetermined coefficients** for power series.

The simpler the coefficient functions in Eq. (1), the easier it is to apply the power-series method. In case all coefficient functions are polynomials, no products of infinite series occur when an infinite power series is substituted into the equation for y. Of course, polynomials are analytic on $(-\infty, \infty)$ since they are represented by finite power series. If the leading coefficient of (1) happens to be a constant, then every real number is an ordinary point of the differential equation and all power-series solutions around such a point converge for all values of x.

EXAMPLE 7

Find a series solution around the origin for the equation

$$y'' + xy' + y = 0 \tag{21}$$

Clearly, $x = 0$, like all other real numbers, is an ordinary point of the differential equation. Therefore, by Theorem 6, every solution around the origin is represented by a power series

$$y = \sum_{n=0}^{\infty} c_n x^n \tag{22}$$

which converges for all values of x. Requiring this series to satisfy Eq. (21), we substitute it for y, then differentiate termwise and find that for all real values of x the equation

$$\sum_{n=2}^{\infty} c_n n(n-1)x^{n-2} + x\sum_{n=1}^{\infty} c_n n x^{n-1} + \sum_{n=0}^{\infty} c_n x^n = 0$$

must hold.

It is our intention to combine these three series into a single series; and for this to be possible, the same power of x must appear in the general term of each series. To achieve this, we move x across the summation sign of the second series and replace the index of summation n by $n - 2$ in the last two series so as to *bring all exponents on x down to the smallest one present in any of the three series.* This strategy accomplishes two important things:

1. It makes the exponent on x the same in all of the series.
2. It makes n the *largest* subscript on c in any of the series.

Thus we have

$$\sum_{n=2}^{\infty} c_n n(n-1)x^{n-2} + \sum_{n=3}^{\infty} c_{n-2}(n-2)x^{n-2} + \sum_{n=2}^{\infty} c_{n-2}x^{n-2} = 0$$

Before we can add different series in powers of x, not only must their general terms involve the same power of x but each series must also have the same range of summation. This is not yet the case with the three series in the last equation since the middle series contains no term corresponding to $n = 2$, as do the other two series.

To obtain a common range, we detach the individual terms that come from the first and last series when $n = 2$, then add them to what is left of all three series. This gives

$$2c_2 + c_0 + \sum_{n=3}^{\infty} c_n n(n-1)x^{n-2} + \sum_{n=3}^{\infty} c_{n-2}(n-2)x^{n-2} + \sum_{n=3}^{\infty} c_{n-2}x^{n-2} = 0$$

Since the last three series have the same range of summation, they can be added term by term. Doing so, then simplifying coefficients of like powers of x, we obtain

$$(2c_2 + c_0) + \sum_{n=3}^{\infty} (n-1)(nc_n + c_{n-2})x^{n-2} = 0$$

Equating the coefficients of each power of x in this power series to zero, we get

(23) $$(a)\ 2c_2 + c_0 = 0 \quad \text{and} \quad (b)\ nc_n + c_{n-2} = 0 \qquad n \geq 3$$

where in writing (23*b*), which is known as the related **recurrence relation,** we have omitted the factor $n - 1$ from the coefficient of x^{n-2} in the last series because $n - 1 \neq 0$ when $n \geq 3$. Solving (23*a*) for c_2, and (23*b*) for the c with the *largest* subscript, we have

(24) $$(a)\ c_2 = -\frac{c_0}{2} \quad \text{and} \quad (b)\ c_n = -\frac{c_{n-2}}{n} \qquad n \geq 3$$

Since the subscripts on c in (24*b*) differ by 2, the subscripts n and $n - 2$ on c must be both odd or both even. In other words, no c with an odd subscript is related to any c whose subscript is even. Starting with (24*a*), then using (24*b*) repeatedly, we list the c_n in two columns, according as n is even or odd, getting in succession

$$c_2 = -\frac{c_0}{2} \qquad c_3 = -\frac{c_1}{3}$$

$$c_4 = -\frac{c_2}{4} \qquad c_5 = -\frac{c_3}{5}$$

$$\cdots\cdots\cdots \qquad \cdots\cdots\cdots$$

$$c_{2k} = -\frac{c_{2k-2}}{2k} \qquad c_{2k+1} = -\frac{c_{2k-1}}{2k+1} \qquad k \geq 1$$

Upon equating the products of the corresponding members in the first column of equations, we obtain

$$c_2 c_4 \cdots c_{2k} = \frac{(-1)^k c_0 c_2 \cdots c_{2k-2}}{2 \cdot 4 \cdots 2k} \qquad k \geq 1$$

Dividing this equation through by the product $c_2 c_4 \cdots c_{2k-2}$ and factoring 2^k out of the denominator, we have

(25) $$c_{2k} = \frac{(-1)^k}{2^k k!} c_0 \qquad k \geq 0,\ c_0 \text{ arbitrary}$$

A similar treatment of the second column of equations yields

$$c_3 c_5 \cdots c_{2k+1} = \frac{(-1)^k c_1 c_3 \cdots c_{2k-1}}{3 \cdot 5 \cdots (2k+1)} \qquad k \geq 1$$

which, when divided through by the product $c_3 c_5 \cdots c_{2k-1}$, gives

$$c_{2k+1} = \frac{(-1)^k}{3 \cdot 5 \cdots (2k+1)} c_1 \qquad k \geq 1$$

To convert the *denominator* of the coefficient of c_1 into a factorial, we multiply both the numerator and denominator of that fraction by $2 \cdot 4 \cdots 2k = 2^k k!$, getting

(26) $$c_{2k+1} = \frac{(-1)^k 2^k k!}{(2k+1)!} c_1 \qquad k \geq 0,\ c_1 \text{ arbitrary}$$

Finally, we write (22) as

$$y = \sum_{k=0}^{\infty} c_{2k} x^{2k} + \sum_{k=0}^{\infty} c_{2k+1} x^{2k+1}$$

and then substitute from (25) and (26) for c_{2k} and c_{2k+1}, respectively, to obtain

(27) $$y = c_0 \sum_{k=0}^{\infty} \frac{(-1)^k}{2^k k!} x^{2k} + c_1 \sum_{k=0}^{\infty} \frac{(-1)^k 2^k k!}{(2k+1)!} x^{2k+1}$$

where c_0 and c_1 are parameters. The functions represented on $(-\infty, \infty)$ by the series

(28) $$(a)\ y_1(x) = \sum_{k=0}^{\infty} \frac{(-1)^k}{2^k k!} x^{2k} \qquad \text{and} \qquad (b)\ y_2(x) = \sum_{k=0}^{\infty} \frac{(-1)^k 2^k k!}{(2k+1)!} x^{2k+1}$$

are not proportional because $y_1(0) = 1$ while $y_2(0) = 0$. They must therefore be fundamental solutions, which makes (27) a complete solution of the given differential equation (21) on $(-\infty, \infty)$.

The formulas (25) and (26) in Example 7 worked out nicely because the coefficient functions in Eq. (21) were not only analytic but also very simple polynomials. Had infinite series been required to represent the coefficient functions, we would have had to multiply those series by the appropriate series for y'', y', and y before collecting coefficients on like powers on x and requiring that they all be zero. Fortunately, in practice, problems this general are not common. Nevertheless, in theory, linear differential equations whose coefficient functions must be expanded as infinite series are of the utmost importance.

Although the series (28*a*) is easily recognized as the Maclaurin expansion of $e^{-x^2/2}$, the function represented by the series (28*b*) is not so easily identified. However, reduction of order gives $y = ue^{-x^2/2}$ as a solution of (21), where $u = \int_0^x e^{t^2/2}\,dt$. Fundamental solutions of (21) are therefore

(29) $$(a)\ e^{-x^2/2} \qquad \text{and} \qquad (b)\ e^{-x^2/2} \int_0^x e^{t^2/2}\,dt$$

We have already observed that $y_2(0) = 0$. Differentiating the series (28*b*) termwise, we find that $y_2'(0) = 1$. The solution represented by (29*b*) also satisfies the initial conditions $y(0) = 0$ and $y'(0) = 1$. Hence, by Theorem 1, Sec. 2.1,

$$y_2(x) = e^{-x^2/2} \int_0^x e^{t^2/2}\,dt$$

Many differential equations have series-type solutions which represent functions that cannot be expressed in terms of the elementary functions of calculus. Historically, new functions defined by such series, whose underlying differential equations occur in a variety of applications, have been given names, studied in detail, and eventually tabulated. This procedure of introducing new functions reflects the fundamental fact that each particular solution of a differential equation can be thought of as a function which under suitable conditions is defined by that equation and its accompanying auxiliary conditions.

EXAMPLE 8

Find a complete solution of the equation

$$y'' + xy = 0 \tag{30}$$

Since all real numbers are ordinary points of the differential equation, every solution around $x = 0$ is represented by a power series

$$y = \sum_{n=0}^{\infty} c_n x^n \qquad |x| < \infty \tag{31}$$

Proceeding as in Example 1 but doing much of the reasoning mentally, we substitute (31) into (30), getting

$$\sum_{n=2}^{\infty} n(n-1)c_n x^{n-2} + \sum_{n=0}^{\infty} c_n x^{n+1} = 0$$

and then

$$\sum_{n=2}^{\infty} n(n-1)c_n x^{n-2} + \sum_{n=3}^{\infty} c_{n-3} x^{n-2} = 0$$

which may be written

$$2c_2 + \sum_{n=3}^{\infty} [n(n-1)c_n + c_{n-3}]\, x^{n-2} = 0$$

Because these equations are to hold for all real values of x, the property (17) applies to the last equation, giving

$$2c_2 = 0 \qquad \text{and} \qquad n(n-1)c_n + c_{n-3} = 0 \qquad n \geq 3$$

Solving for c_2 and c_n, we have

$$(a)\ c_2 = 0 \qquad \text{and} \qquad (b)\ c_n = -\frac{c_{n-3}}{n(n-1)} \qquad n \geq 3 \tag{32}$$

Whereas the subscripts on c in (24b) differed by 2, in (32b) they differ by 3. Thus we display the c_n in three columns according as n is 0, 1, or 2 more than a multiple of 3.

$$c_3 = -\frac{c_0}{3 \cdot 2} \qquad c_4 = -\frac{c_1}{4 \cdot 3} \qquad c_5 = -\frac{c_2}{5 \cdot 4}$$

$$c_6 = -\frac{c_3}{6 \cdot 5} \qquad c_7 = -\frac{c_4}{7 \cdot 6} \qquad c_8 = -\frac{c_5}{8 \cdot 7}$$

$$\cdots\cdots \qquad \cdots\cdots \qquad \cdots\cdots$$

$$c_{3k} = -\frac{c_{3k-3}}{3k(3k-1)} \qquad c_{3k+1} = -\frac{c_{3k-2}}{(3k+1)(3k)} \qquad c_{3k+2} = -\frac{c_{3k-1}}{(3k+2)(3k+1)}$$

Equating the products of corresponding members in each column, we find that for $k \geq 1$,

$$(a)\ c_{3k} = \frac{(-1)^k c_0}{3^k k![2 \cdot 5 \cdots (3k-1)]} \qquad (b)\ c_{3k+1} = \frac{(-1)^k c_1}{3^k k![4 \cdot 7 \cdots (3k+1)]} \qquad (c)\ c_{3k+2} = 0 \tag{33}$$

Making allowance for each type of subscript, we now write (31) as

$$y = \sum_{k=0}^{\infty} c_{3k} x^{3k} + \sum_{k=0}^{\infty} c_{3k+1} x^{3k+1} + \sum_{k=0}^{\infty} c_{3k+2} x^{3k+2}$$

and then delete the final series because, for $k \geq 0$, $c_{3k+2} = 0$. Writing the first terms of the remaining series and substituting from (33), we obtain

$$y = c_0\left\{1 + \sum_{k=1}^{\infty} \frac{(-1)^k x^{3k}}{3^k k![2 \cdot 5 \cdots (3k-1)]}\right\} + c_1\left\{x + \sum_{k=1}^{\infty} \frac{(-1)^k x^{3k+1}}{3^k k![4 \cdot 7 \cdots (3k+1)]}\right\} \tag{34}$$

as a complete solution of Eq. (30) on $(-\infty, \infty)$.

For all values of x, the kth term of each series representing a particular solution of the differential equations in Examples 7 and 8 tends to zero as $k \to \infty$, and the series itself converges. However, as we know, a power series is most useful for computational purposes when x is near its center of expansion.

EXAMPLE 9

Find a fifth-degree polynomial approximation to the solution of the initial-value problem

$$(x-1)y'' + xy' + y = 0 \qquad y(0) = 2,\ y'(0) = -1 \tag{35}$$

We first note where the initial values of y and y' are prescribed and in accordance therewith require

$$y = \sum_{n=0}^{\infty} c_n x^n \tag{36}$$

to satisfy the given differential equation around the ordinary point $x = 0$. This gives us

$$\sum_{n=2}^{\infty} n(n-1)c_n x^{n-1} - \sum_{n=2}^{\infty} n(n-1)c_n x^{n-2} + \sum_{n=1}^{\infty} nc_n x^n + \sum_{n=0}^{\infty} c_n x^n = 0$$

Bringing all exponents on x down to $n-2$ and making the usual provisions for adding series over a common range of summation, we obtain

$$c_0 - 2c_2 + \sum_{n=3}^{\infty} [nc_n - (n-2)c_{n-1} - c_{n-2}](1-n)x^{n-2} = 0$$

and then

$$(a)\ c_0 - 2c_2 = 0 \qquad \text{and} \qquad (b)\ nc_n - (n-2)c_{n-1} - c_{n-2} = 0 \qquad n \geq 3 \tag{37}$$

Solving (37a) for c_2 and the recurrence relation (37b) for c_n yields

$$(a)\ c_2 = \frac{c_0}{2} \qquad \text{and} \qquad (b)\ c_n = \frac{(n-2)c_{n-1} + c_{n-2}}{n} \qquad n \geq 3 \tag{38}$$

Setting $n = 3$ in (38b), then substituting from (38a), we get

$$c_3 = \frac{c_2 + c_1}{3} = \frac{1}{3}\frac{c_0}{2} + \frac{1}{3}c_1 = \frac{1}{6}c_0 + \frac{1}{3}c_1$$

In like manner, we find

$$c_4 = \frac{2c_3 + c_2}{4} = \frac{5}{24}c_0 + \frac{1}{6}c_1 \qquad \text{and} \qquad c_5 = \frac{3c_4 + c_3}{5} = \frac{19}{120}c_0 + \frac{1}{6}c_1$$

These are all of the coefficients of (36) we need in order to find a polynomial of the kind asked for. In fact, by expanding (36) as

$$y = c_0 + c_1x + c_2x^2 + c_3x^3 + c_4x^4 + c_5x^5 + \cdots$$

and then using the preceding formulas to express c_2 through c_5 in terms of c_0 and c_1, we have, after collecting coefficients on c_0 and c_1,

$$y = c_0\left(1 + \frac{x^2}{2} + \frac{x^3}{6} + \frac{5x^4}{24} + \frac{19x^5}{120} + \cdots\right) + c_1\left(x + \frac{x^3}{3} + \frac{x^4}{6} + \frac{x^5}{6} + \cdots\right) \tag{39}$$

which gives the terms up through x^5 of a complete power-series solution in powers of x to the original differential equation. The radius of convergence of each particular solution, indicated within the parentheses of (39), is not less than 1 because each of the coefficient functions

$$p(x) = \frac{x}{x-1} \qquad \text{and} \qquad q(x) = \frac{1}{x-1}$$

in the standard form of the differential equation in (35) has a Maclaurin series that converges for $|x| < 1$ (see Exercises 24 and 70).

Imposing the initial conditions on the complete solution (39) and its derivative, we get

$$y(0) = 2 = c_0 \qquad \text{and} \qquad y'(0) = -1 = c_1$$

Using these values of c_0 and c_1 in (39) and then collecting coefficients of like powers of x, we finally arrive at

$$y = 2 - x + x^2 + \frac{x^4}{4} + \frac{3x^5}{20} \tag{40}$$

as a fifth-degree polynomial approximation to the solution of the initial-value problem.

A recurrence relation between the undetermined coefficients c_n of a proposed power-series solution which involves exactly m different subscripts on c is sometimes called an **m-term recurrence relation.** Thus, those of Examples 7 and 8 and that of Example 9 are, respectively, *two-term* and *three-term* recurrence relations.

EXERCISES

In Exercises 1–4, evaluate each convolution on the range 0 to n.

1. $f * f$, where $f(k) = e^{kx}$, $0 \le k$

2. $f * g$, where $f(k) = 1$ and $g(k) = k$, $0 \le k$

3. $g * g$, where $g(k) = k$, $0 \le k$. *Hint:* $\sum_{k=1}^{n} k^2 = \frac{k(k+1)(2k+1)}{6}$.

4. $f * g$, where $g(k) = \begin{cases} \frac{1}{n+2-k} & 0 \le k \le n \\ 0 & n \le k \end{cases}$ and $f(k) = \frac{1}{k+1}$

Hint: The convolution telescopes.

5. Prove that for $1 \le k \le n$, $\binom{n}{k} + \binom{n}{k-1} = \binom{n+1}{k}$

6. Show that for $0 \le k \le m+n$, $\sum_{j=0}^{k} \binom{m}{j}\binom{n}{k-j} = \binom{m+n}{k}$. *Hint:* Compare the coefficients of t^k in each member of $(1+t)^m(1+t)^n = (1+t)^{m+n}$.

7. Evaluate $\sum_{k=0}^{n} \binom{2n}{2k} - \sum_{k=0}^{n-1} \binom{2n}{2k+1}$

8. Show that for $n \ge 1$, $\sum_{k=0}^{n} \binom{2n}{2k} = 2^{2n-1}$.

9. Multiply $\sum_{n=0}^{\infty} \frac{a^n}{n!}$ and $\sum_{n=0}^{\infty} \frac{(b-a)^n}{n!}$; then use (12) to show that $e^a e^{b-a} = e^b$.

Using power series and Cauchy products, verify each of the following identities.

10. $2 \sinh x \cosh x = \sinh 2x$
11. $e^{ax}e^{bx} = e^{(a+b)x}$
12. $2\cos^2 x = 1 + \cos 2x$. *Hint:* Use Exercise 8.
13. $e^x e^{2x} e^{3x} = e^{6x}$
14. $\cos^2 x + \sin^2 x = 1$. *Hint:* Use Exercise 7.
15. $\cosh^2 x - \sinh^2 x = 1$

Using power series and termwise differentiation, verify each of the derivative formulas.

16. $\frac{d}{dx}e^{ax} = ae^{ax}$
17. $\frac{d}{dx}\sin ax = a\cos ax$
18. $\frac{d}{dx}\cosh ax = a\sinh ax$
19. $\frac{d}{dx}e^{x^3} = 3x^2 e^{x^3}$
20. Define the convolution of two *finite* sequences of functions

$$u_k = u_k(x) \qquad \text{and} \qquad v_k = v_k(x) \qquad 0 \le k \le n,\ x \text{ in } I$$

21. In Example 7, evaluate the wronskian of $y_1(x)$ and $y_2(x)$ at $x = 0$.
22. By direct substitution, show that both $e^{-x^2/2}$ and the series (28*a*) satisfy Eq. (21) on $(-\infty, \infty)$ and the initial conditions $y(0) = 1$, $y'(0) = 0$. What does this imply?
23. Using reduction of order, derive (29*b*) as a solution of Eq. (21).
24. Write Maclaurin series representing $x/(x-1)$ and $1/(x-1)$ on $(-1, 1)$.
25. Show that the Cauchy product of the two power series in x representing $1/(1+x)$ and $1/(1-x)$ reduces to the Maclaurin expansion of $1/(1-x^2)$.

$$Hint: \sum_{k=0}^{n} (-1)^k = \begin{cases} 1 & n \text{ even} \\ 0 & n \text{ odd} \end{cases}$$

26. Expand $1/x$ as a power series in powers of $x - 1$; then integrate termwise to find a power-series representation of $\ln x$.
27. Expand $1/(1+x^2)$ as a power series in terms of x; then integrate termwise to find a power-series representation of $\text{Tan}^{-1} x$.
28. Expand $e^{x^2/2}$ as a power series in x; then integrate termwise to find a power series representing $\int_0^x e^{t^2/2}\,dt$.

Find the radius of convergence of each of the power series.

29. $\sum_{n=0}^{\infty} x^n/(n+3)^2$
30. $\sum_{n=0}^{\infty} n^2(x-1)^n/2^n$
31. $\sum_{n=0}^{\infty} n!\,x^n/n^n$
32. $\sum_{n=0}^{\infty} \left[\frac{n+2}{n+1}\right]^{(n+1)^2} x^{n+1}$

Using the power-series method, find complete solutions around the origin for each of the equations.

33. $y' - \lambda y = 0$
34. $y'' + \lambda^2 y = 0$
35. $y'' - \lambda^2 y = 0$
36. **(a)** Solve the initial-value problem $y'' - xy = 0$; $y(0) = 1$, $y'(0) = 0$, by repeatedly differentiating the equation to obtain $y'''(0), y^{\text{iv}}(0), y^{\text{v}}(0), \ldots$, and then using the Maclaurin formula $c_n = y^{(n)}(0)/n!$.
(b) Using the method suggested in Part **(a)**, find the solution of $y'' - xy = 0$ for which $y(0) = 0$ and $y'(0) = 1$.
37. Using the method suggested in Exercise 36, find solutions of the equation $y'' - 2y' + y = 0$ for which **(a)** $y(0) = 1$ and $y'(0) = 0$, and **(b)** $y(0) = 0$ and $y'(0) = 1$.

Find all power-series solutions around the origin for each of the equations.

38. $y'' - 2xy' = 0$
39. $y'' + xy' + 2y = 0$
40. $y'' - x^2 y = 0$
41. $(25 - x^2)y'' + 2y = 0$
42. $(x^2 + 4)y'' + 6xy' + 4y = 0$
43. $(1 - x^2)y'' - 5xy' - 3y = 0$
44. $(x^2 + 9)y'' - 4xy' + 6y = 0$
45. $(16 - x^4)y'' + 12x^2 y = 0$
46. $y'' + x^2 y' + 3xy = 0$
47. $(1 + x^2)y'' - 2y = 0$
48. $(1 - x^3)y'' - x^2 y' + 9xy = 0$
49. $(1 + x^2)y'' + xy' - y = 0$
50. $y''' + x^2 y'' + 4xy' + 2y = 0$
51. $y''' + x^2 y'' + 5xy' + 3y = 0$

Using the power-series method, solve each of the initial-value problems.

52. $(x^2 - 2)y'' - 2xy' + 2y = 0$; $y(0) = 4$, $y'(0) = 3$
53. $(1 + x^2)y'' - 4xy' + 6y = 0$; $y(0) = 1$, $y'(0) = 6$
54. $y'' + x^2 y' + 2xy = 0$; $y(0) = y'(0) = 1$
55. $(1 - x^2)y'' + xy' - y = 0$; $y(0) = 1$, $y'(0) = -1$
56. Using the *product* notation $\Pi_{m=1}^k a_k = a_1 a_2 \cdots a_k$, show that (33*a*) and (33*b*), Example 8, can be written

$$c_{3k} = \left[\frac{(-1)^k}{(3k)!} \prod_{m=1}^{k} (3m - 2)\right] c_0$$

$$c_{3k+1} = \left[\frac{(-1)^k}{(3k+1)!} \prod_{m=1}^{k} (3m - 1)\right] c_1$$

57. Find the Cauchy product of the Maclaurin series for $e^{-x^2/2}$ and $\int_0^x e^{t^2/2}\,dt$. *Hint:* See Exercise 28.
58. Show that for every nonnegative integer n,

$$\sum_{k=0}^{n} \frac{(-1)^k}{(2k+1)} \binom{n}{k} = \frac{2^{2n}(n!)^2}{(2n+1)!}. \qquad Hint: \text{ Equate the}$$

Cauchy product of Exercise 57 to the series (28*b*), then adjust the indices of summation and apply (18).

Find the first four nonzero terms of two linearly independent power-series solutions around the origin for each of the equations.

59. $(2x + 1)y'' + y' + 2y = 0$
60. $y'' + xy' + (3x + 2)y = 0$

61. $(x^2 - 1)y'' + 3xy' + xy = 0$
62. $(1 - x^3)y'' - xy' + 12xy = 0$
63. $y'' + xy' + e^x y = 0$
64. $y'' + y' + [1/(1 - x)]y = 0$

Using the power-series method, find a complete solution for each of the following equations about the indicated point x_0.

65. $(3 - 4x + x^2)y'' - 6y = 0;\ x_0 = 2$
66. $y'' - (x + 1)^2 y' + 2(x + 1)y = 0;\ x_0 = -1$
67. $y'' - (x + 3)y' - 3y = 0;\ x_0 = -3$
68. $(2x - x^2)y'' - 5(x - 1)y' - 4y = 0;\ x_0 = 1$
69. Extend Definition 4 to include complex singular points.
70. In Theorem 6, a *lower* bound for the radius of convergence R is given by the distance d from x_0 to the nearest (complex) singular point (Exercise 69) of the differential equation. Using this criterion, find a lower bound for the radius of convergence of all series solutions of $(1 + x^2)y'' - 2xy' + 3y = 0$ about the ordinary point (**a**) $x_0 = 2$, and (**b**) $x_0 = -\sqrt{3}$.
71. Work Example 9 by substituting into Eq. (35) the most general fifth-degree polynomial satisfying the given initial conditions.

2.10 THE METHOD OF FROBENIUS

All real numbers except the singular point $x_0 = 0$ are ordinary points of the second-order Euler-Cauchy equation

$$x^2y'' + axy' + by = 0$$

whose normal form involves

$$p(x) = \frac{ax}{x^2} \quad \text{and} \quad q(x) = \frac{b}{x^2}$$

as coefficient functions, both of which are undefined and therefore discontinuous at $x = 0$. Each of the functions defined by $xp(x)$ and $x^2q(x)$ also has a discontinuity at $x = 0$, but in each case it is *removable.*

DEFINITION 1 A function f has a **removable discontinuity** at $x = x_0$ if and only if $\lim_{x \to x_0} f(x) = L$ and either $f(x_0) \neq L$ or else $f(x_0)$ is undefined.

Such a discontinuity of f can be removed by defining $f(x_0) = L$. Thus the functions represented by $xp(x)$ and $x^2q(x)$, if given the respective values

$$a = \lim_{x \to 0} xp(x) \quad \text{and} \quad b = \lim_{x \to 0} x^2q(x)$$

at $x = 0$, become continuous there. For convenience, we shall repeatedly refer to and treat functions, all of whose discontinuities are removable, as continuous functions since for all practical purposes they behave as though they were continuous.

Now, let I_0 be an open interval on which each coefficient function of

$$a_0(x)y'' + a_1(x)y' + a_2(x)y = 0 \tag{1}$$

is analytic and suppose the real number x_0 in I_0 is a singular point of (1). The corresponding standard equation

$$y'' + p(x)y' + q(x)y = 0 \tag{2}$$

has coefficient functions

$$(a)\ p(x) = \frac{a_1(x)}{a_0(x)} \quad \text{and} \quad (b)\ q(x) = \frac{a_2(x)}{a_0(x)} \tag{3}$$

both of which are undefined at $x = x_0$, as are the two functions represented by

$$(4) \qquad (a)\ (x - x_0)p(x) \quad \text{and} \quad (b)\ (x - x_0)^2 q(x)$$

whose discontinuities at x_0 may or may not be simultaneously removable. Depending on the behavior of (4*a*) and (4*b*) near x_0, the singular point itself will be one or the other of two possible kinds.

DEFINITION 2 A singular point x_0 of Eq. (1) is **regular** if and only if the functions represented by (4*a*) and (4*b*) both have removable discontinuities at x_0 and become analytic there when these discontinuities are removed.

DEFINITION 3 A singular point x_0 of Eq. (1) is **irregular** if and only if it is not regular, in which case at least one of the products (4*a*) and (4*b*) cannot be used to define a function that is analytic at x_0.

EXAMPLE 1

Find and classify the singular points of $x(x - 1)^3 y'' + 2(x - 1)^3 y' + 3y = 0$.

By definition, the singular points of the equation are the zeros $x = 0$ and $x = 1$ of the leading coefficient function $a_0(x) = x(x - 1)^3$. From the normal form of the differential equation

$$y'' + \frac{2(x-1)^3}{x(x-1)^3} y' + \frac{3}{x(x-1)^3} y = 0$$

we have

$$p(x) = \frac{2(x-1)^3}{x(x-1)^3} \quad \text{and} \quad q(x) = \frac{3}{x(x-1)^3}$$

The products $xp(x)$ and $x^2 q(x)$ have removable discontinuities at $x = 0, 1$ and at $x = 0$, respectively. Removing these, we obtain

$$xp(x) = 2 \quad \text{and} \quad x^2 q(x) = \frac{3x}{(x-1)^3}$$

each of which represents an analytic function at $x = 0$. Hence $x = 0$ is a regular singular point.

The function defined by $(x - 1)p(x)$ has a removable discontinuity at $x = 1$. When assigned the value 0 there, it becomes analytic at that point and is represented by

$$(x-1)p(x) = \frac{2(x-1)}{x} \qquad x \neq 0$$

On the other hand, the product

$$(x-1)^2 q(x) = \frac{3}{x(x-1)}$$

has no limit as $x \to 1$; consequently, its discontinuity at $x = 1$ cannot be removed. Since a function must be continuous wherever it is analytic, it follows that $x = 1$ is an irregular singular point.

The singular point $x = -1$ of

$$(5) \qquad (x^2 - 1)y'' + (x + 1)y = 0$$

can be removed by dividing the equation through by $x + 1$. Equation (5) is equivalent to the pair of equations

$$(x - 1)y'' + y = 0 \qquad \text{and} \qquad x + 1 = 0$$

The solution $x = -1$ of the algebraic equation is of no interest as a solution of (5) because the unknown in (5) is the dependent variable y, not x.

In general, it is said that a singular point x_0 of Eq. (1) is **removable** if it can be eliminated by dividing the differential equation through by a positive integral power of $x - x_0$. Hereafter we shall suppose Eq. (1) has no removable singular points or that they have all been removed from the equation to start with. Conditions and procedures for solving Eq. (1) near or about an irregular singular point are quite complicated† and are not needed in our later work. Thus we shall continue our discussion of linear differential equations under the assumption that x_0 is a *regular* singular point of (1). For simplicity, x_0 will be taken to be 0; as was explained in Sec. 2.9, a translation reduces the general case to this one.

The second-order Euler-Cauchy equation is an instructive prototype of all such differential equations. We have seen (Sec. 2.8) that it has a nontrivial particular solution of the form

$$y = |x|^r f(x) \qquad r \text{ real} \tag{6}$$

corresponding to each root of the related indicial equation, where $f(x) \equiv 1$ or else $f(x)$ involves $\ln |x|$. None of these solutions can be a power of x unless $f(x) \equiv 1$. Even then, the regular singular point $x = 0$ need not be in any solution interval. If it is not, initial conditions at the origin cannot be satisfied. It is customary to refer to a solution of Eq. (1) over an interval which has 0 as either an interior or a boundary point as a solution **near the origin.**

Clearly, power series alone are inadequate for the solution of (1) near a regular singular point. However, in view of (6) and our experience with power series, we might anticipate that (1) has a solution represented by a more general series

$$\sum_{n=0}^{\infty} a_n x^{n+r} = a_0 x^r + a_1 x^{1+r} + a_2 x^{2+r} + a_3 x^{3+r} + \cdots \qquad a_0 \neq 0 \tag{7}$$

near the origin and where $x > 0$. The number r need not be zero or a positive integer; but unless it is, the solution, if it exists, will not have a Maclaurin series. The restriction $a_0 \neq 0$ simply means that x^r is the *highest* power of x that can be factored out of each term of the series. Were $a_0 = 0$, a still higher power of x could be removed as a factor. Should a solution of (1) be required where $x < 0$, we could set $t = -x$, use a series (7) in powers of t to solve the resulting differential equation where $t > 0$, then replace t by $-x$ to obtain a series

$$\sum_{n=0}^{\infty} a_n(-x)^{n+r} = a_0(-x)^r + a_1(-x)^{1+r} + a_2(-x)^{2+r} + \cdots \qquad a_0 \neq 0 \tag{8}$$

representing the required solution. The two series (7) and (8) can be consolidated as one. To do so, we set $a_n = c_n$ in (7) and factor out x^r; likewise, we set $(-1)^n a_n = c_n$ in (8) and factor out $(-x)^r$. Both results are given by

$$|x|^r \sum_{n=0}^{\infty} c_n x^n \qquad c_0 \neq 0 \tag{9}$$

in which r and the c_n are parameters. By name, (9) is known as a **Frobenius series.**

---

†See, for instance, E. L. Ince, *Ordinary Differential Equations,* pp. 417–437, Dover, New York, 1956.

The strategy for determining r and all the c_n such that (9) will satisfy Eq. (1) is the same whether we are working where $x < 0$ or $x > 0$. We first replace $|x|$ in (9) by either $-x$ or x, according as x is negative or positive. Then to find a solution of (1), we proceed to

1. Substitute the appropriate form of (9) into (1) for y.
2. Perform the indicated differentiations.
3. Equate to zero the coefficient of each power of x.
4. Determine numerical values for r and the c_n.

This procedure is called the **method of Frobenius.**† Let us look at some of the important details involved. Nothing essential to the process is lost by assuming, as we now suppose, that x is positive.

Since $x = 0$ is a regular singular point of (1), Definition (2) tells us that both $xp(x)$ and $x^2q(x)$ are analytic at the origin and therefore have Maclaurin expansions.

$$(10) \qquad (a)\ xp(x) = \sum_{n=0}^{\infty} p_n x^n \quad \text{and} \quad (b)\ x^2q(x) = \sum_{n=0}^{\infty} q_n x^n$$

Our proposed solution

$$(11) \qquad y = \sum_{n=0}^{\infty} c_n x^{n+r} \qquad c_0 \neq 0$$

of Eq. (1) must also satisfy Eq. (2). Multiplying (2) through by x^2 and substituting the series (10a) and (10b) for $xp(x)$ and $x^2q(x)$, we have

$$(12) \qquad x^2y'' + \left(\sum_{n=0}^{\infty} p_n x^n\right)xy' + \left(\sum_{n=0}^{\infty} q_n x^n\right)y = 0$$

Using (11) and the pair of relations

$$(13) \quad (a)\ x^2y'' = \sum_{n=0}^{\infty} (n+r)(n+r-1)c_n x^{n+r} \quad \text{and} \quad (b)\ xy' = \sum_{n=0}^{\infty} (n+r)c_n x^{n+r}$$

we rewrite (12) as

$$\sum_{n=0}^{\infty} (n+r)(n+r-1)c_n x^{n+r} + \left(\sum_{n=0}^{\infty} p_n x^n\right)\left[\sum_{n=0}^{\infty} (n+r)c_n x^{n+r}\right] + \left(\sum_{n=0}^{\infty} q_n x^n\right)\left(\sum_{n=0}^{\infty} c_n x^{n+r}\right) = 0$$

We next replace each of the last two series products by their respective Cauchy products, to obtain

$$\sum_{n=0}^{\infty} (n+r)(n+r-1)c_n x^{n+r} + \sum_{n=0}^{\infty}\left[\sum_{k=0}^{n} (k+r)c_k p_{n-k}\right]x^{n+r} + \sum_{n=0}^{\infty}\left(\sum_{k=0}^{n} c_k q_{n-k}\right)x^{n+r} = 0$$

Finally, collecting coefficients on x^{n+r}, then canceling the nonzero factor x^r, we have

$$(14) \qquad \sum_{n=0}^{\infty}\left[(n+r)(n+r-1)c_n + \sum_{k=0}^{n} (k+r)c_k p_{n-k} + \sum_{k=0}^{n} c_k q_{n-k}\right]x^n = 0$$

†Named after the German mathematician Ferdinand Georg Frobenius (1849–1917).

which will hold on a subinterval of $(0, \infty)$ if and only if, for $n \geq 0$,

$$(n + r)(n + r - 1)c_n + \sum_{k=0}^{n} [(k + r)p_{n-k} + q_{n-k}]c_k = 0 \tag{15}$$

Since $c_0 \neq 0$, it follows from the first of these equations that

$$r^2 + (p_0 - 1)r + q_0 = 0 \tag{16}$$

This quadratic equation in r is known as the **indicial equation** of (1) relative to the singular point $x = 0$, and its roots r_1 and r_2, which might be equal, are known as the **exponents** of the regular singular point. To distinguish exponents, we suppose that $\text{Re}(r_1) \geq \text{Re}(r_2)$. In practice we use the limits

$$(a)\ p_0 = \lim_{x \to 0} xp(x) \qquad \text{and} \qquad (b)\ q_0 = \lim_{x \to 0} x^2 q(x) \tag{17}$$

to write Eq. (16). If (11) is to yield a solution of (1), the value of r must be a root of (16). In other words, r must be a zero of the function

$$h(r) = r^2 + (p_0 - 1)r + q_0 \tag{18}$$

Now, returning to Eq. (15), we add the term of the series for which k equals the upper limit of summation n to the first term, then collect coefficients on c_n, while retaining the rest of the series, to write

$$[(n + r)^2 + (p_0 - 1)(n + r) + q_0]c_n + \sum_{k=0}^{n-1} [(k + r)p_{n-k} + q_{n-k}]c_k = 0 \qquad n \geq 1 \tag{19}$$

This is the **recurrence relation.** The coefficient of c_n is the function $h(n + r)$ and has the value zero if and only if $n + r = r_1$ or $n + r = r_2$. Neither of these conditions is satisfied if $r = r_1$ and n is a positive integer. Hence $h(n + r_1) \neq 0$ and, with $r = r_1$, (19) can be solved for c_n. Doing this, we get

$$c_n = \frac{-1}{h(n + r_1)} \sum_{k=0}^{n-1} [(k + r_1)p_{n-k} + q_{n-k}]c_k \qquad n \geq 1 \tag{20}$$

which ultimately determines every coefficient c_n of (11), from c_1 on, in terms of c_0. However, at present, we have no guarantee that the series thus derived converges on a subinterval $0 < x < R$ of $(0, \infty)$. If it does, then on that subinterval the series (11), with $r = r_1$ and the c_n expressed in terms of c_0, represents a solution of Eq. (1). This same series yields a solution of (1) on $-R < x < 0$ when x^{r_1} is replaced by $-x^{r_1}$ (see Exercise 12). With the c_n determined by (20), both solutions are given by

$$y = |x|^{r_1} \sum_{n=0}^{\infty} c_n x^n \qquad 0 < |x| < R \tag{21}$$

and are generally referred to as a **solution of** (1) **on the deleted neighborhood** $0 < |x| < R$ **of the origin.**

DEFINITION 4 A **deleted neighborhood** $0 < |x - x_0| < R$ **of** x_0 consists of all numbers in the interval $(x_0 - R, x_0 + R)$ except $x = x_0$.

To find a solution of (1) that is linearly independent of (21), our first thought might be to use the second root of the indicial equation r_2 in the recurrence relation (19) (but with new parameters d_n) and then proceed with the Frobenius method by applying it to

$$h(n + r_2)d_n + \sum_{k=0}^{n-1} [(k + r_2)p_{n-k} + q_{n-k}]d_k = 0 \qquad n \geq 1 \tag{22}$$

If $r_1 - r_2 = 0$, so that Eq. (16) has a double root, the method simply yields the general solution defined by (21). If $r_1 - r_2 = m$, where m is a positive integer, then $h(m + r_2) = h(r_1) = 0$ and, for $n = m$, Eq. (22) reads

$$0 \cdot d_m = \sum_{k=0}^{m-1} [(k + r_2)p_{m-k} + q_{m-k}]d_k \tag{23}$$

If this relation holds, the use of (22) leads to a complete solution of (1) involving d_0 and d_m as parameters. But if the sum of the series in (23) is not zero, the method of Frobenius cannot be carried through to a solution.

If $r_1 - r_2 \neq m$, i.e., *whenever the roots of the indicial equation do not differ by an integer,* (22) can be solved for d_n and a solution of (1) linearly independent of (21) found near $x = 0$. Setting $c_0 = 1$ in (21) and $d_0 = 1$ in the general solution obtained when $r_1 - r_2 \neq m$ then gives too linearly independent particular solutions y_1 and y_2, and the linear combination

$$y = c_0 y_1 + d_0 y_2 \qquad c_0, d_0 \text{ parameters} \tag{24}$$

gives a complete solution of Eq. (1) near $x = 0$.

Once y_1 is known, no matter what the value of $r_1 - r_2$ may be, a linearly independent solution y_2 can be found by the method of Sec. 2.4, that is, by setting $y_2 = uy_1$ and then determining u such that y_2 will satisfy Eq. (1). Since the roots of (16) need not be real, the solutions y_1 and y_2 may be complex. Nevertheless, a study of linear differential equations in the complex domain confirms that (24) still represents a complete solution of (1) near the origin. Such a study also yields the following remarkable result† which details a procedure for finding a basis for all solutions of (1) on a deleted neighborhood of the origin in each possible case: $r_1 - r_2 \neq m$, $r_1 - r_2 = m$, and $r_1 - r_2 = 0$.

THEOREM 1 If the differential equation

$$a_0(x)y'' + a_1(x)y' + a_2(x)y = 0 \tag{1}$$

whose standard form is

$$y'' + p(x)y' + q(x)y = 0 \tag{2}$$

has analytic coefficient functions and a regular singular point at $x = 0$, so that $xp(x)$ and $x^2q(x)$ have valid Maclaurin series

$$xp(x) = \sum_{n=0}^{\infty} p_n x^n \quad |x| < R_1 \qquad \text{and} \qquad x^2q(x) = \sum_{n=0}^{\infty} q_n x^n \quad |x| < R_2$$

†See, for instance, Earl D. Rainville, *Intermediate Differential Equations,* 2d ed., pp. 82–116, Macmillan, New York, 1964.

and if the roots r_1 and r_2 of the indicial equation

$$r^2 + (p_0 - 1)r + q_0 = 0$$

are identified such that $\text{Re}(r_1) \geq \text{Re}(r_2)$, then Eq. (1) has a complete solution

$$y = ay_1 + by_2 \qquad a, b \text{ parameters}$$

on a deleted neighborhood $0 < |x| < R$ of the origin, where R is not less than the smaller of R_1 and R_2, and y_1 and y_2 are linearly independent solutions which have the following mathematical structure.

CASE 1 If $r_1 - r_2$ is not an integer, then

$$y_1 = |x|^{r_1} \sum_{n=0}^{\infty} c_n x^n \qquad c_0 \neq 0$$

$$y_2 = |x|^{r_2} \sum_{n=0}^{\infty} d_n x^n \qquad d_0 \neq 0$$

CASE 2 If $r_1 - r_2$ is a positive integer, then

$$y_1 = |x|^{r_1} \sum_{n=0}^{\infty} c_n x^n \qquad c_0 \neq 0$$

$$y_2 = |x|^{r_2} \sum_{n=0}^{\infty} d_n x^n + cy_1(x) \ln |x| \qquad d_0 \neq 0$$

CASE 3 If $r_1 - r_2 = 0$, so that $r_1 = r_2 = r$, then

$$y_1 = |x|^{r} \sum_{n=0}^{\infty} c_n x^n \qquad c_0 \neq 0$$

$$y_2 = |x|^{r} \sum_{n=1}^{\infty} d_n x^n + y_1(x) \ln |x|$$

The parameters in each expression for y_1 and y_2 can be determined to within an arbitrary constant by requiring that expression to satisfy Eq. (1).

In all three cases, Eq. (1) has a Frobenius series solution y_1 involving r_1. In Case 1, the differential equation also has a Frobenius series solution y_2 involving r_2. This possibility also exists in Case 2, since c is unrestricted and thus can be zero. In Case 3, y_2 is never a Frobenius series and d_1 need not be nonzero. Should $d_n = 0$, for $n \geq 1$, we have a so-called **purely logarithmic case** in which $y_2 = y_1(x) \ln |x|$. Although Theorem 1 does not say so, the regular singular point $x = 0$ will sometimes lie in the solution intervals of both y_1 and y_2, and when it does, Eq. (1) has a complete solution not only near but also about the origin. Of course, this never happens in Case 3, nor in Case 2 if $c \neq 0$. Usually the relevant solution intervals are found in practice by directly testing the series involved for convergence.

We next consider three illustrative examples, each involving a different case of Theorem 1. In each example, we first verify that the given differential equation has the origin as a regular singular point. Then, working where $x > 0$, we apply the method of Frobenius to obtain a solution of the equation. The relevant indicial equation and recurrence relation are derived directly rather than by means of equations (16) and (19). This procedure is more fundamental and often yields a simplified form of the recurrence relation more readily than does Eq. (19). To complete the solution process, we utilize the appropriate case of Theorem 1.

EXAMPLE 2

Solve the differential equation $2x^2y'' - x(x-1)y' - y = 0$ near the origin.

Both $p(x) = (1-x)/2x$ and $q(x) = -1/2x^2$ in the standard form of the given equation are undefined at $x = 0$. However, the functions $xp(x) = (1-x)/2$ and $x^2q(x) = -\frac{1}{2}$, being polynomials, are analytic everywhere. Hence the origin is a regular singular point of the differential equation, and by Theorem 1, the equation has at least one Frobenius series solution

$$y = \sum_{n=0}^{\infty} c_n x^{n+r} \qquad c_0 \neq 0$$

on an interval $0 < x < R$. Substituting this series and its first two derivatives into the given equation, then canceling the factor x^r, we get

$$\sum_{n=0}^{\infty} 2(n+r)(n+r-1)c_n x^n - \sum_{n=0}^{\infty} (n+r)c_n x^{n+1} + \sum_{n=0}^{\infty} (n+r)c_n x^n - \sum_{n=0}^{\infty} c_n x^n = 0$$

In the second series we replace n by $n-1$, then collect and factor the coefficients of x^0 and x^n, $n \geq 1$, to obtain

$$(r-1)(2r+1)c_0 + \sum_{n=1}^{\infty} (n+r-1)[(2n+2r+1)c_n - c_{n-1}]x^n = 0$$

Equating the coefficient of c_0 to zero gives the indicial equation

$$(r-1)(2r+1) = 0$$

whose roots are $r_1 = 1$ and $r_2 = -\frac{1}{2}$. We could have written this equation earlier by setting

$$p_0 = \lim_{x\to 0} xp(x) = \lim_{x\to 0} \frac{1-x}{2} = \frac{1}{2} \qquad \text{and} \qquad q_0 = \lim_{x\to 0} x^2q(x) = \lim_{x\to 0}\left(-\frac{1}{2}\right) = -\frac{1}{2}$$

in Eq. (16). Since the difference $r_1 - r_2 = \frac{3}{2}$ of the roots r_1 and r_2 is not an integer, Theorem 1, Case 1, tells us there are two linearly independent Frobenius solutions. To find such a pair, we return to the last series and equate the coefficient of x^n to zero, omitting the factor $n + r - 1$ as we do because it cannot be zero when $n \geq 1$ and r is either 1 or $-\frac{1}{2}$. This gives the recurrence relation

$$(2n+2r+1)c_n - c_{n-1} = 0 \qquad n \geq 1$$

or, solving for c_n,

$$c_n = \frac{c_{n-1}}{2n+2r+1} \qquad n \geq 1$$

Substituting the values 1 and $-\frac{1}{2}$ for r and introducing new coefficients d_n to go with $r_2 = -\frac{1}{2}$, we obtain two recurrence relations, namely,

$$c_n = \frac{c_{n-1}}{2n+3} \qquad \text{and} \qquad d_n = \frac{d_{n-1}}{2n} \qquad n \geq 1$$

We use these to list the c_n and the d_n in two separate columns, getting in succession

$$c_1 = \frac{c_0}{5} \qquad d_1 = \frac{d_0}{2}$$

$$c_2 = \frac{c_1}{7} \qquad d_2 = \frac{d_1}{4}$$

$$c_3 = \frac{c_2}{9} \qquad d_3 = \frac{d_2}{6}$$

$$\cdots\cdots \qquad \cdots\cdots$$

$$c_n = \frac{c_{n-1}}{2n+3} \qquad d_n = \frac{d_{n-1}}{2n}$$

Equating the product of the left-hand members to the product of the right-hand members of the equations in each column, then canceling common factors, we have

$$c_n = \frac{c_0}{5 \cdot 7 \cdot 9 \cdots (2n+3)} \quad \text{and} \quad d_n = \frac{d_0}{2 \cdot 4 \cdot 6 \cdots 2n} \qquad n \geq 1$$

which when written as

$$c_n = \frac{(n+1)!2^{n+1}}{(2n+3)!} 3c_0 \quad \text{and} \quad d_n = \frac{1}{n!2^n} d_0$$

hold for $n \geq 0$. Taking $3c_0 = 1$, $d_0 = 1$, and substituting for c_n and d_n in the two series of Theorem 1, Case 1, we obtain

$$y_1 = x \sum_{n=0}^{\infty} \frac{(n+1)!2^{n+1}}{(2n+3)!} x^n \quad \text{and} \quad y_2 = x^{-1/2} \sum_{n=0}^{\infty} \frac{1}{n!2^n} x^n \qquad 0 < x$$

Consistent with the fact that $xp(x)$ and $x^2q(x)$ are polynomials, the ratio test gives

$$\lim_{n \to \infty} \left| \frac{c_n}{c_{n+1}} \right| = \infty \quad \text{and} \quad \lim_{n \to \infty} \left| \frac{d_n}{d_{n+1}} \right| = \infty$$

which confirms that $R_1 = R_2 = R = \infty$; hence the power series in the expressions for y_1 and y_2 converge on $(-\infty, \infty)$, and y_1 is a solution there. The function y_2 is not defined on $(-\infty, 0]$ because of the factor $x^{-1/2}$. However, by replacing this factor by $|x|^{-1/2}$ we obtain a function which when taken in linear combination with y_1 gives a complete solution

$$y = ay_1 + b|x|^{-1/2} \sum_{n=0}^{\infty} \frac{1}{n!2^n} x^n \qquad a, b \text{ parameters}$$

of the differential equation on the deleted neighborhood $0 < |x| < \infty$ of the origin. This solution happens to be expressible in terms of familiar functions because the power series multiplied by $|x|^{-1/2}$ is the familiar function $e^{x/2}$, and the corresponding particular solution is represented by $|x|^{-1/2}e^{x/2}$. Reduction of order, when applied to $u|x|^{-1/2}e^{x/2}$, leads to a complete solution (see Exercise 55)

$$y = |x|^{-1/2}e^{x/2}\left(a \int_0^x |x|^{1/2}e^{-x/2}\,dx + b\right) \qquad 0 < |x| < \infty$$

EXAMPLE 3

Solve the equation $9x^2y'' + 3xy' + 2(x-4)y = 0$ near the origin.

The coefficient functions in the standard form of this equation give the polynomials

$$xp(x) = \frac{1}{3} \quad \text{and} \quad x^2q(x) = \frac{2(x-4)}{9}$$

which are analytic everywhere. Hence the origin is a regular singular point of the differential equation, and there is at least one Frobenius series solution

$$y = \sum_{n=0}^{\infty} c_n x^{n+r} \qquad c_0 \neq 0$$

on $0 < x < \infty$. Substituting this series and its first two derivatives into the given equation, then canceling the factor x^r, we get

$$\sum_{n=0}^{\infty} 9(n+r)(n+r-1)c_n x^n + \sum_{n=0}^{\infty} 3(n+r)c_n x^n + \sum_{n=0}^{\infty} 2c_n x^{n+1} - \sum_{n=0}^{\infty} 8c_n x^n = 0$$

or collecting coefficients on like powers of x and doing some factoring,

$$(3r-4)(3r+2)c_0 + \sum_{n=1}^{\infty} [(3n+3r-4)(3n+3r+2)c_n + 2c_{n-1}]x^n = 0$$

Equating the coefficient of c_0 to zero gives the indicial equation

$$(3r-4)(3r+2) = 0$$

whose roots are $r_1 = \frac{4}{3}$ and $r_2 = -\frac{2}{3}$. This equation is given by (16) with $p_0 = \frac{1}{3}$ and $q_0 = -\frac{8}{9}$. Since $r_1 - r_2 = 2$ is a positive integer, we are working with a differential equation whose solutions are described by Theorem 1, Case 2.

Returning to the last series and equating the coefficients of x^n to zero, we get the recurrence relation

$$(3n+3r-4)(3n+3r+2)c_n + 2c_{n-1} = 0 \qquad n \geq 1$$

Now we know that the differential equation has a Frobenius series solution corresponding to the larger exponent $r_1 = \frac{4}{3}$ of the ordinary singular point $x = 0$. Substituting this value for r in the recurrence relation, then solving for c_n and giving n successive integral values, we get

$$c_1 = \frac{-2c_0}{9 \cdot 1 \cdot 3} \qquad c_2 = \frac{-2c_1}{9 \cdot 2 \cdot 4} \qquad c_3 = \frac{-2c_2}{9 \cdot 3 \cdot 5} \qquad c_4 = \frac{-2c_3}{9 \cdot 4 \cdot 6} \qquad \cdots \qquad c_n = \frac{-2c_{n-1}}{9(n)(n+2)}$$

Equating the product of the left-hand members of these equations to the product of the right-hand members and canceling common factors leads to the formula

$$c_n = \frac{(-1)^n 2^n (2c_0)}{9^n n!(n+2)!} \qquad n \geq 0$$

Setting $2c_0 = 1$ and substituting into the series for y, we get the particular solution

$$y_1 = \sum_{n=0}^{\infty} \frac{(-1)^n 2^n x^{n+4/3}}{9^n n!(n+2)!} \qquad 0 < x < \infty$$

Returning to the recurrence relation, we next replace r by $-\frac{2}{3}$, introduce new coefficients d_n, and simplify to obtain

$$9(n-2)nd_n + 2d_{n-1} = 0$$

which for $n = 1, 2$ gives $d_1 = \frac{2}{9}d_0$ and $d_1 = 0$, respectively. But $d_0 \neq 0$ in the solution y_2, Theorem 1, Case 2, whether c is zero or not. Thus we seek a solution of the form

$$y_2 = \sum_{n=0}^{\infty} d_n x^{n-2/3} + cy_1 \ln x \qquad d_0 \neq 0,\ 0 < x < \infty$$

To facilitate substituting this expression into the differential equation, we first find

$$-8y_2 = -\sum_{n=0}^{\infty} 8d_n x^{n-2/3} - 8cy_1 \ln x$$

$$2xy_2 = \sum_{n=0}^{\infty} 2d_n x^{n+1/3} + 2xcy_1 \ln x$$

$$3xy_2' = \sum_{n=0}^{\infty} (3n-2)d_n x^{n-2/3} + c(3y_1 + 3xy_1' \ln x)$$

$$9x^2y_2'' = \sum_{n=0}^{\infty} (3n-2)(3n-5)d_n x^{n-2/3} + c(-9y_1 + 18xy_1' + 9x^2y_1'' \ln x)$$

To ensure that y_2 will satisfy the differential equation, we equate the sum of the right-hand members of these equations to zero. In doing so, we write the sum of the four series as

$$0 \cdot d_0 + \sum_{n=1}^{\infty} [9n(n-2)d_n + 2d_{n-1}]x^{n-2/3}$$

We also note that

$$c[9x^2y_1'' + 3xy_1' + 2(x-4)y_1] \ln x = 0$$

because y_1 is a solution of the given equation, and we use the series for y_1 and y_1' to replace $18xy_1' - 6y_1$ by the series

$$\sum_{n=2}^{\infty} \frac{(-1)^n 2^{n-1}(n-1)x^{n-2/3}}{9^{n-3}(n-2)!n!}$$

Thus we have that

$$0 \cdot d_0 + \sum_{n=1}^{\infty} [9n(n-2)d_n + 2d_{n-1}]x^{n-2/3} + c\sum_{n=2}^{\infty} \frac{(-1)^n 2^{n-1}(n-1)x^{n-2/3}}{9^{n-3}(n-2)!n!} = 0$$

must hold for $0 < x < \infty$. As expected, d_0 is arbitrary but not zero.

Taking $n = 1$, we get $[9(-1)d_1 + 2d_0]x^{1/3} = 0$, which yields $d_1 = 2d_0/9$. Taking $n = 2$, we get $[(0 \cdot d_2 + 2d_1) + c \cdot 9]x^{4/3} = 0$ which leaves d_2 arbitrary and yields $c = -2^2 d_0/9^2$. Equating coefficients of all higher powers of x to zero, we have the recurrence relation

$$9n(n-2)d_n + 2d_{n-1} - \frac{(-1)^n 2^{n+1}(n-1)}{9^{n-1}(n-2)!n!} d_0 = 0 \qquad n \ge 3$$

or, solving for d_n,

$$d_n = \frac{2}{9(n-2)(n)}\left[-d_{n-1} + \frac{(-1)^n 2^n (n-1)}{9^{n-1}(n-2)!n!} d_0\right] \qquad n \ge 3$$

For $n = 3$, we have

$$d_3 = \frac{2}{9 \cdot 1 \cdot 3}\left[-d_2 + \frac{(-1)^3 2^3 (2)}{9^2 1! 3!} d_0\right]$$

Since all that is needed to complete the solution process is a particular solution linearly independent of y_1, let us set $d_2 = 0$ and $d_0 = 1$. Then our formula for d_n becomes

$$d_n = \frac{2}{9(n-2)(n)}\left[-d_{n-1} + \frac{(-1)^n 2^n (n-1)}{9^{n-1}(n-2)!n!}\right] \qquad n \ge 3$$

while d_3 reduces to

$$d_3 = \frac{(-1)^3 2^4}{1!3!9^3}\left(\frac{2}{1\cdot 3}\right)$$

We next have

$$d_4 = \frac{2}{9\cdot 2\cdot 4}\left[-d_3 + \frac{(-1)^4 2^4\cdot 3}{9^3 2!4!}\right] = \frac{2}{9\cdot 2\cdot 4}\left[\frac{(-1)^4 2^4\cdot 2}{1!3!9^3\cdot 1\cdot 3} + \frac{(-1)^4 2^4\cdot 3}{9^3 2!4!}\right]$$
$$= \frac{(-1)^4 2^5}{2!4!9^4}\left(\frac{2}{1\cdot 3} + \frac{3}{2\cdot 4}\right)$$

and the same procedure gives

$$d_5 = \frac{(-1)^5 2^6}{3!5!9^5}\left(\frac{2}{1\cdot 3} + \frac{3}{2\cdot 4} + \frac{4}{3\cdot 5}\right)$$

By induction, there follows

$$d_n = \frac{(-1)^n 2^{n+1}}{(n-2)!n!9^n}\sum_{k=1}^{n-2}\frac{k+1}{k(k+2)} \qquad n \ge 3$$

Using partial fractions we get

$$\sum_{k=1}^{n-2}\frac{k+1}{k(k+2)} = \frac{1}{2}\sum_{k=1}^{n-2}\left(\frac{1}{k} + \frac{1}{k+2}\right) = \frac{1}{2}\left(\sum_{k=1}^{n-2}\frac{1}{k} + \sum_{k=3}^{n}\frac{1}{k}\right) = \frac{1}{2}\left(H_{n-2} + H_n - H_2\right)$$

where

(26) $$H_n = \sum_{k=1}^{n}\frac{1}{k} = 1 + \frac{1}{2} + \frac{1}{3} + \cdots + \frac{1}{n}$$

denotes the **nth partial sum of the harmonic series.**

Substituting $d_0 = 1$, $d_1 = \frac{2}{9}$, $d_2 = 0$, $c = -2^2/9^2$, and

$$d_n = \frac{(-1)^n 2^n}{(n-2)!n!9^n}\left(H_{n-2} + H_n - H_2\right) \qquad n \ge 3$$

into our original expression for y_2, we obtain the solution

$$y_2 = x^{-2/3} + \frac{2}{9}x^{1/3} + \sum_{n=3}^{\infty}\frac{(-1)^n 2^n}{(n-2)!n!9^n}\left(H_{n-2} + H_n - H_2\right)x^{n-2/3} - \frac{2^2}{9^2}y_1 \ln x$$

of the differential equation on $0 < x < \infty$. Since $|x|^{4/3} = x^{4/3}$ and $|x|^{-2/3} = x^{-2/3}$, Theorem 1, Case 2, tells us that in terms of the power series

$$f_1(x) = \sum_{n=0}^{\infty}\frac{(-1)^n 2^n x^n}{n!(n+2)!9^n} \qquad \text{and} \qquad f_2(x) = 1 + \frac{2}{9}x + \sum_{n=3}^{\infty}\frac{(-1)^n 2^n (H_{n-2} + H_n - H_2)x^n}{(n-2)!n!9^n}$$

a complete solution on the deleted neighborhood $0 < |x| < \infty$ of the origin is

$$y = ax^{4/3}f_1(x) + b[x^{-2/3}f_2(x) - (\tfrac{2}{9})^2 x^{4/3}f_1(x)\ln|x|] \qquad a, b \text{ parameters}$$

EXAMPLE 4

Solve the differential equation $xy'' + y' + y = 0$ near the origin.

At all points of the real axis the polynomials

$$xp(x) = x\left(\frac{1}{x}\right) = 1 \qquad \text{and} \qquad x^2q(x) = x^2\left(\frac{1}{x}\right) = x$$

are analytic. Hence the origin is a regular singular point of the differential equation, and there is at least one Frobenius series solution

$$y = \sum_{n=0}^{\infty} c_n x^{n+r} \qquad c_0 \neq 0$$

on $0 < x < \infty$. Substituting this series and its first two derivatives into the given equation, then canceling the factor x^r and collecting coefficients on like powers of x, we are lead to the equation

$$r^2 c_0 x^{-1} + \sum_{n=1}^{\infty} [(n + r)^2 c_n + c_{n-1}]x^{n-1} = 0$$

which is to hold for $0 < x < \infty$. The indicial equation $r^2 = 0$ has $r = 0$ as a double root and is given by (16) when $p_0 = 1$ and $q_0 = 0$. Since the indicial equation has just one root, the differential equation has two linearly independent solutions characterized by Theorem 1, Case 3.

In the last series, after setting $r = 0$, we equate the coefficient of x^{n-1} to zero, getting the recurrence relation

$$n^2 c_n + c_{n-1} = 0 \qquad n \geq 1$$

which, when solved for c_n, gives

$$c_n = \frac{-c_{n-1}}{n^2} \qquad n \geq 1$$

There follows in succession

$$c_1 = \frac{-c_0}{1^2} \qquad c_2 = \frac{-c_1}{2^2} \qquad c_3 = \frac{-c_2}{3^2} \qquad c_4 = \frac{-c_3}{4^2} \qquad \cdots \qquad c_n = \frac{-c_{n-1}}{n^2}$$

Multiplying respective members of these equations, then equating the products thus obtained and canceling common factors, we obtain

$$c_n = \frac{(-1)^n c_0}{(n!)^2} \qquad n \geq 0$$

Taking $c_0 = 1$ and substituting into the series for y, we get the particular solution

$$y_1 = \sum_{n=0}^{\infty} \frac{(-1)^n}{(n!)^2} x^n$$

To find a solution linearly independent of y_1, we set

$$y_2 = \sum_{n=1}^{\infty} d_n x^n + y_1 \ln x$$

and require y_2 to satisfy the differential equation by adding the right-hand members of the equations

$$y_2 = \sum_{n=1}^{\infty} d_n x^n + y_1 \ln x$$

$$y_2' = \sum_{n=1}^{\infty} n d_n x^{n-1} + \left(\frac{y_1}{x} + y_1' \ln x\right)$$

$$x y_2'' = \sum_{n=2}^{\infty} n(n-1) d_n x^{n-1} + \left(\frac{-y_1}{x} + 2y_1' + x y_1'' \ln x\right)$$

and equating their sum to zero. At the same time, we write the sum of the three series as

$$d_1 + \sum_{n=2}^{\infty} (n^2 d_n + d_{n-1}) x^{n-1}$$

we note that

$$(x y_1'' + y_1' + y_1) \ln x = 0$$

since y_1 satisfies the differential equation, and we replace $2y_1'$ by

$$-2 + 2 \sum_{n=2}^{\infty} \frac{(-1)^n n x^{n-1}}{(n!)^2}$$

Thus we have that the equation

$$d_1 - 2 + \sum_{n=2}^{\infty} \left[n^2 d_n + d_{n-1} + \frac{2n(-1)^n}{(n!)^2} \right] x^{n-1} = 0$$

must hold for $0 < x < \infty$. Setting coefficients of all powers of x equal to zero yields $d_1 = 2$ and the recurrence relation

$$n^2 d_n + d_{n-1} + \frac{2n(-1)^n}{(n!)^2} = 0 \qquad n \geq 2$$

or solving for d_n,

$$d_n = -\left[\frac{d_{n-1}}{n^2} + \frac{2(-1)^n}{n(n!)^2} \right] \qquad n \geq 2$$

There follows

$$d_2 = -\left[\frac{d_1}{2^2} + \frac{2}{2(2!)^2} \right] = -\left[\frac{2}{(2!)^2} + \frac{2}{2(2!)^2} \right] = \frac{2(-1)}{(2!)^2} \left(1 + \frac{1}{2} \right)$$

$$d_3 = -\left[\frac{d_2}{3^2} + \frac{2(-1)}{3(3!)^2} \right] = -\left[\frac{-2}{3^2(2!)^2} \left(1 + \frac{1}{2} \right) + \frac{-2}{3(3!)^2} \right] = \frac{2(-1)^2}{(3!)^2} \left(1 + \frac{1}{2} + \frac{1}{3} \right)$$

and induction, coupled with $d_1 = 2$, gives

$$d_n = \frac{2(-1)^{n-1}}{(n!)^2} H_n \qquad n \geq 1$$

where, as in Example 3, $H_n = \sum_{k=1}^{n}(1/k)$. Substituting for d_n in our original expression for y_2, we obtain

$$y_2 = 2\sum_{n=1}^{\infty}\frac{(-1)^{n-1}}{(n!)^2}H_n x^n + y_1 \ln x$$

as a solution of the differential equation on $0 < x < \infty$. The power series

$$g(x) = 2\sum_{n=1}^{\infty}\frac{(-1)^{n-1}}{(n!)^2}H_n x^n$$

like that defining y_1, converges for all values of x. A complete solution of the given equation is represented by

$$y(x) = ay_1(x) + b[g(x) + y_1(x)\ln|x|] \qquad a,\ b \text{ parameters}$$

on the deleted neighborhood $0 < |x| < \infty$ of the origin.

EXERCISES

1. Find the singular points of each of the following equations and determine whether they are regular or irregular.
(a) $y'' + xy' + y = 0$
(b) $x^2y'' - \lambda^2 y = 0$
(c) $(1 - x^2)y'' + y' + y = 0$
(d) $x^2y'' + y' + y = 0$
(e) $x^3(2x + 1)y'' + x^2y' - y = 0$
(f) $(x^4 - 3x^3 + 4x)y'' + x^2y' + (x^3 - 2x - 1)y = 0$

2. Work Exercise 1 for the equations:
(a) $e^xy'' + 2y' + xy = 0$
(b) $x^2(1 - x)y'' + (1 - x)y' + y = 0$
(c) $xy'' + y' + x^2y = 0$
(d) $x^2y'' + y' + xy = 0$
(e) $x(3x - 2)^2y'' + 7y' - xy = 0$
(f) $x^2(x^3 + x^2 - x - 1)y'' - 5y' + x^2y = 0$

3. Under what conditions, if any, might one or the other of the functions $p(x)$ and $q(x)$ in Eq. (2) be defined at a singular point of Eq. (1)?

4. In Example 1, **(a)** verify that $(x - 1)p(x) = 2(x - 1)/x$ is analytic at the point $x = 1$ by obtaining its Taylor-series expansion in powers of $x - 1$, and **(b)** verify that $x^2q(x) = 3x/(x - 1)^3$ is analytic at the origin by obtaining its Maclaurin expansion.

Find the indicial equation corresponding to each regular singular point of the following equations. *Hint:* Use Eq. (16).

5. $2xy'' + (1 + x)y' - 5y = 0$
6. $4xy'' - 10(2x - 1)y' - 9y = 0$
7. $3x^2y'' + xy' + (x - 1)y = 0$
8. $(x^2 - 1)^2y'' + 8(x - 1)y' + x^2y = 0$
9. $(2x^4 + x^3 - x^2)y'' + y' - 6xy = 0$
10. $x(1 - x^2)y'' + (x - 1)y' - (x + 1)y = 0$

11. Use Eq. (19) to find the recurrence relation involving the coefficients of the Frobenius series solution of Example 2 in which **(a)** $r = r_1 = 1$, and **(b)** $r = r_2 = -\frac{1}{2}$.

12. Show that if $y_1(x) = x^r\sum_{n=0}^{\infty}c_nx^n$ is a solution of Eq. (1) on $0 < x < R$, then $y_2(x) = (-x)^r\sum_{n=0}^{\infty}c_nx^n$ is a solution on $-R < x < 0$. *Hint:* Substitute y_2, xy_2', and x^2y_2'' into (12), cancel the factor $(-x)^r$, and reduce the result to (14).

13. If (inefficiently) a Frobenius series solution $y = \sum_{n=0}^{\infty}c_nx^{n+r}$ is assumed near an ordinary point of Eq. (1), show that the roots of the related indicial equation are always $r = 0$ and $r = 1$. What does this imply?

14. Show that when the tentative solution y_2, Theorem 1, Case 2 or Case 3, is substituted into Eq. (1), the logarithmic terms drop out and only powers of x remain.

Find two linearly independent solutions near the origin for each of the following equations and determine the values of x for which each solution is valid.

15. $2x^2y'' + (2x^2 + x)y' - y = 0$
16. $2x^2y'' + (x^2 - x)y' + y = 0$
17. $2x^2y'' + xy' - y = 0$
18. $9x^2y'' + 9(x^2 + x)y' + (12x - 1)y = 0$
19. $4x^2y'' + (4x + 1)y = 0$
20. $xy'' + y' + xy = 0$
21. $x(x - 1)y'' + (3x - 1)y' + y = 0$
22. $5x^2y'' + 5xy' + (x - 5)y = 0$
23. $x^2y'' + (x - x^2)y' - y = 0$
24. $x^2y'' + 3xy' + 5y = 0$
25. $2x(x - 1)y'' + 3(x - 1)y' - y = 0$
26. $2x^2y'' - xy' + y = 0$
27. $4x^2y'' - 8x^2y' + (1 + 4x^2)y = 0$
28. $xy'' + (1 - 2x)y' + (x - 1)y = 0$
29. $xy'' + (x - 1)y' - y = 0$
30. $x(x - 1)y'' + 2(x - 1)y' - 2y = 0$
31. $2x^2y'' - x(1 + 2x)y' + (1 + 4x)y = 0$
32. $x^2y'' + (x^2 - x)y' + y = 0$
33. $xy'' + (1 + x)y' + y = 0$

34. $3x^2y'' + (3x^2 + 5x)y' + (6x - 1)y = 0$
35. $2x^2y'' + (2x^2 - 3x)y' + (x + 2)y = 0$
36. $x^2y'' - xy' + (1 + x)y = 0$
37. $x^2y'' + 3xy' + (1 + x)y = 0$
38. $4x^2y'' + y = 0$
39. $xy'' - y' + y = 0$
40. $x^2y'' + (x - 2)y = 0$
41. $4x^2y'' - 2x(x - 2)y' - (3x + 1)y = 0$
42. $x(x - 2)y'' - 2(1 - x)y' - 2y = 0$
43. $2x^2y'' + x(2x + 3)y' + (3x - 1)y = 0$
44. $x(x^2 + 1)y'' - 2(2x^2 + 1)y' + 6xy = 0$
45. $2x^2y'' + 3xy' - y = 0$
46. $4x^2y'' + 4xy' + (4x^2 - 1)y = 0$
47. $x(2x - 3)y'' + 6(1 - x)y' + 6y = 0$
48. $2x(x + 1)y'' + (1 - x)y' + y = 0$
49. $x(1 - x)y'' + xy' - y = 0$
50. $xy'' + (1 - x)y' + y = 0$
51. Find two elementary functions which form a basis for all solutions of the equation in Exercise 23.
52. Find two elementary functions which form a basis for all solutions of the equation in Exercise 29.
53. Show that fundamental solutions of the equation in Exercise 30 are given by

$$y_1 = -2 + 2x$$

and

$$y_2 = y_1 \ln|x| + |x|^{-1}\left[1 + x - 5x^2 + \sum_{n=3}^{\infty} \frac{2x^n}{(n-2)(n-1)}\right]$$

54. **(a)** Using series solutions of the equation in Exercise 46, derive a complete solution on $0 < |x| < \infty$ and express it in terms of elementary functions.
(b) Show that every nontrivial solution becomes unbounded as $x \to 0$ and tends to zero as $x \to \infty$.
55. Show that the final equation of Example 2 is a complete solution of the given differential equation by using reduction of order and the particular solution $y_2 = |x|^{-1/2}e^{x/2}$.
56. Use the method of Frobenius to find linearly independent solutions of the equation

$$x^2y'' - 2axy' + [a(a + 1) + x^2]y = 0$$

57. Use the method of Frobenius to find linearly independent solutions of the equation

$$x^2y'' - 2axy' + [a(a + 1) - x^2]y = 0$$

Find two linearly independent solutions near $x = 1$, and where $1 < x$, for each of the following equations.

58. $3(1 - x)y'' + (x - 3)y' + 2y = 0$
59. $(x - 1)^2y'' + (x^2 - 3x + 2)y' + (2 - x)y = 0$
60. $6(x - 1)^2y'' + (x - 1)y' + y = 0$
61. $x(x - 1)y'' + 2(x - 1)y' - 2y = 0$
62. **(a)** Determine whether $x = 0$ is a regular or an irregular singular point of the equation $x^3y'' + 5xy' - 5y = 0$.
(b) Find a Frobenius series solution of the equation, then use reduction of order to find a complete solution on $0 < |x| < \infty$.
63. Find all Frobenius series solutions of the equation $x^3y'' + y = 0$.
64. Equation (1) is said to have an **ordinary** or a **singular point at infinity,** i.e., at $x = \infty$, according as the equation obtained from (1) under the substitution $x = 1/u$ has an ordinary point or a singular point at $u = 0$. Show that under this transformation the standard equation

$$y'' + p(x)y' + q(x)y = 0$$

becomes the equation

$$u^4\frac{d^2y}{du^2} + \left[2u^3 - u^2p\left(\frac{1}{u}\right)\right]\frac{dy}{du} + q\left(\frac{1}{u}\right)y = 0$$

and use this result to determine the nature of the point at infinity for the equation $(x^2 + 1)y'' + y' + y = 0$.
65. Verify that under the change of dependent variable defined by the substitution

$$y = z \exp\left[-\tfrac{1}{2}\int p(x)\,dx\right]$$

the differential equation $y'' + p(x)y' + q(x)y = 0$ becomes

$$\frac{d^2z}{dx^2} + w(x)z = 0$$

where $\quad w(x) = q(x) - \dfrac{1}{2}p'(x) - \dfrac{1}{4}p^2(x)$

66. Using the results of Exercise 65, determine conditions on the parameters a_1, b_1, a_2, b_2, and c_2 which will ensure that the equation

$$y'' + (a_1 + b_1x)y' + (a_2 + b_2x + c_2x^2)y = 0$$

can be solved in terms of elementary functions.

Using the results of Exercise 65, solve each of the following equations.

67. $y'' + \dfrac{2}{1 + x}y' + y = 0$

68. $y'' + y' \tan\dfrac{x}{2} + \dfrac{5}{4}y = 0$

69. What do you think would be appropriate definitions for the terms *ordinary point, regular singular point, irregular singular point,* and *indicial equation* relative to the following standard third-order linear differential equation?

$$y''' + P(x)y'' + Q(x)y' + R(x)y = 0$$

70. Assuming $xP(x)$, $x^2Q(x)$, and $x^3R(x)$ are all analytic at $x = 0$, what is the indicial equation relative to the origin for the equation

$$y''' + P(x)y'' + Q(x)y' + R(x)y = 0$$

For each of the following equations, find all the power-series solutions near the origin.

71. $xy''' + y = 0$
72. $x^2y''' + y = 0$
73. $y''' + x^2y'' + 4xy' + 2y = 0$
74. $x^2y''' + xy' + 3y = 0$
75. $x^2y''' + xy' + y = 0$
76. $y''' + x^2y'' + 5xy' + 3y = 0$

2.11 APPLICATIONS OF LINEAR DIFFERENTIAL EQUATIONS WITH CONSTANT COEFFICIENTS

Since ordinary linear differential equations with variable coefficients can rarely be solved in terms of elementary functions, they must usually be studied case by case and this can be a major project. Thus we shall defer applications involving such equations to a later chapter (Chap. 12). Linear differential equations with constant coefficients find their most important applications in the study of electric circuits and vibrating mechanical systems. So useful to engineers and scientists are the results of this analysis that we shall devote an entire chapter (Chap. 7) to its major features. Of course, there are many other applications of considerable interest. Following a brief introduction to vibrating mechanical systems, we shall present two typical illustrative examples from mechanics. Electric circuits will be postponed until Chap. 7.

To make sure that certain basic ideas pertaining to vibrating mechanical systems are clearly understood, let us consider an object hanging from a spring, as shown in Fig. 2.3*b*. The weight w of the object is the magnitude of the force of gravity acting on it. The mass m of the object is related to w by the formula $w = mg$. As is customary, we suppose that the object is concentrated in a single ''point mass'' and is not of appreciable dimensions. We also assume that the spring obeys **Hooke's law:**† *Force is proportional to displacement.* This means that if the spring is known to be stretched or compressed a distance s (the magnitude of the displacement) from its neutral, or unstretched, length, then the magnitude of the force F exerted on the spring is given by the formula

$$F = ks \tag{1}$$

In this equation the constant k, whose physical dimensions are clearly force per unit length, is called the **modulus** of the spring. Because Eq. (1) is a linear relation between the force and the corresponding displacement, a spring which obeys Hooke's law is often called a **linear spring.**

†Named for the English physicist Robert Hooke (1635–1703).

Figure 2.3
A simple mass-spring system.

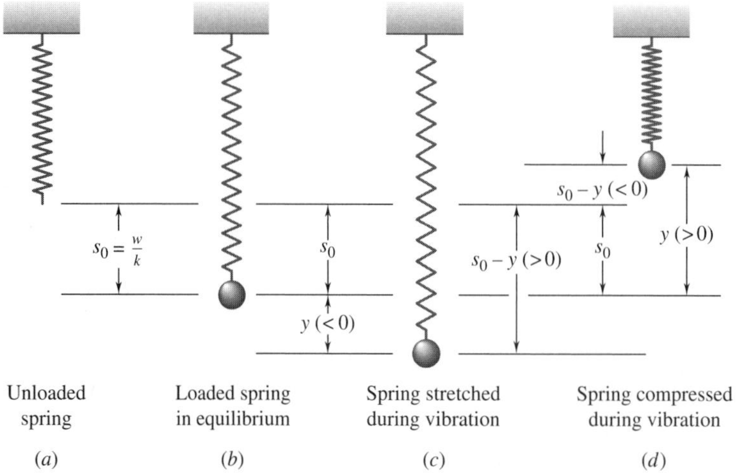

When the weight w hangs from the spring in static (i.e., motionless) equilibrium, the spring is of course stretched and the weight is the force F which stretches it. Hence the distance s_0 which the spring is stretched can be found from Eq. (1):

$$w = ks_0 \qquad \text{or} \qquad s_0 = \frac{w}{k}$$

Now suppose that instead of hanging in static equilibrium, the weight is set in motion in such a way that it moves in a purely vertical direction. As a coordinate to describe the motion of the weight, let us use the distance y from the equilibrium position of the weight to its instantaneous position, the positive direction of y being upward (Fig. 2.3*c* and *d*). When $y = s_0$, the upward movement of the weight has just restored the length of the spring to its neutral, unstretched length, and the force exerted by the spring is instantaneously zero. In general, at the instant when the displacement of the weight from its equilibrium position is y, the change in the length of the spring is $s_0 - y$ and the force the spring exerts on the weight is $k(s_0 - y)$. If $s_0 - y < 0$, the spring is compressed and it exerts a force on the weight in the downward or negative direction. If $s_0 - y > 0$, the spring is stretched and it exerts a force on the weight in the upward or positive direction. Hence in each case the force instantaneously applied to the weight by the spring is given in magnitude and direction by the expression

$$k(s_0 - y)$$

the so-called **elastic force.** Of course, the downward gravitational force $-w = -mg$ also acts on the mass.

A realistic analysis of the vertical motion of the mass would take into account not only the elastic and gravitational forces but also the effects of friction and all other forces that act externally on the suspended mass. In this section these effects are neglected since they are discussed in detail in Chap. 7.

Observing that $m = w/g$ and applying Newton's law, then noting that $ks_0 - w = 0$, we have

$$\frac{w}{g}\frac{d^2y}{dt^2} = k(s_0 - y) - w = -ky$$

or

$$\frac{d^2y}{dt^2} + \frac{kg}{w}y = 0 \tag{2}$$

This is an equation of harmonic motion which has

$$y = A \cos \sqrt{\frac{kg}{w}}t + B \sin \sqrt{\frac{kg}{w}}t \tag{3}$$

as a complete solution. Regardless of the values of A and B, that is, regardless of how the system is set in motion, Eq. (3) describes periodic motion with period

$$2\pi\sqrt{\frac{w}{kg}}$$

or frequency

$$\frac{1}{2\pi}\sqrt{\frac{kg}{w}}$$

If w is measured in kilograms, k in kilograms per centimeter, and g in centimeters per second per second, the period is given in seconds and the frequency in cycles per second.† Whether there is friction in the system or not, the quantity $(1/2\pi)\sqrt{kg/w}$ is called the *natural frequency* of the system because this is the frequency at which the *spring-mass system* would vibrate naturally if no frictional or nonelastic forces, other than gravity, were present.

Of course, there are numerous physical systems whose behavior can be described by a mathematical model which, under appropriate simplifying assumptions, reduces to a differential equation of harmonic oscillation

$$\frac{d^2y}{dt^2} + \omega^2 y = 0$$

involving a time variable t. In such a case, the quantity

$$f = \frac{\omega}{2\pi}$$

is called the **natural frequency** of the system, and its reciprocal

$$p = \frac{2\pi}{\omega}$$

is the corresponding **period.** Frequently, oscillations within a system are undesirable, and it is important to know the natural frequencies at which oscillations can take place in order to avoid or compensate for external influences that might prove destructive to the system. The important part of such a frequency calculation is the formulation of a governing differential equation. For, clearly, the **circular frequency**

$$\omega \text{ radians per unit time}$$

as well as the natural frequency $\omega/2\pi$ Hz can be read from a differential equation of harmonic motion just as well as from any of its solutions, general or particular. Whenever the intended meaning is clear, either of these frequencies might simply be called the *frequency* of the system.

EXAMPLE 1

A SPRING-SUSPENDED WEIGHT

A weight of 7 kg is suspended from a spring of modulus $\frac{36}{35}$ kg/cm. At $t = 0$, while the weight is hanging in static equilibrium, it is suddenly given an initial velocity of 48 cm/s in the downward, or negative, direction. Taking the acceleration of gravity g to be 980 cm/s^2, find the vertical displacement of the weight as a function of t. What are the period and frequency of the subsequent motion? Through what amplitude does the weight move? At what times does the weight reach its extreme displacements above and below its equilibrium position?

The differential equation to be solved in this problem is

$$\frac{7}{980}\frac{d^2y}{dt^2} + \frac{36}{35}y = 0 \qquad \text{or} \qquad \frac{d^2y}{dt^2} + 144y = 0$$

and a complete solution is

$$y = A\cos 12t + B\sin 12t \tag{4}$$

†The unit *cycles per second* is customarily given the name **hertz** (abbreviated Hz) in honor of the German physicist Heinrich Hertz (1857–1894).

The period of the motion is therefore

$$\frac{2\pi}{12} = \frac{\pi}{6}\ \text{s}$$

and the frequency is

$$\frac{6}{\pi}\ \text{Hz}$$

To determine A and B, we use the fact that $y(0) = 0$ (since the weight starts to move from its equilibrium position) and $v(0) = -48$ (since this is the velocity imparted to the weight at $t = 0$). Substituting the initial displacement condition into Eq. (4), we find $A = 0$. Substituting the initial velocity condition into the velocity equation

$$v = 12B \cos 12t$$

we find that $B = -4$. Thus the motion of the weight for $t \geq 0$ is described by the equation

$$y = -4 \sin 12t \tag{5}$$

From Eq. (5) it is clear that the weight alternately rises and falls to a maximum of 4 cm above and below its equilibrium position. In other words, the amplitude of its motion is 4 cm. The extrema ($y = \pm 4$) occur when $\sin 12t = \pm 1$, that is, when

$$12t = \frac{\pi}{2} + n\pi \qquad \text{or} \qquad t = \left(\frac{\pi}{24} + \frac{n\pi}{12}\right)\text{s} \qquad n = 0, 1, 2, \ldots$$

Odd values of n correspond to maxima, even values to minima.

Newton's law is not always the most convenient way to set up the differential equation that describes the behavior of a mechanical system. Sometimes this is most easily done by using the **principle of the conservation of energy**: *if no energy is lost through friction or other irreversible conversions of energy, then in a mechanical system the sum of the instantaneous kinetic and potential energies of the system must remain constant.* A system whose total energy remains constant is called a **conservative system.** To formulate the **equation of motion** for such a system, i.e., a differential equation characterizing the motion, we incorporate the preceding principle into the **energy method** which consists of

a. Finding an equation for the total energy of the system, the so-called **energy equation,** and
b. Differentiating the energy equation to obtain the equation of motion.

To use this method it is of course necessary to have formulas for the various energy forms that may be involved. The following are the ones most commonly encountered.

1. The kinetic energy (KE) of a body of mass m moving with velocity v is given by the formula $\text{KE} = \frac{1}{2}mv^2$.
2. The kinetic energy of a body of moment of inertia I rotating with angualr velocity ω is given by the formula $\text{KE} = \frac{1}{2}I\omega^2$.
3. The change in the potential energy (PE) stored in a spring of modulus k when it is stretched or compressed from an initial length s_0 to a final length s is given by the formula $\text{PE} = \int_{s_0}^{s} ky\,dy = \frac{1}{2}k(s^2 - s_0^2)$.
4. The change in the potential energy of a weight w when it is moved from an initial height h_0 in the earth's gravitational field to a final height h is given by the formula $\text{PE} = w(h - h_0)$ provided that h and h_0 are small in comparison with the radius of the earth.

EXAMPLE 2

A WEIGHT-PULLEY SYSTEM

A weight W_2 is suspended from a pulley of weight W_1 and radius R, as shown in Fig. 2.4*a*. Constraints, which need not be specified, prevent any swinging of the system and permit it to move only in the vertical direction. If a spring of modulus k is inserted in the otherwise inextensible cable which supports the pulley, find the frequency with which the system will vibrate in the vertical direction if it is displaced slightly from its equilibrium position. Friction between the cable and the pulley prevents any slippage and results in negligible energy loss, but all other frictional effects are to be neglected.

To formulate the differential equation governing this problem we first find the total energy of the system so as to apply the energy method. The potential energy consists of two parts: (*a*) the potential energy of the weights W_1 and W_2 due to their position in the gravitational field and (*b*) the potential energy stored in the stretched spring. Taking the equilibrium position of the system as the reference level, with y positive downward and θ positive counterclockwise, for potential energy we have for part (*a*)

(6) $$(\text{PE})_a = -(W_1 + W_2)y$$

the minus sign indicating that a positive y corresponds to a lowering of the weights, hence a decrease in the potential energy. Now the length of both the cable and spring in Fig. 2.4*b* differs from that in Fig. 2.4*a* by an amount $2y$, and because the cable does not stretch, the potential energy in the spring is simply the amount of work required to stretch the spring from its equilibrium elongation, say δ, to its instantaneous elongation $\delta + 2y$. Since the force in the spring at any time is

$$F = \text{elongation} \times \text{force per unit elongation} = sk$$

we have for the potential energy of type (*b*)

(7) $$(\text{PE})_b = \int_{s_1}^{s_2} F\,ds = \int_{\delta}^{\delta+2y} ks\,ds = k\left.\frac{s^2}{2}\right|_{\delta}^{\delta+2y} = 2ky^2 + 2k\delta y$$

The kinetic energy likewise consists of two parts: (*a*) the energy of translation of the weights W_1 and W_2, namely,

(8) $$(\text{KE})_a = \frac{1}{2}\frac{W_1 + W_2}{g}(\dot{y})^2$$

and (*b*) the energy of rotation of the pulley, namely,

$$(\text{KE})_b = \tfrac{1}{2}I(\dot{\theta})^2$$

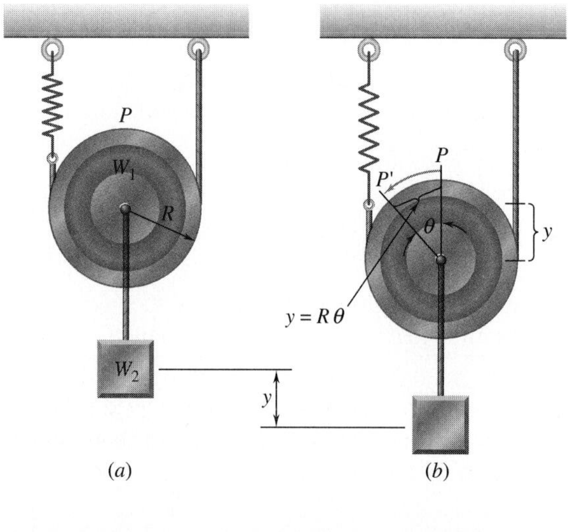

FIGURE 2.4
An unusual spring-suspended weight in equilibrium and after vertical displacement.

or, recalling that the polar moment of inertia of a circular disk is

$$I = M\frac{R^2}{2} = \frac{W_1}{g}\frac{R^2}{2}$$

(9) $$(\text{KE})_b = \frac{1}{2}\frac{W_1}{g}\frac{R^2}{2}(\dot{\theta})^2 = \frac{1}{2}\frac{W_1}{g}\frac{R^2}{2}\left(\frac{\dot{y}}{R}\right)^2 = \frac{W_1}{4g}(\dot{y})^2$$

The conservation of energy now requires that

$$\text{Kinetic energy} + \text{potential energy} = \text{constant}$$

or, substituting from Eqs. (6)–(9),

$$\frac{W_1}{4g}(\dot{y})^2 + \frac{W_1 + W_2}{2g}(\dot{y})^2 + (2ky^2 + 2k\,\delta y) - (W_1 + W_2)y = C$$

Differentiating this energy equation with respect to time, we have

$$\frac{W_1}{2g}\dot{y}\ddot{y} + \frac{W_1 + W_2}{g}\dot{y}\ddot{y} + 4ky\dot{y} + 2k\,\delta\dot{y} - (W_1 + W_2)\dot{y} = 0$$

Dividing out $\dot{y}$ (which surely cannot be identically zero when the system is in motion) and collecting terms gives

$$\frac{3W_1 + 2W_2}{2g}\ddot{y} + 4ky = (W_1 + W_2) - 2k\,\delta = 0$$

the terms on the right equaling zero since the elongation δ of the spring in its equilibrium position, when it is supporting one-half the total weight $W_1 + W_2$, is

$$\delta = \frac{W_1 + W_2}{2k}$$

The differential equation describing the vertical movement of the system is therefore

$$\ddot{y} + \frac{8kg}{3W_1 + 2W_2}y = 0$$

From this, as pointed out before, we can immediately read the natural frequency of the system, namely,

$$\frac{1}{2\pi}\sqrt{\frac{8kg}{3W_1 + 2W_2}} \text{ cycles/unit time}$$

We next consider a mechanical problem that does not involve vibratory motion but which nevertheless leads to a linear differential equation with constant coefficients.

EXAMPLE 3

A CABLE SLIPPING OVER A PEG

A perfectly flexible cable of length $2L$ centimeters, weighing w grams per centimeter, hangs over a frictionless peg of negligible diameter. At $t = 0$, the cable is released from rest in a position in which the portion of the cable on one side is a centimeters longer than that on the other. Find the equation of motion of the cable as it slips over the peg and discuss the motion.

At any time t, let y be the distance that the short end of the cable has risen from its initial position. Then at any instant, the cable on the long side is $(a + 2y)$ cm longer than the cable on the short side (Fig. 2.5). The unbalanced weight of this much cable, namely, $(a + 2y)w$ g, is the force which acts to make the cable move. Since the weight of the entire cable is $2Lw$ g, we then have from Newton's law

$$\frac{2Lw}{g}\frac{d^2y}{dt^2} = (a + 2y)w \qquad \text{or} \qquad \frac{d^2y}{dt^2} - \frac{g}{L}y = \frac{ag}{2L}$$

Using hyperbolic functions, we can write the complementary function of the last equation in the form

$$y_c = c_1 \cosh \sqrt{\frac{g}{L}}t + c_2 \sinh \sqrt{\frac{g}{L}}t$$

By inspection, a particular integral is $Y = -a/2$. Hence a complete solution for y is

$$y = c_1 \cosh \sqrt{\frac{g}{L}}t + c_2 \sinh \sqrt{\frac{g}{L}}t - \frac{a}{2}$$

and, differentiating,

$$v = \sqrt{\frac{g}{L}}\left(c_1 \sinh \sqrt{\frac{g}{L}}t + c_2 \cosh \sqrt{\frac{g}{L}}t\right)$$

Under the conditions of the problem, the cable starts to move from rest in a position in which $y = 0$. Therefore, substituting $y = 0, t = 0$ into the equation for y, and $v = 0, t = 0$ into the equation for v, we obtain $c_1 = a/2$ and $c_2 = 0$. Hence

$$y = \frac{a}{2}\left(\cosh \sqrt{\frac{g}{L}}t - 1\right)$$

This is valid until the short end of the cable reaches the peg, that is, until $y = L - a/2$.

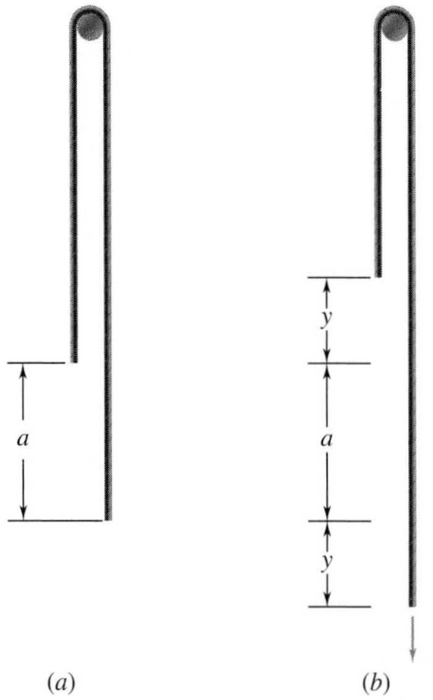

Figure 2.5
A flexible cable slipping over a smooth peg.

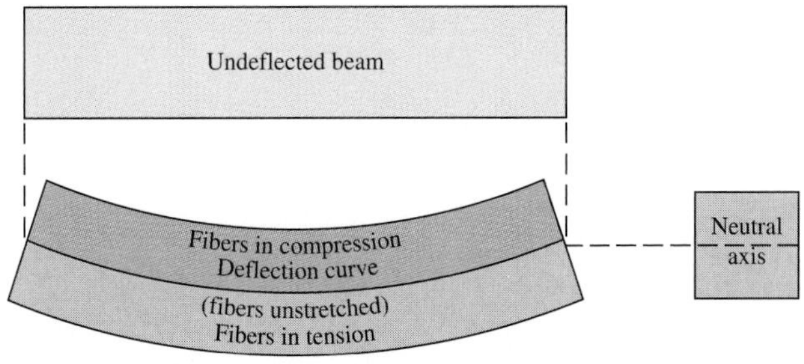

Figure 2.6
A beam before and after bending.

Another important field in which linear differential equations often arise is the bending of beams. Consider a beam which in its undeflected position extends in the direction of the positive x axis. When the beam is bent, it is obvious that the fibers near the concave surface of the beam are compressed, whereas those near the convex surface are stretched. Somewhere between these regions of compression and tension there must, from considerations of continuity, be a surface whose fibers are neither compressed nor stretched. This is known as the **neutral surface** of the beam, and the curve of any particular fiber in this surface is known as the **elastic curve,** or **deflection curve,** of the beam and is taken as the idealized beam itself. The line in which the neutral surface is cut by any plane cross section of the beam is known as the **neutral axis** *of that cross section* (Fig. 2.6).

The loads which cause a beam to bend may be of two sorts: they may be continuously distributed with a density $w(x)$ known as the **load per unit length,** or they may be concentrated at one or more points along the beam. A concentrated load is of course a mathematical fiction which cannot be realized physically since any nonzero load concentrated at a single point would imply a force of infinite intensity which would immediately cut through the beam. Nonetheless, the use of concentrated loads in analyzing various physical systems, such as beams and strings, is both common and fruitful.

A **transverse force** is one whose direction is perpendicular to the length of the beam. The **shear** $V(x)$ at a point x along the beam is the resultant, or algebraic sum, of all transverse forces which act on the beam on the positive side of x. The **moment** $M(x)$ at x is defined as the total moment produced at x by all forces, transverse or not, which act along the beam on one side, or else the other, of the point in question. We shall consider transverse loads and shearing forces to be positive if they act in the direction of the negative y axis (the direction in which loads usually act on a beam). The moment we shall take to be positive if it acts to bend the beam so that it is concave toward the positive y axis (Fig. 2.7). With these conventions of sign (which are not universally adopted), it is shown in the study of the strength of materials that the deflection curve of the beam satisfies the second-order differential equation

$$EIy'' = M \tag{10}$$

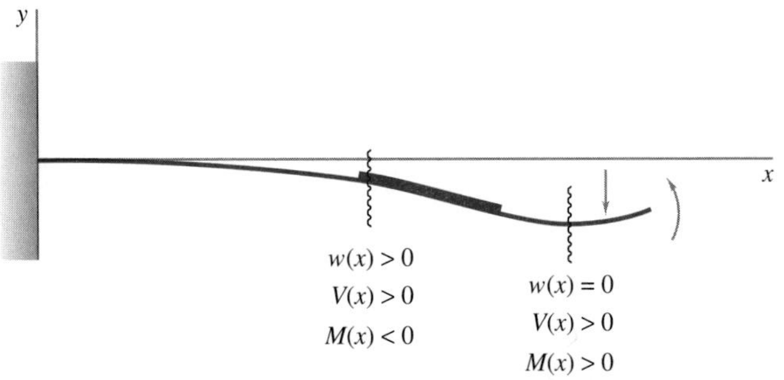

Figure 2.7
The conventions for the signs of the moment, shear, and load per unit length at a general point of a beam.

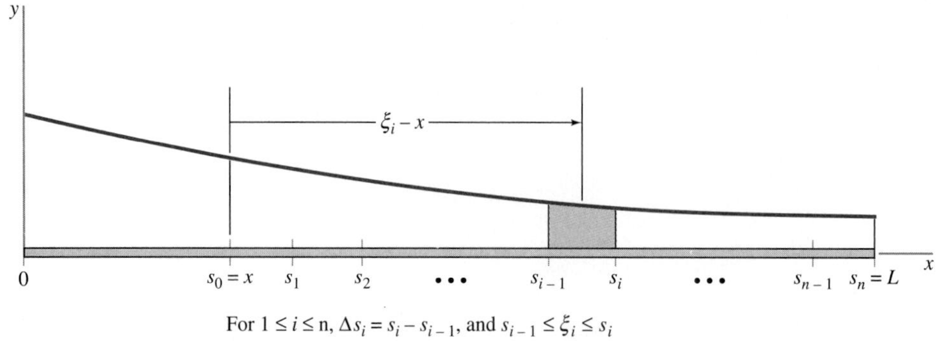

Figure 2.8
A subdivision of the segment from x to L of a beam bearing only a continuously distributed transverse load of density $w(x)$. The load on the subsegment from s_{i-1} to s_i is $w(\xi_i)\,\Delta s_i$.

where E is the modulus of elasticity of the material of the beam, and I, which may be a function of x, is the moment of inertia of the cross-sectional area of the beam about the neutral axis of the cross section at x. If the beam bears only a continuously distributed transverse load of linear density $w(x)$, the shear and the moment at x due to the elementary load $w(\xi_i)\,\Delta s_i$ acting at a distance $\xi_i - x$ to the right of x (Fig. 2.8) are given by

$$\Delta V_i = w(\xi_i)\,\Delta s_i \qquad \text{and} \qquad \Delta M_i = -(\xi_i - x)w(\xi_i)\,\Delta s_i$$

respectively. Summing these from i equals 1 to n and taking limits as the norm of the subdivision of $[x, L]$ goes to zero, we obtain

$$V(x) = \lim_{\|\Delta\|\to 0} \sum_{i=1}^{n} w(\xi_i)\,\Delta s_i = \int_x^L w(s)\,ds$$

and

$$M(x) = -\lim_{\|\Delta\|\to 0} \sum_{i=1}^{n} (\xi_i - x)w(\xi_i)\,\Delta s_i = -\int_x^L (s - x)w(s)\,ds$$

Differentiating the integral for $V(x)$, we get

$$\frac{dV}{dx} = -w(x)$$

Using this result, our integral formula for $V(x)$, integration by parts, and the fact that $V(L) = 0$, we simplify the integral for $M(x)$ as follows:

$$M(x) = -\int_x^L sw(s)\,ds + x\int_x^L w(s)\,ds = \int_x^L sV'(s)\,ds + xV(x)$$

$$= \Big[sV(s)\Big]_x^L - \int_x^L V(s)\,ds + xV(x) = -\int_x^L V(s)\,ds$$

and there follows by differentiation

$$\frac{dM}{dx} = V(x)$$

Recalling Eq. (10) and consolidating results, we have

$$\frac{dM}{dx} = \frac{d(EIy'')}{dx} = V \tag{11}$$

$$\frac{d^2M}{dx^2} = \frac{dV}{dx} = \frac{d^2(EIy'')}{dx^2} = -w \tag{12}$$

In most elementary applications the moment M is an explicit function of x; hence Eq. (10) can be solved and the deflection $y(x)$ determined simply by performing two integrations. However, in problems in which the load has a component in the direction of the length of the beam, M depends on y, and Eq. (10) can be solved only through the use of techniques from the field of differential equations. An interesting example of this sort is provided by the classic problem of the buckling of a slender column.

EXAMPLE 4

BUCKLING OF A COLUMN

A long slender column of length L and uniform cross section whose ends are constrained to remain in the same vertical line but are otherwise free (i.e., able to turn) is compressed by a load F. Determine the possible deflection curves of the column and the loads required to produce each one.

Let coordinates be chosen as shown in Fig. 2.9. Then, clearly, the moment arm of the load F about a general point P on the deflection curve of the beam is y; hence Eq. (10) becomes

$$EIy'' = -Fy \tag{13}$$

the minus sign indicating that when y is positive (as shown), the moment is negative since it has produced a deflection curve which is convex toward the positive y axis.

By hypothesis, the column is of uniform cross section; hence the moment of inertia I is a constant. Therefore (13) is a constant-coefficient differential equation which, when written as $y'' + (F/EI)y = 0$, is seen to have

$$y = A \cos \sqrt{\frac{F}{EI}}x + B \sin \sqrt{\frac{F}{EI}}x \tag{14}$$

as a complete solution.

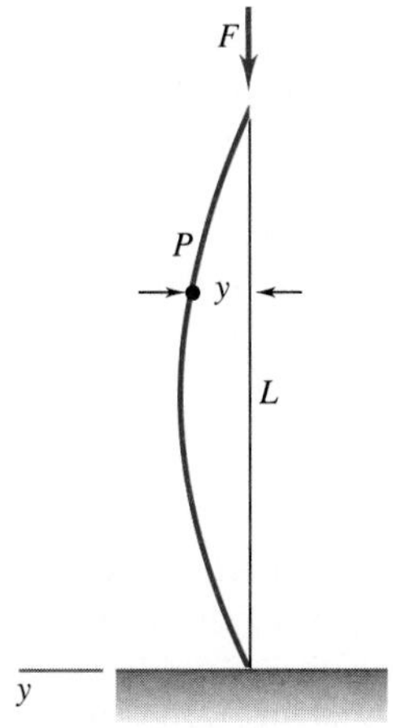

Figure 2.9
A slender column buckling under a vertical load.

To determine the constants A and B, we have the information that $y = 0$ when $x = 0$ and also when $x = L$. Substituting the first of these into (14), we see at once that $A = 0$. Substituting the second, we obtain

$$0 = B \sin \sqrt{\frac{F}{EI}} L$$

Since $\sin \sqrt{(F/EI)}L$ in general is not equal to zero, it follows that $B = 0$, which, since we have already found $A = 0$, means that $y \equiv 0$. However, if the load F has just the right value to make $\sqrt{(F/EI)}L = n\pi$, then the last equation will be satisfied without B being 0 and equilibrium is then possible in a deflected position defined by

$$y = B \sin \frac{n\pi x}{L}$$

Since n can take on any of the values 1, 2, 3, . . . , there are thus infinitely many different critical loads

$$F_n = \left(\frac{n\pi}{L}\right)^2 EI$$

each with its own particular deflection curve. For values of F below the lowest critical load, the column will remain in its undeflected vertical position, or if displaced slightly from it, will return to it as an equilibrium configuration. For values of F above the lowest critical load and different from the higher critical loads, the column can theoretically remain in a vertical position, but the equilibrium is unstable, and if the column is deflected slightly, it will not return to a vertical position but will continue to deflect until it collapses. Thus only the lowest critical load is of much practical significance.

In solving the last example, we evaluated the constants of integration in our complete solution (14) of (13) by imposing suitable end conditions. Were we to be given the load per unit length $w(x)$, the deflection curve might best be found by solving Eq. (12) and then determining the constants of integration in a like manner. For a uniform beam, at a *simply supported,* or *hinged, end* both y and y'' are zero; at a horizontally *built-in end* both y and y' are zero; if the beam extends from 0 to a *free end* at $x = L$, both $y''(L)$ and $y'''(L)$ are zero because the moment and shear are zero there.

EXERCISES

1. Find the modulus of a spring for which a spring-mass system in which $w = 5$ kg will have a circular frequency of 14 rad/s.

For each of the following spring-mass systems, find the displacement y as a function of time. Take $g = 32$ ft/s^2, $g = 384$ in/s^2, or $g = 980$ cm/s^2, as appropriate.

2. $w = 40$ lb, $k = 20$ lb/ft, $y_0 = 2$ ft, $v_0 = -8$ ft/s

3. $w = 15$ g, $k = 3$ g/cm, $y_0 = -1$ cm, $v_0 = 7$ cm/s

4. $w = 45$ kg, $k = 9$ kg/cm, $y_0 = 0$, $v_0 = 28$ cm/s

5. $w = 96$ lb, $k = 16$ lb/in, $y_0 = 10$ in, $v_0 = 0$

6. At $t = 0$, a weight w is suddenly attached to the end of a hanging spring of modulus k. Assuming that friction is negligible, find the subsequent displacement of the weight as a function of time.

7. **(a)** A weight W hangs from a spring of modulus k. Assuming that friction is negligible, derive the differential equation describing the motion of the weight by using the principle of the conservation of energy.

(b) In Part **(a),** suppose that $W = 16$ lb, $k = 6$ lb/in, and that the weight is pulled down 2 in below its equilibrium position and released from rest at that point. Find the equation describing its subsequent motion.

(c) Work Part **(b)** if the weight begins to move from its equilibrium position with an initial velocity of 3 in/s in the positive direction.

(d) Work Part **(b)** if the weight begins to move with initial velocity 8 in/s in the positive direction from a point 4 in above its equilibrium position.

8. A circular cylinder of radius r and height h, made of material weighing w lb/in^3, floats in water in such a way that its axis is always vertical. Neglecting all forces except gravity and the buoyant force of the water, as given by the principle of Archimedes, determine the period with which the cylinder will vibrate in the vertical direction if it is depressed slightly from its equilibrium position and released.

9. A cylinder weighing 50 lb floats in water with its axis vertical. When depressed slightly and released, it vibrates with period 2 s. Neglecting all frictional effects, find the diameter of the cylinder.

10. A straight hollow tube rotates with constant angular velocity ω about a vertical axis which is perpendicular to the tube at its midpoint. A pellet of mass m slides without friction in the interior of the tube. Find the equation describing the radial motion of the pellet until it emerges from the tube, assuming that it starts from rest at a radial distance a from the midpoint of the tube.

11. In Exercise 10, find the magnitude of the force exerted on the pellet by the tube.

12. A straight hollow tube rotates with constant angular velocity ω about a horizontal axis which is perpendicular to the tube at its midpoint. Show that if the initial conditions are suitably chosen, a pellet sliding without friction in the tube will never be ejected but will execute simple harmonic motion within the tube.

13. A beam one of whose ends is free and the other built-in (usually perpendicular to the support) is called a **cantilever beam.** Find an equation for the deflection curve and the **tip deflection,** i.e., the deflection of the free end, of a uniform cantilever beam if a weight $2w$ is hung at its center $x = L/2$. *Hint:* Between the free end of the beam at $x = L$ and the point of loading the moment is identically zero, but this is not the case between the fixed end and the midpoint $x = L/2$. Hence two differential equations are required to obtain the deflection curve $y = y(x)$, and at $x = L/2$ both y and y' must be continuous.

14. Neglecting the effect of its own weight, find the deflection of a uniform cantilever beam of length L at the point x due to a unit downward load at the point s and show that it equals the deflection at s due to a unit positive load at x.

15. A long slender column of uniform cross section is built-in rigidly at its base. Its upper end, which is free to move out of line, bears a vertical load F. Determine the possible deflection curves of the column and the load required to produce each one. *Hint:* Unlike Example 4, it is convenient here to take the origin at the free end of the column. Then $y(0) = 0$ and $y'(L) = 0$.

16. Find an equation for the deflection curve of a uniform cantilever beam of length L bearing a constant load per unit length w.

17. A uniform cantilever beam of length L is subjected to an oblique tensile force F at the free end. Find the deflection as a function of x and the angle θ between the direction of the force and the initial direction of the beam. What is the tip deflection?

A uniform cantilever beam of length L is built-in at $x = 0$. Find the tip deflection if the load per unit length is given by

18. $w(x) = 2L - x$ **19.** $w(x) = L - x$

20. Find an equation for the deflection curve of a uniform beam of length $2L$ whose ends are hinged if the load per unit length is $w(x) = k[u(x) - u(x - L)]$, where k is constant and $u(x)$ is the unit step function.

21. A uniform beam hinged at $x = 0$ and built-in at $x = 1$ bears a load per unit length $w(x) = a + b \sin \pi x$, a, b constants. Find an equation for its deflection curve.

22. A cantilever beam has the shape of a solid of revolution whose radius varies as $\sqrt{x}$, where x is the distance from the free end of the beam. A tensile force F inclined downward at an angle of 45° with the initial direction of the beam is applied at the free end. Find the deflection curve of the beam.

23. A cantilever beam of uniform cross section and length L bears a concentrated load P at its free end. A tensile force F also acts at the free end in the direction of the undeflected beam. By choosing coordinates as shown in Fig. 2.10, find the equation of the deflection curve of the beam and the deflection of the free end relative to the fixed end.

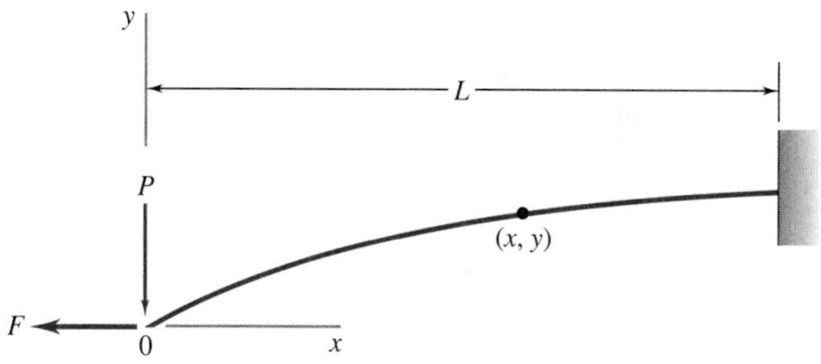

Figure 2.10

24. Under the assumption of very small motions (so that the end of the spring and the weight W may be considered to move in a purely vertical direction) and neglecting friction, determine the natural frequency of the system shown in Fig. 2.11 if the bar is of uniform cross section, absolutely rigid, and of weight w. *Hint:* Use the energy method to obtain the differential equation of the system, recalling that the moment of inertia of a uniform bar of length l and mass m about its midpoint is $\frac{1}{12}ml^2$.

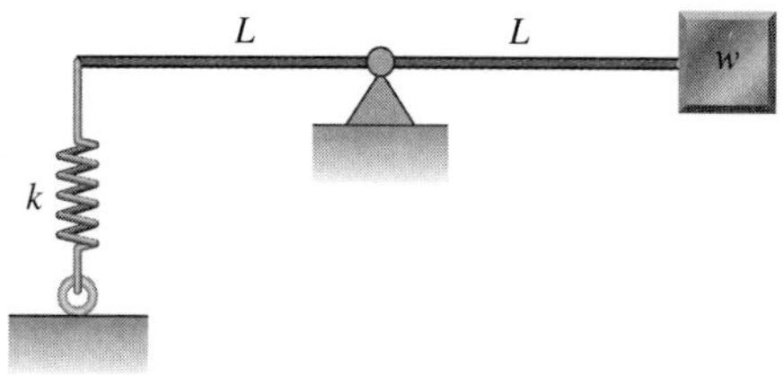

Figure 2.11

25. Work Exercise 24 if the bar is of weight $9W$.

26. Work Exercise 24 if the bar is of weight W.

27. Work Exercise 24 if the bar is of weight W and the end load is connected above and below to vertical springs of modulus k which are neither stretched nor compressed when the bar is horizontal.

28. Under the assumption of very small motions and neglecting friction, determine the natural frequency of the system shown in Fig. 2.12 if the bar is of uniform cross section, absolutely rigid, and of weight w.

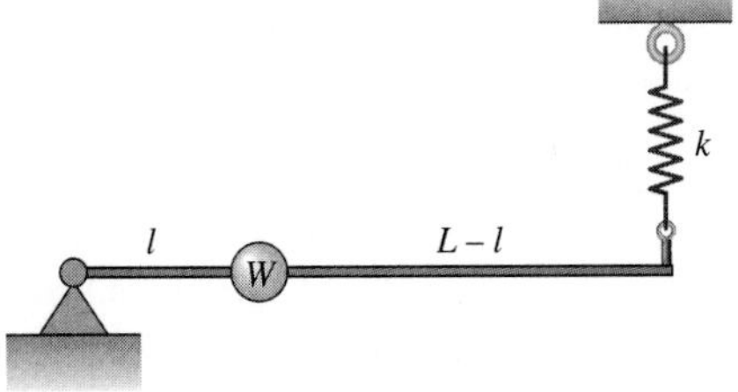

Figure 2.12

29. A weight W hangs by an inextensible cord from the circumference of a pulley of radius R and moment of inertia I. The pulley is prevented from rotating freely by a spring of modulus k, attached as shown in Fig. 2.13. Considering only displacements so small that the departure of the spring from the horizontal can be neglected and neglecting all friction, determine the natural frequency of the oscillations that occur when the system is slightly disturbed.

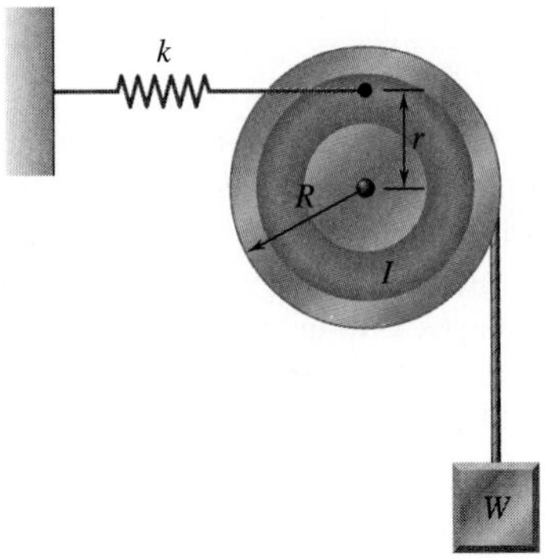

Figure 2.13

30. Under assumptions similar to those of Example 2, find the natural frequency of the mechanical system shown in Fig. 2.14. *Hint:* As in Eq. (9), the kinetic energy of each pulley can be expressed in terms of W and $\dot{y}$ so that the radii of the pulleys need not be specified.

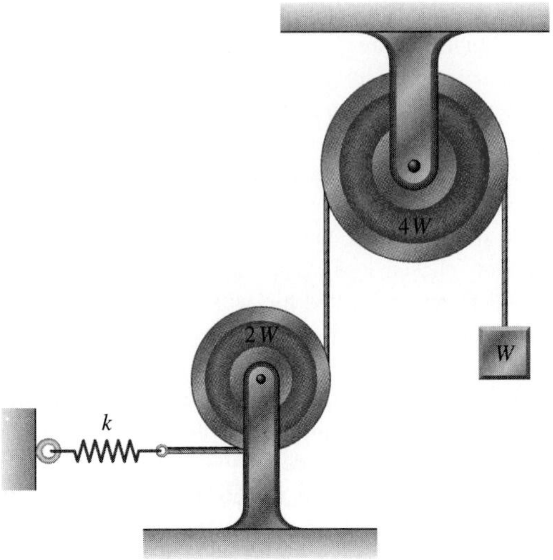

Figure 2.14

31. Work Exercise 30 if the lower pulley is of weight W, the upper pulley of weight $3W$, and a weight $14W$ is suspended from the cable.

32. Find the natural frequency of the system shown in Fig. 2.15.

33. A perfectly flexible cable of length L, weighing w lb/ft, hangs over a pulley as shown in Fig. 2.16a. The radius of the pulley is R, and its moment of inertia is I. Friction between the cable and the pulley prevents any relative slipping, although the pulley is free to turn without appreciable friction. At $t = 0$ the cable is released from rest in a position in which the portion hanging on one side is a ft longer than that hanging on the other. Determine the mo-

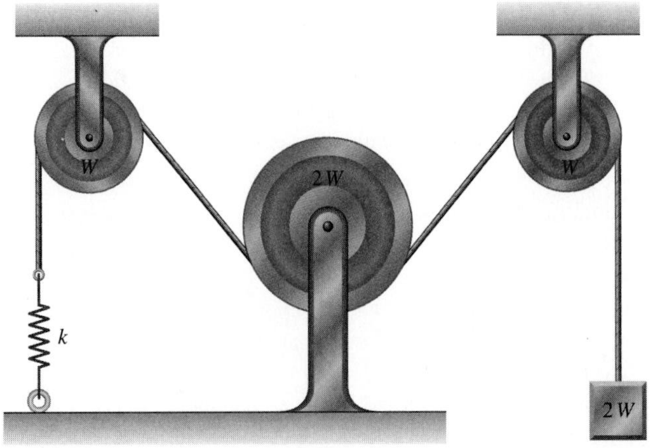

Figure 2.15

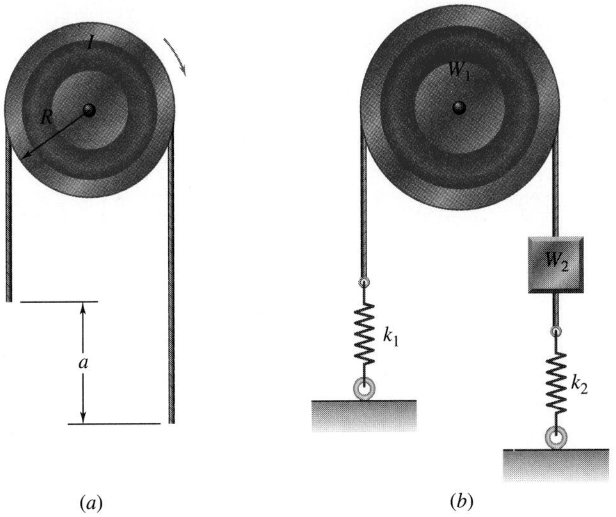

Figure 2.16

tion of the cable until the short end first makes contact with the pulley.

34. A spring of modulus k_1 is attached to one end of an inextensible cable which passes over a pulley of weight W_1, and the other end of the cable is attached to a weight W_2 which is fastened to a spring of modulus k_2, as shown in Fig. 2.16*b*. Neglecting the weight of the cable and springs, find the natural frequency of the system.

35. A perfectly flexible cable of length L ft, weighing w lb/ft, lies in a straight line on a frictionless table top, a ft of the cable hanging over the edge. At $t = 0$ the cable is released and begins to slide off the edge of the table. Assuming that the height of the table is greater than L, determine the motion of the cable until it leaves the table top.

36. Work Exercise 35 if a weight W is attached at the end of the overhang.

37. Work Exercise 35 if $L = 4$ ft, $w = 8$, $a = 1$ ft, and a 32-lb weight is attached at the end of the overhang. *Hint:* Let y denote the instantaneous length of the overhang.

38. A weight w on top of a frictionless surface is connected to a weight W by an inextensible cable which passes over a pulley of weight P as shown in Fig. 2.17. The cable is of length L and its linear density is k. Determine the motion of W until w reaches the edge of the surface if $y(0) = a$ and $\dot{y}(0) = 0$.

39. Work Exercise 38 if $w = 4$ lb, $W = 10$ lb, $P = 4$ lb, $L = 16$ ft, $k = 2$ lb/ft, and $a = 3$ ft.

40. Figure 2.18*a* portrays two concentrated weights W_1 and W_2 attached to a perfectly flexible cable of length $2L$ weighing w g/cm and hanging over a frictionless peg of negligible diameter. At $t = 0$, the system is at rest in a position where W_1 is a units above W_2.

(a) Choosing y as in Example 3, find a differential equation of motion for the system.

(b) Find a complete solution of the equation of motion and impose the initial conditions $y(0) = \dot{y}(0) = 0$.

(c) Show that the results of Example 3 follow from the unique solution of Part **(b).**

(d) Under what conditions, if any, does the system remain motionless?

(e) Discuss the motion of the system if $aw + W_2 - W_1$ is positive, or else negative.

41. A perfectly flexible 60-ft-long cable, weighing $\frac{1}{2}$ lb/ft, has weights of $w = 21$ lb and $W = 29$ lb attached at its

Figure 2.17

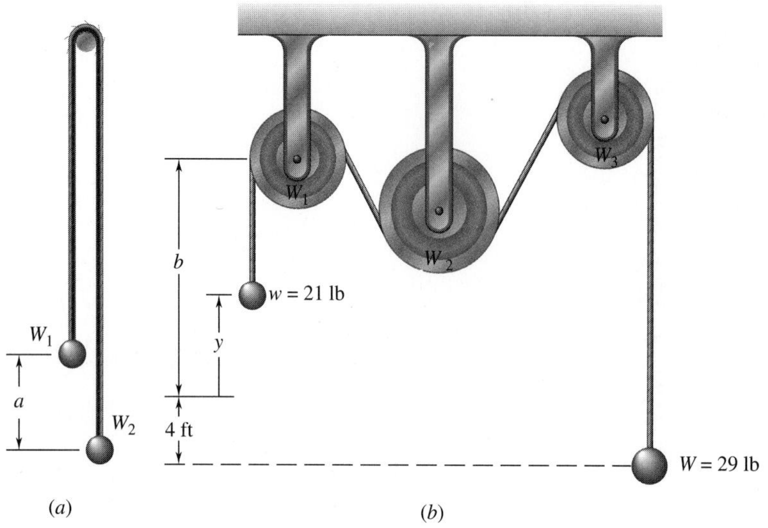

Figure 2.18

ends. The cable passes over two pulleys and under a third as shown in Fig. 2.18*b*. Given that $W_1 = 32$ lb, $W_2 = 48$ lb, and $W_3 = 16$ lb, determine the upward displacement y of w from its initial position assuming w starts from rest while 4 ft above W. *Hint:* Neglect friction and compute potential energies relative to the horizontal through the center of the highest pulley.

42. A uniform shaft of length L rotates about the x axis with constant angular velocity ω. The ends of the shaft are held in bearings which are free to swing out of line, as shown in Fig. 2.19, if the shaft deflects from its neutral position. Show that there are infinitely many critical speeds at which the shaft can rotate in a deflected position, and find these speeds and the associated deflection curves. *Hint:* As a consequence of its rotation, the shaft experiences a load per unit length in the radial direction equal to

$$w(x) = -\frac{\rho A \omega^2}{g} y$$

where A is the cross-sectional area of the shaft and ρ is the density of the material of the shaft. Substitute this into Eq. (12), solve the resulting differential equation, and then impose the conditions that at $x = 0$ and at $x = L$ both the deflection of the shaft and the moment are zero. It will be convenient to use hyperbolic rather than exponential functions in taking account of the real roots of the characteristic equation.

43. Work Exercise 42 if the bearings are fixed in position and cannot swing out of line.

44. Neglecting friction and assuming angular displacements θ so small that θ is a satisfactory approximation to $\sin\theta$ and $\theta^2/2$ is a satisfactory approximation to $1 - \cos\theta$, find the natural frequency of the system shown in Fig. 2.20 if the bar is of uniform cross section, absolutely rigid, and of weight w.

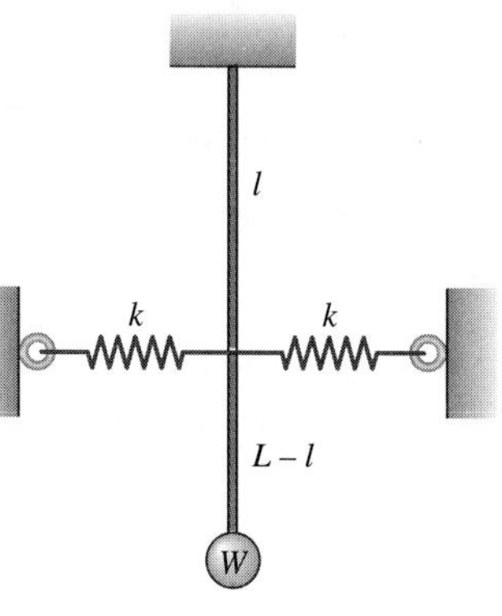

Figure 2.20

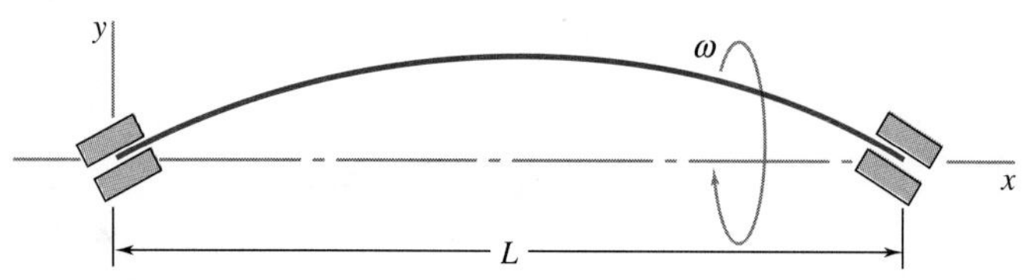

Figure 2.19

45. Neglecting friction and assuming angular displacements θ so small that θ is a satisfactory approximation to $\sin\theta$ and $\theta^2/2$ is a satisfactory approximation to $1 - \cos\theta$, find the natural frequency of the system shown in Fig. 2.21 if the bar is of uniform cross section, absolutely rigid, and of weight w. In what significant respect does this system differ from that in Exercise 44?

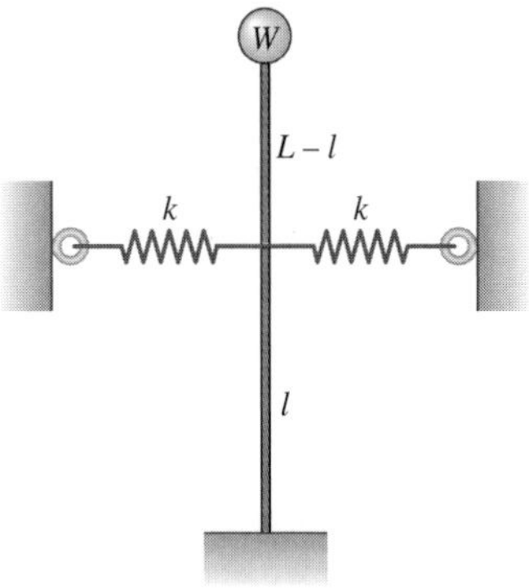

Figure 2.21

46. **(a)** Two disks each of moment of inertia I are connected by an elastic shaft of modulus k, that is, a shaft which requires k units of torque to twist one end through an angle of 1 rad with respect to the other end. The system is mounted in frictionless bearings, as shown in Fig. 2.22. Neglecting the moment of inertia of the shaft, find the natural frequency with which the disks will oscillate if they are twisted through equal but opposite angles and then released.

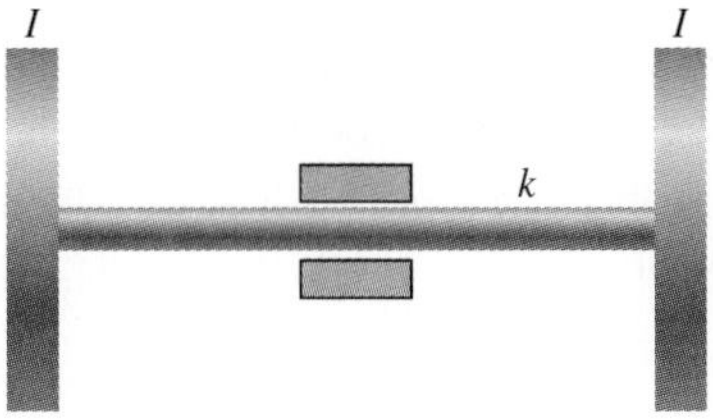

Figure 2.22

(b) What is the natural frequency of the system if the moments of inertia of the disks are, respectively, I_1 and I_2? *Hint:* Not only does the total energy of the system remain constant, but so does the total angular momentum.

47. A pendulum consisting of a mass m at the end of an inextensible cord of length l and negligible mass swings between maximum angular displacements of $\pm\alpha$. If θ is the instantaneous angular displacement of the pendulum from the vertical, use the energy method to show that

$$(\dot{\theta})^2 = 2\frac{g}{l}(\cos\theta - \cos\alpha)$$

From this, determine the period of the pendulum if it swings through an angle small enough to make θ a satisfactory approximation to $\sin\theta$.

48. **(a)** In the expression for $\dot{\theta}$ from Exercise 47, use the half-angle formula $\cos u = 1 - 2\sin^2(u/2)$ to show that

$$(\dot{\theta})^2 = \frac{4g}{l}\left(\sin^2\frac{\alpha}{2} - \sin^2\frac{\theta}{2}\right)$$

Integrate this differential equation assuming that $\theta = 0$ when $t = 0$.

(b) In the integral obtained in Part **(a)** change the variable of integration from θ to ϕ by the substitution

$$\sin\frac{\theta}{2} = \sin\frac{\alpha}{2}\sin\phi$$

and show that

$$t = \sqrt{\frac{l}{g}}\int_0^{\phi}\frac{d\phi}{\sqrt{1 - k^2\sin^2\phi}} \qquad \text{where } k^2 = \sin^2\frac{\alpha}{2}$$

This integral is known as an **elliptic integral of the first kind** and is commonly denoted $F(\phi, k)$. The function $F(\phi, k)$ is tabulated in most elementary handbooks.

49. Using the results of Exercises 47 and 48, together with tables of $F(\phi, k)$, compare the true period of a pendulum swinging through an angle of 90° on each side of the vertical with the period computed under the simplifying assumption that $\sin\theta \doteq \theta$.

2.12 GREEN'S FUNCTIONS

The mathematical objects known as *Green's functions*† serve as mathematical characterizations of important physical concepts. They are of fundamental importance in many practical applications and in the theory of differential equations. They can be introduced either in a purely mathematical way or via their physical counterparts. However, the motivation for their study is more striking, perhaps, when they are developed in a physical setting. So to begin with, we select a simple problem from mechanics whose solution will lead us quickly to a special instance of a Green's function.

†Named for the English mathematical physicist, George Green (1793–1841).

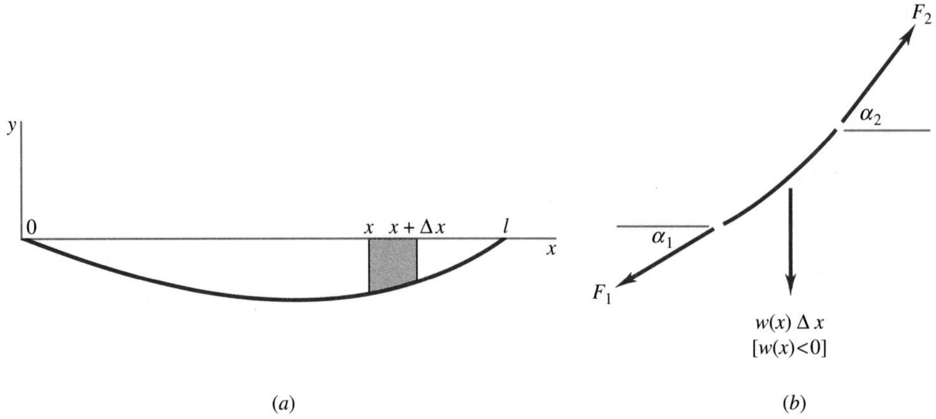

Figure 2.23
A stretched string deflected by a distributed load.

Consider a perfectly flexible elastic string stretched to a length l under tension T. We assume, as a first possibility, that the string bears a distributed load per unit length $w(x)$ which includes the weight of the string itself. Furthermore, we shall suppose that the static deflections produced by this load are all perpendicular to the original, undeflected position of the string and are all in the same plane (Fig. 2.23*a*). Hence, given any two values of x in $[0, l]$, the load acting on the portion of the string between these points is the same before and after the string deflects.

On an arbitrary element of the string we then have the forces shown in Fig. 2.23*b*. As the figure indicates, unlike loadings on a beam, the *positive* direction of $w(x)$ in the case of a string has been chosen *upward.* Since the deflected string is in equilibrium, the net horizontal force and the net vertical force on the element must both be zero. Hence

(1) $$F_1 \cos \alpha_1 = F_2 \cos \alpha_2$$

(2) $$F_2 \sin \alpha_2 = F_1 \sin \alpha_1 - w(x)\,\Delta x$$

The first of these equations tells us that the horizontal component of the force in the string is a constant, and we shall further assume that the deflections are so small that this constant horizontal component does not differ appreciably from the tension T in the string before it is loaded. Then, dividing the respective terms in Eq. (2) by the ''equal'' quantities $F_2 \cos \alpha_2$, $F_1 \cos \alpha_1$, and T, we have

(3) $$\tan \alpha_2 = \tan \alpha_1 - \frac{w(x)\,\Delta x}{T}$$

Now $\tan \alpha_2$ is the slope of the deflection curve at the point $x + \Delta x$, and $\tan \alpha_1$ is the slope of the deflection curve at the point x. Hence Eq. (3) can be rewritten in the form

$$\frac{y'(x + \Delta x) - y'(x)}{\Delta x} = -\frac{w(x)}{T}$$

In the limit as $\Delta x \to 0$, we thus obtain

(4) $$Ty'' = -w(x)$$

as the differential equation satisfied by the deflection curve of the string.

Let us now determine the deflection of the string under the influence of a concentrated rather than a distributed load. Of course, this presumes a weightless string which, like the concept of a concentrated load, is another mathematical fiction. At the outset, we note from Eq. (4) that y'' is zero at all points of the string where there is no distributed load. Hence, since $y'' = 0$ implies that y is a linear function, it follows that the deflection curve of the string under the influence of a single concentrated load P consists of two linear segments, as shown in Fig. 2.24. As in our earlier discussion, the equilibrium of the string implies the following conditions:

$$F_1 \cos \alpha_1 = F_2 \cos \alpha_2 = T$$

$$F_1 \sin \alpha_1 + F_2 \sin \alpha_2 = -P$$

From these we obtain, as before, the equation

$$\tan \alpha_1 + \tan \alpha_2 = \frac{-P}{T}$$

or

$$\frac{-\delta}{s} + \frac{-\delta}{l-s} = \frac{-P}{T} \qquad \text{and} \qquad \delta = \frac{P(l-s)s}{Tl}$$

where δ denotes the transverse deflection of the point of the string that was initially a distance s from the origin.

With the deflection δ known, it is a simple matter to use similar triangles to find the deflection of the string at any point x, The results are

$$y(x, s) = \begin{cases} \dfrac{P(l-s)x}{Tl} & 0 \le x \le s \\ \dfrac{P(l-x)s}{Tl} & s \le x \le l \end{cases} \tag{5}$$

The notation $y(x, s)$, rather than just $y(x)$, is used of course to indicate that the deflection y depends on both the point s where the load is applied and the point x where the deflection is observed. These formulas give the deflection which a load P applied to the string at a point s produces at a point x. Conversely, if the roles of x and s are interchanged, these formulas give the deflection which a load

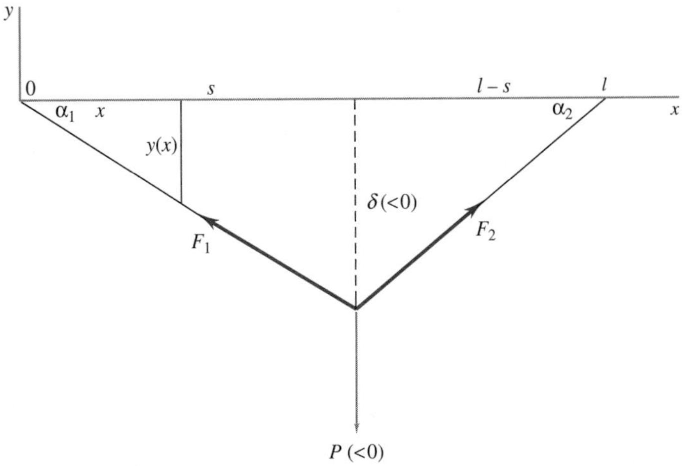

Figure 2.24
A stretched string deflected by a concentrated load.

P applied at the point x produces at the point s. For instance, if $x < s$, the deflection at x due to a load P at s is given by the first of the formulas (5) and is simply

$$\frac{P(l-s)x}{Tl}$$

If the roles of x and s are interchanged, the load is now at x and the deflection is measured at s ($>x$). Its value is therefore given by the second formula in (5) with x and s interchanged of course, hence is

$$\frac{P(l-s)x}{Tl}$$

as before. Thus it is clear that *the deflection of a string at a point x due to a concentrated load P applied at a point s is the same as the deflection produced at the point s by an equal load applied at the point x.* When P is a unit load, it is customary to use the notation $g(x, s)$ as a new name for the corresponding function $y(x, s)$ defined by (5). The preceding observation then asserts that $g(x, s)$ is symmetric in the two variables x and s; that is,

$$g(x, s) = g(s, x)$$

Frequently, $g(x, s)$ is referred to as an **influence function** since it describes the *influence* a unit load concentrated at the point s has at any point x of the string. For definiteness, and to accord with most standard treatments of Green's functions, we shall always express $g(x, s)$ in terms of a unit *positive* load. Thus, in the present problem, we take

$$(6) \qquad g(x, s) = \begin{cases} \dfrac{(l-s)x}{Tl} & 0 \le x \le s \\[2ex] \dfrac{(l-x)s}{Tl} & s \le x \le l \end{cases}$$

to be our influence function, or **Green's function,** for the string. The symmetry of $g(x, s)$ is an illustration of the important *Maxwell-Rayleigh reciprocity law,*† which holds in many physical systems, both mechanical and electrical.

It is interesting and important to note that by means of the influence function $g(x, s)$ an expression for the deflection of a string under an arbitrary distributed load can be found without solving Eq (4). To do this, we reason as follows. Let the interval $[0, l]$ be subdivided into n subintervals by the points $0 = s_0 < s_1 < s_2 < \cdots < s_n = l$, let $\Delta s_i = s_i - s_{i-1}$, and let ξ_i be an arbitrary point of the subinterval $[s_{i-1}, s_i]$. Further, let the portion of the distributed load acting on the segment of the beam from s_{i-1} to s_i, namely, $\Delta w(\xi_i)\,\Delta s_i$, be regarded as concentrated at the point $s = \xi_i$. The deflection produced at the point x by this load is the product of the load and the deflection produced at x by a *unit* load at the point $s = \xi_i$, namely,

$$[w(\xi_i)\,\Delta s_i]g(x, \xi_i)$$

†James Clerk Maxwell (1831–1879) was a great Scottish mathematical physicist. Lord Rayleigh (1842–1919) was an English mathematical physicist.

If we add up all the deflections produced at the point x by the various small concentrated forces which together approximate the actual distributed load, we obtain the sum

$$\sum_{i=1}^{n} w(\xi_i)g(x, \xi_i)\, \Delta s_i$$

In the limit, as each $\Delta s_i \to 0$, this sum becomes an integral, and for the deflection at an arbitrary point x we have the formula

$$y(x) = \int_0^l w(s)g(x, s)\, ds \tag{7}$$

Thus, once the function $g(x, s)$ is known, the deflection of the string under *any* piecewise continuous distributed load is given immediately by the integral (7).

We have already observed that the influence function (6) associated with the stretched string with fixed ends is a symmetric function of the two arguments x and s. Other properties of $g(x, s)$ are also worth noting. In the first place, it is obvious that $g(x, s)$ satisfies the boundary conditions of the problem; i.e., just as the string satisfies the conditions that its deflection is zero when $x = 0$ and when $x = l$, so it is also true that $g(0, s) = g(l, s) = 0$ for all values of s such that $0 \le s \le l$. It is also easy to verify that $g(x, s)$ is a continuous function of x on the interval $[0, l]$. This is obvious, except possibly at the point $x = s$, where we have for the left- and right-hand limits of $g(x, s)$

$$\lim_{x\to s^-} g(x, s) = \lim_{x\to s^-} \frac{(l-s)x}{Tl} = \frac{(l-s)s}{Tl}$$

$$\lim_{x\to s^+} g(x, s) = \lim_{x\to s^+} \frac{(l-x)s}{Tl} = \frac{(l-s)s}{Tl}$$

and these are clearly equal and equal to $g(s, s)$. On the other hand, the derivative of $g(x, s)$ with respect to x is discontinuous at the point $x = s$.

We now proceed to show that $g_x(x, s)$ has a *downward* jump of $-1/T$ at $x = s$. To do this, we note first that $g(x, s)$ is obviously differentiable at all points of the interval $[0, l]$ except possibly at $x = s$. There we observe that

$$\lim_{x\to s^-} g_x(x, s) = \lim_{x\to s^-} \frac{l-s}{Tl} = \frac{l-s}{Tl}$$

$$\lim_{x\to s^+} g_x(x, s) = \lim_{x\to s^+} \left(-\frac{s}{Tl}\right) = -\frac{s}{Tl}$$

These limiting values are not equal, and their difference is

$$g_x(s^+, s) - g_x(s^-, s) = -\frac{s}{Tl} - \frac{l-s}{Tl} = -\frac{1}{T}$$

as asserted. Finally, we note that since $g(x, s)$ consists of two linear expressions, it satisfies the homogeneous differential equation $Ty'' = 0$ at all points of the interval $[0, l]$ except at $x = s$. In fact, at $x = s$ the second derivative $g_{xx}(x, s)$ does not exist since, as we have just observed, $g_x(x, s)$ is discontinuous at that point.

The properties we have just noted are not accidental characteristics of the influence function of one particular problem. Instead, as the following definition makes clear, they identify an important class of functions associated with linear differential equations with variable as well as constant coefficients.

DEFINITION 1 Consider a normal differential equation on $[a, b]$

$$(8) \qquad a_0(x)y'' + a_1(x)y' + a_2(x)y = 0$$

and the homogeneous boundary conditions $\alpha_1 y(a) = \alpha_2 y'(a)$, $\beta_1 y(b) = \beta_2 y'(b)$, where α_1 and α_2 are not both zero and β_1 and β_2 are not both zero. A function $g(x, s)$ with the property that

1. $g(x, s)$ satisfies the differential equation for $a \le x < s$ and for $s < x \le b$.
2. $\alpha_1 g(a, s) = \alpha_2 g_x(a, s)$, $\beta_1 g(b, s) = \beta_2 g_x(b, s)$ for $a \le s \le b$.
3. $g(x, s)$ is a continuous function of x for $a \le x \le b$.
4. $g_x(s, x)$ is continuous for $a \le x < s$ and for $s < x \le b$ but has a jump $-1/a_0(s)$ at $x = s$.

is called the **Green's function** of the problem defined by the given differential equation and its boundary conditions†.

EXAMPLE 1

Using Definition 1, construct the Green's function for the differential equation $y'' + k^2 y = 0$ with the boundary conditions $y(0) = y(b) = 0$, $k \neq 0$.

Since any solution of the equation $y'' + k^2 y = 0$ is of the form $y = A \cos kx + B \sin kx$, it follows from Property 1 that the required function $g(x, s)$ must be defined by expressions of the form

$$g(x, s) = \begin{cases} A_1 \cos kx + B_1 \sin kx & 0 \le x \le s \\ A_2 \cos kx + B_2 \sin kx & s \le x \le b \end{cases}$$

In order for the left-hand boundary condition to be met, as required by Property 2, it is necessary that $A_1 = 0$. Similarly, in order for the boundary condition at $x = b$ to be satisfied by $g(x, s)$, it is necessary that $A_2 \cos kb + B_2 \sin kb = 0$, which will be the case if we take $A_2 = C \sin kb$ and $B_2 = -C \cos kb$, where C is arbitrary. Thus $g(x, s)$ is restricted to the form

$$g(x, s) = \begin{cases} B_1 \sin kx & 0 \le x \le s \\ C(\sin kb \cos kx - \cos kb \sin kx) = C \sin k(b - x) & s \le x \le b \end{cases}$$

Further, in order for $g(x, s)$ to be continuous at $x = s$, as required by Property 3, it is necessary that $B_1 \sin ks = C \sin k(b - s)$, from which it follows that $B_1 = E \sin k(b - s)$ and $C = E \sin ks$, where E is arbitrary. Thus $g(x, s)$ is further reduced to the form

$$g(x, s) = \begin{cases} E \sin k(b - s) \sin kx & 0 \le x \le s \\ E \sin ks \sin k(b - x) & s \le x \le b \end{cases}$$

Finally, to satisfy Property 4, we must have

$$\lim_{x \to s^+} g_x(x, s) - \lim_{x \to s^-} g_x(x, s) = -1 \qquad [\text{Since } a_0(x) = 1]$$

†Questions concerning the existence and uniqueness of Green's functions are discussed in Chap. 14. The property of symmetry, i.e., the property that $g(x, s) = g(s, x)$, is not a part of the definition of a Green's function but is a consequence of the other properties when the differential system has a special form. (See Exercises 28 and 29.)

or

$$\lim_{x\to s^+}[-kE\sin ks\cos k(b-x)] - \lim_{x\to s^-}[kE\sin k(b-s)\cos kx] = -1$$

$$-kE[\sin ks\cos k(b-s) + \sin k(b-s)\cos ks] = -1$$

$$-kE\sin kb = -1$$

$$E = \frac{1}{k\sin kb}$$

With E known, the Green's function $g(x, s)$ is completely determined, and we have

$$g(x,s) = \begin{cases} \dfrac{\sin kx \sin k(b-s)}{k\sin kb} & 0 \le x \le s \\ \dfrac{\sin ks \sin k(b-x)}{k \sin kb} & s \le x \le b \end{cases}$$

provided of course that $kb \neq n\pi$†. It is interesting to note that in this example $g(x, s) = g(s, x)$ even though we did not impose this condition in the course of our derivation.

Green's functions not only are closely related to the influence functions which arise in many practical problems but also have much in common with the results of the method of variation of parameters, discussed in Sec. 2.4. To explore this matter, let us consider again the differential equation

$$a_0(x)y'' + a_1(x)y' + a_2(x)y = f(x) \tag{9}$$

or, equivalently,

$$y'' + \frac{a_1(x)}{a_0(x)}y' + \frac{a_2(x)}{a_0(x)}y = \frac{f(x)}{a_0(x)} \qquad a_0(x) \neq 0 \text{ for } a \le x \le b$$

If $y_1(x)$ and $y_2(x)$ are two linearly independent solutions of the related homogeneous equation, then, according to the method of variation of parameters [Eqs. (6), Sec. 2.4], $Y = u_1y_1 + u_2y_2$ will be a solution of the given nonhomogeneous equation provided

$$u_1' = -\frac{y_2}{y_1y_2' - y_1'y_2}\frac{f}{a_0} \qquad u_2' = \frac{y_1}{y_1y_2' - y_1'y_2}\frac{f}{a_o}$$

†It is worth noting that if $kb = n\pi$, there is a nontrivial function meeting the boundary conditions $y(0) = y(b) = 0$ and satisfying the equation $y'' + k^2y = 0$ at *all* points of the interval $[0, b]$. In fact, beginning with the general solution

$$y = A\cos kx + B\sin kx$$

it is clear that the conditions $y(0) = y(b) = 0$ will be met if and only if $A = 0$ and $B\sin kb = 0$. Since $kb = n\pi$, the second condition is satisfied for all values of B; that is, B need not be zero. This illustrates the important general result that *for equations containing a parameter* (such as k in the equation $y'' + k^2y = 0$) *Green's function fails to exist for any value of the parameter for which there is a nontrivial solution of the differential equation which satisfies the boundary conditions of the problem.*

Let us particularize u_1 and u_2 by integrating their derivatives from a to x and from x to b, respectively. Then, recalling that $y_1y_2' - y_1'y_2$ is the wronskian W of the two solutions y_1 and y_2, we have, using s as a dummy variable of integration,

$$u_1 = -\int_a^x \frac{y_2(s)f(s)}{W(s)a_0(s)}\,ds \qquad u_2 = -\int_x^b \frac{y_1(s)f(s)}{W(s)a_0(s)}\,ds$$

and our particular integral is

$$Y = u_1y_1 + u_2y_2 = -y_1(x)\int_a^x \frac{y_2(s)f(s)}{W(s)a_0(s)}\,ds - y_2(x)\int_x^b \frac{y_1(s)f(s)}{W(s)a_0(s)}\,ds$$

or, moving $y_1(x)$ and $y_2(x)$ into the respective integrands,

$$Y = -\left[\int_a^x \frac{y_1(x)y_2(s)}{W(s)a_0(s)}f(s)\,ds + \int_x^b \frac{y_1(s)y_2(x)}{W(s)a_0(s)}f(s)\,ds\right]$$

which is of the form

$$Y = -\int_a^b k(x, s)\,f(s)\,ds \tag{10}$$

where

$$k(x, s) = \begin{cases} \dfrac{y_1(s)y_2(x)}{W(s)a_0(s)} & x \le s \le b, \text{ that is, } a \le x \le s \\[2ex] \dfrac{y_1(x)y_2(s)}{W(s)a_0(s)} & a \le s \le x, \text{ that is, } s \le x \le b \end{cases} \tag{11}$$

For each value of s the function $k(x, s)$ is a continuous function of x for $a \le x \le b$ because y_1 and y_2 are continuous on $[a, b]$ and at $x = s$ the right- and left-hand limits $k(s^+, s)$ and $k(s^-, s)$ are equal. Equation (11) gives us

$$k_x(x, s) = \begin{cases} \dfrac{y_1(s)y_2'(x)}{W(s)a_0(s)} & a \le x < s \\[2ex] \dfrac{y_1'(x)y_2(s)}{W(s)a_0(s)} & s < x \le b \end{cases} \tag{12}$$

From the properties of y_1 and y_2 and the discontinuity in $k_x(x, s)$ given by

$$\begin{aligned} k_x(s^+, s) - k_x(s^-, s) &= \frac{y_1'(s)y_2(s)}{W(s)a_0(s)} - \frac{y_1(s)y_2'(s)}{W(s)a_0(s)} \\ &= -\frac{y_1(s)y_2'(s) - y_1'(s)y_2(s)}{W(s)a_0(s)} = -\frac{1}{a_0(s)} \end{aligned}$$

it follows that $k_x(x, s)$ is continuous on $a \le x < s$ and on $s < x \le b$ and has the characteristic jump of a Green's function at $x = s$. Since on $[a, b]$, both y_1 and y_2 are solutions of the homogeneous equation

$$a_0(x)y'' + a_1(x)y' + a_2(x) = 0 \tag{8}$$

corresponding to (9), so is $k(x, s)$ except at $x = s$, where $k_x(x, s)$ is discontinuous. In other words, $k(x, s)$ satisfies Eq. (8) on $a \le x < s$ and $s < x \le b$. Thus the so-called **kernel** $k(x, s)$ of the particular integral (10), defined by (11) and obtained by variation of parameters, has three of the four definitive properties of a Green's function set forth in Definition 1.

Now, let us see if $k(x, s)$ can be made to also satisfy the remaining condition of that definition, i.e., the boundary conditions

(13) $$(a)\ \alpha_1 y(a) - \alpha_2 y'(a) = 0 \qquad \alpha_1^2 + \alpha_2^2 \ne 0 \qquad (b)\ \beta_1 y(b) - \beta_2 y'(b) = 0 \qquad \beta_1^2 + \beta_2^2 \ne 0$$

Substituting $k(x, s)$ into the left-hand members of these equations for y, then factoring, we obtain the two relations

(14a) $$\alpha_1 k(a, s) - \alpha_2 k_x(a, s) = [\alpha_1 y_2(a) - \alpha_2 y_2'(a)]\frac{y_1(s)}{W(s)a_0(s)}$$

(14b) $$\beta_1 k(b, s) - \beta_2 k_x(b, s) = [\beta_1 y_1(b) - \beta_2 y_1'(b)]\frac{y_2(s)}{W(s)a_0(s)}$$

The right-hand members here will be zero for each value of s if and only if y_1 and y_2 are linearly independent solutions of (8) such that

(15) $$(a)\ \alpha_1 y_2(a) - \alpha_2 y_2'(a) = 0 \qquad \text{and} \qquad (b)\ \beta_1 y_1(b) - \beta_2 y_1'(b) = 0$$

This being the case, and assuming the Green's function defined by Definition 1 exists and is unique, we have the following important result.

THEOREM 1 The Green's function of the completely homogeneous differential system

$$\begin{aligned} a_0(x)y'' + a_1(x)y' + a_2(x)y &= 0 \\ \alpha_1 y(a) - \alpha_2 y'(a) &= 0 \qquad \alpha_1^2 + \alpha_2^2 \ne 0 \\ \beta_1 y(b) - \beta_2 y'(b) &= 0 \qquad \beta_1^2 + \beta_2^2 \ne 0 \end{aligned}$$

is given by

$$g(x, s) = \begin{cases} \dfrac{y_1(s)y_2(x)}{W(s)a_0(s)} & a \le x \le s \\ \dfrac{y_1(x)y_2(s)}{W(s)a_0(s)} & s \le x \le b \end{cases}$$

where y_1 and y_2 may be any two linearly independent solutions of the differential equation on $[a, b]$ such that

$$\alpha_1 y_2(a) - \alpha_2 y_2'(a) = 0 \qquad \text{and} \qquad \beta_1 y_1(b) - \beta_2 y_1'(b) = 0$$

Any two solutions y_1 and y_2 of (8) suitable for the construction of $g(x, s)$, in accordance with Theorem 1, may be used to write a complete solution of Eq. (9):

(16) $$y = c_1 y_1 + c_2 y_2 - \int_a^b g(x, s)\, f(s)\, ds \qquad c_1, c_2 \text{ parameters}$$

Substitution of this solution for y in the boundary conditions (13a) and (13b) leads to the following important conclusion.

THEOREM 2 The nonhomogeneous differential system

$$a_0(x)y'' + a_1(x)y' + a_2(x)y = f(x)$$

$$\alpha_1 y(a) - \alpha_2 y'(a) = 0 \qquad \alpha_1^2 + \alpha_2^2 \neq 0$$

$$\beta_1 y(b) - \beta_2 y'(b) = 0 \qquad \beta_1^2 + \beta_2^2 \neq 0$$

has the unique solution

$$y = -\int_a^b g(x, s)\, f(s)\, ds \tag{17}$$

on $[a, b]$, where $g(x, s)$ is the Green's function corresponding to the completely homogeneous system in Theorem 1.

PROOF Imposing the boundary conditions (13) on the complete solution (16) of (9), then collecting coefficients on c_1 and c_2 and combining integrals, we have

$$[\alpha_1 y_1(a) - \alpha_2 y_1'(a)]c_1 + [\alpha_1 y_2(a) - \alpha_2 y_2'(a)]c_2 - \int_a^b [\alpha_1 g(a, s) - \alpha_2 g_x(a, s)] f(s)\, ds = 0$$

$$[\beta_1 y_1(b) - \beta_2 y_1'(b)]c_1 + [\beta_1 y_2(b) - \beta_2 y_2'(b)]c_2 - \int_a^b [\beta_1 g(b, s) - \beta_2 g_x(b, s)] f(s)\, ds = 0$$

Since the boundary conditions (13) are satisfied by $g(x, s)$ and because y_2 and y_1 satisfy the conditions (15), the last two equations simplify to the pair

$$[\alpha_1 y_1(a) - \alpha_2 y_1'(a)]c_1 = 0 \qquad [\beta_1 y_2(b) - \beta_2 y_2'(b)]c_2 = 0$$

The coefficients of c_1 and c_2 in these equations must both be nonzero (Exercise 31), hence $c_1 = c_2 = 0$ and (16) reduces to the solution (17) of the theorem. This solution is unique if, as in Example 1, the related completely homogeneous differential system is **incompatible** (Exercise 60), i.e., has only the trivial solution on $[a, b]$. ◗

Theorem 2 tells us that the solution of a *boundary-value problem* comprised of a *general second-order linear differential equation* and *boundary conditions* like those of Definition 1 is given by a formula just like (7), which we derived for the loaded string whose differential equation (4) had $-w(x)$ rather than $f(x)$ as its nonhomogeneous term. In Theorem 1, there is no guarantee that $g(x, s) = g(s, x)$, as was the case for (6) and the Green's function of Example 1.

EXAMPLE 2

Solve the boundary-value problem $x^2y'' + xy' + y = f(x)$, $y'(1) = y(e^{\pi}) = 0$.

The differential equation is an Euler-Cauchy equation and has

$$y_c = c_1 \cos \ln x + c_2 \sin \ln x \qquad 1 \leq x \leq e^{\pi}$$

as a complementary function. In the notation of Theorem 1,

$$a_0(x) = x^2 \qquad \alpha_1 = 0 \qquad a = 1 \qquad \alpha_2 = -1 \qquad \beta_1 = 1 \qquad b = e^{\pi} \qquad \beta_2 = 0$$

Imposing the condition

$$\alpha_1 y_2(a) - \alpha_2 y_2'(a) \equiv 0 \cdot y_2(1) - (-1)y_2'(1) = 0$$

on the complementary function, we obtain

$$-c_1 \sin \ln 1 + c_2 \cos \ln 1 = c_2 = 0$$

Thus we take $y_2 = \cos \ln x$ (see Exercise 33). Imposing the condition

$$\beta_1 y_1(b) - \beta_2 y_1'(b) \equiv 1 \cdot y_1(e^{\pi}) - 0 \cdot y_1'(e^{\pi}) = 0$$

on the complementary function, but with new arbitrary constants d_1 and d_2 in place of c_1 and c_2, we obtain

$$d_1 \cos \ln e^{\pi} + d_2 \sin \ln e^{\pi} = -d_1 = 0$$

Thus we take $y_1 = \sin \ln x$. A simple computation yields $W(x) = -1/x$ for the value of the wronskian of y_1 and y_2.

Substituting into the formula for $g(x, s)$, Theorem 1, we obtain the Green's function

$$g(x, s) = \begin{cases} \dfrac{-\sin \ln s \cos \ln x}{s} & 1 \le x \le s \\[2ex] \dfrac{-\sin \ln x \cos \ln s}{s} & s \le x \le e^{\pi} \end{cases}$$

of the completely homogeneous system. Theorem 2 now gives

$$y = -\int_1^{e^{\pi}} g(x, s)f(s)\, ds = \sin \ln x \int_1^{x} \frac{f(s) \cos \ln s}{s}\, ds + \cos \ln x \int_x^{e^{\pi}} \frac{f(s) \sin \ln s}{s}\, ds$$

as the required solution. Note that in this problem $g(x, s) \neq g(s, x)$.

As a final example of a more sophisticated application of a Green's function, let us consider the following problem. A flexible elastic string of weight per unit length $\rho(x)$ is stretched under tension T between two points a distance l apart. Determine the frequencies at which the string can perform free vibrations and the shape of the curve of maximum deflection corresponding to each natural frequency.

This problem is very much like the one we discussed earlier except that now the transverse load per unit length, instead of being a known static load, is an unknown dynamic load which can be determined as follows. Consider an element of the string of length Δx and mass $[\rho(x)\, \Delta x]/g$. In the absence of frictional effects or other external forces impressed directly on the element, the instantaneous force exerted on the element by the portions of the string to either side of it must equal $[\rho(x)\, \Delta x/g]\ddot{y}$, where $\ddot{y}$ denotes the acceleration of an arbitrary point of the element. According to **Newton's third law,** which says that *to every action there is an equal and opposite reaction,* the element itself exerts a force $-[\rho(x)\, \Delta x/g]\ddot{y}$ on the adjoining segments of the string. Dividing by Δx and letting $\Delta x \to 0$, we obtain $-[\rho(x)/g]\ddot{y}$ for the dynamic load per *unit length* acting on the string.

Now if the string is vibrating at a single natural frequency, all points of the string move harmonically with that frequency and with phase differences which are either 0 or 180°. In other words, during the vibration, the curve of the string is defined by an equation of the form

$$y(x, t) = \phi(x) \sin \omega t$$

where ω is the (as yet unknown) frequency of the vibrations and $\phi(x)$ is the (as yet unknown) curve of maximum displacements. Thus the load per unit length of the string is of the form

$$w(x, t) = -\frac{\rho(x)}{g}[-\omega^2 \phi(x) \sin \omega t]$$

and Eq. (7) gives us

$$y(x, t) \equiv \phi(x) \sin \omega t = \int_0^l g(x, s)\omega^2 \frac{\rho(s)}{g} \phi(s) \sin \omega t \, ds$$

The factor $\sin \omega t$ is independent of the variable of integration s, hence can be removed from the integral and canceled from the equation. Thus the amplitude function $\phi(x)$ satisfies the *integral equation*†

$$\phi(x) = \frac{\omega^2}{g} \int_0^l g(x, s)\rho(s)\phi(s) \, ds$$

In a typical problem, this equation would probably have to be solved approximately in the following way. Make some reasonable guess as to the nature of the function $\phi(x)$, say $\phi_1(x)$; substitute this into the integrand; and compute the integral as a function of x. If the result is proportional to the ''input'' $\phi_1(x)$, we have a solution and the unknown frequency ω can be inferred from the proportionality constant. Of course, it is highly unlikely that we would ever hit upon the solution by guessing, and so the integrated result, or ''output,'' say $\phi_2(x)$, will not be proportional to $\phi_1(x)$ and we shall not have found a solution. However, the process can be repeated, $\phi_2(x)$ can be substituted into the integral to give a new output $\phi_3(x)$, and so on. It can be shown that in a large class of cases the sequence of functions $\{\phi_n(x)\}$ determined by this procedure will converge to a function $\phi(x)$ which is a solution, and that the ratio $\phi_{n-1}(x)/\phi_n(x)$ will approach a value which is independent of x. Thus by repeating the iteration a sufficient number of times, $\phi(x)$ and ω can be approximated with satisfactory accuracy.

In particular, if $\rho(x)$ is a constant, say ρ, then for a uniform string of length l we have the integral equation

$$\phi(x) = \frac{\omega^2 \rho}{g} \int_0^l g(x, s)\phi(s) \, ds$$

or, substituting the Green's function for a string from Eq. (6),

$$\begin{aligned}\phi(x) &= \frac{\omega^2 \rho}{g} \int_0^x \frac{(l - x)s}{Tl} \phi(s) \, ds + \frac{\omega^2 \rho}{g} \int_x^l \frac{(l - s)x}{Tl} \phi(s) \, ds \\ &= \frac{\omega^2 \rho (l - x)}{gTl} \int_0^x s\phi(s) \, ds + \frac{\omega^2 \rho x}{gTl} \int_x^l (l - s)\phi(s) \, ds\end{aligned}$$

A reasonable guess for $\phi(x)$ might be $\phi_1(x) = A \sin (n\pi x/l)$, since this is a simple function with the property that for each value of n it is zero for $x = 0$ and for $x = l$, as the deflection of a string with fixed ends should be. Using this, we have for the integrals on the right.

$$\begin{aligned}&\frac{A\omega^2 \rho (l - x)}{gTl} \int_0^x s \sin \frac{n\pi s}{l} \, ds + \frac{A\omega^2 \rho x}{gTl} \int_x^l (l - s) \sin \frac{n\pi s}{l} \, ds \\ &\qquad = \frac{A\omega^2 \rho (l - x)}{gTl} \left[\frac{l^2}{n^2\pi^2} \sin \frac{n\pi s}{l} - \frac{ls}{n\pi} \cos \frac{n\pi s}{l} \right]_0^x \\ &\qquad + \frac{A\omega^2 \rho x}{gTl} \left[-\frac{(l - s)l}{n\pi} \cos \frac{n\pi s}{l} - \frac{l^2}{n^2\pi^2} \sin \frac{n\pi s}{l} \right]_x^l \\ &\qquad = \frac{A\omega^2 \rho}{gT} \frac{l^2}{n^2\pi^2} \sin \frac{n\pi x}{l}\end{aligned}$$

†An equation involving the integral of an unknown function is called an **integral equation.** As this problem illustrates, Green's functions play an important role in connecting the theory of differential equations with the theory of integral equations.

This will be equal to the input, $A \sin (n\pi x/l)$, if and only if

$$\frac{\omega^2 \rho}{gT} \frac{l^2}{n^2\pi^2} = 1$$

Thus there are infinitely many natural frequencies, given by the formula

$$\omega_n = \frac{n\pi}{l} \sqrt{\frac{gT}{\rho}}$$

with corresponding deflection curves

$$y_n(x, t) = A_n \sin \frac{n\pi x}{l} \sin \omega_n t$$

It is interesting to note that we shall obtain these results by an entirely different method in Sec. 11.5.

EXERCISES

1. Use Theorem 1 to construct the Green's function for a string with fixed ends and of length l.

2. Using Theorem 1, derive and thus verify the Green's function of Example 1.

3. Determine the deflection curve of a string with fixed ends and of length l bearing equal concentrated loads P at $x = l/4$ and at $x = l/2$.

4. Work Exercise 3 if the loads act at $x = l/2$ and at $x = 3l/4$.

5. Sketch the deflection curve of Exercise 3 if $P = T = l = 1$.

6. Sketch the deflection curve of Exercise 4 if $T = l = 1$ and $P = -1$.

7. Determine the deflection curve of a string with fixed ends and of length l bearing concentrated loads $-2P$ at $l/4$ and P at $l/2$.

8. Determine the deflection curve of a string with fixed ends and of length l bearing concentrated loads P at $x = l/4$, $-2P$ at $x = l/2$, and $3P$ at $x = 3l/4$.

9. Find the deflection curve of a string with fixed ends and of length l bearing a load per unit length $w(x) = -x$, first by solving the differential equation (4) and then by using the Green's function for the string.

10. Work Exercise 9 if the load per unit length is

$$w(x) = \begin{cases} 0 & 0 \le x < l/2 \\ -1 & l/2 < x \le l \end{cases}$$

11. Find the deflection curve of the loaded string in Exercise 9 if a concentrated load P is added at $x = l/4$.

12. Work Exercise 11 if P is moved to $x = l/2$. Also find the deflection at $x = l/4$ and at $x = 3l/4$.

13. Use the related Green's function to find the deflection curve of a string stretched under a unit tension, bearing a load per unit length $w(x) = x(x - 1)$, and having fixed ends at $x = 0$ and $x = 1$.

14. Work Exercise 9 if $w(x) = \sin \pi x/l$.

15. **(a)** Determine the deflection curve of the string in Exercise 14 if in addition to the distributed load a concentrated load equal to the tension acts at $x = l/3$.
(b) Sketch the deflection curve when only the concentrated load is present.

16. Find the Green's function of the boundary-value problem $y'' - k^2y = 0$, $k \ne 0$; $y(0) = 0$, $y(1) = 0$ **(a)** using Definition 1, and **(b)** using Theorem 1.

Find the Green's function of each of the following boundary-value problems.

17. $y'' + y' = 0$; $y(0) = 0$, $y'(\pi) = 0$

18. $y'' - y' = 0$; $y(0) = 0$, $y'(1) = 0$

19. $y'' + y = 0$; $y(0) = 0$, $y'(\pi) = 0$

20. $y'' = 0$; $y(0) = 0$, $y(1) = 0$

21. $y'' = 0$; $y(0) = 0$, $y'(l) = 0$

22. $y'' + y' - 2y = 0$; $y(0) = 0$, $y'(1) - y(1) = 0$

23. $y'' + k\ y = 0$, $k \ne 2n\pi$; $y(0) = y(1)$, $y'(0) = y'(1)$

24. $y'' + k^2y = 0$, $k \ne 0$, if the boundary conditions are
(a) $y(0) = y'(b) = 0$ **(b)** $y'(0) = y(b) = 0$
(c) $y'(a) = y'(b) = 0$ **(d)** $y(a) = y'(a)$, $y(b) = 0$

25. $y'' - k^2y = 0$, $k \ne 0$, and boundary conditions as in Parts **(a)**, **(b)**, **(c)**, and **(d)** of Exercise 24.

26. $x^2y'' - 2xy' + 2y = 0$; $y(1) = 0$, $y(2) = 0$

27. $(x^2 - x)y'' - (3x^2 - 2)y' + (9x - 6)y = 0$; $y(-2) - y'(-2) = 0$, $y(-1) - y'(-1) = 0$

28. Construct the Green's function for the equation $y'' + 2y' + 2y = 0$ with the boundary conditions $y(0) = 0$, $y(\pi/2) = 0$. Is this Green's function symmetric? What is the Green's function if the differential equation is $e^{2x}y'' + 2e^{2x}y' + 2e^{2x}y = 0$? Is this Green's function symmetric?

29. **(a)** Show that every normal second-order differential equation

$$a_0(x)y'' + a_1(x)y' + a_2(x)y = 0$$

is equivalent to an equation of the form $[r(x)y']' + p(x)y = 0$.
Hint: Multiply the given equation by $[1/a_0(x)]e^{\int[a_1(x)/a_0(x)]\,dx}$.

(b) Show that the Green's function for a differential system of the form $[r(x)y']' + p(x)y = 0$ with boundary conditions of the type prescribed by Definition 1 is symmetric. *Hint:* Recall Abel's formula for the wronskian of two solutions of a second-order differential equation.

30. Find all Green's functions corresponding to the equation $x^2y'' - 2xy' + 2y = 0$ with the boundary conditions $y(0) = y(1) = 0$. Why does this differential system have infinitely many Green's functions? *Hint:* Where is the differential equation normal?

31. Under the conditions of Theorem 1, show that neither $\alpha_1 y_1(a) - \alpha_2 y_1'(a)$ nor $\beta_1 y_2(b) - \beta_2 y_2'(b)$ can be zero.

32. In Theorem 1, show that if a_0 and W are constants, then the Green's function is symmetric.

33. In Theorem 1, show that the same Green's function is defined by $g(x, s)$ when y_1 and y_2 are particular solutions as when either of the solutions is general.

Solve each of the following boundary-value problems by finding the Green's function of the related homogeneous system and then utilizing Theorem 2.

34. $4y'' + y = 4\cos(x/2)$; $y(0) - 2y'(0) = 0$, $y(2\pi) = 0$

35. $4y'' - y = 4e^{x/2}$; $y(0) = 0$, $y(2) - 2y'(2) = 0$

36. $x^2y'' + 2xy' - 2y = 42$; $y(1) = 0$, $y(2) = 0$

37. $y'' + 4y' + 5y = 3e^{-2x}\sin 2x$; $2y(0) - y'(0) = 0$, $y(\pi) = 0$

38. $4x^2y'' + 8xy' + y = 8x^{-1/2}$; $y(1) = 0$, $y(4) + 8y'(4) = 0$

39. $33x^2y'' - 242xy' + 462y = 1331x^6$; $6y(1) - y'(1) = 0$, $7y(2) - 6y'(2) = 0$

40. The angle of twist θ produced in a uniform shaft of length l by a torque T is given by the formula $\theta = Tl/E_sJ$, where E_s is the modulus of elasticity in shear of the material of the shaft and J is the polar moment of inertia of the cross-sectional area of the shaft about the center of gravity of the cross section. Using this formula, find the influence function which gives the angle of twist at a point x due to a unit torque applied at a point s of a shaft rigidly clamped at the left end and free at the right end.

41. **(a)** By considering the limit of the angle of twist produced in a shaft of infinitesimal length by a torque applied to its ends, show that the torque transmitted *through* any cross section of a twisted shaft is given by the formula

$$T = E_sJ\frac{d\theta}{dx}$$

(b) Using the result of Part **(a)**, show that the angle of twist produced in a uniform shaft by a torque per unit length equal to $t(x)$ satisfies the differential equation

$$E_sJ\theta'' = -t(x)$$

42. Using the result of Part **(b)** of Exercise 41, find the Green's function of a uniform shaft of length l if each end of the shaft is clamped so that it cannot twist.

43. Using the result of Exercise 40, set up the integral equation satisfied by the maximum-deflection function $\phi(x)$ for the free vibrations of a uniform shaft of length l which is clamped at the left end and free at the right. Show that $\phi(x) = A\sin[(2n + 1)\pi x/2l]$ is a solution of this equation for each value of n and find the corresponding natural frequencies.

44. Using the result of Exercise 42, set up the integral equation satisfied by the maximum-deflection function $\phi(x)$ for the free torsional vibrations of a uniform shaft of length l which is clamped at both ends. Show that $\phi(x) = A\sin(nx\pi/l)$ is a solution of this equation for each value of n and find the corresponding natural frequencies.

45. Find the influence function which gives the deflection of a uniform cantilever beam at a point x due to a unit load applied at the point s. Verify that this influence function $g(x, s)$ has the following properties, the primes denoting differentiation with respect to x:

(a) $g(x, s)$ satisfies the differential equation $EIy^{\text{iv}} = 0$ for $0 \le x < s$ and $s < x \le l$.

(b) $g(0, s) = g'(0, s) = g''(l, s) = g'''(l, s) = 0$; that is, $g(x, s)$ satisfies the end conditions for a cantilever beam.

(c) $g(x, s)$, $g'(x, s)$, and $g''(x, s)$ are continuous for $0 \le x \le l$.

(d) $g'''(x, s)$ has a jump of $-1/EI$ at $x = s$.

Is the Green's function, i.e., the influence function, $g(x, s)$ a symmetric function of x and s?

46. Using the influence function obtained in Exercise 45, find the deflection curve of a uniform cantilever beam bearing a concentrated load P at $x = l/2$ and at $x = l$.

47. A concentrated load P acts downward at the midpoint of a uniform cantilever beam of length l. **(a)** Find the tip deflection. **(b)** Where must the load be applied if the tip deflection is to equal $-4Pl^3/81EI$?

48. A uniform cantilever beam of length l bears two concentrated loads, one of magnitude P acting downward at $x = l/2$, the other of magnitude Q acting upward at the free end $x = l$. **(a)** Find the deflection curve. If $P = 160$, what must be the value of Q to make: **(b)** the deflection of the beam zero at $x = l/2$? **(c)** the tip deflection zero? **(d)** If $Q = 0$, what is the nature of the deflection curve from $l/2$ to l?

49. Find the deflection curve of a uniform cantilever beam of length l bearing a constant load w per unit length, first by solving the differential equation $EIy^{\text{iv}} = -w$ and imposing appropriate boundary conditions and then by using the influence function of Exercise 45.

50. Work Exercise 49 if the load per unit length is $w(x) = x$.

51. Work Exercise 49 if the load per unit length is $w(x) = l - x$; also find the tip deflection and the moment at $x = 0$.

52. A uniform cantilever beam of unit length bears a constant load w per unit length and a concentrated load P at its midpoint. Determine the deflection curve of the beam and a relation between w and P for which the tip deflection is zero.

53. A cantilever beam of length l has constant cross-sectional area; however, its weight per unit length at a distance x from the built-in end is given by $\rho(x) = 2l - x$. Use the Green's function of Exercise 45 to find the tip deflection of the beam.

54. Taking the following properties, suggested in Exercise 45, as the definition of the Green's function for a fourth-order differential equation whose leading coefficient is a_0, find the Green's function for the equation $EIy^{\text{iv}} - y = 0$, with the boundary conditions $y(0) = y''(0) = y(l) = y''(l) = 0$:

(a) $g(x, s)$ satisfies the given differential equation except at $x = s$.

(b) $g(x, s)$ satisfies the boundary conditions accompanying the differential equation.

(c) $g(x, s)$, $g'(x, s)$, and $g''(x, s)$ are continuous at all points of the interval of the problem.

(d) $g'''(x, s)$ has a jump of $-1/a_0$ at the point s.

Using the properties set forth in Exercise 54, find the Green's function of each of the following boundary-value problems.

55. $EIy^{\text{iv}} = 0$; $y(0) = y''(0) = y(l) = y''(l) = 0$
(Simply supported beam)

56. $EIy^{\text{iv}} = 0$; $y(0) = y'(0) = 0$, $y(l) = y'(l) = 0$
(Beam with embedded ends)

57. $EIy^{\text{iv}} = 0$; $y(0) = y'(0) = 0$, $y(l) = y''(l) = 0$
(Beam with one end embedded and one simply supported)

58. Find an equation for the deflection curve of a uniform simply supported beam of length $2l$ bearing a load per unit length $w(x) = [u(x) - u(x - l)]k$, k constant.

59. **(a)** Use variation of parameters to find a particular integral

$$Y = -\int_0^1 \kappa(x, s) f(s)\, ds$$

of the nonhomogeneous differential system

$$y'' - k^2 y = f(x) \qquad y(0) = y(1) = 0,\ k \neq 0$$

and verify that the kernel $\kappa(x, s)$ of the particular integral automatically has three of the four definitive properties of the Green's function $g(x, s)$ of the related completely homogeneous differential system.

(b) Derive $g(x, s)$ by imposing the given boundary conditions on $\kappa(x, s)$. *Hint:* Note that if $\sinh ks \cosh k = 0$, then $\cosh ks = [\sinh k(1 - s)]/\sinh k$.

60. Show that if the completely homogeneous differential system related to the differential system of Theorem 2 is incompatible, then the solution (17) of the nonhomogeneous system is unique. *Hint:* What system does the difference of any two solutions of the nonhomogeneous system satisfy?

CHAPTER 3

COMPLEX NUMBERS AND LINEAR ALGEBRA

◖ Complex numbers, vectors, and matrices, with the appropriate rules of operation, are examples of the very general and very important concept of a *linear vector space.* In Chap. 13 we shall discuss linear vector spaces in general; here we study three examples which together form a major part of what is called *linear algebra.*

The elementary properties of complex numbers (Sec. 3.1) will be useful to us again and again in our work, as they were in Sec. 2.5. The important but more difficult topic of *functions of a complex variable* will be treated in detail in Chaps. 17–20.

In Sec. 3.2, we discuss vectors as algebraic objects. In Chap. 15, we will study vectors as geometric objects and develop what might be called the *calculus of vectors.*

Matrices, that is, rectangular arrays of quantities, usually numbers, are introduced in Sec. 3.3, and special forms of these are discussed in Sec. 3.4.

Students usually encounter determinants before they learn about matrices but, properly understood, a determinant is just a function of a square matrix. The properties of determinants are established in Sec. 3.5.

Sections 3.6 and 3.7 deal with the very important topic of systems of linear algebraic equations and their solution by reduction to the *Gauss* or *Gauss-Jordan canonical form.* Certain special methods of solution, such as *Cramer's rule,* are also discussed in Sec. 3.7.

In Sec. 3.8, square matrices are interpreted as transformations which send vectors into vectors. This leads to the concept of the *characteristic values* of a square matrix as the values of a certain parameter for which a vector and its image have the same direction. The characteristic values of a square matrix have many important applications, especially in determining the natural frequencies of vibrating systems, and we shall study them in detail in Secs. 14.6 and 14.7. Characteristic values also arise in what are called *boundary-value problems* in partial differential equations, and much of Chaps. 11 and 12 will be devoted to them and their physical applications.

Prerequisites for this chapter: college algebra and analytic geometry. ◗

3.1 COMPLEX NUMBERS

We have already seen that complex numbers are very useful in solving homogeneous second-order linear differential equations with constant coefficients (Case 3, Sec. 2.5). In particular, we have that the characteristic equation of the differential equation

$$y'' + y = 0$$

is the quadratic equation

$$m^2 + 1 = 0$$

whose solutions as given by the quadratic formula are

$$m = \frac{0 \pm \sqrt{0 - 4(1)(1)}}{2} = \pm\frac{1}{2}\sqrt{-4} = \pm\sqrt{-4/4} = \pm\sqrt{-1}$$

In the notation of Euler, who was the first to set $i = \sqrt{-1}$, we have

$$m = \pm i$$

Since there is no real number whose square is -1, and since -1 is not in the domain $[0, \infty)$ of the square root function (Example 1, Sec. 1.1), it is little wonder that such roots, when first encountered, were discarded as "impossible" or considered to be "imaginary."† A precise definition of the mysterious so-called *imaginary unit* i and a clear understanding of the nature of *complex numbers* of the **standard form**

(1) $$a + bi \quad \text{or} \quad a + ib \qquad a, b \text{ real numbers}$$

were a long time in coming. Several instructive ways of defining complex numbers have been proposed, the most popular of which is to define a **complex number** $z = x + iy$ as the ordered pair

(2) $$z = (x, y)$$

of real numbers x and y subject to a set of axioms among which are the following rules.‡

(a) $(a, b) = (c, d)$ if and only if $a = c$ and $b = d$

(b) $k(a, b) = (ka, kb)$, k real

(c) $(a, b) + (c, d) = (a + c, b + d)$

(d) $(a, b) \times (c, d) = (ac - bd, ad + bc)$

A pair of the form $(x, 0)$ is identified as the real number x, and a number iy with y real and not zero is called a **pure imaginary** number. Since

(3) $$x + iy = (x, y)$$

each complex number corresponds to exactly one ordered pair of real numbers, and conversely. This **one-to-one** correspondence tells us that the set of all complex numbers and the set of all

---

† For a brief history of complex numbers including names and notable achievements of early contributors to the subject, see E. T. Bell, *Development of Mathematics,* 2d ed., pp. 175–181, McGraw-Hill, New York, 1945.

‡ The first axiom system for complex numbers was published by the American mathematician and logician E. V. Huntington (1874–1952).

ordered pairs of real numbers may both be interpreted as models of the same axiom system.† Thus properties of complex numbers may be established by studying the behavior of pairs of real numbers. Of course, the axioms employed must be chosen carefully if all the familiar properties of complex numbers we learned in algebra and calculus are to hold.‡ Several problems involving complex numbers as ordered pairs will be found among the exercises at the end of this section.

In spite of the traditional adjective *"imaginary,"* complex numbers play an essential role in a great many very *real* problems. So true is this that the last four chapters in this book are devoted to the study of functions of a complex variable. At this point in our work, however, our needs are more modest. All that we require is the assurance that operations on complex numbers of the form $x + iy$ follow the familiar rules of the elementary algebra of real numbers, with the additional stipulation that whenever a power of i occurs, it is to be reduced according to the relation $i^2 = -1$. Now that we have a clear understanding of complex numbers and of the symbol i itself (Exercises 45 and 48), we are confident of this. The primary purpose of this section is to highlight some of the most useful algebraic and geometric properties of complex numbers.

A further review of the terminology and notation previously mentioned in Sec. 2.5 will also be helpful.

1. The **real part** of a complex number $z = a + ib$ is a and is denoted by Re z [or Re(z) if z is a complicated expression].
2. The **imaginary part** of $z = a + ib$ is the *real* number b and is denoted by Im z [or by Im(z) if z is a complicated expression].
3. The **conjugate** of $z = a + ib$ is $\bar{z} = a - ib$.

In terms of complex numbers, the rules **a**, **c**, and **d** for ordered pairs of real numbers read

4. The complex numbers $z_1 = a + ib$ and $z_2 = c + id$ are **equal** if and only if $a = c$ and $b = d$, that is, if and only if Re z_1 = Re z_2 and Im z_1 = Im z_2.
5. The **sum** of $z_1 = a + bi$ and $z_2 = c + di$ is $z_1 + z_2 = (a + c) + i(b + d)$.
6. The **product** of $z_1 = a + ib$ and $z_2 = c + id$ is $z_1z_2 = (ac - bd) + i(ad + bc)$.

From **4** and **6**, and then from **5** and **6** together, we have

7. A complex number $z = a + ib$ is **zero** if and only if $a = b = 0$, that is, if and only if Re z = Im z = 0.
8. The **negative** of $z = a + ib$ is $-z = (-1)z = -a - ib$.
9. If $z_2 = c + id$ is subtracted from $z_1 = a + ib$, the **difference** is $z_1 - z_2 = z_1 + (-z_2) = (a - c) + i(b - d)$.

With multiplication defined by **6**, division of complex numbers can be defined as the inverse of multiplication; that is, the quotient $(a + ib)/(c + id)$ is the complex number $x + iy$ such that

$$(c + id)(x + iy) = a + ib$$

Performing the indicated multiplication, we obtain the equation

$$(cx - dy) + i(dx + cy) = a + ib$$

†For a lucid exposition of these ideas, see Alfred Tarski, *Introduction to Logic and to the Methodology of Deductive Sciences,* 3d ed., pp. 117–152, a Galaxy Book, Oxford, New York, 1970.

‡For a step-by-step development of the most important properties of complex numbers as ordered pairs, see Edward Landau, *Foundations of Analysis,* 3d ed., pp. 92–111, Chelsea, New York, 1966.

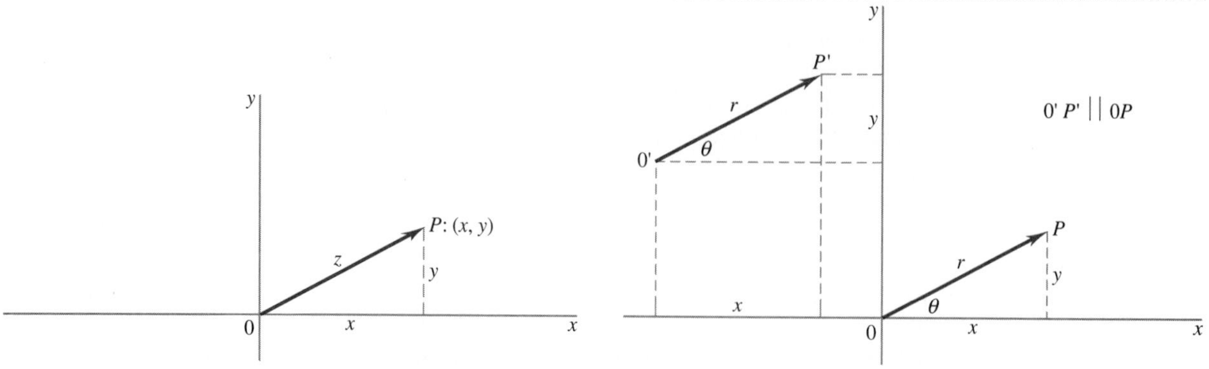

The point P: (x, y) and the vector $\overrightarrow{0P}$ representing the complex number $z = x + iy$.

(a)

Two equivalent vector representations of the complex number $z = x + iy$.

(b)

FIGURE 3.1
Geometric representations of the complex number $z = x + iy$.

which according to **4** implies that

$$cx - dy = a \qquad \text{and} \qquad dx + cy = b$$

Solving these linear equations for x and y gives

$$x = \frac{ac + bd}{c^2 + d^2} \qquad \text{and} \qquad y = \frac{bc - ad}{c^2 + d^2}$$

Hence

$$\text{(4}a\text{)} \qquad \frac{a + ib}{c + id} = \frac{ac + bd}{c^2 + d^2} + i\frac{bc - ad}{c^2 + d^2}$$

provided of course that $c^2 + d^2 \neq 0$; that is, provided that the divisor $c + id$ is not equal to zero. In practice, the quotient of two complex numbers is usually found by multiplying both numerator and denominator by the conjugate of the denominator:

$$\text{(4}b\text{)} \qquad \frac{a + ib}{c + id} = \frac{a + ib}{c + id} \cdot \frac{c - id}{c - id} = \frac{(ac + bd) + i(bc - ad)}{c^2 + d^2} \qquad c + id \neq 0$$

To represent complex numbers geometrically, we first establish a rectangular xy coordinate system in a plane. With each complex number $x + iy$ we associate either the point P: (x, y) or the directed line segment, or vector, $\overrightarrow{OP}$ from the origin O to the image point P (Fig. 3.1a). This sets up a one-to-one correspondence between complex numbers and points of the plane. The x axis consists of the real numbers $x + 0i = x$ and is called the **real axis**. We call the y axis the **imaginary axis** because each of its points corresponds to a complex number of the form iy, y real. A cartesian plane used for graphing complex numbers in this fashion is called an **Argand diagram**† and is commonly referred to simply as the **complex plane** or the z **plane**.

†Named for the French mathematician J. R. Argand (1768–1822).

Since the x and y components of a vector remain the same when the vector is moved parallel to itself (Fig. 3.1*b*), any parallel displacement of $\overrightarrow{OP}$ is also considered to be a geometric representation of the complex number $z = x + iy$. For conciseness of expression, we shall often speak of a complex number and any of its geometric representations as though they were the same thing.

The vector $\overrightarrow{OP}$ which represents the complex number $z = x + iy$ has two important geometric attributes besides its **components**, $x = \text{Re } z$ and $y = \text{Im } z$. The first of these is its length (Fig. 3.1*b*)

$$r = \sqrt{x^2 + y^2} \tag{5}$$

called the **absolute value** of z, written $|z|$, or the **modulus** of z, written mod z. The other is its direction angle θ. Interpreting r and θ as polar coordinates of the point P: (x, y) representing z, we have

$$x = r \cos \theta \qquad \text{and} \qquad y = r \sin \theta \tag{6}$$

and then the formulas

$$z = r \cos \theta + ir \sin \theta = r(\cos \theta + i \sin \theta) \tag{7}$$

Either of these is known as the **polar**, or **trigonometric**, form of z.

Although the modulus r of each complex number is unique, in general, there are many polar angles θ which satisfy Eqs. (6) for the same z. In fact, if θ is such an angle, so are all of the angles

$$\ldots,\ \theta - 4\pi,\ \theta - 2\pi,\ \theta,\ \theta + 2\pi,\ \theta + 4\pi,\ \ldots$$

Each of these angles is called an **amplitude** of z or, more commonly, an **argument** of z, written arg z. If necessary, we can restrict the values of θ so that each $z \neq 0$ will have a single argument. Usually this is done by choosing θ such that

$$(a)\ -\pi < \theta \leq \pi \qquad \text{or else} \qquad (b)\ 0 \leq \theta < 2\pi \tag{8}$$

In either case, arg z becomes single-valued, that is, a function Arg z known as the **principal value** of arg z. Sometimes the formula

$$\theta = \tan^{-1} \frac{y}{x} \tag{9}$$

is used to find an argument of z. When this is done, it is important to note that $\tan^{-1}(y/x)$ denotes two sets of angles in opposite quadrants, the angles of one set being angles of z, the others not. Hence in using Eq. (9), one must be careful to select angles in the proper quadrant, as determined by the signs of x and y.

The addition of two complex numbers has an interesting geometric interpretation which is shown in Fig. 3.2*a*. By drawing one complex number from the terminus of the other, a complex number is determined which has the same initial point as the first number drawn, the same terminal point as the second number drawn, and components which are precisely those of the sum $z_1 + z_2$ of the original numbers. Figure 3.2*b* shows the construction for the difference of two complex numbers, i.e., for the sum $z_1 + (-z_2)$. Evidently, $z_1 - z_2$ is identical in length and direction with the vector drawn from the terminus of z_2 to the terminus of z_1. Both the sum and difference of two complex numbers can be described in terms of the parallelogram having the given numbers for adjacent sides, for the sum is simply the diagonal which passes through the common origin of the vectors representing z_1 and z_2, while the difference is just the other diagonal, properly directed.

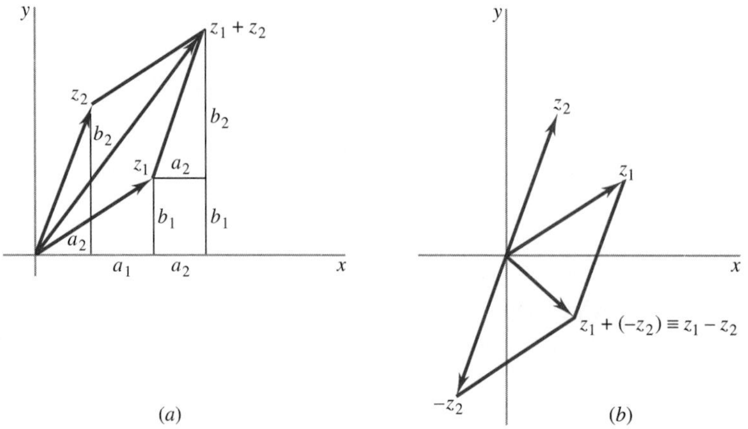

FIGURE 3.2
The sum and difference of the complex numbers $z_1 = a_1 + ib_1$ and $z_2 = a_2 + ib_2$.

Since the length of any side of a triangle must be equal to or less than the sum of the lengths of the other two sides, it follows from the geometric addition of complex numbers (Fig. 3.2*a*) that

$$|z_1 + z_2| \le |z_1| + |z_2| \tag{10}$$

This exceedingly important result is known as the **triangle inequality**. The triangle inequality can readily be extended to three terms, for we have

$$|z_1 + z_2 + z_3| = |z_1 + (z_2 + z_3)| \le |z_1| + |z_2 + z_3| \le |z_1| + |z_2| + |z_3|$$

The important extension to n terms is

$$\left|\sum_{k=1}^{n} z_k\right| \le \sum_{k=1}^{n} |z_k| \tag{11}$$

The components, the conjugate(s), and the absolute value(s) of one or more complex numbers are connected by a number of simple though important relations which we list here for reference. Their proofs are easy, and we leave them as exercises.

(12)

(a) $\operatorname{Re} z = \dfrac{z + \bar{z}}{2}$ **(b)** $\operatorname{Im} z = \dfrac{z - \bar{z}}{2i}$

(c) $|z| = \sqrt{\operatorname{Re}^2 z + \operatorname{Im}^2 z}$ **(d)** $|z| = |\bar{z}|$

(e) $z\bar{z} = |z|^2$ **(f)** $|z_1 z_2| = |z_1| \cdot |z_2|$

(g) $\left|\dfrac{z_1}{z_2}\right| = \dfrac{|z_1|}{|z_2|}$ **(h)** $\overline{z_1 \pm z_2} = \bar{z}_1 \pm \bar{z}_2$

(i) $\overline{z_1 z_2} = \bar{z}_1 \cdot \bar{z}_2$ **(j)** $\overline{\left(\dfrac{z_1}{z_2}\right)} = \dfrac{\bar{z}_1}{\bar{z}_2}$

EXAMPLE 1

What is the absolute value of $(2 + 3i)/(3 - 4i)$?

By definition [Eq. (5)], the absolute value of the complex number is the square root of the sum of the squares of the real and imaginary components of the number. Hence, reducing $(2 + 3i)/(3 - 4i)$ to standard complex form by rationalizing the denominator, we have

$$\frac{2+3i}{3-4i} = \frac{2+3i}{3-4i} \cdot \frac{3+4i}{3+4i} = \frac{-6+17i}{25}$$

and

$$\left|\frac{2+3i}{3-4i}\right| = \sqrt{\left(-\frac{6}{25}\right)^2 + \left(\frac{17}{25}\right)^2} = \frac{\sqrt{13}}{5}$$

The absolute value of $(2+3i)/(3-4i)$ can also be found by using Eq. (12**g**):

$$\left|\frac{2+3i}{3-4i}\right| = \frac{|2+3i|}{|3-4i|} = \frac{\sqrt{2^2+3^2}}{\sqrt{3^2+(-4)^2}} = \frac{\sqrt{13}}{5}$$

as before. Still another method for finding the absolute value of a complex quantity is to use Eq. (12**e**). This is illustrated in a somewhat more complicated setting in the next example.

EXAMPLE 2

If $w = (z+i)/(iz+1)$, show that the restriction $\operatorname{Im} z \le 0$ implies the restriction $|w| \le 1$.

Since we are asked to establish a certain property of $|w|$, our first step is to compute this quantity. This can be done in various ways, but it is probably most convenient to construct the product

$$w\bar{w} = |w|^2 = \frac{z+i}{iz+1}\overline{\left(\frac{z+i}{iz+1}\right)}$$

Since the conjugate of a quotient is the quotient of the conjugates, this can be written as

$$|w|^2 = \frac{z+i}{iz+1}\,\frac{\overline{z+i}}{\overline{iz+1}}$$

Moreover, the conjugate of a sum is the sum of the conjugates; hence we have further

$$|w|^2 = \frac{z+i}{iz+1}\,\frac{\bar{z}+\bar{i}}{\overline{iz}+1}$$

Finally, since $\bar{i} = -i$ and $\overline{iz} = \bar{i}\bar{z} = -i\bar{z}$, we have

$$|w|^2 = \frac{z+i}{iz+1}\,\frac{\bar{z}-i}{-i\bar{z}+1} = \frac{(z\bar{z}+1) - i(z-\bar{z})}{(z\bar{z}+1) + i(z-\bar{z})} = \frac{z\bar{z}+1+2\operatorname{Im} z}{z\bar{z}+1-2\operatorname{Im} z}$$

Since $z\bar{z}+1$ is a positive quantity, it is clear that if $\operatorname{Im} z \le 0$, as given, then the numerator of the last fraction is equal to or less than the denominator. Thus $|w|^2$, hence $|w|$, is at most equal to 1 under the given conditions.

Since the restriction $\operatorname{Im} z \le 0$ implies that z lies on or below the real axis in the plane in which z is plotted, and since $|w| \le 1$ implies that w lies on or inside the unit circle in the plane in which w is plotted, it follows that the given relation

$$w = \frac{z+i}{iz+1}$$

can be thought of as a transformation, or mapping, which sends the lower half of the z plane, point by point, into the region consisting of the unit circle and its interior in the w plane. Mappings of this sort are of considerable importance in applied mathematics, and in Chap. 20 we shall examine their properties in more detail.

Multiplication and division of complex numbers is especially easy to carry out when the numbers are given in polar form. In particular, if we have two complex numbers z_1 and z_2 given in polar form, their product is

$$\begin{aligned} z_1 z_2 &= [r_1(\cos\theta_1 + i\sin\theta_1)][r_2(\cos\theta_2 + i\sin\theta_2)] \\ &= r_1 r_2[(\cos\theta_1\cos\theta_2 - \sin\theta_1\sin\theta_2) + i(\sin\theta_1\cos\theta_2 + \cos\theta_1\sin\theta_2)] \\ &= r_1 r_2[\cos(\theta_1+\theta_2) + i\sin(\theta_1+\theta_2)] \end{aligned} \tag{13}$$

and their quotient can be written

$$\begin{aligned} \frac{z_1}{z_2} &= \frac{r_1(\cos\theta_1 + i\sin\theta_1)}{r_2(\cos\theta_2 + i\sin\theta_2)} \\ &= \frac{r_1(\cos\theta_1 + i\sin\theta_1)(\cos\theta_2 - i\sin\theta_2)}{r_2(\cos\theta_2 + i\sin\theta_2)(\cos\theta_2 - i\sin\theta_2)} \\ &= \frac{r_1}{r_2}\,\frac{(\cos\theta_1\cos\theta_2 + \sin\theta_1\sin\theta_2) + i(\sin\theta_1\cos\theta_2 - \cos\theta_1\sin\theta_2)}{\cos^2\theta_2 + \sin^2\theta_2} \\ &= \frac{r_1}{r_2}[\cos(\theta_1-\theta_2) + i\sin(\theta_1-\theta_2)] \qquad r_2 \neq 0 \end{aligned} \tag{14}$$

In words, then, we have the following alternative definitions:

1. The **product of two complex numbers** is a complex number whose absolute value is the product of the absolute values of the two factors and whose amplitude is the sum of the amplitudes of the two factors.
2. The **quotient of two complex numbers** is a complex number whose absolute value is the quotient of the absolute values of the numbers and whose amplitude is the difference of their amplitudes.

The behavior of the angles of complex numbers when the numbers are multiplied or divided is concisely expressed by the formulas

$$\arg z_1 z_2 = \arg z_1 + \arg z_2 \tag{15}$$

$$\arg\frac{z_1}{z_2} = \arg z_1 - \arg z_2 \tag{16}$$

Equation (15) is to be interpreted as saying that any argument of z_1 plus any argument of z_2 is an argument of the product $z_1 z_2$; and, conversely, every argument of $z_1 z_2$ can be expressed as the sum of any argument of z_1 and some argument of z_2. Equation (16) has an analogous interpretation.

The striking resemblance of (15) and (16) to the corresponding logarithmic and exponential formulas was to be expected, following our informal discovery in Sec. 2.5 of Euler's formula

$$e^{i\theta} = \cos\theta + i\sin\theta \tag{17}$$

since (17) shows that the angle θ of a complex number $z = r(\cos\theta + i\sin\theta)$ is indeed an exponent, and yields the so-called **exponential form** $re^{i\theta}$ of z.

Although we cannot give a rigorous justification of Euler's formula until our discussion of analytic functions of complex variables in Sec. 17.2, two of its consequences will be so useful in much of our work that it is appropriate to introduce them at this time: if we replace θ by $-\theta$ in (17), getting

$$e^{-i\theta} = \cos\theta - i\sin\theta \tag{18}$$

and then add and subtract (17) and (18), we obtain the formulas

$$\cos\theta = \frac{e^{i\theta} + e^{-i\theta}}{2} \tag{19}$$

$$\sin\theta = \frac{e^{i\theta} - e^{-i\theta}}{2i} \tag{20}$$

These will be especially useful in our study of Fourier series.

Returning to Eq. (13), it should be clear that it can be extended at once to products of more than two factors, giving

$$z_1 z_2 \cdots z_n = r_1 r_2 \cdots r_n[\cos(\theta_1 + \theta_2 + \cdots + \theta_n) + i\sin(\theta_1 + \theta_2 + \cdots + \theta_n)]$$

In particular, if all the z's are the same, we have the important result

$$z^n = r^n(\cos n\theta + i\sin n\theta) = r^n e^{in\theta} \tag{21}$$

the last equality following from Euler's formula. If $r = 1$, Eq. (21) is known as **De Moivre's theorem.**† Since the law of division in polar form, (14), gives for the numbers 1 and z the quotient

$$\frac{1}{z} = \frac{1}{r}[\cos(0 - \theta) + i\sin(0 - \theta)] = \frac{1}{r}[\cos(-\theta) + i\sin(-\theta)]$$

which is just the content of Eq. (21) for $n = -1$, it is clear that if $z \neq 0$, Eq. (21) is valid for all integral values of n including $n = 0$.

EXAMPLE 3

Using De Moivre's theorem and the binomial expansion, express $\cos 4\theta$ and $\sin 4\theta$ in terms of powers of $\cos\theta$ and $\sin\theta$.

To do this, we consider $(\cos\theta + i\sin\theta)^4$ and expand it first by De Moivre's theorem and then by the binomial expansion. This gives the identity

$$\begin{aligned}\cos 4\theta + i\sin 4\theta &= \cos^4\theta + 4i\cos^3\theta\sin\theta + 6i^2\cos^2\theta\sin^2\theta + 4i^3\cos\theta\sin^3\theta + i^4\sin^4\theta \\ &= (\cos^4\theta - 6\cos^2\theta\sin^2\theta + \sin^4\theta) + i(4\cos^3\theta\sin\theta - 4\cos\theta\sin^3\theta)\end{aligned}$$

Equating real and imaginary parts of these equal complex expressions, we obtain the required formulas:

$$\cos 4\theta = \cos^4\theta - 6\cos^2\theta\sin^2\theta + \sin^4\theta$$
$$\sin 4\theta = 4(\cos^3\theta\sin\theta - \cos\theta\sin^3\theta)$$

†Named for the French mathematician Abraham De Moivre (1667–1754).

The extension of Eq. (21) to roots of integral order is now an easy matter. In fact, an nth root of $z = r(\cos\theta + i\sin\theta)$ is defined to be any number $w = R(\cos\phi + i\sin\phi)$ such that $w^n = z$, that is,

$$R^n(\cos n\phi + i\sin n\phi) = r(\cos\theta + i\sin\theta) \tag{22}$$

Since two complex numbers which are equal must have the same modulus, it follows from (22) that

$$R^n = r \qquad \text{or} \qquad R = r^{1/n}$$

It should be noted that only real numbers are involved in the determination of R since $r^{1/n}$ is the *real* nth root of the positive quantity r and can always be found by an ordinary logarithmic calculation. Furthermore, the angles of equal complex numbers must either be equal or differ by an integral multiple of 2π. Hence, from (22),

$$n\phi = \theta + 2k\pi \qquad \text{or} \qquad \phi = \frac{\theta + 2k\pi}{n}$$

For $k = 0, 1, \ldots, n-1$, these values of ϕ define n distinct angles which identify n different complex numbers. But as k takes on the values $n, n+1, \ldots$, or the values $-1, -2, \ldots$, the same angles are repeated again and again, each time with an irrelevant difference of 2π in their measures. Thus *there are exactly n distinct values of* $w = z^{1/n}$:

$$w = z^{1/n} = r^{1/n}\left(\cos\frac{\theta + 2k\pi}{n} + i\sin\frac{\theta + 2k\pi}{n}\right) \qquad k = 0, 1, \ldots, n-1 \tag{23}$$

In the complex plane these are represented by radii of the circle with center at the origin and radius $r^{1/n}$, spaced at equal angular intervals of $2\pi/n$, beginning with the radius whose angle is θ/n.

With integral powers and roots defined, the general rational power of a complex number can be defined at once. In fact,

$$z^{p/q} = (z^{1/q})^p = \left[r^{1/q}\left(\cos\frac{\theta + 2k\pi}{q} + i\sin\frac{\theta + 2k\pi}{q}\right)\right]^p$$

$$= r^{p/q}\left[\cos\frac{p}{q}(\theta + 2k\pi) + i\sin\frac{p}{q}(\theta + 2k\pi)\right] \qquad k = 0, 1, \ldots, q-1 \tag{24}$$

The definition of z^α when α is not a rational number, however, must be postponed until Sec. 17.3.

EXAMPLE 4

Find the fourth roots of $-8i$.

To do this, we first note that an angle, or amplitude, of $-8i$ is $3\pi/2$ and the length, or modulus, is 8. Hence, in standard polar form,

$$-8i = 8\left(\cos\frac{3\pi}{2} + i\sin\frac{3\pi}{2}\right)$$

From this, by applying Eq. (23), we find that the four fourth roots of $-8i$ are given by the expression

$$8^{1/4}\left[\cos\frac{1}{4}\left(\frac{3\pi}{2} + 2k\pi\right) + i\sin\frac{1}{4}\left(\frac{3\pi}{2} + 2k\pi\right)\right] \qquad k = 0, 1, 2, 3$$

or, explicitly,

$$r_1 = 8^{1/4}\left(\cos\frac{3\pi}{8} + i\sin\frac{3\pi}{8}\right) \qquad k = 0$$

$$r_2 = 8^{1/4}\left(\cos\frac{7\pi}{8} + i\sin\frac{7\pi}{8}\right) \qquad k = 1$$

$$r_3 = 8^{1/4}\left(\cos\frac{11\pi}{8} + i\sin\frac{11\pi}{8}\right) \qquad k = 2$$

$$r_4 = 8^{1/4}\left(\cos\frac{15\pi}{8} + i\sin\frac{15\pi}{8}\right) \qquad k = 3$$

The coefficient $8^{1/4}$ is of course the *real* fourth root of 8, the value of which is found by a simple logarithmic calculation to be 1.682, to three decimal places.

The n nth roots of 1 are especially significant since they occur in a number of important applications, in particular in what is called the *fast Fourier transform,* which we shall discuss in Sec. 14.2. From (23), their values are given by the formula

$$\rho_k = \cos\frac{2k\pi}{n} + i\sin\frac{2k\pi}{n} = e^{2i\pi k/n} \qquad k = 0, 1, 2, \ldots, n-1$$

In the complex plane, they are represented by n (distinct) radii of the unit circle, spaced at angular intervals of $2\pi/n$, beginning with the radius which falls along the positive real axis. The root corresponding to $k = 1$, that is, the root with the smallest nonzero angle, is called the **primitive** nth **root of** 1 and is usually denoted by the symbol w_n, although when the value of n is clear from the context, n is often dropped as a subscript. Among the various properties of w_n, the following are especially useful:

1. Every integral power of w_n is an nth root of unity.
2. If n is even, $w_n^{n/2} = -1$ and $w_n^2 = w_{n/2}$.
3. If n is a multiple of 4, $w_n^{n/4} = i$, $w_n^{3n/4} = -i$, and $w_n^4 = w_{n/2}^2 = w_{n/4}$.
4. $w_n^k = w_n^l$ if and only if $l - k$ is an integral multiple of n.
5. $\displaystyle\sum_{k=0}^{n-1} (w_n^a)^k = \begin{cases} 0 & a \neq mn \\ n & a = mn \end{cases} \qquad a, m \text{ integers}$
6. If a and b are integers,

$$\sum_{k=0}^{n-1} (w_n^a)^k (w_n^b)^k = \begin{cases} 0 & a + b \neq mn \\ n & a + b = mn \end{cases} \qquad m \text{ an integer}$$

The proofs of Properties **1–3** we leave as exercises. To prove Property **4**, we assume l and k are real. Then since $w_n^k = (e^{i2\pi/n})^k = e^{i2k\pi/n} \neq 0$, $w_n^k = w_n^l$ if and only if

$$1 = w_n^{l-k} = e^{i(l-k)2\pi/n} = \cos\frac{(l-k)2\pi}{n} + i\sin\frac{(l-k)2\pi}{n}$$

Hence $w_n^k = w_n^l$ if and only if

$$1 = \cos\frac{(l-k)2\pi}{n} \qquad \text{and} \qquad 0 = \sin\frac{(l-k)2\pi}{n}$$

which hold if and only if

$$\frac{(l-k)2\pi}{n} = m\,2\pi \qquad \text{or} \qquad l-k = m\cdot n \qquad m \text{ an integer}$$ ◗

To prove Property **5**, we note that the sum is a finite geometric progression with first term 1 and common ratio w_n^a. Hence its value is

$$\frac{1-(w_n^a)^n}{1-w_n^a} = \frac{1-(w_n^n)^a}{1-w_n^a} = 0 \qquad w_n^a \neq 1$$

The condition that $w_n^a \neq 1$ is always satisfied unless a is an integral multiple of n. If this is the case, each term in the original sum is 1 and its value is therefore n, as asserted. ◗

To prove Property **6**, we note that the given sum can be written

$$\sum_{k=0}^{n-1} (w_n^a)^k (w_n^b)^k = \sum_{k=0}^{n-1} (w_n^{a+b})^k$$

It is thus a finite geometric progression with first term 1 and common ratio w_n^{a+b}. Therefore by Property **5**, its value is

$$\begin{cases} 0 & a+b \neq mn \\ n & a+b = mn \end{cases} \qquad m \text{ an integer}$$

as asserted. ◗

As we conclude this introductory discussion of complex numbers, there is one further property that should be carefully noted: *the field of complex numbers cannot be ordered;* that is, the relations ''greater than'' and ''less than'' cannot be consistently defined for complex numbers. Thus whatever our first impression might be, such statements as

$$353 + 294i > 1 + i \qquad \text{and} \qquad -31 - 26i < 0$$

are completely meaningless. The only meaningful inequalities involving complex numbers are those which relate their absolute values (see Exercise 40).

EXERCISES

1. What is the geometric relation between a complex number and its conjugate? Between the number and its negative?

2. If $z_1 = 2 + 3i$, $z_2 = -3 + i$, and $z_3 = 4 - 5i$, reduce each of the following to the form $a + ib$: **(a)** $z_1 z_3$ **(b)** z_2^2 **(c)** $1/z_3$ **(d)** z_3/z_2 **(e)** $z_2 z_3/z_1$ **(f)** $z_3/z_1 z_2$

3. If $z_1 = 3 + 4i$, $z_2 = -2 - 4i$, and $z_3 = 5 - 4i$, reduce each of the following to the form $a + ib$: **(a)** $1/z_1 z_2$ **(b)** z_3^3 **(c)** $(z_1 + z_2)/z_3$ **(d)** $z_1 z_2 z_3$ **(e)** $z_2^2 z_3$ **(f)** $z_1^2 + z_2^2$

4. In Exercise 2, what is **(a)** $\arg z_1$? **(b)** $\arg z_1 z_2$? **(c)** $\arg (z_2 + z_3)$?

5. In Exercise 3, what is **(a)** $\text{Im}(z_1/z_2)$? **(b)** $\text{Re}(z_1 z_2/z_3)$? **(c)** $\text{Im}(z_1 + z_3)/z_2$? **(d)** $|z_3/(z_1 + z_2)|$? **(e)** $|z_1 z_2 z_3|$? **(f)** $\bar{z}_1 \bar{z}_2 \bar{z}_3$? **(g)** $z_1 \bar{z}_2 z_3$?

6. Solve for x and y if $(x^2 - 2x - y) + i(2x - y - 3) = 0$.

7. Prove Parts **a**, **b**, and **c** of (12).

8. Prove Parts **d**, **e**, **f**, and **g** of (12).

9. Prove Parts **h**, **i**, and **j** of (12).

10. Find all the fifth roots of $32(1 - i)$ and express them in both polar form and exponential form.

11. Find all the sixth roots of $8i$ and express them in both polar form and exponential form.

12. Show that $i^i = e^{-\pi/2}$.

13. Show that $[(1 + i)/\sqrt{2}]^i = e^{-\pi/4}$.

14. What is $[(1 - i)/\sqrt{2}]^i$?

15. What is $(-1)^i$?

16. Are the answers in Exercises 12 and 13 unique? What other answers are possible?

17. Prove that $z_1 = a + ib$ and $z_2 = c + id$ are perpendicular if and only if $ac + bd = 0$.

18. Prove that z_1 and z_2 are perpendicular if and only if $z_1\bar{z}_2 + \bar{z}_1 z_2 = 0$.

19. Prove that if both $z_1 + z_2$ and $z_1 z_2$ are real, then either z_1 and z_2 are both real or $z_1 = \bar{z}_2$.

20. Show that $|x + iy| \geq (|x| + |y|)/\sqrt{2}$. Under what conditions will the equality sign hold?

21. Prove that if the product of two complex numbers is zero, at least one of the numbers must be zero.

22. If $f(z)$ is a polynomial in z with real coefficients and $f(2 + 3i) = 1 - i$, what is $f(2 - 3i)$? Is $f(a - ib)$ determined by a knowledge of $f(a + ib)$ if the coefficients of $f(z)$ are not all real?

23. If $B\bar{B} > (A + \bar{A})(C + \bar{C})$, show that the equation $(A + \bar{A})z\bar{z} + Bz + \bar{B}\bar{z} + (C + \bar{C}) = 0$ represents a real circle and find its center and radius.

24. Using De Moivre's theorem, express $\cos 5\theta$ and $\sin 5\theta$ in terms of powers of $\cos\theta$ and $\sin\theta$.

25. If $z_1 = 1 - i$, $z_2 = -2\sqrt{3} + 2i$, $z_3 = -3 - 3\sqrt{3}i$, $z_4 = 4 - 3i$, express each of the following in exponential form without performing the indicated multiplications: **(a)** z_1z_2 **(b)** $z_1z_2z_3$ **(c)** $z_2z_3z_4$ **(d)** $z_1z_2z_3z_4$

26. If $z_1 = 2(1 - i\sqrt{3})$, $z_2 = 3(-1 + i)$, $z_3 = -3 + 4i$, $z_4 = 5 + 12i$, express each of the following in exponential form without performing the indicated multiplications and divisions: **(a)** z_1/z_2 **(b)** $z_1^5 z_3$ **(c)** z_1z_2/z_4 **(d)** $1/z_4^3$ **(e)** z_3/z_4

27. Using Eqs. (19) and (20), verify that $\cos^2 z + \sin^2 z = 1$.

28. Using Eqs. (19) and (20), verify that $\sin 2z = 2\sin z\cos z$ and $\cos 2z = \cos^2 z - \sin^2 z$.

29. Prove algebraically that $|z_1 + z_2| \leq |z_1| + |z_2|$. *Hint:* Consider the identity $|z_1 + z_2|^2 = (z_1 + z_2)(\bar{z}_1 + \bar{z}_2)$.

30. Prove Properties **1**, **2**, and **3** for w_n.

31. Prove Properties **5** and **6** for w_n using the exponential form of w_n.

32. What is the generalization of Properties **2** and **3** for w_n to the case where n is an arbitrary positive integral power of 2?

33. By applying the root-coefficient relations for polynomial equations to the equation $z^n - 1 = 0$, prove that

$$\sum_{k=0}^{n-1} w_n^k = 0 \quad \text{and} \quad \prod_{k=0}^{n-1} w_n^k = (-1)^{n+1}$$

34. If $w = (3z + i)/(i - z)$, show that $\operatorname{Re} z \geq 0$ implies $\operatorname{Im} w \leq 0$.

35. If $w = i(1 - z)/(1 + z)$, show that $|z| < 1$ implies $\operatorname{Im} w > 0$.

36. **(a)** Show that z_1, z_2, z_3 are the vertices of an equilateral triangle if and only if

$$\frac{z_3 - z_1}{z_2 - z_1} = \frac{z_1 - z_2}{z_3 - z_2}$$

Hint: Consider the amplitudes and moduli of the respective fractions.

(b) Using the result of Part **(a)**, show that z_1, z_2, z_3 are the vertices of an equilateral triangle if and only if

$$(z_1 - z_2)^2 + (z_2 - z_3)^2 + (z_3 - z_1)^2 = 0$$

37. Show that three points z_1, z_2, z_3 in the complex plane are collinear if and only if there exist real numbers p, q, r, not all zero, such that $p + q + r = 0$ and $pz_1 + qz_2 + rz_3 = 0$.

38. Show that the triangle whose vertices are the points z_1, z_2, z_3 and the triangle whose vertices are the points z_4, z_5, z_6 are similar if and only if

$$\begin{vmatrix} z_1 & z_4 & 1 \\ z_2 & z_5 & 1 \\ z_3 & z_6 & 1 \end{vmatrix} = 0$$

39. If $z_1, z_2, \ldots, z_n$ and $w_1, w_2, \ldots, w_n$ are complex numbers, prove that

$$\left|\sum_{k=1}^{n} z_k w_k\right|^2 \leq \sum_{k=1}^{n} |z_k|^2 \cdot \sum_{k=1}^{n} |w_k|^2$$

This result is known as **Cauchy's inequality**. *Hint:* Consider the discriminant of the quadratic equation in λ.

$$\sum_{k=1}^{n} (|z_k|\lambda - |w_k|)^2 = 0$$

40. A number system is said to be **ordered** if it contains a subset of numbers P, called **positive numbers**, with the following properties:

1. For any two numbers p_1 and p_2 in P, both $p_1 + p_2$ and p_1p_2 are in P, that is, are positive numbers.
2. For any number q in the system, exactly one of the following possibilities holds: $q = 0$, q is in P, or $-q$ is in P.

If a number system is ordered, it is possible to define the relation **greater than**, or $>$, by saying that $a > b$ means that $a - b$ is a number in P, that is, is positive. Prove that the complex numbers cannot be ordered and that it is therefore meaningless to say that one complex number is greater than another. *Hint:* Note first that $i \neq 0$ and then show that no matter how the subset P is defined, both the assumption that i is in P and the assumption that $-i$ is in P lead at once to a contradiction.

In the following exercises, complex numbers are to be regarded as ordered pairs.

41. Show that (0, 0) is the additive identity. What is the multiplicative identity? What pair corresponds to the number i?

42. Show that "real" × "real" = "real" and that (pure) "imaginary" × "imaginary" = "real".

43. Show that every nonzero number has a multiplicative inverse.

44. Show that both addition and multiplication are associative.

45. Show that $(x, 0) + (0, 1)(y, 0) = (x, y)$. What does this say about $x + iy$?

46. Show that $c(x, y) = (c, 0)(x, y)$, so that multiplication by a real number is consistent with complex multiplication. What does this imply about Rules **b** and **d** for ordered pairs of real numbers?

47. Show that the set of real numbers is a proper subset of the set of all complex numbers.

48. **(a)** Show that $(0, 1) \cdot (0, 1) = -1$. What does this say about i?

(b) Interpret $(0, -1) \cdot (0, -1) = -1$ in terms of complex numbers.

(c) What do Parts **(a)** and **(b)** give as square roots of -1?

49. Show that for all complex numbers z_1, z_2, z_3, the **commutative laws**

$$z_1 + z_2 = z_2 + z_1 \qquad z_1 z_2 = z_2 z_1$$

the **associative laws**

$$(z_1 + z_2) + z_3 = z_1 + (z_2 + z_3) \qquad (z_1 z_2) z_3 = z_1(z_2 z_3)$$

and the **distributive laws**

$$z_1(z_2 + z_3) = z_1 z_2 + z_1 z_3 \qquad (z_1 + z_2) z_3 = z_1 z_3 + z_2 z_3$$

are valid.

50. Show that $z + \bar{z} = 2 \operatorname{Re} z$ and $z - \bar{z} = 2i \operatorname{Im} z$.

3.2 THE ALGEBRA OF VECTORS

Many practical problems give rise to systems of differential equations which are to be solved simultaneously. The methods we developed in Chap. 2 for solving a single linear differential equation can be extended to methods for solving such systems provided all the equations of the system are linear. However, to effect this extension and to make the solution procedure clear and efficient, we will need the material on complex numbers, vectors, matrices, determinants, and systems of linear algebraic equations presented in this chapter. Together these ideas comprise a modest but very important part of the field of mathematics known a *linear algebra.* With these tools available, we will then study linear differential systems in Chap. 4.

In engineering and applied science, physical quantities such as acceleration, force, and velocity, which possess both magnitude and direction, are represented geometrically by directed line segments called **vectors**. Chapter 15 is devoted to classical vector analysis developed from this traditional geometric point of view. By extending the notion of a vector from a geometric object to an algebraic object, it is possible to evolve more general concepts which provide a broader theory unifying a great variety of problems and their solution procedures. In Chaps. 13 and 14 we shall develop such a theory and investigate some of its applications. The present chapter is designed to motivate that work as well as to facilitate our study of linear differential systems in the next chapter.

Physical quantities such as mass, temperature, and volume possess only a magnitude, and no direction. They are commonly known as **scalars**. In a broader sense, a scalar may be thought of as any real or complex number.† To distinguish vectors from scalars, the former will be printed in boldface type **v**. This is a common notation; when handwritten, a vector is usually indicated by putting an arrow above the symbol, thus $\vec{v}$.

Having taken scalars to be numbers, we now define an n-dimensional vector.

DEFINITION 1 An ***n*-dimensional vector v** is an ordered set

$$\mathbf{v} = (v_1, v_2, \ldots, v_n)$$

of n scalars; the scalar v_i, $i = 1, 2, \ldots, n$, is called the ith **component** of **v**.

†In a still broader sense, scalars may be taken to be elements of any mathematical system of the type known in abstract algebra as a *field.* However, real and complex scalars will suffice for our purposes.

Sometimes an n-dimensional vector is called an **ordered n-tuple**. Most vectors which occur in elementary applications have the property that all of their components are real numbers. However, there are important applications, especially in mathematical physics and quantum mechanics, which involve vectors whose components are not real. For this reason, as in Definition 1, many of our basic definitions and fundamental results concerning n-dimensional vectors will be formulated in terms of complex-valued scalars.

For all n-dimensional vectors $\mathbf{u} = (u_1, u_2, \ldots, u_n)$ and $\mathbf{v} = (v_1, v_2, \ldots, v_n)$, $\mathbf{u} = \mathbf{v}$ if and only if $u_i = v_i$ for $i = 1, 2, \ldots, n$. This definition of **vector equality** requires equal vectors to have the same components, and it guarantees the equality of vectors whose corresponding components are equal. Vectors of different dimensions cannot be equal.

Vector addition of n-dimensional vectors is defined by

$$\text{(1)} \qquad \mathbf{u} + \mathbf{v} = (u_1, u_2, \ldots, u_n) + (v_1, v_2, \ldots, v_n) = (u_1 + v_1, u_2 + v_2, \ldots, u_n + v_n)\dagger$$

Clearly, the **sum $\mathbf{u} + \mathbf{v}$** of $\mathbf{u}$ and $\mathbf{v}$ is itself an n-dimensional vector. This fact is expressed by saying that *the set of all n-dimensional vectors is* ***closed*** *under vector addition.*

Multiplication of an n-dimensional vector $\mathbf{v}$ by a scalar k is defined by

$$\text{(2)} \qquad k\mathbf{v} = k(v_1, v_2, \ldots, v_n) = (kv_1, kv_2, \ldots, kv_n) = \mathbf{v}k$$

Equation (2) ensures the equality of the **products** $k\mathbf{v}$ and $\mathbf{v}k$. Thus, when a vector is multiplied by a scalar on the left, the product is the same as when the vector is multiplied by that scalar on the right. The product $k\mathbf{v}$ is of course an n-dimensional vector whose components are the respective components of $\mathbf{v}$ each multiplied by the number k. Since $k\mathbf{v}$ is an n-dimensional vector, it is said that *the set of all n-dimensional vectors is* ***closed*** *under multiplication by a scalar.*

The particular n-dimensional vector $(0, 0, \ldots, 0)$ is called the ***n*-dimensional zero vector** and is usually denoted by $\mathbf{0}$, or more specifically by $\mathbf{0}_n$. To every n-dimensional vector $\mathbf{v} = (v_1, v_2, \ldots, v_n)$, there corresponds a vector $-\mathbf{v} = (-v_1, -v_2, \ldots, -v_n)$ each of whose components is the negative of the corresponding component of $\mathbf{v}$. It is called the **negative** of $\mathbf{v}$, the **additive inverse** of $\mathbf{v}$, or simply the **inverse** of $\mathbf{v}$.

Since every n-dimensional vector has an additive inverse, it is easy to define subtraction for vectors. Given any n-dimensional vectors $\mathbf{u}$ and $\mathbf{v}$, we define the **difference $\mathbf{u} - \mathbf{v}$** to be

$$\text{(3)} \qquad \mathbf{u} - \mathbf{v} = \mathbf{u} + (-\mathbf{v})$$

Obviously, $-\mathbf{v} = (-1)\mathbf{v}$; hence this definition of subtraction is equivalent to taking $\mathbf{u} - \mathbf{v} = \mathbf{u} + (-1)\mathbf{v}$.

EXAMPLE 1

Solve for the vector $\mathbf{v}$ if $2\mathbf{v} - 3(3, 2, -3) = (3, 2, 9)$

By first using (2), then adding the vector $(9, 6, -9)$ to each side of the equation and using (1), we obtain

$$2\mathbf{v} = (12, 8, 0)$$

Finally, multiplying each side of this vector equation by $\frac{1}{2}$, according to (2), we have

$$\mathbf{v} = (6, 4, 0)$$

---

†Since addition of vectors is not the same operation as addition of scalars, it appears that if + is to denote addition of scalars, then some symbol other than + ought to be used in the first two members of (1). The use of + to stand for either operation is conventional, however, and should cause no confusion: the *positioning* of + between two vectors will always signify vector addition, whereas the *positioning* of + between two scalars always indicates scalar addition. The use of + between a vector and a scalar will never occur. Once this is realized, all ambiguity disappears from (1) without any need to proliferate symbols.

The symbol C^n is often used to designate the set of all n-dimensional vectors having complex numbers as components and for which vector addition and multiplication of its elements by a complex number are defined by (1) and (2), respectively. Note carefully that C^n is not merely a set but a set *supplied with two operations,* one of which operates on pairs of vectors in C^n while the other operates on pairs consisting of a vector of C^n and a complex number. Our immediate interest in C^n is to identify its principal algebraic properties.

We have already observed that the operations of *vector addition,* defined by (1), and of *scalar multiplication,* defined by (2), imply the following for each positive integer n.

1. C^n is **closed under vector addition.**
2. C^n is **closed under multiplication by complex numbers.**

These two properties are necessarily locked into C^n because of its two built-in operations.

Other important algebraic properties of C^n are the following: For all **u**, **v**, **w** in C^n, and for all scalars k and l,

(i) $\mathbf{u} + (\mathbf{v} + \mathbf{w}) = (\mathbf{u} + \mathbf{v}) + \mathbf{w}$	**Associative law of vector addition.**
(ii) $\mathbf{u} + \mathbf{v} = \mathbf{v} + \mathbf{u}$	**Commutative law of vector addition.**
(iii) $\mathbf{v} + \mathbf{0} = \mathbf{v}$	**Existence of 0 as an additive identity.**
(iv) $\mathbf{v} + (-\mathbf{v}) = \mathbf{0}$	**Existence of additive inverses.**
(v) $k(\mathbf{u} + \mathbf{v}) = k\mathbf{u} + k\mathbf{v}$	**Scalars distribute over vector addition.**
(vi) $(k + l)\mathbf{v} = k\mathbf{v} + l\mathbf{v}$	**Vectors distribute over scalar addition.**
(vii) $(kl)\mathbf{v} = k(l\mathbf{v})$	**Associative law for multiplication by scalars.**
(viii) $1\mathbf{v} = \mathbf{v}$	**Number 1 is a multiplicative identity.**

Properties **(i)** through **(iv)** follow immediately from (1) and the corresponding properties of complex numbers. For instance, using (1) and scalar inverses, we obtain

$$\begin{aligned}\mathbf{v} + (-\mathbf{v}) &= (v_1, v_2, \ldots, v_n) + (-v_1, -v_2, \ldots, -v_n)\\ &= (v_1 + \{-v_1\}, v_2 + \{-v_2\}, \ldots, v_n + \{-v_n\})\\ &= (0, 0, \ldots, 0) = \mathbf{0}\end{aligned}$$

which establishes **(iv)**.

To prove Properties **(v)** through **(viii)**, (2) is also needed. In particular, to show that **(v)** holds, we use (1) and (2), together with the distributive law of numbers, as follows:

$$\begin{aligned}k(\mathbf{u} + \mathbf{v}) &= k(u_1 + v_1, u_2 + v_2, \ldots, u_n + v_n)\\ &= (k\{u_1 + v_1\}, k\{u_2 + v_2\}, \ldots, k\{u_n + v_n\})\\ &= (ku_1 + kv_1, ku_2 + kv_2, \ldots, ku_n + kv_n) = k\mathbf{u} + k\mathbf{v}\end{aligned}$$

Proofs of the other six properties we leave as exercises.

Of course, C^n has many more algebraic properties. Our reason for calling attention to the few just enumerated is that they serve to characterize completely what is known in mathematics as a **vector space** or **linear space**. Thus, for each positive integer n, C^n is an example of such a space, as is also the set R^n of all n-dimensional vectors, with exclusively *real* components, for which vector addition, and multiplication by *real* numbers, are defined by (1) and (2), repectively.

Many systems in applied mathematics involve two operations with the properties described by Relations **(i)** through **(viii)**. Such systems appear as specific models of the concept of a linear space, which serves to unify them in a single abstract theory. Beginning in Sec. 13.1, we shall take a much closer look at vector spaces and some important practical examples.

Although we can now multiply a scalar and a vector, none of our present operations involves multiplication or division of vectors by vectors. Division of n-dimensional vectors is left undefined.

On the other hand, they are multiplied not only by scalars but also in other ways for various purposes. For instance, it is well known from analytic geometry that if $\mathbf{a} = (a_1, a_2, a_3)$ and $\mathbf{b} = (b_1, b_2, b_3)$ are vectors whose components are real numbers, then

1. $a_1b_1 + a_2b_2 + a_3b_3$ is the *scalar product* of **a** and **b**.
2. **a** and **b** are *perpendicular* if and only if their scalar product is zero.
3. The scalar product of **a** with itself, $a_1^2 + a_2^2 + a_3^2$, is the square of the *length* of **a**.

By analogy, and as a natural extension of these ideas to n-dimensional vectors,

$$\mathbf{u} \cdot \mathbf{v} = u_1v_1 + u_2v_2 + \cdots + u_nv_n = \sum_{k=1}^{n} u_kv_k \tag{4}$$

is said to be the **scalar**, **dot**, or **inner product** of any pair of vectors

$$\mathbf{u} = (u_1, u_2, \ldots, u_n) \qquad \text{and} \qquad \mathbf{v} = (v_1, v_2, \ldots, v_n) \qquad \text{in } R^n$$

The vectors **u** and **v** are **orthogonal** if and only if $\mathbf{u} \cdot \mathbf{v} = 0$. A *nonempty* set of mutually orthogonal *nonzero* vectors is called an **orthogonal set** of vectors. The square root of the inner product of a vector with itself

$$(\mathbf{v} \cdot \mathbf{v})^{1/2} = (v_1^2 + v_2^2 + \cdots + v_n^2)^{1/2} = \left(\sum_{k=1}^{n} v_k^2\right)^{1/2} \tag{5}$$

is called the **length**, or **norm**, of **v**, commonly denoted by $\|\mathbf{v}\|$. Vectors of unit length are called **unit** or **normalized vectors**. A *nonzero* vector can always be reduced to a unit vector by dividing each of its components by the length of the vector. An orthogonal set of unit vectors is said to be an **orthonormal set** of vectors.

EXAMPLE 2

The length of the vector $\mathbf{v} = (1, -2, 2)$ is $\sqrt{\mathbf{v} \cdot \mathbf{v}} = \sqrt{1 + 4 + 4} = 3$. The vector itself is conventionally represented in a three-dimensional cartesian† coordinate system by a directed line segment, or arrow, three units long, extending from the origin to the point $(1, -2, 2)$, as shown in Fig. 3.3*a*. Any arrow having the same

FIGURE 3.3
Geometric representations of the vector $(1, -2, 2)$.

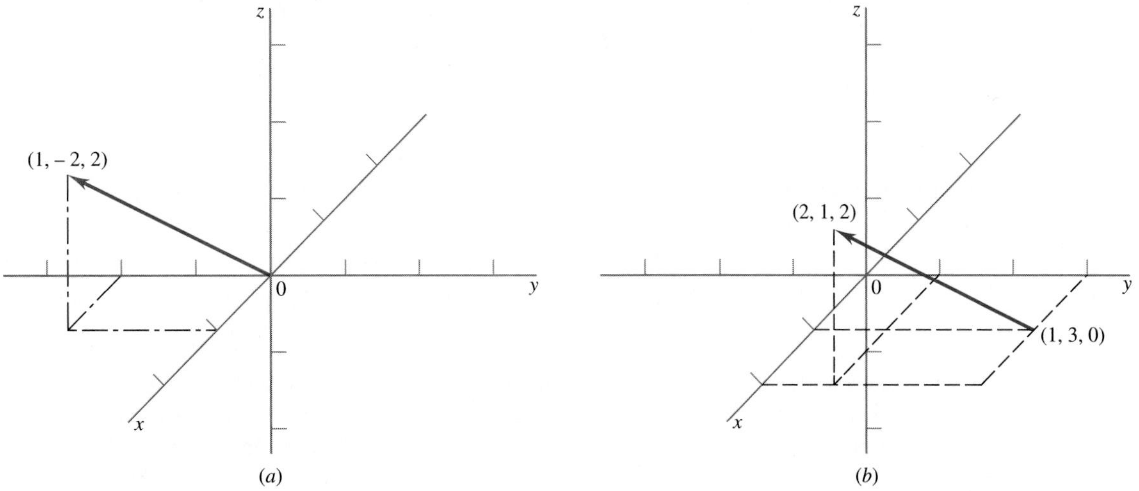

†Named for the French mathematician and philosopher René Descartes (1596–1650).

length and pointing in the same direction is also taken as a geometric representation of the vector **v**. The motivation for this is that all such arrows have perpendicular projections, or components, along the x, y, and z axes which are equal, respectively, to 1, -2, and 2. For instance, Fig. 3.3*b* shows that the (directed) projections of the arrow drawn from (1, 3, 0) to (2, 1, 2) are also 1, -2, 2. Since a vector is uniquely determined by its components, it is natural to agree that all arrows having the same set of projections represent the same vector.

A vector may also be represented geometrically by the point whose cartesian coordinates are the components of the vector. In this case of course the vector has a unique geometric representation rather than an infinite family of representations. As we saw in the last section, these are also the modes of representation for two-dimensional vectors represented arithmetically by complex numbers, $x + iy$.

EXAMPLE 3

Find all vectors of the family $(x, x^2, x + 3, x - 8)$ which are orthogonal to the vector $(1, -1, 4, 3)$ and then normalize each of these vectors.

To find the vectors of the family $(x, x^2, x + 3, x - 8)$ which are orthogonal to the vector $(1, -1, 4, 3)$ we form the inner product of these two vectors and set it equal to zero, getting the equation

$$x - x^2 + 4(x + 3) + 3(x - 8) = -x^2 + 8x - 12 = 0$$

The roots of this equation are $x = 2, 6$. The required vectors are therefore

$$\mathbf{v}(2) = (2, 4, 5, -6) \qquad \text{and} \qquad \mathbf{v}(6) = (6, 36, 9, -2)$$

The length of $\mathbf{v}(2)$ is $(2^2 + 4^2 + 5^2 + \{-6\}^2)^{1/2} = 9$, and the norm, i.e., the length of $\mathbf{v}(6)$ is $(6^2 + \{36\}^2 + 9^2 + \{-2\}^2)^{1/2} = \sqrt{1417}$. Multiplying each vector by the reciprocal of its length, we obtain the respective unit vectors

$$\left(\frac{2}{9}, \frac{4}{9}, \frac{5}{9}, -\frac{6}{9}\right) \qquad \text{and} \qquad \left(\frac{6}{\sqrt{1417}}, \frac{36}{\sqrt{1417}}, \frac{9}{\sqrt{1417}}, \frac{-2}{\sqrt{1417}}\right)$$

When working with systems of linear algebraic or differential equations, it is often necessary to multiply n-dimensional vectors whose components are complex numbers. For any given positive integer n, Eq. (4) defines the **scalar**, **dot**, or **inner product** of any pair of vectors **u** and **v** in C^n, just as it does for vectors of R^n. An inner product of still another kind is used to extend the concepts of *length* and *orthogonality* to vectors with complex components. *Hermitian*† *inner products* are introduced for this purpose.

Recalling that the conjugate of a complex number $z = x + iy$ is the complex number $\bar{z} = x - iy$ and defining the **conjugate of a vector v** in C^n by

$$\bar{\mathbf{v}} = (\bar{v}_1, \bar{v}_2, \ldots, \bar{v}_n) \tag{6}$$

the **hermitian inner product of u into v** is defined as the scalar product of $\bar{\mathbf{u}}$ and **v**; thus the hermitian product of **u** into **v** is given by

$$\bar{\mathbf{u}} \cdot \mathbf{v} = \mathbf{v} \cdot \bar{\mathbf{u}} = \bar{u}_1 v_1 + \bar{u}_2 v_2 + \cdots + \bar{u}_n v_n = \sum_{k=1}^{n} \bar{u}_k v_k \tag{7}$$

In general, the complex number $\bar{\mathbf{u}} \cdot \mathbf{v}$ is *not* equal to its conjugate $\mathbf{u} \cdot \bar{\mathbf{v}}$ because only real numbers are equal to their conjugates. In other words, the hermitian product of **u** into **v** is not the same as the hermitian product of **v** into **u**, except for vectors **u** and **v** whose hermitian products are real.

†Named after the French mathematician Charles Hermite (1822–1901).

The **hermitian length** l_h of a vector **v** in C^n is the nonnegative real number

$$l_h(\mathbf{v}) = (\bar{\mathbf{v}} \cdot \mathbf{v})^{1/2} = \left(\sum_{k=1}^{n} \bar{v}_k v_k\right)^{1/2} \tag{8}$$

and **u** and **v** are **orthogonal in the hermitian sense** if and only if $\bar{\mathbf{u}} \cdot \mathbf{v} = 0$ or $\mathbf{u} \cdot \bar{\mathbf{v}} = 0$.

Several properties of inner products of vectors in R^n are given in Exercises 23 and 24. Here are some important properties of inner products and hermitian products of vectors in C^n, which can be verified directly using (4) or (7).

For all vectors **u**, **v**, and **w** of C^n, and for every scalar k,

1. $\mathbf{u} \cdot \mathbf{v} = \mathbf{v} \cdot \mathbf{u}$
2. $(\mathbf{u} + \mathbf{v}) \cdot \mathbf{w} = \mathbf{u} \cdot \mathbf{w} + \mathbf{v} \cdot \mathbf{w}$
3. $(k\mathbf{u}) \cdot \mathbf{v} = k\mathbf{u} \cdot \mathbf{v} = \mathbf{u} \cdot k\mathbf{v}$
4. $\bar{\mathbf{u}} \cdot \mathbf{u} \geq 0$, the equality sign holding if and only if $\mathbf{u} = \mathbf{0}$

EXAMPLE 4

If $\mathbf{u} = (1, i)$, $\mathbf{v} = (1 - i, 2i)$, and $\mathbf{w} = (1 - i, -i)$, find the inner product of **u** with itself, the hermitian length of **u**, the inner product of **v** and **w**, and show that **v** and **w** are orthogonal in the hermitian sense.

For $\mathbf{u} \cdot \mathbf{u}$ we have $1 + i^2 = 1 - 1 = 0$. For the hermitian length of **u** we have $l_h(\mathbf{u}) = \sqrt{\overline{(1, i)} \cdot (1, i)} = \sqrt{(1, -i) \cdot (1, i)} = \sqrt{1 + 1} = \sqrt{2}$. For $\mathbf{v} \cdot \mathbf{w}$ we have $(1 - i)(1 - i) + (2i)(-i) = 2 - 2i$. The hermitian orthogonality of **v** and **w** follows from the fact that

$$\bar{\mathbf{v}} \cdot \mathbf{w} = (1 + i, -2i) \cdot (1 - i, -i) = 2 - 2 = 0$$

EXERCISES

Find all values of the unknowns which satisfy the following equations.

1. $(x, y, z) - 3(-1, 2, i) = (0, 0, 0)$
2. $7(2, x, x^2) - 5(2, -x, x^2) = (1, 12x, x^2) + 3(1, 0, 1)$
3. $(1, -1, 2) - (x, -1, y) = 7(-2, 0, 2)$
4. $2(z, 1 + i, 1 - z) - (2 + 2i, 2z, -z^2) = \mathbf{0}$
5. $(z^4, z^2, 1) \cdot (1, 1, -z^2) = \tan(\pi/4)$
6. $(x, y, x^2, z) - (y, -x, 1, i) = (1, 1, 0, 0)$
7. $(1, x, x^2, x) \cdot (x, 1, x, x^2) = x^3 + 6x$
8. $(2x^2, x, -2, 1) \cdot (x, -x^2, x, -1) = 1 - x^2$
9. $(\sin x, \sin 2x, \ln e) = (\sin \pi, 2 \sin x, \sec x) \cos x$
10. $2(z^2 + iz, -i, 2z - 2, i) + i(iz - 2, 3 - i, -iz^3, 1) = (z^2, 1 + i, z^2, 3i)$

For the given vectors **u** and **v**, find the value of $\mathbf{u} \cdot \mathbf{v}$ and normalize $\mathbf{u} + \mathbf{v}$.

11. $\mathbf{u} = (1, 1, 3)$, $\mathbf{v} = (3, 1, 1)$
12. $\mathbf{u} = (2, -7, 1)$, $\mathbf{v} = (3, -7, 1)$
13. $\mathbf{u} = (2, 7, -1, 20)$, $\mathbf{v} = (3, -1, 9, -10)$
14. $\mathbf{u} = (-1, 3, -2, 10)$, $\mathbf{v} = (10, 2, 5, -1)$
15. $\mathbf{u} = (3, -3, 1, -3, 3)$, $\mathbf{v} = (4, -1, 1, -1, 3)$
16. $\mathbf{u} = (4, 3, -2, 2, 1)$, $\mathbf{v} = (3, 2, 6, 1, 0)$

In Exercises 17–21, determine whether the given sets of vectors $\{\mathbf{u}, \mathbf{v}, \mathbf{w}\}$ are orthonormal, orthogonal but not orthonormal, or neither.

17. $\mathbf{u} = \dfrac{1}{\sqrt{3}}(1, 1, 1)$, $\mathbf{v} = \dfrac{1}{\sqrt{2}}(1, -1, 0)$, $\mathbf{w} = \dfrac{1}{\sqrt{6}}(-1, -1, 2)$
18. $\mathbf{u} = (\frac{1}{3}, \frac{2}{3}, \frac{2}{3})$, $\mathbf{v} = (-\frac{2}{3}, \frac{2}{3}, -\frac{1}{3})$, $\mathbf{w} = (\frac{2}{3}, \frac{1}{3}, -\frac{2}{3})$
19. $\mathbf{u} = (0, 0, 0)$, $\mathbf{v} = (2, -3, 6)$, $\mathbf{w} = (3, -2, -2)$
20. $\mathbf{u} = (10, -2, 11)$, $\mathbf{v} = (-2, 1, 2)$, $\mathbf{w} = (2, 2, 1)$
21. $\mathbf{u} = (2, -4, 5, -2)$, $\mathbf{v} = (1, -1, -2, -2)$, $\mathbf{w} = (-2, 4, 2, -5)$
22. Establish the following algebraic properties of C^n:
    - **(a)** **(i)**, **(ii)**, and **(iii)**.
    - **(b)** **(vi)**, **(vii)**, and **(viii)**.
    - **(c)** The zero vector of C^n is unique.
    - **(d)** Each vector of C^n has a unique inverse.
    - **(e)** For any **v** in C^n, and any scalar k, $k\mathbf{v} = \mathbf{0}$ if and only if $\mathbf{v} = \mathbf{0}$ or $k = 0$.
23. Given that **v** is a vector of R^n, show that $\mathbf{v} \cdot \mathbf{v} \geq 0$ and that $\mathbf{v} \cdot \mathbf{v} = 0$ if and only if $\mathbf{v} = \mathbf{0}$.
24. Given that **u**, **v**, and **w** are vectors of R^n and that k is a real number, prove that
    - **(a)** $\mathbf{u} \cdot \mathbf{v} = \mathbf{v} \cdot \mathbf{u}$.
    - **(b)** If **v** has length l, then $k\mathbf{v}$ has length kl.
    - **(c)** $(k\mathbf{u}) \cdot \mathbf{v} = k\mathbf{u} \cdot \mathbf{v} = \mathbf{u} \cdot (k\mathbf{v})$.
    - **(d)** $(\mathbf{u} + \mathbf{v}) \cdot (\mathbf{u} + \mathbf{v}) = \mathbf{u} \cdot \mathbf{u} + 2\mathbf{u} \cdot \mathbf{v} + \mathbf{v} \cdot \mathbf{v}$.
25. For all real values of t, **u** and **v** are the vectors of R^4 given by

$$\mathbf{u} = (1, 5t, -4t, t^3)$$

and

$$\mathbf{v} = (20 - t, 4t - 4, 5t^2, t^2)$$

Find all real numbers t for which **u** and **v** are orthogonal.

For the vectors **u** and **v** given in Exercises 26–30, find $\mathbf{u} \cdot \mathbf{v}$, the hermitian products **u** into **v**, and **v** into **u**, the hermitian lengths of **u** and **v**, and determine whether or not **u** and **v** are orthogonal in the hermitian sense.

26. $\mathbf{u} = (3 - 2i, 6)$, $\mathbf{v} = (2, 4 + 4i)$
27. $\mathbf{u} = (1 - i, 1 + i)$, $\mathbf{v} = (5 + i, -3 - i)$
28. $\mathbf{u} = (2 - 2i, 2 - 2i)$, $\mathbf{v} = (1, -1)$
29. $\mathbf{u} = (i, 1 + i, 1)$, $\mathbf{v} = (5 + i, -2 + 2i, -1 + i)$
30. $\mathbf{u} = (i, 1 + i, 1)$, $\mathbf{v} = (-2i, -2, 4 + 5i)$
31. Is it possible to determine x so that the vector $(x^2, x, 1)$ is perpendicular to each of the vectors $(1, 3, 2)$ and $(1, 1, -2)$?
32. Is it possible to determine x so that the vector $(x^2, x, 1)$ is perpendicular to each of the vectors $(1, -5, 6)$ and $(1, 3, -4)$?
33. Which vectors, if any, of the family $(1, 2, 3) - k(2, 3, 4)$ are perpendicular to the vector $(3, 4, 5)$?
34. What is the most general vector which is orthogonal to each of the vectors $(1, 1, 0, 1)$ and $(0, 1, 0, -1)$?
35. **(a)** Is there a vector perpendicular to each of the vectors $(1, 2, 1)$, $(-4, 1, 2)$, $(1, -2, 3)$?
(b) Is there a vector perpendicular to each of the vectors $(1, 2, 1)$, $(-4, 1, 2)$, $(-1, 1, 1)$?
36. **(a)** Is there a vector orthogonal to each of the vectors $(1, 1, 1, 1)$, $(0, 1, -1, 0)$, $(1, 0, 0, -1)$, $(1, -1, -1, 1)$?
(b) Is there a vector orthogonal to each of the vectors $(1, 1, 1, 1)$, $(0, 1, -1, 0)$, $(1, 0, 0, -1)$, $(3, 2, 1, 0)$?
37. **(a)** Normalize the vector $\mathbf{v} = (2, -4, 5, -6)$.
(b) Find all vectors of the family $(3x^3, 6x^2, 6x, 2)$ which are orthogonal to **v**.
38. If $\mathbf{u} = (-1 + i, -1 - i, -2, 1 + i)$ and $\mathbf{v} = (-i, 1, 1 - i, -1)$, **(a)** find all numbers z such that the vector $(z, z^2, z + z^2, z - z^2 + z^3) = \frac{1}{2}(1 - i)\mathbf{u}$, and **(b)** subtract **v** from $\frac{1}{2}(i - 1)\mathbf{u}$.
39. Find all vectors of the real parameter family $(2 - xi, 4i, 2 - 5i, xi)$ that are of hermitian length
(a) 11 **(b)** 25 **(c)** 7
40. Find all vectors of the real parameter family $(2x, \pi + 6i, 2 - \pi i, x^2)$ that are orthogonal to $(1, 1 - i, 1 + i, i)$ in the hermitian sense.

3.3 THE ALGEBRA OF MATRICES

Thus far we have displayed n-dimensional vectors

(1) $$\mathbf{v} = (v_1, v_2, \ldots, v_n)$$

by writing their components in a *row*. For many purposes, it is convenient to write their components in a *column*

(2) $$\mathbf{v} = \begin{bmatrix} v_1 \\ v_2 \\ \vdots \\ v_n \end{bmatrix}$$

To distinguish n-dimensional vectors written in these two ways, we shall call (1) a **row vector** and (2) a **column vector**.

An $m \times n$, or m by n, rectangular array of numbers

(3) $$\begin{bmatrix} a_{11} & a_{12} & \cdots & a_{1n} \\ a_{21} & a_{22} & \cdots & a_{2n} \\ \cdots & \cdots & \cdots & \cdots \\ a_{m1} & a_{m2} & \cdots & a_{mn} \end{bmatrix}$$

enclosed within brackets, and arranged in m rows and n columns, may be thought of as an ordered set of m row vectors

(4) $$\mathbf{u}_i = (a_{i1}, a_{i2}, \ldots, a_{in}) \qquad i = 1, 2, \ldots, m$$

each of dimension n, or as an ordered set of n column vectors

$$(5) \qquad \mathbf{v}_j = \begin{bmatrix} a_{1j} \\ a_{2j} \\ \vdots \\ a_{mj} \end{bmatrix} \qquad j = 1, 2, \ldots, n$$

each of dimension m. Such an array is called a **matrix of size** $m \times n$. A matrix consisting of a single row is called a **row matrix**. A matrix consisting of a single column is called a **column matrix**. Of course, both row matrices and column matrices are vectors. Hence the concept of matrix generalizes the concept of vector.

The **entries** $a_{11}, \ldots, a_{mn}$ in the matrix (3) are called the **elements** of the matrix. With i and j given, the element a_{ij} in the ith row and the jth column is said to have **row index** i and **column index** j. Elements whose row and column indices are the same are called **diagonal** elements.

When there is no possibility of confusion, matrices are often represented by single capital letters. More explicitly, a rectangular array (3), abbreviated $[a_{ij}]$, is used to display some or all of the elements in a matrix A. When distinct matrices are distinguished by different capital letters, corresponding lowercase letters with double subscripts are used to designate their elements. The subscripts need not be i and j, but whatever they are, the *first* subscript identifies the row, and the *second* subscript the column, of the matrix in which the element is found. Elements of a matrix need not be numbers, and later on we shall work with generalized matrices whose entries are themselves matrices.

EXAMPLE 1

For $1 \le i \le 2$ and $1 \le j \le 4$, the elements of a matrix A are determined by the formula $a_{ij} = j[1 + (-1)^{i+j}]/2$. Write A as a rectangular array and identify its diagonal elements.

Computing a_{ij} for the given values of i and j, we obtain

$$A = \begin{bmatrix} 1 & 0 & 3 & 0 \\ 0 & 2 & 0 & 4 \end{bmatrix}$$

which has 1 and 2 as its diagonal elements.

Examples of matrices can be found in many fields. For instance, if m students were given a battery of n different tests, the resulting scores would very probably be displayed in a table containing m rows, one for each student, and n columns, one for each test. The resulting array would of course be an $m \times n$ matrix in which the general element a_{ij} was the score the ith student made on the jth test. Matrices of this sort are of fundamental importance in the branch of mathematical psychology known as *factor analysis*. Similarly, if we had an electric network containing n branches, we might, either experimentally or analytically, determine the current which would flow in the ith branch as a result of inserting a unit voltage in the jth branch. A tabular array of these quantities would also constitute a matrix, the so-called **admittance matrix**, which is of fundamental importance in the theory of electric circuits. Still another example of a matrix is provided by an array of **transition probabilities**: suppose that a system S can exist in any one of n states, say $S_1, S_2, \ldots, S_n$, and that the probability of the system passing from the state S_i to the state S_j by some well-defined random process is p_{ij}. Clearly, the p_{ij}'s can be displayed as an $n \times n$ matrix. Such matrices are of great importance in the theory of probability and its physical applications.

Equality of $m \times n$ matrices closely resembles equality of n-dimensional vectors in that a matrix $A = [a_{ij}]$ **equals** a matrix $B = [b_{ij}]$ if and only if their corresponding elements are equal; that is, $A = B$ if and only if $a_{ij} = b_{ij}$, for $i = 1, 2, \ldots, m$ and $j = 1, 2, \ldots, n$. Matrices of different sizes cannot be equal.

By appropriately defining addition and multiplication of matrices, an algebra of matrices very much like that of vectors can be developed.

Matrices having the same number of rows and columns are added by adding corresponding elements. Thus for all $m \times n$ matrices A and B, their **sum** A + B is a matrix $C = [c_{ij}]$ such that $c_{ij} = a_{ij} + b_{ij}$, for $i = 1, 2, \ldots, m$ and $j = 1, 2, \ldots, n$. Since the sum of A and B is itself an $m \times n$ matrix, it is said that *the set of all* $m \times n$ *matrices is closed under matrix addition.* Addition of matrices of different sizes is undefined.

Multiplication of a matrix by any scalar is accomplished by multiplying every element of the matrix by that scalar. Thus a matrix $D = [d_{ij}]$ is the **product** $kA = Ak$ of an $m \times n$ matrix $A = [a_{ij}]$ and a scalar k if and only if $d_{ij} = ka_{ij}$, for $i = 1, 2, \ldots, m$ and $j = 1, 2, \ldots, n$. Since the product of k and A must have the same size as A, *the set of all* $m \times n$ *matrices is closed under multiplication by a scalar.*

An $n \times n$ matrix has the same number of rows as columns and is called a **square matrix of order** n. A matrix, whether square or not, in which every element is zero is called a **zero matrix** and is denoted by the symbol O. More specifically, a square zero matrix of order n may be designated by O_n, and a general $m \times n$ zero matrix by O_{mn}.

To every $m \times n$ matrix $A = [a_{ij}]$ there corresponds an **additive inverse matrix** $-A = [-a_{ij}]$ each of whose elements is the negative of the corresponding element of A. Since every matrix has an additive inverse, the **difference** A − B of matrices A and B of the same size is defined as the matrix A + (−B), which is a matrix whose elements are the differences of the respective elements of A and B.

By utilizing familiar algebraic laws of numbers, it is easy to verify that matrices have essentially the same algebraic properties as were enumerated in Sec. 3.2 for n-dimensional vectors. Specifically, for all positive integers m and n, for all scalars k and l, and for all $m \times n$ matrices A, B, and C,

(i) $A + (B + C) = (A + B) + C$	**(ii)** $A + B = B + A$
(iii) $A + O = A$	**(iv)** $A + (-A) = O$
(v) $k(A + B) = kA + kB$	**(vi)** $(k + l)A = kA + lA$
(vii) $(kl)A = k(lA)$	**(viii)** $1A = A$

Since addition of $m \times n$ matrices is both associative and commutative, an indicated sum A + B + C of such matrices stands for a single matrix no matter how its terms are associated or in what order the matrices A, B, and C appear. For instance,

$$A + B + C = A + (B + C) = (C + A) + B$$

EXAMPLE 2

Given the 2×3 matrices

$$A = \begin{bmatrix} 1 & 1 & -2 \\ 1 & 1 & 0 \end{bmatrix} \quad B = \begin{bmatrix} 1 & 1 & -1 \\ 1 & 1 & 1 \end{bmatrix} \quad C = \begin{bmatrix} 1 & 2 & 3 \\ 2 & -1 & 1 \end{bmatrix} \quad D = \begin{bmatrix} 1 & -1 & -9 \\ -1 & 4 & 1 \end{bmatrix}$$

find the matrix 2A − 3B + C. Do there exist numbers x and y such that $xB + yC = D$?

Using the definition of the product of a matrix and a scalar and the definitions of matrix addition and subtraction, we have

$$2A - 3B + C = \begin{bmatrix} 2 & 2 & -4 \\ 2 & 2 & 0 \end{bmatrix} - \begin{bmatrix} 3 & 3 & -3 \\ 3 & 3 & 3 \end{bmatrix} + \begin{bmatrix} 1 & 2 & 3 \\ 2 & -1 & 1 \end{bmatrix} = \begin{bmatrix} 0 & 1 & 2 \\ 1 & -2 & -2 \end{bmatrix}$$

Similarly,

$$xB + yC = \begin{bmatrix} x & x & -x \\ x & x & x \end{bmatrix} + \begin{bmatrix} y & 2y & 3y \\ 2y & -y & y \end{bmatrix} = \begin{bmatrix} x+y & x+2y & -x+3y \\ x+2y & x-y & x+y \end{bmatrix}$$

According to the definition of the equality of two matrices, the last matrix will be equal to the matrix D if and only if the following six conditions are met:

$$\begin{array}{ccc} x+y=1 & x+2y=-1 & -x+3y=-9 \\ x+2y=-1 & x-y=4 & x+y=1 \end{array}$$

The two equations in the first column suffice to determine the values of x and y, namely, $x = 3$, $y = -2$. With these values, three of the remaining four conditions *happen* to be fulfilled, but the second equation in the second column is *not* satisfied. Hence, although there are values of x and y for which *five* of the six elements in $xB + yC$ are equal to the corresponding elements in D, there are no values of x and y for which $xB + yC = D$.

As defined by Eq. (4), Sec. 3.2, the inner product of two vectors is a scalar, *not* a vector. The product of two matrices, however, is defined in such a way that it is also a matrix. The multiplication of matrices is based on the inner product of n-dimensional vectors and requires that the matrices being multiplied be *conformable.* Since we have defined an inner product only for vectors with real or complex components, our present discussion of matrix multiplication will be confined to matrices whose elements are either real or complex numbers.

Two matrices A and B are said to be **conformable in the order** AB if and only if the number of columns in A is equal to the number of rows in B. In other words, if A is an $m \times n$ matrix and if B is a $q \times p$ matrix, A and B are conformable in the order AB if and only if $n = q$.

As (4) and (5) point out, any $m \times n$ matrix A and any $n \times p$ matrix B may be thought of, respectively, as ordered sets

$$A = \begin{bmatrix} \mathbf{u}_1 \\ \mathbf{u}_2 \\ \vdots \\ \mathbf{u}_m \end{bmatrix} \qquad B = [\mathbf{v}_1\ \mathbf{v}_2 \cdots \mathbf{v}_p]$$

of m row vectors and of p column vectors, where each vector of either matrix is in C^n. The inner product of any row vector of A with any column vector of B is therefore defined. With this observation and the idea of conformable matrices in mind, we are now prepared to define *row into column multiplication of matrices.*

DEFINITION 1 If A is an $m \times n$ matrix and B an $n \times p$ matrix, so that A and B are conformable in the order AB, then the product C = AB is the $m \times p$ matrix in which the element c_{ij} in the ith row and jth column is the scalar product of the ith row vector $\mathbf{u}_i$ of A and the jth column vector $\mathbf{v}_j$ of B; that is, C = AB is the matrix for which

$$c_{ij} = \mathbf{u}_i \cdot \mathbf{v}_j = \sum_{k=1}^{n} a_{ik} b_{kj}$$

Multiplication is not defined for matrices that are not conformable.

EXAMPLE 3

Find the product AB of

$$A = \begin{bmatrix} 2 & 3 \\ 1 & -1 \\ 0 & 4 \end{bmatrix} \quad \text{and} \quad B = \begin{bmatrix} 5 & -2 & 4 & 7 \\ -6 & 1 & -3 & 0 \end{bmatrix}$$

Applying Definition 1, we have

$$AB = \begin{bmatrix} (2, 3) \cdot (5, -6) & (2, 3) \cdot (-2, 1) & (2, 3) \cdot (4, -3) & (2, 3) \cdot (7, 0) \\ (1, -1) \cdot (5, -6) & (1, -1) \cdot (-2, 1) & (1, -1) \cdot (4, -3) & (1, -1) \cdot (7, 0) \\ (0, 4) \cdot (5, -6) & (0, 4) \cdot (-2, 1) & (0, 4) \cdot (4, -3) & (0, 4) \cdot (7, 0) \end{bmatrix}$$

$$= \begin{bmatrix} 2(5) + 3(-6) & 2(-2) + 3(1) & 2(4) + 3(-3) & 2(7) + 3(0) \\ 1(5) + (-1)(-6) & 1(-2) + (-1)(1) & 1(4) + (-1)(-3) & 1(7) + (-1)(0) \\ 0(5) + 4(-6) & 0(-2) + 4(1) & 0(4) + 4(-3) & 0(7) + 4(0) \end{bmatrix}$$

$$= \begin{bmatrix} -8 & -1 & -1 & 14 \\ 11 & -3 & 7 & 7 \\ -24 & 4 & -12 & 0 \end{bmatrix}$$

For the matrices of this example, BA is not defined because B has more columns than A has rows.

As we have just seen, matrix multiplication is usually noncommutative. Even for matrices A and B which are square and of the same order, AB and BA are *not* necessarily equal.

EXAMPLE 4

For the matrices

$$A = \begin{bmatrix} 1 & 2 \\ 3 & 4 \end{bmatrix} \quad \text{and} \quad B = \begin{bmatrix} 1 & 1 \\ 4 & 1 \end{bmatrix}$$

we have

$$AB = \begin{bmatrix} 1 & 2 \\ 3 & 4 \end{bmatrix}\begin{bmatrix} 1 & 1 \\ 4 & 1 \end{bmatrix} = \begin{bmatrix} 9 & 3 \\ 19 & 7 \end{bmatrix} \quad \text{and} \quad BA = \begin{bmatrix} 1 & 1 \\ 4 & 1 \end{bmatrix}\begin{bmatrix} 1 & 2 \\ 3 & 4 \end{bmatrix} = \begin{bmatrix} 4 & 6 \\ 7 & 12 \end{bmatrix}$$

Since matrix multiplication is not, in general, commutative, it is desirable to be able to describe concisely the order in which two conformable matrices are to be multiplied. This we shall do by adopting the following terminology. In the product AB we shall say that A **premultiplies** B or B is **premultiplied** by A and that B **postmultiplies** A or A is **postmultiplied** by B.

Most familiar properties of multiplication which we know to be true for scalars carry over to matrix multiplication. In particular, for all scalars c and all suitably conformable and additive matrices A, B, C, and D,

1. $c(AB) = (cA)B = A(cB)$	**Associativity of scalar and matrix multiplication**
2. $A(B + C) = AB + AC$	**Distributive law for premultiplication**
3. $(B + C)D = BD + CD$	**Distributive law for postmultiplication**
4. $A(BC) = (AB)C$	**Associative law of matrix multiplication**

A proof of the first property is quite straightforward. Assuming that A is an $m \times n$ matrix and B is an $n \times p$ matrix, we use Definition 1, together with the definition of multiplication of matrices by a scalar, to find the element in the ith row and jth column of each of the $m \times p$ matrices $c(\mathrm{AB})$, $(c\mathrm{A})\mathrm{B}$, and $\mathrm{A}(c\mathrm{B})$. Displaying these elements within brackets and using the usual notation, we have

$$c(\mathrm{AB}) = \left[c\left(\sum_{k=1}^{n} a_{ik}b_{kj}\right)\right] \qquad (c\mathrm{A})\mathrm{B} = \left[\sum_{k=1}^{n} (ca_{ik})b_{kj}\right] \qquad \mathrm{A}(c\mathrm{B}) = \left[\sum_{k=1}^{n} a_{ik}(cb_{kj})\right]$$

Since each expression within brackets completely determines the element in the ith row and jth column of the corresponding matrix, and since all three of these expressions are equal, the matrices themselves must be equal. ◗

Properties **2** and **3** are very much alike so we shall prove only the first of the two. Supposing A to be $m \times n$, and C as well as B to be $n \times p$, we use Definition 1 and the definition of matrix addition to compute the element in the ith row and jth column of each of the $m \times p$ matrices $\mathrm{A}(\mathrm{B} + \mathrm{C})$ and $\mathrm{AB} + \mathrm{AC}$. Displaying these elements within brackets, we have

$$\mathrm{A}(\mathrm{B} + \mathrm{C}) = \left[\sum_{k=1}^{n} a_{ik}(b_{kj} + c_{kj})\right] \qquad \mathrm{AB} + \mathrm{AC} = \left[\sum_{k=1}^{n} a_{ik}b_{kj} + \sum_{k=1}^{n} a_{ik}c_{kj}\right]$$

Using familiar properties of numbers, it is easy to show that the two expressions within brackets are equal, which confirms the equality of the matrices. ◗

To establish Property **4**, we assume the sizes of A, B, and C to be $m \times n$, $n \times p$, and $p \times q$, respectively. Then A(BC) and (AB)C are both $m \times q$ matrices. For all i, j such that $1 \le i \le m$, $1 \le j \le q$, the element in the ith row and jth column of A(BC) is

$$\sum_{k=1}^{n} a_{ik}\left(\sum_{r=1}^{p} b_{kr}c_{rj}\right) = \sum_{k=1}^{n}\sum_{r=1}^{p} a_{ik}b_{kr}c_{rj}$$

and the element in the ith row and jth column of (AB)C is

$$\sum_{r=1}^{p}\left(\sum_{k=1}^{n} a_{ik}b_{kr}\right)c_{rj} = \sum_{r=1}^{p}\sum_{k=1}^{n} a_{ik}b_{kr}c_{rj}$$

The order of summation in either double sum may be reversed because the limits of summation are finite and constant. Hence corresponding elements of A(BC) and (AB)C are equal, and so the matrices themselves must be equal. ◗

Property **4** allows us to drop parentheses and write either A(BC) or (AB)C as ABC. In fact, if A_k is of size $m_k \times m_{k+1}$ and successive pairs of matrices in the product $\mathrm{A}_1\mathrm{A}_2 \cdots \mathrm{A}_n$ are conformable, then the product is unambiguous and its size $m_1 \times m_{n+1}$ can be obtained by writing the size of each A_k in succession and canceling common interior integers:

$$m_1 \times \cancel{m_2}, \cancel{m_2} \times \cancel{m_3}, \ldots, \cancel{m_{n-1}} \times \cancel{m_n}, \cancel{m_n} \times m_{n+1} \rightarrow m_1 \times m_{n+1}$$

EXAMPLE 5

Given the matrices

$$\mathrm{A} = \begin{bmatrix} 2 & 1 & -2 \\ 4 & 2 & -3 \end{bmatrix} \qquad \mathrm{B} = \begin{bmatrix} -2 & 0 & -2 \\ 5 & 2 & 4 \\ 0 & 1 & 1 \end{bmatrix} \qquad \mathrm{C} = \begin{bmatrix} 3 & -2 & 1 & 2 \\ 2 & 4 & 2 & 1 \\ 1 & -2 & -1 & -1 \end{bmatrix}$$

determine the size and the elements of ABC.

Writing the sizes of A, B, and C in succession and canceling common interior integers, we have $2 \times \cancel{3}, \cancel{3} \times \cancel{3}, \cancel{3} \times 4 \to 2 \times 4$. Hence ABC is a 2×4 matrix. Computing AB and then the product (AB)C, we obtain

$$\text{ABC} = \begin{bmatrix} 2 & 1 & -2 \\ 4 & 2 & -3 \end{bmatrix} \begin{bmatrix} -2 & 0 & -2 \\ 5 & 2 & 4 \\ 0 & 1 & 1 \end{bmatrix} \begin{bmatrix} 3 & -2 & 1 & 2 \\ 2 & 4 & 2 & 1 \\ 1 & -2 & -1 & -1 \end{bmatrix}$$

$$= \begin{bmatrix} 1 & 0 & -2 \\ 2 & 1 & -3 \end{bmatrix} \begin{bmatrix} 3 & -2 & 1 & 2 \\ 2 & 4 & 2 & 1 \\ 1 & -2 & -1 & -1 \end{bmatrix} = \begin{bmatrix} 1 & 2 & 3 & 4 \\ 5 & 6 & 7 & 8 \end{bmatrix}$$

EXERCISES

1. For each of the following matrices, identify the third row vector, the second column vector, and find the sum of the diagonal elements:

(a) $\begin{bmatrix} 1 & i & 2 & -3 \\ 2i & 0 & 1 & -i \\ 5 & 2 & i & -1 \end{bmatrix}$ **(b)** $\begin{bmatrix} 1 & 4 \\ 9 & 16 \\ -16 & -9 \\ -4 & -1 \end{bmatrix}$

(c) $\begin{bmatrix} 1 & i & -1 \\ 2 & 0 & 2 \\ -i & 1 & i \end{bmatrix}$ **(d)** $\begin{bmatrix} i & 0 & 1 & 0 \\ 0 & i & 0 & 1 \\ 1 & 0 & i & 0 \\ 0 & 1 & 0 & i \end{bmatrix}$

2. For $1 \le i \le 2$ and $1 \le j \le 3$, let a_{ij} denote the element in the ith row and jth column of a matrix A. Display A as a rectangular array if a_{ij} is given by

(a) $i^2 - j^2$ **(b)** ij **(c)** i^j **(d)** $\cos \pi i \cos \pi j$

In Exercises 3–6, find those numbers z, if any, such that $Az^3 + Bz^2 + Cz = D$.

3. $\text{A} = \begin{bmatrix} 2 & 1 \\ 1 & 2 \end{bmatrix}$ $\text{B} = \begin{bmatrix} -3 & 0 \\ -2 & -5 \end{bmatrix}$

$\text{C} = \begin{bmatrix} -2 & -4 \\ 0 & 2 \end{bmatrix}$ $\text{D} = \begin{bmatrix} 0 & 0 \\ 0 & 0 \end{bmatrix}$

4. $\text{A} = \begin{bmatrix} 1 & 1 \\ 1 & 4 \end{bmatrix}$ $\text{B} = \begin{bmatrix} -3 & -5 \\ 0 & -11 \end{bmatrix}$

$\text{C} = \begin{bmatrix} 1 & 9 \\ -11 & 0 \end{bmatrix}$ $\text{D} = 5\begin{bmatrix} -1 & 1 \\ -4 & -5 \end{bmatrix}$

5. $\text{A} = \begin{bmatrix} 2 & 4 & 2 \\ 1 & 3 & 1 \end{bmatrix}$ $\text{B} = \begin{bmatrix} -1 & 1 & 1 \\ 1 & -2 & -1 \end{bmatrix}$

$\text{C} = \begin{bmatrix} 2 & 4 & 2 \\ 1 & 3 & 1 \end{bmatrix}$ $\text{D} = \begin{bmatrix} 1 & -1 & -1 \\ -1 & 2 & 0 \end{bmatrix}$

6. $\text{A} = \begin{bmatrix} 1 & 0 & 1 \\ 1 & 1 & 1 \end{bmatrix}$

$\text{B} = \begin{bmatrix} i & 1 & -2 + 2i \\ -2 & -2 + i & -4 + i \end{bmatrix}$

$\text{C} = \begin{bmatrix} -4 & -2 + i & -1 - 4i \\ 1 & -2i & 4 - 4i \end{bmatrix}$

$\text{D} = 2\begin{bmatrix} 2i & i & -1 \\ 1 & 0 & -i \end{bmatrix}$

7. Let $f(x)$ be the sum of the diagonal elements of $\begin{bmatrix} x^3 & 2 \\ 3 & -6x^2 \end{bmatrix}$. Find the matrix defined by

$$16\begin{bmatrix} 1 & -1 \\ -1 & -1 \end{bmatrix} x^2 - 4\begin{bmatrix} 5 & -2 \\ -2 & 4 \end{bmatrix} x + \begin{bmatrix} 6 & 13 \\ 19 & 47 \end{bmatrix}$$

when x is the number for which $f(x)$ has

(a) a relative maximum
(b) a relative minimum
(c) an inflection point

8. If $\text{A} = \begin{bmatrix} 1 & 2 \\ 3 & 4 \end{bmatrix}$, $\text{B} = \begin{bmatrix} 1 & -1 \\ 2 & 3 \end{bmatrix}$, and

$\text{C} = \begin{bmatrix} 1 & 4 \\ -1 & 3 \end{bmatrix}$, verify directly that

(a) A(BC) = (AB)C
(b) A(B + C) = AB + AC
(c) B(A − C) = BA − BC
(d) 3(ABC) = A(3B)C

If possible, find a single matrix equal to the following.

9. (a) $\begin{bmatrix} 2 & -3 \\ 3 & 2 \end{bmatrix} \begin{bmatrix} 1 & -2 & 3 \\ -1 & 5 & 4 \end{bmatrix}$

(b) $\begin{bmatrix} 2 & -6 & 3 \\ -5 & 0 & 1 \end{bmatrix} \begin{bmatrix} 2 \\ 1 \\ 0 \end{bmatrix} - \begin{bmatrix} 3 \\ 11 \end{bmatrix}$

10. (a) $\begin{bmatrix} 1 & 2 \\ 2 & -1 \end{bmatrix} \begin{bmatrix} 1 & 0 & 1 \\ 0 & -1 & 0 \end{bmatrix} \begin{bmatrix} 1 & 2 \\ 3 & 4 \end{bmatrix}$

(b) $\begin{bmatrix} 1 & 2 \\ 2 & -1 \end{bmatrix}\begin{bmatrix} 1 & 2 \\ 3 & 4 \end{bmatrix}\begin{bmatrix} 1 & 0 & 1 \\ 0 & -1 & 0 \end{bmatrix}$

11. (a) $[-1 \quad 2 \quad -2 \quad 1]\begin{bmatrix} -1 \\ 2 \\ -2 \\ 1 \end{bmatrix} - [10]$

(b) $\begin{bmatrix} 1 & 0 & -1 \\ 0 & 1 & 0 \\ -1 & 0 & 1 \end{bmatrix}\begin{bmatrix} 2 & 1 \\ 1 & 2 \\ 1 & 1 \end{bmatrix}\begin{bmatrix} 1 & -2 \\ 2 & -1 \end{bmatrix}$

12. (a) $2\begin{bmatrix} 2 & -1 & 4 \\ -1 & 0 & 3 \\ 2 & 1 & 0 \end{bmatrix}\begin{bmatrix} -1 & 5 \\ 4 & 2 \\ -3 & 1 \end{bmatrix} + 4\begin{bmatrix} 9 & -6 \\ 4 & 1 \\ -1 & -6 \end{bmatrix}$

(b) $\begin{bmatrix} 3 & 1 \\ -1 & 0 \\ 2 & 5 \end{bmatrix}\begin{bmatrix} 1 & 0 & 0 \\ 1 & 1 & 0 \\ 1 & 1 & 1 \end{bmatrix}\begin{bmatrix} -1 \\ 0 \\ 1 \end{bmatrix}$

13. State conditions under which a matrix is conformable with itself.

14. (a) Give a sufficient condition for the matrix equation $(A + B)(A - B) = AA - BB$ to hold.

(b) Give an example of two square matrices for which the relation of Part **(a)** is false.

15. Work Exercise 14 for the matrix equation $(A + B)(A + B) = AA + 2AB + BB$.

16. Find the most general matrix that commutes with $\begin{bmatrix} 0 & 1 \\ -1 & 1 \end{bmatrix}$.

17. Find the most general 2×2 matrix A such that $AA = O$.

18. Prove the distributive law under postmultiplication (Property **3**).

Find values (if any) of the unknowns x and y that satisfy the following matrix equations.

19. $\begin{bmatrix} y & -4 & -5 \\ x & 0 & 2 \end{bmatrix}\begin{bmatrix} 3 & 0 \\ x & -2 \\ 0 & 1 \end{bmatrix} = \begin{bmatrix} 5 & 3 \\ y & 2 \end{bmatrix}$

20. $\begin{bmatrix} -4 & -3 \\ 0 & 5 \end{bmatrix}\begin{bmatrix} y & x & -1 \\ -1 & -x & 1 \end{bmatrix} = \begin{bmatrix} 2x & -x & 1 \\ -5 & y & 5 \end{bmatrix}$

21. $\begin{bmatrix} 3 & -2 \\ 3 & 0 \\ 2 & 4 \end{bmatrix}\begin{bmatrix} y & y \\ x & x \end{bmatrix} = \begin{bmatrix} 3 & 3 \\ 3y & 3y \\ 10 & 10 \end{bmatrix}$

22. $\begin{bmatrix} y & -x \\ x & -y \end{bmatrix}\begin{bmatrix} 2 & 0 & 1 \\ 2 & 3 & -1 \end{bmatrix} = \begin{bmatrix} 0 & -3x & 0 \\ 0 & -3y & 0 \end{bmatrix}$

23. (a) Describe the effect of premultiplying

$$A = \begin{bmatrix} a_{11} & a_{12} & a_{13} \\ a_{21} & a_{22} & a_{23} \\ a_{31} & a_{32} & a_{33} \end{bmatrix} \text{ by } B = \begin{bmatrix} 1 & 0 & 0 & 0 \\ 0 & 0 & 1 & 0 \\ 0 & 1 & 0 & 0 \\ 0 & 0 & 0 & 1 \end{bmatrix}.$$

(b) Describe the effect of postmultiplying A by B.

If possible, find a single matrix equal to each of the following.

24. (a) $\begin{bmatrix} -3 & 1+i \\ -i & 2 \end{bmatrix}\begin{bmatrix} 1-i \\ i \end{bmatrix}[1+i \quad 1-i]$

(b) $\begin{bmatrix} 10 & 2-i \\ 3-2i & i \end{bmatrix}\begin{bmatrix} -1 & i & 1 \\ i & 1 & -i \end{bmatrix} + \begin{bmatrix} 1 & 2 \\ 3 & 4 \end{bmatrix}$

25. (a) $\begin{bmatrix} 1-i & 0 & 1 \\ 0 & 2-i & -2 \\ 1 & -4 & 3-i \end{bmatrix}\begin{bmatrix} 1 \\ i \\ -1+i \end{bmatrix} + \begin{bmatrix} 1 \\ -2 \\ 2 \end{bmatrix}$

(b) $[1-i \quad 1+i]\begin{bmatrix} 1 & -i \\ i & 1 \end{bmatrix}\begin{bmatrix} 1+i \\ 1-i \end{bmatrix}$

26. Evaluate $[z \quad \bar{z}]\begin{bmatrix} 1 & -1 & -3 \\ -1 & 1 & -3 \end{bmatrix}\begin{bmatrix} \bar{z} \\ z \\ 1 \end{bmatrix}$ where $z = x + iy$ and x and y are real numbers for which $z\bar{z} + (z + \bar{z}) - 2i(z - \bar{z}) - 2$ has a relative minimum.

Find all values of the unknowns which satisfy the following matrix equations.

27. (a) $\begin{bmatrix} 1 & 0 & 1 & -3 \\ 0 & 1 & -2 & 1 \\ 0 & 0 & 1 & -5 \\ 0 & 0 & 0 & 1 \end{bmatrix}\begin{bmatrix} x \\ y \\ z \\ w \end{bmatrix} = \begin{bmatrix} 1 \\ 2 \\ 5 \\ -1 \end{bmatrix}$

(b) $\begin{bmatrix} i & 1 & 0 \\ 0 & 2i & 1 \\ 0 & 0 & 3i \end{bmatrix}\begin{bmatrix} x \\ y \\ z \end{bmatrix} = \begin{bmatrix} 2+i \\ 2i \\ 6 \end{bmatrix}$

28. If $A = \begin{bmatrix} 3 & 2 \\ 7 & 5 \end{bmatrix}$ and $X = \begin{bmatrix} u & v \\ x & y \end{bmatrix}$, solve each of the following equations:

(a) $AX = \begin{bmatrix} 1 & 0 \\ 0 & 1 \end{bmatrix}$ **(b)** $XA = \begin{bmatrix} 1 & 0 \\ 0 & 1 \end{bmatrix}$

29. If $A = \begin{bmatrix} 1 & 2 \\ 2 & 5 \end{bmatrix}$ and $B = \begin{bmatrix} 3 & 4 \\ 2 & 3 \end{bmatrix}$, solve each of the following equations.

(a) $AX = B$ **(b)** $XA = B$

(c) $BX = A$ **(d)** $XB = A$

30. If w_3 and w_4 are, respectively, the primitive third and fourth roots of unity, evaluate each of the following products:

(a) $\begin{bmatrix} 1 & 1 & 1 \\ 1 & w_3 & w_3^2 \\ 1 & w_3^2 & w_3^4 \end{bmatrix}\begin{bmatrix} 1 & 1 & 1 \\ 1 & w_3^{-1} & w_3^{-2} \\ 1 & w_3^{-2} & w_3^{-4} \end{bmatrix}$

(b) $\begin{bmatrix} 1 & 1 & 1 & 1 \\ 1 & w_4 & w_4^2 & w_4^3 \\ 1 & w_4^2 & w_4^4 & w_4^6 \\ 1 & w_4^3 & w_4^6 & w_4^9 \end{bmatrix}\begin{bmatrix} 1 & 1 & 1 & 1 \\ 1 & w_4^{-1} & w_4^{-2} & w_4^{-3} \\ 1 & w_4^{-2} & w_4^{-4} & w_4^{-6} \\ 1 & w_4^{-3} & w_4^{-6} & w_4^{-9} \end{bmatrix}$

Infer and then prove the corresponding property for w_5.

3.4 SPECIAL MATRICES

Of the many kinds of matrices which arise in various problems, certain ones occur so often that they are given special names. Several of these will now be introduced and some of their most important properties identified.

The **transpose** of an $m \times n$ matrix A is the $n \times m$ matrix, denoted by the symbol A^T, whose elements a_{ji}^T are determined by the relation

$$a_{ji}^T = a_{ij} \qquad 1 \le j \le n \qquad 1 \le i \le m \tag{1}$$

Note that the *first* subscript of a_{ji}^T is j and as such it is the *row* index of the element. Thus, a_{ji}^T occurs in the jth row and ith column of the array $[a_{ji}^T]$ representing A^T, while a_{ij} occurs in the ith row and jth column of the array $[a_{ij}]$ representing A. The jth row of A^T is therefore the jth column of A, and the ith column of A^T is the ith row of A. In other words, A^T is the matrix obtained from a matrix A by writing its successive rows as columns, or its successive columns as rows.

For all matrices A, B, and any scalar c, the following four transpose properties hold.

1. $(A^T)^T = A$.
2. $(cA)^T = cA^T$.
3. If $A + B$ is defined, $(A + B)^T = A^T + B^T$.
4. If AB is defined, $(AB)^T = B^T A^T$.

The first three properties are quite obvious. To prove the fourth one, assume that $A = [a_{ij}]$ is an $m \times n$ matrix and that $B = [b_{ij}]$ is an $n \times p$ matrix, so that $C = AB$ is of size $m \times p$. Then $C^T = (AB)^T$ is of size $p \times m$, and the element in its jth row and ith column is

$$c_{ji}^T = c_{ij} = \sum_{k=1}^{n} a_{ik} b_{kj} \qquad 1 \le j \le p \qquad 1 \le i \le m$$

Now, B^T is of size $p \times n$, A^T is of size $n \times m$, and

$$b_{ji}^T = b_{ij} \qquad 1 \le j \le p \qquad 1 \le i \le n$$

and

$$a_{ji}^T = a_{ij} \qquad 1 \le j \le n \qquad 1 \le i \le m$$

It follows that the element in the jth row and ith column of the $p \times m$ matrix $B^T A^T$ is

$$\sum_{k=1}^{n} b_{jk}^T a_{ki}^T = \sum_{k=1}^{n} b_{kj} a_{ik} = \sum_{k=1}^{n} a_{ik} b_{kj} = c_{ji}^T \qquad 1 \le j \le p \qquad 1 \le i \le m$$

This shows that corresponding elements of the two $p \times m$ matrices $(AB)^T$ and $B^T A^T$ are equal, hence the matrices themselves are equal as asserted. ◗

The **conjugate** of a matrix A is the matrix $\overline{A}$ whose elements are, respectively, the conjugates of the elements of A. A matrix A is **real** if and only if $A = \overline{A}$. A matrix A is **imaginary** if and only if $A = -\overline{A}$. Clearly, real matrices contain just real numbers, and imaginary matrices contain only *pure* imaginary numbers, as elements. The matrix defined by

$$A^* = (\overline{A})^T = \overline{A^T} \tag{2}$$

is called the **tranjugate**, **associate**, or **hermitian conjugate** of A. If A is $m \times n$, A^* is of course $n \times m$. Properties of the tranjugate are very much like those of the transpose (Exercise 8).

If A is an $m \times n$ matrix and if k and l are integers such that $0 < k \leq m$ and $0 < l \leq n$, then the array of elements common to any k rows and any l columns of A is called a $k \times l$ (k by l) **submatrix** of A. In particular, this definition allows a matrix to be a submatrix of itself, for we may let $k = m$ and $l = n$.

EXAMPLE 1

The matrix $A = \begin{bmatrix} 1 & 0 \\ -2 & 4 \\ 3 & 5 \end{bmatrix}$ has $A^T = \begin{bmatrix} 1 & -2 & 3 \\ 0 & 4 & 5 \end{bmatrix}$ as its transpose. Both A and A^T are real matrices because all of their elements are real numbers. As this example illustrates, the diagonal elements of a matrix (1 and 4 in this case) are always the same as the diagonal elements of its transpose. The submatrix of A made up of the elements common to Rows 2 and 3 and Column 1 is the 2×1 matrix $B = \begin{bmatrix} -2 \\ 3 \end{bmatrix}$ whose transpose is $B^T = [-2 \quad 3]$. An easy computation confirms that

$$(AB)^T = B^T A^T = [-2 \quad 16 \quad 9]$$

EXAMPLE 2

For the square matrix

$$A = \begin{bmatrix} 2i & 1+i & 3i \\ -i & 2 & -i \\ i & 3-2i & 0 \end{bmatrix}$$

A transpose, A conjugate, and A tranjugate are given by

$$A^T = \begin{bmatrix} 2i & -i & i \\ 1+i & 2 & 3-2i \\ 3i & -i & 0 \end{bmatrix} \qquad \overline{A} = \begin{bmatrix} -2i & 1-i & -3i \\ i & 2 & i \\ -i & 3+2i & 0 \end{bmatrix}$$

$$A^* = \begin{bmatrix} -2i & i & -i \\ 1-i & 2 & 3+2i \\ -3i & i & 0 \end{bmatrix}$$

Clearly, $A^* = \overline{A}^T = \overline{A^T}$. The submatrix of A consisting of the elements common to its three rows and Columns 1 and 3 is the 3×2 imaginary† matrix

$$B = \begin{bmatrix} 2i & 3i \\ -i & -i \\ i & 0 \end{bmatrix}$$

whose tranjugate is

$$B^* = \begin{bmatrix} -2i & i & -i \\ -3i & i & 0 \end{bmatrix}$$

If we compute AB, take its conjugate, and transpose the result, we obtain (AB)*. Then, postmultiplying B* by A*, we can verify directly that

$$(AB)^* = B^*A^* = \begin{bmatrix} -6+i & 3+2i & -4+3i \\ -5+i & 3+2i & -5+3i \end{bmatrix}$$

†The number $0 = 0i$ is considered to be both real and pure imaginary.

So far, the definitions of this section apply to matrices of arbitrary (but finite) size. The rest of the special matrices we shall introduce here will be square. We have already defined the diagonal elements of a general matrix as those elements whose row and column indices are the same. For *square* matrices, the array of these elements is called the *diagonal,* or *principal diagonal* of the matrix.

DEFINITION 1 The **principal diagonal** of a square matrix $[a_{ij}]$ of order n is the array of those elements whose row and column indices are the same. The **secondary diagonal** is the array of elements extending upward from a_{n1} to a_{1n} and for which $i + j = n + 1$.

Various special square matrices are defined in terms of their diagonals: a matrix A is **upper triangular** if and only if it is square and $a_{ij} = 0$ if $i > j$; in other words, *in an upper triangular matrix, every element below the diagonal is zero.* A matrix A is **lower triangular** if and only if it is square and $a_{ij} = 0$ if $i < j$; that is, *in a lower triangular matrix, every element above the diagonal is zero.* A matrix D is **diagonal** if and only if it is square and $d_{ij} = 0$ if $i \neq j$; in other words, *in a diagonal matrix, every nondiagonal element is zero.* To emphasize that a matrix D is a diagonal matrix with diagonal elements $d_1, d_2, \ldots, d_n$, we shall sometimes write $\mathrm{D} = \mathrm{Dg}(d_1, d_2, \ldots, d_n)$. A **scalar matrix** is a diagonal matrix $\mathrm{Dg}(c, c, \ldots, c)$ whose diagonal elements are all equal. A **unit matrix**, or **identity matrix**, is a diagonal matrix $\mathrm{Dg}(1, 1, \ldots, 1)$ in which each diagonal element is 1. An identity matrix is usually denoted by the symbol I or, more specifically, by the symbol I_n if it is an identity matrix of order n.

EXAMPLE 3

Of the five matrices

$$\begin{bmatrix} 1 & 6 & -4 \\ 0 & -1 & 3 \\ 0 & 0 & 2 \end{bmatrix} \quad \begin{bmatrix} 2 & 0 & 0 \\ 0 & 3 & 0 \\ -1 & -5 & 0 \end{bmatrix} \quad \begin{bmatrix} 3 & 0 & 0 \\ 0 & -2 & 0 \\ 0 & 0 & 6 \end{bmatrix} \quad \begin{bmatrix} 7 & 0 & 0 \\ 0 & 7 & 0 \\ 0 & 0 & 7 \end{bmatrix} \quad \begin{bmatrix} 1 & 0 & 0 \\ 0 & 1 & 0 \\ 0 & 0 & 1 \end{bmatrix}$$

the first, third, fourth, and fifth are upper triangular. All but the first are lower triangular. The last three are diagonal and may be written as $\mathrm{Dg}(3, -2, 6)$, $\mathrm{Dg}(7, 7, 7)$, and $\mathrm{Dg}(1, 1, 1)$, respectively. The last two matrices are scalar matrices. Only the last matrix is an identity matrix, I_3 to be precise.

To indicate that a diagonal matrix D is of order n, we sometimes affix a subscript to the letter D and write D_n. A diagonal matrix D can be expressed concisely in our compact array notation by using the **Kronecker delta**† δ_{ij}, defined by

(3) $$\delta_{ij} = \begin{cases} 0 & \text{if } i \neq j \\ 1 & \text{if } i = j \end{cases}$$

Thus

(4) $$\mathrm{D} = [d_i\, \delta_{ij}] = [\delta_{ij}\, d_j]$$

In particular, an identity matrix I may be written as

(5) $$\mathrm{I} = [\delta_{ij}]$$

†Named for the German mathematician Leopold Kronecker (1823–1891).

and a scalar matrix $S = Dg(c, c, \ldots, c)$ is expressible as

(6) $$S = [c\,\delta_{ij}] = cI$$

If $A = [a_{ij}]$ is an arbitrary square matrix of order n, then the diagonal matrix whose only nonzero elements are those along the diagonal of A is given by

(7) $$Dg(a_{11}, a_{22}, \ldots, a_{nn}) = [a_{ij}\,\delta_{ij}]$$

Our next example brings out two important multiplicative properties of diagonal matrices.

EXAMPLE 4

Given the matrices $D_3 = Dg(1, 2, 3)$, $D_4 = Dg(4, -3, 2, -1)$, and

$$A = \begin{bmatrix} 9 & -8 & 7 & -6 \\ 1 & -2 & 3 & -4 \\ 1 & -2 & 2 & -1 \end{bmatrix}$$

find D_3A and AD_4. Carrying out the required multiplications we obtain

$$D_3A = \begin{bmatrix} 1 & 0 & 0 \\ 0 & 2 & 0 \\ 0 & 0 & 3 \end{bmatrix}\begin{bmatrix} 9 & -8 & 7 & -6 \\ 1 & -2 & 3 & -4 \\ 1 & -2 & 2 & -1 \end{bmatrix} = \begin{bmatrix} 9 & -8 & 7 & -6 \\ 2 & -4 & 6 & -8 \\ 3 & -6 & 6 & -3 \end{bmatrix}$$

and

$$AD_4 = \begin{bmatrix} 9 & -8 & 7 & -6 \\ 1 & -2 & 3 & -4 \\ 1 & -2 & 2 & -1 \end{bmatrix}\begin{bmatrix} 4 & 0 & 0 & 0 \\ 0 & -3 & 0 & 0 \\ 0 & 0 & 2 & 0 \\ 0 & 0 & 0 & -1 \end{bmatrix} = \begin{bmatrix} 36 & 24 & 14 & 6 \\ 4 & 6 & 6 & 4 \\ 4 & 6 & 4 & 1 \end{bmatrix}$$

Clearly, each row of D_3A can be found by simply multiplying the elements of the same row of A by the diagonal element in the corresponding row of D_3, and each column of AD_4 is equal to the corresponding column of A multiplied by the diagonal element in the same column of D_4.

Evidently, diagonal matrices have the following important properties.

THEOREM 1 If A is an $m \times n$ matrix, *premultiplication* of A by a diagonal matrix D_m can be accomplished by multiplying the ith row of A by the ith diagonal element d_i of D_m, $1 \le i \le m$, and *postmultiplication* of A by a diagonal matrix D_n can be accomplished by multiplying the jth column of A by the jth diagonal element of D_n, $1 \le j \le n$.

PROOF To prove Theorem 1, we first compute the element in the ith row and jth column of the $m \times n$ matrix D_mA. Displaying typical elements within brackets and noting that $\delta_{ik} = 0$, if $k \neq i$, while $\delta_{ii} = 1$, we have

$$D_mA = [d_i\,\delta_{ij}][a_{ij}] = \left[\sum_{k=1}^{m} d_i\,\delta_{ik}\,a_{kj}\right] = [d_i\,a_{ij}]$$

From this we see that, for $1 \le i \le m$, every element in the ith row of D_mA is equal to d_i times the corresponding element of A. In other words, the ith row of D_mA is d_i times the ith row of A.

Similarly, we find

$$\mathrm{AD}_n = [a_{ij}][\delta_{ij}\, d_j] = \left[\sum_{k=1}^{n} a_{ik}\,\delta_{kj}\, d_j\right] = [a_{ij}\, d_j]$$

which shows that, for $1 \le j \le n$, the jth column vector of the $m \times n$ matrix AD_n is the jth column vector of A multiplied by d_j, which completes the proof. ◗

As immediate corollaries of Theorem 1, we have

COROLLARY 1 If $\mathrm{S}_m = [k\,\delta_{ij}]$ and $\mathrm{S}_n = [l\,\delta_{ij}]$, then $\mathrm{S}_m\mathrm{A} = k\mathrm{A}$ and $\mathrm{AS}_n = \mathrm{A}l$.

COROLLARY 2 $\mathrm{I}_m\mathrm{A} = \mathrm{A}$ and $\mathrm{AI}_n = \mathrm{A}$.

Corollary 1 states that the product of a matrix A and any conformable scalar matrix $\mathrm{S} = c\mathrm{I}$ is equal to the product of A and the scalar c. The properties of I_m and I_n embodied in Corollary 2 account for why such matrices are called *unit* matrices.

Any square matrix A can be multiplied by itself. The product AA is referred to as the **square** or **second power** of A and is denoted by the symbol A^2. The associative law of matrix multiplication guarantees that for every integer $r \ge 3$, the product $\mathrm{AA}\cdots\mathrm{A}$ of A times itself r times is unambiguous. Thus for every positive integer r, the rth **power** of A is defined by

(8) $$\mathrm{A}^r = \mathrm{AA}\cdots\mathrm{A}$$

where the right-hand member contains r factors. With A^0 defined as the identity matrix I of the same order as A, it is obvious that for any nonnegative integers r and s the familiar laws of exponents

(9) $$\mathrm{A}^r\mathrm{A}^s = \mathrm{A}^{r+s} \qquad \text{and} \qquad (\mathrm{A}^r)^s = \mathrm{A}^{rs}$$

hold for multiplication of square matrices.

A matrix which is not square cannot be multiplied by itself, nor can it be equal to its transpose. The transpose is used to define two important special kinds of square matrices: those which are equal to their transpose and those which are equal to the negative of their transpose.

A matrix A such that $\mathrm{A} = \mathrm{A}^T$, i.e., a square matrix A of order n for which $a_{ij} = a_{ji}$, $1 \le i, j \le n$, is said to be **symmetric** since elements *symmetrically* located with respect to the principal diagonal are equal. A matrix A such that $\mathrm{A} = -\mathrm{A}^T$, i.e., a square matrix A of order n such that $a_{ij} = -a_{ji}$ for $1 \le i,\ j \le n$, and in which therefore $a_{ii} = 0$, is said to be **skew-symmetric**. In applications, real symmetric and skew-symmetric matrices are far more useful than those with complex entries. There are, however, special matrices with complex elements which are of both practical and theoretical interest. The tranjugate, instead of the transpose, is used to define two of these.

A matrix H such that $\mathrm{H} = \mathrm{H}^*$, i.e., a square matrix H of order n such that for $1 \le i, j \le n$, $h_{ij} = \overline{h}_{ji}$, is said to be **hermitian**. Elements of a hermitian matrix which are symmetrically located with respect to the principal diagonal are conjugate complex numbers, and all entries on the principal diagonal are real numbers. A matrix K such that $\mathrm{K} = -\mathrm{K}^*$ is said to be **skew-hermitian**. Clearly, a real symmetric matrix is just a hermitian matrix which is real, and a real skew-symmetric matrix is just a skew-hermitian matrix which is real. Hence, as is to be expected, many fundamental results concerning real symmetric matrices can be derived efficiently as corollaries to more general theorems about hermitian matrices.

EXAMPLE 5

Given the matrix

$$A = \begin{bmatrix} -4 & -3 & -1 \\ 2 & 1 & 1 \\ 4 & -2 & 4 \end{bmatrix}$$

find the matrix $4I - 4A - A^2 + A^3$.

Computing the individual terms of $4I - 4A - A^2 + A^3$, working from left to right, we find

$$4I = \begin{bmatrix} 4 & 0 & 0 \\ 0 & 4 & 0 \\ 0 & 0 & 4 \end{bmatrix} \qquad -4A = \begin{bmatrix} 16 & 12 & 4 \\ -8 & -4 & -4 \\ -16 & 8 & -16 \end{bmatrix}$$

$$-A^2 = \begin{bmatrix} -6 & -11 & 3 \\ 2 & 7 & -3 \\ 4 & 22 & -10 \end{bmatrix} \qquad A^3 = \begin{bmatrix} -14 & -1 & -7 \\ 6 & -7 & 7 \\ 12 & -30 & 22 \end{bmatrix}$$

When we add these, we find that $4I - 4A - A^2 + A^3 = O$, where of course O is the third-order zero matrix.

EXAMPLE 6

The first of the four matrices

$$\begin{bmatrix} 1 & 6 & -2 \\ 6 & 0 & 4 \\ -2 & 4 & -5 \end{bmatrix} \qquad \begin{bmatrix} 0 & 3 & 7 \\ -3 & 0 & -2 \\ -7 & 2 & 0 \end{bmatrix}$$

$$\begin{bmatrix} 4 & 1-i & 3+2i \\ 1+i & 10 & 5i \\ 3-2i & -5i & -3 \end{bmatrix} \qquad \begin{bmatrix} 0 & -5-i & -2+3i \\ 5-i & -3i & -6 \\ 2+3i & 6 & 6i \end{bmatrix}$$

is real, symmetric, and hermitian. Its principal diagonal contains the numbers 1, 0, and -5; its secondary diagonal is made up of the numbers -2, 0, and -2. The second matrix, whose principal diagonal contains three zeros, is real, skew-symmetric, and skew-hermitian. All diagonal elements of the third matrix, which is hermitian, are real numbers, while all diagonal elements of the fourth matrix, which is skew-hermitian, are pure imaginary numbers.

There are three very important types of simple manipulations on the rows of a matrix which we shall use again and again. They are given a common name and characterized individually as follows.

DEFINITION 2 The **elementary row operations** on a matrix are†

1. Interchanging two rows.
2. Multiplying a row by a nonzero number.
3. Adding a number times one row vector to any other row vector.

† Similar operations on the columns of a matrix are discussed in Sec. 14.3.

It is interesting and important to note that each of these three operations on the rows of a matrix A can be accomplished by premultiplying A by a unit matrix on whose rows the same elementary operation has been performed. Unit matrices modified in this way are given a special name.

DEFINITION 3 An **elementary row matrix** P is a unit matrix to which a single elementary row operation has been applied.

EXAMPLE 7

Starting with I_4, display three different types of elementary row matrices.

Elementary row matrices of each of the three types are exemplified, respectively, by

$$\begin{bmatrix} 0 & 0 & 1 & 0 \\ 0 & 1 & 0 & 0 \\ 1 & 0 & 0 & 0 \\ 0 & 0 & 0 & 1 \end{bmatrix} \quad \begin{bmatrix} 1 & 0 & 0 & 0 \\ 0 & 1 & 0 & 0 \\ 0 & 0 & c & 0 \\ 0 & 0 & 0 & 1 \end{bmatrix} \quad \begin{bmatrix} 1 & 0 & 0 & 0 \\ 0 & 1 & k & 0 \\ 0 & 0 & 1 & 0 \\ 0 & 0 & 0 & 1 \end{bmatrix}$$

Just how elementary row operations can be carried out by matrix multiplication is explained more fully in the next three theorems.

THEOREM 2 If A is an arbitrary $m \times n$ matrix and P_m is the elementary row matrix obtained from the identity matrix I_m by interchanging its ith and jth rows, then the product P_mA is identical with A except for the ith and jth rows, which are interchanged.

A proof of this is easy to give using the definition of matrix multiplication, and we leave it as an exercise.

EXAMPLE 8

If $A = \begin{bmatrix} a & b & c & d \\ 1 & 2 & 3 & 4 \\ u & v & x & y \end{bmatrix}$, compute P_3A, where P_3 is the matrix obtained from I_3 by interchanging its first and third rows.

Using the definition of matrix multiplication, we have

$$P_3A = \begin{bmatrix} 0 & 0 & 1 \\ 0 & 1 & 0 \\ 1 & 0 & 0 \end{bmatrix} \begin{bmatrix} a & b & c & d \\ 1 & 2 & 3 & 4 \\ u & v & x & y \end{bmatrix} = \begin{bmatrix} u & v & x & y \\ 1 & 2 & 3 & 4 \\ a & b & c & d \end{bmatrix}$$

By inspection, it is easy to see that the same changes that were made in I_3 have been effected in P_3A.

THEOREM 3 If A is an $m \times n$ matrix and P_m is the matrix obtained from the unit matrix I_m by replacing the element 1 in its ith row by a number c, then the product P_mA is identical with A except that the elements in the ith row are now multiplied by c. If and only if $c \neq 0$, P_m is an elementary matrix and premultiplication of A by P_m performs an elementary row operation on A of Type 2.

This theorem follows as another corollary of Theorem 1 with due regard for Operation **2** of Definition 2.

THEOREM 4 If A is an $m \times n$ matrix and if P_m is the elementary row matrix obtained from I_m by adding to the elements of the jth row of I_m, c times the corresponding elements in the ith row, then the product P_mA is identical with A except in the jth row, which consists of the elements of the jth row of A plus c times the corresponding elements in the ith row of A.

PROOF Let $P_m = [p_{ik}]$ be the matrix obtained from the unit matrix I_m by adding c times its ith row to its jth row and let $P_mA = [\alpha_{ij}]$. Then in every row of P_m except the jth, we have

$$p_{ik} = \delta_{ik} \qquad k = 1, 2, \dots, m;\ i \neq j$$

and in the jth row we have, for $l = 1, 2, \dots, m$,

$$p_{jl} = \begin{cases} 1 & l = j \\ c & l = i \\ 0 & l \neq i, j \end{cases} \tag{10}$$

Then for $i \neq j$, we have

$$\alpha_{ij} = \sum_{k=1}^{m} p_{ik}a_{kj} = \sum_{k=1}^{m} \delta_{ik}\, a_{kj} = a_{ij}$$

which shows that every row in P_mA except the jth is identical with the corresponding row in A. For the jth row in P_mA we have, using (10),

$$\alpha_{jk} = \sum_{l=1}^{m} p_{jl}a_{lk} = a_{jk} + ca_{ik}$$

This shows that the jth row of the product P_mA is equal to the jth row of A *plus* c times the ith row of A, as asserted.

EXAMPLE 9

By what elementary matrix must $A = \begin{bmatrix} 2 & 0 & 1 & 4 \\ 3 & 4 & -1 & 2 \\ 6 & 1 & -2 & 5 \end{bmatrix}$ be premultiplied to obtain a matrix identical with A except that -2 times the second row has been added to the third row?

According to Theorem 4, the required matrix P_3 is obtained from I_3 by adding -2 times the second row of I_3 to its third row; that is,

$$P_3 = \begin{bmatrix} 1 & 0 & 0 \\ 0 & 1 & 0 \\ 0 & -2 & 1 \end{bmatrix}$$

Using this, we have

$$P_3A = \begin{bmatrix} 1 & 0 & 0 \\ 0 & 1 & 0 \\ 0 & -2 & 1 \end{bmatrix}\begin{bmatrix} 2 & 0 & 1 & 4 \\ 3 & 4 & -1 & 2 \\ 6 & 1 & -2 & 5 \end{bmatrix} = \begin{bmatrix} 2 & 0 & 1 & 4 \\ 3 & 4 & -1 & 2 \\ 0 & -7 & 0 & 1 \end{bmatrix}$$

in which the third row is now the required combination of the second and third rows of A.

In Example 9, the operation to be performed on A was so simple that there was no reason to carry it out by first determining P_3 and then performing the necessary premultiplication. However, in more complicated problems, where combinations of the operations described in Theorems 2–4 must be carried out on large, computer-generated matrices, the appropriate transformation matrix may be very important. Our next example is a simple illustration of the kind of rearrangement that must be made on vectors in using what is known as the *fast Fourier transform* (see Sec. 14.2).

EXAMPLE 10

Find a matrix P_6 which, as a premultiplier, will rearrange the vector

$$\mathbf{U} = \begin{bmatrix} a_0 \\ a_1 \\ a_2 \\ a_3 \\ a_4 \\ a_5 \end{bmatrix} \quad \text{into the vector} \quad \mathbf{V} = \begin{bmatrix} a_0 \\ a_2 \\ a_4 \\ a_1 \\ a_3 \\ a_5 \end{bmatrix}$$

To construct P_6 from I_6, we need only keep in mind the definition of matrix multiplication. We are not asked to interchange pairs of elements in **U**, so Theorem 2 cannot be explicitly invoked. However, reflecting for a moment on Theorem 4, we soon see that to move an element from the ith row in **U** to the jth row in **V**, the only nonzero element in the jth row of P_6 must be a 1 in the ith column. To achieve the required rearrangement of **U**, we must

move a_0 from Row 1 to Row 1
move a_1 from Row 2 to Row 4
move a_2 from Row 3 to Row 2
move a_3 from Row 4 to Row 5
move a_4 from Row 5 to Row 3
move a_5 from Row 6 to Row 6

Hence, recalling the remark we just made, it follows that

$$P_6 = \begin{bmatrix} 1 & 0 & 0 & 0 & 0 & 0 \\ 0 & 0 & 1 & 0 & 0 & 0 \\ 0 & 0 & 0 & 0 & 1 & 0 \\ 0 & 1 & 0 & 0 & 0 & 0 \\ 0 & 0 & 0 & 1 & 0 & 0 \\ 0 & 0 & 0 & 0 & 0 & 1 \end{bmatrix}$$

The fact that each row and each column of P_6 contains only a single 1 confirms that we have not carelessly moved two elements into the same row. It is now only a trivial exercise to verify that $P_6\mathbf{U} = \mathbf{V}$.

EXERCISES

1. Write the transpose and identify the diagonal elements of the following matrices:

(a) $[6 \quad -5 \quad 2 \quad 4 \quad 0]$

(b) $\begin{bmatrix} 9 & -1 & 6 & 7 \\ -5 & 3 & 0 & 4 \end{bmatrix}$

(c) $\begin{bmatrix} 2 & 8 \\ -6 & 0 \\ 7 & -1 \end{bmatrix}$

(d) $\begin{bmatrix} 1 & -3 & 4 \\ -3 & 0 & 2 \\ 4 & 2 & 5 \end{bmatrix}$

2. For the given matrices A and B, verify directly that $(AB)^T = B^T A^T$.

(a) $A = \begin{bmatrix} -6 & 2 & 0 & 3 \\ 0 & 4 & 5 & -1 \end{bmatrix}$

$$B = \begin{bmatrix} 1 & 0 \\ 4 & 3 \\ -3 & 0 \\ 2 & -2 \end{bmatrix}$$

(b) $A = \begin{bmatrix} 8 & 2 \\ 10 & 3 \\ 5 & 2 \end{bmatrix}$

$$B = \begin{bmatrix} -1 & 2 & 0 & -3 \\ 5 & -6 & 4 & 6 \end{bmatrix}$$

3. Let S_1, S_2, and S_3 be the submatrices of

$$A = \begin{bmatrix} 10 & 0 & -3 & 7 & 0 \\ 0 & 1 & 2 & 8 & 6 \\ 17 & 6 & -5 & 6 & 4 \\ -1 & -2 & 1 & -9 & 5 \end{bmatrix}$$

whose respective arrays are the elements common to the following rows and columns of A:

S_1: Rows 2 and 3; Columns 3 and 4
S_2: Rows 1 and 4; Columns 3, 4, and 5
S_3: Rows 1, 2, and 4; Columns 1 and 2

(a) Write the submatrices S_1, S_2, and S_3.
(b) Find the product $S_1S_2S_3$ and show that $(S_1S_2S_3)^T = S_3^TS_2^TS_1^T$.
(c) Are any of the matrices $(S_1 + S_2)^T$, $(S_1 + S_3)^T$, and $(S_2 + S_3)^T$ defined?

4. Prove that if A + B is defined, then $(A + B)^T = A^T + B^T$.

5. Prove that the transpose of the product of any number of conformable matrices is equal to the product of the transposed matrices taken in reverse order, i.e.,

$$(A_1A_2 \cdots A_n)^T = A_n^T \cdots A_2^TA_1^T$$

6. Find the transpose, conjugate, and tranjugate of each of the following matrices:

(a) $\begin{bmatrix} 5 & -i & 3 & 6i \\ -2 & 2-3i & 4 & -1 \\ 0 & 4i & 1+i & 3i \end{bmatrix}$

(b) $\begin{bmatrix} 0 & 0 & 4i & 3 \\ 6-5i & -1 & -i & 7 \\ 5+6i & 8i & 2i & 0 \end{bmatrix}$

(c) $\begin{bmatrix} 8 & 7 & 1 & -4 & 2i \\ 2i & -i & -3 & 1-i & 6i \\ 3i & -2 & 2+i & 5 & i \end{bmatrix}$

7. For the given matrices A and B, verify directly that $(AB)^* = B^*A^*$.

(a) $A = \begin{bmatrix} 2-i & 3+2i \\ 1+i & 2i \end{bmatrix}$ $\quad B = \begin{bmatrix} -i \\ 1 \end{bmatrix}$

(b) $A = \begin{bmatrix} 3-2i & 2+3i \\ 5 & 4+i \end{bmatrix}$ $\quad B = \begin{bmatrix} 1-i & 0 \\ -i & 1+i \end{bmatrix}$

8. Prove the following four properties of tranjugates:
(a) $(A^*)^* = A$.
(b) $(cA)^* = \bar{c}A^*$, c a scalar.
(c) If A + B is defined, $(A + B)^* = A^* + B^*$.
(d) If AB is defined, $(AB)^* = B^*A^*$.

In Exercises 9–14, state whether or not the given matrix is (*a*) upper triangular, (*b*) lower triangular, (*c*) diagonal, (*d*) scalar, or (*e*) an identity matrix.

9. $\begin{bmatrix} 1 & 5 & -3 \\ 0 & -2 & 0 \\ 0 & 0 & 6 \end{bmatrix}$ **10.** $\begin{bmatrix} 0 & 0 & 0 \\ 0 & 0 & 0 \\ 0 & 0 & 0 \end{bmatrix}$

11. $\begin{bmatrix} 5 & 0 & 0 \\ 0 & 12 & 0 \\ 0 & 0 & 31 \end{bmatrix}$ **12.** $\begin{bmatrix} 1 & 0 & 0 \\ 0 & \cos 2\pi & 0 \\ 0 & 0 & \ln e \end{bmatrix}$

13. $\begin{bmatrix} 0 & i & 0 & 1 & 1+i \\ 0 & 0 & 2 & 0 & i \\ 0 & 0 & 1 & 1 & -1 \\ 0 & 0 & 0 & i & 2 \end{bmatrix}$ **14.** $\begin{bmatrix} 0 & 0 & 0 & 0 \\ -1 & 2 & 0 & 0 \\ 7 & 0 & 7 & 0 \\ i & 3 & 0 & 1 \end{bmatrix}$

15. Write the following matrices as rectangular arrays:
(a) $Dg(8, -6, 10)$ **(b)** $Dg(\delta_{33}, \delta_{15}, \delta_{24})$
(c) $Dg(i, 1 - i, \delta_{11}, 1 + i)$ **(d)** I_4

In Exercises 16–19, find DA and AD, if possible, where $D = Dg(3, 0, 2, -1)$ and A is the given matrix.

16. $\begin{bmatrix} -2 & 1 & 0 & 3 \\ 9 & 8 & -5 & 9 \\ 0 & -2 & 3 & -4 \\ 7 & -6 & 4 & -1 \end{bmatrix}$ **17.** $\begin{bmatrix} 1 & -2 \\ -4 & 8 \\ 3 & -1 \\ -2 & 7 \end{bmatrix}$

18. $\begin{bmatrix} 1 & 4 & -2 & 3 \\ 2 & -1 & 5 & 0 \end{bmatrix}$ **19.** $\begin{bmatrix} 2i & 0 & 1 & 3i \\ 0 & 1+i & 8 & 0 \\ -i & 1 & 0 & i \\ 1 & 1 & 1 & 1 \end{bmatrix}$

20. If A and O are square matrices of the same order as I, show that multiplication of any two of them is commutative.

21. Prove that multiplication of any two diagonal matrices of the same order is commutative.

22. Show that if A commutes with a diagonal matrix D whose diagonal elements are all distinct, then A is a diagonal matrix. Is A necessarily diagonal if the diagonal elements of D are not distinct?

23. If $A = \begin{bmatrix} 1 & 2 \\ 2 & 4 \end{bmatrix}$, find a nonzero square matrix X of order 2 such that AX is a zero matrix. Is XA = O? If $A = \begin{bmatrix} 1 & 2 \\ 2 & 3 \end{bmatrix}$, is it possible to find a square matrix X such that AX is a zero matrix?

24. Prove that the product of two nonzero matrices may be a zero matrix; i.e., show that AB = O does not imply either A = O or B = O.

25. If $A = \begin{bmatrix} 1 & 2 \\ 3 & 4 \end{bmatrix}$, $B = \begin{bmatrix} 3 & 1 \\ 1 & 0 \end{bmatrix}$, $C = \begin{bmatrix} 1 & 2 \\ 2 & 4 \end{bmatrix}$, and

$X = \begin{bmatrix} x_1 & x_2 \\ x_3 & x_4 \end{bmatrix}$, solve each of the following equations:

(a) AX = B − I **(b)** (B − I)X = A
(c) AX = CI **(d)** (B − I)X = IC
(e) CX = A

26. Evaluate the **matrix polynomial** $X^3 - 4X^2 - X + 4I$ for each of the following matrices.

(a) $X = \begin{bmatrix} 1 & -1 \\ 2 & 0 \end{bmatrix}$ **(b)** $X = \begin{bmatrix} 1 & 1 & 2 \\ 1 & 2 & 1 \\ 2 & 1 & 1 \end{bmatrix}$

(c) $X = \begin{bmatrix} 0 & 1 & 1 \\ -1 & 0 & 1 \\ -1 & -1 & 0 \end{bmatrix}$

27. Verify that $(X - 3I)(X - 2I) = (X - 2I)(X - 3I) = X^2 - 5X + 6I$ for

$$X = \begin{bmatrix} 1 & 2 \\ 2 & -1 \end{bmatrix} \quad \text{and for} \quad X = \begin{bmatrix} 1 & 2 & 0 \\ 0 & 3 & 0 \\ 0 & 0 & 4 \end{bmatrix}$$

28. If $A = \begin{bmatrix} 1 & 2 \\ -1 & 3 \end{bmatrix}$, verify that $A^2 - 4A + 5I = O$ and use this fact to compute A^3, A^4, and A^5 by first expressing them in terms of the matrices A and I. Check these results by calculating A^3, A^4, and A^5 directly from A.

29. Show that $\begin{bmatrix} \cos\theta & \sin\theta \\ -\sin\theta & \cos\theta \end{bmatrix}^n = \begin{bmatrix} \cos n\theta & \sin n\theta \\ -\sin n\theta & \cos n\theta \end{bmatrix}$.

30. Show that $\begin{bmatrix} \cosh\theta & \sinh\theta \\ \sinh\theta & \cosh\theta \end{bmatrix}^n = \begin{bmatrix} \cosh n\theta & \sinh n\theta \\ \sinh n\theta & \cosh n\theta \end{bmatrix}$.

31. Under what conditions, if any, is $(AB)^2 = A^2B^2$?

32. If A and B are square matrices which commute, i.e., square matrices such that AB = BA, prove that A^2 and B^2 also commute. Is this true for general positive integral powers of A and B?

33. Classify each of the following matrices according as it is (*a*) real, (*b*) symmetric, (*c*) skew-symmetric, (*d*) hermitian, or (*e*) skew-hermitian, and identify its principal and secondary diagonals:

(a) $\begin{bmatrix} 2i & -6-9i & 13 \\ 6-9i & 0 & 4i \\ -13 & 4i & 3i \end{bmatrix}$

(b) $\begin{bmatrix} 0 & 1 & 2 \\ -1 & 0 & -3 \\ -2 & 3 & 0 \end{bmatrix}$

(c) $\begin{bmatrix} 1 & 0 & -i \\ 0 & -2 & 4-i \\ i & 4+i & 3 \end{bmatrix}$

(d) $\begin{bmatrix} 7 & 0 & 4 \\ 0 & -2 & 10 \\ 4 & 10 & 5 \end{bmatrix}$

34. Prove that if A is any square matrix, then both AA^T and A^TA are symmetric.

35. If A and B are symmetric matrices of the same order, prove that the product AB is symmetric if and only if AB = BA.

36. If A and B are square matrices which are not symmetric, is it possible for their product to be symmetric?

37. Prove that for all $n \times n$ matrices A and B, and for every scalar k, if A and B are symmetric, then kA and A + B are symmetric.

38. Show that the diagonal elements of every skew-hermitian matrix must be pure imaginary numbers.

39. By the **derivative of a matrix** A we mean the matrix whose elements are the derivatives of the elements of A. Assuming that the elements of each matrix are differentiable functions of x, use this definition to show that

$$\frac{d(AB)}{dx} = \frac{dA}{dx}B + A\frac{dB}{dx}$$

What is $d(ABC)/dx$? Is $d(A^2)/dx = 2A\,dA/dx$?

40. If $A = \begin{bmatrix} e^{\lambda_1 t} & & & \mathbf{0} \\ & e^{\lambda_2 t} & & \\ & & \ddots & \\ \mathbf{0} & & & e^{\lambda_n t} \end{bmatrix}$,† show that $\dfrac{dA}{dt} = DA$, where D is the diagonal matrix $Dg(\lambda_1, \lambda_2, \ldots, \lambda_n)$. What is d^2A/dt^2?

41. If $A = \begin{bmatrix} \sin\lambda_1 t & & & \mathbf{0} \\ & \sin\lambda_2 t & & \\ & & \ddots & \\ \mathbf{0} & & & \sin\lambda_n t \end{bmatrix}$, what is $\dfrac{dA}{dt}$? What is $\dfrac{d^2A}{dt^2}$?

42. If $A = \begin{bmatrix} \cos\lambda_1 t & & & \mathbf{0} \\ & \cos\lambda_2 t & & \\ & & \ddots & \\ \mathbf{0} & & & \cos\lambda_n t \end{bmatrix}$, what is $\dfrac{dA}{dt}$? What is $\dfrac{d^2A}{dt^2}$?

43. Prove Theorem 2.

44. Prove Theorem 3 directly by using the definition of matrix multiplication.

45. By what matrix must $\mathbf{U} = [u_0\ u_1\ u_2\ u_3]^T$ be premultiplied to obtain

(a) $\mathbf{V} = [u_0\ u_2\ u_1\ u_3]^T$?
(b) $\mathbf{V} = [u_3\ u_2\ u_1\ u_0]^T$?
(c) $\mathbf{V} = [au_1\ bu_3\ cu_2\ du_0]^T$?

---

†The boldface zero is a device for indicating that all elements not on the principal diagonal are zero. Diagonal matrices are also written without the boldface zero, i.e., simply with a blank space in all off-diagonal positions.

46. By what matrix must $\mathbf{U} = [u_0\ u_1\ v_0\ v_1\ w_0\ w_1]^T$ be premultiplied to obtain
(a) $\mathbf{V} = [u_0\ v_0\ w_0\ u_1\ v_1\ w_1]^T$?
(b) $\mathbf{V} = [w_1\ u_0\ v_1\ v_0\ w_0\ u_1]^T$?

47. Find a matrix P_5 which can be used as a premultiplier of an arbitrary $5 \times n$ matrix A to effect the following changes in A: Row 1 → Row 2, Row 2 → Row 5, Row 3 → Row 1, Row 4 → Row 3, Row 5 → Row 4.

48. If P_1, P_2, and P_3, are elementary matrices obtained from I_4 by interchanging Rows 1 and 4, multiplying Row 2 by c, and adding k times Row 3 to Row 1, respectively, find $P_1P_2P_3$.

3.5 DETERMINANTS

In a restricted sense at least, the concept of a determinant is already familiar from elementary algebra, where in solving systems of two and three simultaneous linear equations we found it convenient to introduce *determinants of the second and third order.* Determinants of higher order arise in connection with systems of more than three linear equations. Other applications not immediately associated with solving equations also give rise to determinants. It is therefore important that we understand what determinants are and know how to evaluate them. This section is devoted to an investigation of their most important properties.

Later on in our work, we shall have occasion to consider various functions whose domains are sets of matrices. Such is the nature of a determinant. *A determinant is a scalar-valued function whose domain is a set of square matrices.* As soon as the necessary notation and preliminary definitions have been given, we shall describe determinants more precisely.

For every matrix of **order** n, that is, for every $n \times n$ matrix A, the corresponding value of the determinant function is denoted by either $|A|$ or det A. We call $|A|$† the **determinant of A**. As with functions in general, a determinant function and its value are commonly referred to interchangeably. In particular, when some or all of the elements of a matrix

$$\text{(1)} \qquad A = \begin{bmatrix} a_{11} & a_{12} & \cdots & a_{1n} \\ a_{21} & a_{22} & \cdots & a_{2n} \\ \cdots & \cdots & \cdots & \cdots \\ a_{n1} & a_{n2} & \cdots & a_{nn} \end{bmatrix}$$

are displayed explicitly, we write

$$\text{(2)} \qquad |A| = \begin{vmatrix} a_{11} & a_{12} & \cdots & a_{1n} \\ a_{21} & a_{22} & \cdots & a_{2n} \\ \cdots & \cdots & \cdots & \cdots \\ a_{n1} & a_{n2} & \cdots & a_{nn} \end{vmatrix}$$

and refer to (2) as the determinant of A. A determinant $|A|$ is of **order** n if and only if A is of order n. The elements, rows, columns, and principal and secondary diagonals of A are called in turn the **elements**, **rows**, **columns**, and **principal** and **secondary diagonals** of $|A|$.

The determinant $|M|$ formed by the m^2 elements common to any m rows and any m columns of an nth-order determinant $|A|$ is said to be an ***m*th-order minor** of $|A|$. The determinant of order $n - m$ formed by the array of elements which remains when the m rows and m columns containing an mth-order minor $|M|$ are deleted from $|A|$ is called the **complementary minor** of $|M|$. If the numbers of the rows and columns of $|A|$ which contain an mth-order minor $|M|$ are, respectively,

$$i_1, i_2, \ldots, i_m \qquad \text{and} \qquad j_1, j_2, \ldots, j_m$$

---

†The use of vertical bars both in the notation for a determinant and in the notation for the absolute value of a number, while perhaps unfortunate, is widespread. Which meaning is intended in any particular case should always be clear from the context.

then $(-1)^{i_1+i_2+\cdots+i_m+j_1+j_2+\cdots+j_m}$ times the complementary minor of |M| is called the **algebraic complement** of |M|. The first-order minors of |A| are of course just the elements of |A|. Their complementary minors are customarily referred to simply as **minors**, and their algebraic complements are almost universally referred to as **cofactors**. We shall denote the minor of the element a_{ij} by the symbol M_{ij}, and its cofactor by the symbol A_{ij}; thus

$$A_{ij} = (-1)^{i+j} M_{ij}$$

Similarly, we shall use the symbols $M_{ij,kl}$ and $A_{ij,kl}$ to denote, respectively, the complementary minor and the algebraic complement of the second-order minor contained in the ith and jth rows and the kth and lth columns of a determinant |A|; thus

$$A_{ij,kl} = (-1)^{i+j+k+l} M_{ij,kl}$$

The generalization of this notation is obvious.

EXAMPLE 1

In the fifth-order determinant

$$|\mathrm{A}| = \begin{vmatrix} a_{11} & a_{12} & a_{13} & a_{14} & a_{15} \\ a_{21} & a_{22} & a_{23} & a_{24} & a_{25} \\ a_{31} & a_{32} & a_{33} & a_{34} & a_{35} \\ a_{41} & a_{42} & a_{43} & a_{44} & a_{45} \\ a_{51} & a_{52} & a_{53} & a_{54} & a_{55} \end{vmatrix}$$

the minor of the element a_{43} is the fourth-order determinant formed by the elements which remain when the fourth row and third column are deleted from |A|, namely,

$$M_{43} = \begin{vmatrix} a_{11} & a_{12} & a_{14} & a_{15} \\ a_{21} & a_{22} & a_{24} & a_{25} \\ a_{31} & a_{32} & a_{34} & a_{35} \\ a_{51} & a_{52} & a_{54} & a_{55} \end{vmatrix}$$

The cofactor A_{43} of the element a_{43} is equal to this minor times $(-1)^{4+3}$; that is,

$$A_{43} = -M_{43}$$

Similarly, the complementary minor of the second-order minor

$$\begin{vmatrix} a_{31} & a_{34} \\ a_{51} & a_{54} \end{vmatrix}$$

contained in the third and fifth rows and the first and fourth columns of |A| is the third-order determinant formed by the elements which remain when these rows and columns are deleted from |A|:

$$M_{35,14} = \begin{vmatrix} a_{12} & a_{13} & a_{15} \\ a_{22} & a_{23} & a_{25} \\ a_{42} & a_{43} & a_{45} \end{vmatrix}$$

The algebraic complement $A_{35,14}$ of the given second-order minor is equal to the complementary minor $M_{35,14}$ times $(-1)^{3+5+1+4}$; that is,

$$A_{35,14} = -M_{35,14}$$

We are now in a position to give a general definition of a determinant. This can be done in direct fashion, but since the result is unsuited to the practical evaluation of determinants, we choose to give an inductive definition.

DEFINITION 1 The determinant of a matrix with a single element is that element. For every matrix A of order $n \geq 2$,

$$\det \mathrm{A} = a_{11}A_{11} + a_{12}A_{12} + \cdots + a_{1n}A_{1n} = \sum_{k=1}^{n} a_{1k}A_{1k} \tag{3}$$

In words, the **expansion** (3) says that the value of any determinant of order 2 or more is equal to the sum of the products of the elements of its first row and their respective cofactors. Clearly, this makes the value of a determinant of order n depend, in general, upon n determinants of order $n - 1$, each of which in turn depends upon $n - 1$ determinants of order $n - 2$, and so on, until finally the expansion involves only numbers, i.e., determinants of order 1. For this reason, Definition 1 is called an inductive definition.

When applied to a general second-order matrix

$$\mathrm{A} = \begin{bmatrix} a_{11} & a_{12} \\ a_{21} & a_{22} \end{bmatrix}$$

Eq. (3) gives

$$\det \mathrm{A} = a_{11}a_{22} - a_{12}a_{21}$$

The right-hand member of this equation is just the difference between the product of the elements of the principal diagonal and the product of the elements of the secondary diagonal of $|\mathrm{A}|$ as illustrated by the diagram

$$\begin{matrix} (+) & & \\ & \begin{vmatrix} a_{11} & a_{12} \\ a_{21} & a_{22} \end{vmatrix} & = a_{11}a_{22} - a_{12}a_{21} \\ (-) & & \end{matrix} \tag{4}$$

A general third-order determinant can be expanded using Eq. (3) in conjunction with (4):

$$\begin{vmatrix} a_{11} & a_{12} & a_{13} \\ a_{21} & a_{22} & a_{23} \\ a_{31} & a_{32} & a_{33} \end{vmatrix} = a_{11}\begin{vmatrix} a_{22} & a_{23} \\ a_{32} & a_{33} \end{vmatrix} - a_{12}\begin{vmatrix} a_{21} & a_{23} \\ a_{31} & a_{33} \end{vmatrix} + a_{13}\begin{vmatrix} a_{21} & a_{22} \\ a_{31} & a_{32} \end{vmatrix} \tag{5}$$

$$\begin{aligned} &= a_{11}[a_{22}a_{33} - a_{23}a_{32}] - a_{12}[a_{21}a_{33} - a_{23}a_{31}] \\ &\quad + a_{13}[a_{21}a_{32} - a_{22}a_{31}] \\ &= a_{11}a_{22}a_{33} + a_{12}a_{23}a_{31} + a_{13}a_{21}a_{32} - a_{13}a_{22}a_{31} \\ &\quad - a_{11}a_{23}a_{32} - a_{12}a_{21}a_{33} \end{aligned}$$

This expansion can also be obtained by diagonal multiplication, by repeating on the right the first two columns of the determinant and then adding the signed products of the elements on the various diagonals in the resulting array:

$$\begin{matrix} (+) & (+) & (+) & & \\ \left| a_{11} \right. & a_{12} & a_{13} & \left| a_{11} \right. & a_{12} \\ \left| a_{21} \right. & a_{22} & a_{23} & \left| a_{21} \right. & a_{22} \\ \left| a_{31} \right. & a_{32} & a_{33} & \left| a_{31} \right. & a_{32} \\ (-) & (-) & (-) & & \end{matrix} \tag{6}$$

The diagonal scheme of expanding determinants indicated by (4) and (6) is correct *only* for determinants of the second and third orders and generally gives incorrect results when applied to determinants of higher order.

Were we to expand a general fourth-order determinant by means of (3) and then utilize (5), we would obtain 24 terms in the expansion. Fifth-order determinants lead to 120 terms, and nth-order determinants to $n!$ terms. Obviously, if determinants of any appreciable order are to be evaluated efficiently, the expansion process based on repeated applications of (3) must somehow be simplified. There are some very remarkable properties of determinants that expedite their evaluation. A number of these will now be given for determinants of order n (>1).

PROPERTY 1 For every matrix A of order n and for each i such that $1 \leq i \leq n$,

$$\det \mathrm{A} = \sum_{k=1}^{n} a_{ik} A_{ik} \tag{7}$$

and for each j such that $1 \leq j \leq n$,

$$\det \mathrm{A} = \sum_{k=1}^{n} a_{kj} A_{kj} \tag{8}$$

This property shows that if the elements of *any* row, not just the first, or of any column of a determinant are multiplied by their respective cofactors and then added, the sum is the same for all rows and for all columns.† Formula (7) gives **Laplace's expansion**‡ of $|\mathrm{A}|$ by the ith row, and (8) gives the Laplace expansion by the jth column. Observe that for $i = 1$, (7) reduces to (3) of Definition 1.

EXAMPLE 2

Evaluate the fourth-order determinant

$$|\mathrm{A}| = \begin{vmatrix} 1 & 2 & 3 & 4 \\ 4 & 3 & 2 & 1 \\ 0 & -1 & 2 & 3 \\ 1 & 6 & 4 & -2 \end{vmatrix}$$

We shall obtain the value of this determinant using (7) of Property 1. Expanding by the third row because of the presence of the zero element and taking due account of the alternating signs of the successive cofactors, we have

$$|\mathrm{A}| = (0)\begin{vmatrix} 2 & 3 & 4 \\ 3 & 2 & 1 \\ 6 & 4 & -2 \end{vmatrix} - (-1)\begin{vmatrix} 1 & 3 & 4 \\ 4 & 2 & 1 \\ 1 & 4 & -2 \end{vmatrix} + (2)\begin{vmatrix} 1 & 2 & 4 \\ 4 & 3 & 1 \\ 1 & 6 & -2 \end{vmatrix} - (3)\begin{vmatrix} 1 & 2 & 3 \\ 4 & 3 & 2 \\ 1 & 6 & 4 \end{vmatrix}$$

or, expanding the third-order determinants by the diagonal method,

$$|\mathrm{A}| = 0 + 75 + 180 - 105 = 150$$

The same result could have been found just as easily using (8) to expand $|\mathrm{A}|$ by elements of the first column.

Since the same number is obtained whether we expand a determinant in terms of the elements of an arbitrary row or an arbitrary column, we have the following obvious consequence of Property 1:

† For a proof of Property 1, see C. Ray Wylie, *Advanced Engineering Mathematics,* 4th ed., pp. 457–459, McGraw-Hill, New York, 1975.

‡ Named for the French mathematician Pierre Simon de Laplace (1749–1827).

PROPERTY 2 For every square matrix A, det A^T = det A.

EXAMPLE 3

For the matrix

$$A = \begin{bmatrix} 1 & 0 & 0 \\ 5 & -2 & 0 \\ 9 & 14 & 3 \end{bmatrix}$$

show that det A^T = det A.

Expanding successive determinants by elements of their first rows, we find

$$\det A = \begin{vmatrix} 1 & 0 & 0 \\ 5 & -2 & 0 \\ 9 & 14 & 3 \end{vmatrix} = (1)\begin{vmatrix} -2 & 0 \\ 14 & 3 \end{vmatrix} = (1)(-2)(3) = -6$$

Expanding successive determinants by elements of their first columns, we find

$$\det A^T = \begin{vmatrix} 1 & 5 & 9 \\ 0 & -2 & 14 \\ 0 & 0 & 3 \end{vmatrix} = (1)\begin{vmatrix} -2 & 14 \\ 0 & 3 \end{vmatrix} = (1)(-2)(3) = -6 = \det A$$

as was to be shown.

Both A and A^T, in Example 3, are triangular matrices, and the determinant of either matrix is equal to the product of its diagonal elements. This suggests another property of determinants.

PROPERTY 3 If A is a triangular matrix of order n, then det A = $a_{11}a_{22}\cdots a_{nn}$; that is, the determinant of a triangular matrix is equal to the product of its diagonal elements.

PROOF If A is lower triangular, successive expansions by elements of the first row give

$$\det A = \begin{vmatrix} a_{11} & 0 & \cdots & 0 \\ a_{21} & a_{22} & \cdots & 0 \\ \cdots & \cdots & \cdots & \cdots \\ a_{n1} & a_{n2} & \cdots & a_{nn} \end{vmatrix} = a_{11}\begin{vmatrix} a_{22} & \cdots & 0 \\ \cdots & \cdots & \cdots \\ a_{n2} & \cdots & a_{nn} \end{vmatrix} = \cdots = a_{11}a_{22}\cdots a_{nn}$$

If A is upper triangular, successive expansions by elements of the first column confirm the given property.

PROPERTY 4 If a square matrix A has either a zero row or a zero column, then det A = 0.

PROOF If we expand det A, according to Property 1, in terms of the row or column of zero elements, each term in the expansion contains a zero factor. Hence the entire expansion is zero, as asserted.

EXAMPLE 4

Applying Laplace's expansion by elements of the second column, we find that

$$\begin{vmatrix} 5 & 0 & 3 \\ -6 & 0 & 6 \\ 4 & 0 & 2 \end{vmatrix} = -(0)\begin{vmatrix} -6 & 6 \\ 4 & 2 \end{vmatrix} + (0)\begin{vmatrix} 5 & 3 \\ 4 & 2 \end{vmatrix} - (0)\begin{vmatrix} 5 & 3 \\ -6 & 6 \end{vmatrix} = 0$$

PROPERTY 5 If each element in one row (column)† of a determinant is multiplied by a number c, the value of the determinant is multiplied by c.

PROOF Suppose $|B|$ has been obtained from $|A|$ by multiplying row i (column j) of $|A|$ by c. If we expand $|B|$ in terms of row i (column j), each term in the expansion contains c as a factor. If c is then factored from the expansion, the result is just c times the expansion of $|A|$ by the same row (column).

EXAMPLE 5

Evaluate

$$\begin{vmatrix} 3 & 9 & 5 \\ 4 & 6 & 0 \\ -1 & -3 & 2 \end{vmatrix}$$

Factoring 2 from Row 2 and 3 from Column 2, then expanding by Row 2, we find

$$\begin{vmatrix} 3 & 9 & 5 \\ 4 & 6 & 0 \\ -1 & -3 & 2 \end{vmatrix} = 2 \cdot 3 \begin{vmatrix} 3 & 3 & 5 \\ 2 & 1 & 0 \\ -1 & -1 & 2 \end{vmatrix} = 6\left(-2\begin{vmatrix} 3 & 5 \\ -1 & 2 \end{vmatrix} + \begin{vmatrix} 3 & 5 \\ -1 & 2 \end{vmatrix} \right)$$

$$= -6\begin{vmatrix} 3 & 5 \\ -1 & 2 \end{vmatrix} = -6(6+5) = -66$$

PROPERTY 6 If $A = [\mathbf{v}_1 \cdots \mathbf{f}_j + \mathbf{g}_j \cdots \mathbf{v}_n]$ is a square matrix expressed in terms of its column vectors, then

$$\det A = \det [\mathbf{v}_1 \cdots \mathbf{f}_j \cdots \mathbf{v}_n] + \det [\mathbf{v}_1 \cdots \mathbf{g}_j \cdots \mathbf{v}_n]$$

An analogous result holds for A as an array of row vectors if one of them has binomial components.

What this property says is that if each element in one column (row) of a determinant is expressed as a binomial, the determinant can be written as the sum of two determinants. Specifically,

$$(9)\quad \begin{vmatrix} a_{11} & \cdots & f_{1j}+g_{1j} & \cdots & a_{1n} \\ a_{21} & \cdots & f_{2j}+g_{2j} & \cdots & a_{2n} \\ \cdots & \cdots & \cdots & \cdots & \cdots \\ a_{n1} & \cdots & f_{nj}+g_{nj} & \cdots & a_{nn} \end{vmatrix} = \begin{vmatrix} a_{11} & \cdots & f_{1j} & \cdots & a_{1n} \\ a_{21} & \cdots & f_{2j} & \cdots & a_{2n} \\ \cdots & \cdots & \cdots & \cdots & \cdots \\ a_{n1} & \cdots & f_{nj} & \cdots & a_{nn} \end{vmatrix} + \begin{vmatrix} a_{11} & \cdots & g_{1j} & \cdots & a_{1n} \\ a_{21} & \cdots & g_{2j} & \cdots & a_{2n} \\ \cdots & \cdots & \cdots & \cdots & \cdots \\ a_{n1} & \cdots & g_{nj} & \cdots & a_{nn} \end{vmatrix}$$

A like result holds for determinants containing a row of elements which are binomials.

PROOF Using Property 1, we expand det A in terms of the column (row) which contains the binomial elements. If $\mathbf{v}_j = \mathbf{f}_j + \mathbf{g}_j$, (8) gives

$$\det A = \sum_{k=1}^{n} a_{kj}A_{kj} = \sum_{k=1}^{n} (f_{kj} + g_{kj})A_{kj} = \sum_{k=1}^{n} f_{kj}A_{kj} + \sum_{k=1}^{n} g_{kj}A_{kj}$$

† Statements like this, which contain words or phrases in parentheses, are really two statements in one. One version is obtained by consistently omitting the word, or words, in parentheses (the word *column* in this case); the other is obtained by consistently retaining the words in parentheses and omitting the words to which they are alternatives (the word *row*, in this case).

Since the last two sums are, respectively, the expansions of det $[\mathbf{v}_1 \ldots \mathbf{f}_j \ldots \mathbf{v}_n]$ and det $[\mathbf{v}_1 \ldots \mathbf{g}_j \ldots \mathbf{v}_n]$, the first part of Property 6 is established. If some row of A, say $\mathbf{u}_i$, is such that $\mathbf{u}_i = \mathbf{w}_i + \mathbf{z}_i$, (7) yields the analogous result for rows. ◗

EXAMPLE 6

Determine a number k such that

$$\begin{vmatrix} 1 & 2 & -3 \\ -k & 1+3k & 3-k \\ 0 & -6 & 5 \end{vmatrix} = 36$$

With $-k$ regarded as the binomial $0 - k$, the elements of the second row of the given determinant become binomials; hence

$$\begin{vmatrix} 1 & 2 & -3 \\ -k & 1+3k & 3-k \\ 0 & -6 & 5 \end{vmatrix} = \begin{vmatrix} 1 & 2 & -3 \\ 0 & 1 & 3 \\ 0 & -6 & 5 \end{vmatrix} + \begin{vmatrix} 1 & 2 & -3 \\ -k & 3k & -k \\ 0 & -6 & 5 \end{vmatrix}$$

$$= 23 + k\begin{vmatrix} 1 & 2 & -3 \\ -1 & 3 & -1 \\ 0 & -6 & 5 \end{vmatrix} = 23 + k$$

which equals 36 if and only if $k = 13$.

PROPERTY 7 If B is a matrix obtained by interchanging any two rows (columns) of a square matrix A, then det B = −det A.

◖ **PROOF** The property is readily verified for matrices of order 2. Proceeding by induction, we assume that it holds for matrices of order $n - 1$. Let B be a matrix obtained by interchanging two rows of a matrix A of order n (>2) and suppose row i is not one of those interchanged. Using (7) to expand det A and det B by their ith rows and recalling how cofactors were defined, we have

$$\det A = \sum_{k=1}^{n} a_{ik}(-1)^{i+k}M_{ik} \qquad \det B = \sum_{k=1}^{n} a_{ik}(-1)^{i+k}N_{ik}$$

where N_{ik} can be derived by interchanging two rows of the minor M_{ik} of a_{ik} in $|A|$. Since these minors are of order $n - 1$, the induction hypothesis gives $N_{ik} = -M_{ik}$. Substituting this result into the sum for det B, we find

$$\det B = \sum_{k=1}^{n} a_{ik}(-1)^{i+k}(-M_{ik}) = -\sum_{k=1}^{n} a_{ik}(-1)^{i+k}M_{ik} = -\det A$$

By virtue of Property 2, the same proof applies to an interchange of columns. ◗

EXAMPLE 7

The matrix B obtained by interchanging the first and third columns of

$$A = \begin{bmatrix} 1 & 6 & 1 \\ -3 & 9 & -3 \\ 4 & 8 & 4 \end{bmatrix}$$

is the same matrix as A. Consequently, det B = det A. On the other hand, Property 7 gives det B = −det A. It follows that det A = −det A or 2 det A = 0; hence $|A| = 0$.

By interchanging rows (columns) and reasoning as in the last example, it is easy to show that the value of any determinant having two identical rows (columns) is zero. A slightly more general result is this.

PROPERTY 8 If one row (column) vector of a square matrix A is equal to a number c times some other row (column) vector, then $|A| = 0$.

Expressed another way, this property says that if corresponding elements of two rows (columns) of a determinant are proportional, the value of the determinant is zero.

◖ **PROOF** Because of Property 2, it will suffice to establish the foregoing property in the case of proportional rows. So suppose one row vector of a square matrix A is equal to c times some other row vector. By Property 4, if $c = 0$, $|A| = 0$. By Property 5, if $c \neq 0$, $|A| = c|B|$ where $|B| = 0$ because $|B|$ has two identical rows. It follows that $|A| = 0$, whether c is zero or not. ◗

EXAMPLE 8

Evaluate the determinant

$$|A| = \begin{vmatrix} 2 & -3 & -2 \\ 3 & -2 & 5 \\ -6 & 9 & 6 \end{vmatrix}$$

It is apparent that each element of the third row is equal to -3 times the corresponding element of the first row; hence $|A| = 0$.

PROPERTY 9 If a matrix B is obtained from a square matrix A by adding to one row (column) vector of A a number c times a different row (column) vector, then $\det B = \det A$.

◖ **PROOF** If $c = 0$, $B = A$ and $\det B = \det A$. If $c \neq 0$, by applying Property 6 to det B, we obtain two determinants one of which is det A and the other of which contains two proportional rows (or columns). By Property 8, the second determinant is equal to zero, and the stated property is established. ◗

Property 9 is especially useful in the practical evaluation of determinants, for by its repeated application one can reduce to zero a number of the elements in a chosen row (or column) of a given determinant. Then, when the determinant is expanded by elements of this row (or column), most of the products involved will be zero and the computation will be appreciably shortened.

EXAMPLE 9

Find the value of the determinant

$$\begin{vmatrix} 3 & 1 & -1 & 2 & 1 \\ 0 & 3 & 1 & 4 & 2 \\ 1 & 4 & 2 & 3 & 1 \\ 5 & -1 & -3 & 2 & 5 \\ -1 & 1 & 2 & 3 & 2 \end{vmatrix}$$

Here, in an attempt to introduce as many zeros as possible into some row, let us add the third column to the second and to the fifth, and let us add twice the third column to the fourth and 3 times the third column to the first. This gives the new but equal determinant

$$\begin{vmatrix} 0 & 0 & -1 & 0 & 0 \\ 3 & 4 & 1 & 6 & 3 \\ 7 & 6 & 2 & 7 & 3 \\ -4 & -4 & -3 & -4 & 2 \\ 5 & 3 & 2 & 7 & 4 \end{vmatrix}$$

Expanding this in terms of the first row, according to Definition 1, we have

$$(-1)(-1)^{1+3}\begin{vmatrix} 3 & 4 & 6 & 3 \\ 7 & 6 & 7 & 3 \\ -4 & -4 & -4 & 2 \\ 5 & 3 & 7 & 4 \end{vmatrix}$$

Now, adding twice the last column to each of the first three, we obtain the equal determinant

$$-\begin{vmatrix} 9 & 10 & 12 & 3 \\ 13 & 12 & 13 & 3 \\ 0 & 0 & 0 & 2 \\ 13 & 11 & 15 & 4 \end{vmatrix}$$

or expanding in terms of the third row,

$$-(2)(-1)^{3+4}\begin{vmatrix} 9 & 10 & 12 \\ 13 & 12 & 13 \\ 13 & 11 & 15 \end{vmatrix}$$

We can now simplify this by further row or column manipulations or, since it is of the third order, we can expand it by the diagonal method. The result is -166.

PROPERTY 10 For every determinant $|A| = |a_{ij}|$ of order n,

$$\sum_{k=1}^{n} a_{ik}A_{jk} = \delta_{ij}|A| \tag{10}$$

and

$$\sum_{k=1}^{n} a_{kj}A_{ki} = \delta_{ij}|A| \tag{11}$$

where δ_{ij} is the Kronecker delta.

Recalling that the first (second) subscript of an element always signifies the row (column) index, we recognize a_{ik} as the entry of $|A|$ in row i and column k, and A_{jk} as the cofactor of the element a_{jk} of $|A|$ appearing in row j and column k. With these observations in mind, we see that when $i \neq j$, (10) asserts that the sum of the products formed by multiplying the elements of one row of a determinant by the cofactors of the corresponding elements of another row is zero. If $i = j$, (10) is simply Laplace's expansion of $|A|$ by elements of the ith row.

On the other hand, since a_{kj} is in row k and column j of $|A|$ and since A_{ki} is the cofactor of a_{ki} (which is the element in row k and column i), we see that when $i \neq j$, (11) asserts that the sum of the products formed by multiplying the elements of one column of a determinant by the cofactors of the corresponding elements of another column is zero. Of course, if $i = j$, (11) is Laplace's expansion of $|A|$ by elements of the jth column.

By name, Eqs. (10) and (11) are referred to as **alien cofactor laws.**

PROOF By virtue of Property 2, (10) and (11) imply each other and Laplace's expansion holds by Property 1. Hence all we need to prove is (10) in case $i \neq j$. In this case, Property 1 clearly allows us to interpret the sum

$$\sum_{k=1}^{n} a_{ik}A_{jk}$$

as the expansion of a determinant whose jth row is

$$a_{i1}\ a_{i2}\ \ldots\ a_{in}$$

and whose other rows are identical with the corresponding rows of $|A|$. In this new determinant the ith and jth rows are therefore the same, hence by Property 8 the determinant is equal to zero as stated by (10). ◗

EXAMPLE 10

If we multiply the elements in the first row of the determinant

$$\begin{vmatrix} a_{11} & a_{12} & a_{13} \\ a_{21} & a_{22} & a_{23} \\ a_{31} & a_{32} & a_{33} \end{vmatrix}$$

by the cofactors of the corresponding elements in the third row, say, we obtain the sum

$$a_{11}\begin{vmatrix} a_{12} & a_{13} \\ a_{22} & a_{23} \end{vmatrix} - a_{12}\begin{vmatrix} a_{11} & a_{13} \\ a_{21} & a_{23} \end{vmatrix} + a_{13}\begin{vmatrix} a_{11} & a_{12} \\ a_{21} & a_{22} \end{vmatrix}$$

and this is clearly the expansion of the determinant

$$\begin{vmatrix} a_{11} & a_{12} & a_{13} \\ a_{21} & a_{22} & a_{23} \\ a_{11} & a_{12} & a_{13} \end{vmatrix}$$

according to the third row. Since this determinant has two identical rows, it is therefore equal to zero.

Many applied problems lead to determinants some or all of whose elements are functions. In analyzing the behavior of functions represented by determinants, it is often helpful to expand a determinant in terms of **generalized Laplace expansions.** Such expansions are described by the following property.

PROPERTY 11 Let any m rows (columns) be selected from a determinant $|A|$. Then $|A|$ is equal to the sum of the products of all the mth-order minors contained in the chosen rows (columns) each multiplied by its algebraic complement.

No proof of this property will be given, but here is an example to illustrate its use.

EXAMPLE 11

The generalized Laplace expansion of the determinant

$$|A| = \begin{vmatrix} 1 & 2 & 3 & 4 \\ 4 & 3 & 2 & 1 \\ 0 & -1 & 2 & 3 \\ 1 & 6 & 4 & -2 \end{vmatrix}$$

in terms of its first two rows is given by

$$\begin{aligned} |A| = {} & \begin{vmatrix} 1 & 2 \\ 4 & 3 \end{vmatrix} \cdot \begin{vmatrix} 2 & 3 \\ 4 & -2 \end{vmatrix} + \begin{vmatrix} 1 & 3 \\ 4 & 2 \end{vmatrix}\left(-\begin{vmatrix} -1 & 3 \\ 6 & -2 \end{vmatrix}\right) + \begin{vmatrix} 1 & 4 \\ 4 & 1 \end{vmatrix} \cdot \begin{vmatrix} -1 & 2 \\ 6 & 4 \end{vmatrix} \\ & + \begin{vmatrix} 2 & 3 \\ 3 & 2 \end{vmatrix} \cdot \begin{vmatrix} 0 & 3 \\ 1 & -2 \end{vmatrix} + \begin{vmatrix} 2 & 4 \\ 3 & 1 \end{vmatrix}\left(-\begin{vmatrix} 0 & 2 \\ 1 & 4 \end{vmatrix}\right) + \begin{vmatrix} 3 & 4 \\ 2 & 1 \end{vmatrix} \cdot \begin{vmatrix} 0 & -1 \\ 1 & 6 \end{vmatrix} \\ = {} & (-5)(-16) + (-10)(16) + (-15)(-16) \\ & + (-5)(-3) + (-10)(2) + (-5)(1) = 150 \end{aligned}$$

which agrees with the value of $|A|$ found in Example 2.

As a final property of determinants we include one which is reminiscent of matrix multiplication.

PROPERTY 12 If A and B are matrices of the same order, then

(12) $$(\det \mathrm{A})(\det \mathrm{B}) = \det (\mathrm{AB})$$

This property shows that the product of two determinants of the same order is a determinant of like order in which the element in its ith row and jth column is the scalar product of the ith row vector of the first determinant and the jth column vector of the second determinant.

PROOF For simplicity we shall prove this theorem only for determinants of the second order, although for these, direct verification is easier and more natural than the method we shall actually use. The virtue of our proof is that it can be extended immediately to the general case of determinants of any order. We begin by observing that if

$$|\mathrm{A}| = \begin{vmatrix} a_{11} & a_{12} \\ a_{21} & a_{22} \end{vmatrix} \quad \text{and} \quad |\mathrm{B}| = \begin{vmatrix} b_{11} & b_{12} \\ b_{21} & b_{22} \end{vmatrix}$$

then, by Property 11,

$$|\mathrm{A}| \cdot |\mathrm{B}| = \begin{vmatrix} a_{11} & a_{12} \\ a_{21} & a_{22} \end{vmatrix} \cdot \begin{vmatrix} b_{11} & b_{12} \\ b_{21} & b_{22} \end{vmatrix} = \begin{vmatrix} a_{11} & a_{12} & 0 & 0 \\ a_{21} & a_{22} & 0 & 0 \\ c_{11} & c_{12} & b_{11} & b_{12} \\ c_{21} & c_{22} & b_{21} & b_{22} \end{vmatrix}$$

where c_{11}, c_{12}, c_{21}, and c_{22} are completely arbitrary. In particular, it is convenient to take $c_{11} = c_{22} = -1$ and $c_{12} = c_{21} = 0$, so that we have

$$|\mathrm{A}| \cdot |\mathrm{B}| = \begin{vmatrix} a_{11} & a_{12} & 0 & 0 \\ a_{21} & a_{22} & 0 & 0 \\ -1 & 0 & b_{11} & b_{12} \\ 0 & -1 & b_{21} & b_{22} \end{vmatrix}$$

Now if we multiply the elements in the first column by b_{11} and the elements in the second column by b_{21} and then add them to the corresponding elements of the third column, we obtain, by Property 9, the equal determinant

$$|\mathrm{A}| \cdot |\mathrm{B}| = \begin{vmatrix} a_{11} & a_{12} & a_{11}b_{11} + a_{12}b_{21} & 0 \\ a_{21} & a_{22} & a_{21}b_{11} + a_{22}b_{21} & 0 \\ -1 & 0 & 0 & b_{12} \\ 0 & -1 & 0 & b_{22} \end{vmatrix}$$

In the same way, if we multiply the elements in the first column by b_{12} and the elements in the second column by b_{22} and then add them to the corresponding elements in the fourth column, we obtain from the last determinant the equal determinant

$$|\mathrm{A}| \cdot |\mathrm{B}| = \begin{vmatrix} a_{11} & a_{12} & a_{11}b_{11} + a_{12}b_{21} & a_{11}b_{12} + a_{12}b_{22} \\ a_{21} & a_{22} & a_{21}b_{11} + a_{22}b_{21} & a_{21}b_{12} + a_{22}b_{22} \\ -1 & 0 & 0 & 0 \\ 0 & -1 & 0 & 0 \end{vmatrix}$$

If we now expand this determinant by Property 11, applied to the last two rows, we obtain

$$|A| \cdot |B| = \begin{vmatrix} a_{11}b_{11} + a_{12}b_{21} & a_{11}b_{12} + a_{12}b_{22} \\ a_{21}b_{11} + a_{22}b_{21} & a_{21}b_{12} + a_{22}b_{22} \end{vmatrix} = |AB|$$

which is the result asserted by (12). ◗

EXAMPLE 12

For the matrices

$$A = \begin{bmatrix} 1 & 0 & -3 \\ 2 & -5 & 4 \\ -2 & 3 & -1 \end{bmatrix} \quad \text{and} \quad B = \begin{bmatrix} -3 & -2 & 12 \\ 4 & 1 & -6 \\ 2 & 3 & -10 \end{bmatrix}$$

verify directly that det (AB) = det A det B.

After adding 3 times the first column of det A to its third column, we have

$$\det A = \begin{vmatrix} 1 & 0 & 0 \\ 2 & -5 & 10 \\ -2 & 3 & -7 \end{vmatrix} = \begin{vmatrix} -5 & 10 \\ 3 & -7 \end{vmatrix} = 35 - 30 = 5$$

By adding 2 times Row 2 to Row 1, and -3 times Row 2 to Row 3, of det B, we find

$$\det B = \begin{vmatrix} 5 & 0 & 0 \\ 4 & 1 & -6 \\ -10 & 0 & 8 \end{vmatrix} = 5 \cdot 8 = 40$$

We next compute the matrix product AB from which we obtain

$$\det(AB) = \begin{vmatrix} -9 & -11 & 42 \\ -18 & 3 & 14 \\ 16 & 4 & -32 \end{vmatrix} = 4\begin{vmatrix} -9 & -11 & 42 \\ -18 & 3 & 14 \\ 4 & 1 & -8 \end{vmatrix} = 4\begin{vmatrix} 35 & -11 & -46 \\ -30 & 3 & 38 \\ 0 & 1 & 0 \end{vmatrix}$$

$$= (-4)\begin{vmatrix} 35 & -46 \\ -30 & 38 \end{vmatrix} = -(4 \cdot 5 \cdot 2)\begin{vmatrix} 7 & -23 \\ -6 & 19 \end{vmatrix} = -40(133 - 138) = 200$$

$$= \det A \det B$$

EXERCISES

1. Find the value of each of the following determinants:

(a) $\begin{vmatrix} 1 & 2 & 3 & 4 \\ 2 & 1 & 4 & 3 \\ 3 & 4 & 2 & 1 \\ 4 & 3 & 1 & 2 \end{vmatrix}$ **(b)** $\begin{vmatrix} 1 & 2 & 3 & 4 \\ 4 & 3 & 2 & 1 \\ 2 & 1 & 4 & 3 \\ 3 & 4 & 1 & 2 \end{vmatrix}$

(c) $\begin{vmatrix} 0 & 1 & 2 & 3 \\ -1 & 0 & 1 & 2 \\ -2 & -1 & 0 & 3 \\ -3 & -2 & -3 & 0 \end{vmatrix}$

(d) $\begin{vmatrix} 8 & 9 & 2 & 4 \\ -7 & 6 & -1 & 3 \\ 3 & 4 & 0 & 0 \\ 1 & -2 & 0 & 0 \end{vmatrix}$

(e) $\begin{vmatrix} 6 & 4 & -6 & -2 \\ 7 & 0 & 0 & 4 \\ 8 & -3 & 5 & -7 \\ 9 & 0 & 0 & 8 \end{vmatrix}$ **(f)** $\begin{vmatrix} 0 & 4 & 8 & 0 \\ 23 & 13 & 17 & 31 \\ -6 & 11 & 7 & 9 \\ 0 & 3 & 6 & 0 \end{vmatrix}$

2. (a) Find the value(s) of a, if any, for which the diagonal method of expansion yields the correct value for the determinant

$$\begin{vmatrix} -1 & 2 & 3 & 4 \\ -1 & 2 & 0 & 3 \\ 2 & 0 & a & 1 \\ 1 & 4 & -9 & a \end{vmatrix}$$

(b) Show that there is no value of a for which the diagonal method of expansion yields the correct value for the determinant

$$\begin{vmatrix} 1 & 2 & 1 & 1 \\ -1 & 1 & 3 & 2 \\ -1 & 3 & 9 & 1 \\ 2 & 1 & 1 & a \end{vmatrix}$$

3. Show that the number of terms in the expansion of a general determinant of order n is $n!$

4. If $|A|$ is a second-order determinant each of whose elements is a binomial, how many determinants are there in the sum to which $|A|$ is reduced by repeated applications of Property 6? How many are there if $|A|$ is a third-order determinant? If $|A|$ is an nth-order determinant?

5. Calculate the product of each pair of the following determinants, using Property 12, and check your results by multiplying the values of the individual determinants:

(a) $\begin{vmatrix} 1 & 1 & -1 \\ 2 & 1 & 3 \\ 1 & 0 & 1 \end{vmatrix}$ (b) $\begin{vmatrix} -2 & 1 & 1 \\ 3 & 1 & 0 \\ -1 & 2 & 4 \end{vmatrix}$

(c) $\begin{vmatrix} 1 & 3 & 4 \\ 2 & -1 & 0 \\ 0 & 1 & 3 \end{vmatrix}$

6. Prove the following generalization of Property 12: the product of two determinants of the same order is another determinant of that order in which the element in the ith row and jth column is the sum of the products of corresponding elements in the ith row *or* column of the first determinant and the jth row *or* column of the second determinant, a consistent choice of row or column being maintained for all values of i and j. *Hint:* Use Property 2.

7. If $|A| = |a_{ij}|$ is a determinant of order n with the property $a_{ij} = -a_{ji}$ for all values of i and j such that $1 \le i, j \le n$, prove that $|A| = (-1)^n|A|$. What further conclusion can be drawn if n is odd? *Hint:* Use Properties 2 and 5.

8. (a) Show that $|A| = \begin{vmatrix} b^2+ac & bc & c^2 \\ ab & 2ac & bc \\ a^2 & ab & b^2+ac \end{vmatrix} =$

$4a^2b^2c^2$. *Hint:* Verify first that $|A| = \begin{vmatrix} b & c & 0 \\ a & 0 & c \\ 0 & a & b \end{vmatrix}^2$.

(b) Find the value of the determinant

$$\begin{vmatrix} b^2+c^2 & ab & ca \\ ab & c^2+a^2 & bc \\ ca & bc & a^2+b^2 \end{vmatrix}$$

9. Find all values of k which satisfy each of the following equations:

(a) $\begin{vmatrix} k & 3+k & -10 \\ 1-k & 2-k & 5 \\ 2 & 4+k & -k \end{vmatrix} = 48$

(b) $\begin{vmatrix} -1 & 3 & k \\ 2k-3 & 1-k & 3k+1 \\ 2 & k & -2 \end{vmatrix} = 9k - 28$

(c) $\begin{vmatrix} 1 & k & k+2 & k-2 & 100 \\ 0 & k & k-2 & k+2 & 100 \\ 0 & 0 & k+2 & k-2 & 100 \\ 0 & 0 & 0 & k-2 & k+2 \\ 0 & 0 & 0 & 0 & 100 \end{vmatrix} = 0$

10. Without expanding, evaluate $\begin{vmatrix} y+z & z+y & y+x \\ x & y & z \\ 1 & 1 & 1 \end{vmatrix}$.

11. (a) Prove that

$$\begin{vmatrix} 1+a_1 & a_2 & a_3 & \cdots & a_n \\ a_1 & 1+a_2 & a_3 & \cdots & a_n \\ a_1 & a_2 & 1+a_3 & \cdots & a_n \\ \cdots & \cdots & \cdots & \cdots & \cdots \\ a_1 & a_2 & a_3 & \cdots & 1+a_n \end{vmatrix} = 1 + a_1 + a_2 + a_3 + \cdots + a_n$$

(b) Show that the nth-order determinant

$$\begin{vmatrix} a & b & \cdots & b & b \\ b & a & \cdots & b & b \\ \cdots & \cdots & \cdots & \cdots & \cdots \\ b & b & \cdots & a & b \\ b & b & \cdots & b & a \end{vmatrix}$$

is equal to $(a-b)^{n-1}[a+(n-1)b]$.

(c) Prove that

$$\begin{vmatrix} 0 & a_1-a_2 & a_1-a_3 & \cdots & a_1-a_n \\ a_1-a_2 & 0 & a_2-a_3 & \cdots & a_2-a_n \\ a_1-a_3 & a_2-a_3 & 0 & \cdots & a_3-a_n \\ \cdots & \cdots & \cdots & \cdots & \cdots \\ a_1-a_n & a_2-a_n & a_3-a_n & \cdots & 0 \end{vmatrix}$$

$$= (-1)^n 2^{n-2}(a_n - a_1)\prod_{i=1}^{n-1}(a_i - a_{i+1}) \qquad n \ge 2\dagger$$

†Whereas the symbol Σ is used to denote a sum of terms, the symbol Π is used to denote a product of factors. Thus

$$\prod_{i=1}^{n-1}(a_i - a_{i+1}) = (a_1-a_2)(a_2-a_3)\cdots(a_{n-1}-a_n)$$

12. If D_n is the nth-order determinant

$$\begin{vmatrix} 1+x^2 & x & 0 & \cdots & 0 & 0 \\ x & 1+x^2 & x & \cdots & 0 & 0 \\ 0 & x & 1+x^2 & \cdots & 0 & 0 \\ \cdots & \cdots & \cdots & \cdots & \cdots & \cdots \\ 0 & 0 & 0 & \cdots & 1+x^2 & x \\ 0 & 0 & 0 & \cdots & x & 1+x^2 \end{vmatrix}$$

show that $D_n = (1+x^2)D_{n-1} - x^2 D_{n-2}$. Using this relation, determine the value of D_{10} if $x = 1$; if $x = -1$. Is the value of D_n independent of x?

13. If D_n is the nth-order determinant in which each element on the principal diagonal is a, each element immediately above the principal diagonal is b, each element immediately below the principal diagonal is c, and all other elements are zero, obtain a recurrence relation expressing D_n in terms of D_{n-1} and D_{n-2}. Use this relation to infer the value of D_n if $a = 3$, $b = 2$, $c = 1$.

14. Show that the area of the triangle whose vertices are the points (x_1, y_1), (x_2, y_2), (x_3, y_3) is

$$A = \pm\frac{1}{2}\begin{vmatrix} x_1 & y_1 & 1 \\ x_2 & y_2 & 1 \\ x_3 & y_3 & 1 \end{vmatrix}$$

where the plus or the minus sign is to be chosen according as the vertices of the triangle are numbered consecutively in the counterclockwise or the clockwise direction.

15. If P_1: (x_1, y_1), P_2: (x_2, y_2), and P_3: (x_3, y_3) are three points no two of which lie on the same vertical line, show that the equation of the parabola of the family $y = a + bx + cx^2$ which passes through P_1, P_2, and P_3 can be written in the form

$$\begin{vmatrix} y & 1 & x & x^2 \\ y_1 & 1 & x_1 & x_1^2 \\ y_2 & 1 & x_2 & x_2^2 \\ y_3 & 1 & x_3 & x_3^2 \end{vmatrix} = 0$$

Is this result correct if P_1, P_2, and P_3 are collinear?

16. Show that the equation of the circle which passes through the three points P_1: (x_1, y_1), P_2: (x_2, y_2), and P_3: (x_3, y_3) can be written in the form

$$\begin{vmatrix} x^2+y^2 & x & y & 1 \\ x_1^2+y_1^2 & x_1 & y_1 & 1 \\ x_2^2+y_2^2 & x_2 & y_2 & 1 \\ x_3^2+y_3^2 & x_3 & y_3 & 1 \end{vmatrix} = 0$$

Is this result correct if the three points are collinear?

17. Show that

$$\begin{vmatrix} a & -b & -a & b \\ b & a & -b & -a \\ c & -d & c & -d \\ d & c & d & c \end{vmatrix} = 4(a^2+b^2)(c^2+d^2)$$

18. Evaluate the determinant

$$\begin{vmatrix} 0 & 1 & 2 & 3 & \cdots & n-1 \\ 1 & 0 & 1 & 2 & \cdots & n-2 \\ 2 & 1 & 0 & 1 & \cdots & n-3 \\ 3 & 2 & 1 & 0 & \cdots & n-4 \\ \cdots & \cdots & \cdots & \cdots & \cdots & \cdots \\ n-1 & n-2 & n-3 & n-4 & \cdots & 0 \end{vmatrix}$$

19. **(a)** If

$$l_1\colon a_{11}x + a_{12}y + a_{13} = 0$$
$$l_2\colon a_{21}x + a_{22}y + a_{23} = 0$$
$$l_3\colon a_{31}x + a_{32}y + a_{33} = 0$$

are three lines no two of which are parallel, show that l_1, l_2, and l_3 are concurrent if and only if

$$|\text{A}| = \begin{vmatrix} a_{11} & a_{12} & a_{13} \\ a_{21} & a_{22} & a_{23} \\ a_{31} & a_{32} & a_{33} \end{vmatrix} = 0$$

(b) Show that the area of the triangle determined by the lines l_1, l_2, and l_3 is equal to the absolute value of the expression

$$\frac{1}{2A_{13}A_{23}A_{33}}\begin{vmatrix} A_{11} & A_{12} & A_{13} \\ A_{21} & A_{22} & A_{23} \\ A_{31} & A_{32} & A_{33} \end{vmatrix}$$

where A_{ij} is the cofactor of a_{ij} in $|\text{A}|$.

20. Show that

$$\begin{vmatrix} 1 & 1 & 1 & 1 \\ a_1 & a_2 & a_3 & a_4 \\ a_1^2 & a_2^2 & a_3^2 & a_4^2 \\ a_1^3 & a_2^3 & a_3^3 & a_4^3 \end{vmatrix}$$
$$= (a_1-a_2)(a_1-a_3)(a_1-a_4)(a_2-a_3)(a_2-a_4)(a_3-a_4)$$

What is the generalization of this result to determinants of this type of order n? [Determinants of this form are usually referred to as **Vandermonde determinants,** after the French mathematician A. T. Vandermonde (1735–1796).]

21. If $p_1, p_2, \ldots, p_n$ are polynomials, show that the nth-order determinant

$$\begin{vmatrix} p_1(x_1) & p_1(x_2) & \cdots & p_1(x_n) \\ p_2(x_1) & p_2(x_2) & \cdots & p_2(x_n) \\ \cdots & \cdots & \cdots & \cdots \\ p_n(x_1) & p_n(x_2) & \cdots & p_n(x_n) \end{vmatrix}$$

is evenly divisible by

$$\prod_{1\le i<j\le n} (x_i - x_j)$$

22. If $|\text{A}|$ is the nth-order determinant, $n > 1$,

$$\begin{vmatrix} a_{11} & a_{12} & \cdots & a_{1n} \\ a_{21} & a_{22} & \cdots & a_{2n} \\ \cdots & \cdots & \cdots & \cdots \\ a_{n1} & a_{n2} & \cdots & a_{nn} \end{vmatrix}$$

show that

$$\begin{vmatrix} a_{11} & a_{12} & \cdots & a_{1n} & x_1 \\ a_{21} & a_{22} & \cdots & a_{2n} & x_2 \\ \cdots & \cdots & \cdots & \cdots & \cdots \\ a_{n1} & a_{n2} & \cdots & a_{nn} & x_n \\ y_1 & y_2 & \cdots & y_n & 0 \end{vmatrix}$$

is equal to $-\Sigma_{i,j=1}^{n} A_{ij}x_i y_j$, where A_{ij} is the cofactor of a_{ij} in $|A|$.

23. If $|A| = |a_{ij}|$, show that $\partial|A|/\partial a_{ij} = A_{ij}$.

24. If each element of a determinant $|A|$ of order n is a differentiable function of t, show that the derivative of $|A|$ with respect to t is equal to the sum of n determinants, the ith one of which is identical with $|A|$ except for the ith row, which consists of the derivatives of the elements of the ith row of $|A|$. *Hint:* Proceed inductively, using Property 1.

25. If $f_1, f_2, \ldots, f_n$ are suitably differentiable functions of t, show that

$$\frac{d}{dt}\begin{vmatrix} f_1 & \cdots & f_n \\ f_1' & \cdots & f_n' \\ \cdots & \cdots & \cdots \\ f_1^{(n-2)} & \cdots & f_n^{(n-2)} \\ f_1^{(n-1)} & \cdots & f_n^{(n-1)} \end{vmatrix} = \begin{vmatrix} f_1 & \cdots & f_n \\ f_1' & \cdots & f_n' \\ \cdots & \cdots & \cdots \\ f_1^{(n-2)} & \cdots & f_n^{(n-2)} \\ f_1^{(n)} & \cdots & f_n^{(n)} \end{vmatrix}$$

Hint: Use the result of exercise 24.

26. Find $f'(t)$ if $f(t)$ is given by the determinant:

(a) $\begin{vmatrix} e^t & e^{-t} & 1 \\ e^t & -e^{-t} & 0 \\ e^t & -e^{-t} & t \end{vmatrix}$

(b) $\begin{vmatrix} \cos t & \sin t & \ln|t| \\ -\sin t & \cos t & 1/t \\ -\cos t & -\sin t & -1/t^2 \end{vmatrix}$

(c) $\begin{vmatrix} t^2 & t & 0 & 0 \\ 2t & 1 & 0 & 2 \\ 2 & 0 & 1 & 2t \\ 0 & 0 & t & t^2 \end{vmatrix}$

(d) $\begin{vmatrix} \cosh t & \sinh t & e^t & e^{-t} \\ \sinh t & \cosh t & e^t & -e^{-t} \\ e^t & -e^{-t} & \sinh t & \cosh t \\ e^t & e^{-t} & \cosh t & \sinh t \end{vmatrix}$

27. If $P_{ij} = \begin{vmatrix} a_i & a_j \\ b_i & b_j \end{vmatrix}$, $i, j = 1, 2, 3, 4$, $i < j$, show that $P_{12}P_{34} - P_{13}P_{24} + P_{14}P_{23} \equiv 0$. *Hint:* Consider the determinant

$$\begin{vmatrix} a_1 & a_2 & a_3 & a_4 \\ b_1 & b_2 & b_3 & b_4 \\ a_1 & a_2 & a_3 & a_4 \\ b_1 & b_2 & b_3 & b_4 \end{vmatrix}$$

3.6 SYSTEMS OF LINEAR ALGEBRAIC EQUATIONS

Both the terminology and the properties of vectors, matrices, and determinants find immediate application in the solution of linear algebraic equations. For instance, a **solution** of the equation

$$(1) \qquad 2x - y = 5$$

is a *vector* (x_1, y_1) in R^2 such that $2x_1 - y_1 = 5$. There are infinitely many solutions of (1). In the cartesian plane, they may be visualized as the points of the line which is the graph of (1).

A **solution** of the simultaneous equations

$$(2) \qquad \begin{aligned} 2x - y &= 5 \\ x - 2y &= 4 \end{aligned}$$

is a *vector* (x_1, y_1) of R^2 which is a solution of both equations. There is only one solution of this system, namely, $(2, -1)$. This solution is clearly the point of intersection of the two lines defined by Eqs. (2).

Of course, column vectors may be used interchangeably with row vectors to identify solutions of equations involving several variables; thus the solution $(2, -1)$ of (2) can also be written $\begin{bmatrix} 2 \\ -1 \end{bmatrix}$. The system of equations (2) is equivalent to the matrix equation

$$(3a) \qquad \begin{bmatrix} 2 & -1 \\ 1 & -2 \end{bmatrix}\begin{bmatrix} x \\ y \end{bmatrix} = \begin{bmatrix} 5 \\ 4 \end{bmatrix}$$

or

$$\begin{bmatrix} 2x - y \\ x - 2y \end{bmatrix} = \begin{bmatrix} 5 \\ 4 \end{bmatrix} \tag{3b}$$

in the sense that $\begin{bmatrix} 2 \\ -1 \end{bmatrix}$ is the only solution of each system. Since the unknown vector $\begin{bmatrix} x \\ y \end{bmatrix}$ in (3a) is a column vector, it is preferable to write the solution vector in the form $\begin{bmatrix} 2 \\ -1 \end{bmatrix}$, rather than $(2, -1)$. However, the row notation is easier to print, and so we shall specify vector solutions of equations by whichever notation seems most convenient.

Solutions of equations involving complex constants or more than three variables are difficult if not impossible to interpret geometrically, so naturally the analytic aspects of such solutions claim preeminence. Our interest at the moment is to learn how to solve general systems of linear algebraic equations.

A polynomial equation

$$a_1x_1 + a_2x_2 + \cdots + a_nx_n = b \tag{4}$$

of the first degree in the variables $x_1, x_2, \ldots, x_n$ is called a **linear algebraic equation.** If $b \neq 0$, the equation is said to be **nonhomogeneous** and b is called the **nonhomogeneous term.** If $b = 0$, the equation is said to be **homogeneous.** A homogeneous equation has no nonhomogeneous term.

Now suppose we have m simultaneous linear equations

$$\begin{array}{c} a_{11}x_1 + a_{12}x_2 + \cdots + a_{1n}x_n = b_1 \\ a_{21}x_1 + a_{22}x_2 + \cdots + a_{2n}x_n = b_2 \\ \cdots\cdots\cdots\cdots\cdots\cdots\cdots\cdots \\ a_{m1}x_1 + a_{m2}x_2 + \cdots + a_{mn}x_n = b_m \end{array} \tag{5}$$

To identify these equations by name, we shall refer to them collectively as a **system of linear algebraic equations** or a **linear system.** The m numbers $b_1, b_2, \ldots, b_m$ are called the **constants** of the system. If these constants are all zero, the system is said to be **homogeneous.** A linear system is **nonhomogeneous** if and only if at least one of its equations has a nonhomogeneous term.

A **solution** of the ith equation of (5) is an n-dimensional vector $(k_1, k_2, \ldots, k_n)$, or a column vector with the same respective components, such that

$$a_{i1}k_1 + a_{i2}k_2 + \cdots + a_{in}k_n = b_i$$

A solution of System (5) is an n-dimensional vector whose components simultaneously satisfy every equation of the system. Two linear systems are **equivalent** if and only if every solution of either system is a solution of the other.

The elements of the matrix

$$\mathrm{A} = \begin{bmatrix} a_{11} & a_{12} & \cdots & a_{1n} \\ a_{21} & a_{22} & \cdots & a_{2n} \\ \cdots & \cdots & \cdots & \cdots \\ a_{m1} & a_{m2} & \cdots & a_{mn} \end{bmatrix} \tag{6}$$

are called the **coefficients** of (5), and (6) itself is called the **coefficient matrix** of the system. The matrix

$$\begin{bmatrix} a_{11} & a_{12} & \cdots & a_{1n} & b_1 \\ a_{21} & a_{22} & \cdots & a_{2n} & b_2 \\ \cdots & \cdots & \cdots & \cdots & \cdots \\ a_{m1} & a_{m2} & \cdots & a_{mn} & b_n \end{bmatrix} \tag{7}$$

obtained by adjoining the respective constants of (5) to the elements of the coefficient matrix as an $(n + 1)$st column, is known as the **augmented matrix** of the system. If we define **x** and **b** as the column vectors

$$\mathbf{x} = \begin{bmatrix} x_1 \\ x_2 \\ \vdots \\ x_n \end{bmatrix} \qquad \mathbf{b} = \begin{bmatrix} b_1 \\ b_2 \\ \vdots \\ b_m \end{bmatrix} \tag{8}$$

System (5) can be written in the compact matrix form

$$\mathbf{Ax} = \mathbf{b} \tag{9}$$

Obviously, the augmented matrix of any given linear system can be written down by inspection. On the other hand, given the augmented matrix of a linear system, the system itself can immediately be constructed. These simple observations allow us to work with augmented matrices as though they were the systems to which they correspond.

Every *homogeneous* linear system in n variables is clearly satisfied by the n-dimensional zero vector $\mathbf{0}_n$, which is called the **trivial solution** of the system. But, as we shall see, there are nonhomogeneous systems which have no solution. If no solution exists, a system is said to be **inconsistent;** otherwise, it is said to be **consistent.** Every consistent linear system can be solved by systematically performing at most three types of elementary operations upon the equations of the system. If a given system is inconsistent, the process clearly exposes the inconsistency. Each operation has the desirable feature that it always transforms one system into an equivalent system. The three types of operations are these.

1. Interchanging two equations
2. Multiplying an equation through by a nonzero number
3. Adding to one equation a multiple of (i.e., any number times) some other equation.

That a linear system is transformed into an equivalent system by operations of the first two kinds is quite apparent. To prove that the third type of operation has the same effect, we set

$$f_i = f_i(x_1, x_2, \ldots, x_n) = a_{i1}x_1 + a_{i2}x_2 + \cdots + a_{in}x_n \qquad 1 \le i \le \mathrm{m}$$

and write (5) as

$$\begin{aligned} f_1 &= b_1 \\ f_2 &= b_2 \\ &\cdots\cdots \\ f_m &= b_m \end{aligned} \tag{5a}$$

Adding h times the rth equation of this system to the ith equation, we obtain

$$\begin{aligned} f_1 &= b_1 \\ &\cdots\cdots\cdots \\ f_{i-1} &= b_{i-1} \\ hf_r + f_i &= hb_r + b_i \qquad r \ne i \\ f_{i+1} &= b_{i+1} \\ &\cdots\cdots\cdots \\ f_m &= b_m \end{aligned} \tag{10}$$

Clearly, every solution of (5a) is a solution of (10). Conversely, any solution $(k_1, k_2, \ldots, k_n)$ of (10) is evidently a solution of every equation of (5a) except possibly $f_i = b_i$. But, substitution of the components of $(k_1, k_2, \ldots, k_n)$ into the ith equation of (10) gives

$$hf_r(k_1, k_2, \ldots, k_n) + f_i(k_1, k_2, \ldots, k_n) = hb_r + b_i \tag{11}$$

However, because $r \neq i$, the equation $f_r(k_1, k_2, \ldots, k_n) = b_r$ is known to hold. Hence (11) simplifies to

$$f_i(k_1, k_2, \ldots, k_n) = b_i$$

which is the only equation of System (5*a*) which needed to be separately checked. Every solution of (10) is therefore a solution of (5*a*). This shows that (5*a*) and (10) are equivalent systems, and our proof that adding a constant times one equation to another equation does not alter the solution set of a linear system is complete. ◗

We are now ready to use the elementary operations we have just described to systematically eliminate variables from the equations of any particular linear system until the system is either solved or is shown to be inconsistent. This process is carried out most efficiently by operating on the rows of the augmented matrix of the system rather than on the equations themselves. The operations we shall use for this purpose are precisely the three row operations on matrices we described in Definition 2, Sec. 3.4, namely,

1. Interchanging two rows
2. Multiplying a row by a nonzero number
3. Adding a number times the elements of one row to the corresponding elements of another row.

Each of these operations when applied to the rows of a matrix signifies a corresponding elementary operation upon the equations of a related linear system, and all of the linear systems represented by successive matrices in the process are equivalent. **Warning:** Although at times one may be tempted to modify the matrix of a linear system by performing column operations as well as row operations, this must *never* be done. Column operations amount to altering one or more of the coefficients in each equation and do not lead to an equivalent system.

Now comes the crucial question of how a given linear system is actually solved by applying row operations to its augmented matrix. In Sec. 3.4 we saw that each of the elementary row operations on a matrix could be accomplished by premultiplying the matrix by a suitable elementary row matrix. However, as we pointed out, although this is an important stratagem in some problems, when dealing with linear systems it is usually simpler and more natural to work directly with the augmented matrix. A practical procedure is suggested by the observation that a linear system whose coefficient matrix is either triangular or diagonal is easy to solve. We shall soon see why. Thus our method of solution will be to eliminate variables from a given linear system by using elementary row operations to transform the augmented matrix of the system into a matrix having only zeros below, and perhaps certain zeros above, its diagonal elements. This new matrix will enable us to write an equivalent system which can be solved at once. Here are some examples to illustrate the process.

EXAMPLE 1

Solve the system of linear equations

$$\begin{aligned} 3x_1 + 2x_2 &= 4 \\ 2x_1 - x_2 &= -9 \\ x_1 + 3x_2 &= 13 \end{aligned}$$

The augmented matrix of the system is

$$\begin{bmatrix} 3 & 2 & 4 \\ 2 & -1 & -9 \\ 1 & 3 & 13 \end{bmatrix}$$

To begin the elimination process, we move the number 1 in the first column of this matrix to the first row by interchanging Rows 1 and 3. The new matrix

$$\begin{bmatrix} 1 & 3 & 13 \\ 2 & -1 & -9 \\ 3 & 2 & 4 \end{bmatrix}$$

indicates that the first and third equations of the given system of equations have been interchanged. Working with this new matrix, we add -2 times Row 1 to Row 2 and then add -3 times Row 1 to Row 3, obtaining the augmented matrix

$$\begin{bmatrix} 1 & 3 & 13 \\ 0 & -7 & -35 \\ 0 & -7 & -35 \end{bmatrix} \quad \text{of the system} \quad \begin{aligned} x_1 + 3x_2 &= 13 \\ -7x_2 &= -35 \\ -7x_2 &= -35 \end{aligned}$$

which contains x_1 in only the first equation. Next, we add -1 times Row 2 of the preceding matrix to its third row and subsequently multiply Row 2 by $-\frac{1}{7}$, which gives the augmented matrix

$$\begin{bmatrix} 1 & 3 & 13 \\ 0 & 1 & 5 \\ 0 & 0 & 0 \end{bmatrix} \quad \text{of the system} \quad \begin{aligned} x_1 + 3x_2 &= 13 \\ x_2 &= 5 \\ 0 &= 0 \end{aligned}$$

Of course, this system can be solved at once by substitution. But let us carry the elimination process one step further. Upon adding -3 times Row 2 to Row 1 of the last matrix, we obtain the augmented matrix

$$\begin{bmatrix} 1 & 0 & -2 \\ 0 & 1 & 5 \\ 0 & 0 & 0 \end{bmatrix} \quad \text{of the system} \quad \begin{aligned} x_1 &= -2 \\ x_2 &= 5 \\ 0 &= 0 \end{aligned}$$

Since this reduced system is equivalent to the given system, the original system of linear equations has

$$\begin{bmatrix} x_1 \\ x_2 \end{bmatrix} = \begin{bmatrix} -2 \\ 5 \end{bmatrix}$$

as its only solution.

EXAMPLE 2

Find all solutions of the system of linear equations

$$\begin{aligned} 2x_1 + 3x_2 - x_3 &= 1 \\ 3x_1 - 4x_2 + 3x_3 &= -1 \\ 2x_1 - x_2 + x_3 &= -3 \\ 3x_1 + x_2 - 2x_3 &= 4 \end{aligned}$$

The augmented matrix of this system is

$$\begin{bmatrix} 2 & 3 & -1 & 1 \\ 3 & -4 & 3 & -1 \\ 2 & -1 & 1 & -3 \\ 3 & 1 & -2 & 4 \end{bmatrix}$$

We could multiply Row 1 by $\frac{1}{2}$ to obtain a matrix having the number 1 in the first row, first column position. However, this would introduce fractions, which are awkward to work with. So, instead, we subtract Row 3 from Row 2 and then interchange the new second row and Row 1. This results in the matrix

$$\begin{bmatrix} 1 & -3 & 2 & 2 \\ 2 & 3 & -1 & 1 \\ 2 & -1 & 1 & -3 \\ 3 & 1 & -2 & 4 \end{bmatrix}$$

Operating on the rows of this matrix, we add -2 times Row 1 to Rows 2 and 3, and we add -3 times Row 1 to Row 4, which gives

$$\begin{bmatrix} 1 & -3 & 2 & 2 \\ 0 & 9 & -5 & -3 \\ 0 & 5 & -3 & -7 \\ 0 & 10 & -8 & -2 \end{bmatrix}$$

By multiplying Row 2 of this matrix by $\frac{1}{9}$, we could get a matrix having the number 1 in the second row and second column. The same thing can be accomplished, however, by subtracting Row 2 from Row 4 and then interchanging the new fourth row with Row 2. In this way fractions are avoided and we get

$$\begin{bmatrix} 1 & -3 & 2 & 2 \\ 0 & 1 & -3 & 1 \\ 0 & 5 & -3 & -7 \\ 0 & 9 & -5 & -3 \end{bmatrix}$$

To transform this matrix into one having only zeros below the number 1 in column 2, we add -5 times its second row to Row 3, and -9 times Row 2 to Row 4. Then we multiply the new third and fourth rows by $\frac{1}{12}$ and $\frac{1}{2}$, respectively. This results in the matrix

$$\begin{bmatrix} 1 & -3 & 2 & 2 \\ 0 & 1 & -3 & 1 \\ 0 & 0 & 1 & -1 \\ 0 & 0 & 11 & -6 \end{bmatrix}$$

which, upon adding -11 times its third row to the last row, gives the augmented matrix

$$\begin{bmatrix} 1 & -3 & 2 & 2 \\ 0 & 1 & -3 & 1 \\ 0 & 0 & 1 & -1 \\ 0 & 0 & 0 & 5 \end{bmatrix} \quad \text{of the system} \quad \begin{aligned} x_1 - 3x_2 + 2x_3 &= 2 \\ x_2 - 3x_3 &= 1 \\ x_3 &= -1 \\ 0 &= 5 \end{aligned}$$

Since the last equation of this system cannot hold and since this system is equivalent to the given system, the original system of linear equations is inconsistent.

In the next example, we condense the solution process further. The row operations described to the right of each matrix explain how that matrix is transformed into the one that succeeds it. As in the last example, all systems of linear equations intermediate to the given system and the one having the final matrix of the elimination process as its augmented matrix are omitted.

EXAMPLE 3

Determine all solutions of the system of linear algebraic equations

$$\begin{aligned} 4x_1 - 8x_2 + 11x_3 - 11x_4 + 6x_5 &= 4a \\ 3x_1 - 6x_2 + 8x_3 - 9x_4 + 5x_5 &= 3a \\ 2x_1 - 4x_2 + 9x_3 + 5x_4 - 4x_5 &= 2a \\ x_1 - 2x_2 + 3x_3 - 2x_4 + x_5 &= a \end{aligned}$$

where a denotes an arbitrary number.

Starting with the augmented matrix of this system, we apply elementary row operations in succession to one matrix after another until we arrive at the augmented matrix of a linear system that is easy to solve.

$$\begin{bmatrix} 4 & -8 & 11 & -11 & 6 & 4a \\ 3 & -6 & 8 & -9 & 5 & 3a \\ 2 & -4 & 9 & 5 & -4 & 2a \\ 1 & -2 & 3 & -2 & 1 & a \end{bmatrix}$$

Add -4 times Row 4 to Row 1.
Add -3 times Row 4 to Row 2.
Add -2 times Row 4 to Row 3.
Interchange the *new* row 1 and Row 4.

$$\begin{bmatrix} 1 & -2 & 3 & -2 & 1 & a \\ 0 & 0 & -1 & -3 & 2 & 0 \\ 0 & 0 & 3 & 9 & -6 & 0 \\ 0 & 0 & -1 & -3 & 2 & 0 \end{bmatrix}$$

Add $\frac{1}{3}$ times Row 3 to Rows 2 and 4.
Subtract Row 3 from Row 1.
Multiply Row 3 by $\frac{1}{3}$.
Interchange the *new* rows 2 and 3.

$$\begin{bmatrix} 1 & -2 & 0 & -11 & 7 & a \\ 0 & 0 & 1 & 3 & -2 & 0 \\ 0 & 0 & 0 & 0 & 0 & 0 \\ 0 & 0 & 0 & 0 & 0 & 0 \end{bmatrix}$$

This is the augmented matrix of the system

$$\begin{aligned} x_1 - 2x_2 \quad - 11x_4 + 7x_5 &= a \\ x_3 + 3x_4 - 2x_5 &= 0 \\ 0 &= 0 \\ 0 &= 0 \end{aligned}$$

This reduced system can be solved at once by setting $x_2 = c_1$, $x_4 = c_2$, and $x_5 = c_3$, where c_1, c_2, c_3 are arbitrary constants, provided we take $x_1 = 2c_1 + 11c_2 - 7c_3 + a$ and $x_3 = -3c_2 + 2c_3$. For all numbers c_1, c_2, and c_3, this gives the solution

$$(x_1, x_2, x_3, x_4, x_5) = (2c_1 + 11c_2 - 7c_3 + a, c_1, -3c_2 + 2c_3, c_2, c_3)$$

which of course is also a solution of the original system of linear equations. Every specific solution of that system can be obtained by assigning suitable values to c_1, c_2, and c_3. Notice that the number of parameters in the solution is equal to the number of *unknowns* in the given system minus the number of 1s appearing as leading nonzero elements in the rows of the augmented matrix of the reduced linear system. In terms of column vectors, the preceding solution can be written as

$$\begin{bmatrix} x_1 \\ x_2 \\ x_3 \\ x_4 \\ x_5 \end{bmatrix} = c_1 \begin{bmatrix} 2 \\ 1 \\ 0 \\ 0 \\ 0 \end{bmatrix} + c_2 \begin{bmatrix} 11 \\ 0 \\ -3 \\ 1 \\ 0 \end{bmatrix} + c_3 \begin{bmatrix} -7 \\ 0 \\ 2 \\ 0 \\ 1 \end{bmatrix} + \begin{bmatrix} a \\ 0 \\ 0 \\ 0 \\ 0 \end{bmatrix} \tag{12}$$

If $a = 0$, the original linear system is homogeneous, and all of its solutions, including the particular solutions (2, 1, 0, 0, 0), (11, 0, −3, 1, 0), and (−7, 0, 2, 0, 1), are given by (12) with the final vector, which is then **0**, deleted. If $a \neq 0$, the given system is nonhomogeneous and has (a, 0, 0, 0, 0) as an obvious solution. Thus the most general solution of the nonhomogeneous system is a sum of one of its particular solutions and the most general solution of the homogeneous system. To fully understand the striking resemblance this result bears to the manner in which complete solutions of nonhomogeneous linear differential equations are constructed (Sec. 2.3), we need to become familiar with linear transformations. These are discussed in Chap. 13.

As Examples 1–3 illustrate, our elimination process for solving a linear system starts with the augmented matrix of the system. The process continues until it has led to a matrix in which every element below the diagonal elements is zero. Such a matrix is said to be **gaussian,**† and the process whereby elementary row operations are used to reduce a matrix to gaussian form is called **gaussian elimination.** As soon as the augmented matrix of a consistent linear system has been reduced to a gaussian matrix, the system corresponding to the gaussian matrix can be solved by **back substitution,** as illustrated in the next example.

EXAMPLE 4

The nonhomogeneous linear system corresponding to the gaussian matrix

$$\begin{bmatrix} 1 & -2 & 1 & 1 & 2 \\ 0 & 0 & 1 & 1 & 3 \\ 0 & 0 & 0 & 1 & -4 \end{bmatrix} \quad \text{can be written as} \quad \begin{aligned} x_1 - 2x_2 + x_3 + x_4 &= 2 \\ x_3 + x_4 &= 3 \\ x_4 &= -4 \end{aligned}$$

To solve this system, we substitute -4 from the third equation into the second equation for x_4 and solve for x_3, obtaining $x_3 = 7$. Substituting the known values of x_3 and x_4 into the first equation, assigning x_2 an arbitrary value c, and then solving for x_1, we find $x_1 = 2c - 1$. All solutions of the linear system are now determined by $(x_1, x_2, x_3, x_4) = (2c - 1, c, 7, -4) = c(2, 1, 0, 0) + (-1, 0, 7, -4)$.

In Examples 1 and 3, we did not terminate the elimination process at the stage where gaussian matrices were first obtained. Instead, we further simplified the linear systems by continuing the process until the augmented matrices of the original linear systems were brought into *row* (or *row-reduced*) *echelon form.* This form of a matrix is characterized by the following definition.

DEFINITION 1 An $m \times n$ matrix is in **row echelon form** if and only if

(a) The first nonzero entry in any row is 1, and all other entries in the column in which it lies are zero.

(b) Every zero row vector has a higher row index than any nonzero row vector; that is, each zero row vector lies below every nonzero row vector.

(c) If there are r nonzero row vectors, and if for $1 \leq i \leq r$, the first 1 in Row i occurs in Column j_i, then $j_1 < j_2 < \cdots < j_r$ (whence the name *echelon*).

It can be shown that elementary row operations suffice to reduce a general $m \times n$ matrix to row echelon form and that for any given matrix the form is unique.‡ Of course, this implies that r, too, is unique. This process of obtaining the row echelon form is known as **Gauss-Jordan elimination,** conamed for the French mathematician Camille Jordan (1838–1922). In general, a Gauss-Jordan elimination takes more steps than gaussian elimination does, hence is not suitable for computer applications. However, since the final matrix to which every Gauss-Jordan elimination process leads is unique, it is of theoretical significance as a *standard,* or *canonical* form to which an arbitrary matrix can be reduced by means of elementary row operations. On the other hand, gaussian elimination may lead to a variety of matrices, though of course each will yield all solutions of the initial linear system. Both gaussian elimination and Gauss-Jordan elimination can be applied to either homogeneous or nonhomogeneous systems.

---

†Karl Friedrich Gauss (1777–1855) was a great German mathematician.

‡See, for instance, Charles G. Cullen, *Linear Algebra and Differential Equations,* pp. 391–392, Prindle, Weber, and Schmidt, Boston, 1979.

EXAMPLE 5

Specify elementary row operations which will reduce each of the following matrices to row echelon form.

$$\begin{bmatrix} 1 & 1 & -2 & 0 & 0 \\ 0 & 0 & 0 & -1 & 0 \\ 0 & 0 & 0 & 0 & 1 \end{bmatrix} \quad \begin{bmatrix} 0 & 0 & 0 & 0 \\ 1 & -2 & 0 & 1 \\ 0 & 0 & 1 & 0 \\ 0 & 0 & 0 & 0 \end{bmatrix} \quad \begin{bmatrix} 0 & 1 & -2 & 0 & 6 \\ 0 & 0 & 0 & 1 & 3 \\ 0 & 0 & 0 & 0 & 0 \end{bmatrix}$$

Multiplication of Row 2 by -1 reduces the first matrix to row echelon form. The second matrix can be put into the same form by interchanging Rows 1 and 2, and then the *new* Row 2 and Row 3. The third matrix already has the required form.

By *the* **complete solution** of a system of linear algebraic equations we mean, as usual, the set of all possible solutions of the system. By a complete solution, we mean any expression defining all possible solutions. Using the fact that the row echelon form of a given matrix is unique, we can now establish several important results about the existence and number of solutions of linear systems, both homogeneous and nonhomogeneous. Our first result concerns homogeneous systems.

THEOREM 1 A complete solution of a homogeneous system of m equations in n unknowns, $A\mathbf{x} = \mathbf{0}$, always contains $n - r$ arbitrary constants, where r is the number of nonzero rows in the row echelon form of A.

PROOF Let J denote the row echelon form of A, let S denote the linear system corresponding to J, and let $x_{j_1}, x_{j_2}, \ldots, x_{j_r}$ be the unknowns whose indices are the successive column indices of the r leading 1s in J. From **a** of Definition 1, none of these unknowns can occur in two different equations of S, nor can two of them occur in the same equation. Thus S consists of r equations from which $x_{j_1}, x_{j_2}, \ldots, x_{j_r}$ can be expressed as linear combinations of the remaining $n - r$ unknowns. Since these unknowns can be assigned arbitrary values, the proof is complete.

EXAMPLE 6

The row echelon form of the coefficient matrix A of a homogeneous linear system $A\mathbf{x} = \mathbf{0}$ is

$$J = \begin{bmatrix} 1 & 0 & 0 & 1 & -8 & 0 & 2 \\ 0 & 1 & 0 & -3 & 6 & 0 & 0 \\ 0 & 0 & 1 & -1 & 0 & 0 & 2 \\ 0 & 0 & 0 & 0 & 0 & 1 & -3 \\ 0 & 0 & 0 & 0 & 0 & 0 & 0 \\ 0 & 0 & 0 & 0 & 0 & 0 & 0 \\ 0 & 0 & 0 & 0 & 0 & 0 & 0 \\ 0 & 0 & 0 & 0 & 0 & 0 & 0 \end{bmatrix}$$

Find a complete solution of the system and verify that it contains the number of arbitrary constants guaranteed by Theorem 1.

It is easy to check that the requirements of Definition 1 are met and that J is in row echelon form. From the form of the four nonzero rows, it is clear that each of the variables x_1, x_2, x_3, and x_6 occur in one and only one of the equations making up the reduced linear system corresponding to J. Thus, from these four equations, x_1, x_2, x_3, and x_6 can be expressed as linear combinations of the remaining $3 = 7 - 4 = n - r$ variables x_4, x_5, and x_7. For arbitrary values of x_4, x_5, and x_7, these expressions determine a solution of the reduced, hence of the original linear system. Specifically, setting $x_4 = c_1$, $x_5 = c_2$, and $x_7 = c_3$, we have, from the first four rows of J,

$$\begin{aligned}
x_1 &= -x_4 + 8x_5 - 2x_7 = -c_1 + 8c_2 - 2c_3\\
x_2 &= 3x_4 - 6x_5 = 3c_1 - 6c_2\\
x_3 &= x_4 - 2x_7 = c_1 - 2c_3\\
x_6 &= 3x_7 = 3c_3
\end{aligned}$$

Finally, including the definitive equations

$$\begin{aligned}
x_4 &= c_1\\
x_5 &= c_2\\
x_7 &= c_3
\end{aligned}$$

and writing our complete solution in vector form, we have

$$\begin{bmatrix} x_1\\ x_2\\ x_3\\ x_4\\ x_5\\ x_6\\ x_7 \end{bmatrix} = c_1\begin{bmatrix} -1\\ 3\\ 1\\ 1\\ 0\\ 0\\ 0 \end{bmatrix} + c_2\begin{bmatrix} 8\\ -6\\ 0\\ 0\\ 1\\ 0\\ 0 \end{bmatrix} + c_3\begin{bmatrix} -2\\ 0\\ -2\\ 0\\ 0\\ 3\\ 1 \end{bmatrix}$$

Since three arbitrary constants appear in this solution, Theorem 1 is confirmed for the particular system $\mathrm{A}\mathbf{x} = \mathbf{0}$ having J as the row echelon form of A.

According as a linear system has at least as many equations as unknowns, or more unknowns than equations, Theorem 1 in turn implies that:

COROLLARY 1 If $m \geq n$ and $r = n$, then the only solution of a homogeneous linear system of m equations in n unknowns is the trivial solution $\mathbf{x} = \mathbf{0}_n$.

PROOF Since $n - r = n - n = 0$, there are no arbitrary constants in any complete solution. In fact, the linear equations corresponding to the $r = n$ nonzero rows in the row echelon form of the $m \times n$ coefficient matrix are $x_1 = 0, x_2 = 0, \ldots, x_n = 0$.

COROLLARY 2 If $n > m$, a homogeneous linear system of m equations in n unknowns has solutions containing at least $n - m$ arbitrary constants. If $r = m$, the number of parameters in a complete solution is $n - m$.

PROOF Let $\mathrm{A}\mathbf{x} = \mathbf{0}$ denote the linear system. Since the number r of nonzero rows in the row echelon form of A cannot exceed the number of rows in A, it follows that $r \leq m$, and so $n - r \geq n - m$. Hence, by Theorem 1, a complete solution of $\mathrm{A}\mathbf{x} = \mathbf{0}$ must contain at least $n - m$ arbitrary constants. If $r = m$, then the number of parameters in a complete solution is $n - r = n - m$.

As Examples 3 and 6 illustrate, the $n - r$ arbitrary constants which appear in a complete solution of a homogeneous linear system of m equations in n unknowns occur as scalar coefficients of $n - r$ particular solution vectors. A clue to the reason for this is to be found in the fundamental ideas of linear dependence and linear independence which, as defined for functions, were so important to our study of complete solutions of ordinary linear differential equations in Sec. 2.2. These ideas, when expressed in terms of vectors, read:

DEFINITION 2 Vectors $\mathbf{v}_1, \mathbf{v}_2, \ldots, \mathbf{v}_n$ of the same dimension are **linearly dependent** if and only if there exist constants $c_1, c_2, \ldots, c_n$, at least one of which is *not* zero, such that

$$c_1\mathbf{v}_1 + c_2\mathbf{v}_2 + \cdots + c_n\mathbf{v}_n = \mathbf{0} \tag{13}$$

DEFINITION 3 Vectors $\mathbf{v}_1, \mathbf{v}_2, \ldots, \mathbf{v}_n$ of the same dimension are **linearly independent** if and only if they are not linearly dependent, that is, if and only if the relation $c_1\mathbf{v}_1 + c_2\mathbf{v}_2 + \cdots + c_n\mathbf{v}_n = \mathbf{0}$ implies that $c_1 = c_2 = \cdots = c_n = 0$.

If $\mathbf{v}_1 = [v_{1j}]$, $\mathbf{v}_2 = [v_{2j}], \ldots, \mathbf{v}_n = [v_{nj}]$, $j = 1, 2, \ldots, m$, are n row vectors of the same dimension, it follows from the definitions of multiplication of a vector by a scalar, the addition of vectors, and the equality of vectors, that the vector equation (13) implies m scalar equations (one for each of the m components) of the form

$$c_1v_{1j} + c_2v_{2j} + \cdots + c_nv_{nj} = 0 \qquad j = 1, 2, \ldots, m \tag{14}$$

To determine whether the vectors $\mathbf{v}_1, \mathbf{v}_2, \ldots, \mathbf{v}_n$ are linearly dependent or independent requires that we solve the system of homogeneous linear equations (14) and determine whether or not they have a solution other than the trivial solution $\mathbf{c} = (c_1, c_2, \ldots, c_n) = \mathbf{0}$. If the only solution of (14) is the trivial solution $\mathbf{c} = \mathbf{0}$, the vectors are linearly independent. If (14) has a nontrivial solution $\mathbf{c} = (c_1, c_2, \ldots, c_n)$, at least one of whose components is different from zero, the vectors are linearly dependent.

EXAMPLE 7

Show that the vectors $\mathbf{v}_1 = \begin{bmatrix} 1 \\ 2 \\ -1 \end{bmatrix}$, $\mathbf{v}_2 = \begin{bmatrix} 4 \\ -3 \\ 2 \end{bmatrix}$, $\mathbf{v}_3 = \begin{bmatrix} 1 \\ -9 \\ 5 \end{bmatrix}$ are linearly dependent.

The equation $c_1\mathbf{v}_1 + c_2\mathbf{v}_2 + c_3\mathbf{v}_3 = \mathbf{0}$ leads at once to the three scalar equations

$$\begin{aligned} c_1 + 4c_2 + c_3 &= 0 \\ 2c_1 - 3c_2 - 9c_3 &= 0 \\ -c_1 + 2c_2 + 5c_3 &= 0 \end{aligned} \tag{15}$$

To solve this linear system, we use Gauss-Jordan elimination to reduce the coefficient matrix of the system to its row echelon form:

$$\begin{bmatrix} 1 & 0 & -3 \\ 0 & 1 & 1 \\ 0 & 0 & 0 \end{bmatrix}$$

which yields

$$(c_1, c_2, c_3) = \lambda(3, -1, 1) \qquad \lambda \text{ a parameter}$$

as a complete solution of the system (15). Hence, taking $\lambda = 1$, we have $3\mathbf{v}_1 - \mathbf{v}_2 + \mathbf{v}_3 = \mathbf{0}$ which shows that $\mathbf{v}_1$, $\mathbf{v}_2$, and $\mathbf{v}_3$ are linearly dependent, according to Definition 2.

EXAMPLE 8

Show that the vectors $\mathbf{v}_1 = \begin{bmatrix} 2 \\ 3 \\ 4 \\ 0 \\ 0 \end{bmatrix}$, $\mathbf{v}_2 = \begin{bmatrix} 5 \\ 0 \\ -1 \\ 0 \\ 4 \end{bmatrix}$, $\mathbf{v}_3 = \begin{bmatrix} 7 \\ 0 \\ 3 \\ 2 \\ 0 \end{bmatrix}$ are linearly independent.

In this case, the equation $c_1\mathbf{v}_1 + c_2\mathbf{v}_2 + c_3\mathbf{v}_3 = \mathbf{0}$ leads at once to the scalar equations

$$\begin{aligned} 2c_1 + 5c_2 + 7c_3 &= 0 \\ 3c_1 \qquad\qquad &= 0 \\ 4c_1 - c_2 + 3c_3 &= 0 \\ 2c_3 &= 0 \\ 4c_2 \qquad &= 0 \end{aligned}$$

From the second, fourth, and fifth equations it follows that $c_1 = c_2 = c_3 = 0$. Hence, according to Definition 3, $\mathbf{v}_1$, $\mathbf{v}_2$, and $\mathbf{v}_3$ are linearly independent.

Example 8 suggests the following special, but often useful, result.

THEOREM 2 If $\mathbf{v}_1, \mathbf{v}_2, \ldots, \mathbf{v}_n$ are n vectors of dimension m, where $m \geq n$, and if for n *distinct* values of i such that $1 \leq i \leq m$,

(a) the ith component of every vector but one is zero, and

(b) no two of the n nonzero components corresponding to these values of i occur in the same vector.

then the vectors are linearly independent.

PROOF Among the m scalar equations implied by the vector equation $c_1\mathbf{v}_1 + c_2\mathbf{v}_2 + \cdots + c_n\mathbf{v}_n = \mathbf{0}$, consider the n equations corresponding to the n values of i identified in the statement of the theorem. By **a**, each of these equations consists of a single-term expression of the form $c_j v_{ij} = 0$, where $v_{ij} \neq 0$. By **b**, no two of these equations involve the same value of j. These equations thus imply that each of the c's is equal to zero. Therefore, according to Definition 3, the given vectors are linearly independent.

Our intuition now tells us that:

THEOREM 3 Every homogeneous linear system of m equations in n unknowns has $n - r$ *linearly independent* particular solution vectors, and every solution of the system is a linear combination of them.

PROOF From the proof of Theorem 1, we know that every linear system $\mathrm{A}\mathbf{x} = \mathbf{0}$ of m equations in n unknowns has $n - r$ particular solution vectors $\mathbf{v}_1, \mathbf{v}_2, \ldots, \mathbf{v}_{n-r}$, and that every solution of the system is a linear combination of them. To prove that these $n - r$ solution vectors are linearly independent, let $x_{k_1}, x_{k_2}, \ldots, x_{k_{n-r}}$ be the $n - r$ unknowns to which arbitrary values are assigned. Then in the reduced linear system S corresponding to the row echelon form J of A, there are $n - r$ distinct definitive equations (see Example 6) of the form

$$x_{k_i} = c_i \qquad i = 1, 2, \ldots, n - r$$

Thus in the n by $n - r$ matrix $\mathrm{V} = [\mathbf{v}_1 \quad \mathbf{v}_2 \quad \cdots \quad \mathbf{v}_{n-r}]$, there are $n - r$ rows (corresponding to the x_k's) each containing a single 1 and $n - r - 1$ zeros. Furthermore, since the x_k's are distinct, no two of the 1s in these rows can occur in the same column of V. Hence these $n - r$ rows constitute an $n - r$ by $n - r$ submatrix of V which has the same row vectors as I_{n-r}, possibly interchanged. Therefore, by Exercise 38, the $n - r$ particular solution vectors are linearly independent.

For a nonhomogeneous linear system of m equations in n unknowns $\mathrm{A}\mathbf{x} = \mathbf{b}$, whose augmented matrix $[\mathrm{A}\,|\,\mathbf{b}]$ has $[\mathrm{J}\,|\,\mathbf{d}]$ as its row echelon form, there are several possibilities. If the first n rows of J are those of I_n, that is, if $m \geq n$ and $r = n$, and the last $m - n$ rows of $[\mathrm{J}\,|\,\mathbf{d}]$ are zero vectors, then the nth equation of the linear system S corresponding to the nth row of $[\mathrm{J}\,|\,\mathbf{d}]$ can always be solved for a unique value of x_n. Subsequently, back substitution will give a unique value for each of the other unknowns. On the other hand, if there are one or more zero rows in J, the system will be inconsistent unless $\mathbf{d}$ has zero components corresponding to each of the zero rows of J (see Example 2). When this is the case, the linear system will have a complete solution containing $n - r$ arbitrary constants. Summarizing these observations, we have:

THEOREM 4 If $[\mathrm{J}\,|\,\mathbf{d}]$ is the row echelon form of the augmented matrix $[\mathrm{A}\,|\,\mathbf{b}]$ of a nonhomogeneous linear system $\mathrm{A}\mathbf{x} = \mathbf{b}$ of m equations in n unknowns, then

(a) The system has a unique solution if and only if $m \geq n$ and the first n rows of J are those of I_n, whereas the last $m - n$ rows of $[\mathrm{J}\,|\,\mathbf{d}]$ are zero vectors.

(b) If J contains one or more zero rows that are not zero rows of $[\mathrm{J}\,|\,\mathbf{d}]$, the linear system is inconsistent.

(c) If every zero row of J is also a zero row of $[\mathrm{J}\,|\,\mathbf{d}]$, the linear system has a complete solution containing $n - r$ arbitrary constants, where r is the number of nonzero rows in $[\mathrm{J}\,|\,\mathbf{d}]$.

Finally, we have the following theorem, whose proof we leave as an exercise (see Exercise 21). Example 3 is an illustration of this theorem.

THEOREM 5 If $\mathbf{x}_h$ is a complete solution of the homogeneous linear system $\mathrm{A}\mathbf{x} = \mathbf{0}$, and $\mathbf{x}_p$ is any particular solution of the nonhomogeneous system $\mathrm{A}\mathbf{x} = \mathbf{b}$, then a complete solution of the nonhomogeneous system is $\mathbf{x} = \mathbf{x}_h + \mathbf{x}_p$.

EXERCISES

1. If possible, find a complete solution of each of the following linear systems and interpret each solution geometrically.

(a) $\begin{aligned} 3x + 2y &= 3 \\ 4x + 3y &= 6 \end{aligned}$

(b) $\begin{aligned} 4x - 6y &= 11 \\ -2x + 3y &= 10 \end{aligned}$

(c) $\begin{aligned} 5x - 6y - 7 &= 0 \\ 12y - 10x + 14 &= 0 \end{aligned}$

(d) $\begin{aligned} 13x - 10y + 7z &= -4 \\ -4x + 3y - 2z &= 1 \\ 2x + y - 4z &= 7 \end{aligned}$

(e) $\begin{aligned} 3x - 2y + 2z &= 1 \\ 2x + y + 20z &= 3 \\ 4x - 3y &= 2 \end{aligned}$

(f) $\begin{aligned} 2x - y + 2z &= 1 \\ x - 2y + z &= 2 \\ 3x - 2y + 2z &= 1 \end{aligned}$

2. Find the row echelon form of each of the following matrices:

(a) $\begin{bmatrix} 5 & 2 & 1 & 5 \\ 7 & 2 & 2 & 9 \\ 8 & 4 & 1 & 6 \end{bmatrix}$

(b) $\begin{bmatrix} 1 & 2 & -1 & -2 & -1 \\ 2 & 1 & 1 & -1 & 4 \\ 1 & -1 & 2 & 1 & 5 \\ 1 & 3 & -2 & -3 & -3 \end{bmatrix}$

(c) $\begin{bmatrix} 2 & 3 & 2 & 1 & 0 & 0 \\ 3 & 1 & -2 & 0 & 1 & 0 \\ -1 & 0 & 1 & 0 & 0 & 1 \end{bmatrix}$

(d) $\begin{bmatrix} 2 & 1 & 2 & -1 & 7 \\ 1 & 2 & 1 & 1 & 8 \\ 0 & 2 & 0 & 1 & 5 \\ 1 & 1 & 1 & -1 & 4 \end{bmatrix}$

In Exercises 3 through 16, find a complete solution of each linear system which is consistent.

3. $\begin{aligned} x + 2y + 4z - w &= 3 \\ 3x + 4y + 5z - w &= 7 \\ x + 3y + 4z + 5w &= 4 \end{aligned}$

4. $\begin{aligned} x + 2y + 4z + w &= 3 \\ 2x - y + z + 3w &= 7 \\ -4x + 7y + 5z - 7w &= 4 \end{aligned}$

5. $\begin{aligned} x + 2y + 4z + w &= 0 \\ 2x - y + z + 3w &= 0 \\ -4x + 7y + 5z - 7w &= 0 \end{aligned}$

6. $\begin{aligned} x + 4y - z + 2w &= 0 \\ 3x + 5y - 2z - w &= 0 \\ x + 5y + 2z + w &= 0 \end{aligned}$

7. $\begin{aligned} 2x + 5y + 2z - 3w &= 3 \\ 3x + 6y + 5z + 2w &= 2 \\ 4x + 5y + 14z + 14w &= 11 \\ 5x + 10y + 8z + 4w &= 4 \end{aligned}$

8. $\begin{aligned} x - y + z - 2w &= -1 \\ -2x + 2y - z + 2w &= 3 \\ 3x - 3y + 2z - 4w &= -4 \\ -4x + 4y - 3z + 6w &= 5 \end{aligned}$

9. $\begin{aligned} 8x - 4y + 12z + 5w &= 18 \\ 7x - 6y + 13z + 5w &= 17 \\ 5x - 6y + 11z + 4w &= 13 \\ 4x - 4y + 8z + 3w &= 10 \\ 3x - 2y + 5z + 2w &= 7 \end{aligned}$

10. $\begin{aligned} 2x + y - z + w &= 5 \\ x + 2y + w &= 3 \\ 2x + 2y - z + 3w &= 9 \\ x + y - z + w &= 4 \\ 2x + 3y + 3w &= 8 \end{aligned}$

11. $\begin{aligned} 2x - 2y + 4z + 3w &= 9 \\ x - y + 2z + 2w &= 6 \\ 2x - 2y + z + 2w &= 3 \\ x - y + w &= 2 \end{aligned}$

12. $\begin{aligned} -x + 3y - 2z \phantom{{}- 3w} &= 1 \\ 2x - 3y + 4z - 3w &= 4 \\ 3x - 4y - z + 5w &= 6 \\ 4x - 2y - 3z + 5w &= 13 \end{aligned}$

13. $\begin{bmatrix} 1 & -1 & 2 & 3 & 0 \\ 0 & 0 & 1 & 0 & -2 \\ 0 & 0 & 0 & 1 & -4 \end{bmatrix} \begin{bmatrix} x_1 \\ x_2 \\ x_3 \\ x_4 \\ x_5 \end{bmatrix} = \begin{bmatrix} 14 \\ 2 \\ 3 \end{bmatrix}$

14. $\begin{bmatrix} 6 & 0 & 3 & 0 & -5 \\ 0 & 5 & -4 & -3 & -2 \\ 0 & 0 & -4 & 1 & 1 \\ 0 & 0 & 0 & 3 & -6 \\ 0 & 0 & 0 & 0 & -2 \end{bmatrix} \begin{bmatrix} x_1 \\ x_2 \\ x_3 \\ x_4 \\ x_5 \end{bmatrix} = \begin{bmatrix} -4 \\ 4 \\ 2 \\ 3 \\ 2 \end{bmatrix}$

15. $\begin{aligned} x_1 + 2x_2 + 4x_3 - x_4 + 2x_5 &= 3 \\ 3x_1 + 4x_2 + 5x_3 - x_4 - 2x_5 &= 7 \\ x_1 + 3x_2 + 4x_3 + 5x_4 - x_5 &= 4 \end{aligned}$

16. $\begin{aligned} x_1 + x_2 + x_3 - x_4 - x_5 &= 2 \\ x_1 + 2x_2 + 4x_3 - x_4 + 5x_5 &= 3 \\ 3x_1 + 4x_2 + 5x_3 - x_4 - 2x_5 &= 7 \\ x_1 + 3x_2 + 4x_3 + 5x_4 - x_5 &= 4 \\ 2x_1 + 5x_2 + 8x_3 + 4x_4 + x_5 &= 7 \\ x_1 - x_2 - 2x_3 - 7x_4 - x_5 &= 0 \end{aligned}$

17. Find all values of k for which the given linear system is consistent:

(a) $\begin{aligned} x - 3y + 2z &= 1 \\ 2x - 2y &= k^2 \\ 3x - 5y + z &= 0 \\ -2x + 8y + 4z &= 49 \end{aligned}$

(b) $\begin{aligned} 3x - 6y + 2z - 5w &= 3 - k \\ 2x - y + 2z &= 1 - k \\ 2x - y + z - 2w &= 1 \\ x - 2y + z - 2w &= 1 \\ x + 2y - 2z + w &= -2 \end{aligned}$

(c) $\begin{bmatrix} 1 & 2 & 3 & 1 \\ 3 & 2 & 1 & 4 \\ 2 & 6 & 10 & 3 \\ 1 & 1 & 1 & 1 \end{bmatrix} \begin{bmatrix} x \\ y \\ z \\ w \end{bmatrix} = \begin{bmatrix} 3 \\ 7 \\ 7 \\ k \end{bmatrix}$

18. Show that the number of solutions a system of linear equations can have is either *none, one,* or else *infinite.*

19. Show that a homogeneous linear system with more equations than unknowns can have a nontrivial solution.

20. Show that if $\mathbf{x}_1, \mathbf{x}_2, \ldots, \mathbf{x}_n$ are n solutions of the homogeneous linear system $\mathrm{A}\mathbf{x} = \mathbf{0}$, then for all values of the constants c_k the vector sum $c_1\mathbf{x}_1 + c_2\mathbf{x}_2 + \cdots + c_n\mathbf{x}_n = \Sigma_{k=1}^{n} c_k\mathbf{x}_k$ is also a solution of $\mathrm{A}\mathbf{x} = \mathbf{0}$.

21. Prove Theorem 5. *Hint:* Compare this theorem with the statement and proof of Theorem 1, Sec. 2.3.

22. Discuss the solution of a nonhomogeneous linear system with more unknowns than equations.

23. Show that although the gaussian reduction of a linear system may lead to different final forms, the solution obtained in each (solvable) case is a linear combination of the same (or proportional) vectors.

24. Show that a homogeneous linear system with more unknowns than equations always has nontrivial solutions.

In Exercises 25–27, determine whether the given vectors are linearly dependent or independent. If they are dependent, express each vector as a linear combination of the other vectors in the set.

25. $\mathbf{v}_1 = \begin{bmatrix} 1 \\ 2 \\ 3 \end{bmatrix}, \mathbf{v}_2 = \begin{bmatrix} 3 \\ -1 \\ 2 \end{bmatrix}, \mathbf{v}_3 = \begin{bmatrix} 1 \\ -1 \\ 0 \end{bmatrix}$

26. $\mathbf{v}_1 = \begin{bmatrix} 1 \\ 2 \\ 3 \end{bmatrix}, \mathbf{v}_2 = \begin{bmatrix} 2 \\ -4 \\ 1 \end{bmatrix}, \mathbf{v}_3 = \begin{bmatrix} 1 \\ 1 \\ 1 \end{bmatrix}$

27. $\mathbf{v}_1 = [3\ 1\ 3\ {-2}\ 1\ 2]$, $\mathbf{v}_2 = [-2\ 4\ 1\ 1\ 2\ {-2}]$, $\mathbf{v}_3 = [8\ {-2}\ 5\ {-5}\ 0\ 6]$

28. Prove that if the vectors $\mathbf{v}_1, \mathbf{v}_2, \ldots, \mathbf{v}_n$ are mutually orthogonal, they are linearly independent. *Hint:* Consider the equation $c_1\mathbf{v}_1 + c_2\mathbf{v}_2 + \cdots + c_n\mathbf{v}_n = \mathbf{0}$ and form the scalar product of each side with $\mathbf{v}_k$ for $k = 1, 2, \ldots, n$. Is the converse of this statement true?

29. What is meant by the statement that k $m \times n$ matrices are linearly dependent? When are k $m \times n$ matrices linearly independent?

30. Show that the matrices $\mathrm{M}_1 = \begin{bmatrix} 2 & 4 \\ 3 & 1 \end{bmatrix}$, $\mathrm{M}_2 = \begin{bmatrix} 1 & 2 \\ 4 & 3 \end{bmatrix}$, and $\mathrm{M}_3 = \begin{bmatrix} 1 & 2 \\ -6 & -7 \end{bmatrix}$ are linearly dependent.

31. Determine whether the matrices $\mathrm{M}_1 = \begin{bmatrix} 1 & 2 \\ 3 & 4 \end{bmatrix}$, $\mathrm{M}_2 = \begin{bmatrix} 4 & 3 \\ 2 & 1 \end{bmatrix}$, $\mathrm{M}_3 = \begin{bmatrix} 3 & 2 \\ 1 & 4 \end{bmatrix}$ are linearly dependent or independent.

32. For what values of a, b, c, d are the matrices $\begin{bmatrix} 1 & 2 \\ 3 & 4 \end{bmatrix}$, $\begin{bmatrix} 2 & 3 \\ 1 & 4 \end{bmatrix}$, and $\begin{bmatrix} a & b \\ c & d \end{bmatrix}$ linearly dependent?

33. For what value of a, b, c, d are the matrices $\begin{bmatrix} 1 & 2 \\ 3 & 4 \end{bmatrix}$, $\begin{bmatrix} 2 & 1 \\ 3 & 4 \end{bmatrix}$, $\begin{bmatrix} 3 & 1 \\ 4 & 2 \end{bmatrix}$, and $\begin{bmatrix} a & b \\ c & d \end{bmatrix}$ linearly dependent?

34. Show that five or more 2×2 matrices are always linearly dependent.

35. What is the smallest value of k for which k $m \times n$ matrices are always linearly dependent?

36. Show that if the row echelon reduction of a square matrix A of order n has no zero row, then that reduction is the identity matrix I_n.

37. If A and B are conformable matrices, show that

$$AB = [Ab_1 \quad Ab_2 \quad \cdots \quad Ab_n]$$

where $\mathbf{b}_j$ is the jth column vector of B.

38. Justify the following restatement of Theorem 2: if $\mathbf{v}_1, \mathbf{v}_2, \ldots, \mathbf{v}_n$ are n vectors of dimension m, where $m \geq n$, and if the matrix $[\mathbf{v}_1 \quad \mathbf{v}_2 \quad \cdots \quad \mathbf{v}_n]$ contains an $n \times n$ submatrix which can be obtained from I_n by using only elementary row operations of types **1** and **2**, then $\mathbf{v}_1, \mathbf{v}_2, \ldots, \mathbf{v}_n$ are linearly independent.

3.7 SPECIAL LINEAR SYSTEMS, INVERSES, ADJOINTS, AND CRAMER'S RULE

All square matrices fall into one or the other of two mutually exclusive sets, according as their determinants are zero or nonzero. A square matrix whose determinant is zero is said to be **singular.** A square matrix whose determinant is not zero is said to be **nonsingular.** Properties 5, 7, and 9, Sec. 3.5, clearly imply that under elementary row operations, singular matrices are transformed into singular matrices and nonsingular matrices are transformed into nonsingular matrices. In particular, since every gaussian reduction of a square matrix A yields an upper triangular matrix M whose determinant is the product of its diagonal elements, it follows that if A, and hence M, is nonsingular, then none of the diagonal elements of M can be zero. Therefore, by dividing each row of M by the diagonal element in that row, every diagonal element can be made 1. Subsequently, using only row operations of Type **3**, every element above the diagonal in M can be reduced to zero, yielding a final form which is just I. Thus, summarizing, *every nonsingular* (square) *matrix of order n can be reduced to I_n by row operations only,* or *the row echelon form of every nonsingular* $(n \times n)$ *matrix is I_n.*

For solutions of a system of n linear equations in n unknowns

$$(1) \qquad a_{i1}x_1 + a_{i2}x_2 + \cdots + a_{in}x_n = b_i \qquad 1 \leq i \leq n$$

whose matrix form

$$(2) \qquad A\mathbf{x} = \mathbf{b}$$

contains a coefficient matrix $A = [a_{ij}]$ of order n, we have the following existence and uniqueness theorem.

THEOREM 1 If A is a nonsingular matrix, the linear system $A\mathbf{x} = \mathbf{b}$ has exactly one solution.

◖ **PROOF** If A is nonsingular, so that $\det A \neq 0$, elementary row operations can be used to reduce the augmented matrix $[A\mathbf{b}]$ of $A\mathbf{x} = \mathbf{b}$ to a matrix of the form $[I\mathbf{c}]$ (see Exercise 36, Sec. 3.6). Since the linear system $A\mathbf{x} = \mathbf{b}$ and $I\mathbf{x} = \mathbf{x} = \mathbf{c}$ are equivalent, $A\mathbf{x} = \mathbf{b}$ has $\mathbf{c}$ as a solution. Suppose that a column vector $\mathbf{d}$ is also a solution of $A\mathbf{x} = \mathbf{b}$. Then $A\mathbf{c} = \mathbf{b}$ and $A\mathbf{d} = \mathbf{b}$ and, by subtraction, $A(\mathbf{c} - \mathbf{d}) = \mathbf{0}$. Hence $\mathbf{c} - \mathbf{d}$ must be a solution of the homogeneous linear system $A\mathbf{x} = \mathbf{0}$, which is clearly equivalent to $I\mathbf{x} \equiv \mathbf{x} = \mathbf{0}$. This shows that $A\mathbf{x} = \mathbf{0}$ has only a trivial solution, and it follows that $\mathbf{c} = \mathbf{d}$. We have thus proved that $A\mathbf{x} = \mathbf{b}$ has exactly one solution whether $\mathbf{b}$ is a zero vector or not ◗

COROLLARY 1 If A is a nonsingular matrix, the homogeneous linear system $A\mathbf{x} = \mathbf{0}$ has only a trivial solution.

EXAMPLE 1

Given the matrix

$$A = \begin{bmatrix} 1 & 1 & 1 & 1 \\ 1 & 0 & -1 & 0 \\ 0 & 1 & 1 & -1 \\ 2 & 0 & -1 & -3 \end{bmatrix}$$

and the column vector

$$\mathbf{b} = \begin{bmatrix} 5 \\ 10 \\ -5 \\ -10 \end{bmatrix}$$

solve the linear systems $A\mathbf{x} = \mathbf{b}$ and $A\mathbf{x} = \mathbf{0}$.

Using elementary row operations, we find the row echelon form of the augmented matrix $[A\mathbf{b}]$ of the nonhomogeneous system $A\mathbf{x} = \mathbf{b}$ to be

$$[I\mathbf{c}] = \begin{bmatrix} 1 & 0 & 0 & 0 & -2 \\ 0 & 1 & 0 & 0 & 13 \\ 0 & 0 & 1 & 0 & -12 \\ 0 & 0 & 0 & 1 & 6 \end{bmatrix}$$

where A has been reduced to I and **b** has been reduced to the final column vector

$$\mathbf{c} = \begin{bmatrix} -2 \\ 13 \\ -12 \\ 6 \end{bmatrix}$$

Obviously, A is nonsingular, **c** is the unique solution of $A\mathbf{x} = \mathbf{b}$, and the homogeneous system $A\mathbf{x} = \mathbf{0}$ has only the trivial solution $\mathbf{x} = \mathbf{0}_4$.

We now know that every homogeneous linear system has a trivial solution, that a homogeneous system with a nonsingular coefficient matrix has *only* a trivial solution, and that every homogeneous system with more unknowns than equations has infinitely many solutions (Exercise 24, Sec. 3.6). Many exceedingly important practical applications make use of the next theorem which in essence combines Corollary 1 of Theorem 1 and its converse.

THEOREM 2 A homogeneous system of n linear equations in n unknowns $A\mathbf{x} = \mathbf{0}$ has a nontrivial solution if and only if $\det A = 0$.

PROOF Corollary 1, Theorem 1, is logically (i.e., contrapositively) equivalent to the statement that if $A\mathbf{x} = \mathbf{0}$ has a nontrivial solution, then $|A| = 0$. This comprises the *only if* part of the theorem. It remains to prove that if $|A| = 0$, then $A\mathbf{x} = \mathbf{0}$ has a nontrivial solution.

To do this, we observe that by means of elementary row operations, A can be reduced to an upper triangular matrix

$$B = \begin{bmatrix} b_{11} & b_{12} & \cdots & b_{1n} \\ 0 & b_{22} & \cdots & b_{2n} \\ \cdots & \cdots & \cdots & \cdots \\ 0 & 0 & \cdots & b_{nn} \end{bmatrix}$$

in which each b_{ii} is either 0 or 1. Property 3, Sec. 3.5, implies that $\det B = b_{11}b_{22}\cdots b_{nn}$, and Properties 5, 7, and 9 imply that det B is a nonzero number times det A. But we are given that $\det A = 0$; therefore at least one diagonal element of B is zero. Let b_{ii} be the *zero* diagonal element of B with the largest row index. If $i = n$, the nth row of B is identically zero. If $i < n$, it follows that $b_{ij} = 0$ for $j = 1, 2, \ldots, i$, and $b_{jj} = 1$ for $j = i + 1, i + 2, \ldots, n$. This being the case, the last $n - i$ columns can be reduced to the last $n - i$ columns of I_n by elementary row operations of Type **3**. Thus we may suppose that at the outset some row of B was made to be $\mathbf{0}_n$.

The equation of the linear system $\mathrm{B}\mathbf{x} = \mathbf{0}$ which corresponds to any particular zero row of the augmented matrix $[\mathrm{B} \,|\, \mathbf{0}]$ of this system has the form $0 = 0$, hence has no influence on the solutions of the remaining equations. Since the remaining equations constitute a homogeneous linear system with fewer equations than unknowns, it has nontrivial solutions, as does the equivalent system $\mathrm{A}\mathbf{x} = \mathbf{0}$. ◗

An *upper triangular* matrix such that

1. Every diagonal element is either 0 or 1,
2. All elements in any row containing a zero diagonal element are zero,
3. All other elements in any column containing a unit diagonal element are zero,

is said to be in **Hermite canonical form.**

EXAMPLE 2

Determine all solutions of the homogeneous linear system

$$\begin{aligned} x_1 - 2x_2 + 3x_3 + \ \ 9x_4 &= 0 \\ 2x_1 - 4x_2 + 3x_3 + 12x_4 &= 0 \\ 3x_1 - 6x_2 + 7x_3 + 23x_4 &= 0 \\ 4x_1 - 8x_2 + 9x_3 + 30x_4 &= 0 \end{aligned}$$

Using elementary row operations, we transform the coefficient matrix of the system into its Hermite canonical form

$$\mathrm{B} = \begin{bmatrix} 1 & -2 & 0 & 3 \\ 0 & 0 & 0 & 0 \\ 0 & 0 & 1 & 2 \\ 0 & 0 & 0 & 0 \end{bmatrix}$$

Since $\det \mathrm{B} = 0$, the linear system has infinitely many solutions. To find them, we set $x_4 = c_1$ and $x_2 = c_2$ in the equivalent linear system

$$\begin{aligned} x_1 - 2x_2 \quad + 3x_4 &= 0 \\ x_3 + 2x_4 &= 0 \end{aligned}$$

and solve for x_1 and x_3 in terms of the arbitrary constants c_1 and c_2. This gives

$$(x_1, x_2, x_3, x_4) = c_1(-3, 0, -2, 1) + c_2(2, 1, 0, 0)$$

as a complete solution of the given linear system.

If the coefficient matrix A of a homogeneous linear system of n equations in n unknowns $\mathrm{A}\mathbf{x} = \mathbf{0}$ is *singular,* a complete solution of the system can sometimes be found by simply evaluating cofactors of an appropriate row of elements in det A in accordance with the following theorem.

THEOREM 3 If A is a singular matrix of order n such that at least one entry in the kth row of det A has a nonzero cofactor, then a complete solution of the linear system $\mathrm{A}\mathbf{x} = \mathbf{0}$ is given by the column vector

$$\mathbf{x} = c[\mathrm{A}_{k1} \quad \mathrm{A}_{k2} \quad \cdots \quad \mathrm{A}_{kn}]^T \qquad c \neq 0 \tag{3}$$

where A_{kj} is the cofactor of the element a_{kj} in det A; i.e., the unknowns are proportional to the cofactors of their coefficients in the kth row of $|\mathrm{A}|$.

In the next two examples we make use of this theorem. It is proven later on (see Theorem 11, Sec. 14.3).

EXAMPLE 3

Find a complete solution of the linear system

$$\begin{aligned} x_1 + 2x_2 - x_3 &= 0 \\ 3x_1 + 2x_2 + 5x_3 &= 0 \\ 5x_1 + 6x_2 + 3x_3 &= 0 \end{aligned}$$

It is readily verified that det A = 0 and that the cofactor of the element $a_{33} = 3$ in det A, namely,

$$\begin{vmatrix} 1 & 2 \\ 3 & 2 \end{vmatrix}$$

is different from zero. Hence, by Theorem 3,

$$x_1 = c\begin{vmatrix} 2 & -1 \\ 2 & 5 \end{vmatrix} = 12c \qquad x_2 = -c\begin{vmatrix} 1 & -1 \\ 3 & 5 \end{vmatrix} = -8c \qquad x_3 = c\begin{vmatrix} 1 & 2 \\ 3 & 2 \end{vmatrix} = -4c$$

Replacing the parameter c by a new parameter $C = -4c$, we obtain

$$\mathbf{x} = C[-3 \quad 2 \quad 1]^T$$

as a complete solution of the given linear system.

Our next example, which illustrates the use of Theorems 2 and 3, is not quite a routine application since its objective is not to solve a particular set of linear algebraic equations but rather to determine for what values of certain parameters the system has a solution.

EXAMPLE 4

MIXING FERTILIZERS TO ORDER

Commercial fertilizers are usually mixtures containing specified amounts of three components, potassium (K), phosphorus (P), and nitrogen (N). The A-1 Garden Supply Company carries three standard blends, F_1, F_2, and F_3, with the compositions indicated in the following table (matrix):

	K	P	N
F_1	10%	10%	30%
F_2	30%	10%	20%
F_3	20%	20%	30%

Occasionally, a customer with some special need will request a fertilizer whose composition is different from that of F_1, F_2, or F_3, and the company will try to prepare it by mixing suitable amounts of F_1, F_2, and F_3. What mixtures can they produce in this way, and what amounts of each fertilizer must be blended to produce a mixture with the required proportions?

Let the desired percentages of K, P, and N be k, p, and n, respectively. At the outset it is clear that in the final mixture the percentage of each element must be between the maximum and minimum values in F_1, F_2, and F_3. Hence we have the obvious restrictions

$$\begin{aligned} 10 &\le k \le 30 \\ 10 &\le p \le 20 \\ 20 &\le n \le 30 \end{aligned} \tag{4}$$

If x_1 pounds of F_1, x_2 pounds of F_2, and x_3 pounds of F_3 are mixed, the resulting mixture will contain

$$\begin{aligned} &0.10x_1 + 0.30x_2 + 0.20x_3 \text{ pounds of K} \\ &0.10x_1 + 0.10x_2 + 0.20x_3 \text{ pounds of P} \\ &0.30x_1 + 0.20x_2 + 0.30x_3 \text{ pounds of N} \end{aligned}$$

The total amount of the new mixture is $x_1 + x_2 + x_3$, and of this $k\%$ is to be K, $p\%$ is to be P, and $n\%$ is to be N. Hence we have the three conditions

$$0.10x_1 + 0.30x_2 + 0.20x_3 = \frac{k}{100}(x_1 + x_2 + x_3)$$

$$0.10x_1 + 0.10x_2 + 0.20x_3 = \frac{p}{100}(x_1 + x_2 + x_3)$$

$$0.30x_1 + 0.20x_2 + 0.30x_3 = \frac{n}{100}(x_1 + x_2 + x_3)$$

or, multiplying through by 100 and collecting terms,

$$\begin{aligned} (10 - k)x_1 + (30 - k)x_2 + (20 - k)x_3 &= 0 \\ (10 - p)x_1 + (10 - p)x_2 + (20 - p)x_3 &= 0 \\ (30 - n)x_1 + (20 - n)x_2 + (30 - n)x_3 &= 0 \end{aligned} \tag{5}$$

This is a system of three homogeneous linear equations from which we seek solutions other than the trivial solution $x_1 = x_2 = x_3 = 0$. Hence, according to Theorem 2, the determinant of the coefficients in (5) must be zero; that is, we must have

$$\begin{vmatrix} 10 - k & 30 - k & 20 - k \\ 10 - p & 10 - p & 20 - p \\ 30 - n & 20 - n & 30 - n \end{vmatrix} = 0$$

To simplify this equation, we use Property 9, Sec. 3.5, and subtract the first column from the third column and from the second column, getting

$$\begin{vmatrix} 10 - k & 20 & 10 \\ 10 - p & 0 & 10 \\ 30 - n & -10 & 0 \end{vmatrix} = 0$$

Expanding this in terms of the elements in the first column, we have

$$100(10 - k) - 100(10 - p) + 200(30 - n) = 0$$

or

$$k - p + 2n = 60 \tag{6}$$

Thus only mixtures whose percentages satisfy (6), and of course (4), can be produced by mixing F_1, F_2, and F_3.

The amounts of F_1, F_2, and F_3 required to produce a mixture with realizable percentages of K, P, and N can be found most conveniently by using Theorem 3, applied to the third row of the determinant of the matrix of the system (5). This gives us

$$x_1 = \begin{vmatrix} 30 - k & 20 - k \\ 10 - p & 20 - p \end{vmatrix} = 400 - 10k - 10p$$

$$x_2 = -\begin{vmatrix} 10 - k & 20 - k \\ 10 - p & 20 - p \end{vmatrix} = -10k + 10p$$

$$x_3 = \begin{vmatrix} 10 - k & 30 - k \\ 10 - p & 10 - p \end{vmatrix} = -200 + 20p$$

Since the x's are determined only to within an arbitrary nonzero proportionality constant, we have for the final values

$$x_1 = \lambda(40 - k - p) \qquad x_2 = \lambda(p - k) \qquad x_3 = \lambda(2p - 20)$$

where λ is to be determined by the actual amount of the new mixture to be produced.

In most cases, the use of elementary row operations is by far the most efficient way of solving systems of linear algebraic equations although, as the last example illustrates, Theorem 3 is often useful when the coefficient matrix is singular. There are also other special methods which are easy to apply when the coefficient matrix of the system is nonsingular. One of these methods makes use of an idea much like that of a reciprocal for numbers.

It is a fact of elementary algebra that any nonzero quantity Q has a reciprocal $Q^{-1} = 1/Q$ with the property that

$$Q^{-1}Q = QQ^{-1} = 1$$

As is well known, the familiar operation of division is essentially nothing but multiplication involving the reciprocal, or multiplicative inverse, of the divisor as one factor. In matrix algebra, although we do not define division as such, we can in an important class of cases define a reciprocal, or inverse, of a matrix. With inverses defined, multiplication then serves to accomplish all that we might properly expect to do by division.

We shall define an inverse only for a matrix A which is square. What we want of course is a matrix A^{-1} such that

$$A^{-1}A = AA^{-1} = I$$

Now if a matrix B can be found such that BA = I, it can be shown that AB = I and vice versa, and that B is unique (see Exercise 24). Thus if B exists, it does what we want A^{-1} to do. Accordingly, we define the inverse of a matrix as follows.

DEFINITION 1 A matrix A^{-1} is the **inverse** of a matrix A if and only if $A^{-1}A = AA^{-1} = I$.

If $A^{-1}A = I$, then A^{-1} and A clearly have the same order. Moreover, by Property 12, Sec. 3.5, $\det(A^{-1}A) = \det A^{-1} \det A = \det I = 1$. Hence both $\det A^{-1}$ and $\det A$ must be different from zero; that is, if a matrix has an inverse, both the matrix and its inverse must be nonsingular. Thus the *only if* part of the following theorem is certainly true.

THEOREM 4 The inverse A^{-1} of a matrix A exists if and only if A is nonsingular.

To establish the *if* part of the theorem, and therefore the theorem itself, we must still prove that every nonsingular matrix has an inverse.

PROOF Let A be a nonsingular matrix of order n and let $\mathbf{e}_j$ be the jth column of the identity matrix I_n. Then by Theorem 1 the equation

$$A\mathbf{x} = \mathbf{e}_j \qquad j = 1, 2, \ldots, n$$

has a unique solution, say $\mathbf{b}_j$. Since A is the same for each value of j, we can solve these n equations at the same time simply by working with the $n \times 2n$ matrix [AI] and using elementary row operations to reduce it to the form [IB]. Direct computation confirms that $B = A^{-1}$. In fact, we have

$$AB = A[\mathbf{b}_1 \quad \cdots \quad \mathbf{b}_j \quad \cdots \quad \mathbf{b}_n] = [A\mathbf{b}_1 \quad \cdots \quad A\mathbf{b}_j \quad \cdots \quad A\mathbf{b}_n] = [\mathbf{e}_1 \quad \cdots \quad \mathbf{e}_j \quad \cdots \quad \mathbf{e}_n] = I$$

In addition to proving that A has an inverse, we have thus presented a practical procedure for finding it. ◗

Inverses are ideally suited to the problem of solving linear systems with nonsingular coefficient matrices. To see why, consider such a system

$$\mathrm{A}\mathbf{x} = \mathbf{b}$$

and multiply through by A^{-1}. This gives

$$\mathrm{A}^{-1}\mathrm{A}\mathbf{x} = \mathrm{A}^{-1}\mathbf{b} \qquad \mathrm{I}\mathbf{x} = \mathrm{A}^{-1}\mathbf{b} \qquad \mathbf{x} = \mathrm{A}^{-1}\mathbf{b}$$

and, except for the determination of A^{-1}, the system is solved.

EXAMPLE 5

Find the inverse of the coefficient matrix of the linear system

$$\begin{aligned} 2x_1 - x_2 + 3x_3 &= 3 \\ -x_1 + 2x_2 &= 1 \\ 3x_1 - 5x_2 + 2x_3 &= 1 \end{aligned}$$

and then use the inverse to solve the system.

The coefficient matrix of the system is

$$\mathrm{A} = \begin{bmatrix} 2 & -1 & 3 \\ -1 & 2 & 0 \\ 3 & -5 & 2 \end{bmatrix}$$

and det A = 3. Therefore A is nonsingular and A^{-1} exists. To find A^{-1}, we form the matrix [AI], which is the first of the following 3 × 6 matrices, and commence performing elementary row operations as indicated to transform each matrix into the succeeding one.

$$\begin{bmatrix} 2 & -1 & 3 & 1 & 0 & 0 \\ -1 & 2 & 0 & 0 & 1 & 0 \\ 3 & -5 & 2 & 0 & 0 & 1 \end{bmatrix}$$

Add 2 times Row 2 to Row 1.
Add 3 times Row 2 to Row 3.
Multiply Row 2 by −1.
Interchange the new rows 1 and 2.

$$\begin{bmatrix} 1 & -2 & 0 & 0 & -1 & 0 \\ 0 & 3 & 3 & 1 & 2 & 0 \\ 0 & 1 & 2 & 0 & 3 & 1 \end{bmatrix}$$

Add −3 times Row 3 to Row 2.
Add $\frac{2}{3}$ times the new Row 2 to Row 3.
Interchange the new rows 2 and 3.

$$\begin{bmatrix} 1 & -2 & 0 & 0 & -1 & 0 \\ 0 & 1 & 0 & \frac{2}{3} & -\frac{5}{3} & -1 \\ 0 & 0 & -3 & 1 & -7 & -3 \end{bmatrix}$$

Add 2 times Row 2 to Row 1.
Multiply Row 3 by $-\frac{1}{3}$.

$$\begin{bmatrix} 1 & 0 & 0 & \frac{4}{3} & -\frac{13}{3} & -2 \\ 0 & 1 & 0 & \frac{2}{3} & -\frac{5}{3} & -1 \\ 0 & 0 & 1 & -\frac{1}{3} & \frac{7}{3} & 1 \end{bmatrix}$$

From the final matrix we have

$$\mathrm{A}^{-1} = \begin{bmatrix} \frac{4}{3} & -\frac{13}{3} & -2 \\ \frac{2}{3} & -\frac{5}{3} & -1 \\ -\frac{1}{3} & \frac{7}{3} & 1 \end{bmatrix} = \frac{1}{3}\begin{bmatrix} 4 & -13 & -6 \\ 2 & -5 & -3 \\ -1 & 7 & 3 \end{bmatrix}$$

and it is readily verified by direct multiplication that $A^{-1}A = =AA^{-1} = I$. With $\mathbf{b} = \begin{bmatrix} 3 \\ 1 \\ 1 \end{bmatrix}$ we compute $A^{-1}\mathbf{b}$ to obtain the unique solution

$$\begin{bmatrix} x_1 \\ x_2 \\ x_3 \end{bmatrix} = \frac{1}{3}\begin{bmatrix} 4 & -13 & -6 \\ 2 & -5 & -3 \\ -1 & 7 & 3 \end{bmatrix}\begin{bmatrix} 3 \\ 1 \\ 1 \end{bmatrix} = \frac{1}{3}\begin{bmatrix} -7 \\ -2 \\ 7 \end{bmatrix} = \begin{bmatrix} -\frac{7}{3} \\ -\frac{2}{3} \\ \frac{7}{3} \end{bmatrix}$$

of the given system.

According to Properties 1 and 10, Sec. 3.5, if A_{ij} is the cofactor of a_{ij} in the determinant of a matrix $A = [a_{ij}]$ of order $n \geq 2$, then

$$(7) \qquad \begin{bmatrix} a_{11} & a_{12} & \cdots & a_{1n} \\ a_{21} & a_{22} & \cdots & a_{2n} \\ \cdots & \cdots & \cdots & \cdots \\ a_{n1} & a_{n2} & \cdots & a_{nn} \end{bmatrix}\begin{bmatrix} A_{11} & A_{21} & \cdots & A_{n1} \\ A_{12} & A_{22} & \cdots & A_{n2} \\ \cdots & \cdots & \cdots & \cdots \\ A_{1n} & A_{2n} & \cdots & A_{nn} \end{bmatrix} = \begin{bmatrix} |A| & 0 & \cdots & 0 \\ 0 & |A| & \cdots & 0 \\ \cdots & \cdots & \cdots & \cdots \\ 0 & 0 & \cdots & |A| \end{bmatrix} = |A|I$$

Let $C = [A_{ij}]$ be the matrix whose entries are the respective cofactors of the elements in det A. The matrix C^T, i.e., the transpose of C, is called the **adjoint of** A and is denoted by Adj A. In this notation, (7) becomes

$$(8) \qquad A \text{ Adj } A = |A|I$$

This relation holds for any square matrix A, singular or not.

In particular, if A is nonsingular, so that $|A| \neq 0$, then

$$(9) \qquad A\frac{\text{Adj } A}{|A|} = I$$

Hence A^{-1} is determined by

$$(10) \qquad A^{-1} = \frac{\text{Adj } A}{|A|}$$

This relation provides us with an alternative method for finding the inverse of a nonsingular matrix.

EXAMPLE 6

If

$$A = \begin{bmatrix} 1 & 2 & 4 \\ -1 & 0 & 3 \\ 3 & 1 & -2 \end{bmatrix}$$

then the determinant of A is

$$\begin{vmatrix} 1 & 2 & 4 \\ -1 & 0 & 3 \\ 3 & 1 & -2 \end{vmatrix} = 7$$

The adjoint of A is the transpose of the cofactor matrix

$$\begin{bmatrix} \begin{vmatrix} 0 & 3 \\ 1 & -2 \end{vmatrix} & -\begin{vmatrix} -1 & 3 \\ 3 & -2 \end{vmatrix} & \begin{vmatrix} -1 & 0 \\ 3 & 1 \end{vmatrix} \\ -\begin{vmatrix} 2 & 4 \\ 1 & -2 \end{vmatrix} & \begin{vmatrix} 1 & 4 \\ 3 & -2 \end{vmatrix} & -\begin{vmatrix} 1 & 2 \\ 3 & 1 \end{vmatrix} \\ \begin{vmatrix} 2 & 4 \\ 0 & 3 \end{vmatrix} & -\begin{vmatrix} 1 & 4 \\ -1 & 3 \end{vmatrix} & \begin{vmatrix} 1 & 2 \\ -1 & 0 \end{vmatrix} \end{bmatrix} \quad \text{that is} \quad \begin{bmatrix} -3 & 8 & 6 \\ 7 & -14 & -7 \\ -1 & 5 & 2 \end{bmatrix}$$

The inverse of A is therefore

$$A^{-1} = \frac{1}{7}\begin{bmatrix} -3 & 8 & 6 \\ 7 & -14 & -7 \\ -1 & 5 & 2 \end{bmatrix}$$

and

$$A^{-1}A = \frac{1}{7}\begin{bmatrix} -3 & 8 & 6 \\ 7 & -14 & -7 \\ -1 & 5 & 2 \end{bmatrix}\begin{bmatrix} 1 & 2 & 4 \\ -1 & 0 & 3 \\ 3 & 1 & -2 \end{bmatrix} = \frac{1}{7}\begin{bmatrix} 7 & 0 & 0 \\ 0 & 7 & 0 \\ 0 & 0 & 7 \end{bmatrix} = \begin{bmatrix} 1 & 0 & 0 \\ 0 & 1 & 0 \\ 0 & 0 & 1 \end{bmatrix}$$

We have already observed that a linear system $A\mathbf{x} = \mathbf{b}$, in which A is nonsingular, has as its unique solution

$$\mathbf{x} = A^{-1}\mathbf{b} \tag{11}$$

Substituting for A^{-1} from (10), we can write (11) as

$$\begin{bmatrix} x_1 \\ x_2 \\ \vdots \\ x_n \end{bmatrix} = \frac{1}{|A|}\begin{bmatrix} A_{11} & A_{21} & \cdots & A_{n1} \\ A_{12} & A_{22} & \cdots & A_{n2} \\ \cdots & \cdots & \cdots & \cdots \\ A_{1n} & A_{2n} & \cdots & A_{nn} \end{bmatrix}\begin{bmatrix} b_1 \\ b_2 \\ \vdots \\ b_n \end{bmatrix} = \frac{1}{|A|}\begin{bmatrix} b_1A_{11} + b_2A_{21} + \cdots + b_nA_{n1} \\ b_1A_{12} + b_2A_{22} + \cdots + b_nA_{n2} \\ \cdots \\ b_1A_{1n} + b_2A_{2n} + \cdots + b_nA_{nn} \end{bmatrix}$$

For $1 \le j \le n$, we recognize the sum

$$b_1A_{1j} + b_2A_{2j} + \cdots + b_nA_{nj} = \sum_{k=1}^{n} b_kA_{kj}$$

as Laplace's expansion by the jth column of the determinant of the matrix D_j, which is obtained from A by replacing the jth column of A by the column vector **b**. With this observation, we arrive at the following theorem, which is called **Cramer's rule.**

THEOREM 5 If det $A \neq 0$, a system $A\mathbf{x} = \mathbf{b}$ of n linear equations in n unknowns has a unique solution whose jth component is given by

$$x_j = \frac{|D_j|}{|A|} \qquad 1 \le j \le n \tag{12}$$

where the matrix D_j has **b** as its jth column and the corresponding columns of A as its other columns.

EXAMPLE 7

Use Cramer's rule to solve the linear system

$$\begin{aligned} x_1 + x_2 + x_3 + x_4 &= 3 \\ 2x_1 - 2x_2 - x_3 + 2x_4 &= 0 \\ 3x_1 - x_2 + 2x_3 + 2x_4 &= 2 \\ x_1 - x_2 - 2x_3 + x_4 &= 0 \end{aligned}$$

Since

$$|A| = \begin{vmatrix} 1 & 1 & 1 & 1 \\ 2 & -2 & -1 & 2 \\ 3 & -1 & 2 & 2 \\ 1 & -1 & -2 & 1 \end{vmatrix} = 6$$

the determinant of the coefficient matrix is nonzero and Cramer's rule can be applied. Further computations give

$$|D_1| = \begin{vmatrix} 3 & 1 & 1 & 1 \\ 0 & -2 & -1 & 2 \\ 2 & -1 & 2 & 2 \\ 0 & -1 & -2 & 1 \end{vmatrix} = 3 \qquad |D_2| = \begin{vmatrix} 1 & 3 & 1 & 1 \\ 2 & 0 & -1 & 2 \\ 3 & 2 & 2 & 2 \\ 1 & 0 & -2 & 1 \end{vmatrix} = 9$$

$$|D_3| = \begin{vmatrix} 1 & 1 & 3 & 1 \\ 2 & -2 & 0 & 2 \\ 3 & -1 & 2 & 2 \\ 1 & -1 & 0 & 1 \end{vmatrix} = 0 \qquad |D_4| = \begin{vmatrix} 1 & 1 & 1 & 3 \\ 2 & -2 & -1 & 0 \\ 3 & -1 & 2 & 2 \\ 1 & -1 & -2 & 0 \end{vmatrix} = 6$$

Therefore, by (12), the solution of the system is

$$(x_1, x_2, x_3, x_4) = (\tfrac{1}{2}, \tfrac{3}{2}, 0, 1)$$

Now suppose we are required to solve an equation of the type

$$\text{(13)} \qquad AX = B$$

where A is a matrix of size $m \times n$ and B is a matrix of size $m \times p$. Then we must find all **solutions** of (13), i.e., all $n \times p$ matrices K such that AK = B, or else show that no solutions exist. **General matrix equations** of this sort can be written as

$$A[\mathbf{x}_1 \quad \mathbf{x}_2 \quad \cdots \quad \mathbf{x}_p] = [\mathbf{b}_1 \quad \mathbf{b}_2 \quad \cdots \quad \mathbf{b}_p]$$

and treated as a set of p simultaneous linear systems

$$\text{(14)} \qquad A\mathbf{x}_j = \mathbf{b}_j \qquad 1 \le j \le p$$

Every augmented matrix corresponding to a system of (14) is a submatrix of the m by $n + p$ matrix [AB]. Hence, by applying elementary row operations to this matrix, which reduce A to row echelon form, it is possible to find all solutions of (13) or else prove there are none. In particular, if A is nonsingular, the process culminates in the following obvious theorem.

THEOREM 6 If A is a nonsingular $n \times n$ matrix, if B is an $n \times m$ matrix, and if C is an $m \times n$ matrix, the equation AX = B has the unique solution $X = A^{-1}B$ and the equation XA = C has the unique solution $X = CA^{-1}$.

Perhaps it should be mentioned that an equation of the type XA = C can also be handled by applying the elimination process to $A^TX^T = C^T$.

EXAMPLE 8

Solve the matrix equation AX = B, where

$$A = \begin{bmatrix} 1 & -2 & -1 & 1 \\ -2 & 3 & 2 & -3 \\ -1 & 1 & 1 & -2 \\ 0 & -1 & 0 & -1 \\ 2 & -4 & -2 & 1 \end{bmatrix} \quad \text{and} \quad B = \begin{bmatrix} 2 & -1 & 1 \\ 1 & 0 & 1 \\ 3 & -1 & 2 \\ 5 & -2 & 3 \\ 6 & 0 & 5 \end{bmatrix}$$

We begin the solution process by forming the 5 × 7 matrix

$$[AB] = \begin{bmatrix} 1 & -2 & -1 & 1 & 2 & -1 & 1 \\ -2 & 3 & 2 & -3 & 1 & 0 & 1 \\ -1 & 1 & 1 & -2 & 3 & -1 & 2 \\ 0 & -1 & 0 & -1 & 5 & -2 & 3 \\ 2 & -4 & -2 & 1 & 6 & 0 & 5 \end{bmatrix}$$

which we transform into

$$\begin{bmatrix} 1 & 0 & -1 & 0 & -2 & 9 & 4 \\ 0 & 1 & 0 & 0 & -3 & 4 & 0 \\ 0 & 0 & 0 & 1 & -2 & -2 & -3 \\ 0 & 0 & 0 & 0 & 0 & 0 & 0 \\ 0 & 0 & 0 & 0 & 0 & 0 & 0 \end{bmatrix}$$

by applying elementary row operations which reduce A to row echelon form. Interpreting the first five columns of the derived matrix as the augmented matrix of a linear system, we find that the first column of the 4 × 3 solution matrix

$$X = \begin{bmatrix} x_{11} & x_{12} & x_{13} \\ x_{21} & x_{22} & x_{23} \\ x_{31} & x_{32} & x_{33} \\ x_{41} & x_{42} & x_{43} \end{bmatrix}$$

is determined by

$$\begin{bmatrix} x_{11} \\ x_{21} \\ x_{31} \\ x_{41} \end{bmatrix} = c_1 \begin{bmatrix} 1 \\ 0 \\ 1 \\ 0 \end{bmatrix} + \begin{bmatrix} -2 \\ -3 \\ 0 \\ -2 \end{bmatrix}$$

where c_1 is an arbitrary constant. Determining the remaining two columns of X similarly, we obtain

$$X = c_1 \begin{bmatrix} 1 & 0 & 0 \\ 0 & 0 & 0 \\ 1 & 0 & 0 \\ 0 & 0 & 0 \end{bmatrix} + c_2 \begin{bmatrix} 0 & 1 & 0 \\ 0 & 0 & 0 \\ 0 & 1 & 0 \\ 0 & 0 & 0 \end{bmatrix} + c_3 \begin{bmatrix} 0 & 0 & 1 \\ 0 & 0 & 0 \\ 0 & 0 & 1 \\ 0 & 0 & 0 \end{bmatrix} + \begin{bmatrix} -2 & 9 & 4 \\ -3 & 4 & 0 \\ 0 & 0 & 0 \\ -2 & -2 & -3 \end{bmatrix}$$

as a complete solution of AX = B. The three matrices multiplied by the parameters c_1, c_2, and c_3 satisfy the equation AX = O, and the final matrix is a specific solution of AX = B.

As is evident from the next theorem, an equation of the type ABX = I, in which both A and B are nonsingular matrices of order n, has the solution $X = B^{-1}A^{-1}$.

THEOREM 7 If A and B are nonsingular $n \times n$ matrices, then $(AB)^{-1} = B^{-1}A^{-1}$.

PROOF Since A and B are nonsingular, so is the product AB, and $(AB)^{-1}$ is unique. But $(B^{-1}A^{-1})(AB) = B^{-1}(A^{-1}A)B = B^{-1}IB = B^{-1}B = I$; hence $(AB)^{-1} = B^{-1}A^{-1}$.

COROLLARY 1 If $A_1, A_2, \ldots, A_r$ are nonsingular matrices of order n, then $(A_1 A_2 \cdots A_r)^{-1} = A_r^{-1} \cdots A_2^{-1} A_1^{-1}$.

Obviously, if A^{-1} is the inverse of a matrix A, then A is the inverse of A^{-1}, that is, $(A^{-1})^{-1} = A$. Moreover, since $\det(AA^{-1}) = \det A \det A^{-1} = 1$, it is clear that $\det A^{-1} = 1/\det A$.

With the inverse of a nonsingular matrix defined, it is possible to define negative integral powers of any nonsingular matrix.

DEFINITION 2 If A is a nonsingular matrix and if r is a positive integer, then $A^{-r} = (A^{-1})^r$.

Negative powers of singular matrices are not defined. But for all nonsingular matrices A and for all integral values of r and s, $A^rA^s = A^{r+s}$ and $(A^r)^s = A^{rs}$.

EXERCISES

1. Determine which of the following matrices are singular and which are nonsingular.

(a) $\begin{bmatrix} 2 & 4 & 1 \\ 4 & 2 & -1 \\ 7 & 1 & -3 \end{bmatrix}$

(b) $\begin{bmatrix} 0 & 2 & 1 & 4 \\ 1 & 1 & 2 & 1 \\ 1 & -3 & 1 & 4 \\ 3 & 1 & 1 & 6 \end{bmatrix}$

(c) $\begin{bmatrix} 3 & 4 & 5 & 4 \\ 4 & 3 & 3 & 3 \\ 2 & 2 & 3 & 2 \\ 3 & 4 & 6 & 5 \end{bmatrix}$

(d) $\begin{bmatrix} 1 & 0 & 1 & 0 & 1 \\ 2 & 1 & 2 & 1 & 2 \\ -1 & 0 & -1 & 0 & -1 \\ 3 & 2 & 1 & 5 & 4 \end{bmatrix}$

(e) $\begin{bmatrix} 1 & 2 & 3 & 4 \\ 2 & 3 & 4 & 5 \\ 3 & 4 & 5 & 6 \\ 4 & 5 & 6 & 7 \end{bmatrix}$

2. For what values of λ, if any, do the following matrices have inverses?

(a) $\begin{bmatrix} \lambda - 1 & 2 \\ 3 & \lambda \end{bmatrix}$ **(b)** $\begin{bmatrix} \lambda - 1 & \lambda - 2 \\ \lambda - 3 & \lambda - 4 \end{bmatrix}$

(c) $\begin{bmatrix} 3 & 1 & 0 \\ -4 & 2 & 5 \\ \lambda^2 & \lambda & 1 \end{bmatrix}$

(d) $\begin{bmatrix} \lambda - 1 & \lambda & \lambda + 1 \\ 2 & -1 & 3 \\ \lambda + 3 & \lambda - 2 & \lambda + 7 \end{bmatrix}$

(e) $\begin{bmatrix} 2 - \lambda & -1 & 2\lambda \\ 1 & 2 & 3 \\ 2 & 2 & 1 \end{bmatrix}$

3. Find all values of k for which the following linear systems have nontrivial solutions.

(a) $\begin{bmatrix} k + 1 & -1 & 2 \\ -2 & 1 & 1 \\ k^2 & 0 & 3 \end{bmatrix} \begin{bmatrix} x \\ y \\ z \end{bmatrix} = \begin{bmatrix} 0 \\ 0 \\ 0 \end{bmatrix}$

(b) $\begin{bmatrix} 5 & 4 & 6 \\ 4 & 3 & 5 \\ 1 & 3 & 2 \end{bmatrix} \begin{bmatrix} x \\ y \\ z \end{bmatrix} = \begin{bmatrix} \tan \pi \\ \sin k\pi \\ \ln 1 \end{bmatrix}$

(c) $\begin{bmatrix} 5 & 4 & 4 & 5 & k^4 \\ 4 & 3 & 3 & 4 & k^3 \\ 3 & 2 & 2 & 3 & k^2 \\ 2 & 1 & 1 & 2 & k \end{bmatrix} \begin{bmatrix} r \\ s \\ t \\ u \\ v \end{bmatrix} = \begin{bmatrix} 0 \\ 0 \\ 0 \\ 0 \end{bmatrix}$

(d) $\begin{bmatrix} 3 & 0 & -1 & 0 \\ -2 & 1 & -4 & 2 \\ -6 & -1 & 2 & 0 \\ 1 & k & 2 & -1 \end{bmatrix} \begin{bmatrix} w \\ x \\ y \\ z \end{bmatrix} = \begin{bmatrix} 0 \\ 0 \\ 0 \\ 0 \end{bmatrix}$

4. Determine the values of λ, if any, for which the following systems have nontrivial solutions, and find such solutions when they exist.

(a) $(1 - \lambda)x_1 + 2x_2 = 0$
$3x_1 + (2 - \lambda)x_2 = 0$

(b) $(2-\lambda)x_1 + (1-\lambda)x_2 = 0$
$(6-\lambda)x_1 - (5-2\lambda)x_2 = 0$

(c) $\lambda x_1 - 2x_2 + x_3 = 0$
$\lambda x_1 + (1-\lambda)x_2 + x_3 = 0$
$2x_1 - x_2 + 2\lambda x_3 = 0$

(d) $(5-\lambda)x_1 + 4x_2 - 2x_3 = 0$
$4x_1 + (5-\lambda)x_2 - 2x_3 = 0$
$-2x_1 - 2x_2 + (3-2\lambda)x_3 = 0$

5. Justify Eq. (14).

6. Using elementary row operations, reduce each of the following matrices to Hermite canonical form and display the row echelon form of each matrix.

(a) $\begin{bmatrix} 1 & 1 & 1 & 1 \\ 2 & 3 & 3 & 3 \\ 3 & 5 & 6 & 6 \\ 4 & 7 & 8 & 8 \end{bmatrix}$

(b) $\begin{bmatrix} 1 & 1 & -1 & 1 \\ 3 & 3 & -3 & 2 \\ 4 & 4 & -4 & 2 \\ 5 & 5 & -5 & 4 \end{bmatrix}$

(c) $\begin{bmatrix} 1 & 3 & 5 & 7 \\ 2 & 4 & 6 & 8 \\ 5 & 7 & 9 & 11 \\ 4 & 3 & 2 & 8 \end{bmatrix}$

(d) $\begin{bmatrix} 5 & -3 & 2 & -2 & -3 \\ 9 & 4 & 3 & 0 & 2 \\ 0 & 3 & 0 & 1 & 1 \\ 2 & 4 & -3 & -5 & 4 \\ 2 & 1 & 0 & 2 & -1 \end{bmatrix}$

7. Find the adjoint of each of the following matrices; when it exists, find the inverse.

(a) $\begin{bmatrix} 1 & 2 \\ 3 & 4 \end{bmatrix}$ **(b)** $\begin{bmatrix} 2 & -1 & 3 \\ 4 & 0 & -1 \\ 3 & 3 & 2 \end{bmatrix}$

(c) $\begin{bmatrix} 1 & 1 & 1 \\ 1 & 2 & 3 \\ 3 & 2 & 1 \end{bmatrix}$ **(d)** $\begin{bmatrix} 2 & 3 & 1 \\ 1 & -1 & 2 \\ 1 & 9 & -4 \end{bmatrix}$

8. Find and use the inverse of the coefficient matrix to solve each of the following linear systems.

(a) $x_1 - x_2 + 2x_3 = 1$
$2x_1 - x_3 = 2$
$x_1 + x_2 + x_3 = 3$

(b) $3x_1 - 5x_2 + 7x_3 = 2$
$x_1 - 2x_2 + 3x_3 = 17$
$3x_1 - 2x_2 - x_3 = 11$

(c) $2w - 2x + 3y + 4z = 33$
$3w - 5x - 7y + 3z = 11$
$4w + 3x - 4y - 5z = 3$
$5w + 4x + 6y - 2z = 45$

(d) $7w + 2x + 11y - 3z = 42$
$10w - 3x + 6y + 4z = 40$
$4w + 5x - 4y + 2z = 28$
$5w + 2x + 9y - 5z = 34$

9. In Exercise 8, solve each of the linear systems by Cramer's rule.

10. Using Cramer's rule, solve each of the following systems.

(a) $x - y + 2z = -5$
$-x + 3z = 0$
$2x + y = 1$

(b) $x - y + 2z + w = 0$
$-x + 3z + 2w = 2$
$2x + y - w = 1$
$2x + 2y + z + 3w = 14$

11. If, for $1 \le i, j \le n$, the elements $a_{ij}(t)$ of a nonsingular matrix $\mathrm{A} = [a_{ij}(t)]$ and the components $b_i(t)$ of a column vector $\mathbf{b} = [b_i(t)]$ are continuous functions on an interval I, show that the solution of the system $\mathrm{A}(t)\mathbf{x}(t) = \mathbf{b}(t)$ has continuous components over I. *Hint:* Use Cramer's rule.

12. Determine all solutions of the general matrix equation AX = B, given that

(a) $\mathrm{A} = \begin{bmatrix} 1 & 2 & 3 \\ 3 & -2 & 5 \\ 5 & -6 & 7 \end{bmatrix}$ $\mathrm{B} = \begin{bmatrix} -1 & 8 \\ 1 & 8 \\ 3 & 8 \end{bmatrix}$

(b) $\mathrm{A} = \begin{bmatrix} 2 & -1 & 2 \\ 1 & 0 & 1 \\ 3 & -2 & 3 \end{bmatrix}$ $\mathrm{B} = \begin{bmatrix} 1 & 0 & 1 \\ -2 & 1 & -2 \\ 0 & 1 & 0 \end{bmatrix}$

(c) $\mathrm{A} = \begin{bmatrix} 1 & 2 \\ -2 & -1 \\ 1 & 2 \\ 0 & 1 \end{bmatrix}$ $\mathrm{B} = \begin{bmatrix} 2 & -3 & 4 \\ -1 & 0 & 1 \\ 2 & -3 & 4 \\ 1 & -2 & 3 \end{bmatrix}$

(d) $\mathrm{A} = \begin{bmatrix} 2 & -2 & 5 & 2 \\ 1 & -1 & 0 & 1 \\ 1 & -1 & 2 & 1 \\ 1 & -1 & 3 & 1 \\ 2 & -2 & 3 & 2 \end{bmatrix}$

$\mathrm{B} = \begin{bmatrix} 4 & 3 \\ -3 & 4 \\ 1 & 2 \\ 3 & 1 \\ 0 & 5 \end{bmatrix}$

13. For each of the following pairs of matrices, verify that $(\mathrm{AB})^{-1} = \mathrm{B}^{-1}\mathrm{A}^{-1}$.

(a) $\mathrm{A} = \begin{bmatrix} 1 & 1 \\ 1 & 2 \end{bmatrix}$ $\mathrm{B} = \begin{bmatrix} 1 & 2 \\ 2 & 5 \end{bmatrix}$

(b) $\mathrm{A} = \begin{bmatrix} 1 & 0 & -1 \\ 0 & 2 & 0 \\ 1 & 1 & 3 \end{bmatrix}$ $\mathrm{B} = \begin{bmatrix} 1 & 1 & 1 \\ 1 & 2 & 1 \\ 1 & 3 & 2 \end{bmatrix}$

14. Solve each of the following equations.

(a) $\begin{bmatrix} x_{11} & x_{12} \\ x_{21} & x_{22} \end{bmatrix}\begin{bmatrix} 1 & 3 \\ 2 & 4 \end{bmatrix} = \begin{bmatrix} 4 & 6 \\ 3 & 9 \end{bmatrix}$

(b) $\begin{bmatrix} x_{11} & x_{12} & x_{13} \\ x_{21} & x_{22} & x_{23} \\ x_{31} & x_{32} & x_{33} \\ x_{41} & x_{42} & x_{43} \end{bmatrix}\begin{bmatrix} 1 & 2 & 3 \\ 1 & 3 & 4 \\ 1 & 3 & 5 \end{bmatrix} = \begin{bmatrix} 1 & 0 & 3 \\ 0 & 2 & 0 \\ -1 & 0 & -3 \\ 0 & -2 & 0 \end{bmatrix}$

15. Verify the relations $|A^{-1}| = 1/|A|$ and $(A^{-1})^{-1} = A$ for each of the following matrices.

(a) $\begin{bmatrix} 1 & 1 \\ 2 & 3 \end{bmatrix}$ **(b)** $\begin{bmatrix} 2 & 3 \\ 4 & 5 \end{bmatrix}$ **(c)** $\begin{bmatrix} 1 & 1 & 0 \\ 2 & 2 & 1 \\ 1 & 2 & 3 \end{bmatrix}$

16. For the matrix $A = \begin{bmatrix} 4 & -4 & 5 \\ -2 & 3 & -3 \\ 3 & -3 & 4 \end{bmatrix}$, find

(a) A^{-1} **(b)** A^2 **(c)** A^{-2} **(d)** A^3
(e) A^{-3}

17. **(a)** Under what conditions, if any, does AB = AC imply B = C?
(b) If A is a nonsingular matrix, show that AB = O implies B = O.

18. If D is a nonsingular diagonal matrix, prove that D^{-1} is also a diagonal matrix and that each element on the principal diagonal of D^{-1} is the reciprocal of the corresponding element in D. Does a similar result hold for nonsingular triangular matrices?

19. If A is a singular matrix, prove that the product of A and its adjoint is a zero matrix.

20. If A is a nonsingular matrix which commutes with a matrix B, prove that A^{-1} also commutes with B. If B is also nonsingular, do A^{-1} and B^{-1} commute?

21. Prove Corollary 1, Theorem 7.

22. **(a)** If A is a nonsingular matrix, show that the determinant of the adjoint of A is equal to the $(n-1)$st power of the determinant of A.
(b) If A is a nonsingular matrix, show that the adjoint of the adjoint of A is equal to A multiplied by the $(n-2)$nd power of the determinant of A.

23. Establish the following result: if $A = [a_{ij}]$ is a nonsingular matrix and if α_{ji} is the cofactor of the element in the jth row and ith column in the determinant of A^{-1}, then

$$|A|\alpha_{ji} = a_{ij}$$

24. Show that, if A is a square matrix of the same order as I and if B is a matrix such that BA = I, then both A and B are nonsingular, AB = I, and B is unique. *Hint:* To prove AB = I, set C = AB and note that BC = B(AB) = (BA)B = IB = BI so that B(C − I) = O. Then use Corollary 1, Theorem 1, to prove C − I = O.

25. In Example 3, verify that a complete solution can be found by using the cofactors of the elements in either the first row or the second row.

26. Using Theorem 3, solve each of the following systems.

(a) $\begin{aligned} 3x_1 + 4x_2 + 5x_3 &= 0 \\ 2x_1 - 5x_2 + 3x_3 &= 0 \\ 5x_1 - x_2 + 8x_3 &= 0 \end{aligned}$ **(b)** $\begin{aligned} x_1 + 2x_2 + 3x_3 &= 0 \\ 3x_1 + 2x_2 + x_3 &= 0 \\ 5x_1 + 6x_2 + 7x_3 &= 0 \end{aligned}$

(c) $\begin{aligned} x_1 + 2x_2 + 3x_3 + 4x_4 &= 0 \\ 2x_1 + 3x_2 + 4x_3 + x_4 &= 0 \\ 3x_1 + 4x_2 + x_3 + 2x_4 &= 0 \\ 7x_1 + 11x_2 + 11x_3 + 11x_4 &= 0 \end{aligned}$

27. If $A = \begin{bmatrix} 1 & 2 & 3 \\ 2 & 3 & 1 \\ 3 & 1 & 2 \end{bmatrix}$, verify that $(A^2)^{-1} = (A^{-1})^2$.

28. Verify that the matrix $A = \begin{bmatrix} 1 & 2 \\ 3 & 4 \end{bmatrix}$ satisfies the equation $A^2 - 5A - 2I = 0$. How can A^{-1} be found from this relation?

29. In Example 4 show that the same solution is obtained if x_1, x_2, and x_3 are determined from the cofactors of the entries in the first row or the second row of the determinant of the matrix of the system (5).

30. Work Example 4 if the mixtures are given by the table

	K	P	N
F_1	5%	10%	20%
F_2	30%	20%	10%
F_3	20%	20%	30%

31. Work Example 4 if in addition to supplies of F_1, F_2, and F_3 the company has an unlimited supply of the inert material used in each fertilizer.

32. Work Example 4 if the company carries just two fertilizers with compositions

	K	P	N
F_1	10%	10%	30%
F_2	5%	30%	10%

33. Work Exercise 32 if the company has plenty of the inert material used in F_1 and F_2.

34. Show that by means of elementary row operations every square matrix can be reduced to Hermite canonical form.

35. Sometimes the requirement that the Hermite canonical form of a matrix be upper triangular is changed to *lower triangular,* the other three defining properties remaining the same. Is the Hermite canonical form of a matrix the same under either definition? Explain.

36. Using elementary row operations find the Hermite canonical form of the matrix

$$\begin{bmatrix} 2 & -4 & -2 & 3 \\ 1 & -1 & 3 & -1 \\ 3 & -4 & 5 & 0 \\ 1 & -2 & -1 & 3 \end{bmatrix}$$

using both definitions of Exercise 35.

37. Why can Theorem 3 be applied to the third row of the determinant of the matrix of the system (5)?

38. Why can Theorem 3 be applied to the first row or the second row (Exercise 29) of the determinant of the matrix of the system (5)?

39. Given that

$$A = \begin{bmatrix} 1 & 2 & 2 \\ 2 & 2 & 3 \\ 1 & -1 & 3 \end{bmatrix}$$

$$C = \begin{bmatrix} 2 & 1 & 1 \\ 2 & 2 & 1 \\ 1 & 1 & 1 \end{bmatrix}$$

$$D = \begin{bmatrix} 10 \\ 13 \\ 9 \end{bmatrix}$$

and that $C\mathbf{b} = D$, solve the linear system $A\mathbf{x} = \mathbf{b}$.

40. Work Exercise 39 if

$$A = \begin{bmatrix} 2 & 0 & 1 \\ 1 & -2 & 2 \\ 2 & 2 & 0 \end{bmatrix}$$

$$C^{-1} = \frac{1}{12}\begin{bmatrix} 5 & -1 & 3 \\ -8 & 4 & 0 \\ -1 & 5 & -3 \end{bmatrix}$$

$$D = \begin{bmatrix} -2 \\ 17 \\ 25 \end{bmatrix}$$

3.8 CHARACTERISTIC-VALUE PROBLEMS

Many problems, in fields ranging from mechanical vibrations and electric circuit theory to pure geometry, are logically equivalent in that they all lead to the following abstract problem.

> Given a function or transformation T whose domain and range are both subsets of a vector space V, determine those vectors in V, if any, which have the same or opposite direction as their images under T.

We shall encounter problems of this sort again and again in our work, and in this section we begin our study of them by considering a geometric example.

EXAMPLE 1

COMPOSITION OF REFLECTIONS AND ROTATIONS IN THE PLANE

In a euclidean plane, let T_1 denote rotation about the origin O through a fixed angle α and let T_2 denote reflection in the y axis. If $T = T_2T_1$ is the composition of T_1 and T_2, and if P_2 is the image of an arbitrary point P under T, what vectors $\overrightarrow{OP}$ have the same or opposite direction as their images $\overrightarrow{OP_2}$?

If P: $(x, y) = (r\cos\theta, r\sin\theta)$, then under the rotation T_1, P becomes

$$P_1 = T_1P = [r\cos(\theta+\alpha), r\sin(\theta+\alpha)]$$

and, under the reflection T_2, P_1 becomes

$$P_2: (x_2, y_2) = T_2P_1 = T_2(T_1P) = [-r\cos(\theta+\alpha), r\sin(\theta+\alpha)]$$

If we expand the coordinates of P_2 and replace $r\cos\theta$ by x and $r\sin\theta$ by y, we obtain

$$x_2 = -r(\cos\theta\cos\alpha - \sin\theta\sin\alpha) = -x\cos\alpha + y\sin\alpha$$

and

$$y_2 = r(\sin\theta\cos\alpha + \cos\theta\sin\alpha) = y\cos\alpha + x\sin\alpha$$

The equations of the transformation T are thus

$$\begin{aligned} x_2 &= -x\cos\alpha + y\sin\alpha \\ y_2 &= x\sin\alpha + y\cos\alpha \end{aligned} \qquad (1)$$

or in matrix notation

$$(2) \qquad \begin{bmatrix} -\cos\alpha & \sin\alpha \\ \sin\alpha & \cos\alpha \end{bmatrix}\begin{bmatrix} x \\ y \end{bmatrix} = \begin{bmatrix} x_2 \\ y_2 \end{bmatrix} = \begin{bmatrix} 1 & 0 \\ 0 & 1 \end{bmatrix}\begin{bmatrix} x_2 \\ y_2 \end{bmatrix}$$

The vectors $\overrightarrow{OP}$ and $\overrightarrow{OP_2}$ will have the same direction or be directed oppositely if and only if their slopes are equal, that is, if and only if

$$(3) \qquad \frac{y}{x} = \frac{y_2}{x_2} \qquad \text{or equivalently} \qquad x_2 = \lambda x \text{ and } y_2 = \lambda y,\ \lambda \neq 0$$

Making these substitutions in (1) and then collecting terms, we obtain the linear system

$$(4) \qquad \begin{bmatrix} -\lambda - \cos\alpha & \sin\alpha \\ \sin\alpha & -\lambda + \cos\alpha \end{bmatrix}\begin{bmatrix} x \\ y \end{bmatrix} = \begin{bmatrix} 0 \\ 0 \end{bmatrix}$$

From Theorem 2, Sec. 3.7, we know that this system has a nontrivial solution if and only if the determinant of its coefficients is zero. Hence we are led to the equation

$$(5) \qquad \begin{vmatrix} -\lambda - \cos\alpha & \sin\alpha \\ \sin\alpha & -\lambda + \cos\alpha \end{vmatrix} = \lambda^2 - \cos^2\alpha - \sin^2\alpha = \lambda^2 - 1 = 0$$

Thus the system (4) has a nontrivial solution if and only if $\lambda = \pm 1$. For all other values of λ the system has only the trivial solution.

Solving the system (4) when $\lambda = 1$, we obtain the solution vector

$$(6a) \qquad \mathbf{v}_1 = \begin{bmatrix} x \\ y \end{bmatrix} = \begin{bmatrix} \sin\alpha \\ 1 + \cos\alpha \end{bmatrix}$$

For $\lambda = -1$, we obtain the solution vector

$$(6b) \qquad \mathbf{v}_2 = \begin{bmatrix} x \\ y \end{bmatrix} = \begin{bmatrix} -\sin\alpha \\ 1 - \cos\alpha \end{bmatrix}$$

Using (3), it is easy to verify that when $\lambda = 1$ the vectors $\overrightarrow{OP}$ and $\overrightarrow{OP_2}$ extend in the same direction, and when $\lambda = -1$ they extend in opposite directions. It is interesting to note that

$$\mathbf{v}_1 \cdot \mathbf{v}_2 = (\sin\alpha)(-\sin\alpha) + (1 + \cos\alpha)(1 - \cos\alpha) = 0$$

for this indicates that the two solution vectors are perpendicular, or *orthogonal.* This is an illustration of the following general result, which we shall prove in Sec. 14.6: *if* A *is a symmetric matrix, the solution vectors of the equation* $(\mathrm{A} - \lambda\mathrm{I})\mathbf{x} = \mathbf{0}$ *are pairwise orthogonal.* This is an important result since in most physical applications, as in this example [see Eq. (2)], the matrix A is symmetric. We will make use of the property of orthogonality many times in other parts of this book.

It will be helpful now to identify the essential features of Example 1 and restate them in general terms: from one source or another, we have a system of n linear equations in n unknowns, $\mathrm{A}\mathbf{x} = \mathbf{x}'$, which in a linear space of n dimensions (usually R^n) defines a transformation sending a vector $\mathbf{x}$ into an image vector $\mathbf{x}'$. For reasons that vary from problem to problem, we are interested in those vectors $\mathbf{x}$ which are proportional to their images $\mathbf{x}'$. Sometimes, as in Example 1, the nature of the problem requires that we consciously make $\mathbf{x}'$ proportional to $\mathbf{x}$ by writing $\mathbf{x}' = \lambda\mathbf{x}$. In physical problems it is usually the case that the problem itself brings in $\lambda\mathbf{x}$ for $\mathbf{x}'$ from the outset. In either case, we have a linear system which in scalar form appears as

$$\begin{aligned} a_{11}x_1 + a_{12}x_2 + \cdots + a_{1n}x_n &= \lambda x_1 \\ a_{21}x_1 + a_{22}x_2 + \cdots + a_{2n}x_n &= \lambda x_2 \\ \cdots\cdots\cdots\cdots\cdots\cdots\cdots\cdots & \\ a_{n1}x_1 + a_{n2}x_2 + \cdots + a_{nn}x_n &= \lambda x_n \end{aligned} \tag{7}$$

and in matrix form appears as

$$\mathrm{A}\mathbf{x} = \lambda\mathbf{x} \qquad \text{where } \lambda \text{ is a parameter} \tag{8}$$

We can use the $n \times n$ identity matrix I to write (8) as $\mathrm{A}x = \lambda\mathrm{I}x$, or, equivalently, as

$$(a)\ (\mathrm{A} - \lambda\mathrm{I})\mathbf{x} = \mathbf{0} \qquad \text{or} \qquad (b)\ (\lambda\mathrm{I} - \mathrm{A})\mathbf{x} = \mathbf{0} \tag{9}$$

Equation (9a) is preferable in theoretical discussions, while Eq. (9b) is preferable if the equation is to be expanded and used for computational purposes since the coefficient of the highest power of the parameter λ is positive. According to Theorem 2, Sec. 3.7, a homogeneous system of this sort has a nontrivial solution if and only if

$$|\lambda\mathrm{I} - \mathrm{A}| = \begin{vmatrix} \lambda - a_{11} & -a_{12} & \cdots & -a_{1n} \\ -a_{21} & \lambda - a_{22} & \cdots & -a_{2n} \\ \cdots & \cdots & \cdots & \cdots \\ -a_{n1} & -a_{n2} & \cdots & \lambda - a_{nn} \end{vmatrix} = 0 \tag{10}$$

When expanded, the determinant

$$|\lambda\mathrm{I} - \mathrm{A}| = p(\lambda) \tag{11}$$

becomes a polynomial of degree n in the parameter λ. Equation (10) is known as the **characteristic equation** of the matrix A, and $p(\lambda)$ is called the **characteristic polynomial** of A.

For values of λ satisfying (10), and for these values only, the matrix equation (9) has nontrivial solution vectors. The n roots of $p(\lambda) = 0$, which need be neither distinct nor real, are called the **characteristic values** of the matrix A, and the corresponding *nontrivial* solution vectors of Systems (7), (8), or (9) are called the **characteristic vectors**† of A. The problem of finding the characteristic values and characteristic vectors of a square matrix is referred to as a **characteristic-value problem.** In the *characteristic-value problem* discussed in Example 1, Eq. (5) is the *characteristic equation,* $\lambda^2 - 1$ is the *characteristic polynomial,* $\lambda = \pm 1$ are the *characteristic values,* and $\mathbf{v}_1$ and $\mathbf{v}_2$ are the *characteristic vectors.* We shall have occasion to study other kinds of characteristic-value problems in Chaps. 11, 12, and 14.

If A is a real matrix, it should be clear that all coefficients in its characteristic polynomial must be real. Hence complex characteristic values must occur in conjugate pairs. The following theorem asserts that the same is true of complex characteristic vectors.

THEOREM 1 If $\mathbf{x}$ is a characteristic vector corresponding to a complex characteristic value $a + ib$, $b \neq 0$, of a *real* matrix A, then $\bar{\mathbf{x}}$ is a characteristic vector corresponding to the characteristic value $a - ib$.

†Some writers graft the German word *eigen,* meaning "own," "peculiar," or "proper," onto the words *values* and *vectors* and use the hybrid terms *eigenvalues* and *eigenvectors* rather than the terms *characteristic values* and *characteristic vectors.*

PROOF By hypothesis, $A\mathbf{x} = (a + ib)\mathbf{x}$, $b \neq 0$. Taking conjugates, we have $\overline{A\mathbf{x}} = \overline{(a + ib)\mathbf{x}}$ or, since A is real, $A\overline{\mathbf{x}} = (a - ib)\overline{\mathbf{x}}$. This shows that $\overline{\mathbf{x}}$ is a characteristic vector of A corresponding to $a - ib$.

We now turn our attention to several formal characteristic-value problems; that is, we are going to find the characteristic values and characteristic vectors of several specific matrices while leaving to later chapters a discussion of the physical origins of such problems.

EXAMPLE 2

Find the characteristic values and the corresponding characteristic vectors of the matrix

$$A = \begin{bmatrix} 4 & -5 \\ 1 & -2 \end{bmatrix}$$

The characteristic equation of A is

$$\det(\lambda I - A) = \begin{vmatrix} \lambda - 4 & 5 \\ -1 & \lambda + 2 \end{vmatrix} = \lambda^2 - 2\lambda - 3 = (\lambda + 1)(\lambda - 3) = 0$$

and so the characteristic values of A are $\lambda_1 = -1$ and $\lambda_2 = 3$. If $\lambda = -1$, the equation $(\lambda I - A)\mathbf{x} = \mathbf{0}$ is equivalent to the linear system

$$\begin{aligned} -5x_1 + 5x_2 &= 0 \\ -x_1 + x_2 &= 0 \end{aligned} \qquad \text{or to} \qquad -x_1 + x_2 = 0$$

An obvious solution of the last equation is $\mathbf{x}_1 = \begin{bmatrix} 1 \\ 1 \end{bmatrix}$, and of course this vector multiplied by any number is also a solution. Thus, for all *nonzero* numbers c_1, $\mathbf{x} = c_1\begin{bmatrix} 1 \\ 1 \end{bmatrix}$ is a characteristic vector of A corresponding to $\lambda_1 = -1$.

If $\lambda_2 = 3$, the equation $(\lambda I - A)\mathbf{x} = \mathbf{0}$ is equivalent to $-x_1 + 5x_2 = 0$, which has $\mathbf{x}_2 = \begin{bmatrix} 5 \\ 1 \end{bmatrix}$ as an obvious solution. Thus, for every *nonzero* number c_2, $\mathbf{x} = c_2\begin{bmatrix} 5 \\ 1 \end{bmatrix}$ is a characteristic vector of A corresponding to $\lambda_2 = 3$. As a check on the characteristic values and vectors just found, we observe that

$$[\lambda_1 I - A]\begin{bmatrix} 1 \\ 1 \end{bmatrix} = \begin{bmatrix} -5 & 5 \\ -1 & 1 \end{bmatrix}\begin{bmatrix} 1 \\ 1 \end{bmatrix} = \begin{bmatrix} 0 \\ 0 \end{bmatrix} \qquad \text{and that} \qquad [\lambda_2 I - A]\begin{bmatrix} 5 \\ 1 \end{bmatrix} = \begin{bmatrix} -1 & 5 \\ -1 & 5 \end{bmatrix}\begin{bmatrix} 5 \\ 1 \end{bmatrix} = \begin{bmatrix} 0 \\ 0 \end{bmatrix}$$

It is worth noting that the characteristic vectors $\mathbf{x}_1$ and $\mathbf{x}_2$ are not orthogonal since $\mathbf{x}_1 \cdot \mathbf{x}_2 = (1)(5) + (1)(1) \neq 0$. On the basis of our observation in Example 1, orthogonality was not to be expected in this case since A is not a symmetric matrix.

EXAMPLE 3

Find the characteristic values and the corresponding characteristic vectors of the matrix

$$A = \begin{bmatrix} -2 & 2 & 1 \\ 2 & 1 & 2 \\ 1 & 2 & 6 \end{bmatrix}$$

and check the characteristic vectors for pairwise orthogonality.

The characteristic equation of A is $|\lambda I - A| = 0$, or

$$\begin{vmatrix} \lambda + 2 & -2 & -1 \\ -2 & \lambda - 1 & -2 \\ -1 & -2 & \lambda - 6 \end{vmatrix} = \lambda^3 - 5\lambda^2 - 17\lambda + 21 = (\lambda - 1)(\lambda + 3)(\lambda - 7) = 0$$

and so the characteristic values of A are $\lambda = 1, -3, 7$.

When $\lambda = 1$, the equation $(\mathrm{I} - \mathrm{A})\mathbf{x} = \mathbf{0}$ appears in scalar form as

$$\begin{aligned} 3x_1 - 2x_2 - x_3 &= 0 \\ -2x_1 \qquad - 2x_3 &= 0 \\ -x_1 - 2x_2 - 5x_3 &= 0 \end{aligned} \quad \text{which is equivalent to the system} \quad \begin{aligned} x_1 + 2x_2 + 5x_3 &= 0 \\ x_2 + 2x_3 &= 0 \end{aligned}$$

Setting $x_3 = -c_1$, we find $x_2 = 2c_1$ and $x_1 = c_1$. All characteristic vectors of A corresponding to $\lambda = 1$ are therefore given by $\mathbf{x}_1 = c_1\begin{bmatrix} 1 \\ 2 \\ -1 \end{bmatrix}$, where c_1 can be any nonzero number.

When $\lambda = -3$, the equation $(-3\mathrm{I} - \mathrm{A})\mathbf{x} = \mathbf{0}$ becomes the scalar system

$$\begin{aligned} -x_1 - 2x_2 - x_3 &= 0 \\ -2x_1 - 4x_2 - 2x_3 &= 0 \\ -x_1 - 2x_2 - 9x_3 &= 0 \end{aligned} \quad \text{which is equivalent to the system} \quad \begin{aligned} x_1 + 2x_2 + x_3 &= 0 \\ x_3 &= 0 \end{aligned}$$

Setting $x_2 = -c_2$, we find $x_1 = 2c_2$. All characteristic vectors of A corresponding to $\lambda = -3$ are therefore of the form $\mathbf{x}_2 = c_2\begin{bmatrix} 2 \\ -1 \\ 0 \end{bmatrix}$, $c_2 \neq 0$.

When $\lambda = 7$, the equation $(7\mathrm{I} - \mathrm{A})\mathbf{x} = \mathbf{0}$ becomes the system

$$\begin{aligned} 9x_1 - 2x_2 - x_3 &= 0 \\ -2x_1 + 6x_2 - 2x_3 &= 0 \\ -x_1 - 2x_2 + x_3 &= 0 \end{aligned} \quad \text{which is equivalent to the system} \quad \begin{aligned} x_1 + 2x_2 - x_3 &= 0 \\ 5x_2 - 2x_3 &= 0 \end{aligned}$$

Setting $x_3 = 5c_3$, we find $x_2 = 2c_3$, $x_1 = c_3$, and $\mathbf{x}_3 = c_3\begin{bmatrix} 1 \\ 2 \\ 5 \end{bmatrix}$, $c_3 \neq 0$

In this case, since the matrix A is symmetric, we expect that the characteristic vectors will be orthogonal, and indeed we have $\mathbf{x}_1 \cdot \mathbf{x}_2 = (1)(2) + (2)(-1) + (-1)(0) = 0$, $\mathbf{x}_2 \cdot \mathbf{x}_3 = (2)(1) + (-1)(2) + (0)(5) = 0$, $\mathbf{x}_3 \cdot \mathbf{x}_1 = (1)(1) + (2)(2) + (5)(-1) = 0$.

EXAMPLE 4

A factored form of the characteristic equation of the matrix

$$\mathrm{A} = \begin{bmatrix} 3 & 2 & 1 & 1 \\ -1 & 1 & 6 & 6 \\ -1 & 2 & 2 & -2 \\ 1 & -2 & 2 & 6 \end{bmatrix}$$

can be obtained by applying elementary properties of determinants to det $(\lambda\mathrm{I} - \mathrm{A})$. We first add the third row of det $(\lambda\mathrm{I} - \mathrm{A})$ to its fourth row, which permits us to factor $\lambda - 4$ out of the new fourth row. In the remaining determinant, we subtract Column 4 from Column 3 and then expand. This gives us

$$\begin{vmatrix} \lambda - 3 & -2 & -1 & -1 \\ 1 & \lambda - 1 & -6 & -6 \\ 1 & -2 & \lambda - 2 & 2 \\ -1 & 2 & -2 & \lambda - 6 \end{vmatrix} = \begin{vmatrix} \lambda - 3 & -2 & -1 & -1 \\ 1 & \lambda - 1 & -6 & -6 \\ 1 & -2 & \lambda - 2 & 2 \\ 0 & 0 & \lambda - 4 & \lambda - 4 \end{vmatrix}$$

$$= (\lambda - 4)\begin{vmatrix} \lambda - 3 & -2 & 0 & -1 \\ 1 & \lambda - 1 & 0 & -6 \\ 1 & -2 & \lambda - 4 & 2 \\ 0 & 0 & 0 & 1 \end{vmatrix}$$

$$= (\lambda - 4)^2[(\lambda - 2)^2 + 1] = 0$$

From the last of these equations, we see that $\lambda_1 = 4$, $\lambda_2 = 2 + i$, and $\lambda_3 = 2 - i$ are the characteristic values of A.

Using Gauss-Jordan elimination to solve the linear system $(4\mathrm{I} - \mathrm{A})\mathbf{x} = \mathbf{0}$, we find without difficulty that the characteristic vectors of A corresponding to $\lambda_1 = 4$ are given by $\mathbf{x} = c_1(0, 0, 1, -1)$, $c_1 \neq 0$.

To find the characteristic vectors which correspond to $\lambda_2 = 2 + i$, we reduce the matrix

$$[(2+i)\mathrm{I} - \mathrm{A}] = \begin{bmatrix} -1+i & -2 & -1 & -1 \\ 1 & 1+i & -6 & -6 \\ 1 & -2 & i & 2 \\ -1 & 2 & -2 & -4+i \end{bmatrix}$$

to row echelon form and then write the corresponding linear system, which we find to be

$$\begin{aligned} x_1 + \qquad\qquad & x_4 = 0 \\ x_2 + \quad \tfrac{1}{2}(-1+i)&x_4 = 0 \\ x_3 + \qquad & x_4 = 0 \end{aligned}$$

With $x_4 = -2c_2$, we find from these equations that the characteristic vectors of A corresponding to $\lambda_2 = 2 + i$ are given by $\mathbf{x} = c_2(2, -1 + i, 2, -2)$, $c_2 \neq 0$. Having found these vectors, we know from Theorem 1 that the characteristic vectors corresponding to the characteristic value $\lambda_3 = 2 - i$ of A are the vectors $\mathbf{x} = c_3(2, -1 - i, 2, -2)$, $c_3 \neq 0$.

EXERCISES

1. Find the characteristic values and the corresponding characteristic vectors of each of the following matrices.

(a) $\begin{bmatrix} 4 & 6 & 6 \\ 1 & 3 & 2 \\ -1 & -5 & -2 \end{bmatrix}$ **(b)** $\begin{bmatrix} 7 & -2 & -4 \\ 3 & 0 & -2 \\ 6 & -2 & -3 \end{bmatrix}$

(c) $\begin{bmatrix} 2 & 4 & -6 \\ 4 & 2 & -6 \\ -6 & -6 & -15 \end{bmatrix}$ **(d)** $\begin{bmatrix} 11 & -4 & -7 \\ 7 & -2 & -5 \\ 10 & -4 & -6 \end{bmatrix}$

(e) $\begin{bmatrix} 4 & 6 & 6 \\ 1 & 3 & 2 \\ -1 & -4 & -3 \end{bmatrix}$ **(f)** $\begin{bmatrix} -4 & 5 & 5 \\ -5 & 6 & 5 \\ -5 & 5 & 6 \end{bmatrix}$

For which of these, if any, are the characteristic vectors orthogonal?

2. Find the characteristic values and the corresponding characteristic vectors of each of the following matrices.

(a) $\begin{bmatrix} 2 & 1 & 1 \\ 2 & 3 & 2 \\ 1 & 1 & 2 \end{bmatrix}$ **(b)** $\begin{bmatrix} 1 & 1 & -1 \\ -1 & 3 & -1 \\ -1 & 1 & 1 \end{bmatrix}$

(c) $\begin{bmatrix} 0 & 1 & 0 \\ 1 & 2 & 2 \\ 0 & -1 & 0 \end{bmatrix}$ **(d)** $\begin{bmatrix} 1 & 0 & 0 \\ 0 & 1 & 0 \\ 0 & 0 & 1 \end{bmatrix}$

(e) $\begin{bmatrix} 1 & 0 & e \\ 0 & 1 & -\pi \\ -\pi & -e & 1 \end{bmatrix}$

(f) $\begin{bmatrix} h & 0 & 0 \\ 0 & h & 0 \\ \cosh\phi & k\cosh\phi & h \end{bmatrix}$

(g) $\begin{bmatrix} 0 & -1 & 2 & 0 \\ 1 & 0 & 0 & 4 \\ -2 & -2 & 0 & 0 \\ -1 & -1 & 0 & 0 \end{bmatrix}$

(h) $\begin{bmatrix} 3 & -5 & 0 & -7 \\ 1 & -1 & 0 & 0 \\ -2 & 0 & -1 & 1 \\ 0 & -2 & -1 & -3 \end{bmatrix}$

(i) $\begin{bmatrix} \frac{3}{2} & -1 & -2 & 0 \\ 1 & -\frac{1}{2} & -2 & -2 \\ 0 & 0 & \frac{1}{2} & 1 \\ -3 & 0 & -1 & -\frac{3}{2} \end{bmatrix}$

3. **(a)** Find all real values of k for which the matrix

$$\mathrm{A} = \begin{bmatrix} 1 & k & 3 \\ -k & 2 & -k \\ 1 & k & 3 \end{bmatrix}$$

has real characteristic values.

(b) Find the characteristic vectors of A which correspond to each real characteristic value (k real).

(c) With k real, determine the largest possible characteristic value of A and the corresponding characteristic vectors.

(d) Find the real values of k for which A has only two real characteristic values. Find these two characteristic values and the corresponding characteristic vectors.

4. Show that, if there is a nonsingular matrix P such that $P^{-1}AP = B$, then $\det(\lambda I - A) = \det(\lambda I - B)$. *Hint:* $I = P^{-1}IP$.

5. Show that the constant term of the characteristic polynomial $p(\lambda)$ of a square matrix A is either det A or $-\det$ A.

6. Show that A is singular if and only if $\lambda = 0$ is a characteristic value of A. *Hint:* Use Exercise 5.

7. Show that the characteristic equation of every triangular matrix $A = [a_{ij}]$ has the form

$$p(\lambda) = (\lambda - a_{11})(\lambda - a_{22})\cdots(\lambda - a_{nn}) = 0.$$

8. In Example 1, show that the slope of $\mathbf{v}_1$ is $\cot(\alpha/2)$ and that the slope of $\mathbf{v}_2$ is $-\tan(\alpha/2)$. Interpret these results geometrically.

9. Work Example 1 if the order of the transformations T_1 and T_2 is reversed, i.e., if $T = T_1T_2$.

10. **(a)** Work Example 1 if T_2 is reflection in the x axis and $T = T_2T_1$.

(b) Work Part **(a)** if $T = T_1T_2$.

11. **(a)** Work Example 1 if T_2 is reflection in the line $y = x$ and $T = T_2T_1$.

(b) Work Part **(a)** if $T = T_1T_2$.

12. Find the characteristic values and corresponding characteristic vectors of each of the following matrices.

(a) $\begin{bmatrix} 3 & 4 \\ 2 & 1 \end{bmatrix}$ **(b)** $\begin{bmatrix} 6 & 4 \\ -3 & -1 \end{bmatrix}$

(c) $\begin{bmatrix} 2 & -3 & 2 \\ 3 & 12 & -6 \\ -2 & -6 & 7 \end{bmatrix}$ **(d)** $\begin{bmatrix} 2 & -8 & 3 \\ 1 & -1 & 1 \\ 2 & 1 & 2 \end{bmatrix}$

13. Show that characteristic vectors corresponding to different characteristic values cannot be proportional.

14. Show that the three characteristic vectors in Example 3 are linearly independent.

15. Under what conditions, if any, can a characteristic-value problem of the form $(\lambda B - A)\mathbf{x} = \mathbf{0}$ be reduced to a problem of the form $(\lambda I - C)\mathbf{x} = \mathbf{0}$?

16. If $A = \begin{bmatrix} 1 & 2 \\ 3 & 4 \end{bmatrix}$ and $B = \begin{bmatrix} 1 & 1 \\ 4 & 2 \end{bmatrix}$, for what values of λ does the equation $(\lambda A - B)\mathbf{x} = \mathbf{0}$ have a nontrivial solution vector? What are these solution vectors?

CHAPTER 4

SIMULTANEOUS LINEAR DIFFERENTIAL EQUATIONS

In general, as we shall see in Sec. 7.5, the behavior of a physical system of n degrees of freedom is described by a set of n linear differential equations. It is thus an important matter to investigate the properties of simultaneous linear differential equations and how to solve them.

In most applications, the differential equations have constant coefficients, but since the proofs of the basic theorems are no harder when the coefficients are variable, we shall develop the theory for the general case. As we do this, it may be helpful to note how closely the basic results resemble those we established in Secs. 2.1–2.4 for a single linear differential equation.

We shall develop two methods for solving such systems. The first, which is valid whether the coefficients are constant or variable, is analogous to the reduction process we used in Sec. 3.6 for solving systems of linear algebraic equations: three elementary operations (Sec. 4.2) are used to reduce the system until it contains a single equation in just one unknown. Then back substitution completes the solution process.

The second method, valid only when the coefficients are constants, parallels the procedure in Secs. 2.5–2.7 for solving a single linear constant-coefficient differential equation. The set of solutions is regarded as a vector, the so-called *solution vector,* the system is written in matrix notation, and a complementary function is constructed by means of the characteristic values of the matrix of coefficients. If the system is nonhomogeneous, a particular integral can be constructed very much as we did for a single differential equation in Sec. 2.7.

Prerequisite for this chapter: Chaps. 2 and 3.

4.1 SOLUTIONS, CONSISTENCY, AND EQUIVALENCE OF LINEAR DIFFERENTIAL SYSTEMS

In many problems in applied mathematics there are not one but several dependent variables, each a function of a single independent variable, usually time. The formulation of such a problem in mathematical terms frequently leads to a system of simultaneous differential equations. Often these equations are nonlinear and exceedingly difficult, if not impossible, to solve, even with the aid of a computer. In certain important cases, however, they are linear in the dependent variables.

Quite aside from their usefulness from a practical standpoint, linear differential systems are important because in a sense they are to analysis what linear systems are to algebra. Indeed, when each coefficient a_{ij} is replaced by a linear differential operator L_{ij} on an interval I (Definition 1, Sec. 2.7), and when each constant b_i is replaced by a function f_i that is continuous on I, a simultaneous system of m linear algebraic equations in n unknowns

(1) $$a_{i1}x_1 + a_{i2}x_2 + \cdots + a_{in}x_n = b_i \qquad 1 \le i \le m$$

becomes a simultaneous system consisting of m linear differential equations

(2) $$L_{i1}x_1 + L_{i2}x_2 + \cdots + L_{in}x_n = f_i \qquad 1 \le i \le m$$

in n unknown functions $x_1, x_2, \ldots, x_n$. Such a system is called a **linear differential system.** A linear system (2) is **homogeneous** if and only if all of the functions $f_1, f_2, \ldots, f_m$ are trivial on I, and the system is **nonhomogeneous** otherwise.

A matrix

(3) $$\mathrm{F}(t) = \begin{bmatrix} f_{11}(t) & f_{12}(t) & \cdots & f_{1n}(t) \\ \cdots & \cdots & \cdots & \cdots \\ f_{m1}(t) & f_{m2}(t) & \cdots & f_{mn}(t) \end{bmatrix}$$

each of whose elements is a scalar-valued function on an interval I, is called a **matrix function.** The derivative of a matrix function $\mathrm{F}(t)$, if it exists, is a matrix $\mathrm{F}'(t)$ whose elements are the derivatives of the corresponding elements of $\mathrm{F}(t)$ (Exercise 39, Sec. 3.4). A matrix function with only one row, or just one column, is said to be a **vector function.**

A **solution** of a linear differential system (2) on an interval I is a vector function which satisfies each equation of the system for every t in I. A linear differential system is **consistent** or **inconsistent** on I according as it has, or does not have, a solution on I, and one system is **equivalent** to another if and only if every solution of either system is a solution of the other.

EXAMPLE 1

Verify that on $(-\infty, \infty)$ the vector function

$$\mathbf{x}(t) = \begin{bmatrix} t \sinh t \\ \cosh t - \cos t \\ t - \sin t \end{bmatrix}$$

is a solution of the nonhomogeneous linear differential system

$$\begin{aligned} x_1'' - x_1 - x_2'' \qquad - x_2 + x_3'' \qquad &= \sin t \\ x_1'' \qquad - 2x_2'' - tx_2' \qquad + tx_3'' \qquad &= -2\cos t \\ x_1'' - x_1 - 2x_2'' \qquad\qquad - 2x_3' &= -2 \end{aligned}$$

Substituting $t \sinh t$ for x_1, $\cosh t - \cos t$ for x_2, $t - \sin t$ for x_3, and performing the indicated differentiations, we obtain the identities

$$\begin{aligned}(t \sinh t + 2 \cosh t) - (t \sinh t) - (\cosh t + \cos t) - (\cosh t - \cos t) + \sin t &\equiv \sin t \\ (t \sinh t + 2 \cosh t) - 2(\cosh t + \cos t) - t(\sinh t + \sin t) + t(\sin t) &\equiv -2 \cos t \\ (t \sinh t + 2 \cosh t) - (t \sinh t) - 2(\cosh t + \cos t) - 2(1 - \cos t) &\equiv -2\end{aligned}$$

which hold for all real values of t. This confirms that the vector function $\mathbf{x}(t)$ is a solution of the given differential system on $(-\infty, \infty)$.

If the operator coefficients in (2) are very general, the task of solving such a system, or of establishing its consistency or inconsistency, can be formidable even if all scalars and functions involved are real, as we shall generally suppose they are. Nevertheless, there are some exceedingly important special cases of (2) that are relatively easy to solve, notably those cases in which all of the linear operators in the linear differential system have constant coefficients, i.e., are polynomial operators in D (Definition 2, Sec. 2.7). Many physical problems involve differential systems of this kind, and there are various ways of solving them.

In one, which bears a strong resemblance to the solution of systems of simultaneous algebraic equations, the system is reduced by successive elimination of the unknowns until an equivalent system is obtained which has as one of its equations a linear equation in a single dependent variable. This is solved, and then working backward, the other variables are solved for one by one until the problem is completed. A second method, which amounts to considering the system as a single matrix differential equation, generalizes the ideas of complementary function and particular integral and utilizes the characteristic values and vectors of a matrix. This method determines all the unknowns at the same time. Finally, use of the Laplace transformation provides a straightforward operational procedure for solving systems of linear differential equations with constant coefficients which in many applications is preferable to either of the other methods.

In this chapter we shall consider the first two methods, leaving the third to Chap. 10, where we discuss the Laplace transformation and its applications in detail.

EXERCISES

1. Describe each of the following differential systems, stating whether it is linear or nonlinear. If the system is linear, state whether it is homogeneous or not.

(a) $$\begin{aligned} t^2x_1'' - (\cos t)x_2' + x_3'' &= t^3 \\ x_1'' + tx_2' - e^tx_3' &= t \\ 2x_1' - 3t^2x_2'' + (\sin t)x_3 &= 0 \end{aligned}$$

(b) $$\begin{aligned} x_1'' - 2x_2' + x_3'' &= 0 \\ x_1' + x_2 - x_3'' &= 1 \\ x_1 + 2x_2 + x_3'' &= 1/x_1 \end{aligned}$$

2. Write each of the following linear differential systems in operator form.

(a) $$\begin{aligned} x_1'' + 3x_2'' + x_3'' - x_1 &= x_2' \\ 2x_3 - 6x_2' + 5x_1'' - x_1 &= x_3'' \\ x_2' + x_1'' - x_3'' + x_3 &= 2x_1' \end{aligned}$$

(b) $$\begin{aligned} x_1''' - x_2'' + tx_3' &= t + x_3 \\ 2tx_1'' + x_2' + x_3''' + 1 &= x_1 \\ e^tx_1' - x_1'' + x_3'' &= 6t^2x_2 \end{aligned}$$

3. Write each of the following systems in derivative form.

(a) $$\begin{aligned} D(D+1)x_1 - D^2x_2 + 3Dx_3 &= 0 \\ (D^2-1)x_1 + (2D+1)x_2 - 3Dx_3 &= 0 \\ (D+1)^2x_1 - Dx_2 - 3x_3 &= 0 \end{aligned}$$

(b) $$\begin{aligned} D^2(e^tx_1) - D(e^tx_2) + D^2(2e^tx_3) &= e^t \\ D(D-1)x_1 + D^2x_2 + D(x_3 + e^{-t}) &= 0 \\ D(e^{-t}x_1) - 3D^2(e^{-t}x_2) + (D+1)(e^{-t}x_3) &= 0 \end{aligned}$$

4. State which of the following linear operators are, or can be written as, polynomial operators.

(a) $3D^2 - 2D + 1$ **(b)** $3t^2D^2 - 2tD + 5$

(c) $\pi D^2 - eD + \sqrt{2}$

5. (a) If $A(t) = \begin{bmatrix} \cos t & -\sin t \\ \sin t & \cos t \end{bmatrix}$, find $A''(t) + A(t)$.

(b) If $A(t) = \begin{bmatrix} t & e^t \\ 1 & e^{-t} \end{bmatrix}$, find $A^{iv}(t) - A''(t)$.

(c) Given that b is a real parameter and that

$A(t) = \begin{bmatrix} \cos t & \cosh 2t \\ \sin t & \sinh 2t \end{bmatrix} + \begin{bmatrix} t & b \\ b & t \end{bmatrix}$, find $A^{iv}(t) - 3A''(t) - 4A(t)$.

(d) Given that a is a constant and that $\phi(t)$ and $\theta(t)$ are differentiable functions on $(-\infty, \infty)$, show that the vector function $\mathbf{v}(t) = \begin{bmatrix} a \sin \phi(t) \cos \theta(t) \\ a \sin \phi(t) \sin \theta(t) \\ a \cos \phi(t) \end{bmatrix}$ is always orthogonal to its derivative $\dot{\mathbf{v}}(t)$.

In Exercises 6–8, determine whether or not the given vector function is a solution, on the interval specified, of the given differential system.

6. $\mathbf{x}(t) = (1 - t, t)$, $(-\infty, \infty)$,

$$(2D + 1)x_1 + (D + 5)x_2 = 4t$$
$$(D + 2)x_1 + (D + 2)x_2 = 2$$

7. $\mathbf{x}(t) = \begin{bmatrix} t - t \ln |t + 1| \\ t - (t + 1) \ln |t + 1| \end{bmatrix} e^t$, $(-1, \infty)$,

$$x_1' - x_2 = e^t/(t + 1)$$
$$x_1 + x_2' - 2x_2 = 0$$

8. $\mathbf{x}(t) = \begin{bmatrix} \ln |t| \\ t^2 \\ e^{2t} + t \end{bmatrix}$, $(0, \infty)$,

$$2tx_1' - x_2'' + x_3'' - 4x_3 = -4t$$
$$2t^2x_1'' + x_2'' + x_3'' - 2x_3' = -2$$
$$t^2x_1'' + x_2'' + x_3'' = 4e^{2t}$$

9. Show that on every interval I, the differential system

$$x_1'' + 2x_2' - 3x_3'' = t$$
$$2x_1'' - 5x_2' + 4x_3'' = \cos t$$
$$3x_1'' + 6x_2' - 9x_3'' = 2t$$

is inconsistent.

10. Work Exercise 9 for the differential system

$$(D + 2)x + (2D + 1)y + (D - 1)z = t$$
$$(3D + 5)x + (5D + 4)y + (3D + 1)z = 1 + 2t$$
$$(D + 1)x + (D + 2)y + (D + 3)z = 2$$

4.2 THE REDUCTION OF A DIFFERENTIAL SYSTEM TO AN EQUIVALENT SYSTEM

When all the operators in a linear differential system are polynomial operators, the system can either be solved, or else be shown to be inconsistent, in much the same way that linear algebraic systems were dealt with in Sec. 3.6. At most, three types of elementary operations need be performed upon the equations of the differential system to transform it into an equivalent system which is easier to solve. Two of these operations are just like the first two we applied to linear algebraic systems, and the other closely resembles the third operation for linear systems. Specifically, in the case of **linear differential systems with constant coefficients,** i.e., differential systems involving only polynomial operators, the three types of **elementary operations** are these:

1. Interchanging two equations
2. Multiplying an equation through by a nonzero constant
3. Operating on *both* members of one equation with a polynomial operator and adding the result to another equation

Here, as with linear algebraic equations, it is clear that a linear differential system is transformed into an equivalent system by elementary operations of the first two kinds. The third type of operation has the same effect, assuming of course that the functions to which the polynomial operators are applied are sufficiently differentiable. This can be established by an argument almost identical with the one given in Sec. 3.6 for the third of the elementary operations considered there (Exercise 37). Thus, summarizing, we have the following theorem.

THEOREM 1 Each of the elementary operations **1, 2,** and **3,** and hence any sequence of them, transforms a linear differential system into an equivalent system.

Part of the strategy employed to eliminate variables from general linear algebraic systems carries over to linear differential systems with constant coefficients, and it is an interesting mathematical problem to study such systems in detail. However, the linear differential systems which arise in practice usually involve the same number of equations as unknown functions. Moreover, the fundamental existence and uniqueness theorem upon which the theory of linear differential systems is ordinarily based (Sec. 4.3) deals exclusively with systems of this sort. Hence we shall restrict our investigations accordingly. In particular, the constant coefficient systems we consider will all be expressible in the form

$$
\begin{aligned}
p_{11}(D)x_1 + p_{12}(D)x_2 + \cdots + p_{1n}(D)x_n &= f_1(t)\\
p_{21}(D)x_1 + p_{22}(D)x_2 + \cdots + p_{2n}(D)x_n &= f_2(t)\\
\cdots\cdots\cdots\cdots\cdots\cdots\cdots\cdots\cdots\cdots\cdots\cdots\cdots\\
p_{n1}(D)x_1 + p_{n2}(D)x_2 + \cdots + p_{nn}(D)x_n &= f_n(t)
\end{aligned} \tag{1}
$$

where p_{ij} is a polynomial operator in $D = d/dt$, for $1 \le i, j \le n$.

To solve a system like this, or to determine its inconsistency, we try to find elementary operations that will reduce it to an equivalent system of the form

$$
\begin{aligned}
q_{11}(D)x_1 + q_{12}(D)x_2 + \cdots + q_{1n}(D)x_n &= g_1(t)\\
q_{22}(D)x_2 + \cdots + q_{2n}(D)x_n &= g_2(t)\\
\cdots\cdots\cdots\cdots\cdots\cdots\cdots\cdots\cdots\cdots\cdots\cdots\cdots\\
q_{nn}(D)x_n &= g_n(t)
\end{aligned} \tag{2}
$$

where to attain this form it may be necessary to renumber the unknowns. If none of the operators q_{ii}, $1 \le i \le n$, is the **zero operator,** i.e., if none of these operators is the operator all of whose coefficients are zero, the nth equation can be solved for x_n by the methods of Chap. 2, then x_{n-1} can be found from the $(n-1)$st equation, and so forth, until finally all of the unknowns have been determined and the system solved. Here are two examples to illustrate the method.

EXAMPLE 1

Find a complete solution of the linear differential system

$$
\begin{aligned}
\frac{dx}{dt} + 2x + \frac{dy}{dt} + 6y &= 2e^t\\
2\frac{dx}{dt} + 3x + 3\frac{dy}{dt} + 8y &= -1
\end{aligned}
$$

As a rule, the steps in a reduction of a constant-coefficient linear differential system to Form (2) are easier to detect when each system in the reduction process is written in operator form rather than in derivative form. The operator form of the given system is

$$
\begin{aligned}
(D+2)x + (D+6)y &= 2e^t\\
(2D+3)x + (3D+8)y &= -1
\end{aligned}
$$

Our first step in the elimination process is to subtract twice the first equation from the second, after which we interchange equations and have

$$
\begin{aligned}
-x + (D-4)y &= -1 - 4e^t\\
(D+2)x + (D+6)y &= 2e^t
\end{aligned}
$$

Working with this new system, we operate on both members of the first equation with $D + 2$ and add the result to the second equation. Then we simplify the *new* second equation and multiply the first equation through by -1. This gives us

$$
\begin{aligned}
x - (D-4)y &= 1 + 4e^t\\
(D^2 - D - 2)y &= -2 - 10e^t
\end{aligned}
$$

Of course, this system is equivalent to the one we started with, and it is in the form of (2). Since its second equation is an equation in y alone, it is a simple matter to solve for y by the methods of Chap. 2, and we find without difficulty

$$y = c_1 e^{-t} + c_2 e^{2t} + 5e^t + 1$$

With y now completely determined, we substitute into the first equation of the final reduced system to obtain

$$x - (D - 4)(c_1 e^{-t} + c_2 e^{2t} + 5e^t + 1) = 1 + 4e^t$$

Solving this equation for x and simplifying, we find

$$x = -5c_1 e^{-t} - 2c_2 e^{2t} - 11e^t - 3$$

Thus a complete solution of the original differential system is

$$\begin{bmatrix} x \\ y \end{bmatrix} = c_1 \begin{bmatrix} -5 \\ 1 \end{bmatrix} e^{-t} + c_2 \begin{bmatrix} -2 \\ 1 \end{bmatrix} e^{2t} + \begin{bmatrix} -11 \\ 5 \end{bmatrix} e^t + \begin{bmatrix} -3 \\ 1 \end{bmatrix}$$

It is readily verified that both of the vector functions

$$\mathbf{x}_1(t) = \begin{bmatrix} -5 \\ 1 \end{bmatrix} e^{-t} = \begin{bmatrix} -5e^{-t} \\ e^{-t} \end{bmatrix} \qquad \text{and} \qquad \mathbf{x}_2(t) = \begin{bmatrix} -2 \\ 1 \end{bmatrix} e^{2t} = \begin{bmatrix} -2e^{2t} \\ e^{2t} \end{bmatrix}$$

satisfy the homogeneous system corresponding to the given system, which is

$$\frac{dx}{dt} + 2x + \frac{dy}{dt} + 6y = 0$$

$$2\frac{dx}{dt} + 3x + 3\frac{dy}{dt} + 8y = 0$$

and that the vector function

$$\mathbf{v}(t) = \begin{bmatrix} -11 \\ 5 \end{bmatrix} e^t + \begin{bmatrix} -3 \\ 1 \end{bmatrix} = \begin{bmatrix} -11e^t - 3 \\ 5e^t + 1 \end{bmatrix}$$

satisfies the nonhomogeneous system itself. Thus in the complete solution of the given system, the sum $c_1\mathbf{x}_1(t) + c_2\mathbf{x}_2(t)$ appears to play the role of *complementary function* while $\mathbf{v}(t)$ plays the role of *particular integral*. We shall explore these ideas in more detail in Secs. 4.3 and 4.4, and in more general terms when we study linear transformations and linear operator equations in Chap. 13.

EXAMPLE 2

Find a complete solution of the differential system

$$(3) \qquad \begin{aligned} (2D^2 + 3D - 9)x + (D^2 + 7D - 14)y &= 4 \\ (D + 1)x + (D + 2)y &= -8e^{2t} \end{aligned}$$

During the first stage of a reduction process, it is sometimes convenient to eliminate an unknown other than the leading one from all but one equation of a reduced system. To indicate what happens when this is done, we shall eliminate y instead of x from one equation of the present system, although there is no particular advantage in doing so.

To begin the elimination process, we operate on the second equation of (3) with $(-D-5)$,† add the result to the first equation, and simplify. Then we multiply the second equation by 24. This gives us

$$\begin{aligned}(D^2-3D-14)x-24y&=4+56e^{2t}\\ 24(D+1)x+24(D+2)y&=-192e^{2t}\end{aligned}$$

Next we operate on the first equation of this new system with $D+2$, add the result to the second equation, and simplify, obtaining

$$\begin{aligned}(D^2-3D-14)x-24y&=4+56e^{2t}\\ (D^3-D^2+4D-4)x\qquad&=8+32e^{2t}\end{aligned}\tag{4}$$

As it stands, this system is not of the form (2). However, by merely interchanging the x and y terms in the first equation, the system takes on that form. Of course, there is no need to actually make this interchange because the second equation of (4) can be solved at once. The roots of its characteristic equation are $\pm 2i$, 1. Hence a complementary function is $c_1\cos 2t+c_2\sin 2t+c_3e^t$. It is easy to see that $-2+4e^{2t}$ is a particular integral, and therefore

$$x=c_1\cos 2t+c_2\sin 2t+c_3e^t-2+4e^{2t}$$

With x completely determined, a solution for y can be found by substituting the last expression for x into the first equation of (4):

$$(D^2-3D-14)(c_1\cos 2t+c_2\sin 2t+c_3e^t-2+4e^{2t})-24y=4+56e^{2t}$$

Performing the indicated differentiations, collecting like terms, simplifying, and solving for y, we have

$$y=-\tfrac{1}{4}(3c_1+c_2)\cos 2t+\tfrac{1}{4}(c_1-3c_2)\sin 2t-\tfrac{2}{3}c_3e^t+1-5e^{2t}$$

Thus, in terms of the parameters $k_1=c_1/4$, $k_2=c_2/4$, and $k_3=c_3/3$, a complete solution of System (4) is given by

$$\begin{bmatrix}x\\ y\end{bmatrix}=k_1\begin{bmatrix}4\cos 2t\\ -3\cos 2t+\sin 2t\end{bmatrix}+k_2\begin{bmatrix}4\sin 2t\\ -\cos 2t-3\sin 2t\end{bmatrix}+k_3\begin{bmatrix}3\\ -2\end{bmatrix}e^t+\begin{bmatrix}-2\\ 1\end{bmatrix}+\begin{bmatrix}4\\ -5\end{bmatrix}e^{2t}\tag{5}$$

But (3) and (4) are equivalent systems; hence (5) is also a complete solution of the original system.

Before considering another example, let us note an interesting feature of Examples 1 and 2. Although each involves a system of two differential equations in two dependent variables, the complete solution in Example 1 contains two arbitrary constants, whereas the complete solution in Example 2 contains three. What is the explanation? To answer this question, let us first return to System (1) and observe that if D were a number, instead of an operator, then

$$\begin{vmatrix}p_{11}(D) & p_{12}(D) & \cdots & p_{1n}(D)\\ p_{21}(D) & p_{22}(D) & \cdots & p_{2n}(D)\\ \cdots & \cdots & \cdots & \cdots\\ p_{n1}(D) & p_{n2}(D) & \cdots & p_{nn}(D)\end{vmatrix}\tag{6}$$

†The choice of $-D-5$ as an operational multiplier is not just a clever guess. Actually, $D+5$ is the quotient when $D^2+7D-14$ is divided by $D+2$; hence subtracting $(D+5)(D+2)y=(D^2+7D+10)y$ from $(D^2+7D-14)y$ will eliminate both D^2y and Dy, which is one of our objectives.

would be a determinant. Of course, the polynomial operators $p_{ij}(D)$ usually are not numbers. Nevertheless, (6) is known as the **determinant of the operational coefficients** of (1) and can be expanded formally as though each $p_{ij}(D)$ were an algebraic rather than an operational symbol. The result of course is a polynomial in D. As the next theorem tells us, the degree of this polynomial is the answer to our question about the number of arbitrary constants in a complete solution of a linear differential system.

THEOREM 2 If the determinant of the operational coefficients of a system of n linear differential equations with constant coefficients is not identically zero, then the total number of independent arbitrary constants in any complete solution of the system is equal to the degree of the determinant of the operational coefficients, regarded as a polynomial in D. In particular cases in which the determinant of the operational coefficients is identically zero, the system may have no solution or it may have solutions containing any number of independent constants.†

Applying this theorem to System (3), Example 2, we note that the determinant of the operational coefficients is

$$\begin{vmatrix} 2D^2 + 3D - 9 & D^2 + 7D - 14 \\ D + 1 & D + 2 \end{vmatrix}$$

The expanded form of this determinant is the *third*-degree polynomial $D^3 - D^2 + 4D - 4$ in D. Hence, according to Theorem 2, there must be exactly three arbitrary constants in any complete solution of the given system. This explains why the complete solution (5) of System (3) contains *three* arbitrary constants, even though the system has only *two* equations and *two* unknowns.

EXAMPLE 3

Solve the differential system

$$\begin{aligned} Dx + (D-1)y + (D+2)z &= 2e^t \\ (D-1)x + \qquad Dy + (D-2)z &= ae^t \\ (D+1)x + (D-2)y + (D+6)z &= e^t \end{aligned}$$

It is readily verified that the determinant of the operational coefficients of this system is identically zero. Hence, according to Theorem 2, the system may have no solution or it may have solutions containing any number of arbitrary constants. Let us apply elementary operations to the system and find out what the case may be.

By subtracting the second equation from both the first and third equations, and then adding -2 times the *new* first equation to the *new* third equation, we get

$$\begin{aligned} x - y + \qquad 4z &= (2-a)e^t \\ (D-1)x + Dy + (D-2)z &= ae^t \\ 0 &= (a-3)e^t \end{aligned}$$

Of course, this reduced system is equivalent to the given one. Clearly, unless $a = 3$, neither system is consistent.

On the other hand, if $a = 3$, we can operate on the first equation of the reduced system with $-D + 1$, add the result to the second equation, simplify, and omit the third equation which is simply $0 = 0$, to obtain

(7)
$$\begin{aligned} x - \qquad y + \qquad 4z &= -e^t \\ (2D-1)y - (3D-2)z &= 3e^t \end{aligned}$$

†For a proof of this result see, for instance, E. L. Ince, *Ordinary Differential Equations*, pp. 144–150, Dover, New York, 1944.

This system is of the form (2) and is equivalent to the original system. From the second equation of (7), y can be found in terms of $z = z(t)$; then, from the first equation, x too can be found in terms of z. Since z is subject only to the restriction that it be differentiable, it may contain any number of arbitrary constants.

Thus, consistent with the possibilities allowed by Theorem 2, we have found that if $a \neq 3$, the original differential system has no solution; but if $a = 3$, the system may have solutions containing any number of arbitrary constants.

The reduction of a constant-coefficient linear differential system to an equivalent system of the form (2), or to some other equivalent pseudotriangular system, may be carried out most efficiently by applying elementary operations to the rows of the corresponding coefficient matrix of operators or, should the given differential system be nonhomogeneous, to the corresponding augmented matrix. Once this has been done and the particular equation of the equivalent triangular system which contains only one dependent variable has been solved, the system itself can be solved either by back or forward substitution.

EXAMPLE 4

Solve the linear differential system

$$\begin{aligned}(D+1)x + (D-1)y &= \sec t\\ (4D+5)x + (5D-4)y &= 5\sec t\end{aligned}$$

The augmented matrix is

$$\begin{bmatrix} D+1 & D-1 & \sec t \\ 4D+5 & 5D-4 & 5\sec t \end{bmatrix}$$

Subtracting 5 times Row 1 from Row 2, we obtain the augmented matrix

$$\begin{bmatrix} D+1 & D-1 & \sec t \\ -D & 1 & 0 \end{bmatrix}$$

of an equivalent differential system. In this new matrix, we subtract $D - 1$ operating on Row 2 from Row 1. This gives

$$\begin{bmatrix} D^2+1 & 0 & \sec t \\ -D & 1 & 0 \end{bmatrix}$$

From the first row of this matrix we have $(D^2 + 1)x = \sec t$. Noting that the corresponding homogeneous equation is harmonic, and using variation of parameters to find a particular integral, we find that a complete solution of the last equation is

$$x = c_1 \cos t + c_2 \sin t + (\ln|\cos t|)\cos t + t \sin t$$

The second row of the last matrix gives $y = Dx$ which we use, with x now known, to obtain

$$\begin{bmatrix} x \\ y \end{bmatrix} = c_1\begin{bmatrix} \cos t \\ -\sin t \end{bmatrix} + c_2\begin{bmatrix} \sin t \\ \cos t \end{bmatrix} + \ln|\cos t|\begin{bmatrix} \cos t \\ -\sin t \end{bmatrix} + t\begin{bmatrix} \sin t \\ \cos t \end{bmatrix}$$

as a complete solution of the given system on each of the intervals

$$(2n-1)\frac{\pi}{2} < t < (2n+1)\frac{\pi}{2} \qquad n \text{ an integer}$$

EXERCISES

With the understanding that $D \equiv d/dt$, find a complete solution of each of the following systems of equations.

1. $(D+5)x + (D+4)y = e^{-t}$
   $(D+2)x + (D+1)y = 3$
2. $(D+5)x + (D+3)y = e^{-t}$
   $(D+2)x + (D+1)y = 3$
3. $(D+5)x + (D+3)y = e^{-t}$
   $(2D+1)x + (D+1)y = 3$
4. $(D-1)x + (D-2)y = 0$
   $(D-5)x + (2D-7)y = e^{-t}$
5. $(D-1)x + (D+9)y = t$
   $(D-2)x + (2D+9)y = 4$
6. $(2D+5)x - (2D+3)y = t$
   $(D-2)x + (D+2)y = 0$
7. $(2D+3)x + (D+4)y = -\cos t$
   $(D+1)x + (D+2)y = 2 \sin t$
8. $(2D^2+1)x + (D+2)y = 5$
   $(D^2-16)x + (D-4)y = 4$
9. $(D^2+1)x + (D^2+3)y = 0$
   $(3D+1)x + (2D+6)y = 0$
10. $(9D^2+8)x + (3D^2+4)y = 0$
    $(2D^2+1)x + (D^2+2)y = 120 \cos 3t$
11. $(D-1)x - y = t$
    $-2x + (D-1)y - z = 0$
    $-2y + (D-1)z = e^{2t}$
12. $(D+1)x + y + 2z = 0$
    $x + (D+2)y + z = e^{-t}$
    $5x + y + (D-2)z = 5e^{-t}$
13. $(D+1)x + (D+5)y + (2D+5)z = 15e^t$
    $(2D+1)x + (D+2)y + (3D+1)z = 10e^t$
    $(D+3)x + (3D+4)y + (4D+6)z = 21e^t$
14. $(D+1)x + (D+3)y + (2D+3)z = e^t$
    $(2D+1)x + (D+2)y + (3D+1)z = 0$
    $(D+3)x + (3D+11)y + (4D+13)z = 0$
15. $(D+1)x + (D+1)y + (2D+3)z = 0$
    $(2D+1)x + (D+2)y + (3D+5)z = 0$
    $(D+3)x + (3D+1)y + (4D+5)z = 0$
16. $(2D^2-D-1)x + (D-1)y = 0$
    $(D^2-1)x + (D-1)y = 0$
17. $(3D^2+3D+2)x + (D^2+2D+3)y = 0$
    $(2D^2-D-2)x + (D^2+D+1)y = 8$

Find the solution of each of the following systems which satisfies the indicated conditions.

18. $(2D+1)x + (D+2)y = 0$
    $(D-1)x + (D+3)y = 5$
    (a) $x_0 = 0,\ y_0 = 7$
    (b) $x_0 = 7,\ y_0' = 8$
19. $(2D^2+3D-9)x + (D^2+7D-14)y = 4$
    $(D+1)x + (D+2)y = 0$
    (a) $x_0 = -2,\ y_0 = 0,\ y_0' = -6$
    (b) $x_0 = 3,\ y_0 = -3,\ x_0' = 7$
20. Find a system of simultaneous first-order linear constant-coefficient differential equations having

$$x = c_1e^t + c_2e^{2t} \qquad y = c_1e^t + 3c_2e^{2t}$$

as components of a complete solution. *Hint:* First solve the given equations for c_1e^t and c_2e^{2t}. Then apply the operator D to each equation and again solve for c_1e^t and c_2e^{2t}.

21. Work Exercise 20 if $x = c_1e^{2t} + 2c_2e^{-3t}$ and $y = 2c_1e^{2t} - 3c_2e^{-3t}$.
22. If $x = c_1e^{-t} + c_2e^{2t}$ and $y = c_1e^{-t} - 2c_2e^{2t}$, work Exercise 20 by determining conditions on a, b, c, d so that $x = e^{-t}, y = e^{-t}$ and $x = e^{2t}, y = -2e^{2t}$ will satisfy the equation

$$(aD+b)x + (cD+d)y = 0$$

23. **(a)** Work Exercise 20 if $x = c_1 \cos t + c_2 \sin t$ and $y = c_1 \sin t - c_2 \cos t$.
    (b) Show that there is no system of equations of the type described in Exercise 20 having $x = c_1 \cos t + c_2 \sin t$ and $y = c_1 \sin t - 2c_2 \cos t$ as components of a complete solution.
24. **(a)** Using the method of Exercise 22, find a system of simultaneous linear first-order differential equations having

$$x = c_1e^t + c_2e^{-t} + c_3$$
$$y = c_1e^t + c_2e^{-t} + c_3$$
$$z = c_1e^t + c_2e^{-t}$$

as components of a complete solution.

(b) Work Part **(a)** if

$$x = c_1e^t + c_2e^{-t} + c_3e^{2t}$$
$$y = c_1e^t + c_2e^{-t} + c_3e^{2t}$$
$$z = -c_1e^t + c_2e^{-t} + c_3e^{2t}$$

25. Find a system of simultaneous linear constant-coefficient differential equations having $x = c_1e^{-t} + c_2e^t + c_3e^{2t}$ and $y = c_1e^{-t} - c_2e^t + 2c_3e^{2t}$ as components of a complete solution. *Hint:* Begin with a single linear differential equation having x as a complete solution. Then consider the equation $y = (aD^2 + bD + c)x$.
26. If (x_1, y_1) and (x_2, y_2) are two solutions of the system

$$P_{11}(D)x + P_{12}(D)y = 0$$
$$P_{21}(D)x + P_{22}(D)y = 0$$

prove that $(c_1x_1 + c_2x_2, c_1y_1 + c_2y_2)$ is also a solution of the system for all values of the constants c_1 and c_2.

27. In Exercise 17, determine (operational) multiples of the two equations which, when added, will yield an equation

expressing y directly in terms of x and its various derivatives. Can this be done for the system of equations in Exercise 19? Do you think that this can be done in general?

28. A system consists of two tanks each containing V gal of brine. The brine in the first tank initially contains s_1 lb of salt per gallon, the brine in the second tank initially contains s_2 lb of salt per gallon. Fresh brine containing s lb of salt per gallon flows into the first tank at the rate of a gal/min and the mixture, kept uniform by stirring, runs into the second tank at the same rate. From the second tank, the mixture, kept uniform by stirring, runs out at the same rate. Find the amounts of salt in each tank as functions of time. Under what conditions, if any, will the amount of salt in the second tank reach a relative maximum or minimum value?

29. Two tanks are connected as shown in Fig. 4.1. The first tank contains 100 gal of pure water; the second contains 100 gal of brine containing 2 lb of salt per gal. Liquid circulates through the tanks at a constant rate of 5 gal/min. If the brine in each tank is kept uniform by stirring, find the amount of salt in each tank as a function of time.

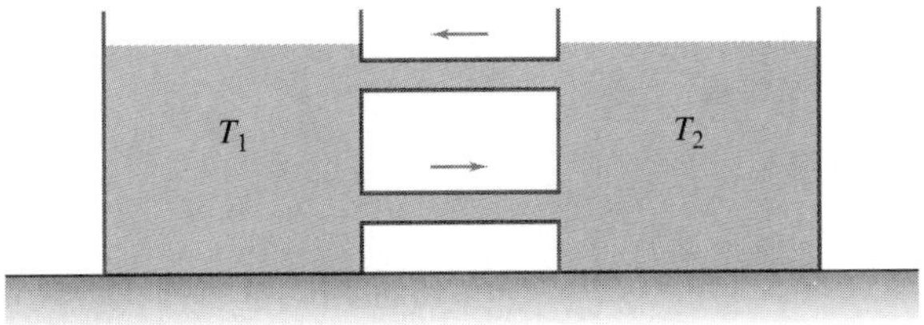

FIGURE 4.1

30. Three tanks are connected as shown in Fig. 4.2. The first tank contains 100 gal of pure water; the second contains 100 gal of brine containing 1 lb of salt per gallon; the third contains 100 gal of brine containing 2 lb of salt per gallon. Liquid circulates through the tanks at a constant rate of 5 gal/min. If the brine in each tank is kept uniform by stirring, find the amount of salt in each tank as a function of time.

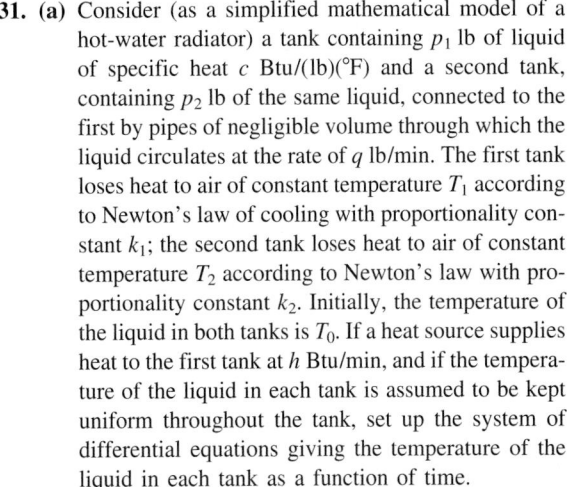

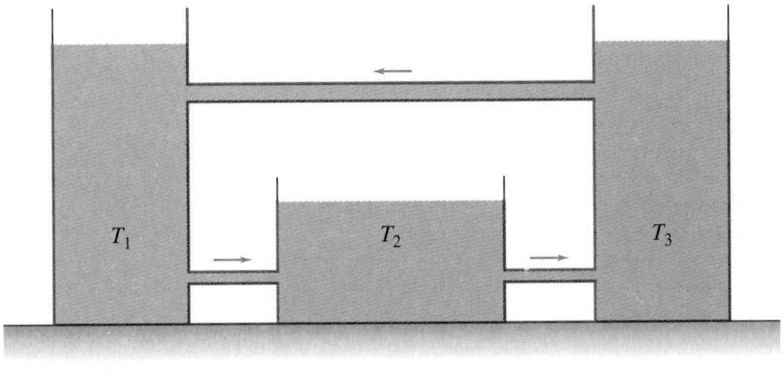

FIGURE 4.2

31. (a) Consider (as a simplified mathematical model of a hot-water radiator) a tank containing p_1 lb of liquid of specific heat c Btu/(lb)(°F) and a second tank, containing p_2 lb of the same liquid, connected to the first by pipes of negligible volume through which the liquid circulates at the rate of q lb/min. The first tank loses heat to air of constant temperature T_1 according to Newton's law of cooling with proportionality constant k_1; the second tank loses heat to air of constant temperature T_2 according to Newton's law with proportionality constant k_2. Initially, the temperature of the liquid in both tanks is T_0. If a heat source supplies heat to the first tank at h Btu/min, and if the temperature of the liquid in each tank is assumed to be kept uniform throughout the tank, set up the system of differential equations giving the temperature of the liquid in each tank as a function of time.

(b) Solve the differential equations derived in Part **(a)** if $c = 1$ Btu/(lb)(°F), $p_1 = 900$ lb, $p_2 = 100$ lb, $k_1 = 3$, $k_2 = 11$, $q = 15$ lb/min, $T_0 = 50°F$, $T_1 = 80°F$, $T_2 = 70°F$, and $h = 100$ Btu/min.

(c) Solve the differential equations derived in Part **(a)** if $c = 1$ Btu/(lb)(°F), $p_1 = p_2 = 200$ lb, $k_1 = 2$, $k_2 = 50$, $q = 10$ lb/min, $T_0 = 60°F$, $T_1 = 90°F$, $T_2 = 70°F$, and $h = 150$ Btu/min.

32. (a) When x is eliminated from the equations

$$(a_{11}D + b_{11})x + (a_{12}D + b_{12})y = f_1$$
$$(a_{21}D + b_{21})x + (a_{22}D + b_{22})y = f_2$$

by applying the operator $a_{21}D + b_{21}$ to the first equation and the operator $a_{11}D + b_{11}$ to the second equation and then subtracting the first of these new equations from the second, the result is

$$[(a_{11}D + b_{11})(a_{22}D + b_{12}) - (a_{12}D + b_{12})(a_{21}D + b_{21})]y = (a_{11}D + b_{11})f_2 - (a_{21}D + b_{21})f_1$$

Show that this equation, together with the equation

$$(a_{11}b_{21} - a_{21}b_{11})x + [a_{11}(a_{22}D + b_{22}) - a_{21}(a_{12}D + b_{12})]y = a_{11}f_2 - a_{21}f_1$$

form a system equivalent to the given system. *Hint:* Derive the last two equations from the first pair by a sequence of elementary operations.

(b) Obtain a second system equivalent to the given system by first eliminating y.

33. Show that if the determinant of the operational coefficients of

$$p_{11}(D)x + p_{12}(D)y = f_1(t)$$
$$p_{21}(D)x + p_{22}(D)y = f_2(t)$$

is not identically zero, then

$$\begin{vmatrix} p_{11}(D) & p_{12}(D) \\ p_{21}(D) & p_{22}(D) \end{vmatrix} x = -\begin{vmatrix} p_{12}(D) & f_1(t) \\ p_{22}(D) & f_2(t) \end{vmatrix}$$

and

$$\begin{vmatrix} p_{11}(D) & p_{12}(D) \\ p_{21}(D) & p_{22}(D) \end{vmatrix} y = \begin{vmatrix} p_{11}(D) & f_1(t) \\ p_{21}(D) & f_2(t) \end{vmatrix}$$

which is precisely what Cramer's rule would yield if applied to the given system as though it were purely algebraic.

34. **(a)** Do the two differential equations in Exercise 33 involving x alone and y alone have the same characteristic equations?

(b) When the two differential equations in x, and in y, of Exercise 33 are solved, are the complementary functions in the solutions the same except for the arbitrary constants involved?

(c) How can the number of parameters in the complementary functions mentioned in Part **(b)** be reduced to the number required by Theorem 2?

(d) Can the adaptation of Cramer's rule described in Exercise 33 be extended to System (1) if $n \geq 3$?

35. If the system

$$(2D^2 - D - 1)x + (2D^2 + 4D - 6)y = 0$$
$$(D^2 + 2D - 3)x + (D^2 + 7D - 8)y = 0$$

is solved by determining both x and y from the differential equations obtained by using Cramer's rule, show that x and y must be substituted into *each* of the given equations to obtain the necessary relations between the arbitrary constants.

36. If each of the operational coefficients in the system

$$P_{11}(D)x + P_{12}(D)y = 0$$
$$P_{21}(D)x + P_{22}(D)y = 0$$

contains $D - a$ as a simple factor and if the determinant of the operational coefficients contains $D - a$ only as a double factor, show that neither x nor y contains a term of the form te^{at} even though the characteristic equation of the differential equations for both x and y has a as a double root. *Hint:* Factor $D - a$ from each operational coefficient, set $(D - a)x = u$ and $(D - a)y = v$, and note that neither u nor v can contain a term of the form e^{at}.

37. Prove that an elementary transformation of Type **3** transforms a linear differential system with constant coefficients into an equivalent system.

38. Solve for x and y in Example 3 if $a = 3$ and **(a)** $z = 3e^t$, **(b)** $z = e^{2t/3}$, **(c)** $z = b + ct + e^{2t/3}$.

39. For what values of a, b, c, if any, will the system

$$(D^2 + D + 1)x + (D + 2)y = 0$$
$$(D^2 + aD + b)x + (D + c)y = 0$$

have a complete solution containing **(a)** exactly two arbitrary constants? **(b)** Exactly one arbitrary constant? **(c)** No arbitrary constants? **(d)** Any number of arbitrary constants?

40. Work Exercise 39 for the system

$$(D^2 + aD + b)x + (D + 1)y = 0$$
$$(D^2 + cD + e)x + (D - 1)y = 0$$

Using the method of Example 4, solve each of the following differential systems wherein a prime indicates differentiation with respect to t.

41. **(a)** $x' - 12x + 5y = 0$
$y' - 30x + 13y = 0$

(b) $x' + 10x - 9y = 0$
$y' + 12x - 11y = 0$

42. **(a)** $x' + y' - x + 2y = 18$
$x' + y' - 4x - y = 9e^{-t}$

(b) $x' + x + y' - 2y = 1$
$2x' + x + 2y' = e^t$

43. **(a)** $2x' + 2y' + x + 4y = 3e^{-2t}$
$x'' + y' - x + 2y = 6$

(b) $x' + 2x + y' + 6y = 2e^t$
$2x' + 3x + 3y' + 8y = -1$

44. **(a)** $x' - x - 2y = 6e^{2t}$
$y' - 3x - 2y = 6e^t$

(b) $x' + y' + 5x + 3y = e^{-t}$
$x' + 2y + y' + y = 3$

45. **(a)** $x' - y - t = 0$
$y' + 2y - 3y + 1 = 0$

(b) $x' - y - 2t = 0$
$y' + x - 2y + 3 = 0$

46. **(a)** $x' - x + y' - 2y = 0$
$y' - 4x - 5y = e^{-t}$

(b) $30x' + 4x - y = 60$
$30y' - 4x + 4y = 0$

47. **(a)** $x' + y' + 2y = \sin t$
$x' + y' - x - y = 0$

(b) $x' - x + y - \cos t = 0$
$y' - x - y - \cos t = 0$

48. **(a)** $x' - 3x - 2y = 0$
$y' + 2x + y = \sin t$

(b) $x' + 3x - 4y = \sin t$
$y' + 2x - 3y = -2$

49. **(a)** $x' + y' - x + 9y = \tan 3t$
$2x' + 3y' - 3x + 18y = 3\tan 3t$

(b) $x' + x + y' - y = \tan t$
$3x' + 4x + 4y' - 3y = 4 \tan t$

50. **(a)** $x' - 2x + y + 4z = 0$
$y' - 2y + 4z = 0$
$z' - y + 2z = 0$

(b) $x' - 2x + y + 4z = 0$
$y' - 2y + 4z = 0$
$z' - 2z = 0$

51. **(a)** $y' - 2y + 2z - w = 0$
$z' - 4z + 4w = 0$
$w' - 4z + 4w = 0$
$x' - 2x = 0$

(b) $x' - 2x - y + 2z = e^t$
$y' + 3x - 4z = e^t$
$z' + 2x + y - 4z = 0$

52. Let $L_1 = 0, L_2 = 0, \ldots, L_6 = 0$ be distinct linear first-order constant-coefficient differential equations in two unknowns, and let S_{ij} be the system consisting of the equations $L_i = 0$ and $L_j = 0$.

(a) If S_{34} and S_{56} are each equivalent to S_{12}, is S_{34} equivalent to S_{56}?

(b) Under what further conditions, if any, will S_{46} be equivalent to S_{12}?

4.3 FUNDAMENTAL CONCEPTS AND THEOREMS CONCERNING FIRST-ORDER SYSTEMS

The methods of solution and central concepts associated with a single ordinary linear differential equation can be extended to linear differential systems. This is most effectively carried out using matrix functions and some ideas of linear algebra, and by working with differential systems that either are or can be reduced to a system of first-order equations of the **standard form**

$$
\begin{aligned}
x_1' &= a_{11}(t)x_1 + a_{12}(t)x_2 + \cdots + a_{1n}(t)x_n + f_1(t) \\
x_2' &= a_{21}(t)x_1 + a_{22}(t)x_2 + \cdots + a_{2n}(t)x_n + f_2(t) \\
&\cdots\cdots\cdots\cdots\cdots\cdots\cdots\cdots\cdots\cdots \\
x_n' &= a_{n1}(t)x_1 + a_{n2}(t)x_2 + \cdots + a_{nn}(t)x_n + f_n(t)
\end{aligned} \tag{1}
$$

where the **coefficient functions** $a_{ij}(t)$, and the **functions of the system** $f_i(t)$, are all real-valued functions. A differential system written in the form (1) is called a **first-order linear differential system,** or simply a **first-order system,** in n unknowns. A first-order system is **normal** on an interval I if and only if its coefficient functions, and the functions of the system are all continuous on I. The problem of finding a solution of a first-order system that satisfies n **initial conditions**

$$x_1(t_0) = k_1, \quad x_2(t_0) = k_2, \quad \ldots, \quad x_n(t_0) = k_n \tag{2}$$

is called an **initial-value problem.**

In terms of the vector functions

$$\mathbf{x} = \begin{bmatrix} x_1 \\ \vdots \\ x_n \end{bmatrix} \qquad \mathbf{f} = \begin{bmatrix} f_1 \\ \vdots \\ f_n \end{bmatrix}$$

and the **coefficient matrix** $\mathrm{A} = [a_{ij}]$, System (1) can be written as the compact **matrix differential equation**

$$\mathbf{x}' = \mathrm{A}(t)\mathbf{x} + \mathbf{f}(t) \tag{3}$$

The **associated homogeneous equation** is of course

$$\mathbf{x}' = \mathrm{A}(t)\mathbf{x} \tag{4}$$

Quite often a first-order differential system and its corresponding matrix equation are spoken of interchangeably. A concise way of writing the initial conditions (2) is obviously $\mathbf{x}(t_0) = \mathbf{k}$, where $\mathbf{k} = \begin{bmatrix} k_1 \\ \vdots \\ k_n \end{bmatrix}$.

EXAMPLE 1

The first-order system

$$\begin{aligned} x_1' &= (\cos t)x_1 + (\sec t)x_2 - \sin t \\ x_2' &= (\sec t)x_1 - (\cos t)x_2 + 1 \end{aligned}$$

has a continuous coefficient matrix

$$\mathrm{A}(t) = \begin{bmatrix} \cos t & \sec \mathrm{t} \\ \sec t & -\cos \mathrm{t} \end{bmatrix}$$

on

$$-\frac{\pi}{2} < t < \frac{\pi}{2}$$

The functions of the system are the components of the vector function

$$\mathbf{f}(t) = \begin{bmatrix} -\sin t \\ 1 \end{bmatrix}$$

which is continuous for all values of t. It is easy to verify that a solution of the initial-value problem

$$\mathbf{x}' = \mathrm{A}(t)\mathbf{x} + \mathbf{f}(t) \qquad \mathbf{x}(0) = \begin{bmatrix} 0 \\ 1 \end{bmatrix}$$

on $-\pi/2 < t < \pi/2$ is $\mathbf{x} = \begin{bmatrix} \tan t \\ \sec t \end{bmatrix}$.

We shall now show that every normal nth-order linear differential equation

$$y^{(n)} + a_1(t)y^{(n-1)} + \cdots + a_{n-1}(t)y' + a_n(t)y = f(t) \tag{5}$$

can be solved by solving a related first-order system. To find a first-order system whose complete solution will yield a complete solution of (5), we set

$$x_1 = y \qquad x_2 = y' \qquad x_3 = y'' \qquad \cdots \qquad x_n = y^{(n-1)} \tag{6}$$

These relations obviously imply all but the last equation of the system

$$\begin{aligned} x_1' &= x_2 \\ x_2' &= x_3 \\ &\vdots \\ x_{n-1}' &= x_n \\ x_n' &= -a_n(t)x_1 - a_{n-1}(t)x_2 - \cdots - a_1(t)x_n + f(t) \end{aligned} \tag{7}$$

The last equation comes upon substituting from (6) into (5), and so it is said that Eq. (5) has been **transformed into,** or **reduced to,** System (7). Given any solution y of Eq. (5), the relations (6) define a corresponding solution $(y, y', \ldots, y^{(n-1)})$ of System (7). Conversely, if $(x_1, x_2, \ldots, x_n)$ is a solution of (7), its first component x_1 is a solution of (5). A complete solution of (7) thus yields a complete solution of (5).

Clearly, (7) is a special case of System (1). By using the first $n - 1$ equations of (7) to eliminate every variable except x_1 from the last equation and then setting $x_1 = y$, it is possible to recover Eq. (5). It is not true, however, that a general linear first-order system can be converted into a single equation in x_1 $(=y)$ from which the other variables of the system can be found by successive differentiation (Exercise 14). Thus (1) may be regarded as a definite generalization of a single nth-order linear differential equation. With these observations in mind, it is natural to expect the theory of first-order systems to include the theory of a single linear equation of order n.

Usually it is much easier to solve a single nth-order equation directly than it is to construct and solve a first-order system whose solution vector will give the solution of the original equation. There are also many differential systems that can be solved without first reducing them to a first-order system. For instance, we saw in the preceding section that elementary operations can be systematically used to solve differential systems with polynomial operators even if their equations are not of the first order. However, this method is prohibitively complicated, at least without the aid of a computer, if either the coefficients in the polynomial operators have several significant digits or the number of unknowns is large. No matter what the equations to be solved may be, it is always desirable to have alternative solution methods available.

After considering a few more basic ideas and proving some general underlying theorems in this section and the next, we shall, in Sec. 4.5, present a purely algebraic method for solving systems of homogeneous linear differential equations with constant coefficients. Our derivation of the method bears a striking resemblance to the procedure we used in Chap. 2 to solve a single homogeneous linear differential equation with constant coefficients. Once the method is established, neither differentiations nor integrations enter into the solution process. Because the method is entirely algebraic, it is well-adapted to the use of computers. Just as in the case of a single equation, it so happens that, in general, a knowledge of a complete solution of a homogeneous linear differential system will enable us to solve any related nonhomogeneous system the functions of which are integrable.

No matter how we propose to solve a linear differential system, it is always essential to know if the system has a solution and, if so, how many. Most systems encountered in practice can be transformed into first-order linear systems (Exercise 11). For first-order linear systems, or any system which can be so expressed, the existence and uniqueness of a solution are guaranteed under the conditions of the next theorem.†

THEOREM 1 Let the first-order linear differential system (1) be normal on an interval I, let t_0 be contained in I, and let $k_1, k_2, \ldots, k_n$ be arbitrary real numbers. Then, over I there exists exactly one solution $\mathbf{x}(t) = [x_1(t), \ldots, x_n(t)]$ of (1) such that $x_1(t_0) = k_1$, $x_2(t_0) = k_2, \ldots, x_n(t_0) = k_n$.

In the compact notation of matrices, the theorem says that if a first-order system $\mathbf{x}' = \mathrm{A}(t)\mathbf{x} + \mathbf{f}(t)$ in n unknowns is normal on an interval I, if t_0 is in I, and if $\mathbf{k}$ is a vector of R^n, then over I there exists exactly one solution $\mathbf{x}(t)$ of the initial-value problem $\mathbf{x}' = \mathrm{A}(t)\mathbf{x} + \mathbf{f}(t)$, $\mathbf{x}(t_0) = \mathbf{k}$. An obvious and important corollary of Theorem 1 is

COROLLARY 1 If a homogeneous first-order system in n dependent variables $\mathbf{x}' = \mathrm{A}(t)\mathbf{x}$ is normal on an interval I, if t_0 is a point of I, and if $\mathbf{x}(t)$ is a solution of $\mathbf{x}' = \mathrm{A}(t)\mathbf{x}$ for which $\mathbf{x}(t_0) = \mathbf{0}$, then $\mathbf{x}(t) \equiv \mathbf{0}$ on I.

---

†This theorem is a special case of a more general fundamental existence and uniqueness theorem. See E. L. Ince, *Ordinary Differential Equations,* pp. 71–72, Dover, New York, 1944.

EXAMPLE 2

The linear system

$$x_1' = (\operatorname{csch} t)x_1 + (1 - t^2)x_2 + \ln|t - 1|$$
$$x_2' = \sqrt{t + 2}\,x_1 + (1 + e^t)x_2 + \operatorname{Tan}^{-1} t$$

is not normal on any interval which contains 0, 1, or any number less than -2. For $\operatorname{csch} t$, $\ln|t - 1|$ and $\sqrt{t + 2}$ are undefined at $t = 0$, $t = 1$, and for $t < -2$, respectively, so that the coefficient matrix

$$\mathrm{A}(t) = \begin{bmatrix} \operatorname{csch} t & 1 - t^2 \\ \sqrt{t + 2} & 1 + e^t \end{bmatrix}$$

fails to be continuous at $t = 0$ and for $t < -2$, while the vector function

$$\mathbf{f}(t) = \begin{bmatrix} \ln|t - 1| \\ \operatorname{Tan}^{-1} t \end{bmatrix}$$

is discontinuous at $t = 1$. On the half-open interval $[-2, 0)$ and over each of the open intervals $(0, 1)$ and $(1, \infty)$, the differential system is normal. Hence, if t_0 is any arbitrary point within any of these intervals, say $\frac{1}{3}$, and if $\mathbf{k}$ is any vector of R^2, say $\begin{bmatrix} e \\ \pi \end{bmatrix}$, Theorem 1 asserts that on the interval $(0, 1)$ there is exactly one solution of the initial-value problem consisting of the given first-order system and the initial condition $\mathbf{x}(\frac{1}{3}) = \begin{bmatrix} e \\ \pi \end{bmatrix}$. The solution is of course a differentiable vector function, and therefore a continuous vector function, at each point of $(0, 1)$. By Corollary 1, the only solution on $(0, 1)$ of the initial-value problem $\mathbf{x}' = \mathrm{A}(t)\mathbf{x}$, $\mathbf{x}(\frac{1}{3}) = \mathbf{0}$ is the trivial solution $\mathbf{x}(t) \equiv \begin{bmatrix} 0 \\ 0 \end{bmatrix}$.

Just as linear combinations of solutions figured prominently in our work with single nth-order equations in Chap. 2, they will also prove indispensable in our study of first-order systems. Of course, solutions of differential systems in more than one unknown will usually be expressed as vector rather than scalar functions, and they will then be called **solution vectors.**

If $\mathbf{v}_1, \mathbf{v}_2, \ldots, \mathbf{v}_m$ are all vector functions with a common domain and the same number of components, a linear combination of them is simply a sum of the form

$$c_1\mathbf{v}_1 + c_2\mathbf{v}_2 + \cdots + c_m\mathbf{v}_m = \sum_{k=1}^{m} c_k\mathbf{v}_k \tag{8}$$

where $c_1, c_2, \ldots, c_m$ are arbitrary constants. The additive vector functions $\mathbf{v}_1, \mathbf{v}_2, \ldots, \mathbf{v}_m$ are **linearly dependent** if and only if there exist numbers $c_1, c_2, \ldots, c_m$, not all zero, such that $c_1\mathbf{v}_1 + c_2\mathbf{v}_2 + \cdots + c_m\mathbf{v}_m = \mathbf{0}$. On the other hand, these vector functions are **linearly independent** if and only if they are not linearly dependent; i.e., if and only if $c_1\mathbf{v}_1 + c_2\mathbf{v}_2 + \cdots + c_m\mathbf{v}_m = \mathbf{0}$ implies $c_1 = c_2 = \cdots = c_m = 0$.

EXAMPLE 3

Determine whether the vector functions defined on $(0, \infty)$ by

$$\mathbf{v}_1(t) = \begin{bmatrix} t + \ln t \\ e^t - t \end{bmatrix} \qquad \mathbf{v}_2(t) = \begin{bmatrix} e^t - t \\ t + t^2 \end{bmatrix} \qquad \mathbf{v}_3(t) = \begin{bmatrix} e^t + 2t + 3\ln t \\ 3e^t - 2t + t^2 \end{bmatrix}$$

are linearly dependent or linearly independent.

Multiplying these three vector functions by c_1, c_2, and c_3, respectively, then adding and equating the vector sum to $\mathbf{0}$, we obtain

$$c_1\mathbf{v}_1 + c_2\mathbf{v}_2 + c_3\mathbf{v}_3 = \begin{bmatrix} c_1(t + \ln t) + c_2(e^t - t) + c_3(e^t + 2t + 3\ln t) \\ c_1(e^t - t) + c_2(t + t^2) + c_3(3e^t - 2t + t^2) \end{bmatrix} = \begin{bmatrix} 0 \\ 0 \end{bmatrix}$$

This vector equation will be satisfied if and only if

$$\begin{aligned} (c_1 - c_2 + 2c_3)t + (c_1 + 3c_3)\ln t + (c_2 + c_3)e^t = 0 \\ (-c_1 + c_2 - 2c_3)t + (c_1 + 3c_3)e^t + (c_2 + c_3)t^2 = 0 \end{aligned}$$

Since t, t^2, e^t, and $\ln t$ are linearly independent *functions,* these two scalar equations will hold if and only if each coefficient of these four functions is zero, i.e., if and only if

$$c_1 - c_2 + 2c_3 = 0 \qquad c_1 + 3c_3 = 0 \qquad c_2 + c_3 = 0$$

We easily find that for every real number k, $(c_1, c_2, c_3) = k(3, 1, -1)$ is a solution of this linear system. It follows that $3\mathbf{v}_1 + \mathbf{v}_2 - \mathbf{v}_3 = \mathbf{0}$, and so the vector functions $\mathbf{v}_1$, $\mathbf{v}_2$, $\mathbf{v}_3$ are linearly dependent.

EXERCISES

1. Find all intervals over which the given first-order system is normal.

(a) $x_1' = \sqrt{t}x_1 - 100x_2 - 1/t$
$x_2' = 10x_1 + \sqrt{1 - t}x_2 + \tan t$

(b) $x_1' = 2e^t x_1 + t^{1/3}x_2 + \ln|t + 2|$
$x_2' = e^{-t}x_1 - 6t^3x_2 - \coth t$

2. Under what conditions is a matrix equation $\mathbf{x}' = \mathrm{A}(t)\mathbf{x} + \mathbf{f}(t)$ normal on an interval I?

3. Write out the first-order system which has the given coefficient matrix $\mathrm{A}(t)$ and whose functions are the respective components of the vector function $\mathbf{f}(t)$.

(a) $\mathrm{A}(t) = \begin{bmatrix} 4 & -1 & 5 \\ 3 & 0 & 2 \\ 0 & 1 & 6 \end{bmatrix} \quad \mathbf{f}(t) = \begin{bmatrix} t \\ 2e^{-t} \\ 0 \end{bmatrix}$

(b) $\mathrm{A}(t) = \begin{bmatrix} 0 & 1 + t & -2t \\ t - 2 & 0 & t^2 \\ 3t & t - 3 & 0 \end{bmatrix} \quad \mathbf{f}(t) = 7t\begin{bmatrix} e^t \\ -1 \\ t \end{bmatrix}$

4. Reduce each equation or system to a first-order system and write the first-order system as a matrix equation.

(a) $y'' - 2ty' + y = t^2$

(b) $y'' + 4y' - 8y = 5\cosh 2t$

(c) $(\cosh t)y''' + (\sinh t)y'' - (\sinh 2t)y = te^t$

(d) $e^t y''' + 2(\cosh t)y'' - 4(\sinh t)y' - e^{2t}y = 7e^{3t}$

(e) $x_1' - 2x_2' + x_1 = t$
$x_2' - 2x_1 + x_2 = t^2$

(f) $x_1'' - 2x_1 + x_2' = \cos t$
$x_1' - x_2' + 3x_2 = \sin t$

5. Reduce each initial-value problem to one involving a first-order system.

(a) $2y'' - 6y' - 8y = \ln|t|;\ y(1) = 1,\ y'(1) = -1$

(b) $y''' - 17ty'' + 13e^t y' = 11 + 7e^{-t};$
$y(0) = 1,\ y'(0) = 2,\ y''(0) = 3$

6. If

$$\mathbf{x}(t_0) = \begin{bmatrix} k_1 \\ \vdots \\ k_n \end{bmatrix}$$

is an initial-condition for System (7), what initial conditions go with Eq. (5)?

7. If $y(t_0) = k_1, y'(t_0) = k_2, \ldots, y^{(n-1)}(t_0) = k_n$ are initial conditions for Eq. (5), what initial conditions go with System (7)?

8. Write an nth-order linear differential equation that can be transformed into the given matrix equation.

(a) $\begin{bmatrix} x_1 \\ x_2 \end{bmatrix}' = \begin{bmatrix} 0 & 1 \\ -e^{2t} & -e^t \end{bmatrix} + \begin{bmatrix} 0 \\ 1 \end{bmatrix} e^{3t}$

(b) $\begin{bmatrix} x_1 \\ x_2 \\ x_3 \end{bmatrix}' = \begin{bmatrix} 0 & 1 & 0 \\ 0 & 0 & 1 \\ 1 & t & t^2 \end{bmatrix}\begin{bmatrix} x_1 \\ x_2 \\ x_3 \end{bmatrix} + \begin{bmatrix} 0 \\ 0 \\ 1 \end{bmatrix} \ln|t|$

9. **(a)** Find a complete solution of $y'' - y' = t$ by the methods of Chap. 2.

(b) Reduce $y'' - y' = t$ to a first-order system.

(c) Using a complete solution of $y'' - y' = t$, solve the system of Part **(b)**.

(d) Using a complete solution of the system of Part **(b)**, solve $y'' - y' = t$.

10. Work Exercise 9 for the equation $y'' - y' = e^t$.

11. **(a)** Let each of the linear operators defined by

$$L_{ij} = a_{ij}(t)D^2 + b_{ij}(t)D + c_{ij}(t) \qquad 1 \le i, j \le 2$$

have continuous coefficient functions $a_{ij}(t)$, $b_{ij}(t)$, $c_{ij}(t)$ on an interval I. Also let $f_1(t)$ and $f_2(t)$ be continuous on this interval. Show that if

$$a_{11}(t)a_{22}(t) - a_{12}(t)a_{21}(t) \neq 0$$

everywhere on I, then the linear differential system

$$L_{11}x_1 + L_{12}x_2 = f_1(t)$$
$$L_{21}x_1 + L_{22}x_2 = f_2(t)$$

can be reduced to a normal first-order system on I.

(b) Extend the definitions, assumptions, and conclusion of Part **(a)** to a system of three linear differential equations in three unknowns

$$L_{i1}x_1 + L_{i2}x_2 + L_{i3}x_3 = f_i(t) \qquad 1 \le i \le 3$$

(c) Show that a similar extension can be made to a system of n linear differential equations in n (>3) unknowns

$$\sum_{j=1}^{n} L_{ij}x_j = f_i(t) \qquad 1 \le i \le n$$

12. Determine whether or not Theorem 1 implies that the given initial-value problem has a unique solution. If it does, specify how large an interval the solution is valid over.

(a) $x_1' = 5x_1 - 4x^2 + \sin t$
$x_2' = 3x_1 + 2x_2 - e^t$
$x_i(-1) = 9,\ x_2(-1) = 19$

(b) $x_1' = tx_1 - t^2x_2 + \tan t$
$x_2' = t^2x_1 + tx_2 + \sec t$
$x_1(0) = x_2(0) = 0$

(c) $x_1' = x_1 + \sqrt{t}x_2 + e^t$
$x_2' = x_2 + t^2x_2 - \sqrt{-t}$
$x_1(0) = -1,\ x_2(0) = -6$

(d) $x_1'' - tx_1' + x_2' = \ln t$
$t^2x_1' + x_2' - 2x_1 = \sqrt{1-t}$
$x_1(1) = 0,\ x_1'(0) = 1,\ x_2(0) = 2$

13. Determine whether the given vector functions are linearly dependent or linearly independent.

(a) $\mathbf{v}_1(t) = \begin{bmatrix} t-2 \\ 3t \end{bmatrix},\ \mathbf{v}_2(t) = \begin{bmatrix} t-1 \\ 2t \end{bmatrix},\ -\infty < t < \infty$

(b) $\mathbf{v}_1(t) = \begin{bmatrix} \cosh t \\ e^t \end{bmatrix},\ \mathbf{v}_2(t) = \begin{bmatrix} e^{-t} \\ \sinh t \end{bmatrix},\ -\infty < t < \infty$

(c) $\mathbf{v}_1(t) = \begin{bmatrix} \cos^2 t \\ \cosh^2 t - 1 \end{bmatrix},\ \mathbf{v}_2(t) = \begin{bmatrix} 1 - \sin^2 t \\ \sinh^2 t \end{bmatrix},$
$-\infty < t < \infty$

(d) $\mathbf{v}_1(t) = \begin{bmatrix} e^t \\ -2t \\ e^{-t} \end{bmatrix},\ \mathbf{v}_2(t) = \begin{bmatrix} e^{-t} \\ 1-t \\ 2e^t \end{bmatrix},$

$\mathbf{v}_3(t) = \begin{bmatrix} 4e^{-t} - 2\sinh t \\ 5 - 3t \\ 9e^t + 2\sinh t \end{bmatrix},\ -\infty < t < \infty$

(e) $\mathbf{v}_1(t) = \begin{bmatrix} 1 + \ln|t| \\ 0 \\ \sqrt{1-t} \end{bmatrix},\ \mathbf{v}_2(t) = \begin{bmatrix} \sin \pi t \\ t^2 - t \\ \cos \pi t \end{bmatrix},$

$\mathbf{v}_3(t) = \begin{bmatrix} \ln|t| \\ 1 \\ \tan t \end{bmatrix},\ 0 < t \le 1$

14. Find a first-order system that cannot be converted into a single linear differential equation. *Hint:* Begin with the first-order system

$$x_1' = ax_1 + x_2 \qquad a, c, d \text{ constants}$$
$$x_2' = cx_1 + dx_2$$

4.4 THEOREMS FOR GENERAL LINEAR DIFFERENTIAL SYSTEMS

The only linear differential systems for which general procedures are available for finding exact solutions are those with constant coefficients. Fortunately, most elementary applications lead to such systems, and naturally these will be of particular interest to us. However, although systems with variable coefficients can rarely be solved analytically, several key theorems describing properties of their solutions can be proved as easily in the setting of general first-order systems as they can be proved in the constant-coefficient case. Accordingly, in this section, we shall focus our attention on first-order systems and establish a number of important results which hold for the general case as well. Our first theorem generalizes a familiar property of a single linear differential equation (Theorem 1, Sec. 2.2) and tells us how to form a family of solutions from a set of particular solutions.

THEOREM 1 If $\mathbf{x}_1, \mathbf{x}_2, \ldots, \mathbf{x}_m$ are m specific solutions of a homogeneous first-order system $\mathbf{x}' = A(t)\mathbf{x}$, then for all values of the constants $c_1, c_2, \ldots, c_m$,

$$\mathbf{x} = c_1\mathbf{x}_1 + c_2\mathbf{x}_2 + \cdots + c_m\mathbf{x}_m = \sum_{k=1}^{m} c_k\mathbf{x}_k$$

is also a solution.

PROOF Let I denote any interval over which $\mathbf{x}_1, \mathbf{x}_2, \ldots, \mathbf{x}_m$ are solutions of $\mathbf{x}'(t) = \mathrm{A}(t)\mathbf{x}$. For each t *in* I

$$\mathbf{x}'(t) = \left[\sum_{k=1}^{m} c_k\mathbf{x}_k(t)\right]' = \sum_{k=1}^{m} c_k\mathbf{x}_k'(t) = \sum_{k=1}^{m} c_k\mathrm{A}(t)\mathbf{x}_k(t) = \mathrm{A}(t)\left[\sum_{k=1}^{m} c_k\mathbf{x}_k(t)\right]$$
$$= \mathrm{A}(t)\mathbf{x}(t)$$

This shows that $\mathbf{x} = \Sigma_{k=1}^{m} c_k\mathbf{x}_k$ satisfies the homogeneous system $\mathbf{x}' = \mathrm{A}(t)\mathbf{x}$ on I, and the theorem is proved.

The essence of this theorem is that every finite linear combination of a number of specific solutions of a homogeneous first-order system is itself a solution, whether the specific solutions are distinct or not.

Our next theorem extends the scope of Theorem 1 and is reminiscent of another familiar result (Theorem 2, Sec. 2.2) pertaining to a single linear differential equation.

THEOREM 2 On any interval I over which a homogeneous first-order system in n unknowns is normal, the system has n linearly independent solutions $\mathbf{x}_1, \mathbf{x}_2, \ldots, \mathbf{x}_n$. Any particular solution of the system on I can be expressed as a linear combination of these n linearly independent solutions.

Except for notation, all essential features involved in a general proof of this theorem are covered by the following proof for a homogeneous system in three unknowns

(1) $$x_i' = a_{i1}(t)x_1 + a_{i2}(t)x_2 + a_{i3}(t)x_3 \qquad 1 \le i \le 3$$

assumed to be normal on an interval I.

PROOF Let $\mathbf{x}_1, \mathbf{x}_2, \mathbf{x}_3$ be the solutions of (1) guaranteed by Theorem 1, Sec. 4.3, which at some point t_0 of I, have the values

(2) $$\mathbf{x}_1(t_0) = \begin{bmatrix} 1 \\ 0 \\ 0 \end{bmatrix} \qquad \mathbf{x}_2(t_0) = \begin{bmatrix} 0 \\ 1 \\ 0 \end{bmatrix} \qquad \mathbf{x}_3(t_0) = \begin{bmatrix} 0 \\ 0 \\ 1 \end{bmatrix}$$

We first prove that $\mathbf{x}_1, \mathbf{x}_2, \mathbf{x}_3$ are linearly independent. To do so we consider an identity of the form $c_1\mathbf{x}_1(t) + c_2\mathbf{x}_2(t) + c_3\mathbf{x}_3(t) \equiv \mathbf{0}$ on I. If such an identity is to hold, then $c_1\mathbf{x}_1(t_0) + c_2\mathbf{x}_2(t_0) + c_3\mathrm{x}_3(t_0) = \mathbf{0}$ because t_0 is contained in I. But, from (2), we see that this condition implies $c_1 = c_2 = c_3 = 0$. Hence, by definition, $\mathbf{x}_1$, $\mathbf{x}_2$, and $\mathbf{x}_3$ are three linearly independent solutions of System (1) on I.

Now suppose $\mathbf{x}$ is any particular solution of (1) over I and that its value at t_0 is

$$\mathbf{x}(t_0) = \begin{bmatrix} k_1 \\ k_2 \\ k_3 \end{bmatrix}$$

Since $\mathbf{x}_1, \mathbf{x}_2, \mathbf{x}_3$, and $\mathbf{x}$ are all solutions of (1) over I, so is the vector function $\mathbf{v}$ given by the linear combination

(3) $$\mathbf{v} = \mathbf{x} - k_1\mathbf{x}_1 - k_2\mathbf{x}_2 - k_3\mathbf{x}_3$$

Evaluating $\mathbf{v}$ at t_0 and again recalling (2), we have

$$\mathbf{v}(t_0) = \mathbf{x}(t_0) - k_1\mathbf{x}_1(t_0) - k_2\mathbf{x}_2(t_0) - k_3\mathbf{x}_3(t_0) = \mathbf{0}$$

It follows, from Corollary 1, Sec. 4.3, that $\mathbf{v}(t) \equiv \mathbf{0}$, hence

$$\mathbf{x}(t) \equiv k_1\mathbf{x}_1(t) + k_2\mathbf{x}_2(t) + k_3\mathbf{x}_3(t) \qquad \text{on } I$$

This shows that $\mathbf{x}$ is expressible as a linear combination of $\mathbf{x}_1$, $\mathbf{x}_2$, and $\mathbf{x}_3$. In the case of System (1), Theorem 2 is now established. ◗

There is a simple criterion, much like the wronskian criterion implicit in Theorems 4 and 5, Sec. 2.2, that can be used to test n particular solutions of a homogeneous system in n unknowns for linear dependence or independence. It involves determinants of matrix functions whose columns are vector functions with a common domain. That is, the determinants are of the form

$$w = \det\,[\mathbf{w}_1\,\mathbf{w}_2 \cdots \mathbf{w}_n] = \begin{vmatrix} w_{11} & w_{12} & \cdots & w_{1n} \\ w_{21} & w_{22} & \cdots & w_{2n} \\ \cdots & \cdots & \cdots & \cdots \\ w_{n1} & w_{n2} & \cdots & w_{nn} \end{vmatrix} \tag{4}$$

where, for $1 \le j \le n$, each component of the vector function $\mathbf{w}_j = \begin{bmatrix} w_{1j} \\ \vdots \\ w_{nj} \end{bmatrix}$ in the jth column of w is defined on an interval I. In other words, the domain of every element of the matrix function $\mathrm{W} = [w_{ij}]$ is I. Of course, $w(t)$ denotes the value of $w = \det\,[\mathbf{w}_1\mathbf{w}_2 \cdots \mathbf{w}_n]$ at any point t in I.

Obviously, every nth-order wronskian is a determinant of type (4). On the other hand, since we defined a wronskian in Sec. 2.2 as a determinant each of whose row vectors (after the first) was the derivative of the preceding row vector, it is clear that not all determinants of the type (4) are wronskians in the sense of Sec. 2.2. Nonetheless, (4) is called the **wronksian of the vector functions** $\mathbf{w}_1, \mathbf{w}_2, \ldots, \mathbf{w}_n$. In particular, when $\mathbf{w}_1, \mathbf{w}_2, \ldots, \mathbf{w}_n$ are solution vectors of a system of first-order equations which a single nth-order linear differential equation has been reduced to, then (4) is identical with the wronskian of the corresponding solutions of the nth-order equation (see Exercise 3). A very important property of wronskians whose columns are solutions of a homogeneous first-order system in the following.

THEOREM 3 On any interval over which a homogeneous first-order system in n unknowns is normal, the wronskian of n particular solutions either is identically zero, or else its value is never zero.

This time we give a proof only for the case $n = 2$ since it clearly indicates how a general proof can be written.

◖ **PROOF** Let $\mathbf{u} = \begin{bmatrix} u_1 \\ u_2 \end{bmatrix}$ and $\mathbf{v} = \begin{bmatrix} v_1 \\ v_2 \end{bmatrix}$ be particular solutions of the first-order system

$$\mathbf{x}' = \begin{bmatrix} a_{11}(t) & a_{12}(t) \\ a_{21}(t) & a_{22}(t) \end{bmatrix} \mathbf{x} \tag{5}$$

over an interval I on which the system is normal. Then for every t in I the four equations

$$\begin{aligned} u_1'(t) &= a_{11}(t)u_1(t) + a_{12}(t)u_2(t) \qquad & v_1'(t) &= a_{11}(t)v_1(t) + a_{12}(t)v_2(t) \\ u_2'(t) &= a_{21}(t)u_1(t) + a_{22}(t)u_2(t) \qquad & v_2'(t) &= a_{21}(t)v_1(t) + a_{22}(t)v_2(t) \end{aligned} \tag{6}$$

must hold, and the wronskian of **u** and **v** has the value

$$w(t) = \begin{vmatrix} u_1(t) & v_1(t) \\ u_2(t) & v_2(t) \end{vmatrix}$$

According to Exercise 24, Sec. 3.5,

$$w'(t) = \begin{vmatrix} u_1'(t) & v_1'(t) \\ u_2(t) & v_2(t) \end{vmatrix} + \begin{vmatrix} u_1(t) & v_1(t) \\ u_2'(t) & v_2'(t) \end{vmatrix}$$

Substituting for $u_1'(t)$, $v_1'(t)$, $u_2'(t)$, and $v_2'(t)$ from (6), we have

$$w'(t) = \begin{vmatrix} a_{11}(t)u_1(t) + a_{12}(t)u_2(t) & a_{11}(t)v_1(t) + a_{12}(t)v_2(t) \\ u_2(t) & v_2(t) \end{vmatrix}$$
$$+ \begin{vmatrix} u_1(t) & v_1(t) \\ a_{21}(t)u_1(t) + a_{22}(t)u_2(t) & a_{21}(t)v_1(t) + a_{22}(t)v_2(t) \end{vmatrix}$$

By using Property 9, Sec. 3.5, to eliminate the terms containing $a_{12}(t)$ and $a_{21}(t)$ and then using Property 5, Sec. 3.5, to factor $a_{11}(t)$ and $a_{22}(t)$ from the resulting determinants, the expression for $w'(t)$ becomes

(7) $$w'(t) = [a_{11}(t) + a_{22}(t)]\begin{vmatrix} u_1(t) & v_1(t) \\ u_2(t) & v_2(t) \end{vmatrix} = [a_{11}(t) + a_{22}(t)]w(t)$$

Since both a_{11} and a_{22} are continuous, w satisfies a normal first-order linear differential equation on I. Hence, according to Corollary 1, Sec. 2.1, if the value of w is zero at even one point of I, then $w(t) \equiv 0$. Thus $w(t)$ is either identically zero, or else never zero, as the theorem asserts. ◗

For a homogeneous first-order system in n unknowns, (7) generalizes to

(8) $$w'(t) = [a_{11}(t) + a_{22}(t) + \cdots + a_{nn}(t)]w(t)$$

Of course, the coefficient matrix $A(t) = [a_{ij}(t)]$ of the system is of order n and w is the wronskian of n solution vectors.

We are now prepared to establish a useful criterion for testing n particular solutions of a homogeneous first-order system in n unknowns for linear independence.

THEOREM 4 Any n particular solutions of a homogeneous first-order system in n unknowns, over an interval I on which the system is normal, are linearly independent if and only if their wronskian has a nonzero value at each point of I.

Here again a proof is given only for the case $n = 2$ because it typifies a general proof in all essential details.

◖ **PROOF** We shall first show that if **u** and **v** are linearly independent particular solutions of (5) on I, the value of their wronskian w is never zero there.

Assuming contrariwise that there is a point t_0 of I such that

$$w(t_0) = \begin{vmatrix} u_1(t_0) & v_1(t_0) \\ u_2(t_0) & v_2(t_0) \end{vmatrix} = 0$$

leads to the conclusion that the homogeneous system of linear algebraic equations

(9) $$\begin{aligned} u_1(t_0)c + v_1(t_0)d &= 0 \\ u_2(t_0)c + v_2(t_0)d &= 0 \end{aligned}$$

has a *nontrivial* solution $(c, d) = (k_1, k_2)$. By Theorem 1, the linear combination

$$\mathbf{x} = k_1\mathbf{u} + k_2\mathbf{v}$$

is a solution of (5). Moreover, since (k_1, k_2) is a solution of Eqs. (9), it follows from the last equation that $\mathbf{x}(t_0) = \mathbf{0}$. According to Corollary 1, Sec. 4.3, this implies that

$$\mathbf{x}(t) = k_1\mathbf{u}(t) + k_2\mathbf{v}(t) \equiv \mathbf{0} \qquad \text{on } I$$

where either $k_1 \neq 0$ or $k_2 \neq 0$. This proves that if $w(t)$ is zero at even one point of I, then **u** and **v** must be linearly dependent. The contrapositive of this implication asserts that if **u** and **v** are linearly independent, then $w(t)$ is never zero on I, as we set out to prove.

Now suppose that $w(t)$ is never zero and that $c_1\mathbf{u}(t) + c_2\mathbf{v}(t) \equiv \mathbf{0}$ on I. Then at any fixed point t_0 of I, Eqs. (9) hold and $w(t_0) \neq 0$. Hence both c_1 and c_2 must be zero. This shows that if $w(t)$ is never zero on I, then **u** and **v** are linearly independent, and the theorem is proved. ◗

An equivalent way of stating Theorem 4 is the following.

THEOREM 5 Any n particular solutions of a homogeneous first-order system in n unknowns over an interval I on which the system is normal are linearly dependent if and only if at some point of I the value of their wronskian is zero.

Before going on to another theorem concerning solutions of homogeneous first-order systems, let us pause and see how some of the theorems we have already learned relate to a specific example.

EXAMPLE 1

From Example 1, Sec. 4.2, we know that on $(-\infty, \infty)$ the vector functions $\mathbf{v}_1$ and $\mathbf{v}_2$ defined by

$$\mathbf{v}_1(t) = \begin{bmatrix} -5 \\ 1 \end{bmatrix} e^{-t} = \begin{bmatrix} -5e^{-t} \\ e^{-t} \end{bmatrix} \qquad \text{and} \qquad \mathbf{v}_2(t) = \begin{bmatrix} -2 \\ 1 \end{bmatrix} e^{2t} = \begin{bmatrix} -2e^{2t} \\ e^{2t} \end{bmatrix}$$

are solutions of the homogeneous linear differential system

$$\begin{aligned} x' + 2x + y' + 6y &= 0 \\ 2x' + 3x + 3y' + 8y &= 0 \end{aligned} \tag{10}$$

Transposing the terms of these equations in x and y, then solving algebraically for x' and y', we obtain

$$x' = \frac{\begin{vmatrix} -2x - 6y & 1 \\ -3x - 8y & 3 \end{vmatrix}}{\begin{vmatrix} 1 & 1 \\ 2 & 3 \end{vmatrix}} = -3x - 10y \qquad \text{and} \qquad y' = \frac{\begin{vmatrix} 1 & -2x - 6y \\ 2 & -3x - 8y \end{vmatrix}}{\begin{vmatrix} 1 & 1 \\ 2 & 3 \end{vmatrix}} = x + 4y$$

In this way, (10) is transformed into the first-order system

$$\begin{aligned} x' &= -3x - 10y \\ y' &= x + 4y \end{aligned} \tag{11}$$

It is readily verified that $\mathbf{v}_1$ and $\mathbf{v}_2$ are solutions of (11). Theorem 1 tells us that every linear combination

$$\mathbf{v} = c_1\mathbf{v}_1 + c_2\mathbf{v}_2 = \begin{bmatrix} -5c_1e^{-t} - 2c_2e^{2t} \\ c_1e^{-t} + c_2e^{2t} \end{bmatrix}$$

satisfies (11) also. Of course, this is easy to confirm directly.

The coefficient matrix of (11) is $A = \begin{bmatrix} -3 & -10 \\ 1 & 4 \end{bmatrix}$, and $a_{11} = -3$, $a_{22} = 4$ are its diagonal elements. Thus, without solving (11), we know from (7) that the wronskian w of any pair of particular solutions must satisfy the first-order linear differential equation $w' = w$. Hence $w(t)$ must be an expression of the form $w(t) = ke^t$, k constant. Computing the wronskian of $\mathbf{v}_1$ and $\mathbf{v}_2$ we get

$$\begin{vmatrix} -5e^{-t} & -2e^{2t} \\ e^{-t} & e^{2t} \end{vmatrix} = -3e^t$$

so, clearly, $k = -3$ and $w(t) = -3e^t$. Since e^t is never zero, neither is $w(t)$. Thus we know from Theorem 4 that $\mathbf{v}_1$ and $\mathbf{v}_2$ are linearly independent solutions of (11) on $(-\infty, \infty)$.

From Theorem 2, we know that a homogeneous first-order system

(12) $$x_i' = a_{i1}(t)x_1 + a_{i2}(t)x_2 + \cdots + a_{in}(t)x_n \qquad 1 \le i \le n$$

which is normal on an interval I, has at least one set of linearly independent solutions in terms of which *any* particular solution can be expressed as a linear combination. However, Theorem 2 does not say that an arbitrary solution of (12) on I can be expressed as a linear combination of *every* set of n linearly independent solutions of (12). This is the assertion of the next theorem, which tells us in effect that System (12) can have no more than n linearly independent solutions on I (Exercise 6).

THEOREM 6 If $\mathbf{x}_1, \mathbf{x}_2, \ldots, \mathbf{x}_n$ are any n linearly independent particular solutions of (12) on an interval I over which the system is normal, any particular solution of (12) on I is a linear combination of $\mathbf{x}_1, \mathbf{x}_2, \ldots, \mathbf{x}_n$.

◖ **PROOF FOR THE CASE** $N = 2$ Let $\mathbf{u}$ and $\mathbf{v}$ be linearly independent particular solutions of (5) on an interval I over which that system is normal, and let $\mathbf{x}$ be an arbitrary particular solution on I. Since $\mathbf{u}$ and $\mathbf{v}$ are linearly independent, the value $w(t)$ of their wronskian w is never zero on I. Hence, for any fixed number t_0 in I, the linear combination

(13) $$\mathbf{y} = \mathbf{x} - \frac{\begin{vmatrix} x_1(t_0) & v_1(t_0) \\ x_2(t_0) & v_2(t_0) \end{vmatrix}}{w(t_0)}\mathbf{u} - \frac{\begin{vmatrix} u_1(t_0) & x_1(t_0) \\ u_2(t_0) & x_2(t_0) \end{vmatrix}}{w(t_0)}\mathbf{v}$$

of $\mathbf{x}$, $\mathbf{u}$, and $\mathbf{v}$ is a well-defined solution of (5), and it is easy to show that $\mathbf{y}(t_0) = 0$ by putting the right-hand side of (13) over $w(t_0)$ as a common denominator and then expanding the three determinants in the numerator. By Corollary 1, Sec. 4.3, this implies that $\mathbf{y}(t) \equiv \mathbf{0}$ on I, and it follows from (13) that $\mathbf{x}$ is a linear combination of $\mathbf{u}$ and $\mathbf{v}$. The proof in the general case follows the same pattern. ◗

Since, in general, every solution of (12) can be expressed as a linear combination of n linearly independent solutions and since any linear combination of solutions of (12) is itself a solution, an arbitrary linear combination

$$\mathbf{x} = c_1\mathbf{x}_1 + c_2\mathbf{x}_2 + \cdots + c_n\mathbf{x}_n$$

of any n linearly independent solutions is called a **complete solution** of (12).

EXAMPLE 2

In Example 1, we found the vector functions $\mathbf{v}_1$ and $\mathbf{v}_2$ to be linearly independent solutions of (11) on $(-\infty, \infty)$. Thus, a complete solution of (11) is

(14) $$\begin{bmatrix} x \\ y \end{bmatrix} = c_1\begin{bmatrix} -5 \\ 1 \end{bmatrix}e^{-t} + c_2\begin{bmatrix} -2 \\ 1 \end{bmatrix}e^{2t}$$

As is clear from Example 1, Sec. 4.2, Eq. (14) also gives a complete solution of (10). The systems (10) and (11) are therefore equivalent.

We now introduce two important definitions that make it easy to describe complete solutions of linear differential systems.

DEFINITION 1 Any n linearly independent particular solutions $\mathbf{x}_1, \mathbf{x}_2, \ldots, \mathbf{x}_n$ of a homogeneous first-order system in n unknowns, on an interval over which the system is normal, are called **fundamental solutions** or a **basis for all solutions,** of the system on that interval.

DEFINITION 2 A matrix $X = [\mathbf{x}_1\mathbf{x}_2 \cdots \mathbf{x}_n]$ whose column vectors $\mathbf{x}_j$ constitute a basis for a homogeneous first-order system in n unknowns is called a **fundamental matrix of the system.**

Clearly, a matrix X is a fundamental matrix if and only if its column vectors are fundamental solutions of some homogeneous linear differential system. Of course, here as elsewhere, it is to be understood that suitable conditions of normality are satisfied by our systems when nothing else is said.

Now, let us consider a general nonhomogeneous system

$$x_i' = a_{i1}(t)x_1 + a_{i2}(t)x_2 + \cdots + a_{in}(t)x_n + f_i(t) \qquad 1 \le i \le n \tag{15}$$

and see how to construct a complete solution of it. As the next theorem clearly indicates, this can be accomplished by using the same procedure that gave us a complete solution of a single nonhomogeneous linear differential equation.

THEOREM 7 If $\mathbf{v}$ is any specific solution of (15), and if $c_1\mathbf{x}_1 + c_2\mathbf{x}_2 + \cdots + c_n\mathbf{x}_n$ is a complete solution of the associated homogeneous system, then a complete solution of the nonhomogeneous system is $\mathbf{x} = c_1\mathbf{x}_1 + c_2\mathbf{x}_2 + \cdots + c_n\mathbf{x}_n + \mathbf{v}$.

PROOF Let $\mathbf{x}$ denote an arbitrary solution and let $\mathbf{v}$ be any specific solution of (15). Then the two matrix equations

$$\mathbf{x}'(t) = A(t)\mathbf{x}(t) + \mathbf{f}(t)$$
$$\mathbf{v}'(t) = A(t)\mathbf{v}(t) + \mathbf{f}(t)$$

must hold. Subtracting the second of these equations memberwise from the first, and rearranging, we obtain

$$\frac{d}{dt}[\mathbf{x}(t) - \mathbf{v}(t)] = \mathrm{A}(t)[\mathbf{x}(t) - \mathbf{v}(t)]$$

This shows that $\mathbf{x} - \mathbf{v}$ is a solution of the homogeneous system $\mathbf{x}' = \mathrm{A}(t)\mathbf{x}$ corresponding to (15). Thus if $\mathbf{x}_1, \mathbf{x}_2, \ldots, \mathbf{x}_n$ is any basis of this homogeneous system, Theorem 6 implies

$$\mathbf{x} - \mathbf{v} = c_1\mathbf{x}_1 + c_2\mathbf{x}_2 + \cdots + c_n\mathbf{x}_n$$

Transposing $\mathbf{v}$, we get a complete solution of (15)

$$\mathbf{x} = c_1\mathbf{x}_1 + c_2\mathbf{x}_2 + \cdots + c_n\mathbf{x}_n + \mathbf{v}$$

as asserted by the theorem.

We are now prepared to extend the concepts of *particular integral* and *complementary function* from a single differential equation to a system of equations. We call the term $\mathbf{v}$ in a complete solution $\mathbf{x} = c_1\mathbf{x}_1 + c_2\mathbf{x}_2 + \cdots + c_n\mathbf{x}_n + \mathbf{v}$ of (15) a **particular integral** of the nonhomogeneous system. The linear combination $c_1\mathbf{x}_1 + c_2\mathbf{x}_2 + \cdots + c_n\mathbf{x}_n$, which is a complete solution of the associated homogeneous system, is called a **complementary function** of the nonhomogeneous system.

The same three-step procedure we used to solve nonhomogeneous linear differential equations can be adapted as follows to nonhomogeneous linear systems in n unknowns.

1. Find n particular solution vectors $\mathbf{x}_1, \mathbf{x}_2, \ldots, \mathbf{x}_n$ of the related homogeneous system which have a nonvanishing wronskian. Then form a *complementary function* $c_1\mathbf{x}_1 + c_2\mathbf{x}_2 + \cdots + c_n\mathbf{x}_n$ of the given system.
2. Find one particular solution $\mathbf{v}$ of the nonhomogeneous system.
3. Add the *particular integral* $\mathbf{v}$ of Step 2 to the complementary function of Step 1 to obtain a complete solution $\mathbf{x} = c_1\mathbf{x}_1 + c_2\mathbf{x}_2 + \cdots + c_n\mathbf{x}_n + \mathbf{v}$ of the given system.

Frequently the nonhomogeneous term $\mathbf{f}(t)$ of a nonhomogeneous system $\mathbf{x}' = \mathrm{A}(t)\mathbf{x} + \mathbf{f}(t)$ will be expressible as a sum of other vector functions. In this event, the following theorem can be used to find a particular integral by stages.

THEOREM 8 If, for $k = 1, 2, \ldots, m$, $\mathbf{v}_k$ is a particular integral of $\mathbf{x}' = \mathrm{A}(t)\mathbf{x} + \mathbf{f}_k(t)$, then $\mathbf{v}_1 + \mathbf{v}_2 + \cdots + \mathbf{v}_m$ is a particular integral of $\mathbf{x}' = \mathrm{A}(t)\mathbf{x} + \mathbf{f}_1(t) + \cdots + \mathbf{f}_m(t)$.

PROOF Since, by hypothesis, $\mathbf{v}_k'(t) = \mathrm{A}(t)\mathbf{v}_k(t) + \mathbf{f}_k(t)$ for $1 \le k \le m$, summation over k gives

$$\left[\sum_{k=1}^{m} \mathbf{v}_k(t)\right]' = \sum_{k=1}^{m} \mathbf{v}_k'(t) = \mathrm{A}(t)\left[\sum_{k=1}^{m} \mathbf{v}_k(t)\right] + \sum_{k=1}^{m} \mathbf{f}_k(t)$$

and this is what the theorem asserts.

In general, it is impossible to find a complete solution of a nonhomogeneous first-order system, for the system may be so complicated that neither a complementary function nor a particular integral can be determined. However, *if* a complementary function can be found, so can a particular integral, for variation of parameters (see Sec. 2.4) works as well on systems as it does on a single equation. Not unexpectedly, when the method of variation of parameters is applied to a system, it becomes necessary to integrate vector functions. This is done by integrating each component of the vector. Similarly, a matrix function $A(t)$ is integrated by integrating each element $a_{ij}(t)$, that is, by definition,

$$\int \mathrm{A}(t)\,dt = \left[\int a_{ij}(t)\,dt\right] \qquad 1 \le i \le m,\ 1 \le j \le n \tag{16}$$

where, if occasion demands, a definite integral may be used instead of an indefinite integral. Of course, every element must be integrable.

EXAMPLE 3

A matrix function X and a vector function $\mathbf{f}$ are defined by

$$\mathrm{X}(t) = \begin{bmatrix} e^t & te^t \\ e^t & (t+1)e^t \end{bmatrix} \qquad -\infty < t < \infty$$

$$\mathbf{f}(t) = \begin{bmatrix} e^t(3t-1)/(t+1) \\ e^t \end{bmatrix} \qquad 0 < |t+1|$$

Disregarding all constants of integration, evaluate $X(t) \int X^{-1}(t)\mathbf{f}(t)\,dt$ for $t \neq -1$.

With Eq. (10), Sec. 3.7, for X^{-1} in mind, we compute

$$|X(t)| = \begin{vmatrix} e^t & te^t \\ e^t & (t+1)e^t \end{vmatrix} = e^{2t} \qquad \text{Adj } X(t) = \begin{bmatrix} (t+1)e^t & -e^t \\ -te^t & e^t \end{bmatrix}^T = \begin{bmatrix} t+1 & -t \\ -1 & 1 \end{bmatrix} e^t$$

and then

$$X^{-1}(t) = \frac{\text{Adj } X(t)}{|X(t)|} = \begin{bmatrix} t+1 & -t \\ -1 & 1 \end{bmatrix} e^{-t}$$

Postmultiplying $X^{-1}(t)$ by $\mathbf{f}(t)$ and rearranging, we get

$$X^{-1}(t)\mathbf{f}(t) = \begin{bmatrix} 2t-1 \\ \dfrac{4}{t+1} - 2 \end{bmatrix} \qquad t \neq -1$$

Integrating this vector function, in accordance with (16), and dropping constants of integration, we have

$$\int X^{-1}(t)\mathbf{f}(t)\,dt = \begin{bmatrix} t^2 - t \\ 4 \ln|t+1| - 2t \end{bmatrix} \qquad t \neq -1$$

Premultiplying this result by $X(t)$ and combining terms, we find that

$$X(t) \int X^{-1}(t)\mathbf{f}(t)\,dt = \begin{bmatrix} 4t \ln|t+1| - t - t^2 \\ 4(t+1)\ln|t+1| - 3t - t^2 \end{bmatrix} e^t \qquad t \neq -1$$

Now suppose we are to solve a nonhomogeneous first-order system (15) and that a fundamental matrix $X = [\mathbf{x}_1\mathbf{x}_2 \cdots \mathbf{x}_n]$ of the associated homogeneous system is known on an interval I. Then, $\mathbf{x}_j' = A\mathbf{x}_j$, for $1 \leq j \leq n$. Clearly, these n vector equations imply, and are implied by, the matrix equation

$$[\mathbf{x}_1'\mathbf{x}_2' \cdots \mathbf{x}_n'] = [A\mathbf{x}_1 A\mathbf{x}_2 \cdots A\mathbf{x}_n] \tag{17}$$

Since $[\mathbf{x}_1'\mathbf{x}_2' \cdots \mathbf{x}_n'] = [\mathbf{x}_1\mathbf{x}_2 \cdots \mathbf{x}_n]' = X'$ and $[A\mathbf{x}_1 A\mathbf{x}_2 \cdots A\mathbf{x}_n] = A[\mathbf{x}_1\mathbf{x}_2 \cdots \mathbf{x}_n] = AX$, Eq. (17) can be written as

$$X' = AX \tag{18}$$

This equation is simply a concise statement of the fact that on I each column vector of X is a solution vector of the homogeneous system $\mathbf{x}' = A\mathbf{x}$ corresponding to the nonhomogeneous differential system (15) whose matrix form is

$$\mathbf{x}' = A\mathbf{x} + \mathbf{f} \tag{19}$$

If $c_1, c_2, \ldots, c_n$ are arbitrary constants,

$$c_1\mathbf{x}_1 + c_2\mathbf{x}_2 + \cdots + c_n\mathbf{x}_n \tag{20}$$

is a complementary function of (19). Letting $\mathbf{c}$ be the constant column vector with successive components $c_1, c_2, \ldots, c_n$, we can write (20) as

$$\mathbf{x}_1c_1 + \mathbf{x}_2c_2 + \cdots + \mathbf{x}_nc_n = [\mathbf{x}_1\mathbf{x}_2 \cdots \mathbf{x}_n]\mathbf{c} = X\mathbf{c} \tag{21}$$

At this point we vary the parameters $c_1, c_2, \ldots, c_n$; that is, we replace these parameters by n unknown functions $u_1, u_2, \ldots, u_n$, assumed to be differentiable on I. Then we require

$$u_1\mathbf{x}_1 + u_2\mathbf{x}_2 + \cdots + u_n\mathbf{x}_n$$

to be a solution of (19). In other words, we replace the constant vector $\mathbf{c}$ by a vector function $\mathbf{u}$ and seek to determine a particular integral of (19) of the type

$$\mathbf{v} = \mathrm{X}\mathbf{u} \tag{22}$$

Substituting $\mathbf{v}$ into (19) for $\mathbf{x}$ and noting that $\mathbf{v}' = (\mathrm{X}\mathbf{u})' = \mathrm{X}'\mathbf{u} + \mathrm{X}\mathbf{u}'$ (see Exercise 39, Sec. 3.4), we obtain

$$\mathrm{X}'\mathbf{u} + \mathrm{X}\mathbf{u}' = \mathrm{A}\mathrm{X}\mathbf{u} + \mathbf{f}$$

From (18), $\mathrm{X}' = \mathrm{A}\mathrm{X}$, so the first terms on each side of this equation cancel, leaving

$$\mathrm{X}\mathbf{u}' = \mathbf{f} \tag{23}$$

Since X is a fundamental matrix of $\mathbf{x}' = \mathrm{A}\mathbf{x}$ on I, the wronskian of $\mathbf{x}_1, \mathbf{x}_2, \ldots, \mathbf{x}_n$ is never zero. Hence, for every t in I, $\det \mathrm{X}(t) \neq 0$, $\mathrm{X}(t)$ is nonsingular, and $\mathrm{X}^{-1}(t)$ exists. It follows from (23) that

$$\mathbf{u}'(t) = \mathrm{X}^{-1}(t)\mathbf{f}(t) \qquad \text{and} \qquad \mathbf{u}(t) = \int \mathrm{X}^{-1}(t)\mathbf{f}(t)\,dt$$

With $\mathbf{u}$ thus determined, a particular integral (22) of (19) is represented by

$$\mathbf{v}(t) = \mathrm{X}(t)\int \mathrm{X}^{-1}(t)\mathbf{f}(t)\,dt \tag{24}$$

where all constants of integration may be omitted because any particular solution of (19), no matter how special, will serve as a particular integral.

EXAMPLE 4

Find a complete solution of the differential system

$$x_1' = x_1 - x_2 + \frac{e^{-t}}{1+t^2}$$

$$x_2' = 2x_1 - 2x_2 + \frac{2e^{-t}}{1+t^2}$$

We could solve this system directly using elementary row operations as in Sec. 4.2. Instead, let us solve only the associated homogeneous system that way and then find a particular integral by means of (24).

The homogeneous system is obviously equivalent to

$$\begin{aligned} 2(D-1)x_1 + \qquad 2x_2 &= 0 \\ -2x_1 + (D+2)x_2 &= 0 \end{aligned}$$

Operating on the second of these equations with $D - 1$, adding the result to the first equation, and then simplifying, we have

$$\begin{aligned} (D^2 + D)x_2 &= 0 \\ -2x_1 + (D+2)x_2 &= 0 \end{aligned}$$

Solving the first of these equations for x_2, then substituting into the second equation and solving for x_1, we get the complementary function

$$\begin{bmatrix} x_1 \\ x_2 \end{bmatrix} = c_1\begin{bmatrix} 1 \\ 1 \end{bmatrix} + c_2\begin{bmatrix} 1 \\ 2 \end{bmatrix}e^{-t}$$

From this it is evident that a fundamental matrix of the homogeneous system is $X(t) = \begin{bmatrix} 1 & e^{-t} \\ 1 & 2e^{-t} \end{bmatrix}$. Without difficulty, we find $|X(t)| = e^{-t}$, $\text{Adj } X(t) = \begin{bmatrix} 2e^{-t} & -e^{-t} \\ -1 & 1 \end{bmatrix}$, and $X^{-1}(t) = \begin{bmatrix} 2 & -1 \\ -e^{t} & e^{t} \end{bmatrix}$. Noting that $\mathbf{f}(t) = \begin{bmatrix} e^{-t}/(1+t^2) \\ 2e^{-t}/(1+t^2) \end{bmatrix}$, we next compute

$$X^{-1}(t)\mathbf{f}(t) = \begin{bmatrix} 2 & -1 \\ -e^{t} & e^{t} \end{bmatrix}\begin{bmatrix} 1 \\ 2 \end{bmatrix}\frac{e^{-t}}{1+t^2} = \begin{bmatrix} 0 \\ \dfrac{1}{1+t^2} \end{bmatrix}$$

which we integrate to obtain

$$\int X^{-1}(t)\mathbf{f}(t)\,dt = \begin{bmatrix} 0 \\ \text{Tan}^{-1} t \end{bmatrix}$$

Using (24), we find

$$\mathbf{v}(t) = X(t)\int X^{-1}(t)\mathbf{f}(t)\,dt = \begin{bmatrix} 1 & e^{-t} \\ 1 & 2e^{-t} \end{bmatrix}\begin{bmatrix} 0 \\ \text{Tan}^{-1} t \end{bmatrix} = \begin{bmatrix} 1 \\ 2 \end{bmatrix}e^{-t}\,\text{Tan}^{-1} t$$

to be a particular integral of the given system. Of course, a complete solution is

$$\begin{bmatrix} x_1 \\ x_2 \end{bmatrix} = c_1\begin{bmatrix} 1 \\ 1 \end{bmatrix} + c_2\begin{bmatrix} 1 \\ 2 \end{bmatrix}e^{-t} + \begin{bmatrix} 1 \\ 2 \end{bmatrix}e^{-t}\,\text{Tan}^{-1} t$$

EXERCISES

1. (a) Determine whether or not

$$\mathbf{x}_1 = \begin{bmatrix} \cos t \\ -\sin t \\ -\cos t \end{bmatrix},\quad \mathbf{x}_2 = \begin{bmatrix} \sin t \\ \cos t \\ -\sin t \end{bmatrix},\quad \mathbf{x}_3 = \begin{bmatrix} t \\ 1 \\ 0 \end{bmatrix}$$

are fundamental solutions of the system

$$\begin{aligned} x_1' &= x_2 \\ x_2' &= x_3 \\ x_3' &= \frac{1}{t}x_1 - x_2 + \frac{1}{t}x_3 \end{aligned}$$

(b) Show that the wronskian w of $\mathbf{x}_1$, $\mathbf{x}_2$, and $\mathbf{x}_3$ has a zero value but that $w(t) \not\equiv 0$. Does this contradict Theorem 3? Explain.

(c) Express $\mathbf{x}_3$ in terms of $w(t)$ and $w'(t)$.

(d) Which, if any, of the following vector functions are solutions of the given differential system?

(i) $\begin{bmatrix} 2t - \cos t \\ 2 + \sin t \\ \cos t \end{bmatrix}$ **(ii)** $\begin{bmatrix} t\cos t \\ -\sin t \\ 0 \end{bmatrix}$

(iii) $\begin{bmatrix} t - 2\sin t \\ 1 - 2\cos t \\ 2\sin t \end{bmatrix}$ **(iv)** $\begin{bmatrix} \cos t + \sin t \\ \cos t - \sin t \\ \sin t - \cos t \end{bmatrix}$

2. (a) Determine whether or not the vector functions

$$\mathbf{x}_1 = \begin{bmatrix} t \\ 1 \\ 1 \end{bmatrix}\quad \mathbf{x}_2 = \begin{bmatrix} 1 \\ t \\ t^2 \end{bmatrix}\quad \mathbf{x}_3 = \begin{bmatrix} -1 \\ 1 \\ t \end{bmatrix}$$

constitute a basis for the system

$$x_1' = \frac{-t}{1-t}x_2 + \frac{1}{1-t}x_3$$

$$x_2' = \frac{1}{1+t}x_1 + \frac{1+t^2}{1-t^2}x_2 - \frac{1}{1-t}x_3$$

$$x_3' = \frac{t}{1+t}x_1 + \frac{t^3+2t+1}{1-t^2}x_2 - \frac{t^2+2t+1}{1-t^2}x_3$$

(b) Show that the wronskian w of $\mathbf{x}_1$, $\mathbf{x}_2$, and $\mathbf{x}_3$ has a zero value but that $w(t) \not\equiv 0$. Does this contradict Theorem 3?

(c) Find all values of t for which $\mathbf{x}_1'$, $\mathbf{x}_2'$, and $\mathbf{x}_3'$ are mutually orthogonal.

(d) Which, if any, of the following vector functions are solutions of the given differential system?

(i) $\begin{bmatrix} t \\ t+2 \\ t^2 \end{bmatrix}$ **(ii)** $\begin{bmatrix} 2 \\ t-1 \\ t(t-1) \end{bmatrix}$

(iii) $\begin{bmatrix} 0 \\ 1+t \\ t^2 \end{bmatrix}$ **(iv)** $\begin{bmatrix} t+1 \\ 0 \\ 1-t \end{bmatrix}$

3. The substitutions $x_1 = y, x_2 = y', \ldots, x_n = y^{(n-1)}$ transform the differential equation $y^{(n)} + a_1(t)y^{(n-1)} + \cdots + a_n(t)y = 0$ into a first-order system $\mathbf{x}' = \mathrm{A}(t)\mathbf{x}$ [see Eqs. (5)–(7), Sec. 4.3]. Show that the wronskian of any n solution vectors of the system coincides with the wronskian of the corresponding solutions of the nth-order equation.

4. In Eq. (13), verify that $\mathbf{y}(t_0) = 0$. *Hint:* Recall Properties 1 and 8, Sec. 3.5.

5. Let $\mathbf{x}' = \mathrm{A}(t)\mathbf{x}$ be a normal homogeneous first-order system in n unknowns over an interval I and let $w = \det[\mathbf{x}_1\mathbf{x}_2 \cdots \mathbf{x}_n]$ be the wronskian of any n solutions of this equation over I.

(a) Derive Eq. (8).

(b) Use (8) to show that if t_0 is any fixed point of I, then for every t in I,

$$w(t) = w(t_0) \exp \int_{t_0}^{t} [a_{11}(s) + a_{22}(s) + \cdots + a_{nn}(s)]\, ds$$

(c) Use the formula for $w(t)$ of Part **(b)** to prove that either $w(t) \equiv 0$, or else $w(t)$ is never zero, on I.

6. Prove that a homogeneous first-order system in n unknowns has at most n linearly independent solutions on an interval over which the system is normal.

7. If a matrix function $\mathrm{X} = [\mathbf{x}_1\mathbf{x}_2 \cdots \mathbf{x}_n]$ is differentiable and nonsingular, show that $\mathbf{x}' = \mathrm{X}'\mathrm{X}^{-1}\mathbf{x}$ is a first-order system which has X as a fundamental matrix. *Hint:* Eliminate $\mathbf{c}$ from $\mathbf{x} = \mathrm{X}\mathbf{c}$ and $\mathbf{x}' = \mathrm{X}'\mathbf{c}$.

8. Find a homogeneous first-order system which has the given matrix as a fundamental matrix. *Hint:* See Exercise 7.

(a) $\begin{bmatrix} \cosh t & \sinh t \\ \sinh t & \cosh t \end{bmatrix}$ **(b)** $\begin{bmatrix} \cos t & -\sin t \\ \sin t & \cos t \end{bmatrix}$

(c) $\begin{bmatrix} t & 1 \\ -1 & t^2 \end{bmatrix}$ **(d)** $\begin{bmatrix} t & 1 & 1 \\ 0 & t & t \\ 0 & 0 & t^2 \end{bmatrix}$

(e) $\begin{bmatrix} \sec t & \sec t & \sec t \\ 0 & \sec t & \tan t \\ 0 & \tan t & \sec t \end{bmatrix}$ **(f)** $\begin{bmatrix} 1 & e^{-t} & 0 \\ t & t & t \\ e^t & 0 & e^t \end{bmatrix}$

9. (a) If $\mathbf{x}_1$ is a solution vector of the equation $\mathbf{x}' = \mathrm{A}\mathbf{x}$, show that $t\mathbf{x}_1$ is a particular integral of the equation $\mathbf{x}' = \mathrm{A}\mathbf{x} + \mathbf{x}_1$.

(b) If $\mathbf{x}_1$ is a solution vector of the equation $\mathbf{x}' = \mathrm{A}\mathbf{x}$, find a particular integral of the equation $\mathbf{x}' = \mathrm{A}\mathbf{x} + t\mathbf{x}_1$.

10. (a) If $\mathrm{A} = \begin{bmatrix} 3 & -2 \\ 4 & -3 \end{bmatrix}$, find a complete solution of the equation $\mathbf{x}' = \mathrm{A}\mathbf{x}$.

Find a particular integral of the equation $\mathbf{x}' = \mathrm{A}\mathbf{x} + \mathbf{f}$ if

(b) $\mathbf{f} = \begin{bmatrix} e^{2t} \\ 1 \end{bmatrix}$ **(c)** $\mathbf{f} = \begin{bmatrix} e^{2t} \\ e^t \end{bmatrix}$

(d) $\mathbf{f} = \begin{bmatrix} t \\ 1 \end{bmatrix}$ **(e)** $\mathbf{f} = \begin{bmatrix} \sin t \\ 0 \end{bmatrix}$

11. (a) If $\mathrm{A} = \begin{bmatrix} 2 & 1 \\ -9 & -4 \end{bmatrix}$, find a complete solution of the equation $\mathbf{x}' = \mathrm{A}\mathbf{x}$.

Find a particular integral of the equation $\mathbf{x}' = \mathrm{A}\mathbf{x} + \mathbf{f}$ if

(b) $\mathbf{f} = \begin{bmatrix} 1 \\ -3 \end{bmatrix}$ **(c)** $\mathbf{f} = \begin{bmatrix} e^{2t} \\ -3e^{2t} \end{bmatrix}$

(d) $\mathbf{f} = \begin{bmatrix} e^{-t} \\ 2e^{-t} \end{bmatrix}$ **(e)** $\mathbf{f} = \begin{bmatrix} t \\ -3t \end{bmatrix}$

12. If $\mathrm{A} = \begin{bmatrix} 0 & 1 \\ -2/t^2 & 2/t \end{bmatrix}$, find a complete solution of the equation $\mathbf{x}' = \mathrm{A}\mathbf{x} + \mathbf{f}$ if

(a) $\mathbf{f} = \begin{bmatrix} 0 \\ 1 \end{bmatrix}$ **(b)** $\mathbf{f} = \begin{bmatrix} 0 \\ t \end{bmatrix}$ **(c)** $\mathbf{f} = \begin{bmatrix} t \\ 0 \end{bmatrix}$

Hint: Write one component of the homogeneous equations as an Euler equation (Sec. 2.8).

In Exercises 13–16, a fundamental matrix of the homogeneous system is given. Find a particular integral of the nonhomogeneous system.

13. $\begin{bmatrix} \cos t & \cos 2t \\ -\cos t & \cos 2t \end{bmatrix}$,

$$\begin{bmatrix} x_1 \\ x_2 \end{bmatrix}' = \begin{bmatrix} -\frac{1}{2}\tan t - \tan 2t & \frac{1}{2}\tan t - \tan 2t \\ \frac{1}{2}\tan t - \tan 2t & -\frac{1}{2}\tan t - \tan 2t \end{bmatrix}\begin{bmatrix} x_1 \\ x_2 \end{bmatrix} + \begin{bmatrix} \sec 2t + \tan t \\ \sec 2t - \tan t \end{bmatrix}$$

14. $\begin{bmatrix} e^{-t} & e^t \\ e^{-t} & 2e^t \end{bmatrix}$, $\begin{aligned} x_1' &= -3x_1 + 2x_2 + e^t \sec^2 t \\ x_2' &= -4x_1 + 3x_2 + 2e^t \sec^2 t \end{aligned}$

15. $\begin{bmatrix} -1 & 1/t \\ 2t & -1 \end{bmatrix}$, $\begin{aligned} x_1' &= -(2/t)x_1 - (1/t^2)x_2 + 1 - (1/t) \\ x_2' &= 2x_1 + (2/t)x_2 + 2 - t \end{aligned}$

16. $\begin{bmatrix} t & t & 1 \\ 0 & 1 & 0 \\ 0 & 1 & t \end{bmatrix}$, $\begin{aligned} x_1' &= \frac{1}{t}x_1 + \frac{1}{t^2}x_2 - \frac{1}{t^2}x_3 + \frac{t}{1+t^2} \\ x_2' &= 3t^2 \\ x_3' &= -\frac{1}{t}x_2 + \frac{1}{t}x_3 + t^2 \end{aligned}$

Solve each of the following initial-value problems.

17. $\begin{aligned} x_1' &= -3x_1 - 4x_2 \\ x_2' &= 2x_1 + x_2 \end{aligned}$ $\quad \mathbf{x}(0) = \begin{bmatrix} \sin \pi \\ \cos \pi \end{bmatrix}$

18. $\begin{aligned} x_1' &= -3x_1 + x_2 \\ x_2' &= -2x_1 \end{aligned}$ $\quad \mathbf{x}(0) = \begin{bmatrix} 0 \\ 1 \end{bmatrix}$

19. $\begin{aligned} x_1' &= 2x_1 - 2x_2 - 1/t \\ x_2' &= 2x_1 - 2x_2 \end{aligned}$ $\quad \mathbf{x}(1) = \begin{bmatrix} 2 \\ 2 \end{bmatrix}$

20. $\begin{aligned} x_1' &= -3x_1 - 4x_2 \\ x_2' &= 2x_1 + x_2 \end{aligned}$ $\quad \mathbf{x}(\pi) = 7e^{-\pi}\begin{bmatrix} 1 \\ -1 \end{bmatrix}$

21. $\begin{aligned} x_1' &= 3x_1 + x_2 - 2\sin t \\ x_2' &= 4x_1 + 3x_2 + 6\cos t \end{aligned}$ $\quad \mathbf{x}(0) = \begin{bmatrix} 3 \\ 1 \end{bmatrix}$

22. $\begin{aligned} x_1' &= x_1 - x_2 + \sin t \\ x_2' &= x_1 + 3x_2 \end{aligned}$ $\quad \mathbf{x}(0) = \frac{1}{25}\begin{bmatrix} -9 \\ 4 \end{bmatrix}$

23. $\begin{aligned} x_1' &= -10x_1 + 9x_2 + 3e^{2t}/t \\ x_2' &= -12x_1 + 11x_2 + 4e^{2t}/t \end{aligned}$ $\quad \mathbf{x}(1) = \begin{bmatrix} 3 \\ 4 \end{bmatrix}e^2$

24. $\begin{aligned} x_1' &= -2x_1 + x_2 + 1 \\ x_2' &= -3x_1 + 2x_2 + 2\sin t \end{aligned}$ $\quad \mathbf{x}(0) = \begin{bmatrix} 7 \\ 11 \end{bmatrix}$

25. $\begin{aligned} x_1' &= -3x_1 + 2x_2 + e^t \sec^2 t \\ x_2' &= -4x_1 + 3x_2 + 2e^t \sec^2 t \end{aligned}$ $\quad \mathbf{x}(0) = \begin{bmatrix} 1 \\ 2 \end{bmatrix}$

26. $\begin{aligned} x_1' &= x_1 - 3x_2 + 3e^t \sec 3t \\ x_2' &= 3x_1 + x_2 \end{aligned}$ $\quad \mathbf{x}(0) = \begin{bmatrix} 0 \\ 0 \end{bmatrix}$

27. $\begin{aligned} x_1' &= x_1 + 3x_2 + 3e^{-2t}t^{1/2} \\ x_2' &= 3x_1 + x_2 - 3e^{-2t}t^{1/2} \end{aligned}$ $\quad \mathbf{x}(0) = \begin{bmatrix} 1 \\ 1 \end{bmatrix}$

28. $\begin{aligned} x_1' &= -x_1 + x_2 \\ x_2' &= -6x_1 + 4x_2 \\ x_3' &= x_2 - x_3 \end{aligned}$ $\quad \mathbf{x}(0) = \begin{bmatrix} 1 \\ 2 \\ 3 \end{bmatrix}$

29. $\begin{aligned} x_1' &= 2x_1 - x_2 - 4x_3 + 2e^{-t} \\ x_2' &= 2x_2 - 4x_3 + 3e^{-t} \\ x_3' &= x_2 - 2x_3 + e^{-t} \end{aligned}$ $\quad \mathbf{x}(0) = \begin{bmatrix} 1 \\ -1 \\ 0 \end{bmatrix}$

30. $\begin{aligned} x_1' &= 3x_1 - 4x_2 + 4x_3 \\ x_2' &= 4x_1 - 5x_2 + 4x_3 \\ x_3' &= 4x_1 - 4x_2 + 3x_3 \end{aligned}$ $\quad \mathbf{x}(0) = \begin{bmatrix} 2 \\ 1 \\ -1 \end{bmatrix}$

31. $\begin{aligned} x_1' &= x_1 + x_3 - 1/(1+t^2) \\ x_2' &= x_2 + 2x_3 - 2/(1+t^2) \\ x_3' &= x_1 + 2x_2 + 5x_3 + 1/(1+t^2) \end{aligned}$ $\quad \mathbf{x}(0) = \begin{bmatrix} 0 \\ 0 \\ 0 \end{bmatrix}$

32. $\begin{aligned} x_1' &= x_1 - x_2 + 4x_3 + e^{3t}/t \\ x_2' &= 3x_1 + 2x_2 - x_3 + 2e^{3t}/t \\ x_3' &= 2x_1 + x_2 - x_3 + e^{3t}/t \end{aligned}$ $\quad \mathbf{x}(1) = \begin{bmatrix} 6e^3 \\ 12e^3 \\ 6e^3 \end{bmatrix}$

4.5 LINEAR DIFFERENTIAL SYSTEMS WITH CONSTANT COEFFICIENTS

We shall now consider, as an alternative to the use of elementary operations, a method of solving linear differential systems with *constant* coefficients which resembles the way in which we solved single linear constant-coefficient differential equations. A significant feature of the method we are about to present is that it puts to use what we already know about characteristic-value problems (see Sec. 3.8). Let us first investigate the method as it applies to first-order systems.

As in the case of a single equation, to solve the nonhomogeneous first-order system

$$\mathbf{x}' = \mathrm{A}\mathbf{x} + \mathbf{f}(t) \tag{1}$$

where A is a *constant* matrix, we begin by looking for a complete solution of the associated homogeneous system

$$\mathbf{x}' = \mathrm{A}\mathbf{x} \tag{2}$$

Guided by our experience in solving single equations, we attempt to find a solution of (2) of the form

$$\mathbf{x} = \mathbf{k}e^{\lambda t} \tag{3}$$

where **k** is a constant vector and λ is a scalar. Substituting (3) into (2), dividing out $e^{\lambda t}$, writing $\mathbf{k}\lambda$ as $\lambda \mathrm{I}\mathbf{k}$, and then collecting coefficients on **k**, we see that (3) yields a solution of (2) if and only if

$$(\lambda \mathrm{I} - \mathrm{A})\mathbf{k} = \mathbf{0} \tag{4}$$

Thus (3) will be a *nontrivial* solution of (2) if and only if λ is a characteristic value of A and **k** is a corresponding characteristic vector. The remaining steps required to find a complementary function of (1) differ somewhat according as

1. All characteristic values of A are real and distinct.
2. All characteristic values of A are real, but some are repeated roots of the characteristic equation det $(\lambda \mathrm{I} - \mathrm{A}) = 0$.
3. Complex characteristic values of A occur.

Perhaps the best way of learning how to handle these three cases, or a combination of the same, is by means of specific examples. Our first example involves a first-order system whose coefficient matrix has only distinct real characteristic values.

EXAMPLE 1

Find a complete solution of the system

$$\begin{aligned} x_1' &= \quad\; - x_2 + x_3 \\ x_2' &= 4x_1 - x_2 - 4x_3 \\ x_3' &= -3x_1 - x_2 + 4x_3 \end{aligned} \tag{5}$$

As (3) suggests, we seek solutions of (5) of the type

$$\begin{bmatrix} x_1 \\ x_2 \\ x_3 \end{bmatrix} = \begin{bmatrix} a \\ b \\ c \end{bmatrix} e^{\lambda t} \qquad a, b, c \text{ constants}$$

Substituting $ae^{\lambda t}$, $be^{\lambda t}$, and $ce^{\lambda t}$ into the given system for x_1, x_2, and x_3, respectively, then dividing out $e^{\lambda t}$, transposing, and collecting terms, we obtain the characteristic-value problem

$$\begin{aligned} \lambda a + \quad\quad\; b - \quad\quad\;\; c &= 0 \\ -4a + (\lambda + 1)b + \quad\;\; 4c &= 0 \\ 3a + \quad\quad\; b + (\lambda - 4)c &= 0 \end{aligned} \tag{6}$$

The corresponding characteristic equation, det $(\lambda \mathrm{I} - A) = 0$, may be expressed in factored form by using familiar properties of determinants, for instance,

$$\begin{vmatrix} \lambda & 1 & -1 \\ -4 & \lambda + 1 & 4 \\ 3 & 1 & \lambda - 4 \end{vmatrix} = \begin{vmatrix} 0 & 0 & -1 \\ 4(\lambda - 1) & \lambda + 5 & 4 \\ (\lambda - 1)(\lambda - 3) & \lambda - 3 & \lambda - 4 \end{vmatrix} = -(\lambda - 1)(\lambda - 3)\begin{vmatrix} 4 & \lambda + 5 \\ 1 & 1 \end{vmatrix}$$

$$= (\lambda + 1)(\lambda - 1)(\lambda - 3) = 0$$

Clearly, $\lambda_1 = -1$, $\lambda_2 = 1$, and $\lambda_3 = 3$ are the characteristic values of the coefficient matrix of (5), i.e., of $\mathrm{A} = \begin{bmatrix} 0 & -1 & 1 \\ 4 & -1 & -4 \\ -3 & -1 & 4 \end{bmatrix}$. If $\lambda = \lambda_1 = -1$, System (6) becomes the linear system

$$\begin{aligned} -a + b - c &= 0 \\ -4a + \quad\; 4c &= 0 \\ 3a + b - 5c &= 0 \end{aligned} \qquad \text{which is equivalent to} \qquad \begin{aligned} a - \quad\; c &= 0 \\ b - 2c &= 0 \end{aligned}$$

Setting $c = 1$, we find that $\begin{bmatrix} 1 \\ 2 \\ 1 \end{bmatrix}$ is a characteristic vector of A corresponding to $\lambda_1 = -1$. By the same procedure, we find $\begin{bmatrix} 1 \\ 0 \\ 1 \end{bmatrix}$ and $\begin{bmatrix} 1 \\ -1 \\ 2 \end{bmatrix}$ to be characteristic vectors of A corresponding to the respective characteristic values $\lambda_2 = 1$ and $\lambda_3 = 3$. It follows that

$$\begin{bmatrix} 1 \\ 2 \\ 1 \end{bmatrix} e^{-t} \qquad \begin{bmatrix} 1 \\ 0 \\ 1 \end{bmatrix} e^{t} \qquad \text{and} \qquad \begin{bmatrix} 1 \\ -1 \\ 2 \end{bmatrix} e^{3t}$$

are particular solutions of (5). Since for every real number t the wronskian w of these three solutions has the nonzero value $w(t) = -2e^{3t}$, a complete solution of the given first-order system is

$$\mathbf{x} = c_1 \begin{bmatrix} 1 \\ 2 \\ 1 \end{bmatrix} e^{-t} + c_2 \begin{bmatrix} 1 \\ 0 \\ 1 \end{bmatrix} e^{t} + c_3 \begin{bmatrix} 1 \\ -1 \\ 2 \end{bmatrix} e^{3t}$$

Our next example involves a simple first-order system whose coefficient matrix has a repeated characteristic value. Although the system has only two unknowns, it serves to illustrate how complete solutions of more general systems can be found when repeated characteristic values occur.

EXAMPLE 2

Find a complete solution of the system

$$\begin{aligned} x_1' &= x_1 - 2x_2 \\ x_2' &= 2x_1 - 3x_2 \end{aligned} \tag{7}$$

The coefficient matrix of this system is $A = \begin{bmatrix} 1 & -2 \\ 2 & -3 \end{bmatrix}$. Since we are seeking solutions of (7) of the type $\begin{bmatrix} a \\ b \end{bmatrix} e^{\lambda t}$, where a and b are constants, the characteristic-value problem to be solved is

$$\begin{bmatrix} \lambda - 1 & 2 \\ -2 & \lambda + 3 \end{bmatrix} \begin{bmatrix} a \\ b \end{bmatrix} = \begin{bmatrix} 0 \\ 0 \end{bmatrix} \tag{8}$$

and the characteristic equation is

$$\begin{vmatrix} \lambda - 1 & 2 \\ -2 & \lambda + 3 \end{vmatrix} = \lambda^2 + 2\lambda + 1 = (\lambda + 1)^2 = 0$$

Hence $\lambda_1 = -1$ is a repeated characteristic value of A. Setting $\lambda = -1$ in (8), we obtain the linear system

$$\begin{aligned} -2a + 2b &= 0 \\ -2a + 2b &= 0 \end{aligned} \qquad \text{which is equivalent to} \qquad a - b = 0$$

Clearly, $\begin{bmatrix} 1 \\ 1 \end{bmatrix}$ is a characteristic vector of A corresponding to $\lambda_1 = -1$, and

$$\begin{bmatrix} 1 \\ 1 \end{bmatrix} e^{-t} \tag{9}$$

is a particular solution of (7).

Strict analogy with a single differential equation having -1 as a double root of its characteristic equation would suggest that a second solution of (7) might be found by multiplying (9) by t. This is not the case, however, since it is easy to verify that $\begin{bmatrix} 1 \\ 1 \end{bmatrix} te^{-t}$ does not satisfy (7). Moreover, the wronskian of this vector function and (9) is identically zero for all values of t. Thus a second solution of (7) which will with (9) provide a basis for all solutions of (7) must be obtained by other means. Fortunately, variation of parameters, which enabled us to cope with either repeated or complex characteristic roots of a single differential equation, can also be employed here. To apply this method, we take $\lambda = -1$, as we must if $\begin{bmatrix} a \\ b \end{bmatrix} e^{\lambda t}$ is to be a solution of (7). Then we vary the parameters a and b and attempt to find a solution of (7) of the form

$$\begin{bmatrix} u \\ v \end{bmatrix} e^{-t} \tag{10}$$

where u and v are functions of t. Of course, we want $\begin{bmatrix} e^{-t} & ue^{-t} \\ e^{-t} & ve^{-t} \end{bmatrix}$ to be a fundamental matrix of (7).

Substituting ue^{-t} and ve^{-t} into (7) for x_1 and x_2, respectively, then canceling all occurrences of e^{-t} and regrouping, we find that (10) will be a solution of (7) if and only if u and v satisfy the equations

$$\begin{aligned} u' &= 2(u - v) \\ v' &= 2(u - v) \end{aligned} \tag{11}$$

Subtracting the second of these equations from the first, we have

$$u' - v' = (u - v)' = 0 \qquad \text{which implies} \qquad u = v + k_1 \tag{12}$$

Replacing u by $v + k_1$ in the second of Eqs. (11), we get $v' = 2k_1$, which integrates into $v = 2k_1t + k_2$. With v thus determined, the last of Eqs. (12) gives $u = 2k_1t + k_2 + k_1$, and (10) becomes

$$\begin{bmatrix} k_1 + k_2 + 2k_1t \\ k_2 + 2k_1t \end{bmatrix} e^{-t} \tag{13}$$

The wronskian of this vector function and (9) has the value

$$\begin{vmatrix} k_1 + k_2 + 2k_1t & 1 \\ k_2 + 2k_1t & 1 \end{vmatrix} e^{-2t} = k_1e^{-2t}$$

Hence (9) and (13) are fundamental solutions of (7), provided $k_1 \neq 0$. In particular, if $k_1 = 1$ and $k_2 = 0$, (9) and (13) give us

$$\begin{bmatrix} e^{-t} & (1 + 2t)e^{-t} \\ e^{-t} & 2te^{-t} \end{bmatrix}$$

as a fundamental matrix and

$$\mathbf{x} = c_1 \begin{bmatrix} 1 \\ 1 \end{bmatrix} e^{-t} + c_2 \begin{bmatrix} 1 + 2t \\ 2t \end{bmatrix} e^{-t}$$

as a complete solution of the original differential system.

A closer look at (13) in the form

$$2k_1 \begin{bmatrix} 1 \\ 1 \end{bmatrix} te^{-t} + \begin{bmatrix} k_1 + k_2 \\ k_2 \end{bmatrix} e^{-t}$$

coupled with the observation that

$$2k_1 \begin{bmatrix} 1 \\ 1 \end{bmatrix} e^{-t} \qquad k_1 \neq 0$$

is itself a nontrivial solution of (7), suggests the following as a shorter method of finding a basis for all solutions of a system like (7) with a repeated characteristic root $\lambda = r$: as soon as a first solution $\begin{bmatrix} a \\ b \end{bmatrix} e^{rt}$ analogous to (9) has been found, determine constants c and d such that

$$\begin{bmatrix} a \\ b \end{bmatrix} te^{rt} + \begin{bmatrix} c \\ d \end{bmatrix} e^{rt} \tag{14}$$

will be a solution of the given system. As this example shows, it is important to note that in general the characteristic vector in the solution corresponding to (9) will *not* be equal to the vector coefficient of e^{rt} but may be chosen as the vector coefficient of te^{rt} in the trial solution (14).

As one further introductory example, let us consider a simple first-order system whose coefficient matrix has complex characteristic values.

EXAMPLE 3

Find a complete solution of the system

$$\begin{aligned} x_1' &= 2x_1 - 3x_2 \\ x_2' &= 3x_1 + 2x_2 \end{aligned} \tag{15}$$

This system will have a solution of the form

$$\begin{bmatrix} a \\ b \end{bmatrix} e^{\lambda t} \tag{16}$$

if and only if

$$\begin{bmatrix} \lambda - 2 & 3 \\ -3 & \lambda - 2 \end{bmatrix} \begin{bmatrix} a \\ b \end{bmatrix} = \begin{bmatrix} 0 \\ 0 \end{bmatrix} \tag{17}$$

Solving the characteristic equation

$$\begin{vmatrix} \lambda - 2 & 3 \\ -3 & \lambda - 2 \end{vmatrix} = \lambda^2 - 4\lambda + 13 = [\lambda - (2 + 3i)][\lambda - (2 - 3i)] = 0$$

we obtain $\lambda_1 = 2 + 3i$ and $\lambda_2 = 2 - 3i$ as conjugate complex characteristic values of the coefficient matrix $A = \begin{bmatrix} 2 & -3 \\ 3 & 2 \end{bmatrix}$. Replacing λ by $2 + 3i$ in the characteristic-value problem (17) and dividing out a factor of 3 from every term, we get the linear system

$$\begin{aligned} ia + b &= 0 \\ -a + bi &= 0 \end{aligned} \qquad \text{which is obviously equivalent to} \qquad ia + b = 0$$

Taking a equal to the coefficient of b, and b equal to minus the coefficient of a, in the last equation, we obtain $\begin{bmatrix} 1 \\ -i \end{bmatrix}$ as a characteristic vector of A corresponding to the characteristic value $\lambda_1 = 2 + 3i$. Substituting these values of a, b, and λ into (16) gives us the complex vector function

$$\begin{bmatrix} 1 \\ -i \end{bmatrix} e^{(2+3i)t} \tag{18}$$

which formally satisfies system (15). Of course, what we want are *real* fundamental solutions of (15). To find these, we use Euler's identity [Eq. (26), Sec. 2.5] to write (18) as

$$\begin{bmatrix} 1 \\ -i \end{bmatrix} e^{2t}e^{i3t} = \begin{bmatrix} e^{i3t} \\ -ie^{i3t} \end{bmatrix} e^{2t} = \begin{bmatrix} \cos 3t + i \sin 3t \\ -i(\cos 3t + i \sin 3t) \end{bmatrix} e^{2t} = \begin{bmatrix} e^{2t} \cos 3t \\ e^{2t} \sin 3t \end{bmatrix} + i \begin{bmatrix} e^{2t} \sin 3t \\ -e^{2t} \cos 3t \end{bmatrix}$$

It is now easy to verify that the real vector functions

$$\begin{bmatrix} \cos 3t \\ \sin 3t \end{bmatrix} e^{2t} \quad \text{and} \quad \begin{bmatrix} \sin 3t \\ -\cos 3t \end{bmatrix} e^{2t}$$

form a basis for all real solutions of system (15). Thus we are able to write a complete solution

$$\mathbf{x} = c_1 \begin{bmatrix} \cos 3t \\ \sin 3t \end{bmatrix} e^{2t} + c_2 \begin{bmatrix} \sin 3t \\ -\cos 3t \end{bmatrix} e^{2t}$$

of the given differential system using only one of the two conjugate characteristic values of A.

It is not always necessary, nor advantageous, to reduce a linear differential system with constant coefficients

$$p_{i1}(D)x_1 + p_{i2}(D)x_2 + \cdots + p_{in}(D)x_n = f_i(t) \qquad 1 \le i \le n \tag{19}$$

to a first-order system before solving, for a system like (19) can usually be solved in the same manner as a first-order system. The concepts of *complementary function* and *particular integral* carry over and, as we shall see in Chap. 13, can even be extended to far more general linear systems.

If we define a **matrix operator** P(D) and vector functions **x** and **f** by

$$\mathrm{P}(D) = \begin{bmatrix} p_{11}(D) & p_{12}(D) & \cdots & p_{1n}(D) \\ p_{21}(D) & p_{22}(D) & \cdots & p_{2n}(D) \\ \cdots & \cdots & \cdots & \cdots \\ p_{n1}(D) & p_{n2}(D) & \cdots & p_{nn}(D) \end{bmatrix} \qquad \mathbf{x} = \begin{bmatrix} x_1 \\ x_2 \\ \vdots \\ x_n \end{bmatrix} \qquad \mathbf{f} = \begin{bmatrix} f_1 \\ f_2 \\ \vdots \\ f_n \end{bmatrix}$$

where the p_{ij}'s are polynomial operators in D, which of course have constant coefficients, System (19) can be written in the compact matrix form

$$\mathrm{P}(D)\mathbf{x} = \mathbf{f}(t) \tag{20}$$

The associated homogeneous equation is

$$\mathrm{P}(D)\mathbf{x} = \mathbf{0} \tag{21}$$

As with first-order systems, a complete solution of (21) is called a **complementary function** of (20), and any particular solution of (20) is called a **particular integral** of that system. To find a complementary function, we assume that solutions of the homogeneous system exist in the form

$$\mathbf{x} = \mathbf{k}e^{\lambda t}$$

just as we did for first-order systems. Since $D^r(e^{\lambda t}) = \lambda^r e^{\lambda t}$, it follows that if we substitute $\mathbf{x} = \mathbf{k}e^{\lambda t}$ into (21), we obtain $\mathrm{P}(\lambda)\mathbf{k}e^{\lambda t} = \mathbf{0}$, or dividing out the scalar factor $e^{\lambda t}$,

$$\mathrm{P}(\lambda)\mathbf{k} = \mathbf{0} \tag{22}$$

Now, by Theorem 2, Sec. 3.7, Eq. (22) will have a nontrivial solution if and only if

$$|P(\lambda)| = 0 \tag{23}$$

This equation is called the **characteristic equation** of both the algebraic system (22) and the original differential system. It is nothing but the determinant of the operational coefficients of (19) equated to zero, with D replaced by λ.

For each root λ_j of the characteristic equation (23) there will be a solution vector $\mathbf{k}_j$ of (22) determined to within an arbitrary scalar factor c_j. If (23) is a polynomial equation in λ of degree N, and if its root $\lambda_1, \lambda_2, \ldots, \lambda_N$ are all distinct real numbers, a complete solution of (21), that is, a complementary function of (20), is then

$$\mathbf{x} = c_1\mathbf{k}_1 e^{\lambda_1 t} + c_2\mathbf{k}_2 e^{\lambda_2 t} + \cdots + c_N\mathbf{k}_N e^{\lambda_N t}$$

EXAMPLE 4

Find a complete solution of the differential system

$$\begin{aligned} (D+1)x + (D+2)y + (D+3)z &= -e^{-t} \\ (D+2)x + (D+3)y + (2D+3)z &= e^{-t} \\ (4D+6)x + (5D+4)y + (20D-12)z &= 7e^{-t} \end{aligned} \tag{24}$$

To find a complementary function of this system, we observe that solutions $\begin{bmatrix} a \\ b \\ c \end{bmatrix} e^{\lambda t}$ of the homogeneous system exist if and only if

$$\begin{aligned} (\lambda+1)a + (\lambda+2)b + (\lambda+3)c &= 0 \\ (\lambda+2)a + (\lambda+3)b + (2\lambda+3)c &= 0 \\ (4\lambda+6)a + (5\lambda+4)b + (20\lambda-12)c &= 0 \end{aligned} \tag{25}$$

The characteristic equation of this system is

$$\begin{vmatrix} \lambda+1 & \lambda+2 & \lambda+3 \\ \lambda+2 & \lambda+3 & 2\lambda+3 \\ 4\lambda+6 & 5\lambda+4 & 20\lambda-12 \end{vmatrix} = 0$$

or, expanding, collecting terms, and factoring,

$$-(\lambda-1)(\lambda-2)(\lambda-3) = 0$$

Substituting the roots $\lambda_1 = 1$, $\lambda_2 = 2$, and $\lambda_3 = 3$ of this equation into (25) one at a time, we obtain the three linear systems

$$\begin{aligned} 2a + 3b + 4c &= 0 \\ 3a + 4b + 5c &= 0 \\ 10a + 9b + 8c &= 0 \end{aligned} \qquad \begin{aligned} 3a + 4b + 5c &= 0 \\ 4a + 5b + 7c &= 0 \\ 14a + 14b + 28c &= 0 \end{aligned} \qquad \begin{aligned} 4a + 5b + 6c &= 0 \\ 5a + 6b + 9c &= 0 \\ 18a + 19b + 48c &= 0 \end{aligned}$$

When the coefficient matrices of these systems are reduced to row echelon form, the systems themselves are transformed, respectively, into

$$\begin{aligned} a - c &= 0 \\ b + 2c &= 0 \end{aligned} \qquad \begin{aligned} a + 3c &= 0 \\ b - c &= 0 \end{aligned} \qquad \begin{aligned} a + 9c &= 0 \\ b - 6c &= 0 \end{aligned}$$

From these systems we see that (1, −2, 1), (3, −1, −1), and (9, −6, −1) are nontrivial solutions of (25) corresponding to the *distinct* real values 1, 2, and 3, respectively, of the parameter λ. A complementary function of (24) is therefore

$$(26)\qquad c_1\begin{bmatrix}1\\-2\\1\end{bmatrix}e^t + c_2\begin{bmatrix}3\\-1\\-1\end{bmatrix}e^{2t} + c_3\begin{bmatrix}9\\-6\\-1\end{bmatrix}e^{3t}$$

To complete the problem we now need to find a particular integral of the given differential system. Since the scalar factor e^{-t} in the nonhomogeneous term $\mathbf{f}(t) = \begin{bmatrix}-1\\1\\7\end{bmatrix}e^{-t}$ of (24) and the scalar factors e^t, e^{2t}, and e^{3t} in the terms of the complementary function are linearly independent functions, we choose a trial solution of (24)

$$(27)\qquad \mathbf{v} = \begin{bmatrix}a\\b\\c\end{bmatrix}e^{-t}$$

exactly as we did in Sec. 2.7, except that the undetermined constant coefficient of e^{-t} is now a vector instead of a scalar. Substituting (27) into (24), collecting terms, and canceling a factor of e^{-t}, we find

$$\begin{aligned} b + 2c &= -1\\ a + 2b + c &= 1\\ 2a - b - 32c &= 7 \end{aligned} \qquad \text{which is equivalent to} \qquad \begin{aligned} a &= 3\\ b &= -1\\ c &= 0 \end{aligned}$$

Hence a particular integral of (24) is $\begin{bmatrix}3\\-1\\0\end{bmatrix}e^{-t}$, and a complete solution of the original system is

$$\mathbf{x} = c_1\begin{bmatrix}1\\-2\\1\end{bmatrix}e^t + c_2\begin{bmatrix}3\\-1\\-1\end{bmatrix}e^{2t} + c_3\begin{bmatrix}9\\-6\\-1\end{bmatrix}e^{3t} + \begin{bmatrix}3\\-1\\0\end{bmatrix}e^{-t}$$

If the characteristic equation $|\mathrm{P}(\lambda)| = 0$ of (19) has a double root, say $\lambda = r$, and if $\mathbf{a}$ is a specific characteristic vector of the equation $\mathrm{P}(r)\mathbf{k} = \mathbf{0}$, then of course

$$(28a)\qquad \mathbf{a}e^{rt}$$

is a nontrivial solution of (21). To obtain a second independent solution, we proceed as indicated at the end of Example 2. That is, we determine a constant vector $\mathbf{b}$ such that

$$(28b)\qquad \mathbf{a}te^{rt} + \mathbf{b}e^{rt}$$

and $\mathbf{a}e^{rt}$ will be linearly independent solutions of Eq. (21). In forming (28*b*), it is important that the term $\mathbf{b}e^{rt}$ be retained in the matrix case. This might have been anticipated had we realized that in general the vector $\mathbf{b}$ will not be a scalar multiple of $\mathbf{a}$; hence in constructing a complete solution of (21), the term involving $\mathbf{b}e^{rt}$ cannot be absorbed in the term involving $\mathbf{a}e^{rt}$, as would necessarily be the case with the term be^{rt} were it included in a trial solution of the form $ate^{rt} + be^{rt}$ for a single scalar differential equation. It can be shown, however (Exercises 44 and 45), that the characteristic vector $\mathbf{a}$ of (28*a*) may be taken as the vector coefficient of te^{rt} in (28*b*).

Similar results hold for roots of (23) of higher multiplicity. Thus, for an m-fold root r, there will be linearly independent particular solutions of (21) of the form

$$\begin{array}{l} \mathbf{a}e^{rt} \\ \mathbf{a}te^{rt} + \mathbf{b}e^{rt} \\ \mathbf{a}t^2e^{rt} + \mathbf{c}_1te^{rt} + \mathbf{c}_2e^{rt} \\ \cdots\cdots\cdots\cdots\cdots\cdots\cdots\cdots\cdots\cdots \\ \mathbf{a}t^{m-1}e^{rt} + \mathbf{k}_1t^{m-2}e^{rt} + \mathbf{k}_2t^{m-3}e^{rt} + \cdots + \mathbf{k}_{m-2}te^{rt} + \mathbf{k}_{m-1}e^{rt} \end{array} \tag{29}$$

As with first-order systems and single equations, if the set of roots $\{\lambda_j\}$ of the characteristic equation (23) includes one or more pairs of conjugate complex numbers, it is desirable to convert all corresponding complex exponential solutions into purely real solutions. To see how this can be accomplished, let $p \pm qi$ be a pair of conjugate complex roots of (23) and let $\mathbf{k}$ be a particular nontrivial solution vector of (22) corresponding to the root $\lambda = p + qi$, so that

$$\mathrm{P}(\lambda)\mathbf{k} = \mathrm{P}(p + qi)\mathbf{k} = \mathbf{0}$$

Then, since all the coefficients in (22) are real, it follows by taking conjugates throughout the last equation that $\overline{\mathbf{k}}$ is a solution vector of (22) corresponding to $\overline{\lambda} = p - qi$. Therefore, in a complex sense, both

$$\mathbf{k}e^{(p+iq)t} \quad \text{and} \quad \overline{\mathbf{k}}e^{(p-iq)t}$$

are particular solutions of (21). By combining these as follows and applying Euler's identity, we obtain the two independent real solutions

$$\begin{aligned} \frac{\mathbf{k}e^{(p+iq)t} + \overline{\mathbf{k}}e^{(p-iq)t}}{2} &= e^{pt}\left(\frac{\mathbf{k} + \overline{\mathbf{k}}}{2}\cos qt - \frac{\mathbf{k} - \overline{\mathbf{k}}}{2i}\sin qt\right) \\ &= e^{pt}[\mathcal{R}(\mathbf{k})\cos qt - \mathcal{I}(\mathbf{k})\sin qt] \end{aligned} \tag{30a}$$

$$\begin{aligned} \frac{\mathbf{k}e^{(p+iq)t} - \overline{\mathbf{k}}e^{(p-iq)t}}{2i} &= e^{pt}\left(\frac{\mathbf{k} - \overline{\mathbf{k}}}{2i}\cos qt + \frac{\mathbf{k} + \overline{\mathbf{k}}}{2}\sin qt\right) \\ &= e^{pt}[\mathcal{I}(\mathbf{k})\cos qt + \mathcal{R}(\mathbf{k})\sin qt] \end{aligned} \tag{30b}$$

where $\mathcal{R}(\mathbf{k})$ and $\mathcal{I}(\mathbf{k})$ denote the column vectors whose components are, respectively, the real parts of the components of $\mathbf{k}$ and the imaginary parts of the components of $\mathbf{k}$.

Our next example illustrates both the treatment of a pair of conjugate complex roots and a pair of repeated real roots.

EXAMPLE 5

Find a complete solution of the system

$$\begin{aligned} (D^2 + D + 8)x_1 + (D^2 + 6D + 3)x_2 &= 0 \\ (D + 1)x_1 + \qquad (D^2 + 1)x_2 &= 0 \end{aligned}$$

In this case the characteristic equation (23) is

$$\begin{vmatrix} \lambda^2 + \lambda + 8 & \lambda^2 + 6\lambda + 3 \\ \lambda + 1 & \lambda^2 + 1 \end{vmatrix} = \lambda^4 + 2\lambda^2 - 8\lambda + 5 = (\lambda - 1)^2(\lambda^2 + 2\lambda + 5) = 0$$

with roots $1, 1, -1 \pm 2i$. For the root $-1 + 2i$, Eq. (22) simplifies to

$$\begin{bmatrix} 4 - 2i & -6 + 8i \\ 2i & -2 - 4i \end{bmatrix}\begin{bmatrix} k_1 \\ k_2 \end{bmatrix} = \begin{bmatrix} 0 \\ 0 \end{bmatrix} \quad \text{which is equivalent to} \quad ik_1 - (1 + 2i)k_2 = 0$$

Taking $k_1 = 1 + 2i$ and $k_2 = i$, we have $\mathbf{k} = \begin{bmatrix} 1+2i \\ i \end{bmatrix}$, $\mathcal{R}(\mathbf{k}) = \begin{bmatrix} 1 \\ 0 \end{bmatrix}$, and $\mathcal{I}(\mathbf{k}) = \begin{bmatrix} 2 \\ 1 \end{bmatrix}$. Thus, from Eqs. (30), two particular solutions of the given differential system are

$$\mathbf{x}_1 = e^{-t}\left\{\begin{bmatrix} 1 \\ 0 \end{bmatrix}\cos 2t - \begin{bmatrix} 2 \\ 1 \end{bmatrix}\sin 2t\right\}$$

$$\mathbf{x}_2 = e^{-t}\left\{\begin{bmatrix} 2 \\ 1 \end{bmatrix}\cos 2t + \begin{bmatrix} 1 \\ 0 \end{bmatrix}\sin 2t\right\}$$

For the repeated root $\lambda = 1$, and a trial solution $\mathbf{a}e^t$, (22) becomes

$$\mathrm{P}(1)\mathbf{a} = \begin{bmatrix} 10 & 10 \\ 2 & 2 \end{bmatrix}\begin{bmatrix} a_1 \\ a_2 \end{bmatrix} = \begin{bmatrix} 0 \\ 0 \end{bmatrix}$$

so we can take

$$\mathbf{a} = \begin{bmatrix} a_1 \\ a_2 \end{bmatrix} = \begin{bmatrix} 1 \\ -1 \end{bmatrix}$$

From (28b), another trial solution is

$$\mathbf{a}te^t + \mathbf{b}e^t$$

or

$$\begin{bmatrix} 1 \\ -1 \end{bmatrix}te^t + \begin{bmatrix} b_1 \\ b_2 \end{bmatrix}e^t$$

Substituting this into the original system, we obtain two equations each of which reduces to

$$2b_1 + 2b_2 = 1$$

Hence we can take $b_1 = 0$ and $b_2 = \frac{1}{2}$. Two solutions associated with the double root $\lambda = 1$ are therefore

$$\mathbf{x}_3 = \mathbf{a}e^t = \begin{bmatrix} 1 \\ -1 \end{bmatrix}e^t \quad \text{and} \quad \mathbf{x}_4 = \mathbf{a}te^t + \mathbf{b}e^t = \begin{bmatrix} 1 \\ -1 \end{bmatrix}te^t + \begin{bmatrix} 0 \\ \frac{1}{2} \end{bmatrix}e^t$$

A complete solution of the original system is now given by

$$\mathbf{x} = c_1\mathbf{x}_1 + c_2\mathbf{x}_2 + c_3\mathbf{x}_3 + c_4\mathbf{x}_4$$

To find a particular integral of a nonhomogeneous system (20), or (1), in the usual case in which $\mathbf{f}(t)$ is a vector function having only a finite number of linearly independent derivatives, we proceed very much as in the case of a single scalar differential equation. At the outset, it is convenient to identify the independent functions $\phi_1(t), \phi_2(t), \ldots, \phi_m(t)$ which appear in the components of $\mathbf{f}(t)$ and then express $\mathbf{f}(t)$ in the form

$$\mathbf{f}(t) = \mathbf{f}_1\phi_1(t) + \mathbf{f}_2\phi_2(t) \cdots + \mathbf{f}_m\phi_m(t) \tag{31}$$

where the **f**'s with subscripts are appropriate constant column vectors. This expression is then compared with a complementary function of the system being solved

$$\mathbf{x}_h(t) = \mathbf{h}_1u_1(t) + \mathbf{h}_2u_2(t) + \cdots + \mathbf{h}_nu_n(t) \tag{32}$$

in which the $\mathbf{h}$'s are constant vectors and $u_1(t), u_2(t), \ldots, u_n(t)$ are known linearly independent scalar functions. For each term of (31) in which neither the ϕ factor nor any of its derivatives is proportional to any of the u's in (32), a particular integral can be constructed as described in Table 2.2, Sec. 2.7, provided that the arbitrary scalar constants appearing in the entries of the table are replaced by undetermined constant vectors. The trial solution is then substituted into the nonhomogeneous system, and the arbitrary components of the coefficient vectors are determined to make the resulting equations hold identically. For any term of (31) in which the ϕ factor, or any of its derivatives, is proportional to one of the u's in (32), the rules of Table 2.2 are still valid with one additional provision: *Not only must the usual choice for a trial particular integral be multiplied by the lowest positive integral power of the independent variable which will eliminate the duplication, but the products of the normal choice and all lower nonnegative integral powers of the independent variable must also be included in the actual choice.* Finally, once a particular integral for each ϕ term in (31) has been found, a particular integral for the nonhomogeneous system having $\mathbf{f}(t)$ as its nonhomogeneous term can be formed by addition as prescribed by Theorem 8, Sec. 4.4. An example should clarify the details of the procedure.

EXAMPLE 6

Find a particular integral of the nonhomogeneous system of equations obtained by adding $2e^t$ and $2e^t + e^{-2t}$ to the right-hand members of the respective equations of Example 5.

At the outset, let us express the nonhomogeneous term of the equivalent matrix equation in the form

$$\mathbf{f}(t) = \begin{bmatrix} 0 \\ 1 \end{bmatrix} e^{-2t} + \begin{bmatrix} 2 \\ 2 \end{bmatrix} e^t \tag{33}$$

and then write the complementary function, found in Example 5, as

$$\mathbf{x}_h = \begin{bmatrix} c_1 + 2c_2 \\ c_2 \end{bmatrix} e^{-t}\cos 2t + \begin{bmatrix} -2c_1 + c_2 \\ -c_1 \end{bmatrix} e^{-t}\sin 2t + \begin{bmatrix} c_3 \\ -c_3 + \frac{1}{2}c_4 \end{bmatrix} e^t + \begin{bmatrix} c_4 \\ -c_4 \end{bmatrix} te^t \tag{34}$$

Since neither e^{-2t} nor any of its derivatives is proportional to any of the linearly independent functions

$$e^{-t}\cos 2t \qquad e^{-t}\sin 2t \qquad e^t \qquad te^t$$

appearing in the terms of $\mathbf{x}_h$, we assume as a trial particular integral, corresponding to the first term $\begin{bmatrix} 0 \\ 1 \end{bmatrix} e^{-2t}$ of $\mathbf{f}(t)$, simply

$$\mathbf{v}_1 = \mathbf{a}e^{-2t} = \begin{bmatrix} a_1 \\ a_2 \end{bmatrix} e^{-2t} \qquad a_1, a_2 \text{ constants}$$

Then, substituting, we have

$$\mathrm{P}(D)\mathbf{v}_1 = \mathrm{P}(D)\mathbf{a}e^{-2t} = \mathrm{P}(-2)\mathbf{a}e^{-2t} = \begin{bmatrix} 10 & -5 \\ -1 & 5 \end{bmatrix}\begin{bmatrix} a_1 \\ a_2 \end{bmatrix} e^{-2t} = \begin{bmatrix} 0 \\ 1 \end{bmatrix} e^{-2t}$$

or, dividing out e^{-2t},

$$\begin{bmatrix} 10 & -5 \\ -1 & 5 \end{bmatrix}\begin{bmatrix} a_1 \\ a_2 \end{bmatrix} = \begin{bmatrix} 0 \\ 1 \end{bmatrix} \qquad \text{and this is equivalent to} \qquad \begin{aligned} 10a_1 - 5a_2 &= 0 \\ -a_1 + 5a_2 &= 1 \end{aligned}$$

It follows that $a_1 = \frac{1}{9}$, $a_2 = \frac{2}{9}$, and $\mathbf{a} = \frac{1}{9}\begin{bmatrix} 1 \\ 2 \end{bmatrix}$. Thus, $\frac{1}{9}\begin{bmatrix} 1 \\ 2 \end{bmatrix} e^{-2t}$ is one term in the particular integral we are seeking.

To find a particular integral corresponding to the second term $\begin{bmatrix} 2 \\ 2 \end{bmatrix} e^t$ of $\mathbf{f}(t)$, we note that since both e^t and te^t occur in terms of the complementary function, the normal choice for a trial particular integral, namely, $\mathbf{b}e^t$, is to be modified by multiplying it by t^2 *and including the terms* $\mathbf{c}te^t$ and $\mathbf{d}e^t$, where

$$\mathbf{b} = \begin{bmatrix} b_1 \\ b_2 \end{bmatrix} \qquad \mathbf{c} = \begin{bmatrix} c_1 \\ c_2 \end{bmatrix} \qquad \mathbf{d} = \begin{bmatrix} d_1 \\ d_2 \end{bmatrix}$$

are constant vectors to be determined. Then, substituting

$$\mathbf{v}_2 = \mathbf{b}t^2e^t + \mathbf{c}te^t + \mathbf{d}e^t$$

into the equation $\mathrm{P}(D)\mathbf{x} = \begin{bmatrix} 2 \\ 2 \end{bmatrix} e^t$ and using the results of Exercises 44 and 46, we obtain

$$\begin{aligned} \mathrm{P}(D)(\mathbf{b}t^2e^t + \mathbf{c}te^t + \mathbf{d}e^t) &= \mathrm{P}(1)\mathbf{b}t^2e^t + 2\mathrm{P}'(1)\mathbf{b}te^t + P''(1)\boldsymbol{b}e^t \\ &\quad + \mathrm{P}(1)\mathbf{c}te^t + \mathrm{P}'(1)\mathbf{c}e^t + \mathrm{P}(1)\mathbf{d}e^t \\ &= \begin{bmatrix} 2 \\ 2 \end{bmatrix} e^t \end{aligned}$$

Hence, equating the coefficients of like functions on the two sides of this equation, we find that

$$\begin{aligned} \mathrm{P}(1)\mathbf{b} &= \mathbf{0} \\ \mathrm{P}(1)\mathbf{c} + 2\mathrm{P}'(1)\mathbf{b} &= \mathbf{0} \\ \mathrm{P}(1)\mathbf{d} + \mathrm{P}'(1)\mathbf{c} + \mathrm{P}''(1)\mathbf{b} &= \begin{bmatrix} 2 \\ 2 \end{bmatrix} \end{aligned}$$

Expanding these equations and simplifying slightly, we obtain the linear system

$$\begin{aligned} b_1 + b_2 &= 0 \\ b_1 + b_2 &= 0 \\ 3b_1 + 8b_2 + 5c_1 + 5c_2 &= 0 \\ b_1 + 2b_2 + c_1 + c_2 &= 0 \\ 2b_1 + 2b_2 + 3c_1 + 8c_2 + 10d_1 + 10d_2 &= 2 \\ 2b_2 + c_1 + 2c_2 + 2d_1 + 2d_2 &= 2 \end{aligned} \tag{35}$$

When reduced to row echelon form, the augmented matrix of this system becomes

$$\begin{bmatrix} 1 & 0 & 0 & 0 & 0 & 0 & -1 \\ 0 & 1 & 0 & 0 & 0 & 0 & 1 \\ 0 & 0 & 1 & 0 & -2 & -2 & -2 \\ 0 & 0 & 0 & 1 & 2 & 2 & 1 \\ 0 & 0 & 0 & 0 & 0 & 0 & 0 \\ 0 & 0 & 0 & 0 & 0 & 0 & 0 \end{bmatrix} \quad \text{which gives us} \quad \begin{bmatrix} b_1 \\ b_2 \\ c_1 \\ c_2 \\ d_1 \\ d_2 \end{bmatrix} = k_1 \begin{bmatrix} 0 \\ 0 \\ 2 \\ -2 \\ 1 \\ 0 \end{bmatrix} + k_2 \begin{bmatrix} 0 \\ 0 \\ 2 \\ -2 \\ 0 \\ 1 \end{bmatrix} + \begin{bmatrix} -1 \\ 1 \\ -2 \\ 1 \\ 0 \\ 0 \end{bmatrix}$$

as a complete solution of (35). In particular, taking $k_1 = k_2 = 0$, we have

$$\mathbf{b} = \begin{bmatrix} -1 \\ 1 \end{bmatrix} \qquad \mathbf{c} = \begin{bmatrix} -2 \\ 1 \end{bmatrix} \qquad \mathbf{d} = \begin{bmatrix} 0 \\ 0 \end{bmatrix}$$

Finally, putting our results together, we get the entire particular solution

$$\mathbf{v} = \frac{1}{9}\begin{bmatrix} 1 \\ 2 \end{bmatrix} e^{-2t} + \begin{bmatrix} -1 \\ 1 \end{bmatrix} t^2 e^t + \begin{bmatrix} -2 \\ 1 \end{bmatrix} te^t$$

EXERCISES

Find a complete solution of each of the following first-order systems. Also, find the solution that satisfies the given initial condition when one is given.

1. $x_1' = -2x_1 - 13x_2$, $x_2' = x_1 + 4x_2$

2. $x_1' = -3x_1 - 4x_2$, $x_2' = 2x_1 + x_2$

3. $x_1' = -\frac{3}{2}x_1 - 2x_2$, $x_2' = 2x_1 + \frac{5}{2}x_2$

4. $x_1' = 3x_1 + 2x_2$, $x_2' = -2x_1 - x_2$

5. $x_1' = -3x_1 - 2x_2$, $x_2' = 2x_1 + x_2$

6. $x_1' = 3x_1 + 4x_2$, $x_2' = -2x_1 - x_2$

7. $x_1' = 3x_1 - 2x_2$, $x_2' = 5x_1 - 3x_2$

8. $x_1' = x_1 - x_2$, $x_2' = x_1 + 3x_2$

9. $x_1' = 4x_1 - 2x_2$, $x_2' = 25x_1 - 10x_2$, $\mathbf{x}(0) = \begin{bmatrix} 10 \\ 40 \end{bmatrix}$

10. $x_1' = -2x_1 + x_2$, $x_2' = -3x_1 + 2x_2 + 2\sin t$, $\mathbf{x}(0) = \begin{bmatrix} 3 \\ 4 \end{bmatrix}$

11. $x_1' = 2x_1 - x_2 + t$, $x_2' = 3x_1 - 2x_2 + 2t$

12. $x_1' = -2x_2 - 2\sec 2t$, $x_2' = 2x_1 + 2\csc 2t$

13. $x_1' = 3x_1 - 2x_3$, $x_2' = -x_1 + 2x_2 + x_3$, $x_3' = 4x_1 - 3x_3$, $\mathbf{x}(0) = \begin{bmatrix} 3 \\ 1 \\ 3 \end{bmatrix}$

14. $x_1' = 4x_1 - 2x_2 - 10x_3$, $x_2' = x_2$, $x_3' = 2x_1 - x_2 - 5x_3$, $\mathbf{x}(0) = \begin{bmatrix} -3 \\ -2 \\ -1 \end{bmatrix}$

15. $x_1' = 3x_1 - x_3$, $x_2' = -2x_1 + 2x_2 + x_3$, $x_3' = 8x_1 - 3x_3$, $\mathbf{x}(0) = \begin{bmatrix} -1 \\ 2 \\ -8 \end{bmatrix}$

16. $x_1' = \frac{1}{2}x_1 - x_2 - \frac{3}{2}x_3 + \dfrac{2t}{1+t^2}$, $x_2' = \frac{1}{2}x_1 - \frac{1}{2}x_3 - \dfrac{2t}{1+t^2}$, $x_3' = -\frac{1}{2}x_1 - x_2 - \frac{1}{2}x_3 + \dfrac{2t}{1+t^2}$, $\mathbf{x}(0) = \begin{bmatrix} 1 \\ -1 \\ 1 \end{bmatrix}$

17. $x_1' = x_1 - 2x_2 - x_3$, $x_2' = x_1 + 2x_2 + e^{-t}$, $x_3' = -x_1 - 3x_2 - 1$

18. Derive (13) by using elementary operations to solve System (11).

Find a complete solution of each of the following systems.

19. $(D + 2)x + (D + 4)y = 1$, $(D + 1)x + (D + 5)y = 2$

20. $(2D + 1)x + (D + 2)y = 0$, $(D + 3)x + (D + 6)y = -3e^t$

21. $(D + 1)x + (4D - 2)y = t - 1$, $(D + 2)x + (5D - 2)y = 2t - 1$

22. $(D + 5)x + (D + 7)y = 4e^{2t}$, $(2D + 1)x + (3D + 1)y = 0$

23. $(2D + 1)x + (D + 2)y = 6e^t$, $(D + 2)x + (D + 4)y = 4e^{-t}$

24. $(2D + 1)x + (D - 1)y = -3\cos t$, $(D + 2)x + (D + 3)y = 5\sin t$

25. $(2D + 1)x + (D + 2)y = 8e^{-t}$, $(D^2 + D + 9)x + (D^2 - 2D + 12)y = 6$

26. $(2D + 1)x + (D^2 + 6D + 1)y = 0$, $(D + 2)x + (D^2 + 2D + 5)y = 6e^{2t}$

27. $(2D^2 + 5)x + (D^2 + 3)y = -8\sin 3t$, $(D^2 + 7)x + (D^2 + 5)y = 8\sin 3t$

28. $(2D + 1)x + (D + 1)y = 0$, $(D - 2)x + (D - 1)y = 0$

29. $(2D + 11)x + (D + 3)y + (D - 2)z = 14e^t$, $(D - 2)x + (D - 1)y + Dz = -2e^t$, $(D + 1)x + (D - 3)y + (2D - 4)z = 4e^t$

30. $Dx + (D - 2)y = -\ln|t|$, $Dx + (2D - 3)y = 1 - \ln|t|$

31. $(3D + 1)x + (D + 7)y = e^{-t}$, $(2D + 1)x + (D + 5)y = e^{-t}$

32. $(D + 5)x + (2D + 1)y = e^{-t} + e^t$, $(D + 7)x + (3D - 1)y = 0$

33. $(D + 5)x + (D + 7)y = 2e^t$, $(2D + 1)x + (3D + 1)y = e^t$

34. $(D + 2)x + (D + 3)y = -4$, $(2D - 6)x + (3D - 4)y = 2$

35. $(D + 1)x + (D + 2)y = -e^t$, $(3D + 1)x + (4D + 7)y = -7e^t$

36. $(D + 1)x + (D + 2)y = -t + 1$, $(5D + 1)x + (6D + 3)y = -2t + 1$

37. $(2D + 1)x + (D + 2)y = e^{-t}$, $(3D - 7)x + (3D + 1)y = 0$

38. $(2D + 1)x + (D + 2)y = \sin t$, $(3D + 1)x + (3D + 5)y = \cos t$

39. Two particles, each of weight w, are attached to a perfectly flexible, weightless, elastic string stretched under tension T as shown in Fig. 4.3. The particles vibrate in a direction perpendicular to the length of the string through amplitudes so small that

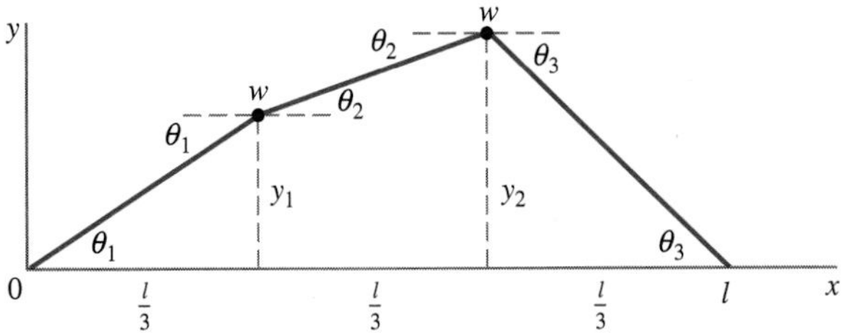

FIGURE 4.3

(a) The tension in the string remains constant.

(b) The angles shown in Fig. 4.3 are so small that their sines can with satisfactory accuracy by approximated by their tangents.

Neglecting all forces but the elastic forces supplied by the string, set up the differential equations describing the behavior of the system, find the natural frequencies of the system and the ratios of the amplitudes of the two particles at each frequency.

40. Work Exercise 39 for the system of three particles shown in Fig. 4.4.

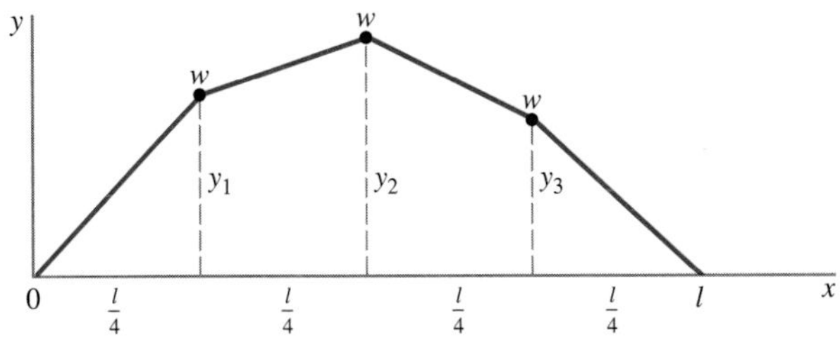

FIGURE 4.4

41. Verify that the characteristic equation of the system

$$\begin{aligned}(2D+1)x+(D+1)y&=0\\(D-4)x+(D-3)y&=0\end{aligned}$$

has the repeated root $\lambda = 1$. Show, further, that the system has solutions of the form $\begin{bmatrix}x\\y\end{bmatrix} = \begin{bmatrix}a\\b\end{bmatrix}e^t$ but not of the form $\begin{bmatrix}x\\y\end{bmatrix} = \begin{bmatrix}c\\d\end{bmatrix}te^t$, where a, b, c, and d are constants. Find a complete solution of the system.

42. Verify that in the system of equations

$$\begin{aligned}(2D+1)x+(D+1)y&=e^t\\(D-7)x+(D-5)y&=0\end{aligned}$$

the nonhomogeneous term e^t duplicates a term in the complementary function of the system. Verify, further, that the system has a particular integral of the form $\begin{bmatrix}x\\y\end{bmatrix} = \begin{bmatrix}a\\b\end{bmatrix}te^t + \begin{bmatrix}c\\d\end{bmatrix}e^t$ but not of the simplistic form $\begin{bmatrix}x\\y\end{bmatrix} = \begin{bmatrix}a\\b\end{bmatrix}te^t$. Find a complete solution of the system.

43. Show that

$$\frac{1}{9}\begin{bmatrix}1\\2\end{bmatrix}e^{-2t} + \begin{bmatrix}-1\\1\end{bmatrix}t^2e^t + \begin{bmatrix}-1\\0\end{bmatrix}te^t + \frac{1}{2}\begin{bmatrix}1\\0\end{bmatrix}e^t$$

is a particular integral of the system in Example 6.

44. Show that $D^m(te^{\lambda t}) = \lambda^m te^{\lambda t} + m\lambda^{m-1}e^{\lambda t}$. Hence show that

$$p(D)te^{\lambda t} = p(\lambda)te^{\lambda t} + p'(\lambda)e^{\lambda t}$$

and

$$\mathrm{P}(D)te^{\lambda t} = \mathrm{P}(\lambda)te^{\lambda t} + \mathrm{P}'(\lambda)e^{\lambda t}$$

where $p(D)$ is a polynomial in the operator D and $\mathrm{P}(D)$ is a matrix whose elements are polynomials in D.

45. Using the results of Exercise 44, show that if $\lambda = r$ is a double root of the characteristic equation $|\mathrm{P}(\lambda)| = 0$, then $\mathbf{x}_1 = \mathbf{a}e^{rt}$ and $\mathbf{x}_2 = \mathbf{b}_1te^{rt} + \mathbf{b}_2e^{rt}$ are two linearly independent solutions of the system $\mathrm{P}(D)\mathbf{x} = \mathbf{0}$ provided the vectors $\mathbf{a}$, $\mathbf{b}_1$, and $\mathbf{b}_2$, where $\mathbf{a} \neq \mathbf{0}$ and $\mathbf{b}_1 \neq \mathbf{0}$, satisfy the equations

$$\mathrm{P}(r)\mathbf{a} = \mathbf{0} \qquad \mathrm{P}(r)\mathbf{b}_1 = \mathbf{0} \qquad \mathrm{P}(r)\mathbf{b}_2 + \mathrm{P}'(r)\mathbf{b}_1 = \mathbf{0}$$

46. Show that

$$D^m(t^2e^{\lambda t}) = \lambda^m t^2e^{\lambda t} + 2m\lambda^{m-1}te^{\lambda t} + m(m-1)\lambda^{m-2}e^{\lambda t}.$$

Hence show that

$$p(D)t^2e^{\lambda t} = p(\lambda)t^2e^{\lambda t} + 2p'(\lambda)te^{\lambda t} + p''(\lambda)e^{\lambda t}$$

and

$$\mathrm{P}(D)t^2e^{\lambda t} = \mathrm{P}(\lambda)t^2e^{\lambda t} + 2\mathrm{P}'(\lambda)te^{\lambda t} + \mathrm{P}''(\lambda)e^{\lambda t}$$

where $p(D)$ is a polynomial in the operator D and $\mathrm{P}(D)$ is a matrix whose elements are polynomials in D.

47. Using the results of Exercise 46, show that if $\lambda = r$ is a triple root of the characteristic equation $|\mathrm{P}(\lambda)| = 0$, then $\mathbf{x}_1 = \mathbf{a}e^{rt}$, $\mathbf{x}_2 = \mathbf{b}_1te^{rt} + \mathbf{b}_2e^{rt}$, and $\mathbf{x}_3 = \mathbf{c}_1t^2e^{rt} + \mathbf{c}_2te^{rt} + \mathbf{c}_3e^{rt}$ are three linearly independent solutions of the system $\mathrm{P}(D)\mathbf{x} = \mathbf{0}$ provided $\mathbf{a}$, $\mathbf{b}_1$, $\mathbf{b}_2$, $\mathbf{c}_1$, $\mathbf{c}_2$, and $\mathbf{c}_3$ are suitably chosen vectors which satisfy the equations

$$\begin{gathered}\mathrm{P}(r)\mathbf{a} = \mathbf{0} \qquad \mathrm{P}(r)\mathbf{b}_1 = \mathbf{0} \qquad \mathrm{P}(r)\mathbf{b}_2 + \mathrm{P}'(r)\mathbf{b}_1 = \mathbf{0}\\ \mathrm{P}(r)\mathbf{c}_1 = \mathbf{0} \qquad \mathrm{P}(r)\mathbf{c}_2 + 2\mathrm{P}'(r)\mathbf{c}_1 = \mathbf{0}\\ \mathrm{P}(r)\mathbf{c}_3 + \mathrm{P}'(r)\mathbf{c}_2 + \mathrm{P}''(r)\mathbf{c}_1 = \mathbf{0}\end{gathered}$$

48. Generalize the results of Exercises 44 and 46 to the function $t^3e^{\lambda t}$.

49. If $\mathbf{a}$, $\mathbf{b}$, and $\mathbf{c}$ are constant column vectors and $\mathrm{P}(D)$ is a matrix whose elements are polynomials in the operator D, show that

(a) $\mathrm{P}(D)\mathbf{a} = \mathrm{P}(0)\mathbf{a}$

(b) $\mathrm{P}(D)\mathbf{b}t = \mathrm{P}(0)\mathbf{b}t + \mathrm{P}'(0)\mathbf{b}$

(c) $\mathrm{P}(D)\mathbf{c}t^2 = \mathrm{P}(0)\mathbf{c}t^2 + 2\mathrm{P}'(0)\mathbf{c}t + \mathrm{P}''(0)\mathbf{c}$

50. In Exercise 45, verify that if $\lambda = r$ is a double root of the characteristic equation of a system of two simultaneous linear differential equations with constant coefficients, then the equations

$$\mathrm{P}(r)\mathbf{b}_1 = \mathbf{0} \quad \text{and} \quad \mathrm{P}(r)\mathbf{b}_2 + \mathrm{P}'(r)\mathbf{b}_1 = \mathbf{0}$$

are always solvable for nontrivial vectors $\mathbf{b}_1$ and $\mathbf{b}_2$.

51. In Example 3, verify that using the root $\lambda_2 = 2 - 3i$ leads to the same particular solutions that were obtained from $\lambda_1 = 2 + 3i$.

52. In Example 3, verify by direct substitution that

$$\begin{bmatrix} \cos 3t \\ \sin 3t \end{bmatrix} e^{2t} \quad \text{and} \quad \begin{bmatrix} \sin 3t \\ -\cos 3t \end{bmatrix} e^{2t}$$

are solutions of the system (15).

53. In Example 3, verify that the wronskian of the two particular solutions is never zero.

Find a complete solution of each of the following systems.

54.
$$\begin{aligned} x_1' &= \quad\ - x_2 + x_3 \\ x_2' &= 4x_1 - x_2 - 4x_3 \\ x_3' &= x_1 - 5x_2 \end{aligned}$$

55.
$$\begin{aligned} x_1' &= \quad\ - 2x_2 + 3x_3 \\ x_2' &= x_1 + 3x_2 - x_3 \\ x_3' &= -x_1 - 2x_2 \end{aligned}$$

56. **(a)** Show that the differential system in Example 4 can be written in the form

$$(\mathrm{A}D - \mathrm{B})\mathbf{x} = \mathbf{f}$$

where A and B are constant matrices.

(b) Show that if variation of parameters is applied to the equation $(\mathrm{A}D - \mathrm{B})\mathbf{x} = \mathbf{f}$, with A nonsingular, the result is the particular integral

$$\mathbf{v}(t) = \mathrm{X}\int [\mathrm{A}\mathrm{X}(t)]^{-1}\mathbf{f}(t)\,dt = \mathrm{X}\int \mathrm{X}^{-1}(t)\mathrm{A}^{-1}\,\mathbf{f}(t)\,dt$$

where $\mathrm{X}(t)$ is a fundamental matrix of the system. Why must A be nonsingular?

(c) Use the formula of Part **(b)** to find a particular integral of the system

$$\begin{aligned} (D-1)x_1 + (D-2)x_2 &= 5e^{4t} \\ (D-5)x_1 + (2D-7)x_2 &= 0 \end{aligned}$$

57. Using the formula of Part **(b)**, Exercise 56, find a particular integral of the system $\mathrm{A}\mathbf{x}' - \mathrm{B}\mathbf{x} = \mathbf{f}(t)$ if $A = \begin{bmatrix} 1 & 0 \\ 0 & 2 \end{bmatrix}$ $\mathrm{B} = \begin{bmatrix} 4 & 1 \\ -4 & 2 \end{bmatrix}$ and

(a) $\mathbf{f}(t) = \begin{bmatrix} e^t \\ 2e^t \end{bmatrix}$ **(b)** $\mathbf{f}(t) = \begin{bmatrix} e^{2t} \\ 0 \end{bmatrix}$

(c) $\mathbf{f}(t) = \begin{bmatrix} e^{3t} \\ -e^{3t} \end{bmatrix}$ **(d)** $\mathbf{f}(t) = \begin{bmatrix} 0 \\ te^{2t} \end{bmatrix}$

58. Discuss the possibility of using annihilator operators to find a particular integral for a nonhomogeneous differential system such as the first-order system (1) or the more general system (20).

59. Find a matrix operator with nonzero determinant which will annihilate each of the following vectors:

(a) $\begin{bmatrix} 1 \\ 2 \end{bmatrix}$ **(b)** $\begin{bmatrix} e^t \\ 2e^t \end{bmatrix}$ **(c)** $\begin{bmatrix} 0 \\ e^{2t} \end{bmatrix}$

(d) $\begin{bmatrix} e^{-t} \\ 0 \end{bmatrix}$ **(e)** $\begin{bmatrix} e^{-t} \\ e^{2t} \end{bmatrix}$ **(f)** $\begin{bmatrix} 1 \\ e^{2t} \end{bmatrix}$

(g) $\begin{bmatrix} \sin t \\ 2\sin t \end{bmatrix}$ **(h)** $\begin{bmatrix} \sin 2t \\ \cos 2t \end{bmatrix}$

Find a complete solution of each of the following systems by using annihilator operators to obtain particular solutions.

60. **(a)**
$$\begin{aligned} x' + x &= 8e^t \\ y' - y &= 3e^t \end{aligned}$$
(b)
$$\begin{aligned} x' - 3x - y &= 4 + 2t \\ y' - 4x - 3y &= 1 + t \end{aligned}$$

61. **(a)**
$$\begin{aligned} x' - x + y + z &= e^{-t} \\ y' + y + z &= 1 \\ z' - z &= 2e^{-t} \end{aligned}$$
(b)
$$\begin{aligned} x' - x + z &= e^t \\ y' - 2y &= e^{2t}\sin t \\ z' - x - z &= 0 \end{aligned}$$

CHAPTER 5

NUMERICAL METHODS

The complexity of today's engineering problems makes the use of high-speed computers a necessity; and much current research is directed to developing even faster, more accurate, and more convenient methods for their use. Nonetheless, there are still basic ideas and methods of numerical analysis with which an engineer should be familiar. This chapter attempts to cover such topics.

Section 5.2 presents essentially classical material on *finite differences* and their properties. In Sec. 5.3, these are used to develop a number of important interpolation formulas. In Sec. 5.4 these are used to obtain formulas for numerical differentiation and integration. Then in Sec. 5.6 these formulas are used to develop a variety of practical methods for the numerical solution of ordinary differential equations. (Numerical methods for solving partial differential equations are given in Sec. 11.9.) The method of least squares is discussed in Sec. 5.5.

Difference equations are discussed in Sec. 5.7. Roughly speaking, these are the finite-difference equivalent of differential equations; that is, they are equations connecting the differences rather than the derivatives of a function. The basic theorems on difference equations and differential equations are strikingly similar; and when the coefficients are constant, the solution processes are very much alike. Difference equations are particularly useful in studying physical systems consisting of a number of identical subsystems identically connected, such as a train of elastically coupled boxcars or the crankshaft of an in-line engine (Sec. 7.6). Their use in studying the stability and instability of formulas for the numerical solution of differential equations is discussed in Sec. 5.8.

Section 5.9 treats the relatively new subject of the G and Z transforms of sequences. These provide an alternative method for solving difference equations and are useful in such fields as signal processing.

Prerequisite for this chapter: Chaps. 1 and 2.

5.1 INTRODUCTION

In calculus we spent a great deal of time practicing the operations of differentiation and integration on a wide variety of functions. For an applied scientist, this is a necessary but by no means sufficient exposure to these topics. More often than not, the functions an engineer must differentiate are tabulated from experimental observations and are not given by familiar formulas. Similarly, in practical applications, functions to be integrated are usually given in tabular form or by expressions for which no elementary antiderivatives exist. Likewise, although in the last four chapters we have learned to solve many important differential equations, it is nonetheless true that equations for which no exact solution can be found arise more and more frequently in present-day applications. As a consequence, applied mathematicians and engineers must be prepared to carry out by approximate numerical methods the processes they learned to do exactly in elementary courses. With this in mind, we shall devote this chapter to a discussion of such topics as numerical differentiation and integration, interpolation, curve fitting, the numerical solution of ordinary differential equations, and difference equations, beginning with the important notion of the *differences of a function.*

5.2 THE DIFFERENCES OF A FUNCTION

Let $y = f(x)$ be a function given in tabular form for a sequence of distinct values of x.

x	$f(x)$
x_0	$f(x_0)$
x_1	$f(x_1)$
x_2	$f(x_2)$
x_3	$f(x_3)$
...	...

If $f(x_i)$ and $f(x_j)$ are any two values of $f(x)$, then the **first,** or **first-order, divided differences** of $f(x)$ are defined by the formula

(1) $$f(x_i, x_j) = \frac{f(x_i) - f(x_j)}{x_i - x_j}$$

Usually, though not necessarily, the subscripts of the arguments x_i and x_j are consecutive integers. Similarly, if $f(x_i, x_j)$ and $f(x_j, x_k)$ are two first divided differences of $f(x)$ having one argument x_j in common, then the **second,** or **second-order, divided differences** of $f(x)$ are defined by the formula

(2) $$f(x_i, x_j, x_k) = \frac{f(x_i, x_j) - f(x_j, x_k)}{x_i - x_k}$$

Proceeding inductively, we define a divided difference of any order as the difference of two divided differences of the next lower order, overlapping in all but one of their arguments, divided by the difference between the extreme, or nonoverlapping, arguments appearing in those differences. From these definitions it is clear that divided differences have the following properties.

PROPERTY 1 Any divided difference of the sum (or difference) of two functions is equal to the sum (or difference) of the divided differences of the individual functions.

PROPERTY 2 Any divided difference of a constant times a function is equal to the constant times the divided difference of the function.

Though obvious only for differences of the first order, it is true (see Exercises 11 and 12) that *divided differences of all orders are symmetric functions of their arguments.* Thus, for example,

$$f(x_i, x_j, x_k) = f(x_i, x_k, x_j) = f(x_j, x_i, x_k) = \cdots$$

In many applications it is convenient to have the divided differences of a function prominently displayed. This is usually done by constructing a **difference table** in which each difference is entered in the appropriate column, midway between the elements in the preceding column from which it has been constructed.

x	$f(x)$			
x_0	$f(x_0)$			
		$f(x_0, x_1)$		
x_1	$f(x_1)$		$f(x_0, x_1, x_2)$	
		$f(x_1, x_2)$		$f(x_0, x_1, x_2, x_3)$
x_2	$f(x_2)$		$f(x_1, x_2, x_3)$	
		$f(x_2, x_3)$		$\cdots$
x_3	$f(x_3)$		$\cdots$	
		$\cdots$		
$\cdots$	$\cdots$			

or in a specific numerical example,

x	x^3				
0	0				
		1			
1	1		4		
		13		1	
3	27		8		0
		37		1	
4	64		14		0
		93		1	
7	343		20		
		193			
9	729				

Usually the values of x in a table of data are equally spaced, and the differences of the function based on sets of consecutive functional values. When this is the case, the denominators in the divided differences of any given order are all equal, and it is customary to omit them. This leads to a modified set of quantities known simply as the **differences** of the function. If the constant difference between successive values of x is h, so that the general value of x in the table is

$$x_k = x_0 + kh \qquad k = \ldots, -2, -1, 0, 1, 2, \ldots,$$

and the corresponding functional value is

$$y_k = f(x_k) = f(x_0 + kh) = f_k$$

then the **first differences** of f are defined by the formula

$$\Delta f_k = f_{k+1} - f_k \tag{3}$$

Differences of higher order are defined in the same way, the **second differences** being

$$\Delta^2 f_k = \Delta(\Delta f_k) = \Delta f_{k+1} - \Delta f_k \tag{4}$$

and, in general, for all positive integral values of n,

$$\Delta^n f_k = \Delta(\Delta^{n-1} f_k) = \Delta^{n-1} f_{k+1} - \Delta^{n-1} f_k \tag{5}$$

These differences are also displayed in difference tables just like divided differences.

Clearly, the **difference operator** Δ has the characteristic properties of a linear operator, that is,

$$\Delta(af_k \pm bg_k) = a\,\Delta f_k \pm b\,\Delta g_k \qquad a,\ b \text{ constants}$$

Moreover, since any nth difference of the mth differences of a function f is obviously an $(n+m)$th difference of f, it is clear that for positive integral exponents, Δ obeys the familiar law of exponents

$$\Delta^n(\Delta^m f_k) = \Delta^{n+m} f_k$$

When the values of the independent variable are equally spaced, the divided differences of a function can easily be expressed in terms of the ordinary differences of the function, and vice versa. Specifically, if the tabular interval is h,

$$f(x_0, x_1) = \frac{f(x_0) - f(x_1)}{x_0 - x_1} = \frac{f_0 - f_1}{-h} = \frac{\Delta f_0}{h}$$

$$f(x_0, x_1, x_2) = \frac{f(x_0, x_1) - f(x_1, x_2)}{x_0 - x_2} = \frac{1}{-2h}\left(\frac{\Delta f_0}{h} - \frac{\Delta f_1}{h}\right) = \frac{\Delta^2 f_0}{2!h^2}$$

and, in general,

$$f(x_0, x_1, \ldots, x_n) = \frac{\Delta^n f_0}{n!h^n} \tag{6}$$

Likewise, if the $n+1$ points used in constructing an nth divided difference are the $n+1$ equally spaced points between $x_{-k} = x_0 - kh$ and $x_{n-k} = x_0 + (n-k)h$, inclusive, rather than the $n+1$ points running from x_0 to x_n, then it is easy to show that

$$f(x_{-k}, x_{-k+1}, \ldots, x_{n-k}) = \frac{\Delta^n f_{-k}}{n!h^n} \tag{7}$$

The Δ symbolism for the differences of a function is known as the **forward-difference** or **advancing-difference, notation.** In some applications, however, another notation known as the **central-difference notation** is more convenient. In this, the symbol δ, called the **central difference operator,** is used instead of Δ, and by definition

$$\delta f_{k+1/2} = f_{k+1} - f_k \qquad \text{for every integer } k$$

The subscript appearing in the symbol for any higher-order central difference is always the average of the subscripts assigned by the preceding definition to the central differences of one lower order which are subtracted in forming that higher-order difference. Thus

$$\Delta f_k = f_{k+1} - f_k = \delta f_{k+1/2} \qquad \Delta f_{k+1} = f_{k+2} - f_{k+1} = \delta f_{k+3/2}$$

$$\Delta^2 f_k = \Delta f_{k+1} - \Delta f_k = \delta f_{k+3/2} - \delta f_{k+1/2} = \delta^2 f_{k+1}$$

The following difference tables show the relation between the advancing- and central-difference notations.

x	f				
x_0	f_0				
		Δf_0			
x_1	f_1		$\Delta^2 f_0$		
		Δf_1		$\Delta^3 f_0$	
x_2	f_2		$\Delta^2 f_1$		$\Delta^4 f_0$
		Δf_2		$\Delta^3 f_1$	
x_3	f_3		$\Delta^2 f_2$		
		Δf_3			
x_4	f_4				

x	f				
x_0	f_0				
		$\delta f_{1/2}$			
x_1	f_1		$\delta^2 f_1$		
		$\delta f_{3/2}$		$\delta^3 f_{3/2}$	
x_2	f_2		$\delta^2 f_2$		$\delta^4 f_2$
		$\delta f_{5/2}$		$\delta^3 f_{5/2}$	
x_3	f_3		$\delta^2 f_3$		
		$\delta f_{7/2}$			
x_4	f_4				

In the first table, elements with the same subscript lie on lines sloping downward, or *advancing* into the table. In the second, elements with the same subscript lie on lines extending horizontally, or *centrally,* into the table.

Closely associated with Δ and δ is the **shift operator** E, defined as the operator that increases the argument of a function by one tabular interval. Thus, by definition,

$$Ef_k = Ef(x_k) = f(x_k + h) = f(x_{k+1}) = f_{k+1}$$

Applying E a second time again increases the argument of f by h; that is,

$$E^2 f_k = E^2 f(x_k) = E[Ef(x_k)] = Ef(x_k + h) = f(x_k + 2h) = f(x_{k+2}) = f_{k+2}$$

and, in general, we define

(8) $$E^r f_k = E^r f(x_k) = f(x_k + rh) = f(x_{k+r}) = f_{k+r}$$

for *any* real number r. Clearly, the operator E has the properties

(9) $$E(af_r \pm bg_r) = aEf_r \pm bEg_r \qquad a, b \text{ constants}$$

(10) $$E^r(E^s f_k) = E^{r+s} f_k \qquad r, s \text{ arbitrary real numbers}$$

Two operators with the property that when they are applied to the same function they yield the same result are said to be **operationally equivalent.** Now from the definition of Δf_k, we have

$$\Delta f_k = f_{k+1} - f_k = Ef_k - f_k = (E - 1)f_k$$

Hence we have the operational equivalences

(11) $$\Delta = E - 1$$

(12) $$E = 1 + \Delta$$

(13) $$E - \Delta = 1$$

Moreover, by definition,

$$\Delta f_k = \delta f_{k+1/2} = \delta E^{1/2} f_k$$

Hence we have the further equivalences

$$\Delta = \delta E^{1/2} \tag{14}$$

$$\delta = \Delta E^{-1/2} \tag{15}$$

Also, substituting from (14) into (11) and then solving for δ, we have

$$\delta = E^{1/2} - E^{-1/2} \tag{16}$$

Equations (11) through (16) are useful in carrying out operational, or formal, derivations of various important formulas in numerical analysis. When deemed necessary, these can then be established rigorously by other means.

By means of (11) we can express the various differences of a function in terms of successive entries in the table of the function. For, using (11), we can write

$$\Delta^n f_k = (E - 1)^n f_k$$

Then, using the binomial expansion, we have

$$\begin{aligned}\Delta^n f_k &= \left[E^n - \binom{n}{1}E^{n-1} + \binom{n}{2}E^{n-2} + \cdots + (-1)^{n-1}\binom{n}{n-1}E + (-1)^n\binom{n}{n}\right]f_k\dagger \\ &= E^n f_k - nE^{n-1}f_k + \frac{n(n-1)}{2}E^{n-2}f_k + \cdots + (-1)^{n-1}nEf_k + (-1)^n f_k \\ &= f_{k+n} - nf_{k+n-1} + \frac{n(n-1)}{2}f_{k+n-2} + \cdots + (-1)^{n-1}nf_{k+1} + (-1)^n f_k \qquad (17)\end{aligned}$$

Specifically, taking $k = 0$ and $n = 1, 2, 3, 4, \ldots$, we have

$$\begin{aligned}\Delta f_0 &= f_1 - f_0 \\ \Delta^2 f_0 &= f_2 - 2f_1 + f_0 \\ \Delta^3 f_0 &= f_3 - 3f_2 + 3f_1 - f_0 \qquad (17a) \\ \Delta^4 f_0 &= f_4 - 4f_3 + 6f_2 - 4f_1 + f_0 \\ &\cdots\cdots\end{aligned}$$

The fact that the first divided difference of a function is precisely the difference quotient whose limit defines the derivative of the function suggests that in some respects the properties of the differences of a function and the properties of the derivatives of a function may be analogous. This is actually the case, and among other interesting results we have the following theorem.

THEOREM 1 The nth divided differences of a polynomial of degree n are constant.

PROOF To prove this theorem, it is clearly sufficient to establish the asserted property for the special polynomial x^n. To do this, we observe that for x^n the first divided difference is simply

$$f(x_i, x_j) = \frac{x_i^n - x_j^n}{x_i - x_j} = x_i^{n-1} + x_i^{n-2}x_j + \cdots + x_i x_j^{n-2} + x_j^{n-1}$$

---

†The quantities $\binom{n}{j}$ are the so-called **binomial coefficients,** defined by the formula

$$\binom{n}{j} = \frac{n!}{j!\,(n-j)!}$$

which is a homogeneous and symmetric function of x_i and x_j of degree $n - 1$. For the second divided difference we have of course

$$f(x_i, x_j, x_k) = \frac{f(x_i, x_j) - f(x_j, x_k)}{x_i - x_k}$$

Moreover, since divided differences of all orders are symmetric functions of the arguments (see Exercises 11 and 12), it follows that the numerator of the last fraction vanishes when $x_i = x_k$. Hence it must contain $x_i - x_k$ as a factor and therefore, as we verified explicitly for the first divided difference, the indicated division is exact. Thus the second divided difference of x^n is a homogeneous and symmetric expression of degree $n - 2$ in x_i, x_j, x_k. Continuing in this way, it is evident that after differencing n times, the degree of the resultant expression will be zero; that is, x^n will have been reduced to a constant, independent of $x_i, x_j, x_k, \ldots, x_n$, as asserted. ◗

Since ordinary differences are proportional to the corresponding divided differences, it is clear that Theorem 1 also holds for these differences, whether they are expressed in terms of the advancing- or central-difference notation. Thus we have the following corollary of Theorem 1.

COROLLARY 1 The nth (ordinary) differences of a polynomial of degree n are constant.

EXERCISES

1. Prove Properties 1 and 2.

2. Construct the divided difference table for each of the following functions on the indicated points.
(a) $x^3 - 2x^2 + 3$; $x = 0.0, 1.0, 1.5, 2.5, 3.0, 3.5$
(b) $2x^3 - x + 3$; $x = 1.0, 1.5, 2.5, 3.0, 4.0, 6.0$

3. Construct the divided difference table for the function $x^4 - 6x^2 + 8x + 1$ on the points $x = 0.0, 0.5, 2.0, 3.0, 3.5, 6.0$.

4. Construct the difference table for **(a)** the function $x^4 - 3x^2 + 1$ evaluated at the points $x = 0, 1, 2, 3, 4, 5, 6$, and **(b)** the function $2x^4 - 3x^3 + x - 5$ evaluated at intervals of $h = 0.2$ from $x = 0$ to $x = 2$, inclusive.

5. For an arbitrary tabular interval h, find the successive differences of **(a)** $x^2 + 3x + 1$ and **(b)** $x^3 - x^2 + 2x$ as functions of x.

6. If $h = 1$, show that

(a) $\Delta \cos(ax + b) = -2 \sin \dfrac{a}{2} \sin\left(ax + b + \dfrac{a}{2}\right)$

(b) $\Delta \sin(ax + b) = 2 \sin \dfrac{a}{2} \cos\left(ax + b + \dfrac{a}{2}\right)$

What function has $\sin ax$ for its first difference? What function has $\cos ax$ for its first difference?

7. For arbitrary h, what are the successive differences of a^x?

8. **(a)** Show that the divided differences of a function are unchanged if the arguments are all increased by the same constant c while the corresponding functional values are left unchanged.
(b) How are the divided differences of a function changed if each argument is multiplied by the same constant c while the corresponding functional values are left unchanged?

9. If $h = 1$, show that for all values of the constants a and b, each of the following functions satisfies the indicated relation:
(a) $y = a2^x + b3^x$; $(E^2 - 5E + 6)y = 0$
(b) $y = a2^x + bx2^x$; $(E^2 - 4E + 4)y = 0$
(c) $y = a3^x + b(-2)^x$; $(\Delta^2 + \Delta - 6)y = 0$
Is there a result similar to this in the theory of differential equations?

10. If we define

$$f(x_0, x_1, \ldots, x_{n-1}, x_n, x_n) = \lim_{x \to x_n} f(x_0, x_1, \ldots, x_{n-1}, x_n, x)$$

show that

$$f(x_0, x_1, \ldots, x_{n-1}, x_n, x_n) = \left.\frac{df(x_0, x_1, \ldots, x_{n-1}, x)}{dx}\right|_{x=x_n}$$

11. Show that

$$f(x_0, x_1) = \frac{f(x_0)}{x_0 - x_1} + \frac{f(x_1)}{x_1 - x_0}$$

and

$$f(x_0, x_1, x_2) = \frac{f(x_0)}{(x_0 - x_1)(x_0 - x_2)} + \frac{f(x_1)}{(x_1 - x_0)(x_1 - x_2)} + \frac{f(x_2)}{(x_2 - x_0)(x_2 - x_1)}$$

What is the generalization of these results to differences of higher order?

12. Show that

$$f(x_0, x_1) = \frac{\begin{vmatrix} f(x_0) & f(x_1) \\ 1 & 1 \end{vmatrix}}{\begin{vmatrix} x_0 & x_1 \\ 1 & 1 \end{vmatrix}}$$

and

$$f(x_0, x_1, x_2) = \frac{\begin{vmatrix} f(x_0) & f(x_1) & f(x_2) \\ x_0 & x_1 & x_2 \\ 1 & 1 & 1 \end{vmatrix}}{\begin{vmatrix} x_0^2 & x_1^2 & x_2^2 \\ x_0 & x_1 & x_2 \\ 1 & 1 & 1 \end{vmatrix}}$$

13. Show that $\lim_{h\to 0} f(x_0 - \phi h, x_0, x_0 + \theta h) = \frac{1}{2}f''(x_0)$. *Hint:* Use the result of Exercise 11.

Factorial Polynomials

14. The product $x^{(n)} = x(x-1)(x-2)\cdots(x-n+1)$ is called a **factorial polynomial.** Show that $\Delta x^{(n)} = nx^{(n-1)}$. What is $\Delta^k x^{(n)}$?

15. What are the difference formulas for the product

$$x_h^{(n)} = x(x-h)(x-2h)\cdots(x-\overline{n-1}h)?$$

16. Of what function is $x^{(n)}$ the first difference? Of what function is it the second difference?

17. Show that if $p(x)$ is a polynomial of degree n, then $p(x)$ can be written in the form

$$p(x) = a_0 + a_1x^{(1)} + a_2x^{(2)} + \cdots + a_nx^{(n)}$$

where $a_j = 1/j!\ \Delta^j p(x)|_{x=0}$. *Hint:* Use the formulas of Exercise 14 to form the successive differences of the expression assumed for $p(x)$. Then evaluate these differences for $x = 0$.

Discuss the relation of this procedure to the usual method of constructing the Maclaurin series of a function.

18. Using the result of Exercise 17, find the polynomial of minimum degree which yields the following data.

(a)

x	y				
0	−1				
		−1			
1	−2		6		
		5		6	
2	3		12		0
		17		6	
3	30		18		
		3			
4	55				

(b)

x	y					
0	6					
		−5				
1	1		2			
		−3		−6		
2	−2		−4		24	
		−7		18		0
3	−9		14		24	
		7		42		
4	−2		56			
		63				
5	61					

19. If $p(x)$ is a polynomial of degree n, show that $p(x)$ can be written in the form

$$p(x) = r_0 + r_1x^{(1)} + r_2x^{(2)} + \cdots + r_nx^{(n)}$$

where the r's are determined from the relations

$$\begin{aligned} p(x) &= r_0 + xq_0(x) \\ q_0(x) &= r_1 + (x-1)q_1(x) \\ q_1(x) &= r_2 + (x-2)q_2(x) \\ &\cdots\cdots \\ q_{n-1}(x) &= r_n \end{aligned}$$

Devise a procedure for calculating the r's through the repeated use of synthetic division.

20. Using the result of Exercise 19, express each of the following polynomials in terms of factorial polynomials.

(a) $x^3 - 3x^2 + 4x - 5$

(b) $2x^4 + x^3 - 4x^2 - x + 1$

(c) $x^5 - 1$

21. The fraction $x^{(-n)} = 1/(x+1)(x+2)\cdots(x+n)$ is called a **reciprocal factorial polynomial.** Show that $\Delta x^{(-n)} = -nx^{(-n-1)}$. What is $\Delta^k x^{(-n)}$?

22. What are the difference formulas for the fraction

$$x_h^{(-n)} = \frac{1}{(x+h)(x+2h)(x+3h)\cdots(x+nh)}?$$

23. Of what function is $x^{(-n)}$ the first difference? Of what function is it the second difference?

24. Express the fraction $1/(x+1)(x+2)(x+4)$ in terms of reciprocal factorial polynomials. *Hint:* Multiply the numerator and denominator by $x+3$, and then note that in the numerator $x+3 = (x+4) - 1$.

25. Express the fraction $1/(x+1)(x+4)$ in terms of reciprocal factorial polynomials. *Hint:* Multiply the numerator and denominator by $(x+2)(x+3)$. Then in the numerator determine a, b, and c so that $(x+2)(x+3) = a(x+3)(x+4) + b(x+4) + c$ is an identity.

26. Express each of the following fractions in terms of reciprocal factorial polynomials.

(a) $\dfrac{x}{(x+1)(x+2)(x+3)}$

(b) $\dfrac{x^2}{(x+1)(x+2)(x+3)(x+4)}$

(c) $\dfrac{1}{(x+1)(x+5)}$

Summation of Series

27. Show that $\Sigma_{k=1}^n y_k = (E^n - 1)/(E-1)\, y_1$, and then, by putting $E - 1 = \Delta$, show that

$$\sum_{k=1}^{n} y_k = \left[n + \frac{n(n-1)}{2!}\Delta + \frac{n(n-1)(n-2)}{3!}\Delta^2 + \cdots\right] y_1$$

28. Using the result of Exercise 27, evaluate

(a) $\sum\limits_{k=1}^{n} k^2$ **(b)** $\sum\limits_{k=1}^{n} k^3$ **(c)** $\sum\limits_{k=1}^{n} k^4$

29. If $F(x)$ is an **antidifference** of $f(x)$, that is, if $\Delta F(x) = f(x)$, show that $\Sigma_{i=1}^n f(x_i) = F(x_{n+1}) - F(x_1)$. Discuss the resemblance of this procedure to the process of definite integration.

30. Work Exercise 28 using the procedure discussed in Exercise 29. *Hint:* First use the result of Exercise 19.

31. Using Exercise 29, evaluate

(a) $\sum_{k=1}^{n} \sin ka$ **(b)** $\sum_{k=1}^{n} \cos ka$

Hint: Use the results of Exercise 6.

32. Evaluate each of the following sums:

(a) $\sum_{k=1}^{n} \frac{1}{(k+1)(k+2)}$

(b) $\sum_{k=1}^{n} \frac{1}{(k+1)(k+3)}$

(c) $\sum_{k=1}^{\infty} \frac{1}{(k+1)(k+2)(k+3)}$

(d) $\sum_{k=1}^{n} \frac{k}{(k+1)(k+2)(k+3)}$

(e) $\sum_{k=1}^{\infty} \frac{k}{(k+1)(k+2)(k+4)}$

33. (a) Show that $\Delta(u_k v_k) = v_{k+1}\,\Delta u_k + u_k\,\Delta v_k$.

(b) Show that $\Delta^{-1}(u_k\,\Delta v_k) = u_k v_k - \Delta^{-1}(v_{k+1}\,\Delta u_k)$, where Δ^{-1} is the symbol of antidifferencing.

(c) Using **(b)** and the result of Exercise 29, show that

$$\sum_{1}^{n} u_k\,\Delta v_k = [u_k v_k]_1^{n+1} - [\Delta^{-1}(v_{k+1}\,\Delta u_k)]_1^{n+1}$$

This important result is known as **summation by parts** and is clearly the analog of integration by parts in calculus.

34. Use the result of Exercise 33**(c)** to evaluate each of the following sums:

(a) $\sum_{k=1}^{n} k2^k$. *Hint:* Let $u_k = k$ and $\Delta v_k = 2^k$.

(b) $\sum_{k=1}^{n} k^2 3^k$ **(c)** $\sum_{k=1}^{n} k \cos ak$ **(d)** $\sum_{k=1}^{n} k \sin ak$

35. Justify Eq. (7).

5.3 INTERPOLATION FORMULAS

One of the most important applications of finite differences has to do with how they are used to determine a function of specified form which takes on a given set of consecutive values in a table of data. Sometimes such an approximating function is needed so that values of the actual function between consecutive tabular values can be found, or *interpolated.* Sometimes the approximating function is needed so that derivatives at particular points and integrals over particular ranges can be calculated for the tabulated function. In particular, formulas for approximating derivatives are of great importance in the numerical solution of ordinary differential equations. In this section we shall be primarily concerned with developing interpolation formulas, leaving their application to numerical differentiation and integration and the numerical solution of differential equations to later sections.

In courses such as algebra and trigonometry, where tables of the elementary functions must occasionally be used, it is customary to obtain values between adjacent entries by the method of *proportional parts* or *linear interpolation.* As is well known, this procedure amounts to replacing the arc of the tabulated function over one tabular interval by its chord and then reading the required functional value from the chord rather than from the arc itself (Fig. 5.1*a*). In this case the formula for the interpolated value turns out to be

$$f(x_0 + rh) = f(x_0) + r[f(x_0 + h) - f(x_0)] = f_0 + r\,\Delta f_0 \tag{1}$$

Obviously, if h is relatively large, or if the graph of $f(x)$ is changing direction rapidly, the chord may not be a good approximation to the arc and linear interpolation may involve a substantial error. One way to overcome this difficulty would be to approximate the graph of $f(x)$ by some curve which would "fit" the true arc more closely than a straight line could and then read the interpolated value from this approximating arc rather than from the chord (Fig. 5.1*b*). If, specifically, the graph of $f(x)$ is approximated over two successive tabular intervals by a parabola of the form $y = a + bx + cx^2$ chosen to pass through the three points

$$[x_0, f(x_0)] \qquad [x_0 + h, f(x_0 + h)] \qquad [x_0 + 2h, f(x_0 + 2h)]$$

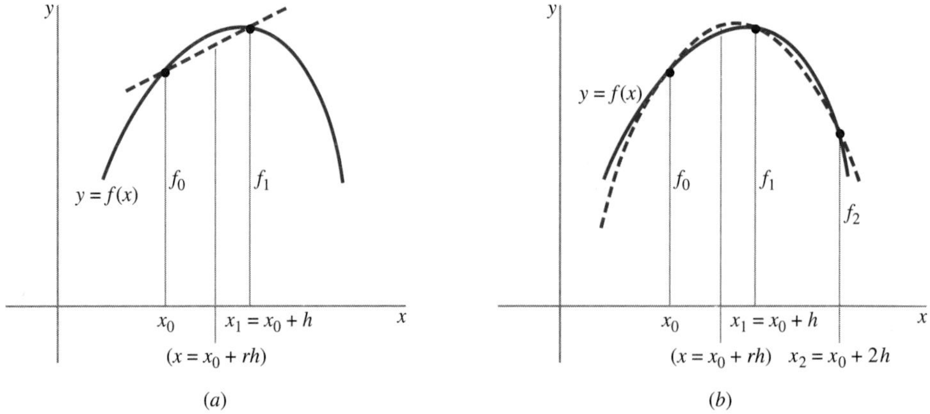

FIGURE 5.1
Straight-line and parabolic approximations to a given function.

the formula for the interpolated value is found without difficulty to be

$$f(x_0 + rh) = f(x_0) + r[f(x_0 + h) - f(x_0)] + \frac{r(r-1)}{2!}[f(x_0 + 2h) - 2f(x_0 + h) + f(x_0)]$$

$$= f_0 + r\,\Delta f_0 + \frac{r(r-1)}{2!}\,\Delta^2 f_0 \tag{2}$$

Proceeding in this fashion, fitting polynomial curves of higher and higher order to more and more consecutive points on the graph of $f(x)$, one could derive a succession of interpolation formulas involving higher and higher differences of the tabulated function and providing, presumably, higher and higher accuracy in the tabulated values. In this section we shall obtain several important interpolation formulas, though we shall derive them by methods more general than the geometric approach we have just suggested.

Probably the most fundamental interpolation formula is **Newton's divided-difference formula**

$$\begin{aligned}(3)\quad f(x) = f(x_0) + (x - x_0)f(x_0, x_1) + (x - x_0)(x - x_1)f(x_0, x_1, x_2) + \cdots \\ + (x - x_0)(x - x_1)\cdots(x - x_{n-1})f(x_0, x_1, \ldots, x_n) \\ + (x - x_0)(x - x_1)\cdots(x - x_n)f(x, x_0, x_1, \ldots, x_n)\end{aligned}$$

From this all the other interpolation formulas of interest to us can easily be derived by suitably specializing the points $x_0, x_1, \ldots, x_n$, which need not be equally spaced or taken in consecutive order. For convenience in establishing (3) we shall restrict our discussion to some special, though adequately typical, value of n, say $n = 2$. Then, beginning with the third difference we have, for arbitrary x,

$$f(x, x_0, x_1, x_2) = \frac{f(x, x_0, x_1) - f(x_0, x_1, x_2)}{x - x_2}$$

and solving for $f(x, x_0, x_1)$, we have

$$f(x, x_0, x_1) = f(x_0, x_1, x_2) + (x - x_2)f(x, x_0, x_1, x_2) \tag{4}$$

But

$$f(x, x_0, x_1) = \frac{f(x, x_0) - f(x_0, x_1)}{x - x_1}$$

and substituting this into (4) and solving for $f(x, x_0)$, we find

(5) $$f(x, x_0) = f(x_0, x_1) + (x - x_1)f(x_0, x_1, x_2) + (x - x_1)(x - x_2)f(x, x_0, x_1, x_2)$$

Finally, since

$$f(x, x_0) = \frac{f(x) - f(x_0)}{x - x_0}$$

we have, on substituting this into (5) and solving for $f(x)$

$$f(x) = f(x_0) + (x - x_0)f(x_0, x_1) + (x - x_0)(x - x_1)f(x_0, x_1, x_2) + (x - x_0)(x - x_1)(x - x_2)f(x, x_0, x_1, x_2)$$

which is precisely (3) in the special case $n = 2$. The extension of the preceding argument to any value of n is obvious.

Once x is given, every term in (3) except the last is completely determined. The last term, however, differs from all the others in that the divided difference which appears in it contains x as one of its arguments, hence is not to be found among the entries in the difference table of $f(x)$. For this reason the last term is usually referred to as the **remainder after** $n + 1$ **terms** or simply as the **error term,** and the interpolation series is often written in the form

(6) $$f(x) = p_n(x) + r_{n+1}(x)$$

where of course $p_n(x)$ is the nth-degree polynomial

(7) $$f(x_0) + (x - x_0)f(x_0, x_1) + (x - x_0)(x - x_1)f(x_0, x_1, x_2) + \cdots + (x - x_0)(x - x_1)\cdots(x - x_{n-1})f(x_0, x_1, \ldots, x_n)$$

and $r_{n+1}(x)$ is the function

(8) $$(x - x_0)(x - x_1)\cdots(x - x_n)f(x, x_0, x_1, \ldots, x_n)$$

The error term $r_{n+1}(x)$ is important for theoretical purposes but is of limited practical value, and we shall leave further discussion of the error term to texts on numerical analysis.†

EXAMPLE 1

Neglecting the error term, use (3) to find $f(5)$ from the following data.

x	$f(x)$	$f(x_i, x_j)$	$f(x_i, x_j, x_k)$	$f(x_i, x_j, x_k, x_l)$	$f(x_i, x_j, x_k, x_l, x_m)$
4.0	1.58740				
		0.12712			
4.5	1.65096		−0.00861		
		0.11421		0.00087	
5.5	1.76517		−0.00687		−0.00009
		0.10390		0.00059	
6.0	1.81712		−0.00539		
		0.09581			
7.0	1.91293				

†See, for instance, F. B. Hildebrand, *Introduction to Numerical Analysis,* 2d ed., pp. 62–68. McGraw-Hill, New York, 1974.

The construction of the difference table of the given function [in this case $f(x) = \sqrt[3]{x}$] presents no problem, and using Newton's formula with $x_0 = 4$ and the error term omitted, we have at once

$$\begin{aligned} f(5) &= 1.58740 + (5.0 - 4.0)(0.12712) + (5.0 - 4.0)(5.0 - 4.5)(-0.00861) \\ &\quad + (5.0 - 4.0)(5.0 - 4.5)(5.0 - 5.5)(0.00087) \\ &\quad + (5.0 - 4.0)(5.0 - 4.5)(5.0 - 5.5)(5.0 - 6.0)(-0.00009) \\ &= 1.70998 \end{aligned}$$

To six decimal places, the correct answer is 1.709975, which rounds off to the interpolated value through the fifth decimal place. Incidentally, we note that the ordinary process of linear interpolation yields the value 1.70806.

Closely associated with Newton's divided-difference formula is **Lagrange's interpolation formula**†

$$(9)\quad f(x) = \frac{(x - x_1)(x - x_2)\cdots(x - x_n)}{(x_0 - x_1)(x_0 - x_2)\cdots(x_0 - x_n)} f(x_0) + \frac{(x - x_0)(x - x_2)\cdots(x - x_n)}{(x_1 - x_0)(x_1 - x_2)\cdots(x_1 - x_n)} f(x_1) + \cdots + \frac{(x - x_0)(x - x_1)\cdots(x - x_{n-1})}{(x_n - x_0)(x_n - x_1)\cdots(x_n - x_{n-1})} f(x_n)$$

Like Newton's divided-difference formula, this formula provides the equation of a polynomial of degree n (or less) which takes on the $n + 1$ prescribed functional values $f(x_0), f(x_1), \ldots, f(x_n)$ when x takes on the values $x_0, x_1, \ldots, x_n$. Equation (9) can easily be derived from Eq. (3), but it is simpler merely to verify its properties. Clearly, it is a polynomial of degree n (or less), since each term on the right is a polynomial of degree n. Moreover, when $x = x_0$, every fraction except the first vanishes because of the factor $x - x_0$, and at the same time the first fraction reduces to 1, leaving just $f(x) = f(x_0)$, as required when $x = x_0$. In the same way, when $x = x_1$, every fraction except the second becomes zero, and we have $f(x) = f(x_1)$. Similarly, we can verify without difficulty that $f(x)$ reduces to $f(x_2), f(x_3), \ldots, f(x_n)$ when $x = x_2, x_3, \ldots, x_n$, as required.

When the points $x_0, x_1, \ldots, x_n$ on which Newton's divided-difference formula is based are regularly spaced with tabular interval h, say, it is generally more convenient to express Formula (3) in terms of ordinary differences. To do this we observe that if

$$x = x_0 + rh \qquad \text{and} \qquad x_k = x_0 + kh$$

then

$$x - x_k = h(r - k) \qquad k = 0, 1, 2, \ldots, n$$

and

$$(10)\quad (x - x_0)(x - x_1)\cdots(x - x_j) = h^{j+1} r(r - 1)\cdots(r - j)$$

Also, from Eq. (6), Sec. 5.2, we have

$$(11)\quad f(x_0, x_1, \ldots, x_{j+1}) = \frac{\Delta^{j+1} f_0}{(j + 1)!\,h^{j+1}}$$

Hence, substituting from (10) and (11) into (3), we find

$$(12)\quad \begin{aligned} f(x) &\equiv f(x_0 + rh) \\ &= f_0 + r\,\Delta f_0 + \frac{r(r - 1)}{2!}\Delta^2 f_0 + \frac{r(r - 1)(r - 2)}{3!}\Delta^3 f_0 + \cdots \end{aligned}$$

†Named for the French mathematician Joseph Louis Lagrange (1736–1813).

which is known as the **forward Gregory-Newton interpolation formula.**† Obviously this is a direct generalization of the formulas of linear and parabolic interpolation [Eqs. (1) and (2)]. Of course, the error term in (3) can be transformed into a corresponding error term for the series (12), but we shall leave this to more advanced texts.

EXAMPLE 2

Compute $\sqrt[3]{4.4}$ from the following table of $\sqrt[3]{x}$.

x	$\sqrt[3]{x}$	Δ	Δ^2	Δ^3	Δ^4
4.0	1.58740				
		0.06356			
4.5	1.65096		−0.00454		
		0.05902		0.00071	
5.0	1.70998		−0.00383		−0.00012
		0.05519		0.00059	
5.5	1.76517		−0.00324		
		0.05195			
6.0	1.81712				

Since the values of $\sqrt[3]{x}$ are given at equally spaced intervals, it is convenient to use the forward Gregory-Newton formula to perform the required interpolation. Clearly, the tabular interval is $h = 0.5$. Hence, taking $x_0 = 4.0$ and noting that $4.4 = 4 + 0.8(0.5)$, we find that $r = 0.8$. Substituting into Eq. (12) then gives us

$$\sqrt[3]{4.4} = 1.58740 + 0.8(0.06356) + \frac{(0.8)(-0.2)}{2}(-0.00454)$$

$$+ \frac{(0.8)(-0.2)(-1.2)}{6}(0.00071) + \frac{(0.8)(-0.2)(-1.2)(-2.2)}{24}(-0.00012) = 1.63863$$

To five places, the correct answer is 1.63864. Linear interpolation involves just the first two terms in the last expression and gives the less accurate value 1.63825.

EXAMPLE 3

Fit a polynomial in x of minimum degree to the following data.

x	$f(x)$	Δ	Δ^2	Δ^3	Δ^4
1.0	2.000				
		1.375			
1.5	3.375		2.250		
		3.625		0.750	
2.0	7.000		3.000		0.000
		6.625		0.750	
2.5	13.625		3.750		
		10.375			
3.0	24.000				

The construction of the difference table presents no problem, and from the constancy of the third differences we know that the required polynomial will be a cubic. If interpolation were our only object, we would simply evaluate the Gregory-Newton formula (12) for the relevant values of r. However, we are asked to find a formula for the function in terms of x, not just its values for certain x's, identified by the corresponding

†Conamed for the Scottish mathematician James Gregory (1638–1675).

values of r. This means that in Eq. (12) we must express r in terms of x, then carry out the indicated multiplications and collect terms. Clearly, for the given table of data, $h = 0.5$ and we have

$$x = x_0 + rh = 1 + \frac{r}{2} \qquad \text{and therefore} \qquad r = 2x - 2$$

Replacing r in Eq. (12) by $2x - 2$ gives

$$f(x_0 + rh) \equiv f(x) = 2 + (2x-2)1.375 + \frac{(2x-2)(2x-3)}{2}2.250 + \frac{(2x-2)(2x-3)(2x-4)}{6}0.750$$

which simplifies at once to $f(x) = x^3 - 2x + 3$.

The forward Gregory-Newton formula expresses $f(x_0 + rh)$ in terms of entries in the difference table of $f(x)$ which lie on a line sloping downward to the right. These differences in turn involve entries in the table which *follow* $f(x_0)$. Sometimes, as in interpolating near the lower end of a table of limited extent, it is desirable to have a formula that uses entries in the difference table which lie on a line sloping upward to the right and therefore involve values of $f(x)$ which *precede* $f(x_0)$. Such a formula can easily be derived by choosing the points $x_0, x_1, x_2, \ldots, x_n$ used in the divided difference formula (3) (which can be any $n + 1$ distinct points) to be, respectively, the points

$$x_0 = x_0 \qquad x_1 = x_0 - h \qquad x_2 = x_0 - 2h \qquad \cdots \qquad x_k = x_0 - kh \qquad \cdots \qquad x_n = x_0 - nh$$

Then

$$x - x_k = (x_0 + rh) - (x_0 - kh) = h(r + k) \qquad k = 0, 1, 2, \ldots, n$$

and

$$(x - x_0)(x - x_1)\cdots(x - x_j) = h^{j+1}r(r+1)\cdots(r+j) \tag{13}$$

In this case, the typical divided difference $f(x_0, x_1, \ldots, x_{j+1})$ is actually $f(x_0, x_{-1}, \ldots, x_{-j-1})$, and from the symmetry of divided differences this is equal to $f(x_{-j-1}, x_{-j}, \ldots, x_{-1}, x_0)$. Hence, using Eq. (7), Sec. 5.2 (with $n = k = j + 1$), we have for our current choice of points

$$f(x_0, x_1, \ldots, x_{j+1}) = \frac{\Delta^{j+1}f_{-j-1}}{(j+1)!h^{j+1}} \tag{14}$$

Finally, substituting from (13) and (14) into (3), we find

$$f(x) \equiv f(x_0 + rh) = f_0 + r\,\Delta f_{-1} + \frac{r(r+1)}{2!}\Delta^2 f_{-2} + \frac{r(r+1)(r+2)}{3!}\Delta^3 f_{-3} + \cdots \tag{15}$$

which is known as the **backward Gregory-Newton interpolation formula.**

There are various ways of obtaining central-difference interpolation formulas. For instance, we can choose the points to be used in Newton's divided difference formula as alternately following and preceding x_0; that is, we can take

$$x_0 = x_0 \qquad x_1 = x_0 + h \qquad x_2 = x_0 - h \qquad x_3 = x_0 + 2h \qquad x_4 = x_0 - 2h \qquad \cdots$$

Then, proceeding very much as we did in deriving the backward Gregory-Newton formula, we find

$$\begin{aligned} f(x) &\equiv f(x_0 + rh) \\ &= f_0 + r\,\Delta f_0 + \frac{r(r-1)}{2!}\Delta^2 f_{-1} + \frac{r(r-1)(r+1)}{3!}\Delta^3 f_{-1} \\ &\quad + \frac{r(r-1)(r+1)(r-2)}{4!}\Delta^4 f_{-2} + \cdots \end{aligned} \tag{16}$$

or, introducing the central-difference operator δ by means of the operational equivalence $\Delta = \delta E^{1/2}$ [Eq. (14), Sec. 5.2].

$$\begin{aligned} f(x_0 + rh) &= f_0 + r\,\delta f_{1/2} + \frac{r(r-1)}{2!}\delta^2 f_0 + \frac{(r+1)r(r-1)}{3!}\delta^3 f_{1/2} \\ &\quad + \frac{(r+1)r(r-1)(r-2)}{4!}\delta^4 f_0 + \cdots \end{aligned} \tag{16a}$$

This is known as the **forward Newton-Gauss interpolation formula.**

In exactly the same way, by choosing the points $x_0, x_1, x_2, \ldots,$ in the order

$$x_0 = x_0 \qquad x_1 = x_0 - h \qquad x_2 = x_0 + h \qquad x_3 = x_0 - 2h \qquad x_4 = x_0 + 2h \qquad \cdots$$

and again substituting into (3) we obtain, after introducing the central-difference notation,

$$\begin{aligned} f(x_0 + rh) &= f_0 + r\,\delta f_{-1/2} + \frac{(r+1)r}{2!}\delta^2 f_0 + \frac{(r+1)r(r-1)}{3!}\delta^3 f_{-1/2} \\ &\quad + \frac{(r+2)(r+1)r(r-1)}{4!}\delta^4 f_0 + \cdots \end{aligned} \tag{17}$$

which is usually referred to as the **backward Newton-Gauss interpolation formula.**

If we take the average of Eqs. (16a) and (17), we obtain what is known as **Stirling's interpolation formula**†

$$\begin{aligned} f(x_0 + rh) &= f_0 + \frac{r}{1!}\,\frac{\delta f_{1/2} + \delta f_{-1/2}}{2} + \frac{r^2}{2!}\delta^2 f_0 \\ &\quad + \frac{r(r^2-1)}{3!}\,\frac{\delta^3 f_{1/2} + \delta^3 f_{-1/2}}{2} + \frac{r^2(r^2-1)}{4!}\delta^4 f_0 + \cdots \end{aligned} \tag{18}$$

Each of the formulas we have developed in this section provides a polynomial approximation P for a function f which takes on the values of f at certain selected base points. As the base points are chosen closer and closer together, it is natural to expect that the accuracy of the approximation will become better and better, but this is not necessarily the case. In fact there are examples‡ which show that if the base points, or *nodes,* of the interpolation are taken closer and closer together, and the degree of the approximating polynomial P is correspondingly increased, then between succes-

†Named for the Scottish mathematician James Stirling (1692–1770).

‡See, for instance, E. Isaacson and H. B. Keller, *Analysis of Numerical Methods,* pp. 275–279, John Wiley, New York, 1966.

sive nodes the difference between f and P may increase without bound. One way to avoid this possibility is to use the relatively new technique of *spline approximation.* We do not have the space to discuss the theory of splines, but we shall conclude this section with a brief explanation of what they are.†

Rather than being a *single* polynomial approximating a function f over an interval $[a, b]$, a **spline function,** or **spline,** is a *piecewise* polynomial approximation $S(x)$ to $f(x)$ such that at each node x_i of the interval of approximation $[a, b]$, $S(x_i) = f(x_i) = f_i$. Thus the graph of S contains the data point, or *knot,* (x_i, f_i) and S is continuous on $[a, b]$.

DEFINITION 1 A spline on $[a, b]$ is of **order** n if and only if

(a) Its first $n - 1$ derivatives exist at each interior node of $[a, b]$.

(b) At least one of the polynomials in terms of which the spline is defined is of degree n, and they are all of degree n or less.

The graph of each defining polynomial of a first-order spline is a line segment connecting a pair of knots on the graph of the function the spline approximates. Although first-order, or **linear,** splines are continuous on $[a, b]$; they are not smooth. **Quadratic,** i.e., second-order, splines are required to have a derivative at each node interior to the interval of approximation $[a, b]$. As a working hypothesis, each polynomial involved in the piecewise definition of such a spline is assumed to be of second degree. Quadratic splines are smooth. The most important splines are **cubic,** or third-order, splines which are required to have both a first and second derivative at each interior node. In general, but not always (see Exercise 23), a cubic spline is defined between each successive pair of nodes by a polynomial of third degree.

To obtain a cubic spline approximation S for a function f on an interval $[a, b]$ whose values f_i are known on a discrete set of base points

$$a = x_0 < x_1 < x_2 < x_3 < \cdots < x_{n-1} < x_n = b \tag{19}$$

not necessarily equally spaced, we begin with the assumption that S is represented by a third-degree polynomial

$$P_i(x) = a_i + b_i x + c_i x^2 + d_i x^3 \qquad 1 \le i \le n \tag{20}$$

on each of the n subintervals of $[a, b]$ determined by the $n + 1$ nodes (19). The $4n$ unknown coefficients in (20) remain to be evaluated consistent with all other defining properties of a cubic spline. Suitable conditions for finding these unknowns may be formulated as follows.

Since the graph of S must contain the end knots (a, f_0) and (b, f_n), we must have

$$S(a) = P_1(a) = f_0 \qquad \text{and} \qquad S(b) = P_n(b) = f_n \qquad \text{(2 conditions)}$$

Because S is to be continuous at each interior node of $[a, b]$, we must have

$$S(x_i) = P_i(x_i) = P_{i+1}(x_i) = f_i \qquad 1 \le i \le n - 1 \qquad (2n - 2 \text{ conditions})$$

†For a further discussion of splines see, for instance, S. C. Chapra and R. P. Canale, *Numerical Methods for Engineers,* 2d ed., pp. 387–399, McGraw-Hill, New York, 1988; F. B. Hildebrand, *Introduction to Numerical Analysis,* loc. cit, pp. 478ff.

The defining properties which require the first and second derivatives of S to exist at each node interior to $[a, b]$ yield the additional relations

$$P_i'(x_i) = P_{i+1}'(x_i) \qquad 1 \le i \le n-1 \qquad (n-1 \text{ conditions})$$

and

$$P_i''(x_i) = P_{i+1}''(x_i) \qquad 1 \le i \le n-1 \qquad (n-1 \text{ conditions})$$

It turns out (Exercises 18 and 19) that the preceding $4n - 2$ conditions are sufficient to determine all but two of the $4n$ coefficients in the cubic expressions of (20) which are to define the spline S. Two further conditions must then be prescribed before the spline is completely determined. Usually these are one or the other of the following pairs of conditions.

1. $S'(a) = P_1'(a) = f'(a)$; $S'(b) = P_n'(b) = f'(b)$
2. $S''(a) = P_1''(a) = 0$; $S''(b) = P_n''(b) = 0$

A cubic spline which satisfies Condition **2** is called a **natural spline** and has the interesting minimum property that if $y(x)$ is any curve which passes through the points $[x_i, f(x_i)]$ and has continuous first and second derivatives on $[a, b]$, then (see Exercise 20)

$$\int_a^b [S''(x)]^2\,dx \le \int_a^b [y''(x)]^2\,dx$$

the equality sign holding if and only if $y(x) \equiv S(x)$.

EXERCISES

1. Establish Eq. (2) by finding the equation of the approximating parabola and evaluating it at $x = x_0 + rh$. *Hint:* Take x_0, x_1, x_2 to be 0, h, $2h$, respectively.

2. Compute **(a)** $f(1.3)$ and **(b)** $f(1.95)$ from the following data:

x	1.1	1.2	1.5	1.7	1.8	2.0
$f(x)$	1.112	1.219	1.636	2.054	2.323	3.011

3. Compute **(a)** $\sqrt{50.2}$ and **(b)** $\sqrt{55.9}$ from the following data.

x	$\sqrt{x}$
50	7.07107
51	7.14143
52	7.21110
53	7.28011
54	7.34847
55	7.41620
56	7.48331

4. Fit a polynomial of minimum degree to the data of Example 1.

5. Fit a polynomial of minimum degree to the following data:

x	−1	1	2	4	5
$f(x)$	13	15	13	33	67

6. (a) Supply the details required to complete the derivation of Eq. (16).
(b) Supply the details required to complete the derivation of Eq. (17).

7. Give an operational derivation of the forward Gregory-Newton interpolation formula by observing that $f(x) \equiv f(x_0 + rh) = E^r f_0$ and then replacing E by its operational equivalent $1 + \Delta$.

8. Give an operational derivation of the backward Gregory-Newton interpolation formula by observing that $f(x) \equiv f(x_0 + rh) = E^r f_0$ and then replacing E by its operational equivalent

$$\frac{E}{E - \Delta} = \frac{1}{1 - \Delta E^{-1}} = (1 - \Delta E^{-1})^{-1}$$

9. If y_0, y_1, y_2, y_3 are the values of a function at the equally spaced values x_0, x_1, x_2, x_3, show that the best estimate of the value of y corresponding to the value of x midway between x_1 and x_2 is

$$\frac{y_1 + y_2}{2} + \frac{(y_1 + y_2) - (y_0 + y_3)}{16}$$

10. Three readings are taken at equally spaced points $x = 0$, h, $2h$ near the maximum (or minimum) of a function $y = f(x)$. Show that the abscissa of the maximum (or minimum) is approximately

$$\left(\frac{1}{2} - \frac{\Delta y_0}{\Delta^2 y_0}\right)h$$

and that the maximum (or minimum) ordinate is approximately

$$y_1 - \frac{(\Delta y_1 + \Delta y_0)^2}{8\Delta^2 y_0}$$

11. Work Exercise 10, given that the three points where the readings are taken are not equally spaced.

12. Derive Lagrange's interpolation formula from Eq. (3) for the case $n = 2$. (*Hint:* Use the results of Exercise 11, Sec. 5.2.)

13. Derive Lagrange's interpolation formula by using partial fractions to expand

$$\frac{f(x)}{(x - x_0)(x - x_1)\cdots(x - x_n)}$$

14. Using the forward Gregory-Newton interpolation formula, show that for all nonnegative values of x,

$$2^x = 1 + \frac{x}{1!} + \frac{x(x-1)}{2!} + \cdots + \frac{x(x-1)\cdots(x-n+1)}{n!} + \cdots$$

15. Using the backward Gregory-Newton interpolation formula, obtain a series for e^x which is valid for all nonpositive values of x.

16. Show that for nonnegative values of x,

$$\frac{1}{1+x} = 1 - \frac{x}{2!} + \frac{x(x-1)}{3!} - \cdots + (-1)^n \frac{x(x-1)\cdots(x-n+1)}{(n+1)!} + \cdots$$

17. Let $f(x)$ be a function such that $f(0) = 1$ and $f(n) = 0$, $n = \pm 1, \pm 2, \ldots$. Using the forward Gregory-Newton interpolation formula, show that $f(x)$ has the formal expansion

$$f(x) \sim 1 - \frac{x}{1} + \frac{x(x-1)}{2!} - \cdots + (-1)^n \frac{x(x-1)\cdots(x-\overline{n-1})}{n!} + \cdots$$

18. Show that besides the conditions which define a general cubic spline, two other conditions must be prescribed before the spline is completely determined. *Hint:* Show that the definition of such a spline imposes four conditions on each polynomial segment except the first and the last, while three conditions are imposed on each of these segments.

19. If the interval $[a, b]$ is divided into n subintervals of respective lengths $x_i - x_{i-1} = h_i$, show that

$$P_i(x) = m_{i-1}\frac{(x_i - x)^2(x - x_{i-1})}{h_i^2} - m_i\frac{(x - x_{i-1})^2(x_i - x)}{h_i^2} + f(x_{i-1})\frac{(x_i - x)^2[2(x - x_{i-1}) + h_k]}{h_i^3} + f(x_i)\frac{(x - x_{i-1})^2[2(x_i - x) + h_i]}{h_i^3}$$

is the equation of a cubic curve such that

$$P_i(x_{i-1}) = f(x_{i-1}) \qquad P_i(x_i) = f(x_i)$$

and

$$P_i'(x_{i-1}) = m_{i-1} \qquad P_i'(x_i) = m_i$$

20. Let y be any function which has continuous first and second derivatives on $[a, b]$ and which has the value $f(x_i)$ at each point x_i in $[a, b]$, and let S be an arbitrary spline on $[a, b]$. Beginning with the identity

$$\int_a^b (y'')^2\,dx - \int_a^b (S'')^2\,dx = \int_a^b (y'' - S'')^2\,dx + 2\int_a^b S''(y'' - S'')\,dx$$

apply integration by parts to the last integral over successive subintervals (x_{i-1}, x_i) and show that

$$2\int_a^b S''(y'' - S'')\,dx = 2\sum_{i=0}^{n-1}\left\{[S''(y' - S')]_{x_i}^{x_{i+1}} - \int_{x_i}^{x_{i-1}} S'''(y' - S')\,dx\right\} = 2[S''(y' - S')]_a^b$$

Under what conditions can we now infer the following minimum property?

$$\int_a^b (S'')^2\,dx \le \int_a^b (y'')^2\,dx$$

21. **(a)** Fit the data of the table

x	3.0	4.5	7.0	9.0
$f(x)$	2.5	1.0	2.5	0.5

with a linear spline.

(b) What is $f(5)$?

22. Can all line segments defined by the polynomials of a linear spline be horizontal? Explain.

23. **(a)** Fit a cubic spline S for which $S'(0) = 0$, $S'(6) = -12$, to the data of the table

x	0	2	4	6
$f(x)$	1	9	41	41

(b) Are all three polynomials involved in the definition of S of degree 3?

24. Fit the data of Exercise 21 with a quadratic spline.

25. **(a)** Fit the data of the table

x	-1	0	1
$f(x)$	-2	0	2

with a cubic spline S such that $S''(-1) = S''(1) = 4k$, $k \neq 0$.

(b) Does the spline of Part **(a)** become a natural spline if $k = 0$? Explain.

5.4. NUMERICAL DIFFERENTIATION AND INTEGRATION

Any of the interpolation formulas we obtained in the last section can be used to find the derivative of a tabular function. For instance, if we consider the forward Gregory-Newton formula

$$f(x_0 + rh) = f_0 + r\,\Delta f_0 + \frac{r(r-1)}{2!}\,\Delta^2 f_0 + \frac{r(r-1)(r-2)}{3!}\,\Delta^3 f_0 + \frac{r(r-1)(r-2)(r-3)}{4!}\,\Delta^4 f_0 + \cdots$$

and differentiate with respect to r, we find

$$hf'(x_0 + rh) = \Delta f_0 + \frac{2r-1}{2}\,\Delta^2 f_0 + \frac{3r^2 - 6r + 2}{6}\,\Delta^3 f_0 + \frac{2r^3 - 9r^2 + 11r - 3}{12}\,\Delta^4 f_0 + \cdots \tag{1}$$

$$h^2 f''(x_0 + rh) = \Delta^2 f_0 + (r-1)\,\Delta^3 f_0 + \frac{6r^2 - 18r + 11}{12}\,\Delta^4 f_0 + \cdots \tag{2}$$

$$h^3 f'''(x_0 + rh) = \Delta^3 f_0 + \frac{2r-3}{2}\,\Delta^4 f_0 + \cdots \tag{3}$$

$$h^4 f^{\text{iv}}(x_0 + rh) = \Delta^4 f_0 + \cdots \tag{4}$$

Specifically, if we put $r = 0$, we find for the successive derivatives at the tabular point x_0,

$$f'(x_0) = \frac{1}{h}\left(\Delta f_0 - \frac{1}{2}\,\Delta^2 f_0 + \frac{1}{3}\,\Delta^3 f_0 - \frac{1}{4}\,\Delta^4 f_0 + \cdots\right) \tag{5}$$

$$f''(x_0) = \frac{1}{h^2}\left(\Delta^2 f_0 - \Delta^3 f_0 + \frac{11}{12}\,\Delta^4 f_0 - \cdots\right) \tag{6}$$

$$f'''(x_0) = \frac{1}{h^3}\left(\Delta^3 f_0 - \frac{3}{2}\,\Delta^4 f_0 + \cdots\right) \tag{7}$$

$$f^{\text{iv}}(x_0) = \frac{1}{h^4}(\Delta^4 f_0 - \cdots) \tag{8}$$

Similarly, from the backward Gregory-Newton formula we obtain

$$(9)\quad hf'(x_0 + rh) = \Delta f_{-1} + \frac{2r+1}{2}\Delta^2 f_{-2} + \frac{3r^2+6r+2}{6}\Delta^3 f_{-3} + \frac{2r^3+9r^2+11r+3}{12}\Delta^4 f_{-4} + \cdots$$

$$(10)\quad h^2 f''(x_0 + rh) = \Delta^2 f_{-2} + (r+1)\,\Delta^3 f_{-3} + \frac{6r^2+18r+11}{12}\Delta^4 f_{-4} + \cdots$$

$$(11)\quad h^3 f'''(x_0 + rh) = \Delta^3 f_{-3} + \frac{2r+3}{2}\Delta^4 f_{-4} + \cdots$$

$$(12)\quad h^4 f^{\text{iv}}(x_0 + rh) = \Delta^4 f_{-4} + \cdots$$

and, at the point x_0,

$$(13)\quad f'(x_0) = \frac{1}{h}\left(\Delta f_{-1} + \frac{1}{2}\Delta^2 f_{-2} + \frac{1}{3}\Delta^3 f_{-3} + \frac{1}{4}\Delta^4 f_{-4} + \cdots\right)$$

$$(14)\quad f''(x_0) = \frac{1}{h^2}\left(\Delta^2 f_{-2} + \Delta^3 f_{-3} + \frac{11}{12}\Delta^4 f_{-4} + \cdots\right)$$

$$(15)\quad f'''(x_0) = \frac{1}{h^3}\left(\Delta^3 f_{-3} + \frac{3}{2}\Delta^4 f_{-4} + \cdots\right)$$

$$(16)\quad f^{\text{iv}}(x_0) = \frac{1}{h^4}(\Delta^4 f_{-4} + \cdots)$$

For a complete development, an error term analogous to Eq. (8), Sec. 5.3, should be found for any formula of numerical differentiation. This can be done, but the results are of little use in routine calculations, and we shall not take them into account. However, it should be borne in mind that unless we are dealing with a polynomial, and have extended to the last nonzero difference the interpolation formula we are differentiating, numerical differentiation may involve errors of considerable magnitude, the errors increasing significantly as derivatives of higher order are computed.

EXAMPLE 1

Find the first and second derivatives of $\sqrt{x}$ at $x = 2.5$ from the table

x	$\sqrt{x}$	Δ	Δ^2
2.50	1.58114		
		0.01573	
2.55	1.59687		−0.00015
		0.01558	
2.60	1.61245		−0.00015
		0.01543	
2.65	1.62788		−0.00014
		0.01529	
2.70	1.64317		−0.00015
		0.01514	
2.75	1.65831		

Using Eqs. (5) and (6) with $x_0 = 2.50$ and $h = 0.05$, we find at once

$$f'(2.5) \doteq \frac{1}{0.05}\left[0.01573 - \frac{1}{2}(-0.00015)\right] = 0.3160$$

$$f''(2.5) \doteq \frac{1}{(0.05)^2}(-0.00015) = -0.0600$$

The correct values to four decimal places are of course

$$f'(2.5) = \left.\frac{-1}{2\sqrt{x}}\right|_{x=2.5} \doteq 0.3162$$

$$f''(2.5) = \left.\frac{-1}{4x\sqrt{x}}\right|_{x=2.5} \doteq -0.0632$$

There are many formulas for numerical integration, but only the *trapezoidal rule* and *Simpson's rule*† are of much importance in elementary applications. To develop the trapezoidal rule for the evaluation of a definite integral

(17) $$\int_a^b f(x)\,dx$$

we let $x_0 = a$ and set $x_n = x_0 + nh = b$, so that $h = (b - a)/n$. Then we approximate (17) over one tabular interval h by integrating the linear approximation to $f(x)$ provided by the first two terms in the forward Gregory-Newton interpolation formula. The result is

$$\begin{aligned}\int_{x_0}^{x_1} f(x)\,dx &= h\int_0^1 f(x_0 + rh)\,dr \doteq h\int_0^1 (f_0 + r\,\Delta f_0)\,dr = h\left(f_0 + \frac{1}{2}\,\Delta f_0\right)\\ &= h\left(\frac{1}{2}f_0 + \frac{1}{2}f_1\right)\end{aligned}$$

Then we apply this formula over the successive intervals (x_0, x_1), (x_1, x_2), (x_2, x_3), . . . , (x_{n-1}, x_n), getting

$$\begin{aligned}\int_{x_0}^{x_n} f(x)\,dx &\doteq h\left(\frac{1}{2}f_0 + \frac{1}{2}f_1\right) + h\left(\frac{1}{2}f_1 + \frac{1}{2}f_2\right) + \cdots + h\left(\frac{1}{2}f_{n-1} + \frac{1}{2}f_n\right)\\ &= h\left(\frac{1}{2}f_0 + f_1 + f_2 + \cdots + f_{n-1} + \frac{1}{2}f_n\right)\end{aligned}$$

which is the **trapezoidal rule.**

Denoting this *approximation* to (17) by a_t and setting

(18) $$\int_a^b f(x)\,dx = a_t + e_t$$

---

†Named for the English mathematician Thomas Simpson (1710–1761).

it is natural to call e_t the **error** committed in evaluating (17) by the trapezoidal rule. If f'' exists everywhere on $a \le x \le b$, it can be shown† that there is a number c such that

$$e_t = -\frac{h^2}{12}(b-a)f''(c) \qquad \text{where } a \le c \le b$$

This formula can be used to determine lower and upper bounds for $|e_t|$ even though an exact value of c is unknown. In particular, if a smallest value m, and a largest value M, of $|f''|$ over the interval of integration can be found, then $km \le |e_t| \le kM$ where $k = (b-a)^3/12n^2$. In our work we shall make no use of this fact.

To obtain Simpson's rule, we take $x_0 = a$, $x_n = x_0 + nh = b$, and $h = (b-a)/n$, as before, except that n must now be *even*. Then, by integrating over *two* successive tabular intervals the parabolic approximation to $f(x)$ provided by the first three terms in the Gregory-Newton formula, we get

$$\begin{aligned}\int_{x_0}^{x_2} f(x)\,dx &= h\int_0^2 f(x_0 + rh)\,dr \doteq h\int_0^2 \left(f_0 + r\,\Delta f_0 + \frac{r^2 - r}{2}\,\Delta^2 f_0\right) dr \\ &= h\left(2f_0 + 2\Delta f_0 + \frac{1}{3}\,\Delta^2 f_0\right) = h\left[2f_0 + 2(f_1 - f_0) + \frac{1}{3}(f_2 - 2f_1 + f_0)\right] \\ &= h\left(\frac{1}{3}f_0 + \frac{4}{3}f_1 + \frac{1}{3}f_2\right)\end{aligned}$$

Application of this formula to the successive pairs of intervals (x_0, x_2), (x_2, x_4), . . . , (x_{n-2}, x_n), gives

$$\begin{aligned}\int_{x_0}^{x_n} f(x)\,dx &\doteq h\left(\frac{1}{3}f_0 + \frac{4}{3}f_1 + \frac{1}{3}f_2\right) + h\left(\frac{1}{3}f_2 + \frac{4}{3}f_3 + \frac{1}{3}f_4\right) + \cdots + h\left(\frac{1}{3}f_{n-2} + \frac{4}{3}f_{n-1} + \frac{1}{3}f_n\right) \\ &= \frac{h}{3}(f_0 + 4f_1 + 2f_2 + 4f_3 + 2f_4 + \cdots + 2f_{n-2} + 4f_{n-1} + f_n) \qquad n \text{ even}\end{aligned}$$

which is **Simpson's rule.**

Denoting this *approximation* to (17) by a_s, we have

(19) $$\int_a^b f(x)\,dx = a_s + e_s$$

where e_s is the **error** committed in evaluating (17) by means of Simpson's rule. If $f^{(iv)}$ is defined at each point of $a \le x \le b$, there is a number c such that

(20) $$e_s = -\frac{h^4}{180}(b-a)f^{(iv)}(c) \qquad \text{where } a \le c \le b$$

Although we shall make no use of the fact, this formula can be used to underestimate, and overestimate, $|e_s|$.

---

†See, for instance, Tom M. Apostol, *Calculus,* vol. 2, pp. 435–436, Blaisdell, New York, 1962.

EXAMPLE 2 VELOCITIES AS VALUES OF A RUNNING INTEGRAL

At $t = 0$, a body of unit mass, initially at rest at the origin, begins to move along the x axis under the influence of a force $F = 1/\sqrt{1+t^3}$. Find the velocity of the body at intervals of 0.2 from $t = 0$ to $t = 2$.

From Newton's law we see at once that

$$v = \int_0^t \frac{du}{\sqrt{1+u^3}}$$

However, this integral cannot be evaluated in terms of elementary functions,† and so the required values of v must be obtained by numerical integration. To do this we shall use the trapezoidal rule because it is especially well suited to the computation of the values of a **running integral,** that is, an integral from a fixed lower limit to a series of equally spaced upper limits. Our first step of course is to evaluate the integrand at the points $t = 0.0, 0.2, \ldots, 2.0$. These values are listed in the second column of the accompanying table. Next, we adjoin to the given data a column containing the averages of successive pairs of values of the integrand. Then we add the entries in this column, beginning at the top, and record the partial total after each addition. From the derivation of the trapezoidal rule, it is clear that these entires, each multiplied by h, are the required values of the integral. The details of this process should be clear from an inspection of the last three columns in the table.

t	$f = \dfrac{1}{\sqrt{1+t^3}}$	$\dfrac{1}{2}(f_{i-1}+f_i)$	$\sum \dfrac{1}{2}(f_{i-1}+f_i)$	$\int_0^t \dfrac{du}{\sqrt{1+u^3}} = v$
0.0	1.0000	. . .	0.0000	0.0000
0.2	0.9960	0.9980	0.9980	0.1996
0.4	0.9695	0.9828	1.9808	0.3962
0.6	0.9068	0.9382	2.9190	0.5838
0.8	0.8132	0.8600	3.7790	0.7558
1.0	0.7071	0.7602	4.5392	0.9078
1.2	0.6054	0.6562	5.1954	1.0391
1.4	0.5168	0.5611	5.7565	1.1513
1.6	0.4430	0.4799	6.2364	1.2473
1.8	0.3826	0.4128	6.6492	1.3298
2.0	0.3333	0.3580	7.0072	1.4014

EXERCISES

1. From the data in the following table compute the first three derivatives of $\ln x$ at $x = 200$ and at $x = 205$ and check them against the exact values:

x	$\ln x$
200	5.29831737
201	5.30330491
202	5.30826770
203	5.31320598
204	5.31811999
205	5.32300998

2. Using the data of Exercise 1, compute the first three derivatives of $\ln x$ at $x = 200.5$ and at $x = 204.5$ and check them against the exact values.

3. From the data in the following table compute $f'(x)$ at $x = 0$ and at $x = 10$:

x	$f(x)$
0	1.000
1	1.221
3	1.822
4	2.226
7	4.055
10	7.389

†Actually, the integral for v is an example of what is known as an *elliptic integral,* one standard form of which we encountered in Exercises 47–49, Sec. 2.11. Tables of such integrals have been constructed, but since we have not learned how to reduce an integral like the present one to standard form, the tables are of no use to us here.

4. Explain how the integration process suggested in Example 2 can be adapted to the evaluation of an integral of the form $\int_x^{x_0} f(t)\,dt$ for a series of equally spaced values of x.
5. Using the trapezoidal rule, compute $\int_0^x e^{-t^2}\,dt$ for $x = n/10$, $1 \le n \le 10$.
6. Using the trapezoidal rule, compute $\int_x^1 (\sin t)/t\,dt$ for values of x at intervals of $h = 0.1$ from 0 to 1.
7. By making two applications of Simpson's rule, establish the following formula for the numerical evaluation of double integrals:

$$\int_{y_0}^{y_2}\int_{x_0}^{x_2} f(x,y)\,dx\,dy = \frac{hk}{9}[(f_{00}+f_{02}+f_{20}+f_{22}) + 4(f_{01}+f_{10}+f_{12}+f_{21}) + 16f_{11}]$$

where h and k are, respectively, the intervals at which x and y are tabulated, and for $0 \le i, j \le 2$, $f_{ij} = f(x_0 + ih, y_0 + jk)$. Give the generalization of this result to integrals of the form

$$\int_{y_0}^{y_{2n}}\int_{x_0}^{x_{2m}} f(x,y)\,dx\,dy$$

8. By restating Maclaurin's expansion,

$$f(x+h) = f(x) + hf'(x) + \frac{h^2}{2!}f''(x) + \frac{h^3}{3!}f'''(x) + \cdots$$

in operational form, establish the operational equivalence $E = e^{hD}$.
9. By differentiating the Lagrange interpolation formula Eq. (9), Sec. 5.3, based on the three points $x_0 = x_0$, $x_1 = x_0 + h$, $x_2 = x_0 + 2h$ and then evaluating the result at x_0, x_1, and x_2, derive the formulas

$$f_0' = \frac{1}{2h}(-3f_0 + 4f_1 - f_2)$$

$$f_1' = \frac{1}{2h}(-f_0 + f_2)$$

$$f_2' = \frac{1}{2h}(f_0 - 4f_1 + 3f_2)$$

Are these equivalent to the results obtained from Eq. (1)?
10. By differentiating the Lagrange interpolation formula based on the five points $x_0 = x_0$, $x_1 = x_0 + h$, $x_2 = x_0 + 2h$, $x_3 = x_0 + 3h$, $x_4 = x_0 + 4h$, obtain formulas similar to those in Exercise **9**.

5.5 THE METHOD OF LEAST SQUARES

The problem of fitting a curve to a set of points admits of two somewhat different interpretations. In the first place, we may ask for the equation of a curve of prescribed type which passes exactly through each point of a given set. If the prescribed curve is to be a polynomial, this is most easily accomplished through the use of interpolation formulas such as those we developed in Sec. 5.3. On the other hand, we may weaken these requirements and ask for some simpler curve whose equation contains too few parameters to permit it to pass exactly through each, or any, of the given points but which is to come "as close as possible" to each point. For instance, given a set of points as in Fig. 5.2*a*, a straight line passing as close as possible to each point may well be more useful than some

FIGURE 5.2
The approximate fitting of a straight line to a set of points.

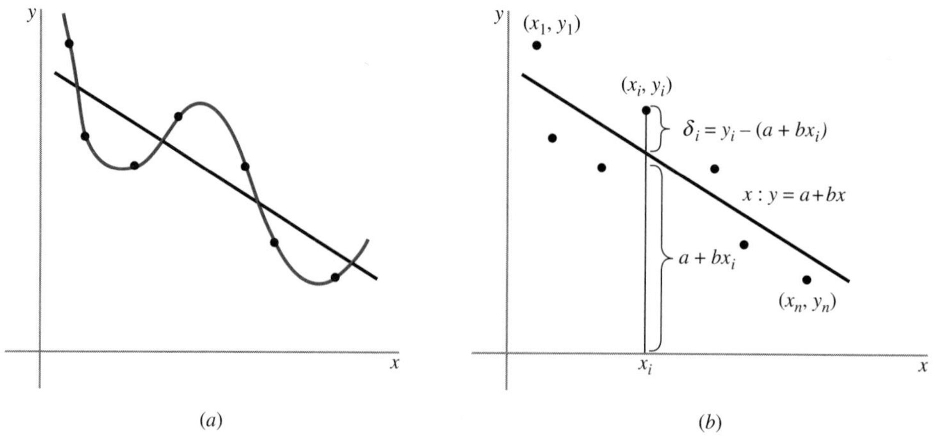

complicated curve passing exactly through each point. This will certainly be the case with experimental data which theoretically should fall along a straight line but which fail to do so because of errors of observation. For most purposes, the necessary measure of "as close as possible" is taken to be the *least square criterion,*† and the process of applying this criterion is known as the **method of least squares,** which we shall now develop.

Let us begin by supposing that we wish to fit a straight line l whose equation is

$$y = a + bx \tag{1}$$

to the n points $(x_1, y_1), (x_2, y_2), \ldots, (x_n, y_n)$. Since two points completely determine a straight line, it will in general be impossible for the required line to pass through more than two of the given points, and it may not pass through any. Hence the coordinates of the general point (x_i, y_i) will not satisfy Eq. (1). That is, when we substitute x_i into Eq. (1), we get not y_i but the ordinate of l which, as we see from Fig. 5.2b, differs from y_i by δ_i. In other words,

$$y_i - (a + bx_i) = \delta_i \neq 0 \tag{2}$$

If we compute the discrepancy δ_i for each point of the set and form the sum of the squares of these quantities (in order to prevent large positive and large negative δ's from canceling each other and thereby giving an unwarranted impression of accuracy), we obtain

$$E = \sum_{i=1}^{n} \delta_i^2 = \sum_{i=1}^{n} (y_i - a - bx_i)^2 \tag{3}$$

The quantity E is obviously a measure of how well the line fits the set of points as a whole. For E will be zero if and only if each of the points lies on l, and the larger E is, the farther the points are, on the average, from l. The **least squares criterion** is now simply this: *The parameters a and b should be chosen so as to make the sum of the squares of the deviations, E, as small as possible.*

To do this, we apply the usual conditions for minimizing a function of several variables and equate to zero the first derivatives of E with respect to a and b. This gives us the two equations

$$\frac{\partial E}{\partial a} = \sum_{i=1}^{n} 2(y_i - a - bx_i)(-1) = 0$$

$$\frac{\partial E}{\partial b} = \sum_{i=1}^{n} 2(y_i - a - bx_i)(-x_i) = 0$$

or, dividing by 2 and collecting terms on the unknown coefficients a and b,

$$na + b\sum_{i=1}^{n} x_i = \sum_{i=1}^{n} y_i \tag{4}$$

$$a\sum_{i=1}^{n} x_i + b\sum_{i=1}^{n} x_i^2 = \sum_{i=1}^{n} x_i y_i \tag{5}$$

Equations (4) and (5) are two simultaneous linear equations whose solution for a and b presents no difficulty.

---

†A brief discussion of the reason for this will be found in A. M. Mood, *Introduction to the Theory of Statistics,* p. 311, McGraw-Hill, New York, 1950.

If we reconsider Eq. (2) from a purely algebraic point of view, it appears that as i varies from 1 to n, it defines a system of n equations in two unknowns, a and b, which should ideally be satisfied but which actually are not. Moreover, minimizing E, as given by (3), is nothing more than minimizing the sum of the squares of the amounts by which these n equations fail to be satisfied. These observations suggest a somewhat more general interpretation of the method of least squares, namely, that it is simply a process for finding the best possible values for a set of m *unknowns, say* $z_1, z_2, \ldots, z_m$, connected by n linear equations

$$\begin{aligned} c_{11}z_1 + c_{12}z_2 + \cdots + c_{1m}z_m &= d_1 \\ c_{21}z_1 + c_{22}z_2 + \cdots + c_{2m}z_m &= d_2 \\ &\cdots\cdots \\ c_{n1}z_1 + c_{n2}z_2 + \cdots + c_{nm}z_m &= d_n \end{aligned} \tag{6}$$

where the c's are known coefficients and $n > m$.

Since the number of equations in the set (6) exceeds the number of unknowns, the system presumably does not admit of an exact solution; i.e., there is no set of values for $z_1, z_2, \ldots, z_m$ for which each equation is exactly satisfied. Hence, letting δ_i be the amount by which the ith equation fails to be satisfied, we consider the discrepancies

$$\delta_i = c_{i1}z_1 + c_{i2}z_2 + \cdots + c_{im}z_m - d_i \qquad i = 1, 2, \ldots, n$$

and attempt to find values of $z_1, z_2, \ldots, z_m$ which will make

$$E = \sum_{i=1}^{n} \delta_i^2 = \sum_{i=1}^{n} (c_{i1}z_1 + c_{i2}z_2 + \cdots + c_{im}z_m - d_i)^2$$

as small as possible.

To minimize E we must equate to zero each of the first partial derivatives

$$\frac{\partial E}{\partial z_1}, \quad \frac{\partial E}{\partial z_2}, \quad \ldots, \quad \frac{\partial E}{\partial z_m}$$

For $\partial E/\partial z_1$ this gives the equation

$$\frac{\partial E}{\partial z_1} = \sum_{i=1}^{n} 2(c_{i1}z_1 + c_{i2}z_2 + \cdots + c_{im}z_m - d_i)c_{i1} = 0$$

or, dividing by 2 and rearranging,

$$z_1 \sum_{i=1}^{n} c_{i1}c_{i1} + z_2 \sum_{i=1}^{n} c_{i2}c_{i1} + \cdots + z_m \sum_{i=1}^{n} c_{im}c_{i1} = \sum_{i=1}^{n} c_{i1}d_i$$

and similarly, for the other derivatives,

$$\begin{aligned} z_1 \sum_{i=1}^{n} c_{i1}c_{i2} + z_2 \sum_{i=1}^{n} c_{i2}c_{i2} + \cdots + z_m \sum_{i=1}^{n} c_{im}c_{i2} &= \sum_{i=1}^{n} c_{i2}d_i \\ &\cdots\cdots \\ z_1 \sum_{i=1}^{n} c_{i1}c_{im} + z_2 \sum_{i=1}^{n} c_{i2}c_{im} + \cdots + z_m \sum_{i=1}^{n} c_{im}c_{im} &= \sum_{i=1}^{n} c_{im}d_i \end{aligned}$$

We have thus obtained a system of m linear equations in m unknowns $z_1, z_2, \ldots, z_m$ whose solution is now a routine matter. As a practical detail, it is worthy of note that these minimizing conditions, or **normal equations** as they are usually called, can be written down at once according to the following rule.

RULE 1 If each of the n linear equations in the system (6) is multiplied by the coefficient of z_i in that equation, the sum of the resulting equations is the ith normal equation in the least-squares solution of the system.

EXAMPLE 1

By the method of least squares, fit a parabolic equation $y = a + bx + cx^2$ to the data

x	-3	-2	0	3	4
y	18	10	2	2	5

Substituting these pairs of values into the equation $y = a + bx + cx^2$, we find that a, b, and c (which play the roles of z_1, z_2, and z_3 in the preceding general discussion) should (but do not) satisfy the five conditions

$$\begin{aligned} a - 3b + 9c &= 18 \\ a - 2b + 4c &= 10 \\ a &= 2 \\ a + 3b + 9c &= 2 \\ a + 4b + 16c &= 5 \end{aligned}$$

In general, three unknowns cannot be made to satisfy more than three conditions; hence the most we can do is to determine values of a, b, and c which will satisfy these five conditions as nearly as possible.

To set up the first of the three normal equations required by the method of least squares, we must multiply each of the equations by the coefficient of a in that equation and then add, getting in this case simply the sum of the five equations

$$5a + 2b + 38c = 37$$

To set up the second normal equation, we multiply each equation by the coefficient of b in that equation and add, getting

$$\begin{aligned} -3a + 9b - 27c &= -54 \\ -2a + 4b - 8c &= -20 \\ 0 + 0 + 0 &= 0 \\ 3a + 9b + 27c &= 6 \\ 4a + 16b + 64c &= 20 \\ \hline 2a + 38b + 56c &= -48 \end{aligned}$$

In the same way, multiplying each equation by the coefficient of c in that equation and adding, we get the third normal equation

$$\begin{aligned} 9a - 27b + 81c &= 162 \\ 4a - 8b + 16c &= 40 \\ 0 + 0 + 0 &= 0 \\ 9a + 27b + 81c &= 18 \\ 16a + 64b + 256c &= 80 \\ \hline 38a + 56b + 434c &= 300 \end{aligned}$$

The solution of the three normal equations is a simple matter, and we find

$$a \doteq 1.82 \qquad b \doteq -2.65 \qquad c \doteq 0.87$$

The required solution is therefore

$$y = 1.82 - 2.65x + 0.87x^2$$

The method of least squares is not limited in its application to problems in which the equations to be satisfied are linear. Sometimes by a suitable transformation the problem can be converted into one in which the parameters do enter linearly. For instance, to fit an equation of the important type $y = ae^{bx}$, we can take the natural logarithm of each side, getting

$$\ln y = \ln a + bx$$

Then, considering x and $\ln y$ as new variables, say X and Y, and $\ln a$ and b as new parameters, say A and B, we can regard the problem as requiring the determination of A and B such that the *linear* equation

$$Y = A + BX$$

gives the best possible fit to the known pairs of values $X(=x)$ and $Y(=\ln y)$. Once A has been found it is of course a simple matter to find the actual parameter a since $A = \ln a$.

Similarly, fitting a function $y = kx^n$ can be reduced to a linear problem by first taking logarithms (preferably to the base 10), getting

$$\log y = \log k + n \log x$$

This equation is linear in the parameters $K = \log k$ and $N = n$. Hence the determination of the parameters can be carried out as outlined before.

On the other hand, it is not possible to make a rigorous linearization of general systems of nonlinear equations of condition. But if a reasonable approximation to a solution of such a system is available, an approximate linearization of the problem can be achieved in the following way.

Let the equations to be satisfied (as nearly as possible) be

$$(7) \qquad f_1(x, y) = 0 \qquad f_2(x, y) = 0 \qquad \cdots \qquad f_n(x, y) = 0$$

and suppose that (x_0, y_0) is known, by inspection or otherwise, to be an approximate solution of the system. Then each function can be expanded in a generalized Taylor series about the point (x_0, y_0), getting, for $i = 1, 2, \ldots, n$,

$$f_i(x, y) = f_i(x_0, y_0) + \left[\left.\frac{\partial f_i}{\partial x}\right|_{x_0, y_0}(x - x_0) + \left.\frac{\partial f_i}{\partial y}\right|_{x_0, y_0}(y - y_0)\right]$$
$$+ \frac{1}{2}\left[\left.\frac{\partial^2 f_i}{\partial x^2}\right|_{x_0, y_0}(x - x_0)^2 + 2\left.\frac{\partial^2 f_i}{\partial x\, \partial y}\right|_{x_0, y_0}(x - x_0)(y - y_0) + \left.\frac{\partial^2 f_i}{\partial y^2}\right|_{x_0, y_0}(y - y_0)^2\right] + \cdots$$

Now if (x_0, y_0) is a reasonable approximation to the required solution, the quantities $x - x_0$ and $y - y_0$ will be small; hence their squares, products, and higher powers will be negligible in comparison with the quantities themselves. Omitting these quantities thus reduces (7) to the system

$$(8) \quad f_i(x, y) = f_i(x_0, y_0) + \left.\frac{\partial f_i}{\partial x}\right|_{x_0, y_0}(x - x_0) + \left.\frac{\partial f_i}{\partial y}\right|_{x_0, y_0}(y - y_0) = 0 \qquad i = 1, 2, \ldots, n$$

which is linear in the unknown corrections $x - x_0$ and $y - y_0$. The method of least squares can now be applied in a straightforward way to the system (8), following which the initial estimate (x_0, y_0) can be appropriately corrected. Of course, if desired, the functions $f_i(x, y)$ can be expanded about the corrected solution (x_1, y_1) and the process repeated. The extension to systems with more than two unknowns

$$f_1(x, y, z, \ldots) = 0 \qquad f_2(x, y, z, \ldots) = 0 \qquad \cdots \qquad f_n(x, y, z, \ldots) = 0$$

is immediate.

A number of interesting variations on the method of least squares will be found in the exercises.

EXERCISES

1. Show that the least-squares approximation to a set of numbers is the average of the numbers.

2. **(a)** By the method of least squares, fit a line $y = a + bx$ to the points

x	-1	0	2	3	5
y	-1	2	9	10	14

(b) Show that the sum of the vertical distances from the points to the required line is zero.
(c) What is the average squared error for the line of best fit?

3. **(a)** By the method of least squares, fit a curve of the form $y = ax + bx^2$ to the points

x	-1	0	1	2	3
y	-3	1	3	2	-1

(b) Is the sum of the vertical distances from the points to the required curve equal to zero?
(c) What is the average squared error for the curve of best fit?
(d) Four of the five points lie below the curve of best fit. Does this seem reasonable? Is there an explanation for this behavior?

4. **(a)** Fit an equation of the form $y = a + bx + cx^2$ to the data

x	-1	0	2	3	5
y	-4	4	8	9	7

(b) Is the sum of the vertical distances from the points to the required line equal to zero?
(c) What is the average squared error for the line of best fit?

5. If an equation of the form $y = a_1 f_1(x) + a_2 f_2(x) + \cdots + a_n f_n(x)$ is fitted to a set of points by the method of least squares, under what conditions, if any, will the algebraic sum of the vertical distances from the points to the curve be zero?

6. Fit a straight line to the data

x	1	3	6	7	9
y	1	5	6	10	12

(a) By minimizing the sum of the squares of the vertical distances from the points to the line.
(b) By minimizing the sum of the squares of the horizontal distances from the points to the line.
What do you think of the relative merits of the two methods?

7. Find the most plausible values of x and y from the following system of equations

$$x + y = 2$$
$$2x - 3y = 9$$
$$20x + 16y = 4$$

(a) Without dividing out the factor 4 from the last equation.
(b) After dividing the last equation by 4. Explain.

8. Fit equations of each of the forms
(a) $ax + by - 1 = 0$, **(b)** $ax + y - c = 0$,
(c) $x + by - c = 0$, to the data

x	0	1	2	3
y	1.1	1.9	3.0	3.9

by minimizing the sum of the squares of the amounts by which each of the equations in turn fails to be satisfied. Compare the results and explain the differences.

9. Fit an equation of the form $y = kx^n$ to the data

x	1	2	3	4
y	0.10	0.80	4.00	13.00

10. Fit an equation of the form $y = Ae^{ax}$ to the data

x	1	2	3	4
y	1.65	2.70	4.50	7.35

11. Find a linear approximation to $y = e^x$ on $[0, 1]$ by minimizing the integral from 0 to 1 of the square of the difference between $y = e^x$ and $y = a + bx$. *Hint:* Apply the minimizing conditions before integrating.

12. Find the equation of the line which is the least-squares approximation to $y = \sin x$ on $[0, \pi/2]$. *Hint:* Proceed as suggested in Exercise 11.

13. Find the equation of the line which is the least-squares approximation to $y = \cos x$ on $[0, \pi/2]$. *Hint:* Proceed as suggested in Exercise 11. Is this line also the least-squares approximation to $y = \cos x$ on $[0, \pi]$?

14. Approximate the solution of the differential equation $y'' + y = 0$ for which $y(0) = 1$ and $y'(0) = 0$ by assuming $y = 1 - ax^2$ and minimizing the integral from 0 to 1 of the square of the amount by which this function fails to satisfy the differential equation.

15. Using the method of Exercise 14, approximate the solution of the differential equation $y'' + y = 0$ for which $y(0) = 0$ and $y'(0) = 1$ by assuming $y = x - ax^3$.

16. Work Exercise 14 by assuming $y = 1 - bx^2 + cx^4$.

17. An equation of the form $y = Ae^{ax}$ is to be fitted to a set of points $(x_1, y_1), (x_2, y_2), \ldots, (x_n, y_n)$, where the x's are equally spaced with tabular interval h. This means that the following equations should (but presumably do not) hold.

$$y_1' - ay_1 = 0 \qquad y_2' - ay_2 = 0 \qquad \cdots \qquad y_n' - ay_n = 0$$

Explain how a can be found by the method of least squares. After a has been found by this method, how can A be determined? *Hint:* Recall Eq. (5), Sec. 5.4.

18. It is desired to fit a curve of the form $y = Ae^{ax}$ to a set of points $(x_1, y_1), (x_2, y_2), \ldots, (x_n, y_n)$ where the x's are equally spaced with tabular interval h. Show that this means that the following equations should (but presumably do not) hold.

$$y_2 - e^{ah}y_1 = 0 \qquad y_3 - e^{ah}y_2 = 0 \qquad \cdots \qquad y_n - e^{ah}y_{n-1} = 0$$

Explain how the method of least squares can be used to find a. After a has been found, how can A be determined?

19. Discuss the merits of the procedures suggested in Exercises 17 and 18 relative to the method of linearizing by taking logarithms and the use of Taylor's series when a curve of the form $y = Ae^{ax}$ is to be fitted to a set of equally spaced points.

20. A circular arc is to be fitted to a set of points $(x_1, y_1), (x_2, y_2), \ldots, (x_n, y_n)$. Discuss the relative merits of doing this by minimizing the sum of the squares of the vertical distances from the points to the circular arc and by taking the equation of the circle in the form $x^2 + y^2 + ax + by + c = 0$ and minimizing the sum of the squares of the amounts by which the coordinates of the points fail to satisfy this equation.

21. Show that Eqs. (4) and (5) can always be solved for a and b by showing that the determinant of the coefficients is always different from zero. *Hint:* Consider the discriminant of the equation

$$(\lambda x_1 + 1)^2 + (\lambda x_2 + 1)^2 + \cdots + (\lambda x_n + 1)^2 = 0$$

thought of as an equation in λ with no real roots.

5.6 THE NUMERICAL SOLUTION OF DIFFERENTIAL EQUATIONS

One of the most important applications of finite differences is to the numerical solution of differential equations which because of their complexity cannot be solved by exact methods. Many procedures are available for doing this,† some of considerable generality, others especially adapted to equations of a particular form. Of the many methods which have been devised, we shall present only the *Runge-Kutta* method‡ and *Milne's* method.§ These can be applied to simultaneous differential equations as well as to single equations of any order, and are therefore adequate for almost any problem one is likely to encounter.

The fundamental problem in the numerical solution of ordinary differential equations is the solution of the first-order equation

$$\frac{dy}{dx} = f(x, y) \tag{1}$$

subject to the initial condition that $y = y_0$ when $x = x_0$. We do not of course expect to find y as an explicit function of x. Instead, our objective is to obtain satisfactory approximations to the values of the solution $y(x)$ on a specified set of x values, $x_1, x_2, x_3, \ldots$. In our discussion of methods for doing this, we shall denote approximations to the exact values $y(x_1), y(x_2), y(x_3), \ldots$, by $y_1, y_2, y_3, \ldots$, respectively; and we shall designate later refinements in these approximations by $(y_i)_2, (y_i)_3, \ldots$.

The first step in the pointwise solution of the initial-value problem $dy/dx = f(x, y)$, $y(x_0) = y_0$ is to approximate the value of y at $x_1 = x_0 + \Delta x$, that is, $y(x_1) = y_0 + \Delta y$. The simplest way to do this is to approximate Δy by the usual differential estimate for the true increment Δy, namely,

$$\Delta y \doteq dy = y'(x_0)\,\Delta x \tag{2}$$

†See, for instance, J. L. Buchanan and P. R. Turner, "Numerical Methods and Analysis," McGraw-Hill, New York, 1992; H. Levy and E. A. Baggott, *Numerical Studies in Differential Equations,* vol. 1, Watts, London, 1934; W. E. Milne, *Numerical Solutions of Differential Equations,* John Wiley, New York, 1953.

‡Conamed for the German mathematicians Carl David Tolmé Runge (1856–1927) and Wilhelm Kutta (1867–1944).

§Named for the American mathematician William E. Milne (1890–1971).

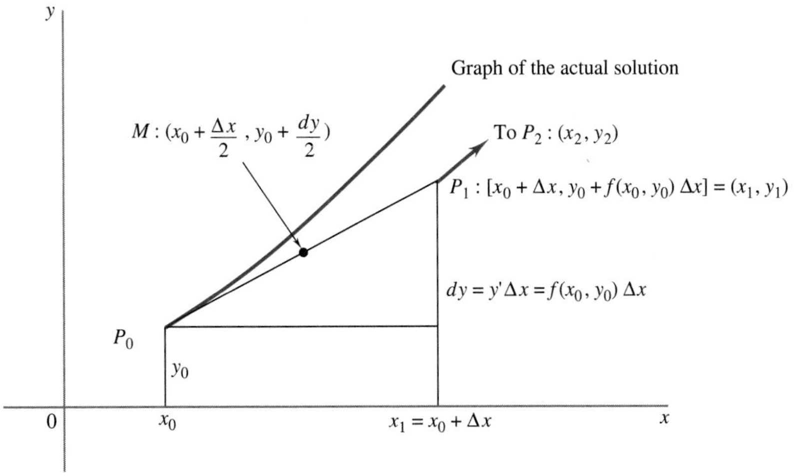

FIGURE 5.3
The Euler method of solving $dy/dx = f(x, y)$.

The differential equation itself gives us the value of the derivative at the point (x_0, y_0) on the required solution curve, for we have from (1)

$$y'(x_0) = f(x_0, y_0)$$

Hence, from (2),

$$\Delta y \doteq f(x_0, y_0)\,\Delta x$$

and therefore

$$y(x_1) = y_0 + \Delta y \doteq y_0 + f(x_0, y_0)\,\Delta x = y_1 \tag{3}$$

Once y_1 has been obtained as an approximation to $y(x_1)$, the same procedure can be repeated at (x_1, y_1) to give

$$y(x_2) = y(x_1) + \Delta y \doteq y_1 + f(x_1, y_1)\,\Delta x = y_2$$

and so on, as far as required. The geometric interpretation of this process, which is known as **Euler's method,** is shown in Fig. 5.3.

On the other hand, having obtained y_1 as a first approximation to $y(x_1)$ by Euler's method, one can use the differential equation (1) to compute y' at the new point P_1: (x_1, y_1) and then use the average of the derivatives at P_0: (x_0, y_0) and P_1: (x_1, y_1) to obtain a (presumably) more accurate estimate of Δy and hence of $y(x_1)$ before attempting to approximate $y(x_2)$. This method yields the value

$$\Delta y \doteq \tfrac{1}{2}[y'(x_0) + y'(x_1)]\,\Delta x$$

and from this, using (1) to obtain the necessary values of the derivatives, we find

$$y_0 + \Delta y \doteq y_0 + \tfrac{1}{2}[f(x_0, y_0) + f(x_1, y_1)]\,\Delta x = (y_1)_2 \tag{4}$$

as a second approximation to $y(x_1)$. This process is known as the **modified Euler method.**

Still another possibility, after determining y_1 as a first estimate of $y(x_1)$ by Euler's method, is to reapproximate Δy and $y(x_1)$ using not the average of the derivatives at P_0: (x_0, y_0) and P_1: (x_1, y_1), as we do in the modified Euler method, but the derivative at the midpoint of the segment P_0P_1 (Fig. 5.3), namely,

$$M\colon \left(\frac{x_0 + x_1}{2}, \frac{y_0 + y_1}{2}\right)$$

This gives the (presumably) improved estimate

$$\Delta y \doteq f\left(\frac{x_0 + x_1}{2}, \frac{y_0 + y_1}{2}\right) \Delta x = f\left[x_0 + \frac{\Delta x}{2}, y_0 + \frac{1}{2} f(x_0, y_0)\,\Delta x\right] \Delta x$$

and, as a third approximation to $y(x_1)$,

$$(5) \qquad (y_1)_3 = y_0 + f\left[x_0 + \frac{\Delta x}{2}, y_0 + \frac{1}{2} f(x_0, y_0)\,\Delta x\right] \Delta x$$

The solution process based on Formula (5) is known as **Runge's method.**

Neither the Euler method, based on Eq. (3), nor the modified Euler method, based on Eq. (4), nor Runge's method, based on Eq. (5), is an effective way of solving the differential equation (1) because unless Δx is very small, the errors implicit in the various approximations for Δy build up significantly as successive values of y are calculated. The *Runge-Kutta* method is essentially a generalization or combination of these three simple procedures in which at each stage three or more preliminary estimates of Δy are calculated. The value then used for Δy to compute the next value of y is a linear combination of these estimates in which the constants of combination are chosen to make the error as small as possible.

Specifically, in **Kutta's third-order method,** the three estimates of Δy are

$$(6) \qquad (\Delta y)_1 = f(x_0, y_0)\,\Delta x$$

which is just the estimate used in Euler's method,

$$(7) \qquad (\Delta y)_2 = f[x_0 + p\,\Delta x, y_0 + p(\Delta y)_1]\,\Delta x \qquad 0 < p < 1$$

which is just like the estimate used in Runge's method except that instead of being evaluated at the midpoint of the segment P_0P_1 (i.e., with $p = \frac{1}{2}$), the derivative is evaluated at a point $[x_0 + p\,\Delta x, y_0 + p(\Delta y)_1]$ yet to be determined; and

$$(8) \qquad (\Delta y)_3 = f[x_0 + q\,\Delta x, y_0 + r(\Delta y)_2 + (q - r)(\Delta y)_1]\,\Delta x \qquad 0 < q, r < 1$$

where q and r are yet to be determined. Finally the value actually used for Δy in the calculation of y_1 is taken to be

$$(9) \qquad (\Delta y)_4 = a(\Delta y)_1 + b(\Delta y)_2 + c(\Delta y)_3$$

where a, b, c are parameters which, like the parameters p, q, r, are to be chosen to ensure the highest possible accuracy in estimating Δy.

The details of the determination of the values of a, b, c, p, q, r need not occupy us here,† but a brief outline of the general procedure should be of some interest, if only to indicate why there is not a *unique* Runge-Kutta method but actually a *family* of such methods.

†See, for instance, C. R. Wylie, *Advanced Engineering Mathematics,* 3d ed., pp. 112–114, McGraw-Hill, New York, 1966.

As a first step, Δy is expanded in a power series in Δx:

$$\Delta y = y(x_1) - y_0 = y'(x_0)\,\Delta x + \frac{1}{2!}y''(x_0)(\Delta x)^2 + \frac{1}{3!}y'''(x_0)(\Delta x)^3 + \cdots \tag{10}$$

using implicit differentiation of the given equation (1) to compute the derivatives of y which must be evaluated to obtain the coefficients in the series. Then, in a similar fashion, the approximations $(\Delta y)_1$, $(\Delta y)_2$, and $(\Delta y)_3$, which are functions of Δx defined by Eqs. (6), (7), and (8), respectively, are expressed as power series in Δx. These three expansions are then substituted into (9), so that $(\Delta y)_4$, the final estimate of Δy, is also expressed as a power series in Δx. Finally, the constants of combination a, b, c, and the parameters p, q, r are chosen to make the series for Δy and the series for $(\Delta y)_4$ identical as far as the terms involving $(\Delta x)^3$. When this is accomplished, the difference between the true value of Δy, as given by the series (10), and the final estimate $(\Delta y)_4$, as obtained from (9), is of the order of $(\Delta x)^4$. In other words, the error inherent in the method is proportional to the fourth power of the tabular interval Δx.

When the coefficients of Δx and $(\Delta x)^2$ in the two series are equated, we obtain the conditions

$$a + b + c = 1 \qquad pb + qc = \tfrac{1}{2} \tag{11}$$

Two more conditions,

$$p^2 b + q^2 c = \tfrac{1}{3} \qquad prc = \tfrac{1}{6} \tag{12}$$

arise when the coefficients of $(\Delta x)^3$ are equated. Equations (11) and the first of Eqs. (12) can easily be solved for a, b, and c in terms of p and q. The second equation in (12) can then be used to express r in terms of p and q also. The results are

$$\begin{aligned} a &= \frac{6pq - 3(p+q) + 2}{6pq} & b &= \frac{2 - 3q}{6p(p-q)} \\ c &= \frac{2 - 3p}{6q(q-p)} & r &= \frac{q(q-p)}{p(2-3p)} \end{aligned} \tag{13}$$

Since p and q are arbitrary, except for the restriction that each should be between 0 and 1, we thus have a two-parameter family of formulas which can be used for the step-by-step solution of the equation $dy/dx = f(x, y)$ with an error on the order of $(\Delta x)^4$.

Two special cases of Kutta's third-order method are worthy of note. In listing them, we shall for convenience introduce the following conventional notation:

$$\Delta x = h \qquad (\Delta y)_1 = k_1 \qquad (\Delta y)_2 = k_2 \qquad (\Delta y)_3 = k_3$$

CASE 1 $\quad a = \tfrac{1}{4} \qquad b = 0 \qquad c = \tfrac{3}{4} \qquad p = \tfrac{1}{3} \qquad q = r = \tfrac{2}{3}$

$$\Delta y \doteq (\Delta y)_4 = \tfrac{1}{4}(k_1 + 3k_3)$$

where

$$\begin{aligned} k_1 &= f(x_0, y_0)h \\ k_2 &= f(x_0 + \tfrac{1}{3}h, y_0 + \tfrac{1}{3}k_1)h \\ k_3 &= f(x_0 + \tfrac{2}{3}h, y_0 + \tfrac{2}{3}k_2)h \end{aligned}$$

CASE 2 $\quad a = \tfrac{1}{4} \qquad b = c = \tfrac{3}{8} \qquad p = q = r = \tfrac{2}{3}$

$$\Delta y \doteq (\Delta y)_4 = \tfrac{1}{8}(2k_1 + 3k_2 + 3k_3)$$

where
$$k_1 = f(x_0, y_0)h$$
$$k_2 = f(x_0 + \tfrac{2}{3}h, y_0 + \tfrac{2}{3}k_1)h$$
$$k_3 = f(x_0 + \tfrac{2}{3}h, y_0 + \tfrac{2}{3}k_2)h$$

The values of the parameters in Case 2 cannot be obtained from Eqs. (13) since $p = q$, but can easily be checked directly in Eqs. (11) and (12).

The preceding discussion can be extended without difficulty (except in detail!) to yield step-by-step solution procedures in which the error is of the order of $h^5 = (\Delta x)^5$. In particular, the following two sets of formulas are quite useful.

CASE 3

$$\Delta y \doteq (\Delta y)_5 = \tfrac{1}{6}(k_1 + 2k_2 + 2k_3 + k_4)$$

where
$$k_1 = f(x_0, y_0)h$$
$$k_2 = f(x_0 + \tfrac{1}{2}h, y_0 + \tfrac{1}{2}k_1)h$$
$$k_3 = f(x_0 + \tfrac{1}{2}h, y_0 + \tfrac{1}{2}k_2)h$$
$$k_4 = f(x_0 + h, y_0 + k_3)h$$

CASE 4

$$\Delta y \doteq (\Delta y)_5 = \tfrac{1}{8}(k_1 + 3k_2 + 3k_3 + k_4)$$

where
$$k_1 = f(x_0, y_0)h$$
$$k_2 = f(x_0 + \tfrac{1}{3}h, y_0 + \tfrac{1}{3}k_1)h$$
$$k_3 = f(x_0 + \tfrac{2}{3}h, y_0 - \tfrac{1}{3}k_1 + k_2)h$$
$$k_4 = f(x_0 + h, y_0 + k_1 - k_2 + k_3)h$$

The solution process based on Case 3 is usually referred to specifically as *the* **Runge-Kutta method.**

EXAMPLE 1

Tabulate the solution of $y' = x^2 + y$ at $x = 0.1, 0.2, 0.3, 0.5$, and 1.0, given that $y = -1$ when $x = 0$.

Using the Runge-Kutta formulas of Case 3 for the first increment, we have

$$k_1 = f(x_0, y_0)h = [(0)^2 - 1](0.1) = -0.10000$$
$$k_2 = f(x_0 + \tfrac{1}{2}h, y_0 + \tfrac{1}{2}k_1)h = [(0.05000)^2 - 1.05000](0.1) = -0.10475$$
$$k_3 = f(x_0 + \tfrac{1}{2}h, y_0 + \tfrac{1}{2}k_2)h = [(0.05000)^2 - 1.05238](0.1) = -0.10500$$
$$k_4 = f(x_0 + h, y_0 + k_3)h = [(0.10000)^2 - 1.10500](0.1) = -0.10950$$

and

$$\begin{aligned}\Delta y &\doteq (\Delta y)_5 \\ &= \tfrac{1}{6}(k_1 + 2k_2 + 2k_3 + k_4) \\ &= \tfrac{1}{6}(-0.10000 - 0.20950 - 0.21000 - 0.10950) \\ &= -0.10483\end{aligned}$$

Hence

$$y_1 = y_0 + (\Delta y)_5 = -1.10483$$

For the second increment, we have similarly

$$k_1 = -0.10948 \qquad k_2 = -0.11371 \qquad k_3 = -0.11392 \qquad k_4 = -0.11788$$

and

$$(\Delta y)_5 = -0.11377$$

Hence

$$y_2 = y_1 + (\Delta y)_5 = -1.21860$$

For the third increment, we have

$$k_1 = -0.11786 \qquad k_2 = -0.12150 \qquad k_3 = -0.12169 \qquad k_4 = -0.12503$$

and

$$(\Delta y)_5 = -0.12154$$

Hence

$$y_3 = y_2 + (\Delta y)_5 = -1.34014$$

For the fourth increment (noting that this time $h = \Delta x = 0.2$), we have

$$k_1 = -0.25003 \qquad k_2 = -0.26103 \qquad k_3 = -0.26213 \qquad k_4 = -0.27045$$

and

$$(\Delta y)_5 = -0.26113$$

Hence

$$y_4 = y_3 + (\Delta y)_5 = -1.60127$$

For the fifth increment (noting that now $h = \Delta x = 0.5$), we have

$$k_1 = -0.67564 \qquad k_2 = -0.68830 \qquad k_3 = -0.69146 \qquad k_4 = -0.64636$$

and

$$(\Delta y)_5 = -0.68025$$

Hence

$$y_5 = y_4 + (\Delta y)_5 = -2.28152$$

The differential equation $y' = x^2 + y$ is so simple that it can be solved exactly without recourse to numerical methods, and by the methods of Chap. 1 or Chap. 2 we find at once that the required solution is

$$y = e^x - x^2 - 2x - 2$$

For $x = 0.1, 0.2, 0.3, 0.5$, and 1.0, this gives us the values, correct to five decimal places,

$$\begin{array}{ll} y(x_1) = y(0.1) \doteq -1.10483 & y(x_2) = y(0.2) \doteq -1.21860 \\ y(x_3) = y(0.3) \doteq -1.34014 & y(x_4) = y(0.5) \doteq -1.60128 \\ y(x_5) = y(1.0) \doteq -2.28172 & \end{array}$$

The values we computed for y_1, y_2, and y_3 agree with these to five decimal places; the value we computed for y_4 differs from the correct value by 1 in the fifth place; and our value for y_5 differs from the correct value by only 20 in the fifth place, i.e., by 2 in the fourth place.

Among other things, this example illustrates the important fact that the value of $h \equiv \Delta x$ need not be constant throughout the process but may be changed as time and circumstances may dictate.

Any of the Runge-Kutta formulas can be used to solve simultaneous differential equations. For instance, using Case 3, we can tabulate the solution of the initial-value problem

$$\frac{dy}{dx} = f(x, y, z) \qquad \frac{dz}{dx} = g(x, y, z) \qquad y = y_0,\ z = z_0, \text{ when } x = x_0$$

at intervals of $\Delta x = h$ (either constant or variable) by first computing not only the k's (as in Example 1) but also the corresponding quantities for z, namely,

$$l_1 = (\Delta z)_1 \qquad l_2 = (\Delta z)_2 \qquad l_3 = (\Delta z)_3 \qquad l_4 = (\Delta z)_4$$

according to the formulas

$$\begin{aligned}
k_1 &\equiv (\Delta y)_1 = f(x_0, y_0, z_0)h & l_1 &\equiv (\Delta z)_1 = g(x_0, y_0, z_0)h\\
k_2 &\equiv (\Delta y)_2 = f(x_0 + \tfrac{1}{2}h, y_0 + \tfrac{1}{2}k_1, z_0 + \tfrac{1}{2}l_1)h & l_2 &\equiv (\Delta z)_2 = g(x_0 + \tfrac{1}{2}h, y_0 + \tfrac{1}{2}k_1, z_0 + \tfrac{1}{2}l_1)h\\
k_3 &\equiv (\Delta y)_3 = f(x_0 + \tfrac{1}{2}h, y_0 + \tfrac{1}{2}k_2, z_0 + \tfrac{1}{2}l_2)h & l_3 &\equiv (\Delta z)_3 = g(x_0 + \tfrac{1}{2}h, y_0 + \tfrac{1}{2}k_2, z_0 + \tfrac{1}{2}l_2)h\\
k_4 &\equiv (\Delta y)_4 = f(x_0 + h, y_0 + k_3, z_0 + l_3)h & l_4 &\equiv (\Delta z)_4 = g(x_0 + h, y_0 + k_3, z_0 + l_3)h
\end{aligned}$$

and then using the formulas

$$\Delta y \doteq (\Delta y)_5 = \tfrac{1}{6}(k_1 + 2k_2 + 2k_3 + k_4)$$
$$\Delta z \doteq (\Delta z)_5 = \tfrac{1}{6}(l_1 + 2l_2 + 2l_3 + l_4)$$

to compute the increments that yield the next pair of (y, z) values.

Since any differential equation of the form

$$y^{(n)} = f(x, y, y', \ldots, y^{(n-1)}) \qquad n \geq 2$$

and any equation of the form $g(x, y, y', \ldots, y^{(n-1)}, y^{(n)}) = 0$ which can be solved for $y^{(n)}$ can be written as a system of simultaneous first-order differential equations, Runge-Kutta methods also suffice for the solution of such equations. For example, under the substitution $dy/dx \equiv y' = z$, the initial-value problem

$$\frac{d^2y}{dx^2} = g(x, y, y') \qquad y = y_0,\ y' = y_0', \text{ when } x = x_0$$

becomes

$$\frac{dy}{dx} = z \qquad \frac{dz}{dx} = g(x, y, z) \qquad y = y_0,\ z = y_0', \text{ when } x = x_0$$

which is just like the system we last discussed with $f(x, y, z) \equiv z$.

To develop **Milne's method** we begin with Eq. (1), Sec. 5.4, written in terms of y rather than f and evaluate it for $r = 1, 2, 3$, and 4, getting

$$y_1' = \frac{1}{h}(\Delta y_0 + \tfrac{1}{2}\Delta^2 y_0 - \tfrac{1}{6}\Delta^3 y_0 + \tfrac{1}{12}\Delta^4 y_0 + \cdots)$$

$$y_2' = \frac{1}{h}(\Delta y_0 + \tfrac{3}{2}\Delta^2 y_0 + \tfrac{1}{3}\Delta^3 y_0 - \tfrac{1}{12}\Delta^4 y_0 + \cdots)$$

$$y_3' = \frac{1}{h}(\Delta y_0 + \tfrac{5}{2}\Delta^2 y_0 + \tfrac{11}{6}\Delta^3 y_0 + \tfrac{1}{4}\Delta^4 y_0 + \cdots)$$

$$y_4' = \frac{1}{h}(\Delta y_0 + \tfrac{7}{2}\Delta^2 y_0 + \tfrac{13}{3}\Delta^3 y_0 + \tfrac{25}{12}\Delta^4 y_0 + \cdots)$$

or, neglecting differences beyond the fourth and replacing the remaining differences by their equivalent expressions in terms of the successive functional values [Eqs. (17*a*), Sec. 5.2]

$$
(14)\qquad
\begin{aligned}
y_1' &= \frac{1}{12h}(-3y_0 - 10y_1 + 18y_2 - 6y_3 + y_4)\\
y_2' &= \frac{1}{12h}(y_0 - 8y_1 + 8y_3 - y_4)\\
y_3' &= \frac{1}{12h}(-y_0 + 6y_1 - 18y_2 + 10y_3 + 3y_4)\\
y_4' &= \frac{1}{12h}(3y_0 - 16y_1 + 36y_2 - 48y_3 + 25y_4)
\end{aligned}
$$

Now if we subtract the second equation in the set (14) from twice the sum of the first and third equations and solve the result for y_4, we obtain

$$y_4 = y_0 + \frac{4h}{3}(2y_1' - y_2' + 2y_3')$$

or, in more general terms,

$$(15)\qquad y_{n+1} = y_{n-3} + \frac{4h}{3}(2y_{n-2}' - y_{n-1}' + 2y_n')$$

If we know the values of y and y' up to and including their values at x_n, Eq. (15) enables us to ''reach out'' one step further and compute y_{n+1}. With y_{n+1} known, we can then return to the given differential equation (1) and compute y_{n+1}'. Then using Eq. (15) again, with n increased by 1 throughout, we can find y_{n+2}, and so on, step by step, until the solution has been extended over the desired range. All that remains is to devise a means of finding enough y's and y''s to get the process started.

One way to obtain the values of y and y' needed to begin the use of Milne's method is to use the Runge-Kutta method, which is ''self-contained'' and needs no independently determined starting values. Another possibility is to begin the tabulation of y by expanding it in a Taylor series about the point $x = x_0$:

$$(16)\qquad y = y_0 + y_0'(x - x_0) + \frac{y_0''}{2!}(x - x_0)^2 + \frac{y_0'''}{3!}(x - x_0)^3 + \cdots$$

The value of y_0 is of course given. The value of y_0' can be found at once by substituting x_0 and y_0 into the given differential equation (1). To find the second derivative, we need only differentiate the given equation, getting

$$(17)\qquad y'' = \frac{\partial f}{\partial x} + \frac{\partial f}{\partial y}y'$$

Since $f(x, y)$ is a given function, its partial derivatives are known and become definite numbers when x_0 and y_0 are substituted into them. Moreover, the value of y' at (x_0, y_0) has already been found, and thus (17) furnishes the value of y_0''. Similarly, differentiating (17) and evaluating the result at (x_0, y_0) will give y''', and so on, assuming that the appropriate derivatives of $f(x, y)$ exist at (x_0, y_0). In this way the first few terms in the expansion of y around $x = x_0$ can be constructed. In especially favorable cases, an explicit formula for the general term of the series (16) can be found and the interval of convergence established. When this happens, (16) is the required solution of Eq.

(1), and we need look no further. In general, however, successive differentiation of $f(x, y)$ becomes too complicated to continue, or the resulting series converges too slowly to be of practical value in computing y at tabular points relatively far from x_0.

On the other hand, for tabular values of x relatively close to x_0 and with (16) available as a representation of y in some neighborhood of $x = x_0$, we can set $x = x_0 + h \equiv x_1$ and calculate y_1. Similarly, setting $x = x_0 + 2h$ and $x_0 + 3h$, we can find y_2 and y_3. Then substituting (x_1, y_1), (x_2, y_2), and (x_3, y_3) into the given differential equation, we can compute y'_1, y'_2, and y'_3 without difficulty. With these values we are then in a position to begin the step-by-step solution of the differential equation by means of Eq. (15).

From the preceding discussion it is clear that Eq. (15) is in general adequate for the step-by-step solution of $y' = f(x, y)$. However, as a precaution against errors of various kinds, it is desirable to have a second, independent formula into which y_{n+1} can be substituted as a check. To obtain such an equation we return to (14) and add 4 times the third equation to the sum of the second and fourth and solve the resulting equation for y_4, getting

$$y_4 = y_2 + \frac{h}{3}(y'_2 + 4y'_3 + y'_4)$$

or, in more general terms,

$$(18) \qquad y_{n+1} = y_{n-1} + \frac{h}{3}(y'_{n-1} + 4y'_n + y'_{n+1})$$

This formula cannot be used as a formula of extrapolation since it involves y'_{n+1}, which cannot be found unless y_{n+1} is already known. However, after y_{n+1} has been calculated by means of (15), y'_{n+1} can be calculated using Eq. (1), and enough information is then available to permit the use of (18) as another means of computing y_{n+1}. If the value of y_{n+1} as given by (18) agrees with the value found from (15), we are ready to move on to the calculation of y_{n+2}. On the other hand, if the two values of y_{n+1} do not agree, we then use the second value of y_{n+1} to compute a new value of y'_{n+1}, which we substitute into (18) to find still another value for y_{n+1}. This process is continued until two successive values of y_{n+1} are in agreement. Once this happens, we are ready to continue the tabulation of y by returning to Eq. (15) and determining an initial estimate of y_{n+2}.

Formulas like (15), which express a new value exclusively in terms of quantities already found, are known as **open formulas** or **predictor formulas.** Those like (18), which express a new value in terms of one or more additional new quantities and which therefore can be used only for purposes of checking and refining, are known as **closed formulas** or **corrector formulas.**

The method of Milne is readily extended to the solution of simultaneous and higher-order equations. For instance, if we have the equations

$$(19) \qquad y' = f(x, y, z) \qquad \text{and} \qquad z' = g(x, y, z)$$

with the initial conditions $y = y_0$, $z = z_0$ when $x = x_0$, and if by independent means we have calculated (y_1, y_2, y_3), (z_1, z_2, z_3), and the related quantities (y'_1, y'_2, y'_3) and (z'_1, z'_2, z'_3), then, using (15) and an identical version of it with z replacing y, we can compute y_4 and z_4. After that, we can compute y'_4 and z'_4 from the given differential equations and again use (15) to obtain y_5 and z_5, and so on, as far as desired. Of course, the closed formula (18) can be used to check and correct both y_{n+1} and z_{n+1} if and when this is deemed necessary.

The application of Milne's method to equations of higher order is immediate since (as we have several times observed) such an equation can always be replaced by a system of simultaneous first-order equations. For instance, $y'' = g(x, y, y')$ is equivalent to the system

$$y' = z \qquad z' = g(x, y, z)$$

which is just a special case, with $f(x, y, z) \equiv z$, of the general problem of two simultaneous first-order equations.

Finally, it should be noted that since Milne's method is based on Eqs. (14), which were derived from the forward Gregory-Newton formula, it follows that in Milne's method, unlike the Runge-Kutta method, the tabular interval h must remain constant.

EXAMPLE 2

Tabulate the solution of $y'' + y^2 = x$ at intervals of $h = 0.1$ from $x = 0$ to $x = 0.6$ if $y = 0$ and $y' = 1$ when $x = 0$.

If we are to use either the Runge-Kutta method or Milne's method, it is first necessary to convert the given second-order equation into a pair of first-order equations by putting $z = y'$. This gives us the related *nonlinear* first-order system

$$(20) \qquad y' = z \qquad z' = x - y^2 \qquad y = 0, z = 1 \qquad \text{when } x = 0$$

Beginning with this information, the Runge-Kutta method can be used to find $y_1, y_2, y_3, z_1, z_2, z_3$, and the corresponding derivatives which we must know before Milne's method can be applied. However, to illustrate the use of Taylor series, we shall obtain the necessary starting values from the expansions of y and z in terms of powers of x.

With a general pair of simultaneous first-order differential equations such as (19), it is necessary to construct the series expansion of y and, independently, the series expansion of z. However, when a second-order equation of the form $y'' = f(x, y, y')$ is reduced to a pair of first-order equations, one of the equations is always $y' = z$. Hence in such cases the series for z can be found by simply differentiating the series for y after the latter has been constructed.

To determine the successive terms in the power series expansion of y in a neighborhood of $x = 0$, we need the values of the first few derivatives of y at $x = 0$. The values $y_0 = 0$ and $y_0' = 1$ are given. That $y_0'' = 0$ follows immediately from the given differential equation. The values of $y_0''', y_0^{iv}, y_0^{v}, y_0^{vi}, \ldots$ can be obtained by repeated differentiation of the given equation and evaluation of the results:

$$\begin{aligned}
y''' &= 1 - 2yy' \\
y^{iv} &= -2yy'' - 2(y')^2 \\
y^{v} &= -2yy''' - 2y'y'' - 4y'y'' = -2yy''' - 6y'y'' \\
y^{vi} &= -2yy^{iv} - 2y'y''' - 6y'y''' - 6(y'')^2 = -2yy^{iv} - 8y'y''' - 6(y'')^2
\end{aligned}$$

Evaluating these derivatives at $x = x_0 = 0$, remembering that $y_0 = 0$, $y_0' = 1$, and $y_0'' = 0$, we obtain

$$y_0''' = 1 \qquad y_0^{iv} = -2 \qquad y_0^{v} = 0 \qquad y_0^{vi} = -8 \qquad \cdots$$

Hence, substituting these values into Eq. (16), we find

$$\begin{aligned}
y &= x + \frac{1}{3!}x^3 - \frac{2}{4!}x^4 - \frac{8}{6!}x^6 + \cdots \\
(21) \qquad &= x + \frac{1}{6}x^3 - \frac{1}{12}x^4 - \frac{1}{90}x^6 + \cdots
\end{aligned}$$

and by differentiating Eq. (21),

$$(22) \qquad z = y' = 1 + \tfrac{1}{2}x^2 - \tfrac{1}{3}x^3 - \tfrac{1}{15}x^5 + \cdots$$

When the series (21) and (22) are evaluated for $x_1 = 0.1$, $x_2 = 0.2$, $x_3 = 0.3$ and the corresponding derivatives of y and z computed from (20), we obtain the following starting values:

$$\begin{array}{llll}
y_0 = 0.0000 & y_0' = 1.0000 & z_0 = 1.0000 & z_0' = 0.0000 \\
y_1 \doteq 0.1002 & y_1' \doteq 1.0047 & z_1 \doteq 1.0047 & z_1' \doteq 0.0900 \\
y_2 \doteq 0.2012 & y_2' \doteq 1.0173 & z_2 \doteq 1.0173 & z_2' \doteq 0.1595 \\
y_3 \doteq 0.3038 & y_3' \doteq 1.0358 & z_3 \doteq 1.0358 & z_3' \doteq 0.2077
\end{array}$$

From these, using Eq. (15) (and its counterpart with y replaced by z), we find

$$y_4 \doteq 0.4085 \quad \text{and} \quad z_4 \doteq 1.0581$$

Corresponding to these values, we find from (20) that

$$y_4' \doteq 1.0581 \quad \text{and} \quad z_4' \doteq 0.2331$$

When we recompute y_4 and z_4 using the closed formula (18), we obtain the same values for y_4 and z_4, so we accept them as correct.

Continuing the process, we return to Eq. (15) and compute y_5 and z_5, getting

$$y_5 \doteq 0.5155 \quad \text{and} \quad z_5 \doteq 1.0817$$

From these we find that

$$y_5' \doteq 1.0817 \quad \text{and} \quad z_5' \doteq 0.2343$$

Checking the values of y_5 and z_5 by recalculating them from Eq. (18), we obtain

$$y_5 \doteq 0.5155 \quad \text{and} \quad z_5 \doteq 1.0816$$

A second application of Eq. (18) repeats these values, which we therefore accept as correct.

Another application of Eq. (15) leads to the values

$$y_6 \doteq 0.6248 \quad \text{and} \quad z_6 \doteq 1.1041$$

which are again confirmed by Eq. (18). By continuing in this fashion, the solution of the given initial-value problem can be tabulated as far as desired.

EXERCISES

1. In Example 1, calculate $y(0.4)$ and $y(0.6)$.
2. For the problem of Example 1, find $y(0.1)$, $y(0.2)$, and $y(0.3)$ using Euler's method.
3. For the problem of Example 1, find $y(0.1)$, $y(0.2)$, and $y(0.3)$ using the modified Euler method.
4. For the problem of Example 1, find $y(0.1)$, $y(0.2)$, and $y(0.3)$ using Runge's method.
5. Using Kutta's third-order approximation Case 1, compute $y(1.1)$, $y(1.2)$, and $y(1.3)$ if $y' = x - y$ and $y = 1$ when $x = 1$. Carry four decimal places in your work and compare your answers with the exact solution.
6. Using Kutta's third-order approximation Case 2, compute $y(0.1)$, $y(0.2)$, and $y(0.3)$ if $y' = x + y$ and $y = 1$ when $x = 0$. Carry four decimal places in your work and compare your answers with the exact solution.
7. Using the Runge-Kutta method 3, tabulate the function $y = e^{-x^2}$ for $x = 0.0, 0.1, 0.2, 0.3, 0.4$, and 0.5. How do your answers compare with the exact values of this function? *Hint:* Find a differential equation satisfied by y.
8. Using the Runge-Kutta method 3, evaluate $\int_0^x e^{-t^2}\,dt$ for $x = 0.0, 0.2, 0.4, 0.6, 0.8$, and 1.0.
9. Using the Runge-Kutta method 3, tabulate the solution of the system

$$\frac{dy}{dx} = x + z \qquad \frac{dz}{dx} = x - y \qquad y = 0,\ z = 1 \text{ when } x = 0$$

at intervals of $h = 0.1$ from $x = 0.0$ to $x = 0.5$.

10. Set up the Kutta third-order approximation corresponding to the values $p = \frac{1}{2}$, $q = 1$ and show that it reduces to Simpson's rule when $f(x, y)$ is independent of y.
11. Given the equation $y' = x - y$ and the starting values

x	y
0.0	1.0000
0.1	0.9097
0.2	0.8375
0.3	0.7816

find $y(0.4)$, $y(0.5)$, and $y(0.6)$ **(a)** using only Eq. (15) and **(b)** using Eq. (18) to correct the values found from Eq. (15). Compare your answers in each case with those given by the exact solution.

12. Work Exercise 11, given the equation $y' = x^2 - y$ and the starting values

x	y
0.0	0.0000
0.1	0.0003
0.2	0.0025
0.3	0.0084

13. Using Milne's method and the Maclaurin series for y to obtain the necessary starting values, tabulate the solution of the equation $y' = x + y^2$ at intervals of $h = 0.1$ from $x = 0$ to $x = 0.5$ if $y_0 = 0$.

14. Show that the use of Maclaurin's expansion alone is sufficient to solve the initial-value problem $y'' = xy$; $y_0 = 1$, $y'_0 = 0$.

15. Work Exercise 11 using the predictor formula

$$y_{n+1} = y_n + \frac{h}{12}(23y'_n - 16y'_{n-1} + 5y'_{n-2})$$

and the corrector formula

$$y_{n+1} = y_n + \frac{h}{12}(5y'_{n+1} + 8y'_n - y'_{n-1})$$

(These equations constitute the so-called **Adams-Moulton method** for the numerical solution of differential equations.)

16. Assuming the values of y_0, y_1, and y_2 of Example 2, continue the solution using the predictor formula

$$y_{n+1} = 2y_n - y_{n-1} + h^2(y''_n + \tfrac{1}{12}\Delta^2 y''_{n-2})$$

and the corrector formula

$$y_{n+1} = 2y_n - y_{n-1} + h^2(y''_n + \tfrac{1}{12}\Delta^2 y''_{n-1})$$

17. Using the data of Example 1, rounded to four places, together with the answers to Exercise 1, compute $y(0.7)$ and $y(0.8)$ using the open formula

$$y_{n+1} = y_n + h(y'_n + \tfrac{1}{2}\Delta y'_{n-1} + \tfrac{5}{12}\Delta^2 y'_{n-2} + \tfrac{3}{8}\Delta^3 y'_{n-3} + \tfrac{251}{720}\Delta^4 y'_{n-4})$$

and the closed formula

$$y_{n+1} = y_n + h(y'_{n+1} - \tfrac{1}{2}\Delta y'_n - \tfrac{1}{12}\Delta^2 y'_{n-1} - \tfrac{1}{24}\Delta^3 y'_{n-2} - \tfrac{19}{720} y'_{n-3})$$

These equations constitute the so-called **Adams-Bashforth method** for the numerical solution of differential equations.

18. Using Milne's method, tabulate the solution of the system

$$\frac{dy}{dx} = y^2 + zx \qquad \frac{dz}{dx} = x^2 + yz$$

at intervals of $h = 0.1$, given $y_0 = 0$, $z_0 = 1$.

19. Show that the error in Milne's predictor formula is $\frac{14}{45}h^5 y_0^{\text{v}}$. In other words, show that the dominant term in the difference between y_4, as given by its Maclaurin expansion, and y_4, as determined by combining the Maclaurin series for y'_1, y'_2, and y'_3 according to Milne's predictor formula, is $\frac{14}{45}h^5 y_0^{\text{v}}$.

20. What is the error in Milne's corrector formula?

21. Find the equation of the polynomial of minimum degree for which y and y' take on prescribed values (y_0, y'_0) at $x = 0$ and (y_1, y'_1) at $x = h$. What is the value of $y_2 = y(2h)$ given by this polynomial? How might this result be used to carry out the step-by-step integration of a differential equation of the form $y' = f(x, y)$? How might an accompanying closed formula be obtained?

22. **(a)** What is the principal part of the error in the predictor formula obtained in Exercise 21? (*Hint:* Recall the explanation in Exercise 19.)
(b) Can a more accurate formula be obtained by starting with the relation $y_2 = ay_0 + by'_0 + cy_1 + dy'_1$ and choosing a, b, c, and d so as to make the error as small as possible?

23. Find the equation of the polynomial of minimum degree for which y and y'' take on prescribed values (y_0, y''_0) at $x = 0$ and (y_1, y''_1) at $x = h$. What is the value of $y_2 = y(2h)$ given by this polynomial? How might this result be used to carry out the step-by-step integration of a differential equation of the form $y'' = f(x, y)$? How might an accompanying closed formula be obtained?

24. **(a)** What is the principal part of the error in the predictor formula obtained in Exercise 23?
(b) Can a more accurate formula be obtained by starting with the relation $y_2 = ay_0 + by''_0 + cy_1 + dy''_1$ and choosing a, b, c, and d so as to make the error as small as possible?

25. **(a)** What is the principal part of the error in the predictor formula given in Exercise 15?
(b) What is the principal part of the error in the closed formula given in Exercise 15?

5.7 DIFFERENCE EQUATIONS

The similarities between the difference operator Δ, introduced in Sec. 5.2, and the derivative operator D, and the fact that derivatives can be approximated by expressions involving differences, suggest that there may be a theory of *difference equations* roughly paralleling the theory of differential equations; and this is indeed the case. However, in the study of difference equations, we do not ordinarily consider equations of the form

$$f(\Delta)y = \phi(x)$$

as might be expected by analogy with the differential equation

$$f(D)y = \phi(x)$$

but rather equations of the form

$$F(E)y = \phi(x)$$

where E is the shift operator, discussed in Sec. 5.2. This of course is simply a matter of notational convenience since, by using the operational equivalence $\Delta = E - 1$ [Eq. (11), Sec. 5.2], any function of Δ can be transformed at once into a function of E, and vice versa.

By a **linear constant-coefficient difference equation** we mean an equation of the form

$$(1) \qquad (a_0E^r + a_1E^{r-1} + \cdots + a_{r-1}E + a_r)y = \phi(x) \qquad a_0, a_1, \ldots, a_r \text{ constants}$$

Since the substitution $t = hx$ will transform a function of t tabulated at intervals of h into a function of x tabulated at unit intervals, it is clearly no restriction to assume $h = 1$, so that invariably $Ef(x) = f(x + 1)$, and we shall do this throughout this section.

In Eq. (1), if both a_0 and a_r are different from zero, as we shall henceforth suppose, the positive integer r is called the **order** of the equation. If $\phi(x)$ is identically zero, Eq. (1) is said to be **homogeneous.**† If $\phi(x)$ is not identically zero, Eq. (1) is said to be **nonhomogeneous.** By a **solution** of Eq. (1) we mean a function of x with the property that when substituted into (1), it reduces the equation to an identity for $x = x_0 \pm n$, $n = 0, 1, 2, \ldots$, x_0 arbitrary.

EXAMPLE 1

Verify that for all values of the constants c_1 and c_2, the expression $y = c_1 + c_2 2^x$ is a solution of the equation

$$(E^2 - 3E + 2)y = 0$$

Since $(E^2 - 3E + 2)y$ means $E^2y - 3Ey + 2y$, substitution of $y = c_1 + c_2 2^x$ yields

$$\begin{aligned}(c_1 + c_2 2^{x+2}) - 3(c_1 + c_2 2^{x+1}) + 2(c_1 + c_2 2^x) &= c_1(1 - 3 + 2) + c_2 2^x(2^2 - 3 \cdot 2 + 2)\\ &= c_1 \cdot 0 + c_2 \cdot 0 = 0\end{aligned}$$

as asserted.

It is interesting to note that the function $y = c_1 + c_2 2^x$ is by no means the only solution of the equation in Example 1. In fact, since $Ef(x) = f(x + 1) = f(x)$ for any function $f(x)$ of period 1, it follows that if $f_1(x)$ and $f_2(x)$ are arbitrary functions of period 1, then both $c_1 f_1(x)$ and $c_2 f_2(x)$ behave as constants as far as the operator E is concerned. Therefore

$$y = c_1 f_1(x) + c_2 f_2(x) 2^x$$

is also a solution of the equation $(E^2 - 3E + 2)y = 0$. In practical problems, however, one is seldom interested in x and y as continuous variables connected by an equation like (1), and in our work we shall attempt no more than the determination of y as a function of x on the domain set $x = \ldots, -2, -1, 0, 1, 2, \ldots$. Hence we shall consistently ignore the possibility of including functions of period 1 in the coefficients of solutions such as $c_1 + c_2 2^x$. Moreover, we shall henceforth emphasize the discrete character of the domain set $\ldots, -2, -1, 0, 1, 2, \ldots$, by using n rather than x as our independent variable.

†A homogeneous linear difference equation with constant coefficients is sometimes referred to as a **linear recurrence formula** since it is simply a repeating, or *recurring,* relation between the values of y at *any* $r + 1$ consecutive values of x.

It is important to note that when suitable starting values of y are given, the exact solution of any homogeneous constant-coefficient linear difference equation can be tabulated in a step-by-step fashion over any desired set of consecutive integers. In fact, given the equation

$$(2) \qquad (a_0E^r + a_1E^{r-1} + \cdots + a_{r-1}E + a_r)y = 0 \qquad a_0 \cdot a_r \neq 0$$

we can divide by a_0 and rename the coefficients, getting

$$(E^r + p_1E^{r-1} + \cdots + p_{r-1}E + p_r)y = 0 \qquad p_k = \frac{a_k}{a_0}$$

Then, setting $E^ky = y_k$, $k = 0, 1, \ldots, r$, we may solve for $E^ry \equiv y_r$, getting

$$(3) \qquad y_r = -p_1y_{r-1} - p_2y_{r-2} - \cdots - p_{r-1}y_1 - p_ry_0$$

Thus if we are given the values of $y(n)$ for $n = 0, 1, 2, \ldots, r-1$,† say $y_0, y_1, y_2, \ldots, y_{r-1}$, we can compute y_r immediately. Then, applying the operator E to Eq. (3), that is, recognizing that (3) is simply a recurrence relation connecting the values of y on *any* set of $r+1$ consecutive integers, we can advance the subscript of each y by 1 and thus obtain y_{r+1}:

$$y_{r+1} = -p_1y_r - p_2y_{r-1} - \cdots - p_{r-1}y_2 - p_ry_1$$

Since y_r has just been found, this equation gives us the value of y_{r+1}. By continuing in this fashion, the solution for y can be tabulated as far as desired. Similarly, since $a_r \neq 0$, we can divide Eq. (2) by a_r, set $q_k = a_k/a_r$, then apply the operator E^{-1} to the resulting equation and solve for $E^{-1}y_0 = y_{-1}$, getting

$$y_{-1} = -q_{r-1}y_0 - q_{r-2}y_1 - \cdots - q_1y_{r-2} - q_0y_{r-1}$$

This equation gives the value of y_{-1} in terms of known quantities. Continuing in this way, the values of $y_{-2}, y_{-3}, \ldots,$ can also be determined.

The tabulation process we have just described for constant-coefficient difference equations can also be applied when the coefficients in Eq. (2) are nonconstant functions of n. In this case, values of y_n can be obtained out to, but not including, the first positive integral value of n for which $a_0(n) = 0$, and back to, but not including, the first negative integral value of n for which $a_r(n) = 0$.

EXAMPLE 2

If $(E^2 - E - 6)y = 0$ and $y_0 = 0$, $y_1 = 1$, determine $y_2, y_3, \ldots, y_8$ and y_{-1}, y_{-2}, y_{-3}.

Clearly, the given equation is equivalent to the recurrence relation

$$y_{n+2} - y_{n+1} - 6y_n = 0 \qquad n = \ldots, -3, -2, -1, 0, 1, 2, 3, \ldots$$

Hence, beginning with $n = 0$ and substituting the given starting values, we have

$$y_2 = y_1 + 6y_0 = 1 + 6 \cdot 0 = 1$$

Then, taking $n = 1$, we have

$$y_3 = y_2 + 6y_1 = 1 + 6 \cdot 1 = 7$$

†This of course is analogous to being given the values of $y_0, y_0', y_0'', \ldots, y_0^{(r-1)}$ in an initial-value problem involving a linear differential equation of order r.

and, continuing in this fashion,

$$y_4 = 13 \qquad y_5 = 55 \qquad y_6 = 133 \qquad y_7 = 463 \qquad y_8 = 1{,}261$$

Similarly, writing the equation as

$$y_n = \tfrac{1}{6}(y_{n+2} - y_{n+1})$$

and taking $n = -1, -2, -3$, we find

$$y_{-1} = \tfrac{1}{6} \qquad y_{-2} = -\tfrac{1}{36} \qquad y_{-3} = \tfrac{7}{216}$$

From the preceding discussion and the accompanying example, it is evident that the *tabulation* of the solution of a linear difference equation is a simple matter. What we really want, however, is a closed *formula* for a complete solution containing arbitrary constants which can be determined to fit prescribed starting values. For linear difference equations with constant coefficients, the determination of a complete solution is not difficult and, in fact, is similar to the methods we have learned for the solution of linear differential equations with constant coefficients. For second-order equations, the process is based on the following theorems, the first two of which remind us of Theorems 1, 4, and 6, Sec. 2.2, while the third is reminiscent of Theorem 1, Sec. 2.3.

THEOREM 1 If $y_1(n)$ and $y_2(n)$ are any two solutions of the homogeneous equation

$$(a_0E^2 + a_1E + a_2)y = 0$$

then $c_1y_1(n) + c_2y_2(n)$, where c_1 and c_2 are arbitrary constants, is also a solution.

THEOREM 2 If $y_1(n)$ and $y_2(n)$ are two solutions of the homogeneous equation

$$(a_0E^2 + a_1E + a_2)y = 0$$

for which

$$C[y_1(n), y_2(n)]\dagger = \begin{vmatrix} y_1(n) & y_2(n) \\ Ey_1(n) & Ey_2(n) \end{vmatrix} \neq 0$$

then any solution $y(n)$ of the homogeneous equation can be written in the form $y(n) = c_1y_1(n) + c_2y_2(n)$, where c_1 and c_2 are suitable constants.

As a consequence of Theorem 2, the expression $c_1y_1(n) + c_2y_2(n)$ is called a **complete solution** of the homogeneous equation (2) when the particular solutions $y_1(n)$ and $y_2(n)$ satisfy the condition $C[y_1(n), y_2(n)] \neq 0$, and $r = 2$.

THEOREM 3 If $Y(n)$ is any particular solution of the nonhomogeneous equation

$$(a_0E^2 + a_1E + a_2)y = \phi(n)$$

and if $c_1y_1(n) + c_2y_2(n)$ is a complete solution of the homogeneous equation obtained from this by deleting the term $\phi(n)$, then any solution of the nonhomogeneous equation can be written in the form

$$y(n) = c_1y_1(n) + c_2y_2(n) + Y(n)$$

where c_1 and c_2 are suitable constants.

†The function $C[y_1(n), y_2(n)]$ is customarily referred to as **Casorati's determinant,** after the Italian mathematician Felice Casorati (1835–1890). Its resemblance to the wronskian $W[y_1(x), y_2(x)]$ (see Sec. 2.2) is apparent.

As in the case of differential equations, the homogeneous equation which results when $\phi(n)$ is deleted from the nonhomogeneous equation is called the **related homogeneous equation,** and a complete solution of the related homogeneous equation is called a **complementary function** of the nonhomogeneous equation. By analogy with differential equations, the particular solution $Y(n)$ is often referred to as a **particular integral** of the nonhomogeneous equation even though no integration is involved in its determination. The extension of these theorems and definitions to difference equations of order greater than 2 is obvious.

The proofs of Theorems 1 and 3 and the proof of Theorem 2 when the coefficients are constants are simple, and we shall leave them as exercises. The proof of Theorem 2 when the coefficients are nonconstant functions of n we leave to more complete texts on difference equations.†

To find particular nontrivial solutions of the homogeneous equation

$$(a_0E^2 + a_1E + a_2)y = 0 \tag{4}$$

where the coefficients a_0, a_1, a_2 are constants, we might try, as with the analogous differential equation, the substitution

$$y = e^{mn} \tag{5}$$

where m is a constant to be determined. However, it is more convenient to assume

$$y = M^n \qquad M \neq 0 \tag{6}$$

which is clearly equivalent to (5) with $M = e^m$. Substituting this into Eq. (4) and recalling our agreement that $Ef(n) = f(n + 1)$, we obtain

$$a_0M^{n+2} + a_1M^{n+1} + a_2M^n = 0$$

or, dividing out M^n, which is not zero,

$$a_0M^2 + a_1M + a_2 = 0 \tag{7}$$

Naturally enough, (7) is called the **characteristic equation** of the difference equation (4).

If the roots M_1 and M_2 of (7) are distinct, then

$$C(M_1^n, M_2^n) = \begin{vmatrix} M_1^n & M_2^n \\ M_1^{n+1} & M_2^{n+1} \end{vmatrix} = M_1^nM_2^n(M_2 - M_1) \neq 0‡$$

hence, by Theorem 2, a complete solution of Eq. (4) is

$$y = C_1M_1^n + c_2M_2^n \tag{8}$$

If M_1 and M_2 are real, this is a completely acceptable form of the solution. However, if M_1 and M_2 are complex, then (8) is inconvenient for many purposes, and it is desirable that we reduce it to a more useful form. To do this, let the roots of the characteristic equation be

$$M_1, M_2 = p \pm iq = re^{\pm i\theta}$$

†See, for instance, L. Brand, "Differential and Difference Equations," pp. 361–363, Wiley, New York, 1966; L. M. Milne-Thompson, *The Calculus of Finite Differences,* pp. 354–355, Macmillan, London, 1933.

‡Since $a_2 \neq 0$ [or else the difference equation (4) would be of order less than 2, contrary to hypothesis], it follows from (7) that neither M_1 nor M_2 can be zero.

where

$$r = \sqrt{p^2 + q^2} \qquad \text{and} \qquad \tan\theta = \frac{q}{p}\dagger$$

Then we can write

$$\begin{aligned} y &= c_1(re^{i\theta})^n + c_2(re^{-i\theta})^n \\ &= r^n(c_1e^{i\theta n} + c_2e^{-i\theta n}) \\ &= r^n[c_1(\cos\theta n + i\sin\theta n) + c_2(\cos\theta n - i\sin\theta n)] \\ &= r^n[(c_1 + c_2)\cos\theta n + i(c_1 - c_2)\sin\theta n] \end{aligned}$$

or, renaming the constants,

$$(9) \qquad y = r^n(A\cos\theta n + B\sin\theta n)$$

If $M_1 = M_2$, clearly $C(M_1^n, M_2^n) = 0$, and we must find a second, independent solution before we can construct a complete solution of Eq. (4). Again, by analogy with differential equations, we are led to try

$$y = nM_1^n$$

By direct substitution we find that this is indeed a solution when the characteristic equation (7) has equal roots. In fact, substituting $y = nM_1^n$ into the difference equation (4) gives

$$a_0(n+2)M_1^{n+2} + a_1(n+1)M_1^{n+1} + a_2nM_1^n = nM_1^n(a_0M_1^2 + a_1M_1 + a_2) + M_1^{n+1}(2a_0M_1 + a_1)$$

On the right-hand side, the coefficient of nM_1^n is zero because, in any event, M_1 satisfies the characteristic equation (7). Furthermore, the coefficient of M_1^{n+1} is zero because when the characteristic equation has equal roots, their common value is $M_1 = -a_1/2a_0$. Moreover, for the solutions $y_1 = M_1^n$ and $y_2 = nM_1^n$, we have

$$C(M_1^n, nM_1^n) = M_1^n(n+1)M_1^{n+1} - nM_1^nM_1^{n+1} = M_1^{2n+1} \neq 0$$

Hence, according to Theorem 2, a complete solution of Eq. (4) when the characteristic equation has equal roots is

$$(10) \qquad y = c_1M_1^n + c_2nM_1^n$$

The results of the preceding discussion are summarized in Table 5.1.

EXAMPLE 3

Find a complete solution of the difference equation

$$(E^2 + 2E + 4)y = 0$$

The characteristic equation in this case is $M^2 + 2M + 4 = 0$, and its roots are $M_1, M_2 = -1 \pm i\sqrt{3}$.

---

†For a discussion of the exponential form of a complex number, see Secs. 3.1 and 17.4.

TABLE 5.1

Difference equation $(a_0E^2 + a_1E + a_2)y = 0 \qquad a_0, a_2 \neq 0$
Characteristic equation: $a_0M^2 + a_1M + a_2 = 0$

Nature of the roots of the characteristic equation	Condition on the coefficients of the characteristic equation	Complete solution of the difference equation
Real and unequal $M_1 \neq M_2$	$a_1^2 - 4a_0a_2 > 0$	$y = c_1M_1^n + c_2M_2^n$
Real and equal $M_1 = M_2$	$a_1^2 - 4a_0a_2 = 0$	$y = c_1M_1^n + c_2nM_1^n$
Conjugate complex $M_1 = p + iq$ $M_2 = p - iq$	$a_1^2 - 4a_0a_2 < 0$	$y = r^n(A\cos\theta n + B\sin\theta n)$ $r = \sqrt{p^2+q^2}$ $\tan\theta = q/p$

Since

$$r = \sqrt{(-1)^2 + (\sqrt{3})^2} = 2 \qquad \text{and} \qquad \theta = \tan^{-1}\frac{\sqrt{3}}{-1} = \frac{2\pi}{3}$$

we have as a complete solution

$$y = 2^n\left(A\cos\frac{2\pi n}{3} + B\sin\frac{2\pi n}{3}\right)$$

EXAMPLE 4

Find a closed formula for the particular solution of the equation $(E^2 - E - 6)y = 0$ that was tabulated in Example 2.

The characteristic equation of the difference equation $(E^2 - E - 6)y = 0$ is

$$M^2 - M - 6 = 0$$

and its roots are $M_1 = -2$ and $M_2 = 3$. Hence a complete solution of the equation is

$$y = c_1(-2)^n + c_2 3^n$$

Since we are given the starting values $y_0 = 0$ and $y_1 = 1$, the coefficients c_1 and c_2 must satisfy the conditions

$$\begin{aligned} c_1 + c_2 &= 0 \\ -2c_1 + 3c_2 &= 1 \end{aligned}$$

Hence $c_1 = -\frac{1}{5}$, $c_2 = \frac{1}{5}$, and the required solution is

$$y = -\tfrac{1}{5}(-2)^n + \tfrac{1}{5}3^n$$

The next example deals with a difference equation which arises in a number of important applications. In particular, we will need the results of this example in Sec. 7.6 when we investigate the behavior of certain electrical and mechanical systems with many degrees of freedom.

EXAMPLE 5

Show that the equation $(E^2 - 2\lambda E + 1)y = 0$ has the indicated solutions in each of the following cases.

(a) $\lambda < -1$	$y = (-1)^n(A \cosh \mu n + B \sinh \mu n)$	$\cosh \mu = -\lambda$
(b) $\lambda = -1$	$y = (-1)^n(A + Bn)$	
(c) $-1 < \lambda < 1$	$y = A \cos \mu n + B \sin \mu n$	$\cos \mu = \lambda$
(d) $\lambda = 1$	$y = A + Bn$	
(e) $\lambda > 1$	$y = A \cosh \mu n + B \sinh \mu n$	$\cosh \mu = \lambda$

Following our general procedure, we first set up the characteristic equation

(11) $$M^2 - 2\lambda M + 1 = 0$$

and determine its roots

(12) $$M_1, M_2 = \lambda \pm \sqrt{\lambda^2 - 1}$$

The structure of this formula makes it clear why we have to consider the five cases indicated in the statement of the problem. The individual cases are handled as follows.

(a) Since $\lambda < -1$ implies $-\lambda > 1$, it follows that it is possible to define a quantity μ by the equation

$$\cosh \mu = -\lambda$$

Under this substitution the roots (12) become

$$\begin{aligned} M_1, M_2 &= -\cosh \mu \pm \sqrt{\cosh^2 \mu - 1} = -\cosh \mu \pm \sinh \mu \\ &= -(\cosh \mu + \sinh \mu), -(\cosh \mu - \sinh \mu) \\ &= -e^{\mu}, -e^{-\mu} \end{aligned}$$

Therefore

$$\begin{aligned} y &= c_1(-e^{\mu})^n + c_2(-e^{-\mu})^n = (-1)^n(c_1 e^{\mu n} + c_2 e^{-\mu n}) \\ &= (-1)^n[c_1(\cosh \mu n + \sinh \mu n) + c_2(\cosh \mu n - \sinh \mu n)] \\ &= (-1)^n[(c_1 + c_2)\cosh \mu n + (c_1 - c_2)\sinh \mu n] \\ &= (-1)^n(A \cosh \mu n + B \sinh \mu n) \end{aligned}$$

where $A = c_1 + c_2$ and $B = c_1 - c_2$.

(b) In this case $\lambda = -1$, so that $M_1 = M_2 = -1$ and therefore

$$y = A(-1)^n + Bn(-1)^n = (-1)^n(A + Bn)$$

(c) Since $-1 < \lambda < 1$, it is possible to define a quantity μ by the equation

$$\cos \mu = \lambda \qquad 0 < \mu < \pi$$

Under this substitution the roots (12) become

$$\begin{aligned} M_1, M_2 &= \cos \mu \pm \sqrt{\cos^2 \mu - 1} = \cos \mu \pm \sqrt{-\sin^2 \mu} \\ &= \cos \mu + i \sin \mu, \cos \mu - i \sin \mu \\ &= e^{i\mu}, e^{-i\mu} \end{aligned}$$

Therefore

$$\begin{aligned} y &= c_1(e^{i\mu})^n + c_2(e^{-i\mu})^n \\ &= c_1(\cos \mu n + i \sin \mu n) + c_2(\cos \mu n - i \sin \mu n) \\ &= (c_1 + c_2)\cos \mu n + i(c_1 - c_2)\sin \mu n \\ &= A \cos \mu n + B \sin \mu n \end{aligned}$$

where $A = c_1 + c_2$ and $B = i(c_1 - c_2)$.

(d) In this case $\lambda = 1$, so that $M_1 = M_2 = 1$ and therefore

$$y = A + Bn$$

(e) Since $\lambda > 1$, it is possible to define a quantity μ by the equation

$$\cosh \mu = \lambda$$

Under this substitution the roots (12) become

$$M_1, M_2 = \cosh \mu \pm \sqrt{\cosh^2 \mu - 1} = \cosh \mu \pm \sinh \mu = e^{\mu}, e^{-\mu}$$

Therefore

$$\begin{aligned} y &= c_1(e^{\mu})^n + c_2(e^{-\mu})^n \\ &= c_1(\cosh \mu n + \sinh \mu n) + c_2(\cosh \mu n - \sinh \mu n) \\ &= (c_1 + c_2)\cosh \mu n + (c_1 - c_2)\sinh \mu n \\ &= A\cosh \mu n + B \sinh \mu n \end{aligned}$$

where, as in Case **a**, $A = c_1 + c_2$ and $B = c_1 - c_2$.

To solve the nonhomogeneous equation

$$(a_0E^2 + a_1E + a_2)y = \phi(n) \tag{4a}$$

by means of Theorem 3, we must add a particular solution of (4*a*) to a complete solution of the related homogeneous equation. To find the necessary particular solution Y, we use the method of undetermined coefficients very much as we did in solving nonhomogeneous differential equations in Sec. 2.7. And here, as in Sec. 2.7, this method is able to handle only equations in which $\phi(n)$ is a linear combination of terms or products of terms of the form†

$$k^n \qquad \cos kn \qquad \sin kn \qquad k \text{ a constant}$$

and

$$n^k \qquad k \text{ a nonnegative integer}$$

We begin by assuming for Y an arbitrary linear combination of all the terms which arise from $\phi(n)$ by repeatedly applying the operator E. As in the case of differential equations, if any term in the initial choice for Y duplicates a term in the complementary function, it and all terms associated with it in Y (see Table 5.2) must be multiplied by the lowest positive integral power of n that will eliminate all duplications. When the form of Y is thus determined, Y is substituted into the difference equation and the arbitrary constants in Y are chosen to make the resulting equation an identity. The procedure is summarized in Table 5.2.

EXAMPLE 6

Find a complete solution of the equation $(E^2 + 2E - 8)y = 5n + 3(2^n)$.

The characteristic equation in this case is $M^2 + 2M - 8 = 0$, and from its roots $M_1 = 2$ and $M_2 = -4$ we can immediately construct the complementary function

$$c_1 2^n + c_2(-4)^n$$

†In Sec. 2.7 we identified these functions as those that possess only a finite number of linearly independent derivatives. Here they appear as the functions which yield only a finite number of linearly independent terms under repeated applications of the operator E.

TABLE 5.2

Difference equation: $(a_0E^2 + a_1E + a_2)y = \phi(n)$

$\phi(n)$†	Necessary choice for particular solution Y‡
1. α (constant)	A
2. αn^k (k a positive integer)	$A_0n^k + A_1n^{k-1} + \cdots + A_{k-1}n + A_k$
3. αk^n	Ak^n
4. $\alpha \cos kn$ 5. $\alpha \sin kn$	$A \cos kn + B \sin kn$
6. $\alpha n^k l^n \cos mn$ 7. $\alpha n^k l^n \sin mn$	$(A_0n^k + \cdots + A_{k-1}n + A_k)l^n \cos mn$ $+ (B_0n^k + \cdots + B_{k-1}n + B_k)l^n \sin mn$

†When $\phi(n)$ consists of a sum of several terms, the appropriate choice for Y is the sum of the Y expressions corresponding to these terms individually.
‡Whenever a term in any of the Y's listed in this column duplicates a term in the complementary function, all terms in that Y expression must be multiplied by the lowest positive integral power of n sufficient to eliminate all such duplications.

According to Lines 2 and 3 in Table 5.2, the particular integral we would normally try is

$$Y = An + B + C2^n$$

However, since the term $C2^n$ duplicates, i.e., is linearly dependent upon, the term c_12^n in the complementary function, we must modify it, according to the second footnote in Table 5.2, by multiplying it by n before including it in our trial solution. Thus we substitute

$$Y = An + B + Cn2^n$$

into the given equation, getting

$$[A(n+2) + B + C(n+2)2^{n+2}] + 2[A(n+1) + B + C(n+1)2^{n+1}] - 8[An + B + Cn2^n] = 5n + 3(2^n)$$

or, collecting terms,

$$-5An + (4A - 5B) + 12C2^n = 5n + 3(2^n)$$

This will be an identity if and only if

$$-5A = 5 \qquad 4A - 5B = 0 \qquad 12C = 3$$

Hence

$$A = -1 \qquad B = -\tfrac{4}{5} \qquad C = \tfrac{1}{4}$$

Therefore $Y = -n - \frac{4}{5} + \frac{1}{4}n2^n$, and the required solution is

$$y = c_12^n + c_2(-4)^n - n - \tfrac{4}{5} + n2^{n-2}$$

There are numerous problems in which the values of a function defined on a set of integers can be connected by a linear recurrence relation; and when this is the case, a formula for the function can be found by solving the corresponding difference equation. Our next three examples are applications of this sort.

EXAMPLE 7

Find a formula for the sum of the series $s_n = \sum_{k=1}^{n} k^2$.

Clearly, the difference between s_{n+1} and s_n is just the last term in s_{n+1}, namely, $(n+1)^2$. Hence s_n satisfies the nonhomogeneous difference equation

$$s_{n+1} - s_n = (n+1)^2 \qquad \text{or} \qquad (E-1)s_n = n^2 + 2n + 1$$

Moreover, although Tables 5.1 and 5.2 appear to be concerned only with second-order equations, the procedures they outline are correct for difference equations of order 1 as well as for equations of higher order. Hence we set up the characteristic equation $M - 1 = 0$, and from its root we construct the complementary function

$$c_1(1)^n = c_1$$

To find a particular integral we would ordinarily try

$$S_n = An^2 + Bn + C$$

but because the term C duplicates the term c_1 in the complementary function we must multiply our normal choice by n before proceeding. Thus we substitute $S_n = An^3 + Bn^2 + Cn$ into the equation $(E-1)s_n = n^2 + 2n + 1$, getting

$$[A(n+1)^3 + B(n+1)^2 + C(n+1)] - [An^3 + Bn^2 + Cn] = n^2 + 2n + 1$$

or, collecting terms,

$$3An^2 + (3A + 2B)n + (A + B + C) = n^2 + 2n + 1$$

This will be an identity if and only if

$$3A = 1 \qquad 3A + 2B = 2 \qquad A + B + C = 1$$

Hence

$$A = \frac{1}{3} \qquad B = \frac{1}{2} \qquad C = \frac{1}{6} \qquad S_n = \frac{n^3}{3} + \frac{n^2}{2} + \frac{n}{6} = \frac{2n^3 + 3n^2 + n}{6}$$

A complete solution is therefore

$$s_n = c_1 + \frac{2n^3 + 3n^2 + n}{6}$$

To determine c_1 we use the obvious fact that $s_1 = 1$. Thus we must have $1 = c_1 + 1$ or $c_1 = 0$, and the required formula is

$$s_n = \frac{2n^3 + 3n^2 + n}{6} = \frac{n(n+1)(2n+1)}{6}$$

EXAMPLE 8

If D_n is the nth-order determinant

$$\begin{vmatrix} 2 & 2 & 0 & 0 & \cdots & 0 & 0 \\ 1 & 2 & 2 & 0 & \cdots & 0 & 0 \\ 0 & 1 & 2 & 2 & \cdots & 0 & 0 \\ 0 & 0 & 1 & 2 & \cdots & 0 & 0 \\ \cdots & \cdots & \cdots & \cdots & \cdots & \cdots & \cdots \\ 0 & 0 & 0 & 0 & \cdots & 2 & 2 \\ 0 & 0 & 0 & 0 & \cdots & 1 & 2 \end{vmatrix}$$

show that $D_n = 2D_{n-1} - 2D_{n-2}$ and determine D_n as a function of n.

The form of the given recurrence relation suggests that we should begin by expanding D_n in terms of the elements in the first column (or row). Doing this, we obtain

$$D_n = 2\begin{vmatrix} 2 & 2 & 0 & \cdots & 0 & 0 \\ 1 & 2 & 2 & \cdots & 0 & 0 \\ 0 & 1 & 2 & \cdots & 0 & 0 \\ \cdots & \cdots & \cdots & \cdots & \cdots & \cdots \\ 0 & 0 & 0 & \cdots & 2 & 2 \\ 0 & 0 & 0 & \cdots & 1 & 2 \end{vmatrix} - \begin{vmatrix} 2 & 0 & 0 & \cdots & 0 & 0 \\ 1 & 2 & 2 & \cdots & 0 & 0 \\ 0 & 1 & 2 & \cdots & 0 & 0 \\ \cdots & \cdots & \cdots & \cdots & \cdots & \cdots \\ 0 & 0 & 0 & \cdots & 2 & 2 \\ 0 & 0 & 0 & \cdots & 1 & 2 \end{vmatrix}$$

The first of the two determinants on the right is identical in structure to D_n except that it contains only $n - 1$ rows and $n - 1$ columns; in other words, it is just D_{n-1}. The second determinant on the right does not have the form of D_n, but if we expand it in terms of the elements in its first row, we obtain just 2 times a determinant of order $n - 2$ which is of the same structure as D_n, that is, is just D_{n-2}. Thus we have verified the asserted recurrence relation

$$D_n = 2D_{n-1} - 2D_{n-2}$$

or, equivalently,

$$(E^2 - 2E + 2)D_n = 0$$

The characteristic equation of this difference equation is $M^2 - 2M + 2 = 0$, and from its roots $M_1, M_2 = 1 \pm i = \sqrt{2}e^{\pm i\pi/4}$ we can immediately construct the complete solution

$$D_n = (\sqrt{2})^n\left(A \cos\frac{n\pi}{4} + B \sin\frac{n\pi}{4}\right)$$

To determine A and B we return to the definition of D_n and note that $D_1 = 2$ and $D_2 = \begin{vmatrix} 2 & 2 \\ 1 & 2 \end{vmatrix} = 2$. Hence A and B must satisfy the conditions

$$2 = \sqrt{2}\left(A\frac{1}{\sqrt{2}} + B\frac{1}{\sqrt{2}}\right) = A + B$$

$$2 = 2(A \cdot 0 + B) = 2B$$

Thus $A = B = 1$, and the required formula is

$$D_n = 2^{n/2}\left(\cos\frac{n\pi}{4} + \sin\frac{n\pi}{4}\right)$$

EXAMPLE 9

POTENTIALS AT POINTS OF A LADDER-TYPE NETWORK

In the system shown in Fig. 5.4, the point P_0 is kept at the constant potential $V = V_0$ with respect to the ground which is assumed to be at the potential zero. The point P_N is connected to the ground by a wire of zero resistance, so that it remains at the potential $V_N = 0$. What is the potential $V_1, V_2, \ldots, V_{N-1}$ at the respective points $P_1, P_2, \ldots, P_{N-1}$?

According to Kirchoff's first law, the sum of the currents flowing toward any junction in a network must equal the sum of the currents flowing away from that junction. Hence at a general point P_{n+1} (Fig. 5.4*b*) we have

$$i_n = i_{n+1} + I_{n+1}$$

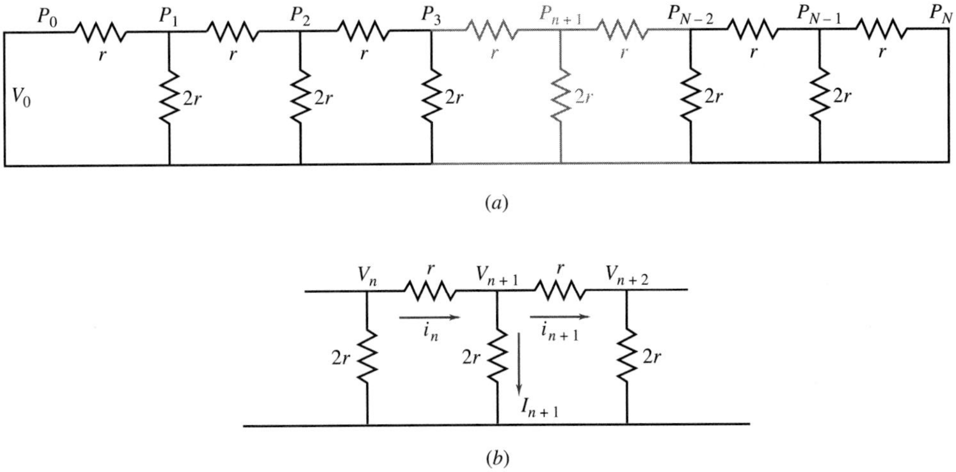

FIGURE 5.4
A ladder-type network with identical loops. (Although the network shown in Fig. 5.4*a* appears to contain exactly seven loops, the number of loops is actually indefinite. This is implied by the fact that the central portion of the figure is drawn with lighter lines; this convention will be used throughout this book to suggest a configuration of indefinite extent.)

If we now replace each current by its equivalent according to Ohm's law, $I = V/R$, where V is the potential difference between the ends of the wire through which the current I is flowing, we obtain

$$\frac{V_n - V_{n+1}}{r} = \frac{V_{n+1} - V_{n+2}}{r} + \frac{V_{n+1} - 0}{2r}$$

or, simplifying and collecting terms,

(13) $$V_{n+2} - \tfrac{5}{2}V_{n+1} + V_n = 0 \qquad n = 0, 1, 2, \ldots, N-2$$

Equation (13) constitutes a system of $N - 1$ linear algebraic equations from which the unknown potentials $V_1, V_2, \ldots, V_{N-1}$ can be found by completely elementary, though very tedious, steps for any particular value of N. However, it is much simpler and much more elegant to regard Eq. (13) as a second-order difference equation

(14) $$(E^2 - \tfrac{5}{2}E + 1)V_n = 0$$

subject to the end conditions $V(0) = V_0$ and $V(N) = 0$, which will serve to determine the values of the arbitrary constants appearing in any complete solution of (14).

Taking this point of view, we first set up the characteristic equation of Eq. (14), namely,

$$M^2 - \tfrac{5}{2}M + 1 = 0$$

From its roots, $M_1 = \frac{1}{2}$ and $M_2 = 2$, we then construct a complete solution of Eq. (14):

(15) $$V_n = A(\tfrac{1}{2})^n + B(2^n)$$

When the boundary conditions at $n = 0$ and $n = N$ are imposed on the complete solution (15), we obtain the equations

$$A + B = V_0 \qquad \text{and} \qquad A(\tfrac{1}{2})^N + B(2^N) = 0$$

When these are solved simultaneously, we find that

$$A = \frac{2^{2N}}{2^{2N} - 1} V_0 \qquad \text{and} \qquad B = -\frac{1}{2^{2N} - 1} V_0$$

The final solution is therefore

$$V_n = \left(\frac{2^{2N}}{2^n} - 2^n \right) \frac{V_0}{2^{2N} - 1} \qquad n = 0, 1, 2, \ldots, N$$

EXERCISES

Express each of the following equations as an equation of the form $F(E)y = \phi(x)$.

1. $\Delta^2 y - 3\,\Delta y + 2y = 0$
2. $\Delta^3 y - 6\,\Delta^2 y + 11\,\Delta y - 6y = x^2$
3. $(\Delta^2 + 2\,\Delta + 2)y = 3^x$
4. $y_{n+2} = y_{n+1} + y_n$
5. If $F(E)$ is the expression obtained from $f(\Delta)$ by the substitution $\Delta = E - 1$, what is the relation between the roots of the equations $f(\Delta) = 0$ and $F(E) = 0$?

Verify that each of the following equations has the indicated solution.

6. $(E^2 + 7E + 12)y = 0 \qquad y = c_1(-3)^n + c_2(-4)^n$
7. $(E^3 + 2E^2 - 5E - 6)y = 0$
$y = c_1(-1)^n + c_2 2^n + c_3(-3)^n$
8. $(E^2 + E + 1)y = 0$
$y = c_1 \cos(2\pi n/3) + c_2 \sin(2\pi n/3)$

Tabulate from $n = 0$ to $n = 5$ the solution of each of the following equations determined by the given starting values.

9. $(E^2 - 2E + 2)y = 0;\ y_0 = 1,\ y_1 = -1$
10. $(E^3 - E^2 + E - 1)y = 0;\ y_0 = 1,\ y_1 = 0,\ y_2 = 2$
11. $(E^2 - E + 2)y = 0;\ y_0 = 1,\ \Delta y_0 = 3$
12. $(E^2 - 2E + 2n)y = 0;\ y_0 = 0,\ y_1 = 1$
13. $(E^2 - nE + 2)y = 0,\ y_0 = y_1 = 1$
14. $(E^2 - 2E + 2)y = n,\ y_0 = y_1 = 0$
15. Tabulate the solution of $y_{n+2} - 3y_{n+1} + 2y_n = 0$ from $n = -4$ to $n = 4$ if $y_0 = 1$ and $y_1 = -1$.
16. Tabulate as far as possible the solution of $(n - 4)y_{n+2} - y_{n+1} + (n + 3)y_n = 0$ if $y_0 = 1$ and $y_1 = 2$.

Find a complete solution of each of the following equations.

17. $(E^2 - 3E - 10)y = 3(2^n)$
18. $(E^2 - 6E + 9)y = 6$
19. $(\Delta^2 - 7\Delta + 12)y = 3^n$
20. $(E^3 + 2E^2 - 4E - 8)y = 3^n$
21. $(E^2 - E - 2)y = 2^{2n}$
22. $(E^2 + 2E - 3)y = 2$
23. $(E^2 - E - 6)y = 6n^2$
24. $(E^3 - 4E^2 + 6E - 4)y = n + 1$
25. $(E^2 + 2E + 1)y = 1$
26. $(E^2 - 3E + 2)y = 2^n + 2^{-n}$
27. $(E^2 + 4)y = \cos n$
28. $(E^2 - 4E + 4)y = 2^n$
29. $(E^4 + 8E^2 - 9)y = 20$
30. $(E^4 + 10E^2 + 9)y = \cos \dfrac{\pi n}{4}$
31. Find a formula for each of the solutions tabulated in Exercises 9–11.
32. If $y_1(x)$ and $y_2(x)$ are any two solutions of the general linear second-order difference equation

$$[a_0(x)E^2 + a_1(x)E + a_2(x)]y = 0$$

show that Casorati's determinant $C[y_1(x), y_2(x)]$ satisfies the relation

$$[a_0(x)E - a_2(x)]C = 0$$

Hint: Write down the conditions that both $y_1(x)$ and $y_2(x)$ satisfy the given equation; then eliminate the terms in $Ey_1(x)$ and $Ey_2(x)$.
33. Prove Theorem 1.
34. Prove Theorem 2. *Hint:* Set up Casorati's determinant for (y_1, y_3) and (y_2, y_3) and use the result of Exercise 32.
35. Prove Theorem 3.
36. Express each of the following nth-order determinants as an explicit function of n:

$$\text{(a)}\ \begin{vmatrix} 5 & 2 & 0 & \cdots & 0 & 0 \\ 3 & 5 & 2 & \cdots & 0 & 0 \\ 0 & 3 & 5 & \cdots & 0 & 0 \\ \cdots & \cdots & \cdots & \cdots & \cdots & \cdots \\ 0 & 0 & 0 & \cdots & 5 & 2 \\ 0 & 0 & 0 & \cdots & 3 & 5 \end{vmatrix} \qquad \text{(b)}\ \begin{vmatrix} 5 & 1 & 0 & \cdots & 0 & 0 \\ 6 & 5 & 1 & \cdots & 0 & 0 \\ 0 & 6 & 5 & \cdots & 0 & 0 \\ \cdots & \cdots & \cdots & \cdots & \cdots & \cdots \\ 0 & 0 & 0 & \cdots & 5 & 1 \\ 0 & 0 & 0 & \cdots & 6 & 5 \end{vmatrix}$$

$$\text{(c)}\ \begin{vmatrix} 4 & 4 & 0 & \cdots & 0 & 0 \\ 1 & 4 & 4 & \cdots & 0 & 0 \\ 0 & 1 & 4 & \cdots & 0 & 0 \\ \cdots & \cdots & \cdots & \cdots & \cdots & \cdots \\ 0 & 0 & 0 & \cdots & 4 & 4 \\ 0 & 0 & 0 & \cdots & 1 & 4 \end{vmatrix} \qquad \text{(d)}\ \begin{vmatrix} 2 & 1 & 0 & \cdots & 0 & 0 \\ 4 & 2 & 1 & \cdots & 0 & 0 \\ 0 & 4 & 2 & \cdots & 0 & 0 \\ \cdots & \cdots & \cdots & \cdots & \cdots & \cdots \\ 0 & 0 & 0 & \cdots & 2 & 1 \\ 0 & 0 & 0 & \cdots & 4 & 2 \end{vmatrix}$$

37. Find a formula for the sum of each of the following series:

(a) $\sum_{i=1}^{n} i$ **(b)** $\sum_{i=1}^{n} (2i-1)^2$ **(c)** $\sum_{i=1}^{n} i^3$

38. Find a formula for the sum of each of the following series:

(a) $\sum_{k=1}^{n} kr^k$ **(b)** $\sum_{k=1}^{n} \sin kr$ **(c)** $\sum_{k=1}^{n} \cos kr$

where r is a constant.

39. Show that the nth-order determinant

$$D_n = \begin{vmatrix} a & 1 & 0 & \cdots & 0 & 0 \\ 1 & a & 1 & \cdots & 0 & 0 \\ 0 & 1 & a & \cdots & 0 & 0 \\ \cdots & \cdots & \cdots & \cdots & \cdots & \cdots \\ 0 & 0 & 0 & \cdots & a & 1 \\ 0 & 0 & 0 & \cdots & 1 & a \end{vmatrix}$$

satisfies the difference equation $(E^2 - aE + 1)D = 0$. Hence show that when $a > 2$,

$$D_n = \frac{\sinh (n+1)\mu}{\sinh \mu} \qquad \text{where } \cosh \mu = \frac{a}{2}$$

What is D_n if $a = 2$? $-2 < a < 2$? $a = -2$? $a < -2$?

40. Work Example 9 if there is a resistor of resistance $2r$ in the wire connecting P_N to the ground.

41. Work Example 9 if both P_0 and P_N are maintained at the potential V_0.

42. Work Example 9 if the common value of the resistance in the vertical branches is kr, $k > 0$.

43. Show that the integral

$$I_n(\lambda) = \int_0^{\pi} \frac{\cos nt - \cos n\lambda}{\cos t - \cos \lambda}\, dt \qquad n = 0, 1, 2, 3, \ldots$$

satisfies the equation $[E^2 - (2 \cos \lambda)E + 1]I_n = 0$. Solve this equation and find I_n as a function of n.

44. A system consists of N spring-connected masses, as shown in Fig. 5.5. What is the displacement of each mass from its original position when the system is again in equilibrium after a force F_0 is applied to the right-hand end?

45. Work Exercise 44 if the direction of the force F_0 is reversed.

46. Verify that when the characteristic equation (7) has complex roots, the particular solutions $y_1 = r^n \cos \theta n$ and $y_2 = r^n \sin \theta n$ are such that $C(y_1, y_2) \neq 0$.

47. Solve Example 9 using the results of Example 5 to solve Eq. (14).

48. Extend the procedure suggested in Exercise 17, Sec. 5.5, to the problem of fitting a curve of the form $y = Ae^{ax} + Be^{bx}$ to a set of y values at equally spaced values of x. Can the procedure be extended to the fitting of a curve of the form $y = Ae^{ax} + Be^{bx} + \cdots + Ke^{kx}$?

49. If $f(x)$ has the known Maclaurin expansion $f(x) = \sum_{n=0}^{\infty} a_n x^n$ and if $f(x)/(A + Bx + Cx^2) = \sum_{n=0}^{\infty} b_n x^n$, show that the b's satisfy the difference equation $Ab_{n+2} + Bb_{n+1} + Cb_n = a_{n+2}$, with the initial conditions $b_0 = a_0/A$ and $b_1 = (Aa_1 - Ba_0)/A^2$.

50. Using the result of Exercise 49, find the Maclaurin expansion of

(a) $\dfrac{1}{2 - 2x + x^2}$ **(b)** $\dfrac{1}{1 - (2\cos\theta)x + x^2}$

51. Solve the system of difference equations

$$\begin{aligned} y_{n+1} + 3y_n - z_n &= 10 \\ -2y_{n+1} + z_{n+1} + 2z_n &= -2 \end{aligned} \qquad y_0 = -3,\ z_0 = 2$$

Hint: Express each equation in terms of the operator E, then formally eliminate one of the unknowns.

52. Solve the system of difference equations

$$\begin{aligned} y_{n+1} - 4y_n - z_n &= 0 \\ y_n + z_{n+1} - 2z_n &= 2^n \end{aligned} \qquad y_0 = 1,\ z_0 = 2$$

53. To each function ϕ whose domain is a set of consecutive integers S there corresponds an **indefinite sum**

$$\theta(n) = \sum_{k=j}^{n-1} \phi(k) + C \qquad C \text{ a parameter}$$

(a) Show that $\Delta\theta(n) = \phi(n)$, j and $n - 1$ in S.
(b) What property is characteristic of Δ^{-1}?

54. If $y_k = au_k + bv_k$ is a complementary function of the nonhomogeneous second-order difference equation

FIGURE 5.5
(See explanation of convention used in Fig. 5.4.)

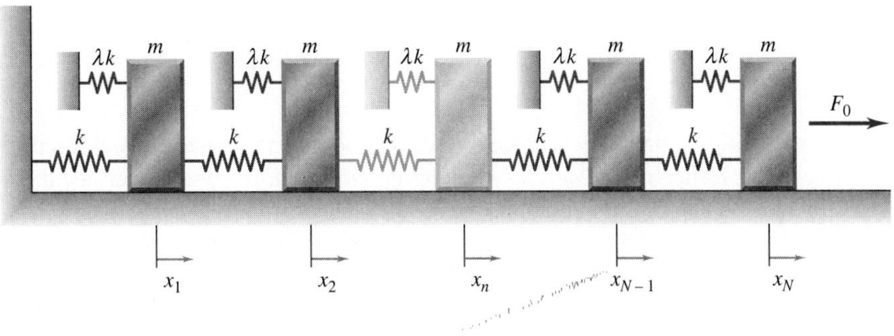

$p_0(k)y_{k+2} + p_1(k)y_{k+1} + p_2(k)y_k = f_k$ on a set of consecutive integers S, and if variation of parameters is used to replace a by A_k and b by B_k in the complementary function, show that the new function thus obtained

$$Y_k = A_k u_k + B_k v_k$$

will be a particular solution of the nonhomogeneous equation if A_k and B_k are functions of k such that

$$\Delta A_k u_{k+1} + \Delta B_k v_{k+1} = 0$$

$$\Delta A_k u_{k+2} + \Delta B_k v_{k+2} = \frac{f_k}{p_0(k)}$$

Hint: Revise the reasoning of Sec. 2.4 starting with the observation that

$$Y_{k+1} = A_k u_{k+1} + B_k v_{k+1} + \Delta A_k u_{k+1} + \Delta B_k v_{k+1}$$

55. In Exercise 54, assume $p_0(k) \equiv 1$ and use the system of equations in ΔA_k and ΔB_k to express A_k and B_k as indefinite sums. *Hint:* See Exercise 53.

56. Use variation of parameters (Exercises 54 and 55) to find a complete solution of the nonhomogeneous difference equation $y_{k+2} - 7y_{k+1} + 6y_k = k$.

57. Reduce the particular solution of the difference equation in Exercise 56, as given in terms of series by variation of parameters, to a specific polynomial in k. *Hint:* Use Exercise 38(**a**).

Find a complementary function for each of the following systems of difference equations.

58. $\begin{aligned} y_{k+1} + y_k + z_k &= f_k \\ 5y_{k+1} - 4y_k + z_{k+1} &= g_k \end{aligned}$

59. $\begin{aligned} y_{k+1} - 4y_k + z_{k+1} - 6z_k &= \ln c(k+1) \\ 2y_{k+1} + y_k + 3z_{k+1} - 2z_k &= 0 \end{aligned}$

Hint: Seek solutions of the form $\mathbf{u}_k = \begin{bmatrix} y_k \\ z_k \end{bmatrix} = \begin{bmatrix} a \\ b \end{bmatrix} M^k$ (see Sec. 4.5).

60. Use variation of parameters to find a particular solution of the system in Exercise 59 of the form $\mathbf{Y}_k = A_k\mathbf{u}_k + B_k\mathbf{v}_k$. (First work Exercise 59.)

5.8 DIFFERENCE EQUATIONS AND THE NUMERICAL SOLUTION OF DIFFERENTIAL EQUATIONS

One of the many uses of difference equations is in analyzing the accuracy of methods for the numerical solution of differential equations. Although a detailed discussion of this topic must be left to more advanced texts on numerical analysis, we can make an introductory investigation of this problem by considering the solution of the simple equation

$$(1) \qquad y' = Ay \qquad A \text{ a constant}$$

by the predictor-corrector formulas used in Milne's method.

Milne's method, as we know, makes use of the predictor formula [Eq. (15), Sec. 5.6]

$$y_{n+1} = y_{n-3} + \frac{4h}{3}(2y'_{n-2} - y'_{n-1} + 2y'_n)$$

to obtain a first estimate of y_{n+1} and then iterates the corrector formula [Eq. (18), Sec. 5.6]

$$y_{n+1} = y_{n-1} + \frac{h}{3}(y'_{n-1} + 4y'_n + y'_{n+1})$$

to determine the value of y_{n+1} which is finally accepted. Applied to the particular differential equation (1), Milne's corrector formula becomes

$$y_{n+1} = y_{n-1} + \frac{h}{3}(Ay_{n-1} + 4Ay_n + Ay_{n+1})$$

which is simply a linear recurrence relation connecting the values of y at three arbitrary consecutive values of x. Hence *the numerical solution given by Milne's method is just a particular solution of the second-order difference equation*

$$(2) \qquad \left(1 - \frac{Ah}{3}\right)y_{n+1} - \frac{4Ah}{3}y_n - \left(1 + \frac{Ah}{3}\right)y_{n-1} = 0 \qquad 1 \pm \frac{Ah}{3} \neq 0$$

The restriction $1 \pm Ah/3 \neq 0$ is not serious, since our concern is with the properties of Milne's method when h is sufficiently small, say $h < 3/|A|$. If, for convenience, we put $Ah/3 = \lambda$, the characteristic equation of Eq. (2) becomes

$$(1 - \lambda)M^2 - 4\lambda M - (1 + \lambda) = 0$$

and its roots are

$$M_1, M_2 = \frac{2\lambda \pm \sqrt{1 + 3\lambda^2}}{1 - \lambda} \qquad \lambda \neq 1$$

Hence the tabular function y_n, which is our approximation to the solution of (1), is contained in the complete solution

$$y_n = c_1 M_1^n + c_2 M_2^n$$

and as $h \to 0$, this particular solution should presumably approach the exact solution of (1), namely,

(3) $$y = y_0 e^{A(x - x_0)}$$

To determine under what conditions, if any, this will be the case, it is convenient to expand M_1 and M_2 in powers of λ, using the binomial theorem, and then retain only the lowest power of λ in the expansion of each root. Doing this, noting that each of the expansions we need is convergent if $3\lambda^2 < 1$, that is, if

$$3\left(\frac{Ah}{3}\right)^2 < 1 \qquad \text{or} \qquad 0 < h < \frac{\sqrt{3}}{|A|}$$

we have

$$\begin{aligned} M_1, M_2 &= [2\lambda \pm (1 + 3\lambda^2)^{1/2}](1 - \lambda)^{-1} \\ &= [2\lambda \pm (1 + \tfrac{3}{2}\lambda^2 + \cdots)](1 + \lambda + \lambda^2 + \cdots) \\ &= (2\lambda \pm 1 + \cdots)(1 + \lambda + \cdots) \end{aligned}$$

and

$$M_1 = 1 + 3\lambda + \cdots \qquad M_2 = -1 + \lambda + \cdots$$

Hence

$$y_n \doteq c_1(1 + 3\lambda)^n + c_2(-1 + \lambda)^n$$

or, finally, since $\lambda = Ah/3$ and $n = (x_n - x_0)/h$,

(4) $$y_n \doteq c_1(1 + Ah)^{(x_n - x_0)/h} + c_2(-1)^n\left(1 - \frac{Ah}{3}\right)^{(x_n - x_0)/h}$$

As $h \to 0$, each term in (4) contains an indeterminate of the form 1^∞. Evaluating these by L'Hospital's rule, or recalling from calculus that

$$\lim_{z \to 0} (1 + z)^{b/z} = e^b$$

we see that for small values of h we have the approximation

(5) $$y_n \doteq c_1 e^{A(x_n - x_0)} + c_2(-1)^n e^{-A(x_n - x_0)/3}$$

A comparison of Eqs. (5) and (3) shows that if our initial data were perfectly accurate and if there were no round-off errors in our calculations, the correct values of c_1 and c_2 would be y_0 and 0, respectively, and (5) would reduce to the exact solution (3) of the differential equation $y' = Ay$ with initial condition $y(x_0) = y_0$. However, because of inevitable errors of various kinds, c_2 will in general not be zero. Thus the numerical solution we obtain is actually the sum of an approximation to the actual solution (3) plus a so-called **parasitic solution** which for small values of h is approximately $c_2(-1)^n e^{-A(x_n - x_0)/3}$.

If $A > 0$, the parasitic solution in (5) is a decreasing exponential function of x_n which soon becomes vanishingly small. In this case, the procedure is said to be **numerically stable.** On the other hand, if $A < 0$, the numerical value of the parasitic solution increases exponentially (while continually alternating in sign), and sooner or later, depending on the size of c_2 and A, it becomes the principal part of the numerical solution we obtain. In this case, the procedure is said to be **numerically unstable.**

Essentially the same analysis, leading to the same conclusions, can be made for the nonhomogeneous equation $y' = Ay + f(x)$. The only difference in this case is that the difference equation arising from Milne's corrector formula is nonhomogeneous rather than homogeneous.

When A is a constant, the equation $y' = Ay$ is so simple that numerical methods for solving it are completely unnecessary, and therefore a discussion of their stability is irrelevant. On the other hand, for the general first-order differential equation

$$y' = f(x, y) \tag{6}$$

numerical methods of solution are often required, and the question of their stability is important. Unfortunately, when Milne's corrector formula is applied to Eq. (6), the resulting difference equation is, in general, nonlinear, and the preceding analysis is not immediately applicable. However, the problem can be approximated by one which does involve a linear difference equation if we first replace $f(x, y)$ by its Taylor expansion around the point (x_0, y_0), namely,

$$\begin{aligned} y' &= f(x, y) \\ &= f(x_0, y_0) + \left[\frac{\partial f}{\partial x}\bigg|_{x_0, y_0} (x - x_0) + \frac{\partial f}{\partial y}\bigg|_{x_0, y_0} (y - y_0) \right] \\ &\quad + \frac{1}{2!}\left[\frac{\partial^2 f}{\partial x^2}\bigg|_{x_0, y_0} (x - x_0)^2 + 2\frac{\partial^2 f}{\partial x\, \partial y}\bigg|_{x_0, y_0} (x - x_0)(y - y_0) + \frac{\partial^2 f}{\partial y^2}\bigg|_{x_0, y_0} (y - y_0)^2 \right] + \cdots \end{aligned}$$

If we now neglect powers of $x - x_0$ and $y - y_0$ higher than the first, then in a suitable neighborhood of (x_0, y_0), the differential equation $y' = f(x, y)$ becomes approximately

$$y' \equiv (y - y_0)' = \frac{\partial f}{\partial y}\bigg|_{x_0, y_0} (y - y_0) + \left[f(x_0, y_0) + \frac{\partial f}{\partial x}\bigg|_{x_0, y_0} (x - x_0) \right]$$

which is a nonhomogeneous differential equation of the form we have just considered. By the preceding discussion, the solution given by Milne's corrector formula will be sable if the coefficient of $y - y_0$, namely,

$$\frac{\partial f}{\partial y}\bigg|_{x_0, y_0}$$

is positive, and unstable if it is negative.

If any of the other predictor-corrector methods we described in Sec. 5.6 (see Exercises 15 and 17, Sec. 5.6) is applied to Eq. (1), the corresponding corrector formula becomes a homogeneous constant-coefficient linear difference equation and the preceding analysis can be repeated. The chief difference between the various difference equations thus obtained is that some may be of order greater than 2. This means that a characteristic equation may be encountered which has more than two roots, so that the related complementary function has the form

$$y_n = c_1 M_1^n + c_2 M_2^n + \cdots + c_k M_k^n$$

As with Milne's method, one of these terms will approach the exact solution of Eq. (1) as $h \to 0$; the others will be parasitic solutions. If the absolute values of the extraneous roots from which the parasitic solutions arise are all less than 1, then as n or x increases, each will decay exponentially and the process will yield a meaningful approximation to the exact solution. On the other hand, if the absolute value of even one of the extraneous roots is greater than 1, then the corresponding parasitic solution will increase exponentially and the process will be numerically unstable.

Since numerical instability is commonly observed in predictor-corrector methods, most workers prefer the Runge-Kutta method, which involves no numerical instability, at least for sufficiently small values of the tabular interval h.

EXERCISES

1. **(a)** Show that Euler's method for the numerical solution of differential equations [Eq. (3), Sec. 5.6] is numerically stable for the differential equation $y' = Ay$ for all values of A.
(b) Verify that as $h \to 0$, the general solution of the difference equation involved in Part **(a)** converges to the general solution of $y' = Ay$ for all values of A.

2. Work Exercise 1 for the modified Euler method [Eq. (4), Sec. 5.6].

3. **(a)** Discuss the stability of the closed formula given in Exercise 15, Sec. 5.6, in relation to the equation $y' = Ay$.
(b) Verify that as $h \to 0$, one of the solutions of the difference equation involved in Part **(a)** converges to the general solution of $y' = Ay$.

4. Show that the corrector formula

$$y_{n+1} = y_n + \frac{h}{24}(9y'_{n+1} + 19y'_n - y'_{n-1} + y'_{n-2})$$

is stable in relation to the equation $y' = Ay$ if $\gamma \equiv Ah/24$ is a sufficiently small negative number. *Hint:* Note that from the characteristic equation γ can easily be plotted as a function of M.

5. Discuss the stability of Milne's predictor formula [Eq. (15), Sec. 5.6] in relation to the equation $y' = Ay$. *Hint:* Note the hint given in Exercise 4.

6. **(a)** Show that the closed formula given in Exercise 16, Sec. 5.6, is numerically stable for the equation $y'' = Ay$ for all values of A.
(b) Verify that as $h \to 0$, the general solution of the difference equation involved in Part **(a)** converges to the general solution of $y'' = Ay$.

5.9 GENERATING FUNCTIONS AND THE G AND Z TRANSFORMATIONS

In Sec. 5.7 we developed methods for solving homogeneous and nonhomogeneous linear constant-coefficient difference equations. However, as we might expect, those methods are not the only ones available, and in this section we shall discuss an alternative procedure that is sometimes preferable. The central idea behind this new method is the notion of a *generating function.*

DEFINITION 1 Let f be a function defined on the set of nonnegative integers.† Then for all values of s for which it converges, the infinite series

$$G(f_n) = \sum_{n=0}^{\infty} f_n s^n$$

is called the **generating function** of the sequence f_n.

†In the important field of *signal processing,* the values of f are usually the values of a continuous signal sampled at regularly spaced intervals of t.

The operation of constructing the generating function $G(f_n)$ of a sequence f_n we shall call the G **transformation.** This casts G in the role of a functional whose value at f_n is a function $F(s)$ called the G **transform** of f_n. Conversely, f_n is called the **inverse transform** of $F(s)$. The domain of F is the set of s values for which the series representation of G converges. In general, if a sequence is denoted by a lowercase letter, we shall denote its G transform by the corresponding capital letter. Thus, to indicate that the G transform of f_n is $F(s)$, we write

$$G(f_n) = F(s)$$

and to indicate that f_n is the inverse G transform of $F(s)$, we write

$$f_n = G^{-1}[F(s)]$$

EXAMPLE 1

If $f_n = 1$, for $n = 0, 1, 2, \ldots$, then

$$G(1) = 1 + s + s^2 + \cdots + s^n + \cdots$$

and by the familiar formula for the sum of an infinite geometric series, we have

$$G(1) = \frac{1}{1-s} \qquad |s| < 1$$

EXAMPLE 2

If $f_n = a^n$ for $n = 0, 1, 2, \ldots$, then

$$G(a^n) = 1 + as + a^2s^2 + \cdots + a^ns^n + \cdots = \frac{1}{1-as} \qquad |s| < \frac{1}{|a|}$$

EXAMPLE 3

Express the G transform of f_{n+1}, $n = 0, 1, 2, \ldots$, in terms of the G transform of f_n.

By definition,

$$G(f_{n+1}) = f_1 + f_2 s + \cdots + f_{n+1}s^n + \cdots$$

If we multiply and divide the series on the right by s and then add and subtract f_0 in the numerator of the resulting fraction, we obtain the required transform:

$$G(f_{n+1}) = \frac{-f_0 + (f_0 + f_1 s + f_2 s^2 + \cdots + f_{n+1}s^{n+1} + \cdots)}{s} = \frac{G(f_n) - f_0}{s}$$

EXAMPLE 4

What is the G transform of f_{n+2}?

By definition,

$$G(f_{n+2}) = f_2 + f_3 s + \cdots + f_{n+2}s^n \cdots$$

This time, to build up $G(f_n)$, we multiply and divide the right-hand side by s^2 and then add and subtract $f_0 + f_1 s$ in the numerator of the resulting fraction, getting

$$G(f_{n+2}) = \frac{-f_0 - f_1 s + (f_0 + f_1 s + f_2 s^2 + \cdots + f_{n+2}s^{n+2} + \cdots)}{s^2}$$

$$= \frac{G(f_n) - f_0 - f_1 s}{s^2}$$

Examples 3 and 4 both suggest and outline the proof of the very important result contained in the following theorem.

THEOREM 1 $G(f_{n+r}) = \dfrac{1}{s^r}\left[G(f_n) - \sum_{i=0}^{r-1} f_i s^i\right]$

To use the G transformation to solve difference equations we must be able to apply it to linear combinations of terms; hence it is important to know that it is a linear operator. The next theorem assures us that this is indeed the case.

THEOREM 2 $G(af_n \pm bh_n) = aG(f_n) \pm bG(h_n)$

PROOF By definition,

(1) $$G(af_n \pm bh_n) = \sum_{n=0}^{\infty} (af_n \pm bh_n)s^n$$

Now every power series converges absolutely within its region of convergence (Theorem 2, Sec. 2.9). Moreover, the terms of an absolutely convergent series can be rearranged at will without altering the sum of the series. Hence we can group the f terms in (1) in one series and the h terms in another and then factor a and b from the respective series, getting

$$G(af_n \pm bh_n) = a\sum_{n=0}^{\infty} f_n s^n \pm b\sum_{n=0}^{\infty} h_n s^n = aG(f_n) \pm bG(h_n)$$

With Theorem 2 and the results of Examples 1–4, we can now illustrate the general procedure for using generating functions to solve linear constant-coefficient difference equations.

EXAMPLE 5

Find the solution of the difference equation $y_{n+2} + y_{n+1} - 6y_n = 1$ for which $y_0 = 0$ and $y_1 = 5$.

We begin by using Theorem 2 to take the G transform of each member of the given equation

$$G(y_{n+2} + y_{n+1} - 6y_n) = G(1)$$

and from this,

$$G(y_{n+2}) + G(y_{n+1}) - 6G(y_n) = G(1)$$

Next, using the results of Examples 1, 3, and 4, together with the given initial conditions, we have

$$\frac{G(y_n) - 0 - 5s}{s^2} + \frac{G(y_n) - 0}{s} - 6G(y_n) = \frac{1}{1-s}$$

From this, by collecting terms and solving for $G(y_n)$ we obtain

$$(1 + s - 6s^2)G(y_n) = 5s + \frac{s^2}{1-s}$$

and

$$G(y_n) = \frac{5s}{(1-2s)(1+3s)} + \frac{s^2}{(1-s)(1-2s)(1+3s)} = F(s)$$

If we now had a suitable table of sequences f_n and their corresponding G transforms, $G(f_n) = F(s)$ (analogous to a table of numbers and their logarithms), we could simply look up $F(s)$ and immediately read the sequence f_n of which it was the G transform. Since at this stage we do not have such a table, we must do the next best thing and attempt to separate $F(s)$ into simpler terms whose inverses we do know. Using the method of partial fractions to accomplish this, we find that

$$G(y_n) = \left(\frac{1}{1-2s} - \frac{1}{1+3s}\right) + \frac{1}{20}\left(-\frac{5}{1-s} + \frac{4}{1-2s} + \frac{1}{1+3s}\right)$$

The inverse of each of the terms which now appear in the expression for $G(y_n)$ can be found immediately by using the results of Examples 1 and 2, and we have

$$\begin{aligned} y_n &= [2^n - (-3)^n] + \tfrac{1}{20}[-5 + 4 \cdot 2^n + (-3)^n] \\ &= -\tfrac{1}{4} + \tfrac{6}{5}2^n - \tfrac{19}{20}(-3)^n \end{aligned}$$

It is easy to verify that the last expression does satisfy the initial conditions that $y_0 = 0$ and $y_1 = 5$.

Example 5 illustrates the important fact that the use of G transforms to solve difference equations introduces the initial conditions at the outset and does not require that we first find a complete solution and then particularize its arbitrary constants.

The following theorem gives us another important property of the G transformation.

THEOREM 3 $G(nf_n) = s\dfrac{d}{ds}[G(f_n)]$

PROOF By definition,

$$\begin{aligned} G(nf_n) &= 0 + f_1 s + 2f_2 s^2 + \cdots + nf_n s^n + \cdots \\ &= s(f_1 + 2f_2 s + \cdots + nf_n s^{n-1} + \cdots) \\ &= s\frac{d}{ds}(f_0 + f_1 s + f_2 s^2 + \cdots + f_n s^n + \cdots) \\ &= s\frac{d}{ds}[G(f_n)] \end{aligned}$$

By applying Theorem 3 r times to the sequence $n^r f_n$, we obtain the following corollary of Theorem 3.

COROLLARY 1 $G(n^r f_n) = \left(s\dfrac{d}{ds}\right)^r G(f_n)$

EXAMPLE 6

What are $G(n)$ and $G(n^2)$?

Since $G(n) = G(n \cdot 1)$, we have, by Theorem 3 and Example 1,

$$G(n) = s\frac{d}{ds}[G(1)] = s\frac{d}{ds}\left(\frac{1}{1-s}\right) = \frac{s}{(1-s)^2}$$

Similarly, since $G(n^2) = G(n \cdot n)$ we have

$$G(n^2) = s\frac{d}{ds}[G(n)] = s\frac{d}{ds}\left[\frac{s}{(1-s)^2}\right] = \frac{s(1+s)}{(1-s)^3}$$

It should be clear from Example 5 that the use of G transforms to solve linear constant-coefficient difference equations leads inevitably to transforms containing terms of the form

$$F(s) = \frac{A + Bs}{C + Ds + Es^2} \qquad E \neq 0 \tag{2}$$

These can always be broken up into partial fractions whose inverses can be written down immediately, but it would be desirable to have formulas that could be applied without first resorting to partial fractions. In doing this there are three cases to consider, according as the zeros of the denominator in (2) are

1. Real and distinct
2. Conjugate complex
3. Real and repeated.

In no case can a root of $Es^2 + Ds + C = 0$ be $s = 0$ because in Definition 1 the series representing G must be *infinite.* Hence, in discussing cases, we may without loss of generality assume that $C = 1$ and then consider separately the two cases $A = 1$, $B = 0$ and $A = 0$, $B = 1$. The next theorem lists the required results.

THEOREM 4 If r_1 and r_2 are the roots of the equation $x^2 + Dx + E = 0$, then

1. If r_1 and r_2 are real and distinct,

(a) $$G^{-1}\left[\frac{1}{(1 - r_1 s)(1 - r_2 s)}\right] = \frac{r_1^{n+1} - r_2^{n+1}}{r_1 - r_2}$$

(b) $$G^{-1}\left[\frac{s}{(1 - r_1 s)(1 - r_2 s)}\right] = \frac{r_1^n - r_2^n}{r_1 - r_2}$$

2. If r_1 and r_2 are conjugate complex numbers such that $r_1, r_2 = \rho e^{\pm i\phi}$,

(a) $$G^{-1}\left[\frac{1}{(1 - r_1 s)(1 - r_2 s)}\right] = \rho^n \frac{\sin (n + 1)\phi}{\sin \phi}$$

(b) $$G^{-1}\left[\frac{s}{(1 - r_1 s)(1 - r_2 s)}\right] = \rho^{n-1} \frac{\sin n\phi}{\sin \phi}$$

3. If $r_1 = r_2 = r$,

(a) $$G^{-1}\left[\frac{1}{(1 - rs)^2}\right] = (n + 1)r^n$$

(b) $$G^{-1}\left[\frac{s}{(1 - rs)^2}\right] = nr^{n-1}$$

PROOF In Cases **1** and **2** it is easy to verify that

$$\frac{1}{(1 - r_1 s)(1 - r_2 s)} = \frac{1}{r_1 - r_2}\left(\frac{r_1}{1 - r_1 s} - \frac{r_2}{1 - r_2 s}\right)$$

Hence, by Example 2,

$$G^{-1}\left[\frac{1}{(1 - r_1 s)(1 - r_2 s)}\right] = \left(\frac{r_1}{r_1 - r_2}\right) r_1^n - \left(\frac{r_2}{r_1 - r_2}\right) r_2^n = \frac{r_1^{n+1} - r_2^{n+1}}{r_1 - r_2}$$

This establishes Formula **1a** of Case **1.** Moreover, if $r_1 = \rho e^{i\phi}$ and $r_2 = \rho e^{-i\phi}$, then the last fraction becomes

$$\frac{(\rho e^{i\phi})^{n+1} - (\rho e^{-i\phi})^{n+1}}{\rho e^{i\phi} - \rho e^{-i\phi}} = \rho^n \frac{e^{i(n+1)\phi} - e^{-i(n+1)\phi}}{e^{i\phi} - e^{-i\phi}} = \rho^n \frac{\sin (n+1)\phi}{\sin \phi}$$

which establishes Formula **2a** of Case **2.**

In Cases **1** and **2,** it is easy to verify that

$$\frac{s}{(1 - r_1 s)(1 - r_2 s)} = \frac{1}{r_1 - r_2}\left(\frac{1}{1 - r_1 s} - \frac{1}{1 - r_2 s}\right)$$

Hence, again using Example 2, we have

$$G^{-1}\left[\frac{s}{(1 - r_1 s)(1 - r_2)}\right] = \frac{r_1^n - r_2^n}{r_1 - r_2}$$

which establishes Formula **1b.** If $r_1 = \rho e^{i\phi}$ and $r_2 = \rho e^{-i\phi}$, then the last fraction becomes

$$\frac{(\rho e^{i\phi})^{n} - (\rho e^{-i\phi})^{n}}{\rho e^{i\phi} - \rho e^{-i\phi}} = \rho^{n-1} \frac{e^{in\phi} - e^{-in\phi}}{e^{i\phi} - e^{-i\phi}} = \rho^{n-1} \frac{\sin n\phi}{\sin \phi}$$

which establishes Formula **2b.**

In Case **3** we can write

$$\frac{1}{(1 - rs)^2} = \frac{rs + (1 - rs)}{(1 - rs)^2} = \frac{rs}{(1 - rs)^2} + \frac{1}{1 - rs} = s\frac{d}{ds}\left(\frac{1}{1 - rs}\right) + \frac{1}{1 - rs}$$

Hence, by Theorem 3 and Example 2, we have

$$G^{-1}\left[\frac{1}{(1 - rs)^2}\right] = nr^n + r^n = (n+1)r^n$$

which establishes Formula **3a.** Finally, since

$$\frac{s}{(1 - rs)^2} = \frac{s}{r}\frac{r}{(1 - rs)^2} = \frac{s}{r}\frac{d}{ds}\left(\frac{1}{1 - rs}\right)$$

we have, by Theorem 3,

$$G^{-1}\left[\frac{s}{(1 - rs)^2}\right] = \frac{1}{r} nr^n = nr^{n-1}$$

which is Formula **3b.** ◗

The use of G transforms can readily be extended to the solution of systems of simultaneous linear difference equations. The only new feature is that instead of a single linear equation in one unknown transform, there is now a system of several linear equations in several unknown transforms, as many equations as there are unknown sequences.

EXAMPLE 7

Solve for y_n and z_n if

$$\begin{aligned} y_{n+1} + 2y_n + 2z_n &= 0 \\ 6y_n - z_{n+1} + 4z_n &= 0 \end{aligned} \qquad y_0 = 1,\ z_0 = -2$$

Taking the G transform of each term in each equation, using the result of Example 3, we have

$$\frac{Y(s)-1}{s}+2Y(s)+2Z(s)=0 \qquad\qquad (1+2s)Y(s)+2sZ(s)=1$$

or

$$6Y(s)-\frac{Z(s)+2}{s}+4Z(s)=0 \qquad\qquad 6sY(s)+(4s-1)Z(s)=2$$

Solving simultaneously the last two equations, we obtain

$$Y(s)=\frac{\begin{vmatrix}1 & 2s\\ 2 & 4s-1\end{vmatrix}}{\begin{vmatrix}1+2s & 2s\\ 6s & 4s-1\end{vmatrix}}=\frac{1}{1-2s+4s^2}$$

$$Z(s)=\frac{\begin{vmatrix}1+2s & 1\\ 6s & 2\end{vmatrix}}{\begin{vmatrix}1+2s & 2s\\ 6s & 4s-1\end{vmatrix}}=\frac{2s-2}{1-2s+4s^2}$$

The roots of the equation $x^2-2x+4=0$ are $r_1, r_2=1\pm i\sqrt{3}$. To convert these to the exponential form $\rho e^{\pm i\phi}$, we note that $\rho=\sqrt{1+(\sqrt{3})^2}=2$ and $\phi=\tan^{-1}(\sqrt{3}/1)=\pi/3$. Hence $r_1, r_2=2e^{\pm i\pi/3}$. Therefore, by Formula **2a** of Theorem 4, we have

$$y_n=2^n\frac{\sin(n+1)\pi/3}{\sin\pi/3}=\frac{2^{n+1}}{\sqrt{3}}\sin\frac{(n+1)\pi}{3}$$

Likewise, by separating the expression for $Z(s)$ into two fractions and then using Formulas **2a** and **2b** of Theorem 4, we have

$$z_n=2\left(2^{n-1}\frac{\sin n\pi/3}{\sin\pi/3}\right)-2\left(2^n\frac{\sin(n+1)\pi/3}{\sin\pi/3}\right)$$
$$=\frac{2^{n+1}}{\sqrt{3}}\left[\sin\frac{n\pi}{3}-2\sin\frac{(n+1)\pi}{3}\right]$$

It is easy to verify that the initial conditions $y_0=1$ and $z_0=-2$ are satisfied.

So far, our discussion has dealt with the G transform of single functions of n and the inverses of such transforms. There is, however, a result of considerable importance, known as the **convolution theorem,** which deals with the product of two transforms, and with this we conclude the theory of this section.

THEOREM 5 If $G(u_n)$ and $G(v_n)$ are the generating functions of the sequences $\{u_n\}$ and $\{v_n\}$, respectively, then

$$G(u_n)G(v_n)=G\left(\sum_{k=0}^{n}u_kv_{n-k}\right)=G\left(\sum_{k=0}^{n}u_{n-k}v_k\right)$$

PROOF By definition,

$$G(u_n)G(v_n)=\left(\sum_{k=0}^{\infty}u_ks^k\right)\left(\sum_{k=0}^{\infty}v_ks^k\right)$$

Forming the Cauchy product of these two series (see Definition 6, Sec. 2.9), we have

$$G(u_n)G(v_n)=\sum_{n=0}^{\infty}\sum_{k=0}^{n}(u_ks^k)(v_{n-k}s^{n-k})=\sum_{n=0}^{\infty}\left[\sum_{k=0}^{n}u_kv_{n-k}\right]s^n$$

Clearly, this expression is simply the statement that $G(u_n)G(v_n)$ is the generating function, or G transform, of the quantity in brackets, which is just the first assertion of the theorem. Since u_n and v_n enter the problem in a completely symmetric way, the second assertion of the theorem follows without further proof. ◗

As in Sec. 2.9, each of the sums $\Sigma_{k=0}^{n} u_k v_{n-k}$ and $\Sigma_{k=0}^{n} u_{n-k} v_k$ is called the **convolution** (on the range 0 to n) of the sequences $\{u_n\}$ and $\{v_n\}$, and each is denoted by either of the symbols $u_n * v_n$ or $v_n * u_n$. We will encounter the convolution of two functions again in our study of Fourier transforms in Chap. 9 and in the study of Laplace transforms in Chap. 10.

Although the G transformation is primarily useful for solving linear constant-coefficient difference equations, it can sometimes be used effectively to solve difference equations which also involve derivatives. The next example illustrates how this leads to a differential equation, rather than an algebraic equation, from which the unknown transform and eventually its inverse is to be found.

EXAMPLE 8

Solve the difference-differential equation

$$\frac{dy_{n+1}}{dt} = y_n(t)$$

given that when $t = 0$, y is the sequence $y_n(0) = h_n$, $n = 0, 1, 2, \ldots$, and when $n = 0$, y is the function $y_0(t) = f(t)$.

In this case, the G transform, or generating function, of the required solution will be a function of the variable t as well as the transform parameter s, say $Y(s, t)$. Thus, using the result of Example 3 to transform the given equation, we have†

$$\frac{d}{dt}\left[\frac{Y(s, t) - y_0(t)}{s}\right] = Y(s, t)$$

or

(3)
$$\frac{dY(s, t)}{dt} - sY(s, t) = y_0'(t) = f'(t)$$

Equation (3) is a linear first-order differential equation in the unknown transform $Y(s, t)$. To solve it, we first compute the integrating factor $e^{-s\int dt} = e^{-st}$ and then proceed as we learned to do in Sec. 1.11. The result, after introducing the dummy variable u, is

$$Y(s, t)e^{-st} = \int_0^t e^{-su} f'(u)\, du + c$$

and

$$Y(s, t) = e^{st}\int_0^t e^{-su} f'(u)\, du + ce^{st} = \int_0^t e^{s(t-u)} f'(u)\, du + ce^{st}$$

When $t = 0$, the integral is zero, and we have $Y(s, 0) = c$. Moreover, by definition,

$$Y(s, 0) = G[y_n(0)] = G(h_n)$$

Hence

$$Y(s, t) = \int_0^t e^{s(t-u)} f'(u)\, du + e^{st} G(h_n)$$

†This assumes that the operations of taking the G transform and differentiating with respect to t are commutative.

The difficult part of the problem now is to recover $y(s, t)$ from its transform. To do this we first replace $e^{s(t-u)}$ and e^{st} by their series equivalents and then recognize that

$$e^{st} = \sum_{n=0}^{\infty} \frac{t^n}{n!} s^n$$

is just the generating function, or G transform, of the sequence $t^n/n!$. This allows us to write

$$Y(s, t) = \int_0^t \sum_{n=0}^{\infty} \frac{(t-u)^n}{n!} s^n f'(u)\, du + G\left(\frac{t^n}{n!}\right) G(h_n) \tag{4}$$

By the convolution theorem (Theorem 5), the last term in (4) can be written as a single transform

$$G\left[\sum_{j=0}^{n} \frac{t^j}{j!} h_{n-j}\right] = \sum_{n=0}^{\infty} \left[\sum_{j=0}^{n} \frac{t^j}{j!} h_{n-j}\right] s^n$$

If we make this substitution in (4) and then interchange the order of integration and summation in the first term, we have further

$$Y(s, t) = \sum_{n=0}^{\infty} \left[\int_0^t \frac{(t-u)^n}{n!} s^n f'(u)\, du\right] + \sum_{n=0}^{\infty} \left[\sum_{j=0}^{n} \frac{t^j}{j!} h_{n-j}\right] s^n$$

Finally, if we combine the two sums and factor out s^n, we have

$$Y(s, t) = \sum_{n=0}^{\infty} \left[\int_0^t \frac{(t-u)^n}{n!} f'(u)\, du + \sum_{j=0}^{n} \frac{t^j}{j!} h_{n-j}\right] s^n$$

The structure of the right-hand side of this equation shows that it is just the G transform of the expression in brackets. Thus the required solution is

$$y_n(t) = \int_0^t \frac{(t-u)^n}{n!} f'(u)\, du + \sum_{j=0}^{n} \frac{t^j}{j!} h_{n-j} \tag{5}$$

Two special cases of (5) are worthy of note. If $f(t) = 1$, then $f'(t) = 0$ and we have simply

$$y(t) = \sum_{j=0}^{n} \frac{t^j}{j!} h_{n-j}$$

Similarly, if $f(t) = t$, the integral in (5) becomes

$$\int_0^t \frac{(t-u)^n}{n!} du = -\frac{(t-u)^{n+1}}{(n+1)!}\bigg|_0^t = \frac{t^{n+1}}{(n+1)!}$$

and

$$y(t) = \frac{t^{n+1}}{(n+1)!} + \sum_{j=0}^{n} \frac{t^j}{j!} h_{n-j}$$

If, further, $h_n = 1$, the last expression reduces to $y_n(t) = \sum_{j=0}^{n+1} (t^j/j!)$.

As we have seen, the G transformation is a convenient alternative to the methods of Sec. 5.7 for solving linear constant-coefficient difference equations. However, in certain other applications it is more useful if it is expressed in terms of $z = 1/s$ rather than s itself. When this is done, we have what is called the **Z transformation.** This has proved useful in signal processing techniques, and we shall return to it briefly in Sec. 10.12 where we shall investigate its relation to the Laplace transformation. Meanwhile, as a summary of our present discussion, we list, both as G transforms and as Z transforms, the results we have obtained in this section, together with a few others whose derivations we leave as exercises.

TABLE 5.3

Sequence	G transform	Z transform
1. f_n	$\sum^{\infty} f_n s^n$	$\sum^{\infty} f_n z^{-n}$
2. 1	$\dfrac{1}{1-s}$	$\dfrac{z}{z-1}$
3. a^n	$\dfrac{1}{1-as}$	$\dfrac{z}{z-a}$
4. $\dfrac{a^n}{n!}$	e^{as}	$e^{a/z}$
5. f_{n+1}	$\dfrac{F(s)-f_0}{s}$	$z[F(z)-f_0]$
6. f_{n+2}	$\dfrac{F(s)-f_0-f_1 s}{s^2}$	$z^2[F(z)-f_0-f_1 z^{-1}]$
7. f_{n+r}	$\dfrac{F(s)-\sum_{j=0}^{r-1} f_j s^j}{s^r}$	$z^r\left[F(z)-\sum_{j=0}^{r-1} f_j z^{-j}\right]$
8. $\begin{cases} f_{n-r} & n=r, r+1, \ldots \\ 0 & n=0, 1, \ldots, r-1 \end{cases}$	$s^r F(s)$	$z^{-r}F(z)$
9. n	$\dfrac{s}{(1-s)^2}$	$\dfrac{z}{(z-1)^2}$
10. nf_n	$s\dfrac{d}{ds}F(s)$	$-z\dfrac{d}{dz}F(z)$
11. $n^r f_n$	$\left(s\dfrac{d}{ds}\right)^r F(s)$	$\left(-z\dfrac{d}{dz}\right)^r F(z)$
12. $\dfrac{r_1^{n+1}-r_2^{n-1}}{r_1-r_2}$	$\dfrac{1}{(1-r_1 s)(1-r_2 s)}$	$\dfrac{z^2}{(z-r_1)(z-r_2)}$
13. $\dfrac{r_1^n-r_2^n}{r_1-r_2}$	$\dfrac{s}{(1-r_1 s)(1-r_2 s)}$	$\dfrac{z}{(z-r_1)(z-r_2)}$
14. $(n+1)r^n$	$\dfrac{1}{(1-rs)^2}$	$\dfrac{z^2}{(z-r)^2}$
15. nr^{n-1}	$\dfrac{s}{(1-rs)^2}$	$\dfrac{z}{(z-r)^2}$
16. $\rho^n\dfrac{\sin(n+1)\phi}{\sin\phi}$	$\dfrac{1}{1-2\rho s\cos\phi+\rho^2 s^2}$	$\dfrac{z^2}{z^2-2\rho z\cos\phi+\rho^2}$
17. $\rho^n\dfrac{\sin n\phi}{\sin\phi}$	$\dfrac{s}{1-2\rho s\cos\phi+\rho^2 s^2}$	$\dfrac{z}{z^2-2\rho z\cos\phi+\rho^2}$
18. $\begin{cases} \sum_{j=0}^{n} f_j g_{n-j} \\ \sum_{j=0}^{n} f_{n-j} g_j \end{cases}$	$F(s)G(s)$	$F(z)G(z)$

EXERCISES

Find the indicated solutions of each of the following equations or systems of equations.

1. $y_{n+2} - 6y_{n+1} + 8y_n = 0;\ y_0 = 1,\ y_1 = -1$
2. $y_{n+2} - 4y_n = 1;\ y_0 = 1,\ y_1 = 0$
3. $y_{n+2} - 6y_{n+1} + 9y_n = 0;\ y_0 = 0,\ y_1 = 1$
4. $y_{n+2} + 3y_{n+1} + 9y_n = 0;\ y_0 = 2,\ y_1 = 1$
5. $(E^3 - 6E^2 + 11E - 6)y_n = 0;\ y_0 = 0,\ y_1 = 1,\ y_2 = 0$
6. $y_{n+2} + 2y_{n+1} - 8y_n = 3^n;\ y_0 = 0,\ y_1 = 1$
7. $\begin{aligned} y_{n+1} + 3y_n - z_n &= 0 \\ -2y_n + z_{n+1} + 2z_n &= 0 \end{aligned} \qquad y_0 = 0,\ z_0 = 1$
8. $\begin{aligned} 3y_{n+1} + 16y_n + z_{n+1} + 7z_n &= 0 \\ 2y_{n+1} + y_n + z_{n+1} + z_n &= 0 \end{aligned} \qquad y_0 = 1,\ z_0 = -3$
9. $\begin{aligned} 2y_{n+1} + y_n + 5z_{n+1} + 8z_n &= 0 \\ y_{n+1} + y_n + 2z_{n+1} + 4z_n &= 0 \end{aligned} \qquad y_0 = 2,\ z_0 = -1$
10. $\begin{aligned} y_{n+1} + y_n + 4z_n &= 0 \\ 2y_{n+1} + y_n + z_{n+1} + 13z_n &= 0 \end{aligned} \qquad y_0 = 4,\ z_0 = -1$
11. What is f_n if **(a)** $G(f_n) = \dfrac{s}{1-s}$? **(b)** $G(f_n) = \dfrac{s^2}{1-s}$?
12. What is f_n if **(a)** $G(f_n) = \dfrac{s}{1-as}$? **(b)** $G(f_n) = \dfrac{s^2}{1-as}$?

Using the convolution theorem, verify each of the following entries in Table 5.3.

13. Entry 12. *Hint:* Note that $\dfrac{1}{(1-r_1s)(1-r_2s)} = G(r_1^n)G(r_2^n)$.
14. Entry 13.
15. Entry 14
16. Entry 15
17. Prove Entry 8 in Table 5.3.
18. Justify the Z forms of Entries 10 and 11 in Table 5.3.
19. What is $G(\sin na)$? *Hint:* use Entry 17 in Table 5.3.
20. What is $G(\cos na)$? *Hint:* Take the transform of the exponential form of $\cos na$
21. What is $G[\sin (n+1)a]$?
22. What is $G[\cos (n+1)a]$?
23. What is $G[\sin (na+b)]$?
24. What is $G[\cos (na+b)]$?
25. What is $G^{-1}\left[\dfrac{1}{(1-s)^3}\right]$? *Hint:* Use the convolution theorem.
26. If $F(s)$ is the G transform of y_n, what is the inverse of $F(s)/(1-s)$?
27. Solve the difference-differential equation $dy_{n+1}/dt + y_n = 0$ subject to the conditions $y_0(t) = f(t)$, $y_n(0) = h_n$, $n = 0, 1, 2, \ldots$.
28. Solve the difference-differential equation $dy_{n+1}/dt = ay_n$ subject to the conditions $y_0(t) = f(t)$, $y_n(0) = h_n$, $n = 0, 1, 2, \ldots$.
29. Solve the difference-differential equation $dy_{n+1}/dt = 2ty_n$ subject to the conditions $y_0(t) = f(t)$, $y_n(0) = h_n$, $n = 0, 1, 2, \ldots$.
30. If the continuous signal e^{at} is instantaneously sampled at intervals of $\Delta t = T$, beginning at $t = 0$, what are the G and Z transforms of the sequence of sample values?
31. What is $G(e^{at}\cos bt)$?
32. What is $G(e^{at}\sin bt)$?
33. What is $G[(n+1)f_n]$?
34. What is $G[(n+r)f_n]$?
35. If $f_n = n^{(r)} \equiv n(n-1)(n-2)\cdots(n-r+1)$, show that

$$G(f_n) = \frac{r!s^r}{(1-s)^{r+1}}$$

CHAPTER 6

THE DESCRIPTIVE THEORY OF ORDINARY DIFFERENTIAL EQUATIONS

By the *descriptive theory of differential equations* we mean the study of such questions as "When does an equation have periodic solutions?" and "When do its solutions remain bounded and when do they become unbounded as the independent variable becomes infinite?"

In this chapter we restrict our attention to the system of first-order equations

$$\frac{dx}{dt} = F(x, y)$$

$$\frac{dy}{dt} = G(x, y)$$

which includes the important case of a single second-order equation

$$\frac{d^2x}{dt^2} = G\left(x, \frac{dx}{dt}\right)$$

if we put $dx/dt = y$ so that y itself becomes $F(x, y)$.

Our procedure is to begin with the linear system

$$\frac{dx}{dt} = ax + by$$

$$\frac{dy}{dt} = cx + ey$$

which is easily solvable by the methods of Sec. 4.5. With these results to guide us, we are then able to develop corresponding results for the general case.

Prerequisites: Chaps. 1–4.

6.1 INTRODUCTION

Up to this point, most of our work has been concerned with devising methods for solving differential equations. For several types of first-order equations and for the very important class of linear constant-coefficient equations of all orders, we are now able to obtain solutions in finite form. For many linear differential equations with variable coefficients, we have learned how to find solutions in the form of infinite series. And for equations whose complexity makes exact solution difficult or impossible, we have developed methods for obtaining accurate numerical approximations to their solutions. Given sufficient time and computer resources, it is probably correct to say that all solutions of any solvable differential equation, or system of differential equations, can be found with acceptable accuracy. However, many questions phrased in terms of differential equations can be adequately answered without actually solving the equations. Frequently, only some descriptive property, such as the stability or periodicity of solutions, is to be investigated. In this chapter we shall undertake an introductory discussion of these matters for systems which can be described by two first-order differential equations.

6.2 THE PHASE PLANE AND CRITICAL POINTS

As a starting point, let us consider the initial-value problem

$$(1) \qquad \frac{d^2x}{dt^2} = f\left(x, \frac{dx}{dt}\right) \qquad x = x_0, \frac{dx}{dt} = x_0' \text{ when } t = 0$$

where f has continuous first partial derivatives with respect to x and x' for all values of x. As we have pointed out several times in our previous work, if we put $dx/dt = y$, this equation can be replaced by the system

$$(2) \qquad \begin{aligned} \frac{dx}{dt} &= y \\ \frac{dy}{dt} &= f(x, y) \end{aligned} \qquad x = x_0,\ y = y_0\ (= x_0') \text{ when } t = 0$$

By a solution of this system we mean, as usual, a pair of differentiable functions $[x = x(t), y = y(t)]$ which on some interval containing $t = 0$ reduce Eqs. (2) to identities and are such that $x(0) = x_0$ and $y(0) = y_0$. In our study of the initial-value problem (1), we shall work exclusively with the related system (2).

Although they have different physical interpretations, the variables x and y $(= x')$ play the same role as dependent variables in the two equations. In other words, the displacement x and the velocity y have the same formal or mathematical significance, and we shall so regard them.

If we choose, we can think of the solutions $x = x(t)$ and $y = y(t)$ as the parametric equations of an arc in the xy plane which passes through the point (x_0, y_0). From this point of view, the xy plane is called the **phase plane** of the original equation (1) or of the system (2), and the arc defined parametrically by the equations $x = x(t)$ and $y = y(t)$ is called a **path** or **orbit** or **trajectory** of either (1) or (2). When the points of a trajectory are in one-to-one correspondence with the values of the parameter t (which need not be the case, as Exercise 16 shows), the direction in which t increases is said to be the **positive direction** on the trajectory.

If the values $x = x_0$ and $y = y_0$ $(= x_0')$ are assigned when $t = t_0$ rather than when $t = 0$, we obtain a different solution but the same trajectory since the equations

$$x = x(t) \qquad y = y(t) \qquad \alpha < t < \beta$$

define the same arc as the equations

$$x = x(t - t_0) \qquad y = y(t - t_0) \qquad \alpha + t_0 < t < \beta + t_0$$

Thus *solution* and *trajectory* are not synonomous terms.

If the parameter t is eliminated between the equations $x = x(t)$ and $y = y(t)$, we obtain the xy equation of a curve, which may or may not be the corresponding trajectory. For example, if we eliminate the parameter t between the equations

$$x = e^t \qquad \text{and} \qquad y = e^{2t} \qquad -\infty < t < \infty \tag{3}$$

we obtain $y = x^2$. However, the parabola represented by this equation is not the trajectory defined by (3) because these equations define only the portion of the parabola to the right of the y axis; that is, the trajectory is only the (open) right half of the parabola $y = x^2$.

The xy equation of a curve containing the trajectory through (x_0, y_0) can also be found by first eliminating t (in the form of the differential dt) from the system (2) by dividing the second equation by the first, getting

$$\frac{dy}{dx} = \frac{f(x, y)}{y} \qquad (y \neq 0),\ y = y_0 \text{ when } x = x_0$$

and then solving this equation in the variables x and y.

More generally, the preceding observations can all be applied to systems of the form

$$\begin{aligned} \frac{dx}{dt} &= g(x, y) \\ \frac{dy}{dt} &= f(x, y) \end{aligned} \qquad x = x_0,\ y = y_0 \text{ when } t = 0 \tag{4}$$

where both f and g are functions possessing continuous first partial derivatives. In this case, unless $g(x, y) \equiv y$, x and y are not the displacement and velocity of some particle. Nonetheless, the xy plane is still called the **phase** plane, and the arc defined parametrically by any solution

$$x = x(t) \qquad y = y(t)$$

is called a **path** or **orbit** or **trajectory** of the system. Systems such as (4) in which the independent variable t does not appear explicitly in either f or g are said to be **autonomous.** In this chapter we shall be concerned exclusively with autonomous systems.

EXAMPLE 1

Find the trajectories of the system

$$\begin{aligned} \frac{dx}{dt} &= 3x + y \\ \frac{dy}{dt} &= x + 3y \end{aligned} \tag{5}$$

Using the methods of Chap. 4, we seek solutions of the form

$$\begin{bmatrix} x \\ y \end{bmatrix} = \begin{bmatrix} A \\ B \end{bmatrix} e^{mt}$$

Substituting into (5) and dividing by e^{mt} yields the equations

$$\begin{aligned}(m-3)A - B &= 0\\ -A + (m-3)B &= 0\end{aligned} \tag{6}$$

This system of algebraic equations will have a nontrivial solution if and only if

$$\begin{vmatrix} m-3 & -1 \\ -1 & m-3 \end{vmatrix} = m^2 - 6m + 8 = 0$$

that is, if and only if $m = 2$ or $m = 4$. If $m = 2$, then, from (6), $-A - B = 0$, and we have the particular solution

$$\begin{bmatrix} x \\ y \end{bmatrix} = \begin{bmatrix} 1 \\ -1 \end{bmatrix} e^{2t}$$

If $m = 4$, then, from (6), $A - B = 0$, and we have the particular solution

$$\begin{bmatrix} x \\ y \end{bmatrix} = \begin{bmatrix} 1 \\ 1 \end{bmatrix} e^{4t}$$

These two solutions are linearly independent, hence a complete solution is

$$\begin{bmatrix} x \\ y \end{bmatrix} = c_1 \begin{bmatrix} 1 \\ -1 \end{bmatrix} e^{2t} + c_2 \begin{bmatrix} 1 \\ 1 \end{bmatrix} e^{4t}$$

The trajectories of the system are thus defined parametrically by the equations

$$\begin{aligned} x &= c_1 e^{2t} + c_2 e^{4t} \\ y &= -c_1 e^{2t} + c_2 e^{4t} \qquad -\infty < t < \infty \end{aligned} \tag{7}$$

To eliminate the parameter t between the equations in (7), we first add these equations and then subtract them, getting

$$\begin{aligned} x + y &= 2c_2 e^{4t} \\ x - y &= 2c_1 e^{2t} \end{aligned}$$

Finally, eliminating the exponentials between these equations, we have

$$\left(\frac{x-y}{2c_1}\right)^2 = e^{4t} = \frac{x+y}{2c_2}$$

or

$$(x-y)^2 = k(x+y) \qquad k = \frac{2c_1^2}{c_2} \tag{8}$$

If $k = 0$, this yields the line $y = x$, corresponding to $c_1 = 0$. The line $y = -x$, corresponding to $c_2 = 0$, is not included in (8) unless k is allowed to become infinite.

Alternatively, if we divide the second of the equations (5) by the first, we obtain

$$\frac{dy}{dx} = \frac{x + 3y}{3x + y}$$

This is a homogeneous first-order equation which can easily be solved by means of the substitution $y = ux$, as we saw in Sec. 1.10. The result, of course, is Eq. (8).

Equation (8) defines a family of parabolas (Fig. 6.1) which, with one exception, are all tangent at the origin to the line $x + y = 0$ and have the line $x - y = 0$ as axis. The one exception is the degenerate parabola

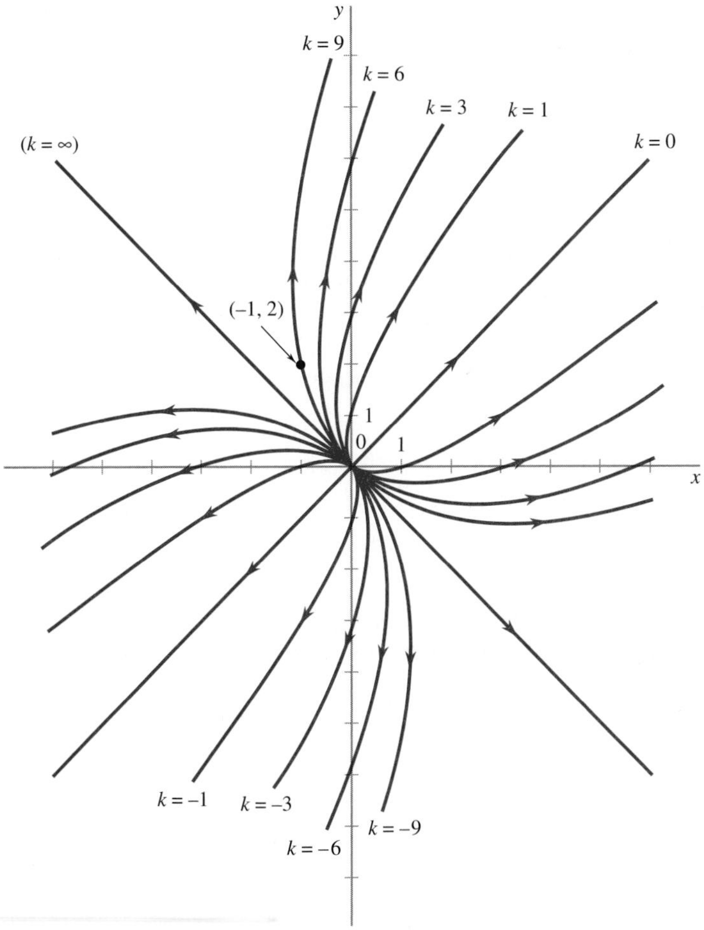

FIGURE 6.1
Typical trajectories of the system $dx/dt = 3x + y$, $dy/dt = x + 3y$.

consisting of the repeated line $x - y = 0$, corresponding to the value $k = 0$. These curves are *not* the trajectories of the given system. For instance, if $x = -1$ and $y = 2$ when $t = 0$, then, from (7), c_1 and c_2 must satisfy the conditions

$$\begin{aligned} c_1 + c_2 &= -1 \\ -c_1 + c_2 &= 2 \end{aligned}$$

Hence $c_1 = -\frac{3}{2}$, $c_2 = \frac{1}{2}$, and

$$\begin{aligned} x &= -\tfrac{3}{2}e^{2t} + \tfrac{1}{2}e^{4t} \\ y &= \tfrac{3}{2}e^{2t} \quad + \tfrac{1}{2}e^{4t} \qquad -\infty < t < \infty \end{aligned}$$

The arc, or trajectory, corresponding to these equations is only the (open) half of the parabola $(x - y)^2 = 9(x + y)$ extending from the origin (which is excluded) through the point $(-1, 2)$. This is clear from the fact that, according to the last formula, y is always greater than 0. Since c_1 appears as a square in the parameter k in Eq. (8), it follows that the other (open) half of the parabola $(x - y)^2 = k(x + y)$ is the trajectory

$$\begin{aligned} x &= \tfrac{3}{2}e^{2t} + \tfrac{1}{2}e^{4t} \\ y &= -\tfrac{3}{2}e^{2t} + \tfrac{1}{2}e^{4t} \end{aligned}$$

obtained by putting $c_1 = \frac{3}{2}$ and $c_2 = \frac{1}{2}$ in Eqs. (7). More generally, each parabola of the family (8) (with the origin excluded) is the union of two trajectories. Similarly, each of the lines $y = \pm x$ (with the origin excluded) is also the union of two trajectories.

In general, the equation

$$\frac{dy}{dx} = \frac{f(x, y)}{g(x, y)}$$

obtained from (4) by dividing the second equation by the first, gives the slope at the point (x, y) of the trajectory of (4) which passes through that point. However, if (x_0, y_0) is a point where $f(x, y)$ and $g(x, y)$ are simultaneously zero, then the slope at that point is an indeterminate of the form 0/0. Such a point is called a **critical point** or **equilibrium point** of the system (4). If a critical point has the property that there exists a circle which contains it but no other critical point, then that critical point is said to be **isolated.** Throughout this chapter we shall be concerned only with isolated critical points.

In the next section we shall investigate the nature of the trajectories of the system

$$\text{(9)} \qquad \begin{aligned} \frac{dx}{dt} &= ax + by \\ \frac{dy}{dt} &= cx + ey \qquad a, b, c, e \text{ constants} \end{aligned}$$

in the neighborhood of the obvious isolated critical point (0, 0). Since the equations in (9) are linear, the system can always be solved explicitly. Moreover, since the related first-order equation

$$\frac{dy}{dx} = \frac{cx + ey}{ax + by}$$

is homogeneous, it too can be solved in every case. However, a knowledge of the possible configurations of the trajectories of (9) in the neighborhood of (0, 0) is fundamental for the descriptive study of the system (4) in the general case when at least one of the functions f, g is nonlinear.

EXERCISES

Determine the critical points of each of the following systems.

1. $dy/dt = x + 2y - 3$
$dx/dt = 3x + y + 1$

2. $dy/dt = xy + x - 2y + 4$
$dx/dt = 3x - y + 2$

3. $dy/dt = 9x^2 + 16y^2 - 25$
$dx/dt = 16x^2 + 9y^2 - 25$

4. $dy/dt = x - y$
$dx/dt = 3x - y - 3xy + x^3$

Find parametric equations for the trajectories of each of the following equations. In each case, find the xy equation of the family of curves containing the trajectories.

5. $x'' - 3x' + 2x = 0$ **6.** $x'' + 3x' + 2x = 0$

7. $x'' + x = 0$ **8.** $x'' + 4x = 0$

Find parametric equations for the trajectories of each of the following systems. In each case, find the xy equation of the family of curves containing the trajectories.

9. $dy/dt = x$
$dx/dt = y$

10. $dy/dt = x$
$dx/dt = -y$

11. $dy/dt = y$
$dx/dt = -x + 2y$

12. $dy/dt = 3x + y$
$dx/dt = -x + y$

13. Find the xy equation of the family of curves containing the trajectories of the equation $x'' + x + 2x^3 = 0$. Can parametric equations be found for the trajectories themselves? What are the trajectories?

14. For an autonomous system, how many trajectories pass through a particular point which is not a critical point? Why?

15. Find the solutions of the nonautonomous system

$$\frac{dx}{dt} = y \qquad \frac{dy}{dt} = 6t$$

which satisfy the conditions **(a)** $x = 1$, $y = 1$ when $t = 1$ and **(b)** $x = 1$, $y = 1$ when $t = 2$. How many different trajectories pass through the point (1, 1)? How does this behavior compare with that of an autonomous system?

16. Determine A and B so that $x = At^2$, $y = Bt^2$ will be a solution of the system

$$\frac{dx}{dt} = 8\sqrt{y} \qquad \frac{dy}{dt} = \sqrt{x}$$

Describe the trajectory defined by this solution and show that the parameter t increases in each direction on this path.

6.3 CRITICAL POINTS AND THE TRAJECTORIES OF LINEAR SYSTEMS

In this section we shall examine the possible configurations of the trajectories of the system

$$\text{(1)} \qquad \begin{aligned} \frac{dx}{dt} &= ax + by \\ \frac{dy}{dt} &= cx + ey \end{aligned} \qquad a, b, c, e, \text{ real constants; } ae - bc \neq 0$$

or of the corresponding single equation

$$\text{(1}a\text{)} \qquad \frac{dy}{dx} = \frac{cx + ey}{ax + by}$$

in the neighborhood of the critical point (0, 0). The restriction $ae - bc \neq 0$ is necessary, for otherwise every point of the line $ax + by = 0$ is a critical point of (1) and (0, 0) is not an isolated critical point, as we have supposed. To see this, assume (for definiteness) that $a \neq 0$ and consider the equations

$$\text{(2)} \qquad \begin{aligned} ax + by &= 0 \\ cx + ey &= 0 \end{aligned}$$

If c times the first equation is subtracted from a times the second, we obtain $(ae - bc)y = 0$. Hence (2) is equivalent to the system

$$\text{(3)} \qquad \begin{aligned} ax + by &= 0 \\ (ae - bc)y &= 0 \end{aligned}$$

If $ae - bc \neq 0$, the only solution of (3), and hence of (2), is $x = y = 0$, and (0, 0) is the only critical point of the system (1). On the other hand, if $ae - bc = 0$, then the system (2) and the single equation $ax + by = 0$ have identical solutions; that is, every point of the line $ax + by = 0$ is a critical point of the system (1) and the equation (1a).

As usual, to solve the system (1), we assume $x = Ae^{mt}$, $y = Be^{mt}$, substitute, and obtain at once the characteristic equation

$$\text{(4)} \qquad \begin{vmatrix} m - a & -b \\ -c & m - e \end{vmatrix} = m^2 - (a + e)m + ae - bc = 0$$

As we learned in Chap. 4, the roots of this equation determine the form of the solutions for x and y, and these in turn determine the nature of the trajectories of (1) and (1a). There are three distinct cases to consider, according as the discriminant

$$\Delta = (a + e)^2 - 4(ae - bc)$$

of (4) is greater than, equal to, or less than zero. However, there are several subcases which must be distinguished. For instance, when the roots of (4) are real and distinct, the trajectories are significantly different when the roots are of like sign and when they are of unlike sign. Furthermore, when the roots are complex, the configuration of the trajectories is significantly different when the roots are pure imaginaries and when they are not.

The various possibilities are described in Table 6.1. Each of these cases can be investigated in complete generality by solving either Eqs. (1) or Eq (1a) under the appropriate restrictions on the coefficients. However, for simplicity, we shall explain the general nature of the trajectories in each case by considering only a suitable prototype.

TABLE 6.1

Nature of the characteristic roots	Conditions on the coefficients in the characteristic equation $[\Delta \equiv (a+e)^2 - 4(ae - bc)]$
1. Real, unequal, and of like sign	$\Delta > 0$, $ae - bc > 0$†
2. Real, unequal, and of unlike sign	$\Delta > 0$, $ae - bc < 0$
3a. Real and equal	$\Delta = 0$; b, c not both zero
3b. Real and equal	$\Delta = 0$, $b = c = 0$ $\quad (a = e)$
4. Pure imaginary	$\Delta < 0$, $a + e = 0$ $\quad (ae - bc > 0)$
5. Complex but not pure imaginary	$\Delta < 0$, $a + e \neq 0$

†From the elementary theory of quadratic equations, the constant term $ae - bc$ in the characteristic equation (4) is equal to the product of the roots. Hence the roots are of like sign when $ae - bc > 0$ and of unlike sign when $ae - bc < 0$. Since $ae - bc \neq 0$, the characteristic equation can never have zero as a root.

In discussing the various cases, it will be convenient to have the following definitions.

DEFINITION 1 A trajectory T defined by the equations $x = x(t)$ and $y = y(t)$ is said to **approach** the critical point $(0, 0)$ as $t \to +\infty$ if and only if

$$\lim_{t\to+\infty} x(t) = 0 \qquad \text{and} \qquad \lim_{t\to+\infty} y(t) = 0$$

T is said to **approach** the critical point $(0, 0)$ as $t \to -\infty$ if and only if

$$\lim_{t\to-\infty} x(t) = 0 \qquad \text{and} \qquad \lim_{t\to-\infty} y(t) = 0$$

DEFINITION 2 A trajectory T defined by the equations $x = x(t)$ and $y = y(t)$ is said to **enter** the critical point $(0, 0)$ as $t \to +\infty$ if and only if T approaches $(0, 0)$ as $t \to +\infty$ and

$$\lim_{t\to+\infty} \frac{y(t)}{x(t)}$$

either exists or is $\pm\infty$. T is said to **enter** the critical point $(0, 0)$ as $t \to -\infty$ if and only if T approaches $(0, 0)$ as $t \to -\infty$ and

$$\lim_{t\to-\infty} \frac{y(t)}{x(t)}$$

either exists or is $\pm\infty$.

According to Definition 2, if a trajectory enters a critical point, it must not only approach the critical point as $t \to \infty$ or as $t \to -\infty$, but also must approach it in a definite limiting direction.

DEFINITION 3 A trajectory T defined by the equations $x = x(t)$ and $y = y(t)$ is said to **recede indefinitely** from the critical point (0, 0) as $t \to +\infty$ (or $t \to -\infty$) if and only if at least one of the function $x(t)$ and $y(t)$ becomes infinite as $t \to +\infty$ (or $t \to -\infty$).

CASE 1 For the system

$$\frac{dx}{dt} = \lambda x \qquad \frac{dy}{dt} = 2\lambda y \qquad \lambda \neq 0 \tag{5}$$

we have $a = \lambda$, $b = c = 0$, $e = 2\lambda$. Hence $\Delta = \lambda^2$, $ae - bc = 2\lambda^2$, and thus the characteristic roots λ, 2λ are real, unequal, and of like sign. In this case the related equation (1a) is

$$\frac{dy}{dx} = \frac{2y}{x}$$

and its solution is

$$y = kx^2 \tag{6}$$

This is the equation of a family of parabolas each tangent to the line $y = 0$ at the critical point (0, 0) and each having $x = 0$ as axis.

If the parametric equations of the trajectories are found by solving for x and y from Eqs. (5), we find

$$x = Ae^{\lambda t} \qquad y = Be^{2\lambda t} \tag{7}$$

Eliminating t between these equation leads of course to the family of curves (6), although in every case the trajectories are only portions of the corresponding curves. From Eqs. (7) it is clear that as $t \to +\infty$, every trajectory, including the open halves of the line $x = 0$ corresponding to $A = 0$, recedes indefinitely from the critical point if $\lambda > 0$ and approaches the critical point in a well-defined direction if $\lambda < 0$. Figure 6.2 shows the configuration of the trajectories when $\lambda > 0$ and both characteristic roots are positive. When $\lambda < 0$ and the characteristic roots are negative, the direction of each arrow must be reversed.

For this case, the significant feature of the configuration of the trajectories around the critical point is that there is a neighborhood of the critical point such that all trajectories in that neighborhood either

1. Approach and enter the critical point as $t \to +\infty$
2. Approach and enter the critical point as $t \to -\infty$

A critical point around which the trajectories have this property is called a **node.** The origin in Example 1, Sec. 6.2, is another example of a node.

CASE 2 For the system

$$\frac{dx}{dt} = \lambda x \qquad \frac{dy}{dt} = -\lambda y \qquad \lambda \neq 0 \tag{8}$$

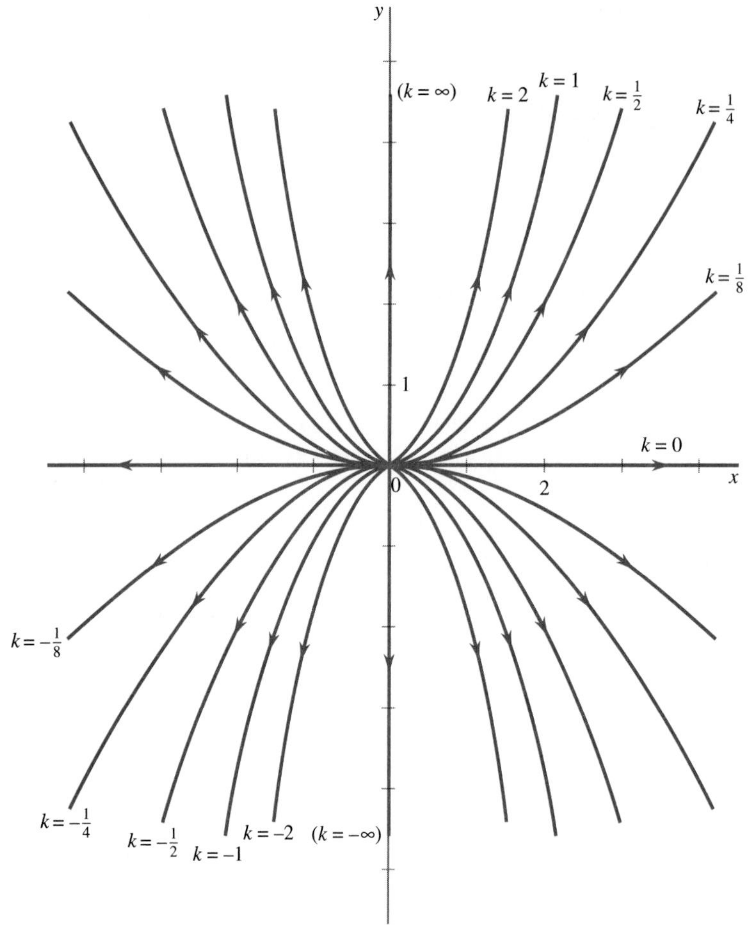

FIGURE 6.2
Typical trajectories around the node of the system $dx/dt = \lambda x$, $dy/dt = 2\lambda y$ $(\lambda > 0)$.

we have $a = \lambda$, $b = c = 0$, $e = -\lambda$. Hence $\Delta = 4\lambda^2$, $ae - bc = -\lambda^2$, and the characteristic roots λ and $-\lambda$ are real, unequal, and of opposite sign. In this case the related equation (1*a*) is

$$\frac{dy}{dx} = -\frac{y}{x}$$

and its solution is

(9) $$xy = k$$

This is the equation of a family of curves which, for $k \neq 0$, are hyperbolas having $x = 0$ and $y = 0$ as asymptotes. For $k = 0$ the corresponding curve consists of the asymptotes themselves. From Eqs. (8) we obtain at once the parametric equations of the trajectories

(10) $$x = Ae^{\lambda t} \qquad y = Be^{-\lambda t}$$

Elimination of t between these leads immediately to Eq. (9). From Eqs. (10) it is clear that whether λ is positive or negative, each hyperbolic trajectory recedes indefinitely from the critical point

(0, 0) both as $t \to +\infty$ and as $t \to -\infty$. As $t \to +\infty$, the trajectories on $x = 0$ approach and enter the critical point if $\lambda > 0$ and recede indefinitely from the critical point if $\lambda < 0$. As $t \to +\infty$, the trajectories on $y = 0$ recede indefinitely from the critical point if $\lambda > 0$ and approach and enter the critical point if $\lambda < 0$. Figure 6.3 shows the trajectories for $\lambda > 0$. If $\lambda < 0$, the direction of each arrow must be reversed.

A critical point around which the configuration of trajectories resembles that shown in Fig. 6.3 is called a **saddle point.** A saddle point is characterized by the existence of a neighborhood of the critical point in which

1. There are two trajectories which approach and enter the critical point from opposite directions as $t \to +\infty$, and there are two other trajectories which approach and enter the critical point from opposite directions as $t \to -\infty$.
2. All other trajectories recede indefinitely from the critical point both as $t \to +\infty$ and as $t \to -\infty$.

FIGURE 6.3
Typical trajectories around the saddle point of the system $dx/dt = \lambda x$, $dy/dt = -\lambda y$ ($\lambda > 0$).

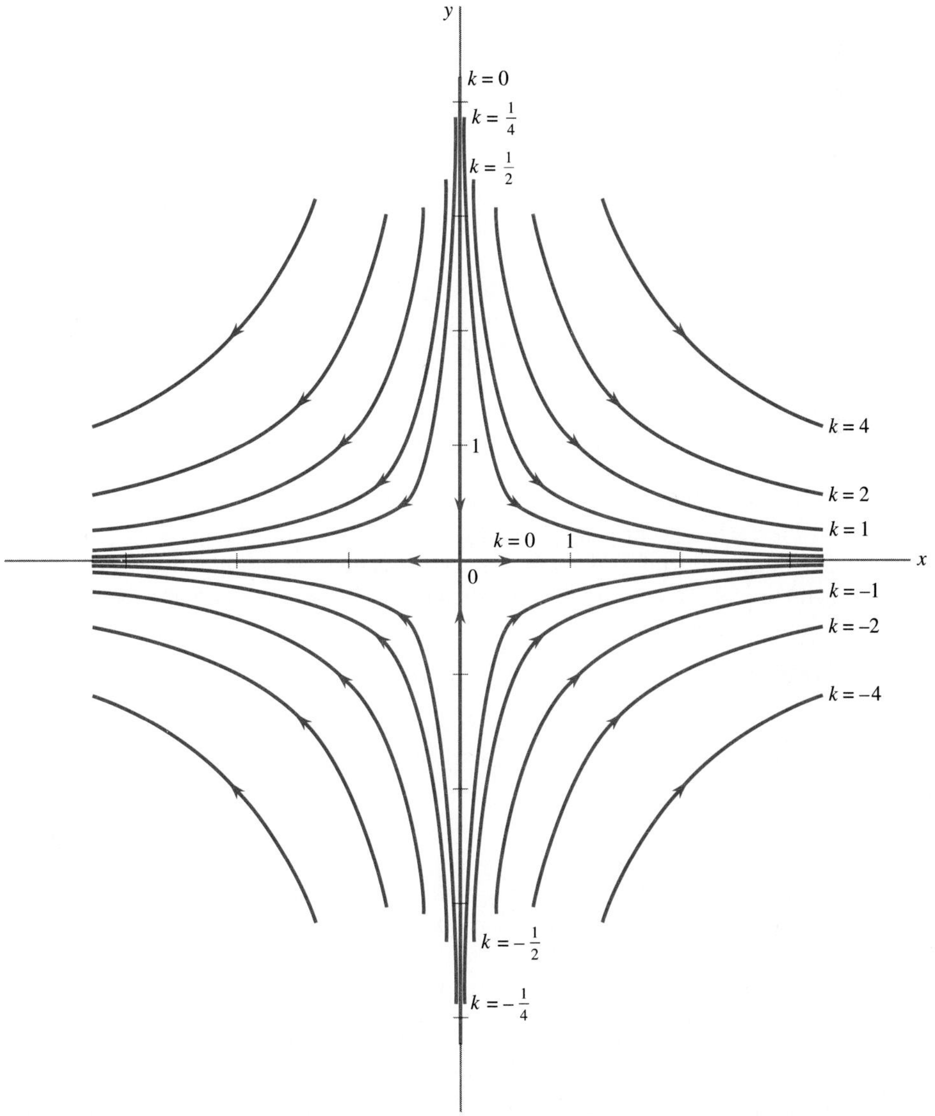

CASE 3A For the system

$$\text{(11)} \qquad \frac{dx}{dt} = \lambda x \qquad \frac{dy}{dt} = \lambda x + \lambda y \qquad \lambda \neq 0$$

we have $a = \lambda$, $b = 0$, $c = e = \lambda$. Hence $\Delta = 0$, $ae - bc = \lambda^2$, and the characteristic roots λ, λ are real and equal. In this case the related equation (1a) is

$$\frac{dy}{dx} = \frac{x + y}{x}$$

and its solution (obtained by solving it either as a homogeneous or a linear equation) is

$$\text{(12)} \qquad y = x \ln |cx| \qquad c \neq 0$$

Since

$$\lim_{x \to 0} (x \ln |cx|) = 0$$

and since

$$\frac{dy}{dx} = \frac{d}{dx}(x \ln |cx|) = 1 + \ln |cx|$$

becomes infinite as $x \to 0$, it follows that (12) is the equation of a family of transcendental curves each of which approaches the critical point (0, 0) with limiting slope ∞ as $x \to 0$. From (11) we find that the parametric equations of the trajectories are

$$x = Ae^{\lambda t}$$
$$y = Be^{\lambda t} + A\lambda te^{\lambda t}$$

Thus if $\lambda > 0$, each trajectory recedes indefinitely from the critical point as $t \to +\infty$, while if $\lambda < 0$, each trajectory approaches and enters the critical point at $t \to +\infty$. Therefore the critical point in this case is also a node. Figure 6.4 shows the configuration of the trajectories when $\lambda > 0$ and the (equal) roots are positive. When $\lambda < 0$, the curves are the same, but the direction of each arrow must be reversed.

CASE 3B For the system

$$\frac{dx}{dt} = \lambda x \qquad \frac{dy}{dt} = \lambda y$$

we have $a = e = \lambda$, $b = c = 0$, $\Delta = 0$, and the equal characteristic roots are λ, λ. The parametric equations of the trajectories in this case are

$$x = Ae^{\lambda t} \qquad y = Be^{\lambda t}$$

and the associated family of curves are the family of lines

$$y = kx$$

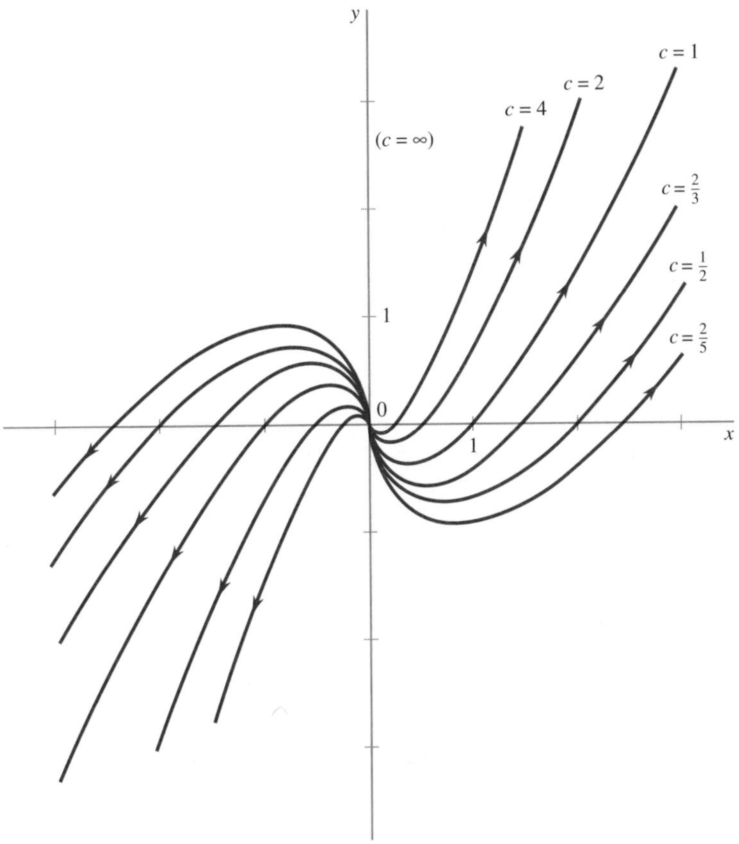

FIGURE 6.4
Typical trajectories around the node of the system $dx/dt = \lambda x$, $dy/dt = \lambda x + \lambda y$ $(\lambda > 0)$.

If $\lambda > 0$, every trajectory recedes indefinitely from the critical point as $t \to +\infty$; and if $\lambda < 0$, every trajectory approaches and enters the critical point as $t \to +\infty$. Figure 6.5 shows the configuration of trajectories when $\lambda > 0$ and the (equal) roots are positive. When $\lambda < 0$ and the roots are negative, the direction of the arrows must be reversed. The critical point in this example is also a node, although it differs from the patterns of Cases 1 and 3*a* in Table 6.1 in that the curves do not have a common tangent at the critical point. A node of this type is sometimes called a **proper node,** while nodes of the type occurring in Cases 1 and 3*a* are called **improper nodes.**

CASE 4 For the system

$$\frac{dx}{dt} = \lambda y \qquad (13)$$

$$\frac{dy}{dt} = -\lambda x \qquad \lambda \neq 0$$

we have $a = e = 0$, $b = \lambda$, $c = -\lambda$. Hence $\Delta = -4\lambda^2$, $a + e = 0$, and the characteristic roots $\pm i\lambda$ are pure imaginaries. The related equation (1*a*) is

$$\frac{dy}{dx} = -\frac{x}{y} \qquad (14)$$

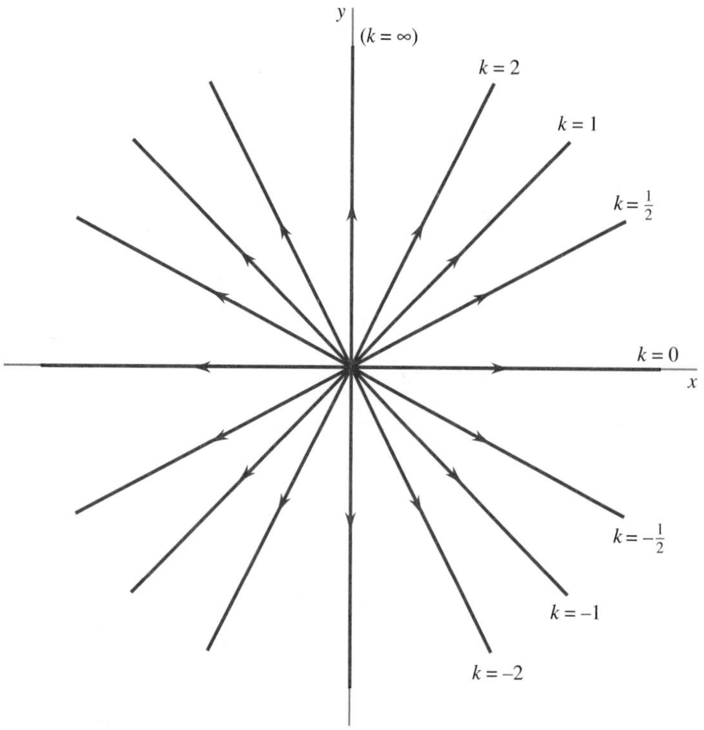

FIGURE 6.5
Typical trajectories around the node of the system $dx/dt = \lambda x$, $dy/dt = \lambda y$ ($\lambda > 0$).

and its solution is $x^2 + y^2 = r^2$. This is the equation of a family of circles each having the critical point (0, 0) as center. Solving Eqs. (13), we find that the parametric equations of the trajectories are

$$\begin{aligned} x &= A \cos \lambda t + B \sin \lambda t \\ y &= -A \sin \lambda t + B \cos \lambda t \end{aligned} \tag{15}$$

and these we recognize as parametric equations of the circles of the family $x^2 + y^2 = r^2$ which we obtained from (14). From Eqs. (15) it follows (see Exercise 13) that as $t \to +\infty$, all trajectories are traversed repeatedly in the clockwise direction if $\lambda > 0$, and in the counterclockwise direction if $\lambda < 0$. Figure 6.6 shows the configuration of the trajectories for $\lambda > 0$.

A critical point around which the configuration of trajectories resembles the configuration shown in Fig. 6.6 is called a **center.** Its significant characteristics are:

1. There exists a neighborhood of the critical point containing an infinite set of closed trajectories each of which contains the critical point in its interior.
2. For every $\epsilon > 0$, there are trajectories in this neighborhood whose maximum chord is less than ϵ in length.

CASE 5 For the system

$$\begin{aligned} \frac{dx}{dt} &= \lambda x - y \\ \frac{dy}{dt} &= x + \lambda y \qquad \lambda \neq 0 \end{aligned} \tag{16}$$

we have $a = e = \lambda$, $b = -1$, $c = 1$. Hence $\Delta = -4$, $a + e = 2\lambda$, and (since $\lambda \neq 0$) the characteristic roots $\lambda \pm i$ are complex numbers which are not pure imaginaries. The related equation (1*a*) is

$$\frac{dy}{dx} = \frac{x + \lambda y}{\lambda x - y}$$

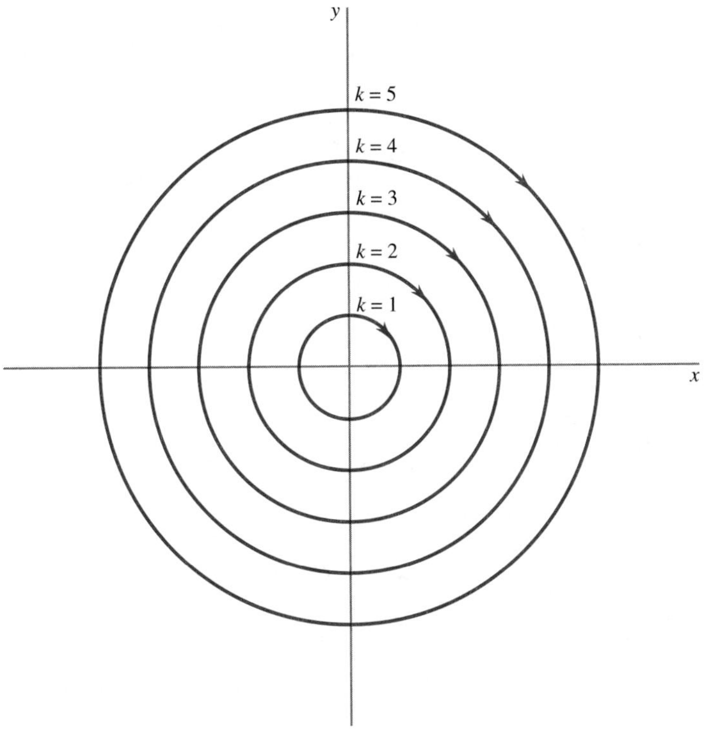

FIGURE 6.6
Typical trajectories around the center of the system $dx/dt = \lambda y$, $dy/dt = -\lambda x$ $(\lambda > 0)$.

and its solution (found by solving it as a homogeneous equation) is

$$x^2 + y^2 = K^2 e^{2\lambda \tan^{-1}(y/x)} \qquad K \neq 0$$

or, transforming to polar coordinates,

$$r^2 = K^2 e^{2\lambda\theta}$$

or finally, taking square roots and letting k be either positive or negative,

$$r = ke^{\lambda\theta}$$

This is the equation of a family of spirals, each of which winds around the critical point (0, 0) as a limit. From Eqs. (16) we find that the parametric equations of the trajectories are

$$x = e^{\lambda t}(A \cos t + B \sin t)$$
$$y = e^{\lambda t}(A \sin t - B \cos t)$$

From these it is evident that if $\lambda > 0$, every trajectory recedes indefinitely from the critical point as $t \to +\infty$, and if $\lambda < 0$, every trajectory approaches the critical point as $t \to +\infty$. Figure 6.7 shows the configuration of the trajectories for $\lambda > 0$.

A critical point around which the configuration of trajectories resembles the configuration shown in Fig. 6.7 is called a **spiral point** or **focal point.** Its significant characteristics are:

1. There exists a neighborhood of the critical point such that every trajectory in this neighborhood approaches the critical point either as $t \to +\infty$ or as $t \to -\infty$.
2. As each trajectory approaches the critical point, it winds around the critical point an infinite number of times.

An examination of the families of trajectories appearing in Figs. 6.2–6.7 indicates, first of all, that only around a center are there periodic solutions of the given system, for only in this case does a trajectory containing a point (x_0, y_0) ever return to that point and hence (by the fundamental existence and uniqueness theorem) repeat the behavior it began previously at that point. The question of stability, which we shall soon consider, cannot be answered solely by an inspection of the family of trajectories because (except around a saddle point) trajectories may approach, or recede from, the critical point as t becomes infinite, depending on the directions established on the trajectories by the roots of the characteristic equation.

To aid us in discussing stability, we need both the concept of a *stable critical point* and the concept of an *asymptotically stable critical point.* To formulate these definitions, let C: (x_0, y_0) be an isolated critical point of the system

$$\frac{dx}{dt} = g(x, y)$$

$$\frac{dy}{dt} = f(x, y)$$

let Γ: $\{x = x(t),\ y = y(t)\}$ be an arbitrary trajectory of the system, and let

$$D(t) = \sqrt{[x(t) - x_0]^2 + [y(t) - y_0]^2}$$

be the distance of an arbitrary point on Γ from the critical point.

DEFINITION 1 The isolated critical point C is said to be **stable** if and only if for every $\epsilon > 0$ there exists a $\delta > 0$ such that on any trajectory which contains a point $[x(t^*), y(t^*)]$ for which $D(t^*) < \delta$ the distance $D(t)$ exists and is less than ϵ for all $t \geq t^*$.

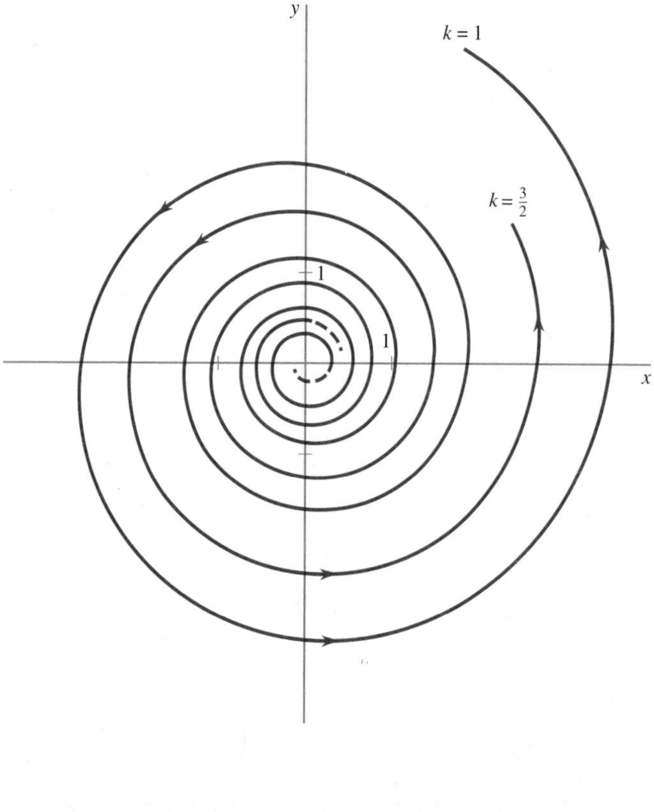

FIGURE 6.7
Typical trajectories around the spiral point of the system $dx/dt = \lambda x - y$, $dy/dt = x + \lambda y$ $(\lambda > 0)$.

Stated less formally, Definition 1 says that a critical point is stable if and only if for every distance $\epsilon > 0$ there is a distance δ (necessarily equal to or less than ϵ) such that any trajectory that once comes within δ of the critical point remains within ϵ of the critical point for all subsequent values of t. Figure 6.8*a* illustrates this behavior.

DEFINITION 2 The isolated critical point C is said to be **asymptotically stable** if and only if, first, it is stable and, second, there exists a $\delta^* > 0$ such that if $D(t^*) < \delta^*$, then

$$\lim_{t \to +\infty} x(t) = x_0 \quad \text{and} \quad \lim_{t \to +\infty} y(t) = y_0$$

Less formally, Definition 2 says that a critical point C is asymptotically stable if and only if it is stable and every trajectory that comes sufficiently close to C actually approaches C. Figure 6.8*b* illustrates this behavior.

Naturally, to complete our vocabulary, we also have the following definition.

DEFINITION 3 A critical point which is not stable is said to be **unstable.**

Table 6.2 summarizes the stability properties of the various critical points we have discussed.

The preceding results are stated still more concisely in the following theorem.

THEOREM 1 Given the linear system $dx/dt = ax + by$, $dy/dt = cx + ey$ with $ae - bc \neq 0$, so that $(0, 0)$ is the only critical point of the system. If the roots of the characteristic equation $m^2 - (a + e)m + (ae - bc) = 0$ are real and negative or complex with negative real part, then the critical point is asymptotically stable. If the roots of the characteristic equation are real and positive or complex with positive real part, then the critical point is unstable. If the roots of the characteristic equation are pure imaginary, then the critical point is stable but not asymptotically stable. Only when the characteristic roots are pure imaginaries does the given system have periodic solutions.

FIGURE 6.8
(*a*) A stable trajectory; (*b*) an asymptotically stable trajectory.

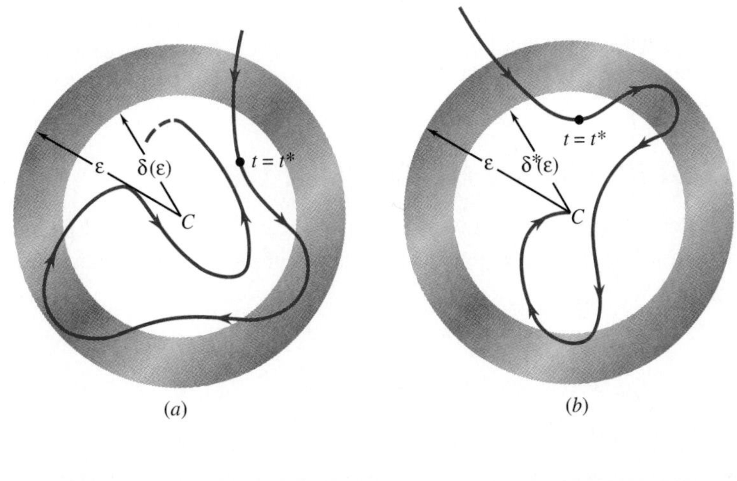

TABLE 6.2 Stability properties of the critical point (0, 0) of the linear system $dx/dt = ax + by$, $dy/dt = cx + ey$, whose characteristic equation is

$$m^2 - (a + e)m + (ae - bc) = 0$$

Nature of the characteristic roots	Nature of the critical point	Stability of the critical point
1. Real, unequal, and of like sign	Node (improper)	Asymptotically stable if roots are negative; unstable if roots are positive
2. Real, unequal, and of unlike sign	Saddle point	Unstable
3. Real and equal	Node (proper or improper)	Asymptotically stable if roots are negative; unstable if roots are positive
4. Pure imaginary	Center	Stable but not asymptotically stable
5. Complex but not pure imaginary	Spiral point	Asymptotically stable if real part of roots is negative; unstable if real part of roots is positive

EXERCISES

Determine the nature of the critical point (0, 0) for each of the following systems and tell whether it is stable, asymptotically stable, or unstable.

1. $dx/dt = 2x + 5y$
$dy/dt = x - 2y$

2. $dx/dt = 2x + 5y$
$dy/dt = -x + 5y$

3. $dx/dt = -4x + 3y$
$dy/dt = -2x + y$

4. $dx/dt = 2x + y$
$dy/dt = -x + 2y$

5. $dx/dt = x - 4y$
$dy/dt = x + 5y$

6. $dx/dt = -3x + y$
$dy/dt = -x - 3y$

7. Is the following an acceptable definition of asymptotic stability? A critical point C is asymptotically stable if every trajectory that comes sufficiently close to C actually approaches C. Justify your answer.

8. The motion of a mass-spring system is governed by the equation $m(d^2x/dt^2) + c(dx/dt) + kx = 0$. Convert this to a system of two first-order equations, show that this system has a single critical point, identify its type, and determine its stability properties.

9. Solve the system (11) and derive Eq. (12) by eliminating the parameter t between the functions thus obtained.

10. Work Exercise 9 for the system used to illustrate Case **4.**

11. Work Exercise 9 for the system used to illustrate Case **5.**

12. Show that if a trajectory of the system $dx/dt = g(x, y)$, $dy/dt = f(x, y)$ starts at a point which is not a critical point, it cannot reach a critical point in a finite length of time. *Hint:* Let $x = x(t)$, $y = y(t)$ be equations of the given trajectory, let (x_0, y_0) be the critical point, and suppose that there is a time $t = t_0$ for which $x(t_0) = x_0$ and $y(t_0) = y_0$. Then note that $x = x_0$, $y = y_0$ is a solution of the initial-value problem for which $x = x_0$ and $y = y_0$ when $t = t_0$.

13. Either by using the polar coordinate relation $\theta = \tan^{-1}(y/x)$ or by expressing x and y as single trigonometric functions, verify that as $t \to +\infty$, the trajectories described by (15) are traversed in the clockwise direction if $\lambda > 0$ and in the counterclockwise direction if $\lambda < 0$.

6.4 CRITICAL POINTS OF SYSTEMS THAT ARE APPROXIMATELY LINEAR

Having completed our preparatory investigation of the critical points and paths of linear systems, we now turn our attention to the general nonlinear autonomous system

$$\frac{dx}{dt} = F(x, y) \qquad (1)$$

$$\frac{dy}{dt} = G(x, y)$$

As a starting point, we assume that this system has an isolated critical point at the origin, which implies of course that $F(0, 0) = G(0, 0) = 0$. In this section we shall suppose further that in some neighborhood of the origin, F and G can be written in the form

$$\begin{aligned} F(x, y) &= ax + by + f(x, y) \\ G(x, y) &= cx + ey + g(x, y) \end{aligned} \tag{2}$$

where

$$ae - bc \neq 0 \tag{3}$$

and f and g are nonlinear functions which are small in comparison with $r = \sqrt{x^2 + y^2}$. More explicitly, the last requirement means that

$$\lim_{(x,y)\to(0,0)} \frac{f(x, y)}{\sqrt{x^2 + y^2}} = 0 \quad \text{and} \quad \lim_{(x,y)\to(0,0)} \frac{g(x, y)}{\sqrt{x^2 + y^2}} = 0 \tag{4}$$

In particular, these requirements will be met if, as we shall suppose, F and G possess MacLaurin expansions and

$$\left(\frac{\partial F}{\partial x}\bigg|_{0,0}\right)\left(\frac{\partial G}{\partial y}\bigg|_{0,0}\right) - \left(\frac{\partial F}{\partial y}\bigg|_{0,0}\right)\left(\frac{\partial G}{\partial x}\bigg|_{0,0}\right) \neq 0 \tag{5}$$

When this is the case, we can write

$$F(x, y) = F(0, 0) + \left[\frac{\partial F}{\partial x}\bigg|_{0,0} x + \frac{\partial F}{\partial y}\bigg|_{0,0} y\right] + \frac{1}{2}\left[\frac{\partial^2 F}{\partial x^2}\bigg|_{0,0} x^2 + 2\frac{\partial^2 F}{\partial x\,\partial y}\bigg|_{0,0} xy + \frac{\partial^2 F}{\partial y^2}\bigg|_{0,0} y^2\right] + \cdots \tag{6}$$

and

$$G(x, y) = G(0, 0) + \left[\frac{\partial G}{\partial x}\bigg|_{0,0} x + \frac{\partial G}{\partial y}\bigg|_{0,0} y\right] + \frac{1}{2}\left[\frac{\partial^2 G}{\partial x^2}\bigg|_{0,0} x^2 + 2\frac{\partial^2 G}{\partial x\,\partial y}\bigg|_{0,0} xy + \frac{\partial^2 G}{\partial y^2}\bigg|_{0,0} y^2\right] + \cdots \tag{7}$$

In these series, $F(0, 0)$ and $G(0, 0)$ are zero, by hypothesis. Hence if we make the identifications

$$a = \frac{\partial F}{\partial x}\bigg|_{0,0} \qquad b = \frac{\partial F}{\partial y}\bigg|_{0,0} \qquad c = \frac{\partial G}{\partial x}\bigg|_{0,0} \qquad e = \frac{\partial G}{\partial y}\bigg|_{0,0} \tag{8}$$

and take $f(x, y)$ and $g(x,y)$ to be the portions of the respective series consisting of the terms in x and y of the second degree and higher, (6) and (7) have the form of the right-hand members of Eqs. (2).

With a, b, c, e defined by (8), Condition (3) is satisfied because of the assumption (5). Moreover, the definitions we have given for $f(x, y)$ and $g(x, y)$ make it clear that in sufficiently small neighborhoods of the origin, Condition (4) is satisfied; that is, $f(x, y)$ and $g(x, y)$ are negligibly small in comparison with x and y. This suggests that around the origin the system defined by (1) and (2) behaves essentially like the linear system we studied in the last section, that is, is **almost linear.** This is true in some cases but false in others, as the following theorem, which we cite without proof, makes clear.

THEOREM 1 Consider the nonlinear autonomous system

$$\frac{dx}{dt} = ax + by + f(x, y)$$

$$\frac{dy}{dt} = cx + ey + g(x, y)$$

where $ae - bc \neq 0$ and

$$\lim_{(x,y)\to(0,0)} \frac{f(x, y)}{\sqrt{x^2 + y^2}} = 0 \qquad \lim_{(x,y)\to(0,0)} \frac{g(x, y)}{\sqrt{x^2 + y^2}} = 0$$

and the corresponding linear system

$$\frac{dx}{dt} = ax + by$$

$$\frac{dy}{dt} = cx + ey$$

Let $(0, 0)$ be an isolated critical point for each system and let m_1 and m_2 be the roots of the characteristic equation of the latter system:

$$m^2 - (a + e)m + (ae - bc) = 0$$

Then the critical points of the two systems are related as follows.

1. If m_1 and m_2 are real, unequal, and of like sign, then the critical point $(0, 0)$ is a node of both the linear system and the nonlinear system.
2. If m_1 and m_2 are real, unequal, and of unlike sign, then the critical point $(0, 0)$ is a saddle point of both the linear system and the nonlinear system, although the trajectories which are the asymptotes may be curves rather than straight lines.
3. If m_1 and m_2 are conjugate complex numbers which are not pure imaginaries, then the critical point $(0, 0)$ is a spiral point of both the linear system and the nonlinear system.
4. If m_1 and m_2 are real and equal, then the critical point $(0, 0)$ is a node of both the linear system and the nonlinear system except when simultaneously $a = e \neq 0$ and $b = c = 0$.
5. If m_1 and m_2 are real and equal and if simultaneously $a = e \neq 0$ and $b = c = 0$, then although $(0, 0)$ is a node of the linear system, it may be either a node or a spiral point of the nonlinear system.
6. If m_1 and m_2 are pure imaginaries, then although $(0, 0)$ is a center of the linear system, it may be either a center or a spiral point of the nonlinear system.

EXAMPLE 1

Determine the nature of the critical point $(0, 0)$ for the system

$$\frac{dx}{dt} = x + 2y + x \cos y$$

$$\frac{dy}{dt} = -y - \sin y$$

Our first step is to determine the related linear system. This requires that the linear terms in $x \cos y$ and $-\sin y$ be found and combined with the linear terms originally present. Hence we replace $x \cos y$ and $-\sin y$ by their Maclaurin expansions and then collect terms, getting

$$\frac{dx}{dt} = x + 2y + x\left(1 - \frac{y^2}{2!} + \frac{y^4}{4!} - \cdots\right) = 2x + 2y + x\left(-\frac{y^2}{2} + \frac{y^4}{24} - \cdots\right)$$

$$\frac{dy}{dt} = -y - \left(y - \frac{y^3}{3!} + \frac{y^5}{5!} - \cdots\right) = -2y - \left(-\frac{y^3}{6} + \frac{y^5}{120} - \cdots\right)$$

The linear system we must investigate in order to apply Theorem 1 is therefore

$$\frac{dx}{dt} = 2x + 2y$$

$$\frac{dy}{dt} = -2y$$

The characteristic equation of this system is $m^2 - 4 = 0$. Since its roots, $m = \pm 2$, are real, unequal, and of unlike sign, the critical point $(0, 0)$ is a saddle point of the linear system. Therefore, by Observation 2 of Theorem 1, it is also a saddle point of the given nonlinear system.

The equation obtained by dividing the two given equations is

$$\frac{dy}{dx} = \frac{-y - \sin y}{x + 2y + x \cos y}$$

or

$$(y + \sin y)\, dx + (x + 2y + x \cos y)\, dy = 0$$

This equation happens to be exact, hence (atypically) it can easily be solved. The solution, obtained by inspection, is

$$xy + y^2 + x \sin y = k$$

Figure 6.9*a* shows the configuration of the trajectories for the given nonlinear system, and Fig. 6.9*b* shows the trajectories of the related linear system. In this case one of the asymptotes of the trajectories of the nonlinear system is a curve and not a straight line.

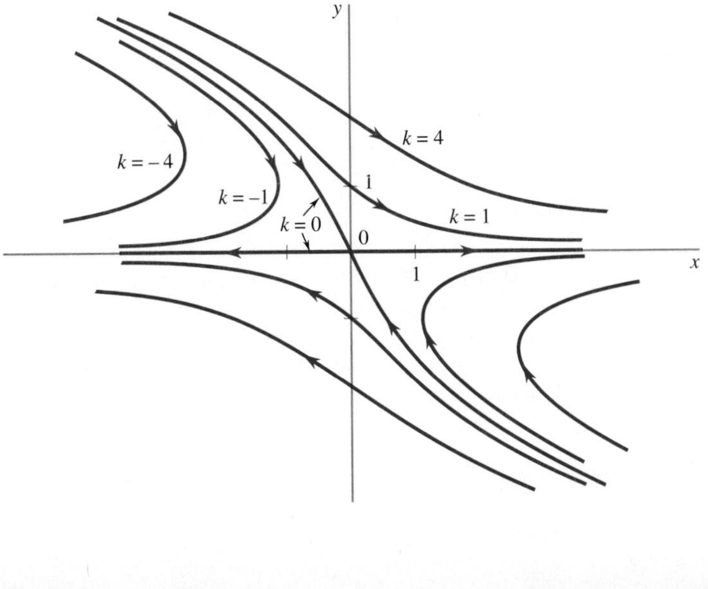

FIGURE 6.9*a*

Typical trajectories around the critical point of the nonlinear system $dx/dt = x + 2y + x \cos y$, $dy/dt = -y - \sin y$.

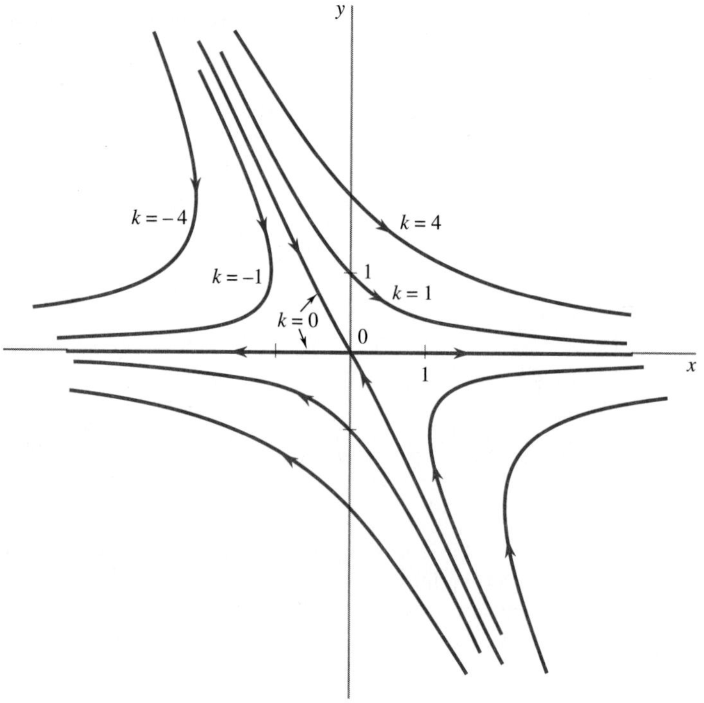

FIGURE 6.9b
Typical trajectories around the critical point of the system of Fig. 6.9*a* for its related linear system $dx/dt = 2x + 2y$, $dy/dt = -2y$.

EXAMPLE 2

Determine the nature of the critical point (0, 0) for the system

$$\frac{dx}{dt} = y$$

$$\frac{dy}{dt} = -x - y^2$$

Since $\lim_{(x,y)\to(0,0)} y^2/\sqrt{x^2+y^2} = 0$, the related linear system is

$$\frac{dx}{dt} = y$$

$$\frac{dy}{dt} = -x$$

and its characteristic equation is $m^2 + 1 = 0$. Since the characteristic roots are pure imaginaries, the critical point (0, 0) is a center for the linear system. Therefore, according to Observation 6 of Theorem 1, (0, 0) is either a center or a spiral point of the given nonlinear system, but the theorem does not tell us which.

In this case, however, the nature of the critical point can be determined, since the trajectories can be found explicitly. In fact, from the original pair of equations we find

$$\frac{dy}{dx} = -\frac{x}{y} - y \qquad \text{or} \qquad \frac{dy}{dx} + y = -\frac{x}{y}$$

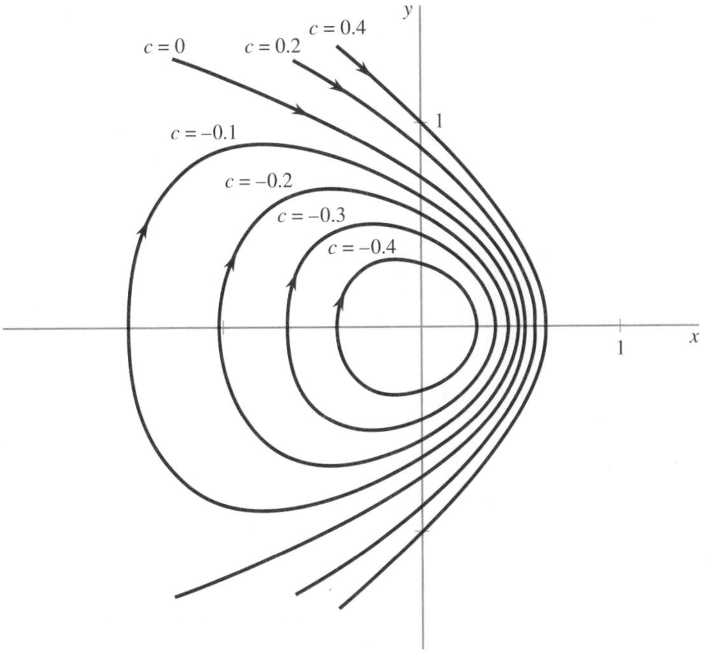

FIGURE 6.10
Typical trajectories around the center of the system $dx/dt = y$, $dy/dt = -x - y^2$.

which is a Bernoulli equation (Sec. 1.12) with $n = -1$. We therefore multiply the last equation by $2y$ and then put $z = y^2$, getting

$$\frac{dz}{dx} + 2z = -2x$$

Solving this as a linear first-order equation (Sec. 1.11), we find

$$z = y^2 = -x + \tfrac{1}{2} + Ce^{-2x}$$

If $C = 0$, this is the equation of a parabola. For $C > 0$, the trajectories are open curves somewhat resembling parabolas. For $-\frac{1}{2} < C < 0$, the trajectories are simple closed curves containing the critical point in their interiors. For $C = -\frac{1}{2}$, the path reduces to the single point (0, 0), and for $C < -\frac{1}{2}$, no path curves exist. From the plot of the paths shown in Fig. 6.10, it is clear that (0, 0) is a center of the given system.

EXAMPLE 3

Determine the nature of the critical point (0, 0) for the system

$$\frac{dx}{dt} = y$$

$$\frac{dy}{dt} = -x - y^3$$

By inspection, the related linear system is

$$\frac{dx}{dt} = y$$

$$\frac{dy}{dt} = -x$$

which is the same approximating system we found in Example 2. Therefore, again, (0, 0) is a center of the related linear system and either a center or a spiral point of the given nonlinear system, but Theorem 1 cannot tell us which.

In this case, neither the given system nor the related equation

$$\frac{dy}{dx} = -\frac{x + y^3}{y} \tag{9}$$

can be solved exactly by any method with which we are familiar. However, from a purely descriptive consideration of the last equation, it is possible to argue convincingly that (0, 0) is a spiral point. To do this, we observe first that if we substitute m for dy/dx in Eq. (9), the resulting curves

$$\{C_m\}\colon m = -\frac{x + y^3}{y} \quad \text{or} \quad x = -my - y^3 \qquad -\infty < m < +\infty \tag{10}$$

are the loci of points at which the solutions of Eq. (9); i.e., the trajectories of the given system have slope m. In other words, at every point where a trajectory intersects a curve C_m of the family (10), the trajectory has slope m. Curves which have this property with respect to the solutions of a differential equation are said to be **isoclines,** or "curves of equal slope" for that equation. In particular, $x = -y^3$ is the locus of points at which the trajectories have slope zero, $y = 0$ is the locus of points at which their slope is infinite, $x = -y - y^3$ is the locus of points at which their slope is 1, and $x = y - y^3$ is the locus of points at which their slope is -1. These observations are illustrated in Fig. 6.11a, where short segments of the appropriate inclination indicate the slope associated with each isocline. In the second quadrant, the intersections of a vertical line with the curves (10) correspond to values of m which decrease monotonically from $+\infty$ when $y = 0$ to $-\infty$ as $y \to \infty$. In the fourth quadrant, the intersections of a vertical line and the curves (10) correspond to values of m which decrease monotonically from $+\infty$ when $y = 0$ to $-\infty$ as $y \to -\infty$.

Now consider an arbitrary circle $\Gamma\colon x^2 + y^2 = r^2$, where r is sufficiently small that in the first and third quadrants Γ intersects each isocline in a single point (see Exercise 19). Let A and B be the ends of the horizontal diameter of Γ and let C and D be the points where the isocline C_0 intersects Γ (Fig. 6.11b). From the order relations among the isoclines shown in Fig. 6.11a, it follows that

1. Along the arc AC, m increases monotonically from $-\infty$ to 0.
2. Along the arc CB, m increases monotonically from 0 to $+\infty$.
3. Along the arc BD, m increases monotonically from $-\infty$ to 0.
4. Along the arc DA, m increases monotonically from 0 to $+\infty$.

Now let P: $(r\cos\theta, r\sin\theta)$ be an arbitrary point on Γ distinct from A and B. At P the slope of the circle is

$$m_\gamma = -\frac{\cos\theta}{\sin\theta}$$

and at P the slope assigned by Eq. (9) to the trajectory which passes through P is

$$m_t = -\frac{\cos\theta + r^2\sin^3\theta}{\sin\theta} = -\frac{\cos\theta}{\sin\theta} - r^2\sin^2\theta$$

Since P is neither A nor B, the term $r^2\sin^2\theta$ is different from zero; hence from the last equation it follows that $m_t < m_\gamma$. The relative directions of the tangent to the circle at P and the tangent to the trajectory through P are therefore as shown in Fig. 6.11b.

Thus at all points on any sufficiently small circle $x^2 + y^2 = r^2$ (except at the ends of the horizontal diameter, where the tangent to the trajectory is also tangent to the circle), the tangent to the trajectory through that point is directed into (or out of) the interior of the circle. This strongly suggests that the trajectories must be spirals approaching (or receding from) the critical point as a point traverses them in the clockwise (or counterclockwise) direction. Moreover, from the equation $dx/dt = y$, it follows that the trajectories are traversed in the clockwise direction as $t \to +\infty$.

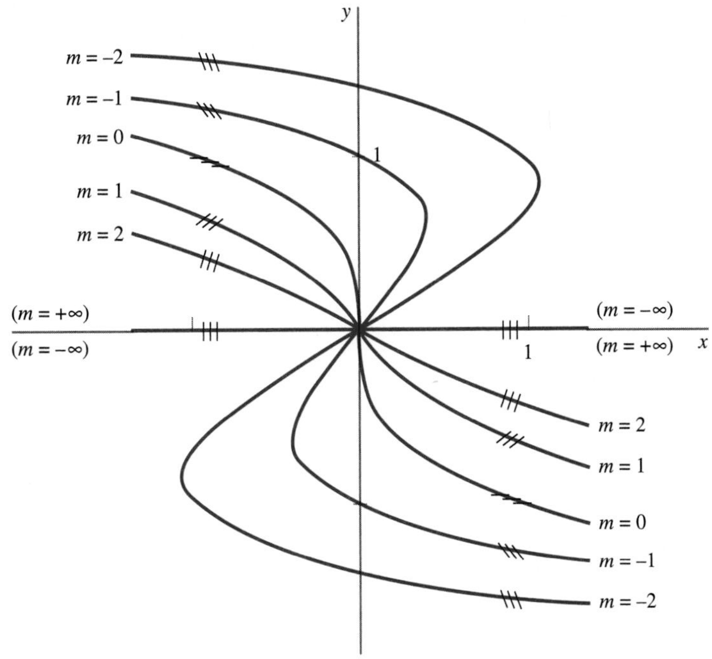

FIGURE 6.11a
Typical isoclines of the system $dx/dt = y$, $dy/dt = -x - y^3$.

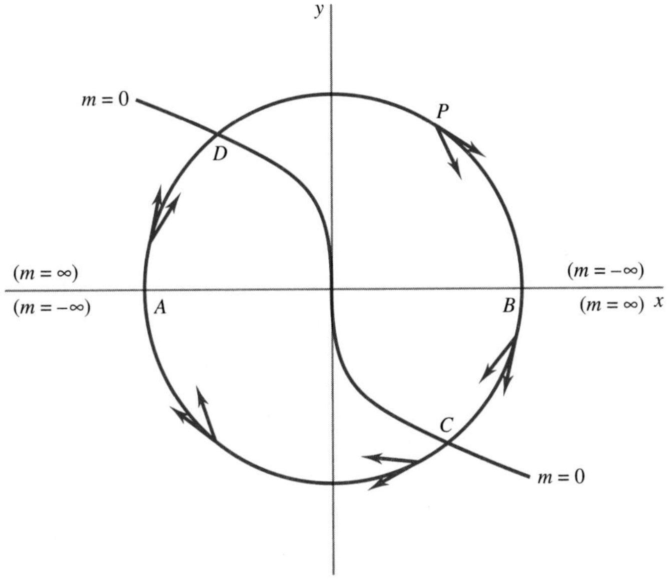

FIGURE 6.11b

When the Taylor expansions of $F(x, y)$ and $G(x, y)$ each contain linear terms, Theorem 1 gives us information about the nature of the critical points of the system

$$\frac{dx}{dt} = F(x, y)$$

$$\frac{dy}{dt} = G(x, y)$$

but it does not give us any information about the stability of the critical point. The next theorem, which we also cite without proof, provides us with useful information of this sort.

THEOREM 2 Consider the nonlinear autonomous system

$$\frac{dx}{dt} = ax + by + f(x, y)$$

$$\frac{dy}{dt} = cx + ey + g(x, y)$$

where $ae - bc \neq 0$ and

$$\lim_{(x,y)\to(0,0)} \frac{f(x, y)}{\sqrt{x^2 + y^2}} = 0 \qquad \lim_{(x,y)\to(0,0)} \frac{g(x, y)}{\sqrt{x^2 + y^2}} = 0$$

and the corresponding linear system

$$\frac{dx}{dt} = ax + by$$

$$\frac{dy}{dt} = cx + ey$$

Let $(0, 0)$ be an isolated critical point of each system, and let m_1 and m_2 be the roots of the characteristic equation of the linear system:

$$m^2 - (a + e)m + (ae - bc) = 0$$

Then the stability of the two systems at $(0, 0)$ is determined as follows.

1. If m_1 and m_2 are real and negative or conjugate complex with negative real part, then $(0, 0)$ is an asymptotically stable critical point of both the linear system and the given nonlinear system.
2. If m_1 and m_2 are pure imaginary, then although $(0, 0)$ is a stable critical point of the linear system, it is not necessarily a stable critical point of the nonlinear system. In fact, the critical point $(0, 0)$ of the given nonlinear system may be asymptotically stable, stable but not asymptotically stable, or unstable.
3. If either m_1 or m_2 is real and positive, or if m_1 and m_2 are conjugate complex with positive real parts, then $(0, 0)$ is an unstable critical point for both the linear system and the given nonlinear system.

Reconsidering Example 1 in the light of Theorem 2, we note that since the characteristic roots of the related linear system are ± 2, therefore, by Observation 3 of Theorem 2, the critical point $(0, 0)$ is unstable for both the linear system and the nonlinear system. We confirmed this of course by determining and plotting the path curves (Fig. 6.9). In both Examples 2 and 3 the characteristic roots of the related linear system were the conjugate imaginaries $\pm i$. Hence, according to Observation 2 of Theorem 2, the critical point $(0, 0)$ while stable for the linear system may be asymptotically stable, stable but not asymptotically stable, or unstable for the given nonlinear system. From our analysis of the path curves, we found the critical point $(0, 0)$ to be stable but not asymptotically stable in Example 2 and asymptotically stable in Example 3.

EXERCISES

Determine the nature and stability of the critical point (0, 0) of each of the following systems.

1. $dx/dt = 3x + 4y + x^2$
$dy/dt = 4x - 3y - 2xy$

2. $dx/dt = 6x + 10y - x^2$
$dy/dt = -4x - 6y + 2xy$

3. $dx/dt = -x - x\cos y$
$dy/dt = y + \sin y$

4. $dx/dt = x + 2y + 2\sin y$
$dy/dt = -3y - xe^x$

5. $dx/dt = 1 + y - e^{-x}$
$dy/dt = y - \sin x$

6. $dx/dt = e^{-x+y} - \cos x$
$dy/dt = \sin(x - 3y)$

7. $dx/dt = x + x^2 - 3xy$
$dy/dt = -2x + y + 3y^2$

8. $dx/dt = -\sin(x - y)$
$dy/dt = 1 - 5y - e^x$

9. Find the equation of the family of trajectories in Exercise 1.

10. Find the equation of the family of trajectories in Exercise 3 and sketch several members of the family.

Determine the critical points of each of the following systems and discuss their nature and stability.

11. $dx/dt = 2y + x^2$
$dy/dt = -2x - 4y$

Hint: Translate axes by the substitutions $X = x + h$, $Y = y + k$, where h and k are chosen so that the critical point becomes (0, 0).

12. $dx/dt = -3y + 3xy$
$dy/dt = 2x - y - 3$

13. $dx/dt = x + y^2$
$dy/dt = x + y$

14. $dx/dt = 1 - xy$
$dy/dt = x - y^3$

15. $dx/dt = 1 - y$
$dy/dt = x^2 - y^2$

16. If the equation for dy/dx obtained from the system

$$\frac{dx}{dt} = ax + by + f(x, y) \qquad \frac{dy}{dt} = cx + ey + g(x, y)$$

is exact (Sec. 1.7), show that the critical point (0, 0) is never asymptotically stable.

17. Where does the geometric argument used in Example 3 to prove that the trajectories are spirals break down when applied to the system in Example 2?

18. Justify the directions indicated by the arrows on the trajectories **(a)** in Fig. 6.9 and **(b)** in Fig. 6.10.

19. In Example 3, show that the argument requires that $r < 3\sqrt{3}/2$. *Hint:* Note that the circle Γ must not intersect any trajectory in more than one point in the first quadrant.

The following exercises illustrate the use of the phase plane in the study of conservative dynamical systems with a single degree of freedom.

20. According to Newton's law, the motion of a particle of mass m under the influence of a force $F = F(x)$ is governed by the differential equation $m\,(d^2x/dt^2) = F(x)$. Express this equation as a pair of simultaneous equations in x and y $(= dx/dt)$. Given that $x = x_0$ and y $(= dx/dt) = y_0$ when $t = 0$, show that the equation of the family of trajectories is $my^2 - my_0^2 = 2[\int_0^x F(x)\,dx - \int_0^{x_0} F(x)\,dx]$.

21. Recalling that the kinetic energy of a moving particle is $\frac{1}{2}m(dx/dt)^2$ and that the potential energy of the particle is $V(x) = -\int_0^x F(x)\,dx$, show that the equation of the trajectories obtained in Exercise 20 can be written in the form

$$\frac{1}{2}my^2 + V(x) = h \qquad \text{or} \qquad y = \pm\sqrt{\frac{2}{m}[h - V(x)]}$$

What is the significance of the constant h?

22. Show that $V'(x) = 0$ at every critical point of the system discussed in Exercise 20.

23. If $V(x_0)$ is a minimum at a critical point $(x_0, 0)$ of the system discussed in Exercise 20, show that the critical point is a center and therefore stable. *Hint:* Let $h_0 = V(x_0)$ be the minimum value of $V(x)$ and let a and b, with $a < b$, be the values of x at which $V(x) = h_1 > h_0$. Then using the formula for y obtained in Exercise 21, show that the trajectory corresponding to $h = h_1$ is a closed curve enclosing the point $(x_0, 0)$ and extending from $x = a$ to $x = b$.

24. If $V(x_0)$ is a maximum at a critical point $(x_0, 0)$ of the system discussed in Exercise 20, show that the critical point is a saddle point and therefore unstable. *Hint:* As in Exercise 23, let $h_0 = V(x_0)$ be the maximum value of $V(x)$, let $h_1 < h_0$ be an arbitrary value of V, and let a and b, with $a < b$, be the values of x at which $V(x) = h_1$. Then using the formula for y obtained in Exercise 21, show that the trajectory corresponding to $h = h_1$ is a hyperbola-like curve opening to the left from $x = a$ and to the right from $x = b$. If $h_2 > h_0$, show similarly that the trajectory corresponding to $h = h_2$ is a hyperbola-like curve opening upward and downward from vertices directly above and below $x = x_0$.

25. What is the trajectory corresponding to $h = h_0$ **(a)** in Exercise 23 and **(b)** in Exercise 24?

26. Discuss the case in which $V(x)$ has a horizontal inflection at $x = x_0$.

27. Assuming $m = 2$, discuss the configuration of the trajectories around the critical point (0, 0) if $F(x) = -2x - 4x^3$.

28. Work Exercise 27 if $F(x) = 2x + 4x^3$.

29. If $m = 2$ and $F(x) = -2x + 4x^3$, show that the system has three critical points and determine the nature of each. Discuss the particular trajectory corresponding to $h = \frac{1}{4}$.

30. If $m = 2$ and $F(x) = 2x - 4x^3$, show that the system has three critical points and determine the nature of each. Discuss the particular trajectories corresponding to $h = 0$ and $h = -\frac{1}{4}$.

31. If $m = 2$, discuss the nature of each critical point of the system if **(a)** $F(x) = 2x - 3x^2$ and **(b)** $F(x) = -2x + 3x^2$.

6.5 SYSTEMS THAT ARE NOT APPROXIMATELY LINEAR

Up to this point we have been concerned exclusively with autonomous systems

$$\frac{dx}{dt} = F(x, y) \qquad \frac{dy}{dt} = G(x, y)$$

in which the Taylor expansions of both $F(x, y)$ and $G(x, y)$ around a critical point contained linear terms; and we have seen that from the system resulting when only the linear terms were retained, certain properties of the actual nonlinear system could be inferred. If the expansions of $F(x, y)$ and $G(x, y)$ around a critical point contain no linear terms, Theorems 1 and 2 of the last section are clearly inapplicable, and an investigation of the trajectories and their stability is too difficult to attempt in an elementary treatment such as ours. Hence we shall content ourselves with an example or two and then cite, without proof, several fundamental theorems that pertain to these more general autonomous systems.

As an illustration of the complicated configurations of path curves which may occur, let us consider first the system

$$\frac{dx}{dt} = y^2 - x^2 \qquad \frac{dy}{dt} = 2xy \qquad \text{or} \qquad \frac{dy}{dx} = \frac{2xy}{y^2 - x^2}$$

Clearly, $(0, 0)$ is an isolated critical point of this system, but since neither $y^2 - x^2$ nor $2xy$ contains any linear terms, the theorems of the last section cannot be applied to help us determine its nature or stability. In this case, however, the equation for dy/dx is easy to solve, for it can be written in the form

$$3y^2\, dy = 3x^2\, dy + 6xy\, dx$$

and then integrated by inspection. This gives us

$$y^3 = 3x^2y + k$$

as the equation of the family of trajectories. Several of these curves are shown in Fig. 6.12. Evidently, the critical point $(0, 0)$ is a kind of ''super'' saddle point with three asymptotes. Each of the trajectories is a cubic curve consisting of three branches located in alternate sectors formed by the three asymptotes.

As a second example, let us consider the system

$$\frac{dx}{dt} = x^2 \qquad \frac{dy}{dt} = 2y^2 - xy$$

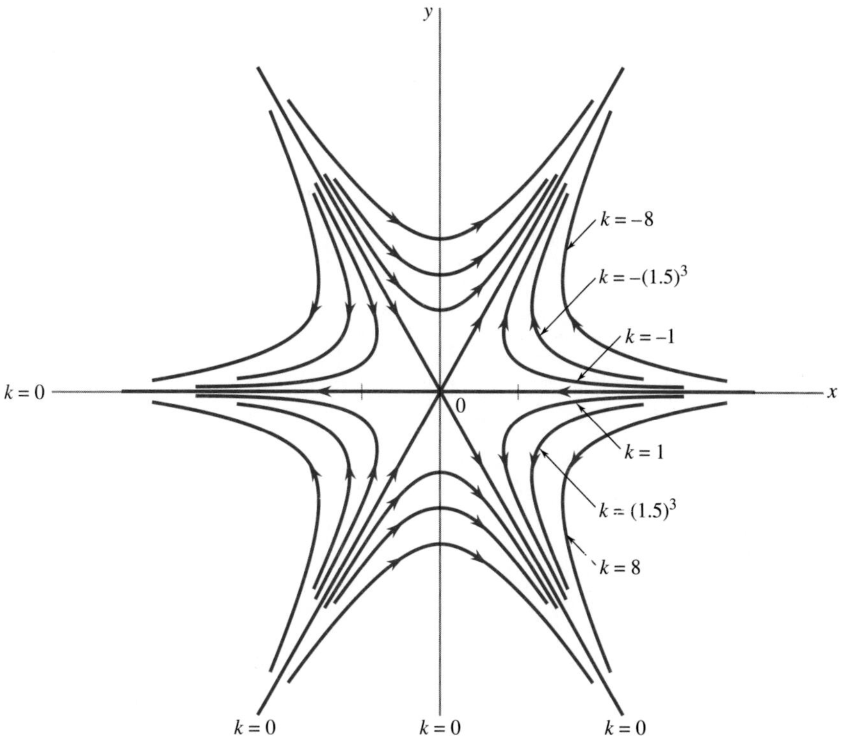

FIGURE 6.12
Typical trajectories around the critical point (0, 0) of the system $dx/dt = y^2 - x^2$, $dy/dt = 2xy$.

Since the associated equation

$$\frac{dy}{dx} = \frac{2y^2 - xy}{x^2}$$

is homogeneous, it is possible in this case also to solve for the equation of the family of trajectories. Thus, making the substitution $y = ux$ (Sec. 1.10), we obtain the separable equation

$$x\,du = 2(u^2 - u)\,dx$$

Separating variables, integrating, and then replacing u by y/x, we obtain finally

$$y = \frac{x}{1 - kx^2}$$

Several curves of this family are shown in Fig. 6.13. In this case the critical point (0, 0) appears to be a combination of a node and a saddle point.

Since there are infinitely many possibilities for the configuration of trajectories around a critical point of a general nonlinear system, no attempt is made to describe and classify the critical points of such systems. Instead, studies of these systems are concerned primarily with questions of stability and periodicity. We shall conclude this section by first introducing a necessary definition and then stating without proof several of the basic theorems dealing with these matters.

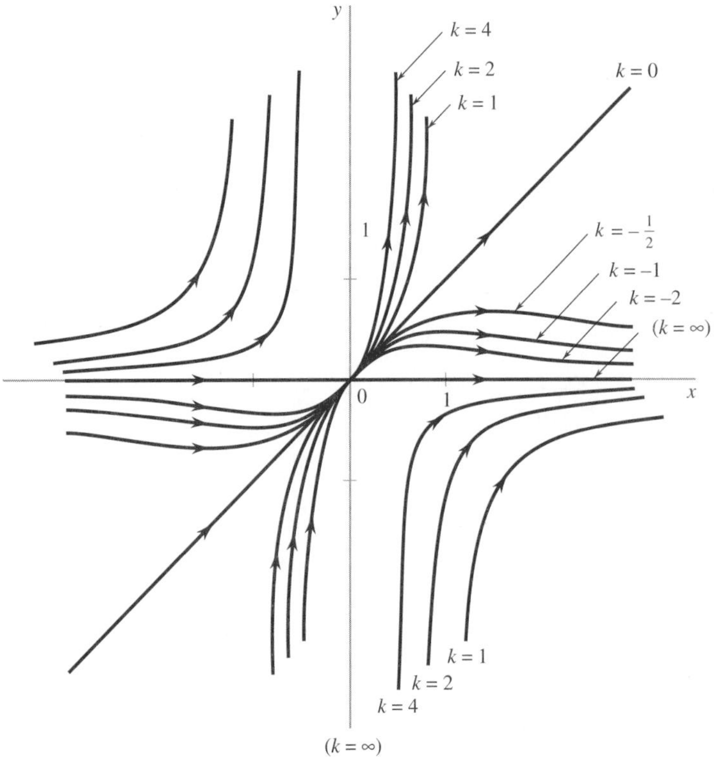

FIGURE 6.13
Typical trajectories around the critical point $(0, 0)$ of the system $dx/dt = x^2$, $dy/dt = 2y^2 - xy$.

We begin by introducing the concept of *definiteness* for general continuous functions. This is an important property which we will study at greater length in our discussion of quadratic forms and matrices in Chap. 14.

DEFINITION 1 Let D be a domain containing the point (x_0, y_0) in its interior and let V be a function defined throughout D.

1. If $V(x_0, y_0) = 0$ and if $V(x, y) > 0$ at all other points of D, then V is said to be **positive definite** throughout D.
2. If $V(x_0, y_0) = 0$ and if $V(x, y) \geq 0$ at all other points of D, with the equality holding for at least one point other than (x_0, y_0), then V is said to be **positive semidefinite** throughout D.
3. If $V(x_0, y_0) = 0$ and if $V(x, y) < 0$ at all other points of D, then V is said to be **negative definite** throughout D.
4. If $V(x_0, y_0) = 0$ and if $V(x, y) \leq 0$ at all other points of D, with the equality holding for at least one point other than (x_0, y_0), then V is said to be **negative semidefinite** throughout D.

EXAMPLE 1

If (x_0, y_0) is the point $(0, 0)$ and if D is the interior of the circle $x^2 + y^2 = 1$, then

1. The function $V = (x + y)^2 + y^4$ is positive definite in D since it is nonnegative throughout D and zero only at $(0, 0)$.
2. The function $V = y^2(x^2 + y^2)$ is positive semidefinite in D since it is nonnegative throughout D but vanishes in D not only at $(0, 0)$ but also at all points of the horizontal diameter of D.
3. The function $V = (x^2 + y^2)/(x^2 + y^2 - 2)$ is negative definite in D since it is nonpositive throughout D and is zero only at $(0, 0)$.

4. The function $V = x^2/(x^2 + y^2 - 2)$ is negative semidefinite in D since it is nonpositive throughout D but vanishes in D not only at $(0, 0)$ but also at all points of the vertical diameter of D.

Suppose, now, that we have a trajectory

$$\Gamma: x = x(t),\ y = y(t)$$

of the system

$$\frac{dx}{dt} = F(x, y) \qquad \frac{dy}{dt} = G(x, y) \tag{1}$$

and a function V defined and differentiable at all points of Γ. This means that on the curve Γ, V is a function of t,

$$V = V[x(t), y(t)]$$

and can be differentiated with respect to t. Thus, from the familiar chain rule, we have

$$\frac{dV}{dt} \equiv \dot{V} = \frac{\partial V}{\partial x}\frac{dx}{dt} + \frac{\partial V}{\partial y}\frac{dy}{dt}$$

or, using the values of dx/dt and dy/dt provided by (1),

$$\frac{dV}{dt} \equiv \dot{V} = \frac{\partial V}{\partial x}F(x, y) + \frac{\partial V}{\partial y}G(x, y) \tag{2}$$

The function $\dot{V}$ defined by (2) is called the **derivative of V with respect to the system** (1).

We are now in a position to state two theorems, due to the Russian mathematician A. M. Liapunov (1857–1918), which deal, respectively, with the stability and the instability of the general system (1).

THEOREM 1 If the autonomous system

$$\frac{dx}{dt} = F(x, y)$$

$$\frac{dy}{dt} = G(x, y)$$

has an isolated critical point at the origin, if F and G have continuous first partial derivatives in some domain D containing the origin, and if there exists a function $V = V(x, y)$ with continuous first partial derivatives which is positive definite in D, then

1. If the derivative of V with respect to the given system is negative definite in D, the origin is an asymptotically stable critical point.
2. If the derivative of V with respect to the given system is negative semidefinite in D, the origin is at least a stable critical point.

THEOREM 2 Let the origin be an isolated critical point of the autonomous system

$$\frac{dx}{dt} = F(x, y)$$

$$\frac{dy}{dt} = G(x, y)$$

Let F and G have continuous first partial derivatives in some domain D containing the origin and let V be a function which has continuous first partial derivatives in D and takes on the value 0 at the critical point (0, 0). Then the origin is an unstable critical point if either of the following conditions is satisfied.

1. In every neighborhood of the origin there is at least one point where V is positive and the derivative of V with respect to the system is positive definite in D.
2. In every neighborhood of the origin there is at least one point where V is negative and the derivative of V with respect to the system is negative definite in D.

A function $V = V(x, y)$ having either the properties described in Theorem 1 or the properties described in Theorem 2 is said to be a **Liapunov function** for the corresponding autonomous system. Unfortunately, Theorems 1 and 2 contain no clues to help us determine an appropriate function V, and it usually takes considerable ingenuity to construct one, if indeed it exists. However, the following theorem from elementary algebra is useful in identifying a class of functions from which a suitable Liapunov function $V(x, y)$ can sometimes be selected.

THEOREM 3 The function $V(x, y) = Ax^2 + Bxy + Cy^2$

1. Is positive definite if and only if $A > 0$ and $B^2 - 4AC < 0$
2. Is negative definite if and only if $A < 0$ and $B^2 - 4AC < 0$
3. Is positive semidefinite if and only if $A > 0$ and $B^2 - 4AC = 0$
4. Is negative semidefinite if and only if $A < 0$ and $B^2 - 4AC = 0$

EXAMPLE 2

Show that the origin is an asymptotically stable critical point of the system

$$\frac{dx}{dt} = y - 2x^3$$

$$\frac{dy}{dt} = -2x - 3y^5$$

It is not difficult to verify that the origin is the only critical point of the given system. To determine its nature, let us try, tentatively, to find a Liapunov function of the form

$$V(x, y) = Ax^2 + Cy^2$$

Using this, we see at once that

$$\begin{aligned}\dot{V} &= 2Ax(y - 2x^3) + 2Cy(-2x - 3y^5)\\ &= -4Ax^4 + (2A - 4C)xy - 6Cy^6\end{aligned}$$

If we take $A = 2$ and $C = 1$, this becomes

$$\dot{V} = -8x^4 - 6y^6$$

which is clearly a negative definite function. Moreover, with $A = 2$ and $C = 1$, the function V becomes

$$V = 2x^2 + y^2$$

which is obviously positive definite. Hence Condition 1 of Theorem 1 is satisfied, and therefore the origin is an asymptotically stable critical point, as asserted.

EXAMPLE 3

Show that the origin is an asymptotically stable critical point of the system

$$\frac{dx}{dt} = -x^5 - y^3$$

$$\frac{dy}{dt} = 3x^3 - y^3$$

Clearly, the origin is the only critical point of the system. Again, to investigate it, let us begin with the tentative Liapunov function

$$V = Ax^2 + Cy^2$$

From this we find at once that

$$\begin{aligned}\dot{V} &= 2Ax(-x^5 - y^3) + 2Cy(3x^3 - y^3)\\ &= -2Ax^6 + (-2Axy^3 + 6Cx^3y) - 2Cy^4\end{aligned}$$

As in Example 2, the first and last terms in $\dot{V}$ form a negative definite function if A and C are positive, but in this case the components of the middle term do not cancel and it is thus not clear whether $\dot{V}$ is negative definite or not.

Reconsidering the middle term in $\dot{V}$, we note that it could be eliminated by proper choice of A and C if the factor x in the first term were x^3 and the factor y in the second term were y^3. Moreover, these replacements can be accomplished by beginning with the tentative Liapunov function

$$V = Ax^4 + Cy^4$$

In fact, for this function we have

$$\begin{aligned}\dot{V} &= 4Ax^3(-x^5 - y^3) + 4Cy^3(3x^3 - y^3)\\ &= -4Ax^8 + (-4A + 12C)x^3y^3 - 4Cy^6\end{aligned}$$

From this, by choosing $A = 3$ and $C = 1$, we obtain

$$\dot{V} = -12x^8 - 4y^6$$

and

$$V = 3x^4 + y^4$$

Thus our second choice for V is positive definite, and the related derivative $\dot{V}$ is negative definite. Hence, by Theorem 1, the origin is an asymptotically stable critical point of the given system, as asserted.

EXAMPLE 4

Determine the nature of the critical point $(0, 0)$ for the system

$$\frac{dx}{dt} = 2x^3 - 2xy^2$$

$$\frac{dy}{dt} = -5x^2y - 4y^3$$

Again trying

$$V = Ax^2 + Cy^2$$

we find

$$\begin{aligned}\dot{V} &= 2Ax(2x^3 - 2xy^2) + 2Cy(-5x^2y - 4y^3)\\ &= 4Ax^4 + (-4A - 10C)x^2y^2 - 8Cy^4\end{aligned}$$

Here, if we eliminate the middle term by taking $A = 5$ and $C = -2$, we are left with

$$\dot{V}=20x^4 + 16y^4$$

and

$$V = 5x^2 - 2y^2$$

Obviously $\dot{V}$ is positive definite; hence Theorem 1 cannot be applied. However, it is clear that $V(0, 0) = 0$ and that $V(x, y) > 0$ at points arbitrarily close to $(0, 0)$ since in fact V is positive at all points of the x axis except the origin. Hence Condition 1 of Theorem 2 is satisfied, and therefore $(0, 0)$ is an unstable critical point of the system.

EXERCISES

Using Liapunov functions of the form $Ax^2 + By^2$, verify that for each of the following systems the origin is a critical point of the indicated type.

1. $dx/dt = -x^3 + xy^2$ $\quad dy/dt = -4x^2y - y^3$
Asymptotically stable

2. $dx/dt = -x^3 + 3xy^2$ $\quad dy/dt = x^2y - 4y^3$
At least stable

3. $dx/dt = 2xy + x^3$ $\quad dy/dt = 2x^2 - y^3$ $\quad$ Unstable

4. $dx/dt = -x - x^3 + 4xy^2$ $\quad dy/dt = -2x - y - 4y^3$
Asymptotically stable

Using Liapunov functions of the form $V = Ax^2 + By^2$, determine the nature of the critical point $(0, 0)$ for each of the following systems.

5. $dx/dt = -x^3 + 3y^2$ $\quad dy/dt = -y^3 - 2xy$

6. $dx/dt = -x^3 - 8xy^2$ $\quad dy/dt = -2x^2y + 9y^3$

7. $dx/dt = x - 4y + x^3$ $\quad dy/dt = -4y - y^3 + 2x^2y$

8. $dx/dt = -x + 2y - x^3$ $\quad dy/dt = -y + 2x^2y - y^3$

9. Using the Liapunov function $V = xy$, determine the nature of the critical point $(0, 0)$ for the system

$$\frac{dx}{dt} = -2xy - y^3 \qquad \frac{dy}{dt} = -x + 2y^2$$

10. Using a Liapunov function of the form $V = Ax^2 + By^4$, determine the nature of the critical point $(0, 0)$ for the system

$$\frac{dx}{dt} = -x^3 - 3xy^4 \qquad \frac{dy}{dt} = x^2y - 2y^3 - y^5$$

11. Show that the qualification "at least" in Conclusion 2 of Theorem 1 is necessary by exhibiting a positive definite Liapunov function whose derivative with respect to the system

$$\frac{dx}{dt} = -x^3 + xy^2 \qquad \frac{dy}{dt} = -3x^2y - y^3$$

is negative semidefinite and another positive definite Liapunov function whose derivative with respect to this system is negative definite.

12. Using a Liapunov function of the form $V = Ax^2 + By^2$, show that the critical point $(0, 0)$ is asymptotically stable for the system

$$\frac{dx}{dt} = a^2y - xf(x, y) \qquad \frac{dy}{dt} = -b^2x - yf(x, y)$$

if $f(x, y)$ is positive throughout some neighborhood of the origin and is unstable if $f(x, y)$ is negative throughout some neighborhood of the origin.

13. Show that $(0, 0)$ is an asymptotically stable critical point of the system

$$\frac{dx}{dt} = y^3 - x^3f(x, y) \qquad \frac{dy}{dt} = -x^3 - y^3f(x, y)$$

if $f(x, y)$ is positive throughout some neighborhood of the origin and is an unstable critical point if $f(x, y)$ is negative throughout some neighborhood of the origin.

14. For each of the following functions, determine whether in some neighborhood of the origin it is positive definite, positive semidefinite, negative definite, negative semidefinite, or none of these.

(a) $x^2 - 4xy + 5y^2$ $\quad$ **(b)** $x^2 + x^4 + 4xy + 4y^2$
(c) $1 - e^{-xy}$ $\quad$ **(d)** $x^4 - 2x^2y^4 + y^8$
(e) $x^4 - 3x^2y^4 + y^8$ $\quad$ **(f)** $x^4 - x^2y^2 + y^8$
(g) $2x^4 - 4x^3y + 4x^2y^2 - 4xy^3 + 2y^4$
(h) $\sin(xy)$

6.6 PERIODIC SOLUTIONS AND LIMIT CYCLES

In Sec. 6.3 we observed that if the roots of the characteristic equation

$$m^2 - (a + e)m + (ae - bc) = 0$$

of the linear system

$$(1) \qquad \frac{dx}{dt} = ax + by \qquad \frac{dy}{dt} = cx + ey$$

were pure imaginaries, then the critical point (0, 0) was a center. This meant that the family of trajectories was a continuous infinity of nested simple closed curves each containing the critical point (0, 0) in its interior (Fig. 6.6). In this case the system had infinitely many periodic solutions, but in no other case did it have any periodic solutions at all.

In Sec. 6.4 we saw further that if the roots of the characteristic equation of the linear system associated with the approximately linear system

$$(2) \qquad \frac{dx}{dt} = ax + by + f(x, y) \qquad \frac{dy}{dt} = cx + ey + g(x, y)$$

were pure imaginaries, then the singular point (0, 0) might be a center or it might be a spiral point. This seems to suggest that for certain nonlinear systems it is also true that there are either infinitely many periodic solutions or no periodic solutions. This is not the case, however, and it is not difficult to imagine why. For all we know, the trajectories winding outward from a spiral point instead of spiraling to infinity may wind ever and ever nearer to some simple closed curve C enclosing the critical point and forming a barrier, so to speak, to the outward progress of the spirals. At the same time, paths outside C may spiral in from infinity and wind ever more closely around C on the outside. If this is possible, then there might be one, or perhaps several, periodic solutions forming the boundaries of annular regions around the critical point in which the members of the infinite family of spiral paths would lie.

The simplest example of a system exhibiting this behavior is probably

$$(3) \qquad \frac{dx}{dt} = y + x - x(x^2 + y^2) \qquad \frac{dy}{dt} = -x + y - y(x^2 + y^2)$$

and we shall soon see that it does indeed have a single periodic solution toward which the other solutions all spiral. As we prepare to investigate this system, it should be clear that since we are concerned with spiral behavior, it will probably be convenient to find dr/dt and $d\theta/dt$ and work in polar coordinates rather than work with Eqs. (3). In particular, from the sign of dr/dt we should be able to tell whether r increases indefinitely, decreases to zero, or approaches a limiting value $r = r_0$ as t becomes infinite. Accordingly, as a preliminary step, we begin with the familiar polar coordinate relations and compute the derivatives of r and θ with respect to t. From $r^2 = x^2 + y^2$, $r > 0$, we find at once

$$r\frac{dr}{dt} = x\frac{dx}{dt} + y\frac{dy}{dt} \tag{4}$$

Then from the relation $\theta = \tan^{-1}(y/x)$ we find

$$\frac{d\theta}{dt} = \frac{1}{1 + (y/x)^2} \cdot \frac{x(dy/dt) - y(dx/dt)}{x^2} = -\frac{1}{x^2 + y^2}\left(y\frac{dx}{dt} - x\frac{dy}{dt}\right)$$

or

$$-r^2\frac{d\theta}{dt} = y\frac{dx}{dt} - x\frac{dy}{dt} \tag{5}$$

Returning to Eqs. (3), if we multiply the first equation by x and the second by y and add, we obtain

$$x\frac{dx}{dt} + y\frac{dy}{dt} = x^2 + y^2 - (x^2 + y^2)^2$$

or, using (4),

$$r\frac{dr}{dt} = r^2 - r^4$$

or finally

$$\frac{dr}{dt} = r(1 - r^2) \tag{6}$$

Similarly, if we multiply the first of Eqs. (3) by y and the second by x and subtract, we get

$$y\frac{dx}{dt} - x\frac{dy}{dt} = x^2 + y^2$$

or, using (5),

$$-r^2\frac{d\theta}{dt} = r^2$$

or finally

$$\frac{d\theta}{dt} = -1 \tag{7}$$

From (6) it is clear that $dr/dt > 0$ if $0 < r < 1$; that is, as t increases, r increases by (6), and the path curves spiral outward as long as they lie within the circle $r = 1$. It also follows from (6) that $dr/dt < 0$ if $r > 1$; that is, r decreases, and again by (6) the path curves spiral inward as long as they lie outside the circle $r = 1$. Moreover, the circle $r = 1$ is itself a trajectory since Eqs. (6) and (7) are satisfied by

$$r = 1 \qquad \theta = -t + t_0 \qquad t_0 \text{ arbitrary}$$

More explicitly, if we separate variables in Eq. (6) and then use partial-fraction techniques, we have

$$\frac{dr}{r(1-r^2)} = dt$$

$$\frac{1}{2}\left(\frac{2}{r} + \frac{1}{1-r} - \frac{1}{1+r}\right) dr = dt$$

$$\ln \frac{r^2}{|1-r^2|} = 2t - \ln|k| \qquad k \neq 0$$

$$r^2 = \frac{e^{2t}}{k+e^{2t}} = \frac{1}{1+ke^{-2t}}$$

or

(8) $$r = \frac{1}{\sqrt{1+ke^{-2t}}}$$

and, from (7), $\theta = -t + t_0$.

The value $k = 0$ was excluded in the derivation of (8). However, if it is now used in (8), it yields the circle $r = 1$ which, as we have noted, is one of the trajectories. If $k > 0$, the corresponding trajectories spiral toward this circle from its interior as $t \to +\infty$. If $k < 0$, the corresponding curves spiral toward the circle from the outside as $t \to +\infty$. Figure 6.14 shows the circle $r = 1$ and several of these spirals.

A simple closed curve in the phase plane which has nonclosed paths spiraling toward it, either as $t \to \infty$ or as $t \to -\infty$, is called a **limit cycle.** If all trajectories which start sufficiently close to a limit cycle, both inside and outside, spiral toward it as $t \to +\infty$, the limit cycle is said to be **stable.** If trajectories on one side of the limit cycle spiral toward it while those on the other side spiral away from it as $t \to +\infty$, the limit cycle is said to be **semistable.** If trajectories on both sides spiral away from it as $t \to +\infty$, the limit cycle is said to be **unstable.**

FIGURE 6.14
The limit cycle of the system $dx/dt = x + y - x(x^2+y^2)$, $dy/dt = -x + y - y(x^2+y^2)$.

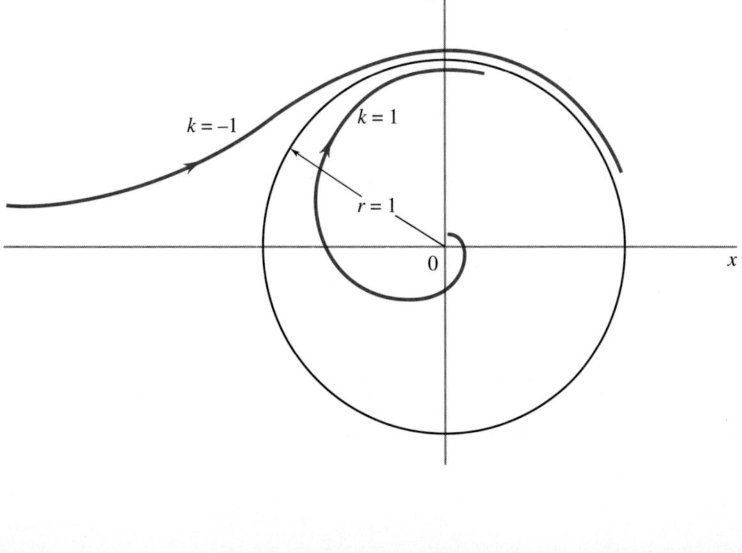

In the example we have just discussed, the existence of a limit cycle was established by actually solving the given system of equations. Usually, however, the equations are too difficult to solve, and we must fall back on other methods to establish the existence or nonexistence of a limit cycle, that is, a periodic solution. The fundamental results are contained in the following pair of theorems, one giving conditions sufficient for the existence of a limit cycle, the other giving conditions sufficient for the nonexistence of a limit cycle.

THEOREM 1† Consider the autonomous system

$$\frac{dx}{dt} = F(x, y)$$

$$\frac{dy}{dt} = G(x, y)$$

where F and G have continuous first partial derivatives in some domain D. Let D_1 be a bounded subdomain in D, let R be the region consisting of D_1 and its boundary, and suppose that R contains no critical points of the given system. If there exists a solution $x = x(t)$, $y = y(t)$ of the system and a value t_0 such that the trajectory defined by $x = x(t)$, $y = y(t)$ remains in R for all $t \geq t_0$, then either

1. $x = x(t)$, $y = y(t)$ is a closed trajectory, i.e., a periodic solution.
2. $x = x(t)$, $y = y(t)$ spirals toward a closed trajectory in R as $t \to \infty$.

In either case, the given system has a periodic solution.

THEOREM 2‡ Let the functions F and G in the autonomous system

$$\frac{dx}{dt} = F(x, y)$$

$$\frac{dy}{dt} = G(x, y)$$

have continuous first partial derivatives in a simply connected domain D.§

If $\partial F/\partial x + \partial G/\partial y$ has the same sign throughout D, then there is no periodic solution of the given system which lies entirely in D.

If we reconsider the system (3) in the light of Theorem 1, we must first find a closed region R containing no critical points of the system, and then we must show that there is a trajectory which remains in R for all sufficiently large values of t. Since R must contain no critical points, it cannot be the interior of a circle with center at the origin. However, since (0, 0) is the only critical point, R can be the closed annular region defined by two concentric circles having (0, 0) as center, say $r = \frac{1}{3}$ and $r = 3$. Using the solution (8) which we found for r, it is clear that there is a solution, say the one for which $k = 3$, which remains in R for all $t \geq 0$. Had we not obtained the explicit solution (8), we could have inferred the existence of the required trajectory in R by noting from (6) that for any solution starting in R, the radius r is a decreasing function of t if $1 < r < 3$ and an increasing

†This is customarily referred to as the **Poincaré-Bendixon theorem,** after the French mathematician Henri Poincaré (1854–1912) and the Swedish mathematician Ivar Bendixon (1861–1935).
‡This theorem is usually known as **Bendixon's theorem.**
§See p. 1064.

function of t if $\frac{1}{3} < r < 1$. Hence any solution which starts in R must remain in R. Thus the hypotheses of Theorem 1 are satisfied, and the system (3) has a periodic solution in R, namely,

$$\begin{aligned} r &= 1 \\ \theta &= -t + t_0 \end{aligned}$$

as we observed earlier.

To illustrate Theorem 2, consider the system

$$\begin{aligned} \frac{dx}{dt} &= x + 2xy + x^3 \\ \frac{dy}{dt} &= -y^2 + x^2 y \end{aligned}$$

Clearly, the functions $F = x + 2xy + x^3$ and $G = -y^2 + x^2y$ have continuous first partial derivatives at all points of the xy plane. Furthermore,

$$\frac{\partial F}{\partial x} + \frac{\partial G}{\partial y} = (1 + 2y + 3x^2) + (-2y + x^2) = 1 + 4x^2$$

Since $1 + 4x^2$ is positive for all values of x (and y), Theorem 2 guarantees that the given system has no periodic solutions anywhere in the phase plane.

One of the most important applications of the ideas we have been discussing is to nonlinear second-order differential equations, in particular to the equation

$$\frac{d^2x}{dt^2} + p(x)\frac{dx}{dt} + q(x) = 0 \tag{9}$$

which, as we have seen, is equivalent to the system

$$\begin{aligned} \frac{dx}{dt} &= y \\ \frac{dy}{dt} &= -q(x) - p(x)y \end{aligned}$$

For particular instances of Eq. (9) that arise in important technical problems, Theorems 1 and 2 are often unable to establish either the existence or the nonexistence of periodic solutions. In such cases, the following theorem, due to the American mathematician Norman Levinson (1912–1975), is frequently applicable.

THEOREM 3 Consider the second-order differential equation

$$\frac{d^2x}{dt^2} + p(x)\frac{dx}{dt} + q(x) = 0$$

and let

$$P(x) = \int_0^x p(s)\,ds \qquad \text{and} \qquad Q(x) = \int_0^x q(s)\,ds$$

If

1. $p(x)$ is an even, continuous function for all values of x
2. $q(x)$ is an odd function which is positive for $x > 0$ and has a continuous derivative for all values of x
3. There exists a number x_0 such that $P(x) < 0$ for $0 < x < x_0$ and $P(x) > 0$ for $x > x_0$
4. $P(x)$ increases monotonically for $x > x_0$ and becomes infinite as $x \to +\infty$
5. $Q(x)$ becomes infinite as $x \to +\infty$

then the given equation has a unique closed trajectory toward which all other trajectories (except the trivial solution $x = y \equiv dx/dt = 0$) spiral as $t \to +\infty$.

EXAMPLE 1

Investigate the existence of periodic solutions of the so-called **van der Pol equation,**†

$$\frac{d^2x}{dt^2} + \lambda(x^2 - 1)\frac{dx}{dt} + x = 0 \qquad \lambda > 0$$

As a first step, let us convert this equation into the equivalent system

$$\frac{dx}{dt} = y$$
$$\frac{dy}{dt} = -x - \lambda(x^2 - 1)y \tag{10}$$

and attempt to apply Theorems 1 and 2. Considering Theorem 2 first (since it is a little simpler), we note that

$$F = y \qquad G = -x - \lambda(x^2 - 1)y$$

and therefore

$$\frac{\partial F}{\partial x} + \frac{\partial G}{\partial y} = -\lambda(x^2 - 1)$$

This is positive at all points in the simply connected infinite vertical strip between the lines $x = -1$ and $x = 1$. Hence, by Theorem 2, we conclude that the given equation has no closed trajectory which lies entirely in this portion of the xy plane. Moreover, $-\lambda(x^2 - 1)$ is of constant negative sign throughout the half-plane to the left of $x = -1$ and throughout the half-plane to the right of $x = 1$. Hence there are no closed trajectories in these regions of the phase plane. However, for all we know from Theorem 2, there may be solutions of the system (10) which define closed trajectories lying partly in one of these regions and partly in another.

In an attempt to apply Theorem 1, we next compute

$$x\frac{dx}{dt} + y\frac{dy}{dt} = r\frac{dr}{dt} = -\lambda(x^2 - 1)y^2$$

from which we conclude that

$$\frac{dr}{dt} = -\lambda(r^2\cos^2\theta - 1)r\sin^2\theta$$

This shows that if $0 < r < 1$, the radius r is an increasing function of t and the path curves spiral outward. If we could show similarly that r is a decreasing function of t on paths that start sufficiently far from the origin [as we did in our discussion of the system (3)], we could conclude that there was an annular region in which a

†Balthasar van der Pol (1889–1959) was a Dutch radio engineer.

trajectory would have to remain, and we could apply Theorem 1. However, for $r > 1$, dr/dt is a decreasing function of t where $\cos^2\theta > 1/r^2$, but an increasing function of t where $\cos^2\theta < 1/r^2$, and so our attempt to invoke Theorem 1 fails.

As a final effort, let us try to apply Theorem 3. Clearly, $p(x) = \lambda(x^2 - 1)$ is an even, continuous function, and $q(x) = x$ is an odd function with a continuous derivative, as required by the theorem. Furthermore,

$$Q(x) = \int_0^x s\, ds = \frac{x^2}{2}$$

becomes infinite as $x \to \infty$. Also,

$$P(x) = \int_0^x \lambda(s^2 - 1)\, ds = \lambda\left(\frac{x^3}{3} - x\right)$$

and if we take $x_0 = \sqrt{3}$, we have

$$P(x) < 0 \text{ for } 0 < x < x_0 = \sqrt{3} \qquad \text{and} \qquad P(x) > 0 \text{ for } x > x_0 = \sqrt{3}$$

Finally, since $P'(x) = \lambda(x^2 - 1) > 0$ for $x > 1$, it follows that P is a monotonically increasing function of x for $x > x_0 = \sqrt{3}$ and P becomes infinite as $x \to +\infty$. Thus all the hypotheses of Theorem 3 are satisfied, and we can conclude that van der Pol's equation has a unique closed trajectory toward which all other trajectories spiral.

EXERCISES

For each of the following autonomous systems, find all limit cycles and determine their stability.

1. $dr/dt = r(4 - r^2)$ $\quad d\theta/dt = 1$
2. $dr/dt = r(1 - r)^2$ $\quad d\theta/dt = -1$
3. $dr/dt = r(r - 1)(r - 2)$ $\quad d\theta/dt = 1$
4. $dr/dt = r(r^2 + r - 6)$ $\quad d\theta/dt = 1$
5. $dr/dt = r(r - 1)(r - 2)^2(r - 3)$ $\quad d\theta/dt = -1$
6. $dr/dt = \sin \pi r$ $\quad d\theta/dt = -1$
7. Find the equation of the family of trajectories in **(a)** Exercise 1, **(b)** Exercise 2, and **(c)** Exercise 3.

Show that none of the following systems has a limit cycle.

8. $dx/dt = x + 2y + x^3$ $\quad dy/dt = -x^2y + 2y^3$
9. $dx/dt = -2x - x\sin y$ $\quad dy/dt = -x^2y^3$
10. $dx/dt = x + x^2 + 2y\cos x$ $\quad dy/dt = -2xy + y^2\sin x + y^3$
11. $dx/dt = y\sinh xy$ $\quad dy/dt = y - x^2 + y\cosh x$
12. Show that the system

$$\frac{dx}{dt} = -y + \frac{x}{r}f(r) \qquad \frac{dy}{dt} = x + \frac{y}{r}f(r)$$

has limit cycles corresponding to the zeros of $f(r)$. What is the direction of motion on these curves?

Show that each of the following systems has a limit cycle.

13. $dx/dt = -2y + 4x - x(x^2 + 4y^2)$
    $dy/dt = 2x + 4y - y(x^2 + 4y^2)$
14. $dx/dt = x + 3y - x(x^2 + 9y^2)$
    $dy/dt = -x + 2y - y(x^2 + 9y^2)$

Show that each of the following equations has a periodic solution.

15. $\dfrac{d^2x}{dt^2} + (x^2 - 1)\dfrac{dx}{dt} + x + \sin x = 0$
16. $\dfrac{d^2x}{dt^2} + (x^4 - 1)\dfrac{dx}{dt} + x^3 = 0$
17. $\dfrac{d^2x}{dt^2} + (x^4 - 4x^2)\dfrac{dx}{dt} + x + x^3 = 0$
18. $\dfrac{d^2x}{dt^2} + (4x^2 - 1)\dfrac{dx}{dt} + \sinh x = 0$
19. $\dfrac{d^2x}{dt^2} - (4 + 9x^2 - 5x^4)\dfrac{dx}{dt} - x + x^3 = 0$
20. For what values of λ, if any, does the equation

$$\frac{d^2x}{dt^2} + (\cosh x - \lambda)\frac{dx}{dt} + \sinh x = 0$$

have a periodic solution?

21. What is the difference, if any, between a closed trajectory and a periodic solution?

CHAPTER 7

MECHANICAL SYSTEMS AND ELECTRIC CIRCUITS

The most important application of linear constant-coefficient differential equations is to vibrating mechanical systems and electric circuits. This chapter begins (Sec. 7.2) with the derivation of the basic differential equations for each type of system, emphasizing their mathematical identity, through a simple change of vocabulary.

Mechanical systems with one degree of freedom are investigated in Sec. 7.3, and such fundamental concepts as *natural frequency, critical damping, free* and *forced motion, transient* and *steady-state motion, magnification ratio* and *phase shift, resonance,* and *beats* are studied.

Section 7.4 discusses electric circuits with one degree of freedom, emphasizing the important concept of *impedance,* both *real* and *complex.*

In Sec. 7.5, the material on simultaneous linear differential equations which we studied in Chap. 4 is used to study the behavior of mechanical and electrical systems with several degrees of freedom.

In Sec. 7.6, through the use of difference equations (Sec. 5.7) combined with differential equations, systems with an arbitrarily large but finite number of degrees of freedom are studied. Particular applications include *wave propagation, wave filters,* and *wave traps.*

In Sec. 7.7, we learn how, through the use of appropriate scale factors, the mathematical identity between the differential equations of mechanical and electrical systems can be made a numerical identity also.

Prerequisites for this chapter: Chaps. 2–4, Sec. 5.7, and a calculus-based physics course.

7.1 INTRODUCTION

An examination of the application of differential equations to mechanical systems and electric circuits is valuable for at least two reasons. In the first place, such a study will give us useful information about a variety of physical systems of considerable practical interest. Second, it will provide us with a striking example of the role mathematics plays in unifying the behavior of widely differing real-world phenomena. Specifically, we shall find that the motion of a system of spring-connected masses is described by differential equations which are mathematically identical with those which arise in the analysis of an electric network containing interconnected resistors, capacitors, and inductors. As a result, various vibrating mechanical systems can be simulated, or modeled, by electric circuits whose currents or voltages, as we choose, will give the *exact* values of the displacements in the corresponding mechanical systems when suitable scale factors are introduced. Moreover, a computer program for solving a problem in one field is automatically a program for solving similar problems in either field, depending on the physical interpretations we give to the abstract variables in the program. This aspect of mathematics is pervasive, but especially prominent in the material we shall study in this chapter.

7.2 SYSTEMS WITH ONE DEGREE OF FREEDOM

A system which can be described by one dependent variable, or coordinate, i.e., by one physical datum such as a displacement, an angle, a current, or a voltage, is called a **system of one degree of freedom.** A system requiring more than one coordinate for its complete description is called a **system of several degrees of freedom.** The **number of degrees of freedom** a system has is equal to the *minimum number of dependent variables* required to specify the state of the system. A single differential equation suffices for the mathematical description of a system with one degree of freedom. A set of simultaneous differential equations, as many equations as there are degrees of freedom, is necessary for the analysis of systems of more than one degree of freedom. We shall begin our investigations by considering, as prototypes, each of the systems shown in Fig. 7.1. In each case we assume that all the elements of the system are concentrated, or **lumped.** In other words, such things as the distributed mass of the springs in Fig. 7.1*a*, the distributed moment of inertia of the shaft in Fig. 7.1*b*, and the distributed resistance of the leads in Fig. 7.1*c* and *d* we assume to be either negligible or taken into account through suitable corrections added to the corresponding major elements.† (See Exercise 3.)

In Fig. 7.1*a* we assume that the weight is guided, so that only vertical motion without swinging is possible. As the dashpot indicates, the effects of friction are not negligible, nor are other external forces neglected, as they were in Sec. 2.11. Instead, we suppose that a retarding force proportional to the velocity acts at all times. Friction of this sort is known as **viscous friction,** or **viscous damping,** and the proportionality constant c, which must be nonnegative, is called the **damping coefficient,** or simply the **damping.** The existence of this kind of friction is well established for moderate velocities, although for large velocities the resistance may be more nearly proportional to the square, or even the cube, of the velocity.

The analysis of this system is based on **Newton's law:**

$$\text{Mass} \times \text{acceleration} = \text{force} \tag{1}$$

Measuring the instantaneous displacement y of the weight from its equilibrium position and substituting w/g for the mass, we have

†In many problems these assumptions are not sufficiently accurate, and the continuous distribution of the components of the system must be considered. As we shall see in Chap. 11, this leads to *partial* rather than *ordinary* differential equations.

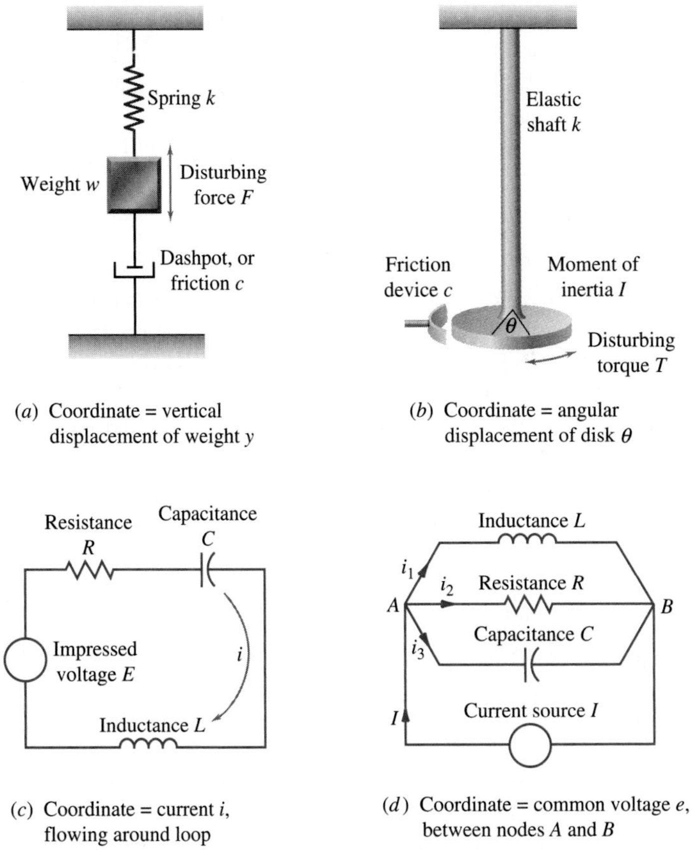

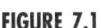

FIGURE 7.1
Four simple systems of one degree of freedom: (a) translational mechanical; (b) torsional mechanical; (c) series electrical; (d) parallel electrical.

(2) $$\text{Mass} \times \text{acceleration} = \frac{w}{g}\frac{d^2y}{dt^2}$$

The most obvious force acting on the mass is the attraction of gravity:

(3) $$\text{Gravitational force} = -w$$

the minus sign indicating that this force acts downward. As we observed in Sec. 2.11, the force the spring exerts on the mass is

(4) $$\text{Elastic force} = k\left(\frac{w}{k} - y\right) = w - ky$$

where w/k is the amount the spring is stretched when the weight hangs in equilibrium. To determine the frictional force, we observe that the velocity of the weight is dy/dt; hence, from the assumption of viscous damping,

(5) $$\text{Frictional force} = -c\frac{dy}{dt}$$

the minus sign indicating that the resistance always acts in opposition to the velocity. Finally, through some external agency, a disturbing, or impressed force, often periodic, may act on the weight, upsetting its condition of equilibrium. In this chapter we shall consider in detail only the simple, though highly important cases

(6*a*) $\qquad$ Impressed force $= F_0 \cos \omega t$

(6*b*) $\qquad$ Impressed force $= F_0 \sin \omega t$ $\qquad F_0 > 0,\ \omega \geq 0$, constants

Of course, if $\omega = 0$, Eq. (6*a*) represents a constant force of magnitude F_0. Once we have learned how to find the effect of these simple periodic forces and have learned a little about Fourier series (Chap. 8), we will be able to determine the effect of almost any periodic force however complicated.

Considering specifically the case in which the impressed force is cosinusoidal, and equating the right-hand member of Eq. (2) to the sum of the forces (3) through (6*a*), we obtain the **equation of motion**

$$\frac{w}{g}\frac{d^2y}{dt^2} = -w + (w - ky) - c\frac{dy}{dt} + F_0 \cos \omega t$$

or

(7)
$$\frac{w}{g}\frac{d^2y}{dt^2} + c\frac{dy}{dt} + ky = F_0 \cos \omega t$$

Again, as in Sec. 2.11, we note that the gravitational force on the weight is canceled by that part of the elastic force due to the initial elongation of the spring. Therefore, in analyzing problems like this in the future, we shall from the outset neglect both the gravitational force and the initial, or static, elongation of the spring which it produces.

Equation (7) is a typical nonhomogeneous linear differential equation of the second order with constant coefficients whose complete solution we can easily find by the methods of Chap. 2. Presumably, it will be accompanied by given initial conditions of displacement and velocity,

$$y(0) = y_0 \qquad \text{and} \qquad \left.\frac{dy}{dt}\right|_{t=0} = \dot{y}_0$$

and by using these, the constants in any complete solution for the displacement y can be determined. However, before continuing with the solution of Eq. (7) we shall derive the equations governing the other systems shown in Fig. 7.1.

The analysis of the system shown in Fig. 7.1*b* is based on Newton's law in torsional form:

(8) $\qquad$ Moment of inertia $\times$ angular acceleration $=$ torque

In this case the various torques are

(9) $\qquad$ Elastic torque due to twisting of the shaft $= -k\theta$

where k is now the **torsional modulus** of the shaft.

(10)
$$\text{Viscous damping torque} = -c\frac{d\theta}{dt}$$

and one or the other of the following external torques

(11*a*) Impressed torque $= T_0 \cos \omega t$

(11*b*) Impressed torque $= T_0 \sin \omega t$

$T_0 > 0,\ \omega \geq 0$, constants

Noting that the angular acceleration is $d^2\theta/dt^2$ and assuming a cosinusoidal impressed torque, we have on substituting into the torsional version of Newton's law [Eq. (8)]

$$I\frac{d^2\theta}{dt^2} = -k\theta - c\frac{d\theta}{dt} + T_0 \cos \omega t$$

or

$$I\frac{d^2\theta}{dt^2} + c\frac{d\theta}{dt} + k\theta = T_0 \cos \omega t \tag{12}$$

This too is a completely familiar differential equation, and when it is accompanied by initial conditions of angular displacement and velocity,

$$\theta(0) = \theta_0 \qquad \text{and} \qquad \left.\frac{d\theta}{dt}\right|_{t=0} = \dot{\theta}_0$$

it can easily be solved for the function describing the angle of twist of any particular system.

The analysis of the series, or one-loop, electric circuit (Fig. 7.1*c*) is based on **Kirchoff's second law:**† the *algebraic sum of the potential differences around any closed loop in an electric network is zero,* or *the voltage impressed on a closed loop is equal to the sum of the voltage drops in the rest of the loop.* Using well-known electrical laws, we have

(13) Voltage drop across the resistor $= iR$

(14) Voltage drop across the capacitor $= \dfrac{1}{C}\displaystyle\int^t i\, dt$

(15) Voltage drop across the inductor $= L\dfrac{di}{dt}$

Thus, considering the important case in which

(16) Impressed voltage $= E_0 \cos \omega t$ $\quad E_0 > 0,\ \omega \geq 0$, constants

we have, on substituting from Eqs. (13)–(16) into Kirchoff's second law,

$$L\frac{di}{dt} + iR + \frac{1}{C}\int^t i\, dt = E_0 \cos \omega t \tag{17}$$

An equation such as this is called an **integrodifferential equation** since an integral as well as a derivative of the dependent variable appears in it. Given the initial current, the operational methods we shall develop in Chap. 10 will handle Eq. (17) directly, but before we can apply the

†Named for the German physicist Gustav Robert Kirchoff (1824–1877).

techniques available at this stage, we must convert (17) into a pure differential equation. There are several ways of doing this. The first is to regard not i but $\int^t i\,dt$ as the dependent variable of the problem. This is not merely a mathematical stratagem, for the quantity

$$Q = \int^t i\,dt$$

i.e., the integrated flow of current into the capacitor, is precisely the quantity of electricity, or electric charge, on the capacitor at time t. In terms of Q, we have

$$i = \frac{dQ}{dt} \qquad \text{and} \qquad \frac{di}{dt} = \frac{d^2Q}{dt^2}$$

and Eq. (17) becomes

$$L\frac{d^2Q}{dt^2} + R\frac{dQ}{dt} + \frac{1}{C}Q = E_0 \cos \omega t \tag{18a}$$

subject of course to given initial conditions of charge and current

$$Q(0) \equiv \int^{t=0} i\,dt = Q_0 \qquad \text{and} \qquad \left.\frac{dQ}{dt}\right|_{t=0} = i(0) = i_0$$

On the other hand, we can also convert Eq. (17) into a differential equation simply by differentiating it with respect to time, getting

$$L\frac{d^2i}{dt^2} + R\frac{di}{dt} + \frac{1}{C}i = -\omega E_0 \sin \omega t \tag{18b}$$

The initial conditions required for an equation of this form are

$$i(0) = i_0 \qquad \text{and} \qquad \left.\frac{di}{dt}\right|_{t=0} = \dot{i}_0$$

The first of these will be given. If Q_0 is also prescribed, the second can be found from Eq. (17) since

$$\frac{di}{dt} = \frac{1}{L}\left(E_0 \cos \omega t - iR - \int^t i\,dt\right)$$

and the right-hand side will be completely known at $t = 0$.

To establish the differential equation describing the behavior of the parallel, or one-node-pair, electric circuit shown in Fig. 7.1d, we must use **Kirchoff's first law.** *The algebraic sum of the currents flowing toward any point in an electrical network is zero.* However, before we can do this, we must solve Eqs. (13), (14), and (15) for i in terms of the unknown potential difference e between the points A and B. This gives us, respectively,

$$\text{Current through the resistor} = \frac{e}{R} \tag{19}$$

(20) $$\text{Current (apparently) through the capacitor} = C\frac{de}{dt}$$

(21) $$\text{Current through the inductor} = \frac{1}{L}\int^t e\,dt$$

Thus, assuming the important case of a current source such that

(22) $$\text{Impressed current} = I_0 \cos \omega t \qquad I_0 > 0,\ \omega \geq 0,\ \text{constants}$$

we have, on substituting into Kirchoff's first law,

(23) $$C\frac{de}{dt} + \frac{1}{R}e + \frac{1}{L}\int^t e\,dt = I_0 \cos \omega t$$

Again our derivation has led us to an integrodifferential equation. To convert it to a pure differential equation we can consider $\int^t e\,dt = U$, say, as a new dependent variable, getting

(24*a*) $$C\frac{d^2U}{dt^2} + \frac{1}{R}\frac{dU}{dt} + \frac{1}{L}U = I_0 \cos \omega t$$

subject to initial conditions of the form

$$U(0) \equiv \int^{t=0} e\,dt = U_0 \qquad \text{and} \qquad \left.\frac{dU}{dt}\right|_{t=0} = e_0$$

On the other hand, we can simply differentiate Eq. (23) with respect to time, getting

(24*b*) $$C\frac{d^2e}{dt^2} + \frac{1}{R}\frac{de}{dt} + \frac{1}{L}e = -\omega I_0 \sin \omega t$$

subject to initial conditions of the form

$$e(0) = e_0 \qquad \text{and} \qquad \left.\frac{de}{dt}\right|_{t=0} = \dot{e}_0$$

When we collect the differential equations we have derived,

(7) $$\frac{w}{g}\frac{d^2y}{dt^2} + c\frac{dy}{dt} + ky = F_0 \cos \omega t$$ **(Translational-mechanical)**

(12) $$I\frac{d^2\theta}{dt^2} + c\frac{d\theta}{dt} + k\theta = T_0 \cos \omega t$$ **(Torsional-mechanical)**

(18*a*) $$L\frac{d^2Q}{dt^2} + R\frac{dQ}{dt} + \frac{1}{C}Q = E_0 \cos \omega t$$

(18*b*) $$L\frac{d^2i}{dt^2} + R\frac{di}{dt} + \frac{1}{C}i = -\omega E_0 \sin \omega t$$

(Series-electrical)

$$\left.\begin{aligned} &(24a) \qquad C\frac{d^2U}{dt^2} + \frac{1}{R}\frac{dU}{dt} + \frac{1}{L}U = I_0 \cos \omega t \\ &(24b) \qquad C\frac{d^2e}{dt^2} + \frac{1}{R}\frac{de}{dt} + \frac{1}{L}e = -\omega I_0 \sin \omega t \end{aligned}\right\} \textbf{(Parallel-electrical)}$$

their essential mathematical identity becomes apparent. Moreover, we can see the possibility of various physical analogies, i.e., vocabulary correlations. For instance, upon comparing the translational-mechanical and the series-electrical equations, we notice the correspondences

$$\begin{aligned} \text{Mass } \frac{w}{g} &\longleftrightarrow \text{inductance } L \\ \text{Damping } c &\longleftrightarrow \text{resistance } R \\ \text{Spring modulus } k &\longleftrightarrow \text{elastance } \frac{1}{C} \\ \text{Impressed force } F &\longleftrightarrow \begin{cases} \text{impressed voltage } E & \text{[Using (18}a\text{)]} \\ dE/dt & \text{[Using (18}b\text{)]} \end{cases} \\ \text{Displacement } y &\longleftrightarrow \begin{cases} \text{charge } Q & \text{[Using (18}a\text{)]} \\ \text{current } i & \text{[Using (18}b\text{)]} \end{cases} \end{aligned}$$

and, if we compare the translational-mechanical and the parallel-electrical equations, we have the correspondences

$$\begin{aligned} \text{Mass } \frac{w}{g} &\longleftrightarrow \text{capacitance } C \\ \text{Damping } c &\longleftrightarrow \text{conductance } \frac{1}{R} \\ \text{Spring modulus } k &\longleftrightarrow \text{susceptance } \frac{1}{L} \\ \text{Impressed force } F &\longleftrightarrow \begin{cases} \text{impressed current } I & \text{[Using (24}a\text{)]} \\ dI/dt & \text{[Using (24}b\text{)]} \end{cases} \\ \text{Displacement } y &\longleftrightarrow \begin{cases} \int^t e\,dt & \text{[Using (24}a\text{)]} \\ \text{voltage } e & \text{[Using (24}b\text{)]} \end{cases} \end{aligned}$$

However, we must not go so far as to assume that a differential equation derived for a mechanical system, with numerical coefficients determined by the components of that system, will necessarily be the differential equation of some *realizable* electrical model. For instance, the numerical value of w/g in Eq. (7) may be much larger than the numerical value of any realizable inductance L in Eqs. (18) or any attainable capacitance C in Eqs. (24). As a consequence, for such purposes as computer simulation, and modeling a circuit to run empirical tests on a proposed mechanical configuration, it would be desirable if the related equations were not only identical in structure but in an exact quantitative sense as well. Such a mathematical identity between equations can easily be accomplished by reducing each equation to dimensionless form through the introduction of suitable scale factors. These ideas are discussed in greater detail in the final section of this chapter.

In the next two sections we shall undertake an extensive investigation of translational-mechanical systems and series-electrical circuits with one degree of freedom. Then we shall study both mechanical systems and electric circuits with several degrees of freedom. Following this, through the use of both difference equations and differential equations, we shall investigate systems with many degrees of freedom.

EXERCISES

1. A weight w hangs from a spring of modulus k_1, which in turn hangs from a spring of modulus k_2. What is the modulus k of the suspension formed by the two springs together? *Hint:* The deflection of an equivalent spring must be the sum of the deflections of the two given springs.
2. **(a)** A weight w hangs from two springs of equal length but different moduli, k_1 and k_2, each of which is connected directly to the weight. What is the modulus k of the suspension formed by the two springs together?
   (b) A weight w hangs from a spring of modulus k_1 and at the same time is supported from below by a spring of modulus k_2. What is the modulus of the suspension formed by the two springs together?
   Hint: In both Parts **(a)** and **(b)**, the force exerted by an equivalent spring must be the sum of the forces exerted by the two given springs.
3. One reasonable way to take into account the (relatively small) moment of inertia of the elastic shaft shown in Fig. 7.1*b* is to compute the kinetic energy of the shaft as it vibrates with the disk and then add to the moment of inertia of the disk a correction sufficient to increase its kinetic energy by an amount equal to the kinetic energy of the shaft. Following this procedure, show that if the polar moment of inertia of the shaft is I_s, the correction to be added to the moment of inertia of the disk is $\frac{1}{3}I_s$. *Hint:* Determine the angular velocity of the shaft as a function of the distance from the point of support and then integrate the kinetic energy of infinitesimal elements of the shaft along the length of the shaft. What correction should be added to the mass in the system shown in Fig. 7.1*a* to take into account the distributed mass of the spring?
4. In Fig. 7.2*a* the unstretched lengths of the springs are l_1 and l_2, respectively. What is the equilibrium position of the weight? Set up the differential equation describing the motion of the weight and show that it is independent of the initial elongations of the springs.
5. In Fig. 7.2*b* the unstretched lengths of the springs are l_1 and l_2, respectively. What is the equilibrium position of the weight? Set up the differential equation describing the motion of the weight and show that it is independent of the initial elongation of the springs. (Because of this result and the result of Exercise 4, it is customary in setting up mass-spring problems to assume that all springs are initially unstretched.)

FIGURE 7.2

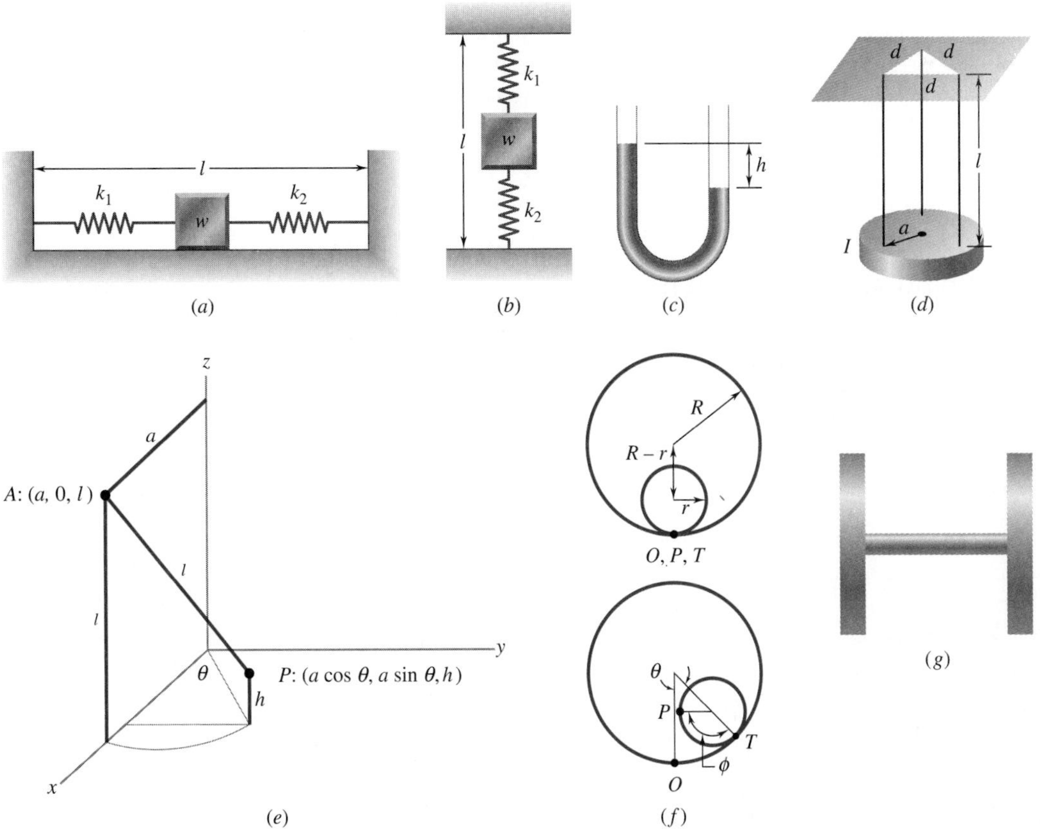

6. Show that the differential equation of the series electric circuit can be written in the form

$$LC\frac{d^2e_c}{dt^2} + RC\frac{de_c}{dt} + e_c = E_0 \cos \omega t$$

where e_c is the potential difference across the capacitor, and also in the form

$$LC\frac{d^2e_r}{dt^2} + RC\frac{de_r}{dt} + e_r = -\omega RCE_0 \sin \omega t$$

where e_r is the potential difference across the resistor. What is the form of the equation if the dependent variable is the potential difference e_i across the inductor?

7. What is the form of the differential equation of the parallel electric circuit if the dependent variable is the current through the inductor? The current through the resistor? The current (apparently) through the capacitor?

In working the following problems, it may be helpful to recall the material on the energy method that was presented in Sec. 2.11.

8. **(a)** Show that the potential energy stored in a shaft of torsional modulus k when one end is twisted through an angle θ with respect to the other end is equal to $\frac{1}{2}k\theta^2$.
   (b) Using the principle of the conservation of energy, derive the differential equation describing the motion of the disk shown in Fig. 7.1*b* if friction is negligible and there is no impressed torque.

9. The U tube shown in Fig. 7.2*c* is of uniform cross-sectional area A, and the column of liquid in the tube is of total length l. Assuming that friction is negligible, use the energy method to obtain the differential equation which describes the oscillation of the liquid in the tube. *Hint:* Note that in the position shown, a column of liquid of length $h/2$ has been lifted a distance $h/2$.

10. A circular disk of moment of inertia I is suspended by three identical inextensible cords of length l as shown in Fig. 7.2*d*. Assuming that friction is negligible, use the energy method to obtain the differential equation which defines the torsional vibrations of the system if it twists without tilting or swinging. *Hint:* As the disk rotates, it rises by a very small amount h. Calculate h by computing the distance between A and P in Fig. 7.2*e* and setting it equal to l. Assume an angle of twist so small that $\sin \theta$ and $\cos \theta$ can be approximated by θ and $1 - \frac{1}{2}\theta^2$, respectively, and h^2 can be neglected in comparison with h.

11. A cylinder of weight w and radius r rolls without slipping in a cylindrical depression of radius R as shown in Fig. 7.2*f*. Using the energy method, obtain the differential equation which defines the oscillation of the cylinder about its lowest point. *Hint:* In calculating the portion of the kinetic energy due to the rotation of the cylinder, note that the angle through which the cylinder has rotated is not ϕ but $\phi - \theta$. Why?

12. Work Exercise 11 if the rolling object is a sphere of radius r and not a cylinder. *Hint:* Figure 7.2*f* serves equally well for a cylinder and a sphere. The only difference is that the moment of inertia of a sphere about a diameter is $\frac{2}{5}mr^2$.

13. The torsional system shown in Fig. 7.2*g* appears to have two degrees of freedom since angular coordinates θ_1 and θ_2 are required to specify the positions of the respective disks. If one disk is rotated with respect to the other and the system is released from rest in this twisted position, and if friction is negligible, use the energy method to show that a single differential equation in the dependent variable $\theta = \theta_1 - \theta_2$ suffices to describe the motion of one disk relative to the other. *Hint:* Recall that in a frictionless system the total angular momentum remains constant.

7.3 THE TRANSLATIONAL-MECHANICAL SYSTEM

In the last section we saw that the displacement y of the weight in the translational system (Fig. 7.1*a*), driven by a cosinusoidal impressed force, satisfies the differential equation

$$\frac{w}{g}\frac{d^2y}{dt^2} + c\frac{dy}{dt} + ky = F_0 \cos \omega t \tag{1}$$

In accordance with the general theory of Chap. 2, y must therefore consist of two parts. One is the *complementary function,* obtained by solving Eq. (1) when the term representing the impressed force $F_0 \cos \omega t$ is deleted; the other is the *particular integral,* corresponding to the term $F_0 \cos \omega t$. The complementary function describes the motion of the weight in the absence of any external disturbance. This intrinsic or natural behavior of the system is called the **free motion.** The particular integral† describes the response of the system to a specific influence external to the system, the

†From our work in Chap. 2, we know that a nonhomogeneous linear differential equation has infinitely many particular integrals. When we speak here of *the* particular integral, we mean the unique solution containing no terms that are also in the complementary function, that is, the solution obtained by the method of undetermined coefficients.

driving force $F_0 \cos \omega t$ in the case we are considering here. The behavior it represents is called the **forced motion.**

The nature of the free motion of the system will depend upon the roots of the related characteristic equation

$$\frac{w}{g}m^2 + cm + k = 0$$

namely,

$$m_1, m_2 = -\frac{cg}{2w} \pm \frac{g}{2w}\sqrt{c^2 - \frac{4kw}{g}}$$

Since g, w, and k are all positive and c is nonnegative, and since the radical, when real, is certainly less than c, it follows that the roots m_1 and m_2, if real, are always negative and, if complex, have real parts which are negative or zero. We must now consider three possibilities:

$$c^2 - \frac{4kw}{g}\begin{cases} > 0 \\ = 0 \\ < 0 \end{cases}$$

If $c^2 - 4kw/g > 0$, or $c^2 > 4kw/g$, there is a relatively large amount of friction and, naturally enough, the system or its motion is said to be **overdamped.** In this case the roots of the characteristic equation are real and unequal, and the free motion (i.e., the motion described by the complementary function) is given by

$$y = Ae^{m_1 t} + Be^{m_2 t}$$

where, as we have already pointed out, both m_1 and m_2 are negative. Thus y approaches zero as time increases indefinitely. This of course is perfectly consistent with the familiar observation that if a system upon which no external force is acting is displaced from its equilibrium position, it will eventually return to that position as friction causes the motion to subside.

If we set $y = 0$, we obtain the equation

$$Ae^{m_1 t} + Be^{m_2 t} = 0 \qquad \text{or} \qquad e^{(m_1 - m_2)t} = -\frac{B}{A} \qquad A \neq 0$$

If A and B, which will of course be determined by the initial conditions of the problem, are of opposite sign, there is one and only one value of t that satisfies the last equation. On the other hand, since a real exponential function must always be positive, it follows that when A and B have the same sign or when one or the other of them is zero, there is no time when $y = 0$. A plot of the displacement y during the free motion of an overdamped system must therefore resemble one of the curves shown in Fig. 7.3 or the reflection of one of these curves in the t axis. Figures 7.3a, b, and

FIGURE 7.3
Displacement-time plots for free, overdamped, and critically damped motion.

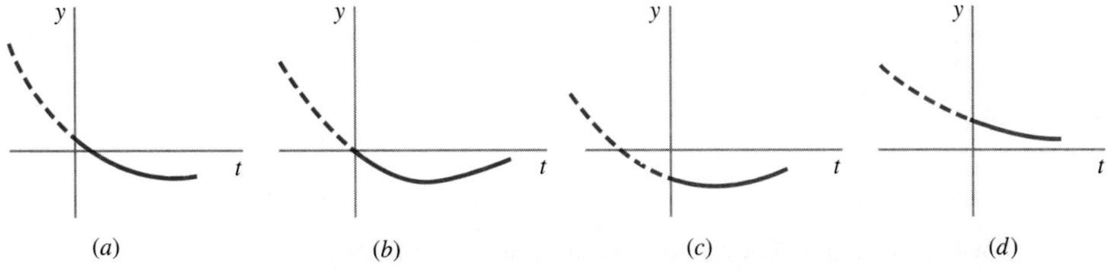

c illustrate the possibilities when A and B are of opposite sign and y vanishes once and only once. Assuming that the weight starts its motion when $t = 0$, the zero of y may occur at $t = 0$ or in the physically irrelevant interval $-\infty < t < 0$. Figure 7.3d illustrates both the case when A and B are of like sign and the case when either A or B is zero and y can never vanish.

If $c^2 - 4kw/g = 0$, we have the borderline case in which the roots of the characteristic equation are real and equal:

$$m_1 = m_2 = m = -\frac{cq}{2w}$$

When this occurs, the motion is said to be **critically damped,** and the exact value of the damping that produces it, namely,

$$c_c = 2\sqrt{\frac{kw}{g}} \tag{2}$$

is known as the **critical damping.** In this case the free motion is given by

$$y = Ae^{mt} + Bte^{mt} = (A + Bt)e^{mt}$$

If we set $y = 0$, we obtain

$$(A + Bt)e^{mt} = 0 \qquad \text{or} \qquad t = -\frac{A}{B} \qquad B \neq 0$$

If $B = 0$, there is no value of t for which $y = 0$, but in all other cases there is one and only one value of t for which $y = 0$. This may be in the physically irrelevant interval $-\infty < t < 0$, however, and so it is possible that y will not vanish during the actual motion, even when $B \neq 0$. Clearly, there is no essential difference in the character of the motion in the overdamped and critically damped cases, and the possible plots of the displacement y in the critically damped case are also represented by the curves in Fig. 7.3 and their reflections in the t axis.

If $c^2 - 4kw/g < 0$, in which case $c < c_c$, the motion is said to be **underdamped.** The roots of the characteristic equation are then the conjugate complex numbers

$$m_1, m_2 = -\frac{cg}{2w} \pm i\frac{g}{2w}\sqrt{\frac{4kw}{g} - c^2} = -\frac{cg}{2w} \pm i\sqrt{\frac{kg}{w} - \frac{c^2g^2}{4w^2}} = -p \pm iq$$

where

$$p = \frac{cg}{2w} \qquad \text{and} \qquad q = \sqrt{\frac{kg}{w} - \frac{c^2g^2}{4w^2}} \tag{3}$$

The free motion is therefore described by

$$y = e^{-pt}(A \cos qt + B \sin qt) \tag{4a}$$

or equally well by

$$y = Ge^{-pt} \cos (qt - H) \tag{4b}$$

or by

$$y = Ke^{-pt} \sin (qt - L) \tag{4c}$$

where A, B, G, H, K, and L are arbitrary constants.

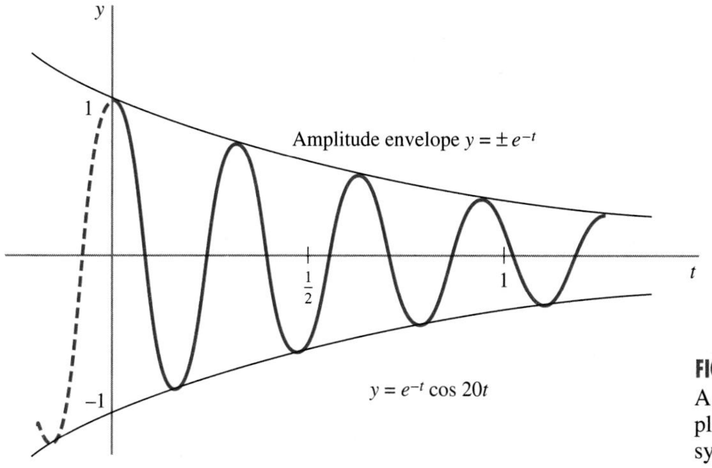

FIGURE 7.4
A typical displacement-time plot for an underdamped system.

When $0 < c < c_c$, the motion described by (4a), (4b), or (4c) is known as a **damped oscillation,** and its general appearance is shown in Fig. 7.4. It is not periodic, since the factor e^{-pt} that multiplies the trigonometric terms is continuously decreasing. However, there are regularly spaced passages through the equilibrium position at intervals of π/q. In fact, from the description of the motion provided by Eq. (4b), it is clear that $y = 0$ whenever

$$\cos(qt - H) = 0$$

i.e., when $qt - H = (\pi/2) + n\pi$ or

$$t = \frac{1}{q}\left(H + \frac{\pi}{2}\right) + \frac{n\pi}{q} \qquad n = 0, 1, 2, \ldots$$

Thus we can speak of the **pseudoperiod,** or **period with damping,**

$$\text{(5a)} \qquad \mathrm{p}_d = \frac{2\pi}{q} \text{ units of time}$$

of the **pseudofrequency,** or **frequency with damping,**

$$\text{(5b)} \qquad \mathrm{f}_d = \frac{1}{\mathrm{p}_d} = \frac{q}{2\pi} = \frac{1}{2\pi}\sqrt{\frac{kg}{w} - \frac{c^2g^2}{4w^2}} \text{ cycles per unit time}$$

and of the **pseudo circular frequency**

$$\text{(5c)} \qquad \omega_d = \frac{2\pi}{\mathrm{p}_d} = q = \sqrt{\frac{kg}{w} - \frac{c^2g^2}{4w^2}} \text{ radians per unit time}$$

If $c = 0$, that is, if there is no damping in the system, the homogeneous differential equation corresponding to Eq. (1) is harmonic, so the motion is strictly periodic and (5b) particularizes to

$$\mathrm{f}_n = \frac{1}{2\pi}\sqrt{\frac{kg}{w}} \text{ cycles per unit time}$$

which of course is the natural frequency of the system, as we found in Sec. 2.11. This explains why the natural frequency is sometimes called the **undamped frequency** of a vibrating system, and why *friction is always neglected in natural frequency calculations.*

From Eq. (5*b*), it is evident that the frequency when damping is present is always less than the natural frequency. The ratio of these two frequencies equals

$$\frac{\omega_d}{\omega_n} = \frac{\sqrt{kg/w - c^2g^2/4w^2}}{\sqrt{kg/w}} = \sqrt{1 - \frac{c^2g}{4kw}} = \sqrt{1 - \frac{c^2}{c_c^2}} \tag{6}$$

since, from Eq. (2), $c_c^2 = 4kw/g$. Figure 7.5 shows a plot of ω_d/ω_n vs. c/c_c. Evidently, if the actual damping is only a small fraction of the critical damping, as it often is, $\omega_d/\omega_n \doteq 1$ and the effect of friction on the frequency of the motion is very small.

Still using Eq. (4*b*), it is clear that the extreme values of y occur when

$$\frac{dy}{dt} = G[-pe^{-pt}\cos(qt - H) - qe^{-pt}\sin(qt - H)] = 0$$

i.e., when $\tan(qt - H) = -p/q$ or, finally, when

$$t = \frac{H}{q} - \frac{1}{q}\operatorname{Tan}^{-1}\frac{p}{q} + \frac{n\pi}{q} = T + \frac{n\pi}{q}$$

where T denotes the constant $H/q - (1/q)\operatorname{Tan}^{-1}(p/q)$.

The ratio of successive extreme displacements on the same side of the equilibrium position is a quantity of considerable importance. Its value is

$$\begin{aligned}\frac{y_n}{y_{n+2}} &= \frac{y\left(T + \dfrac{n\pi}{q}\right)}{y\left[T + \dfrac{(n+2)\pi}{q}\right]} \\ &= \frac{G\exp\left[-p\left(T + \dfrac{n\pi}{q}\right)\right]\cos\left[q\left(T + \dfrac{n}{q}\pi\right) - H\right]}{G\exp\left\{-p\left[T + \dfrac{(n+2)\pi}{q}\right]\right\}\cos\left[q\left(T + \dfrac{n+2}{q}\pi\right) - H\right]} \\ &= e^{2\pi p/q}\frac{\cos(qT + n\pi - H)}{\cos(qT + n\pi - H + 2\pi)} \\ &= e^{2\pi p/q}\end{aligned} \tag{7}$$

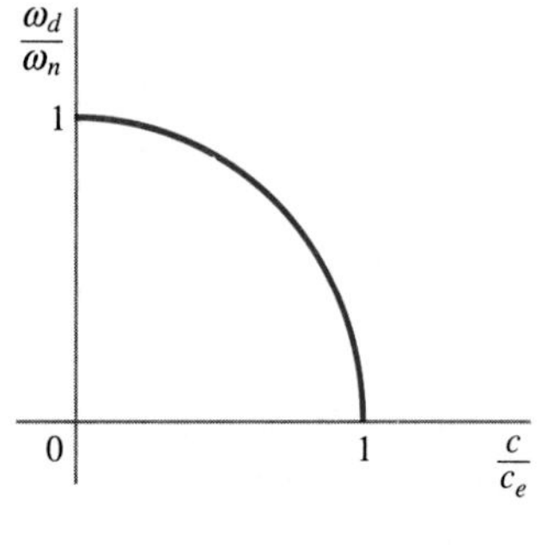

FIGURE 7.5
The effect of friction on frequency in an underdamped system.

Since this result depends only on the parameters of the system and not on n, we have thus established the following remarkable result.

THEOREM 1 The ratio of successive maximum (or minimum) displacements remains constant throughout the entire free motion of an underdamped system.

Equation (7) provides a means for determining the amount of damping in an underdamped system from observable data. To see how this can be done, we begin by taking natural logarithms in Eq. (7), getting

$$\ln \frac{y_n}{y_{n+2}} = \frac{2\pi p}{q} \tag{8}$$

This quantity, known as the **logarithmic decrement** δ, is a convenient measure, in **nepers per cycle,** of the rate at which the motion dies away.† Substituting for p and q from (3) into (8), we find

$$\delta = \frac{2\pi p}{q} = 2\pi \frac{cg/2w}{(g/2w)\sqrt{4kw/g - c^2}} = 2\pi \frac{c}{\sqrt{c_c^2 - c^2}}$$

Solved for c/c_c, this becomes

$$\frac{c}{c_c} = \frac{\delta}{\sqrt{\delta^2 + 4\pi^2}} \tag{9}$$

Since y_n and y_{n+2} are quantities which can be read from the output of various recording devices, δ can easily be computed from Eq. (8). Then from Eq. (9) the fraction of critical damping in the system can be found at once. If ω_d has been obtained experimentally, and if ω_n can be determined from an analysis of the system, the value of c/c_c can also be found from Eq. (6) which, when solved for c/c_c, becomes

$$\frac{c}{c_c} = \sqrt{1 - \left(\frac{\omega_d}{\omega_n}\right)^2}$$

Now that we have investigated the free motion of the translational-mechanical system in the overdamped, critically damped, and underdamped cases, it remains for us to consider the forced motion. To do this we must of course find a particular integral of Eq. (1):

$$\frac{w}{g}\frac{d^2y}{dt^2} + c\frac{dy}{dt} + ky = F_0 \cos \omega t \tag{1}$$

Assuming, as usual,

$$Y = A \cos \omega t + B \sin \omega t$$

†Equivalently, though less conventionally, the rate of attenuation can be expressed in **decibels per cycle** by means of the definition

$$\text{Decibels} = 20 \ln \frac{y_n}{y_{n+2}}$$

and substituting into (1), collecting terms, and equating the coefficients of cos ωt and sin ωt on each side of the equation, we obtain the two conditions

$$\left(k - \omega^2 \frac{w}{g}\right)A + \omega c B = F_0$$

$$-\omega c A + \left(k - \omega^2 \frac{w}{g}\right)B = 0$$

from which we find immediately

$$A = \frac{k - \omega^2(w/g)}{[k - \omega^2(w/g)]^2 + (\omega c)^2} F_0$$

$$B = \frac{\omega c}{[k - \omega^2(w/g)]^2 + (\omega c)^2} F_0$$

and

$$Y = F_0 \frac{[k - \omega^2(w/g)] \cos \omega t + \omega c \sin \omega t}{[k - \omega^2(w/g)]^2 + (\omega c)^2}$$

The fact that both Y and F_0/k have the dimensions of *length* suggests that if we factor k out of the numerator of the last compound fraction and k^2 out of the denominator, Y will be expressed as the product of F_0/k and a necessarily dimensionless quantity, namely,

$$\frac{[1 - \omega^2(w/kg)] \cos \omega t + (\omega c/k) \sin \omega t}{[1 - \omega^2(w/kg)]^2 + (\omega c/k)^2}$$

From the familiar relations $\omega_n = \sqrt{kg/w}$ and $c_c = 2\sqrt{kw/g}$, we have

$$1 - \omega^2(w/kg) = 1 - \frac{\omega^2}{kg/w} = 1 - \frac{\omega^2}{\omega_n^2} \qquad \omega c/k = \frac{2\omega}{\sqrt{kg/w}} \frac{c}{2\sqrt{kw/g}} = 2\frac{\omega}{\omega_n}\frac{c}{c_c}$$

Hence

$$Y = \frac{F_0}{k} \frac{\left(1 - \frac{\omega^2}{\omega_n^2}\right) \cos \omega t + 2\frac{\omega}{\omega_n}\frac{c}{c_c} \sin \omega t}{\left(1 - \frac{\omega^2}{\omega_n^2}\right)^2 + \left(2\frac{\omega}{\omega_n}\frac{c}{c_c}\right)^2}$$

If the last fraction is expressed in terms of the cosine function, the displacement Y will involve the same function as the disturbing force that produced it. This will make it easy to determine the phase relation between the **response** Y and the **excitation** $F_0 \cos \omega t$. Referring to the triangle in Fig. 7.6, it is evident that Y can be written in the form

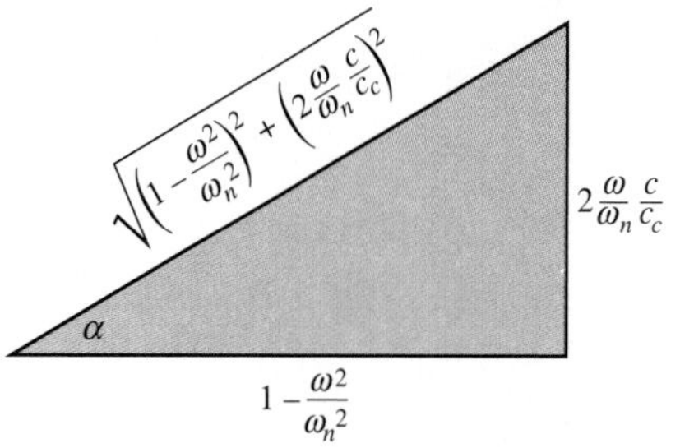

FIGURE 7.6
The triangle defining the phase angle appearing in Eqs. (10*a*) and (10*b*).

$$Y = \frac{F_0}{k}\frac{1}{\sqrt{\left(1-\frac{\omega^2}{\omega_n^2}\right)^2+\left(2\frac{\omega}{\omega_n}\frac{c}{c_c}\right)^2}}(\cos\alpha\cos\omega t+\sin\alpha\sin\omega t)$$

$$= \frac{F_0}{k}\frac{1}{\sqrt{\left(1-\frac{\omega^2}{\omega_n^2}\right)^2+\left(2\frac{\omega}{\omega_n}\frac{c}{c_c}\right)^2}}\cos(\omega t-\alpha)$$

For a disturbing force $F = F_0 \sin \omega t$, an identical derivation, in which Fig. 7.6 remains relevant, yields a sinusoidal formula for the response (see Exercise 30). Combining results for both kinds of excitation, we obtain

(10*a*) $$Y = \delta_{st} M \cos(\omega t - \alpha) \qquad F = F_0 \cos \omega t$$

(10*b*) $$Y = \delta_{st} M \sin(\omega t - \alpha) \qquad F = F_0 \sin \omega t$$

where $\delta_{st} = F_0/k$ is the **static deflection** a constant force of magnitude F_0 would produce in a spring of modulus k.

The quantity

(11) $$M = \frac{1}{\sqrt{\left(1-\frac{\omega^2}{\omega_n^2}\right)^2+\left(2\frac{\omega}{\omega_n}\frac{c}{c_c}\right)^2}}$$

is called the **magnification ratio.** It is the factor by which the static deflection produced in a spring of modulus k by a steady force F_0 must be multiplied in order to give the amplitude of the vibrations which result when the same force acts dynamically with frequency ω. It is assumed of course that F_0 is restricted so that the static deflection it would produce does not exceed the elastic limit of the spring. Curves of the magnification ratio M plotted against the **frequency ratio** ω/ω_n for various values of the **damping ratio** c/c_c are shown in Fig. 7.7. An inspection of Fig. 7.7 reveals the following interesting facts.

(a) $M = 1$, regardless of the amount of damping, if $\omega/\omega_n = 0$.

(b) If $c/c_c \geq 1/\sqrt{2}$, the magnification ratio decreases steadily as ω/ω_n increases from 0.

(c) For all values of c/c_c, M approaches zero as the impressed frequency is raised indefinitely above the natural frequency of the system.

(d) If $0 < c/c_c < 1/\sqrt{2}$, M rises to a maximum as ω/ω_n increases from 0, the peak value of M occurring in all cases before the impressed frequency reaches the natural frequency ω_n.

(e) If $0 < c/c_c < 1/\sqrt{2}$, the smaller the amount of friction, the larger the maximum of M, until for $c/c_c = 0$, M becomes infinite as ω/ω_n tends to 1.

For any particular system, ω_n, c, and c_c are fixed, and therefore the magnification ratio M depends only on the frequency of the excitation ω. Unless the amplitude of the excitation F_0 also depends on ω (see Exercises 46 and 47), this means that the amplitude of the *response,* given by $A = \delta_{st} M$, is also a maximum for the same frequency that maximizes M. Observations **a** and **b** clearly imply that if $c/c_c \geq 1/\sqrt{2}$, the maximum of M is 1, the maximum of A is δ_{st}, and these maxima occur when $\omega = 0$.

When the response of a translational-mechanical system has an amplitude A greater than the static deflection δ_{st}, i.e., when $M > 1$, the system is said to be **resonating.** Clearly, by Observations **b** and **d**, a system can resonate only if $0 \leq c/c_c < 1/\sqrt{2}$. For each value of c/c_c in this interval, there is a corresponding band of frequencies at which resonance occurs, that is, for which $M > 1$. To identify this band as a function of c/c_c, we need only determine the values of ω for which the

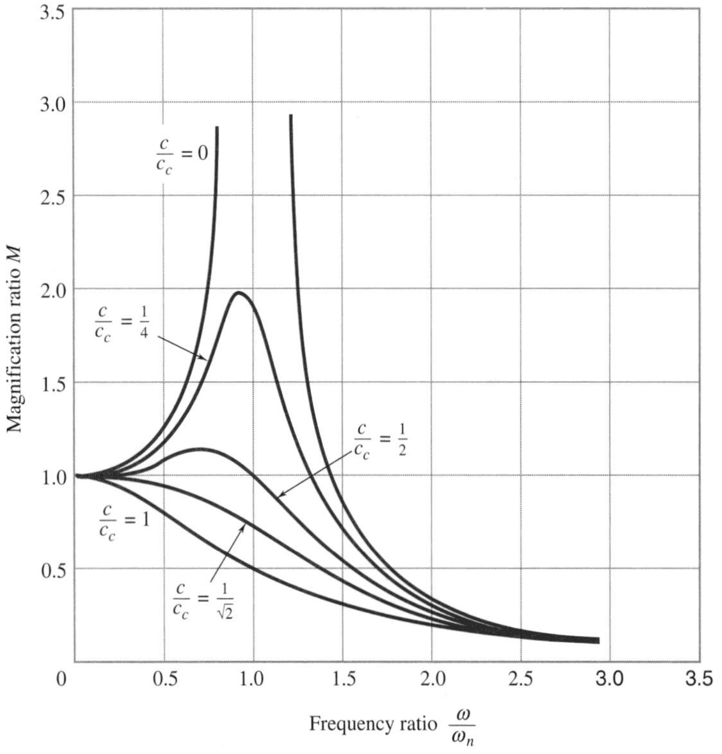

FIGURE 7.7
Curves of the magnification ratio M as a function of the impressed frequency ratio ω/ω_n for various amounts of damping.

denominator of M in (11) is less than 1. Solving this inequality, we find that for each value of $c/c_c < 1/\sqrt{2}$, resonance occurs when and only when

$$(12a) \qquad 0 < \omega < \omega_n\sqrt{2[1 - 2(c/c_c)^2]}$$

This set of frequencies is called the **resonant frequency band** of the system.

Within each resonant frequency band, the frequency at which the maxima of M and A occur can be found by regarding M as a function of $x = \omega/\omega_n$, with $a = c/c_c$ as a parameter. By equating dM/dx to zero and solving for x, we find that the frequency at which M is a maximum is (see Exercise 45)

$$(12b) \qquad \omega_r = \omega_n\sqrt{1 - 2(c/c_c)^2} \qquad 0 \le c/c_c < 1/\sqrt{2}$$

The frequency ω_r defined by (12b) is spoken of as *the* **resonant frequency** of the system, and when the system is being driven at this frequency it is said to be in **true resonance** with the excitation. As c/c_c increases toward $1/\sqrt{2}$, ω_r tends to zero but ω_n remains fixed. Thus, the resonant frequency ω_r is decidedly different from the natural frequency ω_n unless c/c_c is close to zero.

Theoretically, if $c/c_c = 0$, then according to (12b) $\omega_r = \omega_n$, and according to (11) $M = \infty$. Such a condition, of what might be called **pure resonance,** cannot occur in the real world because every physical system must contain some friction. Therefore M must remain finite. However, if c/c_c is sufficiently close to zero and the impressed frequency ω is sufficiently close to ω_n, M may be very large and oscillations of considerable magnitude may build up. In mechanical systems (though not necessarily in electrical networks), large amplitudes of the response are almost always to be avoided. Not only are they noisy, but they also produce alternating stresses that can quickly disable or even destroy a system. In dealing with such problems, a vibration engineer must first of all determine the resonant frequency band(s) of the system, and then take steps to prevent or control resonance effects at those frequencies. In all but the simplest cases, ω_r must be found experimentally or, alternatively, ω_n must be calculated approximately from design data and then checked

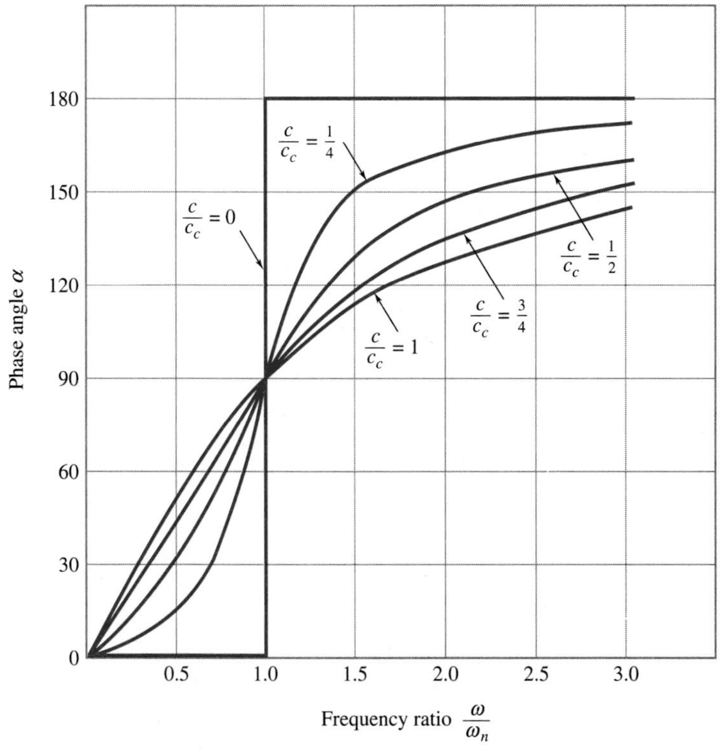

FIGURE 7.8
Curves of the phase angle α as a function of the impressed frequency ratio ω/ω_n for various amounts of damping.

experimentally. Subsequently, avoiding resonance effects will usually involve one or more of the following steps.

1. Eliminating exciting forces whose frequencies fall in a resonant frequency band.
2. Increasing the damping in the system through the use of friction dampers or dynamic dampers.
3. Making design changes in the system that will alter the resonant frequency band(s).

The preceding observations apply equally well, and even more urgently, to systems with more than one degree of freedom (Secs. 7.5 and 7.6), where there are several natural frequencies, and resonant frequency bands, at which a system may be in resonance with a periodic excitation.

The angle

$$\alpha = \tan^{-1} \frac{2(\omega/\omega_n)(c/c_c)}{1 - (\omega/\omega_n)^2} \tag{13}$$

which appears in Eqs. (10) and is shown in Fig. 7.6, is known as the **phase angle** or **angle of lag** of the response. It is important to note that α is *not* to be read from the principal-value branch of the arctangent relation, for it is evident from Fig. 7.6 that $\sin \alpha$ is always nonnegative, whereas $\cos \alpha$ can be positive, negative, or zero, according as ω is less than, greater than, or equal to ω_n. Hence $0 \le \alpha \le \pi$, and so α cannot be a negative angle in the principal-value range $(-\pi/2, \pi/2)$. Plots of α versus the frequency ratio ω/ω_n for various values of the damping ratio c/c_c are shown in Fig. 7.8.

The physical significance of α is shown in Fig. 7.9 for a typical excitation and response defined by Eqs. (10). Both the excitation $F = F_0 \cos \omega t$ and the response $Y = \delta_{st} M \cos (\omega t - \alpha)$ are of period $2\pi/\omega$. The displacement Y reaches its maxima when $\omega t - \alpha = 2n\pi$ or when $t = \alpha/\omega + 2n\pi/\omega$, whereas the excitation F reaches its maxima when $\omega t = 2n\pi$ or when $t = 2n\pi/\omega$. Thus the maxima of the response occur α/ω units of time, or a fraction of a cycle equal to $\alpha/2\pi$, *after* or *later* than the driving force reaches its corresponding peak. The same conclusions apply to

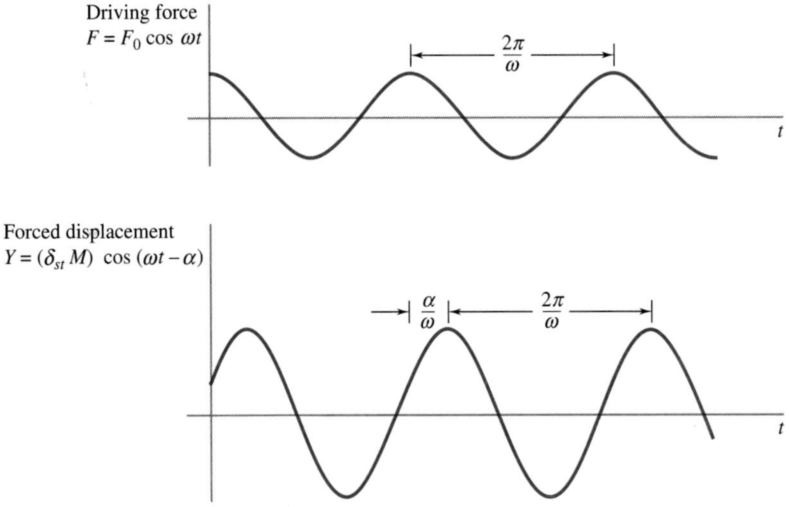

FIGURE 7.9
The significance of the phase angle as a measure of the time by which the response lags the excitation in a mechanical system.

Eqs. (10*b*). When the frequency of the disturbing force is well below the natural frequency of the system, α is small and the forced vibrations lag only slightly behind the driving force. When the impressed frequency is equal to the natural frequency, $\alpha = \pi/2$ and the response of the system lags the excitation by one quarter-cycle. As ω increases indefinitely, the lag of the response approaches a half-cycle or, in other words, the response becomes 180° out of phase with respect to the driving force.

The results of our detailed study of the vibrating weight can now be summarized. The complete motion of the system consists of two parts. The first is described by the complementary function of the underlying differential equation and may be oscillatory or nonoscillatory, according as the amount of friction in the system is less than or at least as large as the critical damping figure for the system. In any case, however, this part of the solution contains terms that decay exponentially, and it soon becomes vanishingly small. For this reason it is known as the **transient.** The complete expression for the transient contains two arbitrary constants which, after a complete solution has been constructed, must be determined to fit initial conditions of displacement and velocity. The second part of the solution is described by the particular integral. In the highly important case in which the system is acted upon by a pure harmonic disturbing force, this term represents a harmonic displacement of the same frequency as the excitation but lagging behind the latter. The amplitude of this displacement is a determinate multiple of the steady deflection that would be produced in the system by a constant force of the same magnitude as the actual alternating force. This factor of magnification M, like the angle of lag α, depends on only two dimensionless parameters, ω/ω_n, which is the ratio of the impressed frequency to the natural frequency of the system, and c/c_c, which is the ratio of the actual amount of damping to the critical damping of the system. The motion described by the particular integral does not decay as time goes on but continues its periodic behavior indefinitely provided resonance effects are nondestructive. For this reason, although it is obviously not independent of the time, it is known as the **steady state.**

EXAMPLE 1

IMPRESSED FREQUENCY EFFECTS

A 50 lb weight is suspended from a spring of modulus 20 lb/in. When the system is vibrating freely, it is observed that in consecutive cycles the maximum displacement decreases by 40 percent. If a force equal to 10 cos ωt lb acts upon the system, find the amplitude and phase lag of the resultant steady-state motion if **(a)** $\omega = 6$, **(b)** $\omega = 12$, and **(c)** $\omega = 18$ rad/s.

The first step here is to determine the amount of damping present in the system. From the given data it is clear that

$$y_{n+2} = 0.60 y_n$$

and thus that

$$\delta = \ln \frac{y_n}{y_{n+2}} = \ln \frac{1}{0.60} \doteq 0.511$$

Hence, by Eq. (9),

$$\frac{c}{c_c} = \frac{\delta}{(\sqrt{\delta^2 + 4\pi^2})} \doteq \frac{0.511}{\sqrt{(0.511)^2 + 4\pi^2}} \doteq 0.081$$

Next we must compute the natural frequency of the system

$$\omega_n = \sqrt{\frac{kg}{w}} = \sqrt{\frac{20 \times 384}{50}} \doteq 12.4 \text{ rad/s}$$

Finally, it is clear that a 10 lb force acting statically will stretch a spring of modulus 20 lb/in a distance

$$\delta_{st} = \tfrac{10}{20} = 0.5 \text{ in}$$

Knowing c/c_c, ω_n, and δ_{st}, we can now use Eq. (11) to compute the magnification ratio, the equation $A = \delta_{st} M$ to find the amplitude, Eq. (13) to compute the phase angle, and the first of Eqs. (10*a*) to obtain the steady-state motion for $\omega = 6$, 12, and 18 rad/s. Direct substitution gives the accompanying table.

ω	M	A	α	Y
6	1.30	0.65 in	0.10	$0.65 \cos(6t - 0.10)$ in
12	5.94	2.97 in	1.19	$2.97 \cos(12t - 1.19)$ in
18	0.88	0.44 in	2.93	$0.44 \cos(18t - 2.93)$ in

The amplitude corresponding to the impressed frequency $\omega = 12$ is much larger than either of the others because this frequency very nearly coincides with the resonant frequency of the system $\omega_r \doteq 12.3$ rad/s determined by (12*b*). One further calculation, using Eq. (12*a*), gives $0 < \omega < 17.4$ as the resonant frequency band. Of course, it contains both $\omega = 6$ and $\omega = 12$, but not $\omega = 18$, as the corresponding values of M would imply.

EXAMPLE 2

AMPLITUDE MODULATION AND BEATS

A system containing a negligible amount of damping is disturbed from its equilibrium position by the sudden application at $t = 0$ of a force equal to $F_0 \sin \omega t$. Discuss, qualitatively, the subsequent motion of the system if ω is close to the natural frequency ω_n and if $\omega = \omega_n$.

The differential equation to be solved here is

$$\frac{w}{g}\frac{d^2y}{dt^2} + ky = F_0 \sin \omega t$$

The complementary function is clearly

$$A \cos \sqrt{\frac{kg}{w}}t + B \sin \sqrt{\frac{kg}{w}}t = A \cos \omega_n t + B \sin \omega_n t$$

and it is easy to verify that a particular integral is

$$Y = \frac{F_0}{k - \omega^2(w/g)} \sin \omega t = \frac{\omega_n^2 \delta_{st}}{\omega_n^2 - \omega^2} \sin \omega t \qquad \omega \neq \omega_n$$

Hence a complete solution can be written

$$y = A \cos \omega_n t + B \sin \omega_n t + \frac{\omega_n^2 \delta_{st}}{\omega_n^2 - \omega^2} \sin \omega t$$

Since $y = 0$ when $t = 0$, we must have $A = 0$, leaving

(14) $$y = B \sin \omega_n t + \frac{\omega_n^2 \delta_{st}}{\omega_n^2 - \omega^2} \sin \omega t$$

and

$$v = \frac{dy}{dt} = \omega_n B \cos \omega_n t + \frac{\omega_n^2 \delta_{st}}{\omega_n^2 - \omega^2} \omega \cos \omega t$$

Substituting $v = 0$ and $t = 0$ in the last equation, we obtain

$$0 = \omega_n B + \omega \frac{\omega_n^2 \delta_{st}}{\omega_n^2 - \omega^2} \qquad \text{or} \qquad B = -\frac{\omega \omega_n \delta_{st}}{\omega_n^2 - \omega^2}$$

Hence, substituting this into (14), we find for the required solution

(15) $$y = \frac{\omega_n \delta_{st}}{\omega^2 - \omega_n^2}(\omega \sin \omega_n t - \omega_n \sin \omega t) \qquad 0 \leqq t,\ \omega \neq \omega_n$$

Since we are concerned with the behavior of the system when the impressed frequency ω is close to the natural frequency ω_n, it is natural to put $\omega - \omega_n = 2\epsilon$, where ϵ is an arbitrarily small quantity. After doing this and rearranging slightly, Eq. (15) becomes

$$y = \frac{\omega_n \delta_{st}}{4\epsilon(\omega_n + \epsilon)}[\omega_n \sin \omega_n t + 2\epsilon \sin \omega_n t - \omega_n \sin (\omega_n + 2\epsilon)t]$$

For descriptive purposes, we may at this stage neglect the second term in brackets and the second term in the binomial in the denominator. Making these approximations and then converting the difference of the remaining sine terms into a product, we obtain as an approximate description of the motion

$$-\frac{\omega_n \delta_{st}}{2\epsilon} \cos (\omega_n + \epsilon)t \sin \epsilon t \doteq y$$

from which, by again neglecting ϵ in comparison with ω_n, we obtain the further approximation

(16) $$y_a = -\frac{\omega_n \delta_{st}}{2\epsilon} \cos \omega_n t \sin \epsilon t \doteq y$$

To help us visualize this approximation to the exact solution y, we note first that the period of the factor $\cos \omega_n t$, namely, $2\pi/\omega_n$, is much smaller than the period of the factor $\sin \epsilon t$, namely, $2\pi/\epsilon$. Specifically, there are $(2\pi/\epsilon)/(2\pi/\omega_n) = \omega_n/\epsilon$ cycles of $\cos \omega_n t$ in every period of $\sin \epsilon t$. In other words, the graph of $\cos \omega_n t$ fluctuates much more rapidly than the graph of $\sin \epsilon t$ does. This suggests that in the expression (16) for y_a, we focus our attention on the factor

$$-\frac{\omega_n \delta_{st}}{2\epsilon} \cos \omega_n t \equiv -A \cos \omega_n t$$

of amplitude $A = \omega_n \delta_{st}/2\epsilon$ and period $2\pi/\omega_n$ and express y_a as

(17) $$(a)\ y_a = a(t) \cos \omega_n t \qquad \text{where} \qquad (b)\ a(t) = -A \sin \epsilon t$$

By definition, the **amplitude** of y_a equals half the difference between the greatest and least values of y_a on each successive period of cos $\omega_n t$. As t increases, this amplitude changes from cycle to cycle as the factor sin ϵt in (17*b*) slowly varies between -1 and 1. Because this changing amplitude is due solely to the presence of this factor, we say that in the representation (16) of y_a the function $-A \cos \omega_n t$ is **amplitude-modulated** by sin ϵt.

Just as the graph of $A \cos \omega_n t$ lies on or between (i.e., is contained or *enveloped* by) the horizontal lines $y = \pm A$, touching them only when $\cos \omega_n t = \pm 1$, so the curves

$$y = \pm A \sin \epsilon t$$

enclose or *envelop* the graph of y_a. These two curves, to which y_a periodically rises and falls, touching them only when $\cos \omega_n t = \pm 1$, are said to be the **amplitude envelop,** or **envelope** of the graph of y_a, i.e, of the graph of $-A \cos \omega_n t$ amplitude-modulated by sin ϵt.

Figure 7.10*a* summarizes our geometric observations about the general character of the motion when the impressed frequency is close to, but not equal to, the natural frequency.

The behavior we have just described is most vividly perceived when sound waves are involved. In this case, our ears would hear a tone whose *pitch* was determined by the frequency $\omega_n/2\pi$ of the factor cos $\omega_n t$ and whose *intensity* or *loudness* varied between 0 (or momentary silence) at the times when $\sin \epsilon t = 0$ and $\omega_n \delta_{st}/2\epsilon$ at the times when $\sin \epsilon t = \pm 1$. This low-frequency variation in the intensity of a high-frequency signal is called **beating,** and each individual maximum of the signal intensity is called a **beat.** Since, as Fig. 7.10*a* illustrates, there are two extremes of $|\sin \epsilon t|$ in each period of sin ϵt, the number of beats per unit time is

$$2\left(\frac{|\epsilon|}{2\pi}\right) = \frac{|\omega - \omega_n|}{2\pi} = \left|\frac{\omega}{2\pi} - \frac{\omega_n}{2\pi}\right|$$

In other words, the number of beats per unit time is equal to the absolute value of the difference between the two frequencies that are involved.

FIGURE 7.10
The phenomena of beats and pure resonance.

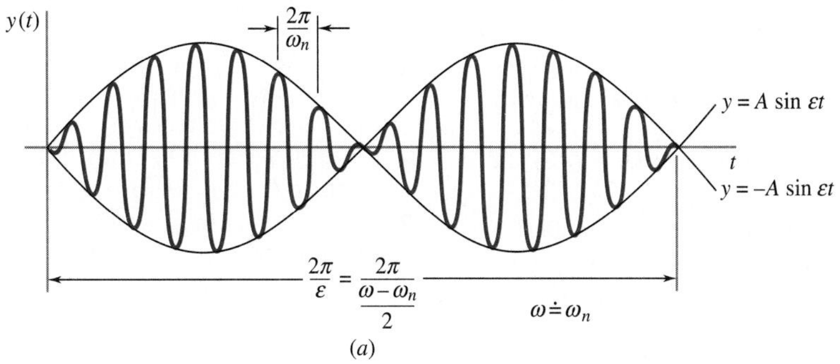

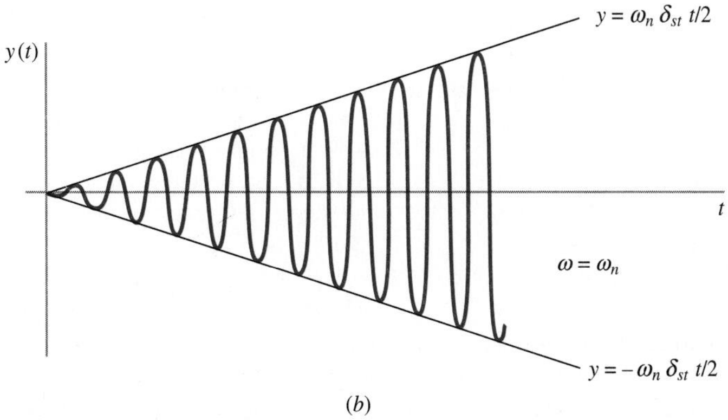

The phenomenon of beats should be anticipated whenever damping effects are small and

(a) An impressed frequency is close to a natural frequency of a system, or

(b) Two disturbances of slightly different frequencies but approximately equal amplitudes are impressed on a system, regardless of what its natural frequencies may be; or, in more complicated systems,

(c) Two natural frequencies of a system are close together and the system is vibrating with nearly equal amplitudes at those frequencies.

To investigate what happens as $\omega \to \omega_n$ or, equivalently, as $\epsilon = (\omega - \omega_n)/2 \to 0$, it is convenient to multiply and divide the expression (16) for y_a by t and write

$$y_a = -\frac{t\omega_n\,\delta_{st}}{2}\cos\omega_n t\,\frac{\sin\epsilon t}{\epsilon t}$$

Now, as $\epsilon \to 0$, the last fraction approaches 1, and we have

$$\lim_{\epsilon\to 0} y_a = -\frac{\omega_n\,\delta_{st}}{2}t\cos\omega_n t$$

The function $-(\omega_n\,\delta_{st}/2)\cos\omega_n t$ is now amplitude-modulated by the function t, and when thus modulated has the pair of lines

$$y = \pm\frac{\omega_n\,\delta_{st}}{2}t$$

as its envelop. In this limiting case, the amplitude of the system increases beyond all bounds, as expected under conditions of pure resonance. Figure 7.10*b* shows a plot of the response when $\omega = \omega_n$.

EXERCISES

In the following problems, take the acceleration of gravity to be 384 in/sec^2 or 32 ft/sec^2, as appropriate.

1. A weight of 48 lb hangs from a spring of modulus 18 lb/in. At $t = 0$ the weight is pulled down a distance of 2 in and released from that position with an upward velocity of 2 in/sec. Find an equation describing its subsequent motion.
2. The displacement x of a particle that moves along a straight line satisfies $\ddot{x} + 2\dot{x} + 4x = bx$. For what values of b is the motion oscillatory, and for what value of b is the natural frequency 1 Hz?
3. A unit mass particle moves along a horizontal straight line. When it is at a distance x from a fixed point O on the line, it experiences a restoring force equal to $-x$, a force of repulsion equal to $2dx/dt$, and an impressed force equal to e^t/t. All other forces are negligible. Write the equation of motion of the particle and find its complete solution.
4. A spring is stretched 3 in by a force of 2 lb. A 5 lb weight is attached and released from rest 1 ft above its equilibrium position. If $c = 3$ lb/ft/sec, determine the displacement of the weight and when, if ever, it passes through its equilibrium position.
5. A spring is stretched 6 in by a 90 lb force. A 10 lb weight is attached and undergoes damped vertical motion. Find its displacement y given that $y(0) = -0.5$ ft, $\dot{y}(0) = 14$ ft/s, and $c = 15$ lb/ft/s. When, if ever, is the displacement zero?
6. A weight of 96 lb hangs from a spring of modulus 25 lb/in. The damping in the system is 60 percent of critical. Find the displacement of the weight as a function of t if it is pulled down 1 in from its equilibrium position and given an upward velocity of 2 in/s.
7. A 64 lb weight suspended from a spring of modulus $k = \frac{5}{6}$ lb/in is being driven by a force $F = 34\sin 2t$ lb. The coefficient of damping is 4 lb/ft/s, $y(0) = -4$ ft, and $\dot{y}(0) = 2$ ft/s. Find the displacement of the weight when $t = \pi/2$ s.
8. A weight of 24 lb hangs from a spring of modulus 6 lb/in. The weight is retarded in its motion by a frictional force whose damping is equal to $\frac{5}{4}$ lb/in/s. At $t = 0$ the weight is moving through the point $y = 2$ in with velocity 3 in/s. Find an equation describing its subsequent motion.
9. A 24 lb weight is suspended from a spring of modulus 16 lb/in. The damping in the system is $\frac{5}{12}$ of critical. Find the steady-state motion and resonant frequency if the disturbing force is $F = 40\cos 24t$ lb.
10. A weight of 60 lb hangs from a spring of modulus 40 lb/in. Find the steady-state motion and resonant frequency of this system if the damping equals $\frac{3}{2}$ lb/in/s and $F = 5\cos 20t$ lb.
11. A 32 lb weight hangs from a spring of modulus 49 lb/in

and is acted upon by a force $F = 7\sqrt{2}\sin 21t$ lb. The damping coefficient equals $\frac{7}{12}$ lb/in/s. Find the steady-state motion and the resonant frequency of the system.

12. A weight of 48 lb hangs from a spring of modulus 72 lb/in. If the damping equals 0.06 lb/in/s and a disturbing force $F = 12\cos 24t$ lb acts on the weight, find the steady-state motion and the resonant frequency of the system.

13. A weight of 75 lb hangs in equilibrium from a spring of modulus 18 lb/in. Find the displacement of the weight as a function of time if a force of 45 lb acting downward is suddenly applied to the weight.

14. What is the minimum amount of friction that would have to be introduced in the system in Exercise 13 to keep it from oscillating? Find the displacement of the weight as a function of time if this much friction is present.

15. A weight of 48 lb hangs in equilibrium from a spring of modulus 32 lb/in. If the damping equals 5 lb/in/s, find the deflection of the weight as a function of time if an excitation $F_0 \sin 16t$ lb is suddenly applied at $t = 0$.

16. In Example 1, if the applied frequency must be 12 rad/s but if the resultant amplitude must be no more than 2 in, to what figure must the value of c/c_c be increased?

17. A weight of 128 lb hangs from a spring of modulus 27 lb/in. The damping in the system is 1 lb/in/s. Determine the steady-state response of the system to a force equal to $12\sin\omega t$ lb if **(a)** $\omega = 5$, **(b)** $\omega = 10$, and **(c)** $\omega = 15$, rad/s.

18. A weight of 1600 lb is connected to a wall by a spring of modulus 720 lb/ft. At $t = 0$ a weight of 800 lb traveling at 15 ft/s collides with the first weight and thereafter moves with it as a single body. Neglecting friction, find the equation describing the motion after impact of the combined mass. *Hint:* By the principle of the conservation of momentum, the momentum of the system before impact is the same as its momentum after impact.

19. What is the minimum amount of friction the system in Exercise 18 must contain if the motion after impact is to be nonoscillatory? What is the equation of motion of the combined mass in this case?

20. A weight w is suspended from a spring of modulus k but is temporarily supported at a level where the spring is unstretched. At $t = 0$ the support is removed, the weight falls, and the system begins vertical oscillations. Neglecting friction, what is the maximum distance to which the weight falls? About what point does the weight oscillate? If y is measured from the initial position of the weight, what is the equation of its subsequent motion? *Hint:* Recall that the energy stored in a spring of modulus k stretched a distance y is equal to $ky^2/2$. Then note that this energy, reduced by the potential energy lost by the weight in moving to its lowest position, must be equal to the potential energy of the system in its initial position. Why?

21. What is the value of the magnification ratio if friction in a system is just equal to c_c?

22. If friction is neglected, show that the natural frequency of a system consisting of a mass on an elastic suspension is approximately equal to $3.12/\sqrt{\delta_{st}}$ Hz, where δ_{st} is the deflection (in inches) produced in the suspension when the mass hangs on it in static equilibrium.

23. A motor of unknown weight is set on a felt mounting pad of unknown spring constant. What is the natural frequency of the system if the motor is observed to compress the pad $\frac{1}{16}$ in?

24. Show that the logarithmic decrement δ can also be computed by the formula

$$\delta = \frac{1}{k}\ln\frac{y_n}{y_{n+2k}} \qquad k = 1, 2, 3, \ldots$$

25. A weight of 54 lb hangs from a spring of modulus 36 lb/in. During the free motion of the system it is observed that the maximum displacement of the weight decreases to one-tenth of its value in five complete cycles of the motion. Find the equation describing the steady-state motion produced by a force equal to $6\sin 15t$ lb.

26. A weight of 96 lb hangs from a spring of modulus 25 lb/in. Damping in the system is 60 percent of critical. Determine the motion of the weight if it is pulled downward 1 in from its equilibrium position and released with an upward velocity of 2 in/s.

27. Work Exercise 26 if a constant force of 50 lb is suddenly applied to the system when it is at rest in its equilibrium position.

28. A weight of 128 lb hangs from a spring of modulus 75 lb/in. Damping in the system is 28 percent of critical. Determine the motion of the weight if it is pulled downward 2 in from its equilibrium position and suddenly released.

29. Work Exercise 28 if a force equal to F_0e^{-10t} lb is suddenly applied to the system when it is at rest in its equilibrium position.

30. Investigate the motion of a weight hanging on a spring when the disturbing force is equal to $F_0\sin\omega t$ instead of $F_0\cos\omega t$. In particular, show that Eqs. (11) and (13) for the magnification ratio and phase shift, respectively, are still the same.

31. For a given value of c/c_c, determine the minimum number of cycles required to produce a reduction of at least 50 percent in the maxima of a damped oscillation.

32. Prove that the logarithmic decrement δ is equal to the natural logarithm of the ratio of *any* nonzero displacement to the displacement one full cycle later.

33. If c/c_c is small, show that the logarithmic decrement is approximately

$$\delta = \frac{y_n - y_{n+2}}{y_n} = \frac{\Delta y_n}{y_n}$$

34. Show that the energy dissipated during the nth cycle of a damped oscillation is equal to $(k/2)(y_n^2 - y_{n+2}^2)$. Hence, using Exercise 33, show that when c/c_c is small, the energy loss per cycle during the nth cycle is approximately $k\delta y_n^2$.

35. If the roots of the characteristic equation in the overdamped case are $m = -r \pm s$, show that in general the complementary function can be written

$$y = Ae^{-rt} \cosh (st + B)$$

or

$$y = Ce^{-rt} \sinh (st + D)$$

according as y has no real zero or one real zero. Are there any exceptions?

36. Show that a complete description of the motion of an underdamped system acted upon by a force $F_0 \cos \omega t$ can be written in the form

$$y = \exp [(-c/c_c)\omega_n t][A \cos \omega t + B \sin \omega t] + \delta_{st} M \cos (\omega t - \alpha)$$

Hint: Recall Eq. (6).

37. **(a)** Obtain a result comparable to that of Exercise 36 if the system is critically damped.

(b) Obtain a result comparable to that of Exercise 36 if the system is overdamped.

38. An overdamped system begins to move from its equilibrium position with velocity v_0. Show that its maximum displacement occurs when

$$t = \frac{1}{\omega_n \sqrt{(c/c_c)^2 - 1}} \tanh^{-1} \sqrt{1 - \left(\frac{c}{c_c}\right)^2}$$

Hint: Use the result of Exercise 35.

39. In Exercise 38, show that the maximum displacement is

$$y_{\max} = \frac{v_0}{\omega_n}\left(\tan \frac{\gamma}{2}\right)^{\sec \gamma} \qquad \text{where } \gamma = \mathrm{Sin}^{-1} \frac{c}{c_c}$$

40. Investigate the answers to Exercises 38 and 39 in the limit when c/c_c approaches 1. Check your conclusions by working directly with the equation for the transient in the critically damped case.

41. Show that the maximum displacements during the free motion of an underdamped system do not occur midway between the zeros of the displacement but precede the midpoints by the constant amount

$$\frac{\mathrm{Sin}^{-1} (c/c_c)}{\omega_n \sqrt{1 - (c/c_c)^2}}$$

42. Show that Eq. (1) can be written in the form

$$\frac{d^2y}{dt^2} + 2\frac{c}{c_c}\omega_n \frac{dy}{dt} + \omega_n^2 y = \delta_{st}\omega_n^2 \cos \omega t$$

43. In the critically damped case, show that the common value of the roots of the characteristic equation of Eq. (1) is $-\omega_n$.

44. A weight of 16 lb is hanging at rest in its equilibrium position from a spring whose elongation is $\frac{8}{3}$ in. At $t = 0$ the weight begins to move under the action of a disturbing force $F = \sqrt{216} \sin \omega t$ lb. The motion is damped by a frictional force for which $c/c_c = 3\sqrt{2}/10$. Find the steady-state motion of the weight when the forcing function is in resonance with the system.

45. Show that the maxima of the curves of the magnification ratio versus the frequency ratio occur when

$$\frac{\omega}{\omega_n} = \sqrt{1 - 2\left(\frac{c}{c_c}\right)^2} \qquad 0 \le \frac{c}{c_c} < \frac{1}{\sqrt{2}}$$

What are the maximum values?

46. In many applications involving forces arising from rotating parts that have become unbalanced, the amplitude of the sinusoidal disturbing force acting on a system is not constant but varies as the square of the frequency. If a weight suspended from a spring is acted upon by a force of this character, determine its steady-state motion. In particular, determine the form of the magnification ratio and the formula for the angle of lag.

47. Show that the maxima of the curves of the magnification ratio versus the frequency ratio under the conditions of Exercise 46 always occur at values of the impressed frequency ω which are greater than the natural frequency of the system ω_n.

48. A uniform bar of length l and weight w rests on two parallel rollers which rotate about fixed horizontal axes as shown in Fig. 7.11*a*. Friction between the bar and each roller is assumed to be "dry" or **coulomb,** i.e., proportional to the normal force between the bar and the roller, the proportionality constant being the so-called **coefficient of friction** μ. When the bar, which always remains in a line perpendicular to the axes of the rollers, is displaced slightly from a symmetric position, it executes small oscillations in the horizontal direction. (Why?) Determine the period of these oscillations and show how the value of μ can thus be found experimentally. *Hint:* The vertical components of the forces which the rollers exert on the bar vary as the position of the bar changes, but their sum is always numerically equal to the weight of the bar. The value of the vertical force exerted on the

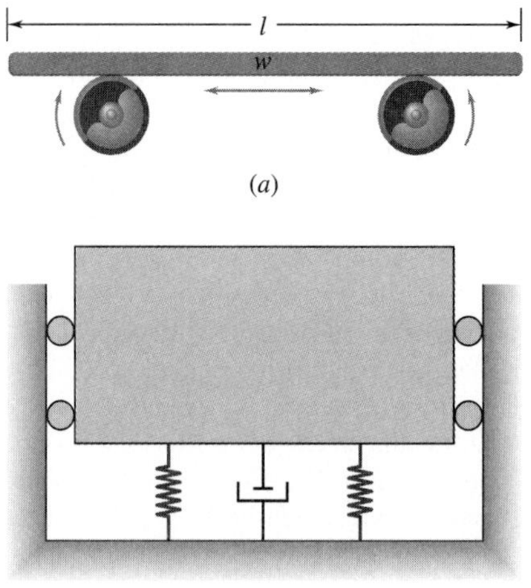

(*a*)

(*b*)

FIGURE 7.11

bar by either of the rollers can be found by taking moments about the point of contact of the bar and the other roller and equating to zero the sum of these moments.

49. What is the natural frequency of the system shown in Fig. 7.2*d* and discussed in Exercise 10, Sec. 7.2? Explain how such a system can be used to determine experimentally the polar moment of inertia of a large, irregular object.

50. A particle of weight w moves along the x axis under the influence of a force equal to $-kx$. Friction in the system is assumed to be dry, rather than viscous; i.e., it is proportional to the normal force between the particle and the surface on which it moves and does not depend on the velocity. Show that the motion of the particle is governed by the differential equations

$$\frac{w}{g}\frac{d^2x}{dt^2} + kx = \begin{cases} \mu w & \text{(Particle moving to the left)} \\ -\mu w & \text{(Particle moving to the right)} \end{cases}$$

If the particle starts from rest at the point $x = x_0$, find x as a function of t. What is the decrease in amplitude per cycle? When will the particle come to rest?

51. If y_0 and v_0 are, respectively, the initial displacement and velocity with which an overdamped system begins its motion, show that

$$\frac{w}{g}v_0^2 + cv_0y_0 + ky_0^2 > 0$$

is the condition that the complementary function, i.e., the transient, should have a real zero. What further condition is necessary to ensure that the zero of the complementary function will be nonnegative, i.e., will occur during the actual motion?

52. In Example 2, find the exact solution y of the differential equation when $\omega = \omega_n$ and let y_0 be defined by $y_0 = \lim_{\epsilon\to 0} y_a$. In what respect does y differ from y_0? What part of y is given by y_0?

53. In Exercise 52, show that for $t > 0$ the zeros of y_0 are given by $t = (2m + 1)\pi/2\omega_n$ and that the extremes of y_0 occur when $\tan \omega_n t = 1/\omega_n t$. Also verify that the zeros of y satisfy the equation $\tan \omega_n t = \omega_n t$ and that y has an extremum when $t = m\pi/\omega_n$.

54. Work Example 2 if the impressed force is $F_0 \cos \omega t$.

55. A weight of 384 lb hangs from a spring of modulus 169 lb/in. The damping is 10 times the critical value divided by $26\sqrt{2}$. A disturbing force equal to $170 \cos \omega t$ lb acts on the weight. Find the pseudoperiod of the transient, the resonant frequency, the steady state when $\omega = \omega_r$, and the value of ω for which the phase angle of the response equals 135°.

56. A weight W is suspended from a spring and is vibrating in resonance with a periodic driving force $F = 3\sqrt{2} \cos \omega_r t$ lb. The natural frequency of the system is $2/\pi$ Hz and the amplitude of the steady state is 6 in. Given that $c/c_c = \omega_r/\omega_n$, find the magnification ratio M, the spring constant k, the weight W, and the steady-state motion of the system. *Hint:* Use Eq. (12*b*) to simplify the formula for M.

57. A 48 lb weight hangs from a spring of modulus 50 lb/in. In 10 cycles of the motion it is observed that the maximum displacement decreases by 50 percent. Determine the steady-state motion of the system if it is acted upon simultaneously by forces equal to $F_0 \cos 15t$ and $F_0 \cos 16t$. Do you think that these two forces will produce the phenomenon of beats? Why?

58. A system is acted upon by two forces

$$F_1 \sin \omega_1 t \qquad \text{and} \qquad F_2 \sin \omega_2 t$$

Friction, though present in the system, is so small that it can be neglected in determining the forced motion. Discuss the steady-state behavior of the system if ω_1 and ω_2 are nearly equal but neither is close to the natural frequency of the system. In particular, show that the response consists of terms of frequency $(\omega_1 + \omega_2)/2$ whose amplitudes are modulated by factors of frequency $(\omega_1 - \omega_2)/2$. *Hint:* Note first that the assumption of at least a little friction in the system means that only the particular integrals contribute to the steady-state motion. (Why?) Then, after the particular integrals have been determined, note that the expression $K_1 \sin \omega_1 t + K_2 \sin \omega_2 t$ can be written in the form

$$\frac{K_1 + K_2}{2}(\sin \omega_1 t + \sin \omega_2 t) + \frac{K_1 - K_2}{2}(\sin \omega_1 t - \sin \omega_2 t)$$

Finally, convert these sums and differences into products.

59. Using the result of Exercise 45, find the value of the phase angle α when the amplitude of the forced steady-state vibration is a maximum.

60. How much work is done by a force $F = F_0 \cos \omega t$ during one cycle of forced steady-state vibrations? *Hint:* Observe that $dW = F\,ds = Fv\,dt$.

61. How much energy is dissipated as a result of friction during one cycle of a forced steady-state vibration? *Hint:* Remember that with viscous damping, the force of friction is proportional to the velocity.

62. Using the results of Exercises 60 and 61, show that when c/c_c is small, the maximum amplitude of the steady-state vibrations produced by a force $F = F_0 \cos \omega t$ is approximately $A_{\max} = F_0/c\omega$.

63. Because it is slightly unbalanced, an electric motor is subject to a disturbing force equal to $F_0\omega^2 \cos \omega t$. To reduce the force transmitted to the foundation, the motor is mounted on a combination of springs and a dashpot, as shown in Fig. 7.11*b*. Show that the maximum value of the vertical force transmitted to the foundation is

$$F_0\omega^2\frac{\sqrt{1 + \left(2\dfrac{\omega}{\omega_n}\dfrac{c}{c_c}\right)^2}}{\sqrt{\left(1 - \dfrac{\omega^2}{\omega_n^2}\right)^2 + \left(2\dfrac{\omega}{\omega_n}\dfrac{c}{c_c}\right)^2}}$$

Hint: The elastic force ky and the friction force $-c\dot{y}$, each of which affects the foundation, are 90° out of phase and their resultant must be considered.

64. For a given value of c/c_c, what is the range of values of ω/ω_n for which $M \geq 1$?

7.4 THE SERIES ELECTRIC CIRCUIT

All the results we obtained in the last section can, after a suitable change in terminology, be applied to any of the other systems we have considered. However, the concepts central in one field are not always of equal importance in related fields, and it seems desirable to illustrate the minor differences in the application of our general theory to various classes of systems by considering one of the electric circuits in some detail.

For the simple series circuit with an alternating impressed voltage, we derived (among several equivalent forms) the equation

$$L\frac{d^2Q}{dt^2} + R\frac{dQ}{dt} + \frac{1}{C}Q = E_0 \cos \omega t \tag{1}$$

and on comparing this with the differential equation of the vibrating weight

$$\frac{w}{g}\frac{d^2y}{dt^2} + c\frac{dy}{dt} + ky = F_0 \cos \omega t$$

we noted the correspondences

$$\begin{aligned}
\text{Mass } \frac{w}{g} &\longleftrightarrow \text{inductance } L\\
\text{Damping } c &\longleftrightarrow \text{resistance } R\\
\text{Spring modulus } k &\longleftrightarrow \text{elastance } \frac{1}{C}\\
\text{Impressed force } F &\longleftrightarrow \text{impressed voltage } E\\
\text{Displacement } y &\longleftrightarrow \text{charge } Q\\
\text{Velocity } v &\longleftrightarrow \text{current } i
\end{aligned}$$

Extending these correspondences to the derived results by making the appropriate substitutions, we infer from the natural frequency of the mechanical system

$$\omega_n = \sqrt{\frac{kg}{w}}$$

that the electric circuit has a **natural frequency**

$$\Omega_n = \sqrt{\frac{1}{LC}}$$

Furthermore, the concept of critical damping

$$c_c = 2\sqrt{\frac{kw}{g}}$$

leads to the concept of **critical resistance**

$$R_c = 2\sqrt{\frac{L}{C}}$$

which determines whether the free behavior of the electrical system will be oscillatory or nonoscillatory.

The notion of magnification ratio can also be extended to the electrical case, but it is not customary to do so because the extension would relate to Q (the analog of the displacement y), whereas in most series circuits it is not Q but i that is the variable of interest. In working with series circuits, a more useful concept arises by considering not the charge Q but the current i produced by an impressed voltage $E_0 \cos \omega t$. This is easily done by converting the derivative

$$\dot{y} = -\delta_{st} M\omega \sin(\omega t - \alpha) \qquad \alpha = \tan^{-1} \frac{2(\omega/\omega_n)(c/c_c)}{1 - (\omega/\omega_n)^2}$$

of the particular integral Y given by Eq. (10*a*), Sec. 7.3, into its electrical equivalent. Direct substitution of electrical for corresponding mechanical parameters gives

$$i = \frac{-E_0 C\omega \sin(\omega t - \alpha)}{\sqrt{(1 - LC\omega^2)^2 + (RC\omega)^2}} \qquad \alpha = \tan^{-1} \frac{RC\omega}{1 - LC\omega^2}$$

Dividing numerator and denominator of the two fractional expressions by $C\omega$ and setting $\alpha = \pi/2 + \delta$ so that the current i and the impressed voltage E will be expressed in terms of the same trigonometric function, we obtain, after straightforward simplifications, the **steady-state** current

(2) $$i = \frac{E_0 \cos(\omega t - \delta)}{\sqrt{R^2 + (\omega L - 1/\omega C)^2}}$$

where, since $\tan \delta = -1/\tan \alpha$,

(3) $$\delta = \operatorname{Tan}^{-1} \frac{\omega L - 1/\omega C}{R}$$

the principal-value designation being appropriate in (3) because δ increases from $-\pi/2$ to $\pi/2$ as α increases from 0 to π.

From Eq. (2) we infer that the steady-state current produced by an alternating voltage is of the same frequency as the voltage but differs from it in phase by

$$\frac{\delta}{\omega} \text{ units of time} \qquad \text{or} \qquad \frac{\delta/\omega}{2\pi/\omega} = \frac{\delta}{2\pi} \text{ cycles}$$

If δ is positive, the steady-state current *lags* the voltage; if δ is negative, the steady-state current *leads* the voltage.

Furthermore, from Eq. (2) we see that the amplitude of the steady-state current is obtained by dividing the amplitude of the impressed voltage E_0 by the expression

(4) $$\sqrt{R^2 + \left(\omega L - \frac{1}{\omega C}\right)^2}$$

By analogy with **Ohm's law,**† $I = E/R$, the quantity (4) thus appears as a generalized resistance, although it is actually called the **impedance** of the circuit. While not the analog of the magnification ratio, the impedance is clearly a similar concept. Since impedance is defined as

$$\frac{\text{Voltage}}{\text{Current}}$$

†Georg Simon Ohm (1787–1854) was a German physicist.

the mechanical quantity corresponding to this is the ratio

$$\frac{\text{Force}}{\text{Velocity}}$$

This is called the **mechanical impedance** by some writers, and in certain mechanical problems has proved a useful notion.†

There is another approach to the problem of determining the steady-state current produced by a harmonic voltage that is well worth investigating. Suppose that given *either* $E_0 \cos \omega t$ *or* $E_0 \sin \omega t$, we write the basic differential equation (1) in the form

$$(5) \qquad L\frac{d^2Q}{dt^2} + R\frac{dQ}{dt} + \frac{1}{C}Q = E_0 e^{j\omega t} = E_0(\cos \omega t + j \sin \omega t)‡$$

This includes both possibilities for the voltage, and if the real and imaginary terms retain their identity throughout the analysis, then the real part of the particular integral corresponding to $e^{j\omega t}$ will be the particular integral for $E_0 \cos \omega t$ and the imaginary part will be the particular integral for $E_0 \sin \omega t$.

To see that this is actualıy the case, we must first find a particular integral of Eq. (5). Although $e^{j\omega t}$ is not real, we nevertheless assume that there is a particular integral of the form

$$Q = Ae^{j\omega t}$$

and substitute it into the differential equation. This gives

$$L(-\omega^2 Ae^{j\omega t}) + R(j\omega Ae^{j\omega t}) + \frac{1}{C}(Ae^{j\omega t}) = E_0 e^{j\omega t}$$

which will be an identity if and only if

$$A = \frac{E_0}{-\omega^2 L + j\omega R + 1/C}$$

Hence

$$Q = \frac{E_0}{j\omega R - \omega^2 L + 1/C} e^{j\omega t}$$

From this, by differentiation, we find that

$$\frac{dQ}{dt} = i = \frac{j\omega E_0}{j\omega R - \omega^2 L + 1/C} e^{j\omega t} = \frac{E_0}{R + j(\omega L - 1/\omega C)} e^{j\omega t}$$

To find the real and imaginary parts of this expression, it is convenient to use the fact (Sec. 3.1) that any complex number $a + jb$ can be written in the form $a + jb = re^{j\delta}$, where the magnitude r

---

†See, for instance, T. von Karman and M. A. Biot, *Mathematical Methods in Engineering,* pp. 370–378, McGraw-Hill, New York, 1940.

‡To avoid confusing $i = \sqrt{-1}$ with i = current, we shall throughout the rest of this chapter follow the practice, standard in electrical engineering, of writing $j = \sqrt{-1}$.

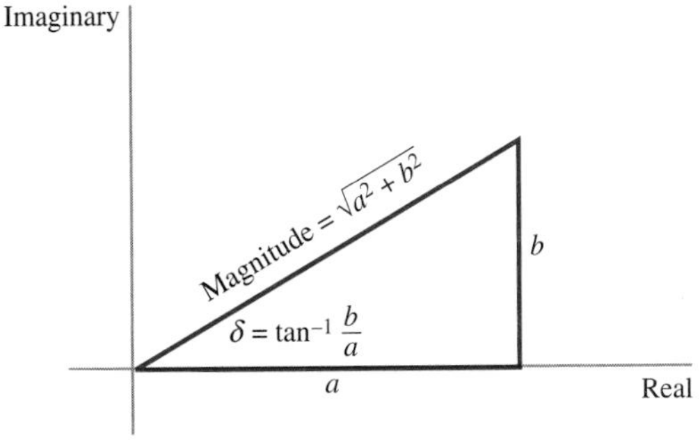

FIGURE 7.12
The relations among the magnitude, angle, and components of a general complex number $a + jb$.

and the angle δ of the complex number are related to the components a and b as shown in Fig. 7.12. Applied to the denominator of the second expression for i, this gives

$$R + j\left(\omega L - \frac{1}{\omega C}\right) = \sqrt{R^2 + \left(\omega L - \frac{1}{\omega C}\right)^2}\, e^{j\delta}$$

where

$$\delta = \text{Tan}^{-1}\frac{\omega L - 1/\omega C}{R}$$

Hence we can rewrite i in the form

$$\begin{aligned} i &= \frac{E_0}{\sqrt{R^2 + [\omega L - 1/(\omega C)]^2}\, e^{j\delta}}\, e^{j\omega t} \\ &= \frac{E_0}{\sqrt{R^2 + [\omega L - 1/(\omega C)]^2}}\, e^{j(\omega t - \delta)} \\ &= E_0 \frac{\cos(\omega t - \delta) + j\sin(\omega t - \delta)}{\sqrt{R^2 + [\omega L - 1/(\omega C)]^2}} \end{aligned}$$

Comparing this with Eqs. (2) and (3) makes it clear that the real part here is exactly the particular integral corresponding to $E_0 \cos \omega t$, as we derived it directly. Similarly, had we taken the trouble to work it out explicitly, we would have found for the particular integral corresponding to $E_0 \sin \omega t$ precisely the imaginary part of the last expression. Since it is much easier to find the particular integral corresponding to an exponential term than it is to find the particular integral corresponding to a cosine or sine term, the advantage of using $E_0 e^{j\omega t}$ in place of $E_0 \cos \omega t$ or $E_0 \sin \omega t$ is obvious.

The expression

$$(6) \qquad R + j\left(\omega L - \frac{1}{\omega C}\right) \quad \text{or} \quad j\omega L + R + \frac{1}{j\omega C}$$

is called the **complex impedance** Z. Its magnitude is the quantity (4) which we referred to simply as the impedance. Its angle δ is the **phase shift.** The real part of Z is clearly a **resistance.** The imaginary part of Z is called the **reactance.** The reciprocal of Z is called the **admittance.** The real part of the admittance is called the **conductance,** and the imaginary part is called the **susceptance.** The **inductive reactance** X_L is defined by $X_L = \omega L$ and the **capacitive reactance** X_C by $X_C = 1/\omega C$. In a circuit containing just an inductor, just a resistor, or just a capacitor, the complex impedance $Z = R + j(X_L - X_C)$ reduces to $Z_L = j\omega L$, $Z_R = R$, and $Z_C = 1/j\omega C$, respectively. Thus the successive terms in the second form of (6) identify in turn the complex impedance of an inductor $j\omega L$, the impedance of a resistor R, and the complex impedance of a capacitor $1/j\omega C$. This

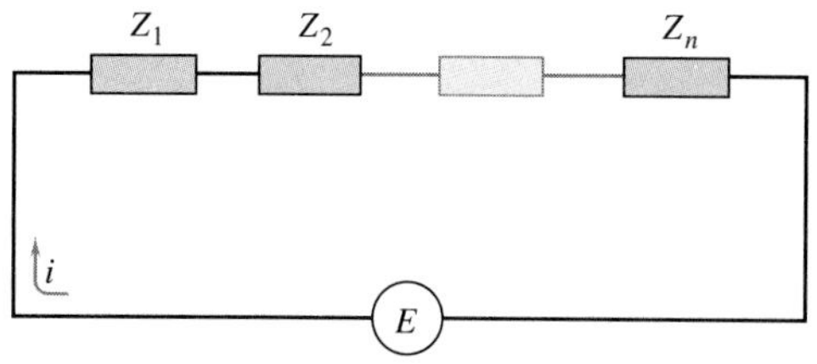

FIGURE 7.13
Impedances connected in series.

feature makes the latter form of (6) the preferred expression for the complex impedance of a basic series circuit. Different series combinations are usually identified by simply indicating the kind of elements they contain. For instance, an *RC* circuit contains a resistor and a capacitor but no inductor, whereas an *RLC* circuit contains one element of each kind.

The property that makes complex impedances so useful in circuit analysis is that when any electrical elements are connected in series or in parallel, the corresponding impedances combine just as simple resistances do. More specifically,

(a) The steady-state current through a series of impedances $Z_1, Z_2, \ldots, Z_n$ (Fig. 7.13) can be found by dividing the impressed (complex) voltage by the single impedance

$$Z = Z_1 + Z_2 + \cdots + Z_n$$

(see Exercise 17).

(b) The steady-state current through a set of impedances $Z_1, Z_2, \ldots, Z_n$ connected in parallel (Fig. 7.14) can be found by dividing the (complex) impressed voltage by the single impedance Z defined by the expression

$$\frac{1}{Z} = \frac{1}{Z_1} + \frac{1}{Z_2} + \cdots + \frac{1}{Z_n}$$

(see Exercise 17).

These properties make it unnecessary to use differential equations in determining the *steady-state* behavior of an electric network (or of a mechanical system, if the concept of mechanical impedance is used). For the *transient* behavior, however, this is not true until the impedance concept is generalized through the use of the Laplace transform (Chap. 10).

EXAMPLE 1

AN IMPEDANCE CALCULATION

What is the single impedance equivalent to the combination of impedances shown in Fig. 7.15?

FIGURE 7.14
Impedances connected in parallel.

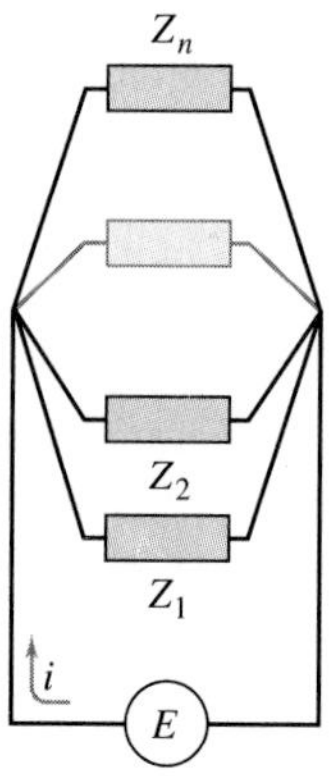

FIGURE 7.15
A mixed combination of impedances.

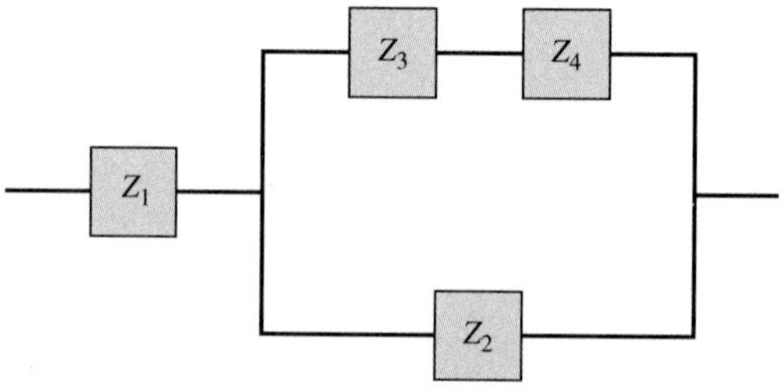

By Property **a**, the series combination of Z_3 and Z_4 can be replaced by the single impedance $Z_5 = Z_3 + Z_4$. Next, the parallel combination of Z_2 and Z_5 can be replaced according to Property **b** by the single impedance Z_6 defined by

$$\frac{1}{Z_6} = \frac{1}{Z_2} + \frac{1}{Z_5} = \frac{1}{Z_2} + \frac{1}{Z_3 + Z_4} \qquad \text{or} \qquad Z_6 = \frac{Z_2(Z_3 + Z_4)}{Z_2 + Z_3 + Z_4}$$

Finally, the impedance of the entire combination is

$$Z = Z_1 + Z_6 = Z_1 + \frac{Z_2(Z_3 + Z_4)}{Z_2 + Z_3 + Z_4}$$

The use of this result in a particular problem would entail only a straightforward application of the algebra of complex numbers.

EXAMPLE 2

A SERIES CIRCUIT WITH A PERIODIC VOLTAGE

A series circuit contains the elements $L = 1$ H, $R = 1{,}000\ \Omega$, $C = 6.25 \times 10^{-6}$ F. Find the steady-state current produced in this circuit by an impressed voltage $E = 100 \cos 200t$ V. By what fraction of a cycle does current lag or lead the voltage E?

This problem can of course be solved by finding the particular integral of either of the related differential equations

$$\frac{d^2Q}{dt^2} + 1{,}000\frac{dQ}{dt} + \frac{1}{6.25 \times 10^{-6}}Q = 100 \cos 200t \tag{7}$$

or

$$\frac{d^2i}{dt^2} + 1{,}000\frac{di}{dt} + \frac{1}{6.25 \times 10^{-6}}i = -20{,}000 \sin 200t$$

However, since we are asked only for the steady-state current, we may disregard the complementary function of this equation, for it gives only the transient current. To find the steady-state current, we shall use the impedance concept since it leads most directly to the amplitude and phase relations that we are asked to determine. Accordingly, we first imagine that the actual voltage $E = 100 \cos 200t$ in Eq. (7) is replaced by the complex voltage

$$100e^{j200t}$$

Next, we compute the complex impedance

$$Z = R + j\left(\omega L - \frac{1}{\omega C}\right) = 1{,}000 + j\left(200 - \frac{160{,}000}{200}\right) = 1{,}000 - 600j$$

and convert it to exponential form, getting

$$\begin{aligned} Z &= \sqrt{1{,}000^2 + 600^2}e^{j\delta} \\ &= 200\sqrt{34}e^{j\delta} \qquad \text{where } \delta = \text{Tan}^{-1}\left(-\frac{600}{1{,}000}\right) \doteq -0.29 \text{ rad} \end{aligned}$$

Then we divide the complex voltage by the complex impedance to get the complex current:

$$\begin{aligned} \frac{E_0 e^{j\omega t}}{Z} = \frac{100e^{j200t}}{200\sqrt{34}e^{-0.29j}} &= \frac{\sqrt{34}}{68}e^{j(200t+0.29)} \\ &\doteq 0.086[\cos (200t + 0.29) + j \sin (200t + 0.29)] \end{aligned}$$

The real part of the last expression is the required steady-state current:

$$i_{ss} \doteq 0.086 \cos (200t + 0.29) \text{ A}$$

The amplitude of the steady-state current is 0.086 A. The phase angle of 0.29 rad represents $0.29/(2\pi) \doteq 0.046$ cycle, and because it is positive here, i.e., negative in the standard form (2), the current leads the voltage by this amount.

EXAMPLE 3

A SERIES CIRCUIT WITH A CONSTANT VOLTAGE

If a constant voltage $E = 24$ V is suddenly switched into the circuit described in Example 2 and if at that instant both the charge and current are zero, find the peak value of the resultant current.

The differential equation we must solve is

$$\frac{d^2Q}{dt^2} + 1{,}000\frac{dQ}{dt} + \frac{Q}{6.25 \times 10^{-6}} = 24$$

subject to the conditions that $Q = i = 0$ when $t = 0$. The characteristic equation in this case is

$$m^2 + 1{,}000m + 160{,}000 = 0$$

and its roots are $m_1 = -200$, $m_2 = -800$. Hence the complementary function is

$$c_1 e^{-200t} + c_2 e^{-800t}$$

To find a particular integral, we assume $Q = A$ and substitute into the differential equation, getting immediately

$$A = 150 \times 10^{-6}$$

A complete solution is therefore

$$Q = c_1 e^{-200t} + c_2 e^{-800t} + 150 \times 10^{-6}$$

Differentiation gives

$$\frac{dQ}{dt} = i = -200c_1 e^{-200t} - 800c_2 e^{-800t}$$

Substituting the initial conditions for Q and i gives the pair of equations

$$c_1 + c_2 + 150 \times 10^{-6} = 0 \qquad \text{and} \qquad c_1 + 4c_2 = 0$$

from which we find at once

$$c_1 = -200 \times 10^{-6} \qquad c_2 = 50 \times 10^{-6}$$

and

$$i = 0.04(e^{-200t} - e^{-800t})$$

To find the time when i is a maximum, we must equate to zero the time derivative of i:

$$0.04(-200e^{-200t} + 800e^{-800t}) = 0$$

Dividing out $0.04 \times 800e^{-200t}$ and transposing, we have $e^{-600t} = \frac{1}{4}$. From this, taking logarithms, we find

$$-600t = -\ln 4 \doteq -1.386 \qquad \text{and} \qquad t_{max} \doteq 0.0023 \text{ s}$$

The maximum value of i can now be found by substituting the value of t_{max} into the general expression for i. The result is

$$i_{max} \doteq 0.019 \text{ A}$$

EXERCISES

1. In Example 3, find the potential difference across each element as a function of time.
2. How much energy is dissipated in one cycle of the steady-state current in Example 2? *Hint:* When I is in amperes and R in ohms, the power delivered by a current I flowing through a resistance R under a voltage E is equal to EI or I^2R watts.
3. An open series circuit contains the elements $L = 0.01$ H, $R = 250\ \Omega$, $C = 10^{-6}$ F. At $t = 0$, with the capacitor charged to the value $Q_0 = 10^{-5}$ C, the circuit is closed. Find the resultant current as a function of time.
4. Work Exercise 3, given the circuit elements $L = 6.4 \times 10^{-3}$ H, $R = 160\ \Omega$, $C = 10^{-6}$ F.
5. Work Exercise 3, given the circuit elements $L = 0.01$ H, $R = 120\ \Omega$, $C = 10^{-6}$ F.
6. A series circuit in which $Q_0 = i_0 = 0$ contains the elements $L = 0.15$ H, $R = 800\ \Omega$, $C = 4 \times 10^{-6}$ F. If a constant voltage $E = 25$ V is suddenly switched into the circuit, find the resultant current as a function of time.
7. Work Exercise 6, given the circuit elements $L = 0.16$ H, $R = 800\ \Omega$, $C = 10^{-6}$ F.
8. Work Exercise 7, given that $E = e^{-500t}$ V.
9. A series circuit in which $Q_0 = i_0 = 0$ contains the elements $L = 1$ H, $R = 1{,}000\ \Omega$, $C = 4 \times 10^{-6}$ F. At $t = 0$ a voltage $E = 110 \sin 50\pi t$ is suddenly switched into the circuit. Find the resultant current as a function of time.
10. Work Exercise 9 in case $L = 1$ H, $R = 800\ \Omega$, $C = 4 \times 10^{-6}$ F, $E = 120 \cos 600t$ V.
11. A series circuit in which $Q_0 = i_0 = 0$ contains the elements $L = 0.02$ H, $R = 250\ \Omega$, $C = 2 \times 10^{-6}$ F. A constant voltage $E = 28$ V is suddenly switched into the circuit. Find the time it takes for the potential difference across the capacitor to build up to one-half its final value.
12. A capacitor, resistor, and inductor for which $C = 4 \times 10^{-6}$ F, $R = 250\ \Omega$, and $L = 1$ H, are connected in parallel. A current source delivering a constant current $I = 0.01$ A is suddenly connected across the common terminals of the elements. Find the resultant voltage as a function of time.
13. Find the steady-state voltage in Exercise 12 if $I = 0.01 \sin 150t$ A.
14. For the series electric circuit, what is the analog of the static deflection δ_{st} in a mechanical system?
15. Show that Eq. (1) can be written in the form

$$\frac{d^2Q}{dt^2} + 2\frac{R}{R_c}\Omega_n\frac{dQ}{dt} + \Omega_n^2 Q = Q_{st}\Omega_n^2 \cos \omega t$$

where Q_{st} is the quantity identified in Exercise 14 as the analog of the static deflection δ_{st}.

16. Find the particular integral of Eq. (1) corresponding to the impressed voltage $E_0 \sin \omega t$ and verify that it produces a current equal to the imaginary part of the particular integral corresponding to the complex voltage $E_0 e^{j\omega t}$.
17. **(a)** Prove that if a set of elements with impedance Z_1 is connected in series with a set of elements with impedance Z_2, the impedance of the resulting combination is $Z_1 + Z_2$. *Hint:* Note that the potential difference across $Z_1 + Z_2$ is equal to the difference across Z_1 plus the difference across Z_2.

    (b) Prove that if a set of elements with impedance Z_1 is connected in parallel with a set of elements with impedance Z_2, the impedance Z of the resultant combination is given by

$$\frac{1}{Z} = \frac{1}{Z_1} + \frac{1}{Z_2}$$

    How does this result generalize to more than two impedances connected in parallel? *Hint:* Note that the current flowing through the combination is equal to the sum of the currents flowing through Z_1 and Z_2.

18. Express in terms of the individual impedances, the overall impedance of the combination shown in **(a)** Fig. 7.16*a*, **(b)** Fig. 7.16*b*, **(c)** Fig. 7.16*c*, and **(d)** Fig. 7.16*d*.
19. A constant voltage is suddenly switched into a nonoscillatory *RLC* series circuit in which $Q_0 = i_0 = 0$. Show that the potential difference across the capacitor can never overshoot its final value.
20. If the frequency of the voltage $E_0 \cos \omega t$ impressed on a series circuit is the same as the natural frequency of the circuit, show that the steady-state potential difference across the inductor and the capacitor are each equal to $E_0 R_c/2R$.
21. For what value(s) of ω is the impedance

$$\sqrt{R^2 + (\omega L - 1/\omega C)^2}$$

a minimum? How does this compare with the corresponding property of the magnification ratio? Explain.

22. Instead of using the ratio R/R_c as a dimensionless parameter in circuit analysis, it is customary to use the so-called **quality factor** Q (not to be confused with the charge Q), defined to be $R_c/2R$. Express the impedance and the phase angle of a simple *RLC* series circuit in terms of the resistance R, the frequency ratio Ω/Ω_n, and the quality factor Q.

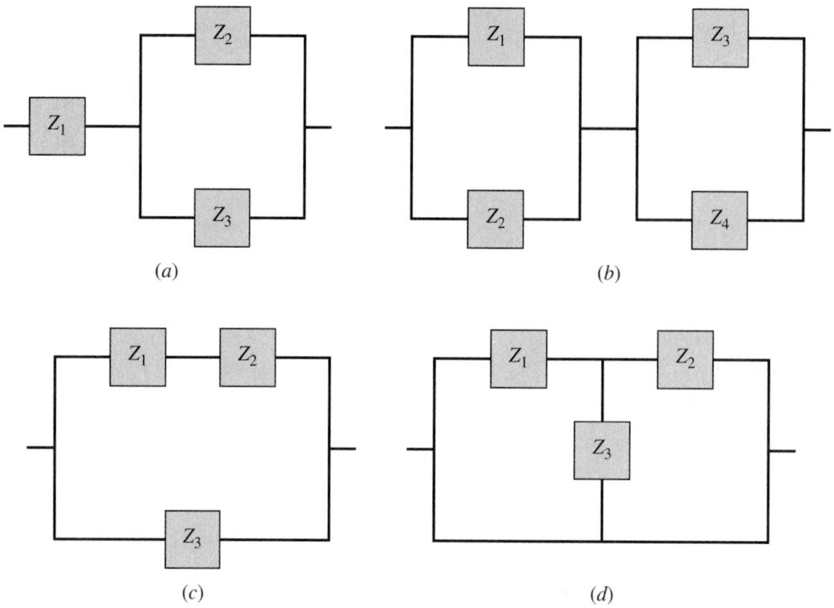

FIGURE 7.16

23. Find the steady-state voltage between the terminals A and B of the parallel network shown in Fig. 7.1d by solving either Eq. (24a) or Eq. (24b), Sec. 7.2, and verify that your answer is equal to the steady-state voltage calculated directly from the impedances $j\omega L$, R, and $1/j\omega C$ of the three branches of the network.

24. At $t = 0$ the switch in an RL circuit is closed and current begins to flow due to a battery of constant voltage E in the circuit. Find the limiting value of the current i_∞ as $t \to \infty$. When does the current have the value $(1-e^{-1})i_\infty$?

25. An RLC electric circuit containing the elements $R = \sqrt{3} \times 10^3\ \Omega$, $L = 50/3\pi$ H, $C = 25 \times 10^{-6}/3\pi$ F, is activated by an impressed voltage $E = 120 \cos 60\pi t$ V. Find the steady-state current, the natural frequency, and the resonant frequency of the circuit.

7.5 SYSTEMS WITH SEVERAL DEGREES OF FREEDOM

The laws of Newton and Kirchoff, together with the theory of simultaneous differential equations developed in Chap. 4, form the basis for the solution of a variety of problems involving systems with more than one degree of freedom. The details of such applications can best be made clear through examples.

EXAMPLE 1

A MASS-SPRING SYSTEM WITH TWO DEGREES OF FREEDOM

Assuming friction to be negligible, find the natural frequencies of the mass-spring system shown in Fig. 7.17. What are the relative amplitudes with which the two masses vibrate at each of the natural frequencies? If a force $F = 40 \sin 3t$ is suddenly applied to M_1 when the system is in static equilibrium, find the displacements of M_1 and M_2 as functions of time.

As usual, we suppose the masses to be guided, by constraints which need not be specified, so that they can move only in the vertical direction. We shall use the instantaneous displacements of the masses from their equilibrium positions as coordinates to describe the system, displacements above the equilibrium positions being considered positive. Since friction is assumed to be negligible, the only forces acting on the masses besides the attraction of gravity are those transmitted to them by the attached springs. Moreover, as suggested by the derivation of Eq. (7), Sec. 7.2, and confirmed for systems with two degrees of freedom by Exercise 14 at the end of this section, the force of gravity can be neglected provided we also neglect the initial elongation of the springs and assume that each is unstretched when the system is in equilibrium.

When the displacements of the masses M_1 and M_2 are y_1 and y_2, respectively, the upper spring is changed in length by the amount y_1 and the lower spring is changed in length by the amount $y_1 - y_2$. Because of these changes in length, the springs exert forces of magnitudes

$$|8y_1| \quad \text{and} \quad |4(y_1 - y_2)|$$

respectively. Hence, applying Newton's law to each mass and taking due account of the direction of the forces applied to each mass by the attached springs, we have

$$(1) \qquad 4\frac{d^2y_1}{dt^2} = -8y_1 - 4(y_1 - y_2) \qquad 2\frac{d^2y_2}{dt^2} = 4(y_1 - y_2)$$

or

$$(2) \qquad \begin{aligned} (D^2 + 3)y_1 - \qquad\quad y_2 &= 0 \\ -2y_1 + (D^2 + 2)y_2 &= 0 \end{aligned}$$

Operating on the first of Eqs. (2) with $D^2 + 2$ and adding the result to the second equation, we obtain an equivalent differential system which simplifies to

$$(3) \qquad \begin{aligned} (D^2 + 3)y_1 - y_2 &= 0 \\ (D^2 + 1)(D^2 + 4)y_1 &= 0 \end{aligned}$$

Solving the second of these equations for y_1, then substituting into the first equation and solving for y_2, we find that a complete solution of System (1) is

$$(4) \qquad \begin{bmatrix} y_1 \\ y_2 \end{bmatrix} = \begin{bmatrix} 1 \\ 2 \end{bmatrix}(c_1 \cos t + c_2 \sin t) + \begin{bmatrix} 1 \\ -1 \end{bmatrix}(c_3 \cos 2t + c_4 \sin 2t)$$

From (4) we see that the system has two natural frequencies, $\omega_1 = 1$ and $\omega_2 = 2$. If the system is vibrating at the single frequency $\omega_1 = 1$, that is, if $c_3 = c_4 = 0$, it follows from the coefficient matrix $\begin{bmatrix} 1 \\ 2 \end{bmatrix}$ that M_1 and M_2 move up and down together but M_2 vibrates through twice the amplitude of M_1. Similarly, if the system is vibrating at the single frequency $\omega_2 = 2$, that is, if $c_1 = c_2 = 0$, it follows from the coefficient matrix $\begin{bmatrix} 1 \\ -1 \end{bmatrix}$ that M_1 and M_2 vibrate with equal amplitudes but in opposite directions. Since (4) is a complete solution of (1), ω_1 and ω_2 are the only natural frequencies of the system.

To determine the motion when the force $F = 40 \sin 3t$ acts on M_1, we must first find a particular integral for the differential system

$$(1a) \qquad \begin{aligned} (D^2 + 3)y_1 - \qquad\quad y_2 &= 10 \sin 3t \\ -2y_1 + (D^2 + 2)y_2 &= 0 \end{aligned}$$

obtained from (1) by adding the term 40 sin $3t$ to the right-hand side of the first equation and then simplifying.

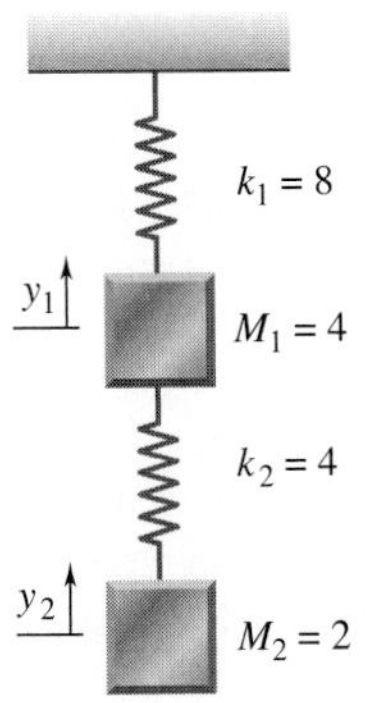

FIGURE 7.17
A simple mass-spring system.

Since only derivatives of even order occur in (1*a*) (because of our assumption that friction is negligible), we need only assume

$$\begin{bmatrix} Y_1 \\ Y_2 \end{bmatrix} = \begin{bmatrix} A_1 \\ A_2 \end{bmatrix} \sin 3t$$

Then substituting into (1*a*), dividing out the common factor sin 3*t*, and simplifying, we obtain the conditions

$$\begin{aligned} -6A_1 - \ \ A_2 &= 10 \\ -2A_1 - 7A_2 &= 0 \end{aligned}$$

from which we find $A_1 = -\frac{7}{4}$ and $A_2 = \frac{2}{4}$. The required particular integral is then

$$\begin{bmatrix} Y_1 \\ Y_2 \end{bmatrix} = \begin{bmatrix} -\frac{7}{4} \\ \frac{1}{2} \end{bmatrix} \sin 3t \tag{5}$$

Adding (5) to (4) gives us a complete solution of the nonhomogeneous system:

$$\begin{bmatrix} y_1 \\ y_2 \end{bmatrix} = \begin{bmatrix} 1 \\ 2 \end{bmatrix}(c_1 \cos t + c_2 \sin t) + \begin{bmatrix} 1 \\ -1 \end{bmatrix}(c_3 \cos 2t + c_4 \sin 2t) + \begin{bmatrix} -\frac{7}{4} \\ \frac{1}{2} \end{bmatrix} \sin 3t$$

Imposing now the first of our initial conditions, namely, $y_1 = y_2 = 0$ when $t = 0$, we find

$$\begin{bmatrix} 0 \\ 0 \end{bmatrix} = \begin{bmatrix} 1 \\ 2 \end{bmatrix} c_1 + \begin{bmatrix} 1 \\ -1 \end{bmatrix} c_3 \qquad \text{or} \qquad \begin{aligned} c_1 + c_3 &= 0 \\ 2c_1 - c_3 &= 0 \end{aligned}$$

from which it follows that $c_1 = c_3 = 0$. Differentiating what remains of $\begin{bmatrix} y_1 \\ y_2 \end{bmatrix}$ to obtain $\begin{bmatrix} \dot{y}_1 \\ \dot{y}_2 \end{bmatrix}$ and then imposing the condition that $\dot{y}_1 = \dot{y}_2 = 0$ when $t = 0$, we obtain similarly the equation

$$\begin{bmatrix} 0 \\ 0 \end{bmatrix} = \begin{bmatrix} 1 \\ 2 \end{bmatrix} c_2 + \begin{bmatrix} 1 \\ -1 \end{bmatrix} 2c_4 + 3\begin{bmatrix} -\frac{7}{4} \\ \frac{1}{2} \end{bmatrix} \qquad \text{or} \qquad \begin{aligned} c_2 + 2c_4 &= \tfrac{21}{4} \\ 2c_2 - 2c_4 &= -\tfrac{6}{4} \end{aligned}$$

whence $c_2 = \frac{5}{4}$ and $c_4 = 2$. The required solution for y_1 and y_2, when M_1 is acted upon by the force $F = 40 \sin 3t$ and the system begins to move from rest in its equilibrium position, is thus given by

$$\begin{bmatrix} y_1 \\ y_2 \end{bmatrix} = \begin{bmatrix} \frac{5}{4} \\ \frac{5}{2} \end{bmatrix} \sin t + \begin{bmatrix} 2 \\ -2 \end{bmatrix} \sin 2t + \begin{bmatrix} -\frac{7}{4} \\ \frac{1}{2} \end{bmatrix} \sin 3t$$

From this solution it appears that the terms from the complementary function, i.e., the terms describing vibrations at the natural frequencies $\omega_1 = 1$ and $\omega_2 = 2$, persist indefinitely, and this is correct under our assumption that there is no friction in the system. However, any amount of friction, no matter how small, would actually cause these terms to die away, and only the term of frequency $\omega = 3$ would appear in the steady state.

EXAMPLE 2

A VIBRATING SPRING-SUPPORTED BAR

A uniform bar 4 ft long and weighing 16 lb/ft is supported as shown in Fig. 7.18 on springs of moduli 24 and 15 lb/in, respectively. If the springs are guided so that only vertical displacement of the center of the bar is possible, and if friction is neglected, find the (natural) frequencies at which the system would *begin* to vibrate if disturbed slightly from its equilibrium position.

This is a system with two degrees of freedom because two coordinates suffice to specify it completely, for instance, the vertical displacement y of the center of gravity of the bar (with y positive upward) and the angle of rotation θ of the bar about its center of gravity (with θ positive in the counterclockwise direction). Under the requirements of the problem, we are to find the frequencies at which vibrations of arbitrarily small amplitudes can occur. In particular, this means that the bar turns through angles so small that the horizontal displacements of its ends can be neglected and each end can be assumed to move in a purely vertical line.

From Fig. 7.18*b*, it is clear that the instantaneous deflections of the left- and right-hand springs are, respectively,

$$y - 24 \sin \theta \qquad \text{and} \qquad y + 24 \sin \theta$$

or, if we make the usual small-angle approximation $\sin \theta \doteq \theta$,

$$y - 24\theta \qquad \text{and} \qquad y + 24\theta$$

Hence the unbalanced forces the springs apply to the ends of the bar are†

$$-24(y - 24\theta) \qquad \text{and} \qquad -15(y + 24\theta)$$

Newton's law applied to the translation of the center of gravity of the bar therefore gives the equation

$$\frac{64}{384}\frac{d^2y}{dt^2} = -24(y - 24\theta) - 15(y + 24\theta)$$

or

$$\frac{d^2y}{dt^2} + 234y - 1{,}296\theta = 0 \tag{6}$$

Recalling our assumption that θ is so small that the ends of the bar move in a purely vertical direction, it is clear that the moment arm of each spring force about the center of gravity of the bar is 24 in. Hence, computing the torques applied to the ends of the bar by the spring forces and then applying Newton's law in torsional form to the rotation of the bar about its center of gravity, we have (using the fact that the moment of inertia of a uniform bar of length l about its midpoint is $ml^2/12$)

$$\frac{64}{384}\frac{(48)^2}{12}\frac{d^2\theta}{dt^2} = 24[24(y - 24\theta)] - 24[15(y + 24\theta)]$$

or

$$\frac{d^2\theta}{dt^2} - \frac{27}{4}y + 702\theta = 0 \tag{7}$$

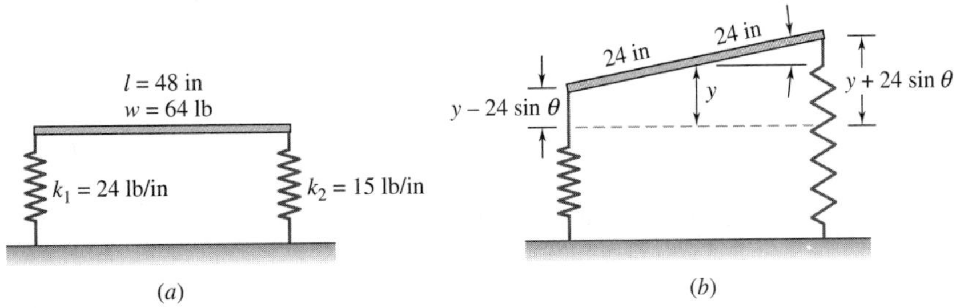

FIGURE 7.18
A simple spring-supported bar.

†As we have pointed out repeatedly, we can neglect the gravitational force on the bar and the portion of the elastic forces due to the compression of the springs in their equilibrium position. An equivalent point of view is to imagine that the motion takes place not in a vertical plane but on a frictionless horizontal plane, so that gravitational effects are irrelevant.

Since we are only asked to find the natural frequencies of the mechanical configuration, there is no need to solve the simultaneous differential equations (6) and (7), for the required frequencies are determined by the roots of the corresponding characteristic equation

$$\begin{vmatrix} m^2 + 234 & -1{,}296 \\ -\frac{27}{4} & m^2 + 702 \end{vmatrix} = m^4 + 936m^2 + 155{,}520 = (m^2 + 216)(m^2 + 720) = 0$$

The roots of this equation are $\pm 6\sqrt{6}j$ and $\pm 12\sqrt{5}j$, and for these imaginary values of m, and these only, there are nontrivial (real) trigonometric vector solutions of the two equations of motion. In other words, for all other values of m the only solution of Eqs. (6) and (7) is the trivial solution $\begin{bmatrix} y \\ \theta \end{bmatrix} \equiv \begin{bmatrix} 0 \\ 0 \end{bmatrix}$.

The components y and θ of every nontrivial solution corresponding to the characteristic roots $\pm 6\sqrt{6}j$ are linear combinations of $\cos 6\sqrt{6}t$ and $\sin 6\sqrt{6}t$ which oscillate with frequency $\omega_1 = 6\sqrt{6}$ rad/s. Likewise, the components of every nontrivial solution corresponding to the characteristic roots $\pm 12\sqrt{5}j$ are linear combinations of $\cos 12\sqrt{5}t$ and $\sin 12\sqrt{5}t$ which oscillate with frequency $\omega_2 = 12\sqrt{5}$ rad/s. Thus ω_1 and ω_2 are the natural frequencies of the given configuration under the assumption of small-amplitude vibrations. Converted from radians per second, they are

$$f_1 = \frac{6\sqrt{6}}{2\pi} \doteq 2.34 \text{ Hz} \qquad \text{and} \qquad f_2 = \frac{12\sqrt{5}}{2\pi} \doteq 4.27 \text{ Hz}$$

EXAMPLE 3

A TWO-LOOP ELECTRIC CIRCUIT

In the circuit shown in Fig. 7.19, find the current in each loop as a function of time, given that all charges and currents are zero when the switch is closed at $t = 0$.

We take as variables the currents i_1 and i_2 flowing in the respective loops, the positive direction of current flow being in the clockwise direction. This gives us $i_1 - i_2$ for the current through the common branch when referred to the left-hand loop. Relative to the right-hand loop the current in this branch is $i_2 - i_1$. Applying Kirchoff's second law to each loop, we obtain the equations

$$0.5\frac{di_1}{dt} + 200(i_1 - i_2) = 50$$

$$300i_2 + 200(i_2 - i_1) + \frac{1}{50 \times 10^{-6}} \int^t i_2 \, dt = 0$$

or, setting $D = d/dt$, letting $Q_2 = \int^t i_2 \, dt$, and rearranging,

$$\begin{aligned} (D + 400)i_1 - 400\, DQ_2 &= 100 \\ -2i_1 + (5D + 200)Q_2 &= 0 \end{aligned} \tag{8}$$

This differential system can easily be solved by the method used in Example 1, that is, by adding $(D + 400)$ "times" the second equation to twice the first equation and then solving the resulting equation for Q_2. It can also be solved by changing it into an equivalent first-order system. Choosing the second alternative, we add 80 times the second equation to the first, then rearrange terms to obtain the nonhomogeneous system

$$\begin{aligned} \dot{i}_1 &= -240i_1 - 16{,}000Q_2 + 100 \\ \dot{Q}_2 &= \tfrac{2}{5}i_1 - 40Q_2 \end{aligned} \tag{9}$$

As usual (see Example 1, Sec. 4.5), we first seek solutions of the form

$$\begin{bmatrix} i_1 \\ Q_2 \end{bmatrix} = \begin{bmatrix} a \\ b \end{bmatrix} e^{\lambda t}$$

for the related homogeneous system. This leads at once to the characteristic-value problem

$$\begin{bmatrix} \lambda + 240 & 16{,}000 \\ -\frac{2}{5} & \lambda + 40 \end{bmatrix} \begin{bmatrix} a \\ b \end{bmatrix} = \begin{bmatrix} 0 \\ 0 \end{bmatrix} \tag{10}$$

The corresponding characteristic equation is

$$\begin{vmatrix} \lambda + 240 & 16{,}000 \\ -\frac{2}{5} & \lambda + 40 \end{vmatrix} = \lambda^2 + 280\lambda + 16{,}000 = (\lambda + 80)(\lambda + 200) = 0$$

and the characteristic numbers are $\lambda_1 = -80$ and $\lambda_2 = -200$. It is readily verified that corresponding solution vectors of (10) are $\begin{bmatrix} -100 \\ 1 \end{bmatrix}$ and $\begin{bmatrix} -400 \\ 1 \end{bmatrix}$. Hence

$$c_1 \begin{bmatrix} -100 \\ 1 \end{bmatrix} e^{-80t} + c_2 \begin{bmatrix} -400 \\ 1 \end{bmatrix} e^{-200t} \tag{11}$$

is a complementary function.

To obtain a particular integral of (9) we assume $\begin{bmatrix} i_1 \\ Q_2 \end{bmatrix} = \begin{bmatrix} A \\ B \end{bmatrix}$ and substitute, getting the linear system

$$\begin{aligned} 0 &= -240A - 16{,}000B + 100 \\ 0 &= \tfrac{2}{5}A - 40B \end{aligned}$$

whose solution is

$$\begin{bmatrix} A \\ B \end{bmatrix} = \begin{bmatrix} \frac{1}{4} \\ \frac{1}{400} \end{bmatrix}$$

A complete solution of (9) is therefore

$$\begin{bmatrix} i_1 \\ Q_2 \end{bmatrix} = c_1 \begin{bmatrix} -100 \\ 1 \end{bmatrix} e^{-80t} + c_2 \begin{bmatrix} -400 \\ 1 \end{bmatrix} e^{-200t} + \begin{bmatrix} \frac{1}{4} \\ \frac{1}{400} \end{bmatrix}$$

Since $i_1 = 0$ and $Q_2 = 0$ when $t = 0$, we must have

$$\begin{aligned} 0 &= -100c_1 - 400c_2 + \tfrac{1}{4} \\ 0 &= c_1 + c_2 + \tfrac{1}{400} \end{aligned}$$

From these we find at once that $c_1 = -\frac{1}{240}$ and $c_2 = \frac{1}{600}$. The required currents are therefore

$$i_1 = \tfrac{5}{12}e^{-80t} - \tfrac{2}{3}e^{-200t} + \tfrac{1}{4} \qquad \text{A}$$

$$i_2 = \frac{dQ_2}{dt} = \tfrac{1}{3}e^{-80t} - \tfrac{1}{3}e^{-200t} \qquad \text{A}$$

Evidently, $i_2 = 0$ when $t = 0$, as required.

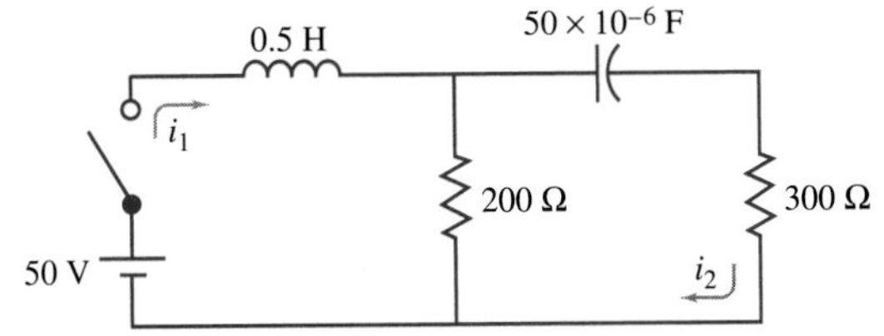

FIGURE 7.19
A simple two-loop electric circuit.

EXERCISES

1. If $M_1 = 1$, $M_2 = 2$, $k_1 = 1$, $k_2 = k_3 = 2$ for the system shown in Fig. 7.20*a*, find the natural frequencies of the system. What are the relative amplitudes of the two masses when the system is vibrating at each of these frequencies? *Hint:* When only natural frequencies and corresponding amplitude ratios are required, the simplest way to proceed is the following:

(a) Set up the appropriate differential equations.

(b) Assume $x_1 = A_1 \cos \omega t$, $x_2 = A_2 \cos \omega t$, substitute into the differential equations, and simplify.

(c) Determine the values of ω for which there are nontrivial values of A_1 and A_2 by equating to zero the determinant of the coefficients of the linear equations obtained in Step **b**.

(d) Solve for the values of A_1 and A_2 corresponding to each natural frequency.

2. Work Exercise 1 if $M_1 = 1$, $M_2 = 3$, $k_1 = 1$, $k_2 = k_3 = 3$.

3. Work Exercise 1 if

(a) $M_1 = M_2 = 1$, $k_1 = 1$, $k_2 = 3$, $k_3 = 9$

(b) $M_1 = M_2 = 2$, $k_1 = k_3 = 8$, $k_2 = 5$

(c) $M_1 = 2$, $M_2 = 1$, $k_1 = 4$, $k_2 = k_3 = 1$

(d) $M_1 = 4$, $M_2 = 9$, $k_1 = 1$, $k_2 = 3$, $k_3 = 6$

(e) $M_1 = 4$, $M_2 = 2$, $k_1 = 8$, $k_2 = 4$, $k_3 = 0$

4. **(a)** Find the natural frequencies and relative amplitudes for the system shown in Fig. 7.20*b*.

(b) Work Part **a** if $M_1 = M_2 = M_3 = 1$, $k_1 = 1$, $k_2 = k_3 = 2$, $k_4 = 3$.

5. Find the natural frequencies and relative amplitudes for the torsional system shown in **(a)** Figure 7.20*c* and **(b)** Figure 7.20*d*.

6. In Exercise 1, find the displacements of M_1 and M_2 as functions of t if the system starts to move from rest when $x_1 = 1$ and $x_2 = 0$.

7. In Exercise 2, find the displacements of M_1 and M_2 as functions of t if the system starts to move from rest when $x_1 = 1$ and $x_2 = -1$.

8. In Example 1, find the displacements of M_1 and M_2 as functions of t if the force $F = 40 \sin 3t$ is suddenly applied to M_2 when the system is at rest in its equilibrium position.

9. In Example 1, find the displacements of M_1 and M_2 as functions of t if the force $F = 40$ is suddenly applied to M_2 when the system is at rest in its equilibrium position.

10. In Example 2, find the relation between y and θ when the system is vibrating at each of its natural frequencies.

11. Find the steady-state motion produced in the system discussed in Example 1 by a force $F = F_0 \cos 3t$ acting on M_1 if a frictional force equal to $-4dy_1/dt$ acts on M_1 and a frictional force equal to $-2dy_2/dt$ acts on M_2. What are the phase relations between M_1 and M_2 and the driving force? *Hint:* Replace $F_0 \cos 3t$ by $F_0 e^{j3t}$.

12. Find the steady-state motion produced in the system in Exercise 1 by a force $F = F_0 \sin 3t$ acting on M_2 if a frictional force equal to $-2dx_1/dt$ acts on M_1 and a frictional force equal to $-dx_2/dt$ acts on M_2. What are the phase relations between M_1 and M_2 and the driving force? *Hint:* Replace $F_0 \sin 3t$ by $F_0 e^{j3t}$.

13. Prove that for no values of the parameters M_1, M_2, k_1, k_2, k_3 can the two natural frequencies of the system shown in Fig. 7.20*a* be equal.

14. If the system shown in Fig. 7.20*a* is turned so that the masses move (without swinging) in the vertical rather than the horizontal direction, set up the differential equations governing the motion, taking into account the force of gravity and the forces due to the initial elongations of the springs, and verify that these cancel each other out identically.

15. A mass M_1 hanging from a spring of modulus k_1 consti-

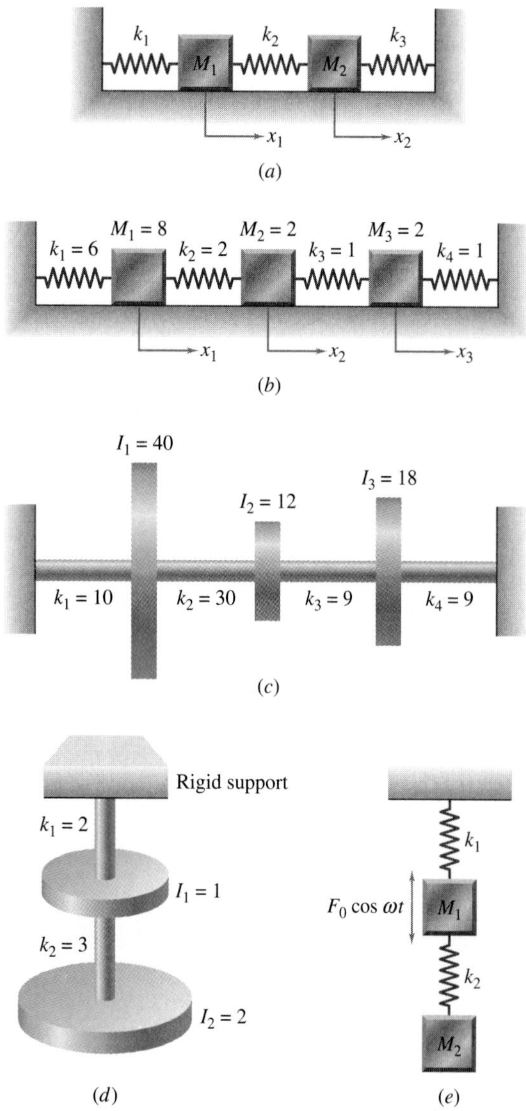

FIGURE 7.20

tutes a system with natural frequency $\omega = \sqrt{k_1/M_1}$. If a mass M_2 hangs from M_1 by a spring of modulus k_2, prove that the two natural frequencies of the resulting system are such that one is always greater than ω and one is always less than ω. Is this true if M_2 is connected to the ground by a spring of modulus k_3?

16. In the system shown in Fig. 7.20*e* the parameters M_1, k_1, and ω are assumed to be known. Determine k_2 and M_2 so that in the steady-state forced motion of the system the mass M_1 will remain at rest. What is the resultant amplitude of the mass M_2? If this amplitude is deemed to be too large, can viscous friction be introduced into the **dynamic damper,** i.e., the (M_2, k_2) subsystem, to reduce it?

17. Find the natural frequencies of the system shown in Fig. 7.20*a* if $k_1 = k_3$ and $M_1 = M_2$. What are the relative amplitudes of the two masses when the system is vibrating at each of these frequencies? How do the quantities $x_1 + x_2$ and $x_1 - x_2$ vary with time? Under what conditions, if any, would you expect this system to exhibit the phenomenon of beats?

18. Neglecting friction, find the equations of motion of the system shown in Fig. 7.20*a* if $M_1 = M_2 = \frac{1}{9}$, $k_1 = k_3 = 9$, and $k_2 = \frac{20}{9}$ and if the system starts from rest in the position where $x_1 = 2$ and $x_2 = 0$. Analyze this motion for the existence of beats.

19. Discuss Example 2 if the 15 lb/in spring is replaced by another 24 lb/in spring.

20. In Example 1 find the particular integral associated with a force equal to $F_0 \sin \omega t$ acting on M_1 and discuss the corresponding steady-state motion as a function of ω.

21. Discuss the steady-state behavior of the system in Exercise 2 if

(a) A force $F = \cos t$ is applied to M_1.

(b) A force $F = -\cos t$ is applied to M_2.

(c) A force $F = \cos t$ is applied to M_1 and a force $F = -\cos t$ is applied to M_2.

Find the steady-state motion of the system if a force $F = \cos \omega t$ is applied to M_1 and a force $F = -\cos \omega t$ is applied to M_2 and discuss the limit of your solution as ω approaches 1. Do infinite amplitudes always occur when a system is driven at one of its natural frequencies?

22. Discuss the steady-state behavior of the system in Example 1 if

(a) A force $F = \sin 2t$ is applied to M_1.

(b) A force $F = \sin 2t$ is applied to M_2.

(c) A force $F = \sin 2t$ is applied to M_1 and M_2.

Find the steady-state motion of the system if a force $F = \sin \omega t$ is applied to both M_1 and M_2 and discuss the limit of your solution as ω approaches 2. Do infinite amplitudes always occur when a system is driven at one of its natural frequencies?

23. In the network shown in Fig. 7.21 the current and the charge on the capacitor in the closed loop are both zero, but the capacitor in the open loop bears a charge Q_0. Find the current in each loop as a function of time after the switch is closed.

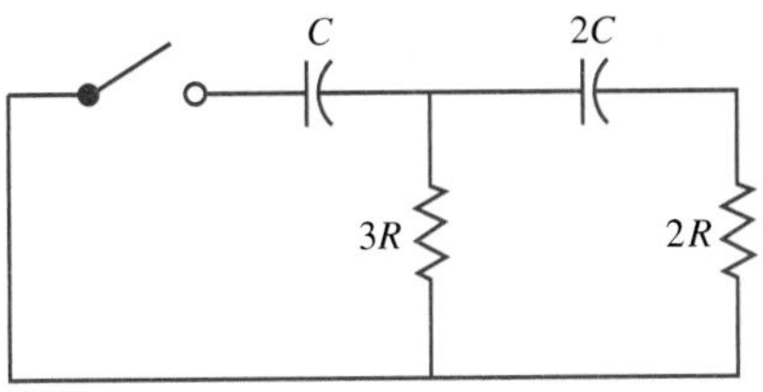

FIGURE 7.21

24. Work Exercise 23 for the hookup shown in Fig. 7.22.

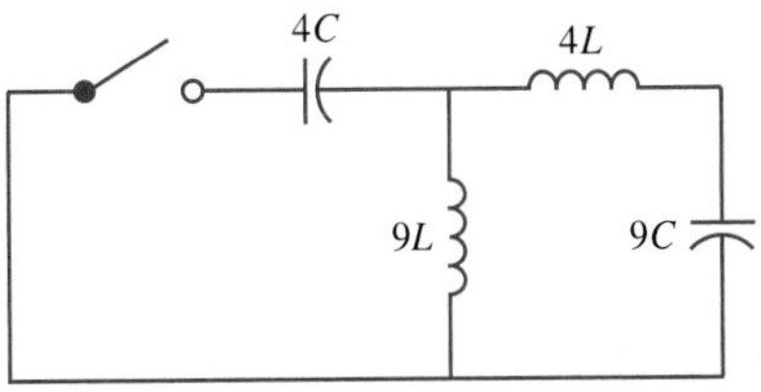

FIGURE 7.22

25. Find the current in each loop of the network shown in Fig. 7.23 if the switch is closed at an instant when all currents are zero.

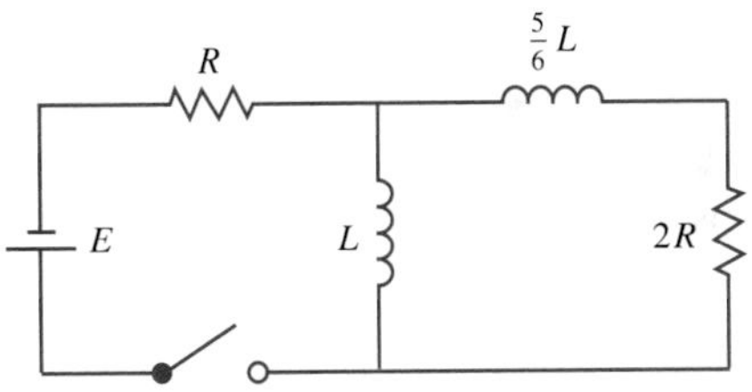

FIGURE 7.23

26. (a) What is the amperage of each current in Exercise 25 if $R = 5\ \Omega$, $L = 1$ H, and $E = 85$ V? As $t \to \infty$, what happens to each current?

(b) Work Exercise 25 for the hookup shown in Fig. 7.24 if all currents and charges are zero when the switch is closed.

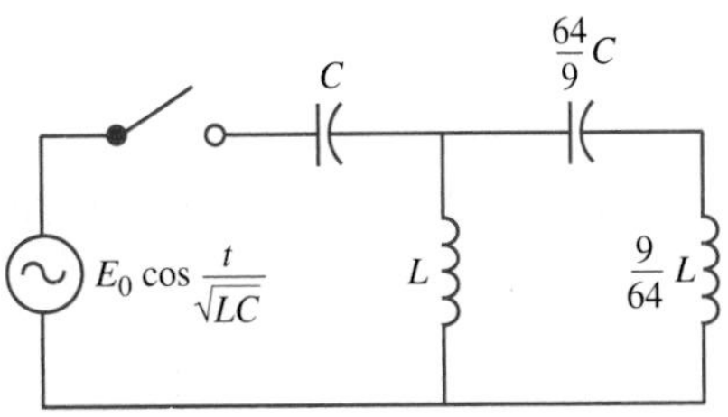

FIGURE 7.24

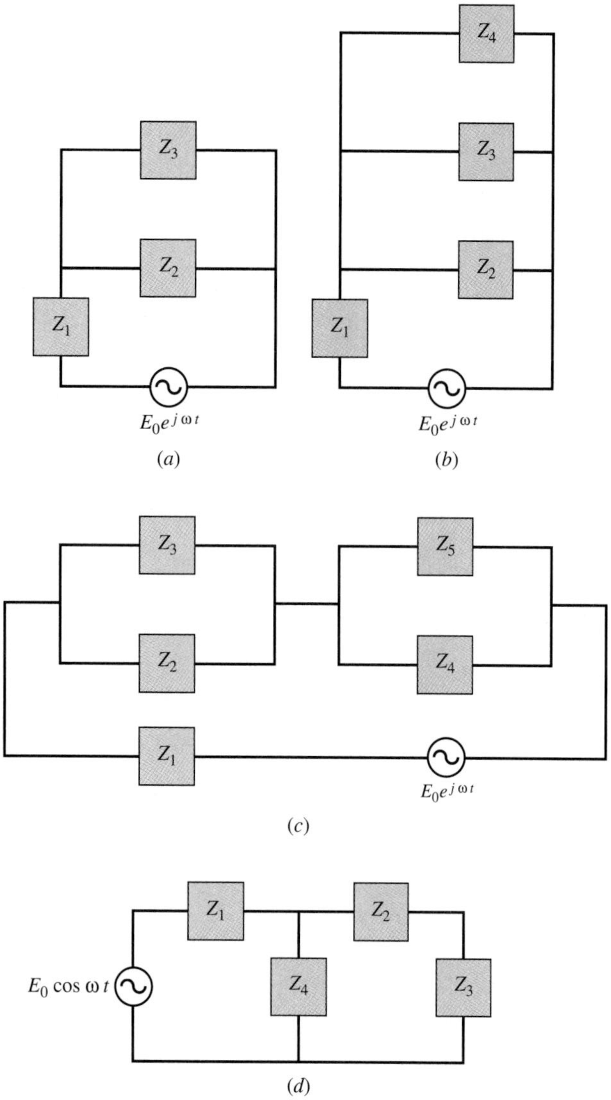

FIGURE 7.25

27. For the circuit shown in Fig. 7.25*a*, express the steady-state current through each impedance in terms of the Z's.

28. For the circuit shown in Fig. 7.25*b*, express the steady-state current through each impedance in terms of the Z's.

29. For the circuit shown in Fig. 7.25*c*, express the steady-state current through each impedance in terms of the Z's.

30. In Exercise 27, if $Z_1 = R$, $Z_2 = j\omega L$, and $Z_3 = 1/j\omega C$, find the steady-state current through Z_1 produced by **(a)** the voltage $E_0 \cos \omega t$ and **(b)** the voltage $E_0 \sin \omega t$.

By first finding the impedance of the system shown in Fig. 7.25*d* (see Example 1, Sec. 7.4) and then finding the values of ω for which the impedance is zero, determine for each of the following sets of data the frequencies at which the system can oscillate.

31. $Z_1 = j\omega L$, $Z_2 = \frac{2}{3} j\omega L$, $Z_3 = 4/3j\omega C$, $Z_4 = 1/j\omega C$

32. $Z_1 = j\omega L$, $Z_2 = \frac{1}{6} j\omega L$, $Z_3 = 3/8j\omega C$, $Z_4 = 1/j\omega C$

33. $Z_1 = j\omega L$, $Z_2 = \frac{4}{9} j\omega L$, $Z_3 = 4/9j\omega C$, $Z_4 = 1/j\omega C$

34. $Z_1 = j\omega L$, $Z_2 = \frac{3}{4} j\omega L$, $Z_3 = 5/3j\omega C$, $Z_4 = 1/j\omega C$

7.6 SYSTEMS WITH MANY DEGREES OF FREEDOM

As we saw in the last section, the analysis of both mechanical and electrical networks with only a few degrees of freedom can easily be carried out through the use of simultaneous differential equations. On the other hand, if the number of degrees of freedom is even moderately large, the computations are in general so extensive that computer assistance becomes necessary. There is, however, one class of problems where, regardless of the number of degrees of freedom, solutions can readily be found. These are problems involving either mechanical or electrical networks consisting of a number of identical segments identically connected. Such problems arise, for instance, in the study of longitudinal vibrations of a train of boxcars, torsional vibrations of the crankshaft of an in-line engine, a chain of electrical insulators, a transmission line, or an electrical wave filter. As expected, problems of this sort are usually formulated in terms of differential equations, or in some cases in terms of impedances. But then, because of the identical character of the segments of the system, difference equations are used to complete the solution. The details of this approach can best be made clear through the use of examples.

EXAMPLE 1 A CHAIN OF SPRING-CONNECTED MASSES

Figure 7.26 shows a chain of n equal masses connected by springs of equal modulus k. Find the natural frequencies of the system, and for each frequency find the relative amplitudes with which each mass vibrates.

Since friction is to be neglected, the only forces acting on the masses are those due to the extension and compression of the attached springs. If we take the displacements $\{x_j\}$ of the respective masses from their equilibrium positions as coordinates, the instantaneous length of the spring between the jth and $(j+1)$st masses in $x_{j+1} - x_j$ and the force exerted by this spring on the adjacent masses is $\pm k(x_{j+1} - x_j)$. Newton's law now gives us the following set of differential equations:

$$\begin{aligned} M\ddot{x}_1 &= -k(x_1 - x_2) \\ M\ddot{x}_j &= k(x_{j-1} - x_j) - k(x_j - x_{j+1}) \qquad j = 2, 3, \ldots, n-1 \\ M\ddot{x}_n &= k(x_{n-1} - x_n) \end{aligned}$$

Since we are concerned only with the possibility of harmonic vibrations, it is no specialization to assume a solution of the form

$$x_j = A_j \sin \omega t \qquad j = 1, 2, \ldots, n \qquad \omega > 0$$

where the A_j's are constants yet to be determined. Substituting these into the equations of motion, dividing out $\sin \omega t$, and collecting terms, we obtain the set of algebraic equations

$$A_2 - \left(1 - \omega^2 \frac{M}{k}\right) A_1 = 0 \tag{1a}$$

$$A_{j+1} - 2\left(1 - \omega^2 \frac{M}{2k}\right) A_j + A_{j-1} = 0 \qquad j = 2, 3, \ldots, n-1 \tag{1b}$$

$$\left(1 - \omega^2 \frac{M}{k}\right) A_n - A_{n-1} = 0 \tag{1c}$$

Equation (1b) is a linear second-order difference equation, and the dissimilar equations (1a) and (1b) are the boundary conditions a complete solution of Eq. (1b) must satisfy.

If free vibrations are to take place, as we assume they do, we know from Example 5, Sec. 5.7, that $1 - \omega^2(M/2k)$ must be between -1 and 1 exclusive. Hence we can put

$$1 - \omega^2 \frac{M}{2k} = \cos \mu$$

FIGURE 7.26
A system of spring-connected masses.

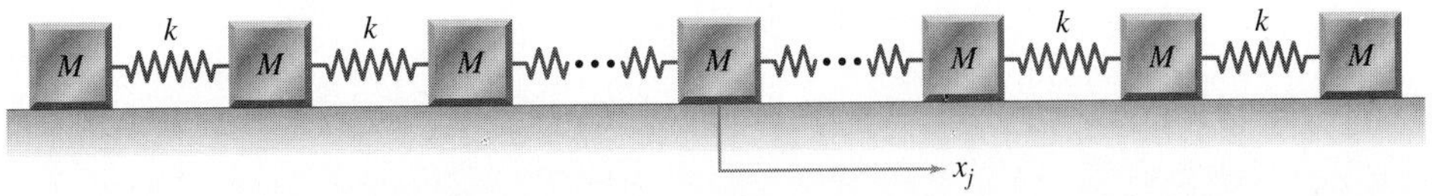

and according to the example just cited, a complete solution of the difference equation (1b) is then

(2) $$A_j = a \cos j\mu + b \sin j\mu$$

Noting that $1 - \omega^2(M/k) = 2 \cos \mu - 1$, we find that the boundary conditions (1a) and (1c) become, respectively,

$$A_2 - (2 \cos \mu - 1)A_1 = 0 \qquad \text{and} \qquad (2 \cos \mu - 1)A_n - A_{n-1} = 0$$

Imposing these conditions on the solution (2), we obtain the equations

(3) $$(a \cos 2\mu + b \sin 2\mu) - (2 \cos \mu - 1)(a \cos \mu + b \sin \mu) = 0$$

(4) $$(2 \cos \mu - 1)(a \cos n\mu + b \sin n\mu) - [a \cos (n - 1)\mu + b \sin (n - 1)\mu] = 0$$

Using the familiar double-angle formulas for the sine and cosine, we reduce Eq. (3) to

(5) $$a(1 - \cos \mu) - b \sin \mu = 0$$

Collecting terms in (4), we obtain

$$a[2 \cos \mu \cos n\mu - \cos n\mu - \cos (n - 1)\mu] + b[2 \cos \mu \sin n\mu - \sin n\mu - \sin (n - 1)\mu] = 0$$

to which we apply the formulas

$$2 \cos u \cos v = \cos (u + v) + \cos (u - v)$$
$$2 \cos u \sin v = \sin (u + v) - \sin (u - v)$$

thus reducing Eq. (4) to

(6) $$a[\cos n\mu - \cos (n + 1)\mu] + b[\sin n\mu - \sin (n + 1)\mu] = 0$$

Equations (5) and (6) are two homogeneous linear algebraic equations in a and b; there is a nontrivial solution for a and b, and therefore for A_j [Eq.(2)] if and only if the determinant of the coefficients in Eqs. (5) and (6) is equal to zero:

$$\begin{vmatrix} 1 - \cos \mu & -\sin \mu \\ \cos n\mu - \cos (n + 1)\mu & \sin n\mu - \sin (n + 1)\mu \end{vmatrix} = 0$$

To simplify this equation, we apply the formulas $1 - \cos u = 2 \sin^2 (u/2)$, $\sin u = 2 \sin (u/2) \cos (u/2)$ and the identities

$$\cos u - \cos v = 2 \sin \frac{u + v}{2} \sin \frac{v - u}{2}$$

$$\sin u - \sin v = 2 \sin \frac{u - v}{2} \cos \frac{u + v}{2}$$

to the elements of the determinant, getting

$$\begin{vmatrix} 2 \sin^2 \dfrac{\mu}{2} & -2 \sin \dfrac{\mu}{2} \cos \dfrac{\mu}{2} \\ 2 \sin \dfrac{(2n + 1)\mu}{2} \sin \dfrac{\mu}{2} & -2 \sin \dfrac{\mu}{2} \cos \dfrac{(2n + 1)\mu}{2} \end{vmatrix} = -4 \sin^2 \frac{\mu}{2} \begin{vmatrix} \sin \dfrac{\mu}{2} & \cos \dfrac{\mu}{2} \\ \sin \dfrac{(2n + 1)\mu}{2} & \cos \dfrac{(2n + 1)\mu}{2} \end{vmatrix}$$

$$= -2(1 - \cos \mu) \sin \left[\frac{\mu}{2} - \frac{(2n + 1)\mu}{2}\right] = 2(1 - \cos \mu) \sin n\mu = 0$$

Since $1 - \cos \mu = \omega^2 M/2k > 0$, the algebraic system comprised of Eqs. (5) and (6) will have a nontrivial solution for a and b, and Eq. (2) a nontrivial solution for A_j if and only if

$$\sin n\mu = 0$$

This equation is called the **frequency equation** of the mechanical system because its roots determine all possible frequencies at which the system can vibrate freely. In fact, from $\cos \mu = 1 - \omega^2 M/2k$, we have

$$\omega = \sqrt{\frac{k}{M}}\sqrt{2(1 - \cos \mu)} = \sqrt{\frac{k}{M}}\sqrt{4 \sin^2 \frac{\mu}{2}} = 2\sqrt{\frac{k}{M}}\left|\sin \frac{\mu}{2}\right|$$

For any one of the roots of the frequency equation, namely,

$$\mu_N = \frac{N\pi}{n} \qquad N \text{ an integer}$$

the last expression for ω becomes

$$2\sqrt{\frac{k}{M}}\left|\sin \frac{\mu_N}{2}\right| = 2\sqrt{\frac{k}{M}}\left|\sin \frac{N\pi}{2n}\right|$$

Only positive integers N need be considered because $\omega > 0$ and $|\sin (N\pi/2n)| = |\sin (-N\pi/2n)|$. For $N = 1, 2, \ldots, n - 1$, the last formula takes on $n - 1$ distinct nonzero values. As N continues to increase, these values simply repeat again and again. Hence all the natural frequencies of the system are given by the formula

(7) $$\omega_N = 2\sqrt{\frac{k}{M}} \sin \frac{N\pi}{2n} \qquad N = 1, 2, \ldots, n - 1$$

It is interesting that although the system contains n masses, it has only $n - 1$ natural frequencies. This is characteristic of systems completely isolated from their surroundings (see Exercise 13, Sec. 7.2).

To find the relative amplitudes of the respective masses when the system is vibrating at one of its natural frequencies, say ω_N, we use the fact that when $\mu = \mu_N$, Eqs. (5) and (6) are equivalent and a and b can be found equally well from either equation. Choosing Eq. (5) because it seems a little simpler, we have

$$a(1 - \cos \mu_N) - b \sin \mu_N = 0$$

whence

$$a_N = \lambda_N \sin \mu_N \qquad \text{and} \qquad b_N = \lambda_N(1 - \cos \mu_N)$$

where, for each N, λ_N is an arbitrary constant. Substituting these values into Eq. (2), we get

(8) $$\begin{aligned}(A_j)_N &= \lambda_N[\sin \mu_N \cos j\mu_N + (1 - \cos \mu_N) \sin j\mu_N] \\ &= \lambda_N[\sin j\mu_N - \sin (j - 1)\mu_N] \\ &= 2\lambda_N \sin \tfrac{1}{2}\mu_N \cos (j - \tfrac{1}{2})\mu_N \\ &= \beta_N \cos (j - \tfrac{1}{2})\mu_N\end{aligned}$$

where, for each N, the coefficient $\beta_N = 2\lambda_N \sin \frac{1}{2}\mu_N$ is just another arbitrary constant since it is independent of j, i.e., does not vary from mass to mass when the system is vibrating at a particular natural frequency ω_N. Thus the successive amplitudes of the masses $|(A_j)_N|$ are proportional to $|\cos (j - \frac{1}{2})\mu_N|$, that is, since $\mu_N = N\pi/n$,

$$|(A_j)_N| \propto \left|\cos \frac{(2j - 1)N\pi}{2n}\right| \qquad \begin{cases} N = 1, 2, \ldots, n - 1 \\ j = 1, 2, \ldots, n \end{cases} \quad \text{for each } N$$

Without the absolute value signs $\cos [(2j - 1)N\pi]/2n$ gives the relative maximum displacements of the successive masses, some of which will be positive and some negative.

EXAMPLE 2

A SYSTEM OF ROTATING DISKS

In the system shown in Fig. 7.27, the moment of inertia of the end disks is $I/2$, and the moment of inertia of the other disks is I. The torsional modulus of each connecting shaft is c. Neglecting friction, determine the motion of the system if a periodic torque $T = T_0 \sin \omega t$; T_0, $\omega > 0$, is applied to the disk at the left end.

As coordinates to describe the system, we shall use the angles of rotation θ_k of the successive disks, and we shall use the torsional version of Newton's law to set up the appropriate differential equations in terms of these coordinates. In doing this, we note that the angle of twist of the shaft between the kth and $(k + 1)$st disks is $\theta_{k+1} - \theta_k$, hence the torque exerted by this shaft on the adjoining disks is $\pm c(\theta_{k+1} - \theta_k)$. Since friction is to be neglected, these are the only torques acting on the disks except for the torque $T = T_0 \sin \omega t$ applied to the leftmost disk. Reasoning in this way, we find that the equation of motion of the first disk is

$$\frac{I}{2}\ddot{\theta}_0 = c(\theta_1 - \theta_0) + T_0 \sin \omega t \tag{9}$$

Similarly, the equation of the kth disk is

$$I\ddot{\theta}_k = c(\theta_{k+1} - \theta_k) - c(\theta_k - \theta_{k-1}) \tag{10}$$

and clearly this is valid for all disks except the first and last, that is, for all values of k except $k = 0$ and $k = n$. Finally, the equation of motion of the disk at the right end is

$$\frac{I}{2}\ddot{\theta}_n = c(\theta_{n-1} - \theta_n) \tag{11}$$

Since friction is assumed to be negligible, each disk will either move in phase with the applied torque or be 180° out of phase with it (Sec. 7.3). Hence we assume $\theta_k = A_k \sin \omega t$, where ω is the frequency of the applied torque and the A_k's are constants to be determined. Substituting these expressions into Eqs. (9), (10), and (11), dividing out $c \sin \omega t$, setting $\omega^2 I/c = \alpha^2$, and collecting terms, we obtain the equations

$$A_1 - \left(1 - \frac{\alpha^2}{2}\right)A_0 = -\frac{T_0}{2} \tag{12}$$

$$A_{k+1} - 2\left(1 - \frac{\alpha^2}{2}\right)A_k + A_{k-1} = 0 \qquad k = 1, 2, \ldots, n - 1 \tag{13}$$

$$\left(1 - \frac{\alpha^2}{2}\right)A_n - A_{n-1} = 0 \tag{14}$$

Equation (13) is a homogeneous second-order difference equation just like Eq. (1*b*), Example 1, and the equation we studied in Example 5, Sec. 5.7, where now $\lambda = 1 - \alpha^2/2$. Equations (12) and (14) are of course the two end conditions which a complete solution of Eq. (13) must satisfy. With Example 5, Sec. 5.7, in mind, it appears that there are five cases we must investigate:

$$1 - \frac{\alpha^2}{2} > 1 \qquad \text{(i.e., } \alpha^2 < 0\text{)}$$

$$1 - \frac{\alpha^2}{2} = 1 \qquad \text{(i.e., } \alpha^2 = 0\text{)}$$

$$-1 < 1 - \frac{\alpha^2}{2} < 1 \qquad \text{(i.e., } 0 < \alpha^2 < 4\text{)}$$

$$1 - \frac{\alpha^2}{2} = -1 \qquad \text{(i.e., } \alpha^2 = 4\text{)}$$

$$1 - \frac{\alpha^2}{2} < -1 \qquad \text{(i.e., } \alpha^2 > 4\text{)}$$

The first two cases are impossible, however, since ω^2, I, and c are intrinsically positive physical quantities, hence $\alpha^2 = \omega^2 I/c$ must be positive.

In the third case, guided by the results of Example 5, Sec. 5.7, we put $1 - \alpha^2/2 = \cos \mu$, getting the difference equation

$$A_{k+1} - 2A_k \cos \mu + A_{k-1} = 0 \tag{15}$$

from Eq. (13), and the boundary conditions

$$A_1 - A_0 \cos \mu = \frac{-T_0}{c} \tag{16}$$

and

$$A_n \cos \mu - A_{n-1} = 0 \tag{17}$$

from Eqs. (12) and (14), respectively. In Sec. 5.7, we found that a complete solution of Eq. (15) is

$$A_k = a \cos k\mu + b \sin k\mu \tag{18}$$

and it remains now to determine the coefficients a and b. Imposing the first boundary condition, (16), we obtain

$$(a \cos \mu + b \sin \mu) - a \cos \mu = \frac{-T_0}{c}$$

whence

$$b = -\frac{T_0}{c \sin \mu}$$

Imposing the second boundary condition, (17), we obtain

$$(a \cos n\mu + b \sin n\mu) \cos \mu - [a \cos (n-1)\mu + b \sin (n-1)\mu] = 0$$

This simplifies remarkably if we expand the last two terms as, respectively, the cosine and sine of the difference of $n\mu$ and μ:

$$a \cos n\mu \cos \mu + b \sin n\mu \cos \mu - a(\cos n\mu \cos \mu + \sin n\mu \sin \mu) - b(\sin n\mu \cos \mu - \cos n\mu \sin \mu) = 0$$

When simplified, this gives us

$$a = \frac{\cos n\mu}{\sin n\mu} \quad b = \frac{-T_0 \cos n\mu}{c \sin \mu \sin n\mu}$$

FIGURE 7.27
A system of rotating disks connected by identical shafts.

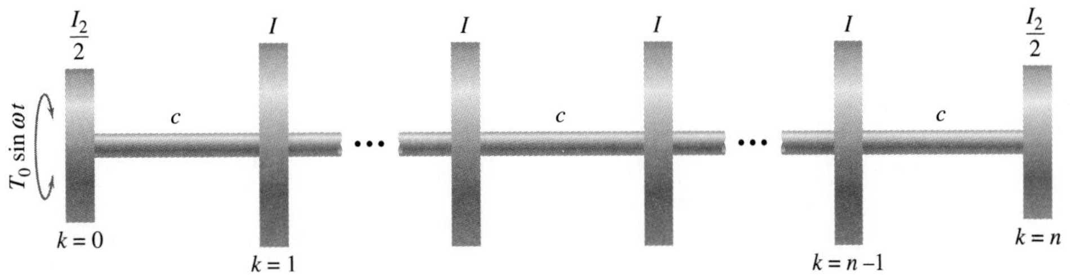

With the values of a and b now determined, we have from (18),

$$A_k = \frac{-T_0}{c \sin \mu}\left(\frac{\cos n\mu}{\sin n\mu}\cos k\mu + \sin k\mu\right)$$

$$= \frac{-T_0}{c \sin \mu}\,\frac{\cos n\mu \cos k\mu + \sin n\mu \cos k\mu}{\sin n\mu}$$

(19)
$$= \frac{-T_0 \cos (n-k)\mu}{c \sin \mu \sin n\mu}$$

and

$$\theta_k = A_k \sin \omega t = -\frac{T_0 \cos (n-k)\mu}{c \sin \mu \sin n\mu} \sin \omega t \qquad \sin n\mu \neq 0$$

Incidentally, since $|A_k|$ becomes infinite as μ tends to any zero of $\sin n\mu$, it follows that when the applied torque has any one of these zeros as its frequency, the system is being driven at pure resonance. In other words, as a by-product of our work thus far, we have discovered that the natural, or resonant, frequencies of the torsional system shown in Fig. 7.27 are determined by the roots of the equation $\sin n\mu = 0$. Thus we have found the **frequency equation** of the system. Since the roots of this equation are $\mu = N\pi/n$, where N is an integer, and since $\cos \mu = 1 - \alpha^2/2 = 1 - \frac{1}{2}(\omega^2 I/c)$, the natural frequencies themselves can be found by solving for ω from the equation

$$\cos \mu = 1 - \frac{1}{2}\frac{\omega^2 I}{c}$$

The result is

$$\omega_N = \sqrt{\frac{2c}{I}}\sqrt{1 - \cos\frac{N\pi}{n}} = 2\sqrt{\frac{c}{I}}\left|\sin\frac{N\pi}{2n}\right| \qquad N = 1, 2, \ldots, n$$

The inclusion of n among the values that N can assume merits a word of explanation. When $N = n$, $\mu_N = \pi$ and $\cos \mu_N = -1$, contrary to our current assumption that $-1 < \cos \mu < 1$. However, by reasoning independently of the use of difference equations (see Exercise 18), it can be shown that when $1 - \alpha^2/2 = -1$, the frequency $\omega_n = 2\sqrt{c/I}$ is actually one of the natural frequencies of the system. This is consistent with one of our observations in Example 1 since, without the torque $T_0 \sin \omega t$ the present system is isolated from its surroundings, has $n + 1$ disks, and has $(n + 1) - 1 = n$ natural frequencies. Values of N greater than n need not be considered because they only repeat the frequencies already found (see Exercises 19 and 20).

If $\alpha^2 = 4$, which is the fourth case we must consider, Eq. (13) reduces to $A_{k+1} + 2A_k + A_{k-1} = 0$, and the corresponding characteristic equation has -1 as a repeated root. A complete solution of the difference equation is therefore $A_k = (-1)^k(a + bk)$. Imposing the boundary condition (12), we get $b = T_0/c$; then using (14) we obtain $b = 0$. But this is impossible since, by hypothesis, $T_0 \neq 0$. Consequently, Case 4 is also ruled out.

Finally, if $1 - \alpha^2/2 < -1$, we can put $1 - \alpha^2/2 = -\cosh \mu$, getting from Eq. (13), the difference equation

(20)
$$A_{k+1} + 2A_k \cosh \mu + A_{k-1} = 0$$

and from (12) and (14), the boundary conditions

(21)
$$A_1 + A_0 \cosh \mu = \frac{-T_0}{c}$$

and

(22)
$$A_n \cosh \mu + A_{n-1} = 0$$

According to Example 5, Sec. 5.7, a complete solution of the difference equation (20) is

(23)
$$A_k = (-1)^k(a \cosh k\mu + b \sinh k\mu)$$

Imposing the boundary condition (21), we have

$$-(a \cosh \mu + b \sinh \mu) + a \cosh \mu = \frac{-T_0}{c}$$

and

$$b = \frac{T_0}{c \sinh \mu}$$

Imposing the boundary condition (22), we obtain

$$(-1)^n(a \cosh n\mu + b \sinh n\mu) \cosh \mu + (-1)^{n-1}[a \cosh (n-1)\mu + b \sinh (n-1)\mu] = 0$$

Expanding the last two terms and simplifying, we have

$$-(a \cosh n\mu \cosh \mu + b \sinh n\mu \cosh \mu) + a(\cosh n\mu \cosh \mu - \sinh n\mu \sinh \mu) \\ + b(\sinh n\mu \cosh \mu - \cosh n\mu \sinh \mu) = 0$$

and

$$a = -\frac{\cosh n\mu}{\sinh n\mu} \quad b = -\frac{T_0 \cosh n\mu}{c \sinh \mu \sinh n\mu}$$

With a and b now determined, we can return to Eq. (23) and find A_k:

$$\begin{aligned} A_k &= (-1)^{k+1} \frac{T_0}{c \sinh \mu} \left(\frac{\cosh n\mu}{\sinh n\mu} \cosh k\mu - \sinh k\mu \right) \\ &= (-1)^{k+1} \frac{T_0}{c \sinh \mu} \left(\frac{\cosh n\mu \cosh k\mu - \sinh n\mu \sinh k\mu}{\sinh n\mu} \right) \\ &= (-1)^{k+1} \frac{T_0 \cosh (n-k)\mu}{c \sinh \mu \sinh n\mu} \end{aligned} \tag{24}$$

Finally,

$$\theta_k = A_k \sin \omega t = (-1)^{k+1} \frac{T_0 \cosh (n-k)\mu}{c \sinh \mu \sinh n\mu} \sin \omega t$$

As we shall soon see, a very important feature of the last example is that the response of the system shown in Fig. 7.27 to an applied harmonic torque $T = T_0 \sin \omega t$ depends on which of two sets of frequencies contain the impressed frequency ω. One set is defined by the condition

$$-1 < 1 - \frac{\alpha^2}{2} < 1 \qquad \text{or} \qquad 0 < \alpha^2 < 4 \qquad \text{or} \qquad 0 < \omega^2 \frac{I}{c} < 4$$

or finally

$$0 < \omega < 2\sqrt{\frac{c}{I}}$$

The other set is defined by the condition

$$1 - \frac{\alpha^2}{2} < -1 \qquad \text{or} \qquad \alpha^2 > 4 \qquad \text{or} \qquad \frac{\omega^2 I}{c} > 4$$

or finally

$$\omega > 2\sqrt{\frac{c}{I}}$$

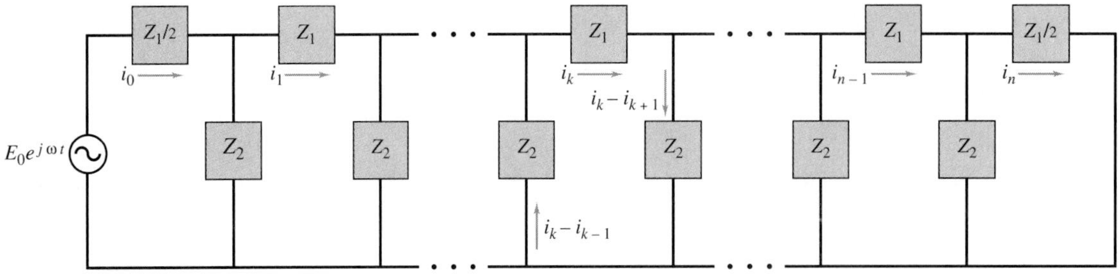

FIGURE 7.28
A general filter network.

In the first case, the ratio of the amplitude of the last disk to the amplitude of the first disk is, from (19), $|A_n|/|A_0| = 1/|\cos n\mu|$, which is equal to or greater than 1 for all values of n. In the second case, the ratio of these amplitudes is, from (24), $1/\cosh n\mu$, and for any value of μ, this fraction can be made arbitrarily small by taking n sufficiently large. In other words, if $0 < \omega < 2\sqrt{c/I}$, a harmonic torque $T_0 \sin \omega t$ applied to the first disk produces a harmonic oscillation at the last disk whose amplitude is at least as great as the amplitude of the first disk; but if $\omega > 2\sqrt{c/I}$, then, as n increases, a harmonic torque acting on the first disk produces only a vanishingly small oscillation at the last disk. To put it still differently, we can say that the system of disks shown in Fig. 7.27 acts as a **wave filter,** passing waves whose frequencies are less than $2\sqrt{c/I}$ but attenuating to extinction waves whose frequencies are greater than $2\sqrt{c/I}$. This characteristic suggests the much more important possibility of constructing an analogous electric network which will serve as an **electrical filter,** passing waves of certain frequencies while annihilating waves of all other frequencies.

To explore this possibility, let us consider the network shown in Fig. 7.28 in which Z_1 and Z_2 are unspecified complex impedances containing arbitrary reactances X_1 and X_2 but no resistances; that is, $Z_1 = jX_1$ and $Z_2 = jX_2$ are *pure imaginary* impedances. If we take the positive direction of current flow in each loop to be clockwise, and apply Kirchoff's second law to each of the dissimilar end loops and to a typical one of the identical intermediate loops, we obtain for the first loop, the equation

$$i_0\frac{Z_1}{2} + (i_0 - i_1)Z_2 = E_0e^{j\omega t}$$

or

(25) $$i_1 - \left(1 + \frac{Z_1}{2Z_2}\right)i_0 = -\frac{E_0e^{j\omega t}}{Z_2}$$

for the kth loop, the equation

$$i_kZ_1 + (i_k - i_{k+1})Z_2 + (i_k - i_{k-1})Z_2 = 0$$

or

(26) $$i_{k+1} - 2\left(1 + \frac{Z_1}{2Z_2}\right)i_k + i_{k-1} = 0 \qquad k = 1, 2, \ldots, n-1$$

and for the last loop, the equation

$$i_n\frac{Z_1}{2} + (i_n - i_{n-1})Z_2 = 0$$

or

$$\left(1 + \frac{Z_1}{2Z_2}\right)i_n - i_{n-1} = 0 \tag{27}$$

Equations (25), (26), and (27) are almost identical with Eqs. (12), (13), and (14), respectively. Hence, from the results of Example 5, Sec. 5.7, as illustrated in Example 2, it is clear that the network shown in Fig. 7.28 will allow a wave to pass through without attenuation if and only if

$$\left|1 + \frac{Z_1}{2Z_2}\right| < 1$$

Since both Z_1 and Z_2 are pure imaginary impedances, it follows that Z_1/Z_2 is a real quantity, so that the last inequality can be written in the form

$$-1 < 1 + \frac{Z_1}{2Z_2} < 1 \tag{28}$$

Once Z_1 and Z_2 are specified, this becomes a condition on the frequency ω, which is contained in Z_1 and Z_2 as a matter of definition. There are now four cases to consider.

(a) If $Z_1 = j\omega L_1$ and $Z_2 = j\omega L_2$, that is, if each Z represents an inductor, then

$$1 + \frac{Z_1}{2Z_2} = 1 + \frac{L_1}{2L_2}$$

Since the last quantity is always greater than 1, a system with these impedances will allow no wave to pass through. In other words, it serves as a **wave trap.** Similarly, if $Z_1 = 1/j\omega C_1$ and $Z_2 = 1/j\omega C_2$, that is, if both Z_1 and Z_2 represent capacitors, then

$$1 + \frac{Z_1}{2Z_2} = 1 + \frac{C_2}{2C_1} > 1$$

and again the system functions as a wave trap at all frequencies.

(b) If $Z_1 = j\omega L$ and $Z_2 = 1/j\omega C$, then

$$1 + \frac{Z_1}{2Z_2} = 1 + \frac{j\omega L}{2/j\omega C} = 1 - \frac{\omega^2 LC}{2}$$

and (28) becomes

$$-1 < 1 - \frac{\omega^2 LC}{2} < 1 \qquad \text{or} \qquad 0 < \omega^2 < \frac{4}{LC}$$

This arrangement of impedances thus constitutes a **low-pass filter** which passes waves of all frequencies from 0 up to $2/\sqrt{LC}$ and, depending on the number of loops n, attenuates waves of all higher frequencies to arbitrarily small amplitudes.

(c) If $Z_1 = 1/j\omega C$ and $Z_2 = j\omega L$, then

$$1 + \frac{Z_1}{2Z_2} = 1 + \frac{1/j\omega C}{2j\omega L} = 1 - \frac{1}{2\omega^2 LC}$$

and (28) becomes

$$-1 < 1 - \frac{1}{2\omega^2 LC} < 1 \qquad \text{or} \qquad \omega^2 > \frac{1}{4LC}$$

This arrangement of impedances thus serves as a **high-pass filter** since it passes waves of all frequencies greater than $1/2\sqrt{LC}$ but attenuates waves of all lower frequencies to arbitrarily small amplitudes.

(d) If $Z_1 = j\omega L_1 + 1/j\omega C_1$ and $Z_2 = j\omega L_2 + 1/j\omega C_2$, then

$$1 + \frac{Z_1}{2Z_2} = 1 + \frac{j\omega L_1 + 1/j\omega C_1}{2(j\omega L_2 + 1/j\omega C_2)} = 1 + \frac{\omega^2 L_1 - 1/C_1}{2(\omega^2 L_2 - 1/C_2)}$$

and (28) becomes

$$-1 < 1 + \frac{\omega^2 L_1 - 1/C_1}{2(\omega^2 L_2 - 1/C_2)} < 1$$

or

(29) $$-4 < \frac{\omega^2 L_1 - 1/C_1}{\omega^2 L_2 - 1/C_2} < 0$$

Waves whose frequencies satisfy the inequality (29) are passed by the network; waves of all other frequencies are annihilated. Therefore in this case the system functions as a **band-pass filter.** If the inequality signs in the last expression are replaced by equality signs and the resulting equations solved for ω, we find that the *endpoints* of the passband are

$$\omega_1 = \frac{1}{\sqrt{L_1 C_1}} \qquad \text{and} \qquad \omega_2 = \frac{1}{\sqrt{L_1 C_1}}\sqrt{\frac{1 + 4C_1/C_2}{1 + 4L_2/L_1}}$$

If $L_1 C_1 > L_2 C_2$, the interval is

$$\frac{1}{\sqrt{L_1 C_1}} < \omega < \frac{1}{\sqrt{L_1 C_1}}\sqrt{\frac{1 + 4C_1/C_2}{1 + 4L_2/L_1}}$$

If $L_1 C_1 < L_2 C_2$, the interval is

$$\frac{1}{\sqrt{L_1 C_1}}\sqrt{\frac{1 + 4C_1/C_2}{1 + 4L_2/L_1}} < \omega < \frac{1}{\sqrt{L_1 C_1}}$$

By suitably choosing C_1, C_2, L_1, L_2, the passband can be made very narrow or very wide, as desired.

To determine the effect these filters have on the waves they allow to pass, we must, as in Example 2, solve the general difference equation (26) and impose the boundary conditions (25) and (27) on the solution. The work is almost exactly like the corresponding part of Example 2 and is the same for each of the three types of filters. For each type of filter we have

$$-1 < 1 + \frac{Z_1}{2Z_2} < 1$$

Hence in each case we can set $1 + Z_1/2Z_2 = \cos\mu$, getting from Eq. (26) the familiar difference equation

$$i_{k+1} - 2i_k \cos\mu + i_{k-1} = 0$$

whose solution we know is

$$i_k = a\cos k\mu + b\sin k\mu$$

When we impose the first boundary condition (25) we find immediately that

$$b = \frac{-E_0}{Z_2 \sin\mu} e^{j\omega t}$$

Imposing the second boundary condition (27), we obtain

$$(a\cos n\mu + b\sin n\mu)\cos\mu - [a\cos(n-1)\mu + b\sin(n-1)\mu] = 0$$

Expanding the last two terms and simplifying, exactly as we did in Example 2, we find that

$$a = \frac{-E_0 \cos n\mu}{Z_2 \sin\mu \sin n\mu} e^{j\omega t}$$

and finally

$$i_k = -\frac{\cos(n-k)\mu}{Z_2 \sin\mu \sin n\mu} E_0 e^{j\omega t}$$

From this formula it is easy to see that

$$\frac{|i_n|}{|i_0|} = \frac{1}{|\cos n\mu|} \tag{30}$$

Since the filter networks we have just considered contained only reactances, the waves that were passed suffered no phase shifts, although according to Eq. (30) their amplitudes were changed. If the impedances contained resistances, both the amplitude of a wave *and* its phase with respect to $E_0 e^{j\omega t}$ would be changed. A perfect filter would presumably be one in which the waves that were passed suffered no change in either amplitude or phase. This ideal situation is impossible to achieve, however, for the amplitude characteristics and the phase characteristics of a filter cannot be independently specified. In fact, in U.S. Patent 2,123,178, H. W. Bode of Bell Laboratories has shown that if amplitudes are specified over any portion of the passband, the phase shifts for waves in that range of frequencies are determined. Conversely, if phase shifts are specified over a certain range of frequencies, the resultant amplitudes are determined.

EXERCISES

1. In Example 1, show that $(A_j)_N = (-1)^N (A_{n-j+1})_N$. What is the physical interpretation of this?
2. In Example 1, why is the possibility that $N = n$ excluded in Eq. (7)?
3. In Example 2, do adjacent disks necessarily rotate in opposite directions?
4. In Example 2, if $I = 45$, $c = 475$, and $\omega = 5$, what is the ratio of θ_n to θ_0 for $n = 5$, 10, and 20?
5. Work Exercise 4 if $c = 250$ instead of 475.
6. Verify the two expressions for the endpoints of the passband that were obtained from Eq. (29).

Note: In Exercises 7–11 the difference equation to be considered is $A_{k+1} - 2A_k \cos\mu + A_{k-1} = 0$, with μ suitably defined, and the work consists mainly in determining the appropriate end conditions and imposing them on the general solution of the difference equation.

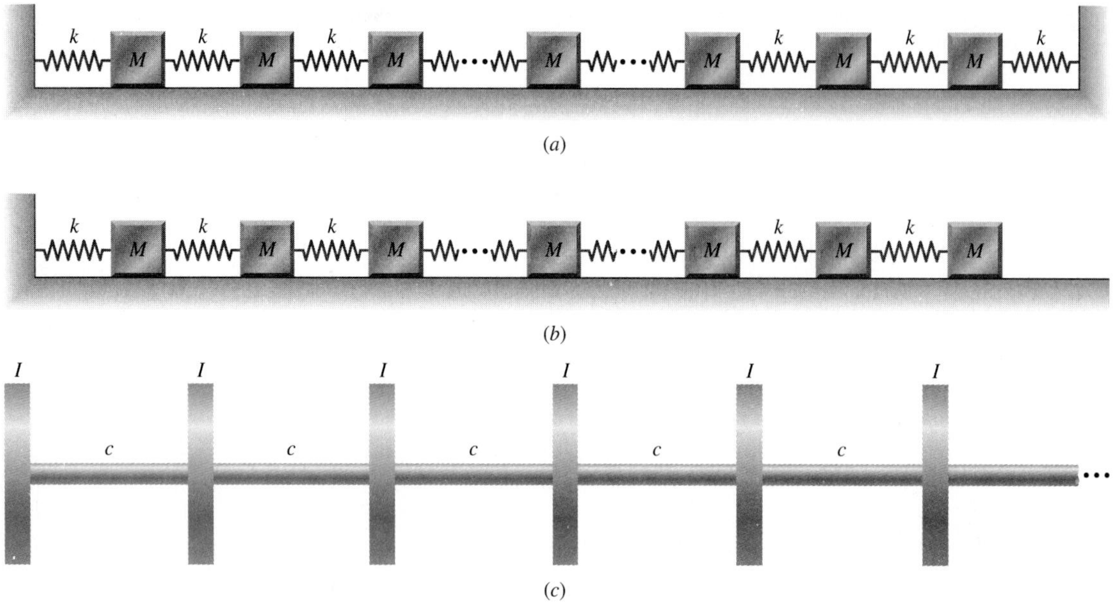

FIGURE 7.29

7. Work Example 1 for the system of n spring-connected masses shown in Fig. 7.29*a*.
8. Work Example 1 for the system of n spring-connected masses shown in Fig. 7.29*b*.
9. Find the natural frequencies of the system shown in Fig. 7.27 and for each frequency find the relative amplitude of each vibrating disk in the absence of the applied torque.
10. Work Exercise 9 if the disk at the right-hand end is replaced by one of moment of inertia I.
11. Work Exercise 9 if the disk at the right-hand end is clamped so that it cannot rotate.
12. Discuss the behavior of the system in Example 2 if an additional restoring torque proportional to the angle of rotation is applied to each disk. In particular, show that the system can now transmit waves only in a band between two frequencies, and find these frequencies.
13. If ω is a frequency such that $0 < \omega^2 I/c < 4$ and $\cos\mu = 1 - \omega^2 I/2c$, show that $\cos\mu k \cos\omega t$, $\cos\mu k \sin\omega t$, $\sin\mu k \cos\omega t$, and $\sin\mu k \sin\omega t$ are four particular solutions of the difference equation

$$\theta_{k+1} - 2\theta_k \cos\mu + \theta_{k-1} = 0$$

satisfied by the angular displacements of three consecutive disks in the infinite chain shown in Fig. 7.29*c* when $0 < \omega^2 I/c < 4$. Hence show that

$$\begin{matrix} \cos(\omega t - \mu k) & & \cos(\omega t + \mu k) \\ \sin(\omega t - \mu k) & \text{and} & \sin(\omega t + \mu k) \end{matrix}$$

are also particular solutions.

If the leftmost disk in the chain is forced to move according to the law $\theta_0 = A\cos\omega t$, show that $\theta_k = A\cos(\omega t - \mu k)$ satisfies this condition and describes a rotational wave which progresses from one disk to the next in μ/ω units of time. If the distance between adjacent disks is l, what is the velocity of the wave? What is the length of the wave? Does the function $\theta_k = \cos(\omega t + \mu k)$ describe a possible wave of the system?
14. In Exercise 13, discuss the velocity and length of the wave as μ approaches zero and as μ becomes infinite.
15. If ω is a frequency such that $\omega^2 I/c > 4$ and $\cosh\mu = \omega^2 I/2c - 1$, show that $(-1)^k \cosh\mu k \cos\omega t$, $(-1)^k \cosh\mu k \sin\omega t$, $(-1)^k \sinh\mu k \cos\omega t$, and $(-1)^k \sinh\mu k \sin\omega t$ are four particular solutions of the difference equation

$$\theta_{k+1} + 2\theta_k \cosh\mu + \theta_{k-1} = 0$$

which is satisfied by the angular displacements of three consecutive disks in the infinite chain shown in Fig.

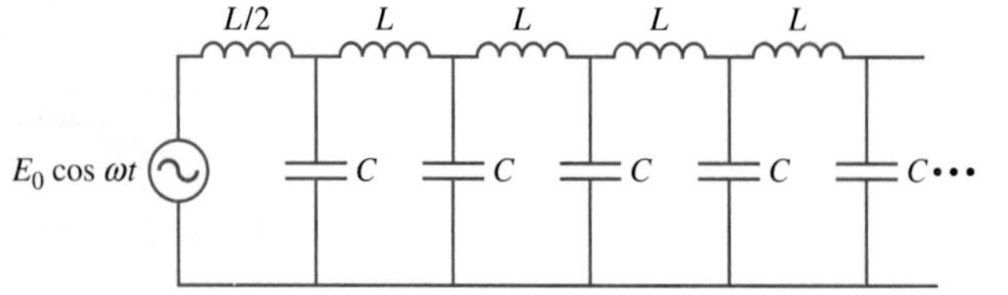

FIGURE 7.30

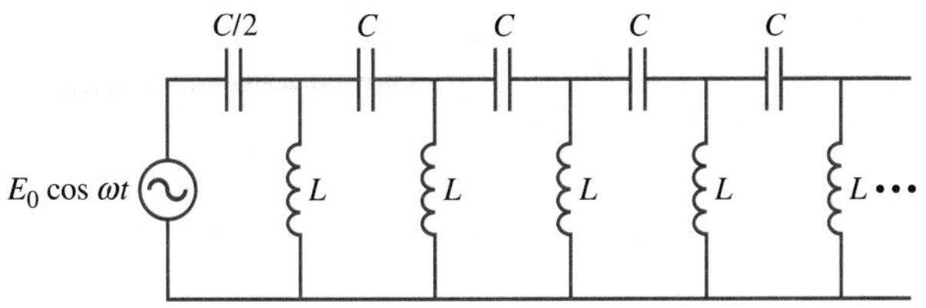

FIGURE 7.31

7.29*c* when $\omega^2 > 4c/I$. Hence show that

$$(-1)^k e^{\mu k} \cos \omega t \qquad (-1)^k e^{\mu k} \sin \omega t$$
$$(-1)^k e^{-\mu k} \cos \omega t \qquad (-1)^k e^{-\mu k} \sin \omega t$$

are also particular solutions. Discuss the motion of the system if the leftmost disk is forced to move according to the law $\theta_0 = A \cos \omega t$.

16. Discuss the response of the infinite low-pass network shown in Fig. 7.30 to a voltage $E_0 \cos \omega t$ in the first loop.

17. Discuss the response of the infinite high-pass network shown in Fig. 7.31 to a voltage $E_0 \cos \omega t$ in the first loop.

18. Show that $\omega_n = 2\sqrt{c/I}$ is one of the natural frequencies of the system discussed in Example 2. *Hint:* Note that $\omega = 2\sqrt{c/I}$ implies that $1 - \alpha^2/2 = -1$. Then, beginning with Eq. (12) with the right-hand side set equal to zero, solve for A_1 in terms of A_0. Then continuing through Eqs. (13), solve for A_2 in terms of A_1, A_3 in terms of A_2, and so on, until all the equations are shown to be consistent.

19. Show that if n is a positive integer, then all values of

$$\left|\sin \frac{N\pi}{2n}\right| \qquad N \text{ an integer}$$

are given by

$$\left|\sin \frac{N\pi}{2n}\right| \qquad N = 0, 1, 2, \ldots, n$$

Hint: Consider values of $N\pi/2n$ on the interval $[0, \pi/2]$.

20. Show that if n is a positive integer, then for every integer k,

$$\left|\sin \frac{(n+k)\pi}{2n}\right| = \left|\sin \frac{(n-k)\pi}{2n}\right|$$

What does this imply about the values

$$\left|\sin \frac{N\pi}{2n}\right| \qquad N \text{ an integer}$$

7.7 ELECTROMECHANICAL ANALOGIES

Although we have already noted the mathematical similarity between the differential equations which model simple translational-mechanical systems, torsional-mechanical systems, series-electrical circuits, and parallel-electrical circuits (Sec. 7.2), we have not yet made such correspondences exact so that whether we are primarily interested in a mechanical system or electric circuit, all the conclusions we derive for either become applicable to the other. The strategy for this is to reduce each equation to dimensionless form so that they are no longer encumbered with units peculiar to mechanical systems or electric circuits. To see how this can be done, let us consider as an illustration the equation

$$(1) \qquad \frac{w}{g}\frac{d^2y}{dt^2} + c\frac{dy}{dt} + ky = F_0 \cos \omega t$$

for the translational-mechanical system [Eq. (7), Sec. 7.2], and the equation

$$(2) \qquad L\frac{d^2Q}{dt^2} + R\frac{dQ}{dt} + \frac{1}{C}Q = E_0 \cos \omega t$$

for the series circuit [Eq. (18*a*), Sec. 7.2].

To reduce Eq. (1) to dimensionless form, let ν be an arbitrary frequency, let σ be an arbitrary distance, and let Z and T be dimensionless variables defined by the relations

$$Z = \frac{y}{\sigma} \qquad \text{and} \qquad T = \nu t \qquad \sigma, \nu \text{ constants}$$

Then,

$$\frac{dy}{dt} = \frac{d(\sigma Z)}{dT}\frac{dT}{dt} = \sigma\nu\frac{dZ}{dT}$$

and in a similar fashion,

$$\frac{d^2y}{dt^2} = \sigma\nu^2\frac{d^2Z}{dT^2}$$

Under these substitutions, Eq. (1) becomes

$$\frac{w}{g}\sigma\nu^2\frac{d^2Z}{dT^2} + c\sigma\nu\frac{dZ}{dT} + k\sigma Z = F_0\cos\frac{\omega}{\nu}T$$

or, dividing through by $w\sigma\nu^2/g$,

(3)
$$\frac{d^2Z}{dT^2} + \frac{cg}{w\nu}\frac{dZ}{dT} + \frac{kg}{w\nu^2}Z = \frac{F_0 g}{w\sigma\nu^2}\cos\frac{\omega}{\nu}T$$

Equation (3) is entirely dimensionless, for not only are the variables Z and T dimensionless, but, as is easy to verify (Exercise 1), so too are the coefficients

$$\frac{cg}{w\nu} \qquad \frac{kg}{w\nu^2} \qquad \text{and} \qquad \frac{F_0 g}{w\sigma\nu^2}$$

To reduce Eq. (2) to dimensionless form, let q be an arbitrary charge and let ν be an arbitrary frequency. Then, in terms of the dimensionless variables

$$Z = \frac{Q}{q} \qquad \text{and} \qquad T = \nu t \qquad q,\ \nu \text{ constants}$$

we have

$$\frac{dQ}{dt} = q\nu\frac{dZ}{dT} \qquad \text{and} \qquad \frac{d^2Q}{dt^2} = q\nu^2\frac{d^2Z}{dT^2}$$

These substitutions reduce Eq. (2) to the form

$$Lq\nu^2\frac{d^2Z}{dT^2} + Rq\nu\frac{dZ}{dT} + \frac{q}{C}Z = E_0\cos\frac{\omega}{\nu}T$$

or, dividing through by $Lq\nu^2$,

(4)
$$\frac{d^2Z}{dT^2} + \frac{R}{L\nu}\frac{dZ}{dT} + \frac{1}{LC\nu^2}Z = \frac{E_0}{Lq\nu^2}\cos\frac{\omega}{\nu}T$$

Again we have achieved a dimensionless form, for not only are the variables Z and T dimensionless but so too are the coefficients (Exercise 2)

$$\frac{R}{L\nu} \qquad \frac{1}{LC\nu^2} \qquad \text{and} \qquad \frac{E_0}{Lq\nu^2}$$

By suitably choosing the arbitrary parameters σ, q, and ν (which need not be the same for each system), the differential equations describing two quite different physical systems can be reduced to the same dimensionless equation. Subsequently, by applying the appropriate scale factors, the solution of the dimensionless equation can be interpreted as the solution of either physical system.

Suppose now that we have a translational-mechanical system for which we wish to construct an exact series circuit in which the charge will represent displacement and current will represent velocity. To begin, we replace the frequencies ω and ν in (4) by ω_e and ν_e, respectively, and set $T = \nu_e t_e$, so that the frequencies for the mechanical system and electric circuit, as well as their time scales, can be distinguished. With this change of notation, Eq. (4) becomes

$$\text{(4}a\text{)} \qquad \frac{d^2Z}{dT^2} + \frac{R}{L\nu_e}\frac{dZ}{dT} + \frac{1}{LC\nu_e^2}Z = \frac{E_0}{Lq\nu_e^2}\cos\frac{\omega_e}{\nu_e}T$$

Once we know the mechanical parameters w/g, c, k, F_0, and ω, we can choose, as convenient or necessary, the electrical parameters L, R, C, E_0, and ω_e and the **scalar factors** σ, ν, q, and ν_e. Comparing Eqs. (3) and (4a), it is clear that they will be identical provided only that the corresponding groups of dimensionless parameters are numerically equal:

$$\text{(5)} \qquad \begin{aligned} \frac{cg}{w\nu} &= \frac{R}{L\nu_e} \\ \frac{kg}{w\nu^2} &= \frac{1}{LC\nu_e^2} \\ \frac{F_0 g}{w\sigma\nu^2} &= \frac{E_0}{Lq\nu_e^2} \\ \frac{\omega}{\nu} &= \frac{\omega_e}{\nu_e} \end{aligned}$$

Of the nine quantities apparently at our disposal, namely,

$$L,\ R,\ C,\ E_0,\ \omega_e,\ \sigma,\ \nu,\ q,\ \text{and}\ \nu_e$$

only seven are essentially arbitrary, because only the ratios

$$\frac{\sigma}{q} \quad \text{and} \quad \frac{\nu}{\nu_e}$$

are significant. However, since these quantities need satisfy only the four relations (5), it is clear that we have considerable freedom to suit our convenience (e.g., availability of electrical components in the laboratory) in assembling the required model.

Suppose now that the electrical counterpart of the mechanical system has been built and that the appropriate initial conditions for the electric circuit have been determined from the relations

$$\text{(6)} \qquad \frac{y}{\sigma} = Z = \frac{Q}{q} \quad \text{and} \quad \frac{1}{\sigma\nu}\frac{dy}{dt} = \frac{dZ}{dT} = \frac{1}{q\nu_e}\frac{dQ}{dt} = \frac{i}{q\nu_e}$$

by evaluating them when $t = 0$, at which time $t_e = 0$ and $T = 0$ also. If the appropriate observations are now made on the electric circuit, we obtain either graphs of Q and i versus t_e or tables of values of Q and i for various values of t_e. To convert these electrical data to corresponding data for the mechanical system, they must be multiplied by the proper conversion factors. The values of these factors are evident from the dual interpretations of Z and dZ/dT shown in (6) and from the dual interpretation of T,

$$\nu t = T = \nu_e t_e$$

These imply, respectively, that

$$y = \left(\frac{\sigma}{q}\right)Q \qquad \frac{dy}{dt} \equiv v = \left(\frac{\sigma\nu}{q\nu_e}\right)i \qquad \text{and} \qquad t = \left(\frac{\nu_e}{\nu}\right)t_e$$

By using the conversion factors appearing inside the parentheses of these relations, data for the mechanical system can be read directly from the graphs or tables of the electrical data.

All of the preceding ideas apply equally well to the analogy between the mechanical system described by Eq. (1) and the parallel electric circuit of Sec. 7.2 described by

$$C\frac{d^2e}{dt^2} + \frac{1}{R}\frac{de}{dt} + \frac{1}{L}e = -\omega I_0 \sin \omega t \tag{7}$$

Equation (7) can be reduced to dimensionless form much as Eq. (2) was. Then its dimensionless parameters can be matched with those arising from Eq. (1), and numerical data for the mechanical system can be obtained by applying the proper conversion factors to data obtained from the parallel circuit. The only difference is that in converting Eq. (7) to dimensionless form, an arbitrary voltage rather than an arbitrary charge must be introduced.

A mechanical system with a single degree of freedom is so simple to analyze mathematically that, except for demonstration purposes, there is no practical reason for constructing an electric model of it. This is not the case, however, when there are several degrees of freedom, especially if the behavior of the system for a number of values of the parameters is to be determined. In such cases the ease with which electrical components can be connected and disconnected and the ease with which voltages and currents can be measured often make it convenient to study a complicated mechanical system by making experimental measurements on an electrical model, i.e., on an electrical analog of the system.

The theory of this procedure is a direct extension of that which we have just discussed. However, in laying out equivalent circuits a certain amount of ingenuity is required, which only practice can supply. Lacking space for this, we shall conclude our discussion with a few general observations and an example or two.

When we first consider making an electrical model of a mechanical system, we always have two distinct possibilities. The model may be based on the analogy between the mechanical system and the series circuit described by Eq. (2), or it may be based on the analogy with the parallel circuit described by Eq. (7). If we choose the series analogy, we have the correspondences

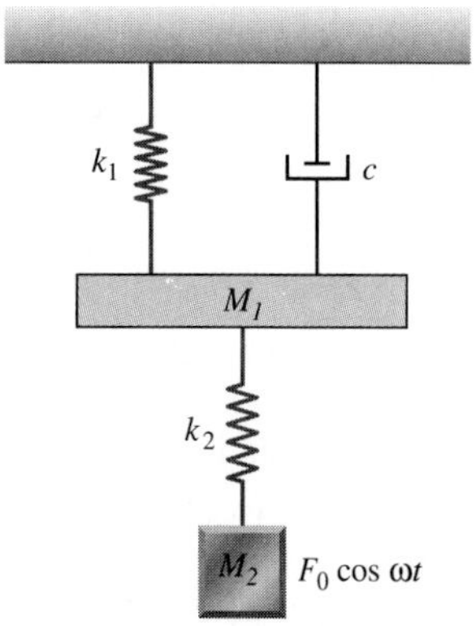

FIGURE 7.32
A translational mechanical system with two springs and masses.

$$\text{Mass } \frac{w}{g} \text{ (or moment of inertia } I) \longleftrightarrow \text{inductance } L$$

$$\text{Damping coefficient } c \longleftrightarrow \text{resistance } R$$

$$\text{Spring (or shaft) modulus } k \longleftrightarrow \text{elastance } \frac{1}{C}$$

$$\text{Force } F \text{ (or torque } T) \longleftrightarrow \text{voltage } E$$

and depending upon the point of view we take toward the electrical model

$$\left.\begin{array}{r}\text{Displacement}\\ \text{Velocity}\end{array}\right\} \longleftrightarrow \text{current}$$

where for a torsional-mechanical system both the displacement and the velocity will be angular.

From the last set of corresponding quantities, we see that if components of a mechanical system experience the same displacement, or velocity, difference between their terminals, then their electrical images must be elements through which the same current flows, that is, elements in series. Thus in Fig. 7.32, both the spring k_1 and the dashpot c experience the same displacement between their respective terminals as is experienced by the mass M_1; hence their analogs must be connected in series, as shown in Fig. 7.33*a*. Similarly, M_2 and the impressed force $F_0 \cos \omega t$ undergo the same displacement and so must have images which are in series (Fig. 7.33*b*). If a mechanical component experiences a displacement (or displacement difference) equal to the sum of displacements (or displacement differences) of other components, then the current through its electrical analog must be the sum of the currents through the electrical images of those other components. In other words, all electric elements must lie in branches having a common endpoint. Thus in Fig. 7.32 the displacement of M_2 is equal to the displacement of M_1 plus the displacement difference across the the spring k_2; hence their electrical analogs must be connected as shown in Fig. 7.33*c*. A network meeting these requirements is shown in Fig. 7.33*d*. The corresponding mechanical components are indicated in parentheses.

Since the coordinates in a mechanical system are the displacements of the various masses, it follows that in the series electrical model the coordinates will be the currents through the various inductors; that is, there must be a single named current flowing through each inductor. This is most conveniently arranged if the inductors are placed in branches which do not appear to be common to two adjacent loops. On the other hand, unless one terminal of a spring or dashpot is grounded, they experience displacements which are the differences of other displacements. Hence their electrical

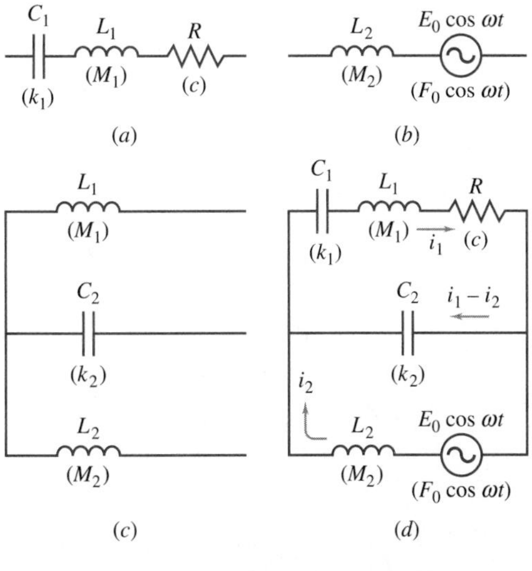

FIGURE 7.33
The assemblage of a series electrical model of the system shown in Fig. 6.32.

images should be placed in branches common to loops in which named currents are flowing. Figure 7.33*d* illustrates these observations.

If we choose the parallel network analogy, we have the correspondences

$$\text{Mass } \frac{w}{g} \text{ (or moment of inertia } I) \longleftrightarrow \text{capacitance } C$$

$$\text{Damping coefficient } c \longleftrightarrow \text{conductance } \frac{1}{R}$$

$$\text{Spring (or shaft) modulus } k \longleftrightarrow \text{susceptance } \frac{1}{L}$$

$$\text{Force } F \text{ (or torque } T) \longleftrightarrow \text{current } I$$

and, as we choose,

$$\left.\begin{matrix}\text{Displacement}\\ \text{Velocity}\end{matrix}\right\} \longleftrightarrow \text{voltage}$$

From the last set of corresponding quantities, it is clear that components in *mechanical parallel,* i.e., constituent parts whose terminals experience the same displacement or velocity differences, correspond to electrical elements across which there is the same voltage difference, in other words, to elements in *electrical parallel.* Moreover, components that are in *mechanical series,* and hence experience displacement and velocity differences totaling the difference across the combination as a whole, correspond to electrical elements whose voltage differences add to the voltage difference across the entire combination, i.e., to elements in *electrical series.* As a result of this direct correspondence, a mechanical system has essentially the same structure as its parallel-electrical analog, and all that is necessary to convert one into the other is to replace each mechanical component by its electrical counterpart, as indicated in the last tabulation. In implementing this process, the simplest way to diagram the mechanical system is first to indicate that all masses are grounded and then to connect the terminals of all mechanical elements that have the same displacement or velocity. Figure 7.34 shows a parallel-electrical analog of the mechanical system shown in Fig. 7.32 diagrammed in this way.

To make exact the correspondence between mechanical systems with more than one degree of freedom and their electrical models we must, as in the case of systems with a single degree of freedom, reduce the underlying differential equations to dimensionless form and then determine the physical makeup of the model so that corresponding dimensionless groups of parameters will be numerically equal. The procedure is exactly the same as for a single equation except that now arbitrary scale factors are to be introduced for *each* of the dependent variables.

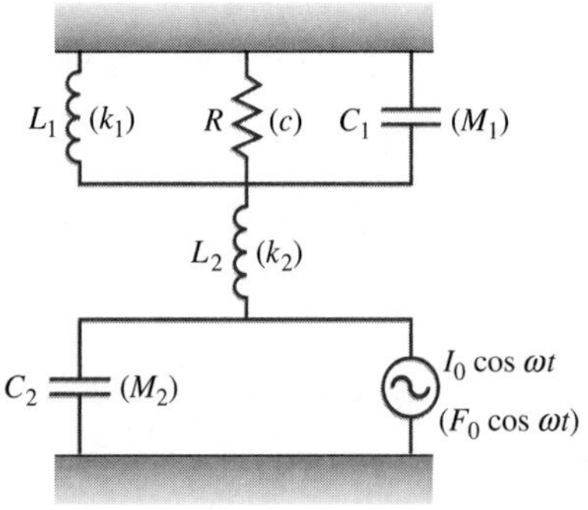

FIGURE 7.34
A parallel electrical model of the system shown in Fig. 7.32.

EXERCISES

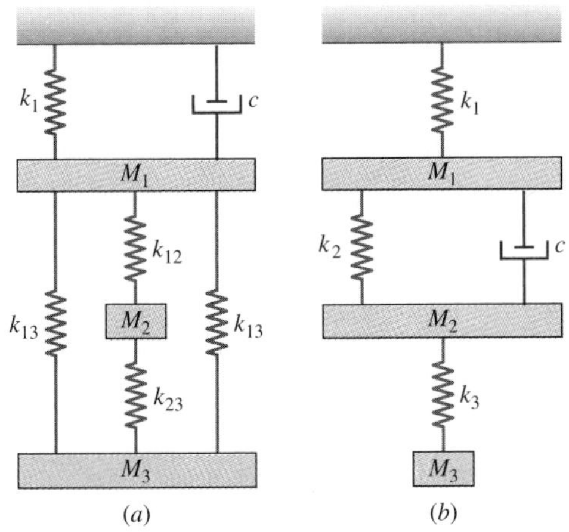

FIGURE 7.35

1. Verify that the coefficients in Eq. (3) are dimensionless.
2. Verify that the coefficients in Eq. (4) are dimensionless.
3. Reduce Eq. (7) to dimensionless form and verify that the coefficients in the resulting equation are dimensionless.
4. Reduce Eq. (12), Sec. 7.2, for the torsional vibrations of a disk, to dimensionless form and verify that the coefficients in the resulting equation are dimensionless.
5. Set up the differential equations describing the behavior of the system shown in Fig. 7.32 and reduce them to dimensionless form.
6. Set up the differential equations describing the behavior of the series analog (Fig. 7.33*d*) of the system shown in Fig. 7.32 and reduce them to dimensionless form.
7. Set up the differential equations describing the behavior of the parallel analog (Fig. 7.34) of the system shown in Fig. 7.32 and reduce them to dimensionless form.
8. If the dashpot shown in Fig. 7.32 is placed between M_1 and M_2 and if a third spring k_3 is added to connect M_2 to the ground, diagram **(a)** the series analog of the resulting system and **(b)** the parallel analog of the resulting system.
9. Diagram **(a)** the series analog of the system shown in Fig. 7.35*a* and **(b)** the parallel analog of the system shown in Fig. 7.35*a*.
10. Diagram **(a)** the series analog of the system shown in Fig. 7.35*b* and **(b)** the parallel analog of the system shown in Fig. 7.35*b*.
11. Figure 7.36*a* represents a freight train with $n - 2$ boxcars, an engine, and a caboose. Diagram the series analog of the system. Set up the differential equations describing the behavior of the train if sudden braking should cause longitudinal vibrations to occur. Set up the differential equations describing the behavior of the analog and reduce each set of equations to dimensionless form.
12. Figure 7.36*b* represents the crankshaft of a four-cylinder in-line engine with a flywheel at one end and a fan at the other. Diagram both the series analog and the parallel analog for this system.
13. Diagram an electric circuit which will be the series analog of the system discussed in Example 2, Sec. 7.5, if the bar is of length $2l$, mass M, and moment of inertia J and if the springs are of moduli k_1 and k_2. Discuss the special case when $k_1 = k_2$. *Hint:* Set up the differential equations for the general case, considering $l\theta$ as one of the dependent variables. Then rearrange the right-hand members until they represent the elastic forces acting on two spring-connected masses.

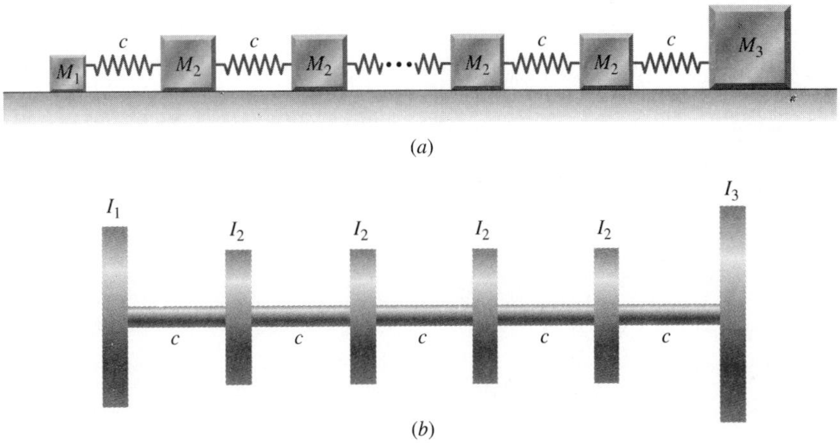

FIGURE 7.36

CHAPTER 8

FOURIER SERIES

Fourier series are infinite series of sines and cosines which, unlike Taylor series, are capable of representing almost any periodic function whether continuous or not. Looking back to the last chapter, this means that if we have a system, either mechanical or electrical, acted on by a periodic force which is not a simple sine or cosine wave, such a force can be represented by a Fourier series, each term of which we learned to handle in Secs. 7.3 and 7.4.

In Sec. 8.2, we obtain the so-called *Euler formulas* for the coefficients in the Fourier series of a given function. These are definite integrals which are usually complicated enough to require integration by parts for their evaluation. To avoid this, in Sec. 8.3 integration by parts is incorporated in these formulas, and recurrence relations are obtained which often are simpler to use than the Euler formulas. These have the additional virtue that they enable one to infer useful information about the continuity of a function from its Fourier coefficients, and vice versa.

In Sec. 8.4 we develop what are called *half-range series.* These are expansions in terms of sines only or cosines only, either of which can represent a finite segment of almost any function whether continuous or not. As we shall see repeatedly in Chap. 11, this is crucial in making solutions of partial differential equations fit the initial conditions or boundary conditions of a physical problem.

In Sec. 8.5, we develop several alternative forms of Fourier series. One of these involves complex exponentials instead of sines and cosines and is especially useful when the complex impedance is used to study an electric circuit.

Examples in Sec. 8.6 illustrate applications to such things as *forced vibrations that exhibit subharmonic resonance, power factors* and *RMS values for electric circuits,* the *least-squares approximation properties of the partial sums of Fourier series,* and the *deflection of cantilever beams under arbitrary loads.*

Prerequisites for this chapter: Secs. 2.5–2.7, 3.1, 5.5, Chap. 7, single-variable calculus.

8.1 PERIODIC FUNCTIONS

In Chap. 2 we learned how to solve ordinary linear constant-coefficient differential equations containing nonhomogeneous terms of the form

$$A \cos \omega t + B \sin \omega t + C \tag{1}$$

in which A, B, C, and ω are constants. Then in Chap. 7, we discovered that such equations are fundamental in the study of physical systems subjected to certain periodic excitations. In many cases, however, the forces, torques, voltages, or currents which act on a system, although periodic, are by no means so simple that they can be represented by an expression like (1). For instance, the voltage impressed on an electric circuit might consist of a series of pulses, as shown in Fig. 8.1*a*, or the disturbing influence acting on a mechanical system might be a force of constant magnitude whose direction is periodically and instantaneously reversed, as shown in Fig. 8.1*b*. The functions represented by these graphs are examples of what are known as *periodic functions.*

DEFINITION 1 A function f is **periodic** if and only if there exists a *positive* number $2p$ such that for every t in the domain of f, $f(t + 2p) = f(t)$. The number $2p$ is called a **period** of f.

If $2p$ is a period of a function f having t in its domain, it follows that

$$f(t) = f(t + 2p) = f[(t + 2p) + 2p] = f(t + 4p)$$

Hence $4p$ is also period of f; and by an obvious extension we conclude that *if $2p$ is a period of the function f, then for every positive integer n, $2np$ is also a period of f.* It is thus clear that every periodic function has arbitrarily large periods and contains arbitrarily *large* numbers in its domain. We call $2p$ the **fundamental period,** or simply *the* **period,** if it is the smallest period of f. Conversely, when we say that f is a function **of period** $2p$, we mean that $2p$ is its fundamental period. Not all periodic functions have a fundamental or smallest period.

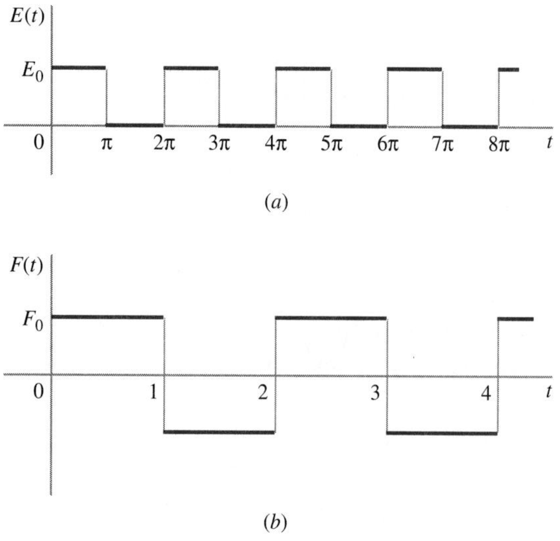

FIGURE 8.1
Typical periodic functions.

EXAMPLE 1

Let f be a constant function defined for all real values of t by $f(t) = c$. Then, for every positive number $2p$,

$$f(t + 2p) = c = f(t)$$

Hence $2p$ is a period of f. Since there is no smallest positive number, f has no fundamental period.

EXAMPLE 2

From the familiar fact that 2π is the fundamental period of both $\cos\theta$ and $\sin\theta$ it follows, upon taking $\theta = \omega t$, that 2π is the smallest positive number for which

$$\cos \omega t = \cos(\omega t + 2\pi) = \cos \omega\left(t + \frac{2\pi}{\omega}\right)$$

and

$$\sin \omega t = \sin(\omega t + 2\pi) = \sin \omega\left(t + \frac{2\pi}{\omega}\right)$$

Thus we see that if $\omega > 0$, the fundamental period of both $\cos \omega t$ and $\sin \omega t$ is $2\pi/\omega$. In particular, if $\omega = n\pi/p$, then $2\pi/\omega = 2\pi/(n\pi/p) = 2p/n$ is the fundamental period of both $\cos(n\pi t/p)$ and $\sin(n\pi t/p)$. Now, as we noted before, any positive integral multiple of the fundamental period of a function is also a period. Hence $n(2p/n) = 2p$ is also a period of both $\cos(n\pi t/p)$ and $\sin(n\pi t/p)$ for every positive integer n.

EXAMPLE 3

A periodic function need not be defined for all values of its independent variable. For example, $\tan t$ is undefined for the values $t = (\pi/2) + n\pi$. Nevertheless, $\tan t$ is a periodic function because for every t *in its domain of definition* $\tan(t + \pi) = \tan t$. Thus π is a period of $\tan t$, the fundamental period, in fact.

Many of the functions we shall encounter in our study of Fourier series will have isolated points of discontinuity where the value of the function is undefined or, if it exists, is given arbitrarily. Much of the usefulness of Fourier series lies in their ability to represent discontinuous periodic functions like those shown in Fig. 8.1.

From Examples 1 and 2 it is clear that every term in the series

$$\begin{aligned}(2)\qquad &\frac{1}{2}a_0{}^{\dagger} + a_1 \cos\frac{\pi t}{p} + a_2 \cos\frac{2\pi t}{p} + \cdots + a_n \cos\frac{n\pi t}{p} + \cdots\\ &+ b_1 \sin\frac{\pi t}{p} + b_2 \sin\frac{2\pi t}{p} + \cdots + b_n \sin\frac{n\pi t}{p} + \cdots\end{aligned}$$

has $2p$ as one of its periods. Since $2p$ is the smallest period of either $\cos(\pi t/p)$ or $\sin(\pi t/p)$, it follows that if either a_1 or b_1 is nonzero, the series as a whole has $2p$ as its fundamental period. This naturally raises the question of whether the coefficients a_0, a_n, b_n $(n = 1, 2, 3, \ldots)$ can be determined so that the series (2) will represent an arbitrary function of period $2p$ at every point of its domain.

If this is the case, then in systems which can be modeled by linear differential equations, the steady-state response of the system to a periodic disturbance represented by $f(t)$ can be found by extending to the series for $f(t)$ the procedure we use when (1) is the nonhomogeneous term in the

†The introduction of the factor $\frac{1}{2}$ is a conventional device to render more symmetric the final formulas for the coefficients.

relevant differential equation: First, we apply to each term of the series the amplitude change and phase shift corresponding to a single harmonic disturbance of that particular frequency. For a mechanical system these will be given by the magnification ratio and the phase-shift formulas [Eqs. (11) and (13), Sec. 7.3]. For an electrical system these will be given by the complex impedance [Eq. (6), Sec. 7.4]. Then we sum these responses to the individual terms in the trigonometric expansion of $f(t)$ to obtain the series giving the response of the system to $f(t)$ itself. The possibility of such expansions, their determination when they exist, and their applications are the subject matter of *Fourier*† *analysis* to which this chapter is devoted.

EXERCISES

1. The following functions are defined on $(-\infty, \infty)$. Find their fundamental periods.

(a) $1 + \cos t + \cos 2t$ **(b)** $\sin \pi t + \sin 3\pi t$
(c) $\sin (5t + \pi)$ **(d)** $\cos \frac{3}{2}(\pi - t)$
(e) $\ln (1 + |\sin t|)$ **(f)** $\cos (\sin t)$
(g) $\sum_{n=1}^{10} \left(\cos \frac{n\pi t}{p} + \sin \frac{n\pi t}{p}\right)$ **(h)** $\cos \pi t \sin 9\pi t$
(i) $\cosh (\cos t + \sin 2t)$ **(j)** $e^{\sin (\pi t/2)}$

2. According to the recording thermograph, the temperature T in a thermostatically controlled missile silo is given by the function $T(t) = 65 + 3 \sin (\pi t/3)$, where $T(t)$ is measured in degrees Fahrenheit‡ and t is in hours.

(a) Between what extreme temperatures does $T(t)$ fluctuate?
(b) What is the period of T?
(c) For what values of t does T have its maximum values?

3. Given that a function f is continuous on $(-\infty, \infty)$ and of period $2p$, find all values of d for which

$$\int_d^{d+2p} f(t)\, dt = \int_0^{2p} f(t)\, dt$$

4. Let $2b$ be the period of a periodic function $f(t)$ whose domain is $(-\infty, \infty)$. Find a positive number k such that $f(kt)$ will be of period $2p$.

5. If $2p$ is a period and D is the domain of both f and g, show that $2p$ is also a period of $f \pm g$. Is $2p$ also a period of $f \times g$?

6. If $2p$ is the fundamental period and D the domain of both f and g, is $2p$ also the fundamental period of $f \pm g$? Is $2p$ the fundamental period of $f \times g$?

7. What is the fundamental period of the function defined on $(-\infty, \infty)$ by **(a)** $e^{in\pi t/p}$ and **(b)** $e^{-in\pi t/p}$ if n is a positive integer?

8. The constant vector $\mathbf{c} = (c_1, c_2, \ldots, c_n)$ has nonzero components and, for t real,

$$\mathbf{v}(t) = [\cos (\pi t/p), \cos (2\pi t/p), \ldots, \cos (n\pi t/p)].$$

What is the fundamental period of $\mathbf{c} \cdot \mathbf{v}$?

9. Work Exercise 8 if $c_1 = c_2 = 0$, $c_3 \neq 0$.

10. Justify the assertion that if a_1 and b_1 are not both zero, the series (2) has $2p$ as its fundamental period.

11. What is the period of the matrix function

$$\mathrm{A}(t) = \begin{bmatrix} \sin 3t & \cos 5t & -\cos 2t \\ \cos 3t & \sin 5t & -\sin 2t \\ 3 & 5 & 2 \end{bmatrix}?$$

12. The position vector of a charged particle whirling about a fixed axis is

$$\mathbf{r}(t) = (3 \cos 2000\pi t, 2 \sin 2000\pi t, \cos 1000\pi t).$$

What is the period of the motion, i.e., of the vector function $\mathbf{r}$?

13. Over a prolonged period of sampling, it is found that the average monthly population of legal-size trout in a tributary of the Missouri drainage is approximated by the formula

$$\overline{P}(n) = 10{,}000 - 2800n + 2090n^2 - 530n^3 + 55n^4 - 2n^5 \qquad 1 \le n \le 12$$

where n is the number of the month. The **periodic extension** P of $\overline{P}$ to all positive integers is defined by

$$P(n) = \overline{P}(n) \qquad n = 1, 2, \ldots, 12$$
$$P(n + 12) = P(n)$$

(a) Find $P(38)$ and identify the average population it represents.
(b) Evaluate $P(153) - P(74)$ and interpret the result in terms of trout populations.
(c) During any given year, or any 2 years, what is the maximum fluctuation in the average monthly trout populations? *Hint:* The critical values of $\overline{P}(x)$ would

†Named for J. B. J. Fourier (1758–1830), a French mathematician and sometime confidant of Napoleon who first undertook the systematic study of such expansions in a memorable monograph *Theorie analytique de la chaleur,* published in 1822. The use of such series in particular problems, however, dates from the time of Daniel Bernoulli (1700–1782), who used them to solve certain problems concerning vibrating strings.
‡Named for the German physicist Gabriel D. Fahrenheit (1686–1736).

all have been integers had $1 \leqq x \leqq 12$ been the domain.

(d) The third largest and largest values of $\overline{P}$ correspond to the spawning seasons of *rainbow* and *brown* trout, respectively. Find these values and the months when they occur.

14. A function E defined on the positive critical values of $\sin \pi t - \pi t \cos \pi t$ has the value 1 or -1 according as the corresponding extremum is a maximum or a minimum. What is **(a)** the domain of E? **(b)** the product of $E(2)E(5)$? **(c)** the period of E?

15. If f is defined on $(-\infty, \infty)$ by the formula

$$f(t) = \begin{cases} 0 & t \text{ a rational number} \\ 1 & t \text{ an irrational number} \end{cases}$$

show that f is periodic and find the set of all its periods.

8.2 THE EULER FORMULAS

To obtain formulas for the so-called **Fourier coefficients** a_n and b_n in the expansion

$$\begin{aligned} f(t) &= \frac{1}{2}a_0 + a_1 \cos\frac{\pi t}{p} + a_2 \cos\frac{2\pi t}{p} + \cdots + a_n \cos\frac{n\pi t}{p} + \cdots \\ &\quad + b_1 \sin\frac{\pi t}{p} + b_2 \sin\frac{2\pi t}{p} + \cdots + b_n \sin\frac{n\pi t}{p} + \cdots \\ &= \frac{1}{2}a_0 + \sum_{n=0}^{\infty}\left(a_n \cos\frac{n\pi t}{p} + b_n \sin\frac{n\pi t}{p}\right) \end{aligned} \tag{1}$$

of a function f of period $2p$, assuming of course that the expansion exists and represents $f(t)$, we shall need the following definite integrals, which are valid for all values of d, provided m and n are integers satisfying the given restrictions:

$$\int_d^{d+2p} \cos\frac{n\pi t}{p}\,dt = 0 \qquad n \neq 0 \tag{2}$$

$$\int_d^{d+2p} \sin\frac{n\pi t}{p}\,dt = 0 \tag{3}$$

$$\int_d^{d+2p} \cos\frac{m\pi t}{p}\cos\frac{n\pi t}{p}\,dt = 0 \qquad m \neq n \tag{4}$$

$$\int_d^{d+2p} \cos^2\frac{n\pi t}{p}\,dt = p \qquad n \neq 0 \tag{5}$$

$$\int_d^{d+2p} \cos\frac{m\pi t}{p}\sin\frac{n\pi t}{p}\,dt = 0 \tag{6}$$

$$\int_d^{d+2p} \sin\frac{m\pi t}{p}\sin\frac{n\pi t}{p}\,dt = 0 \qquad m \neq n \tag{7}$$

$$\int_d^{d+2p} \sin^2\frac{n\pi t}{p}\,dt = p \qquad n \neq 0 \tag{8}$$

With these integrals available, the determination of a_n and b_n proceeds as follows.

To find a_0, we assume that the series (1) can legitimately be integrated term by term from $t = d$ to $t = d + 2p$.† Then,

$$\int_d^{d+2p} f(t)\,dt = \frac{a_0}{2}\int_d^{d+2p} dt + a_1\int_d^{d+2p}\cos\frac{\pi t}{p}\,dt + \cdots + a_n\int_d^{d+2p}\cos\frac{n\pi t}{p}\,dt + \cdots$$
$$+ b_1\int_d^{d+2p}\sin\frac{\pi t}{p}\,dt + \cdots + b_n\int_d^{d+2p}\sin\frac{n\pi t}{p}\,dt + \cdots$$

The integral on the left can always be evaluated since $f(t)$ is a known function which is assumed to be integrable. At worst, some method of approximate integration will be required. The first term on the right is simply

$$\frac{1}{2}a_0 t\Big|_d^{d+2p} = pa_0$$

By Eq. (2) all integrals with a cosine in the integrand vanish, and by Eq. (3) all integrals containing a sine vanish. Hence the integrated result reduces to

$$\int_d^{d+2p} f(t)\,dt = pa_0$$

or

$$a_0 = \frac{1}{p}\int_d^{d+2p} f(t)\,dt \tag{9}$$

To find a_n $(n = 1, 2, 3, \ldots)$, we multiply each side of (1) by $\cos(n\pi t/p)$ and then integrate from d to $d + 2p$, assuming again that term-by-term integration is justified. This gives

$$\int_d^{d+2p} f(t)\cos\frac{n\pi t}{p}\,dt = \frac{1}{2}a_0\int_d^{d+2p}\cos\frac{n\pi t}{p}\,dt + a_1\int_d^{d+2p}\cos\frac{\pi t}{p}\cos\frac{n\pi t}{p}\,dt + \cdots$$
$$+ a_n\int_d^{d+2p}\cos^2\frac{n\pi t}{p}\,dt + \cdots$$
$$+ b_1\int_d^{d+2p}\sin\frac{\pi t}{p}\cos\frac{n\pi t}{p}\,dt + \cdots + b_n\int_d^{d+2p}\sin\frac{n\pi t}{p}\cos\frac{n\pi t}{p}\,dt + \cdots$$

Again, the integral on the left is completely determined. By Eqs. (2) and (4), all integrals on the right containing only cosine terms vanish except the one involving $\cos^2(n\pi t/p)$, which by Eq. (5) is equal to p. Finally, by Eq. (6), every integral which contains a sine is zero. Hence

$$\int_d^{d+2p} f(t)\cos\frac{n\pi t}{p}\,dt = pa_n$$

or

$$a_n = \frac{1}{p}\int_d^{d+2p} f(t)\cos\frac{n\pi t}{p}\,dt \tag{10}$$

---

†A sufficient condition for a series of integrable functions to be integrable term by term is that it be uniformly convergent. See Definition 2, Sec. 18.1.

To determine b_n, we continue essentially the same procedure. We multiply (1) by sin $(n\pi t/p)$ and then integrate from d to $d + 2p$, getting

$$\int_d^{d+2p} f(t) \sin \frac{n\pi t}{p}\, dt = \frac{1}{2}a_0 \int_d^{d+2p} \sin \frac{n\pi t}{p}\, dt + a_1 \int_d^{d+2p} \cos \frac{\pi t}{p} \sin \frac{n\pi t}{p}\, dt + \cdots$$

$$+ a_n \int_d^{d+2p} \cos \frac{n\pi t}{p} \sin \frac{n\pi t}{p}\, dt + \cdots$$

$$+ b_1 \int_d^{d+2p} \sin \frac{\pi t}{p} \sin \frac{n\pi t}{p}\, dt + \cdots + b_n \int_d^{d+2p} \sin^2 \frac{n\pi t}{p}\, dt + \cdots$$

As before, every integral on the right vanishes but one, leaving

$$\int_d^{d+2p} f(t) \sin \frac{n\pi t}{p}\, dt = pb_n$$

or

$$b_n = \frac{1}{p} \int_d^{d+2p} f(t) \sin \frac{n\pi t}{p}\, dt \tag{11}$$

Formulas (9), (10), and (11) for the Fourier coefficients are known as the **Euler** or **Euler-Fourier formulas,** and the series (1), when its coefficients have these values, is known as the **Fourier series** of $f(t)$. In most applications, the interval over which the coefficients are computed is either $(-p, p)$ or $(0, 2p)$; so the value of d in the Euler formulas is usually either $-p$ or 0. Actually, Formula (9) for a_0 need not be listed, or remembered, since it can be obtained from Formula (10) for a_n by putting $n = 0$. It was to achieve this that we wrote the constant term as $\frac{1}{2}a_0$ in the original expansion.

The fact that the formula for a_0 can be obtained by simply putting $n = 0$ in the general formula for a_n does *not* mean that the value of a_0 in a particular problem can be obtained by putting $n = 0$ in the *integrated* formula for a_n. For instance, in Example 2, the integrated formula for a_n is undefined when $n = 0$, although a_0 is well defined by the integral that results when n is set equal to zero *before* integration. A similar situation arises if, for any other value of n, the integrated formula for either a_n or b_n is undefined. For instance, when $n = 1$ in Example 1, the integrated formulas for a_n and b_n are both undefined, and so a_1 and b_1 must be found by setting $n = 1$ before integrating the pertinent formulas.

If a function of period $2p$ is defined everywhere on an interval $d \le t \le d + 2p$ except possibly at a finite number of points, the *interval* is often referred to as **one period** or a **period interval** of the function. It is obvious that the integrations to be performed in Eqs. (2)–(11) range over one period of f.

We must be careful at this stage not to delude ourselves with the belief that we have proved that ever periodic function $f(t)$ has a Fourier expansion that converges to it. What our analysis has shown is merely that *if* a function $f(t)$ has an expansion of the form (1) for which term-by-term integration is valid, *then* the coefficients in that series must be given by the Euler formulas. Questions concerning the convergence of Fourier series and (if they converge) the conditions under which they will represent the functions that generated them are many and difficult. These problems are primarily of theoretical interest, however, for almost any conceivable practical application is covered by the famous **theorem of Dirichlet.**†

†Named for the German mathematician Peter Gustave Lejeune Dirichlet (1805–1859). For a proof of this theorem see, for instance, Philip Franklin, "A Simple Discussion of the Representation of Functions by Fourier Series," in *Selected Papers on Calculus,* pp. 357–361, Mathematical Association of America, Washington, 1969.

THEOREM 1 If $f(t)$ is a bounded periodic function which in any one period has at most a finite number of local maxima and minima and a finite number of points of discontinuity, then the Fourier series of $f(t)$ converges to $f(t)$ at all points where $f(t)$ is continuous and converges to the average of the right- and left-hand limits of $f(t)$ at each point where $f(t)$ is discontinuous.

The conditions of Theorem 1 under which a periodic function possesses a valid Fourier expansion are referred to collectively as the Dirichlet conditions.

DEFINITION 1 A function satisfies the **Dirichlet conditions** on an interval I if and only if the function is bounded and has at most a finite number of local maxima and minima and a finite number of discontinuities on I.

Theorem 1 makes it clear that a function need not be continuous in order to have a valid Fourier series. This means that a periodic function may have a graph each period of which consists of a number of disjointed arcs of different curves, each defined by a different formula, and still be represented by a Fourier series. Thus, in using the Euler formulas to find the coefficients in the expansion of such a function, it will be necessary to break up the range of integration $[d, d+2p]$ to correspond to the various segments of the function. For instance, in Fig. 8.2 the function $f(t)$ is defined by three different expressions $f_1(t)$, $f_2(t)$, and $f_3(t)$ on successive portions of the period interval $d \le t \le d+2p$. Hence it is necessary to write the Euler formulas as

$$
\begin{aligned}
a_n &= \frac{1}{p}\int_d^{d+2p} f(t)\cos\frac{n\pi t}{p}\,dt \\
&= \frac{1}{p}\int_d^{r} f_1(t)\cos\frac{n\pi t}{p}\,dt + \frac{1}{p}\int_r^{s} f_2(t)\cos\frac{n\pi t}{p}\,dt + \frac{1}{p}\int_s^{d+2p} f_3(t)\cos\frac{n\pi t}{p}\,dt \\
b_n &= \frac{1}{p}\int_d^{d+2p} f(t)\sin\frac{n\pi t}{p}\,dt \\
&= \frac{1}{p}\int_d^{r} f_1(t)\sin\frac{n\pi t}{p}\,dt + \frac{1}{p}\int_r^{s} f_2(t)\sin\frac{n\pi t}{p}\,dt + \frac{1}{p}\int_s^{d+2p} f_3(t)\sin\frac{n\pi t}{p}\,dt
\end{aligned}
$$

Incidentally, according to Theorem 1, the Fourier series of the function shown in Fig. 8.2 will converge to the average values, indicated by dots, at the discontinuities at d, r, and $d+2p$, regardless of the definition (or lack of definition) of the function at these points.

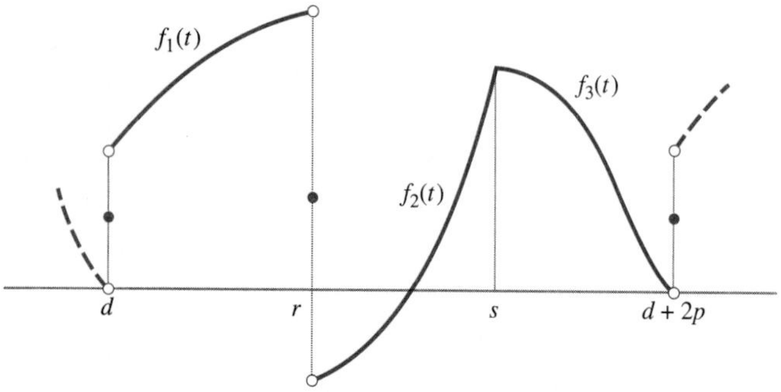

FIGURE 8.2
A periodic function defined by different formulas over different portions of a period.

EXAMPLE 1

What is the Fourier expansion of the periodic function whose definition in one period is

$$f(t) = \begin{cases} 0 & -\pi \le t \le 0 \\ \sin t & 0 \le t \le \pi \end{cases}$$

In this case the half-period of the given function is $p = \pi$. Hence, taking $d = -\pi$ in the Euler formulas, we have

$$a_n = \frac{1}{\pi}\int_{-\pi}^{\pi} f(t)\cos nt\,dt = \frac{1}{\pi}\int_{-\pi}^{0} 0\cos nt\,dt + \frac{1}{\pi}\int_{0}^{\pi}\sin t\cos nt\,dt$$

$$= \frac{1}{\pi}\left[-\frac{1}{2}\left\{\frac{\cos(1-n)t}{1-n} + \frac{\cos(1+n)t}{1+n}\right\}\right]_0^{\pi}$$

$$= -\frac{1}{2\pi}\left[\frac{\cos(\pi - n\pi)}{1-n} + \frac{\cos(\pi + n\pi)}{1+n} - \left(\frac{1}{1-n} + \frac{1}{1+n}\right)\right]$$

$$= -\frac{1}{2\pi}\left(\frac{-\cos n\pi}{1-n} + \frac{-\cos n\pi}{1+n} - \frac{2}{1-n^2}\right)$$

$$= \frac{1+\cos n\pi}{\pi(1-n^2)} \qquad n \ne 1$$

$$a_1 = \frac{1}{\pi}\int_0^{\pi}\sin t\cos t\,dt = \left.\frac{\sin^2 t}{2\pi}\right|_0^{\pi} = 0$$

$$b_n = \frac{1}{\pi}\int_{-\pi}^{\pi} f(t)\sin nt\,dt = \frac{1}{\pi}\int_{-\pi}^{0} 0\sin nt\,dt + \frac{1}{\pi}\int_{0}^{\pi}\sin t\sin nt\,dt$$

$$= \frac{1}{\pi}\left[\frac{1}{2}\left\{\frac{\sin(1-n)t}{1-n} - \frac{\sin(1+n)t}{1+n}\right\}\right]_0^{\pi} = 0 \qquad n \ne 1$$

$$b_1 = \frac{1}{\pi}\int_0^{\pi}\sin^2 t\,dt = \frac{1}{\pi}\left[\frac{t}{2} - \frac{\sin 2t}{4}\right]_0^{\pi} = \frac{1}{2}$$

Hence, evaluating the coefficients for $n = 0, 1, 2, \ldots,$ we have

$$f(t) = \frac{1}{\pi} + \frac{\sin t}{2} - \frac{2}{\pi}\left(\frac{\cos 2t}{3} + \frac{\cos 4t}{15} + \frac{\cos 6t}{35} + \frac{\cos 8t}{63} + \cdots\right)$$

Plots showing the accuracy with which the first n terms of this series represent the given function are shown on the next page in Fig. 8.3 for $n = 1, 2, 3$. For $n = 4, 5, 6, \ldots,$ the graphs of the partial sums are almost indistinguishable from the graph of $f(t)$.

EXAMPLE 2

What is the Fourier expansion of the periodic function whose definition in one period is

$$f(t) = 4 - t^2 \qquad -2 \le t \le 2$$

In this case the half-period of the given function is $p = 2$. Hence, taking $d = -2$ and using integration by parts to evaluate the integrals in the Euler formulas, we have

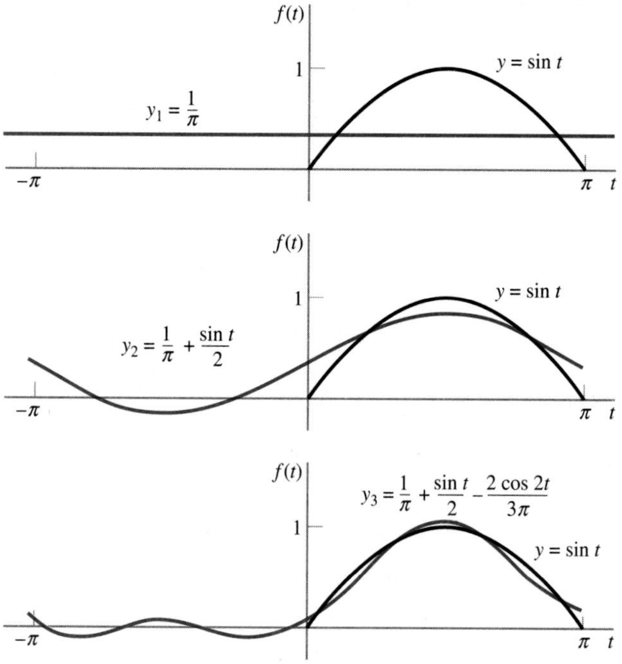

FIGURE 8.3
The approximation of a function by the first few terms of its Fourier expansion.

$$a_n = \frac{1}{2}\int_{-2}^{2} (4 - t^2) \cos \frac{n\pi t}{2}\, dt$$

$$= \frac{1}{2}\left[\frac{8}{n\pi} \sin \frac{n\pi t}{2}\right]_{-2}^{2} - \frac{1}{2}\left[\frac{8t}{n^2\pi^2} \cos \frac{n\pi t}{2} + \left(\frac{2t^2}{n\pi} - \frac{16}{n^3\pi^3}\right) \sin \frac{n\pi t}{2}\right]_{-2}^{2}$$

$$= -\frac{16}{n^2\pi^2} \cos n\pi = \frac{16}{n^2\pi^2}(-1)^{n+1} \qquad n \neq 0$$

$$a_0 = \frac{1}{2}\int_{-2}^{2} (4 - t^2)\, dt = \frac{1}{2}\left[4t - \frac{t^3}{3}\right]_{-2}^{2} = \frac{16}{3}$$

$$b_n = \frac{1}{2}\int_{-2}^{2} (4 - t^2) \sin \frac{n\pi t}{2}\, dt$$

$$= \frac{1}{2}\left[-\frac{8}{n\pi} \cos \frac{n\pi t}{2}\right]_{-2}^{2} - \frac{1}{2}\left[\frac{8t}{n^2\pi^2} \sin \frac{n\pi t}{2} - \left(\frac{2t^2}{n\pi} - \frac{16}{n^3\pi^3}\right) \cos \frac{n\pi t}{2}\right]_{-2}^{2}$$

$$= 0$$

Substituting these coefficients into the series (1), we obtain

$$f(t) = \frac{8}{3} + \frac{16}{\pi^2}\left(\cos \frac{\pi t}{2} - \frac{1}{2^2} \cos \frac{2\pi t}{2} + \frac{1}{3^2} \cos \frac{3\pi t}{2} - \frac{1}{4^2} \cos \frac{4\pi t}{2} + \cdots\right)$$

Interesting numerical series can often be obtained from Fourier series by evaluating them for particular values of the independent variable. For instance, if we set $t = 0$ in the series we have just obtained, we find

$$f(0) = 4 = \frac{8}{3} + \frac{16}{\pi^2}\left(1 - \frac{1}{2^2} + \frac{1}{3^2} - \frac{1}{4^2} + \cdots\right)$$

or, solving for the series

$$1 - \frac{1}{2^2} + \frac{1}{3^2} - \frac{1}{4^2} + \cdots = \frac{\pi^2}{12}$$

EXERCISES

Find the Fourier expansions of the periodic functions whose definitions on one period are

1. $f(t) = \begin{cases} 1 & 0 < t < 1 \\ 0 & 1 < t < 2 \end{cases}$

2. $f(t) = \begin{cases} 1 & 0 < t < \pi \\ -1 & \pi < t < 2\pi \end{cases}$

3. $f(t) = \begin{cases} 1 & 0 < t < 1 \\ 0 & 1 < t < 4 \end{cases}$

4. $f(t) = \begin{cases} 0 & 0 < t < 3\pi/4 \\ 1 & 3\pi/4 < t < 2\pi \end{cases}$

5. $f(t) = \begin{cases} 1 & 0 < t < 1 \\ 0 & 1 < t < 3 \\ -1 & 3 < t < 4 \end{cases}$

6. $f(t) = \begin{cases} 0 & 0 < t < 2\pi/3 \\ 1 & 2\pi/3 < t < 4\pi/3 \\ 2 & 4\pi/3 < t < 2\pi \end{cases}$

7. $f(t) = t \qquad -1 < t < 1$

8. $f(t) = |t| \qquad -2 \le t \le 2$

9. $f(t) = \begin{cases} t & 0 \le t < 1 \\ 0 & 1 < t \le 2 \end{cases}$

10. $f(t) = \begin{cases} t & 0 \le t \le 2 \\ 4 - t & 2 \le t \le 4 \end{cases}$

11. $f(t) = \sin t \qquad 0 \le t \le \pi$

12. $f(t) = \cos t \qquad -\pi/2 \le t \le \pi/2$

13. $f(t) = e^{-t} \qquad 0 < t < 1$

14. $f(t) = t^2 \qquad 0 < t < 1$

15. $f(t) = a^2 - t^2 \qquad -a \le t \le a$

16. $f(t) = t(a - t) \qquad 0 \le t \le a$

17. $f(t) = \begin{cases} 0 & -2 \le t \le -1 \\ \cos \dfrac{\pi t}{2} & -1 \le t \le 1 \\ 0 & 1 \le t \le 2 \end{cases}$

18. $f(t) = \begin{cases} 0 & -2 \le t < -1 \\ \sin \dfrac{\pi t}{2} & -1 < t < 1 \\ 0 & 1 < t \le 2 \end{cases}$

19. $f(t) = \sin^3 t;\ 0 \le t \le 2\pi$. *Hint:* You do not need the Euler formulas to solve this problem.

20. $f(t) = \sin^5 t;\ 0 \le t \le 2\pi$

21. Establish the following numerical results.

$$1 + \frac{1}{2^2} + \frac{1}{3^2} + \frac{1}{4^2} + \frac{1}{5^2} + \cdots + \frac{1}{n^2} + \cdots = \frac{\pi^2}{6}$$

$$1 + \frac{1}{3^2} + \frac{1}{5^2} + \frac{1}{7^2} + \cdots + \frac{1}{(2n-1)^2} + \cdots = \frac{\pi^2}{8}$$

Hint: Use the results of Example 2.

22. Show that the answer to Example 1 can be written in the form

$$f(t) = \frac{1}{\pi} + \frac{\sin t}{2} - \frac{2}{\pi} \sum_{n=1}^{\infty} \frac{\cos 2nt}{4n^2 - 1}$$

Hence show that

$$\frac{\pi - 2}{4} = \sum_{n=1}^{\infty} \frac{(-1)^{n+1}}{(2n-1)(2n+1)}$$

23. **(a)** Find the Fourier expansion of the periodic function whose definition in one period is

$$f(t) = 1 + t \qquad -1 < t < 1$$

(b) Use the Fourier series of f to find the sum of the series

$$\sum_{n=0}^{\infty} \frac{(-1)^n}{2n+1}$$

(c) If F denotes the function defined on $(-\infty, \infty)$ by the Fourier series of f, find $F(1) + F(-5) - 3F(0)$.

24. Find the Fourier expansion of the periodic function whose definition in one period is

$$f(t) = e^t \qquad -\pi < t < \pi$$

Use your result to find the sum of the series

$$\sum_{n=2}^{\infty} \frac{(-1)^n}{1 + n^2}$$

25. Show by an example that a function may satisfy the Dirichlet conditions while its derivative does not.

8.3 ALTERNATIVE FORMULAS FOR THE FOURIER COEFFICIENTS

In many problems, as we saw in Example 2 of the last section, integration by parts is required for evaluation of the integrals that appear in the Euler formulas. This suggests that in certain instances the calculation of the Fourier coefficients in the Fourier expansion of a given periodic function can perhaps be simplified by using results derived from the Euler formulas by integrating the formulas themselves by parts. We shall devote this section to a discussion of these results and some important topics related to them.

Preparatory to this, let us first recall that a function f has a **jump** J_k at t_k if and only if

$$J_k = f(t_k^+) - f(t_k^-) \tag{1}$$

where $f(t_k^+)$ and $f(t_k^-)$ are the right- and left-hand limits of $f(t)$ as t tends toward t_k (Definition 1, Sec. 1.1).

Since the Dirichlet conditions imply the existence of right- and left-hand limits at all points, the functions we will be considering will all have well-defined jumps at every point. If f has a finite discontinuity at t_k, then J_k measures the abrupt change in $f(t)$ at t_k. If f is continuous or has a removable discontinuity at t_k, then of course $J_k = 0$. According as J_k is positive or negative, the graph of f jumps upward or downward as t increases through t_k.

Let us now apply integration by parts to the Euler formula for a_n. We assume, for simplicity, that f is discontinuous at the endpoint d and at a single interior point t_k of the period interval $[d, d + 2p]$. Because of these discontinuities, the integral for a_n is improper and, as Fig. 8.4 indicates, is to be interpreted as

$$a_n = \lim_{\epsilon\to 0} \frac{1}{p}\int_{d+\epsilon}^{t_k-\epsilon} f(t)\cos\frac{n\pi t}{p}\,dt + \lim_{\epsilon\to 0}\frac{1}{p}\int_{t_k+\epsilon}^{d+2p-\epsilon} f(t)\cos\frac{n\pi t}{p}\,dt\dagger$$

With $n \neq 0$, and

$$u = f(t) \qquad dv = \cos\frac{n\pi t}{p}\,dt$$

$$du = f'(t)\,dt \qquad v = \frac{p}{n\pi}\sin\frac{n\pi t}{p}$$

the formula for a_n becomes

$$\begin{aligned} a_n = &\lim_{\epsilon\to 0}\frac{1}{p}\left[\frac{p}{n\pi}f(t)\sin\frac{n\pi t}{p}\Bigg|_{d+\epsilon}^{t_k-\epsilon} - \frac{p}{n\pi}\int_{d+\epsilon}^{t_k-\epsilon} f'(t)\sin\frac{n\pi t}{p}\,dt\right] \\ &+ \lim_{\epsilon\to 0}\frac{1}{p}\left[\frac{p}{n\pi}f(t)\sin\frac{n\pi t}{p}\Bigg|_{t_k+\epsilon}^{d+2p-\epsilon} - \frac{p}{n\pi}\int_{t_k+\epsilon}^{d+2p-\epsilon} f'(t)\sin\frac{n\pi t}{p}\,dt\right] \end{aligned}$$

†Since only finite discontinuities are involved, the same infinitesimal, ϵ, can be used on each side of each discontinuity.

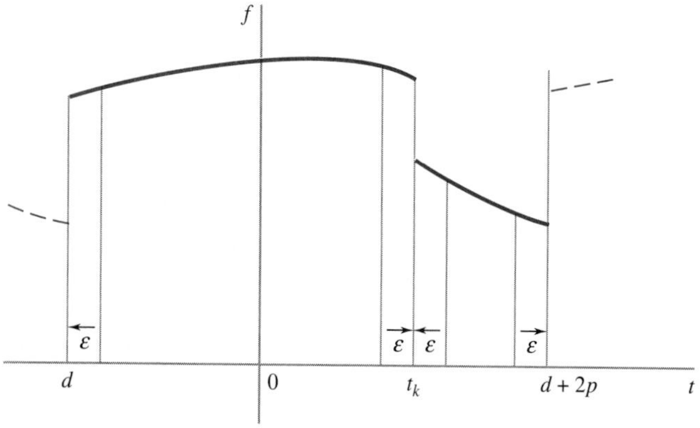

FIGURE 8.4
The figure used in the proof of Theorem 1.

Evaluating and grouping terms appropriately, noting that $f(d + 2p - \epsilon) = f(d - \epsilon)$, this becomes

$$a_n = -\frac{1}{n\pi} \lim_{\epsilon \to 0} \left[f(t_k + \epsilon) \sin \frac{n\pi(t_k + \epsilon)}{p} - f(t_k - \epsilon) \sin \frac{n\pi(t_k - \epsilon)}{p} \right]$$
$$- \frac{1}{n\pi} \lim_{\epsilon \to 0} \left[f(d + \epsilon) \sin \frac{n\pi(d + \epsilon)}{p} - f(d - \epsilon) \sin \frac{n\pi(d - \epsilon)}{p} \right]$$
$$- \frac{1}{n\pi} \lim_{\epsilon \to 0} \left[\int_{d+\epsilon}^{t_k-\epsilon} f'(t) \sin \frac{n\pi t}{p}\, dt + \int_{t_k+\epsilon}^{d+2p-\epsilon} f'(t) \sin \frac{n\pi t}{p}\, dt \right]$$

Since $\sin (n\pi t/p)$ is everywhere continuous, the limits of the first two bracketed quantities are, respectively,

$$f(t_k^+) \sin \frac{n\pi t_k}{p} - f(t_k^-) \sin \frac{n\pi t_k}{p} = J_k \sin \frac{n\pi t_k}{p}$$

and

$$f(d^+) \sin \frac{n\pi d}{p} - f(d^-) \sin \frac{n\pi d}{p} = J_d \sin \frac{n\pi d}{p}$$

Furthermore, the limit of the sum of the two integrals inside the last set of brackets is just

$$\int_d^{d+2p} f'(t) \sin \frac{n\pi t}{p}\, dt = pb_n'$$

where b_n' is the coefficient of $\sin (n\pi t/p)$ in the Fourier expansion of $f'(t)$. Thus, for the special case when f has nonzero jumps only at the two points d and t_k, we have the alternative formula

$$(2) \qquad a_n = -\frac{1}{n\pi}\left(J_d \sin \frac{n\pi d}{p} + J_k \sin \frac{n\pi t_k}{p} \right) - \frac{p}{n\pi} b_n' \qquad n \neq 0$$

Should f have nonzero jumps other than J_d and J_k on the half-open interval $d \le t < d + 2p$, each of them would contribute a term to the right-hand member of (2) having exactly the same structure as the terms that arose from the jumps in f at d and t_k. However, *and this is important,* a jump at $d + 2p$, which must equal the jump at d, does *not* yield an additional term because it has already been taken into account by the term coming from the jump at d. Of course, f need not have a nonzero jump at d, but it might.

Assuming that f has m nonzero jumps on $[d, d + 2p)$ and treating k as a running index, these observations give us the following useful result.

THEOREM 1 If f is a function of period $2p$ which satisfies the Dirichlet conditions and has nonzero jumps $J_1, J_2, \ldots, J_m$ at the respective points $t_1 < t_2 < \cdots < t_m$ of $[d, d + 2p)$, then

$$(3) \qquad a_n = -\frac{p}{n\pi} b_n' - \frac{1}{n\pi} \sum_{k=1}^{m} J_k \sin \frac{n\pi t_k}{p} \qquad n \neq 0$$

Since we had to assume that $n \neq 0$, it follows that, in general, a_0 must still be calculated from the original Euler formula.

Identical reasoning applied to the Euler formula for b_n leads to the following companion theorem in which a_n' denotes the coefficient of $\cos (n\pi t/p)$ in the Fourier expansion of $f'(t)$.

THEOREM 2 If f is a function of period $2p$ which satisfies the Dirichlet conditions and has nonzero jumps $J_1, J_2, \ldots, J_m$ at the respective points $t_1 < t_2 < \cdots < t_m$ of $[d, d+2p)$, then

(4) $$b_n = \frac{p}{n\pi} a_n' + \frac{1}{n\pi} \sum_{k=1}^{m} J_k \cos \frac{n\pi t_k}{p}$$

A single application of the **alternative formulas** (3) and (4) will give the Fourier coefficients of a periodic function whose graph over a period interval consists of horizontal straight-line segments.

EXAMPLE 1

Find the coefficients in the Fourier expansion of the periodic function whose definition over one period is

$$f(t) = \begin{cases} -1 & -2 < t < -1 \\ 0 & -1 < t < 0 \\ 1 & 0 < t < 1 \\ 0 & 1 < t < 2 \end{cases}$$

Noting that $p = 2$, we first compute $a_0 = \frac{1}{2}[\int_{-2}^{-1} (-1)\,dt + \int_0^1 (1)\,dt] = 0$. Next we graph f over the given period interval and indicate its periodic extension immediately to the left of $t = -2$ so that its jump at $t = -2$ can be identified (Fig. 8.5). Using the graph, we tabulate the points of discontinuity and corresponding jumps of f on $-2 \le t < 2$:

k	1	2	3	4
t_k	-2	-1	0	1
J_k	-1	1	1	-1

Substituting $p = 2$ and these values of t_k and J_k into (3) and (4), we obtain

$$a_n = \frac{-2}{n\pi} b_n' - \frac{1}{n\pi}\left[(-1)\sin\frac{n\pi(-2)}{2} + (1)\sin\frac{n\pi(-1)}{2} + (1)\sin\frac{n\pi(0)}{2} + (-1)\sin\frac{n\pi(1)}{2}\right]$$

and

$$b_n = \frac{2}{n\pi} a_n' + \frac{1}{n\pi}\left[(-1)\cos\frac{n\pi(-2)}{2} + (1)\cos\frac{n\pi(-1)}{2} + (1)\cos\frac{n\pi(0)}{2} + (-1)\cos\frac{n\pi(1)}{2}\right]$$

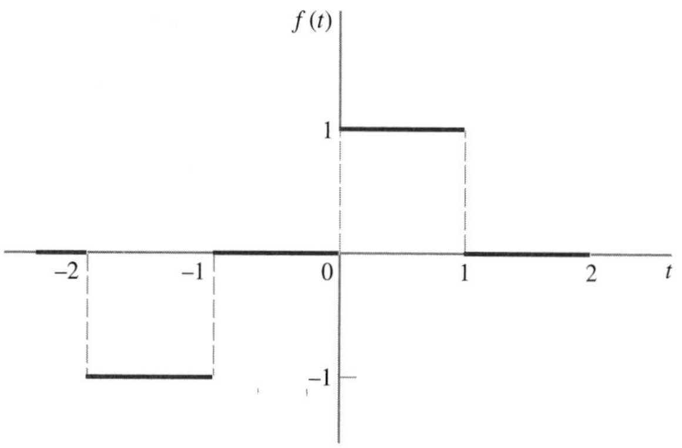

FIGURE 8.5
The periodic function in Example 1.

For $n \geq 1$, $a'_n = b'_n = 0$ because $f'(t) \equiv 0$. Hence the Fourier coefficients in the expansion of f simplify to

$$a_n = \frac{2}{n\pi}\sin\frac{n\pi}{2} = \begin{cases} 0 & n \text{ even} \\ \dfrac{2}{n\pi}(-1)^{(n-1)/2} & n \text{ odd} \end{cases}$$

and

$$b_n = \frac{1}{n\pi}[(-1)^{n+1} + 1] = \begin{cases} 0 & n \text{ even} \\ \dfrac{2}{n\pi} & n \text{ odd} \end{cases}$$

Since all Fourier coefficients for which n is even are zero, the only nontrivial terms in the Fourier expansion of f are those for which n is odd. Setting $n = 2N - 1$ to take these into account, we have

$$f(t) = \frac{2}{\pi}\left[\sum_{N=1}^{\infty} \frac{(-1)^{N-1}}{2N-1}\cos\frac{(2N-1)\pi t}{2} + \sum_{N=1}^{\infty} \frac{1}{2N-1}\sin\frac{(2N-1)\pi t}{2}\right]$$

as the Fourier expansion of f.

It is important to realize that as long as successive derivatives of a periodic function f satisfy the Dirichlet conditions, so that their Fourier coefficients exist, Eqs. (3) and (4) serve as *iteration formulas* in that they express the coefficients in the Fourier series of any particular derivative of f in terms of the nonzero jumps of that derivative on a half-open period interval and the Fourier coefficients of the next higher-order derivative of f. For example, the coefficients a'_n and b'_n can be expressed in terms of the Fourier coefficients a''_n and b''_n of f'' and the nonzero jumps $J'_1, J'_2, \ldots,$ J'_{m_1} at the points of discontinuity $t'_1 < t'_2 < \cdots < t'_{m_1}$ of f' on $d \leq t < d + 2p$. When substituted for a'_n in (4) and for b'_n in (3) these expressions will give new alternative formulas for a_n and b_n in terms of a''_n, b''_n, and the nonzero jumps of both f and f' on $[d, d + 2p)$.

EXAMPLE 2

Find the Fourier expansion of the periodic function whose definition on one period is $f(t) = t$, $-1 < t < 1$.

Graphs of f and f' are shown in Fig 8.6, from which it is evident that $p = 1$, that the only nonzero jump of $f(t)$ on $-1 \leq t < 1$ is $J_1 = -2$ at $t_1 = -1$, and that $f'(t)$ has no nonzero jump. For $n \geq 1$, $a''_n = b''_n = 0$ because all values of f'' are zero. Using the original formula for a_0 and applying (3) and (4) through one iteration to find the other coefficients, we thus have

FIGURE 8.6
The function expanded in Example 2 and its derivative.

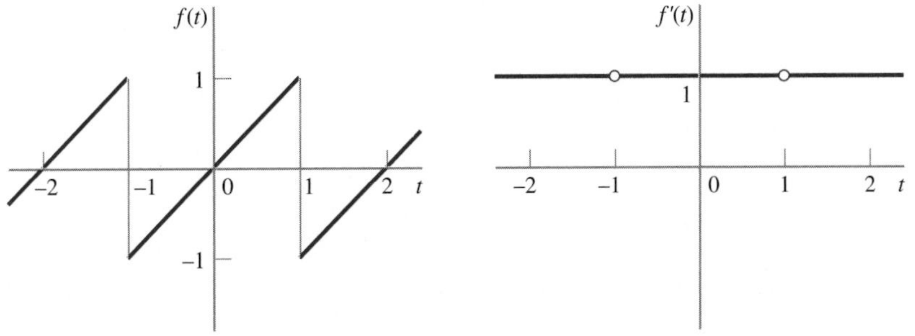

$$a_0 = \frac{1}{1}\int_{-1}^{1} t\,dt = \frac{t^2}{2}\bigg|_{-1}^{1} = 0$$

$$a_n = -\frac{1}{n\pi}b_n' - \frac{1}{n\pi}[(-2)\sin n\pi(-1)] = -\frac{1}{n\pi}\left[\frac{1}{n\pi}a_n''\right] = 0 \qquad n \neq 0$$

$$b_n = \frac{1}{n\pi}a_n' + \frac{1}{n\pi}[(-2)\cos n\pi(-1)] = \frac{1}{n\pi}\left[\frac{-1}{n\pi}b_n''\right] + \frac{2(-1)^{n+1}}{n\pi} = \frac{2(-1)^{n+1}}{n\pi}$$

Hence, for every t in the domain of f,

$$f(t) = \frac{2}{\pi}\sum_{n=1}^{\infty}\frac{(-1)^{n+1}\sin n\pi t}{n}$$

The iterative procedure employed in Example 2 can be readily extended to periodic functions defined by several polynomial expressions over a period interval. When applied to functions of this kind, the formulas for a_n and b_n eventually involve just the jumps of f and its derivatives over a half-open period interval. No unknown Fourier coefficients remain because all derivatives of a polynomial of order higher than its degree are identically zero, and a zero function of any period has only zero Fourier coefficients.

EXAMPLE 3

Find the Fourier expansion of the periodic function whose definition over one period is

$$f(t) = \begin{cases} (t-1)^2 & 0 \le t \le 2 \\ 2t - 3 & 2 \le t \le 3 \\ 9 - 2t & 3 \le t \le 4 \end{cases}$$

Our first step is to graph f, f', and f''. This we have done in Fig. 8.7. From these graphs it is evident that on the period interval $0 \le t < 4$, $f(t)$ has no nonzero jumps, whereas

$$f'(t) = \begin{cases} 2(t-1) & 0 \le t \le 2 \\ 2 & 2 \le t < 3 \\ -2 & 3 < t \le 4 \end{cases} \qquad \text{and} \qquad f''(t) = \begin{cases} 2 & 0 < t < 2 \\ 0 & 2 < t < 3 \\ 0 & 3 < t < 4 \end{cases}$$

FIGURE 8.7
The function expanded in Example 3 and its first two derivatives.

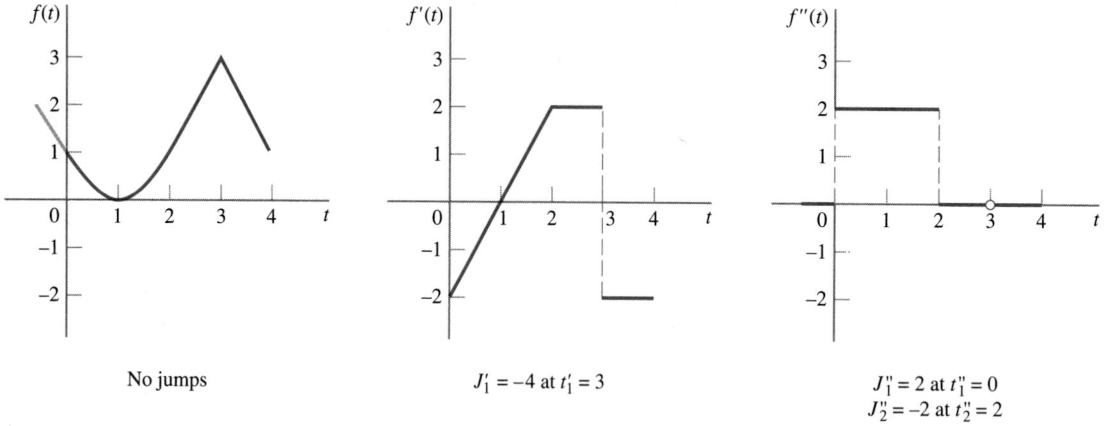

have the discontinuities and corresponding jumps indicated. For $n \geq 1$, $a_n''' = b_n''' = 0$ because $f'''(t) \equiv 0$. With this information in mind and with $p = 2$, we use the original Euler formula to compute a_0 and apply (3) and (4) through two iterations to obtain

$$a_0 = \frac{1}{2}\left[\int_0^2 (t-1)^2\,dt + \int_2^3 (2t-3)\,dt + \int_3^4 (9-2t)\,dt\right] = \frac{7}{3}$$

$$a_n = -\frac{2}{n\pi}b_n' = \frac{-2}{n\pi}\left[\frac{2}{n\pi}a_n'' + \frac{1}{n\pi}(-4)\cos\frac{3n\pi}{2}\right]$$

$$= \frac{-4}{n^2\pi^2}\left\{\frac{-2}{n\pi}b_n''' - \frac{1}{n\pi}[(2)\sin 0 + (-2)\sin n\pi]\right\} + \frac{8}{n^2\pi^2}\cos\frac{3n\pi}{2}$$

$$= \frac{8}{n^2\pi^2}\cos\frac{3n\pi}{2} = \begin{cases} \dfrac{2(-1)^N}{N^2\pi^2} & n = 2N \\ 0 & n = 2N-1 \end{cases} \qquad N = 1, 2, 3, \ldots$$

$$b_n = \frac{2}{n\pi}a_n' = \frac{2}{n\pi}\left[\frac{-2}{n\pi}b_n'' - \frac{1}{n\pi}(-4)\sin\frac{3n\pi}{2}\right]$$

$$= \frac{-4}{n^2\pi^2}\left\{\frac{2}{n\pi}a_n''' + \frac{1}{n\pi}[(2)\cos 0 + (-2)\cos n\pi]\right\} + \frac{8}{n^2\pi^2}\sin\frac{3n\pi}{2}$$

$$= \frac{-8}{n^3\pi^3}[1-(-1)^n] + \frac{8}{n^2\pi^2}\sin\frac{3n\pi}{2} = \begin{cases} 0 & n = 2N \\ \dfrac{8[(-1)^N(2N-1)\pi - 2]}{(2N-1)^3\pi^3} & n = 2N-1 \end{cases}$$

Our final formulas for a_n and b_n tell us that $a_{2N}\cos N\pi t$ is the Nth nontrivial cosine term and that $b_{2N-1}\sin[(2N-1)\pi t/2]$ is the Nth nontrivial sine term in the Fourier expansion of f. Substituting our known values of a_{2N} and b_{2N-1} into these expressions, then summing each of them over N and adding $a_0/2 = \frac{7}{6}$ to the resultant series, we get

$$f(t) = \frac{7}{6} + \frac{2}{\pi^2}\sum_{N=1}^{\infty}\frac{(-1)^N}{N^2}\cos N\pi t + \frac{8}{\pi^3}\sum_{N=1}^{\infty}\frac{(-1)^N(2N-1)\pi - 2}{(2N-1)^3}\sin\frac{(2N-1)\pi t}{2}$$

as the required Fourier series representation of f on $(-\infty, \infty)$.

Formulas (3) and (4) can also be used to determine the Fourier coefficients of a number of periodic functions even though they have no derivative that vanishes identically. This is true, in particular, if the function is proportional to one of its derivatives over a period interval. One such case is the function of Example 1 in the last section, which is a sine wave that has passed through a half-wave rectifier which eliminates the negative half of each wave. For comparison, let us find its Fourier coefficients by our iterative method.

EXAMPLE 4

Use (3) and (4) to find the Fourier coefficients of the periodic function whose definition on one period is

$$f(t) = \begin{cases} 0 & -\pi \leq t \leq 0 \\ \sin t & 0 \leq t \leq \pi \end{cases}$$

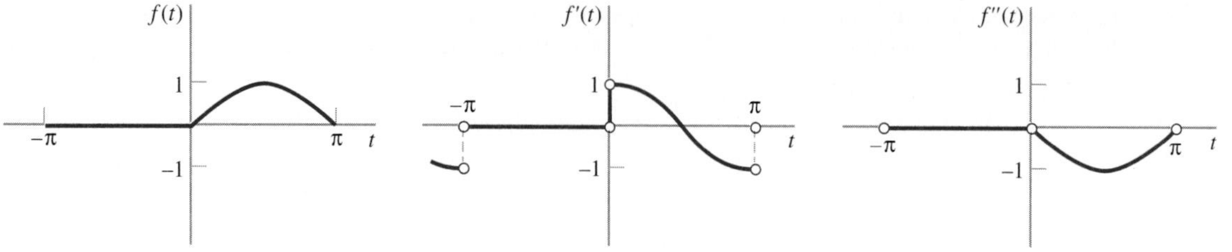

FIGURE 8.8
The periodic function in Example 4 and its first two derivatives.

Graphs of the given function and its first two derivatives are shown in Fig. 8.8. From these we conclude that f is everywhere continuous, that $f'(t)$ has just two nonzero jumps, $J_1' = 1$ at $t_1' = -\pi$ and $J_2' = 1$ at $t_2' = 0$, on $[-\pi, \pi)$, and that $f''(t) = -f(t)$ at all points in the domain of f'', which implies that $a_n'' = -a_n$ and $b_n'' = -b_n$. Thus, with $p = \pi$, our iterative procedure gives

$$a_n = -\frac{1}{n}b_n' = \frac{-1}{n}\left\{\frac{1}{n}a_n'' + \frac{1}{n\pi}[1 \cdot \cos(-n\pi) + 1 \cdot \cos 0]\right\} = \frac{1}{n^2}a_n - \frac{1}{n^2\pi}(\cos n\pi + 1)$$

and

$$b_n = \frac{1}{n}a_n' = \frac{1}{n}\left\{-\frac{1}{n}b_n'' - \frac{1}{n\pi}[1 \cdot \sin(-n\pi) + 1 \cdot \sin 0]\right\} = \frac{1}{n^2}b_n$$

By equating the first and fourth members in each of these chains of equations, we obtain two equations, one involving a_n and the other b_n, which upon collecting terms on a_n and b_n, respectively, can be simplified to yield

$$\pi(n^2 - 1)a_n = -(\cos n\pi + 1) \qquad \text{and} \qquad (n^2 - 1)b_n = 0$$

Solving these equations, we obtain

$$a_n = \frac{1 + \cos n\pi}{\pi(1 - n^2)} \qquad b_n = 0 \qquad n \neq 1$$

which are the same values we obtained by integration in Example 1, Sec. 8.2. The values of a_1 and b_1 must be found from the Euler formulas by integration, but since this was done in the last section, there is no need to repeat the work here.

As we have seen, new pairs of alternative formulas for a_n and b_n can be derived by using (3) and (4) as iteration formulas. Repeated indefinitely, this simple process, coupled with an obvious extension of notation, leads to the following general results:

$$(5)\quad a_n = -\frac{1}{n\pi}\sum_{k=1}^{m} J_k \sin\frac{n\pi t_k}{p} - \frac{p}{n^2\pi^2}\sum_{k=1}^{m_1} J_k' \cos\frac{n\pi t_k'}{p} + \frac{p^2}{n^3\pi^3}\sum_{k=1}^{m_2} J_k'' \sin\frac{n\pi t_k''}{p}$$
$$+ \frac{p^3}{n^4\pi^4}\sum_{k=1}^{m_3} J_k''' \cos\frac{n\pi t_k'''}{p} - \Sigma - \Sigma + \Sigma + \Sigma \cdots$$

$$(6)\quad b_n = \frac{1}{n\pi}\sum_{k=1}^{m} J_k \cos\frac{n\pi t_k}{p} - \frac{p}{n^2\pi^2}\sum_{k=1}^{m_1} J_k' \sin\frac{n\pi t_k'}{p} - \frac{p^2}{n^3\pi^3}\sum_{k=1}^{m_2} J_k'' \cos\frac{n\pi t_k''}{p}$$
$$+ \frac{p^3}{n^4\pi^4}\sum_{k=1}^{m_3} J_k''' \sin\frac{n\pi t_k'''}{p} + \Sigma - \Sigma - \Sigma + \Sigma + \Sigma \cdots$$

The pattern of the signs of successive terms in (5) and (6) is unusual. In the formula for a_n, *pairs* of negative terms alternate with *pairs* of positive terms from the very beginning. In the formula for b_n, the first term is positive but thereafter pairs of negative terms alternate with pairs of positive terms.

From our earlier discussion, we know that the series appearing in (5) and (6) can be evaluated one after another as long as successive derivatives of the periodic function f satisfy the Dirichlet conditions. Formulas (5) and (6) are useful primarily when

(a) f is defined by a finite number of polynomial expressions on a period interval and both (5) and (6) terminate because after a certain point the derivatives of f are identically zero.

(b) f is proportional to one of its derivatives and explicit formulas for a_n and b_n are desired.

In most cases, the use of (3) and (4) as iteration formulas is at least as effective as the use of (5) and (6).

Nevertheless, the series in (5) and (6) provide important information about the rate of convergence of the Fourier series of $f(t)$. Specifically, if f is discontinuous, then, because of the jumps in $f(t)$, the first term in either (5) or (6), and usually both first terms, will be different from zero. This means that the most slowly decreasing term in a_n and b_n is of the order of $1/n$ and therefore at least one of the coefficients can decrease no faster than c/n.† As a result, although Theorem 1, Sec. 8.2, guarantees that the Fourier series for $f(t)$ converges, it does so very slowly. On the other hand, if $f(t)$ is continuous but $f'(t)$ is discontinuous, then $f(t)$ has no jumps but $f'(t)$ has. In this case, the leading terms in both (5) and (6) are zero, but at least one of the second terms is not. This means that the most slowly decreasing term in the formulas for a_n and b_n is of the order of $1/n^2$, and thus at least one of the coefficients a_n and b_n can decrease no faster than c/n^2. In general, if $f(t)$ and its first $k - 1$ derivatives are everywhere continuous but the kth derivative is discontinuous, then the first k terms in both (5) and (6) are zero, but at least one of the terms in the $(k + 1)$st series is different from zero. Hence the coefficients are of the order $1/n^{k+1}$. Conversely, if we know the Fourier expansion of a function, then even though we do not know the function itself, we can draw conclusions about its continuity and the continuity of its derivatives by observing the rate at which the coefficients a_n and b_n approach zero. These observations are summarized in the following theorem.

THEOREM 3 As n becomes infinite, the coefficients a_n and b_n in the Fourier expansion of a periodic function satisfying the Dirichlet conditions always approach zero at least as rapidly as c/n, where c is a constant independent of n. If the function has one or more points of discontinuity, then either a_n or b_n, and in general both, can decrease no faster than this. In general, if a function $f(t)$ and its first k derivatives satisfy the Dirichlet conditions and are everywhere continuous, then as n becomes infinite, the coefficients a_n and b_n in the Fourier series of $f(t)$ tend to zero at least as rapidly as c/n^{k+2}. If, in addition, the $(k + 1)$st derivative of $f(t)$ is not everywhere continuous, then either a_n or b_n, and in general both, can tend to zero no faster than c/n^{k+2}.

More concisely, though less accurately, Theorem 3 asserts that the smoother the function, the faster its Fourier expansion converges. Conversely, by observing the rate at which the coefficients

†This does not mean that the coefficients are proportional to $1/n$. What it does mean is that there exists a number c, independent of n, such that both a_n and b_n are less than c/n for all sufficiently large values of n. For example, $2n/(n^2 + 9)$ approaches zero at least as rapidly as c/n because

$$\frac{2n}{n^2 + 9} < \frac{2n}{n^2} = \frac{2}{n} \qquad \text{for all values of } n$$

but it does not approach zero as rapidly as c/n^2 because there is no value of c such that

$$\frac{2n}{n^2 + 9} < \frac{c}{n^2}$$

for all sufficiently large values of n.

in the Fourier expansion of an otherwise unknown function approach zero, one can obtain useful information about the degree of smoothness of the function.

EXAMPLE 5

Without finding the coefficients in the Fourier expansion of the periodic function whose definition on one period is

$$f(t) = \begin{cases} \sin \pi t & -1 \leqq t \leqq 0 \\ \pi(t - t^2) & 0 \leqq t \leqq 1 \end{cases}$$

determine how rapidly they will decrease as n becomes infinite.

Clearly, the given function is everywhere continuous. Moreover, its derivative is also continuous everywhere on the given period, except possibly at the points -1, 0, 1, where the representations of the function change. Actually, at $t = 0$ the slopes of $y = \sin \pi t$ and $y = \pi(t - t^2)$ are each equal to π, and at $t = -1$ and $t = 1$ and slope of each of the adjoining curves is $-\pi$. Hence $f'(t)$ is everywhere continuous. However, $f''(t)$ is discontinuous at the origin since when $t = 0$ the second derivative of $\sin t$ is equal to zero while the second derivative of $\pi(t - t^2)$ is equal to -2π. Thus, according to Theorem 3, both a_n and b_n tend to zero at least as fast as c/n^3 while at least one of them can decrease no faster than c/n^3.

EXAMPLE 6

A periodic function $f(t)$ is known to have Fourier coefficients given by the formulas

$$a_n = \frac{n\pi}{n^4 + \pi^4} \qquad b_n = \frac{\pi}{n^2 + n\pi + \pi^2}$$

What can be said about the continuity of f and its derivatives?

By inspection, we see that a_n tends to zero as c/n^3, while b_n tends to zero only as c/n^2. Since continuity determines, and in turn is determined by, the more slowly decreasing one of the pair (a_n, b_n), it follows from Theorem 3 that $f(t)$ is continuous everywhere but $f'(t)$ has at least one point of discontinuity in each period. In other words, in each period the graph of $f(t)$ must have at least one **corner point** where the tangent changes direction abruptly.

We conclude this section with two theorems on the termwise intergration and differentiation of Fourier series. In view of the behavior of the Fourier coefficients described in Theorem 3, they appear highly plausible, but we must leave their proof to more advanced texts.

THEOREM 4 The integral of any periodic function which satisfies the Dirichlet conditions can be found by term-by-term integration of the Fourier series of the function.†

THEOREM 5 If $f(t)$ is a periodic function that satisfies the Dirichlet conditions and is everywhere continuous, and if $f'(t)$ also satisfies the Dirichlet conditions, then wherever it exists, $f'(t)$ can be found by term-by-term differentiation of the Fourier series of $f(t)$.‡

EXAMPLE 7

Beginning with the series obtained in Example 2, for the periodic function whose definition in one period is $f(t) = t$, $-1 < t < 1$, use Theorem 4 to obtain the Fourier series for the periodic function whose definition in one period is $f_1(t) = t^2$, $-1 \leq t \leq 1$.

Integrating the series obtained in Example 2 termwise, with t temporarily confined to the interval $-1 < t < 1$, we have

$$\int t\,dt = \frac{t^2}{2} + c = \frac{2}{\pi}\sum_{n=1}^{\infty}\frac{(-1)^{n+1}}{n}\int \sin n\pi t\,dt = \frac{2}{\pi^2}\sum_{n=1}^{\infty}\frac{(-1)^n}{n^2}\cos n\pi t$$

†See, for instance, E. C. Titchmarsh, *Theory of Functions,* pp. 419–421, Oxford, New York, 1939.
‡See, for instance, E. T. Whittaker and G. M. Watson, *Modern Analysis,* pp. 168–169, Cambridge, New York, 1963.

Putting $t = 0$ in order to solve for c, we get

$$c = \frac{2}{\pi^2}\sum_{n=1}^{\infty}\frac{(-1)^n}{n^2} = -\frac{2}{\pi^2}\left(1 - \frac{1}{2^2} + \frac{1}{3^2} - \frac{1}{4^2} + \cdots\right) = -\frac{2}{\pi^2}\frac{\pi^2}{12} = -\frac{1}{6}$$

the value of the numerical series being obtained from Example 2, Sec. 8.2. Hence, substituting for c and solving for t^2,

$$t^2 = \frac{1}{3} + \frac{4}{\pi^2}\sum_{n=1}^{\infty}\frac{(-1)^n \cos n\pi t}{n^2} \qquad -1 \le t \le 1$$

where the last interval may be closed, as is confirmed by the fact that the coefficients in the series for t^2 decrease as c/n^2. This of course indicates that the periodic function f_1 itself is continuous everywhere, and thus is represented by the preceding series for all real values of t, but has a graph with at least one corner point on each period interval.

EXERCISES

Use Formulas (3) and (4), iteratively if necessary, to find the Fourier expansions of the periodic functions whose definitions on one period are

1. $f(t) = \begin{cases} 1 & 0 < t < \pi \\ 2 & \pi < t < 2\pi \end{cases}$

2. $f(t) = \begin{cases} 1 & 0 < t < 1 \\ 0 & 1 < t < 2p \end{cases}$

3. $f(t) = \begin{cases} 0 & 0 < t \le 1 \\ t - 1 & 1 \le t < 2 \end{cases}$

4. $f(t) = \begin{cases} t - 1 & -1 < t < 0 \\ t + 1 & 0 < t < 1 \end{cases}$

5. $f(t) = 2 - t \qquad 0 < t < 2$

6. $f(t) = t - t^2 \qquad -1 < t < 1$

7. $f(t) = 1 + |t| \qquad -p \le t \le p$

8. $f(t) = 3 - t^2 \qquad -2 \le t \le 2$

9. $f(t) = \begin{cases} 1 + t^2 & 0 \le t \le 1 \\ 3 - t & 1 \le t \le 2 \end{cases}$

10. $f(t) = \begin{cases} 0 & -1 < t < 0 \\ 1 + t^2 & 0 < t < 1 \end{cases}$

11. $f(t) = \begin{cases} t^2 & -1 \le t \le 0 \\ t^3 & 0 \le t \le 1 \end{cases}$

12. $f(t) = \begin{cases} t - 2 & 0 < t \le 2 \\ (t - 2)(t - 4) & 2 \le t < 4 \end{cases}$

13. $f(t) = e^{kt} \qquad 0 < t < 2p$

14. $f(t) = t^4 \qquad -\pi \le t \le \pi$

15. $f(t) = \cos t \qquad 0 < t < \pi$

16. $f(t) = \sin t \qquad 0 \le t \le \pi$

17. $f(t) = \cosh t \qquad -a \le t \le a$

18. $f(t) = \sinh at \qquad -1 < t < 1$

19. $f(t) = \begin{cases} \cosh t & -1 < t \le 0 \\ e^t & 0 \le t < 1 \end{cases}$

20. $f(t) = \begin{cases} \cos t & 0 < t < \pi \\ 0 & \pi < t < 2\pi \end{cases}$

21. $f(t) = \begin{cases} 2 & -2\pi \le t \le 0 \\ 1 + \cos t & 0 \le t \le 2\pi \end{cases}$

22. $f(t) = \begin{cases} \sin t & 0 \le t \le \pi/2 \\ 1 & \pi/2 \le t \le \pi \\ -\cos t & \pi \le t \le 3\pi/2 \\ 0 & 3\pi/2 \le t \le 2\pi \end{cases}$

Hint: In Exercises 21 and 22, choose functions f_1 and f_2 such that $f_1 + f_2 = f$ and add their Fourier coefficients to obtain those of f.

Without finding the Fourier coefficients of the following functions having the given periods, determine how fast the coefficients will decrease as n becomes infinite.

23. $f(t) = t - t^3 \qquad -1 \le t \le 1$

24. $f(t) = (1 - t^2)^2 \qquad -1 \le t \le 1$

25. $f(t) = \begin{cases} 0 & -\pi \le t \le 0 \\ 1 - \cos t & 0 \le t \le \pi \end{cases}$

26. $f(t) = \begin{cases} 0 & -2 \le t \le 0 \\ t^3(2 - t)^2 & 0 \le t \le 2 \end{cases}$

27. $f(t) = \begin{cases} \cos t & -\pi \le t \le 0 \\ \cos 2t & 0 \le t \le \pi/2 \end{cases}$

28. $f(t) = \begin{cases} (1/\pi)\sin \pi t & -1 \le t \le 0 \\ t - 2t^3 + t^4 & 0 \le t \le 1 \end{cases}$

29. Determine the value of a so that the Fourier coefficients of the periodic function $f(t) = at + t^3$, $-1 \le t \le 1$, will all be of the form c/n^3. Can an interesting numerical series be obtained from the expansion?

30. Determine the values of a and b so that the Fourier coefficients of the periodic function $f(t) = at^2 + bt^3 + t^4$, $0 \le t \le 1$, will all be of the form c/n^4. Can any interesting numerical series be obtained from the expansion?

31. A periodic function is known to have Fourier coefficients given by the formulas

$$a_n = \frac{2n}{\sqrt{(n^2-1)^2+4n^2}} \qquad b_n = \frac{2}{\sqrt{(n^2-1)^2+4n^2}}$$

What can be said about the continuity of the function and its derivatives?

32. Work Exercise 31 if

$$a_n = \frac{1}{(2n-1)\sqrt{n^2+1}} \qquad b_n = \frac{1}{n^2\sqrt{n^2+1}}$$

33. If $f(t)$ is the periodic function whose definition in one period is

$$f(t) = \begin{cases} 0 & -1 < t < 0 \\ 1 & 0 < t < 1 \end{cases}$$

use Theorem 4 to find the Fourier expansions of

$$\int_{-1}^{t} f(s)\,ds \qquad \text{and} \qquad \int_{-1}^{t}\int_{-1}^{r} f(s)\,ds\,dr$$

A 128-lb weight is suspended from a spring of modulus 75 lb/in, and the weight vibrates vertically under the influence of an impressed periodic force $f(t)$. Assuming negligible friction, determine which, if any, of the following possibilities for f will produce a condition of pure resonance.

34. $f(t) = \begin{cases} t & 0 \le t \le \pi \\ 2\pi - t & \pi \le t \le 2\pi \end{cases}$

35. $f(t) = \begin{cases} 2 & 0 \le t < \pi \\ 2t/\pi - 2 & \pi \le t \le 2\pi \end{cases}$

36. $f(t) = \begin{cases} a & 0 \le t < \pi \\ -a & \pi \le t < 2\pi \end{cases}$

37. $f(t) = \begin{cases} \cos t & 0 \le t < \pi \\ -\cos t & \pi < t < 2\pi \end{cases}$

38. A periodic force defined by the function of Exercise 16 acts on the unit mass in the mechanical system of Exercise 1, Sec. 7.5. Is pure resonance to be expected? Explain.

8.4 HALF-RANGE EXPANSIONS

When a periodic function possesses certain symmetry properties, the coefficients in its Fourier series become especially simple. This was illustrated in Example 2, Sec. 8.2, and in Example 2 of the last section. In the first of these two examples the given function, which was defined over one period by $f(t) = 4 - t^2$, $-2 \le t \le 2$, was symmetric in the y axis and its expansion contained only cosine terms, i.e., only terms which themselves were symmetric in the y axis. In the other example the given function was defined by $f(t) = t$ in the period $-1 < t < 1$ and was symmetric in the origin. Its expansion contained only sine terms, i.e., only terms which themselves were symmetric in the origin. In this section we shall investigate in detail just what effect the symmetry of $f(t)$ has on the coefficients in its Fourier series.

Taking $d = -p$ in the Euler formulas [Eqs. (10) and (11), Sec. 8.2] for a_n and b_n, we can write

$$a_n = \frac{1}{p}\int_{-p}^{p} f(t)\cos\frac{n\pi t}{p}\,dt \tag{1}$$

$$b_n = \frac{1}{p}\int_{-p}^{p} f(t)\sin\frac{n\pi t}{p}\,dt \tag{2}$$

Since the product of two even or two odd functions is even, whereas the product of an odd and an even function is odd [Exercise 14(**b**), Sec. 1.1], and since the cosine function is even while the sine function is odd, it follows that if $f(t)$ is an even function, the integrand of (1) is even and the integrand of (2) is odd. Consequently, (assuming the Dirichlet conditions) Theorems 1 and 2, Sec. 1.1, give the following useful result.

THEOREM 1 If $f(t)$ is an even periodic function which satisfies the Dirichlet conditions, the coefficients in the Fourier series of $f(t)$ are given by the formulas

$$a_n = \frac{2}{p}\int_{0}^{p} f(t)\cos\frac{n\pi t}{p}\,dt \qquad b_n \equiv 0 \qquad \text{where } 2p \text{ is the period of } f(t).$$

Similarly, if f is an odd function, the integrand of (1) is odd and the integrand of (2) is even. Thus we get the companion result:

THEOREM 2 If $f(t)$ is an odd periodic function which satisfies the Dirichlet conditions, the coefficients in the Fourier series of $f(t)$ are given by the formulas

$$a_n \equiv 0 \qquad b_n = \frac{2}{p}\int_0^p f(t)\sin\frac{n\pi t}{p}\,dt$$

where $2p$ is the period of $f(t)$.

The observations we have just made about the Fourier coefficients of odd and even functions serve to reduce by half the labor of expanding such functions. However, their chief value is that they allow us to meet the requirements of certain problems† in which expansions containing *only* cosine terms or expansions containing *only* sine terms must be constructed.

Suppose, for instance, that the conditions of a problem require us to consider the values of a function f *only* on the closed interval $[0, p]$. In other words, conditions of periodicity are irrelevant to the problem. This being the case, we can, if it suits our purposes, extend f into a periodic function F, defined for all t and of period $2p$, in the following way: on $[0, p]$ set $F(t) = f(t)$, over $(-p, 0)$ let F have arbitrary values $\phi(t)$ (subject to the Dirichlet conditions, of course), and for all other values of t let F be defined by the periodicity condition $F(t + 2p) = F(t)$. Thus

$$F(t)\ddagger = \begin{cases} \phi(t) & -p < t < 0 \\ f(t) & 0 \le t \le p \\ F(t+2p) & t \text{ real} \end{cases}$$

By the Dirichlet theorem, this periodic continuation of f has a Fourier expansion which converges to $F(t)$ wherever F is continuous. Thus, *regardless of the extension* ϕ, the Fourier series of F likewise converges to $f(t)$ between 0 and p since on $[0, p]$, $F(t) = f(t)$.

In particular, if we extend the function f from 0 to $-p$ by reflecting its graph between 0 and p in the vertical axis, so that $\phi(t) = f(-t)$, the periodic continuation F is an *even* function with domain $(-\infty, \infty)$; hence its Fourier expansion will contain only cosine terms whose coefficients, according to Theorem 1, are

$$a_n = \frac{2}{p}\int_0^p F(t)\cos\frac{n\pi t}{p}\,dt = \frac{2}{p}\int_0^p f(t)\cos\frac{n\pi t}{p}\,dt$$

Of course, the constant term $\frac{1}{2}a_0 = \frac{1}{2}a_0 \cos(0\pi t/p)$ will also be included.

On the other hand, we may extend the same function f from 0 to $-p$ by reflecting its graph between 0 and p in the origin, so that $\phi(t) = -f(-t)$. But unless $f(0) = f(p) = 0$, the periodic continuation F with domain $(-\infty, \infty)$ will not be strictly *odd.* (Why?) Nevertheless, its Fourier expansion will contain only sine terms whose coefficients are those of Theorem 2, namely,

$$b_n = \frac{2}{p}\int_0^p F(t)\sin\frac{n\pi t}{p}\,dt = \frac{2}{p}\int_0^p f(t)\sin\frac{n\pi t}{p}\,dt$$

because neither the coefficients $\{b_n\}$ nor the convergence of the Fourier series of F is affected by a finite change in the value of f at either 0 or p. In fact, the series remains unchanged even if 0 or p is deleted from the original domain of f.

Putting together the preceding results, we reach the remarkable conclusion that simply by *imagining* the appropriate extension of a function f defined only from 0 to p, we can obtain

---

†Examples of such problems will be found in Sec. 11.5.

‡Functions initially defined on open or half-open intervals from 0 to p can be extended similarly by paying due attention to where equality signs are inserted to identify the domain of ϕ.

expansions representing f and containing only cosine terms or only sine terms, as we please. Furthermore, the formulas for the coefficients in these expansions involve only the values of f on the *original interval* from 0 to p and do not depend in any way upon the extensions we used to obtain those formulas. Such series are called **half-range cosine series** if they contain only cosine terms; they are called **half-range sine series** if they contain only sine terms. Collectively, we speak of such series as **half-range expansions.**

EXAMPLE 1

Find the Fourier coefficients in the half-range sine expansion of the function $f(t) = t^2$, $0 \le t < 1$.

In Example 7, Sec. 8.3, without realizing it, we obtained the Fourier expansion of the *even* function shown in Fig. 8.9*a*. Now we are asked to determine the half-range sine expansion of t^2, that is, the Fourier series of the *odd* function shown in Fig. 8.9*b*. According to Theorem 2, we know that for this function

$$a_n \equiv 0 \qquad \text{and} \qquad b_n = \frac{2}{1}\int_0^1 t^2 \sin\frac{n\pi t}{1}\,dt$$

and it would not be difficult to find b_n by evaluating the last integral. It is simpler, however, to evaluate b_n using the jumps of the odd periodic continuation F of f and its derivatives on a half-open period interval.

We already know that $p = 1$. Sketching graphs of F' and F'' to go with that of F, we see at once that in terms of our established notation, the relevant nonzero jumps on $[-1, 1)$ are

$$J_1 = -2 \text{ at } t_1 = -1 \qquad J_1'' = -4 \text{ at } t_1'' = -1 \qquad J_2'' = 4 \text{ at } t_2'' = 0$$

There are no nonzero jumps of the type J_k''' because $F''' \equiv 0$. Substituting these values into Eq. (6), Sec. 8.3, which arose from our iterative process, we have

$$b_n = \frac{1}{n\pi}(-2)\cos(-n\pi) - \frac{1}{n^3\pi^3}[-4\cos(-n\pi) + 4\cos 0]$$

$$= \frac{2(-1)^{n+1}}{n\pi} - \frac{4}{n^3\pi^3}[(-1)^{n+1} + 1] = \begin{cases} \dfrac{2}{n\pi} - \dfrac{8}{n^3\pi^3} & n \text{ odd} \\[2ex] \dfrac{-2}{n\pi} & n \text{ even} \end{cases}$$

FIGURE 8.9
Even and odd periodic continuations of a function.

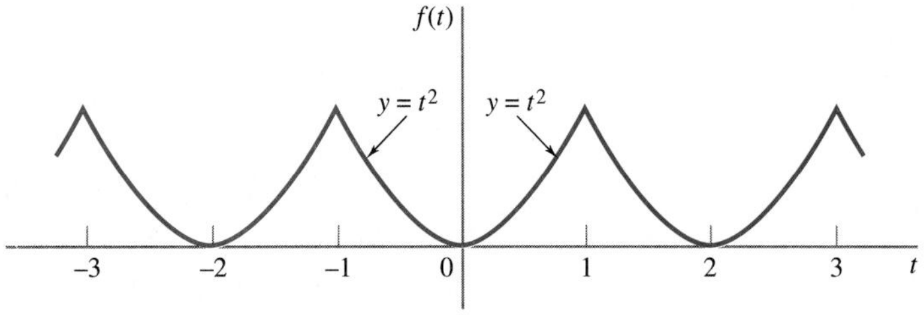

(*a*) The even periodic continuation of the function $f(t) = t^2$, $0 \le t \le 1$.

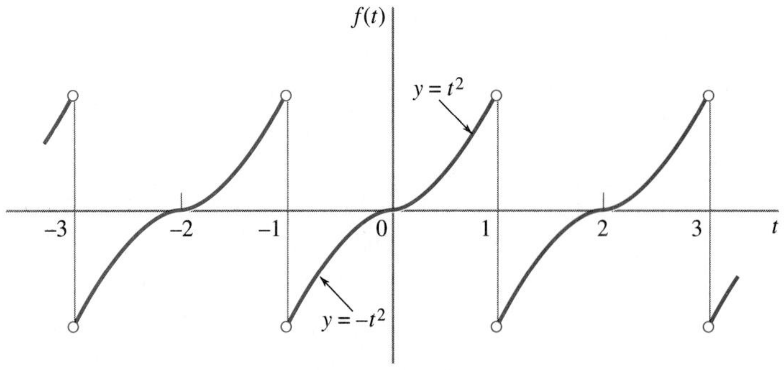

(*b*) The odd periodic continuation of the function $f(t) = t^2$, $0 \le t < 1$.

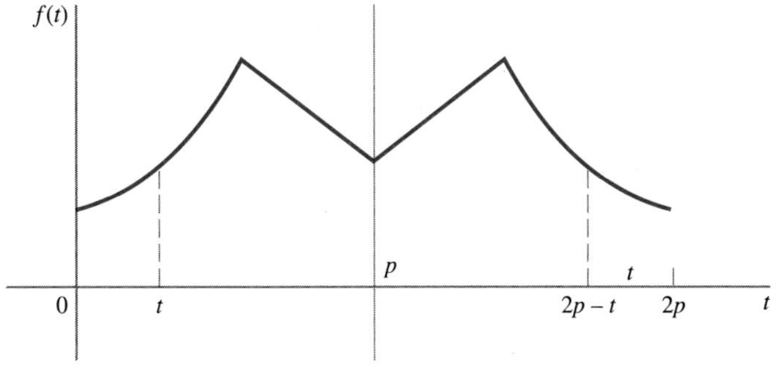

FIGURE 8.10
The extended function discussed in Theorem 3.

Closely associated with half-range cosine and sine series is another special expansion obtained in the following way: given a function f defined originally only for $0 \le t \le p$, let us extend it from p to $2p$ by reflecting it in the line $t = p$. As Fig. 8.10 illustrates, the extended function F is defined by

$$F(t) = \begin{cases} f(t) & 0 \le t \le p \\ f(2p - t) & p \le t \le 2p \end{cases}$$

For the half-range sine expansion of the extended function F we have, by Theorem 2,

$$b_n = \frac{2}{2p}\int_0^{2p} F(t)\sin\frac{n\pi t}{2p}\,dt$$

(3) $$= \frac{1}{p}\int_0^{p} f(t)\sin\frac{n\pi t}{2p}\,dt + \frac{1}{p}\int_p^{2p} f(2p - t)\sin\frac{n\pi t}{2p}\,dt$$

If we make the substitutions $s = 2p - t$ and $ds = -dt$ in the second of the integrals in (3), we obtain

(4) $$\frac{1}{p}\int_p^{0} f(s)\sin\left[\frac{n\pi}{2p}(2p - s)\right](-ds) = \frac{1}{p}\int_0^{p} f(s)\sin\left(n\pi - \frac{n\pi s}{2p}\right)ds$$

Now

$$\sin\left(n\pi - \frac{n\pi s}{2p}\right) = \sin n\pi\cos\frac{n\pi s}{2p} - \cos n\pi\sin\frac{n\pi s}{2p} = -(-1)^n\sin\frac{n\pi s}{2p}$$

Hence the integral (4) becomes

$$-(-1)^n\frac{1}{p}\int_0^{p} f(s)\sin\frac{n\pi s}{2p}\,ds$$

Except for the dummy variable s, this is identical with the first integral in (3) if n is odd and is the negative of the first integral in (3) if n is even. Thus, combining the integrals in the two cases, we have established the following theorem.

THEOREM 3 If $f(t)$, originally defined for $0 \le t \le p$, is extended over the interval $(p, 2p)$ by reflection in the line $t = p$, then for the half-range sine expansion of the extended function,

$$b_n = \begin{cases} 0 & n \text{ even} \\ \dfrac{2}{p}\displaystyle\int_0^{p} f(t)\sin\frac{n\pi t}{2p}\,dt & n \text{ odd} \end{cases}$$

The special series described in Theorem 3 is important because there are problems† in which a function given on an interval $(0, p)$ must be expanded in terms of sines whose angles must all be *odd* multiples of a fundamental angle of the form $\pi t/2p$.

EXERCISES

Use Theorems 1 and 2 to find the half-range cosine and sine expansions of each of the following functions.

1. $f(t) = \begin{cases} 1 & 0 \le t \le \pi \\ 0 & \pi < t \le 2\pi \end{cases}$

2. $f(t) = \begin{cases} 1 & 0 \le t \le \pi \\ 0 & \pi < t \le 4\pi \end{cases}$

3. $f(t) = \begin{cases} 1 - t & 0 \le t \le p \\ 0 & p < t \le 2p \end{cases}$

4. $f(t) = \begin{cases} 1 - t & 0 \le t \le p \\ 0 & p < t \le 3p \end{cases}$

5. A periodic function is defined over one period by

$$f(t) = \begin{cases} 0 & -\pi < t < 0 \\ 2a & 0 < t < \pi \end{cases}$$

Verify that $g(t) = f(t) - a$ is an odd function and use this fact to obtain the Fourier expansion of f.

6. What is the geometric significance, if any, of a Fourier expansion in which $a_n = 0$ for $n = 1, 2, 3, \ldots$, but $a_0 \ne 0$?

7. Prove Theorem 2.

Find the Fourier expansions of the periodic functions whose definitions on one period are as follows.

8. $f(t) = \begin{cases} -a & -p < t < 0 \\ a & 0 < t < p \end{cases}$

9. $f(t) = \begin{cases} a & 0 \le |t| < \pi \\ b & \pi < |t| \le 2\pi \end{cases}$

10. $f(t) = \begin{cases} e^{-t} & -1 \le t \le 0 \\ e^{t} & 0 \le t \le 1 \end{cases}$

11. $f(t) = \begin{cases} -\cos \pi t & -1 < t < 0 \\ \cos \pi t & 0 < t < 1 \end{cases}$

Hint: In Exercises 12–15, f is the sum of an odd and an even function.

12. $f(t) = t + t^2 \qquad -1 < t < 1$

13. $f(t) = |t| + \sinh t \qquad -1 < t < 1$

14. $f(t) = t + \cos^4 \pi t \qquad -1 < t < 1$

15. $f(t) = t^2 + \sin^3 \pi t \qquad -1 \le t \le 1$

Find the half-range sine and cosine expansions of each of the following functions.

16. $f(t) = 1 \qquad 0 < t < p$

17. $f(t) = t \qquad 0 \le t < l$

18. $f(t) = \cos t \qquad 0 < t < 2\pi$

19. $f(t) = t(l - t) \qquad 0 \le t \le l$

20. $f(t) = \begin{cases} 0 & 0 \le t < 1 \\ 1 & 1 \le t < 2 \end{cases}$

21. $f(t) = \begin{cases} at & 0 \le t \le l/2 \\ a(l - t) & l/2 \le t \le l \end{cases}$

22. $f(t) = e^{-t} \qquad 0 < t < p$

23. $f(t) = \sin t \qquad 0 \le t \le 2\pi$

24. $f(t) = \cos at \qquad 0 < t < \pi$, a not an integer

25. $f(t) = \sin at \qquad 0 \le t < \pi$, a not an integer

26. By setting $t = \pi$ in the half-range cosine expansion in Exercise 24, show that

$$\cot a\pi = \frac{1}{a\pi} - \frac{2a}{\pi}\sum_{n=1}^{\infty}\frac{1}{n^2 - a^2} = \frac{1}{\pi}\sum_{-\infty}^{\infty}\frac{1}{n + a}$$

27. Using the results of Exercises 25 and 26, show that

$$\csc a\pi = \frac{1}{a\pi} - \frac{2a}{\pi}\sum_{n=1}^{\infty}\frac{(-1)^n}{n^2 - a^2} = \frac{1}{\pi}\sum_{-\infty}^{\infty}\frac{(-1)^n}{n + a}$$

Obtain an expansion containing only terms of the form $\sin (n\pi t/2p)$, where n is odd, for each of the following functions.

28. $f(t) = e^{-t} \qquad 0 < t \le p$

29. $f(t) = p - t \qquad 0 < t \le p$

Without calculating the coefficients, determine how fast the coefficients in the half-range expansions of the following functions will decrease.

30. $f(t) = t(1 - t)^2 \qquad 0 \le t \le 1$

31. $f(t) = t^2(1 - t^2)^2 \qquad 0 \le t \le 1$

32. $f(t) = \begin{cases} t^4 - 3t^3 + 2t^2 + t & 0 \le t \le 1 \\ 2t - t^2 & 1 \le t \le 2 \end{cases}$

33. $f(t) = \dfrac{1}{2 + \cos t} \qquad 0 \le t \le 2\pi$

34. Extend the function $f(t) = t$, $0 \le t \le 1$, from $t = 1$ to $t = 2$ in such a way that the half-range sine expansion of the extended function will have coefficients decreasing as c/n^4.

35. Determine how $f(t)$, originally defined only for $0 \le t \le p$, must be extended from p to $2p$ if the half-range cosine expansion of the extended function is to contain no terms

†Examples of such problems will be found in Sec. 11.5.

of the form $\cos(n\pi t/p)$, n even. Derive a formula for the nonzero coefficients.

36. Using the results of Example 1, show that

$$1 - \frac{1}{3^3} + \frac{1}{5^3} - \frac{1}{7^3} + \cdots = \frac{\pi^3}{32}.$$

37. Show that $1 + \frac{1}{3^4} + \frac{1}{5^4} + \frac{1}{7^4} + \cdots = \frac{\pi^4}{96}$. *Hint:* First integrate the result of Example 1.

38. During the two-second interval $0 \le t \le 2$ a voltage $E(t) = 2t - t^2$ is impressed upon an electric circuit. Find the half-range sine expansion of E.

39. A horizontal beam of length l bears a load $w(x)$ equal to

$$w(x) = \begin{cases} 0 & 0 < x < l/4 \\ 100 & l/4 < x < l/2 \\ 0 & l/2 < x < 3l/4 \\ 100 & 3l/4 < x < l \end{cases}$$

Find the half-range Fourier sine and cosine series that represent the load w on $0 < x < l$.

40. Find the half-range cosine expansion of the function $f(t) = \sin a(\pi/2 - x)$, $0 < x < \pi$, a not an integer, and from it show that

$$\frac{\pi \sin a\left(\frac{\pi}{2} - x\right)}{4a \cos \frac{a\pi}{2}} = \sum_{n=1}^{\infty} \frac{\cos(2n-1)x}{(2n-1)^2 - a^2}$$

41. In Exercise 40, if the function defined by the series is called $g(a, x)$, show that

$$\frac{g(a_1, x) - g(a_2, x)}{a_1^2 - a_2^2} = \frac{g(a_1, x)}{a_1^2 - a_2^2} + \frac{g(a_2, x)}{a_2^2 - a_1^2}$$

$$= \sum_{n=1}^{\infty} \frac{\cos(2n-1)x}{[(2n-1)^2 - a_1^2][(2n-1)^2 - a_2^2]}$$

42. Extend Exercise 41 by showing that

$$\frac{1}{a_1^2 - a_3^2}\left[\frac{g(a_1, x) - g(a_2, x)}{a_1^2 - a_2^2} - \frac{g(a_2, x) - g(a_3, x)}{a_2^2 - a_3^2}\right]$$

$$= \frac{g(a_1, x)}{(a_1^2 - a_2^2)(a_1^2 - a_3^2)} + \frac{g(a_2, x)}{(a_2^2 - a_1^2)(a_2^2 - a_3^2)} + \frac{g(a_3, x)}{(a_3^2 - a_1^2)(a_3^2 - a_2^2)}$$

$$= \sum_{n=1}^{\infty} \frac{\cos(2n-1)x}{[(2n-1)^2 - a_1^2][(2n-1)^2 - a_2^2][(2n-1)^2 - a_3^2]}$$

8.5 ALTERNATIVE FORMS OF FOURIER SERIES

The original form of the Fourier series of a function, as derived in Sec. 8.2, can be converted into several other trigonometric forms and into one in which imaginary exponentials appear instead of real trigonometric functions. For instance, in the original **standard form**

$$f(t) = \frac{a_0}{2} + \sum_{n=1}^{\infty}\left(a_n \cos\frac{n\pi t}{p} + b_n \sin\frac{n\pi t}{p}\right) \tag{1}$$

we can apply to each pair of terms of the same frequency the usual procedure for reducing the sum of a sine and a cosine of the same angle to a single term:

$$f(t) = \frac{a_0}{2} + \sum_{n=1}^{\infty} \sqrt{a_n^2 + b_n^2}\left(\frac{a_n}{\sqrt{a_n^2 + b_n^2}} \cos\frac{n\pi t}{p} + \frac{b_n}{\sqrt{a_n^2 + b_n^2}} \sin\frac{n\pi t}{p}\right)$$

If we now define the angles γ_n and δ_n from the triangle shown in Fig. 8.11 and set

$$A_0 = \frac{a_0}{2} \quad \text{and} \quad A_n = \sqrt{a_n^2 + b_n^2} \tag{2}$$

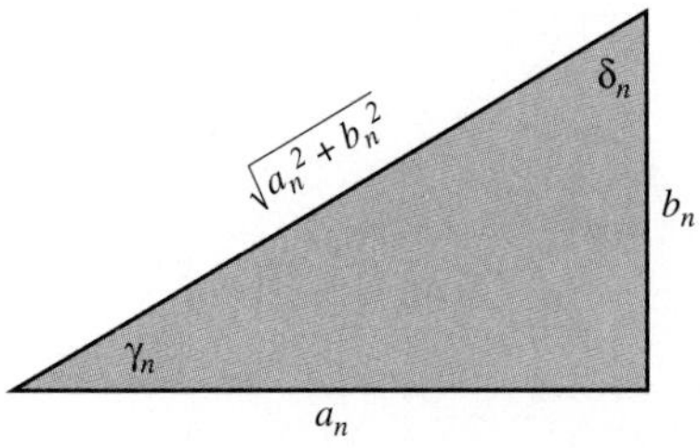

FIGURE 8.11
The triangle defining the phase angles γ_n and δ_n for the resultant of the terms of frequency $n\pi/p$ in a standard Fourier series.

the last series can be written

$$f(t) = A_0 + \sum_{n=1}^{\infty} A_n\left(\cos\frac{n\pi t}{p}\cos\gamma_n + \sin\frac{n\pi t}{p}\sin\gamma_n\right)$$

(3) $$= A_0 + \sum_{n=1}^{\infty} A_n \cos\left(\frac{n\pi t}{p} - \gamma_n\right)$$

or, equally well,

$$f(t) = A_0 + \sum_{n=1}^{\infty} A_n\left(\cos\frac{n\pi t}{p}\sin\delta_n + \sin\frac{n\pi t}{p}\cos\delta_n\right)$$

(4) $$= A_0 + \sum_{n=1}^{\infty} A_n \sin\left(\frac{n\pi t}{p} + \delta_n\right)$$

In the **harmonic cosine series** (3), the **nth harmonic** is $\cos(n\pi t/p - \gamma_n)$; in the **harmonic sine series** (4) it is $\sin(n\pi t/p + \delta_n)$. In either series, the quantity $A_n = \sqrt{a_n^2 + b_n^2}$ is the **resultant amplitude** of the components of frequency $n\pi/p$ in (1), that is, the **amplitude of the nth harmonic** in either harmonic series. The phase angles

(5) $$\gamma_n = \tan^{-1}\frac{b_n}{a_n} \qquad \text{and} \qquad \delta_n = \tan^{-1}\frac{a_n}{b_n} = \frac{\pi}{2} - \gamma_n$$

measure the lag or lead of the nth harmonic with reference to a standard cosine or standard sine wave of the same frequency.

If γ_n is positive, the nth harmonic lags (i.e., reaches its maxima after) the corresponding cosine function $\cos(n\pi t/p)$ by the time interval $\gamma_n p/n\pi$. If γ_n is negative, the nth harmonic leads $\cos(n\pi t/p)$ by this amount. If δ_n is positive, the nth harmonic leads the corresponding sine function $\sin(n\pi t/p)$ by the time interval $\delta_n p/n\pi$. If δ_n is negative, the nth harmonic lags $\sin(n\pi t/p)$ by this amount.

The standard Fourier series (1) is converted into a complex exponential form by substituting their exponential equivalents for the cosine and sine terms:

$$f(t) = \frac{a_0}{2} + \sum_{n=1}^{\infty}\left(a_n\frac{e^{ni\pi t/p} + e^{-ni\pi t/p}}{2} + b_n\frac{e^{ni\pi t/p} - e^{-ni\pi t/p}}{2i}\right)$$

Collecting terms on the various exponentials and noting that $1/i = -i$, we obtain

$$f(t) = \frac{a_0}{2} + \sum_{n=1}^{\infty}\left(\frac{a_n - ib_n}{2}e^{ni\pi t/p} + \frac{a_n + ib_n}{2}e^{-ni\pi t/p}\right)$$

If we now define

(6) $$c_0 = \frac{a_0}{2} \qquad c_n = \frac{a_n - ib_n}{2} \qquad c_{-n} = \frac{a_n + ib_n}{2}$$

the last series can be written in the more symmetric form

(7) $$f(t) = \sum_{n=-\infty}^{\infty} c_n e^{ni\pi t/p}$$

When it is used at all, this exponential form is used as a basic form in its own right; i.e., it is not obtained by transformation from the trigonometric form but is constructed directly from the given function. To do this requires that expressions be available for the *direct* evaluation of the coefficients c_n. These can easily be found from the definitions of c_0, c_n, and c_{-n}. For

$$c_0 = \frac{1}{2}a_0 = \frac{1}{2p}\int_d^{d+2p} f(t)\,dt$$

$$c_n = \frac{a_n - ib_n}{2} = \frac{1}{2}\left[\frac{1}{p}\int_d^{d+2p} f(t)\cos\frac{n\pi t}{p}\,dt - i\frac{1}{p}\int_d^{d+2p} f(t)\sin\frac{n\pi t}{p}\,dt\right]$$

$$= \frac{1}{2p}\int_d^{d+2p} f(t)\left(\cos\frac{n\pi t}{p} - i\sin\frac{n\pi t}{p}\right)dt$$

$$= \frac{1}{2p}\int_d^{d+2p} f(t)e^{-ni\pi t/p}\,dt$$

$$c_{-n} = \frac{a_n + ib_n}{2} = \frac{1}{2}\left[\frac{1}{p}\int_d^{d+2p} f(t)\cos\frac{n\pi t}{p}\,dt + i\frac{1}{p}\int_d^{d+2p} f(t)\sin\frac{n\pi t}{p}\,dt\right]$$

$$= \frac{1}{2p}\int_d^{d+2p} f(t)\left(\cos\frac{n\pi t}{p} + i\sin\frac{n\pi t}{p}\right)dt$$

$$= \frac{1}{2p}\int_d^{d+2p} f(t)e^{ni\pi t/p}\,dt$$

Clearly, whether the index n is positive, negative, or zero, c_n is correctly given by the single formula

$$c_n = \frac{1}{2p}\int_d^{d+2p} f(t)e^{-ni\pi t/p}\,dt \tag{8}$$

As usual, d will almost always be either $-p$ or 0. A series of the form (7), with its coefficients determined by (8), is called a **complex exponential Fourier series.**

Because the summation in Eq. (7) extends over all integers from $-\infty$ to ∞, it seems plausible that the signs in the exponentials in (7) and (8) could be reversed without invalidating the formulas, since every term in the series (7) would still be included with the proper coefficient. In fact, if we replace n by $-n$, (7) and (8) become, respectively,

$$f(t) = \sum_{n=-\infty}^{\infty} c_{-n}e^{-ni\pi t/p} \qquad \text{and} \qquad c_{-n} = \frac{1}{2p}\int_d^{d+2p} f(t)e^{ni\pi t/p}\,dt$$

In these formulas, c_{-n} is just an arbitrary name that can be replaced by any other symbol, say c_n, without altering its meaning or validity. Making this replacement, we obtain the alternative formulas

$$f(t) = \sum_{n=-\infty}^{\infty} c_n e^{-ni\pi t/p} \tag{7a}$$

$$c_n = \frac{1}{2p}\int_d^{d+2p} f(t)e^{ni\pi t/p}\,dt \tag{8a}$$

which tell us that (7) and (8) do, indeed, *remain valid if the signs in their exponents are reversed.*

Despite the obvious differences between Eqs. (7) and (8) [or between (7*a*) and (8*a*)], their definitive roles and mode of composition are much alike. Given f, a function c such that $c(n) = c_n$ is defined over the integers by Eq. (8) [or (8*a*)]; given a function c such that $c(n) = c_n$, the function f is determined by Eq. (7) [or (7*a*)]. Equation (8) involves the continuous variable t and so, perforce, integration is involved; Eq. (7) involves the discrete variable n and so, necessarily, summation rather than integration is required. In (8), $f(t)$ is multiplied by $e^{-ni\pi t/p}$; in (7), c_n is multiplied by the reciprocal factor $e^{ni\pi t/p}$. Taken together, Eqs. (7) and (8) constitute two characterizations of the function f: Eq. (7) represents f in the time domain; Eq. (8) characterizes f (through its Fourier coefficients) in the discrete domain of integers. Later on, in our study of Fourier integrals and Laplace transforms, we shall discuss in detail other reciprocal pairs of functions analogous to the relation established between $f(t)$ and c_n by Eqs. (7) and (8).

EXAMPLE 1

Find the complex form of the Fourier series of the periodic function whose definition in one period is $f(t) = e^{-t}$, $-1 < t < 1$.

Since $p = 1$, we have from (8), taking $d = -1$,

$$c_n = \frac{1}{2}\int_{-1}^{1} e^{-t}e^{-ni\pi t}\,dt = \frac{1}{2}\left[\frac{e^{-(1+ni\pi)t}}{-(1+ni\pi)}\right]_{-1}^{1}$$

$$= \frac{e^{-(1+ni\pi)} - e^{(1+ni\pi)}}{-2(1+ni\pi)}$$

$$= \frac{ee^{ni\pi} - e^{-1}e^{-ni\pi}}{2(1+ni\pi)}$$

Now $e^{i\pi} = \cos\pi + i\sin\pi = -1$, and thus $e^{ni\pi} = e^{-ni\pi} = (-1)^n$. Hence

$$c_n = \frac{(-1)^n}{(1+ni\pi)}\,\frac{e - e^{-1}}{2} = \frac{(-1)^n(1-ni\pi)\sinh 1}{1+n^2\pi^2}$$

The expansion of $f(t)$ is therefore

(9) $$f(t) = \sum_{n=-\infty}^{\infty} (-1)^n \frac{(1-ni\pi)\sinh 1}{1+n^2\pi^2} e^{ni\pi t}$$

The relations (6) can easily be solved for a_n and b_n. Specifically, upon listing the value of a_0, then adding and next subtracting the expressions for c_n and c_{-n}, we find that

(10) $$a_0 = 2c_0 \qquad a_n = c_n + c_{-n} \qquad b_n = i(c_n - c_{-n})$$

These formulas are useful when it is desired to revert from the complex exponential form of a Fourier series to a real trigonometric form.

EXAMPLE 2

Convert the exponential series (9) of Example 1 into a real trigonometric form.

Using the coefficient of $e^{ni\pi t}$ in (9) to substitute for c_0, c_n, and c_{-n} in Eqs. (10), we get

$$a_0 = 2\sinh 1$$

$$a_n = \frac{(-1)^n(1-ni\pi)\sinh 1}{1+n^2\pi^2} + \frac{(-1)^n(1+ni\pi)\sinh 1}{1+n^2\pi^2} = \frac{(-1)^n\, 2\sinh 1}{1+n^2\pi^2}$$

$$b_n = i\left[\frac{(-1)^n(1-ni\pi)\sinh 1}{1+n^2\pi^2} - \frac{(-1)^n(1+ni\pi)\sinh 1}{1+n^2\pi^2}\right] = \frac{(-1)^n 2n\pi\sinh 1}{1+n^2\pi^2}$$

Hence a real form of (9) is

$$f(t) = \sinh 1\left[1 + 2\sum_{n=1}^{\infty}\frac{(-1)^n}{1+n^2\pi^2}(\cos n\pi t + n\pi \sin n\pi t)\right] \tag{11}$$

In some problems, chiefly in electrical engineering, it is convenient to have the **complex Fourier coefficients** c_n and c_{-n} expressed in exponential form. To do this, we recall from Sec. 3.1 that any complex number $z = x + iy$ can be written in the exponential form $z = re^{i\theta}$, where $r = |z| = \sqrt{x^2+y^2}$ and $\theta = \arg z = \tan^{-1}(y/x)$. Then from (2), (6), and (5) we have

$$c_0 = A_0 \qquad |c_n| = |c_{-n}| = \tfrac{1}{2}\sqrt{a_n^2+b_n^2} = \tfrac{1}{2}A_n$$

$$\arg c_{-n} = \tan^{-1}\left(\frac{b_n}{a_n}\right) = \gamma_n \qquad \text{and} \qquad \arg c_n = \tan^{-1}\left(\frac{-b_n}{a_n}\right) = -\gamma_n$$

Finally, combining these results, we have

$$c_0 = A_0 \qquad c_n = \tfrac{1}{2}A_n e^{-i\gamma_n} \qquad c_{-n} = \tfrac{1}{2}A_n e^{i\gamma_n} \tag{12}$$

EXERCISES

What is the resultant amplitude and what are the phase angles of the terms of frequency $n\pi/p$ in the harmonic cosine and sine series of each of the following functions with periods as indicated?

1. $f(t) = \begin{cases} 1 & 0 < t < 1 \\ 0 & 1 < t < 4 \end{cases}$

2. $f(t) = \begin{cases} 0 & -p \le t \le 0 \\ t & 0 \le t < p \end{cases}$

3. $f(t) = \begin{cases} 0 & -p < t < 0 \\ e^{-t} & 0 < t < p \end{cases}$

4. $f(t) = \begin{cases} 0 & -p \le t \le 0 \\ pt - t^2 & 0 \le t \le p \end{cases}$

5. $f(t) = e^t \quad -p < t < p$

6. $f(t) = t + t^2 \quad -1 < t < 1$

7. Verify the assertions made in the first paragraph after Eqs. (5) concerning the significance of the phase angles γ_n and δ_n.

Find the complex exponential Fourier series of the periodic functions whose definitions in one period are

8. $f(t) = \begin{cases} 1 & 0 < t < 1 \\ 0 & 1 < t < 2 \end{cases}$

9. $f(t) = \begin{cases} 1 & 0 < t < 1 \\ -1 & 1 < t < 2 \end{cases}$

10. $f(t) = t \quad 0 < t < 1$

11. $f(t) = e^{-t} \quad -p < t < p$

12. $f(t) = \cos t \quad -\pi/2 \le t \le \pi/2$. *Hint:* Use the fact that $\cos t = \frac{1}{2}(e^{it} + e^{-it})$.

13. $f(t) = \sin t \quad 0 \le t \le \pi$

14. $f(t) = \sinh t \quad -1 < t < 1$

15. $f(t) = \cosh t \quad -1 \le t \le 1$

16. $f(t) = \begin{cases} 2 & 0 < t < 2 \\ 1 & 2 < t < 4 \end{cases}$

Convert each of the following complex exponential Fourier series into a real trigonometric series.

17. The complex series of Exercise 16.

18. $\displaystyle\sum_{n=-\infty}^{\infty}\frac{-4i}{\pi^3(2n-1)^3}e^{(2n-1)i\pi t}$

19. $\displaystyle\frac{\pi}{4} + \sum_{n=-\infty}^{\infty}\left(\frac{\cos n\pi - 1}{2\pi n^2} + i\frac{3\cos n\pi}{2n}\right)e^{int}$.

Note: $c_0 = \dfrac{\pi}{4}$ is the term for which $n = 0$.

20. Write the coefficients in the series in Exercise 18 in exponential form.

21. Write the coefficients in the series in Exercise 19 in exponential form.

22. Show that the amplitude of the nth harmonic in the harmonic cosine or sine series of a periodic function $f(t)$ is equal to $2\sqrt{c_n c_{-n}}$.

23. Find the complex exponential form of the Fourier series of the periodic function

$$F(t) = \begin{cases} 0 & -p/2 \le t < -a/2 \\ 1 & -a/2 < t < a/2 \\ 0 & a/2 < t \le p/2 \end{cases}$$

This important function is called a **pulse train of width** a **and period** p.

8.6 APPLICATIONS OF FOURIER SERIES

Fourier series play an important role in the solution of many problems in partial differential equations, and we shall see numerous examples of this in Chap. 11. There are many other applications of course and in this section we shall consider several that are relevant to the present stage of our work. We begin with two examples that involve the analysis of physical systems subjected to general periodic disturbances. Though the use of Fourier series is new, the techniques applied are not, for they entail, in effect, just straightforward extensions of the ideas of Chap. 7.

EXAMPLE 1

A SPRING-MASS SYSTEM DRIVEN BY AN ALTERNATING SQUARE-WAVE FORCE

Determine the steady-state forced vibrations of the system shown in Fig. 8.12*a* if the applied force $F(t)$ is that shown in Fig. 8.12*b*.

Since the concepts we developed in our study of forced vibrations in Sec. 7.3 are applicable only to periodic functions that are simple sines and cosines, our first step must be to express the driving force $F(t)$ in terms of such functions; i.e., our first step must be to determine the Fourier expansion of $F(t)$. Clearly, an odd extension of the function is suggested by the graph of $F(t)$. Since no cosine terms are ever present in a half-range sine expansion, we need only compute b_n:

$$b_n = \frac{2}{\frac{1}{2}}\int_0^{1/2} 20 \sin\frac{n\pi t}{\frac{1}{2}}\,dt = 80\left[-\frac{\cos 2n\pi t}{2n\pi}\right]_0^{1/2} = 40\frac{1-\cos n\pi}{n\pi}$$

$$= \begin{cases} 0 & n \text{ even} \\ \dfrac{80}{n\pi} & n \text{ odd} \end{cases}$$

Hence

$$F(t) = \frac{80}{\pi}\left(\sin 2\pi t + \frac{\sin 6\pi t}{3} + \frac{\sin 10\pi t}{5} + \frac{\sin 14\pi t}{7} + \cdots\right) \text{ lb}$$

and this is the expression that would appear on the right-hand side of the differential equation describing the motion of the system if we were to set up the equation. Now we are asked only to find the *steady-state* forced motion of the system; hence we need to determine only the particular integral corresponding to $F(t)$. Moreover, since the differential equation (even though we have not set it up) is obviously linear, the required particular integral can be found very simply by using the ideas of Sec. 7.3. In fact, all we need to do is apply the proper magnification and phase shift to each component of the driving force $F(t)$ and add the results. Preparatory to this, we must determine the static deflections that would be produced in the system by steady forces having the magnitudes of the various terms of $F(t)$. These are equal to

$$(\delta_{st})_n = \frac{80/n\pi \text{ lb}}{100 \text{ lb/in}} = \frac{4}{5n\pi} \text{ in} \qquad n \text{ odd}$$

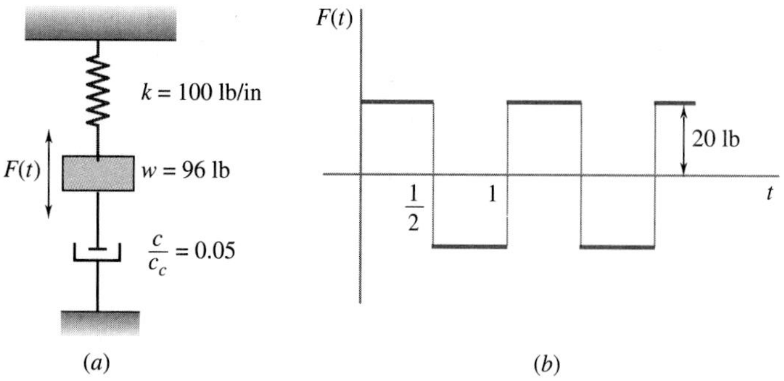

FIGURE 8.12
A spring-mass system acted upon by an alternating square-wave force.

TABLE 8.1

Term	δ_{st}	$\dfrac{\omega}{\omega_N}$	$M = \dfrac{1}{\sqrt{\left(1-\dfrac{\omega^2}{\omega_N^2}\right)^2+\left(2\dfrac{c}{c_c}\dfrac{\omega}{\omega_N}\right)^2}}$	$\alpha = \tan^{-1}\dfrac{2\dfrac{c}{c_c}\dfrac{\omega}{\omega_N}}{1-\dfrac{\omega^2}{\omega_N^2}}$	Steady-state term = $\delta_{st} M \sin(\omega t - \alpha)$
1	$\dfrac{4}{5\pi}$	$\dfrac{2\pi}{20}$	1.11	0.035 rad	$0.28 \sin(2\pi t - 0.035)$
2	$\dfrac{4}{15\pi}$	$\dfrac{6\pi}{20}$	6.83	0.701 rad	$0.58 \sin(6\pi t - 0.701)$
3	$\dfrac{4}{25\pi}$	$\dfrac{10\pi}{20}$	0.68	3.035 rad	$0.03 \sin(10\pi t - 3.035)$
4	$\dfrac{4}{35\pi}$	$\dfrac{14\pi}{20}$	0.26	3.084 rad	$0.01 \sin(14\pi t - 3.084)$
...	...	...	...	...	...

Then we must calculate the natural frequency of the system:

$$\omega_N\dagger = \sqrt{\frac{kg}{w}} = \sqrt{\frac{100 \times 384}{96}} = 20 \text{ rad/s}$$

The rest of the work can best be presented in tabular form (see Table 8.1). Figure 8.13 shows one period of the steady-state motion of the system plotted as a function of time.

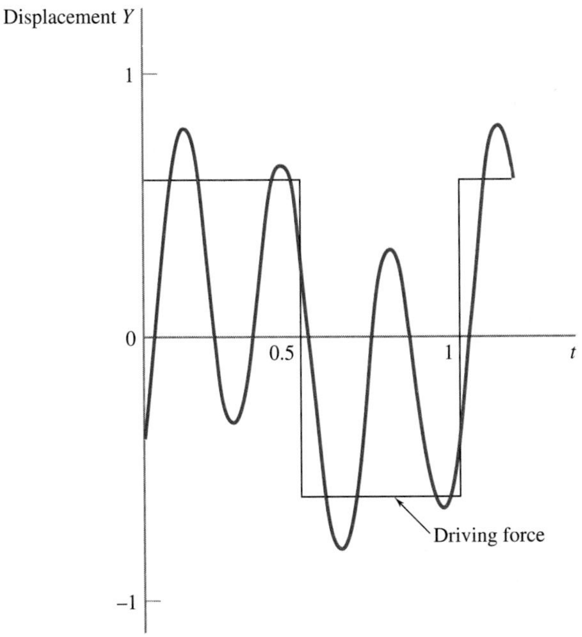

FIGURE 8.13
A response of apparent frequency greater than that of the excitation producing it.

†We must remember that here the subscript N in ω_N stands for *natural* and is in no way connected with the parameter n which identifies the general term in the Fourier expansion of $F(t)$.

This example illustrates an exceedingly important but sometimes misunderstood characteristic of forced vibrations. In general, the frequency of the response will be the same as the frequency of the excitation. Thus despite appearances in Fig. 8.13, the actual frequency of the displacement Y is one cycle per second because the period of the driving force is 1 s. Nonetheless, as the figure shows, an observer would see the weight moving through three pronounced and seemingly complete cycles during each cycle of the driving force. The reason for this is the following: if the excitation is periodic but is not a pure sine or cosine function, its Fourier expansion will contain terms whose frequencies are above the fundamental or apparent frequency of the excitation. Moreover, each of these terms, as a component of the driving force, has its own magnification ratio, phase shift, and corresponding steady-state term, as indicated in Table 8.1. If the frequency of one of these harmonics is close to the resonant frequency [Eq. (12*b*), Sec. 7.3]. and if the amount of friction is small, the corresponding magnification ratio will be large and its value may offset many times the smaller amplitude and corresponding static deflection of that harmonic. As a result, the steady-state term it contributes to the entire response (i.e., to the sum of all the steady-state terms) will be the dominant part of the response, and the system will appear to vibrate at a frequency which is an integral multiple of the frequency of the excitation. For instance, in the present example, the second harmonic of the excitation is $(80/3\pi)\sin 6\pi t$, its frequency of 6π rad/s is close to the resonant frequency

$$\omega_r = [20\sqrt{1 - 2(0.05)^2} \doteq 19.95] \text{ rad/s} \doteq 3.2 \text{ Hz}$$

of the mechanical system, its magnification ratio of 6.83 is more than six times larger than that of any other harmonic, and as Fig. 8.13 clearly shows, although the driving force alternates only once per second, the weight is seen to move up and down three times per second. Behavior such as this is sometimes called **harmonic resonance.** The related phenomenon in which one cycle of the response extends over *more* than one cycle of the excitation is known as **subharmonic resonance** and can occur only in nonlinear systems.

It is interesting to note that although the driving force in this example is discontinuous, both the displacement and velocity it produces are continuous. This is suggested by the plot of the displacement shown in Fig. 8.13 and confirmed by an application of Theorem 3, Sec. 8.3. In fact, since the circular frequency of the general term in the Fourier expansion of the driving force $F(t)$ is $(2n - 1)2\pi \sim 4n\pi$, it follows by neglecting all but the highest power of n in the denominator of the magnification ratio M, that for n sufficiently large, M is arbitrarily close to

$$\frac{1}{\omega^2/\omega_N^2} \sim \frac{1}{(4n\pi)^2/(20)^2} = \frac{25}{n^2\pi^2}$$

Therefore, since the static deflection corresponding to the amplitude of the general term in the expansion of $F(t)$ is

$$(\delta_{st})_n = \frac{4}{5(2n - 1)\pi} \sim \frac{2}{5n\pi}$$

it follows that as n becomes infinite, the coefficient of the general term in the expansion of the steady-state displacement, namely, $(\delta_{st})_n M$, tends to zero as

$$\frac{2}{5n\pi}\frac{25}{n^2\pi^2} = \frac{10}{n^3\pi^3}$$

Thus, according to Theorem 3, Sec. 8.3, the steady-state displacement $Y(t)$ and the steady-state velocity $\dot{Y}(t)$ are continuous, but the acceleration $\ddot{Y}(t)$ is discontinuous.

Of course, independent of this reasoning, the acceleration must be discontinuous since, from Newton's law,

$$\text{Acceleration} = \frac{\text{force}}{\text{mass}}$$

and in this problem the given force is discontinuous.

EXAMPLE 2 A SERIES CIRCUIT WITH A SQUARE-WAVE VOLTAGE

Find the steady-state current produced in the circuit shown in Fig. 8.14*a* by the periodic voltage shown in Fig. 8.14*b*.

As in Example 1, our first step here is to determine the Fourier expansion of the impressed voltage. However, since we plan to use the complex impedance to find the steady-state current, we use the complex exponential rather than the real trigonometric form of the Fourier series. Hence we compute

$$c_n = \frac{1}{0.01}\int_0^{0.005} E_0 e^{-ni\pi t/0.005}\,dt = 100E_0 \frac{e^{-ni\pi t/0.005}}{-ni\pi/0.005}\bigg|_0^{0.005}$$

$$= E_0 \frac{1-e^{-ni\pi}}{2ni\pi}$$

$$= \begin{cases} 0 & n \text{ even } n \neq 0 \\ \dfrac{E_0}{ni\pi} = -\dfrac{iE_0}{n\pi} & n \text{ odd} \end{cases}$$

$$c_0 = \frac{1}{0.01}\int_0^{0.005} E_0\,dt = \frac{E_0}{2}$$

Therefore

$$E(t) = E_0\left(\cdots + \frac{ie^{-600i\pi t}}{3\pi} + \frac{ie^{-200i\pi t}}{\pi} + \frac{1}{2} - \frac{ie^{200i\pi t}}{\pi} - \frac{ie^{600i\pi t}}{3\pi} - \cdots\right)$$

In Sec. 7.4 we showed that the steady-state current produced by a voltage of the form $Ae^{i\omega t}$ can be found simply by dividing the voltage by the complex impedance

$$Z(\omega) = R + i\left(\omega L - \frac{1}{\omega C}\right)$$

Using the data of the present problem, we have

$$Z(\omega) = 250 + i\left(0.02\omega - \frac{10^6}{2\omega}\right)$$

or, since

$$\omega = \omega_n = 200n\pi \qquad n \text{ odd}$$

we have

$$Z(\omega_n) \equiv Z_n = 250 + i\left(4n\pi - \frac{2{,}500}{n\pi}\right) \qquad n \text{ odd}$$

FIGURE 8.14
A series circuit driven by a square-wave voltage.

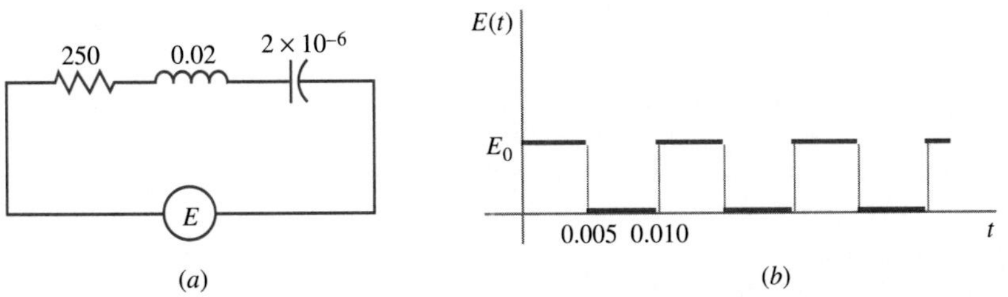

Hence, dividing each term in the expansion of the voltage $E(t)$ by the value of Z for the corresponding frequency (i.e., the corresponding value of n), we find

$$I(t) = \sum_{n=-\infty}^{\infty} D_n e^{200ni\pi t} \qquad n \text{ odd}†$$

where

$$D_n = \frac{c_n}{Z_n} = -\frac{iE_0}{n\pi}\frac{1}{250 + i(4n\pi - 2{,}500/n\pi)} = \frac{-iE_0}{250n\pi + i(4n^2\pi^2 - 2{,}500)}$$

If we want the real trigonometric form of this expansion, namely,

$$I(t) = \frac{a_0}{2} + a_1 \cos 200\pi t + a_3 \cos 600\pi t + \cdots + b_1 \sin 200\pi t + b_3 \sin 600\pi t + \cdots$$

we have at once

$$a_n = D_n + D_{-n} = -iE_0\left[\frac{1}{250n\pi + i(4n^2\pi^2 - 2{,}500)} + \frac{1}{-250n\pi + i(4n^2\pi^2 - 2{,}500)}\right]$$

$$= -\frac{2E_0(4n^2\pi^2 - 2{,}500)}{(250n\pi)^2 + (4n^2\pi^2 - 2{,}500)^2} \qquad n \text{ odd}$$

$$b_n = i(D_n - D_{-n}) = E_0\left[\frac{1}{250n\pi + i(4n^2\pi^2 - 2{,}500)} + \frac{1}{-250n\pi + i(4n^2\pi^2 - 2{,}500)}\right]$$

$$= \frac{500n\pi E_0}{(250n\pi)^2 + (4n^2\pi^2 - 2{,}500)^2} \qquad n \text{ odd}$$

EXAMPLE 3

POWER DELIVERED BY A PERIODIC CURRENT

If f and f^* are two periodic functions with the same fundamental period $2p$, express the average value of their product in terms of the amplitudes and phase angles of the terms in the harmonic cosine series of the two functions. Using this result, find a constant direct current which will deliver the same power through a resistance R as a given periodic current.

We begin by letting

$$f(t) = \frac{1}{2}a_0 + \sum_{n=1}^{\infty}\left(a_n \cos\frac{n\pi t}{p} + b_n \sin\frac{n\pi t}{p}\right)$$

$$= A_0 + \sum_{n=1}^{\infty} A_n \cos\left(\frac{n\pi t}{p} - \gamma_n\right) = \sum_{n=-\infty}^{\infty} c_n e^{ni\pi t/p}$$

$$f^*(t) = \frac{1}{2}a_0^* + \sum_{n=1}^{\infty}\left(a_n^* \cos\frac{n\pi t}{p} + b_n^* \sin\frac{n\pi t}{p}\right)$$

$$= A_0^* + \sum_{n=1}^{\infty} A_n^* \cos\left(\frac{n\pi t}{p} - \gamma_n^*\right) = \sum_{n=-\infty}^{\infty} c_n^* e^{ni\pi t/p}$$

be the Fourier expansions of f and f^*, relations between the coefficients and phase angles in the various equivalent representations being given by Eqs. (2), (5), (6), and (10), Sec. 8.5. Since f and f^* have the same fundamental period, it is clearly sufficient to find the average of their product over a single period. This we do

†Because of the presence of the capacitor, the impedance Z_0 for the dc component, or component of zero frequency, is infinite. Hence the term $E_0/2$ in the expansion of $E(t)$ makes no contribution to the steady-state current.

as usual, by integrating ff^* over one period and dividing by the length of a period. Choosing the particular period $(0, 2p)$ and using the complex form of each series, we have for the **average value of** ff^*,

$$P = \frac{1}{2p}\int_0^{2p} f(t)f^*(t)\,dt = \frac{1}{2p}\int_0^{2p}\left(\sum_{n=-\infty}^{\infty} c_n e^{ni\pi t/p}\right)\left(\sum_{m=-\infty}^{\infty} c_m^* e^{mi\pi t/p}\right) dt$$

where the index of summation in the second series has been changed from n to m to avoid confusion when the two series are multiplied.

When the product of the series is formed, each term of either series is multiplied by every term of the other series. In other words, the general term in the product is

$$(c_n e^{ni\pi t/p})(c_m^* e^{mi\pi t/p}) = c_n c_m^* e^{(m+n)i\pi t/p}$$

and the product itself is a double series

$$f(t)f^*(t) = \sum_{m=-\infty}^{\infty}\sum_{n=-\infty}^{\infty} c_n c_m^* e^{(m+n)i\pi t/p}$$

The integral for the average value thus becomes

$$P = \frac{1}{2p}\int_0^{2p}\left(\sum_{m=-\infty}^{\infty}\sum_{n=-\infty}^{\infty} c_n c_m^* e^{(m+n)i\pi t/p}\right) dt$$

or, integrating term by term,

$$P = \frac{1}{2p}\sum_{m=-\infty}^{\infty}\sum_{n=-\infty}^{\infty}\left(c_n c_m^* \int_0^{2p} e^{(m+n)i\pi t/p}\,dt\right)$$

For $m + n \neq 0$, we have

$$\int_0^{2p} e^{(m+n)i\pi t/p}\,dt = \left.\frac{e^{(m+n)i\pi t/p}}{(m+n)i\pi/p}\right|_0^{2p} = \frac{e^{(m+n)2i\pi} - 1}{(m+n)i\pi/p} = 0$$

since $e^{2i\pi} = \cos 2\pi + i\sin 2\pi = 1$. If $m + n = 0$, i.e., when $m = -n$, the integral is simply

$$\int_0^{2p} dt = 2p$$

Since nonzero terms arise only when $m = -n$, the last double sum becomes effectively a single sum, and the expression for P is just

$$P = \sum_{n=-\infty}^{\infty} c_n c_{-n}^* = c_0 c_0^* + \sum_{n=1}^{\infty} (c_n c_{-n}^* + c_{-n} c_n^*)$$

To express P in terms of the amplitudes and phase angles of the cosine harmonics of f and f^*, we next use the exponential forms of c_n and c_n^* as given by Eqs. (12), Sec. 8.5. This gives

$$P = A_0 A_0^* + \sum_{n=1}^{\infty}\left[\left(\frac{1}{2}A_n e^{-i\gamma_n}\right)\left(\frac{1}{2}A_n^* e^{i\gamma_n^*}\right) + \left(\frac{1}{2}A_n e^{i\gamma_n}\right)\left(\frac{1}{2}A_n^* e^{-i\gamma_n^*}\right)\right]$$

$$= A_0 A_0^* + \frac{1}{2}\sum_{n=1}^{\infty} A_n A_n^* \frac{e^{i(\gamma_n - \gamma_n^*)} + e^{-i(\gamma_n - \gamma_n^*)}}{2}$$

$$(1) \qquad = A_0 A_0^* + \frac{1}{2}\sum_{n=1}^{\infty} A_n A_n^* \cos(\gamma_n - \gamma_n^*)$$

for the period average P.

In the usual applications in electrical engineering, one of the functions, say f^*, is a voltage E, the other f is a current I, and P stands for the **average power per cycle** delivered by the current. As a consequence, P is called the **power product** of the voltage and current. The factor $\cos(\gamma_n - \gamma_n^*)$ in the nth term of (1) is called the **power factor** of the nth harmonic. Clearly, if $\gamma_n - \gamma_n^* = \pi/2$, then $\cos(\gamma_n - \gamma_n^*) = 0$ and no power is delivered by the nth harmonic of the current. When this occurs, the nth harmonics of the voltage and current are said to be in **quadrature.**

If the impedance through which the current flows is a pure resistance R, then $E(t) = RI(t)$ and it follows that the integral for the power product is simply

$$P = R\left[\frac{1}{2p}\int_0^{2p} I^2(t)\,dt\right]$$

Now the **root mean square value** of an integrable function g over an interval (a, b) in the domain of g, abbreviated g_{rms}, is defined by

$$g_{\text{rms}} = \left[\frac{1}{b-a}\int_a^b g^2(t)\,dt\right]^{1/2}$$

Therefore, with I_{rms} computed over $(0, 2p)$, the power product is given by

$$P = RI_{\text{rms}}^2$$

On the other hand, Since $E = IR$, we have $A_n^* = RA_n$ and $\gamma_n^* = \gamma_n$. Thus (1) becomes

$$P = R\left(A_0^2 + \frac{1}{2}\sum_{n=1}^{\infty} A_n^2\right) = R\left[\frac{a_0^2}{4} + \frac{1}{2}\sum_{n=1}^{\infty}(a_n^2 + b_n^2)\right]$$

Equating RI_{rms}^2 to the last expression for P, then solving for I_{rms}, we get

$$I_{\text{rms}} = \frac{1}{2}\sqrt{a_0^2 + 2\sum_{n=1}^{\infty}(a_n^2 + b_n^2)}$$

This gives the value of a constant, or dc, current which over any number of complete cycles would deliver the same power through a resistor as the actual periodic current I.

As Fig. 8.3 suggests, and as we certainly expect, the Nth partial sum of a Fourier series

(2) $$S_N = \frac{1}{2}a_0 + \sum_{n=1}^{N}\left(a_n \cos\frac{n\pi t}{p} + b_n \sin\frac{n\pi t}{p}\right)$$

is a good approximation, even for moderate values of N, to the function f represented by the series. Moreover, as N increases, S_N provides a better and better approximation to f and in the limit, according to Theorem 1, Sec. 8.2, S_N converges to $f(t)$ at every point where f is continuous. Nonetheless, there remains the question of whether or not some other (finite) sum of the same trigonometric form but with different coefficients, say

(3) $$\bar{S}_n(t) = \frac{1}{2}\bar{a}_0 + \sum_{n=1}^{N}\left(\bar{a}_n \cos\frac{n\pi t}{p} + \bar{b}_n \sin\frac{n\pi t}{p}\right)$$

might not be a still better approximation. Before we can decide this matter, we must first agree on how we are going to measure the accuracy with which S_N and $\bar{S}_N$ approximate f. Naturally, we want a measure that reflects goodness of fit over an entire period, say $(0, 2p)$.

One such measure is the *least-square criterion,* which arises as a natural extension of our discussion of the method of least squares in Sec. 5.5. It is almost universally accepted and is the one we shall use.

THE LEAST-SQUARE CRITERION The trigonometric partial sum $\overline{S}_N$ which best approximates a periodic function f over one full period is the one for which the **total squared error**

$$\overline{E} = \int_0^{2p} [f(t) - \overline{S}_N(t)]^2 \, dt \tag{4}$$

is a minimum.

EXAMPLE 4

LEAST-SQUARE PROPERTIES OF FOURIER SERIES

Show that among all trigonometric sums (3), the Nth partial sum (2) of the Fourier series of a periodic function f is, for every value of N the best least-square approximation to f over one period (and hence over any number of complete cycles).

Our first step is to compute $\overline{E}$ and then compare it with the error

$$E = \int_0^{2p} [f(t) - S_N(t)]^2 \, dt \tag{5}$$

associated with the particular sum $S_N(t)$.

From (4), we have

$$\overline{E} = \int_0^{2p} [f^2(t) - 2f(t)\overline{S}_N(t) + \overline{S}_N^2(t)] \, dt$$

$$= \int_0^{2p} f^2(t) \, dt - 2\int_0^{2p} f(t)\left[\frac{\overline{a}_0}{2} + \sum_{n=1}^{N}\left(\overline{a}_n \cos\frac{n\pi t}{p} + \overline{b}_n \sin\frac{n\pi t}{p}\right)\right] dt \tag{6}$$

$$+ \int_0^{2p}\left[\frac{\overline{a}_0}{2} + \sum_{n=1}^{N}\left(\overline{a}_n \cos\frac{n\pi t}{p} + \overline{b}_n \sin\frac{n\pi t}{p}\right)\right]^2 dt$$

Recalling from Eqs. (9)–(11), Sec. 8.2, the Euler formulas for the Fourier coefficients of a function f, it is clear that for $n \geq 1$ the integrals in the second term of (6) are, respectively, $p\overline{a}_n a_n$ and $p\overline{b}_n b_n$ while for $n = 0$ the integral is $p\overline{a}_0 a_0/2$. Hence the entire second term simplifies to

$$-2p\left[\frac{\overline{a}_0 a_0}{2} + \sum_{n=1}^{N} (\overline{a}_n a_n + \overline{b}_n b_n)\right]$$

Likewise, recalling Eqs. (2)–(8), Sec. 8.2, it follows that after the sum in the third term of (6) is squared, the integral of each cross-product is zero while the integrals of the squared terms are $p\overline{a}_n^2$ and $p\overline{b}_n^2$ if $n \geq 1$ and $p\overline{a}_0^2/2$ if $n = 0$. Hence the entire third term reduces to

$$p\left[\frac{\overline{a}_0^2}{2} + \sum_{n=1}^{N} (\overline{a}_n^2 + \overline{b}_n^2)\right]$$

The entire expression for $\overline{E}$ thus becomes

$$\overline{E} = \int_0^{2p} f^2(t) \, dt - 2p\left[\frac{\overline{a}_0 a_0}{2} + \sum_{n=1}^{N} (\overline{a}_n a_n + \overline{b}_n b_n)\right] + p\left[\frac{\overline{a}_0^2}{2} + \sum_{n=1}^{N} (\overline{a}_n^2 + \overline{b}_n^2)\right]$$

From this we can find E immediately by putting $\bar{a}_0 = a_0$, $\bar{a}_n = a_n$, and $\bar{b}_n = b_n$. The result is

$$E = \int_0^{2p} f^2(t)\,dt - 2p\left[\frac{a_0^2}{2} + \sum_{n=1}^{N}(a_n^2 + b_n^2)\right] + p\left[\frac{a_0^2}{2} + \sum_{n=1}^{N}(a_n^2 + b_n^2)\right]$$

(7) $$= \int_0^{2p} f^2(t)\,dt - p\left[\frac{a_0^2}{2} + \sum_{n=1}^{N}(a_n^2 + b_n^2)\right]$$

If we now subtract E from $\bar{E}$ and make the obvious regroupings, we have

$$\bar{E} - E = \frac{p}{2}(\bar{a}_0^2 - 2\bar{a}_0 a_0 + a_0^2) + p\sum_{n=1}^{N}[(\bar{a}_n^2 + \bar{b}_n^2) - 2(\bar{a}_n a_n + \bar{b}_n b_n) + (a_n^2 + b_n^2)]$$

(8) $$= p\left\{\left(\frac{\bar{a}_0 - a_0}{\sqrt{2}}\right)^2 + \sum_{n=1}^{N}[(\bar{a}_n - a_n)^2 + (\bar{b}_n - b_n)^2]\right\}$$

Since (8) is the product of the positive number p and a sum consisting exclusively of squared real numbers, its value must be nonnegative. Hence, for every possible set of coefficients $\{\bar{a}_n, \bar{b}_n\}$, we have

$$\bar{E} \geq E$$

Moreover, the equality sign can hold if and only if each squared difference in (8) is zero, that is, if and only if $\bar{a}_0 = a_0$, $\bar{a}_n = a_n$, $\bar{b}_n = b_n$, and therefore $\bar{S}_N(t) = S_N(t)$. The best approximation property of $S_N(t)$ stated at the beginning of this example is now established.

To continue, we note the following consequence of (7): since E is always equal to or greater than zero, as is evident from the nonnegative integrand of (5), it follows that

(9) $$\frac{1}{p}\int_0^{2p} f^2(t)\,dt \geq \frac{1}{2}a_0^2 + \sum_{n=1}^{\infty}(a_n^2 + b_n^2)$$

This important result is known as **Bessel's inequality.** It holds for the Fourier coefficients of any periodic function f for which the integral on the left exists, even though f may not satisfy the Dirichlet conditions. We shall encounter another form of this inequality in a more advanced setting in Sec. 11.6.

If $a_0^2/4$ is subtracted from both members of (9), what remains on the right is the sum of the squares of the Fourier coefficients of f. This sum clearly has

$$\frac{1}{p}\int_0^{2p} f^2(t)\,dt - \frac{a_0^2}{4}$$

as an upper bound.

In our next example we encounter for the first time the use of Fourier series to represent a function which in the context of our problem is not periodic. The problem concerns the static deflection of a simply supported beam of length l. The beam is not vibrating, so there is no periodic motion involved. Moreover, there is nothing periodic about the beam itself, which extends only from $x = 0$ to $x = l$. However, as we saw in Sec. 8.4, we can *imagine* a periodic continuation of a function given only from 0 to l and then use the Fourier series of that extended function to represent the original nonperiodic function on its interval of definition $[0, l]$.

EXAMPLE 5

DEFLECTION OF A SIMPLY SUPPORTED BEAM

A uniform beam of length l, simply supported at each end, bears an arbitrary load per unit length given by the function $w(x)$. Neglecting the weight of the beam, find the deflection curve of the beam.

Using Formula 12, Sec. 2.11, it is possible to find the required deflection curve by solving the simple differential equation

$$EIy^{\text{iv}} = -w(x) \tag{10}$$

and then imposing the end conditions of a simply supported beam, namely,

$$\begin{aligned} y(0) &= y(l) = 0 && \text{(No deflection at either end)} \\ y''(0) &= y''(l) = 0 && \text{(No moment at either end)} \end{aligned} \tag{11}$$

on the solution. The problem can also be solved by using the Green's function of the beam (Exercise 55, Sec. 2.12). However, for simply supported beams, it is also possible to determine the deflection curve by an application of Fourier series. This method is often preferable to direct integration of (10) when the distributed load is given by different analytic expressions over different portions of the beam.

Our objective is to obtain the deflection curve of the beam as a Fourier series satisfying the differential equation (10) and the boundary conditions (11). Since a Fourier series will vanish at $x = 0$ and at $x = l$ only if each of its terms does also, it is clear that we must assume for the deflection y a half-range sine expansion†

$$y = \sum_{n=1}^{\infty} b_n \sin\frac{n\pi x}{l}$$

in which the b_n's have yet to be determined. Guided by this, our next step is to expand the given load function $w(x)$ in a half-range sine series, getting, by now familiar steps,

$$w(x) = \sum_{n=1}^{\infty} B_n \sin\frac{n\pi x}{l} \qquad B_n = \frac{2}{l}\int_0^l w(x)\sin\frac{n\pi x}{l}\,dx$$

With $w(x)$ thus expressed, the differential equation (10) becomes

$$EIy^{\text{iv}} = -\sum_{n=1}^{\infty} B_n \sin\frac{n\pi x}{l}$$

If we now substitute the series we assumed for y, we obtain

$$EI\frac{\pi^4}{l^4}\sum_{n=1}^{\infty} n^4 b_n \sin\frac{n\pi x}{l} = -\sum_{n=1}^{\infty} B_n \sin\frac{n\pi x}{l}$$

For this equation to hold identically on $0 < x < l$, the coefficients of like terms on each side must be equal. Hence for every $n \geq 1$,

$$EI\frac{\pi^4}{l^4}n^4 b_n = -B_n$$

and

$$b_n = -\frac{l^4}{\pi^4 n^4 Ei}B_n$$

†Other types of beams, cantilevers for instance, cannot be handled by this method since none of the other forms of Fourier series have terms that can satisfy the relevant end conditions. An extension of this method, using series other than Fourier series, is possible, however, and is suggested in Example 4 and in Exercise 47, Sec. 11.6.

With B_n determined by the given load function $w(x)$, b_n is now completely determined and our problem is solved. Because of the factor n^4 in the denominator of the formula for b_n and because B_n must be at least of the order c/n, the series we have found for the deflection y converges very rapidly.

The primary objective of applied mathematics is to formulate and solve mathematical problems that adequately describe and solve problems from the real world. Usually specific numerical answers are required for such problems. On the other hand, there are times when our purpose is merely to understand or explain something we have observed. Our next example is of this sort.

EXAMPLE 6

MODELING RADIATION INTENSITY

When a switch is opened or closed in a wire carrying electric current, the magnetic field around the wire changes abruptly and observable effects may be produced; for instance, a sudden "click" may be heard in a nearby radio receiver. A team of engineers, interested in investigating the radio interference produced in this way by the periodic opening and closing of a switch, began by measuring the intensity of the radiation as a function of radiation frequency. To do this, they used a band-pass filter that screened out all frequencies except those in a range so narrow that in effect only one term in the Fourier expansion of the radiation intensity $f(t)$ was recorded. By varying the location of the passband along the frequency scale, they thus obtained the value of the resultant amplitude $A_n = \sqrt{a_n^2 + b_n^2}$ for various frequencies of the emitted radiation. When they plotted their data as $\ln A_n$ versus $\ln \omega_n$, they obtained a set of points that fell rather closely along a line of slope -1. This suggested a relation of the form

$$\ln A_n = -\ln \omega_n + \ln k \qquad \text{or} \qquad A_n = \frac{k}{\omega_n}$$

or, since $\omega_n = n\pi/p$,

$$A_n = \frac{pk/\pi}{n}$$

To construct a mathematical model of their experiment, the engineers naturally assumed that the switch opened and closed instantaneously and that they were dealing with a discontinuous phenomenon. More specifically, by using suitable scale factors, they set up a discontinuous pulse pattern of period $2p = 4$ with intensity $f(t) = 1$ over half the period, as shown in Fig. 8.15*a*. This seemed to fit their observations with satisfactory accuracy because, by Theorem 3, Sec. 8.3, the coefficients in the Fourier expansion of a discontinuous function do decrease as c/n.

However, when they subsequently repeated the experiment with a filter capable of reaching much higher frequencies, they encountered a curious thing. For frequencies just beyond the capability of their first filter, they obtained readings that scattered badly and showed no definite trend. Then as they reached still higher frequencies, to their surprise, regularity reappeared, but this time the points they plotted fell very nearly along a line of slope -2.

An explanation for all this is quite simple, given an appropriate model. Clearly, the discontinuous model of the radiation intensity suggested by Fig. 8.15*a* must be modified. Now the extended readings imply a relation of the form $A_n = c/n^2$ which, according to Theorem 3, Sec. 8.3, cannot hold for a discontinuous function. Hence a continuous model of f with periodic pulses like those shown in Fig. 8.15*b* is suggested, a being sufficiently close to 1 to represent an extremely rapid but still not instantaneous closing of the switch.

To verify that such a model is consistent with all of the observed data, it is necessary to find the Fourier expansion of the newly postulated intensity function. Since the new model, like the first, is an even function, every b_n is zero. To calculate a_n we could use the appropriate Euler formula; but since it is easier, we shall use the alternate formulas of Sec. 8.3.

We note first, from Fig. 8.15*b*, that f is now continuous and that f'' is a trivial function. The value of a_n

therefore depends only on the jumps of f' on $[-2, 2)$ which, to be precise, are upward jumps of $-1/(a-1)$ at $x = \pm 1$ and downward jumps of $1/(a-1)$ at $x = \pm a$. Therefore, for $n \geq 1$, and with $p = 2$,

$$\begin{aligned} a_n &= -\frac{2}{n\pi} b_n' = -\frac{2}{n\pi}\left[\frac{1}{n\pi}\left(2\frac{-1}{a-1}\cos\frac{n\pi}{2} + 2\frac{1}{a-1}\cos\frac{n\pi a}{2}\right)\right] \\ &= \frac{4}{n^2\pi^2(a-1)}\left[\cos\frac{n\pi}{2} - \cos\frac{n\pi a}{2}\right] \\ &= \frac{4}{n^2\pi^2(a-1)}\left[-2\sin\frac{n\pi(1+a)}{4}\sin\frac{n\pi(1-a)}{4}\right] \\ &= \frac{2}{n\pi}\sin\frac{n\pi(1+a)}{4}\left[\sin\frac{n\pi(1-a)}{4}\bigg/\frac{n\pi(1-a)}{4}\right] \end{aligned}$$

For fixed n, no matter how large, the quotient

(12) $$\sin\frac{n\pi(1-a)}{4}\bigg/\frac{n\pi(1-a)}{4}$$

has the indeterminate form 0/0 when $a = 1$, and its limiting value as $a \to 1$ is 1. Equivalently, if a, $0 < a < 1$, is fixed and $1 - a$ is very small, then until n has increased enough so that $n(1-a)$ is no longer small, the quotient (12) is close to 1. Clearly, for this range of n values, that is, for the corresponding range of frequencies, the last expression for a_n has the approximate form c/n, as was observed during the first phase of the experiment. However, for values of n beyond this point, the quotient (12) is no longer near 1. The numerator $\sin[n\pi(1-a)/4]$ is bounded between 1 and -1, but the denominator increases as the first power of n. Hence eventually a_n behaves like c/n^2, as observed during the second phase of the experiment, and as expected of the radiation resulting from a switch that does not open and close instantaneously.

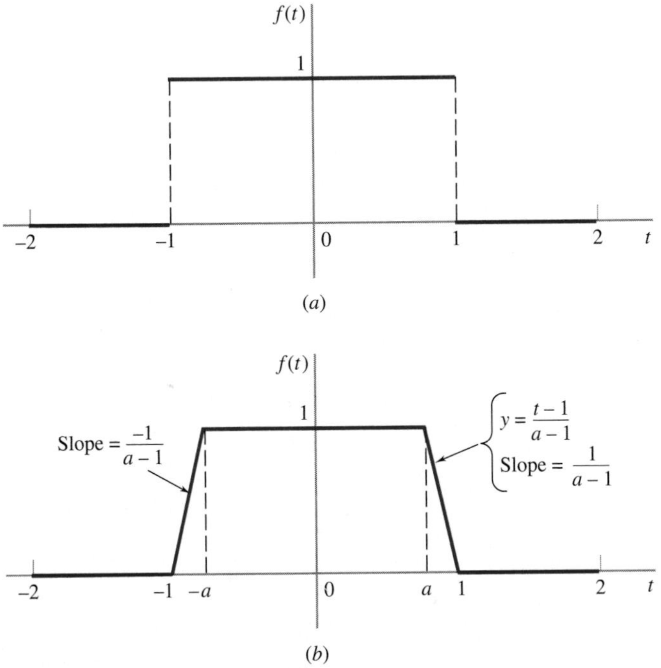

FIGURE 8.15
Two models of the pulses discussed in Example 6.

EXERCISES

Work Example 1 with all numerical data the same except for the changes indicated.

1. $c = 0$ **2.** $k = 81$

3. $k = 144$; $c/c_c = 0.1$

4. $k = 64$; $F(t) = \begin{cases} F_0 & 0 \le t < 1 \\ 0 & 1 \le t < 2 \\ F(t+2) & 0 \le t \end{cases}$

5. $k = 144$; $F(t) = \begin{cases} t & 0 \le t < \frac{1}{2} \\ t - 1 & \frac{1}{2} \le t < 1 \\ F(t+1) & 0 \le t \end{cases}$

6. Work Example 1 if $k = 3$ lb/in, $w = 8$ lb, $c/c_c = 0.1$, and over one period

$$F(t) = \begin{cases} F_0 \text{ lb} & 0 \le t < 1 \\ 0 & 1 \le t < 2 \end{cases}$$

7. Work Exercise 6 if $k = 60$ lb/in, $w = 160$ lb, and $c/c_c = 0.04$.

8. Work Example 1 if $k = 40$ lb/in, $w = 100$ lb, $c/c_c = 0.1$, and over one period

$$F(t) = \begin{cases} F_0 t \text{ lb} & 0 \le t \le \frac{1}{2} \\ F_0(t-1) \text{ lb} & \frac{1}{2} < t \le 1 \end{cases}$$

9. Work Exercise 8 if the amount of friction is reduced to $c/c_c = 0.02$.

10. In Example 1, discuss the problem of determining the complete motion, transient as well as steady-state.

11. Determine the steady-state motion produced in the system shown in Fig. 7.17 if **(a)** the force $F(t)$ shown in Fig. 8.12*b* acts on M_1 and **(b)** the force shown in Fig. 8.12*b* acts on M_2.

12. Work Example 2 if the capacitor and the inductor are removed from the circuit.

13. Work Example 2 if the resistor is removed from the circuit.

14. Work Example 2 if over one period

$$E(t) = \begin{cases} E_0 & 0 < t < 0.005 \\ -E_0 & 0.005 < t < 0.01 \end{cases}$$

15. Work Example 2 if $R = 600$, $L = 1$, $C = 4 \times 10^{-6}$, and E is the periodic voltage whose definition over the period $0 \le t < 0.01$ is

(a) $E(t) = E_0 t$ **(b)** $E(t) = 100 \sin 50\pi t$

16. Work Example 2 if $R = 100$, $L = 0.4$, $C = 10^{-5}$, and E is defined over the period $0 \le t < 0.02$ by

(a) $E(t) = E_0 t$ **(b)** $E(t) = 100 \sin 50\pi t$

17. If $F(t)$ is the periodic function whose definition in one period is

$$F(t) = \begin{cases} 1 & 0 < t < \pi \\ 0 & \pi < t < 2\pi \end{cases}$$

find the solution of each of the following differential equations which satisfies the indicated conditions.

(a) $y'' - y = F(t)$; $y(0) = y'(0) = 0$

(b) $y'' + y = F(t)$; $y(0) = y'(0) = 0$

(c) $y'' - y = F(t)$; $y(0) = 1$, $y'(0) = 0$

(d) $y'' + y = F(t)$; $y(0) = 1$, $y'(0) = 0$

(e) $y'' - 3y' + 2y = F(t)$; $y(0) = y'(0) = 0$

Hint: First expand $F(t)$ in a Fourier series and then find a particular integral for $F(t)$ term by term. Finally, add this particular integral to the complementary function and impose the initial conditions on the complete solution.

18. If $F(t)$ is the periodic function whose definition over one period is

$$F(t) = |t| \qquad -\pi \le t \le \pi$$

find the solution of each of the following differential equations which satisfies the given initial conditions.

(a) $y'' - y = F(t)$; $y(0) = y'(0) = 0$

(b) $y'' - y = F(t)$; $y(0) = 0$, $y'(0) = 1$

(c) $y'' + 4y = F(t)$; $y(0) = y'(0) = 0$

(d) $y'' + 9y = F(t)$; $y(0) = y'(0) = 0$

19. What error is made in approximating each of the following periodic functions over the specified period by the sum of the first three nonzero terms in its Fourier series?

(a) $f(t) = e^{-t}$ $\quad 0 < t < 1$

(b) $f(t) = e^{t}$ $\quad 0 < t < 1$

(c) $f(t) = \begin{cases} t & 0 < t < 1 \\ 0 & 1 < t < 2 \end{cases}$

(d) $f(t) = \begin{cases} 1 & 0 < t < 1 \\ 0 & 1 < t < 4 \end{cases}$

20. Determine the values of a and b which make the line $y = a + bx$ the best least-square approximation to e^x for $0 < x < 1$. *Hint:* Using the least-square criterion, the error to be minimized is

$$E(a, b) = \int_0^1 [e^x - (a + bx)]^2 \, dx$$

To minimize this, recall from calculus that a necessary condition for a minimum is that

$$\frac{\partial E}{\partial a} = 0 \qquad \text{and} \qquad \frac{\partial E}{\partial b} = 0$$

Then obtain these partial derivatives by differentiating with respect to a and b inside the integral sign using Leibniz' rule (see footnote, p. 661). Finally, solve for a and b from the linear equations that result from the subsequent integrations.

21. Find the value of a that makes each of the following functions the best least-square approximation of its type to $\cos x$ for $-\pi/2 \le x \le \pi/2$. *Hint:* Recall the hint for Exercise 20.

(a) $f(x) = a(\pi^2 - 4x^2)$ **(b)** $f(x) = 1 - ax^2$

(c) $f(x) = a - x^2/4$

What is the property of each of these functions that makes it a reasonable family of one-parameter approximations to $\cos x$?

22. The function $y = \cos x$ is the unique solution of the differential equation $y'' + y = 0$ for which $y_0 = 1$ and $y_0' = 0$. The function $y_1 = 1 - bx^2$ also satisfies the conditions $y_0 = 1$ and $y_0' = 0$, but it does not satisfy the differential equation. In some sense, then, y_1 might be considered a good approximation to $y = \cos x$ over an interval $(0, x_1)$ if the amount by which it failed to satisfy the equation $y'' + y = 0$ over $(0, x_1)$ were as small as possible. Using the least-square criterion, this would require

$$E(b) = \int_0^{x_1} (y_1'' + y_1)^2 \, dx$$

to be a minimum. Determine the minimizing value of b if
(a) $x_1 = 1$ **(b)** $x_1 = \pi/2$ **(c)** $x_1 = 2$
What values do Parts **(a)** and **(c)** give for the first positive intercept of $y = \cos x$?

23. Let $\bar{S}_n = \dfrac{\bar{a}_0}{2} + \bar{a}_1 \cos x + \cdots + \bar{a}_n \cos nx + \cdots + \bar{b}_1 \sin x + \cdots + \bar{b}_n \sin nx + \cdots$ where the $\bar{a}_n$'s and $\bar{b}_n$'s are arbitrary coefficients, and define $\bar{E}(\bar{a}_0, \bar{a}_1, \ldots, \bar{a}_n, \ldots, \bar{b}_1, \ldots, \bar{b}_n, \ldots) = \int_0^{2\pi} [f(t) - \bar{S}_n]^2 \, dx$ where $f(t)$ is an arbitrary function satisfying the Dirichlet conditions. Extending the hint given in Exercise 20, show that the values of the $\bar{a}_n$'s and $\bar{b}_n$'s which make $\bar{E}$ a minimum are the Fourier coefficients of $f(t)$. *Observation:* This problem is much simpler than Exercise 20.

24. In Example 5, find the coefficients in the Fourier expansion of the deflection y if $w(x)$ is a constant.

25. Work Exercise 24 if $w(x) = kx$.

26. Work Exercise 24 if

$$w(x) = \begin{cases} 2P_0x/l & 0 \le x \le l/2 \\ 2P_0(l - x)/l & l/2 \le x \le l \end{cases}$$

For the beam of Example 5, find the Fourier series representation of the deflection y due to each of the following loads.

27. $w(x) = 2P_0 \sin(2\pi x/l) \cos(\pi x/l)$

28. $w(x) = 1 + \sin(\pi x/l)$

29. $w(x) = \begin{cases} 0 & 0 \le x < l/2 - a \\ P_0/2a & l/2 - a \le x \le l/2 + a \\ 0 & l/2 + a < x \le l \end{cases}$

30. By letting $a \to 0$ in the expression for the deflection obtained in Exercise 29, find the Fourier expansion of the deflection y produced by a concentrated load P_0 acting at the midpoint of the beam.

31. In Example 5, determine the deflection curve of the beam if $w(x) = x$ by integrating the equation $EIy^{\text{iv}} = -x$ and imposing the end conditions on the result. Compute the value of the deflection y at $l/2$ and compare your answer with that given by **(a)** the first term in the Fourier series obtained in Exercise 25 and **(b)** the sum of the first two terms obtained in Exercise 25.

32. What do you think the engineers in Example 6 would observe if they repeated their experiment using a filter capable of reaching frequencies several orders of magnitude higher than those they had previously reached?

33. Could an experiment such as that described in Example 6 be used to obtain information about the speed with which a given switch opens and closes? How?

34. If

$$f(t) = \begin{cases} 2 & 0 < t < 1 - 1/2a \\ 1 - \sin \pi at & 1 - 1/2a < t < 1 + 1/2a \\ 0 & 1 + 1/2a < t < 2 \end{cases}$$

and if a is a large even integer, discuss the behavior of the coefficients in the half-range cosine expansion of $f(t)$ for small and medium values of n as well as for $n \to \infty$.

35. For c units of time water flows through a turbine at a constant rate b. During the next $2(\pi - c)$ units of time the constant rate of flow is a. Then for c more units of time the flow rate is b again. Thereafter water discharges through the turbine in the same way periodically.
(a) Find a half-range cosine expansion for the periodic rate $r(t)$ at which water flows through the turbine.
(b) Particularize the series of Part **(a)** for the cases $a = b$, $c = 0$, and $c = \pi$.
(c) Using the series of Part **(a)**, obtain a half-range sine series expansion for the function f defined on $0 < c \le \pi$ by $f(c) = \pi - c$. *Hint:* Take $t = 0$.
(d) Use the series of Part **(a)** to obtain a half-range sine series expansion for the function g defined on $0 \le c < \pi$ by $g(c) = c$. *Hint:* Take $t = \pi$.
(e) Find the root mean square value of $r(t)$.
(f) Particularize r_{rms} for the cases $a = b$, $c = 0$, and $c = \pi$.

According to Eq. (4), Sec. 2.12, the static deflection y of a uniform horizontal string stretched under tension T and bearing a vertical load per unit length $w(x)$ satisfies the equation

$$T\frac{d^2y}{dx^2} = -w(x)$$

provided the deflections are small. Find the static deflection of a string stretched between $x = 0$ and $x = l$ due to each of the following loads.

36. $w(x) = \begin{cases} 0 & 0 < x < l/2 - a \\ P_0/2a & l/2 - a < x < l/2 + a \\ 0 & l/2 + a < x < l \end{cases}$

Does your solution for y have a limit as $a \to 0$? What is the significance of the limit if there is one?

37. $w(x) = x(l - x)$; $0 \le x \le l$

38. $w(x) = \begin{cases} P_0x & 0 \le x \le l/2 \\ P_0(l - x) & l/2 \le x \le l \end{cases}$

39. How is the solution of Exercise 37 affected if the load per unit length is $w(x) = l^2 - 4x^2$, $-l/2 \le x \le l/2$?

40. Show that the Nth partial sum of the Fourier series of a periodic function is not necessarily the best least-square trigonometric approximation to the function over an interval which is not a full period. *Hint:* Construct a counterexample.

CHAPTER 9

FOURIER INTEGRALS AND FOURIER TRANSFORMS

Fourier series, by their very nature, are limited to the representation of periodic functions. On the other hand, many important functions, such as an isolated pulse or a decaying exponential, for instance, are nonperiodic, and an effective representation for such functions is often required. The *Fourier integral,* which arises from the Fourier series of a periodic function as the period becomes infinite, provides such a representation. Section 9.1 discusses this transition from the periodic case, when the frequencies of the various terms form an arithmetic progression, to the nonperiodic case, when the frequency is a continuous variable.

Properly interpreted, the Fourier integral can be thought of as a transformation which transforms a nonperiodic function of time t, say, into a function of a continuous frequency variable ω. This transformation is useful in many fields. Its most important properties and some typical applications are given in Secs. 9.3 and 9.4.

Section 9.5 introduces what are called *singularity functions,* whose physical counterparts include such things as an impulsive force of zero duration or a load concentrated at a single point, and determines their Fourier transforms.

Section 9.6 attempts to motivate the work of the next chapter on the Laplace transformation by tracing its evolution from the Fourier transformation we have studied in this chapter.

Prerequisite for this chapter: Chap. 8.

9.1 THE FOURIER INTEGRAL AS THE LIMIT OF A FOURIER SERIES

The properties of Fourier series developed thus far are adequate to accomplish the expansion of any periodic function satisfying the Dirichlet conditions and, in conjunction with the theory of Chap. 7, enable us to find the response of numerous mechanical and electrical systems to general periodic disturbances. On the other hand, in many problems the impressed force or voltage is nonperiodic rather than periodic—a single unrepeated pulse, for instance. Functions of this sort cannot be handled directly through the use of Fourier series, since such series necessarily define only periodic functions. However, by investigating the limit (if any) which is approached by a Fourier series as the period of the given function becomes infinite, a suitable representation for nonperiodic functions can perhaps be obtained. An example is probably the best way to introduce the theory of this procedure.

Consider, then, the function $f_p(t)$ shown (with unequal vertical and horizontal scales) in Fig. 9.1, defined on one period by

$$f_p(t) = \begin{cases} 0 & -p < t < -1 \\ 1 & -1 < t < 1 \\ 0 & 1 < t < p \end{cases}$$

in the limit as $p \to \infty$, as suggested by Fig. 9.2. This function is clearly even, and thus its Fourier expansion contains only cosine terms, i.e.,

$$(1) \qquad f_p(t) = \frac{a_0}{2} + a_1 \cos\frac{\pi t}{p} + a_2 \cos\frac{2\pi t}{p} + \cdots + a_n \cos\frac{n\pi t}{p} + \cdots$$

where

$$(2a) \qquad a_0 = \frac{2}{p}\int_0^1 1\,dt = \frac{2}{p}$$

and

$$(2b) \qquad a_n = \frac{2}{p}\int_0^1 1 \cos\frac{n\pi t}{p}\,dt = \frac{2}{p}\left[\frac{\sin(n\pi t/p)}{n\pi/p}\right]_0^1 = \frac{2}{p}\,\frac{\sin(n\pi/p)}{n\pi/p}$$

In the Fourier series of $f_p(t)$, the frequency of the nth term is

$$\omega_n = \frac{n\pi}{p}$$

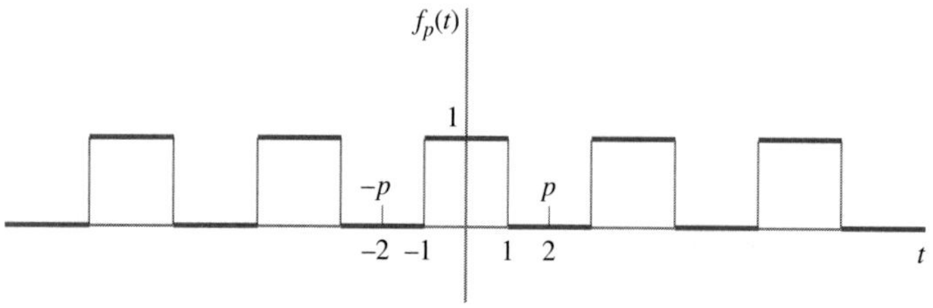

FIGURE 9.1
A periodic function of period $2p = 4$.

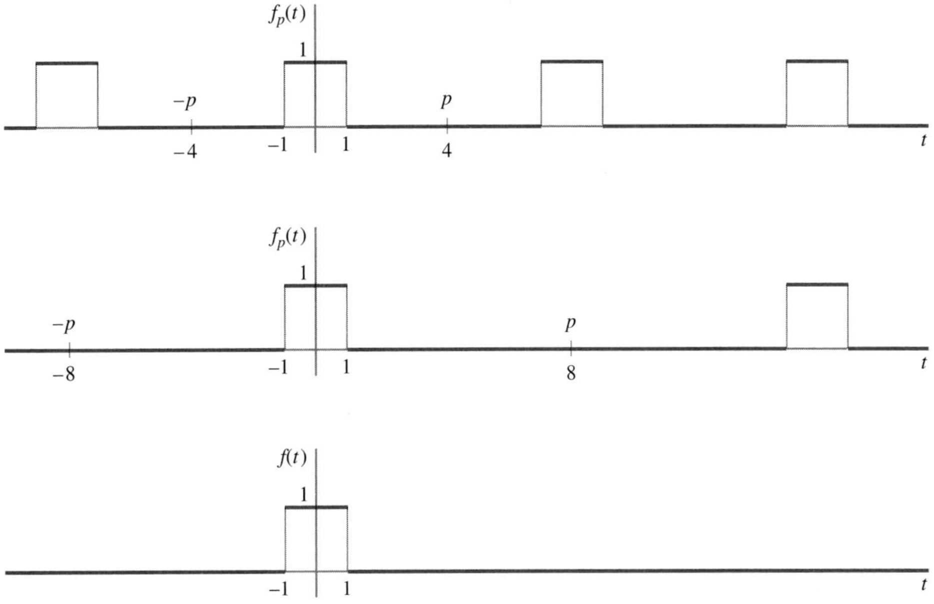

FIGURE 9.2
The nonperiodic limit of a sequence of periodic functions whose periods become infinite.

and the difference in frequency between successive terms is

$$\Delta\omega = \omega_{n+1} - \omega_n = \frac{(n+1)\pi}{p} - \frac{n\pi}{p} = \frac{\pi}{p}$$

Hence it is possible to write the coefficient of the nth term in the form

$$a_n = \frac{2}{p}\frac{\sin\omega_n}{\omega_n} = \frac{2}{\pi}\frac{\sin\omega_n}{\omega_n}\frac{\pi}{p} = \frac{2}{\pi}\frac{\sin\omega_n}{\omega_n}\Delta\omega \tag{3}$$

If we define what we might call the **amplitude function**

$$A(\omega) = \begin{cases} \dfrac{2}{\pi} & \omega = 0 \\ \dfrac{2}{\pi}\dfrac{\sin\omega}{\omega} & \omega > 0 \end{cases}$$

it follows from (2a) and (3) that for all values of p, the successive Fourier coefficients in (1) are proportional, respectively, to the ordinates of the amplitude curve $y = A(\omega)$ which start at $\omega = 0$ and are spaced thereafter at equal intervals of $\Delta\omega = \pi/p$. As p increases, these ordinates move closer together, but the amplitude curve remains unchanged since it, unlike $\Delta\omega$, is independent of p. The geometric significance of these observations is shown in Fig. 9.3. Clearly, for all values of p, the magnitude of a_n is equal to the area of the nth rectangle

$$|A(\omega_n)|\,\Delta\omega = \begin{cases} \dfrac{2}{\pi}\Delta\omega & n = 0 \\ \dfrac{2}{\pi}\dfrac{|\sin\omega_n|}{\omega_n}\Delta\omega & n \geq 1 \end{cases}$$

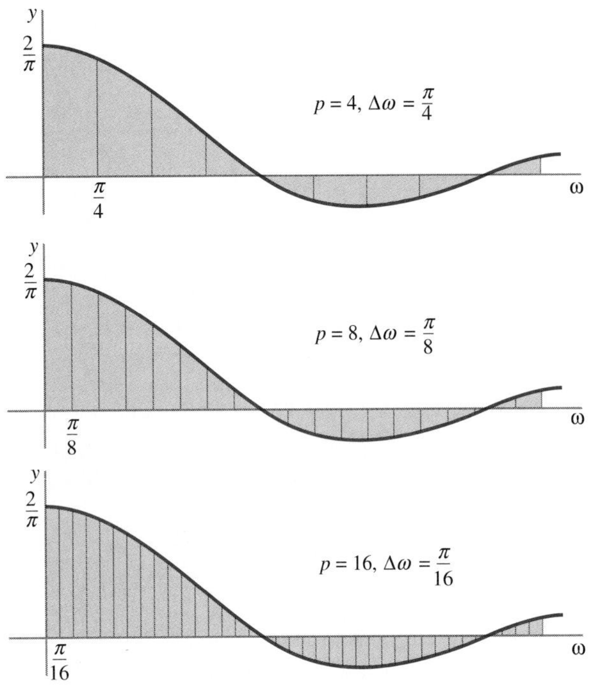

FIGURE 9.3

The amplitude function $y = \frac{2}{\pi}\frac{\sin\omega}{\omega} = A(\omega)$ which defines $a_n = A(\omega)\,\Delta\,\omega$.

and approaches zero as p becomes infinite. If the value of a_n given by (3) is substituted into (1), we obtain

$$f_p(t) = \frac{1}{p} + \frac{2}{\pi}\sum_{n=1}^{\infty}\frac{\sin\omega_n}{\omega_n}\cos\omega_n t\,\Delta\omega \tag{4}$$

which, for every finite value of p, is still a periodic function of t.

The nth term of the series in (4) involves the function $[(\sin\omega)/\omega]\cos\omega t$ evaluated at a point of the nth subinterval $[\omega_{n-1}, \omega_n]$ (the right-hand endpoint, in fact), whose length is $\Delta\omega = \pi/p$. For every value of p the set of all these subintervals covers the entire positive half-axis. Moreover, as p increases, the first term $1/p$ of (4) tends to zero. This suggests that as $p \to \infty$ and $\Delta\omega = \pi/p \to 0$, the right-hand member of (4) becomes an improper integral, presumably

$$\frac{2}{\pi}\int_0^{\infty}\frac{\sin\omega\cos\omega t}{\omega}\,d\omega \tag{5}$$

and that this integral then represents the nonperiodic limit of $f_p(t)$. Although we have not proved it, this conjecture is indeed correct.

Guided by the preceding example, let us now consider the question of whether a *general* nonperiodic function can be represented by a trigonometric *integral* in a manner analogous to the representation of a general periodic function by a trigonometric *series.*

To begin, we let f be a nonperiodic function with domain $(-\infty, \infty)$ and such that

(a) On every finite interval, f satisfies the Dirichlet conditions.

(b) $\int_{-\infty}^{\infty} |f(t)|\,dt$ exists.

We then define a periodic function f_p of period $2p$ in terms of f as follows:

$$f_p(t) = \begin{cases} f(t) & -p \le t < p \\ f_p(t + 2p) & t \text{ real} \end{cases}$$

Clearly, $f(t)$ is the limit of $f_p(t)$ as $p \to \infty$. Because f satisfies the Dirichlet conditions on every finite interval, it follows that for every value of p the function f_p satisfies those same conditions in each period and hence possesses a valid Fourier series. Using the complex exponential form of the Fourier series of an arbitrary function [Eqs. (7) and (8), Sec. 8.5] and noting that on $[-p, p)$, $f_p(t) = f(t)$, we can therefore write

$$f_p(t) = \sum_{n=-\infty}^{\infty} c_n e^{ni\pi t/p} \qquad \text{where} \qquad c_n = \frac{1}{2p}\int_{-p}^{p} f(t)e^{-ni\pi t/p}\,dt$$

Substituting c_n into the expression for $f_p(t)$ gives

$$\begin{aligned} f_p(t) &= \sum_{n=-\infty}^{\infty} \left[\frac{1}{2p}\int_{-p}^{p} f(t)e^{-ni\pi t/p}\,dt\right]e^{ni\pi t/p} \\ &= \sum_{n=-\infty}^{\infty} \left[\frac{1}{2\pi}\int_{-p}^{p} f(t)e^{-ni\pi t/p}\,dt\right]e^{ni\pi t/p}\,\frac{\pi}{p} \end{aligned}$$

Now, as before, let us denote the frequency of the general term by

$$\omega_n = \frac{n\pi}{p}$$

and the difference in frequency between successive terms by

$$\Delta\omega = \frac{\pi}{p}$$

Then $f_p(t)$ can be written

$$(6) \qquad f_p(t) = \sum_{n=-\infty}^{\infty} \left[\frac{1}{2\pi}\int_{-p}^{p} f(t)e^{-i\omega_n t}\,dt\right]e^{i\omega_n t}\,\Delta\omega$$

If we now set

$$(7) \qquad C_p(\omega) = \frac{1}{2\pi}\int_{-p}^{p} f(t)e^{-i\omega t}\,dt$$

and for each p define a function F_p by

$$(8) \qquad F_p(\omega) = C_p(\omega)e^{i\omega t}$$

Eq. (6) becomes simply

$$(9) \qquad f_p(t) = \sum_{n=-\infty}^{\infty} F_p(\omega_n)\,\Delta\omega$$

where ω_n is now the right-hand endpoint of the nth subinterval $[\omega_{n-1}, \omega_n]$ of a set of intervals which covers the entire real axis and $\Delta\omega \to 0$ as $p \to \infty$. This suggests that as $p \to \infty$, (9) becomes an improper integral over $(-\infty, \infty)$. Of course, the elementary theory of the definite integral is not applicable to (9) because the limits of summation are not finite, and the function F_p being evaluated at the points ω_n changes, in general, with p.

However, since f, by hypothesis, is absolutely integrable over $(-\infty, \infty)$, and because $|f(t)e^{-i\omega t}| = |f(t)|$, the integral

$$C(\omega) = \lim_{p\to\infty} C_p(\omega) = \frac{1}{2\pi}\int_{-\infty}^{\infty} f(t)e^{-i\omega t}\,dt$$

exists. Therefore it follows from (8) that

$$F(\omega) = \lim_{p\to\infty} F_p(\omega) = \lim_{p\to\infty} C_p(\omega)e^{i\omega t} = C(\omega)e^{i\omega t}$$

exists. Thus there is good reason to believe that for large values of p, (9) approximates

$$\sum_{n=-\infty}^{\infty} F(\omega_n)\,\Delta\omega$$

and that as $p \to \infty$, the nonperiodic limit $f(t)$ of $f_p(t)$ is correctly given by the formula

$$f(t) = \int_{-\infty}^{\infty} F(\omega)\,d\omega = \int_{-\infty}^{\infty} C(\omega)e^{i\omega t}\,d\omega$$

$$= \int_{-\infty}^{\infty}\left[\frac{1}{2\pi}\int_{-\infty}^{\infty} f(t)e^{-i\omega t}\,dt\right]e^{i\omega t}\,d\omega \qquad (10)$$

Though our proof† of it has been far from complete, we do have the following fundamental theorem.

THEOREM 1 If on every finite interval, f satisfies the Dirichlet conditions and if the improper integral $\int_{-\infty}^{\infty} |f(t)|\,dt$ exists, then the **Fourier integral**

$$f(t) = \int_{-\infty}^{\infty}\left[\frac{1}{2\pi}\int_{-\infty}^{\infty} f(t)e^{-i\omega t}\,dt\right]e^{i\omega t}\,d\omega \qquad (10)$$

gives the value of f at every point where f is continuous and gives the average of the right- and left-hand limits of $f(t)$ at every point where f is discontinuous.

The Fourier integral (10) can be written as the integral

$$f(t) = \int_{-\infty}^{\infty} C(\omega)e^{i\omega t}\,d\omega \qquad (11)$$

in which C is the **coefficient function**

$$C(\omega) = \frac{1}{2\pi}\int_{-\infty}^{\infty} f(t)e^{-i\omega t}\,dt \qquad (12)$$

This **complex Fourier integral pair** bears a striking resemblance to the Fourier series formulas (7) and (8) of Sec. 8.5. Thus the expression (11), with $C(\omega)$ given by (12), is naturally called a **complex exponential Fourier integral.**

---

†For a rigorous proof of Theorem 1, see, for instance, J. W. Brown and R. V. Churchill, *Fourier Series and Boundary Value Problems,* 5th ed., pp. 217–226, McGraw-Hill, New York, 1993.

Specification of the coefficient function $C(\omega)$ is tantamount to specifying $f(t)$ since, when $C(\omega)$ is known, $f(t)$ is completely determined through (11). In effect, we thus have two different ways of characterizing the function of our discussion: $f(t)$ in the time domain and $C(\omega)$ in the frequency domain. By analogy with the theory of light, $C(\omega)$ is often called the **spectrum** of $f(t)$, since it provides a measure of the **intensity** of f in the frequency interval between ω_0 and $\omega_0 + \Delta\omega$; that is, for every t in the domain of f, $C(\omega)$ determines how much of the value of f at t is attributable to values of ω between ω_0 and $\omega_0 + \Delta\omega$.

EXAMPLE 1

Find the complex exponential Fourier integral representation of the function

$$f(t) = \begin{cases} 0 & t < 0 \\ e^{-at} & t > 0 \end{cases} \qquad a > 0$$

Using Eq. (12), with the range of integration reduced to $(0, \infty)$ since $f(t) \equiv 0$ on $(-\infty, 0)$, we have

$$C(\omega) = \frac{1}{2\pi}\int_0^\infty e^{-at}e^{-i\omega t}\,dt = \frac{1}{2\pi}\int_0^\infty e^{-(a+i\omega)t}\,dt$$

$$= \frac{\lim_{b\to\infty}[e^{-(a+i\omega)t}]_0^b}{-2\pi(a+i\omega)} = \frac{1 - \lim_{b\to\infty} e^{-(a+i\omega)b}}{2\pi(a+i\omega)}$$

Now, for $a > 0$,

$$\lim_{b\to\infty}\left|e^{-(a+i\omega)b}\right| = \lim_{b\to\infty}\left|e^{-ab}e^{-i\omega b}\right| = \lim_{b\to\infty} e^{-ab} = 0$$

and this implies $\lim_{b\to\infty} e^{-(a+i\omega)b} = 0$; hence $C(\omega) = 1/2\pi(a + i\omega)$. Substituting the coefficient function $C(\omega)$ into (11) gives

$$f(t) = \frac{1}{2\pi}\int_{-\infty}^{\infty} \frac{1}{a+i\omega} e^{i\omega t}\,d\omega$$

as the required Fourier integral representation of f.

Just as in the case of Fourier series (see Sec. 8.5), the complex exponential integral (10) can be expressed in alternative real forms. To do this, we change the dummy variable of integration from t to τ in the inner integral in (10) and then move $e^{i\omega t}$ across the inner integral sign, which we can do because it does not involve τ. This gives

$$f(t) = \frac{1}{2\pi}\int_{-\infty}^{\infty}\int_{-\infty}^{\infty} f(\tau)e^{-i\omega(\tau-t)}\,d\tau\,d\omega \tag{13}$$

In this, we can replace the exponential by its trigonometric equivalent, getting

$$f(t) = \frac{1}{2\pi}\int_{-\infty}^{\infty}\int_{-\infty}^{\infty} f(\tau)[\cos\omega(\tau-t) - i\sin\omega(\tau-t)]\,d\tau\,d\omega$$

If we break this up into two integrals, we get

$$f(t) = \frac{1}{2\pi}\int_{-\infty}^{\infty}\int_{-\infty}^{\infty} f(\tau)\cos\omega(\tau-t)\,d\tau\,d\omega$$

$$- \frac{i}{2\pi}\int_{-\infty}^{\infty}\int_{-\infty}^{\infty} f(\tau)\sin\omega(\tau-t)\,d\tau\,d\omega$$

The fact that sin $\omega(\tau - t)$ is an odd function of ω makes it seem plausible that the second integral is always zero, and indeed this must be the case since by hypothesis $f(t)$ is purely real. Thus we obtain the real trigonometric representation

$$(14a) \qquad f(t) = \frac{1}{2\pi}\int_{-\infty}^{\infty}\int_{-\infty}^{\infty} f(\tau)\cos\omega(\tau - t)\,d\tau\,d\omega$$

Since the integrand of (14a) is an even function of ω, we need perform the ω integration only between 0 and ∞, provided we multiply the result by 2. This gives us the modified form

$$(14b) \qquad f(t) = \frac{1}{\pi}\int_{0}^{\infty}\int_{-\infty}^{\infty} f(\tau)\cos\omega(\tau - t)\,d\tau\,d\omega$$

which, when cos $\omega(\tau - t)$ is expanded, becomes

$$\begin{aligned} f(t) &= \frac{1}{\pi}\int_{0}^{\infty}\int_{-\infty}^{\infty} f(\tau)[\cos\omega\tau\cos\omega t + \sin\omega\tau\sin\omega t]\,d\tau\,d\omega \\ &= \int_{0}^{\infty}\left\{\left[\frac{1}{\pi}\int_{-\infty}^{\infty} f(\tau)\cos\omega\tau\,d\tau\right]\cos\omega t + \left[\frac{1}{\pi}\int_{-\infty}^{\infty} f(\tau)\sin\omega\tau\,d\tau\right]\sin\omega t\right\}d\omega \end{aligned}$$

Upon reverting back to t (instead of τ) as dummy variable in the two inner integrals and defining the (real) **coefficient functions,**

$$(15) \qquad (a)\ A(\omega) = \frac{1}{\pi}\int_{-\infty}^{\infty} f(t)\cos\omega t\,dt \qquad (b)\ B(\omega) = \frac{1}{\pi}\int_{-\infty}^{\infty} f(t)\sin\omega t\,dt$$

we arrive at the **standard Fourier integral representation** of f

$$(16) \qquad f(t) = \int_{0}^{\infty}[A(\omega)\cos\omega t + B(\omega)\sin\omega t]\,d\omega$$

Clearly, Eq. (16) is a direct analog of the standard Fourier series given by Eq. (1), Sec. 8.5, with Eqs. (15a) and (15b) as counterparts of the Euler formulas when $d = -p$. In working with Fourier expansions we frequently wrote them as the sum of a cosine and sine series. Likewise, (16) can be written in the real **trigonometric** form

$$(17) \qquad f(t) = \int_{0}^{\infty} A(\omega)\cos\omega t\,d\omega + \int_{0}^{\infty} B(\omega)\sin\omega t\,d\omega$$

thus displaying the Fourier integral representation of a function f as the sum of a cosine and sine integral.

EXAMPLE 2

For the function f of Example 1, Eqs. (15a) and (15b) read

$$A(\omega) = \frac{1}{\pi}\int_{0}^{\infty} e^{-at}\cos\omega t\,dt \qquad \text{and} \qquad B(\omega) = \frac{1}{\pi}\int_{0}^{\infty} e^{-at}\sin\omega t\,dt$$

These coefficient functions can be evaluated simultaneously by equating $A(\omega)$ to the real part and $B(\omega)$ to the imaginary part of

$$\frac{1}{\pi}\int_0^\infty e^{-at}\,e^{i\omega t}\,dt$$

whose value, by comparison with $C(\omega)$ of the last example, is

$$\frac{1}{\pi(a - i\omega)} = \frac{a + i\omega}{\pi(a^2 + \omega^2)}$$

Hence

$$A(\omega) = \frac{a}{\pi(a^2 + \omega^2)} \qquad \text{and} \qquad B(\omega) = \frac{\omega}{\pi(a^2 + \omega^2)}$$

and (17) gives as a Fourier integral representation of f

$$f(t) = \frac{a}{\pi}\int_0^\infty \frac{\cos \omega t}{a^2 + \omega^2}\,d\omega + \frac{1}{\pi}\int_0^\infty \frac{\omega \sin \omega t}{a^2 + \omega^2}\,d\omega$$

which, upon combining integrals, yields the standard Fourier integral representation of f.

If f is an even function, the integrands of (15a) and (15b) are, respectively, even and odd functions of t. In this case, $B(\omega) = 0$ and both (16) and (17) can be written

$$f(t) = \frac{2}{\pi}\int_0^\infty \int_0^\infty f(\tau)\cos \omega\tau \cos \omega t\,d\tau\,d\omega \qquad f \text{ even} \tag{18}$$

This is the so-called **Fourier cosine integral** of f and is analogous to the half-range cosine expansion of an even periodic function.

If the function f is odd, the integrands of (15a) and (15b) are, respectively, odd and even functions of t. In this case, $A(\omega) = 0$ and both (16) and (17) can be written

$$f(t) = \frac{2}{\pi}\int_0^\infty \int_0^\infty f(\tau)\sin \omega\tau \sin \omega t\,d\tau\,d\omega \qquad f \text{ odd} \tag{19}$$

This is the **Fourier sine integral** of f, the analog of the half-range sine expansion of an odd periodic function.

Thus, by merely imagining an even or odd extension of a function known to satisfy the conditions of Theorem 1 only on the positive half-axis, we can obtain either a Fourier cosine or a Fourier sine integral representation of the function on $(0, \infty)$. This is an important property, for there are problems, in partial differential equations for instance, which require half-range integral representations of a particular type.

When working with even and odd functions, it is convenient to have the cosine integral (18) split into the **Fourier cosine integral pair**

$$f(t) = \int_0^\infty A(\omega)\cos \omega t\,d\omega \tag{20a}$$

$$A(\omega) = \frac{2}{\pi}\int_0^\infty f(t)\cos \omega t\,dt \tag{20b}$$

f even

and the sine integral (19) expressed as the **Fourier sine integral pair**

$$f(t) = \int_0^\infty B(\omega)\sin\omega t\, d\omega \tag{21a}$$

$$B(\omega) = \frac{2}{\pi}\int_0^\infty f(t)\sin\omega t\, dt \tag{21b}$$

f odd

As soon as f is given, according as it is even or odd, the coefficient function $A(\omega)$ or $B(\omega)$ can immediately be evaluated. Substitution of this value into (20a) or (21a), whichever is appropriate, then gives the Fourier integral of f.

If f is a function such that $f(t) = 0$ for $t < 0$, the expression for $A(\omega)$ given by (20b) is twice that given by (15a). Likewise, in this case, the expression for $B(\omega)$ given by (21b) is twice that of (15b). Hence, comparing Eqs. (20a) and (21a) with Eq. (17), we have the following useful result.

THEOREM 2 If for $t < 0$, $f(t) = 0$, then the Fourier cosine integral (18) and the Fourier sine integral (19) are, respectively, just twice the first and second terms in the trigonometric Fourier integral (17) of f.

EXAMPLE 3

Find the Fourier cosine and sine integral representations of

$$f(t) = e^{-at} \qquad 0 < t < \infty,\ a > 0$$

Applying Theorem 2 to the last equation of Example 2, we see at once that the required cosine integral is

$$f(t) = \frac{2a}{\pi}\int_0^\infty \frac{\cos\omega t}{a^2+\omega^2}\, d\omega$$

and the sine integral is

$$f(t) = \frac{2}{\pi}\int_0^\infty \frac{\omega\sin\omega t}{a^2+\omega^2}\, d\omega$$

It is interesting to note the behavior of the coefficient functions

$$A(\omega) = \frac{2}{\pi}\frac{a}{a^2+\omega^2} \qquad \text{and} \qquad B(\omega) = \frac{2}{\pi}\frac{\omega}{a^2+\omega^2}$$

in Example 3. As $\omega \to \infty$, the cosine coefficient function $A(\omega)$ approaches zero as c/ω^2, whereas the sine coefficient function $B(\omega)$ approaches zero as c/ω. Recalling Theorem 3, Sec. 8.3, this should have been expected, since $A(\omega)$ relates to the even function shown in Fig. 9.4a which becomes everywhere continuous upon removal of the discontinuity at $t = 0$, while $B(\omega)$ relates to the odd function shown in Fig. 9.4b whose discontinuity at $t = 0$ is nonremovable. Theorems 1 and 2, Sec. 8.3, can also be extended to Fourier coefficient functions, but we shall leave this to the exercises.

In the second paragraph of this section, we introduced a family of periodic functions $f_p(t)$ for which the limiting function, as $p \to \infty$, was the nonperiodic isolated pulse

$$f(t) = \begin{cases} 1 & |t| < 1 \\ 0 & |t| > 1 \end{cases} \tag{22}$$

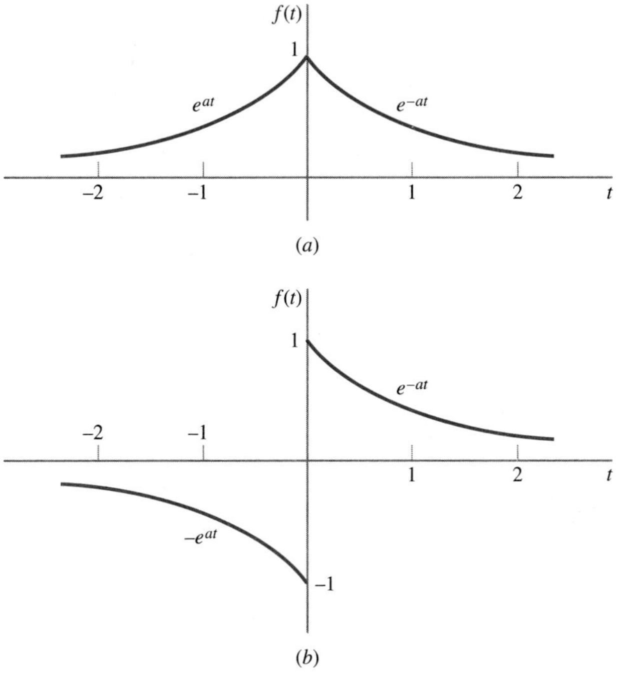

FIGURE 9.4
Graphs of the functions discussed in Example 3.

Since this function is clearly even, its Fourier integral can be found using (20*b*), and we have

$$A(\omega) = \frac{2}{\pi}\int_0^1 \cos \omega t \, dt = \frac{2}{\pi}\left[\frac{\sin \omega t}{\omega}\right]_0^1 = \frac{2}{\pi}\frac{\sin \omega}{\omega}$$

Using this, (20*a*) then gives us

$$f(t) = \frac{2}{\pi}\int_0^\infty \frac{\sin \omega \cos \omega t}{\omega}\, d\omega \tag{23}$$

in agreement with the previously predicted limit (5).

Although it is impossible to find an elementary antiderivative of the integrand in (23), we do know from Theorem 1 that

$$\frac{2}{\pi}\int_0^\infty \frac{\sin \omega \cos \omega t}{\omega}\, d\omega = \begin{cases} 1 & |t| < 1 \\ \frac{1}{2} & |t| = 1 \\ 0 & |t| > 1 \end{cases} \tag{24}$$

In particular, setting $t = 0$, we get

$$\int_0^\infty \frac{\sin \omega}{\omega}\, d\omega = \frac{\pi}{2} \tag{25}$$

which gives the limit, as $x \to \infty$, of the so-called **sine integral function,** denoted by Si and defined on $(-\infty, \infty)$ by

$$\mathrm{Si}(x) = \int_0^x \frac{\sin \omega}{\omega}\, d\omega \tag{26}$$

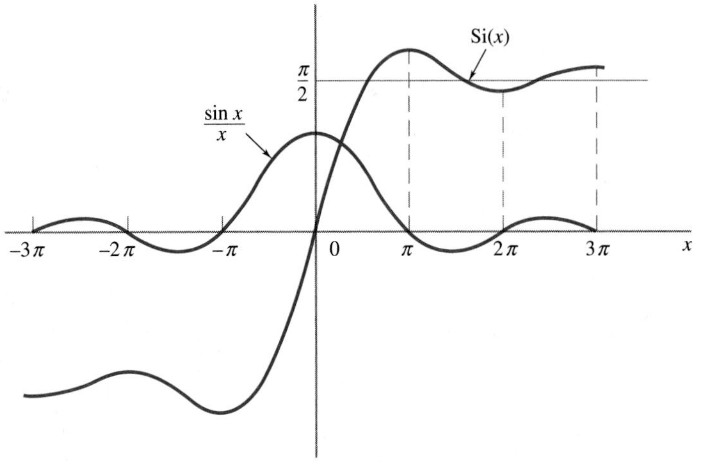

FIGURE 9.5
Graphs of Si(x) and its integrand.

As we shall see, this function plays a very important role in the study of Fourier integrals. A portion of its graph is shown in Fig. 9.5. Values of Si(x) are tabulated in numerous handbooks.†

EXERCISES

1. Each of the following representations identifies a typical member f_p of a family of periodic functions whose periods $2p$ increase with p. Define an amplitude function for each family and in each case express $\lim_{p\to\infty} f_p(t)$ as an integral.

(a) $f_p(t) = \begin{cases} 0 & -p < t < -1 \\ -1 & -1 < t < 0 \\ 1 & 0 < t < 1 \\ 0 & 1 < t < p \end{cases}$

(b) $f_p(t) = \begin{cases} 0 & -p < t < -1 \\ t & -1 < t < 1 \\ 0 & 1 < t < p \end{cases}$

(c) $f_p(t) = \begin{cases} 0 & -p < t \le -1 \\ 1+t & -1 \le t \le 0 \\ 1-t & 0 \le t \le 1 \\ 0 & 1 \le t < p \end{cases}$

(d) $f_p(t) = \begin{cases} 0 & -p < t < -1 \\ -1-t & -1 < t < 0 \\ 1-t & 0 < t < 1 \\ 0 & 1 < t < p \end{cases}$

2. Work Exercise 1 for the two families of periodic functions characterized by

(a) $f_p(t) = \begin{cases} 0 & -p < t < -1 \\ e^t & -1 < t \le 0 \\ e^{-t} & 0 \le t < 1 \\ 0 & 1 < t < p \end{cases}$

(b) $f_p(t) = \begin{cases} e^t & -p \le t \le 0 \\ e^{-t} & 0 \le t \le p \end{cases}$

In what significant ways do the results of Part **(a)** differ from the results of Part **(b)**?

3. If in each period, $f_p(t)$ consists of a single pulse whose shape and duration remain fixed as $p \to \infty$, show that the summand of (9) is independent of p.

With f_p defined in terms of f by

$$f_p(t) = \begin{cases} f(t) & -p \le t < p \\ f_p(t+2p) & t \text{ real} \end{cases}$$

sketch two periods of f_1 and f_2 for each of the following functions.

4. $f(t) = \begin{cases} 0 & t < 0 \\ e^{-t} & t > 0 \end{cases}$

5. $f(t) = \begin{cases} 0 & -\infty < t < 0 \\ 1 & 0 < t < 2 \\ 0 & 2 < t < \infty \end{cases}$

6. $f(t) = \begin{cases} 0 & -\infty < t \le 0 \\ \sin(\pi t/2) & 0 \le t \le 4 \\ 0 & 4 \le t < \infty \end{cases}$

7. $f(t) = \begin{cases} 0 & -\infty < t < 0 \\ 2-t & 0 < t \le 2 \\ 0 & 2 \le t < \infty \end{cases}$

Find the complex exponential Fourier integral of each of the following functions and reduce the integral to a purely real form.

8. $f(t) = \begin{cases} 0 & -\infty < t < 0 \\ 1 & 0 < t < 1 \\ 0 & 1 < t < \infty \end{cases}$

†See, for instance, E. Jahnke, F. Emde, and F. Lösch, *Tables of Higher Functions,* 6th ed., McGraw-Hill, New York, 1960.

9. $f(t) = \begin{cases} 0 & -\infty < t < 0 \\ 2^{-t} & 0 < t < \infty \end{cases}$

10. $f(t) = \begin{cases} 0 & -\infty < t \le -1 \\ 1+t & -1 \le t \le 0 \\ 1 & 0 \le t < 1 \\ 0 & 1 < t < \infty \end{cases}$

11. $f(t) = \begin{cases} 0 & -\infty < t \le -1 \\ 1-t^2 & -1 \le t \le 0 \\ \cos t & 0 \le t \le \pi/2 \\ 0 & \pi/2 \le t < \infty \end{cases}$

12. $f(t) = \begin{cases} 0 & |t| > p \\ \cosh t & |t| < p \end{cases}$

13. $f(t) = \begin{cases} 0 & |t| > p \\ \sinh t & |t| < p \end{cases}$

Use the formulas for the coefficient functions of the standard Fourier integral or one of the formulas (20*b*) or (21*b*) to find the Fourier integral of each of the following functions.

14. $f(t) = \begin{cases} \pi & -1 < t < 0 \\ -\pi & 0 < t < 1 \\ 0 & 1 < |t| \end{cases}$

15. $f(t) = \begin{cases} b & -1 < t < 0 \\ a & 0 < t < 1 \\ 0 & 1 < |t| \end{cases}$

16. $f(t) = \begin{cases} -t & |t| < \pi \\ 0 & |t| > \pi \end{cases}$

17. $f(t) = \begin{cases} 1+t & |t| < 1 \\ 0 & |t| > 1 \end{cases}$

18. $f(t) = \begin{cases} \sin t & |t| \le \pi \\ 0 & |t| \ge \pi \end{cases}$

19. $f(t) = \begin{cases} \cos t & |t| \le \pi/2 \\ 0 & |t| \ge \pi/2 \end{cases}$

20. $f(t) = \begin{cases} a & -2 < t < -1 \\ b & -1 < t < 1 \\ a & 1 < t < 2 \\ 0 & 2 < |t| \end{cases}$

21. $f(t) = \begin{cases} 0 & -\infty < t < -1 \\ 1-t & -1 < t < 0 \\ t & 0 < t < 1 \\ 0 & 1 < t < \infty \end{cases}$

22. $f(t) = \begin{cases} 0 & |t| > 1 \\ t^2 & |t| < 1 \end{cases}$

23. $f(t) = \begin{cases} 1-t^2 & |t| \le 1 \\ 0 & |t| \ge 1 \end{cases}$

24. $f(t) = \begin{cases} |\sin \pi t| & |t| \le 1 \\ 0 & |t| \ge 1 \end{cases}$

25. $f(t) = \begin{cases} 0 & 1 < |t| \\ 1+at & -1 < t \le 0 \\ 1+bt & 0 \le t < 1 \end{cases}$

26. $f(t) = \begin{cases} 0 & -\infty < t < b \\ e^{-at} & b < t < \infty,\ 0 < a \end{cases}$

27. $f(t) = \begin{cases} e^{a(t-b)} & -\infty < t < b,\ 0 < a \\ 0 & b < t < \infty \end{cases}$

Find the Fourier cosine and sine integral representations of each of the following functions.

28. $f(t) = \begin{cases} 0 & 0 \le t < 1 \\ -1 & 1 < t < 2 \\ 0 & 2 < t < \infty \end{cases}$

29. $f(t) = \begin{cases} 0 & 0 \le t < a \\ d & a < t < b \\ k & b < t < c \\ 0 & c < t < \infty \end{cases}$

30. $f(t) = \begin{cases} t-t^2 & 0 \le t \le 1 \\ 0 & 1 \le t < \infty \end{cases}$

31. $f(t) = \begin{cases} (2-t)/e & 0 < t \le 1 \\ e^{-t} & 1 \le t < \infty \end{cases}$

32. $f(t) = \begin{cases} 1+\sin t & 0 < t \le \pi \\ 2 - t/\pi & \pi \le t \le 2\pi \\ 0 & 2\pi \le t < \infty \end{cases}$

33. $f(t) = \begin{cases} 1+\cos t & 0 \le t \le \pi \\ 0 & \pi \le t < \infty \end{cases}$

Without finding any Fourier coefficient functions, discuss the relation between the continuity of each of the following functions and the rate at which its spectrum tends to zero as $\omega \to \infty$.

34. $f(t) = \begin{cases} 0 & -\infty < t < 0 \\ 1/n & n-1 < t < n,\ n \le 10 \\ 0 & 10 < t < \infty \end{cases}$

35. $f(t) = \begin{cases} 1+t/2 & -2 \le t \le -1 \\ 1/(1+t^2) & |t| \le 1 \\ 1-t/2 & 1 \le t \le 2 \\ 0 & |t| \ge 2 \end{cases}$

36. $f(t) = \begin{cases} 0 & -\infty < t \le -\pi/2 \\ 1+\sin t & -\pi/2 \le t \le 3\pi/2 \\ 0 & 3\pi/2 \le t < \infty \end{cases}$

37. $f(t) = \begin{cases} \dfrac{4(t+e)}{5(e-\ln 2)} & -e \le t \le -\ln 2 \\ \operatorname{sech} t & |t| \le \ln 2 \\ \dfrac{4(e-t)}{5(e-\ln 2)} & \ln 2 \le t \le e \\ 0 & |t| \ge e \end{cases}$

Using Theorem 1 in conjunction with the Fourier integral of the function given in the exercise indicated, establish each of the following integration formulas.

38. $\displaystyle\int_0^\infty \frac{(1-\cos\omega)\sin\omega}{\omega}\,d\omega = \frac{\pi}{4}$ (Exercise 14)

39. $\displaystyle\int_0^\infty \frac{\sin^2 \pi\omega}{1-\omega^2}\,d\omega = 0$ (Exercise 18)

40. $\displaystyle\int_0^\infty \frac{1-\cos 2\pi\omega - \pi\omega\sin 2\pi\omega}{\omega^2}\,d\omega = \frac{\pi^2}{2}$ (Exercise 16)

41. $\displaystyle\int_0^\infty \frac{[(b-a)+2a\cos\omega]\sin 2\omega}{\omega}\,d\omega = \frac{(a+b)\pi}{2}$ (Exercise 20)

42. $\displaystyle\int_0^\infty \frac{\sin \pi\omega}{\omega(1-\omega^2)}\,d\omega = \pi$ (Exercise 33)

43. $\displaystyle\int_0^\infty \frac{\sin 2\pi\omega}{\omega(1-\omega^2)}\,d\omega = 0$ (Exercise 33)

44. $\displaystyle\int_0^\infty \frac{1-\cos\omega}{\omega^2}\,d\omega = \frac{\pi}{2}$ (Exercise 25)

45. $\displaystyle\int_0^\infty \left(\frac{\sin\omega}{\omega}\right)^2 d\omega = \frac{\pi}{2}$ (Exercise 25)

46. $\displaystyle\int_0^\infty \frac{(1+b)\omega\sin\omega - b(1-\cos\omega)}{\omega^2}\,d\omega = \frac{\pi}{2}$ (Exercise 25)

47. $\displaystyle\int_0^\infty \frac{[(1-b)\omega\cos\omega + b\sin\omega]\sin\omega}{\omega^2}\,d\omega = \frac{(1+b)\pi}{4}$ (Exercise 25)

48. Verify Eq. (25) using integration by parts in Exercise 44.

49. Find the Fourier integral of the function defined on $(-\infty,\infty)$ by $f(t) = e^{-t^2/2}$ and use it to evaluate $\int_0^\infty e^{-\omega^2/2}\,d\omega$. *Hint:* Use a table of integrals to find $A(\omega)$.

50. Find the Fourier integral of the function defined on $(-\infty,\infty)$ by $f(t) = te^{-t^2/2}$ and use it to evaluate $\int_0^\infty \omega e^{-\omega^2/2}\sin\omega\,d\omega$. *Hint:* Use integration by parts and a table of integrals to find $B(\omega)$.

51. Find the Fourier integral of the function defined on $(-\infty,\infty)$ by $f(t) = 1/(1+t^2)$ and then evaluate $\int_0^\infty\int_0^\infty e^{-\omega}\cos\omega t\,d\omega\,dt$. *Hint:* Use a table of integrals to find $A(\omega)$.

Express the spectrum $C(\omega)$ of each of the following functions in the polar form $C(\omega) = r(\omega)e^{i\theta(\omega)}$, where $r(\omega)$ is the magnitude, or absolute value, of $C(\omega)$ and $\theta(\omega)$ is the argument, or phase angle, of $C(\omega)$. *Hint:* $|u/v| = |u|/|v|$.

52. $f(t) = \begin{cases} 0 & -\infty < t < b \\ e^{-a(t-b)} & b < t < \infty \end{cases}$

53. $f(t) = \begin{cases} 0 & -\infty < t < -b \\ e^{-a(t+b)} & -b < t < \infty \end{cases}$

54. $f(t) = \begin{cases} 0 & -\infty < t < 0 \\ 1 & 0 < t < 1 \\ 0 & 1 < t < \infty \end{cases}$

55. $f(t) = \begin{cases} 0 & -\infty < t < 1 \\ 1 & 1 < t < 2 \\ 0 & 2 < t < \infty \end{cases}$

56. Suppose f has a nonzero jump J_k at the point t_k and that the number of such jumps is either finite or infinite. Letting $C'(\omega)$ denote the spectrum of f' and assuming the existence of both $C(\omega)$ and $C'(\omega)$, use integration by parts just as we did in deriving Formulas (3) and (4), Sec. 8.3, to show that

$$C(\omega) = \frac{1}{i\omega}C'(\omega) + \frac{1}{2\pi i\omega}\sum_k J_k e^{-i\omega t_k}$$

where the summation takes place over all integral values of k used to identify the jumps of f on $(-\infty,\infty)$.

57. Show that the standard coefficient functions (15*a*) and (15*b*) appearing in Eqs. (16) and (17) are given in terms of the coefficient functions $A'(\omega)$ and $B'(\omega)$ in the standard Fourier integral of f' by

$$(27a)\quad A(\omega) = -\frac{1}{\omega}B'(\omega) - \frac{1}{\pi\omega}\sum_k J_k \sin\omega t_k$$

$$(27b)\quad B(\omega) = \frac{1}{\omega}A'(\omega) + \frac{1}{\pi\omega}\sum_k J_k \cos\omega t_k$$

Hint: Follow the procedure and use the jump notation of Exercise 56.

58. Use Formulas (27*a*) and (27*b*) iteratively to express $A'(\omega)$ and $B'(\omega)$ in terms of the Fourier coefficient functions $A''(\omega)$ and $B''(\omega)$ of f'' and the jumps J_k' of f'.

59. Use the results of Exercises 57 and 58 to express $A(\omega)$ and $B(\omega)$ in terms of $A'(\omega)$, $B'(\omega)$, $A''(\omega)$, $B''(\omega)$, and the jumps of f and f'.

Using (27*a*) and (27*b*) as iteration formulas, find Fourier integrals of each of the following functions.

60. $f(t) = \begin{cases} 0 & -\infty < t < -1 \\ k & -1 < t < b \\ 0 & b < t < \infty \end{cases}$

61. $f(t) = \begin{cases} \pi + t & -\pi \le t \le 0 \\ \pi - t & 0 \le t \le \pi \\ 0 & |t| \ge \pi \end{cases}$

62. $f(t) = \begin{cases} 0 & -\infty < t \le 0 \\ \sin t & 0 \le t \le \pi \\ 0 & \pi \le t < \infty \end{cases}$

63. $f(t) = \begin{cases} \pi t & |t| < \pi \\ 0 & |t| > \pi \end{cases}$

64. $f(t) = \begin{cases} t^2 - 1 & |t| \le 1 \\ 0 & |t| \ge 1 \end{cases}$

65. $f(t) = \begin{cases} \sin \pi t & |t| \le 1 \\ 0 & |t| \ge 1 \end{cases}$

66. $f(t) = ke^{-|t|} \qquad -\infty < t < \infty$

67. $f(t) = \begin{cases} t & -1 < t \le 0 \\ t^2 & 0 \le t < 1 \\ 0 & |t| > 1 \end{cases}$

Using (27*a*) and (27*b*), find Fourier cosine and sine integral representations for each of the following functions.

68. $f(t) = \begin{cases} 0 & 0 \le t < 1 \\ 1 & 1 < t < b \\ 0 & b < t < \infty \end{cases}$

69. $f(t) = \begin{cases} 2t - 4 & 0 < t \le 2 \\ 0 & 2 \le t < \infty \end{cases}$

70. $f(t) = \begin{cases} e^t & 0 < t < 1 \\ 0 & 1 < t < \infty \end{cases}$

71. $f(t) = \begin{cases} 0 & 0 \le t < 1 \\ \sin(\pi t/2) & 1 < t \le 2 \\ 0 & 2 \le t < \infty \end{cases}$

72. $f(t) = \begin{cases} 1 + t + t^2/2 & 0 < t < 1 \\ 0 & 1 < t < \infty \end{cases}$

Hint: Add the integrals of three functions whose sum is the even (odd) extension of *f*.

Express each of the following definite integrals in terms of the sine integral function.

73. $\int_a^b \frac{\sin x^n}{x}\,dx$. *Hint:* Let $x^n = u$.

74. $\int_a^b \sin e^{\lambda x}\,dx$

75. $\int_a^b \frac{\sin x \cos x}{x}\,dx$

76. $\int_0^{27} \frac{\sin x^{1/3}}{x}\,dx$

77. $\int_a^b \frac{\cos x}{x^2}\,dx$. *Hint:* First integrate by parts.

78. $\int_a^b \frac{\sin x}{x^3}\,dx$. *Hint:* Use integration by parts.

79. $\int_1^2 \frac{\sin x \sin 2x \sin 3x}{x}\,dx$

Hint: Convert the product of two factors into a sum or difference, then repeat the process.

The **cosine integral function** Ci(x) is defined by the formula

$$\text{Ci}(x) = -\int_x^\infty \frac{\cos u}{u}\,du \qquad 0 < x$$

Evaluate the following definite integrals in terms of Ci(x).

80. $\int_1^2 \frac{\cos x^n}{x}\,dx$

81. $\int_a^b \cos e^{\lambda x}\,dx$

82. $\int_a^b \frac{\cos x \cos 2x}{x}\,dx$

83. $\int_a^b \frac{\sin x \sin 2x}{x}\,dx$

The **exponential integral function** Ei(x) is defined by the formula

$$\text{Ei}(x) = \int_x^\infty \frac{e^{-u}}{u}\,du \qquad 0 < x$$

Evaluate the following definite integrals in terms of Ei(x).

84. $\int_2^\infty \frac{e^{-3x}}{x}\,dx$

85. $\int_1^\infty \frac{e^{-x}}{x+1}\,dx$

86. $\int_2^\infty \frac{e^{-x}}{x^2}\,dx$. *Hint:* First use integration by parts.

87. $\int_1^\infty \frac{e^{-x}}{x^2 + x}\,dx$. *Hint:* First use partial fractions.

88. If $f(x, y)$ is a function satisfying suitable conditions, show that *f* has a **two-dimensional Fourier coefficient function** given by the formula

$$G(\xi, \eta) = \frac{1}{2\pi}\int_{-\infty}^{\infty}\int_{-\infty}^{\infty} f(x, y)e^{-i(\xi x + \eta y)}\,dx\,dy$$

Hint: First use Eq. (12) to find the spectrum $C(\xi, y)$ of *f*, thought of as a function of the single variable *x*. Then find the coefficient function of $C(\xi, y)$, thought of as a function of *y*.

89. If $G(\xi, \eta)$ as given in Exercise 88 is the two-dimensional coefficient function of $f(x, y)$, show that

$$f(x, y) = \frac{1}{2\pi}\int_{-\infty}^{\infty}\int_{-\infty}^{\infty} G(\xi, \eta)e^{i(x\xi + y\eta)}\,d\xi\,d\eta$$

9.2 FOURIER INTEGRAL APPROXIMATIONS AND THE GIBBS PHENOMENON

In the Fourier series representation of a periodic function *f*, it was a matter of some interest to determine how well the first few terms of the expansion represented the function (Fig. 8.3). The corresponding problem in the nonperiodic case is to investigate how well the Fourier integral represents a function when only the components in the lower portion of the (continuous) frequency range are taken into account. For the Fourier integral, this can be done in surprisingly general terms.

Let $P(\omega_0, t)$ be the approximation to a function *f* provided by that portion of its Fourier integral involving only frequencies between 0 and ω_0, if the Fourier cosine or sine integral is used, or between $-\omega_0$ and ω_0, if the standard or complex integral is used. Choosing the complex form and using Eq. (10), Sec. 9.1, we have, upon changing the dummy variable in the inner integral to τ and then moving $e^{i\omega t}$ across the inner integral sign,

$$P(\omega_0, t) = \frac{1}{2\pi}\int_{-\omega_0}^{\omega_0}\int_{-\infty}^{\infty} f(\tau)e^{-i\omega(\tau - t)}\,d\tau\,d\omega$$

Interchanging the order of integration, we get

$$\begin{aligned} P(\omega_0, t) &= \frac{1}{2\pi}\int_{-\infty}^{\infty} f(\tau)\int_{-\omega_0}^{\omega_0} e^{-i\omega(\tau-t)}\,d\omega\,d\tau \\ &= \frac{1}{2\pi}\int_{-\infty}^{\infty} f(\tau)\left[\frac{e^{-i\omega(\tau-t)}}{-i(\tau-t)}\right]_{-\omega_0}^{\omega_0} d\tau \\ &= \frac{1}{\pi}\int_{-\infty}^{\infty} f(\tau)\frac{e^{i\omega_0(\tau-t)}-e^{-i\omega_0(\tau-t)}}{2i(\tau-t)}\,d\tau \end{aligned}$$

(1) $$= \frac{1}{\pi}\int_{-\infty}^{\infty} f(\tau)\frac{\sin\omega_0(\tau-t)}{\tau-t}\,d\tau$$

This remarkable result is valid for any function f that has a Fourier integral representation and plays an important role in the theory of Fourier integrals and many more advanced applications.

To illustrate its use, let us apply it to the isolated pulse of Eq. (22), Sec. 9.1. For this function, Eq. (1) becomes

$$P(\omega_0, t) = \frac{1}{\pi}\int_{-1}^{1}\frac{\sin\omega_0(\tau-t)}{\tau-t}\,d\tau$$

or, making the substitution $\omega_0(\tau - t) = -x$, we get $d\tau = -dx/\omega_0$ and

$$\begin{aligned} P(\omega_0, t) &= \frac{1}{\pi}\int_{\omega_0(t-1)}^{\omega_0(t+1)}\frac{\sin x}{x}\,dx \\ &= \frac{1}{\pi}\int_{0}^{\omega_0(t+1)}\frac{\sin x}{x}\,dx - \frac{1}{\pi}\int_{0}^{\omega_0(t-1)}\frac{\sin x}{x}\,dx \end{aligned}$$

(2) $$= \frac{1}{\pi}\mathrm{Si}[\omega_0(t+1)] - \frac{1}{\pi}\mathrm{Si}[\omega_0(t-1)]$$

where $\mathrm{Si}(x)$ is the sine integral function we introduced in Eq. (26), Sec. 9.1. Figure 9.6 shows this approximation for $\omega_0 = 4$, 8, and 16. Physically speaking, these curves represent the output of an ideal low-pass filter, cutting off all frequencies above ω_0 when the input signal is an isolated rectangular pulse.

An examination of the highest and lowest points on the curves shown in Fig. 9.6 reveals that, as ω_0 increases, these points move closer and closer to the discontinuities of f at $t = \pm 1$, but they do not settle down on the graph of the function. This illustrates an interesting and important characteristic of both Fourier integrals and Fourier series. As ω_0 (or the number of terms) increases and the approximation to $f(t)$ becomes better and better, it appears that the vertical amount by which the approximating curve overshoots the graph of the function on the right and left of each discontinuity does *not* approach zero, as we might expect, but is wiped out in the limit by being pinched into a vanishingly small interval on each side of the discontinuity. This behavior is somewhat apparent from Eq. (2) because, as ω_0 increases, configurations are compressed in the direction of the t axis inversely as ω_0.

To better understand these ideas as they relate to a general nonperiodic function, let us now consider a function f with a single jump J which, without loss of generality, we may suppose is positive at the origin, that is,

$$J = f(0^+) - f(0^-) > 0$$

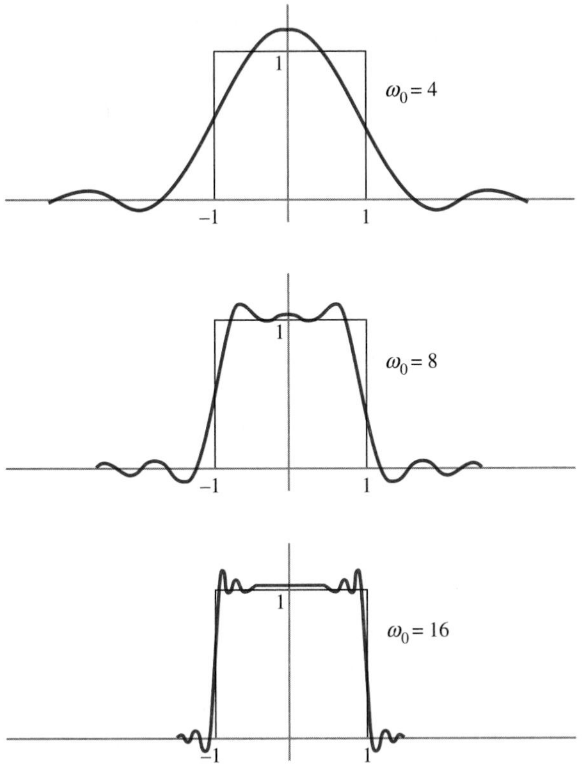

FIGURE 9.6
The approximation of a function by its Fourier integral taken only over frequencies less than ω_0.

Clearly, if the right half of the graph of f, where $t > 0$, is translated downward a distance J and then joined to the left half (Fig. 9.7), the resultant graph will be that of a function ϕ which is continuous on $(-\infty, \infty)$. To compensate for this displacement we must add to ϕ a step function of height J whose point of discontinuity occurs at $t = 0$, that is,

$$f(t) = \phi(t) + Ju(t)$$

Applying Eq. (1) to f, written in the last form, we obtain

$$P(\omega_0, t) = \frac{1}{\pi}\int_{-\infty}^{\infty} \phi(\tau)\frac{\sin \omega_0(\tau - t)}{\tau - t}\,d\tau + \frac{J}{\pi}\int_{-\infty}^{\infty} u(\tau)\frac{\sin \omega_0(\tau - t)}{\tau - t}\,d\tau$$

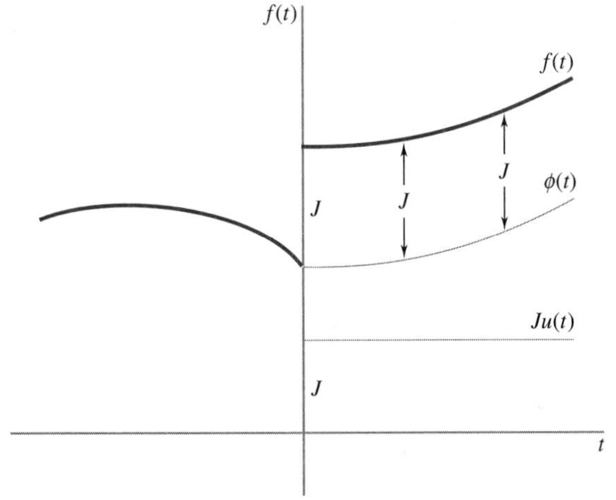

FIGURE 9.7
A function made continuous by the translation of one segment.

As $\omega_0 \to \infty$, the first term in this formula for $P(\omega_0, t)$ converges to the Fourier integral of ϕ, which for all t equals $\phi(t)$ because ϕ is everywhere continuous. Hence we must look to the second term of the formula for an explanation of the overshoot of $P(\omega_0, t)$ on either side of $t = 0$. This term is simply

$$\frac{J}{\pi}\int_0^\infty \frac{\sin \omega_0(\tau - t)}{\tau - t}\, d\tau$$

and under the substitution $x = -\omega_0(\tau - t)$ it becomes

$$\frac{J}{\pi}\int_{-\infty}^{\omega_0 t} \frac{\sin x}{x}\, dx = J\left(\frac{1}{\pi}\int_{-\infty}^{0} \frac{\sin x}{x}\, dx + \frac{1}{\pi}\int_0^{\omega_0 t} \frac{\sin x}{x}\, dx\right)$$

Since $(\sin x)/x$ is an even function, the first integral on the right is the same as the integral from 0 to ∞, which we have already seen to be equal to $\pi/2$ [see Eq. (25), Sec. 9.1]. The entire right-hand side thus becomes

$$J\left[\frac{1}{2} + \frac{1}{\pi}\mathrm{Si}(\omega_0 t)\right] \tag{3}$$

and for each t this is the amount that must be added to the approximation for ϕ to obtain the approximation $P(\omega_0, t)$ to f itself, for any particular value of ω_0.

To investigate this in more detail it is convenient to consider the graph of the related function

$$s(z) = \frac{1}{2} + \frac{1}{\pi}\mathrm{Si}(z)$$

shown in Fig. 9.8. It is easy to verify that for positive values of z, the first, and largest, maximum of this function occurs where $z = \pi$. Now $z = \omega_0 t = \pi$ implies that $t = \pi/\omega_0$. Hence the only effect of ω_0 in (3), as ω_0 increases, is to compress the horizontal scale without altering the ordinates of the graph. Therefore the amount of overshoot to the right of $t = 0$, which is equal to

$$J\left[\frac{1}{2} + \frac{1}{\pi}\mathrm{Si}(\pi)\right] - J = J\left[\frac{1}{\pi}\mathrm{Si}(\pi) - \frac{1}{2}\right] \doteq 0.09\text{ J}$$

remains constant as the t scale is increasingly compressed. Since $\mathrm{Si}(x)$ is an odd function, the overshoot to the left of $t = 0$ is

$$J\left[\frac{1}{2} - \frac{1}{\pi}\mathrm{Si}(\pi)\right] \doteq -0.09\text{ J}$$

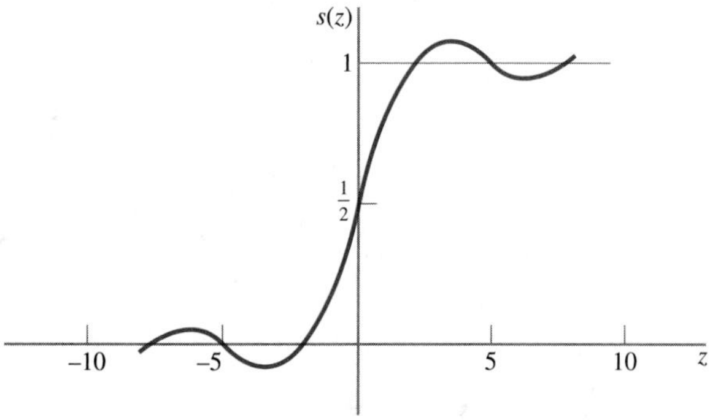

FIGURE 9.8
The graph of $s(z) = \frac{1}{2} + (1/\pi)\,\mathrm{Si}\,(z)$.

and it too is independent of ω_0. Both overshoots disappear in the limit not because they shrink to zero but because their bases are pinched to zero as $\omega_0 \to \infty$ so that the t scale is increasingly compressed. This curious behavior is known as the **Gibbs phenomenon,** after the American mathematical physicist Josiah Willard Gibbs (1839–1903), who brought it to the attention of the scientific world in a letter to the journal *Nature* in 1899. Actually, it was first noticed in 1848 by the mathematician Henry Wilbraham and first explained by the American mathematician Maxime Bocher (1867–1918) in 1906.

We are now prepared to show why the Fourier integral of a function f converges to the average of the right- and left-hand limits at each discontinuity of f. Since $\mathrm{Si}(0) = 0$, it follows from (3) that for every value of ω_0 the approximation to $J \cdot u(t)$ at $t = 0$ is $\frac{1}{2}J$. For each value of ω_0, this is the amount that must be added to the approximation to ϕ at $t = 0$ in order to obtain the approximation $P(\omega_0, 0)$ to f at $t = 0$. Since, as we observed earlier, the approximation to $\phi(t)$ converges to $\phi(t)$ for all values of t and since, by definition, $\phi(0) = f(0^-)$, it follows that as $\omega_0 \to \infty$ we have

$$\lim_{\omega_0 \to \infty} P(\omega_0, 0) = \phi(0) + \tfrac{1}{2}J = f(0^-) + \tfrac{1}{2}[f(0^+) - f(0^-)] = \tfrac{1}{2}[f(0^+) + f(0^-)]$$

The preceding argument can also be applied to explain the Gibbs phenomenon and the average-value property of Fourier series, although, curiously, the details are somewhat more complicated.

EXERCISES

For each of the following functions, use Eq. (1) to express $P(\omega_0, t)$ in terms of sine integral functions.

1. $f(t) = \begin{cases} 0 & -\infty < t < -1 \\ -1 & -1 < t < 0 \\ 1 & 0 < t < 1 \\ 0 & 1 < t < \infty \end{cases}$

2. $f(t) = \begin{cases} 0 & -\infty < t < -p \\ 1 & -p < t < 0 \\ -1 & 0 < t < p \\ 0 & p < t < \infty \end{cases}$

3. $f(t) = \begin{cases} 0 & -\infty < t < -1 \\ 1 & -1 < t < 2 \\ 0 & 2 < t < \infty \end{cases}$

4. $f(t) = \begin{cases} 0 & |t| < 1 \\ 1 & 1 < t^2 < 4 \\ 0 & |t| > 2 \end{cases}$

5. $f(t) = \begin{cases} t & |t| < 1 \\ 0 & |t| > 1 \end{cases}$

6. $f(t) = \begin{cases} |t| & |t| < 1 \\ 0 & |t| > 1 \end{cases}$

7. Work Exercise 2 using the Fourier sine integral pair, Eqs. (21*a*) and (21*b*), Sec. 9.1.

8. Work Exercise 4 using the Fourier cosine integral pair, Eqs. (20*a*) and (20*b*), Sec. 9.1.

9. Work Exercise 5 using the Fourier sine integral pair.

10. Work Exercise 6 using the Fourier cosine integral pair.

11. Using the Fourier cosine integral pair, express $P(\omega_0, t)$ in terms of sine integral functions for

$$f(t) = \begin{cases} 0 & -\infty < t \le -1 \\ 1 + t & -1 \le t \le 0 \\ 1 - t & 0 \le t \le 1 \\ 0 & 1 \le t < \infty \end{cases}$$

12. Verify the assertion made on p. 560 that the first positive maximum of the function $s(z) = \frac{1}{2} + (1/\pi)\ \mathrm{Si}(z)$ is $\frac{1}{2} + (1/\pi)\ \mathrm{Si}(\pi)$. *Hint:* What is the derivative of the integral defining $\mathrm{Si}(z)$?

9.3 PROPERTIES OF FOURIER TRANSFORMS

So far, except for making use of the alternative formulas (27*a*) and (27*b*), Sec. 9.1, the only method we have for finding the Fourier integral of a function is by direct integration using suitable combinations of integral formulas as stipulated in Sec. 9.1. However, if the members of a Fourier transform *pair* [Eqs. (11) and (12), (20*a*) and (20*b*), or (21*a*) and (21*b*), Sec. 9.1] are thought of primarily as functionals, or *transformations,* they are found to have properties which in many cases enable us to find the Fourier integral of a function without integration. In this section we shall develop some of these properties and illustrate some of their uses.

In the complex integral pair [Eqs. (11) and (12), Sec. 9.1], the factor $1/2\pi$ was associated with the integral for the coefficient function $C(\omega)$ because that was where it arose most naturally in the transition from Fourier series to Fourier integrals. However, by incorporating the factor $1/2\pi$ in the integral for $f(t)$ and rewriting the transform pair as

$$f(t) = \frac{1}{2\pi}\int_{-\infty}^{\infty} F(\omega)e^{i\omega t}\,d\omega \tag{1}$$

$$F(\omega) = \int_{-\infty}^{\infty} f(t)e^{-i\omega t}\,dt \tag{2}$$

a simpler coefficient function $F(\omega) = 2\pi C(\omega)$ is introduced which, since $1/2\pi$ is not a factor of $F(\omega)$, is easier to write and tabulate than $C(\omega)$. Thus in dealing with *complex* Fourier transform pairs, for convenience, we shall henceforth use Formulas (1) and (2). Even when $f(t)$ is real, the function $F(\omega)$ is generally complex.

The process of multiplying $f(t)$ by $e^{-i\omega t}$ and then integrating from $t = -\infty$ to $t = \infty$ assigns a single function F to each function f that satisfies the conditions of Theorem 1, Sec. 9.1. Thus this process, together with its domain of functions, is itself a function, called the **Fourier transformation,** whose value at f is the function F. Denoting this transformation by $\mathcal{F}$, we write

$$F(\omega) = \mathcal{F}[f(t)] \tag{3}$$

which emphasizes the dependence of $F(\omega)$ on $f(t)$. When interpreted as a value of $\mathcal{F}$, the coefficient function $F(\omega)$ is called the **Fourier transform** of $f(t)$. The function $f(t)$ is in turn referred to as the **inverse Fourier transform** of $F(\omega)$ and is denoted by

$$\mathcal{F}^{-1}[F(\omega)] = f(t) \tag{4}$$

Together, f and F, as given by (1) and (2), are commonly referred to as the **complex Fourier transform pair.**

The Fourier cosine and sine integral pairs [Eqs. (20) and (21), Sec. 9.1] can also be interpreted as pairs of mutually inverse transforms. For reasons of symmetry, this is usually done by splitting the numerical factor $2/\pi$ in $A(\omega)$ and $B(\omega)$ into two equal factors and assigning one factor to each integral in the pair. The cosine integral pair then becomes the so-called **Fourier cosine transform pair**

$$f(t) = \sqrt{\frac{2}{\pi}}\int_0^{\infty} F_c(\omega)\cos\omega t\,d\omega \tag{5a}$$

$$F_c(\omega) = \sqrt{\frac{2}{\pi}}\int_0^{\infty} f(t)\cos\omega t\,dt \tag{5b}$$

f even

and the sine integral pair is converted into the **Fourier sine transform pair**

$$f(t) = \sqrt{\frac{2}{\pi}}\int_0^{\infty} F_s(\omega)\sin\omega t\,d\omega \tag{6a}$$

$$F_s(\omega) = \sqrt{\frac{2}{\pi}}\int_0^{\infty} f(t)\sin\omega t\,dt \tag{6b}$$

f odd

where

$$(a)\ F_c(\omega) = \sqrt{\pi/2}\,A(\omega) \qquad (b)\ F_s(\omega) = \sqrt{\pi/2}\,B(\omega) \tag{7}$$

In both the cosine transform pair (5) and Eq. (7*a*), the function $F_c(\omega)$ is known as the **Fourier cosine transform** of $f(t)$, and $f(t)$ is the **inverse Fourier cosine transform** of $F_c(\omega)$; that is, in terms of the Fourier cosine transformation $\mathcal{F}_c$,

$$(a)\ \mathcal{F}_c[f(t)] = F_c(\omega) \qquad (b)\ \mathcal{F}_c^{-1}[F_c(\omega)] = f(t) \tag{8}$$

The function $F_s(\omega)$ in the sine transform pair (6) and Eq. (7*b*) is the **Fourier sine transform** of $f(t)$, and $f(t)$ is the **inverse Fourier sine transform** of $F_s(\omega)$. In terms of the Fourier sine transformation $\mathcal{F}_s$,

$$(a)\ \mathcal{F}_s[f(t)] = F_s(\omega) \qquad (b)\ \mathcal{F}_s^{-1}[F_s(\omega)] = f(t) \tag{9}$$

Each of the last two transform pairs is *completely symmetric* in that an interchange of the functional variables and of the scalar variables in either equation results in the other equation of the pair. For example, when f and F_s, as well as t and ω, are interchanged in (6*a*), the result is (6*b*). This symmetry makes both pairs easy to remember.

The various forms of the Fourier integral that we have encountered and the minor differences between their limits, coefficients, and notation may sometimes seem confusing. Hence, before concentrating in this section on properties of the Fourier transformations $\mathcal{F}$, $\mathcal{F}_c$, and $\mathcal{F}_s$, it may be well to bring together all the forms we have discussed.

TABLE 9.1 The Fourier Integral and Its Variants

The complex exponential Fourier integral, Eq. (10), Sec. 9.1:

$$f(t) = \int_{-\infty}^{\infty} \left[\frac{1}{2\pi}\int_{-\infty}^{\infty} f(t)e^{-i\omega t}\,dt\right]e^{i\omega t}\,d\omega$$

The complex Fourier integral pair, Eqs. (11) and (12), Sec. 9.1:

$$f(t) = \int_{-\infty}^{\infty} C(\omega)e^{i\omega t}\,d\omega$$

$$C(\omega) = \frac{1}{2\pi}\int_{-\infty}^{\infty} f(t)e^{-i\omega t}\,dt$$

[Compare Eqs. (7) and (8), Sec. 8.5.]

The standard Fourier integral, Eqs. (15*a*), (15*b*), and (16), Sec. 9.1:

$$f(t) = \int_0^{\infty} [A(\omega)\cos\omega t + B(\omega)\sin\omega t]\,d\omega$$

$$A(\omega) = \frac{1}{\pi}\int_{-\infty}^{\infty} f(t)\cos\omega t\,dt \quad \text{and} \quad B(\omega) = \frac{1}{\pi}\int_{-\infty}^{\infty} f(t)\sin\omega t\,dt$$

[Compare Eqs. (1), (10), and (11), with $d = -p$, Sec. 8.2.]

The Fourier cosine integral, Eq. (18), Sec. 9.1:

$$f(t) = \frac{2}{\pi}\int_0^{\infty}\int_0^{\infty} f(\tau)\cos\omega\tau\cos\omega t\,d\tau\,d\omega \qquad f \text{ even}$$

The Fourier cosine integral pair, Eqs. (20*a*) and (20*b*), Sec. 9.1:

$$f(t) = \int_0^{\infty} A(\omega)\cos\omega t\,d\omega$$

$$A(\omega) = \frac{2}{\pi}\int_0^{\infty} f(t)\cos\omega t\,dt$$

f even (Compare Theorem 1, Sec. 8.4.)

The Fourier sine integral, Eq. (19), Sec. 9.1:

$$f(t) = \frac{2}{\pi}\int_0^\infty \int_0^\infty f(\tau)\sin\tau\omega\sin\omega t\,d\tau\,d\omega \qquad f\text{ odd}$$

The Fourier sine integral pair, Eqs. (21*a*) and (21*b*), Sec. 9.1:

$$\begin{aligned} f(t) &= \int_0^\infty B(\omega)\sin\omega t\,d\omega \\ B(\omega) &= \frac{2}{\pi}\int_0^\infty f(t)\sin\omega t\,dt \end{aligned} \qquad f\text{ odd} \qquad \text{(Compare Theorem 2, Sec. 8.4.)}$$

The complex Fourier transform pair, Eqs. (1) and (2):

$$f(t) = \frac{1}{2\pi}\int_{-\infty}^\infty F(\omega)e^{i\omega t}\,d\omega$$

$$F(\omega) = \int_{-\infty}^\infty f(t)e^{-i\omega t}\,dt$$

The Fourier cosine transform pair, Eqs. (5*a*) and (5*b*):

$$\begin{aligned} f(t) &= \sqrt{\frac{2}{\pi}}\int_0^\infty F_c(\omega)\cos\omega t\,d\omega \\ F_c(\omega) &= \sqrt{\frac{2}{\pi}}\int_0^\infty f(t)\cos\omega t\,dt \end{aligned} \qquad f\text{ even}$$

The Fourier sine transform pair, Eqs. (6*a*) and (6*b*):

$$\begin{aligned} f(t) &= \sqrt{\frac{2}{\pi}}\int_0^\infty F_s(\omega)\sin\omega t\,d\omega \\ F_s(\omega) &= \sqrt{\frac{2}{\pi}}\int_0^\infty f(t)\sin\omega t\,dt \end{aligned} \qquad f\text{ odd}$$

In general, the complex Fourier transformation and the Fourier cosine and sine transformations convert operations on a function in either the t domain or the frequency domain into simpler operations on a corresponding function in the other domain. By studying properties of the transformations themselves, we gain, among other things, the advantage of being able to find, without integration, the transforms of many functions from known transforms of related functions. This is very useful in constructing tables of transforms† and in certain applications, especially in communications engineering, where Fourier integrals of numerous complicated functions are encountered.

In the remainder of this section we shall establish a number of properties of our transformations and illustrate their use. Although t need not be a time variable, in most applications it is, and so we shall treat it as such as we attribute names to our principal results.

Several important transforms are listed in Table 9.2. The first entry is just 2π times the coefficient function $C(\omega)$ found in Example 1, Sec. 9.1. Entries **b** and **c** can be obtained similarly by direct integration and subsequent rationalization of the complex form of the transform $F(\omega)$. Entries **d** and **e** follow from Example 2 and Theorem 2, Sec. 9.1, after the coefficient $2/\pi$ is equally split between the relevant transform pairs. Entry **f** is easily verified by direct integration.

†See, in particular, A. Erdelyi, W. Magnus, F. Oberhettinger, and F. Tricomi, *Tables of Integral Transforms,* 2 vols., McGraw-Hill, New York, 1954.

TABLE 9.2

a. $f(t) = \begin{cases} 0 & t < 0 \\ e^{-at} & 0 < t,\ a > 0 \end{cases}$			$F(\omega) = \dfrac{1}{a + i\omega}$
b. $f(t) = \begin{cases} e^{at} & t \le 0 \\ e^{-at} & 0 \le t \end{cases}$		$a > 0$	$F(\omega) = \dfrac{2a}{a^2 + \omega^2}$
c. $f(t) = \begin{cases} -e^{at} & t < 0 \\ e^{-at} & 0 < t \end{cases}$		$a > 0$	$F(\omega) = \dfrac{-2i\omega}{a^2 + \omega^2}$
d. $f(t) = e^{-at}$	$0 < t$	$a > 0$	$F_c(\omega) = \sqrt{\dfrac{2}{\pi}}\dfrac{a}{a^2 + \omega^2}$
e. $f(t) = e^{-at}$	$0 < t$		$F_s(\omega) = \sqrt{\dfrac{2}{\pi}}\dfrac{\omega}{a^2 + \omega^2}$
f. $f(t) = \begin{cases} 0 & -\infty < t < -k \\ a & -k < t < 0 \\ b & 0 < t < l \\ 0 & l < t < \infty \end{cases}$			$F(\omega) = \dfrac{1}{i\omega}[(b - a) + ae^{i\omega k} - be^{-i\omega l}]$

Our first objective will be to show that the Fourier transformation distributes over sums of functions and commutes with scalars. In particular, we have

THEOREM 1 **(Linearity)** If the Fourier transforms of f_1 and f_2 exist, then

$$\mathcal{F}(a_1 f_1 + a_2 f_2) = a_1\mathcal{F}(f_1) + a_2\mathcal{F}(f_2) \qquad a_1,\ a_2 \text{ constants}$$

PROOF From the definition of the Fourier transformation $\mathcal{F}$ and familiar properties of integrals, we have

$$\begin{aligned}\mathcal{F}(a_1 f_1 + a_2 f_2) &= \int_{-\infty}^{\infty} [a_1 f_1(t) + a_2 f_2(t)]e^{-i\omega t}\,dt \\ &= a_1\int_{-\infty}^{\infty} f_1(t)e^{-i\omega t}\,dt + a_2\int_{-\infty}^{\infty} f_2(t)e^{-i\omega t}\,dt \\ &= a_1\mathcal{F}(f_1) + a_2\mathcal{F}(f_2)\end{aligned}$$

The property ascribed to the Fourier transformation by Theorem 1 is also valid for the Fourier cosine and sine transformations and, in each case, its extension to a linear combination of more than two functions is immediate. In other words, *all three Fourier transformations are linear operators,* as are their inverses.

THEOREM 2 **(Symmetry)** If $F(\omega)$ is the Fourier transform of $f(t)$, then $2\pi f(-\omega)$ is the transform of $F(t)$.

PROOF By hypothesis,

$$F(\omega) = \int_{-\infty}^{\infty} f(t)e^{-i\omega t}\,dt \qquad \text{and inversely} \qquad f(t) = \frac{1}{2\pi}\int_{-\infty}^{\infty} F(\omega)e^{i\omega t}\,d\omega$$

If in the latter integral we replace t by $-t$, it becomes

$$2\pi f(-t) = \int_{-\infty}^{\infty} F(\omega)e^{-i\omega t}\,d\omega$$

Finally, by interchanging the symbols ω and t, we obtain

$$2\pi f(-\omega) = \int_{-\infty}^{\infty} F(t)e^{-i\omega t}\,dt$$

which is just the statement that $2\pi f(-\omega)$ is the Fourier transform of $F(t)$. ◗

THEOREM 3 **(Change of Time Scale)** If a is a positive real constant,

$$\mathcal{F}[f(at)] = \frac{1}{a}F\left(\frac{\omega}{a}\right)$$

◖ **PROOF** By definition, $\mathcal{F}[f(at)] = \int_{-\infty}^{\infty} f(at)e^{-i\omega t}\,dt$
Under the substitution $at = z$ we have $dt = dz/a$, and the integral becomes

$$\frac{1}{a}\int_{-\infty}^{\infty} f(z)e^{-i\omega z/a}\,dz = \frac{1}{a}\mathcal{F}[f(z)]_{\omega\to\omega/a} = \frac{1}{a}F\left(\frac{\omega}{a}\right)$$

as asserted. ◗

The change of scale property is also valid for both the cosine and sine transformations and is established in the same way.

THEOREM 4 **(Time Shifting)**

$$\mathcal{F}[f(t-t_0)] = e^{-i\omega t_0}\mathcal{F}[f(t)] = e^{-i\omega t_0}F(\omega)$$

◖ **PROOF** By definition, $\mathcal{F}[f(t-t_0)] = \int_{-\infty}^{\infty} f(t-t_0)e^{-i\omega t}\,dt$
If we put $t - t_0 = z$ and $dt = dz$, we have

$$\begin{aligned}\mathcal{F}[f(t-t_0)] &= \int_{-\infty}^{\infty} f(z)e^{-i\omega(z+t_0)}\,dz = e^{-i\omega t_0}\int_{-\infty}^{\infty} f(z)e^{-i\omega z}\,dz \\ &= e^{-i\omega t_0}\mathcal{F}[f(z)] = e^{-i\omega t_0}F(\omega)\end{aligned}$$ ◗

In words, the time-shifting theorem says that translating a function an amount t_0 in the time domain results in the transform of the function being multiplied by $e^{-i\omega t_0}$. Conversely, as a tool for recovering a function from its transform, Theorem 4 asserts that if a transform contains a factor of the form $e^{-i\omega t_0}$, its inverse can be found by first deleting the factor $e^{-i\omega t_0}$, then finding the inverse of the remaining factor, and finally translating that inverse an amount equal to t_0.

Since the transform $F(\omega)$ can be written in the complex form

$$F(\omega) = r(\omega)e^{i\theta(\omega)}$$

where $r(\omega)$ is the magnitude and $\theta(\omega)$ is the polar angle of $F(\omega)$, it follows that translating $f(t)$ an amount t_0 in the time domain adds $-\omega t_0$ to the argument of $F(\omega)$; that is,

$$\begin{aligned}\mathcal{F}[f(t-t_0)] &= F(\omega)e^{-i\omega t_0} = [r(\omega)e^{i\theta(\omega)}]e^{-i\omega t_0} \\ &= r(\omega)e^{i[\theta(\omega)-\omega t_0]}\end{aligned} \tag{10}$$

The cosine and sine transformations have no comparably simple time-shifting property.

THEOREM 5 **(Frequency Shifting)** If ω_0 is a real constant, then

$$\mathcal{F}[e^{i\omega_0 t}f(t)] = \mathcal{F}[f(t)]_{\omega\to\omega-\omega_0} = F(\omega-\omega_0)$$

PROOF By definition,

$$\mathcal{F}[e^{i\omega_0 t}f(t)] = \int_{-\infty}^{\infty}[f(t)e^{i\omega_0 t}]e^{-i\omega t}\,dt = \int_{-\infty}^{\infty} f(t)e^{-i(\omega-\omega_0)t}\,dt$$
$$= \mathcal{F}[f(t)]_{\omega\to\omega-\omega_0} = F(\omega-\omega_0)$$

In words, the frequency-shifting theorem says that multiplying a function by $e^{i\omega_0 t}$ in the time domain results in the transform being shifted an amount ω_0 in the frequency domain. Conversely, as a tool for determining a function when its transform is known, Theorem 5 asserts that if the transform of a function is shifted an amount ω_0 and the inverse of the resulting function of ω is found, then $e^{-i\omega_0 t}$ times that inverse is the inverse of the original transform before translation.

The cosine and sine transformations have no comparably simple frequency-shifting property.

EXAMPLE 1

What is the Fourier transform of the product $f(t)\cos at$?

Using Theorem 5 and the complex exponential form of the cosine, we have

$$\mathcal{F}[f(t)\cos at] = \mathcal{F}\left[f(t)\frac{e^{iat}+e^{-iat}}{2}\right] = \mathcal{F}\left[\frac{f(t)e^{iat}}{2}\right] + \mathcal{F}\left[\frac{f(t)e^{-iat}}{2}\right]$$
$$= \tfrac{1}{2}[F(\omega-a)+F(\omega+a)]$$

If we think of $f(t)\cos at$ as a carrier signal, $\cos at$, modulated by $f(t)$, this transform allows us to express the transform of the modulated signal in terms of the transform F of the modulating function f, which determines the amplitude envelope of the signal.

EXAMPLE 2

Find the inverse of the Fourier transform $F(\omega) = \dfrac{1}{4a^4+\omega^4}$, $a > 0$.

We first express the denominator of $F(\omega)$ as the product of two quadratic factors with real coefficients by using the four fourth roots of $-4a^4$, namely,

$$z_1 = a(1+i) \qquad z_2 = a(-1+i) \qquad z_3 = a(-1-i) = \bar{z}_2 \qquad z_4 = a(1-i) = \bar{z}_1$$

to write

$$4a^4+\omega^4 = [(\omega-z_1)(\omega-\bar{z}_1)][(\omega-z_2)(\omega-\bar{z}_2)] = [(\omega-a)^2+a^2][(\omega+a)^2+a^2]$$

The method of partial fractions then gives

$$F(\omega) = \frac{1}{8a^3}\left[\frac{2a-\omega}{a^2+(\omega-a)^2} + \frac{2a+\omega}{a^2+(\omega+a)^2}\right] = \frac{1}{8a^3}\left[\frac{a-(\omega-a)}{a^2+(\omega-a)^2} + \frac{a+(\omega+a)}{a^2+(\omega+a)^2}\right]$$
$$= \frac{1}{8a^3}\left[\frac{1}{2}\frac{2a}{a^2+(\omega-a)^2} + \frac{1}{2}\frac{2a}{a^2+(\omega+a)^2} + \frac{1}{2i}\frac{-2i(\omega-a)}{a^2+(\omega-a)^2} - \frac{1}{2i}\frac{-2i(\omega+a)}{a^2+(\omega+a)^2}\right]$$

According to **b**, Table 9.2, and Theorem 5, the inverse of the first two terms inside the last pair of brackets is

$$g(t) = \begin{cases} \tfrac{1}{2}e^{at}e^{iat} + \tfrac{1}{2}e^{at}e^{-iat} = e^{at}\cos at & t \le 0 \\ \tfrac{1}{2}e^{-at}e^{iat} + \tfrac{1}{2}e^{-at}e^{-iat} = e^{-at}\cos at & 0 \le t \end{cases}$$

and by **c**, Table 9.2, and Theorem 5, the inverse of the last two terms in brackets is

$$h(t) = \begin{cases} \dfrac{1}{2i}(-e^{at}e^{iat}) - \dfrac{1}{2i}(-e^{at}e^{-iat}) = -e^{at}\sin at & t \le 0 \\ \dfrac{1}{2i}e^{-at}e^{iat} - \dfrac{1}{2i}e^{-at}e^{-iat} = e^{-at}\sin at & 0 \le t \end{cases}$$

Thus

$$f(t) = \mathcal{F}^{-1}[F(\omega)] = \frac{1}{8a^3}[g(t) + h(t)] = \begin{cases} (1/8a^3)e^{at}(\cos at - \sin at) & t \le 0 \\ (1/8a^3)e^{-at}(\cos at + \sin at) & 0 \le t \end{cases}$$

THEOREM 6 **(Time Differentiation)** If $f(t)$ is continuous and $f'(t)$ is at least piecewise continuous on $(-\infty, \infty)$, and if $\int_{-\infty}^{\infty} |f(t)|\, dt$ and $\int_{-\infty}^{\infty} |f'(t)|\, dt$ exist, then

$$\mathcal{F}[f'(t)] = i\omega\mathcal{F}[f(t)] = i\omega F(\omega)$$

PROOF By definition, $\mathcal{F}[f'(t)] = \int_{-\infty}^{\infty} f'(t)e^{-i\omega t}\, dt$. Integrating this by parts, with $u = e^{-i\omega t}$ and $dv = f'(t)\, dt$, we have

$$\mathcal{F}[f'(t)] = [e^{-i\omega t}f(t)]_{-\infty}^{\infty} + i\omega\int_{-\infty}^{\infty} f(t)e^{-i\omega t}\, dt$$

Since $f(t)$ is assumed to be absolutely integrable, $f(t)$ must vanish at both $-\infty$ and ∞. Hence the integrated portion of the last expression is equal to zero. What remains is just $i\omega F(\omega)$.

As long as successive derivatives of $f(t)$ are continuous and absolutely integrable on $(-\infty, \infty)$, the process of time differentiation can be continued, leading to the formulas

$$\mathcal{F}[f''(t)] = i\omega\mathcal{F}[f'(t)] = (i\omega)^2F(\omega)$$
$$\mathcal{F}(f'''(t)] = i\omega\mathcal{F}[f''(t)] = (i\omega)^3F(\omega)$$
$$\cdots\cdots\cdots\cdots$$

(11) $$\mathcal{F}[f^{(n)}(t)] = (i\omega)^nF(\omega)$$

EXAMPLE 3

Find the Fourier transform of the function shown in Fig. 9.9*a* whose definition is

$$f(t) = \begin{cases} 0 & -\infty < t \le -1 \\ 1 + t & -1 \le t \le 0 \\ 1 - t & 0 \le t \le 1 \\ 0 & 1 \le t < \infty \end{cases}$$

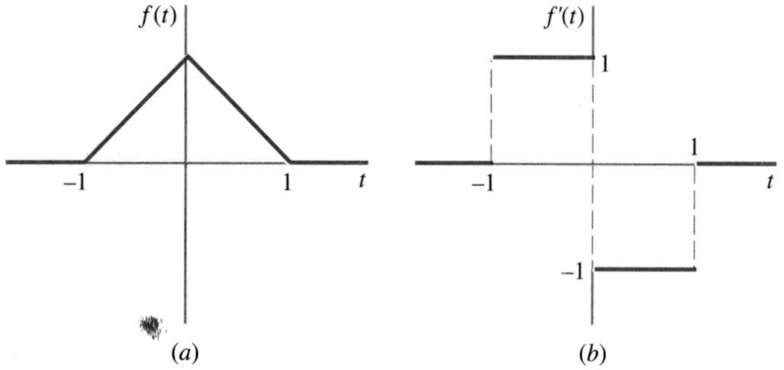

FIGURE 9.9
Graphs of the functions $f(t)$ and $f'(t)$ discussed in Example 3.

The transform of this function can be found of course by direct integration, but the required integration can be reduced, or even eliminated, by using Theorem 6. Clearly, the derivative of $f(t)$ is the step function shown in Fig. 9.9*b*, whose Fourier transform can be found by a much simpler integration, or by taking $a = k = l = -b = 1$ in the transform **f** of Table 9.2 The latter method gives

$$\mathcal{F}[f'(t)] = \frac{1}{i\omega}(-2 + e^{i\omega} + e^{-i\omega}) = \frac{2(-1 + \cos\omega)}{i\omega}$$

Hence, reversing Theorem 6, we have

$$F(\omega) = \mathcal{F}[f(t)] = \frac{1}{i\omega}\left[\frac{2(-1 + \cos\omega)}{i\omega}\right] = \frac{2(1 - \cos\omega)}{\omega^2}$$

Both the cosine and sine transformations have time differentiation properties comparable to Theorem 6. We state these properties and two important extensions of them as corollaries to Theorem 6.

COROLLARY 1 **(Cosine Transforms of Derivatives)** If $f(t)$ satisfies the conditions of Theorem 6 on $[0, \infty)$, then

$$\mathcal{F}_c[f'(t)] = \omega\mathcal{F}_s[f(t)] - \sqrt{\frac{2}{\pi}}f(0) = \omega F_s(\omega) - \sqrt{\frac{2}{\pi}}f(0)$$

In this corollary, the right-hand limit $f(0^+)$ takes the place of $f(0)$ should the domain of f be the open interval $(0, \infty)$.

COROLLARY 2 **(Sine Transforms of Derivatives)** If $f(t)$ satisfies the conditions of Theorem 6 on $[0, \infty)$, then

$$\mathcal{F}_s[f'(t)] = -\omega\mathcal{F}_c[f(t)] = -\omega F_c(\omega)$$

If on $[0, \infty)$, f' and f'' satisfy the respective conditions prescribed for f and f' by Theorem 6, we can continue by applying each corollary to f' rather than f, getting

COROLLARY 3 **(Cosine Transforms of Second Derivatives)** If f' and f'' satisfy the conditions of Theorem 6 on $[0, \infty)$, then

$$\mathcal{F}_c[f''(t)] = -\omega^2\mathcal{F}_c[f(t)] - \sqrt{\frac{2}{\pi}}f'(0) = -\omega^2 F_c(\omega) - \sqrt{\frac{2}{\pi}}f'(0)$$

COROLLARY 4 **(Sine Transforms of Second Derivatives)** If f' and f'' satisfy the conditions of Theorem 6 on $[0, \infty)$, then

$$\mathcal{F}_s[f''(t)] = -\omega^2\mathcal{F}_s[f(t)] + \sqrt{\frac{2}{\pi}}\omega f(0) = -\omega^2 F_s(\omega) + \sqrt{\frac{2}{\pi}}\omega f(0)$$

While Corollaries 1 and 2 each involve transformations of both types, Corollaries 3 and 4 have the distinct advantage of involving only one kind of transformation.

EXAMPLE 4

Using Corollaries 3 and 4, find the cosine and sine transforms of

$$f(t) = e^{-at} \qquad t \geq 0,\ a > 0$$

Differentiating f twice we get $f''(t) = a^2 f(t)$. Taking the cosine transformation of each side of this equation and then using Corollary 3, we have

$$\mathscr{F}_c[f''(t)] = a^2 \mathscr{F}_c[f(t)]$$

$$-\omega^2 F_c(\omega) - \sqrt{\frac{2}{\pi}} f'(0) = a^2 F_c(\omega)$$

Noting that $f'(0) = -a$ and solving for $F_c(\omega)$, we obtain

$$F_c(\omega) = \sqrt{\frac{2}{\pi}} \frac{a}{a^2 + \omega^2}$$

which is the transform **d** given in Table 9.2. Similarly, equating the sine transforms of $f''(t)$ and $a^2 f(t)$, then using Corollary 4 and setting $f(0) = 1$, we have

$$\mathscr{F}_s[f''(t)] = a^2 \mathscr{F}_s[f(t)]$$

$$-\omega^2 F_s(\omega) + \sqrt{\frac{2}{\pi}} \omega f(0) = a^2 F_s(\omega)$$

$$F_s(\omega) = \sqrt{\frac{2}{\pi}} \frac{\omega}{a^2 + \omega^2}$$

which agrees with the transform **e**, Table 9.2.

As we should now expect, the time differentiation theorem has a counterpart in the frequency domain which is

THEOREM 7 **(Frequency Differentiation)** If $F(\omega)$ is the Fourier transform of $f(t)$, then

$$F'(\omega) = \mathscr{F}[-itf(t)]$$

PROOF By definition,

$$F(\omega) = \int_{-\infty}^{\infty} f(t) e^{-i\omega t}\, dt \qquad \text{and} \qquad F'(\omega) = \frac{d}{d\omega}\left[\int_{-\infty}^{\infty} f(t) e^{-i\omega t}\, dt\right]$$

Since $f(t)$ is known to be absolutely integrable on $(-\infty, \infty)$, the indicated differentiation can be carried out inside the integral sign (see Theorems 1 and 4, Sec. 10.3). Doing this, we have

$$F'(\omega) = \int_{-\infty}^{\infty} [-itf(t)] e^{-i\omega t}\, dt = \mathscr{F}[-itf(t)]$$

In words, this theorem says that differentiating a transform in the frequency domain amounts to multiplying its inverse by $-it$ in the time domain. Conversely, as a tool for finding inverses, this theorem says that if an antiderivative of a transform has a known inverse $f(t)$, then $-it$ times that inverse is the inverse of the original transform.

If we choose, we can multiply both sides of the conclusion in Theorem 7 by i, getting the equivalent formula

$$\mathscr{F}[tf(t)] = iF'(\omega)$$

As long as the products of $f(t)$ and successive integral powers of t are absolutely integrable on $(-\infty, \infty)$, the preceding process can be continued, leading to the formula

$$\mathcal{F}[t^n f(t)] = i^n F^{(n)}(\omega) \tag{12}$$

EXAMPLE 5

For $a < 0$, b real, $z = a + bi$, and $k \geqq 0$ an integer, find the Fourier transform of the function

$$g(t) = \begin{cases} 0 & t < 0 \\ t^k e^{zt} & 0 < t \end{cases}$$

For $a < 0$, the transform of the function

$$f(t) = \begin{cases} 0 & t < 0 \\ e^{zt} & 0 < t \end{cases}$$

is given by

$$F(\omega) = \int_0^\infty e^{zt} e^{-i\omega t}\, dt = \left. \frac{e^{[a+(b-\omega)i]t}}{z - i\omega} \right|_0^\infty = \frac{1}{i\omega - z}$$

Clearly, $g = t^k f$. Hence, using (12) and the easily derived relation

$$F^{(k)}(\omega) = \frac{(-1)^k i^k k!}{(i\omega - z)^{k+1}}$$

we have

$$\mathcal{F}[g(t)] = \mathcal{F}[t^k f(t)] = i^k F^{(k)}(\omega) = \frac{(-1)^k i^{2k} k!}{(i\omega - z)^{k+1}} = \frac{k!}{(i\omega - z)^{k+1}}$$

In Sec. 2.9 we considered a particular combination of two functions called their *convolution*. This concept, when extended to functions with more general domains, has many practical applications and is of fundamental importance to our further study of the Fourier transformation.

DEFINITION 1 The function h defined by the integral

$$h(t) = \int_a^{t-a} f(\lambda) g(t - \lambda)\, d\lambda$$

is called the **convolution** of the functions f and g over the interval from a to $t - a$.

It is customary to denote h by $f * g$, and it is easily proved, by making the change of variable $t - \lambda = \mu$ in the convolution integral, that $f * g = g * f$.

In many problems, $a = 0$, in which case we have the **unilateral convolution** of f and g from 0 to t given by

$$(f * g)(t) = \int_0^t f(\lambda) g(t - \lambda)\, d\lambda \tag{13}$$

If $a = -\infty$, we obtain the **bilateral convolution** over $(-\infty, \infty)$ given by

$$(f * g)(t) = \int_{-\infty}^{\infty} f(\lambda)g(t - \lambda)\,d\lambda \tag{14}$$

The unilateral convolution (13) plays an important role in the theory and applications of Laplace transformations (see Sec. 10.11). Both the unilateral convolution (13) and the bilateral convolution (14) are important in working with Fourier transformations.

We already know (Sec. 2.9) that the Cauchy product of two *series* is a sum of convolutions, thus it is natural to ask if the product of two *integrals* yields a similar result. In the case of Fourier integrals, which are the kind we are presently concerned with, we find that if two functions f and g are represented by

$$f(t) = \frac{1}{2\pi}\int_{-\infty}^{\infty} F(\omega)e^{i\omega t}\,d\omega \qquad \text{and} \qquad g(t) = \frac{1}{2\pi}\int_{-\infty}^{\infty} G(\omega)e^{i\omega t}\,d\omega$$

then, upon replacing the dummy variable ω in the first integral by λ and in the second integral by μ, the product of f and g has the representation

$$f(t)g(t) = \frac{1}{4\pi^2}\int_{-\infty}^{\infty} F(\lambda)e^{i\lambda t}\,d\lambda \int_{-\infty}^{\infty} G(\mu)e^{i\mu t}\,d\mu$$

Because the dummy variables are now different, we can write the product of the integrals as the repeated integral

$$f(t)g(t) = \frac{1}{4\pi^2}\int_{-\infty}^{\infty}\int_{-\infty}^{\infty} F(\lambda)G(\mu)e^{i(\lambda+\mu)t}\,d\lambda\,d\mu$$

The change of variable $\mu = \omega - \lambda$ gives $d\mu = d\omega$, and

$$\begin{aligned} f(t)g(t) &= \frac{1}{4\pi^2}\int_{-\infty}^{\infty}\int_{-\infty}^{\infty} F(\lambda)G(\omega - \lambda)e^{i\omega t}\,d\lambda\,d\omega \\ &= \frac{1}{2\pi}\int_{-\infty}^{\infty}\left[\frac{1}{2\pi}\int_{-\infty}^{\infty} F(\lambda)G(\omega - \lambda)\,d\lambda\right]e^{i\omega t}\,d\omega \end{aligned}$$

The structure of the last expression shows that the quantity in brackets, namely, $1/2\pi$ times the bilateral convolution of the transforms $F(\omega)$ and $G(\omega)$, is the Fourier transform of the product $f(t)g(t)$ on the left; that is,

$$\mathcal{F}[f(t)g(t)] = \frac{1}{2\pi}\int_{-\infty}^{\infty} F(\lambda)G(\omega - \lambda)\,d\lambda = \frac{1}{2\pi}F(\omega) * G(\omega)$$

We have thus established the important result

THEOREM 8 **(Frequency Convolution)** If f and g satisfy the Dirichlet conditions on every finite interval and are absolutely integrable on $(-\infty, \infty)$, then

$$\mathcal{F}[f(t)g(t)] = \frac{1}{2\pi}F(\omega) * G(\omega)$$

With this theorem established, we are prepared to prove

THEOREM 9 **(Parseval's Modulus Theorem)** If a real function $f(t)$ with domain $(-\infty, \infty)$ has a Fourier transform whose polar form is $F(\omega) = r(\omega)e^{i\theta(\omega)}$, then

$$\int_{-\infty}^{\infty} f^2(t)\,dt = \frac{1}{2\pi}\int_{-\infty}^{\infty} r^2(\omega)\,d\omega \tag{15}$$

PROOF We first note from the definitive integral, written with ω replaced by τ,

$$F(\tau) = \int_{-\infty}^{\infty} f(t)e^{-i\tau t}\,dt$$

that when $f(t)$ is real, both i and the real variable τ enter only through the product $i\tau$. Hence, if we take the conjugate of $F(\tau)$ by replacing i by $-i$, and subsequently replace τ by $-\tau$, we have left the transform unchanged. In other words, whatever the argument may be, $F(\tau) = \overline{F}(-\tau)$. In particular, then, $F(\tau - \omega) = \overline{F}(\omega - \tau)$. The convolution of $F(\tau)$ with itself can therefore be written

$$F(\tau) * F(\tau) = \int_{-\infty}^{\infty} F(\omega)F(\tau - \omega)\,d\omega = \int_{-\infty}^{\infty} F(\omega)\overline{F}(\omega - \tau)\,d\omega$$

Now from this result and the frequency convolution theorem, we have

$$\mathcal{F}[f^2(t)] = \frac{1}{2\pi}F(\tau) * F(\tau) = \frac{1}{2\pi}\int_{-\infty}^{\infty} F(\omega)\overline{F}(\omega - \tau)\,d\omega$$

From this, by putting $\tau = 0$, we obtain

$$\int_{-\infty}^{\infty} f^2(t)\,dt = \frac{1}{2\pi}\int_{-\infty}^{\infty} F(\omega)\overline{F}(\omega)\,d\omega$$

But the product of any number, real or complex, and its conjugate is the square of the absolute value of the number. Hence the last equation yields

$$\int_{-\infty}^{\infty} f^2(t)\,dt = \frac{1}{2\pi}\int_{-\infty}^{\infty} r^2(\omega)\,d\omega$$

as asserted by the theorem.

In the case of a signal $f(t)$, the square of the modulus of $F(\omega)$, namely, $r^2(\omega)$, is called the **energy spectrum** of the signal because $r^2(\omega)\,\Delta\omega$ represents the amount of energy in any $\Delta\omega$ segment of the frequency spectrum, and the integral of $r^2(\omega)$ over $(-\infty, \infty)$ gives the **total energy** in the signal. Parseval's theorem is also true if $f(t)$ is complex, but we shall leave the proof of this as an exercise.

EXAMPLE 6

Verify Parseval's theorem for the function

$$f(t) = \begin{cases} 0 & t < 0 \\ e^{-t} & t > 0 \end{cases}$$

In this particular case

$$\int_{-\infty}^{\infty} f^2(t)\,dt = \int_{0}^{\infty} (e^{-t})^2\,dt = \frac{1}{2}e^{-2t}\Big|_{\infty}^{0} = \frac{1}{2}$$

The transform $F(\omega)$ of f is given by **a**, Table 9.2. To find the square of its modulus, we first express $F(\omega)$ in standard complex form

$$F(\omega) = \frac{1}{1 + i\omega} = \frac{1}{1 + i\omega}\,\frac{1 - i\omega}{1 - i\omega} = \frac{1}{1 + \omega^2} + i\frac{-\omega}{1 + \omega^2}$$

From this we get

$$r^2(\omega) = \left(\frac{1}{1 + \omega^2}\right)^2 + \left(\frac{-\omega}{1 + \omega^2}\right)^2 = \frac{1}{1 + \omega^2}$$

and

$$\frac{1}{2\pi}\int_{-\infty}^{\infty} r^2(\omega)\,d\omega = \frac{1}{2\pi}\int_{-\infty}^{\infty} \frac{d\omega}{1 + \omega^2} = \frac{1}{\pi}\int_0^{\infty} \frac{d\omega}{1 + \omega^2} = \frac{1}{\pi}\operatorname{Tan}^{-1}\omega\Big|_0^{\infty} = \frac{1}{2}$$

Hence Parseval's theorem holds for the given function f.

In essentially the same way that we proved the frequency convolution theorem, we can multiply the integrals of two transforms to establish the following companion theorem.

THEOREM 10 **(Time Convolution)** The inverse of the product of two transforms is the convolution of their inverses; that is,

$$\mathcal{F}^{-1}[F(\omega)G(\omega)] = \int_{-\infty}^{\infty} f(\lambda)g(t - \lambda)\,d\lambda = f(t) * g(t)$$

This theorem is often useful for finding the inverses of transforms which can be resolved into factors whose inverses we know. However, when the inverses of the individual factors are functions which are identically zero for $t < 0$, the bilateral convolution appearing in the statement of Theorem 10 must be replaced by the unilateral convolution.

To see why this is so, we first note that if f and g are functions which are trivial for $t < 0$, then their bilateral convolution reduces to

$$f(t) * g(t) = \int_0^{\infty} f(\lambda)g(t - \lambda)\,d\lambda$$

Moreover, when $t - \lambda < 0$ in the course of the integration, the second factor in the integrand becomes zero and the integration effectively terminates at $\lambda = t$. In other words, the bilateral convolution becomes the unilateral convolution

$$f(t) * g(t) = \int_0^{t} f(\lambda)g(t - \lambda)\,d\lambda$$

This observation we formalize in the following corollary to Theorem 10.

COROLLARY 1 **(Unilateral Time Convolution)** If $F(\omega)$ and $G(\omega)$ are the Fourier transforms of two functions $f(t)$ and $g(t)$, which are identically zero for $t < 0$, then

$$\mathcal{F}^{-1}[F(\omega)G(\omega)] = \int_0^{t} f(\lambda)g(t - \lambda)\,d\lambda$$

EXAMPLE 7

Find the inverse of the Fourier transform $F(\omega) = 1/(6 + 5i - \omega^2)$.

By writing $-\omega^2$ as $i^2\omega^2$, the denominator becomes easy to factor, thus

$$F(\omega) = \frac{1}{6 + 5i + (i\omega)^2} = \frac{1}{2 + i\omega}\,\frac{1}{3 + i\omega}$$

By **a**, Table 9.2, the two factors of the last product are, respectively, the Fourier transforms of the functions

$$g(t) = \begin{cases} 0 & t < 0 \\ e^{-2t} & 0 < t \end{cases} \quad \text{and} \quad h(t) = \begin{cases} 0 & t < 0 \\ e^{-3t} & 0 < t \end{cases}$$

both of which are identically zero for $t < 0$. Hence, by Corollary 1, Theorem 10, the required inverse is

$$f(t) = \int_0^t e^{-2\lambda} e^{-3(t-\lambda)}\, d\lambda = e^{-3t} \int_0^t e^{\lambda}\, d\lambda = \begin{cases} 0 & t \le 0 \\ e^{-2t} - e^{-3t} & 0 \le t \end{cases}$$

EXAMPLE 8

Find the inverse $f(t)$ of the Fourier transform $F(\omega) = [4a \sin k\omega]/\omega(a^2 + \omega^2)$, $a > 0$.

Guided by Theorem 10, we express the given transform as the product

$$F(\omega) = \left(\frac{2 \sin k\omega}{\omega}\right)\left(\frac{2a}{a^2 + \omega^2}\right) = \left(\frac{e^{ik\omega} - e^{-ik\omega}}{i\omega}\right)\left(\frac{2a}{a^2 + \omega^2}\right)$$

With $l = k$ and $a = b = 1$, we see from **f**, Table 9.2, that the inverse of the first factor in $F(\omega)$ is

$$g(t) = \begin{cases} 0 & |t| > k \\ 1 & |t| < k \end{cases}$$

From **b**, Table 9.2, the inverse of the second factor is

$$h(t) = \begin{cases} e^{at} & t \le 0 \\ e^{-at} & t \ge 0 \end{cases}$$

Hence, by Theorem 10,

$$f(t) = \int_{-\infty}^{\infty} g(\lambda) h(t - \lambda)\, d\lambda = \int_{-k}^{k} h(t - \lambda)\, d\lambda$$

To evaluate the last integral, we observe that

$$h(t - \lambda) = \begin{cases} e^{a(t-\lambda)} & t \le \lambda \\ e^{-a(t-\lambda)} & \lambda \le t \end{cases}$$

Thus, depending on the relation of t to the limits on λ, we have the following three cases to consider.

1. $t \le -k$: $f(t) = \displaystyle\int_{-k}^{k} e^{a(t-\lambda)}\, d\lambda = \frac{e^{at}}{-a}[e^{-a\lambda}]_{-k}^{k} = \frac{2 \sinh ak}{a} e^{at}$

2. $|t| \le k$: $f(t) = \displaystyle\int_{-k}^{t} e^{-a(t-\lambda)}\, d\lambda + \int_{t}^{k} e^{a(t-\lambda)}\, d\lambda = \frac{e^{-at}}{a}[e^{a\lambda}]_{-k}^{t} - \frac{e^{at}}{a}[e^{-a\lambda}]_{t}^{k}$

$$= \frac{2(1 - e^{-ak} \cosh at)}{a}$$

3. $t \ge k$: $f(t) = \displaystyle\int_{-k}^{k} e^{-a(t-\lambda)}\, d\lambda = \frac{e^{-at}}{a}[e^{a\lambda}]_{-k}^{k} = \frac{2 \sinh ak}{a} e^{-at}$

Finally, combining cases, we have for the required inverse transform

$$f(t) = \begin{cases} \dfrac{2 \sinh ak}{a} e^{at} & -\infty < t \le -k \\ \dfrac{2(1 - e^{-ak} \cosh at)}{a} & -k \le t \le k \\ \dfrac{2 \sinh ak}{a} e^{-at} & k \le t < \infty \end{cases}$$

EXERCISES

1. Given the result of Example 3, use the theorems of this section to find the Fourier transforms of the functions shown in Fig. 9.10.

Using the transforms and theorems of this section, find the inverse of each of the following Fourier transforms.

2. **(a)** $\dfrac{e^{-2i\omega}}{3(1+i\omega)}$ **(b)** $\dfrac{e^{3i\omega}}{\pi(2+i\omega)}$

(c) $\dfrac{1-e^{-3i\omega}}{2+i\omega}$ **(d)** $\dfrac{e^{-i\omega}}{2\pi(1+i\omega)(2+i\omega)}$

(e) $\dfrac{e^{-(2+i\omega)}}{2+i\omega}$ **(f)** $\dfrac{e^{-3i\omega}}{3+2i\omega}$

(g) $\dfrac{i\sin\omega}{a+i\omega}$ **(h)** $\dfrac{e^{-2i\omega}}{-\omega^2+4i\omega+3}$

3. **(a)** $\dfrac{1}{(i\omega+1)(i\omega+2)}$ **(b)** $\dfrac{i\omega}{(i\omega+1)(i\omega+2)}$

(c) $\dfrac{1}{i\omega+1-i}$ **(d)** $\dfrac{e^{-i\omega}}{i\omega+2}$

(e) $\dfrac{e^{-2i\omega}}{(i\omega+3)^2}$

4. **(a)** $\dfrac{\sin(\omega-2)}{\omega-2}$ **(b)** $\dfrac{\sin(\omega/2)}{\omega}$

(c) $\dfrac{1}{(i\omega+1)^3}$ *Hint:* What is the antiderivative of the transform?

(d) $\dfrac{e^{-i\omega}}{1+\omega^2}$ *Hint:* Factor the denominator.

5. **(a)** If $F(\omega) = F_r(\omega) + iF_i(\omega)$ is the Fourier transform of a function $f(t)$ which is identically zero before $t = 0$, show that $F_r(\omega)$ is the transform of one-half the even extension of $f(t)$ and that $iF_i(\omega)$ is the transform of one-half the odd extension of $f(t)$.

(b) What are the functions represented, respectively, by the real and the imaginary terms in the transform $1/(a + i\omega)$?

6. **(a)** Show that the Fourier transform of the function

$$f(t) = \begin{cases} 0 & -\infty < t \le 0 \\ \sin t & 0 \le t \le \pi \\ 0 & \pi \le t < \infty \end{cases}$$

is

$$F(\omega) = \frac{1}{2\pi}\,\frac{1+e^{-i\pi\omega}}{1-\omega^2}$$

(b) Using the result of Part **a**, find the Fourier transform of each of the following functions

FIGURE 9.10

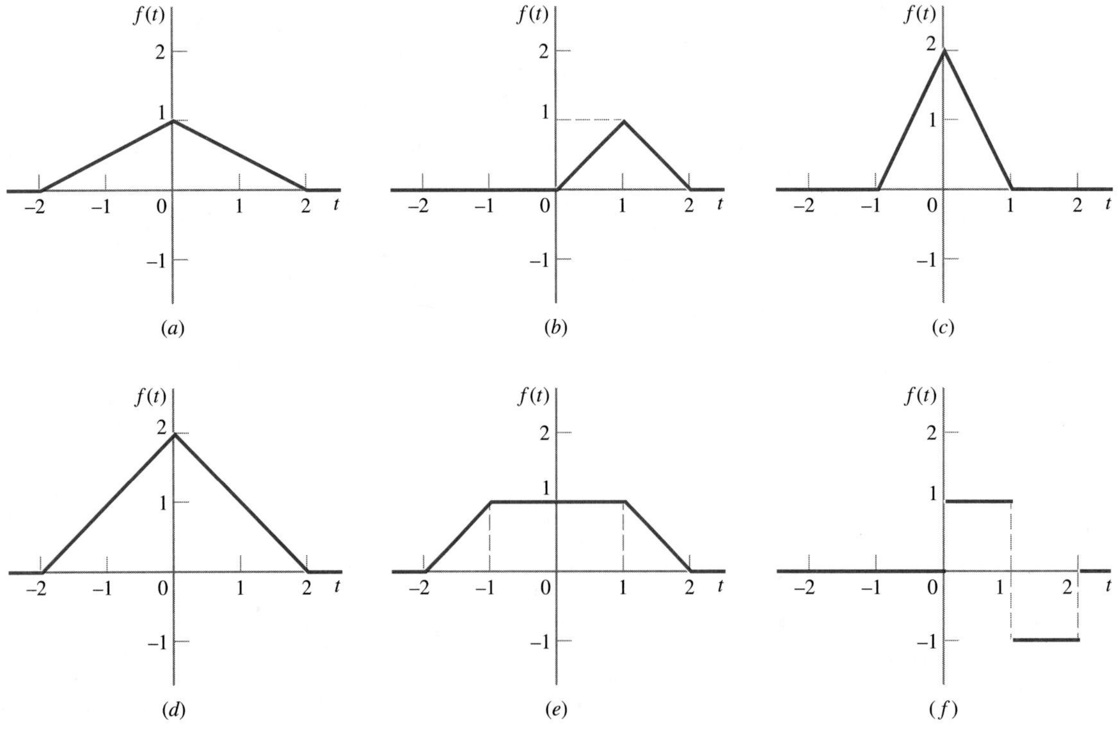

$$f(t) = \begin{cases} 0 & -\infty < t \leq -\pi \\ \sin t & -\pi \leq t \leq \pi \\ 0 & \pi \leq t < \infty \end{cases}$$

and $$f(t) = \begin{cases} 0 & -\infty < t \leq -\pi \\ |\sin t| & -\pi \leq t \leq \pi \\ 0 & \pi \leq t < \infty \end{cases}$$

7. Use partial fractions and Theorem 6 to find the inverse of the Fourier transform $F(\omega)$ of Example 2.

8. Use the identity in a and ω given by $\dfrac{1}{4a^4+\omega^4} = \dfrac{1}{[a^2+(\omega-a)^2][a^2+(\omega+a)^2]}$ and Theorem 10 to find the inverse of the Fourier transform $F(\omega)$ of Example 2.

9. Show that

(a) $$\int_{-\infty}^{\infty} \frac{dz}{(a+iz)(a-iz+i\omega)} = \frac{2\pi}{2a+i\omega}$$

(b) $$\int_{-\infty}^{\infty} \frac{dz}{(a-iz)(a-iz+i\omega)} = 0$$

Hint: Interpret each integral as a convolution and then use Theorem 8.

10. If $F(\omega)$ and $G(\omega)$ are, respectively, the Fourier transforms of $f(t)$ and $g(t)$, show that

(a) $\displaystyle\int_{-\infty}^{\infty} f(t)g(t)\,dt = 2\pi\int_{-\infty}^{\infty} F(\tau)G(-\tau)\,d\tau$. *Hint:* Suitably specialize Theorem 8.

(b) $\displaystyle\int_{-\infty}^{\infty} f(t)g(-t)\,dt = 2\pi\int_{-\infty}^{\infty} F(\tau)G(\tau)\,d\tau$. *Hint:* Suitably specialize Theorem 8.

11. Show that $\displaystyle\int_{-\infty}^{\infty} \frac{\sin z}{z}\frac{\sin(\omega-z)}{\omega-z}\,dz = \frac{\pi\sin\omega}{\omega}$.

12. Using Theorem 10, find the inverse of each of the following Fourier transforms.

(a) $\dfrac{1}{(a+i\omega)^2}$ **(b)** $\dfrac{1}{(a+i\omega)^3}$

13. Using Theorem 10, find the inverse of $F(\omega) = \dfrac{e^{-3i\omega}}{(1+i\omega)(2+i\omega)}$. Check your result by using a different factorization of $F(\omega)$.

14. Find the inverse of each of the following transforms.

(a) $\dfrac{3+2i\omega}{(1+i\omega)(2+i\omega)^2}$ **(b)** $\dfrac{8+4i\omega}{(1+i\omega)^2(3+i\omega)^2}$

15. (a) If $f(t)$ is identically zero for $t<0$ and continuous on $[0,\infty)$ and if $f'(t)$ is at least piecewise continuous on $[0,\infty)$, show that Theorem 6 becomes

$$\mathscr{F}[f'(t)] = i\omega\mathscr{F}[f(t)] - f(0)$$

(b) Under the appropriate conditions, what is $\mathscr{F}[f''(t)]$?

16. (a) Prove Corollary 1, Theorem 6.
(b) Prove Corollary 2, Theorem 6.
(c) Prove Corollary 3, Theorem 6.
(d) Prove Corollary 4, Theorem 6.

17. If $F(\omega)$ is the Fourier transform of $f(t)$, show that $\mathscr{F}[\int_{-\infty}^{t} f(s)\,ds] = F(\omega)/i\omega$.

18. Show that Parseval's theorem is also true when f is complex, i.e., when $f(t) = f_r(t) + if_i(t)$. *Hint:* First show that if

$$\mathscr{F}[f(t)] = F(\omega) \qquad \text{then} \qquad \mathscr{F}[\bar{f}(t)] = \bar{F}(-\omega)$$

Then apply Theorem 8 to the product $f(t)\bar{f}(t) = |f|^2(t)$.

19. Let p be a unit pulse function between -1 and 1, so that

$$p(t) = u(t+1) - u(t-1) = \begin{cases} 0 & |t| > 1 \\ 1 & |t| < 1 \end{cases}$$

Show by repeated differentiation of $\mathscr{F}[p(t)]$ with respect to ω that

$$\mathscr{F}[t^n p(t)] = \frac{i^n}{\pi}\frac{d^n S(\omega)}{d\omega^n}$$

where $S(\omega) = (\sin\omega)/\omega$. Explain how this result can be used to obtain the Fourier transform of a single arbitrary pulse $f(t)p(t)$ defined between -1 and 1 by a convergent power series.†

20. Using the definitions of Exercise 19, show that

$$\mathscr{F}[e^{ni\pi t}p(t)] = \frac{1}{\pi}S(\omega - n\pi)$$

Explain how this result can be used to obtain the Fourier transform of a single pulse defined between -1 and 1 by a Fourier series in either complex exponential or real trigonometric form.

21. If $f(t)$ is a pulse defined between -1 and 1 by either of the equivalent series

$$\sum_{n=0}^{\infty}(a_n\cos n\pi t + b_n\sin n\pi t) \qquad \text{or} \qquad \sum_{n=-\infty}^{\infty} c_n e^{ni\pi t}$$

use the results of Exercise 20 to show that **(a)** $a_n = \pi F(n\pi) + F(-n\pi)$, **(b)** $b_n = i\pi F(n\pi) - F(-n\pi)$, and **(c)** $c_n = \pi F(n\pi)$, where $F(\omega)$ is the Fourier transform of the pulse.

22. Is the Fourier transform of the function $f(t)$ given by **b**, Table 9.2, the same as the Fourier cosine transform of the following function? Why?

$$g(t) = e^{-at} \qquad t \geqq 0$$

23. Find the Fourier transform of the finite wave train

$$f(t) = \begin{cases} 0 & |t| > T \\ \sin\omega_0 t & |t| < T \end{cases}$$

and show that the larger the value of T, the more $F(\omega)$ concentrates around the frequencies $-\omega_0$ and ω_0, that is,

†In particular problems, *Tables of the Function (sin u)/u and Its First Eleven Derivatives,* Harvard University Press, Cambridge, Mass., will be of considerable help.

the more accurately ω_0 can be determined by scanning the signal through a narrow bandpass filter.

24. Verify that the Fourier transform of the function $f(t)$ of Example 8 is

$$F(\omega) = \frac{4a \sin k\omega}{\omega(a^2 + \omega^2)}$$

25. **(a)** Assuming that f has domain $(0, \infty)$ and the Fourier sine transform $F_s(\omega)$, express the transform $F(\omega)$ of the odd extension of f in terms of $F_s(\omega)$.

(b) Use the required relation of Part **a**, and the procedure of Example 8, to find the inverse of the Fourier transform $F(\omega) = \dfrac{e^{-ik\omega} - 1}{2\pi^2(a^2 + \omega^2)}$

26. Show that if one or the other of the functions $f(t)$ or $g(t)$ is absolutely integrable and the other bounded on $(-\infty, \infty)$, then their bilateral convolution exists.

9.4 APPLICATIONS OF FOURIER INTEGRALS AND TRANSFORMS

The most important applications of Fourier integrals and Fourier transforms are probably to be found in fields such as signal analysis, communications engineering, optics, and spectroscopy, where wave phenomena are conspicuously involved. On the other hand, there are also applications to problems formulated in terms of differential equations, both ordinary and partial, where wave phenomena really play no part. Many such applications are extensions of the use of Fourier series, and the chief difference is simply the difference between summation and integration. This is well illustrated in the solution of ordinary differential equations where disturbing influences must be represented in some easily handled way. Thus, in both Examples 1 and 2 of Sec. 8.6 there was a periodic forcing function which was replaced by a Fourier series. Then each term in the Fourier series was treated as though it were the only term involved, and finally all these partial results were added to form the actual solution. Had the forcing function been nonperiodic, it could not have been represented by a Fourier series, but under suitable conditions a Fourier integral representation could have been found for it. In such a case there are no discrete terms to be handled individually and then summed. Instead, general infinitesimal segments of the continuous-frequency representation are treated as though they were individual terms, and then the total effect of their infinitesimal contributions is found by integration. Our first example illustrates this process in detail in a routine mathematical exercise.

EXAMPLE 1

Find a particular integral of the equation

$$y'' + 3y' + 2y = f(t) \qquad \text{if} \qquad f(t) = \begin{cases} 1 & |t| < 1 \\ 0 & |t| > 1 \end{cases}$$

In this problem, $f(t)$ is the pulse function we studied in detail in Sec 9.1, and its Fourier integral representation, as given by Eq. (23), Sec. 9.1, is

$$f(t) = \frac{2}{\pi}\int_0^\infty \frac{\sin \omega}{\omega} \cos \omega t \, d\omega$$

Hence the given equation can be written in the form

$$y'' + 3y' + 2y = \frac{2}{\pi}\int_0^\infty \frac{\sin \omega}{\omega} \cos \omega t \, d\omega \tag{1}$$

We cannot identify individual terms in the Fourier integral representation of f, so instead we consider the infinitesimal portion df of $f(t)$ corresponding to or arising from the frequencies in a general infinitesimal segment $d\omega$ of the continuous-frequency scale:

$$df = \frac{2}{\pi}\frac{\sin\omega}{\omega}\cos\omega t\, d\omega$$

This is an expression of the familiar form $A\cos\omega t$, where $A = \frac{2}{\pi}\frac{\sin\omega}{\omega}\,d\omega$; and a particular integral for it can be found, as usual, by the method of undetermined coefficients. Thus, to find the infinitesimal contribution dY which the infinitesimal df makes to the total particular integral Y, we form the linear combination

$$dY = a\cos\omega t + b\sin\omega t$$

and substitute dY for y in (1), with the right-hand member replaced by df. Collecting terms on $\cos\omega t$ and $\sin\omega t$ and subsequently equating coefficients of like terms on both sides, we obtain the equations

$$(2-\omega^2)a + 3\omega b = \frac{2}{\pi}\frac{\sin\omega}{\omega}\,d\omega$$

$$-3\omega a + (2-\omega^2)b = 0$$

Solving these for a and b and observing that $(2-\omega^2)^2 + 9\omega^2 = (1+\omega^2)(4+\omega^2)$, we find

$$a = \frac{2-\omega^2}{(1+\omega^2)(4+\omega^2)}\frac{2}{\pi}\frac{\sin\omega}{\omega}\,d\omega \qquad b = \frac{3\omega}{(1+\omega^2)(4+\omega^2)}\frac{2}{\pi}\frac{\sin\omega}{\omega}\,d\omega$$

Hence

$$dY = \frac{(2-\omega^2)\cos\omega t + 3\omega\sin\omega t}{(1+\omega^2)(4+\omega^2)}\frac{2}{\pi}\frac{\sin\omega}{\omega}\,d\omega$$

Finally, ''adding up'' all these infinitesimal contributions to Y by integration over the entire range of ω, we get the Fourier integral representation

$$Y(t) = \frac{2}{\pi}\int_0^\infty \frac{(2-\omega^2)\cos\omega t + 3\omega\sin\omega t}{(1+\omega^2)(4+\omega^2)}\frac{\sin\omega}{\omega}\,d\omega$$

of the required particular integral Y.

Our next example is a variation of Example 5, Sec. 8.6, in which Fourier integrals, but not Fourier series, may be used to determine the deflection of a beam.

EXAMPLE 2

DEFLECTION OF AN INFINITE BEAM AS A FOURIER INTEGRAL

A beam of infinite length rests on an elastic foundation which applies to the beam a distributed force proportional to the deflection but opposite in sign. Neglecting the weight of the beam, find an equation of its deflection curve if the beam bears a total load $P_0 > 0$, uniformly distributed between $x = -l$ and $x = l$, so that the load per unit length $p(x)$ is given by

$$p(x) = \begin{cases} P_0/2l & |x| < l \\ 0 & |x| > l \end{cases}$$

Using Eq. (12), Sec. 2.11, and taking into account both the elastic load ky, which acts upward when y is negative, and the static load $p(x)$, which acts downward, we obtain

$$EI\frac{d^4y}{dx^4} = -ky - p(x) \tag{2}$$

as the equation to be solved to find the deflection $y(x)$.

In this problem, since the beam extends to infinity in both directions and since neither the load nor the deflection is periodic, we cannot use Fourier series to represent y and $p(x)$ as we did in Example 5, Sec. 8.6, but we can use Fourier integrals. Because of the symmetry of the loading, the deflection y will be an even function. Hence we shall utilize the cosine transform pair. Thus, using Eq. (5a), Sec. 9.3, we can write

$$y(x) = \sqrt{\frac{2}{\pi}} \int_0^\infty Y_c(\omega) \cos \omega x \, d\omega \tag{3}$$

where $Y_c(\omega)$ is the Fourier cosine transform of y, yet to be determined. Similarly, we have for $p(x)$

$$p(x) = \sqrt{\frac{2}{\pi}} \int_0^\infty P_c(\omega) \cos \omega x \, d\omega$$

where

$$P_c(\omega) = \sqrt{\frac{2}{\pi}} \int_0^\infty p(x) \cos \omega x \, dx = \sqrt{\frac{2}{\pi}} \int_0^l \frac{P_0}{2l} \cos \omega x \, dx = \frac{P_0}{\sqrt{2\pi}} \frac{\sin \omega l}{l\omega}$$

Thus

$$p(x) = \frac{P_0}{\pi l} \int_0^\infty \frac{\sin \omega l}{\omega} \cos \omega x \, d\omega$$

Now, differentiating (3) with respect to x to find y^{iv} and then substituting into (2), we obtain

$$EI\sqrt{\frac{2}{\pi}} \int_0^\infty \omega^4 Y_c(\omega) \cos \omega x \, d\omega = -k\sqrt{\frac{2}{\pi}} \int_0^\infty Y_c(\omega) \cos \omega x \, d\omega - \frac{P_0}{\pi l} \int_0^\infty \frac{\sin \omega l}{\omega} \cos \omega x \, d\omega$$

or, multiplying through by $\sqrt{\frac{\pi}{2}}$ and combining the integrals,

$$\int_0^\infty \left[(EI\omega^4 + k) Y_c(\omega) + \frac{P_0}{\sqrt{2\pi} l} \frac{\sin \omega l}{\omega} \right] \cos \omega x \, d\omega = 0$$

Since this must be an identity for all values of x, it follows that the coefficient of $\cos \omega x$ in the integrand must be zero. Hence

$$(EI\omega^4 + k) Y_c(\omega) + \frac{P_0}{\sqrt{2\pi} l} \frac{\sin \omega l}{\omega} = 0$$

and

$$Y_c(\omega) = -\frac{P_0}{\sqrt{2\pi} l} \frac{\sin \omega l}{\omega(EI\omega^4 + k)}$$

Thus, finally

$$y(x) = -\frac{P_0}{\pi l} \int_0^\infty \frac{\sin \omega l}{\omega(EI\omega^4 + k)} \cos \omega x \, d\omega \tag{4}$$

It is informative to consider what happens to the value of y as $l \to 0$, that is, as the total load, while remaining constant, is concentrated on a shorter and shorter segment of the beam. Moving the factor l into the integrand and noting that

$$\lim_{l \to 0} \frac{\sin \omega l}{\omega l} = 1$$

we see that $y(x)$ approaches the limit

$$-\frac{P_0}{\pi}\int_0^\infty \frac{\cos \omega x}{EI\omega^4 + k}\,d\omega \tag{5}$$

which is the deflection at x produced in the beam by a concentrated downward load P_0 applied at the origin.

In Examples 1 and 2, the differential equations involved were ordinary nonhomogeneous linear differential equations with constant coefficients. In each case, a Fourier integral representation of a particular solution was found. Now let us see how Fourier transforms may be used to find particular solutions of such equations.

Proceeding formally, we apply the Fourier transformation to both members of the equation

$$a_0\frac{d^n y}{dt^n} + \cdots + a_n y = f(t) \tag{6}$$

Making use of Eq. (11), Sec. 9.3, but with $Y(\omega) = \mathcal{F}[y(t)]$, and assuming $f(t)$ has the transform $F(\omega)$, we obtain

$$[a_0(i\omega)^n + \cdots + a_n]Y(\omega) = F(\omega)$$

Solving this equation for $Y(\omega)$ and introducing the so-called **transfer function**

$$W(s) = \frac{1}{a_0 s^n + \cdots + a_n} \tag{7}$$

we have for the Fourier transform of $y(t)$,

$$Y(\omega) = W(i\omega)F(\omega) \tag{8}$$

For ω real, $W(i\omega)$ is called the **frequency response function.** If Eq. (6) is **stable,** i.e., if all characteristic roots have negative real parts, then $W(i\omega)$ has an inverse (see Exercise 14). Hence, by Theorem 10, Sec. 9.3, assuming $W(i\omega) = \mathcal{F}[w(t)]$,

$$y(t) = w(t) * f(t) \tag{9}$$

This result, although derived under rather stringent assumptions, can be used under much more general conditions because the convolution of the **weight function** $w(t)$ and $f(t)$ may exist even though the differential equation is unstable or if f has no transform (see Example 5).

EXAMPLE 3

Use Fourier transforms to find a particular integral of the differential equation of Example 1.

Writing Fourier transforms of both members of the equation, we have

$$[(i\omega)^2 + 3i\omega + 2]Y(\omega) = F(\omega)$$

Solving for $Y(\omega)$, then using partial fractions, we get

$$Y(\omega) = \left[\frac{1}{(i\omega)^2 + 3i\omega + 2}\right]F(\omega) = \left[\frac{1}{1 + i\omega} - \frac{1}{2 + i\omega}\right]F(\omega) \equiv G(\omega)F(\omega)$$

By Theorem 10, Sec. 9.3, and from the definition of $f(t)$,

$$y(t) = \int_{-\infty}^{\infty} g(t-\lambda)f(\lambda)\,d\lambda = \int_{-1}^{1} g(t-\lambda)\,d\lambda$$

where the inverse transform $g(t)$ of $G(\omega)$ is given by

$$g(t) = \begin{cases} 0 & t \le 0 \\ e^{-t} - e^{-2t} & 0 \le t \end{cases} \qquad \text{and} \qquad g(t-\lambda) = \begin{cases} 0 & t \le \lambda \\ e^{-(t-\lambda)} - e^{-2(t-\lambda)} & \lambda \le t \end{cases}$$

Depending on how t relates to the limits of integration on λ, we have

$$y(t) = \begin{cases} \displaystyle\int_{-1}^{1} 0\,d\lambda & t \le -1 \\ \displaystyle\int_{-1}^{t} (e^{-(t-\lambda)} - e^{-2(t-\lambda)})\,d\lambda = e^{-t}\int_{-1}^{t} e^{\lambda}\,d\lambda - e^{-2t}\int_{-1}^{t} e^{2\lambda}\,d\lambda & -1 \le t \le 1 \\ \displaystyle\int_{-1}^{1} (e^{-(t-\lambda)} - e^{-2(t-\lambda)})\,d\lambda = e^{-t}\int_{-1}^{1} e^{\lambda}\,d\lambda - e^{-2t}\int_{-1}^{1} e^{2\lambda}\,d\lambda & 1 \le t \end{cases}$$

Finally, performing the indicated integrations and simplifying, we obtain

$$y(t) = \begin{cases} 0 & t \le -1 \\ \frac{1}{2} - e^{-(t+1)} + \frac{1}{2}e^{-2(t+1)} & -1 \le t \le 1 \\ 2e^{-t}\sinh 1 - e^{-2t}\sinh 2 & 1 \le t \end{cases}$$

which gives us a trivial solution of the differential equation on $(-\infty, -1)$ and nontrivial solutions on $(-1, 1)$ and $(1, \infty)$. These particular solutions are also represented by the Fourier integral $Y(t)$ of Example 1 because, as might be expected, $y(t)$ has $Y(t)$ as its Fourier integral representation (see Exercise 15). Thus, $y(t)$ *provides an evaluation of* $Y(t)$. The structure of $y(t)$ shows that e^{-t} and e^{-2t} form a basis for the homogeneous differential equation and that $y_p(t) = \frac{1}{2}$ is a particular solution on the normal interval $(-1, 1)$ (see Exercise 16).

By checking limiting values at $t = -1$ and $t = 1$, we find that the function y represented by $y(t)$ is continuous and that y' can be made continuous on $(-\infty, \infty)$. But, at these same two points, y'' has nonremovable discontinuities. This behavior is consistent with the fact that as $\omega \to \infty$, the *transform* $Y(\omega)$ of $y(t)$ [not to be confused with the *Fourier integral* $Y(t)$ of Example 1] tends to zero as c/ω^3.

EXAMPLE 4

DEFLECTION OF AN INFINITE BEAM IN EXPLICIT FORM

Use Fourier transforms to find the deflection of the beam described in Example 2.

From Eq. (2), the deflection $y(x)$ satisfies the differential equation

$$EIy^{\text{iv}} + ky = -p(x)$$

on $(-\infty, \infty)$. Equating Fourier transforms of both sides of this equation, then solving for $Y(\omega)$, we obtain

$$Y(\omega) = -\frac{1}{EI}W(\omega)P(\omega) \qquad \text{where} \qquad W(\omega) = \frac{1}{4a^4 + \omega^4} \qquad a = \left(\frac{k}{4EI}\right)^{1/4}$$

and $P(\omega)$ is the transform of $p(x)$. The inverse transform of $Y(\omega)$ can now be written as a convolution with a reduced range of integration by noting that the definition of $p(x)$ implies

$$p(x-\lambda) = \begin{cases} P_0/2l & |x-\lambda| < l \\ 0 & |x-\lambda| > l \end{cases}$$

so that

$$y(x) = -\frac{1}{EI}w(x) * p(x) = -\frac{1}{EI}\int_{-\infty}^{\infty} w(\lambda)p(x-\lambda)\,d\lambda = -\frac{P_0}{2EIl}\int_{x-l}^{x+l} w(\lambda)\,d\lambda$$

Example 2, Sec. 9.3, gives

$$w(x) = \begin{cases} \dfrac{1}{8a^3} e^{ax}(\cos ax - \sin ax) & x \le 0 \\ \dfrac{1}{8a^3} e^{-ax}(\cos ax + \sin ax) & 0 \le x \end{cases}$$

as the inverse transform of $W(\omega)$. Thus, having due regard for how x is related to the limits of integration on λ and setting $b = P_0/16EIla^3$, we have

$$y(x) = \begin{cases} -b\displaystyle\int_{x-l}^{x+l} e^{a\lambda}(\cos a\lambda - \sin a\lambda)\,d\lambda & x \le -l \\ -b\left[\displaystyle\int_{x-l}^{0} e^{a\lambda}(\cos a\lambda - \sin a\lambda)\,d\lambda + \int_0^{x+l} e^{-a\lambda}(\cos a\lambda + \sin a\lambda)\,d\lambda\right] & -l \le x \le l \\ -b\displaystyle\int_{x-l}^{x+l} e^{-a\lambda}(\cos a\lambda + \sin a\lambda)\,d\lambda & l \le x \end{cases}$$

Performing the indicated integrations, then replacing the common coefficient b/a in each resultant representation of y by $P_0/4kl$ and simplifying, we find the deflection of the beam to be given by

$$y(x) = \begin{cases} \dfrac{-P_0}{4kl}[e^{a(x+l)} \cos a(x+l) - e^{a(x-l)} \cos a(x-l)] & x \le -l \\ \dfrac{-P_0}{4kl}[2 - e^{a(x-l)} \cos a(x-l) - e^{-a(x+l)} \cos a(x+l)] & -l \le x \le l \\ \dfrac{-P_0}{4kl}[e^{-a(x-l)} \cos a(x-l) - e^{-a(x+l)} \cos a(x+l)] & l \le x \end{cases}$$

A change from x to $-x$ in the argument of $y(x)$ shows that $y(-x) \equiv y(x)$, and so the deflection is an even function of x, as was inferred in Example 2. As with the deflection y, all discontinuities in y', and in the moment and shear, are removable because as $\omega \to \infty$ the transform $Y(\omega)$ tends to zero as c/ω^5 [see Eq. (4)]. From the representations of y on $(-\infty, -l]$ and $[l, \infty)$, we see that $\lim_{x\to\pm\infty} y(x) = 0$.

EXAMPLE 5

Find a particular solution for $2y'' - 5y' - 3y = 20e^{2t}$ by means of Fourier transforms.

From the transfer function

$$W(s) = \frac{1}{2s^2 - 5s - 3} = \frac{1}{(2s+1)(s-3)} = -\frac{1}{7}\left(\frac{2}{2s+1} + \frac{-1}{s-3}\right)$$

we get the frequency response function

$$W(i\omega) = -\frac{1}{7}\left(\frac{1}{\frac{1}{2} + i\omega} + \frac{-1}{-3 + i\omega}\right)$$

Entry **a**, Table 9.2, and the relation $\int_{-\infty}^{0} e^{at}e^{-i\omega t}\,dt = -1/(-a + i\omega)$, $a > 0$, give us in turn

$$\mathcal{F}^{-1}\left[\frac{1}{\frac{1}{2} + i\omega}\right] = \begin{cases} 0 & t < 0 \\ e^{-t/2} & 0 < t \end{cases} \quad \text{and} \quad \mathcal{F}^{-1}\left[\frac{-1}{-3 + i\omega}\right] = \begin{cases} e^{3t} & t < 0 \\ 0 & 0 < t \end{cases}$$

Hence the inverse transform of $W(i\omega)$ is the weight function

$$w(t) = \begin{cases} -e^{3t}/7 & t < 0 \\ -e^{-t/2}/7 & 0 < t \end{cases}$$

The differential equation is unstable, and its nonhomogeneous term $f(t) = 20e^{2t}$ has no transform; nevertheless, a formal application of Eq. (9) gives

$$y(t) = w(t) * f(t) = -\frac{20}{7}\left[\int_{-\infty}^{0} e^{3\lambda}e^{2(t-\lambda)}\,d\lambda + \int_{0}^{\infty} e^{-\lambda/2}e^{2(t-\lambda)}\,d\lambda\right] = -4e^{2t}$$

as a particular solution of the given equation on $(-\infty, \infty)$. The factored form of the denominator in the transfer function $W(s)$ identifies

$$c_1e^{-t/2} + c_2e^{3t}$$

as a complementary function. A complete solution of the differential equation on $(-\infty, \infty)$ is therefore

$$c_1e^{-t/2} + c_2e^{3t} - 4e^{2t}$$

We next consider an important topic in information theory which deals with the analysis of signals and involves an interesting interplay between Fourier series and Fourier integrals.

Let f denote a continuous signal defined for all values of t. If $f(t)$ is observed, or *sampled,* only at certain times, then presumably some information in the signal is lost. It is remarkable, however, that under certain conditions a function is completely determined if its values are known on an appropriate set of points. Specifically, we have the following result, due to the American applied mathematician Claude Elwood Shannon (b. 1916).

THEOREM 1 **(The Time-Sampling Theorem)** If the Fourier transform of a function $f(t)$ is **band-limited,** that is, if for some smallest positive number ω_0,

$$F(\omega) = 0 \qquad \text{for} \qquad |\omega| > \omega_0$$

then $f(t)$ is completely determined by the values

$$f_n = f(n\pi/\omega_0)$$

and

$$f(t) = \sum_{n=-\infty}^{\infty} f_n \frac{\sin(\omega_0 t - n\pi)}{\omega_0 t - n\pi}$$

PROOF Because of the band-limited hypothesis on $F(\omega)$, the Fourier integral (1), Sec. 9.3, for $f(t)$ becomes

$$f(t) = \frac{1}{2\pi}\int_{-\omega_0}^{\omega_0} F(\omega)e^{i\omega t}\,d\omega \tag{10}$$

Evaluating this integral for $t = n\pi/\omega_0$, we obtain the sample values

$$f_n = f(n\pi/\omega_0) = \frac{1}{2\pi}\int_{-\omega_0}^{\omega_0} F(\omega)e^{i\omega n\pi/\omega_0}\,d\omega \tag{11}$$

Now, working in the frequency domain, let us extend the *restriction of F to* $(-\omega_0, \omega_0)$ into a periodic function G of period $2\omega_0$, as indicated in Fig. 9.11b, by defining

$$G(\omega) = \begin{cases} F(\omega) & -\omega_0 < \omega < \omega_0 \\ G(\omega + 2\omega_0) = G(\omega) & \omega \quad \text{real} \end{cases}$$

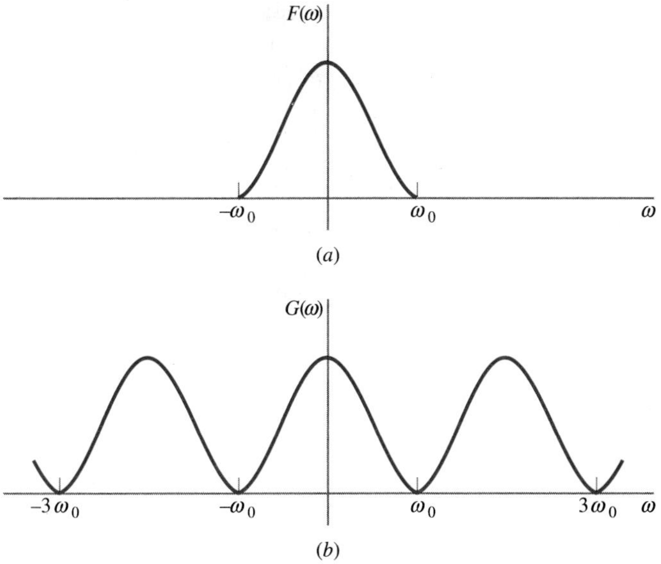

FIGURE 9.11
The restriction of $F(\omega)$ to $(-\omega_0, \omega_0)$ and its periodic extension $G(\omega)$ discussed in the proof of Theorem 1.

$G(\omega)$ can of course be represented by a complex Fourier series, and we have [see Eqs. (7*a*) and (8*a*), Sec. 8.5]

$$G(\omega) = \sum_{n=-\infty}^{\infty} c_n e^{-ni\pi\omega/\omega_0} \tag{12}$$

where

$$c_n = \frac{1}{2\omega_0}\int_{-\omega_0}^{\omega_0} F(\omega)e^{ni\pi\omega/\omega_0}\,d\omega \tag{13}$$

Comparing Eqs. (11) and (13), we conclude that $c_n = (\pi/\omega_0)f(n\pi/\omega_0)$ and therefore, from (12),

$$G(\omega) = \sum_{n=-\infty}^{\infty} \frac{\pi}{\omega_0} f_n e^{-ni\pi\omega/\omega_0}$$

Now, by definition, $G(\omega) \equiv F(\omega)$ on the interval $-\omega_0 < \omega < \omega_0$. Hence, multiplying $G(\omega)$ by the pulse function

$$p(\omega) = \begin{cases} 1 & |\omega| < \omega_0 \\ 0 & |\omega| > \omega_0 \end{cases}$$

in essence gives $F(\omega)$ everywhere on the ω axis, that is,

$$F(\omega) = p(\omega)\sum_{n=-\infty}^{\infty} \frac{\pi}{\omega_0} f_n e^{-ni\pi\omega/\omega_0} = \sum_{n=-\infty}^{\infty} \frac{\pi}{\omega_0} f_n p(\omega) e^{-ni\pi\omega/\omega_0} \tag{14}$$

The Fourier transform of $p(t)$, as given by **f**, Table 9.2, with $k = l = \omega_0$ and $a = b = 1$, is

$$P(\omega) = \frac{1}{i\omega}(e^{i\omega_0\omega} - e^{-i\omega_0\omega}) = \frac{2\sin\omega_0\omega}{\omega}$$

By the symmetry theorem (Theorem 2, Sec. 9.3), the transform of $P(t)$ is $2\pi p(-\omega) = 2\pi p(\omega)$. Therefore the inverse of $p(\omega)$ is $(1/2\pi)P(t) = (\sin \omega_0 t)/\pi t$. Using this result, together with the time-shifting theorem (Theorem 4, Sec. 9.3) to take the inverse of the second series in (14) term by term, we have

$$\mathcal{F}^{-1}[F(\omega)] = f(t) = \sum_{n=-\infty}^{\infty} \frac{\pi}{\omega_0} f_n \left(\frac{\sin \omega_0 t}{\pi t}\right)_{t \to t - \frac{n\pi}{\omega_0}} = \sum_{n=-\infty}^{\infty} f_n \frac{\sin (\omega_0 t - n\pi)}{\omega_0 t - n\pi}$$

as asserted. ◗

Although Fourier transforms worked well in solving the differential equations of Examples 1, 4, and 5, such applications are limited by the number of suitable transform pairs available and by the failure of certain integrals to converge. The Fourier transformation is best suited to the solution of differential equations over infinite domains. Under appropriate symmetry conditions, the cosine and sine transformations are also well tailored to such problems. Such transformations can sometimes be used to solve initial-value problems. However, as we shall see in the next chapter, the Laplace transformation is much more efficient for this purpose; hence, except for several exercises, we shall not pursue the use of Fourier transforms to this end.

The Laplace transformation, like the three Fourier transformations, converts ordinary differential equations into algebraic equations and reduces a partial differential equation to a differential equation with one less independent variable. This feature and the inversion process associated with each transformation are what make transform pairs so effective in solving initial-value or boundary-value problems.

EXERCISES

1. Using the method and function $f(t)$ of Example 1, find a particular solution of the equation $y'' + ay' + by = f(t)$.

Work Exercise 1 for each of the following functions.

2. $f(t) = \begin{cases} 0 & -\infty < t \le -1 \\ 1+t & -1 \le t \le 0 \\ 1-t & 0 \le t \le 1 \\ 0 & 1 \le t < \infty \end{cases}$

3. $f(t) = \begin{cases} t & |t| < 1 \\ 0 & |t| > 1 \end{cases}$

4. $f(t) = \begin{cases} \sin \dfrac{\pi t}{2} & |t| < 1 \\ 0 & |t| > 1 \end{cases}$

5. Work Example 2 if the load per unit length is

$$p(x) = \begin{cases} 0 & -\infty < x \le -l \\ P_0(l+x)/l^2 & -l \le x \le 0 \\ P_0(l-x)/l^2 & 0 \le x \le l \\ 0 & l \le x < \infty \end{cases}$$

and find the limit of the deflection $y(x)$ as $l \to 0$.

Work Exercise 5 for each of the following loadings.

6. $p(x) = \begin{cases} 0 & -\infty < x < b-l \\ P_0/2l & b-l < x < b+l \\ 0 & b+l < x < \infty \end{cases}$

7. $p(x) = \begin{cases} 0 & -\infty < x < -l \\ -P_0/l & -l < x < 0 \\ P_0/l & 0 < x < l \\ 0 & l < x < \infty \end{cases}$

8. Find the Fourier integral representing the deflection $y(x)$ of the beam in Example 2 if the load per unit length is

$$p(x) = \begin{cases} e^{cx} & -\infty < x \le 0 \\ e^{-cx} & 0 \le x < \infty \end{cases} \qquad c > 0$$

Use Fourier transforms to find the deflection of the beam in Example 2 if the load per unit length is that of

9. Exercise 5 **10.** Exercise 6

11. Exercise 7 **12.** Exercise 8

13. Discuss the possibility of using the method of Example 2 to determine the deflection of a semi-infinite beam resting on an elastic support.

14. Show that if Eq. (6) is stable, then its frequency response function $W(i\omega)$ has an inverse Fourier transform. *Hint:* Use partial fractions to write $W(i\omega)$ as a linear combination of expressions of the type $1/(i\omega - z)^{k+1}$, $k \ge 0$ an integer, and observe that Example 5, Sec. 9.3, guarantees an inverse transform for each such expression.

15. **(a)** Write the real Fourier integral representation of $Y(t)$, Example 1, as a complex exponential Fourier integral and identify the Fourier transform of $Y(t)$.

(b) Find the Fourier transform $F(\omega)$ of the pulse function

$f(t)$, Example 1, then form the product of $G(\omega) = 1/[(i\omega)^2 + 3i\omega + 2]$ and $F(\omega)$ to derive the Fourier transform of the function $y(t)$, Example 3.

(c) Verify that the transform of $Y(t)$ in Part **a** and that of $y(t)$ in Part **b** are the same. What does this imply about $Y(t)$ and $y(t)$?

(d) Find the Fourier transform of $y(t)$ directly and use it to derive the Fourier integral representation of $y(t)$.

16. Find a complete solution of the differential equation in Example 1.

17. Show that the function $y(t)$ found in Example 3 is unique in that for $s = -1, 1$, it satisfies the differential equation of Example 1 on the normal intervals $(-\infty, -1), (-1, 1), (1, \infty)$ and the auxiliary conditions $y(s^-) = y(s^+)$, $y'(s^-) = y'(s^+)$, $\lim_{t \to -\infty} y(t) = 0$.

Use Fourier transforms to find a particular integral of each of the following differential equations, assuming that in each case the right-hand side is identically zero before $t = 0$.

18. $y'' + 3y' + 2y = e^{-3t}$ **19.** $y'' + 2y' + y = e^{-t}$

20. $y' + y = \begin{cases} 1 & 0 < t < 1 \\ 0 & 1 < t < \infty \end{cases}$

21. $y' + y = \begin{cases} e^t & 0 < t < 1 \\ 0 & 1 < t < \infty \end{cases}$

Apply Eq. (9) to find a particular integral of each of the following equations.

22. $y'' + 3y' + 2y = 6e^t$

23. $2y'' + 7y' + 3y = 100e^{2t}$

24. $y' + y = \sin t$ **25.** $y' + 4y = 25 \cos 3t$

26. $y' + y = e^{-t}$ **27.** $3y' - 2y = e^{t/3}$

28. $y'' + 3y' + 2y = \begin{cases} 1 & |t - \frac{1}{2}| < \frac{1}{2} \\ 0 & |t - \frac{1}{2}| > \frac{1}{2} \end{cases}$

29. $2y'' - 7y' + 3y = \begin{cases} a & |t| < b \\ 0 & |t| > b \end{cases}$

30. Find the solution of $6y'' + 5y' + y = 1/(1 + t^2)$ determined by Eq. (9) and show that its values all lie between 0 and 1.

Solve each of the following initial-value problems by taking Fourier transforms as indicated in Exercise 15, Sec. 9.3.

31. $y'' + 3y' + 2y = e^{-t}$; $y(0) = 0$, $y'(0) = 1$

32. $y'' + 3y' + 2y = e^{-t}$; $y(0) = 1$, $y'(0) = 0$

33. $3y'' + 10y' + 3y = 64e^{-3t}$; $y(0) = 1$, $y'(0) = -3$

34. $y'' + 2ay' + (a^2 + b^2)y = f(t)$; $y(0) = -1$, $y'(0) = 0$, where $f(t) = \begin{cases} k & 0 < t < 1 \\ 0 & 1 < t < \infty \end{cases}$, the equation is stable, and $b \neq 0$.

35. Use the formula $\mathcal{F}[y'(t)] = i\omega Y(\omega) - y(0)$ to solve the initial-value problem

$$y' + ay = 0; \qquad y(0) = k,\ a > 0$$

on $[0, \infty)$. Is the solution valid on $(-\infty, \infty)$? Explain.

36. Write a formula, analogous to that of Exercise 35, which may be used to solve the initial-value problem $y' - ay = 0$; $y(0) = k$, $a > 0$, on $(-\infty, 0]$. Use the formula to find the solution. Is the solution valid on $(-\infty, \infty)$? Explain.

37. **(a)** Find the transfer, frequency response, and weight functions corresponding to the differential equation $y'' - (a + b)y' + aby = e^{mt}$, $0 < a < b$, m real.

(b) For what values of m does $w(t) * e^{mt}$ converge?

(c) Use Eq. (9) to find a particular integral of the differential equation. For what values of m is the solution valid?

38. Using the real form of the Fourier integral given by Eq. (14*a*), Sec. 9.1, and the concepts of magnification ratio and phase shift developed in Sec. 7.3, obtain a formula for the steady-state response of a mechanical system of one degree of freedom to a nonperiodic driving force.

39. Show that the steady-state response of an electric circuit of impedance $Z(\omega)$ to a nonperiodic voltage $E(t) = \int_{-\infty}^{\infty} g(\omega)e^{i\omega t}\, d\omega$ can be written

$$I(t) = 2\int_0^{\infty} \{\text{Re}[g(\omega)/Z(\omega)] \cos \omega t - \text{Im}[g(\omega)/Z(\omega)] \sin \omega t\}\, d\omega$$

where $\text{Re}[g(\omega)/Z(\omega)]$ and $\text{Im}[g(\omega)/Z(\omega)]$ denote, respectively, the real and imaginary parts of the complex expression $g(\omega)/Z(\omega)$.

40. Show that if $t_0 > 0$ and if $f(t) = 0$ for all values of t such that $|t| > t_0$, then the Fourier transform of f is completely determined if the values of $F(\omega)$ are known on the set of points $\omega_n = n\pi/t_0$, $n = \ldots, -2, -1, 0, 1, 2, \ldots$.

9.5 SINGULARITY FUNCTIONS AND THEIR FOURIER TRANSFORMS

In Sec. 2.11 we introduced the notion of a concentrated load. If one were asked to explain what is meant by a concentrated load, the explanation would probably go something like this: imagine that over an interval $(x_0, x_0 + a)$ a beam bears a uniformly distributed load whose magnitude per unit length is P_0/a. Then no matter how small a may be, the total load on the beam is $(P_0/a) \times a = P_0$. As $a \to 0$, the idea of a **concentrated load** emerges from this as the limiting form of a set of distributed loads. A similar example is the so-called **impulse function,** conceived as the limit of a force of magnitude F_0/a applied to a system during a vanishingly small interval of time $(t_0, t_0 + a)$.

Both a concentrated load and an impulse function are closely related to what is known as the *delta function,* often called the *Dirac delta function,* after the English theoretical physicist P. A. M.

Dirac (1902–1984). As suggested by our discussion of concentrated loads and impulses, the delta function, written $\delta(t)$, is often described (though hardly defined!) by the formulas

$$\delta(t) = \begin{cases} 0 & t \neq 0 \\ \infty & t = 0 \end{cases} \tag{1a}$$

$$\int_{-\infty}^{\infty} \delta(t)\, dt = 1 \tag{1b}$$

It turns out (see Exercise 6) that in discussing the properties of concentrated loads and impulses, neither the load density $P(x)$ nor the applied force $F(t)$ need be uniformly distributed over the respective intervals $(x_0, x_0 + a)$ and $(t_0, t_0 + a)$. All that is required is that

$$\int_{x_0}^{x_0+a} P(x)\, dx \qquad \text{and} \qquad \int_{t_0}^{t_0+a} F(t)\, dt$$

remain constant as $a \to 0$. This leads to a second description of the delta function as the limit of a sequence of nonnegative functions

$$\delta(t) = \lim_{n\to\infty} f_n(t) \tag{2a}$$

such that

$$\int_{-\infty}^{\infty} f_n(t)\, dt = 1 \qquad \text{and} \qquad \lim_{n\to\infty} f_n(t) = 0 \qquad \text{for } t \neq 0 \tag{2b}$$

The delta function is also described by the so-called **sampling property:** for all functions ϕ which are continuous around the origin,

$$\int_{-\infty}^{\infty} \delta(t)\phi(t)\, dt = \phi(0) \tag{3}$$

To make this seem plausible, suppose that the functions of the sequence in (2a) are identically zero for $|t| > 1/n$, so that, from the first of Eqs. (2b),

$$\int_{-\infty}^{\infty} f_n(t)\, dt = \int_{-1/n}^{1/n} f_n(t)\, dt = 1$$

Since $\phi(t)$ is assumed to be continuous around the origin, we can, without loss of generality, suppose that n is sufficiently large that ϕ is continuous over the interval $[-1/n, 1/n]$. Let ϕ_M and ϕ_m be, respectively, the largest and smallest values of ϕ on $[-1/n, 1/n]$. Then,

$$\phi_m = \phi_m \int_{-1/n}^{1/n} f_n(t)\, dt \le \int_{-1/n}^{1/n} f_n(t)\phi(t)\, dt \le \phi_M \int_{-1/n}^{1/n} f_n(t)\, dt = \phi_M$$

As n becomes infinite, both ϕ_M and ϕ_m approach $\phi(0)$, and according to (2a), $f_n(t)$ tends to $\delta(t)$. Hence the middle integral in the continued inequality, whose limiting value is $\int_{-\infty}^{\infty} \delta(t)\phi(t)\, dt$, approaches $\phi(0)$, as prescribed by (3). In view of the unusual properties suggested by Eqs. (1a) and

(1*b*), (2*a*) and (2*b*), and by Eq. (3), it is not surprising that $\delta(t)$ is called a *singularity function.* Another example of a singularity function is introduced in Exercise 10.

None of the three descriptions of $\delta(t)$ that we have given is logically satisfying, but each is a guide to understanding the delta function. Of the three, Eq. (3) is the most enlightening. Thought of as a *functional* which assigns the value $\phi(0)$ to any function ϕ that is continuous around the origin, the **sampling operator**

$$\int_{-\infty}^{\infty} \delta(t)[\quad]\,dt \tag{4}$$

is an example of what is called a *generalized function* or a *distribution.* It is beyond the scope of this text to discuss the theory of distributions† or even to discuss (4) as a particular distribution. However, in this section we shall present some of the properties and applications of $\delta(t)$ and its Fourier transform. We will also encounter the delta function and several other singularity functions in the next chapter when we study the Laplace transformation. Many of the distinctive properties of the delta function, and of distributions in general, are similar to the formal properties of definite integrals, and we shall use this fact in our discussion of $\delta(t)$.

One notable feature of the sampling operator (4) is that it is a linear functional, that is,

$$\begin{aligned}\int_{-\infty}^{\infty} \delta(t)[a_1\phi_1(t) + a_2\phi_2(t)]\,dt &= a_1\int_{-\infty}^{\infty} \delta(t)\phi_1(t)\,dt + a_2\int_{-\infty}^{\infty} \delta(t)\phi_2(t)\,dt \\ &= a_1\phi_1(0) + a_2\phi_2(0)\end{aligned} \tag{5}$$

Two other features are the **general sampling property**

$$\int_{-\infty}^{\infty} \delta(t - t_0)\phi(t)\,dt = \int_{-\infty}^{\infty} \delta(t)\phi(t + t_0)\,dt = \phi(t_0) \tag{6}$$

and the **change of scale property**

$$\int_{-\infty}^{\infty} \delta(at)\phi(t)\,dt = \frac{1}{a}\int_{-\infty}^{\infty} \delta(t)\phi(t/a)\,dt = \frac{1}{a}\phi(0) \tag{7}$$

which are justified by the changes of variables

$$t - t_0 \to t \qquad \text{and} \qquad at \to t$$

in the initial integrals of (6) and (7), respectively.

The derivative of $\delta(t)$ is characterized by the formula

$$\int_{-\infty}^{\infty} \delta'(t)\phi(t)\,dt = -\phi'(0) \tag{8}$$

†Nelson Dunford and Jacob T. Schwartz, *Linear Operators,* pt. II, pp. 1644–1680, Interscience, New York, 1963.

The appropriateness of (8) becomes apparent if we apply the usual definition of a derivative to $\delta(t)$ and use Eqs. (3) and (6). Doing this, we have

$$\int_{-\infty}^{\infty} \frac{\delta(t+\epsilon)-\delta(t)}{\epsilon}\phi(t)\,dt = \frac{1}{\epsilon}\int_{-\infty}^{\infty} \delta(t+\epsilon)\phi(t)\,dt - \frac{1}{\epsilon}\int_{-\infty}^{\infty} \delta(t)\phi(t)\,dt = -\frac{\phi(-\epsilon)-\phi(0)}{-\epsilon}$$

which in the limit, as ϵ approaches zero, gives

$$\int_{-\infty}^{\infty} \delta'(t)\phi(t)\,dt = -\phi'(0)$$

as asserted. Continuing in this way, assuming the relevant derivatives of ϕ exist and are continuous around the origin, we have the general formula

$$\int_{-\infty}^{\infty} \delta^{(n)}(t)\phi(t)\,dt = (-1)^n\phi^{(n)}(0) \tag{9}$$

The meaning of the product $\delta(t)f(t)$ of $\delta(t)$ and a function $f(t)$ continuous around the origin is given by

$$\int_{-\infty}^{\infty} [\delta(t)f(t)]\phi(t)\,dt = \int_{-\infty}^{\infty} \delta(t)[f(t)\phi(t)]\,dt = f(0)\phi(0) \tag{10}$$

Since it is also true that

$$\int_{-\infty}^{\infty} f(0)\,\delta(t)\phi(t)\,dt = f(0)\int_{-\infty}^{\infty} \delta(t)\phi(t)\,dt = f(0)\phi(0)$$

the sampling operators obtained from (4) by replacing $\delta(t)$ by $\delta(t)f(t)$ or $f(0)\delta(t)$ are the same, that is to say, they are operationally equivalent because they produce the same result when applied to all suitable functions ϕ. This we indicate by simply writing

$$\delta(t)f(t) = f(t)\,\delta(t) = f(0)\,\delta(t) \tag{11}$$

In particular, we have $t\,\delta(t) = 0$.

The product $f(t)\,\delta'(t)$ is given by the formula

$$f(t)\,\delta'(t) = f(0)\,\delta'(t) - f'(0)\,\delta(t) \tag{12}$$

To verify this, we note first that, by (8),

$$\int_{-\infty}^{\infty} [\delta'(t)f(t)]\phi(t)\,dt = \int_{-\infty}^{\infty} \delta'(t)[f(t)\phi(t)]\,dt = -[f(t)\phi(t)]'_{t=0} = -\int_{-\infty}^{\infty} \delta(t)[f(t)\phi(t)]'\,dt$$

$$= -\int_{-\infty}^{\infty} \delta(t)[f(t)\phi'(t) + f'(t)\phi(t)]\,dt = -f(0)\phi'(0) - f'(0)\phi(0)$$

However, we also have

$$\int_{-\infty}^{\infty} [f(0)\,\delta'(t) - f'(0)\,\delta(t)]\phi(t)\,dt = f(0)\int_{-\infty}^{\infty} \delta'(t)\phi(t)\,dt - f'(0)\int_{-\infty}^{\infty} \delta(t)\phi(t)\,dt$$

$$= -f(0)\phi'(0) - f'(0)\phi(0)$$

Thus the two functionals corresponding to

$$\delta'(t)f(t) \qquad \text{and} \qquad f(0)\,\delta'(t) - f'(0)\,\delta(t)$$

produce the same result when applied to all suitable functions ϕ. Hence they are equal, and (12) is confirmed. In particular, we see from (12) that $t\,\delta'(t) = -\delta(t)$.

The delta function clearly fails to satisfy the conditions of Theorem 1, Sec. 9.1. Nevertheless, it does have a Fourier transform. In fact, from Eq. (2), Sec. 9.3, and Eq. (3), extended to complex-valued functions of a real variable, we have

$$\mathscr{F}[\delta(t)] = \int_{-\infty}^{\infty} \delta(t)e^{-i\omega t}\,dt = 1$$

Likewise, for the Fourier transform of $\delta(t - t_0)$, we have, by Eq. (6),

$$\mathscr{F}[\delta(t - t_0)] = \int_{-\infty}^{\infty} \delta(t - t_0)e^{-i\omega t}\,dt = e^{-i\omega t_0}$$

Hence we have the following transform pair

$$\delta(t - t_0) \longleftrightarrow e^{-i\omega t_0} \tag{13}$$

consistent with the time-shifting theorem.

It should be noted that although the transform pair (13) is a correct result, $\delta(t - t_0)$ cannot be recovered from its transform $e^{-i\omega t_0}$ through the use of Eq. (1), Sec. 9.3. In fact, even for the simple case $t_0 = 0$, that equation gives

$$\delta(t) = \frac{1}{2\pi}\int_{-\infty}^{\infty} 1 \cdot e^{i\omega t}\,d\omega = \frac{1}{2\pi}\int_{-\infty}^{\infty} (\cos \omega t + i \sin \omega t)\,d\omega$$

and this integral is meaningless until it is interpreted by the more advanced theory of distributions.

On the other hand, if $F(\omega) = \delta(\omega - \omega_0)$, we have, according to Eq. (1), Sec. 9.3,

$$f(t) = \mathscr{F}^{-1}[\delta(\omega - \omega_0)] = \frac{1}{2\pi}\int_{-\infty}^{\infty} \delta(\omega - \omega_0)e^{i\omega t}\,d\omega = \frac{1}{2\pi}e^{i\omega_0 t}$$

That is, we have the transform pair

$$\frac{1}{2\pi}e^{i\omega_0 t} \longleftrightarrow \delta(\omega - \omega_0) \tag{14a}$$

with

$$\frac{1}{2\pi} \longleftrightarrow \delta(\omega) \tag{14b}$$

as a special case.

EXAMPLE 1

What is the Fourier transform of cos $\omega_0 t$?

Since cos $\omega_0 t$ does not satisfy the conditions of Theorem 1, Sec. 9.1, it is not surprising that Eq. (2), Sec. 9.3, when applied to cos $\omega_0 t$, fails to give a corresponding Fourier transform (see Exercise 17). Nonetheless, it is possible to assign cos $\omega_0 t$ a transform. To do so, we simply use the exponential form of the cosine function, together with Eq. (14*a*), to write

$$\mathcal{F}[\cos \omega_0 t] = \mathcal{F}\left[\frac{e^{i\omega_0 t} + e^{-i\omega_0 t}}{2}\right] = \pi[\delta(\omega - \omega_0) + \delta(\omega + \omega_0)]$$

EXAMPLE 2

What is the Fourier transform of the so-called **signum function**

$$\operatorname{sgn} t = \begin{cases} -1 & t < 0 \\ 1 & 0 < t \end{cases}$$

Like cos $\omega_0 t$ in Example 1, sgn t does not satisfy the conditions of Theorem 1, Sec. 9.1, nor can its transform be found (by methods at our disposal) from Eq. (2), Sec. 9.3. However, it is not difficult to verify that

(15) $$\mathcal{F}[\operatorname{sgn} t] = \frac{2}{i\omega}$$

To do this, we use Eq. (1), Sec. 9.3, which gives

$$\mathcal{F}^{-1}\left[\frac{2}{i\omega}\right] = \frac{1}{2\pi}\int_{-\infty}^{\infty} \frac{2}{i\omega} e^{i\omega t}\, d\omega = \frac{1}{\pi}\int_{-\infty}^{\infty} \frac{\cos \omega t + i \sin \omega t}{i\omega}\, d\omega = \frac{1}{\pi}\int_{-\infty}^{\infty} \frac{\sin \omega t}{\omega}\, d\omega$$

since $(\cos \omega t)/\omega$ is an odd function of ω. The last integral is clearly an odd function of t. Therefore, if we can show that its value for $t > 0$ is equal to 1, its value for $t < 0$ will be -1 and our verification will be complete. To this end, we make the change of variable $\omega t = u$, $t > 0$, and utilize Eq. (25), Sec. 9.1, getting

$$\frac{1}{\pi}\int_{-\infty}^{\infty} \frac{\sin \omega t}{\omega}\, d\omega = \frac{1}{\pi}\int_{-\infty}^{\infty} \frac{\sin u}{u}\, du = \frac{2}{\pi}\int_{0}^{\infty} \frac{\sin u}{u}\, du = \frac{2}{\pi}\cdot\frac{\pi}{2} = 1$$

hence Eq. (15) is confirmed.

EXAMPLE 3

Find the Fourier transform of the unit step function

$$u(t) = \begin{cases} 0 & t < 0 \\ 1 & 0 < t \end{cases}$$

We first note that $u(t) = \frac{1}{2}(1 + \operatorname{sgn} t)$. Then, from (14*b*) and (15), we have

(16) $$\mathcal{F}[u(t)] = \mathcal{F}\left[\frac{1}{2}\right] + \frac{1}{2}\mathcal{F}[\operatorname{sgn} t] = \pi\delta(\omega) + \frac{1}{i\omega}$$

EXAMPLE 4

What is $f(t)$ if its Fourier transform is $F(\omega) = u(\omega)$, where $u(\omega)$ is the unit step function?

From the symmetry theorem, Theorem 2, Sec. 9.3, we know that if $f(t)$ and $F(\omega)$ form a transform pair, then $F(t)$ and $2\pi f(-\omega)$ form another pair. By hypothesis, the function F is the unit step function. Hence, by Example 3,

$$\mathcal{F}[F(t)] = \mathcal{F}[u(t)] = \pi\delta(\omega) + \frac{1}{i\omega}$$

The symmetry theorem, extended to generalized functions, asserts that $\mathcal{F}[F(t)]$ is also equal to $2\pi f(-\omega)$, consequently,

$$2\pi f(-\omega) = \pi\delta(\omega) + \frac{1}{i\omega}$$

This equation determines the structure of the function f. In fact, upon replacing $-\omega$ by t and using the relation $\delta(-t) = \delta(t)$ (see Exercise 3), we have

$$f(t) = \mathcal{F}^{-1}[u(\omega)] = \frac{\delta(t)}{2} - \frac{1}{2\pi i t} = \frac{\delta(t)}{2} + \frac{i}{2\pi t} \tag{17}$$

as the required inverse of $u(\omega)$.

EXERCISES

1. By making the appropriate change of variable, verify Eq. (6).
2. By making the appropriate change of variable, verify Eq. (7).
3. By making the appropriate change of variable, in Eq. (3), show that $\delta(t)$ is an even function.
4. **(a)** What is the Fourier transform $F(\omega)$ of the function

$$f(t) = \begin{cases} 0 & |t| > a \\ 1/2a & |t| < a \end{cases}$$

(b) What is $\lim_{a\to 0} F(\omega)$?

5. **(a)** What is the Fourier transform $F(\omega)$ of the function

$$f(t) = \begin{cases} 0 & t < 0 \\ 1/3a & 0 < t < 3a \\ 0 & 3a < t \end{cases}$$

(b) What is $\lim_{a\to 0} F(\omega)$?

6. **(a)** What is the Fourier transform $F(\omega)$ of the function

$$f(t) = \begin{cases} 0 & |t| > 0 \\ (a+t)/a^2 & -a < t < 0 \\ (a-t)/a^2 & 0 < t < a \end{cases}$$

(b) What is $\lim_{a\to 0} F(\omega)$?

7. **(a)** If $f(t) = \begin{cases} 0 & |t| > a \\ 1/2a & |t| < a \end{cases}$, evaluate the integral

$$I = \int_{-\infty}^{\infty} f(t) \cos 3t \, dt$$

(b) What is $\lim_{a\to 0} I$?

8. Work Exercise 7 if

$$f(t) = \begin{cases} 0 & t < 3\pi/4 \\ 1/a & 3\pi/4 < t < 3\pi/4 + a \\ 0 & 3\pi/4 + a < t \end{cases}$$

9. **(a)** If $f(t) = \begin{cases} 0 & t < 0 \\ \dfrac{2(a-t)}{a^2} & 0 < t < a, \\ 0 & a < t \end{cases}$

show that for all values of a $\int_{-\infty}^{\infty} f(t)\, dt = 1$.

(b) Evaluate $I = \int_{-\infty}^{\infty} \frac{f(t)}{4+t^2}\, dt$

(c) What is $\lim_{a\to 0} I$?

(d) Is this consistent with Eq. (3)?

10. Another singularity function known as the **unit doublet** is often described as the limit as $a \to 0$ of the function

$$f(t) = \begin{cases} 0 & t < 0 \\ 1/a^2 & 0 < t < a \\ -1/a^2 & a < t < 2a \\ 0 & 2a < t \end{cases}$$

(a) Show that $\int_{-\infty}^{\infty} f(t)\, dt = 0$ and $\int_{-\infty}^{\infty} tf(t)\, dt = -1$

(b) What is the Fourier transform $F(\omega)$ of $f(t)$?

(c) What is $\lim_{a\to 0} F(\omega)$?

(d) Evaluate $I = \int_{-\infty}^{\infty} f(t) \sin 2t\, dt$.

(e) What is $\lim_{a\to 0} I$?

(f) How does the unit doublet appear to be related to $\delta'(t)$?

11. Work Exercise 10 if

$$f(t) = \begin{cases} 0 & t < 1 \\ 1/a^2 & 1 < t < 1+a \\ -1/a^2 & 1+a < t < 1+2a \\ 0 & 1+2a < t \end{cases}$$

12. Show that the Fourier transform of the Fourier series

$$f(t) = \sum_{n=-\infty}^{\infty} c_n e^{ni\pi t/p} \text{ is } F(\omega) = 2\pi \sum_{n=-\infty}^{\infty} c_n \delta(\omega - n\pi/p)$$

13. Show that the Fourier transform of the finite sequence of unit impulses $f(t) = \sum_{k=0}^{n-1} \delta(t - kT)$ is

$$F(\omega) = \sum_{k=0}^{n-1} e^{-ik\omega T} = e^{-i(n-1)T\omega/2} \frac{\sin(nT\omega/2)}{\sin(T\omega/2)} \qquad n \geq 1$$

Hint: After taking the transform term by term using Eq. (13), sum the resulting geometric progression. Then factor the appropriate exponential from numerator and denominator.

14. If $F(\omega)$ is the Fourier transform of a real function $f(t)$, so that

$$f(t) = \frac{1}{2\pi}\int_{-\infty}^{\infty} F(\omega)e^{i\omega t}\,d\omega$$

and if $(1/\pi)\int_{-\infty}^{\infty} F(\omega)e^{i\omega t}\,d\omega = f_1(t) + if_2(t)$, show that

$$f_1(t) = f(t) \qquad \text{and} \qquad f_2(t) = \frac{1}{\pi}\int_{-\infty}^{\infty} \frac{f(\tau)}{t-\tau}\,d\tau$$

Hint: Note that the transform of $f_1(t) + if_2(t)$ is equal to $2F(\omega)u(\omega)$. Then use the time convolution theorem and the result of Example 4.

15. **(a)** What is the Fourier transform of the step-modulated signal $f(t) = u(t)\cos\omega_0 t$? *Hint:* Recall Example 1, Sec. 9.3.

(b) What is the Fourier transform of $u(t)\sin\omega_0 t$?

16. Derive Eq. (5), Sec. 9.4, by assuming $p(x) = P_0\delta(x)$ at the outset. *Hint:* Since $\mathcal{F}[\delta(x)] = 1$, Eq. (1), Sec. 9.3, gives $\delta(x) = (1/2\pi)\int_{-\infty}^{\infty} e^{i\omega x}\,d\omega$.

17. **(a)** Show that $\cos\omega_0 t$ is not absolutely integrable on $(-\infty, \infty)$.

(b) Show that if $f(t) = \cos\omega_0 t$, Eq. (2), Sec. 9.3, reduces to

$$F(\cos\omega_0 t) = \int_0^{\infty} [\cos(\omega_0 + \omega)t + \cos(\omega_0 - \omega)t]\,dt$$

and fails to define a Fourier transform.

Use Fourier transforms to find particular solutions of each of the following differential equations.

18. $y'' + 2y' = u(t)e^{-t}$

19. $y'' + 4y' = 40u(t)e^{-5t}$

20. $3y''' + 7y'' + 5y' + y = 4u(t)$

9.6 FROM THE FOURIER INTEGRAL TO THE LAPLACE TRANSFORMATION

As we have already seen, there are numerous occasions when the function to be represented by a Fourier integral is identically zero on $(-\infty, 0)$. In fact, it was this possibility that led us to the important corollary to Theorem 10, Sec. 9.3. When this is the case, the complex Fourier transform pair, given by Eqs. (1) and (2), Sec. 9.3, becomes the so-called **unilateral Fourier transform pair**

$$f(t) = \frac{1}{2\pi}\int_{-\infty}^{\infty} F(\omega)e^{i\omega t}\,d\omega \tag{1a}$$

$$F(\omega) = \int_0^{\infty} f(t)e^{-i\omega t}\,dt \tag{1b}$$

If f is given, $(1a)$ is an integral equation involving F as an unknown function. If f has the Dirichlet properties and is absolutely integrable on $[0, \infty)$, a function $F(\omega)$ is determined by $(1b)$ and, as we know, that function is the solution to Eq. $(1a)$. We call $F(\omega)$ the **unilateral Fourier transform** of f, and we call $f(t)$ the **inverse transform** of F.

Useful as it is in many applications, the unilateral transform pair, Eqs. $(1a)$ and $(1b)$, is so limited by the requirement that f be absolutely integrable on $[0, \infty)$ that it is inadequate to represent such a simple and important function as the unit step function

$$u(t) = \begin{cases} 0 & -\infty < t < 0 \\ 1 & 0 < t < \infty \end{cases}$$

shown in Fig. 9.12d and with which we are now quite familiar. In fact, for the function u we have from $(1b)$,

$$F(\omega) = \int_0^{\infty} u(t)e^{-i\omega t}\,dt = \int_0^{\infty} e^{-i\omega t}\,dt = \left.\frac{e^{-i\omega t}}{-i\omega}\right|_0^{\infty} = \left.\frac{\cos\omega t - i\sin\omega t}{-i\omega}\right|_0^{\infty}$$

and this is completely meaningless, since both the cosine and sine oscillate without limit as their arguments become infinite.

As an artifice to handle this case and others like it, the function e^{-at} is sometimes inserted into the integral for $F(\omega)$ in place of the unit step function. Now, as we have already seen (Example 1,

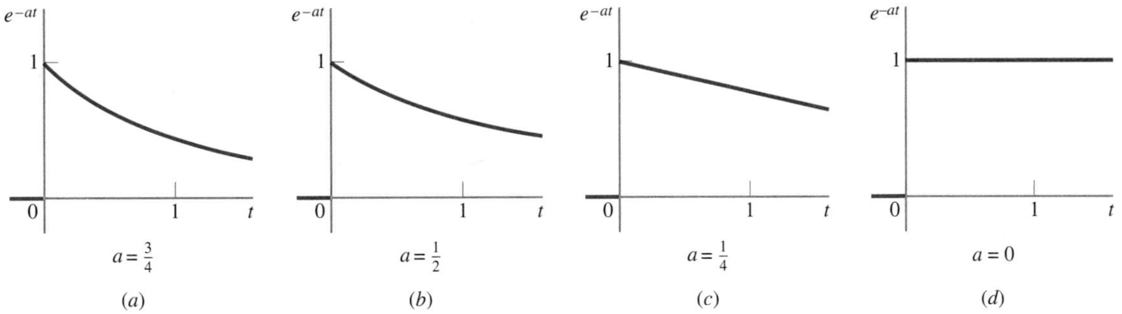

FIGURE 9.12
The approach of e^{-at} $(a, t > 0)$ to the unit step function as a approaches zero.

Sec. 9.1), a function that is trivial on $(-\infty, 0)$ and given by e^{-at} on $(0, \infty)$ has the coefficient function $1/[2\pi(a + i\omega)]$ and hence the unilateral Fourier transform [remembering that $1/2\pi$ has been transferred from $F(\omega)$ to $f(t)$]

$$F(\omega) = \frac{1}{a + i\omega} \qquad a > 0$$

Moreover, when a approaches zero, e^{-at}, considered for $t > 0$, approaches the unit step function, as suggested in Fig. 9.12. Hence it is natural to hope that the order of the operations of letting a approach zero and taking the unilateral transform can be interchanged. If this is the case, then we can postpone letting $a \to 0$ until *after* the transform has been found, and all will be well.

In the present problem the development proceeds as follows. Substituting the transform $1/(a + i\omega)$ for $F(\omega)$ in Eq. (1*a*), we have

$$\begin{aligned} f(t) &= \frac{1}{2\pi}\int_{-\infty}^{\infty} F(\omega)e^{i\omega t}\,d\omega = \frac{1}{2\pi}\int_{-\infty}^{\infty} \frac{e^{i\omega t}}{a + i\omega}\,d\omega \\ &= \frac{1}{2\pi}\int_{-\infty}^{\infty} \frac{\cos\omega t + i\sin\omega t}{a + i\omega}\,\frac{a - i\omega}{a - i\omega}\,d\omega \\ &= \frac{1}{2\pi}\int_{-\infty}^{\infty} \frac{(a\cos\omega t + \omega\sin\omega t) + i(a\sin\omega t - \omega\cos\omega t)}{a^2 + \omega^2}\,d\omega \end{aligned}$$

Now, the imaginary part of the integrand, namely,

$$\frac{a\sin\omega t - \omega\cos\omega t}{a^2 + \omega^2}$$

is an odd function of ω and hence will vanish when integrated between the limits $-\infty$ and ∞. On the other hand, the real part of the integrand is an even function of ω, and thus, integrating over just the positive half of the range and multiplying by 2, we have

$$f(t) = \frac{1}{\pi}\int_0^{\infty} \frac{a\cos\omega t + \omega\sin\omega t}{a^2 + \omega^2}\,d\omega = \frac{1}{\pi}\int_0^{\infty} \frac{a\cos\omega t}{a^2 + \omega^2}\,d\omega + \frac{1}{\pi}\int_0^{\infty} \frac{\omega\sin\omega t}{a^2 + \omega^2}\,d\omega$$

In the first integral in the right member, let $\omega = az$. Then,

$$f(t) = \frac{1}{\pi}\int_0^\infty \frac{\cos azt}{1+z^2}\,dz + \frac{1}{\pi}\int_0^\infty \frac{\omega \sin \omega t}{a^2+\omega^2}\,d\omega$$

We are now in a position to let a approach zero. As this happens,

$$f(t) \equiv e^{-at} \to u(t) \qquad t > 0$$

and thus we obtain

$$u(t) = \frac{1}{\pi}\int_0^\infty \frac{dz}{1+z^2} + \frac{1}{\pi}\int_0^\infty \frac{\sin \omega t}{\omega}\,d\omega = \frac{1}{2} + \frac{1}{\pi}\int_0^\infty \frac{\sin \omega t}{\omega}\,d\omega$$

Incidentally, since $u(t) = 1$ for $t > 0$, it follows that if we put $t = 1$ in this result, we have

$$1 = \frac{1}{2} + \frac{1}{\pi}\int_0^\infty \frac{\sin \omega}{\omega}\,d\omega \qquad \text{and hence} \qquad \int_0^\infty \frac{\sin \omega}{\omega}\,d\omega \equiv \text{Si}(\infty) = \frac{\pi}{2}$$

a result with which we are already familiar.

The factor e^{-at} $(a > 0)$ is called a **convergence factor** since, as we have just seen, when it is inserted in the integrands of certain divergent infinite integrals, it decreases with increasing t at a rate sufficient to make them converge. The use we have just made of this convergence factor is both artificial and clumsy, however, and it would be desirable to make this procedure more systematic. To do this, let us define an auxiliary function

$$g(t) = \begin{cases} 0 & t < 0 \\ e^{-at}f(t) & t > 0 \end{cases}$$

where $f(t)$ is the function of actual interest. Then, applying the unilateral Fourier transformation (1b) to $g(t)$, which surely has a unilateral transform if $f(t)$ does, we have

(2) $$G(\omega) = \int_0^\infty g(t)e^{-i\omega t}\,dt = \int_0^\infty [e^{-at}f(t)]e^{-i\omega t}\,dt = \int_0^\infty f(t)e^{-(a+i\omega)t}\,dt$$

whose inverse transform is

(3) $$g(t) = e^{-at}f(t) = \frac{1}{2\pi}\int_{-\infty}^\infty G(\omega)e^{i\omega t}\,d\omega$$

Multiplying both sides of the last equation here by e^{at}, we obtain

$$f(t) = \frac{e^{at}}{2\pi}\int_{-\infty}^\infty G(\omega)e^{i\omega t}\,d\omega = \frac{1}{2\pi}\int_{-\infty}^\infty G(\omega)e^{(a+i\omega)t}\,d\omega$$

Now from the last form of the expression for $G(\omega)$ in (2), it is clear that ω enters the analysis only through the binomial $a + i\omega$. To emphasize this fact, we shall write $F(a + i\omega)$ instead of $G(\omega)$. Then the equations of the transform pair become

$$f(t) = \frac{1}{2\pi}\int_{-\infty}^{\infty} F(a + i\omega)e^{(a+i\omega)t}\, d\omega$$

$$F(a + i\omega) = \int_0^{\infty} f(t)e^{-(a+i\omega)t}\, dt$$

Finally, let us put $a + i\omega = s$, noting that

$$d\omega = \frac{d(a + i\omega)}{i} = \frac{ds}{i}$$

and that $s = a - i\infty$ when $\omega = -\infty$ and $s = a + i\infty$ when $\omega = \infty$. This gives the pair of equations

$$f(t) = \frac{1}{2\pi i}\int_{a-i\infty}^{a+i\infty} F(s)e^{st}\, ds \qquad F(s) = \int_0^{\infty} f(t)e^{-st}\, dt$$

These equations for $f(t)$ and $F(s)$ constitute a **Laplace transform pair.** The function $F(s)$ is known as the **Laplace transform** of $f(t)$. The integral for $f(t)$ is known as the **complex inversion integral.**

The process of multiplying a function $f(t)$ by e^{-st} and then integrating the product $e^{-st}f(t)$ from 0 to ∞, together with the set of functions for which there exist values of s such that the resultant integral converges, make up a function named the **Laplace transformation.** We have thus naturally and inevitably encountered this transformation through our attempt to provide the unilateral Fourier transformation with a ''built-in'' convergence factor. The Laplace transformation is the foundation of the modern form of the *operational calculus,* which was originated in quite another form by the English electrical engineer Oliver Heaviside (1850–1925) around 1890.

In the next chapter we shall develop an extensive list of formulas for the use of Laplace transforms in solving ordinary differential equations and determining the behavior of linear systems. Many of these formulas will be quite similar to those we developed for Fourier transforms in Sec. 9.3, but because of the effect of the convergence factor e^{-st} they will be applicable to a much larger class of functions. The meaning and use of the complex inversion integral we must leave to the chapters on complex variable theory.

EXERCISES

1. Show that $f(t) = t$ does not have a unilateral Fourier transform but that $g(t) = e^{-at}t$, $a > 0$, does.
2. Does t^2 have a unilateral Fourier transform? Does $e^{-at}t^2$?
3. Does $\cos kt$ have a unilateral Fourier transform? Does $e^{-at}\cos kt$?
4. Does $\sin kt$ have a unilateral Fourier transform? Does $e^{-at}\sin kt$?

CHAPTER 10

THE LAPLACE TRANSFORMATION

The *Laplace transformation* is probably the best tool for solving linear differential equations with constant coefficients, especially if a particular solution satisfying given initial conditions is required. As a process, the Laplace transformation converts a problem in differential equations into one involving one or more linear algebraic equations in which the initial conditions are already incorporated and from which the transform can be found by elementary algebra. Relatively simple operations on the transform correspond to more complicated operations on the function being transformed, much as the addition and subtraction of logarithms correspond to multiplication and division of the numbers themselves. Finally, just as the answer to a numerical problem is found by taking antilogs, so the solution of a differential equation is obtained by taking the inverse or ''anti'' Laplace transform.

To justify the existence of Laplace transforms and to establish their useful properties, a certain amount of theory is required. Accordingly, after the Laplace transformation is defined and illustrated in Sec. 10.2, we take time to discuss such things as the *absolute* and *uniform convergence* of infinite integrals and their *continuity, differentiability,* and *integrability* with respect to a parameter. These put our derivations on a sound foundation and in places give us additional useful information about transforms.

In addition to a wealth of information about the properties and uses of Laplace transforms, this chapter also gives us additional information (Sec. 10.11) about the singularity functions we introduced in Sec. 9.5.

This chapter ends with a useful comparison of Fourier transforms, Laplace transforms, and Z transforms.

Prerequisites for this chapter: single-variable calculus, Chap. 9.

10.1 INTRODUCTION

In Sec. 2.5 we introduced the symbol D to represent the operation of taking the derivative of a function with respect to its independent variable, whatever that variable might be. Subsequently, we made extensive use of the symbol D as a notational convenience, but we made no attempt to assign operational properties to it, though this can be done.

To explore the matter briefly, consider the equation

$$(D - a)y = f(t) \tag{1}$$

A naive student, misled by the algebraic appearance of Eq. (1), might "divide" by $D - a$ and claim that $y = f(t)/(D - a)$ was the required answer; but of course this is just meaningless formalism. On the other hand, a more experienced student might do the same thing and then inquire if there is any way in which $1/(D - a)$ could be interpreted as a meaningful operator in its own right. Pursuing this thought, the student might return to Eq. (1) and solve it by the methods of Sec. 1.11, getting

$$\frac{1}{D - a} f(t) = e^{at} \int e^{-at} f(t)\, dt + c_1 e^{at} \tag{2}$$

Initially, this operational interpretation of $1/(D - a)$ merely provides an alternative way of solving certain types of linear first-order equations, but perhaps it can be extended. Proceeding in a purely formal way, let us apply formula (2) to the equation

$$(D^2 - 3D + 2)y = e^t \tag{3}$$

"Dividing" by $D^2 - 3D + 2$, then using partial fractions, and finally applying Formula (2), we have

$$\begin{aligned} y &= \frac{e^t}{D^2 - 3D + 2} = \frac{e^t}{(D - 1)(D - 2)} = \left(\frac{1}{D - 2} - \frac{1}{D - 1}\right)e^t \\ &= \left(e^{2t} \int e^{-2t} e^t\, dt + c_1 e^{2t}\right) - \left(e^t \int e^{-t} e^t\, dt - c_2 e^t\right) \\ &= [e^{2t}(-e^{-t}) + c_1 e^{2t}] - [e^t(t) - c_2 e^t] \\ &= c_1 e^{2t} + c_2 e^t - te^t - e^t \end{aligned}$$

Surprisingly, perhaps, our formal manipulations have led to the correct answer, for $c_1 e^{2t} + c_2 e^t$ is the complementary function of Eq. (3), and it is easy to verify that $-te^t - e^t$ is a particular integral. Moreover, the particular integral, $-te^t - e^t$, emerged without any special treatment even though the nonhomogeneous term in Eq. (3) duplicated a term in the complementary function!

Although the interpretation of D as an operator goes back to Leibniz (1646–1716), the English engineer Oliver Heaviside (1850–1925) was the first to make effective and extensive use of it. With his *operational calculus,* Heaviside solved a great variety of difficult and important physical problems that classical methods had been unable to handle. Because his work was formal and not rigorous, it was scorned by purists; but because it "worked," it was widely accepted by engineers and applied mathematicians.† As the great power and utility of Heaviside's operational calculus became more and more apparent, mathematicians who had previously scoffed at it became interested in trying to justify it. These efforts continued with varying degrees of success for several

† Reproached by someone because of the lack of a logical justification for his methods, Heaviside is said to have replied, "Shall I refuse to eat my dinner because I do not understand the process of digestion?"

decades until finally, around 1930, it was recognized that an integral transform, originally constructed by Laplace (1749–1827) over a century before, not only furnished an adequate theoretical foundation for Heaviside's work but in fact provided a more systematic alternative to the methods themselves.

In this chapter we shall study the modern form of operational calculus based on the Laplace transformation. As we shall see, this transformation is a powerful tool for anyone using linear differential equations with constant coefficients to investigate either the transient or the steady-state behavior of physical systems because

1. It allows initial conditions to be introduced at the outset so that there is no need to find and then particularize general solutions.
2. It converts *differential* equations in one or more functions of t into linear *algebraic* equations in the transforms of those functions.

The material on Laplace transforms which we will study in this chapter bears a striking resemblance to the material on Fourier transforms we studied in the last chapter; and it should be interesting and helpful to note the numerous similarities. The chief differences, which should also be noted, are

1. Fourier transforms usually involve functions defined on the entire real axis, but only functions that are restricted to the nonnegative real axis have Laplace transforms.
2. For $f(t)$ to have a Fourier transform, $\int_{-\infty}^{\infty} |f(t)|\,dt$ must exist, but it can have a Laplace transform as long as it does not increase faster than some simple exponential function e^{at} as $t \to \infty$.
3. Both direct and inverse Fourier transforms occur regularly in almost all applications, but except in quite advanced problems (which we will not encounter until Chap. 19), only the direct Laplace transform is used.
4. In general, both direct and inverse Fourier transforms involve complex quantities, but in the usual applications of the Laplace transformation, only real quantities occur.

EXERCISES

Using Eq. (2), find a complete solution of each of the following equations.

1. $(D^2 - 5D + 6)y = e^t$
2. $(D^2 + D - 2)y = e^t$
3. $(D^2 + 4D + 3)y = t$
4. $(D^2 - D - 6)y = e^{-2t}$
5. $(D^2 - 2D + 1)y = e^{2t}$. *Hint:* Write the "solution" for y in the form $y = \dfrac{1}{D-1}\left(\dfrac{1}{D-1}e^{2t}\right)$ and use Eq. (2) twice.
6. $(D^2 - 4D + 4)y = e^{2t}$
7. $(D^3 - 6D^2 + 11D - 6)y = 6$
8. $(D^3 + 2D^2 - D - 2)y = e^t$

10.2 DEFINITIONS AND BASIC THEORY

The **Laplace transformation** is an operation, denoted by the symbol $\mathcal{L}$, which associates with each function $f(t)$, satisfying suitable conditions for $t \geqq 0$, a unique function $F(s)$ called the **Laplace transform** of $f(t)$, according to the rule (Sec. 9.6)

$$\mathcal{L}[f(t)] = F(s) = \int_0^\infty f(t)e^{-st}\,dt \tag{1a}$$

Although we have at this point no formula for obtaining $f(t)$ when $F(s)$ is known, we call $f(t)$ the **inverse Laplace transform**, or simply the **inverse**, of $F(s)$ and we write

$$f(t) = \mathcal{L}^{-1}[F(s)] \tag{2a}$$

The variable of integration t in Eq. (1a) is of course a dummy variable and can be replaced at pleasure with any other symbol. From time to time in our work, we shall find it convenient to do this, and we shall usually use $\phi(s)$ as an alternative notation for $\mathcal{L}[f(t)]$. In terms of $\phi(s)$, and with braces in place of brackets, Eq. (1a) reads

$$\mathcal{L}\{f(t)\} = \phi(s) = \int_0^\infty f(t)e^{-st}\,dt \tag{1}$$

and Eq. (2a) becomes

$$f(t) = \mathcal{L}^{-1}\{\phi(s)\} \tag{2}$$

EXAMPLE 1

What is $\mathcal{L}\{e^{-at}\}$?

By Eq. (1),

$$\mathcal{L}\{e^{-at}\} = \int_0^\infty e^{-at}e^{-st}\,dt = \int_0^\infty e^{-(s+a)t}\,dt$$

$$= \frac{e^{-(s+a)t}}{-(s+a)}\bigg|_0^\infty = \frac{1}{s+a}$$

provided s is such that $s + a > 0$. Conversely, we can also write

$$e^{-at} = \mathcal{L}^{-1}\left\{\frac{1}{s+a}\right\}$$

EXAMPLE 2

What is $\mathcal{L}\{\cos at\}$?

By (1),

$$\mathcal{L}\{\cos at\} = \int_0^\infty \cos at\, e^{-st}\,dt$$

$$= \frac{e^{-st}(-s\cos at + a\sin at)}{s^2 + a^2}\bigg|_0^\infty$$

$$= \frac{s}{s^2 + a^2} \qquad s > 0$$

Conversely, we can also write

$$\cos at = \mathcal{L}^{-1}\left\{\frac{s}{s^2 + a^2}\right\}$$

The simplest and most obvious property of the Laplace transformation is that it is linear.

THEOREM 1 If the transforms exist, then for all values of the constants a and b,

$$\mathcal{L}\{af_1(t) + bf_2(t)\} = a\mathcal{L}\{f_1(t)\} + b\mathcal{L}\{f_2(t)\}$$

The proof of this is identical to the proof of Theorem 1, Sec. 9.2, for Fourier transforms and need not be repeated here.

EXAMPLE 3

What is the inverse of the transform $\phi(s) = \dfrac{1}{s^2 + 4s + 3}$?

Although we have (at this stage) no explicit formula for finding the inverse of the given transform, it is often possible to rearrange a transform so that the inverse of the individual terms can be identified from previous work. Thus in the present problem we can write

$$\frac{1}{s^2 + 4s + 3} = \frac{1}{(s + 1)(s + 3)} = \frac{1}{2}\left(\frac{1}{s + 1} - \frac{1}{s + 3}\right)$$

Hence, using Theorem 1 in reverse and the result of Example 1, we have

$$\mathscr{L}^{-1}\left\{\frac{1}{s^2 + 4s + 3}\right\} = \frac{1}{2}\mathscr{L}^{-1}\left\{\frac{1}{s + 1}\right\} - \frac{1}{2}\mathscr{L}^{-1}\left\{\frac{1}{s + 3}\right\}$$
$$= \frac{1}{2}(e^{-t} - e^{-3t})$$

Both Eq. ($1a$), which defined the Laplace transform, and Theorem 1 raise the fundamental question of when a function $f(t)$ has a Laplace transform. The discussion in Sec. 9.6 of the evolution of the Laplace transformation from the unilateral Fourier integral made it appear that for the Laplace transform of $f(t)$ to exist and for $f(t)$ to be recoverable from its transform it is sufficient that

(a) In every finite interval of $[0, \infty)$, $f(t)$ should satisfy the Dirichlet conditions (Definition 1, Sec. 8.2), that is, should be bounded and have at most a finite number of maxima and minima and a finite number of discontinuities.

(b) There should exist a real constant a such that the improper integral $\int_0^\infty e^{-at}|f(t)|\,dt$ is convergent.

Functions that satisfy Condition **a** we shall henceforth describe as **piecewise regular** on $[0, \infty)$.

Conditions **a** and **b** are frequently replaced by the following conditions:

(a′) Every interval of the form $0 \le t_1 \le t \le t_2$ can be divided into a finite number of subintervals such that $f(t)$ is continuous in the interior of each subinterval and approaches finite limits as t approaches either endpoint of the interval from the interior.

(b′) There is a constant α with the property that $e^{-\alpha t}|f(t)|$ remains bounded as t becomes infinite; that is, there are constant α, M, and T such that

$$e^{-\alpha t}|f(t)| < M \qquad \text{for all } t > T$$

Functions that satisfy Condition **a′** are said to be **piecewise continuous** on $[0, \infty)$. Functions that satisfy Condition **b′** are said to be of **exponential order**.

Condition **a′** is weaker, that is, is less restrictive, than Condition **a** because, clearly, it imposes no restriction on the number of maxima and minima $f(t)$ may have in a finite interval. On the other hand, Condition **b′** is stronger, that is, is more restrictive, than Condition **b**. This is hardly obvious, but follows from the fact, illustrated by Exercises 5 and 6, that the integral in **b** may exist even though $f(t)$ is not of exponential order.

For our purposes there is no practical difference between Conditions **a** and **a′**; hence we shall continue to require our functions to be piecewise regular, as we did, necessarily, in working with

Fourier series and Fourier integrals in Chaps. 8 and 9. Condition **b**′, however, is much more convenient for us to use than Condition **b**, and we shall consistently assume that our functions are of exponential order.

The number α required by Condition **b**′ is not unique. For instance, if

$$e^{-2t}|f(t)| < M \qquad \text{for all } t > T$$

then surely

$$e^{-2.1t}|f(t)| \qquad e^{-3.5t}|f(t)| \qquad e^{-5t}|f(t)| \qquad \cdots$$

will also be less than M for all $t > T$ since

$$e^{-2.1t} \qquad e^{-3.5t} \qquad e^{-5t} \qquad \cdots$$

all decrease faster than e^{-2t}. The observation that for a function of exponential order there are infinitely many values of α for which Condition **b**′ holds leads to the following important concept.

DEFINITION 1 If $f(t)$ is a function of exponential order such that

$$\text{(3)} \qquad e^{-\alpha t}|f(t)| < M \qquad \text{for all } t > T$$

then the greatest lower bound α_0 of all values of α for which the inequality (3) holds is called the **abscissa of convergence** of $f(t)$.

From this definition it follows that if a function $f(t)$ is of exponential order with abscissa of convergence α_0, then Condition **b**′ holds for any value of $\alpha > \alpha_0$ but fails to hold for any value of $\alpha < \alpha_0$. The condition may or may not hold for α_0 itself. For instance, if $f(t) = t$, then for every positive α and for no others, the product $e^{-\alpha t}|f(t)| = |t|e^{-\alpha t}$ remains bounded and in fact approaches zero as t becomes infinite. Since the greatest lower bound of the set of all positive numbers is the number zero, it follows that in this case $\alpha_0 = 0$. However, for α_0 itself, the product $|t|e^{-\alpha_0 t} = |t|$ increases beyond all bounds as $t \to \infty$. Thus for the function $f(t) = t$, the abscissa of convergence, namely, $\alpha_0 = 0$, is not one of the α's that can be used in Condition **b**′.

On the other hand, if $f(t) = e^{2t}$, then for any α greater than or *equal* to 2, $e^{-\alpha t}|f(t)| = e^{-\alpha t}e^{2t} = e^{-(\alpha-2)t}$ is bounded as $t \to \infty$. Since the greatest lower bound of all numbers greater than or equal to 2 is 2, it is clear that in this case the abscissa of convergence is 2 and moreover is a value of α that will serve in Condition **b**′.

Since $e^{-\alpha t}|f(t)| < M$ implies only that $|f(t)| < Me^{\alpha t}$, it is obvious that if a function is of exponential order, its absolute value need not remain bounded as $t \to \infty$, but it must not increase more rapidly than some constant multiple of a simple exponential function of t. As the particular function $f(t) = \sin e^{t^2}$ shows, *the derivative of a function of exponential order is not necessarily of exponential order.* On the other hand, it is not difficult to show (see Exercise 9) that *if $f(t)$ is piecewise regular on* $[0, \infty)$ *and of exponential order, then* $\int_a^t f(x)\,dx$, $a \geq 0$, *is continuous on* $[0, \infty)$ *and of exponential order.*

The existence of the Laplace transform of any function of exponential order which is piecewise regular on $[0, \infty)$ follows as a corollary of the following fundamental theorem.

THEOREM 2 **(Absolute Convergence)** If $f(t)$ is piecewise regular on $[0, \infty)$ and of exponential order, then for any value of s which is greater than the abscissa of convergence of $f(t)$, the integral $\int_0^\infty f(t)e^{-st}\,dt$ converges absolutely.

PROOF To establish this theorem we must show that under the hypotheses of the theorem, $\int_0^\infty |f(t)e^{-st}|\,dt$ converges; that is, we must show that

$$\lim_{b\to\infty}\int_0^b |f(t)e^{-st}|\,dt = \lim_{b\to\infty}\int_0^b |f(t)|e^{-st}\,dt \tag{4}$$

exists. To do this, we need to have a bound for $|f(t)|$. The hypothesis that $f(t)$ is of exponential order gives a bound of the form

$$|f(t)| < M_1 e^{\alpha t} \qquad \text{for all } t > T, \text{ and } \alpha > \alpha_0$$

where α_0 is the abscissa of convergence of $f(t)$. Under the assumption that $f(t)$ is also piecewise regular, it can be shown (Exercise 7) that there is a number M such that *over the entire range of integration* $[0, \infty)$,

$$|f(t)| < Me^{\alpha t} \qquad M \geq M_1,\ \alpha > \alpha_0$$

Assuming such a bound, our proof proceeds as follows.

$$\int_0^b |f(t)|e^{-st}\,dt < \int_0^b Me^{\alpha t}e^{-st}\,dt = M\int_0^b e^{-(s-\alpha)t}\,dt$$

$$= \frac{M}{-(s-\alpha)}e^{-(s-\alpha)t}\Big|_0^b = \frac{M}{s-\alpha}[1 - e^{-(s-\alpha)b}]$$

Now if $s > \alpha > \alpha_0$, the exponential in the last expression decreases monotonically and approaches zero as b becomes infinite. The entire expression therefore approaches $M/(s - \alpha)$ as $b \to \infty$, and we have

$$\int_0^b |f(t)|e^{-st}\,dt < \frac{M}{s-\alpha} \qquad s > \alpha > \alpha_0$$

Since the integrand of the last integral is clearly nonnegative, the integral itself is a monotonically increasing function of b. Hence, being bounded above, as we have just seen, it must approach a limit as $b \to \infty$. Since $s > \alpha > \alpha_0$ is equivalent to $s > \alpha_0$, the theorem is established.

In calculus we learned that an infinite series converges if it converges absolutely, and the same thing is true for infinite integrals. Hence, since we have just shown that $\int_0^\infty |f(t)|e^{-st}\,dt$ converges, it follows that $\int_0^\infty f(t)e^{-st}\,dt$ converges, and we have the following corollary.

COROLLARY 1 Sufficient (but not necessary) conditions for a function to have a Laplace transform are that the function be piecewise regular on $[0, \infty)$ and of exponential order.

Since the absolute value of an integral is always equal to or less than the integral of the absolute value of its integrand, it follows from the proof of Theorem 2 that

$$\left|\int_0^b f(t)e^{-st}\,dt\right| \leq \int_0^b |f(t)|e^{-st}\,dt < \frac{M}{s-\alpha}$$

Hence, letting $b \to \infty$, we have the following useful result.

COROLLARY 2 If $f(t)$ is piecewise regular on $[0, \infty)$ and of exponential order with abscissa of convergence α_0, then for all values of s and α such that $s > \alpha > \alpha_0$,

$$|\mathscr{L}\{f(t)\}| \leq \frac{M}{s-\alpha} \qquad \text{where } M \text{ is independent of } s$$

Finally, from Corollary 2 we draw the following interesting conclusions.

COROLLARY 3 If $f(t)$ is piecewise regular on $[0, \infty)$ and of exponential order, then $\mathscr{L}\{f(t)\}$ approaches zero as s becomes infinite.

COROLLARY 4 If $f(t)$ is piecewise regular on $[0, \infty)$ and of exponential order, then $s\mathscr{L}\{f(t)\}$ is bounded as s becomes infinite.

Corollaries 3 and 4 make it clear that not all functions of s are Laplace transforms—or at least not Laplace transforms of functions of the "respectable" type that are piecewise regular on $[0, \infty)$ and of exponential order. For instance, $\phi(s) = s/(s-1)$ does not approach zero as s becomes infinite; hence it is not the Laplace transform of any "respectable" function. Likewise, although $\phi(s) = 1/\sqrt{s}$ does approach zero as s becomes infinite, it too is not the transform of any "respectable" function since $s\phi(s) = \sqrt{s}$ is not bounded as s becomes infinite.

EXERCISES

1. Which of the following functions are of exponential order?
   (a) t^n (b) $\tan t$ (c) e^{t^2}
   (d) $\cosh t$ (e) $1/t$ (f) t^2e^{3t}
2. Show by an example that it is possible for the abscissa of convergence of a function to be negative.
3. What is the abscissa of convergence of each of the following functions?
   (a) $\cos kt$ (b) $\sin kt$ (c) t^2
   (d) $\cosh kt$ (e) $\sinh kt$ (f) $\ln(1+t)$
4. For which functions in Exercise 3 is α_0 a value of α which will serve in Condition **b**′?
5. Prove that if a piecewise regular function satisfies Condition **b**, it does not necessarily satisfy Condition **b**′. *Hint:* consider the function

$$f(t) = \begin{cases} e^{n^2} & t = n \\ 0 & t \neq n \end{cases} \qquad n = 0, 1, 2, \ldots$$

6. Show that the function $f(t) = 2te^{t^2}\cos e^{t^2}$ has a Laplace transform even though $f(t)$ is not of exponential order.
7. Complete the proof of Theorem 2 by showing that if $f(t)$ is piecewise regular and of exponential order, then numbers α and M exist such that

$$|f(t)| < Me^{\alpha t} \qquad \text{for all } t \geq 0$$

Hint: Since $f(t)$ is of exponential order, numbers α, M_1, and T exist such that

$$|f(t)| < M_1e^{\alpha t} \qquad \text{for } t > T$$

Also, since $f(t)$ is piecewise regular, it is bounded on any finite interval, in particular on $0 \leq t \leq T$. Thus there exists a number M_2 such that

$$|f(t)| < M_2 = (M_2e^{-\alpha t})e^{\alpha t} \qquad \text{for } 0 \leq t \leq T$$

Now show that if M is the larger of the numbers

$$(M_1, M_2), \text{ if } \alpha \geq 0 \qquad \text{or} \qquad (M_1, M_2e^{-\alpha T}), \text{ if } \alpha < 0$$

then $|f(t)| < Me^{\alpha t}$ for all $t \geq 0$.

8. Prove that a function $f(t)$ is of exponential order if and only if s can be chosen so that $\lim_{t\to\infty} e^{-st}|f(t)| = 0$. If $f(t)$ is of exponential order, show that its abscissa of convergence α_0 is the greatest lower bound of all values of s such that $\lim_{t\to\infty} e^{-st}|f(t)| = 0$.
9. Prove that if $f(t)$ is piecewise regular on $[0, \infty)$ and of exponential order, then $\int_0^t f(x)\,dx$ is continuous on $[0, \infty)$ and of exponential order. Show further that if α_0 and α_1 are, respectively, the abscissas of convergence of $f(t)$ and $\int_0^t f(x)\,dx$ and if $\alpha_0 \geq 0$, then $\alpha_1 \leq \alpha_0$. Is it necessarily true that $\alpha_1 \leq \alpha_0$ if $\alpha_0 < 0$?
10. Show that each of the following integrals converges absolutely for $s \geq 0$.
    (a) $\displaystyle\int_0^\infty \frac{\sin st}{1+t^2}\,dt$ (b) $\displaystyle\int_0^\infty \frac{\cos st}{1+st+t^2}\,dt$
11. Show that neither of the following integrals converges absolutely.
    (a) $\displaystyle\int_0^\infty \frac{\sin t}{t}\,dt$ (b) $\displaystyle\int_0^\infty \frac{\sin t}{1+st}\,dt$

    Hint: In each case, express the integral as a sum of integrals over $(0, \pi)$, $(\pi, 2\pi)$, $(2\pi, 3\pi)$, ..., $(\overline{n-1}\pi, n\pi)$, ...; then make a suitable overestimate of the denominator of each integral.
12. If both f and g are functions of exponential order, show that the product fg is also a function of exponential order.

10.3 UNIFORM CONVERGENCE AND ITS CONSEQUENCES

In the last section we learned that every function that is piecewise regular on $[0, \infty)$ and of exponential order has a Laplace transform, and we learned something about the behavior of the transforms of such functions as $s \to \infty$. Before the Laplace transformation can become really useful, however, we need to explore one further bit of theory, the *uniform convergence* of the transform integral. Theorem 2, Sec. 10.2, established that the Laplace transform integral converges absolutely. But absolute convergence is not sufficient to justify differentiating and integrating the transform integral with respect to s, as we will need to do to develop some of the important transform formulas. Instead, uniform convergence is required, and to this we now turn our attention.

To help us appreciate the concept of uniform convergence as a refinement of the idea of ordinary convergence, suppose that the integral

$$(1) \qquad \int_a^\infty F(s, t)\, dt$$

converges for some particular value of s, say s_0. This means that

$$\lim_{b\to\infty} \int_a^b F(s_0, t)\, dt$$

exists. This in turn means that given any $\epsilon > 0$, there exists a number B, depending on ϵ and s_0, such that

$$\left| \int_a^b F(s_0, t)\, dt - \int_a^\infty F(s_0, t)\, dt \right| < \epsilon \qquad \text{for all } b > B(\epsilon, s_0)$$

If we combine the two integrals, this becomes the equivalent statement

$$(2) \qquad \left| \int_b^\infty F(s_0, t)\, dt \right| < \epsilon \qquad \text{for all } b > B(\epsilon, s_0)$$

If the integral (1) converges for other values of s, say all the s's in some set S, then for each of these values there must exist a corresponding value of B to serve in the counterpart of Condition (2). In general these B's will all be different, and no one will serve for all the s's in S. However, if it should happen that the set of B's thus determined is bounded, that is, if there is a number $B^*(\epsilon)$ at least as big as any of the B's, then B^* could be used to establish the convergence of (1) for all values of s under consideration. In other words, there would be one value of B that would work equally well, or *uniformly*, for all the s's under consideration. This idea we now formalize in the following definition.

DEFINITION 1 The improper integral $\int_a^\infty F(s, t)\, dt$ is said to **converge uniformly** over a given set S of s values if and only if given any $\epsilon > 0$ there exists a number B, depending on ϵ but not on s, such that

$$\left| \int_b^\infty F(s, t)\, dt \right| < \epsilon \qquad \text{for } b > B \text{ and all } s \text{ in } S$$

Since integrals that converge uniformly have many nice properties, our next theorem is a very important one.

THEOREM 1 **(Uniform Convergence)** If $f(t)$ is piecewise regular on $[0, \infty)$ and of exponential order with abscissa of convergence α_0, then for any number $s_0 > \alpha_0$,

$$\mathcal{L}\{f(t)\} = \int_b^\infty f(t)e^{-st}\,dt$$

converges uniformly for all values of s such that $s \geq s_0$.

PROOF To prove this theorem, we must show that given any $\epsilon > 0$, there exists a number B, depending on ϵ but not on s, such that

$$\left|\int_b^\infty f(t)e^{-st}\,dt\right| < \epsilon \qquad \text{for all } b > B \text{ and all } s \geq s_0 > \alpha_0$$

Now the absolute value of an integral is always equal to or less than the integral of the absolute value of its integrand; hence

$$\left|\int_b^\infty f(t)e^{-st}\,dt\right| \leq \int_b^\infty |f(t)|e^{-st}\,dt$$

Further, we know that for $s > \alpha_0$ the integral on the right approaches zero as b becomes infinite because this is implied by the fact that

$$\int_0^\infty |f(t)|e^{-st}\,dt$$

is convergent for $s > \alpha_0$ (Theorem 2, Sec. 10.2). In other words, given any $\epsilon > 0$ and any $s_0 > \alpha_0$, there exists a number B such that

$$\int_b^\infty |f(t)|e^{-s_0 t}\,dt < \epsilon \qquad \text{for all } b > B$$

Now if $s \geq s_0$, it is obvious that $e^{-st} \leq e^{-s_0 t}$ for all $t \geq 0$. Hence

$$\left|\int_b^\infty f(t)e^{-st}\,dt\right| \leq \int_b^\infty |f(t)|e^{-st}\,dt \leq \int_b^\infty |f(t)|e^{-s_0 t}\,dt$$

and so for any $s \geq s_0$, the integral on the left is less than ϵ for all values of b greater than the particular B that suffices for the integral on the right. This value of B is clearly independent of s (since it arises from the *specific* value $s = s_0$), and so the proof of the theorem is complete.

The important properties of uniformly convergent integrals, which we now know are shared by the Laplace transform integral, are stated in a form suitable for our use in the following three theorems whose proofs we leave to more advanced texts.†

† See, for instance, H. S. Carslaw, *Fourier Series*, pp. 198–201, Dover, New York, 1930.

THEOREM 2 **(Continuity)** If $g(s, t)$ is a continuous function of s and t for $\alpha \le s \le \beta$ and $t \ge a$, if $f(t)$ is at least piecewise regular for $t \ge a$, and if the integral $G(s) = \int_a^\infty f(t)g(s, t)\, dt$ converges uniformly over the interval $\alpha \le s \le \beta$, then $G(s)$ is a continuous function of s for $\alpha \le s \le \beta$.

Since the definitive property of a continuous function is that

$$\lim_{s \to s_0} G(s) = G(s_0)$$

this theorem states, in effect, that under the appropriate conditions the limit of $G(s)$ can be found by taking the limit inside the t-integral sign or, equivalently, that the order of integrating with respect to t and taking the limit with respect to s can be interchanged.

THEOREM 3 **(Integrability)** If $g(s, t)$ is a continuous function of s and t for $\alpha \le s \le \beta$ and $t \ge a$, if $f(t)$ is at least piecewise regular for $t \ge a$, and if the integral $G(s) = \int_a^\infty f(t)g(s, t)\, dt$ converges uniformly over the interval $\alpha \le s \le \beta$, then

$$\int_\alpha^\beta G(s)\, ds \equiv \int_\alpha^\beta \left[\int_a^\infty f(t)g(s, t)\, dt\right] ds = \int_a^\infty \left[\int_\alpha^\beta f(t)g(s, t)\, ds\right] dt$$

In words, Theorem 3 asserts that under the appropriate conditions the integral of $G(s)$ can be found by integrating inside the t-integral sign or, equivalently, that the order of integrating with respect to t and with respect to s can be interchanged.

THEOREM 4 **(Differentiability)** If $g(s, t)$ and $g_s(s, t) \equiv \partial g(s, t)/\partial s$ are continuous functions of s and t for $\alpha \le s \le \beta$ and $t \ge a$, if $f(t)$ is at least piecewise regular for $t \ge a$, if the integral

$$G(s) = \int_a^\infty f(t)g(s, t)\, dt$$

converges, and if $\int_a^\infty f(t)g_s(s, t)\, dt$ converges uniformly over the interval $\alpha \le s \le \beta$, then

$$G'(s) \equiv \frac{d}{ds}\int_a^\infty f(t)g(s, t)\, dt = \int_a^\infty f(t)\frac{\partial g(s, t)}{\partial s}\, dt = \int_a^\infty f(t)g_s(s, t)\, dt$$

In words, Theorem 4 asserts that under the appropriate conditions, the derivative of $G(s)$ can be found by differentiating inside the t-integral sign or, equivalently, that the order of integrating with respect to t and differentiating with respect to s can be interchanged.

Obviously, if we take $g(s, t)$ to be the continuous function e^{-st} and take $a = 0$, the integral $G(s)$ referred to in the last three theorems is precisely the Laplace transform of the function $f(t)$.

EXAMPLE 1

Applying Theorem 4 to the Laplace transform of e^{-at} which we found in Example 1, Sec. 10.2, show that $\mathcal{L}\{te^{-at}\} = 1/(s + a)^2$.

If we differentiate the result of Example 1, Sec. 10.2, we have

$$\frac{d}{ds}\int_0^\infty e^{-at}e^{-st}\, dt = \frac{d}{ds}\left(\frac{1}{s + a}\right) = \frac{-1}{(s + a)^2}$$

According to Theorem 4, the derivative of the integral can also be found by differentiating with respect to s inside the integral sign, and we have, alternatively,

$$\frac{d}{ds}\int_0^\infty e^{-at}e^{-st}\, dt = \int_0^\infty e^{-at}\frac{\partial}{\partial s}(e^{-st})\, dt = -\int_0^\infty (te^{-at})e^{-st}\, dt$$

Hence, equating the two expressions for the derivative of the integral and multiplying by -1, we have

$$\int_0^\infty (te^{-at})e^{-st}\,dt = \frac{1}{(s+a)^2}$$

which is just the assertion that $\mathscr{L}\{te^{-at}\} = 1/(s+a)^2$.

In succeeding sections we shall find that many relatively complicated operations upon a function $f(t)$, such as differentiation and integration, can be replaced by simple algebraic operations, such as multiplication and division by s, upon the transform of $f(t)$. This is analogous to the way in which operations such as the multiplication and division of numbers are replaced by the simpler operations of addition and subtraction when we work not with the numbers themselves but with their logarithms. Our primary purpose in this chapter is to develop rules of transformation and tables of transforms which can be used, like tables of logarithms, to facilitate the manipulation of functions and by means of which we can recover the proper function from its transform at the end of a problem.

EXERCISES

Prove that each of the following integrals is uniformly convergent over the indicated set of s values.

1. $\displaystyle\int_0^\infty \frac{\sin st}{1+t^2}\,dt$; all real values of s

2. $\displaystyle\int_0^\infty \frac{\cos st}{1+2t+t^2}\,dt$; all real values of s

3. $\displaystyle\int_1^\infty \frac{1}{\sqrt{s^4+t^4}}\,dt$; all real values of s

4. $\displaystyle\int_0^\infty e^{-st^2}\,dt$; all values of $s > s_0 > 0$

Hint: Note first that if $b > 0$, it is sufficient to consider the integral only over the range (b, ∞). Then note that if $s > s_0$, we have $st^2 > s_0t^2 > s_0t$ if $t > 1$.

5. $\displaystyle\int_0^\infty e^{-t^2}\cos st\,dt$; all real values of s

6. If $G(s)$ is the function of s defined by the integral in Exercise 1, can $G'(s)$ be found by differentiating with respect to s inside the integral sign? Why?

7. Work Exercise 6 for the function defined by the integral in Exercise 2.

8. Work Exercise 6 for the function defined by the integral in Exercise 3.

9. Work Exercise 6 for the function defined by the integral in Exercise 5.

10. If $G(s)$ is the function defined by the integral in Exercise 5, show that $G'(s) = -\dfrac{s}{2}G(s)$. Then by solving this differential equation and using the fact that $\int_0^\infty e^{-t^2}\,dt = \frac{1}{2}\sqrt{\pi}$, find $G(s)$. *Hint:* After differentiating with respect to s inside the integral, apply integration by parts to the resulting integral.

10.4 THE GENERAL METHOD

We have already seen (Theorem 1, Sec. 10.2) that the Laplace transformation is a linear operator. In addition to this result, the utility of the Laplace transformation is based primarily upon the following two theorems.

THEOREM 1 If $f(t)$ is a continuous function of exponential order on $[0, \infty)$ whose derivative is also of exponential order and at least piecewise regular on $[0, \infty)$, then

$$\mathscr{L}\{f'(t)\} = s\mathscr{L}\{f(t)\} - f(0)$$

PROOF To prove this, let us suppose for definiteness that there is a single point, say $t = t_0$, where although $f(t)$ is continuous, its derivative has a finite jump, as suggested by Fig. 10.1. Then, by definition,

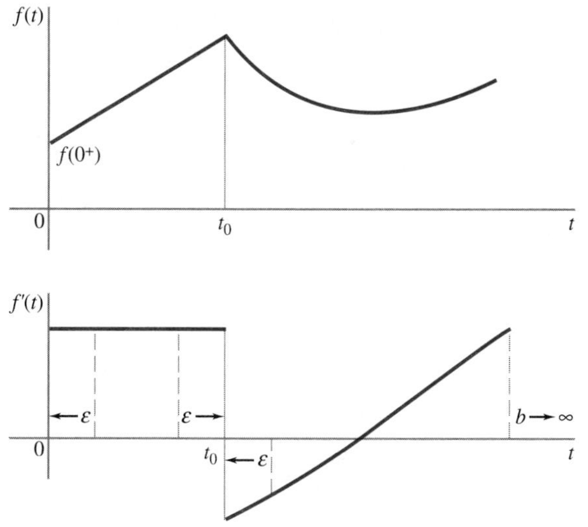

FIGURE 10.1
A continuous function whose derivative has a point of discontinuity.

$$\mathcal{L}\{f'(t)\} = \int_0^\infty f'(t)e^{-st}\,dt$$

$$= \lim_{\substack{\epsilon\to 0\\ b\to\infty}} \left[\int_\epsilon^{t_0-\epsilon} f'(t)e^{-st}\,dt + \int_{t_0+\epsilon}^{b} f'(t)e^{-st}\,dt\right]$$

If we use integration by parts on these integrals, choosing

$$u = e^{-st} \qquad dv = f'(t)\,dt$$
$$du = -se^{-st}\,dt \qquad v = f(t)$$

we have

$$\mathcal{L}\{f'(t)\} = \lim_{\substack{\epsilon\to 0\\ b\to\infty}} \left[e^{-st}f(t)\Big|_\epsilon^{t_0-\epsilon} + s\int_\epsilon^{t_0-\epsilon} f(t)e^{-st}\,dt + e^{-st}f(t)\Big|_{t_0+\epsilon}^{b} + s\int_{t_0+\epsilon}^{b} f(t)e^{-st}\,dt\right]$$

In the limit the two integrals remaining combine to give precisely

$$s\int_0^\infty f(t)e^{-st}\,dt = s\mathcal{L}\{f(t)\}$$

The continuity of f at $t = 0$ implies $f(0^+) = f(0)$. Thus the first evaluated portion gives

$$e^{-st_0}f(t_0^-) - f(0^+) = e^{-st_0}f(t_0^-) - f(0)$$

and the second yields simply

$$0 - e^{-st_0}f(t_0^+)$$

because, since $f(t)$ is of exponential order, s can be chose sufficiently large, i.e., greater than the abscissa of convergence of $f(t)$, for the contribution from the upper limit to be zero. Now $f(t)$ was assumed to be continuous. Hence at t_0 (as at all other points where $t > 0$) its right- and left-hand limits must be equal. Therefore the terms

$$e^{-st_0}f(t_0^-) \qquad \text{and} \qquad -e^{-st_0}f(t_0^+)$$

cancel, leaving finally

$$\mathcal{L}\{f'(t)\} = s\mathcal{L}\{f(t)\} - f(0)$$

as asserted. ◗

The preceding proof is readily extended to functions whose derivatives have more than one finite jump. The extension of the theorem to the relatively unimportant case in which $f(t)$ itself is permitted to have finite jumps is indicated in Exercise 2.

COROLLARY 1 If both $f(t)$ and $f'(t)$ are continuous functions of exponential order on $[0, \infty)$ and if $f''(t)$ is also of exponential order and at least piecewise regular on $[0, \infty)$, then

$$\mathscr{L}\{f''(t)\} = s^2\mathscr{L}\{f(t)\} - sf(0) - f'(0)$$

PROOF This result follows immediately by applying Theorem 1 twice to $f''(t)$:

$$\begin{aligned}\mathscr{L}\{f''(t)\} = \mathscr{L}\{[f'(t)]'\} &= s\mathscr{L}\{f'(t)\} - f'(0)\\ &= s[s\mathscr{L}\{f(t)\} - f(0)] - f'(0)\\ &= s^2\mathscr{L}\{f(t)\} - sf(0) - f'(0)\end{aligned}$$

as asserted.

In the same way, by repeated use of Theorem 1, we can prove the extended result contained in Corollary 2.

COROLLARY 2 If $f(t), f'(t), f''(t), \ldots, f^{(n-1)}(t)$ are continuous and of exponential order on $[0, \infty)$ and if $f^{(n)}(t)$ is of exponential order and at least piecewise regular on $[0, \infty)$, then

$$\mathscr{L}\{f^{(n)}(t)\} = s^n\mathscr{L}\{f(t)\} - s^{n-1}f(0) - s^{n-2}f'(0) - \cdots - f^{(n-1)}(0)$$

Although we need many more formulas before the Laplace transformation can be applied effectively to specific problems, Theorem 1 and its corollaries allow us to outline all the essential steps in the usual application of this method to the solution of differential equations with constant coefficients. Suppose, for instance, that we are given the equation

$$ay'' + by' + cy = f(t) \qquad a,\ b,\ c \text{ constants}$$

Taking the Laplace transform of each side and using the linearity of the transformation, we have

$$a\mathscr{L}\{y''\} + b\mathscr{L}\{y'\} + c\mathscr{L}\{y\} = \mathscr{L}\{f(t)\}$$

Applying Theorem 1 and its first corollary to the first two terms, we have further

$$a[s^2\mathscr{L}\{y\} - sy_0 - y'_0] + b[s\mathscr{L}\{y\} - y_0] + c\mathscr{L}\{y\} = \mathscr{L}\{f(t)\}$$

where y_0 and y'_0 are the given initial values of y and y'. Collecting terms on $\mathscr{L}\{y\}$ and then solving for $\mathscr{L}\{y\}$, we have finally

$$\mathscr{L}\{y\} = \frac{\mathscr{L}\{f(t)\} + (as + b)y_0 + ay'_0}{as^2 + bs + c}$$

Now $f(t)$ is a given function of t; hence its Laplace transform (if it exists) is a perfectly definite function of s, although except for the integral definition provided by Eq. (1), Sec. 10.2, and the results of Examples 1 and 2, Sec. 10.2, we have as yet no formulas for finding it. Moreover, y_0 and y'_0 are definite numbers known from the data of the problem. Hence the transform of y is a completely determined function of s. Thus if we had available a sufficiently extensive table of transforms, we could find in it the function $y(t)$ having the right-hand side of the last equation for its transform, *and this function would be the formal solution to our problem, initial conditions and all.*† If deemed necessary, this formal solution could then be substituted into the differential equation to verify that it is indeed the genuine solution.

This brief discussion illustrates what we have already said about the two great advantages of the Laplace transformation in solving linear differential equations with constant coefficients: it reduces the problem to one in algebra, and it takes care of initial conditions without the necessity of first constructing a complete solution and then specializing the arbitrary constants it contains.

Since we are primarily concerned in this book with *differential* equations, it is natural that we should be interested in a result like Theorem 1, which expresses the Laplace transform of the derivative of a function in terms of the transform of the function itself. On the other hand, there is no obvious reason why we should be concerned about the Laplace transform of the integral of a function. However, if we recall the discussion of the series electrical circuit in Sec. 7.2, we find a reason. There, through the use of Kirchoff's second law, we found that the behavior of the general series circuit is described by the equation

$$L\frac{di}{dt} + iR + \frac{1}{C}Q = E(t) \tag{1}$$

which we discovered could be written in either of the forms

$$L\frac{d^2Q}{dt^2} + R\frac{dQ}{dt} + \frac{1}{C}Q = E(t) \qquad \text{or} \qquad L\frac{di}{dt} + iR + \frac{1}{C}\int^t i\,dt = E(t)$$

With formulas for the Laplace transforms of derivatives available, we can now use transform methods to solve the first of these equations for Q and then find i from the relation $i = dQ/dt$. However, if we knew a formula for the Laplace transform of the integral of a function, we could solve the second of these equations for i without first having to find Q. The formula we need to do this is given in the next theorem.

THEOREM 2 If $f(t)$ is piecewise regular on $[0, \infty)$ and of exponential order, then the Laplace transform of $\int_0^t f(x)\,dx$ is given by the formula

$$\mathcal{L}\left\{\int_0^t f(x)\,dx\right\} = \frac{1}{s}\mathcal{L}\{f(t)\}$$

PROOF Since $f(t)$ is piecewise regular on $[0, \infty)$ and of exponential order, the function $F(t) = \int_0^t f(x)\,dx$ is continuous on $[0, \infty)$ and of exponential order (Exercise 9, Sec. 10.2). Hence Theorem 1 can be applied to the function $F(t)$. Observing that $F'(t) = f(t)$, we thus have

$$\mathcal{L}\{f(t)\} = \mathcal{L}\{F'(t)\} = s\mathcal{L}\{F(t)\} - F(0) = s\mathcal{L}\left\{\int_0^t f(x)\,dx\right\} - \int_0^0 f(x)\,dx$$

Hence, dividing the first and last members in this continued equality by s, we have

$$\mathcal{L}\left\{\int_0^t f(x)\,dx\right\} = \frac{1}{s}\mathcal{L}\{f(t)\}$$

as asserted. The extension of this result to repeated integrals is asked for in Exercise 1.

† This of course assumes the "obvious" theorem that the function having a given function of s as its Laplace transform is unique or, in other words, that if two functions have the same transform, they are identical. This is strictly true if the functions are continuous. If discontinuities are permitted, the most we can say is that two functions with the same transform cannot differ over any interval of positive length, although they may differ at various isolated points. A detailed discussion of this result (Lerch's theorem) would take us too far afield, but we shall assume it repeatedly throughout this chapter. (For a proof of Lerch's theorem, see D. V. Widder, *The Laplace Transform,* pp. 59–63, Princeton University Press, Princeton, N.J., 1941.)

Theorems 1 and 2 are not only useful for finding the transforms of derivatives and integrals of functions but are also of considerable help in finding the inverses of transforms containing the factors s and $1/s$, respectively. The companion result for Theorem 1 is the following.

THEOREM 3 **(s-Multiplied Transforms)** If the inverse of $\phi(s)$ is a function $f(t)$ which is continuous on $[0, \infty)$ and of exponential order, if $f'(t)$ is at least piecewise regular on $[0, \infty)$ and of exponential order, and if $f(0) = 0$, then the inverse of $s\phi(s)$ is $f'(t)$, that is,

$$\mathcal{L}^{-1}\{s\phi(s)\} = f'(t) = \frac{d}{dt}\mathcal{L}^{-1}\{\phi(s)\}$$

PROOF Under the conditions of the theorem, we can apply Theorem 1 to $f(t)$, getting

$$\mathcal{L}\{f'(t)\} = s\mathcal{L}\{f(t)\} - f(0) = s\phi(s)$$

or, equivalently,

$$\mathcal{L}^{-1}\{s\phi(s)\} = f'(t) = \frac{d}{dt}\mathcal{L}^{-1}\{\phi(s)\}$$

which is the assertion of the theorem.

As a working rule for finding inverses, Theorem 3 can be restated in the following way.

If a Laplace transform contains the factor s, the inverse of that transform can be found by suppressing the factor s, determining the inverse of the remaining portion of the transform, verifying that that inverse is zero when $t = 0$, and finally differentiating it with respect to t.

The companion theorem that makes Theorem 2 useful for finding inverses is the following.

THEOREM 4 **(s-Divided Transforms)** If the inverse of $\phi(s)$ is a function $f(t)$ which is piecewise regular on $[0, \infty)$ and of exponential order, then the inverse of $\phi(s)/s$ is $\int_0^t f(x)\,dx$.

PROOF Under the conditions of the theorem we can apply Theorem 2 to $f(t)$, getting

$$\mathcal{L}\left\{\int_0^t f(x)\,dx\right\} = \frac{1}{s}\mathcal{L}\{f(t)\} = \frac{1}{s}\phi(s)$$

or

$$\mathcal{L}^{-1}\left\{\frac{1}{s}\phi(s)\right\} = \int_0^t f(x)\,dx$$

as asserted.

As a working rule for finding inverse transforms, Theorem 4 can be restated in the following way.

If a Laplace transform contains the factor 1/s, the inverse of that transform can be found by suppressing the factor 1/s, determining the inverse of the remaining portion of the transform, and finally integrating that inverse with respect to t from 0 to t.

EXAMPLE 1

Given $\mathcal{L}\{\sin 2t\} = \dfrac{2}{s^2 + 4}$, find $\mathcal{L}^{-1}\left\{\dfrac{s}{s^2 + 4}\right\}$.

Example 2, Sec. 10.2, gives $f(t) = \cos 2t$ as the required inverse. To verify this result, we suppress the factor s, then use the given formula to write the inverse of the remaining portion of the transform, getting

$$\mathscr{L}^{-1}\left\{\frac{1}{s^2+4}\right\} = \frac{1}{2}\mathscr{L}^{-1}\left\{\frac{2}{s^2+4}\right\} = \frac{1}{2}\sin 2t$$

which has the value 0 at $t = 0$. Differentiating this inverse with respect to t and invoking Theorem 3, we get

$$f(t) = \frac{d}{dt}\mathscr{L}^{-1}\left\{\frac{1}{s^2+4}\right\} = \frac{d}{dt}\left(\frac{1}{2}\sin 2t\right) = \cos 2t$$

as before. The usual applications of Theorem 3 are usually not of this trivial character.

EXAMPLE 2

What is $\mathscr{L}^{-1}\left\{\frac{1}{s(s^2+4)}\right\}$?

Suppressing the factor $1/s$, we get $\phi(s) = 1/(s^2+4) = \frac{1}{2}[2/(s^2+4)]$ whose inverse, from Example 1, is $\frac{1}{2}\sin 2t$. Integrating this function from 0 to t, we have by Theorem 4

$$\mathscr{L}^{-1}\left\{\frac{1}{s(s^2+4)}\right\} = \int_0^t \frac{\sin 2t}{2}\,dt = -\frac{\cos 2t}{4}\Big|_0^t = \frac{1-\cos 2t}{4} = \frac{1}{2}\sin^2 t$$

EXERCISES

1. Show that $\mathscr{L}\{\int_0^t \int_0^v f(u)\,du\,dv\} = (1/s^2)\mathscr{L}\{f(t)\}$.
2. If $f(t)$ satisfies all the conditions of Theorem 1 except that it has a jump equal to J_0 at $t = t_0$, show that
$$\mathscr{L}\{f'(t)\} = s\mathscr{L}\{f(t)\} - f_0 - J_0 e^{-st_0}$$
3. Is the result of Theorem 2 valid if $f(t)$ has a jump J_0 at $t = t_0$?
4. Devise a proof of Theorem 2 that does not use Theorem 1. *Hint:* Set up the integral that defines $\mathscr{L}\{\int_0^t f(u)\,du\}$ and then use integration by parts.
5. Show that $\mathscr{L}\{f(at)\} = (1/a)\mathscr{L}\{f(t)\}|_{s\to s/a}$. *Hint:* Set up the integral that defines $\mathscr{L}\{f(at)\}$ and then make the appropriate change of variable. This result is usually referred to as the **change of scale theorem**.
6. **(a)** Given $\mathscr{L}\{\sin t\} = 1/(s^2+1)$, use Theorem 1 to obtain $\mathscr{L}\{\cos t\}$.
   (b) Given $\mathscr{L}\{\cos t\} = s/(s^2+1)$, use Theorem 1 to obtain $\mathscr{L}\{\sin t\}$.
   (c) Given $\mathscr{L}\{\sin t\} = 1/(s^2+1)$, use Theorem 2 to obtain $\mathscr{L}\{\cos t\}$.
   (d) Given $\mathscr{L}\{\cos t\} = s/(s^2+1)$, use Theorem 2 to obtain $\mathscr{L}\{\sin t\}$.
7. **(a)** Given $\mathscr{L}\{\sin t\} = 1/(s^2+1)$, use the result of Exercise 5 to obtain $\mathscr{L}\{\sin at\}$.
   (b) Given $\mathscr{L}\{\cos t\} = s/(s^2+1)$, use the result of Exercise 5 to obtain $\mathscr{L}\{\cos at\}$.
8. Explain how the Laplace transformation can be used to solve a system of simultaneous linear differential equations with constant coefficients. In particular, given that $y = y_0$ and $z = z_0$ when $t = 0$, obtain formulas for the Laplace transforms of y and z if
$$a_1\frac{dy}{dt} + b_1 y + c_1\frac{dz}{dt} + d_1 z = f_1(t)$$
$$a_2\frac{dy}{dt} + b_2 y + c_2\frac{dz}{dt} + d_2 z = f_2(t)$$
9. The function of n defined by the equation
$$S[f(t)] = \int_0^\pi f(t)\sin nt\,dt \qquad n = 1, 2, 3, \ldots$$
is called the **finite sine transform** of $f(t)$. Show that
$$S[f''(t)] = -n^2 S[f(t)] + n[f(0) - (-1)^n f(\pi)]$$
10. The function of n defined by the equation
$$C[f(t)] = \int_0^\pi f(t)\cos nt\,dt \qquad n = 0, 1, 2, 3, \ldots$$
is called the **finite cosine transform** of $f(t)$. Show that
$$C[(f''(t)] = -n^2 C[f(t)] - f'(0) + (-1)^n f'(\pi)$$
11. Obtain formulas expressing $S[f'(t)]$ and $C[f'(t)]$ in terms of $C[f(t)]$ and $S[f(t)]$, respectively.
12. **(a)** Noting that the values of $S[f(t)]$ are just $\pi/2$ times the coefficients in the half-range sine expansion of $f(t)$, defined for $0 < t < \pi$, use the result of Exercise 9 to obtain the half-range sine expansion of $f(t) = t^2$, $0 < t < \pi$, from the half-range sine expansion of $f''(t) = 2$. How does this method compare with the direct calculation of the expansion of t^2 using the formulas of Sec. 8.2? Using the formulas of Sec. 8.4?
    (b) Employing ideas similar to those in Part (a), use the result of Exercise 10 to obtain the half-range cosine

expansion of $f(t) = t^2$, $0 < t < \pi$, from the expansion of $f''(t) = 2$.

13. Let $T[f(t)]$ be a general integral transform

$$T[f(t)] = \int_a^b f(t)K(s, t)\,dt$$

where $K(s, t)$ is the so-called **kernel** of the transformation. Obtain conditions on $K(s, t)$ so that $T[f'(t)]$ and $T[f''(t)]$ contain no terms involving evaluations of f or any of its derivatives. Find at least one kernel satisfying these conditions.

14. If $f(t)$ is of exponential order and continuous on $[0, \infty)$, with $f(0) = 0$, and if $f'(t)$ is at least piecewise regular on $[0, \infty)$ and of exponential order, show that as s becomes infinite, $\mathscr{L}\{f(t)\}$ tends to zero at least as rapidly as c/s^2. Can this result be generalized?

15. How is the initial charge Q_0 on the capacitor taken into account when the Laplace transformation is used to solve the integrodifferential equation corresponding to (1)?

10.5 THE TRANSFORMS OF SPECIAL FUNCTIONS

Among all the functions whose transforms we might now think of tabulating, the most important are the simple ones

$$e^{-at} \qquad \cos bt \qquad \sin bt \qquad t^n$$

and the unit step function

$$u(t) = \begin{cases} 0 & t < 0 \\ 1 & t > 0 \end{cases}$$

which we discussed in Sec. 1.1. Once we know the transforms of these functions, nearly all the formulas we shall need can be obtained through the use of a few additional theorems which we shall establish in later sections of this chapter. The specific results are the following.

FORMULA 1 $$\mathscr{L}\{e^{-at}\} = \frac{1}{s+a}$$

FORMULA 2 $$\mathscr{L}\{\cos bt\} = \frac{s}{s^2 + b^2}$$

FORMULA 3 $$\mathscr{L}\{\sin bt\} = \frac{b}{s^2 + b^2}$$

FORMULA 4 $$\mathscr{L}\{t^n\} = \begin{cases} \dfrac{\Gamma(n+1)}{s^{n+1}} & n > -1 \\[2ex] \dfrac{n!}{s^{n+1}} & n \text{ a nonnegative integer} \end{cases}$$

FORMULA 5 $$\mathscr{L}\{u(t)\} = \frac{1}{s}$$

Formula 1 was derived in Example 1, Sec. 10.2.
Formula 2 was derived in Example 2, Sec. 10.2.
To prove Formula 3, we have

$$\mathscr{L}\{\sin bt\} = \int_0^\infty \sin bt\, e^{-st}\,dt = \left.\frac{e^{-st}(-s \sin bt - b \cos bt)}{s^2 + b^2}\right|_0^\infty$$

$$= \frac{b}{s^2 + b^2} \qquad \text{if } s > 0$$

Before we can prove Formula 4, it will be necessary for us to investigate briefly the so-called **gamma function** or **generalized factorial function** defined by the equation

$$\Gamma(x) = \int_0^\infty e^{-t} t^{x-1}\, dt \tag{1}$$

This improper integral can be shown to be convergent for all $x > 0$.†

To determine the simple properties of the gamma function and its relation to the familiar factorial function

$$n! = n(n-1)(n-2)\cdots 3 \cdot 2 \cdot 1$$

defined in elementary algebra for positive integral values of n, let us apply integration by parts to the definitive integral (1), taking

$$u = e^{-t} \qquad dv = t^{x-1}\, dt$$

$$du = -e^{-t}\, dt \qquad v = \frac{1}{x} t^x$$

Then

$$\Gamma(x) = \frac{t^x e^{-t}}{x}\bigg|_0^\infty + \frac{1}{x}\int_0^\infty e^{-t} t^x\, dt$$

Under the restriction $x > 0$, the integrated portion vanishes at both limits. By comparison with (1), it is clear that the integral that remains is simply $\Gamma(x+1)$. Thus we have established the important recurrence relation

$$\Gamma(x) = \frac{\Gamma(x+1)}{x} \qquad x > 0 \tag{2}$$

or

$$x\Gamma(x) = \Gamma(x+1) \tag{2a}$$

Moreover, we have specifically

$$\Gamma(1) = \int_0^\infty e^{-t}\, dt = -e^{-t}\bigg|_0^\infty = 1$$

Therefore, using (2*a*),

$$\Gamma(2) = 1 \cdot \Gamma(1) = 1$$
$$\Gamma(3) = 2 \cdot \Gamma(2) = 2 \cdot 1 = 2!$$
$$\Gamma(4) = 3 \cdot \Gamma(3) = 3 \cdot 2! = 3!$$

and, in general,

$$\Gamma(n+1) = n! \qquad n = 1, 2, 3, \ldots \tag{3}$$

† See, for instance, P. Franklin, *A Treatise on Advanced Calculus*, p. 559, John Wiley, New York, 1947.

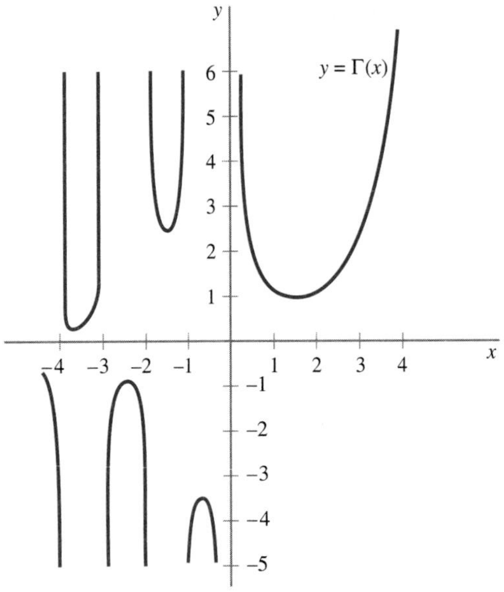

FIGURE 10.2
The function $y = \Gamma(x)$.

The connection between the gamma function and ordinary factorials is now clear. However, the gamma function constitutes an essential extension of the idea of a factorial since its argument x is not restricted to positive integral values but can vary continuously.

From (2) and the fact that $\Gamma(1) = 1$, it is evident that $\Gamma(x)$ becomes infinite as x approaches zero. It is thus clear that $\Gamma(x)$ cannot be defined for $x = 0, -1, -2, \ldots,$ in a way consistent with Eq. (2); hence we shall leave it undefined for these values of x. For all other values of x, however, $\Gamma(x)$ is well defined, the use of the recurrence formula (2a) effectively removing the restriction that x be positive, which the integral definition (1) requires. By methods that need not concern us here, tables of $\Gamma(x)$ have been constructed and can be found, usually as tables of $\log \Gamma(x)$, in most elementary handbooks. Because of the recurrence formula which the gamma function satisfies, these tables ordinarily cover only a unit interval on x, usually the interval $1 \le x \le 2$. A plot of $\Gamma(x)$ is shown in Fig. 10.2.

EXAMPLE 1

What is the value of $I = \int_0^\infty \sqrt{z}e^{-z^3}\,dz$?

This integral is typical of many that can be reduced to the standard form of the gamma function by a suitable substitution. In this case a comparison of the given integral with (1) suggests that we should let

$$z^3 = t \qquad \text{or} \qquad z = t^{1/3} \qquad dz = \tfrac{1}{3}t^{-2/3}\,dt$$

getting

$$I = \int_0^\infty \sqrt{t^{1/3}}e^{-t}\left(\frac{1}{3}t^{-2/3}\right)dt = \frac{1}{3}\int_0^\infty e^{-t}t^{(1/2)-1}\,dt = \frac{1}{3}\Gamma\left(\frac{1}{2}\right)$$

Since $\Gamma(\frac{1}{2})$ cannot be found in the usual table, which lists $\Gamma(x)$ only for $1 \le x \le 2$, it is necessary to use the recurrence relation (2) to bring the argument of the gamma function into this interval:

$$I = \frac{1}{3}\Gamma\left(\frac{1}{2}\right) = \frac{1}{3}\frac{\Gamma(\frac{3}{2})}{\frac{1}{2}} \doteq \frac{2}{3}(0.86623) \doteq 0.59082\dagger$$

† Actually, the value of $\Gamma(\frac{1}{2})$ is known exactly and in fact is equal to $\sqrt{\pi}$ (Exercise 19). Hence in this example $I = \sqrt{\pi}/3$.

Returning now to Formula 4, we have by definition

$$\mathcal{L}\{t^n\} = \int_0^\infty t^n e^{-st}\, dt$$

In an attempt to reduce this integral to the standard form of the gamma function, let us make the substitution

$$t = \frac{z}{s} \qquad dt = \frac{dz}{s}$$

Then, if $s > 0$,

$$\mathcal{L}\{t^n\} = \int_0^\infty \left(\frac{z}{s}\right)^n e^{-z} \frac{dz}{s} = \frac{1}{s^{n+1}} \int_0^\infty e^{-z} z^n\, dz = \frac{\Gamma(n+1)}{s^{n+1}} \qquad n > -1$$

Since $\Gamma(n+1) = n!$ when n is a positive integer, this establishes the second part of Formula 4 also.

It is interesting to note that if $-1 < n < 0$, then

$$s\mathcal{L}\{t^n\} = \frac{\Gamma(n+1)}{s^n}$$

is not bounded as $s \to \infty$. Hence, according to Corollary 4, Theorem 2, Sec. 10.2, $\Gamma(n+1)/s^{n+1}$ cannot be the Laplace transform of a piecewise regular function of exponential order. This of course is not surprising since when n is negative, t^n, although of exponential order (with abscissa of convergence $\alpha_0 = 0$), is not bounded in the neighborhood of the origin and so is not piecewise regular. Although the improper integral defining $\mathcal{L}\{t^n\}$ exists for $n > -1$, it does not exist for $n \leq -1$. The first version of Formula 4 must therefore be qualified by the restriction $n > -1$.

Formula 5 can be obtained immediately by taking $n = 0$ in Formula 4.

We will encounter the gamma function again in Chap. 12 where it is involved in the definition of Bessel functions. It is also of great importance in both pure and applied statistics.

EXAMPLE 2

What is the Laplace transform of sinh bt?

Since $\sinh bt = (e^{bt} - e^{-bt})/2$, we have by Formula 1,

$$\mathcal{L}\{\sinh bt\} = \mathcal{L}\left\{\frac{e^{bt} - e^{-bt}}{2}\right\} = \frac{1}{2}\left(\frac{1}{s-b} - \frac{1}{s+b}\right) = \frac{b}{s^2 - b^2}$$

The analogy with Formula 3 for the transform of sin bt is apparent.

EXAMPLE 3

If $\mathcal{L}\{y\} = \dfrac{s+1}{s^2 + s - 6}$, what is $y(t)$?

None of our formulas yields a transform resembling this one. However, factoring the denominator and then using the method of partial fractions, we have

$$\frac{s+1}{s^2 + s - 6} = \frac{s+1}{(s-2)(s+3)} = \frac{1}{5}\left(\frac{3}{s-2} + \frac{2}{s+3}\right)$$

Formula 1 can now be applied (in reverse) to the individual terms, and we find

$$y(t) = \tfrac{1}{5}(3e^{2t} + 2e^{-3t})$$

EXAMPLE 4

Find the particular solution of the differential equation $y'' - 3y' + 2y = 12e^{-2t}$ for which $y = 2$ and $y' = 6$ when $t = 0$.

Taking the Laplace transform of each side of the given equation, using Theorem 1 and its first corollary from Sec. 10.4 and Formula 1 of this section, we have

$$\mathcal{L}\{y''\} - 3\mathcal{L}\{y'\} + 2\mathcal{L}\{y\} = \mathcal{L}\{12e^{-2t}\}$$

$$[s^2\mathcal{L}\{y\} - 2s - 6] - 3[s\mathcal{L}\{y\} - 2] + 2\mathcal{L}\{y\} = \frac{12}{s+2}$$

and finally

$$\mathcal{L}\{y\} = \frac{2s^2 + 4s + 12}{(s-1)(s-2)(s+2)}$$

By an easy application of the method of partial fractions, we can now express $\mathcal{L}\{y\}$ in the form

$$\mathcal{L}\{y\} = -\frac{6}{s-1} + \frac{7}{s-2} + \frac{1}{s+2}$$

Hence, applying Formula 1 in reverse, we have

$$y = -6e^t + 7e^{2t} + e^{-2t}$$

It is of course an easy matter to verify that y satisfies the given initial-value problem.

EXAMPLE 5

Solve for $y(t)$ from the simultaneous equations

$$y' + 2y + 6\int_0^t z\,dt = -2u(t)$$

$$y' + z' + z = 0 \qquad \text{if } y_0 = -5 \text{ and } z_0 = 6$$

We begin by taking the Laplace transform of each equation term for term, using Theorems 1 and 2, Sec. 10.4, and Formula 5:

$$[s\mathcal{L}\{y\} + 5] + 2\mathcal{L}\{y\} + \frac{6}{s}\mathcal{L}\{z\} = -\frac{2}{s}$$

$$[s\mathcal{L}\{y\} + 5] + [s\mathcal{L}\{z\} - 6] + \mathcal{L}\{z\} = 0$$

Obvious simplifications then lead to the following pair of linear equations in the transforms of the unknown functions $y(t)$ and $z(t)$:

$$(s^2 + 2s)\mathcal{L}\{y\} + 6\mathcal{L}\{z\} = -2 - 5s$$

$$s\mathcal{L}\{y\} + (s+1)\mathcal{L}\{z\} = 1$$

Since we are only asked to find $y(t)$, we solve these simultaneous algebraic equations for $\mathcal{L}\{y\}$, getting

$$\mathcal{L}\{y\} = \frac{\begin{vmatrix} -2-5s & 6 \\ 1 & s+1 \end{vmatrix}}{\begin{vmatrix} s^2+2s & 6 \\ s & s+1 \end{vmatrix}} = \frac{-5s^2 - 7s - 8}{s^3 + 3s^2 - 4s}$$

Applying the method of partial fractions to this expression, we have

$$\mathcal{L}\{y\} = \frac{-5s^2 - 7s - 8}{s^3 + 3s^2 - 4s} = \frac{-5s^2 - 7s - 8}{s(s-1)(s+4)} = \frac{2}{s} - \frac{4}{s-1} - \frac{3}{s+4}$$

Finally, determining the inverse of each of these terms, we find

$$y(t) = 2u(t) - 4e^{t} - 3e^{-4t}$$

EXERCISES

1. Plot each of the following functions.
   (a) $u(t-2)$ (b) $u(2-t)$ (c) $u(t^2)$
   (d) $u(t^2-1)$ (e) $u(\sin 2t)$
   (f) $u(t^3-6t^2+11t-6)$
2. Plot each of the following functions.
   (a) $u(t-2)-u(t-1)$
   (b) $u(t)+u(t-1)+u(t-2)+u(t-3)+\cdots$
   (c) $t^2u(t-1)$ (d) $(t-1)^2u(t-1)$
   (e) $2u(\sin \pi t)-1$ (f) $t[u(t)-u(t-1)]$
3. What are
   (a) $\mathcal{L}\{\cosh kt\}$ (b) $\mathcal{L}\{(t+1)^2\}$
   (c) $\mathcal{L}\{(1-3e^{2t})^2\}$ (d) $\mathcal{L}\{\sin 2t + \cos 2t\}$
4. What is $\mathcal{L}\{\cos(at+b)\}$? *Hint:* First express $\cos(at+b)$ as the difference of two terms.
5. What is $\mathcal{L}\{\cos^2 bt\}$? *Hint:* First express $\cos^2 bt$ as a function of $2bt$.
6. Find the inverse of each of the following transforms.
   (a) $\dfrac{1}{s+3}$ (b) $\dfrac{1}{s^4}$ (c) $\dfrac{1}{s^2+9}$
   (d) $\dfrac{2s+3}{s^2+9}$ (e) $\dfrac{1}{4s+5}$ (f) $\dfrac{1}{4s^2+9}$
   (g) $\dfrac{3s+1}{(s-1)(s+3)}$ (h) $\dfrac{s+6}{s^2-4}$ (i) $\dfrac{s+4}{s^2+5s+6}$
7. Find the solution of each of the following differential equations that satisfies the given conditions.
   (a) $y''+4y'-5y=0;\ y_0=1,\ y_0'=0$
   (b) $y''-4y=0;\ y_0=-1,\ y_0'=1$
   (c) $4y''+y=0;\ y_0=2,\ y_0'=1$
   (d) $y''+4y=u(t);\ y_0=y_0'=0$
8. Find the solution of each of the following equations that satisfies the given conditions.
   (a) $y''+3y'+2y=e^t;\ y_0=1,\ y_0'=0$
   (b) $y'''+6y''+11y'+6y=0;\ y_0=2,\ y_0'=1,\ y_0''=-1$
   (c) $y'+4y+3\displaystyle\int_0^t y\,dt=t;\ y_0=0$
   (d) $y-4\displaystyle\int_0^t y\,dt=e^{-t}$. Should, or can, a value for y_0 be given?
9. Solve for $z(t)$ in Example 5.
10. Find the solution of the following system of equations.

$$y'+y+2z'+3z=e^{-t}$$
$$3y'-y+4z'+z=0 \qquad y_0=-1,\ z_0=0$$

11. Find the solution of the following system of equations.

$$(D+1)y+(2D-1)z=e^{-t}$$
$$(2D+3)y+(4D+1)z=e^{-t} \qquad y_0=2,\ z_0=-1$$

12. Find the solution of the following system of equations.

$$(D+1)y+(2D+3)z=e^{-2t}$$
$$(D-4)y+(3D-8)z=0 \qquad y_0=z_0=0$$

13. Find the solution of the following system of equations.

$$(D-1)y+(D-2)z=0$$
$$(4D+2)y+(5D+2)z=u(t) \qquad y_0=-1,\ z_0=1$$

14. Evaluate each of the following integrals.
   (a) $\displaystyle\int_0^\infty \frac{e^{-x}}{\sqrt{x}}\,dx$
   (b) $\displaystyle\int_0^\infty \exp(-\sqrt{x})\,dx$
   (c) $\displaystyle\int_0^\infty (x+1)^2e^{-x^3}\,dx$
15. Evaluate each of the following integrals.
   (a) $\displaystyle\int_0^\infty \frac{x^c}{c^x}\,dx$. *Hint:* Recall that $c^x=\exp(x\ln c)$.
   (b) $\displaystyle\int_0^1 \frac{dx}{[\ln(1/x)]^{1/2}}$. *Hint:* Let $\ln\dfrac{1}{x}=z$.
   (c) $\displaystyle\int_0^1 x^m\left(\ln\frac{1}{x}\right)^n dx$
16. Show that

$$\int_0^{\pi/2}\cos^{2m-1}\theta\,\sin^{2n-1}\theta\,d\theta=\frac{\Gamma(m)\Gamma(n)}{2\Gamma(m+n)} \qquad m,n>0$$

Hint: First show that $\Gamma(m)=\int_0^\infty 2x^{2m-1}e^{-x^2}\,dx$ and $\Gamma(n)=\int_0^\infty 2y^{2n-1}e^{-y^2}\,dy$. Then multiply these integrals and interpret the resulting repeated integral as a volume over the xy plane. Finally, compute this volume using polar coordinates, noting that the integral in polar coordinates can be written as the product of two integrals, one of which is the required integral and the other is $2\Gamma(m+n)$.

17. By setting $2m-1=k$ and $n=\frac{1}{2}$ in the result of Exercise 16, show that

$$\int_0^{\pi/2}\cos^k\theta\,d\theta=\frac{\sqrt{\pi}\,\Gamma[(k+1)/2]}{2\,\Gamma[(k/2)+1]} \qquad k>-1$$

What is $\int_0^{\pi/2}\sin^k\theta\,d\theta$?

18. Put $x=\cos^2\theta$ in the integral in Exercise 16 and show that

$$\int_0^1 x^{m-1}(1-x)^{n-1}\,dx=\frac{\Gamma(m)\Gamma(n)}{\Gamma(m+n)} \qquad m,n>0$$

This integral, which has important applications in statistics, is usually referred to as the **beta function** of m and n, $B(m, n)$.

19. Show that $\Gamma(\frac{1}{2}) = \sqrt{\pi}$. *Hint:* First show that

$$\Gamma(\tfrac{1}{2}) = 2\int_0^\infty e^{-x^2}\,dx = 2\int_0^\infty e^{-y^2}\,dy$$

Then multiply these integrals and evaluate the resulting repeated integral by changing to polar coordinates.

20. A particle of mass m moves along the x axis under the influence of a force that varies inversely as the distance from the origin. If the particle begins to move from rest at the point $x = a$, find the time it takes to reach the origin. *Hint:* After the equation of motion is set up, recall Case 2, Sec. 1.13.

21. Evaluate each of the following integrals.

(a) $\displaystyle\int_0^{\pi/2} \sqrt{\cos\theta}\,d\theta$ **(b)** $\displaystyle\int_0^{\pi/2} \sqrt{\tan\theta}\,d\theta$

(c) $\displaystyle\int_0^{\pi/2} \cos^{2/3}\theta \sin^{3/4}\theta\,d\theta$

22. Evaluate each of the following integrals.

(a) $\displaystyle\int_0^1 \frac{1}{(1-z^4)^{1/2}}\,dz$. *Hint:* Let $z^4 = x$.

(b) $\displaystyle\int_0^1 (1-z^k)^{1/k}\,dz$ **(c)** $\displaystyle\int_0^a \sqrt{a^n - z^n}\,dz$

23. **(a)** Show that $B(m, n) = B(n, m)$.

(b) Show that $B(m, n) = \displaystyle\int_0^\infty \frac{z^{m-1}}{(1+z)^{m+n}}\,dz$. *Hint:* Transform the integral that defines $B(m, n)$ by the substitution $x = z/(1 + z)$.

(c) Show that $B(m, m) = \dfrac{\sqrt{\pi}\,\Gamma(m)}{2^{2m-1}\Gamma(m+\frac{1}{2})}$ *Hint:* Use the double-angle sine formula in the integral in Exercise 16.

24. Can problems like those in Exercises 10–13 be solved by Laplace transform methods if the initial data are y_0' and z_0' instead of y_0 and z_0? How?

25. How can Laplace transform methods be used to find general rather than particular solutions of differential equations?

10.6 THE SHIFTING THEOREMS

One of the most useful properties of the Laplace transformation is contained in the so-called **first shifting theorem**.

THEOREM 1 **(s-Shifting)** If $f(t)$ is piecewise regular on $[0, \infty)$ and of exponential order, then

$$\mathscr{L}\{e^{-at}f(t)\} = \mathscr{L}\{f(t)\}_{s\to s+a}$$

PROOF By definition,

$$\mathscr{L}\{e^{-at}f(t)\} = \int_0^\infty [e^{-at}f(t)]e^{-st}\,dt = \int_0^\infty f(t)e^{-(s+a)t}\,dt$$

and the last integral is in structure exactly the Laplace transform of $f(t)$ except that $s + a$ takes the place of s.

As a tool for finding transforms, Theorem 1 states that

The transform of e^{-at} times a function of t is equal to the transform of the function itself with s replaced by $s + a$.

Conversely, as a tool for finding inverses, Theorem 1 states that

If $s + a$ is replaced by s in the transform of a function $f(t)$, that is, if s is replaced by $s - a$, then $f(t)$ is equal to e^{-at} times the inverse of the modified transform.

The last observation is formalized in the following corollary to Theorem 1.

COROLLARY 1 $\mathscr{L}^{-1}\{\phi(s)\} = e^{-at}\mathscr{L}^{-1}\{\phi(s-a)\}$

By means of Theorem 1, we can easily derive the following important formulas from Formulas 2–4 of the last section.

FORMULA 1 $\mathscr{L}\{e^{-at}\cos bt\} = \dfrac{s+a}{(s+a)^2 + b^2}$

FORMULA 2 $\mathscr{L}\{e^{-at}\sin bt\} = \dfrac{b}{(s+a)^2+b^2}$

FORMULA 3 $\mathscr{L}\{e^{-at}t^n\} = \begin{cases} \dfrac{\Gamma(n+1)}{(s+a)^{n+1}} & n > -1 \\ \dfrac{n!}{(s+a)^{n+1}} & n \text{ a nonnegative integer} \end{cases}$

EXAMPLE 1

If $\mathscr{L}\{y\} = \dfrac{2s+5}{s^2+4s+13}$, what is y?

By obvious manipulations designed to make all occurrences of s involve the same binomial, we have

$$\mathscr{L}\{y\} = \frac{2(s+2)+1}{(s+2)^2+3^2} = 2\frac{s+2}{(s+2)^2+3^2} + \frac{1}{3}\frac{3}{(s+2)^2+3^2}$$

Hence, by Formulas 1 and 2,

$$y = 2e^{-2t}\cos 3t + \tfrac{1}{3}e^{-2t}\sin 3t$$

EXAMPLE 2

What is the solution of the differential equation

$$y'' + 2y' + y = te^{-t}$$

for which $y_0 = 1$ and $y_0' = -2$?

Transforming both sides of the given equation, using Formula 3 to transform the right-hand side, we have

$$[s^2\mathscr{L}\{y\} - s + 2] + 2[s\mathscr{L}\{y\} - 1] + \mathscr{L}\{y\} = \frac{1}{(s+1)^2}$$

$$(s^2 + 2s + 1)\mathscr{L}\{y\} = \frac{1}{(s+1)^2} + s$$

$$\mathscr{L}\{y\} = \frac{1}{(s+1)^4} + \frac{s}{(s+1)^2}$$

By Formula 3, the inverse of the first fraction in $\mathscr{L}\{y\}$ is

$$\frac{t^3e^{-t}}{3!}$$

To find the inverse of the second fraction, we can write it in the form

$$\frac{s+1-1}{(s+1)^2} = \frac{1}{s+1} - \frac{1}{(s+1)^2}$$

and take the inverse of each term, or we can suppress the factor s, take the inverse of what remains, and then differentiate that inverse, according to Theorem 3, Sec. 10.4. By either method we obtain immediately $e^{-t} - te^{-t}$. Hence

$$y = \frac{t^3e^{-t}}{3!} + e^{-t} - te^{-t}$$

In this example, the characteristic equation of the differential equation has equal roots. Moreover, the term on the right is a part of the complementary function. Nevertheless, neither of these features requires any special treatment in the operational solution of the problem. This is another of the many advantages of the Laplace transform method of solving linear differential equations with constant coefficients.

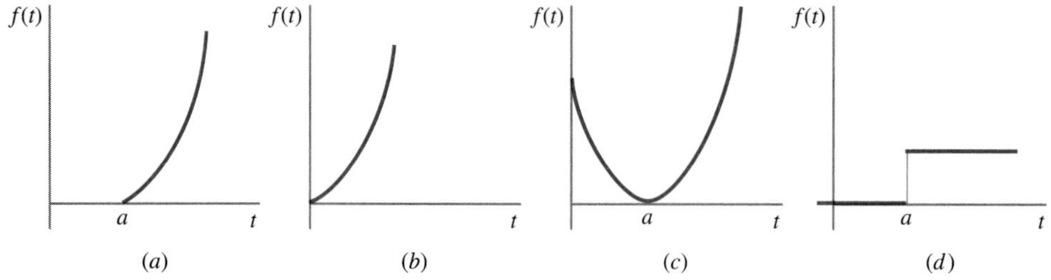

FIGURE 10.3
Plot describing the graph of a function that has been translated and cut off.

In many problems a system that becomes active at $t = 0$ because of some initial disturbance is subsequently acted upon by another disturbance beginning at some later time, say $t = a$. The analytic representation of such delayed functions and the nature of their Laplace transforms are therefore matters of some importance. To illustrate, suppose we wish an expression describing the function whose graph is shown in Fig. 10.3a, the graph being congruent to the right half of the parabola $y = t^2$ shown in Fig. 10.3b. It is not enough to recall the translation formula from analytic geometry and write $f(t) = (t - a)^2$ because this equation, even with the usual qualification that $f(t) \equiv 0$ for $t < 0$, defines the curve shown in Fig. 10.3c and not the required graph. However, if we take the unit step function and translate it a units to the right by writing $u(t - a)$, we obtain the function shown in Fig. 10.3d. Since this vanishes for $t < a$ and is equal to 1 for $t > a$, the product $(t - a)^2 u(t - a)$ will be identically zero for $t < a$ and will be identically equal to $(t - a)^2$ for $t > a$ and hence will define precisely the arc we want. More generally, the expression

$$f(t - a)u(t - a)$$

represents the function obtained by

(a) Translating $f(t)$ a units to the right

(b) Cutting it off, i.e., making it vanish identically to the left of $t = a$

EXAMPLE 3

What is the equation of the function whose graph is shown in Fig. 10.4a?

Clearly, we can regard this function as the sum of the two translated step functions shown in Fig. 10.4b. Hence its equation is

$$u(t - a) - u(t - b)$$

Although the function shown in Fig. 10.4a is not ordinarily given a name, it could appropriately be called a **filter function**; for when any other function is multiplied by this filter function, it is annihilated completely, i.e., reduced identically to zero, outside the "passband" $a < t < b$ and reproduced without any change whatsoever for values of t within the passband.

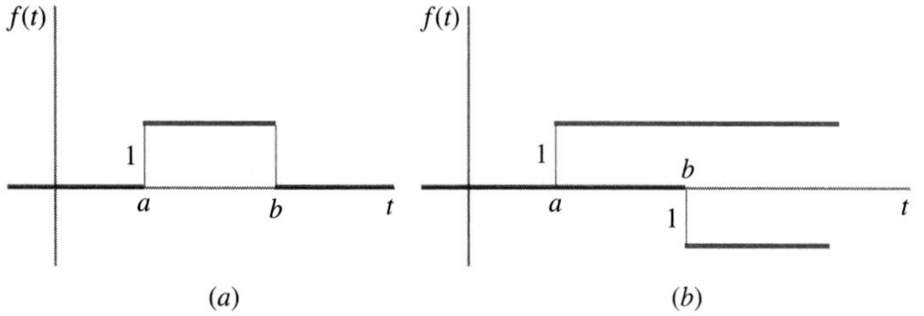

FIGURE 10.4
The construction of a rectangular pulse, or filter function, from two step functions.

EXAMPLE 4

What is the equation of the function whose graph is shown in Fig. 10.5?

To obtain an equation for the segment of this function between 1 and 2, we must multiply the expression $2(t-1)$ by a factor that will be zero to the left of 1, unity between 1 and 2, and zero to the right of 2. By Example 3, such a function is the filter function $u(t-1)-u(t-2)$. Hence

$$2(t-1)[u(t-1)-u(t-2)]$$

defines the segment of the given function between 1 and 2 and vanishes elsewhere. Similarly,

$$(-t+4)[u(t-2)-u(t-4)]$$

defines the segment of the given function between 2 and 4 and vanishes elsewhere. The complete representation of the function is therefore

$$2(t-1)[u(t-1)-u(t-2)]+(-t+4)[u(t-2)-u(t-4)]$$
$$=2(t-1)u(t-1)-3(t-2)u(t-2)+(t-4)u(t-4)$$

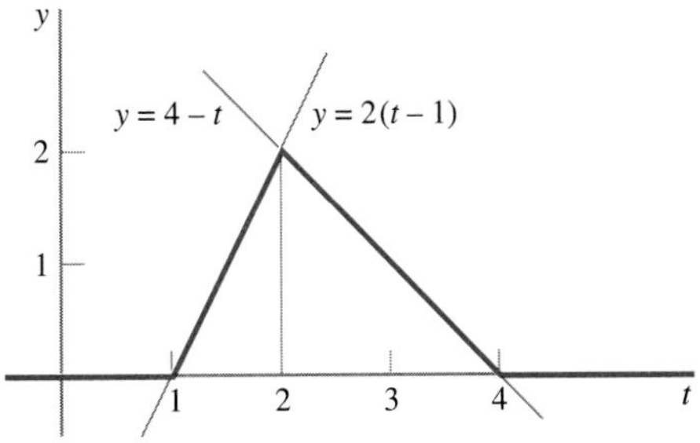

FIGURE 10.5
A graph consisting of straight-line segments.

The transforms of functions that have been translated and cut off are given by the so-called **second shifting theorem**.

THEOREM 2 **(t-Shifting)** $\mathcal{L}\{f(t-a)u(t-a)\}=e^{-as}\mathcal{L}\{f(t)\}$, $a\ge 0$.

PROOF To prove this theorem, we have by definition

$$\mathcal{L}\{f(t-a)u(t-a)\}=\int_0^\infty f(t-a)u(t-a)e^{-st}\,dt=\int_a^\infty f(t-a)e^{-st}\,dt$$

the last step following since the integration effectively starts not at $t=0$ but at $t=a$ because $f(t-a)u(t-a)$ vanishes identically to the left of $t=a$. Now let $t-a=T$ and $dt=dT$. Then the last integral becomes

$$\int_0^\infty f(T)e^{-s(T+a)}\,dT=e^{-as}\int_0^\infty f(T)e^{-sT}\,dT=e^{-as}\mathcal{L}\{f(t)\}$$

as asserted.

Before Theorem 2 can be applied, the function being transformed must be expressed in terms of the same binomial argument that appears in the unit step function. This will not often be the case; so it will frequently be necessary to alter the form of the function, as originally given, before it can be conveniently transformed. In many cases this can be done by inspection. On the other hand, we can always proceed in the following systematic way. Suppose we wish to transform

$$f(t)u(t-a)$$

As it stands, this cannot be handled by Theorem 2, so we rewrite it in the form

$$f[(t-a)+a]u(t-a) \equiv F(t-a)u(t-a)$$

where, by definition, $F(t-a) = f[(t-a)+a] = f(t)$, or

$$F(t) = f(t+a)$$

Theorem 2 can now be applied, and we have

$$\mathcal{L}\{f(t)u(t-a)\} = \mathcal{L}\{F(t-a)u(t-a)\} = e^{-as}\mathcal{L}\{F(t)\} = e^{-as}\mathcal{L}\{f(t+a)\}$$

Thus we have established the following useful result.

COROLLARY 1 $\mathcal{L}\{f(t)u(t-a)\} = e^{-as}\mathcal{L}\{f(t+a)\}$, $a \geq 0$.

As a tool for finding inverses, it is convenient to restate Theorem 2 in the following form.

COROLLARY 2 If $\mathcal{L}^{-1}\{\phi(s)\} = f(t)$, then $\mathcal{L}^{-1}\{e^{-as}\phi(s)\} = f(t-a)u(t-a)$.

In words, Corollary 2 states that

Suppressing the factor e^{-as} in a transform requires that the inverse of what remains be translated a units to the right and cut off to the left of $t = a$.

EXAMPLE 5

What is the transform of the arc of the parabola $y = -t^2 + 3t - 2$ that lies in the first quadrant, as indicated in Fig. 10.6?

The equation of the function to be transformed is clearly

$$g(t) = [-t^2 + 3t - 2][u(t-1) - u(t-2)]$$

According to Corollary 1, Theorem 2, the required transform is

$$e^{-s}\mathcal{L}\{f(t+1)\} - e^{-2s}\mathcal{L}\{f(t+2)\}$$

Easy calculations show that

$$f(t+1) = -(t+1)^2 + 3(t+1) - 2 = t - t^2$$
$$f(t+2) = -(t+2)^2 + 3(t+2) - 2 = -t - t^2$$

Hence the required transform is

$$\mathcal{L}\{g(t)\} = e^{-s}\mathcal{L}\{t - t^2\} - e^{-2s}\mathcal{L}\{-t - t^2\}$$
$$= e^{-s}\left(\frac{1}{s^2} - \frac{2}{s^3}\right) + e^{-2s}\left(\frac{1}{s^2} + \frac{2}{s^3}\right)$$

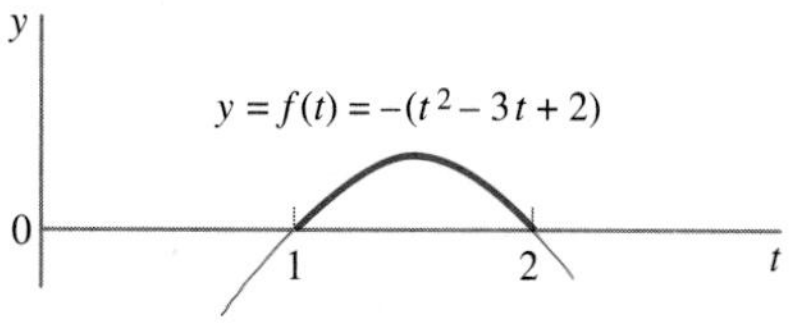

FIGURE 10.6
A parabolic pulse.

EXAMPLE 6

Find the solution of the equation $y' + 3y + 2\int_0^t y\,dt = f(t)$ for which $y_0 = 1$ if $f(t)$ is the delayed function shown in Fig. 10.7.

In this case $f(t) = 2u(t-1) - 2u(t-2)$, and thus the integrodifferential equation can be written

$$y' + 3y + 2\int_0^t y\,dt = 2u(t-1) - 2u(t-2)$$

Taking transforms and using Theorem 2 to transform the right-hand side, we have

$$[s\mathcal{L}\{y\} - 1] + 3\mathcal{L}\{y\} + \frac{2}{s}\mathcal{L}\{y\} = \frac{2e^{-s}}{s} - \frac{2e^{-2s}}{s}$$

or

$$(s^2 + 3s + 2)\mathcal{L}\{y\} = s + 2e^{-s} - 2e^{-2s}$$

and

$$\mathcal{L}\{y\} = \frac{s}{(s+1)(s+2)} + \frac{2e^{-s}}{(s+1)(s+2)} - \frac{2e^{-2s}}{(s+1)(s+2)}$$

The first term in $\mathcal{L}\{y\}$ can be written

$$\frac{2}{s+2} - \frac{1}{s+1}$$

and by inspection its inverse is $2e^{-2t} - e^{-t}$. If the exponential factors are suppressed in the second and third terms in $\mathcal{L}\{y\}$, the remaining algebraic portion can be written

$$2\left(\frac{1}{s+1} - \frac{1}{s+2}\right)$$

and the inverse of this is

$$2e^{-t} - 2e^{-2t} \tag{1}$$

However, because the factor e^{-s} has been suppressed in the second term in $\mathcal{L}\{y\}$, it is necessary to take the inverse (1), translate it one unit to the right, and then cut it off to the left of $t = 1$. Likewise, since the factor e^{-2s} has been suppressed in the third term, it is also necessary to take the inverse (1), translate it two units to the right, and then cut it off to the left of $t = 2$. The entire inverse for y is therefore

$$y = [2e^{-2t} - e^{-t}] + 2[e^{-(t-1)} - e^{-2(t-1)}]u(t-1) - 2[e^{-(t-2)} - e^{-2(t-2)}]u(t-2)$$

Plots of these three terms and of their sum, that is, y itself, are shown in Fig. 10.8.

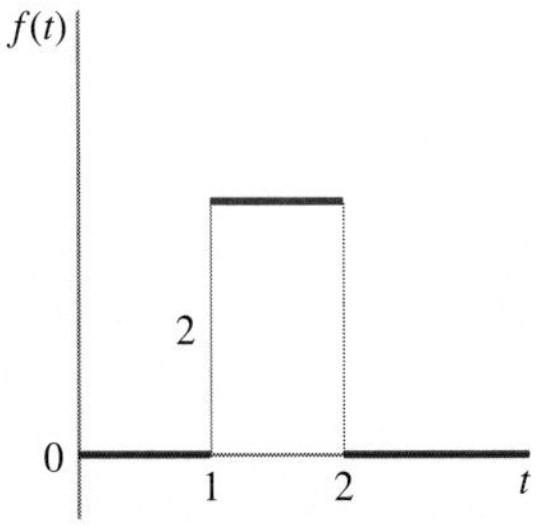

FIGURE 10.7
A rectangular pulse.

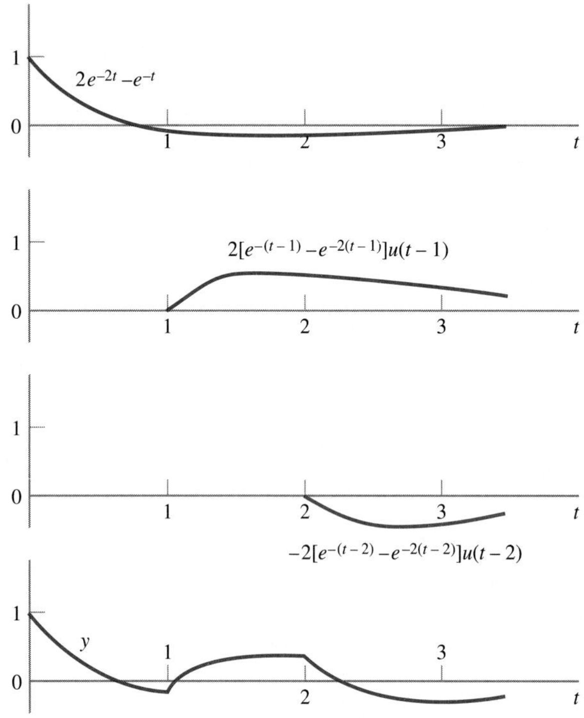

FIGURE 10.8
The solution of Example 6.

EXERCISES

Find the Laplace transform of each of the following functions.

1. **(a)** $e^{2t}\cos 5t$ **(b)** $e^{-t}t^{5/2}$ **(c)** $u(t-a)$

2. **(a)** $t^2\cosh t$ **(b)** $e^t\cosh 2t$ **(c)** $u(t^3-1)$

3. **(a)** $4e^{2t}t^{3/2}$ **(b)** $\cosh t\sinh t$ **(c)** $u(1-e^{-t})$

4. **(a)** $e^t t^{3/2}$ **(b)** $e^{-t}\int_0^t \sin 2x\,dx$
(c) $t^2u(t-2)$

5. **(a)** $e^{-t}\int_0^t xe^x\,dx$ **(b)** $\int_1^t x\,dx$
(c) $(t^2-1)u(t-1)$

6. **(a)** $e^{2t}(t-5)u(t-5)$ **(b)** $e^{-3t}(t^5+\cos t)$
(c) $u(t-1)\cos(t-1)$

7. **(a)** $\frac{1}{2}\sin at\sinh bt$ **(b)** $\int_{\pi/2}^t \sin^2 x\,dx$
(c) $u(t-1)\cos t$

8. **(a)** $f(t) = \begin{cases} \sin t & 0 \le t \le \pi \\ 0 & \pi \le t \end{cases}$

(b) $f(t) = \begin{cases} 0 & 0 < t < 1 \\ t^2 & 1 < t < 2 \\ 1 & 2 < t \end{cases}$

(c) $e^{2t}u(t-2)$

9. **(a)** $\sin^2 t + t^2u(t-1)$
(b) $[u(t) - u(t-\pi)]\sin 2t$
(c) $e^{-t}\cos(t-1)$

10. The function graphed in Fig. 10.9*a*.

11. The function graphed in Fig. 10.9*b*.

12. The function graphed in Fig. 10.9*c*.

13. The function graphed in Fig. 10.9*d*.

Find the inverse of each of the following transforms.

14. **(a)** $\dfrac{s+2}{(s+1)^3}$ **(b)** $\dfrac{1}{s(s^2+1)}$ **(c)** $\dfrac{1}{(s+2)^4}$

15. **(a)** $\dfrac{3s}{(s+5)^4}$ **(b)** $\dfrac{s+1}{s^2+s-6}$ **(c)** $\dfrac{s+2}{s^2+4s+5}$

16. **(a)** $\dfrac{4s}{(s-7)^3}$ **(b)** $\dfrac{\Gamma(3)}{(s-1)^{4/3}}$ **(c)** $\dfrac{s+5}{s^2+4s+13}$

17. **(a)** $\dfrac{s-2}{(s-5)^4}$ **(b)** $\dfrac{s+3}{(s+1)(s-3)}$ **(c)** $\dfrac{s}{(s+2)^3}$

18. **(a)** $\dfrac{\sqrt{\pi}}{s^{3/2}}$ **(b)** $\dfrac{1}{(s+1)(s^2+2s+5)}$ **(c)** $\dfrac{e^{-s}}{(s+2)^5}$

19. **(a)** $\dfrac{s(1+e^{-\pi s})}{s^2+1}$ **(b)** $\dfrac{e^{-s}+e^{-2s}}{s^2-3s+2}$ **(c)** $\dfrac{e^{-2s}}{s^2-9}$

Find the solution of each of the following differential equations that satisfies the indicated conditions.

20. **(a)** $y'' - 2y' + y = te^t$; $y(0) = 1$, $y'(0) = 0$
(b) $y'' + y = 3\sin t$; $y(0) = 1$, $y'(0) = -\frac{3}{2}$
(c) $y'' + 4y' + 5y = u(t-2)$; $y(0) = 1$, $y'(0) = -2$

21. **(a)** $y''' + y' = e^{2t}$; $y(0) = y'(0) = y''(0) = 0$
(b) $y'' - 4y' + 4y = te^{2t} + 25\sin t$; $y(0) = 1$, $y'(0) = 4$
(c) $y'' + 2y' + y = e^{-(t-1)}u(t-1)$; $y(0) = 0$, $y'(0) = 1$

22. **(a)** $y'' + 4y' + 3y = e^{-t}$; $y(0) = y'(0) = 1$

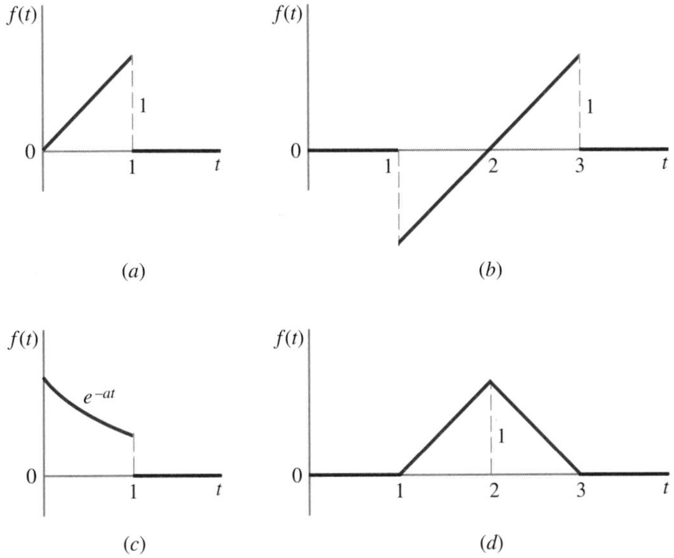

FIGURE 10.9

(b) $y''' + 3y'' + 7y' + 5y = 16e^{-t}\cos 2t$; $\quad y(0) = 2$, $y'(0) = -4$, $y''(0) = -2$

(c) $y'' + 6y' + 10y = u(t-3)$; $y(0) = 0$, $y'(0) = 2$

23. **(a)** $y^{\text{iv}} + 2y'' + y = 0$; $y(0) = 0$, $y'(0) = 1$, $y''(0) = 2$, $y'''(0) = 3$

(b) $y'' + 4y' + 13y = 2e^{-t}$; $y(0) = 0$, $y'(0) = -1$

(c) $y'' + 4y' + 13y = u(t-3)$; $y(0) = y'(0) = 1$

24. **(a)** $y^{\text{iv}} + y''' = \cos t$; $y(0) = y'(0) = y'''(0) = 0$, $y''(0) = 2$

(b) $y'' + 4y' + 3y = e^{3t}$; $y(0) = 0$, $y'(0) = 1$

(c) $y''' - 2y'' - y' + 2y = u(t-2)$; $\quad y(0) = y'(0) = 0$, $y''(0) = 1$

Find the solution of each of the following equations that satisfies such initial conditions as are given.

25. **(a)** $y' + \int_0^t y(x)\,dx + 2y = e^{-t}t^{1/2}$; $y(0) = 0$

(b) $y' + 4y + 3\int_0^t y(x)\,dx = 6 - e^{-t}$; $y(0) = 1$

(c) $y' + \int_0^t y(x)\,dx = tu(t-2)$; $y(0) = 1$

26. **(a)** $y'' - 3y' + 3y - \int_0^t y(x)\,dx = e^t(t-1)$; $\quad y(0) = 0$, $y'(0) = 1$

(b) $y' + \int_0^t y(x)\,dx + 2y = e^{-t}t^5$; $y(0) = 0$

(c) $y = 2(e^t - 1 - t) - \frac{1}{2}t^2 + \int_0^t \int_0^t y(t)\,dt\,dt + u(t - \pi)$

27. Use Laplace transforms to find a complete solution for each of the equations:

(a) $y'' - 2y' + y = 1$ **(b)** $y'' + 2ky' + k^2y = 0$

Find the solution of each of the following systems of equations that satisfies such initial conditions as are given.

28. **(a)** $\begin{aligned} x' - x + 2y &= 0 \\ y' - 2x - y &= 0 \end{aligned}$ $\quad x(0) = 5,\ y(0) = -5$

(b) $\begin{aligned} x' - 3x - 2y &= 0 \\ y' + 2x + y &= \sin t \end{aligned}$ $\quad x(0) = 2,\ y(0) = -\frac{3}{2}$

(c) $\begin{aligned} y' + y + z' &= e^{-3(t-3)}u(t-3) \\ 4y + z' + 9z &= 0 \end{aligned}$ $\quad \begin{bmatrix} y(0) \\ z(0) \end{bmatrix} = \begin{bmatrix} 1 \\ 0 \end{bmatrix}$

29. **(a)** $\begin{aligned} x' + 2x + y' + 6y &= 2e^t \\ 2x' + 3x + 3y' + 8y &= -1 \end{aligned}$ $\quad \begin{bmatrix} x(0) \\ y(0) \end{bmatrix} = \begin{bmatrix} -14 \\ 6 \end{bmatrix}$

(b) $\begin{aligned} x' + 2x + y &= 2e^{-3t} \\ 3x' + x + y' + 3y &= 0 \end{aligned}$ $\quad x(0) = y(0) = 0$

(c) $\begin{aligned} x'' - 3x - 6y &= 6e^{2t} \\ y' - x - 2y &= 0 \end{aligned}$ $\quad x(0) = x'(0) = y(0) = 0$

(d) $\begin{aligned} 4x + 8\int_0^t\int_0^u x(v)\,dv\,du - 2\int_0^t\int_0^u y(v)\,dv\,du &= 0 \\ y - 2\int_0^t\int_0^u x(v)\,dv\,du + 2\int_0^t\int_0^u y(v)\,dv\,du &= 12t^2 \end{aligned}$

30. Find real numbers a and k for which $\int_0^t\int_0^s\int_0^r du\,dr\,ds = at^k$.

31. Evaluate $\int_0^\infty (1 + x)e^{-x^2}\,dx$. *Hint:* Let $x = t^{1/2}$.

32. Evaluate $\displaystyle\int_0^\infty \frac{(\sin tx)}{x}\,dx$, $t > 0$. *Hint:* Interchange the order of integration in the double integral defining the transform of the integral.

33. Given that $\mathcal{L}\{t^{-1/2}\} = (\pi/s)^{1/2}$, find $\mathcal{L}\{t^{1/2}\}$. *Hint:* Integrate the integral for $\mathcal{L}\{t^{1/2}\}$ by parts.

34. Given that $\mathcal{L}\{2t^{1/2}\} = \sqrt{\pi}/s^{3/2}$, find $\mathcal{L}^{-1}\{\sqrt{\pi/s}\}$. *Hint:* $\sqrt{\pi}/\sqrt{s} = s\sqrt{\pi}/s^{3/2}$.

35. Evaluate $\displaystyle\lim_{a\to 0} \frac{1}{a^2}\mathcal{L}\{u(t) - 2u(t-a) + u(t-2a)\}$.

36. Solve the initial-value problem $y''(t) - y(t-1) = b$; $y(0^+) = y'(0^+) = 0$, if $y(t) = 0$ and $b = 0$ for $t \le 0$.

37. A 4-lb weight suspended from a spring of modulus $k = \frac{9}{8}$ lb/ft is acted upon by a force $f(t) = t^2e^{-3t}/8$ lb. The coefficient of friction is $c = \frac{3}{4}$ lb-s/ft. Initially the weight is motionless and has a positive displacement from equilibrium of 3 in. Use Laplace transforms to solve the initial-value problem describing the displacement y of the weight.

38. Use Laplace transforms to find the deflection y of a uniform beam of length l under a constant load per unit length w if both ends are hinged. *Hint:* Take $a = 1/EI$ and $w(x) = w[u(x) - u(x - l)]$ in Eq. (12), Sec. 2.11.

39. Work Exercise 38 if the beam is a cantilever of length $l = 2c$ with its free end at $x = 0$.

40. The angular deflection of a right-cylindrical shaft $\theta(x, t)$ satisfies the partial differential equation

$$\frac{\partial^2\theta(x, t)}{\partial t^2} = a^2\frac{\partial^2\theta(x, t)}{\partial x^2} \qquad a^2 = \frac{E_s}{\delta} \text{ constant}$$

the initial conditions $\theta(x, 0) = \theta_0 x/l$, $\theta_t(x, 0) = 0$, and the boundary conditions $\theta(0, t) = 0$, $\theta_x(l, t) = 0$. Use Laplace transforms (treating x as a constant) to express the deflection of the free end $\theta(l, t)$ in terms of $H(c, t) = \mathcal{L}^{-1}\left\{\frac{1}{s^2}\tanh\frac{cs}{2}\right\}$.

10.7 THE DIFFERENTIATION AND INTEGRATION OF TRANSFORMS

We have already made repeated use of Theorems 1 and 2, Sec. 10.4, on the transforms of derivatives and integrals. On the other hand, it is sometimes convenient or necessary to consider the derivatives and integrals of transforms. The basis for this is contained in the next two theorems.

THEOREM 1 **(Transform Differentiation)** If $f(t)$ is piecewise regular on $[0, \infty)$ and of exponential order, and if $\mathcal{L}\{f(t)\} = \phi(s)$, then $\mathcal{L}\{tf(t)\} = -\phi'(s)$.

PROOF By definition we have $\mathcal{L}\{f(t)\} = \int_0^\infty f(t)e^{-st}\,dt = \phi(s)$, and by differentiating this with respect to s we obtain

$$\frac{d}{ds}\int_0^\infty f(t)e^{-st}\,dt = \phi'(s)$$

If we now interchange the order of differentiation and integration, that is, if we perform the indicated differentiation inside the integral sign, the last expression becomes

$$\int_0^\infty f(t)\frac{\partial}{\partial s}(e^{-st})\,dt = -\int_0^\infty [tf(t)]e^{-st}\,dt = \phi'(s)$$

Clearly, the last equality is simply the statement that

$$\mathcal{L}\{tf(t)\} = -\phi'(s)$$

which is the assertion of the theorem.

The proof of the theorem will be complete if we can show that the interchange of the order of differentiation and integration is justified. To do this, we note that under our usual assumption that $f(t)$ is piecewise regular and of exponential order, the product $tf(t)$ also satisfies these conditions. Hence, by Theorems 1 and 4, Sec. 10.3, differentiation inside the integral sign is legitimate, and the theorem is proved.

By taking inverses in the assertion of Theorem 1 and then solving for $f(t)$, we obtain the following useful result.

COROLLARY 1 If $\mathcal{L}\{f(t)\} = \phi(s)$, then

$$f(t) \equiv \mathcal{L}^{-1}\{\phi(s)\} = -\frac{1}{t}\mathcal{L}^{-1}\{\phi'(s)\}$$

Restated as a tool for finding inverses, Corollary 1 says that

The inverse of a transform is equal to $-1/t$ times the inverse of the derivative of the transform.

Corollary 1 is often helpful when the inverse of a transform cannot conveniently be found but the inverse of the derivative of the transform is known. The extension of Theorem 1 and its corollary to repeated differentiation of transforms is obvious.

EXAMPLE 1

What is the Laplace transform of $f(t) = t \sin at$?

By an immediate application of Theorem 1, we have

$$\mathscr{L}\{t \sin at\} = -\frac{d}{ds}[\mathscr{L}\{\sin at\}] = -\frac{d}{ds}\left[\frac{a}{s^2 + a^2}\right] = \frac{2as}{(s^2 + a^2)^2}$$

EXAMPLE 2

What is y if $\mathscr{L}\{y\} \equiv \phi(s) = \ln \dfrac{s+1}{s-1}$?

Here, in an attempt to obtain a more familiar function of s with which to work, we differentiate $\phi(s)$ and then use Corollary 1, Theorem 1:

$$y = -\frac{1}{t}\mathscr{L}^{-1}\left\{\frac{d}{ds}\left(\ln \frac{s+1}{s-1}\right)\right\} = -\frac{1}{t}\mathscr{L}^{-1}\left\{\frac{1}{s+1} - \frac{1}{s-1}\right\}$$

$$= \frac{e^{-t} - e^{t}}{-t} = \frac{2 \sinh t}{t}$$

As we have stressed, the Laplace transformation is a powerful tool for solving linear differential equations with *constant coefficients.* However, since Theorem 1 allows us to transform products containing a positive integral power of t as one factor, Laplace transform methods are often useful for solving linear differential equations whose coefficients are *polynomials in t.* The general procedure in such cases is to transform each term of the equation, using Theorem 1 wherever necessary, and then perform the indicated differentiations. The result is not an *algebraic* equation the way it is when the coefficients in the differential equation are constants. Instead, it is a linear *differential* equation whose dependent variable is the Laplace transform of the required function and whose coefficients are functions of s. In many cases, this differential equation can be solved for the transform, and the required inverse determined by familiar methods. The general method is illustrated in the next example.

EXAMPLE 3

Solve the differential equation

(1) $$ty'' - 2(t+1)y' + (t+2)y = 0$$

If we denote $\mathscr{L}\{y(t)\}$ by $Y(s)$ as we take Laplace transforms term by term throughout this equation, using Theorem 1 together with Theorem 1, Sec. 10.4, and its first corollary, we obtain

$$-\frac{d}{ds}[s^2Y(s) - sy_0 - y_0'] + 2\frac{d}{ds}[sY(s) - y_0] - 2[sY(s) - y_0] - \frac{d}{ds}Y(s) + 2Y(s) = 0$$

Performing the indicated differentiations and collecting terms then gives us

$$(-s^2 + 2s - 1)Y'(s) + (-4s + 4)Y(s) = -3y_0$$

or, reducing this first-order linear differential equation in Y to standard form,

(2) $$Y' + \frac{4}{s-1}Y = \frac{3y_0}{(s-1)^2}$$

Next, following the theory of Sec. 1.11, we construct the integrating factor

$$e^{\int 4\,ds/(s-1)} = e^{4\ln(s-1)} = (s-1)^4$$

After multiplying each side of Eq. (2) by this factor and then integrating, we have

$$(s-1)^4 Y = y_0(s-1)^3 + c$$

$$Y(s) = \frac{y_0}{s-1} + \frac{c}{(s-1)^4}$$

Finally, using Formula 3, Sec. 10.6, to determine the inverses of the terms on the right, we have for the required solution of (1),

$$y = y_0 e^t + \frac{c}{6} t^3 e^t$$

THEOREM 2 **(Transform Integration)** If $f(t)$ is piecewise regular on $[0, \infty)$ and of exponential order, if $\mathscr{L}\{f(t)\} = \phi(s)$, and if $f(t)/t$ has a limit as t approaches zero from the right, then

$$\mathscr{L}\left\{\frac{f(t)}{t}\right\} = \int_s^\infty \phi(s)\,ds = \int_0^\infty \mathscr{L}\{f(t)\}\,ds$$

PROOF By definition,

$$\phi(s) = \mathscr{L}\{f(t)\} = \int_0^\infty f(t)e^{-st}\,dt$$

Hence, integrating from s to ∞, we obtain

$$\int_s^\infty \phi(s)\,ds = \int_s^\infty \left[\int_0^\infty f(t)e^{-st}\,dt\right] ds$$

Now under the assumption that

$$\lim_{t\to 0^+}\left[\frac{f(t)}{t}\right]$$

exists and that $f(t)$ itself is piecewise regular and of exponential order, it follows from Theorems 1 and 3, Sec. 10.3, that the integration with respect to s can be performed inside the t integral, i.e., that the order of integration in the repeated integral can be reversed. Hence, performing the s integration first, we have

$$\begin{aligned}\int_s^\infty \phi(s)\,ds &= \int_0^\infty \int_s^\infty f(t)e^{-st}\,ds\,dt = \int_0^\infty f(t)\left[\frac{e^{-st}}{-t}\right]_s^\infty dt \\ &= \int_0^\infty \left[\frac{f(t)}{t}\right] e^{-st}\,dt \\ &= \mathscr{L}\left\{\frac{f(t)}{t}\right\}\end{aligned}$$

as asserted.

By taking inverses in the assertion of Theorem 2 and then solving for $f(t)$, we obtain the following useful result.

COROLLARY 1 If $\mathcal{L}\{f(t)\} = \phi(s)$, then

$$f(t) \equiv \mathcal{L}^{-1}\{\phi(s)\} = t\mathcal{L}^{-1}\left\{\int_s^\infty \phi(s)\, ds\right\}$$

Restated as a tool for finding inverses, Corollary 1 says that

The inverse of a transform is equal to t times the inverse of the integral of the transform from s to ∞.

Corollary 1 is often useful in finding inverses when the integral of a transform is simpler to work with than the transform itself.

EXAMPLE 4

What is $\mathcal{L}\left\{\dfrac{\sin kt}{t}\right\}$?

By Theorem 2, we have

$$\mathcal{L}\left\{\frac{\sin kt}{t}\right\} = \int_s^\infty \mathcal{L}\{\sin kt\}\, ds = \int_s^\infty \frac{k}{s^2+k^2}\, ds = \operatorname{Tan}^{-1}\frac{s}{k}\bigg|_s^\infty$$

$$= \frac{\pi}{2} - \operatorname{Tan}^{-1}\frac{s}{k} = \operatorname{Cot}^{-1}\frac{s}{k}$$

EXAMPLE 5

What is y if $\mathcal{L}\{y\} = s/(s^2-1)^2$?

Using Corollary 1, Theorem 2, we have immediately

$$y = t\mathcal{L}^{-1}\left\{\int_s^\infty \frac{s}{(s^2-1)^2}\, ds\right\} = t\mathcal{L}^{-1}\left\{\frac{-1}{2(s^2-1)}\bigg|_s^\infty\right\}$$

$$= t\mathcal{L}^{-1}\left\{\frac{1}{2(s^2-1)}\right\}$$

$$= t\mathcal{L}^{-1}\left\{\frac{1}{4}\left(\frac{1}{s-1} - \frac{1}{s+1}\right)\right\}$$

$$= \frac{t}{4}(e^t - e^{-t}) = \frac{t\sinh t}{2}$$

EXERCISES

Find the Laplace transform of each of the following functions.

1. $t\cos 3t$
2. te^{2t}
3. t^2e^{-t}
4. $te^{-3t}\sin 2t$
5. $t^2\cos 2t$
6. $t^2\sin at$
7. $t\int_0^t e^{-3t}\cos 2t\, dt$
8. $t\int_0^t e^{-3t}\sin 2t\, dt$
9. $e^{-3t}\int_0^t t\sin 2t\, dt$
10. $\int_0^t te^{-3t}\sin 2t\, dt$
11. $\int_0^t t^3e^{-2t}\, dt$
12. $e^{-2t}\int_0^t t^2e^{2t}\, dt$
13. $\dfrac{1-\cos 3t}{t}$
14. $\dfrac{e^{2t}-1}{t}$
15. $\dfrac{e^{-3t}\sin 2t}{t}$
16. $\dfrac{e^t-\cos t}{t}$
17. $e^{-3t}\int_0^t \dfrac{\sin 2t}{t}\, dt$
18. $\int_0^t \dfrac{e^{-at}\cos bt}{t}\, dt$

Find the inverse of each of the following transforms.

19. $\dfrac{s}{(s+2)^4}$ **20.** $\dfrac{1}{s(s+1)^2}$

21. $\dfrac{1}{s^2(s+1)}$ **22.** $\ln\dfrac{s+a}{s+b}$

23. $\ln\dfrac{s^2-1}{s^2}$ **24.** $\dfrac{1}{(s+1)(s^2+2s+5)}$

25. $\ln\dfrac{s^2+1}{s(s+1)}$ **26.** $\mathrm{Tanh}^{-1}\dfrac{1}{s}$

27. $\dfrac{1}{s}\mathrm{Tan}^{-1}\dfrac{1}{s}$

28. $\dfrac{2}{(s^2+4)^2}$. *Hint:* First multiply and divide the transform by s.

29. $s\ln\dfrac{s-1}{s+1}+2$. What role, if any, does the additive constant 2 play in this problem?

30. $\dfrac{s+2}{(s^2+4s+5)^2}$ **31.** $\dfrac{2s+3}{(s^2+3s+2)^2}$

32. $\dfrac{1}{(s^2+2s+2)^2}$

Find the solution of each of the following differential equations that satisfies the indicated conditions.

33. $y''+4y=\cos 2t;\ y_0=y_0'=1$

34. $y''+3y'+2y=u(t-1);\ y_0=0,\ y_0'=1$

35. $y''+4y'+4y=e^{-2(t-1)}u(t-1);\ y_0=1,\ y_0'=-1$

36. $y''+2y'+10y=e^{-t}\cos t;\ y_0=1,\ y_0'=-5$

37. $y^{\text{iv}}+2y''+y=0;\ y_0=y_0'=y_0''=0,\ y_0'''=1$

38. Discuss the following problem as a possible application of the Laplace transformation. The behavior of a certain system is governed by a linear differential equation of the second order with constant, though unknown, coefficients. The response of the system, $y(t)$, to a specific test disturbance $f(t)$, a unit step function, say, can be recorded. From the relation

$$(as^2+bs+c)\mathscr{L}\{y(t)\}=\mathscr{L}\{f(t)\}=\int_0^\infty f(t)e^{-st}\,dt$$

in which $f(t)$ is given and $y(t)$ has been recorded, is it possible, using graphical or numerical integration to evaluate $\mathscr{L}\{y(t)\}$ for various values of s, and thus obtain a set of linear equations from which a, b, and c can be determined?

Solve each of the following differential equations.

39. $ty''+2(t-1)y'+(t-2)y=0$

40. $ty''+2(2t-1)y'+4(t-1)y=0$

41. $ty''-2y'+ty=0$

42. $ty''+2(2t-1)y'-4y=0$

43. Solve the initial-value problem $ty''+y'=te^{-t}$, $y(0^+)=y'(0^+)=0$.

44. Solve the boundary-value problem $ty''-(t+3)y'+4y=t-1;\ y(0)=y(1)=0$.

45. Show that the equation $(at+b)y''+(ct+d)y'+(et+f)y=0$ can be solved in terms of elementary functions if $b=0$, $d=2a\lambda$, and $f=c\lambda$, where λ is an integer greater than 1.

10.8 LIMIT THEOREMS

In Sec. 10.2 we made the following observations about the limits of Laplace transforms.

If $f(t)$ is piecewise regular on $[0,\infty)$ and of exponential order, then

(a) $\lim_{s\to\infty}\mathscr{L}\{f(t)\}=0$ (Corollary 3, Theorem 2, Sec. 10.2).

(b) $s\mathscr{L}\{f(t)\}$ is bounded as $s\to\infty$ (Corollary 4, Theorem 2, Sec. 10.2).

In this section we extend these results by proving several additional limit theorems that enable us to draw conclusions about the limiting behavior of a function $f(t)$ from the limiting behavior of its Laplace transform.

THEOREM 1 If $f(t)$ is continuous on $(0,\infty)$, if $\lim_{t\to0^+}f(t)$ exists, if $f'(t)$ is at least piecewise regular on $[0,\infty)$, and if both $f(t)$ and $f'(t)$ are of exponential order, then

$$\lim_{s\to\infty}s\mathscr{L}\{f(t)\}=\lim_{t\to0^+}f(t)=f(0^+)$$

PROOF From Theorem 1, Sec. 10.4, we have

$$\mathscr{L}\{f'(t)\}=s\mathscr{L}\{f(t)\}-f(0^+)$$

where $f(0)$ has been replaced by $f(0^+)$ since $f(0)$ may not be defined. Hence, taking the limit of each side, we have

(1) $$\lim_{s\to\infty} \mathcal{L}\{f'(t)\} = \lim_{s\to\infty} s\mathcal{L}\{f(t)\} - f(0^+)$$

By hypothesis, $f'(t)$ is piecewise regular on $[0, \infty)$ and of exponential order. Therefore, by Observation **a** above, its Laplace transform must approach zero as s becomes infinite. Hence, from (1),

$$\lim_{s\to\infty} s\mathcal{L}\{f(t)\} = f(0^+)$$

as asserted. ◗

If f' and f'' satisfy the appropriate additional conditions, Theorem 1 can be applied to $f'(t)$ and we have

COROLLARY 1 If $f(t)$ and $f'(t)$ are continuous on $(0, \infty)$, if $\lim_{t\to 0^+} f'(t)$ exists, if $f''(t)$ is at least piecewise regular on $[0, \infty)$, and if $f(t)$, $f'(t)$, and $f''(t)$ are of exponential order, then

$$\lim_{s\to\infty} s[s\mathcal{L}\{f(t)\} - f(0^+)] = f'(0^+)$$

EXAMPLE 1

If $\mathcal{L}\{f(t)\} = \dfrac{s+3}{2s^2+2s+1}$, what are the values of $f(0^+)$ and $f'(0^+)$?

By Theorem 1,

$$f(0^+) = \lim_{s\to\infty} s\mathcal{L}\{f(t)\} = \lim_{s\to\infty} \frac{s^2+3s}{2s^2+2s+1} = \frac{1}{2}$$

Similarly, by Corollary 1,

$$f'(0^+) = \lim_{s\to\infty} s\left[\frac{s^2+3s}{2s^2+2s+1} - \frac{1}{2}\right] = \lim_{s\to\infty} \frac{4s^2-s}{2(2s^2+2s+1)} = 1$$

THEOREM 2 If $f(t)$ is continuous on $(0, \infty)$, if $f'(t)$ is at least piecewise regular on $[0, \infty)$, if both $f(t)$ and $f'(t)$ are of exponential order, and if the abscissa of convergence of $f'(t)$ is negative, then

$$\lim_{s\to 0} s\mathcal{L}\{f(t)\} = \lim_{t\to\infty} f(t)$$

◖ **PROOF** Here, as in the proof of Theorem 1, we begin with the result of Theorem 1, Sec. 10.4, but this time we take limits as $s \to 0$:

$$\lim_{s\to 0} \mathcal{L}\{f'(t)\} = \lim_{s\to 0} s\mathcal{L}\{f(t)\} - f(0^+)$$

or

(2) $$\lim_{s\to 0} s\mathcal{L}\{f(t)\} = \lim_{s\to 0} \mathcal{L}\{f'(t)\} + f(0^+)$$

Now

$$\lim_{s\to 0} \mathcal{L}\{f'(t)\} = \lim_{s\to 0} \int_0^\infty f'(t)e^{-st}\, dt$$

and under the hypotheses of the present theorem we can invoke Theorems 1 and 2, Sec. 10.3, and take the limit on the right inside the integral sign. Thus

$$\lim_{s\to 0} \mathscr{L}\{f'(t)\} = \int_0^\infty f'(t)[\lim_{s\to 0} e^{-st}]\, dt = \int_0^\infty f'(t)\, dt = f(t)\Big|_0^\infty$$

$$= \lim_{t\to\infty} f(t) - f(0^+)$$

Substituting this into (2), we have finally

$$\lim_{s\to 0} s\mathscr{L}\{f(t)\} = [\lim_{t\to\infty} f(t) - f(0^+)] + f(0^+)$$

$$= \lim_{t\to\infty} f(t)$$

as asserted.† ◗

EXAMPLE 2

If y is the solution on $[0, \infty)$ of the initial-value problem

$$y'' + 4y' + 4y = 0 \qquad y(0) = 0,\ y'(0) = a$$

what is $\lim_{t\to\infty} y(t)$?

Taking transforms of the differential equation term by term, having due regard for the initial conditions, then solving for the transform $Y(s)$ of $y(t)$, we get

$$Y(s) = \frac{a}{(s+2)^2}$$

which is bounded for all nonnegative values of s. Hence, by Theorem 2,

$$\lim_{t\to\infty} y(t) = \lim_{s\to 0} sY(s) = 0$$

EXAMPLE 3

If $\mathscr{L}\{f(t)\} = \dfrac{1}{s(s^2 + 2s + 2)}$, what is $\lim_{t\to\infty} f(t)$?

Since $s^2 + 2s + 2$ is zero only for the values $s = -1 \pm i$, $s\mathscr{L}\{f(t)\}$ is bounded for all nonnegative real values of s and all complex values of s with nonnegative real parts. Hence Theorem 2 can be applied, and we have

$$\lim_{t\to\infty} f(t) = \lim_{s\to 0} s\mathscr{L}\{f(t)\} = \lim_{s\to 0} \frac{1}{s^2 + 2s + 2} = \frac{1}{2}$$

† In realistic applications of this theorem, $\mathscr{L}\{f(t)\}$ will be known, but $f(t)$ and its abscissa of convergence will be unknown. Hence it is desirable that conditions for the use of the theorem be expressed in terms of $\mathscr{L}\{f(t)\}$ rather than $f(t)$. This can be done since it is possible to show that Theorem 2 holds if and only if $s\mathscr{L}\{f(t)\}$ is bounded for all nonnegative real values of s and all complex values of s with nonnegative real parts. Thus, for example, even though $\lim_{s\to 0} s/(s^2 + 1)$ exists, Theorem 2 cannot be applied to $\mathscr{L}\{f(t)\} = 1/(s^2 + 1)$ since this becomes infinite for the values $s = \pm i \equiv 0 \pm i$. In this case of course $f(t) = \sin t$, and clearly $\lim_{t\to 0} \sin t$ does not exist.

Our final theorem is one that enables us to evaluate certain infinite integrals for which no elementary antiderivatives exist, provided we know the Laplace transforms of the *integrands* of the integrals.

THEOREM 3 If the abscissa of convergence of $f(t)$ is negative, then

$$\lim_{s\to 0} \mathcal{L}\{f(t)\} = \int_0^\infty f(t)\,dt$$

PROOF By definition,

$$\mathcal{L}\{f(t)\} = \int_0^\infty f(t)e^{-st}\,dt$$

and hence, taking limits as $s \to 0$,

$$\lim_{s\to 0} \mathcal{L}\{f(t)\} = \lim_{s\to 0} \int_0^\infty f(t)e^{-st}\,dt$$

By hypothesis, the abscissa of convergence of $f(t)$ is negative. Therefore, by Theorem 1, Sec. 10.3, the integral converges uniformly on the interval $0 \le s < \infty$, and hence, by Theorem 2, Sec. 10.3, the limit on the right can be taken inside the integral sign. Thus

$$\lim_{s\to 0} \mathcal{L}\{f(t)\} = \int_0^\infty f(t)[\lim_{s\to 0} e^{-st}]\,dt = \int_0^\infty f(t)\,dt$$

as asserted.

EXAMPLE 4

Evaluate $\displaystyle\int_0^\infty \frac{e^{-at}\cos bt - e^{-pt}\cos qt}{t}\,dt \qquad a, p > 0$

The presence of the factors e^{-at} and e^{-pt} makes it clear that the abscissa of convergence of the integrand is negative and Theorem 3 can be applied. The essential step, therefore, is to find the Laplace transform of the integrand of the given integral. According to Theorem 2, Sec. 10.7, this is

$$\int_s^\infty \mathcal{L}\{e^{-at}\cos bt - e^{-pt}\cos qt\}\,ds$$

$$= \int_s^\infty \left[\frac{s+a}{(s+a)^2+b^2} - \frac{s+p}{(s+p)^2+q^2}\right] ds = [\tfrac{1}{2}\ln\{(s+a)^2+b^2\} - \tfrac{1}{2}\ln\{(s+p)^2+q^2\}]_s^\infty$$

$$= \frac{1}{2}\ln\frac{(s+a)^2+b^2}{(s+p)^2+q^2}\Bigg|_s^\infty = \frac{1}{2}\ln\frac{(s+p)^2+q^2}{(s+a)^2+b^2}$$

Finally, using Theorem 3, we take the limit of this transform as $s \to 0$, getting

$$\frac{1}{2}\ln\frac{p^2+q^2}{a^2+b^2}$$

as the value of the given integral.

EXERCISES

Find the values of $f(0^+)$ and of $\lim_{t\to\infty} f(t)$, if it exists, if $\mathscr{L}\{f(t)\}$ is

1. $\dfrac{s^2+1}{s^3+6s^2+11s+6}$

2. $\dfrac{s+3}{2s^3-3s^2-2s}$

3. $\dfrac{s^2+s+1}{s^3-s^2+2}$

4. $\dfrac{s+2}{s^3+3s^2+4s+2}$

5. Show that in Theorem 3 the condition that the abscissa of convergence of the integrand be negative is sufficient but *not* necessary. *Hint:* Consider the integral $\displaystyle\int_0^\infty \frac{\sin t}{t}\,dt$.

Evaluate each of the following definite integrals.

6. $\displaystyle\int_0^\infty \frac{e^{-t}\sin 2t}{t}\,dt$

7. $\displaystyle\int_0^\infty \frac{1-\cos t}{t}e^{-2t}\,dt$

8. $\displaystyle\int_0^\infty \frac{\cos t-\cos 2t}{t}e^{-3t}\,dt$

9. $\displaystyle\int_0^\infty \frac{\sin t-\sin 2t}{t}e^{-3t}\,dt$

10. $\displaystyle\int_0^\infty \frac{e^{-at}\sin bt-e^{-pt}\sin qt}{t}\,dt$

11. $\displaystyle\int_0^\infty \frac{e^{-at}\sin bt+e^{-pt}\sin qt}{t}\,dt$

12. $\displaystyle\int_0^\infty \frac{e^{-2t}\sinh t}{t}\,dt$

13. $\displaystyle\int_0^\infty \frac{e^{-3t}(1-\cosh 2t)}{t}\,dt$

14. Prove Corollary 1, Theorem 1.

15. Show that under the appropriate conditions,

$$f''(0^+) = \lim_{s\to\infty} s[s^2\mathscr{L}\{f(t)\} - sf(0^+) - f'(0^+)]$$

What conditions beyond those of Theorem 1 are necessary for the validity of this result? Can the value of $f^{(n)}(0^+)$ be found by an extension of this result?

16. Show that under the appropriate conditions

$$\lim_{s\to 0} s[s\mathscr{L}\{f(t)\} - f(0^+)] = \lim_{t\to\infty} f'(t)$$

What conditions beyond those of Theorem 2 are necessary for the validity of this result? Can this result be generalized to the determination of $\lim_{t\to\infty} f^{(n)}(t)$ from $\mathscr{L}\{f(t)\}$?

17. Prove Theorem 1 without assuming that $f(t)$ is continuous. *Hint:* Use the result of Exercise 2, Sec. 10.4.

18. Prove Theorem 2 without assuming that $f(t)$ is continuous.

19. Where in the proof of Theorem 3 is use made of the hypothesis that the abscissa of convergence of $f(t)$ is negative?

20. If $Y(s)$ is the Laplace transform of the solution $y = y(t)$ of the initial-value problem $y''' + y' = e^t$; $y(0) = y'(0) = y''(0) = 0$, **(a)** does $\lim_{s\to 0} Y(s)$ exist? **(b)** Can Theorem 2 be applied to find $\lim_{t\to\infty} y(t)$? Explain. **(c)** Find $y(t)$. Does $\lim_{t\to\infty} y(t)$ exist?

10.9 THE HEAVISIDE EXPANSION THEOREMS

The frequent use we have made of partial fractions indicates clearly the importance of this technique in operational calculus. It is therefore highly desirable to have this procedure systematized as much as possible. The following theorems, usually associated with the name of Heaviside, are very useful in this connection.

THEOREM 1 If $f(t) = \mathscr{L}^{-1}\{p(s)/q(s)\}$, where $p(s)$ and $q(s)$ are polynomials and the degree of $q(s)$ is greater than the degree of $p(s)$, then the term in $f(t)$ corresponding to an unrepeated real linear factor $s - a$ of $q(s)$ is equally well

$$\frac{p(a)}{q'(a)}e^{at} \qquad \text{or} \qquad \frac{p(a)}{Q(a)}e^{at}$$

where $Q(s)$ is the product of all the factors of $q(s)$ except $s - a$.

PROOF In the familiar partial fraction decomposition of $p(s)/q(s)$, an unrepeated linear factor $s - a$ of $q(s)$ gives rise to a single fraction of the form $A/(s - a)$. Hence if we denote by $h(s)$ the sum of all the fractions corresponding to the other factors of $q(s)$, we can write

$$\frac{p(s)}{q(s)} = \frac{A}{s-a} + h(s)$$

where, since $s - a$ is an *unrepeated* factor of $q(s)$, the term $h(s)$ does not contain $s - a$ as a factor of its denominator and hence remains finite as s approaches a. Multiplying this identity by $s - a$ then gives

$$\frac{(s-a)p(s)}{q(s)} = \frac{p(s)}{q(s)/(s-a)} = A + (s-a)h(s)$$

If we now let s approach a, the second term on the right vanishes and we have

$$A = \lim_{s\to a} \frac{p(s)}{q(s)/(s-a)}$$

The limit of the numerator here is evidently $p(a)$. The denominator appears as an indeterminate of the form 0/0. However, if we evaluate it as usual according to L'Hospital's rule, by differentiating numerator and denominator with respect to s and then letting s approach a, we obtain just $q'(a)$. Hence

$$A = \frac{p(a)}{q'(a)}$$

On the other hand, we could have eliminated the indeterminacy before passing to the limit simply by canceling $s - a$ into $q(s)$, which by hypothesis contains this factor. Doing this, we obtain the equivalent form of A:

$$A = \frac{p(a)}{Q(a)}$$

Finally, taking inverses, we see that the fraction $A/(s - a)$ gives rise to the term

$$Ae^{at} = \frac{p(a)}{q'(a)}e^{at} = \frac{p(a)}{Q(a)}e^{at}$$

in the inverse $f(t)$, as asserted. ◗

If $q(s)$ contains only unrepeated linear factors, then by applying Theorem 1 to each factor in turn, we obtain the following useful result.

COROLLARY 1 If $f(t) = \mathcal{L}^{-1}\{p(s)/q(s)\}$ and if $q(s)$ is completely factorable into unrepeated real linear factors

$$(s - a_1),\quad (s - a_2),\quad \ldots,\quad (s - a_n)$$

then

$$f(t) = \sum_{i=1}^{n} \frac{p(a_i)}{q'(a_i)} e^{a_i t} = \sum_{i=1}^{n} \frac{p(a_i)}{Q_i(a_i)} e^{a_i t}$$

where $Q_i(s)$ is the product of all the factors of $q(s)$ except the factor $s - a_i$.

EXAMPLE 1

If $\mathcal{L}\{f(t)\} = (s^2 + 2)/s(s + 1)(s + 2)$, what is $f(t)$?

The roots of the denominator in this case are $s = 0, -1, -2$. Hence we must compute the values of

$$p(s) = s^2 + 2 \qquad \text{and} \qquad q'(s) = 3s^2 + 6s + 2$$

for these values of s. The results are

$$\begin{array}{lll} p(0) = 2 & p(-1) = 3 & p(-2) = 6 \\ q'(0) = 2 & q'(-1) = -1 & q'(-2) = 2 \end{array}$$

From the corollary of Theorem 1 we now have at once

$$f(t) = \frac{2}{2}e^{0t} + \frac{3}{-1}e^{-t} + \frac{6}{2}e^{-2t} = 1 - 3e^{-t} + 3e^{-2t}$$

Equally well, of course, we could have obtained the coefficients in the inverse by suppressing in turn each of the factors of the denominator and evaluating the rest of the fraction at the root corresponding to the suppressed factor. In particular, we note that

$$Q_1(0) = (s+1)(s+2)\Big|_{s=0} = 2 = q'(0)$$

$$Q_2(-1) = s(s+2)\Big|_{s=-1} = -1 = q'(-1)$$

$$Q_3(-2) = s(s+1)\Big|_{s=-2} = 2 = q'(-2)$$

THEOREM 2 If $f(t) = \mathcal{L}^{-1}\{p(s)/q(s)\}$, where $p(s)$ and $q(s)$ are polynomials and the degree of $q(s)$ is greater then the degree of $p(s)$, then the terms in $f(t)$ corresponding to a repeated real linear factor $(s-a)^r$ in $q(s)$ are

$$\left[\frac{\phi^{(r-1)}(a)}{(r-1)!} + \frac{\phi^{(r-2)}(a)}{(r-2)!}\frac{t}{1!} + \cdots + \frac{\phi'(a)}{1!}\frac{t^{r-2}}{(r-2)!} + \phi(a)\frac{t^{r-1}}{(r-1)!}\right]e^{at}$$

where $\phi(s)$ is the quotient of $p(s)$ and all the factors of $q(s)$ except $(s-a)^r$.

PROOF From the familiar theory of partial fractions we recall that a repeated linear factor $(s-a)^r$ of $q(s)$ gives rise to the component fractions

$$\frac{A_1}{s-a} + \frac{A_2}{(s-a)^2} + \cdots + \frac{A_{r-1}}{(s-a)^{r-1}} + \frac{A_r}{(s-a)^r}$$

If, as before, we let $h(s)$ denote the sum of the fractions corresponding to all the other factors of $q(s)$, we have

$$\frac{p(s)}{q(s)} \equiv \frac{\phi(s)}{(s-a)^r} = \frac{A_1}{s-a} + \frac{A_2}{(s-a)^2} + \cdots + \frac{A_{r-1}}{(s-a)^{r-1}} + \frac{A_r}{(s-a)^r} + h(s)$$

Multiplying this identity by $(s-a)^r$ gives

$$\phi(s) = A_1(s-a)^{r-1} + A_2(s-a)^{r-2} + \cdots + A_{r-1}(s-a) + A_r + (s-a)^r h(s)$$

If we put $s = a$ in this identity, we obtain

$$\phi(a) = A_r$$

If we now differentiate $\phi(s)$, we have

$$\phi'(s) = A_1(r-1)(s-a)^{r-2} + A_2(r-2)(s-a)^{r-3} + \cdots + A_{r-1} + [r(s-a)^{r-1}h(s) + (s-a)^r h'(s)]$$

Again setting $s = a$, we find this time

$$\phi'(a) = A_{r-1}$$

Continuing this process of differentiation and evaluation and noting that the first $r-1$ derivatives of the product $(s-a)^r h(s)$ will all vanish when $s = a$, we obtain successively

$$\begin{aligned} \phi''(a) &= 2!A_{r-2} \\ \phi'''(a) &= 3!A_{r-3} \\ &\cdots\cdots\cdots \\ \phi^{(r-1)}(a) &= (r-1)!A_1 \end{aligned}$$

or

$$A_{r-k} = \frac{\phi^{(k)}(a)}{k!} \qquad k = 0, 1, 2, \ldots, r-1$$

The terms in the expansion of $p(s)/q(s)$ that correspond to the factor $(s-a)^r$ are therefore

$$\frac{\phi^{(r-1)}(a)}{(r-1)!}\frac{1}{s-a} + \frac{\phi^{(r-2)}(a)}{(r-2)!}\frac{1}{(s-a)^2} + \cdots + \frac{\phi'(a)}{1!}\frac{1}{(s-a)^{r-1}} + \phi(a)\frac{1}{(s-a)^r}$$

When we recall from Formula 3, Sec. 10.6, that

$$\mathcal{L}^{-1}\left\{\frac{1}{(s-a)^n}\right\} = \frac{t^{n-1}e^{at}}{(n-1)!}$$

it is evident that the terms in $f(t)$ which arise from these fractions are

$$\frac{\phi^{(r-1)}(a)}{(r-1)!}e^{at} + \frac{\phi^{(r-2)}(a)}{(r-2)!}\frac{te^{at}}{1!} + \cdots + \frac{\phi'(a)}{1!}\frac{t^{r-2}e^{at}}{(r-2)!} + \phi(a)\frac{t^{r-1}e^{at}}{(r-1)!}$$

Finally, if we factor out e^{at} from this expression, we have precisely the assertion of the theorem. ◗

EXAMPLE 2

If $\mathcal{L}\{f(t)\} = \dfrac{s}{(s+1)^3(s+2)}$, what is $f(t)$?

Considering first the unrepeated factor $s+2$, we identify $p(s) = s$ and $Q(s) = (s+1)^3$. Hence, evaluating these for $s = -2$ and using Theorem 1, we obtain

$$\frac{p(-2)}{Q(-2)}e^{-2t} = \frac{-2}{(-1)^3}e^{-2t} = 2e^{-2t}$$

as one term in the required inverse.

Considering next the repeated factor $(s+1)^3$, we identify $\phi(s) = s/(s+2)$, $\phi'(s) = 2/(s+2)^2$, $\phi''(s) = -4/(s+2)^3$, and therefore $\phi(-1) = -1$, $\phi'(-1) = 2$, $\phi''(-1) = -4$. According to Theorem 2, the terms in the inverse corresponding to the repeated linear factor $(s+1)^3$ are then

$$\frac{-4}{2}e^{-t} + \frac{2}{1\cdot 1}te^{-t} + \frac{-1}{2}t^2e^{-t}$$

The entire inverse is therefore

$$f(t) = 2e^{-2t} - (\tfrac{1}{2}t^2 - 2t + 2)e^{-t}$$

THEOREM 3 If $f(t) = \mathcal{L}^{-1}\{p(s)/q(s)\}$, where $p(s)$ and $q(s)$ are polynomials and the degree of $q(s)$ is greater than the degree of $p(s)$, then the terms in $f(t)$ which correspond to an unrepeated, irreducible quadratic factor $(s+a)^2 + b^2$ of $q(s)$ are

$$\frac{e^{-at}}{b}(\phi_i \cos bt + \phi_r \sin bt)$$

where ϕ_r and ϕ_i are, respectively, the real and the imaginary parts of $\phi(-a+ib)$ and where $\phi(s)$ is the quotient of $p(s)$ and all the factors of $q(s)$ except the factor $(s+a)^2 + b^2$.

PROOF From the familiar theory of partial fractions, we recall that an unrepeated irreducible quadratic factor $(s+a)^2 + b^2$ of $q(s)$ gives rise to a single fraction of the form

$$\frac{As + B}{(s+a)^2 + b^2}$$

in the partial-fraction expansion of $p(s)/q(s)$. If again we let $h(s)$ denote the sum of the fractions corresponding to all the other factors of $q(s)$, we can therefore write

$$\frac{p(s)}{q(s)} \equiv \frac{\phi(s)}{(s+a)^2 + b^2} = \frac{As + B}{(s+a)^2 + b^2} + h(s)$$

Multiplying this identity by $(s+a)^2 + b^2$, we obtain

$$\phi(s) = As + B + [(s+a)^2 + b^2]h(s)$$

Now put $s = -a + ib$. This value of course makes $(s+a)^2 + b^2$ vanish; hence the last product drops out, leaving

$$\phi(-a+ib) = (-a+ib)A + B$$

or, reducing $\phi(-a+ib)$ to its standard complex form $\phi_r + i\phi_i$,

$$\phi_r + i\phi_i = (-aA + B) + ibA$$

Equating real and imaginary terms, respectively, in this equality, we find

$$\phi_r = -aA + B \qquad \text{and} \qquad \phi_i = bA$$

or, solving for A and B,

$$A = \frac{\phi_i}{b} \qquad B = \frac{b\phi_r + a\phi_i}{b}$$

Thus the partial fraction corresponding to the quadratic factor $(s+a)^2+b^2$ is

$$\frac{As+B}{(s+a)^2+b^2}=\frac{1}{b}\frac{\phi_i s+(b\phi_r+a\phi_i)}{(s+a)^2+b^2}=\frac{1}{b}\left[\frac{(s+a)\phi_i}{(s+a)^2+b^2}+\frac{b\phi_r}{(s+a)^2+b^2}\right]$$

By Formulas 1 and 2, Sec. 10.6, the inverse of the last expression is

$$\frac{1}{b}(\phi_i e^{-at}\cos bt+\phi_r e^{-at}\sin bt)$$

Finally, factoring out e^{-at}, we have the assertion of the theorem. ◗

EXAMPLE 3

If $\mathcal{L}\{y\}=\dfrac{s}{(s+2)^2(s^2+2s+10)}$, what is y?

Considering first the repeated linear factor, we identify

$$\phi(s)=\frac{s}{s^2+2s+10}\qquad\text{and}\qquad\phi'(s)=\frac{-s^2+10}{(s^2+2s+10)^2}$$

Evaluating these for the root $s=-2$, we obtain

$$\phi(-2)=-\tfrac{1}{5}\qquad\text{and}\qquad\phi'(-2)=\tfrac{3}{50}$$

Hence, by Theorem 2, the terms in y corresponding to $(s+2)^2$ are

$$\left(\frac{3}{50}-\frac{t}{5}\right)e^{-2t}=\frac{(3-10t)e^{-2t}}{50}$$

For the quadratic factor $s^2+2s+10\equiv(s+1)^2+3^2$, we have

$$\phi(s)=\frac{s}{(s+2)^2}$$

Hence

$$\phi(-a+ib)=\phi(-1+3i)=\frac{-1+3i}{[(-1+3i)+2]^2}$$
$$=\frac{-1+3i}{(1+3i)^2}=\frac{-1+3i}{-8+6i}=\frac{13-9i}{50}$$

and thus $\phi_r=\frac{13}{50}$, $\phi_i=-\frac{9}{50}$. The term in y corresponding to the factor $s^2+2s+10$ is therefore

$$\frac{1}{3}\,\frac{e^{-t}(-9\cos 3t+13\sin 3t)}{50}$$

Adding the two partial inverses, we have finally

$$y=\frac{(3-10t)e^{-2t}}{50}+\frac{e^{-t}(-9\cos 3t+13\sin 3t)}{150}$$

There is a fourth theorem dealing with repeated irreducible quadratic factors, but because of its complexity and limited usefulness, we shall not develop it here. Fortunately, many of the simpler transforms involving repeated quadratic factors can be handled by other means, e.g., the convolution theorem of Sec. 10.11.

EXERCISES

Find the functions that have the following transforms.

1. $\dfrac{s^2 - s + 3}{(s+1)(s+2)(s+3)}$
2. $\dfrac{s^2 + 4}{s^3 + 2s^2 - s - 2}$
3. $\dfrac{s}{(s+1)(s+2)^2}$
4. $\dfrac{1}{(s+2)^2(s+3)^2}$
5. $\dfrac{s+2}{(s+1)(s^2+4)}$
6. $\dfrac{s}{s^4 - 2s^2 + 1}$
7. $\dfrac{s}{(s+2)^2(s^2+1)}$
8. $\dfrac{s+2}{(s^2+1)(s^2+4)}$
9. $\dfrac{2s+3}{(s+2)(s^2+4s+5)}$
10. $\dfrac{s+2}{s^4 + 4s^3 + 4s^2 - 4s - 5}$
11. $\dfrac{s^2+2}{(s^2+4s+5)(s^2+6s+10)}$
12. $\dfrac{s^2}{(s^2+1)(s+2)^3}$

Solve each of the following differential equations.

13. $y''' - 2y'' - y' + 2y = u(t-2)$; $y_0 = y_0' = 0$, $y_0'' = 1$
14. $y''' + 3y'' + 3y' + y = \cosh t$; $y_0 = y_0' = y_0'' = 0$
15. $y^{iv} + 2y''' + 2y'' + 2y' + y = e^{-t}$;
$y_0 = y_0' = y_0'' = y_0''' = 0$

Solve each of the following systems of equations.

16. $\begin{aligned} 3y' + 8y + 2z' + 5z &= e^{-t} \\ y' \quad\;\; + z' + z &= 0 \end{aligned}$ $\quad y_0 = 2,\ z_0 = -2$
17. $\begin{aligned} 4y' + 9y + 3z' + 7z &= 0 \\ y' \quad\;\; + z' + z &= \sin t \end{aligned}$ $\quad y_0 = z_0 = 0$
18. $\begin{aligned} 2y' + 10y + z' + 7z &= e^{-3t} \\ y' + y + z' + 2z &= e^{-3t} \end{aligned}$ $\quad y_0 = z_0 = 0$
19. In the proof of Theorem 3, verify that if the identity
$$\phi(s) = As + B + [(s+a)^2 + b^2]h(s)$$
is evaluated for $s = -a - ib$ instead of for $s = -a + ib$, the same inverse is obtained.
20. Find the inverse of $s/(s+1)(s^2+2s+5)$ by factoring the irreducible quadratic factor $s^2 + 2s + 5$† into the unrepeated linear factors $s + 1 + 2i$ and $s + 1 - 2i$ and then applying Theorem 1 to these factors as well as to the real factor $s + 1$. Does your answer agree with the result obtained by using Theorem 3 to handle the quadratic factor? Do you think that this alternative procedure could be used to handle irreducible quadratic factors in general?
21. Using the procedure suggested in Exercise 20, find the inverse of $s/(s^2+4)^2$. Does your answer agree with the result obtained by using Corollary 1, Theorem 2, Sec. 10.7?
22. If $q(s)$ is a polynomial of degree n containing only unrepeated real linear factors, show that the sum of the numerators of the fractions in the partial fraction decomposition of $p(s)/q(s)$ is equal to the coefficient of s^{n-1} in $p(s)$. Is there a comparable result if $q(s)$ contains only unrepeated real linear factors and unrepeated quadratic factors?

10.10 THE TRANSFORMS OF PERIODIC FUNCTIONS

The application of the Laplace transformation to the important case of general periodic functions is based on the following theorem.

THEOREM 1 If a function f is periodic with period k on $[0, \infty)$ and piecewise regular on the period interval $0 \le t \le k$, then

$$\mathcal{L}\{f(t)\} = \frac{\displaystyle\int_0^k f(t)e^{-st}\,dt}{1 - e^{-ks}} \qquad s > 0$$

†There is no contradiction in factoring an expression previously described as irreducible because *irreducible* means, technically, "having no *real* factors."

PROOF Since f is periodic and piecewise regular over every period, there exists a constant M such that $|f(t)| < M$ for all $t \geq 0$. Hence f is of exponential order with abscissa of convergence $\alpha_0 = 0$, and so $\mathcal{L}\{f(t)\}$ exists if $s > 0$. By definition,

$$\begin{aligned}\mathcal{L}\{f(t)\} &= \int_0^\infty f(t)e^{-st}\,dt \\ &= \int_0^k f(t)e^{-st}\,dt + \int_k^{2k} f(t)e^{-st}\,dt + \int_{2k}^{3k} f(t)e^{-st}\,dt + \cdots\end{aligned}$$

Now, in the second integral, let $t = T + k$; in the third integral let $t = T + 2k$; and, in general, let $t = T + nk$ in the $(n + 1)$st integral. In each case $dt = dT$, and the new limits become 0 and k. Hence

$$\begin{aligned}\mathcal{L}\{f(t)\} &= \int_0^k f(t)e^{-st}\,dt + \int_0^k f(T+k)e^{-s(T+k)}\,dT \\ &\quad + \int_0^k f(T+2k)e^{-s(T+2k)}\,dT + \cdots \\ &= \int_0^k f(T)e^{-sT}\,dT + e^{-ks}\int_0^k f(T+k)e^{-sT}\,dT \\ &\quad + e^{-2ks}\int_0^k f(T+2k)e^{-sT}\,dT + \cdots\end{aligned}$$

However, $f(T + k) = f(T + 2k) = \cdots = f(T + nk) = \cdots = f(T)$ for all values of T since, by hypothesis, $f(t)$ is of period k. Thus we have

$$\begin{aligned}\mathcal{L}\{f(t)\} &= \int_0^k f(T)e^{-sT}\,dT + e^{-ks}\int_0^k f(T)e^{-sT}\,dT + e^{-2ks}\int_0^k f(T)e^{-sT}\,dT + \cdots \\ &= (1 + e^{-ks} + e^{-2ks} + \cdots)\int_0^k f(T)e^{-sT}\,dT\end{aligned}$$

Finally, if the infinite geometric series that multiplies the integral is summed, using the familiar formula $S = 1/(1 - r)$, where the common ratio $r = e^{-ks}$ is less than 1, we have

$$\mathcal{L}\{f(t)\} = \frac{1}{1 - e^{-ks}}\int_0^k f(T)e^{-sT}\,dT$$

which is the assertion of the theorem.

EXAMPLE 1

What is the transform of the rectangular wave shown in Fig. 10.10?

The period of the given function is $2b$. Hence, by Theorem 1,

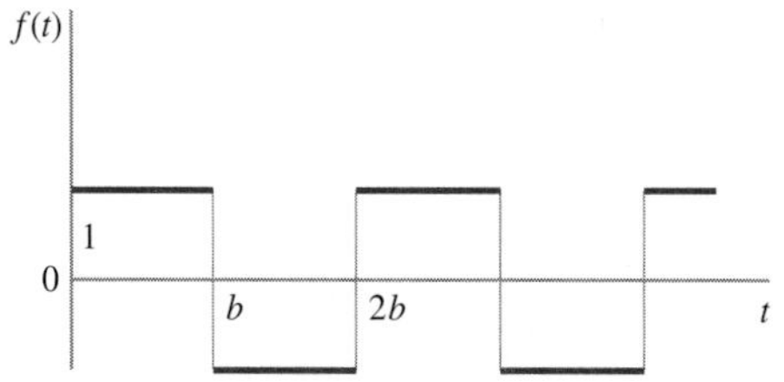

FIGURE 10.10
An alternating rectangular wave.

$$\begin{aligned}\mathcal{L}\{f(t)\} &= \frac{1}{1-e^{-2bs}}\int_0^{2b} f(t)e^{-st}\,dt \\ &= \frac{1}{1-e^{-2bs}}\left(\int_0^b 1e^{-st}\,dt + \int_b^{2b} -1e^{-st}\,dt\right) \\ &= \frac{1}{1-e^{-2bs}}\left(\frac{e^{-st}}{-s}\bigg|_0^b - \frac{e^{-st}}{-s}\bigg|_b^{2b}\right) \\ &= \frac{1}{1-e^{-2bs}}\,\frac{1-2e^{-bs}+e^{-2bs}}{s} = \frac{(1-e^{-bs})^2}{s(1-e^{-bs})(1+e^{-bs})} \\ &= \frac{1-e^{-bs}}{s(1+e^{-bs})} = \frac{e^{bs/2}-e^{-bs/2}}{s(e^{bs/2}+e^{-bs/2})} = \frac{1}{s}\tanh\frac{bs}{2}\end{aligned}$$

EXAMPLE 2

What is the transform of the sawtooth wave shown in Fig. 10.11?

In this case, the period of the function is k. Hence

$$\begin{aligned}\mathcal{L}\{f(t)\} &= \frac{1}{1-e^{-ks}}\int_0^k te^{-st}\,dt = \frac{1}{1-e^{-ks}}\left[\frac{e^{-st}}{s^2}(-st-1)\right]_0^k \\ &= \frac{1-(1+ks)e^{-ks}}{s^2(1-e^{-ks})} = \frac{(1+ks)-(1+ks)e^{-ks}-ks}{s^2(1-e^{-ks})} \\ &= \frac{(1+ks)(1-e^{-ks})-ks}{s^2(1-e^{-ks})} \\ &= \frac{1+ks}{s^2} - \frac{k}{s(1-e^{-ks})}\end{aligned}$$

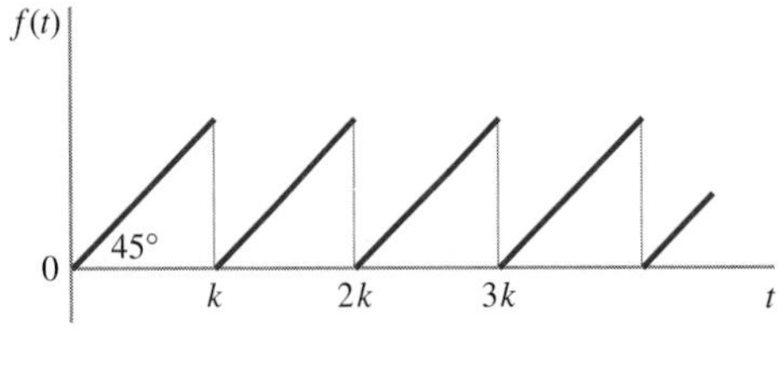

FIGURE 10.11
A sawtooth wave.

EXAMPLE 3

What is the transform of the **staircase function**

$$f(t) = n+1 \qquad nk < t < (n+1)k,\ n = 0, 1, 2, \ldots$$

shown in Fig. 10.12*a*?

The required transform can easily be found by direct calculation. However, it is even simpler to obtain it by considering $f(t)$ to be the difference between the two functions shown in Fig. 10.12*b*. The transform of the linear function $(t+k)/k$ can be found at once by using Formula 4, Sec. 10.5. Furthermore, except for the obvious coefficient $1/k$ needed to give the proper slope to the "teeth" in Fig. 10.12*b*, the transform of the sawtooth function was obtained in the last example. Hence

$$\mathcal{L}\{f(t)\} = \frac{1}{k}\left[\frac{1}{s^2}+\frac{k}{s}\right] - \frac{1}{k}\left[\frac{1+ks}{s^2} - \frac{k}{s(1-e^{-ks})}\right] = \frac{1}{s(1-e^{-ks})}$$

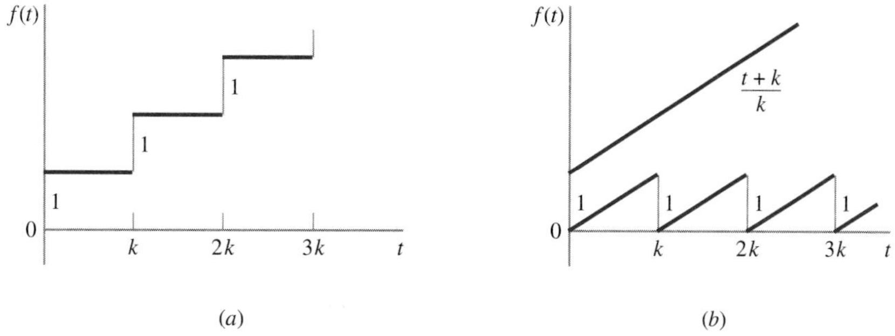

FIGURE 10.12
The staircase function and its synthesis.

EXAMPLE 4

What is $f(t)$ if $\mathcal{L}\{f(t)\} = \dfrac{1}{(s+a)(1-e^{-ks})}$?

Although $\mathcal{L}\{f(t)\}$ somewhat resembles the transform of the staircase function obtained in the last example, the correspondence is not sufficiently close to provide us with the required inverse. Moreover, we cannot successfully employ the result of the last example after first using the corollary of Theorem 1, Sec. 10.6, for after we replace s by $s - a$ (intending to multiply the inverse of the resulting fraction by e^{-at}), the given transform becomes

$$\frac{1}{s(1-e^{-k(s-a)})} = \frac{1}{s(1-e^{ak}e^{-ks})}$$

and now, because of the factor e^{ak}, which is not equal to 1 except in the trivial cases $a = 0$ or $k = 0$, we still do not have the transform of the staircase function. It appears, therefore, that we must make a direct attack upon the problem. To do this, let us reverse the derivation of Theorem 1 and replace $1/(1 - e^{-ks})$ by the infinite geometric series of which it is the sum:

$$\mathcal{L}\{f(t)\} = \frac{1}{(s+a)(1-e^{-ks})} = \frac{1}{s+a}(1 + e^{-ks} + e^{-2ks} + e^{-3ks} + \cdots)$$

$$= \frac{1}{s+a} + \frac{e^{-ks}}{s+a} + \frac{e^{-2ks}}{s+a} + \frac{e^{-3ks}}{s+a} + \cdots$$

Now let us assume that we can take the inverse of this infinite series term by term. If we neglect the exponential factor in, say, the $(n + 1)$st term, the inverse of what remains is obvious, namely, e^{-at}. But having neglected the exponential e^{-nks}, we must, according to Corollary 2 of Theorem 2, Sec. 10.6, translate the function e^{-at} to the right a distance of nk and then cut it off to the left of $t = nk$. When this is done for each term, we have

$$f(t) = e^{-at} + e^{-a(t-k)}u(t-k) + e^{-a(t-2k)}u(t-2k) + e^{-a(t-3k)}u(t-3k) + \cdots$$

When we take into account the cutoff properties of the various translated step functions, it is clear that $f(t)$ is equal to

e^{-at}	over the interval $(0, k)$
$e^{-at} + e^{ak}e^{-at}$	over the interval $(k, 2k)$
$e^{-at} + e^{ak}e^{-at} + e^{2ak}e^{-at}$	*over the interval* $(2k, 3k)$
....	....
$e^{-at} + e^{ak}e^{-at} + e^{2ak}e^{-at} + \cdots + e^{nak}e^{-at}$	over the interval $[nk, (n+1)k]$

In order to obtain a more convenient expression for $f(t)$ over the general interval $nk < t < (n+1)k$, we can sum the finite geometric progression defining $f(t)$ in this range. Since this progression contains $n + 1$ terms and has the common ratio $r = e^{ak}$, it follows that over this interval we have

$$f(t) = e^{-at}(1 + e^{ak} + e^{2ak} + \cdots + e^{nak}) = e^{-at}\frac{(e^{ak})^{n+1} - 1}{e^{ak} - 1}$$

$$= \frac{e^{-a[t-(n+1)k]}}{e^{ak} - 1} - \frac{e^{-at}}{e^{ak} - 1} \qquad nk < t < (n+1)k$$

Now, to achieve a more symmetric form, let us define $\tau = t - (n+1)k$. Clearly, $t = nk$ corresponds to $\tau = -k$, and $t = (n+1)k$ corresponds to $\tau = 0$, so that, for each value of n, the parameter τ ranges from $-k$ to 0 as t ranges from nk to $(n+1)k$. If we make this substitution in the first fraction only, $f(t)$ assumes the form

$$(1) \qquad f(t) = \frac{e^{-a\tau}}{e^{ak} - 1} - \frac{e^{-at}}{e^{ak} - 1} \qquad \begin{matrix} -k < \tau < 0 \\ nk < t < (n+1)k \end{matrix}$$

The second term describes a continuous function which decreases steadily if $a > 0$. The first term is completely independent of n, that is, yields the same set of values over each interval because no matter what n may be, as t ranges from nk to $(n+1)k$, τ always ranges from $-k$ to 0. Moreover, the first term is discontinuous, since at the left end of any interval, where $\tau = -k$, its value is

$$\frac{e^{-a(-k)}}{e^{ak} - 1}$$

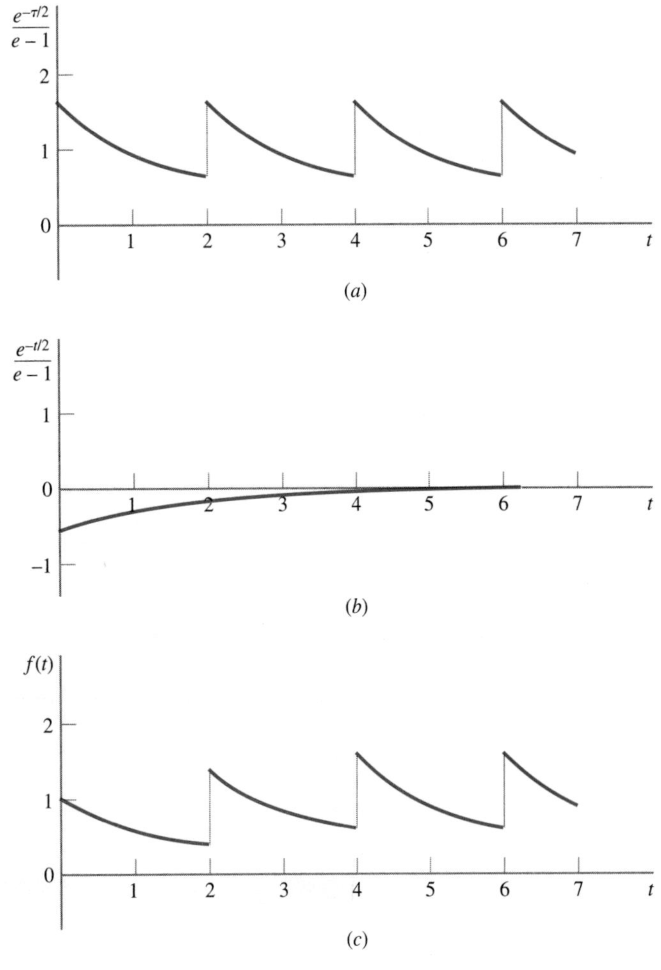

FIGURE 10.13

Plot showing the inverse of $\phi(s) = \dfrac{1}{(s + \frac{1}{2})(1 - e^{-2s})}$.

while at the right end, where $\tau = 0$, its value is

$$\frac{1}{e^{ak} - 1}$$

The periodic function which it represents therefore has an upward jump of

$$\frac{e^{ak}}{e^{ak} - 1} - \frac{1}{e^{ak} - 1} = 1$$

at each of the points $t = k, 2k, 3k, \ldots$.

Figure 10.13a shows the discontinuous periodic function represented by the first fraction in the expression (1) for $f(t)$; Fig. 10.13b shows the continuous transient term represented by the second fraction; and Fig. 10.13c shows a plot of $f(t)$ itself for $a = \frac{1}{2}$ and $k = 2$.

EXAMPLE 5

What is the solution of the equation $y' + 3y + 2\int_0^t y\,dt = f(t)$ if $y_0 = 1$ and if $f(t)$ is the function shown in Fig. 10.11 with $k = 1$?

Taking the transform of each side of the given equation, using the result of Example 2 to transform $f(t)$, we have

$$[s\mathscr{L}\{y\} - 1] + 3\mathscr{L}\{y\} + \frac{2}{s}\mathscr{L}\{y\} = \frac{1+s}{s^2} - \frac{1}{s(1 - e^{-s})}$$

or, multiplying through by s^2 and collecting terms,

$$(s^3 + 3s^2 + 2s)\mathscr{L}\{y\} = (s^2 + s + 1) - \frac{s}{1 - e^{-s}}$$

and finally

$$\mathscr{L}\{y\} = \frac{s^2 + s + 1}{s(s+1)(s+2)} - \frac{1}{(s+1)(s+2)(1 - e^{-s})}$$

The inverse of the first fraction can be found immediately by the corollary of the first Heaviside theorem (Sec. 10.9):

$$\tfrac{1}{2} - e^{-t} + \tfrac{3}{2}e^{-2t}$$

To find the inverse of the second fraction we must write

$$\frac{1}{(s+1)(s+2)(1 - e^{-s})} = \left(\frac{1}{s+1} - \frac{1}{s+2}\right)\frac{1}{1 - e^{-s}}$$

$$= \frac{1}{(s+1)(1 - e^{-s})} - \frac{1}{(s+2)(1 - e^{-s})}$$

and then use the results of Example 4. In this case $k = 1$, and thus the inverse over the interval $n < t < n + 1$ is

$$\left(\frac{e^{-\tau}}{e - 1} - \frac{e^{-t}}{e - 1}\right) - \left(\frac{e^{-2\tau}}{e^2 - 1} - \frac{e^{-2t}}{e^2 - 1}\right)$$

or

$$\left(\frac{e^{-\tau}}{e - 1} - \frac{e^{-2\tau}}{e^2 - 1}\right) - \left(\frac{e^{-t}}{e - 1} - \frac{e^{-2t}}{e^2 - 1}\right) \qquad -1 < \tau < 0$$

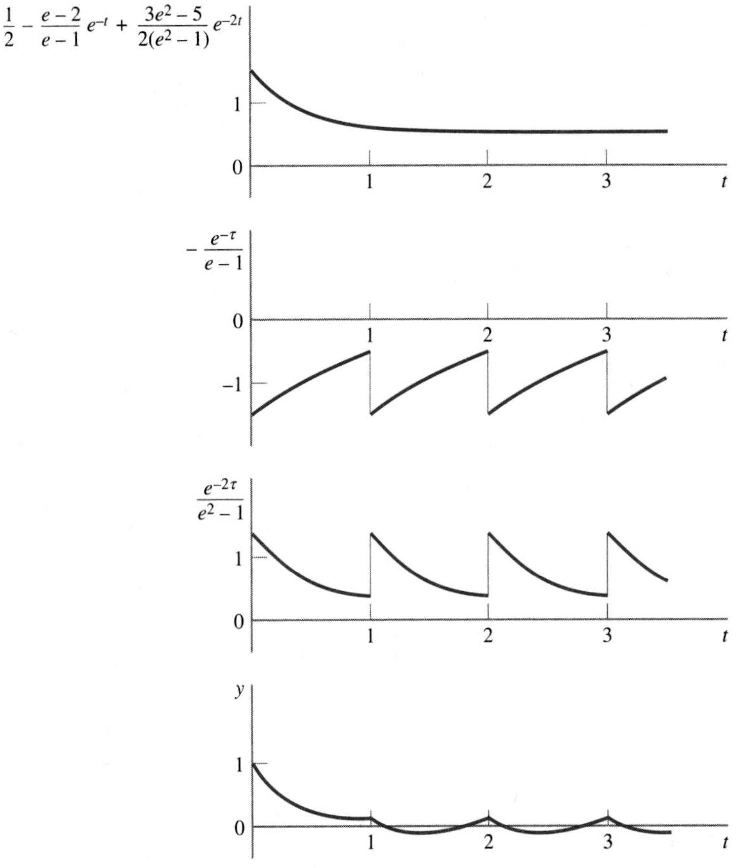

FIGURE 10.14
The solution of Example 5.

The second term is obviously a continuous function of t and is simply an additional contribution to the transient of the system. The periodic function defined by the first term is also continuous in this case because the unit jumps exhibited by each of the fractions at $t = 1, 2, 3, \ldots,$ are of opposite sign and hence cancel each other. The entire solution for y is therefore

$$y = \frac{1 - 2e^{-t} + 3e^{-2t}}{2} + \left(\frac{e^{-t}}{e-1} - \frac{e^{-2t}}{e^2-1}\right) - \left(\frac{e^{-\tau}}{e-1} - \frac{e^{-2\tau}}{e^2-1}\right)$$

$$= \underbrace{\left[-\frac{e-2}{e-1}e^{-t} + \frac{3e^2-5}{2(e^2-1)}e^{-2t}\right]}_{\text{Transient}} + \underbrace{\left(\frac{1}{2} - \frac{e^{-\tau}}{e-1} + \frac{e^{-2\tau}}{e^2-1}\right)}_{\text{Steady-state}} \qquad -1 < \tau < 0$$

Figure 10.14 shows a plot of the component terms and of y itself.

The analysis of equations like the one we encountered in Example 5 is so important that a table of additional results similar to the one we obtained in Example 4 would be highly desirable. Using for the most part only the procedure illustrated in Example 4, we can easily develop such a table, as we shall now show.

To eliminate unnecessary writing, it will be convenient to introduce the functions defined in Table 10.1, where k is an arbitrary positive number, n is a nonnegative integer, and x is a variable which is to be replaced by t or τ, as indicated in Table 10.2. The functions $\phi_1(x, k)$ and $\phi_2(x, k)$ are, respectively, the staircase function and the **Morse dot function.** The functions $\phi_3(x, k)$ and $\phi_4(x, k)$ are the integrals from 0 to x of $\phi_1(x, k)$ and $\phi_2(x, k)$, respectively. The function $\phi_5(x, a, k)$ is precisely the one we encountered in the solution of Examples 4 and 5. The others, though somewhat more complicated, arise in the same way and can be plotted just as easily when the parameters a, b, and k are known.

Table 10.2 lists the inverses of all elementary periodic-type transforms which are likely to be encountered. Of course, as Example 5 illustrated, it is usually necessary to employ the method of partial fractions before the results of Table 10.2 can be applied.

Formulas **1** to **4** are obtained by easy application of Theorem 2, Sec. 10.4. Formula **5** was derived in detail in Example 4, and the derivations of Formulas **6–10** follow almost exactly the same pattern. All that is necessary is to express as complex exponentials the sines and cosines appearing in the inverses of the individual terms after the transform has been converted into an

TABLE 10.1A

Definition of functional symbol	Definition of function over general interval $nk < x < (n+1)k$
$\phi_1(x, k)$	$n + 1$
$\phi_2(x, k)$	$\dfrac{(-1)^n + 1}{2}$
$\phi_3(x, k)$	$(n+1)x - \dfrac{n(n+1)k}{2}$
$\phi_4(x, k)$	$\dfrac{(-1)^n + 1}{2}x + \dfrac{k}{4}[1 - (-1)^n(2n+1)]$

TABLE 10.1B

Definition of functional symbol	Definition of function
$\phi_5(x, a, k)$	$\dfrac{e^{-ax}}{e^{ak} - 1}$
$\phi_6(x, a, k)$	$\dfrac{e^{-ax}}{e^{ak} + 1}$
$\phi_7(x, a, b, k)$	$\dfrac{e^{-ax}\cos b(x+k) - e^{-a(x+k)}\cos bx}{2(\cosh ak - \cos bk)}$
$\phi_8(x, a, b, k)$	$\dfrac{e^{-ax}\cos b(x+k) + e^{-a(x+k)}\cos bx}{2(\cosh ak + \cos bk)}$
$\phi_9(x, a, b, k)$	$\dfrac{e^{-ax}\sin b(x+k) - e^{-a(x+k)}\sin bx}{2(\cosh ak - \cos bk)}$
$\phi_{10}(x, a, b, k)$	$\dfrac{e^{-ax}\sin b(x+k) + e^{-a(x+k)}\sin bx}{2(\cosh ak + \cos bk)}$
$\phi_{11}(x, a, k)$	$\dfrac{(x+k)e^{-ax} - xe^{-a(x+k)}}{2(\cosh ak - 1)}$
$\phi_{12}(x, a, k)$	$\dfrac{(x+k)e^{-ax} + xe^{-a(x+k)}}{2(\cosh ak + 1)}$

TABLE 10.2

Laplace transform		Inverse over general interval $nk < t < (n+1)k$, $-k < \tau < 0$
1. $\dfrac{1}{s(1-e^{-ks})}$		$\phi_1(t, k)$
2. $\dfrac{1}{s(1+e^{-ks})}$		$\phi_2(t, k)$
3. $\dfrac{1}{s^2(1-e^{-ks})}$		$\phi_3(t, k)$
4. $\dfrac{1}{s^2(1+e^{-ks})}$		$\phi_4(t, k)$
5. $\dfrac{1}{(s+a)(1-e^{-ks})}$	$a \neq 0$	$\phi_5(\tau, a, k) - \phi_5(t, a, k)$
6. $\dfrac{1}{(s+a)(1+e^{-ks})}$	$a \neq 0$	$(-1)^n\phi_6(\tau, a, k) + \phi_6(t, a, k)$
7. $\dfrac{s+a}{[(s+a)^2+b^2](1-e^{-ks})}$		$\phi_7(\tau, a, b, k) - \phi_7(t, a, b, k)$†
8. $\dfrac{s+a}{[(s+a)^2+b^2](1+e^{-ks})}$		$(-1)^n\phi_8(\tau, a, b, k) + \phi_8(t, a, b, k)$‡
9. $\dfrac{b}{[(s+a)^2+b^2](1-e^{-ks})}$		$\phi_9(\tau, a, b, k) - \phi_9(t, a, b, k)$†
10. $\dfrac{b}{[(s+a)^2+b^2](1+e^{-ks})}$		$(-1)^n\phi_{10}(\tau, a, b, k) + \phi_{10}(t, a, b, k)$‡
11. $\dfrac{1}{(s+a)^2(1-e^{-ks})}$	$a \neq 0$	$\phi_{11}(\tau, a, k) - \phi_{11}(t, a, k)$
12. $\dfrac{1}{(s+a)^2(1+e^{-ks})}$	$a \neq 0$	$(-1)^n\phi_{12}(\tau, a, k) + \phi_{12}(t, a, k)$

† The possibility that, simultaneously, a is zero and bk is an even multiple of π is to be ruled out.

‡ The possibility that, simultaneously, a is zero and bk is an odd multiple of π is to be ruled out.

infinite series. The expression for $f(t)$ over any interval $nk < t < (n+1)k$ is then, as in Example 4, just a finite geometric progression which can be summed and converted into a purely real form without difficulty.

The derivations of Formulas **11** and **12** are somewhat different because of the repeated factors in the denominators of the transforms. Over the general interval $nk < t < (n+1)k$, these lead to expressions for $f(t)$ which are series of the form

$$\sum_{j=0}^{n} (t - jk)e^{-a(t-jk)} = te^{-at}\sum_{j=0}^{n} (e^{ak})^j - ke^{-at}\sum_{j=0}^{n} j(e^{ak})^j$$

in the case of Formula **11**, and

$$\sum_{j=0}^{n} (-1)^j(t - jk)e^{-a(t-jk)} = te^{-at}\sum_{j=0}^{n} (-e^{-ak})^j - ke^{-at}\sum_{j=0}^{n} j(-e^{-ak})^j$$

in the case of Formula **12**. In each instance, the second series is not a geometric progression and must be summed by other means. Using difference equations for this purpose, much as in Example 7, Sec. 5.7, we find without difficulty that

$$S_n = \sum_{j=0}^{n} jr^j = \frac{r + [n(r-1) - 1]r^{n+1}}{(r-1)^2} \qquad r \neq 1$$

With $r = e^{ak}$ and $r = -e^{ak}$, Formulas **11** and **12** can easily be established.

The transient, or t-evaluated, terms in the inverses in Table 10.2 are all continuous for all $t \geq 0$. This is true of the periodic, or τ-evaluated, terms if and only if the degree of the polynomial part of the denominator of the transform exceeds the degree of the numerator by more than 1. If this is not the case, there is an upward jump of 1 at each of the points $t = k, 2k, 3k, \ldots, nk, \ldots,$ if the denominator of the transform contains $1 - e^{-ks}$ and a jump of $(-1)^n$ if the denominator of the transform contains $1 + e^{-ks}$.

EXAMPLE 6 A SERIES CIRCUIT WITH A MORSE DOT VOLTAGE

A simple series circuit contains the components $R = 400\ \Omega$, $L = 0.2$ H, and $C = 10^{-6}$ F. At $t = 0$, while the circuit is completely passive, a Morse dot voltage of magnitude E_0 and period 0.002 s is switched into the circuit. Find both the transient and the steady-state currents that result.

The initial-value problem to be solved is

$$0.2\frac{di}{dt} + 400i + 10^6 \int_0^t i\, dt = E(t) \equiv E_0\phi_2(t, 0.001) \qquad i(0) = 0$$

Taking Laplace transforms of each side, using Formula 2, Table 10.2, to transform on the right, we obtain

$$0.2s\mathcal{L}\{i\} + 400\mathcal{L}\{i\} + \frac{10^6}{s}\mathcal{L}\{i\} = \frac{E_0}{s(1 + e^{-0.001s})}$$

Hence, multiplying through by $5s$, collecting terms, and solving for $\mathcal{L}\{i\}$,

$$\mathcal{L}\{i\} = \frac{5E_0}{[(s + 1{,}000)^2 + (2{,}000)^2](1 + e^{-0.001s})}$$

The entire solution for $i(t)$ is now given by Formula 10, Table 10.2, with $a = 1000$, $b = 2000$, and $k = 0.001$:

$$i(t) = \frac{E_0[(-1)^n\phi_{10}(\tau, 1{,}000, 2{,}000, 0.001) + \phi_{10}(t, 1{,}000, 2{,}000, 0.001)]}{400}$$

The periodic steady-state current is described by the first term; the transient is described by the second term. The formula in column 2 of row 6, Table 10.1B, gives the explicit description of the steady state in one period:

$$i_{ss} = \frac{E_0(-1)^n[e^{-1{,}000\tau}\sin 2{,}000(\tau + 0.001) + e^{-1{,}000(\tau+0.001)}\sin 2{,}000\tau]}{800(\cosh 1 + \cos 2)} \qquad -0.001 < \tau < 0$$

The operational solution of a problem like Example 6, leading as it does to a relatively simple finite expression for the total response of a system is, in general, to be preferred to the use of Fourier series, which leaves the answer for the steady state in the form of an infinite series and is incapable of providing an expression for the transient.

EXERCISES

1. Using Theorem 1, verify that

(a) $\mathscr{L}\{\sin bt\} = \dfrac{b}{s^2 + b^2}$ **(b)** $\mathscr{L}\{\cos bt\} = \dfrac{s}{s^2 + b^2}$

2. Obtain the Laplace transform of the staircase function, Fig. 10.12*a*, by direct evaluation of the Laplace transform integral.

Find the Laplace transforms of the periodic functions whose definitions in one period are the following.

3. $f(t) = \sin t;\ 0 \le t \le \pi$

4. $f(t) = \begin{cases} \sin t & 0 \le t \le \pi \\ 0 & \pi \le t \le 2\pi \end{cases}$

5. $f(t) = \begin{cases} \sin 2t & 0 \le t \le \pi \\ 0 & \pi \le t \le 2\pi \end{cases}$

6. $f(t) = \begin{cases} 1 & 0 < t < a \\ 0 & a < t < 2a \\ -1 & 2a < t < 3a \\ 0 & 3a < t < 4a \end{cases}$

7. $f(t) = \begin{cases} t & 0 < t < a \\ 0 & a < t < 2a \end{cases}$

Find the inverse of each of the following transforms.

8. $\dfrac{s}{(s+1)(s+2)(1-e^{-2s})}$ **9.** $\dfrac{2s+5}{(s^2+5s+6)(1+e^{-2s})}$

10. $\dfrac{e^{-s}}{s(s^2+2s+5)(1+e^{-s})}$ *Hint:* Add and subtract 1 in the numerator and then break up the fraction.

11. $\dfrac{1}{(s+1)(s+2)^2(1+e^{-s})}$

12. $\dfrac{2-s^2}{(s^4+5s^2+4)(1-e^{-2s})}$

Solve the following differential equations, $f(t)$ being in each case a periodic function defined over one period as indicated.

13. $y'' + y = f(t);\ f(t) = \begin{cases} 1 & 0 < t < 1 \\ 0 & 1 < t < 2 \end{cases}$ $\quad y_0 = y_0' = 0$

14. $y'' + 4y' + 4y = f(t);\ f(t) = \begin{cases} 1 & 0 < t < 1 \\ 0 & 1 < t < 2 \end{cases}$

$y_0 = y_0' = 0$

15. $y' + 4y + 3\displaystyle\int_0^t y\,dt = f(t);\ f(t) = \begin{cases} 1 & 0 < t < 2 \\ -1 & 2 < t < 4 \end{cases}$

$y_0 = 1$

16. Find and plot the inverse of the transform

$$\frac{1+e^{-s}}{s(1-e^{-s})(1+e^{-2s})}$$

Hint: Treating e^{-s} as though it were the simple variable x, break up into partial fractions the fraction that multiplies $1/s$.

17. Work Exercise 16 for the transform

$$\frac{1-e^{-s}}{s(1+e^{-s})(1+e^{-2s})}$$

18. Show that when $a = 0$, $b = 1$, and $k = \pi$, the function $f(t) = \phi_9(\tau, a, b, k) - \phi_9(t, a, b, k)$ is precisely the rectified sine wave given in Exercise 4. What is $f(t)$ if $a = 0$, $b = 2$, and $k = \pi/2$?

19. Describe geometrically the function

$$f(t) = (-1)^n\phi_{10}(\tau, a, b, k) + \phi_{10}(t, a, b, k)$$

when $a = 0$, $b = 2$, and $k = \pi$. What is $f(t)$ if $a = 0$, $b = 4$, and $k = \pi$?

20. If $\phi(s)$ is the transform of the periodic function whose definition in one period is

$$f(t) = \begin{cases} g(t) & 0 < t < k \\ 0 & k < t < 2k \end{cases}$$

show that $\phi(s)\,(1 + e^{-ks})$ is the transform of the periodic function whose definition in one period is

$$F(t) = g(t) \qquad 0 < t < k$$

Restate this observation as a procedure for obtaining $\mathscr{L}\{f(t)\}$ from $\mathscr{L}\{F(t)\}$.

21. Show that if $f(t)$ and $F(t)$ are periodic functions whose definitions in one period are, respectively,

$$\begin{cases} g(t) & 0 < t < k \\ 0 & k < t < 2k \end{cases} \quad \text{and} \quad g(t) \quad 0 < t < k$$

then their Laplace transforms have the respective forms

$$\frac{\phi(s)}{1-e^{-2ks}} \quad \text{and} \quad \frac{\phi(s)}{1-e^{-ks}}$$

Is the converse true, that is, if $\mathscr{L}\{f(t)\}$ and $\mathscr{L}\{F(t)\}$ have the indicated forms, are $f(t)$ and $F(t)$ necessarily of the indicated forms?

22. In the proof of Theorem 1, the geometric series that had to be summed is meaningful only if the absolute value of the common ratio $r = e^{-ks}$ is less than 1. Does this cause any problems?

23. According to the footnotes to Table 10.2 certain values of a, b, and k cannot be allowed to occur simultaneously in Formulas **7** to **10** of Table 10.2 because the formulas become meaningless for these values. Why?

24. Derive the following formulas in Table 10.2:

(a) Formula **6** **(b)** Formula **7** **(c)** Formula **8**
(d) Formula **9** **(e)** Formula **10** **(f)** Formula **11**
(g) Formula **12**

25. Find $\mathscr{L}\{f(t)\}$ if

(a) $f(t) = 2^n;\ n < t < n+1$

(b) $f(t) = (-2)^n;\ n < t < n+1$

26. Find $\mathscr{L}\{f(t)\}$ if $f(t) = (n+1)\sin t;\ n\pi < t < (n+1)\pi$.

27. Find the function whose Laplace transform is $\phi(s) = (1-e^{-s})/s(e-e^{-s})$. *Hint:* First expand $1/(e-e^{-s})$ in powers of $e^{-(s+1)}$; then note that the transform can be written

$$\phi(s) = \sum_{n=0}^{\infty} e^{-(n+1)}\left[\frac{e^{-(n+1)s} - e^{-ns}}{-s}\right]$$

28. Find the function whose Laplace transform is
$$\phi(s) = \frac{e^{-s} + e^{-2s}}{s(1 - e^{-s})^2}.$$

29. Verify the statement made following Table 10.2 about the continuity of the periodic, or τ-evaluated, terms in Table 10.2.

30. What is the inverse of the transform $\phi(s) = 1/(s \sinh s)$? *Hint:* Use the exponential definition of sinh s.

31. What is the inverse of the transform $\phi(s) = 1/(s \cosh s)$?

Let f be a transformable function of fundamental period k and domain $[0, \infty)$ so that for all $t \geq 0$, $f(t + k) = f(t)$. Define functions f_n such that for $t \geq 0$,

$$f_n(t) = [u(t - nk) - u(t - \{n + 1\}k)]f(t)$$

n a nonnegative integer.

32. Express $f(t)$ in terms of $f_n(t)$.

33. Express $\mathcal{L}\{f_n(t)\}$ in terms of $\mathcal{L}\{f(t)\}$.

34. Express $\mathcal{L}\{f(t)\}$ in terms of $\mathcal{L}\{f_0(t)\}$.

35. **(a)** In Exercise 34, set $\phi(s) = \mathcal{L}\{f(t)\}$, $\phi_0(s) = \mathcal{L}\{f_0(t)\}$, and express $\phi(s)$ in terms of $\phi_0(s)$.
(b) How does your result in Part **(a)** compare with the formula for $\mathcal{L}\{f(t)\}$ in Theorem 1?

Use the method outlined in Exercises 32–35 to find the Laplace transform of each of the following functions.

36. $f(t) = |\sin t|;\ t \geq 0$

37. $f(t) = \begin{cases} \cos t & 0 \leq t \leq \pi \\ 0 & \pi < t \leq 2\pi \end{cases}$

$f(t + 2\pi) = f(t),\ t \geq 0$

38. $f(t) = \begin{cases} 1 & 0 < t < 2 \\ -1 & 2 < t < 4 \end{cases} \qquad f(t + 4) = f(t),\ t \geq 0$

39. $f(t) = \begin{cases} 1 & 0 \leq t < 1 \\ 0 & 1 \leq t < 2 \end{cases} \qquad f(t + 2) = f(t),\ t \geq 0$

10.11 CONVOLUTION AND DUHAMEL FORMULAS

In this section we conclude our discussion of the properties of the Laplace transformation by establishing a result of considerable theoretical, as well as practical, interest. As we shall soon see, it involves the notion of the *convolution* of two functions, which we encountered in a somewhat different context in Sec. 9.3.

THEOREM 1
$$\mathcal{L}\{f(t)\}\mathcal{L}\{g(t)\} = \mathcal{L}\left\{\int_0^t f(t - \lambda)g(\lambda)\,d\lambda\right\} = \mathcal{L}\left\{\int_0^t f(\lambda)g(t - \lambda)\,d\lambda\right\}$$

PROOF Working with the term on the right in the first equality, we have by definition

$$(1) \qquad \mathcal{L}\left\{\int_0^t f(t - \lambda)g(\lambda)\,d\lambda\right\} = \int_0^\infty \left[\int_0^t f(t - \lambda)g(\lambda)\,d\lambda\right]e^{-st}\,dt$$

Now

$$u(t - \lambda) = \begin{cases} 1 & \lambda < t \\ 0 & \lambda > t \end{cases}$$

and thus

$$f(t - \lambda)g(\lambda)u(t - \lambda) = \begin{cases} f(t - \lambda)g(\lambda) & \lambda < t \\ 0 & \lambda > t \end{cases}$$

Since this product vanishes for all values of λ greater than the upper limit t, the inner integration in (1) can be extended to infinity if the factor $u(t - \lambda)$ is inserted in the integrand. Hence

$$(2) \qquad \mathcal{L}\left\{\int_0^t f(t - \lambda)g(\lambda)\,d\lambda\right\} = \int_0^\infty \left[\int_0^\infty f(t - \lambda)g(\lambda)u(t - \lambda)\,d\lambda\right]e^{-st}\,dt$$

Now our usual assumptions about the functions we transform are sufficient to permit the order of integration in (2) to be interchanged:

$$(3)\qquad \mathscr{L}\left\{\int_0^t f(t-\lambda)g(\lambda)\,d\lambda\right\} = \int_0^\infty\left[\int_0^\infty f(t-\lambda)g(\lambda)u(t-\lambda)e^{-st}\,dt\right]d\lambda$$

$$= \int_0^\infty g(\lambda)\left[\int_0^\infty f(t-\lambda)u(t-\lambda)e^{-st}\,dt\right]d\lambda$$

Because of the presence of $u(t-\lambda)$, the integrand of the inner integral is identically zero for all $t < \lambda$. Hence the inner integration effectively starts not at $t = 0$ but at $t = \lambda$. Therefore

$$(4)\qquad \mathscr{L}\left\{\int_0^t f(t-\lambda)g(\lambda)\,d\lambda\right\} = \int_0^\infty g(\lambda)\left[\int_\lambda^\infty f(t-\lambda)e^{-st}\,dt\right]d\lambda$$

Now, in the inner integral on the right of (4), let $t - \lambda = \tau$ and $dt = d\tau$. Then

$$\mathscr{L}\left\{\int_0^t f(t-\lambda)g(\lambda)\,d\lambda\right\} = \int_0^\infty g(\lambda)\left[\int_0^\infty f(\tau)e^{-s(\tau+\lambda)}\,d\tau\right]d\lambda$$

$$= \int_0^\infty g(\lambda)e^{-s\lambda}\left[\int_0^\infty f(\tau)e^{-s\tau}\,d\tau\right]d\lambda$$

$$= \left[\int_0^\infty f(\tau)e^{-s\tau}\,d\tau\right]\left[\int_0^\infty g(\lambda)e^{-s\lambda}\,d\lambda\right]$$

$$= \mathscr{L}\{f(t)\}\mathscr{L}\{g(t)\}$$

as asserted. ◗

As we observed in Sec. 9.3, the **convolution** or **Faltung**† integral

$$\int_0^t f(t-\lambda)g(\lambda)\,d\lambda$$

is frequently denoted simply by $f(t) * g(t)$. In this notation Theorem 1 becomes

$$\mathscr{L}\{f(t)\}\mathscr{L}\{g(t)\} = \mathscr{L}\{f(t) * g(t)\}$$

One of the simpler applications of the convolution theorem is its use in solving nonhomogeneous linear differential equations. The next three examples illustrate variations on this theme.

EXAMPLE 1

Find a particular integral of the differential equation

$$y'' + 2ay' + (a^2 + b^2)y = f(t)$$

Taking the Laplace transform of each side of the given equation and assuming $y_0 = y_0' = 0$ since we desire only a *particular* solution, we have

$$\mathscr{L}\{y\} = \frac{1}{(s+a)^2 + b^2}\mathscr{L}\{f(t)\}$$

†German for "folding."

Now

$$\frac{1}{(s+a)^2+b^2} = \mathcal{L}\left\{\frac{e^{-at}\sin bt}{b}\right\}$$

Hence

$$\mathcal{L}\{y\} = \mathcal{L}\left\{\frac{e^{-at}\sin bt}{b}\right\}\mathcal{L}\{f(t)\}$$

and therefore, by the convolution theorem,

$$y = \frac{1}{b}\int_0^t f(t-\lambda)e^{-a\lambda}\sin b\lambda\, d\lambda$$

or, equally well,

$$y = \frac{1}{b}\int_0^t f(\lambda)e^{-a(t-\lambda)}\sin b(t-\lambda)\, d\lambda = \frac{e^{-at}}{b}\int_0^t f(\lambda)e^{a\lambda}\sin b(t-\lambda)\, d\lambda$$

It is interesting to compare this procedure with the method of variation of parameters (Sec. 2.4) for the determination of particular integrals of linear differential equations. The two give identical results in the case of constant-coefficient differential equations.

The next example illustrates how in some cases the convolution theorem serves as an alternative to the fourth Heaviside theorem when finding the inverse of a transform whose denominator contains a repeated quadratic factor.

EXAMPLE 2

Find the solution of the differential equation

$$y'' + 4y' + 13y = \tfrac{1}{3}e^{-2t}\sin 3t$$

for which $y = 1$ and $y' = -2$ when $t = 0$.

Taking the Laplace transform of each side of the given equation, we have

$$(s^2\mathcal{L}\{y\} - s + 2) + 4(s\mathcal{L}\{y\} - 1) + 13\mathcal{L}\{y\} = \frac{1}{(s+2)^2+3^2}$$

and, solving for $\mathcal{L}\{y\}$,

$$\mathcal{L}\{y\} = \frac{s+2}{(s+2)^2+3^2} + \frac{1}{[(s+2)^2+3^2]^2}$$

The inverse of the first term is $e^{-2t}\cos 3t$. To find the inverse of the second term, it is convenient to begin by using the corollary of the first shifting theorem (Theorem 1, Sec. 10.6) to obtain

$$\mathcal{L}^{-1}\left\{\frac{1}{[(s+2)^2+3^2]^2}\right\} = e^{-2t}\mathcal{L}^{-1}\left\{\frac{1}{(s^2+3^2)^2}\right\} \tag{5}$$

Now

$$\frac{1}{(s^2+3^2)^2} = \mathcal{L}\left\{\frac{\sin 3t}{3}\right\}\mathcal{L}\left\{\frac{\sin 3t}{3}\right\}$$

Hence, by the convolution theorem,

$$\begin{aligned}
\mathscr{L}^{-1}\left\{\frac{1}{(s^2+3^2)^2}\right\} &= \frac{1}{9}\int_0^t \sin 3(t-\lambda)\sin 3\lambda\, d\lambda \\
&= \frac{1}{9}\int_0^t \frac{\cos(6\lambda-3t)-\cos 3t}{2}\, d\lambda \\
&= \frac{1}{18}\left[\frac{\sin(6\lambda-3t)}{6} - \lambda\cos 3t\right]_0^t \\
&= \frac{1}{18}\left(\frac{\sin 3t}{3} - t\cos 3t\right)
\end{aligned}$$

Therefore, from (5),

$$\mathscr{L}^{-1}\left\{\frac{1}{[(s+2)^2+3^2]^2}\right\} = \frac{e^{-2t}(\sin 3t - 3t\cos 3t)}{54}$$

and, combining the inverses of the two terms in $\mathscr{L}\{y\}$,

$$y = e^{-2t}\cos 3t + \frac{e^{-2t}(\sin 3t - 3t\cos 3t)}{54}$$

Exercises 36–39 in Sec. 8.6 called for the use of Fourier series to determine the static deflection of a stretched string bearing an arbitrary distributed load. The next example illustrates how the convolution theorem can be used to obtain the deflection as a definite integral rather than as an infinite series.

EXAMPLE 3

DEFLECTION OF AN ELASTIC STRING OF FINITE LENGTH

Using Laplace transform methods, solve the equation

$$\frac{d^2y}{dx^2} = -\frac{w(x)}{T}$$

and find the deflection $y(x)$ of an elastic string stretched under tension T between $x = 0$ and $x = l$ if the string bears a distributed load $w(x)$.

Taking the Laplace transform of each side of the given equation, using the obvious deflection condition $y(0) = 0$ but carrying the unknown slope $y'(0)$ as a parameter to be determined, we have (remembering that now the independent variable is x not t)

$$s^2\mathscr{L}\{y\} - y_0' = -\frac{1}{T}\mathscr{L}\{w(x)\}$$

or

$$\mathscr{L}\{y\} = \frac{1}{s^2}y_0' - \frac{1}{T}\frac{1}{s^2}\mathscr{L}\{w(x)\}$$

Now

$$\frac{1}{s^2} = \mathscr{L}\{x\}$$

Hence

$$\mathscr{L}\{y\} = y_0'\mathscr{L}\{x\} - \frac{1}{T}\mathscr{L}\{x\}\mathscr{L}\{w(x)\}$$

Therefore, taking inverses, using the convolution theorem to obtain the inverse of the second term on the right, we have

$$y = y_0'x - \frac{1}{T}\int_0^x (x - \lambda)w(\lambda)\,d\lambda$$

To find y_0' we use the right-hand end condition, namely, $y(l) = 0$. This gives us

$$0 = y_0'l - \frac{1}{T}\int_0^l (l - \lambda)w(\lambda)\,d\lambda \qquad \text{or} \qquad y_0' = \frac{1}{Tl}\int_0^l (l - \lambda)w(\lambda)\,d\lambda$$

With y_0' determined, the solution is complete:

$$y(x) = \frac{x}{Tl}\int_0^l (l - \lambda)w(\lambda)\,d\lambda - \frac{1}{T}\int_0^x (x - \lambda)w(\lambda)\,d\lambda$$

We have already used the Laplace transformation to solve the integrodifferential equation

$$L\frac{di}{dt} + iR + \frac{1}{C}\int_0^t i\,dt = E(t)$$

which describes the behavior of an *RLC* circuit. Now, with the convolution theorem available, we are able to solve more general integral equations in which the unknown function occurs both inside and outside the integral. Specifically, through the use of convolution we can solve equations of the form

$$y(t) = f(t) + k\int_0^t y(t - \lambda)g(\lambda)\,d\lambda$$

where f and g are known functions of t, and $y(t)$ is to be determined. This is by no means the most general form of an integral equation but is an important type that arises frequently in the study of systems that exhibit *hysteresis,* that is, systems in which the behavior of the system depends explicitly on its state at some previous time. The use of the convolution theorem to solve such equations is illustrated in our next example.

EXAMPLE 4

Solve the integral equation $y(t) = \sin 5t - 6\int_0^t y(t - \lambda)\cos 5\lambda\,d\lambda$.

If we transform each side of the given equation using the convolution theorem to transform the integral on the right, we obtain

$$\begin{aligned}\mathcal{L}\{y(t)\} &= \frac{5}{s^2 + 25} - 6\mathcal{L}\{y(t)\}\mathcal{L}\{\cos 5t\}\\ &= \frac{5}{s^2 + 25} - \frac{6s}{s^2 + 25}\mathcal{L}\{y(t)\}\end{aligned}$$

Solving this for $\mathcal{L}\{y(t)\}$, we have

$$\mathcal{L}\{y(t)\}\left[1 + \frac{6s}{s^2 + 25}\right] = \frac{5}{s^2 + 25}$$

$$\mathcal{L}\{y(t)\} = \frac{5}{s^2 + 6s + 25} = \frac{5}{4}\frac{4}{(s + 3)^2 + 4^2}$$

and finally

$$y(t) = \tfrac{5}{4}e^{-3t}\sin 4t$$

An especially important application of the convolution theorem makes it possible to determine the response of a system to a general excitation if its response to a unit step function is known. To develop this idea we shall need the concepts of *transfer function* and *indicial admittance.*

Any physical system capable of responding to an excitation can be thought of as a device by means of which an input function is transformed into an output function. If we assume that all initial conditions are zero at the moment when a single excitation, or **input**, $f(t)$ begins to act, then by setting up the differential equations describing the system, taking Laplace transformations, and solving for the transform of the **output** $y(t)$, we obtain a relation of the form

$$\mathcal{L}\{y(t)\} = \frac{\mathcal{L}\{f(t)\}}{Z(s)} \tag{6}$$

where $Z(s)$ is a function of s whose coefficients depend solely on the parameters of the system. Moreover, in the usual applications to linear systems, $Z(s)$ will be just the quotient of two polynomials in s.

In electrical problems where the input is an applied voltage $E_0e^{j\omega t}$ and the output is the resultant current, the function $Z(s)$, except for the fact that the frequency variable $j\omega$ is replaced by the Laplace transform parameter s, is just the impedance of the network. However, the importance of $Z(s)$ is not restricted to electric circuits, and for systems of all sorts the function

$$\frac{1}{Z(s)} = \frac{\mathcal{L}\{y(t)\}}{\mathcal{L}\{f(t)\}} = \frac{\mathcal{L}\{\text{output}\}}{\mathcal{L}\{\text{input}\}}$$

is an exceedingly important quantity, usually called the **transfer function**. In particular, after s has been replaced by $j\omega$, the transfer function can be used to determine the effect of any system on the phase and amplitude of a sinusoidal input of arbitrary frequency, just as in the electrical case.

If a unit step function is applied to a system with transfer function $1/Z(s)$, then from (6) we have

$$\mathcal{L}\{y(t)\} = \frac{\mathcal{L}\{u(t)\}}{Z(s)} = \frac{1}{sZ(s)}$$

The response in this particular case is called the **indicial admittance** $A(t)$, that is,

$$\mathcal{L}\{A(t)\} = \frac{1}{sZ(s)} \tag{7}$$

Using (7), we can now rewrite (6) in the form

$$\mathcal{L}\{y(t)\} = \frac{\mathcal{L}\{f(t)\}}{Z(s)} = \frac{s\mathcal{L}\{f(t)\}}{sZ(s)} = s\mathcal{L}\{A(t)\}\mathcal{L}\{f(t)\}$$

Hence, by the convolution theorem,

$$\mathcal{L}\{y(t)\} = s\mathcal{L}\left\{\int_0^t A(t-\lambda)f(\lambda)\,d\lambda\right\} = s\mathcal{L}\left\{\int_0^t A(\lambda)f(t-\lambda)\,d\lambda\right\}$$

Because of the factor s, it follows from Theorem 3, Sec. 10.4, that

$$y(t) = \frac{d}{dt}\left[\int_0^t A(t-\lambda)f(\lambda)\,d\lambda\right] = \frac{d}{dt}\left[\int_0^t A(\lambda)f(t-\lambda)\,d\lambda\right]$$

Therefore, performing the indicated differentiations,† we have equivalently,

(8) $$y(t) = \int_0^t A'(t-\lambda)f(\lambda)\,d\lambda + A(0)f(t)$$

and

(9) $$y(t) = \int_0^t A(\lambda)f'(t-\lambda)\,d\lambda + A(t)f(0)$$

Since $A(t)$ is by definition the response of a system which is initially passive, it follows that $A(0) = 0$. Hence Eq. (8) becomes simply

(10) $$y(t) = \int_0^t A'(t-\lambda)f(\lambda)\,d\lambda$$

Finally, by making the change of variable $\tau = t - \lambda$ in the integrals in (9) and (10), we obtain the related expressions

(11) $$y(t) = \int_0^t A'(\tau)f(t-\tau)\,d\tau$$

(12) $$y(t) = \int_0^t A(t-\tau)f'(\tau)\,d\tau + A(t)f(0)$$

Equations (9)–(12) all serve to express the response of a system to a general driving function $f(t)$ in terms of the experimentally accessible response of the system to a unit step function. They are often referred to collectively as **Duhamel's formulas**, after the French mathematician J. M. C. Duhamel (1797–1872).

It is possible to interpret these integrals in physical terms as follows. Let the driving function $f(t)$ be given and imagine it approximated by a series of step functions, as shown in Fig. 10.15. The first step function is of noninfinitesimal magnitude $f(0)$. All later step functions in the approximation are of infinitesimal magnitude, and their contributions in the limit will have to be taken into account by integration. Specifically, since

$$\frac{\Delta f}{\Delta\lambda} \doteq \left.\frac{df}{dt}\right|_{t=\lambda} = f'(\lambda)$$

we have for the height Δf_i of the general infinitesimal step function the approximate expression

$$\Delta f_i \doteq f'(\lambda_i)\,\Delta\lambda_i$$

† According to **Leibniz' rule**, if $F(t) = \int_{a(t)}^{b(t)} \phi(x, t)\,dx$, where a and b are differentiable functions of t and where $\phi(x, t)$ and $\partial\phi(x, t)/\partial t$ are continuous in x and t, then

$$\frac{dF}{dt} = \int_{a(t)}^{b(t)} \frac{\partial\phi(x, t)}{\partial t}\,dx + \phi[b(t), t]\frac{db(t)}{dt} - \phi[a(t), t]\frac{da(t)}{dt}$$

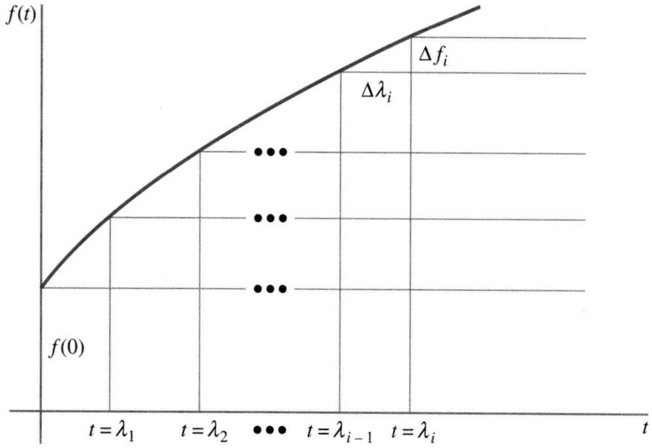

FIGURE 10.15
The synthesis of a general function by means of step functions.

Now if $A(t)$ is the indicial admittance of the system, the first step function $f(0)u(t)$ produces a response equal to

$$f(0)A(t)$$

from the very definition of the indicial admittance as the response of the system to a unit step input beginning to act at $t = 0$. For the second step function $\Delta f_1\, u(t - \lambda_1)$, there is a lag of $t = \lambda_1$ units of time before it begins to act. Hence the infinitesimal response it produces is

$$\Delta f_1[A(t - \lambda_1)] \qquad \text{or} \qquad f'(\lambda_1)\,\Delta\lambda_1[A(t - \lambda_1)]$$

Similarly, the third step function produces the response

$$f'(\lambda_2)\,\Delta\lambda_2[A(t - \lambda_2)]$$

and in general the $(i + 1)$st step function produces the response

$$f'(\lambda_i)\,\Delta\lambda_i[A(t - \lambda_i)]$$

If these contributions to the total response are added, we obtain for the response at a general time t,

$$\begin{aligned} y(t) &= f(0)A(t) + f'(\lambda_1)\,\Delta\lambda_1[A(t - \lambda_1)] + f'(\lambda_2)\,\Delta\lambda_2[A(t - \lambda_2)] + \cdots \\ &\qquad + f'(\lambda_i)\,\Delta\lambda_i[A(t - \lambda_i)] + \cdots \\ &= f(0)A(t) + \Sigma f'(\lambda_i)A(t - \lambda_i)\,\Delta\lambda_i \end{aligned}$$

the summation extending over all step function inputs which have begun to act up to the instant t. In the limit when each $\Delta\lambda$ approaches zero and the height of each step function after the first, $f(0)u(t)$, approaches zero, the sum in the last expression becomes an integral and, except for the dummy variable, we have Eq. (12).

To give a physical interpretation of Eq. (10), we must first determine the significance of the derivative of the indicial admittance $A(t)$. To do this, we will need to use the concept of a unit impulse, or delta function, which we introduced in Sec. 9.5. There, among other ways, we described the delta function $\delta(t) \equiv I(t)$, as $\lim_{a\to 0} f(t)$, where

$$(13) \qquad f(t) = \begin{cases} 0 & t < 0 \\ 1/a & 0 < t < a \\ 0 & a < t \end{cases} \qquad \text{and} \qquad \int_{-\infty}^{\infty} \delta(t)\,dt = 1$$

We also described it in terms of its very important sampling property

$$\int_{-\infty}^{\infty} \delta(t - t_0)\phi(t)\, dt = \phi(t_0) \tag{14}$$

The Fourier transform of the unit impulse, $\delta(t) = I(t)$, was determined [Eq. (13), Sec. 9.5], and related ideas pertaining to the unit doublet $D(t)$ were considered (Exercise 10, Sec. 9.5). Here, since we are now working with the Laplace transformation, we must make a similar determination of the Laplace transforms of both $\delta(t)$ and $D(t)$.

To determine $\mathcal{L}\{\delta(t)\}$, it is convenient to alter the definition of $f(t)$ from that given in Eq. (13) to

$$f(t) = \frac{u(t) - u(t-a)}{a}$$

as suggested in Fig. 10.16. Transforming this expression, we have for all $a > 0$,

$$\mathcal{L}\{f(t)\} = \frac{1}{a}\left(\frac{1}{s} - \frac{e^{-as}}{s}\right)$$

As $a \to 0$, this assumes the indeterminate form 0/0, but evaluating it as usual by l'Hôspital's rule, we obtain immediately the limiting value 1; that is,

$$\mathcal{L}\{\delta(t)\} = 1 \tag{15}$$

In the same way we can show that the Laplace transform of the unit doublet (Fig. 10.17) is

$$\mathcal{L}\{D(t)\} = s \tag{16}$$

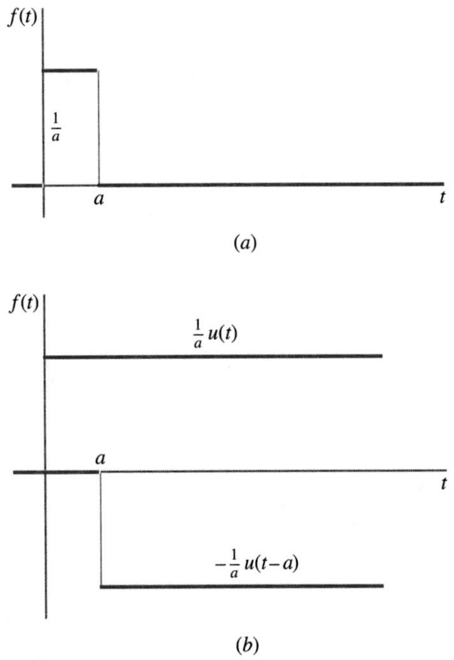

FIGURE 10.16
Plot suggesting the nature of a unit impulse.

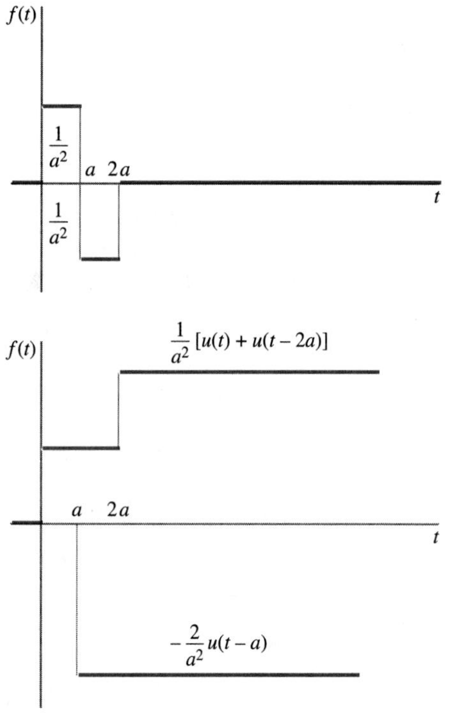

FIGURE 10.17
Plot suggesting the nature of a unit doublet.

and the transforms of the higher singularity functions follow exactly the same pattern (Exercise 14). Since these transforms do not approach zero as s becomes infinite, we know from Corollary 3, Theorem 2, Sec. 10.2, that they are not the transforms of piecewise regular functions of exponential order. This of course is obvious, for although all the singularity functions are of exponential order, they are limiting forms involving unbounded behavior in the neighborhood of the origin, hence are not piecewise regular.

We are now in a position to resume our attempt to give a physical interpretation to Eq. (10). For convenience let us denote by $h(t)$ the response of the system under discussion when the input is a unit impulse. We have already seen [Eq. (6)] that

$$\mathcal{L}\{y(t)\} = \frac{\mathcal{L}\{f(t)\}}{Z(s)}$$

Hence if $f(t)$ is a unit impulse, so that $\mathcal{L}\{f(t)\} = 1$ and $y(t) = h(t)$, we have

$$\mathcal{L}\{h(t)\} = \frac{1}{Z(s)} = s\frac{1}{sZ(s)} = s\mathcal{L}\{A(t)\}$$

Thus, from Theorem 3, Sec. 10.4, it follows that

$$h(t) = \frac{dA(t)}{dt} = A'(t)$$

or in words, *the response of a system to a unit impulse is the derivative of the response of the system to a unit step function.*

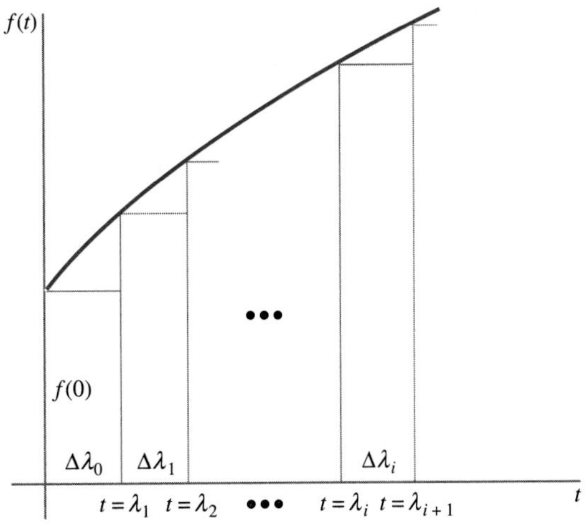

FIGURE 10.18
The synthesis of a general function by means of impulses.

Now let a general input $f(t)$ be approximated by a series of infinitesimal impulses, as suggested in Fig. 10.18. For the first impulse, whose magnitude, by definition, is the product

$$f(0)\,\Delta\lambda_0 \equiv f(\lambda_0)\,\Delta\lambda_0$$

the infinitesimal response is $[f(\lambda_0)\,\Delta\lambda_0]A'(t)$ since $A'(t) \equiv h(t)$ is the response to a unit impulse acting at $t = 0$. The second impulse $f(\lambda_1)\,\Delta\lambda_1$ does not occur until $t = \lambda_1$; hence the response it produces is $[f(\lambda_1)\,\Delta\lambda_1]A'(t - \lambda_1)$, and in general, the response produced by the $(i + 1)$st impulse is

$$[f(\lambda_i)\,\Delta\lambda_i]A'(t - \lambda_i)$$

If these contributions to the total response are added, we obtain for the response at a general time t,

$$y(t) = \Sigma f(\lambda_i)A'(t - \lambda_i)\,\Delta\lambda_i$$

the summation extending over all impulses that have acted on the system up to the time t. In the limit when each $\Delta\lambda$ approaches zero, the last sum becomes an integral, and we have Eq. (10).

EXERCISES

Using the convolution theorem, find the inverse of each of the following transforms.

1. $\dfrac{1}{(s-1)(s-2)}$. How do you think the solution using convolution compares with the solution using the first Heaviside theorem?

2. $\dfrac{1}{s^2(s^2+1)}$. How do you think the solution using convolution compares with the solution using the second Heaviside theorem? How does it compare with the solution using Theorem 4, Sec. 10.4?

3. $\dfrac{1}{(s^2+4)^2}$

4. $\dfrac{s}{(s^2+1)(s^2+4)}$

5. $\dfrac{s}{(s^2+9)^3}$. *Hint:* First apply Theorem 3, Sec. 10.4.

6. $\dfrac{s^2+4s+4}{(s^2+4s+13)^2}$. *Hint:* First apply Corollary 1, Theorem 1, Sec. 10.6.

7. $\dfrac{s^4+2s^2-s}{(s+1)(s^2+1)^2}$. *Hint*: First show that the numerator is equal to $(s^2+1)^2 - (s+1)$.

Using the appropriate singularity functions, find the inverse of each of the following transforms.

8. $\dfrac{s}{s+2}$

9. $\dfrac{s^3+2s+3}{s^2+4}$

10. $\dfrac{se^{-s}}{s+1}$

11. $\dfrac{s^2e^{-s}}{s+1}$

12. Using the convolution theorem, find a particular integral of the equation $y'' + 2ay' + a^2y = f(t)$.

13. Using the convolution theorem, find a particular integral of the equation $y'' + (a + b)y' + aby = f(t)$.

14. **(a)** Verify that the Laplace transform of the unit doublet $D(t)$ is s.
(b) The **unit triplet** T is given by $T(t) = \lim_{a\to 0} [u(t) - 3u(t - a) + 3u(t - 2a) - u(t - 3a)]/a^3$.
Verify that the Laplace transform of T is s^2.

15. According to Theorem 2, Sec. 10.6, the transform of the shifted unit impulse, $I(t - t_0)$, is e^{-st_0}. Verify this by considering $I(t - t_0)$ to be the limit as a approaches zero of the function

$$\frac{u(t - t_0) - u(t - t_0 - a)}{a}$$

and then assuming that the transform of the limit is the limit of the transform.

16. According to Theorem 2, Sec. 10.6, the transform of the shifted unit doublet, $D(t - t_0)$, is se^{-st_0}. Verify this by considering $D(t - t_0)$ to be the limit as a approaches zero of the function

$$\frac{u(t - t_0) - 2u(t - t_0 - a) + u(t - t_0 - 2a)}{a^2}$$

and then assuming that the transform of the limit is the limit of the transform.

17. Find the transform of the triangular pulse shown in Fig. 10.19*a*. What is the limit of the transform of this pulse as $a \to 0$?

18. Determine k so that the parabolic pulse shown in Fig. 10.19*b* will have unit area. What is the transform of this pulse? What is the limit of the transform of this pulse as $a \to 0$?

What is the solution of each of the following integral equations in each of the indicated cases?

19. $y(t) = te^{-t} + k \int_0^t y(t - \lambda)\lambda e^{-\lambda}\, d\lambda$
(a) $k = 1$ **(b)** $k = -1$
(c) $k = 4$ **(d)** $k = -4$

20. $y(t) = t + k \int_0^t y(t - \lambda)e^{-\lambda}\, d\lambda$
(a) $k = 1$ **(b)** $k = -1$ **(c)** $k = 2$

21. $y(t) = e^{-at} + k \int_0^t y(t - \lambda)\lambda e^{-a\lambda}\, d\lambda$
(a) $a = 1, k = 4$ **(b)** $a = 1, k = -4$
(c) $a = 4, k = 1$ **(d)** $a = 4, k = -1$

22. $y(t) = \sinh t + k \int_0^t y(t - \lambda) \sin \lambda\, d\lambda$
(a) $k = 5$ **(b)** $k = -3$

23. $y(t) = t^2 + k \int_0^t y(t - \lambda)\lambda\, d\lambda$
(a) $k = 1$ **(b)** $k = -1$

24. $y(t) = 1 + k \int_0^t y(t - \lambda) \cos a\lambda\, d\lambda$
(a) $a = 2, k = 5$ **(b)** $a = 2, k = -4$
(c) $a = 5, k = 10$ **(d)** $a = 5, k = -8$

25. At $t = 0$, while it is completely passive, a system governed by the differential equation $y'' + 2y' + y = 0$ is activated by a unit impulse $I(t)$. Determine its subsequent behavior. Is it possible to reduce the system to a completely passive state by the application of a suitable impulse $kI(t - t_0)$ at a later time $t = t_0$? Can the system be reduced to a completely passive state by the sudden application of a suitable impulse and a suitable doublet at a later time $t = t_0$?

26. Work Exercise 25 if the differential equation governing the behavior of the system is
(a) $y'' + 3y' + 2y = 0$ **(b)** $y'' + 2y' + 2y = 0$

27. Show that the Laplace transform of the function $f(t) = \sum_{n=0}^{\infty} I(t - nt_0)$ is $\phi(s) = 1/(1 - e^{-t_0 s})$. Discuss this result in the light of the theory of Sec. 10.10.

28. Work Exercise 27 if $f(t) = \sum_{n=0}^{\infty} (-1)^n I(t - nt_0)$.

29. Find $A(t)$ and $h(t)$ for the equation $y'' + 3y' + 2y = 0$. Verify that $h(t) = A'(t)$ and then verify Eqs. (10) and (12) when this equation is driven by the function e^t.

30. **(a)** In Exercise 1, Sec. 7.5, find $A(t)$ and $h(t)$ if the input is applied to M_1 and if the output is the response, i.e., the displacement, of M_2. Verify that $h(t) = A'(t)$. What is the response of M_2 to an arbitrary force applied to M_1 when the system is in equilibrium?
(b) Work Part **(a)** if the input is applied to M_2 and the output is the response of M_1.

FIGURE 10.19

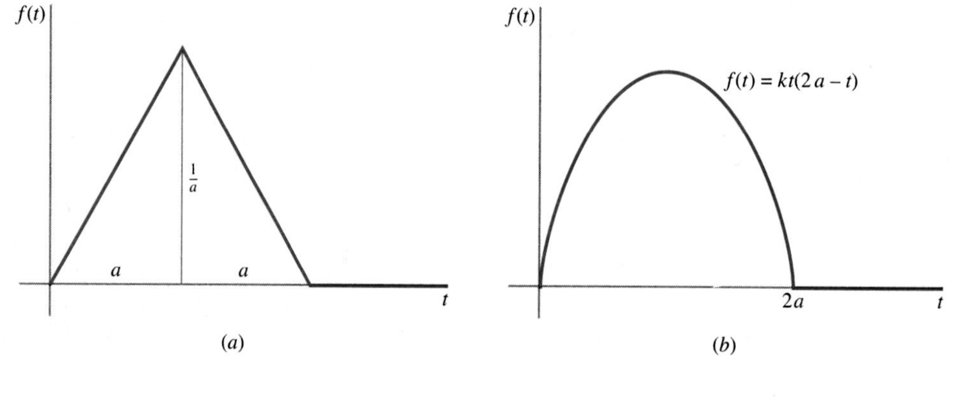

31. Find $A(t)$ and $h(t)$ for the system shown in Fig. 10.20 if the input voltage is applied across the indicated terminals and if the output is the current through R_2. Verify that $h(t) = A'(t)$. What is the current through R_2 due to an arbitrary voltage $E(t)$ across the terminals when all charges and currents in the system are zero?

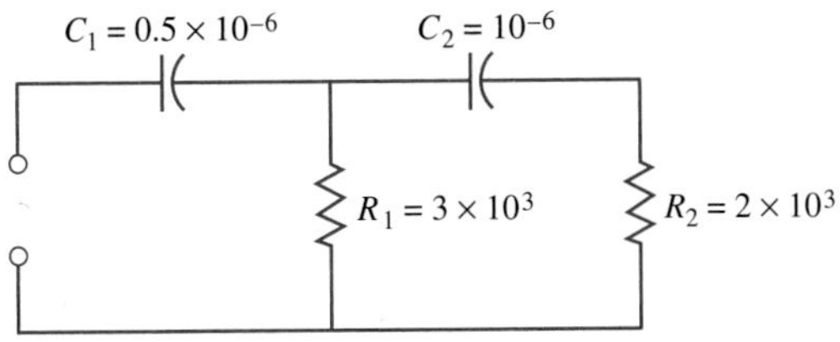

FIGURE 10.20

32. Show that the solution of the equation $ay'' + by' + cy = 0$ when $y_0 = 0$ and $y_0' = 1$ is exactly the same as the solution of the equation $ay'' + by' + cy = aI(t)$ when $y_0 = y_0' = 0$. Does this fact have a physical interpretation? With what combination of singularity functions must an initially passive second-order equation be driven in order to have the same solution as the undriven equation with initial conditions $y(0) = y_0$ and $y'(0) = y_0'$?

33. **(a)** Show that

$$\mathcal{L}^{-1}\left\{\frac{1}{\sqrt{s}(s-1)}\right\} = \frac{2e^t}{\sqrt{\pi}}\int_0^{\sqrt{t}} e^{-\tau^2}\,d\tau$$

Hint: Use the convolution theorem and then in the resulting integral let $\sqrt{\lambda} = \tau$.

(b) What is $\mathcal{L}^{-1}\{1/(s\sqrt{s+1})\}$?

34. What is $\mathcal{L}^{-1}\{1/(s + \sqrt{s})\}$? *Hint:* First rationalize the denominator, then use the result of Exercise 33(**a**).

35. **(a)** What is $\mathcal{L}^{-1}\{1/[\sqrt{s}(s^2+1)]\}$?

(b) What is $\mathcal{L}^{-1}\{s/[\sqrt{s}(s^2+1)]\}$?

36. Show that

$$f(t) * [g(t) \pm h(t)] = [f(t) * g(t)] \pm [f(t) * h(t)]$$

37. Show that

$$f(t) * [g(t) * h(t)] = \int_0^t \int_0^\lambda f(t-\lambda)g(\lambda-\mu)h(\mu)\,d\mu\,d\lambda$$

38. Show that $f(t) * [g(t) * h(t)] = [f(t) * g(t)] * h(t)$.

39. Show that $\mathcal{L}\{f(t)\}\mathcal{L}\{g(t)\}\mathcal{L}\{h(t)\} = \mathcal{L}\{f(t) * g(t) * h(t)\}$.

40. Show that $1 * 1 = t$ and that $1 * 1 * 1 = t^2/2$. What is the generalization of these results to n factors?

41. What is **(a)** $\delta(t-a) * f(t)$ and **(b)** $u(t-a) * f(t)$?

42. If $f(0) = g(0) = 0$, show that $f'(t) * g(t) = f(t) * g'(t)$ and that

$$[f(t) * g(t)]' = \frac{f'(t) * g(t) + f(t) * g'(t)}{2}$$

43. If $f(t)$ and $g(t)$ are given functions, is it possible to solve the equation $f(t) * x(t) = g(t)$ for $x(t)$?

44. What is $t^m * t^n$, m, n nonnegative integers?

Solve each of the following equations.

45. $t^3 = \int_0^t (t-x)^2 y'(x)\,dx$; $y(0) = 0$

46. $2t^{7/2} = \int_0^t (t-x)^{5/2} y'(x)\,dx + 5\pi t^3$; $y(0) = 0$

47. Abel's integral equation $f(t) = \int_0^t (t-\tau)^{-b} y'(\tau)\,d\tau$; $0 < b < 1$

48. $y = 3t + \int_0^t \sin(t-u)y(u)\,du$

49. $512t^{9/2} = 10{,}080 \int_0^t \int_0^\tau (\tau - x)^{3/2} y'(x)\,dx\,d\tau + 945\pi t^4$; $y(0) = 0$

50. Show that

(a) $\mathcal{L}\left\{\frac{f(t)}{t}\right\} = \int_0^\infty \phi(s)\,ds - 1 * \phi(s)$

Hint: $\int_s^\infty = \int_0^\infty - \int_0^s$.

(b) $\mathcal{L}\left\{\frac{f(t)}{t^2}\right\} = s * \phi(s) - \int_0^\infty (s-\tau)f(\tau)\,d\tau$. *Hint:* Express the Laplace transform as a double integral and reverse the order of integration.

10.12 THE $\mathcal{L}$, $\mathcal{F}$, AND Z TRANSFORMATIONS COMPARED

The Laplace transformation, as we have learned in this chapter and Sec. 9.6, is defined by the pair of formulas

(1) $$\mathcal{L}\{f(t)\} \equiv \phi(s) = \int_0^\infty f(t)e^{-st}\,dt$$

(2) $$f(t) = \mathcal{L}^{-1}\{\phi(s)\} = \frac{1}{2\pi i}\int_{a-i\infty}^{a+i\infty} \phi(s)e^{st}\,ds$$

Except for the singularity functions, the functions to which the Laplace transformation can be applied must be zero for $t < 0$ and on $[0, \infty)$ must be piecewise regular and of exponential order. The parameter s in the Laplace transform of a function appears to be a real variable of no particular significance. However, as we learned in Sec. 9.6, s is actually a complex number of the form

$a + i\omega$, where a can be any real number greater than the abscissa of convergence of $f(t)$ and ω is a frequency. If the abscissa of convergence of $f(t)$ is negative, a may be set equal to zero, and in this case the Laplace transform defines the spectrum of $f(t)$, that is, $\phi(s) \equiv \phi(i\omega)$. The inverse transform (2) involves line integration in the complex plane and is of little use in routine applications. In such cases, $f(t)$ is usually recovered from its transform by referring to a table of transforms and theorems relating to transforms.

The Fourier transformation, as we learned in Sec. 9.3, is defined in its general complex form by the pair of formulas

$$\mathcal{F}[f(t)] \equiv F(\omega) = \int_{-\infty}^{\infty} f(t)e^{-i\omega t}\,dt \tag{3}$$

$$f(t) = \mathcal{F}^{-1}[F(\omega)] = \frac{1}{2\pi}\int_{-\infty}^{\infty} F(\omega)e^{i\omega t}\,d\omega \tag{4}$$

Except for singularity functions, the functions to which the Fourier transformation can be applied must satisfy the Dirichlet conditions on every finite interval and must also be absolutely integrable, that is,

$$\int_{-\infty}^{\infty} |f(t)|\,dt$$

must exist. The parameter ω in the Fourier transform of a function $f(t)$ is a frequency, and the Fourier transform itself defines the spectrum of $f(t)$. For functions that have both a Fourier transform and a Laplace transform, the two transforms are identical under the substitution $s \leftrightarrow i\omega$. Unlike the inversion formula (2) for Laplace transforms, the inversion formula (4) is frequently used, in addition to tables of transforms, to recover $f(t)$ from its Fourier transform.

The Z transformation associates a sequence S_n with its generating function, and therefore only sequences, not continuous functions, have Z transforms. In Sec. 5.9, we investigated this transformation, first expressed in terms of the parameter $s = z^{-1}$, as an alternative method of solving difference equations. However, for sequences obtained by sampling continuous signals at regular intervals, the Z transform is just the Laplace transform of the sequence resulting from the sampling. To see this, consider the sequence of values $f(0), f(T), f(2T), \ldots, f(nT), \ldots$, obtained by sampling $f(t)$ at $t = 0, T, 2T, \ldots, nT, \ldots$. The generating function of this sequence, that is, its Z transform, is by definition,

$$f(0) + f(T)z^{-1} + f(2T)z^{-2} + \cdots + f(nT)z^{-n} + \cdots \tag{5}$$

On the other hand, if we consider the sum

$$\sum_{n=0}^{\infty} f(nT)\,\delta(t - nT)$$

and take its Laplace transform term by term, noting that in each term $f(nT)$ is a constant, we obtain, using the formula $\mathcal{L}\{\delta(t - nT)\} = e^{-nTs}$ of Exercise 15, Sec. 10.11,

$$\sum_{n=0}^{\infty} f(nT)e^{-nsT}$$

which is identical with (5) under the substitution $z = e^{sT}$.

EXAMPLE 1

What is the Z transform of $f(t) = e^{-at}$, sampled at intervals of $\Delta t = T$, beginning at $t = 0$?

In terms of z, the generating function of the sequence

$$1, e^{-aT}, e^{-2aT}, \ldots, e^{-naT}, \ldots$$

is

$$1 + e^{-aT}z^{-1} + e^{-2aT}z^{-2} + \cdots + e^{-naT}z^{-n} + \cdots = \frac{1}{1 - e^{-aT}z^{-1}} = \frac{z}{z - e^{-aT}}$$

On the other hand, the Laplace transform of the sequence

$$\sum_{n=0}^{\infty} e^{-anT}\,\delta(t - nT)$$

is

$$\sum_{n=0}^{\infty} e^{-anT}e^{-nsT} = \sum_{n=0}^{\infty} e^{-n(a+s)T} = \frac{1}{1 - e^{-(a+s)T}} = \frac{e^{sT}}{e^{sT} - e^{-aT}}$$

which is just

$$\frac{z}{z - e^{-aT}} \qquad \text{with } z = e^{sT}$$

There are several ways of finding the inverse of a Z transform. In many cases, the Z transform of a sequence will be a rational fractional function, $p(z)/q(z)$. If this fraction is converted into a power series in z^{-1} by long division or the binomial theorem, then, by definition, the coefficients of the successive powers of z^{-1} will be the corresponding terms in the required sequence. Alternatively, $p(z)/q(z)$ can be converted into simple fractions whose inverses can be found from a table of Z transforms or by long division.

EXAMPLE 2

What is the sequence whose Z transform is $z/(z^2 + z - 6)$?

Using the method of partial fractions, it is easy to show that

$$\frac{z}{z^2 + z - 6} = \frac{1}{5}\left(\frac{3}{z + 3} + \frac{2}{z - 2}\right) = \frac{1}{5z}\left(\frac{3}{1 + 3z^{-1}} + \frac{2}{1 - 2z^{-1}}\right)$$

Hence

$$\begin{aligned}\frac{z}{z^2 + z - 6} &= \frac{3}{5z}\left(1 - 3z^{-1} + 3^2z^{-2} - 3^3z^{-3} + \cdots\right)\\ &\quad + \frac{2}{5z}(1 + 2z^{-1} + 2^2z^{-2} + 2^3z^{-3} + \cdots)\\ &= \left(\frac{2 + 3}{5}\right)\frac{1}{z} + \left(\frac{2^2 - 3^2}{5}\right)\frac{1}{z^2} + \left(\frac{2^3 - 3^3}{5}\right)\frac{1}{z^3} + \cdots\end{aligned}$$

The required sequence is therefore

$$u_n = \frac{2^n - (-3)^n}{5} \qquad n = 1, 2, 3, \ldots$$

When the Z transform of a sequence is not of the form $p(z)/q(z)$ and the inverse cannot be found from tables, there is a formula analogous to Eq. (2), but its derivation and applications are beyond the scope of our discussion. It can be found in B. C. Kuo, *Digital Control Systems,* SRL Publishing Co., Champaign, Ill., 1977.

An extended discussion of the relations between the Laplace, Fourier, and Z transformations and their application to signal processing can be found in the book *Signal Processing Techniques* by Russ Roberts, Interstate Electronics Corporation, Anaheim, Calif., 1979. These three transformations are by no means the only ones of importance to an applied scientist. A brief description of 10 such transformations can be found in the article ''Numerical Transforms,'' R. N. Bracewell, *Science,* vol. 248, no. 4956, May 11, 1990, pp. 697–704.

EXERCISES

1. If $Z(z) = u_0 + u_1\frac{1}{z} + u_2\frac{1}{z^2} + \cdots$ is the Z transform of the sequence $u_0, u_1, u_2, \ldots$, what is the Z transform of the function $f(t) = e^{-at}$, sampled at intervals of $\Delta t = T$, beginning at **(a)** $t = 0$ and **(b)** $t = T$?

2. **(a)** Show that the Z transform of the sequence obtained by sampling $f(t) = \sin \omega t$ at intervals of $\Delta t = T$, beginning at $t = 0$, is

$$Z(z) = \frac{z \sin \omega T}{z^2 - 2z \cos \omega T + 1}$$

(b) Verify the formula in Part **(a)** by showing that it reduces to the sequences obtained when $\omega T = \pi, \pi/2, \pi/3$, and $\pi/4$.

3. What is the Z transform of the function $f(t) = \cos \omega t$, sampled at intervals of $\Delta t = T$, beginning at $t = 0$?

4. What is the Z transform of the function $f(t) = t$, sampled at intervals of $\Delta t = T$, beginning at $t = 0$?

5. Work Exercise 4 if $f(t) = t^2$.

6. What is the sequence whose Z transform is $1/(z - 4)$?

7. Work Exercise 6 if $Z(z) = 1/(z^2 - 4)$.

8. **(a)** What is the spectrum of the solution of the equation $y'' + 3y' + 2y = 0$ for which $y_0 = 1$ and $y_0' = -2$?

(b) Write the spectrum in the form $r(\omega)e^{i\phi(\omega)}$ and plot r and ϕ as functions of ω.

9. Work Exercise 8 for the differential equation $y'' + 2y' + y = 0$ with initial conditions $y_0 = 0$, $y_0' = 1$.

10. Can we speak of the spectrum of the solution of the differential equation $y'' + y' - 6y = 0$ for which **(a)** $y_0 = 1$, $y_0' = -3$ and **(b)** $y_0 = -3$, $y_0' = 1$?

CHAPTER 11

PARTIAL DIFFERENTIAL EQUATIONS

An engineer's contact with partial differential equations usually comes in the study of *boundary-value problems,* that is, problems in which a quantity, such as temperature or displacement, is required

(a) To satisfy some partial differential equation at every point of a region R,

(b) To satisfy given conditions at $t = 0$ at every point of R, and

(c) To satisfy permanent conditions given at every point of the boundary of R.

In Sec. 11.2 we derive carefully the partial differential equations we will learn to solve later in this chapter. These deal primarily with heat flow and with vibrating systems such as strings, bars, and beams. Then, following a descriptive study of the *wave equation* in Sec. 11.4, we develop in Sec. 11.5 the general method of solving linear partial differential equations known as the *separation of variables.* This involves reducing a partial differential equation to two or more ordinary differential equations which can then be solved by the methods of Chap. 2. In attempting to make the solutions of one or more of these ordinary differential equations fit the *initial conditions* or the *boundary conditions* of the problem, we will find in our early examples that Fourier series often play an essential role through their ability to represent an arbitrary function over an arbitrary finite interval (Sec. 8.4).

In other problems, however, Fourier series prove inadequate, and we are led to more general functions known as *orthogonal functions,* which include Fourier series as a very important but very special subclass. After investigating the fundamental properties of such functions (Sec. 11.6), we illustrate their use in Sec. 11.7 in a variety of applications in heat flow, vibrating strings and beams, and transmission line problems.

Later, in Sec. 11.8, we discuss the use of Laplace transforms in solving partial differential equations. Finally, in Sec. 11.9 we discuss methods for the numerical solution of boundary-value problems involving partial differential equations.

Prerequisites for this chapter: multivariable calculus, Chaps. 2 and 8, a calculus-based physics course.

Prerequisite for Sec. 11.8: Chap. 10.

Prerequisite for Sec. 11.9: Secs. 5.2–5.4 and 5.6.

11.1 INTRODUCTION

In our previous work, most notably in Chap. 7, we have seen how the analysis of mechanical and electrical systems containing lumped parameters often leads to ordinary differential equations in which the time t is the (only) independent variable. However, the assumption that all masses exist as conceptualized mass points; that all springs are weightless; or that elements of an electric circuit are concentrated in ideal resistors, capacitors, and inductors, rather than continuously distributed, is frequently not sufficiently accurate. In such cases, a more realistic approach must take into account the fact that the dependent variables depend not only on t but also on one or more space variables. Because there is more than one independent variable, the formulation of such problems leads to partial, rather than ordinary, differential equations. In this chapter we shall discuss such equations as they commonly arise in applied mathematics. We shall begin by examining in some detail the derivation from physical principles of a number of important partial differential equations. Then, knowing the forms of most common occurrence, we shall investigate methods of solution and their application to specific problems.

11.2 THE DERIVATION OF EQUATIONS

One of the first physical problems to be attacked through the use of partial differential equations was that of the vibration of a stretched, flexible string. Today, after nearly 250 years, it is still an excellent initial example.†

Let us consider, then, an elastic string stretched under a tension T between two points on the x axis (Fig. 11.1*a*). The weight of the string per unit length after it is stretched, but before it is displaced, we suppose to be a known function $w(x)$. Besides the elastic forces inherent in the system, the string may also be acted upon by a distributed transverse load whose magnitude per unit length we assume to be a known function of x, y, t, and the transverse velocity $\dot{y}$, say $|f(x, y, \dot{y}, t|$. In formulating the problem we assume that

(a) The motion takes place entirely in one plane, and in this plane each particle moves at right angles to the equilibrium position of the string.

(b) The deflection of the string during the motion is so small that the resulting change in length of the string has no appreciable effect on the tension T.

(c) The string is perfectly flexible, i.e., can transmit force only in the direction of its length.

(d) The slope of the deflection curve of the string is at all points and at all times so small that with satisfactory accuracy $\sin\alpha$ can be replaced by $\tan\alpha$, where α is the inclination angle of the tangent to the deflection curve.

FIGURE 11.1
A typical element of a vibrating string.

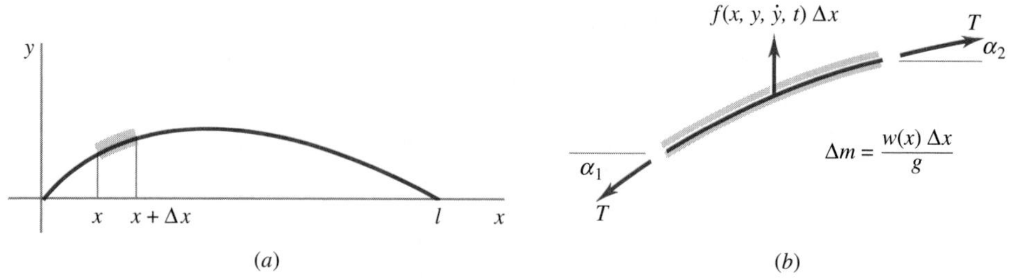

†We have already considered one aspect of this problem in Sec. 2.12. There, through the use of Green's function, we were able to determine the natural frequencies and the corresponding deflection curves of a uniform string vibrating transversely. Now, through the use of partial differential equations, we are able to make an even more detailed study of vibrating strings.

Gravitational and frictional forces, if any, we suppose to be taken into account in the expression for the load per unit length $f(x, y, \dot{y}, t)$.

With these assumptions in mind, let us consider a general infinitesimal segment of the string as a free body (Fig. 11.1*b*). By Assumption **(a)**, the mass of such an element is $\Delta m = w(x)\,\Delta x/g$. By Assumption **(b)**, the forces which act at the ends of the element are the same, namely, T. By Assumption **(c)**, these forces are directed along the respective tangents to the deflection curve; and, by Assumption **(d)**, their transverse components are

$$T \sin \alpha_2 = T \sin \alpha\big|_{x+\Delta x} \doteq T \tan \alpha\big|_{x+\Delta x}$$

and

$$T \sin \alpha_1 = T \sin \alpha\big|_{x} \doteq T \tan \alpha\big|_{x}$$

The acceleration produced in Δm by these forces and by the portion of the distributed load $f(x, y, \dot{y}, t)\,\Delta x$ which acts over the interval Δx, whether $f(x, y, \dot{y}, t)$ is positive (as indicated in Fig. 11.1*b*) or negative, is approximately $\partial^2 y/\partial t^2$, where y is the ordinate of an arbitrary point of the element. The time derivative is here written as a partial derivative because obviously y depends not only upon t but upon x as well. Applying Newton's law to the element, we can thus write

$$\text{(1)} \qquad \frac{w(x)\,\Delta x}{g}\frac{\partial^2 y}{\partial t^2} = T \tan \alpha\big|_{x+\Delta x} - T \tan \alpha\big|_{x} + f(x, y, \dot{y}, t)\,\Delta x$$

or, dividing by Δx,

$$\frac{w(x)}{g}\frac{\partial^2 y}{\partial t^2} = T\left(\frac{\tan \alpha\big|_{x+\Delta x} - \tan \alpha\big|_{x}}{\Delta x}\right) + f(x, y, \dot{y}, t)$$

The fraction on the right consists of the difference between the values of $\tan \alpha$ at $x + \Delta x$ and at x, divided by the difference Δx. In other words, it is precisely the difference quotient for the function $\tan \alpha$. Hence its limit as $\Delta x \to 0$ is the derivative of $\tan \alpha$ with respect to x, that is, $(\partial \tan \alpha)/\partial x$. But since $\tan \alpha = \partial y/\partial x$, this can be written simply as $\partial^2 y/\partial x^2$. Our final result, then, is that the deflection $y(x, t)$ of a stretched string satisfies the partial differential equation†

$$\text{(2)} \qquad \frac{\partial^2 y}{\partial t^2} = \frac{Tg}{w(x)}\frac{\partial^2 y}{\partial x^2} + \frac{g}{w(x)} f(x, y, \dot{y}, t)$$

In most applications the weight of the string per unit length $w(x)$ is a constant and, in comparison with the tension T, the external forces are negligible; i.e., with satisfactory accuracy $f(x, y, \dot{y}, t)$ can be assumed to be identically zero. When this is the case, Eq. (2) reduces to the **one-dimensional wave equation**

$$\text{(3)} \qquad \frac{\partial^2 y}{\partial t^2} = a^2 \frac{\partial^2 y}{\partial x^2} \qquad a^2 = \frac{Tg}{w}$$

†The question of what constitutes a satisfactory derivation of the partial differential equation describing a given physical system is not a simple one. To attempt to give a careful limiting argument is, in effect, ''to strain at a gnat and swallow a camel'' since, being ultimately atomic, no physical system is continuous. Perhaps our purported derivations should be regarded merely as plausibility arguments suggesting that certain partial differential equations be accepted as the axioms of a theoretical or ''rational'' study of applied mathematics whose practical importance, in contrast to its purely mathematical interest, is to be judged by how well its conclusions describe past observations and predict new ones.

The dimensions of a^2 are

$$\frac{\text{Force} \times \text{acceleration}}{\text{Weight/unit length}} = \frac{(ML/T^2)(L/T^2)}{(ML/T^2)(1/L)} = \frac{L^2}{T^2}$$

that is, a has the dimensions of velocity. The significance of this will become apparent in Sec. 11.3 when we discuss the *d'Alembert solution* of the wave equation.

The problem of the vibrating string generalizes naturally to the problem of the transverse vibration of a flexible membrane. With z as the displacement, the analysis, which is called for in Exercise 31, leads to the equation

$$\frac{\partial^2 z}{\partial t^2} = \frac{Tg}{w(x, y)}\left(\frac{\partial^2 z}{\partial x^2} + \frac{\partial^2 z}{\partial y^2}\right) + \frac{g}{w(x, y)} f(x, y, z, \dot{z}, t) \tag{4}$$

where T is the tension per unit length, assumed to be constant throughout the membrane, w is the weight per unit area of the membrane, and $f(x, y, z, \dot{z}, t)$ is the load per unit area due to whatever external or frictional forces may act on the membrane.

If the weight per unit area of the membrane is the same at all points and if there are no external or frictional forces, i.e., if $w(x, y)$ is a constant and if $f(x, y, z, \dot{z}, t) \equiv 0$, then Eq. (4) reduces to the **two-dimensional wave equation**

$$\frac{\partial^2 z}{\partial t^2} = a^2\left(\frac{\partial^2 z}{\partial x^2} + \frac{\partial^2 z}{\partial y^2}\right) \qquad a^2 = \frac{Tg}{w} \tag{5a}$$

Here, as in the one-dimensional wave equation for the vibrations of a uniform string, the parameter a has the dimensions of velocity.

Similarly, the analysis of acoustical or electrical vibrations in three-dimensional regions leads to the **three-dimensional wave equation**

$$\frac{\partial^2 \phi}{\partial t^2} = a^2\left(\frac{\partial^2 \phi}{\partial x^2} + \frac{\partial^2 \phi}{\partial y^2} + \frac{\partial^2 \phi}{\partial z^2}\right) \tag{5b}$$

Here again, the parameter a has the dimensions of velocity.

As another problem leading to a partial differential equation, let us consider a shaft vibrating torsionally (Fig. 11.2a). The material of the shaft we assume to have a modulus of elasticity in shear E_s and to be of uniform weight per unit volume ρ. The cross-sectional area of the shaft at a distance x from one end of the shaft we suppose to be a known function of x, say $A(x)$. The polar moment of inertia $J(x)$ of a general cross section about its center of gravity we also suppose to be

FIGURE 11.2
A typical element of a vibrating shaft.

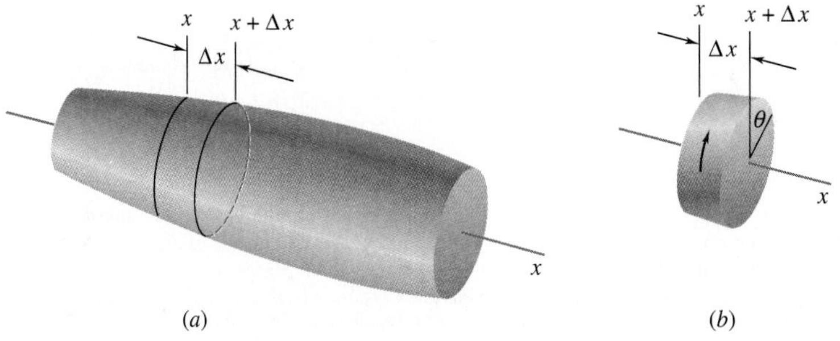

known. In addition to the internal elastic torques, the shaft may also be acted upon by a distributed external torque whose magnitude per unit length is a known function, say $|f(x, \theta, \dot{\theta}, t)|$, where θ is the angle through which a general cross section has rotated from its equilibrium position and $\dot{\theta}$ is the angular velocity with which that cross section rotates while the shaft is vibrating. We assume further that

(a) All cross sections of the shaft remain plane during rotation.

(b) Each cross section rotates about its center of gravity.

(c) The shape of a general cross section does not depart greatly from a circle.

Frictional torques, if any, we suppose to be taken into account in the expression for the distributed torque per unit length $f(x, \theta, \dot{\theta}, t)$.

We begin by considering as a free body an infinitesimal segment of the shaft bounded by two cross sections a distance Δx apart (Fig. 11.2*b*). The mass of such a disk is approximately

$$\Delta m = \frac{\rho A(x)\,\Delta x}{g}$$

and, by definition, the radius of gyration of the cross section at x is

$$k = \sqrt{\frac{J(x)}{A(x)}}$$

Hence the polar moment of inertia of the infinitesimal disk is

$$\Delta I = k^2\,\Delta m = \frac{J(x)}{A(x)} \times \frac{\rho A(x)\,\Delta x}{g} = \frac{J(x)\rho\,\Delta x}{g}$$

The rotation of such an element is produced by the portion of the distributed torque $f(x, \theta, \dot{\theta}, t)\,\Delta x$ which acts on it and by the elastic torques T transmitted to it through its end sections by the adjacent portions of the shaft. Therefore, applying Newton's law in torsional form, we have

$$\frac{J(x)\rho\,\Delta x}{g} \times \frac{\partial^2\theta}{\partial t^2} = T|_{x+\Delta x,t} - T|_{x,t} + f(x, \theta, \dot{\theta}, t)\,\Delta x$$

or, dividing by Δx and then letting $\Delta x \to 0$,

$$\frac{J(x)\rho}{g} \times \frac{\partial^2\theta}{\partial t^2} = \frac{\partial T}{\partial x} + f(x, \theta, \dot{\theta}, t) \tag{6}$$

Now from the strength of materials, it is known that the torque transmitted through any cross section of a twisted shaft is proportional to the twist per unit length, i.e., the slope of the (θ, x) curve, at that cross section:

$$T = \tau \frac{\partial\theta}{\partial x}$$

The proportionality factor τ is known as the **torsional rigidity.** For shafts that are solids of revolution it can be shown that

$$\tau = E_s J(x)$$

and this result can be used with satisfactory accuracy whenever the cross sections of a shaft are approximately circular. Hence in such cases Eq. (6) becomes

$$\frac{J(x)\rho}{g}\frac{\partial^2\theta}{\partial t^2} = \frac{\partial[E_s J(x)\,\partial\theta/\partial x]}{\partial x} + f(x, \theta, \dot{\theta}, t) \tag{7}$$

However, for configurations whose cross sections differ appreciably from circles, e.g., propeller blades or aircraft wings, it is necessary to determine the torsional rigidity τ by experimental means and continue the solution of Eq. (6) by numerical rather than analytical methods.

In most elementary applications the shafts are of uniform circular cross section, and there are no external distributed torques. In such cases $J(x)$ is a constant, $f(x, \theta, \dot{\theta}, t)$ is identically zero, and Eq. (7) reduces to

$$\frac{\partial^2\theta}{\partial t^2} = a^2\frac{\partial^2\theta}{\partial x^2} \qquad a^2 = \frac{E_s g}{\rho} \tag{8}$$

which is, again, just the one-dimensional wave equation.

Another vibration problem of considerable practical interest concerns the transverse vibrations of a beam. To obtain the partial differential equation describing these vibrations, let us first choose a coordinate system such that a neutral fiber of the beam in its undeflected position coincides with a portion of the x axis and the deflections occur in the direction of the y axis. A general cross section of the beam we assume to be of known area $A(x)$ and known moment of inertia $I(x)$ about its neutral axis (see p. 165). The material of the beam we suppose to be of weight per unit volume ρ and modulus of elasticity E. In addition to internal forces, the beam may also be acted upon by a distributed load of known intensity $f(x, y, \dot{y}, t)$. Gravitational forces and frictional forces, if any, we suppose to be included in this distributed load. Finally, we assume that all particles of the beam move in a purely transverse direction, i.e., that the slight rotation of the cross sections as the beam vibrates is negligible.

During vibration all external forces acting on a segment of the beam of length Δx, having the beam cross sections at x and $x + \Delta x$ as interior faces, are given by

$$-f(x, y, \dot{y}, t)\,\Delta x$$

the minus sign being necessary because we have taken loads to be positive downward (Sec. 2.11) but must take forces to be positive upward in applying Newton's second law with the y axis directed upward. Since all motion is assumed to be transverse, only the two shearing forces acting across the interior faces of the segment need be considered. These forces have also been taken positive downward, so they too must be accompanied by a sign change in applying Newton's law. Thus the internal forces are

$$-V(x + \Delta x, t) + V(x, t) = -[V(x + \Delta x, t) - V(x, t)]$$

For the mass of the segment we have $\rho A(x)\,\Delta x/g$, and Newton's law gives

$$\frac{\rho A(x)\,\Delta x}{g}\frac{\partial^2 y}{\partial t^2} = -[V(x + \Delta x, t) - V(x, t)] - f(x, y, \dot{y}, t)\,\Delta x$$

Dividing by Δx and letting $\Delta x \to 0$, we obtain

$$\frac{\rho A(x)}{g}\frac{\partial^2 y}{\partial t^2} = -\frac{\partial V}{\partial x} - f(x, y, \dot{y}, t) \tag{9}$$

Now from the discussion in Sec. 2.11 it is clear that

$$\frac{\partial V}{\partial x} = \frac{\partial^2[EI(x)\,\partial^2 y/\partial x^2]}{\partial x^2}$$

Substituting this expression into (9) and rearranging, we have finally

$$\frac{\partial^2[EI(x)\,\partial^2 y/\partial x^2]}{\partial x^2} = -\frac{\rho}{g}A(x)\frac{\partial^2 y}{\partial t^2} - f(x, y, \dot{y}, t) \tag{10}$$

In many important applications the beam under consideration is of constant cross section, and there is no external load; that is, A and I are constants and $f(x, y, \dot{y}, t) \equiv 0$. Under these conditions Eq. (10) reduces to the simpler form

$$(11) \qquad a^2 \frac{\partial^4 y}{\partial x^4} = -\frac{\partial^2 y}{\partial t^2} \qquad a^2 = \frac{EIg}{A\rho}$$

In this case the parameter a does *not* have the dimensions of velocity.

We encounter an entirely different class of problems leading to partial differential equations when we undertake the study of the flow of heat in thermally conducting regions. The region may of course be one-, two-, or three-dimensional, and the governing differential equation will vary accordingly. The variation, however, is just what we would expect by analogy with the one-, two-, and three-dimensional wave equations. Hence the derivation of the equation for one-dimensional heat flow, which can readily be extended to the other two cases, should be sufficient for our purposes [see Exercises 25 and 26].

Let us consider, then, the thermal conditions in a long thin rod of constant cross section whose surface, except for the ends, is insulated against the flow of heat. If there is heat flow in the rod, the temperature will vary along the rod, but the rod is assumed to be so thin that over each cross section the temperature is constant. We base our analysis on the following experimental laws:

(a) Heat flows in the direction of decreasing temperature.

(b) The rate at which heat flows through an area is proportional to the area and to the temperature gradient, in degrees per unit distance, in the direction perpendicular to the area.

(c) The quantity of heat gained or lost by a body when its temperature changes, i.e., the change in heat content or thermal energy, is proportional to the mass of the body and to the temperature change.

The proportionality constant in **b** is called the **thermal conductivity** of the material k. The proportionality constant in **c** is called the **specific heat** of the material c.

With the rod located along the x axis as shown in Fig. 11.3, let us consider the thermal conditions in the infinitesimal segment of the rod between x and $x + \Delta x$. If A is the cross sectional area of the rod and ρ is the weight per unit volume of the material, then the mass of the segment is

$$\Delta m = \frac{\rho}{g} \Delta V = \frac{\rho A \, \Delta x}{g}$$

Furthermore, if $u = u(x, t)$ is the temperature at any cross section x at any time t and if Δu is the average temperature rise in the infinitesimal segment in the infinitesimal time interval Δt, then by **c** the quantity of heat stored in this segment during this time is

$$\Delta H = c \, \Delta m \, \Delta u = \frac{c\rho}{g} A \, \Delta x \, \Delta u$$

and the rate at which heat is being stored in the segment is approximately

$$(12) \qquad \frac{\Delta H}{\Delta t} = \frac{c\rho}{g} A \, \Delta x \frac{\Delta u}{\Delta t}$$

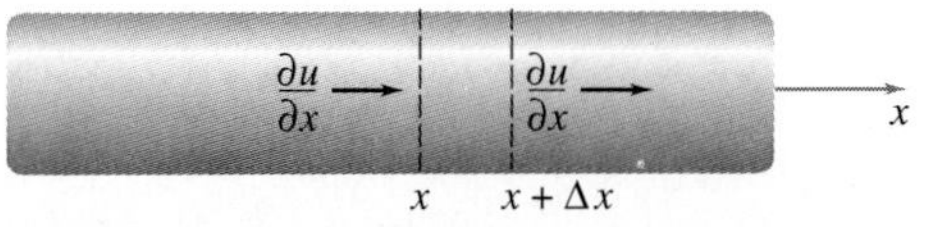

FIGURE 11.3
A typical element in one-dimensional heat flow in a thin rod.

The heat that produces the temperature change Δu comes from two sources. First, heat may be generated throughout the body, by electrical or chemical means, for instance, at a known rate per unit volume, say $f(x, t)$. The rate at which heat is being received by the element from this source is then

$$f(x, t)\,\Delta V = f(x, t)A\,\Delta x \tag{13}$$

Second, the element may also gain heat by virtue of heat transfer through its left and right faces.

Specifically, by **b**, the rate at which heat flows into the segment through its left end is

$$-kA\left.\frac{\partial u}{\partial x}\right|_{x,t}$$

The minus sign is necessary because the segment *gains* heat through its left end if the temperature gradient $\partial u/\partial x$ is *negative*. Similarly, the element gains heat through its right end at the approximate rate

$$kA\left.\frac{\partial u}{\partial x}\right|_{x+\Delta x,t}$$

Now the rate at which heat is being stored in the segment, Eq. (12), must equal the rate at which it is being produced in the segment, Eq. (13), plus the rate at which heat flows into the segment through its ends. Hence we have approximately

$$\frac{c\rho}{g}A\,\Delta x\frac{\Delta u}{\Delta t} = kA\left.\frac{\partial u}{\partial x}\right|_{x+\Delta x,t} - kA\left.\frac{\partial u}{\partial x}\right|_{x,t} + f(x, t)A\,\Delta x$$

Finally, dividing through by $kA\,\Delta x$ and letting Δx and Δt approach zero, we obtain the equation for one-dimensional heat conduction:

$$a^2\frac{\partial u}{\partial t} = \frac{\partial^2 u}{\partial x^2} + \frac{1}{k}f(x, t) \qquad a^2 = \frac{c\rho}{kg} \tag{14}$$

The parameter a in this equation does *not* have the dimensions of velocity.

In many important problems, heat is neither generated nor absorbed in the body. Under these conditions $f(x, t) \equiv 0$ and Eq. (14) becomes the **one-dimensional heat equation**

$$a^2\frac{\partial u}{\partial t} = \frac{\partial^2 u}{\partial x^2} \qquad a^2 = \frac{c\rho}{kg} \tag{14a}$$

For heat flow in two dimensions the heat equation becomes (Exercise 25)

$$a^2\frac{\partial u}{\partial t} = \frac{\partial^2 u}{\partial x^2} + \frac{\partial^2 u}{\partial y^2} + \frac{1}{k}f(x, y, t) \qquad a^2 = \frac{c\rho}{kg} \tag{14b}$$

and in three dimensions it becomes (Exercise 26)

$$a^2\frac{\partial u}{\partial t} = \frac{\partial^2 u}{\partial x^2} + \frac{\partial^2 u}{\partial y^2} + \frac{\partial^2 u}{\partial z^2} + \frac{1}{k}f(x, y, z, t) \qquad a^2 = \frac{c\rho}{kg} \tag{14c}$$

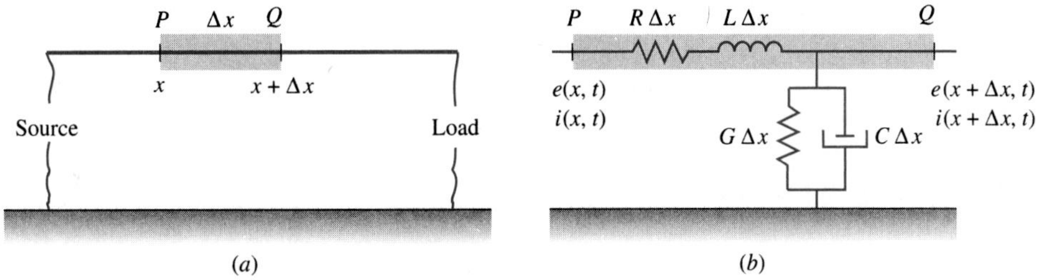

FIGURE 11.4
A typical element of a transmission line.

When heat is neither generated nor absorbed in the body and we are concerned only with the limiting steady-state temperature distribution that exists when all change of temperature with time has ceased, Eq. (14*b*) becomes

(15*a*)
$$\frac{\partial^2 u}{\partial x^2} + \frac{\partial^2 u}{\partial y^2} = 0$$

and Eq. (14*c*) becomes

(15*b*)
$$\frac{\partial^2 u}{\partial x^2} + \frac{\partial^2 u}{\partial y^2} + \frac{\partial^2 u}{\partial z^2} = 0$$

These exceedingly important equations, which arise in many applications besides steady-state heat flow, are known, respectively, as **Laplace's equation in two dimensions** and **Laplace's equation in three dimensions.** Equation (15*b*) is often written in the abbreviated form

(16)
$$\nabla^2 u = 0$$

As a final example of the derivation of partial differential equations from physical principles, we consider the flow of electricity in a long cable or transmission line. We assume the cable to be imperfectly insulated so that there is both capacitance and current leakage to ground (Fig. 11.4). Specifically, let

x = distance from the sending end of the cable
$e(x, t)$ = potential at any point on the cable at any time
$i(x, t)$ = current at any point on the cable at any time
R = resistance of the cable *per unit length*
L = inductance of the cable *per unit length*
G = conductance to ground *per unit length of cable*
C = capacitance to ground *per unit length of cable*

Since the potential at Q is equal to the potential at P minus the drop in potential along the element PQ, we see from the equivalent circuit shown in Fig. 11.4*b* that

$$e(x + \Delta x, t) = e(x, t) - (R\,\Delta x)i - (L\,\Delta x)\frac{\partial i}{\partial t}\dagger$$

---

†See Eqs. (13), (15), (19), and (20), Sec. 7.2, for an explicit statement of the circuit laws used in this derivation.

or

$$e(x + \Delta x, t) - e(x, t) = -(R\,\Delta x)i - (L\,\Delta x)\frac{\partial i}{\partial t}$$

Finally, dividing by Δx and then letting Δx approach zero gives

$$\frac{\partial e}{\partial x} = -Ri - L\frac{\partial i}{\partial t} \tag{17}$$

Likewise, the current at Q is equal to the current at P minus the current loss through leakage to ground and the apparent current loss due to the varying charge stored on the element. Hence, referring again to Fig. 11.4*b*, we have

$$i(x + \Delta x, t) = i(x, t) - (G\,\Delta x)e - (C\,\Delta x)\frac{\partial e}{\partial t}$$

or

$$i(x + \Delta x, t) - i(x, t) = -(G\,\Delta x)e - (C\,\Delta x)\frac{\partial e}{\partial t}$$

Finally, dividing by Δx and letting Δx approach zero gives

$$\frac{\partial i}{\partial x} = -Ge - C\frac{\partial e}{\partial t} \tag{18}$$

If we differentiate Eq. (17) with respect to x and Eq. (18) with respect to t, we obtain

$$\frac{\partial^2 e}{\partial x^2} = -R\frac{\partial i}{\partial x} - L\frac{\partial^2 i}{\partial x\,\partial t}$$

$$\frac{\partial^2 i}{\partial t\,\partial x} = -G\frac{\partial e}{\partial t} - C\frac{\partial^2 e}{\partial t^2}$$

If we eliminate the term $\partial^2 i/(\partial t\,\partial x)$ $[\equiv \partial^2 i/(\partial x\,\partial t)]$ between these two equations and then substitute for $\partial i/\partial x$ from (18), we find

$$\frac{\partial^2 e}{\partial x^2} = LC\frac{\partial^2 e}{\partial t^2} + (RC + GL)\frac{\partial e}{\partial t} + RGe \tag{19}$$

By differentiating Eq. (17) with respect to t and Eq. (18) with respect to x and then eliminating the derivatives of e, we obtain a similar equation for i:

$$\frac{\partial^2 i}{\partial x^2} = LC\frac{\partial^2 i}{\partial t^2} + (RC + GL)\frac{\partial i}{\partial t} + RGi \tag{20}$$

Equations (19) and (20) are known as the **telephone equations.** They assert that e and i satisfy the same partial differential equation. Two special cases of the telephone equations are worthy of note:

(a) If leakage and inductance are negligible, i.e., if $G = L = 0$, as they are, for example, for coaxial cables, Eqs. (19) and (20) reduce, respectively, to

$$\frac{\partial^2 e}{\partial x^2} = RC\frac{\partial e}{\partial t} \tag{21a}$$

$$\frac{\partial^2 i}{\partial x^2} = RC\frac{\partial i}{\partial t} \tag{21b}$$

These are known as the **telegraph equations.** Mathematically, they are identical with the one-dimensional heat equation Eq. (14*a*).

(b) At high frequencies the factor introduced by the time differentiation is large. Hence the terms involving e and $\partial e/\partial t$ or i and $\partial i/\partial t$ are insignificant in comparison with the terms containing the corresponding second derivatives $\partial^2 e/\partial t^2$ and $\partial^2 i/\partial t^2$. In this case Eqs. (19) and (20) reduce, respectively, to

$$\frac{\partial^2 e}{\partial x^2} = LC\frac{\partial^2 e}{\partial t^2} \tag{22a}$$

$$\frac{\partial^2 i}{\partial x^2} = LC\frac{\partial^2 i}{\partial t^2} \tag{22b}$$

Each of these is an example of the one-dimensional wave equation [Eq. (3)], $1/\sqrt{LC}$ having, in fact, the dimensions of velocity. These equations are obtained at any frequency of course if $R = G = 0$.

It is interesting to note that nowhere in the derivation of any of the preceding equations was any use made of boundary conditions. In other words, the same partial differential equation is satisfied by the deflections of a membrane whether the membrane is round or square, the same equation is satisfied by the deflections of a vibrating beam whether the beam is a cantilever, fixed at one end and free at the other, or a highway bridge, held in place at both ends. Likewise, the flow of heat in a rod is described by the same equation whether the ends of the rod are maintained at constant temperatures, insulated, or allowed to cool by radiation into the surrounding air. Of course, had we chosen to use polar rather than rectangular coordinates to study the vibrations of a membrane or the flow of heat in a thin metal sheet, we would have obtained different equations, but again the derivations would not have been influenced by any boundary condition.

In general, as we shall soon see, the role of boundary conditions, e.g., permanent conditions of constraint or of temperature, is to select from the infinite set of *all* solutions of the relevant partial differential equation those which, alone, are suitable for the problem at hand. Subsequent to this, initial conditions of displacement, velocity, or temperature, say, determine the coefficients with which these solutions must be combined to yield a solution to the original problem.

EXAMPLE 1

Show that each of the functions

$$u_1 = \cos \lambda x e^{-\lambda^2 t/a^2} \quad \text{and} \quad u_2 = \sin \lambda x e^{-\lambda^2 t/a^2}$$
$$u_3 = \cosh \lambda x e^{\lambda^2 t/a^2} \quad \text{and} \quad u_4 = \sinh \lambda x e^{\lambda^2 t/a^2}$$

is a solution of the one-dimensional heat equation $\partial^2 u/\partial x^2 = a^2\, \partial u/\partial t$ for all values of the parameter λ.

We first note that in each of the four products x occurs by itself in one of the factors, while t occurs by itself in the other factor. Hence partial differentiation with respect to either x or t is simply ordinary total differentiation of one or the other of the factors. This is a simplification that we will exploit effectively later in this chapter. Considering u_1 first, it is clear that

$$\frac{\partial^2 u_1}{\partial x^2} = (-\lambda^2 \cos \lambda x)e^{-\lambda^2 t/a^2}$$

Similarly, differentiating u_1 with respect to t gives

$$\frac{\partial u_1}{\partial t} = \cos \lambda x\left(-\frac{\lambda^2}{a^2}e^{-\lambda^2 t/a^2}\right)$$

A comparison of these results shows us that no matter what the value of λ may be, $\partial^2 u_1/\partial x^2$ is indeed equal to a^2 times $\partial u_1/\partial t$, and u_1 is thus a solution of the heat equation. Exactly the same procedure shows that u_2, u_3, and u_4 are also solutions of the heat equation (provided that in differentiating u_3 and u_4 we remember that $\cosh \lambda x$ and $\sinh \lambda x$ do not change sign when differentiated).

This simple example may be used to illustrate the remark we made previously that boundary conditions serve to select certain solutions as relevant to a problem while rejecting others. For instance, if we were studying a rod that extended from $x = 0$ to $x = 1$ and had each of its ends maintained at the temperature 0, it is clear that neither u_1 nor u_3 could satisfy even the left-hand end condition since neither $\cos \lambda x$ nor $\cosh \lambda x$ is zero when $x = 0$. Thus even though they are perfectly good solutions of the heat equation, they would be of no use to us.

Furthermore, although u_4 does satisfy the left-hand end condition, if $\lambda \neq 0$ so that u_4 is nontrivial, it cannot satisfy the right-hand end condition since $\sinh \lambda x$ is zero only if $\lambda x = 0$. Thus the boundary conditions have served to reject another solution, and only u_2 remains a possibly useful solution.

In order for the right-hand end condition to be satisfied, it is necessary that

$$e^{-\lambda^2 t/a^2} \sin \lambda x = 0$$

when $x = 1$, and this will be true for all values of t if and only if $\sin \lambda = 0$. Fortunately, unlike $\sinh \lambda$, $\sin \lambda$ can be zero for values of λ other than 0. Hence the boundary conditions have made a further reduction in the relevant solutions, limiting us to those corresponding to values of λ for which $\sin \lambda = 0$, that is, $\lambda = n\pi$. In later sections we shall learn how to combine the solutions that the boundary conditions have left us, namely,

$$\pm e^{-\pi^2 t/a^2} \sin \pi x \qquad \pm e^{-4\pi^2 t/a^2} \sin 2\pi x \qquad \pm e^{-9\pi^2 t/a^2} \sin 3\pi x \qquad \cdots$$

and obtain a solution satisfying both boundary conditions and initial conditions.

EXERCISES

1. **(a)** Show that in Eqs. (5*a*) and (5*b*) the parameter a has the dimensions of velocity.
   (b) Show that in Eqs. (11), (14*a*), (14*b*), and (14*c*) the parameter a does not have the dimensions of velocity.
2. **(a)** A thin insulated rod of uniform cross section extends along the x axis from $x = 0$ to $x = 1$. What do you think is the temperature distribution in the rod after steady-state conditions have been reached if the left end of the rod is maintained at the temperature 0 and the right end is maintained at the temperature 100? Does your ''guess'' satisfy the heat equation under the given conditions?
   (b) If the left and right ends of the rod in Part **(a)** are maintained at the respective temperatures $u = 100$ and $u = 50$, what is the formula for the temperature along the rod after all change with time has ceased? *Hint:* When $\partial u/\partial t = 0$, the resulting equation can be

solved by two simple integrations, following which the constants of integration can be determined to fit the end conditions.

3. Verify that each of the following equations has the indicated solution:

(a) $\dfrac{\partial^2 u}{\partial x\,\partial y} = 0;\ u = f(x) + g(y)$

(b) $u\dfrac{\partial^2 u}{\partial x\,\partial y} = \dfrac{\partial u}{\partial x}\dfrac{\partial u}{\partial y};\ u = f(x)g(y)$

(c) $a\dfrac{\partial u}{\partial x} + b\dfrac{\partial u}{\partial y} = 0;\ u = f(ay - bx)$

(d) $\dfrac{\partial^2 y}{\partial x^2} = \dfrac{\partial^2 y}{\partial t^2};\ y = f(x - t) + g(x + t)$

(e) $\dfrac{\partial^3 u}{\partial x^3} - 6\dfrac{\partial^3 u}{\partial y\,\partial x^2} + 11\dfrac{\partial^3 u}{\partial y^2\,\partial x} - 6\dfrac{\partial^3 u}{\partial y^3} = 0;$
$u = f(x + y) + g(2x + y) + h(3x + y)$

4. Determine under what conditions, if any, each of the functions **(a)** $ax + by$, **(b)** $ax^2 + bxy + cy^2$, **(c)** $ax^3 + bx^2y + cxy^2 + dy^3$, and **(d)** $ax^4 + bx^2y^2 + cy^4$ will be a solution of the equation

$$\frac{\partial^2 u}{\partial x^2} + 2\frac{\partial^2 u}{\partial x\,\partial y} + \frac{\partial^2 u}{\partial y^2} = 0$$

5. Explain how the method of undetermined coefficients can be used to obtain a particular solution of the equation

$$a\frac{\partial^2 u}{\partial x^2} + b\frac{\partial^2 u}{\partial x\,\partial y} + c\frac{\partial^2 u}{\partial y^2} = \phi(x, y)$$

if

(a) $\phi(x, y) = e^{mx+ny}$

(b) $\phi(x, y) = \sin(mx + ny)$

(c) $\phi(x, y) = \cos(mx + ny)$

(d) $\phi(x, y) = px^2 + qxy + ry^2$

(e) $\phi(x, y)$ is a homogeneous polynomial of degree n in x and y.

6. Find a particular solution of each of the following equations.

(a) $\dfrac{\partial^2 u}{\partial x^2} + \dfrac{\partial^2 u}{\partial y^2} = \cos(x + 2y)$

(b) $\dfrac{\partial^2 u}{\partial x^2} - \dfrac{\partial u}{\partial y} = 2e^{2x+3y}$

(c) $\dfrac{\partial^2 u}{\partial x^2} + 3\dfrac{\partial^2 u}{\partial x\,\partial y} + 2\dfrac{\partial^2 u}{\partial y^2} = 2x - y$

(d) $\dfrac{\partial^2 u}{\partial x^2} + \dfrac{\partial^2 u}{\partial y^2} - \dfrac{\partial^2 u}{\partial t^2} = 6(x^2 + y^2)\sin t$

7. Show that if $u_1(x, y)$ and $u_2(x, y)$ are solutions of the equation

$$A\frac{\partial^2 u}{\partial x^2} + B\frac{\partial^2 u}{\partial x\,\partial y} + C\frac{\partial^2 u}{\partial y^2} + D\frac{\partial u}{\partial x} + E\frac{\partial u}{\partial y} + F = 0$$

then for all values of the constants c_1 and c_2, the expression $u = c_1u_1 + c_2u_2$ is also a solution. Is this still true if A, B, C, D, E, and F are functions of x and y?

8. **(a)** Show that for all values of the constants A, B, and λ, the function $u = (A\cos\lambda x + B\sin\lambda x)e^{-\lambda^2 t/a^2}$ is a solution of the equation $\partial^2u/\partial x^2 = a^2\,\partial u/\partial t$. For what values of A, B, and λ, if any, is there a nontrivial solution in this family that satisfies the conditions $u(0, t) = 0$ and $\partial u/\partial x|_{l,t} = 0$?

(b) Show that for all values of A, B, C, D, and λ, $u = (A\cos\lambda x + B\sin\lambda x)(C\cosh\lambda y + D\sinh\lambda y)$ is a solution of the equation $\partial^2u/\partial x^2 + \partial^2u/\partial y^2 = 0$. For what values of the constants A, B, C, D, and λ, if any, is there a nontrivial solution in this family that satisfies the conditions $u(x, 0) = u(x, l) = 0$?

9. **(a)** Show that for all values of A, B, C, D, and λ, $u = [A\cos(\lambda x/a) + B\sin(\lambda x/a)](C\cos\lambda t + D\sin\lambda t)$ is a solution of the equation $a^2\,\partial^2u/\partial x^2 = \partial^2u/\partial t^2$. For what values of the constants A, B, C, D, and λ, if any, is there a nontrivial solution in this family that satisfies the conditions $u(0, t) = u(l, t) = 0$?

(b) Show that for all values of A, B, C, D, and λ, $u = (A\cosh\lambda x + B\sinh\lambda x)(C\cos\lambda y + D\sin\lambda y)$ is a solution of the equation $\partial^2u/\partial x^2 + \partial^2u/\partial y^2 = 0$. For what values of the constants A, B, C, D, and λ, if any, is there a nontrivial solution in this family that satisfies the conditions $\partial u/\partial x|_{0,y} = \partial u/\partial x|_{l,y} = 0$?

10. **(a)** Show that Laplace's equation in three dimensions

$$\frac{\partial u^2}{\partial x^2} + \frac{\partial^2 u}{\partial y^2} + \frac{\partial^2 u}{\partial z^2} = 0$$

is satisfied by the function
$u = 1/[(x - a)^2 + (y - b)^2 + (z - c)^2]^{1/2}$
for all values of the constants a, b, and c.

(b) Determine whether or not Laplace's equation in two dimensions, $\dfrac{\partial^2 u}{\partial x^2} + \dfrac{\partial^2 u}{\partial y^2} = 0$ is satisfied by the function $u = 1/[(x - a)^2 + (y - b)^2]^{1/2}$ for any values of a, b.

11. Show that for all values of the constants a and b, Laplace's equation in two dimensions is satisfied by the function

$$u = \ln[(x - a)^2 + (y - b)^2]$$

12. What is the form of the one-dimensional heat equation if the cross-sectional area of the rod is a function of x?

13. Work Part **(b)** of Exercise (2) if the cross section of the rod varies from $x = 0$ to $x = 1$ according to the formulas **(a)** $A(x) = 1 + x$ and **(b)** $A(x) = 1/(1 + x)$. *Hint:* Use the result of Exercise 12.

14. Using the result of Exercise 12, show that the steady-state temperature in a long, thin rod of variable cross section is always a monotonic function of x. Under what conditions, if any, will the graph of the temperature distribution in the rod be everywhere convex up? Under what conditions will it be everywhere concave up? *Hint:* Recall the second-derivative test for convexity and concavity.

15. Consider the telephone equations in the so-called distortionless case when $RC = LG$ and put $a^2 = RG$ and $v^2 = 1/LC$. Prove that if $e(x, t)$ or, equally well, $i(x, t)$ is written in the form $e(x, t) = \epsilon^{-avt}y(x, t)$, then the function y satisfies the wave equation $v^2\, \partial^2 y/\partial x^2 = \partial^2 y/\partial t^2$. *Note:* To avoid confusion with the voltage, ϵ is here used in place of e to denote the base of natural logarithms.

16. Show that under the substitution $u(x, y) = w(x, y)e^{-(bx+ay)}$ the equation $\partial^2 u/\partial x\, \partial y + a\, \partial u/\partial x + b\, \partial u/\partial y + cu = 0$ becomes $\partial^2 w/\partial x\, \partial y + (c - ab)w = 0$.

17. Derive the differential equation satisfied by the concentration, i.e., fluid per unit volume, u, of a liquid diffusing through a long, porous rod whose lateral surface is impervious to diffusion. Assume that the rod is of constant cross section and that the concentration is the same at all points of each cross section. *Hint:* Set up the fluid balance for the infinitesimal segment of the rod between x and $x + \Delta x$ using the fact that the rate at which fluid diffuses through an area is proportional to the area and to the concentration gradient in the direction perpendicular to the area.

18. A circular disk whose upper and lower faces are insulated is so thin that heat flow in the disk can be considered two-dimensional. Derive the differential equation governing the flow of heat in the disk if the temperature is independent of θ, that is, is a function of r and t only. *Hint:* Since the temperature is independent of θ, it is convenient to base the analysis on the heat balance in the infinitesimal circular ring between r and $r + \Delta r$.

19. The radii of the inner and outer boundaries of an annular ring are 1 and 2, respectively. The upper and lower surfaces of the annulus are perfectly insulated. The entire inner boundary is maintained at the temperature $u = 0$, and the outer boundary is maintained at the temperature $u = 100$. What is the steady-state temperature distribution in the annulus? *Hint:* Use the result of Exercise 18.

20. Derive the differential equation governing the flow of heat in a sphere when the temperature is a function of r and t only. *Hint:* It will be convenient to base the analysis on the heat balance in the infinitesimal spherical shell between r and $r + \Delta r$.

21. The inner and outer radii of a spherical shell are 1 and 6, respectively. The entire inner spherical boundary is maintained at the temperature $u = 100$, and the entire outer surface is maintained at the temperature $u = 50$. What is the steady-state temperature distribution in the shell? *Hint:* Use the result of Exercise 20.

22. **(a)** By considering the resultant horizontal force on the element of the string shown in Fig. 11.1b, show that the assumption that the tension in the string remains constant is not strictly true.

(b) Show that the assumption that the tension in the string remains constant is equivalent to the assumption that $(\partial y/\partial x)(\partial^2 y/\partial x^2) = 0$, that is, that the deflection curve of the string is a straight line.

23. Derive the differential equation satisfied by the temperature in a long, thin rod of constant circular cross section whose lateral surface loses heat according to Newton's law of cooling: *The rate of heat loss per unit area is proportional to the difference between the surface temperature and the temperature of the surrounding air.* Assume that the temperature is constant over each cross section. How does the analysis differ if the cross sections are not circular?

24. If the rod in Exercise 23 is of unit length and if each end is maintained at the temperature 100, what is the steady-state temperature in the rod as a function of x? *Hint:* It is easier to impose the end conditions if hyperbolic rather than exponential functions are used in solving the ordinary differential equation that results when $\partial u/\partial t = 0$.

25. Derive Eq. (14b) for heat flow in two dimensions. *Hint:* Consider the flow of heat through the pairs of opposite sides of the element shown in Fig. 11.5. Evaluate the temperature gradient at the midpoint of each side and apply in each direction the reasoning used to derive Eq. (14).

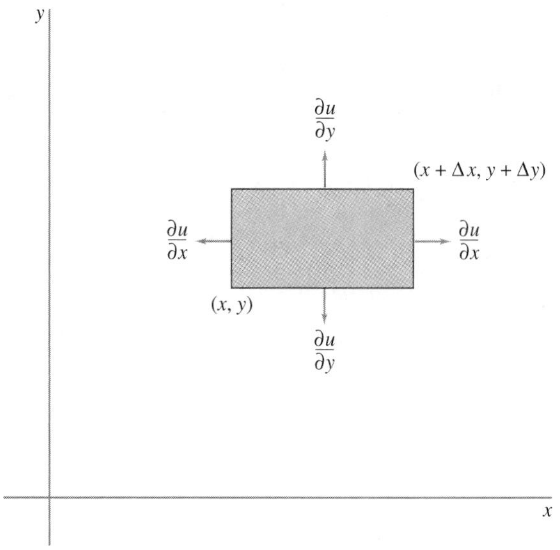

FIGURE 11.5
A typical rectangular coordinate element in a region of two-dimensional heat flow.

26. Derive Eq. (14c) for heat flow in three dimensions. *Hint:* Apply the reasoning suggested in Exercise 25 to the flow of heat through the pairs of opposite faces of a volume element $\Delta V = \Delta x\, \Delta y\, \Delta z$ with one corner at the point (x, y, z) and with ρ as the weight per unit volume of the conducting solid.

27. A thin sheet is not insulated but loses heat through its upper and lower surfaces according to Newton's law of cooling (see Exercise 23). Find the equation satisfied by the temperature distribution in the sheet as a function of x, y, and t. *Hint:* Recall the hint given in Exercise 25.

28. Consider a region of space filled with a moving fluid. Let the density of the fluid at the point (x, y, z) at the time t be $\rho(x, y, z, t)$ and let the particle instantaneously at the point (x, y, z) have velocity components v_x, v_y, v_z, respectively, in the directions of the coordinate axes. By considering the flow through the boundaries of an infinitesimal region of dimensions Δx, Δy, Δz, show that the velocity components satisfy the so-called **equation of continuity**

$$\frac{\partial(\rho v_x)}{\partial x} + \frac{\partial(\rho v_y)}{\partial y} + \frac{\partial(\rho v_z)}{\partial z} + \frac{\partial \rho}{\partial t} = 0$$

29. A thin, circular disk is not insulated but loses heat through its upper and lower surfaces according to Newton's law of cooling (see Exercise 23). If the temperature distribution is independent of θ, derive the equation satisfied by the temperature as a function of r and t.

30. Derive Eq. (3) by approximating the string by a weightless elastic cord bearing n equally spaced mass particles each equal to $(1/n)$th of the mass of the string and then letting n become infinite. *Hint:* Recall the relation between the second difference and the second derivative of a function.

31. Derive the differential equation governing the transverse vibration of an elastic membrane of weight per unit area $w(x, y)$ when stretched in the xy plane and for which the tension per unit length T is the same at all points and in all directions in the membrane. Assume that the membrane is acted on by a transverse force whose magnitude per unit area is $|f(x, y, z, \dot{z}, t)|$ and that during vibration each particle moves in a direction perpendicular to the xy plane. *Hint:* Consider the infinitesimal element of the membrane shown in Fig. 11.6. By hypothesis, the forces applied at the midpoints of the sides A, B, C, D by the adjacent material are

$$A\colon T\,\Delta x \qquad B\colon T\,\Delta y \qquad C\colon T\,\Delta x \qquad D\colon T\,\Delta y$$

Assuming these forces are tangential to the surface element as it vibrates and that deflections $z(x, y, t)$ are so small that sines may be satisfactorily approximated by tangents, the z components of these forces are

$$A\colon -T\,\Delta x \left.\frac{\partial z}{\partial y}\right|_y \qquad B\colon T\,\Delta y \left.\frac{\partial z}{\partial x}\right|_{x+\Delta x}$$

$$C\colon T\,\Delta x \left.\frac{\partial z}{\partial y}\right|_{y+\Delta y} \qquad D\colon -T\,\Delta y \left.\frac{\partial z}{\partial x}\right|_x$$

Taking these forces and the force $f(x, y, z, \dot{z}, t)\,\Delta x\,\Delta y$ into account, now apply Newton's law to the mass of the element, divide by $\Delta x\,\Delta y$, and take the limit as Δx and Δy approach zero.

32. **(a)** By assuming $u(x, t) = \phi(x) \sin t$, find a solution of the equation $\dfrac{\partial^2 u}{\partial x^2} - \dfrac{\partial^2 u}{\partial t^2} = x \sin t$ satisfying the conditions $u(0, t) = u(l, t) = 0$, $l \neq n\pi$.

(b) By assuming $u(x, t) = \phi_1(x) \sin t + \phi_2(x) \cos t + \phi_3(x) t \sin t + \phi_4(x) t \cos t$, find a solution when $l = n\pi$.

33. Derive the polar coordinate form of the two-dimensional heat equation. *Hint:* Apply the reasoning suggested in Exercise 25 to the pairs of opposite sides of the polar coordinate element $\Delta A = r\,\Delta r\,\Delta\theta$ shown in Fig. 11.7.

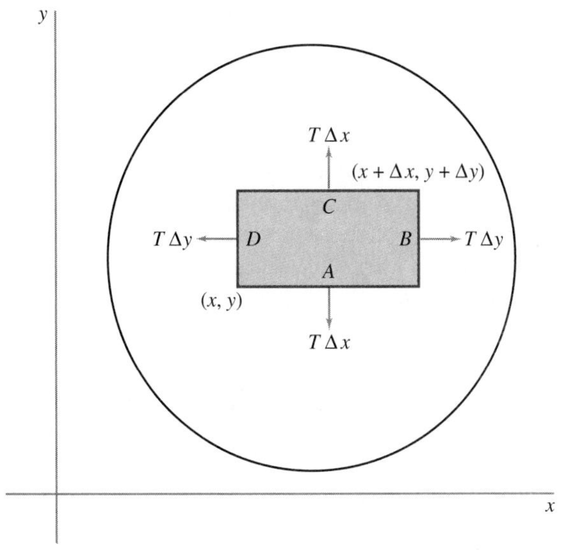

FIGURE 11.6
A typical element of a vibrating membrane.

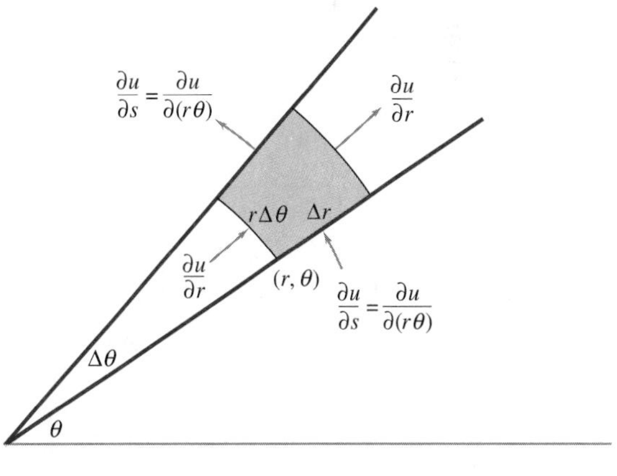

FIGURE 11.7
A typical polar coordinate element in a region of two-dimensional heat flow.

34. Derive the polar coordinate form of the equation describing the transverse vibrations of a plane membrane. *Hint:* Apply the reasoning suggested in Exercise 31 to the element shown in Fig. 11.7.

35. If $u(x, t)$ is the displacement of a general cross section of a bar which is vibrating longitudinally, i.e., in the direction of its length, show that

$$A(x)\frac{\partial^2 u}{\partial t^2} = \frac{Eg}{\rho}\frac{\partial[A(x)\,\partial u/\partial x]}{\partial x}$$

where $A(x)$ = cross-sectional area of the bar, E = modulus of elasticity of the material of the bar, ρ = weight per unit volume of the material of the bar. *Hint:* Use the definition of the modulus of elasticity,

$$E = \frac{\text{stress}}{\text{strain}} = \frac{\text{force per unit area}}{\text{stretch per unit length}}$$

to obtain the expression $F = EA\,\partial u/\partial x$ for the force transmitted through a general cross section of the bar.

36. Derive the cylindrical coordinate form of the three-dimensional heat equation. *Hint:* Apply the reasoning suggested in Exercise 26 to the pairs of opposite faces of the cylindrical coordinate element $\Delta V = r\,\Delta r\,\Delta\theta\,\Delta z$.

37. If u is a function of r and t only, what is the result of transforming

(a) The two-dimensional wave equation

$a^2\left(\frac{\partial^2 u}{\partial x^2} + \frac{\partial^2 u}{\partial y^2}\right) = \frac{\partial^2 u}{\partial t^2}$ by the substitution $w = ru$, where $r = \sqrt{x^2 + y^2}$?

(b) The three-dimensional wave equation

$a^2\left(\frac{\partial^2 u}{\partial x^2} + \frac{\partial^2 u}{\partial y^2} + \frac{\partial^2 u}{\partial z^2}\right) = \frac{\partial^2 u}{\partial t^2}$ by the substitution $w = ru$, where $r = \sqrt{x^2 + y^2 + z^2}$?

38. **(a)** Show that when Laplace's equation in cartesian coordinates

$$\frac{\partial^2 u}{\partial x^2} + \frac{\partial^2 u}{\partial y^2} + \frac{\partial^2 u}{\partial z^2} = 0$$

is transformed into cylindrical coordinates by means of the substitutions $x = r\cos\theta$, $y = r\sin\theta$, $z = z$, it becomes

$$\frac{\partial^2 u}{\partial r^2} + \frac{1}{r}\frac{\partial u}{\partial r} + \frac{1}{r^2}\frac{\partial^2 u}{\partial \theta^2} + \frac{\partial^2 u}{\partial z^2} = 0$$

(b) Show that when Laplace's equation in cartesian coordinates is transformed into spherical coordinates by means of the substitutions $x = r\sin\theta\cos\phi$, $y = r\sin\theta\sin\phi$, $z = r\cos\theta$, it becomes

$$r^2\sin\theta\frac{\partial^2 u}{\partial r^2} + 2r\sin\theta\frac{\partial u}{\partial r} + \sin\theta\frac{\partial^2 u}{\partial \theta^2} + \cos\theta\frac{\partial u}{\partial \theta} + \frac{1}{\sin\theta}\frac{\partial^2 u}{\partial \phi^2} = 0$$

39. A uniform beam of length $2l$ has hinged ends and bears a load per unit length

$$w(x) = [u(x) - u(x - l)]k \qquad k \text{ constant}$$

Find the deflection curve of the beam

(a) By integrating the beam equation $EIy^{\text{iv}} = -w(x)$ and imposing appropriate auxiliary conditions.

(b) Using the related Green's function (Exercise 55, Sec. 2.12).

(c) By means of Fourier series (see Example 5, Sec. 8.6).

(d) Using Laplace transforms.

40. Modify the derivation of Eq. (10) to take into account the rotational inertia of the cross sections of the beam. *Hint:* Consider the portion of the beam between x and $x + \Delta x$ and compute its moment of inertia about the neutral axis of its end section. Then note that as the beam vibrates, this slice of the beam remains perpendicular to the deflection curve of the beam and hence rotates through an angle equal to the inclination angle of the deflection curve at the corresponding point. Next, assuming deflections so small that the inclination angle α can be replaced by its tangent $\partial y/\partial x$, compute the angular acceleration $\ddot{\alpha}$ of the slice and the inertial moment this angular acceleration implies. Then note that between x and $x + \Delta x$ the total moment decreases by an amount equal to the inertial moment of the slice. Finally, include this change in the expression for $\partial M/\partial x$ and complete the derivation as in the text.

11.3 THE D'ALEMBERT SOLUTION OF THE WAVE EQUATION

Each of the homogeneous partial differential equations derived in the last section, i.e, those equations in which $f \equiv 0$, can be solved by a method of considerable generality known as *separation of variables*. For the one-dimensional wave equation, however, there is also an elegant special method of solution due to D'Alembert† which, because of the importance of this equation, we shall examine in some detail before developing more general techniques.

The whole matter is very simple. In fact, if f is a function possessing a second derivative, then by the chain rule,

$$\frac{\partial f(x - at)}{\partial t} = -af'(x - at) \qquad \frac{\partial f(x - at)}{\partial x} = f'(x - at)$$

$$\frac{\partial^2 f(x - at)}{\partial t^2} = a^2 f''(x - at) \qquad \frac{\partial^2 f(x - at)}{\partial x^2} = f''(x - at)$$

†Named for the French mathematician and philosopher Jean Le Rond D'Alembert (1717–1783).

and from these results it is evident that $y = f(x - at)$ satisfies the equation

$$\frac{\partial^2 y}{\partial t^2} = a^2 \frac{\partial^2 y}{\partial x^2} \tag{1}$$

It is an equally simple matter to prove that if g is an arbitrary twice-differentiable function, then $g(x + at)$ is likewise a solution of (1). Hence, since (1) is a linear equation, it follows that the sum

$$y = f(x - at) + g(x + at) \tag{2}$$

is also a solution. In fact, it can be shown (see Exercise 23) that if f and g are arbitrary twice-differentiable functions, then *D'Alembert's solution* (2) is a *complete* solution of (1); that is, *any* solution of (1) can be expressed in the form (2).

This form of the solution of the wave equation is especially useful for revealing the significance of the parameter a and its dimensions of velocity. Suppose, specifically, that we consider the motion of a uniform string† stretching from $-\infty$ to ∞. If its transverse displacement is given by (2), we have in fact two waves traveling in opposite directions along the string, each with velocity a. To see this, consider first the function $f(x - at)$. At $t = 0$, it defines the curve $y = f(x)$, and at any later time $t = t_1$, it defines the curve $y = f(x - at_1)$. But these curves are identical except that the latter is translated to the right a distance equal to at_1. Thus the entire configuration moves along the string without distortion a distance of at_1 in t_1 units of time. The velocity with which the wave is propogated is therefore

$$v = \frac{at_1}{t_1} = a$$

Similarly, the function $g(x + at)$ defines a configuration which moves to the left along the string with constant velocity a. The total displacement of the string is of course the algebraic sum of these two traveling waves.

To carry the solution through in detail, let us suppose that the initial displacement of the string at any point x is given by $\phi(x)$ and that the initial velocity of the string at any point is $\theta(x)$. Then, as conditions to determine the form of f and g, we have, from (2) and its derivative with respect to t,

$$y(x, 0) \equiv \phi(x) = [f(x - at) + g(x + at)]_{t=0} = f(x) + g(x) \tag{3}$$

$$\left.\frac{\partial y}{\partial t}\right|_{x,0} \equiv \theta(x) = [-af'(x - at) + ag'(x + at)]_{t=0} = -af'(x) + ag'(x) \tag{4}$$

Dividing Eq. (4) by a and then integrating with respect to x, we find

$$-f(x) + g(x) = \frac{1}{a}\int_{x_0}^{x} \theta(s)\, ds - f(x_0) + g(x_0)$$

Combining this with (3) by subtraction and addition, and setting $k = \frac{1}{2}[f(x_0) - g(x_0)]$, we obtain

$$f(x) = \frac{1}{2}\left[\phi(x) - \frac{1}{a}\int_{x_0}^{x} \theta(s)\, ds\right] + k \qquad g(x) = \frac{1}{2}\left[\phi(x) + \frac{1}{a}\int_{x_0}^{x} \theta(s)\, ds\right] - k$$

With the forms of f and g known, we can now write

$$\begin{aligned} y &= f(x - at) + g(x + at) \\ &= \left[\frac{\phi(x - at)}{2} - \frac{1}{2a}\int_{x_0}^{x-at} \theta(s)\, ds\right] + \left[\frac{\phi(x + at)}{2} + \frac{1}{2a}\int_{x_0}^{x+at} \theta(s)\, ds\right] \end{aligned}$$

†The use of the string as an illustration is purely a matter of convenience, and *any* quantity satisfying the wave equation possesses the mathematical properties developed for the string.

or, combining the integrals, we have the **initial-value D'Alembert solution** of (1)

(5) $$y(x, t) = \frac{\phi(x - at) + \phi(x + at)}{2} + \frac{1}{2a}\int_{x-at}^{x+at} \theta(s)\, ds$$

for which $y(x, 0) = \phi(x)$ and $y_t(x, 0) = \theta(x)$.

EXAMPLE 1

MOTION OF A TWO-WAY INFINITE STRING

A string stretching to infinity in both directions is given the initial displacement

$$\phi(x) = \frac{1}{1 + 8x^2}\dagger$$

and released from rest. Determine its subsequent motion.

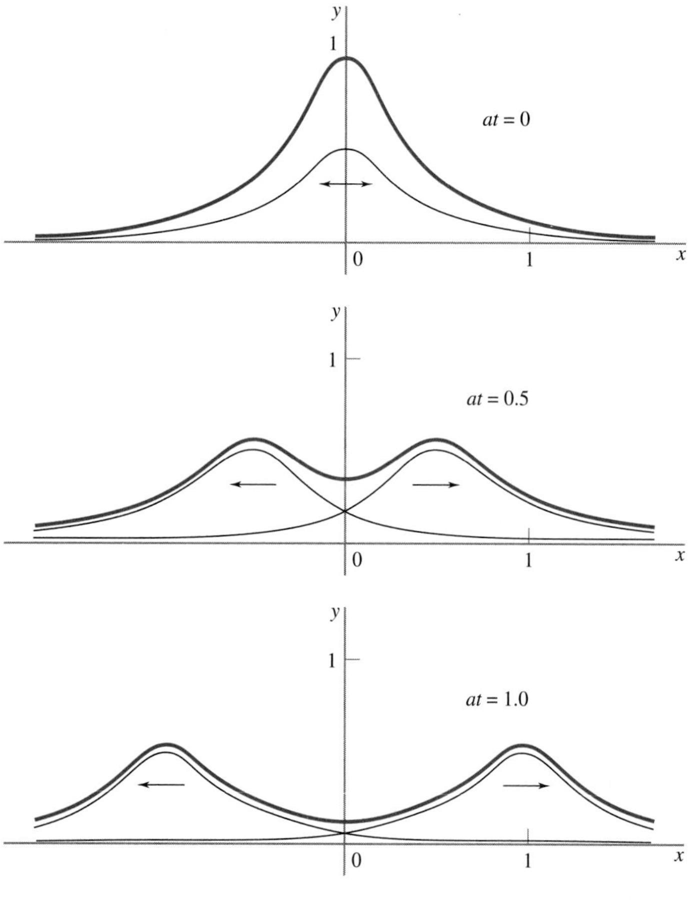

FIGURE 11.8
The propagation of a disturbance along a two-way infinite string.

†The initial deflection curve $y = \phi(x)$ clearly violates Assumption **(d)**, Sec. 11.2, since at $x = -\frac{1}{4}$ (for instance), $\phi'(x) \equiv \tan\alpha = \frac{16}{9} \doteq 1.78$ while $\sin\alpha \doteq 0.87$. This difficulty is easily overcome, however, by assuming instead of $\phi(x)$ a new deflection curve

$$\phi^*(x) = \frac{1}{k}\phi(x)$$

where k is a sufficiently large constant, say $k = 10{,}000$. Using $\phi(x)$ instead of $\phi^*(x)$ in this and similar problems is just a convenient way of eliminating the constant factor $1/k$ at each step of our work.

Since $\theta(x) \equiv 0$, we have from (5) simply

$$y(x, t) = \frac{\phi(x - at) + \phi(x + at)}{2} = \frac{1}{2}\left[\frac{1}{1 + 8(x - at)^2} + \frac{1}{1 + 8(x + at)^2}\right]$$

The deflection of the string when $at = 0.0$, 0.5, and 1.0 is shown in Fig. 11.8.

The motion of a semi-infinite string whose finite end is fixed can be completely determined by considering the motion of one-half of a two-way infinite string having a fixed point, or **node,** located at some point, say the origin. For the motion of either string on $[0, \infty)$ is described by a complete solution of (1) subject to the same boundary and initial conditions. To capitalize on this fact, we need only imagine the actual string, stretching from 0 to ∞, to be extended in the opposite direction to $-\infty$. The initial conditions of velocity and displacement for the new portion of the string we define to be equal in magnitude but opposite in sign to those given for the actual string.† The solution for the resulting two-way infinite string can be written down at once using Eq. (5). In the nature of the extended initial conditions, the displacement at the origin due to the wave traveling to the right from the left half of the string will always be equal but opposite in sign to the displacement at the origin due to the wave traveling to the left from the right half of the string. Hence the string will always remain at rest at the origin, and the solution for the right half of the extended string will be precisely the solution of the original problem.

EXAMPLE 2

MOTION OF A SEMI-INFINITE STRING

A semi-infinite string is given the displacement shown in Fig. 11.9*a* and released from rest. Determine its subsequent motion.

We first imagine the string extended to $-\infty$ and released from rest in the extended initial configuration shown in Fig. 11.9*b*. Since $\theta(x) \equiv 0$, we have, from (5),

$$y(x, t) = \frac{\phi(x - at) + \phi(x + at)}{2}$$

where $\phi(x)$ is the displacement shown in Fig. 11.9*b*.‡ We thus have two displacement waves, each of shape defined by $\frac{1}{2}\phi(x)$, one traveling to the right and one traveling to the left along the string. Plots of these waves are shown in Fig. 11.10. An inspection of these configurations reveals the important fact that a displacement wave is reflected from a fixed end without distortion but with reversal of sign.

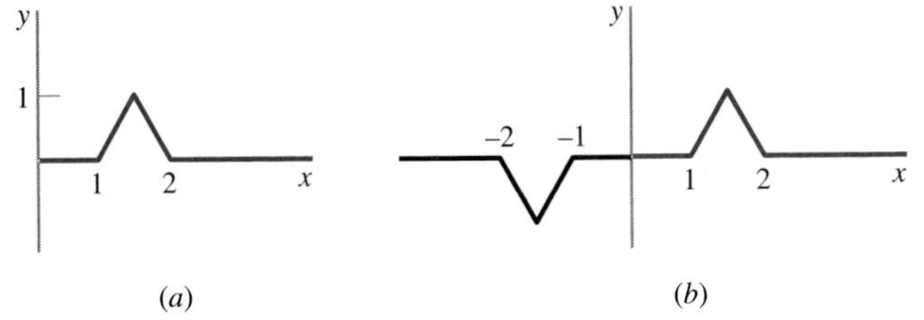

FIGURE 11.9
A semi-infinite string and its conceptual extension.

†This method of extending the initial conditions is sufficient but not necessary (see Exercise 17).

‡If, as suggested by Fig. 11.9*a*, the graph of $\phi(x)$ has one or more corner points, then, strictly speaking, $\phi(x)$ does not describe an admissible initial displacement function. In fact, in the derivation of Eq. (5) both $f(x)$ and $g(x)$ were assumed to be twice-differentiable, and therefore $\phi(x)$ must also be twice-differentiable, which is not the case if there are points where the derivative of $\phi(x)$ is undefined. The apparent solutions obtained from Eq. (5) by overlooking this fact are therefore at best only formal solutions and are to be viewed with suspicion unless and until it is verified directly that they satisfy the given partial differential equation and its accompanying boundary and initial conditions. Questions concerning the existence and uniqueness of solutions of partial differential equations are quite difficult, and in our work we shall be concerned mainly with techniques for obtaining formal solutions. For an extended discussion of the problem of establishing the validity of solutions derived by purely formal means, see, for instance, James Ward Brown and R. V. Churchill, *Fourier Series and Boundary Value Problems,* 5th ed., pp. 129–168, McGraw-Hill, New York, 1993.

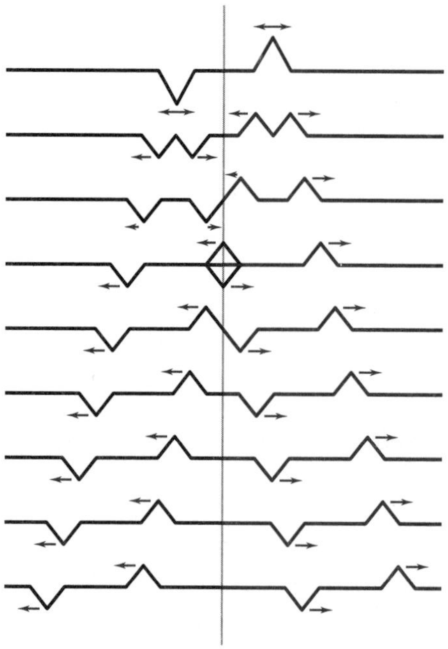

FIGURE 11.10
The propagation of a disturbance along a semi-infinite string.

The motion of a finite string can be obtained as the motion of a segment of an infinite string with suitably defined initial displacement and velocity. If the string has fixed ends and is given between 0 and l, say, we first imagine that it is extended from 0 to $-l$ with initial conditions equal but opposite in sign to those for the actual string. Then we extend the string to infinity in each direction subject to initial conditions which duplicate with period $2l$ the initial conditions between $-l$ and l.† Finally, after we obtain the solution for the infinite string we have thus created, its behavior for $0 \le x \le l$ will be a complete description of the motion of the actual finite string in which we are really interested.

EXAMPLE 3

MOTION OF A STRING OF FINITE LENGTH

A string of length l is given the displacement shown in Fig. 11.11 and released from rest. Determine its subsequent motion.

A suitable extension of the string and one-half cycle of its motion are shown in Fig. 11.12. An inspection of Fig. 11.12 shows that the period of the motion, i.e., the least time for its return to its initial state, is just the time for either of the traveling waves to traverse a distance $2l$. In other words, since the velocity of the waves is a, the period is $2l/a$. The frequency of the vibrations is therefore $a/(2l)$. We shall encounter this formula again when we solve the wave equation by the method of separation of variables.

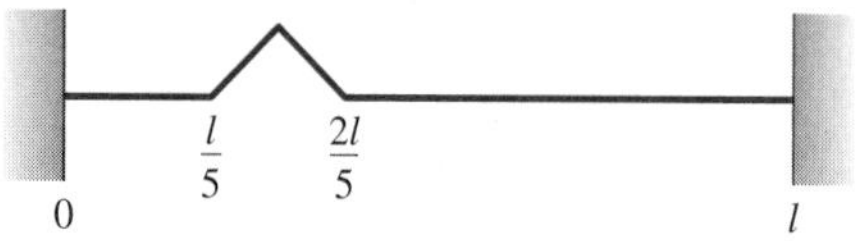

FIGURE 11.11
A finite string with initial displacement.

†This of course is essentially the procedure we used in Sec. 8.4 to obtain the half-range sine expansion of a function originally defined only over a finite portion of the real axis. The relation of Fourier series to the problem of the vibrating string and to the solution of the wave equation in general will become clear in Sec. 11.5 when we develop the method of separation of variables.

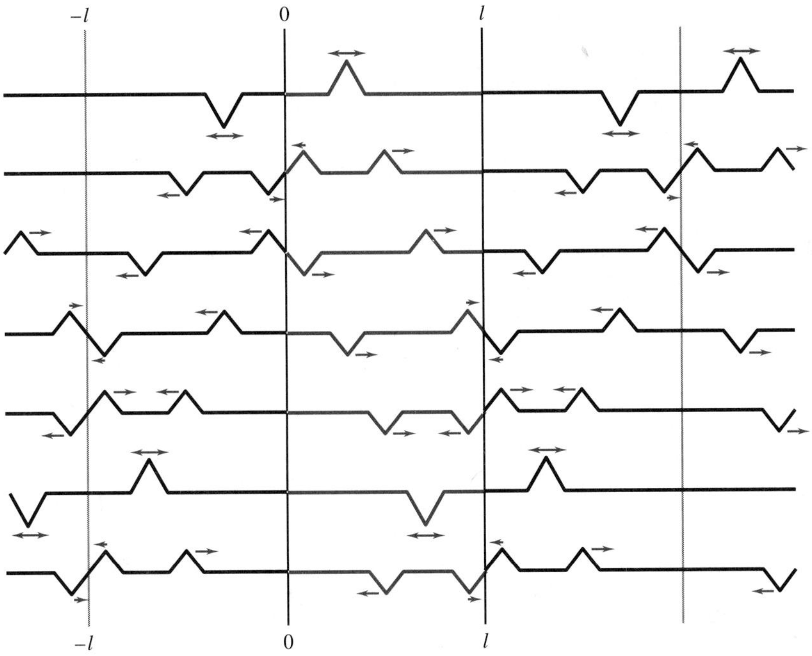

FIGURE 11.12
One half-cycle of the motion of a finite string.

The preceding examples were especially simple because they involved only traveling waves produced by initial displacements, not velocities. Moreover, they were also purely descriptive and did not require any use of the equations of the traveling waves. In our final example we seek precise information about how a semi-infinite string behaves when excited by an initial velocity distributed over a portion of the string. In working problems of this sort, it is frequently helpful to use *filter functions,* such as we introduced in Sec. 1.1 and studied again in Sec. 10.6, to represent velocities restricted to finite sections of the string. If we do this, the use of Eq. (5) to obtain the deflection of the string will require the integration of functions containing a unit step function as one factor. This is not a difficult matter, but with variable limits, such as those appearing in Eq. (5), it is sometimes rather complicated. The following lemma, whose proof we leave as an exercise (see Exercise 22) is usually the most convenient way of performing such integrations.

LEMMA 1 If $f(s)$ is piecewise continuous on $[a, b]$, then

$$\int_a^b u(s-c)f(s)\,ds = [u(s-c)\{F(s)-F(c)\}]_a^b$$

where $F(s)$ is an antiderivative of $f(s)$. Equivalently,

$$\int_a^b u(s-c)f(s)\,ds = [u(s-c)F^*(s)]_a^b$$

where $F^*(s)$ is the antiderivative of $f(s)$ which vanishes at the point $s = c$ where the step function $u(s-c)$ has its finite jump.

EXAMPLE 4

MOTION OF A STRUCK SEMI-INFINITE STRING

At $t = 0$ a uniform string stretched along the x axis from 0 to ∞ is struck in such a way that the portion of the string between $x = 0$ and $x = 1$ is given unit velocity. Find the subsequent displacement of the string and plot it as a function of x for $at = \frac{3}{4}$.

Using filter functions, it is clear that the initial velocity of the string is given by the formula

$$u(x) - u(x-1)$$

Following the discussion that preceded Example 2, we now imagine an extension of the string to $-\infty$, and on this conceptual extension we imagine the skew-symmetric initial velocity condition

$$-u(x+1) + u(x)$$

Thus on the extended string we have

$$y(x, 0) \equiv \phi(x) = 0$$
$$\dot{y}(x, 0) \equiv \theta(x) = -u(x+1) + 2u(x) - u(x-1)$$

Hence, from Eq. (5), we have

$$y(x, t) = \frac{1}{2a}\int_{x-at}^{x+at} [-u(s+1) + 2u(s) - u(s-1)]\, ds$$

In this integral, each term contains a different unit step function. Hence, according to the second version of Lemma 1, a different antiderivative must be used in integrating each term. The necessary choices are obvious, and we have further

$$y(x, t) = \frac{1}{2a}[-(s+1)u(s+1) + 2s\, u(s) - (s-1)\, u(s-1)]_{x-at}^{x+at}$$

$$= \frac{1}{2a}[-(x+at+1)u(x+at+1) + (x-at+1)u(x-at+1) + 2(x+at)u(x+at)$$
$$- 2(x-at)u(x-at) - (x+at-1)u(x+at-1) + (x-at-1)u(x-at-1)]$$

Setting $at = \frac{3}{4}$ in the formula for $y(x, t)$, we obtain at once

$$y\left(x, \frac{3}{4a}\right) = \frac{1}{2a}\left[-\left(x+\frac{7}{4}\right)u\left(x+\frac{7}{4}\right) + \left(x+\frac{1}{4}\right)u\left(x+\frac{1}{4}\right) + 2\left(x+\frac{3}{4}\right)u\left(x+\frac{3}{4}\right)\right.$$
$$\left. - 2\left(x-\frac{3}{4}\right)u\left(x-\frac{3}{4}\right) - \left(x-\frac{1}{4}\right)u\left(x-\frac{1}{4}\right) + \left(x-\frac{7}{4}\right)u\left(x-\frac{7}{4}\right)\right]$$

In the first three terms in this expression, each step function is 1 for all nonnegative values of x. Hence, collectively, these terms contribute

$$\frac{1}{2a}\left[-\left(x+\frac{7}{4}\right) + \left(x+\frac{1}{4}\right) + 2\left(x+\frac{3}{4}\right)\right] = \frac{2x}{2a}$$

to $y(x, 3/4a)$. Noting that the other step functions jump from 0 to 1 at $x = \frac{1}{4}$, $x = \frac{3}{4}$, and $x = \frac{7}{4}$, respectively, and taking into account the resulting contributions to $y(x, 3/4a)$ at each of these points, we have finally

$$0 \le x \le \frac{1}{4}: \qquad y = \frac{2x}{2a}$$

$$\frac{1}{4} \le x \le \frac{3}{4}: \qquad y = \frac{2x}{2a} + \frac{1}{2a}\left[-\left(x-\frac{1}{4}\right)\right] = \frac{1}{2a}\left(x+\frac{1}{4}\right)$$

$$\frac{3}{4} \le x \le \frac{7}{4}: \qquad y = \frac{1}{2a}\left(x+\frac{1}{4}\right) + \frac{1}{2a}\left[-2\left(x-\frac{3}{4}\right)\right] = \frac{1}{2a}\left(-x+\frac{7}{4}\right)$$

$$\frac{7}{4} \le x: \qquad y = \frac{1}{2a}\left(-x+\frac{7}{4}\right) + \frac{1}{2a}\left(x-\frac{7}{4}\right) = 0$$

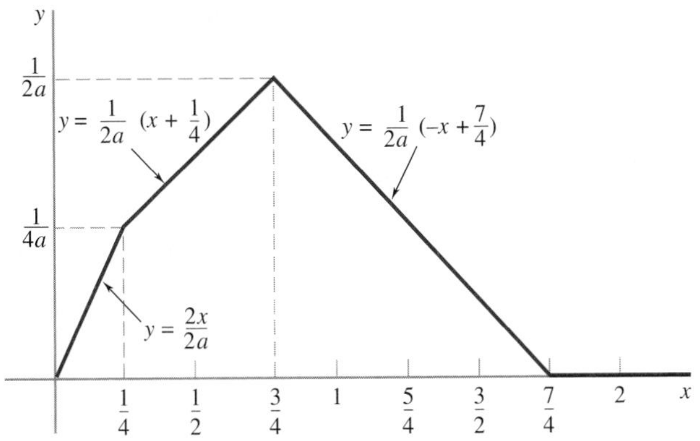

FIGURE 11.13
The plot of $y(x, \frac{3}{4}a)$.

A plot of $y(x, 3/4a)$ is shown in Fig. 11.13.

EXERCISES

1. **(a)** A uniform string stretched from $x = 0$ to $x = \pi$ is given the initial displacement $y(x, 0) = \sin x$ and released from rest in that position. Find the subsequent displacement of the string as a function of x and t.
   (b) Work Part **(a)** if $y(x, 0) = \sin 2x$. *Hint:* It is not necessary to use unit step functions in solving this problem.
2. **(a)** While at rest in its equilibrium position, a uniform string stretched from $x = 0$ to $x = \pi$ is given the initial velocity $\dot{y}(x, 0) = \sin x$. Find the subsequent displacement of the string as a function of x and t.
   (b) Work Part **(a)** if $\dot{y}(x, 0) = \sin 2x$. *Hint:* It is not necessary to use unit step functions in solving this problem.
3. A uniform string stretched from $x = 0$ to $x = \pi$ is given the initial displacement $y(x, 0) = \sin mx$ and set in motion from that position with initial velocity $\dot{y}(x, 0) = \sin nx$. Find the subsequent motion of the string as a function of x and t.
4. **(a)** A uniform string stretched from $x = 0$ to $x = l$ is given the initial displacement $y(x, 0) = \sum_{n=1}^{\infty} B_n \sin(n\pi x/l)$ and released from rest in that position. Find the subsequent displacement of the string as a function of x and t. With what frequencies does the string vibrate?
   (b) Work Part **(a)** if the string is set in motion from its equilibrium position with initial velocity $\dot{y}(x, 0) = \sum_{n=1}^{\infty} B_n \sin(n\pi x/l)$. With what frequencies does the string vibrate?
5. In Example 4, plot the deflection of the string as a function of x for $at = \frac{1}{4}$ and $at = \frac{5}{4}$.
6. In Example 4, plot the deflection of the string as a function of x for $at = 1$ and $at = 2$.
7. While in its equilibrium position, a string stretching from $-\infty$ to ∞ is struck in such a way that the portion of the string between $x = -1$ and $x = 1$ is given the initial velocity 1. Find the subsequent deflection of the string.
8. Work Exercise 7 using the antiderivative s instead of the antiderivatives $s + 1$ and $s - 1$ and verify that the expression for the deflection of the string is discontinuous.
9. In Exercise 7, find the kinetic energy of the string at any time t.
10. While in its equilibrium position, a string stretching from $-\infty$ to ∞ is given the initial velocity

$$\dot{y}(x, 0) = \begin{cases} \cos(\pi x/2) & -1 \le x \le 1 \\ 0 & x^2 \le 1 \end{cases}$$

    Find the subsequent deflection of the string as a function of x and t. Is there any significant difference between the deflection curve in this case and the deflection curve in Exercise 7?
11. At $t = 0$, while at rest in its equilibrium position, the right half of a string stretching from $-\infty$ to ∞ is given the initial velocity $\dot{y}(x, 0) = 1/(1 + x)$. Find the subsequent deflection of the string. Verify that the deflection is continuous for all values of x and t, that it reduces to 0 when $t = 0$, and that it yields the correct initial velocity.
12. Work Exercise 11 if the string stretches only from 0 to ∞.
13. Work Exercise 12 if the initial velocity is $\dot{y}(x, 0) = e^{-x}$, $x > 0$.
14. In Exercise 13, find the total kinetic energy of the string.
15. Find the deflection of a string stretching from ∞ to $-\infty$ that starts from its equilibrium position with initial velocity

$$\dot{y}(x, 0) = \begin{cases} e^{x} & x < 0 \\ e^{-x} & x > 0 \end{cases}$$

16. Work Exercise 15 if the initial velocity is

$$\dot{y}(x,0)=\begin{cases}-e^{x} & x<0\\ e^{-x} & x>0\end{cases}$$

17. A uniform string stretching from 0 to ∞ begins its motion with initial displacement $\phi(x)$ and initial velocity $\theta(x)$. Show that its motion can be found as the motion of the right half of a two-way infinite string provided merely that the initial displacement $\phi(-x)$ and the initial velocity $\theta(-x)$ for the negative extension of the string satisfy the condition

$$\phi(x)+\phi(-x)=-\frac{1}{a}\int_{-x}^{x}\theta(s)\,ds$$

18. If a semi-infinite string begins its motion with initial displacement $\phi(x)=(1/a)\sin x$ and initial velocity $\theta(x)=1$, and if the negative extension of the string is imagined to have the initial displacement $\phi(x)=0$, $x<0$, find the necessary initial velocity on the extended portion of the string if the origin is to remain at rest.

19. A string stretched from $-\infty$ to ∞. Show that there will be no wave traveling to the left along the string if the initial displacement $\phi(x)$ and the initial velocity $\theta(x)$ satisfy the relation $\theta(x)=-a\phi'(x)$. How must $\phi(x)$ and $\theta(x)$ be related if there is to be no wave traveling to the right along the string?

20. The initial displacement of a two-way infinite string is $y(x,0)=\exp(-x^2)$. With what velocity must the string start to move if its subsequent motion is to consist solely of a wave traveling to the right? Of a wave traveling to the left?

21. The initial velocity of a two-way infinite string is

$$\dot{y}(x,0)=\begin{cases}\sin x & x^2<\pi^2\\ 0 & x^2\geq\pi^2\end{cases}$$

From what initial displacement must the string start to move if its subsequent motion is to consist solely of a wave traveling to the right? Of a wave traveling to the left?

22. Prove Lemma 1. *Hint:* Consider the three cases: $c\leq a$, $b\leq c$, and $a<c<b$.

23. Show that under the substitutions $u=x-at$ and $v=x+at$ the equation $\dfrac{\partial^2 y}{\partial t^2}=a^2\dfrac{\partial^2 y}{\partial x^2}$ becomes $\dfrac{\partial^2 y}{\partial u\,\partial v}=0$. Hence show that $y=f(x-at)+g(x+at)$ is the most general solution of the one-dimensional wave equation.

24. **(a)** Discuss the possibility of extending the d'Alembert solution to the two-dimensional wave equation

$$a^2\left(\frac{\partial^2 u}{\partial x^2}+\frac{\partial^2 u}{\partial y^2}\right)=\frac{\partial^2 u}{\partial t^2}$$

(b) Discuss the possibility of finding solutions of the form $u=e^{\lambda x+\mu y}$ for the equation

$$A\frac{\partial^2 u}{\partial x^2}+B\frac{\partial u^2}{\partial x\,\partial y}+C\frac{\partial^2 u}{\partial y^2}+D\frac{\partial u}{\partial x}+E\frac{\partial u}{\partial y}+Fu=0$$

where A, B, C, D, E, and F are constants.

25. A uniform string stretching from 0 to ∞ is initially displaced into the curve $y=xe^{-x}$ and released from rest in that position. Find its subsequent displacement as a function of x and t.

26. At $t=0$ a uniform string stretching along the x axis from 0 to ∞ is struck in such a way that the portion of the string between $x=1$ and $x=4$ is given unit initial velocity. Find the subsequent displacement of the string $y(x,t)$.

27. In Exercise 26, find the deflection at $x=10$ as a function of t.

28. In Exercise 26, find and plot the deflection at $x=1$ as a function of at.

29. A uniform string stretching from $-\infty$ to ∞ is given the initial displacement

$$y(x,0)=\begin{cases}1-|x| & x^2<1\\ 0 & x^2\geq 1\end{cases}$$

and released from rest. Find the displacement of the string as a function of x and t. What is the transverse velocity of the string at $x=0$?

30. In Exercise 29 find the displacement and velocity at $x=3$ as functions of at.

31. Work Exercise 29 if

$$y(x,0)=\begin{cases}\cos x & x^2<\pi^2/4\\ 0 & x^2\geq\pi^2/4\end{cases}$$

32. A uniform string stretching from 0 to ∞ while at rest in its equilibrium position is struck in such a way that the portion of the string between $x=0$ and $x=\pi$ is given the velocity $\sin x$. Find the displacement of the string as a function of x and t.

33. The initial velocity of a two-way infinite string is

$$\dot{y}(x,0)=\begin{cases}a(2x+k\pi\cos\pi x) & |x|<1\\ 0 & |x|>1\end{cases}$$

What must be its initial displacement if its motion is to consist solely of a wave traveling to the left?

34. Work Exercise 33 if

$$\dot{y}(x,0)=\begin{cases}a(2x-5) & 2<x<3\\ 0 & x<2 \quad\text{or}\quad 3<x\end{cases}$$

and the motion is to consist solely of a wave traveling to the right.

35. A uniform semi-infinite string is given the initial displacement

$$y(x,0)=\begin{cases}x^2\sin x & 0\leq x\leq\pi\\ 0 & \pi\leq x\end{cases}$$

and released from rest. Find the displacement and velocity of the string at $x=\pi$ as a function of at.

36. Work Exercise 35 if the initial displacement is

$$y(x,0) = \begin{cases} 1 - \cos 2x & 0 \le x \le \pi \\ 0 & \pi \le x \end{cases}$$

37. Work Exercise 35 for a two-way infinite string having the initial displacement

$$y(x,0) = \begin{cases} 1 - \cos x & |x - \pi| \le \pi \\ 0 & |x - \pi| \ge \pi \end{cases}$$

38. A uniform semi-infinite string is given the initial displacement

$$y(x,0) = \begin{cases} x(1 - x) & 0 \le x \le 1 \\ 0 & 1 \le x \end{cases}$$

and released from rest. Find the displacement and velocity of the string at $x = 1$ as a function of at.

39. A uniform string stretching from 0 to ∞ is given the initial displacement

$$y(x,0) = \begin{cases} \cos x & \pi/2 \le x \le 3\pi/2 \\ 0 & 0 \le x \le \pi/2,\ 3\pi/2 \le x \end{cases}$$

and released from rest in that position. Find the displacement of the string as a function of x and t. What is the velocity of the string at $x = \pi$ as a function of at?

40. A two-way infinite string is given the initial displacement $y(x,0) = 2[u(x - 2\pi) - u(x - 3\pi)] \sin x$ and released from rest.

(a) Find the displacement of the string as a function of x and t.

(b) Sketch the displacement curve when $at = 3\pi$.

(c) What is the transverse velocity of the string at $x = 0$?

(d) When, if ever, is $\dot{y}(0, t)$ negative?

11.4 CHARACTERISTICS AND THE CLASSIFICATION OF PARTIAL DIFFERENTIAL EQUATIONS

The simplicity and elegance of the d'Alembert solution of the wave equation raises the question of whether other partial differential equations can be solved by this method. Let us consider first the possibility of finding solutions of the form $u = f(x + \lambda y)$ for the equation

$$(1) \qquad A\frac{\partial^2 u}{\partial x^2} + 2B\frac{\partial^2 u}{\partial x\,\partial y} + C\frac{\partial^2 u}{\partial y^2} = 0 \qquad A, B, C \text{ constants}$$

Substituting our tentative solution, we have

$$Af''(x + \lambda y) + 2B\lambda f''(x + \lambda y) + C\lambda^2 f''(x + \lambda y) = 0$$

which will be an identity if and only if

$$(2) \qquad C\lambda^2 + 2B\lambda + A = 0$$

Thus there are two, one, or no (real) values of λ for which solutions of the form $f(x + \lambda y)$ exist, according as the discriminant $B^2 - AC$ is greater than, equal to, or less than zero. By analogy with the criterion for a conic to be a hyperbola, a parabola, or an ellipse, Eq. (1) is said to be a **hyperbolic, parabolic,** or **elliptic** equation according as $B^2 - AC > 0$, $B^2 - AC = 0$, or $B^2 - AC < 0$.

In each case, the loci

$$x + \lambda_1 y = c_1 \qquad \text{and} \qquad x + \lambda_2 y = c_2$$

where λ_1 and λ_2 are the roots of Eq. (2), are called the **characteristic curves,** or simply the **characteristics,** of Eq. (1). In particular, when λ_1 and λ_2 are real and distinct, $x + \lambda_1 y = c_1$ and $x + \lambda_2 y = c_2$ define two families of parallel lines in the xy plane. The slopes of the lines of these two families are, respectively,

$$y_1' = \frac{-1}{\lambda_1} \qquad \text{and} \qquad y_2' = \frac{-1}{\lambda_2}$$

and in each case it is easy to verify that y' satisfies the relation

$$(3) \qquad A(y')^2 - 2By' + C = 0$$

The similarity between Eq. (3) and Eq. (1) is striking, but the difference between the signs of the middle terms should be carefully noted.

More generally, the equation

$$(4) \qquad A(x, y)u_{xx} + 2B(x, y)u_{xy} + C(x, y)u_{yy} = g(u, u_x, u_y, x, y)$$

is said to be **hyperbolic, parabolic,** or **elliptic** throughout a region R according as $B^2(x, y) - A(x, y)C(x, y)$ is greater than, equal to, or less than zero at all points of R. Furthermore, generalizing the property expressed by Eq. (3), the curves

$$\phi(x, y) = c_1 \qquad \text{and} \qquad \psi(x, y) = c_2$$

which are the solutions of the differential equation

$$(5) \qquad A(x, y)(y')^2 - 2B(x, y)y' + C(x, y) = 0$$

are said to be the **characteristics** of the partial differential equation (4).

The simplest, and in elementary applications the most important, examples of hyperbolic, parabolic, and elliptic partial differential equations are, respectively,

(a) The wave equation $a^2 \dfrac{\partial^2 u}{\partial x^2} - \dfrac{\partial^2 u}{\partial t^2} = 0$

(b) The heat equation $\dfrac{\partial^2 u}{\partial x^2} - a^2 \dfrac{\partial u}{\partial t} = 0$

(c) Laplace's equation $\dfrac{\partial^2 u}{\partial x^2} + \dfrac{\partial^2 u}{\partial y^2} = 0$

The concept of the characteristics of a partial differential equation is of great importance in the theory of such equations. In particular, through their use it is possible to reduce the general equation (4) to one or the other of the canonical forms described in the following theorem.

THEOREM 1 Consider the equation

$$A(x, y)u_{xx} + 2B(x, y)u_{xy} + C(x, y)u_{yy} = g(u, u_x, u_y, x, y)$$

and let $\phi(x, y) = c_1$ and $\psi(x, y) = c_2$ be independent solutions of the auxiliary equation

$$A(x, y)(y')^2 - 2B(x, y)y' + C(x, y) = 0$$

Then

(a) If the given equation is hyperbolic, the change of variables defined by the substitutions $r = \phi(x, y)$, $s = \psi(x, y)$ will reduce it to the standard form $u_{rs} = G(u, u_r, u_s, r, s)$.

(b) If the given equation is parabolic, the change of variables defined by the substitutions $r = x$, $s = \phi(x, y)$ will reduce it to the standard form $u_{rr} = G(u, u_r, u_s, r, s)$.

(c) If the given equation is elliptic, the change of variables defined by the substitutions $r = \dfrac{\phi(x, y) + \psi(x, y)}{2}$, $s = \dfrac{\phi(x, y) - \psi(x, y)}{2i}$ will reduce it to the standard form $u_{rr} + u_{ss} = G(u, u_r, u_s, r, s)$.

Since the equations we will be considering in our work will almost always come to us in a convenient form, we have no particular need for this theorem, and hence we shall leave its proof to more advanced texts.

EXAMPLE 1

Find the characteristics of the hyperbolic equation

$$x\frac{\partial^2 u}{\partial x^2} - y\frac{\partial^2 u}{\partial x\,\partial y} + \frac{\partial u}{\partial x} = 0$$

Reduce the equation to canonical form and solve it, if possible.

In this case, Eq. (5) becomes

$$x(y')^2 + yy' = 0 \qquad \text{or} \qquad y'(xy' + y) = 0$$

The first factor implies that $y = c_1$. From the second, we have

$$x\frac{dy}{dx} + y = 0$$

$$\frac{dy}{y} + \frac{dx}{x} = 0$$

$$\ln|y| + \ln|x| = \ln|c_2|$$

and finally

$$xy = c_2$$

The characteristics are thus $y = c_1$ and $xy = c_2$.

From the substitutions

$$r = y \qquad \text{and} \qquad s = xy$$

prescribed by **a** of Theorem 1 we have, by the chain rule for partial derivatives,

$$\frac{\partial u}{\partial x} = \frac{\partial u}{\partial r}\frac{\partial r}{\partial x} + \frac{\partial u}{\partial s}\frac{\partial s}{\partial x} = y\frac{\partial u}{\partial s}$$

$$\frac{\partial^2 u}{\partial x^2} = y\left(\frac{\partial^2 u}{\partial r\,\partial s}\frac{\partial r}{\partial x} + \frac{\partial^2 u}{\partial s^2}\frac{\partial s}{\partial x}\right) = y^2\frac{\partial^2 u}{\partial s^2}$$

$$\frac{\partial^2 u}{\partial y\,\partial x} = \frac{\partial u}{\partial s} + y\left(\frac{\partial^2 u}{\partial r\,\partial s}\frac{\partial r}{\partial y} + \frac{\partial^2 u}{\partial s^2}\frac{\partial s}{\partial y}\right) = \frac{\partial u}{\partial s} + y\frac{\partial^2 u}{\partial r\,\partial s} + xy\frac{\partial^2 u}{\partial s^2}$$

Hence, substituting into the given equation, we have

$$x\left(y^2\frac{\partial^2 u}{\partial s^2}\right) - y\left(\frac{\partial u}{\partial s} + y\frac{\partial^2 u}{\partial r\,\partial s} + xy\frac{\partial^2 u}{\partial s^2}\right) + y\frac{\partial u}{\partial s} = 0$$

or

$$-y^2\frac{\partial^2 u}{\partial u\,\partial s} = 0$$

or, finally,

$$\frac{\partial^2 u}{\partial r\,\partial s} = 0$$

The solution of the last equation is simply

$$u(r, s) = f(r) + g(s)$$

and from this we find that the solution of the original equation is

$$u(x, y) = f(y) + g(xy)$$

where f and g are arbitrary, twice-differentiable functions.

The characteristics of parabolic and elliptic equations are of relatively little significance, at least at the level of our work. On the other hand, the two families of characteristics of hyperbolic equations, being real and distinct, are of considerable practical value. In particular, for the one-dimensional wave equation, a consideration of the characteristics can give us a good deal of information about the propagation of disturbances through the medium.

For the wave equation

$$a^2 \frac{\partial^2 u}{\partial x^2} - \frac{\partial^2 u}{\partial t^2} = 0$$

in which t now plays the role of y in the general equation (1), the auxiliary equation (3) is

$$a^2\left(\frac{dt}{dx}\right)^2 - 1 = 0$$

Hence $dt/dx = \pm 1/a$, and the characteristics are defined by

$$x - at = c_1 \qquad \text{and} \qquad x + at = c_2$$

as, in effect, we discovered in our discussion of D'Alembert's solution of the wave equation. Now if $x - at$ and $x + at$ are constant along the characteristics of the respective families, then so too are $f(x - at)$ and $g(x + at)$. In other words, at every point of any characteristic of the family $x - at = c_1$, the value of $f(x - at)$ is its value at the x intercept of that characteristic. Similarly, at any point of any characteristic of the family $x + at = c_2$, the value of $g(x + at)$ is its value at the x intercept of that characteristic. Furthermore, at any point P on the x axis, the values of $u(x, 0)$ and $u_t(x, 0)$ are given as initial conditions of displacement and velocity. Thus they suffice to determine the (constant) values of f and g along the two characteristics that pass through P. The characteristics therefore represent the paths in the xt plane along which disturbances in the medium are propagated. Finally, since the solution of the wave equation is $u(x, t) = f(x - at) + g(x + at)$, it follows that the value of u at any point in the xt plane is the sum of the values of f and g on the respective characteristics that pass through that point.

As an illustration of these ideas, let us consider a two-way infinite string which is set in motion by a disturbance confined to the interval between P_1: $(x_1, 0)$ and P_2: $(x_2, 0)$, and let us attempt to describe the state of motion of the string at any point x at any subsequent time t. This can be done, as we have just said, by tracing back to their x intercepts the two characteristics which pass through a general point Q: (x, t). In doing this it will be helpful to consider the six regions into which the upper half of the xt plane is divided by the characteristics $x - at = x_1$ and $x + at = x_1$ that pass through P_1 and the characteristics $x - at = x_2$ and $x + at = x_2$ that pass through P_2, as shown in Fig. 11.14.

Specifically, at the time and place corresponding to any point in Region I, III, or V the string is at rest, since the characteristics which pass through such a point cannot intersect the x axis in a point between P_1 and P_2. That is, they intersect the line $t = 0$ only in points where the initial disturbance is zero. In Region IV the left-moving wave is currently affecting the string, but the right-moving wave is not, because the characteristic of negative slope through any point in Region IV intersects the x axis in a point between P_1 and P_2 while the characteristic of positive slope

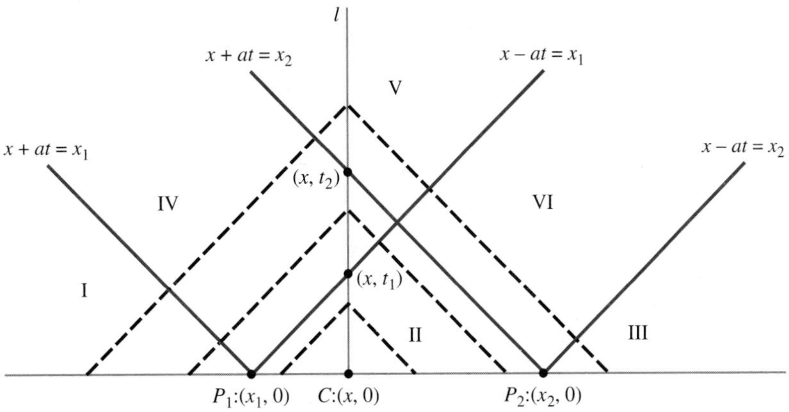

FIGURE 11.14
The descriptive significance of the characteristics of the wave equation.

through such a point does not. Similarly, in Region VI the right-moving wave is currently affecting the string but the left-moving wave is not. In Region II both the right-moving and the left-moving waves are currently affecting the string.

The behavior of a particular point on the string throughout the course of its motion can be determined qualitatively by moving upward along the vertical line which passes through the point in question. For example, as we move upward along the vertical line l through the point C in Fig. 11.14, we see that at C the string is affected by both traveling waves until the time t_1 corresponding to the intersection of l and the characteristic $x - at = x_1$. At that time the right-moving wave passes by, but the left-moving wave continues to affect the string until the time t_2 corresponding to the point where l leaves Region IV and enters Region V. At this time, both waves have passed by and the string has come to rest permanently at the point C.

Observations similar to the preceding can be made for the one-way infinite string and for finite strings, but we shall leave these to the exercises.

EXERCISES

1. Prove that if the equation

$$A\frac{\partial^2 u}{\partial x^2} + 2B\frac{\partial^2 u}{\partial x\, \partial y} + C\frac{\partial^2 u}{\partial y^2} = 0 \qquad A, B, C \text{ constants}$$

is parabolic and if $\lambda = a$ is the repeated root of the equation $C\lambda^2 + 2B\lambda + A = 0$, then both $u = f(x + ay)$ and $u = yf(x + ay)$ are solutions of the given equation.

2. Under the hypotheses of Exercise 1, show that $u = xf(x + ay)$ is also a solution of the given equation.

3. For what values of x and y is each of the following equations hyperbolic? Parbolic? Elliptic?

(a) $(y+1)\dfrac{\partial^2 u}{\partial x^2} + 2x\dfrac{\partial^2 u}{\partial x\, \partial y} + \dfrac{\partial^2 u}{\partial y^2} = x + y$

(b) $(x+1)\dfrac{\partial^2 u}{\partial x^2} + 2x\dfrac{\partial^2 u}{\partial x\, \partial y} + y\dfrac{\partial^2 u}{\partial y^2} = u$

(c) $(1-y)\dfrac{\partial^2 u}{\partial x^2} + 2(1-x)\dfrac{\partial^2 u}{\partial x\, \partial y} + (1+y)\dfrac{\partial^2 u}{\partial y^2} - x\dfrac{\partial u}{\partial y} + y\dfrac{\partial u}{\partial x} = 0$

4. Solve each of the following equations.

(a) $3u_{xx} - 4u_{xy} + u_{yy} = 0$
(b) $2u_{xx} - 5u_{xy} + 2u_{yy} = 0$
(c) $9u_{xx} + 6u_{xy} + u_{yy} = 0$
(d) $4u_{xx} + 4u_{xy} + u_{yy} = \sin(x + 2y)$
(e) $u_{xx} + 2u_{xy} - 3u_{yy} = e^{x+2y}$
(f) $25u_{xx} - 10u_{xy} + u_{yy} = x^2 + y$

5. Discuss the use of characteristics in describing qualitatively the behavior of a one-way infinite string which is set in motion by an initial disturbance restricted to the interval between P_1: $(x_1, 0)$ and P_2: $(x_2, 0)$. *Hint:* Since the left-hand end of the string remains fixed, the sum of the values of u on the characteristics which intersect in a point on the vertical line through the left-hand end of the string must be zero.

6. While a one-way infinite string is at rest in its equilibrium position, its left-hand end begins to move so that subsequently $u(0, t) = \phi(t)$. Using the properties of the characteristics of the wave equation, show that a particle of the string at a distance x_1 from the fixed end moves precisely as did the left-hand end of the string x_1/a units of time earlier.

7. Discuss the use of characteristics in describing qualitatively the behavior of a finite string which is set in motion by a disturbance extending over its entire length. *Hint:* Recall the suggestion made in Exercise 5.

8. Using the properties of the characteristics of the wave equation, show that the period of vibration of a string of length l with fixed ends is $2l/a$.

9. At $t = 0$, while a stretched string of length l is at rest in its equilibrium position, its left-hand end begins to move according to the law $u(0, t) = \sin 2\pi t$. Using the properties of the characteristics of the wave equation and taking $a = 1$, determine the deflection curve of the string for
(a) $0 \le t \le l$ **(b)** $l \le t \le 2l$ **(c)** $2l \le t \le 3l$

10. Find the characteristics of each of the following equations and reduce each equation to the appropriate standard form, using the substitutions given in Theorem 1. Where possible, obtain real solutions for the equation.
(a) $u_{xx} - 4x^2 u_{yy} = \frac{1}{x} u_x$ **(b)** $xu_{xy} + yu_{yy} = 0$
(c) $x^2 u_{xx} + 2xyu_{xy} + y^2 u_{yy} = 0$
(d) $x^2 u_{xx} - 2xyu_{xy} + y^2 u_{yy} = 0$

11. Suppose that u, u_x, and u_y are known at every point of a curve Γ in the xy plane, where u is a solution of the equation $A(x, y)u_{xx} + 2B(x, y)u_{xy} + C(x, y)u_{yy} = G(u, u_x, u_y, x, y)$. Show that the auxiliary equation

$$A(x, y)(y')^2 - 2B(x, y)y' + C(x, y) = 0$$

is the condition that the values of the second derivatives u_{xx}, u_{xy}, and u_{yy} *cannot* be calculated at the points of Γ. *Hint:* Note that $du_x = u_{xx}\,dx + u_{xy}\,dy$ and $du_y = u_{xy}\,dx + u_{yy}\,dy$ and then consider the possibility of solving these two equations simultaneously with the given partial differential equation for u_{xx}, u_{xy}, and u_{yy}.

11.5 SEPARATION OF VARIABLES

We are now ready to consider the solution of partial differential equations by the method of *separation of variables.* Although this method is not universally applicable, it suffices for most of the partial differential equations encountered in elementary applications in engineering and physics and leads directly to the heart of the branch of mathematics dealing with *boundary-value problems,* which is really the central subject of this chapter.

The idea behind the method is the familiar mathematical strategem of reducing a new problem to dependence upon an old one. In this case we attempt to convert the given partial differential equation into several ordinary differential equations, hoping that what we know about the latter will prove adequate for a successful continuation of the search for solutions.

To illustrate the details of the procedure, we shall again consider the one-dimensional wave equation, this time taking as a specific representation a uniform, undamped, torsionally vibrating shaft of finite length, subject to given initial conditions of angular displacement and velocity.† This means that our objective is to find solutions of the equation [Eq. (8), Sec. 11.2]

$$\frac{\partial^2 \theta}{\partial t^2} = a^2 \frac{\partial^2 \theta}{\partial x^2}$$

which are *periodic in time* and satisfy the initial conditions of angular displacement and angular velocity

$$(1) \qquad \theta(x, 0) = f(x) \qquad \text{and} \qquad \left.\frac{\partial \theta}{\partial t}\right|_{x,0} = g(x)$$

We assume, as a working hypothesis, that solutions for the angle of twist θ exist as products of a function of x alone and a function of t alone:

$$\theta(x, t) = X(x)T(t)$$

†We choose the torsionally vibrating shaft as a model here not primarily for the sake of variety but because free end conditions, as well as fixed end conditions, are important to consider and are quite realistic for a shaft, although for a string, while possible (see Example 1, Sec. 11.8), they are less natural. Of course, all the results of the present discussion are relevant to any system satisfying the one-dimensional wave equation.

If this is the case, partial differentiation of θ amounts to total differentiation of one or the other of the factors of θ, and we have

$$\frac{\partial^2\theta}{\partial x^2} = X''T \qquad \text{and} \qquad \frac{\partial^2\theta}{\partial t^2} = XT''$$

Substituting these into the wave equation, we obtain

$$XT'' = a^2X''T$$

Dividing by the product XT then gives

$$\text{(2)} \qquad \frac{T''}{T} = a^2\frac{X''}{X}$$

as a necessary condition that $\theta(x, t) = X(x)T(t)$ should be a solution.

Now the left-hand member of (2) is clearly independent of x. Hence (in spite of its appearance) the right-hand side of (2) must also be independent of x since it is identically equal to the expression on the left. Similarly, each member of (2) must be independent of t. Therefore, being independent of both x and t, each side of (2) must be a constant, say μ, and we can write

$$\frac{T''}{T} = a^2\frac{X''}{X} = \mu$$

Thus the determination of solutions of the original partial differential equation has been reduced to the determination of solutions of the two ordinary differential equations

$$T'' = \mu T \qquad \text{and} \qquad X'' = \frac{\mu}{a^2}X$$

Assuming that we need consider only real values of μ, there are thus three cases to investigate:

$$\mu > 0 \qquad \mu = 0 \qquad \mu < 0$$

If $\mu > 0$, we can write $\mu = \lambda^2$, $\lambda > 0$. In this case the two differential equations and their solutions are

$$T'' = \lambda^2 T \qquad\qquad X'' = \frac{\lambda^2}{a^2}X$$

$$T = Ae^{\lambda t} + Be^{-\lambda t} \qquad X = Ce^{\lambda x/a} + De^{-\lambda x/a}$$

But a solution of the form

$$\theta(x, t) = X(x)T(t) = (Ce^{\lambda x/a} + De^{-\lambda x/a})(Ae^{\lambda t} + Be^{-\lambda t})$$

cannot describe the undamped vibrations of a system because it is not periodic, i.e., does not repeat itself periodically as time increases. Hence, although product solutions of the wave equation exist for $\mu > 0$, they have no significance in relation to the problem we are considering.

If $\mu = 0$, the equations in T and X and their solutions are

$$T'' = 0 \qquad\qquad X'' = 0$$

$$T = At + B \qquad X = Cx + D$$

But again, a solution of the form

$$\theta(x, t) = X(x)T(t) = (Cx + D)(At + B)$$

cannot describe periodic motion, although if $C = 0$, it does describe a uniform rotation of the shaft. For a shaft with even one end fixed, uniform rotation is impossible. Hence for such shafts the possibility that $\mu = 0$ must be rejected. On the other hand, a shaft with both ends free is capable of uniform rotation, and for such shafts $\mu = 0$ remains a possibility.

Finally, if $\mu < 0$, we can write $\mu = -\lambda^2$, $\lambda > 0$. Then the component differential equations and their solutions are

$$T'' = -\lambda^2 T \qquad X'' = -\frac{\lambda^2}{a^2}X$$

$$T = A\cos\lambda t + B\sin\lambda t \qquad X = C\cos\frac{\lambda}{a}x + D\sin\frac{\lambda}{a}x$$

In this case the solution

(3) $$\theta(x, t) = X(x)T(t) = \left(C\cos\frac{\lambda}{a}x + D\sin\frac{\lambda}{a}x\right)(A\cos\lambda t + B\sin\lambda t)$$

is clearly periodic, repeating itself every time t increases by $2\pi/\lambda$. In other words, $\theta(x, t)$, as defined in (3), represents vibratory motion with period $2\pi/\lambda$ or frequency $\lambda/2\pi$.

It remains now to find the value or values of λ and the constants A, B, C, and D. Since the admissible values of λ are determined by the end, or boundary, conditions the shaft must satisfy, the continuation now varies in some respects, depending upon how the shaft is constrained at its ends. We shall discuss in turn the following commonly occurring cases (Fig. 11.15).

(a) Both ends of the shaft are fixed, or built-in, i.e., are constrained so that no twisting can take place.

(b) Both ends of the shaft are free to twist.

(c) One end of the shaft is built-in; the other is free to twist.

If both ends of the shaft are held fixed, we have the following conditions to impose upon the general expression (3) for $\theta(x, t)$, assuming the x axis is chosen along the shaft so that the left end of the shaft is at $x = 0$ and the right end is at $x = l$:

$$\theta(0, t) = \theta(l, t) = 0 \qquad \text{identically in } t$$

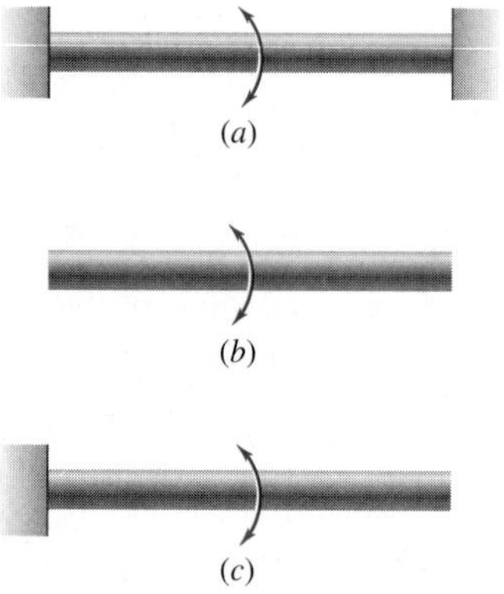

FIGURE 11.15
End conditions for a shaft vibrating torsionally: (*a*) fixed-fixed; (*b*) free-free; (*c*) fixed-free.

Substituting $x = 0$ into the expression (3), we find

$$\theta(0, t) = 0 = C(A \cos \lambda t + B \sin \lambda t)$$

This condition will obviously be fulfilled for all values of t if both A and B are zero. In this case, however, $\theta(x, t)$ is zero at all times, and the shaft remains motionless, a possible but trivial solution in which we have no interest. Hence we are driven to the other alternative, $C = 0$, which reduces (3) to the form

$$\theta(x, t) = \left(D \sin \frac{\lambda}{a} x\right)(A \cos \lambda t + B \sin \lambda t)$$

The second boundary condition, namely, that the right-hand end of the shaft remains motionless at all times, requires that

$$\theta(l, t) = 0 = \left(D \sin \frac{\lambda}{a} l\right)(A \cos \lambda t + B \sin \lambda t)$$

As before, we reject the possibility that $A = B = 0$ since it leads only to a trivial solution. Moreover, we cannot permit $D = 0$, since that too, with C already zero, leads to the trivial case. The only possibility remaining is that

$$\sin \frac{\lambda}{a} l = 0 \qquad \text{or} \qquad \frac{\lambda l}{a} = n\pi$$

From the continuous infinity of values of the positive parameter λ for which periodic product solutions of the wave equation exist, we have thus been forced to reject all but the values

(4) $$\lambda_n = \frac{n\pi a}{l} \qquad n = 1, 2, 3, \ldots$$

To each of these values of λ, *and to no others,* there corresponds a product solution

$$\theta_n(x, t) = \left(\sin \frac{\lambda_n}{a} x\right)(A_n \cos \lambda_n t + B_n \sin \lambda_n t)$$

$$= \sin \frac{n\pi x}{l}\left(A_n \cos \frac{n\pi a t}{l} + B_n \sin \frac{n\pi a t}{l}\right)^\dagger$$

which, in addition to being periodic in time, also satisfies the end, or boundary conditions of the problem we are currently considering. It remains now to construct from these product solutions a solution that will also satisfy the given initial conditions of angular displacement and angular velocity (1).

Clearly, no single product solution $\theta_n(x, t)$ can satisfy the arbitrary initial displacement condition $\theta(x, 0) = f(x)$ or the arbitrary initial velocity condition $\left.\frac{\partial \theta}{\partial t}\right|_{x,0} = g(x)$. However, the wave

†The constants A and B now bear subscripts to indicate that they are not necessarily the same in solutions associated with different values of λ. The constant D can of course be absorbed in the constants A and B and need not be explicitly included.

equation is linear, and thus if we have several solutions, their sum is also a solution. Hence it is natural (though perhaps optimistic in view of the questions of convergence that are raised) to ask if an *infinite* series of *all* the θ_n's, say

$$\theta(x, t) = \sum_{n=1}^{\infty} \theta_n(x, t) = \sum_{n=1}^{\infty} \sin\frac{n\pi x}{l}\left(A_n \cos\frac{n\pi at}{l} + B_n \sin\frac{n\pi at}{l}\right) \tag{5}$$

can be made to yield a solution fitting the given initial conditions of angular displacement and velocity.

This can be done, and in fact in this case the determination of the coefficients A_n and B_n requires nothing more than a simple application of Fourier series, as developed in Chap. 8. For if we set $t = 0$ in $\theta(x, t)$, we obtain from Eq. (5) and the given initial displacement condition (1) the requirement that

$$\theta(x, 0) = f(x) = \sum_{n=1}^{\infty} A_n \sin\frac{n\pi x}{l}$$

The problem of determining the A_n's so that this will be true is nothing but the problem of expanding a given function $f(x)$ in a half-range sine series over the interval $(0, l)$. Using Theorem 2, Sec. 8.4, we have explicitly

$$A_n = \frac{2}{l}\int_0^l f(x) \sin\frac{n\pi x}{l}\,dx$$

To determine the B_n's, we note, further that from (5),

$$\frac{\partial\theta}{\partial t} = \sum_{n=1}^{\infty} \sin\frac{n\pi x}{l}\left(-A_n \sin\frac{n\pi at}{l} + B_n \cos\frac{n\pi at}{l}\right)\frac{n\pi a}{l}$$

Hence, putting $t = 0$, we have, from the initial velocity condition,

$$\left.\frac{\partial\theta}{\partial t}\right|_{x,0} = g(x) = \sum_{n=1}^{\infty}\left(\frac{n\pi a}{l}B_n\right)\sin\frac{n\pi x}{l}$$

This again merely requires that the B_n's be determined so that the quantities

$$\frac{n\pi a}{l}B_n$$

will be the coefficients in the half-range sine expansion of the known function $g(x)$. Thus

$$\frac{n\pi a}{l}B_n = \frac{2}{l}\int_0^l g(x)\sin\frac{n\pi x}{l}\,dx \qquad \text{or} \qquad B_n = \frac{2}{n\pi a}\int_0^l g(x)\sin\frac{n\pi x}{l}\,dx$$

Aside from questions of convergence, our problem is now solved. We know that a uniform shaft with both ends restrained against turning can vibrate torsionally at any of an infinite number of natural frequencies,

$$\frac{\lambda_n}{2\pi} = \frac{na}{2l} \text{ cycles per unit time} \qquad n = 1, 2, 3, \ldots$$

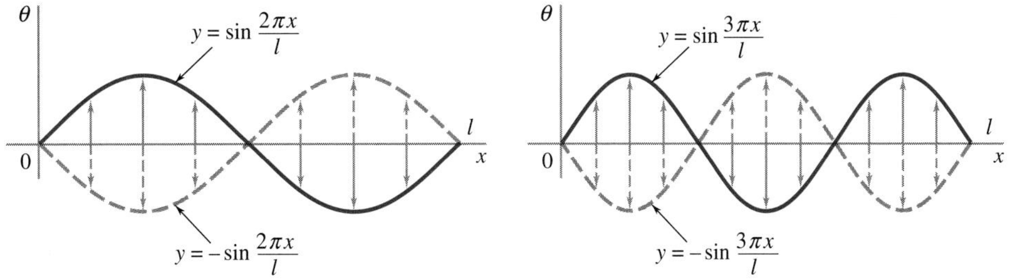

FIGURE 11.16
Typical normal modes of a fixed-fixed shaft.

If and when the shaft vibrates at a *single* one of these frequencies, say $\lambda_n/2\pi$, we know that the motion of the shaft is described by the term

$$\sin \frac{n\pi x}{l}\left(A_n \cos \frac{n\pi at}{l} + B_n \sin \frac{n\pi at}{l}\right)$$

in Eq. (5). From the time factor, we see that at every x the cross section vibrates harmonically with period $2l/na$. From the space factor, $\sin n\pi x/l$, we see that at every x the cross section vibrates harmonically between amplitudes proportional to $\pm\sin(n\pi x/l)$. Typical amplitude curves, for $n = 2$ and $n = 3$, are shown in Fig. 11.16. As in the more familiar case of a vibrating string, a shaft vibrating torsionally at one of its natural frequencies also has **nodes,** that is, points at which the cross section remains stationary and does not vibrate. Specifically, as Fig. 11.16 suggests, it is easy to show that when it is vibrating at its n*th* natural frequency, a fixed-fixed shaft has $n - 1$ equally spaced nodes.

Finally, when we are given the initial angular displacement and angular velocity at each cross section, we know how to determine A_n and B_n by familiar Fourier theory. Thus we are able to express the twisting motion of the shaft under any starting conditions (satisfying the Dirichlet conditions, of course) as a sum of pure harmonic motions at its various natural frequencies.

Before we can discuss the free-free shaft, we must be clear about just how *free* it is. From Fig. 11.15*b*, it appears that it is entirely unrestrained and can move, or be moved, in any number of ways having nothing to do with angular vibration. What we mean here by a free-free shaft is one which is held in place by frictionless bearings so that rotation and angular vibration are possible but all other motion is prevented. With this understood, the analysis of the torsional behavior of such a shaft follows closely the preceding discussion once we realize, as we pointed out previously, that the case $\mu = 0$ with its accompanying solution $\theta(x, t) = (At + B)$ represents uniform angular rotation and may be relevant if we are concerned with torsional vibrations superposed on a uniformly rotating shaft.

Our first step is to obtain the proper formulation of the new end conditions. To do this, we observe that at a free end, although we do not know the amount of twist, we do know that there is no torque acting at the end section since there is no material of the shaft beyond the end section. Recalling, from the discussion in Sec. 11.2, the expression for the torque transmitted through a general cross section of a twisted shaft, we thus find that the free ends are characterized by the requirement that

$$\text{Torque at end section} = E_s J \left.\frac{\partial \theta}{\partial x}\right|_{\text{end}} = 0$$

Since E_s is a nonzero constant of the material of the shaft, and since J cannot vanish for a shaft of uniform cross section such as we are considering, it follows that at a free end $\partial\theta/\partial x = 0$.

Returning to the original product solution (3), we find that

$$\frac{\partial \theta}{\partial x} = \left(-C\frac{\lambda}{a}\sin\frac{\lambda}{a}x + D\frac{\lambda}{a}\cos\frac{\lambda}{a}x\right)(A\cos\lambda t + B\sin\lambda t)$$

Substituting $x = 0$ and equating the result to zero, we obtain the condition

$$\frac{\lambda}{a}D(A\cos\lambda t + B\sin\lambda t) = 0 \qquad \text{for all } t$$

and from this we conclude that $D = 0$. Similarly, imposing the right-hand end condition by substituting $x = l$ and again equating the result to zero, we find

$$-C\frac{\lambda}{a}\sin\frac{\lambda l}{a}(A\cos\lambda t + B\sin\lambda t) = 0 \qquad \text{for all } t$$

Since we cannot permit $C = 0$, we must have

$$\sin\frac{\lambda l}{a} = 0 \qquad \text{or} \qquad \frac{\lambda l}{a} = n\pi$$

Thus, as in the fixed-ends case, to have the end conditions of the problem fulfilled, λ must be restricted to one of the discrete set of values

$$\lambda_n = \frac{n\pi a}{l} \qquad n = 1, 2, 3, \ldots$$

Again we construct the product solution for each admissible value of λ, getting

$$\theta_n(x, t) = \left(\cos\frac{\lambda_n}{a}x\right)(A_n\cos\lambda_n t + B_n\sin\lambda_n t)$$

(6) $$= \cos\frac{n\pi x}{l}\left(A_n\cos\frac{n\pi at}{l} + B_n\sin\frac{n\pi at}{l}\right)$$

and attempt to form an infinite series of these solutions

(7) $$\theta(x, t) = \sum_{n=1}^{\infty}\cos\frac{n\pi x}{l}\left(A_n\cos\frac{n\pi at}{l} + B_n\sin\frac{n\pi at}{l}\right)$$

which will satisfy the initial displacement condition $\theta(x, 0) = f(x)$ and the initial velocity condition $\partial\theta/\partial t|_{x,0} = g(x)$.

To satisfy the initial displacement condition, we set $t = 0$ in (7), getting

(7*a*) $$\theta(x, 0) = f(x) = \sum_{n=1}^{\infty} A_n\cos\frac{n\pi x}{l}$$

which requires that the A_n's be the coefficients in the half-range cosine expansion† of the known function $f(x)$ over the interval $(0, l)$, that is, that

$$A_n = \frac{2}{l}\int_0^l f(x)\cos\frac{n\pi x}{l}\,dx$$

To satisfy the initial velocity condition, we differentiate (7) with respect to t and then set $t = 0$, getting

$$\left.\frac{\partial\theta}{\partial t}\right|_{x,0} = g(x) = \sum_{n=1}^{\infty}\left(\frac{n\pi a}{l}B_n\right)\cos\frac{n\pi x}{l} \tag{7b}$$

which requires that the quantities

$$\frac{n\pi a}{l}B_n$$

be the coefficients in the half-range cosine series for $g(x)$ over the interval $(0, l)$, that is, that

$$\frac{n\pi a}{l}B_n = \frac{2}{l}\int_0^l g(x)\cos\frac{n\pi x}{l}\,dx \qquad \text{or} \qquad B_n = \frac{2}{n\pi a}\int_0^l g(x)\cos\frac{n\pi x}{l}\,dx$$

Since the values of λ for which nontrivial solutions exist are the same for both the fixed-fixed and free-free shafts, it follows that the natural frequencies of the two shaft configurations are the same, namely, $\lambda_n/2\pi = na/2l$. However, when the two shafts are vibrating at a single one of their common frequencies, say the nth, the amplitudes through which corresponding cross sections vibrate are *not* the same. For the fixed-fixed shaft we saw that the amplitudes along the shaft were proportional to $\pm\sin(n\pi x/l)$, whereas from Eq. (6) it is clear that for the free-free shaft the amplitude at any x is proportional to $\pm\cos(n\pi x/l)$. Typical amplitude curves, for $n = 2$ and $n = 3$, are shown in Fig. 11.17. These curves illustrate the fact that, like the fixed-fixed shaft, the free-free shaft also has nodes when it is vibrating at a single natural frequency.

FIGURE 11.17

Typical normal modes of a free-free shaft.

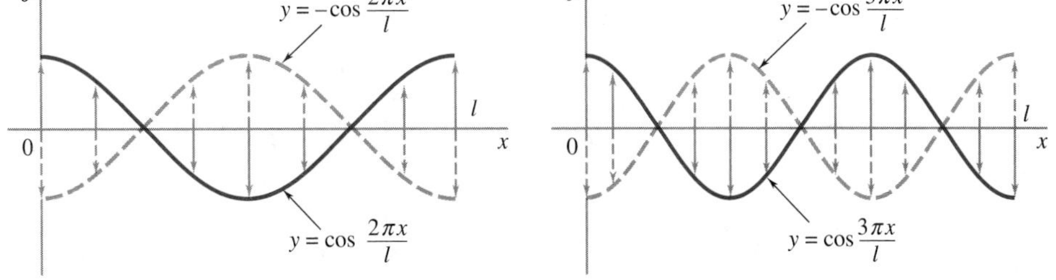

†In general, the half-range cosine expansion of a function begins with a constant term. The absence of that term here means simply that the vibrations we are studying are not superposed on a uniform rotation of the shaft. Had this been the case, Eq. (7) would have had to include the term $\theta_0(x, t) = At + B$, arising from the possibility that $\mu = 0$, and the series in the equation $7a$ involving $\theta(x, 0)$ and $f(x)$ would have begun with the term $\theta_0(x, 0) = A$.

For the free-free shaft, just as for the fixed-fixed shaft, when we are given initial conditions of displacement and velocity (satisfying the Dirichlet conditions, of course) we can determine the coefficients A_n and B_n in Eq. (7) by familiar Fourier theory. Thus again we are able to express the motion of a free-free shaft, twisting under any starting conditions, as a sum of the simple harmonic motions occurring at its various natural frequencies.

The case of the shaft with one end fixed and the other free can be disposed of quickly. Taking the fixed end at $x = 0$ and the free end at $x = l$, we have the two conditions

$$\theta(0, t) = 0 \qquad \text{and} \qquad \left.\frac{\partial \theta}{\partial x}\right|_{l,t} = 0 \qquad \text{for all } t$$

Imposing the first of these upon the general product solution (3) gives

$$C(A \cos \lambda t + B \sin \lambda t) = 0$$

from which it follows that $C = 0$. Imposing the second then gives

$$\frac{\lambda}{a} D \cos \frac{\lambda l}{a}(A \cos \lambda t + B \sin \lambda t) = 0$$

from which we conclude that

$$\cos \frac{\lambda l}{a} = 0 \qquad \frac{\lambda l}{a} = \frac{(2n-1)\pi}{2} \qquad \text{and} \qquad \lambda_n = \frac{(2n-1)\pi a}{2l}$$

The general solution of the problem, formed by adding together the product solutions corresponding to each λ_n is therefore

$$\begin{aligned}\theta(x, t) &= \sum_{n=1}^{\infty} \left(\sin \frac{\lambda_n}{a} x\right)(A \cos \lambda_n t + B_n \sin \lambda_n t) \\ &= \sum_{n=1}^{\infty} \sin \frac{(2n-1)\pi x}{2l}\left[A_n \cos \frac{(2n-1)\pi a t}{2l} + B_n \sin \frac{(2n-1)\pi a t}{2l}\right]\end{aligned}$$

To fit the initial displacement condition $\theta(x, 0) = f(x)$, we must have

$$f(x) = \sum_{n=1}^{\infty} A_n \sin \frac{(2n-1)\pi x}{2l}$$

This is not quite the usual half-range sine expansion problem since the arguments of the various terms are not integral multiples of the fundamental argument $\pi x/l$. It is, however, the half-range sine expansion covered by Theorem 3, Sec. 8.4. There, the formula for the coefficients was shown to be

$$A_n = \frac{2}{l}\int_0^l f(x) \sin \frac{(2n-1)\pi x}{2l}\, dx$$

Similarly, to satisfy the initial velocity condition $\partial\theta/\partial t|_{x,0} = g(x)$, we must have

$$g(x) = \sum_{n=1}^{\infty}\left[\frac{(2n-1)\pi a}{2} B_n\right] \sin \frac{(2n-1)\pi x}{2l}$$

which requires that

$$B_n = \frac{4}{(2n-1)\pi a}\int_0^l g(x) \sin \frac{(2n-1)\pi x}{2l}\, dx$$

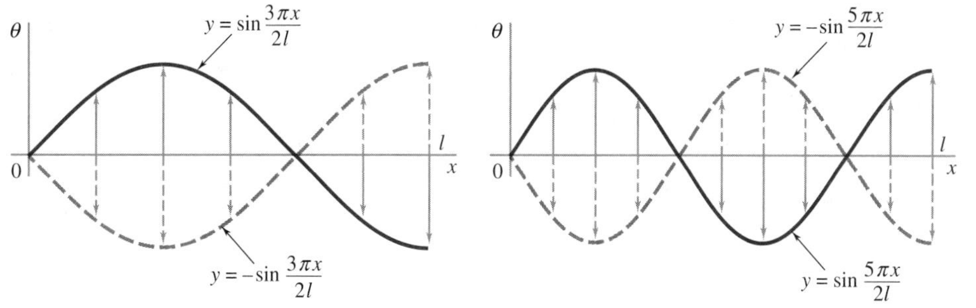

FIGURE 11.18
Typical normal modes of a fixed-free shaft.

The values of λ for which nontrivial solutions exist for the shaft with one end fixed and one end free, namely, $\lambda_n/2\pi = (2n-1)a/4l$, are not the same as those for the fixed-fixed and free-free shafts. Moreover, when a fixed-free shaft is vibrating at a single one of its natural frequencies, the distribution of the amplitudes through which the various cross sections vibrate is different from that for either the fixed-fixed or free-free shaft. For a fixed-free shaft, vibrating at a single natural frequency $\lambda_n/2\pi$, the cross section at any point x twists periodically between amplitudes proportional to $\pm\sin(2n-1)\pi x/2l$. Typical amplitude curves of this family are shown in Fig. 11.18. Again, we note the existence of nodes along the shaft. Finally, using familiar Fourier theory, the response of the shaft to any starting conditions satisfying the Dirichlet conditions can be expressed as a sum of simple harmonic motions whose frequencies are the natural frequencies of the shaft.

Thus far, our discussion has centered on the one-dimensional wave equation as applied to the study of torsionally vibrating shafts. However, our primary concern is neither the wave equation nor vibrating shafts. Instead, our real interest is in understanding the method of separation of variables and the variety of problems to which it can be applied. Accordingly, in concluding this section we shall illustrate the versatility of this method by considering two problems in heat flow, one involving Laplace's equation in two dimensions, the other involving the one-dimensional heat equation, to which the ideas we have been developing can be applied with almost no change.

EXAMPLE 1

THE STEADY-STATE TEMPERATURE OF A UNIT SQUARE

A sheet of metal coincides with the square in the xy plane whose vertices are the points $(0, 0)$, $(1, 0)$ $(1, 1)$, and $(0, 1)$. The two faces of the sheet are insulated, and the sheet is so thin that heat flow in it can be regarded as two-dimensional. The edges parallel to the x axis are insulated, and the left-hand edge is maintained at the constant temperature 0. If the temperature distribution $u(1, y) = f(y)$ is maintained along the right-hand edge, find the steady-state temperature distribution throughout the sheet.

Our first observation is that since we are asked to find the steady-state temperature distribution in the sheet, we may legitimately suppose that the derivative $\partial u/\partial t$ is identically zero and that no heat sources exist. Under these assumptions, the two-dimensional heat equation [Eq. (14b), Sec. 11.2] reduces to Laplace's equation [Eq. (15a), Sec. 11.2]:

$$\frac{\partial^2 u}{\partial x^2} + \frac{\partial^2 u}{\partial y^2} = 0$$

In an attempt to separate variables in this equation and reduce the problem to the solution of two ordinary differential equations, we assume as a tentative solution the product

$$u(x, y) = X(x)Y(y)$$

Then, differentiating, substituting, dividing by the product XY, and transposing, we obtain

$$\frac{X''}{X} = -\frac{Y''}{Y} \tag{8}$$

Exactly as in the case of the wave equation, we conclude that the common value of these two fractions must be a constant, say μ, and we have to consider the possibility that μ is less than, equal to, or greater than zero.

If $\mu < 0$, say $\mu = -\lambda^2$, $\lambda > 0$, we have from the last equation

$$X'' = -\lambda^2 X \qquad Y'' = \lambda^2 Y$$
$$X = A\cos\lambda x + B\sin\lambda x \qquad Y = C\cosh\lambda y + D\sinh\lambda y$$

and

$$u(x, y) = X(x)Y(y) = (A\cos\lambda x + B\sin\lambda x)(C\cosh\lambda y + D\sinh\lambda y) \tag{9}$$

We are given that along the left-hand edge of the sheet, $u(0, y) = 0$. Hence, evaluating (9) for $x = 0$, we must have

$$u(0, y) = A(C\cosh\lambda y + D\sinh\lambda y) = 0$$

for all values of y between $y = 0$ and $y = 1$. This will be the case if $C = D = 0$, but then Y is identically zero and there is only the trivial solution for u. Hence we must have $A = 0$. Thus, absorbing the coefficient B in the arbitrary constants C and D, (9) reduces to

$$u(x, y) = \sin\lambda x(C\cosh\lambda y + D\sinh\lambda y) \tag{10}$$

We must now impose the boundary conditions that hold along the upper and lower edges of the sheet. Since these edges are insulated, it follows that at every point on each of these edges the normal temperature gradient $\partial u/\partial y$ must be zero, for otherwise there would be heat flow across the edges. Thus, differentiating (10) with respect to y, we obtain

$$\frac{\partial u}{\partial y} = \sin\lambda x(\lambda C\sinh\lambda y + \lambda D\cosh\lambda y)$$

and evaluating this for $y = 0$ and $y = 1$ and equating the results to zero, we find

$$\sin\lambda x(\lambda D) = 0 \qquad \text{and} \qquad \sin\lambda x(\lambda C\sinh\lambda + \lambda D\cosh\lambda) = 0$$

These conditions will hold for all x such that $0 < x < 1$ only if $D = 0$ (to satisfy the first equation) and if $\lambda C\sinh\lambda = 0$ (to satisfy the second equation). Since $\sinh\lambda$ cannot be zero for any positive value of λ, it follows that C must be zero. But again, if C and D are both zero, the factor Y is identically zero and there is only the trivial solution for u. Thus the possibility that $\mu < 0$ leads to no solution of interest to us.

If $\mu = 0$, then from (8),

$$X'' = 0 \qquad Y'' = 0$$
$$X = Ax + B \qquad Y = Cy + D$$

and

$$u(x, y) = X(x)Y(y) = (Ax + B)(Cy + D) \tag{11}$$

Imposing the condition $u(0, y) = 0$ on Eq. (11), we find that

$$B(Cy + D) = 0$$

and in order to obtain a nontrivial solution it is necessary that $B = 0$, leaving

$$u(x, y) = x(Cy + D) \tag{12}$$

after the coefficient A is absorbed in the arbitrary constants C and D.

To impose the insulated conditions that hold along the upper and lower edges, we first compute $\partial u/\partial y$, getting

$$\frac{\partial u}{\partial y} = Cx$$

This will be zero by $y = 0$ and at $y = 1$ if and only if $C = 0$. Hence (12) reduces to the very simple, though necessary, product solution $u(x, y) = Dx$ or, replacing D by $\frac{1}{2}C_0$ for later convenience,

$$u_0(x, y) = \tfrac{1}{2}C_0 x \tag{13}$$

Finally, if $\mu > 0$, say $\mu = \lambda^2$, $\lambda > 0$, we have

$$X'' = \lambda^2 x \qquad\qquad Y = -\lambda^2 Y$$
$$X = A \cosh \lambda x + B \sinh \lambda x \qquad Y = C \cos \lambda y + D \sin \lambda y$$

and

$$u(x, y) = X(x)Y(y) = (A \cosh \lambda x + B \sinh \lambda x)(C \cos \lambda y + D \sin \lambda y) \tag{14}$$

From the condition $u(0, y) = 0$, which hold along the left-hand edge, we obtain

$$A(C \cos \lambda y + D \sin \lambda Y) = 0$$

from which we conclude that $A = 0$, leaving (after the coefficient B is absorbed in the arbitrary constants C and D)

$$u(x, y) = \sinh \lambda x(C \cos \lambda y + D \sin \lambda y) \tag{15}$$

To impose the insulated conditions that hold along the upper and lower edges, we compute $\partial u/\partial y$ from Eq. (15), evaluate it at $y = 0$ and $y = 1$, and equate the results to zero, getting

$$\sinh \lambda x(D\lambda) = 0 \qquad \text{and} \qquad \sinh \lambda x(-C\lambda \sin \lambda + D\lambda \cos \lambda) = 0$$

These will hold nontrivially for all x such that $0 < x < 1$ if and only if

$$D = 0 \qquad \text{and} \qquad C\lambda \sin \lambda = 0$$

Since $C = 0$ implies that Y, and hence u, is identically zero, we conclude from the second of these equations that $C \neq 0$ and therefore

$$\sin \lambda = 0 \qquad \text{or} \qquad \lambda_n = n\pi \qquad n = 1, 2, 3, \ldots$$

These values of λ do not have the immediate physical significance of natural frequencies that their counterparts did in the case of the wave equation. However, together with $\lambda = 0$, they are the *only* values of λ for which nontrivial solutions of the form $u(x, y) = X(x)Y(y)$ exist for the present problem. In other words, the solution (13) and the solutions

$$u_n(x, y) = C_n \sinh \lambda_n x \cos \lambda_n y = C_n \sinh n\pi x \cos n\pi y \tag{16}$$

are the only product solutions of Laplace's equation which also satisfy the three boundary conditions

$$u(0, y) = 0 \qquad \left.\frac{\partial u}{\partial y}\right|_{x,0} = 0 \qquad \left.\frac{\partial u}{\partial y}\right|_{x,1} = 0$$

The final boundary condition, namely, that along the right-hand edge of the sheet the temperature distribution $u(1, y) = f(y)$ is maintained, cannot be met by any one of the individual product solutions (16). In fact, if we set $x = 1$ in Eq. (16), we obtain the relation

$$C_n \sinh n\pi \cos n\pi y = f(y)$$

which clearly cannot hold for an arbitrary function $f(y)$. However, Laplace's equation is linear, and therefore sums of solutions are also solutions. Hence, finally, we form an infinite series of all the product solutions, including of course the solution (13) which we found when $\lambda = 0$,

$$u(x, y) = \sum_{n=0}^{\infty} u_n(x, y) = \tfrac{1}{2}C_0 x + \sum_{n=0}^{\infty} C_n \sinh n\pi x \cos n\pi y$$

and attempt to determine the C's so that the function defined by this series will reduce to the given temperature distribution $f(y)$ when $x = 1$. This requires that

$$u(1, y) \equiv f(y) = \tfrac{1}{2}C_0 + \sum_{n=0}^{\infty} (C_n \sinh n\pi) \cos n\pi y$$

and this we recognize as simply the problem of expanding the given function $f(y)$ in a half-range cosine expansion, with coefficients $C_n \sinh n\pi$, over the interval (0, 1). Thus we must have (Theorem 1, Sec. 8.4)

$$C_0 = 2\int_0^1 f(y)\,dy \qquad C_n \sinh n\pi = 2\int_0^1 f(y) \cos n\pi y\,dy$$

and

$$C_n = \frac{2}{\sinh n\pi}\int_0^1 f(y)\cos n\pi y\,dy \qquad n = 1, 2, 3, \ldots$$

With C_n determined, the formal solution of our problem is now complete.

EXAMPLE 2

TEMPERATURES ALONG A THIN ROD OF FINITE LENGTH

A rod of length l has its lateral surface insulated and is so thin that heat flow in the rod can be regarded as one-dimensional. Initially the rod is at the temperature 100 throughout. At $t = 0$ the temperature at the left end of the rod is suddenly reduced to 50 and maintained thereafter at that value, while the right end is maintained at 100. Find the temperature at any point in the rod at any subsequent time.

Here we must solve the one-dimensional heat equation

$$\frac{\partial^2 u}{\partial x^2} = a^2 \frac{\partial u}{\partial t}$$

subject to the boundary, i.e., end, conditions

$$u(0, t) = 50 \qquad u(l, t) = 100$$

and the initial condition

$$u(x, 0) = 100$$

Assuming a product solution $u(x, t) = X(x)T(t)$ and differentiating, substituting, and dividing by the product XT, we get

$$\frac{X''}{X} = a^2 \frac{T'}{T} \tag{17}$$

Again we reason that the common value of these two fractions must be a constant, say μ, and we have to consider the possibility that μ is greater than, equal to, or less than zero.

If $\mu > 0$, say $\mu = \lambda^2$, $\lambda > 0$, we have from the last equation,

$$X'' = \lambda^2 X \qquad T' = \frac{\lambda^2 T}{a^2}$$

$$X = A\cosh\lambda x + B\sinh\lambda x \qquad T = C\exp\frac{\lambda^2 t}{a^2}$$

and

$$u(x, t) = X(x)T(t) = (A\cosh\lambda x + B\sinh\lambda x)C\exp\frac{\lambda^2 t}{a^2}$$

This must be rejected immediately, however, because the positive exponential function of t implies that the temperature increases without limit as t becomes infinite, which is impossible under the conditions of the problem.

If $\mu = 0$, we have from (17),

$$X'' = 0 \qquad T' = 0$$
$$X = Ax + B \qquad T = C$$

and

$$u(x, t) = X(x)T(t) = C(Ax + B)$$

or simply

$$u(x, t) = Ax + B$$

if we absorb the coefficient C in the arbitrary constants A and B. At the left-hand end we have the condition $u(0, t) = 50$, which implies that $B = 50$. At the right-hand end we have the condition $u(l, t) = 100$, which implies that $Al + 50 = 100$ or $A = 50/l$. Thus

$$u_0(x, t) = 50 + \frac{50x}{l} \tag{18}$$

is a solution of the heat equation that satisfies both end conditions of the present problem [though of course it does not satisfy the initial condition that $u(x, 0) = 100$].

If $\mu < 0$, say $\mu = -\lambda^2$, $\lambda > 0$, then from (17),

$$X'' = -\lambda^2 X \qquad T' = \frac{-\lambda^2 T}{a^2}$$
$$X = A\cos\lambda x + B\sin\lambda x \qquad T = C\exp\frac{-\lambda^2 t}{a^2}$$

and

$$u(x, t) = X(x)T(t) = (A\cos\lambda x + B\sin\lambda x)\exp\frac{-\lambda^2 t}{a^2} \tag{19}$$

where again we have absorbed the coefficient C in the arbitrary constants A and B.

If we thoughtlessly attempt to impose the end conditions of the problem on the product solution (19) it appears that we must have

$$50 = A\exp\frac{-\lambda^2 t}{a^2} \qquad \text{and} \qquad 100 = (A\cos\lambda l + B\sin\lambda l)\exp\frac{-\lambda^2 t}{a^2}$$

and, clearly, there are no values of the constants A and B for which these equations can hold for all values of t. A moment's reflection, however, should convince us that we actually want the product solutions given by (19) to be zero at $x = 0$ and at $x = l$. In fact, since we already have the solution $u_0(x, t) = 50 + 50x/l$ [Eq. (18)], which takes the value 50 at $x = 0$ and the value 100 at $x = l$, and since we must eventually form a series of this and all the other product solutions in order to satisfy the given initial temperature condition, it is clear that if the latter are zero at each end of the rod, then at $x = 0$ and at $x = l$ the entire series will reduce to

$$u(0, t) = 50 + 0 + 0 + 0 + \cdots = 50$$

and

$$u(l, t) = 100 + 0 + 0 + 0 + \cdots = 100$$

as required.

Imposing the new conditions,† namely, $u(0, t) = 0$ and $u(l, t) = 0$, which we have found that the general product solutions must satisfy, we obtain

$$A\exp\frac{-\lambda^2 t}{a^2} = 0 \qquad \text{and} \qquad [A\cos\lambda l + B\sin\lambda l]\exp\frac{-\lambda^2 t}{a^2} = 0$$

†These conditions are sometimes called **homogeneous boundary conditions,** in contrast to the nonzero conditions of the problem itself which are called **nonhomogeneous boundary conditions.**

The first of these implies that $A = 0$. The second implies that if we are to avoid the trivial solution arising if $B = 0$, then we must have

$$\sin \lambda l = 0 \qquad \text{or} \qquad \lambda_n = \frac{n\pi}{l} \qquad n = 1, 2, 3, \ldots$$

Using these values of λ, we obtain the family of product solutions

$$u_n(x, t) = B_n \sin \frac{n\pi x}{l} \exp \frac{-n^2\pi^2 t}{a^2 l^2} \qquad n = 1, 2, 3, \ldots$$

Hence, forming an infinite series of all these solutions, together with the solution (18), preparatory to imposing the initial condition that $u(x, 0) = 100$, we have

$$u(x, t) = 50 + \frac{50x}{l} + \sum_{n=1}^{\infty} B_n \sin \frac{n\pi x}{l} \exp \frac{-n^2\pi^2 t}{a^2 l^2} \tag{20}$$

Finally, putting $t = 0$ in this expression, we obtain

$$u(x, 0) \equiv 100 = 50 + \frac{50x}{l} + \sum_{n=1}^{\infty} B_n \sin \frac{n\pi x}{l}$$

Thus the B's are simply the coefficients in the half-range sine expansion of the difference

$$100 - \left(50 + \frac{50x}{l}\right) = 50 - \frac{50x}{l}$$

Therefore

$$B_n = \frac{2}{l} \int_0^l \left(50 - \frac{50x}{l}\right) \sin \frac{n\pi x}{l}\, dx$$

and hence, by an easy integration,

$$B_n = \frac{100}{n\pi}$$

Inserting these values in Eq. (20), we obtain as the solution to our problem

$$u(x, t) = 50 + \frac{50x}{l} + \frac{100}{\pi} \sum_{n=1}^{\infty} \frac{1}{n} \sin \frac{n\pi x}{l} \exp \frac{-n^2\pi^2 t}{a^2 l^2}$$

EXERCISES

1. Describe a physical system that might reasonably be modeled by
   - **(a)** A fixed-fixed shaft, not necessarily of uniform cross section, vibrating torsionally
   - **(b)** A free-free shaft, not necessarily of uniform cross section, vibrating torsionally
   - **(c)** A fixed-free shaft, not necessarily of uniform cross section, vibrating torsionally.
2. Repeat the analysis of the torsional vibrations of a uniform fixed-free shaft if the left end is free and the right end is fixed. Is the frequency equation in this case the same as the one obtained on p. 708? Is this to be expected? How do the amplitude-distribution functions compare?
3. Determine the frequency equation and the amplitude-distribution fuction for a uniform shaft of length $2l$, vibrating torsionally, if the shaft is
   - **(a)** Free at $x = -l$ and at $x = l$.
   - **(b)** Fixed at $x = -l$ and fixed at $x = l$.
   - **(c)** Fixed at $x = -l$ and free at $x = l$.

   In each case, how do the frequencies compare with those

of a comparable shaft of length l with similar end conditions?

4. A uniform string stretched between $x = 0$ and $x = l$ is set in motion with initial displacement $y(x, 0) = f(x)$ and initial velocity $\dot{y}(x, 0) = g(x)$. Applying the method of separation of variables to the wave equation

$$a^2 \frac{\partial^2 y}{\partial x^2} = \frac{\partial^2 y}{\partial t^2}$$

determine the series expansion for $y(x, t)$.

5. Show that the natural frequencies of a uniform string are given by the formula

$$f_n = \frac{n}{2l}\sqrt{\frac{Tg}{w}} \text{ cycles/unit time}$$

where l = length of the string, T = tension under which the string is stretched, w = weight of the string per unit length. How does doubling the tension affect the pitch of the fundamental tone of the string? Why is it that the strings of most stringed instruments are either of different lengths or have their lengths changed by the performer while being played?

6. Verify that the solutions of the wave equation obtained in this section can all be written in the form

$$\theta(x, t) = f(x - at) + g(x + at)$$

7. Which of the following partial differential equations can be reduced to two or more ordinary differential equations by the method of separation of variables?

(a) $a\dfrac{\partial^2 u}{\partial x\, \partial y} + bu = 0$ **(b)** $x^2\dfrac{\partial^2 u}{\partial x^2} + y\dfrac{\partial^2 u}{\partial y^2} = 0$

(c) $a\dfrac{\partial^2 u}{\partial x^2} + b\dfrac{\partial^2 u}{\partial x\, \partial y} + c\dfrac{\partial u}{\partial x} = 0$

(d) $a^2\dfrac{\partial^4 y}{\partial x^4} + \dfrac{\partial^2 y}{\partial t^2} = 0$

(e) $a\dfrac{\partial^2 u}{\partial x^2} + b\dfrac{\partial^2 u}{\partial x\, \partial y} + c\dfrac{\partial^2 u}{\partial y^2} = 0$

(f) $\dfrac{\partial^2 z}{\partial x^2} + xy\dfrac{\partial z}{\partial y} = 0$

(g) $a\dfrac{\partial^2 u}{\partial x^2} + b\dfrac{\partial^2 u}{\partial y^2} + c\dfrac{\partial^2 u}{\partial z^2} = 0$. *Hint:* Assume a three-factor product solution, $u(x, y, z) = X(x)Y(y)Z(z)$, and separate variables in two stages.

(h) $\dfrac{\partial^2 u}{\partial x^2} + \dfrac{\partial^2 u}{\partial y^2} = a^2\dfrac{\partial u}{\partial t}$

(i) $\dfrac{\partial^2 u}{\partial r^2} + \dfrac{1}{r}\dfrac{\partial u}{\partial r} + \dfrac{1}{r^2}\dfrac{\partial^2 u}{\partial \theta^2} = a^2\dfrac{\partial u}{\partial t}$

(j) $J(x)\dfrac{\partial^2 \theta}{\partial t^2} = a^2\dfrac{\partial[J(x)\, \partial\theta/\partial x]}{\partial x}$

(k) $x^2\dfrac{\partial^2 u}{\partial x^2} = a^2\dfrac{\partial[r^2\, \partial u/\partial x]}{\partial r}$

8. A uniform shaft fixed at $x = 0$ and free at $x = l$ is twisted so that each cross section is rotated through an angle proportional to its distance from the fixed end. If the shaft is released from rest in this position, find its subsequent angular displacement as a function of x and t. *Hint:* Recall Exercise 17, Sec. 8.4.

9. A uniform shaft fixed at $x = 0$ and at $x = l$ is twisted so that each cross section is rotated through an angle proportional to $x(l - x)$. If the shaft is released from rest in this position, find its subsequent angular displacmenet as a function of x and t. *Hint:* Recall Exercise 19, Sec. 8.4.

10. A uniform shaft free at $x = 0$ and at $x = l$ is twisted so that each cross section is rotated through an angle proportional to $(2x - l)/2$. If the shaft is supported in frictionless bearings that permit it to twist freely, find its subsequent angular displacement as a function of x and t if it is released from rest in its twisted position.

11. Work Example 1 if the conditions along the edges are

(a) $u(x, 0) = u(0, y) = u(1, y) = 0;\ u(x, 1) = f(x)$

(b) $u(0, y) = u(1, y) = \partial u/\partial y|_{x,0} = 0;\ u(x, 1) = 100$

(c) $\partial u/\partial x|_{0,y} = \partial u/\partial x|_{1,y} = \partial u/\partial y|_{x,1} = 0;\ u(x, 0) = f(x)$

(d) $\partial u/\partial x|_{0,y} = \partial u/\partial y|_{x,0} = u(1, y) = 0;\ u(x, 1) = 100$

(e) $u(x, 0) = u(1, y) = \partial u/\partial x|_{0,y} = 0;$ $u(x, 1) = \cos \pi x/2$

12. A thin sheet of metal bounded by the x axis and the lines $x = 0$ and $x = 1$ and extending to infinity in the y direction has its upper and lower faces insulated and its vertical edges maintained at the constant temperature $u = 0$. Over its lower edge the temperature distribution $u(x, 0) = 100$ is maintained. Find the steady-state temperature at any point in the sheet. *Hint:* Since $u(x, y)$ must remain bounded as $y \to \infty$, it will be convenient to use exponential rather than hyperbolic functions at the appropriate point in the solution for the factor Y.

13. Work Exercise 12 if the boundary conditions are

(a) $\partial u/\partial x|_{0,y} = \partial u/\partial x|_{1,y} = 0,\ u(x, 0) = 100$

(b) $u(0, y) = 0,\ u(1, y) = 100,\ u(x, 0) = 100x$

(c) $u(0, y) = 0,\ \partial u/\partial x|_{1,y} = 0,\ u(x, 0) = 100$

(d) $u(0, y) = u(x, 0) = 0,\ u(1, y) = 100$

(e) $\partial u/\partial x|_{0,y} = \partial u/\partial x|_{1,y} = 0,\ u(x, 0) = \sin \pi x$

14. A uniform string stretched between the points $x = 0$ and $x = l$ is given the following initial displacement and velocity.

$$y(x, 0) = \sin(\pi x/l) \qquad 0 \le x \le l$$

$$\dot{y}(x, 0) = \begin{cases} 0 & 0 \le x < l/4 \\ a & l/4 < x < 3l/4 \\ 0 & 3l/4 < x \le l \end{cases}$$

Find its subsequent displacement as a function of x and t.

15. A uniform string stretched between $x = 0$ and $x = l$ is given the initial displacement

$$y(x, 0) = f(x) = \begin{cases} x & 0 \le x \le l/2 \\ l - x & l/2 \le x \le l \end{cases}$$

and released from rest. Find its subsequent displacement as a function of x and t. *Hint:* Recall the result of Exercise 21, Sec. 8.4.

16. In Exercise 15, show that the displacement of the midpoint of the string, $y(l/2, t)$, varies as a linear function of at between its successive maximum and minimum values. *Hint:* In the cosine expansion of t obtained in Exercise 17, Sec. 8.4, replace t by at, then compare the resulting series with the series for $y(l/2, t)$.

17. While in its equilibrium position, a uniform string stretched between $x = 0$ and $x = l$ is given the initial velocity

$$\dot{y}(x, 0) = g(x) = \begin{cases} ax/l & 0 \le x \le l/2 \\ a(l - x)/l & l/2 \le x \le l \end{cases}$$

Find its subsequent displacement as a function of x and t.

18. In Exercise 17, show that the displacement of the midpoint of the string, $y(l/2, t)$, varies as a quadratic function of at between its maximum and minimum values. *Hint:* In the sine expansion of $t(l - t)$ obtained in Exercise 19, Sec. 8.4, replace t by at, then compare the resulting series with the series for $y(l/2, t)$.

19. While in its equilibrium position, a uniform string stretched between $x = 0$ and $x = l$ is given the initial velocity

$$\dot{y}(x, 0) = g(x) = \begin{cases} 0 & 0 < x < (l - k)/2 \\ a/k & (l - k)/2 < x < (l + k)/2 \\ 0 & (l + k)/2 < x < l \end{cases}$$

Find its subsequent displacement as a function of x and t. Does your answer appear to have a meaningful limit as $k \to 0$? If so, to what problem do you think it is the answer?

20. It was shown in Exercise 17, Sec. 11.2, that the partial differential equation governing the diffusion of a liquid through a porous solid is exactly the same as the heat equation, except that the dependent variable represents the concentration of the diffusing liquid rather than the temperature. Using this fact, find the concentration in a long, thin bar of length l whose lateral surface is impervious to diffusion, if the ends of the bar are maintained at the concentration 0 and if the concentration in the bar at $t = 0$ is $C_0x(l - x)$.

21. Work Exercise 20 if each end of the bar is impervious to diffusion. *Hint:* If there is no diffusion through a surface, then the normal derivative of the concentration must be equal to zero on that surface.

22. Verify the statement that a fixed-fixed shaft, or string, vibrating at its nth natural frequency, has $n - 1$ equally spaced nodes. What is the corresponding result for a shaft that is free at both ends? A shaft that is fixed at one end and free at the other?

23. Discuss the restriction implicitly imposed on $f(x)$ and $g(x)$ by the absence of constant terms in the series in Eqs. (7*a*) and (7*b*). What is the physical significance of these restrictions?

24. Show that the torsional vibrations of a uniform fixed-free shaft of length l vibrating at one of its natural frequencies are always the same as those of the left half of a similar fixed-fixed shaft of length $2l$ vibrating at the same frequency. Is the converse true? That is, does the motion of the left half of a fixed-fixed shaft of length $2l$ always represent a possible motion of a fixed-free shaft of length l?

25. The curved surface of a thin rod of length l is insulated against the flow of heat. Initially the temperature throughout the rod is $u(x, 0) = 100$. Assuming heat flow in the rod is one-dimensional, find the temperature at any point in the rod at any subsequent time if at $t = 0$ the temperature at each end of the rod is suddenly reduced to 0 and maintained at that temperature thereafter.

26. **(a)** Work Exercise 25 if the initial temperature distribution in the rod is $u(x, 0) = 100$, the left end of the rod is insulated, and the right end of the rod is maintained after $t = 0$ at the constant temperature $u(l, t) = 0$. *Hint:* Recall that the temperature gradient through an insulated surface must be zero.

(b) Work Exercise 25 if the initial temperature distribution in the rod is $u(x, 0) = 0$, the left end of the rod is insulated, and the right end of the rod is maintained after $t = 0$ at the constant temperature $u(l, t) = 100$.

27. **(a)** Work Exercise 25 if the initial temperature distribution in the rod is $u(x, 0) = 0$ and if the left- and right-hand ends of the rod are maintained after $t = 0$ at the respective temperatures $u(0, t) = 0$ and $u(l, t) = 100$.

(b) Work Exercise 25 if the initial temperature distribution in the rod is $u(x, 0) = 0$ and if the left- and right-hand ends are maintained after $t = 0$ at the respective temperatures $u(0, t) = 50$ and $u(l, t) = 100$.

28. **(a)** Work Exercise 25 if the initial temperature distribution in the rod is $u(x, 0) = u_0x/l$.

(b) Work Exercise 25 if the initial temperature distribution in the rod is $u(x, 0) = u_0x/l$ and each end of the rod is insulated.

29. In Exercise 25, find the temperature at the midpoint of the rod and show that for arbitrarily small positive values of t it is different from 100. What does this appear to say about the rate at which thermal disturbances are propagated? Do you agree with this conclusion? *Hint:* Consider $u(l/2, t)$ as a power series in the quantity

$$z = \exp \frac{-\pi^2 t}{a^2 l^2}$$

and recall that a power series cannot converge to a constant over any interval unless it converges to that same constant at *every* point of its interval of convergence.

30. **(a)** In Exercise 25, determine the quantity of heat remaining in the rod when $t = \alpha a^2 l^2$ and verify that it is equal to the quantity of heat in the rod at $t = 0$ minus the heat lost through the two ends between $t = 0$ and $t = \alpha a^2 l^2$. *Hint:* Recall the laws of heat flow given in Sec. 11.2 in connection with the derivation of the heat equation. Then, at the appropriate point, use the results of Exercise 21, Sec. 8.2.

(b) In Exercise 26(**a**), determine the quantity of heat remaining in the rod when $t = \alpha a^2 l^2$ and verify that it is equal to the amount of heat in the rod when $t = 0$ minus the heat lost through the right end between $t = 0$ and $t = \alpha a^2 l^2$.

31. **(a)** Use separation of variables to solve $\partial u/\partial t = \alpha\, \partial^2 u/\partial x^2$, α constant, subject to the boundary conditions $u(0, t) = u(\pi, t) = 0$ and the initial condition $u(x, 0) = \sin 2x$.

(b) Use separation of variables to solve Laplace's equation $u_{xx} + u_{yy} = 0$ subject to the homogeneous boundary conditions $u(x, 0) = u_x(\pi, y) = u_x(0, y) = 0$ and the nonhomogeneous boundary condition $u(x, 2) = 4 + 3\cos x - 2\cos 2x$.

32. Work Exercise 25 if $l = 1$, the initial temperature distribution is $u(x, 0) = 0$, and both ends of the rod are maintained at the temperature 100.

33. **(a)** A thin sheet of metal bounded by the y axis and the lines $y = 0$ and $y = 1$ and extending to infinity in the positive x direction has its upper and lower faces insulated. Its horizontal edges along $y = 1$ and $y = 0$ are maintained at the respective temperatures 100 and 0, and along its vertical edge the temperature distribution $u(0, y) = 100(1 - y)$ is maintained. Find the steady-state temperature at any point in the sheet.

(b) Work Part **(a)** if the horizontal boundary lines are $y = 0$ and $y = \pi/2$, if $u(x, \pi/2) = b$, $u_y(x, 0) = 0$, and if $u(0, y) = 2\cos y \cos 2y$.

34. Work Example 1 if the sheet of metal coincides with the rectangle in the xy plane whose vertices are $(0, 0)$, $(\pi, 0)$, $(\pi, 1)$, and $(0, 1)$ and if all edges are maintained at the temperature 0 except for the edge along the x axis where the temperature distribution $u(x, 0) = \pi x - x^2$ is maintained.

35. **(a)** Work Exercise 31(**a**) if the auxiliary conditions are $u(0, t) = u(\pi, t) = 10$ and $u(x, 0) = 10 + 2\sin x \cos x$. *Hint:* First set $U = u - 10$.

(b) The curved surface of a right circular cylinder of radius b and height h is insulated, and its upper and lower bases are maintained at the temperatures 0 and 100, respectively. Find the steady-state temperature in the cylinder. *Hint:* Use Laplace's equation in cylindrical coordinates [Exercise 38(**a**), Sec. 11.2] and note the values of $\partial u/\partial r$ and $\partial u/\partial \theta$.

11.6 ORTHOGONAL FUNCTIONS AND THE GENERAL EXPANSION PROBLEM

The problems we discussed in the last section embody all the significant features of the general boundary-value problem, but they give an exaggerated picture of the role of Fourier series in the final expansion required to fit the initial conditions or some other auxiliary condition of the problem. The situation is somewhat as it was when we were studying vectors in calculus. There we always used unit vectors in the directions of the x, y, and z axes when we wanted to represent other vectors. And, officially at least, we weren't told that a general vector could be expressed equally well in terms of *any* three mutually perpendicular, or *orthogonal,* vectors or, in other words, that the standard unit vectors **(i, j, k)** formed just one of infinitely many sets that could be used.

This is where we are at present. In the last section, the nature of the problems we studied restricted us to one or the other of two particular sets of functions

$$\left\{\cos\frac{n\pi x}{l},\ \sin\frac{n\pi x}{l}\right\} \qquad \text{and} \qquad \left\{\cos\frac{(2n-1)\pi x}{2l},\ \sin\frac{(2n-1)\pi x}{2l}\right\}$$

in terms of which we had to expand certain other functions. It is only natural to ask if there might be other sets of functions that could, or might *have* to, be used for expansion purposes in other kinds of problems. Such is indeed the case. In this section we shall investigate how they arise, extend the concept of *orthogonality* to functions, and show how orthogonal sets of functions over suitable intervals are used in problems where Fourier series of sines and cosines having a common period do not serve our purpose. We begin with an example very much like Example 2 in the last section, in which, perhaps unexpectedly, Fourier series play no part.

EXAMPLE 1 TEMPERATURES ALONG A FINITE ROD WITH A RADIATING END

A thin rod of length l has its lateral surface insulated against the flow of heat. Its left end is maintained at the constant temperature $u(0, t) = 0$, and at $x = l$ its right end radiates freely into air of constant temperature $u = 0$. If the initial temperature distribution in the rod is given by

$$u(x, 0) = f(x)$$

find the temperature at any point of the rod at any subsequent time.

Since the rod is thin and since its lateral surface is insulated, we shall assume as usual that all points of any given cross section are at the same temperature and that the flow of heat in the rod is entirely in the x direction. Since there are no sources of heat in the rod, this means that we have to solve the one-dimensional heat equation [Eq. (14a), Sec. 11.2]

$$\frac{\partial^2 u}{\partial x^2} = a^2 \frac{\partial u}{\partial t} \tag{1}$$

subject to the given end conditions.

At the left end we have a familiar constant temperature condition, $u(0, t) = 0$. At the right end we have the condition of free radiation, which we have not encountered before. To express it in a form we can apply to our problem, we use **Stefan's law:**† *Under conditions of free radiation, the temperature gradient normal to the surface is proportional to the difference between the surface temperature and the ambient temperature, provided this difference is moderately small.* In our problem this becomes

$$-\left.\frac{\partial u}{\partial x}\right|_{l,t} = h[u(l, t) - u_0] = hu(l, t) \tag{2}$$

since we are told that the air temperature u_0 is 0. The minus sign is necessary in this case since for heat to flow to the surface to be radiated, the temperature within the body must be greater than the surface temperature; i.e., the temperature gradient at the surface $\partial u/\partial x$ must be negative. The constant h in Eq. (2) is known as the **surface conductivity.**

As in Example 2 in Sec. 11.5, we begin by assuming a product solution $u(x, t) = X(x)T(t)$ and separating variables, getting

$$\frac{X''}{X} = a^2 \frac{T'}{T} = \mu \qquad \mu \text{ a constant}$$

The possibility that $\mu > 0$ can be rejected immediately since, as we have seen several times, it leads to the conclusion that $T(t)$, and hence the product $u(x, t) = X(x)T(t)$, becomes infinite as t becomes infinite.

If $\mu = 0$, we have simply

$$X'' = 0 \qquad T' = 0$$
$$X = Ax + B \qquad T = C$$

from which, letting $C = 1$, as we can without loss of generality since A and B are arbitrary, we obtain

$$u = XT = Ax + B$$

For this to be relevant to our problem it must reduce to zero when $x = 0$; hence $B = 0$. Moreover, it must satisfy Eq. (2) when $x = l$; hence $-A = hAl$, which implies that $A = 0$, since both h and l are positive constants. Thus $\mu = 0$ leads only to a trivial solution and must also be rejected.

Finally, if $\mu < 0$, say $\mu = -\lambda^2$, $\lambda > 0$, the component differential equations and their solutions are

$$X'' = -\lambda^2 X \qquad T' = -\frac{\lambda^2}{a^2} T$$

$$X = A \cos \lambda x + B \sin \lambda x \qquad T = C \exp(-\lambda^2 t/a^2)$$

and, again letting $C = 1$,

$$u = XT = (A \cos \lambda x + B \sin \lambda x) \exp(-\lambda^2 t/a^2)$$

†Named for the Austrian physicist Josef Stefan (1835–1893).

To satisfy the left end condition, namely, $u(0, t) = 0$, we put $x = 0$ in the last expression, getting $0 = A \exp(-\lambda^2 t/a^2)$. Hence $A = 0$, and u reduces to

$$u = B \exp(-\lambda^2 t/a^2) \sin \lambda x$$

To satisfy the right end condition (2) we must have

$$-B \exp\left(\frac{-\lambda^2 t}{a^2}\right)\lambda \cos \lambda l = hB \exp\left(\frac{-\lambda^2 t}{a^2}\right) \sin \lambda l$$

or, dividing out the exponential and collecting terms,

$$B(h \sin \lambda l + \lambda \cos \lambda l) = 0$$

If $B = 0$, the solution is trivial, since A is already known to be zero. Hence to obtain a meaningful solution we must have

$$h \sin \lambda l + \lambda \cos \lambda l = 0$$

or

$$\tan \lambda l = -\frac{\lambda}{h} = -\frac{\lambda l}{hl}$$

or finally

$$\tan z = -\alpha z$$

where

$$z = \lambda l \qquad \text{and} \qquad \alpha = 1/hl$$

The equation $\tan z = -\alpha z$ is not like the simple equations

$$\sin \lambda l = 0 \qquad \text{and} \qquad \cos \frac{\lambda l}{a} = 0$$

which determined the admissible values of λ in the second example and for the fixed-free shaft in the last section, and its roots cannot be found by inspection. To determine them, it is convenient to consider the graphs of the two functions $y_1 = \tan z$ and $y_2 = -\alpha z$. The abscissas of the points of intersection of these curves (Fig. 11.19), being values of z for which $y_1 = y_2$, are then the solutions of the equation $\tan z = -\alpha z$. Obviously, there are an infinite number of roots z_n. However, unlike the roots of $\sin \lambda l = 0$ and $\cos \lambda l/a = 0$, they are not evenly spaced, although, as Fig. 11.19 indicates, the interval between successive values of z_n *approaches* π as n becomes infinite.

From each root z_n we obtain at once the corresponding value of λ

$$\lambda_n = \frac{z_n}{l}$$

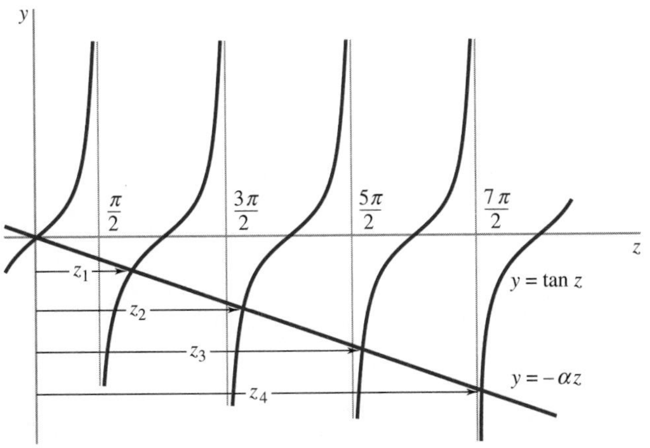

FIGURE 11.19
The graphical solution of the equation $\tan z = -\alpha z$.

and the associated product solution

$$u_n(x, t) = T_n(t)X_n(x) = B_n \exp\left(-\frac{\lambda_n^2 t}{a^2}\right) \sin \lambda_n x$$

Then we form a series of these particular solutions

$$u(x, t) = \sum_{n=1}^{\infty} u_n(x, t) = \sum_{n=1}^{\infty} B_n \exp\left(-\frac{\lambda_n^2 t}{a^2}\right) \sin \lambda_n x \tag{3}$$

and attempt to determine the constants B_n so that the function defined by the series will satisfy the initial condition

$$u(x, 0) = f(x)$$

Finally, putting $t = 0$ in (3), we find that this requires

$$u(x, 0) \equiv f(x) = \sum_{n=1}^{\infty} B_n \sin \lambda_n x \tag{4}$$

Thus, as in the examples in Sec. 11.5, to satisfy the initial conditions we must be able to expand an arbitrary function in an infinite series of known functions determined by a differential equation and a set of boundary conditions. However, although the functions in terms of which the expansion is to be carried out are sines, the values of λ appearing in their arguments are spaced at incommensurable intervals and so the required series is *not* a Fourier series. Clearly, something is involved which includes Fourier series as a special case but is itself more general and more fundamental.

If we review thoughtfully our earlier discussion of Fourier series (Sec. 8.2), it should be apparent that the decisive property of the set of functions $\{\cos (n\pi x/l), \sin (n\pi x/l)\}$ which made it possible to determine one by one the coefficients in the assumed expansion

$$f(x) = \frac{1}{2}a_0 + a_1 \cos \frac{\pi x}{l} + a_2 \cos \frac{2\pi x}{l} + \cdots + b_1 \sin \frac{\pi x}{l} + b_2 \sin \frac{2\pi x}{l} + \cdots$$

was that the integral of the product of any two (different) members of the set taken over the appropriate interval is zero. For it was this which enabled us to multiply the series for $f(x)$ by $\cos (n\pi x/l)$ or $\sin (n\pi x/l)$ and eliminate all but one of the unknown coefficients simply by integrating from d to $d + 2l$.

Now sines and cosines are by no means the only functions from which sets can be constructed having the property that the integral between suitable limits of the product of two distinct members of the set is zero. In fact, the trigonometric functions that appear in Fourier expansions are merely one of the simplest examples of infinitely many such systems of functions, whose existence we shall soon establish.

DEFINITION 1 If a sequence of real functions

$$\{\phi_n(x)\} \qquad n = 1, 2, 3, \ldots$$

which are defined over some interval (a, b), finite or infinite, has the property that

$$\int_a^b \phi_m(x)\phi_n(x)\,dx \begin{cases} =0 & m \neq n \\ \neq 0 & m = n \end{cases}$$

then the functions are said to form an **orthogonal set** on that interval.

DEFINITION 2 If the functions of an orthogonal set $\{\phi_n(x)\}$ have the property that

$$\int_a^b \phi_n^2(x)\,dx = 1 \qquad \text{for all values of } n$$

then the functions are said to be **orthonormal** on the interval (a, b).

Any set of orthogonal functions can easily be converted into an orthonormal set. In fact, if the functions of the set $\{\phi_n(x)\}$ are orthogonal, and if k_n is the (necessarily positive) value of $\int_a^b \phi_n^2(x)\,dx$, then the functions

$$\frac{\phi_1(x)}{\sqrt{k_1}}, \quad \frac{\phi_2(x)}{\sqrt{k_2}}, \quad \frac{\phi_3(x)}{\sqrt{k_3}}, \quad \ldots$$

are clearly orthonormal. It is therefore no specialization to assume that an orthogonal set of functions is also orthonormal.

DEFINITION 3 If a sequence of real functions $\{\phi_n(x)\}$ has the property that over some interval (a, b), finite or infinite,

$$\int_a^b p(x)\phi_m(x)\phi_n(x)\,dx \begin{cases} = 0 & m \neq n \\ \neq 0 & m = n \end{cases} \qquad p(x) > 0 \text{ on } (a, b)$$

then the functions are said to be **orthogonal with respect to the weight function** $p(x)$ on that interval.

Any set of functions orthogonal with respect to a weight function $p(x)$ can be converted into a set of functions orthogonal in the first sense (Definition 1) simply by multiplying each member of the set by $\sqrt{p(x)}$ since, by Definition 3, $p(x) > 0$ on the interval of orthogonality.

With respect to any set of functions $\{\phi_n(x)\}$ orthogonal over an interval (a, b), an arbitrary function $f(x)$ has a formal expansion analogous to a Fourier expansion, for we can write

$$(5) \qquad f(x) = a_1\phi_1(x) + a_2\phi_2(x) + a_3\phi_3(x) + \cdots + a_n\phi_n(x) + \cdots$$

Then, multiplying by $\phi_n(x)$ and integrating formally between the appropriate limits, a and b, we have

$$\int_a^b f(x)\phi_n(x)\,dx = a_1\int_a^b \phi_1(x)\phi_n(x)\,dx + a_2\int_a^b \phi_2(x)\phi_n(x)\,dx + \cdots + a_n\int_a^b \phi_n^2(x)\,dx + \cdots$$

From the property of orthogonality, all integrals on the right are zero except the one that contains a square in its integrand. Hence we can solve at once for a_n as the quotient of two known integrals:

$$a_n = \frac{\displaystyle\int_a^b f(x)\phi_n(x)\,dx}{\displaystyle\int_a^b \phi_n^2(x)\,dx}$$

However, although the orthogonality of the ϕ's makes it possible to determine the coefficients in the expansion (5), this property is not sufficient to guarantee that this series converges to $f(x)$ or even converges at all. Nevertheless, the a_n thus determined and the series (5) are referred to, respectively, as the **Fourier constants** and the **generalized Fourier series** of f corresponding to the orthogonal set of functions $\{\phi_n(x)\}$.

To pursue this matter a little further,† it is convenient to introduce the notion of a *null function.*

DEFINITION 4 A real function $f(x)$ is said to be a **null function** on the interval (a, b) if $\int_a^b f^2(x)\,dx = 0$.

If $f(x)$ is identically zero, it is obviously a null function. However, a null function need not be identically zero. In fact, since the area under a curve is not altered when the ordinate of the curve at one or more isolated points is changed, it is clear that we can have $\int_a^b f^2(x)\,dx = 0$ even though $f(x)$ has nonzero values at a finite or countably infinite number of points between a and b. On the other hand, if there is any subinterval of (a, b) (no matter how short) at all points of which $f(x)$ is different from zero, then $\int_a^b f^2(x)\,dx \neq 0$, and $f(x)$ is not a null function. From this it is not difficult to show that *a null function is zero at every point where it is continuous.*

Clearly, if an arbitrary function is multiplied by a null function, the product is also a null function and hence the integral of such a product is zero. In particular, this means that any null function is orthogonal to every member of an orthogonal set $\{\phi_n(x)\}$. Is it also conceivable that a nonnull function $g(x)$ might be orthogonal to every ϕ, that is, that we might have

$$\int_a^b g(x)\phi_n(x)\,dx = 0 \qquad \text{for all values of } n$$

In such a case, every coefficient in the expansion of $g(x)$ in terms of the ϕ's would be zero, and the series (5) would converge to zero at all points of (a, b) even though $g(x)$ was not a null function. That this is actually possible is easily shown by example. For instance, although the functions $\{\sin nx\}$ are readily shown to be orthogonal over the interval $(-\pi, \pi)$, not every function can be represented on this interval by a series of the form

$$a_1 \sin x + a_2 \sin 2x + \cdots + a_n \sin nx + \cdots$$

In particular, if $g(x) = x^2$, we have for the coefficients in its formal expansion

$$a_n = \frac{\displaystyle\int_{-\pi}^{\pi} x^2 \sin nx\,dx}{\displaystyle\int_{-\pi}^{\pi} \sin^2 nx\,dx}$$

$$= \frac{1}{\pi}\left[\frac{2x}{n^2}\sin nx - \left(\frac{x^2}{n} - \frac{2}{n^3}\right)\cos nx\right]_{-\pi}^{\pi} = 0$$

for all values of n. More generally, since every member of the set $\{\sin nx\}$ is odd, it is clear that no series of these functions can represent *any* nonzero even function on the interval $(-\pi, \pi)$.

Evidently, important as it is, orthogonality is not the whole story, and the functions in our orthogonal systems must possess some further property before the expansion (5) can be used with confidence. What is required is that the set of functions $\{\phi_n(x)\}$, in addition to being orthogonal, should also possess the property of *completeness* described in the following definition.

DEFINITION 5 A set of orthogonal functions $\{\phi_n(x)\}$ is said to be **complete** if the relation $\int_a^b f(x)\phi_n(x)\,dx = 0$ can hold for all values of n only if $f(x)$ is a null function on (a, b).

†An extended discussion of these ideas requires a knowledge of the Lebesgue integral, which we do not assume in this book.

If $\{\phi_n(x)\}$ is a complete orthogonal set, then clearly not all the coefficients in the expansion of a nonnull function $g(x)$ can be zero and thus no nontrivial function can have a trivial expansion. In fact, we have the following theorem.

THEOREM 1 If the formal expansion

$$a_1\phi_1(x) + a_2\phi_2(x) + \cdots + a_n\phi_n(x) + \cdots$$

of a function $g(x)$ in terms of the members of a complete orthonormal set $\{\phi_n(x)\}$ converges and can be integrated term by term, then the sum of the series differs from $g(x)$ by at most a null function; i.e., the sum of the series cannot differ from $g(x)$ over any interval of finite length.

PROOF By hypothesis, the series $\sum_{n=1}^{\infty} a_n\phi_n(x)$ converges to some function; hence it is meaningful to consider the difference

$$h(x) = g(x) - \sum_{n=1}^{\infty} a_n\phi_n(x)$$

If we can prove that $h(x)$ is a null function, the assertion of the theorem will be established. To do this, consider

$$\begin{aligned}\int_a^b \phi_m(x)h(x)\,dx &= \int_a^b \phi_m(x)\left[g(x) - \sum_{n=1}^{\infty} a_n\phi_n(x)\right]dx \\ &= \int_a^b \phi_m(x)g(x)\,dx - \int_a^b \phi_m(x)\left[\sum_{n=1}^{\infty} a_n\phi_n(x)\right]dx \\ &= \int_a^b \phi_m(x)g(x)\,dx - \sum_{n=1}^{\infty} a_n\int_a^b \phi_m(x)\phi_n(x)\,dx\end{aligned}$$

The first integral in the last expression is simply the coefficient a_m. Furthermore, since the ϕ's form an orthonormal set, every integral in the sum is zero except the one for which $n = m$, which is equal to 1. Hence we have finally

$$\int_a^b \phi_m(x)h(x)\,dx = a_m - a_m = 0 \qquad m = 1, 2, 3, \ldots$$

Thus $h(x)$ is orthogonal to every one of the ϕ's. Therefore, since the ϕ's form a complete set, $h(x)$ must be a null function, and the theorem is established.

Closely associated with the concept of completeness is the concept of *closure*,† described in the following definitions.

DEFINITION 6 If $\lim_{n\to\infty} \int_a^b [f(x) - S_n(x)]^2\,dx = 0$, the sequence of functions $\{S_n(x)\}$ is said to **converge in the mean** to $f(x)$.

†What we have called *completeness* some authors call *closure*, and vice versa.

DEFINITION 7 If $S_n(x) = a_1\phi_1(x) + a_2\phi_2(x) + \cdots + a_n\phi_n(x)$ is the nth partial sum of the expansion of $f(x)$ in terms of the members of an orthonormal set $\{\phi_n(x)\}$, and if $\{S_n(x)\}$ converges in the mean to $f(x)$ for every $f(x)$, then the set $\{\phi_n(x)\}$ is said to be **closed.**

One important property of closed orthonormal sets is contained in the following so-called (amplitude) **theorem of Parseval.**

THEOREM 2 If $a_1\phi_1(x) + a_2\phi_2(x) + \cdots$ is the expansion of a function $f(x)$ in terms of the members of a closed orthonormal set $\{\phi_n(x)\}$, then

$$\sum_{n=1}^{\infty} a_n^2 = \int_a^b f^2(x)\,dx$$

PROOF From the definition of closure, we have

$$\lim_{m\to\infty} \int_a^b \left[f(x) - \sum_{n=1}^{m} a_n\phi_n(x)\right]^2 dx = 0$$

or

$$\lim_{m\to\infty} \int_a^b \left[f^2(x) - 2f(x)\sum_{n=1}^{m} a_n\phi_n(x) + \left\{\sum_{n=1}^{m} a_n\phi_n(x)\right\}^2\right] dx = 0$$

If we now perform the indicated integration, remembering that

$$\int_a^b f(x)\phi_n(x)\,dx = a_n$$

and observing that in the integral of the sum in the last term

$$\int_a^b \phi_m(x)\phi_n(x)\,dx = \begin{cases} 0 & m \neq n \\ 1 & m = n \end{cases}$$

we obtain

$$\lim_{m\to\infty} \left[\int_a^b f^2(x)\,dx - 2\sum_{n=1}^{m} a_n^2 + \sum_{n=1}^{m} a_n^2\right] = 0$$

or

$$\sum_{n=1}^{\infty} a_n^2 = \int_a^b f^2(x)\,dx$$

as asserted.

As an immediate consequence of the last theorem we have the following result.

THEOREM 3 A closed orthonormal system $\{\phi_n(x)\}$ is also complete.

PROOF To prove this, let us suppose, contrary to the theorem, that the closed orthonormal system $\{\phi_n(x)\}$ is not complete. This implies that there is at least one nonnull function $f(x)$ which is orthogonal to each of the ϕ's and which therefore has the property that every coefficient in its expansion in terms of the ϕ's is zero. However, since the set $\{\phi_n(x)\}$ is closed, we have from Parseval's theorem,

$$\int_a^b f^2(x)\,dx = \sum_{n=1}^{\infty} a_n^2$$

Hence, since each a_n is zero, as we have just observed, it follows that $f(x)$ is a null function, contrary to our assumption. This contradiction forces us to abandon the supposition that the closed set $\{\phi_n(x)\}$ is incomplete, and the theorem is established.

The converse of Theorem 3 is also true, but the proof of this fact is difficult and we shall not attempt it.

A great deal of important advanced mathematics deals with the properties of special orthogonal systems and with the validity of the formal expansion we have just created. In the next chapter we shall examine in some detail two such systems, the Bessel functions and the Legendre polynomials. Questions concerning the convergence of the generalized Fourier series (5), however, we shall not discuss, and in our work we shall assume not only that all the expansions we obtain converge but also that they actually represent the functions which generated them.

The terminology of Definitions 1–7 also applies to a set of functions each of which is defined on a closed interval $[a, b]$. Whether it is closed or open, an interval on which a set of functions is orthogonal is called a **fundamental interval** of that set. The formula (5*a*) which we found for the coefficients a_n in Eq. (5), and Theorems 1–3, are all valid whether the related fundamental interval is open or closed.

Orthogonal functions arise naturally and inevitably in many problems in pure and applied mathematics. Their existence is guaranteed by our next theorem, the so-called *Sturm-Liouville theorem.* This important result and the boundary-value problem with which it deals are conamed for the Swiss mathematician J. C. F. Sturm (1803–1855) and the French mathematician Joseph Liouville (1809–1882).

THEOREM 4 **(Sturm-Liouville Theorem)** Consider the differential equation

$$\frac{d[r(x)y']}{dx} + [q(x) + \lambda p(x)]y = 0$$

where

(a) $p(x)$ and $r(x)$ are continuous and positive on the closed interval $x_1 \leq x \leq x_2$.
(b) $q(x)$ is continuous at least on the open interval $x_1 < x < x_2$.

If $\lambda_1, \lambda_2, \lambda_3, \ldots,$ are distinct values of the parameter λ for which this equation has nontrivial solutions $y_1, y_2, y_3, \ldots,$ possessing continuous first derivatives and satisfying the boundary conditions

(c) $a_1 y(x_1) - b_1 y'(x_1) = 0$, a_1, b_1 not both zero
(d) $a_2 y(x_2) - b_2 y'(x_2) = 0$, a_2, b_2 not both zero

then the functions $\{y_n(x)\}$ form a system orthogonal with respect to the weight function $p(x)$ over the interval $[x_1, x_2]$.

PROOF To prove this, let y_m and y_n be the solutions associated with two distinct values of λ, say λ_m and λ_n. This means that

$$\frac{d(ry_m')}{dx} + (q + \lambda_m p)y_m = 0$$

$$\frac{d(ry_n')}{dx} + (q + \lambda_n p)y_n = 0$$

If y_m times the second of these equations is subtracted from y_n times the first, we obtain

$$(\lambda_m - \lambda_n)py_m y_n = y_m(ry_n')' - y_n(ry_m')' \tag{6}$$

Now, by performing the indicated differentiations, it is easy to verify that

$$y_m(ry_n')' - y_n(ry_m')' = \frac{d}{dx}[r(y_m y_n' - y_m' y_n)]$$

Hence, substituting the last expression for the right-hand side of Eq. (6) and then integrating from x_1 to x_2, we have

$$(7) \qquad (\lambda_m - \lambda_n)\int_{x_1}^{x_2} py_my_n\,dx = [r(y_my_n' - y_m'y_n)]_{x_1}^{x_2}$$

$$= r(x_2)[W(y_m, y_n)]_{x=x_2} - r(x_1)[W(y_m, y_n)]_{x=x_1}$$

where $W(y_m, y_n)$ denotes, as usual, the wronskian of y_m and y_n. [See Eq. (3), Sec. 2.2].

Now y_m and y_n are not merely solutions of the given differential equation. For all values of the integers m and n, they also satisfy the boundary conditions **c** and **d**. From the boundary condition at $x = x_2$, we have

$$a_2y_m(x_2) - b_2y_m'(x_2) = 0$$

$$a_2y_n(x_2) - b_2y_n'(x_2) = 0$$

in which, by hypothesis, a_2 and b_2 are not both zero. Thus these equations constitute a system of two homogeneous linear equations in two "unknowns" with a nontrivial solution, and therefore the determinant of their coefficients must be zero. Since this determinant is precisely $W(y_m, y_n)|_{x=x_2}$, it follows that the antiderivative in (7) vanishes at the upper limit, $x = x_2$. In exactly the same fashion it follows that the antiderivative in (7) vanishes at $x = x_1$. Thus we have shown that under the conditions of the theorem

$$(\lambda_m - \lambda_n)\int_{x_1}^{x_2} py_my_n\,dx = 0$$

Since λ_m and λ_n were any two *distinct* values of λ, the difference $\lambda_m - \lambda_n$ cannot vanish. Hence

$$\int_{x_1}^{x_2} py_my_n\,dx = 0 \qquad m \neq n$$

and the theorem is established. ◗

In each of the problems we considered in Sec. 11.5 the functions in terms of which we had to expand a function specified by some auxiliary condition satisfied a differential equation and a set of boundary conditions included under Theorem 4. This, and not the coincidental fact that Fourier series were involved, explains why the final expansion could be carried out in each case.

For a **Sturm-Liouville equation** $[r(x)y']' + [q(x) + \lambda p(x)]y = 0$ in which r and p are continuous on $[x_1, x_2]$, q is continuous, and $r(x) > 0$ and $p(x) > 0$ on the open interval (x_1, x_2), Eq. (7) still holds. Moreover, if $r(x_1) = 0$, the boundary condition **c** becomes irrelevant; i.e., the antiderivative in (7) vanishes at the lower limit $x = x_1$ without the need for any boundary condition there on the solutions y_m and y_n. Likewise, if $r(x_2) = 0$, the boundary condition **d** is irrelevant. Thus we have shown that under this slight relaxation of the conditions in Theorem 4, we have the following corollary.

COROLLARY If $\lambda_1 < \lambda_2 < \lambda_3 < \cdots$ are values of λ to which there correspond the nontrivial solutions $y_1, y_2, y_3, \ldots$ of the Sturm-Liouville equation

$$\frac{d[r(x)y']}{dx} = [q(x) + \lambda p(x)]y = 0$$

which satisfy the one boundary condition

(c) $a_1y(x_1) - b_1y'(x_1) = 0$ if $r(x_2) = 0$, a_1, b_1 not both zero

or the one boundary condition

(d) $a_2 y(x_2) - b_2 y'(x_2) = 0$ if $r(x_1) = 0$, a_2, b_2 not both zero

then, in either case,

$$\int_{x_1}^{x_2} p y_m y_n \, dx = 0 \qquad m \neq n$$

If in this corollary the function $r(x)$ vanishes at both x_1 and x_2, it is still possible to have an orthogonal set of solutions $\{y_n(x)\}$ on $[x_1, x_2]$ even though *no boundary conditions* apply. (See Sec. 12.9.)

EXAMPLE 1

TEMPERATURES ALONG A FINITE ROD WITH A RADIATING END (*cont.*)

When we left Example 1 in order to develop the theory necessary to complete its solution, we were faced with the necessity of expanding the initial temperature $u(x, 0) = f(x)$ in a series of the form (4)

$$f(x) = \sum_{n=1}^{\infty} B_n \sin \lambda_n x$$

where the functions in the set $\{\sin \lambda_n x\}$ were solutions of the differential equation

$$X'' + \lambda^2 X = 0$$

which satisfied the conditions

$$X(0) = 0$$
$$hX(l) + X'(l) = 0$$

We now note that this equation and the accompanying boundary conditions are in all respects a special case covered by Theorem 4. In fact, with X and λ^2 written in place of y and λ, respectively, we have

$$r(x) = 1 \qquad q(x) = 0 \qquad p(x) = 1$$
$$x_1 = 0 \qquad x_2 = l$$
$$a_1 = 1 \qquad b_1 = 0 \qquad a_2 = h \qquad b_2 = -1$$

Hence, by Theorem 4, the functions $\{\sin \lambda_n x\}$ form a set orthogonal with respect to the weight function $p(x) = 1$ on the interval $(0, l)$.

To determine B_n we now multiply Eq. (4) by $\sin \lambda_n x$ and integrate term by term from 0 to l. Because of the orthogonality of the functions $\{\sin \lambda_n x\}$, every integral on the right vanishes except the one whose integrand contains $\sin^2 \lambda_n x$. Therefore

$$B_n = \frac{\int_0^l f(x) \sin \lambda_n x \, dx}{\int_0^l \sin^2 \lambda_n x \, dx}$$

or, evaluating the integral in the denominator and recalling that $z_n \equiv \lambda_n l$ satisfies the equation $\sin z_n = -\alpha z_n \cos z_n$, we have

$$B_n = \frac{2}{l(1 + \alpha \cos^2 z_n)} \int_0^l f(x) \sin \lambda_n x \, dx$$

With B_n determined, the formal solution of our heat flow problem is now complete.

Although Theorem 4 guarantees that under suitable conditions the solutions of certain differential equations form orthogonal systems, it does not guarantee that such systems are complete. This means that unless completeness is established in some other way, expansions in terms of such systems are only formal solutions. Since it is difficult to prove that a set of orthogonal functions is complete, we shall simply assume this property and leave to more advanced treatments the validation of our formal solutions.

However, while we cannot prove that an orthogonal system is complete, we may in some instances be able to show that it is incomplete. One way to do this is to invoke Definition 5 and exhibit a nonnull function which is orthogonal to every member of the set. The difficulty here is that such a function is usually not easy to find. Another, less certain, method is to expand some simple function and note the degree to which the resultant series approximates that function. Sometimes the discrepancy is so great that we can be sure that one or more members must be missing from the orthogonal set we are trying to use. The next example alerts us to this possibility.

EXAMPLE 2

Expand $f(x) = x^2$ in terms of the trigonometric functions $\{\cos \lambda x, \sin \lambda x\}$ which satisfy the boundary conditions

$$y(0) = 0 \qquad \text{and} \qquad 2y(1) = y'(1)$$

Clearly, the general term in the required expansion is of the form $y = A \cos \lambda x + B \sin \lambda x$, and our first task is to determine the values of λ for which such a term can satisfy the given boundary conditions. From the requirement that $y(0) = 0$, it is evident that $A = 0$, leaving $y = B \sin \lambda x$. Imposing the second condition, we find that we must have

$$2B \sin \lambda = B\lambda \cos \lambda$$

If $B = 0$ we have only the trivial solution, since A is already known to be zero. Hence we must have

$$2 \sin \lambda = \lambda \cos \lambda \qquad \text{or} \qquad \tan \lambda = \frac{\lambda}{2}$$

an equation very much like the one we encountered in Example 1. The first three roots of this equation are

$$\lambda_1 \doteq 4.275 \qquad \lambda_2 \doteq 7.597 \qquad \lambda_3 \doteq 10.813$$

With the admissible values of λ known (or at least knowable), we now set up the required expansion

$$x^2 = B_1 \sin \lambda_1 x + B_2 \sin \lambda_2 x + \cdots + B_n \sin \lambda_n x + \cdots \tag{8}$$

and proceed to determine the formula for B_n.

Since both the equation satisfied by $\sin \lambda x$ and $\cos \lambda x$, namely, $y'' + \lambda^2 y = 0$, and the given boundary conditions meet the requirements of Theorem 4, it follows that in this case the functions $\{\sin \lambda_n x\}$ are orthogonal with respect to the weight function 1 on the interval (0, 1). Hence we can find B_n, as usual, by multiplying the expansion (8) by $\sin \lambda_n x$ and integrating term by term from $x = 0$ to $x = 1$. Because of the property of orthogonality, every term but one is zero, and we have

$$B_n = \frac{\displaystyle\int_0^1 x^2 \sin \lambda_n x \, dx}{\displaystyle\int_0^1 \sin^2 \lambda_n x \, dx}$$

For the integral in the numerator we have

$$\left[\frac{2x \sin \lambda_n x}{\lambda_n^2} - \left(\frac{x^2}{\lambda_n} - \frac{2}{\lambda_n^3}\right) \cos \lambda_n x\right]_0^1$$

Evaluating this and then simplifying the result by using the fact that λ_n satisfies the characteristic equation $2 \sin \lambda = \lambda \cos \lambda$, we obtain

$$-\frac{2}{\lambda_n^3}(1 - \cos \lambda_n)$$

For the integral in the denominator we have

$$\frac{1}{2}\left[x - \frac{\sin \lambda_n x \cos \lambda_n x}{\lambda_n}\right]_0^1$$

Evaluating this, and again using the characteristic equation to simplify the result, we obtain

$$\tfrac{1}{4}(2 - \cos^2 \lambda_n)$$

Hence

$$B_n = \frac{-2(1 - \cos \lambda_n)/\lambda_n^3}{\frac{1}{4}(2 - \cos^2 \lambda_n)} = \frac{8}{\lambda_n^3}\left[\frac{\cos \lambda_n - 1}{2 - \cos^2 \lambda_n}\right]$$

From this, using the values for λ_1, λ_2, and λ_3 which we listed earlier, we find

$$B_1 \doteq -0.0801 \qquad B_2 \doteq -0.0118 \qquad B_3 \doteq -0.0038$$

In the spirit of our work up to this point it is now natural to feel that with the B's determined, we have obtained the formal solution to the problem and our work is done. However, the size of B_1, B_2, and B_3 together with the fact that $|\sin \lambda_n x| \le 1$ suggest strongly that the series (8) is a very poor approximation to x^2, and this is confirmed by Fig. 11.20 which shows a plot of the first few terms of the series. Surely something is wrong!

The difficulty lies in the way the problem is worded, ''Expand $f(x) = x^2$ in terms of the trigonometric functions $\{\cos \lambda x, \sin \lambda x\}$. . . .'' Actually, we do not expand a function in terms of functions of a particular form; properly, we expand a function in terms of the set of all solutions of a particular differential equation. In the present example, that equation is $y'' + \lambda^2 y = 0$; and so far we have only worked with the solutions that arise when $\lambda^2 > 0$. If we do not consider the possibilities $\lambda^2 = 0$ and $\lambda^2 < 0$, we may miss one or more additional solutions and attempt (as we did) to expand x^2 in terms of functions of an incomplete set.

The possibility that $\lambda^2 = 0$ is easily disposed of. The corresponding family of solutions is $y = Ax + B$, and it is clear that no nontrivial member of this set can satisfy the given boundary conditions. If $\lambda^2 < 0$, the equation becomes $y'' - \lambda^2 y = 0$, whose complete solution is $y = A \cosh \lambda x + B \sinh \lambda x$. The condition that $y(0) = 0$ implies that $A = 0$, leaving $y = B \sinh \lambda x$. Imposing the condition at $x = 1$ gives us the equation $2B \sinh \lambda = \lambda B \cosh \lambda$. Since there is only a trivial solution if $B = 0$, we must have

$$2 \sinh \lambda = \lambda \cosh \lambda \qquad \text{or} \qquad \tanh \lambda = \frac{\lambda}{2}$$

This equation has the single root $\lambda_0 \doteq 1.915$. Hence the function $y = \sinh \lambda_0 x$ must be adjoined to the set $\{\sin \lambda_n x\}$ and inserted as a first term in the series (8):

(8*a*) $$x^2 = B_0 \sinh \lambda_0 x + B_1 \sin \lambda_1 x + B_2 \sin \lambda_2 x + \cdots + B_n \sin \lambda_n x + \cdots$$

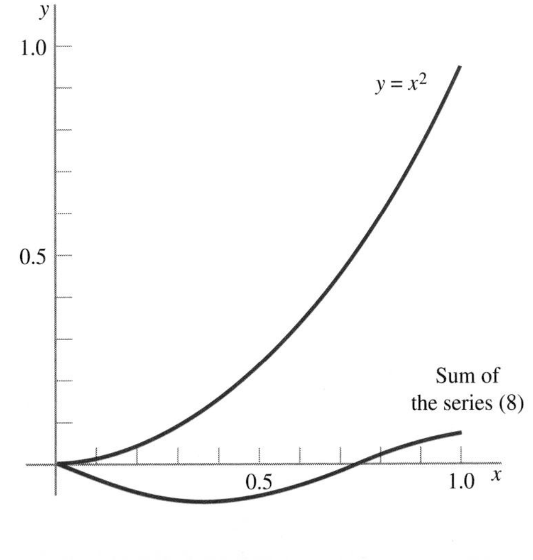

FIGURE 11.20
The approximation of $y = x^2$ by a series of incomplete orthogonal functions.

We have already obtained the values of $B_1, B_2, B_3, \ldots$. For the value of B_0 we have, by now familiar reasoning,

$$B_0 = \frac{\int_0^1 x^2 \sinh \lambda_0 x \, dx}{\int_0^1 \sinh^2 \lambda_0 x \, dx}$$

The evaluation of these integrals is almost identical to the evaluation of the integrals in the formula for B_n. When the result is simplified by using the equation $2 \sinh \lambda_0 = \lambda_0 \cosh \lambda_0$, we obtain the expression

$$B_0 = \frac{8}{\lambda_0^3}\left(\frac{\cosh \lambda_0 - 1}{\cosh^2 \lambda_0 - 2}\right) \doteq 0.2805$$

The graph of the series (8*a*) is almost indistinguishable from the graph of $y = x^2$. In fact, for $0 \le x \le 1$ the maximum difference between x^2 and the sum of the first four terms in (8*a*) is only 0.005. We can thus feel reasonably confident that with the inclusion of $\sinh \lambda_0 x$ we now have a complete orthogonal system.

Other examples of incomplete orthogonal sets of trigonometric functions will be found among the exercises. We shall also encounter this phenomenon in the next chapter when we attempt to expand functions in series involving Bessel functions.

Problems involving second-order differential equations are not the only ones in which orthogonal functions arise. In particular, we have the following important theorem covering fourth-order systems, of which the vibrating beam is a special case.

THEOREM 5 Given the differential equation

$$\frac{d^2[r(x)y'']}{dx^2} + [q(x) + \lambda p(x)]y = 0$$

where $r(x)$ and $p(x)$ are continuous and positive on the closed interval $[a, b]$ and $q(x)$ is continuous at least on the open interval (a, b). If $\lambda_1, \lambda_2, \lambda_3, \ldots$, are distinct values of the parameter λ for which there exist nontrivial solutions of this equation possessing continuous third derivatives and satisfying the boundary conditions

$$a_1 y(a) - \alpha_1 (ry'')'\big|_{x=a} = 0 \qquad a_2 y'(a) - \alpha_2 (ry'')\big|_{x=a} = 0$$
$$b_1 y(b) - \beta_1 (ry'')'\big|_{x=b} = 0 \qquad b_2 y'(b) - \beta_2 (ry'')\big|_{x=b} = 0$$

where neither a_i and α_i nor b_i and β_i are both zero, and if $y_1, y_2, y_3, \ldots$, are nontrivial solutions corresponding to these values of λ, then the functions $\{y_n(x)\}$ form a set orthogonal with respect to the weight $p(x)$ over the interval $[a, b]$.

EXAMPLE 3

A VIBRATING UNIFORM CANTILEVER BEAM

A uniform cantilever of length l begins to vibrate with initial displacement $y(x, 0) = f(x)$ and initial velocity $\partial y/\partial t\big|_{x,0} = g(x)$. Find its displacement at any point at any subsequent time.

For definiteness, let us assume that the beam extends along the positive x-axis with its built-in end at the origin. Then, since the beam is of uniform cross section and bears no external load, we have to solve Eq. (11), Sec. 11.2,

$$a^2 \frac{\partial^4 y}{\partial x^4} = -\frac{\partial^2 y}{\partial t^2}$$

subject to the end conditions that define a cantilever:

$$y(0,t)=0 \qquad \text{Displacement at a built-in end} = 0$$

$$\left.\frac{\partial y}{\partial x}\right|_{0,t}=0 \qquad \text{Slope at a built-in end} = 0$$

$$\left.\frac{\partial^2 y}{\partial x^2}\right|_{l,t}=0 \qquad \text{Moment } EI\frac{\partial^2 y}{\partial x^2} \text{ at a free end} = 0$$

$$\left.\frac{\partial^3 y}{\partial x^3}\right|_{l,t}=0 \qquad \text{Shear } \frac{\partial[EI(\partial^2 y/\partial x^2)]}{\partial x} \text{ at a free end} = 0$$

We could begin, as we have so many times before, by assuming a product solution $Y(x,t)=X(x)T(t)$, substituting, separating variables, and considering in turn the various possibilities for the separation parameter. However, *when the nature of the problem guarantees that the time factor will be harmonic,* it is common practice to bypass the separation step and being with the assumption that $T(t)=(A\cos\lambda t+B\sin\lambda t)$, so that $Y(x,t)$ is of the form

$$Y(x,t)=X(x)(A\cos\lambda t+B\sin\lambda t)$$

Substituting Y in this form effectively separates the partial differential equation, because the factor $(A\cos\lambda t+B\sin\lambda t)$ always divides out, leaving $-\lambda^2$ in its place in the second time derivative. The result is then an ordinary differential equation in which only X appears.

In the present problem, this stratagem gives us at once the equation

$$a^2X^{\text{iv}}=\lambda^2X$$

from which we find that

$$X=C\cos\left(\sqrt{\frac{\lambda}{a}}x\right)+D\sin\left(\sqrt{\frac{\lambda}{a}}x\right)+E\cosh\left(\sqrt{\frac{\lambda}{a}}x\right)+F\sinh\left(\sqrt{\frac{\lambda}{a}}x\right) \tag{9}$$

Imposing the first boundary condition, which like the other three is of course a condition on $X(x)$ alone since it must hold for all values of t, we obtain

$$C+E=0$$

Imposing the second boundary condition by differentiating Eq. (9) and then setting $x=0$, we obtain

$$D+F=0$$

Similarly, imposing the third boundary condition by differentiating Eq. (9) twice and then setting $x=l$, we find

$$-C\cos\left(\sqrt{\frac{\lambda}{a}}l\right)-D\sin\left(\sqrt{\frac{\lambda}{a}}l\right)+E\cosh\left(\sqrt{\frac{\lambda}{a}}l\right)+F\sinh\left(\sqrt{\frac{\lambda}{a}}l\right)=0$$

Finally, imposing the fourth boundary condition by differentiating Eq. (9) three times and then setting $x=l$, we obtain

$$C\sin\left(\sqrt{\frac{\lambda}{a}}l\right)-D\cos\left(\sqrt{\frac{\lambda}{a}}l\right)+E\sinh\left(\sqrt{\frac{\lambda}{a}}l\right)+F\cosh\left(\sqrt{\frac{\lambda}{a}}l\right)=0$$

Hence, setting

$$z=\sqrt{\frac{\lambda}{a}}l \tag{10}$$

we must have

$$-C\cos z - D\sin z + E\cosh z + F\sinh z = 0$$
$$C\sin z - D\cos z + E\sinh z + F\cosh z = 0$$

If we eliminate C and D from these equations by using the conditions $C + E = 0$ and $D + F = 0$, we obtain the system

$$\begin{aligned} E(\cosh z + \cos z) + F(\sinh z + \sin z) &= 0 \\ E(\sinh z - \sin z) + F(\cosh z + \cos z) &= 0 \end{aligned} \tag{11}$$

These equations will have a solution other than the obvious trivial solution $E = F = 0$ if and only if the determinant of the coefficients is equal to zero. Thus we must have

$$\begin{vmatrix} \cosh z + \cos z & \sinh z + \sin z \\ \sinh z - \sin z & \cosh z + \cos z \end{vmatrix} = 0$$

or, expanding and simplifying,

$$\cosh z \cos z = -1$$

The existence of infinitely many roots of this equation, that is, $\cos z = -1/\cosh z$, can be inferred from Fig. 11.21, where the graphs of

$$y = \cos z \quad \text{and} \quad y = -\frac{1}{\cosh z}$$

are plotted.

From these roots, $z_1, z_2, z_3, \ldots$, we can find the relevant values of λ at once from Eq. (10):

$$\lambda_1 = \frac{az_1^2}{l^2}, \quad \lambda_2 = \frac{az_2^2}{l^2}, \quad \lambda_3 = \frac{az_3^2}{l^2}, \quad \cdots$$

When z has any one of the values $z_1, z_2, z_3, \ldots$ (and E and F have the corresponding values E_n and F_n), Eqs. (11) become dependent and we can write either

$$\frac{E_n}{F_n} = -\frac{\sinh z_n + \sin z_n}{\cosh z_n + \cos z_n} \quad \text{or} \quad \frac{E_n}{F_n} = -\frac{\cosh z_n + \cos z_n}{\sinh z_n - \sin z_n}$$

as we choose. Using the former, we have

$$E_n = -C_n = -(\sinh z_n + \sin z_n)K_n$$
$$F_n = -D_n = (\cosh z_n + \cos z_n)K_n$$

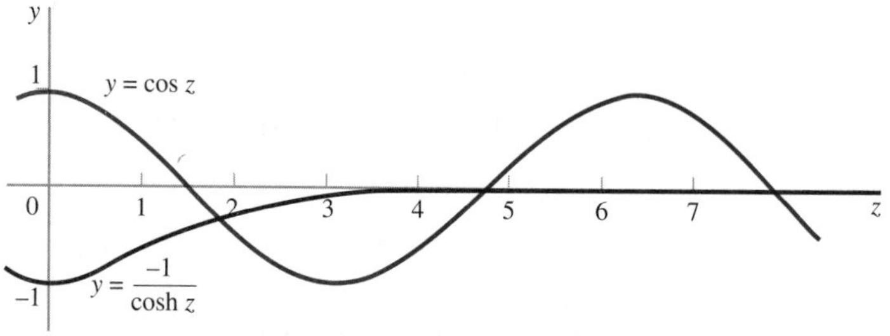

FIGURE 11.21
The graphical solution of the equation $\cos z = -1/\cosh z$.

where K_n is an arbitrary constant. Therefore, substituting into Eq. (9), we have

$$X_n(x) = C_n \cos\left(\sqrt{\frac{\lambda_n}{a}}x\right) + D_n \sin\left(\sqrt{\frac{\lambda_n}{a}}x\right) + E_n \cosh\left(\sqrt{\frac{\lambda_n}{a}}x\right) + F_n \sinh\left(\sqrt{\frac{\lambda_n}{a}}x\right)$$
$$= K_n(\sinh z_n + \sin z_n)\left[\cos\left(z_n\frac{x}{l}\right) - \cosh\left(z_n\frac{x}{l}\right)\right]$$
$$- K_n(\cosh z_n + \cos z_n)\left[\sin\left(z_n\frac{x}{l}\right) - \sinh\left(z_n\frac{x}{l}\right)\right]$$

Hence, absorbing K_n in the (still arbitrary) coefficients A_n and B_n in the product solution for y and redefining X_n to be the completely determined function

$$(12)\quad X_n = (\sinh z_n + \sin z_n)\left[\cos\left(z_n\frac{x}{l}\right) - \cosh\left(z_n\frac{x}{l}\right)\right] - (\cosh z_n + \cos z_n)\left[\sin\left(z_n\frac{x}{l}\right) - \sinh\left(z_n\frac{x}{l}\right)\right]$$

we have

$$y(x, t) = \sum_{n=1}^{\infty} X_n(x)T_n(t) = \sum_{n=1}^{\infty} X_n(x)(A_n \cos \lambda_n t + B_n \sin \lambda_n t)$$

as the formal solution of the partial differential equation which meets the four given boundary conditions.

To satisfy the initial displacement condition, we must have

$$(13)\qquad y(x, 0) \equiv f(x) = \sum_{n=1}^{\infty} A_n X_n(x)$$

and to satisfy the initial velocity condition, we must have

$$(14)\qquad \left.\frac{\partial y}{\partial t}\right|_{x,0} \equiv g(x) = \sum_{n=1}^{\infty} (\lambda_n B_n)X_n(x)$$

Thus again, to satisfy the initial conditions we must be able to expand an arbitrary function in an infinite series of known functions, in this case the functions of the set $\{X_n(x)\}$ defined by Eq. (12). These bear little or no resemblance to the terms of a Fourier series, but the required expansions can easily be carried out using the orthogonality of the X's which is guaranteed by Theorem 5 (and of course their completeness, which, as usual, we must assume). In fact, with $-\lambda^2/a^2$ written in place of λ, our problem is just the special case of Theorem 5 for which

$$\begin{array}{cccc} r(x) = 1 & q(x) = 0 & p(x) = 1 & \\ & a = 0 & b = 1 & \\ a_1 = 1 & \alpha_1 = 0 & b_1 = 0 & \beta_1 = -1 \\ a_2 = 1 & \alpha_2 = 0 & b_2 = 0 & \beta_2 = -1 \end{array}$$

Hence the functions of the set $\{X_n(x)\}$ are orthogonal with respect to the weight function $p(x) = 1$ over the interval $(0, l)$.

With the orthogonality of the X_n's now established, we can determine A_n and B_n immediately by multiplying Eqs. (13) and (14) by $X_n(x)$ and integrating from 0 to l. The results are

$$A_n = \frac{\int_0^l f(x)X_n(x)\,dx}{\int_0^l X_n^2(x)\,dx} \quad \text{and} \quad B_n = \frac{\int_0^l g(x)X_n(x)\,dx}{\lambda_n\int_0^l X_n^2(x)\,dx}$$

We are now in a position to summarize the essential features of a **boundary-value problem:** By assuming that solutions for the dependent variable exist in the form of products of functions of the respective independent variables, the original partial differential equation is broken down into several ordinary differential equations, each involving a parameter λ which ranges over a continuous infinity of values.

When the boundary conditions of the problem are imposed upon the product solutions obtained by solving the component ordinary differential equations, it is necessary, in order to avoid solutions which are identically zero, that the parameter λ satisfy a certain equation. This equation is known as the **characteristic equation** of the problem, and its roots, in general infinite in number, are known as the **characteristic values, eigenvalues,** or **eigenwerte**† of the problem. Only for these values can nontrivial solutions be found satisfying both the partial differential equation and the given boundary conditions. In a vibration problem, the characteristic values determine the natural frequencies of the system, and the characteristic equation is therefore usually called the **frequency equation.** The solutions corresponding to the respective characteristic values are known as the **characteristic functions** or **eigenfunctions** of the problem. In a vibration problem, they are usually called the **normal modes** since they define the relative amplitudes of the extreme positions between which the system oscillates when it is vibrating at a single natural frequency, i.e., in a ''normal'' manner.

To satisfy the initial conditions of the problem it is necessary to be able to express an arbitrary function as an infinite series of the characteristic functions of the problem. This can be done in most cases of interest because under very general conditions the characteristic functions of a boundary-value problem form an orthogonal set over the particular interval related to the problem.

In Example 5, Sec. 8.6, we used Fourier series to obtain the deflection curve of a simply supported beam bearing a distributed load $w(x)$. At that time we pointed out that Fourier series could not be used to find the deflection curves of beams that were not simply supported, but that similar methods using other types of series would become available later in our work. We are now in a position to discuss these more general series, and our next example illustrates how the orthogonal functions obtained in Example 3 can be used to find the deflection curve of a cantilever bearing an arbitrary distributed load.

EXAMPLE 4

DEFLECTION OF A LOADED UNIFORM CANTILEVER

A uniform cantilever extending along the x axis from $x = 0$ to $x = l$ bears a distributed load per unit length $w(x)$. Find the deflection curve of the beam under the given load.

In the last example, we found that the functions $\{X_n\}$ defined by Eq. (12) satisfied the conditions that characterize a cantilever, namely,

$$X(0) = X'(0) = X''(l) = X'''(l) = 0$$

and were orthogonal with respect to the weight function $p(x) = 1$ on the interval [0, 1].‡ Thus, by what should now be a familiar extension of the Fourier procedure, we expand the given load function in a series of the X_n's, obtaining

$$w(x) = A_1X_1(x) + A_2X_2(x) + \cdots + A_nX_n(x) + \cdots \tag{15}$$

†German for *characteristic values.*

‡Admittedly, the X_n's are defined by rather complicated expressions, but the definition is explicit, the individual terms involve only familiar functions, and it would not be difficult to tabulate the X_n's if we desired. Logically, the X_n's are as simple as the functions $\sin n\pi x/l$ of Example 5, Sec. 8.6, and are used in exactly the same way.

where

$$A_n = \frac{\int_0^l w(x)X_n(x)\,dx}{\int_0^l X_n^2(x)\,dx}$$

Furthermore, since the individual X_n's satisfy the end conditions of a cantilever, so too will the function defined by any series of them. Thus we naturally assume that the (unknown) deflection curve we are seeking can also be represented by a series of the X_n's, say

$$y(x) = B_1X_1(x) + B_2X_2(x) + \cdots + B_nX_n(x) + \cdots \tag{16}$$

where the B's are coefficients to be determined.

If, just as we did in Example 5, Sec. 8.6, we substitute Eqs. (15) and (16) into the equation

$$y^{\text{iv}} = -\frac{w(x)}{EI}$$

which describes the static deflection of a cantilever bearing a distributed load, we obtain

$$\sum_{n=1}^{\infty} B_nX_n^{\text{iv}} = -\frac{1}{EI}\sum_{n=1}^{\infty} A_nX_n$$

or, since $X_n^{\text{iv}} = k_n^4X_n$, $(k_n^4 = \lambda_n^2/a^2)$,

$$\sum_{n=1}^{\infty} B_nk_n^4X_n = -\frac{1}{EI}\sum_{n=1}^{\infty} A_nX_n$$

Hence, equating coefficients of like terms on each side of the last equation, we have the formula

$$B_n = -\frac{A_n}{EIk_n^4} = -\frac{1}{EIk_n^4}\,\frac{\int_0^l w(x)X_n(x)\,dx}{\int_0^l X_n^2(x)\,dx}$$

The presence of the factor k_n^4 in the denominator of the expression for B_n means that the series defining the deflection curve will converge quite rapidly so that only a few terms will provide a satisfactory approximation.

We have not yet obtained a formula for the integral

$$\int_0^l X_n^2(x)\,dx$$

which appears in the denominator of the expressions for A_n and B_n, but this can be done and the general formula is given in Exercise 46. In particular, for a uniform cantilever we have the very simple result

$$\int_0^l X_n^2(x)\,dx = \frac{l}{4}X_n^2(l)$$

Sometimes, because of the nature of the boundary conditions, the property of orthogonality cannot be inferred from either Theorem 4 or 5.† In this case, orthogonality (if the solutions have

†Sometimes, in fact, the functions we obtain do not have the property of orthogonality, as we now know it, at all. Exercise 31(**b**) gives an example of this sort.

this property) must be established independently of either theorem. Our next example illustrates this possibility. It involves a problem in heat flow in which two conducting sheets with different thermal properties are joined together to form a single sheet, so that in additional to boundary conditions of the usual sort there are also two conditions of continuity at the interface that have no counterpart in either Theorem 4 or 5.

EXAMPLE 5

TEMPERATURES IN A COMPOSITE RECTANGULAR SHEET

A composite sheet of metal coincides with the rectangle in the xy plane whose vertices are the points $(-a_1, 0)$, $(a_2, 0)$, (a_2, b), $(-a_1, b)$ (Fig. 11.22). The thermal conductivity of the portion of the sheet to the left of the y axis is k_1, the conductivity of the portion to the right of the y axis is k_2. The edges of the sheet parallel to the y axis are maintained at the temperature 0. The lower edge of the sheet is perfectly insulated, and the upper edge is maintained at the temperature T_0. Assuming that heat flow in the sheet is strictly two-dimensional, find the steady-state temperature distribution in the sheet.

Since we are to solve a problem in steady-state heat flow in two dimensions, we begin with Laplace's equation

$$\frac{\partial^2 u}{\partial x^2} + \frac{\partial^2 u}{\partial y^2} = 0$$

Assuming a product solution $u(x, y) = X(x)Y(y)$ and separating variables presents no problem, we find, as usual,

$$\frac{X''}{X} = -\frac{Y''}{Y} = \mu \qquad \mu \text{ a constant}$$

However, because of the different thermal properties of the two sections of the sheet, we must solve the equation for X independently over the two intervals $-a_1 < x < 0$ and $0 < x < a_2$ and then fit these solutions together so that both the temperature and the rate of heat flow are continuous at $x = 0$.

Considering first the possibility that $\mu > 0$, say $u = \lambda^2$, $\lambda > 0$, we have from the equation for X, namely $X'' = \lambda^2 X$, the two solutions

$$X_1(x) = A_1 \cosh \lambda x + B_1 \sinh \lambda x \qquad -a_1 < x < 0$$
$$X_2(x) = A_2 \cosh \lambda x + B_2 \sinh \lambda x \qquad 0 < x < a_2$$

Now the left-hand edge condition $u(-a_1, y) = 0$ implies $X_1(-a_1)Y(y) = 0$, which in turn implies that $X_1(-a_1) = 0$. Hence

$$A_1 \cosh \lambda a_1 - B_1 \sinh \lambda a_1 = 0$$

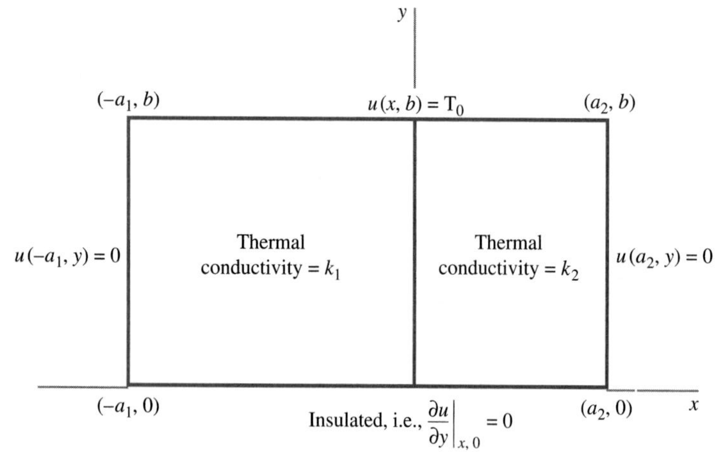

FIGURE 11.22
The composite sheet discussed in Example 5.

so that we may take $A_1 = C_1 \sinh \lambda a_1$ and $B_1 = C_1 \cosh \lambda a_1$, getting

$$X_1(x) = C_1(\sinh \lambda a_1 \cosh \lambda x + \cosh \lambda a_1 \sinh \lambda x) = C_1 \sinh \lambda(a_1 + x)$$

Similarly, from the condition $u(a_2, y) = 0$, we conclude that

$$X_2(x) = C_2 \sinh \lambda(a_2 - x)$$

In order for the temperature to be continuous at $x = 0$, we must have $X_1(0^-)Y(y) = X_2(0^+)Y(y)$, which implies that

$$C_1 \sinh \lambda a_1 = C_2 \sinh \lambda a_2$$

Thus we can take $C_1 = K \sinh \lambda a_2$ and $C_2 = K \sinh \lambda a_1$, getting

$$\begin{aligned} X_1(x) &= K \sinh \lambda a_2 \sinh \lambda(a_1 + x) \\ X_2(x) &= K \sinh \lambda a_1 \sinh \lambda(a_2 - x) \end{aligned}$$

Finally, in order for the rate of heat flow to be continuous at $x = 0$, we must have

$$k_1 X_1'(0^-)Y(y) = k_2 X_2'(0^+)Y(y)$$

which implies that

$$k_1 \lambda K \sinh \lambda a_2 \cosh \lambda a_1 = -k_2 \lambda K \sinh \lambda a_1 \cosh \lambda a_2$$

or

$$\lambda K(k_1 \sinh \lambda a_2 \cosh \lambda a_1 + k_2 \sinh \lambda a_1 \cosh \lambda a_2) = 0$$

Since k_1, k_2, a_1, a_2, and λ are all positive and since the hyperbolic sine and cosine are positive for all positive values of their arguments, it follows from the last equation that $K = 0$. Thus the assumption that $\mu > 0$ leads only to a trivial solution.

In the same way, though we shall leave the details as an exercise, it can be shown that the assumption $\mu = 0$ also leads only to a trivial solution.

If $\mu < 0$, say $\mu = -\lambda^2$, $\lambda > 0$, we have from the equations $X'' = -\lambda^2 X$ and $Y'' = \lambda^2 Y$,

$$\begin{aligned} X_1(x) &= A_1 \cos \lambda x + B_1 \sin \lambda x \qquad -a_1 < x < 0 \\ X_2(x) &= A_2 \cos \lambda x + B_2 \sin \lambda x \qquad 0 < x < a_2 \end{aligned}$$

and in either case

$$Y(y) = D \cosh \lambda y + E \sinh \lambda y$$

From the insulated condition along the lower edge of the sheet, namely,

$$\left.\frac{\partial u}{\partial y}\right|_{x,0} = 0 \qquad \text{or} \qquad X(x)Y'(0) = 0$$

we conclude that $E = 0$, so that $Y(y) = D \cosh \lambda y$.

Now proceeding exactly as we did in the case $\mu = \lambda^2$, we find that the boundary conditions $u(-a_1, y) = u(a_2, y) = 0$ and the condition that the temperature be continuous at $x = 0$ imply that

$$X_1(x) = K \sin \lambda a_2 \sin \lambda(a_1 + x) \qquad \text{and} \qquad X_2(x) = K \sin \lambda a_1 \sin \lambda(a_2 - x)$$

The condition that the rate of heat flow be continuous at $x = 0$ then leads to the equation

$$k_1 K \lambda \sin \lambda a_2 \cos \lambda a_1 = -k_2 K \lambda \sin \lambda a_1 \cos \lambda a_2$$

or

$$\lambda K(k_1 \sin \lambda a_2 \cos \lambda a_1 + k_2 \sin \lambda a_1 \cos \lambda a_2) = 0$$

If $K = 0$, the solution becomes trivial. Hence we must have

$$k_1 \sin \lambda a_2 \cos \lambda a_1 + k_2 \sin \lambda a_1 \cos \lambda a_2 = 0 \qquad \lambda > 0 \tag{17}$$

This is the *characteristic equation* of our problem. For its roots (which are simple and distinct) and for no other values of λ, we have nontrivial solutions of Laplace's equation satisfying all the conditions of the problem except the one along the upper edge of the sheet. Specifically, these solutions are

$$u_n(x, y) = K_n X_n(x) Y_n(y) = K_n X_n(x) \cosh \lambda_n y \tag{18}$$

where the *characteristic functions* X_n are given by

$$X_n(x) = \begin{cases} \sin \lambda_n a_2 \sin \lambda_n(a_1 + x) & -a_1 < x < 0 \\ \sin \lambda_n a_1 \sin \lambda_n(a_2 - x) & 0 < x < a_2 \end{cases} \tag{19}$$

No one of the solutions (18), by itself, can fit the assigned temperature condition $u(x, b) = T_0$ which holds along the upper edge of the sheet. Hence, as usual, we form an infinite series

$$u(x, y) = \sum_{n=1}^{\infty} u_n(x, y) = \sum_{n=1}^{\infty} K_n X_n(x) \cosh \lambda_n y$$

and attempt to determine the coefficients K_n so that when $y = b$, the series

$$\sum_{n=1}^{\infty} (K_n \cosh \lambda_n b) X_n(x)$$

will be the expansion of the function $u(x, b) = T_0$.

To do this requires that the X_n's be orthogonal, but because of the internal conditions at $x = 0$, this cannot be inferred from Theorem 4. However, since the characteristic functions satisfy the Sturm-Liouville equation $(X')' + \lambda^2 X = 0$ on the separate intervals $(-a_1, 0)$ and $(0, a_2)$ over each of which $p(x) \equiv 1$ is *constant,* let us seek constants c_1, c_2 for which (hopefully) the X_n's will be orthogonal on $(-a_1, a_2)$ with respect to a weight function

$$P(x) = \begin{cases} c_1 & -a_1 < x < 0 \\ c_2 & 0 < x < a_2 \end{cases}$$

If we begin with the equations

$$X''_m + \lambda_m^2 X_m = 0$$
$$X''_n + \lambda_n^2 X_n = 0$$

satisfied by any two of the X's, multiply the first equation by $P(x)X_n$ and the second by $P(x)X_m$, subtract the results, and then integrate from $-a_1$ to a_2, we obtain

$$(\lambda_m^2 - \lambda_n^2) \int_{-a_1}^{a_2} P(x) X_m X_n \, dx = \int_{-a_1}^{a_2} P(x)(X''_n X_m - X''_m X_n) \, dx = \int_{-a_1}^{a_2} P(x)(X'_n X_m - X'_m X_n)' \, dx$$

Breaking up the range of integration, substituting the appropriate values of $P(x)$, and integrating, we have

$$\begin{aligned}(\lambda_m^2 - \lambda_n^2) \int_{-a_1}^{a_2} P(x) X_m X_n \, dx &= c_1 [X'_n X_m - X'_m X_n]_{-a_1}^{0} + c_2 [X'_n X_m - X'_m X_n]_0^{a_2} \\ &= c_1[\{X'_n(0^-)X_m(0^-) - X'_m(0^-)X_n(0^-)\} - \{X'_n(-a_1)X_m(-a_1) - X'_m(-a_1)X_n(-a_1)\}] \\ &\quad + c_2[\{X'_n(a_2)X_m(a_2) - X'_m(a_2)X_n(a_2)\} - \{X'_n(0^+)X_m(0^+) - X'_m(0^+)X_n(0^+)\}]\end{aligned}$$

Now from the boundary conditions of the problem, $X_n(-a_1) = X_n(a_2) = 0$ for all values of n. Furthermore, from the internal conditions at $x = 0$, we have $X_n(0^-) = X_n(0^+)$ and $k_1X_n'(0^-) = k_2X_n'(0^+)$ for all values of n. Hence the entire right-hand side of the last equation involving c_1 and c_2 reduces to

$$\left[c_1\left(\frac{k_2}{k_1}\right) - c_2\right][X_n'(0^+)X_m(0^+) - X_m'(0^+)X_n(0^+)]$$

which vanishes for the pairs of values $c_1 = k_1$, $c_2 = k_2$, and $c_1 = 1$, $c_2 = k_2/k_1$. Using the first pair, we have that the functions (19) are orthogonal on $(-a_1, a_2)$ with respect to the weight function

$$P(x) = \begin{cases} k_1 & -a_1 < x < 0 \\ k_2 & 0 < x < a_2 \end{cases} \tag{20}$$

Using this property, the K_n's can easily be found, and we have

$$K_n \cosh \lambda_n b = \frac{\displaystyle\int_{-a_1}^{a_2} P(x)X_n(x)T_0\,dx}{\displaystyle\int_{-a_1}^{a_2} P(x)X_n^2(x)\,dx}$$

or

$$K_n = \frac{T_0}{\cosh \lambda_n b} \frac{\displaystyle\int_{-a_1}^{0} k_1 \sin \lambda_n a_2 \sin \lambda_n(a_1 + x)\,dx + \int_0^{a_2} k_2 \sin \lambda_n a_1 \sin \lambda_n(a_2 - x)\,dx}{\displaystyle\int_{-a_1}^{0} k_1 \sin^2 \lambda_n a_2 \sin^2 \lambda_n(a_1 + x)\,dx + \int_0^{a_2} k_2 \sin^2 \lambda_n a_1 \sin^2 \lambda_n(a_2 - x)\,dx}$$

where the λ's are the roots of the characteristic equation (17).

Performing the integrations and evaluating, using the characteristic equation (17) to simplify the result, we have finally

$$K_n = \frac{2T_0}{\lambda_n \cosh \lambda_n b} \frac{k_2 \sin \lambda_n a_1 + k_1 \sin \lambda_n a_2}{k_2 a_2 \sin^2 \lambda_n a_1 + k_1 a_1 \sin^2 \lambda_n a_2}$$

The basic ideas of this section pertinent to the orthogonality of functions and related expansion problems have interesting counterparts when adapted to vectors (see Sec. 13.1). As the last example indicates, these ideas also enter into the solution of multiple-layer boundary-value problems which upon separation of variables lead to **interface Sturm-Liouville systems** of the type

$$\frac{d}{dx}[r(x)y_k'] + [\lambda p_k(x) + q_k(x)]y_k = 0 \qquad x_{k-1} < x < x_k,\ 1 \le k \le K \tag{21}$$

$$a_1y_1'(a) + b_1y_1(a) = 0 \tag{21a}$$

$$[a_{k1}y_k'(x_k^-) + b_{k1}y_k(x_k^-)] + [c_{k1}y_{k+1}'(x_k^+) + d_{k1}y_{k+1}(x_k^+)] = 0 \tag{k, 1}$$

$$[a_{k2}y_k'(x_k^-) + b_{k2}y_k(x_k^-)] + [c_{k2}y_{k+1}'(x_k^+) + d_{k2}y_{k+1}(x_k^+)] = 0 \tag{k, 2}$$

$$1 \le k \le K - 1$$

$$c_Ky_K'(b) + d_Ky_K(b) = 0 \tag{21b}$$

where $a = x_0$, $x_K = b$, and $2 \le K$. In addition to requiring suitable behavior of r, the p_k and q_k,† to ensure that the **boundary conditions** (21a) and (21b) are both present we require the inequalities

$$(a)\ a_1^2 + b_1^2 \ne 0 \qquad \text{and} \qquad (b)\ c_K^2 + d_K^2 \ne 0 \tag{22}$$

†For details, see the article by L. C. Barrett and D. N. Winslow, "Interlacing Theorems for Interface Sturm-Liouville Systems," *Journal of Mathematical Analysis and Applications,* vol. 129, no. 2, Feb. 1, 1988.

to hold. With regard to the $2(K-1)$ **interface conditions** $(k, 1)$ and $(k, 2)$, we set

$$(23)\qquad (a)\ |AB|_k = \begin{vmatrix} a_{k1} & b_{k1} \\ a_{k2} & b_{k2} \end{vmatrix} \quad \text{and} \quad (b)\ |CD|_k = \begin{vmatrix} c_{k1} & d_{k1} \\ c_{k2} & d_{k2} \end{vmatrix} \qquad 1 \le k \le K-1$$

and assume that

$$(24)\qquad (a)\ |AB|_k|CD|_k \neq 0 \quad \text{and} \quad (b)\ R_k = \frac{|CD|_k}{|AB|_k} > 0 \qquad 1 \le k \le K-1$$

Under these assumptions, an extension of the procedure used to derive the weight function (20) of Example (5) leads to the following general result for interface Sturm-Liouville systems.

THEOREM 6 The set of characteristic functions is an orthogonal set on the fundamental interval (a, b) with respect to the weight function

$$(25)\qquad P(x) = \left(\prod_{j=0}^{k-1} R_j\right)p_k(x) \qquad x_{k-1} < x < x_k,\ 1 \le k \le K$$

where $R_0 = 1$.

EXAMPLE 6

Writing the interface system of Example 5 as

$$\begin{aligned} &[X_k']' + \lambda^2 X_k = 0 \qquad x_{k-1} < x < x_k,\ 1 \le k \le 2 \\ &X_1(-a_1) = 0 \\ &X_1(0^-) - X_2(0^+) = 0 \\ &k_1X_1'(0^-) - k_2X_2'(0^+) = 0 \\ &X_2(a_2) = 0 \end{aligned}$$

we see that in this case there is just one pair of interface conditions. Inspecting coefficients in the first of these, we have

$$a_{11} = 0 \qquad b_{11} = 1 \qquad c_{11} = 0 \qquad d_{11} = -1$$

and from the second of these, we have

$$a_{12} = k_1 \qquad b_{12} = 0 \qquad c_{12} = -k_2 \qquad d_{12} = 0$$

Substituting into (23), we have

$$|AB|_1 = \begin{vmatrix} 0 & 1 \\ k_1 & 0 \end{vmatrix} = -k_1 \qquad |CD|_1 = \begin{vmatrix} 0 & -1 \\ -k_2 & 0 \end{vmatrix} = -k_2$$

With $R_0 = 1$ and now with $R_1 = k_2/k_1$, Eq. (25) gives

$$P(x) = \begin{cases} R_0p_1(x) = 1 & -a_1 < x < 0 \\ R_0R_1p_2(x) = k_2/k_1 & 0 < x < a_2 \end{cases}$$

which is the weight function we would have obtained had we used the second pair of values we found for c_1 and c_2 in Example 5. Of course, the product of any positive number and a valid weight function is itself a valid weight function.

To find the characteristic numbers of an interface Sturm-Liouville system, we must actually solve the related characteristic equation. For this purpose, it is very helpful to first have the following important concept in mind.

DEFINITION 8 The zeros of a function $f(\lambda)$ **interlace** the zeros of a function $g(\lambda)$ on an interval I if and only if on this interval

1. No simple zero of $g(\lambda)$ is a zero of $f(\lambda)$.
2. Each double zero of $g(\lambda)$ is a zero of $f(\lambda)$.
3. All zeros of $g(\lambda)$ are either simple or double and all zeros of $f(\lambda)$ are simple.
4. Between any two consecutive zeros of $g(\lambda)$, whether simple or double, there is exactly one zero of $f(\lambda)$.

If all zeros of $g(\lambda)$ are simple, this definition becomes one for **separation** of the zeros of $g(\lambda)$ by those of $f(\lambda)$ on I.

It is easy to give a geometric proof that if f and g are continuously differentiable functions on an interval I where

$$W(\lambda) = \begin{vmatrix} f(\lambda) & g(\lambda) \\ f'(\lambda) & g'(\lambda) \end{vmatrix} \neq 0$$

then the zeros of f and g separate one another on I (see Exercise 39, Sec. 2.2). Now let $f_1(\lambda), f_2(\lambda), f_3(\lambda)$, and $f_4(\lambda)$ be four continuously differentiable functions on an interval I, define

$$W_{ij}(\lambda) = \begin{vmatrix} f_i(\lambda) & f_j(\lambda) \\ f_i'(\lambda) & f_j'(\lambda) \end{vmatrix} \tag{26}$$

to be the *wronskian* of f_i and f_j, $1 \le i, j \le 4$, and introduce the **auxiliary function**

$$F(\lambda) = f_1(\lambda)f_4(\lambda) - f_2(\lambda)f_3(\lambda) \tag{27}$$

Then the following **fundamental interlacing theorem** applies (Exercise 72).

THEOREM 7 If $W_{12}(\lambda)W_{34}(\lambda) < 0$ on I, the zeros of $F(\lambda)$ interlace the zeros of either of the product functions $f_1(\lambda)f_3(\lambda)$ or $f_2(\lambda)f_4(\lambda)$ on I.

We next show how the characteristic equation of a general interface system can be expressed in the form $F(\lambda) = 0$, where $F(\lambda)$ has the structure (27) and falls within the scope of Theorem 7, thereby making it possible to isolate each successive characteristic number within a specific interval or determine it exactly.

EXAMPLE 7

AN APPLICATION OF INTERLACING

In Example 6, express the characteristic equation of the given interface system in the form $F(\lambda) = f_1(\lambda)f_4(\lambda) - f_2(\lambda)f_3(\lambda) = 0$ where $W_{12}(\lambda)W_{34}(\lambda) < 0$. Subsequently, take $a_1 = 4\pi$, $a_2 = \pi$, then isolate the first four characteristic numbers within disjoint intervals and determine λ_5 exactly.

Imposing the *successive* auxiliary conditions of the system on

$$X_1 = A_1 \cos \lambda x + B_1 \sin \lambda x \qquad \text{and} \qquad X_2 = A_2 \cos \lambda x + B_2 \sin \lambda x$$

and dividing out the factor λ, $\lambda > 0$, from the third equation thus obtained, we have the linear algebraic system in the unknowns A_1, B_1, A_2, and B_2:

$$\begin{aligned} A_1 \cos a_1\lambda - B_1 \sin a_1\lambda &= 0 \\ A_1 - A_2 &= 0 \\ k_1 B_1 - k_2 B_2 &= 0 \\ A_2 \cos a_2\lambda + B_2 \sin a_2\lambda &= 0 \end{aligned}$$

TABLE 11.1

Product function	Zeros on the interval (0, 1]
$f_1(\lambda)f_3(\lambda)$	$\frac{1}{4}$ $\frac{1}{2}$ $\frac{3}{4}$ 1
$f_2(\lambda)f_4(\lambda)$	$\frac{1}{8}$ $\frac{3}{8}$ $\frac{1}{2}$ $\frac{5}{8}$ $\frac{7}{8}$

To avoid a trivial solution we must have

$$\begin{vmatrix} \cos a_1\lambda & -\sin a_1\lambda & 0 & 0 \\ 1 & 0 & -1 & 0 \\ 0 & k_1 & 0 & -k_2 \\ 0 & 0 & \cos a_2\lambda & \sin a_2\lambda \end{vmatrix} = 0$$

The generalized Laplace expansion (Property 11, Sec. 3.5) of this determinant in terms of the first two *columns* is

$$\begin{vmatrix} \cos a_1\lambda & -\sin a_1\lambda \\ 1 & 0 \end{vmatrix}\begin{vmatrix} 0 & -k_2 \\ \cos a_2\lambda & \sin a_2\lambda \end{vmatrix} - \begin{vmatrix} \cos a_1\lambda & -\sin a_1\lambda \\ 0 & k_1 \end{vmatrix}\begin{vmatrix} -1 & 0 \\ \cos a_2\lambda & \sin a_2\lambda \end{vmatrix} = 0$$

Denoting the entire left-hand member of this equation by $F(\lambda)$ and expanding each determinant gives

(28) $$F(\lambda) = (\sin a_1\lambda)(k_2 \cos a_2\lambda) - (k_1 \cos a_1\lambda)(-\sin a_2\lambda) = 0$$

which is the related characteristic equation of the interface system. Setting

$$f_1(\lambda) = \sin a_1\lambda \qquad f_4(\lambda) = k_2 \cos a_2\lambda \qquad f_2(\lambda) = k_1 \cos a_1\lambda \qquad f_3(\lambda) = -\sin a_2\lambda$$

then expanding and subsequently multiplying the wronskians $W_{12}(\lambda)$ and $W_{34}(\lambda)$, we obtain $W_{12}(\lambda)W_{34}(\lambda) = -a_1a_2k_1k_2 < 0$. Hence Eq. (28) gives the characteristic equation in the required form and, according to Theorem 7, the zeros of $F(\lambda)$ interlace the zeros of each of the product functions $f_1(\lambda)f_3(\lambda)$ and $f_2(\lambda)f_4(\lambda)$.

Now the zeros of f_1, f_4, f_2, and f_3 are the same as the zeros of $\sin a_1\lambda$, $\cos a_2\lambda$, $\cos a_1\lambda$, and $\sin a_2\lambda$, respectively. By solving for the zeros of these functions in the special case for which $a_1 = 4\pi$ and $a_2 = \pi$ and taking appropriate unions of zeros, we find the zeros in Table 11.1.

From the interlacing property of Theorem 7, the double zero of $\sin 4\pi\lambda \sin \pi\lambda$ at $\lambda = 0$ is a simple zero of the auxiliary function $F(\lambda)$. Since the next larger zero of $f_1(\lambda)f_3(\lambda)$ is $\lambda = \frac{1}{4}$, the first characteristic value λ_1 must lie in the interval $(0, \frac{1}{4})$. It cannot equal the zero $\lambda = \frac{1}{8}$ of f_2f_4 nor can it be less than $\frac{1}{8}$ because $F(\lambda)$ has just one zero between the consecutive zeros $\lambda = -\frac{1}{8}$ and $\lambda = \frac{1}{8}$ of f_2f_4. Hence

$$\tfrac{1}{8} < \lambda_1 < \tfrac{1}{4}$$

Still keeping the nature of interlacing in mind, we find in like manner

$$\tfrac{3}{8} < \lambda_2 < \tfrac{1}{2} \qquad \tfrac{1}{2} < \lambda_3 < \tfrac{5}{8} \qquad \tfrac{3}{4} < \lambda_4 < \tfrac{7}{8}$$

Finally, the double zero of f_1f_3 at $\lambda = 1$ gives $\lambda_5 = 1$.

EXERCISES

1. Show that $f(x) = 1$ and $g(x) = x$ are orthogonal on $(-1, 1)$ and find constants a and b such that $h(x) = 1 + ax + bx^2$ will be orthogonal to both $f(x)$ and $g(x)$ on $(-1, 1)$.
2. Find constants a and b for which $h(x) = 1 + ax + bx^2$ is orthogonal to both $g(x) = x$ and $q(x) = x^2$ on $(-1, 1)$.
3. **(a)** Find the relation between the constants a and b for which the functions

$$f(x) = \begin{cases} x & -c < x < 0 \\ x^2 & 0 < x < c \end{cases}$$

$$g(x) = \begin{cases} -ax & -c < x < 0 \\ b & 0 < x < c \end{cases}$$

are orthogonal on $(-c, c)$.

(b) Find a weight function having positive values with

respect to which $f(x)$ and $g(x)$ are orthogonal on $(-c, c)$.

4. Show that the set of polynomials $p_0(x) = 1$, $p_1(x) = x$, $p_2(x) = (3x^2 - 1)/2$, $p_3(x) = (5x^2 - 3x)/2$ are mutually orthogonal on $(-1, 1)$ and find a function $f(n)$ such that for $n = 0, 1, 2, 3$,

$$\int_{-1}^{1} p_n^2(x)\, dx = f(n)$$

In Exercises 5–7, find the characteristic values and characteristic functions of the given differential system and verify by direct integration that the characteristic functions form an orthogonal set on the fundamental interval.

5. $y'' + 16\pi^2\lambda^2 y = 0$, $y(-2) = y(-1) = 0$

6. $y'' + 9\pi^2\lambda^2 y = 0$, $y(-1) = y(1) = 0$

7. $9y'' + \lambda^2 y = 0$, $y(6) = y(8) = 0$

8. A thin sheet of metal coincides with the **unit square** in the xy plane [the square whose vertices are (0, 0), (1, 0), (1, 1), and (0, 1)]. Find the steady-state temperature distribution in the sheet if no heat is lost through its faces, the left edge of the sheet is insulated, the right edge radiates into air of constant temperature $u = 0$, the lower edge is maintained at the temperature $u = 0$, and the upper edge is maintained at the temperature $u = 50$.

9. Work Example 1 if the right end of the rod is maintained at the constant temperature $u(l, t) = 0$ and the left end radiates freely into air of constant temperature $u = 0$. *Hint:* At the left end the boundary condition is not Eq. (2) but $\partial u/\partial x = hu$.

10. Work Example 1 if the left end of the rod is insulated. Verify that the characteristic equation has infinitely many roots and that the interval between consecutive roots approaches π as n becomes infinite.

11. Work Example 1 if the right end of the rod radiates freely into air of constant temperature $u = 70$. *Hint:* Let $U = u - 70$ be the independent variable.

12. Work Example 1 if the left end of the rod is maintained at the constant temperature $u(0, t) = 100$. *Hint:* Note, as we pointed out in Example 2, Sec. 11.5, that after we have found one solution which takes on the value $u = 100$ at the left end, all other solutions must take on the value $u = 0$ at that end.

13. Work Example 1 if both ends radiate freely into air of constant temperature $u = 0$. Verify that the characteristic equation has infinitely many roots, that the interval between consecutive roots approaches π as n becomes infinite, and that the characteristic functions are orthogonal.

14. **(a)** A sheet of metal coincides with the unit square in the xy plane. The upper and lower surfaces of the sheet are insulated, and the sheet is so thin that heat flow in it can be considered two-dimensional. Find the steady-state temperature distribution in the sheet if the lower edge radiates freely into air of constant temperature $u = 0$, the left edge and the right edge are maintained at the constant temperature $u = 0$, and the temperature distribution $u(x, 1) = f(x)$ is maintained along the upper edge.

(b) Work Part **(a)** if the upper edge of the sheet radiates freely into air of constant temperature $u = 0$, the left edge and lower edge are maintained at the constant temperature $u = 0$, and the temperature distribution $u(1, y) = f(y)$ is maintained along the right edge.

15. **(a)** A sheet of metal coincides with the rectangle in the xy plane whose vertices are the points $(0, 0)$, $(a, 0)$, (a, b), $(0, b)$. The upper and lower surfaces of the sheet are insulated, and the sheet is so thin that heat flow in it can be regarded as two-dimensional. Find the steady-state temperature distribution in the sheet if the lower edge is maintained at the constant temperature $u = 0$, the left edge is insulated, the right edge radiates freely into air of constant temperature $u = 0$, and the temperature distribution $u(x, b) = f(x)$ is maintained along the upper edge.

(b) Work Part **(a)** if the lower edge of the sheet is maintained at the constant temperature $u = 0$, the left edge is insulated, the upper edge radiates freely into air of constant temperature $u = 0$, and the temperature distribution $u(a, y) = f(y)$ is maintained along the right edge.

16. **(a)** A thin strip of metal extending to infinity in the y direction is bounded by the lines $x = 0$, $x = 1$ and the segment of the x axis between $x = 0$ and $x = 1$. The upper and lower surfaces of the strip are insulated, and heat flow in the strip can be considered two-dimensional. Find the steady-state temperature distribution in the strip if the left edge is maintained at the constant temperature $u = 0$, the right edge radiates freely into air of constant temperature $u = 0$, and the temperature distribution $u(x, 0) = f(x)$ is maintained along the lower edge.

(b) Work Part **(a)** if instead of being held at the constant temperature $u = 0$, the left edge is insulated.

17. **(a)** A thin rod of length l loses heat through its lateral surface according to Newton's law of cooling and through its ends which, like the surrounding air, are maintained at the constant temperature $u(0, t) = u(l, t) = 0$. If the temperature throughout the rod at $t = 0$ is $u = 100$, find the temperature at any point of the rod at any subsequent time. *Hint:* Recall the equation derived in Exercises 23, Sec. 11.2, and when separating this equation, associated the parameter b^2 with the fraction a^2T'/T rather than with the fraction X''/X.

(b) Work Part **(a)** if the left end of the rod is held at the constant temperature $u = 0$, the right end is held at the temperature $u = T_0$, and the temperature throughout the rod at $t = 0$ is $u(x, 0) = 0$.

18. **(a)** A thin rod of length l loses heat through its lateral surface according to Newton's law of cooling, through its left end which is maintained at the constant temperature $u = 0$, and through its right end which radiates freely into air of constant temperature $u = 0$. If the temperature throughout the rod at $t = 0$ is $u(x, 0) = f(x)$, find the temperature at any point in the rod at any subsequent time. *Hint:* Recall the hint given in Exercise 17.

(b) Work Part **(a)** if the left end of the rod is insulated.

19. A thin sheet of metal coincides with the unit square in the xy plane. The surfaces of the sheet are not insulated and as a result lose heat into air of constant temperature $u = 0$ according to Newton's law of cooling. Find the steady-state temperature in the sheet if its left, right, and lower edges are maintained at the constant temperature $u = 0$ and its upper edge is maintained at the constant temperature $u(x, 1) = 100$. *Hint:* Recall the equation derived in Exercise 27, Sec. 11.2, and when separating this equation associate the term $2h/k$ with the fraction Y''/Y rather than the fraction X''/X.

20. Work Exercise 19 if the lower edge of the sheet is insulated.

21. A thin rod whose lateral surface is insulated extends along the x axis from $x = 0$ to $x = a_1 + a_2$. The left end of the rod is maintained at the temperature $u = 0$, and the right end is maintained at the constant temperature $u = 100$. Find the steady-state temperature distribution in the rod if the thermal conductivity of the portion of the rod between $x = 0$ and $x = a_1$ is k_1 while the thermal conductivity of the portion of the rod between $x = a_1$ and $x = a_1 + a_2$ is k_2. *Hint:* Solve the problem in two parts and fit them together by making both the temperature and the temperature gradient continuous at $x = a_1$.

22. Work Exercise 21 if the rod consists of three segments of lengths a_1, a_2, and a_3 whose thermal conductivities are, respectively, k_1, k_2, and k_3.

23. A uniform shaft of length l has its left end fixed and its right end elastically restrained, i.e., attached to a torsional spring which applies a restoring torque proportional to the angle through which the end section has rotated. Find the frequency equation for the shaft and verify that the nth natural frequency of the shaft is between the nth natural frequencies of a fixed-free and a fixed-fixed shaft of the same dimensions. Using Theorem 4, verify also that the normal modes of the shaft are orthogonal.

24. Work Exercise 23 if the left end of the shaft is free rather than fixed, this time showing that the nth natural frequency of the shaft is between the nth natural frequencies of a free-free shaft and a fixed-free shaft of the same dimensions.

25. Work Exercise 23 if both ends of the shaft are elastically restrained, this time showing that the nth natural frequency of the shaft is between the nth natural frequencies of a free-free shaft and a fixed-fixed shaft of the same dimensions.

26. In Exercises 23 and 24, verify the orthogonality of the normal modes by direct integration.

27. What is the lowest natural frequency of a steel shaft of length 100 in and diameter 2 in which is restrained as in **(a)** Exercise 23, **(b)** Exercise 24, and **(c)** Exercise 25 if the modulus of the torsional springs is 1000 in-lb/rad? Take $E_s = 12{,}000{,}000$ lb/in^2 and $\rho = 0.25$ lb/in^3.

28. What value of k will make the lowest natural frequency of the shaft in Exercise 23 halfway between the lowest natural frequencies of a fixed-free shaft and a fixed-fixed shaft of the same dimensions? *Hint:* Use the physical data given in Exercise 27.

29. Find the frequency equation for the torsional vibrations of a shaft of length $2l$ clamped at $x = 0$ and free at $x = 2l$ if the radius of the portion of the shaft between $x = 0$ and $x = l$ is r_1 and the radius of the portion of the shaft between $x = l$ and $x = 2l$ is r_2. *Hint:* Solve the problem separately for the interval $(0, l)$ and the interval $(l, 2l)$ and then, in addition to the two end conditions, impose the condition that at $x = l$ both the angle of twist and the transmitted torque are continuous.

30. Find the frequency equation for a uniform torsional cantilever if a disk of polar moment of inertia I_p is attached to the free end of the shaft. *Hint:* At the free end of the shaft the boundary condition is that the torque, instead of being zero, is equal to the inertia torque of the disk.

31. **(a)** Show that the normal modes of the shaft in Exercise 29 are orthogonal. *Hint:* Specialize Eq. (7) to each segment of the shaft.

(b) Show that the normal modes of the shaft in Exercise 30 are not orthogonal. *Hint:* Try to verify their orthogonality by direct integration.

32. Find the frequency equation and normal modes for the transverse vibrations of a uniform beam of length l whose ends are

(a) Hinged-hinged. *Hint:* A **hinged end** is one where a beam, though constrained so that it cannot deflect, is still free to turn, i.e., an end where both the displacement and the moment are zero at all times. A hinged end is often referred to as a **simply supported end.**

(b) Fixed-fixed **(c)** Free-free

(d) Fixed-hinged **(e)** Free-hinged

Hint: The conditions defining a free end and a fixed end were given in Example 3.

33. Show that for $n \geq 4$, corresponding roots of the equations

$$\cos z \cosh z = -1 \qquad \text{and} \qquad \cos z \cosh z = 1$$

agree to at least three decimal places. Interpret this as a statement about the frequencies of beam vibrations.

34. Find the frequency equation for the transverse vibrations of a uniform cantilever of length l bearing a concentrated mass M at the free end. Discuss the physical significance of the limiting cases when $M = 0$ and when $M = \infty$. *Hint:* At the free end, one boundary condition is that the shear, instead of being zero, is equal to the inertia force of the attached mass M.

35. Find the frequency equation for the transverse vibrations of a uniform hinged-hinged beam bearing a concentrated mass M at its midpoint. Discuss the physical significance of the limiting cases when $M = 0$ and when $M = \infty$. *Hint:* It will be convenient to imagine the beam extending from $x = -l$ to $x = l$, with M attached at the point $x = 0$. First solve the beam equation [Eq. (11), Sec. 11.2] for $-l \leq x < 0$ and for $0 < x \leq l$. Then fit these two solutions together so that at $x = 0$ the displacement, the slope, and the moment are continuous and the shear has a jump equal to the inertia force of the mass M.

36. Show that on the interval $(-\pi, \pi)$ the functions of the set $\{\cos nx\}$, $n = 0, 1, 2, \ldots$, are orthogonal but that the set is not complete.

37. Show that for an orthonormal system $\{\phi_n(x)\}$, whether closed or not, we have **Bessel's inequality**

$$\sum_{n=1}^{\infty} a_n^2 \le \int_a^b f^2(x)\,dx$$

where the a's are the coefficients in the generalized Fourier expansion of $f(x)$ in terms of the ϕ's and (a, b) is the interval of orthogonality. Using this result, show that

$$\lim_{n\to\infty} \int_a^b \phi_n(x) f(x)\,dx = 0$$

38. If $\{\phi_n(x)\}$ is an orthonormal set over the interval (a, b), show that the values of the c's which make

$$\int_a^b [f(x) - c_1\phi_1(x) - c_2\phi_2(x) - \cdots - c_n\phi_n(x)]^2\,dx$$

a minimum are the corresponding coefficients in the generalized Fourier expansion of $f(x)$ in terms of the ϕ's.

39. What is the minimum value of the integral in Exercise 38.

40. Prove that the general linear second-order differential equation

$$p_0(x)y'' + p_1(x)y' + p_2(x)y = -\lambda y$$

can be reduced to an equation of the Sturm-Liouville form (see Theorem 4) by multiplying it by the factor

$$\frac{1}{p_0(x)} \exp\left[\int_{x_0}^{x} \frac{p_1(x)}{p_0(x)}\,dx\right]$$

41. Show that the nontrivial solutions of the equation

$$xy'' + y' + \frac{\mu}{x}y = 0 \qquad \mu > 0$$

satisfying each of the following sets of conditions form an orthogonal set on the indicated interval. In each case solve the equation, impose the given boundary conditions, obtain the characteristic functions, and verify their orthogonality by direct integration.

(a) $y(1) = 0,\ y(e) = 0$

(b) $y'(1) = 0,\ y'(e) = 0$

(c) $y(1) = 0,\ y'(\sqrt{e}) = 0$

(d) $y(1) = 0,\ y(e) = y'(e)$

42. Work Exercise 41 for each of the following boundary-value problems.

(a) $x^2y'' + 2xy' + \mu y = 0 \qquad y(1) = y(e) = 0$

(b) $x^2y'' + 3xy' + \mu y = 0 \qquad y(1) = y'(\sqrt{e}) = 0$

43. Prove that a null function is zero at every point where it is continuous.

44. By considering the sequence of functions

$$S_n(x) = \begin{cases} 0 & 0 \le x < 1/(n+1) \\ 1/[x \ln(n+1)]^{1/2} & 1/(n+1) \le x \le 1 \end{cases}$$

and the limiting function $f(x) = 0$, $0 \le x \le 1$, show that a sequence may converge to a function in the ordinary sense without converging to that function in the mean.

45. Prove Theorem 5. *Hint:* By combining the equations satisfied by y_m and y_n, as in the proof of Theorem 4, and then using integration by parts twice, first obtain the antiderivative

$$(\lambda_m - \lambda_n)\int p y_m y_n\,dx$$
$$= [y_m(ry_n'')' - y_n(ry_m'')'] - [y_m'(ry_n'') - y_n'(ry_m'')]$$

46. (a) Specialize the antiderivative obtained in Exercise 45 to the case of the beam equation $y^{\text{iv}} - k^4 y = 0$ and then, by evaluating the limit as $k_m \to k_n$, show that for beams whose ends are fixed, free, or simply supported, in any combination

$$\int_0^l y^2(k_n x)\,dx =$$
$$\frac{l}{4k_n^4}[y(k_n l)y^{\text{iv}}(k_n l) - 2y'(k_n l)y'''(k_n l) + \{y''(k_n l)\}^2]$$

Hint: Note first that every solution of the beam equation is a function of the product argument $k_n x$. Then before differentiating with respect to k_m in the application of l'Hospital's rule, use the fact that

$$\frac{d^j y(k_n x)}{dx^j} = k_n^j \frac{d^j y(k_n x)}{[d(k_n x)]^j}$$

Then, at the appropriate point, revert to differentiation with respect to x.

(b) Specialize the result of Part **(a)** to a beam fixed at $x = l$, free at $x = l$, and simply supported at $x = l$.

47. Using the method of Example 4 and the results of Exercise 46 obtain a formula for the static deflection an arbitrary distributed load will produce in a uniform beam which is

(a) Fixed at $x = 0$ and at $x = l$

(b) Fixed at $x = 0$ and simply supported at $x = l$

(c) Simply supported at $x = 0$ and at $x = l$.

How do the answers in Parts **(b)** and **(c)** differ?

48. By substituting $\theta(x, t) = X(x)\sin\omega t$ into Eq. (7), Sec. 11.2 (with $f \equiv 0$), dividing out $\sin\omega t$, and integrating the resulting equation between the appropriate limits, show that the characteristic functions for the torsional vibrations of a nonuniform shaft fixed at $x = 0$ and free at $x = l$ satisfy the *integral equation*

$$X(x) = \frac{\omega^2\rho}{E_s g}\int_0^x \frac{1}{J(s)}\int_r^l J(s)X(s)\,ds\,dr$$

49. Find the integral equations satisfied by the characteristic functions for the torsional vibrations of a nonuniform shaft which is

(a) Fixed at $x = 0$ and at $x = l$
(b) Free at $x = 0$ and at $x = l$

50. The integral equations obtained in Exercises 48 and 49 are often used in the following way to approximate the natural frequencies of a nonuniform shaft vibrating torsionally when an exact solution is impossible. Assume a reasonable expression $X_1(x)$ as a first approximation to the characteristic function $X(x)$, substitute this into the right-hand side of the appropriate integral equation, and perform the indicated integrations, getting a second function, say $X_2(x)$. Repeat the process by substituting $X_2(x)$ and integrating to get $X_3(x)$, and so on. If the ratio $X_n(x)/X_{n+1}(x)$ becomes (approximately) the same for all values of x, then this ratio is (approximately) equal to $\omega^2\rho/E_s g$ and $X_n(x)$ is (approximately) the characteristic function, or normal mode, corresponding to the frequency ω determined from the now (approximately) known ratio $\omega^2\rho/E_s g$. This procedure is often referred to as the **method of Stodola.** Prove that this iterative process does indeed converge to the lowest natural frequency of the shaft and the corresponding characteristic function. *Hint:* Express $X_1(x)$ as an infinite series of the characteristic functions $\{\phi_i(x)\}$ of the shaft and observe that $\phi_i(x)$, though not yet known, satisfies the integral equation when ω has the corresponding value ω_i.

51. How can the method of Stodola be modified to obtain other natural frequencies besides the lowest? *Hint:* Observe that after ϕ_1 has been obtained, the orthogonality property of the ϕ's can be used to determine a function $X_1(x)$ in whose expansion ϕ_1 will not appear.

52. Obtain the integral equation for the problem of the torsional vibrations of a nonuniform shaft built-in at $x = 0$ and bearing a disk of polar moment of inertia I at the free end. *Hint:* At the free end of the shaft, the torque transmitted through the end section is equal to the inertia torque of the disk, $-I\partial^2\theta/\partial t^2$, evaluated at $x = l$.

53. Obtain the integral equation for the problem of the transverse vibration of a nonuniform string of length l.

54. Obtain the integral equation for the transverse vibration of a nonuniform cantilever beam of length l and prove that the method of Stodola converges to the lowest natural frequency and corresponding normal mode of the cantilever.

55. Discuss Example 5 in the special case when $a_1 = a_2$.

56. Discuss Example 5 in the special case when $k_1 = k_2$.

57. Work Example 5 if the lower edge of the sheet is also maintained at the temperature 0.

58. Show that the trigonometric functions that are solutions of the equation $y'' + \lambda^2 y = 0$ and satisfy the boundary conditions $y(0) = 0$ and $y(l) - ly'(l) = 0$ form an orthogonal set which is not complete. *Hint:* Try to expand $f(x) = x$.

59. In Exercise 58, show that x should be included, with the trigonometric functions, as one of the characteristic functions of the problem.

60. If $\{y(\lambda_n x)\}$ is the set of all solutions of the equation $y'' + \lambda^2 y = 0$ satisfying the boundary conditions

$$y(0) = 0 \qquad ay(l) - by'(l) = 0 \qquad a, b \text{ of like sign}$$

show that

(a) If $b > al$, the set $\{y(\lambda_n x)\}$ consists exclusively of trigonometric functions.
(b) If $b = al$, then in addition to trigonometric functions, the set $\{y(\lambda_n x)\}$ contains the function $y = x$.
(c) If $b < al$, then in addition to trigonometric functions, the set $\{y(\lambda_n x)\}$ contains one hyperbolic function.

Hint: Consider the possibility that in the given equation the parameter λ^2 may be zero or negative. Also at the appropriate point, recall the graph of the hyperbolic tangent function.

61. In Exercise 60 show that if the condition at $x = 0$ is $y'(0) = 0$, then the set $\{y(\lambda_n x)\}$ always contains one hyperbolic function.

62. In Exercises 60 and 61, show that if a and b are of opposite sign, then the set $\{y(\lambda_n x)\}$ always contains only trigonometric functions. (This is why we were satisfied with the solution we found in Example 1.)

63. Which of the boundary conditions

$$ay(l) - by'(l) = 0 \quad a, b \text{ of like sign} \quad \text{(Exercise 60)}$$

and

$$ay(l) - by'(l) = 0 \quad a, b \text{ of opposite sign} \quad \text{(Exercise 62)}$$

is the more realistic? *Hint:* Recall Stefan's law from Example 1.

64. Discuss Exercise 60 if the boundary conditions are

$$ay(0) - by'(0) = 0 \qquad \text{and} \qquad y(l) = 0$$

65. Discuss Exercise 23 if the torque applied at the end of the shaft acts to increase the angle of twist.

66. If $r(x_1) = r(x_2)$, show that the conclusion of Theorem 4 holds in case the so-called **periodic boundary conditions** $y(x_1) = y(x_2)$ and $y'(x_1) = y'(x_2)$ apply.

67. In Example 5, show that the assumption $\mu = 0$ leads only to a trivial solution.

68. Show that the function $f(x) = \cos x \cosh x + 1$ has infinitely many zeros on $(0, \infty)$. *Hint:* Consider the wronskian of $f(x)$ and $g(x) = \cosh x \sin x$.

69. A thin rod whose lateral surface is perfectly insulated extends along the x axis from $x = -1$ to $x = 1$. The thermal conductivity of the left half of the rod is k_1, the conductivity of the right half is k_2. At $t = 0$, when the temperature throughout the rod is 100, the temperature at each end is suddenly reduced to 0 and maintained thereafter at that value. Find the temperature at any point in the rod at any subsequent time. *Hint:* In separating variables, associate the parameter a^2 with X rather than T.

70. Work Exercise 69 if, while the temperature throughout the rod is 0, the right-hand end is suddenly raised to 100 and maintained thereafter at that value.

71. The frequency equation for the transverse vibrations of a free-hinged uniform beam can be written $\sin z \cosh z - \cos z \sinh z = 0$. Determine whether or not on $0 < z < \infty$

its frequencies separate those of an identical beam whose end conditions and frequency equation are
(a) Fixed-fixed, $\cos z \cosh z = 1$
(b) Hinged-hinged, $\sin z = 0$
Hint: See the hint for Exercise 68.

72. Give a geometric proof of Theorem 7. *Hint:* Consider vectors $\mathbf{u} = [f_1(\lambda), f_2(\lambda)]$ and $\mathbf{v} = [f_3(\lambda), f_4(\lambda)]$ originating at the origin, note how they rotate if $W_{12}(\lambda)W_{34}(\lambda) < 0$, interpret the absolute value of $F(\lambda)$ as the area of the parallelogram having $\mathbf{u}$ and $\mathbf{v}$ as two sides, then plot the zeros of $F(\lambda), f_1(\lambda)f_3(\lambda)$, and $f_2(\lambda)f_4(\lambda)$ along a line as $\mathbf{u}$ and $\mathbf{v}$ rotate.

73. If $f_1(\lambda) = \sin a\lambda$, $f_2(\lambda) = k\cos a\lambda$, $f_3(\lambda) = \cos b\lambda$, $f_4(\lambda) = h\sin b\lambda$, and $F(\lambda) = f_1(\lambda)f_4(\lambda) - f_2(\lambda)f_3(\lambda)$, find a condition which if satisfied by a, b, k, and h, will guarantee that the zeros of $F(\lambda)$ interlace the zeros of both $f_1(\lambda)f_3(\lambda)$ and $f_2(\lambda)f_4(\lambda)$ on $(-\infty, \infty)$.

74. The axis of a composite shaft, consisting of two cylindrical segments bonded together at $x = \beta$, extends along the x axis from $x = 0$ to $x = 1$. The left end of the shaft is free; its other end is fixed. Other shaft properties are given in the following table.

Property	Left segment	Right segment
Shear modulus of elasticity	E_1	E_2
Weight density	ρ_1	ρ_2
Polar moment of inertia of a cross section	J_1	J_2
Cross-sectional radius	r_1	r_2
Angle of twist	$\theta_1(x,t)$	$\theta_2(x,t)$

Find the frequency equation $F(z) = 0$ for the free vibrations of the shaft and define functions $f_1(z), f_2(z), f_3(z)$, and $f_4(z)$ such that the zeros of $F(z)$ interlace the zeros of either $f_1(z)f_3(z)$ or $f_2(z)f_4(z)$ on $(0, \infty)$.

75. A composite cylindrical shaft of uniform material and length 2 is fixed at $x = 0$ and free at $x = 2$. Its radius between $x = 0$ and $x = 1$ is 0.1; between $x = 1$ and $x = 2$ the radius is 0.2. Find the frequency equation $F(z) = 0$ for free torsional vibrations and define functions f_1, f_2, f_3, and f_4 such that the zeros of $F(z)$ interlace the zeros of either $f_1(z)f_3(z)$ or $f_2(z)f_4(z)$ on $(0, \infty)$.

76. A free-fixed composite shaft consisting of two cylindrical segments bonded at $x = 0$ extends from $x = -2$ to $x = 3$. Each segment is uniform, but the radii and material of the two segments are different.
(a) Formulate a boundary-value problem satisfied by the angular deflection $\theta(x, t)$ of the shaft during free torsional vibrations.
(b) Separate variables to obtain a related interface Sturm-Liouville system.
(c) Find a weight function with respect to which the characteristic functions of the interface system are orthogonal on the fundamental interval $(-2, 3)$.

77. Work Exercise 76 for a fixed-free composite shaft consisting of two unlike segments of unit length bonded at $x = 0$. Assuming a^2 has the values $1/16\pi^2$ and $1/\pi^2$ on the respective intervals $(-1, 0)$ and $(0, 1)$, find intervals which contain just the first characteristic number λ_1, just λ_2, and find the exact value of λ_3.

78. A composite shaft like that of Exercise 77 for which $a_1^2 = 1/16\pi^2$ on $(-1, 0)$ and $a_2^2 = 1/\pi^2$ on $(0, 1)$ has the initial angle of twist

$$\theta(x, t) \equiv f(x) = \begin{cases} 0 & -1 \le x \le 0 \\ x & 0 \le x \le 1 \end{cases}$$

when it is released from rest.
(a) Formulate a boundary-value problem with initial conditions whose solution gives the angle of twist $\theta(x, t)$ of the shaft during free torsional vibrations.
(b) Use separation of variables to obtain a related interface Sturm-Liouville system.
(c) Find the frequency equation and natural frequencies of the shaft.
(d) Find the characteristic functions $X_n(x)$ of the interface system and the weight function $P(x)$ with respect to which these functions are orthogonal on $(-1, 1)$. Then verify this orthogonality property by direct integration.
(e) Finally, write a formal series representation for the angle of twist $\theta(x, t)$ in the vibrating shaft.

11.7 FURTHER APPLICATIONS

Many problems in partial differential equations involve features not found in the relatively simple examples we have used to explain the standard elementary theory. We cannot here investigate in detail the variations and extensions of this theory, but as illustrations we shall present several additional examples exhibiting techniques of practical interest.

In our first example we investigate non-steady-state heat flow in a two-dimensional region and see how a partial differential equation involving three rather than two independent variables leads to a *double* series of characteristic functions and two separate expansion problems. Then we consider two problems dealing with vibrating strings. In one we investigate the effect of friction on the behavior of a string; in the other we consider forced vibrations and see how their analysis leads to a nonhomogeneous rather than a homogeneous differential equation. The fourth example deals with an infinitely long transmission line. In it we shall see that, though a partial differential equation may be separable, it may be impossible to make its product solutions fit the boundary conditions and so other methods must be used to solve it. In our last three examples, which again involve heat flow, we shall see how Fourier integrals rather than Fourier series enter into problems where the boundary conditions fail to provide a characteristic equation and λ remains a continuous parameter.

The important matter of the application of Laplace transform methods to the solution of partial differential equations is discussed in Sec. 11.8. Numerical methods of solution will be considered in Sec. 11.9.

EXAMPLE 1

TRANSIENT AND STEADY-STATE TEMPERATURES OF A UNIT SQUARE

A thin sheet of metal whose faces are insulated coincides with the unit square in the xy plane. Initially, the temperature in the sheet is $u(x, y, 0) = g(x, y)$. If there are no sources of heat in the sheet, find the temperature at any point at any subsequent time, given that the upper and lower edges of the sheet are insulated, the left-hand edge is held at the temperature 0, and the temperature distribution $u(1, y, t) = f(y)$ is maintained along the right-hand edge.

In this problem we must solve the non-steady-state heat equation in two dimensions

$$\frac{\partial^2 u}{\partial x^2} + \frac{\partial^2 u}{\partial y^2} = a^2 \frac{\partial u}{\partial t} \tag{1}$$

This is the first time in our work that we have been asked to solve a partial differential equation containing *three* rather than *two* independent variables, and almost certainly we should begin by assuming a three-factor product expression for u,

$$u(x, y, t) = X(x)Y(y)T(t) \tag{2}$$

and attempting to separate variables. However, before we plunge into the details, it will pay us to reflect on the general nature of the problem.

Since no internal heat source exists and none of the edge conditions involve t explicitly, it should be clear that while the temperature will certainly vary with time, it will none the less approach a steady-state condition independent of time. In other words, the solution will consist of a transient part and a steady-state part,

$$u(x, y, t) = u_{\text{tr}} + u_{\text{ss}}$$

Moreover, since the steady state will persist after the transient has become negligible, the steady-state term must by itself satisfy all the boundary conditions of the problem. Specifically, the steady-state solution must assume the values $u = f(y)$ along the right-hand edge of the sheet. But if this is so, then the transient term must assume the value zero along the right-hand edge. For if both the transient and the steady-state terms took on the values $f(y)$, the entire solution would then take on the values $f(y) + f(y) = 2f(y)$ and the condition along the right-hand edge would not be satisfied. (See Example 2, Sec. 11.5, for a discussion of this matter in a different setting.)

Since no internal heat source exists and none of the edge conditions involve t explicitly, it should be clear that while the temperature will certainly vary with time, it will none the less approach a steady-state condition independent of time. In other words, the solution will consist of a transient part and a steady-state part,

$$\begin{array}{ll}
\textit{Transient} & \textit{Steady-state} \\
u(0, y, t) = 0 & u(0, y, t) = 0 \\
\left.\dfrac{\partial u}{\partial y}\right|_{x, 0, t} = 0 & \left.\dfrac{\partial u}{\partial y}\right|_{x, 0, t} = 0 \\
\left.\dfrac{\partial u}{\partial y}\right|_{x, 1, t} = 0 & \left.\dfrac{\partial u}{\partial y}\right|_{x, 1, t} = 0 \\
u(1, y, t) = 0 & u(1, y, t) = f(y)
\end{array} \tag{3}$$

We now begin the solution process by substituting the product solution (2) into Eq. (1), dividing by the product XYT, and isolating the factor X. Since (1) is the time-dependent form of the heat equation, this procedure will give us the transient part of the final solution. The result of this first separation step is

$$\frac{X''}{X} = a^2\frac{T'}{T} - \frac{Y''}{Y} \tag{4}$$

Although y and t enter together on the right-hand side of (4), they are both independent of x, and so each side of this equation must be a constant, say μ. Thus the factor X satisfies the equation

$$X'' = \mu X$$

If $\mu > 0$, say $\mu = \lambda^2$, $\lambda > 0$, we have

$$X = A\cosh\lambda x + B\sinh\lambda x$$

From the first of the boundary conditions (3) we have

$$u(0, y, t) = X(0)Y(y)T(t) = AY(y)T(t) = 0$$

Since this must hold for *all* values of y and t, it follows that $A = 0$. Likewise, from the fourth of the boundary conditions (for the transient), it follows that

$$u(1, y, t) = X(1)Y(y)T(t) = (B\sinh\lambda)Y(y)T(t) = 0$$

and from this we conclude that $B = 0$ also. Hence, when $\mu > 0$, the factor $X(x)$ vanishes identically and only a trivial solution is possible.

If $\mu = 0$, we have $X = Ax + B$, and again the boundary conditions (3) can be satisfied only if $A = B = 0$.

If $\mu < 0$, say $\mu = -\lambda^2$, $\lambda > 0$, we have

$$X = A\cos\lambda x + B\sin\lambda x$$

From the first of the boundary conditions (3) we conclude that $A = 0$. The fourth requires that $(B\sin\lambda)Y(y)T(t) = 0$, and since we cannot permit B to be zero (for then every solution is trivial), we must have

$$\sin\lambda = 0 \qquad \text{and} \qquad \lambda_m = m\pi \qquad m = 1, 2, 3, \ldots$$

Therefore, dropping a nonzero arbitrary factor B_m,

$$X_m(x) = \sin m\pi x \qquad m = 1, 2, 3, \ldots \tag{5}$$

Continuing with the other equation arising from (4), we now have

$$a^2\frac{T'}{T} - \frac{Y''}{Y} = \mu = -\lambda^2 = -m^2\pi^2$$

or

$$\frac{Y''}{Y} = a^2\frac{T'}{T} + m^2\pi^2 \tag{6}$$

Since y and t are independent, each member of the last equation must be a constant, say η. Thus the factor Y satisfies the equation

$$Y'' = \eta Y$$

If $\eta > 0$, say $\eta = \nu^2$, $\nu > 0$, we have

$$Y = C\cosh\nu y + D\sinh\nu y$$

and

$$Y' = \nu C\sinh\nu y + \nu D\cosh\nu y$$

From the second of the boundary conditions (3), namely,

$$\left.\frac{\partial u}{\partial y}\right|_{x,\,0,\,t} = X(x)Y'(0)T(t) = (\sin m\pi x)(\nu D)T(t) = 0$$

it follows that $D = 0$. Likewise, from the third of the conditions (3), namely,

$$\left.\frac{\partial u}{\partial y}\right|_{x,\,1,\,t} = X(x)Y'(1)T(t) = (\sin m\pi x)(\nu C \sinh \nu)T(t) = 0$$

it follows that $C = 0$. Hence, when $\eta > 0$, the factor $Y(y)$ vanishes identically and only a trivial solution is possible.

If $\eta = 0$, we have $Y = Cy + D$, and this time the boundary conditions (3) require that $C = 0$ but do not restrict D. Hence $Y = D$ is a possible solution for the factor Y.

Finally, if $\eta < 0$, say $\eta = -\nu^2$, $\nu > 0$, we have

$$Y = C \cos \nu y + D \sin \nu y$$

and

$$Y' = -\nu C \sin \nu y + \nu D \cos \nu y$$

From the second of the conditions (3) we conclude again that $D = 0$. The third of the conditions (3) requires that

$$X(x)Y'(1)T(t) = (\sin m\pi x)(-\nu C \sin \nu)T(t) = 0$$

and since we cannot permit $C = 0$, we must have

$$\sin \nu = 0 \qquad \text{and} \qquad \nu_n = n\pi \qquad n = 1, 2, 3, \ldots$$

Therefore

$$Y_n(y) = \cos n\pi y \qquad n = 1, 2, 3, \ldots$$

or, including the solution $Y =$ constant obtained when $\eta = 0$,

(7) $$Y_n(y) = \cos n\pi y \qquad n = 0, 1, 2, 3, \ldots$$

From (6) it is now clear that the factor T satisfies the equation

$$T' = -\frac{(m^2 + n^2)\pi^2}{a^2}T$$

and hence that

(8) $$T = E_{mn} \exp\left[-\frac{(m^2 + n^2)\pi^2 t}{a^2}\right]$$

Therefore, combining (5) and (7) with (8), we can write the general product term in the solution for the transient as

$$E_{mn} \sin m\pi x \cos n\pi y \exp\left[-\frac{(m^2 + n^2)\pi^2 t}{a^2}\right]$$

The entire transient, of course, is the sum of the these terms taken over all values of m and n:

$$u_{\text{tr}} = \sum_{m,n} E_{mn} \sin m\pi x \cos n\pi y \exp\left[-\frac{(m^2 + n^2)\pi^2 t}{a^2}\right]$$

In summing the double series for the transient u_{tr}, it is convenient first to add up all the terms corresponding to a particular value of n, getting

$$\sum_{m=1}^{\infty} E_{mn} \sin m\pi x \cos n\pi y \exp\left[-\frac{(m^2+n^2)\pi^2 t}{a^2}\right]$$

and then to add these sums for all values of n, getting finally

$$u_{tr} = \sum_{n=0}^{\infty}\sum_{m=1}^{\infty} E_{mn} \sin m\pi x \cos n\pi y \exp\left[-\frac{(m^2+n^2)\pi^2 t}{a^2}\right]$$

Having found the transient part of the solution, we must now find the steady-state part u_{ss}, add the two, and proceed to determine the coefficients E_{mn} so that the sum $u_{tr} + u_{ss}$ will reduce to the given distribution $u(x, y, 0) = g(x, y)$ when $t = 0$. In general, the determination of u_{ss} is a significant problem in itself, since it involves returning to Eq. (1), deleting the time derivative, and then solving Laplace's equation in two dimensions subject to the appropriate boundary conditions. In the present problem, however, all this was done for us in Example 1, Sec. 11.5. Hence, identifying the solution obtained there as the steady-state expression u_{ss} we now require, we can write

$$u(x, y, t) = u_{ss} + \sum_{n=0}^{\infty}\sum_{m=1}^{\infty} E_{mn} \sin m\pi x \cos n\pi y \exp\left[-\frac{(m^2+n^2)\pi^2 t}{a^2}\right] \tag{9}$$

The final requirement is that Eq. (9) should reduce to $g(x, y)$ when $t = 0$:

$$g(x, y) = u_{ss} + \sum_{n=0}^{\infty}\left(\sum_{m=1}^{\infty} E_{mn} \sin m\pi x\right) \cos n\pi y \tag{10}$$

Now the inner sum in Eq. (10) is a function of n and x only, say $G_n(x)$, if $n \neq 0$, and $\frac{1}{2}G_0(x)$, if $n = 0$. Hence Eq. (10) can be written

$$g(x, y) - u_{ss} = \tfrac{1}{2}G_0(x) + \sum_{n=1}^{\infty} G_n(x) \cos n\pi y$$

But for any particular value of x, this is just the half-range cosine expansion of the known function $g(x, y) - u_{ss}$, thought of as a function of y. Hence, by familiar theory,

$$G_n(x) = 2\int_0^1 [g(x, y) - u_{ss}] \cos n\pi y \, dy \tag{11}$$

But, by definition,

$$G_n(x) = \sum_{m=1}^{\infty} E_{mn} \sin m\pi x \qquad n \neq 0 \qquad \text{and} \qquad \tfrac{1}{2}G_0(x) = \sum_{m=1}^{\infty} E_{m0}(x) \sin m\pi x$$

and these are just the half-range sine expansions of the now known functions $G_n(x)$ and $\frac{1}{2}G_0(x)$. Therefore

$$E_{mn} = \begin{cases} 2\displaystyle\int_0^1 G_n(x) \sin m\pi x \, dx & n \neq 0 \\ 2\displaystyle\int_0^1 \tfrac{1}{2}G_0(x) \sin m\pi x \, dx & n = 0 \end{cases} \tag{12}$$

If we wish, we can now substitute from Eq. (11) into Eq. (12), getting

$$E_{mn} = \begin{cases} 4\int_0^1 \int_0^1 [g(x, y) - u_{ss}] \cos n\pi y \sin m\pi x \, dy \, dx & n \neq 0 \\ 2\int_0^1 \int_0^1 [g(x, y) - u_{ss}] \sin m\pi x \, dy \, dx & n = 0 \end{cases}$$

With the coefficients E_{mn} now known, the solution for $u(x, y, t)$ is now completely determined.

EXAMPLE 2 A VIBRATING STRING ACTED UPON BY FRICTION

A uniform string of length l is acted upon by a distributed frictional force per unit length equal to

$$-\frac{2cw}{g}\frac{\partial y}{\partial t}$$

where c is an arbitrary proportionality constant. Discuss the subsequent motion of the string, given that it starts with initial displacement $y(x, 0) = \phi(x)$ and initial velocity $\dot{y}(x, 0) = \theta(x)$. In particular, show that certain frequencies in the spectrum of the string may be overdamped while others are underdamped and determine which ones are of each type.

Recalling Eq. (2) Sec. 11.2, it is evident that we must solve the equation

$$\frac{\partial^2 y}{\partial t^2} = a^2\frac{\partial^2 y}{\partial x^2} + \frac{g}{w}\left(-\frac{2cw}{g}\frac{\partial y}{\partial t}\right) = a^2\frac{\partial^2 y}{\partial x^2} - 2c\frac{\partial y}{\partial t}$$

subject to the boundary conditions

(13) $$y(0, t) = y(l, t) = 0$$

and the initial conditions

(14) $$y(x, 0) = \phi(x) \qquad \left.\frac{\partial y}{\partial t}\right|_{x,0} = \theta(x)$$

Assuming a product solution $y(x, t) = X(x)T(t)$ and separating variables, as usual, we obtain

$$\frac{T''}{T} + 2c\frac{T'}{T} = a^2\frac{X''}{X} = \mu \qquad \mu \text{ a constant}$$

If $\mu \geq 0$, it is a purely routine matter to show that the only solution satisfying the boundary conditions (13) is the trivial one. If $\mu < 0$, say $\mu = -\lambda^2$, $\lambda > 0$, we have

$$X(x) = A\cos\frac{\lambda}{a}x + B\sin\frac{\lambda}{a}x$$

and from conditions (13), we conclude that

$$A = 0 \qquad \text{and} \qquad B\sin\frac{\lambda}{a}l = 0$$

Therefore

$$\lambda_n = \frac{n\pi a}{l} \qquad \text{and} \qquad X_n(x) = \sin\frac{n\pi x}{l}$$

To determine the corresponding factors T_n, we must solve the equation

$$T'' + 2cT' + \frac{n^2\pi^2a^2}{l^2}T = 0$$

The roots of the auxiliary equation of this differential equation are

$$-c \pm \sqrt{c^2 - \frac{n^2\pi^2a^2}{l^2}}$$

and we recognize two possibilities. If $c \geq n\pi a/l$, the roots are real and negative and the corresponding product solution for y is either critically damped† (if the equality sign holds) or overdamped. If $c < n\pi a/l$, the corresponding solution for y is underdamped.

Restating these observations in terms of n, we can say that for those harmonics for which $n \leq lc/\pi a$ the string behaves in either a critically damped or overdamped fashion, while for those harmonics for which $n > lc/\pi a$ the string behaves in an underdamped fashion. In other words, it is possible that, with damping, the fundamental and the first few harmonics may be missing from the tone of the string because the corresponding solutions for y, being critically damped or overdamped, are not oscillatory.

Specifically, let us suppose that for the first k values of n the corresponding product solutions are overdamped, with real characteristic roots

$$(-r_1, -r_1'), \quad (-r_2, -r_2'), \quad \ldots, \quad (-r_k, -r_k')$$

and that thereafter the solutions are underdamped, with complex characteristic roots

$$-p_{k+1} \pm iq_{k+1}, \quad -p_{k+2} \pm iq_{k+2}, \quad \ldots$$

Then for the product solutions we have

$$y_n(x,t) = \begin{cases} \sin\dfrac{n\pi x}{l}(A_n e^{-r_n t} + B_n e^{-r_n' t}) & n = 1, 2, \ldots, k \\[2ex] \sin\dfrac{n\pi x}{l} e^{-p_n t}(A_n \cos q_n t + B_n \sin q_n t) & n = k+1, k+2, \ldots \end{cases}$$

Forming an infinite series of these product solutions, we obtain for y the expression

$$y(x,t) = \sum_{n=1}^{\infty} y_n(x,t) = \sum_{n=1}^{k} \sin\frac{n\pi x}{l}(A_n e^{-r_n t} + B_n e^{-r_n' t})$$

$$+ \sum_{n=k+1}^{\infty} \sin\frac{n\pi x}{l}[e^{-p_n t}(A_n \cos q_n t + B_n \sin q_n t)]$$

To meet the first of the initial conditions (14), we must have

$$y(x,0) \equiv \phi(x) = \sum_{n=1}^{k}(A_n + B_n)\sin\frac{n\pi x}{l} + \sum_{n=k+1}^{\infty} A_n \sin\frac{n\pi x}{l}$$

Hence, recognizing that this is just a half-range sine expansion, we have

(15) $$A_n + B_n = \frac{2}{l}\int_0^l \phi(x)\sin\frac{n\pi x}{l}\,dx \qquad n = 1, 2, \ldots, k$$

(16) $$A_n = \frac{2}{l}\int_0^l \phi(x)\sin\frac{n\pi x}{l}\,dx \qquad n = k+1, k+2, \ldots$$

†See Sec. 7.3.

To meet the second of the conditions (14), we must have

$$\left.\frac{\partial y}{\partial t}\right|_{x,0} \equiv \theta(x) = \sum_{n=1}^{k} (-r_n A_n - r_n' B_n) \sin \frac{n\pi x}{l} + \sum_{n=k+1}^{\infty} (-p_n A_n + q_n B_n) \sin \frac{n\pi x}{l}$$

which, since each $p_n = c$, implies that

$$(17) \qquad -r_n A_n - r_n' B_n = \frac{2}{l} \int_0^l \theta(x) \sin \frac{n\pi x}{l}\, dx \qquad n = 1, 2, \ldots, k$$

$$(18) \qquad -cA_n + q_n B_n = \frac{2}{l} \int_0^l \theta(x) \sin \frac{n\pi x}{l}\, dx \qquad n = k+1, k+2, \ldots$$

For values of n equal to or less than k, the values of A_n and B_n can be found by solving Eqs. (15) and (17) simultaneously. For values of n greater than k, the coefficient A_n is given explicitly by Eq. (16), and subsequently B_n is given by Eq. (18). With the coefficients A_n and B_n thus determined, the formal solution of the problem is now complete.

EXAMPLE 3 A VIBRATING STRING ACTED UPON BY A PERIODIC FORCE

A uniform string of length l is acted upon by a distributed periodic force per unit length

$$f(x, t) = \frac{w}{g} \phi(x) \sin \omega t$$

If the string is initially at rest in its equilibrium position, determine its subsequent motion given that frictional effects are negligible.

From Eq. (2), Sec. 11.2, it is clear that the deflection of the string satisfies the partial differential equation

$$(19) \qquad \frac{\partial^2 y}{\partial t^2} = a^2 \frac{\partial^2 y}{\partial x^2} + \phi(x) \sin \omega t$$

As in the case of a system with a single degree of freedom, the motion of the string consists of two parts, one described by the solution of the homogeneous equation

$$(19a) \qquad \frac{\partial^2 y}{\partial t^2} = a^2 \frac{\partial^2 y}{\partial x^2}$$

and the other described by a particular solution corresponding to the nonhomogeneous term $\phi(x) \sin \omega t$.

To find the solution of the homogeneous equation (19*a*), that is, to determine the free motion of the string, we assume a product solution

$$y_H(x, t) = X(x)T(t)$$

and proceed *exactly* as we did in solving the wave equation for the torsional vibrations of a fixed-fixed shaft of uniform cross section in Sec. 11.5. The result is

$$(20) \qquad y_H(x, t) = \sum_{n=1}^{\infty} \sin \frac{n\pi x}{l} \left(A_n \cos \frac{n\pi at}{l} + B_n \sin \frac{n\pi at}{l} \right)$$

To find a particular solution of the nonhomogeneous equation (19), we observe that from physical considerations, the motion produced by the applied force must be periodic with the same period as the force. Moreover, since the system is assumed to be frictionless, the motion of the string ought to be in phase with the force. Hence it is reasonable to assume a solution of the form

$$Y(x, t) = \Phi(x) \sin \omega t$$

From this point we can now continue in either of two ways:

a. In the first place, before substituting $Y(x, t) = \Phi(x) \sin \omega t$ into the nonhomogeneous equation (19), we can expand $\phi(x)$ into a series of the characteristic functions of the homogeneous problem; i.e., we can express $\phi(x)$ as a half-range sine series with known coefficients,† say

$$\phi(x) = \sum_{n=1}^{\infty} C_n \sin \frac{n\pi x}{l} \qquad \text{where} \qquad C_n = \frac{2}{l} \int_0^l \phi(x) \sin \frac{n\pi x}{l}\, dx$$

Then, assuming for the corresponding factor $\Phi(x)$ in the particular integral a half-range sine series with undetermined coefficients, say

$$\Phi(x) = \sum_{n=1}^{\infty} D_n \sin \frac{n\pi x}{l}$$

we have, on substituting

$$Y(x, t) = \left(\sum_{n=1}^{\infty} D_n \sin \frac{n\pi x}{l} \right) \sin \omega t$$

into Eq. (19) and then dividing out $\sin \omega t$,

$$-\omega^2 \sum_{n=1}^{\infty} D_n \sin \frac{n\pi x}{l} = -a^2 \sum_{n=1}^{\infty} D_n \left(\frac{n\pi}{l} \right)^2 \sin \frac{n\pi x}{l} + \sum_{n=1}^{\infty} C_n \sin \frac{n\pi x}{l}$$

Making this relation an identity by collecting terms and then equating to zero the coefficient of $\sin (n\pi x/l)$ for each value of n, we find

$$D_n = \frac{C_n}{(n\pi a/l)^2 - \omega^2} = \frac{C_n}{\omega_n^2 - \omega^2}$$

where $\omega_n = n\pi a/l$ is the nth natural frequency of the string.

Hence

$$\Phi(x) = \sum_{n=1}^{\infty} D_n \sin \frac{n\pi x}{l} = \sum_{n=1}^{\infty} \frac{C_n}{\omega_n^2 - \omega^2} \sin \frac{n\pi x}{l}$$

and

(21) $$Y(x, t) = \Phi(x) \sin \omega t = \left(\sum_{n=1}^{\infty} \frac{C_n}{\omega_n^2 - \omega^2} \sin \frac{n\pi x}{l} \right) \sin \omega t$$

Thus the complete formal solution of the nonhomogeneous equation becomes, from (20) and (21),

$$\begin{aligned} y(x, t) &= y_H(x, t) + Y(x, t) \\ &= \sum_{n=1}^{\infty} \sin \frac{n\pi x}{l} \left(A_n \cos \frac{n\pi at}{l} + B_n \sin \frac{n\pi at}{l} \right) + \left(\sum_{n=1}^{\infty} \frac{C_n}{\omega_n^2 - \omega^2} \sin \frac{n\pi x}{l} \right) \sin \omega t \\ &= \sum_{n=1}^{\infty} \sin \frac{n\pi x}{l} \left(A_n \cos \frac{n\pi at}{l} + B_n \sin \frac{n\pi at}{l} + \frac{C_n}{\omega_n^2 - \omega^2} \sin \omega t \right) \end{aligned}$$

†This is essentially what we did in Example 5, Sec. 8.6, and Example 4, Sec. 11.6.

To satisfy the initial conditions, we must have

$$y(x, 0) \equiv 0 = \sum_{n=1}^{\infty} A_n \sin \frac{n\pi x}{l}$$

from which $A_n \equiv 0$; and

$$\dot{y}(x, 0) \equiv 0 = \sum_{n=1}^{\infty} \sin \frac{n\pi x}{l} \left(\frac{n\pi a}{l} B_n + \frac{\omega C_n}{\omega_n^2 - \omega^2} \right)$$

from which

$$\omega_n B_n + \frac{\omega C_n}{\omega_n^2 - \omega^2} = 0 \qquad \text{or} \qquad B_n = -\frac{\omega C_n}{\omega_n(\omega_n^2 - \omega^2)}$$

From the expression for B_n it is clear that the frequency ω of the impressed force must not coincide with any natural frequency ω_n of the string unless the corresponding coefficient C_n is equal to zero, i.e., unless the term

$$\sin \frac{n\pi x}{l} \equiv \sin \omega_n \frac{x}{a}$$

is missing from the half-range expansion of $\phi(x)$. If $\omega = \omega_n$ and $C_n \neq 0$ for some particular value of n, then the string is effectively being driven at a condition of resonance and displacements of arbitrarily large amplitudes will be built up (see Exercise 9).

b. Instead of assuming a half-range sine expansion for $\Phi(x)$, as we have just done, we can, if we choose, determine $\Phi(x)$ directly as follows. We can substitute $Y(x, t)$ into the nonhomogeneous equation (19), divide out the common factor $\sin \omega t$, solve the resulting nonhomogeneous ordinary differential equation, namely,

$$-\omega^2 \Phi = a^2 \Phi'' + \phi(x)$$

and impose upon its solution the boundary conditions

$$Y(0, t) = Y(l, t) = 0$$

i.e., the conditions that,

$$\Phi(0) = \Phi(l) = 0$$

When $\Phi(x)$ has been determined so that these conditions are fulfilled, we can then construct the complete solution

$$y(x, t) = y_H(x, t) + Y(x, t)$$

(22) $$= \sum_{n=1}^{\infty} \sin \frac{n\pi x}{l} \left(A_n \cos \frac{n\pi a t}{l} + B_n \sin \frac{n\pi a t}{l} \right) + \Phi(x) \sin \omega t$$

The initial conditions can now be imposed, giving

(23) $$y(x, 0) \equiv 0 = \sum_{n=1}^{\infty} A_n \sin \frac{n\pi x}{l}$$

and

(24) $$\dot{y}(x, 0) \equiv 0 = \sum_{n=1}^{\infty} \frac{n\pi a}{l} B_n \sin \frac{n\pi x}{l} + \omega \Phi(x)$$

From (23) we conclude that the A's are the coefficients in the half-range sine expansion of 0; hence $A_n = 0$ for all values of n. From (24) we conclude, similarly, that the quantities $\{(n\pi a/l)B_n\}$ are the coefficients in the half-range sine expansion of $-\omega\Phi(x)$; hence

$$B_n = -\frac{2\omega}{n\pi a}\int_0^l \Phi(x)\sin\frac{n\pi x}{l}\,dx$$

provided that ω is not equal to one of the natural frequencies $\{\omega_n\} = \{n\pi a/l\}$ of the system, i.e., provided that the system is not being driven at resonance.

EXAMPLE 4

THE STEADY-STATE POTENTIAL ALONG AN INFINITE TRANSMISSION LINE

Find the steady-state potential at any point of an infinitely long transmission line if a signal voltage $E_0 \cos \omega t$ is applied at the sending end $x = 0$.

Here we have to solve the telephone equation [Eq. (19), Sec. 11.2]

(25) $$\frac{\partial^2 e}{\partial x^2} = LC\frac{\partial^2 e}{\partial t^2} + (RC + GL)\frac{\partial e}{\partial t} + RGe$$

subject to the boundary conditions

(26) $$e(0, t) = E_0 \cos \omega t \qquad e(x, t) \text{ bounded as } x \to \infty$$

If we assume a product solution $e(x, t) = X(x)T(t)$ and separate variables, we obtain

$$\frac{X''}{X} = \frac{LCT'' + (RC + GL)T' + RGT}{T} = \mu$$

Thus the factor T satisfies the differential equation

$$LCT'' + (RC + GL)T' + (RG - \mu)T = 0$$

and hence T must be one or of the other of the forms (see the note in Exercise 15, Sec. 11.2).

$$\epsilon^{pt}(A \cos qt + B \sin qt)$$
$$\epsilon^{pt}(At + B)$$
$$A\epsilon^{p_1 t} + B\epsilon^{p_2 t}$$

Now the steady-state behavior of a system acted upon by a periodic disturbance must be periodic. Under no circumstances, however, can the last two expressions represent periodic behavior. Moreover, the first of the three expressions can represent periodic behavior only if $p = 0$, which, since $p \equiv -(RC + GL)/2LC \neq 0$, is impossible. Hence no product solution of Eq. (25) is capable of describing what we know the steady-state response of the line must be.

If we reconsider the problem in an attempt to find an alternative method of solution, it seems reasonable to expect that under the given conditions the voltage along the line will vary harmonically with time while exhibiting attenuation and phase shift depending on the distance from the sending end. Hence, as an alternative approach, we are led to try an expression of the form

(27) $$e(x, t) = E_0\epsilon^{-ax}\cos(\omega t + bx)$$

If $a > 0$, this obviously satisfies each of the boundary conditions (26), and perhaps the constants a and b can be determined so that it will satisfy the differential equation (25) also.

If we substitute the tentative solution (27) into the telephone equation (25), divide out $E_0\epsilon^{-ax}$, and collect terms, we obtain without difficulty

$$(a^2 - b^2 + LC\omega^2 - RG)\cos(\omega t + bx) + [2ab + \omega(RC + GL)]\sin(\omega t + bx) = 0$$

This will be an identity if and only if

$$a^2 - b^2 = RG - LC\omega^2 \tag{28}$$
$$2ab = -(RC + GL)\omega \tag{29}$$

Now by adding the square of Eq. (29) to the square of Eq. (28), we obtain

$$(a^2 + b^2)^2 = (RG - LC\omega^2)^2 + (RC + GL)^2\omega^2$$

or

$$a^2 + b^2 = \sqrt{(RG - LC\omega^2)^2 + (RC + GL)^2\omega^2} \tag{30}$$

Finally, by solving (28) and (30) simultaneously, we find

$$a^2 = \tfrac{1}{2}[\sqrt{(RG - LC\omega^2)^2 + (RC + GL)^2\omega^2} + (RG - LC\omega^2)]$$
$$b^2 = \tfrac{1}{2}[\sqrt{(RG - LC\omega^2)^2 + (RC + GL)^2\omega^2} - (RG - LC\omega^2)]$$

From the form of these equations it is clear that a^2 and b^2 are both positive. Hence a and b are real, and with their values now determined, Eq. (27) becomes the required solution. In a similar manner of course the steady-state response to a signal voltage of the form $E_0 \sin \omega t$ can be found.

By means of these results it is now possible to find the steady-state voltage corresponding to *any* periodic signal. For if $e(0, t) = f(t)$ is a periodic function with period $2p$, it can be expanded in a Fourier series

$$f(t) = \frac{1}{2}a_0 + a_1 \cos\frac{\pi t}{p} + a_2 \cos\frac{2\pi t}{p} + \cdots + b_1 \sin\frac{\pi t}{p} + b_2 \sin\frac{2\pi t}{p} + \cdots$$

and the steady-state solution for each of these terms can be found. Then, since the telephone equation is linear, the sum of the steady-state responses to each of these terms will be the steady-state response of the line to the entire signal $f(t)$. Moreover, if the input signal is not periodic, the steady-state response can still be found by a similar analysis after first expressing the input signal as a Fourier integral instead of a Fourier series.

EXAMPLE 5

TEMPERATURE IN A SEMI-INFINITE ROD

A slender rod whose curved surface is insulated stretches from $x = 0$ to $x = \infty$. Find the temperature in the rod as a function of x and t if the left end of the rod is maintained at the constant temperature 0°C and if initially the temperature is given by $u(x, 0) = f(x)$.

Exactly as in Example 1, Sec. 11.6, we find that the function

$$u = B \exp\left(\frac{-\lambda^2 t}{a^2}\right) \sin \lambda x$$

satisfies the heat equation

$$\frac{\partial^2 u}{\partial x^2} = a^2 \frac{\partial u}{\partial t}$$

and the boundary condition at the left-hand end of the rod,

$$u(0, t) \equiv 0$$

Lacking a second boundary condition, however, we have no further restriction on λ. Therefore, instead of having an infinite set of *discrete* characteristic values λ_n, with corresponding solutions

$$u_n(x, t) = B_n \exp\left(\frac{-\lambda_n^2 t}{a^2}\right) \sin \lambda_n x$$

we have a *continuous* family of solutions

$$u_\lambda(x, t) = B(\lambda) \exp\left(\frac{-\lambda^2 t}{a^2}\right) \sin \lambda x$$

where the arbitrary constant B is now associated not with n but with the continuous parameter λ, which without loss of generality we can assume to be nonnegative.

We cannot speak of an infinite series of particular solutions in this case. Instead of *adding* the product solutions for each value of n, we therefore try *integrating* them over all values of λ, getting

$$u(x, t) = \int_0^\infty B(\lambda) \exp\left(-\frac{\lambda^2 t}{a^2}\right) \sin \lambda x \, d\lambda \tag{31}$$

By direct substitution it is easily verified that this integral is a solution of the heat equation.

It is now necessary to impose the initial condition $u(x, 0) = f(x)$ on the solution $u(x, t)$. When we set $t = 0$ in Eq. (31), it is clear that this requires that

$$f(x) = \int_0^\infty B(\lambda) \sin \lambda x \, d\lambda$$

This we recognize as just an instance of the Fourier integral considered in Sec. 9.1. There, in discussing what we called the Fourier sine integral, we saw [Eq. (21*b*), Sec. 9.1] that if

$$f(x) = \int_0^\infty B(\lambda) \sin \lambda x \, d\lambda$$

then the coefficient function $B(\lambda)$ is given by the formula

$$B(\lambda) = \frac{2}{\pi} \int_0^\infty f(x) \sin \lambda x \, dx$$

Introducing the dummy variable s for x in the integral defining $B(\lambda)$ and then substituting for $B(\lambda)$ in Eq. (31), we have

$$\begin{aligned} u(x, t) &= \int_0^\infty \exp\left(-\frac{\lambda^2 t}{a^2}\right)\left[\frac{2}{\pi}\int_0^\infty f(s) \sin \lambda s \, ds\right] \sin \lambda x \, d\lambda \\ &= \frac{2}{\pi}\int_0^\infty \int_0^\infty \exp\left(-\frac{\lambda^2 t}{a^2}\right) f(s) \sin \lambda s \sin \lambda x \, ds \, d\lambda \end{aligned}$$

which is the required solution.

In Examples 1 and 2 in Sec. 9.4, we illustrated the use of Fourier integrals in solving ordinary differential equations. Then in Examples 3–5 we illustrated how Fourier transforms could be used for that purpose. Now in the example we have just worked, we have shown that Fourier integrals can also be used to solve problems involving partial differential equations. In the next example, to round out the picture, we will illustrate the use of Fourier transforms in solving partial differential equations. For purposes of comparison, we will do this by using the sine transform to rework the problem we solved in Example 5.

EXAMPLE 6

AN APPLICATION OF FOURIER SINE TRANSFORMS

Work Example 5 using Fourier transforms.

Since our problem involves the semi-infinite interval $[0, \infty)$, it is clear that we must work with either the sine or cosine transform. Furthermore, since the left end condition gives us the value of $u(0, t)$ but no information about the derivative $u_x(0, t)$, it follows, by considering Corollaries 3 and 4 of Theorem 6, Sec. 9.3, that the sine transform is the one that is appropriate here. Accordingly, taking the sine transform of both sides of the heat equation, $a^2u_t = u_{xx}$, *treating x as the variable with respect to which we are transforming,* we have, from Corollary 4, Theorem 6, Sec. 9.3,

$$\begin{aligned} a^2\mathcal{F}_s[u_t(x, t)] &= \mathcal{F}_s[u_{xx}(x, t)] \\ &= -\omega^2\mathcal{F}_s[u(x, t)] + \sqrt{\frac{2}{\pi}}\,\omega u(0, t) \\ &= -\omega^2\mathcal{F}_s[u(x, t)] \qquad \text{since } u(0, t) = 0 \end{aligned}$$

Assuming that the operations of taking the sine transform with respect to x and differentiating with respect to t can be interchanged, the last equation can be rewritten in the form

$$\frac{d}{dt}\mathcal{F}_s[u(x, t)] = -\frac{\omega^2}{a^2}\mathcal{F}_s[u(x, t)]$$

or, displaying the transform explicitly as a function of ω and t,

$$\frac{d}{dt}U_s(\omega, t) = -\frac{\omega^2}{a^2}U_s(\omega, t)$$

This is a very simple separable differential equation whose solution can be written down at once:

$$U_s(\omega, t) = k(\omega)e^{-\omega^2 t/a^2} \tag{32}$$

where of course the integration constant k must be recognized as depending on ω.

Setting $t = 0$ in Eq. (32), it follows that

$$U_s(\omega, 0) = k(\omega)$$

that is, $k(\omega)$ is the sine transform of the given initial temperature function $u(x, 0) = f(x)$, namely, $\mathcal{F}_s[f(x)] = F_s(\omega)$. Thus (32) becomes

$$U_s(\omega, t) = F_s(\omega)e^{-\omega^2 t/a^2} \tag{33}$$

Applying the inversion formula for sine transforms [Eq. (6*a*), Sec. 9.3] to Eq. (33), we have

$$u(x, t) = \sqrt{\frac{2}{\pi}}\int_0^\infty F_s(\omega)e^{-\omega^2 t/a^2}\sin \omega x\, d\omega \tag{34}$$

If we now substitute the integral expression for $F_s(\omega)$ [Eq. (6*b*), Sec. 9.3] into (34), we have finally,

$$\begin{aligned} u(x, t) &= \sqrt{\frac{2}{\pi}}\int_0^\infty \left[\sqrt{\frac{2}{\pi}}\int_0^\infty f(s)\sin \omega s\, ds\right] e^{-\omega^2 t/a^2}\sin \omega x\, d\omega \\ &= \frac{2}{\pi}\int_0^\infty\int_0^\infty f(s)e^{-\omega^2 t/a^2}\sin \omega s \sin \omega x\, ds\, d\omega \end{aligned}$$

which is precisely the answer we obtained in Example 5, except that the dummy variable λ in Example 5 is now ω.

EXAMPLE 7

TEMPERATURES IN A HALF-PLANE

Find the steady-state temperature distribution in a thin, insulated sheet coinciding with the upper half of the xy plane if the portion of the x axis between $x = -a$ and $x = a$ is maintained at the temperature T_0 while the rest of the x axis is maintained at the temperature 0.

In this problem we are to solve the steady-state heat equation, i.e., Laplace's equation, subject to the boundary condition

$$u(x, 0) = f(x) = \begin{cases} T_0 & |x| < a \\ 0 & |x| > a \end{cases}$$

and the further (reasonable) condition that $u(x, y) \to 0$ as $y \to \infty$.

Assuming a product solution $u(x, y) = X(x)Y(y)$, substituting into Laplace's equation, and separating variables gives us, as usual,

$$\frac{X''}{X} = -\frac{Y''}{Y} = \mu \qquad \mu \text{ a constant}$$

If $\mu > 0$, say $\mu = \lambda^2$, $\lambda > 0$, we have $Y = C \cos \lambda y + D \sin \lambda y$, and if u is to approach zero as y becomes infinite, it follows that both C and D must be zero. Thus $\mu > 0$ leads only to the trivial solution. Similarly, if $\mu = 0$, then $Y = Cy + D$, and if u is to approach zero as y becomes infinite, both C and D must be zero. Thus only the trivial solution is possible if $\mu = 0$.

If $\mu < 0$, say $\mu = -\lambda^2$, $\lambda > 0$, then

$$X = A \cos \lambda x + B \sin \lambda x \qquad \text{and} \qquad Y = Ce^{\lambda y} + De^{-\lambda y}$$

Since u must approach zero as y becomes infinite, it follows that $C = 0$. Also, since it is clear from the symmetry of the problem that $u(x, y) = u(-x, y)$, it follows that $B = 0$. There are no further conditions to restrict λ and so, instead of having a countably infinite set of *discrete* characteristic values λ_n with corresponding product solutions

$$u_n(x, y) = A_n e^{-\lambda_n y} \cos \lambda_n x$$

we have a *continuous* family of solutions

$$u_\lambda(x, y) = A(\lambda)e^{-\lambda y} \cos \lambda x$$

where the coefficient A is now associated not with n but with the continuous parameter λ, which may assume any positive value. As in the last example, we cannot speak of an infinite series of the product solutions we have obtained. Instead of *adding* them for each value of n, we must therefore *integrate* them over all values of λ, getting

$$(35) \qquad u(x, y) = \int_0^\infty u_\lambda(x, y)\, d\lambda = \int_0^\infty A(\lambda)e^{-\lambda y} \cos \lambda x \, d\lambda$$

Imposing the boundary conditions that holds along the x axis, it appears that we must have

$$u(x, 0) = \int_0^\infty A(\lambda) \cos \lambda x \, d\lambda = f(x) = \begin{cases} T_0 & |x| < a \\ 0 & |x| > a \end{cases}$$

In other words, the last integral must be the half-range cosine integral of the given function $f(x)$. Therefore, by Eq. (20*b*), Sec. 9.1, the coefficient function $A(\lambda)$ is given by the formula

$$\begin{aligned} A(\lambda) &= \frac{2}{\pi}\int_0^\infty f(x) \cos \lambda x \, dx = \frac{2}{\pi}\int_0^a T_0 \cos \lambda x \, dx \\ &= \frac{2}{\pi\lambda} T_0 \sin \lambda x \Big|_0^a \\ &= \frac{2}{\pi} T_0 \frac{\sin \lambda a}{\lambda} \end{aligned}$$

Thus, returning to Eq. (35) and substituting the value we have just obtained for $A(\lambda)$, we have the answer to our problem:

$$u(x, y) = \frac{2}{\pi} T_0 \int_0^\infty \frac{\sin \lambda a \cos \lambda x}{\lambda} e^{-\lambda y} \, d\lambda$$

The limits on the last integral and the presence of the factor $e^{-\lambda y}$ in the integrand are strongly suggestive of the Laplace transformation. In fact, by pursuing this observation, the formula for $u(x, y)$ can be simplified in a remarkable way. To do this, let us interpret λ as the variable t and y as the variable s in the usual Laplace transform notation. Then $u(x, y)$ is the Laplace transform of

$$\frac{2}{\pi}T_0\frac{\sin \lambda a \cos\lambda x}{\lambda}$$

and this transform, by Theorem 2, Sec. 10.7, is equal to

$$\int_y^{\infty} \mathcal{L}\left[\frac{2}{\pi}T_0 \sin \lambda a \cos \lambda x\right] dy$$

Now $2 \sin \lambda a \cos \lambda x = \sin (x + a)\lambda - \sin (x - a)\lambda$. Hence

$$\begin{aligned}\mathcal{L}\left[\frac{2}{\pi}T_0 \sin \lambda a \cos \lambda x\right] &= \frac{1}{\pi}T_0\mathcal{L}[\sin (x + a)\lambda - \sin (x - a)\lambda] \\ &= \frac{1}{\pi}T_0\left[\frac{x + a}{(x + a)^2 + y^2} - \frac{x - a}{(x - a)^2 + y^2}\right]\end{aligned}$$

Thus

$$\begin{aligned}u(x, y) &= \frac{1}{\pi}T_0\int_y^{\infty}\left[\frac{x + a}{(x + a)^2 + y^2} - \frac{x - a}{(x - a)^2 + y^2}\right] dy \\ &= \frac{1}{\pi}\,T_0\left[-\tan^{-1}\frac{y}{x + a} + \tan^{-1}\frac{y}{x - a}\right]\end{aligned}$$

Now from Fig. 11.23 it is apparent that

$$\tan^{-1}\frac{y}{x - a} = \phi_1 \qquad \text{and} \qquad \tan^{-1}\frac{y}{x + a} = \phi_2$$

Hence $u(x, y) = (1/\pi)T_0(\phi_1 - \phi_2) = (1/\pi)T_0\theta$, by the exterior angle theorem. The isotherms of this problem are thus the curves on which θ is a constant. Now the locus of a point P which varies in such a way that $\angle A_1PA_2$ remains constant is a circular arc having A_1 and A_2 as endpoints. Hence the family of isotherms is the family of circular arcs that pass through A_1 and A_2 and lie in the upper half-plane.

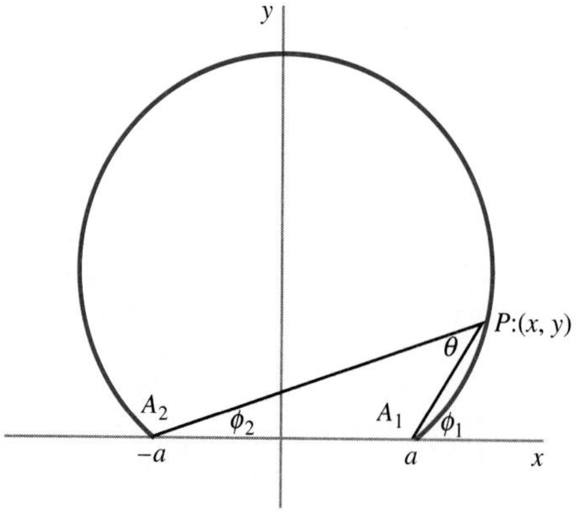

FIGURE 11.23
A typical isotherm in Example 7.

EXERCISES

1. Work Example 1 if $f(y) = 0$ and
   (a) $g(x, y) = 100$ (b) $g(x, y) = xy$
   (c) $g(x, y) = x + y$
2. Work Example 1 if $f(y) = 100$ and
   (a) $g(x, y) = 0$ (b) $g(x, y) = 100y$
   Hint: You should be able to infer u_{ss} by inspection.
3. Work Example 1 if all four edges are maintained at the temperature 0 and
   (a) $g(x, y) = 100$ (b) $g(x, y) = xy$
   (c) $g(x, y) = x + y$
4. Work Example 1 if the left-hand edge is also insulated, the right-hand edge is maintained at 0, and
   (a) $g(x, y) = 100$ (b) $g(x, y) = 100y$
5. In Example 1, show that if all four edges are insulated, the (constant) steady-state temperature is equal to E_{00} regardless of the initial temperature distribution $g(x, y)$. What is $u(x, y, t)$?
6. Work Example 1 if all four edges are insulated and
   (a) $g(x, y) = xy$ (b) $g(x, y) = x + y$
7. Work Example 3 if
   (a) $f(x, t) = \frac{w}{g} \sin \frac{\pi x}{l} \sin \frac{2\pi at}{l}$
   (b) $f(x, t) = \frac{w}{g} \sin \frac{2\pi x}{l} \sin \frac{\pi at}{l}$
   (c) $f(x, t) = \frac{w}{g} \sin \frac{2\pi x}{l} \cos \frac{\pi at}{l}$
   Is it possible for a system to be driven at one of its natural frequencies without resonance occurring?
8. Assuming that the free or natural motion of the string, $y_H(x, t)$, dies away because of small, though otherwise negligible, frictional effects, find the steady-state motion produced in a uniform string of length l produced by a periodic distributed force per unit length equal to
   (a) $\frac{w}{g} x \sin \omega t$ (b) $\frac{w}{g} x(l - x) \cos \omega t$
   What restrictions must be placed on ω?
9. In Example 3, how can the first method of solution be modified to yield a solution when ω is one of the natural frequencies of the string?
10. A uniform string of length l is acted on by a periodic distributed force per unit length $(w/g)\phi(x) \sin \omega t$, where

$$\phi(x) = \begin{cases} 0 & 0 < x < (l - k)/2 \\ 1 & (l - k)/2 < x < (l + k)/2 \\ 0 & (l + k)/2 < x < l \end{cases}$$

   Using the first method of Example 3, find the steady-state motion of the string and discuss the possibility of resonance as k varies from 0 to l.
11. In Example 3, where does the second method of solution break down when ω is one of the natural frequencies of the string?
12. If a uniform string of length l is being driven by a force $f(x, t) = (w/g)\phi(x) \sin \omega t$, show that resonance will not occur if $\int_0^l \phi(x) \sin (n\pi x/l)\, dx = 0$.
13. Discuss how each of the methods of Example 3 could be applied to the forced torsional vibrations of a uniform shaft of length l (**a**) with both ends fixed, (**b**) with both ends free, and (**c**) with one end fixed and one end free.
14. Compare methods **a** and **b** of Example 3 with regard to
    (a) The relative difficulty of carrying them out for a given forcing function $\phi(x) \sin \omega t$
    (b) The information they provide about conditions of resonance
    (c) The ease with which they provide the steady-state solution, if that is all that is required
    (d) The ease with which they can handle problems in which $\phi(x)$ has corner points or jumps.
15. Can the procedures described in Example 3 be modified to obtain a description of the motion of the string when the impressed force is of the form $f(x, t) = \phi(x)\theta(t)$ and
    (a) $\theta(t)$ is a general periodic function? How?
    (b) $\theta(t)$ is nonperiodic? How?
16. A uniform string of length l is acted upon by an elastic restoring force per unit length equal to $(w/g)c^2y$. Find the deflection of the string as a function of x and t if the initial conditions are $y(x, 0) = f(x)$ and $\dot{y}(x, 0) = h(x)$. What are the natural frequencies of the string? Do they form an arithmetic progression? *Hint:* In separating variables, associate c^2 with the factor T, not the factor X.
17. Work Exercise 16 if (**a**) $f(x) = \sin (\pi x/l)$ and $h(x) = 0$, (**b**) $f(x) = 0$ and $h(x) = \sin (\pi x/l)$, (**c**) $f(x) = 0$ and

$$h(x) = \begin{cases} 0 & 0 < x < l/4 \\ 1 & l/4 < x < 3l/4 \\ 0 & 3l/4 < x < l \end{cases}$$

18. In Exercise 16, find the steady-state motion of the string if it is acted upon by a force per unit length equal to $f(x, t) = (w/g) \sin (\pi x/l) \sin (n\pi at/l)$.
19. A string subject to viscous damping as described in Example 2 is acted upon by a force per unit length equal to $(wa^2/gl) \sin (n\pi x/l) \sin \omega t$. Discuss the possibility of finding the steady-state motion by assuming a solution of the form $y(x, t) = \sin (n\pi x/l)(A \sin \omega t + B \cos \omega t)$. Are the concepts of magnification ratio and phase shift relevant here? (See Sec. 7.3.) Can the procedure described here be adapted to the determination of the steady-state response of a string to an impressed force per unit length equal to $(wa^2/gl)\phi(x) \sin \omega t$?
20. A thin sheet of metal coincides with the unit square in the xy plane. The faces of the sheet, instead of being insulated, radiate into air of constant temperature 0. If the edges of the sheet are maintained at the temperature 0 and if the initial temperature distribution in the sheet is $u(x, y, 0) = g(x, y)$, find the temperature in the sheet as a function of x, y, and t. *Hint:* The differential equation to be solved was derived in Exercise 27, Sec. 11.2.

21. Work Exercise 20 if all four edges of the sheet are insulated.

22. **(a)** If an arbitrary temperature distribution is maintained along *each* of the edges of a square sheet of metal, how can the steady-state temperature distribution in the sheet be found?

(b) If an arbitrary temperature distribution is maintained along *each* of the edges of a square sheet of metal and if the initial temperature distribution in the sheet is known, how can the temperature be found as a function of x, y, and t?

23. Work Example 4 by replacing the signal voltage $E_0 \cos \omega t$ by $E_0 \epsilon^{j\omega t}$ and assuming a solution of the form

$$e(x, t) = E_0 \epsilon^{j\omega t + (a+jb)x}$$

24. A slender rod of infinite length has its curved surface insulated. Find the steady-state temperature distribution in the rod if the temperature at the finite end of the rod varies according to the law $u(0, t) = \sin \omega t$. Explain how this result can be used to determine the steady-state temperature distribution produced by an arbitrary periodic temperature condition at the finite end of the rod.

25. Work Example 5 if the left end of the rod is insulated.

26. Work Example 5 if

(a) $f(x) = e^{-x}$

(b) $f(x) = \begin{cases} T_0 & 0 < x < k \\ 0 & k < x < \infty \end{cases}$

27. Work Example 5 if the surface of the rod, instead of being insulated, radiates into air of constant temperature 0. *Hint:* The equation to be solved here was derived in Exercise 23, Sec. 11.2.

28. Assuming that the function defined by Eq. (31) can legitimately be differentiated inside the integral sign, verify that it satisfies the one-dimensional heat equation.

29. A slender rod of length l has its curved surface insulated. Its right-hand end is maintained at the constant temperature $u(l, t) = 0$. At the left-hand end the temperature varies according to the law $u(0, t) = \sin \omega t$. Determine the steady-state temperature distribution in the rod. Explain how this result can be used to determine the steady-state temperature distribution produced by an arbitrary periodic temperature condition at the left-hand end of the rod. *Hint:* Verify that λ can be chosen so that

$$u_1(x, t) = \sin \omega t \cos \frac{\lambda x}{l} \cosh \lambda\left(2 - \frac{x}{l}\right) - \cos \omega t \sin \frac{\lambda x}{l} \sinh \lambda\left(2 - \frac{x}{l}\right)$$

and

$$u_2(x, t) = \sin \omega t \cos \lambda\left(2 - \frac{x}{l}\right) \cosh \frac{\lambda x}{l} - \cos \omega t \sin \lambda\left(2 - \frac{x}{l}\right) \sinh \frac{\lambda x}{l}$$

are solutions of the one-dimensional heat equation. Then determine A_1 and A_2 so that

$$u(x, t) = A_1 u_1(x, t) + A_2 u_2(x, t)$$

satisfies the boundary conditions of the problem.

30. A slender rod of length l has its curved surface insulated. Heat is generated in the rod at a rate per unit length equal to $\phi(x)$. Find the temperature in the rod as a function of x and t if both ends of the rod are maintained at the temperature 0 and if the initial temperature distribution in the rod is $u(x, 0) = g(x)$.

31. Work Exercise 30 if the left-hand end of the rod is insulated and the right-hand end is maintained at the temperature 0.

32. A thin sheet of metal bounded by the x axis and the upper halves of the lines $x = 0$ and $x = 1$ has its upper and lower surfaces insulated. If its edges are maintained at the temperature 0 and if the initial temperature distribution in the sheet is $u(x, y, 0) = g(x, y)$, find the temperature at any point in the sheet at any subsequent time.

33. Work Exercise 32 if the lower edge of the sheet is insulated.

34. Work Exercise 32 if the two vertical edges of the sheet are insulated.

35. Work Exercise 32 if, instead of being insulated, the two faces of the sheet radiate into air of constant temperature 0.

36. A solid coincides with the unit cube in an xyz coordinate system. The temperature distribution $u(x, y, ,0) = f(x, y)$ is maintained over the face of the solid which lies in the xy plane. The other five faces are maintained at the temperature 0. Find the steady-state temperature in the solid as a function of x, y, and z. *Hint:* Assume a product solution $u(x, y, z) = X(x)Y(y)Z(z)$, substitute into the three-dimensional form of Laplace's equation, and separate variables in successive steps.

37. Work Exercise 36 if

(a) The upper horizontal face of the solid is insulated

(b) The four vertical faces of the solid are insulated

(c) The front and back faces of the solid are insulated

38. Work Example 7 if the temperature distribution maintained along the x axis is

(a) $f(x) = T_0 e^{-|x|}$

(b) $f(x) = \begin{cases} T_0(a^2 - x^2) & x^2 \le a^2 \\ 0 & x^2 \ge a^2 \end{cases}$

(c) $f(x) = \begin{cases} T_0(a - x) & |x| < a \\ 0 & |x| > a \end{cases}$

39. A thin sheet of metal coincides with the semi-infinite strip bounded by the x axis and the positive halves of the lines $x = 0$ and $x = 1$. The faces of the sheet are insulated, the lower edge is maintained at the temperature 0, and the temperature distribution $u(1, y) = f(y)$ is maintained along the right-hand edge. Find the steady-state temperature distribution in the sheet if

(a) The left-hand edge is maintained at the temperature 0.

(b) The left-hand edge is insulated.

40. Work both parts of Exercise 39 if in each case the lower edge is maintained at the temperature 100 and

$$f(y) = \begin{cases} 100 & 0 < y < a \\ 0 & a < y < \infty \end{cases}$$

Hint: The required answer is the sum of the answers to two problems. One is the problem solved in Exercise 39 with $f(x)$ the function given here. The other is the problem when $f(y) = 0$ and $u(x, 0) = 100$.

41. A thin sheet of metal coincides with the first quadrant in the xy plane. The faces of the sheet are insulated. Find the steady-state temperature distribution in the sheet if the temperature distribution

$$u(0, y) \equiv f(y) = \begin{cases} 100 & 0 < y < a \\ 0 & a < y < \infty \end{cases}$$

is maintained along the positive y axis and

(a) The lower edge is maintained at the temperature 0.

(b) The lower edge is insulated.

42. Determine the natural frequencies of a square drumhead.

43. A uniform beam of length l simply supported at each end is acted on by a periodic distributed load per unit length equal to $\phi(x) \sin \omega t$. Neglecting friction (but assuming nonetheless that the transient dies away eventually), find the steady-state motion of the beam by first expanding $\phi(x)$ as a series of the characteristic functions of a simply supported beam. Explain how the second method described in Example 3 can also be used to determine the steady-state motion of the beam.

44. Work Exercise 43 if

(a) $\phi(x) = \sin(\pi x/l)$

(b) $\phi(x) = \begin{cases} 0 & 0 < x < l/4 \\ 1 & l/4 < x < 3l/4 \\ 0 & 3l/4 < x < l \end{cases}$

45. Can the methods of Example 3, referred to in Exercise 43, be applied to beams other than simply supported ones? How?

46. Is it possible for an applied force to drive a beam at one of its natural frequencies without resonance occurring? How?

47. Find the steady-state displacement of a uniform cantilever of length l if the free end of the beam is forced to move so that its displacement is equal to $A_0 \sin \omega t$.

48. Find the steady-state motion of a uniform cantilever of length l if a force equal to $F_0 \sin \omega t$ is applied at the free end of the beam.

49. Find the frequency equation of the uniform cantilever beam shown in Fig. 11.24*a*. Discuss the limiting cases when $k = 0$ and $k = \infty$.

50. Assuming that the attached mass is guided so that it can move only in a direction perpendicular to the length of the beam, find the frequency equation of the uniform cantilever shown in Fig. 11.24*b*.

51. Use Fourier transforms to work Example 5 if the left end of the rod is insulated.

52. Use Fourier transforms to work Example 5 if the rod stretches from $x = -\infty$ to $x = \infty$.

53. Work Exercise 41(**a**) if the y axis above $y = a$ is insulated and heat flows into the sheet through the segment $0 < y < a$ at the constant rate q per unit length.

54. A thin sheet of metal coincides with the infinite strip in the xy plane bounded on the left and right by the lines $x = 0$ and $x = 1$. The faces of the sheet are insulated, its initial temperature is 0, heat is generated in the sheet at a rate per unit volume that is proportional to the distance from $x = 0$, and its edges are maintained at the temperature 0. Find the temperature in the sheet as a function of x and t.

55. If the transmission line in Example 4 is initially dead, i.e., if at $t = 0$ the potential and current along the line are identically zero, determine the complete response of the line, transient as well as steady-state, to the signal voltage $E_0 \cos \omega t$. *Hint:* Show that if $-p \pm iq$ are the roots of the equation

$$LCm^2 + (RC + GL)m + (RG + \lambda^2) = 0$$

then

$$e_\lambda = \epsilon^{-pt}[A(\lambda) \cos qt + B(\lambda) \sin qt] \sin \lambda x$$

FIGURE 11.24

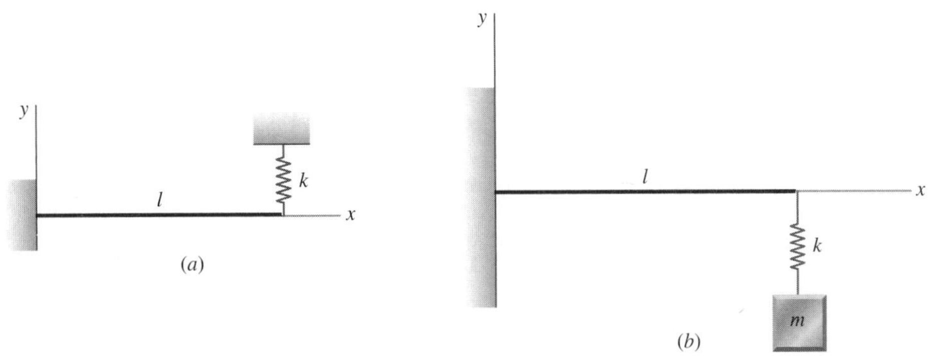

is a solution of the telephone equation which is bounded as $x \to \infty$ and is zero for all values of t when $x = 0$. Then show that the steady-state solution found in Example 4 plus the integral of e_λ over all values of λ is a solution which satisfies both boundary conditions (26). Finally, determine $A(\lambda)$ and $B(\lambda)$ so that both e and $\partial e/\partial t$ are zero when $t = 0$.

11.8 LAPLACE TRANSFORM METHODS

In several examples in Sec. 9.4 we illustrated how Fourier transforms could be used to convert an ordinary constant-coefficient *differential* equation into an *algebraic* equation from which the transform of the dependent variable and, subsequently, the dependent variable could be found. Then in Chap. 10, in much greater detail, we showed how the Laplace transformation could be used similarly to solve ordinary constant-coefficient differential equations by first converting them into algebraic equations in the transform(s) of the dependent variable(s). In Example 6 (and Exercises 51 and 52) in the last section, we extended this technique by showing how Fourier transforms could be used to convert a *partial* differential equation into an *ordinary* differential equation in the transform of the dependent variable. Not unexpectedly, the Laplace transformation can also be applied to partial differential equations to convert them into ordinary differential equations from which the transform of the dependent variable and then the dependent variable itself can be found. Laplace transforms are probably more useful than Fourier transforms for solving partial differential equations, and in this section we shall investigate how they can be used for this purpose. The general procedure is as follows.

The given partial differential equation, with its accompanying boundary conditions and initial conditions, is transformed with respect to one of its independent variables, usually t. Partial derivatives with respect to this variable are of course transformed by the familiar formulas of Theorem 1, Sec. 10.4, and its corollaries. For partial derivatives with respect to the other independent variable, we assume† that the operations of differentiating and taking Laplace transforms can be interchanged. Then if the independent variables are x and t, say, we have

$$\mathcal{L}\left\{\frac{\partial f(x,t)}{\partial x}\right\} = \int_0^\infty \frac{\partial f(x,t)}{\partial x} e^{-st}\,dt = \frac{\partial}{\partial x}\int_0^\infty f(x,t)e^{-st}\,dt$$
$$= \frac{d}{dx}\mathcal{L}\{f(x,t)\}$$

the derivative in the term being a total derivative because $\mathcal{L}\{f(x,t)\}$ is not a function of t. Similar formulas hold of course for x derivatives of higher orders. Thus the result of the transformation is an ordinary differential equation in $\mathcal{L}\{f(x,t)\}$ in which x is the independent variable and s enters as a parameter. Because s occurs in the coefficients of the differential equation, the arbitrary constants appearing in its complete solution will in general be functions of s which must be determined by imposing the transformed boundary conditions on the complete solution of the transformed differential equation. After this has been done, the inverse transformation is carried out and the solution to the original problem is obtained. The details of this process can best be made clear through examples.

EXAMPLE 1

A SEMI-INFINITE STRING WITH A MOVING END

A semi-infinite string is initially at rest in a position coinciding with the positive half of the x axis. At $t = 0$, the left-hand end of the string begins to move along the y axis in a manner described by $y(0, t) = f(t)$, where $f(t)$ is a known function. Find the displacement $y(x, t)$ of the string at any point at any subsequent time.

†This is justified by Theorems 1 and 4, Sec. 10.3.

The partial differential equation to be solved is of course the one-dimensional wave equation

(1) $$\frac{\partial^2 y}{\partial t^2} = a^2 \frac{\partial^2 y}{\partial x^2}$$

subject to the boundary conditions

(2) $$y(0, t) = f(t)$$

(3) $$y(x, t) \text{ bounded as } x \to \infty$$

and the initial conditions

(4) $$y(x, 0) = 0$$

(5) $$\left.\frac{\partial y}{\partial t}\right|_{x,0} = 0$$

If we take Laplace transforms in Eq. (1) *with respect to t,* we obtain

$$s^2\mathcal{L}\{y(x, t)\} - sy(x, 0) - \left.\frac{\partial y}{\partial t}\right|_{x,0} = a^2\mathcal{L}\left\{\frac{\partial^2 y(x, t)}{\partial x^2}\right\} = a^2\frac{d^2}{dx^2}\mathcal{L}\{y(x, t)\}$$

or, using the initial conditions (4) and (5),

(6) $$\frac{d^2\mathcal{L}\{y(x, t)\}}{dx^2} - \frac{s^2}{a^2}\mathcal{L}\{y(x, t)\} = 0$$

Solving this ordinary differential equation for $\mathcal{L}\{y(x, t)\}$, we find without difficulty that

(7) $$\mathcal{L}\{y(x, t)\} = A(s)e^{-(s/a)x} + B(s)e^{(s/a)x}$$

To determine the coefficient functions $A(s)$ and $B(s)$, we observe first that if $y(x, t)$ remains finite as $x \to \infty$ [Condition 3], so must $\mathcal{L}\{y(x, t)\}$. Hence $B(s)$ must be zero. Furthermore, putting $x = 0$ in (7), after $B(s)$ is set equal to zero, we have $\mathcal{L}\{y(0, t)\} = A(s)$, and from the boundary condition (2) we have $\mathcal{L}\{y(0, t)\} = \mathcal{L}\{f(t)\}$. Therefore (7) becomes

$$\mathcal{L}\{y(x, t)\} = \mathcal{L}\{f(t)\}e^{-(s/a)x} = \mathcal{L}\{f(t)\}e^{-(x/a)s}$$

The inverse of this can be found at once by suppressing the exponential factor and using Corollary 2 of Theorem 2, Sec. 10.6. The solution to our problem is therefore

$$y(x, t) = f\left(t - \frac{x}{a}\right)u\left(t - \frac{x}{a}\right)$$

which represents a wave traveling to the right along the string with velocity a. Evidently, the effect of this wave is to give the string at a general point the same displacement that the left-hand end of the string had x/a units of time earlier (see Exercise 6, Sec. 11.4).

EXAMPLE 2 A SEMI-INFINITE STRING WITH A TRAVELING LOAD

A semi-infinite string is initially at rest in a position coinciding with the positive half of the x axis. Its left-hand end is fixed. A concentrated transverse force of magnitude F_0 moves along the string with constant velocity v, beginning at $t = 0$ at the point $x = 0$. Find the displacement $y(x, t)$ of the string at any point at any subsequent time.

In this problem, since there is an external force applied to the string, we must use the nonhomogeneous wave equation [Eq. (2), Sec. 11.2]

$$\frac{\partial^2 y}{\partial t^2} = a^2\frac{\partial^2 y}{\partial x^2} + \frac{g}{w}F(x, t)$$

To obtain $F(x, t)$, we observe that a single concentrated load F_0 acting at the point $x = vt$ corresponds to a load per unit length which is infinite at $x = vt$ and zero everywhere else. Hence, assuming that F_0 acts on the string in the negative y direction, we can write

$$F(x, t) = -F_0\delta\left(t - \frac{x}{v}\right)$$

where $\delta(t - x/v)$ is the unit impulse, or δ function, discussed in Secs. 9.5 and 10.11. Our problem, therefore, is to solve the equation

$$\frac{\partial^2 y}{\partial t^2} = a^2\frac{\partial^2 y}{\partial x^2} - \frac{g}{w}F_0\delta\left(t - \frac{x}{v}\right) \tag{8}$$

subject to the boundary conditions

$$y(0, t) = 0 \tag{9}$$

$$y(x, t) \text{ bounded as } x \to \infty \tag{10}$$

and the initial conditions

$$y(x, 0) = 0 \tag{11}$$

$$\left.\frac{\partial y}{\partial t}\right|_{x,0} = 0 \tag{12}$$

If we take Laplace transforms in Eq. (8) with respect to t and use the initial conditions (11) and (12), we obtain, just as in Example 1,

$$s^2\mathcal{L}\{y(x, t)\} = a^2\frac{d^2}{dx^2}\mathcal{L}\{y(x, t)\} - \frac{g}{w}F_0e^{-(x/v)s}$$

or

$$\frac{d^2}{dx^2}\mathcal{L}\{y(x, t)\} - \frac{s^2}{a^2}\mathcal{L}\{y(x, t)\} = \frac{gF_0}{a^2w}e^{-(s/v)x} \tag{13}$$

The complete solution of this equation by the methods of Chap. 2 presents no difficulty, and we find for the complete solution

$$\mathcal{L}\{y(x, t)\} = A(s)e^{-(s/a)x} + B(s)e^{(s/a)x} + \begin{cases} \dfrac{gv^2F_0}{w(a^2 - v^2)s^2}e^{-(s/v)x} & v \neq a \\ -\dfrac{gF_0}{2was}xe^{-(s/v)x} & v = a \end{cases} \tag{14}$$

In each case we must have $B(s) = 0$ in order that $\mathcal{L}\{y(x, t)\}$ should remain finite as $x \to \infty$. To determine $A(s)$ we have, from the boundary condition (9), the information that when $x = 0$,

$$\mathcal{L}\{y(x, t)\} \equiv \mathcal{L}\{y(0, t)\} = 0$$

Hence, substituting into Eq. (14), we obtain

$$A(s) = \begin{cases} -\dfrac{gv^2F_0}{w(a^2 - v^2)s^2} & v \neq a \\ 0 & v = a \end{cases}$$

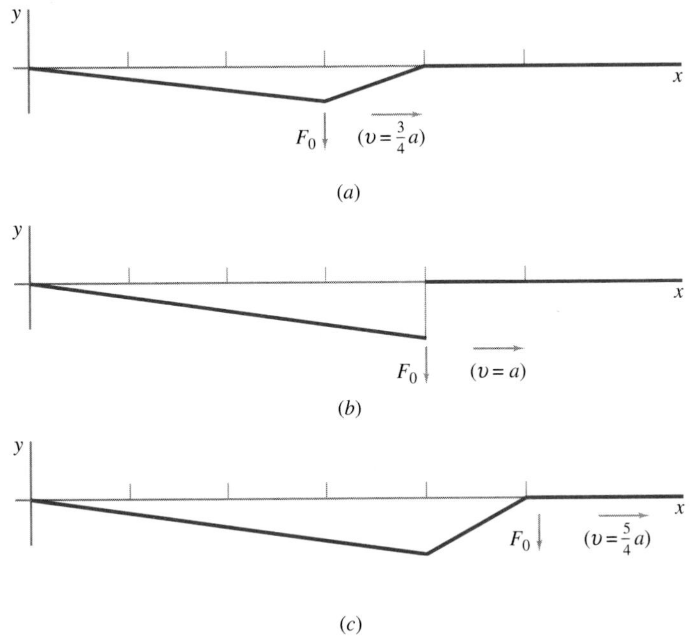

FIGURE 11.25
The displacement of a semi-infinite string produced by a concentrated force moving along the string with a velocity of (a) $\frac{3}{4}$, (b) 1, and (c) $\frac{5}{4}$ times the propagation velocity for the string.

and therefore

$$\mathcal{L}\{y(x,t)\} = \begin{cases} \dfrac{gv^2F_0}{w(a^2 - v^2)s^2}(e^{-(x/v)s} - e^{-(x/a)s}) & v \neq a \\ -\dfrac{gF_0}{2was}xe^{-(x/a)s} & v = a \end{cases}$$

Taking inverses, we have finally

$$y(x,t) = \frac{gv^2F_0}{w(a^2 - v^2)}\left[\left(t - \frac{x}{v}\right)u\left(t - \frac{x}{v}\right) - \left(t - \frac{x}{a}\right)u\left(t - \frac{x}{a}\right)\right] \qquad v \neq a \tag{15}$$

and

$$y(x,t) = -\frac{gF_0}{2wa}xu\left(t - \frac{x}{a}\right) \qquad v = a \tag{16}$$

Plots of (15) in the subsonic case $v = \frac{3}{4}a$ and the supersonic case $v = \frac{5}{4}a$, and of the transonic case $v = a$ described by (16) are shown in Fig. 11.25 for a typical time t. The discontinuity in $y(x,t)$ in the transonic case when the disturbance travels with exactly the propagation velocity a is interesting.

EXAMPLE 3

THE POTENTIAL ALONG A SEMI-INFINITE CABLE

A semi-infinite cable of negligible leakage and inductance is initially dead. At $t = 0$ an arbitrary signal voltage $E(t)$ is suddenly applied at the sending end. Find the potential $e(x,t)$ at any point on the cable at any subsequent time.

In this problem we have to solve the telegraph equation [Eq. (21*a*), Sec. 11.2],

$$\frac{\partial^2 e}{\partial x^2} = a^2\frac{\partial e}{\partial t}\dagger \qquad a^2 = RC \tag{17}$$

†This is identical with the one-dimensional heat equation, and so all our conclusions apply equally well to the flow of heat in a slender, insulated, semi-infinite rod whose left end is maintained at an arbitrary time-dependent temperature.

subject to the boundary conditions

(18) $$e(0, t) = E(t)$$

(19) $$e(x, t) \text{ bounded as } x \to \infty$$

and the initial condition

(20) $$e(x, 0) = 0$$

Taking Laplace transforms in (17) with respect to t and using the initial condition (20), we obtain

$$\frac{d^2}{dx^2}\mathcal{L}\{e(x, t)\} = a^2 s \mathcal{L}\{e(x, t)\}$$

as the ordinary differential equation satisfied by the transform of the potential. Solving this for $\mathcal{L}\{e(x, t)\}$, we find without difficulty that

(21) $$\mathcal{L}\{e(x, t)\} = A(s) \exp(-a\sqrt{s}x) + B(s) \exp(a\sqrt{s}x)$$

Since $e(x, t)$ and hence $\mathcal{L}\{e(x, t)\}$ are to remain finite as $x \to \infty$, it is necessary that $B(s) = 0$. To determine $A(s)$ we observe that when $x = 0$,

$$\mathcal{L}\{e(x, t)\} = \mathcal{L}\{e(0, t)\} = \mathcal{L}\{E(t)\}$$

Hence, substituting into Eq. (21), we find

$$A(s) = \mathcal{L}\{E(t)\}$$

and

(22) $$\mathcal{L}\{e(x, t)\} = \mathcal{L}\{E(t)\} \exp(-ax\sqrt{s})$$

To determine $e(x, t)$ it will be necessary to use the convolution theorem, but before this can be done we must know the inverse of $\exp(-ax\sqrt{s})$. Up to this point in our work we have not encountered a function of t having this function of s for its transform. However, it can be shown (see Exercises 21 and 22) that

$$\mathcal{L}\left\{\frac{b\epsilon^{-b^2/4t}}{2\sqrt{\pi}t^{3/2}}\right\}^{\dagger} = \exp(-b\sqrt{s})$$

Hence, taking $b = ax$ and setting up the convolution integral, we obtain from (22),

(23) $$e(x, t) = \frac{ax}{2\sqrt{\pi}} \int_0^t E(t - \lambda) \frac{\epsilon^{-a^2x^2/4\lambda}}{\lambda^{3/2}} \, d\lambda$$

In particular, if $E(t)$ is a unit step voltage, we have, since $u(t - \lambda) = 1$ for $\lambda < t$ and $u(t - \lambda) = 0$ for $\lambda > t$,

$$e(x, t) = \frac{ax}{2\sqrt{\pi}} \int_0^t \frac{\epsilon^{-a^2x^2/4\lambda}}{\lambda^{3/2}} \, d\lambda$$

---

†Here, as usual, to avoid confusion with the voltage, the symbol ϵ is used in place of e to denote the base of natural logarithms.

If we now change the variable of integration from λ to z by the substitution $a^2x^2/4\lambda = z^2$, then $\lambda = a^2x^2/4z^2$, $d\lambda = -a^2x^2/2z^3\,dz$, and the last integral becomes

$$e(x,t) = \frac{ax}{2\sqrt{\pi}}\int_{\infty}^{ax/2\sqrt{t}} \epsilon^{-z^2}\frac{8z^3}{a^3x^3}\left(-\frac{a^2x^2}{2z^3}\,dz\right)$$

$$= \frac{2}{\sqrt{\pi}}\int_{ax/2\sqrt{t}}^{\infty} \epsilon^{-z^2}\,dz$$

(24) $$= \frac{2}{\sqrt{\pi}}\int_0^{\infty} \epsilon^{-z^2}\,dz - \frac{2}{\sqrt{\pi}}\int_0^{ax/2\sqrt{t}} \epsilon^{-z^2}\,dz$$

Under the substitution $z^2 = v$, the first integral becomes

$$\frac{1}{\sqrt{\pi}}\int_0^{\infty} \epsilon^{-v} v^{(1/2)-1}\,dv = \frac{1}{\sqrt{\pi}}\Gamma(\tfrac{1}{2}) = 1$$

since $\Gamma(\frac{1}{2}) = \sqrt{\pi}$. Hence Eq. (24) can be written

$$e(x,t) = 1 - \frac{2}{\sqrt{\pi}}\int_0^{ax/2\sqrt{t}} \epsilon^{-z^2}\,dz$$

(25) $$= 1 - \operatorname{erf}\frac{ax}{2\sqrt{t}}$$

where

(26) $$\operatorname{erf}\theta = \frac{2}{\sqrt{\pi}}\int_0^{\theta} \epsilon^{-z^2}\,dz$$

This is the so-called **error function,** a tabulated function which can be found in most handbooks of mathematical tables.†

EXERCISES

1. Using Laplace transform methods, determine the motion of a uniform string of length l if its initial displacement is $y(x, 0) = \sin(m\pi x/l)$ and its initial velocity is $\dot{y}(x, 0) = 0$.
2. Work Exercise 1 if the initial displacement of the string is $y(x, 0) = 0$ given that its initial velocity is $\dot{y}(x, 0) = \sin(n\pi x/l)$.
3. Can the results of Exercises 1 and 2 be used to obtain the motion produced in the string by arbitrary initial conditions? How?
4. Using Laplace transform methods, determine the response of a uniform string of length l to a distributed force equal to $f(x, t) = (w/g)\sin(n\pi x/l)\sin\omega t$ if the string is initially at rest in its equilibrium position. Assume $\omega \neq \omega_n = n\pi a/l$.
5. Can the result of Exercise 4 be used to determine the

†Actually, most handbooks list not the error function as here defined and used in physics and engineering but the so-called **probability integral** of mathematical statistics:

$$\Phi(\theta) = \frac{1}{\sqrt{2\pi}}\int_{-\infty}^{0} \epsilon^{-w^2/2}\,dw$$

If the substitution $z = w/\sqrt{2}$ is made in the error function (26), it becomes

$$\frac{2}{\sqrt{2\pi}}\int_0^{\sqrt{2}\theta} \epsilon^{-w^2/2}\,dw$$

and we obtain the relation

$$\operatorname{erf}\theta = 2\Phi(\sqrt{2}\theta) - 1$$

response of the string to a distributed force $f(x, t) = g(x) \sin \omega t$, where $g(x)$ is defined arbitrarily on the interval $0 < x < l$? How? Can the result of Exercise 4 be used to determine the response of the string to a distributed force $f(x, t) = h(t) \sin (n\pi x/l)$, where $h(t)$ is an arbitrary periodic function of t? How?

6. Work Exercise 4 if ω is one of the natural frequencies of the string, $\omega_n = n\pi a/l$.

7. Using Laplace transform methods, determine the torsional motion of a uniform shaft of length l, fixed at $x = 0$ and free at $x = l$, if it is given the initial displacement $\theta(x, 0) = \sin [(2n + 1)\pi x/2l]$ and released from rest.

8. Work Exercise 7 if the shaft is given the initial angular velocity $\dot{\theta}(x, 0) = \sin [(2n + 1)\pi x/2l]$ while in its equilibrium position.

9. Using Laplace transform methods, determine the torsional motion of a uniform shaft of length l, free at both $x = 0$ and $x = l$, if it is given the initial displacement $\theta(x, 0) = \cos (n\pi x/l)$ and released from rest.

10. Work Exercise 9 if the shaft is given the initial angular velocity $\dot{\theta}(x, 0) = \cos (n\pi x/l)$ while in its equilibrium position.

11. In Example 3, what is the response of the line if $E(t)$ is a unit impulse voltage? *Hint:* Recall from Sec. 10.11 the relation between the response of a system to a unit step function and to a unit impulse.

12. Using Eq. (25) and the appropriate Duhamel formula (see Sec. 10.11), obtain a formula different from Eq. (23) for the response of the line in Example 3 to an arbitrary voltage.

13. A semi-infinite string whose weight per unit length is w has its left end fixed at the origin. The infinite end is fastened to a ring which slides without friction along a vertical rod. Initially the string is at rest in a position coinciding with the positive x axis. At $t = 0$ the support that maintained the string in its horizontal position is removed and the string begins to fall freely under the influence of gravity. Determine its subsequent position as a function of x and t.

14. Work Example 2 if the force moving along the string is $-F_0 u(t - x/v)$.

15. Work Example 2 if the force moving along the string is $-F_0 u(x/v - t)$.

16. Work Example 3 if $E(t) = u(t) - u(t - k)$.

17. A shaft of uniform cross section is built-in at $x = 0$ and free at $x = l$. At $t = 0$, while the shaft is at rest in its equilibrium position, a constant torque T_0 is suddenly applied at the free end. Find the Laplace transform of the resultant angular displacement. What is the angular displacement of the free end as a function of time? *Hint:* The end condition at $x = l$ is $E_s J(\partial\theta/\partial x) = T_0$.

18. Work Exercise 17 if the torque applied at the free end is a unit impulse instead of a step function.

19. Work Exercise 17 if the constant torque T_0 is suddenly removed at $t = b$. Discuss the subsequent motion of the end of the shaft if

(a) $b = 2l/a$ **(b)** $b = 4l/a$

20. A semi-infinite string, initially at rest in a position coinciding with the positive x axis, is acted upon by a concentrated force $F_0 \sin \omega t$ applied at the point $x = b$. Find the Laplace transform of the resultant displacement of the string. What is the displacement of the string at the point $x = b$ as a function of time?

21. If

$$f(\lambda) = \int_0^\infty \frac{e^{-z}e^{-\lambda/z}}{\sqrt{z}}\, dz$$

show by means of the substitution $u = \lambda/z$ that

$$f(\lambda) = \sqrt{\lambda} \int_0^\infty \frac{e^{-u}e^{-\lambda/u}}{u^{3/2}}\, du$$

Hence, by differentiating the first expression for $f(\lambda)$, show that

$$f'(\lambda) = -\frac{f(\lambda)}{\sqrt{\lambda}}$$

Solve this differential equation, using the fact that

$$f(0) = \Gamma(\tfrac{1}{2}) = \sqrt{\pi}$$

and show that

$$f(\lambda) = \sqrt{\pi} \exp (-2\sqrt{\lambda})$$

Finally, use this result to show that

$$\mathcal{L}\left\{\frac{e^{-b^2/4t}}{(\pi t)^{1/2}}\right\} = \frac{\exp (-b\sqrt{s})}{\sqrt{s}}$$

22. Use the results of Exercise 21, together with Theorem 2, Sec. 10.7, to show that

$$\mathcal{L}\left\{\frac{be^{-b^2/4t}}{2\sqrt{\pi}t^{3/2}}\right\} = \exp (-b\sqrt{s})$$

11.9 THE NUMERICAL SOLUTION OF PARTIAL DIFFERENTIAL EQUATIONS

The work of this chapter has made it abundantly clear that there are many interesting and important problems involving partial differential equations for which exact solutions (usually in the form of infinite series) can be found. It is equally true, however, that applied scientists are encountering more and more problems in partial differential equations that cannot be solved exactly; and it is the purpose of this section to present an introductory account of the numerical methods by which approximate solutions to such equations can be obtained. In doing this, we will find that solution procedures differ somewhat according as the equation to be solved is elliptic, parabolic, or hyper-

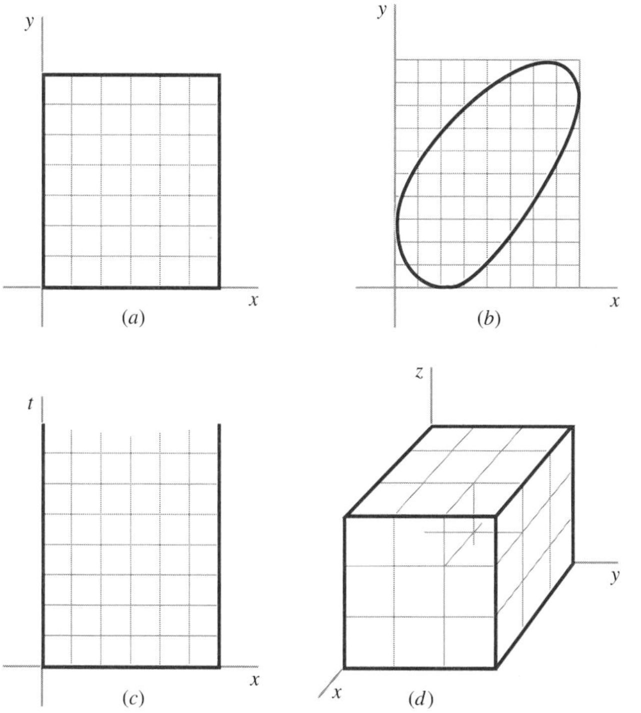

FIGURE 11.26
Rectangular lattices superimposed on various regions.

bolic (see Sec. 11.4); and we shall discuss Laplace's equation, the one-dimensional heat equation, and the one-dimensional wave equation as respective prototypes.

As in the case of ordinary differential equations which must be solved numerically (Sec. 5.6), the objective here is to obtain approximate values for the solution on a suitable set of points. Usually this means that one seeks the values of the solution at the points of a rectangular grid, or lattice,† extending over some portion of the domain of the solution. Figure 11.26 illustrates several possibilities. Figure 11.26*a* shows a rectangular grid covering a rectangular region in the *xy* plane in which we might be attempting to solve Laplace's equation. Figure 11.26*b* shows a rectangular grid superposed on a nonrectangular region in which we might be attempting to solve Laplace's equation. Figure 11.26*c* shows a rectangular grid superposed on an infinite rectangular region in the *xt* plane in which we might be attempting to solve either the one-dimensional heat equation or the one-dimensional wave equation. Finally, Fig. 11.26*d* suggests a rectangular grid superposed on a rectangular region in space in which we might be attempting to solve a problem involving Laplace's equation in three dimensions.

In Fig. 11.26*a*, *c*, and *d*, the outermost points of the lattice all fall on the boundary of the region, where the values of the solution are given as data of the problem. This is not the case for the irregular region shown in Fig. 11.26*b*, however, and the general formulas we shall soon develop must be modified for the lattice points in the region which are adjacent to, but not actually on, the boundary.

The fundamental idea on which the numerical solution of partial differential equations is based is this: each of the partial derivatives appearing in the equation is replaced by a finite-difference approximation. When these differences are evaluated at each of the mesh points, the result is a set of simultaneous equations which can be solved either directly or by various iterative procedures. Of course, if the number of lattice points is even moderately large, hand solution becomes prohibitively time-consuming, and high-speed computers must be used.

†The terms *net* and *mesh* are also used to describe the set of points at which the values of the solution are sought.

Specifically, in a plane grid, if the coordinates of the mesh points (named neutrally for the moment) are $p_i = p_0 + ih$ and $q_j = q_0 + jk$, then from the usual difference quotient approximation to the first derivative we have

$$\text{(1}a\text{)} \qquad \left.\frac{\partial f}{\partial p}\right|_{p_i,q_j} = \frac{f(p_{i+1}, q_j) - f(p_i, q_j)}{h} = \frac{f_{i+1,j} - f_{i,j}}{h}$$

and, similarly,

$$\text{(1}b\text{)} \qquad \left.\frac{\partial f}{\partial q}\right|_{p_i,q_j} = \frac{f_{i,j+1} - f_{i,j}}{k}$$

Furthermore, if we differentiate Stirling's interpolation formula [Eq. (18), Sec. 5.2] twice, set $r = 1$, neglect all differences beyond the second, and identify p_i with x_0, we have

$$\text{(2}a\text{)} \qquad \left.\frac{\partial^2 f}{\partial p^2}\right|_{p_i,q_j} = \frac{f(p_{i+1}, q_j) - 2f(p_i, q_j) + f(p_{i-1}, q_j)}{h^2} = \frac{f_{i+1,j} - 2f_{i,j} + f_{i-1,j}}{h^2}$$

and, similarly,

$$\text{(2}b\text{)} \qquad \left.\frac{\partial^2 f}{\partial q^2}\right|_{p_i,q_i} = \frac{f_{i,j+1} - 2f_{i,j} - f_{i,j-1}}{k^2}$$

We now turn our attention to the application of these general ideas to the solution of elliptic, parabolic, and hyperbolic equations.

CASE 1 **Elliptic Equations (Laplace's Equation in Two Dimensions)** Using Eqs. (2) to approximate each of the partial derivatives in the two-dimensional form of Laplace's equation, namely,

$$\frac{\partial^2 f}{\partial x^2} + \frac{\partial^2 f}{\partial y^2} = 0$$

we obtain, as a difference equation approximating the actual equation,

$$\frac{f_{i+1,j} - 2f_{i,j} + f_{i-1,j}}{h^2} + \frac{f_{i,j+1} - 2f_{i,j} + f_{i,j-1}}{k^2} = 0$$

or, making the natural and convenient assumption that $h = k$ and solving for $f_{i,j}$,

$$\text{(3)} \qquad f_{i,j} = \frac{f_{i+1,j} + f_{i,j+1} + f_{i-1,j} + f_{i,j-1}}{4}$$

This asserts that *the value of f at any mesh point is equal to the average of the values of f at the four adjacent mesh points,* as shown in Fig. 11.27. (Compare this with the result of Exercise 16, Sec. 17.6.) The configuration shown in Fig. 11.27 is often called a **star.**

If Eq. (3) is evaluated at each of the mesh points that are not boundary points, where the value of the solution f is initially given, the result is a system of simultaneous linear equations in the unknown functional values f_{ij}. The number of equations is of course just equal to the number of mesh points at which the value of f is to be calculated; and (at least for rectangular regions) it can

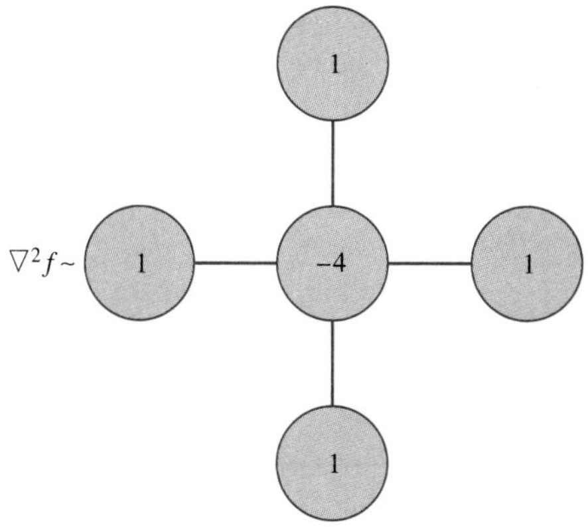

FIGURE 11.27
The finite difference "star" that approximates $\partial^2 f/\partial x^2 + \partial^2 f/\partial y^2$.

be shown that this system of equations always has a unique nontrivial solution.† In important problems, where high accuracy is required, the number of mesh points, hence the number of equations, may be anywhere from several hundred to several thousand. Each equation is very simple, however, for none can contain more than five of the unknown functional values. Large systems of linear equations of this simple structure have been extensively studied, and efficient computer programs are available for their solution.

To illustrate the formulation and solution of such a system, let us attempt to approximate the steady-state temperature distribution in the square region shown in Fig. 11.28, using the grid obtained by dividing each edge into four equal parts. The unknowns in this problem are the temperatures at the nine points of the grid which are not boundary points and at which the temperature is not determined by the given boundary conditions.

At the outset, we note that from symmetry $f_{11} = f_{31}$, $f_{12} = f_{32}$, and $f_{13} = f_{33}$, so that our problem actually involves only six equations in the six unknowns, $f_{11}, f_{12}, f_{13}, f_{21}, f_{22}$, and f_{23}.

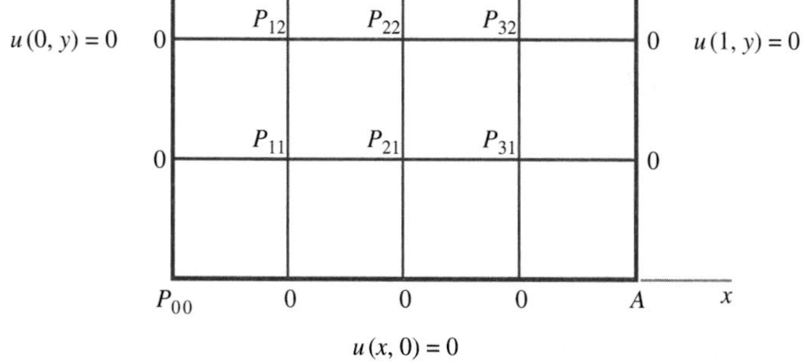

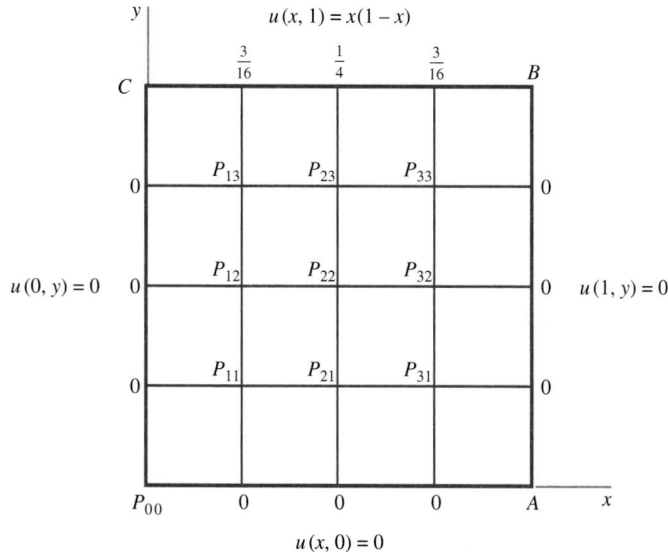

FIGURE 11.28
A typical lattice used in the approximate solution of Laplace's equation in the unit square.

†See, for instance, E. Isaacson and H. B. Keller, *Analysis of Numerical Methods,* pp. 447–448, Wiley, New York, 1966.

Applying formula (3) at each of the six mesh points P_{11}, P_{12}, P_{13}, P_{21}, P_{22}, P_{23} and taking into account the symmetries we have just noted and the known values of f on the boundary, we have at P_{11}

$$4f_{11} - f_{01} - f_{10} - f_{21} - f_{12} = 0$$

or, noting that by hypothesis $f_{01} = f_{10} = 0$,

$$4f_{11} - f_{21} - f_{12} = 0 \tag{4}$$

Similarly, at P_{12}, P_{13}, P_{21}, P_{22}, P_{23} we have, respectively,

$$4f_{12} - f_{11} - f_{22} - f_{13} = 0 \tag{5}$$
$$4f_{13} - f_{12} - f_{23} = \tfrac{3}{16} \tag{6}$$
$$4f_{21} - 2f_{11} - f_{22} = 0 \tag{7}$$
$$4f_{22} - f_{21} - 2f_{12} - f_{23} = 0 \tag{8}$$
$$4f_{23} - f_{22} - 2f_{13} = \tfrac{1}{4} \tag{9}$$

Using Eqs. (4), (5), and (6) to eliminate f_{21}, f_{22}, and f_{23} from Eqs. (7), (8), and (9), we obtain the system

$$\begin{aligned} 15f_{11} - 8f_{12} + f_{13} &= 0 \\ -8f_{11} + 16f_{12} - 8f_{13} &= -\tfrac{3}{16} \\ f_{11} - 8f_{12} + 15f_{13} &= 1 \end{aligned}$$

from which we find at once that

$$f_{11} \doteq 0.0151 \doteq f_{31} \qquad f_{12} \doteq 0.0391 \doteq f_{32} \qquad f_{13} \doteq 0.0865 \doteq f_{33}$$

and from these

$$f_{21} \doteq 0.0212 \qquad f_{22} \doteq 0.0547 \qquad f_{23} \doteq 0.1194$$

The correct values, as determined from the series solution obtained by the method of separation of variables, are

$$\begin{gathered} f_{11} = f_{31} \doteq 0.0137 \qquad f_{12} = f_{32} \doteq 0.0364 \qquad f_{13} = f_{33} \doteq 0.0833 \\ f_{21} \doteq 0.0194 \qquad f_{22} \doteq 0.0513 \qquad f_{23} \doteq 0.1159 \end{gathered}$$

Problems in which the normal derivative, rather than the function itself, is specified along a part, or all, of the boundary can also be handled by the method we have been discussing. Suppose, for instance, that along the edge AB of the region shown in Fig. 11.28 the value of the normal derivative is required to be λ times the value of the function. Then approximating the derivative by its difference quotient, the boundary condition

$$\frac{\partial f}{\partial n} = \lambda f$$

that is,

$$\frac{\partial f}{\partial x} = \lambda f$$

becomes

$$\frac{f_{4j} - f_{3j}}{h} = \lambda f_{4j} \qquad \text{or} \qquad f_{4j} = \frac{f_{3j}}{1 - \lambda h}$$

Now, applying Eq. (3) at P_{33}, for example, we have

$$4f_{33} - f_{43} - f_{34} - f_{23} - f_{32} = 0$$

$$4f_{33} - \frac{f_{33}}{1-\lambda h} - \frac{3}{16} - f_{23} - f_{32} = 0$$

or, finally,

$$\frac{3-4\lambda h}{1-\lambda h} f_{33} - f_{23} - f_{32} = \frac{3}{16}$$

Along an insulated boundary λ is of course equal to zero and, in particular, the last equation becomes simply

$$3f_{33} - f_{23} - f_{32} = \frac{3}{16}$$

In any event, the system of equations thus obtained can be solved just as in the preceding case.

To obtain the proper finite-difference relation at a lattice point adjacent to an irregular boundary, as suggested in Fig. 11.29, it is convenient to use a divided difference approximation for the derivatives. Thus suppose that in Fig. 11.29 the boundary points A and B are at the distances $\theta_A h$ and $\theta_B h$ from E, where θ_A and θ_B are each less than 1. Then, using second divided differences in the x and y directions as approximations to $\partial^2 f/\partial x^2$ and $\partial^2 f/\partial y^2$, we have (see Exercise 13, Sec. 5.2)

$$\left.\frac{\partial^2 f}{\partial x^2}\right|_E = 2\left[\frac{f_D}{h(h+\theta_B h)} + \frac{f_E}{-h(\theta_B h)} + \frac{f_B}{\theta_B h(\theta_B h + h)}\right]$$

and

$$\left.\frac{\partial^2 f}{\partial y^2}\right|_E = 2\left[\frac{f_C}{h(h+\theta_A h)} + \frac{f_E}{-h(\theta_A h)} + \frac{f_A}{\theta_A h(\theta_A h + h)}\right]$$

or

$$\nabla^2 f|_E = \frac{2}{h^2}\left[\frac{f_A}{\theta_A(1+\theta_A)} + \frac{f_B}{\theta_B(1+\theta_B)} + \frac{f_C}{1+\theta_A} + \frac{f_D}{1+\theta_B} - \frac{\theta_A+\theta_B}{\theta_A\theta_B} f_E\right] = 0 \tag{10}$$

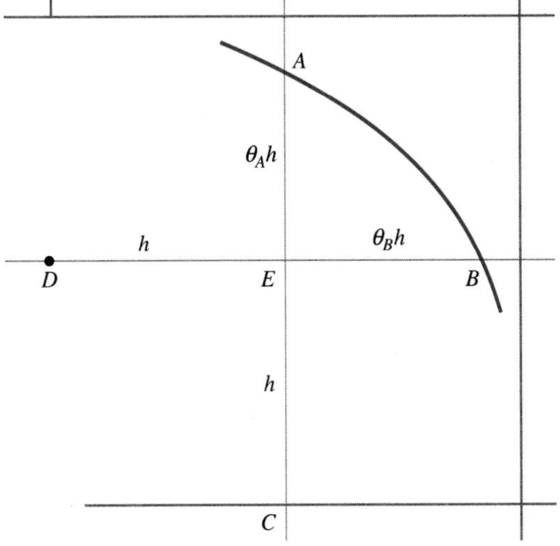

FIGURE 11.29
Geometry of a lattice point adjacent to an irregular boundary.

This is the equation which is to be used at those points of the lattice whose immediate neighbors in the lattice fall outside the boundary of the region.

There is another way in which the finite-difference approximation to the Laplacian can be used to determine the value of the solution at the points of the lattice. It is a simple iterative method which proceeds as follows:† We first recall that the finite-difference approximation to the Laplacian [Eq. (3)] expresses the value of the solution at any mesh point as the average of the values at the four adjacent points. Thus, after an initial estimate for the value of the solution at each mesh point has been made, they can be corrected and improved by systematically moving through the lattice and replacing each value according to Eq. (3). In doing this, each value as soon as it is corrected should be used in all subsequent calculations. Of course, in regions with irregular boundaries, Eq. (10) must be used to correct the values of the solution at those mesh points whose neighbors lie outside the boundary.

As an illustration of this method, let us reconsider the problem we have just worked. Beginning with the estimates shown in Fig. 11.30*a*, we have for the first refinement of f_{13} the value

$$\frac{0.1875 + 0.0000 + 0.1200 + 0.0600}{4} = 0.0919$$

Continuing through the lattice as indicated,‡ using the corrected values as soon as they become available (but taking no advantage of the symmetry of the problem), we obtain the values shown in Fig. 11.30*b*. Values bearing the subscript 1 were obtained by a single iteration; values bearing the subscript 5 were obtained after five iterations.

CASE 2 **Parabolic Equations (The One-Dimensional Heat Equation)** For the one-dimensional heat equation

$$\frac{\partial^2 f}{\partial x^2} = a^2 \frac{\partial f}{\partial t}$$

FIGURE 11.30
Data from an iterative solution of Laplace's equation.

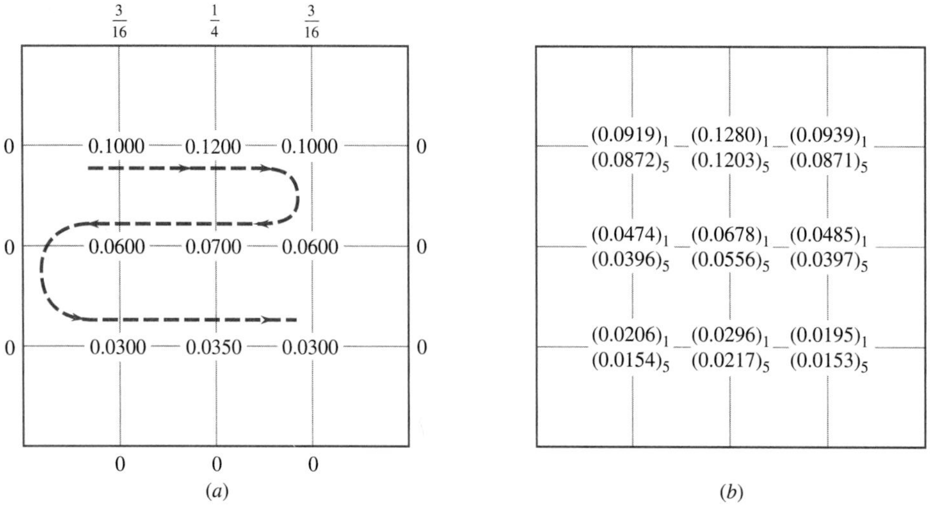

†This procedure is usually known as the **Liebmann method,** after the German scientist who first proposed it in a paper in 1918.

‡Any other path would serve just as well.

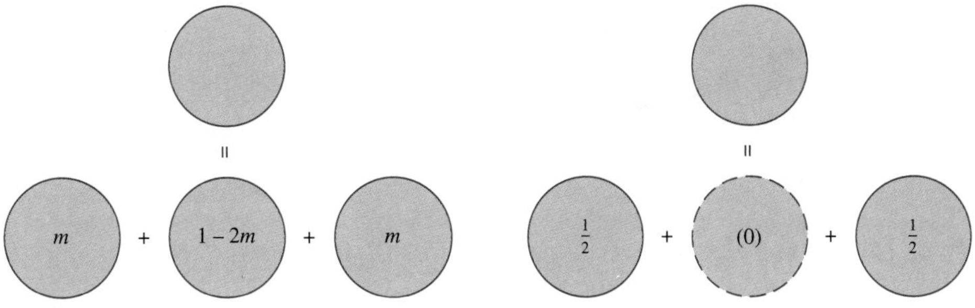

FIGURE 11.31
Iteration formulas for the one-dimensional heat equation.

the region of the xt plane over which a solution is sought is always infinite because of the infinite increase of time. Thus a typical lattice would be that shown in Fig. 11.26c. As a finite-difference approximation to the heat equation we have, using Eqs. (1b) and (2a).

$$\frac{1}{h^2}(f_{i+1,j} - 2f_{i,j} + f_{i-1,j}) = \frac{a^2}{k}(f_{i,j+1} - f_{i,j})$$

or, setting $m = k/a^2h^2$ and solving for $f_{i,j+1}$,

$$f_{i,j+1} = mf_{i+1,j} + (1 - 2m)f_{i,j} + mf_{i-1,j} \tag{11}$$

Clearly, it would be convenient to choose h and k so that the value of m is $\frac{1}{2}$.

The values of f on the boundary are of course provided by the data of the problem. Thus the given initial condition $f(x, 0)$ provides the values of $f_{00}, f_{10}, f_{20}, \ldots$. Similarly, end conditions of the form

$$f(0, t) = g_1(t) \qquad f(l, t) = g_2(t)$$

where g_1 and g_2 are usually, though not necessarily, constant, furnish the values of $f_{01}, f_{02}, f_{03}, \ldots$ and $f_{l1}, f_{l2}, f_{l3}, \ldots$. Insulated end conditions can of course be handled as we outlined before in our discussion of Laplace's equation.

Once the values of f at the lattice points on the boundary have been determined from the conditions of the problem, the determination of the solution over the rest of the lattice proceeds in a straightforward way, using the extrapolation pattern provided by Eq. (11) and shown in Fig. 11.31. First, the values of $f_{11}, f_{21}, \ldots, f_{l-1,1}$ are calculated from the known values of $f_{00}, f_{10}, f_{20}, \ldots, f_{l0}$. Then using these new values and the boundary values f_{01} and f_{l1}, the solution is "marched" forward by calculating the values of f at the lattice points in the third row, and so on as far as desired.

CASE 3 **Hyperbolic Equations (The One-Dimensional Wave Equation)** The solution procedure for the one-dimensional wave equation

$$a^2\frac{\partial^2 f}{\partial x^2} = \frac{\partial^2 f}{\partial t^2}$$

is very much like that for the one-dimensional heat equation. Because of the increase of time, the region of the xt plane over which a solution is sought is always infinite. Hence a solution must be found by "marching" forward a step at a time from known initial values. The finite-difference

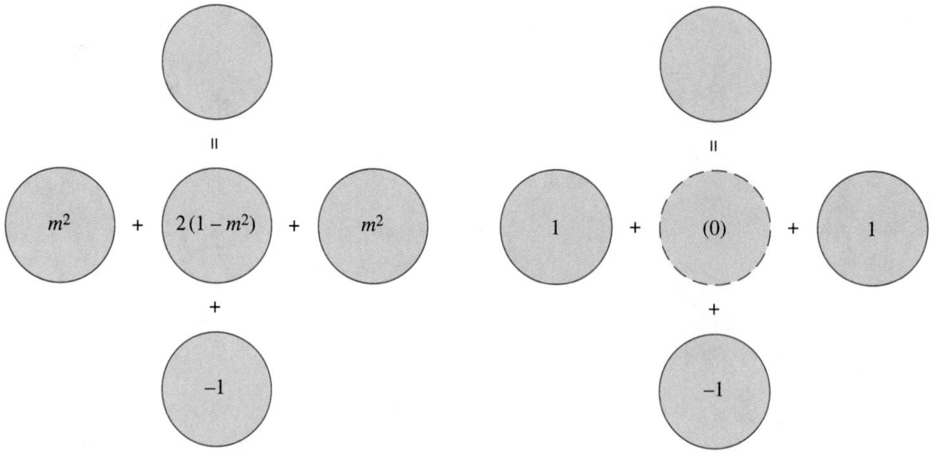

FIGURE 11.32
Iterative formulas for the one-dimensional wave equation.

approximation that provides the necessary extrapolation formula comes immediately from Eqs. (2*a*) and (2*b*) and is

$$a^2\left[\frac{1}{h^2}(f_{i+1,j} - 2f_{i,j} + f_{i-1,j})\right] = \frac{1}{k^2}(f_{i,j+1} - 2f_{i,j} + f_{i,j-1})$$

or, setting $m^2 = a^2k^2/h^2$ and solving for $f_{i,j+1}$,

(12) $$f_{i,j+1} = 2(1 - m^2)f_{i,j} + m^2(f_{i+1,j} + f_{i-1,j}) - f_{i,j-1}$$

If the dimensions of the lattice, h and k, are chosen so that $m = 1$, Eq. (12) assumes the especially convenient form

(13) $$f_{i,j+1} = f_{i+1,j} + f_{i-1,j} - f_{i,j-1}$$

Equations (12) and (13) are shown schematically in Fig. 11.32.

In a particular problem the wave equation would be accompanied by end conditions of the form

$$f(0, t) = g_1(t) \qquad \text{and} \qquad f(l, t) = g_2(t)$$

where usually, though not necessarily, g_1 and g_2 would be identically zero, and by initial conditions of the form

$$f(x, 0) = \phi(x) \qquad \text{and} \qquad \left.\frac{\partial f}{\partial t}\right|_{x,0} = \theta(x)$$

The two end conditions provide the values of f at the mesh points on the left- and right-hand boundaries of the grid. The first of the two initial conditions provides the values of f on the lowest row of mesh points. The second initial condition provides the values of f at the mesh points in the second row of the lattice since the approximation

$$\left.\frac{\partial f}{\partial t}\right|_{x_i,0} = \frac{f_{i1} - f_{i0}}{k} = \theta(x_i)$$

becomes

$$f_{i1} = f_{i0} + k\theta(x_i)$$

With these values known, either formula (12) or (13) allows us to calculate the values of f at the lattice points in the third row, and thus the solution can be ''marched'' forward as far as desired.

There are of course many questions concerning refinements in the solution procedures we have outlined, as well as their accuracy, which we have not discussed. These, however, we must leave to more advanced texts.†

EXERCISES

1. Work the example considered under Case 1, using the Liebmann method but taking advantage of the symmetry of the problem.
2. Work the example considered under Case 1, taking as initial conditions the values obtained by solving Eqs. (4)–(9). Explain.
3. After the Liebmann method had been applied to the lattice shown in Fig. 11.28, it was desired to obtain a more accurate approximation by repeating the process with the lattice interval reduced to half its former size. Explain how reasonable estimates can be found for the starting values at the new lattice points.
4. Work the example considered under Case 1 if the upper edge is insulated, the lower edge is maintained at the temperature 100, and the two vertical edges are maintained at the temperature 0.
5. Taking advantage of all symmetries, at how many points must equations be set up, or independent calculations be made, in the lattices in the following figures:
   (a) Fig. 11.33*a* **(b)** Fig. 11.33*b*
   (c) Fig. 11.33*c* **(d)** Fig. 11.33*d*
6. Set up the minimum number of equations required for the lattices in **(a)** Fig. 11.33*b* and **(b)** Fig. 11.33*c*.
7. Derive Eq. (10).

Find the steady-state temperature distribution in each of the following regions using the indicated lattice.

8. The region shown in Fig. 11.33*a*
9. The region shown in Fig. 11.33*b*
10. The region shown in Fig. 11.33*c*
11. The region shown in Fig. 11.33*d*

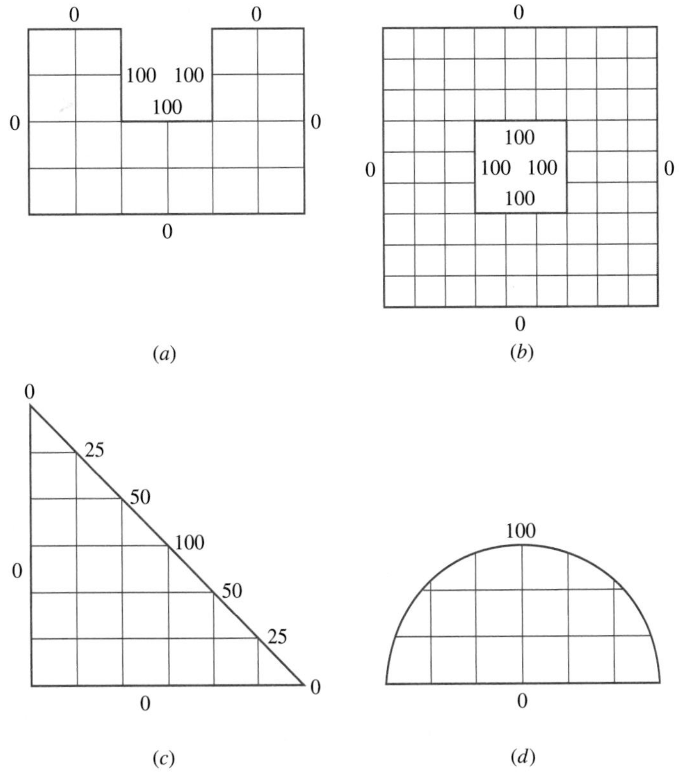

FIGURE 11.33

†See, for example, E. Isaacson and H. B. Keller, *Analysis of Numerical Methods,* Wiley, New York, 1966; K. S. Kunz, *Numerical Analysis,* McGraw-Hill, New York, 1957; S. H. Crandall, *Engineering Analysis,* McGraw-Hill, New York, 1956; J. D. Hoffman, *Numerical Methods for Engineers and Scientists,* McGraw-Hill, New York, 1992.

12. Derive a finite difference approximation to the three-dimensional Laplacian $u_{xx} + u_{yy} + u_{zz}$.

13. Using the indicated lattice, find the function that satisfies the nonhomogeneous equation $u_{xx} + u_{yy} = -2$ and vanishes on the boundary of the region shown in Fig. 11.28. (This problem is of importance in studying the torsion of prismatic bars.)

14. A thin rod is initially at the temperature $u(x, 0) = 0$. Find its temperature as a function of x and t if at $t = 0$ the left-hand end of the rod is suddenly raised to the temperature 100 and maintained thereafter at that temperature while the right-hand end is maintained at the temperature 0. Assume a lattice based on a division of the rod into 10 equal parts and

(a) $m = \frac{1}{2}$ **(b)** $m = 1$

15. A thin rod of unit length is initially at the temperature distribution $u(x, 0) = 100(1 - 2x)^2$. If each end of the rod is insulated, find the temperature in the rod as a function of x and t. Assume a lattice based on a division of the rod into 10 equal segments and $m = \frac{1}{2}$.

16. A thin rod is initially at the temperature $u(x, 0) = 0$. At $t = 0$, while the right-hand end is maintained at the temperature 0, the left-hand end is subjected to the periodic temperature condition $u(0, t) = \sin(\pi t/12)$. Assuming a lattice based on a division of the rod into 10 equal parts, and time intervals $\Delta t = 1$, find the temperature in the rod as a function of x and t. Assume $m = \frac{1}{2}$.

17. A string of length l is displaced into the initial shape $y = \sin(\pi x/l)$ and released from rest in that position. Assuming a lattice based on a division of the string into 12 equal segments, find the deflection of the string as a function of x and t. Assume $m = 1$.

18. Derive a finite difference approximation for

(a) The time-dependent heat equation in two dimensions

$$u_{xx} + u_{yy} = a^2 u_t$$

(b) The two-dimensional wave equation

$$z_{tt} = a^2(z_{xx} + z_{yy})$$

19. Can the approximate solution of the wave equation for a string of variable density be carried out by the methods of this section? How? Can the approximate solution of the heat equation be carried out by the methods of this section if the specific heat and thermal conductivity vary along the rod? How?

CHAPTER 12

BESSEL FUNCTIONS AND LEGENDRE POLYNOMIALS

In Secs. 2.9 and 2.10, we pointed out that when a differential equation with variable coefficients is solved through the use of infinite series, the resulting series does not usually represent any of the familiar elementary functions. However, if such an equation proves to be an important one, its series solutions are often taken as the definitions of new functions to be studied in detail and eventually tabulated. In this chapter, we investigate the solutions of two such equations, each of which arises as one of the ordinary differential equations resulting from the separation of variables in certain boundary-value problems. One of these is *Bessel's equation,* which occurs frequently in problems involving circular symmetry. The other is *Legendre's equation,* which arises in problems involving spherical symmetry.

In Secs. 12.2 and 12.3, we obtain solutions for Bessel's equation and several closely related equations. Then, after identifying a large class of differential equations whose solutions can be expressed in terms of Bessel functions (Sec. 12.4), we investigate the properties of Bessel functions and their numerous identities. Then, after verifying that under suitable conditions these functions have the property of orthogonality (Sec. 11.6), we examine a number of typical applications to heat flow. One example leads to an important Laplace transform formula, and a number of such formulas appear among the exercises.

In Sec. 12.9, we undertake a similar study of Legendre's equation. This leads us to *Legendre polynomials,* a number of their properties, including orthogonality, and some typical applications in heat flow and potential theory.

Prerequisite for this chapter: Chaps. 2 and 11, a calculus-based course in physics.

12.1 INTRODUCTION

In the last chapter, we solved a number of significant physical problems by the method of separation of variables. We were able to do this because the separation process led to ordinary differential equations with *constant* coefficients, which could then be solved by the methods of Chap. 2. This is usually the case when the underlying partial differential equation is expressed in cartesian coordinates, although even in this case, if the geometric or physical properties of the body vary from point to point, as in Eqs. (2), (4), (7), and (10), Sec. 11.2, then equations with variable coefficients will necessarily arise. On the other hand, when an equation such as the wave equation, the heat equation, or Laplace's equation is expressed in cylindrical or spherical coordinates, the separation process always leads to differential equations with variable coefficients, and these can rarely be solved in terms of familiar elementary functions. Exercises 33–35, Sec. 11.2, led to equations of this nature. When dealing with linear differential equations with variable coefficients, the usual procedure is to obtain solutions in the form of infinite series. Historically, in many instances such series have been taken as the definition of new functions which were subsequently studied in detail and eventually tabulated.

Series solutions of differential equations with variable coefficients are usually obtained by the method of Frobenius, which we studied in Sec. 2.10. In this chapter, we shall use this method to solve **Bessel's equation,**

$$x^2y'' + xy' + (\lambda^2x^2 - \nu^2)y = 0$$

which is named for the German mathematician and astronomer Friedrich Wilhelm Bessel (1784–1846), and **Legendre's equation,**

$$(1 - x^2)y'' - 2xy' + n(n + 1)y = 0$$

which is named for the French mathematician Adrien Marie Legendre (1752–1833).

Finally, after we have obtained series solutions for each of these equations, we shall take these series as the definitions of new functions, known, respectively, as **Bessel functions** and **Legendre polynomials,** and study in some detail their properties and their applications.

EXERCISES

1. Assuming $u(r, t) = R(r)T(t)$, separate variables in **(a)** the equation derived in Exercise 18, Sec. 11.2, and **(b)** the equation derived in Exercise 20, Sec. 11.2.
2. Assuming in each case that the function f is identically zero, assume the appropriate product solution and separate variables in **(a)** Eq. (2), Sec. 11.2; **(b)** Eq. (7), Sec. 11.2; **(c)** Eq. (10), Sec. 11.2.

12.2 THE SERIES SOLUTION OF BESSEL'S EQUATION

One of the most important of all variable-coefficient differential equations is

$$(1) \qquad x^2\frac{d^2y}{dx^2} + x\frac{dy}{dx} + (\lambda^2x^2 - \nu^2)y = 0$$

which is known as **Bessel's equation of order ν with a parameter λ.** This arises in a great variety of problems, including almost all applications involving partial differential equations, such as the wave equation or the heat equation, in regions possessing circular symmetry.

As a preliminary step to make the series solution of (1) a little simpler, it is convenient to eliminate the parameter λ (temporarily) by the change of independent variable defined by the substitution

$$(2) \qquad t = \lambda x$$

Since $dy/dx = \lambda\, dy/dt$ and $d^2y/dx^2 = \lambda^2\, d^2y/dt^2$, this substitution changes Eq. (1) into the equation

$$t^2 \frac{d^2y}{dt^2} + t\frac{dy}{dt} + (t^2 - \nu^2)y = 0 \tag{3}$$

which is known simply as **Bessel's equation of order** ν.

For Eq. (3) it is clear that in the notation of Sec. 2.10,

$$p(t) = \frac{1}{t} \qquad \text{and} \qquad q(t) = \frac{t^2 - \nu^2}{t^2}$$

Hence the origin is a regular singular point of the equation, and all other values of t are ordinary points. At the origin, around which we propose to obtain series solutions of (3), the indicial equation [Eq. (16), Sec. 2.10] is $r^2 - \nu^2 = 0$, and therefore, by the theory of Sec. 2.10, we are led to try a series solution of the form

$$y_\nu = \sum_{k=0}^{\infty} a_k t^{\nu+k} \tag{4}$$

corresponding to the root $r = \nu$, $\nu \geqq 0$; the root $r = -\nu$ we shall consider later.

Substituting (4) into Eq. (3), we obtain

$$t^2 \sum_{k=0}^{\infty} a_k(\nu + k)(\nu + k - 1)t^{\nu+k-2} + t \sum_{k=0}^{\infty} a_k(\nu + k)t^{\nu+k-1} + (t^2 - \nu^2) \sum_{k=0}^{\infty} a_k t^{\nu+k} = 0$$

or, bringing the coefficients into the respective sums and then combining terms,

$$\sum_{k=0}^{\infty} a_k[(\nu + k)(\nu + k - 1) + (\nu + k) - \nu^2]t^{\nu+k} + \sum_{k=0}^{\infty} a_k t^{\nu+k+2} = 0$$

or

$$\sum_{k=1}^{\infty} a_k k(2\nu + k)t^{\nu+k} + \sum_{k=0}^{\infty} a_k t^{\nu+k+2} = 0$$

If the term corresponding to $k = 1$ is detached from the first sum and if the index of summation in the second sum is changed from k to $k - 2$, the two series can be combined into one:

$$\left[a_1(2\nu + 1)t^{\nu+1} + \sum_{k=2}^{\infty} a_k k(2\nu + k)t^{\nu+k}\right] + \sum_{k=2}^{\infty} a_{k-2} t^{\nu+k}$$

$$= a_1(2\nu + 1)t^{\nu+1} + \sum_{k=2}^{\infty} [k(2\nu + k)a_k + a_{k-2}]t^{\nu+k} = 0$$

The last equation will be satisfied identically if and only if the coefficient of each power of t is zero, that is, if and only if

$$a_1(2\nu + 1) = 0 \tag{5}$$

and the a's satisfy the recurrence relation

$$k(2\nu + k)a_k + a_{k-2} = 0 \qquad \text{or} \qquad a_k = -\frac{a_{k-2}}{k(2\nu + k)} \qquad k = 2, 3, \ldots \tag{6}$$

From (5) and the restriction that $\nu \geq 0$, it is clear that $a_1 = 0$. From (6) it then follows that $a_3 = a_5 = \cdots = a_{2m+1} = \cdots = 0$. Likewise, from (6) it follows that

$$a_2 = -\frac{a_0}{2(2\nu+2)} = -\frac{a_0}{2^2 \cdot 1!(\nu+1)}$$

$$a_4 = -\frac{a_2}{4(2\nu+4)} = -\frac{a_2}{2^2 \cdot 2(\nu+2)} = \frac{a_0}{2^4 \cdot 2!(\nu+2)(\nu+1)}$$

$$a_6 = \frac{a_4}{6(2\nu+6)} = -\frac{a_4}{2^2 \cdot 3(\nu+3)} = -\frac{a_0}{2^6 \cdot 3!(\nu+3)(\nu+2)(\nu+1)}$$

and in general

$$a_{2m} = \frac{(-1)^m a_0}{2^{2m} m!(\nu+m)(\nu+m-1)\cdots(\nu+3)(\nu+2)(\nu+1)} \qquad m = 1, 2, 3, \ldots$$

Now a_{2m} is the coefficient of $t^{\nu+2m}$ in the series (4) for y_ν. Hence it would probably be convenient if a_{2m} contained the factor $2^{\nu+2m}$ in its denominator instead of just 2^{2m}. To achieve this, we multiply and divide a_{2m} by 2^ν, getting

$$a_{2m} = \frac{(-1)^m}{2^{\nu+2m} m!(\nu+m)\cdots(\nu+2)(\nu+1)}(2^\nu a_0)$$

Furthermore, the factors $(\nu+m)\cdots(\nu+2)(\nu+1)$ in the denominator of a_{2m} suggest a factorial. In fact, if ν were a positive integer, a factorial could be created by multiplying the numerator and denominator of a_{2m} by $(\nu!)$. However, since ν is not necessarily an integer, we must use not $(\nu!)$ but its generalization, $\Gamma(\nu+1)$ (see Sec. 10.5) for this purpose. Doing this, we can write

$$a_{2m} = \frac{(-1)^m}{2^{\nu+2m} m!(\nu+m)\cdots(\nu+2)(\nu+1)\Gamma(\nu+1)}[2^\nu \Gamma(\nu+1) a_0]$$

Since the gamma function satisfies the recurrence relation

$$(\nu+j)\Gamma(\nu+j) = \Gamma(\nu+j+1)$$

the factors $(\nu+1), (\nu+2), \ldots, (\nu+m)$ can be successively telescoped into the gamma function, and the expression for a_{2m} can be written

$$a_{2m} = \frac{(-1)^m}{2^{\nu+2m} m!\Gamma(\nu+m+1)}[2^\nu \Gamma(\nu+1) a_0]$$

In this formula a_0 is still arbitrary, and since we are looking only for particular solutions, it is convenient to take

$$a_0 = \frac{1}{2^\nu \Gamma(\nu+1)}$$

so that finally

$$a_{2m} = \frac{(-1)^m}{2^{\nu+2m} m!\Gamma(\nu+m+1)} \qquad m = 0, 1, 2, \ldots$$

With a_k thus determined for even values of k and $a_k = 0$ for odd values of k, substitution into (4) gives a solution y_ν for each $\nu \geq 0$. For each ν, the function y_ν is called a **Bessel function of the first kind of order** ν and is denoted by the symbol $J_\nu(t)$. Thus

$$J_\nu(t) = t^\nu\left[\frac{1}{2^\nu\Gamma(\nu+1)} - \frac{t^2}{2^{\nu+2}\Gamma(\nu+2)} + \frac{t^4}{2^{\nu+4}2!\Gamma(\nu+3)} - \cdots\right]$$

$$= \sum_{m=0}^{\infty} \frac{(-1)^m t^{\nu+2m}}{2^{\nu+2m} m!\Gamma(\nu+m+1)} \tag{7}$$

Since Bessel's equation of order ν has no finite singular points except the origin, it follows from Theorem 1, Sec. 2.10, that the series (7) converges for all $t \geq 0$. To ensure convergence to a real-valued function for *all* values of t, the factor t^ν in (7) must of course be replaced by $|t|^\nu$. Graphs of $J_0(t)$ and $J_1(t)$ are shown in Fig. 12.1. Their resemblance to the graphs of $\cos t$ and $\sin t$ is interesting. In particular, they illustrate the important fact that for each value of ν the equation $J_\nu(t) = 0$ has infinitely many real roots.

Let us now consider the series arising from the other root of the indicial equation, namely, $r = -\nu$. We could of course begin with a series analogous to (4) and determine its coefficients one by one, just as we did for $J_\nu(t)$, but there is no need to go to this trouble. In fact, since ν enters Bessel's equation only in the form of a square, it follows that the series obtained from (7) by replacing ν by $-\nu$ will satisfy Bessel's equation provided only that the gamma functions appearing in the denominators of the various terms are all defined. This is necessarily the case unless ν is an integer; hence when ν is not an integer, the function

$$J_{-\nu}(t) = \sum_{m=0}^{\infty} \frac{(-1)^m t^{-\nu+2m}}{2^{-\nu+2m} m!\Gamma(-\nu+m+1)} \tag{8}$$

is a second particular solution of Bessel's equation of order ν. Moreover, since $J_{-\nu}(t)$ contains negative powers of t while $J_\nu(t)$ does not, it is obvious that in the neighborhood of the origin $J_{-\nu}(t)$ is unbounded while $J_\nu(t)$ remains finite. Hence, when ν is not an integer, $J_\nu(t)$ and $J_{-\nu}(t)$ are two linearly independent solutions of Bessel's equation. According to Theorem 6, Sec. 2.2 a complete solution of Bessel's equation when ν is not an integer is then

$$y(t) = c_1 J_\nu(t) + c_2 J_{-\nu}(t) \tag{9}$$

From the symmetric way in which ν enters into Eq. (9), it is evident that our earlier restriction that $\nu \geq 0$ can now be removed.

For many purposes it is convenient to take the linear combination

$$Y_\nu(t) = \frac{\cos \nu\pi J_\nu(t) - J_{-\nu}(t)}{\sin \nu\pi} \qquad \nu \text{ is not an integer} \tag{10}$$

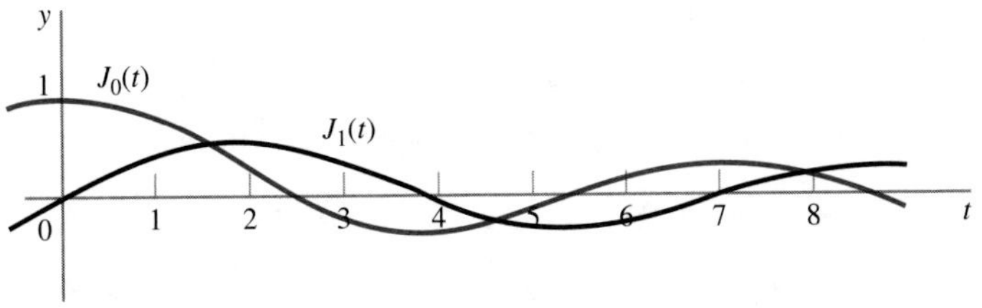

FIGURE 12.1
The Bessel functions of the first kind $J_0(t)$ and $J_1(t)$.

instead of $J_{-\nu}(t)$ as a second, independent solution of Bessel's equation. Using $Y_\nu(t)$, which is known as the **Bessel function of the second kind of order** ν, we can thus write a complete solution of Bessel's equation in the alternative form

$$y(t) = c_1 J_\nu(t) + c_2 Y_\nu(t) \qquad \nu \text{ not an integer} \tag{11}$$

In some applications it is convenient to use still another form of a complete solution of Bessel's equation based on the two particular solutions

$$\begin{aligned} H_\nu^{(1)}(t) &= J_\nu(t) + iY_\nu(t) \\ H_\nu^{(2)}(t) &= J_\nu(t) - iY_\nu(t) \end{aligned} \qquad \nu \text{ not an integer} \tag{12}$$

These are known as **Hankel functions**† or **Bessel functions of the third kind of order** ν, and in terms of them a complete solution of Eq. (3) can be written

$$y(t) = c_1 H_\nu^{(1)}(t) + c_2 H_\nu^{(2)}(t) \tag{13}$$

It is interesting to note that (9), (11), and (13) are correct expressions for a complete solution of Eq. (3) even when ν is an odd multiple of $\frac{1}{2}$ and the roots of the indicial equation $r^2 - \nu^2 = 0$ differ by an integer. In Sec. 2.10, we pointed out that when this happens, a second, independent series solution of the form (4) usually does not exist. It *may* exist, however, and this is one of the instances when it actually does.

If ν is an integer, say $\nu = n$, the situation is somewhat different. Again the roots of the indicial equation differ by an integer, namely, $2n$, and it is to be expected that a second solution of the form (4) does not exist. In fact, when we consider $J_{-n}(t)$ as the limit of $J_\nu(t)$ as ν approaches $-n$ and remember that the value of the gamma function becomes infinite when its argument approaches any nonpositive integer, then it follows that as ν approaches $-n$, the first n terms in the series (7) approach zero and the series effectively begins with the term for which $m = n$:

$$J_{-n}(t) = \sum_{m=n}^{\infty} \frac{(-1)^m t^{-n+2m}}{2^{-n+2m} m!\Gamma(-n+m+1)}$$

In this, let the variable of summation be changed from m to j by the substitution $m = j + n$. Then,

$$\begin{aligned} J_{-n}(t) &= \sum_{j=0}^{\infty} \frac{(-1)^{j+n} t^{-n+2(j+n)}}{2^{-n+2(j+n)}(j+n)!\Gamma[-n+(j+n)+1]} \\ &= \sum_{j=0}^{\infty} \frac{(-1)^n(-1)^j t^{n+2j}}{2^{n+2j}(j+n)!\Gamma(j+1)} \\ &= (-1)^n \sum_{j=0}^{\infty} \frac{(-1)^j t^{n+2j}}{2^{n+2j}\Gamma(n+j+1)j!} = (-1)^n J_n(t) \end{aligned} \tag{14}$$

Thus, when ν is an integer, the function $J_{-\nu}(t)$ is proportional to $J_\nu(t)$. These two solutions are therefore not independent, and the linear combination $c_1 J_\nu(t) + c_2 J_{-\nu}(t)$ is no longer a complete solution of Bessel's equation. Moreover, without additional definitions, neither (11) nor (13) provides a complete solution since $Y_\nu(t)$ as defined by (10) assumes the indeterminate form 0/0 when ν is an integer.

---

†Named for the German mathematician Hermann Hankel (1839–1873).

A complete solution when ν is an integer can be found in any of several ways. One is to use the method developed in Sec. 2.4 for finding a second solution of a linear second-order differential equation when one solution is known. Another is to use the method suggested in Sec. 2.10. Still another procedure is to evaluate the limit of $Y_\nu(t)$ as $\nu \to n$. This limit, which can be proved to exist and which turns out to be independent of $J_n(t)$ for all values of n, is commonly denoted by $Y_n(t)$; that is,

$$(15) \qquad Y_n(t) = \lim_{\nu \to n} Y_\nu(t) = \lim_{\nu \to n} \frac{\cos \nu\pi J_\nu(t) - J_{-\nu}(t)}{\sin \nu\pi}$$

More explicitly, this limiting process leads to the following expression for $Y_n(t)$:

$$(15a) \quad Y_n(t) = \frac{2}{\pi}\left(\ln\frac{t}{2} + \gamma\dagger\right)J_n(t) - \frac{1}{\pi}\sum_{k=0}^{n-1}\frac{2^{n-2k}(n-k-1)!\ddagger}{t^{n-2k}k!}$$
$$-\frac{1}{\pi}\sum_{k=0}^{\infty}\frac{(-1)^k t^{n+2k}}{2^{n+2k}k!(n+k)!} \times \left[\left(1+\frac{1}{2}+\cdots+\frac{1}{k}\right)\S + \left(1+\frac{1}{2}+\cdots+\frac{1}{n+k}\right)\right]$$

The corresponding specializations of the Hankel functions are defined in the obvious way in terms of $Y_n(t)$:

$$(16) \qquad H_n^{(1)}(t) = J_n(t) + iY_n(t) \qquad \text{and} \qquad H_n^{(2)}(t) = J_n(t) - iY_n(t)$$

With Formulas (15) and (16), we can now eliminate from (11) and (13) the restriction that ν is not an integer and use these results for all values of ν, integral as well as nonintegral. Plots of $Y_0(t)$ and $Y_1(t)$ are shown in Fig. 12.2. Among other things, they illustrate the important fact that for all values of ν, $Y_\nu(t)$ is unbounded in the neighborhood of the origin and, like $J_\nu(t)$ and $J_{-\nu}(t)$, has infinitely many real zeros.

Reversing the transformation (2) which we used to eliminate the parameter λ from the general form of Bessel's equation, we can now summarize the results of the preceding discussion in the following theorem.

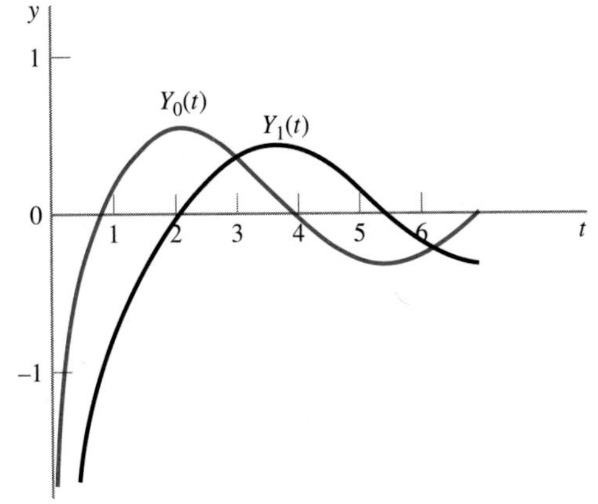

FIGURE 12.2
The Bessel functions of the second kind $Y_0(t)$ and $Y_1(t)$.

†The symbol γ denotes the following limit:

$$\lim_{m\to\infty}\left(1+\frac{1}{2}+\frac{1}{3}+\cdots+\frac{1}{m} - \ln m\right) = 0.5772\cdots$$

this important number is known as **Euler's constant.**
‡When $n = 0$, this sum is to be taken equal to zero.
§When $k = 0$, this sum is to be taken equal to zero.

THEOREM 1 For all values of ν, a complete solution of Bessel's equation of order ν with a parameter λ,

$$x^2y'' + xy' + (\lambda^2x^2 - \nu^2)y = 0$$

can be written in either of the forms

$$y(x) = c_1J_\nu(\lambda x) + c_2Y_\nu(\lambda x)$$

or

$$y(x) = c_1H_\nu^{(1)}(\lambda x) + c_2H_\nu^{(2)}(\lambda x)$$

If ν is not an integer, a complete solution can also be written in the form

$$y(x) = c_1J_\nu(\lambda x) + c_2J_{-\nu}(\lambda x)$$

$J_\nu(\lambda x)$, $J_{-\nu}(\lambda x)$, and $Y_\nu(\lambda x)$ all have infinitely many real zeros. If $\nu \geq 0$, $J_\nu(\lambda x)$ is finite for all values of x, but $J_{-\nu}(\lambda x)$ and $Y_\nu(\lambda x)$ are unbounded in the neighborhood of the origin. $H_\nu^{(1)}(\lambda x)$ and $H_\nu^{(2)}(\lambda x)$ are complex-valued functions when x is real.

EXERCISES

1. Explain the role of $\cos \nu\pi$ and $\sin \nu\pi$ in Eq. (10).
2. In Exercise 1(**a**), Sec. 12.1, show that the equation obtained for the factor $R(r)$ is Bessel's equation of order zero.
3. A thin circular sheet of metal of radius a, whose upper and lower faces are insulated, is initially at the temperature 100. At $t = 0$ the temperature around the circumference is suddenly reduced to zero and maintained thereafter at that value. Using the result of Exercise 2, proceed as far as possible toward the determination of the temperature in the sheet as a function of r and t. What is the characteristic equation of the problem? What are the characteristic functions? What is the expansion required to fit the initial temperature distribution in the sheet? What more do we need to know before we can determine the coefficients in that expansion? *Hint:* Note first that in the general solution $R(r) = AJ_0(\lambda r) + BY_0(\lambda r)$ of Bessel's equation of order zero, the term $Y_0(\lambda r)$ is infinite at $r = 0$ and hence cannot be part of the required solution. Then recall the summary of boundary-value problems given on p. 734.
4. Using the ratio test, show that the $J_\nu(t)$ series converges for all values of t.
5. By termwise differentiation, show that $J_0'(t) = -J_1(t)$.
6. Show that under the change of dependent variable defined by the substitution $y = u/\sqrt{t}$, Bessel's equation of order ν becomes

$$\frac{d^2u}{dt^2} + \left(1 + \frac{1 - 4\nu^2}{4t^2}\right)u = 0$$

Hence show that for large values of t, solutions of Bessel's equation are described approximately by expressions of the form

$$c_1\frac{\sin t}{\sqrt{t}} + c_2\frac{\cos t}{\sqrt{t}}$$

7. It can be shown that for large values of t, $J_0(t)$ is approximately equal to

$$\frac{2}{\pi t}\cos\left(t - \frac{\pi}{4}\right)$$

Using this expression, approximate the first four zeros of $J_0(t)$.

8. If y_1 and y_2 are any two solutions of Bessel's equation of order ν, show that $y_1y_2' - y_1'y_2 = c/t$, where c is a suitable constant. *Hint:* Recall Abel's formula from Sec. 2.2.
9. If y_1 and y_2 are two linearly independent solutions of Bessel's equation of order ν, show that there is no value of t for which y_1 and y_2 are simultaneously zero. *Hint:* Use the result of Exercise 8.
10. Show that with the possible exception of the origin, no nontrivial solution of Bessel's equation can have a double zero. *Hint:* Use the result of Exercise 8.
11. By determining the coefficient of $1/x$ on the left-hand side, show that if ν is not an integer,

$$J_\nu(x)J_{-\nu}'(x) - J_\nu'(x)J_{-\nu}(x) = -\frac{2}{\pi x}\sin \nu\pi$$

Is this result true if ν is an integer? *Hint:* Use the result of Exercise 8 and the fact that $\Gamma(x)\Gamma(1 - x) = \pi/\sin \pi x$ if x is not an integer.

12. Using Eq. (10) and the result of Exercise 11, show that

$$J_\nu(x)Y_\nu'(x) - J_\nu'(x)Y_\nu(x) = \frac{2}{\pi x}$$

13. Show that $\int \dfrac{dx}{xJ_\nu^2(x)}\,dx = \dfrac{\pi Y_\nu(x)}{2J_\nu(x)} + c$. What is $\int \dfrac{dx}{xY_\nu^2(x)}$? *Hint:* Divide the result of Exercise 12 by $J_\nu^2(x)$ and $Y_\nu^2(x)$ in turn and integrate.

14. What is $\dfrac{d}{dx}\left[\ln \dfrac{Y_\nu(x)}{J_\nu(x)}\right]$?

15. If ν is not an integer, show that the expression

$$\frac{\pi}{2\sin\nu\pi}\left[J_\nu(x)\int_a^x f(s)J_{-\nu}(s)\,ds - J_{-\nu}(x)\int_a^x f(s)J_\nu(s)\,ds\right]$$

is a particular solution of the nonhomogeneous Bessel equation

$$x^2y'' + xy' + (x^2 - \nu^2)y = xf(x)$$

Using the result of Exercise 12, obtain a similar expression that is valid when ν is an integer.

16. Show that in every interval of length π on the t axis there is at least one zero of $J_0(t)$ and at least one zero of $Y_0(t)$. *Hint:* Apply the comparison theorem of Exercise 59, Sec. 2.2, to the equations $u'' + (1 + 1/4t^2)u = 0$ and $u'' + u = 0$.

17. Show that the smallest positive root of the equation $J_\nu(x) = 0$ is equal to or greater than $\frac{1}{2}\sqrt{4\nu^2 - 1}$, $\nu > \frac{1}{2}$. *Hint:* Recall the result of Exercise 60, Sec. 2.2.

18. Show that if $\nu_1 > \nu_2$, then between any two consecutive zeros of $J_{\nu_1}(t)$ there is at least one zero of $J_{\nu_2}(t)$. *Hint:* Recall the comparison theorem established in Exercise 59, Sec. 2.2, and apply it to the transformed equation obtained in Exercise 6.

19. Obtain the recurrence relation satisfied by the coefficients in the power-series solution of Bessel's equation of order zero about the point $t = 1$. Using this, determine the first four nonzero terms in two linearly independent solutions about $t = 1$.

20. Use variation of parameters to find a solution to Bessel's equation of order n that is linearly independent of $J_n(t)$. *Hint:* Recall Eq. (10), Sec. 2.4.

Solve each of the following initial-value problems.

21. $xy'' + y' + xy = 0$; $y(0) = 13$, $y'(0) = 0$

22. $xy'' + y' + xy = 0$; $y(0) = 3$, $y'(0) = 0$

23. $4x^2y'' + 4xy' - (9 - 4x^2)y = 0$; $y(0) = y'(0) = 0$

24. Write a complete solution on $(0, \infty)$ of the equation $x^2y'' + xy' + (\lambda^2x^2 - 1)y = 0$ and find all nontrivial solutions that are bounded on $[0, 1]$ and vanish at $x = 1$.

25. **(a)** Find all nontrivial solutions on $(0, \infty)$ of Bessel's equation of order $\nu = 1$ with parameter λ which are bounded on $[0, 10]$ and vanish at $x = 10$.
(b) Particularize the three solutions corresponding to the first three positive zeros of $J_1(z)$: $z_1 \doteq 3.832$, $z_2 \doteq 7.016$, and $z_3 \doteq 10.174$.

26. A tapered shaft of length l, modulus of elasticity E, and density ρ, and with square cross sections of area $A(x) = kx$, undergoes free longitudinal vibrations. Its pointed end is free; its other end is fixed. Express the natural frequencies of the shaft in terms of the positive zeros of a Bessel function. *Hint:* See Exercise 35, Sec. 11.2.

12.3 MODIFIED BESSEL FUNCTIONS

Certain equations closely resembling Bessel's equation occur so often that their solutions are also named and studied as functions in their own right. The most important of these is

$$x^2y'' + xy' - (x^2 + \nu^2)y = 0 \tag{1}$$

which is known as the **modified Bessel equation of order** ν. Since this can be written in the form

$$x^2y'' + xy' + (i^2x^2 - \nu^2)y = 0$$

it is evident that this is nothing but Bessel's equation of order ν with the imaginary parameter $\lambda = i$. However, in actual applications, to write the complete solution of (1) in the form

$$y = c_1J_\nu(ix) + c_2Y_\nu(ix)$$

and retain the imaginaries would be about as awkward as taking the solution of

$$y'' - y = 0$$

to be

$$y = c_1\cos ix + c_2\sin ix$$

and use this complex expression instead of resorting to real exponentials or hyperbolic functions.

Accordingly, we seek modifications of $J_\nu(ix)$ and $Y_\nu(ix)$ which will be real functions of real variables.

Now,

$$J_\nu(ix) = \sum_{k=0}^{\infty} \frac{(-1)^k(ix)^{\nu+2k}}{2^{\nu+2k}k!\Gamma(\nu+k+1)} = i^\nu \sum_{k=0}^{\infty} \frac{x^{\nu+2k}}{2^{\nu+2k}k!\Gamma(\nu+k+1)}$$

Moreover, $J_\nu(ix)$ multiplied by any constant will also be a solution of the equation we are considering. Hence, in particular, we can multiply it by $i^{-\nu}$, getting

$$i^{-\nu}J_\nu(ix) = \sum_{k=0}^{\infty} \frac{x^{\nu+2k}}{2^{\nu+2k}k!\Gamma(\nu+k+1)}$$

This is a completely real function, identical with $J_\nu(x)$ except that its terms, instead of alternating in sign, are all positive. This new function, which is related to $J_\nu(x)$ in the same way that $\cosh x$ and $\sinh x$ are related to $\cos x$ and $\sin x$, is known as the **modified Bessel function of the first kind of order** ν and is customarily denoted by $I_\nu(x)$, although a few authors and some handbooks continue to use the original designation $i^{-\nu}J_\nu(ix)$ for this function. If ν is not an integer, the function $I_{-\nu}(x)$ obtained from $I_\nu(x)$ by replacing ν by $-\nu$ throughout is a second, independent solution of Eq. (1), whose complete solution can therefore be written

$$y(x) = c_1 I_\nu(x) + c_2 I_{-\nu}(x) \qquad \nu \text{ not an integer}$$

On the other hand, instead of using $I_{-\nu}(x)$, many writers take the second solution of the modified Bessel equation to be the linear combination

$$K_\nu(x) = \frac{\pi}{2}\frac{I_{-\nu}(x) - I_\nu(x)}{\sin \nu\pi} \qquad \nu \text{ not an integer} \tag{2}$$

which is known as the **modified Bessel function of the second kind of order** ν. If ν is not an integer, this is a well-defined solution which is clearly independent of $I_\nu(x)$. If ν is an integer n, this assumes the indeterminate form 0/0, but a tedious evaluation by L'Hospital's rule leads to a limiting expression

$$K_n(x) = \lim_{\nu\to n} K_\nu(x) = \lim_{\nu\to n} \frac{\pi}{2}\frac{I_{-\nu}(x) - I_\nu(x)}{\sin \nu\pi}$$

$$= (-1)^{n+1}\left(\ln\frac{x}{2} + \gamma\dagger\right)I_n(x) + \frac{1}{2}\sum_{k=0}^{n-1}\frac{(-1)^k(n-k-1)!2^{n-2k}}{k!x^{n-2k}}\ddagger$$

$$+\frac{(-1)^n}{2}\sum_{k=0}^{\infty}\frac{x^{n+2k}}{k!(n+k)!2^{n+2k}}\left\{\left[1+\frac{1}{2}+\cdots+\frac{1}{k}\right]\S+\left[1+\frac{1}{2}+\cdots+\frac{1}{n+k}\right]\right\}$$

which is a solution independent of $I_n(x)$. This is a useful result because, as we might expect, $I_\nu(x)$ and $I_{-\nu}(x)$ are not independent when ν is an integer. In fact, when $\nu = n$, we have the identity [see Eq. (14), Sec. 12.2]

$$(-1)^n J_{-n}(ix) = J_n(ix)$$

†The symbol γ denotes Euler's constant.
‡When $n = 0$, this sum is to be taken equal to zero.
§When $k = 0$, this sum is to be taken equal to zero.

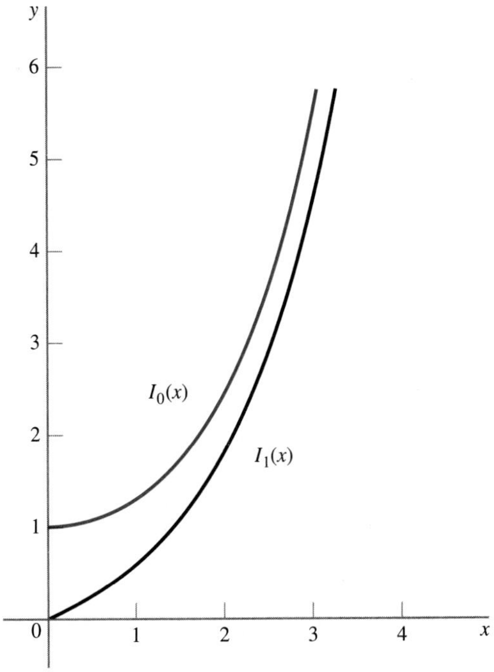

FIGURE 12.3
The modified Bessel functions of the first kind $I_0(x)$ and $I_1(x)$.

and then by obvious steps,

$$(i^2)^n J_{-n}(ix) = J_n(ix)$$
$$i^n J_{-n}(ix) = i^{-n} J_n(ix)$$
$$I_{-n}(x) = I_n(x)$$

Plots of $I_0(x)$ and $I_1(x)$ are shown in Fig. 12.3; plots of $K_0(x)$ and $K_1(x)$ are shown in Fig. 12.4. As these graphs illustrate, the modified Bessel functions have no real zeros except possibly at $x = 0$. They also illustrate that for $\nu \geq 0$, $I_\nu(x)$ is finite at the origin but $K_\nu(x)$, like $I_{-\nu}(x)$, becomes infinite as x approaches zero.

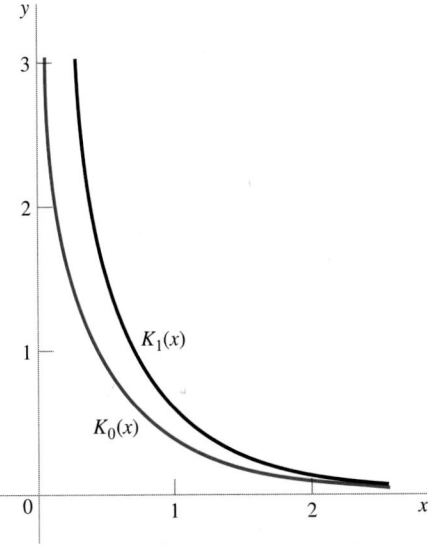

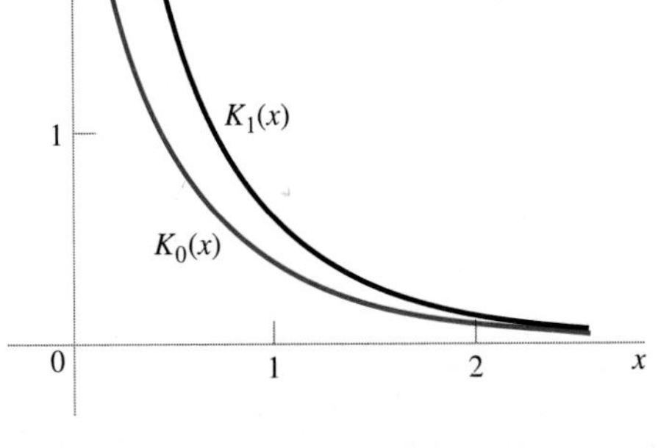

FIGURE 12.4
The modified Bessel functions of the second kind $K_0(x)$ and $K_1(x)$.

Like the ordinary Bessel equation, the modified Bessel equation frequently occurs in a form containing a parameter λ:

$$x^2y'' + xy' - (\lambda^2x^2 + \nu^2)y = 0 \tag{3}$$

A complete solution of this is of course

$$y = c_1I_\nu(\lambda x) + c_2K_\nu(\lambda x) \qquad \nu \text{ unrestricted}$$

If ν is not an integer, we have the alternative form

$$y = c_1I_\nu(\lambda x) + c_2I_{-\nu}(\lambda x)$$

In the last chapter we found it sufficient to assume that when we separated variables in a partial differential equation, only real separation constants had to be considered. This is not always the case, however, for as we shall see in Example 5, Sec. 12.8, there are important problems that require the separation constant to be imaginary. In particular, in a number of instances, one of the equations resulting from the separation process is still another modification of Bessel's equation, namely,

$$x^2y'' + xy' + (-ix^2 - \nu^2)y = 0 \tag{4}$$

This can be written in the form

$$x^2y'' + xy'[(\sqrt{-i})^2x^2 - \nu^2]y = 0$$

and regarded as Bessel's equation of order ν with parameter $\sqrt{-i}$, or it can be written in the form

$$x^2y'' + xy' - [(\sqrt{i})^2x^2 + \nu^2]y = 0$$

and regarded as the modified Bessel equation of order ν with parameter $\sqrt{i}$. From the first point of view, a complete solution can be written

$$y = c_1J_\nu(\sqrt{-i}x) + c_2Y_\nu(\sqrt{-i}x)$$

From the second point of view, a complete solution can be written

$$y = c_1I_\nu(\sqrt{i}x) + c_2K_\nu(\sqrt{i}x)$$

A complete solution can of course be constructed from *any* pair of independent particular solutions, and it is customary in studying Eq. (4) to select $J_\nu(\sqrt{-i}x)$ and $K_\nu(\sqrt{i}x)$ for this purpose. Making this choice and recalling that $-i = i^3$ and therefore

$$\sqrt{-i} = (-i)^{1/2} = i^{3/2}$$

we have for the complete solution of Eq. (4)

$$y = c_1J_\nu(i^{3/2}x) + c_2K_\nu(i^{1/2}x)$$

Now

$$J_\nu(i^{3/2}x) = \sum_{k=0}^{\infty} \frac{(-1)^k(i^{3/2}x)^{\nu+2k}}{2^{\nu+2k}k!\Gamma(\nu + k + 1)}$$

$$= i^{3\nu/2} \sum_{k=0}^{\infty} \frac{(-1)^k i^{3k} x^{\nu+2k}}{2^{\nu+2k}k!\Gamma(\nu + k + 1)}$$

Moreover, i^{3k} can take on only one of the four values

$$\begin{array}{rl} 1 & k = 0, 4, 8, \ldots \\ -i & k = 1, 5, 9, \ldots \\ -1 & k = 2, 6, 10, \ldots \\ i & k = 3, 7, 11, \ldots \end{array}$$

Hence the first, third, fifth, . . . , terms in the series for $J_\nu(i^{3/2}x)$ are real, and the second, fourth, sixth, . . . , are imaginary. The real terms, for which $k = 0, 2, 4, \ldots$, can be identified equally well by putting $k = 2j$ and letting j take on the values $0, 1, 2, \ldots$. The imaginary terms, for which $k = 1, 3, 5, \ldots$, can be identified similarly by putting $k = 2j + 1$ and again letting j take on the values $0, 1, 2, \ldots$. Moreover, when $k = 2j$, then,

$$(-1)^k i^{3k} = (-1)^{2j} i^{6j} = i^{10j} = (-1)^j$$

and when $k = 2j+1$, then

$$(-1)^k i^{3k} = (-1)^{2j+1} i^{6j+3} = i^{10j+5} = (-1)^j i$$

Hence, using these observations to separate the last series into its real and imaginary parts, we have

$$\begin{aligned} J_\nu(i^{3/2}x) &= i^{3\nu/2}\left[\sum_{j=0}^{\infty} \frac{(-1)^j x^{\nu+4j}}{2^{\nu+4j}(2j)!\Gamma(\nu + 2j + 1)} + i \sum_{j=0}^{\infty} \frac{(-1)^j x^{\nu+2+4j}}{2^{\nu+2+4j}(2j + 1)!\Gamma(\nu + 2j + 2)}\right] \\ &= i^{3\nu/2}(\Sigma_R + i\Sigma_I) \end{aligned}$$

Moreover, since $i = e^{i\pi/2}$, we have

$$i^{3\nu/2} = (e^{i\pi/2})^{3\nu/2} = e^{i3\nu\pi/4} = \cos\frac{3\nu\pi}{4} + i \sin\frac{3\nu\pi}{4}$$

and therefore

$$\begin{aligned} J_\nu(i^{3/2}x) &= \left(\cos\frac{3\nu\pi}{4} + i\sin\frac{3\nu\pi}{4}\right)(\Sigma_R + i\,\Sigma_I) \\ &= \left(\cos\frac{3\nu\pi}{4}\Sigma_R - \sin\frac{3\nu\pi}{4}\Sigma_I\right) + i\left(\cos\frac{3\nu\pi}{4}\Sigma_I + \sin\frac{3\nu\pi}{4}\Sigma_R\right) \end{aligned}$$

$J_\nu(i^{3/2}x)$ thus consists of one purely real series plus i times a second purely real series. The series forming the real part of this expression defines the function $\mathbf{ber}_\nu\, x$. The series forming the imaginary part defines the function $\mathbf{bei}_\nu\, x$. The letters *be* suggest the relation between these new functions and the Bessel functions themselves. The terminal letters r and i of course suggest the adjectives *real* and *imaginary*. For the important case $\nu = 0$, we have explicitly

$$\text{ber}_0\, x \equiv \text{ber}\, x = \sum_{j=0}^{\infty} \frac{(-1)^j x^{4j}}{2^{4j}[(2j)!]^2}$$

$$\text{bei}_0\, x \equiv \text{bei}\, x = \sum_{j=0}^{\infty} \frac{(-1)^j x^{4j+2}}{2^{4j+2}[(2j+1)!]^2}$$

Plots of ber x and bei x are shown in Fig. 12.5. The graphs oscillate with increasing amplitudes.

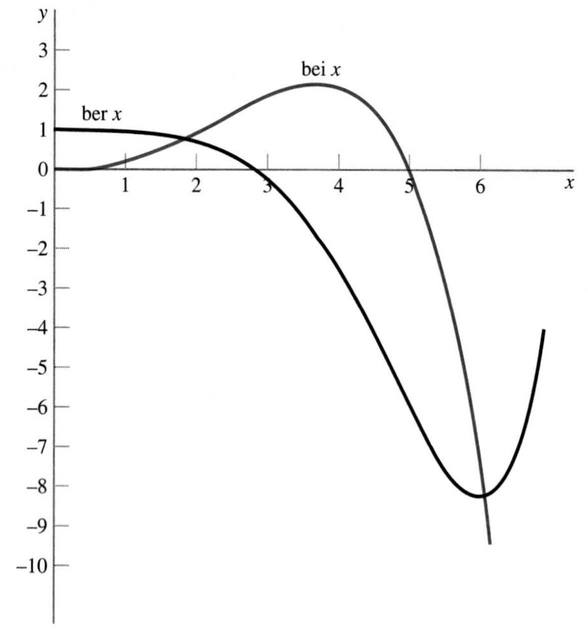

FIGURE 12.5
The functions ber x and bei x.

In a similar way, $K_\nu(i^{1/2}x)$ can be expressed as a real series plus i times another real series. These series are taken as the definitions of the new functions **ker**$_\nu\, x$ and **kei**$_\nu\, x$,† respectively. A complete solution of Eq. (4) can thus be written

$$y = c_1(\mathrm{ber}_\nu\, x + i\,\mathrm{bei}_\nu\, x) + c_2(\mathrm{ker}_\nu\, x + i\,\mathrm{kei}_\nu\, x)$$

The function $\mathrm{ber}_\nu\, x + i\,\mathrm{bei}_\nu\, x$ is finite at the origin but becomes infinite as x becomes infinite; $\mathrm{ker}_\nu\, x + i\,\mathrm{kei}_\nu\, x$ is infinite at the origin but approaches zero as x becomes infinite.

For real values of x the expression $\mathrm{ber}_\nu\, x + i\,\mathrm{bei}_\nu\, x$ is a complex number and, like any complex number, has a characteristic length, or modulus, $M_\nu(x)$ and a characteristic angle, or amplitude, $\theta_\nu(x)$. Clearly (Fig. 12.6),

$$M_\nu(x) = \sqrt{\mathrm{ber}_\nu^2\, x + \mathrm{bei}_\nu^2\, x} \qquad \text{and} \qquad \theta_\nu(x) = \tan^{-1}\frac{\mathrm{bei}_\nu\, x}{\mathrm{ber}_\nu\, x}$$

In many applications, it is convenient to replace the expression $\mathrm{ber}_\nu\, x + i\,\mathrm{bei}_\nu\, x$ by its exponential equivalent

$$M_\nu(x) \exp\,[i\theta_\nu(x)]$$

and for this reason the functions $M_\nu(x)$ and $\theta_\nu(x)$, like the other Bessel functions, have been tabulated for certain important values of ν.

Similarly, of course, the expression $\mathrm{ker}_\nu\, x + i\,\mathrm{kei}_\nu\, x$ can be replaced by an equivalent exponential form

$$N_\nu(x) \exp\,[i\phi_\nu(x)]$$

where

$$N_\nu(x) = \sqrt{\mathrm{ker}_\nu^2\, x + \mathrm{kei}_\nu^2\, x} \qquad \text{and} \qquad \phi_\nu(x) = \tan^{-1}\frac{\mathrm{kei}_\nu\, x}{\mathrm{ker}_\nu\, x}$$

These functions have also been tabulated.

†The letters *ke* are derived from the name of the British mathematical physicist Lord Kelvin (1824–1907).

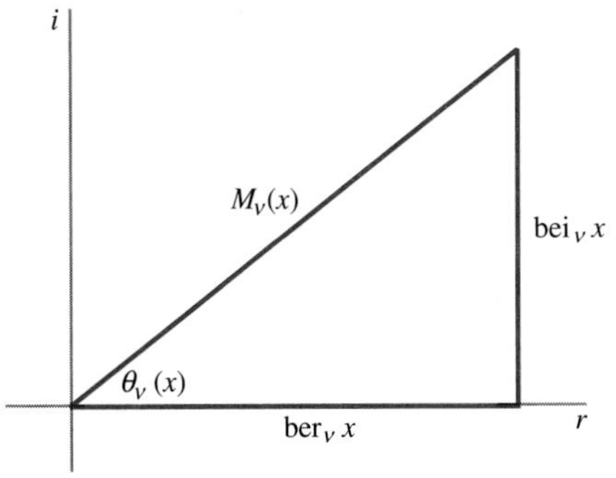

FIGURE 12.6
The polar representation of $\text{ber}_\nu x + \text{i bei}_\nu x$.

EXERCISES

1. What is the role of $\sin \nu\pi$ in Eq. (2)?
2. By termwise differentiation, show that $I_0'(x) = I_1(x)$.
3. Show that under the change of dependent variable defined by the substitution $y = u/\sqrt{x}$, the modified Bessel equation of order ν becomes

$$\frac{d^2u}{dx^2} - \left(1 + \frac{4\nu^2 - 1}{4x^2}\right)u = 0$$

Hence show that for large values of x, solutions of the modified Bessel equation are described approximately by expressions of the form

$$c_1\frac{e^x}{\sqrt{x}} + c_2\frac{e^{-x}}{\sqrt{x}}$$

More precisely, it can be shown that

$$I_\nu(x) \sim \frac{e^x}{\sqrt{2\pi x}} \qquad \text{and} \qquad K_\nu(x) \sim \sqrt{\frac{\pi}{2x}}e^{-x}$$

4. If y_1 and y_2 are any two solutions of the modified Bessel equation of order ν, show that $y_1y_2' - y_1'y_2 = c/x$, where c is a suitable constant.
5. By determining the coefficient of $1/x$ on the left-hand side, show that if ν is not an integer,

$$I_\nu(x)I_{-\nu}'(x) - I_\nu'(x)I_{-\nu}(x) = -\frac{2}{\pi x}\sin \nu\pi$$

Hint: Use the result of Exercise 4 and the hint for Exercise 11, Sec. 12.2.

6. Using Eq. (2) and the result of Exercise 5, show that

$$I_\nu(x)K_\nu'(x) - I_\nu'(x)K_\nu(x) = \frac{-1}{x}$$

7. Show that $\displaystyle\int \frac{dx}{xI_\nu^2(x)} = -\frac{K_\nu(x)}{I_\nu(x)} + c$. *Hint:* Divide the result of Exercise 6 by $I_\nu^2(x)$ and integrate.
8. What is $\displaystyle\int \frac{dx}{xK_\nu^2(x)}\,dx$?
9. If the temperature distribution in a thin circular disk is independent of θ and if the disk loses heat through its faces into air of constant temperature 0, it was shown in Exercise 29, Sec. 11.2, that the temperature satisfies the equation

$$\frac{\partial(r\;\partial u/\partial r)}{\partial r} = b^2ru + a^2r\frac{\partial u}{\partial t}$$

Show that the steady-state temperature in the disk satisfies the modified Bessel equation of order 0.

10. Using the result of Exercise 9, find the steady-state temperature in the disk if its circumference is maintained at the temperature 100.
11. If the disk in Exercises 9 and 10 is an annular ring of inner radius 1 and outer radius 2, find the steady-state temperature in the annulus if the inner boundary is maintained at the temperature 0 and the outer boundary is maintained at the temperature 100.
12. Show that $J_0(i^{1/2}x) = \text{ber}\, x - i\,\text{bei}\, x$
13. Show that $(x\,\text{ber}'\,x)' = -x\,\text{bei}\,x$ and that $(x\,\text{bei}'\,x)' = x\,\text{ber}\,x$.
14. Verify that $\displaystyle \text{ber}_\nu\,(x) = \sum_{k=0}^{\infty}\frac{(-1)^k x^{\nu+2k}\cos \frac{3}{4}(\nu + 2k)\pi}{2^{\nu+2k}k!\Gamma(\nu + k + 1)}$.
15. Verify that $\displaystyle \text{bei}_\nu\,(x) = \sum_{k=0}^{\infty}\frac{(-1)^k x^{\nu+2k}\sin \frac{3}{4}(\nu + 2k)\pi}{2^{\nu+2k}k!\Gamma(\nu + k + 1)}$.

12.4 EQUATIONS SOLVABLE IN TERMS OF BESSEL FUNCTIONS

There are many differential equations whose solutions can be expressed in terms of Bessel functions, in particular, the large and important family described in the following theorem.

THEOREM 1 If $(1 - a)^2 \geq 4c$ and if neither d, p, nor q is zero, then, except in the obvious special cases when it reduces to Euler's equation,† the differential equation

†Equation (1), Sec. 2.8.

$$x^2y'' + x(a + 2bx^p)y' + [c + dx^{2q} + b(a + p - 1)x^p + b^2x^{2p}]y = 0$$

has as a complete solution

$$y = x^{\alpha}e^{-\beta x^p}[c_1 J_\nu(\lambda x^q) + c_2 Y_\nu(\lambda x^q)]$$

where

$$\alpha = \frac{1-a}{2} \qquad \beta = \frac{b}{p} \qquad \lambda = \frac{\sqrt{|d|}}{q} \qquad \nu = \frac{\sqrt{(1-a)^2 - 4c}}{2q}$$

If $d < 0$, J_ν and Y_ν are to be replaced by I_ν and K_ν, respectively. If ν is not an integer, Y_ν and K_ν can be replaced by $J_{-\nu}$ and $I_{-\nu}$ if desired.

The proof of this theorem, while straightforward, is lengthy and involved, and we shall not present it here. It consists in transforming the given equation by means of the substitutions

$$y = x^{(1-a)/2}e^{-(b/p)x^p}Y \qquad \text{and} \qquad x = \left(\frac{qX}{\sqrt{|d|}}\right)^{1/q}$$

and verifying that when the parameters are properly identified, the result is precisely Bessel's equation in terms of the new variables X and Y.

One special case of Theorem 1 is useful enough to be stated as a corollary.

COROLLARY 1 If $(1 - r)^2 \geq 4b$, then except in the special cases $a = 0$ (and $s = r - 2$) when it reduces to Euler's equation, the differential equation

$$(x^r y')' + (ax^s + bx^{r-2})y = 0$$

has as a complete solution

$$y = x^{\alpha}[c_1 J_\nu(\lambda x^{\gamma}) + c_2 Y_\nu(\lambda x^{\gamma})]$$

where

$$\alpha = \frac{1-r}{2} \qquad \gamma = \frac{2-r+s}{2} \qquad \lambda = \frac{2\sqrt{|a|}}{2-r+s} \qquad \nu = \frac{\sqrt{(1-r)^2 - 4b}}{2-r+s}$$

If $a < 0$, J_ν and Y_ν are to be replaced by I_ν and K_ν, respectively. If ν is not an integer, Y_ν and K_ν can be replaced by $J_{-\nu}$ and $I_{-\nu}$ if desired.

EXAMPLE 1

Find a complete solution of the equation

$$x^2y'' + x(4x^4 - 3)y' + (4x^8 - 5x^2 + 3)y = 0$$

Clearly, this is a special case of the equation of Theorem 1 with $a = -3$, $b = 2$, $p = 4$, $c = 3$, $d = -5$, and $q = 1$. Hence

$$\alpha = 2 \qquad \beta = \tfrac{1}{2} \qquad \lambda = \sqrt{|-5|} = \sqrt{5} \qquad \nu = 1$$

A complete solution is therefore

$$y = x^2e^{-x^4/2}[c_1 I_1(\sqrt{5}x) + c_2 K_1(\sqrt{5}x)]$$

EXAMPLE 2

What is a complete solution of the equation $y'' + y = 0$?

Obviously, one possibility is

$$y = c_1 \cos x + c_2 \sin x$$

However, $y'' + y = 0$ is also a special case of the equation of Corollary 1, with $r = 0$, $s = 0$, $a = 1$, and $b = 0$. Hence

$$\alpha = \tfrac{1}{2} \qquad \gamma = 1 \qquad \lambda = 1 \qquad \nu = \tfrac{1}{2}$$

and so we can also write

$$y = \sqrt{x}[d_1 J_{1/2}(x) + d_2 J_{-1/2}(x)]$$

It follows, therefore, from Theorem 6, Sec. 2.2, that for proper choice of the constants c_1 and c_2, each of the particular solutions

$$\sqrt{x}J_{1/2}(x) \qquad \text{and} \qquad \sqrt{x}J_{-1/2}(x)$$

must be expressible in the form $c_1 \cos x + c_2 \sin x$.

Now, since $\Gamma(\frac{3}{2}) = \frac{1}{2}\Gamma(\frac{1}{2}) = \frac{1}{2}\sqrt{\pi}$, the series for $J_{1/2}(x)$ begins with the term

$$\frac{x^{1/2}}{2^{1/2}\Gamma(\frac{3}{2})} = \sqrt{\frac{2x}{\pi}}$$

Hence the series for $\sqrt{x}J_{1/2}(x)$ begins with the term $\sqrt{2/\pi}x$. Therefore if we write

$$\sqrt{x}J_{1/2}(x) = \sqrt{\frac{2}{\pi}}x - \cdots = c_1 \cos x + c_2 \sin x$$

$$= c_1\left(1 - \frac{x^2}{2} + \cdots\right) + c_2\left(x - \frac{x^3}{6} + \cdots\right)$$

and put $x = 0$ in this identity, we find $c_1 = 0$. Subsequently, by equating the coefficients of x, we find

$$c_2 = \sqrt{\frac{2}{\pi}}$$

We have thus established the interesting and important result that

$$\sqrt{x}J_{1/2}(x) = \sqrt{\frac{2}{\pi}}\sin x \qquad \text{or} \qquad J_{1/2}(x) = \sqrt{\frac{2}{\pi x}}\sin x$$

In a similar fashion it can be shown that

$$J_{-1/2}(x) = \sqrt{\frac{2}{\pi x}}\cos x$$

EXAMPLE 3

Find a complete solution of the equation $x^2y'' - xy' + (1 + x)y = 0$.

The relatively simple form of the given equation suggests that it may be solvable through the use of Corollary 1. However, before Corollary 1 can be used it is necessary that the terms involving the first and second derivatives be combined into a single term which is the derivative of a product of the form $x^r y'$. In general the derivative terms in an equation of the form

$$p_0(x)y'' + p_1(x)y' + p_2(x)y = 0$$

will not constitute the derivative of a product, but multiplication by a suitable factor will always accomplish this. Sometimes the necessary factor will be evident by inspection, but in any case it can always be found by using the result of Exercise 40, Sec. 11.6.

Thus, identifying $p_0(x)$ as x^2 and $p_1(x)$ as $-x$, we compute

$$\frac{1}{p_0(x)} \exp\left[\int \frac{p_1(x)}{p_0(x)}\,dx\right] = \frac{1}{x^2}\exp\left(\int -\frac{dx}{x}\right) = \frac{1}{x^2}\exp(-\ln x) = \frac{1}{x^3}$$

Multiplying the given equation by $1/x^3$ thus gives us

$$\frac{y''}{x} - \frac{y'}{x^2} + \left(\frac{1}{x^3} + \frac{1}{x^2}\right)y = 0$$

Knowing that the derivative terms now do form the derivative of a product, it is easy to identify the product, and we recognize that the last equation can be written as

$$(x^{-1}y')' + (x^{-2} + x^{-3})y = 0$$

Clearly, this is an equation covered by Corollary 1, with $r = -1$, $s = -2$, $a = b = 1$. Hence, verifying that $\alpha = 1$, $\gamma = \frac{1}{2}$, $\lambda = 2$, and $\nu = 0$, we have as the required solution

$$y = x[c_1 J_0(2\sqrt{x}) + c_2 Y_0(2\sqrt{x})]$$

EXERCISES

1. Verify that Corollary 1 is a special case of Theorem 1.

Solve each of the following equations in terms of Bessel functions.

2. $y'' - x^2 y = 0$
3. $xy'' - y' + y = 0$
4. $x^2y'' + (x - 2)y = 0$
5. $4x^2y'' + (4x + 1)y = 0$
6. $xy'' + y' + xy = 0$
7. $x^2y'' - xy' + (1 + x)y = 0$
8. $x^2y'' - 2xy' + (1 + x)y = 0$
9. $x^2y'' + 3xy' + (1 + x)y = 0$
10. $xy'' + y = 0$
11. $xy'' + y' + x^2y = 0$
12. $xy'' + 2y' + 4xy = 0$
13. $xy'' - y' + 4x^5y = 0$
14. $x^2y'' - x(2 - x)y' + (2 - x)y = 0$
15. $x^2y'' - x(4 - x)y' + (6 - 2x)y = 0$
16. $x^2y'' - 2xy' + (2 - x^2)y = 0$
17. $x^2y'' - 2xy' + (2 - x^3)y = 0$
18. $x^2y'' + (2x^2 + x)y' + (x^2 + 3x - 1)y = 0$
19. $x^2y'' + 2xy' + (x^4 + x^2 - 2)y = 0$
20. $(x^2y')' + [x^2 - n(n - 1)]y = 0$

Can the following equations be solved in terms of Bessel functions?

21. $2x^2y'' + (2x^2 + x)y' - y = 0$
22. $9x^2y'' + 9(x^2 + x)y' + (12x - 1)y = 0$
23. Solve the equation $[(a + bx)y']' - y = 0$. *Hint:* Introduce a new variable via the substitution $t = a + bx$.
24. By making the appropriate change of independent variable, solve the equation $[(1 + x)^2y']' + (1 + 3x)y = 0$.
25. Solve $y'' + ae^{mx}y = 0$. *Hint:* Introduce a new independent variable via the substitution $t = e^{mx}$.
26. Show that $I_{1/2}(x) = \sqrt{2/\pi x}\,\sinh x$ and that $I_{-1/2}(x) = \sqrt{2/\pi x}\,\cosh x$.
27. What relations must exist among the coefficients in
$x^2y'' + x(b_0 + b_1x)y' + (c_0 + c_1x + c_2x^2 + c_3x^3)y = 0$
in order for this equation to be solvable in terms of Bessel functions?
28. Use Theorem 1 to solve the initial differential equation in Example 3.

By first introducing a new dependent variable via the substitution described in Exercise 65, Sec. 2.10, find a complete solution of each of the following equations.

29. $y'' - (1 + x)y' + \dfrac{x^2 - 1}{4}y = 0$
30. $y'' + \dfrac{2}{x + 3}y' + x^4y = 0$
31. $y'' + 2x^2y' + (x^4 + x - 2/x^2)y = 0$
32. $y'' - (2\tan x)y' + (x - 1)y = 0$
33. Show that any solution of

$$(x^{m-1}y')' + kx^{m-2}y = 0 \text{ or } (x^{m-1}y')' - kx^{m-2}y = 0$$

will also satisfy the equation $(x^m y'')'' - k^2x^{m-2}y = 0$.

Using the results of Exercise 33, find a complete solution of each of the following equations.

34. $(x^2y'')'' - 9y = 0$
35. $x^2y^{iv} - 4y = 0$
36. $x^2y^{iv} + 8xy''' + 12y'' - 9y = 0$
37. Show that if $b = \frac{1}{4} - \nu^2$, the function $y = \sqrt{x}J_\nu(x)$ satisfies the differential equation $y'' + (1 + b/x^2)y = 0$. Hence, applying the result of Exercise 59, Sec. 2.2, to this equation and the equation $y'' + k^2y = 0$, for suitable k, show that $J_\nu(x)$ has infinitely many zeros.
38. **(a)** Use the particular solution $y = x$ and reduction of order to find a complete solution of the differential equation in Exercise 14.

(b) Use Theorem 1 to write a complete solution of that equation involving Bessel functions.
(c) Use these complete solutions to verify the identities of Exercise 26.

39. Find a complete solution of the differential equations:
(a) $x^5y'' + 5x^4y' - y = 0$
(b) $x^2y'' + 12(x^4 - x)y' + [11 - x^2 + (6x^3 - 5)^2]y = 0$

40. Prove that the zeros of $J_{1/2}(x)$ and $J_{-1/2}(x)$ separate one another on $(0, \infty)$. *Hint:* Use the representations for $J_{1/2}$ and $J_{-1/2}$ found in Example 2.

Solve each of the initial-value problems.

41. $t^2y'' - 2ty' + (t^2 + 2)y = 0$; $y(\pi) = 0$, $y'(\pi) = -5\pi$
42. $xy'' + y' - xy = 0$; $y(0) = 10$, $y'(0) = 0$
43. Work Exercise 26, Sec. 12.2, if $A(x) = kx^2$.

12.5 IDENTITIES FOR THE BESSEL FUNCTIONS

The Bessel functions are related by an amazing array of identities. Fundmental among these are the results of the next two theorems and their consequences.

THEOREM 1 For all values of ν, $\dfrac{d[x^\nu J_\nu(x)]}{dx} = x^\nu J_{\nu-1}(x)$.

PROOF To prove this theorem, we first multiply the series for $J_\nu(x)$ by x^ν, then differentiate the resulting series term by term, cancel common factors, and finally remove x^ν from each term:

$$J_\nu(x) = \sum_{k=0}^{\infty} \frac{(-1)^k x^{\nu+2k}}{2^{\nu+2k}k!\Gamma(\nu + k + 1)}$$

$$x^\nu J_\nu(x) = \sum_{k=0}^{\infty} \frac{(-1)^k x^{2\nu+2k}}{2^{\nu+2k}k!\Gamma(\nu + k + 1)}$$

$$\begin{aligned}\frac{d[x^\nu J_\nu(x)]}{dx} &= \sum_{k=0}^{\infty} \frac{(-1)^k 2(\nu + k)x^{2\nu+2k-1}}{2^{\nu+2k}k!(\nu + k)\Gamma(\nu + k)} \\ &= \sum_{k=0}^{\infty} \frac{(-1)^k x^\nu x^{\nu+2k-1}}{2^{\nu-1+2k}k!\Gamma(\nu + k)} \\ &= x^\nu \sum_{k=0}^{\infty} \frac{(-1)^k x^{\nu-1+2k}}{2^{\nu-1+2k}k!\Gamma(\nu-1 + k + 1)} \\ &= x^\nu J_{\nu-1}(x)\end{aligned}$$

THEOREM 2 For all values of ν, $\dfrac{d[x^{-\nu} J_\nu(x)]}{dx} = -x^{-\nu}J_{\nu+1}(x)$.

PROOF This theorem can be proved in essentially the same manner as Theorem 1, but it is easier and perhaps more instructive to proceed as follows. By applying Corollary 1, Theorem 1, Sec. 12.4, to the equation

$$\frac{d(x^{-1-2\nu}y')}{dx} + x^{-1-2\nu}y = 0$$

with $r = s = -1 - 2\nu$, $a = 1$, $b = 0$, and therefore $\alpha = 1 + \nu$, $\gamma = 1$, $\lambda = 1$, ν (for the corollary) $= 1 + \nu$, it is clear that $y = x^{1+\nu}J_{1+\nu}(x)$ is a particular solution of the equation we are considering. Hence, substituting it for y, we have

$$\frac{d[x^{-1-2\nu}\{x^{1+\nu}J_{1+\nu}(x)\}']}{dx} + x^{-1-2\nu}[x^{1+\nu}J_{1+\nu}(x)] = 0$$

Now, using Theorem 1 to compute the derivative of the quantity $x^{1+\nu}J_{1+\nu}(x)$ in the first term, we have further

$$\frac{d[x^{-1-2\nu}\{x^{1+\nu}J_\nu(x)\}]}{dx} = -x^{-\nu}J_{\nu+1}(x)$$

or, finally,

$$\frac{d[x^{-\nu}J_\nu(x)[}{dx} = -x^{-\nu}J_{\nu+1}(x)$$ ◗

Using Theorems 1 and 2, we can now show that the Bessel function of the second kind, $Y_\nu(x)$, satisfies the same identities that $J_\nu(x)$ satisfies. Specifically, we have the following pair of theorems whose proofs are almost identical.

THEOREM 3 For all values of ν, $\dfrac{d[x^\nu Y_\nu(x)]}{dx} = x^\nu Y_{\nu-1}(x)$.

◖ **PROOF** By definition [Eq. (10), Sec. 12.2],

$$Y_\nu(x) = \frac{\cos \nu\pi J_\nu(x) - J_{-\nu}(x)}{\sin \nu\pi}$$

Hence, multiplying this identity by x^ν and then differentiating with respect to x, we have

$$\frac{d[x^\nu Y_\nu(x)]}{dx} = \frac{\cos \nu\pi}{\sin \nu\pi}\frac{d[x^\nu J_\nu(x)]}{dx} - \frac{1}{\sin \nu\pi}\frac{d[x^\nu J_{-\nu}(x)]}{dx}$$

Applying Theorems 1 and 2 to the respective derivatives on the right, we have further,

$$\frac{d[x^\nu Y_\nu(x)]}{dx} = \frac{\cos \nu\pi}{\sin \nu\pi}[x^\nu J_{\nu-1}(x)] - \frac{1}{\sin \nu\pi}[-x^\nu J_{-(\nu-1)}(x)]$$

Now $\cos \nu\pi = -\cos(\nu - 1)\pi$ and $\sin \nu\pi = -\sin(\nu - 1)\pi$. Hence the last expression can be written in the form

$$\frac{d[x^\nu Y_\nu(x)]}{dx} = x^\nu\left[\frac{\cos(\nu - 1)\pi J_{\nu-1}(x) - J_{-(\nu-1)}(x)}{\sin(\nu - 1)\pi}\right]$$

$$= x^\nu Y_{\nu-1}(x)$$

The exceptional case when ν is an integer is handled by recalling that $Y_n(x)$ is defined to be $\lim_{\nu\to n} Y_\nu(x)$. ◗

THEOREM 4 For all values of ν, $\dfrac{d[x^{-\nu}Y_\nu(x)]}{dx} = -x^{-\nu}Y_{\nu+1}(x)$.

Since $J_\nu(x)$, $J_{-\nu}(x)$, and $Y_\nu(x)$ satisfy the same identities, and since any solution of Bessel's equation can be written as a linear combination of either $J_\nu(x)$ and $J_{-\nu}(x)$ or of $J_\nu(x)$ and $Y_\nu(x)$, the following theorem, proved in the same way that Theorem 3 was proved, extends the results of Theorems 1–4 to *all* solutions of Bessel's equation.

THEOREM 5 If $y_\nu(x)$ is any solution of Bessel's equation of order ν, then

$$\frac{d[x^\nu y_\nu(x)]}{dx} = x^\nu y_{\nu-1}(x) \qquad \text{and} \qquad \frac{d[x^{-\nu}y_\nu(x)]}{dx} = -x^{-\nu}y_{\nu+1}(x)$$

For the modified Bessel functions $I_\nu(x)$ and $K_\nu(x)$, arguments similar to those we have just used lead to the following theorems.

THEOREM 6 For all values of ν,

$$\frac{d[x^\nu I_\nu(x)]}{dx} = x^\nu I_{\nu-1}(x) \quad \text{and} \quad \frac{d[x^\nu K_\nu(x)]}{dx} = -x^\nu K_{\nu-1}(x)$$

THEOREM 7 For all values of ν,

$$\frac{d[x^{-\nu} I_\nu(x)]}{dx} = x^{-\nu} I_{\nu+1}(x) \quad \text{and} \quad \frac{d[x^{-\nu} K_\nu(x)]}{dx} = -x^{-\nu} K_{\nu+1}(x)$$

If we return to the identities of Theorems 1 and 2 and perform the indicated differentiations, we obtain, respectively,

$$x^\nu J_\nu'(x) + \nu x^{\nu-1} J_\nu(x) = x^\nu J_{\nu-1}(x)$$
$$x^{-\nu} J_\nu'(x) - \nu x^{-\nu-1} J_\nu(x) = -x^{-\nu} J_{\nu+1}(x)$$

Dividing the first of these by x^ν and multiplying the second by x^ν and solving for $J'(x)$ in each case gives

(1) $$J_\nu'(x) = J_{\nu-1}(x) - \frac{\nu}{x} J_\nu(x)$$

(2) $$J_\nu'(x) = \frac{\nu}{x} J_\nu(x) - J_{\nu+1}(x)$$

Adding these and dividing by 2, we obtain a third formula for $J_\nu'(x)$:

(3) $$J_\nu'(x) = \frac{J_{\nu-1}(x) - J_{\nu+1}(x)}{2}$$

Subtracting (2) from (1) gives the important recurrence formula

$$J_{\nu-1}(x) + J_{\nu+1}(x) = \frac{2\nu}{x} J_\nu(x)$$

Written as

(4) $$J_{\nu+1}(x) = \frac{2\nu}{x} J_\nu(x) - J_{\nu-1}(x)$$

this formula serves to express Bessel functions of higher orders in terms of functions of lower orders, frequently a useful manipulation. Written as

(5) $$J_{\nu-1}(x) = \frac{2\nu}{x} J_\nu(x) - J_{\nu+1}(x)$$

it serves similarly to express Bessel functions of large negative orders (for instance) in terms of Bessel functions whose orders are numerically smaller.

EXAMPLE 1

Express $J_4(ax)$ in terms of $J_0(ax)$ and $J_1(ax)$.

Taking $\nu = 3$ in Eq. (4), we first have

$$J_4(ax) = \frac{6}{ax} J_3(ax) - J_2(ax)$$

Applying (4) again to $J_3(ax)$ and then to $J_2(ax)$, we have further

$$J_4(ax) = \frac{6}{ax}\left[\frac{4}{ax}J_2(ax) - J_1(ax)\right] - J_2(ax) = \left(\frac{24}{a^2x^2} - 1\right)J_2(ax) - \frac{6}{ax}J_1(ax)$$

$$= \left(\frac{24}{a^2x^2} - 1\right)\left[\frac{2}{ax}J_1(ax) - J_0(ax)\right] - \frac{6}{ax}J_1(ax) = \left(\frac{48}{a^3x^3} - \frac{8}{ax}\right)J_1(ax) - \left(\frac{24}{ax^2} - 1\right)J_0(ax)$$

EXAMPLE 2

Show that $d[xJ_\nu(x)J_{\nu+1}(x)]/dx = x[J_\nu^2(x) - J_{\nu+1}^2(x)]$.

Performing the indicated differentiation, we have

$$\frac{d[xJ_\nu(x)J_{\nu+1}(x)]}{dx} = J_\nu(x)J_{\nu+1}(x) + xJ'_\nu(x)J_{\nu+1}(x) + xJ_\nu(x)J'_{\nu+1}(x)$$

Then, substituting for $xJ'_\nu(x)$ from (2) and for $xJ'_{\nu+1}$ from (1), we have

$$\begin{aligned}\frac{d[xJ_\nu(x)J_{\nu+1}(x)]}{dx} &= J_\nu(x)J_{\nu+1}(x) + J_{\nu+1}(x)[\nu J_\nu(x) - xJ_{\nu+1}(x)] \\ &\quad + J_\nu(x)[xJ_\nu(x) - (\nu+1)J_{\nu+1}(x)] \\ &= x[J_\nu^2(x) - J_{\nu+1}^2(x)]\end{aligned}$$

The basic differentiation identities of Theorems 1 and 2, when written as integration formulas

(6) $$\int x^\nu J_{\nu-1}(x)\,dx = x^\nu J_\nu(x) + c$$

(7) $$\int x^{-\nu} J_{\nu+1}(x)\,dx = -x^{-\nu} J_\nu(x) + c$$

suffice for the integration of numerous simple expressions involving Bessel functions. For example, taking $\nu = 1$ in (6), we have

$$\int xJ_0(x)\,dx = xJ_1(x) + c$$

Similarly, taking $\nu = 0$ in (7), we find

$$\int J_1(x)\,dx = -J_0(x) + c$$

Usually, however, integration by parts must be used in addition to formulas (6) and (7).

EXAMPLE 3

What is $\int J_3(x)\,dx$?

With formula (7) in mind, it seems convenient to multiply and divide the integrand by x^2 before we attempt integration by parts. We thus begin with the integral in the following form:

$$\int x^2[x^{-2}J_3(x)]\,dx$$

Now, integrating by parts with

$$u = x^2 \qquad dv = x^{-2}J_3(x)\,dx$$
$$du = 2x\,dx \qquad v = -x^{-2}J_2(x)$$

we have, using (4) and (7) with $\nu = 1$,

$$\int J_3(x)\,dx = -J_2(x) + 2\int x^{-1}J_2(x)\,dx = -\left[\frac{2}{x}J_1(x) - J_0(x)\right] - 2x^{-1}J_1(x) + c$$
$$= J_0(x) - \frac{4}{x}J_1(x) + c$$

EXAMPLE 4

What is $\int \frac{J_2(3x)}{x^2}\,dx$?

Here it is convenient to multiply the numerator and denominator of the integrand by $9x^2$, getting

$$\frac{1}{9}\int (3x)^2 J_2(3x)\frac{dx}{x^4}$$

Now, integrating by parts with

$$u = (3x)^2 J_2(3x) \qquad dv = \frac{dx}{x^4}$$
$$du = (3x)^2 J_1(3x)3\,dx \qquad v = -\frac{1}{3x^3}$$

we have

$$\int \frac{J_2(3x)}{x^2}\,dx = \frac{1}{9}\left[-\frac{3J_2(3x)}{x} + 3\int 3xJ_1(3x)\frac{dx}{x^2}\right]$$

Again using integration by parts, with

$$u = 3xJ_1(3x) \qquad dv = \frac{dx}{x^2}$$
$$du = 3xJ_0(3x)3\,dx \qquad v = -\frac{1}{x}$$

we have further

$$\int \frac{J_2(3x)}{x^2}\,dx = \frac{1}{9}\left[-\frac{3J_2(3x)}{x} + 3\left\{-3J_1(3x) + 9\int J_0(3x)\,dx\right\}\right]$$
$$= -\frac{J_2(3x)}{3x} - J_1(3x) + 3\int J_0(3x)\,dx$$

The residual integral $\int J_0(3x)\,dx$ cannot be evaluated in finite form in terms of any of the Bessel functions we have encountered.

In general, an integral of the form

$$\int x^m J_n(x)\,dx$$

where m and n are integers such that $m + n \geq 0$, can be completely integrated if $m + n$ is odd but will ultimately depend upon the residual integral $\int J_0(x)\,dx$ if $m + n$ is even. For this reason $\int_0^x J_0(x)\,dx$ has now been tabulated.†

EXAMPLE 5

Show that $\displaystyle\int_0^{\pi/2} J_0(x\cos\phi)\cos\phi\,d\phi = \frac{\sin x}{x}$

As a first step, let us replace $J_0(x\cos\phi)$ by its equivalent infinite series and then perform the indicated integration term by term. This gives us

$$\int_0^{\pi/2} J_0(x\cos\phi)\cos\phi\,d\phi = \int_0^{\pi/2}\sum_{k=0}^{\infty}\left[\frac{(-1)^k(x\cos\phi)^{2k}}{2^{2k}(k!)^2}\right]\cos\phi\,d\phi$$
$$= \sum_{k=0}^{\infty}\left[\frac{(-1)^k x^{2k}}{2^{2k}(k!)^2}\int_0^{\pi/2}\cos^{2k+1}\phi\,d\phi\right]$$

The integral that occurs in the general term is given immediately by the first of **Wallis' formulas,**‡ namely,

$$\int_0^{\pi/2}\cos^{2k+1}\phi\,d\phi = \frac{2\cdot 4\cdot 6\cdots(2k)}{1\cdot 3\cdot 5\cdots(2k+1)} = \frac{2^{2k}(k!)^2}{(2k+1)!}$$

Using this, the last series becomes

$$\sum_{k=0}^{\infty}\left[\frac{(-1)^k x^{2k}}{2^{2k}(k!)^2}\cdot\frac{2^{2k}(k!)^2}{(2k+1)!}\right] = \sum_{k=0}^{\infty}\frac{(-1)^k x^{2k}}{(2k+1)!} = \frac{1}{x}\sum_{k=0}^{\infty}\frac{(-1)^k x^{2k+1}}{(2k+1)!}$$

Since the last series is just the Maclaurin expansion of $\sin x$, the given integral is equal to $(\sin x)/x$, as asserted.

EXERCISES

1. What is $d[x^{\nu}J_{\nu}(\lambda x)]/dx$? What is $d[x^{-\nu}J_{\nu}(\lambda x)]/dx$?
2. Work Example 2 by first writing it in the form
$$\frac{d}{dx}[\{x^{-\nu}J_{\nu}(x)\}\{x^{\nu+1}J_{\nu+1}(x)\}]$$
3. Show that $d[x^2J_0(x)]/dx = 2xJ_0(x) - x^2J_1(x)$. What is $d[x^2J_0(\lambda x)]/dx$?
4. **(a)** Express $J_3(x)$ in terms of $J_1(x)$ and $J_0(x)$.
   (b) Express $J_3(\lambda x^2)$ in terms of $J_1(\lambda x^2)$ and $J_0(\lambda x^2)$.
5. Express $J_{3/2}(x)$ and $J_{-3/2}(x)$ in terms of $\sin x$ and $\cos x$.
6. Show that $d[x^4J_2(x)]/dx = (x^4 + 4x^2)J_1(x) - 2x^3J_0(x)$. What is $d[x^4J_2(\lambda x)]/dx$?
7. Show that $d[x^3J_1(\lambda x)J_2(\lambda x)]/dx = \lambda x^3[J_1^2(\lambda x) + J_0(\lambda x)J_2(\lambda x)]$.
8. Show that $d[x^2J_3(\lambda x)]/dx = 5xJ_3(\lambda x) - \lambda x^2J_4(\lambda x) = -xJ_3(\lambda x) + \lambda x^2J_2(\lambda x)$ and verify that the two answers are equivalent.
9. Differentiate each of the following functions:
   (a) $x^2J_1(\lambda x^2)$ **(b)** $x^2J_1^2(\lambda x)$ **(c)** $xJ_0(\lambda x^2)$
10. Differentiate each of the following functions:
   (a) $x^4J_1(\lambda x)J_3(\lambda x)$ **(b)** $x^4J_3(\lambda x)$
   (c) $J_1(\lambda x)/xJ_2(\lambda x)$
11. Show that each of the following functions satisfies the indicated equation:
   (a) $y = xJ_1(2x)$, $(x^{-1}y')' + 4x^{-1}y = 0$
   (b) $y = x^4J_2(2x^2)$, $(x^{-7}y')' + 16x^{-5}y = 0$
12. Perform each of the following integrations:
   (a) $\int x^2J_1(2x)\,dx$ **(b)** $\int J_0(x)J_1(x)\,dx$
13. Perform each of the following integrations:
   (a) $\int x^2J_0(x)J_1(x)\,dx$ **(b)** $\int x^5J_1(x)\,dx$
14. Show that
   (a) $\int xJ_0(x)\,dx = xJ_1(x) + c$
   (b) $\int x^2J_0(x)\,dx = x^2J_1(x) + xJ_0(x) - \int J_0(x)\,dx + c$
   (c) $\int x^3J_0(x)\,dx = (x^3 - 4x)J_1(x) + 2x^2J_0(x) + c$

†*Handbook of Mathematical Functions,* Superintendent of Documents, Government Printing Office, Washington, D.C., 1965.

‡Named for the English mathematician John Wallis (1616–1703). Wallis' second formula is given in the hint for Exercise 27.

(d) $\int x^4 J_0(x)\,dx = (x^4 - 9x^2)J_1(x) + (3x^3 - 9x)J_0(x) + 9\int J_0(x)\,dx + c$

15. Show that

(a) $\displaystyle\int \frac{1}{x}J_1(x)\,dx = -J_1(x) + \int J_0(x)\,dx + c$

(b) $\int J_1(x)\,dx = -J_0(x) + c$

(c) $\int xJ_1(x)\,dx = -xJ_0(x) + \int J_0(x)\,dx + c$

(d) $\int x^2 J_1(x)\,dx = 2xJ_1(x) - x^2 J_0(x) + c$

(e) $\int x^3 J_1(x)\,dx = 3x^2 J_1(x) - (x^3 - 3x)J_0(x) - 3\int J_0(x)\,dx + c$

(f) $\int x^4 J_1(x)\,dx = (4x^3 - 16x)J_1(x) - (x^4 - 8x^2)J_0(x) + c$

16. What is **(a)** $\int xJ_0(\sqrt{x})\,dx$? **(b)** $\int xJ_2(\sqrt{x})\,dx$?

17. What is **(a)** $\int xJ_2(1 - x)\,dx$? **(b)** $\int x \ln x J_0(\lambda x)\,dx$?

18. Show that

(a) $I_\nu'(x) = I_{\nu-1}(x) - \dfrac{\nu}{x}I_\nu(x)$

(b) $I_\nu'(x) = \dfrac{\nu}{x}I_\nu(x) + I_{\nu+1}(x)$

(c) $I_\nu'(x) = \frac{1}{2}[I_{\nu+1}(x) + I_{\nu-1}(x)]$

(d) $I_{\nu-1}(x) - I_{\nu+1}(x) = \dfrac{2\nu}{x}I_\nu(x)$

19. What is

(a) $\int xI_0(x)\,dx$ **(b)** $\int x^2 I_0(x)\,dx$

(c) $\int xI_1(x)\,dx$ **(d)** $\int x^2 I_1(x)\,dx$

20. Show that

(a) $4J_\nu''(x) = J_{\nu-2}(x) - 2J_\nu(x) + J_{\nu+2}(x)$

(b) $8J_\nu'''(x) = J_{\nu-3}(x) - 3J_{\nu-1}(x) + 3J_{\nu+1}(x) - J_{\nu+3}(x)$

21. If $y_\nu(\lambda x)$ is any solution of Bessel's equation of order ν with parameter λ, show that

$$\frac{d}{dx}[x^2 y_{\nu-1}(\lambda x)y_{\nu+1}(\lambda x)] = 2x^2 y_\nu(\lambda x)\frac{d}{dx}[y_\nu(\lambda x)]$$

22. If $y_\nu(\lambda x)$ is any solution of Bessel's equation of order ν with parameter λ, show that

$$\int xy_\nu^2(\lambda x)\,dx = \tfrac{1}{2}x^2[y_\nu^2(\lambda x) - y_{\nu-1}(\lambda x)y_{\nu+1}(\lambda x)] + c$$

Hint: After integrating by parts, with $u = y_\nu^2(\lambda x)$, use the result of Exercise 21.

23. Recall the result of Exercise 11, Sec. 12.2, to show that $J_\nu(x)J_{-(\nu+1)}(x) + J_{\nu+1}(x)J_{-\nu}(x) = -(2\sin\nu\pi)/\pi x$.

24. Show that $J_\nu(x)Y_{\nu+1}(x) + J_{\nu+1}(x)Y_\nu(x) = -2/\pi x$. *Hint:* Recall the result of Exercise 12, Sec. 12.2.

25. What is $I_\nu(x)I_{-(\nu+1)}(x) - I_{\nu+1}(x)I_{-\nu}(x)$? *Hint:* Recall the result of Exercise 5, Sec. 12.3.

26. What is $I_\nu(x)K_{\nu+1}(x) + I_{\nu+1}(x)K_\nu(x)$? *Hint:* Recall the result of Exercise 6, Sec. 12.3.

27. By expressing the exponential as an infinite series and then integrating term by term, show that

$$I_0(x) = \frac{1}{\pi}\int_0^\pi \exp(x\cos\phi)\,d\phi$$

Hint: It will be helpful to use **Wallis' second formula**

$$\int_0^{\pi/2}\cos^{2k}\phi\,d\phi = \frac{1\cdot 3\cdot 5\cdots(2k-1)}{2\cdot 4\cdot 6\cdots(2k)}\cdot\frac{\pi}{2} = \frac{(2k)!}{2^{2k}(k!)^2}\cdot\frac{\pi}{2}$$

28. Show that

$$\int J_0(x)\,dx = J_1(x) + \int\frac{1}{x}J_1(x)\,dx$$
$$= J_1(x) + \frac{1}{x}J_2(x) + 1\cdot 3\int\frac{1}{x^2}J_2(x)\,dx$$
$$= J_1(x) + \frac{1}{x}J_2(x) + \frac{1\cdot 3}{x^2}J_3(x) + 1\cdot 3\cdot 5\int\frac{1}{x^3}J_3(x)\,dx$$
$$\cdots\cdots$$
$$= J_1(x) + \frac{1}{x}J_2(x) + \frac{1\cdot 3}{x^2}J_3(x) + \cdots + \frac{(2n-2)!}{2^{n-1}(n-1)!x^{n-1}}J_n(x) + \frac{(2n)!}{2^n n!}\int\frac{1}{x^n}J_n(x)\,dx$$

Hint: Use repeated integration by parts, each time taking $dv = x^{k+1}J_k(x)\,dx$.

29. Show that $\int_0^{\pi/2} J_1(x\cos\phi)\,d\phi = (1 - \cos x)/x$. *Hint:* The work proceeds very much as in Example 5.

30. Show that $\int_0^{\pi/2} J_2(x\cos\phi)\cos\phi\,d\phi = 2(1 - \cos x)/x^2 - (\sin x)/x$. *Hint:* First express $J_2(x\cos\phi)$ in terms of $J_1(x\cos\phi)$ and $J_0(x\cos\phi)$.

31. Show that $\int x[J_{\nu-1}^2(\lambda x) - J_{\nu+1}^2(\lambda x)]\,dx = 2(\nu/\lambda^2)J_\nu^2(\lambda x) + c$. *Hint:* Begin by factoring the integrand. Is this result true for all solutions of Bessel's equation of order ν with parameter λ?

32. **(a)** Show that $\int J_0(x)\cos x\,dx = xJ_0(x)\cos x + xJ_1(x)\sin x + c$.

(b) What is $\int J_0(x)\sin x\,dx$?

33. **(a)** Show that $\int J_1(x)\cos x\,dx = xJ_1(x)\cos x - J_0(x)[x\sin x + \cos x] + c$

(b) What is $\int J_1(x)\sin x\,dx$?

34. What is **(a)** $\int xJ_0(x)\cos x\,dx$? **(b)** $\int xJ_1(x)\sin x\,dx$? *Hint:* Integrate by parts in Part **(a)**, taking $u = x$ and using Exercise 32**(a)** to determine v. Then integrate by parts taking $u = \cos x$. Finally, solve these two equations for the integrals required in Parts **(a)** and **(b)**.

35. What is **(a)** $\int xJ_0(x)\sin x\,dx$? **(b)** $\int xJ_1(x)\cos x\,dx$?

36. Show that $x^2 J_2(x) = \frac{1}{2}\int_0^x t(x^2 - t^2)J_0(t)\,dt$. *Hint:* Observe that $x^2 J_2(x) = \int_0^x t^2 J_1(t)\,dt = \int_0^x t\int_0^t sJ_0(s)\,ds\,dt$ and then change the order of integration in the iterated integral.

37. Show that $x^3 J_3(x) = (1/2^2 2!)\int_0^x t(x^2 - t^2)^2 J_0(t)\,dt$

38. Verify by induction the following generalization of Exercises 36 and 37:

(8) $x^{n+1}J_{n+1}(x) = \dfrac{1}{2^n n!}\int_0^x t(x^2 - t^2)^n J_0(t)\,dt$

39. By substituting $x \sin\theta$ for t in Formula (8), show that $J_{n+1}(x) = (x^{n+1}/2^n n!)\int_0^{\pi/2} \sin\theta \cos^{2n+1}\theta\, J_0(x\sin\theta)\,d\theta$.

40. **(a)** Verify that $x^2\left(\dfrac{1}{x}\dfrac{d}{dx}\right)\left(\dfrac{1}{x}\dfrac{d}{dx}\right)J_0(x) = J_2(x)$.

(b) What is $x^3\left(\dfrac{1}{x}\dfrac{d}{dx}\right)\left(\dfrac{1}{x}\dfrac{d}{dx}\right)\left(\dfrac{1}{x}\dfrac{d}{dx}\right)J_0(x)$?

41. Using mathematical induction, prove the following generalization of Exercise 40:

$$J_n(x) = (-1)^n x^n \left(\frac{1}{x}\frac{d}{dx}\right)^n J_0(x)$$

42. Verify that $y_1 = J_0(\lambda x)$, $y_2 = Y_0(\lambda x)$, $y_3 = xJ_1(\lambda x)$, and $y_4 = xY_1(\lambda x)$ are four particular solutions of the equation

$$\left(\frac{d^2}{dx^2} + \frac{1}{x}\frac{d}{dx} + \lambda^2\right)\left(\frac{d^2}{dx^2} + \frac{1}{x}\frac{d}{dx} + \lambda^2\right)y = 0$$

Hint: Note that y_1 and y_2 are solutions of this equation because they are annihilated by the second-order differential operator. Then verify that

$$\left(\frac{d^2}{dx^2} + \frac{1}{x}\frac{d}{dx} + \lambda^2\right)y_3 = 2\lambda J_0(\lambda x)$$

and

$$\left(\frac{d^2}{dx^2} + \frac{1}{x}\frac{d}{dx} + \lambda^2\right)y_4 = 2\lambda Y_0(\lambda x)$$

43. Find a complete solution of the equation

$$\left(\frac{d^2}{dx^2} + \frac{1}{x}\frac{d}{dx} - \lambda^2\right)\left(\frac{d^2}{dx^2} + \frac{1}{x}\frac{d}{dx} - \lambda^2\right)y = 0$$

44. With Exercise 42 in mind, it should be obvious that $y_1 = J_0(\lambda x)$, $y_2 = Y_0(\lambda x)$, $y_3 = xJ_1(\lambda x)$, and $y_4 = xY_1(\lambda x)$ are solutions of the equation

$$\left(\frac{d^2}{dx^2} + \frac{1}{x}\frac{d}{dx} + \lambda^2\right)^3 y = 0$$

Infer and then verify two additional solutions of this equation which are linearly independent of y_1, y_2, y_3, and y_4.

45. How should the equation in Exercise 42 be modified if $y_1 = J_\nu(\lambda x)$, $y_2 = Y_\nu(\lambda x)$, $y_3 = xJ_{\nu+1}(\lambda x)$, and $y_4 = xY_{\nu+1}(\lambda x)$ are to be its solutions? Verify your conjecture.

46. Show that $\int_0^\infty e^{-ax} J_0(\lambda x)\,dx = 1/\sqrt{a^2+\lambda^2}$. *Hint:* Replace $J_0(\lambda x)$ by its infinite series and then integrate term by term, recognizing that in each term the integral defines a gamma function with integral argument. Then show that the resulting series is identical with the binomial expansion of $1/\sqrt{a^2+\lambda^2}$. It is interesting (and important) to note that this exercise gives us the Laplace transform of $J_0(\lambda x)$.

47. What is $\int_0^\infty xe^{-ax}J_1(\lambda x)\,dx$? *Hint:* Differentiate both sides of the formula of Exercise 46 with respect to λ.

48. What is $\int_0^\infty xe^{-ax}J_0(\lambda x)\,dx$? *Hint:* Use integration by parts on the result of Exercise 47.

49. What is $\int_0^\infty e^{-ax}J_1(\lambda x)\,dx$? *Hint:* Use integration by parts on the result of Exercise 46.

50. What is $\int_0^\infty xe^{-ax}J_2(\lambda x)\,dx$? *Hint:* Express $J_2(\lambda x)$ in terms of $J_1(\lambda x)$ and $J_0(\lambda x)$.

51. What is $\int_0^\infty x^2e^{-ax}J_1(\lambda x)\,dx$?

52. Show that $\int xJ_n^2(x)\,dx = (x^2/2)[J_n^2(x) - J_{n-1}(x)J_{n+1}(x)] + c$. *Hint:* After integrating by parts, the result of Exercise 21 may be helpful.

53. What is **(a)** $\int J_0(\sqrt{x})\,dx$? **(b)** $\int J_2(\sqrt{x})\,dx$?

54. Given $a > \lambda$. **(a)** What is $\int_0^\infty e^{-ax}I_0(\lambda x)\,dx$? **(b)** What is $\int_0^\infty e^{-ax}I_1(\lambda x)\,dx$? *Hint:* Proceed as in Exercise 46 and at the appropriate point use the result of Exercise 3, Sec. 12.3.

55. Express **(a)** $J_4(x)$, **(b)** $J_5(x)$, **(c)** $\int J_3(x)\,dx$, and **(d)** $\int_0^x \lambda^5 J_0(\lambda)\,d\lambda$ in terms of $J_0(x)$ and $J_1(x)$.

56. Show that $J_\nu''(x) = \left[\dfrac{\nu(\nu+1)}{x^2} - 1\right]J_\nu(x) - \dfrac{1}{x}J_{\nu-1}(x)$

57. **(a)** Find the solution $y = y(x)$ on $(0, \infty)$ of the initial-value problem

$$4x^2y'' + 4xy' - (25 - 4x^2)y = 0$$
$$y(0) = 0,\ y(\pi) = 3\sqrt{2}/\pi$$

(b) What is the limit of $y(x)$ as $x \to 0$? **(c)** as $x \to \infty$? **(d)** What is $\lim_{n\to\infty}(x_n - n\pi)$, where x_n denotes the nth zero of $y(x)$?

58. What is **(a)** $\int xJ_2(x)\cos x\,dx$? **(b)** $\int xJ_2(x)\sin x\,dx$?

12.6 THE GENERATING FUNCTION FOR $\{J_n(x)\}$

In addition to the identities we derived in the last section, there are others that are obtained most readily from the so-called *generating function* for the sequence $\{J_n(x)\}$.

DEFINITION 1 **(Generating Functions)** If $\{f_n(x)\}$ is an infinite sequence of functions and if $\Phi(x, t)$ is a function with the property that when it is expanded in terms of powers of t the coefficient of t^n is $f_n(x)$, then (whether the series converges or not) $\Phi(x, t)$ is said to be a **generating function** for the sequence $\{f_n(x)\}$.

Clearly, the generating function of a sequence of functions is a generalization of the idea of a generating function for a sequence of numerical values, which we encountered in our discussion of S and Z transforms in Sec. 5.9.

The generating function for the Bessel functions of the first kind of integral order is given by the following theorem.

THEOREM 1 $$\exp\frac{x}{2}\left(t-\frac{1}{t}\right) = J_0(x) + \sum_{n=1}^{\infty} J_n(x)[t^n + (-1)^n t^{-n}] = \sum_{n=-\infty}^{\infty} J_n(x)t^n$$

PROOF Our first step of course must be to express the generating function as a power series in t. To do this, it is convenient to factor it, getting

$$e^{xt/2}e^{-x/2t}$$

and then replace each factor by its Maclaurin series. This gives us

$$\left(\sum_{j=0}^{\infty}\frac{1}{j!}\frac{x^j t^j}{2^j}\right)\left(\sum_{k=0}^{\infty}\frac{(-1)^k}{k!}\frac{x^k t^{-k}}{2^k}\right)$$

When these series are multiplied together, we obtain a term containing t^n, $n \geq 0$, when and only when the general term in the second series (i.e., the term containing t^{-k}) is multiplied by the term in the first series which contains t^{n+k} (i.e., the term for which $j = n + k$). Therefore, taking into account all possible values of k, we find that the total coefficient of t^n in the product of the two series is

$$\sum_{k=0}^{\infty}\left[\frac{1}{(n+k)!}\frac{x^{n+k}}{2^{n+k}}\right]\left[\frac{(-1)^k}{k!}\frac{x^k}{2^k}\right] = \sum_{k=0}^{\infty}\frac{(-1)^k x^{n+2k}}{2^{n+2k}k!\Gamma(n+k+1)} = J_n(x)$$

Similarly, a term containing t^{-n} arises when and only when the general term in the first series (i.e., the term containing t^j), is multiplied by the term in the second series which contains t^{-n-j} (i.e., the term for which $k = n + j$). Therefore, taking into account all possible values of j, we find the total coefficient of t^{-n} in the product of the two series is

$$\sum_{j=0}^{\infty}\left[\frac{1}{j!}\frac{x^j}{2^j}\right]\left[\frac{(-1)^{n+j}}{(n+j)!}\frac{x^{n+j}}{2^{n+j}}\right] = (-1)^n\sum_{j=0}^{\infty}\frac{(-1)^j x^{n+2j}}{2^{n+2j}j!\Gamma(n+j+1)}$$
$$= (-1)^n J_n(x) = J_{-n}(x)$$

Hence, summing all powers of t, positive, negative, and zero, we have

(1) $$\exp\frac{x}{2}\left(t-\frac{1}{t}\right) = J_0(x) + \sum_{n=1}^{\infty} J_n(x)[t^n + (-1)^n t^{-n}] = \sum_{n=-\infty}^{\infty} J_n(x)t^n$$

Numerous identities can now be obtained by evaluating the generating function for specific values of t or by manipulating it in various ways.

EXAMPLE 1

If Eq. (1) is evaluated for $t = 1$, the left-hand side becomes $e^0 = 1$, while on the right if n is odd, say $n = 2k - 1$, the terms cancel, and if n is even, say $n = 2k$, the terms combine. Hence

$$1 = J_0(x) + 2\sum_{k=1}^{\infty} J_{2k}(x)$$

EXAMPLE 2

If Eq. (1) is evaluated for $t = i$, the left-hand side becomes

$$\exp \frac{x}{2}\left(i - \frac{1}{i}\right) = e^{ix} = \cos x + i \sin x$$

On the right, since $(-1)^n i^{-n} = (-1/i)^n = i^n$, we have

$$J_0(x) + \sum_{n=1}^{\infty} J_n(x) 2i^n = J_0(x) + 2[iJ_1(x) - J_2(x) - iJ_3(x) + J_4(x) + \cdots]$$

Hence, equating real and imaginary terms on the two sides, we have

(2) $$\cos x = J_0(x) - 2J_2(x) + 2J_4(x) - \cdots$$

(3) $$\sin x = 2J_1(x) - 2J_3(x) + 2J_5(x) - \cdots$$

EXAMPLE 3

If Eq. (1) is evaluated for $t = e^{i\phi}$, the left-hand side becomes

$$\exp \frac{x}{2}(e^{i\phi} - e^{-i\phi}) = e^{ix \sin \phi}$$

On the right when n is even, say $n = 2k$, we have

$$t^n + (-1)^n t^{-n} = t^{2k} + (-1)^{2k} t^{-2k} = e^{i2k\phi} + e^{-i2k\phi} = 2 \cos 2k\phi$$

Similarly, when n is odd, say $n = 2k - 1$, we have

$$\begin{aligned} t^n + (-1)^n t^{-n} = t^{2k-1} + (-1)^{2k-1} t^{-2k+1} &= e^{i(2k-1)\phi} - e^{-i(2k-1)\phi} \\ &= 2i \sin (2k - 1)\phi \end{aligned}$$

Thus, by grouping odd terms and even terms, Eq. (1) can be written

(4) $$e^{ix \sin \phi} \equiv \cos (x \sin \phi) + i \sin (x \sin \phi) = J_0(x) + 2 \sum_{k=1}^{\infty} J_{2k}(x) \cos 2k\phi + 2i \sum_{k=1}^{\infty} J_{2k-1}(x) \sin (2k - 1)\phi$$

Equating real and imaginary parts in the last two members, we obtain the identities

(5) $$\cos (x \sin \phi) = J_0(x) + 2 \sum_{k=1}^{\infty} J_{2k}(x) \cos 2k\phi$$

(6) $$\sin (x \sin \phi) = 2 \sum_{k=1}^{\infty} J_{2k-1}(x) \sin (2k - 1)\phi$$

The series on the right in (5) and (6) are of course just the Fourier expansions of the functions on the left.

If we multiply both sides of (5) by $\cos n\phi$ and both sides of (6) by $\sin n\phi$ and integrate each identity with respect to ϕ from 0 to π, remembering that

$$\int_0^{\pi} \cos m\phi \cos n\phi \, d\phi = \int_0^{\pi} \sin m\phi \sin n\phi \, d\phi = 0 \qquad m \neq n$$

$$\int_0^{\pi} \cos^2 n\phi \, d\phi = \int_0^{\pi} \sin^2 n\phi \, d\phi = \frac{\pi}{2}$$

we obtain

$$\int_0^\pi \cos n\phi \cos(x\sin\phi)\,d\phi = \begin{cases} \pi J_n(x) & n \text{ even} \\ 0 & n \text{ odd} \end{cases}$$

$$\int_0^\pi \sin n\phi \sin(x\sin\phi)\,d\phi = \begin{cases} 0 & n \text{ even} \\ \pi J_n(x) & n \text{ odd} \end{cases}$$

If we add these two expressions and divide by π, we have for all integral values of n,

$$J_n(x) = \frac{1}{\pi}\int_0^\pi [\cos n\phi\cos(x\sin\phi) + \sin n\phi\sin(x\sin\phi)]\,d\phi$$

since for every value of n, one or the other of the integrals vanishes while the remaining one contributes $J_n(x)$. Finally, using the formula for the cosine of the difference of two quantities, we have

$$(7) \qquad J_n(x) = \frac{1}{\pi}\int_0^\pi \cos(n\phi - x\sin\phi)\,d\phi$$

Formulas (2), (3) and (5)–(7) are useful in the analysis of wave forms that are phase-modulated.

EXERCISES

1. What identity results when Eq. (1) is evaluated for $t = -1$?

2. What identity is obtained when Eq. (1) is evaluated for $t = -i$?

3. What identity is obtained when Eq. (1) is evaluated for $t = ie^{i\phi}$?

4. **(a)** Show that $(1/2\pi)\int_0^{2\pi} e^{ix\sin\phi}\,d\phi = J_0(x)$. *Hint:* Integrate Eq. (4).

(b) What is $\int_0^{2\pi} e^{ix\sin\phi}\cos n\phi\,d\phi$?

(c) What is $\int_0^{2\pi} e^{ix\sin\phi}\sin n\phi\,d\phi$?

5. **(a)** Show that $(1/2\pi)\int_0^{2\pi} e^{\pm ix\cos\phi}\,d\phi = J_0(x)$. *Hint:* Use the result of Exercise 3.

(b) What is $\int_0^{2\pi} e^{ix\cos\phi}\cos n\phi\,d\phi$?

(c) What is $\int_0^{2\pi} e^{ix\cos\phi}\sin n\phi\,d\phi$?

6. By differentiating Eq. (1) with respect to x, derive Eq. (3), Sec. 12.5.

7. By differentiating Eq. (1) with respect to t, derive Eq. (4), Sec. 12.5.

8. Explain how the results of Exercises 6 and 7 can be used to obtain Theorems 1 and 2, Sec. 12.5.

9. If the generating function (1) is multiplied by

$$\frac{x}{2}\left(1 + \frac{1}{t^2}\right) = \frac{d}{dt}\left[\frac{x}{2}\left(t - \frac{1}{t}\right)\right]$$

and integrated with respect to t, what identity is obtained by comparing the coefficients of like powers of t?

10. By integrating Eq. (1) with respect to x, show that $\int J_0(x)\,dx = 2[J_1(x) + J_3(x) + J_5(x) + \cdots]$. *Hint:* On the left-hand side, after integrating, express $1/\frac{1}{2}[t - (1/t)]$ as a power series in $1/t$. What is $\int J_1(x)\,dx$? $\int J_2(x)\,dx$? $\int J_3(x)\,dx$?

11. Show that $\exp(x/2)(t + 1/t)$ is the generating function for $\{I_n(x)\}$.

12. What identity is obtained from the generating function of $\{I_n(x)\}$ given in Exercise 11 if **(a)** $t = 1$? **(b)** $t = -1$?

13. What identity is obtained if t is set equal to $e^{i\phi}$ in the generating function for $\{I_n(x)\}$ given in Exercise 11?

14. What identity is obtained if t is set equal to $ie^{i\phi}$ in the generating function for $\{I_n(x)\}$ given in Exercise 11?

15. Show that $\sum_{n=1}^{\infty}(2n-1)J_{2n-1}(x) = x/2$. *Hint:* Differentiate Eq. (6) and then set $\phi = 0$.

16. By suitably differentiating and evaluating Eq. (5), show that

(a) $\displaystyle\sum_{n=1}^{\infty}(2n)^2 J_{2n}(x) = \frac{1}{2}x^2$

(b) $\displaystyle\sum_{n=1}^{\infty}(-1)^{n-1}(2n)^2 J_{2n}(x) = \frac{x\sin x}{2}$

17. By multiplying the generating functions for the sequences $\{J_n(x)\}$ and $\{J_n(y)\}$, show that

$J_0(x+y) = \displaystyle\sum_{k=-\infty}^{\infty} J_k(x)J_{-k}(y)$. What is $J_n(x+y)$?

18. Using the result of Exercise 17, show that

(a) $\displaystyle J_0(2x) = J_0^2(x) + 2\sum_{k=1}^{\infty}(-1)^k J_k^2(x)$

(b) $\displaystyle 1 = J_0^2(x) + 2\sum_{k=1}^{\infty} J_n^2(x)$

19. Show that

$$\exp\left[-\frac{x}{2t}\left(k - \frac{1}{k}\right)\right]\cdot\sum_{n=-\infty}^{\infty} J_n(x)(kt)^n = \sum_{n=-\infty}^{\infty} J_n(kx)t^n$$

20. By putting $k = i$ in the result of Exercise 19, show that

$$I_n(x) = \sum_{k=0}^{\infty}\frac{x^k}{k!}J_{n+k}(x)$$

12.7 THE ORTHOGONALITY OF THE BESSEL FUNCTIONS

As we shall see in the next section, Bessel functions are useful in a variety of applications. Among these, the most important are probably boundary-value problems in which Bessel functions enter as the characteristic functions in terms of which some other function is to be expanded. To carry out such expansions, it is necessary, as we learned in Sec. 11.6, that the characteristic functions form an orthogonal set. The following theorem guarantees that under suitable conditions the Bessel functions do form an orthogonal set.

THEOREM 1 If $\lambda_1 < \lambda_2 < \lambda_3 < \cdots$ are values of λ to which there correspond nontrivial solutions $J_\nu(\lambda_1 x)$, $J_\nu(\lambda_2 x)$, $J_\nu(\lambda_3 x)$, . . . , of Bessel's equation $x^2y'' + xy' + (\lambda^2x^2 - \nu^2)y = 0$ which satisfy conditions of the form

(a) $a_1 y(x_1) - b_1 y'(x_1) = 0 \quad a_1, b_1$ not both zero

(b) $a_2 y(x_2) - b_2 y'(x_2) = 0 \quad a_2, b_2$ not both zero

where $0 < x_1 < x_2$, then the Bessel functions $\{J_\nu(\lambda_n x)\}$ form an orthogonal set with respect to the weight function x on the interval $x_1 < x < x_2$; that is,

$$\int_{x_1}^{x_2} xJ_\nu(\lambda_m x)J_\nu(\lambda_n x)\,dx = 0 \qquad m \neq n$$

Solutions of Bessel's equation which are bounded in the neighborhood of the origin and satisfy Condition **(b)** are orthogonal with respect to the weight function x on the interval $0 < x < x_2$.

PROOF If Bessel's equation in its usual form is divided by x, we obtain

$$xy'' + y' + \left(-\frac{\nu^2}{x} + \lambda^2 x\right)y = 0 \qquad \text{or} \qquad (xy')' + \left(-\frac{\nu^2}{x} + \lambda^2 x\right)y = 0$$

Clearly, the last equation is a special case of the equation covered by the Sturm-Liouville theorem (Theorem 4, Sec. 11.6) with

$$p(x) = x \qquad q(x) = -\frac{\nu^2}{x} \qquad r(x) = x$$

and λ^2 written in place of λ. Moreover, Conditions **a** and **b** are precisely the conditions imposed on the solutions by the Sturm-Liouville theorem. Hence, without further discussion, it follows that the first assertion of the theorem is true.

If $x_1 = 0$, then $r(x_1) = 0$ and, according to the corollary of the Sturm-Lioville theorem, a boundary condition of the form **a** will not be needed (and will not be available). In this case the only requirement for orthogonality in addition to **b** is that the solution y be bounded in the neighborhood of the origin. This verifies the second assertion of the theorem.

The modified Bessel equation

$$x^2y'' + xy' + (-\lambda^2x^2 - \nu^2)y = 0 \qquad \text{or} \qquad (xy')' + \left(-\frac{\nu^2}{x} - \lambda^2 x\right)y = 0$$

is also an equation of the form covered by Theorem 4, Sec. 11.6. Its solutions are nonoscillatory. In general, none of them, and in particular cases (see Exercises 6–16) at most one modified Bessel function, can satisfy auxiliary conditions like those of Theorem 1. Consequently, neither the hypothesis concerning the existence of infinitely many characteristic functions nor the concluding orthogonality condition of the Sturm-Liouville theorem applies. In fact, we have the following theorem.

THEOREM 2 The solutions of a Sturm-Liouville system whose differential equation is a modified Bessel equation do not form an orthogonal set with respect to any weight function over any interval of the x axis.

The conclusions of Theorem 1 can be extended significantly by noting that the equation of Corollary 1, Theorem 1, Sec. 12.4, namely,

$$(x^r y')' + (bx^{r-2} + ax^s)y = 0$$

is also an equation of the form covered by Theorem 4, Sec. 11.6, provided we make the identifications

$$r(x) = x^r \qquad p(x) = x^s \qquad \lambda = a \qquad q(x) = bx^{r-2}$$

Hence if suitable necessary conditions on the coefficient functions are met, then solutions satisfying the appropriate boundary conditions will be orthogonal. The first of these conditions is that $r(x) = x^r$ and $p(x) = x^s$ be continuous on the closed interval $x_1 \le x \le x_2$. This is automatically true if $x_1 > 0$. If $x_1 = 0$, it will be true only if the exponents r and s are nonnegative. The second condition is that $q(x) = bx^{r-2}$ be continuous at least on the open interval $x_1 < x < x_2$, and this is always the case. Thus we have the following theorem.

THEOREM 3 If $0 < x_1 < x_2$, solutions of the equation

$$(x^r y')' + (bx^{r-2} + ax^s)y = 0$$

which satisfy the conditions

(**a**) $a_1 y(x_1) - b_1 y'(x_1) = 0 \quad a_1,\ b_1$ not both zero
(**b**) $a_2 y(x_2) - b_2 y'(x_2) = 0 \quad a_2,\ b_2$ not both zero

and correspond to distinct values of the parameter a, are orthogonal with respect to the weight function x^s over the interval (x_1, x_2).

If $x_1 = 0$ and $r,\ s \ge 0$, solutions which are bounded in the neighborhood of the origin and satisfy Condition **b** are orthogonal with respect to the weight function x^s over the interval $(0, x_2)$.

As we learned in Sec. 11.6, the orthogonality of the characteristic functions of a boundary-value problem guarantees that (at least formally) an arbitrary integrable function can be expanded in an infinite series of the characteristic functions. However, for practical purposes the numerical values of the coefficients in such expansions must be known; and before these can be found, the integral of the weighted square of the general characteristic function,

$$\int_{x_1}^{x_2} p(x) y_n^2(x)\, dx$$

must be determined. This is sometimes a difficult matter, but for the Bessel functions covered by Theorem 1, the required antiderivative was obtained in Exercise 22, Sec. 12.5. However, the antiderivative we found there involves not only $y_\nu(\lambda x)$ but also $y_{\nu-1}(\lambda x)$ and $y_{\nu+1}(\lambda x)$ and hence is inconvenient to use with the usual boundary conditions, which involve only $y_\nu(\lambda x)$ and $y_\nu'(\lambda x)$. The following theorem restates the result of that exercise in a form more suited to our needs.

THEOREM 4 If $y_\nu(\lambda x)$ is any solution of Bessel's equation of order ν with a parameter λ, then,

$$\int x y_\nu^2(\lambda x)\, dx = \frac{1}{2\lambda^2}\left[(\lambda^2 x^2 - \nu^2) y_\nu^2(\lambda x) + x^2 \left\{\frac{dy_\nu(\lambda x)}{dx}\right\}^2\right] + c$$

PROOF We begin with the result of Exercise 22, Sec. 12.5, namely,

$$\int xy_\nu^2(\lambda x)\,dx = \frac{1}{2}x^2y_\nu^2(\lambda x) - \frac{1}{2}x^2y_{\nu-1}(\lambda x)y_{\nu+1}(\lambda x) + c \tag{1}$$

and in the second term on the right we replace $y_{\nu-1}(\lambda x)$ and $y_{\nu+1}(\lambda x)$ by their equivalents as given, respectively, by Eqs. (1) and (2), Sec. 12.5. This term then becomes

$$\frac{-1}{2}x^2\left[\frac{dy_\nu(\lambda x)}{d(\lambda x)} + \frac{\nu}{\lambda x}y_\nu(\lambda x)\right]\left[\frac{\nu}{\lambda x}y_\nu(\lambda x) - \frac{dy_\nu(\lambda x)}{d(\lambda x)}\right] = \frac{-1}{2}x^2\left[\frac{\nu^2}{\lambda^2x^2}y_\nu^2(\lambda x) - \frac{1}{\lambda^2}\left\{\frac{dy_\nu(\lambda x)}{dx}\right\}^2\right]$$

Substituting this into Eq. (1), we have finally

$$\int xy_\nu^2(\lambda x)\,dx = \frac{1}{2}x^2y_\nu^2(\lambda x) - \frac{1}{2}x^2\left[\frac{\nu^2}{\lambda^2x^2}y_\nu^2(\lambda x) - \frac{1}{\lambda^2}\left\{\frac{dy_\nu(\lambda x)}{dx}\right\}^2\right] + c$$

$$= \frac{1}{2\lambda^2}\left[(\lambda^2x^2 - \nu^2)y_\nu^2(\lambda x) + x^2\left\{\frac{dy_\nu(\lambda x)}{dx}\right\}^2\right] + c \tag{2}$$

In the applications we have in mind, the parameter λ in Eq. (2) will be one of the set of characteristic values $\{\lambda_m\}$ determined by imposing boundary conditions of the form

$$a_iy_\nu(\lambda_m x_i) - b_i\left.\frac{dy_\nu(\lambda_m x)}{dx}\right|_{x=x_i} = 0 \qquad \begin{cases} i = 1, 2 \\ a_i, b_i \text{ not both zero}\end{cases} \tag{3}$$

on the general solution of Bessel's equation. There are thus six possible cases to consider. In the first place, if $x_1 \neq 0$, each of the conditions (3) is relevant and there are four possibilities according as b_1 and b_2 are, independently, equal to zero or different from zero. If $x_1 = 0$, the requirement at $x_1 = 0$ is simply that the solution be bounded in the neighborhood of the origin, and only the condition at $x = x_2$ is relevant. In this case there are just two possibilities: $b_2 = 0$ or $b_2 \neq 0$. The value of the integral (2) in these six cases is given in the next theorem.

THEOREM 5 For solutions of Bessel's equation of order ν which satisfy boundary conditions of the form

$$a_iy_\nu(\lambda_m x_i) - b_i\left.\frac{dy_\nu(\lambda_m x_i)}{dx}\right|_{x=x_i} = 0 \qquad i = 1,\ 2;\ x_1 \neq 0;\ a_i,\ b_i \text{ not both zero}$$

the value of the integral

$$\int_{x_1}^{x_2} xy_\nu^2(\lambda_m x)\,dx$$

is

(i) $\dfrac{x_2^2}{2}y_{\nu+1}^2(\lambda_m x_2) - \dfrac{x_1^2}{2}y_{\nu+1}^2(\lambda_m x_1);\ b_1 = b_2 = 0$

(ii) $\dfrac{x_2^2}{2}y_{\nu+1}^2(\lambda_m x_2) - \dfrac{y_\nu^2(\lambda_m x_1)}{2\lambda_m^2}\left[(\lambda_m x_1)^2 - \nu^2 + \left(\dfrac{x_1a_1}{b_1}\right)^2\right];\ b_1 \neq 0,\ b_2 = 0$

(iii) $\dfrac{y_\nu^2(\lambda_m x_2)}{2\lambda_m^2}\left[(\lambda_m x_2)^2 - \nu^2 + \left(\dfrac{x_2a_2}{b_2}\right)^2\right] - \dfrac{x_1^2}{2}y_{\nu+1}^2(\lambda_m x_1);\ b_1 = 0,\ b_2 \neq 0$

(iv) $\dfrac{y_\nu{}^2(\lambda_m x_2)}{2\lambda_m^2}\left[(\lambda_m x_2)^2 - \nu^2 + \left(\dfrac{x_2 a_2}{b_2}\right)^2\right] - \dfrac{y_\nu^2(\lambda_m x_1)}{2\lambda_m^2}\left[(\lambda_m x_1)^2 - \nu^2 + \left(\dfrac{x_1 a_1}{b_1}\right)^2\right]; \begin{matrix} b_1 \neq 0 \\ b_2 \neq 0 \end{matrix}$

If $x_1 = 0$, the value of the integral is

(v) $\dfrac{x_2^2}{2} y_{\nu+1}^2(\lambda_m x_2);\ b_2 = 0$

(vi) $\dfrac{y_\nu^2(\lambda_m x_2)}{2\lambda_m^2}\left[(\lambda_m x_2)^2 - \nu^2 + \left(\dfrac{x_2 a_2}{b_2}\right)^2\right];\ b_2 \neq 0$

PROOF To prove this theorem there are just two calculations to make: the evaluation of the antiderivative in Theorem 4 when $b_i = 0$, and its evaluation when $b_i \neq 0$. Each of the six cases can then be obtained by suitably pairing these two results.

If $b_i = 0$, the boundary condition at $x = x_i$ is simply $y_\nu(\lambda_m x_i) = 0$, and the antiderivative given by Theorem 4 becomes

$$\frac{1}{2\lambda_m^2}\left(x_i \left.\frac{dy_\nu(\lambda_m x)}{dx}\right|_{x=x_i}\right)^2 \tag{4}$$

This can be further simplified by recalling from the last section that all solutions of Bessel's equation J_ν, $J_{-\nu}$, Y_ν, $H_\nu^{(1)}$, and $H_\nu^{(2)}$, as well as arbitrary linear combinations of these functions, satisfy the identity [Eq. (2), Sec. 12.5],

$$t\frac{dy_\nu(t)}{dt} = \nu y_\nu(t) - t y_{\nu+1}(t)$$

or

$$x\frac{dy_\nu(\lambda_m x)}{dx} = \nu y_\nu(\lambda_m x) - \lambda_m x y_{\nu+1}(\lambda_m x)$$

Evaluating this at $x = x_i$ and again using the fact that when $b_i = 0$, the boundary condition becomes simply $y_\nu(\lambda_m x_i) = 0$, we find

$$x\left.\frac{dy_\nu(\lambda_m x_i)}{dx}\right|_{x=x_i} = -\lambda_m x_i y_{\nu+1}(\lambda_m x_i)$$

Hence the value of the antiderivative (4) is just

$$\frac{x_i^2}{2} y_{\nu+1}^2(\lambda_m x_i)$$

If $b_i \neq 0$, we can solve for the derivative term in the boundary condition. Then, substituting it into the antiderivative given in Theorem 4, we have at $x = x_i$

$$\frac{1}{2\lambda_m^2}\left[x_i^2\left\{\frac{a_i}{b_i} y_\nu(\lambda_m x_i)\right\}^2 + \{(\lambda_m x_i)^2 - \nu^2\} y_\nu^2(\lambda_m x_i)\right]$$

or, factoring out $y_\nu^2(\lambda_m x_i)$,

$$\frac{1}{2\lambda_m^2} y_\nu^2(\lambda_m x_i)\left[(\lambda_m x_i)^2 - \nu^2 + \left(\frac{a_i x_i}{b_i}\right)^2\right]$$

By suitably pairing the two results we have now obtained, the six assertions of the theorem can be established. ◗

For the more general functions shown to be orthogonal by Theorem 3, it is possible to prove the following result, analogous to Theorem 4. (See Exercise 4.)

THEOREM 6 If y is any solution of the equation

$$(x^r y')' + (\lambda^2 x^s + b x^{r-2})y = 0 \qquad r, s \geq 0 \tag{5}$$

then

$$\int x^s y^2\, dx = \frac{x^r}{(2-r+s)\lambda^2}\left[x(y')^2 + (r-1)yy' + \left(\frac{b}{x} + \lambda^2 x^{1-r+s}\right)y^2\right] + c$$

When solutions of Eq. (5) satisfy the boundary conditions of Theorem 3, it is a straightforward matter to evaluate the antiderivative given by Theorem 6 and compute the definite integral

$$\int_{x_1}^{x_2} x^s y^2\, dx$$

which must be known before a given function can be expanded in terms of these solutions.

EXAMPLE 1

Expand the function $f(x) = 4x - x^3$ over the interval [0, 2] in terms of the Bessel functions of order 1 which are bounded at the origin and satisfy the boundary condition $y(2) = 0$.

A complete solution of Bessel's equation of order 1 with a parameter is

$$y = c_1 J_1(\lambda x) + c_2 Y_1(\lambda x)$$

Since $Y_1(\lambda x)$ is unbounded in the neighborhood of the origin, it follows that $c_2 = 0$. The boundary condition at $x = 2$ then requires that $c_1 J_1(2\lambda) = 0$. If $c_1 = 0$, the solution is identically zero and of no use to us. Hence the parameter λ must satisfy the equation

$$J_1(2\lambda) = 0$$

This is the characteristic equation of the problem. Now the positive roots of the equation $J_1(z) = 0$ are†

$$z_1 \doteq 3.832 \qquad z_2 \doteq 7.016 \qquad z_3 \doteq 10.174$$

and so forth. Hence, since $z = 2\lambda$, the characteristic values of the problem are

$$\lambda_1 \doteq 1.916 \qquad \lambda_2 \doteq 3.508 \qquad \lambda_3 \doteq 5.087$$

and so forth. The characteristic functions are then

$$\begin{aligned} y_1 &= J_1(\lambda_1 x) \doteq J_1(1.916x) \\ y_2 &= J_1(\lambda_2 x) \doteq J_1(3.508x) \\ y_3 &= J_1(\lambda_3 x) \doteq J_1(5.087x) \\ &\cdots\cdots\cdots \end{aligned}$$

and we must now determine the values of the coefficients in the proposed expansion

$$4x - x^3 = C_1 J_1(\lambda_1 x) + C_2 J_1(\lambda_2 x) + C_3 J_1(\lambda_3 x) + \cdots + C_n J_1(\lambda_n x) + \cdots \tag{6}$$

†See, for instance, Eugene Jahnke, Fritz Emde, and Friedrich Losch, *Tables of Higher Functions,* 6th ed., p. 192, McGraw-Hill, New York, 1960; or *Handbook of Mathematical Functions,* p. 409. Government Printing Office, Washington, D.C., 1965.

As we observed before, when they satisfy suitable conditions on an interval I: $x_1 \le x \le x_2$, Bessel functions in general, and the functions $\{J_1(\lambda_n x)\}$ in particular, are orthogonal with respect to the weight function x on the interval I. Therefore to find C_n in the series (6) we multiply the series by $xJ_1(\lambda_n x)$ and then integrate term by term from $x = 0$ to $x = 2$, that is, over the interval determined by the boundary conditions. From the orthogonality of the Bessel functions $\{J_1(\lambda_n x)\}$, it follows that every term on the right except the nth is equal to zero, and we have

$$\int_0^2 (4x - x^3)xJ_1(\lambda_n x)\,dx = C_n \int_0^2 xJ_1^2(\lambda_n x)\,dx$$

From Part **(v)**, of Theorem 5 (i.e., $x_1 = 0$ and $b_2 = 0$), the integral on the right is equal to $2J_2^2(2\lambda_n)$. Hence

$$C_n = \frac{\displaystyle\int_0^2 (4x^2 - x^4)J_1(\lambda_n x)\,dx}{2J_2^2(2\lambda_n)}$$

For the first term in the numerator of C_n we have, from Eq. (6), Sec. 12.5,

$$4\int_0^2 x^2 J_1(\lambda_n x)\,dx = \frac{4}{\lambda_n^3}\int_0^2 (\lambda_n x)^2 J_1(\lambda_n x)\,d(\lambda_n x) = \frac{4}{\lambda_n^3}(\lambda_n x)^2 J_2(\lambda_n x)\Big|_0^2$$
$$= \frac{16}{\lambda_n} J_2(2\lambda_n)$$

For the second integral in the numerator of C_n we have, from Exercise 15**(f)**, Sec. 12.5,

$$\int_0^2 x^4 J_1(\lambda_n x)\,dx = \frac{1}{\lambda_n^5}\int_0^2 (\lambda_n x)^4 J_1(\lambda_n x)\,d(\lambda_n x)$$
$$= \frac{1}{\lambda_n^5}[\{4(\lambda_n x)^3 - 16(\lambda_n x)\}J_1(\lambda_n x) - \{(\lambda_n x)^4 - 8(\lambda_n x)^2\}J_0(\lambda_n x)]_0^2$$
$$= \left(-\frac{16}{\lambda_n} + \frac{32}{\lambda_n^3}\right)J_0(2\lambda_n)$$

Hence

$$C_n = \frac{\dfrac{16}{\lambda_n}J_2(2\lambda_n) - \left(-\dfrac{16}{\lambda_n} + \dfrac{32}{\lambda_n^3}\right)J_0(2\lambda_n)}{2J_2^2(2\lambda_n)}$$

This can be simplified significantly by using Eq. (4), Sec. 12.5, to express J_2 in terms of J_1 and J_0. Thus

$$J_2(2\lambda_n) = \frac{2}{2\lambda_n}J_1(2\lambda_n) - J_0(2\lambda_n) = -J_0(2\lambda_n)$$

since the characteristic equation of this problem is $J_1(2\lambda) = 0$. Making this substitution for $J_2(2\lambda_n)$ and simplifying, we obtain as the final form of C_n

$$C_n = -\frac{16}{\lambda_n^3 J_0(2\lambda_n)}$$

The required expansion is therefore

$$4x - x^3 = -16\sum_{n=1}^{\infty} \frac{1}{\lambda_n^3 J_0(2\lambda_n)} J_1(\lambda_n x)$$

Plots showing the degree to which the first term and first two terms of this series approximate the graph of $4x - x^3$ are shown in Fig. 12.7.

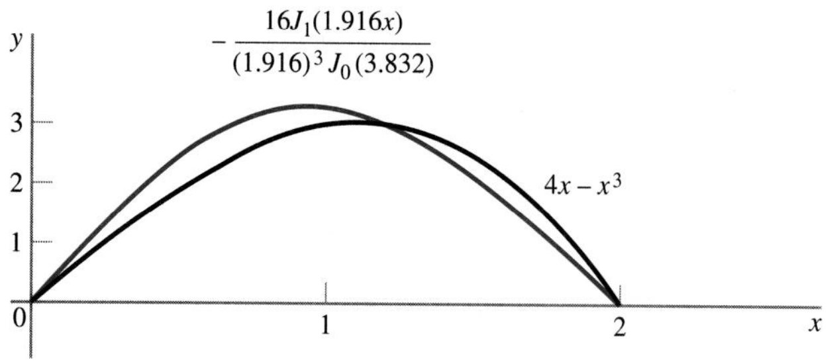

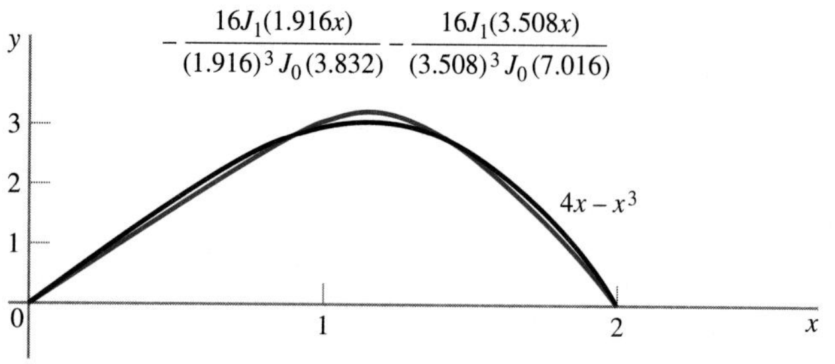

FIGURE 12.7
The approximation of a function by the first two terms of a Bessel function expansion.

EXERCISES

1. Using as a model the flow of heat in a thin circular disk or a thin circular annulus where the temperature is a function of r only, identify physical boundary conditions corresponding to each of the six cases of Theorem 5.
2. Work Exercise 1 using as a model the flow of heat in a sphere or a spherical shell where the temperature is a function of r only.
3. Verify by differentiation that the antiderivative obtained in Theorem 4 is correct.
4. Verify by differentiation that the antiderivative given in Theorem 6 is correct.
5. Verify by differentiation that

$$\int xI_\nu^2(\lambda x)\,dx = \frac{1}{2\lambda^2}[(\lambda^2x^2+\nu^2)I_\nu(\lambda x) - x^2\{I_\nu'(\lambda x)\}^2]$$

6. **(a)** Show that the Bessel functions of order zero which are bounded in the neighborhood of the origin and satisfy the boundary condition $y'(x_2) = 0$ form an incomplete set of orthogonal functions with respect to the weight function x on the interval $(0, x_2)$. *Hint:* Try to expand the function $f(x) = 1$.
   (b) In Part **(a)**, should the function $y = 1$ be adjoined to the set of Bessel functions as one of the characteristic functions of the problem? *Hint:* Show that there is a value of λ for which $y = 1$ is a solution of Bessel's equation of order zero which satisfies the boundary condition at $x = x_2$.
7. **(a)** Show that the Bessel functions of order ν which are bounded in the neighborhood of the origin and satisfy the boundary condition $\nu y(x_2) - x_2 y'(x_2)$ form an incomplete set of orthogonal functions with respect to the weight function x on the interval $(0, x_2)$. *Hint:* Try to expand the function $f(x) = x^\nu$.
   (b) In Part **(a)**, should $y = x^\nu$ be adjoined to the set of Bessel functions as one of the characteristic functions of the problem? *Hint:* Show that there is a value of λ for which $y = x^\nu$ is a solution of Bessel's equation of order ν which satisfies the boundary condition at $x = x_2$.
8. Are the observations of Exercise 7 valid if the boundary condition is $\nu y(x_2) + x_2 y'(x_2) = 0$?
9. If $\{y(\lambda_n x)\}$ is the set of all solutions of the equation $x^2y'' + xy' + (\lambda^2x^2 - \nu^2)y = 0$ which are bounded in the neighborhood of the origin and satisfy the boundary condition

$$ay(x_2) - by'(x_2) = 0 \qquad a,\, b \text{ positive}$$

   show that
   (a) If $ax_2 - b\nu < 0$, the set $\{y(\lambda_n x)\}$ consists exclusively of Bessel functions of the first kind of order ν.
   (b) If $ax_2 - b\nu = 0$, then in addition to Bessel functions of the first kind of order ν, the set $\{y(\lambda_n x)\}$ contains $y = x^\nu$.
   (c) If $ax_2 - b\nu > 0$, then in addition to Bessel functions of the first kind of order ν, the set $\{y(\lambda_n x)\}$ contains at least one function of the form $I_\nu(\lambda x)$.
   Hint: Consider the possibility that in the given equation the parameter λ^2 may be zero or negative and that the resulting solution(s) may satisfy the boundary condition.
10. In Exercise 9, show that if a and b are of opposite sign, the set $\{y(\lambda_n x)\}$ consists exclusively of Bessel functions of the first kind of order ν.

11. If λ_n is *any* root of the equation

$$J_0(\lambda) - \left.\frac{dJ_0(\lambda x)}{dx}\right|_{x=1} = J_0(\lambda) + \lambda J_1(\lambda) = 0$$

and if μ is *the* root of the equation

$$I_0(\mu) - \left.\frac{dI_0(\mu x)}{dx}\right|_{x=1} = I_0(\mu) - \mu I_1(\mu) = 0$$

Show that $I(\mu x)$ is orthogonal to every member of the set $\{J_0(\lambda_n x)\}$ with respect to the weight function x on the interval $(0, 1)$. *Hint:* Integrate $\int_0^1 xI_0(\mu x)J_0(\lambda_n x)\,dx$ by parts, using two different choices for u and dv and add the results. Then add and subtract the appropriate quantity in the antiderivative.

12. Generalize Exercise 11 by showing that if λ_n is *any* root of the equation

$$aJ_\nu(\lambda x_2) - b\left.\frac{dJ_\nu(\lambda x)}{dx}\right|_{x=x_2} = 0 \qquad a, b \text{ positive} \qquad ax_2 - b\nu > 0$$

and if μ is *the* root of the equation

$$aI_\nu(\mu x) - b\left.\frac{dI_\nu(\mu x)}{dx}\right|_{x=x_2} = 0$$

then $I_\nu(\mu x)$ is orthogonal to every member of the set $\{J_\nu(\lambda_n x)\}$ with respect to the weight function x on the interval $(0, x_2)$. *Hint:* In the integral $\int_0^{x_2} xJ_\nu(\lambda_n x)I_\nu(\mu x)\,dx$ put $x = x^{-\nu}x^{\nu+1}$, then integrate using two different choices for u and dv and combine the results.

13. **(a)** What is $\int xI_\nu(\lambda_1 x)I_\nu(\lambda_2 x)\,dx$? *Hint:* Notice the hint in Exercise 12.

(b) Using the result of Part **(a)**, show that the equation

$$aI_\nu(\lambda x_2) - b\left.\frac{dI_\nu(\lambda x)}{dx}\right|_{x=x_2} = 0 \text{ has at most one root.}$$

Hint: Note that the integral in Part **(a)** can never be zero. [*Note:* This result shows that in Exercise **9c** there is only one function of the form $I_\nu(\lambda x)$. It also justifies the italicized word *the* in Exercises 11 and 12.]

14. Determine the value of λ for which $I_1(\lambda x)$ satisfies the boundary condition $2y(\lambda x) - d[y(\lambda x)]/dx = 0$ at $x = 1$. *Hint:* Show that the given condition can be written in either of the forms $I_1(\lambda) - \lambda I_2(\lambda) = 0$ or $\lambda I_0(\lambda) - 3I_1(\lambda) = 0$; then use tables of modified Bessel functions to solve one of these equations by trial and error.

15. Determine the value of λ for which $I_1(\lambda x)$ satisfies the boundary condition $2y(\lambda x) - 3d[y(\lambda x)]/dx = 0$ at $x = 2$. Can this condition be satisfied at $x = 3$? At $x = 1$?

16. Determine the value of λ for which $I_2(\lambda x)$ satisfies the boundary condition $3y(\lambda x) - d[y(\lambda x)]/dx = 0$ at $x = 1$.

In Exercises 17 through 30, expand the indicated function over the given interval in terms of the set of all solutions of Bessel's equation of the indicated order that are bounded in the neighborhood of the origin and satisfy the prescribed boundary condition. *Hint:* Keep in mind the observations made in Exercise 9, and in particular exercises recall the results of Exercises 14 and 15, Sec. 12.5.

17. $f(x) = 1$; $\nu = 0$, $y(3) = 0$
18. $f(x) = 1$; $\nu = 0$, $y'(3) = 0$
19. $f(x) = 1$; $\nu = 0$, $y(3) - y'(3) = 0$
20. $f(x) = 1$; $\nu = 0$, $y(3) + y'(3) = 0$
21. $f(x) = x$; $\nu = 1$, $y(3) = 0$
22. $f(x) = x$; $\nu = 1$, $y'(3) = 0$
23. $f(x) = x$; $\nu = 1$, $y(3) - 4y'(3) = 0$
24. $f(x) = x$; $\nu = 1$, $y(3) - 3y'(3) = 0$
25. $f(x) = x$; $\nu = 1$, $y(3) - 2y'(3) = 0$
26. $f(x) = 4x - x^3$; $\nu = 1$, $y'(2) = 0$
27. $f(x) = 4x - x^3$; $\nu = 1$, $y(2) + y'(2) = 0$
28. $f(x) = x^2$; $\nu = 2$, $y(2) = 0$
29. $f(x) = x^2$; $\nu = 2$, $y'(2) = 0$
30. $f(x) = x^2$; $\nu = 2$, $y(2) - y'(2) = 0$

31. Expand the function $f(x) = \sqrt{x}$ over the interval $(0, \pi)$ in terms of the solutions of Bessel's equation of order $\frac{1}{2}$ which satisfy the condition $y(\pi) = 0$ and are bounded in the neighborhood of the origin. Is this a Fourier series?

32. Work Exercise 31 if $\nu = \frac{3}{2}$.

33. If solutions of Bessel's equation of order ν are to satisfy the boundary conditions $y(1) = y(2) = 0$, what is the characteristic equation of the problem?

34. Work Exercise 33 if the boundary conditions are $y(1) = y'(2) = 0$.

35. Work Exercise 33 if the boundary conditions are $y'(1) = y'(2) = 0$.

36. Expand $f(x) = \sqrt{x}$ over the interval $(0, 1)$ in terms of the solutions of the equation $y'' + (\lambda^2 x^2 + 1/4x^2)y = 0$ which are bounded in the neighborhood of the origin and satisfy the boundary condition $y(1) = 0$.

37. If $\lambda_1, \lambda_2, \lambda_3, \ldots,$ are the zeros of $J_0(x)$, show that

$$\sum_{i=1}^{\infty}\frac{1}{\lambda_i^2} = \frac{1}{4}$$

Hint: Assume that $J_0(x)$ can be written in the factored form

$$J_0(x) = \left(1 - \frac{x}{\lambda_1}\right)\left(1 + \frac{x}{\lambda_1}\right)\cdots\left(1 - \frac{x}{\lambda_i}\right)\left(1 + \frac{x}{\lambda_i}\right)\cdots$$

$$= \prod_{i=1}^{\infty}\left(1 - \frac{x^2}{\lambda_i^2}\right)$$

and note that the required sum is the coefficient of x^2 in the infinite product.

38. What is the sum of the reciprocals of the squares of the nonzero roots of $J_1(x) = 0$?

39. Find the value of the sum $\Sigma_{i=1}^{\infty}(1/\lambda_i^4)$ where, as in Exercise 37, the λ's are the zeros of $J_0(x)$. *Hint:* Consider the infinite product expansions of both $J_0(x)$ and $J_0^2(x)$.

40. Expand the function $f(x) = x^2$ on $[0, 2]$ in terms of the solutions of the equation $[xy']' + \lambda^2 xy = 0$ for which $y(0^+)$ is finite and $y'(2) = 0$.

41. Expand the function $f(x) = 1$ on $[0, 1]$ in terms of the solutions of the equation $[xy']' + (\lambda^2 x - 1/x)y = 0$ for which $y(0^+)$ is finite and $y(1) = 0$.

12.8 APPLICATIONS OF BESSEL FUNCTIONS

Bessel functions occur in a great many applied problems. In principle they are always to be expected when partial differential equations are used in the study of configurations possessing circular symmetry. On the other hand, they also arise in numerous problems where neither circular symmetry nor partial differential equations are involved. In this section we shall conclude our study of Bessel functions by discussing a variety of problems where their use is required.

EXAMPLE 1

NATURAL FREQUENCIES OF A HANGING CABLE

A uniform, perfectly flexible cable of length l and weight per unit length w hangs by one end from a frictionless support, as shown in Fig. 12.8. Neglecting friction, find the natural frequencies of the cable.

This is essentially the problem of the vibrating string which we discussed in Secs. 11.2 and 11.3 and again in Examples 2 and 3, Sec. 11.7, with two important differences: Here, instead of being constant, the tension at a general point of the cable is equal to the weight wx of the cable below that point, and now one end of the cable is free, whereas before both ends were fixed. Hence, because the tension wx is variable, Eq. (1), Sec. 11.2, becomes in the limit

$$\frac{w}{g}\frac{\partial^2 y}{\partial t^2} = \frac{\partial(wx\,\partial y/\partial x)}{\partial x} \tag{1}$$

As usual, we assume a product solution $y(x, t) = X(x)T(t)$ and attempt to separate variables. Thus, substituting, we have

$$\frac{w}{g}T''X = \frac{\partial(wxTX')}{\partial x} = wT(xX')' \qquad \text{or} \qquad \frac{(xX')'}{X} = \frac{T''}{gT}$$

The common value of the last two fractions must be a negative constant, say $-\lambda^2$, $\lambda > 0$, for otherwise T will not be a periodic function of the time t, as we know it must. Hence for the factor T we have

$$T'' = -\lambda^2 gT \qquad \text{and} \qquad T = C\cos\lambda\sqrt{g}t + D\sin\lambda\sqrt{g}t \tag{2}$$

For the factor X we have the differential equation

$$(xX')' + \lambda^2 X = 0 \tag{3}$$

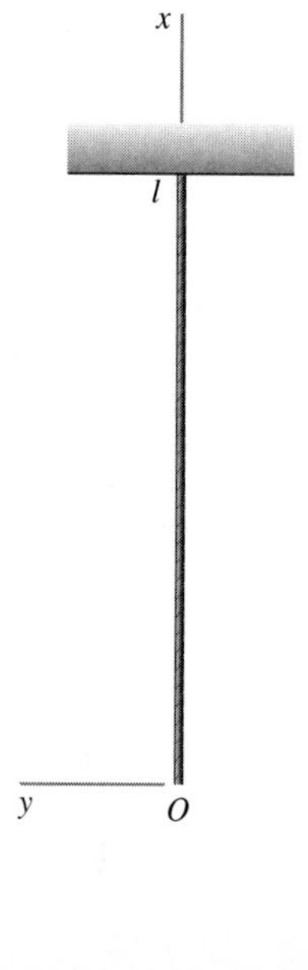

FIGURE 12.8
A hanging cable.

Using Corollary 1, Theorem 1, Sec. 12.4, we find at once that the solution for X is

$$X = AJ_0(2\lambda\sqrt{x}) + BY_0(2\lambda\sqrt{x})$$

Since the deflection $y(0, t) = X(0)T(t)$ at the free end of the cable obviously remains finite, whereas $Y_0(2\lambda\sqrt{x})$ becomes infinite as x approaches zero, it is clear that B must be zero. Moreover, for all values of t, the displacement y is zero when $x = l$. Hence $X(l) = 0$, that is,

$$J_0(2\lambda\sqrt{l}) = 0 \tag{4}$$

Mathematically speaking, (4) is the characteristic equation of the problem. If $z_1, z_2, z_3, \ldots$, are the roots of the equation $J_0(z) = 0$, then

$$\lambda_1 = \frac{z_1}{2\sqrt{l}} \qquad \lambda_2 = \frac{z_2}{2\sqrt{l}} \qquad \lambda_3 = \frac{z_3}{2\sqrt{l}}$$

etc., are the characteristic values of the problem. For these values *and for these values only,* Eq. (1) has periodic solutions which are bounded in the neighborhood of $x = 0$ and are zero at $x = l$. Physically speaking, (4) is also the frequency equation of the problem, for, from the time factor T given by Eq. (2), it is clear that the natural frequencies of the cable are given by the formula

$$\frac{\omega_n}{2\pi} = \frac{\lambda_n\sqrt{g}}{2\pi} = \frac{z_n}{4\pi}\sqrt{\frac{g}{l}} \quad \text{cycles per unit time}$$

It is interesting to note that if the flexible cable were replaced by a rigid bar of the same length and weight per unit length, the latter would swing about one end with frequency

$$\frac{1}{2\pi}\sqrt{\frac{3g}{2l}}$$

The ratio of the lowest frequency of the cable to the single frequency of an equivalent bar is thus (since $z_1 \doteq 2.4048$)

$$\frac{z_1}{2}\sqrt{\frac{g}{l}} \Big/ \sqrt{\frac{3g}{2l}} \doteq \frac{1.2024}{1.2247} \doteq 0.9818$$

EXAMPLE 2 NATURAL FREQUENCIES OF AN ANNULAR MEMBRANE

An elastic membrane spans an annular region whose inner and outer radii are r_1 and r_2, respectively. What are the natural frequencies of the membrane for modes of vibration which possess circular symmetry, i.e., are independent of the polar coordinate θ?

From the statement of the problem it is clear that we must solve the two-dimensional wave equation expressed in polar coordinates (see Exercise 38(**a**), Sec. 11.2) in the special case in which $\partial y/\partial\theta$ is identically zero. Thus we assume a product solution $y(r, t) = R(r)T(t)$ and attempt to separate variables in the equation

$$a^2\left(\frac{\partial^2 y}{\partial r^2} + \frac{1}{r}\frac{\partial y}{\partial r}\right) = \frac{\partial^2 y}{\partial t^2}$$

The result is

$$\frac{R''}{R} + \frac{1}{r}\frac{R'}{R} = \frac{1}{a^2}\frac{T''}{T}$$

The common value of the two members of this equation must be a negative constant, say $-\lambda^2$, $\lambda > 0$, for otherwise T will not be a periodic function of the time t, as we know it must. Hence for the factor T we have

$$T'' = -\lambda^2 a^2 T \qquad \text{and} \qquad T = C\cos\lambda at + D\sin\lambda at$$

For the factor R we have the equation

$$rR'' + R' + \lambda^2 rR = 0$$

which is precisely Bessel's equation of order zero. Hence

$$R(r) = AJ_0(\lambda r) + BY_0(\lambda r)$$

In this problem, since the origin does not lie in the given configuration, there is no condition of boundedness in the neighborhood of $x = 0$. Instead, the relevant boundary conditions are $y(r_1, t) = y(r_2, t) = 0$. Hence we must have

$$R(r_1)T(t) = [AJ_0(\lambda r_1) + BY_0(\lambda r_1)]T(t) = 0$$

and

$$R(r_2)T(t) = [AJ_0(\lambda r_2) + BY_0(\lambda r_2)]T(t) = 0$$

Thus, since $T(t) \not\equiv 0$, A and B must satisfy the pair of equations

$$AJ_0(\lambda r_1) + BY_0(\lambda r_1) = 0$$
$$AJ_0(\lambda r_2) + BY_0(\lambda r_2) = 0$$

If $A = B = 0$, we have only the trivial solution corresponding to the membrane remaining permanently at rest. Hence, in order that there should be a nontrivial solution for A and B, the determinant of the coefficients of the last system must be zero; that is,

$$\begin{vmatrix} J_0(\lambda r_1) & Y_0(\lambda r_1) \\ J_0(\lambda r_2) & Y_0(\lambda r_2) \end{vmatrix} = J_0(\lambda r_1)Y_0(\lambda r_2) - Y_0(\lambda r_1)J_0(\lambda r_2) = 0$$

and this is the characteristic equation, or frequency equation, of the problem. If its roots are $\lambda_1, \lambda_2, \lambda_3, \ldots,$† then from the expression for T it follows that the required natural frequencies are

$$\frac{\omega_n}{2\pi} = \frac{\lambda_n a}{2\pi} = \frac{\lambda_n}{2\pi}\sqrt{\frac{\tau g}{w}} \quad \text{cycles per unit time}$$

where τ is the tension per unit length in the membrane and w is its weight per unit area.

EXAMPLE 3

TEMPERATURES IN A COOLING FIN

A metal fin of triangular cross section is attached to a wall to help carry off heat from the latter. Assuming dimensions and coordinates as shown in Fig. 12.9, find the steady-state temperature distribution from the base of the fin to its tip if the wall temperature is u_w and if the fin cools freely into air of constant temperature u_0.

We shall base our analysis on a unit length of the fin and shall assume that the fin is so thin that temperature variations parallel to the base can be neglected. In other words, we assume that the temperature is the same at all points of any particular cross section parallel to the wall. The physical laws we will need for the formulation of the problem are **Fourier's law of heat conduction**:

> *The rate at which heat flows through an area is proportional to the area and to the temperature gradient in the direction perpendicular to the area.*

and the following form of **Newton's law of cooling**:

> *The rate at which heat is lost from an area on the surface of a body is proportional to the area and to the difference between the surface temperature of the body and the temperature of the surrounding medium.*

---

† The first five roots of this equation for selected values of the ratio r_1/r_2 are given in *Handbook of Mathematical Functions*, p. 415, Superintendent of Documents, Government Printing Office, Washington, D.C., 1965.

The proportionality constants in these laws are known, respectively, as the **thermal conductivity** and the **surface conductivity**.

Now consider the heat balance in the infinitesimal portion of the fin between x and $x + \Delta x$. This element gains heat by internal flow through its right-hand face (the face nearest to the wall) and loses heat by internal flow through its left-hand face and also by cooling through its upper and lower surfaces. Through the right-hand face, the gain of heat per unit time is, by Fourier's law,

$$\text{Area} \times \text{thermal conductivity} \times \text{temperature gradient}$$

or

$$\left[\left(1\frac{bx}{a}\right)k\frac{du}{dx}\right]_{x+\Delta x} = \left(\frac{bkx}{a}\frac{du}{dx}\right)_{x+\Delta x}$$

Similarly, through the left-hand face the element loses heat at the rate

$$\left(\frac{bkx}{a}\frac{du}{dx}\right)_x$$

By Newton's law, the element loses heat through its upper and lower surfaces at the rate

$$\text{Area} \times \text{surface conductivity} \times (\text{surface temperature} - \text{air temperature})$$

or

$$2\left(1\frac{\Delta x}{\cos\theta}\right)h(u - u_0) = \frac{2h(u - u_0)\,\Delta x}{\cos\theta}$$

Under steady-state conditions the rate of gain of heat must equal the rate of loss, and thus we have

$$\left(\frac{bkx}{a}\frac{du}{dx}\right)_{x+\Delta x} = \left(\frac{bkx}{a}\frac{du}{dx}\right)_x + \frac{2h(u - u_0)\,\Delta x}{\cos\theta}$$

Writing this as

$$\frac{(x\,du/dx)_{x+\Delta x} - (x\,du/dx)_x}{\Delta x} - \frac{2ah}{bk\cos\theta}(u - u_0) = 0$$

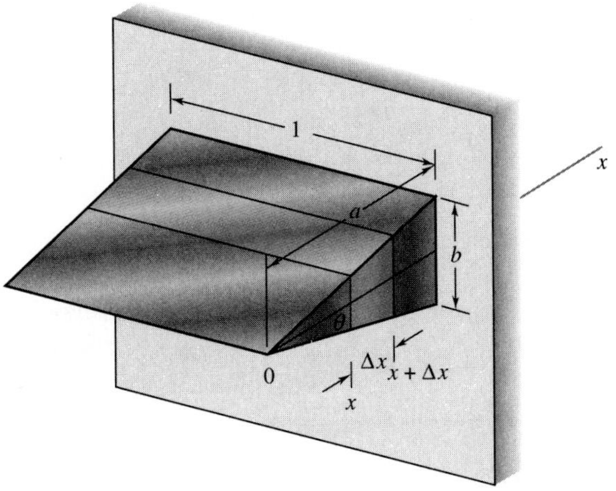

FIGURE 12.9
A portion of a triangular cooling fin attached to a flat wall.

and letting $\Delta x \to 0$, we obtain the differential equation

$$\frac{d(xu')}{dx} - \frac{2ah}{bk \cos \theta}(u - u_0) = 0$$

If we set $U = u - u_0$ and $\alpha^2 = 2ah/(bk \cos \theta) = [2h/(k \sin \theta)]$, this becomes

$$\frac{d(xU')}{dx} - \alpha^2 U = 0$$

This can be solved immediately by means of the corollary of Theorem 1, Sec. 12.4, and we have

$$U = u - u_0 = c_1 I_0(2\alpha\sqrt{x}) + c_2 K_0(2\alpha\sqrt{x})$$

Since $K_0(2\alpha\sqrt{x})$ is infinite when $x = 0$, c_2 must be zero, leaving

$$u - u_0 = c_1 I_0(2\alpha\sqrt{x})$$

Furthermore, $u = u_w$ when $x = a$; hence

$$u_w - u_0 = c_1 I_0(2\alpha\sqrt{a}) \qquad \text{or} \qquad c_1 = \frac{u_w - u_0}{I_0(2\alpha\sqrt{a})}$$

Therefore

$$u = u_0 + (u_w - u_0)\frac{I_0(2\alpha\sqrt{x})}{I_0(2\alpha\sqrt{a})}$$

EXAMPLE 4

TEMPERATURES IN A HALF-CYLINDER

A solid consists of one-half of a right circular cylinder of radius b and height h (Fig. 12.10). The lower base, the curved surface, and the vertical plane face are maintained at the constant temperature $u = 0$. Over the upper base the temperature is a known function of position, that is, $u(r, \theta, h) = f(r, \theta)$. Assuming that there are no sources or sinks of heat in the body, find the steady-state temperature at any point in the body.

Because of the nature of the boundaries of the solid, it is essentially impossible to use the three-dimensional heat equation in the cartesian form appearing in Eq. (14*c*), Sec. 11.2, namely,

$$a^2 \frac{\partial u}{\partial t} = \frac{\partial^2 u}{\partial x^2} + \frac{\partial^2 u}{\partial y^2} + \frac{\partial^2 u}{\partial z^2} + \frac{1}{k} f(x, y, z, t)$$

Instead we must use it as expressed in cylindrical coordinates by the change of variables

$$x = r \cos \theta \qquad y = r \sin \theta \qquad z = z$$

namely,

$$a^2 \frac{\partial u}{\partial t} = \frac{\partial^2 u}{\partial r^2} + \frac{1}{r}\frac{\partial u}{\partial r} + \frac{1}{r^2}\frac{\partial^2 u}{\partial \theta^2} + \frac{\partial^2 u}{\partial z^2} + \frac{1}{k} g(r, \theta, z, t)$$

In the present problem there are neither sources nor sinks of heat; hence $g(r, \theta, z, t) \equiv 0$. Likewise, we are asked to find only the steady-state temperature distribution. Hence $\partial u/\partial t \equiv 0$ and therefore the equation we must solve is

$$\frac{\partial^2 u}{\partial r^2} + \frac{1}{r}\frac{\partial u}{\partial r} + \frac{1}{r^2}\frac{\partial^2 u}{\partial \theta^2} + \frac{\partial^2 u}{\partial z^2} = 0 \tag{5}$$

Our first step is to assume a product solution $u(r, \theta, z) = R(r)\Theta(\theta)Z(z)$ and substitute it into (5) in an attempt to separate the variables. This gives

$$R''\Theta Z + \frac{1}{r}R'\Theta Z + \frac{1}{r^2}R\Theta''Z + R\Theta Z'' = 0$$

or, multiplying by r^2 and dividing by the product $R\Theta Z$,

$$r^2\frac{R''}{R} + r\frac{R'}{R} + r^2\frac{Z''}{Z} = -\frac{\Theta''}{\Theta} = \mu_1$$

where the common value μ_1 is necessarily a constant, since the variables appearing on the respective sides of the equation are independent of each other.

If $\mu_1 < 0$, say $\mu_1 = -\nu^2$, $\nu > 0$, then $\Theta''/\Theta = \nu^2$ and

$$\Theta = A \cosh \nu\theta + B \sinh \nu\theta \tag{6}$$

Now, by hypothesis,

$$u(r, 0, z) = R(r)\Theta(0)Z(z) = 0 \qquad \text{and} \qquad u(r, \pi, z) = R(r)\Theta(\pi)Z(z) = 0$$

and these can hold for all values of r and z, as required by the given boundary conditions, only if $\Theta(0) = \Theta(\pi) = 0$. From (6) we see that the condition $\Theta(0) = 0$ will be satisfied only if $A = 0$. To satisfy the condition $\Theta(\pi) = 0$, it is necessary that

$$B \sinh \nu\pi = 0$$

which, since $\nu \neq 0$, is possible only if $B = 0$. Thus the possibility $\mu_1 < 0$ leads only to a trivial solution and hence must be rejected.

If $\mu_1 = 0$, then $\Theta'' = 0$ and

$$\Theta = A + B\theta$$

Again imposing the conditions $\Theta(0) = \Theta(\pi) = 0$, we find, as before, that $A = B = 0$. Hence the possibility $\mu_1 = 0$ must also be rejected, since it, too, leads only to a trivial solution.

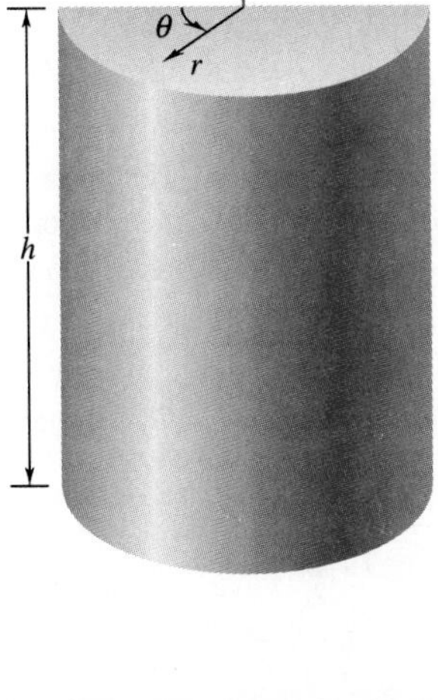

FIGURE 12.10
A half-cylinder in which heat flow occurs because of surface temperature conditions.

Finally, if $\mu_1 > 0$, say $\mu_1 = \nu^2$, $\nu > 0$, we have

$$\frac{\Theta''}{\Theta} = -\nu^2 \qquad \text{and} \qquad \Theta = A \cos \nu\theta + B \sin \nu\theta$$

For this to vanish when $\theta = 0$, we must have $A = 0$. For it to vanish where $\theta = \pi$, it is necessary that

$$B \sin \nu\pi = 0$$

Since we cannot permit B to be zero, because that would again lead to a trivial solution, we must have

$$\sin \nu\pi = 0$$

Hence

$$\nu = 1, 2, 3, \ldots, n, \ldots$$

and so for Θ we have the family of solutions

$$\Theta_n(\theta) = \sin n\theta$$

With $\mu_1 = \nu^2$ now known to be n^2, the differential equation for R and Z becomes

$$r^2\frac{R''}{R} + r\frac{R'}{R} + r^2\frac{Z''}{Z} = n^2$$

or, rearranging,

$$\frac{Z''}{Z} = \frac{n^2}{r^2} - \frac{R''}{R} - \frac{1}{r}\frac{R'}{R} = \mu_2$$

where again, since r and z are independent variables, it follows that the common value μ_2 must be a constant.

If $\mu_2 < 0$, say $\mu_2 = -\lambda^2$, $\lambda > 0$, we have†

$$\frac{R''}{R} + \frac{1}{r}\frac{R'}{R} - \lambda^2 - \frac{n^2}{r^2} = 0 \qquad \text{or} \qquad r^2R'' + rR' - (\lambda^2 r^2 + n^2)R = 0$$

which is precisely the modified Bessel equation of order n. Hence

$$R = CI_n(\lambda r) + DK_n(\lambda r)$$

Now $K_n(\lambda r)$ is infinite when $r = 0$; hence, to keep the temperature finite on the axis of the cylinder, it is necessary that $D = 0$. Also, from the given boundary conditions,

$$u(b, \theta, z) = R(b)\Theta(\theta)Z(z) = 0$$

and this is possible for all values of θ and z only if

$$R(b) = CI_n(\lambda b) = 0$$

But the modified Bessel function I_n is never zero except possibly at the origin. Therefore the last condition can hold only if $C = 0$. But with C and D both zero, the solution is trivial, and so the possibility that $\mu_2 < 0$ must be rejected.

---

† Since the group of terms involving Z is much simpler than the group of terms involving R, one might think that it would be better to work next with the equation $Z''/Z = \mu_2$. However, it is not the simplicity of the terms themselves but the nature of the boundary conditions the variables must satisfy that determines the continuation. Clearly, the condition $R(b) = 0$ is one that can be imposed on R once it is found; whereas the condition $u = f(r, \theta)$ when $z = h$ cannot be imposed on the factor Z by itself but must finally be satisfied by a series expansion involving all the product solutions.

If $\mu_2 = 0$, then

$$\frac{R''}{R} + \frac{1}{r}\frac{R'}{R} - \frac{n^2}{r^2} = 0 \qquad \text{or} \qquad r^2R'' + rR' - n^2R = 0$$

This is not a Bessel-type equation but is instead an example of the Euler equation (Sec. 2.8). By the usual change of independent variable

$$r = e^v \qquad \text{or} \qquad v = \ln r$$

it becomes

$$\frac{d^2R}{dv^2} - n^2R = 0$$

so that

$$R = Ce^{nv} + De^{-nv} = Cr^n + Dr^{-n}$$

To keep the temperature finite on the axis of the cylinder, where $r = 0$, it is necessary that $D = 0$. To keep the temperature zero when $r = b$, it is necessary that

$$R(b) \equiv Cb^n = 0$$

which will be the case only if $C = 0$. This means that again the solution is trivial, and $\mu_2 = 0$ must also be rejected.

Finally, if $\mu_2 > 0$, say $\mu_2 = \lambda^2$, $\lambda > 0$, we have

$$\frac{R''}{R} + \frac{1}{r}\frac{R'}{R} + \lambda^2 - \frac{n^2}{r^2} = 0 \qquad \text{or} \qquad r^2R'' + rR' + (\lambda^2r^2 - n^2)R = 0$$

and

$$R = CJ_n(\lambda r) + DY_n(\lambda r)$$

Since $Y_n(\lambda r)$ is infinite when $r = 0$, we must have $D = 0$. To keep the temperature zero on the curved surface of the cylinder, we must have

$$R(b) \equiv CJ_n(\lambda b) = 0$$

Since $C = 0$ leads to a trivial solution, it is thus necessary that

$$J_n(\lambda b) = 0$$

that is, λ is restricted to the set of values

$$\left\{\frac{\rho_{nm}}{b}\right\}$$

where ρ_{nm} is the mth one of the roots of the equation $J_n(x) = 0$. Thus, for every value of n, there are infinitely many particular solutions for R, namely,

$$R_{nm}(r) = J_n(\lambda_{nm}r)$$

Now that we know that $\mu_2 = \lambda_{nm}^2$, it is an easy matter to solve for Z, and we have

$$\frac{Z''}{Z} = \lambda_{nm}^2 \qquad \text{and} \qquad Z = E\cosh\lambda_{nm}z + F\sinh\lambda_{nm}z$$

Since $u(r, \theta, 0) \equiv R(r)\Theta(\theta)Z(0) = 0$, it follows that $Z(0) = 0$, from which we conclude that $E = 0$. The solution for Z associated with R_{nm} is therefore

$$Z_{nm}(z) = \sinh\lambda_{nm}z$$

For *each* n we therefore have infinitely many product solutions consisting of the same factor $\Theta_n(\theta) = \sin n\theta$ multiplied by the product of any pair of corresponding R's and Z's:

$$u_{nm} = A_{nm} J_n(\lambda_{nm} r) \sinh \lambda_{nm} z \sin n\theta$$

In other words, we have a double array of product solutions,

$$\begin{array}{l} u_{11}, u_{12}, u_{13}, \ldots, u_{1m}, \ldots \\ u_{21}, u_{22}, u_{23}, \ldots, u_{2m}, \ldots \\ \cdots\cdots\cdots\cdots\cdots\cdots \\ u_{n1}, u_{n2}, u_{n3}, \ldots, u_{nm}, \ldots \\ \cdots\cdots\cdots\cdots\cdots\cdots \end{array}$$

Since none of the product solutions by itself is capable of representing the given temperature distribution $f(r, \theta)$ on the upper base, we must construct an infinite series of the u_{nm}'s and try to make it fit the temperature condition when $z = h$. To build up a series for u we first add up all the product solutions associated with a particular value of n, getting

$$u_n = \sum_{m=1}^{\infty} u_{nm} = \sin n\theta \sum_{m=1}^{\infty} A_{nm} J_n(\lambda_{nm} r) \sinh \lambda_{nm} z$$

This of course amounts to forming the sums of the elements in each of the rows in the above array. Next we add up all these series for every value of n:

$$u(r, \theta, z) = \sum_{n=1}^{\infty} u_n = \sum_{n=1}^{\infty} \left[\sin n\theta \sum_{m=1}^{\infty} A_{nm} J_n(\lambda_{nm} r) \sinh \lambda_{nm} z \right] \tag{7}$$

The final step now is to determine the A's so that this double series will reduce to $f(r, \theta)$ when $z = h$:

$$u(r, \theta, h) = f(r, \theta) = \sum_{n=1}^{\infty} \left[\sin n\theta \sum_{m=1}^{\infty} A_{nm} J_n(\lambda_{nm} r) \sinh \lambda_{nm} h \right] \tag{8}$$

To carry out this expansion, let us imagine that r is held constant and that θ is allowed to vary over the range of the problem $(0, \pi)$. Under these conditions the inner sum in (8) is effectively a constant depending on n, say G_n, or more explicitly $G_n(r)$. That is,

$$f(r, \theta) = \sum_{n=1}^{\infty} G_n \sin n\theta$$

But the determination of the G's is a familiar problem—nothing more in fact than the Fourier sine-expansion problem—and we can write immediately

$$G_n \equiv G_n(r) = \frac{2}{\pi} \int_0^{\pi} f(r, \theta) \sin n\theta \, d\theta \tag{9}$$

Thus $G_n(r)$ is a *known* function of r. But, by definition, $G_n(r)$ was the inner sum in (8); that is,

$$G_n(r) = \sum_{m=1}^{\infty} (A_{nm} \sinh \lambda_{nm} h) J_n(\lambda_{nm} r)$$

Hence it is clear that the A's must be such that the products $A_{nm} \sinh \lambda_{nm} h$ are the coefficients in a Bessel function expansion of the now known function $G_n(r)$. Therefore, recalling that the λ's were determined by the condition $J_n(\lambda b) = 0$, we have from the theory of Sec. 12.7 (in particular, Part **(v)** Theorem 5),

$$A_{nm} \sinh \lambda_{nm} h = \frac{\displaystyle\int_0^b r G_n(r) J_n(\lambda_{nm} r) \, dr}{(b^2/2) J_{n+1}^2(\lambda_{nm} b)}$$

Hence

$$A_{nm} = \frac{\int_0^b rG_n(r)J_n(\lambda_{nm}r)\,dr}{(b^2/2)\sinh\lambda_{nm}h\,J_{n+1}^2(\lambda_{nm}b)}$$

where $G_n(r)$ is given by (9). With the coefficients in the series (7) now determined, the problem is solved.

EXAMPLE 5

TEMPERATURES IN A CIRCULAR PLATE

A thin circular plate of radius b has its upper and lower faces insulated against the flow of heat. Around the circumference of the plate the time-dependent temperature distribution $u(b, \theta, t) = f(\theta)\cos\omega t$ is maintained. Find the steady-state temperature distribution throughout the plate.

In this problem, the steady-state temperature distribution we are asked to find is not one that is *independent* of time but is instead the limiting *periodic* distribution the temperature approaches as t increases indefinitely and all transients die away. Thus, even though we are seeking a description of the steady state of the system, we must retain the term $\partial u/\partial t$ in the heat equation. On the other hand, since the plate is very thin and has perfectly insulated faces, we may legitimately assume that the heat flow is two-dimensional, i.e., is independent of z. Hence the equation we must solve is

$$\frac{\partial^2 u}{\partial r^2} + \frac{1}{r}\frac{\partial u}{\partial r} + \frac{1}{r^2}\frac{\partial^2 u}{\partial \theta^2} = a^2\frac{\partial u}{\partial t}$$

If, as usual, we assume a product solution

$$u(r, \theta, t) = R(r)\Theta(\theta)T(t)$$

substitute, and divide by $R\Theta T$, we obtain

$$\frac{R''}{R} + \frac{1}{r}\frac{R'}{R} + \frac{1}{r^2}\frac{\Theta''}{\Theta} = a^2\frac{T'}{T} = \mu_1$$

Here, however, we are faced with a situation we have not previously encountered. For whether μ_1 is positive, negative, or zero, the solution of the equation

$$a^2\frac{T'}{T} = \mu_1$$

cannot describe a nonconstant periodic function of t, as we know T must be. The only possible continuation is to assume that T is a *complex* periodic function of period ω, namely,

$$T(t) = e^{i\omega t} = \cos\omega t + i\sin\omega t$$

This means that

$$a^2\frac{T'}{T} = a^2\frac{i\omega e^{i\omega t}}{e^{i\omega t}} = a^2\omega i$$

which implies that $\mu_1 = ia^2\omega$. Thus, for the first time, we have a problem in which a complex separation constant is required.

Using this value of μ_1 and separating again, we have

$$r^2\frac{R''}{R} + r\frac{R'}{R} - ia^2\omega r^2 = -\frac{\Theta''}{\Theta} = \mu_2$$

Clearly, the boundary condition is a periodic function of θ of period 2π. Hence $\mu_2 = n^2$, and therefore from the equation $-\Theta''/\Theta = \mu_2 = n^2$, we have

$$\Theta_n(\theta) = A_n\cos n\theta + B_n\sin n\theta \qquad n = 0, 1, 2, \ldots$$

The factor R is now to be determined from the equation

$$r^2\frac{R''}{R} + r\frac{R'}{R} - ia^2\omega r^2 = n^2$$

or

$$r^2R'' + rR' - (ia^2\omega r^2 + n^2)R = 0$$

A complete solution of this equation is, from Sec. 12.3,

$$R(r) = C(\text{ber}_n\, a\sqrt{\omega}r + i\,\text{bei}_n\, a\sqrt{\omega}r) + D(\text{ker}_n\, a\sqrt{\omega}r + i\,\text{kei}_n\, a\sqrt{\omega}r)$$

Since $\text{ker}_n\, a\sqrt{\omega}r + i\,\text{kei}_n\, a\sqrt{\omega}r$ is infinite when $r = 0$, whereas the temperature is obviously finite at the center of the plate, it is necessary that $D = 0$, leaving

$$R(r) = C(\text{ber}_n\, a\sqrt{\omega}r + i\,\text{bei}_n\, a\sqrt{\omega}r) = CM_n(a\sqrt{\omega}r)\exp\,[i\vartheta_n(a\sqrt{\omega}r)]$$

where M_n denotes the modulus and ϑ_n the amplitude, respectively, of $\text{ber}_n\, a\sqrt{\omega}r + i\,\text{bei}_n\, a\sqrt{\omega}r$.

We thus have a family of product solutions,

$$u_n(r, \theta, t) = M_n(a\sqrt{\omega}r)\exp\,[i\vartheta_n(a\sqrt{\omega}r)](A_n \cos n\theta + B_n \sin n\theta)e^{i\omega t}$$

Since none of these by itself can satisfy the boundary condition, we form an infinite series of them

$$u(r, \theta, t) = \sum_{n=0}^{\infty} M_n(a\sqrt{\omega}r)\exp\,[i\vartheta_n(a\sqrt{\omega}r)](A_n \cos n\theta + B_n \sin n\theta)e^{i\omega t} \tag{10}$$

and attempt to determine A_n and B_n so that this series will reduce to the appropriate condition on the boundary where $r = b$. However, before we can do this we must modify the boundary condition as given. Since we have been forced to take T to be the complex exponential $e^{i\omega t}$, we must also change the boundary condition from $f(\theta)\cos\omega t$ to the complex form

$$f(\theta)e^{i\omega t} = f(\theta)(\cos\omega t + i\sin\omega t)$$

Then, when we have solved the problem for this modified boundary condition, the answer to the actual problem will be just the real part of the resulting complex solution.†

Putting $r = b$ and $u(b, \theta, t) = f(\theta)e^{i\omega t}$ in (10), we have, after dividing out the factor $e^{i\omega t}$,

$$f(\theta) = \sum_{n=0}^{\infty} M_n(a\sqrt{\omega}b)\exp\,[i\vartheta_n(a\sqrt{\omega}b)](A_n \cos n\theta + B_n \sin n\theta)$$

From this, by familiar Fourier theory, we have

$$M_n(a\sqrt{\omega}b)\exp\,[i\vartheta_n(a\sqrt{\omega}b)]A_n = \frac{1}{\pi}\int_0^{2\pi} f(\theta)\cos n\theta\, d\theta = \frac{1}{\pi}\int_0^{2\pi} f(s)\cos ns\, ds$$

or

$$A_n = \frac{\exp\,[-i\vartheta_n(a\sqrt{\omega}b)]}{M_n(a\sqrt{\omega}b)}\,\frac{1}{\pi}\int_0^{2\pi} f(s)\cos ns\, ds$$

---

†This stratagem is similar to the one we introduced in Sec. 7.4 when we replaced the voltages $E_0\cos\omega t$ and $E_0\sin\omega t$ by the complex voltage $E_0e^{i\omega t}$, solved for the resulting current, and then identified the real and imaginary components of this complex current as the responses to $E_0\cos\omega t$ and $E_0\sin\omega t$, respectively.

and

$$M_n(a\sqrt{\omega}b)\exp[i\vartheta_n(a\sqrt{\omega}b)]B_n = \frac{1}{\pi}\int_0^{2\pi} f(\theta)\sin n\theta\, d\theta = \frac{1}{\pi}\int_0^{2\pi} f(s)\sin ns\, ds$$

or

$$B_n = \frac{\exp[-i\vartheta_n(a\sqrt{\omega}b)]}{M_n(a\sqrt{\omega}b)}\frac{1}{\pi}\int_0^{2\pi} f(s)\sin ns\, ds$$

Substituting these into the series in (10) and combining the exponentials, we obtain

$$\frac{1}{\pi}\sum_{n=0}^{\infty}\frac{M_n(a\sqrt{\omega}r)}{M_n(a\sqrt{\omega}b)}\exp\{i[\omega t + \vartheta_n(a\sqrt{\omega}r) - \vartheta_n(a\sqrt{\omega}b)]\}\left[\cos n\theta\int_0^{2\pi} f(s)\cos ns\, ds + \sin n\theta\int_0^{2\pi} f(s)\sin ns\, ds\right]$$

or, consolidating the integrals,

$$\frac{1}{\pi}\sum_{n=0}^{\infty}\frac{M_n(a\sqrt{\omega}r)}{M_n(a\sqrt{\omega}b)}\exp\{i[\omega t + \vartheta_n(a\sqrt{\omega}r) - \vartheta_n(a\sqrt{\omega}b)]\}\int_0^{2\pi} f(s)\cos n(\theta - s)\, ds$$

Finally, by retaining only the real part of this series, we find that the solution to our problem is

$$u(r,\theta,t) = \frac{1}{\pi}\sum_{n=0}^{\infty}\frac{M_n(a\sqrt{\omega}r)}{M_n(a\sqrt{\omega}b)}\cos[\omega t + \vartheta_n(a\sqrt{\omega}r) - \vartheta_n(a\sqrt{\omega}b)]\int_0^{2\pi} f(s)\cos n(\theta - s)\, ds$$

From this, it is evident that although at every point the temperature varies periodically with frequency ω, there is a phase difference between the temperature at an arbitrary radius and the temperature on the boundary.

EXAMPLE 6

What is the Laplace transform of $J_0(at)$?

The most natural way to determine the Laplace transform of $J_0(at)$ is probably to transform the infinite series for $J_0(at)$ term by term, using Formula 4, Sec. 10.5, and then identify the resulting series. Doing this, we have

$$\mathcal{L}\{J_0(at)\} = \mathcal{L}\left\{\sum_{n=0}^{\infty}\frac{(-1)^n(at)^{2n}}{2^{2n}n!n!}\right\} = \sum_{n=0}^{\infty}\frac{(-1)^n a^{2n}}{2^{2n}n!n!}\mathcal{L}\{t^{2n}\}$$

$$= \sum_{n=0}^{\infty}\frac{(-1)^n a^{2n}}{2^{2n}n!n!}\cdot\frac{(2n)!}{s^{2n+1}} = \frac{1}{s}\left[1 + \sum_{n=1}^{\infty}\frac{(-1)^n a^{2n}(2n)!}{2^{2n}n!n!s^{2n}}\right]$$

If in the last summand we now remove the even factors

$$2\cdot 4\cdot 6\cdots 2n = 2^n n!$$

from $(2n)!$, cancel these against the corresponding factors in the denominator, and then pair the remaining odd factors in the numerator with the remaining 2's in the denominator, we have

$$\mathcal{L}\{J_0(at)\} = \frac{1}{s}\left[1 + \sum_{n=1}^{\infty}\frac{(-\frac{1}{2})(-\frac{3}{2})(-\frac{5}{2})\cdots[-(2n-1)/2]}{n!}\left(\frac{a}{s}\right)^{2n}\right]$$

and this we recognize as simply the binomial expansion of

$$\frac{1}{(s^2 + a^2)^{1/2}} = \frac{1}{s}\left(1 + \frac{a^2}{s^2}\right)^{-1/2}$$

Thus we have established the important formula

$$\mathcal{L}\{J_0(at)\} = \frac{1}{(s^2 + a^2)^{1/2}} \tag{11}$$

Moreover, since $I_0(at)$ differs from $J_0(at)$ only in having just positive terms in its series expansion, and since $1/(s^2 - a^2)^{1/2}$ differs from $1/(s^2 + a^2)^{1/2}$ only in having just positive terms in its series expansion, we have also established the formula

$$\mathcal{L}\{I_0(at)\} = \frac{1}{(s^2 - a^2)^{1/2}} \tag{11.2}$$

Numerous other transforms follow readily from Eq. (11). For instance, since

$$\frac{dJ_0(at)}{dt} = -aJ_1(at)$$

it follows from Theorem 1, Sec. 10.4, that

$$\begin{aligned}\mathcal{L}\{J_1(at)\} &= -\frac{1}{a}\mathcal{L}\left\{\frac{dJ_0(at)}{dt}\right\} = -\frac{1}{a}[s\mathcal{L}\{J_0(at)\} - J_0(0)] \\ &= -\frac{1}{a}\left[\frac{s}{(s^2 + a^2)^{1/2}} - 1\right] = \frac{1}{a}\frac{(s^2 + a^2)^{1/2} - s}{(s^2 + a^2)^{1/2}} \\ &= \frac{a}{(s^2 + a^2)^{1/2}[s + (s^2 + a^2)^{1/2}]}\end{aligned}$$

A number of other transforms will be found among the exercises.

In some of the boundary-value problems we have previously considered (see, for instance, Examples 5–7, Sec. 11.7) we found that the boundary conditions were not sufficient to provide us with a characteristic equation. As a result, the separation parameter remained a continuous variable and we were unable to identify a countably infinite set of product solutions. In such cases, instead of constructing an *infinite series* of the characteristic functions, we had to form an *infinite integral* of them and use the Fourier integral, or a Fourier transform pair, rather than Fourier series to obtain our final solution. This situation may also occur in problems where functions other than the familiar sines and cosines of Fourier theory are involved. In particular, the Fourier transform has an analog known as the *Hankel transform* which often arises in problems involving Bessel functions. We cannot undertake an extended investigation of the Hankel transform, but we shall conclude this section by first indicating the connection between the Fourier transform and the Hankel transform and then discussing one typical application.

Our starting point is the two-dimensional Fourier transform pair we obtained in Exercises 88 and 89, Sec. 9.1, namely,

$$f(x, y) = \frac{1}{2\pi}\int_{-\infty}^{\infty}\int_{-\infty}^{\infty} G(\xi, \eta)e^{i(x\xi + y\eta)}\, d\xi\, d\eta \tag{12}$$

where

$$G(\xi, \eta) = \frac{1}{2\pi}\int_{-\infty}^{\infty}\int_{-\infty}^{\infty} f(x, y)e^{-i(\xi x+\eta y)}\,dx\,dy \tag{13}$$

For our purposes it is necessary to specialize these relations by supposing that $f(x, y)$ is a function of x and y only through the combination $r = \sqrt{x^2 + y^2}$; that is, we shall suppose that $f(x, y)$ is actually a function of the single variable r, say $f(x, y) = F(r)$. Thus, changing to polar coordinates via the substitutions

$$x = r\cos\theta \qquad y = r\sin\theta \qquad \xi = \rho\cos\phi \qquad \eta = \rho\sin\phi$$

noting that

$$\xi x + \eta y = r\rho\,(\cos\theta\cos\phi + \sin\theta\sin\phi) = r\rho\cos(\theta - \phi)$$

and recalling that the differential of area in polar coordinates is

$$dA = r\,dr\,d\theta$$

the integral (13) becomes

$$\begin{aligned} G(\xi, \eta) &= \frac{1}{2\pi}\int_0^{\infty}\int_0^{2\pi} F(r)e^{-ir\rho\cos(\theta-\phi)}(r\,dr\,d\theta) \\ &= \frac{1}{2\pi}\int_0^{\infty} rF(r)\left[\int_0^{2\pi} e^{-ir\rho\cos(\theta-\phi)}\,d\theta\right]dr \end{aligned} \tag{14}$$

Now as a function of θ, $e^{-ir\rho\cos(\theta-\phi)}$ is of period 2π. Hence if we put $\theta - \phi = u$ in the inner integral, we obtain

$$\int_{-\phi}^{2\pi-\phi} e^{-ir\rho\cos u}\,du$$

and this is equal to

$$\int_0^{2\pi} e^{-ir\rho\cos u}\,du$$

since the integral of a periodic function over one full period is the same no matter where the integration begins. Moreover, as we saw in Exercise 5(**a**), Sec. 12.6, the last integral is equal to $2\pi J_0(\rho r)$. Therefore, substituting, Eq. (14) becomes

$$G(\xi, \eta) = \int_0^{\infty} rF(r)J_0(\rho r)\,dr$$

Since the last integral is clearly a function of the single variable ρ, this shows that $G(\xi, \eta)$ is actually a function of ξ and η only through the combination $\rho = \sqrt{\xi^2 + \eta^2}$; that is, $G(\xi, \eta) = \mathcal{H}(\rho)$, say, and we can write

$$\mathcal{H}(\rho) = \int_0^{\infty} rF(r)J_0(\rho r)\,dr \tag{15}$$

Conversely, if we begin with the inversion formula (12), make the same substitutions, and integrate over the $\rho\phi$ plane, we have

$$\begin{aligned} f(x, y) \equiv F(r) &= \frac{1}{2\pi}\int_0^\infty \int_0^{2\pi} G(\xi, \eta) e^{i\rho r \cos(\theta-\phi)} (\rho\, d\rho\, d\phi) \\ &= \frac{1}{2\pi}\int_0^\infty \rho \mathcal{H}(\rho)\left[\int_0^{2\pi} e^{i\rho r \cos(\phi-\theta)}\, d\phi\right] d\rho \\ &= \int_0^\infty \rho\mathcal{H}(\rho) J_0(\rho r)\, d\rho \end{aligned} \tag{16}$$

Equations (15) and (16) constitute the **Hankel transform pair for Bessel functions of order** $\nu = 0$. More generally, it can be shown† that *if* $F(r)$ *is piecewise continuous and if* $\int_0^\infty |F(r)|\, dr$ *exists, then*

$$F(r) = \int_0^\infty \rho\mathcal{H}(\rho) J_\nu(\rho r)\, d\rho$$

where

$$\mathcal{H}(\rho) = \int_0^\infty rF(r) J_\nu(\rho r)\, dr$$

EXAMPLE 7

TEMPERATURES IN A HALF-SPACE

Find the steady-state temperature distribution in the half-space $z > 0$ if the temperature distribution $u(r, 0) = F(r)$ is maintained on the bounding plane face $z = 0$. Assume that $F(r)$ satisfies the conditions for the existence of its Hankel transform.

From the statement of the problem it is clear that the temperature u is independent of the polar angle θ. Hence we begin with the steady-state heat equation in the form

$$\frac{\partial^2 u}{\partial r^2} + \frac{1}{r}\frac{\partial u}{\partial r} + \frac{\partial^2 u}{\partial z^2} = 0$$

Assuming a product solution $u(r, z) = R(r)Z(z)$ and separating variables, we obtain at once

$$\frac{R''}{R} + \frac{1}{r}\frac{R'}{R} = -\frac{Z''}{Z} = \mu$$

If $\mu = \lambda^2$, $\lambda > 0$, the equation for the factor R becomes

$$\frac{R''}{R} + \frac{1}{r}\frac{R'}{R} - \lambda^2 = 0 \qquad \text{or} \qquad r^2R'' + rR' - \lambda^2 r^2 R = 0$$

which is the modified Bessel equation of order $\nu = 0$ whose solution is

$$R(r) = A_1 I_0(\lambda r) + A_2 K_0(\lambda r)$$

Since u, and hence R, must be bounded in the neighborhood of $r = 0$, it follows that $A_2 = 0$. Similarly, since u must also be bounded as r becomes infinite, whereas $\lim_{r\to\infty} I_0(\lambda r) = \infty$, it follows that $A_1 = 0$. Thus $\mu = \lambda^2$ leads only to a trivial solution and must be rejected.

† See, for instance, Ian N. Sneddon, *Fourier Transforms,* pp. 48–70, McGraw-Hill, New York, 1951.

If $\mu = 0$, the equation for R is

$$\frac{R''}{R} + \frac{1}{r}\frac{R'}{R} = 0 \qquad \text{or} \qquad r^2R'' + rR' = 0$$

This is an instance of Euler's equation, and its solution is readily found to be

$$R(r) = A_1 + A_2 \ln r$$

Again, boundedness in the neighborhood of the origin requires that $A_2 = 0$. Also, since the implied condition

$$\int_0^\infty |F(r)|\, dr < \infty$$

requires that $\lim_{r\to\infty} F(r) = 0$, it follows that $A_1 = 0$. Thus the possibility that $\mu = 0$ also leads only to the trivial solution.

If $\mu = -\lambda^2$, $\lambda > 0$, the equations for R and Z become, respectively,

$$r^2R'' + rR' + \lambda^2r^2R = 0 \qquad \text{and} \qquad Z'' = \lambda^2 Z$$

The first of these is Bessel's equation of order $\nu = 0$, and hence

$$R(r) = A_1J_0(\lambda r) + A_2Y_0(\lambda r)$$

The requirement that u, and hence R, be bounded in the neighborhood of $r = 0$ implies that $A_2 = 0$, leaving

$$R(r) = A_1J_0(\lambda r)$$

From the equation for Z we conclude that

$$Z(z) = B_1e^{-\lambda z} + B_2e^{\lambda z}$$

Since u, and hence Z, must be bounded as $z \to \infty$, it follows that $B_2 = 0$, leaving

$$Z(z) = B_1e^{-\lambda z}$$

Since we have no other boundary condition to impose, we have no further restriction on λ, which therefore remains a continuous variable. Thus we have a *continuous* family of product solutions

$$u_\lambda(r, z) = A(\lambda)\lambda J_0(\lambda r)e^{-\lambda z}$$

where, for later convenience in using the Hankel transform, we have written the still-arbitrary coefficient $A_1(\lambda)B_1(\lambda)$ as $A(\lambda)\lambda$.

We now *integrate* all these product solutions in an attempt to construct an expression which can be made to satisfy the boundary condition $u(r, 0) = F(r)$. The result is

$$u(r, z) = \int_0^\infty u_\lambda(r, z)\, d\lambda = \int_0^\infty \lambda A(\lambda)e^{-\lambda z}J_0(\lambda r)\, d\lambda$$

Putting $z = 0$ in this solution, we obtain

$$u(r, 0) \equiv F(r) = \int_0^\infty \lambda A(\lambda)J_0(\lambda r)\, d\lambda$$

which we recognize as one of the members of a Hankel transform pair. Hence, by Eq. (15), the coefficient function $A(\lambda)$ must be given by

$$A(\lambda) = \int_0^\infty rF(r)J_0(\lambda r)\, dr$$

Thus, finally,

$$u(r,z) = \int_0^\infty \lambda e^{-\lambda z}\left[\int_0^\infty rF(r)J_0(\lambda r)\,dr\right]J_0(\lambda r)\,d\lambda$$

In particular, if

$$F(r) = \begin{cases} u_0 & r < a \\ 0 & r > a \end{cases}$$

we have

$$A(\lambda) = \int_0^a ru_0 J_0(\lambda r)\,dr = \frac{u_0}{\lambda} rJ_1(\lambda r)\Big|_0^a = \frac{au_0}{\lambda}J_1(\lambda a)$$

and

$$u(r,z) = au_0\int_0^\infty e^{-\lambda z}J_1(\lambda a)J_0(\lambda r)\,d\lambda$$

EXERCISES

1. In Example 1, if the weight per unit length of the cable varies according to the law $w(x) = (k+1)x^k$, what is the frequency equation?
2. At $t = 0$, while the cable in Example 1 is at rest in the vertical position, a uniform horizontal velocity v is imparted to the portion of the cable between $x = 0$ and $x = \alpha l$. find the expression describing the subsequent motion of the cable.
3. If the cable in Example 1 is released from rest at $t = 0$ in a position extending along the line $y = \alpha(l - x)$, find the expression describing its subsequent motion.
4. Determine the natural frequencies of a uniform circular membrane when it is vibrating in a shape that is independent of the polar angle θ.
5. At $t = 0$ while a uniform circular membrane of radius B is at rest in its undeflected position, it is struck in such a way that all points within a circle of radius b, concentric with the membrane, are given initial velocity v. Find the expression describing the subsequent motion of the membrane.
6. At $t = 0$ a uniform circular membrane of radius b is released from rest in a deflected position described by the equation $z = b^2 - r^2$. Find the expression describing its subsequent motion.
7. Determine the natural frequencies of a uniform circular membrane of radius b for modes of vibration that are not necessarily independent of the polar angle θ. What are the associated nodal lines?
8. Find the expression describing the steady-state vibrations of a uniform circular membrane of radius b driven by a force per unit area equal to $F_0 \cos \omega t$.
9. What are the natural frequencies of a uniform elastic membrane spanning a semicircle of radius b?
10. In Example 3, verify that all the heat that enters the fin is lost from its surface. What fraction of the heat entering the fin is lost from the section between $x = 0$ and $x = \epsilon a$?
11. Work Example 3 if the fin is of rectangular rather than triangular cross section and if heat loss from the end of the fin is neglected. *Hint:* In addition to the temperature condition at the base of the fin, it will be necessary to use the fact that the rate at which heat enters the fin is equal to the rate at which heat is lost from the fin.
12. Show that the radial temperature distribution in a thin fin of rectangular cross section and outer radius B that completely encircles a heated cylinder of radius b satisfies the differential equation

$$\frac{d(x\,du/dx)}{dx} - \frac{2hx(u - u_0)}{kw} = 0$$

where x is measured radially outward from the axis of the cylinder, w is the thickness of the fin, and the other parameters have the same significance as in Example 3.

13. Work Example 4 if the curved surface of the solid is insulated.
14. Work Example 4 if the upper base is maintained at the temperature $u(r, \theta, h) = 0$ while the temperature distribution $u(b, \theta, z) = g(\theta, z)$ is maintained on the curved surface.
15. Could Example 4 be worked if an arbitrary temperature distribution $f(r, \theta)$ is maintained on the upper base *and* an arbitrary temperature distribution $g(\theta, z)$ is maintained on the curved surface? How?
16. A right circular cylinder of radius b and height h has its upper and lower bases maintained at the constant temperature $u = 0$. The curved surface is maintained at the temperature distribution $u(b, \theta, z) = f(z)$. Find the steady-state temperature distribution throughout the cylinder.

17. Work Exercise 16 if the upper and lower bases of the cylinder are insulated.

18. A right circular cylinder of radius b and height h has its lower base maintained at the constant temperature $u = 100$ and its upper base maintained at the constant temperature $u = 0$. If the curved surface is insulated, find the steady-state temperature distribution in the cylinder.

19. Work Exercise 18 if the curved surface instead of being insulated is maintained at the temperature $u = 50$.

20. A right circular cylinder of radius b and height h has its lower base and curved surface maintained at the temperature $u = 0$. If the upper base is maintained at the temperature $u = 100$, find the steady-state temperature distribution throughout the cylinder.

21. Work Exercise 20 if the temperature distribution $u = b^2 - r^2$ is maintained over the upper base.

22. A thin circular disk of radius b whose upper and lower surfaces are insulated is initially at the temperature $u(r, \theta) = f(r)$. If the temperature around the circumference of the disk is suddenly reduced to 0 and maintained at that value, find the temperature in the disk as a function of r and t.

23. A two-dimensional sheet whose surfaces are insulated has the shape of a semicircle of radius b. At $t = 0$, while the temperature throughout the sheet is $u = 100$, the temperature around the entire boundary is suddenly reduced to 0 and maintained thereafter at that value. Find the temperature in the sheet as a function of r, θ, and t.

24. Work Exercise 23 if the sheet has the shape of a quarter-circle.

25. Work Exercise 23 if the sheet is an annular region of inner radius b and outer radius B.

26. A circular disk of radius b has its surfaces insulated. At $t = 0$ the temperature throughout the disk is given by the formula $u(r, \theta, 0) = 100(1 - r^2/b^2)$. Find the temperature distribution in the disk if at $t = 0$ the temperature around the circumference is suddenly raised to 100 and maintained thereafter at that value.

27. A cylinder of height h and radius b is initially at temperature $u = 100$ throughout. At $t = 0$ the temperature over both ends and the curved surface of the cylinder is suddenly reduced to 0 and held at that value thereafter. Find the temperature at any point in the cylinder at any subsequent time.

28. A right circular cylinder of radius b and infinite length has its curved surface maintained at the temperature $u = 0$. If the base of the cylinder is maintained at the temperature $u = 100$, find the steady-state temperature distribution in the cylinder.

29. Show that $\mathcal{L}\{tJ_0(at)\} = s/(s^2 + a^2)^{3/2}$. *Hint:* Recall Theorem 1, Sec. 10.7.

30. Show that $\mathcal{L}\{tJ_1(at)\} = a/(s^2 + a^2)^{3/2}$. *Hint:* Recall Eq. (6), Sec. 12.5, and Theorem 2, Sec. 10.7.

31. Show that $\mathcal{L}\{t^2J_1(at)\} = 3as/(s^2 + a^2)^{5/2}$.

32. Show that $\mathcal{L}\{t^2J_2(at)\} = 3a^2/(s^2 + a^2)^{5/2}$.

33. Starting with the results of Exercises 29 and 31, prove by mathematical induction that

$$\mathcal{L}\{t^nJ_{n-1}(at)\} = \frac{(2n)!a^{n-1}s}{2^n n!(s^2 + a^2)^{n+1/2}}$$

34. Starting with the results of Exercises 30 and 32, prove by mathematical induction that

$$\mathcal{L}\{t^nJ_n(at)\} = \frac{(2n)!a^n}{2^n n!(s^2 + a^2)^{n+1/2}}$$

35. Show that $\mathcal{L}\{J_2(at)\} = \dfrac{a^2}{(s^2 + a^2)^{1/2}[s + (s^2 + a^2)^{1/2}]^2}$.
Hint: Recall from Eq. (3), Sec. 12.5, that

$$J_2(at) = J_0(at) - 2\frac{dJ_1(at)}{a\,dt}$$

36. From Exercise 35 and the last result in Example 6, infer the formula for $\mathcal{L}\{J_n(at)\}$ and prove it by mathematical induction.

37. Show that $\int_0^\infty J_0(at)\,dt = 1/a$. *Hint:* Consider the integral that defines the Laplace transform of $J_0(at)$.

38. What is **(a)** $\int_0^\infty J_1(at)\,dt$? **(b)** $\int_0^\infty tJ_0(at)\,dt$? **(c)** $\int_0^\infty tJ_1(at)\,dt$?

39. Show that $\mathcal{L}\{I_1(at)\} = \dfrac{a}{(s^2 - a^2)^{1/2}[s + (s^2 - a^2)^{1/2}]}$.

40. What is **(a)** $\mathcal{L}\{tI_0(at)\}$? **(b)** $\mathcal{L}\{tI_1(at)\}$? **(c)** $\mathcal{L}\{t^2I_0(at)\}$?

41. If $\mathcal{L}\{f(t)\} = 1/\sqrt{s^2 + 4s + 13}$, what is $f(t)$? *Hint:* Write the radicand as the sum of two squares and then use Corollary 1, Theorem 1, Sec. 10.6.

42. If $\mathcal{L}\{f(t)\} = 1/(s^2 + 2s + 10)^{5/2}$, what is $f(t)$? *Hint:* Note the hint in Exercise 41, then use the result of Exercise 32.

43. What is $\mathcal{L}^{-1}\{1/\sqrt{s(s-4)}\}$?

44. Prove: $\mathcal{L}^{-1}\{1/(s + a)\sqrt{s^2 + b^2}\} = e^{-at}\int_0^t e^{ax}J_0(bx)\,dx$. *Hint:* Use Corollary 1, Theorem 1, Sec. 10.6, and Theorem 4, Sec. 10.4.

45. What is $\mathcal{L}^{-1}\{1/\sqrt{s^4 - a^4}\}$? *Hint:* Factor the radicand into two quadratic factors and then use Theorem 1, Sec. 10.11.

46. What is $\mathcal{L}^{-1}\{1/\sqrt{s^4 + 5s^2 + 4}\}$?

47. Show that $\int_0^t J_0(\lambda)J_0(t - \lambda)\,d\lambda = \sin t$. *Hint:* Use Theorem 1, Sec. 10.11.

48. Find a particular integral of the equation $y'' + y = J_0(t)$. *Hint:* Take the Laplace transform of each side of this equation and then use the result of Exercise 30.

49. Show that $\int_0^t \sin(t - \lambda)J_0(\lambda)\,d\lambda = tJ_1(t)$.

50. What is $\int_0^t \sin(t - \lambda)J_1(\lambda)\,d\lambda$?

51. Show that $\int_0^t \cos(t - \lambda)J_0(\lambda)d\lambda = tJ_0(\lambda)$.

52. Show that $\int_0^t J_0(t - \lambda)J_1(\lambda)\,d\lambda = J_0(t) - \cos t$.

53. What is $\int_0^t J_1(t - \lambda)J_1(\lambda)\,d\lambda$?

54. Show that $I_0(t) = \dfrac{e^{-t}}{\pi}\displaystyle\int_0^t \frac{e^{2\lambda}}{\sqrt{\lambda(t - \lambda)}}\,d\lambda$. *Hint:* Combine Formula 4, Sec. 10.5, for the case $n = -\frac{1}{2}$ with Theorem 1, Sec. 10.6, and then apply Theorem 1, Sec. 10.11, to the factored form of the Laplace transform of $I_0(t)$.

55. Show that $\mathcal{L}\{J_0(2\sqrt{t})\} = e^{-1/s}/s$. *Hint:* Show first that $y = J_0(2\sqrt{t})$ is a solution of the equation $(ty')' + y = 0$ for which $y(0) = 1$ and $y'(0) = 0$. Then by taking the Laplace transform of each side of this equation, show that the transform of the solution for which $y(0) = 1$ and $y'(0) = 0$ is $Ce^{-1/s}/s$. Finally, determine C by comparing the leading terms in the expansions of $\mathcal{L}\{J_0(2\sqrt{t})\}$ and $Ce^{-1/s}/s$ in terms of powers of s.

56. What is $\mathcal{L}\{I_0(2\sqrt{t})\}$?

57. Show that $\mathcal{L}\{J_n(at)/t\} = a^n/n(s + \sqrt{s^2 + a^2})^n$. *Hint:* Recall the result of Exercise 36. Then apply Theorem 2, Sec. 10.7, to $J_n(at)/t$, noting that

$$\frac{d(s + \sqrt{s^2 + a^2})}{ds} = \frac{s + \sqrt{s^2 + a^2}}{(s^2 + a^2)^{1/2}}$$

58. Show that

$$J_{m+n}(t) = n\int_0^t J_m(t-\lambda)J_n(\lambda)\frac{d\lambda}{\lambda}$$

$$= m\int_0^t J_m(t-\lambda)J_n(\lambda)\frac{d\lambda}{t-\lambda}$$

Hint: Factor the result of Exercise 36 in the appropriate way and use Theorem 1, Sec. 10.11.

59. The response of a certain network of tuned circuits to a voltage e_0 applied in the first circuit is described by the set of differential equations†

$$LC\frac{d^2e_1}{dt^2} + RC\frac{de_1}{dt} + e_1 = e_0$$

$$LC\frac{d^2e_2}{dt^2} + RC\frac{de_2}{dt} + e_2 = e_1$$

$$\cdots\cdots\cdots\cdots\cdots\cdots$$

$$LC\frac{d^2e_n}{dt^2} + RC\frac{de_n}{dt} + e_n = e_{n-1}$$

If e_0 is a unit impulse, find e_n as a function of t. *Hint:* Use Laplace transform methods.

60. When alternating current flows through a long cylindrical conductor, the current density is not uniform over the cross section but increases from the center toward the surface. If the actual current, whether of the form cos ωt or sin ωt, is replaced by the complex current $e^{i\omega t}$, as in Example 5, it can be shown‡ that the resulting current density σ satisfies the differential equation

$$\frac{d^2\sigma}{dr^2} + \frac{1}{r}\frac{d\sigma}{dr} - ik^2\sigma = 0$$

where $k^2 = 4\pi\omega/\rho$ and ρ is the resistivity of the conductor. Solve this equation and find the current density as a function of r in a cylindrical conductor of outer radius a if the current density at the surface is σ_0.

61. In Exercise 60, show that the fraction of the current carried by the portion of the conductor between $r = 0$ and $r = \lambda a$ is, in magnitude,

$$\frac{I_{\lambda a}}{I_a} = \lambda\frac{M_1(\lambda ka)}{M_1(ka)}$$

where $M_1(x)$ is the magnitude of the complex quantity

$$J_1(i^{3/2}x) = \text{ber}_1\, x + i\,\text{bei}_1\, x$$

Hint: Take $dA = 2\pi r\,dr$ when integrating the current density and remember that $\int xJ_0(\alpha x)\,dx = (x/\alpha)J_1(\alpha x)$ for complex as well as real values of α.

62. Find the first two natural frequencies of a steel shaft 100 in long vibrating torsionally if the shaft is built-in at one end and free at the other and if its radius at a distance x from the free end is $r(x) = (x/100)^{1/4}$. Steel weighs 0.285 lb/in^3, and its modulus of elasticity in sheer is $E_s = 12 \times 10^6$ lb/in^2.

63. Work Exercise 62 if the radius of the shaft varies according to the law

(a) $r(x) = \left(\frac{x}{100}\right)^{1/2}$ **(b)** $r(x) = \left(\frac{x}{100}\right)^{3/4}$

(c) $r(x) = \frac{x}{100}$

64. In Exercise 62, find the instantaneous deflection of the shaft if it starts to vibrate from rest with initial displacement $\theta(x, 0) = (100 - x)/100$.

65. An elastic string whose weight per unit length is $w(x) = w_0(1 + \alpha x)$, where x is the distance from one end of the string, is stretched under tension T between two points a distance l apart. Find the equation defining the natural frequencies of the string.

66. Work Exercise 65 if the weight per unit length of the string is $w(x) = w_0e^{\alpha x}$.

67. A body whose mass varies according to the law $m(t) = m_0e^{-\alpha t}$ moves along the x axis under the influence of an attractive force equal to $F(x) = -k^2x$. Find the equation of motion of the body if it starts from rest at the point $x = x_0$. *Hint:* Recall the general form of Newton's second law, $d(mv)/dt = F$. Then, at the appropriate point introduce a new independent variable by the substitution $z = e^{\alpha t/2}$.

68. Work Exercise 67 if $F(x) = k^2x$.

† See Samuel Sabaroff, "Impulse Excitation of a Cascade of Series Tuned Circuits," *Proceedings of the IRE,* December 1944, pp. 758–760.

‡ See, for example, N. W. McLachlan, *Bessel Functions for Engineers,* pp. 134–137, Oxford University Press, London, 1934.

69. Work Exercise 67 is $m(t) = m_0(1 + \alpha t)^{-1}$. *Hint:* At the appropriate point, make the substitution $1 + \alpha t = z^{2/3}$.

70. Work Exercise 67 if the mass $m(t) = m_0(1 + \alpha t)^{-1}$ and $F(x) = k^2x$.

71. A cantilever beam of length l and rectangular cross section has its upper surface horizontal. The width of the beam is b, and its thickness d varies directly as the cube root of the distance x from the free end. If an oblique tensile force F whose direction makes an angle θ with the horizontal acts at the free end of the beam, find the equation of the deflection curve of the beam. *Hint:* The static deflection of a beam satisfies the differential equation $EIy'' = M(x, y)$, where $M(x, y)$ is the moment at the point, (x, y) produced by the forces acting on the beam. The moment of inertia of a rectangle about its axis of symmetry is $bd^3/12$.

72. Work Exercise 71 if the force F is an oblique compressive force.

73. In Example 7, verify that $\int_0^\infty \lambda A(\lambda)e^{-\lambda z}J_0(\lambda r)\, d\lambda$ is a solution of the equation

$$\frac{\partial^2 u}{\partial r^2} + \frac{1}{r}\frac{\partial u}{\partial r} + \frac{\partial^2 u}{\partial z^2} = 0$$

74. Work Example 7 if $F(r) = e^{-r}$. *Hint:* Recall Exercise 48, Sec. 12.5.

75. Work Example 7 if

$$F(r) = \begin{cases} a^2 - r^2 & r < a \\ 0 & r > a \end{cases}$$

76. Find the steady-state temperature distribution in the region of space between the plane $z = 0$ and the plane $z = 1$ if on each plane the temperature distribution

$$u(r) = \begin{cases} u_0 & r < a \\ 0 & r > a \end{cases}$$

is maintained.

77. A thin sheet of metal whose upper and lower surfaces are insulated extends over the entire xy plane. A disk of radius a with center at the origin is removed from the sheet, heated until its temperature is u_0 throughout, and then, at $t = 0$, put back into the sheet. Assuming perfect thermal contact between the disk and the rest of the sheet, find the temperature as a function of r and t throughout the sheet, including the disk.

78. Show that the function $\phi(x) = \int_0^\infty \exp(-x \cosh \theta)\, d\theta$, $x > 0$, satisfies the equation $x\phi'' + \phi' - x\phi = 0$. Hence show that $\phi(x)$ is of the form $CK_0(x)$. (It can be shown that the constant C has the value 1.)

79. Work Exercise 16 if $b = 1$ and $h = \pi$.

80. Work Exercise 11 if heat loss through the end of the fin is not neglected.

81. A solid cylinder bounded by the surfaces $r = 1$, $z = 0$, and $z = 1$ has its lower base insulated. Find its steady-state temperature if

(a) The curved surface and upper base are maintained at the respective temperatures 0 and 1.

(b) The curved surface is maintained at the temperature 1 and the temperature distribution $u(r, \theta, 1) = r^2$ is maintained over the upper base.

82. Work Exercise 62 if the length of the shaft is 20 in and $r(x)$ equals

(a) $\left(\dfrac{x}{20}\right)^{1/4}$ **(b)** $\left(\dfrac{x}{20}\right)^{1/2}$ **(c)** $\left(\dfrac{x}{20}\right)^{3/4}$ **(d)** $\dfrac{x}{20}$

83. Work Example 3 if the fin, instead of running out to a point at $x = 0$, is cut off at $x = c$ and if heat loss through the end of the fin is neglected. *Hint:* Let u_c be the initially unknown temperature at the end where $x = c$, use u_c, with u_w, to determine the integration constants c_1 and c_2, then determine u_c by equating the heat entering the fin to the heat lost from its surface.

84. Neglecting heat loss through the exposed cylindrical surface of the thin fin, solve the equation of Exercise 12 and find the temperature distribution in the fin if the cylinder temperature is u_c.

85. The lower end of a long thin rod of uniform cross section is clamped so that the rod is vertical. Determine the values of the parameters of the rod for which buckling will occur if the upper end of the rod is displaced slightly from its neutral position. *Hint:* Choosing axes as in Example 4, Sec. 2.11, the problem can be solved by using the relation $(EIy'')' = V$, where V is the transverse component of the weight of the portion of the rod above a general point x, and then noting that $y'' \doteq d\theta/dx$ for inclination angles θ small enough that $\tan\theta \doteq \sin\theta \doteq \theta$.

86. Find the frequency equation for the transverse vibrations of a cantilever whose width is constant but whose depth varies directly as the distance from the free end. *Hint:* To solve the differential equation defining the normal modes of the beam, recall Exercise 33, Sec. 12.4.

87. Find the frequency equation for the transverse vibrations of a cantilever which is a solid of revolution whose radius varies directly as the distance from the free end.

88. A bar has the shape of a truncated right circular cone of length l, the radii of the bases being r and R. Find the frequency equation for the torsional vibrations of the bar assuming both ends of the bar free.

89. Determine the limiting form of the frequency equation in Exercise 88 when $r \to R$. Check by comparing your result with the frequency equation derived directly for a uniform bar. *Hint:* Express the Bessel functions in terms of sines and cosines.

90. Find the steady-state temperature distribution in a two-dimensional region having the shape of a semicircle of radius b if the diametral boundary is maintained at the constant temperature $u = u_0$ and the curved boundary is maintained at the temperature distribution $u(b, \theta) = f(\theta)$.

91. A two-dimensional region having the shape of a semicircle of radius b is initially at the uniform temperature $u = 100°C$. At $t = 0$ the temperature along the bounding diameter is reduced to 0°C and maintained thereafter at that temperature. The curved boundary is maintained at the temperature $u = 100$°C. Find the temperature at any point of the region at any subsequent time.

92. Find the steady-state temperature distribution in the region between two concentric circles of radii r_1 and r_2 if the portion of the inner boundary between $\theta = 0$ and $\theta = \pi$ is maintained at the temperature $u = 100$°C while the remainder of the inner boundary and the entire outer boundary are maintained at the temperature $u = 0$°C.

93. The region between two concentric circles of radii r_1 and r_2 is initially at the uniform temperature $u = 0$°C. At $t = 0$ the temperature around the entire inner boundary is suddenly raised to 100°C and maintained thereafter at that value. Find the temperature at any point of the region at any subsequent time if the outer boundary is maintained at the temperature 0°C.

94. If the temperature throughout the cylinder in Exercise 16 at $t = 0$ is given by the known function $\phi(r, z)$, find the temperature distribution at any subsequent time.

95. A free fixed composite shaft of unit length consists of two segments bonded together at $x = \beta$, $0 < \beta < 1$. The radius of the cross section of the left segment a distance x from the free end is $r(x) = bx$. The right segment is a right circular cylinder of radius r_0 and length $1 - \beta$. Find the frequency equation for free torsional vibrations of the shaft. *Hint:* With ρ_1, ρ_2 and G_1, G_2 as the weight densities and moduli of elasticity in shear of the tapered and cylindrical segments, respectively, parallel the procedure of Examples 5 and 7, Sec. 11.6.

12.9 LEGENDRE POLYNOMIALS

In Example 4, Sec. 12.8, we found that when we solved the steady-state heat equation, i.e., Laplace's equation, in cylindrical coordinates, one of the ordinary differential equations arising from the separation of variables was Bessel's equation. In very much the same way, it turns out that when we apply the method of separation of variables to Laplace's equation in spherical coordinates, one of the ordinary differential equations that results is *Legendre's equation.*

If the expression

$$\nabla^2 F \equiv \frac{\partial^2 F}{\partial x^2} + \frac{\partial^2 F}{\partial y^2} + \frac{\partial^2 F}{\partial z^2}$$

is transformed from cartesian coordinates to spherical coordinates by means of the relations (Fig. 12.11)

$$x = r \sin\theta \cos\phi \qquad y = r \sin\theta \sin\phi \qquad z = r\cos\theta$$

we obtain, after a lengthy but straightforward reduction,

$$\nabla^2 F = \frac{1}{r^2 \sin\theta}\left(r^2 \sin\theta \frac{\partial^2 F}{\partial r^2} + 2r\sin\theta \frac{\partial F}{\partial r} + \sin\theta \frac{\partial^2 F}{\partial \theta^2} + \cos\theta \frac{\partial F}{\partial \theta} + \frac{1}{\sin\theta}\frac{\partial^2 F}{\partial \phi^2}\right)$$

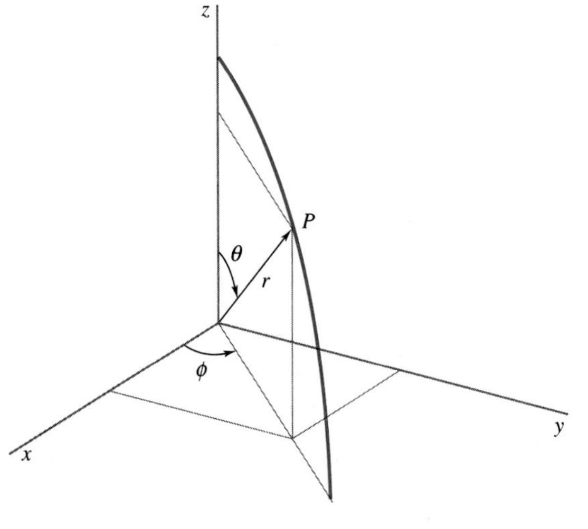

FIGURE 12.11
The relation between rectangular and spherical coordinates.

Hence, when Laplace's equation $\nabla^2 F = 0$ is expressed in spherical coordinates, it becomes

$$(1) \qquad r^2 \sin\theta \frac{\partial^2 F}{\partial r^2} + 2r\sin\theta \frac{\partial F}{\partial r} + \sin\theta \frac{\partial^2 F}{\partial \theta^2} + \cos\theta \frac{\partial F}{\partial \theta} + \frac{1}{\sin\theta}\frac{\partial^2 F}{\partial \phi^2} = 0$$

Any solution $F(r, \theta, \phi)$ of this equation is known as a **spherical harmonic**.

In an attempt to solve Eq. (1), we assume a product solution $F(r, \theta, \phi) = R(r)\Theta(\theta)\Phi(\phi)$. Then, substituting into (1), we have

$$r^2 \sin\theta\, R''\Theta\Phi + 2r\sin\theta\, R'\Theta\Phi + \sin\theta\, R\Theta''\Phi + \cos\theta\, R\Theta'\Phi + \frac{1}{\sin\theta} R\Theta\Phi'' = 0$$

Dividing through by $R\Theta\Phi \sin\theta$ and rearranging gives us

$$r^2\frac{R''}{R} + 2r\frac{R'}{R} = -\left(\frac{\Theta''}{\Theta} + \frac{\cos\theta}{\sin\theta}\frac{\Theta'}{\Theta} + \frac{1}{\sin^2\theta}\frac{\Phi''}{\Phi}\right)$$

This relation can hold only if the common value of these two expressions is a constant. For later convenience we write this constant as $n(n+1)$. Thus we are led to the two equations

$$(2) \qquad r^2R'' + 2rR' - n(n+1)R = 0$$

$$(3) \qquad \frac{\Theta''}{\Theta} + \frac{\cos\theta}{\sin\theta}\frac{\Theta'}{\Theta} + \frac{1}{\sin^2\theta}\frac{\Phi''}{\Phi} + n(n+1) = 0$$

Separating variables in Eq. (3), we have further

$$(4) \qquad \sin^2\theta\frac{\Theta''}{\Theta} + \cos\theta\sin\theta\frac{\Theta'}{\Theta} + n(n+1)\sin^2\Theta = -\frac{\Phi''}{\Phi}$$

Again, we reason that the two members of this equation must have a common constant value which, for convenience, we take to be m^2. This gives us, finally, the two equations

$$(5) \qquad \Phi'' + m^2\Phi = 0$$

$$(6) \qquad \sin^2\theta\,\Theta'' + \cos\theta\sin\theta\,\Theta' + [n(n+1)\sin^2\theta - m^2)\Theta = 0$$

Two of the three equations into which Eq. (1) has now been separated are familiar and easy to solve. Equation (2) is an Euler equation whose complete solution is

$$R(r) = c_1r^n + c_2\frac{1}{r^{n+1}}$$

The solution of Eq. (5) is

$$\Phi(\phi) = c_3\cos m\phi + c_4\sin m\phi$$

Equation (6), however, is unfamiliar, and we will have to solve it by the method of Frobenius. It is known as the **associated Legendre equation**,† although it is usually studied in the form obtained by setting $x = \cos\theta$.

---

†Named for the French mathematician Adrien Marie Legendre, (1752–1833).

If $x = \cos\theta$, then

$$\frac{d\Theta}{d\theta} = \frac{d\Theta}{dx}\frac{dx}{d\theta} = -\sin\theta\frac{d\Theta}{dx}$$

$$\frac{d^2\Theta}{d\theta^2} = \frac{d}{d\theta}\left(-\sin\theta\frac{d\Theta}{dx}\right) = -\cos\theta\frac{d\Theta}{dx} - \sin\theta\frac{d^2\Theta}{dx^2}\frac{dx}{d\theta}$$

$$= -\cos\theta\frac{d\Theta}{dx} + \sin^2\theta\frac{d^2\Theta}{dx^2}$$

Hence, substituting these expressions into (6), we obtain

$$\sin^2\theta\left(-\cos\theta\frac{d\Theta}{dx} + \sin^2\theta\frac{d^2\Theta}{dx^2}\right) + \sin\theta\cos\theta\left(-\sin\theta\frac{d\Theta}{dx}\right) + [n(n+1)\sin^2\theta - m^2]\Theta = 0$$

Dividing out $\sin^2\theta$, substituting $x = \cos\theta$ in the coefficients, and simplifying gives

$$(1 - x^2)\frac{d^2\Theta}{dx^2} - 2x\frac{d\Theta}{dx} + \left[n(n+1) - \frac{m^2}{1-x^2}\right]\Theta = 0 \tag{7}$$

This is the **algebraic form of the associated Legendre equation**. If $m = 0$, that is, if the solution of the original equation (1) is independent of the longitude angle ϕ, then Eq. (7) reduces to

$$(1 - x^2)\frac{d^2\Theta}{dx^2} - 2x\frac{d\Theta}{dx} + n(n+1)\Theta = 0 \tag{8}$$

which is known simply as **Legendre's equation**.

To solve Eq. (8), we use the method of Frobenius and assume a series solution of the form

$$\Theta(x) = \sum_{k=0}^{\infty} a_k x^{k+c}$$

whence

$$\Theta'(x) = \sum_{k=0}^{\infty} a_k(k+c)x^{k+c-1}$$

and

$$\Theta''(x) = \sum_{k=0}^{\infty} a_k(k+c)(k+c-1)x^{k+c-2}$$

Then, substituting, we have, after moving x^2 and x into the sums they multiply,

$$\sum_{k=0}^{\infty} a_k(k+c)(k+c-1)x^{k+c-2} - \sum_{k=0}^{\infty} a_k(k+c)(k+c-1)x^{k+c}$$
$$- 2\sum_{k=0}^{\infty} a_k(k+c)x^{k+c} + n(n+1)\sum_{k=0}^{\infty} a_k x^{k+c} = 0$$

In this identity, the coefficients of the two lowest powers of x, namely, the first two terms in the first sum, are respectively,

$$a_0 c(c-1) \qquad \text{and} \qquad a_1(c+1)c$$

If $c = 0$, each of these is zero without restrictions on a_0 and a_1; hence the choice $c = 0$ will lead to two independent solutions of Eq. (8). Of course, this is to be expected since $x = 0$, the point about which we are expanding the solution, is an ordinary point of Eq. (8). Putting $c = 0$ and then combining the last three sums, we obtain

$$\sum_{k=2}^{\infty} a_k k(k-1)x^{k-2} - \sum_{k=0}^{\infty} a_k[k(k-1) + 2k - n(n+1)]x^k = 0$$

If we now change the variable of summation in the first sum from k to $k + 2$ and again combine the sums, we have, further,

$$\sum_{k=0}^{\infty} [a_{k+2}(k+1)(k+2) + a_k\{-k(k+1) + n(n+1)\}]x^k = 0$$

or

$$\sum_{k=0}^{\infty} [a_{k+2}(k+1)(k+2) + a_k(n-k)(n+k+1)]x^k = 0$$

This will be an identity if and only if

$$(k+1)(k+2)a_{k+2} + (n-k)(n+k+1)a_k = 0$$

and therefore the recurrence relation connecting the coefficients is

$$a_{k+2} = -\frac{(n-k)(n+k+1)}{(k+1)(k+2)}a_k \qquad k \geqq 0$$

and, specifically,

$$\begin{aligned}
a_0 &= a_0 & a_1 &= a_1 \\
a_2 &= -\frac{n(n+1)}{2!}a_0 & a_3 &= -\frac{(n-1)(n+2)}{3!}a_1 \\
a_4 &= \frac{n(n-2)(n+1)(n+3)}{4!}a_0 & a_5 &= \frac{(n-1)(n-3)(n+2)(n+4)}{5!}a_1 \\
&\cdots\cdots\cdots & &\cdots\cdots\cdots
\end{aligned}$$

Hence a complete solution of Eq. (8) can be written

$$(9) \quad \Theta(x) = a_0\left[1 - \frac{n(n+1)}{2!}x^2 + \frac{n(n-2)(n+1)(n+3)}{4!}x^4 - \cdots\right] + a_1\left[x - \frac{(n-1)(n+2)}{3!}x^3 + \frac{(n-1)(n-3)(n+2)(n+4)}{5!}x^5 - \cdots\right]$$

These infinite series define what are known as **Legendre functions of the second kind**. Since $x = \pm 1$ are the only singular points of Eq. (8), it follows from Theorem 6, Sec. 2.9, that the radius of convergence of each of these series is at least 1. It can be shown, however (see Exercise 1), that (unless one or the other terminates because n is an integer) neither of these series converges at either of the points $x = \pm 1$. In other words, the interval of convergence of each series in (9) is $-1 < x < 1$.

In many applications the parameter n is a positive integer. If it is odd, then clearly the second series in (9) contains only a finite number of terms; if it is even, then the first series contains only a finite number of terms. In either case, the series that reduces to a finite sum is known as a **Legendre polynomial** or **zonal harmonic of order** n. The usual standard form for the Legendre polynomials is obtained by assigning to a_0 and a_1 values which will make the coefficient of the highest power of x in each series equal to

$$\frac{(2n)!}{2^n(n!)^2}$$

These values are

$$a_0 = (-1)^{n/2}\frac{1\cdot 3\cdot 5\cdots(n-1)}{2\cdot 4\cdot 6\cdots n} = (-1)^{n/2}\frac{n!}{2^n[(n/2)!]^2}$$

$$a_1 = (-1)^{(n-1)/2}\frac{1\cdot 3\cdot 5\cdots n}{2\cdot 4\cdot 6\cdots(n-1)} = (-1)^{(n-1)/2}\frac{(n+1)!}{2^n[(n-1)/2]![(n+1)/2]!}$$

and the resulting general formula is

$$P_n(x) = \sum_{k=0}^{N}(-1)^k\frac{(2n-2k)!}{2^n k!(n-k)!(n-2k)!}x^{n-2k} \qquad \begin{matrix} N = n/2,\ n \text{ even} \\ N = (n-1)/2,\ n \text{ odd}\end{matrix} \tag{10}$$

Specifically, this formula gives

$$P_0(x) = 1 \qquad P_1(x) = x$$

$$P_2(x) = \tfrac{1}{2}(3x^2 - 1) \qquad P_3(x) = \tfrac{1}{2}(5x^3 - 3x)$$

$$P_4(x) = \tfrac{1}{8}(35x^4 - 30x^2 + 3) \qquad P_5(x) = \tfrac{1}{8}(63x^5 - 70x^3 + 15x)$$

$$P_6(x) = \tfrac{1}{16}(231x^6 - 315x^4 + 105x^2 - 5) \qquad P_7(x) = \tfrac{1}{16}(429x^7 - 693x^5 + 315x^3 - 35x)$$

As these particular results illustrate, and as we shall soon prove, $P_n(1) = 1$ and $P_n(-1) = (-1)^n$ for all values of n. Since, as we observed earlier, the infinite series in (9) diverge when $x = \pm 1$, we have the following important result.

THEOREM 1 To within an arbitrary constant multiplier, $P_n(x)$ is the only solution of Legendre's equation which is finite on the closed interval $-1 \le x \le 1$.

Like the Bessel functions, the Legendre polynomials have a generating function (see Definition 1, Sec. 12.6) from which various important properties can be derived. For the Legendre polynomials, the generating function is $f(x, z) = (1 - 2xz + z^2)^{-1/2}$, as the following theorem makes clear.

THEOREM 2

$$\frac{1}{\sqrt{1 - 2xz + z^2}} = P_0(x) + P_1(x)z + P_2(x)z^2 + \cdots + P_n(x)z^n + \cdots$$

PROOF To prove this, we rearrange the radical slightly and then expand it by the binomial theorem, getting

$$[1 - z(2x - z)]^{-1/2} = 1 + \frac{1}{2}z(2x - z) + \frac{1\cdot 3}{2^2 2!}z^2(2x - z)^2 + \cdots$$

$$+ \frac{1\cdot 3\cdots(2n-3)}{2^{n-1}(n-1)!}z^{n-1}(2x - z)^{n-1} + \frac{1\cdot 3\cdots(2n-1)}{2^n n!}z^n(2x - z)^n + \cdots$$

Now z^n can occur only in the terms out to and including the one containing $z^n(2x - z)^n$. In the last of these terms it arises when the coefficient z^n is multiplied by the first term in the expansion of $(2x - z)^n$, namely, $(2x)^n$. In the preceding term it arises when the coefficient z^{n-1} is multiplied by the second term in the expansion of $(2x - z)^{n-1}$, namely, $\frac{n-1}{1!}(2x)^{n-2}z$. Proceeding from term to term in this fashion, we find that the total coefficient of z^n is

$$\frac{1 \cdot 3 \cdots (2n-1)}{2^n n!}(2x)^n - \frac{1 \cdot 3 \cdots (2n-3)}{2^{n-1}(n-1)!} \cdot \frac{n-1}{1!}(2x)^{n-2} + \frac{1 \cdot 3 \cdots (2n-5)}{2^{n-2}(n-2)!} \cdot \frac{(n-2)(n-3)}{2!}(2x)^{n-4} - \cdots$$

Multiplying and dividing by the factors needed to complete the factorials in the respective numerators gives

$$\frac{(2n)!}{2^n n! n!}x^n - \frac{(2n-2)!}{2^n 1!(n-1)!(n-2)!}x^{n-2} + \frac{(2n-4)!}{2^n 2!(n-2)!(n-4)!}x^{n-4} - \cdots$$

as the total coefficient of z^n in the expansion of $(1 - 2xz + z^2)^{-1/2}$. Since this series is precisely the expanded form of $P_n(x)$, as given by (10), we have thus shown that

$$(1 - 2xz + z^2)^{-1/2} = \sum_{n=0}^{\infty} P_n(x)z^n$$

as asserted. ◗

EXAMPLE 1

What is $P_n(1)$? What is $P_n(-1)$?

To answer the first of these questions, let us return to the assertion of Theorem 2 and set $x = 1$. This makes the radicand in the generating function a perfect square, and we have

$$\frac{1}{1-z} = \sum_{n=0}^{\infty} P_n(1)z^n$$

By the binomial theorem,

$$\frac{1}{1-z} = 1 + z + z^2 + z^3 + \cdots + z^n + \cdots$$

Hence, comparing coefficients, it follows that $P_n(1) = 1$ for every value of n.

To answer the second question, we put $x = -1$ in the generating function, getting

$$\frac{1}{1+z} = 1 - z + z^2 - z^3 + \cdots + (-1)^n z^n + \cdots$$

Again, comparing coefficients with the series $\sum_{n=0}^{\infty} P_n(-1)z^n$, it follows that $P_n(-1) = (-1)^n$ for every value of n.

EXAMPLE 2

Establish the recurrence relation

$$(n+1)P_{n+1}(x) = x(2n+1)P_n(x) - nP_{n-1}(x)$$

If we let $g(x, z) = \dfrac{1}{(1 - 2xz + z^2)^{1/2}}$, then

$$\frac{\partial g(x, z)}{\partial z} = -\frac{1}{2}\frac{-2x + 2z}{(1 - 2xz + z^2)^{3/2}} = \frac{x - z}{(1 - 2xz + z^2)(1 - 2xz + z^2)^{1/2}}$$

or

$$(1 - 2xz + z^2)\frac{\partial g(x, z)}{\partial z} = (x - z)g(x, z)$$

If we now substitute $\Sigma_{n=0}^{\infty} P_n(x)z^n$ for $g(x, z)$, according to Theorem 2, we have

$$(1 - 2xz + z^2)[\cdots + P_{n-1}(x)(n - 1)z^{n-2} + P_n(x)nz^{n-1} + P_{n+1}(x)(n + 1)z^n + \cdots]$$
$$= (x - z)[\cdots P_{n-1}(x)z^{n-1} + P_n(x)z^n + \cdots]$$

Equating the coefficients of z^n on each side of this equation, we have

$$P_{n+1}(x)(n + 1) - 2xP_n(x)n + P_{n-1}(x)(n - 1) = xP_n(x) - P_{n-1}(x)$$

Collecting terms and solving for $(n + 1)P_{n+1}(x)$ now gives us the assertion of the problem,

$$(n + 1)P_{n+1}(x) = (2n + 1)xP_n(x) - nP_{n-1}(x)$$

In many applications the algebraic form of the Legendre polynomials is the most useful. There are problems, however, in which it is essential that they be expressed in terms of θ, the colatitude angle of the spherical coordinate system with which our discussion began. This can easily be done by reversing the transformation $x = \cos\theta$ which led from the trigonometric to the algebraic form of Legendre's equation. However, replacing x by $\cos\theta$ in $P_n(x)$ leads to expressions that are quite inconvenient because of the powers of $\cos\theta$ they contain. Fortunately, using the generating function provided by Theorem 2, we can easily derive more useful forms in which cosines of multiples of θ take the place of powers of $\cos\theta$.

To do this, let us substitute

$$x = \cos\theta = \frac{e^{i\theta} + e^{-i\theta}}{2}$$

into the generating function, getting

$$[1 - z(e^{i\theta} + e^{-i\theta}) + z^2]^{-1/2} = [(1 - ze^{i\theta})(1 - ze^{-i\theta})]^{-1/2} = \sum_{n=0}^{\infty} P_n(x)z^n$$

If we use the binomial theorem to expand each of the factors in the middle of this continued identity, we obtain

$$(1 - ze^{i\theta})^{-1/2} = \sum_{n=0}^{\infty} a_n z^n e^{ni\theta} \qquad \text{and} \qquad (1 - ze^{-i\theta})^{-1/2} = \sum_{n=0}^{\infty} a_n z^n e^{-in\theta}$$

where

$$a_n = \frac{1 \cdot 3 \cdot 5 \cdots (2n - 1)}{2 \cdot 4 \cdot 6 \cdots (2n)} \qquad n \geq 1,\ a_0 = 1 \tag{11a}$$

To find the coefficient of z^n in the product of these two series, we form their Cauchy product (Definition 6, Sec. 2.9), getting

$$\sum_{n=0}^{\infty}\sum_{k=0}^{n}(a_{n-k}z^{n-k}e^{(n-k)i\theta})(a_k z^k e^{-ki\theta}) = \sum_{n=0}^{\infty}\left(z^n \sum_{k=0}^{n} a_k a_{n-k} e^{(n-2k)i\theta}\right) \tag{11b}$$

In the inner sum, which is the sum of interest to use, it is easy to verify that for $n \geq 1$, the terms corresponding to $k = j$ and $k = n - j$ are identical except that the exponents on e are of opposite sign. Hence these terms can be paired, and we have for the coefficient of z^n,

$$\begin{aligned} P_n(\cos\theta) &= a_0 a_n(e^{ni\theta} + e^{-ni\theta}) + a_1 a_{n-1}(e^{(n-2)i\theta} + e^{-(n-2)i\theta}) + \cdots \\ &= 2[a_0 a_n \cos n\theta + a_1 a_{n-1}\cos(n-2)\theta + \cdots] \end{aligned} \tag{11c}$$

If n is odd, the number of terms is even and each has a place in one of the pairs. In this case, the last term in the sum is

$$a_{(n-1)/2}a_{(n+1)/2}\cos\theta$$

If n is even, the number of terms is odd and the middle term, perforce, remains unpaired. In this case, the series (11c) for $P_n(\cos\theta)$ ends with the constant term

$$a_{n/2}a_{n/2}$$

Using (11a) to compute values of the a_n, we find from the unit coefficient of z^0 in (11b) and from Eq. (11c), whether n is odd or even, the specific expressions

$$\begin{aligned} P_0(\cos\theta) &= 1 \\ P_1(\cos\theta) &= \cos\theta \\ P_2(\cos\theta) &= \tfrac{1}{4}(3\cos 2\theta + 1) \\ P_3(\cos\theta) &= \tfrac{1}{8}(5\cos 3\theta + 3\cos\theta) \\ P_4(\cos\theta) &= \tfrac{1}{64}(35\cos 4\theta + 20\cos 2\theta + 9) \\ P_5(\cos\theta) &= \tfrac{1}{128}(63\cos 5\theta + 35\cos 3\theta + 30\cos\theta) \\ P_6(\cos\theta) &= \tfrac{1}{512}(231\cos 6\theta + 126\cos 4\theta + 105\cos 2\theta + 50) \end{aligned}$$

Another fundamental identity involving Legendre polynomials is **Rodrigues' formula,**† contained in the following theorem.

THEOREM 3

$$P_n(x) = \frac{1}{2^n n!}\frac{d^n(x^2-1)^n}{dx^n} \qquad n = 0, 1, 2, \ldots$$

PROOF We shall establish this result by actually carrying out the indicated differentiations, using the familiar formula in which m, n are nonnegative integers,

$$D^n x^m = m(m-1)(m-2)\cdots(m-n+1)x^{m-n} = \begin{cases} \dfrac{m!}{(m-n)!}x^{m-n} & n \leq m \\ 0 & n > m \end{cases}$$

†Named for the French economist and mathematician Olinde Rodrigues (1794–1851).

We begin by replacing $(x^2 - 1)^n$ by its binomial expansion, getting

$$\frac{1}{2^n n!} D^n \left[\sum_{k=0}^{n} (-1)^k \frac{n!}{k!(n-k)!} (x^2)^{n-k} \right] = \frac{1}{2^n n!} \sum_{k=0}^{n} (-1)^k \frac{n!}{k!(n-k)!} D^n x^{2n-2k}$$

In the last series, the only nonzero terms are those for which $2n - 2k \geq n$, and so the series terminates when k, which is integral-valued, reaches the largest integer N contained in $n/2$. If n is even, $N = n/2$; if n is odd, $N = (n-1)/2$. Hence if we move the constant factor $1/2^n n!$ inside the last summation sign, cancel $n!$ from numerator and denominator, change the upper limit of summation to N, and replace $D^n x^{2n-2k}$ by $\dfrac{(2n-2k)!}{(n-2k)!} x^{n-2k}$, the result is precisely $P_n(x)$, as given by Eq. (10), and the theorem is established. ◗

Various consequences of Rodrigues' formula are indicated in the exercises.

It would be surprising if our interest in Legendre polynomials did not stem, at least in part, from the expectation that they form an orthogonal set, and such is indeed the case. In fact, since Legendre's equation [Eq. (8)] can be written in the form

$$\frac{d[(1 - x^2)y']}{dx} + n(n+1)y = 0$$

it is clear that it is a special case, with

$$p(x) = 1 \qquad q(x) = 0 \qquad r(x) = 1 - x^2 \qquad \lambda = n(n+1)$$

of the equation covered by Theorem 4, Sec. 11.6. Hence if solutions of Legendre's equation corresponding to distinct values of n satisfy suitable boundary conditions at the ends of an interval, then over that interval they are orthogonal with respect to the weight function $p(x) = 1$. In particular, for the important interval $[-1, 1]$, no boundary conditions are necessary since $r(x) \equiv 1 - x^2$ vanishes at each endpoint. For the Legendre polynomials this gives us the following fundamental result,

$$(12) \qquad \int_{-1}^{1} P_m(x) P_n(x)\, dx = 0 \qquad m \neq n$$

Before the property of orthogonality can be used to expand an arbitrary function in terms of Legendre polynomials, we must of course know the value of the integral of the square of the general Legendre polynomial. This can be found in various ways, but perhaps the simplest is to use the generating function provided by Theorem 2. If we square the identity

$$\frac{1}{(1 - 2xz + z^2)^{1/2}} = P_0(x) + P_1(x)z + \cdots + P_n(x)z^n + \cdots$$

and integrate with respect to x from -1 to 1, we obtain

$$\int_{-1}^{1} \frac{dx}{1 - 2xz + z^2} = \int_{-1}^{1} [P_0(x) + P_1(x)z + \cdots + P_n(x)z^n + \cdots]^2\, dx$$

The integral on the left is easily evaluated. On the right, all integrals involving the product of two different P's are zero because of the orthogonality property (12). Hence

$$(13) \quad -\frac{1}{2z} \ln (1 - 2xz + z^2)\Big|_{-1}^{1} = \int_{-1}^{1} P_0^2(x)\, dx + z^2 \int_{-1}^{1} P_1^2(x)\, dx + \cdots + z^{2n} \int_{-1}^{1} P_n^2(x)\, dx + \cdots$$

Evaluation of the left-hand member leads at once to

$$-\frac{1}{2z}[\ln(1-z)^2 - \ln(1+z)^2] = \frac{1}{z}[\ln(1+z) - \ln(1-z)]$$

If we now replace the logarithms by their respective power series, (Example 6, Sec. 2.9), we obtain

$$\frac{1}{z}\left(z - \frac{z^2}{2} + \frac{z^3}{3} - \cdots - \frac{z^{2n}}{2n} + \frac{z^{2n+1}}{2n+1} - \cdots\right)$$
$$- \frac{1}{z}\left(-z - \frac{z^2}{2} - \frac{z^3}{3} - \cdots - \frac{z^{2n}}{2n} - \frac{z^{2n+1}}{2n+1} - \cdots\right)$$
$$= 2\left(1 + \frac{z^2}{3} + \frac{z^4}{5} + \cdots + \frac{z^{2n}}{2n+1} + \cdots\right)$$

Hence, comparing coefficients in this series and in the right-hand member of (13), we obtain the desired result:

$$\int_{-1}^{1} P_n^2(x)\, dx = \frac{2}{2n+1} \tag{14}$$

By means of the substitution $x = \cos\theta$, Eq. (14) can be transformed at once into the corresponding results for Legendre polynomials in trigonometric form. Hence, summarizing the preceding discussion, we have the following important theorem.

THEOREM 4 The Legendre polynomials in algebraic form satisfy the orthogonality relations

$$\int_{-1}^{1} P_m(x)P_n(x)\, dx = \begin{cases} 0 & m \neq n \\ 2/(2n+1) & m = n \end{cases}$$

In trigonometric form, the Legendre polynomials satisfy the orthogonality relations

$$\int_{0}^{\pi} P_m(\cos\theta)P_n(\cos\theta)\sin\theta\, d\theta = \begin{cases} 0 & m \neq n \\ 2/(2n+1) & m = n \end{cases}$$

EXAMPLE 3

TEMPERATURES IN A SPHERE

If the known temperature distribution $u = f(\theta)$ is maintained over the entire surface of a sphere of radius b, what is the steady-state temperature at any point in the sphere?

In this problem we are to solve the steady-state heat equation, i.e., Laplace's equation, in spherical coordinates when the temperature distribution is independent of the longitude angle ϕ. We have already carried out the process of separating variables in this equation and obtained Eqs. (2), (5), and (8) for the respective factors $R(r)$, $\Phi(\phi)$, and $\Theta(\theta)$. Because in the present problem the temperature does not depend on ϕ, Eq. (5) is irrelevant and we are concerned only with solutions of Eqs. (2) and (8).

We have already found the solutions of Eq. (2), namely,

$$R(r) = c_1 r^n + c_2 \frac{1}{r^{n+1}}$$

However, since the temperature must remain finite at the center of the sphere, where $r = 0$, the term c_2/r^{n+1} must be rejected, leaving $R(r) = c_1 r^n$.

The solutions of Eq. (8) are Legendre functions. Of these, according to Theorem 1, to within a constant multiplier the only ones that are finite over the θ interval $[0, \pi]$ (corresponding to the x interval $[-1, 1]$) are the Legendre polynomials $\{P_n(\cos\theta)\}$. Hence our product solutions are all of the form

$$u_n(r, \theta) = R(r)\Theta(\theta) = C_n r^n P_n(\cos\theta)$$

We must now construct an infinite series of these products

$$u(r, \theta) = \sum_{n=0}^{\infty} u_n(r, \theta) = \sum_{n=0}^{\infty} C_n r^n P_n(\cos\theta) \tag{15}$$

and attempt to determine the coefficients so that when $r = b$ the temperature becomes $u(b, \theta) = f(\theta)$:

$$f(\theta) = \sum_{n=0}^{\infty} C_n b^n P_n(\cos\theta)$$

To find C_n we proceed as usual when working with orthogonal functions; that is, we multiply the last equation by $\sin\theta\, P_m(\cos\theta)$ and integrate from $\theta = 0$ to $\theta = \pi$. By virtue of the orthogonality properties of the P's (Theorem 4), all integrals on the right except the one for which $m = n$ become zero and we have

$$\int_0^{\pi} f(\theta) \sin\theta\, P_n(\cos\theta)\, d\theta = C_n b^n \frac{2}{2n+1}$$

or

$$C_n = \frac{2n+1}{2b^n} \int_0^{\pi} f(\theta) \sin\theta\, P_n(\cos\theta)\, d\theta$$

With the coefficients in the series (15) determined, the formal solution of the problem is now complete.

Legendre polynomials are especially useful in what is known as *potential theory*. We shall investigate this subject in more detail in the chapter on vector analysis, but by anticipating just a little we can set the stage for an example of the application of Legendre polynomials to a boundary-value problem whose solution does not involve the property of orthogonality.

According to Newton's law of universal gravitation, two particles of mass m_1 and m_2 attract each other with a force directly proportional to the product of their masses and inversely proportional to the square of the distance d between them, i.e.,

$$F = k\frac{m_1 m_2}{d^2}$$

If m_1 is a point mass but m_2 is a distributed mass, a disk or a sphere, for example, both the distance d and the direction of the gravitational attraction at m_1 due to the mass elements into which m_2 may be subdivided will vary from point to point in m_2. Consequently, the resultant force m_2 exerts on m_1 will have to be found by integration over all the mass elements in m_2. This complicated integration process can be avoided, however, and replaced by differentiation if we know what is called the *gravitational potential* of m_2. The following definition makes this idea more precise.

DEFINITION 1 The **gravitational potential** of a distributed mass M is a function of position, $V(x, y, z)$ or $V^*(r, \theta, \phi)$, with the property that the directional derivative of V or V^* in any direction at a point P is the component in that direction of the attraction of M on a unit mass located at P.

It can be shown, and in Chapter 15 we shall do this, that every distributed mass M has an associated gravitational potential.

The determination of the potential function for a mass M is clearly an important problem. Moreover, it is a problem that, in some cases at least, we are currently prepared to handle, since as the following theorem shows, it involves solving a familiar partial differential equation.

THEOREM 5 In empty space, the gravitational potential of any mass M satisfies Laplace's equation.

Actually, Theorem 5 can be extended remarkably because not only gravitational force fields but also electrostatic and electromagnetic fields and velocity fields in hydrodynamics have potential functions that satisfy Laplace's equation.

Focusing now on the solution process and leaving the deeper physical relations to a later chapter, let us consider the following example.

EXAMPLE 4

THE GRAVITATIONAL POTENTIAL OF A RING

What is the gravitational potential $V(r, \theta, \phi)$ of a thin circular ring of radius c and mass M if the potential on the axis of the ring at a distance r from the center of the ring is $M/\sqrt{c^2 + r^2}$?

The statement of the problem implies that we should work in a spherical coordinate system whose origin is the center of the ring and whose polar axis (i.e., the line $\theta = 0$) is the axis of the ring (Fig. 12.12). Clearly, the gravitational effect of the ring is symmetric about the axis of the ring; i.e., it is independent of the longitude angle ϕ. Hence, according to Theorem 5, we are to solve Laplace's equation in spherical coordinates when it is independent of ϕ. Since we have already carried out the separation process for this case of Laplace's equation and obtained Eqs. (2) and (8), we know that our product solutions are of the form

$$V_n = \left(A_n r^n + B_{n+1}\frac{1}{r^{n+1}}\right)P_n(\cos\theta)$$

From this we must, as usual, construct an infinite series

(16) $$V = \sum_{n=0}^{\infty}\left(A_n r^n + B_{n+1}\frac{1}{r^{n+1}}\right)P_n(\cos\theta)$$

which will reduce to the given boundary value when $\theta = 0$:

(17) $$\frac{M}{\sqrt{c^2 + r^2}} = \sum_{n=0}^{\infty}\left(A_n r^n + B_{n+1}\frac{1}{r^{n+1}}\right)P_n(\cos 0)$$

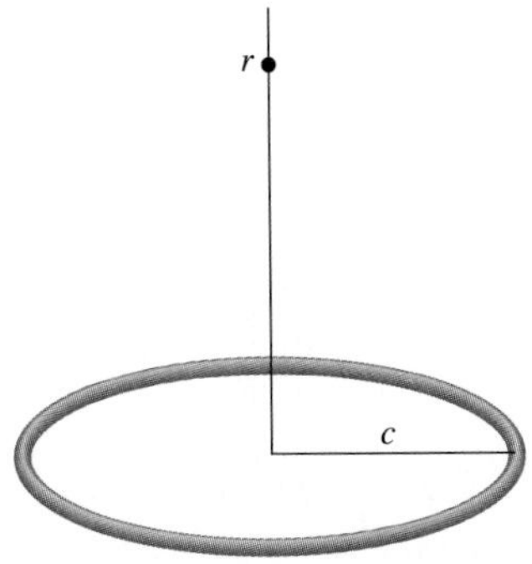

FIGURE 12.12
A thin ring whose gravitational potential is to be calculated.

The property of orthogonality is irrelevant here because after evaluation at $\theta = 0$, the P's are no longer variable functions of θ. In fact, as we discovered in Example 1, $P_n(\cos 0) = P_n(1) = 1$ for all values of n. Hence Eq. (17) becomes simply

$$\text{(18a)} \qquad \frac{M}{\sqrt{c^2 + r^2}} = \sum_{n=0}^{\infty} \left(A_n r^n + B_{n+1} \frac{1}{r^{n+1}} \right)$$

However, in this case the A's and the B's can be determined easily by comparing coefficients in the series on the right of Eq. (18a) with the coefficients in the binomial expansion of the term on the left. In doing this, there are two cases to be considered, according as $r < c$ or $r > c$.

If $r < c$, we have

$$\text{(18b)} \qquad \frac{M}{\sqrt{c^2 + r^2}} = \frac{M}{c}\left(1 + \frac{r^2}{c^2}\right)^{-1/2} = \frac{M}{c}\left(1 - \frac{1}{2}\frac{r^2}{c^2} + \frac{1 \cdot 3}{2 \cdot 4}\frac{r^4}{c^4} - \frac{1 \cdot 3 \cdot 5}{2 \cdot 4 \cdot 6}\frac{r^6}{c^6} + \frac{1 \cdot 3 \cdot 5 \cdot 7}{2 \cdot 4 \cdot 6 \cdot 8}\frac{r^8}{c^8} - \cdots\right)$$

Since this series contains only positive even powers of r, it is clear, by comparing it with (18a), that

$B_{n+1} = 0$ for all values of n

$A_n = 0$ if n is odd

$A_n =$ coefficient of r^n in (18b) if n is even

With these values for the coefficients, Eq. (16) becomes

$$\text{(19)} \qquad V(r, \theta) = \frac{M}{c}\left[P_0(\cos\theta) - \frac{1}{2}\frac{r^2}{c^2}P_2(\cos\theta) + \frac{1 \cdot 3}{2 \cdot 4}\frac{r^4}{c^4}P_4(\cos\theta) - \frac{1 \cdot 3 \cdot 5}{2 \cdot 4 \cdot 6}\frac{r^6}{c^6}P_6(\cos\theta) + \frac{1 \cdot 3 \cdot 5 \cdot 7}{2 \cdot 4 \cdot 6 \cdot 8}\frac{r^8}{c^8}P_8(\cos\theta) - \cdots\right] \qquad r < c$$

If $r > c$, we have

$$\text{(18c)} \qquad \frac{M}{\sqrt{c^2 + r^2}} = \frac{M}{r}\left(1 + \frac{c^2}{r^2}\right)^{-1/2}$$
$$= \frac{M}{r}\left(1 - \frac{1}{2}\frac{c^2}{r^2} + \frac{1 \cdot 3}{2 \cdot 4}\frac{c^4}{r^4} - \frac{1 \cdot 3 \cdot 5}{2 \cdot 4 \cdot 6}\frac{c^6}{r^6} + \frac{1 \cdot 3 \cdot 5 \cdot 7}{2 \cdot 4 \cdot 6 \cdot 8}\frac{c^8}{r^8} - \cdots\right)$$
$$= \frac{M}{c}\left(\frac{c}{r} - \frac{1}{2}\frac{c^3}{r^3} + \frac{1 \cdot 3}{2 \cdot 4}\frac{c^5}{r^5} - \frac{1 \cdot 3 \cdot 5}{2 \cdot 4 \cdot 6}\frac{c^7}{r^7} + \frac{1 \cdot 3 \cdot 5 \cdot 7}{2 \cdot 4 \cdot 6 \cdot 8}\frac{c^9}{r^9} - \cdots\right)$$

Since this series contains only odd negative powers of r, it is clear, by comparing it with (18a), that

$A_n = 0$ for all values of n

$B_{n+1} = 0$ if $n + 1$ is even

$B_{n+1} =$ coefficient of $\dfrac{1}{r^{n+1}}$ in Eq. (18c) if $n + 1$ is odd

With these values for the coefficients, Eq. (16) becomes

$$\text{(20)} \qquad V(r, \theta) = \frac{M}{c}\left[\frac{c}{r}P_0(\cos\theta) - \frac{1}{2}\frac{c^3}{r^3}P_2(\cos\theta) + \frac{1 \cdot 3}{2 \cdot 4}\frac{c^5}{r^5}P_4(\cos\theta) - \frac{1 \cdot 3 \cdot 5}{2 \cdot 4 \cdot 6}\frac{c^7}{r^7}P_6(\cos\theta) + \frac{1 \cdot 3 \cdot 5 \cdot 7}{2 \cdot 4 \cdot 6 \cdot 8}\frac{c^9}{r^9}P_8(\cos\theta) - \cdots\right] \qquad r > c$$

Equations (19) and (20) together constitute the required solution for the gravitational potential of the ring.

EXERCISES

1. Use the ratio test to prove that when the series in Eq. (9) do not terminate, the interval of convergence of each is $-1 < x < 1$.

2. Show that $P_{2n}(x)$ is an even function and $P_{2n+1}(x)$ is an odd function. Hence show that

$$\int_{-1}^{1} P_{2n}(x)\,dx = 2\int_{0}^{1} P_{2n}(x)\,dx \qquad \int_{-1}^{1} P_{2n+1}(x)\,dx = 0$$

3. Show that $P_{2n+1}(0) = 0$, and

$$P_{2n}(0) = (-1)^n \frac{1\cdot 3\cdot 5\cdots(2n-1)}{2^n n!} = (-1)^n \frac{(2n)!}{2^{2n}(n!)^2}$$

Hint: Set $x = 0$ in the generating function in Theorem 2.

4. Show that both the Legendre polynomials with even subscripts and the Legendre polynomials with odd subscripts form orthogonal sets over the interval (0, 1).

5. By differentiating the generating function for the Legendre polynomials with respect to x and then setting $x = 0$, show that $P'_{2n}(0) = 0$. What is $P'_{2n+1}(0)$?

6. Letting $g(x, z)$ denote the generating function for the Legendre polynomials, show that

$$z\frac{\partial g}{\partial z} = (x - z)\frac{\partial g}{\partial x}$$

Using this, establish the recurrence relation

$$nP_n(x) - xP'_n(x) + P'_{n-1}(x) = 0$$

7. Using Rodrigues' formula for $P_n(x)$ and repeated integration by parts, show that

$$\int_{-1}^{1} f(x)P_n(x)\,dx = \frac{(-1)^n}{2^n n!}\int_{-1}^{1} f^{(n)}(x)(x^2-1)^n\,dx$$

Hence show that

$$\int_{-1}^{1} x^m P_n(x)\,dx = 0 \qquad \text{if } m < n$$

Hint: Recall that if a function has a repeated factor $(x-a)^k$, then all derivatives of that function through the $(k-1)$st will vanish at $x = a$.

8. Use the results of Exercise 7 to prove that

$$\int_{-1}^{1} P_m(x)P_n(x)\,dx = 0 \qquad \text{if } m \neq n$$

Hint: If $m \neq n$, the order of one of the factors in the integrand must be greater than the highest power of x in the other.

9. Express x^2, x^3, and x^4 in terms of Legendre polynomials.

10. Using Rodrigues' formula, prove that

$$P'_{n+1}(x) - P'_{n-1}(x) = (2n+1)P_n(x)$$

Hence show that

$$\int_{x}^{1} P_n(x)\,dx = \frac{1}{2n+1}[P_{n-1}(x) - P_{n+1}(x)]$$

Hint: After Rodrigues' formula has been used on the left member, factor D^n from the resulting expression and then simplify the remaining portion, or simply use Exercise 6.

11. Show that

$$\int_{-1}^{1} P_n(x)P'_{n+1}(x)\,dx = 2$$

Hint: Use the result of Exercise 10.

12. Show that

$$\int_{-1}^{1} xP_n(x)P_{n-1}(x)\,dx = \frac{2n}{4n^2-1}$$

Hint: Use the result of Example 2.

13. It is desired to approximate a function $f(x)$ over the interval $(-1, 1)$ by a polynomial $P(x)$ of degree n which will make the integral

$$\int_{-1}^{1} [f(x) - P(x)]^2\,dx$$

a minimum. Show that $P(x)$ is the nth partial sum of the expansion of $f(x)$ over the interval $(-1, 1)$ in terms of Legendre polynomials. *Hint:* First express $P(x)$ as an arbitrary linear combination of the first n Legendre polynomials.

14. Consider the functions x^i, $i = 0, 1, 2, 3, \ldots,$ and let $f_n(x) = \sum_{i=0}^{n} a_{ni}x^i$, where the a's are coefficients to be determined so that $a_{nn} > 0$ and

$$\int_{-1}^{1} f_m(x)f_n(x)\,dx = \begin{cases} 0 & m \neq n \\ 2/(2n+1) & m = n \end{cases}$$

Calculate the a's for $n = 0, 1, 2, 3$ and show that, at least for these values of n, $f_n(x) = P_n(x)$.

15. Show that

$$\int xP_n(x)\,dx = \frac{n+1}{(2n+1)(2n+3)}P_{n+2}(x) + \frac{1}{(2n-1)(2n+3)}P_n(x) - \frac{n}{(2n-1)(2n+1)}P_{n-2}(x) + c$$

Hint: Use the results of Example 2 and Exercise 10.

16. Show that

$$\int xP_n(x)\,dx = -\frac{1}{(2n+1)(2n+3)}P_{n+2}(x) + \frac{x}{2n+1}P_{n+1}(x) + \frac{2}{(2n-1)(2n+3)}P_n(x) - \frac{x}{2n+1}P_{n-1}(x) - \frac{1}{(2n-1)(2n+1)}P_{n-2}(x) + c$$

Hint: Begin with integration by parts, then use the result of Exercise 10.

17. **(a)** If $f(x)$ is an odd function, show that

$$\int_{-1}^{1} f(x)P_n(x)\,dx = \begin{cases} 0 & n \text{ even} \\ 2\int_0^1 f(x)P_n(x)\,dx & n \text{ odd} \end{cases}$$

(b) If $f(x)$ is an even function, show that

$$\int_{-1}^{1} f(x)P_n(x)\,dx = \begin{cases} 2\int_0^1 f(x)P_n(x)\,dx & n \text{ even} \\ 0 & n \text{ odd} \end{cases}$$

18. Expand $f(x) = \begin{cases} -1 & -1 < x < 0 \\ 1 & 0 < x < 1 \end{cases}$ in a series of Legendre polynomials.

19. Expand $f(x) = \begin{cases} 0 & -1 < x < 0 \\ 1 & 0 < x < 1 \end{cases}$ in a series of Legendre polynomials.

20. Expand $f(x) = \begin{cases} -x & -1 < x < 0 \\ x & 0 < x < 1 \end{cases}$ in a series of Legendre polynomials.

21. Expand $f(x) = \begin{cases} 0 & -1 < x < 0 \\ x & 0 < x < 1 \end{cases}$ in a series of Legendre polynomials.

22. Show that $P_m(x)P_n(x)\Big|_{-1}^{1} = \begin{cases} 2 & m+n \text{ odd} \\ 0 & m+n \text{ even} \end{cases}$

23. Show that

$$\int_{-1}^{1} P_m(x)P_n'(x)\,dx = \begin{cases} 0 & m+n \text{ even or } n \le m \\ 2 & m+n \text{ odd and } n > m \end{cases}$$

Hint: Integrate by parts and use the result of Exercise 22.

24. Show that the expansion of $P_n'(x)$ in terms of Legendre polynomials is the finite series

$$P_n'(x) = (2n-1)P_{n-1}(x) + (2n-5)P_{n-3}(x) + (2n-9)P_{n-5}(x) + \cdots$$

Hint: Use Theorem 4 and the results of Exercise 23.

25. The upper half of the surface of a sphere of radius b is maintained at the temperature $u = 100$, and the lower half is maintained at the temperature $u = 0$. Find the steady-state temperature distribution in the sphere. *Hint:* Express the integral for A_n (Example 4) in terms of x when evaluating it, then use the result of Exercise 10.

26. Work Exercise 25 if the upper half of the surface is maintained at the temperature $u = 50$ and the lower half of the surface is maintained at the temperature $u = -50$.

27. Work Exercise 25 if the upper half of the surface is maintained at the temperature $u = 50\cos\theta$ and the lower half of the surface is maintained at the temperature $u = -50\cos\theta$. *Hint:* Recall the result of Exercise 15.

28. The curved surface of a solid hemisphere is maintained at the temperature $u = 100$, and the base is maintained at the temperature $u = 0$. Find the steady-state temperature distribution in the hemisphere.

29. In Example 3, if the temperature throughout the sphere is initially $u = 0$, find the temperature at any point in the sphere at any subsequent time. *Hint:* The equation to be solved is Eq. (1) with the term involving ϕ deleted and the term $a^2\,\partial u/\partial t$ inserted on the right.

30. Find the steady-state temperature distribution in a spherical shell of inner radius b_1 and outer radius b_2 if the temperature distributions $u(b_1, \theta) = f_1(\theta)$ and $u(b_2, \theta) = f_2(\theta)$ are maintained over the inner and outer surfaces, respectively.

31. What is the gravitational potential of a thin circular disk of radius a and mass M if the potential on the axis of the disk at a distance r from the center of the disk is

$$\frac{2M}{a^2}(\sqrt{r^2 + a^2} - r)?$$

32. If an electric charge M is placed on a thin disk of radius c, it distributes itself so that the electrostatic potential on the axis of the disk at a distance r from the center of the disk is

$$\frac{M}{2c}\cos^{-1}\frac{r^2 - c^2}{r^2 + c^2}$$

Given this boundary condition, find the electrostatic potential of the charged disk at any point in space. *Hint:* Differentiate the expression for the potential on the axis of the disk, express the derivative as a power series in r, integrate the series and determine the integration constant(s), and then proceed as in Example 4.

33. Introduce a new independent variable in the associated Legendre equation in trigonometric form by means of the substitution $\theta = x/n$ and show that as $n \to \infty$, the resulting equation approaches Bessel's equation.

34. Show that the change of dependent variable defined by the substitution $y = u/\sqrt{\sin\theta}$ transforms Legendre's equation into the equation

$$\frac{d^2u}{d\theta^2} + \left[\left(n + \frac{1}{2}\right)^2 + \frac{1}{4\sin^2\theta}\right]u = 0$$

Hence show that for large values of n, $P_n(\cos\theta)$ is described approximately by the expression

$$P_n(\cos\theta) = A_n \cos\left[\frac{2n+1}{2}\theta + \alpha_n\right]$$

35. Substituting the result of Exercise 34 into the integral

$$\int_0^{\pi} P_n^2(\cos\theta)\sin\theta\,d\theta = \frac{2}{2n+1}$$

and then carrying out the integration, show that A_n approaches $\sqrt{2/n\pi}$ as n becomes infinite.

36. It can be shown that one solution of the associated Legendre equation (7) is the function $P_n^{(m)}(x)$, defined as

$$P_n^{(m)}(x) = (1 - x^2)^{m/2}\frac{d^m P_n(x)}{dx^m}$$

Write out $P_1^{(1)}(x)$, $P_2^{(1)}(x)$, $P_3^{(1)}(x)$, $P_1^{(2)}(x)$, $P_2^{(2)}(x)$, $P_3^{(2)}(x)$

37. Show that $P_n(x)$ has exactly n zeros in the interval $(-1, 1)$. *Hint:* Assume that $x_1, x_2, \ldots, x_k$, where $k < n$, are the only zeros of $P_n(x)$ in the interval $(-1, 1)$, consider the integral

$$\int_{-1}^{1} P_n(x)(x - x_1)(x - x_2)\cdots(x - x_k)\,dx$$

and use the result of Exercise 7.

38. Find power-series solutions of the equation

$$y'' - 2xy' + 2ny = 0 \qquad n \text{ a positive integer}$$

and show that they are orthogonal with respect to the weight function e^{-x^2} over the integral $(-\infty, \infty)$. This equation is known as **Hermite's equation**, and its polynomial solutions are known as **Hermite polynomials**.

39. If $H_n(x)$ is a Hermite polynomial, show that

$$H_n(x) = Ce^{x^2}\frac{d^n e^{-x^2}}{dx^n}$$

where C is a constant.

40. Find power-series solutions of the equation

$$xy'' + (1 - x)y' + ny = 0 \qquad n \text{ a positive integer}$$

and show that they are orthogonal with respect to the weight function e^{-x} over the interval $(0, \infty)$. This equation is known as **Laguerre's equation**, after the French mathematician Edmond Nicolas Laguerre (1834–1886), and its polynomial solutions are known as **Laguerre polynomials**.

41. If $L_n(x)$ is a Laguerre polynomial, show that

$$L_n(x) = Ce^x\frac{d^n(x^n e^{-x})}{dx^n}$$

where C is a constant.

42. In the proof of Theorem 2, what is the first term in the expansion of $[1 - z(2x - z)]^{-1/2}$ that contributes to the coefficient of z^n?

43. Work Exercise 30 if $f_1(\theta) = T_0$ and $f_2(\theta) = T_1$ and compare your answer with the result of Exercise 40, Sec. 1.15.

44. Show that the equation $(1 - x^2)y'' - 2xy' + \lambda y = 0$ has a polynomial solution if and only if $\lambda = n(n + 1)$, where n is an integer.

45. By first putting $x = \cos\theta$ in the generating function for the Legendre polynomials, then setting $\theta = \pi/2$, and finally using the fact that

$$\binom{-\frac{1}{2}}{m} = (-1)^m\frac{(2m)!}{2^{2m}(m!)^2}$$

approaches $(-1)^m(1/\sqrt{m\pi})$ as m increases indefinitely, show that the value of the phase angle α_n in the formula of Exercise 34 is $-\pi/4$.

CHAPTER 13

VECTOR SPACES AND LINEAR TRANSFORMATIONS

In several places (Secs. 4.4 and 5.7) we have noted the similarity between the basic theorems for linear differential equations, systems of such equations formulated as matrix equations, and linear difference equations. This similarity is not accidental but exists because each of these equations consists of a peculiar abstract transformation from one linear space into another. In Secs. 13.1–13.3 we shall investigate linear spaces in further detail. This will draw together the general ideas of linear dependence and independence we have encountered in a variety of special settings (Secs. 2.2, 3.6, and 4.3). It will also give us a more comprehensive view of the important property of orthogonality, both of functions and vectors, and introduce us to the concept of a *function space*. In particular, we will learn (Sec. 13.4) how the *Gram-Schmidt orthogonalization process* allows us to form orthogonal sets from arbitrary sets of vectors or functions, provided they are linearly independent.

In Secs. 13.5 and 13.6, we examine the important concept of linear transformations and their properties. Finally, in Sec. 13.7 we study in general terms the idea of a *linear operator,* which we have encountered in so many special cases (Secs. 2.5, 2.7, 5.2, 5.9, 9.3, and 10.2), and investigate *linear operator equations.*

Prerequisites for this chapter: Chap. 3, Sec. 11.6.

13.1 VECTOR SPACES

Let us now return to the subject of vectors and matrices, introduced in Chap. 3, and resume our study from a more fundamental point of view. Among other things, this should make clear why vectors, matrices, and functions have so many common properties and why so many things about linear systems and linear differential equations are so much alike.

Thus far, we have thought of vectors as ordered n-tuples, and of matrices as rectangular arrays, of scalars. On closer inspection, both vectors and matrices are seen to be functions with rather simple domains. For instance, *an n-dimensional vector* $\mathbf{v} = (v_1, v_2, \ldots, v_n)$ *is just a function* v *whose domain is the set of positive integers* $\{1, 2, \ldots, n\}$ *and whose functional values* $v(1) = v_1$, $v(2) = v_2, \ldots, v(n) = v_n$ *are the components* of $\mathbf{v}$. In particular, if $\mathbf{v}$ is in R^n and n is not too large, it is easy to graph the function v in R^2. Such a graph gives a geometric representation of $\mathbf{v}$ even if $n > 3$ and we cannot visualize $\mathbf{v}$ as either an arrow or a point.

EXAMPLE 1

The vector $\mathbf{v} = (1, -2, 2)$ can be identified as the function v having $\{1, 2, 3\}$ as domain and $v(1) = 1$, $v(2) = -2$, and $v(3) = 2$ as its functional values. A graph of this function (Fig. 13.1a) becomes a geometric representation of $\mathbf{v}$ also. A graph of $\mathbf{w} = (2, 2, 1)$ is plotted in Fig. 13.1b. Since $\mathbf{v} \cdot \mathbf{w} = 0$, $\mathbf{v}$ and $\mathbf{w}$ are of course orthogonal.

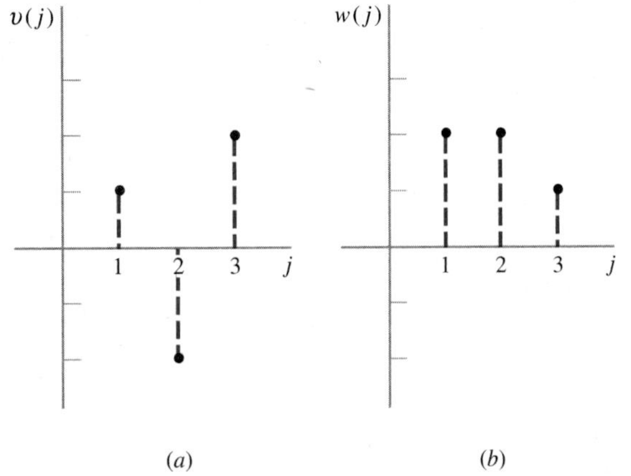

FIGURE 13.1
Graphs of vectors.

The analogy between vectors, thought of as functions of a discrete variable j defined on a set of positive integers $\{1, 2, \ldots, n\}$, and functions of a continuous variable x defined on some interval $[a, b]$ of the real axis, is worth pursuing a little further. To find the inner product of two n-dimensional vectors $\mathbf{u}$ and $\mathbf{v}$, commonly denoted by $(\mathbf{u}, \mathbf{v})$, we multiply corresponding components of the two vectors and add all these products, getting

$$(1a) \qquad (\mathbf{u}, \mathbf{v}) = \sum_{j=1}^{n} u_j v_j$$

Now let us change our point of view and think of $\mathbf{u}$ and $\mathbf{v}$ as the respective functions u and v which have $\{1, 2, \ldots, n\}$ as a common domain, and values defined by the equations $u(j) = u_j$, $v(j) = v_j$. Since u and v identify the same vectors as do $\mathbf{u}$ and $\mathbf{v}$, we may substitute u for $\mathbf{u}$, v for $\mathbf{v}$, and functional values for corresponding components, in Eq. (1a), to rewrite that equation as

$$(1b) \qquad (u, v) = \sum_{j=1}^{n} u(j)v(j)$$

This equation defines the **inner product** of *functions* u and v having a discrete set of consecutive integers $\{1, 2, \ldots, n\}$ as their domain.

To formulate an analogous definition pertaining to pairs of functions for which an interval, instead of a discrete point set, is their domain, we next consider two real-valued functions f and g defined on $[a, b]$. For each value of x belonging to $[a, b]$, the product $f(x)g(x)$ of corresponding functional values of f and g is a real number. Guided by Eq. (1b) which defines (u, v) as a sum of products of the type $u(j)v(j)$, we proceed to *sum* the products $f(x)g(x)$ arising from all different values of x, which we do by integration rather than arithmetic addition because the variable x ranges over all values of the continuum $[a, b]$. This leads us to the following generalization of Eq. (1b):

$$(f, g) = \int_a^b f(x)g(x)\,dx \tag{2}$$

The integral here, if it exists, defines the **inner product,** denoted by (f, g), of a general pair of functions defined on $[a, b]$. This definition also applies to pairs of functions defined on an open interval (a, b).

When we represent a nonzero three-dimensional vector as an appropriate arrow in three-dimensional space, the arrow serves as a geometric image of the vector. A degenerate arrow consisting of the point at the origin is the geometric image of the zero vector $\mathbf{0}_3 = (0, 0, 0)$. Whether interpreted analytically or geometrically, the vanishing of the inner product of two vectors of R^3 means that these vectors are perpendicular, or *orthogonal.* It follows that any two vectors of R^3, at least one of which is $\mathbf{0}_3$, are orthogonal, that is, $\mathbf{0}_3$ is orthogonal to every vector of R^3. This terminology is naturally extended to higher-dimensional vectors whose components are all real (see Sec. 3.2). Since each vector of R^n may be thought of as a real-valued function, we are also led to say that two functions u and v are **orthogonal on a set of consecutive integers** $\{1, 2, \ldots, n\}$ if this set is the domain of both functions and the inner product (u, v), defined by Eq. (1b), is zero. Two functions f and g, having an interval $[a, b]$ as their domain, are said to be **orthogonal on** $[a, b]$ if the inner product (f, g), defined by Eq. (2), is zero. Extending this terminology somewhat, we say

DEFINITION 1 Two functions f and g are **orthogonal on an interval** $[a, b]$ if and only if $\int_a^b f(x)g(x)\,dx = 0$.

This definition relaxes the requirement that functions which are orthogonal on an interval $[a, b]$ must have $[a, b]$ as their domain and is akin to Definition 1, Sec. 11.6. Note carefully, however, that Definition 1, Sec. 11.6, excludes the trivial function on an interval (a, b) from every *orthogonal set* of functions on (a, b). Definition 3, Sec. 11.6, extends the concept of orthogonality to *orthogonality of functions with respect to a weight function.* As a characterization of an interesting parallel concept of *orthogonality of vectors with respect to a weight vector,* we have

DEFINITION 2 For all n-dimensional vectors $\mathbf{u}$, $\mathbf{v}$, and $\mathbf{w}$, the vectors $\mathbf{u}$ and $\mathbf{v}$ are **orthogonal with respect to the weight vector w** if and only if $\sum_{j=1}^n w_j u_j v_j = 0$.

Note how the preceding sum $\sum_{j=1}^n w_j u_j v_j$ of products of corresponding components of $\mathbf{w}$, $\mathbf{u}$, and $\mathbf{v}$ is now cast in the role of the integral $\int_a^b p(x)\phi_m(x)\phi_n(x)\,dx$ in the analogous definition concerning weighted orthogonality of functions. We tacitly assume that $\mathbf{u}$, $\mathbf{v}$, and $\mathbf{w}$ have only real components, and that p and the ϕ's are real-valued functions. Hermitian orthogonality (Sec. 3.2) can be extended to complex-valued functions (Exercise 19).

In Chapter 11, without mentioning the term *inner product,* we used expansions in terms of orthogonal functions in the solution of boundary-value problems. Under suitable conditions, an arbitrary vector can be expanded likewise in terms of orthogonal vectors.

Consider, for instance, an expansion problem in which $\{\mathbf{u}_j\}$, $1 \le j \le m$, is a **set of** $m \le n$ **mutually orthogonal vectors** of R^n, that is,

$$(\mathbf{u}_j, \mathbf{u}_k)\begin{cases} = 0 & j \ne k \\ \ne 0 & j = k \end{cases} \tag{3}$$

Given the integers m and n, many such sets of vectors can be found (Exercise 10). Our problem is to represent an arbitrary vector $\mathbf{v}$ of R^n by a series of the form

$$\sum_{j=1}^{m} a_j\mathbf{u}_j = a_1\mathbf{u}_1 + a_2\mathbf{u}_2 + \cdots + a_m\mathbf{u}_m \tag{4}$$

that is, we are to find real values for the coefficients a_j such that

$$\mathbf{v} = a_1\mathbf{u}_1 + a_2\mathbf{u}_2 + \cdots + a_m\mathbf{u}_m \tag{5}$$

To do this, working directly with vectors, we form the inner product of $\mathbf{u}_k$ and each member of (5). Equating these, we get (Exercise 11)

$$(\mathbf{u}_k, \mathbf{v}) = a_1(\mathbf{u}_k, \mathbf{u}_1) + a_2(\mathbf{u}_k, \mathbf{u}_2) + \cdots + a_m(\mathbf{u}_k, \mathbf{u}_m) \qquad 1 \le k \le m \tag{6}$$

Since, for $1 \le j, k \le m$, $(\mathbf{u}_k, \mathbf{u}_j) = (\mathbf{u}_j, \mathbf{u}_k)$, the conditions (3) imply that every inner product in the right-hand member of Eq. (6), except $(\mathbf{u}_k, \mathbf{u}_k)$, is zero and we have

$$(\mathbf{u}_k, \mathbf{v}) = a_k(\mathbf{u}_k, \mathbf{u}_k) \qquad \text{and} \qquad a_k = \frac{(\mathbf{u}_k, \mathbf{v})}{(\mathbf{u}_k, \mathbf{u}_k)} \qquad 1 \le k \le m \tag{7}$$

Using standard double subscript notation (Sec. 3.3), we set $\mathbf{u}_k = (u_{k1}, u_{k2}, \ldots, u_{kn})$, $1 \le k \le m$, take $\mathbf{v} = (v_1, v_2, \ldots, v_n)$, and expand inner products in our formula for the a_k's, to rewrite it as

$$a_k = \frac{\displaystyle\sum_{i=1}^{n} v_i u_{ki}}{\displaystyle\sum_{i=1}^{n} u_{ki}^2}$$

This equation is analogous to the formula

$$a_k = \frac{\displaystyle\int_a^b f(x)\phi_k(x)\,dx}{\displaystyle\int_a^b \phi_k^2(x)\,dx}$$

for the Fourier constants in a generalized Fourier series expansion

$$f(x) = a_1\phi_1(x) + a_2\phi_2(x) + \cdots + a_k\phi_k(x) + \cdots$$

of a real-valued function $f(x)$ in terms of an orthogonal set of functions $\{\phi_k(x)\}$ whose fundamental interval is (a, b) (Sec. 11.6).

Our derivation of Eqs. (7) started with the assumption that

1. For every vector $\mathbf{v}$ of R^n, values of the coefficients a_j exist for which the vector represented by the right-hand member of (5) is $\mathbf{v}$.

Let us now suppose that

2. Some vector $\mathbf{w} \ne \mathbf{0}_n$ is orthogonal to every vector of the orthogonal set $\{\mathbf{u}_j\}$.

According to **1**, there are values of the coefficients a_j in (4) for which

$$\mathbf{w} = a_1\mathbf{u}_1 + a_2\mathbf{u}_2 + \cdots + a_m\mathbf{u}_m$$

According to **2** and the second of Eqs. (7), all a_j's corresponding to $\mathbf{w}$ are zero. Hence $\mathbf{w} = \mathbf{0}_n$. But this contradicts the premise $\mathbf{w} \ne \mathbf{0}_n$. Thus if **2** is true, then **1** is false. In other words, if a vector other than the zero vector is orthogonal to all vectors of the set $\{\mathbf{u}_j\}$, there is no expansion (4) representing that vector. This being the case, in the language of Sec. 11.6 we would say that the set $\{\mathbf{u}_j\}$ is *incomplete;* in the language of linear algebra we would say more conventionally that the $\mathbf{u}_j$'s fail to *span* (Sec. 13.2), or do not provide a *basis* (Sec. 13.3) for, R^n.

Having examined some of the ways in which vectors and functions defined on intervals resemble each other, it is natural next to regard matrices as generalized vectors (Sec. 3.3) and study them in greater detail. In particular, matrices can be identified as functions defined on appropriate sets of ordered pairs of positive integers, and from this point of view are analogous to functions of two variables defined on some (rectangular) region of R^2.

DEFINITION 3 An $m \times n$ matrix is a function of two variables whose domain is the set of all ordered pairs (i, j), where $1 \le i \le m$ and $1 \le j \le n$.

In keeping with our practice of identifying a function by one of its representations, we shall not distinguish between a matrix and the array of its elements. Of course, if $A = [a_{ij}]$ represents a matrix, then the element a_{ij} in the ith row and jth column is simply the value $A(i, j)$ of A corresponding to the ordered pair of integers (i, j). To graph a matrix with real elements, we simply plot its values in R^3.

EXAMPLE 2

The matrix A defined by $A(i, j) = a_{ij} = 3[1 + \sin(i\pi/2)] + i(i - j - 1)$, for $1 \le i \le 3$ and $1 \le j \le 4$, is represented by the array $\begin{bmatrix} 5 & 4 & 3 & 2 \\ 3 & 1 & -1 & -3 \\ 3 & 0 & -3 & -6 \end{bmatrix}$. A graph of this matrix is shown in Fig. 13.2.

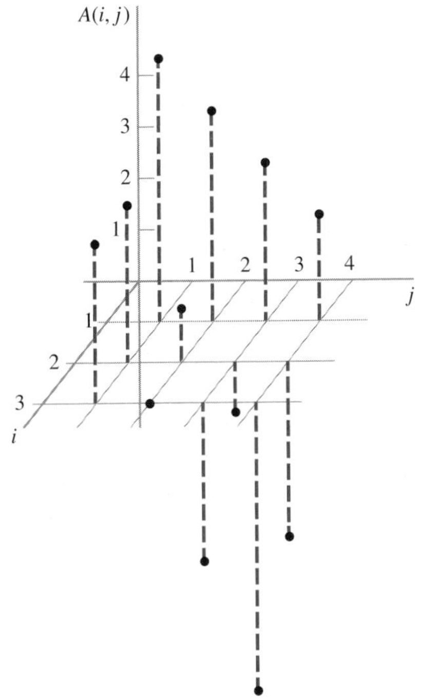

FIGURE 13.2
Graph of a matrix.

The fact that we can add vectors only if they have the same number of components and can add matrices only if they have the same size now appears as simply an illustration of the fact that two functions can be added only if they have the same domain.

Nothing in Definition 3 rules out the possibility that the elements of a matrix are themselves matrices. In fact, it is often convenient to subdivide, or **partition,** a matrix into submatrices and then regard the original matrix as a new matrix having these submatrices as elements. In Chaps. 3 and 4 we did this several times when, in particular, we wrote an $m \times n$ matrix A as a row matrix $[\mathbf{v}_1, \mathbf{v}_2 \cdots \mathbf{v}_n]$ having the successive column vectors of A as elements. Similarly, it is sometimes helpful to regard A as a column matrix $\begin{bmatrix} \mathbf{u}_1 \\ \vdots \\ \mathbf{u}_m \end{bmatrix}$ whose elements are the respective row vectors of A.

EXAMPLE 3

Among numerous possibilities, we can write

$$A = \left[\begin{array}{ccc:c} a_{11} & a_{12} & a_{13} & a_{14} \\ a_{21} & a_{22} & a_{23} & a_{24} \\ \hdashline a_{31} & a_{32} & a_{33} & a_{34} \end{array}\right] = \begin{bmatrix} A_{11} & A_{12} \\ A_{21} & A_{22} \end{bmatrix}$$

where

$$A_{11} = \begin{bmatrix} a_{11} & a_{12} & a_{13} \\ a_{21} & a_{22} & a_{23} \end{bmatrix} \quad A_{12} = \begin{bmatrix} a_{14} \\ a_{24} \end{bmatrix} \quad A_{21} = [a_{31} \quad a_{32} \quad a_{33}] \quad A_{22} = [a_{34}]$$

or, equally well,

$$A = [A_{11} \quad A_{12} \quad A_{13} \quad A_{14}]$$

where now

$$A_{11} = \begin{bmatrix} a_{11} \\ a_{21} \\ a_{31} \end{bmatrix} \quad A_{12} = \begin{bmatrix} a_{12} \\ a_{22} \\ a_{32} \end{bmatrix} \quad A_{13} = \begin{bmatrix} a_{13} \\ a_{23} \\ a_{33} \end{bmatrix} \quad A_{14} = \begin{bmatrix} a_{14} \\ a_{24} \\ a_{34} \end{bmatrix}$$

In constructing the product of two matrices it is sometimes convenient to partition them before performing the multiplication. This can be done in many ways, but it is of course necessary that the given matrices be conformable and that the various submatrices which must be multiplied together also be conformable. This requirement imposes no restriction on the horizontal partitioning of the first matrix or on the vertical partitioning of the second matrix. It does require, however, that the columns of the first matrix be partitioned into groups such that the number of columns in each group is equal to the number of rows in the corresponding groups into which the rows of the second matrix are partitioned. Matrices for which this is the case are said to be **conformably partitioned.**

EXAMPLE 4

By direct multiplication we have

$$\begin{bmatrix} 1 & 1 & 1 \\ 2 & -1 & 0 \\ -1 & 0 & 2 \end{bmatrix} \begin{bmatrix} 1 & 2 & 3 & -1 \\ 3 & -1 & 1 & 0 \\ 0 & 0 & -2 & 1 \end{bmatrix} = \begin{bmatrix} 4 & 1 & 2 & 0 \\ -1 & 5 & 5 & -2 \\ -1 & -2 & -7 & 3 \end{bmatrix}$$

On the other hand we can write, among various other possibilities,

$$\left[\begin{array}{cc:c} 1 & 1 & 1 \\ 2 & -1 & 0 \\ \hdashline -1 & 0 & 2 \end{array}\right] = \begin{bmatrix} A_{11} & A_{12} \\ A_{21} & A_{22} \end{bmatrix} \quad \text{and} \quad \left[\begin{array}{cccc} 1 & 2 & 3 & -1 \\ 3 & -1 & 1 & 0 \\ \hdashline 0 & 0 & -2 & 1 \end{array}\right] = \begin{bmatrix} B_{11} \\ B_{21} \end{bmatrix}$$

and, from this point of view, the product of the two matrices is

$$\begin{bmatrix} A_{11} & A_{12} \\ A_{21} & A_{22} \end{bmatrix} \begin{bmatrix} B_{11} \\ B_{21} \end{bmatrix} = \begin{bmatrix} A_{11}B_{11} + A_{12}B_{21} \\ A_{21}B_{11} + A_{22}B_{21} \end{bmatrix}$$

or, performing the indicated multiplications and additions of the submatrices,

$$\begin{bmatrix} \begin{bmatrix} 4 & 1 & 4 & -1 \\ -1 & 5 & 5 & -2 \end{bmatrix} + \begin{bmatrix} 0 & 0 & -2 & 1 \\ 0 & 0 & 0 & 0 \end{bmatrix} \\ [\,-1 \quad -2 \quad -3 \quad 1\,] + [\,0 \quad 0 \quad -4 \quad 2\,] \end{bmatrix} = \begin{bmatrix} 4 & 1 & 2 & 0 \\ -1 & 5 & 5 & -2 \\ -1 & -2 & -7 & 3 \end{bmatrix}$$

as before.

The observation that both vectors and matrices can be regarded as functions having many properties in common suggests that there may be other kinds of functions which share the same properties. Indeed, when appropriate basic concepts are identified and restated in terms of unspecified objects, an abstract structure emerges which has numerous mathematical systems, including n-dimensional vectors and $m \times n$ matrices with their respective algebras, as particular models. A study of this abstract structure then enables us to investigate simultaneously the properties of many specific systems of practical interest.

Let us begin by considering two nonempty sets, one a set S of scalars, the other a set V of unspecified entities. To prevent any confusion, we shall use boldface letters to denote members of V and italics to designate scalars. Guided by results we have found to be true for vectors and matrices, we assume that any two elements in V can be combined to give a unique member of V, and that any element in V can also be combined with any scalar in S to give a unique member of V. Whatever the actual nature of these operations may be, we shall refer to them, respectively, as *addition* and *multiplication*. Thus $\mathbf{u} + \mathbf{v}$ will indicate the sum of a pair of elements in V, and $c\mathbf{u} = \mathbf{u}c$ will indicate the product of $\mathbf{u}$ and a scalar c. A set V whose members can be multiplied by scalars, as well as added, and for which all statements in the following definition hold, is called a **vector space** or a **linear space.** The members of V are called **vectors.** Statements **I** and **II** require that every vector space be closed under its two intrinsic operations. The other eight axioms simply extend to vector spaces of all kinds the fundamental algebraic properties of vectors and matrices specified by **(i)** through **(viii),** Secs. 3.2 and 3.3.

DEFINITION 4 A nonempty set V, any pair of whose elements can be added and each of whose elements can be multiplied by an arbitrary scalar of a set S, is a vector space over S if and only if for all $\mathbf{u}$, $\mathbf{v}$, and $\mathbf{w}$ in V,

I. $\mathbf{u} + \mathbf{v}$ is a member of V.	**V is closed under vector addition.**
(i) $\mathbf{u} + (\mathbf{v} + \mathbf{w}) = (\mathbf{u} + \mathbf{v}) + \mathbf{w}$	**Associative law of vector addition**
(ii) $\mathbf{u} + \mathbf{v} = \mathbf{v} + \mathbf{u}$	**Commutative law of vector addition**
There exists a unique vector $\mathbf{0}$ in V such that for every $\mathbf{u}$ in V	
(iii) $\mathbf{u} + \mathbf{0} = \mathbf{u}$.	**Existence of an additive identity called the zero vector**
To each $\mathbf{u}$ in V there corresponds a vector $-\mathbf{u}$ in V such that	
(iv) $\mathbf{u} + (-\mathbf{u}) = \mathbf{0}$.	**Existence of additive inverses**
For all $\mathbf{u}$, $\mathbf{v}$ in V and all a, b in S,	
II. $a\mathbf{u} = \mathbf{u}a$ is a member of V.	**V is closed under multiplication by scalars.**
(v) $a(\mathbf{u} + \mathbf{v}) = a\mathbf{u} + a\mathbf{v}$	**Scalars distribute over vector addition.**
(vi) $(a + b)\mathbf{u} = a\mathbf{u} + b\mathbf{u}$	**Vectors distribute over scalar addition.**
(vii) $(ab)\mathbf{u} = a(b\mathbf{u})$	**Associative law of multiplication by scalars.**
(viii) $1\mathbf{u} = \mathbf{u}$	**The multiplicative identity for scalars is a multiplicative identity for multiplication of vectors by scalars.**

If S is the set of all complex numbers, the definition just given is said to define a **vector space over the complex numbers,** or a **complex vector space.** On the other hand, if S is the set of all real numbers, the space defined is called a **real vector space,** or a **vector space over the real numbers.** Both complex and real vector spaces are often referred to simply as **vector spaces,** or **linear spaces,** the adjectives *complex* and *real* usually being dropped when the intended meaning is clear.

By using corresponding properties of complex or real numbers, as in Sec. 3.2, it is a straightforward matter to verify that C^n and R^n are complex and real vector spaces, respectively. With addition, and multiplication by a scalar, defined as in Sec. 3.3, the set of all $m \times n$ matrices with complex elements constitutes another complex linear space, while the set of all $m \times n$ matrices with real elements is a real linear space. Real linear spaces are the ones we shall be primarily interested in. Especially important among these are those whose vectors are real-valued functions.

Consider, for example, **a set $F(D)$ consisting of all real-valued functions with a common domain D.** Given any pair of functions f, g in $F(D)$, we may take their **sum** $f + g$ to be the function which, for every x in D, has the value

$$(f + g)(x) = f(x) + g(x) \tag{8}$$

This definition of **function addition** assigns a real value to $f + g$ at each x in D, which is equal to the ordinary sum of the corresponding values of f and g. Hence $f + g$ is in $F(D)$, and so this set is closed under function addition.

For any real number c and any f in $F(D)$, we may take their **product** $cf = fc$ to be the function which for every x in D, has the value

$$(cf)(x) = c[f(x)] = cf(x) \tag{9}$$

This definition of **multiplication by reals** assigns a real value to cf at each x in D, which is equal to the numerical product of the real number c and the corresponding value of f. Hence cf is in $F(D)$, and so this set is closed under multiplication by real numbers. We proceed to show that by incorporating the two operations just defined into $F(D)$, a real vector space is obtained.

As we have just observed, $F(D)$ has the properties of closure required by **I** and **II** of Definition 4. Let f, g, and h be arbitrary functions of $F(D)$ and let a and b denote any real numbers. Because of closure, each member of the next six equations [Eqs. (10)–(15)] is a function with domain D. The argument right after each equation shows that for every x in D, the equated functions have identical values. Hence all six equations hold. Notice how each argument makes use of (8) or (9) or an associative, commutative, or distributive law of real numbers.

(10) $f + (g + h) = (f + g) + h$

$$\begin{aligned}[f + (g + h)](x) &= f(x) + (g + h)(x) = f(x) + [g(x) + h(x)] \\ &= [f(x) + g(x)] + h(x) = (f + g)(x) + h(x) \\ &= [(f + g) + h](x)\end{aligned}$$

(11) $f + g = g + f$

$$(f + g)(x) = f(x) + g(x) = g(x) + f(x) = (g + f)(x)$$

(12) $a(f + g) = af + ag$

$$\begin{aligned}[a(f + g)](x) &= a(f + g)(x) = a[f(x) + g(x)] = af(x) + ag(x) \\ &= (af)(x) + (ag)(x) = (af + ag)(x)\end{aligned}$$

(13) $(a + b)f = af + bf$

$$\begin{aligned}[(a + b)f](x) &= (a + b)f(x) = af(x) + bf(x) = (af)(x) + (bf)(x) \\ &= (af + bf)(x)\end{aligned}$$

(14) $(ab)f = a(bf)$

$$[(ab)f](x) = (ab)f(x) = a[bf(x)] = a(bf)(x) = [a(bf)](x)$$

(15) $1f = f$

$$(1f)(x) = 1f(x) = f(x)$$

These results confirm that $F(D)$ has all of the defining properties of a vector space except those stipulated by **(iii)** and **(iv)** of Definition 4. Clearly, the trivial function with domain D is the additive identity in $F(D)$, and for every f in $F(D)$, the function $-f$ such that at each x in D, $(-f)(x) = -f(x)$, is obviously an additive inverse of f. Hence $F(D)$ possesses all requisite properties of a real vector space and we have proved

THEOREM 1 A set that consists of all real-valued functions having a common domain, is a real vector space.

EXAMPLE 5

Is the set of all vectors in R^3 with second component 1 a real vector space?

Let the given set, together with the two operations previously defined for R^3, be denoted by W. Both of the vectors (0, 1, 0) and (0, 1, 1) are in W, but their sum (0, 2, 1) is not. This shows that W is not closed under addition; nor is W closed under multiplication by real numbers because $(0, 2, 0) = 2(0, 1, 0)$ is not in W. Since W is not closed under its intrinsic operations, W is not a real vector space.

EXAMPLE 6

For any fixed integer $n \geq 2$, let V denote the set of all $n \times n$ symmetric matrices, with multiplication of such matrices by complex numbers, and their addition, defined as for $m \times n$ matrices. Is V a vector space?

It is readily verified that V is closed under its two accompanying operations (see Exercise 37, Sec. 3.4). The zero matrix O_n is the additive identity of V. All other properties of a vector space are possessed by V because they hold for the set of all matrices of order n. In particular, -1 times any matrix in V gives the corresponding additive inverse matrix which is also in V because of closure. Hence V is a complex vector space.

The same reasoning shows that the set W of all $n \times n$ symmetric matrices, together with their usual addition but with scalar multiplication of the elements of W restricted to real numbers, is a real vector space.

EXAMPLE 7

An **nth-degree polynomial in x with real coefficients** is a function p represented on $(-\infty, \infty)$ by an expression of the form

$$p(x) = p_0 + p_1x + \cdots + p_nx^n$$

where $p_0, p_1, \ldots, p_n$ are real numbers and $p_n \neq 0$. Let q be such a polynomial of degree m represented by

$$q(x) = q_0 + q_1x + \cdots + q_mx^m$$

and suppose for definiteness that $m \geq n$. If $m > n$, the sum $p + q$ of p and q is the mth-degree polynomial in x with real coefficients represented by

$$(p + q)(x) = (p_0 + q_0) + (p_1 + q_1)x + \cdots + (p_m + q_m)x^m$$

where $p_k = 0$, for $n + 1 \leq k \leq m$. If $m = n$, it may happen that for some integer N, $p_N + q_N \neq 0$, while for every integer k such that $0 \leq N < k \leq n$, $p_k + q_k = 0$. In this case, $p + q$ is a polynomial in x of degree N. On the other hand, for $0 \leq k \leq n$, we could have $p_k + q_k = 0$. In this case, $p + q$ is trivial on $(-\infty, \infty)$ and is called the **zero polynomial.** By convention the zero polynomial has no degree.

The product of p by any real number $c \neq 0$ is the nth-degree polynomial in x with real coefficients represented by

$$(cp)(x) = cp_0 + cp_1x + \cdots + cp_nx^n$$

If $c = 0$, then cp is the zero polynomial. Clearly the zero polynomial is an additive identity in the **set P of all polynomials in x with real coefficients.**

As we have just verified, P is closed under function addition and multiplication by reals. The remaining properties of a vector space are also possessed by P because, by Theorem 1, they hold for the set of all real-valued functions having $(-\infty, \infty)$ as their domain. The set P of all polynomials in x with real coefficients is therefore a real vector space.

Once a set V, having two operations, is known to be a vector space, V itself is commonly referred to by the same name. For example, such sets as the set of real numbers, the set of complex numbers, the set of all ordered pairs of real numbers, and the set of all polynomials in x with real coefficients, are often called vector spaces without any mention of their implicit operations. Although the members of any vector space are called vectors, different kinds of vectors such as n-tuples, matrices, and functions are distinguished by more specific names. As with n-dimensional vectors, vectors in general are often, but not always, denoted by boldface letters.

If V is any vector space over a set of scalars S, and $\mathbf{v}_1$, $\mathbf{v}_2$, and $\mathbf{v}_3$ are any vectors of V, then, by the associative law of vector addition, $\mathbf{v}_1 + (\mathbf{v}_2 + \mathbf{v}_3) = (\mathbf{v}_1 + \mathbf{v}_2) + \mathbf{v}_3$. Since both members of this equation denote the same vector, parentheses are usually omitted, that is, by definition,

$$\mathbf{v}_1 + \mathbf{v}_2 + \mathbf{v}_3 = \mathbf{v}_1 + (\mathbf{v}_2 + \mathbf{v}_3) = (\mathbf{v}_1 + \mathbf{v}_2) + \mathbf{v}_3 \tag{16}$$

Now, let c_1, c_2, and c_3 be arbitrary scalars of S. Then, because of closure, $c_1\mathbf{v}_1$, $c_2\mathbf{v}_2$, $c_3\mathbf{v}_3$, $c_1\mathbf{v}_1 + c_2\mathbf{v}_2$, and $c_1\mathbf{v}_1 + c_2\mathbf{v}_2 + c_3\mathbf{v}_3$ are all vectors in V. It follows, by induction, that for all vectors $\mathbf{v}_1, \mathbf{v}_2, \ldots, \mathbf{v}_n$ in V, and for arbitrary scalars $c_1, c_2, \ldots, c_n$ of S,

$$c_1\mathbf{v}_1 + c_2\mathbf{v}_2 + \cdots + c_n\mathbf{v}_n = \sum_{k=1}^{n} c_k\mathbf{v}_k \tag{17}$$

is a vector in V. We call (17) a **linear combination** of the vectors $\mathbf{v}_1, \mathbf{v}_2, \ldots, \mathbf{v}_n$ whether or not these vectors are distinct. The commutative law of vector addition implies that any linear combination of vectors obtained by rearranging terms of (17) still designates the same vector as (17).

Many algebraic properties of vector spaces are not to be found among their defining properties. Some of these properties are important enough to deserve special attention. For instance, every operation called *vector addition* is assumed to be **well defined** with respect to the equals relation. This simply means that, for all $\mathbf{u}$, $\mathbf{v}$, $\mathbf{w}$, and $\mathbf{x}$ in V:

$$\text{If} \quad \mathbf{u} = \mathbf{v} \quad \text{and} \quad \mathbf{w} = \mathbf{x} \quad \text{then} \quad \mathbf{u} + \mathbf{w} = \mathbf{v} + \mathbf{x} \tag{18}$$

In particular, if $\mathbf{u} + \mathbf{w} = \mathbf{v} + \mathbf{w}$, then $\mathbf{u} + \mathbf{w} + (-\mathbf{w}) = \mathbf{v} + \mathbf{w} + (-\mathbf{w})$, i.e., $\mathbf{u} + \mathbf{0} = \mathbf{v} + \mathbf{0}$ or $\mathbf{u} = \mathbf{v}$. This proves the **right cancelation law:**

$$\text{If} \quad \mathbf{u} + \mathbf{w} = \mathbf{v} + \mathbf{w} \quad \text{then} \quad \mathbf{u} = \mathbf{v} \tag{19}$$

The **left cancelation law:**

$$\text{If} \quad \mathbf{w} + \mathbf{u} = \mathbf{w} + \mathbf{v} \quad \text{then} \quad \mathbf{u} = \mathbf{v} \tag{20}$$

also holds (Exercise 59), and so like terms can be canceled on the left too.

For any scalar c in S, $c\mathbf{0} + \mathbf{0} = c\mathbf{0} = c(\mathbf{0} + \mathbf{0}) = c\mathbf{0} + c\mathbf{0}$ so that, by cancelation,

$$c\mathbf{0} = \mathbf{0} \tag{21}$$

Since the zero element $\mathbf{0}$ of V is unique, and since $\mathbf{v} + 0\mathbf{v} = 1\mathbf{v} + 0\mathbf{v} = (1 + 0)\mathbf{v} = 1\mathbf{v} = \mathbf{v}$, it follows that

$$0\mathbf{v} = \mathbf{0} \tag{22}$$

The inverse of any element $\mathbf{v}$ in V is also unique, i.e.,

$$\text{If} \quad \mathbf{v} + \mathbf{w} = \mathbf{0} \quad \text{then} \quad \mathbf{w} = -\mathbf{v} \tag{23}$$

for if $\mathbf{v} + \mathbf{w} = \mathbf{0}$, then, since $\mathbf{v} + (-\mathbf{v}) = \mathbf{0}$, clearly $\mathbf{v} + \mathbf{w} = \mathbf{v} + (-\mathbf{v})$ which implies $\mathbf{w} = -\mathbf{v}$.

Observing that $\mathbf{v} + (-1)\mathbf{v} = 1\mathbf{v} + (-1)\mathbf{v} = [1 + (-1)]\mathbf{v} = 0\mathbf{v} = \mathbf{0}$, we see that

$$-\mathbf{v} = (-1)\mathbf{v} \tag{24}$$

To subtract a vector $\mathbf{v}$ from a vector $\mathbf{u}$, we form the **difference**

$$\mathbf{u} - \mathbf{v} = \mathbf{u} + (-\mathbf{v}) = \mathbf{u} + (-1)\mathbf{v} \tag{25}$$

and call the operation of adding to $\mathbf{u}$ the additive inverse $-\mathbf{v}$ of $\mathbf{v}$ **subtraction.**

As a final elementary property, we have (Exercise 60)

$$c\mathbf{v} = \mathbf{0} \quad \text{if and only if} \quad c = 0 \quad \text{or} \quad \mathbf{v} = \mathbf{0} \tag{26}$$

None of the preceding properties of vector spaces are very surprising, and this makes them easy to remember, as we must, because they are usually used without any comment.

Just as inner products and **norms,** i.e., lengths, play an important role in many three-dimensional geometric problems, more general inner products and norms prove essential to the analysis of numerous applied problems involving real vector spaces. We encountered several examples of this when we took up Fourier series and partial differential equations. A real vector space R^n in which every pair of vectors has an inner product and each individual vector has a norm [Eqs. (4) and (5), Sec. 3.2] is called a **euclidean** $\boldsymbol{n}$**-space.**

A common notation for the *norm* of a function f is $\|f\|$. A continuous real-valued function f with domain $[a, b]$ has a **norm** defined by

$$\|f\| = \left[\int_a^b f^2(x)\,dx\right]^{1/2} \tag{27}$$

Alternatively, we often use the compact notation $\|f\| = (f, f)^{1/2}$. For complex-valued functions f and g with a common domain $[a, b]$, the **hermitian inner product** of f and g is defined by

$$(f, g) = \int_a^b \overline{f(x)}g(x)\,dx \tag{28}$$

provided the integral here exists. A continuous function w having complex values $w(x) = u(x) + iv(x)$ and an interval $[a, b]$ as domain, has a **hermitian norm** defined by

$$\|w\| = \left[\int_a^b \overline{w(x)}w(x)\,dx\right]^{1/2} = \left[\int_a^b \{u^2(x) + v^2(x)\}\,dx\right]^{1/2} \tag{29}$$

In terms of the hermitian inner product (w, w), $\|w\| = (w, w)^{1/2}$.

EXERCISES

1. Identify the function whose values are given by $\tan j\pi$, $1 \le j \le 4$.
2. Two functions u and v are defined by

$$u(j) = \sin(j\pi/2) \qquad v(j) = \cos(j\pi/2) \qquad 1 \le j \le 5$$

(a) Evaluate the inner product (u, v).
(b) Where, if at all, are u and v orthogonal?

3. Find all orthogonal pairs of vectors that can be formed from the vectors:

$$\mathbf{u} = (1, 2, 3) \qquad \mathbf{v} = (4, 5) \qquad \mathbf{w} = (0, 6, 0, 7)$$
$$\mathbf{r} = (-1, 0) \qquad \mathbf{s} = (-2, 0, -1, 0)$$
$$\boldsymbol{\rho} = (1, -1, 0, 1, -1)$$

4. Determine whether or not the following pairs of functions are orthogonal on $(-\infty, \infty)$.
(a) $\sqrt{x}$, $\cosh x$ (b) e^x, e^{-x}
(c) $(1 + x^2)^{-2}$, $2x$

5. Are the functions f and g defined by

$$f(x) = \begin{cases} -1 & -2 \le x \le 0 \\ 1 & 0 < x \le 2 \end{cases}$$

and $$g(x) = \begin{cases} 2x + 4 & -2 \le x \le 0 \\ 1 & 0 < x \le 2 \end{cases}$$

(a) Orthogonal on $[-2, 2]$?
(b) Orthogonal on $[-\frac{1}{2}, \frac{7}{4}]$?

6. Verify that the vectors $\mathbf{u} = (3, -2, 1)$ and $\mathbf{v} = (1, 2, 1)$ are orthogonal and find a vector $\mathbf{w}$ such that $\mathbf{u}$, $\mathbf{v}$, and $\mathbf{w}$ are mutually orthogonal.
7. (a) Show that the vectors $\mathbf{b}_1 = (1, 1, 1)$, $\mathbf{b}_2 = (1, 0, -1)$, and $\mathbf{b}_3 = (1, -2, 1)$ are mutually orthogonal.
(b) Find the length of each given vector.
(c) Use Part **(a)** to express $(1, 0, 5)$ as a linear combination of $\mathbf{b}_1$, $\mathbf{b}_2$, and $\mathbf{b}_3$.
8. Show that $f_1(x) = x$ and $f_2(x) = 3x - 2$ are orthogonal on the interval $(0, 1)$ and find a quadratic function $f_3(x)$ with the value $f_3(0) = 1$ that is orthogonal to both $f_1(x)$ and $f_2(x)$ on $(0, 1)$.
9. (a) Prove that

$$\mathbf{u}_1 = (1, 1, -1, 1) \qquad \mathbf{u}_2 = (1, -1, 3, 3)$$
$$\mathbf{u}_3 = (-2, 1, 0, 1) \qquad \mathbf{u}_4 = (1, 4, 3, -2)$$

form an orthogonal set of four-dimensional vectors.
(b) Expand each of the following vectors in terms of the vectors of Part **(a)**.

$$\mathbf{v}_1 = (1, 0, 0, 0) \qquad \mathbf{v}_2 = (1, 0, 2, 3)$$
$$\mathbf{v}_3 = (1, 1, 1, 1) \qquad \mathbf{v}_4 = (0, 1, 2, 5)$$

10. For $1 \le j \le n$, let $\mathbf{e}_j$ be the vector of R^n whose jth component is 1 and whose other components are all 0 and let $c_j \ne 0$ be a real number. Prove that every set $\{c_j\mathbf{e}_j\}$ of two or more vectors is an orthogonal set of vectors of R^n.
11. By replacing $\mathbf{u}$'s by f's and $\mathbf{v}$'s by g's, convert each of the following algebraic laws for inner products of vectors into a corresponding law for inner products of functions.

(I) $(\mathbf{u}, \mathbf{v}) = (\mathbf{v}, \mathbf{u})$ **Commutative law**

(II) $(c\mathbf{u}, \mathbf{v}) = (\mathbf{u}, c\mathbf{v}) = c(\mathbf{u}, \mathbf{v})$ **Scalar positioning**

(III) $(\mathbf{v}, \mathbf{u}_1 + \mathbf{u}_2) = (\mathbf{v}, \mathbf{u}_1) + (\mathbf{v}, \mathbf{u}_2)$ **Left distributive law**

(IV) $(\mathbf{v}_1 + \mathbf{v}_2, \mathbf{u}) = (\mathbf{v}_1, \mathbf{u}) + (\mathbf{v}_2, \mathbf{u})$ **Right distributive law**

How can this substitution process be justified and the new laws pertaining to inner products of functions thus verified?

12. Prove that the vectors

$$\mathbf{u}_1 = (9, 4, -6) \qquad \mathbf{u}_2 = (2, -3, 1) \qquad \mathbf{u}_3 = (2, 3, 5)$$

form a complete orthogonal set for the vectors of R^3. *Hint:* Show that an arbitrary vector of R^3 is a linear combination of $\mathbf{u}_1$, $\mathbf{u}_2$, and $\mathbf{u}_3$.

13. Prove that the vectors

$$\mathbf{u}_1 = (1, 1, 1, 0) \qquad \mathbf{u}_2 = (1, 0, -1, 1)$$
$$\mathbf{u}_3 = (-2, 1, 1, 3)$$

form an incomplete set of orthogonal vectors. Find a fourth vector $\mathbf{u}_4$ which together with the given vectors form a complete orthogonal set of vectors of R^4.

14. Prove that the vectors

$$\mathbf{u}_1 = (2, 3, 3) \qquad \mathbf{u}_2 = (1, 0, -1) \qquad \mathbf{u}_3 = (2, -5, 3)$$

form an orthogonal set with respect to the weight vector $\mathbf{w} = (3, 2, 2)$. Explain how an arbitrary vector of R^3 can be expanded in terms of these vectors and illustrate by expanding the vector $\mathbf{v} = (1, -2, 3)$.

15. Show that the functions $P_0(x) = 1$, $P_1(x) = x$, $P_2(x) = \frac{1}{2}(3x^2 - 1)$ are mutually orthogonal on $(-1, 1)$. Using this orthogonality property, expand $f(x) = x^2$ in a series of the form $x^2 = a_0P_0(x) + a_1P_1(x) + a_2P_2(x)$.
16. Show that the functions P_0, P_1, and P_3, defined in Exercise 15 form a complete orthogonal set in the space of all quadratic functions by expanding $a + bx + cx^2$ in terms of $P_0(x)$, $P_1(x)$, and $P_2(x)$.
17. Show that the function $g(x) = \frac{1}{2}(5x^3 - 3x)$ has only a trivial expansion in terms of the functions P_0, P_1, and P_2 of Exercise 15. Does this prove that the P's form an incomplete orthogonal set for the space of all cubic functions?
18. Expand the function $h(x) = x^4$ in terms of the functions P_0, P_1, and P_2 of Exercise 15. Does this expansion represent x^4? Does this prove that the P's form an incomplete orthogonal set for the space of all quartic functions?
19. Modify Definition 1 so as to extend hermitian orthogonality, as defined for vectors (Sec. 3.2), to a pair of complex-valued functions f and g defined on an interval $[a, b]$. *Hint:* Make use of the hermitian product (28).
20. Evaluate the inner product of the functions f and g defined by

$$f(x) = e^{2\pi ix} \qquad g(x) = e^{-2\pi ix} \qquad -1 \le x \le 1$$

Are these functions orthogonal in the hermitian sense on $[-1, 1]$?

21. Evaluate the inner product of the functions w_1 and w_2 defined by

$$w_1(x) = 2i \cos x \qquad w_2(x) = i \sin x \qquad 0 \le x \le 3\pi/2$$

Is w_1 orthogonal to w_2 on $[0, 3\pi/2]$ in the hermitian sense?

22. Are the functions w_n defined, for $n = 0, \pm 1, \pm 2, \ldots$, by

$$w_n(x) = e^{inx} \qquad -\pi < x < \pi$$

mutually orthogonal on $(-\pi, \pi)$ in the hermitian sense? Is one of these functions the trivial function on $(-\pi, \pi)$?

23. Find the norm of each of the following functions.
(a) $f(x) = \sqrt{x} \sin x;\ 0 \le x \le 2\pi$
(b) $g(x) = \ln x \quad 1 \le x \le e$
(c) $h(x) = (1 + x^2)^{-1/2};\ -\infty < x < \infty$
(d) $q(x) = \sqrt{x}\, J_{1/2}(x) \quad -\pi \le x \le \pi$

24. Find the hermitian norm of each of the following functions.
(a) $w(x) = \dfrac{1}{\sqrt{x}} - i \ln x \quad 1 \le x \le e$
(b) Each w_n of Exercise 22

25. Identify the array $\begin{bmatrix} 2 & -3 & 4 & -5 \\ -3 & 4 & -5 & 6 \\ 4 & -5 & 6 & -7 \end{bmatrix}$ as a function A defined on pairs of integers.

26. Using the indicated partitioning, multiply the following pairs of matrices and check by multiplying without regard to the partitioning.

(a) $\left[\begin{array}{cc|c} 1 & 2 & -1 \\ \hline 3 & 0 & 2 \end{array}\right]$ and $\left[\begin{array}{c|c} 3 & 1 \\ 1 & 3 \\ \hline 2 & 0 \end{array}\right]$

(b) $\left[\begin{array}{cc|c|c} 1 & 0 & -2 & 3 \\ 2 & -1 & 0 & 1 \\ \hline 1 & 1 & 2 & 2 \end{array}\right]$ and $\left[\begin{array}{cc} 1 & -2 \\ -3 & 0 \\ \hline 1 & 1 \\ \hline 2 & -2 \end{array}\right]$

27. **(a)** Does the definition of A^T given by (1), Sec. 3.4, hold if the a_{ij}'s are submatrices of A?
(b) If A and B are conformable matrices, show that the ith row vector in the product AB is $\mathbf{u}_i$ B where $\mathbf{u}_i$ is the ith row vector of A. What is the jth column vector in the product AB?
(c) Represent the matrix $B = \begin{bmatrix} 0 & 2 & 0 \\ 2 & 2 & 2 \end{bmatrix}$ as a function $B(i, j) = a + b \cos ij\pi$ defined on pairs of positive integers by finding suitable values for a and b.

28. **(a)** Verify that the set of all real numbers is a vector space over itself.
(b) Verify that the set of all complex numbers is a complex vector space.

29. Show that the set of all linear combinations of the three-dimensional vectors $\mathbf{e}_1 = (1, 0, 0)$, $\mathbf{e}_2 = (0, 1, 0)$, and $\mathbf{e}_3 = (0, 0, 1)$ is a real vector space and identify the space.

30. Give a geometric description of R^2 and R^3.

31. For the given vectors $\mathbf{u}$ and $\mathbf{v}$ and the corresponding number c, write the sum $\mathbf{u} + \mathbf{v}$ and the product $c(\mathbf{u} + \mathbf{v})$ as vectors of R^2 or R^3. Illustrate each sum and product geometrically.
(a) $\mathbf{u} = (-1, 1)$, $\mathbf{v} = (1, 1)$, $c = -\frac{1}{2}$
(b) $\mathbf{u} = (2, 1)$, $\mathbf{v} = (3, -1)$, $c = \frac{1}{5}$
(c) $\mathbf{u} = (2, 3)$, $\mathbf{v} = (-3, 1)$, $c = -\frac{1}{4}$
(d) $\mathbf{u} = (1, 2)$, $\mathbf{v} = (2, 1)$, $c = -1$
(e) $\mathbf{u} = (-1, -2, 0)$, $\mathbf{v} = (2, 1, 1)$, $c = 2$
(f) $\mathbf{u} = (1, -4, 1)$, $\mathbf{v} = (-3, 10, 1)$, $c = -\frac{1}{2}$

32. Determine a polynomial p such that $p = aq + br$, where
(a) $q(x) = x^3 - x + 1$, $r(x) = x^2 + 2x - 5$, $a = -\frac{5}{3}$, $b = -\frac{1}{3}$
(b) $q(x) = x^4 - x^2 + 2x$, $r(x) = 3x^4 - x^3 + 6x$, $a = 3$, $b = -1$
(c) $q(x) = x - 2x^3 + 3x^4$, $r(x) = 3 - x + 3x^2 + 2x^3 - 3x^4$, $a = 3$, $b = 3$
(d) $q(x) = 2 - 3x + \frac{9}{2}x^2 + 7x^3$, $r(x) = 1 - 12x + 2x^3$, $a = \frac{2}{3}$, $b = -\frac{1}{3}$

33. Specify, if possible, a domain over which the given formulas for $f(x)$ and $g(x)$ represent two real-valued functions which can be added.
(a) $f(x) = \sqrt{x^2 - 4}$, $g(x) = \sqrt{36 - 9x^2}$
(b) $f(x) = 1 + x^{3/2}$, $g(x) = \sqrt{1 - x^2}$
(c) $f(x) = \sqrt{5x^2 - x^4 - 4}$, $g(x) = \sqrt{\cos[\pi(x - 2)/2]}$
(d) $f(x) = 1 + \ln x$, $g(x) = 10 + (-x)^{1/4}$

34. **(a)** Find a domain D for which the set of all real-valued functions $F(D)$ defined on D and the set of all $m \times n$ matrices with real elements are the same linear space.
(b) Find a domain D for which $F(D)$ and R^n are the same linear space.

35. Determine which, if any, of the following statements are true.
(a) R^n is a linear space over the complex numbers.
(b) Every complex vector space is also a real vector space.
(c) If $\mathbf{u}$ and $\mathbf{v}$ are members of the same vector space, their scalar product is real.
(d) There are vector spaces for which subtraction cannot be defined.
(e) Some vector spaces contain a single element.

In each of Exercises 36–54, the set containing the elements specified also possesses its relevant operations of multiplication over reals and vector addition. Is the set a real vector space?

36. All vectors of R^3 with second component 0

37. The zero vector of R^5

38. All vectors of R^3 whose three successive components are in arithmetic progression

39. All vectors of R^3 whose three successive components are in harmonic progression

40. All 3×3 nonsingular matrices with real elements

41. All 3×3 singular matrices with real elements

42. All 3×3 skew-symmetric matrices

43. All polynomials of even degree

44. All polynomials of degree $n \leq 3$ with real coefficients

45. All real-valued functions continuous on an interval I

46. All real-valued functions defined on $-1 \leq x \leq 1$ such that $f(-1) = f(1)$

47. All real-valued functions defined on $-1 \leq x \leq 1$ such that $f(0) = 1$

48. All solutions of the differential equation $y'' - 5y' + 6y = 0$

49. All solutions of the differential equation $y''' - 6y'' + 11y' - 6y = 0$

50. All solutions of $a_0 y'' + a_1 y' + a_2 y = 0$ on an interval I over which the equation is normal

51. All solutions of $\mathrm{A}\mathbf{x} = \mathbf{0}$ where A is an $m \times n$ matrix

52. All infinite sequences $\mathbf{v} = (v_1, v_2, \ldots)$ of real numbers with multiplication of such sequences by a real number c, and their addition, defined by $c\mathbf{v} = (cv_1, cv_2, \ldots)$ and $\mathbf{u} + \mathbf{v} = (u_1 + v_1, u_2 + v_2, \ldots)$

53. All convergent infinite series of real numbers

54. All real-valued functions f defined on a rectangle R: $|x - x_0| \leq a$, $|y - y_0| \leq b$

55. If possible express the vector $\mathbf{v}$ as a linear combination of $\mathbf{v}_1$, $\mathbf{v}_2$, and $\mathbf{v}_3$.

(a) $\mathbf{v} = (3, 2, 2)$, $\mathbf{v}_1 = (2, -1, 4)$, $\mathbf{v}_2 = (0, 1, 0)$, $\mathbf{v}_3 = (-2, 0, 4)$

(b) $\mathbf{v} = (0, 0, -1)$, $\mathbf{v}_1 = (2, -1, 1)$, $\mathbf{v}_2 = (1, 0, 1)$, $\mathbf{v}_3 = (1, -1, 0)$

(c) $\mathbf{v} = (0, 0, 0)$, $\mathbf{v}_1 = (3, -1, 2)$, $\mathbf{v}_2 = (-2, 1, -1)$, $\mathbf{v}_3 = (3, -2, 1)$

56. Give a geometric description of the set of all linear combinations of the vectors $\mathbf{u}$ and $\mathbf{v}$.

(a) $\mathbf{u} = (0, 0)$, $\mathbf{v} = (1, 1)$

(b) $\mathbf{u} = (1, -2)$, $\mathbf{v} = (-2, 4)$

(c) $\mathbf{u} = (0, 1, 1)$, $\mathbf{v} = (0, -1, 1)$

(d) $\mathbf{u} = (1, -1, 0)$ $\mathbf{v} = (-1, 1, 0)$

(e) $\mathbf{u} = (1, 0, 1)$, $\mathbf{v} = (0, 1, 1)$

(f) $\mathbf{u} = (-1, -1, 1)$, $\mathbf{v} = (1, 1, 1)$

57. **(a)** If the scalar product of two matrices of the same size is defined to be the sum of the products of the corresponding elements in the two matrices, show that

$$\mathrm{B}_1 = \begin{bmatrix} 1 & 1 \\ 1 & 0 \end{bmatrix}, \quad \mathrm{B}_2 = \begin{bmatrix} 1 & 0 \\ -1 & 1 \end{bmatrix}, \quad \mathrm{B}_3 = \begin{bmatrix} -1 & 1 \\ 0 & 1 \end{bmatrix},$$

$\mathrm{B}_4 = \begin{bmatrix} 0 & -2 \\ 2 & 2 \end{bmatrix}$ are mutually orthogonal matrices.

(b) What is the norm or ''length'' of each of these matrices?

(c) Using the orthogonality property of Part **(a)**, express $\mathrm{I} = \begin{bmatrix} 1 & 0 \\ 0 & 1 \end{bmatrix}$ and $\mathrm{A} = \begin{bmatrix} 1 & 2 \\ 3 & 4 \end{bmatrix}$ as linear combinations of B_1, B_2, B_3, and B_4.

(d) Express A as a linear combination of the orthonormal set of matrices $\mathrm{E}_k = \begin{bmatrix} \delta_{1k} & \delta_{2k} \\ \delta_{3k} & \delta_{4k} \end{bmatrix}$, $k = 1, 2, 3, 4$.

58. Using the definition of Exercise 57, verify that $M_1 = \begin{bmatrix} 1 & 0 \\ 0 & 1 \end{bmatrix}$ and $M_2 = \begin{bmatrix} 1 & 2 \\ 0 & -1 \end{bmatrix}$ are orthogonal and find matrices M_3 and M_4 such that M_1, M_2, M_3, and M_4 are mutually orthogonal.

59. Prove the left cancelation law (20): If $\mathbf{w} + \mathbf{u} = \mathbf{w} + \mathbf{v}$, then $\mathbf{u} = \mathbf{v}$.

60. Prove (26), that is, show that $c\mathbf{v} = \mathbf{0}$ if and only if $c = 0$ or $\mathbf{v} = \mathbf{0}$.

13.2 SUBSPACES, LINEAR DEPENDENCE, AND LINEAR INDEPENDENCE

A set U each of whose elements is a member of a set W is called a **subset** of W. Examples 6 and 7 of the last section, as well as several of the exercises there, illustrate the fact that a vector space V can include a nonempty subset which itself is a vector space relative to the basic operations and set of scalars S associated with V. Such a subset is called a **subspace** of V. Since every pair of vectors in a vector space V can be added, the same is true of every pair of vectors in any nonempty subset U of V. Clearly, each vector in U can also be multiplied by any scalar in S. But there is no guarantee that an arbitrary subset U of V is a vector space. Of course, any given U can be tested step by step to see if it has all the defining properties of a vector space. However, in dealing with a set U known to be a nonempty subset of a vector space, it is much easier to determine if U is a subspace by means of the following criterion.

THEOREM 1 If U is a nonempty subset of a vector space V over a set of scalars S, and if for all vectors $\mathbf{u}$ and $\mathbf{v}$ in U and for every scalar c in S both $\mathbf{u} + \mathbf{v}$ and $c\mathbf{u}$ are in U, then U is a subspace of V.

In terms of closure, this theorem says that if a nonempty subset of a vector space V is closed under the operations of V, then that subset is a subspace of V.

PROOF Under the hypotheses of the theorem, U is closed under the operations of V, every element in U is in V, and there is at least one element in U. Hence, for every $\mathbf{u}$ in U, $(-1)\mathbf{u} = -\mathbf{u}$ is in U, as in $(-1)\mathbf{u} + \mathbf{u} = \mathbf{0}$. This shows that the inverse of every element of U is in U, as is the zero vector of V. All other properties of a vector space hold for U since they hold for V which includes U.

Now let us consider several examples pertaining to subspaces and to the use of Theorem 1.

EXAMPLE 1

Every vector space is a subspace of itself. Let $\mathbf{0}$ be the zero element of a vector space V. Since $\mathbf{0} + \mathbf{0} = \mathbf{0}$ and $c\mathbf{0} = \mathbf{0}$, the subset $\{\mathbf{0}\}$ of V is closed under addition and under multiplication by a scalar. Hence $\{\mathbf{0}\}$ is a subspace of V. We call $\{\mathbf{0}\}$ the **trivial subspace** of V. If $V \neq \{\mathbf{0}\}$, V has at least two subspaces. But if $V = \{\mathbf{0}\}$, the trivial subspace of V is V itself so that V has just one subspace.

EXAMPLE 2

For $1 \leq k \leq n$, let $\mathbf{e}_k$ be the vector of R^n whose kth component is 1 and whose other components are all zero. Let V_k denote the set of all linear combinations of $\mathbf{e}_k$. It is readily verified that V_k is a subset of R^n which is closed under the operations of addition and multiplication by reals defined for R^n. Hence $V_1, V_2, \ldots, V_n$ are all subspaces of R^n. If $j \neq k$, no vector of V_j except the zero vector belongs to V_k.

If $n = 1$, $V_1 = R$ is the set of all real numbers together with its operations of addition and multiplication. This vector space has just two subspaces, namely, R itself and the trivial subspace $\{0\}$.

If $n = 2$, V_1 is the subspace of R^2 which contains all vectors of the type $(x, 0)$, where x is a real number. The subspace V_2 of R^2 contains all vectors of the type $(0, y)$, where y is real. The trivial subspace of R^2 is $\{(0, 0)\}$. For each nonzero real number m, the set of all linear combinations of the vector $(1, m)$ is another subspace of R^2. Finally, the set of all linear combinations of $\mathbf{e}_1 = (1, 0)$ and $\mathbf{e}_2 = (0, 1)$ is R^2. Clearly, R^2 has infinitely many subspaces.

Our geometric intuition tells us that R^2 is a cartesian plane, that V_1 and V_2 are coordinate axes, that $\{(0, 0)\}$ is the origin, and that for each m, the set of all linear combinations of $(1, m)$ is a line through the origin with slope m. These observations indicate how the subspaces of R^n may be analyzed when $n \geq 3$.

EXAMPLE 3

We have already shown that with function addition and multiplication by reals defined by Eqs. (8) and (9), Sec. 13.1, the set of all real-valued functions $F(D)$ having D as domain is a real vector space. If D is an interval I, the corresponding vector space $F(I)$ contains as elements just those real-valued functions which are defined on I.

Let $\mathscr{C}^{(n)}(I)$ denote **the set of all real-valued functions which are continuously differentiable n times on I**. Obviously, $\mathscr{C}^{(n)}(I)$ is a subset of $F(I)$. For any real number c, and for all functions f and g whose nth-order derivatives are continuous over I, both $f + g$ and cf are n times continuously differentiable on I. In other words, $\mathscr{C}^{(n)}(I)$ is closed under function addition and multiplication by reals. Hence, by Theorem 1, for every positive integer n, $\mathscr{C}^{(n)}(I)$ is a linear space. If $n = 0$, $\mathscr{C}^{(0)}(I)$ is interpreted to be **the set $\mathscr{C}(I)$ of all continuous functions on** I, which is also a linear space (Exercise 45, Sec. 13.1).

For integers m and n such that $0 \leq m < n$, every function that is continuously differentiable n times is certainly continuously differentiable m times on I. On the other hand, there are functions which are continuously differentiable m times which are not continuously differentiable n times on I. Hence, $\mathscr{C}^{(n)}(I)$ is a subspace of $\mathscr{C}^{(m)}(I)$ and these vector spaces are distinct. Since there are functions that can be differentiated any number of times on I, it follows that $F(I)$ has infinitely many subspaces, including its trivial subspace whose only member is the trivial function with domain I.

Theorem 1 provides a convenient way of testing whether or not any subset of a vector space V is a subspace. Our next theorem points out how linear combinations and subspaces are connected.

THEOREM 2 The set U of all linear combinations of n arbitrary vectors $\mathbf{v}_1, \mathbf{v}_2, \ldots, \mathbf{v}_n$ of a vector space V is a subspace of V. Each of the vectors $\mathbf{v}_1, \mathbf{v}_2, \ldots, \mathbf{v}_n$ is in U.

PROOF Let $\mathbf{u}$ and $\mathbf{w}$ be arbitrary elements of U. Then $\mathbf{u}$ and $\mathbf{w}$ are both linear combinations of $\mathbf{v}_1, \mathbf{v}_2, \ldots, \mathbf{v}_n$; that is, there exist scalars a_k and b_k, $1 \le k \le n$, such that $\mathbf{u} = \Sigma_{k=1}^n a_k\mathbf{v}_k$ and $\mathbf{w} = \Sigma_{k=1}^n b_k\mathbf{v}_k$. Since all elements of U are vectors of V, they can be added, and they can also be multiplied by any scalar c. In particular,

$$\mathbf{u} + \mathbf{w} = \sum_{k=1}^{n} (a_k + b_k)\mathbf{v}_k \qquad \text{and} \qquad c\mathbf{u} = \sum_{k=1}^{n} (ca_k)\mathbf{v}_k$$

This shows that for every $\mathbf{u}$ and $\mathbf{w}$ in U and for every scalar c, $\mathbf{u} + \mathbf{w}$ and $c\mathbf{u}$ are linear combinations of $\mathbf{v}_1, \mathbf{v}_2, \ldots, \mathbf{v}_n$. Hence U is closed under the operations of V and is therefore a subspace of V. Since for $1 \le j \le n$,

$$\mathbf{v}_j = \sum_{k=1}^{n} \delta_{jk}\mathbf{v}_k$$

each of the vectors $\mathbf{v}_1, \mathbf{v}_2, \ldots, \mathbf{v}_n$ is in U.

To signify that every vector of a linear space V is a linear combination of n specific vectors $\mathbf{v}_1, \mathbf{v}_2, \ldots, \mathbf{v}_n$, we say that V is **spanned** by these vectors or that these vectors **span** V. We also say that the set $\{\mathbf{v}_1, \mathbf{v}_2, \ldots, \mathbf{v}_n\}$ **spans** V.

In Example 2, we observed that R^2 is spanned by the vectors (1, 0) and (0, 1), and we saw that each of the vectors (0, 0), (1, 0), (0, 1), and $(1, m)$, with $m \neq 0$, spans a subspace of R^2. A geometric interpretation of each subspace was also given. Here are two more examples to illustrate these ideas.

EXAMPLE 4

Find the subspace U of R^3 spanned by the vectors $\mathbf{v}_1 = (1, -2, 1)$, $\mathbf{v}_2 = (-2, 0, 3)$, and $\mathbf{v}_3 = (3, -2, -2)$. Are the vectors $(4, -4, -1)$ and $(6, -6, -2)$ in U?

Every vector $\mathbf{v} = (x, y, z)$ in U must be a linear combination of $\mathbf{v}_1$, $\mathbf{v}_2$, and $\mathbf{v}_3$; hence there are numbers a, b, and c for which

$$\mathbf{v} = a\mathbf{v}_1 + b\mathbf{v}_2 + c\mathbf{v}_3 \qquad \text{or} \qquad (x, y, z) = (a - 2b + 3c, -2a - 2c, a + 3b - 2c)$$

The last equation holds if and only if the linear system in a, b, and c

$$\begin{aligned} a - 2b + 3c &= x \\ -2a \qquad - 2c &= y \\ a + 3b - 2c &= z \end{aligned} \qquad \text{or the equivalent system} \qquad \begin{aligned} a - 2b + 3c &= x \\ b - c &= x + y + z \\ 0 &= 6x + 5y + 4z \end{aligned}$$

is consistent, and this is the case if and only if $6x + 5y + 4z = 0$. This equation defines a plane in an xyz cartesian 3-space, which plane we may interpret as a geometric representation of the required subspace U. Clearly, the point which represents the vector $(4, -4, -1)$ lies in the plane, but the point depicting $(6, -6, -2)$ does not.

EXAMPLE 5

Let I be an interval over which a homogeneous nth-order linear differential equation

$$(1) \qquad a_0(x)y^{(n)} + a_1(x)y^{(n-1)} + \cdots + a_{n-1}(x)y' + a_n(x)y = 0$$

is normal. Then, every solution of (1) on I has a continuous nth-order derivative over I. Moreover, every solution of (1) over I can be expressed as a linear combination of any n linearly independent particular solutions $y_1, y_2, \ldots, y_n$ of (1) on I. In the language of linear spaces, these results say that every solution of (1) is a vector of $\mathscr{C}^{(n)}(I)$ and that any n fundamental solution vectors, i.e., a basis, of (1) on I spans a single subspace of $\mathscr{C}^{(n)}(I)$ which subspace is of course *the* complete solution of (1) over I. That the complete solution y of (1) on I is indeed a subspace of $\mathscr{C}^{(n)}(I)$ is a direct consequence of Theorem 2.

In Example 4, every linear combination of $\mathbf{v}_1$, $\mathbf{v}_2$, and $\mathbf{v}_3$ is a linear combination of just $\mathbf{v}_1$ and $\mathbf{v}_2$ because $\mathbf{v}_3 = \mathbf{v}_1 - \mathbf{v}_2$. Likewise, in Example 5, more than n solutions of (1) over I still span some subspace of the complete solution y. These observations suggest that if V is any vector space spanned by a *finite* number of vectors, then there is a unique integer n such that V can be spanned by n vectors, but not by less than n vectors, of the space. A proof of this result is given in the next section.

Preparatory thereto, we now extend the definitions of *linear dependence* and *linear independence,* first introduced in Sec. 2.2, to vectors of an arbitrary linear space.

DEFINITION 1 A finite set of vectors $\{\mathbf{v}_1, \mathbf{v}_2, \ldots, \mathbf{v}_n\}$ of a vector space V is linearly dependent if and only if there exist scalars $c_1, c_2, \ldots, c_n$, at least one of which is not zero, such that $c_1\mathbf{v}_1 + c_2\mathbf{v}_2 + \cdots + c_n\mathbf{v}_n = \mathbf{0}$.

DEFINITION 2 A finite set of vectors $\{\mathbf{v}_1, \mathbf{v}_2, \ldots, \mathbf{v}_n\}$ of a vector space V is linearly independent if and only if it is not linearly dependent, i.e., if and only if $c_1\mathbf{v}_1 + c_2\mathbf{v}_2 + \cdots + c_n\mathbf{v}_n = \mathbf{0}$ implies $c_1 = c_2 = \cdots = c_n = 0$.

The scalars $c_1, c_2, \ldots, c_n$ in these two definitions are to be real or complex according as V is real or complex. This guarantees that the individual products $c_k\mathbf{v}_k$, as well as sums of such products, are vectors of V. According as a set of vectors $\{\mathbf{v}_1, \mathbf{v}_2, \ldots, \mathbf{v}_n\}$ is linearly dependent, or linearly independent, we also say that the vectors $\mathbf{v}_1, \mathbf{v}_2, \ldots, \mathbf{v}_n$, themselves, are **linearly dependent,** or **linearly independent.** An important property of every linearly dependent set containing two or more vectors is

PROPERTY 1 A set of vectors $\{\mathbf{v}_1, \mathbf{v}_2, \ldots, \mathbf{v}_n\}$ is linearly dependent if and only if some vector of the set is a linear combination of the other vectors of the set.

PROOF Let $U = \{\mathbf{v}_1, \mathbf{v}_2, \ldots, \mathbf{v}_n\}$. If some vector $\mathbf{v}_k$ of U is a linear combination $\mathbf{v}_k = a_1\mathbf{v}_1 + \cdots + a_{k-1}\mathbf{v}_{k-1} + a_{k+1}\mathbf{v}_{k+1} + \cdots + a_n\mathbf{v}_n$ of the other vectors of U, then $\Sigma_{j=1}^{n} a_j\mathbf{v}_j = \mathbf{0}$, where $a_k = -1 \neq 0$. Hence, by definition, U is linearly dependent. Conversely, if U is linearly dependent, there are scalars $c_1, c_2, \ldots, c_n$, at least one of which, say c_j, is nonzero such that $c_1\mathbf{v}_1 + c_2\mathbf{v}_2 + \cdots + c_n\mathbf{v}_n = \mathbf{0}$. Dividing through by c_j and transposing, we obtain $\mathbf{v}_j$ as a linear combination

$$\mathbf{v}_j = -\frac{c_1}{c_j}\mathbf{v}_1 - \cdots - \frac{c_{j-1}}{c_j}\mathbf{v}_{j-1} - \frac{c_{j+1}}{c_j}\mathbf{v}_{j+1} - \cdots - \frac{c_n}{c_j}\mathbf{v}_n$$

of the other vectors of U.

EXAMPLE 6

Determine whether the vectors

$$\mathbf{v}_1 = (1, 1, -2, -2) \quad \mathbf{v}_2 = (2, -3, 0, 2) \quad \mathbf{v}_3 = (-2, 0, 2, 2) \quad \mathbf{v}_4 = (3, -3, -2, 2)$$

are linearly dependent or linearly independent in R^4. Show that the first three vectors are linearly independent.

Direct computation gives

$$\sum_{k=1}^{4} c_k\mathbf{v}_k = (c_1 + 2c_2 - 2c_3 + 3c_4, c_1 - 3c_2 - 3c_4, -2c_1 + 2c_3 - 2c_4, -2c_1 + 2c_2 + 2c_3 + 2c_4)$$

and this vector will be the zero vector $(0, 0, 0, 0)$ of R^4 if and only if (c_1, c_2, c_3, c_4) is a solution of the homogeneous linear system

$$\begin{aligned} c_1 + 2c_2 - 2c_3 + 3c_4 &= 0 \\ c_1 - 3c_2 \qquad\quad - 3c_4 &= 0 \\ -2c_1 \qquad + 2c_3 - 2c_4 &= 0 \\ -2c_1 + 2c_2 + 2c_3 + 2c_4 &= 0 \end{aligned}$$

Reducing the augmented matrix of this system to row echelon form we obtain the equivalent system

$$\begin{aligned} c_1 \qquad\quad + 3c_4 &= 0 \\ c_2 \quad + 2c_4 &= 0 \\ c_3 + 2c_4 &= 0 \end{aligned}$$

which has $(c_1, c_2, c_3, c_4) = (3, 2, 2, -1)$ as a nontrivial solution. Hence $3\mathbf{v}_1 + 2\mathbf{v}_2 + 2\mathbf{v}_3 - \mathbf{v}_4 = \mathbf{0}$, and from this it follows that the four given vectors are linearly dependent. On the other hand, $c_1\mathbf{v}_1 + c_2\mathbf{v}_2 + c_3\mathbf{v}_3 = \mathbf{0}$ if and only if (c_1, c_2, c_3) is a solution of the linear system

$$\begin{aligned} c_1 + 2c_2 - 2c_3 &= 0 \\ c_1 - 3c_2 \qquad &= 0 \\ -2c_1 \qquad + 2c_3 &= 0 \\ -2c_1 + 2c_2 + 2c_3 &= 0 \end{aligned} \qquad \text{which is equivalent to the system} \qquad \begin{aligned} c_1 &= 0 \\ c_2 &= 0 \\ c_3 &= 0 \end{aligned}$$

Hence, $\mathbf{v}_1$, $\mathbf{v}_2$ and $\mathbf{v}_3$ are linearly independent vectors. These three vectors span the same subspace of R^4 as the four given vectors because $\mathbf{v}_4 = 3\mathbf{v}_1 + 2\mathbf{v}_2 + 2\mathbf{v}_3$.

Another important property of finite sets of vectors is the following.

PROPERTY 2 If each of n vectors $\mathbf{v}_1, \mathbf{v}_2, \ldots, \mathbf{v}_n$ is a linear combination of m vectors $\mathbf{u}_1, \mathbf{u}_2, \ldots, \mathbf{u}_m$ of a vector space V and if $m < n$, then the vectors $\mathbf{v}_1, \mathbf{v}_2, \ldots, \mathbf{v}_n$ are linearly dependent.

PROOF By hypothesis, scalars $a_{1j}, a_{2j}, \ldots, a_{mj}$ exist such that for $1 \leq j \leq n$,

$$\mathbf{v}_j = a_{1j}\mathbf{u}_1 + a_{2j}\mathbf{u}_2 + \cdots + a_{mj}\mathbf{u}_m = \sum_{i=1}^{m} a_{ij}\mathbf{u}_i$$

Substituting this expression for $\mathbf{v}_j$ into the equation

$$c_1\mathbf{v}_1 + c_2\mathbf{v}_2 + \cdots + c_n\mathbf{v}_n = \sum_{j=1}^{n} c_j\mathbf{v}_j = \mathbf{0}$$

then interchanging the order of summation, we obtain

$$\sum_{j=1}^{n} c_j \sum_{i=1}^{m} a_{ij}\mathbf{u}_i = \sum_{i=1}^{m} \left(\sum_{j=1}^{n} a_{ij}c_j\right)\mathbf{u}_i = \mathbf{0}$$

The last equation clearly holds if for $1 \leq i \leq m$,

$$\sum_{j=1}^{n} a_{ij}c_j = a_{i1}c_1 + a_{i2}c_2 + \cdots + a_{in}c_n = 0$$

Since $m < n$, this is a homogeneous linear system with fewer equations than unknowns. It must therefore have a nontrivial solution (Exercise 24, Sec. 3.6). Thus constants $c_1, c_2, \ldots, c_n$, not all zero, exist such that $\Sigma_{j=1}^{n} c_j\mathbf{v}_j = \mathbf{0}$. This proves that the set $\{\mathbf{v}_1, \mathbf{v}_2, \ldots, \mathbf{v}_n\}$ is linearly dependent.

As an application of Property 2 to vectors of R^n, consider the following example.

EXAMPLE 7

In terms of the Kronecker delta [Eq. (3), Sec. 3.4], the vectors $\mathbf{e}_k$ described in Example 2 are given by

$$\mathbf{e}_k = (\delta_{k1}, \delta_{k2}, \ldots, \delta_{kn}) \qquad 1 \le k \le n \tag{2}$$

A finite number of multiplications by scalars and vector additions confirms that any vector $\mathbf{x} = (x_1, x_2, \ldots, x_n)$ of R^n is a linear combination $\mathbf{x} = x_1\mathbf{e}_1 + x_2\mathbf{e}_2 + \cdots + x_n\mathbf{e}_n$ of the n vectors $\mathbf{e}_k$. Hence, by Property 2, every set $\{\mathbf{v}_1, \mathbf{v}_2, \ldots, \mathbf{v}_m\}$ containing *more* than n vectors of R^n is linearly dependent. If $c_1\mathbf{e}_1 + c_2\mathbf{e}_2 + \cdots + c_n\mathbf{e}_n = (c_1, c_2, \ldots, c_n) = \mathbf{0}$, then $c_1 = c_2 = \cdots = c_n = 0$. Thus the vectors $\mathbf{e}_1, \mathbf{e}_2, \ldots, \mathbf{e}_n$ are linearly independent.

With $m = n$, $\mathbf{u}_k = \mathbf{e}_k^T$, $\mathbf{v}_j = \begin{bmatrix} a_{1j} \\ a_{2j} \\ \vdots \\ a_{nj} \end{bmatrix}$ and $\mathbf{c} = \begin{bmatrix} c_1 \\ c_2 \\ \vdots \\ c_n \end{bmatrix}$, the homogeneous system arrived at in the proof of Property 2 particularizes to $\mathrm{A}\mathbf{c} = \mathbf{0}$, where $\mathrm{A} = [\mathbf{v}_1 \ \mathbf{v}_2 \ \cdots \ \mathbf{v}_n]$ is an $n \times n$ matrix with real elements. This linear system has a nontrivial solution if and only if A is singular (Theorem 2, Sec. 3.7). Hence, any n vectors $\mathbf{v}_1, \mathbf{v}_2, \ldots, \mathbf{v}_n$ of the space R^n are linearly dependent if and only if $\det [\mathbf{v}_1 \ \mathbf{v}_2 \ \cdots \ \mathbf{v}_n] = 0$, and of course the set $\{\mathbf{v}_1, \mathbf{v}_2, \ldots, \mathbf{v}_n\}$ is linearly independent if and only if $\det [\mathbf{v}_1 \ \mathbf{v}_2 \ \cdots \ \mathbf{v}_n] \neq 0$.

A function f which possesses a derivative of order n at each point of an interval is said to be **differentiable n times on that interval.** For any interval I, the **set $F^{(n)}(I)$ of all real-valued functions which are differentiable n times on I** is a vector space. In particular, $F^{(0)}(I)$ is the space $F(I)$ of Example 3. A necessary condition for the linear dependence of n functions belonging to $F^{(n-1)}(I)$ is provided by the following theorem.

THEOREM 3 Let the functions $f_1, f_2, \ldots, f_n$ be differentiable $n - 1$ times on an interval I. If the functions are linearly dependent, then, for every x in I, their wronskian $W(f_1, f_2, \ldots, f_n)$ has the value $W(x) = 0$.

PROOF Because the functions $f_1, f_2, \ldots, f_n$ are linearly dependent over I, there are numbers $c_1, c_2, \ldots, c_n$, not all zero, such that, for every x in I,

$$c_1 f_1(x) + c_2 f_2(x) + \cdots + c_n f_n(x) = 0 \tag{3}$$

Since $c_1 f_1 + c_2 f_2 + \cdots + c_n f_n$ is the trivial function with domain I, its first $n - 1$ derivatives are also the zero function. Hence, for all x in I,

$$\begin{gathered} c_1 f_1'(x) + c_2 f_2'(x) + \cdots + c_n f_n'(x) = 0 \\ \cdots\cdots\cdots\cdots\cdots\cdots\cdots\cdots \\ c_1 f_1^{(n-1)}(x) + c_2 f_2^{(n-1)}(x) + \cdots + c_n f_n^{(n-1)}(x) = 0 \end{gathered} \tag{4}$$

For each x in I, Eq. (3) and the $n - 1$ equations (4) constitute a system of n homogeneous linear equations in the variables $c_1, c_2, \ldots, c_n$. Since the system has a nontrivial solution, the determinant of its coefficients must be zero; hence

$$W(x) = \begin{vmatrix} f_1(x) & f_2(x) & \cdots & f_n(x) \\ f_1'(x) & f_2'(x) & \cdots & f_n'(x) \\ \cdots & \cdots & \cdots & \cdots \\ f_1^{(n-1)}(x) & f_2^{(n-1)}(x) & \cdots & f_n^{(n-1)}(x) \end{vmatrix} \equiv 0$$

By restating Theorem 3 in its logically equivalent contrapositive form, we obtain the following useful result.

COROLLARY 1 Let the functions $f_1, f_2, \ldots, f_n$ be differentiable $n-1$ times on an interval I. If for some x in I, the wronskian $W(f_1, f_2, \ldots, f_n)$ has a value $W(x) \neq 0$, then the functions $f_1, f_2, \ldots, f_n$ are linearly independent.

The converse of Theorem 3 is not true. To show this, it is sufficient to exhibit two functions which are linearly independent over an interval I but whose wronskian vanishes identically on I. One such pair, defined on $(-\infty, \infty)$, is the following.

$$f_1 = x^2 \qquad f_2 = \begin{cases} -x^2 & x \leq 0 \\ x^2 & x \geq 0 \end{cases}$$

Clearly, $f_1' = 2x$ for all values of x, while

$$f_2' = \begin{cases} -2x & x \leq 0 \\ 2x & x \geq 0 \end{cases}$$

Therefore, for $x \leq 0$, the wronskian of f_1 and f_2 is

$$\begin{vmatrix} x^2 & -x^2 \\ 2x & -2x \end{vmatrix}$$

and for $x \geq 0$, the wronskian is

$$\begin{vmatrix} x^2 & x^2 \\ 2x & 2x \end{vmatrix}$$

and each of these determinants is identically zero. However, f_1 and f_2 are linearly independent over any interval containing the origin. In fact, if $c_1 f_1(x) + c_2 f_2(x) \equiv 0$, then

$$c_1 f_1(1) + c_2 f_2(1) = c_1 + c_2 = 0$$

and

$$c_1 f_1(-1) + c_2 f_2(-1) = c_1 - c_2 = 0$$

Hence $c_1 = c_2 = 0$, and f_1 and f_2 are linearly independent, as asserted.

It is important to realize that neither Theorem 3 nor Corollary 1 requires the functions f_1, $f_2, \ldots, f_n$ to be solutions of a differential equation. The functions involved may be any arbitrary functions of $F^{(n-1)}(I)$. Two similar results apply to arbitrary vector functions defined over an interval I (see Exercises 47 and 48).

EXERCISES

1. Let U be the set of all vectors in R^2 of the form $(-a, a)$.
(a) Is U closed under addition of its elements?
(b) Is U closed under multiplication of its elements by real numbers?
(c) Is U a subspace of R^2?
(d) Give a geometric interpretation of U.

2. Let Q be the set containing the zero polynomial and all polynomials of even degree in x with real coefficients.
(a) If p and q are in Q, is $p + q$ in Q?
(b) If p is in Q and c is a real number, is cp in Q?
(c) Is Q a subspace of the vector space P containing all polynomials in x with real coefficients?

3. Answer *yes* or *no* according as the specified set is or is not a subspace of the indicated vector space. Justify your answer.
(a) All vectors (a, b, c) in R^3 such that $b^2 = a + c$
(b) All vectors in R^3 of the form $(a + b, -a, 2b)$
(c) All vectors in R^4 of the form $(-a, c, -b, c)$
(d) All vectors (a, b) in R^2 such that $a^2 = b^2$
(e) All vectors (a, b, c) in R^3 for which $3a - 2b + c = 0$
(f) All vectors in C^2 of the form $(z, \bar{z})$
(g) The subset of all two-dimensional vectors of the form $(z, \bar{z})$ with multiplication by scalars restricted to real numbers

(h) All vectors of R^4 contained in R^5
(i) All upper triangular matrices of order n
(j) All hermitian matrices of order n

4. Let P denote the vector space of all polynomials in x with real coefficients. Answer *yes* or *no* according as the specified subset of P is or is not a subspace. Justify your answer.
(a) The zero polynomial and all polynomials of degree less than 6
(b) All polynomials with even coefficients
(c) The zero polynomial and all fourth-degree polynomials
(d) The zero polynomial and all polynomials of odd degree
(e) All polynomials having x^2 as a factor
(f) All polynomials whose successive coefficients are in arithmetic progression

5. Answer *yes* or *no* according as the specified set of real-valued functions is or is not a subspace of the vector space containing all real functions defined on the given interval. Justify your answer.
(a) All functions f defined on $(-3, 3)$ such that $f(-2) = 3f(2)$
(b) All functions f defined on $(0, 10]$ such that $\sum_{k=1}^{10} f(k) = 0$
(c) All differentiable functions f on $(-6, 6)$ such that $f'(0) = 1$
(d) All nonincreasing functions on $[a, b]$
(e) All continuous functions f on $[-a, a]$ such that $\int_{-a}^{a} [f(x) + x]\, dx = 0$
(f) All solutions on I of a homogeneous nth-order linear differential equation which is normal over I

6. Show that if **0** is the zero vector of a vector space V, then the trivial space $\{\mathbf{0}\}$ is a subspace of every subspace of V.

7. Prove that a nonempty subset U of a vector space V over a set of scalars S is a subspace of V if and only if for all **u** and **w** in U and for all a and b in S, the linear combination $a\mathbf{u} + b\mathbf{w}$ is contained in U.

8. Determine the subspace of R^3 spanned by $(1, 1, 1)$, $(2, 1, 0)$ and $(1, 1, -1)$.

9. Find all real values of t for which $(1 + t, 2t, t^2 - 1)$ and $(-2, t + 1, t)$ are orthogonal vectors and, for each such value of t, write the corresponding pair of vectors; also determine the subspace of R^3 which they span.

10. The set $U^{\perp}$ of all vectors in R^n which are orthogonal to every vector in a subspace U of R^n is called the **orthogonal complement** of U. Show that $U^{\perp}$ is a subspace of R^n. *Hint:* **v** is in $U^{\perp}$ if and only if for every **u** in U, $\mathbf{v} \cdot \mathbf{u} = 0$.

11. Prove that a linear system $\mathrm{A}\mathbf{x} = \mathbf{b}$ is consistent if and only if the vector **b** is in the linear space spanned by the column vectors of the matrix A.

12. Find an equation, or equations, defining the subspace of R^3 which is spanned by the given vector, or vectors, and interpret the subspace geometrically.
(a) $(2, 1, -2), (5, 2, -2)$ **(b)** $(2, 3, 4)$
(c) $(1, -1, 0), (1, 0, 2)$ **(d)** $(1, -3, 2), (-2, 0, 3)$
(e) $(1, -2, 5), (-2, 4, -10)$
(f) $(-1, 8, 3), (2, -8, 2)$
(g) $(1, -2, 1), (-2, 4, -2), (3, -6, 3)$
(h) $(1, -3, 2), (-2, 1, 2), (-3, -1, 6)$

13. Determine whether or not the following vectors are in the subspace of R^4 spanned by $(3, -1, 0, 3)$, $(4, 0, 1, -4)$ and $(5, 1, 2, -11)$.
(a) $(10, -1, 6, -4)$ **(b)** $(-9, 6, 0, 3)$
(c) $(0, 0, a, 0)$, a real **(d)** $(-10, 3, -7, -7)$
(e) $(13, -2, 13, -1)$
(f) $(7a, -a, b, -a)$, a, b real

14. The vectors $(4, -1, 3)$ and $(2, 1, -1)$ span a subspace U, and the vectors $(3, 3, 4)$ and $(5, 4, 11)$ span a subspace W, of R^3. Are U and W the same?

15. Find the subspace of P spanned by the given polynomials.
(a) $1, x, x^2$ **(b)** $x, x^3, 3x - 2x^3$
(c) $x - 1, x^2 - 1, x^3 - 1$

16. Determine which of the following polynomials are in the subspace of P spanned by $1 - x$, $x + x^3$, and x^4.
(a) $x^4 - x^3 - 1$ **(b)** $4x^4 + 3x^3 + 2x + 1$
(c) $4x^4 + 3x^3 + 2$ **(d)** $x^4 - x^3 + 1$
(e) $6 + 13x + 19x^3 + 47x^4$
(f) $a + ax^3 + bx^4$, a, b real

17. Determine which of the following functions are in the linear space spanned by 1, $\cos 2x$, and $\sin 2x$.
(a) $\cos^2 x$ **(b)** $5 - \sin x \cos x$
(c) $\cos x[\cos x + \sin x]$ **(d)** $1 + \cos x$
(e) $1 + \sin^2 x$ **(f)** $3 - 2 \sin x$

18. Determine whether the given vectors of the indicated space are linearly dependent or linearly independent.
(a) $(-2, 5), (3, -8)$ in R^2
(b) $(1 - i, 2 - 3i), (1 - 3i, 1 - 8i)$ in C^2
(c) $(1, 0, 0), (1, 1, 0), (1, 1, 1)$ in R^3
(d) $(2, 3, -1), (3, 2, 2), (4, 4, -1)$ in R^3
(e) $(3, 4), (4, 5), (5, 6)$ in R^2
(f) $(6, -1, 6), (4, -4, -1), (2, -1, 1)$ in R^3
(g) $(1, 2, -1, 2), (-2, -5, 3, 0), (3, 5, -2, 10)$ in R^4
(h) $(2, -1, -2, 1), (3, -2, 0, 1), (4, -1, 2, 1)$ in R^4
(i) $(2, i, -1), (1, 1 - i, 0), (i, -1, 1 + i)$ in C^3
(j) $(-i, 0, i), (1 - i, i, 2 + i), (2 - 3i, 2i, 4 + 3i)$ in C^3

In the statements of Exercises 19–28, V denotes a nontrivial vector space and W is a nonempty subset of V containing a finite number of vectors. Prove each statement.

19. If W contains the zero vector of V, then W is linearly dependent.

20. If W is linearly independent, W does not contain the zero vector of V.

21. If some subset of W is linearly dependent, so is W.

22. If W is linearly independent, then every subset of W is linearly independent. (Assume that the empty set is linearly independent.)

23. A linearly dependent set W can have a linearly independent subset.

24. If W contains at least one nonzero vector and if V_1 is the linear space spanned by the vectors of W, then there exists a linearly independent subset of W whose vectors span V_1.

25. If both X and Y are linearly independent finite subsets of V, the set containing all vectors in X and Y is not necessarily linearly independent.

26. If $\mathbf{x}$ and $\mathbf{y}$ are linearly independent vectors of V and if $\mathbf{u} = a\mathbf{x} + b\mathbf{y}$ and $\mathbf{v} = c\mathbf{x} + d\mathbf{y}$, then $\mathbf{u}$ and $\mathbf{v}$ are linearly independent if and only if $ad - bc \neq 0$.

27. If $\mathbf{u}_1, \mathbf{u}_2, \ldots, \mathbf{u}_n$ are linearly independent vectors of V and if for $1 \leq j \leq n$,

$$\mathbf{v}_j = \sum_{i=1}^{n} a_{ij}\mathbf{u}_i$$

the vectors $\mathbf{v}_1, \mathbf{v}_2, \ldots, \mathbf{v}_n$ are linearly independent if and only if $\det \mathrm{A} \neq 0$.

28. Each vector $\mathbf{v}$ in the linear space spanned by n linearly independent vectors $\mathbf{u}_1, \mathbf{u}_2, \ldots, \mathbf{u}_n$ of V is expressible as a unique linear combination of $\mathbf{u}_1, \mathbf{u}_2, \ldots, \mathbf{u}_n$.

29. Prove that a correct converse to Theorem 3, for the case $n = 2$, is the following: If for every x in I the wronskian of f_1 and f_2 vanishes identically and if one of the functions is never zero, then the functions are linearly dependent.

30. Determine whether the given vectors of P, or of $F(I)$ where I is as indicated, are linearly dependent or linearly independent.

(a) $x^2 - 1,\ 1 + x,\ 2x - 1$
(b) $1,\ \sec x,\ \tan x;\quad 0 \leq x < \pi/2$
(c) $1,\ \mathrm{Cos}^{-1} x,\ \mathrm{Sin}^{-1} x;\quad |x| \leq 1$
(d) $\ln x,\ x \ln x,\ \ln (x + x^2);\quad x > 0$
(e) $\sin 2x,\ \cos 2x,\ \cos^2 x;\quad (-\infty, \infty)$
(f) $1, x, x^2, \ldots, x^n$
(g) $2x^2 - 1,\ x + 1,\ x - 1,\ 3x - 2x^2$
(h) $\cos x,\ \sin x,\ \cosh x,\ \sinh x;\quad (-\infty, \infty)$
(i) $x^{1/2}, x^{1/3}, x^{1/4};\quad x \geq 0$
(j) $1 + x^3,\ 1 + |x|^3;\quad (-\infty, \infty)$

31. Prove that a polynomial in x with real coefficients $p_0 + p_1x + \cdots + p_nx^n$ is the zero polynomial if and only if, for $0 \leq k \leq n$, $p_k = 0$.

32. If possible, express the polynomial $3x - 5x^2 + x^3$ as a linear combination of the polynomials $1 - x + x^2$, $2 - x^2$, and $x^3 - 2x$.

33. Show that the polynomials $2x^2 - 1$, $x + 1$, $x - 1$, and $3x - 2x^2$ are linearly dependent by finding nonzero values of c_1, c_2, c_3, and c_4 for which

$$c_1(2x^2 - 1) + c_2(x + 1) + c_3(x - 1) + c_4(3x - 2x^2) \equiv 0$$

34. Show that any set containing at least $n + 2$ real polynomials whose degrees are all at most n is linearly dependent.

35. Show that, for every positive integer n, the functions 1, $\cos x$, $\cos 2x, \ldots, \cos nx$ are linearly independent vectors in $\mathscr{C}(-\infty, \infty)$. *Hint:* Multiply $\sum_{k=0}^{n} a_k \cos kx$ by $\cos jx$, $0 \leq j \leq n$, and integrate from $-\pi$ to π.

36. Any m vectors $\mathbf{v}_1, \mathbf{v}_2, \ldots, \mathbf{v}_m$ of R^n are **mutually orthogonal** if and only if, for $1 \leq j,\ k \leq m$, and $j \neq k$, $\mathbf{v}_j \cdot \mathbf{v}_k = 0$. Prove that every set of nontrivial mutually orthogonal vectors of R^n is linearly independent.

37. Let $f_1, f_2, \ldots, f_n$ be mutually orthogonal nontrivial continuous functions on $[a, b]$ so that, for $1 \leq j,\ k \leq n$, and $j \neq k$, $\int_a^b f_j(x)f_k(x)\,dx = 0$. Show that $f_1, f_2, \ldots, f_n$ are linearly independent.

38. **(a)** Using the definition recalled in Exercise 37, determine if the functions defined by $1, x, \frac{1}{2}(3x^2 - 1)$, and $\frac{1}{2}(5x^3 - 3x)$ on $[-1, 1]$ are mutually orthogonal there.
(b) Do the same expressions define linearly independent polynomials of P?

39. What conditions must a, b, c, and d satisfy for the matrices $\begin{bmatrix} 1 & 2 \\ -1 & 0 \end{bmatrix}$, $\begin{bmatrix} 2 & 3 \\ -2 & 1 \end{bmatrix}$, and $\begin{bmatrix} a & b \\ c & d \end{bmatrix}$ to be linearly dependent?

40. Show that five or more 2×2 matrices are always linearly dependent. How many 3×3 matrices must we have before we can be sure that they are linearly dependent?

41. If A is a square matrix, and if the equation $\mathrm{A}\mathbf{x} = \mathbf{0}$ has k linearly independent solutions, show that the same is true of the equation $\mathrm{A}^T\mathbf{x} = \mathbf{0}$. Is this result true if A is not a square matrix?

42. Show that if $\mathbf{v}_1, \mathbf{v}_2, \ldots, \mathbf{v}_m$ are m vectors of R^n, where $m \leq n$, and if for $i = 1, 2, \ldots, m$, the first nonzero component of $\mathbf{v}_i$ is the ith, then $\mathbf{v}_1, \mathbf{v}_2, \ldots, \mathbf{v}_m$ are linearly independent.

43. Show that if $\mathbf{v}_1, \mathbf{v}_2, \ldots, \mathbf{v}_m$ are m vectors of R^n, where $m \leq n$, and if for $i = 1, 2, \ldots, m$, the last nonzero component of $\mathbf{v}_i$ is the $(n - m + i)$th, then $\mathbf{v}_1, \mathbf{v}_2, \ldots, \mathbf{v}_m$ are linearly independent.

44. Let $\mathbf{v}_1, \mathbf{v}_2, \ldots, \mathbf{v}_m$ be vectors of C^n which are mutually orthogonal in the hermitian sense so that, for $1 \leq j, k \leq m$, and $j \neq k$, $\overline{\mathbf{v}}_j \cdot \mathbf{v}_k = 0$. Prove that if none of the vectors $\mathbf{v}_1, \mathbf{v}_2, \ldots, \mathbf{v}_m$ is the zero vector, these vectors are linearly independent.

45. Prove that n vectors $\mathbf{v}_1, \mathbf{v}_2, \ldots, \mathbf{v}_n$ of C^n are linearly dependent if and only if the so-called **gram determinant,** or **grammian,** of $\mathbf{v}_1, \mathbf{v}_2, \ldots, \mathbf{v}_n$,

$$G = \begin{vmatrix} \overline{\mathbf{v}}_1 \cdot \mathbf{v}_1 & \overline{\mathbf{v}}_1 \cdot \mathbf{v}_2 & \cdots & \overline{\mathbf{v}}_1 \cdot \mathbf{v}_n \\ \overline{\mathbf{v}}_2 \cdot \mathbf{v}_1 & \overline{\mathbf{v}}_2 \cdot \mathbf{v}_2 & \cdots & \overline{\mathbf{v}}_2 \cdot \mathbf{v}_n \\ \cdots & \cdots & \cdots & \cdots \\ \overline{\mathbf{v}}_n \cdot \mathbf{v}_1 & \overline{\mathbf{v}}_n \cdot \mathbf{v}_2 & \cdots & \overline{\mathbf{v}}_n \cdot \mathbf{v}_n \end{vmatrix}$$

is equal to zero.

46. Particularize Exercise 45 to n vectors of R^n.

Let $\mathbf{f}_1, \mathbf{f}_2, \ldots, \mathbf{f}_n$ be vector functions with n components and defined on an interval I.

47. Show that if $\mathbf{f}_1, \mathbf{f}_2, \ldots, \mathbf{f}_n$ are linearly dependent, then for every t in I, their wronskian $w(\mathbf{f}_1, \mathbf{f}_2, \ldots, \mathbf{f}_n)$ has the value $w(t) = 0$.

48. Show that if for some t in I, the wronskian

$w(\mathbf{f}_1, \mathbf{f}_2, \ldots, \mathbf{f}_n)$ has a value $w(t) \neq 0$, then $\mathbf{f}_1, \mathbf{f}_2, \ldots, \mathbf{f}_n$ are linearly independent.

49. Determine whether or not the given vector functions, defined on $(-\infty, \infty)$ are linearly dependent or linearly independent.

(a) $\mathbf{f}_1(t) = \begin{bmatrix} |t| \\ |\sin \pi t| \end{bmatrix}$, $\mathbf{f}_2(t) = \begin{bmatrix} 1 - |t| \\ |\cos \pi t| \end{bmatrix}$

(b) $\mathbf{f}_1(t) = \begin{bmatrix} t^3 \\ t|t| \end{bmatrix}$, $\mathbf{f}_2(t) = \begin{bmatrix} |t|^3 \\ t^2 \end{bmatrix}$

(c) $\mathbf{f}_1(t) = \begin{bmatrix} 3e^{2t} \\ 2e^{-t} \\ -e^t \end{bmatrix}$, $\mathbf{f}_2(t) = \begin{bmatrix} 14e^{2t} \\ -9e^{-t} \\ -13e^t \end{bmatrix}$, $\mathbf{f}_3(t) = \begin{bmatrix} -e^{2t} \\ 3e^{-t} \\ 2e^t \end{bmatrix}$

(d) $\mathbf{f}_1(t) = \begin{bmatrix} -e^{2t} \\ 2e^{-t} \\ 3e^t \end{bmatrix}$, $\mathbf{f}_2(t) = \begin{bmatrix} 5e^{2t} \\ -8e^{-t} \\ -12e^t \end{bmatrix}$, $\mathbf{f}_3(t) = \begin{bmatrix} 2e^{2t} \\ -5e^{-t} \\ -5e^t \end{bmatrix}$

50. Show that the vector functions defined on $(-\infty, \infty)$ by

$$\mathbf{f}_1(t) = \begin{bmatrix} t^2 \\ t \\ 2 \end{bmatrix} \qquad \mathbf{f}_2(t) = \begin{bmatrix} 3t^3 \\ -2t^2 \\ t \end{bmatrix} \qquad \mathbf{f}_3(t) = \begin{bmatrix} 2t^2e^t \\ te^t \\ 3e^t \end{bmatrix}$$

are linearly independent and find the value $w(t)$ of their wronskian $w(\mathbf{f}_1, \mathbf{f}_2, \mathbf{f}_3)$.

13.3 BASES AND DIMENSION

In Example 7 of Sec. 13.2, we observed that every set containing more than n vectors of R^n is linearly dependent and that R^n is spanned by the n linearly independent vectors $\mathbf{e}_k = (\delta_{k1}, \delta_{k2}, \ldots, \delta_{kn})$, $1 \leq k \leq n$. Just as n linearly independent solutions of an nth-order homogeneous linear differential equation or of a homogeneous first-order system in n unknowns were called a basis for a complete solution, the set of vectors $\{\mathbf{e}_1, \mathbf{e}_2, \ldots, \mathbf{e}_n\}$ is called a **basis** for R^n. The concept of a basis carries over to more general vector spaces in consequence of the following theorem in which the number m is finite.

THEOREM 1 Every nontrivial vector space V, spanned by m vectors, is spanned by a set of n $(\leq m)$ linearly independent vectors.† All linearly independent sets of vectors which span V contain the same number of vectors. Any linearly dependent set of vectors which spans V contains more than n vectors.

PROOF By assumption, V is spanned by a finite number of vectors $\mathbf{v}_1, \mathbf{v}_2, \ldots, \mathbf{v}_m$, at least one of which is not $\mathbf{0}$. If these m vectors are linearly dependent, it follows from Property 1, Sec. 13.2, that there are n linearly independent vectors which span V, where $n < m$ (Exercise 24, Sec. 13.2). Thus every vector space that is spanned by m vectors is spanned by n $(\leq m)$ linearly independent vectors. Suppose $\{\mathbf{u}_1, \mathbf{u}_2, \ldots, \mathbf{u}_n\}$ and $\{\mathbf{w}_1, \mathbf{w}_2, \ldots, \mathbf{w}_k\}$ are both linearly independent sets of vectors which span V. Then each vector in either set is a linear combination of the vectors in the other set, and Property 2, Sec. 13.2, implies $n = k$, i.e., n is unique. Hence all linearly independent sets of vectors which span V contain the same number of vectors. We have already noted that every dependent set whose vectors span V must contain more than n vectors.

We are now prepared to say what constitutes a *basis* of any vector space spanned by a finite number of vectors.

DEFINITION 1 A finite set of vectors $B = \{\mathbf{b}_1, \mathbf{b}_2, \ldots, \mathbf{b}_n\}$ is a **basis** of a vector space V if and only if B is linearly independent and V is spanned by B.

Sometimes the vectors $\mathbf{b}_1, \mathbf{b}_2, \ldots, \mathbf{b}_n$ of a basis are themselves called a **basis.** According to Theorem 1, the number of vectors in any particular basis of a linear space is the same as the number of vectors in any other basis. This fact is used to define what is meant by the *dimension* of a vector space.

†This theorem also holds for a trivial linear space $\{\mathbf{0}\}$ provided it is assumed that the empty set spans $\{\mathbf{0}\}$.

DEFINITON 2 A vector space V has **dimension** n if and only if V has a basis containing n vectors.

To signify that a vector space is n-dimensional, we write

$$\dim V = n \tag{1}$$

A trivial space has dimension 0.

If $\{\mathbf{b}_1, \mathbf{b}_2, \ldots, \mathbf{b}_n\}$ is a basis of a linear space V and if $a_1, a_2, \ldots, a_n$ are nonzero scalars, it is readily verified that $\{a_1\mathbf{b}_1, a_2\mathbf{b}_2, \ldots, a_n\mathbf{b}_n\}$ is a basis for V. Thus every vector space of dimension n (≥ 1) has infinitely many bases. Our next theorem tells us that any n linearly independent vectors of a linear space of dimension n form a basis of the space.

THEOREM 2 Every set of n linearly independent vectors $\{\mathbf{v}_1, \mathbf{v}_2, \ldots, \mathbf{v}_n\}$ of an n-dimensional vector space V is a basis of V.

PROOF Each of the vectors $\mathbf{v}_1, \mathbf{v}_2, \ldots, \mathbf{v}_n$ is a linear combination of these n vectors because every vector is obviously equal to itself. For any vector $\mathbf{v}$ of V that is *not* one of the vectors $\mathbf{v}_1, \mathbf{v}_2, \ldots, \mathbf{v}_n$, the set $\{\mathbf{v}_1, \mathbf{v}_2, \ldots, \mathbf{v}_n, \mathbf{v}\}$ is linearly dependent because each vector of this set is a linear combination of any n basis vectors. Constants $c_1, c_2, \ldots, c_n, c$ therefore exist, at least one of which is not zero, such that $c_1\mathbf{v}_1 + c_2\mathbf{v}_2 + \cdots + c_n\mathbf{v}_n + c\mathbf{v} = \mathbf{0}$. Clearly, $c \neq 0$, since $\mathbf{v}_1, \mathbf{v}_2, \ldots, \mathbf{v}_n$ are linearly independent. The preceding equation can thus be solved for $\mathbf{v}$. This proves that every vector of V is a linear combination of $\mathbf{v}_1, \mathbf{v}_2, \ldots, \mathbf{v}_n$. Since these n vectors span V and are linearly independent, they are a basis of V.

We now consider a theorem that brings out an important relation between the vectors and bases of a linear space.

THEOREM 3 A set of vectors $\{\mathbf{b}_1, \mathbf{b}_2, \ldots, \mathbf{b}_n\}$ is a basis of a vector space V if and only if each vector in V is uniquely expressible as a linear combination of $\mathbf{b}_1, \mathbf{b}_2, \ldots, \mathbf{b}_n$.

PROOF If $\{\mathbf{b}_1, \mathbf{b}_2, \ldots, \mathbf{b}_n\}$ is a basis for V, any vector $\mathbf{v}$ in V is some linear combination $\mathbf{v} = c_1\mathbf{b}_1 + c_2\mathbf{b}_2 + \cdots + c_n\mathbf{b}_n$ of $\mathbf{b}_1, \mathbf{b}_2, \ldots, \mathbf{b}_n$. Suppose that $\mathbf{v}$ is also represented by $\mathbf{v} = a_1\mathbf{b}_1 + a_2\mathbf{b}_2 + \cdots + a_n\mathbf{b}_n$. Then termwise subtraction of the last two equations gives $(a_1 - c_1)\mathbf{b}_1 + (a_2 - c_2)\mathbf{b}_2 + \cdots + (a_n - c_n)\mathbf{b}_n = \mathbf{0}$. Since $\mathbf{b}_1, \mathbf{b}_2, \ldots, \mathbf{b}_n$ are linearly independent, we have $a_k - c_k = 0$, for $1 \leq k \leq n$, i.e., $a_k = c_k$. This proves that every vector of V is uniquely expressible as a linear combination of $\mathbf{b}_1, \mathbf{b}_2, \ldots, \mathbf{b}_n$.

Conversely, if every vector of V is represented by a unique linear combination of $\mathbf{b}_1, \mathbf{b}_2, \ldots, \mathbf{b}_n$, these vectors span V. In particular, the zero vector has the unique representation

$$\mathbf{0} = 0\mathbf{b}_1 + 0\mathbf{b}_2 + \cdots + 0\mathbf{b}_n$$

Since this representation of $\mathbf{0}$ is unique, if

$$\mathbf{0} = c_1\mathbf{b}_1 + c_2\mathbf{b}_2 + \cdots + c_n\mathbf{b}_n$$

then $c_k = 0$, for $1 \leq k \leq n$. Hence, $\mathbf{b}_1, \mathbf{b}_2, \ldots, \mathbf{b}_n$ are linearly independent. Since the set $\{\mathbf{b}_1, \mathbf{b}_2, \ldots, \mathbf{b}_n\}$ spans V and is linearly independent, it is a basis for V. Theorem 3 is now established.

Now, let us consider several examples involving the concepts of basis, dimension, and the unique representation of vectors by means of linear combinations.

EXAMPLE 1

Since $\begin{vmatrix} 0 & 1 & 1 \\ 1 & 0 & 1 \\ 1 & 1 & 0 \end{vmatrix} = 2$, the vectors $(0, 1, 1)$, $(1, 0, 1)$, and $(1, 1, 0)$ are a basis for R^3. Of course, $\{(1, 0, 0), (0, 1, 0), (0, 0, 1)\}$ is another basis. All bases of R^3 contain exactly three vectors; hence dim $R^3 = 3$. Every subset of R^3 containing four or more vectors is linearly dependent.

EXAMPLE 2

A linear space V is spanned by the vectors $\mathbf{u}_1 = (1, 2, 1, -2)$, $\mathbf{u}_2 = (2, 3, 2, -3)$, $\mathbf{u}_3 = (2, 5, 2, -5)$, $\mathbf{u}_4 = (3, 4, 3, -4)$, and $\mathbf{u}_5 = (3, 5, 3, -5)$. Find a basis for V and the dimension of V.

Since we are given five vectors of R^4, they must be linearly dependent. To find a linearly independent subset of $\{\mathbf{u}_1, \mathbf{u}_2, \mathbf{u}_3, \mathbf{u}_4, \mathbf{u}_5\}$ that spans V, we form the matrix having the given vectors as its respective rows and then apply elementary row operations, as indicated, to reduce the matrix to row echelon form:

$$\begin{bmatrix} 1 & 2 & 1 & -2 \\ 2 & 3 & 2 & -3 \\ 2 & 5 & 2 & -5 \\ 3 & 4 & 3 & -4 \\ 3 & 5 & 3 & -5 \end{bmatrix} = \begin{bmatrix} \mathbf{u}_1 \\ \mathbf{u}_2 \\ \mathbf{u}_3 \\ \mathbf{u}_4 \\ \mathbf{u}_5 \end{bmatrix} \qquad \begin{bmatrix} 1 & 2 & 1 & -2 \\ 0 & -1 & 0 & 1 \\ 0 & 2 & 0 & -2 \\ 0 & -2 & 0 & 2 \\ 0 & 1 & 0 & -1 \end{bmatrix} = \begin{bmatrix} \mathbf{u}_1 \\ \mathbf{u}_2 - 2\mathbf{u}_1 \\ \mathbf{u}_3 - \mathbf{u}_2 \\ \mathbf{u}_4 - 3\mathbf{u}_1 \\ \mathbf{u}_5 - \mathbf{u}_4 \end{bmatrix}$$

$$\begin{bmatrix} 1 & 0 & 1 & 0 \\ 0 & 1 & 0 & -1 \\ 0 & 0 & 0 & 0 \\ 0 & 0 & 0 & 0 \\ 0 & 0 & 0 & 0 \end{bmatrix} = \begin{bmatrix} -3\mathbf{u}_1 + 2\mathbf{u}_2 \\ 2\mathbf{u}_1 - \mathbf{u}_2 \\ \mathbf{u}_3 + \mathbf{u}_2 - 4\mathbf{u}_1 \\ \mathbf{u}_4 - 3\mathbf{u}_1 + \mathbf{u}_3 - \mathbf{u}_2 \\ \mathbf{u}_5 - \mathbf{u}_4 + \mathbf{u}_2 - 2\mathbf{u}_1 \end{bmatrix}$$

From the final equation it is clear that

$$\begin{aligned} \mathbf{u}_3 + \mathbf{u}_2 - 4\mathbf{u}_1 &= \mathbf{0} \\ \mathbf{u}_4 - 3\mathbf{u}_1 + \mathbf{u}_3 - \mathbf{u}_2 &= \mathbf{0} \\ \mathbf{u}_5 - \mathbf{u}_4 + \mathbf{u}_2 - 2\mathbf{u}_1 &= \mathbf{0} \end{aligned} \quad \text{and from these equations we find} \quad \begin{aligned} \mathbf{u}_3 &= 4\mathbf{u}_1 - \mathbf{u}_2 \\ \mathbf{u}_4 &= -\mathbf{u}_1 + 2\mathbf{u}_2 \\ \mathbf{u}_5 &= \mathbf{u}_1 + \mathbf{u}_2 \end{aligned}$$

Using these relations it is easy to show that every linear combination of the five given vectors is a linear combination of $\mathbf{u}_1$ and $\mathbf{u}_2$ alone. Hence $\mathbf{u}_1$ and $\mathbf{u}_2$ span V.

If $c_1\mathbf{u}_1 + c_2\mathbf{u}_2 = \mathbf{0}$, then (c_1, c_2) must be a solution of the linear system

$$\begin{aligned} c_1 + 2c_2 &= 0 \\ 2c_1 + 3c_2 &= 0 \\ c_1 + 2c_2 &= 0 \\ -2c_1 - 3c_2 &= 0 \end{aligned} \quad \text{which is equivalent to} \quad \begin{bmatrix} c_1 \\ c_2 \end{bmatrix} = \begin{bmatrix} 0 \\ 0 \end{bmatrix}$$

Thus, $\mathbf{u}_1$ and $\mathbf{u}_2$ are linearly independent vectors which span V. The set $\{\mathbf{u}_1, \mathbf{u}_2\}$ is therefore a basis for V and dim $V = 2$. The sets $\{\mathbf{u}_3, \mathbf{u}_4\}$, $\{\mathbf{u}_3, \mathbf{u}_5\}$, and $\{\mathbf{u}_4, \mathbf{u}_5\}$ are also bases of V (see Exercise 27, Sec. 13.2, or apply the definition of linear independence).

By extending the procedure of this example in an obvious way (Exercise 30), it is possible to find a basis of any space spanned by a nontrivial finite subset of R^n or C^n.

In many applications involving matrices, the notion of *rank* plays an important role.

DEFINITION 3 The **(row) rank of a matrix** A with scalar elements is the maximum number of linearly independent row vectors of A.

Insofar as rank is concerned, only nonzero matrices are of much importance.† So in general we shall assume that all of our matrices are of positive rank. The rank r of a matrix A, as we have just defined it, is obviously equal to the dimension of the linear space spanned by the row vectors of A. As the preceding example illustrates, r is also equal to the number of nonzero rows in the row echelon form of A (Exercise 29). Hence Theorems 1 and 5, Sec. 3.6, imply the following theorem.

THEOREM 4 Every complete solution of a consistent system of linear algebraic equations $A\mathbf{x} = \mathbf{b}$ in n unknowns contains $n - r$ arbitrary constants, where r is the rank of A.

A vector space which has no finite basis is said to be an **infinite dimensional** space. Many of the most important linear spaces which arise in practice are of this kind.

EXAMPLE 3

Show that the vector space P of all polynomials in x with real coefficients is infinite dimensional.

For every positive integer n,

$$\begin{vmatrix} 1 & x & \cdots & x^{n-1} \\ 0 & 1 & \cdots & (n-1)x^{n-2} \\ \cdots & \cdots & \cdots & \cdots \\ 0 & 0 & \cdots & (n-1)! \end{vmatrix} = 0!1!\cdots(n-1)! = \prod_{k=0}^{n-1} k! \neq 0$$

Hence by Corollary 1, Theorem 3, Sec. 13.2, $1, x, \ldots, x^{n-1}$ are linearly independent polynomials which span the subspace P_n of P **whose elements are the zero polynomial and all polynomials of degree less than** n. By definition, $\{1, x, \ldots, x^{n-1}\}$ is a basis for P_n and dim $P_n = n$. No matter how large n may be, there is a polynomial of degree $m > n$. Thus P has no finite basis. On the other hand, Theorem 3 guarantees that each polynomial p in P_n is represented by a unique linear combination

$$p(x) = p_0 + p_1 x + \cdots + p_n x^{n-1}$$

of $1, x, \ldots, x^{n-1}$.

Of course, no vector space which has P as a subspace can have a finite basis. So obviously the space $\mathscr{C}(-\infty, \infty)$ of all real-valued functions continuous on $(-\infty, \infty)$ and the space $F(-\infty, \infty)$ of all real-valued functions on $(-\infty, \infty)$ are both infinite dimensional spaces.

EXAMPLE 4

As we consider spaces R^n and imagine n increasing without bound, we are led inevitably to the idea of an infinite dimensional space R^∞. This space contains all vectors $\mathbf{v} = (v_1, v_2, v_3, \ldots)$ which have an infinite sequence of real components. The natural extension of the inner product of two vectors $\mathbf{u} = (u_1, u_2, u_3, \ldots)$ and $\mathbf{v} = (v_1, v_2, v_3, \ldots)$ of R^∞, denoted by $(\mathbf{u}, \mathbf{v})$ and defined by

(2) $$(\mathbf{u}, \mathbf{v}) = \sum_{j=1}^{\infty} u_j v_j$$

fails to exist for many pairs of vectors of R^∞ (Exercise 34) because the infinite series (2) diverges. There are also many vectors of R^∞ for which the series

(3) $$(\mathbf{v}, \mathbf{v}) = \sum_{j=1}^{\infty} v_j^2$$

diverges (Exercise 35). For each such vector, the familiar formula $\|\mathbf{v}\| = (\mathbf{v}, \mathbf{v})^{1/2}$ fails to define a finite length, or norm, for that vector. Excluding these troublesome vectors and keeping all those for which the series (3) is convergent still leaves an infinite dimensional vector space known as a **Hilbert space.**‡

†Since a zero matrix has no linearly independent row vectors, the rank of every zero matrix is 0.
‡Named for David Hilbert (1862–1943), a famous German mathematician.

We leave the study of Hilbert and other abstract spaces to books devoted to their study and to texts on functional analysis. Our next example deals with vectors and bases of a finite dimensional vector space R^n.

EXAMPLE 5

According to Theorem 3, each vector $\mathbf{x} = (x_1, x_2, \ldots, x_n)$ of R^n is represented by a unique linear combination

$$\mathbf{x} = x_1\mathbf{e}_1 + x_2\mathbf{e}_2 + \cdots + x_n\mathbf{e}_n \tag{4}$$

of the basis vectors $\mathbf{e}_k = (\delta_{k1}, \delta_{k2}, \ldots, \delta_{kn})$, $1 \le k \le n$. The orthonormal basis $\{\mathbf{e}_1, \mathbf{e}_2, \ldots, \mathbf{e}_n\}$ is called the **standard basis** of R^n, and the coefficient x_k of $\mathbf{e}_k$ in (4) is said to be the kth **standard component** of $\mathbf{x}$. Whenever a vector $\mathbf{x}$ of R^n is specified by enclosing its successive components within parentheses, the components are to be interpreted as the standard components of $\mathbf{x}$. For instance, the standard basis vectors of R^4 are $\mathbf{e}_1 = (1, 0, 0, 0)$, $\mathbf{e}_2 = (0, 1, 0, 0)$, $\mathbf{e}_3 = (0, 0, 1, 0)$, and $\mathbf{e}_4 = (0, 0, 0, 1)$. In terms of these vectors, the four-dimensional vector $\mathbf{v} = (2, -3, 4, -5)$ is given by *the* linear combination

$$\mathbf{v} = 2\mathbf{e}_1 - 3\mathbf{e}_2 + 4\mathbf{e}_3 - 5\mathbf{e}_4$$

Each vector $\mathbf{x}$ of R^n is also represented by a unique linear combination

$$\mathbf{x} = a_1\mathbf{b}_1 + a_2\mathbf{b}_2 + \cdots + a_n\mathbf{b}_n \tag{5}$$

of any other basis $\mathbf{b}_1, \mathbf{b}_2, \ldots, \mathbf{b}_n$ of R^n. The vectors $\mathbf{b}_1 = (0, 0, 0, 1)$, $\mathbf{b}_2 = (0, 0, 1, 1)$, $\mathbf{b}_3 = (0, 1, 1, 1)$, and $\mathbf{b}_4 = (1, 1, 1, 1)$, for example, are basis vectors of R^4 because the determinant with these vectors as its rows is nonzero. We find that *the* linear combination of these vectors which represents $\mathbf{v}$ is

$$\mathbf{v} = -9\mathbf{b}_1 + 7\mathbf{b}_2 - 5\mathbf{b}_3 + 2\mathbf{b}_4$$

The standard components of $\mathbf{v}$ are *not* equal to the respective coefficients of $\mathbf{b}_1$, $\mathbf{b}_2$, $\mathbf{b}_3$, and $\mathbf{b}_4$ in this linear combination. This does not imply, however, that there are no vectors in R^n for which the coefficients in (4) and (5) are equal (Exercise 20). But it can be shown that if, for every $\mathbf{x}$ in R^n,

$$x_1\mathbf{e}_1 + x_2\mathbf{e}_2 + \cdots + x_n\mathbf{e}_n = x_1\mathbf{b}_1 + x_2\mathbf{b}_2 + \cdots + x_n\mathbf{b}_n$$

then $\mathbf{b}_k = \mathbf{e}_k$ for $1 \le k \le n$ (Exercise 25).

Occasionally it is convenient to work with a nonstandard basis. The coefficients $a_1, a_2, \ldots, a_n$ of (5) are then called **components of x relative to the basis** $\{\mathbf{b}_1, \mathbf{b}_2, \ldots, \mathbf{b}_n\}$. To designate $\mathbf{x}$ by means of these components we enclose them within a pair of special grouping symbols; thus $\langle a_1, a_2, \ldots, a_n\rangle$. For example, $\langle -9, 7, -5, 2\rangle$ designates the vector $\mathbf{v}$ and displays its components relative to the basis (0, 0, 0, 1), (0, 0, 1, 1), (0, 1, 1, 1), and (1, 1, 1, 1), whereas $(2, -3, 4, -5)$ displays the standard components of $\mathbf{v}$.

Vectors of R^n are by no means the only vectors that have components. Consider a general but finite dimensional space V having $\mathbf{b}_1, \mathbf{b}_2, \ldots, \mathbf{b}_n$ as basis vectors. Each vector $\mathbf{v}$ in V is represented by a unique linear combination

$$\mathbf{v} = c_1\mathbf{b}_1 + c_2\mathbf{b}_2 + \cdots + c_n\mathbf{b}_n \tag{6}$$

of these basis vectors. The scalars $c_1, c_2, \ldots, c_n$ are called the **components,** or **coordinates,** of $\mathbf{v}$ relative to the basis $\{\mathbf{b}_1, \mathbf{b}_2, \ldots, \mathbf{b}_n\}$. The basis itself is said to form a **coordinate system** for V, and each subspace of V spanned by a single basis vector $\mathbf{b}_k$, $1 \le k \le n$, is called a **coordinate axis** of that coordinate system.

EXERCISES

1. Find all values of k for which the set of vectors $\{(3, 1, 2), (-2, k, 5), (19k, 18, 19k)\}$ is *not* a basis of R^3.

2. Find all values of θ for which the set of vectors $\{(2, 3, 0, 2), (1, 1, \sin\theta, 1), (1, 2, 0, 1), (1, 2, \sin\theta, 2)\}$ is *not* a basis of R^4.

3. Is $\{(1, 0, 1, 0), (0, 1, 0, -1)\}$ a basis for the linear space V of Example 2? Explain.

4. If a is a real number, is $\{1, x - a, (x - a)^2, \ldots, (x - a)^{n-1}\}$ a basis for the linear space P_n of Example 3? Explain.

5. Find a basis for, and the dimension of, C^n.

6. Determine whether or not the given set of vectors is a basis of the indicated space.

(a) $\{(2, -1, 1), (2, 0, 3), (1, 1, -2)\}$; R^3

(b) $\{(9, 0, 13), (2, 4, 3), (-1, 16, -1)\}$; R^3

(c) $\{(2, 5, 3, 3), (1, 1, 0, 1), (1, 4, 3, 2), (-1, 2, 3, 0)\}$; R^4

(d) $\{(3, 5, 1, -1), (-2, 3, 1, 2), (1, 3, 0, 4), \mathbf{u}\}$; R^4 where $\mathbf{u} = (2, -1, 4, 1)$

(e) $\{(5, -5, 2, 4), (-4, 3, 0, -4), (3, -3, 1, 2), \mathbf{v}\}$; R^4 where $\mathbf{v} = (2, -2, 1, 2)$

(f) $\{3x - 2, x, x + 1\}$; P_2

(g) $\{x^2 - 1, x^2 + x + 1, x^2\}$; P_3

(h) $\{x^2, x^2 + x, x^2 - x\}$; P_3

(i) $\{(i, 1 - i, 2 + i), (2 - i, i, -1 - i), \mathbf{w}\}$; C^3 where $\mathbf{w} = (2 + i, 2 - i, 3 + i)$

(j) $\left\{\begin{bmatrix} 1 & 0 \\ 0 & 1 \end{bmatrix}, \begin{bmatrix} 0 & 1 \\ 1 & 0 \end{bmatrix}\right\}$; all matrices of order 2

(k) $\left\{\begin{bmatrix} 0 & 1 \\ 0 & 0 \end{bmatrix}, \begin{bmatrix} 1 & 0 \\ 0 & 0 \end{bmatrix}, \begin{bmatrix} 0 & 0 \\ 0 & 1 \end{bmatrix}\right\}$; all upper triangular matrices of order 2

(l) $\{(1, 0, 3), (0, 1, -6)\}$; the linear space spanned by all solutions of the equation

$$3x - 6y - z = 0$$

7. Find a basis, and the dimension, of the vector space V spanned by the given vectors.

(a) $(3, -1, 2), (7, 4, -2), (8, 1, 5)$

(b) $(4, -1, 7), (3, 3, 4), (2, 1, 3), (1, -1, 2)$

(c) $(3, 9, 3, 5), (4, 12, 4, 5), (2, 6, 1, 0), (5, 15, 3, 2)$

(d) $(7, -7, -14, 6), (5, -5, -10, 4), (2, -2, -4, 1), (3, -3, -6, 2), (9, -9, -18, 7)$

(e) $1 - x, 3 + x^2, 2 - x + x^2$

(f) $1 - \frac{1}{2}x^2, 1 + 3x + \frac{9}{2}x^2, 1 + \frac{3}{2}x + 2x^2$

(g) $x^2 - 4, x - 1, 2x + 3$

(h) $1 - x^3, 2x + x^3, x - 1, x^2 + x$

(i) $\begin{bmatrix} -1 & 1 \\ 2 & 0 \end{bmatrix}, \begin{bmatrix} 0 & 1 \\ -1 & 1 \end{bmatrix}, \begin{bmatrix} -2 & 0 \\ 1 & -1 \end{bmatrix}, \begin{bmatrix} 1 & 1 \\ 1 & 1 \end{bmatrix}$

(j) $\begin{bmatrix} 1 & 0 \\ 0 & 0 \end{bmatrix}, \begin{bmatrix} 0 & 1 \\ 0 & 0 \end{bmatrix}, \begin{bmatrix} 0 & 0 \\ 1 & 0 \end{bmatrix}, \begin{bmatrix} 0 & 0 \\ 0 & 1 \end{bmatrix}$

(k) $1, \cos^2 x, \sin^2 x, \cos 2x$

(l) $1, \cosh^2 x, \sinh^2 x, \cosh 2x, e^{2x}, e^{-2x}$

8. Find a basis, and the dimension, of the vector space spanned by all solutions of the given equation(s).

(a) $3x_1 - 2x_2 + 5x_3 = 0$

(b)
$$\begin{aligned} 4x_1 - 12x_2 + 3x_3 &= 0 \\ 3x_1 - 9x_2 + 2x_3 &= 0 \\ 2x_1 - 6x_2 + x_3 &= 0 \end{aligned}$$

(c) $y''' - 6y'' + 11y' - 6y = 0$

(d) $y^{\text{iv}} - y = 0$; $y(0) = y'(0) = 0$

9. Show that if U is a subspace of a finite dimensional vector space V, then $\dim U \leq \dim V$. If $\dim U = \dim V$, show that $U = V$.

10. Show that the linear space spanned by the functions e^x, $e^{2x}, \ldots, e^{nx}$ has dimension n.

11. Show that the linear space spanned by the infinite sequence of functions $e^{r_1 x}, e^{r_2 x}, e^{r_3 x}, \ldots,$ where $r_1, r_2, r_3, \ldots,$ are distinct real numbers, is infinite dimensional.

12. Is the space of all real-valued functions defined on $[0, 1]$ infinite dimensional?

13. Is the linear space spanned by the infinite sequence of functions $\sin x, \sin 2x, \sin 3x, \ldots,$ finite dimensional?

14. If θ is a fixed real number and if $0 \leq \theta < 2\pi$, is $\{(\cos\theta, \sin\theta), (-\sin\theta, \cos\theta)\}$ a basis for R^2?

15. Show that the standard basis of R^n also serves as a basis for C^n.

16. Find a basis, and the dimension, of the vector space of all matrices of order 2.

17. Find a basis, and the dimension, of the vector space of all matrices of order 3.

18. Verify that a basis for R^3 is given by $(1, 0, 2)$, $(3, 0, 1)$, and $(1, 2, 3)$. Express each of the following vectors as a linear combination of these basis vectors.

(a) $(1, 0, 0)$ **(b)** $(0, 1, 0)$ **(c)** $(0, 0, 1)$
(d) $(1, -2, -4)$ **(e)** $(3, -2, 0)$ **(f)** $(-5, 2, 1)$

19. Verify that $2 - x^2$, $x^3 - x$, $2 - 3x^2$, and $3 - x^3$ form a basis for P_4. Express each of the following polynomials as a linear combination of this basis.

(a) x^2 **(b)** $1 + x$ **(c)** x^3
(d) $x + x^2$

20. Verify that the vector $(1, 2)$ has coordinates $\langle 1, 2\rangle$ relative to the basis $\{(-3, -2), (2, 2)\}$ of R^2.

21. Find the coordinates of the vector $(2, -3, 4, -5)$ relative to the basis $(0, 1, 0, -1)$, $(1, 0, -1, 1)$, $(0, 0, -1, 0)$, $(-1, 1, -1, 0)$ of R^4.

22. Find coordinates of the following vectors relative to the basis $(3, -2, 1)$, $(1, -2, 3)$, $(1, 0, 1)$ of R^3.

(a) $(4, 4, 4)$ **(b)** $(0, 4, 0)$ **(c)** $(1, 1, 0)$
(d) $(0, 1, 1)$ **(e)** $(2, -2, -4)$ **(f)** $(20, 0, 0)$

23. The components of the following vectors are given relative to the basis of Exercise 22. Find the standard components of each vector.

(a) $\langle 2, -1, 0\rangle$ **(b)** $\langle 1, 0, 1\rangle$ **(c)** $\langle 1, 1, 1\rangle$
(d) $\langle 0, 0, 0\rangle$ **(e)** $\langle 1, 2, 3\rangle$ **(f)** $\langle 1, -1, 1\rangle$

24. Find two different bases for P_3 such that the coordinates of the polynomial $1 - x + x^2$ relative to either basis will be the same as the coordinates of the polynomial relative to the basis $\{1, x, x^2\}$.

25. Let $\{\mathbf{b}_1, \mathbf{b}_2, \ldots, \mathbf{b}_n\}$ be a basis for R^n. Show that if, for every vector $\mathbf{x}$ in R^n,

$$x_1\mathbf{e}_1 + x_2\mathbf{e}_2 + \cdots + x_n\mathbf{e}_n = x_1\mathbf{b}_1 + x_2\mathbf{b}_2 + \cdots + x_n\mathbf{b}_n$$

then $\mathbf{b}_k = \mathbf{e}_k$ for $1 \le k \le n$.

26. Let $\mathbf{b}_1, \mathbf{b}_2, \ldots, \mathbf{b}_m$ be linearly independent vectors of an n-dimensional vector space V, where $m < n$. Show that there are $n - m$ vectors $\mathbf{u}_{m+1}, \mathbf{u}_{m+2}, \ldots, \mathbf{u}_n$ in V such that $\{\mathbf{b}_1, \mathbf{b}_2, \ldots, \mathbf{b}_m, \mathbf{u}_{m+1}, \ldots, \mathbf{u}_n\}$ is a basis for V.

27. Construct a basis for R^4 having $(2, 0, -6, 1)$ and $(3, -1, 0, 6)$ as two of its elements.

28. Construct a basis for P_4 having $1 - x^3$ and $2x + 3x^2$ as two of its elements.

29. Prove that a set of row vectors $\{\mathbf{u}_1, \mathbf{u}_2, \ldots, \mathbf{u}_m\}$ of C^n spans a subspace U of C^n of dimension r if and only if r is the number of 1's appearing as leading *nonzero* elements in the rows of the row echelon form of the matrix $\mathrm{A} = \begin{bmatrix} \mathbf{u}_1 \\ \vdots \\ \mathbf{u}_m \end{bmatrix}$.

30. Show that the set of nonzero row vectors in the row echelon form of a matrix A is a basis for the vector space spanned by the row vectors of A.

31. Using Definition 3, find the rank of the matrix $\mathrm{A} = \begin{bmatrix} 1 & 2 & -1 & 3 \\ 3 & 4 & 0 & -1 \\ -1 & 0 & -2 & 7 \end{bmatrix}$. Also find the dimension of the subspace U of R^4 and of the subspace V of R^3, spanned by the row and column vectors of A, respectively.

32. If the rows of a matrix are linearly dependent, are the columns necessarily linearly dependent?

33. Identify a basis of R^∞.

For all positive integers j, the jth component of vectors $\mathbf{u}$, $\mathbf{v}$, and $\mathbf{w}$ of R^∞ are $u_j = 1/j$, $v_j = 1/\sqrt{j}$, and $w_j = 1/\sqrt[3]{j}$, respectively.

34. Determine which of the inner products $(\mathbf{u}, \mathbf{v})$, $(\mathbf{u}, \mathbf{w})$ and $(c\mathbf{v}, \mathbf{w})$, $c \neq 0$, are defined by (2).

35. Which, if any, of the vectors $\mathbf{u}$, $\mathbf{v}$, and $\mathbf{w}$ belong to the Hilbert space of Example 4?

13.4 ANGLES, PROJECTIONS, AND THE GRAM-SCHMIDT ORTHOGONALIZATION PROCESS

We have repeatedly used inner products to find lengths of vectors and to identify orthogonal vectors. In calculus, we also used inner products of two- and three-dimensional vectors to find angles between lines or planes and to find projections of vectors onto lines or planes. In spaces R^n of dimension $n > 3$, no significant change in the use of inner products is required to solve similar problems involving angles and projections. To show that this is so, we begin with the premise that three distinct points not all on the same line, say the vectors

$$\mathbf{u} = (u_1, u_2, \ldots, u_n) \qquad \mathbf{v} = (v_1, v_2, \ldots, v_n) \qquad \mathbf{0}_n = (0, 0, \ldots, 0)$$

regarded as geometric points of R^n, determine a plane and a triangle Δ in that plane.

Picturing the nonzero vectors $\mathbf{u}$ and $\mathbf{v}$ as geometric arrows emanating from $\mathbf{0}_n$, we recognize $\mathbf{u}$, $\mathbf{v}$, and $\mathbf{v} - \mathbf{u}$ as sides of Δ. The lengths of the sides of Δ which form the angle θ with vertex $\mathbf{0}_n$ are $\|\mathbf{u}\|$ and $\|\mathbf{v}\|$. The side opposite θ has length $\|\mathbf{v} - \mathbf{u}\|$. Applying the law of cosines to Δ, we obtain

$$(1) \qquad \|\mathbf{v} - \mathbf{u}\|^2 = \|\mathbf{u}\|^2 + \|\mathbf{v}\|^2 - 2\|\mathbf{u}\|\,\|\mathbf{v}\| \cos\theta$$

or

$$(\mathbf{v} - \mathbf{u}) \cdot (\mathbf{v} - \mathbf{u}) = \mathbf{u} \cdot \mathbf{u} + \mathbf{v} \cdot \mathbf{v} - 2\|\mathbf{u}\|\,\|\mathbf{v}\| \cos\theta$$

Expanding the scalar product in the left-hand member of the last equation, then simplifying, we get

$$(2) \qquad \mathbf{u} \cdot \mathbf{v} = \|\mathbf{u}\|\,\|\mathbf{v}\| \cos\theta$$

If θ is a right angle, (1) becomes the pythagorean theorem† and (2) reduces to the orthogonality condition $\mathbf{u} \cdot \mathbf{v} = 0$. If $\mathbf{v} = c\mathbf{u}$, $c \neq 0$, the points $\mathbf{0}_n$, $\mathbf{u}$, and $\mathbf{v}$ are colinear and our derivation breaks down; nevertheless, (2) still holds with $\theta = 0$ or $\theta = \pi$ according as $c > 0$ or $c < 0$ (Exercise 9).

†Pythagoras of Samos (c. 580–c. 500 B.C., Greek) is credited with having introduced *proof* into mathematics.

Unless **u** and **v** are linearly dependent, their *geometric arrows* originating at $\mathbf{0}_n$ form two angles with vertex $\mathbf{0}_n$. If the measure of the smaller angle is θ, the other angle measures $2\pi - \theta$.

DEFINITION 1 **The angle between two nonzero vectors u and v of** R^n, $n > 1$, is the *smaller* angle of intersection of their representative arrows originating at $\mathbf{0}_n$ if **u** and **v** are linearly independent, but is 0 or π according as $c > 0$ or $c < 0$ if $\mathbf{v} = c\mathbf{u}$.

An angle between two vectors must belong to the interval $0 \leqq \theta \leqq \pi$ and is an **undirected angle,** i.e., an angle that is *not* assigned a direction, having neither an initial nor a terminal side.

Returning to (2) and solving for $\cos\theta$, we obtain

THEOREM 1 The cosine of the angle θ between any two nonzero vectors **u** and **v** of a vector space R^n, $n > 1$, is given by

$$\cos\theta = \frac{\mathbf{u}\cdot\mathbf{v}}{\|\mathbf{u}\|\,\|\mathbf{v}\|} \tag{3}$$

With (3) available, we are now able to compute not only lengths and distances, but also angles, by means of inner products, when working with vectors of R^n.

The orthogonal projection of a point P **onto a line** l is either P or else the foot of the perpendicular from P to l, according as P is or is not on l. Extending this concept to vectors, we say that **the orthogonal projection of a vector onto a line** l is a vector whose initial and terminal points are the projections onto l of the initial and terminal points, respectively, of the original vector. Assuming $n > 1$ and that **u** and **v** are linearly independent vectors of R^n, the subspace of R^n spanned by **u** is a line l through $\mathbf{0}_n$ in the plane of **u** and **v**. We now proceed to find a formula for the orthogonal projection of **v** onto l.

With **v** visualized as an arrow originating at $\mathbf{0}_n$, its orthogonal projection **p** onto l has the same initial point because l passes through $\mathbf{0}_n$. Since **u** spans l and **p** is directed along l, a real number c exists such that $\mathbf{p} = c\mathbf{u}$. From a geometric sketch of **v**, **p**, and the vector $\mathbf{v} - \mathbf{p}$ coincident with the perpendicular from l to the tip of **v**, it is evident that $\mathbf{v} - c\mathbf{u}$ and **u** are orthogonal or, more generally, that *the vector obtained by subtracting from* **v** *its orthogonal projection onto the line* l *spanned by* **u** *is a vector perpendicular to* **u**. Hence

$$(\mathbf{v} - c\mathbf{u})\cdot\mathbf{u} = 0 \qquad \text{or} \qquad \mathbf{v}\cdot\mathbf{u} - c\mathbf{u}\cdot\mathbf{u} = 0 \qquad \text{or} \qquad c = \frac{\mathbf{v}\cdot\mathbf{u}}{\|\mathbf{u}\|^2}$$

With c thus determined, the vector $\mathbf{p} = c\mathbf{u}$ yields the vector defined by

$$\mathbf{p} = \frac{\mathbf{v}\cdot\mathbf{u}}{\|\mathbf{u}\|^2}\mathbf{u} \tag{4}$$

as **the orthogonal projection of v onto the subspace** l **of** R^n **spanned by u.** This vector is also called **the vector projection of v in the direction of,** or **along the line spanned by, u.** A simpler name for it is **the u component of v.** Using (2) to substitute for $\mathbf{v}\cdot\mathbf{u}$ in (4), we get

$$\mathbf{p} = \frac{\|\mathbf{v}\|\,\|\mathbf{u}\|\cos\theta}{\|\mathbf{u}\|^2}\mathbf{u} = \|\mathbf{v}\|\cos\theta\frac{\mathbf{u}}{\|\mathbf{u}\|} \qquad \text{where } \theta \text{ is the angle between } \mathbf{u} \text{ and } \mathbf{v}.$$

The final term here expresses **p** as a unit vector $\mathbf{u}/\|\mathbf{u}\|$ in the direction of **u** multiplied by the scalar quantity

(5) $\|\mathbf{v}\|\cos\theta$ called **the scalar projection of v in the direction of,** or **along, u.**

As a rule, an orthonormal basis of a subspace V of either R^n or C^n is much easier to work with than a basis whose members are merely linearly independent. If dim $V = m$ and $m > 1$, then by a procedure known as the **Gram-Schmidt† orthogonalization process,** it is always possible to find linear combinations of m linearly independent vectors in V which determine an orthogonal basis of V. Subsequent normalization of these orthogonal vectors yields an orthonormal basis. The process can be extended in a straightforward way to general **inner product,** or **pre-Hilbert, spaces,** i.e., to linear spaces for which an inner product is defined. However, we shall center our attention on vectors of R^n. We continue to use the notation $\|\mathbf{v}\|$ to denote the *norm,* or length, of a vector $\mathbf{v}$ in R^n.

The idea behind the general process is easy to grasp when successive steps of the procedure are developed by visualizing them in the familiar setting of three-dimensional space.

EXAMPLE 1

ORTHONORMAL BASES

Find an orthonormal basis for the subspace V of R^3 spanned by the vectors $\mathbf{v}_1 = (2, 4, -4)$ and $\mathbf{v}_2 = (-3, 6, 0)$. Also find an orthonormal basis for R^3, other than the standard basis, by using the fact that $\mathbf{v}_1$, $\mathbf{v}_2$, and $\mathbf{v}_3 = (7, 2, 1)$ are linearly independent.

The vectors $\mathbf{v}_1$ and $\mathbf{v}_2$ are drawn in Fig. 13.3. Clearly, dim $V = 2$. To find an orthogonal basis $\{\mathbf{u}_1, \mathbf{u}_2\}$ for V, we first set

$$\mathbf{u}_1 = \mathbf{v}_1 = (2, 4, -4)$$

FIGURE 13.3
Orthogonal vectors $\mathbf{u}_1$, $\mathbf{u}_2$, $\mathbf{u}_3$ derived from the vectors $\mathbf{v}_1$, $\mathbf{v}_2$, $\mathbf{v}_3$ by the Gram-Schmidt orthogonalization process.

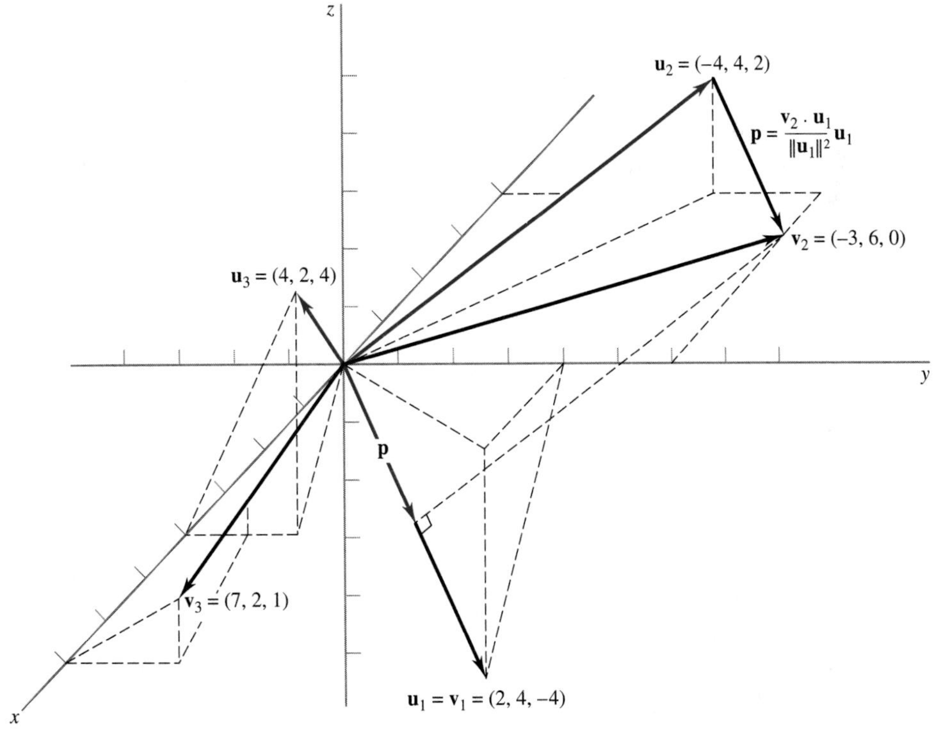

†Named jointly for the Danish actuary Jørgen Pedersen Gram (1850–1916) and the German mathematician Erhardt Schmidt (1876–1959).

Then we project $\mathbf{v}_2$ onto $\mathbf{u}_1$. This *vector projection* is the vector $\mathbf{p}$ lying along $\mathbf{u}_1$ in Fig. 13.3. Substituting $\mathbf{v}_2$ for $\mathbf{v}$ and $\mathbf{u}_1$ for each $\mathbf{u}$ in Eq. (4), we find that the vector projection $\mathbf{p}$ of $\mathbf{v}_2$ in the direction of $\mathbf{u}_1$ is given by

$$\mathbf{p} = \frac{\mathbf{v}_2 \cdot \mathbf{u}_1}{\|\mathbf{u}_1\|^2}\mathbf{u}_1$$

Subtracting $\mathbf{p}$ from $\mathbf{v}_2$, we obtain the vector

$$\mathbf{u}_2 = \mathbf{v}_2 - \frac{\mathbf{v}_2 \cdot \mathbf{u}_1}{\|\mathbf{u}_1\|^2}\mathbf{u}_1 = (-3, 6, 0) - \frac{(-3, 6, 0) \cdot (2, 4, -4)}{(2, 4, -4) \cdot (2, 4, -4)}(2, 4, -4) = (-4, 4, 2)$$

Obviously, every linear combination of $\mathbf{u}_1$ and $\mathbf{u}_2$ is a linear combination of $\mathbf{v}_1$ and $\mathbf{v}_2$, and vice versa. Moreover, $\mathbf{u}_1 \cdot \mathbf{u}_2 = (2, 4, -4) \cdot (-4, 4, 2) = -8 + 16 - 8 = 0$; hence $\{\mathbf{u}_1, \mathbf{u}_2\}$ is an orthogonal basis for V. Normalizing $\mathbf{u}_1$ and $\mathbf{u}_2$, we obtain the orthonormal basis $\mathbf{b}_1 = \frac{1}{3}(1, 2, -2)$, $\mathbf{b}_2 = \frac{1}{3}(-2, 2, 1)$ of V.

Continuing the process, we next project $\mathbf{v}_3$ onto $\mathbf{u}_1$ and $\mathbf{u}_2$ and then subtract these vector projections from $\mathbf{v}_3$, thus removing from $\mathbf{v}_3$ its components along $\mathbf{u}_1$ and $\mathbf{u}_2$. This gives

$$\begin{aligned}\mathbf{u}_3 &= \mathbf{v}_3 - \frac{\mathbf{v}_3 \cdot \mathbf{u}_1}{\|\mathbf{u}_1\|^2}\mathbf{u}_1 - \frac{\mathbf{v}_3 \cdot \mathbf{u}_2}{\|\mathbf{u}_2\|^2}\mathbf{u}_2 \\ &= (7, 2, 1) - \frac{(7, 2, 1) \cdot (2, 4, -4)}{36}(2, 4, -4) - \frac{(7, 2, 1) \cdot (-4, 4, 2)}{36}(-4, 4, 2) = (4, 2, 4)\end{aligned}$$

It is now a simple matter to check that $\mathbf{u}_1$, $\mathbf{u}_2$, and $\mathbf{u}_3$ are mutually orthogonal and thus form a basis for R^3. Normalizing $\mathbf{u}_3$, we obtain the unit vector $\mathbf{b}_3 = \frac{1}{3}(2, 1, 2)$ which, together with $\mathbf{b}_1$ and $\mathbf{b}_2$, gives us an orthonormal basis for R^3.

Had we begun the orthogonalization process by taking $\mathbf{u}_1 = \mathbf{v}_2 = (-3, 6, 0)$, we would have found $\mathbf{u}_2$ to be $(\frac{16}{5}, \frac{8}{5}, -4)$. Normalization of $\mathbf{u}_1$ and $\mathbf{u}_2$ as thus determined gives a second orthonormal basis for V. Of course, by continuing the process, another orthonormal basis of R^3 can be found.

The Gram-Schmidt orthogonalization process is by no means limited to the problem of finding orthogonal bases for subspaces of R^3.

THEOREM 2 A subspace V of R^n spanned by m (>1) linearly independent vectors $\mathbf{v}_1, \mathbf{v}_2, \ldots, \mathbf{v}_m$ has an orthogonal basis $\{\mathbf{u}_1, \mathbf{u}_2, \ldots, \mathbf{u}_m\}$, where

$$\begin{aligned}\mathbf{u}_1 &= \mathbf{v}_1 \\ \mathbf{u}_2 &= \mathbf{v}_2 - \frac{\mathbf{v}_2 \cdot \mathbf{u}_1}{\|\mathbf{u}_1\|^2}\mathbf{u}_1 \\ \mathbf{u}_3 &= \mathbf{v}_3 - \frac{\mathbf{v}_3 \cdot \mathbf{u}_1}{\|\mathbf{u}_1\|^2}\mathbf{u}_1 - \frac{\mathbf{v}_3 \cdot \mathbf{u}_2}{\|\mathbf{u}_2\|^2}\mathbf{u}_2 \\ &\cdots\cdots\cdots \\ \mathbf{u}_m &= \mathbf{v}_m - \frac{\mathbf{v}_m \cdot \mathbf{u}_1}{\|\mathbf{u}_1\|^2}\mathbf{u}_1 - \frac{\mathbf{v}_m \cdot \mathbf{u}_2}{\|\mathbf{u}_2\|^2}\mathbf{u}_2 - \cdots - \frac{\mathbf{v}_m \cdot \mathbf{u}_{m-1}}{\|\mathbf{u}_{m-1}\|^2}\mathbf{u}_{m-1}\end{aligned} \tag{6}$$

PROOF As in Example 1, we set $\mathbf{u}_1 = \mathbf{v}_1$. Since the vectors $\mathbf{v}_1, \mathbf{v}_2, \ldots, \mathbf{v}_m$ are linearly independent, none of them is the zero vector of R^n. Hence $\|\mathbf{u}_1\| \neq 0$ and $\mathbf{u}_2$, as given by Formulas (6), is defined. Clearly, $\mathbf{u}_1$ and $\mathbf{u}_2$ are linear combinations of $\mathbf{v}_1$ and $\mathbf{v}_2$, and conversely. Moreover,

$$\mathbf{u}_2 \cdot \mathbf{u}_1 = \mathbf{v}_2 \cdot \mathbf{u}_1 - \frac{\mathbf{v}_2 \cdot \mathbf{u}_1}{\|\mathbf{u}_1\|^2}\|\mathbf{u}_1\|^2 = 0$$

Therefore, $\{\mathbf{u}_1, \mathbf{u}_2\}$ is an orthogonal basis of the linear space spanned by $\{\mathbf{v}_1, \mathbf{v}_2\}$. Of course, $\|\mathbf{u}_2\| \neq 0$ because $\mathbf{u}_1 = \mathbf{v}_1$, and $\mathbf{v}_2$ is not a linear combination of $\mathbf{v}_1$.

If $m > 2$, $\mathbf{u}_3$ is defined by Formulas (6). We see at once that $\mathbf{u}_1$, $\mathbf{u}_2$, and $\mathbf{u}_3$ are linear combinations of $\mathbf{v}_1$, $\mathbf{v}_2$, and $\mathbf{v}_3$, and vice versa. Since $\mathbf{u}_2 \cdot \mathbf{u}_1 = 0$, we have

$$\mathbf{u}_3 \cdot \mathbf{u}_1 = \mathbf{v}_3 \cdot \mathbf{u}_1 - \frac{\mathbf{v}_3 \cdot \mathbf{u}_1}{\|\mathbf{u}_1\|^2}\|\mathbf{u}_1\|^2 - \frac{\mathbf{v}_3 \cdot \mathbf{u}_2}{\|\mathbf{u}_2\|^2}\mathbf{u}_2 \cdot \mathbf{u}_1 = 0$$

and

$$\mathbf{u}_3 \cdot \mathbf{u}_2 = \mathbf{v}_3 \cdot \mathbf{u}_2 - \frac{\mathbf{v}_3 \cdot \mathbf{u}_1}{\|\mathbf{u}_1\|^2}\mathbf{u}_1 \cdot \mathbf{u}_2 - \frac{\mathbf{v}_3 \cdot \mathbf{u}_2}{\|\mathbf{u}_2\|^2}\|\mathbf{u}_2\|^2 = 0$$

It follows that $\{\mathbf{u}_1, \mathbf{u}_2, \mathbf{u}_3\}$ is an orthogonal basis of the linear space spanned by $\{\mathbf{v}_1, \mathbf{v}_2, \mathbf{v}_3\}$ and of course $\|\mathbf{u}_3\| \neq 0$ because $\mathbf{v}_3$ is not a linear combination of $\mathbf{v}_1$ and $\mathbf{v}_2$.

If $m > 3$, by the same kind of reasoning, we find that $\mathbf{u}_4, \mathbf{u}_5, \ldots, \mathbf{u}_m$, as determined by Formulas (6), are indeed defined, and that $\mathbf{u}_1, \mathbf{u}_2, \ldots, \mathbf{u}_m$ are linear combinations of $\mathbf{v}_1, \mathbf{v}_2, \ldots, \mathbf{v}_m$, and conversely. We also find that $\mathbf{u}_i \cdot \mathbf{u}_j = 0$ if $j < i$ and $i = 2, 3, \ldots, m$. This gives us $\{\mathbf{u}_1, \mathbf{u}_2, \ldots, \mathbf{u}_m\}$ as an orthogonal basis of the subspace of R^n spanned by $\{\mathbf{v}_1, \mathbf{v}_2, \ldots, \mathbf{v}_m\}$, as asserted by the theorem. ◗

In dealing with inner product spaces other than R^n, it is sometimes more convenient to develop the Gram-Schmidt orthogonalization process along the lines suggested in Exercise 18.

Now, let us reexamine the problem of expanding an arbitrary vector $\mathbf{v}$ of an n-dimensional inner product space V as a linear combination

$$\mathbf{v} = \sum_{j=1}^{n} c_j\mathbf{b}_j = c_1\mathbf{b}_1 + c_2\mathbf{b}_2 + \cdots + c_n\mathbf{b}_n \tag{7}$$

of an orthogonal set of basis vectors $\{\mathbf{b}_1, \mathbf{b}_2, \ldots, \mathbf{b}_n\}$ for V. Theorem 3, Sec. 13.3, tells us the expansion (7) is unique. To determine the coordinates c_j, we use inner products as in Sec. 13.1, and the fact that the basis vectors are mutually orthogonal, to obtain

$$(a)\ (\mathbf{v}, \mathbf{b}_j) = c_j(\mathbf{b}_j, \mathbf{b}_j) \quad \text{or} \quad (b)\ c_j = \frac{(\mathbf{v}, \mathbf{b}_j)}{\|\mathbf{b}_j\|^2} \qquad 1 \le j \le n \tag{8}$$

Substituting these values into (7), we have

$$\mathbf{v} = \sum_{j=1}^{n} \frac{(\mathbf{v}, \mathbf{b}_j)}{\|\mathbf{b}_j\|^2}\mathbf{b}_j = \frac{(\mathbf{v}, \mathbf{b}_1)}{\|\mathbf{b}_1\|^2}\mathbf{b}_1 + \frac{(\mathbf{v}, \mathbf{b}_2)}{\|\mathbf{b}_2\|^2}\mathbf{b}_2 + \cdots + \frac{(\mathbf{v}, \mathbf{b}_n)}{\|\mathbf{b}_n\|^2}\mathbf{b}_n \tag{9}$$

The terms of this expansion remind us of Eq. (4) which, upon changing notation by writing the dot product $\mathbf{v} \cdot \mathbf{u}$ as the inner product $(\mathbf{v}, \mathbf{u})$, gives the formula

$$\mathbf{p} = \frac{(\mathbf{v}, \mathbf{u})}{\|\mathbf{u}\|^2}\mathbf{u} \tag{10}$$

for the projection $\mathbf{p}$ of $\mathbf{v}$ in the direction of $\mathbf{u}$. This expression for $\mathbf{p}$ has the very same structure as each term of (9). Thus, independent of the nature of the vectors of V, the terms of (9) may be interpreted as *projections.* So, naturally, the jth term

$$\frac{(\mathbf{v}, \mathbf{b}_j)}{\|\mathbf{b}_j\|^2}\mathbf{b}_j \qquad 1 \le j \le n \tag{11}$$

of (9) is called **the projection of v,** or **the component of v, along** b_j and Eq. (9) shows that **v** is the sum of all its projections along the basis vectors.

The preceding discussion should strike us as remarkably similar to the expansion problems we encountered in Chaps. 11 and 12, and it is interesting and important now to restate them as problems in a linear vector space in which vectors are interpreted as functions. Doing this should make clear the logical identity of certain ideas in vector analysis with similar ideas in the study of orthogonal functions, and should help each subject to clarify the other.

A **function space** is a linear space with its vectors identified as functions, all with a common domain of definition. With respect to an orthogonal set on an interval $[a, b]$ of infinitely many functions $\{\phi_1(x), \phi_2(x), \ldots\}$, we have seen (Secs. 11.6 and 13.1) that an arbitrary function f of a space of functions defined on $[a, b]$ has a formal generalized Fourier series expansion

$$f(x) = \sum_{k=1}^{\infty} a_k\phi_k(x) = a_1\phi_1(x) + a_2\phi_2(x) + \cdots + a_k\phi_k(x) + \cdots \tag{12}$$

for which the coefficient formula is

$$a_k = \frac{\int_a^b f(x)\phi_k(x)\,dx}{\int_a^b \phi_k^2(x)\,dx} \qquad 1 \le k \tag{13a}$$

Rewriting this formula in terms of inner products and norms of functions [Eqs. (2) and (27), Sec. 13.1], we have

$$a_k = \frac{(f, \phi_k)}{\|\phi_k\|^2} \qquad 1 \le k \tag{13b}$$

Substituting these values into (12) and subsequently replacing functional values by the functions themselves, we obtain

$$f = \sum_{k=1}^{\infty} \frac{(f, \phi_k)}{\|\phi_k\|^2}\phi_k = \frac{(f, \phi_1)}{\|\phi_1\|^2}\phi_1 + \frac{(f, \phi_2)}{\|\phi_2\|^2}\phi_2 + \cdots + \frac{(f, \phi_k)}{\|\phi_k\|^2}\phi_k + \cdots \tag{14}$$

The general term of this expansion

$$\frac{(f, \phi_k)}{\|\phi_k\|^2}\phi_k \qquad 1 \le k \tag{15}$$

closely resembles both (10) and (11). This analogy explains why (15) is referred to as **the projection of the function f onto the function ϕ_k** and is sometimes called **the ϕ_k component of f.** With the meaning of the term *projection* thus enlarged to apply to functions as well as vectors, the expansion (14) of f is seen to be just the sum of all projections of f onto the infinite sequence of mutually orthogonal nontrivial functions $\phi_1, \phi_2, \phi_3, \ldots$, on the interval $[a, b]$. The scalar quantities of (13b) which define the a_k are the **scalar projections of f onto the same sequence of ϕ's.**

EXAMPLE 2

FOURIER COEFFICIENTS AS SCALAR PROJECTIONS

For all positive integers n, the number 2π is a period of $\cos nt$ and $\sin nt$. The constant function $f(t) \equiv 1$ also has 2π as a period. Taken all together, these functions form an orthogonal set of functions

$$\{1, \cos nt, \sin nt\} \qquad n = 1, 2, 3, \ldots \tag{16}$$

on the interval $[0, 2\pi]$. Euler formulas (see Sec. 8.2) for the coefficients occurring in the series expansion

$$y(t) = \frac{1}{2}a_0 + \sum_{n=1}^{\infty} (a_n \cos nt + b_n \sin nt) \tag{17}$$

of a function $y(t)$ defined at each point of $[0, 2\pi]$ and of period 2π, are given by

$$(a)\ a_n = \frac{1}{\pi}\int_0^{2\pi} y(t) \cos nt\, dt \qquad (b)\ b_n = \frac{1}{\pi}\int_0^{2\pi} y(t) \sin nt\, dt \tag{18}$$

Focusing our attention on the *restriction* Y of the function y to the period interval $[0, 2\pi]$ and on the *restrictions* c_n and s_n of all functions of (16) to the same interval, we note that c_n and s_n are defined for $0 \leq t \leq 2\pi$ by

$$(a)\ c_n(t) = \cos nt \quad n = 0, 1, 2, 3, \ldots \qquad (b)\ s_n(t) = \sin nt \quad n = 1, 2, 3, \ldots \tag{19}$$

With the understanding that Y and each c_n and s_n belong to a function space whose functions have $[0, 2\pi]$ as their common domain, and with the identity $Y(t) \equiv y(t)$ on $[0, 2\pi]$ in mind, along with (19*a*) and (19*b*), we write the inner products

$$(a)\ (Y, c_n) = \int_0^{2\pi} y(t) \cos nt\, dt \qquad (b)\ (Y, s_n) = \int_0^{2\pi} y(t) \sin nt\, dt \tag{20}$$

By definition,

$$\|c_n\|^2 = \int_0^{2\pi} \cos^2 nt\, dt \quad \text{and} \quad \|s_n\|^2 = \int_0^{2\pi} \sin^2 nt\, dt$$

Evaluating these integrals, we have

$$(a)\ \|c_0\|^2 = 2\pi \qquad (b)\ \|c_n\|^2 = \pi \qquad (c)\ \|s_n\|^2 = \pi \qquad n = 1, 2, 3, \ldots \tag{21}$$

Substituting from Eqs. (20) and (21) into (18*a*) and (18*b*), we obtain

$$(a)\ \frac{a_0}{2} = \frac{(Y, c_0)}{\|c_0\|^2} \qquad (b)\ a_n = \frac{(Y, c_n)}{\|c_n\|^2} \qquad (c)\ b_n = \frac{(Y, s_n)}{\|s_n\|^2} \qquad n = 1, 2, 3, \ldots \tag{22}$$

With its Fourier coefficients thus determined, Eq. (17) becomes

$$y(t) = \sum_{n=0}^{\infty} \frac{(Y, c_n)}{\|c_n\|^2} \cos nt + \sum_{n=1}^{\infty} \frac{(Y, s_n)}{\|s_n\|^2} \sin nt \tag{23}$$

which expresses the function $y(t)$ as the sum of its projections onto every one of the functions

$$\cos nt \quad n = 0, 1, 2, 3 \ldots \quad \text{and} \quad \sin nt \quad n = 1, 2, 3, \ldots$$

which belong to the orthogonal set of functions (16).

Now let us consider a periodic function y for which $[d, d + 2p]$ is a period interval. From Sec. 8.2, we know that

$$\left\{1, \cos\frac{n\pi t}{p}, \sin\frac{n\pi t}{p}\right\} \qquad n = 1, 2, 3, \ldots \tag{24}$$

is an orthogonal set of functions on $[d, d+2p]$ and that each function of (24) has $2p$ as a period. The function Y defined by

$$Y(t) = y(t) \qquad d \le t \le d + 2p \tag{25}$$

is the **restriction of** y **to** $[d, d+2p]$, and for every *nonnegative* integer n, the function c_n defined by

$$c_n(t) = \cos\frac{n\pi t}{p} \qquad d \le t \le d + 2p \tag{26}$$

is the **restriction of** $\cos(n\pi t/p)$ **to** $[d, d+2p]$. Likewise, for every *positive* integer n, the function s_n defined by

$$s_n(t) = \sin\frac{n\pi t}{p} \qquad d \le t \le d + 2p \tag{27}$$

is the **restriction of** $\sin(n\pi t/p)$ **to** $[d, d+2p]$. Assuming the relevant integrals exist, the inner products (Y, c_n) and (Y, s_n) are defined by

$$(Y, c_n) = \int_d^{d+2p} y(t)\cos\frac{n\pi t}{p}\,dt \qquad n = 0, 1, 2, 3, \ldots \tag{28a}$$

$$(Y, s_n) = \int_d^{d+2p} y(t)\sin\frac{n\pi t}{p}\,dt \qquad n = 1, 2, 3, \ldots \tag{28b}$$

The squares of the norms of the restricted trigonometric functions c_n and s_n, found by three elementary integrations, are

$$(a)\ \|c_0\|^2 = 2p \qquad (b)\ \|c_n\|^2 = p \qquad (c)\ \|s_n\|^2 = p \qquad n = 1, 2, 3, \ldots \tag{29}$$

We are now prepared to prove the following important result.

THEOREM 3 The coefficients of the Fourier series representation

$$y(t) = \sum_{n=0}^{\infty} a_n \cos\frac{n\pi t}{p} + \sum_{n=1}^{\infty} b_n \sin\frac{n\pi t}{p} \tag{30}$$

of a function y of period $2p$ are the *scalar* projections

$$(a)\ a_n = \frac{(Y, c_n)}{\|c_n\|^2} \qquad n = 0, 1, 2, 3, \ldots \qquad (b)\ b_n = \frac{(Y, s_n)}{\|s_n\|^2} \qquad n = 1, 2, 3, \ldots \tag{31}$$

of the restriction Y of y, to an *arbitrary* period interval $[d, d+2p]$ of y, onto the restrictions c_n and s_n of $\cos(n\pi t/p)$ and $\sin(n\pi t/p)$, respectively, to the same period interval $[d, d+2p]$.

PROOF First of all, we suppose that y is integrable over some period interval. Each trigonometric function of Eq. (30), as well as y, has $2p$ as a period. Hence, for all relevant values of n, the product functions $y\cos(n\pi t/p)$ and $y\sin(n\pi t/p)$ have $2p$ as a period also and are integrable over whatever period interval y itself is. But if the integral over one period interval of a periodic function has a finite value, its integral over any other period interval has the same value (Exercise 23). It follows that for every number d such that $[d, d+2p]$ is a period interval of y, the inner products (Y, c_n) and (Y, s_n) are well defined by Eqs. (28) and the square of the norms c_n and s_n are well defined by

$$(a)\ \|c_n\|^2 = \int_d^{d+2p} \cos^2\frac{n\pi t}{p}\,dt \qquad (b)\ \|s_n\|^2 = \int_d^{d+2p} \sin^2\frac{n\pi t}{p}\,dt \tag{32}$$

With t restricted to the arbitrary period interval $[d, d + 2p]$ and with index of summation k, Eq. (30) reads

$$Y(t) = \sum_{k=0}^{\infty} a_k c_k(t) + \sum_{k=1}^{\infty} b_k s_k(t) \qquad d \leq t \leq d + 2p$$

Taking inner products of both members of this equation with c_n, then with s_n, and using the orthogonality of the members of the set of functions (24) on the interval $[d, d + 2p]$, we are left with the relations

$$(Y, c_n) = a_n\|c_n\|^2 \qquad n = 0, 1, 2, 3, \ldots \qquad \text{and} \qquad (Y, s_n) = b_n\|s_n\|^2 \qquad n = 1, 2, 3, \ldots$$

Solving these for a_n *and* b_n, we arrive at Eqs. (31). ◗

In addition to being easy to remember and apply in finding Fourier series of functions, the Formulas (31) indicate and emphasize the breadth and importance of projections in general expansion problems involving vectors or functions of various linear spaces. To represent the function y of Theorem 3 as the sum of all its projections onto the functions

$$\cos\frac{n\pi t}{p} \qquad n = 0, 1, 2, 3, \ldots \qquad \text{and} \qquad \sin\frac{n\pi t}{p} \qquad n = 1, 2, 3, \ldots$$

we merely substitute the scalar projections (31) into Eq. (30), getting

$$(33) \qquad y(t) = \sum_{n=0}^{\infty} \frac{(Y, c_n)}{\|c_n\|^2} \cos\frac{n\pi t}{p} + \sum_{n=1}^{\infty} \frac{(Y, s_n)}{\|s_n\|^2} \sin\frac{n\pi t}{p}$$

and we have such a sum.

For practical purposes, numerical values of the scalar projections are needed. As was pointed out in the proof of Theorem 3, if the period interval $[d, d + 2p]$ to which y and the other functions of (30) are restricted is changed to another period interval, the projections (31) are *not* affected in value. Consequently, the number d in the limits of the integrals defining the inner product and norm of each scalar projection may be chosen arbitrarily, consistent with the requirement that $[d, d + 2p]$ must be a period interval of y. In particular, by using appropriate analytic representations of y, we may take $d = -p$ or $d = 0$, whichever choice proves more convenient.

Since the constant term of the Fourier series (30) is a_0, instead of $a_0/2$ as in Eq. (17), the value of a_0 given by (31*a*) with $n = 0$ is *not* the value given by the Euler formula for a_0 (Exercise 26). There is no need to divide a_0 by 2 when its numerical value, as computed using (31*a*), is substituted into

$$(34) \qquad y(t) = a_0 + \sum_{n=1}^{\infty} \left(a_n \cos\frac{n\pi t}{p} + \sin\frac{n\pi t}{p}\right)$$

as the leading term of the Fourier series.

EXERCISES

1. Find the cosine of the angle θ between each of the following pairs of vectors.
(a) (2, 6), (4, 4) **(b)** (3, 4), (−6, −8)
(c) (3, −2), (6, 3) **(d)** (2, −2, 1), (−6, 2, 3)
(e) (2, 1, −2), (1, −2, 2)
(f) (2, −4, 1), (3, 1, −2)
(g) (1, 0, 2, 0, 1), (3, −2, 1, −1, 1)
(h) (1, 3, 4, −6, 7, 8, 9), (5, 3, 3, 3, 2, −2, 2)

2. Justify writing the solution of Eq. (3) for θ as $\theta = \text{Cos}^{-1}\dfrac{\mathbf{u}\cdot\mathbf{v}}{\|\mathbf{u}\|\,\|\mathbf{v}\|}$ in terms of the *principal* inverse cosine function.

3. Find the angle θ between each of the following pairs of vectors.
(a) (3, −1), (2, 1) **(b)** (−2, −3), (6, 9)
(c) $(2\sqrt{3}, 2)$, $(-\sqrt{3}, 1)$

(d) $(-3, 4, -5)$, $(3, -4, 0)$
(e) $(1, 1, 2)$, $(2, -1, 1)$
(f) $(1, 2, 2)$, $(4, 0, -3)$
(g) $(1, 2, 3, 2)$, $(4, -3, 3, 4)$
(h) $(3, -3, 2, 1, -4, 5, 6)$, $(2, 1, 1, 2, 1, 1, 2)$

4. Find k if the vectors $(3, 2)$ and $(2, k)$ are **(a)** orthogonal and **(b)** parallel.

5. Find k if the angle between the vectors $(1, k)$ and $(1, -1)$ is $\pi/3$.

6. Might the angle between two vectors be undefined? Explain.

7. Given the angle α between $\mathbf{u}$ and $\mathbf{v}$, find the angle between
(a) $\mathbf{u}$ and $-\mathbf{v}$ **(b)** $-\mathbf{u}$ and $\mathbf{v}$
(c) $-\mathbf{u}$ and $-\mathbf{v}$

8. Express Eq. (3) in terms of the angle $2\pi - \theta$ and inner products involving $\mathbf{u}$ or $\mathbf{v}$.

9. Show that Eq. (2) holds for vectors $\mathbf{u}$ and $\mathbf{v}$ of R^n, $n > 1$, such that $\mathbf{v} = c\mathbf{u}$, $c \neq 0$, if $\theta = 0$ and $c > 0$ or else if $\theta = \pi$ and $c < 0$. *Hint:* $\sqrt{c^2} = |c|$.

10. Prove that (3) is dimensionally correct by showing that the value of $\cos\theta$ is unchanged if $\mathbf{u}$ is replaced by $c\mathbf{u}$ and $\mathbf{v}$ is replaced by $k\mathbf{v}$, where $c, k > 0$.

11. Show that the **scalar component of v in the direction of u,** that is, the scalar projection (5), is given by $(\mathbf{v}\cdot\mathbf{u})/\|\mathbf{u}\|$.

12. For each of the following pairs of vectors, find the *scalar* component of $\mathbf{v}$ in the direction of $\mathbf{u}$.
(a) $\mathbf{v} = (2, -1)$, $\mathbf{u} = (1, 1)$
(b) $\mathbf{v} = (5, 1)$, $\mathbf{u} = (3, -2)$
(c) $\mathbf{v} = (5, -14)$, $\mathbf{u} = (12, -5)$
(d) $\mathbf{v} = (2, 2, -1)$, $\mathbf{u} = (4, 5, 3)$
(e) $\mathbf{v} = (2, 1, 1)$, $\mathbf{u} = (2, 4, -5)$
(f) $\mathbf{v} = (5, 1, 3)$, $\mathbf{u} = (5, -7, 4)$
(g) $\mathbf{v} = (-1, 3, 3, 3, 1)$, $\mathbf{u} = (12, 4, 2, 2, 1)$
(h) $\mathbf{v} = (1, -3, 2, \pi, 1, -1, 3)$, $\mathbf{u} = (1, 2, 3, 0, 5, 7, 9)$

13. For each of the following pairs of vectors, find the *vector* component of $\mathbf{v}$ in the direction of $\mathbf{u}$.
(a) $\mathbf{v} = (10, 5)$, $\mathbf{u} = (3, 4)$
(b) $\mathbf{v} = (1, 1)$, $\mathbf{u} = (1, -7)$
(c) $\mathbf{v} = (\pi^3, e^3)$, $\mathbf{u} = (\pi, -e)$
(d) $\mathbf{v} = (3, -3, 3)$, $\mathbf{u} = (1, -2, 4)$
(e) $\mathbf{v} = (5, 3, -9)$, $\mathbf{u} = (2, -3, 2)$
(f) $\mathbf{v} = (4, 20, -3)$, $\mathbf{u} = (2, -1, 2)$
(g) $\mathbf{v} = (3, 3, 7, 5, 3)$, $\mathbf{u} = (1, -2, -7, -2, 1)$
(h) $\mathbf{v} = (12, 5, 1, 3, 2\pi, 7, 6\pi)$,
$\mathbf{u} = (1, 0, -2, 0, 2\pi, 0, \pi)$

14. Show that the square of *the* distance d from a vector $\mathbf{v}$ (interpreted as a point) to a line l (spanned by a nonzero vector $\mathbf{u}$) of R^n, $n > 1$, is given by

$$d^2 = \left\|\mathbf{v} - \frac{\mathbf{v}\cdot\mathbf{u}}{\|\mathbf{u}\|^2}\mathbf{u}\right\|^2$$

15. In Exercise 14, show that the formula for d^2 can be written

$$d^2 = \frac{(\mathbf{v}\cdot\mathbf{v})(\mathbf{u}\cdot\mathbf{u}) - (\mathbf{v}\cdot\mathbf{u})^2}{\mathbf{u}\cdot\mathbf{u}}$$

16. **(a)** Show that for all vectors $\mathbf{u}$ and $\mathbf{v}$ of a space R^n with inner product (,), the **Schwarz inequality**†

$$|(\mathbf{u}\cdot\mathbf{v})| \le \|\mathbf{u}\|\,\|\mathbf{v}\|$$

holds. *Hint:* In Exercise 15, impose the condition $d^2 \ge 0$.
(b) What is a necessary and sufficient condition for equality between the members of the Schwarz inequality?

17. Prove that the Schwarz inequality is *equivalent* to the **triangle inequality**

$$\|\mathbf{u} + \mathbf{v}\| \le \|\mathbf{u}\| + \|\mathbf{v}\|$$

Hint: Show that each inequality implies the other.

18. Let $\mathbf{v}_1, \mathbf{v}_2, \ldots, \mathbf{v}_m$ be linearly independent vectors of R^n. Choose one of them, say $\mathbf{v}_1$, and set $\mathbf{u}_1 = \mathbf{v}_1$. Next, choose any of the original vectors except $\mathbf{v}_1$, say $\mathbf{v}_2$, and set $\mathbf{u}_2 = \mathbf{v}_2 - c_1\mathbf{u}_1$, where the real number c_1 is to be determined so that $\mathbf{u}_2$ is orthogonal to $\mathbf{u}_1$. This of course requires that

$$0 = \mathbf{u}_2\cdot\mathbf{u}_1 = \mathbf{v}_2\cdot\mathbf{u}_1 - c_1\|\mathbf{u}_1\|^2 \qquad \text{or} \qquad c_1 = \frac{\mathbf{v}_2\cdot\mathbf{u}_1}{\|\mathbf{u}_1\|^2}$$

so that
$$\mathbf{u}_2 = \mathbf{v}_2 - \frac{\mathbf{v}_2\cdot\mathbf{u}_1}{\|\mathbf{u}_1\|^2}\mathbf{u}_1$$

By continuing this process, prove Theorem 2 for $m > 2$.

19. Using the Gram-Schmidt process, find an orthonormal basis for the subspace of R^2, R^3, or R^4 spanned by the given vectors.
(a) $(3, 4)$, $(-1, 1)$ **(b)** $(2, 3, 6)$, $(7, 12, 8)$
(c) $(2, -14, 5)$, $(6, -12, 9)$
(d) $(1, 1, 0)$, $(1, 0, 1)$, $(0, 1, 1)$
(e) $(1, 2, 2)$, $(1, 4, 0)$, $(2, 0, 1)$
(f) $(1, 1, 1, 1)$, $(0, 1, 2, 2)$, $(0, 0, 1, 1)$
(g) $(1, -1, 1, -1)$, $(3, -1, 5, 1)$, $(1, 2, 0, 2)$
(h) $(2, -8, 2, 3)$, $(9, -12, 9, 10)$, $(9, -4, 5, 7)$

20. Apply the orthogonalization process described in Exercise 18 to the functions 1, x, x^2, and x^3 on the interval $[-1, 1]$. How do your results compare with the first four Legendre polynomials (Sec. 12.9)? What do you think is the generalization of this result to the infinite set of functions $\{1, x, x^2, x^3, x^4, \ldots\}$?

21. Apply the orthogonalization process described in Exercise 18 to the functions 1, x, x^2, and x^3 on the interval $[0, 1]$.

22. Express each of the following vectors as a linear combination of the orthogonal basis vectors $\mathbf{b}_1 = (1, 0, 1, 2)$,

---

†Named for the German mathematician Hermann Amandus Schwarz (1843–1921).

$\mathbf{b}_2 = (0, 1, 2, -1)$, $\mathbf{b}_3 = (0, 5, -2, 1)$, and $\mathbf{b}_4 = (5, 0, -1, -2)$ of R^4.
(a) $(1, 1, 1, 1)$ (b) $(5, 7, 1, 3)$
(c) $(6, 6, 0, 0)$ (d) $(1, -2, 3, -4)$
(e) $(1, -1, -1, 3)$ (f) $(5, 0, 5, 4)$

23. Show that if a periodic function f is integrable on a period interval $[d, d + 2p]$, then for all k such that $[k, k + 2p]$ is a period interval of f,

$$\int_k^{k+2p} f(t)\, dt = \int_d^{d+2p} f(t)\, dt$$

24. Answer *yes* or *no* according as the specified set is or is not a function space. Justify each *no* answer.
(a) All polynomials of degree $n \le 5$
(b) All polynomials of even degree
(c) All solutions of $y'' - y = 0$ on $(-\infty, \infty)$
(d) All linear combinations of $\cos 2x$ and $\sin 2x$
(e) $\{1, \cos x, \sin x\}$
(f) All twice-differentiable functions on $[0, 1]$
(g) All linear combinations of x and $\ln x$
(h) $\{\sqrt{x}, x, 1\}$

25. A periodic function y is represented on one period interval by

$$y(t) = 2\pi + t \qquad -2\pi < t < -\pi$$

Use scalar projections to find values of the coefficients in the Fourier expansion of y and represent y as the sum of its projections onto cosine and sine functions.

26. Assuming $[d, d + 2p]$ is a period interval of a function y, compare the value of a_0 as given by the relevant Euler formula with the value of a_0 determined by Eq. (31*a*).

27. Work Exercise 25 for the periodic function y_1 represented on one period by

$$y_1(t) = \begin{cases} 0 & -2\pi < t < -\pi \\ \pi & -\pi < t < 0 \end{cases}$$

and use the alternative formulas [Eqs. (3) and (4), Sec. 8.3] to verify the Fourier coefficients found by means of scalar projections.

28. Let Y be a restriction of the periodic function y, Exercise 25, to two adjacent period intervals of y. Find the scalar projections of Y onto the restrictions of the functions

$$\cos nt \quad n = 0, 1, 2, 3, \ldots \quad \text{and} \quad \sin nt \quad n = 1, 2, 3, \ldots$$

to the same set of points and represent y as the sum of all its projections onto the preceding cosine and sine functions.

29. Take Y_1 to be the restriction of the periodic function y, Exercise 25, to the period interval $[0, \pi]$ and evaluate the scalar projections of Y_1 onto the restrictions of the functions

$$\cos 2nt \quad n = 0, 1, 2, 3, \ldots \quad \text{and} \quad \sin 2nt \quad n = 1, 2, 3, \ldots$$

to the same period interval. Compare the values of these scalar projections with the values of the scalar projections of Y in Exercise 28.

30. Represent the periodic function y_2 defined on one period interval by

$$y_2(t) = 2\pi + t \qquad -2\pi < t < 0$$

as the sum of its projections onto the functions

$$\cos nt \quad n = 0, 1, 2, 3, \ldots \quad \text{and} \quad \sin nt \quad n = 1, 2, 3, \ldots$$

and show that the projection of y_2 onto each of these functions is the sum of the projections of y and y_1, Exercises 27 and 28, onto the same functions.

13.5 LINEAR TRANSFORMATIONS

Functions are often called **transformations, operators,** or **mappings.** A common notation for a function f, which assigns exactly one element of a set B to each element of a set A, is

$$f\colon A \to B \tag{1}$$

We may think of f as a rule, or operator, which when applied to any element of A, transforms, or maps, that element into an element of B. Thus (1) is said to denote a **transformation** (**operator,** or **mapping**) ***f* from *A* into *B*.**

Many engineering and scientific applications of mathematics involve functions defined on vector spaces whose values are elements of a vector space. Epsecially important among such functions are those known as *linear transformations.* Although we have seldom mentioned the fact, as we shall soon realize, we have already made extensive use of transformations of this kind. Let us see just what it is that makes a transformation linear.

We have already said that the domain of a linear operator must be a vector space and that its functional values must be vectors of a vector space. More precisely, if $\mathbf{v}$ is any vector in the domain V of a linear transformation T, and if the corresponding value $T(\mathbf{v})$ of T, usually written $T\mathbf{v}$, is a vector of W, then both V and W are to be real, or else both are to be complex, vector spaces. In other words, V and W must be vector spaces over the same set of scalars. With this stipulation in force, a linear transformation T from V into W, may be defined as follows.

DEFINITION 1 A function T from a vector space V into a vector space W is a **linear transformation** if and only if, for all **u**, **v** in V and for every scalar c,

$$T(\mathbf{u} + \mathbf{v}) = T(\mathbf{u}) + T(\mathbf{v}) = T\mathbf{u} + T\mathbf{v} \tag{2}$$

and

$$T(c\mathbf{v}) = cT(\mathbf{v}) = cT\mathbf{v} \tag{3}$$

Linear transformations are also called **linear operators,** or **linear mappings.**

Our insistence that V and W be vector spaces over the same set of scalars guarantees that if $c\mathbf{v}$ is a vector of V, then $cT\mathbf{v}$ is a vector of W, and vice versa. Unless V and W are the same space, the two operations associated with V will differ from those of W, and the zero vector of V may differ from that of W. Since both V and W are closed under their operations of addition, and of multiplication by scalars, (2) and (3) are meaningful statements for all **u**, **v** in V, and for every scalar c.

The value of T at **v**, i.e., $T\mathbf{v}$, is frequently referred to as the **image** of **v** under T. In terms of images, (2) simply says that the sum $\mathbf{u} + \mathbf{v}$ of any two vectors of V must have an image $T(\mathbf{u} + \mathbf{v})$ which is the sum $T\mathbf{u} + T\mathbf{v}$ of the individual images of **u** and **v**. Similarly, (3) requires the product $c\mathbf{v}$ of any scalar c and a vector **v** of V to have an image $T(c\mathbf{v})$ given by the product $cT\mathbf{v}$ of c and the image $T\mathbf{v}$ of **v**. These two properties imply (Exercise 19) that for every finite linear combination $\Sigma_{k=1}^{n} c_k\mathbf{v}_k$ of vectors of V,

$$T\left(\sum_{k=1}^{n} c_k\mathbf{v}_k\right) = \sum_{k=1}^{n} c_kT\mathbf{v}_k \tag{4}$$

This equation tells us that the image of a finite linear combination of vectors in the domain of a linear transformation can be found by simply replacing each vector of the linear combination by its image. In particular, if $n = 2$, Eq. (4) becomes

$$T(c_1\mathbf{v}_1 + c_2\mathbf{v}_2) = c_1T\mathbf{v}_1 + c_2T\mathbf{v}_2 \tag{5}$$

Since this equation holds if and only if T is linear (Exercise 18), a linear transformation is sometimes defined by requiring all vectors $\mathbf{v}_1$, $\mathbf{v}_2$ of V and arbitrary scalars c_1, c_2 to satisfy (5).

EXAMPLE 1

Show that the function z defined on R^2 by $z(x, y) = 3x - 5y$ is a linear transformation from R^2 into R.

First of all, we need to clear up a matter of notation. Given any vector $\mathbf{v} = (x, y)$ in R^2, it is to be understood that

$$z(\mathbf{v}) = z[(x, y)] = z(x, y)$$

From this it is evident that the domain of z is R^2 and of course the values of z are all real numbers. In other words, z is a transformation from R^2 into R. For all $\mathbf{v}_1 = (x_1, y_1)$, $\mathbf{v}_2 = (x_2, y_2)$ in R^2,

$$\begin{aligned} z(\mathbf{v}_1 + \mathbf{v}_2) &= z[(x_1, y_1) + (x_2, y_2)] = z[(x_1 + x_2, y_1 + y_2)] = z(x_1 + x_2, y_1 + y_2) \\ &= 3(x_1 + x_2) - 5(y_1 + y_2) = (3x_1 - 5y_1) + (3x_2 - 5y_2) = z(x_1, y_1) + z(x_2, y_2) \\ &= z(\mathbf{v}_1) + z(\mathbf{v}_2) \end{aligned}$$

Hence z satisfies Condition (2) of Definition 1.

For every vector $\mathbf{v} = (x, y)$ in R^2 and for any real number c,

$$\begin{aligned} z(c\mathbf{v}) &= z[c(x, y)] = z[(cx, cy)] = z(cx, cy) = 3cx - 5cy = c(3x - 5y) = cz(x, y) \\ &= cz(\mathbf{v}) \end{aligned}$$

Hence z satisfies Condition (3) of Definition 1. Since R^2 and R are both real vector spaces, this proves that z is a linear transformation from R^2 into R.

EXAMPLE 2

Let the vector space $\mathscr{C}[0, 1]$ of all real-valued continuous functions on [0, 1] be the domain of an integral operator T which transforms each function f in $\mathscr{C}[0, 1]$ into the function Tf represented on $0 \le x \le 1$ by

$$Tf(x) = \int_0^x f(s)\, ds$$

Since the derivative of Tf with respect to x is f, the function Tf belongs to the real vector space $\mathscr{C}^{(1)}[0, 1]$ each of whose members is a real-valued continuously differentiable function on [0, 1]. Thus T is a transformation from the real space $\mathscr{C}[0, 1]$ into $\mathscr{C}^{(1)}[0, 1]$. Now, if f_1 and f_2 are arbitrary functions of $\mathscr{C}[0, 1]$ and if c_1 and c_2 are any real numbers, then, for $0 \le x \le 1$,

$$\begin{aligned} T(c_1f_1 + c_2f_2)(x) &= \int_0^x (c_1f_1 + c_2f_2)(s)\, ds = \int_0^x [(c_1f_1)(s) + (c_2f_2)(s)]\, ds \\ &= \int_0^x [c_1f_1(s) + c_2f_2(s)]\, ds = \int_0^x c_1f_1(s)\, ds + \int_0^x c_2f_2(s)\, ds \\ &= c_1 \int_0^x f_1(s)\, ds + c_2 \int_0^x f_2(s)\, ds = c_1Tf_1(x) + c_2Tf_2(x) \end{aligned}$$

so that

$$T(c_1f_1 + c_2f_2) = c_1Tf_1 + c_1Tf_2$$

Comparing this result with (5), we see that T is a linear transformation from $\mathscr{C}[0, 1]$ into $\mathscr{C}^{(1)}[0, 1]$.

EXAMPLE 3

Let A be an $m \times n$ matrix with complex elements. Let C^n be the space of all n-dimensional *column* vectors with complex components. If $\mathbf{v}$ is a vector of C^n, then A$\mathbf{v}$ is a vector in C^m. Thus A is a transformation from the complex vector space C^n into the complex vector space C^m. Using familiar distributive and associative laws for matrices, we find that for all $\mathbf{v}_1$, $\mathbf{v}_2$ in C^n and for arbitrary complex numbers c_1 and c_2,

$$\mathrm{A}(c_1\mathbf{v}_1 + c_2\mathbf{v}_2) = \mathrm{A}(c_1\mathbf{v}_1) + \mathrm{A}(c_2\mathbf{v}_2) = c_1\mathrm{A}\mathbf{v}_1 + c_2\mathrm{A}\mathbf{v}_2$$

Comparing this result with (5), we see that A is a linear transformation from C^n into C^m.

EXAMPLE 4

Definition 1, Sec. 2.7, defines an expression of the form

$$L = a_0D^n + a_1D^{n-1} + \cdots + a_{n-1}D + a_n$$

to be an nth-order linear differential operator. It is assumed of course that $n \ge 1$, that the real-valued functions $a_0, a_1, \ldots, a_n$ are defined on an interval I, and that $a_0(x) \not\equiv 0$ on I. Show that L is indeed a linear operator.

If y is any function in the space $F^{(n)}(I)$ of all real-valued functions which are differentiable n times on I, then, by definition,

$$Ly = (a_0D^n + a_1D^{n-1} + \cdots + a_{n-1}D + a_n)y = a_0y^{(n)} + a_0y^{(n-1)} + \cdots + a_{n-1}y' + a_ny \tag{6}$$

From this we infer that L transforms functions of the real space $F^{(n)}(I)$ into functions of the real space $F(I)$.

Using elementary properties of derivatives, we find that if y_1, y_2 have nth-order derivatives on I, and if c_1 and c_2 are real numbers, then

$$\begin{aligned}L(c_1y_1 + c_2y_2) &= a_0(c_1y_1 + c_2y_2)^{(n)} + a_1(c_1y_1 + c_2y_2)^{(n-1)} + \cdots + a_n(c_1y_1 + c_2y_2)\\ &= a_0(c_1y_1^{(n)} + c_2y_2^{(n)}) + a_1(c_1y_1^{(n-1)} + c_2y_2^{(n-1)}) + \cdots + a_n(c_1y_1 + c_2y_2)\\ &= c_1(a_0y_1^{(n)} + a_1y_1^{(n-1)} + \cdots + a_ny_1) + c_2(a_0y_2^{(n)} + a_1y_2^{(n-1)} + \cdots + a_ny_2)\\ &= c_1Ly_1 + c_2Ly_2\end{aligned}$$

which shows that L is a linear transformation from $F^{(n)}(I)$ into $F(I)$.

EXAMPLE 5

Is the transformation g from R^2 into R defined by $g(x, y) = \sqrt{x^2 + y^2}$ a linear transformation?

For g to be a linear transformation, the equation $g(\mathbf{v}_1 + \mathbf{v}_2) = g(\mathbf{v}_1) + g(\mathbf{v}_2)$ must hold for all $\mathbf{v}_1$, $\mathbf{v}_2$ in R^2. Both $\mathbf{v}_1 = (1, 0)$ and $\mathbf{v}_2 = (0, 1)$ are in R^2, and $g(\mathbf{v}_1) = g(\mathbf{v}_2) = 1$. But

$$g(\mathbf{v}_1 + \mathbf{v}_2) = g(1, 1) = \sqrt{2} \neq 2 = g(\mathbf{v}_1) + g(\mathbf{v}_2)$$

and so g is not a linear transformation.

It is important to notice that every linear transformation T: $V \to W$ maps the zero vector $\mathbf{o}$ of V into the zero vector $\mathbf{0}$ of W. To see this, set $c = 0$ in (3) and recall that any vector multiplied by the number 0 is a zero vector. This gives

$$T(\mathbf{o}) = \mathbf{0} \tag{7}$$

Of course, if V is a subspace of W, or vice versa, then $\mathbf{o} = \mathbf{0}$. Should W be a subspace of V, it is said that T is a linear transformation **on** V.

Especially important in the study of linear transformations is *the set K_T of all elements of V which T maps into* $\mathbf{0}$. We have already noted that $\mathbf{o}$ is contained in K_T. Suppose $\mathbf{v}_1$ and $\mathbf{v}_2$ are arbitrary vectors of K_T so that $T(\mathbf{v}_1) = \mathbf{0}$ and $T(\mathbf{v}_2) = \mathbf{0}$. Then

$$T(c_1\mathbf{v}_1 + c_2\mathbf{v}_2) = T(c_1\mathbf{v}_1) + T(c_2\mathbf{v}_2) = c_1T(\mathbf{v}_1) + c_2T(\mathbf{v}_2) = c_1\mathbf{0} + c_2\mathbf{0} = \mathbf{0}$$

This proves that K_T is a subspace of V (Exercise 7, Sec. 13.2). We call K_T the **kernel,** or **null space,** of the linear transformation T.

As with functions in general, *the set R_T of all functional values of T* is called the **range** of T. This set is a subset of W. If $\mathbf{w}_1$ and $\mathbf{w}_2$ are vectors of R_T, there must be vectors $\mathbf{v}_1$ and $\mathbf{v}_2$ in V such that $\mathbf{w}_1 = T\mathbf{v}_1$ and $\mathbf{w}_2 = T\mathbf{v}_2$. Since T is linear,

$$\mathbf{w}_1 + \mathbf{w}_2 = T\mathbf{v}_1 + T\mathbf{v}_2 = T(\mathbf{v}_1 + \mathbf{v}_2) \qquad \text{and} \qquad c\mathbf{w}_1 = cT(\mathbf{v}_1) = T(c\mathbf{v}_1)$$

By closure, the vectors $\mathbf{v}_1 + \mathbf{v}_2$ and $c\mathbf{v}_1$ are in V, and they have $\mathbf{w}_1 + \mathbf{w}_2$ and $c\mathbf{w}_1$ as their respective images under T. Hence R_T is closed under the operations of W and is therefore a subspace of W. The range of T need not be all of W. On the other hand, it can be. To describe the case when $R_T = W$, we say that T is a transformation from V **onto** W. Every transformation from V onto W is a transformation from V into W, but not conversely.

A linear transformation T: $V \to W$ is **one-to-one** if and only if for all $\mathbf{v}_1$ and $\mathbf{v}_2$ in V, if $T\mathbf{v}_1 = T\mathbf{v}_2$ then $\mathbf{v}_1 = \mathbf{v}_2$. If T is one-to-one, each vector in the range of T is the image of exactly one vector in the domain of T. No vector in the **codomain** of T, namely W, can be the image of more than one vector in V. Some vectors in W might not be in R_T. In other words, a one-to-one linear transformation is *not* necessarily a transformation from its domain onto it codomain. To express the fact that a linear transformation is *both* one-to-one and onto, it is said to be an **isomorphism.** For reasons which will soon be apparent, such transformations are also called **invertible,** or **nonsingular,** transformations.

A very useful theorem concerning linear transformations is the following.

THEOREM 1 A linear transformation $T: V \to W$ is one-to-one if and only if $K_T = \{\mathbf{o}\}$. In other words, T is not one-to-one if and only if its kernel K_T is not the trivial subspace $\{\mathbf{o}\}$ of V.

PROOF We know that $T\mathbf{o} = \mathbf{0}$. Assume $T\mathbf{v} = \mathbf{0}$. Then $T\mathbf{v} = T\mathbf{o}$. If T is one-to-one, $T\mathbf{v} = T\mathbf{o}$ implies $\mathbf{v} = \mathbf{o}$, and so $K_T = \{\mathbf{o}\}$.

On the other hand, since T is linear, $T\mathbf{v}_2 = T\mathbf{v}_1$ if and only if $T\mathbf{v}_2 - T\mathbf{v}_1 = T\mathbf{v}_2 + T(-\mathbf{v}_1) = T(\mathbf{v}_2 - \mathbf{v}_1) = \mathbf{0}$. Hence, if $K_T = \{\mathbf{o}\}$, then $T\mathbf{v}_2 = T\mathbf{v}_1$ implies $\mathbf{v}_2 = \mathbf{v}_1$, so that T is one-to-one.

Now, let us see how all of these ideas relate to real systems of linear algebraic equations.

EXAMPLE 6

Reasoning as in Example 3, we find that every *real* $m \times n$ matrix is a linear transformation from the space R^n of all *real* n-dimensional column vectors into R^m. When used as transformations, matrices sometimes are called **matrix operators.**

Let A be an arbitrary $m \times n$ real matrix. The set of all solutions of the homogeneous linear system

$$\mathrm{A}\mathbf{x} = \mathbf{0} \tag{8}$$

is the kernel of A. Thus we speak of K_A as the **solution space** of (8), and of course the zero vector of R^n is in this space. Since A is a linear operator, every finite linear combination of solutions of (8) is itself a solution (Exercise 20, Sec. 3.6). In fact, if $\{\mathbf{x}_1, \mathbf{x}_2, \ldots, \mathbf{x}_N\}$ is a basis for K_A, every solution of (8) can be expressed as a linear combination of these basis vectors. The range of A is the subspace R_A of R^m containing all vectors $\mathbf{y}$ for which the nonhomogeneous linear system

$$\mathrm{A}\mathbf{x} = \mathbf{y} \tag{9}$$

has a solution. If $\mathbf{y}$ is not in R_A, this system is inconsistent.

Now, let A be nonsingular and of order n. Then to every vector $\mathbf{y}$ in R^n there corresponds the unique solution $\mathbf{x} = \mathrm{A}^{-1}\mathbf{y}$ of (9). The range of A is therefore R^n, and A maps R^n onto R^n. In particular, if $\mathbf{y} = \mathbf{0}$, (9) reduces to (8), and the corresponding unique solution is $\mathbf{x} = \mathbf{0}$. The kernel of A is therefore $\{\mathbf{0}\}$. From Theorem 1, it follows that A is a one-to-one transformation. Finally, since the domain and codomain of A are the same, A is a matrix operator on R^n which is both onto and one-to-one, that is, A is an isomorphism.

EXERCISES

1. State whether or not the given transformation is linear and prove or disprove its linearity.

(a) $f: R^2 \to R$ where $f(x, y) = 5x - 6y$

(b) $f: R^2 \to R^2$ where $f(x, y) = (2x + 3y, -x + 6y)$

(c) $f: R^3 \to R$ where $f(x, y, z) = 4xyz$

(d) $f: C \to C$ where $f(z) = 10 - 3z$

(e) $f: P \to P$ where $f[p(x)] = 2p(x - 1) + p(0)$

(f) $f: R^3 \to R$ where $f = [-1 \quad 0 \quad 2]$

(g) $f: \mathscr{C}[0, 1] \to R$ where $f(g) = \int_0^1 g(s)\, ds$

(h) $f: C^n \to R$ where $f(\mathbf{v}) = \bar{\mathbf{v}} \cdot \mathbf{v}$

(i) $f: R^2 \to R^3$ where $f(x, y) = (x, 2y, x - y)$

(j) $f: R^2 \to \mathscr{C}(-\infty, \infty)$ where $f(x, y) = |\cos x| + |\sin y|$

In Exercises 2–5, each matrix is a linear transformation on R^2. **(a)** Find the image of the given vector $\mathbf{v}$ under each corresponding transformation. **(b)** Find the kernel of each transformation. **(c)** Is the transformation one-to-one? Explain. **(d)** Does the transformation map R^2 onto itself? If not, determine the range of the transformation. **(e)** Which of the vectors $\mathbf{v}_1 = (3, -2)$, $\mathbf{v}_2 = (-6, -2)$, $\mathbf{v}_3 = (3, 1)$, and $\mathbf{v}_4 = (6, -4)$ are in the kernel, and which are in the range, of each transformation?

2. $\mathrm{A} = \dfrac{1}{2}\begin{bmatrix} \sqrt{3} & 1 \\ -1 & \sqrt{3} \end{bmatrix}$; $\mathbf{v} = \dfrac{1}{2}\begin{bmatrix} -1 \\ \sqrt{3} \end{bmatrix}$

3. $A = \frac{1}{\sqrt{2}}\begin{bmatrix} 1 & 1 \\ -1 & 1 \end{bmatrix}$; $\mathbf{v} = \sqrt{2}\begin{bmatrix} 2 \\ 1 \end{bmatrix}$

4. $A = \begin{bmatrix} 3 & -9 \\ -2 & 6 \end{bmatrix}$; $\mathbf{v} = \begin{bmatrix} -5 \\ -2 \end{bmatrix}$

5. $A = \begin{bmatrix} 3 & 1 \\ 5 & 2 \end{bmatrix}$; $\mathbf{v} = \begin{bmatrix} 1 \\ -1 \end{bmatrix}$

6. The matrix $A = \begin{bmatrix} 1 & -2 & 1 \\ 0 & 1 & 2 \end{bmatrix}$ is a linear transformation from R^3 into R^2.

(a) What is the image of $\begin{bmatrix} 5 \\ 2 \\ -1 \end{bmatrix}$ under A?

(b) Find the kernel of A.

(c) Is A a one-to-one transformation? Explain.

(d) Does A map R^3 onto R^2?

(e) Which, if any, of the vectors $\mathbf{v}_1 = (1, 1, 1)$, $\mathbf{v}_2 = (-10, -4, 2)$, $\mathbf{v}_3 = (0, 0, 0)$, and $\mathbf{v}_4 = (5, 2, 1)$ are in the kernel of A?

7. A linear transformation T on R^3 is given by $T(x, y, z) = (-z, x, x + z)$.

(a) What is the image of $(-3, 10, 5)$ under T?

(b) Find the kernel of T and interpret it geometrically.

(c) Is T a one-to-one transformation? Explain.

8. Let T be a linear transformation on R^2 such that $T\mathbf{e}_1 = (-1, 5)$ and $T\mathbf{e}_2 = (5, 2)$, where $\mathbf{e}_1 = (1, 0)$ and $\mathbf{e}_2 = (0, 1)$. If $\mathbf{v} = (-2, 2)$, find $T\mathbf{v}$.

9. Let T be a linear transformation from R^3 into R^4. If $\mathbf{v}_1 = (2, -1, 2)$, $\mathbf{v}_2 = (1, 3, 1)$, $\mathbf{v}_3 = (4, -9, 4)$, $T\mathbf{v}_1 = (1, 0, 1, 0)$, and $T\mathbf{v}_2 = (2, -1, 1, -2)$, find $T\mathbf{v}_3$.

10. (a) Verify that the transformations T_1 and T_2 defined by

$$T_1(x_1, x_2, x_3, \ldots) = (x_1, x_3, x_5, \ldots)$$
$$T_2(x_1, x_2, x_3, \ldots) = (0, x_1, x_2, \ldots)$$

are linear transformations on the space R^∞ of all infinite sequences of real numbers.

(b) Find the kernels of T_1 and T_2.

(c) Do T_1 and T_2 map R^∞ onto itself?

(d) Are T_1 and T_2 one-to-one transformations on R^∞?

11. Which, if any, of the following functions are in the null space of the operator $D^2(D + 1)$?

(a) $\sin x$ **(b)** x **(c)** 1
(d) $\cos x$ **(e)** e^{-x}

12. Which, if any, of the following functions are in the null space of the operator $D^4 - 1$?

(a) e^{-x} **(b)** $\cosh x$ **(c)** x
(d) $\sin x$ **(e)** $x^4/4!$

13. Which, if any, of the following functions are in the kernel of the linear transformation from $\mathscr{C}[-1, 1]$ into R defined by $Tf = \int_{-1}^{1} f(s)\, ds$?

(a) $x^5 \cos 5x$ **(b)** $\tan x$ **(c)** $\sin^9 x$
(d) $\sqrt{1 + x^{10}}$ **(e)** $\sec x$

14. Find all values of m for which x^m is in the kernel of the operator $x^2D^2 - 3xD + 3$.

15. Match the following functions and operators so that on an appropriate interval the function is in the kernel of the operator.

(a) $x^2D^2 - xD - 8$ **(b)** $D^2 - 2D + 5$
(c) $x^3D^3 - 2xD$
(d) $(x^2 - 2)D^2 + 3xD - 15$
(e) $\ln |x|$ **(f)** $x^3 - x$
(g) x^4 **(h)** $e^x \cos x$

16. Describe the kernel of an nth-order linear differential operator L which is normal on an interval I.

17. Is it possible to have a linear transformation from C^n into R^n? Explain.

18. Show that a transformation $T: V \to W$ is linear if and only if for all scalars c_1, c_2 and for all $\mathbf{v}_1$, $\mathbf{v}_2$ in V, $T(c_1\mathbf{v}_1 + c_2\mathbf{v}_2) = c_1T\mathbf{v}_1 + c_2T\mathbf{v}_2$.

19. Establish Eq. (4).

In Exercises 20–23, **(a)** write the given transformation as a matrix operator, **(b)** determine a basis for its range, and **(c)** give the dimension of the range.

20. $f(x_1, x_2) = (2x_1 - x_2, x_1 + x_2)$

21. $f(x_1, x_2, x_3) = (x_1 + x_2, x_1 - x_3, x_2 + x_3, x_1 - x_2 + x_3)$

22. $f(w, x, y) = (3w + x + 5y, -2w + 4x + 6y)$

23. $f(w, x, y, z) = (w + 2x + y + 3z, 2w - x + y + z, w + y + z)$

24. (a) Does the transformation $T(x, y, z) = (x + 2y + z, x + 3y - 2z)$ map R^3 onto R^2?

(b) Is T an isomorphism? Explain.

25. Let V be the space of all piecewise regular functions on $[0, \infty)$ which are of exponential order with an abscissa of convergence α_0. Show that the Laplace transformation $\mathscr{L}$ is a linear operator from V into $F(\alpha_0, \infty)$.

26. Let $F(n)$ denote the space of all real-valued functions defined over the integers. Show that the shift operator E (Sec. 5.2) is a linear transformation on $F(n)$.

27. In Example 6, how is N related to the rank r of A?

28. Express the dimension of the null space of an $m \times n$ matrix A in terms of n and the rank r of A.

29. If a function transforms R^n into R^m, must it be linear?

30. Let $T: V \to W$ be a linear transformation, where $\dim V = n$ and $\dim W = m$.

(a) Show that if $m > n$, then T cannot be a transformation from V onto W.

(b) If $m < n$, show that T cannot be one-to-one.

13.6 SUMS, PRODUCTS, AND INVERSES OF LINEAR TRANSFORMATIONS

A remarkable property of the set of all linear transformations from any particular vector space into another is that such a set becomes a vector space itself, once appropriate operations of addition and of multiplication of transformations by scalars have been defined. Since transformations are functions, **equal** transformations must have the same domain and identical values.

Function addition is extended to linear transformations by simply requiring that they satisfy Eq. (8), Sec. 13.1. Thus, given any pair of linear transformations T_1, T_2 from V into W, their **sum** $T_1 + T_2$ is defined as the transformation which, for every $\mathbf{v}$ in V, has the value

(1) $$(T_1 + T_2)(\mathbf{v}) = T_1(\mathbf{v}) + T_2(\mathbf{v}) = T_1\mathbf{v} + T_2\mathbf{v}$$

Since $T_1\mathbf{v}$ and $T_2\mathbf{v}$ are in W and since W is closed under addition of its vectors, the sum $T_1 + T_2$ itself is a transformation from V into W.

If S is the set of scalars over which V and W are vector spaces, the **product** cT of a scalar c of S and a linear transformation $T: V \to W$ is defined to be the transformation which, for every $\mathbf{v}$ in V, has the value

(2) $$(cT)(\mathbf{v}) = c[T(\mathbf{v})] = cT(\mathbf{v}) = cT\mathbf{v}$$

Clearly, cT is also a transformation from V into W.

A verification that the transformation $T_1 + T_2$ is linear can be carried out by replacing $\mathbf{v}$ in the first and last members of (1) by a linear combination $c_1\mathbf{v}_1 + c_2\mathbf{v}_2$ of vectors in V and then using the fact that both T_1 and T_2 are linear transformations. This gives

$$\begin{aligned}(T_1 + T_2)(c_1\mathbf{v}_1 + c_2\mathbf{v}_2) &= T_1(c_1\mathbf{v}_1 + c_2\mathbf{v}_2) + T_2(c_1\mathbf{v}_1 + c_2\mathbf{v}_2)\\ &= c_1T_1\mathbf{v}_1 + c_2T_1\mathbf{v}_2 + c_1T_2\mathbf{v}_1 + c_2T_2\mathbf{v}_2\\ &= c_1(T_1 + T_2)(\mathbf{v}_1) + c_2(T_1 + T_2)(\mathbf{v}_2)\end{aligned}$$

Comparing this result with Eq. (5), Sec. 13.5, we see that $T_1 + T_2$ is indeed linear. The same substitution can be used to confirm the linearity of cT.

That the set of all linear transformations from V into W, together with its two built-in operations, is a linear space is easy to show. All we need to do is reason with the transformations of such a set exactly as we did with the functions of $F(D)$ in Sec. 13.1.

The transformation O which maps every vector of V onto the zero vector of W is not only unique but is also a linear transformation from V into W (Exercise 24). It is called the **zero transformation,** no matter what vector spaces V and W stand for. Of course, for every $T: V \to W$, the linear transformation $-T$ such that

(3) $$(-T)(\mathbf{v}) = -T\mathbf{v} \qquad \text{for all } \mathbf{v} \text{ in } V$$

is the **additive inverse** of T.

If W is a subspace of U, a linear transformation from V into W is also a linear transformation from V into U. On the other hand, if W is a subspace of V, a linear transformation from V into W is a linear transformation on V. The set of all linear transformations on a vector space V contains a transformation $\mathbf{I}$ called the **identity transformation.** It has the special property that

(4) $$\mathbf{I}(\mathbf{v}) = \mathbf{v} \qquad \text{for all } \mathbf{v} \text{ in } V$$

The restriction of $\mathbf{I}$ to any subspace of V is the identity transformation of that subspace. In other words, an identity transformation is a one-to-one linear transformation of a vector space onto itself. But this is not all. If V is a subspace of U, the identity transformation on V may be regarded as a linear transformation from V into U.

EXAMPLE 1

Let I be an arbitrary interval. If $0 \le k \le n$, the space $F^{(n)}(I)$ of all n times differentiable functions on I is a subspace of $F^{(k)}(I)$. Furthermore, the differential operator D^k is a linear transformation from $F^{(n)}(I)$ into $F^{(n-k)}(I)$ since, after being differentiated k times, every function in $F^{(n)}(I)$ can be differentiated at least $n - k$ more times. Thus D^k is a linear transformation from $F^{(n)}(I)$ into $F(I)$ since $F^{(n-k)}(I)$ is a subspace of $F(I)$.

Clearly, multiplication of any function f in $F(I)$ by 1 gives f. We may therefore interpret multiplication of functions by 1 as the identity transformation on $F(I)$. The restriction of this transformation to $F^{(n)}(I)$ is the identity transformation on $F^{(n)}(I)$. This restricted transformation can also be interpreted as a linear transformation from $F^{(n)}(I)$ into $F(I)$. Let us denote it by **I**. Then, for every f in $F^{(n)}(I)$, $\mathbf{I}f = f$, and for every real number c,

$$c\mathbf{I}f = cf \tag{5}$$

Because each of the operators $D^n, D^{n-1}, \ldots, D, \mathbf{I}$ belongs to the set of all linear transformations from $F^{(n)}(I)$ into $F(I)$, they can be multiplied by scalars and added. Thus any linear combination

$$a_0D^n + a_1D^{n-1} + \cdots + a_{n-1}D + a_n\mathbf{I} \tag{6}$$

of these operators is a linear transformation from $F^{(n)}(I)$ into $F(I)$. In particular, if $a_j = 0$, for $0 \le j \le n$, (6) becomes the zero transformation $\mathbf{O}$ which transforms every function in $F^{(n)}(I)$ into the trivial function on I.

If $a_0 \neq 0$, (6) is an nth-order linear differential operator with constant coefficients. In view of (5), we may write (6) as

$$L = a_0D^n + a_1D^{n-1} + \cdots + a_{n-1}D + a_n \qquad a_0 \neq 0 \tag{7}$$

It should now be clear why this operator transforms functions of $F^{(n)}(I)$ into functions of $F(I)$ and why

$$Lf = (a_0D^n + a_1D^{n-1} + \cdots + a_{n-1}D + a_n)f = a_0f^{(n)} + a_1f^{(n-1)} + \cdots + a_{n-1}f' + a_nf \tag{8}$$

We now know that by adding suitable linear transformations and by multiplying them by scalars, it is possible to produce more and more linear transformations. Another way of constructing linear transformations is by *composition* or, as we shall say, by *multiplication*. The circumstances under which linear transformations can be multiplied are these.

Let U, V, and W be vector spaces, all real or else all complex. Let T_1 and T_2 be linear transformations from U into V, and from V into W, respectively. Then, for every $\mathbf{u}$ in U, $T_1\mathbf{u}$ is a vector of V. Hence $T_2(T_1\mathbf{u})$ is defined and is a vector of W. Thus, to each vector $\mathbf{u}$ of U, there is made to correspond a unique vector $T_2(T_1\mathbf{u})$ of W. We may think of this correspondence, brought about by successive applications of T_1 and T_2, as a single transformation T_2T_1 from U directly into W. The transformation T_2T_1 is said to be the **product** of T_1 and T_2 **in that order** and is defined as follows.

DEFINITION 1 The **product** T_2T_1 of any two linear transformations $T_1\colon U \to V$ and $T_2\colon V \to W$ is the transformation from U into W such that

$$T_2T_1(\mathbf{u}) = T_2(T_1\mathbf{u}) \qquad \text{for all } \mathbf{u} \text{ in } U \tag{9}$$

The transformations $T_1\colon U \to V$, $T_2\colon V \to W$, and $T_2T_1\colon U \to W$ are portrayed pictorially in Fig. 13.4. Note that the vector $T_1\mathbf{u}$ in V is the image of the vector $\mathbf{u}$ in U under T_1, while the vector $T_2(T_1\mathbf{u})$ in W is the image of the vector $T_1\mathbf{u}$ in V under T_2 but is the image $T_2T_1(\mathbf{u})$ of the vector $\mathbf{u}$ in U under T_2T_1. Since functions are defined only over their domains, the product T_2T_1 can exist only if the range of T_1 is included in the domain of T_2.

By substituting an arbitrary linear combination $c_1\mathbf{u}_1 + c_2\mathbf{u}_2$ into (9) for $\mathbf{u}$ and then using the linearity of T_1 and T_2, it is readily verified that the product T_2T_1 is a linear transformation from U into W. To emphasize this fact, parentheses are dropped and $T_2T_1(\mathbf{u})$ is written $T_2T_1\mathbf{u}$. If T_1 and T_2 are both linear transformations on V, then T_2T_1 and T_1T_2 are also linear transformations on V. However, the transformations T_2T_1 and T_1T_2 will usually be different, even when their domains are the same.

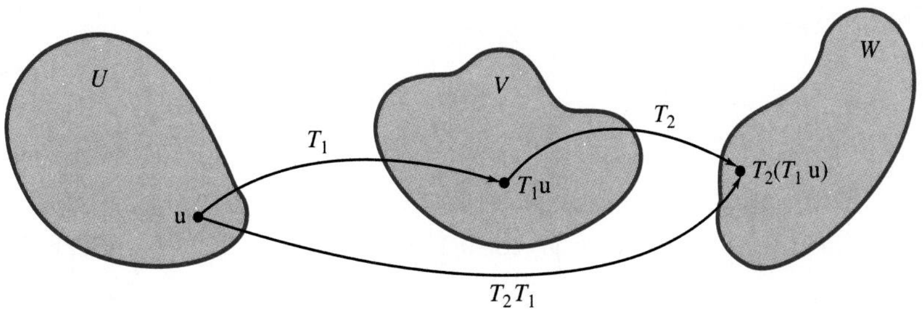

FIGURE 13.4
Images of a vector **u** under successive linear transformations T_1 and T_2 and under their product T_2T_1.

Clearly, products of more than two linear transformations can be formed provided the domain of each transformation in the product $T_nT_{n-1}\cdots T_2T_1$ includes the range of the transformation immediately to the right of it. All finite products of this sort turn out to be linear transformations.

EXAMPLE 2

Since an $m \times n$ matrix A with complex elements is a linear transformation from C^n into C^m, while an $l \times m$ matrix B is a linear transformation from C^m into C^l, the matrix product BA must be a linear transformation from C^n into C^l.†

In a descriptive way, we can think of A as a transformation which sends a column vector **x** of C^n into an m-component column vector $\mathbf{y} = \mathrm{A}\mathbf{x}$ of C^m. Likewise, B sends **y** into a column vector $\mathbf{z} = \mathrm{B}\mathbf{y}$ of C^l. Of course BA sends **x** directly into **z**, as is evident in the special case of matrix transformations from the relations

$$\mathbf{z} = \mathrm{B}\mathbf{y} = \mathrm{B}(\mathrm{A}\mathbf{x}) = \mathrm{BA}(\mathbf{x}) = \mathrm{BA}\mathbf{x}$$

Finally, BA sends a linear combination $c_1\mathbf{x}_1 + c_2\mathbf{x}_2$ of vectors in C^n into the linear combination $c_1\mathrm{BA}\mathbf{x}_1 + c_2\mathrm{BA}\mathbf{x}_2$ of vectors in C^l because the transformation BA is linear.

If $l \neq n$, the domain of A does not include the range of B, and so AB is undefined. If $l = n$, the domain of AB is C^m while the domain of BA is C^n. Thus, if $m \neq n$, AB $\neq$ BA. In fact, even if $l = m = n$, AB $\neq$ BA unless A and B commute.

EXAMPLE 3

The differential operator $D = d/dx$ is a linear transformation from the space $\mathscr{C}^{(1)}[0, 1]$ of all continuously differentiable functions on $0 \leq x \leq 1$ into the space $\mathscr{C}[0, 1]$ of continuous functions on [0, 1]. Indeed,

$$Df = f' \qquad \text{for every } f \text{ in } \mathscr{C}^{(1)}[0, 1]$$

The integral operator T such that, for every continuous function f on $0 \leq x \leq 1$, Tf is represented by

$$(Tf)(x) = Tf(x) = \int_0^x f(s)\, ds \qquad 0 \leq x \leq 1$$

is a linear transformation from $\mathscr{C}[0, 1]$ into $\mathscr{C}^{(1)}[0, 1]$. Moreover, the restriction T_0 of T to $\mathscr{C}^{(1)}[0, 1]$ is a linear transformation from $\mathscr{C}^{(1)}[0, 1]$ into $\mathscr{C}^{(2)}[0, 1]$, and hence into $\mathscr{C}[0, 1]$ since $\mathscr{C}^{(2)}[0, 1]$ is a subspace of $\mathscr{C}[0, 1]$. Linear combinations of D and T_0 are therefore linear transformations from $\mathscr{C}^{(1)}[0, 1]$ into $\mathscr{C}[0, 1]$. In particular, $D + T_0$ transforms the continuously differentiable function g, given by

$$g(x) = 10x^{3/2} \qquad 0 \leq x \leq 1$$

†Historically, the definition of a matrix product was introduced by the English mathematician Arthur Cayley (1821–1895) as a result of his investigations on linear transformations involving matrices.

into the continuous function $(D + T_0)g$ defined on $0 \le x \le 1$ by

$$(D + T_0)g(x) = (D + T_0)(10x^{3/2}) = 10(Dx^{3/2} + T_0x^{3/2}) = 15x^{1/2} + 10\int_0^x s^{3/2}\,ds$$
$$= 15x^{1/2} + 4x^{5/2}$$

Since the domain of D, namely, $\mathscr{C}^{(1)}[0, 1]$, includes the range of T, the product DT is defined and is a linear transformation on $\mathscr{C}[0, 1]$. For any continuous function f over $0 \le x \le 1$, DTf is the continuous function defined by

$$DTf(x) = D\int_0^x f(s)\,ds = f(x) \qquad 0 \le x \le 1$$

From this we see that DT is the identity transformation on $\mathscr{C}[0, 1]$.

The product TD is defined because the domain of T, namely, $\mathscr{C}[0, 1]$, includes the range of D. This product is a linear transformation on $\mathscr{C}^{(1)}[0, 1]$. For any function f in $\mathscr{C}^{(1)}[0, 1]$, TDf is the continuously differentiable function

$$TDf(x) = Tf'(x) = \int_0^x f'(s)\,ds = f(x) - f(0)$$

Clearly, DT and TD have different domains. Furthermore, only those functions in $\mathscr{C}^{(1)}[0, 1]$ which have the value zero at $x = 0$ are their own images under TD. So obviously, $DT \neq TD$.

Although multiplication of linear transformations is generally noncommutative, most of the properties satisfied by ordinary products of numbers carry over to products of linear transformations (Exercise 23).

In Section 13.5, we saw that a linear transformation $T\colon V \to W$ is one-to-one and onto if and only if the kernel of T is the trivial subspace of V and the range of T is W, i.e., T is an isomorphism if and only if

(10) $$K_T = \{\mathbf{o}\} \qquad \text{and} \qquad R_T = W$$

Linear transformations of this kind are especially important because they have an *inverse,* as do real-valued functions that are one-to-one and onto. For, if conditions (10) hold, then each vector in W is the image under T of a single vector in V or, to put it another way, to each vector $\mathbf{w}$ in W there corresponds a *unique* vector of V, namely, that vector $\mathbf{v}$ in V which T maps into $\mathbf{w}$. Thus we have a function defined whose domain is W. This function is denoted by T^{-1}, read "T inverse," and is called the **inverse** of T. Since T transforms every vector of V into a vector of W, every vector in V must be a value of T^{-1}. The range of T^{-1} is therefore V. Thus, with T in mind, we see that T^{-1} is *the* function with domain W and range V such that, for every $\mathbf{w}$ in W,

(11) $$T^{-1}(\mathbf{w}) = \mathbf{v} \qquad \text{if and only if } T\mathbf{v} = \mathbf{w}$$

Interestingly enough, T^{-1} is a *linear* transformation from W onto V. Summarizing, we have

THEOREM 1 If a linear transformation $T\colon V \to W$ is an isomorphism, its inverse transform T^{-1} from W onto V is linear.

PROOF Let $\mathbf{w}_1 = T\mathbf{v}_1$ and $\mathbf{w}_2 = T\mathbf{v}_2$ be arbitrary vectors of W, and let c_1 and c_2 be arbitrary scalars. In accordance with (11), and since T is linear,

(12) $$T^{-1}(c_1\mathbf{w}_1 + c_2\mathbf{w}_2) = \mathbf{v} \qquad \text{if and only if} \qquad T\mathbf{v} = c_1\mathbf{w}_1 + c_2\mathbf{w}_2 = c_1T\mathbf{v}_1 + c_2T\mathbf{v}_2 = T(c_1\mathbf{v}_1 + c_2\mathbf{v}_2)$$

Now, T is one-to-one and so $\mathbf{v} = c_1\mathbf{v}_1 + c_2\mathbf{v}_2$. Moreover, $\mathbf{v}_1 = T^{-1}(\mathbf{w}_1)$ and $\mathbf{v}_2 = T^{-1}(\mathbf{w}_2)$; hence $\mathbf{v} = c_1T^{-1}(\mathbf{w}_1) + c_2T^{-1}(\mathbf{w}_2)$. Substituting this expression for $\mathbf{v}$ into the first of Eqs. (12), we have

$$T^{-1}(c_1\mathbf{w}_1 + c_2\mathbf{w}_2) = c_1T^{-1}(\mathbf{w}_1) + c_2T^{-1}(\mathbf{w}_2)$$

which shows that T^{-1} is linear.

Not only is the transformation T^{-1}: $W \rightarrow V$ onto, but it is one-to-one as well. So T^{-1} has an inverse, and of course $(T^{-1})^{-1} = T$.

Since under assumptions (10) the domain W of T^{-1} is the range R_T of T, the product $T^{-1}T$ is defined and has V both as its domain and range. In fact, $T^{-1}T\mathbf{v} = \mathbf{v}$ for every $\mathbf{v}$ in V, and so $T^{-1}T$ is the identity transformation $\mathbf{I}_v$ on V; that is, $T^{-1}T = \mathbf{I}_v$.

On the other hand, the domain V of T is the range of T^{-1}. So the product TT^{-1} is also defined and has W as both its domain and range. Since $TT^{-1}\mathbf{w} = \mathbf{w}$ for every $\mathbf{w}$ in W, TT^{-1} is the identity transformation $\mathbf{I}_w$ on W; that is, $TT^{-1} = \mathbf{I}_w$.

EXAMPLE 4

VECTOR ROTATION BY A MATRIX

For all real values of α, the matrix $\mathrm{A} = \begin{bmatrix} \cos\alpha & -\sin\alpha & 0 \\ \sin\alpha & \cos\alpha & 0 \\ 0 & 0 & 1 \end{bmatrix}$ is nonsingular because $\det A = 1$. It follows from Example 6, Sec. 13.5, that A is a one-to-one linear transformation from R^3 onto R^3, and so A is invertible.

To determine the geometric effects of A on the vectors of R^3, let us regard these vectors as points in three-dimensional space. Expressing a general vector $\mathbf{v}$ in terms of both rectangular coordinates x, y, z and cylindrical coordinates r, θ, z (Fig. 13.5a) and using column notation, we have

$$\mathbf{v} = \begin{bmatrix} x \\ y \\ z \end{bmatrix} = \begin{bmatrix} r\cos\theta \\ r\sin\theta \\ z \end{bmatrix}$$

Applying A to $\mathbf{v}$, we obtain

$$\mathrm{A}\mathbf{v} = \begin{bmatrix} \cos\alpha & -\sin\alpha & 0 \\ \sin\alpha & \cos\alpha & 0 \\ 0 & 0 & 1 \end{bmatrix} \begin{bmatrix} r\cos\theta \\ r\sin\theta \\ z \end{bmatrix} = \begin{bmatrix} r\cos\theta\cos\alpha - r\sin\theta\sin\alpha \\ r\cos\theta\sin\alpha + r\sin\theta\cos\alpha \\ z \end{bmatrix}$$

$$= \begin{bmatrix} r\cos(\theta+\alpha) \\ r\sin(\theta+\alpha) \\ z \end{bmatrix} = \begin{bmatrix} x' \\ y' \\ z' \end{bmatrix} = \mathbf{v}'$$

Of course, $\mathbf{v}'$ is the image of $\mathbf{v}$ under A. Since $\mathbf{v}$ and $\mathbf{v}'$ have the same third component and since $(x')^2 + (y')^2 = r^2 = x^2 + y^2$, $\mathbf{v}$ and $\mathbf{v}'$ lie in a plane perpendicular to the z axis, and $\mathbf{v}'$ is at the same distance from the z axis as $\mathbf{v}$. Now, the difference in the polar angle $\theta + \alpha$ of $\mathbf{v}'$ and the polar angle θ of $\mathbf{v}$ is α. Thus, A rotates each point in space about the z axis through the angle α (Fig. 13.5b). In particular, the orthonormal vectors $(1, 0, 0)$ and $(0, 1, 0)$ rotate into the respective orthonormal vectors $(\cos\alpha, \sin\alpha, 0)$ and $(-\sin\alpha, \cos\alpha, 0)$, while $(0, 0, 1)$ is left unchanged. If $\alpha = \pi/4$, A rotates the point $(\sqrt{2}, 0, 2)$ into the point $(1, 1, 2)$ as shown in Fig. 13.5c. Should $\alpha = 0$, A becomes the identity matrix I which acts as the identity transformation on R^3.

Geometric considerations clearly indicate that for any given angle α, A^{-1} ought to rotate every point in space about the z axis through the angle $-\alpha$. This suggests that A^{-1} can be obtained from A by simply replacing α by $-\alpha$, which gives

$$\mathrm{A}^{-1} = \begin{bmatrix} \cos\alpha & \sin\alpha & 0 \\ -\sin\alpha & \cos\alpha & 0 \\ 0 & 0 & 1 \end{bmatrix}$$

With A^{-1} thus determined, it is readily verified that $\mathrm{A}^{-1}\mathrm{A} = \mathrm{A}\mathrm{A}^{-1} = \mathrm{I}$. It is also apparent that

$$\mathrm{A}^T = \mathrm{A}^{-1} \tag{13}$$

This is not a mere coincidence. Any matrix A whose transpose is its inverse is said to be **orthogonal.** Although we cannot go into details, it is worth noting that orthogonal matrices derive their name from the fact that the matrix of a transformation which is a rotation of mutually perpendicular, or orthogonal, axes in two, three, or more dimensions is always orthogonal.

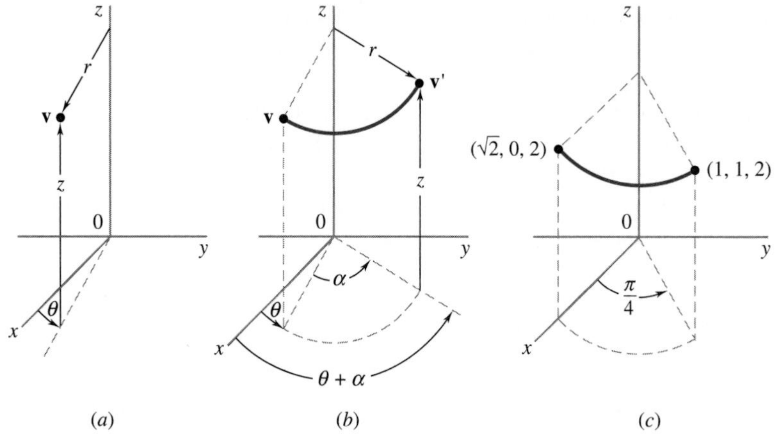

FIGURE 13.5
Rotation of vectors by an orthogonal matrix.

TABLE 13.1

A

$\overline{A}$ = conjugate of A
A^T = transpose of A
$\overline{A}^T$ = tranjugate of A
A^{-1} = inverse or reciprocal of A (A nonsingular)

Condition on A	Type
$A = \overline{A}$	Real
$A = -\overline{A}$	Imaginary
$A = A^T$	Symmetric
$A = -A^T$	Skew-symmetric
$A = \overline{A}^T$	Hermitian
$A = -\overline{A}^T$	Skew-hermitian
$A = (A^T)^{-1}$; that is, $A^{-1} = A^T$ or $AA^T = I$	Orthogonal
$A = (\overline{A}^T)^{-1}$; that is, $A^{-1} = \overline{A}^T$ or $A\overline{A}^T = I$	Unitary

As a generalization of the concept of an orthogonal matrix, a matrix A with complex elements and such that

$$\overline{A}^T = A^{-1} \tag{14}$$

is called a **unitary, or hermitian orthogonal,** matrix. In Table 13.1 we bring together a number of the special matrices introduced in Sec. 3.4, as well as the two new types just defined.

EXERCISES

1. **(a)** Verify that both $A = \begin{bmatrix} 2 & -3 & 1 \\ 1 & 4 & 2 \end{bmatrix}$ and $B = \begin{bmatrix} 1 & -2 & 1 \\ 3 & 2 & 2 \end{bmatrix}$ are linear transformations from R^3 onto R^2.

(b) Are 3A, −B, A + 2B, and 7A − 4B linear transformations? Explain.

(c) Find the range of A − B and show that A − B is not a linear transformation from R^3 onto R^2.

(d) Verify directly that the range of A − B is a linear space.

(e) Identify all vectors in R^3 which have $\begin{bmatrix} -1 \\ 5 \end{bmatrix}$ as their image under 2B − A.

2. Let T be the transformation from R^3 into R^2 defined by $T(x, y, z) = \begin{bmatrix} x - y \\ z + x \end{bmatrix}$.
(a) Express T as a matrix and thus show that T is linear.
(b) Express $-T$ as a matrix.
(c) Verify directly that $T - T$ is a zero transformation.
(d) Find the kernel of $2T + A$ where $A = \begin{bmatrix} 0 & 4 & 2 \\ 0 & 2 & 0 \end{bmatrix}$.

3. Let T_1 be the integral operator which transforms each function f of $\mathscr{C}[-2, 2]$ into the continuously differentiable function $T_1 f$ defined by

$$T_1 f(x) = \int_0^x f(s)\, ds \qquad -2 \le x \le 2$$

Also let $T_2 = D^2$ and let $T_3 = xD + 6$, where $D = d/dx$.
(a) Show that $T_1 + T_2 + T_3$ is a linear transformation from $\mathscr{C}^{(2)}[-2, 2]$ into $\mathscr{C}[-2, 2]$.
State whether or not the following functions are in the domain of $T_1 + T_2 + T_3$ and apply this operator to those that are.
(b) $\cos x$ **(c)** $\text{Tan}^{-1} x$ **(d)** $\tan x$
(e) $\sinh x$ **(f)** $\ln |x|$

4. The set of all elements which a number of sets contain in common is called the **intersection** of those sets. If T_1: $V_1 \to W_1$ and T_2: $V_2 \to W_2$ are linear transformations, if the intersection $V_1 \cap V_2$ of V_1 and V_2 is not empty, and if c is a scalar, are all of the transformations cT_1, cT_2, and $T_1 + T_2$ necessarily linear? Explain.

5. The set of all elements contained in at least one of a number of sets is called the **union** of those sets. If T: $V \to W$ is a linear transformation, if V_1 and V_2 are subspaces of V, and if $V_1 \cup V_2$ is the union of V_1 and V_2, is the transformation T: $V_1 \cup V_2 \to W$ necessarily linear? Explain.

In Exercises 6–11 find, if possible, T_1T_2 and T_2T_1. If both products exist, determine whether or not $T_1T_2 = T_2T_1$.

6. $T_1(x, y) = (-x, y)$, $T_2(x, y) = (-x, -y)$, for all (x, y) in R^2
7. $T_1(x, y) = (0, y)$, $T_2(x, y) = (x, 0)$, for all (x, y) in R^2
8. $T_1(x, y) = (x - y, 2x + y)$, $T_2(x, y) = (7x + 5y, 7y - 10x)$, for all (x, y) in R^2
9. $T_1(x, y) = (x - 3y, x + 2y)$, $T_2(x, y) = (2x - 2y, x - y)$, for all (x, y) in R^2
10. $T_1(x, y, z) = (z, x - y)$, for all (x, y, z) in R^3; $T_2(x, y) = (x, x + y, -y)$, for all (x, y) in R^2
11. $T_1(w, x, y, z) = (w - z, y + z, x - y)$, for all (w, x, y, z) in R^4; $T_2(x, y, z) = (x + y, y - z)$, for all (x, y, z) in R^3
12. Show that there are no linear transformations T_1: $R^3 \to R^2$ and T_2: $R^2 \to R^3$ such that T_2T_1 is the identity transformation on R^3 while T_1T_2 is the identity transformation on R^2. *Hint:* Express T_1 and T_2 as matrices; see Exercise 2.

In Exercises 13–17, T_1 and T_2 are linear transformations on the space of all polynomials P. Determine T_1T_2 and T_2T_1 and apply each product to the given polynomial.

13. $T_1 p(x) = 4xp'(x)$, $T_2 p(x) = p(x) - xp'(x)$, $9 + 6x - x^2$
14. $T_1 p(x) = p(x) - p(x - h)$, $T_2 p(x) = p(x) - p(x + h)$, $1 + x^2$
15. $T_1 p(x) = p'(x)$, $T_2 p(x) = xp(x)$, $1 + x^3$
16. $T_1 = xD - 2$, $T_2 = D + 3$, $1 - 6x$
17. $T_1 = axD + h$, $T_2 = bxD + k$, x^{10}, with a, h, b, k scalars
18. Let T_1 and T_2 be linear integral operators defined by

$$T_1 f(x) = \int_{-1}^x f(s)\, ds \qquad \text{and} \qquad T_2 f(x) = \int_x^1 f(s)\, ds$$

where $-1 \le x \le 1$ and f is in $\mathscr{C}[-1, 1]$.
(a) Find T_1T_2 and T_2T_1. **(b)** Is $T_1T_2 = T_2T_1$?
(c) Apply T_1T_2 and T_2T_1 to $1/(x + 2)$.
(d) Find $T_1 + T_2$.
(e) What is the range of $T_1 + T_2$?
(f) Apply $T_1 + T_2$ to $\text{Sin}^{-1} x$.

19. Let T_1 and T_2 be the linear integral operators defined on the space of all polynomials P by

$$T_1 p(x) = \int_0^x \int_0^t p(s)\, ds\, dt \qquad T_2 p(x) = \int_0^x \int_t^x p(s)\, ds\, dt$$

(a) Reduce the double integrals defining T_1 and T_2 to single integrals.
(b) Find T_1T_2 and T_2T_1. **(c)** Is $T_1T_2 = T_2T_1$?
(d) Apply T_1T_2 and T_2T_1 to $30x$.
(e) Find $T_1 + T_2$.
(f) Apply $T_1 + T_2$ to $30x$. **(g)** Find $T_2(T_1 + T_2)$.
(h) Apply $T_2(T_1 + T_2)$ to $30x$.

20. If possible, find inverses of the following linear transformations on R^2 or R^3.
(a) $T(x, y) = (-y, y - x)$
(b) $T(x, y) = (3x - y, y - 2x)$
(c) $T(x, y) = (x \cos\theta - y \sin\theta, x \sin\theta + y \cos\theta)$
(d) $T(x, y, z) = (-y, z, y + z)$
(e) $T(x, y, z) = (x - y - z, 2x + z, 3x - y)$
(f) $T(x, y, z) = (x - y, y - z, z)$
(g) $A = \begin{bmatrix} 0 & 1 & -1 \\ 0 & 0 & 1 \\ 1 & -1 & 0 \end{bmatrix}$ **(h)** $A = \begin{bmatrix} 1 & 0 & -1 \\ 1 & -1 & -1 \\ 1 & -1 & -2 \end{bmatrix}$

21. Determine which, if any, of the following matrices are orthogonal and which are unitary.
(a) $\begin{bmatrix} \dfrac{1 - i}{\sqrt{3}} & \dfrac{1}{\sqrt{2}} \\ \dfrac{-1}{\sqrt{3}} & \dfrac{1 + i}{\sqrt{2}} \end{bmatrix}$ **(b)** $\begin{bmatrix} \dfrac{\sqrt{2}(1 + i)}{6} & \dfrac{2(1 + i)}{3} \\ \dfrac{-2\sqrt{2}}{3} & \dfrac{1}{3} \end{bmatrix}$

(c) $\begin{bmatrix} \dfrac{2}{3} & \dfrac{1}{\sqrt{2}} & \dfrac{1}{3\sqrt{2}} \\ \dfrac{2}{3} & \dfrac{-1}{\sqrt{2}} & \dfrac{1}{3\sqrt{2}} \\ \dfrac{-1}{3} & 0 & \dfrac{4}{3\sqrt{2}} \end{bmatrix}$ **(d)** $\begin{bmatrix} 0 & 1 & 0 \\ \dfrac{4}{5} & 0 & \dfrac{3}{5} \\ \dfrac{3}{5} & 0 & \dfrac{-4}{5} \end{bmatrix}$

22. Prove that the set of all linear transformations from one vector space into another is itself a vector space, assuming the transformations can be added, and multiplied by scalars, in accordance with Eqs. (1) and (2).

23. Let T_1, T_2, and T_3 be linear transformations and let c be an arbitrary scalar. Assuming that all of the indicated products exist, show that
(a) $T_1(T_2T_3) = (T_1T_2)T_3$
(b) $(T_1 + T_2)T_3 = T_1T_3 + T_2T_3$
(c) $T_1(T_2 + T_3) = T_1T_2 + T_1T_3$
(d) $(cT_1)T_2 = T_1(cT_2) = c(T_1T_2)$
(e) If T_1: $V \to W$, then $T_1\mathbf{I}_v = T_1$ and $\mathbf{I}_wT_1 = T_1$

24. Show that a transformation from a vector space V into a vector space W which maps every vector of V onto the zero vector of W is a unique linear transformation from V into W.

25. Given the transformation T: $R^2 \to R^2$, where $T(x, y) = (ax + cy, cx + by)$, find $T^2(x, y)$. *Hint:* $T^2 = TT$.

26. If two linear transformations T_1: $R^3 \to R^3$ and T_2: $R^3 \to R^3$ are isomorphisms, are T_1T_2 and T_2T_1 isomorphisms? Explain.

27. **(a)** In Exercise 25, write T as a matrix operator A.
(b) If $a = \cos\theta$, $b = -\cos\theta$, and $c = \sin\theta$, is A symmetric? Orthogonal?

28. Prove that if T_1 and T_2 are invertible linear transformations on a linear space V, then both T_1T_2 and T_2T_1 are invertible and $(T_1T_2)^{-1} = T_2^{-1}T_1^{-1}$.

29. Prove that the determinant of any orthogonal matrix is either 1 or -1. Is the converse true?

30. Prove that a real matrix is orthogonal if and only if its column vectors are unit vectors which are mutually orthogonal.

31. Prove that if the column vectors of a real matrix A are mutually orthogonal unit vectors, so are the row vectors of A.

32. Show that a square matrix is unitary if and only if its columns are orthonormal vectors in the hermitian sense.

33. Prove that the product of two orthogonal matrices is an orthogonal matrix.

34. Prove that the inverse of an orthogonal matrix is an orthogonal matrix.

35. Show that a matrix U which is both unitary and hermitian must satisfy the equation $X^2 = I$.

13.7 LINEAR OPERATOR EQUATIONS

Given a transformation f: $A \to B$, then for any b in B, we can formulate a condition

$$f(x) = b \tag{1}$$

and ask which, if any, members of A satisfy this condition. Such a condition is called an **operator equation.** An entity a is a **solution** of (1) if and only if it **satisfies** (1), i.e., if and only if a is in A and $f(a) = b$. The **complete solution** of an operator equation is the set of all its solutions. Of course, this set might be empty. For (1) to have a solution, b must be in the range of f. If b is not in the range of f, there is no solution of (1).

The definitions of onto and of one-to-one transformations, given in Sec. 13.5, apply to general functions as well as to linear transformations. If f maps A onto B, then, given any b in B, Eq. (1) has a solution. If f is also one-to-one, the solution is unique and f has an inverse. Theorems that specify conditions under which an operator equation has a solution are called **existence theorems.** They save us from looking for solutions when there are none. Theorems stipulating conditions under which an operator equation has *at most one* solution are called **uniqueness theorems.** If one solution has been found, they say when there are no others.

EXAMPLE 1 THE PRINCIPAL INVERSE SINE

The familiar *sine* function transforms the set of all real numbers R into itself. As we know, the range of $\sin x$ is the set of all real numbers y such that $|y| \le 1$. For any real number b, if $|b| \le 1$, the operator equation

$$\sin x = b$$

has infinitely many solutions. In particular, the complete solution of

$$\sin x = 0$$

is the set of all real numbers x such that $x = n\pi$, where n is an integer. If $|b| > 1$, $\sin x = b$ has no solution. Clearly, the sine function does not map R onto R, but it does map R onto $[-1, 1]$. The mapping is not one-to-one, however. By restricting the domain of sin x to $-\pi/2 \leq x \leq \pi/2$, we obtain a function Sin x which is a one-to-one transformation from $[-\pi/2, \pi/2]$ onto $[-1, 1]$. Thus the equation Sin $x = b$, with $|b| \leq 1$, has a unique solution $x = \text{Sin}^{-1} b$, where the function $\text{Sin}^{-1} x$ is the inverse of Sin x. By name, $\text{Sin}^{-1} x$ is called the **principal arc,** or **inverse, sine function.** Its domain is $[-1, 1]$; its range is $[-\pi/2, \pi/2]$.

Now let us replace f, x and b in (1) by T, **v**, and **w**, respectively, where T stands for a *linear* transformation from V into W, where **v** is a variable which may be replaced by any vector of V, and where **w** is a prescribed vector of W. The resulting equation

$$T\mathbf{v} = \mathbf{w} \tag{2}$$

in the variable **v**, is called a **linear operator equation.** According as **w** is or is not the zero vector **0** of W, Eq. (2) is said to be **homogeneous** or **nonhomogeneous.** If $\mathbf{w} \neq \mathbf{0}$, so that (2) is nonhomogeneous,

$$T\mathbf{v} = \mathbf{0} \tag{3}$$

is referred to as the **homogeneous linear operator equation corresponding to** (2).

Clearly, the complete solution of (3) is the kernel K_T of the transformation T. In addition to containing all solutions of (3), K_T is a subspace of V. For this reason, K_T is frequently called the **solution space** of (3). Although Eq. (2) may have no solutions, (3) always has at least one solution, namely, $\mathbf{v} = \mathbf{o}$, because K_T must contain the zero vector **o** of V.

If $c_1, c_2, \ldots, c_m$ are arbitrary scalars, while $\mathbf{v}_1, \mathbf{v}_2, \ldots, \mathbf{v}_m$ are any m solutions of (3), then

$$T\left(\sum_{j=1}^{m} c_j\mathbf{v}_j\right) = \sum_{j=1}^{m} c_jT(\mathbf{v}_j) = \mathbf{0}$$

This shows that any finite linear combination of solutions of a homogeneous linear operator equation is itself a solution of that equation.

In case (2) has a solution $\mathbf{v}_p$, it can be added to any solution $\mathbf{v}_h$ of (3) and the sum $\mathbf{v}_h + \mathbf{v}_p$ is also a solution of (2). To verify this, simply observe that

$$T(\mathbf{v}_h + \mathbf{v}_p) = T(\mathbf{v}_h) + T(\mathbf{v}_p) = \mathbf{0} + \mathbf{w} = \mathbf{w}$$

Carrying this result further, we have

THEOREM 1 If $\mathbf{v}_h$ represents the complete solution of Eq. (3) and if $\mathbf{v}_p$ is any *particular* solution of Eq. (2), then $\mathbf{v}_h + \mathbf{v}_p$ represents the complete solution of (2).

PROOF If **v** is an arbitrary solution of (2), both

$$T\mathbf{v} = \mathbf{w} \qquad \text{and} \qquad T\mathbf{v}_p = \mathbf{w}$$

hold. These equations imply that

$$T\mathbf{v} = T\mathbf{v}_p \qquad \text{or} \qquad T\mathbf{v} - T\mathbf{v}_p = \mathbf{0}$$

Since T is linear, it follows that

$$T(\mathbf{v} - \mathbf{v}_p) = \mathbf{0}$$

From this we see that $\mathbf{v} - \mathbf{v}_p$ is represented by $\mathbf{v}_h$, and we have

$$\mathbf{v} - \mathbf{v}_p = \mathbf{v}_h \qquad \text{or} \qquad \mathbf{v} = \mathbf{v}_h + \mathbf{v}_p$$

as asserted.

As another useful observation, we note that if $\mathbf{v}_k$ is a solution of a nonhomogeneous equation $T\mathbf{v} = \mathbf{w}_k$, for $1 \leq k \leq m$, then

$$\sum_{k=1}^{m} \mathbf{w}_k = \sum_{k=1}^{m} T\mathbf{v}_k = T\left(\sum_{k=1}^{m} \mathbf{v}_k\right)$$

and it follows that $\mathbf{v}_1 + \mathbf{v}_2 + \cdots + \mathbf{v}_m$ is a solution of $T\mathbf{v} = \mathbf{w}_1 + \mathbf{w}_2 + \cdots + \mathbf{w}_m$.

An operator equation is **consistent,** or **solvable,** if and only if it has a solution. Otherwise it is **inconsistent.** If Eq. (2) is consistent and $\mathbf{v}_p$ is some particular solution, then a complete solution $\mathbf{v}_h + \mathbf{v}_p$ and $\mathbf{v}_p$ will be identical if and only if $\mathbf{v}_h = \mathbf{o}$. But, $\mathbf{v}_h = \mathbf{o}$ if and only if $K_T = \{\mathbf{o}\}$. Thus, if (2) has a solution at all, it is unique if and only if T has a trivial null space. This is not surprising since T is then one-to-one.

From Sec. 13.3, we know that every vector in the kernel K_T of a linear transformation $T: V \rightarrow W$ is expressible as a linear combination of the members of a finite linearly independent set of n vectors of K_T if and only if some basis of K_T contains n vectors.† This being the case, K_T has dimension n, where n is a unique nonnegative integer. In particular, $\dim K_T = 0$ if and only if $K_T = \{\mathbf{o}\}$, in which event T is one-to-one, the empty set serves as a basis for K_T, and the only solution of (3) is the so-called **trivial solution** $\mathbf{v} = \mathbf{o}$. Clearly, K_T is a nontrivial space if and only if $\dim K_T = n > 0$.

Since vectors in K_T are solutions of (3) and, conversely, every solution of (3) can be expressed as a linear combination of any n basis vectors of K_T, a linear combination of any n linearly independent particular solutions of (3) is called a **complete solution** of (3) and a **complementary function** of (2).

The preceding remarks clearly explain why so many ideas and results concerning linear differential equations, linear difference equations, linear algebraic equations, and Laplace transformations, not to mention other important problems involving linear transformations, bear such a striking resemblance to one another.

In working with operator equations, it is often necessary to find solutions which satisfy certain auxiliary conditions. Of course the operator equations to be solved, as well as the auxiliary conditions, may differ from problem to problem. Equation (1) gives the form of a general operator equation. Assuming that (1) is consistent, its complete solution X is a nonempty subset of the domain A of f. Clearly, only auxiliary conditions pertaining to solutions of (1), and hence to elements of A, are of any interest to us. Each such condition specifies a unique subset of A.‡ Let A_0 be the subset of A specified by a given condition. Then, for every x in A, x satisfies the condition in question if and only if $x \in A_0$. A condition

(4) $$x \in A_0$$

[read "x is an element (or member) of A_0"] is called a **set theoretic condition.** By simply adjoining such a condition to Eq. (1), we obtain an **operator system**

(5*a*) $$f(x) = b$$
(5*b*) $$x \in A_0$$

where A_0 may be any subset of A. A complete solution of such a system, which may be the empty set, will contain all elements of A in the intersection of X and A_0.

†The vector **o** in V is considered to be a linear combination of the vectors of the empty set $\varnothing$.
‡This assertion follows from the *axiom of specification* of *set theory*.

A **linear set theoretic condition** $\mathbf{v} \in U$ is one in which U is a vecto · space. By adjoining such a condition to Eq. (2), we obtain a **linear operator system**

$$T\mathbf{v} = \mathbf{w} \qquad (6a)$$
$$\mathbf{v} \in U \qquad (6b)$$

where U may be any subspace of V. The **homogeneous system** corresponding to (6) is

$$T\mathbf{v} = \mathbf{0} \qquad (7a)$$
$$\mathbf{v} \in U \qquad (7b)$$

Of course, a vector $\mathbf{v}_1$ of V is a **solution** of System (7) if and only if $T\mathbf{v}_1 = \mathbf{0}$ and $\mathbf{v}_1 \in U$.

Sometimes, such as when T is a normal linear differential operator and the condition $\mathbf{v} \in U$ takes the form of a set of linear boundary conditions, the total number of linearly independent solutions of (7) is called the **index of compatibility** of the system.

The fact that every solution of (7) must be in the intersection of K_T and U is sufficient to ensure that $K_T \cap U$ is a subspace of both K_T and U. Therefore

$$\dim (K_T \cap U) \leq \dim K_T \qquad \text{and} \qquad \dim (K_T \cap U) \leq \dim U$$

This suggests that the index of compatibility of a general homogeneous linear operator system be defined as follows:

DEFINITION 1 The **index of compatibility of System** (7) is k if and only if $k = \dim (K_T \cap U)$.

DEFINITION 2 **System** (7) **is incompatible** if and only if $\dim (K_T \cap U) = 0$.

In case no auxiliary condition is imposed, System (7) reduces to Eq. (3), and we say that **the index of compatibility of** (3) is equal to the dimension of K_T. Finally, Eq. (3) is said to be **incompatible** if and only if $K_T = \{\mathbf{o}\}$, so that the only solution of (3) is the trivial solution.

A definition of the index of compatibility, and of incompatibility, for more general systems involving several linear operator or set theoretic conditions may be readily formulated along the lines of Definitions 1 and 2. To further clarify ideas, we now consider two specific examples.

EXAMPLE 2

THE INDEX OF COMPATIBILITY OF A LINEAR ALGEBRAIC SYSTEM

To a system of m simultaneous, homogeneous, linear algebraic equations in n unknowns let there be adjoined a system of p more such equations. The result is a linear operator system

$$\mathrm{A}\mathbf{x} = \mathbf{0} \qquad (8a)$$
$$\mathrm{B}\mathbf{x} = \mathbf{0} \qquad (8b)$$

where A and B are in turn $m \times n$ and $p \times n$ matrices. Let us suppose A and B are real. Since $\mathrm{B}\mathbf{x} = \mathbf{0}$ if and only if $x \in K_{\mathrm{B}}$, we see that this system is a special case of (7) in which $V = R^n$, $T = \mathrm{A}$, and $U = K_{\mathrm{B}}$. On the other hand, (8) is equivalent to the linear system $\begin{bmatrix} \mathrm{A} \\ \mathrm{B} \end{bmatrix}\mathbf{x} = \mathbf{0}$. Hence, by Theorem 4, Sec. 13.3, the index of compatibility of (8) is equal to $n - r$, where r is the rank of the $(m + p)$ by n matrix $\begin{bmatrix} \mathrm{A} \\ \mathrm{B} \end{bmatrix}$.

Consider the specific operator system

$$\begin{bmatrix} 4 & -2 & 0 \\ 2 & 2 & -2 \end{bmatrix}\begin{bmatrix} x_1 \\ x_2 \\ x_3 \end{bmatrix} = \begin{bmatrix} 0 \\ 0 \end{bmatrix} \qquad \begin{bmatrix} 3 & 0 & -1 \\ -7 & 1 & 2 \end{bmatrix}\begin{bmatrix} x_1 \\ x_2 \\ x_3 \end{bmatrix} = \begin{bmatrix} 0 \\ 0 \end{bmatrix}$$

The rank of the matrix $\begin{bmatrix} 4 & -2 & 0 \\ 2 & 2 & -2 \\ 3 & 0 & -1 \\ -7 & 1 & 2 \end{bmatrix}$ is 3 because its row echelon form $\begin{bmatrix} 1 & 0 & 0 \\ 0 & 1 & 0 \\ 0 & 0 & 1 \\ 0 & 0 & 0 \end{bmatrix}$ has three nonzero rows. Therefore $k = 3 - 3 = 0$, and the system is incompatible.

EXAMPLE 3

THE INDEX OF COMPATIBILITY OF A BOUNDARY-VALUE PROBLEM

Let D be an $n \times n$ diagonal matrix in which each diagonal element is the differential operator d/dt, let $\mathrm{F} = [f_{ij}(t)]$ be an $n \times n$ matrix function continuous on an interval $[a, b]$, and let A and B be $m \times n$ constant matrices. Then the two-point boundary-value problem

(9*a*) $$(\mathrm{D} + \mathrm{F})\mathbf{x} = \mathbf{0}$$
(9*b*) $$\mathrm{A}\mathbf{x}(a) + \mathrm{B}\mathbf{x}(b) = \mathbf{0}$$

is a linear operator system in which the vector space V of (7) is now the set of all continuously differentiable n-component vector functions on $[a, b]$, T is the matrix operator D + F, and U is the subspace of V containing all vectors $\mathbf{x}$ such that $\mathrm{A}\mathbf{x}(a) + \mathrm{B}\mathbf{x}(b) = \mathbf{0}$. By Definition 1, the index of compatibility of (9) is given by $k = \dim (K_{\mathrm{D+F}} \cap U)$. From Sec. 4.4, the kernel $K_{\mathrm{D+F}}$ of D + F has dimension n.

Assuming X is a fundamental matrix of (9*a*), so that its n columns $\mathbf{x}_j$ constitute a basis for $K_{\mathrm{D+F}}$, a complete solution of (9*a*) is

(10) $$\mathbf{x}_h = \sum_{j=1}^{n} c_j\mathbf{x}_j$$

The requirement that a vector $\mathbf{x}$ shall be in $K_{\mathrm{D+F}} \cap U$ amounts to requiring that $\mathbf{x}_h$, which is already in $K_{\mathrm{D+F}}$, shall also be in U. This means that

(11) $$\mathrm{A}\mathbf{x}_h(a) + \mathrm{B}\mathbf{x}_h(b) = [\mathrm{A}, \mathrm{B}]\begin{bmatrix} \mathbf{x}_h(a) \\ \mathbf{x}_h(b) \end{bmatrix} = [\mathrm{A}, \mathrm{B}]\begin{bmatrix} \mathrm{X}(a) \\ \mathrm{X}(b) \end{bmatrix}\mathbf{c} = \mathbf{0}$$

must hold. Hence if the $m \times n$ matrix $\mathrm{AX}(a) + \mathrm{BX}(b)$ is of rank r, then $k = n - r$ is the index of compatibility of System (9).

EXERCISES

1. **(a)** Onto what subset of R does $\cos x$ map R?
   (b) The restriction $\mathrm{Cos}\, x$ of $\cos x$ to $[0, \pi]$ is a one-to-one mapping from $[0, \pi]$ onto what subset of R?
   (c) Identify the inverse of $\mathrm{Cos}\, x$, giving both its domain and range.
2. **(a)** What is the range of $\tan x$?
   (b) The restriction $\mathrm{Tan}\, x$ of $\tan x$ to $(-\pi/2, \pi/2)$ is a one-to-one transformation from $(-\pi/2, \pi/2)$ onto what subset of R?
   (c) Identify the inverse of $\mathrm{Tan}\, x$, giving both its domain and range.
3. Specify whether or nor the following operator equations are linear:
   (a) $[(1 + \cos^2 x)D^2 + x^{1/3}]y = \sin 2x$
   (b) $[k^2\Delta^2 - (k + 1)\Delta]f_k = k!$
   (c) $8x - 6y + 4z = 5$
   (d) $|x + y| - 19 = 0$
   (e) $3y'' - x \sin y = 0$
   (f) $\int_0^x (t^2 \sin t - t)y(t)\, dt = t^3y$
   (g) $\begin{bmatrix} 1 & 2 & 4 \\ -3 & 0 & 2 \end{bmatrix}\begin{bmatrix} x \\ y \end{bmatrix} = \begin{bmatrix} 2 \\ 7 \end{bmatrix}$

(h) $\begin{bmatrix} x \\ y \end{bmatrix}' = \begin{bmatrix} t & e^t \\ 1 & \cosh t \end{bmatrix}\begin{bmatrix} x \\ y \end{bmatrix} + \begin{bmatrix} \sec t \\ \tan t \end{bmatrix}$

(i) $\dfrac{\partial}{\partial x}\left[(6x)^{1/2}\dfrac{\partial y}{\partial x}\right] = \dfrac{\partial^2 y}{\partial t^2}$

(j) $r^2\dfrac{\partial^2 u}{\partial r^2} + r\dfrac{\partial u}{\partial r} = u\dfrac{\partial u}{\partial t}$

(k) $w\nabla^2 w + 9w = x + y + z$

(l) $\nabla^2 w + w_x + w_y + w_z = 3xyz$

4. Let $L(y) = a_0 y^{(n)} + a_1 y^{(n-1)} + \cdots + a_{n-1}y' + a_n y = 0$ be normal on an interval $[a, b]$, let A and B be $m \times n$ constant matrices, and let $\mathbf{y} = [y\, y' \cdots y^{(n-1)}]^T$.

(a) Verify that the two-point boundary-value problem

(12) **(i)** $L(y) = 0$ **(ii)** $\mathrm{A}\mathbf{y}(a) + \mathrm{B}\mathbf{y}(b) = \mathbf{0}$

is a special case of the linear operator system (7).

(b) Show that if $y_1, y_2, \ldots, y_n$ form a basis for all solutions on $[a, b]$ of (12**i**), then the matrix function $\mathrm{W}(x) = [y_j^{(i-1)}(x)]$ is nonsingular on $[a, b]$.

(c) Show that a complete solution $y = \sum_{j=1}^{n} c_j y_j$ of (12**i**) satisfies (12**ii**) if and only if

$$[\mathrm{AW}(a) + \mathrm{BW}(b)]\mathbf{c} = \mathbf{0}$$

where $\mathbf{c} = [c_1 \;\; c_2 \;\; \cdots \;\; c_n]^T$.

(d) Show that $\mathrm{AW}(a) + \mathrm{BW}(b) = [Y_i(y_j)]$; $i = 1, 2, \ldots, m$; $j = 1, 2, \ldots, n$; where

$$Y_i(y) = a_{i1}y(a) + \cdots + a_{in}y^{(n-1)}(a) + b_{i1}y(b) + \cdots + b_{in}y^{(n-1)}(b) = 0$$

is the ith boundary condition of System (12).

(e) Prove that if the matrix $\mathrm{Y} = [Y_i(y_j)]$ is of rank r, then the index of compatibility k of (12) is given by $k = n - r$, in which case the system admits of $n - r$ linearly independent solutions on $[a, b]$. Of course, more than $n - r$ solutions must be linearly dependent.

5. **(a)** Use the results of Exercise 4 to find the index of compatibility of the boundary value problem

(13) **(i)** $xy''' - y'' + xy' - y = 0$
(ii) $y(\pi) + y'(3\pi) = 0$, $y'(\pi) + y(3\pi) = 0$

Hint: One solution of (13**i**) can be found by inspection. The reduced equation (Sec. 2.4) can be solved using Theorem 1, Sec. 12.4, together with appropriate Bessel identities. An alternative, but less challenging procedure, is to seek solutions of the types x^m and e^{mx}.

(b) Convert (13) into a boundary-value problem involving a first-order linear system and use Example 3 to find the index of compatibility of the problem.

(c) Find a complete solution of System (13).

In Exercises 6–10, find the index of compatibility of the linear operator system consisting of Eqs. **(a)** and **(b).**

6. **(a)** $\begin{bmatrix} 1 & -2 & 0 & 3 \\ 2 & 0 & 1 & 1 \\ 3 & 1 & 2 & 0 \end{bmatrix}\begin{bmatrix} w \\ x \\ y \\ z \end{bmatrix} = \begin{bmatrix} 0 \\ 0 \\ 0 \\ 0 \end{bmatrix}$

(b) $\begin{bmatrix} -2 & 1 & 1 & 2 \\ -3 & 0 & -2 & 1 \end{bmatrix}\begin{bmatrix} w \\ x \\ y \\ z \end{bmatrix} = \begin{bmatrix} 0 \\ 0 \\ 0 \\ 0 \end{bmatrix}$

7. **(a)** $\begin{bmatrix} 2 & 2 & 3 & 1 \\ 1 & 5 & 3 & 2 \\ 2 & -2 & -1 & -1 \end{bmatrix}\begin{bmatrix} w \\ x \\ y \\ z \end{bmatrix} = \begin{bmatrix} 0 \\ 0 \\ 0 \\ 0 \end{bmatrix}$

(b) $\begin{bmatrix} 3 & 1 & 5 & 1 \\ 1 & -1 & 2 & 0 \end{bmatrix}\begin{bmatrix} w \\ x \\ y \\ z \end{bmatrix} = \begin{bmatrix} 0 \\ 0 \\ 0 \\ 0 \end{bmatrix}$

8. **(a)** $y^{\text{iv}} + y'' = 0$

(b) $y(0) + y''(0) + y'(\pi) - y'''(\pi) = 0$
$y'(0) - y'''(0) + y''(\pi) = 0$
$y''(0) + y'''(0) - y'(\pi) - 2y'''(\pi) = 0$

9. **(a)** $x^3y''' - 3x^2y'' + 6xy' - 6y = 0$

(b) $y'(1) - y''(1) + y(2) = 0$
$y(1) - y''(2) = 0$
$8y(1) - 8y''(1) - 13y(2) + 18y'(2) - 6y''(2) = 0$

10. **(a)** $\begin{bmatrix} x \\ y \\ z \end{bmatrix}' = \dfrac{1}{t}\begin{bmatrix} 0 & -1 & 1 \\ -t & 0 & t \\ -t & -1 & 1+t \end{bmatrix}\begin{bmatrix} x \\ y \\ z \end{bmatrix}$

Hint: $\mathrm{X} = \begin{bmatrix} 1 & t & 0 \\ 1 & 0 & e^t \\ 1 & t & e^t \end{bmatrix}$

(b) $2y(1) - 2z(1) - y(2) + z(2) = 0$
$x(1) + x(2) + y(2) = 0$

CHAPTER 14

APPLICATIONS AND FURTHER PROPERTIES OF MATRICES

This chapter is a balanced presentation of a number of applications of matrices, some fairly simple and some quite complex, and an extension of the theory introduced in Secs. 3.3–3.8. In Sec. 14.1 the applications are straightforward, involving such things as communication networks and probability processes and could probably have been handled in Chap. 3. Section 14.2 deals with the use of matrices in the relatively new field of the *fast Fourier transform,* which is useful in *signal processing,* where the trigonometric representation of discrete rather than continuous functions is required.

Section 14.3 introduces the important concept of the *rank* of a matrix and its role in the transformation of matrices to one or another canonical form by means of elementary row or column operations.

In Sec. 14.4, the idea of a *Green's function,* which we introduced in Sec. 2.12 for a single linear differential equation, is extended to linear systems of differential equations, an extension that would be all but impossible were it not for matrix notation and certain basic properties of matrices.

Section 14.5 deals with *quadratic forms,* that is, homogeneous second-degree expressions in n variables, and their intimate relation to square matrices. Section 14.6 is a major extension of the ideas of the characteristic values and characteristic vectors of a square matrix introduced in Sec. 3.8. Particular emphasis is placed on the linear dependence and independence and the orthogonality of the characteristic vectors. Section 14.7 deals with transformations of square matrices, not by row and column operations but by transformation of the variables in the related quadratic form. In both Secs. 14.6 and 14.7, these ideas and the related concept of the *normal coordinates* of a dynamical system are illustrated in a series of examples dealing with the vibration of a spring-connected, three-mass system.

Section 14.8 discusses functions of a square matrix, including such forms as polynomials, rational fractional functions, and infinite series.

Prerequisites for this chapter: Sec. 2.12, Chap. 3, Sec. 7.6.

14.1 MISCELLANEOUS APPLICATIONS

The most common application of matrices is probably their use in solving systems of linear equations, which today may contain thousands of equations and involve extensive use of high-speed computers. On the other hand, matrices are also used for the orderly display and manipulation of data of various kinds. As an example of this sort, we begin our discussion with what sociologists call a **communication network,** that is, a group of individuals in which only certain pairs can communicate, some in both directions, some perhaps in only one direction. One example might be a group of people with unlisted phone numbers, some of which are known to only a few others in the group. Another example might be a group of nations which, for political reasons, might not allow planes from certain countries to land at their airfields. Such relations can easily be displayed in what is called an **accessibility matrix,** using the following conventions: if there are n individuals, $P_1, P_2, \ldots, P_n$, construct a matrix $\mathrm{P} = [p_{ij}]$ in which

$$p_{ij} = \begin{cases} 1 & \text{if } P_i \text{ can communicate with } P_j,\ i \neq j \\ 0 & \text{if } P_i \text{ cannot communicate with } P_j,\ i \neq j \\ 0 & \text{if } i = j, \text{ since communication with oneself is trivial} \end{cases}$$

Clearly, if P_i and P_j can communicate with each other, $p_{ij} = p_{ji} = 1$.

Now consider the square of the matrix P. The general entry in P^2 is

$$p_{ij}^{(2)} = \sum_{k=1}^{n} p_{ik}p_{kj} \tag{1}$$

If either p_{ik} or p_{kj} is zero, the product $p_{ik}p_{kj}$ is zero, which means that P_i does not have a two-step communication path to P_j via P_k. The product $p_{ik}p_{kj}$ is 1 if and only if $p_{ik} = p_{kj} = 1$, that is, if and only if P_i has a two-step communication path to P_j via P_k. Thus, for each pair (i, j), the sum in (1) is the number of different ways in which P_i can communicate with P_j via *some* other party. In particular, P_i receives personal feedback, i.e., has only one two-step communication path back to P_i, if and only if $p_{ii}^{(2)} = 1$. Higher powers of P have similar interpretations, but we leave these as exercises.

EXAMPLE 1

COMMUNICATION PATHS

Figure 14.1 shows graphically the one-step communication paths existing among five individuals. P_i can communicate with P_j if and only if a directed line segment extends from the point P_i to the point P_j. Construct the matrix P that displays these relations, then compute and interpret P^2.

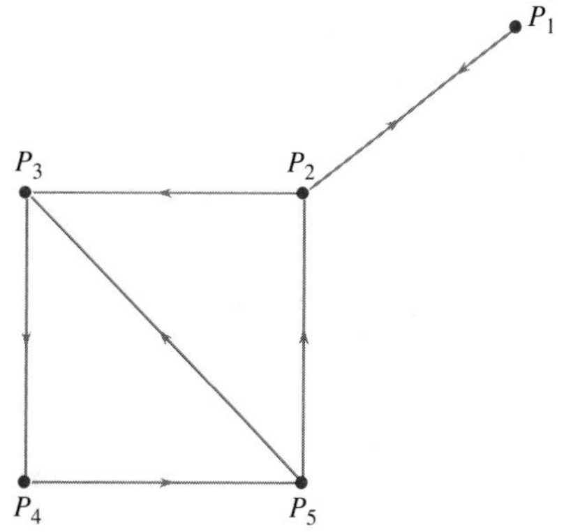

FIGURE 14.1
Graph showing communication paths.

Inspection of Fig. 14.1 leads at once to the accessibility matrix

$$P = \begin{bmatrix} 0 & 1 & 0 & 0 & 0 \\ 1 & 0 & 1 & 0 & 0 \\ 0 & 0 & 0 & 1 & 0 \\ 0 & 0 & 0 & 0 & 1 \\ 0 & 1 & 1 & 0 & 0 \end{bmatrix}$$

and a routine calculation gives

$$P^2 = \begin{bmatrix} 1 & 0 & 1 & 0 & 0 \\ 0 & 1 & 0 & 1 & 0 \\ 0 & 0 & 0 & 0 & 1 \\ 0 & 1 & 1 & 0 & 0 \\ 1 & 0 & 1 & 1 & 0 \end{bmatrix}$$

Since each nonzero element of this matrix is 1, no two individuals are connected by more than one two-step path. The row and column in which each 1 occurs identify the respective source and destination of a corresponding two-step communication path. Hence both P_1 and P_2 get personal feedback, P_1 has a two-step path to P_3, and P_2 has one to P_4. Also, P_3, has a two-step path to just P_5, and P_4 has two-step paths to only P_2 and P_3. Finally, P_5 has two-step paths to three other individuals, P_1, P_3, and P_4.

At a more technical level, *transition probabilities* are another example of the use of matrices to display and manipulate data. As in Sec. 3.3, suppose that a system S can exist in any of n states, S_1, S_2, . . . , S_n, and that by some random process the system may pass directly from the ith to the jth state with probability $p_{ij}^{(1)}$ (i, $j = 1, 2, \ldots, n$). Naturally, the system may also pass from the ith to the jth state by first passing through some intermediate state, say the kth; and the calculation of these two-step transition probabilities is a matter of some importance. Now the probability that the system will pass from the ith to the jth state via the kth state is the product of the probability $p_{ik}^{(1)}$ that it will pass in one step from S_i to S_k and the probability $p_{kj}^{(1)}$ that it will subsequently pass in one step from S_k to S_j. Furthermore, since in any two-step transition from S_i to S_j, the system must pass through *some* intermediate state (including of course S_i and S_j themselves), the probability that the system will pass in exactly two steps from S_i to S_j can be found by adding the products $p_{ik}^{(1)}p_{kj}^{(1)}$ for all possible intermediate steps, i.e., for all values of k. Thus

$$p_{ij}^{(2)} = \sum_{k=1}^{n} p_{ik}^{(1)}p_{kj}^{(1)}$$

But the sum on the right in this formula is precisely the element in the ith row and jth column of the square of the matrix of one-step transition probabilities, $P = [p_{ij}^{(1)}]$. In other words, *the matrix of two-step transition probabilities for any system S is the square of the matrix of one-step transition probabilities for S.* A similar argument shows that the matrix of three-step transition probabilities for S is the cube of the matrix of one-step transition probabilities, and so on.

EXAMPLE 2

TRANSITION PROBABILITIES

Let S be the system consisting of two players A and B who begin with \$2 apiece and match coins until one or the other of them has no more money. If the states of the system are defined by the number of dollars in A's possession, specifically if the system is in the state S_{i+1} whenever A has i dollars ($i = 0, 1, 2, 3, 4$), find the matrix of one-step transition probabilities. Then, by raising this matrix to the second, third, and fourth powers, find the matrices containing the two-, three-, and four-step transition probabilities for S. What is the probability that A will be ruined in at most four turns? What is the probability that A will be ruined in exactly four turns?

Clearly, unless A or B is bankrupt, A must either win a dollar or lose a dollar on each turn, and the probability of each of these events is $\frac{1}{2}$. Hence, if A has i dollars ($i = 1, 2, 3$), that is, if the system is in the state $S_{i+1}(i = 1, 2, 3)$, the probability of a one-step transition to S_i is $\frac{1}{2}$, the probability of a one-step transition to S_{i+2} is $\frac{1}{2}$, and the probability of a one-step transition to any other state is 0. On the other hand, if the system is in the state S_1, that is, if A is bankrupt, the system remains in that state; so the probability of a one-step transition from S_1 to S_1 is 1, and the probability of any other transition from S_1 is 0. Similarly, if the system is in the state S_5, that is, if B is bankrupt, the system remains in that state; hence the probability of a one-step transition from S_5 to S_5 is 1, and the probability of any other transition is 0. Thus the matrix of one-step transition probabilities is

$$\mathrm{P} = \begin{bmatrix} 1 & 0 & 0 & 0 & 0 \\ \frac{1}{2} & 0 & \frac{1}{2} & 0 & 0 \\ 0 & \frac{1}{2} & 0 & \frac{1}{2} & 0 \\ 0 & 0 & \frac{1}{2} & 0 & \frac{1}{2} \\ 0 & 0 & 0 & 0 & 1 \end{bmatrix}$$

By multiplying this matrix by itself, we find at once that the matrix of two-step transition probabilities is

$$\mathrm{P}^2 = \begin{bmatrix} 1 & 0 & 0 & 0 & 0 \\ \frac{1}{2} & \frac{1}{4} & 0 & \frac{1}{4} & 0 \\ \frac{1}{4} & 0 & \frac{1}{2} & 0 & \frac{1}{4} \\ 0 & \frac{1}{4} & 0 & \frac{1}{4} & \frac{1}{2} \\ 0 & 0 & 0 & 0 & 1 \end{bmatrix}$$

Similarly, by computing P^3 and P^4, we find the matrices of three-step and four-step transition probabilities, respectively, to be

$$\mathrm{P}^3 = \begin{bmatrix} 1 & 0 & 0 & 0 & 0 \\ \frac{5}{8} & 0 & \frac{1}{4} & 0 & \frac{1}{8} \\ \frac{1}{4} & \frac{1}{4} & 0 & \frac{1}{4} & \frac{1}{4} \\ \frac{1}{8} & 0 & \frac{1}{4} & 0 & \frac{5}{8} \\ 0 & 0 & 0 & 0 & 1 \end{bmatrix} \quad \text{and} \quad \mathrm{P}^4 = \begin{bmatrix} 1 & 0 & 0 & 0 & 0 \\ \frac{5}{8} & \frac{1}{8} & 0 & \frac{1}{8} & \frac{1}{8} \\ \frac{3}{8} & 0 & \frac{1}{4} & 0 & \frac{3}{8} \\ \frac{1}{8} & \frac{1}{8} & 0 & \frac{1}{8} & \frac{5}{8} \\ 0 & 0 & 0 & 0 & 1 \end{bmatrix}$$

The probability that A is ruined in at most four turns is simply the probability of a four-step transition from S_3 to S_1, namely, $p_{31}^{(4)} = \frac{3}{8}$, since among such transitions are included those in which the system reaches S_1 in less than four steps and then remains there. The probability that A is ruined in four turns and not before is the probability that S reaches S_1 in four steps but does not reach it in three steps or less, namely, the difference $p_{31}^{(4)} - p_{31}^{(3)} = \frac{3}{8} - \frac{1}{4} = \frac{1}{8}$.

As a third matrix application, we choose one involving a physical illustration of a matrix and its inverse. Both matrices have the desirable feature that the physical significance of each of their elements is not only easy to identify but also has immediate practical value.

EXAMPLE 3

STIFFNESS AND FLEXIBILITY MATRICES

Let us consider the mass-spring system shown in Fig. 14.2 and determine the forces acting on each of the masses as a result of arbitrary displacements x_1, x_2, and x_3 of the respective masses.

The modulus of each spring is the indicated value of k; that is, the force required to stretch each spring a unit distance is the corresponding value of k. If the masses are displaced by the respective amounts x_1, x_2, and x_3, the changes in the lengths of the various springs, and the forces represented by these changes are as given in Table 14.1, a positive force indicating that the spring is stretched and a negative force indicating that the spring is compressed. Hence, taking due account of the direction of the force applied to each mass by each spring attached to it, we find that the forces f_1, f_2, and f_3 acting on the respective masses are

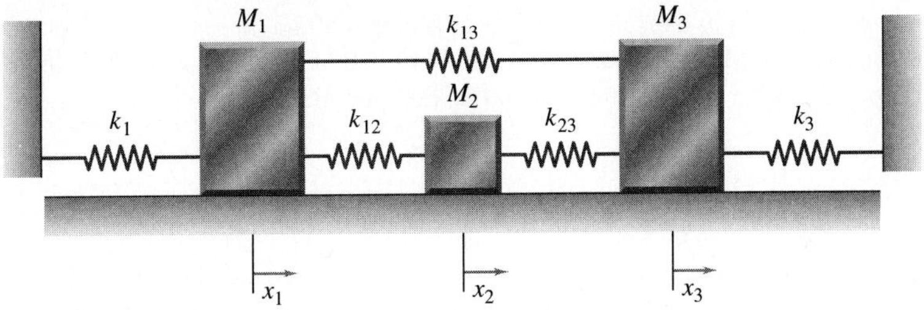

FIGURE 14.2
A typical system of spring-connected masses.

TABLE 14.1

Spring modulus	Change in spring length	Force
k_1	x_1	$k_1 x_1$
k_{12}	$x_2 - x_1$	$k_{12}(x_2 - x_1)$
k_{13}	$x_3 - x_1$	$k_{13}(x_3 - x_1)$
k_{23}	$x_3 - x_2$	$k_{23}(x_3 - x_2)$
k_3	$-x_3$	$k_3(-x_3)$

(2)
$$\begin{aligned} f_1 &= -k_1 x_1 + k_{12}(x_2 - x_1) + k_{13}(x_3 - x_1) \\ f_2 &= -k_{12}(x_2 - x_1) + k_{23}(x_3 - x_2) \\ f_3 &= -k_{13}(x_3 - x_1) - k_{23}(x_3 - x_2) - k_3 x_3 \end{aligned}$$

or, collecting terms and rewriting in matrix notation,

(3)
$$\mathbf{f} = \mathrm{K}\mathbf{x}$$

where

$$\mathbf{f} = \begin{bmatrix} f_1 \\ f_2 \\ f_3 \end{bmatrix} \quad \mathbf{x} = \begin{bmatrix} x_1 \\ x_2 \\ x_3 \end{bmatrix} \quad \mathrm{K} = \begin{bmatrix} -(k_1 + k_{12} + k_{13}) & k_{12} & k_{13} \\ k_{12} & -(k_{12} + k_{23}) & k_{23} \\ k_{13} & k_{23} & -(k_{13} + k_{23} + k_3) \end{bmatrix}$$

Evaluating the first of Eqs. (2) for $x_1 = 1$ and $x_2 = x_3 = 0$, it is clear that $-(k_1 + k_{12} + k_{13})$ is the force applied to the first mass as a result of a unit displacement of that mass. A similar evaluation shows that, in general, the element in the ith row and jth column of K is the force applied to the ith mass as a result of a unit displacement of the jth mass. Because of this property, the matrix K is usually referred to as the **stiffness matrix** of the system.

It can easily be verified that for all positive values of the k's the matrix K is nonsingular. Hence, for the physical system shown in Fig. 14.2, K^{-1} exists, and we can solve Eq. (3) for **x**, getting

$$\mathbf{x} = \mathrm{K}^{-1}\mathbf{f}$$

Now, evaluating the right-hand side of this equation for a force vector **f** whose jth component is 1 and whose other components are all zero, it follows that the element of K^{-1} is the ith row and jth column is the displacement produced in the ith mass as a result of a unit force applied to the jth mass. Because of this property, the matrix K^{-1} is usually referred to as the **elasticity** or **flexibility matrix** of the system. Our discussion has thus illustrated the important fact that for any elastic system, the flexibility matrix is the inverse of the stiffness matrix, and vice versa. The symmetry of the stiffness and elasticity matrices, which asserts, for instance, that the force acting on the ith mass as a result of a unit displacement of the jth mass is equal to the force acting on the jth mass as a result of a unit displacement of the ith mass, is an illustration of the so-called **reciprocity theorem** of Maxwell, Rayleigh, and Betti.† It is interesting in this connection to recall the symmetry properties of the Green's function and its interpretation as an influence function, discussed in Sec. 2.12.

In Sec. 3.4, we saw that each of the three elementary row operations on a matrix A (Definition 2, Sec. 3.4) could be accomplished by premultiplying A by a conformable unit matrix on which the corresponding row operation had been performed. Clearly, any sequence of row operations on A can be accomplished by premultiplying A by the appropriate modified unit matrices. However, in many cases it is possible to construct a *single* matrix which will carry out a specified sequence of elementary row operations on A. The following example illustrates this possibility, which we will find useful in the next section in our discussion of the fast Fourier transform.

EXAMPLE 4

Given the vectors $\mathbf{U} = \begin{bmatrix} a_0 \\ a_2 \\ a_4 \end{bmatrix}$ and $\mathbf{V} = \begin{bmatrix} a_1 \\ a_3 \\ a_5 \end{bmatrix}$, devise a process of matrix multiplication with which the vector $\mathbf{Z} = \begin{bmatrix} a_0 + a_1 \\ a_2 + 2a_3 \\ a_4 + 3a_5 \\ a_0 - a_1 \\ a_2 - 2a_3 \\ a_4 - 3a_5 \end{bmatrix}$ can be obtained from $\mathbf{U}$ and $\mathbf{V}$.

It should be clear at the outset that we must first combine $\mathbf{U}$ and $\mathbf{V}$ into the six-vector $\mathbf{W} = \begin{bmatrix} \mathbf{U} \\ \mathbf{V} \end{bmatrix} = \begin{bmatrix} a_0 \\ a_2 \\ a_4 \\ a_1 \\ a_3 \\ a_5 \end{bmatrix}$. Then, in addition to shifting some of the elements of $\mathbf{W}$ from one row to another, as we did in Example 10, Sec. 3.4, we must also add given multiples of certain elements to other elements, very much as we did in Theorem 4, Sec. 3.4. To construct a matrix P_6 which will accomplish these two objectives, we observe that to shift the element in the ith row of $\mathbf{W}$ into the jth row and at the same time add c times the element in the kth row to it requires that the nonzero elements in the jth row of P_6 consist of a 1 in the ith column and a c in the kth column. With these observations in mind, we must now construct P_6 to accomplish the following manipulations:

Move a_0 from Row 1 to Row 1 and add a_1 from Row 4
Move a_2 from Row 2 to Row 2 and add $2a_3$ from Row 5
Move a_4 from Row 3 to Row 3 and add $3a_5$ from Row 6
Move a_0 from Row 1 to Row 4 and add $-a_1$ from Row 4
Move a_2 from Row 2 to Row 5 and add $-2a_3$ from Row 5
Move a_4 from Row 3 to Row 6 and add $-3a_5$ from Row 6

†James Clerk Maxwell (1831–1879) and J. W. S. Rayleigh (1842–1919) were English mathematical physicists. Enrico Betti (1823–1892) was an Italian mathematician.

The result is

$$P_6 = \begin{bmatrix} 1 & 0 & 0 & 1 & 0 & 0 \\ 0 & 1 & 0 & 0 & 2 & 0 \\ 0 & 0 & 1 & 0 & 0 & 3 \\ 1 & 0 & 0 & -1 & 0 & 0 \\ 0 & 1 & 0 & 0 & -2 & 0 \\ 0 & 0 & 1 & 0 & 0 & -3 \end{bmatrix}$$

and it is easy to verify that $P_6\mathbf{W} = \mathbf{Z}$.

EXERCISES

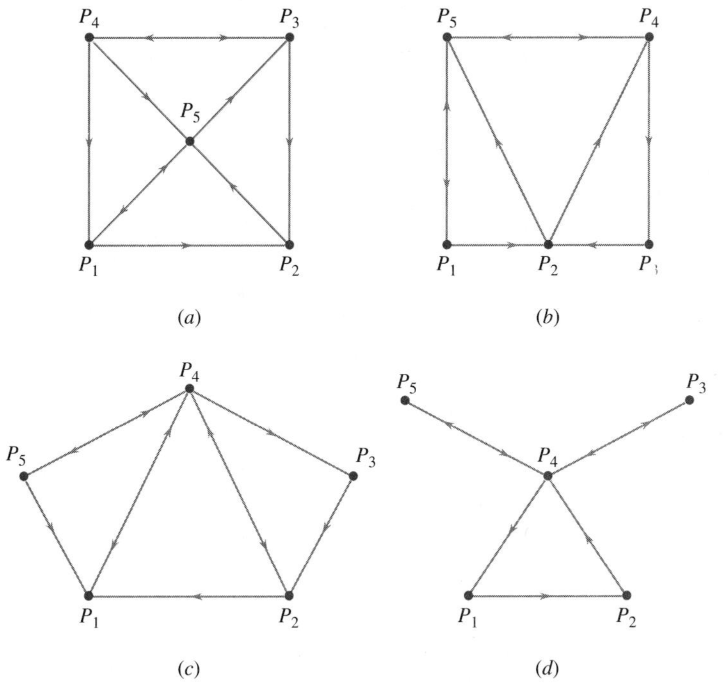

FIGURE 14.3

1. Construct the accessibility matrix P associated with each of the following graphs:
   (a) Fig. 14.3*a* **(b)** Fig. 14.3*b* **(c)** Fig. 14.3*c*
   (d) Fig. 14.3*d*
   In each case compute P^2 and identify each of the various two-step paths in the graph.
2. Discuss the significance of higher powers of the accessibility matrix.
3. Justify the statement that the cube and fourth power of a transition matrix are, respectively, the matrices of the probabilities of three- and four-step transitions.
4. *A* and *B* play the following game: *A* throws a die, and if either a 1 or a 2 shows, *B* pays *A* \$1; if a 3 or a 4 shows, *A* pays *B* \$1; and if a 5 or a 6 shows, the turn passes to *B* without any payment. If *A* and *B* each start with \$2 and if the states of the system are S_{i+1} $(i = 0, 1, 2, 3, 4)$, where i is the number of dollars in *A*'s possession, set up the transition matrix P and compute P^2, P^3, and P^4. What is the probability that *A* will be ruined in exactly four turns? What is the probability that the game will continue for more than four turns?
5. Work Exercise 4 if *A* starts with \$1 and *B* starts with \$2 and the states of the system are suitably redefined.
6. Show that in any matrix of transition probabilities, the sum of the entries in any row is 1.
7. Consider the system *S* consisting of four boxes B_1, B_2, B_3, and B_4 and a single ball and let the system be in the state S_i $(i = 1, 2, 3, 4)$ if the ball is in box B_i. Transitions from one state to another take place in the following manner: a die is thrown, and if a 1, 2, or 3 turns up, the ball is taken from whichever box it is in and placed in the box bearing the number showing on the die. If 4, 5, or 6 turns up, the

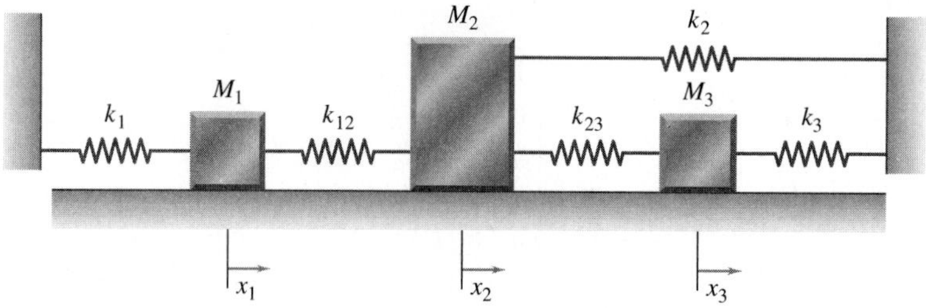

FIGURE 14.4

ball is taken from wherever it is and placed in box B_4. Find the matrix of one-, two-, and three-step transition probabilities for the system.

8. Consider the system consisting of three boxes B_1, B_2, and B_3 and a single ball and let the system be in the state S_i ($i = 1, 2, 3$) if the ball is in the box B_i. Transitions from one state to another take place in the following manner: three coins are tossed. If no heads turn up, the ball is not moved; if one or more heads turn up, the ball is taken from its box and placed in the box corresponding to the number of heads showing. Find the matrix of one-, two-, three-, and four-step transition probabilities for the system.
9. Show that for all positive values of the k's, the matrix K in Eq. (3) is nonsingular.
10. Find the stiffness matrix for the system shown in Fig. 14.4.
11. In mechanics it is shown that if a cantilever beam bears a concentrated load P at a distance s from the free end, then the deflection y at a distance x from the fixed end is given by the formula

$$y = \begin{cases} \dfrac{Px^2(x - 3s)}{6EI} & x \leq s \\ \dfrac{Ps^2(s - 3x)}{6EI} & x \geq s \end{cases}$$

where E and I are physical constants of the beam. Using this formula, obtain the stiffness and elasticity matrices relating the forces and deflections at the positions $s = L/3, 2L/3, L$ and $x = L/3, 2L/3, L$. In what respect, if any, does this problem differ significantly from the example discussed in the text?
12. Assuming that all springs are initially unstretched, compute the potential energy of the system shown in Fig. 14.2 when the three masses have arbitrary displacements x_1, x_2, and x_3 and show that the force acting on the ith mass is the negative of the partial derivative of the potential energy with respect to x_i.
13. Let K be the stiffness matrix for a frictionless mass-spring system S, let M be the diagonal matrix whose diagonal elements are the respective masses of S, and suppose that **x** is a vector such that **Kx** is proportional to **Mx**. Show that the square root of the negative of this proportionality constant is one of the natural frequencies of the system. *Hint:* Use Newton's law to set up the equations of motion of the respective masses and then substitute $x_i = a_i \cos \omega t$.
14. Guided by analogy with the mechanical case, what do you think is the reciprocity theorem for electric circuits?
15. By what matrix must $\mathbf{U} = [u_0\ u_1\ u_2\ u_3]^T$ be premultiplied to obtain
    (a) $\mathbf{V} = [u_0\ u_2\ u_1\ u_3]^T$? **(b)** $\mathbf{V} = [u_3\ u_2\ u_1\ u_0]^T$?
    (c) $\mathbf{V} = [au_1\ bu_3\ cu_2\ du_0]^T$?
16. By what matrix must $[u_0\ u_1\ v_0\ v_1\ w_0\ w_1]^T$ be premultiplied to obtain $\mathbf{V} = [u_0 + au_1\ \ u_0 - au_1\ \ v_0 + bv_1\ \ v_0 - bv_1\ \ w_0 + cw_1\ \ w_0 - cw_1]^T$?
17. Find a matrix P_5 which can be used as a premultiplier of an arbitrary $5 \times n$ matrix A to effect the following changes in A:

    Row 1 becomes Row 3 + 2 Row 5
    Row 2 becomes Row 5 + Row 1
    Row 3 becomes 2 Row 2 + 3 Row 4
    Row 4 becomes Row 1 + Row 5
    Row 5 becomes Row 2 + Row 3 + Row 4

18. Find a matrix P_5 which can be used as a premultiplier of an arbitrary $5 \times n$ matrix A to effect the following changes in A: Row 1 → Row 2, Row 2 → Row 5, Row 3 → Row 1, Row 4 → Row 3, Row 5 → Row 4.
19. Consider the set of all complex numbers, the set of all real 2×2 matrices of the form $\mathrm{M} = \begin{bmatrix} x & y \\ -y & x \end{bmatrix}$, and the one-to-one correspondence between these two sets defined by the association $x + iy \leftrightarrow \begin{bmatrix} x & y \\ -y & x \end{bmatrix}$. Using the usual definitions for the sum and product of two complex numbers and the sum and product of two matrices, show that if $z_j \leftrightarrow \mathrm{M}_j$ and $z_k \leftrightarrow \mathrm{M}_k$, then $z_j + z_k \leftrightarrow \mathrm{M}_j + \mathrm{M}_k$ and $z_j z_k \leftrightarrow \mathrm{M}_j\mathrm{M}_k$. If $z_j \leftrightarrow \mathrm{M}_j$, show that the determinant of M_j is equal to the square of the absolute value of z_j. What is the significance of the adjoint of M_j? The inverse of M_j? What are the characteristic values of M_j? (A one-to-one relation-preserving correspondence like this one is called an **isomorphism**.)

14.2 THE DISCRETE AND FAST FOURIER TRANSFORMS

In Sec. 8.2 we learned that any periodic function f which satisfies the Dirichlet conditions can be expanded in an infinite series of cosines and sines. We also learned that the Fourier series of a function f converges to $f(t)$ at every point t where f is continuous. After extending these ideas to aperiodic functions on $(-\infty, \infty)$ and Fourier integrals, in Sec. 9.3 we expressed the complex Fourier integral representation of a function f as the complex Fourier transform pair

$$f(t) = \frac{1}{2\pi}\int_{-\infty}^{\infty} F(\omega)e^{i\omega t}\,d\omega$$

$$F(\omega) = \int_{-\infty}^{\infty} f(t)e^{-i\omega t}\,dt$$

and called $F(\omega)$ the Fourier transform of $f(t)$. Important as these results are, there are an increasing number of problems of current interest to which neither infinite Fourier series nor Fourier integrals apply. Specifically, in many applications the values of a function can be ascertained only for a finite number of equally spaced points. For instance, such a set of values might be the sequence obtained by instantaneously sampling a continuous signal at a set of equally spaced times. Then the aim is to find a *finite* Fourier series whose sum at each recorded point in the domain of f equals the corresponding recorded value of the function at that point. More precisely, the problem is this:

Let f be a function whose values

$$f_0, f_1, f_2, \ldots, f_j, \ldots, f_{n-1}$$

are known at n equally spaced points. By the introduction of a suitable scalar, if necessary (see Exercise 12), we assume that these are the n points

$$t_0 = 0, \quad t_1 = \frac{2\pi}{n}, \quad t_2 = 2\left(\frac{2\pi}{n}\right), \quad \ldots, \quad t_j = j\left(\frac{2\pi}{n}\right), \quad \ldots, \quad t_{n-1} = (n-1)\left(\frac{2\pi}{n}\right)$$

of the interval $[0, 2\pi)$. We are to find a finite Fourier series which will give the values of f at each of these n points.

One way of finding a series representation of f of this sort is known as the *discrete Fourier transform method.* Since powers of e are so much easier to work with algebraically than sines and cosines, we assume a finite complex exponential Fourier series representation of f:

$$f(t) = a_0 + a_1e^{it} + a_2e^{2it} + \cdots + a_ke^{kit} + \cdots + a_{n-1}e^{(n-1)it} \tag{1}$$

and attempt to determine the n coefficients $a_0, a_1, a_2, \ldots, a_{n-1}$ so that (1) will be satisfied by each of the pairs of values

$$(t_j, f_j) = \left(\frac{2j\pi}{n}, f_j\right) \qquad j = 0, 1, 2, \ldots, n-1$$

Substituting these data points into (1), we obtain the n equations

$$f_j = \sum_{k=0}^{n-1} a_ke^{kit_j} \qquad j = 0, 1, 2, \ldots, n-1 \tag{2}$$

To find the coefficient a_k of the general term in the series (1), we multiply both sides of (2) by e^{-iht_j}, getting

$$f_je^{-iht_j} = \sum_{k=0}^{n-1} a_ke^{i(k-h)t_j} \qquad j = 0, 1, 2, \ldots, n-1$$

Equating the sum of the left-hand members of these n equations to the sum of their right-hand members, then reversing the order of summation on the right and remembering that $t_j = 2j\pi/n$, we have

$$\sum_{j=0}^{n-1} f_j e^{-iht_j} = \sum_{j=0}^{n-1}\sum_{k=0}^{n-1} a_k e^{i(k-h)t_j} = \sum_{k=0}^{n-1} a_k \sum_{j=0}^{n-1} e^{i(k-h)2j\pi/n} \tag{3}$$

The last series in (3) is a finite geometric series which we rewrite and sum as follows:

$$\sum_{j=0}^{n-1} e^{i(k-h)2j\pi/n} = \sum_{j=0}^{n-1} r^j = \frac{1-r^n}{1-r} \qquad r \neq 1, \text{ where } r = e^{i(k-h)2\pi/n} \tag{4}$$

If h is an integer, as we shall now suppose, and $h \neq k$, then $r \neq 1$ and (recall Euler's formula, Sec. 3.1)

$$r^n = e^{i(k-h)2\pi} = \cos(k-h)2\pi = 1$$

so that

$$1 - r^n = 0 \qquad \text{and} \qquad \sum_{j=0}^{n-1} r^j = 0$$

On the other hand, if $h = k$, then

$$r = e^0 = 1 \qquad \text{and} \qquad \sum_{j=0}^{n-1} r^j = n$$

Combining cases and remembering the definitions of r and t_j, we rewrite (4) as

$$\sum_{j=0}^{n-1} e^{i(k-h)t_j} = \begin{cases} n & h = k \\ 0 & h \neq k \end{cases} \tag{5}$$

This tells us that the last member of (3) is zero unless $k = h$, in which case it reduces to the product $a_h n$; hence Eq. (3) can be written

$$\sum_{j=0}^{n-1} f_j e^{-iht_j} = a_h n$$

Solving this for a_h and replacing h by k, we find the coefficient of the general term in (1) to be

$$a_k = \frac{1}{n}\sum_{j=0}^{n-1} f_j e^{-kit_j} \qquad k = 0, 1, 2, \ldots, n-1 \tag{6}$$

These n equations and the n equations (2) can be expressed in especially elegant vector-matrix notation. To do so, we define the n-dimensional column vectors

$$\mathbf{F} = \begin{bmatrix} f_0 \\ f_1 \\ \vdots \\ f_{n-1} \end{bmatrix} \qquad \mathbf{A} = \begin{bmatrix} a_0 \\ a_1 \\ \vdots \\ a_{n-1} \end{bmatrix}$$

and recall that the primitive nth root of unity is

$$w_n = e^{2i\pi/n}$$

which, for convenience we shall denote simply by w. (See Sec. 3.1 for properties of w_n.) With $t_j = 2j\pi/n$, the n equations (2) become

$$f_j = \sum_{k=0}^{n-1} a_k(e^{2i\pi/n})^{kj} = \sum_{k=0}^{n-1} w^{jk}a_k \qquad j = 0, 1, 2, \ldots, n-1$$

which in vector-matrix form reads

$$\mathbf{F} = \mathbf{MA} \tag{7}$$

where, since $w^{jk} = w^{kj}$, the so-called **Fourier matrix**

$$\mathrm{M} = [w^{jk}] = \begin{bmatrix} 1 & 1 & 1 & \cdot & 1 \\ 1 & w & w^2 & \cdot & w^{n-1} \\ 1 & w^2 & w^4 & \cdot & w^{2(n-1)} \\ \cdots & \cdots & \cdots & \cdots & \cdots \\ 1 & w^{n-1} & w^{2(n-1)} & \cdot & w^{(n-1)^2} \end{bmatrix} \tag{7a}$$

is symmetric and of order n, *with the rows and columns numbered from 0 to $n-1$* as it is natural to do with Fourier matrices.

To write the n equations (6) as a vector-matrix equation, we shall use the fundamental property of Fourier matrices set forth in the following theorem.

THEOREM 1 The inverse of the coefficient matrix

$$\mathrm{M} = [w^{jk}] \qquad j, k = 0, 1, 2, \ldots, n-1$$

of the system (2) is the matrix

$$\mathrm{M}^{-1} = \frac{1}{n}[w^{-jk}] \qquad j, k = 0, 1, 2, \ldots, n-1$$

PROOF For $j, k = 0, 1, 2, \ldots, n-1$, the entry in the jth row and kth column of $\mathrm{M}^{-1}\mathrm{M}$ as given by the scalar product of the jth row vector of $(1/n)[w^{-jk}]$ and the kth column vector of $[w^{jk}]$ equals the dot product

$$\frac{1}{n}(w^0, w^{-j}, w^{-2j}, \ldots, w^{-(n-1)j}) \cdot (w^0, w^k, w^{2k}, \ldots, w^{(n-1)k}) =$$

$$\frac{1}{n}[1 + w^{k-j} + w^{2(k-j)} + \cdots + w^{(n-1)(k-j)}] = \frac{1}{n}\sum_{m=0}^{n-1} w^{m(k-j)}$$

From Formula (5), with index of summation m, $t_m = 2m\pi/n$, and $h = j$, we have

$$\sum_{m=0}^{n-1} e^{i(k-j)(2m\pi/n)} = \sum_{m=0}^{n-1} w^{m(k-j)} = \begin{cases} n & j = k \\ 0 & j \neq k \end{cases}$$

Hence all diagonal elements of $\mathrm{M}^{-1}\mathrm{M}$ are 1, and all off-diagonal elements are 0. This proves that

$$\mathrm{M}^{-1}\mathrm{M} = \frac{1}{n}[w^{-jk}] \times [w^{jk}] = I_n$$

and shows that $\mathrm{M}^{-1} = (1/n)[w^{-jk}]$, as asserted.

Returning to (6), we substitute $2j\pi/n$ for t_j, introduce powers of w, and take $1/n$ across the summation sign to write

$$a_k = \frac{1}{n}\sum_{j=0}^{n-1} f_j e^{-ki2j\pi/n} = \frac{1}{n}\sum_{j=0}^{n-1} f_j w^{-jk} = \sum_{j=0}^{n-1} \frac{1}{n} w^{-jk} f_j \qquad k = 0, 1, 2, \ldots, n-1$$

which in view of Theorem 1 gives

$$\mathbf{A} = \mathbf{M}^{-1}\mathbf{F} \tag{8}$$

as the matrix form of the n equations (6). Equations (2) and (6) or, equivalently, the two matrix equations (7) and (8) are known as **discrete Fourier transform pairs.**

By incorporating the scalar $1/n$ in (7) and rewriting (7) and (8) as the transform pair

$$\mathbf{F} = \frac{1}{n}[w^{jk}]\mathbf{G} \tag{9a}$$

$$\mathbf{G} = [w^{-jk}]\mathbf{F} \tag{9b}$$

a discrete analog of the familiar complex Fourier transform pair is obtained. The vector **G** is called the **discrete Fourier transform** of **F**, and **F** is said to be the **inverse discrete Fourier transform** of **G**. Once a finite complex Fourier series has been found whose sum equals the value of a function f at each of n equally spaced points of some interval, a finite trigonometric series representation of the function over the same n points can be obtained by replacing exponentials by linear combinations of cosines and sines.

EXAMPLE 1

FINITE TRIGONOMETRIC REPRESENTATIONS

Find a finite trigonometric representation of the function f whose values are 0, 1, 2, and 3, respectively, when $t = 0, \pi/2, \pi, 3\pi/2$.

With $n = 4$, $w = w_4$, and M^{-1} given by Theorem 1, we have from (8), the given data, the symmetry of M^{-1}, and the fact that $w^{-1} = w^{-9} = -i$, $w^{-2} = w^{-6} = -1$, $w^{-3} = i$, $w^{-4} = 1$,

$$\begin{bmatrix} a_0 \\ a_1 \\ a_2 \\ a_3 \end{bmatrix} = \frac{1}{4}\begin{bmatrix} w^0 & w^0 & w^0 & w^0 \\ w^0 & w^{-1} & w^{-2} & w^{-3} \\ w^0 & w^{-2} & w^{-4} & w^{-6} \\ w^0 & w^{-3} & w^{-6} & w^{-9} \end{bmatrix}\begin{bmatrix} 0 \\ 1 \\ 2 \\ 3 \end{bmatrix} = \frac{1}{4}\begin{bmatrix} 1 & 1 & 1 & 1 \\ 1 & -i & -1 & i \\ 1 & -1 & 1 & -1 \\ 1 & i & -1 & -i \end{bmatrix}\begin{bmatrix} 0 \\ 1 \\ 2 \\ 3 \end{bmatrix} = \begin{bmatrix} \frac{3}{2} \\ (-1+i)/2 \\ -\frac{1}{2} \\ (-1-i)/2 \end{bmatrix}$$

Substituting the values of a_0, a_1, a_2, and a_3 thus determined into (1) yields

$$f(t) = \tfrac{1}{2}[3 + (-1+i)e^{it} - e^{2it} - (1+i)e^{3it}]$$

Reverting to real trigonometric functions through the use of Euler's formula and rearranging, we obtain the trigonometric representation

$$f(t) = \tfrac{3}{2} - \tfrac{1}{2}\cos t - \tfrac{1}{2}\cos 2t - \tfrac{1}{2}\cos 3t - \tfrac{1}{2}\sin t + \tfrac{1}{2}\sin 3t$$
$$+\, i(\tfrac{1}{2}\cos t - \tfrac{1}{2}\cos 3t - \tfrac{1}{2}\sin t - \tfrac{1}{2}\sin 2t - \tfrac{1}{2}\sin 3t)$$

The fact that the imaginary terms in this expression for f do not cancel identically may at first seem surprising since f is supposedly a purely real function. However, the only purpose of this expression is to serve as a representation of f at each of the data points $t = 0, \pi/2, \pi, 3\pi/2$, and by direct substitution it is easy to verify that this is the case. The complex values of f corresponding to other values of t are entirely irrelevant to the specified, or empirical, data. As we should expect, it can be shown (see Exercise 18) that for all values of

t equally spaced at a distance d over a finite interval, the imaginary terms in the complex exponential series (1) representing f are identically zero. Thus the real terms in (1) constitute the required trigonometric representation of f, and this representation is periodic with period 2π. In the present problem this means that we can write simply

$$f(t) = \tfrac{3}{2} - \tfrac{1}{2}\cos t - \tfrac{1}{2}\cos 2t - \tfrac{1}{2}\cos 3t - \tfrac{1}{2}\sin t + \tfrac{1}{2}\sin 3t$$

Either of the discrete Fourier transform pairs [(2), (6)] or [(7), (8)] gives us two elegant, and essentially familiar, formulas for the representation of a sequence of data as a finite exponential Fourier series. However, the labor involved in using these formulas may be prohibitively great, even for a computer, if n is large, as it usually is. Specifically, to compute one of the coefficients in (1) through the use of Eq. (6) requires n multiplications, and since (1) contains n terms, the total number of multiplications necessary to determine the series is $n \times n = n^2$. Disregarding the factor $1/n$, the same number of multiplications is required if the column vector $\mathbf{A}$ of coefficients in (1) is calculated by constructing the matrix product $\mathbf{A} = \mathrm{M}^{-1}\mathbf{F}$. Any modification of the usual row-into-column way of taking matrix-vector products like $\mathrm{M}^{-1}\mathbf{F}$ and $\mathrm{M}\mathbf{A}$ that would serve to reduce significantly the number of multiplications would be highly desirable. The **fast Fourier transform** (FFT), to which we now turn our attention, is just such a modification comprised of an expeditious rearrangement of vector-matrix products each containing a Fourier matrix M or its inverse. It can be shown that if M is of order n, through the use of the FFT the number of multiplications is reduced from n^2 to approximately $n \log_2 n$.

As usual in Fourier analysis, the fast Fourier transform addresses either of two related problems: **(1)** Given the values of a discrete function, determine the coefficients in its Fourier representation, and **(2)** Given the coefficients in the Fourier representation of a discrete function, determine the values of the function. From the structure of the discrete Fourier transform pair [(7), (8)], it is clear that the two problems are almost identical, and for definiteness, we shall explain the procedure by considering the second, that is, by applying a FFT to determine the values of a discrete function assuming the coefficients in its finite complex exponential Fourier series are known.

Let us suppose, then, that for some high value of n we know the n coefficients $a_0, a_1, a_2, \ldots, a_{n-1}$ and wish to find the values $f_0, f_1, f_2, \ldots, f_{n-1}$ of the sequence they determine. In other words, we seek to determine $\mathbf{F}$ from the equation

(7) $$\mathbf{F} = \mathrm{M}\mathbf{A} \qquad \text{or} \qquad \mathbf{F} = [w^{jk}]\mathbf{A}$$

To avoid carrying out the n^2 multiplications involved in calculating the product $\mathrm{M}\mathbf{A}$, the fast Fourier transform reduces the problem successively to two simpler, but mathematically identical, problems each one-half the size, then four problems one-fourth the size, eight problems one-eighth the size, and so on until at the last stage there are $n/2$ problems each involving only the 2×2 matrix

$$\mathrm{M}_2 = [w_2^{jk}] = \begin{bmatrix} 1 & 1 \\ 1 & -1 \end{bmatrix}$$

Clearly, for this sort of reduction to be possible, n must be a power of 2, say $n = 2^l$, and in sampling a continuous signal, say, n is almost always so chosen, although theoretically n could be a power of some other integer.

To accomplish this reduction, the process proceeds as follows: from the matrix equation $\mathbf{F} = [w^{jk}]\mathbf{A}$ we infer the n scalar relations

(10) $$f_j = \sum_{k=0}^{n-1} w^{jk} a_k \qquad j = 0, 1, 2, \ldots, n-1$$

which is just the system of equations (2) with $t_j = 2\pi j/n$. Letting $n/2 = m$, each series of (10) can be separated into two simpler series, one containing the m even coefficients a_{2k}, the other containing the m odd coefficients a_{2k+1}. Breaking up the series (10) in this fashion and attaching the subscript n to w, so that in subsequent steps we can distinguish the mth roots of 1 from the nth roots of 1, we have

$$f_j = \sum_{k=0}^{m-1} w_n^{j(2k)} a_{2k} + \sum_{k=0}^{m-1} w_n^{j(2k+1)} a_{2k+1}$$

$$= \sum_{k=0}^{m-1} w_n^{2(jk)} a_{2k} + w_n^j \sum_{k=0}^{m-1} w_n^{2(jk)} a_{2k+1}$$

(11) $$= \sum_{k=0}^{m-1} w_m^{jk} a_{2k} + w_n^j \sum_{k=0}^{m-1} w_m^{jk} a_{2k+1} \qquad j = 0, 1, 2, \ldots, n-1$$

since $w_n^2 = w_{n/2} = w_m$ (see Exercise 30, Sec. 3.1).

It is now convenient to break up the list of f's, given by (11), into two lists, one containing the m values from f_0 to f_{m-1}, the other containing the m values from f_m to f_{n-1}, and then to index the second half-list from $j = 0$ to $j = m - 1$, like the first half-list, by replacing j by $j + m$. This gives us

$$f_j = \sum_{k=0}^{m-1} w_m^{jk} a_{2k} + w_n^j \sum_{k=0}^{m-1} w_m^{jk} a_{2k+1} \qquad j = 0, 1, \ldots, m-1$$

$$f_{j+m} = \sum_{k=0}^{m-1} w_m^{(j+m)k} a_{2k} + w_n^{j+m} \sum_{k=0}^{m-1} w_m^{(j+m)k} a_{2k+1} \qquad j = 0, 1, \ldots, m-1$$

The formula for f_{j+m} can be simplified by noting that

$$w_m^{(j+m)k} = (w_m^m)^k w_m^{jk} = w_m^{jk} \qquad \text{and that} \qquad w_n^{j+m} = w_n^j w_n^{n/2} = -w_n^j$$

With these changes, Eq. (11) can be written as the pair of equations

(12) $$\begin{aligned} f_j &= \sum_{k=0}^{m-1} w_m^{jk} a_{2k} + w_n^j \sum_{k=0}^{m-1} w_m^{jk} a_{2k+1} \\ f_{j+m} &= \sum_{k=0}^{m-1} w_m^{jk} a_{2k} - w_n^j \sum_{k=0}^{m-1} w_m^{jk} a_{2k+1} \end{aligned} \qquad j = 0, 1, \ldots, m-1$$

in which the expressions for f_j and f_{j+m} differ only in the sign before the second sum in each equation.

The structural similarity between the two equations in (12) becomes even more striking if we assign simple names to the two sums. Since the first sum in each of Eqs. (12) contains the first, third, fifth, . . . , coefficients, we shall designate it by the symbol $f_j^{(o)}$, using the superscript (o) to suggest the word *odd*, even though the a's in the sum have even subscripts. Thus we define

(13) $$f_j^{(o)} = \sum_{k=0}^{m-1} w_m^{jk} a_{2k} \qquad j = 0, 1, 2, \ldots, m-1$$

and, naturally,

$$\mathbf{F}^{(o)} = \begin{bmatrix} f_0^{(o)} \\ f_1^{(o)} \\ \vdots \\ f_{m-1}^{(o)} \end{bmatrix} \tag{14}$$

In the same way, since the second sum in each of Eqs. (12) contains the second, fourth, sixth, ... , coefficients, we shall denote it by the symbol $f_j^{(e)}$, using the superscript (e) to suggest the word *even,* even though the coefficients themselves bear odd subscripts. Thus we shall write

$$f_j^{(e)} = \sum_{k=0}^{m-1} w_m^{jk} a_{2k+1} \qquad j = 0, 1, 2, \ldots, m-1 \tag{15}$$

and

$$\mathbf{F}^{(e)} = \begin{bmatrix} f_0^{(e)} \\ f_1^{(e)} \\ \vdots \\ f_{m-1}^{(e)} \end{bmatrix} \tag{16}$$

Using Eqs. (13) and (15), we can rewrite (12) in the form

$$\begin{aligned} f_j &= f_j^{(o)} + w_n^j f_j^{(e)} \\ f_{j+m} &= f_j^{(o)} - w_n^j f_j^{(e)} \end{aligned} \qquad j = 0, 1, 2, \ldots, m-1. \tag{17}$$

These relations are displayed in Fig. 14.5 in what is commonly called a **butterfly pattern,** where the multipliers on the arrows, known as **twiddle factors,** are the coefficients of $f_j^{(o)}$ and $f_j^{(e)}$ in the formulas for the f's to which the arrows point. In the formula for f_j, for instance, 1 and w_n^j are the coefficients of $f_j^{(o)}$ and $f_j^{(e)}$. The entire content of Eqs. (17) can be displayed by a superposition of butterfly patterns called a **flow graph.** In particular, for $n = 8$, the superposed pattern is shown in Fig. 14.6.

As soon as we know the $f_j^{(o)}$'s and the $f_j^{(e)}$'s, that is, $\mathbf{F}^{(o)}$ and $\mathbf{F}^{(e)}$, we can obtain the ordered sequence of the f_j's, that is, $\mathbf{F}$, by interweaving the components of $\mathbf{F}^{(o)}$ and $\mathbf{F}^{(e)}$ according to Eqs. (17). This can always be done by premultiplying the vector

$$\begin{bmatrix} \mathbf{F}^{(o)} \\ \mathbf{F}^{(e)} \end{bmatrix}$$

by a matrix which can be written down at once using Eqs. (17) as a pattern. (Recall Example 4 in the last section.) For $n = 8$, the required matrix and the indicated multiplication is the following.

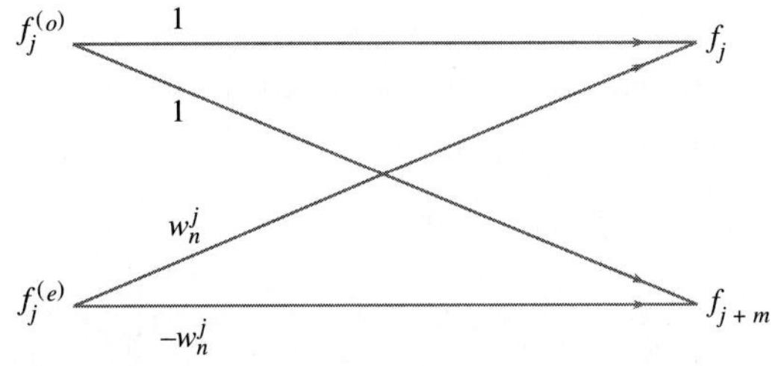

FIGURE 14.5
A typical butterfly pattern.

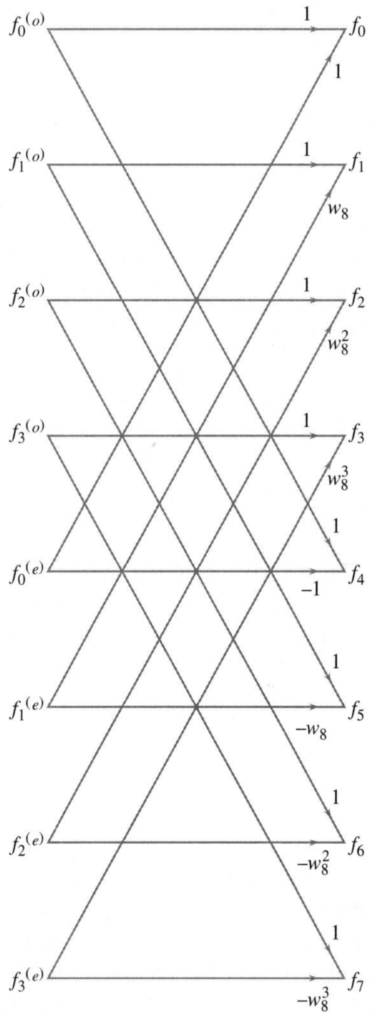

FIGURE 14.6
The complete butterfly pattern for Eqs. (17) when $n = 8$.

(18)
$$\begin{bmatrix} 1 & 0 & 0 & 0 & 1 & 0 & 0 & 0 \\ 0 & 1 & 0 & 0 & 0 & w_8 & 0 & 0 \\ 0 & 0 & 1 & 0 & 0 & 0 & w_8^2 & 0 \\ 0 & 0 & 0 & 1 & 0 & 0 & 0 & w_8^3 \\ 1 & 0 & 0 & 0 & -1 & 0 & 0 & 0 \\ 0 & 1 & 0 & 0 & 0 & -w_8 & 0 & 0 \\ 0 & 0 & 1 & 0 & 0 & 0 & -w_8^2 & 0 \\ 0 & 0 & 0 & 1 & 0 & 0 & 0 & -w_8^3 \end{bmatrix} \begin{bmatrix} f_0^{(o)} \\ f_1^{(o)} \\ f_2^{(o)} \\ f_3^{(o)} \\ f_0^{(e)} \\ f_1^{(e)} \\ f_2^{(e)} \\ f_3^{(e)} \end{bmatrix} = \begin{bmatrix} f_0 \\ f_1 \\ f_2 \\ f_3 \\ f_4 \\ f_5 \\ f_6 \\ f_7 \end{bmatrix}$$

[The 8×8 matrix in (18) we shall later refer to as R_3.]

The next step, now, is to determine the $f_j^{(o)}$'s and the $f_j^{(e)}$'s, that is $\mathbf{F}^{(o)}$ and $\mathbf{F}^{(e)}$. To do this, we note first that Eqs. (13) and (15) are just the general scalar equations implied by the respective matrix equations

(19)
$$\mathbf{M}_{n/2}\mathbf{A}_o = \mathbf{F}^{(o)} \quad \text{or} \quad [w_m^{jk}]\mathbf{A}_o = \mathbf{F}^{(o)} \qquad \mathbf{A}_o = \begin{bmatrix} a_0 \\ a_2 \\ \vdots \\ a_{n-2} \end{bmatrix}$$

and

$$(20) \qquad M_{n/2}\mathbf{A}_e = \mathbf{F}^{(e)} \qquad \text{or} \qquad [w_m^{jk}]\mathbf{A}_e = \mathbf{F}^{(e)} \qquad \mathbf{A}_e = \begin{bmatrix} a_1 \\ a_3 \\ \vdots \\ a_{n-1} \end{bmatrix}$$

$$j, k = 0, 1, 2, \ldots, m-1$$

Clearly these are mathematically identical with Eq. (7), except that each is just half the size. Moreover, (19) and (20) can be combined into a single $n \times n$ matrix equation, namely,

$$(21) \qquad \begin{bmatrix} M_{n/2} & 0 \\ 0 & M_{n/2} \end{bmatrix} \begin{bmatrix} \mathbf{A}_o \\ \mathbf{A}_e \end{bmatrix} = \begin{bmatrix} \mathbf{F}^{(o)} \\ \mathbf{F}^{(e)} \end{bmatrix}$$

The solution process for each of Eqs. (19) and (20) parallels exactly the procedure that led from Eq. (7) to (19) and (20), and our biggest problem is devising a notation to describe it. For the purpose of our exposition, we shall introduce the following symbols: when the sum $f_j^{(o)}$ in (13) is broken up into two sums consisting of alternate elements of $\mathbf{A}_o$, we shall (very much as we did in defining $f_j^{(o)}$ and $f_j^{(e)}$) use $f_j^{(oo)}$ to name the sum containing the first, third, fifth, . . . , coefficients in $\mathbf{A}_o$, and $f_j^{(oe)}$ to name the sum containing the second, fourth, sixth, . . . , coefficients in $\mathbf{A}_o$. Similarly, we shall use $f_j^{(eo)}$ and $f_j^{(ee)}$ to name the corresponding sums into which $f_j^{(e)}$, as given by (15), is broken up. Thus, at the next stage after (17), we now have the *four* equations

$$(22) \qquad \begin{aligned} f_j^{(o)} &= f_j^{(oo)} + w_m^j f_j^{(oe)} \\ f_{j+m/2}^{(o)} &= f_j^{(oo)} - w_m^j f_j^{(oe)} \\ f_j^{(e)} &= f_j^{(eo)} + w_m^j f_j^{(ee)} \\ f_{j+m/2}^{(e)} &= f_j^{(eo)} - w_m^j f_j^{(ee)} \end{aligned} \qquad j = 0, 1, 2, \ldots, m/2 - 1$$

If, in analogy with Eqs. (14), (16), and the vectors $\mathbf{A}_o$ and $\mathbf{A}_e$ in (19) and (20), we introduce the vectors

$$\mathbf{F}^{(oo)} = \begin{bmatrix} f_0^{(oo)} \\ f_1^{(oo)} \\ \vdots \\ f_{m/2-1}^{(oo)} \end{bmatrix} \qquad \mathbf{F}^{(oe)} = \begin{bmatrix} f_0^{(oe)} \\ f_1^{(oe)} \\ \vdots \\ f_{m/2-1}^{(oe)} \end{bmatrix} \qquad \mathbf{F}^{(eo)} = \begin{bmatrix} f_0^{(eo)} \\ f_1^{(eo)} \\ \vdots \\ f_{m/2-1}^{(eo)} \end{bmatrix} \qquad \mathbf{F}^{(ee)} = \begin{bmatrix} f_0^{(ee)} \\ f_1^{(ee)} \\ \vdots \\ f_{m/2-1}^{(ee)} \end{bmatrix}$$

and

$$\mathbf{A}_{oo} = \begin{bmatrix} a_0 \\ a_4 \\ \vdots \\ a_{n-4} \end{bmatrix} \qquad \mathbf{A}_{oe} = \begin{bmatrix} a_2 \\ a_6 \\ \vdots \\ a_{n-2} \end{bmatrix} \qquad \mathbf{A}_{eo} = \begin{bmatrix} a_1 \\ a_5 \\ \vdots \\ a_{n-3} \end{bmatrix} \qquad \mathbf{A}_{ee} = \begin{bmatrix} a_3 \\ a_7 \\ \vdots \\ a_{n-1} \end{bmatrix}$$

it is clear that just as (13) and (15) were typical scalar equations implied by the respective matrix equations (19) and (20), so the equations obtained by setting $f_j^{(oo)}, f_j^{(oe)}, f_j^{(eo)}$, and $f_j^{(ee)}$ equal to the series we have designated for them are typical scalar equations implied by the respective matrix equations

$$M_{n/4}\mathbf{A}_{oo} = \mathbf{F}^{(oo)} \qquad M_{n/4}\mathbf{A}_{oe} = \mathbf{F}^{(oe)} \qquad M_{n/4}\mathbf{A}_{eo} = \mathbf{F}^{(eo)} \qquad M_{n/4}\mathbf{A}_{ee} = \mathbf{F}^{(ee)}$$

where $M_{n/4} = [w_{n/4}^{jk}]$, $j, k = 0, 1, 2, \ldots, n/4 - 1$. Furthermore, these four equations can be combined to form a partitioned $n \times n$ matrix equation analogous to (21):

$$
\begin{bmatrix} M_{n/4} & 0 & 0 & 0 \\ 0 & M_{n/4} & 0 & 0 \\ 0 & 0 & M_{n/4} & 0 \\ 0 & 0 & 0 & M_{n/4} \end{bmatrix} \begin{bmatrix} \mathbf{A}_{oo} \\ \mathbf{A}_{oe} \\ \mathbf{A}_{eo} \\ \mathbf{A}_{ee} \end{bmatrix} = \begin{bmatrix} \mathbf{F}^{(oo)} \\ \mathbf{F}^{(oe)} \\ \mathbf{F}^{(eo)} \\ \mathbf{F}^{(ee)} \end{bmatrix} \tag{23}
$$

After $\mathbf{F}^{(oo)}$, $\mathbf{F}^{(oe)}$, $\mathbf{F}^{(eo)}$, and $\mathbf{F}^{(ee)}$ have been found, their components will have to be interwoven to yield the vector

$$
\begin{bmatrix} \mathbf{F}^{(o)} \\ \mathbf{F}^{(e)} \end{bmatrix}
$$

from which eventually the vector **F** will be found. [Recall Eq. (18).] This rearrangement can be accomplished by premultiplying the vector on the right in (23) by a matrix that can easily be written down, using Eqs. (22) as a pattern. For $n = 8$, the required matrix and the indicated multiplication is the following.

$$
\begin{bmatrix} 1 & 0 & 1 & 0 & 0 & 0 & 0 & 0 \\ 0 & 1 & 0 & i & 0 & 0 & 0 & 0 \\ 1 & 0 & -1 & 0 & 0 & 0 & 0 & 0 \\ 0 & 1 & 0 & -i & 0 & 0 & 0 & 0 \\ 0 & 0 & 0 & 0 & 1 & 0 & 1 & 0 \\ 0 & 0 & 0 & 0 & 0 & 1 & 0 & i \\ 0 & 0 & 0 & 0 & 1 & 0 & -1 & 0 \\ 0 & 0 & 0 & 0 & 0 & 1 & 0 & -i \end{bmatrix} \begin{bmatrix} f_0^{(oo)} \\ f_1^{(oo)} \\ f_0^{(oe)} \\ f_1^{(oe)} \\ f_0^{(eo)} \\ f_1^{(eo)} \\ f_0^{(ee)} \\ f_1^{(ee)} \end{bmatrix} = \begin{bmatrix} f_0^{(o)} \\ f_1^{(o)} \\ f_2^{(o)} \\ f_3^{(o)} \\ f_0^{(e)} \\ f_1^{(e)} \\ f_2^{(e)} \\ f_3^{(e)} \end{bmatrix} \tag{24}
$$

[The 8×8 matrix in (24) we shall later refer to as R_2.]

If $n = 2^l$, this process of dividing the problem into simpler problems of half the size can be carried out $l - 1$ times, leading finally to a stage in which the matrix

$$
\begin{bmatrix} M_2 & & & \mathbf{O} \\ & M_2 & & \\ & & \ddots & \\ \mathbf{O} & & & M_2 \end{bmatrix} \qquad M_2 = \begin{bmatrix} 1 & 1 \\ 1 & -1 \end{bmatrix} \tag{25}
$$

premultiplies a suitable rearrangement of the given vector of coefficients **A** and converts that vector into the earliest or most "primitive" of the **F** vectors. Thus, in summary, the fast Fourier transform obtains the vector of functional values **F** from the vector of coefficients **A** by $l + 1$ successive premultiplications by modifications of I_n. The first of these matrices is the one that rearranges the vector of coefficients **A** into the form required in the use of (25). The second is the matrix appearing in (25). The remaining $l - 1$ are the interweaving matrices required as we go from stage to stage back to the vector of functional values **F**. By this process we have, in effect, factored the matrix $M(=M_n) = [w_n^{jk}]$ in Eq. (7) into $l + 1$ matrices, each of which consists largely of zeros, a very desirable computational feature.

When $n = 8$, the last equation in the chain is Eq. (23) which, when written at length becomes

$$
\begin{bmatrix} 1 & 1 & 0 & 0 & 0 & 0 & 0 & 0 \\ 1 & -1 & 0 & 0 & 0 & 0 & 0 & 0 \\ 0 & 0 & 1 & 1 & 0 & 0 & 0 & 0 \\ 0 & 0 & 1 & -1 & 0 & 0 & 0 & 0 \\ 0 & 0 & 0 & 0 & 1 & 1 & 0 & 0 \\ 0 & 0 & 0 & 0 & 1 & -1 & 0 & 0 \\ 0 & 0 & 0 & 0 & 0 & 0 & 1 & 1 \\ 0 & 0 & 0 & 0 & 0 & 0 & 1 & -1 \end{bmatrix} \begin{bmatrix} a_0 \\ a_4 \\ a_2 \\ a_6 \\ a_1 \\ a_5 \\ a_3 \\ a_7 \end{bmatrix} = \begin{bmatrix} f_0^{(oo)} \\ f_1^{(oo)} \\ f_0^{(oe)} \\ f_1^{(oe)} \\ f_0^{(eo)} \\ f_1^{(eo)} \\ f_0^{(ee)} \\ f_1^{(ee)} \end{bmatrix} \tag{26}
$$

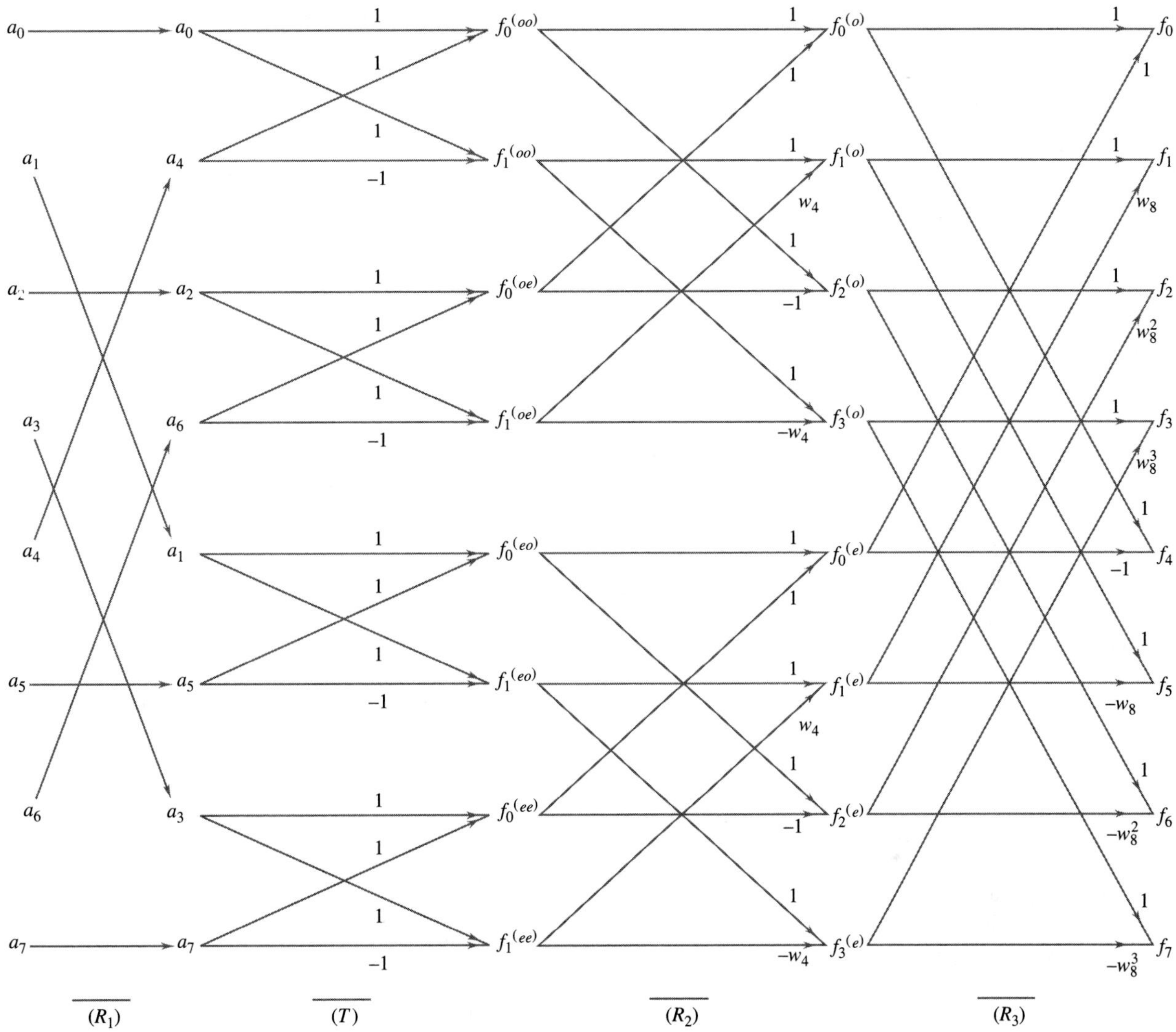

FIGURE 14.7
Flowchart for the fast Fourier transform when $n = 8$.

The matrix in (26), which transforms the rearranged vector **A** into the first of the **F** vectors, we shall call T. The matrix required to transform **A** into the vector appearing on the left in (26) is easily verified to be

$$R_1 = \begin{bmatrix} 1 & 0 & 0 & 0 & 0 & 0 & 0 & 0 \\ 0 & 0 & 0 & 0 & 1 & 0 & 0 & 0 \\ 0 & 0 & 1 & 0 & 0 & 0 & 0 & 0 \\ 0 & 0 & 0 & 0 & 0 & 0 & 1 & 0 \\ 0 & 1 & 0 & 0 & 0 & 0 & 0 & 0 \\ 0 & 0 & 0 & 0 & 0 & 1 & 0 & 0 \\ 0 & 0 & 0 & 1 & 0 & 0 & 0 & 0 \\ 0 & 0 & 0 & 0 & 0 & 0 & 0 & 1 \end{bmatrix}$$

The subsequent interweavings that finally give the vector **F** are accomplished by the matrix R_2, given in (24), and the matrix R_3, given in (18). Thus, finally, for the case $n = 8 = 2^3$, we can write

$$\mathbf{F} = (R_3R_2TR_1)\mathbf{A}$$

The complete flowchart for the case $n = 8$ is shown in Fig. 14.7.

EXERCISES

1. (a) What is $|M_4|$?
   (b) What is M_4^2?
2. Write out M_8 in terms of w_8, expressing each entry as the lowest possible positive power of w_8.
3. Write out M_8^{-1} in terms of w_8, expressing each entry as the lowest possible positive power of w_8.
4. Explain how the fast Fourier transform can be used to solve the equation $\mathbf{A} = M_n^{-1}\mathbf{F}$, $n = 4, 8, 16, \ldots$.
5. Write out the matrices in the fast Fourier transform solution of the equation $\mathbf{F} = M_4\mathbf{A}$ and verify that their product is M_4.
6. Write out the matrices in the fast Fourier transform solution of the equation $\mathbf{A} = M_4^{-1}\mathbf{F}$ and verify that their product is M_4^{-1}.
7. If $\mathbf{A} = [1\ 1\ 0\ 1]^T$, find **F** from the equation $\mathbf{F} = M_4\mathbf{A}$ **(a)** by performing the indicated multiplication and **(b)** by using the fast Fourier transform. Write **F** in trigonometric form.
8. Work Exercise 7 if $\mathbf{A} = [2\ 2\ 1\ -1]^T$.
9. If **A** is a real vector, what conditions must its components satisfy in order that **(a)** $\mathbf{F} = M_4\mathbf{A}$ will be a real vector and **(b)** $\mathbf{F} = M_4\mathbf{A}$ will be a pure imaginary vector?
10. If **A** is a real vector, show that the vector $\mathbf{F} = M_8\mathbf{A}$ is real if and only if $a_1 = a_7$, $a_2 = a_6$, $a_3 = a_5$. *Hint:* Recall that an expression is real if and only if it is equal to its conjugate.
11. If **A** is a real vector, under what conditions, if any, will the vector $\mathbf{F} = M_8\mathbf{A}$ be a pure imaginary vector?
12. How can a trigonometric representation be found for a function defined on n equally spaced points in an interval of the form $[0, 2p)$? *Hint:* Recall the nature of the terms used in the Fourier expansion of a function of period $2p$.
13. Use the fast Fourier transform to obtain a trigonometric representation of the vector $\mathbf{F} = [0\ 1\ 2\ 3]^T$ on the interval $[0, 2\pi)$.
14. Work Exercise 13 if $\mathbf{F} = [1\ 1\ 2\ 1]$ on the interval $[0, 4)$.
15. If $\mathbf{A} = [1\ 0\ 1\ 0\ 1\ 0\ 1\ 0]^T$, use the fast Fourier transform to find **F** **(a)** on the interval $[0, 2\pi)$ and **(b)** on the interval $[0, 4\pi)$.
16. Work Exercise 15 if $\mathbf{A} = [0\ 1\ 0\ 1\ 0\ 1\ 0\ 1]^T$.
17. A vector **v** of dimension $n = 4m$ is said to be **odd** if $v_{2m} = v_0$ and $v_{n-j} = -v_j$, $j = 1, 2, \ldots, 2m - 1$. Show that in the equation $\mathbf{F} = M_n\mathbf{A}$, **F** is odd if and only if **A** is odd.
18. If the values of f are f_0, f_1, f_2, f_3, show that the imaginary terms in the finite complex exponential Fourier expansion of f vanish at each point in the domain of f.

14.3 RANK AND THE EQUIVALENCE OF MATRICES

In Chap. 3, we introduced the ideas of vectors and matrices and investigated a number of their elementary properties which, like those of Chap. 13, proved to be very useful. In the remainder of this chapter, we continue our study of matrices and present some more applications, beginning in this section with *rank* and the *equivalence of matrices*.

If B is a square matrix, any square submatrix M of B whose principal diagonal is a part of the principal diagonal of B is called a **principal submatrix** of B, and $|M| = \det M$ is said to be a **principal minor** of B. The determinants of the square submatrices of any matrix A, whether A is square or not, are called the **minors** of A.

One of the most important characteristics of a matrix with real or complex elements is its *rank*. We have already said what is meant by the *row rank* of a matrix (Definition 3, Sec. 13.3). Similarly, the maximum number of linearly independent column vectors in a matrix is called its **column rank.** Still another definition of the rank of a matrix is the following.

DEFINITION 1 The **rank** of a zero matrix is 0. The **rank** of a nonzero matrix A is the largest integer r for which there exists an rth-order minor of A whose value is not zero.

This, the so-called *classical* definition of rank, is often referred to more specifically as the **determinant rank** of A. As it turns out (Exercise 2), these three ranks are all equal. Thus, without ambiguity, we may speak of *the* rank of a matrix. In practice, zero matrices are relatively unimportant. We shall therefore suppose in the theorems of this section that all matrices are of positive rank, even though many of the theorems hold for zero matrices.

EXAMPLE 1

The matrix $\begin{bmatrix} 1 & 2 & -1 & 3 \\ 3 & 4 & 0 & -1 \\ -1 & 0 & -2 & 7 \end{bmatrix}$ is of rank 2 since each of the third-order submatrices

$$\begin{bmatrix} 2 & -1 & 3 \\ 4 & 0 & -1 \\ 0 & -2 & 7 \end{bmatrix} \quad \begin{bmatrix} 1 & -1 & 3 \\ 3 & 0 & -1 \\ -1 & -2 & 7 \end{bmatrix} \quad \begin{bmatrix} 1 & 2 & 3 \\ 3 & 4 & -1 \\ -1 & 0 & 7 \end{bmatrix} \quad \begin{bmatrix} 1 & 2 & -1 \\ 3 & 4 & 0 \\ -1 & 0 & -2 \end{bmatrix}$$

is singular, while not all second-order minors are zero. Specifically, the determinant of the 2×2 submatrix in the upper left-hand corner is different from zero.

As an immediate consequence of Property 12, Sec. 3.5, we have the following result.

THEOREM 1 If A and B are $n \times n$ matrices of rank n, then both AB and BA are of rank n.

In Sec. 3.4, we introduced three types of elementary operations for manipulating the rows of a matrix. For some purposes, it is necessary to consider the effect of performing **elementary column operations** on a matrix. These are defined analogously to be the following.

1. Interchanging two columns
2. Multiplying a column by a nonzero number
3. Adding a constant times the elements of one column to the corresponding elements of another column.

For convenience, we shall speak of elementary row and column operations collectively as **elementary operations.** One of their most important properties is the following.

THEOREM 2 The rank of a matrix is not altered by any sequence of elementary operations.

PROOF Let A be an arbitrary matrix and let r be its rank. To prove the theorem, it is clearly sufficient to prove that no single elementary operation can change the rank of A.

After an operation of Type 1, every submatrix of A exists *somewhere* in the resulting matrix with at most two rows or two columns interchanged. By Property 7, Sec. 3.5, an interchange of rows or columns does not affect the vanishing or nonvanishing of a determinant. Hence no operation of Type 1 can alter the rank of A.

By Property 5, Sec. 3.5, an operation of Type 2 does not affect the vanishing or nonvanishing of any minor of A; hence it cannot alter the rank of A.

Finally, consider an operation of Type 3 in which a number c times each element of the ith row of A is added to the corresponding element of the jth row ($i \neq j$) and let B denote the resulting matrix. (A Type-3 column operation can be handled similarly.) If $c = 0$, then B = A, and no proof is required. By Property 9, Sec. 3.5, the value of any minor of A involving neither the ith nor the jth row, both the ith and the jth rows, or the ith but not the jth row is unaffected by the operation. By Property 6, Sec. 3.5, any minor $|M|$ of B involving the jth row but not the ith row can be written in the form

$$|M| = |S_1| + c|S_2| \qquad c \neq 0$$

where S_1 and S_2 are square submatrices of A of the same order as M. Obviously, if both $|S_1| = 0$ and $|S_2| = 0$, then $|M| = 0$. Contrapositively, if $|M| \neq 0$, either $|S_1| \neq 0$ or $|S_2| \neq 0$. Hence no transformation of Type 3 can produce a matrix B whose rank exceeds the rank of A. On the other hand, no operation of this kind can yield a matrix whose rank is less than the rank of A, either. For if this were the case, the operation consisting of adding $-c$ times the elements of the ith row to the corresponding elements of the jth row in the matrix B would be a Type-3 operation which restored the matrix B to its original form A and hence increased its rank to the original value r; and we have just proved this to be impossible. Thus the rank of A, being neither increased nor decreased by an operation of Type 3, is invariant under this operation also, and our proof is complete. ◗

In Sec. 3.4 we observed that any elementary row operation on a matrix A can be accomplished by premultiplying A by a conformable unit matrix on whose rows the same operation has been performed. Likewise, it is easy to show that any column operation on A can be accomplished by postmultiplying A by a conformable unit matrix on whose columns the same operation has been performed. More precisely, we have the following theorems which extend Theorems 2–4 of Sec. 3.4 to include column operations also.

THEOREM 3 If A is an arbitrary $m \times n$ matrix and if $P_m(Q_n)$,† is the matrix obtained from the identity matrix $I_m(I_n)$ by interchanging its ith and jth rows (columns), then the product $P_mA(AQ_n)$ is identical with A except for the ith and jth rows (columns) which are interchanged.

THEOREM 4 If A is an arbitrary $m \times n$ matrix and if $P_m(Q_n)$ is the matrix obtained from the identity matrix $I_m(I_n)$ by multiplying the elements in the ith row (column) by c, then the product $P_mA(AQ_n)$ is identical with A except for the ith row (column) which consists of the elements of the ith row (column) of A each multiplied by c.

THEOREM 5 If A is an $m \times n$ matrix and if $P_m(Q_n)$ is the matrix obtained from the identity matrix $I_m(I_n)$ by adding to the elements of the jth row (column) c times the corresponding elements in the ith row (column), then the product $P_mA(AQ_n)$ is identical with A except for the jth row (column), which consists of the elements of the jth row (column) of A plus c times the corresponding elements in the ith row (column) of A.

Any matrix obtained by performing elementary row (column) operations on a unit matrix is called an **elementary row (column) matrix.** By Theorem 2, every elementary matrix is nonsingular.

From Theorems 3–5 it is clear that a sequence of elementary operations $E_1, E_2, \ldots, E_k$ ($E_1, E_2, \ldots, E_l$) on the rows (columns) of an $m \times n$ matrix A can be accomplished by premultiplying (postmultiplying) A by a sequence of elementary row (column) matrices $P_1, P_2, \ldots, P_k$ ($Q_1, Q_2, \ldots, Q_l$). The product $P_k \cdots P_2 P_1$ ($Q_1 Q_2 \cdots Q_l$) of the matrices by which A is premultiplied (postmultiplied) can of course be expressed as a single matrix P (Q) which, in virtue of

†See the footnote on p. 232.

Theorem 1, is necessarily of rank m (n). To express the fact that a matrix B can be obtained by performing elementary operations on a matrix A, so that $B = P_k \cdots P_2P_1AQ_1Q_2 \cdots Q_l = PAQ$, we say that B is **equivalent** to A. More concisely, we have the following definition.

DEFINITION 2 A matrix B is **equivalent** to a matrix A if and only if nonsingular matrices P and Q exist such that B = PAQ.

As is to be expected, for $m \times n$ matrices A, B, and C,

(a) A is equivalent to A, since $A = I_mAI_n$.

(b) If B is equivalent to A, then A is equivalent to B because $A = P^{-1}BQ^{-1}$.

(c) If C is equivalent to B and B is equivalent to A, then C is equivalent to A since C = RBS and B = PAQ together imply C = (RP)B(QS) with RP and QS nonsingular.

The word *equivalent* is aptly used in Definition 2 since Properties A, B, and C assert, respectively, that the relation is *reflexive, symmetric,* and *transitive,* and these are the definitive characteristics of any equivalence relation.

In Sec. 3.7 we observed that row operations alone suffice to reduce any nonsingular matrix of order n to the identity matrix I_n. It turns out that the same is true of elementary column operations (Exercise 11). Thus we have the following theorem.

THEOREM 6 Any nonsingular $n \times n$ matrix can be reduced to the identity matrix I_n by elementary row operations or, equally well, by elementary column operations.

It is natural to inquire whether, conversely, every nonsingular matrix can be obtained from the identity matrix of the same order by means of elementary row or column operations. Our next theorem answers this question.

THEOREM 7 Any nonsingular matrix of order n can be obtained by performing elementary row operations or, equally well, elementary column operations on I_n.

PROOF Let B be an arbitrary nonsingular $n \times n$ matrix. Theorem 6 assures us that a sequence of elementary row operations can be found which will reduce B to I_n. Hence, by Theorems 3–5, there exist elementary row matrices $P_1, P_2, \ldots, P_k$ such that

(1)
$$I_n = P_k \cdots P_2 P_1 B$$

Since every elementary matrix is nonsingular, we may premultiply this equation by $(P_k \cdots P_2P_1)^{-1}$, which gives $P_1^{-1}P_2^{-1} \cdots P_k^{-1}I_n = B$. This shows that B can be obtained from I_n by applying, in reverse order, the inverses of the successive elementary row operations by which B is converted into I_n. A similar argument confirms that B can also be expressed as a product of elementary column matrices and hence is derivable from I_n by a sequence of elementary column operations.

Incidentally, since the inverse of a matrix must be unique, it follows from (1) that

(2)
$$B^{-1} = P_k \cdots P_2 P_1$$

This formula provides another method of determining the inverse of a matrix, which we shall now illustrate.

EXAMPLE 2

Find a sequence of elementary row operations $E_1, E_2, \ldots,$ which will reduce the matrix

$$B = \begin{bmatrix} 1 & 2 & 0 \\ 2 & 3 & -1 \\ -1 & -1 & 2 \end{bmatrix}$$

to I_3, determine the elementary row matrices $P_1, P_2, \ldots$, corresponding to these operations, and use these matrices to compute B^{-1}.

By inspection it is clear that B can be reduced to I_3 by the following sequence of row operations.

$$E_1:\ \text{Row 2} - 2\ \text{Row 1}\quad P_1 = \begin{bmatrix} 1 & 0 & 0 \\ -2 & 1 & 0 \\ 0 & 0 & 1 \end{bmatrix} \qquad P_1B = \begin{bmatrix} 1 & 2 & 0 \\ 0 & -1 & -1 \\ -1 & -1 & 2 \end{bmatrix}$$

$$E_2:\ \text{Row 3} + \ \text{Row 1}\quad P_2 = \begin{bmatrix} 1 & 0 & 0 \\ 0 & 1 & 0 \\ 1 & 0 & 1 \end{bmatrix} \qquad P_2P_1B = \begin{bmatrix} 1 & 2 & 0 \\ 0 & -1 & -1 \\ 0 & 1 & 2 \end{bmatrix}$$

$$E_3:\ -1\ \text{Row 2}\quad P_3 = \begin{bmatrix} 1 & 0 & 0 \\ 0 & -1 & 0 \\ 0 & 0 & 1 \end{bmatrix} \qquad P_3P_2P_1B = \begin{bmatrix} 1 & 2 & 0 \\ 0 & 1 & 1 \\ 0 & 1 & 2 \end{bmatrix}$$

$$E_4:\ \text{Row 3} - \ \text{Row 2}\quad P_4 = \begin{bmatrix} 1 & 0 & 0 \\ 0 & 1 & 0 \\ 0 & -1 & 1 \end{bmatrix} \qquad P_4P_3P_2P_1B = \begin{bmatrix} 1 & 2 & 0 \\ 0 & 1 & 1 \\ 0 & 0 & 1 \end{bmatrix}$$

$$E_5:\ \text{Row 2} - \ \text{Row 3}\quad P_5 = \begin{bmatrix} 1 & 0 & 0 \\ 0 & 1 & -1 \\ 0 & 0 & 1 \end{bmatrix} \qquad P_5P_4P_3P_2P_1B = \begin{bmatrix} 1 & 2 & 0 \\ 0 & 1 & 0 \\ 0 & 0 & 1 \end{bmatrix}$$

$$E_6:\ \text{Row 1} - 2\ \text{Row 2}\quad P_6 = \begin{bmatrix} 1 & -2 & 0 \\ 0 & 1 & 0 \\ 0 & 0 & 1 \end{bmatrix} \qquad P_6P_5P_4P_3P_2P_1B = \begin{bmatrix} 1 & 0 & 0 \\ 0 & 1 & 0 \\ 0 & 0 & 1 \end{bmatrix}$$

The inverse B^{-1} of the given matrix can now be found either by using (2), namely,

$$B^{-1} = P_6P_5P_4P_3P_2P_1$$

or by performing on I_3 the same sequence of row operations used to reduce B to I_3, very much as we did in proving Theorem 4, Sec. 3.7,

$$B^{-1} = E_6E_5E_4E_3E_2E_1I_3$$

The result by either method is $B^{-1} = \begin{bmatrix} -5 & 4 & 2 \\ 3 & -2 & -1 \\ -1 & 1 & 1 \end{bmatrix}$.

By first using elementary row operations to reduce an $m \times n$ matrix A to row echelon form and then using elementary column operations, if necessary, it is possible to prove the following theorem.

THEOREM 8 Any $m \times n$ matrix A of rank r can be reduced by elementary operations, which in general will involve both rows and columns, to an $m \times n$ matrix M in which, for $i = 1, 2, \ldots, r$, $m_{ii} = 1$ and all other elements are zero.

From Theorem 2 it is clear that equivalent matrices have the same rank. In view of Theorem 8, given any two $m \times n$ matrices of the same rank r, each can be reduced by elementary operations to an $m \times n$ matrix M all of whose elements are zero except for the first r diagonal elements, each of which is a 1. Hence either of the two matrices can be converted into the other, via M, by elementary operations. Thus we have established the following important result.

TABLE 14.2

If P, Q are arbitrary nonsingular matrices	$B = PAQ$	is an equivalence transformation and B is equivalent to A
If $P = Q^{-1}$	$B = Q^{-1}AQ$	is a similarity transformation and B is similar to A
If $P = Q^T$	$B = Q^TAQ$	is a congruence transformation and B is congruent to A
If $P = Q^T = Q^{-1}$	$B = Q^TAQ = Q^{-1}AQ$	is an orthogonal transformation and B is orthogonally similar to A
If $P = \overline{Q}^T = Q^{-1}$	$B = \overline{Q}^TAQ = Q^{-1}AQ$	is a unitary transformation and B is unitarily similar to A

THEOREM 9 Two $m \times n$ matrices are equivalent if and only if they have the same rank.

The equivalence relation $B = PAQ$, where P and Q are nonsingular, is a very general one. It can be thought of as a transformation of a matrix A into a matrix B, and the usual terminology reflects this point of view. For instance, the matrix M of Theorem 8 is known as the **canonical form** of an $m \times n$ matrix of rank r **under equivalence transformations.** Many applications involve special cases of equivalence transformations in which P and Q have further properties. Table 14.2 summarizes the various cases of particular interest.

Theorem 4 of Sec. 3.6 gave us a useful criterion for testing the consistency of a nonhomogeneous linear system of equations and for determining the nature of a complete solution should solutions exist. That test, which was based on the row echelon form of the augmented matrix of the system, can now be restated in terms of the ranks of the coefficient matrix and the augmented matrix of the system alternatively as follows.

THEOREM 10 A system of m simultaneous linear algebraic equations in n unknowns, $A\mathbf{x} = \mathbf{b}$, has a solution if and only if the coefficient matrix A and the augmented matrix $[A\mathbf{b}]$ have the same rank r. When solutions exist, the number of arbitrary constants in any complete solution is $n - r$.

PROOF Let r_1 be the rank of $[A\mathbf{b}]$ and r the rank of A. Clearly, $r_1 \geq r$. Theorem 2 guarantees that r_1 is equal to the rank of the row echelon form of $[A\mathbf{b}]$ and that r is equal to the rank of the row echleon form of A. If $r_1 > r$, the system of linear equations corresponding to the row echelon form of $[A\mathbf{b}]$ contains at least one equation of the form $0 = k$, $k \neq 0$. This implies that the linear system $A\mathbf{x} = \mathbf{b}$ is inconsistent. It follows that $A\mathbf{x} = \mathbf{b}$ is consistent if and only if $r_1 = r$, in which case the number of arbitrary constants in any complete solution is $n - r$, as we pointed out in Sec. 3.6. ◗

Another result that warrants restatement in terms of *rank* is Theorem 3, Sec. 3.7, which translates into the following.

THEOREM 11 If the coefficient matrix A of a homogeneous system of n linear equations in n unknowns is of rank $n - 1$ and if the submatrix obtained from A by omitting the kth row is also of rank $n - 1$, then a complete solution of $A\mathbf{x} = \mathbf{0}$ is given by $\mathbf{x} = c\mathbf{v}$, where $\mathbf{v} = [A_{k1}\ A_{k2} \cdots A_{kn}]^T$ is the column matrix whose components are the cofactors of the elements of the kth row of det A and c is an arbitrary constant.

PROOF Since the rank of the $(n-1) \times n$ matrix remaining after the kth row is deleted from A is $n-1$, at least one of the cofactors A_{ki} of the elements of the kth row of $|A|$ is different from zero. Hence $\mathbf{v} \neq \mathbf{0}$. Applying Eq. (10), Property 10, Sec. 3.5, with k and j interchanged in that equation, and noting that $|A| = 0$ by hypothesis, we find by direct computation

$$A\mathbf{v} = [\delta_{ik}|A|] \qquad 1 \leq i \leq n \qquad \text{or} \qquad A\mathbf{v} = \mathbf{0}$$

which shows that $\mathbf{v}$ is a solution of $A\mathbf{x} = \mathbf{0}$. The rank of both A and $[A\mathbf{0}]$ is $r = (n-1)$. Moreover, $n - (n-1) = 1$, and $\mathbf{v}$ is nontrivial. Therefore, by Theorem 10, $c\mathbf{v}$ is a complete solution, as asserted.

Here is a useful corollary of Theorem 11, whose proof we leave as an exercise.

COROLLARY 1 If the coefficient matrix A of a homogeneous system of $n-1$ linear equations in n unknowns $A\mathbf{x} = \mathbf{0}$ is of rank $n-1$, then a complete solution of the system is given by $c\mathbf{v}$, where $\mathbf{v} = [|M_1|, -|M_2|, \ldots, (-1)^{n-1}|M_n|]^T$ and M_j is the $(n-1) \times (n-1)$ submatrix obtained from A by deleting the jth column of A.

EXAMPLE 3

Use Theorem 11 to find a complete solution of the system

$$\begin{aligned} x_1 - 2x_2 + x_3 + 3x_4 &= 0 \\ 2x_1 + 2x_2 - x_3 + x_4 &= 0 \\ -x_1 - x_2 + 3x_3 + 2x_4 &= 0 \\ x_1 - 8x_2 - x_3 + 3x_4 &= 0 \end{aligned}$$

It is easy to verify that the coefficient matrix A of this system is singular but that the 3×3 submatrix in the upper left-hand corner is nonsingular. Thus A and the submatrix remaining when the last row is deleted from A are both of rank 3. Hence, according to the last theorem, the successive components of a complete solution $\mathbf{x}$ are proportional to the respective cofactors of the last row of $|A|$. Thus we have

$$x_1 = -c\begin{vmatrix} -2 & 1 & 3 \\ 2 & -1 & 1 \\ -1 & 3 & 2 \end{vmatrix} = -20c \qquad x_2 = c\begin{vmatrix} 1 & 1 & 3 \\ 2 & -1 & 1 \\ -1 & 3 & 2 \end{vmatrix} = 5c$$

$$x_3 = -c\begin{vmatrix} 1 & -2 & 3 \\ 2 & 2 & 1 \\ -1 & -1 & 2 \end{vmatrix} = -15c \qquad x_4 = c\begin{vmatrix} 1 & -2 & 1 \\ 2 & 2 & -1 \\ -1 & -1 & 3 \end{vmatrix} = 15c$$

or, setting $-5c = k$ and writing $\mathbf{x}$ as a row vector,

$$\mathbf{x} = k(4, -1, 3, -3)$$

We conclude this section with two theorems concerning the rank of a product of matrices.

THEOREM 12 If A and B are conformable matrices of rank r and ρ, respectively, the rank of the product AB is equal to or less than the smaller of r and ρ.

PROOF Let A be an $m \times n$ matrix, let B be an $n \times p$ matrix, and let the row vectors of A be $\mathbf{u}_1, \mathbf{u}_2, \ldots, \mathbf{u}_m$. Then the ith row vector in the product AB is [Exercise 27(**b**), Sec. 13.1]

$$\mathbf{u}_i B \qquad i = 1, 2, \ldots, m$$

Since A is of rank r, the maximum number of linearly independent row vectors of A is r. Without loss of generality, we may suppose that $\mathbf{u}_1, \mathbf{u}_2, \ldots, \mathbf{u}_r$ are linearly independent. Then all other row vectors of A are linear combinations of these r vectors:

$$\mathbf{u}_i = c_{1i}\mathbf{u}_1 + c_{2i}\mathbf{u}_2 + \cdots + c_{ri}\mathbf{u}_r \qquad i = r+1, r+2, \ldots, m$$

Substituting into (3), we find that for $i = r + 1, r + 2, \ldots, m$, the ith row vector of the product AB is

$$\mathbf{u}_i\mathrm{B} = (c_{1i}\mathbf{u}_1 + c_{2i}\mathbf{u}_2 + \cdots + c_{ri}\mathbf{u}_r)\mathrm{B} = c_{1i}\mathbf{u}_1\mathrm{B} + c_{2i}\mathbf{u}_2\mathrm{B} + \cdots + c_{ri}\mathbf{u}_r\mathrm{B}$$

But this shows that each row of AB after the rth is a linear combination of the first r rows, which in turn proves that AB contains at most r linearly independent row vectors and hence is at most of rank r. A similar argument using a column partition of B shows that the rank of AB is at most equal to ρ. Therefore, the rank of AB is at most equal to the smaller of r and ρ, as asserted. If A and B are also conformable in the order BA, it is clear that the rank of BA is also equal to or less than the smaller of the pair of positive integers (r, ρ). ◗

The estimate for the rank of the product AB provided by the last theorem can be supplemented with the following result.†

THEOREM 13 If **A** is an $m \times n$ matrix of rank r and if **B** is an $n \times p$ matrix of rank ρ, the rank of the product **AB** is equal to or greater than $r + \rho - n$.

EXERCISES

1. The linear space $U(V)$ spanned by the row (column) vectors of a matrix A is called the **row (column) space** of A.
  (a) Show that A is of (row) rank r if and only if $r = \dim U$.
  (b) Show that A is of (column) rank r if and only if $r = \dim V$.

2. Show that, for every matrix A, the row rank, the column rank, and the determinant rank are all equal.

3. Show that if A and B are $n \times n$ matrices of rank n, then both AB and BA are of rank n.

4. If a matrix is of rank r, is it possible that for some value of ρ less than r all minors of order ρ are equal to zero? Why?

5. Determine the rank of each of the following matrices.

(a) $\begin{bmatrix} 3 & -1 & 2 & 4 \\ 6 & 2 & -4 & -8 \end{bmatrix}$ **(b)** $\begin{bmatrix} 2 & -1 & 3 \\ 1 & -2 & 3 \\ 5 & 0 & 3 \end{bmatrix}$

(c) $\begin{bmatrix} 1 & 2 & 3 & 4 \\ 2 & 1 & 4 & 3 \\ 3 & 0 & 5 & -10 \end{bmatrix}$ **(d)** $\begin{bmatrix} 1 & 2 & 3 & 4 & 5 \\ 2 & 3 & 4 & 5 & 6 \\ 3 & 4 & 5 & 6 & 7 \\ 4 & 5 & 6 & 7 & 8 \end{bmatrix}$

6. Determine the rank of each of the following matrices as a function of λ.

(a) $\begin{bmatrix} 8(1-\lambda) & -2 & 0 \\ -2 & 3-2\lambda & -1 \\ 0 & -1 & 2(1-\lambda) \end{bmatrix}$

(b) $\begin{bmatrix} 1-\lambda & 1 & 1 \\ 1 & 3-\lambda & 3 \\ 2 & 1 & 4-\lambda \end{bmatrix}$

(c) $\begin{bmatrix} 5-\lambda & 4 & -2 \\ 4 & 5-\lambda & -2 \\ -2 & -2 & 3-2\lambda \end{bmatrix}$

(d) $\begin{bmatrix} 3-\lambda & -1 & 1 \\ 6 & -2-\lambda & 3 \\ 2 & -1 & 2-\lambda \end{bmatrix}$

7. Show by an example that even when A and B are conformable in either order, the rank of AB is in general not equal to the rank of BA.

8. Prove Theorem 3.

9. Prove Theorem 4.

10. Prove Theorem 5.

11. Prove that elementary column operations alone suffice to reduce a nonsingular matrix of order n to I_n. *Hint:* First reduce the nonsingular matrix to lower triangular form.

12. If A is an $m \times 1$ matrix and B is a $1 \times n$ matrix, neither of which is a zero matrix, what is the rank of AB?

13. Prove Theorem 8.

14. Work Example 2 using only elementary column operations.

15. Work Example 2 if B is the matrix $\begin{bmatrix} 2 & 1 & 2 \\ 0 & 1 & 1 \\ 1 & 1 & 1 \end{bmatrix}$.

†Both this result and Theorem 12 are due to the English mathematician J. J. Sylvester (1814–1897) and are known together as **Sylvester's law of nullity.** A proof of Theorem 13 can be found in L. Mirsky, *Linear Algebra,* p. 162, Oxford, New York, 1955.

16. Find a sequence of elementary operations which will reduce the matrix

$$\begin{bmatrix} 0 & 1 & 1 & 3 & 2 \\ 1 & 2 & -1 & 0 & 1 \\ 0 & 1 & 2 & 2 & 1 \\ 1 & 0 & -4 & -5 & -2 \end{bmatrix}$$

to the standard form described in Theorem 8. What is the rank of this matrix?

17. Show that $A = \begin{bmatrix} 0 & 1 & 0 \\ 1 & 2 & 1 \end{bmatrix}$ and $B = \begin{bmatrix} 1 & 0 & 0 \\ 0 & 1 & 1 \end{bmatrix}$ are equivalent and find nonsingular matrices P and Q such that $B = PAQ$.

18. Work Exercise 17 for the matrices

$$A = \begin{bmatrix} 0 & 1 & 0 \\ 1 & 2 & 1 \\ 1 & 1 & 2 \end{bmatrix} \quad \text{and} \quad B = \begin{bmatrix} 1 & 1 & 0 \\ 0 & 1 & 1 \\ 2 & 1 & 1 \end{bmatrix}$$

19. Work Exercise 17 for the matrices

$$A = \begin{bmatrix} 2 & 1 & 0 \\ 1 & 2 & 1 \\ 4 & -1 & -2 \end{bmatrix} \quad \text{and} \quad B = \begin{bmatrix} 2 & 1 & 0 \\ 1 & 1 & 2 \\ 0 & -1 & -4 \end{bmatrix}$$

20. Using Theorem 11 and its corollary, solve each of the following systems.

(a)
$$\begin{aligned} x_1 - 2x_2 + 3x_3 &= 0 \\ 2x_1 + 3x_2 - x_3 &= 0 \\ 4x_1 - x_2 + 5x_3 &= 0 \end{aligned}$$

(b)
$$\begin{aligned} x_1 - 2x_2 + x_3 + 3x_4 &= 0 \\ 2x_1 + x_2 - 3x_3 + x_4 &= 0 \\ 3x_1 + 3x_2 - 2x_3 + x_4 &= 0 \end{aligned}$$

21. Show that if in each row of a determinant the absolute value of the element on the principal diagonal is greater than the sum of the absolute values of the remaining elements in that row, then the value of the determinant is different from zero. *Hint:* Assume the conclusion is false and apply Theorem 2, Sec. 3.7, to obtain a contradiction.

22. **(a)** Find the regions in the xy plane in which Exercise 21 guarantees that

$$\begin{vmatrix} x & 1 & 1 \\ 1 & y & 1 \\ 1 & 1 & 3 \end{vmatrix} \neq 0$$

(b) Expand the determinant in Part **(a)**, plot the locus of the pairs of values for which it is equal to zero, and verify that this locus has no point in common with the regions identified in Part **(a)**.

23. Prove that if an $n \times (n+1)$ matrix A contains a column of elements that are not all zero and if every nth-order determinant in A which contains this column vanishes, then the rank of A is less than n. *Hint:* Expand each of the vanishing determinants in terms of the elements in their common column, consider the determinant of the resulting system of equations, and use the result of Exercise 22, Sec. 3.7.

24. If A is a square matrix, p a positive integer, and $\mathbf{x}$ a vector such that $A^p\mathbf{x} \neq \mathbf{0}$ but $A^{p+1}\mathbf{x} = \mathbf{0}$, show that the vectors $\mathbf{x}, A\mathbf{x}, A^2\mathbf{x}, \ldots, A^p\mathbf{x}$ are linearly independent.

25. Let A and B be matrices conformable in the order AB. Prove that the rank of AB is equal to the rank of B if and only if $B\mathbf{x} = \mathbf{0}$ for every vector $\mathbf{x}$ such that $AB\mathbf{x} = \mathbf{0}$.

26. If A is a square matrix and if p is a positive integer such that A^p and A^{p+1} have the same rank r, show that A^{p+2}, A^{p+3}, A^{p+4}, . . . , are also of rank r.

27. Prove that if A is an $n \times n$ matrix, then A^n, A^{n+1}, A^{n+2}, . . . , all have the same rank.

28. Show that if A is an $m \times n$ matrix of rank $r \geq m$, then $r = m$.

29. Show that if A is a matrix of order n, then a complete solution of the system $A\mathbf{x} = \mathbf{0}$ contains the same number of parameters as a complete solution of $A^T\mathbf{x} = \mathbf{0}$.

30. Prove Corollary 1, Theorem 11.

31. By what matrix must $\mathbf{u} = [u_0\ u_1\ u_2\ u_3]^T$ be premultiplied to obtain each of the following?

(a) $\mathbf{v} = [u_0\ u_2\ u_1\ u_3]^T$ **(b)** $\mathbf{v} = [u_3\ u_2\ u_1\ u_0]^T$

(c) $\mathbf{v} = [au_2\ bu_3\ cu_1\ du_0]^T$

32. By what matrix must $\mathbf{u} = [u_0\ u_1\ u_2\ u_3]$ be postmultiplied to obtain each of the following?

(a) $\mathbf{v} = [u_2\ u_1\ u_0\ u_3]$ **(b)** $\mathbf{v} = [u_1\ u_2\ u_3\ u_0]$

(c) $\mathbf{v} = [2u_3\ 3u_2\ 4u_1\ 5u_0]$

33. By what matrices must the matrix $\begin{bmatrix} a_{11} & a_{12} & a_{13} & a_{14} \\ a_{21} & a_{22} & a_{23} & a_{24} \\ a_{31} & a_{32} & a_{33} & a_{34} \\ a_{41} & a_{42} & a_{43} & a_{44} \end{bmatrix}$ be pre- and postmultiplied to obtain the matrix

$$\begin{bmatrix} a_{34} & a_{32} & a_{33} & a_{31} \\ a_{44} & a_{42} & a_{43} & a_{41} \\ a_{14} & a_{12} & a_{13} & a_{11} \\ a_{24} & a_{22} & a_{23} & a_{21} \end{bmatrix}?$$

34. Show that A and A^T have the same principal minors.

35. **(a)** Give an example in which the rank of AB is equal to the smaller of the ranks of A and B.

(b) Give an example in which the rank of AB is less than the smaller of the ranks of A and B.

14.4 THE EXISTENCE OF GREEN'S FUNCTIONS FOR GENERAL LINEAR DIFFERENTIAL SYSTEMS

In Sec. 2.12 we defined Green's functions for both second- and fourth-order differential systems and used them to solve several applied problems. Because Green's functions are of considerable importance, both in theory and in practice, we shall now give a brief account of how the underlying ideas and results of Sec. 2.12 can be extended to more general boundary-value problems.

Consider a nonhomogeneous **two-point boundary-value problem** consisting of a normal nth-order differential equation

(1*a*) $$L(y) = a_0 y^{(n)} + a_1 y^{(n-1)} + \cdots + a_{n-1} y' + a_n y = f$$

on an interval $[a, b]$ and n **two-point boundary conditions**

(1*b*) $$a_{i1} y(a) + a_{i2} y'(a) + \cdots + a_{in} y^{(n-1)}(a) + b_{i1} y(b) + b_{i2} y'(b) + \cdots + b_{in} y^{(n-1)}(b) = c_i \qquad i = 1, 2, \ldots, n$$

A more concise form of this system is

(2*a*) $$L(y) = f$$

(2*b*) $$\mathbf{A}\mathbf{y}(a) + \mathbf{B}\mathbf{y}(b) = \mathbf{C}$$

where $\mathrm{A} = [a_{ij}]$ and $\mathrm{B} = [b_{ij}]$ are constant $n \times n$ matrices, $\mathbf{y}(x) = [y(x)\ y'(x) \cdots y^{(n-1)}(x)]^T$, and $\mathbf{C} = [c_1\ c_2 \cdots c_n]^T$ is a constant-column vector. The **completely homogeneous system** corresponding to (2) is

(3*a*) $$L(y) = 0$$

(3*b*) $$\mathbf{A}\mathbf{y}(a) + \mathbf{B}\mathbf{y}(b) = \mathbf{0}$$

In our earlier work, we saw that the differential system of Example 1, Sec. 2.12, had a Green's function if and only if the system was incompatible, that is, had only a trivial solution, in which case the Green's function was unique. The same is true of the System (3).

For a function to be a solution of (3) it must satisfy the system and be continuous together with its first $n - 1$ derivatives on $[a, b]$. The Green's function $g(x, s)$ of (3), when it exists, also satisfies the boundary conditions (3*b*). Moreover, as Definition 1 and Exercise 54, Sec. 2.12, indicate, $g(x, s)$ is to be continuous together with its first $n - 2$ derivatives with respect to x on $[a, b]$. On the other hand, its $(n - 1)$st derivative $g^{(n-1)}(x, s)$ with respect to x is to have a jump of $-1/a_0(s)$ at the point $x = s$ of $[a, b]$, and this prevents $g(x, s)$ from being a solution of (3). Our first definition specifies the precise nature of $g(x, s)$.

DEFINITION 1 The **Green's function** of System (3) is a function $g(x, s)$ such that for any fixed number s in $[a, b]$,

1. $g(x, s)$ is a solution of $L(y) = 0$ on $a \le x < s$ and on $s < x \le b$.
2. $g(x, s)$ satisfies the boundary conditions (3*b*), that is, $\mathbf{A}\mathbf{g}(a, s) + \mathbf{B}\mathbf{g}(b, s) = \mathbf{0}$, where

$$\mathbf{g}(x, s) = [g(x, s)\ g'(x, s) \cdots g^{(n-1)}(x, s)]^T$$

3. $g(x, s), g'(x, s), \ldots, g^{(n-2)}(x, s)$ are continuous on $[a, b]$.
4. $g^{(n-1)}(x, s)$ is continuous on $[a, b]$ except at $x = s$ where it has a jump of $-1/a_0(s)$.

Theorem 2, Sec. 2.12, suggests the following generalization whose proof appears incidentally in the proof of Theorem 4 later on.

THEOREM 1 If the Green's function of (3) exists, then the nonhomogeneous system

(4*a*) $$L(y) = f$$

(4*b*) $$\mathbf{A}\mathbf{y}(a) + \mathbf{B}\mathbf{y}(b) = \mathbf{0}$$

has the function represented by

(5) $$y(x) = -\int_a^b g(x, s) f(s)\, ds \qquad a \le x \le b$$

as its unique solution.

Green's functions can also be used to solve a nonhomogeneous boundary-value problem of the form (2) when $\mathbf{C} \neq \mathbf{0}$. However, it is usually easier to solve such a system by using the substitutions (6) of Sec. 4.3 to transform it into a first-order differential system whose solution yields a solution of the original system. In fact, by developing a Green's function theory for general first-order systems, it is possible to obtain Theorem 1 together with existence and uniqueness theorems pertaining to the Green's function of a homogeneous system (3) as corollaries of the more general theory. For this reason, we shall now concentrate on first-order linear differential systems of the form

$$[\mathrm{D} + \mathrm{F}]\mathbf{y} = \mathbf{f} \tag{6a}$$

$$\mathrm{A}\mathbf{y}(a) + \mathrm{B}\mathbf{y}(b) = \mathbf{c} \tag{6b}$$

where D is an $n \times n$ diagonal matrix in which each diagonal element is d/dx, $\mathrm{F} = [f_{ij}(x)]$ is an nth-order matrix function continuous on an interval $[a, b]$, $\mathbf{f}$ is a continuous vector function on $[a, b]$, A and B are constant matrices of order n, and $\mathbf{c}$ is a constant vector. The completely homogeneous system corresponding to (6) is of course

$$[\mathrm{D} + \mathrm{F}]\mathbf{y} = \mathbf{0} \tag{7a}$$

$$\mathrm{A}\mathbf{y}(a) + \mathrm{B}\mathbf{y}(b) = \mathbf{0} \tag{7b}$$

For $\mathbf{y} = \mathbf{y}(x)$ to be a solution of this completely homogeneous two-point boundary-value problem $\mathbf{y}(x)$ must be a continuous *vector function* which reduces (7*a*) to an identity on $[a, b]$ and which satisfies (7*b*). In contrast, a Green's function of System (7) is a *matrix function* of order n.

DEFINITION 2 The **Green's function** of System (7) is an $n \times n$ matrix function $\mathrm{G}(x, s) = [g_{ij}(x, s)]$ such that for any fixed number s in $[a, b]$,

1. The columns of $\mathrm{G}(x, s)$ are solutions of (7*a*) on $a \leq x < s$ and on $s < x \leq b$, that is, $[\mathrm{D} + \mathrm{F}]\mathrm{G}(x, s) = \mathrm{O}$ on $[a, b]$ except at $x = s$.
2. The columns of $\mathrm{G}(x, s)$ satisfy (7*b*), that is, $\mathrm{AG}(a, s) + \mathrm{BG}(b, s) = \mathrm{O}$.
3. The off-diagonal elements $g_{ij}(x, s)$, $i \neq j$, of $\mathrm{G}(x, s)$ are continuous on $[a, b]$.
4. The diagonal elements $g_{ii}(x, s)$, $i = 1, 2, \ldots, n$, of $\mathrm{G}(x, s)$ have an upward unit jump at $x = s$, so that $\mathrm{G}(s^+, s) - \mathrm{G}(s^-, s) = \mathrm{I}$.†

Before this definition can be accepted and used, we must show that it is possible for System (7) to have a Green's function and that when it does, the Green's function is unique. This we shall do by proving the following theorem.

THEOREM 2 A Green's function of System (7) exists and is unique if and only if the system is incompatible.

PROOF Let s be any fixed number in $[a, b]$ and let $\mathrm{G}(x, s)$ be any $n \times n$ matrix function defined on $a \leq x \leq b$. Then a necessary and sufficient condition for the matrix equation $[\mathrm{D} + \mathrm{F}(x)]\mathrm{G}(x, s) = \mathrm{O}$ to hold on the two separate intervals $a \leq x < s$ and $s < x \leq b$ is that the columns of $\mathrm{G}(x, s)$ be linear combinations of the column vectors of a fundamental matrix Y of (7*a*) (Definition 2, Sec. 4.4). In other words, $\mathrm{G}(x, s)$ satisfies the first condition of Definition 2 if and only if $\mathrm{G}(x, s)$ has the form

$$\mathrm{G}(x, s) = \begin{cases} \mathrm{Y}(x)\mathrm{R}(s) & a \leq x < s \\ \mathrm{Y}(x)\mathrm{T}(s) & s < x \leq b \end{cases} \tag{8}$$

†If, in (7*a*), D is replaced by a diagonal matrix $\mathrm{Dg}[q_1 d/dx, q_2 d/dx, \ldots, q_n d/dx]$, where $\Pi_{i=1}^{n} q_i(x) \neq 0$ on $[a, b]$, the matrix I is to be replaced by $\mathrm{Dg}\,[1/q_1(s), 1/q_2(s), \ldots, 1/q_n(s)]$.

where R and T are arbitrary $n \times n$ matrices which may depend on the parameter s but not on the variable x. Condition **2** of Definition 2 is satisfied by $G(x, s)$ if and only if

$$AY(a)R(s) + BY(b)T(s) = O \tag{9}$$

Since Y is continuous on $[a, b]$, $Y(s^+) = Y(s^-) = Y(s)$. Hence the two remaining conditions of Definition 2 are fulfilled by $G(x, s)$ if and only if

$$Y(s)[T(s) - R(s)] = I \tag{10}$$

This equation can be solved for $T(s)$ because $Y(s)$ is nonsingular, and we have at once

$$T(s) = R(s) + Y^{-1}(s) \tag{11}$$

Substituting this expression for $T(s)$ in (9) and rearranging, we obtain

$$[AY(a) + BY(b)]R(s) = -BY(b)Y^{-1}(s)$$

A necessary and sufficient condition for $R(s)$ to be uniquely determined by this equation is that $[AY(a) + BY(b)]$ be nonsingular, in which case

$$R(s) = -[AY(a) + BY(b)]^{-1}BY(b)Y^{-1}(s) \tag{12}$$

and System (7) is incompatible (Example 3, Sec. 13.7). It follows that the Green's function of the homogeneous system (7) exists and is uniquely determined by (8), (11), and (12) if and only if (7) is incompatible. ◗

Although both $R(s)$ and $T(s)$ are defined for $s = a$ and $s = b$, it is clear from (8) that $G(x, a)$ cannot be defined in terms of R, and $G(x, b)$ cannot be defined in terms of T, since s cannot equal a in R and cannot equal b in T. However, we can use the following definitions

$$G(x, a) = Y(x)T(a) \tag{13a}$$
$$G(x, b) = Y(x)R(b) \tag{13b}$$

In the preceding proof we pointed out that the system (7) is incompatible if and only if $AY(a) + BY(b) = [A, B]\begin{bmatrix} Y(a) \\ Y(b) \end{bmatrix}$ is nonsingular. Since Y stands for a fundamental matrix, the $2n \times n$ matrix $\begin{bmatrix} Y(a) \\ Y(b) \end{bmatrix}$ must be of rank n. According to Theorem 12, Sec. 14.3, the rank of $[AY(a) + BY(b)]$ is less than or at most equal to the smaller of the ranks of $[A, B]$ and $\begin{bmatrix} Y(a) \\ Y(b) \end{bmatrix}$. Therefore, if $[AY(a) + BY(b)]$ is nonsingular, the rank of $[A, B]$ must also be n. This proves the following theorem.

THEOREM 3 If the system (7) is incompatible, the $n \times 2n$ matrix $[A, B]$ is of rank n.

Throughout the remainder of this section we shall assume that the homogeneous system (7) corresponding to System (6) is incompatible. Then $[A, B]$, being of rank n, must have at least one nonsingular $n \times n$ submatrix C whose successive columns are, say, those which in $[A, B]$ have column indices $k_1, k_2, \ldots, k_n$. If M is the $2n \times n$ matrix in which the submatrix whose *row* indices are $k_1, k_2, \ldots, k_n$ is I_n and in which all other entries are zero, then it is clear that $C = [A, B]M$. With C and M in mind, we can now state our main result.

THEOREM 4 The solution of the nonhomogeneous system

$$[\mathrm{D} + \mathrm{F}]\mathbf{y} = \mathbf{f} \tag{6a}$$
$$\mathrm{A}\mathbf{y}(a) + \mathrm{B}\mathbf{y}(b) = \mathbf{c} \tag{6b}$$

is given by

$$\mathbf{y}(x) = \int_a^b G(x, s)\mathbf{f}(s)\, ds + [G(x, a) - G(x, b)]\mathrm{MC}^{-1}\mathbf{c} \tag{14}$$

PROOF At the outset we observe that if (14) is a solution of the system (6), it must be unique since by hypothesis the system (7) is incompatible. To prove that (14) is indeed a solution of (6) we first show that

$$\mathbf{y}(x) = \int_a^b G(x, s)\mathbf{f}(s)\, ds \tag{15}$$

is a solution when $\mathbf{c} = \mathbf{0}$. Substituting for $\mathbf{y}$ from (15) into the left-hand member of (6a), writing F as F(x), recalling the definition of G(x, s) from (8), using Leibniz' rule (p. 661) to differentiate the integrals with respect to x, subsequently observing that (10) implies that $\mathrm{Y}(x)[\mathrm{T}(x) - \mathrm{R}(x)] = \mathrm{I}$, and finally noting from Property **1**, Definition 2, that $[\mathrm{D} + \mathrm{F}(x)]\mathrm{G}(x, s) = \mathrm{O}$ at all points of $[a, b]$ except $x = s$, we have

$$\begin{aligned}
[\mathrm{D} + \mathrm{F}(x)]\mathbf{y}(x) &= \mathrm{D}\int_a^b \mathrm{G}(x, s)\mathbf{f}(s)\, ds + \mathrm{F}(x)\int_a^b \mathrm{G}(x, s)\mathbf{f}(s)\, ds \\
&= \mathrm{D}\int_a^x \mathrm{Y}(x)\mathrm{T}(s)\mathbf{f}(s)\, ds + \mathrm{D}\int_x^b \mathrm{Y}(x)\mathrm{R}(s)\mathbf{f}(s)\, ds \\
&\quad + \int_a^b F(x)\mathrm{G}(x, s)\mathbf{f}(s)\, ds \\
&= \mathrm{Y}(x)\mathrm{T}(x)\mathbf{f}(x) + \int_a^x \mathrm{DY}(x)\mathrm{T}(s)\mathbf{f}(s)\, ds - \mathrm{Y}(x)\mathrm{R}(x)\mathbf{f}(x) \\
&\quad + \int_x^b DY(x)\mathrm{R}(s)\mathbf{f}(s)\, ds + \int_a^b \mathrm{F}(x)\mathrm{G}(x, s)\mathbf{f}(s)\, ds \\
&= \mathrm{Y}(x)[\mathrm{T}(x) - \mathrm{R}(x)]\mathbf{f}(x) + \int_a^b [\mathrm{D} + \mathrm{F}(x)]\mathrm{G}(x, s)\mathbf{f}(s)\, ds \\
&= \mathbf{If}(x) + \mathbf{0} = \mathbf{f}(x)
\end{aligned}$$

This shows that (15) is a solution of (6a) on $[a, b]$. Moreover, because of the similiarity between (15) and (5), the preceding argument is essentially a proof of Theorem 1, as well.

To verify that (15) also satisfies (7b), which of course is (6b) with $\mathbf{c} = \mathbf{0}$, we substitute (15) into the left-hand member of (6b), combine terms, and utilize the second property of Definition 2, getting

$$\begin{aligned}
\mathrm{A}\mathbf{y}(a) + \mathrm{B}\mathbf{y}(b) &= \mathrm{A}\int_a^b \mathrm{G}(a, s)\mathbf{f}(s)\, ds + \mathrm{B}\int_a^b \mathrm{G}(b, s)\mathbf{f}(s)\, ds \\
&= \int_a^b [\mathrm{AG}(a, s) + \mathrm{BG}(b, s)]\mathbf{f}(s)\, ds = \mathbf{0}
\end{aligned}$$

Having proven (15) to be the solution of System (6) when $\mathbf{c} = \mathbf{0}$, we next show that

$$\mathbf{y}(x) = [\mathrm{G}(x, a), -\mathrm{G}(x, b)]\mathrm{M}\mathrm{C}^{-1}\mathbf{c} \tag{16}$$

is a solution of (6) when $\mathbf{f}(x) \equiv \mathbf{0}$ on $[a, b]$. Substituting (16) into the left-hand member of (6a), using (13), and noting that Y is a fundamental matrix of (7a), we obtain

$$\begin{aligned}\{\mathrm{D} + \mathrm{F}(x)\}\mathbf{y}(x) &= \{\mathrm{D} + \mathrm{F}(x)\}[\mathrm{G}(x, a), -\mathrm{G}(x, b)]\mathrm{M}\mathrm{C}^{-1}\mathbf{c} \\ &= [\{\mathrm{D} + \mathrm{F}(x)\}\mathrm{Y}(x)\mathrm{T}(a), -\{\mathrm{D} + \mathrm{F}(x)\}\mathrm{Y}(x)\mathrm{R}(b)]\mathrm{M}\mathrm{C}^{-1}\mathbf{c} \\ &= [\mathrm{O}, \mathrm{O}]\mathrm{M}\mathrm{C}^{-1}\mathbf{c} = \mathbf{0}\end{aligned}$$

Hence (16) is a solution of the homogeneous first-order system corresponding to (6a).

To show that (16) also satisfies (6b), we note that in view of (8), the second property of Definition 2 can be written

$$\mathrm{AG}(a, s) + \mathrm{BG}(b, s) = \mathrm{AY}(a)\mathrm{R}(s) + \mathrm{BY}(b)\mathrm{T}(s) = 0$$

from which, by letting $s = b$ and $s = a$, we obtain

$$-\mathrm{AY}(a)\mathrm{R}(b) = \mathrm{BY}(b)\mathrm{T}(b) \qquad \text{and} \qquad \mathrm{BY}(b)\mathrm{T}(a) = -\mathrm{AY}(a)\mathrm{R}(a)$$

Now substituting (16) into the left-hand member of (6b) and making use of these relations, together with (13), (10), and the definition of M, we have

$$\begin{aligned}\mathrm{A}\mathbf{y}(a) + \mathrm{B}\mathbf{y}(b) &= \mathrm{A}[\mathrm{G}(a, a), -\mathrm{G}(a, b)]\mathrm{M}\mathrm{C}^{-1}\mathbf{c} + \mathrm{B}[\mathrm{G}(b, a), -\mathrm{G}(b, b)]\mathrm{M}\mathrm{C}^{-1}\mathbf{c} \\ &= [\mathrm{AY}(a)\mathrm{T}(a), -\mathrm{AY}(a)\mathrm{R}(b)]\mathrm{M}\mathrm{C}^{-1}\mathbf{c} \\ &\quad + [\mathrm{BY}(b)\mathrm{T}(a), -\mathrm{BY}(b)\mathrm{R}(b)]\mathrm{M}\mathrm{C}^{-1}\mathbf{c} \\ &= [\mathrm{AY}(a)\mathrm{T}(a) + \mathrm{BY}(b)\mathrm{T}(a), -\mathrm{AY}(a)\mathrm{R}(b) - \mathrm{BY}(b)\mathrm{R}(b)]\mathrm{M}\mathrm{C}^{-1}\mathbf{c} \\ &= [\mathrm{AY}(a)\mathrm{T}(a) - \mathrm{AY}(a)\mathrm{R}(a), \mathrm{BY}(b)\mathrm{T}(b) - \mathrm{BY}(b)\mathrm{R}(b)]\mathrm{M}\mathrm{C}^{-1}\mathbf{c} \\ &= [\mathrm{AY}(a)\{\mathrm{T}(a) - \mathrm{R}(a)\}, \mathrm{BY}(b)\{\mathrm{T}(b) - \mathrm{R}(b)\}]\mathrm{M}\mathrm{C}^{-1}\mathbf{c} \\ &= [\mathrm{A}, \mathrm{B}]\mathrm{M}\mathrm{C}^{-1}\mathbf{c} = \mathrm{C}\mathrm{C}^{-1}\mathbf{c} = \mathbf{c}\end{aligned}$$

This shows that (16) satisfies (6b). Finally, since D + F, A, and B are linear operators, it follows that when $\mathbf{y}$, as given by (14), is substituted into Eqs. (6), the first term yields $\mathbf{f}$ in (6a) and $\mathbf{0}$ in (6b), while the second term yields $\mathbf{0}$ in (6a) and $\mathbf{c}$ in (6b). Thus their sum (14), the sum of (15) and (16), gives the solution of System (6) as the theorem asserts. ◗

By way of illustration, we now apply Theorem 4 to solve a specific nonhomogeneous two-point boundary-value problem.

EXAMPLE 1

Find the solution of the nonhomogeneous differential system

$$x^2u'' - 2xu' + 2u = 2x^3 \tag{17a}$$

$$\begin{aligned}2u(1) - u'(1) &= -1 \\ u(1) - u'(1) + u(2) - u'(2) &= 4\end{aligned} \tag{17b}$$

The homogeneous differential equation corresponding to (17a) can be solved at once as an Euler-Cauchy equation or, in this case, one solution can be found by inspection and another by variation of parameters. Either method gives $u_1 = x$ and $u_2 = x^2$ as fundamental solutions on [1, 2]. Although the Green's function $g(x, s)$ of the completely homogeneous system corresponding to (17) exists and is easy to construct, at present we are unable to express the solution of (17) in terms of $g(x, s)$ because the boundary conditions are nonhomogeneous. Accordingly, following the preceding discussion, we first transform (17) into a first-order system to which Theorem 4 applies. The usual substitutions $y_1 = u$ and $y_2 = u'$ give us

(18*a*)
$$\begin{bmatrix} d/dx & -1 \\ 2/x^2 & d/dx - 2/x \end{bmatrix} \begin{bmatrix} y_1 \\ y_2 \end{bmatrix} = \begin{bmatrix} 0 \\ 2x \end{bmatrix}$$

(18*b*)
$$\begin{bmatrix} 2 & -1 \\ 1 & -1 \end{bmatrix} \begin{bmatrix} y_1(1) \\ y_2(1) \end{bmatrix} + \begin{bmatrix} 0 & 0 \\ 1 & -1 \end{bmatrix} \begin{bmatrix} y_1(2) \\ y_2(2) \end{bmatrix} = \begin{bmatrix} -1 \\ 4 \end{bmatrix}$$

A fundamental matrix of the homogeneous system corresponding to (18*a*) is

$$\mathrm{Y}(x) = \begin{bmatrix} x^2 & x \\ 2x & 1 \end{bmatrix} \text{ and its inverse is } \mathrm{Y}^{-1}(x) = \begin{bmatrix} -1/x^2 & 1/x \\ 2/x & -1 \end{bmatrix}$$

With $\mathrm{A} = \begin{bmatrix} 2 & -1 \\ 1 & -1 \end{bmatrix}$ and $\mathrm{B} = \begin{bmatrix} 0 & 0 \\ 1 & -1 \end{bmatrix}$, we compute $\mathrm{AY}(1) + \mathrm{BY}(2) = \begin{bmatrix} 0 & 1 \\ -1 & 1 \end{bmatrix}$. Since this matrix is nonsingular, the completely homogeneous system corresponding to System (18) is incompatible and therefore has a Green's function $\mathrm{G}(x, s)$. Using (12) and (11) to compute

$$\mathrm{R}(s) = \begin{bmatrix} 2/s & -1 \\ 0 & 0 \end{bmatrix} \quad \text{and} \quad \mathrm{T}(s) = \begin{bmatrix} (2s-1)/s^2 & (1-s)/s \\ 2/s & -1 \end{bmatrix}$$

and then utilizing (8), we find that $\mathrm{G}(x, s)$ is given by

$$\mathrm{G}(x, s) = \begin{cases} \begin{bmatrix} 2x^2/s & -x^2 \\ 4x/s & -2x \end{bmatrix} & 1 \le x < s \\ \begin{bmatrix} [2s(x^2+x) - x^2]/s^2 & [x^2 - s(x^2+x)]/s \\ [2s(2x+1) - 2x]/s^2 & [2x - s(2x+1)]/s \end{bmatrix} & s < x \le 2 \end{cases}$$

In this particular example, Eqs. (13) become

(13*a*)
$$\mathrm{G}(x, 1) = \begin{bmatrix} x^2 + 2x & -x \\ 2x + 2 & -1 \end{bmatrix}$$

(13*b*)
$$\mathrm{G}(x, 2) = \begin{bmatrix} x^2 & -x^2 \\ 2x & -2x \end{bmatrix}$$

Before (14) can be used to solve System (18), we must decide on a choice for C and then identify the related matrix M. Selecting $\mathrm{C} = \begin{bmatrix} -1 & 0 \\ -1 & 1 \end{bmatrix}$ as the matrix made up of the second and third columns of [A, B] and observing that $\mathrm{C} = \mathrm{C}^{-1}$, we obtain

$$\begin{bmatrix} y_1(x) \\ y_2(x) \end{bmatrix} = \int_1^x \begin{bmatrix} [2s(x^2+x) - x^2]/s^2 & [x^2 - s(x^2+x)]/s \\ [2s(2x+1) - 2x]/s^2 & [2x - s(2x+1)]/s \end{bmatrix} \begin{bmatrix} 0 \\ 2s \end{bmatrix} ds + \int_x^2 \begin{bmatrix} 2x^2/s & -x^2 \\ 4x/s & -2x \end{bmatrix} \begin{bmatrix} 0 \\ 2s \end{bmatrix} ds$$
$$+ \begin{bmatrix} x^2 + 2x & -x & -x^2 & x^2 \\ 2x + 2 & -1 & -2x & 2x \end{bmatrix} \begin{bmatrix} 0 & 0 \\ 1 & 0 \\ 0 & 1 \\ 0 & 0 \end{bmatrix} \begin{bmatrix} -1 & 0 \\ -1 & 1 \end{bmatrix} \begin{bmatrix} -1 \\ 4 \end{bmatrix}$$

which eventually leads to

$$\begin{bmatrix} y_1(x) \\ y_2(x) \end{bmatrix} = \begin{bmatrix} x^3 - 10x^2 \\ 3x^2 - 20x \end{bmatrix}$$

The solution of System (17) is of course $u = x^3 - 10x^2$.

Returning to the homogeneous system (3), let us replace y by u and then make the substitutions $y_1 = u,\ y_2 = u',\ \ldots,\ y_n = u^{(n-1)}$. This leads to the differential system

$$\text{(19a)}\quad \begin{aligned} y_i' - y_{i+1} &= 0 \qquad 1 \le i \le n-1 \\ a_0 y_n' + a_1 y_{n-1} + \cdots + a_n y_1 &= 0 \end{aligned}$$

$$\text{(19b)}\quad \mathbf{A}\mathbf{y}(a) + \mathbf{B}\mathbf{y}(b) = \mathbf{0}$$

Clearly, the Green's function $\mathrm{G}(x, s) = [g_{ij}(x, s)]$ of this system, if it exists, has the property that for $1 \le i, j \le n$, the element $g_{ij}(x, s)$ of $\mathrm{G}(x, s)$ is the derivative with respect to x of order $i - 1$ of $g_{1j}(x, s)$. To allow for the possibility that $a_0 \not\equiv 1$ on $[a, b]$, in which case (19a) is not what we called a first-order system, the final diagonal element $g_{nn}(x, s)$ of $\mathrm{G}(x, s)$ is to have a jump of $1/a_0(s)$ at $x = s$ in accordance with the footnote to Definition 2. These properties, together with the other defining properties of $\mathrm{G}(x, s)$ and $g(x, s)$, imply that the Green's function $g(x, s)$ of system (3) is given by $-g_{1n}(x, s)$ (Exercise 12). Obviously, $g_{1n}(x, s)$ itself would become the Green's function of System (3) were we to alter Definition 1 by requiring $g^{(n-1)}(x, s)$ to have $1/a_0(s)$ rather than $-1/a_0(s)$ as its jump at $x = s$. Of course, this would change the physical meaning of $g(x, s)$ in any given application.

EXERCISES

1. **(a)** Find the Green's function of the homogeneous boundary-value problem $u'' + u = 0,\ u(0) = 0,\ u'(\pi) = 0$.

(b) Use Theorem 1 and the Green's function of Part **(a)** to solve the system
$u'' + u = -3 \sin 2x,\ u(0) = 0,\ u'(\pi) = 0$.

(c) Use Theorem 4 to solve the system
$u'' + u = -3 \sin 2x,\ u(0) = 0,\ u'(\pi) = 2$

2. **(a)** Verify the solution of Example 1 by selecting C to be the matrix made up of the first and second columns in [A, B].

(b) Can C be made up of any two distinct columns of [A, B]? Explain.

Utilize Theorem 4 to solve each of the following boundary-value problems.

3. $u'' - (1/x)u' = 3x$
$u'(1) = 3,\ u'(1) - 2u(2) + 2u'(2) = 11$

4. $u''' - (1/x)u'' = 12x$
$u(1) = 1,\ u'(1) = 4,\ u''(2) - u'(2) = 16$

5. $u''' + (\tan x)u'' = \sin 2x$
$u(0) = 1,\ u'(0) + u'(\pi/4) = 0,\ u''(0) = 0$

6. $u''' + (3/x)u'' = 6/x$
$2u(1) - u(2) = -1,\ 2u'(1) - u'(2) = 0,$
$u''(1) - u''(2) = 0$

7. $u''' - (1/x)u'' = 30x,\ u(1) = 1,\ u(2) = 0,\ u''(1) = 0$

8. $x^3u''' - 6x^2u'' + 18xu' - 24u = -3x^9,$
$u(1) = u'(1) = u''(1) + u(2) = 0$

9. $\dfrac{d\mathbf{y}}{dx} = \begin{bmatrix} 0 & 1 \\ -1 & 2 \end{bmatrix}\mathbf{y} + \begin{bmatrix} e^x/(x+1) \\ 0 \end{bmatrix},$

$\begin{bmatrix} 0 & -1 \\ -1 & 2 \end{bmatrix}\mathbf{y}(0) + \begin{bmatrix} 1 & 0 \\ 0 & 1 \end{bmatrix}\mathbf{y}(1) = \begin{bmatrix} 1 & -\ln 2 \\ 1 & -\ln 4 \end{bmatrix} e$

10. $\dfrac{d\mathbf{y}}{dx} + \begin{bmatrix} 10 & -6 \\ 12 & -7 \end{bmatrix}\mathbf{y} = \begin{bmatrix} 10 \\ 18 \end{bmatrix} e^{-3x}$

$\begin{bmatrix} 1/e^2 & 1/e^2 \\ -1/e & -1/e \end{bmatrix}\mathbf{y}(0) + \begin{bmatrix} 3 & -4 \\ -3 & 4 \end{bmatrix}\mathbf{y}(1) = \begin{bmatrix} 7/e^2 \\ -14/e \end{bmatrix}$

11. If in Definition 1, $g^{(n-1)}(x, s)$ is required to have $1/a_0(s)$ instead of $-1/a_0(s)$ as its jump at $x = s$, what effect does this have on Theorem 1?

12. Show that if System (19) has the Green's function $\mathrm{G}(x, s) = [g_{ij}(x, s)]$, then $-g_{1n}(x, s)$ is the Green's function $g(x, s)$ of System (3).

14.5 QUADRATIC FORMS

In this section we continue our study of matrices by introducing what are known as *quadratic forms, hermitian forms,* and *bilinear forms.*

By a **quadratic form** we mean a homogeneous second-degree expression Q in n variables $x_1, x_2, \ldots, x_n$ of the form

$$\text{(1)}\quad \begin{aligned} Q = a_{11}x_1^2 &+ 2a_{12}x_1x_2 + \cdots + 2a_{1n}x_1x_n \\ &+ \quad a_{22}x_2^2 + \cdots + 2a_{2n}x_2x_n \\ &+ \cdots\cdots\cdots\cdots\cdots\cdots \\ &\qquad\qquad + a_{nn}x_n^2 \end{aligned}$$

In order that matrix notation can be applied to quadratic forms, it is customary to separate each of the cross-products into two equal terms and introduce additional coefficients for the new terms by the definition $a_{ji} = a_{ij}$. When this has been done, Q can be expressed in the more symmetric form

$$\begin{aligned} Q = \;& a_{11}x_1^2 + a_{12}x_1x_2 + \cdots + a_{1n}x_1x_n \\ &+ a_{21}x_2x_1 + a_{22}x_2^2 + \cdots + a_{2n}x_2x_n \\ &+ \cdots\cdots\cdots\cdots\cdots\cdots\cdots\cdots \\ &+ a_{n1}x_nx_1 + a_{n2}x_nx_2 + \cdots + a_{nn}x_n^2 \end{aligned} \qquad a_{ji} = a_{ij}$$

If we now define $\mathbf{x} = \begin{bmatrix} x_1 \\ x_2 \\ \vdots \\ x_n \end{bmatrix}$, $A = \begin{bmatrix} a_{11} & a_{12} & \cdots & a_{1n} \\ a_{21} & a_{22} & \cdots & a_{2n} \\ \cdots & \cdots & \cdots & \cdots \\ a_{n1} & a_{n2} & \cdots & a_{nn} \end{bmatrix}$, and write Q as $Q(\mathbf{x})$ to denote the dependence of Q on $\mathbf{x}$,† it is clear from the definition of matrix multiplication that Q can be written in the compact form

$$Q(\mathbf{x}) = \mathbf{x}^T A\mathbf{x} \tag{2}$$

where, since $a_{ji} = a_{ij}$, the **matrix of the quadratic form** A is necessarily symmetric.

If a quadratic form with real coefficients has the property that it is equal to or greater than zero for all real values of its variables, it is said to be **positive**. A positive form which is zero *only* for the values $x_1 = x_2 = \cdots x_n = 0$ is said to be **positive-definite**. A positive form which is zero for real values other than $x_1 = x_2 = \cdots = x_n = 0$ is said to be **positive-semidefinite**. A quadratic form with real coefficients which is equal to or less than zero for all real values of its variables is said to be **negative**. A negative form which is zero *only* for the values $x_1 = x_2 = \cdots = x_n = 0$ is said to be **negative-definite**. A negative form which is zero for real values other than $x_1 = x_2 = \cdots = x_n = 0$ is said to be **negative-semidefinite**. Clearly, a negative-definite or negative-semidefinite form can be converted into a positive form of corresponding type by multiplying it by -1. A quadratic form which can take on both positive and negative values for real values of its variables is said to be **indefinite**. Examples of quadratic forms of each type are shown in Table 14.3.

The matrix A of the quadratic form $Q(\mathbf{x}) = \mathbf{x}^T A\mathbf{x}$ is said to be **positive-definite, negative-definite, positive-semidefinite, negative-semidefinite**, or **indefinite** according to the nature of $Q(\mathbf{x})$. Correspondingly, $Q(\mathbf{x})$ is said to be **singular** or **nonsingular** according as its matrix A is singular or nonsingular, i.e., according as $|A|$ is equal to or different from zero.

TABLE 14.3

Type of quadratic form	Example
Positive-definite	$x_1^2 + x_2^2$
Negative-definite	$-(x_1^2 + x_2^2)$
Positive-semidefinite	$(x_1 - x_2)^2$
Negative-semidefinite	$-(x_1 - x_2)^2$
Indefinite	$x_1^2 - x_2^2$

†Throughout the rest of this chapter we shall find it convenient to regard every vector $\mathbf{v}$ as a column vector unless it is written as $\mathbf{v}^T$, in which case it is always to be interpreted as a row vector.

A quadratic form which is **definite**, that is, is either positive-definite or negative-definite, is necessarily nonsingular. In fact, if we suppose that $Q(\mathbf{x})$ is both definite and singular, we are led at once to a contradiction, as follows. Let us first write $Q(\mathbf{x})$ in the partially factored form

$$\begin{aligned} Q = &\quad (a_{11}x_1 + a_{12}x_2 + \cdots + a_{1n}x_n)x_1 \\ &+ (a_{21}x_2 + a_{22}x_2 + \cdots + a_{2n}x_n)x_2 \\ &+ \cdots\cdots\cdots\cdots\cdots\cdots\cdots\cdots\cdots \\ &+ (a_{n1}x_1 + a_{n2}x_2 + \cdots + a_{nn}x_n)x_n \end{aligned}$$

Then, assuming that det A = 0, it follows from Theorem 2, Sec. 3.7 that there is a nontrivial solution of the system of equations obtained by equating to zero the expressions in parentheses in $Q(\mathbf{x})$. Finally, we observe that for the values of the x's in this (nontrivial) solution, $Q(\mathbf{x})$ itself is equal to zero, contrary to the hypothesis that it is definite.

The converse of the preceding observation is not true. In other words, a nonsingular quadratic form is not necessarily definite. For example, the form $x_1^2 - 4x_1x_2 + 3x_2^2 + 2x_3^2$ is nonsingular since the determinant of its matrix, namely,

$$\begin{vmatrix} 1 & -2 & 0 \\ -2 & 3 & 0 \\ 0 & 0 & 2 \end{vmatrix} = -2$$

is different from zero, yet it is not definite since it is zero for the nontrivial set of values $x_1 = 3$, $x_2 = 1$, $x_3 = 0$.

The complete criterion for the definiteness of a quadratic form is contained in the following theorem, for whose proof we must refer to texts on higher algebra.†

THEOREM 1 A necessary and sufficient condition that the real quadratic form $\mathbf{x}^T\mathrm{A}\mathbf{x}$ be positive-definite (negative-definite) is that the quantities

$$a_{11}, \qquad \begin{vmatrix} a_{11} & a_{12} \\ a_{21} & a_{22} \end{vmatrix}, \qquad \begin{vmatrix} a_{11} & a_{12} & a_{13} \\ a_{21} & a_{22} & a_{23} \\ a_{31} & a_{32} & a_{33} \end{vmatrix}, \qquad \cdots \qquad, \begin{vmatrix} a_{11} & \cdots & a_{1n} \\ \cdots & \cdots & \cdots \\ a_{n1} & \cdots & a_{nn} \end{vmatrix}$$

should all be positive (alternate in sign with a_{11} negative).

Clearly, sets of necessary and sufficient conditions for the definiteness of $\mathbf{x}^T\mathrm{A}\mathbf{x}$ equivalent to those given in Theorem 1 can be obtained by first permuting the variables in $\mathbf{x}^T\mathrm{A}\mathbf{x}$ and then applying Theorem 1. Recalling the definition of *principal minors* given in Sec. 14.3, this yields the following more general results.

THEOREM 2 A necessary and sufficient condition that the real quadratic form $\mathbf{x}^T\mathrm{A}\mathbf{x}$ be positive-definite is that every principal minor of A be positive.

THEOREM 3 A necessary and sufficient condition that the real quadratic form $\mathbf{x}^T\mathrm{A}\mathbf{x}$ be negative-definite is that every principal minor of A of odd order be negative and every principal minor of even order be positive.

†See, for instance, W. L. Farrar, *Algebra,* pp. 138–141, Oxford, New York, 1941.

EXAMPLE 1

The quadratic form

$$[x_1 \quad x_2 \quad x_3]\begin{bmatrix} 1 & 2 & -2 \\ 2 & 5 & -4 \\ -2 & -4 & 5 \end{bmatrix}\begin{bmatrix} x_1 \\ x_2 \\ x_3 \end{bmatrix} = \begin{matrix} x_1^2 + 2x_1x_2 - 2x_1x_3 \\ +2x_2x_1 + \ 5x_2^2 \ - 4x_2x_3 \\ -2x_3x_1 - 4x_3x_2 + 5x_3^2 \end{matrix}$$

is positive-definite, since the three quantities

$$1 \qquad \begin{vmatrix} 1 & 2 \\ 2 & 5 \end{vmatrix} = 1 \qquad \text{and} \qquad \begin{vmatrix} 1 & 2 & -2 \\ 2 & 5 & -4 \\ -2 & -4 & 5 \end{vmatrix} = 1$$

are all positive. Moreover, all the other principal minors, namely, the diagonal elements

$$a_{22} = 5 \qquad \text{and} \qquad a_{33} = 5$$

and the second-order determinants

$$\begin{vmatrix} a_{11} & a_{13} \\ a_{31} & a_{33} \end{vmatrix} = \begin{vmatrix} 1 & -2 \\ -2 & 5 \end{vmatrix} = 1 \qquad \text{and} \qquad \begin{vmatrix} a_{22} & a_{23} \\ a_{32} & a_{33} \end{vmatrix} = \begin{vmatrix} 5 & -4 \\ -4 & 5 \end{vmatrix} = 9$$

are also positive, in accordance with Theorem 2. In this case the quadratic form can be written equivalently as

$$(x_1 + 2x_2 - 2x_3)^2 + x_2^2 + x_3^2$$

which, being a sum of squares, can vanish only if

$$x_1 + 2x_2 - 2x_3 = 0 \qquad x_2 = 0 \qquad x_3 = 0$$

and these in turn can hold simultaneously only if $x_1 = x_2 = x_3 = 0$.

On the other hand, the quadratic form

$$[x_1 \quad x_2 \quad x_3]\begin{bmatrix} 1 & 2 & -2 \\ 2 & 3 & -4 \\ -2 & -4 & 5 \end{bmatrix}\begin{bmatrix} x_1 \\ x_2 \\ x_3 \end{bmatrix} = \begin{matrix} x_1^2 + 2x_1x_2 - 2x_1x_3 \\ +2x_2x_1 + \ 3x_2^2 \ - 4x_2x_3 \\ -2x_3x_1 - 4x_3x_2 + 5x_3^2 \end{matrix}$$

is not definite since the three quantities

$$1 \qquad \begin{vmatrix} 1 & 2 \\ 2 & 3 \end{vmatrix} = -1 \qquad \begin{vmatrix} 1 & 2 & -2 \\ 2 & 3 & -4 \\ -2 & -4 & 5 \end{vmatrix} = -1$$

do not fulfill either of the conditions of Theorem 1. In fact, this quadratic form can be written as

$$(x_1 + 2x_2 - 2x_3)^2 - x_2^2 + x_3^2$$

and since this expression takes on the value 1 when $x_1 = 2$, $x_2 = 0$, $x_3 = 1$ and takes on the value -1 when $x_1 = -2$, $x_2 = 1$, $x_3 = 0$, it is actually indefinite.

In our definition of a quadratic form, neither the matrix of coefficients A nor the n component vector **x** was restricted to be real. However, in most elementary applications both A and **x** will be real, and only for real-valued quadratic forms are such properties as definiteness and indefiniteness defined. Actually, when complex quantities are involved, quadratic forms as we have defined them are almost always replaced by related expressions known as *hermitian forms.*

DEFINITION 1 If H is a hermitian matrix, the expression $\bar{\mathbf{x}}^T H\mathbf{x}$ is known as a **hermitian form**.

Recalling the definition of a hermitian matrix (Sec. 3.4), it is easy to verify that any hermitian form is equal to its transposed conjugate. Moreover, since such a form is a scalar, i.e., a 1×1 matrix, it is also equal to its transpose. Thus, since its transpose is equal to the conjugate of its transpose, it must be real and we have the following result.

THEOREM 4 The value of a hermitian form is real for all values of its variables.

Because of Theorem 4, positive- and negative-definite, positive- and negative-semidefinite, and indefinite hermitian forms can be defined precisely as the corresponding types of quadratic forms were defined. Moreover, it can be shown that the criteria for definiteness contained in Theorems 1–3 hold without change for hermitian forms.

Closely associated with quadratic forms are what are known as *bilinear forms.*

DEFINITION 2 If A is a symmetric matrix, the expression $\mathbf{y}^T A\mathbf{x}$ is known as a **bilinear form**.

Clearly, if $\mathbf{y} = \mathbf{x}$, the bilinear form $\mathbf{y}^T A\mathbf{x}$ becomes the quadratic form $\mathbf{x}^T A\mathbf{x}$. If the components of $\mathbf{y}$ are thought of as the coordinates of a ''point'' in a hyperspace of the appropriate number of dimensions, the bilinear form $\mathbf{y}^T A\mathbf{x}$ is sometimes called the **polar** of the point $\mathbf{y}$ with respect to the quadratic form $\mathbf{x}^T A\mathbf{x}$.

It is interesting to note that the scalar product of two vectors $\mathbf{y}$ and $\mathbf{x}$, namely, $\mathbf{y}^T\mathbf{x}$, can be thought of as the bilinear form $\mathbf{y}^T I\mathbf{x}$. The condition that $\mathbf{y}$ and $\mathbf{x}$ be orthogonal is, then, just the condition that the bilinear form $\mathbf{y}^T I\mathbf{x}$ be equal to zero. This suggests that the simple notion of orthogonality introduced in Sec. 3.2, by analogy with the familiar results of solid analytic geometry, be extended to include the following concept of *generalized orthogonality.*

DEFINITION 3 Two vectors $\mathbf{x}$ and $\mathbf{y}$ are **orthogonal with respect to a symmetric matrix** A if and only if the bilinear form $\mathbf{y}^T A\mathbf{x}$ is equal to zero.

In the spirit of Definition 3, the notion of the length of a vector can also be generalized. In fact, the definition of the length of a vector $\mathbf{x}$ introduced in Sec. 3.2, namely, $\sqrt{\mathbf{x}^T\mathbf{x}}$, can be rewritten $\sqrt{\mathbf{x}^T I\mathbf{x}}$, and this suggests

$$\sqrt{\mathbf{x}^T A\mathbf{x}} \tag{3}$$

as the generalized length of the vector $\mathbf{x}$ with respect to a symmetric matrix A. This is meaningful, however, only if the quantity under the radical is nonnegative. It is also desirable that the generalized length of $\mathbf{x}$ be zero just in case $\mathbf{x} = \mathbf{0}$. Hence we require A to be positive-definite. A vector whose length with respect to a given symmetric, positive-definite matrix is 1 is said to be **normalized** with respect to that matrix. Any nonzero vector $\mathbf{x}$ can always be normalized with respect to a given symmetric, positive-definite matrix A by dividing it by the positive quantity $\sqrt{\mathbf{x}^T A\mathbf{x}}$.

Clearly, the Schmidt orthogonalization process we discussed in Sec. 13.4 can be carried out equally well using the concepts of generalized orthogonality and generalized length. The notion of orthogonality with respect to a symmetric, nonunit matrix will be of considerable importance in the work of this chapter.

Just as it was convenient in analytic geometry to be able to remove the cross-product term from the equation of a conic, so in many applications involving quadratic forms it is desirable to be able to remove the cross-product terms by a suitable transformation and express the quadratic form as a sum of squares. There are numerous ways of doing this, among which the following, due to Lagrange, is particularly effective. The general idea is first to group together all terms containing one of the variables as a factor, say the terms containing x_1, and then, by suitable manipulations, to convert this part of the expression into a perfect square. Then among the terms not included in this square, those containing one of the remaining variables, say x_2, are converted into a perfect square; and so on, until the process terminates.

To begin the process, let us assume that $a_{11} \neq 0$ and group together all terms containing x_1 as a factor:

$$(a_{11}x_1^2 + 2a_{12}x_1x_2 + \cdots + 2a_{1n}x_1x_n) + \sum_{i,\,j=2}^{n} a_{ij}x_ix_j$$
$$= a_{11}\left(x_1^2 + \frac{2a_{12}}{a_{11}}x_1x_2 + \cdots + \frac{2a_{1n}}{a_{11}}x_1x_n\right) + \phi_1(x_2, x_3, \ldots, x_n)$$

Now, adding and subtracting the appropriate squares and cross-product terms, none of which involves x_1, we have

$$a_{11}\left[\left(x_1 + \frac{a_{12}}{a_{11}}x_2 + \cdots + \frac{a_{1n}}{a_{11}}x_n\right)^2 - \left(\frac{a_{12}}{a_{11}}\right)^2 x_2^2 - \cdots - \left(\frac{a_{1n}}{a_{11}}\right)^2 x_n^2\right.$$
$$\left. -2\frac{a_{12}}{a_{11}}\frac{a_{13}}{a_{11}}x_2x_3 - \cdots - 2\frac{a_{1,n-1}}{a_{11}}\frac{a_{1n}}{a_{11}}x_{n-1}x_n\right] + \phi_1(x_2, x_3, \ldots, x_n)$$
$$= a_{11}\left(x_1 + \frac{a_{12}}{a_{11}}x_2 + \cdots + \frac{a_{1n}}{a_{11}}x_n\right)^2 + \phi_2(x_2, x_3, \ldots, x_n)$$

The linear matrix transformation $T_1\colon \mathbf{y} = \begin{bmatrix} 1 & a_{12}/a_{11} & \cdots & a_{1n}/a_{11} \\ 0 & 1 & \cdots & 0 \\ \cdots & \cdots & \cdots & \cdots \\ 0 & 0 & \cdots & 1 \end{bmatrix}\mathbf{x}$, which is obviously nonsingular, now reduces Q to the form $a_{11}y_1^2 + \phi_2(y_2, y_3, \ldots, y_n)$, where ϕ_2 is of course a quadratic form in the $n-1$ variables $y_2, y_3, \ldots, y_n$ with coefficients b_{ij}, say.

This process can, in general, be applied to ϕ_2, and a second nonsingular transformation $T_2\colon \mathbf{z} = \begin{bmatrix} 1 & 0 & 0 & \cdots & 0 \\ 0 & 1 & b_{23}/b_{22} & \cdots & b_{2n}/b_{22} \\ 0 & 0 & 1 & \cdots & 0 \\ \cdots & \cdots & \cdots & \cdots & \cdots \\ 0 & 0 & 0 & \cdots & 1 \end{bmatrix}\mathbf{y}$ yields the reduction $a_{11}z_1^2 + b_{22}z_2^2 + \phi_3(z_3, z_4, \ldots, z_n)$.

The continuation is now obvious, and the required transformation is finally the product, or composition, of the successive linear transformations $T_1, T_2, \ldots, T_n$.

If at any stage all square terms are missing from the form $\phi_k(u_k, u_{k+1}, \ldots, u_n)$, the process must be modified. If this occurs, either no more terms remain and the reduction is complete or else there is at least one cross-product term with nonzero coefficient, say u_iu_j, where we may suppose that $i < j$. If this is the case, the nonsingular transformation

$$\begin{aligned} u_k &= v_k \qquad k \neq i, j \\ u_i &= v_i + v_j \\ u_j &= v_i - v_j \end{aligned}$$

will clearly introduce the square terms v_i^2 and v_j^2, after which the process can be continued as originally described.

It is important to note that the linear transformation employed at each stage of the reduction process is rank-preserving. Hence, since the rank of a diagonal matrix is equal to the number of its nonzero diagonal elements, it follows that when $\mathbf{x}^T\mathrm{A}\mathbf{x}$ is transformed into a sum of squares by the Lagrange reduction, the number of square terms present in the final result is equal to the rank of the matrix of the original form. It is also clear that when a positive-definite quadratic form is reduced to a sum of squares by the Lagrange reduction, the final result must consist of the square of each variable with a *positive* coefficient.

EXAMPLE 2

Find a transformation which will reduce to a sum of squares the quadratic form $\mathbf{x}^T\mathrm{A}\mathbf{x}$, where

$$\mathrm{A} = \begin{bmatrix} 1 & -1 & 0 & 2 \\ -1 & 4 & 6 & 4 \\ 0 & 6 & 11 & 8 \\ 2 & 4 & 8 & 8 \end{bmatrix}$$

Following the Lagrange procedure, we first group together the terms containing x_1 as a factor and then complete the square on these terms:

$$\begin{aligned}(x_1^2 - 2x_1x_2 + 4x_1x_4) &+ (4x_2^2 + 12x_2x_3 + 8x_2x_4 + 11x_3^2 + 16x_3x_4 + 8x_4^2)\\ &= [(x_1 - x_2 + 2x_4)^2 - x_2^2 + 4x_2x_4 - 4x_4^2]\\ &\quad + (4x_2^2 + 12x_2x_3 + 8x_2x_4 + 11x_3^2 + 16x_3x_4 + 8x_4^2)\\ &= (x_1 - x_2 + 2x_4)^2 + (3x_2^2 + 12x_2x_3 + 12x_2x_4 + 11x_3^2 + 16x_3x_4 + 4x_4^2)\end{aligned}$$

Eventually we will replace $x_1 - x_2 + 2x_4$ by a new variable y_1, and similar expressions in the other x's by y_2, y_3, and y_4; but before it is convenient to do this we must continue the grouping process and incorporate the terms involving x_2, then x_3, and finally x_4 into expressions that are perfect squares. Proceeding in this way, we have

$$\begin{aligned}(x_1 - x_2 + 2x_4)^2 &+ 3(x_2^2 + 4x_2x_3 + 4x_2x_4) + (11x_3^2 + 16x_3x_4 + 4x_4^2)\\ &= (x_1 - x_2 + 2x_4)^2 + [3(x_2 + 2x_3 + 2x_4)^2 - 12x_3^2 - 24x_3x_4 - 12x_4^2]\\ &\quad + (11x_3^2 + 16x_3x_4 + 4x_4^2)\\ &= (x_1 - x_2 + 2x_4)^2 + 3(x_2 + 2x_3 + 2x_4)^2 - (x_3^2 + 8x_3x_4 + 8x_4^2)\\ &= (x_1 - x_2 + 2x_4)^2 + 3(x_2 + 2x_3 + 2x_4)^2 - (x_3^2 + 8x_3x_4 + 16x_4^2) + 8x_4^2\\ &= (x_1 - x_2 + 2x_4)^2 + 3(x_2 + 2x_3 + 2x_4)^2 - (x_3 + 4x_4)^2 + 8(x_4)^2\end{aligned}$$

If we now replace the expressions in the respective parentheses by y_1, y_2, y_3, and y_4, the original quadratic form appears as the sum of squares

$$y_1^2 + 3y_2^2 - y_3^2 + 8y_4^2 \tag{4}$$

The transformation we have thus constructed can conveniently be expressed in matrix form as

$$T^{-1}\colon \mathbf{y} = \mathrm{M}^{-1}\mathbf{x} \qquad \text{where } \mathrm{M}^{-1} = \begin{bmatrix} 1 & -1 & 0 & 2 \\ 0 & 1 & 2 & 2 \\ 0 & 0 & 1 & 4 \\ 0 & 0 & 0 & 1 \end{bmatrix} \tag{5}$$

We have called the transformation T^{-1} rather than T and denoted its matrix by M^{-1} rather than M because we are more interested in the transformation that expresses **x** in terms of **y**. Clearly, M^{-1} is nonsingular and therefore has an inverse M, which turns out to be

$$\mathrm{M} = \begin{bmatrix} 1 & 1 & -2 & 4 \\ 0 & 1 & -2 & 6 \\ 0 & 0 & 1 & -4 \\ 0 & 0 & 0 & 1 \end{bmatrix}$$

Thus, solving (5) for **x**, we obtain the direct transformation from **x** to **y**,

$$(6) \qquad T\colon \mathbf{x} = \mathrm{M}\mathbf{y}$$

Using T, we have

$$\mathbf{x}^T\mathrm{A}\mathbf{x} = (\mathrm{M}\mathbf{y})^T\mathrm{A}(\mathrm{M}\mathbf{y}) = \mathbf{y}^T(\mathrm{M}^T\mathrm{A}\mathrm{M})\mathbf{y}$$

and it is easy to verify that $\mathrm{M}^T\mathrm{AM}$ is the diagonal matrix

$$\mathrm{M}^T\mathrm{AM} = \begin{bmatrix} 1 & 0 & 0 & 0 \\ 0 & 3 & 0 & 0 \\ 0 & 0 & -1 & 0 \\ 0 & 0 & 0 & 8 \end{bmatrix}$$

As this example illustrates, a linear transformation T: $\mathbf{x} = \mathrm{M}\mathbf{y}$ which will reduce a given quadratic form $Q = \mathbf{x}^T\mathrm{A}\mathbf{x}$ to an expression involving only the squares of the components of the new variable **y** can always be found by applying the Lagrange procedure to Q to find M^{-1}, and thence M.

EXERCISES

1. Classify each of the following quadratic forms:
   (a) $3x_1^2 + 3x_2^2 + 6x_3^2 - 2x_1x_2 - 4x_1x_3$
   (b) $-x_1^2 - 3x_2^2 - 5x_3^2 + 2x_1x_2 + 2x_1x_3 + 2x_2x_3$
   (c) $x_1^2 + 4x_2^2 + 4x_3^2 + 4x_1x_2 + 4x_1x_3 + 6x_2x_3$
   (d) $2x_1^2 + 2x_2^2 + x_3^2 + 2x_1x_3 + 2x_2x_3$
   (e) $x_1^2 + 2x_2^2 + 2x_3^2 + 4x_4^2 - 2x_1x_2 + 2x_2x_3 + 6x_3x_4$
2. Find a transformation that will reduce each of the following quadratic forms to a sum of squares:
   (a) $x_1^2 + 5x_2^2 + 2x_3^2 + 4x_1x_2 + 2x_1x_3 + 6x_2x_3$
   (b) $x_1^2 + x_2^2 + 3x_3^2 + 4x_4^2 + 2x_1x_2 + 4x_1x_3 + 2x_2x_3 + 2x_2x_4 + 6x_3x_4$
   (c) $x_1^2 + 5x_2^2 + 5x_3^2 + 2x_4^2 - 2x_1x_2 + 4x_1x_3 + 2x_1x_4 - 6x_2x_4 + 2x_3x_4$
   (d) $x_1^2 + 4x_2^2 + 2x_3^2 + 4x_4^2 + 4x_1x_2 + 2x_1x_3 + 2x_1x_4 + 2x_2x_3 - 2x_2x_4 - 2x_3x_4$
   (e) $x_1x_2 + x_3x_4$
   (f) $x_1x_2 + x_3x_4 + x_4x_5 + x_5x_6$
3. Obtain two more linear transformations each of which will reduce the quadratic form of Example 2 to a sum of squares. Do the sums of squares obtained in the respective cases have any common characteristic?
4. Show that the type of a quadratic form is not altered by the process of reducing it to a sum of squares.
5. Applying the result of Exercise 4 to the results of Exercise 2, determine the type of each of the quadratic forms in Exercise 2.
6. If A is a symmetric matrix, show that $\mathbf{y}^T\mathrm{A}\mathbf{x} = \mathbf{x}^T\mathrm{A}\mathbf{y}$.
7. If $Q(\mathbf{x}) = \mathbf{x}^T\mathrm{A}\mathbf{x}$, where A is a symmetric matrix, show that
$$Q(\lambda\mathbf{x} + \mu\mathbf{y}) = \lambda^2\mathbf{x}^T\mathrm{A}\mathbf{x} + 2\lambda\mu\mathbf{x}^T\mathrm{A}\mathbf{y} + \mu^2\mathbf{y}^T\mathrm{A}\mathbf{y}$$
8. If $\mathbf{x}^T\mathrm{A}\mathbf{x}$ is a positive-definite quadratic form, show that $(\mathbf{x}^T\mathrm{A}\mathbf{y})^2 \le (\mathbf{x}^T\mathrm{A}\mathbf{x})(\mathbf{y}^T\mathrm{A}\mathbf{y})$, the equality sign holding if and only if **x** or **y** is a zero vector or $\mathbf{x} = \mathbf{y}$. *Hint:* Use the result of Exercise 7.
9. Prove that a nonsingular quadratic form cannot be semidefinite. *Hint:* Use the result of Exercise 4.
10. If **v** is a vector for which an indefinite quadratic form $Q = \mathbf{x}^T\mathrm{A}\mathbf{x}$ is equal to zero and if ϵ is an arbitrary (small) positive number, show that there are vectors each of whose components differs from the corresponding component of **v** by less than ϵ for which Q is positive and also such vectors for which Q is negative.
11. Show that the bilinear form $\mathbf{y}^T\mathrm{A}\mathbf{x}$, that is, the polar of **y** with respect to the quadratic form $Q = \mathbf{x}^T\mathrm{A}\mathbf{x}$, can be obtained by applying the so-called **polar operator**
$$\frac{1}{2}\left(y_1\frac{\partial}{\partial x_1} + y_2\frac{\partial}{\partial x_2} + \cdots + y_n\frac{\partial}{\partial x_n}\right)$$
to Q.
12. **(a)** Show that regardless of the character of $\mathbf{x}^T\mathrm{A}\mathbf{x}$ and $\mathbf{x}^T\mathrm{B}\mathbf{x}$, there are always singular quadratic forms in

the family $\lambda\mathbf{x}^T\mathrm{A}\mathbf{x} + \mu\mathbf{x}^T\mathrm{B}\mathbf{x}$. In general, how many singular forms are there in such a family?

(b) If $\mathrm{A} = \begin{bmatrix} 1 & 0 & 0 \\ 0 & 0 & 1 \\ 0 & 1 & 0 \end{bmatrix}$ and $\mathrm{B} = \begin{bmatrix} 0 & 1 & 1 \\ 1 & 0 & -1 \\ 1 & -1 & 0 \end{bmatrix}$, determine the singular quadratic forms in the family $\lambda\mathbf{x}^T\mathrm{A}\mathbf{x} + \mu\mathbf{x}^T\mathrm{B}\mathbf{x}$.

13. Find the potential energy stored in the system shown in Fig. 14.2 as a result of the displacements x_1, x_2, and x_3 and show that it is a positive-definite quadratic form in the x's. What is the relation of the matrix of this form to the matrix K discussed in Sec. 14.1 in connection with this system? *Hint:* Recall that the work required to stretch a spring of modulus k a distance s is equal to $\frac{1}{2}ks^2$.

14. Work Exercise 13 for the system shown in Fig. 14.4.

15. From the vectors

$$\mathbf{v}_1 = \begin{bmatrix} 1 \\ 0 \\ 0 \end{bmatrix} \qquad \mathbf{v}_2 = \begin{bmatrix} 1 \\ 1 \\ 0 \end{bmatrix} \qquad \mathbf{v}_3 = \begin{bmatrix} 1 \\ 1 \\ 1 \end{bmatrix}$$

construct a set of vectors orthonormal with respect to the matrix $\begin{bmatrix} 1 & 0 & 0 \\ 0 & 2 & 0 \\ 0 & 0 & 3 \end{bmatrix}$.

16. Work Exercise 15 for the vectors

$$\mathbf{v}_1 = \begin{bmatrix} 1 \\ 0 \\ 0 \end{bmatrix} \qquad \mathbf{v}_2 = \begin{bmatrix} 0 \\ 1 \\ 0 \end{bmatrix} \qquad \mathbf{v}_3 = \begin{bmatrix} 0 \\ 0 \\ 1 \end{bmatrix}$$

and the matrix $\begin{bmatrix} 1 & 1 & 0 \\ 0 & 2 & 0 \\ 0 & 0 & 2 \end{bmatrix}$.

17. If A is a symmetric $n \times n$ matrix, the locus of the equation $\mathbf{x}^T\mathrm{A}\mathbf{X} = 0$ is called a **quadric hypersurface** S. Show that in R^n every line l intersects S in two points which may be **(a)** real and distinct, **(b)** real and equal, or **(c)** conjugate complex, and determine the criterion for each possibility. *Hint:* Let l be represented parametrically in terms of two of its points, P and Q.

18. Find the coordinates of the intersections of the hyperquadric S: $x_1^2 + 2x_1x_3 + x_2^2 - 4x_2x_4 + x_4^2 = 0$ and the line determined by the points P: $(1, -1, 2, -1)$ and Q: $(1, 1, -4, -1)$.

19. Let S be the hyperquadric $\mathbf{X}^T\mathrm{A}\mathbf{X} = 0$ and let T be the polar of an arbitrary point P with respect to S. Show that if Q is a point common to S and T, then the line PQ is tangent to S, that is, intersects S in two coincident points.

20. Let P be an arbitrary point of the hyperquadric S: $\mathbf{X}^T\mathrm{A}\mathbf{X} = 0$, let T be the polar of P with respect to S, and let Q be any point common to S and T. Show that the line PQ lies entirely on S.

21. Show that if the point P lies in the polar of the point Q with respect to the hyperquadric S: $\mathbf{X}^T\mathrm{A}\mathbf{X} = 0$, then Q lies in the polar of Q with respect to S.

14.6 CHARACTERISTIC VALUES AND CHARACTERISTIC VECTORS OF A MATRIX

In Sec. 3.8 we identified a characteristic value problem as the determination of those vectors, if any, each of which has an image proportional to itself under a linear transformation $\mathbf{y} = \mathrm{A}\mathbf{x}$, where A is an nth-order matrix. Thus, the problem is to find all *nontrivial* vectors $\mathbf{x}$ such that $\mathrm{A}\mathbf{x} = \lambda\mathbf{x}$ or, equivalently, $\mathrm{A}\mathbf{x} = \lambda\mathrm{I}\mathbf{x}$, or finally,

$$(1) \qquad (\mathrm{A} - \lambda\mathrm{I})\mathbf{x} = \mathbf{0} \qquad \text{or} \qquad (\lambda\mathrm{I} - \mathrm{A})\mathbf{x} = \mathbf{0} \qquad \lambda \text{ a parameter}$$

In general, we shall work with the first form of Eq. (1). Although a geometric interpretation of the problem requires that λ and A be real, we shall be interested in (1) whether A or λ is real or complex. We first review several ideas from Sec. 3.8.

Since (1) is equivalent to a homogeneous system of n linear equations in n unknowns and since we are interested only in nontrivial solutions, it follows from Theorem 2, Sec. 3.7, that we must have det $(\mathrm{A} - \lambda\mathrm{I}) = 0$. This equation, namely,

$$(2) \qquad |\mathrm{A} - \lambda\mathrm{I}| = \begin{vmatrix} a_{11} - \lambda & a_{12} & \cdots & a_{1n} \\ a_{21} & a_{22} - \lambda & \cdots & a_{2n} \\ \cdots & \cdots & \cdots & \cdots \\ a_{n1} & a_{n2} & \cdots & a_{nn} - \lambda \end{vmatrix} = 0$$

is a polynomial equation of degree n in the parameter λ with leading coefficient $(-1)^n$, say,

$$(3) \qquad |\mathrm{A} - \lambda\mathrm{I}| = (-1)^n[\lambda^n - \beta_1\lambda^{n-1} + \beta_2\lambda^{n-2} + \cdots + (-1)^{n-1}\beta_{n-1}\lambda + (-1)^n\beta_n] = 0$$

Both this equation and the equivalent equation obtained by dropping the factor $(-1)^n$ are known as the **characteristic equation** of the matrix A, the expression in brackets is known as the **characteristic polynomial** of A, the roots of (3) are known as the **characteristic roots** or **characteristic values** of A, and the solutions of (1) corresponding to these roots are known as the **characteristic vectors** of A. In this section we shall extend the ideas of Sec. 3.8 and consider such topics as the number of characteristic roots and characteristic vectors of a matrix, the orthogonality and linear independence of the characteristic vectors, and the extension of these ideas to the more general equation $(\mathrm{A} - \lambda \mathrm{B})x = 0$.

Since most of the applications we have in mind involve matrices which are either real and symmetric or hermitian, we shall for the most part limit the rest of our discussion to the characteristic values and characteristic vectors of such matrices. We begin, however, with several theorems dealing with the characteristic values and characteristic vectors of arbitrary square matrices.

Since the characteristic equation (3) of a square matrix A is a polynomial equation, its roots, say $\lambda_1, \lambda_2, \ldots, \lambda_n$, are connected with its coefficients $-\beta_1, \beta_2, -\beta_3, \ldots, (-1)^n\beta_n$ by the well-known root-coefficient relations

$$
\begin{aligned}
\beta_1 &= \lambda_1 + \lambda_2 + \cdots + \lambda_n \\
\beta_2 &= \lambda_1\lambda_2 + \lambda_1\lambda_3 + \cdots + \lambda_{n-1}\lambda_n \\
\beta_3 &= \lambda_1\lambda_2\lambda_3 + \cdots + \lambda_{n-2}\lambda_{n-1}\lambda_n \\
&\cdots\cdots\cdots\cdots\cdots\cdots \\
\beta_n &= \lambda_1\lambda_2\lambda_3\cdots\lambda_n
\end{aligned} \tag{4}
$$

Furthermore, if we set $\lambda = 0$ in Eq. (3), we obtain

$$|\mathrm{A}| = (-1)^{2n}\beta_n = \beta_n \tag{5}$$

Hence, from the last of Eqs. (4), we have

$$|\mathrm{A}| = \lambda_1\lambda_2\lambda_3\cdots\lambda_n$$

From this it follows that $|\mathrm{A}|$ is zero if and only if at least one of the λ's is zero. Thus we have established the following theorem.

THEOREM 1 A matrix is singular if and only if at least one of its characteristic values is zero.

Equation (5) is only the first of a series of relations connecting the coefficients in Eq. (3) with the principal minors of A. For instance, if $|\mathrm{A} - \lambda \mathrm{I}|$ is written as the sum of 2^n determinants by repeated use of the addition theorem (Property 6, Sec. 3.5) it is clear that the terms containing the first power of λ are obtained by multiplying the term $-\lambda$ in each diagonal element of $|\mathrm{A} - \lambda \mathrm{I}|$ by the λ-free part of the cofactor of that element. Thus the coefficient of λ in Eq. (3), namely,

$$(-1)^{2n-1}\beta_{n-1} = -\beta_{n-1}$$

is equal to

$$-(A_{11} + A_{22} + \cdots + A_{nn})$$

Hence it follows that

$$\beta_{n-1} = A_{11} + A_{22} + \cdots + A_{nn}$$

Similarly, the terms containing λ^2 in the expansion of $|\mathrm{A} - \lambda \mathrm{I}|$ are found by multiplying the terms containing $-\lambda$ in every pair of diagonal elements by the λ-free part of the algebraic complement of the second-order minor containing those diagonal elements. Thus the coefficient of λ^2 in Eq. (3), namely,

$$(-1)^{2n-2}\beta_{n-2} = \beta_{n-2}$$

is equal to

$$A_{12,12} + A_{13,13} + \cdots + A_{(n-1,n),(n-1,n)}$$

The continuation is obvious, and we therefore have the following theorem.

THEOREM 2 If $\lambda^n - \beta_1\lambda^{n-1} + \cdots + (-1)^{n-1}\beta_{n-1}\lambda + (-1)^n\beta_n = 0$ is the characteristic equation of a square matrix A, then β_i is equal to the sum of all the principal minors of order i in A.

For $i = 1$, we have as a special case of Theorem 2, the relation $\beta_1 = \lambda_1 + \lambda_2 + \cdots + \lambda_n = a_{11} + a_{22} + \cdots + a_{nn}$. The quantity $a_{11} + a_{22} + \cdots + a_{nn}$ is called the **trace** of A.

Since the principal minors of a matrix A and its transpose A^T are identical (see Exercise 34, Sec. 14.3), it follows from the last theorem that A and A^T have the same characteristic equation and therefore the same characteristic values. Thus we have the following corollary of Theorem 2.

COROLLARY 1 If A is a square matrix, A and A^T have the same characteristic values.

EXAMPLE 1

CONSUMER PREFERENCES

A marketing survey was made of a group of affluent people who buy a new car each year. Their choice of cars at the time of the survey is given by the *ownership vector*

$$\mathbf{x} = \begin{bmatrix} x_1 \\ x_2 \\ x_3 \end{bmatrix}$$

in which x_i is the fraction of the group owning a car of make C_i. The cars they will purchase during the year are indicated by the transition matrix

$$A = [p_{ij}] = \begin{bmatrix} 0.7 & 0.3 & 0.1 \\ 0.1 & 0.6 & 0.1 \\ 0.2 & 0.1 & 0.8 \end{bmatrix}$$

in which p_{ij} is the fraction of those owning a car of make C_j who will switch to a car of make C_i, where of course C_i and C_j may be cars of the same make. If these figures remain constant as time goes on, is there a steady-state ownership vector, that is, is there a particular ownership vector which, if achieved, will remain constant year after year? If so, what is that vector?

Let y_i be the fraction of the group who will own a C_i at the end of the year. Then y_i is made up of those who currently own a C_i and are not going to switch, namely,

$$p_{ii}x_i$$

plus the fraction of those who currently own a C_j but are going to switch to a C_i, namely,

$$p_{ij}x_j$$

plus the fraction of those who currently own a C_k but will switch to a C_i. Thus $y_i = p_{ii}x_i + p_{ij}x_j + p_{ik}x_k$. Clearly, this sum is the scalar product of the ownership vector $\mathbf{x}$ and the ith row vector of the transition matrix A. Thus, the new ownership vector $\mathbf{y}$ is obtained from the current ownership vector $\mathbf{x}$ by the transformation

$$A\mathbf{x} = \mathbf{y}$$

The vector $\mathbf{x}$ will be the required steady-state vector if and only if $\mathbf{y}$ is proportional to $\mathbf{x}$, that is, if and only if $\mathbf{x}$ satisfies the equation

$$A\mathbf{x} = \lambda\mathbf{x} \qquad \text{or} \qquad (A - \lambda I)\mathbf{x} = \mathbf{0}$$

Thus $\mathbf{x}$ must be one of the characteristic vectors of A. To find these we must first find the characteristic roots of A by solving the characteristic equation of A, namely,

$$\begin{vmatrix} 0.7-\lambda & 0.3 & 0.1 \\ 0.1 & 0.6-\lambda & 0.1 \\ 0.2 & 0.1 & 0.8-\lambda \end{vmatrix} = -(\lambda^3 - 2.1\lambda^2 + 1.4\lambda - 0.3) = 0$$

getting $\lambda_1 = 1.0$, $\lambda_2 = 0.6$, $\lambda_3 = 0.5$ (see Exercise 25). By substituting each of these in turn into the system

$$\begin{bmatrix} 0.7-\lambda & 0.3 & 0.1 \\ 0.1 & 0.6-\lambda & 0.1 \\ 0.2 & 0.1 & 0.8-\lambda \end{bmatrix}\begin{bmatrix} x_1 \\ x_2 \\ x_3 \end{bmatrix} = \mathbf{0}$$

and solving for $\mathbf{x}$, we find that the characteristic vectors are

$$\lambda_1 = 1\colon \mathbf{x}_1 = \begin{bmatrix} 7 \\ 4 \\ 9 \end{bmatrix} \qquad \lambda_2 = 0.6\colon \mathbf{x}_2 = \begin{bmatrix} 1 \\ 0 \\ -1 \end{bmatrix} \qquad \lambda_3 = 0.5\colon \mathbf{x}_3 = \begin{bmatrix} 2 \\ -1 \\ -1 \end{bmatrix}$$

Since the fraction of the group owning a particular make of car cannot be negative, neither $\mathbf{x}_2$ nor $\mathbf{x}_3$ is relevant to our problem. Hence the required values must come from $\mathbf{x}_1$ and are therefore in the ratio 7:4:9 or, in terms of percentages,

$$x_1 = 35\% \qquad x_2 = 20\% \qquad x_3 = 45\%$$

This example shows that there is a steady-state ownership vector, but it does not show that the process will converge to that vector as time goes on. However, from the results of Exercises 15–17 it follows that this is indeed the case, regardless of the initial vector. (See also Exercise 47.)

The characteristic polynomial of a matrix A, and hence the coefficients $\{\beta_i\}$ and the characteristic roots $\{\lambda_i\}$ have the interesting property of being invariant under any similarity transformation. More precisely, we have the following theorem.

THEOREM 3 If A and B are similar matrices, then A and B have the same characteristic equation.

PROOF Let $|A - \lambda I|$ be the characteristic polynomial of the matrix A and let B be a matrix similar to A; that is, let B be any matrix such that $B = S^{-1}AS$ for some nonsingular matrix S. Then the characteristic polynomial of B is

$$\begin{aligned} |B - \lambda I| &= |S^{-1}AS - \lambda I| = |S^{-1}AS - \lambda S^{-1}IS| \\ &= |S^{-1}(A - \lambda I)S| = |S^{-1}| \cdot |A - \lambda I| \cdot |S| \end{aligned}$$

since the determinant of a product of square matrices is equal to the product of the determinants of the individual matrices. In particular, $|S^{-1}||S| = |S^{-1}S| = |I| = 1$. Hence $|B - \lambda I| = |A - \lambda I|$, as asserted.

The next three theorems also deal with the characteristic values and characteristic vectors of arbitrary square matrices.

THEOREM 4 A characteristic vector of a square matrix cannot correspond to two distinct characteristic values.

PROOF Let λ_1 and λ_2 be distinct characteristic values of a square matrix A and let $\mathbf{x}_1$ be a characteristic vector of A corresponding, if possible, to both λ_1 and λ_2. Then, simultaneously,

$$(A - \lambda_1 I)\mathbf{x}_1 = \mathbf{0} \qquad \text{and} \qquad (A - \lambda_2 I)\mathbf{x}_1 = \mathbf{0}$$

Hence, subtracting,

$$(\lambda_2 - \lambda_1)I\mathbf{x}_1 = (\lambda_2 - \lambda_1)\mathbf{x}_1 = \mathbf{0} \tag{6}$$

However, by hypothesis, $\lambda_1 \neq \lambda_2$. Moreover, a characteristic vector is, by definition, a *nontrivial* solution vector of $(A - \lambda I)\mathbf{x} = \mathbf{0}$. Thus $\mathbf{x}_1 \neq \mathbf{0}$, and therefore Eq. (6) cannot hold. Hence the assumption that a characteristic vector can correspond to each of two distinct characteristic values must be abandoned, and the theorem is established.

THEOREM 5 If $\mathbf{x}_1, \mathbf{x}_2, \ldots, \mathbf{x}_m$, $m \leq n$, are characteristic vectors corresponding, respectively, to the distinct characteristic values $\lambda_1, \lambda_2, \ldots, \lambda_m$ of an $n \times n$ matrix A, then $\mathbf{x}_1, \mathbf{x}_2, \ldots, \mathbf{x}_m$ are linearly independent.

PROOF Let $\mathbf{x}_1, \mathbf{x}_2, \ldots, \mathbf{x}_m$ be characteristic vectors corresponding, respectively, to the distinct characteristic values $\lambda_1, \lambda_2, \ldots, \lambda_m$ of a square matrix A, and let us suppose, contrary to the theorem, that $\mathbf{x}_1, \mathbf{x}_2, \ldots, \mathbf{x}_m$ are dependent. More specifically, let us suppose that the maximum number of linearly independent vectors in the set is k, where $1 \leq k < m$, and for convenience, let them be the first k $\mathbf{x}$'s. Then any relation of the form

$$\alpha_1\mathbf{x}_1 + \alpha_2\mathbf{x}_2 + \cdots + \alpha_k\mathbf{x}_k = \mathbf{0}$$

implies that $\alpha_1 = \alpha_2 = \cdots = \alpha_k = 0$, but there does exist a nontrivial set of γ's, with $\gamma_{k+1} \neq 0$, such that

$$\gamma_1\mathbf{x}_1 + \gamma_2\mathbf{x}_2 + \cdots + \gamma_k\mathbf{x}_k + \gamma_{k+1}\mathbf{x}_{k+1} = \mathbf{0} \tag{7}$$

Now multiply Eq. (7) on the left by the matrix A, getting

$$\gamma_1 A\mathbf{x}_1 + \gamma_2 A\mathbf{x}_2 + \cdots + \gamma_k A\mathbf{x}_k + \gamma_{k+1} A\mathbf{x}_{k+1} = \mathbf{0}$$

However, $A\mathbf{x}_i = \lambda_i\mathbf{x}_i$ for each i. Hence the last equation becomes

$$\gamma_1\lambda_1\mathbf{x}_1 + \gamma_2\lambda_2\mathbf{x}_2 + \cdots + \gamma_k\lambda_k\mathbf{x}_k + \gamma_{k+1}\lambda_{k+1}\mathbf{x}_{k+1} = \mathbf{0} \tag{8}$$

If we now multiply Eq. (7) by λ_{k+1} and subtract the result from Eq. (8), we obtain

$$(\lambda_1 - \lambda_{k+1})\gamma_1\mathbf{x}_1 + (\lambda_2 - \lambda_{k+1})\gamma_2\mathbf{x}_2 + \cdots + (\lambda_k - \lambda_{k+1})\gamma_k\mathbf{x}_k = \mathbf{0} \tag{9}$$

Since $\mathbf{x}_1, \mathbf{x}_2, \ldots, \mathbf{x}_k$ are linearly independent, by hypothesis, it follows that each coefficient in (9) is equal to zero. Hence, since $\lambda_i - \lambda_{k+1} \neq 0$, $i = 1, 2, \ldots, k$, it must be that

$$\gamma_i = 0 \qquad i = 1, 2, \ldots, k$$

But if this is the case, it follows from (7) that

$$\gamma_{k+1}\mathbf{x}_{k+1} = \mathbf{0}$$

which is impossible, since neither the scalar γ_{k+1} nor the vector $\mathbf{x}_{k+1}$ is zero. This contradiction overthrows the possibility that the characteristic vectors $\mathbf{x}_1, \mathbf{x}_2, \ldots, \mathbf{x}_m$ are linearly dependent, and the theorem is established.

In particular, if an $n \times n$ matrix has n distinct characteristic values, the last theorem tells us that it has n linearly independent characteristic vectors. Hence, as a special case of Theorem 2, Sec. 13.3, we have the following result.

COROLLARY 1 If the characteristic values of an $n \times n$ matrix A are all distinct, then A has n linearly independent characteristic vectors, and any vector with n components can be expressed as a linear combination of the characteristic vectors of A.

Since the characteristic equation of an $n \times n$ matrix is always of degree n, it is obvious that if repeated roots are counted the appropriate number of times, such a matrix always has exactly n characteristic roots. With the same convention one might perhaps be able to say that an $n \times n$ matrix always has exactly n characteristic vectors. However, attempting to assign a multiplicity to a characteristic vector associated with a repeated characteristic root is completely artificial and without significance. The decisive consideration is the number of linearly independent characteristic vectors of a given matrix; hence it is of fundamental importance to know when more than one independent characteristic vector is associated with a repeated characteristic root. The next theorem gives us a partial answer to this question.

THEOREM 6 If λ_1 is a characteristic root of multiplicity r of an $n \times n$ matrix A, then the rank of $A - \lambda_1 I$ is equal to or greater than $n - r$.

PROOF If A is an $n \times n$ matrix and if $\lambda = \lambda_1$ is a repeated root of multiplicity r of the characteristic equation $|A - \lambda I| = 0$, then when we write $\lambda = \lambda_1 + w$, so that $\lambda = \lambda_1$ corresponds to $w = w_1 = 0$, it is clear that $w_1 = 0$ is a repeated root of multiplicity r of the equation

$$|A - (\lambda_1 + w)I| = |(A - \lambda_1 I) - wI| = 0$$

Hence the expanded form of the last equation, say

$$(-1)^n[w^n - \sigma_1 w^{n-1} + \sigma_2 w^{n-2} - \cdots + (-1)^{n-1}\sigma_{n-1}w + (-1)^n\sigma_n] = 0$$

must contain w^r as a factor and must therefore reduce to

$$(-1)^n[w^n - \sigma_1 w^{n-1} + \cdots + (-1)^{n-r}\sigma_{n-r}w^r] = 0$$

where $\sigma_{n-r} \neq 0$. Now, by Theorem 2, the coefficient σ_{n-r} is equal to the sum of all principal minors of order $n - r$ of the matrix $A - \lambda_1 I$. Hence, since $\sigma_{n-r} \neq 0$, at least one of these minors must be different from zero. In other words, the rank of $A - \lambda_1 I$ must be at least as great as $n - r$, as asserted.

If, for a particular root λ_1 of multiplicity r of an $n \times n$ matrix A, the equality sign holds in the assertion of Theorem 6, then, according to Theorem 10, Sec. 14.3, there are exactly $n - (n - r) = r$ linearly independent characteristic vectors associated with λ_1. Such a characteristic root is said to be **regular**. However, this is the exception rather than the rule, and in general there will be a single independent characteristic vector associated with a repeated root of any multiplicity. For instance, for the matrix

$$A = \begin{bmatrix} -3 & -7 & -5 \\ 2 & 4 & 3 \\ 1 & 2 & 2 \end{bmatrix}$$

we have

$$|A - \lambda I| = \begin{vmatrix} -3-\lambda & -7 & -5 \\ 2 & 4-\lambda & 3 \\ 1 & 2 & 2-\lambda \end{vmatrix}$$
$$= -\lambda^3 + 3\lambda^2 - 3\lambda + 1 = -(\lambda - 1)^3 = 0$$

Thus A has a single characteristic root, $\lambda = 1$. Moreover, for $\lambda = 1$ the rank of

$$(A - \lambda I)_{\lambda=1} = A - I = \begin{bmatrix} -4 & -7 & -5 \\ 2 & 3 & 3 \\ 1 & 2 & 1 \end{bmatrix}$$

is clearly 2. Hence, according to Theorem 11, Sec. 14.3, the system of equations $(A - I)\mathbf{x} = \mathbf{0}$ has a single independent solution, namely,

$$\mathbf{x} = \begin{bmatrix} -3 \\ 1 \\ 1 \end{bmatrix}$$

and thus A has just one independent characteristic vector.

Later in this section we shall see that for hermitian matrices, and hence for real symmetric matrices, the assertion of the last theorem can be sharpened to a strict equality; in other words, we shall prove that if $\lambda = \lambda_1$ is a characteristic value of multiplicity r of a hermitian matrix or a real symmetric matrix, then the rank of $\mathrm{A} - \lambda_1\mathrm{I}$ is *exactly* $n - r$. Preparatory to this, however, it will be convenient to prove first some other theorems about hermitian matrices.

THEOREM 7 The characteristic values of a hermitian matrix are all real.

PROOF Let H be a hermitian matrix, let λ_1 be any one of its characteristic values, and let $\mathbf{x}_1$ be a characteristic vector corresponding to λ_1. Then,

$$(\mathrm{H} - \lambda_1\mathrm{I})\mathbf{x}_1 = \mathbf{0}$$

or

$$\mathrm{H}\mathbf{x}_1 = \lambda_1\mathbf{x}_1 \tag{10}$$

and from this, by premultiplying by $\overline{\mathbf{x}}_1^T$, we obtain

$$\overline{\mathbf{x}}_1^T\mathrm{H}\mathbf{x}_1 = \lambda_1\overline{\mathbf{x}}_1^T\mathbf{x}_1 \tag{11}$$

Now, from the familiar properties of conjugate complex numbers, $\overline{\mathbf{x}}_1^T\mathbf{x}_1$ is real and in fact positive. Furthermore, from Theorem 4, Sec. 14.5, we know that $\overline{\mathbf{x}}_1^T\mathrm{H}\mathbf{x}_1$ is also real. Hence it follows immediately from Eq. (11) that λ_1 is real, as asserted.

Since, as we observed in Sec. 3.4, a real symmetric matrix is just a special case of a hermitian matrix, we have the following corollary of Theorem 7.

COROLLARY 1 The characteristic values of a real symmetric matrix are all real.

Furthermore, since $i\mathrm{K}$ is hermitian if K is skew-hermitian and since $|\mathrm{K} - \lambda\mathrm{I}| = 0$ implies that $|i\mathrm{K} - i\lambda\mathrm{I}| = 0$, it follows that if λ_1 is a characteristic value of the skew-hermitian matrix K, then $i\lambda_1$ is a characteristic value of the hermitian matrix $i\mathrm{K}$. Hence, by Theorem 7, $i\lambda_1$ is real, and therefore λ_1 is a pure imaginary. Thus we have established the following result.

COROLLARY 2 The characteristic values of a skew-hermitian matrix are all pure imaginary.

Knowing now that the characteristic roots of a hermitian matrix H are all real, we can return to the characteristic equation of H and prove the following result.

THEOREM 8 If $\lambda^n - \beta_1\lambda^{n-1} + \beta_2\lambda^{n-2} - \cdots + (-1)^{n-1}\beta_{n-1}\lambda + (-1)^n\beta_n = 0$ is the characteristic equation of a hermitian matrix H, then the characteristic values of H are all positive if and only if each β is positive.

PROOF If H is a hermitian matrix, it follows from Theorem 7 that the roots of the characteristic equation

$$\lambda^n - \beta_1\lambda^{n-1} + \beta_2\lambda^{n-2} - \cdots + (-1)^{n-1}\beta_{n-1}\lambda + (-1)^n\beta_n = 0$$

are all real. Furthermore, if each β is positive, it follows from Descartes' rule of signs that no root of the characteristic equation can be negative or zero. Hence all the characteristic roots must be positive. Conversely, if the characteristic roots of H are all positive, then from the root-coefficient relations

$$\begin{aligned}\beta_1 &= \lambda_1 + \lambda_2 + \cdots + \lambda_n\\ \beta_2 &= \lambda_1\lambda_2 + \lambda_1\lambda_3 + \cdots + \lambda_{n-1}\lambda_n\\ &\cdots\cdots\cdots\cdots\cdots\cdots\\ \beta_n &= \lambda_1\lambda_2\cdots\lambda_n\end{aligned}$$

it follows at once that each β is positive, as asserted.

COROLLARY 1 If $\lambda^n - \beta_1\lambda^{n-1} + \beta_2\lambda^{n-2} - \cdots + (-1)^{n-1}\beta_{n-1}\lambda + (-1)^n\beta_n = 0$ is the characteristic equation of a real symmetric matrix A, then the characteristic roots of A are all positive if and only if each β is positive.

One of the most important properties of the characteristic vectors of a hermitian matrix is that of orthogonality. More precisely, we have the following theorem.

THEOREM 9 If $\mathbf{x}_i$ and $\mathbf{x}_j$ are characteristic vectors corresponding, respectively, to the distinct characteristic values λ_i and λ_j of a hermitian matrix H, then $\bar{\mathbf{x}}_i^T\mathbf{x}_j = 0$.

PROOF By hypothesis, we have

$$\text{(12)} \qquad \mathrm{H}\mathbf{x}_i = \lambda_i\mathbf{x}_i$$
$$\text{(13)} \qquad \mathrm{H}\mathbf{x}_j = \lambda_j\mathbf{x}_j$$

If in the first of these we take the conjugate and then the transpose of each member, we obtain

$$\bar{\mathbf{x}}_i^T\bar{\mathrm{H}}^T = \bar{\lambda}_i\bar{\mathbf{x}}_i^T$$

or, since $\bar{\mathrm{H}}^T = \mathrm{H}$, by hypothesis, and $\bar{\lambda}_i = \lambda_i$, by Theorem 7,

$$\text{(14)} \qquad \bar{\mathbf{x}}_i^T\mathrm{H} = \lambda_i\bar{\mathbf{x}}_i^T$$

Now, if we premultiply Eq. (13) by $\bar{\mathbf{x}}_i^T$ and postmultiply Eq. (14) by $\mathbf{x}_j$, we obtain, respectively,

$$\bar{\mathbf{x}}_i^T\mathrm{H}\mathbf{x}_j = \lambda_j\bar{\mathbf{x}}_i^T\mathbf{x}_j \qquad \text{and} \qquad \bar{\mathbf{x}}_i^T\mathrm{H}\mathbf{x}_j = \lambda_i\bar{\mathbf{x}}_i^T\mathbf{x}_j$$

Finally, subtracting these equations, we have

$$(\lambda_i - \lambda_j)\bar{\mathbf{x}}_i^T\mathbf{x}_j = 0 \qquad i \neq j$$

or, since $\lambda_i \neq \lambda_j$, by hypothesis,

$$\bar{\mathbf{x}}_i^T\mathbf{x}_j = 0$$

as asserted.

COROLLARY 1 If $\mathbf{x}_i$ and $\mathbf{x}_j$ are characteristic vectors corresponding, respectively, to the distinct characteristic values λ_i and λ_j of a hermitian matrix H, then $\bar{\mathbf{x}}_i^T\mathrm{H}\mathbf{x}_j = 0$.

PROOF Since $\bar{\mathbf{x}}_i^T\mathbf{x}_j = 0$, Corollary 1 is an immediate consequence of either of the equations following (14).

COROLLARY 2 If $\mathbf{x}_i$ and $\mathbf{x}_j$ are characteristic vectors corresponding, respectively, to the distinct characteristic values λ_i and λ_j of a real symmetric matrix A, then $\mathbf{x}_i^T\mathbf{x}_j = \mathbf{x}_i^T\mathrm{A}\mathbf{x}_j = 0$.

We are now in a position to return to the question we raised earlier in this section about the rank of the matrix $\mathrm{H} - \lambda_i\mathrm{I}$ when H is hermitian and λ_i is a characteristic root of H of multiplicity r. As the next theorem shows, every characteristic root of a hermitian matrix is regular; i.e., if H is hermitian, then for every characteristic root λ_i of multiplicity r, the rank of $\mathrm{H} - \lambda_i\mathrm{I}$ drops to the minimum allowed by Theorem 6, namely $n - r$, and there are r linearly independent characteristic vectors of H corresponding to λ_i.

THEOREM 10 If H is a hermitian matrix, then to every r-fold characteristic root of H there correspond exactly r linearly independent characteristic vectors.

PROOF Let λ_1 be a characteristic root of multiplicity r of a hermitian matrix H, let $\mathbf{x}_1$ be any characteristic vector corresponding to λ_1, and let $\mathbf{v}_2, \mathbf{v}_3, \ldots, \mathbf{v}_n$ be $n-1$ vectors such that $\mathbf{x}_1, \mathbf{v}_2, \mathbf{v}_3, \ldots, \mathbf{v}_n$ are linearly independent. By the Schmidt orthogonalization process it is always possible to convert these vectors into a set of orthonormal vectors $\mathbf{u}_1, \mathbf{u}_2, \mathbf{u}_3, \ldots, \mathbf{u}_n$ in which, in particular, $\mathbf{u}_1$ is the normalized form of $\mathbf{x}_1$. The vectors in the set $\{\mathbf{u}_i\}$ thus satisfy the relations

$$\bar{\mathbf{u}}_i^T \mathbf{u}_j = \begin{cases} 1 & i = j \\ 0 & i \neq j \end{cases} \tag{15}$$

Now let U be the $n \times n$ matrix whose columns are, respectively, $\mathbf{u}_1, \mathbf{u}_2, \mathbf{u}_3, \ldots, \mathbf{u}_n$. Then from (15) it is clear that

$$\bar{\mathrm{U}}^T \mathrm{U} = \mathrm{I}$$

Hence $\bar{\mathrm{U}}^T$ is the inverse of U; and therefore from Theorem 3 it follows that the matrix H and the matrix $\bar{\mathrm{U}}^T\mathrm{HU} \equiv \mathrm{U}^{-1}\mathrm{HU}$ have the same characteristic polynomial. In other words, the equations

$$|\bar{\mathrm{U}}^T\mathrm{HU} - \lambda\mathrm{I}| = 0 \qquad \text{and} \qquad |\mathrm{H} - \lambda\mathrm{I}| = 0$$

have the same roots. Now, remembering that $\mathrm{H}\mathbf{u}_1 = \lambda_1\mathbf{u}_1$, since by hypothesis $\mathbf{u}_1$ is a characteristic vector of H corresponding to $\lambda = \lambda_1$, we have

$$\bar{\mathrm{U}}^T\mathrm{HU} = \begin{bmatrix} \bar{\mathbf{u}}_1^T \\ \bar{\mathbf{u}}_2^T \\ \vdots \\ \bar{\mathbf{u}}_n^T \end{bmatrix} \mathrm{H}[\mathbf{u}_1 \quad \mathbf{u}_2 \quad \cdots \quad \mathbf{u}_n] = \begin{bmatrix} \bar{\mathbf{u}}_1^T \\ \bar{\mathbf{u}}_2^T \\ \vdots \\ \bar{\mathbf{u}}_n^T \end{bmatrix} [\mathrm{H}\mathbf{u}_1 \quad \mathrm{H}\mathbf{u}_2 \quad \cdots \quad \mathrm{H}\mathbf{u}_n]$$

$$= \begin{bmatrix} \bar{\mathbf{u}}_1^T \\ \bar{\mathbf{u}}_2^T \\ \vdots \\ \bar{\mathbf{u}}_n^T \end{bmatrix} [\lambda_1\mathbf{u}_1 \quad \mathrm{H}\mathbf{u}_2 \quad \cdots \quad \mathrm{H}\mathbf{u}_n] = \begin{bmatrix} \lambda_1 & \bar{\mathbf{u}}_1^T\mathrm{H}\mathbf{u}_2 & \cdots & \bar{\mathbf{u}}_1^T\mathrm{H}\mathbf{u}_n \\ 0 & \bar{\mathbf{u}}_2^T\mathrm{H}\mathbf{u}_2 & \cdots & \bar{\mathbf{u}}_2^T\mathrm{H}\mathbf{u}_n \\ \cdots & \cdots & \cdots & \cdots \\ 0 & \bar{\mathbf{u}}_n^T\mathrm{H}\mathbf{u}_2 & \cdots & \bar{\mathbf{u}}_n^T\mathrm{H}\mathbf{u}_n \end{bmatrix}$$

The zeros in the first column in the last matrix appear because of the relation $\bar{\mathbf{u}}_i^T\mathbf{u}_1 = 0,\ i \neq 1$, guaranteed by (15). The remaining entries in the last matrix are, in general, not equal to zero since the $\mathbf{u}_i$'s are not orthogonal with respect to the matrix H. However, because H, and therefore $\bar{\mathrm{U}}^T\mathrm{HU}$, is hermitian (see Exercise 7), it follows that since the last $n-1$ elements in the first column are zero, so too are the last $n-1$ elements in the first row. Thus $\bar{\mathrm{U}}^T\mathrm{HU}$ is of the form

$$\begin{bmatrix} \lambda_1 & 0 & 0 & \cdots & 0 \\ 0 & \alpha_{22} & \alpha_{23} & \cdots & \alpha_{2n} \\ \cdots & \cdots & \cdots & \cdots & \cdots \\ 0 & \alpha_{n2} & \alpha_{n3} & \cdots & \alpha_{nn} \end{bmatrix}$$

and

$$[\bar{\mathrm{U}}^T\mathrm{HU} - \lambda\mathrm{I}] = \begin{bmatrix} \lambda_1 - \lambda & 0 & 0 & \cdots & 0 \\ 0 & \alpha_{22} - \lambda & \alpha_{23} & \cdots & \alpha_{2n} \\ 0 & \alpha_{32} & \alpha_{33} - \lambda & \cdots & \alpha_{3n} \\ \cdots & \cdots & \cdots & \cdots & \cdots \\ 0 & \alpha_{n2} & \alpha_{n3} & \cdots & \alpha_{nn} - \lambda \end{bmatrix}$$

Therefore, if $\lambda = \lambda_1$ is a repeated root of

$$|\mathrm{H} - \lambda\mathrm{I}| = |\bar{\mathrm{U}}^T\mathrm{HU} - \lambda\mathrm{I}| = 0$$

then $\lambda_1 - \lambda$ must be a factor of the minor of the element in the first row and first column of $|\overline{U}^T HU - \lambda I|$. But if this minor vanishes when $\lambda = \lambda_1$, then the rank of $\overline{U}^T HU - \lambda_1 I$ is at most $n - 2$, since all other minors of order $n - 1$ obviously contain either a row of zeros or a column of zeros. Furthermore, since

$$\overline{U}^T HU - \lambda_1 I \equiv \overline{U}^T (H - \lambda_1 I)U \qquad \text{and} \qquad H - \lambda_1 I$$

are equivalent matrices and therefore have the same rank, it follows that the rank of $H - \lambda_1 I$ is also at most $n - 2$. Hence, by Theorem 10, Sec. 14.3, $(H - \lambda_1 I)\mathbf{x} = \mathbf{0}$ has at least two linearly independent solution vectors, and H has at least two linearly independent characteristic vectors corresponding to λ_1.

If the multiplicity of λ_1 is more than 2, the preceding argument can be repeated, using this time any unitary matrix U whose first *two* columns are any two orthonormal characteristic vectors of H corresponding to the characteristic value λ_1. In this case we obtain at once the relation

$$\overline{U}^T HU - \lambda I = \begin{bmatrix} \lambda_1 - \lambda & 0 & 0 & 0 & \cdots & 0 \\ 0 & \lambda_1 - \lambda & 0 & 0 & \cdots & 0 \\ 0 & 0 & \gamma_{33} - \lambda & \gamma_{34} & \cdots & \gamma_{3n} \\ 0 & 0 & \gamma_{43} & \gamma_{44} - \lambda & \cdots & \gamma_{4n} \\ \cdots & \cdots & \cdots & \cdots & \cdots & \cdots \\ 0 & 0 & \gamma_{n3} & \gamma_{n4} & \cdots & \gamma_{nn} - \lambda \end{bmatrix}$$

Then, since $\lambda - \lambda_1$ is a characteristic root of multiplicity greater than 2, it follows that $\lambda_1 - \lambda$ must be a factor of the complementary minor of the second-order minor in the first two rows and first two columns of $|\overline{U}^T HU - \lambda I|$. Hence since all other $(n - 2)$nd-order minors obviously vanish, it is evident that when $\lambda = \lambda_1$, the rank of $\overline{U}^T HU - \lambda I$, and therefore the rank of $H - \lambda I$, is not more than $n - 3$ and H has at least three linearly independent characteristic vectors corresponding to the characteristic value $\lambda = \lambda_1$.

Clearly, this procedure can be continued until we reach the conclusion that if λ_1 is an r-fold characteristic root of H, then the rank of $H - \lambda_1 I$ is at most $n - r$, hence H has at least r independent characteristic vectors corresponding to λ_1. But by Theorem 6, H can have at most r linearly independent characteristic vectors corresponding to the r-fold root λ_1. Therefore H must have *exactly* r linearly independent characteristic vectors corresponding to λ_1, as asserted. ◗

Since, as we have repeatedly observed, a real symmetric matrix is a special case of a hermitian matrix, it is clear that we also have the following result.

COROLLARY 1 If A is a real symmetric matrix, then to every r-fold characteristic root of A there correspond exactly r linearly independent characteristic vectors.

We are now in a position to prove the following fundamental theorem.

THEOREM 11 Every $n \times n$ hermitian matrix has n linearly independent characteristic vectors.

◖ **PROOF** Let H be an $n \times n$ hermitian matrix. It may of course possess one or more repeated characteristic roots, but if it does, we know from the last theorem that to each root of multiplicity r there correspond exactly r linearly independent characteristic vectors. Hence H cannot have more than n linearly independent characteristic vectors. Specifically, let the characteristic roots of H be

$$\lambda_1, \lambda_2, \ldots, \lambda_k \qquad 1 \le k \le n$$

let the multiplicity of λ_i be r_i, where $\sum_{i=1}^{k} r_i = n$, and let

$$\mathbf{x}_{i1}, \mathbf{x}_{i2}, \ldots, \mathbf{x}_{ir_i}$$

be r_i independent characteristic vectors corresponding to λ_i. Suppose, now, contrary to the assertion of the theorem, that these n characteristic vectors of H are not linearly independent. Then there exists a relation of the form

(16) $$(c_{11}\mathbf{x}_{11} + \cdots + c_{1r_1}\mathbf{x}_{1r_1}) + (c_{21}\mathbf{x}_{21} + \cdots + c_{2r_2}\mathbf{x}_{2r_2}) + \cdots + (c_{k1}\mathbf{x}_{k1} + \cdots + c_{kr_k}\mathbf{x}_{kr_k}) = \mathbf{0}$$

in which at least one of the c's is different from zero.

Now consider a typical group of terms, say the ith, in the last expression. By Exercise 20, Sec. 3.6, unless the c's in such a group are all zero, the combination defines a characteristic vector corresponding to the characteristic value $\lambda = \lambda_i$. Thus Eq. (16) is simply an expression of the form

$$c_1\mathbf{x}_1 + c_2\mathbf{x}_2 + \cdots + c_i\mathbf{x}_i + \cdots + c_k\mathbf{x}_k = \mathbf{0}$$

in which each c is either 0 or 1 and at least one c is different from zero. But since the $\mathbf{x}$'s now correspond to distinct characteristic values, it follows from Theorem 5 that they are linearly independent and hence that each c must be zero. This contradiction establishes the theorem. ◗

COROLLARY 1 Every real symmetric $n \times n$ matrix has n linearly independent characteristic vectors.

If an $n \times n$ matrix has n linearly independent characteristic vectors, then, by means of the Schmidt orthogonalization process applied to the vectors in each of the sets corresponding to a repeated characteristic root, a set of normalized mutually orthogonal characteristic vectors can always be constructed. Hence we have the following important result.

COROLLARY 2 Every $n \times n$ hermitian or real symmetric matrix has a set of n orthonormal characteristic vectors.

An $n \times n$ matrix whose columns are linearly independent characteristic vectors of an $n \times n$ matrix A is said to be a **modal matrix** of A.

In many applications in physics, chemistry, and engineering, it is necessary to consider matrix equations of the form $(A - \lambda B)\mathbf{x} = \mathbf{0}$ in which A and B are either hermitian or real and symmetric. Such an equation will of course have nontrivial solutions if and only if the determinant of the coefficients is equal to zero. Paralleling our earlier terminology for the special case in which $B = I$, the equation $|A - \lambda B| = 0$ is called the **characteristic equation** of the system, its roots are called the **characteristic roots** or **characteristic values** of the system, and the corresponding nontrivial solutions are called the **characteristic vectors** of the system. As one would expect, the theory of the equation $(A - \lambda B)\mathbf{x} = \mathbf{0}$ resembles closely the theory of the equation $(A - \lambda I)\mathbf{x} = \mathbf{0}$ which we have developed in this section. In particular, we have the following results.

THEOREM 12 The equation $(A - \lambda B)\mathbf{x} = \mathbf{0}$ has zero as a characteristic root if and only if A is singular.

◖ **PROOF** This follows immediately from a consideration of the characteristic equation $|A - \lambda B| = 0$ when the left-hand side is expressed as a polynomial in λ. ◗

THEOREM 13 If A and B are hermitian matrices and if B is definite, then the characteristic values of $(A - \lambda B)\mathbf{x} = \mathbf{0}$ are all real.

◖ **PROOF** Let A and B be hermitian matrices, let B be definite, and let $\mathbf{x}_1$ be a characteristic vector of the equation $(A - \lambda B)\mathbf{x} = \mathbf{0}$ corresponding to an arbitrary characteristic value λ_1. Then,

(17) $$A\mathbf{x}_1 = \lambda_1 B\mathbf{x}_1$$

Hence, premultiplying Eq. (17) by $\bar{\mathbf{x}}_1^T$, we have

(18) $$\bar{\mathbf{x}}_1^T A\mathbf{x}_1 = \lambda_1 \bar{\mathbf{x}}_1^T B\mathbf{x}_1$$

Now from Theorem 4, Sec. 14.5, we know that both $\bar{\mathbf{x}}_1^T A\mathbf{x}_1$ and $\bar{\mathbf{x}}_1^T B\mathbf{x}_1$ are real numbers. Moreover, since B is definite, $\bar{\mathbf{x}}_1^T B\mathbf{x}_1 \neq 0$. Hence it follows from Eq. (18) that λ_1 is a real number, as asserted. ◗

COROLLARY 1 If A and B are real symmetric matrices and if B is definite, then the characteristic values of $(A - \lambda B)\mathbf{x} = \mathbf{0}$ are all real.

By inspection of Eq. (18), the following results are obtained immediately.

COROLLARY 2 If A and B are hermitian (or real symmetric) matrices which are both positive-definite or both negative-definite, then the characteristic values of $(A - \lambda B)\mathbf{x} = \mathbf{0}$ are all positive.

COROLLARY 3 If A and B are hermitian (or real symmetric) matrices and if A is positive-definite and B is negative-definite, or vice versa, then the characteristic values of $(A - \lambda B)\mathbf{x} = \mathbf{0}$ are all negative.

THEOREM 14 If A and B are hermitian matrices and if $\mathbf{x}_1, \mathbf{x}_2, \ldots, \mathbf{x}_k$ are characteristic vectors of the equation $(A - \lambda B)\mathbf{x} = \mathbf{0}$ corresponding, respectively, to the distinct characteristic values $\lambda_1, \lambda_2, \ldots, \lambda_k$, then the $\mathbf{x}$'s satisfy the generalized orthogonality conditions

$$\overline{\mathbf{x}}_i^T B\mathbf{x}_j = 0 \qquad \text{and} \qquad \overline{\mathbf{x}}_i^T A\mathbf{x}_j = 0 \qquad i \neq j$$

PROOF Let A and B be hermitian matrices, let λ_i and λ_j be distinct characteristic values of the equation $(A - \lambda B)\mathbf{x} = \mathbf{0}$, and let $\mathbf{x}_i$ and $\mathbf{x}_j$ be characteristic vectors corresponding, respectively, to λ_i and λ_j. Then

$$A\mathbf{x}_i = \lambda_i B\mathbf{x}_i \qquad \text{and} \qquad A\mathbf{x}_j = \lambda_j B\mathbf{x}_j$$

If we premultiply the first of these equations by $\overline{\mathbf{x}}_j^T$ and the second by $\overline{\mathbf{x}}_i^T$, we obtain, respectively,

$$\overline{\mathbf{x}}_j^T A\mathbf{x}_i = \lambda_i \overline{\mathbf{x}}_j^T B\mathbf{x}_i \tag{19}$$

and

$$\overline{\mathbf{x}}_i^T A\mathbf{x}_j = \lambda_j \overline{\mathbf{x}}_i^T B\mathbf{x}_j \tag{20}$$

Now, if we take the transpose and then the conjugate of each side of Eq. (19), remembering that both A and B are hermitian, we obtain

$$\overline{\mathbf{x}}_i^T A\mathbf{x}_j = \lambda_i \overline{\mathbf{x}}_i^T B\mathbf{x}_j \tag{21}$$

Finally, subtracting Eq. (20) from Eq. (21), we have

$$(\lambda_i - \lambda_j)\overline{\mathbf{x}}_i^T B\mathbf{x}_j = 0$$

Therefore, since $\lambda_i \neq \lambda_j$, by hypothesis, it follows that

$$\overline{\mathbf{x}}_i^T B\mathbf{x}_j = 0$$

as asserted. Furthermore, if $\overline{\mathbf{x}}_i^T B\mathbf{x}_j = 0$, then it follows from Eq. (21) that $\overline{\mathbf{x}}_i^T A\mathbf{x}_j = 0$, and our proof is complete.

COROLLARY 1 If A and B are real symmetric matrices, and if $\mathbf{x}_1, \mathbf{x}_2, \ldots, \mathbf{x}_k$ are characteristic vectors of the equation $(A - \lambda B)\mathbf{x} = \mathbf{0}$, corresponding, respectively, to the distinct characteristic values $\lambda_1, \lambda_2, \ldots, \lambda_k$, then the $\mathbf{x}$'s satisfy the generalized orthogonality conditions

$$\mathbf{x}_i^T B\mathbf{x}_j = 0 \qquad \text{and} \qquad \mathbf{x}_i^T A\mathbf{x}_j = 0 \qquad i \neq j$$

THEOREM 15 If A and B are hermitian matrices, if B is definite, and if $\mathbf{x}_1, \mathbf{x}_2, \ldots, \mathbf{x}_k$ are characteristic vectors of $(A - \lambda B)\mathbf{x} = \mathbf{0}$ corresponding, respectively, to the distinct characteristic values $\lambda_1, \lambda_2, \ldots, \lambda_k$, then $\mathbf{x}_1, \mathbf{x}_2, \ldots, \mathbf{x}_k$ are linearly independent.

PROOF Let A and B be hermitian matrices, let B be definite, and let us suppose, contrary to the theorem, that the characteristic vectors $\mathbf{x}_1, \mathbf{x}_2, \ldots, \mathbf{x}_k$ corresponding, respectively, to the distinct characteristic values $\lambda_1, \lambda_2, \ldots, \lambda_k$ of $(A - \lambda B)\mathbf{x} = \mathbf{0}$ are linearly dependent. Then there exists a relation of the form

$$c_1\mathbf{x}_1 + c_2\mathbf{x}_2 + \cdots + c_k\mathbf{x}_k = \mathbf{0}$$

in which at least one of the c's, say c_i, is different from zero. Now if we multiply the last equation through on the left by $\overline{\mathbf{x}}_i^T B$, we get

$$c_1\overline{\mathbf{x}}_i^T B\mathbf{x}_1 + c_2\overline{\mathbf{x}}_i^T B\mathbf{x}_2 + \cdots + c_i\overline{\mathbf{x}}_i^T B\mathbf{x}_i + \cdots + c_k\overline{\mathbf{x}}_i^T B\mathbf{x}_k = 0$$

However, from the orthogonality guaranteed by Theorem 14, it follows that every term in the last equation except $c_i\overline{\mathbf{x}}_i^T B\mathbf{x}_i$ is equal to zero. Moreover, by hypothesis, B is either positive-definite or negative-definite. Hence $\overline{\mathbf{x}}_i^T B\mathbf{x}_i \neq 0$, and therefore $c_i = 0$, contrary to the assumption of linear dependence. This contradiction shows that the **x**'s must be linearly independent, and the theorem is established.

COROLLARY 1 If A and B are real symmetric matrices, if B is definite, and if $\mathbf{x}_1, \mathbf{x}_2, \ldots, \mathbf{x}_k$ are characteristic vectors of $(A - \lambda B)\mathbf{x} = \mathbf{0}$ corresponding, respectively, to the distinct characteristic values $\lambda_1, \lambda_2, \ldots, \lambda_k$, then $\mathbf{x}_1, \mathbf{x}_2, \ldots, \mathbf{x}_k$ are linearly independent.

Theorem 15 must not be misinterpreted as asserting that if A and B are hermitian (or real symmetric) $n \times n$ matrices and if B is definite, then $(A - \lambda B)\mathbf{x} = \mathbf{0}$ has n linearly independent characteristic vectors. It guarantees that characteristic vectors corresponding to *distinct* characteristic values of $(A - \lambda B)\mathbf{x} = \mathbf{0}$ are linearly independent, but it says nothing about how many distinct characteristic values there are or about how many independent characteristic vectors correspond to a repeated characteristic value. If, because of repeated roots, $(A - \lambda B)\mathbf{x} = \mathbf{0}$ has fewer than n distinct characteristic values, then, for all we know at present, $(A - \lambda B)\mathbf{x} = 0$ may have fewer than n linearly independent characteristic vectors. However, by a proof very much like the proof of Theorem 10, the following result can be established.

THEOREM 16 If A and B are hermitian (or real symmetric) matrices and if B is positive-definite, then to a repeated characteristic value of $(A - \lambda B)\mathbf{x} = \mathbf{0}$ of multiplicity r there correspond exactly r linearly independent characteristic vectors.

With this theorem, it is not difficult to establish the following counterpart of Theorem 11.

THEOREM 17 If A and B are hermitian (or real symmetric) $n \times n$ matrices and if B is positive-definite, then the equation $(A - \lambda B)\mathbf{x} = \mathbf{0}$ has exactly n linearly independent characteristic vectors.

By a straightforward application of the Schmidt orthogonalization process to the n linearly independent characteristic vectors of $(A - \lambda B)\mathbf{x} = \mathbf{0}$ guaranteed by Theorem 17, we can establish the following useful result.

COROLLARY 1 If A and B are hermitian (or real symmetric) matrices and if B is positive-definite, then $(A - \lambda B)\mathbf{x} = \mathbf{0}$ possesses n characteristic vectors which are orthonormal with respect to B.

With Theorem 17 and its corollary available, it is now an easy matter to express an arbitrary vector $\mathbf{c}$ with n components as a linear combination of the characteristic vectors of the equation $(A - \lambda B)\mathbf{x} = \mathbf{0}$, provided that A and B are hermitian or real symmetric and B is positive-definite.

For we can write

$$\mathbf{c} = c_1\mathbf{x}_1 + c_2\mathbf{x}_2 + \cdots + c_n\mathbf{x}_n \tag{22}$$

where the **x**'s are characteristic vectors of $(\mathrm{A} - \lambda\mathrm{B})\mathbf{x} = \mathbf{0}$ mutually orthogonal with respect to B. Then, if we premultiply Eq. (22) by $\overline{\mathbf{x}}_i^T\mathrm{B}$, we obtain

$$\overline{\mathbf{x}}_i^T\mathrm{B}\mathbf{c} = c_1\overline{\mathbf{x}}_i^T\mathrm{B}\mathbf{x}_1 + \cdots + c_i\overline{\mathbf{x}}_i^T\mathrm{B}\mathbf{x}_i + \cdots + c_n\overline{\mathbf{x}}_i^T\mathrm{B}\mathbf{x}_n$$

From the orthogonality of the **x**'s, it follows that every term on the right except $c_i\overline{\mathbf{x}}_i^T\mathrm{B}\mathbf{x}_i$ is equal to zero. Moreover, since B is positive-definite, it follows that $\overline{\mathbf{x}}_i^T\mathrm{B}\mathbf{x}_i \neq 0$. Hence we can solve for c_i, getting

$$c_i = \frac{\overline{\mathbf{x}}_i^T\mathrm{B}\mathbf{c}}{\overline{\mathbf{x}}_i^T\mathrm{B}\mathbf{x}_i} \qquad i = 1, 2, \ldots, n$$

If the **x**'s have been normalized with respect to B, that is, if $\overline{\mathbf{x}}_i^T\mathrm{B}\mathbf{x}_i = 1$, $i = 1, 2, \ldots, n$, the last formula reduces to the simpler expression

$$c_i = \overline{\mathbf{x}}_i^T\mathrm{B}\mathbf{c} \qquad i = 1, 2, \ldots, n$$

The fact that we were able to solve for the coefficients in the expansion (22) without solving any simultaneous equations should make clear the great convenience of working with a set of vectors which are orthogonal.†

The ideas we have developed in this section are especially relevant to physical systems with several degrees of freedom, such as those we studied in Sec. 7.5. Once the differential equations describing such a system have been set up, they can easily be converted into a single matrix equation. The characteristic values of this matrix are then the squares of the natural frequencies of the system, and the corresponding characteristic vectors are the normal modes of the system. Subsequently, the set of linearly independent characteristic vectors (guaranteed by Theorem 17 and its corollary) and their orthogonality properties (guaranteed by Theorem 14 and its corollary) make it a routine matter to determine the response of the system to given initial conditions and to harmonic forcing functions. Our last three examples illustrate these ideas in detail.

EXAMPLE 2

NATURAL FREQUENCIES AND NORMAL MODES OF A THREE-MASS SYSTEM

Find the natural frequencies and the normal modes of the system shown in Fig. 14.8.

Neglecting friction, as we always do in making natural frequency calculations, the only forces acting in the system are those transmitted to the three masses by the springs directly attached to them. When the instantaneous displacements of the masses are x_1, x_2, x_3, the lengths of the springs have changed from their unstretched initial values by the respective amounts

$$x_1, \qquad x_2 - x_1, \qquad x_3 - x_2, \qquad -x_3$$

each amount being positive or negative according as that spring is elongated or compressed. Hence the forces instantaneously exerted by the springs are, respectively,

$$3x_1, \qquad 3(x_2 - x_1), \qquad 3(x_3 - x_2), \qquad -x_3$$

†The similarity between this procedure and the process by which we found the coefficients in the Fourier series of a function in Sec. 8.2 and Sec. 11.5 and the coefficients in the more general expansions in Sec. 11.6 is worthy of note.

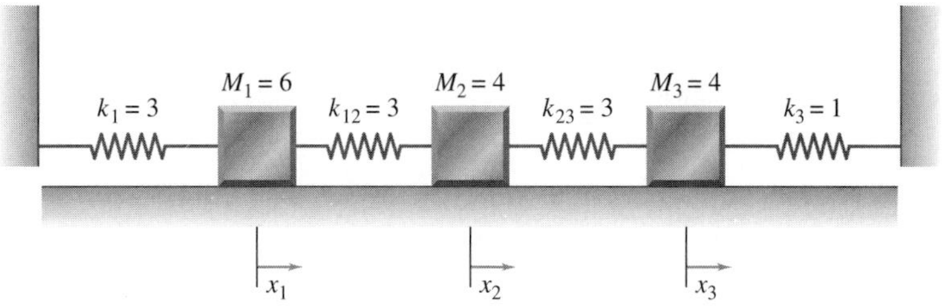

FIGURE 14.8
A simple mass-spring system.

Therefore, applying Newton's law to each mass, taking due account of the signs of the forces applied by the attached springs, we obtain the following three differential equations.

$$\begin{aligned} 6D^2x_1 &= -3x_1 + 3(x_2 - x_1) \\ 4D^2x_2 &= -3(x_2 - x_1) + 3(x_3 - x_2) \\ 4D^2x_3 &= -3(x_3 - x_2) - x_3 \end{aligned}$$

or

(23) $$(\mathrm{A} + \mathrm{B}D^2)\mathbf{x} = \mathbf{0}$$

where

$$\mathrm{A} = \begin{bmatrix} 6 & -3 & 0 \\ -3 & 6 & -3 \\ 0 & -3 & 4 \end{bmatrix} \qquad \mathrm{B} = \begin{bmatrix} 6 & 0 & 0 \\ 0 & 4 & 0 \\ 0 & 0 & 4 \end{bmatrix} \qquad \mathbf{x} = \begin{bmatrix} x_1 \\ x_2 \\ x_3 \end{bmatrix} \qquad D = \frac{d}{dt}$$

Since there is no dissipation of energy through friction, each mass must move harmonically with constant amplitude about its equilibrium position. Hence as a solution we can assume

$$\mathbf{x} = \mathbf{a}\cos\omega t \qquad \mathbf{a} = \begin{bmatrix} a_1 \\ a_2 \\ a_3 \end{bmatrix}$$

(The assumption $\mathbf{x} = \mathbf{a}\sin\omega t$ would serve equally well.) Substituting this into (23), we obtain, after dividing out $\cos\omega t$,

(24) $$(\mathrm{A} - \omega^2\mathrm{B})\mathbf{a} = \mathbf{0}$$

In this equation, ω^2 plays the role of λ in our previous discussion. Since **A** and **B** are symmetric and **B** is clearly positive-definite, Corollary 1 of Theorem 13 assures us that the values of ω^2 satisfying the equation

$$|\mathrm{A} - \omega^2\mathrm{B}| = 0$$

are positive; hence the natural frequencies ω are all real.

Expanding the determinant $|\mathrm{A} - \omega^2\mathrm{B}| = 0$ leads to the equation

$$-6(4\omega^2 - 1)(\omega^2 - 1)(4\omega^2 - 9) = 0$$

and therefore the natural frequencies are

$$\omega_1 = \tfrac{1}{2} \qquad \omega_2 = 1 \qquad \omega_3 = \tfrac{3}{2}$$

For these frequencies and for no others, Eq. (24) has nontrivial solutions, and the physical system can vibrate freely.

Solving the systems of equations

$$(\mathrm{A} - \omega_i^2 \mathrm{B})\mathbf{a} = \mathbf{0} \qquad i = 1, 2, 3$$

and noting that only the ratios of the components of **a** are significant, we find at once that for

$$\omega_1 = \frac{1}{2}, \ \mathbf{a}_1 = \begin{bmatrix} 2 \\ 3 \\ 3 \end{bmatrix} \quad \text{and} \quad \mathbf{x}_1 = \mathbf{a}_1 \cos\frac{t}{2}$$

$$\omega_2 = 1, \ \mathbf{a}_2 = \begin{bmatrix} 1 \\ 0 \\ -1 \end{bmatrix} \quad \text{and} \quad \mathbf{x}_2 = \mathbf{a}_2 \cos t$$

$$\omega_3 = \frac{3}{2}, \ \mathbf{a}_3 = \begin{bmatrix} 2 \\ -5 \\ 3 \end{bmatrix} \quad \text{and} \quad \mathbf{x}_3 = \mathbf{a}_3 \cos\frac{3t}{2}$$

When the system is vibrating at the radian per unit time frequency $\omega_1 = \frac{1}{2}$, the masses all move in the same direction, M_1 moving with an amplitude $\frac{2}{3}$, the common amplitude of M_2 and M_3.

When the system is vibrating with frequency $\omega_2 = 1$, M_2 remains at rest, while M_1 and M_3 move in opposite directions but with equal amplitudes.

When the system is vibrating with frequency $\omega_3 = \frac{3}{2}$, M_1 and M_3 move in the same direction with respective amplitudes $2k$ and $3k$, while M_2 moves in the opposite direction with amplitude $5k$. Thus the characteristic vectors, i.e., the so-called **normal modes**, $\mathbf{a}_1$, $\mathbf{a}_2$, and $\mathbf{a}_3$, give, in magnitude and direction, the relative amplitudes with which the masses move at the respective frequencies ω_1, ω_2, and ω_3.

EXAMPLE 3

FREE MOTION OF A THREE-MASS SYSTEM

If the three masses shown in Fig. 14.8 are initially displaced so that

$$(25) \qquad (x_1)_0 = 2 \qquad (x_2)_0 = -1 \qquad (x_3)_0 = 1 \qquad \text{or} \qquad (\mathbf{x})_0 = \begin{bmatrix} 2 \\ -1 \\ 1 \end{bmatrix}$$

and while in this configuration are given the initial velocities

$$(26) \qquad (\dot{x}_1)_0 = 1 \qquad (\dot{x}_2)_0 = 2 \qquad (\dot{x}_3)_0 = 0 \qquad \text{or} \qquad (\dot{\mathbf{x}})_0 = \begin{bmatrix} 1 \\ 2 \\ 0 \end{bmatrix}$$

determine the subsequent motion of each mass.

In Example 2 we found that the assumption $\mathbf{x} = \mathbf{a} \cos \omega t$ led to the three solutions

$$\mathbf{x}_1 = \mathbf{a}_1 \cos\frac{t}{2} \qquad \mathbf{x}_2 = \mathbf{a}_2 \cos t \qquad \mathbf{x}_3 = \mathbf{a}_3 \cos\frac{3t}{2}$$

In exactly the same way, the assumption $\mathbf{x} = \mathbf{a} \sin \omega t$ would have led to the same three values of ω, the same respective characteristic vectors, and the three additional solutions

$$\mathbf{x}_4 = \mathbf{a}_1 \sin\frac{t}{2} \qquad \mathbf{x}_5 = \mathbf{a}_2 \sin t \qquad \mathbf{x}_6 = \mathbf{a}_3 \sin\frac{3t}{2}$$

Hence, since (23) is a linear equation, a complete solution can be written in the form

$$\mathbf{x} = c_1\mathbf{x}_1 + c_2\mathbf{x}_2 + c_3\mathbf{x}_3 + c_4\mathbf{x}_4 + c_5\mathbf{x}_5 + c_6\mathbf{x}_6$$

(27) $$= c_1\mathbf{a}_1 \cos\frac{t}{2} + c_2\mathbf{a}_2 \cos t + c_3\mathbf{a}_3 \cos\frac{3t}{2} + c_4\mathbf{a}_1 \sin\frac{t}{2} + c_5\mathbf{a}_2 \sin t + c_6\mathbf{a}_3 \sin\frac{3t}{2}$$

where the c's are arbitrary constants to be determined so that the initial conditions (25) and (26) are satisfied.

If we now set $t = 0$ in Eq. (27) and set $\mathbf{x} = (\mathbf{x})_0$, we obtain

(28) $$(\mathbf{x})_0 = c_1\mathbf{a}_1 + c_2\mathbf{a}_2 + c_3\mathbf{a}_3$$

Using the methods developed in Sec. 7.5, we would have to interpret this matrix equation as a system of three linear equations which would have to be solved simultaneously for c_1, c_2, c_3. Now, because of the orthogonality properties of the $\mathbf{a}$'s, we can solve for the c's one at a time, as follows.

If we multiply Eq. (28) through on the left by $\mathbf{a}_1^T\mathrm{B}$, we obtain

$$\mathbf{a}_1^T\mathrm{B}(\mathbf{x})_0 = c_1(\mathbf{a}_1^T\mathrm{B}\mathbf{a}_1) + c_2(\mathbf{a}_1^T\mathrm{B}\mathbf{a}_2) + c_3(\mathbf{a}_1^T\mathrm{B}\mathbf{a}_3)$$

By Corollary 1, Theorem 14, both $\mathbf{a}_1^T\mathrm{B}\mathbf{a}_2$ and $\mathbf{a}_1^T\mathrm{B}\mathbf{a}_3$ are equal to zero. Moreover, since B is positive-definite, $\mathbf{a}_1^T\mathrm{B}\mathbf{a}_1 \neq 0$. Hence we have at once

$$c_1 = \frac{\mathbf{a}_1^T\mathrm{B}(\mathbf{x})_0}{\mathbf{a}_1^T\mathrm{B}\mathbf{a}_1} \quad \text{or in the present case} \quad c_1 = \frac{24}{96} = \frac{1}{4}$$

Similarly, multiplying Eq. (28) through on the left by $\mathbf{a}_2^T\mathrm{B}$ and $\mathbf{a}_3^T\mathrm{B}$ and then solving for c_2 and c_3, we have

$$c_2 = \frac{\mathbf{a}_2^T\mathrm{B}(\mathbf{x})_0}{\mathbf{a}_2^T\mathrm{B}\mathbf{a}_2} = \frac{8}{10} = \frac{4}{5} \quad \text{and} \quad c_3 = \frac{\mathbf{a}_3^T\mathrm{B}(\mathbf{x})_0}{\mathbf{a}_3^T\mathrm{B}\mathbf{a}_3} = \frac{56}{160} = \frac{7}{20}$$

To find c_4, c_5, c_6 we first differentiate Eq. (27) with respect to t and then set $t = 0$ and $\dot{\mathbf{x}} = (\dot{\mathbf{x}})_0$, getting

(29) $$(\dot{\mathbf{x}})_0 = \tfrac{1}{2}c_4\mathbf{a}_1 + c_5\mathbf{a}_2 + \tfrac{3}{2}c_6\mathbf{a}_3$$

Again, we multiply Eq. (29) on the left by $\mathbf{a}_1^T\mathrm{B}$, $\mathbf{a}_2^T\mathrm{B}$, and $\mathbf{a}_3^T\mathrm{B}$ in turn and use the orthogonality of the $\mathbf{a}$'s to determine c_4, c_5, c_6, one at a time. The results are

$$c_4 = \frac{2\mathbf{a}_1^T\mathrm{B}(\dot{\mathbf{x}})_0}{\mathbf{a}_1^T\mathrm{B}\mathbf{a}_1} = \frac{3}{4} \qquad c_5 = \frac{\mathbf{a}_2^T\mathrm{B}(\dot{\mathbf{x}})_0}{\mathbf{a}_2^T\mathrm{B}\mathbf{a}_2} = \frac{3}{5} \qquad c_6 = \frac{2\mathbf{a}_3^T\mathrm{B}(\dot{\mathbf{x}})_0}{\mathbf{a}_3^T\mathrm{B}\mathbf{a}_3} = -\frac{7}{60}$$

Hence the required solution is

$$\mathbf{x} = \frac{1}{4}\mathbf{a}_1 \cos\frac{t}{2} + \frac{4}{5}\mathbf{a}_2 \cos t + \frac{7}{20}\mathbf{a}_3 \cos\frac{3t}{2} + \frac{3}{4}\mathbf{a}_1 \sin\frac{t}{2} + \frac{3}{5}\mathbf{a}_2 \sin t - \frac{7}{60}\mathbf{a}_3 \sin\frac{3t}{2}$$

As we shall see in the next example, there are problems in which it is necessary to solve nonhomogeneous matrix equations of the form

(30) $$(\mathrm{A} - \lambda\mathrm{B})\mathbf{x} = \mathbf{f} \qquad \mathbf{f} \text{ a constant vector}$$

To solve such an equation, it is convenient to assume for $\mathbf{x}$ an arbitrary linear combination of the characteristic vectors of the homogeneous equation $(\mathrm{A} - \lambda\mathrm{B})\mathbf{x} = \mathbf{0}$,

(31) $$\mathbf{x} = \sum_{i=1}^{n} \alpha_i\mathbf{a}_i$$

Then, substituting into (30), we have

$$(A - \lambda B) \sum_{i=1}^{n} \alpha_i \mathbf{a}_i = \mathbf{f}$$

or

$$\sum_{i=1}^{n} \alpha_i (A\mathbf{a}_i - \lambda B\mathbf{a}_i) = \mathbf{f} \tag{32}$$

Now, by definition, the characteristic vector $\mathbf{a}_i$ is a solution of the homogeneous equation $(A - \lambda_i B)\mathbf{x} = \mathbf{0}$. Hence $A\mathbf{a}_i = \lambda_i B\mathbf{a}_i$, and Eq. (32) can be written

$$\sum_{i=1}^{n} \alpha_i (\lambda_i B\mathbf{a}_i - \lambda B\mathbf{a}_i) = \mathbf{f}$$

or

$$\sum_{i=1}^{n} \alpha_i (\lambda_i - \lambda) B\mathbf{a}_i = \mathbf{f} \tag{33}$$

To find α_i, we multiply (33) through on the left by $\mathbf{a}_j^T$, noting that because of the orthogonality of the $\mathbf{a}$'s the only nonzero term in the sum is the one for which $j = i$. Thus we are left with

$$\alpha_i (\lambda_i - \lambda) \mathbf{a}_i^T B\mathbf{a}_i = \mathbf{a}_i^T \mathbf{f}$$

or

$$\alpha_i = \frac{\mathbf{a}_i^T \mathbf{f}}{(\lambda_i - \lambda) \mathbf{a}_i^T B\mathbf{a}_i}$$

The required solution of (30) is therefore

$$\mathbf{x} = \sum_{i=1}^{n} \frac{\mathbf{a}_i^T \mathbf{f}}{(\lambda_i - \lambda) \mathbf{a}_i^T B\mathbf{a}_i} \mathbf{a}_i \tag{34}$$

If the characteristic vectors have been normalized, then $\mathbf{a}_i^T B\mathbf{a}_i = 1$ and Eq. (35) assumes the simpler form

$$\mathbf{x} = \sum_{i=1}^{n} \frac{\mathbf{a}_i^T \mathbf{f}}{\lambda_i - \lambda} \mathbf{a}_i$$

It should be noted that if $\lambda = \lambda_i$, the system (30) is inconsistent unless $\mathbf{a}_i^T \mathbf{f} = 0$.

EXAMPLE 4 FORCED RESPONSE OF A THREE-MASS SYSTEM

Determine the response of the system shown in Fig. 14.8 to the force vector $\mathbf{F}(t) = \mathbf{f} \sin \omega t$ if $\mathbf{f} = \begin{bmatrix} -1 \\ 2 \\ 1 \end{bmatrix}$.

Drawing on the results of Example 2, it appears that we must find a particular integral of the equation $(A + BD^2)\mathbf{x} = \mathbf{f} \sin \omega t$. As usual, we assume $\mathbf{x} = \boldsymbol{\phi} \sin \omega t$. Then substituting and dividing out $\sin \omega t$, we obtain the matrix equation

$$(A - \omega^2 B)\boldsymbol{\phi} = \mathbf{f}$$

This is a nonhomogeneous equation just like Eq. (30) except that λ is replaced by ω^2. Hence we know its solution is given by Eq. (34):

$$\boldsymbol{\phi} = \sum_{i=1}^{3} \frac{\mathbf{a}_i^T\mathbf{f}}{(\omega_i^2 - \omega^2)\mathbf{a}_i^T\mathrm{B}\mathbf{a}_i}\mathbf{a}_i$$

In our particular problem we have $\omega_1 = \frac{1}{2}$, $\omega_2 = 1$, $\omega_3 = \frac{3}{2}$, and

$$\mathbf{a}_1^T\mathbf{f} = 7 \qquad \mathbf{a}_2^T\mathbf{f} = -2 \qquad \mathbf{a}_3^T\mathbf{f} = -9 \qquad \mathbf{a}_1^T\mathrm{B}\mathbf{a}_1 = 96 \qquad \mathbf{a}_2^T\mathrm{B}\mathbf{a}_2 = 10 \qquad \mathbf{a}_3^T\mathrm{B}\mathbf{a}_3 = 160$$

Thus substituting and making obvious numerical simplifications, we have for the required solution

$$\mathbf{x} = \boldsymbol{\phi}\sin\omega t = \left(\frac{7}{24(1-4\omega^2)}\begin{bmatrix}2\\3\\3\end{bmatrix} - \frac{1}{5(1-\omega^2)}\begin{bmatrix}-1\\0\\1\end{bmatrix} - \frac{9}{40(9-4\omega^2)}\begin{bmatrix}2\\-5\\3\end{bmatrix}\right)\sin\omega t$$

It is interesting to note that if $\mathbf{a}_i^T\mathbf{f} = 0$, then $\mathbf{a}_i^T\mathbf{f}/(\omega_i^2 - \omega^2)\mathbf{a}_i^T\mathrm{B}\mathbf{a}_i$ is missing from the particular integral, *regardless of the value of* ω. From this we infer the surprising fact that *a system can be driven at one of its natural frequencies by a force* $\mathbf{f}\sin\omega t$ *without resonating, provided* $\mathbf{a}_i^T\mathbf{f} = 0$, *where* $\mathbf{a}_i$ *is the characteristic vector corresponding to the natural frequency* ω_i.

EXERCISES

1. Show that the characteristic-value problem

$$\begin{aligned}2x_1 - 2x_2 &= \lambda x_1\\ 4x_1 - 2x_2 &= \lambda x_2\end{aligned}$$

has no real nontrivial solutions for any value of λ.

2. Find the characteristic values and the corresponding characteristic vectors for the equation $(\mathrm{A} - \lambda\mathrm{B})\mathbf{x} = \mathbf{0}$ if

(a) $\mathrm{A} = \begin{bmatrix}2 & -1 & 0\\ -1 & 2 & -1\\ 0 & -1 & 2\end{bmatrix} \qquad \mathrm{B} = \begin{bmatrix}3 & 0 & 0\\ 0 & 4 & 0\\ 0 & 0 & 3\end{bmatrix}$

(b) $\mathrm{A} = \begin{bmatrix}6 & -3 & 0\\ -3 & 6 & -3\\ 0 & -3 & 4\end{bmatrix} \qquad \mathrm{B} = \begin{bmatrix}6 & 0 & 0\\ 0 & 4 & 0\\ 0 & 0 & 4\end{bmatrix}$

(c) $\mathrm{A} = \begin{bmatrix}3 & -1 & 0\\ -1 & 1 & -1\\ 0 & -1 & 5\end{bmatrix} \qquad \mathrm{B} = \begin{bmatrix}4 & 0 & 0\\ 0 & 1 & 0\\ 0 & 0 & 4\end{bmatrix}$

In each case, verify all orthogonality relations and, using orthogonality properties, express the vector

$$\mathbf{v} = \begin{bmatrix}1\\2\\3\end{bmatrix}$$

as a linear combination of the characteristic vectors.

3. Find three solution vectors of $(\mathrm{A} - \lambda\mathrm{B})\mathbf{x} = \mathbf{0}$ which are orthonormal with respect to B if

(a) $\mathrm{A} = \begin{bmatrix}7 & 1 & -1\\ 1 & 4 & -1\\ -1 & -1 & 3\end{bmatrix} \qquad \mathrm{B} = \begin{bmatrix}6 & 0 & 0\\ 0 & 3 & 0\\ 0 & 0 & 2\end{bmatrix}$

(b) $\mathrm{A} = \begin{bmatrix}7 & -1 & -1\\ -1 & 4 & 1\\ -1 & 1 & 3\end{bmatrix} \qquad \mathrm{B} = \begin{bmatrix}6 & 0 & 0\\ 0 & 3 & 0\\ 0 & 0 & 2\end{bmatrix}$

Are the characteristic vectors orthogonal with respect to A in the respective cases? Are they orthonormal with respect to A?

4. Prove Theorem 17.

5. In Corollary 1 of Theorem 17 why is it necessary to restrict B to be positive-definite?

6. Prove that iK is hermitian if K is skew-hermitian.

7. If H is hermitian, prove that $\overline{\mathrm{U}}^T\mathrm{HU}$ is also hermitian.

8. Under what conditions, if any, is it possible for every value of λ to be a characteristic value of the equation $(\mathrm{A} - \lambda\mathrm{B})\mathbf{x} = \mathbf{0}$?

9. Show by an example that if A and B are indefinite, the characteristic values of $(\mathrm{A} - \lambda\mathrm{B})\mathbf{x} = \mathbf{0}$ need not be real even though A and B are real and symmetric.

10. Show that, if either A or B is nonsingular, then AB and BA have the same characteristic values. Hence prove that there are no matrices A and B, with either A or B nonsingular, such that AB − BA = I. (These results hold even when both A and B are singular.)

11. Show that the characteristic values of a real skew-symmetric matrix are either zero or pure imaginary.

12. Prove that if a 2 × 2 matrix has characteristic vectors which are orthogonal, it is symmetric.

13. Prove that if every characteristic value of a hermitian or real symmetric matrix is zero, the matrix is a zero matrix. Is this true if the matrix is not symmetric?

14. Show by an example that Corollary 1 of Theorem 10 is false for symmetric matrices which are not real.

15. **(a)** If $(A - \lambda B)\mathbf{x} = \mathbf{0}$, where B is a nonsingular matrix, show that $M\mathbf{x} = \lambda\mathbf{x}$ and $M^k\mathbf{x} = \lambda^k\mathbf{x}$, where $M = B^{-1}A$.

(b) If $\mathbf{c}_1 = \alpha_1\mathbf{x}_1 + \alpha_2\mathbf{x}_2 + \cdots + \alpha_n\mathbf{x}_n$ is an arbitrary linear combination of the characteristic vectors of $(A - \lambda B)\mathbf{x} = \mathbf{0}$ corresponding to the distinct characteristic values $\lambda_1, \lambda_2, \ldots, \lambda_n$, show that $M^k\mathbf{c}_1 = \alpha_1\lambda_1^k\mathbf{x}_1 + \alpha_2\lambda_2^k\mathbf{x}_2 + \cdots + \alpha_n\lambda_n^k\mathbf{x}_n$.

16. In the notation of Exercise 15, let $\mathbf{c}_{k+1} = M^k\mathbf{c}_1$ and let λ_1 be the largest of the characteristic values of $(A - \lambda B)\mathbf{x} = \mathbf{0}$. Prove that the ratios of corresponding components in $\mathbf{c}_{k+1}$ and $\mathbf{c}_k$ approach λ_1 as k becomes infinite.

17. Extend the results of Exercise 16 by showing that the ratios of successive components in $\mathbf{c}_{k+1}$ approach the ratios of the corresponding components in $\mathbf{x}_1$ as k becomes infinite.

18. Using the results of Exercises 16 and 17, approximate the largest characteristic value and the corresponding characteristic vector of the system $(A - \lambda B)\mathbf{x} = \mathbf{0}$, where

$$A = \begin{bmatrix} 3 & -1 & -2 \\ -1 & 2 & -2 \\ -2 & -2 & 6 \end{bmatrix} \quad \text{and} \quad B = \begin{bmatrix} 2 & 0 & 0 \\ 0 & 3 & 0 \\ 0 & 0 & 1 \end{bmatrix}$$

19. Work Exercise 18 for the system

$$\begin{aligned} 3x_1 - x_2 \qquad &= 4\lambda x_1 \\ -x_1 + x_2 - x_3 &= \lambda x_2 \\ -x_2 + 5x_3 &= 4\lambda x_3 \end{aligned}$$

20. If A and B are symmetric matrices, explain how the results of Exercises 16 and 17 can be extended to obtain the other characteristic values and characteristic vectors of the system $(A - \lambda B)\mathbf{x} = \mathbf{0}$. *Hint:* Having found the characteristic vector $\mathbf{x}_1$ corresponding to the largest characteristic value λ_1, show how to determine a vector $\mathbf{c}_1$ which is orthogonal to $\mathbf{x}_1$ and then consider the result of carrying out the iteration procedure which is begun with such a vector.

21. **(a)** Find the remaining characteristic values and the corresponding vectors for the system in Exercise 18.

(b) Find the remaining characteristic values and the corresponding characteristic vectors for the system in Exercise 19.

22. **(a)** Show that the characteristic equation of the matrix $\begin{bmatrix} 0 & 1 \\ -c & -b \end{bmatrix}$ is the quadratic equation $\lambda^2 + b\lambda + c = 0$.

(b) Show that the polynomial equation

$$p(\lambda) = \lambda^n + a_1\lambda^{n-1} + \cdots + a_{n-1}\lambda + a_n = 0$$

is the characteristic equation of the matrix

$$\begin{bmatrix} 0 & 1 & 0 & \cdots & 0 \\ 0 & 0 & 1 & \cdots & 0 \\ \cdots & \cdots & \cdots & \cdots & \cdots \\ 0 & 0 & 0 & \cdots & 1 \\ -a_n & -a_{n-1} & -a_{n-2} & \cdots & -a_1 \end{bmatrix}$$

This matrix is called the **companion matrix of the polynomial** p. *Hint:* Proceed inductively, beginning with the result of Part **(a)**.

23. In Exercise 22, if r_1 is a characteristic root of the companion matrix of a polynomial p, what is the corresponding characteristic vector of the companion matrix?

24. Using the results of Exercises 16 and 17, approximate the root of largest absolute value of each of the following polynomials.

(a) $x^2 - 4x + 2 = 0$

(b) $x^3 - 6x^2 + 11x - 5 = 0$

(c) $x^3 + 4x^2 + x - 5 = 0$

25. A square matrix in which the sum of the elements in each row is 1 is called a **stochastic matrix**. Show that every stochastic matrix has 1 as one of its characteristic values.

26. Show that if the characteristic roots of A are distinct and those of B are also distinct, then A and B commute if and only if A and B have the same characteristic vectors.

27. If $\mathbf{x}_1$ and $\mathbf{x}_2$ are characteristic vectors of A corresponding, respectively, to the characteristic values λ_1 and λ_2, under what conditions, if any, is $\lambda_1\mathbf{x}_1 + \lambda_2\mathbf{x}_2$ a characteristic vector of A?

28. **(a)** In Example 2, verify that the characteristic vectors are orthogonal with respect to the matrix A.

(b) Normalize the characteristic vectors with respect to B.

(c) Normalize the characteristic vectors with respect to A.

29. Determine the response of the system shown in Fig. 14.8 to the force vector $\mathbf{f} = \begin{bmatrix} 3 \\ 0 \\ -2 \end{bmatrix} \sin \omega t$.

30. Determine the behavior of the system shown in Fig. 14.8 if the force vector $\mathbf{f} = \begin{bmatrix} 3 \\ 0 \\ -2 \end{bmatrix}$ is applied to the system when it is at rest in its equilibrium position.

31. Determine the motion of the system shown in Fig. 14.8 if the initial displacement is $(\mathbf{x})_0 = \begin{bmatrix} 1 \\ 1 \\ 1 \end{bmatrix}$ and the initial velocity is $(\dot{\mathbf{x}})_0 = \begin{bmatrix} -3 \\ 1 \\ 1 \end{bmatrix}$.

32. **(a)** For the system shown in Fig. 14.9 find the natural frequencies and the normal modes and verify that the characteristic vectors are orthogonal with respect to A and to B.

(b) If the system begins to move with initial displacement $(\mathbf{x})_0 = \begin{bmatrix} 1 \\ 2 \end{bmatrix}$ and initial velocity $(\dot{\mathbf{x}})_0 = \begin{bmatrix} 2 \\ -1 \end{bmatrix}$, determine its subsequent motion.

(c) If the force vector $\mathbf{f} = \begin{bmatrix} 1 \\ 2 \end{bmatrix} \sin 3t$ is applied to

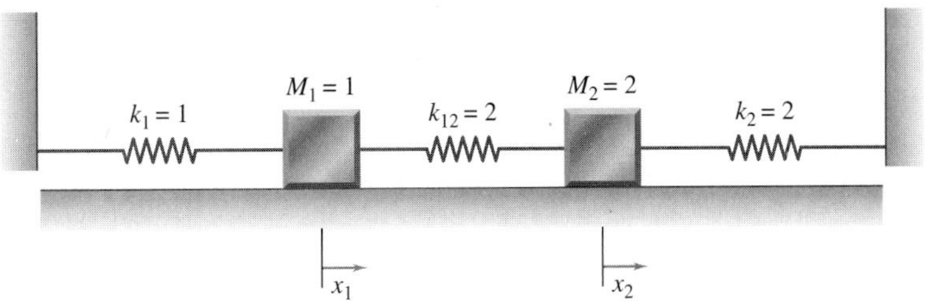

FIGURE 14.9

the system when it is at rest in its equilibrium position, determine its subsequent motion.

33. The system shown in Fig. 14.10 begins to move with initial displacement $(\mathbf{x})_0 = \begin{bmatrix} 1 \\ 1 \\ -1 \end{bmatrix}$ and initial velocity $(\dot{\mathbf{x}})_0 = \begin{bmatrix} 1 \\ 0 \\ 2 \end{bmatrix}$. Determine its subsequent motion.

34. Determine the steady-state motion produced in the system shown in Fig. 14.10 by the force vector $\mathbf{f} = \begin{bmatrix} 0 \\ 0 \\ 1 \end{bmatrix} \cos 2t$.

35. Find the natural frequencies and normal modes of the system shown in Fig. 14.11 if $k_1 = k_2 = k_3 = k_{12} = k_{23} = k_{13} = 1$ and $m_1 = m_2 = m_3 = 1$. Verify all orthogonality relations for the characteristic vectors.

36. Work Exercise 35 if $k_1 = 13$, $k_2 = k_3 = k_{12} = k_{23} = k_{13} = 1$ and $m_1 = 4$, $m_2 = m_3 = 1$.

37. Work Exercise 35 if $k_1 = 1$, $k_2 = 5$, $k_3 = 13$, $k_{12} = k_{23} = k_{13} = 1$ and $m_1 = 4$, $m_2 = 8$, $m_3 = 16$.

Exercises 35–37 show that the statement "a system with n degrees of freedom has n different natural frequencies" is not always true. Investigate this matter further by proving the following results.

38. The system shown in Fig. 14.11 will have a single natural frequency if and only if $k_{12} = k_{23} = k_{13} = 0$ and $k_1/m_1 = k_2/m_2 = k_3/m_3$, that is, if and only if it consists of three uncoupled systems having the same natural frequency.

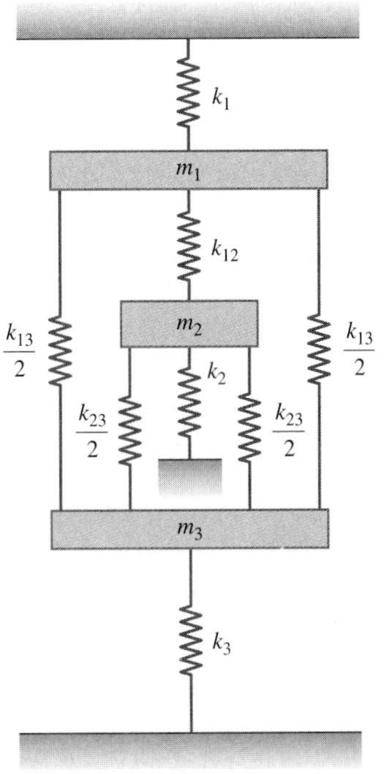

FIGURE 14.11

39. When the system shown in Fig. 14.11 does not consist of two or more uncoupled subsystems, its frequency equation will have a simple root ω_1 and a double root ω_2; that is, the system will have just two natural frequencies, if and only if

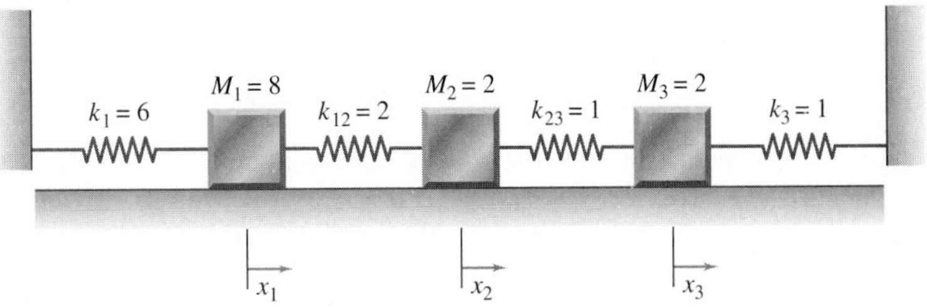

FIGURE 14.10

$$\omega_2^2 m_1 = k_1 + k_{12} + k_{13} + \frac{k_{12}k_{13}}{k_{23}}$$

$$\omega_2^2 m_2 = k_{12} + k_2 + k_{23} + \frac{k_{12}k_{23}}{k_{13}}$$

$$\omega_2^2 m_3 = k_{13} + k_{23} + k_3 + \frac{k_{13}k_{23}}{k_{12}}$$

40. When the system shown in Fig. 14.11 has a repeated frequency ω_2, the components d_1, d_2, d_3 of the normal mode corresponding to the repeated frequency satisfy the relation

$$\frac{d_1}{k_{23}} + \frac{d_2}{k_{13}} + \frac{d_3}{k_{12}} = 0$$

41. When the system shown in Fig. 14.11 has a repeated frequency, the components c_1, c_2, c_3 of the normal mode corresponding to the unrepeated natural frequency are proportional to

$$\frac{k_{12}k_{13}}{m_1} \qquad \frac{k_{12}k_{23}}{m_2} \qquad \frac{k_{13}k_{23}}{m_3}$$

42. When the system shown in Fig. 14.11 has a repeated frequency, the repeated frequency ω_2 and the unrepeated frequency ω_1 are connected by the relation

$$\omega_2^2 - \omega_1^2 = \frac{k_{12}k_{13}}{k_{23}m_1} + \frac{k_{12}k_{23}}{k_{13}m_2} + \frac{k_{13}k_{23}}{k_{12}m_3}$$

43. When the system shown in Fig. 14.11 is vibrating at a repeated frequency, the masses need not move in phase or with phrase differences of 180°. Hence there may be no time when the three masses are simultaneously at rest.

44. According to Corollary 1, Theorem 2, A and A^T have the same characteristic values. Do they therefore have the same characteristic vectors? Why?

45. If B is nonsingular, show that $(A - \lambda B)\mathbf{x} = \mathbf{0}$ and $(B^{-1}A - \lambda I)\mathbf{x} = \mathbf{0}$ have the same characteristic roots.

46. If B is nonsingular, can Theorems 12–17 be established by first multiplying $(A - \lambda B)\mathbf{x} = \mathbf{0}$ on the left by B^{-1} and then applying Theorems 1–11?

47. In Example 1, compute the ownership vectors for the next 5 years, given that the present vector is $\mathbf{x}_0 = [0.4 \ \ 0.2 \ \ 0.4]$.

14.7 THE TRANSFORMATION OF MATRICES

In previous sections we have already encountered the idea of the transformation of matrices. For instance, in Sec. 14.3 we defined equivalent matrices as matrices A and B connected by a relation, or *equivalence* transformation, of the form

$$B = PAQ \qquad P, Q \text{ nonsingular}$$

Again in Example 2, Sec. 14.5, we observed that if the variables in a quadratic form $\mathbf{x}^T A\mathbf{x}$ are subjected to a nonsingular linear transformation $\mathbf{x} = M\mathbf{y}$, then the quadratic form becomes $\mathbf{y}^T B\mathbf{y}$, where B is obtained from A by the *congruence* transformation

$$B = M^T AM \qquad M \text{ nonsingular}$$

In particular, we observed that if a nonsingular matrix M can be found with the property that $B = M^T AM$ is a diagonal matrix, then in terms of the new variables introduced by the substitution $\mathbf{x} = M\mathbf{y}$, the quadratic form $\mathbf{x}^T A\mathbf{x}$ becomes just a sum of squares. In this section we shall consider briefly the question of just when it is possible to transform an $n \times n$ matrix A into a diagonal matrix by multiplying it on the left and on the right by suitable matrices P and Q.

Because an equivalence transformation is simply a composition of elementary transformations, it is clear that for any square matrix A many pairs of nonsingular matrices P and Q can be found such that PAQ is diagonal. In other words, we have the following result, which is little more than a restatement of Theorem 8, Sec. 14.3, for square matrices.

THEOREM 1 Any square matrix is equivalent to a diagonal matrix.

COROLLARY 1 If a matrix A of rank r is equivalent to a diagonal matrix D, then D has exactly r nonzero diagonal elements.

A square matrix cannot in general be transformed into a diagonal matrix by a transformation more restricted than an equivalence transformation. However, in many important special cases this is possible, as the following theorems show.

THEOREM 2 A square matrix is congruent to a diagonal matrix if and only if it is symmetric.

PROOF The proof that if a matrix A is symmetric, then it is congruent to a diagonal matrix was essentially given in our discussion of the Lagrange reduction in Sec. 14.5. For if A is symmetric, then it can be regarded as the matrix of a quadratic form and the Lagrange reduction provides a linear transformation whose matrix Q is a nonsingular matrix with the property that Q^TAQ is diagonal.

On the other hand, if A is congruent to a diagonal matrix D, then there exists a nonsingular matrix Q such that $A = Q^TDQ$. Furthermore, if this is the case, then the transpose of A is

$$A^T = (Q^TDQ)^T = Q^TD^TQ$$

However, $D^T = D$, since any diagonal matrix is obviously symmetric. Hence

$$A^T = Q^TD^TQ = Q^TDQ = A$$

Therefore, A, being equal to its transpose, is symmetric, as asserted.

From the nature of the Lagrange reduction it is evident that a symmetric matrix can be diagonalized by a congruence transformation in many ways; and among these there is always at least one which will simultaneously diagonalize a second given matrix, provided it is positive-definite. More precisely, we have the following important theorem.

THEOREM 3 Let $\lambda_1, \lambda_2, \ldots, \lambda_n$ be the characteristic values of the equation $(A - \lambda B)\mathbf{x} = \mathbf{0}$, where A and B are hermitian (or real symmetric) $n \times n$ matrices and B is positive-definite. Let $\mathbf{x}_1, \mathbf{x}_2, \ldots, \mathbf{x}_n$ be n independent characteristic vectors corresponding to $\lambda_1, \lambda_2, \ldots, \lambda_n$; and let the $\mathbf{x}$'s be orthonormal with respect to B. Let M be the matrix whose columns are the characteristic vectors $\mathbf{x}_1, \mathbf{x}_2, \ldots, \mathbf{x}_n$; and let D be the diagonal matrix whose diagonal elements are the characteristic values $\lambda_1, \lambda_2, \ldots, \lambda_n$. Then $\overline{M}^TBM = I$ and $\overline{M}^TAM = D$ (or $M^TBM = I$ and $M^TAM = D$).

PROOF Let $\lambda_1, \lambda_2, \ldots, \lambda_n$ be the characteristic values of the equation $(A - \lambda B)\mathbf{x} = \mathbf{0}$. Whether or not there are repeated roots among the λ's, we know, from Corollary 1 of Theorem 17, Sec. 14.6, that there exists a set of characteristic vectors $\mathbf{x}_1, \mathbf{x}_2, \ldots, \mathbf{x}_n$ orthonormal with respect to B, that is, such that

$$\overline{\mathbf{x}}_i^TB\mathbf{x}_j = \begin{cases} 1 & i = j \\ 0 & i \neq j \end{cases} \tag{1}$$

Now, writing the matrix M in partitioned form, for convenience, we have

$$M = [\mathbf{x}_1 \quad \mathbf{x}_2 \quad \cdots \quad \mathbf{x}_n] \tag{2}$$

and

$$\overline{M}^TBM = \begin{bmatrix} \overline{\mathbf{x}}_1^T \\ \overline{\mathbf{x}}_2^T \\ \vdots \\ \overline{\mathbf{x}}_n^T \end{bmatrix} [B\mathbf{x}_1 \quad B\mathbf{x}_2 \quad \cdots \quad B\mathbf{x}_n] = \begin{bmatrix} \overline{\mathbf{x}}_1^TB\mathbf{x}_1 & \overline{\mathbf{x}}_1^TB\mathbf{x}_2 & \cdots & \overline{\mathbf{x}}_1^TB\mathbf{x}_n \\ \overline{\mathbf{x}}_2^TB\mathbf{x}_1 & \overline{\mathbf{x}}_2^TB\mathbf{x}_2 & \cdots & \overline{\mathbf{x}}_2^TB\mathbf{x}_n \\ \cdots & \cdots & \cdots & \cdots \\ \overline{\mathbf{x}}_n^TB\mathbf{x}_1 & \overline{\mathbf{x}}_n^TB\mathbf{x}_2 & \cdots & \overline{\mathbf{x}}_n^TB\mathbf{x}_n \end{bmatrix} = I \tag{3}$$

the last equality holding in virtue of (1).

Next, we premultiply Eq. (2) by A, replace $A\mathbf{x}_i$ by its equal $\lambda_iB\mathbf{x}_i$, $1 \leqq i \leqq n$, and then apply

Theorem 1, Sec. 3.4, to obtain

$$\begin{aligned} AM &= [A\mathbf{x}_1 \quad A\mathbf{x}_2 \quad \cdots \quad A\mathbf{x}_n] = [\lambda_1 B\mathbf{x}_1 \quad \lambda_2 B\mathbf{x}_2 \quad \cdots \quad \lambda_n B\mathbf{x}_n] \\ &= B[\lambda_1\mathbf{x}_1 \quad \lambda_2\mathbf{x}_2 \quad \cdots \quad \lambda_n\mathbf{x}_n] \\ &= B[\mathbf{x}_1 \quad \mathbf{x}_2 \quad \cdots \quad \mathbf{x}_n]\mathrm{Dg}(\lambda_1, \lambda_2, \ldots, \lambda_n) = BMD \end{aligned}$$

Therefore, by (3), $\overline{M}^T AM = \overline{M}^T(BMD) = (\overline{M}^T BM)D = D$, which is the second assertion of the theorem. ◗

COROLLARY 1 If A and B are hermitian (or real symmetric) matrices, if B is positive-definite, and if M is a matrix whose columns are characteristic vectors of $(A - \lambda B)\mathbf{x} = \mathbf{0}$ which are orthonormal with respect to B, then the substitution $\mathbf{x} = M\mathbf{y}$ simultaneously reduces the hermitian (or quadratic) forms $\overline{\mathbf{x}}^T A\mathbf{x}$ and $\overline{\mathbf{x}}^T B\mathbf{x}$ (or $\mathbf{x}^T A\mathbf{x}$ and $\mathbf{x}^T B\mathbf{x}$) to $\overline{\mathbf{y}}^T D\mathbf{y}$ and $\overline{\mathbf{y}}^T I\mathbf{y} \equiv \overline{\mathbf{y}}^T\mathbf{y}$ (or $\mathbf{y}^T D\mathbf{y}$ and $\mathbf{y}^T I\mathbf{y} \equiv \mathbf{y}^T\mathbf{y}$), respectively, where D is the diagonal matrix whose diagonal elements are the characteristic values which correspond, respectively, to the column vectors of M.

The conditions under which a square matrix can be diagonalized by a similarity transformation are contained in the next theorem.

THEOREM 4 An $n \times n$ matrix is similar to a diagonal matrix if and only if it has n linearly independent characteristic vectors.

◖ **PROOF** Let A be an $n \times n$ matrix and suppose first that A is similar to a diagonal matrix D; that is, let us suppose that there exists a matrix S such that

$$S^{-1}AS = D = [d_i\delta_{ij}] \qquad 1 \le i, j \le n \tag{4}$$

If, for convenience, we write S in the partitioned form $S = [\mathbf{s}_1 \quad \mathbf{s}_2 \quad \cdots \quad \mathbf{s}_n]$, then, premultiplying Eq. (4) by S, we have $AS = SD$, or

$$\begin{aligned} [A\mathbf{s}_1 \quad A\mathbf{s}_2 \quad \cdots \quad A\mathbf{s}_n] &= [\mathbf{s}_1 \quad \mathbf{s}_2 \quad \cdots \quad \mathbf{s}_n]\mathrm{Dg}(d_1, d_2, \ldots, d_n) \\ &= [d_1\mathbf{s}_1 \quad d_2\mathbf{s}_2 \quad \cdots \quad d_n\mathbf{s}_n] \end{aligned}$$

Hence it follows that

$$A\mathbf{s}_i = d_i\mathbf{s}_i = d_i I\mathbf{s}_i \qquad i = 1, 2, \ldots, n$$

which shows that $\mathbf{x}_i = \mathbf{s}_i$ is a characteristic vector corresponding to the characteristic value $\lambda_i = d_i$ of the equation $(A - \lambda I)\mathbf{x} = \mathbf{0}$. Thus the n columns of the transforming matrix S are characteristic vectors of the matrix A. Moreover, since the inverse of S exists, by hypothesis, S is of rank n. Hence the n columns of S are linearly independent. This proves that the matrix A has n linearly independent characteristic vectors, and the necessity assertion of the theorem is verified.

Suppose next that A has n linearly independent characteristic vectors $\mathbf{x}_1, \mathbf{x}_2, \ldots, \mathbf{x}_n$ corresponding to the (possibly repeated) characteristic values $\lambda_1, \lambda_2, \ldots, \lambda_n$. Then, by hypothesis,

$$A\mathbf{x}_i = \lambda_i I\mathbf{x}_i = \lambda_i\mathbf{x}_i \qquad i = 1, 2, \ldots, n$$

Now, let S be the matrix whose columns are the characteristic vectors $\mathbf{x}_1, \mathbf{x}_2, \ldots, \mathbf{x}_n$; that is, let S be a modal matrix of A. Then, since the characteristic vectors are independent, by hypothesis, S has rank n, S^{-1} exists, and we can write

$$\begin{aligned} S^{-1}AS &= S^{-1}A[\mathbf{x}_1 \quad \mathbf{x}_2 \quad \cdots \quad \mathbf{x}_n] = S^{-1}[A\mathbf{x}_1 \quad A\mathbf{x}_2 \quad \cdots \quad A\mathbf{x}_n] \\ &= S^{-1}[\lambda_1\mathbf{x}_1 \quad \lambda_2\mathbf{x}_2 \quad \cdots \quad \lambda_n\mathbf{x}_n] \\ &= S^{-1}[\mathbf{x}_1 \quad \mathbf{x}_2 \quad \cdots \quad \mathbf{x}_n]\mathrm{Dg}(\lambda_1, \lambda_2, \ldots, \lambda_n) = S^{-1}S\mathrm{Dg}(\lambda_1, \lambda_2, \ldots, \lambda_n) \\ &= \mathrm{Dg}(\lambda_1, \lambda_2, \ldots, \lambda_n) \end{aligned}$$

Hence A is similar to a diagonal matrix, and the sufficiency assertion of the theorem is also verified. ◗

Since every hermitian and every real symmetric matrix has n linearly independent characteristic vectors (Theorem 11, Sec. 14.6), it is clear that the last theorem contains the following important special case.

COROLLARY 1 Every hermitian and every real symmetric matrix is similar to a diagonal matrix.

Using the Schmidt process, it is clear that if a matrix A has n independent characteristic vectors, it has in fact a set of n orthonormal characteristic vectors. Moreover, as we saw in Exercise 30, Sec. 13.6, a matrix whose columns are orthonormal is an orthogonal matrix. Hence, taking the matrix S in Theorem 4 to be a matrix whose columns are orthonormal characteristic vectors of A, we have the following results.

COROLLARY 2 Every real symmetric matrix is orthogonally similar to a diagonal matrix.

COROLLARY 3 If a matrix is orthogonally similar to a diagonal matrix, it is symmetric.

By essentially the same argument, the following companion results for hermitian matrices can be established.

COROLLARY 4 Every hermitian matrix is unitarily similar to a diagonal matrix.

COROLLARY 5 If a matrix is unitarily similar to a diagonal matrix, it is hermitian.

From Theorem 4 it is clear that not every square matrix is similar to a diagonal matrix. However, in more advanced texts† it is shown that every square matrix is similar to a triangular matrix. Furthermore, it can be shown‡ that every square matrix A is similar to an ''almost diagonal'' matrix C whose diagonal elements are the characteristic values of A, whose elements immediately above the principal diagonal are either 0 or 1, and whose remaining elements are all 0. This standard form of a matrix is known as the **classical** or **Jordan canonical** form.

EXAMPLE 1

NORMAL COORDINATES OF A THREE-MASS SYSTEM

If A and B are the matrices defined in Example 2, Sec. 14.6, for the system shown in Fig. 14.8, p. 971, find the transformation which will simultaneously reduce the quadratic forms $\mathbf{x}^T\mathrm{A}\mathbf{x}$ and $\mathbf{x}^T\mathrm{B}\mathbf{x}$ to diagonal form, i.e., to a sum of squares.

Drawing on the results of Examples 2 and 3 in the last section, we first normalize the characteristic vectors of the equation $(\mathrm{A} - \lambda\mathrm{B})\mathbf{x} = \mathbf{0}$ with respect to B, getting the unit vectors

$$\mathbf{u}_1 = \frac{\mathbf{a}_1}{(\mathbf{a}_1^T\mathrm{B}\mathbf{a}_1)^{1/2}} = \frac{1}{\sqrt{96}}\begin{bmatrix} 2 \\ 3 \\ 3 \end{bmatrix} \qquad \mathbf{u}_2 = \frac{\mathbf{a}_2}{(\mathbf{a}_2^T\mathrm{B}\mathbf{a}_2)^{1/2}} = \frac{1}{\sqrt{10}}\begin{bmatrix} 1 \\ 0 \\ -1 \end{bmatrix}$$

$$\mathbf{u}_3 = \frac{\mathbf{a}_3}{(\mathbf{a}_3^T\mathrm{B}\mathbf{a}_3)^{1/2}} = \frac{1}{\sqrt{160}}\begin{bmatrix} 2 \\ -5 \\ 3 \end{bmatrix}$$

According to Theorem 3, $\mathbf{u}_1$, $\mathbf{u}_2$, and $\mathbf{u}_3$ are the respective columns of the matrix M that defines the transformation we are looking for:

$$(5) \qquad \mathbf{x} = \mathrm{M}\mathbf{y} \qquad \text{or} \qquad \begin{bmatrix} x_1 \\ x_2 \\ x_3 \end{bmatrix} = \begin{bmatrix} \frac{2}{\sqrt{96}} & \frac{1}{\sqrt{10}} & \frac{2}{\sqrt{160}} \\ \frac{3}{\sqrt{96}} & 0 & \frac{-5}{\sqrt{160}} \\ \frac{3}{\sqrt{96}} & \frac{-1}{\sqrt{10}} & \frac{3}{\sqrt{160}} \end{bmatrix} \begin{bmatrix} y_1 \\ y_2 \\ y_3 \end{bmatrix}$$

†See, for instance, L. Mirsky, *Linear Algebra,* p. 307, Oxford, New York, 1955.
‡See, for instance, R. A. Rosenbaum, *Projective Geometry and Modern Algebra,* pp. 316–327, Addison-Wesley, Reading, Mass., 1963.

It is now a routine matter of matrix multiplication to verify that under the transformation $\mathbf{x} = \mathrm{M}\mathbf{y}$, we have

$$\mathbf{x}^T\mathrm{A}\mathbf{x} \longrightarrow \mathbf{y}^T\mathrm{M}^T\mathrm{A}\mathrm{M}\mathbf{y} = \tfrac{1}{4}y_1^2 + y_2^2 + \tfrac{9}{4}y_3^2$$

and

$$\mathbf{x}^T\mathrm{B}\mathbf{x} \longrightarrow \mathbf{y}^T\mathrm{M}^T\mathrm{B}\mathrm{M}\mathbf{y} = y_1^2 + y_2^2 + y_3^2$$

Equally well, these transformations can be accomplished by substituting into

$$\mathbf{x}^T\mathrm{A}\mathbf{x} = 6x_1^2 - 6x_1x_2 + 6x_2^2 - 6x_2x_3 + 4x_3^2$$

and

$$\mathbf{x}^T\mathrm{B}\mathbf{x} = 6x_1^2 + 4x_2^2 + 4x_3^2$$

the scalar relations implied by Eq. (5), namely,

$$(6)\qquad x_1 = \frac{2y_1}{\sqrt{96}} + \frac{y_2}{\sqrt{10}} + \frac{2y_3}{\sqrt{160}} \qquad x_2 = \frac{3y_1}{\sqrt{96}} - \frac{5y_3}{\sqrt{160}} \qquad x_3 = \frac{3y_1}{\sqrt{96}} - \frac{y_2}{\sqrt{10}} + \frac{3y_3}{\sqrt{160}}$$

In the present problem it is easy to identify the two quadratic forms $\mathbf{x}^T\mathrm{A}\mathbf{x}$ and $\mathbf{x}^T\mathrm{B}\mathbf{x}$ in terms of physical concepts. In fact, since the energy stored in a spring of modulus k which has been stretched a distance s is $\frac{1}{2}ks^2$, it follows that the instantaneous potential energy of the system shown in Fig. 14.8 is

$$\tfrac{1}{2}[3x_1^2 + 3(x_2 - x_1)^2 + 3(x_3 - x_2)^2 + x_3^2] = \tfrac{1}{2}[6x_1^2 - 6x_1x_2 + 6x_2^2 - 6x_2x_3 + 4x_3^2] = \tfrac{1}{2}\mathbf{x}^T\mathrm{A}\mathbf{x}$$

Also, the kinetic energy of a mass m moving with velocity v is $\frac{1}{2}mv^2$. Hence the instantaneous kinetic energy of the system is

$$\tfrac{1}{2}(6\dot{x}_1^2 + 4\dot{x}_2^2 + 4\dot{x}_3^2) = \tfrac{1}{2}\dot{\mathbf{x}}^T\mathrm{B}\dot{\mathbf{x}}$$

From this it follows that when the system is vibrating at one of its natural frequencies ω_i, its *maximum* kinetic energy is

$$\tfrac{1}{2}\omega_i^2\mathbf{x}^T\mathrm{B}\mathbf{x}$$

Coordinates like y_1, y_2, y_3, in terms of which both the potential energy and the kinetic energy of a system are given by quadratic forms which are sums of squares, are known as **normal coordinates**. In the present problem, either from the inverse relation $\mathbf{y} = \mathrm{M}^{-1}\mathbf{x}$ or by solving Eqs. (6) simultaneously, we have

$$(7)\qquad \begin{aligned} y_1 &= \frac{\sqrt{6}}{2}(x_1 + x_2 + x_3) \\ y_2 &= \frac{2}{\sqrt{10}}(3x_1 - 2x_3) \\ y_3 &= \frac{1}{\sqrt{10}}(3x_1 - 5x_2 + 3x_3) \end{aligned}$$

The significance of the y's becomes clear when we substitute into Eqs. (7) the scalar expressions for the x's derived from the matrix equation

$$\mathbf{x} = \begin{bmatrix} x_1 \\ x_2 \\ x_3 \end{bmatrix} = c_1\begin{bmatrix} 2 \\ 3 \\ 3 \end{bmatrix}\cos\frac{t}{2} + c_2\begin{bmatrix} 1 \\ 0 \\ -1 \end{bmatrix}\cos t + c_3\begin{bmatrix} 2 \\ -5 \\ 3 \end{bmatrix}\cos\frac{3t}{2}$$

For instance, under these substitutions, y_1 becomes

$$
\begin{aligned}
y_1 = \frac{\sqrt{6}}{2}\Big(2c_1 \cos\frac{t}{2} &+ c_2 \cos t + 2c_3 \cos\frac{3t}{2} \\
+ 3c_1 \cos\frac{t}{2} &\qquad\qquad - 5c_3 \cos\frac{3t}{2} \\
+ 3c_1 \cos\frac{t}{2} &- c_2 \cos t + 3c_3 \cos\frac{3t}{2}\Big) = 4\sqrt{6}c_1 \cos\frac{t}{2}
\end{aligned}
$$

Similarly, we find that under these substitutions,

$$y_2 = \sqrt{10}c_2 \cos t \qquad \text{and} \qquad y_3 = 4\sqrt{10}c_3 \cos\frac{3t}{2}$$

Thus, unlike the natural, or ''naive'' coordinates x_1, x_2, x_3, which contain terms involving each of the natural frequencies, each of the normal coordinates varies at a single one of the natural frequencies of the system. It is highly unlikely that one would ever be able to look at a specific system and recognize its normal coordinates when the problem was being formulated. Nonetheless, the existence and properties of normal coordinates are important in theoretical discussions.

EXERCISES

1. For each of the following matrices A, find a pair of matrices (P, Q) such that PAQ is a diagonal matrix.

(a) $\begin{bmatrix} 1 & 2 \\ 3 & 4 \end{bmatrix}$ **(b)** $\begin{bmatrix} 1 & -1 \\ 0 & 3 \end{bmatrix}$

(c) $\begin{bmatrix} 1 & -1 & 1 \\ 2 & 1 & 2 \\ 0 & 1 & 3 \end{bmatrix}$ **(d)** $\begin{bmatrix} 1 & 0 & 3 \\ 1 & -1 & 1 \\ -1 & 3 & 3 \end{bmatrix}$

2. For each of the following matrices A, find two nonsingular matrices Q such that Q^TAQ is a diagonal matrix.

(a) $\begin{bmatrix} 1 & 2 \\ 2 & 3 \end{bmatrix}$ **(b)** $\begin{bmatrix} 1 & -1 \\ -1 & 0 \end{bmatrix}$

(c) $\begin{bmatrix} 1 & 1 & 1 \\ 1 & 2 & 0 \\ 1 & 0 & 3 \end{bmatrix}$ **(d)** $\begin{bmatrix} 1 & 2 & 0 \\ 2 & 5 & 2 \\ 0 & 2 & 4 \end{bmatrix}$

3. For each of the following pairs of matrices (A, B), find a congruence transformation which will simultaneously reduce A and B to diagonal form, and carry out the diagonalization.

(a) $\begin{bmatrix} 3 & -2 \\ -2 & 4 \end{bmatrix}\begin{bmatrix} 1 & 0 \\ 0 & 2 \end{bmatrix}$ **(b)** $\begin{bmatrix} 6 & 2 \\ 2 & 2 \end{bmatrix}\begin{bmatrix} 2 & 0 \\ 0 & 1 \end{bmatrix}$

(c) $\begin{bmatrix} 4 & 3 \\ 3 & 6 \end{bmatrix}\begin{bmatrix} 1 & 0 \\ 0 & 3 \end{bmatrix}$ **(d)** $\begin{bmatrix} 2 & 2 \\ 2 & 3 \end{bmatrix}\begin{bmatrix} 1 & 1 \\ 1 & 2 \end{bmatrix}$

(e) $\begin{bmatrix} 8 & -2 & 0 \\ -2 & 3 & -1 \\ 0 & -1 & 2 \end{bmatrix}\begin{bmatrix} 8 & 0 & 0 \\ 0 & 2 & 0 \\ 0 & 0 & 2 \end{bmatrix}$

(f) $\begin{bmatrix} 3 & -1 & 0 \\ -1 & 1 & -1 \\ 0 & -1 & 5 \end{bmatrix}\begin{bmatrix} 4 & 0 & 0 \\ 0 & 1 & 0 \\ 0 & 0 & 4 \end{bmatrix}$

(g) $\begin{bmatrix} 6 & 0 & 2 \\ 0 & 6 & -4 \\ 2 & -4 & 6 \end{bmatrix}\begin{bmatrix} 3 & 1 & 1 \\ 1 & 3 & -1 \\ 1 & -1 & 3 \end{bmatrix}$

(h) $\begin{bmatrix} 7 & -1 & 0 \\ -1 & 11 & -4 \\ 0 & -4 & 10 \end{bmatrix}\begin{bmatrix} 6 & -2 & -1 \\ -2 & 10 & -5 \\ -1 & -5 & 9 \end{bmatrix}$

4. **(a)** If A and B are hermitian (or real symmetric) matrices, show that there exist congruence transformations which will simultaneously diagonalize A and B even though B is not definite.

(b) Find a congruence transformation which will simultaneously diagonalize $\begin{bmatrix} -2 & 1 \\ 1 & 1 \end{bmatrix}$ and $\begin{bmatrix} 1 & 0 \\ 0 & -2 \end{bmatrix}$.

(c) Find a congruence transformation which will simultaneously diagonalize $\begin{bmatrix} -3 & 3 \\ 3 & 0 \end{bmatrix}$ and $\begin{bmatrix} -7 & 5 \\ 5 & -1 \end{bmatrix}$.

5. Find a similarity transformation which will reduce each of the following matrices to diagonal form.

(a) $\begin{bmatrix} -3 & 2 \\ -10 & 6 \end{bmatrix}$ **(b)** $\begin{bmatrix} 0 & -2 \\ -2 & 0 \end{bmatrix}$ **(c)** $\begin{bmatrix} 2 & 1 \\ 2 & 1 \end{bmatrix}$

(d) $\begin{bmatrix} 5 & -2 & -1 \\ -1 & 4 & -1 \\ 1 & -2 & 3 \end{bmatrix}$ **(e)** $\begin{bmatrix} 2 & -3 & 3 \\ 0 & 3 & -1 \\ 0 & -1 & 3 \end{bmatrix}$

(f) $\begin{bmatrix} 3 & -2 & -2 \\ -1 & 2 & 0 \\ 1 & -1 & 1 \end{bmatrix}$

6. Find the normal coordinates for the system shown in Fig. 14.9. *Hint:* Recall Exercise 32, Sec. 14.6.

7. Find the normal coordinates for the system shown in Fig. 14.4. *Hint:* Recall Exercise 33, Sec. 14.3.

8. If the behavior of a system is described by the vector equation

$$\mathbf{x} = c_1\mathbf{a}_1 \cos \omega_1 t + c_2\mathbf{a}_2 \cos \omega_2 t + c_3\mathbf{a}_3 \cos \omega_3 t$$

where the **a**'s are the characteristic vectors and the c's are arbitrary constants, explain how the normal coordinates can be obtained from the three scalar equations implied by the given vector equation. *Hint:* Recall that each normal coordinate must vary with a single natural frequency.

14.8 FUNCTIONS OF A SQUARE MATRIX

In Sec. 3.4 we were able to define positive integral powers of a square matrix A and to verify that for arbitrary positive integers r and s,

$$\text{(1)} \qquad A^rA^s = A^sA^r = A^{r+s}$$

Moreover, we verified in Sec. 3.7 that if A is a nonsingular matrix, it has an inverse A^{-1} such that $AA^{-1} = A^{-1}A = I$, and we defined negative integral powers of A by the relation

$$A^{-n} = (A^{-1})^n$$

Thus, after we introduced the definition $A^0 = I$, it became clear that for any nonsingular matrix A, Eq. (1) holds for all integral values of r and s. It is now natural to define polynomial functions of a square matrix and, if possible, rational fractional functions.

DEFINITION 1 A **polynomial function** of a square matrix A is a finite linear combination of nonnegative integral powers of A, including $A^0 = I$.

EXAMPLE 1

If $A = \begin{bmatrix} 1 & 2 \\ 3 & -4 \end{bmatrix}$ and $p(x) = x^2 + 5x + 4$, then

$$p(A) = A^2 + 5A + 4I = \begin{bmatrix} 7 & -6 \\ -9 & 22 \end{bmatrix} + 5\begin{bmatrix} 1 & 2 \\ 3 & -4 \end{bmatrix} + 4\begin{bmatrix} 1 & 0 \\ 0 & 1 \end{bmatrix} = \begin{bmatrix} 16 & 4 \\ 6 & 6 \end{bmatrix}$$

It is interesting to note that $p(A)$ can also be evaluated by using the factored forms of $p(x)$, namely,

$$p(x) = (x + 4)(x + 1) = (x + 1)(x + 4)$$

$$p(A) = (A + 4I)(A + I) = \left(\begin{bmatrix} 1 & 2 \\ 3 & -4 \end{bmatrix} + 4\begin{bmatrix} 1 & 0 \\ 0 & 1 \end{bmatrix}\right)\left(\begin{bmatrix} 1 & 2 \\ 3 & -4 \end{bmatrix} + \begin{bmatrix} 1 & 0 \\ 0 & 1 \end{bmatrix}\right)$$

$$= \begin{bmatrix} 5 & 2 \\ 3 & 0 \end{bmatrix}\begin{bmatrix} 2 & 2 \\ 3 & -3 \end{bmatrix} = \begin{bmatrix} 16 & 4 \\ 6 & 6 \end{bmatrix}$$

$$p(A) = (A + I)(A + 4I) = \left(\begin{bmatrix} 1 & 2 \\ 3 & -4 \end{bmatrix} + \begin{bmatrix} 1 & 0 \\ 0 & 1 \end{bmatrix}\right)\left(\begin{bmatrix} 1 & 2 \\ 3 & -4 \end{bmatrix} + 4\begin{bmatrix} 1 & 0 \\ 0 & 1 \end{bmatrix}\right)$$

$$= \begin{bmatrix} 2 & 2 \\ 3 & -3 \end{bmatrix}\begin{bmatrix} 5 & 2 \\ 3 & 0 \end{bmatrix} = \begin{bmatrix} 16 & 4 \\ 6 & 6 \end{bmatrix}$$

In this example it is of course not clear, especially in view of the noncommutative character of matrix multiplication, whether the fact that $p(A)$ can be computed equally well from

$$A^2 + 5A + 4I \qquad (A + 4I)(A + I) \qquad \text{or} \qquad (A + I)(A + 4I)$$

is a result of some special property of A and $p(x)$ or is illustrative of some general principle. Actually the latter is the case. In fact, any identical relation involving sums and products of scalar polynomials is valid for the corresponding matrix polynomials, as the following important, but almost obvious, theorem assures us.

THEOREM 1 Any polynomial identity between scalar polynomials implies a corresponding identity for matrix polynomials.

PROOF Clearly, any polynomial relation between scalar polynomials can be constructed using only the operations of addition and multiplication. For instance,

$$[f_1(x)f_2(x) + f_3(x)]f_4(x) = f_5(x)$$

is completely equivalent to the chain of relations

$$\phi(x)f_4(x) = f_5(x) \qquad \phi(x) = \Psi(x) + f_3(x) \qquad \Psi(x) = f_1(x)f_2(x)$$

Hence to prove the theorem it is sufficient to show that for any polynomials f, g, s, p and any square matrix A,

(a) If $f(x) + g(x) = s(x)$, then $f(A) + g(A) = s(A)$.

(b) If $f(x)g(x) = p(x)$, then $f(A)g(A) = p(A)$.

To prove the first of these, let

$$f(x) = \sum_{i=0}^{m} a_i x^i \qquad g(x) = \sum_{i=0}^{n} b_i x^i \qquad s(x) = \sum_{i=0}^{t} c_i x^i$$

where $t = \max(m, n)$, $c_i = a_i + b_i$, and the coefficients of any powers of x not present are understood to be zero. Then

$$f(\text{A}) + g(\text{A}) = \sum_{i=0}^{m} a_i \text{A}^i + \sum_{i=0}^{n} b_i \text{A}^i = \sum_{i=0}^{t} (a_i + b_i)\text{A}^i = \sum_{i=0}^{t} c_i \text{A}^i = s(\text{A})$$

as asserted.

To prove Part b, let

$$f(x) = \sum_{i=0}^{m} a_i x^i \qquad g(x) = \sum_{j=0}^{n} b_j x^j \qquad p(x) = \sum_{k=0}^{t} c_k x^k$$

where $t = m + n$ and $c_k = \Sigma_{i,j}\, a_i b_j$, the summation extending over all values of i and j such that $i + j = k$ and of course $0 \le i \le m$, $0 \le j \le n$. Then, using the distributive property of matrix multiplication and the associative and commutative properties of matrix addition, we have

$$f(A)g(A) = \left(\sum_{i=0}^{m} a_i \text{A}^i\right)\left(\sum_{j=0}^{n} b_j \text{A}^j\right) = \sum_{i=0}^{m}\sum_{j=0}^{n} (a_i\text{A}^i)(b_j\text{A}^j) = \sum_{i=0}^{m}\sum_{j=0}^{n} a_i b_j \text{A}^{i+j}$$

or, grouping together all terms involving the same power of A,

$$f(\text{A})g(\text{A}) = \sum_{k=0}^{m+n} c_k \text{A}^k \qquad \text{where } k = i + j \text{ and } c_k = \sum_{i,j} a_i b_j$$

$$= p(A)$$

as asserted.

Since $f(x)g(x) = g(x)f(x)$, it follows from Theorem 1 that

$$f(A)g(A) = g(A)f(A) \tag{2}$$

In other words, we have the following result.

COROLLARY 1 Any two polynomials in a matrix A commute with each other.

If $g(A)$ is a nonsingular matrix, then $g^{-1}(A)$ exists and we can premultiply and postmultiply each side of Eq. (2) by $g^{-1}(A)$, getting

$$g^{-1}(A)f(A)g(A)g^{-1}(A) = g^{-1}(A)g(A)f(A)g^{-1}(A)$$

or

$$g^{-1}(A)f(A) = f(A)g^{-1}(A) \tag{3}$$

With this identity, we are now in a position to define rational fractional functions of a square matrix A.

DEFINITION 2 If $f(x)$ and $g(x)$ are scalar polynomials and if A is a square matrix such that $g(A)$ is nonsingular, then either of the equal matrices $g^{-1}(A)f(A)$ and $f(A)g^{-1}(A)$ is called the **quotient** of $f(A)$ by $g(A)$ and is written $f(A)/g(A)$.

It is now relatively easy to prove the following extension of Theorem 1 (see Exercise 3).

THEOREM 2 Any identity between rational fractional functions of a scalar variable implies a corresponding matrix identity, provided all the matrix functions are defined.

With rational functions of a square matrix now defined, it is natural to ask whether the characteristic values of a rational function of a square matrix A can be expressed in terms of the characteristic values of A. This is indeed the case, as the following chain of theorems makes clear.

THEOREM 3 If $\lambda_1, \lambda_2, \ldots, \lambda_n$ are the (possibly repeated) characteristic values of a square matrix A, and if f is any polynomial, then the determinant of $f(A)$ is given by the formula

$$|f(A)| = f(\lambda_1)f(\lambda_2)\cdots f(\lambda_n)$$

PROOF Let the factored form of the characteristic polynomial of the given matrix A be

$$|A - \lambda I| = \prod_{i=1}^{n} (\lambda_i - \lambda) \tag{4}$$

and let the factored form of the given polynomial f be

$$f(t) = c(t - r_1)(t - r_2)\cdots(t - r_k) \tag{5}$$

Then, since Theorem 1 assures us that identities between scalar polynomials imply corresponding matrix identities, we have

$$f(A) = c(A - r_1 I)(A - r_2 I)\cdots(A - r_k I)$$

Furthermore, since the determinant of a product of square matrices is equal to the product of the determinants of the matrix factors and since the scalar factor c incorporated into any one of the matrix factors reappears as the factor c^n in the determinant of that matrix, we have

$$|f(A)| = c^n|A - r_1 I| \cdot |A - r_2 I|\cdots|A - r_k I| = c^n \prod_{j=1}^{k} |A - r_j I| \tag{6}$$

However, $|A - r_j I|$ is just the characteristic polynomial of A evaluated for $\lambda = r_j$. Hence, by (4),

$$|A - r_j I| = \prod_{i=1}^{n} (\lambda_i - r_j)$$

and therefore, substituting into (6), we have

$$|f(A)| = c^n \prod_{j=1}^{k} \prod_{i=1}^{n} (\lambda_i - r_j)$$

Next, interchanging the order in which the products are formed, by first grouping together all the factors corresponding to a given value of i and assigning a single factor c to each such group, we have

$$|f(A)| = c^n \prod_{i=1}^{n} \prod_{j=1}^{k} (\lambda_i - r_j) = \prod_{i=1}^{n} \left[c \prod_{j=1}^{k} (\lambda_i - r_j) \right]$$

Finally we observe that, with the coefficient c, the inner product in the last expression is precisely the evaluation of the factored form (5) of the given polynomial for $t = \lambda_i$. Hence

$$|f(A)| = \prod_{i=1}^{n} f(\lambda_i)$$

as asserted. ◗

THEOREM 4 If $\lambda_1, \lambda_2, \ldots, \lambda_n$ are the characteristic values of a square matrix A, if $f = g/h$ is a rational fractional function, and if $h(A)$ is nonsingular, then the determinant of $f(A)$ is given by the formula

$$|f(A)| = f(\lambda_1)f(\lambda_2)\cdots f(\lambda_n)$$

◖ **PROOF** Since, by definition, $f(A) = g(A)/h(A) = g(A)h^{-1}(A)$ and since the determinant of a product of square matrices is equal to the product of the determinants of the matrix factors, we have

$$|f(A)| = |g(A)h^{-1}(A)| = |g(A)| \cdot |h^{-1}(A)|$$

Moreover, as we observed in Sec. 3.7, $|h^{-1}(A)| = 1/|h(A)|$. Therefore

$$|f(A)| = \frac{|g(A)|}{|h(A)|}$$

However, by Theorem 3, since g and h are polynomials,

$$|g(A)| = g(\lambda_1)g(\lambda_2)\cdots g(\lambda_n) \qquad \text{and} \qquad |h(A)| = h(\lambda_1)h(\lambda_2)\cdots h(\lambda_n)$$

Hence

$$\begin{aligned} |f(A)| &= \frac{g(\lambda_1)g(\lambda_2)\cdots g(\lambda_n)}{h(\lambda_1)h(\lambda_2)\cdots h(\lambda_n)} \\ &= f(\lambda_1)f(\lambda_2)\cdots f(\lambda_n) \end{aligned}$$

as asserted. ◗

THEOREM 5 If $\lambda_1, \lambda_2, \ldots, \lambda_n$ are the characteristic values of a square matrix A, and if $f = g/h$, where g and h are polynomials such that $|h(\mathrm{A})| \neq 0$, then the characteristic values of $f(\mathrm{A})$ are $f(\lambda_1), f(\lambda_2), \ldots, f(\lambda_n)$.

PROOF Let $\phi(x)$ be the function defined by the expression

$$\phi(x) = f(x) - \lambda = \frac{g(x)}{h(x)} - \lambda = \frac{g(x) - \lambda h(x)}{h(x)}$$

Clearly, $g(x) - \lambda h(x)$ is a polynomial, and therefore $\phi(x)$ is a rational fractional function of x. Hence, by the last theorem,

$$|\phi(\mathrm{A})| = \phi(\lambda_1)\phi(\lambda_2)\cdots\phi(\lambda_n)$$

In other words, for all values of λ,

$$|f(\mathrm{A}) - \lambda \mathrm{I}| = [f(\lambda_1) - \lambda][f(\lambda_2) - \lambda]\cdots[f(\lambda_n) - \lambda]$$

The right-hand side of this identity is thus the factored form of the characteristic polynomial $|f(\mathrm{A}) - \lambda \mathrm{I}|$. Hence, by inspection, the roots of the characteristic equation of $f(\mathrm{A})$ are

$$\lambda = f(\lambda_1), f(\lambda_2), \ldots, f(\lambda_n)$$

as asserted.

COROLLARY 1 If the characteristic values of a square matrix A are $\lambda_1, \lambda_2, \ldots, \lambda_n$, then for all integral values of k if A is nonsingular and for all nonnegative integral values of k if A is singular, the characteristic values of A^k are $\lambda_1^k, \lambda_2^k, \ldots, \lambda_n^k$.

In view of Theorem 5, one might expect that not only the characteristic values but also the characteristic vectors of $f(\mathrm{A})$ would depend in some way on the function f. This is not the case, however, and it is not difficult (see Exercise 8) to prove the following rather surprising result.

COROLLARY 2 If $\mathbf{x}_i$ is a characteristic vector corresponding to the characteristic value λ_i of a square matrix A, and if p is a polynomial, then $\mathbf{x}_i$ is also a characteristic vector corresponding to the characteristic value $p(\lambda_i)$ of the matrix $p(\mathrm{A})$.

EXAMPLE 2

As an illustration of Theorem 5, consider the matrix $\mathrm{A} = \begin{bmatrix} 1 & -2 \\ 3 & -4 \end{bmatrix}$ and the function $\phi(x) = x/(x+3)$. The characteristic equation of A is

$$|\mathrm{A} - \lambda \mathrm{I}| = \begin{vmatrix} 1-\lambda & -2 \\ 3 & -4-\lambda \end{vmatrix} = \lambda^2 + 3\lambda + 2 = 0$$

Hence the characteristic roots are $\lambda = -1, -2$. Therefore, according to Theorem 5, the characteristic roots of $\phi(\mathrm{A})$ are

$$\phi(-1) = -\tfrac{1}{2} \qquad \text{and} \qquad \phi(-2) = -2$$

To confirm this, we have, by direct calculation,

$$\phi(\mathrm{A}) = \frac{\mathrm{A}}{\mathrm{A} + 3\mathrm{I}} = \mathrm{A}(\mathrm{A} + 3\mathrm{I})^{-1} = \begin{bmatrix} 1 & -2 \\ 3 & -4 \end{bmatrix}\begin{bmatrix} 4 & -2 \\ 3 & -1 \end{bmatrix}^{-1}$$

$$= \begin{bmatrix} 1 & -2 \\ 3 & -4 \end{bmatrix} \frac{1}{2}\begin{bmatrix} -1 & 2 \\ -3 & 4 \end{bmatrix}$$

$$= \frac{1}{2}\begin{bmatrix} 5 & -6 \\ 9 & -10 \end{bmatrix} = \begin{bmatrix} \frac{5}{2} & -3 \\ \frac{9}{2} & -5 \end{bmatrix}$$

The characteristic roots of $\phi(A)$ are therefore the roots of the equation

$$|\phi(A) - \lambda I| = \begin{vmatrix} \frac{5}{2} - \lambda & -3 \\ \frac{9}{2} & -5 - \lambda \end{vmatrix} = \lambda^2 + \tfrac{5}{2}\lambda + 1 = 0$$

or $-\frac{1}{2}$ and -2, as before.

If p is a polynomial and A is a square matrix, the evaluation of $p(A)$ is a perfectly straightforward matter. However, when A is a matrix similar to a diagonal matrix, the evaluation of $p(A)$ can be appreciably simplified. To establish the result upon which this simplification is based, it is convenient first to prove the following lemmas.

LEMMA 1 If $B = SAS^{-1}$, then $B^n = SA^nS^{-1}$.

PROOF Clearly, the lemma is true for $n = 2$, since

$$B^2 = (SAS^{-1})(SAS^{-1}) = SA(S^{-1}S)AS^{-1} = SA^2S^{-1}$$

Assuming, then, that the lemma is true for $n = k$, we have

$$B^{k+1} = BB^k = (SAS^{-1})(SA^kS^{-1}) = SA(S^{-1}S)A^kS^{-1} = SA^{k+1}S^{-1}$$

which completes the induction and establishes the lemma.

If we now apply Lemma 1 to each term of any polynomial function of B and then use the distributive property of matrix multiplication, we obtain the following result.

LEMMA 2 If $B = SAS^{-1}$ and if p is a polynomial, then $p(B) = Sp(A)S^{-1}$, that is, $p(SAS^{-1}) = Sp(A)S^{-1}$.

Furthermore, by another easy induction we can establish the following property of diagonal matrices.

LEMMA 3 If $D = Dg(d_1, d_2, \ldots, d_n)$, then $D^k = Dg(d_1^k, d_2^k, \ldots, d_n^k)$.

Finally, by applying Lemma 3 to each term of any polynomial function of a diagonal matrix D and then using the definition of matrix addition, we have the following result.

LEMMA 4 If $D = Dg(d_1, d_2, \ldots, d_n)$ and p is any polynomial, then $p(D) = Dg\,[p(d_1), p(d_2), \ldots, p(d_n)]$.

Using Lemmas 2 and 4, we can now prove the following theorem.

THEOREM 6 If A is a matrix similar to a diagonal matrix, that is, if

$$S^{-1}AS = D = \begin{bmatrix} \lambda_1 & & \mathbf{O} \\ & \lambda_2 & \\ \mathbf{O} & & \lambda_n \end{bmatrix}$$

where $\lambda_1, \lambda_2, \ldots, \lambda_n$ are the characteristic values of A, then

$$p(A) = S\begin{bmatrix} p(\lambda_1) & & \mathbf{O} \\ & p(\lambda_2) & \\ \mathbf{O} & & p(\lambda_n) \end{bmatrix}S^{-1}$$

PROOF By Lemma 4, $p(D) = Dg[p(\lambda_1), p(\lambda_2), \ldots, p(\lambda_n)]$. Also, since $S^{-1}AS = D$, it follows that $A = SDS^{-1}$. Hence, using Lemma 2, we have

$$SDg[p(\lambda_1), p(\lambda_2), \ldots, p(\lambda_n)]S^{-1} = Sp(D)S^{-1} = p(SDS^{-1}) = p(A)$$

as asserted.

EXAMPLE 3

EXISTENCE OF A STEADY-STATE OWNERSHIP VECTOR

In Example 1, Sec. 14.6, we found the steady-state solution vector of a problem in consumer preferences defined by the matrix equation

$$\mathrm{A}\mathbf{x} = \mathbf{y} \qquad \text{where } \mathrm{A} = \begin{bmatrix} 0.7 & 0.3 & 0.1 \\ 0.1 & 0.6 & 0.1 \\ 0.2 & 0.1 & 0.8 \end{bmatrix}$$

However, we did not investigate how $\mathbf{x}$ varies from year to year as its limiting value is approached. With Theorem 6 available, we can now answer this question. If $\mathbf{x}_0$ is the ownership vector at a given time t_0, then a year later the ownership vector will be

$$\mathbf{x}_1 = \mathrm{A}\mathbf{x}_0$$

Similarly, after a second year the ownership vector will be

$$\mathbf{x}_2 = \mathrm{A}\mathbf{x}_1 = \mathrm{A}(\mathrm{A}\mathbf{x}_0) = \mathrm{A}^2\mathbf{x}_0$$

The iterative continuation is obvious, and after n years the ownership vector is

$$\mathbf{x}_n = \mathrm{A}^n\mathbf{x}_0 \tag{7}$$

Thus our problem is to compute the nth power of A.

Taking $p(\mathrm{A})$ to be A^n in Theorem 6, we have

$$\mathrm{A}^n = \mathrm{S}\begin{bmatrix} \lambda_1^n & 0 & 0 \\ 0 & \lambda_2^n & 0 \\ 0 & 0 & \lambda_3^n \end{bmatrix}\mathrm{S}^{-1}$$

where the λ's are the characteristic values of A, and S is a modal matrix of A, that is, a matrix whose columns are linearly independent characteristic vectors of A. Using the results of Example 1, Sec. 14.6, we thus have

$$\mathrm{A}^n = \begin{bmatrix} 7 & 1 & 2 \\ 4 & 0 & -1 \\ 9 & -1 & -1 \end{bmatrix}\begin{bmatrix} 1 & 0 & 0 \\ 0 & (0.6)^n & 0 \\ 0 & 0 & (0.5)^n \end{bmatrix}\left(\frac{-1}{20}\begin{bmatrix} -1 & -1 & -1 \\ -5 & -25 & 15 \\ -4 & 16 & -4 \end{bmatrix}\right)$$

With this value for A^n, Eq. (7) gives the value of $\mathbf{x}_n$ for any initial vector $\mathbf{x}_0$.

It is interesting to verify that as n increases indefinitely, $\mathrm{A}^n\mathbf{x}_0$ does approach the solution we obtained in Example 1, Sec. 14.6. Clearly, as $n \to \infty$, $(0.6)^n$ and $(0.5)^n$ approach zero; hence we need only evaluate the product

$$\begin{bmatrix} (x_1)_\infty \\ (x_2)_\infty \\ (x_3)_\infty \end{bmatrix} = \begin{bmatrix} 7 & 1 & 2 \\ 4 & 0 & -1 \\ 9 & -1 & -1 \end{bmatrix}\begin{bmatrix} 1 & 0 & 0 \\ 0 & 0 & 0 \\ 0 & 0 & 0 \end{bmatrix}\left(\frac{-1}{20}\begin{bmatrix} -1 & -1 & -1 \\ -5 & -25 & 15 \\ -4 & 16 & -4 \end{bmatrix}\right)\begin{bmatrix} (x_1)_0 \\ (x_2)_0 \\ (x_3)_0 \end{bmatrix}$$

Routine multiplication yields

$$(x_1)_\infty = \tfrac{7}{20}\{(x_1)_0 + (x_2)_0 + (x_3)_0\}$$

$$(x_2)_\infty = \tfrac{4}{20}\{(x_1)_0 + (x_2)_0 + (x_3)_0\}$$

$$(x_3)_\infty = \tfrac{9}{20}\{(x_1)_0 + (x_2)_0 + (x_3)_0\}$$

which confirms our earlier conclusion that eventually 35 percent of those surveyed will own a car of make C_1, 20 percent will own a car of make C_2, and 45 percent will own a car of make C_3.

After polynomial functions of a square matrix have been defined, it is natural to consider polynomial equations in a matrix variable. In particular, now that we have developed procedures for evaluating $p(A)$, that is, solving the equation $p(A) = X$, we shall consider the problem of solving the nontrivial equation $p(X) = A$, where p is a given polynomial, A is a given square matrix, and X is a matrix variable. By means of examples (see Exercise 1) it is easy to show that there are polynomial equations $p(X) = A$ which have no solution. In one important case, however, the equation $p(X) = A$ can always be solved, as the following theorem makes clear.

THEOREM 7 If A is similar to a diagonal matrix, and if p is a scalar polynomial, then the equation $p(X) = A$ is solvable for X.

PROOF By hypothesis, since A is similar to a diagonal matrix D, there exists a nonsingular matrix S with the property that $S^{-1}AS = D$, or $A = SDS^{-1}$, where, say, $D = Dg(d_1, d_2, \ldots, d_n)$. Now let r_i be any root of the equation $p(x) = d_i$. Then if $R = Dg(r_1, r_2, \ldots, r_n)$ and $X = SRS^{-1}$, we have, by Lemma 2,

$$p(X) = p(SRS^{-1}) = Sp(R)S^{-1}$$

Moreover, by Lemma 4 and the fact that $p(r_i) = d_i$,

$$p(R) = Dg[p(r_1), p(r_2), \ldots, p(r_n)] = Dg(d_1, d_2, \ldots, d_n) = D$$

Therefore $p(X) = Sp(R)S^{-1} = SDS^{-1} = A$, which proves that if A is similar to a diagonal matrix, then $p(X) = A$ has the solution $X = SRS^{-1}$. If the polynomial p is of degree k, the scalar equation $p(x) = d_i$ has, in general, k distinct roots. Hence there are k distinct choices for each of the diagonal elements in R, and therefore $p(X) = A$ has, in general, at least k^n different solutions.

By applying the preceding theorem to the particular equation $X^2 = A$, we obtain the following corollary.

COROLLARY 1 An $n \times n$ matrix with distinct characteristic values has 2^n or 2^{n-1} distinct square roots, according as it is nonsingular or singular.

PROOF Let A be an $n \times n$ matrix with n distinct characteristic values $\lambda_1, \lambda_2, \ldots, \lambda_n$. It follows, then, by Theorem 4, Sec. 14.7, that there exists a nonsingular matrix S such that $A = SDg(\lambda_1, \lambda_2, \ldots, \lambda_n)S^{-1}$. Thus according to the last theorem, for any choice of plus and minus signs $X = SDg(\pm\sqrt{\lambda_1}, \pm\sqrt{\lambda_2}, \ldots, \pm\sqrt{\lambda_n})S^{-1}$ satisfies the equation $X^2 = A$. If A is nonsingular, none of the λ's is zero and there are 2^n combinations of signs each leading to a different matrix X satisfying $X^2 = A$. On the other hand, if A is singular but still has distinct characteristic values, then, by Theorem 1, Sec. 14.6, one and only one of the λ's must be zero, and therefore for one of the diagonal elements there is only a single choice rather than two. Hence in this case there can be only 2^{n-1} distinct square roots, as asserted.

EXAMPLE 4

Solve the equation $X^2 - 4X + 4I = \begin{bmatrix} 4 & 3 \\ 5 & 6 \end{bmatrix}$.

The characteristic equation of the matrix $A = \begin{bmatrix} 4 & 3 \\ 5 & 6 \end{bmatrix}$ is

$$\begin{vmatrix} 4-\lambda & 3 \\ 5 & 6-\lambda \end{vmatrix} = \lambda^2 - 10\lambda + 9 = 0$$

Hence the characteristic values of A are $\lambda_1 = 1$ and $\lambda_2 = 9$, and corresponding characteristic vectors are $\mathbf{x}_1 = \begin{bmatrix} 1 \\ -1 \end{bmatrix}$, $\mathbf{x}_2 = \begin{bmatrix} 3 \\ 5 \end{bmatrix}$. Therefore, by Theorem 4, Sec. 14.7, A is similar to a diagonal matrix; that is,

$$S^{-1}AS = D \qquad \text{or} \qquad A = SAS^{-1}$$

where S is the modal matrix $\begin{bmatrix} 1 & 3 \\ -1 & 5 \end{bmatrix}$, $S^{-1} = \frac{1}{8}\begin{bmatrix} 5 & -3 \\ 1 & 1 \end{bmatrix}$, and $D = \begin{bmatrix} 1 & 0 \\ 0 & 9 \end{bmatrix}$. We must now solve the equations $p(x) = d_i$ for r_i, $i = 1, 2$,

$$\begin{array}{cc} x^2 - 4x + 4 = d_1 = 1 & x^2 - 4x + 4 = d_2 = 9 \\ x = r_1 = 1, 3 & x = r_2 = -1, 5 \end{array}$$

Pairing each possibility for r_1 with each possibility for r_2, we thus obtain four possibilities for the matrix R,

$$R_1 = \begin{bmatrix} 1 & 0 \\ 0 & -1 \end{bmatrix} \quad R_2 = \begin{bmatrix} 1 & 0 \\ 0 & 5 \end{bmatrix} \quad R_3 = \begin{bmatrix} 3 & 0 \\ 0 & -1 \end{bmatrix} \quad R_4 = \begin{bmatrix} 3 & 0 \\ 0 & 5 \end{bmatrix}$$

Then according to Theorem 7, the solutions of the given equation are

$$X_1 = SR_1S^{-1} = \begin{bmatrix} 1 & 3 \\ -1 & 5 \end{bmatrix}\begin{bmatrix} 1 & 0 \\ 0 & -1 \end{bmatrix}\frac{1}{8}\begin{bmatrix} 5 & -3 \\ 1 & 1 \end{bmatrix} = \frac{1}{4}\begin{bmatrix} 1 & -3 \\ -5 & -1 \end{bmatrix}$$

and, similarly,

$$X_2 = SR_2S^{-1} = \frac{1}{2}\begin{bmatrix} 5 & 3 \\ 5 & 7 \end{bmatrix} \qquad X_3 = SR_3S^{-1} = \frac{1}{2}\begin{bmatrix} 3 & -3 \\ -5 & 1 \end{bmatrix}$$

and

$$X_4 = SR_4S^{-1} = \frac{1}{4}\begin{bmatrix} 15 & 3 \\ 5 & 17 \end{bmatrix}$$

EXERCISES

1. Prove that there is no matrix satisfying the equation $X^2 = \begin{bmatrix} 0 & 1 \\ 0 & 0 \end{bmatrix}$.

2. Show that for particular polynomials p and particular matrices A, each of the following cases is possible.
(a) A nonsingular, $p(A)$ nonsingular
(b) A nonsingular, $p(A)$ singular
(c) A singular, $p(A)$ nonsingular
(d) A singular, $p(A)$ singular

3. Prove Theorem 2. *Hint:* Note first that it is sufficient to prove that

$$\frac{f_1(x)}{f_2(x)} + \frac{f_3(x)}{f_4(x)} = \frac{f_5(x)}{f_6(x)} \quad \text{implies} \quad \frac{f_1(A)}{f_2(A)} + \frac{f_3(A)}{f_4(A)} = \frac{f_5(A)}{f_6(A)}$$

$$\frac{f_1(x)}{f_2(x)}\frac{f_3(x)}{f_4(x)} = \frac{f_5(x)}{f_6(x)} \quad \text{implies} \quad \frac{f_1(A)}{f_2(A)}\frac{f_3(A)}{f_4(A)} = \frac{f_5(A)}{f_6(A)}$$

$$\frac{f_1(x)/f_2(x)}{f_3(x)/f_4(x)} = \frac{f_5(x)}{f_6(x)} \quad \text{implies} \quad \frac{f_1(A)/f_2(A)}{f_3(A)/f_4(A)} = \frac{f_5(A)}{f_6(A)}$$

Then, clear of fractions in the scalar identities, use Theorem 1, and multiply the resulting identities by the appropriate inverses.

4. If A is a diagonal matrix and p is any scalar polynomial, show that $p(A)$ is also a diagonal matrix.

5. If A is a diagonal matrix and f is a rational fractional function, is $f(A)$ necessarily a diagonal matrix?

6. Show that I_2 has infinitely many distinct square roots.

7. Prove that an $n \times n$ matrix with distinct characteristic values has no square roots other than those identified by Corollary 1, Theorem 7. *Hint:* Use the result of Exercise 22, Sec. 3.4.

8. Prove Corollary 2, Theorem 5. *Hint:* First prove the assertion for the special polynomials $p(A) = A^k$ by premultiplying the equation $A\mathbf{x}_i = \lambda_i\mathbf{x}_i$ by $A, A^2, \ldots, A^{k-1}$ in turn.

9. By actually constructing an infinite family of solutions, show that each of the following matrix equations is satisfied by infinitely many matrices.
(a) $X^2 - 2X - 3I_2 = O$ **(b)** $X^2 - 4X + 3I_2 = O$
(c) $X^2 - 4X - 5I_2 = O$
(d) $X^3 - 6X^2 + 11X - 6I_3 = O$

10. Without attempting to find the solutions, show that for all values of a and b the equation $X^2 + aX + bI_2 = O$ has infinitely many solutions. What do you think is the generalization of this result to equations in an $n \times n$ matrix variable?

11. Show that the following matrix equations have no solutions.

(a) $X^2 - 2X - 3I_2 = \begin{bmatrix} -4 & 1 \\ 0 & -4 \end{bmatrix}$

(b) $X^2 - 4X + 3I_2 = \begin{bmatrix} -1 & 2 \\ 0 & -1 \end{bmatrix}$

(c) $X^2 - 4X - I_2 = \begin{bmatrix} -9 & 3 \\ 0 & -9 \end{bmatrix}$

(d) $X^2 - 4X + 3I_3 = \begin{bmatrix} 2 & 0 & 1 \\ 1 & -1 & 0 \\ 0 & 0 & -1 \end{bmatrix}$

12. If A and B commute, show that A commutes with any polynomial in B.

13. Verify each of the following identities for $X = \begin{bmatrix} 1 & 2 \\ 1 & 3 \end{bmatrix}$ and $X = \begin{bmatrix} 0 & 2 \\ 0 & -2 \end{bmatrix}$.

(a) $(X - I)^2 = X^2 - 2X + I$

(b) $X^3 - I = (X - I)(X^2 + X + I)$

(c) $\dfrac{2X}{X^2 - I} = \dfrac{I}{X - I} + \dfrac{I}{X + I}$

(d) $\dfrac{X^2}{X - 2I} = X + 2I + \dfrac{4I}{X - 2I}$

14. If $A^k = O$ for some positive integer k, prove that every characteristic value of A is zero.

15. Solve each of the following matrix equations.

(a) $X^2 - 5X + 3I = \begin{bmatrix} 1 & -4 \\ 2 & -5 \end{bmatrix}$

(b) $X^2 + 6X + 9I = \begin{bmatrix} -5 & 9 \\ -6 & 10 \end{bmatrix}$

(c) $X^3 = \begin{bmatrix} -6 & 14 \\ -7 & 15 \end{bmatrix}$

(d) $X^3 = \begin{bmatrix} 8 & -7 & -7 \\ -9 & 10 & 11 \\ 9 & -9 & -10 \end{bmatrix}$

16. If $f(x) = x/(x + 4)$, compute $f(A)$ for each of the following matrices A.

(a) $\begin{bmatrix} 1 & -4 \\ 2 & -5 \end{bmatrix}$ **(b)** $\begin{bmatrix} 4 & -1 \\ 6 & -1 \end{bmatrix}$ **(c)** $\begin{bmatrix} 2 & -1 \\ 4 & -3 \end{bmatrix}$

(d) $\begin{bmatrix} -3 & 1 & 0 \\ 1 & -4 & 0 \\ 0 & 0 & -5 \end{bmatrix}$ **(e)** $\begin{bmatrix} -1 & 2 & 2 \\ 2 & -1 & -2 \\ -2 & 2 & 3 \end{bmatrix}$

17. If $p(x) = x^4 - x^3 - 3x^2 + 4x + 2$, evaluate $p(A)$ for each of the following matrices A.

(a) $\begin{bmatrix} 4 & 1 \\ -3 & 0 \end{bmatrix}$ **(b)** $\begin{bmatrix} -1 & -2 \\ 3 & 4 \end{bmatrix}$ **(c)** $\begin{bmatrix} 4 & 6 \\ -3 & -5 \end{bmatrix}$

(d) $\begin{bmatrix} -4 & -9 & -3 \\ 1 & 4 & 1 \\ 3 & 3 & 2 \end{bmatrix}$ **(e)** $\begin{bmatrix} 2 & 1 & 1 \\ 1 & 4 & 3 \\ -1 & -1 & 0 \end{bmatrix}$

18. Verify that the characteristic values of $f(A)$ are equal to $f(\lambda_i)$ for each of the following functions and each of the given matrices.

(a) $x^2 - 2x + 3$ **(b)** $x^2 - 4x + 3$

(c) $x^3 + x^2 + x + 1$

(i) $\begin{bmatrix} 1 & -4 \\ 2 & -5 \end{bmatrix}$ **(ii)** $\begin{bmatrix} 4 & -1 \\ 6 & -1 \end{bmatrix}$ **(iii)** $\begin{bmatrix} 2 & 1 \\ 1 & 2 \end{bmatrix}$

19. Verify that the characteristic values of $f(A)$ are equal to $f(\lambda_i)$ for each of the following functions and each of the given matrices.

(a) $\dfrac{x}{x^2 + 1}$ **(b)** $\dfrac{x - 2}{x + 2}$ **(c)** $\dfrac{x + 1}{x^2 + x + 1}$

(i) $\begin{bmatrix} -1 & 2 \\ -1 & 2 \end{bmatrix}$ **(ii)** $\begin{bmatrix} 4 & 2 \\ 1 & 3 \end{bmatrix}$ **(iii)** $\begin{bmatrix} 5 & 2 \\ 2 & 2 \end{bmatrix}$

20. If p is a polynomial, is it possible for $p(A)$ to have a characteristic vector which is not a characteristic vector of A? Justify your answer.

21. Show that every $n \times n$ matrix satisfies a polynomial equation of degree at most n^2.

22. Let $P(\lambda) = C_0 + C_1\lambda + \cdots + C_k\lambda^k$ be a polynomial in the scalar variable λ with coefficients $C_0, C_1, \ldots, C_k$ which are square matrices of order n and let A also be an $n \times n$ matrix. Show by an example that in general $C_0 + C_1A + \cdots + C_kA^k \neq C_0 + AC_1 + \cdots + A^kC_k$. If P(A) is defined by the matrix equation $P(A) = C_0 + C_1A + \cdots + C_kA^k$, show by an example that Theorem 1 does not hold for polynomials whose coefficients are matrices.

23. According to the **Cayley-Hamilton theorem**:† *Every square matrix satisfies its own characteristic equation.* Verify this theorem for each of the following matrices.

(a) $\begin{bmatrix} 4 & 3 \\ 5 & 6 \end{bmatrix}$ **(b)** $\begin{bmatrix} -1 & -2 \\ 3 & 4 \end{bmatrix}$

(c) $\begin{bmatrix} 2 & 1 & 1 \\ 1 & 4 & 3 \\ -1 & -1 & 0 \end{bmatrix}$

24. (a) Show by means of the Cayley-Hamilton theorem that any polynomial in a square matrix of order n with scalar coefficients can be reduced to a polynomial of degree at most $n - 1$. Illustrate this fact by reducing the polynomial $A^5 - A^4 + 2A^2 + A + 2I$, where

---

†Named jointly for Arthur Cayley and the Irish mathematician William Rowan Hamilton (1805–1865). For a proof of this theorem and of Sylvester's identity (Exercise 25) see, for instance, C. R. Wylie, *Advanced Engineering Mathematics,* 4th ed., pp. 576–581, McGraw Hill, New York, 1975.

$$A = \begin{bmatrix} -1 & 2 & 2 \\ 2 & -1 & -2 \\ -2 & 2 & 3 \end{bmatrix}$$

to a polynomial of degree less than 3.

(b) Explain how the Cayley-Hamilton theorem can be used to find the inverse of a square matrix and illustrate this fact by finding the inverse of the matrix in Part **(a)**.

25. If A is an $n \times n$ matrix with distinct characteristic values and if $p(x)$ is a scalar polynomial, **Sylvester's identity** asserts that

$$p(\mathrm{A}) = \sum_{k=1}^{n} \left[\frac{p(\lambda_k)}{\prod_{\substack{i=1 \\ i \neq k}}^{n} (\lambda_k - \lambda_i)} \prod_{\substack{i=1 \\ i \neq k}}^{n} (\mathrm{A} - \lambda_i \mathrm{I}) \right]$$

where $\lambda_1, \lambda_2, \ldots, \lambda_n$ are the characteristic values of A. Use this result to carry out the reduction called for in Part **(a)** of Exercise 24 if

$$A = \begin{bmatrix} 2 & 1 & 1 \\ 1 & 4 & 3 \\ -1 & -1 & 0 \end{bmatrix}$$

26. If $A = \begin{bmatrix} 4 & 2 \\ -3 & -1 \end{bmatrix}$, use the Cayley-Hamilton theorem (Exercise 23) to express A^n in terms of A and I.

27. Work Exercise 26 if $A = \begin{bmatrix} 4 & 2 \\ -1 & 1 \end{bmatrix}$. *Hint:* Consider the use of difference equations.

28. In Example 2, find the characteristic vectors of A and verify that they are also the characteristic vectors of $\phi(A)$ corresponding to the characteristic values $\phi(-1) = -\frac{1}{2}$ and $\phi(-2) = -2$.

29. If $A = \begin{bmatrix} 1 & 3 \\ -1 & 5 \end{bmatrix}$ and $\phi(x) = x^2 - 3x + 5$, find the characteristic values and characteristic vectors of A and verify that the characteristic vectors of A are also characteristic vectors of $\phi(A)$.

30. Assuming that Theorem 6 holds for transcendental as well as algebraic functions, evaluate $\phi(A) = e^A$, where $A = \begin{bmatrix} 2 & 6 \\ 1 & 3 \end{bmatrix}$. Verify your result by using the Cayley-Hamilton theorem (Exercise 23) to simplify the series

$$e^A = I + A + \frac{1}{2!}A^2 + \frac{1}{3!}A^3 + \cdots$$

31. Work Exercise 30 if $A = \begin{bmatrix} 2 & 5 \\ 1 & -2 \end{bmatrix}$ and $\phi(A) = \sin A$.

32. **(a)** Work Exercise 30 if $A = \begin{bmatrix} 4 & 2 \\ -6 & -3 \end{bmatrix}$ and $\phi(A) = \sin A$.

(b) Work Exercise 30 if $A = \begin{bmatrix} 4 & 2 \\ -6 & -3 \end{bmatrix}$ and $\phi(A) = \cos A$.

(c) Using the results of Parts **(a)** and **(b)**, verify that $\cos^2 A + \sin^2 A = I$.

CHAPTER 15

VECTOR ANALYSIS

◖ Unlike Sec. 3.2, this chapter deals exclusively with vectors in three dimensions, but otherwise it represents a major extension and redirection of Sec. 3.2. Secs. 15.1–15.3 deal with the algebra of vectors, thought of primarily as geometric rather than algebraic objects, and *dot* and *cross products* are defined.

Sections 15.4 and 15.5 consider vectors whose components are functions of a single scalar variable, the operator ∇ (''del'') is introduced, and the *divergence* and *curl* of a variable vector are defined. Applications of these ideas include motion along curvilinear paths and certain topics from differential geometry such as the *curvature* and *torsion* of twisted curves.

In Sec. 15.6, the elementary calculus idea of a definite integral is extended from integration along a straight line to integration along a curve, over a surface, or throughout a volume. In Sec. 15.7, the ideas of Sec. 15.6 are used to develop a number of important theorems, including *Green's lemma,* the *divergence theorem,* and the *theorems of Gauss, Stokes,* and *Green.* In Sec. 15.8, these are used to show that in empty space the gravitational potential satisfies *Poisson's equation* and to derive *Maxwell's equations* for electric fields.

Prerequisites for this chapter: multivariable calculus, Sec. 3.2, Sec. 12.9, a calculus-based course in physics. ◗

15.1 BASIC ALGEBRAIC AND GEOMETRIC PROPERTIES OF VECTORS IN 3-SPACE

In Sec. 3.2, preparatory to our discussion of matrices and determinants, we introduced the notion of a scalar and defined an n-dimensional vector $\mathbf{v}$ as an ordered set of n scalars, say $\mathbf{v} = (v_1, v_2, \ldots, v_n)$. We then developed the basic algebra of vectors and became familiar with the linear spaces C^n and R^n. By means of the inner product, the concepts of length and orthogonality were defined for vectors of R^n ($n \geq 2$). Subsequently, we gave a geometrical interpretation to the vectors of R^3 by agreeing that each such vector $\mathbf{v}$ could be represented either by the point P whose three-dimensional cartesian coordinates (v_1, v_2, v_3) are the respective components of $\mathbf{v}$ or by any directed line segment having the direction and length of $\overrightarrow{OP}$, where O is the origin of the cartesian coordinate system. These two kinds of geometric representations provide simple and routine images of any vector in 3-space, i.e., in R^3. They also prove useful in finding angles between, and projections of, vectors in a Euclidean n-space, even if $n > 3$ and serious perceptual difficulties are involved (Sec. 13.4).

The definition of a vector as an ordered set of n scalars immediately focuses attention on the essentially algebraic nature of the components of the vector and leaves the geometric interpretation to follow. Historically, however, the concept of a vector arose from physical and geometric considerations of quantities possessing both magnitude and direction out of which the idea of components evolved. Of course, whether they enter through the definition or subsequently, the components of a vector play the same role in both the modern and the traditional development. In particular, the two approaches lead very quickly to the same formulas for the scalar product of two vectors, the length of a vector, and the angle between two vectors.

On the other hand, there are numerous results, for the most part restricted to two and three dimensions, which follow more naturally from the historical rather than the modern approach. These either have significant mathematical content or are of great practical importance, especially in engineering and physics, and are well worth our attention. To facilitate their study, we require some additional vector geometry and algebra and must extend the methods of calculus to vector functions. In this chapter we shall investigate these ideas and study vectors and vector functions following, for the most part, the development and notation of Josiah Willard Gibbs (1839–1903), a great American mathematical physicist and thermodynamicist. Collectively, this subject matter is known as **vector analysis.**

To distinguish vectors from scalars, we shall continue to denote the former by symbols, either uppercase or lowercase letters, set in boldface type thus, $\mathbf{V}$, consistent with the convention of Sec. 3.2. Since a real scalar quantity has magnitude only, it can be adequately represented by a mark on a fixed scale. To represent a general vector geometrically, however, and display both its direction and magnitude, we must use an arrow or directed line segment whose direction is that of the vector and whose length is equal (on some convenient scale) to the magnitude of the vector.

For convenience, we shall often refer to a representative line segment as though it were the vector itself. We shall also call the magnitude, length, or norm of a vector the **absolute value** of the vector and indicate it either by enclosing the symbol for the vector between ordinary absolute-value bars or by simply setting the symbol of the vector in *ordinary* rather than boldface type. Thus, in place of the notation $\|\mathbf{V}\|$ previously used to denote the norm or length of a vector $\mathbf{V}$, either

$$|\mathbf{V}| \quad \text{or} \quad V$$

will henceforth denote the magnitude, or absolute value, of $\mathbf{V}$. Regardless of its direction, a vector whose length, or absolute value, is unity is called a **unit vector.** In the setting of 3-space, with which we are now concerned, a vector is said to be **zero** if its absolute value is zero. Zero vectors will still be denoted by a zero set in boldface type. The direction of a zero vector is undefined.

Vectors of equal length which have the same direction are said to be **equal,** regardless of the points in space where they may originate. In other words, a vector quantity can be represented equally well by any of infinitely many **equivalent** line segments, all having the same length and the same direction. It is therefore customary to say that *a vector can be moved parallel to itself without change.*† If two vectors have the same length but are oppositely directed, either is called the **negative** of the other.

The *sum* $\mathbf{A} + \mathbf{B}$ of two vectors $\mathbf{A}$ and $\mathbf{B}$ is defined geometrically by the familiar **parallelogram law,** which as a definition reads as follows.

DEFINITION 1 If $\mathbf{A}$ and $\mathbf{B}$ are drawn from the same point, or origin, and if the parallelogram having $\mathbf{A}$ and $\mathbf{B}$ as adjacent sides is constructed, then the **sum** $\mathbf{A} + \mathbf{B}$ is the vector whose point of origin is the common origin of $\mathbf{A}$ and $\mathbf{B}$ and which coincides with the diagonal of the parallelogram containing that point (Fig. 15.1*a*).

From this definition it is evident that

$$\mathbf{A} + \mathbf{B} = \mathbf{B} + \mathbf{A}$$

i.e., that *vector addition is commutative,* and that

$$\mathbf{A} + (\mathbf{B} + \mathbf{C}) = (\mathbf{A} + \mathbf{B}) + \mathbf{C}$$

i.e., that *vector addition is associative.* By the **difference** of vectors $\mathbf{A}$ and $\mathbf{B}$, we mean the sum of the first and the negative of the second; i.e.,

$$\mathbf{A} - \mathbf{B} = \mathbf{A} + (-\mathbf{B})$$

(Fig. 15.1*b*). By the **product of a scalar** a **and a vector A** we mean the vector $a\mathbf{A} = \mathbf{A}a$ whose length is equal to the product of $|a|$ and the magnitude of $\mathbf{A}$ and whose direction is the same as the direction of $\mathbf{A}$ if a is positive and opposite to it if a is negative. An important consequence of this definition is that $\mathbf{A}/|\mathbf{A}| = \mathbf{A}/A$ is a unit vector in the direction of $\mathbf{A}$, assuming of course that $\mathbf{A} \neq \mathbf{0}$.

In addition to the product of a scalar and a vector, two other types of products are defined in vector analysis. One of these is the *dot* or *scalar* or *inner product* of two vectors $\mathbf{A}$ and $\mathbf{B}$, uniformly indicated by placing a dot between the two factors.

FIGURE 15.1
The geometric addition and subtraction of vectors.

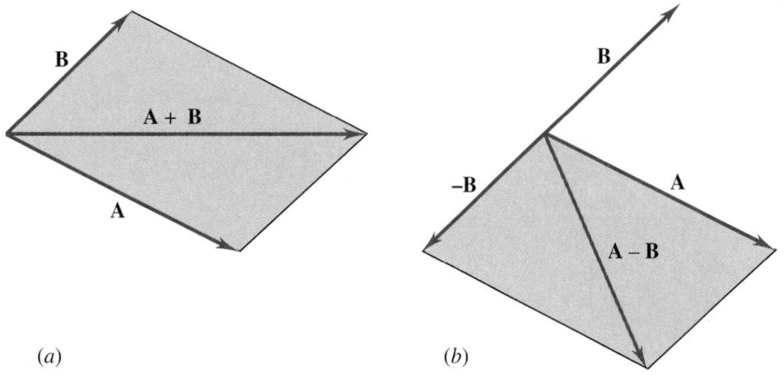

†In some applications, e.g., in dealing with forces whose points of application or lines of action cannot be shifted, it is necessary to think of a vector as fixed or at least restricted in position. Such vectors are usually said to be **bound,** in contrast to unrestricted vectors, which are said to be **free.**

DEFINITION 2 The **dot product** $\mathbf{A} \cdot \mathbf{B}$ of two nonzero vectors is equal to the product of their absolute values and the cosine of the angle θ between their (positive) directions; but if either $\mathbf{A}$ or $\mathbf{B}$ is zero, then $\mathbf{A} \cdot \mathbf{B} = 0$.

This definition gives, as a working formula,

$$\mathbf{A} \cdot \mathbf{B} = |\mathbf{A}|\,|\mathbf{B}| \cos\theta = AB\cos\theta \tag{1}$$

where, if either $\mathbf{A}$ or $\mathbf{B}$ is zero, it is to be remembered that $\mathbf{A} \cdot \mathbf{B} = 0$. In any case, $\mathbf{A} \cdot \mathbf{B}$ is a scalar. Definition 2 may be regarded as a geometric version in 3-space of the broader analytic concept introduced under the same terminology in Sec. 3.2, and Eq. (1) is Eq. (2), Sec. 13.4, now applied to vectors of 3-space.

From the definitive relation (1) it is clear that *dot multiplication is commutative;* i.e.,

$$\mathbf{A} \cdot \mathbf{B} = \mathbf{B} \cdot \mathbf{A} \tag{2}$$

If $\mathbf{A} \neq \mathbf{0}$ and $\mathbf{B} = \mathbf{A}$, then $\theta = 0$ and (1) gives

$$\mathbf{A} \cdot \mathbf{A} = |\mathbf{A}|^2 = A^2 \tag{3}$$

which reduces to $\mathbf{0} \cdot \mathbf{0} = 0$, in case $\mathbf{A}$ is zero. We already know (Sec. 3.2) that *dot multiplication is distributive over addition;* i.e.,

$$\mathbf{A} \cdot (\mathbf{B} + \mathbf{C}) = \mathbf{A} \cdot \mathbf{B} + \mathbf{A} \cdot \mathbf{C} \tag{4}$$

For nonzero vectors $\mathbf{A}$ and $\mathbf{B}$, since $|\mathbf{A}| \cos\theta$ is just the projection of $\mathbf{A}$ in the direction of $\mathbf{B}$, and since $|\mathbf{B}| \cos\theta$ is the projection of $\mathbf{B}$ in the direction of $\mathbf{A}$ (Fig. 15.2), it follows from (1) that *the dot product of two nonzero vectors is equal to the length of either of them multiplied by the projection of the other upon it.* Furthermore, since $\mathbf{A}/A$ is a unit vector in the direction of $\mathbf{A}$, we have in the terminology of Sec. 13.4 that

$$\mathbf{B} \cdot \mathbf{A}/A = B\cos\theta \tag{5}$$

is the **scalar projection of B in the direction of A,** and that

$$(\mathbf{B} \cdot \mathbf{A}/A)\mathbf{A}/A = \left(\frac{\mathbf{B} \cdot \mathbf{A}}{A^2}\right)\mathbf{A} \tag{6}$$

is the **vector projection of B in the direction of A.**

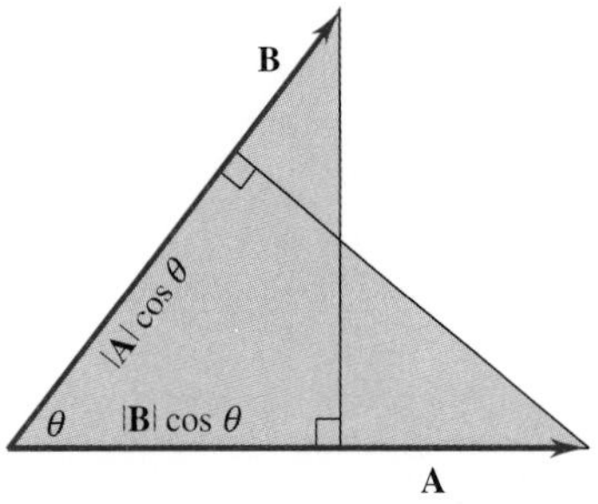

FIGURE 15.2
The geometric projection of either of two nonzero vectors upon the other.

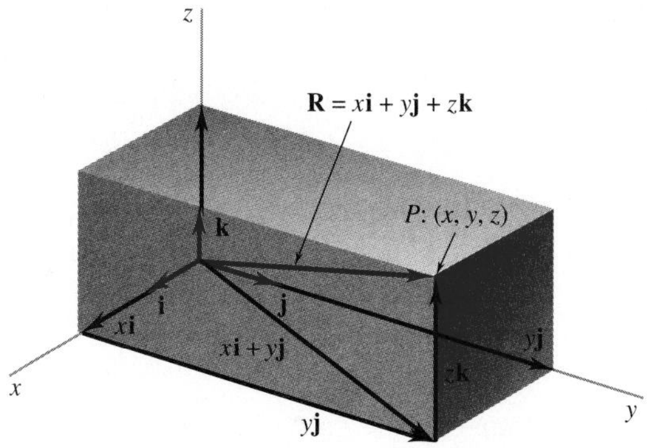

FIGURE 15.3
The representation of a vector as a linear combination of the unit vectors **i**, **j**, **k**.

It is often convenient to be able to refer vector expressions to a cartesian frame of reference. To provide for this we define **i**, **j** and **k** to be vectors of unit length directed, respectively, along the positive x, y, and z axes of a right-handed rectangular coordinate system. Then $x\mathbf{i}$, $y\mathbf{j}$, and $z\mathbf{k}$ represent vectors of lengths $|x|$, $|y|$, and $|z|$ which run parallel to the respective axes; and from the definition of vector addition it is evident that the vector joining the origin to a general point P: (x, y, z) (Fig. 15.3) can be written

$$\mathbf{R} = x\mathbf{i} + y\mathbf{j} + z\mathbf{k} \tag{7}$$

In more general terms, any vector whose components along the coordinate axes are, respectively a_1, a_2, and a_3 can be written

$$\mathbf{A} = a_1\mathbf{i} + a_2\mathbf{j} + a_3\mathbf{k}$$

If, further,

$$\mathbf{B} = b_1\mathbf{i} + b_2\mathbf{j} + b_3\mathbf{k}$$

it follows readily from the parallelogram law for vector addition that

$$\mathbf{A} \pm \mathbf{B} = (a_1 \pm b_1)\mathbf{i} + (a_2 \pm b_2)\mathbf{j} + (a_3 \pm b_3)\mathbf{k}$$

Clearly, as we saw in Sec. 3.2, *two vectors will be equal if and only if their respective components are equal.* Hence any vector equation implies three scalar equations.

Since the dot product of perpendicular vectors is zero, it follows that

$$\mathbf{i} \cdot \mathbf{j} = \mathbf{j} \cdot \mathbf{k} = \mathbf{k} \cdot \mathbf{i} = 0 \tag{8}$$

Moreover, applying (3) to the unit vectors **i**, **j**, **k**, we have

$$\mathbf{i} \cdot \mathbf{i} = \mathbf{j} \cdot \mathbf{j} = \mathbf{k} \cdot \mathbf{k} = 1 \tag{9}$$

In component form

$$\mathbf{i} = (1, 0, 0) \qquad \mathbf{j} = (0, 1, 0) \qquad \mathbf{k} = (0, 0, 1)$$

Hence $\{\mathbf{i}, \mathbf{j}, \mathbf{k}\}$ is the **standard basis of all vectors in** 3-**space** (Example 5, Sec. 13.3). By writing

$$\mathbf{A} \cdot \mathbf{B} = (a_1\mathbf{i} + a_2\mathbf{j} + a_3\mathbf{k}) \cdot (b_1\mathbf{i} + b_2\mathbf{j} + b_3\mathbf{k})$$

and using the fact that dot multiplication is distributive over addition [Eq. (4)] to expand, then simplifying by applying (8) and (9), we obtain the familiar formula

$$\mathbf{A} \cdot \mathbf{B} = a_1b_1 + a_2b_2 + a_3b_3 \tag{10}$$

of analytic geometry. In particular, taking $\mathbf{B} = \mathbf{A}$, we have

$$\mathbf{A} \cdot \mathbf{A} = |\mathbf{A}|^2 = A^2 = a_1^2 + a_2^2 + a_3^2$$

or

$$(11) \qquad |\mathbf{A}| = A = \sqrt{a_1^2 + a_2^2 + a_3^2}$$

On the other hand, assuming neither **A** nor **B** is zero, if we write $\mathbf{A} \cdot \mathbf{B} = |\mathbf{A}|\,|\mathbf{B}| \cos \theta$, then solve for $\cos \theta$, and subsequently utilize (10) and (11), we obtain in succession

$$(12) \qquad (a)\ \cos \theta = \frac{\mathbf{A} \cdot \mathbf{B}}{|\mathbf{A}|\,|\mathbf{B}|} \qquad (b)\ \cos \theta = \frac{a_1 b_1 + a_2 b_2 + a_3 b_3}{\sqrt{a_1^2 + a_2^2 + a_3^2}\sqrt{b_1^2 + b_2^2 + b_3^2}}$$

Eq. (12*b*) is another familiar formula from analytic geometry, where the *a*'s and *b*'s were introduced not as components of two vectors but as direction numbers of two straight lines. Eq. (12*a*) is just Eq. (3), Sec. 13.4, particularized to vectors of 3-space.

By means of vector methods, many important mathematical and practical results can be derived efficiently.

EXAMPLE 1

Using vector methods, derive the law of cosines.

To do this, let the sides of the given triangle be **A**, **B**, and $\mathbf{C} = \mathbf{A} - \mathbf{B}$. Then

$$\mathbf{C} \cdot \mathbf{C} = (\mathbf{A} - \mathbf{B}) \cdot (\mathbf{A} - \mathbf{B}) = \mathbf{A} \cdot \mathbf{A} - 2\mathbf{A} \cdot \mathbf{B} + \mathbf{B} \cdot \mathbf{B}$$

or, using (1) and (3),

$$C^2 = A^2 + B^2 - 2AB \cos \theta$$

which is the law of cosines.

EXAMPLE 2

If (x, y, z) and (x', y', z') are two right-handed coordinate systems having a common origin, obtain by vector methods the transformation equations connecting the two sets of coordinates.

To do this, let **i**, **j**, **k** and $\mathbf{i}', \mathbf{j}', \mathbf{k}'$ be unit vectors in the directions of the respective axes (Fig. 15.4), and let P be a general point in space having coordinates (x, y, z) and (x', y', z') in the respective systems. Now the coordinates (x', y', z') are simply the components of the vector $\overrightarrow{OP}$ along the x', y', z' axes. Hence, if we write

$$\mathbf{R} = \overrightarrow{OP} = x\mathbf{i} + y\mathbf{j} + z\mathbf{k}$$

and observe that the dot products of this vector with the unit vectors $\mathbf{i}', \mathbf{j}'$, and $\mathbf{k}'$ are its components in these directions, we find the required transformation formulas to be

$$\begin{aligned} x' &= \mathbf{R} \cdot \mathbf{i}' = (x\mathbf{i} + y\mathbf{j} + z\mathbf{k}) \cdot \mathbf{i}' = x(\mathbf{i} \cdot \mathbf{i}') + y(\mathbf{j} \cdot \mathbf{i}') + z(\mathbf{k} \cdot \mathbf{i}') \\ y' &= \mathbf{R} \cdot \mathbf{j}' = (x\mathbf{i} + y\mathbf{j} + z\mathbf{k}) \cdot \mathbf{j}' = x(\mathbf{i} \cdot \mathbf{j}') + y(\mathbf{j} \cdot \mathbf{j}') + z(\mathbf{k} \cdot \mathbf{j}') \\ z' &= \mathbf{R} \cdot \mathbf{k}' = (x\mathbf{i} + y\mathbf{j} + z\mathbf{k}) \cdot \mathbf{k}' = x(\mathbf{i} \cdot \mathbf{k}') + y(\mathbf{j} \cdot \mathbf{k}') + z(\mathbf{k} \cdot \mathbf{k}') \end{aligned}$$

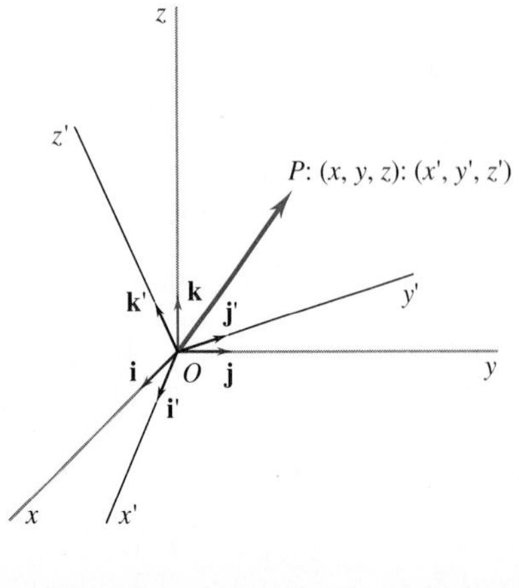

FIGURE 15.4
Two rectangular coordinate systems with the same origin.

From (1), the products $\mathbf{i} \cdot \mathbf{i}', \mathbf{j} \cdot \mathbf{i}', \ldots, \mathbf{k} \cdot \mathbf{k}'$ are just the cosines of the angles between the various axes of the two systems and are known from the data of the problem.

EXERCISES

In Exercises 1–3, for each of the sets of vectors find **(a)** the absolute values of **A**, **B**, and **C**, **(b)** unit vectors in the directions of **A** and **C**, **(c)** $\mathbf{A} \cdot \mathbf{B}$, **(d)** the scalar projection of **C** along **B**, **(e)** the vector projection of **B** along **A**, and **(f)** the angle between **A** and **B**.

1. $\mathbf{A} = 2\mathbf{i} - 2\mathbf{j} + \mathbf{k}$
$\mathbf{B} = \mathbf{i} + 8\mathbf{j} - 4\mathbf{k}$
$\mathbf{C} = 12\mathbf{i} - 4\mathbf{j} - 3\mathbf{k}$

2. $\mathbf{A} = 2\mathbf{i} - 3\mathbf{j} + 6\mathbf{k}$
$\mathbf{B} = 10\mathbf{i} + 2\mathbf{j} + 11\mathbf{k}$
$\mathbf{C} = 2\mathbf{i} - 9\mathbf{j} - 6\mathbf{k}$

3. $\mathbf{A} = 10\mathbf{i} + 10\mathbf{j} + 5\mathbf{k}$
$\mathbf{B} = 5\mathbf{i} - 2\mathbf{j} - 14\mathbf{k}$
$\mathbf{C} = 4\mathbf{i} + 7\mathbf{j} - 4\mathbf{k}$

4. If **A** is a given vector and $\mathbf{A} \cdot \mathbf{X} = \mathbf{A} \cdot \mathbf{Y}$, can we conclude that $\mathbf{X} = \mathbf{Y}$? Explain. What can we conclude?

5. Are two vectors equal if they have equal components in a given direction? In two given directions? In three given directions? In an arbitrary direction?

6. If the initial point of a vector **B** is the **terminus** (terminal point) of **A**, what is the vector from the initial point of **A** to the terminus of **B**? Explain.

7. The **resultant** of a set of vectors is the sum of those vectors. Find the resultant of the seven vectors **j**, $2\mathbf{k}$, $3\mathbf{j} - 5\mathbf{k}$, $2\mathbf{i} - 4\mathbf{j} - 2\mathbf{k}$, $-3\mathbf{i} - 2\mathbf{j} + 3\mathbf{k}$, $4\mathbf{i} + 2\mathbf{j} + 2\mathbf{k}$, and $-3\mathbf{i}$. What is the negative of the resultant?

8. Find all values of a for which the vector $-3\mathbf{i} + a\mathbf{j} + a^2\mathbf{k}$ is perpendicular to $\mathbf{i} - 2\mathbf{j} + \mathbf{k}$.

9. Find a unit vector which is perpendicular to both $\mathbf{i} - 2\mathbf{j} + \mathbf{k}$ and $3\mathbf{i} + \mathbf{j} - 2\mathbf{k}$.

10. For the vectors

$$\mathbf{A} = \mathbf{i} - 2\mathbf{j} + \mathbf{k} \qquad \mathbf{B} = -\mathbf{i} - 2\mathbf{j} + \mathbf{k}$$
$$\mathbf{C} = \mathbf{i} + \mathbf{j} + 2\mathbf{k} \qquad \mathbf{D} = -2\mathbf{i} + \mathbf{j} + 3\mathbf{k}$$

find **(a)** $(\mathbf{A} \cdot \mathbf{C})(\mathbf{B} \cdot \mathbf{D}) - (\mathbf{A} \cdot \mathbf{D})(\mathbf{B} \cdot \mathbf{C})$,
(b) $(\mathbf{C} \cdot \mathbf{A})\mathbf{B} - (\mathbf{C} \cdot \mathbf{B})\mathbf{A}$, and
(c) $(B\mathbf{A} + A\mathbf{B}) \cdot (B\mathbf{A} - A\mathbf{B}) + |3\mathbf{C} - 2\mathbf{D}|$.

11. The vectors **P**, **Q**, and **R** have the same directions as **i**, $-\mathbf{i} + \mathbf{j}$, and $-\mathbf{k}$, respectively, If $|\mathbf{P}| = 2$, $\mathbf{Q} \cdot \mathbf{Q} = 18$, and $\mathbf{R} \cdot \mathbf{R} = 16$, **(a)** find **P**, **Q**, **R**, $\mathbf{Q} - \mathbf{R}$, and **(b)** express $-3\mathbf{i} + 9\mathbf{j} + 8\mathbf{k}$ as a linear combination of **P**, **Q**, and **R**.

12. Find all vectors **v** such that $(\mathbf{v} \cdot \mathbf{v})\mathbf{v} = 169\mathbf{v}$ and $4\mathbf{i} \cdot \mathbf{v} = 3\mathbf{j} \cdot \mathbf{v} = \mathbf{k} \cdot \mathbf{v}$.

13. Given that $\mathbf{U} = \mathbf{i} - 2\mathbf{j} + 3\mathbf{k}$, $\mathbf{V} = 2\mathbf{i} + \mathbf{j} + 4\mathbf{k}$, $\mathbf{W} = \mathbf{i} + 3\mathbf{j} + 3\mathbf{k}$, and $(\mathbf{U} \cdot \mathbf{R} - 10)\mathbf{i} + (\mathbf{V} \cdot \mathbf{R} - 20)\mathbf{j} + (\mathbf{W} \cdot \mathbf{R} - 20)\mathbf{k} = \mathbf{0}$, find **R**.

14. Find the angle between $\mathbf{i} + 2\mathbf{j} + 2\mathbf{k}$ and $\mathbf{i} + \mathbf{k}$.

15. Find the angle between the diagonal of a cube and one of its edges.

16. Find the *acute* angle between the lines lying along the vectors $\mathbf{i} - 2\mathbf{j} + 2\mathbf{k}$ and $-6\mathbf{i} + 6\mathbf{j} + 7\mathbf{k}$.

17. Find the scalar and vector projections of $-7\mathbf{i} + 14\mathbf{j} + 7\mathbf{k}$ along $2\mathbf{i} + 3\mathbf{j} - 6\mathbf{k}$.

18. Prove that for all values of the a's and b's

$$(a_1b_1 + a_2b_2 + a_3b_3)^2 \le (a_1^2 + a_2^2 + a_3^2)(b_1^2 + b_2^2 + b_3^2)$$

This is the special case $n = 3$ of **Cauchy's inequality,**

$$\left(\sum_{i=1}^{n} a_i b_i\right)^2 \le \left(\sum_{i=1}^{n} a_i^2\right)\left(\sum_{i=1}^{n} b_i^2\right)$$

19. Show that if **r** and **s** are nonzero constant vectors and the scalar b is chosen such that $|\mathbf{r} + b\mathbf{s}|$ is a minimum, then $|b\mathbf{s}|^2 + |\mathbf{r} + b\mathbf{s}|^2 = |\mathbf{r}|^2$. *Hint:* Minimize $|\mathbf{r} + b\mathbf{s}|^2$ as a function of b.

20. **(a)** Show that if $\mathbf{A} = a_1\mathbf{i} + a_2\mathbf{j} + a_3\mathbf{k}$ is a constant vector drawn from the origin, the locus of the endpoints of the vectors $\mathbf{R} = x\mathbf{i} + y\mathbf{j} + z\mathbf{k}$ which satisfy the equation $(\mathbf{R} - \mathbf{A}) \cdot \mathbf{A} = 0$ is a plane perpendicular to **A** at its terminal point. What is the locus of the endpoints of the vectors which satisfy **(b)** the equation $(\mathbf{R} - \mathbf{A}) \cdot \mathbf{R} = 0$? **(c)** The equation $(\mathbf{R} - \mathbf{A}) \cdot (\mathbf{R} - \mathbf{A}) = 0$?

Show by vector methods that

21. An angle inscribed in a semicircle is a right angle.

22. The diagonals of a parallelogram bisect each other.

23. The lines joining consecutive midpoints of a quadrilateral form a parallelogram.

24. The diagonals of a parallelogram are perpendicular if and only if the parallelogram is a rhombus.

25. The medians of a triangle meet in a point of trisection of each median.

26. The altitudes of an arbitrary triangle are concurrent.

27. By considering the dot product of the two vectors

$$\mathbf{A} = a_1\mathbf{i} + a_2\mathbf{j} \quad \text{and} \quad \mathbf{B} = b_1\mathbf{i} + b_2\mathbf{j}$$

derive the formula for the cosine of the difference of two angles.

28. Assuming that $\mathbf{V} \cdot \mathbf{V} = |\mathbf{V}|^2$ for any vector **V**, use the law of cosines to show that $\mathbf{A} \cdot \mathbf{B} = |\mathbf{A}|\,|\mathbf{B}| \cos\theta$.

29. Using the relation between dot products and projections, carry through in detail a geometric proof that dot multiplication is distributive over addition.

30. Using vector methods, show that the segment joining the midpoints of two sides of any triangle is parallel to the third side and half as long. *Hint:* Let $\triangle A_1A_2A_3$ be an arbitrary triangle, let A_4 be the midpoint of A_1A_3, let A_5 be the midpoint of A_2A_3, and let $\mathbf{A}_i\mathbf{A}_j$ denote the vector whose initial point is A_i and whose terminal point is A_j. Then verify that $\mathbf{A}_1\mathbf{A}_2 = \mathbf{A}_1\mathbf{A}_3 + \mathbf{A}_3\mathbf{A}_2$, $\mathbf{A}_1\mathbf{A}_4 = \frac{1}{2}\mathbf{A}_1\mathbf{A}_3$, $\mathbf{A}_2\mathbf{A}_5 = \frac{1}{2}\mathbf{A}_2\mathbf{A}_3$, and $\mathbf{A}_1\mathbf{A}_4 + \mathbf{A}_4\mathbf{A}_5 = \mathbf{A}_1\mathbf{A}_2 + \mathbf{A}_2\mathbf{A}_5$. Finally, combine these relations to obtain the conclusion that

$$\mathbf{A}_4\mathbf{A}_5 = \tfrac{1}{2}\mathbf{A}_1\mathbf{A}_2$$

31. If $A_1A_2A_3A_4$ is an arbitrary parallelogram, prove by vector methods that the line joining A_1 to the midpoint of A_2A_3 intersects the diagonal A_2A_4 in one of the trisection points of the diagonal. *Hint:* Let A_5 be the midpoint of A_2A_3, let A_6 be the intersection of A_1A_5 and A_2A_4, let $A_1A_6 = xA_1A_5$, and let $A_2A_6 = yA_2A_4$. Then by obtaining two different representations of $\mathbf{A_1A_6}$, show that

$$(1 - y)\mathbf{A_1A_2} + y\mathbf{A_1A_4} = x(\mathbf{A_1A_2} + \tfrac{1}{2}\mathbf{A_1A_4})$$

and finally show that this implies that $x = \frac{2}{3}$ and $y = \frac{1}{3}$.

32. A person traveling east at 3 mi/h finds that the wind appears to be blowing directly from the north. At twice that velocity the wind appears to come from the northeast. What is the wind velocity? *Hint:* Use only vectors in a plane.

33. A wheel 2 ft in diameter bears a load of 2,000 lb and is at rest on a smooth horizontal surface. What horizontal force directed through the center of the wheel will be required to start the wheel over a rectangular block 6 in high?

15.2 THE CROSS PRODUCT, LINES, AND PLANES

The third type of product with which vector analysis deals is another carryover from a standard calculus with analytic geometry course and is the *vector,* or *cross, product of two vectors,* indicated by putting a cross between the factors.† Unlike the dot product of two vectors, which is always a scalar, a cross product $\mathbf{A} \times \mathbf{B}$ is always a vector. We can define the cross product either geometrically or analytically. Geometrically, we have, with θ as the angle between $\mathbf{A}$ and $\mathbf{B}$, the following definition.

DEFINITION 1 The **cross product** $\mathbf{A} \times \mathbf{B}$ of two nonzero vectors is the vector whose absolute value is the product of the absolute values of $\mathbf{A}$, $\mathbf{B}$, and the sine of θ and which, if $\theta \neq 0$ or π, is perpendicular to both $\mathbf{A}$ and $\mathbf{B}$ and so sensed that a right-handed screw turned from $\mathbf{A}$ toward $\mathbf{B}$ through the angle θ would advance in the direction of $\mathbf{A} \times \mathbf{B}$ (Fig. 15.5*a*). If either $\mathbf{A}$ or $\mathbf{B}$ is zero, then $\mathbf{A} \times \mathbf{B} = \mathbf{0}$.

If θ is the angle between two nonzero vectors, then $0 \leq \theta \leq \pi$ (Sec. 13.4) and by definition we have for the absolute value of $\mathbf{A} \times \mathbf{B}$

$$|\mathbf{A} \times \mathbf{B}| = |\mathbf{A}|\,|\mathbf{B}| \sin\theta = AB \sin\theta \tag{1}$$

where it is to be understood that if either $\mathbf{A}$ or $\mathbf{B}$ is zero, so that θ is undefined, then $|\mathbf{A} \times \mathbf{B}| = 0$.

For any two noncollinear vectors $\mathbf{A}$ and $\mathbf{B}$ drawn from a common point, the projection of $\mathbf{B}$ on a perpendicular to $\mathbf{A}$ is the altitude $|\mathbf{B}| \sin\theta$ of the parallelogram having $\mathbf{A}$ as base and $\mathbf{A}$ and $\mathbf{B}$ as adjacent sides (Fig. 15.5*b*). Hence it follows from (1) that the absolute value of $\mathbf{A} \times \mathbf{B}$ is equal to the area of this parallelogram.

Two nonzero vectors for which the angle between them is either 0 or π are said to be *parallel.* In general,

DEFINITION 2 Two vectors $\mathbf{A}$ and $\mathbf{B}$ are **parallel** if and only if $\mathbf{A} \times \mathbf{B} = \mathbf{0}$.

In particular, the zero vector is parallel to every vector in 3-space. However, if $\mathbf{A} \times \mathbf{B} = \mathbf{0}$, we cannot conclude that either $\mathbf{A}$ or $\mathbf{B}$ is zero, for this product will vanish if $\sin\theta = 0$. Hence, we have

---

†Meaning has also been given to the symbol **AB**, and in fact under the name **dyad** such combinations have been extensively studied, for instance in J. W. Gibbs and E. B. Wilson, ''Vector Analysis,'' Yale, New Haven, 1929. These are actually special cases of what are known as *tensors* and are outside the province of this chapter. For an introductory discussion of *tensor analysis* see, for instance, C. R. Wylie, ''Advanced Engineering Mathematics,'' 4th ed., pp. 696–732, McGraw-Hill, New York, 1975.

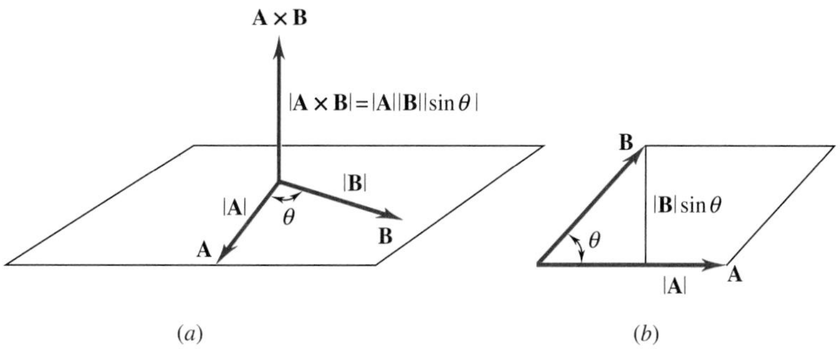

FIGURE 15.5
The geometric interpretation of a general vector product $\mathbf{A} \times \mathbf{B}$ and of its absolute value.

THEOREM 1 If $\mathbf{A} \times \mathbf{B} = \mathbf{0}$, then either at least one of the vectors $\mathbf{A}$, $\mathbf{B}$ is zero or else $\mathbf{A}$ and $\mathbf{B}$ are parallel nonzero vectors.

Any two nonparallel vectors $\mathbf{A}$ and $\mathbf{B}$ taken together with $\mathbf{A} \times \mathbf{B}$, in that order, form a **right-handed** system of vectors. In terms of a unit vector $\mathbf{u}$ in the direction of $\mathbf{A} \times \mathbf{B}$, we may write

$$\mathbf{A} \times \mathbf{B} = AB \sin \theta \, \mathbf{u} \tag{2}$$

From Definition 1, it follows that for all vectors $\mathbf{A}$ and $\mathbf{B}$

$$(a)\ (\mathbf{A} \times \mathbf{B}) \cdot \mathbf{A} = 0 \qquad (b)\ (\mathbf{A} \times \mathbf{B}) \cdot \mathbf{B} = 0 \tag{3}$$

because either $\mathbf{A} \times \mathbf{B} \neq \mathbf{0}$ and therefore is perpendicular to both $\mathbf{A}$ and $\mathbf{B}$ by definition or else $\mathbf{A} \times \mathbf{B} = \mathbf{0}$ and thus is perpendicular to every vector. This proves

THEOREM 2 For all vectors $\mathbf{A}$ and $\mathbf{B}$ in 3-space, $\mathbf{A} \times \mathbf{B}$ is perpendicular, or orthogonal, to both $\mathbf{A}$ and $\mathbf{B}$.

Eq. (1) clearly implies that the magnitude of $\mathbf{B} \times \mathbf{A}$ is equal to the magnitude of $\mathbf{A} \times \mathbf{B}$. However, for nonparallel vectors $\mathbf{A}$ and $\mathbf{B}$, since the direction of $\mathbf{A} \times \mathbf{B}$ is determined by the right-hand rule, interchanging $\mathbf{A}$ and $\mathbf{B}$ reverses the direction, or sign, of the cross product. Therefore the direction of $\mathbf{B} \times \mathbf{A}$ is opposite that of $\mathbf{A} \times \mathbf{B}$. Hence, *cross-multiplication is not commutative.* In fact, for arbitrary vectors $\mathbf{A}$ and $\mathbf{B}$, we have

$$\mathbf{B} \times \mathbf{A} = -\mathbf{A} \times \mathbf{B} \tag{4}$$

Multiplication in which products obey this rule is sometimes said to be **anticommutative.** The foregoing observations point out why we must be especially careful to preserve the proper order of factors in any expression involving cross products, and change signs in accordance with Eq. (4) if the order of the factors is altered.

Using a geometric argument (see Exercise 56), it is not difficult to show that *vector multiplication is distributive over addition;* i.e.,

$$\mathbf{A} \times (\mathbf{B} + \mathbf{C}) = \mathbf{A} \times \mathbf{B} + \mathbf{A} \times \mathbf{C} \tag{5}$$

For the cross products of the unit vectors **i**, **j**, **k**, Eqs. (2) and (4) give at once

$$\begin{aligned} \mathbf{i}\times\mathbf{i} &= \mathbf{j}\times\mathbf{j} = \mathbf{k}\times\mathbf{k} = \mathbf{0} \\ \mathbf{i}\times\mathbf{j} &= -\mathbf{j}\times\mathbf{i} = \mathbf{k} \\ \mathbf{j}\times\mathbf{k} &= -\mathbf{k}\times\mathbf{j} = \mathbf{i} \\ \mathbf{k}\times\mathbf{i} &= -\mathbf{i}\times\mathbf{k} = \mathbf{j} \end{aligned} \tag{6}$$

Hence, using (6) and the fact that cross multiplication is distributive over addition, we obtain for

$$\mathbf{A}\times\mathbf{B} = (a_1\mathbf{i} + a_2\mathbf{j} + a_3\mathbf{k}) \times (b_1\mathbf{i} + b_2\mathbf{j} + b_3\mathbf{k})$$

the expression

$$\mathbf{A}\times\mathbf{B} = (a_2b_3 - a_3b_2)\mathbf{i} - (a_1b_3 - a_3b_1)\mathbf{j} + (a_1b_2 - a_2b_1)\mathbf{k} \tag{7}$$

which provides an *analytic definition* of the **cross product** of two vectors that may be applied to find such a product, in a straightforward way, whenever its factors are expressed in component form. It is much easier to remember and apply Eq. (7) when it is written

$$\mathbf{A}\times\mathbf{B} = \begin{vmatrix} \mathbf{i} & \mathbf{j} & \mathbf{k} \\ a_1 & a_2 & a_3 \\ b_1 & b_2 & b_3 \end{vmatrix} \tag{8}$$

in which the expansion of the symbolic determinant by its first-row entries, in accordance with Laplace's expansion (Sec. 3.5), is precisely the right-hand member of (7). The anticommutative character of vector multiplication thus corresponds to the property that interchanging two rows of a determinant changes the sign of the determinant. Other properties of determinants serve likewise to unveil matching peculiarities of the cross product.

EXAMPLE 1

Property 6, Sec. 3.5, applied to rows gives

$$\begin{vmatrix} \mathbf{i} & \mathbf{j} & \mathbf{k} \\ a_1 + b_1 & a_2 + b_2 & a_3 + b_3 \\ c_1 & c_2 & c_3 \end{vmatrix} = \begin{vmatrix} \mathbf{i} & \mathbf{j} & \mathbf{k} \\ a_1 & a_2 & a_3 \\ c_1 & c_2 & c_3 \end{vmatrix} + \begin{vmatrix} \mathbf{i} & \mathbf{j} & \mathbf{k} \\ b_1 & b_2 & b_3 \\ c_1 & c_2 & c_3 \end{vmatrix}$$

from which we conclude that

$$(\mathbf{A} + \mathbf{B})\times\mathbf{C} = \mathbf{A}\times\mathbf{C} + \mathbf{B}\times\mathbf{C}$$

This result and Eq. (5) tell us that *vector multiplication is distributive over addition from both the right and the left.*

Implicit within Definition 1 is the notion of a *normal vector* to a plane.

DEFINITION 3 A vector **n** is a **normal vector** to a plane if and only if it is nonzero, originates at a point P in the plane, and is perpendicular to any two nonparallel vectors in the plane through P.

With reference to a standard right-handed coordinate system,

DEFINITION 4 The **position vector** of a general point P: (x, y, z) is the vector extending from the origin to P given by $\mathbf{R} = x\mathbf{i} + y\mathbf{j} + z\mathbf{k}$.

The use of position vectors, normal vectors, and vector algebra greatly facilitates the study of many geometric problems; especially those concerned with lines and planes in 3-space.

EXAMPLE 2

Find a vector equation, and a set of parametric scalar equations, for the line L through a point P_0: (x_0, y_0, z_0) in the direction of a vector $\mathbf{A} = a_1\mathbf{i} + a_2\mathbf{j} + a_3\mathbf{k}$.

Subtracting the position vector $\mathbf{R}_0 = x_0\mathbf{i} + y_0\mathbf{j} + z_0\mathbf{k}$ of the point P_0 from the position vector $\mathbf{R} = x\mathbf{i} + y\mathbf{j} + z\mathbf{k}$ of a general point P: (x, y, z) on L, we obtain a vector $\mathbf{R} - \mathbf{R}_0$ which will be parallel to $\mathbf{A}$ if and only if to each point P on L there corresponds a real scalar t such that $\mathbf{R} - \mathbf{R}_0$ is the product of t and $\mathbf{A}$. Hence, a **parametric vector equation** for L is

(9) $$\mathbf{R} = \mathbf{R}_0 + t\mathbf{A} \qquad t \text{ a parameter}$$

Equating components of this equation leads to the **parametric scalar equations** for L

(10) $$x = x_0 + a_1 t \qquad y = y_0 + a_2 t \qquad z = z_0 + a_3 t$$

encountered in analytic geometry. Since by assumption $\mathbf{A}$ has a direction and is therefore nonzero, a_1, a_2, and a_3 cannot all be zero.

In Example 2, if none of the components of $\mathbf{A}$ is zero, each of Eqs. (10) is solvable for t. Equating their solutions gives the familiar **symmetric equations** of L

(11) $$\frac{x - x_0}{a_1} = \frac{y - y_0}{a_2} = \frac{z - z_0}{a_3}$$

special cases of which we leave to the exercises.

EXAMPLE 3

Find vector and scalar equations of a plane p through a point P_0: (x_0, y_0, z_0) for which $\mathbf{n} = a\mathbf{i} + b\mathbf{j} + c\mathbf{k}$ is a normal vector to p originating at P_0. Then show that the points A: (1, 3, 3), B: (3, 5, 0), C: (2, 5, 1) determine a plane, and find an equation for it.

The difference $\mathbf{R} - \mathbf{R}_0$ of the position vector $\mathbf{R}$ of an arbitrary point P: (x, y, z) in p and the position vector $\mathbf{R}_0$ of P_0 is a vector in p containing P_0; hence, $\mathbf{n}$ and $\mathbf{R} - \mathbf{R}_0$ are perpendicular. A **vector equation** of p is therefore

(12) $$(\mathbf{R} - \mathbf{R}_0) \cdot \mathbf{n} = 0$$

which, when expanded, gives

(13) $$a(x - x_0) + b(y - y_0) + c(z - z_0) = 0$$

as a **scalar equation** of p. Carefully note where the *components,* and the *coordinates* of the initial point, of $\mathbf{n}$ appear in (13).

To show that the points A, B, C determine a plane, we shall establish that they are noncollinear by proving that the two vectors originating at A and terminating at B and C are nonparallel. These two vectors are given in terms of the respective position vectors $\mathbf{A}, \mathbf{B}, \mathbf{C}$ of the given points by

$$\mathbf{B} - \mathbf{A} = 2\mathbf{i} + 2\mathbf{j} - 3\mathbf{k} \qquad \mathbf{C} - \mathbf{A} = \mathbf{i} + 2\mathbf{j} - 2\mathbf{k}$$

Their cross product

$$\mathbf{n} = (\mathbf{B} - \mathbf{A}) \times (\mathbf{C} - \mathbf{A}) = \begin{vmatrix} \mathbf{i} & \mathbf{j} & \mathbf{k} \\ 2 & 2 & -3 \\ 1 & 2 & -2 \end{vmatrix} = 2\mathbf{i} + \mathbf{j} + 2\mathbf{k}$$

is not zero; hence, $\mathbf{B} - \mathbf{A}$ and $\mathbf{C} - \mathbf{A}$ are not parallel and A, B, C determine a plane. Substituting the coordinates of A: $(1, 3, 3)$ and the components of the normal $\mathbf{n}$ to the plane at A into (13), we obtain

$$2(x - 1) + (y - 3) + 2(z - 3) = 0$$

or

$$2x + y + 2z = 11$$

as an equation of the plane determined by A, B, and C.

An indicated product of the form $\mathbf{A} \times (\mathbf{B} \times \mathbf{C})$ is meaningful because, by definition, $\mathbf{B} \times \mathbf{C}$ is a vector and hence is an appropriate factor in a vector product with $\mathbf{A}$. However, there are vectors $\mathbf{A}$, $\mathbf{B}$, and $\mathbf{C}$ for which

(14) $$\mathbf{A} \times (\mathbf{B} \times \mathbf{C}) \neq (\mathbf{A} \times \mathbf{B}) \times \mathbf{C}$$

For instance, if $\mathbf{A} = \mathbf{B} = \mathbf{i}$ and $\mathbf{C} = \mathbf{j}$, then $\mathbf{A} \times (\mathbf{B} \times \mathbf{C})$ and $(\mathbf{A} \times \mathbf{B}) \times \mathbf{C}$ reduce to the unequal vectors $-\mathbf{j}$ and $\mathbf{0}$, respectively. Therefore, we can state the following theorem.

THEOREM 3 Vector multiplication is *not* associative.

EXERCISES

In Exercises 1–6, find $\mathbf{A} \times \mathbf{B}$, $|\mathbf{A} \times \mathbf{B}|$, and the angle θ between $\mathbf{A}$ and $\mathbf{B}$.

1. $\mathbf{A} = 2\mathbf{i} - 3\mathbf{j} + \mathbf{k}$, $\mathbf{B} = 3\mathbf{i} + 2\mathbf{j}$

2. $\mathbf{A} = \mathbf{i} + 2\mathbf{j} - 2\mathbf{k}$, $\mathbf{B} = 2\mathbf{i} - \mathbf{j} + 3\mathbf{k}$

3. $\mathbf{A} = \mathbf{i} + \mathbf{j} + \mathbf{k}$, $\mathbf{B} = 2\mathbf{i} + 3\mathbf{j} - \mathbf{k}$

4. $\mathbf{A} = \mathbf{i} + 3\mathbf{j} - 2\mathbf{k}$, $\mathbf{B} = -2\mathbf{i} - 6\mathbf{j} + 4\mathbf{k}$

5. $\mathbf{A} = 3\mathbf{i} + \mathbf{j} - 2\mathbf{k}$, $\mathbf{B} = \mathbf{i} + 2\mathbf{j} + \mathbf{k}$

6. $\mathbf{A} = 2\mathbf{i} + \mathbf{j} + 2\mathbf{k}$, $\mathbf{B} = 2\mathbf{i} - \mathbf{j} + 4\mathbf{k}$

In Exercises 7–12, use both the geometric and the analytic definitions to find $\mathbf{A} \times \mathbf{B}$.

7. $\mathbf{A} = 2\mathbf{i} - 2\mathbf{j} + \mathbf{k}$, $\mathbf{B} = 12\mathbf{i} - 4\mathbf{j} - 3\mathbf{k}$

8. $\mathbf{A} = 2\mathbf{i} - 3\mathbf{j} + 6\mathbf{k}$, $\mathbf{B} = 2\mathbf{i} - 9\mathbf{j} - 6\mathbf{k}$

9. $\mathbf{A} = 10\mathbf{i} + 10\mathbf{j} + 5\mathbf{k}$, $\mathbf{B} = 4\mathbf{i} + 7\mathbf{j} - 4\mathbf{k}$

10. $\mathbf{A} = \mathbf{i} + 2\mathbf{j} + 2\mathbf{k}$, $\mathbf{B} = \mathbf{i} + \mathbf{j}$

11. $\mathbf{A} = \mathbf{i} + \mathbf{j} - \mathbf{k}$, $\mathbf{B} = 3\mathbf{i} - \mathbf{j} + 2\mathbf{k}$

12. $\mathbf{A} = 2\mathbf{i} - \mathbf{j} + 3\mathbf{k}$, $\mathbf{B} = \mathbf{i} - 4\mathbf{j} + \mathbf{k}$

13. Find a unit vector perpendicular to both $\mathbf{i} - 2\mathbf{j} + \mathbf{k}$ and $-5\mathbf{i} + 4\mathbf{j} - 2\mathbf{k}$.

14. Given the vectors $\mathbf{v} = 2\mathbf{i} - 2\mathbf{j} + \mathbf{k}$ and $\mathbf{w} = -2\mathbf{i} + 3\mathbf{j} - 6\mathbf{k}$, find all vectors $\mathbf{u}$ such that $\mathbf{u} \times \mathbf{v} = \mathbf{w}$.

15. Find the area of the parallelogram which has $2\mathbf{i} - \mathbf{j} + \mathbf{k}$ and $-\mathbf{i} + 2\mathbf{j} + 2\mathbf{k}$ as two adjacent sides.

16. Find the area of the triangle whose vertices are the points:
(a) $(1, -2, 1)$, $(-2, 1, -2)$, $(-3, -2, 1)$
(b) $(-1, -2, 1)$, $(0, 0, 4)$, $(-3, 2, -1)$

17. Find the area of every parallelogram which has the points of Exercise 16(**a**) as three of its vertices.

18. A triangle has vertices $(1, 1, 1)$, $(2, 2, 2)$, and $(1, 1, y)$ and area $\csc \pi/4$. Find y.

19. Show that the area of a triangle whose vertices are $(x_1, y_1, 0)$, $(x_2, y_2, 0)$, and $(x_3, y_3, 0)$ equals one-half the absolute value of the determinant

$$\begin{vmatrix} 1 & 1 & 1 \\ x_1 & x_2 & x_3 \\ y_1 & y_2 & y_3 \end{vmatrix}$$

20. The vector $\mathbf{v} = -2\mathbf{i} + 3\mathbf{j} + 6\mathbf{k}$ is projected onto the line through $(1, 0, -1)$ and $(-3, 6, 11)$. Find the length of the projection. Is the line parallel to $\mathbf{v}$?

Using Eq. (8), show that for all vectors $\mathbf{A}$ and $\mathbf{B}$ in 3-space

21. $\mathbf{A} \times \mathbf{B}$ is perpendicular to both $\mathbf{A}$ and $\mathbf{B}$.

22. If either $\mathbf{A}$ or $\mathbf{B}$ is zero, then $\mathbf{A} \times \mathbf{B} = \mathbf{0}$.

23. $\mathbf{A} \times (c\mathbf{B}) = (c\mathbf{A}) \times \mathbf{B} = c(\mathbf{A} \times \mathbf{B})$, c an arbitrary real number

24. $|\mathbf{A} \times \mathbf{B}| = AB \sin \theta$. *Hint:* First show that $|\mathbf{A} \times \mathbf{B}|^2 = A^2B^2 - (\mathbf{A} \cdot \mathbf{B})^2$.

25. Show that for all vectors $\mathbf{A}$ and $\mathbf{B}$ in 3-space, $(\mathbf{A} + \mathbf{B}) \times (\mathbf{A} - \mathbf{B}) = 2(\mathbf{B} \times \mathbf{A})$.

26. Find symmetric equations of the line through $(-2, 5, -1)$ perpendicular to the plane $y + 2z = 7$.

27. **(a)** Find symmetric equations of the line containing (1, 4, 3) which is perpendicular to both of the lines $\frac{x-1}{2} = y + 3 = \frac{z-2}{4}$ and $\frac{x+2}{3} = \frac{y-4}{2} = \frac{z+1}{-2}$.
(b) Find all points on the line the square of whose distance from (1, 4, 3) is 357.

28. Find **(a)** a parametric vector equation and **(b)** symmetric equations for the intersection of the planes $x - 2y + z = 8$ and $2x - y + z = 10$.

29. Find parametric equations for the intersection of the planes $2x - y + z + 2 = 0$ and $x + y + z = 0$.

30. **(a)** Show that the two sets of parametric scalar equations

$$x = 1,\ y = 3 + 2t,\ z = 4 + t$$

and

$$x = 1 + 4s,\ y = 3 + 2s,\ z = 4 + 2s$$

define nonparallel lines.
(b) Find a vector and a scalar equation of the plane containing these lines.

31. **(a)** Find the obtuse angle between the lines defined by

$$x = 1 + s,\ y = 2 + s,\ z = -1$$

and

$$x = 1 + t,\ y = 2 + 2t,\ z = -1 + 2t$$

(b) Find symmetric equations of the line perpendicular to these lines which passes through their point of intersection.

32. Find the point on the y axis that falls on the line through $(2, 1, 2/\sqrt{3})$ and $(1, 2, 1/\sqrt{3})$.

33. Modify Eqs. (11) to suit the special case of Eqs. (10) for which:
(a) $a_1 = 0,\ a_1a_2 \neq 0$ **(b)** $a_1 \neq 0,\ a_2^2 + a_3^2 = 0$

34. Does Eq. (9) remain an equation of the line through P_0: (x_0, y_0, z_0) if **A** is replaced by an arbitrary nonzero vector parallel to **A**? Explain.

35. Three vertices of a parallelogram are A: (1, 1, 1), B: (2, −3, 5), and C: (4, −1, 6). One of its diagonals joins B and C. Find **(a)** the area of the parallelogram, **(b)** its other vertex, and **(c)** symmetric equations of the line lying along its other diagonal.

36. Find an equation of the plane through (0, 0, 3), (1, 1, 1) and (−1, 1, 2).

37. Find an equation of the plane perpendicular to $\mathbf{i} + 2\mathbf{j} + 3\mathbf{k}$ which passes through the point (1, 2, 3).

38. Find an equation of the plane through (1, −2, 3) perpendicular to the line $\mathbf{R} - (2\mathbf{i} - \mathbf{j} - 2\mathbf{k}) = t(2\mathbf{i} + \mathbf{j} + \mathbf{k})$.

39. Find an equation of the plane containing the line $\frac{x+2}{3} = \frac{y}{-2} = z$ and the point (1, 1, 2).

40. In Eq. (12), if **n** is normalized, a **normal vector equation**

$$(15) \qquad (\mathbf{R} - \mathbf{R}_0) \cdot \frac{\mathbf{n}}{n} = 0$$

of the plane p, Example 3, is obtained. Show that a corresponding **normal scalar equation** is

$$(16) \qquad \frac{a(x - x_0) + b(y - y_0) + c(z - z_0)}{\sqrt{a^2 + b^2 + +c^2}} = 0$$

41. **(a)** Show that if **n** is a normal vector to a plane p at any point P_0: (x_0, y_0, z_0) of p, then the distance d from an arbitrary point P_1: (x_1, y_1, z_1) to p is given by

$$(17) \qquad d = \left|(\mathbf{R}_1 - \mathbf{R}_0) \cdot \frac{\mathbf{n}}{n}\right|$$

Hint: Project $\mathbf{R}_1 - \mathbf{R}_0$ onto **n**.
(b) Express (17) in scalar form.

42. Find the distance from the point (6, 2, 1) to the plane through (1, 2, 3) perpendicular to $2\mathbf{i} + 2\mathbf{j} + \mathbf{k}$.

43. How far is the point (5, 4, 5) from the plane $x + 2y + 3z = 14$?

44. How far is the origin from the plane $3x + 4y - z = 26$?

45. Find an equation for each plane that is parallel to, and a unit distance from, the plane $2x - y + 2z = 5$.

46. Find an equation of each plane that is parallel to, and at a distance 3 from, the plane $x - 2y - 2z = 10$.

47. Show that if P_1: (x_1, y_1, z_1) and P_2: (x_2, y_2, z_2) are any two distinct points on a line L, then the distance d from an arbitrary point A: (a_1, a_2, a_3) to L is given by

$$(18) \qquad d = \left|\frac{(\mathbf{A} - \mathbf{R}_1) \times (\mathbf{R}_2 - \mathbf{R}_1)}{|\mathbf{R}_2 - \mathbf{R}_1|}\right|$$

Hint: What is the projection of $\mathbf{A} - \mathbf{R}_1$ onto a line perpendicular to $\mathbf{R}_2 - \mathbf{R}_1$?

48. Find the distance from the point (5, 0, −4) to the line $x - 1 = \frac{y+2}{-2} = \frac{z+1}{3}$.

49. Find the distance between the plane $11x - 3y + 7z = 33$ and the plane through the points (1, 0, 4), (2, −1, 2), and (4, 4, 1).

50. Find a unit vector which is parallel to the plane of $\mathbf{i} + \mathbf{j} - 2\mathbf{k}$ and $3\mathbf{i} - 2\mathbf{j} + \mathbf{k}$ and perpendicular to $2\mathbf{i} + 2\mathbf{j} - \mathbf{k}$.

51. Show that the lines $\mathbf{R} = \mathbf{R}_0 + t\mathbf{A}$ and $\mathbf{R} = \mathbf{R}_1 + s\mathbf{B}$ intersect if and only if $(\mathbf{R}_0 - \mathbf{R}_1) \cdot (\mathbf{A} \times \mathbf{B}) = 0$.

52. Show that if $\mathbf{A} + \mathbf{B} + \mathbf{C} = \mathbf{0}$, then $\mathbf{A} \times \mathbf{B} = \mathbf{B} \times \mathbf{C} = \mathbf{C} \times \mathbf{A}$. Is the converse true?

53. Prove that two nonzero vectors are linearly dependent if and only if they are parallel.

54. By considering the cross product of $\mathbf{A} = a_1\mathbf{i} + a_2\mathbf{j}$ and $\mathbf{B} = b_1\mathbf{i} + b_2\mathbf{j}$, derive the formula for the sine of the difference of two angles.

55. Three noncollinear points L, M, and N lie in a plane p. Prove that if **L**, **M**, and **N** are the vectors to these points from an arbitrary origin, then the vector $(\mathbf{L} \times \mathbf{M}) + (\mathbf{M} \times \mathbf{N}) + (\mathbf{N} \times \mathbf{L})$ is perpendicular to p.

56. Show that the product $\mathbf{A} \times \mathbf{B}$ can be obtained by the following steps: First, project **B** onto a plane perpendicular to **A**, then rotate this projection through 90° about **A**, and finally multiply this rotated vector by $|\mathbf{A}|$. Show, further, that these steps always transform the sides of a triangle into the sides of a second triangle. Applying this result to an arbitrary triangle with sides **B**, **C**, and $\mathbf{B} + \mathbf{C}$, show that vector multiplication is distributive over addition; that is, show that

$$\mathbf{A} \times (\mathbf{B} + \mathbf{C}) = (\mathbf{A} \times \mathbf{B}) + (\mathbf{A} \times \mathbf{C})$$

and

$$(\mathbf{B} + \mathbf{C}) \times \mathbf{A} = (\mathbf{B} \times \mathbf{A}) + (\mathbf{C} \times \mathbf{A})$$

57. In mechanics the **moment** M **of a force F about a point** O is defined as the magnitude of **F** times the perpendicular distance from the point O to the line of action of **F**. If the **vector moment M** is defined as the vector whose magnitude is M and whose direction is perpendicular to the plane of O and **F**, show that $\mathbf{M} = \mathbf{R} \times \mathbf{F}$, where **R** is the vector from O to any point on the line of action of **F**. Would $\mathbf{M} = \mathbf{F} \times \mathbf{R}$ be an equally acceptable definition? Explain.

58. Let B be a rigid body which is rotating with uniform angular velocity ω about an axis l, and let the motion of B be described by a vector **w** of length ω lying along l and so directed that the rotation of B would cause a right-handed screw to advance in the direction of **w**. If **R** is the vector drawn from a fixed origin on l to an arbitrary point P in B, show that the linear velocity of P is $\mathbf{w} \times \mathbf{R}$.

15.3 COMPOSITION OF PRODUCTS INVOLVING VECTORS

Products involving three rather than merely one or two vectors offer the following possibilities:

$$(\mathbf{A} \cdot \mathbf{B})\mathbf{C} \qquad \mathbf{A} \cdot (\mathbf{B} \times \mathbf{C}) \qquad \mathbf{A} \times (\mathbf{B} \times \mathbf{C})$$

The first of these is very easy to analyze. In fact, $\mathbf{A} \cdot \mathbf{B}$ is just a scalar, and thus $(\mathbf{A} \cdot \mathbf{B})\mathbf{C}$ is simply a vector whose length is $|\mathbf{A} \cdot \mathbf{B}|$ times the length of **C** and whose direction, if $\mathbf{C} \neq \mathbf{0}$, is the same as that of **C** or opposite it, according to whether $\mathbf{A} \cdot \mathbf{B}$ is positive or negative.

The second combination of products $\mathbf{A} \cdot (\mathbf{B} \times \mathbf{C})$ is known as the **scalar triple product** of **A**, **B**, and **C**. In this product, the parentheses enclosing the vector product $\mathbf{B} \times \mathbf{C}$ are superfluous because there is only one alternative interpretation of $\mathbf{A} \cdot \mathbf{B} \times \mathbf{C}$, namely, $(\mathbf{A} \cdot \mathbf{B}) \times \mathbf{C}$, and this is meaningless, since both factors in a cross product must be vectors, whereas $\mathbf{A} \cdot \mathbf{B}$ is a scalar. Thus no meaning but the intended one can be sensibly attributed to an expression $\mathbf{A} \cdot \mathbf{B} \times \mathbf{C}$, and so it is customary to omit parentheses and interpret such a product as the dot product of **A** and $\mathbf{B} \times \mathbf{C}$. Of course, the value of every scalar triple product is a scalar.

To evaluate $\mathbf{A} \cdot \mathbf{B} \times \mathbf{C}$ in terms of the components of **A**, **B**, and **C**, we first write the determinant representing $\mathbf{B} \times \mathbf{C}$ and then expand the determinant using Laplace's expansion by entries of the first row, to obtain

$$\mathbf{B} \times \mathbf{C} = \begin{vmatrix} \mathbf{i} & \mathbf{j} & \mathbf{k} \\ b_1 & b_2 & b_3 \\ c_1 & c_2 & c_3 \end{vmatrix} = \begin{vmatrix} b_2 & b_3 \\ c_2 & c_3 \end{vmatrix} \mathbf{i} - \begin{vmatrix} b_1 & b_3 \\ c_1 & c_3 \end{vmatrix} \mathbf{j} + \begin{vmatrix} b_1 & b_2 \\ c_1 & c_2 \end{vmatrix} \mathbf{k}$$

From this it follows that

$$\mathbf{A} \cdot \mathbf{B} \times \mathbf{C} = a_1 \begin{vmatrix} b_2 & b_3 \\ c_2 & c_3 \end{vmatrix} - a_2 \begin{vmatrix} b_1 & b_3 \\ c_1 & c_3 \end{vmatrix} + a_3 \begin{vmatrix} b_1 & b_2 \\ c_1 & c_2 \end{vmatrix}$$

which is just the relation

$$(1) \qquad \mathbf{A} \cdot \mathbf{B} \times \mathbf{C} = \begin{vmatrix} a_1 & a_2 & a_3 \\ b_1 & b_2 & b_3 \\ c_1 & c_2 & c_3 \end{vmatrix}$$

with the determinant in the right-hand member written as its Laplace's expansion by the first row.

By writing Laplace's expansion by the first row of the determinant representing **A × B** and dotting it into **C**, we obtain

$$\mathbf{A} \times \mathbf{B} \cdot \mathbf{C} = \begin{vmatrix} a_2 & a_3 \\ b_2 & b_3 \end{vmatrix} c_1 - \begin{vmatrix} a_1 & a_3 \\ b_1 & b_3 \end{vmatrix} c_2 + \begin{vmatrix} a_1 & a_2 \\ b_1 & b_2 \end{vmatrix} c_3$$

which is Laplace's expansion by the third row of the determinant in (1). Hence

$$\mathbf{A} \times \mathbf{B} \cdot \mathbf{C} = \mathbf{A} \cdot \mathbf{B} \times \mathbf{C} \tag{2}$$

because the value of a determinant is given by Laplace's expansion by any of its rows. In words, Eq. (2) asserts

THEOREM 1 In any scalar triple product, the dot and the cross can be interchanged without altering the value of the product.

The scalar triple product of **A**, **B**, and **C** is usually written simply as [**ABC**] and, because of the appearance of this notation, is often called their **box product,** although a more convincing reason is found in Theorem 3.

Since the value of a determinant is reversed in sign by interchanging any two rows [Property 7, Sec. 3.5], from (1) or the anticommutative character of cross products it follows that $\mathbf{A} \cdot \mathbf{B} \times \mathbf{C} = -\mathbf{A} \cdot \mathbf{C} \times \mathbf{B}$ and, more generally,

$$[\mathbf{ABC}] = [\mathbf{BCA}] = [\mathbf{CAB}] = -[\mathbf{ACB}] = -[\mathbf{BAC}] = -[\mathbf{CBA}] \tag{3}$$

A **cyclic permutation** is the retreating of each member of a finite ordered set one position, the first member taking the place of the last. In particular, a cyclic permutation of **ABC** is **BCA**. According to Eqs. (3)

THEOREM 2 Any cyclic permutation of the factors in a scalar triple product leaves the value of the product unchanged, whereas any permutation which reverses the original cyclic order changes the sign of the product.

Geometrically, three noncoplanar vectors **A**, **B**, and **C**, all drawn from the same point, determine a parallelepiped having these vectors as concurrent edges. If we regard the parallelogram having **B** and **C** as adjacent sides as the base of the parallelepiped, then **B × C** is a vector whose direction is perpendicular to the base and whose magnitude is equal to the area of the base. Moreover, the altitude $h = \mathbf{A} \cos \theta$ of the parallelepiped (Fig. 15.6a) is the projection of **A** along

FIGURE 15.6
Geometric interpretations of the scalar triple product.

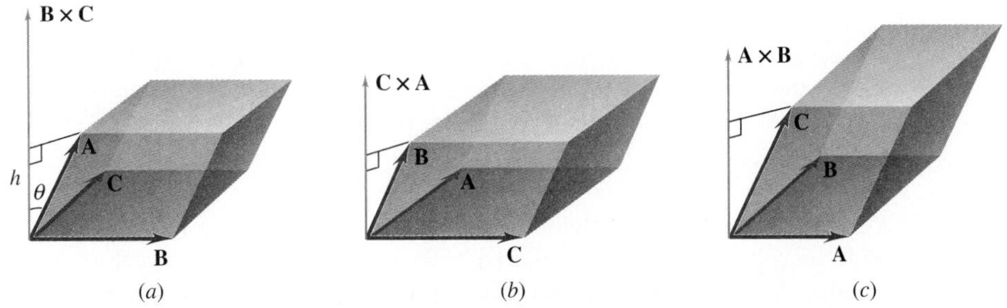

$\mathbf{B} \times \mathbf{C}$. Hence $\mathbf{A} \cdot \mathbf{B} \times \mathbf{C}$, whose value is just the magnitude of $\mathbf{B} \times \mathbf{C}$ multiplied by the projection of $\mathbf{A}$ on $\mathbf{B} \times \mathbf{C}$, is numerically equal to the volume of the parallelepiped. As an important geometric interpretation of the scalar triple product, we thus have

THEOREM 3 The absolute value of the scalar triple product $\mathbf{A} \cdot \mathbf{B} \times \mathbf{C}$ of three noncoplanar vectors is the volume of the parallelepiped having **A**, **B**, and **C** as concurrent edges.

If the vectors **A**, **B**, and **C** are coplanar or parallel to the same plane, they determine a degenerate parallelepiped of zero volume, and conversely. Hence

(4) $[\mathbf{ABC}] = 0$ if and only if **A**, **B**, and **C** are parallel to one and the same plane.

In particular, if two factors of a box product have the same direction, the product is zero.

EXAMPLE 1

If **A**, **B**, and **C** are three vectors which are not parallel to the same plane, show that any vector **V** can be expressed as a linear combination of **A**, **B**, and **C**.

If we write

$$\mathbf{V} = a\mathbf{A} + b\mathbf{B} + c\mathbf{C} \tag{5}$$

where a, b, and c are scalar constants to be determined, and form the cross product of each member with the vector **B**, we obtain

$$\mathbf{V} \times \mathbf{B} = a\mathbf{A} \times \mathbf{B} + b\mathbf{B} \times \mathbf{B} + c\mathbf{C} \times \mathbf{B} = a\mathbf{A} \times \mathbf{B} + c\mathbf{C} \times \mathbf{B}$$

where the term $b\mathbf{B} \times \mathbf{B}$ is $\mathbf{0}$ because it is the product of a scalar and a cross product whose factors are identical. Now if we form the dot product of the last result with the vector **C**, we have

$$\mathbf{V} \times \mathbf{B} \cdot \mathbf{C} = a\mathbf{A} \times \mathbf{B} \cdot \mathbf{C} + c\mathbf{C} \times \mathbf{B} \cdot \mathbf{C} = a\mathbf{A} \times \mathbf{B} \cdot \mathbf{C}$$

where the term $c\mathbf{C} \times \mathbf{B} \cdot \mathbf{C}$ vanishes because it is the product of a scalar and a scalar triple product with two factors identical. By hypothesis, **A**, **B**, and **C** are not parallel to the same plane. Hence $\mathbf{A} \times \mathbf{B} \cdot \mathbf{C}$ is different from zero, and we can solve for a, getting

$$a = \frac{[\mathbf{VBC}]}{[\mathbf{ABC}]}$$

In the same way we can obtain the remaining constants in the required linear combination:

$$b = \frac{[\mathbf{AVC}]}{[\mathbf{ABC}]} \qquad c = \frac{[\mathbf{ABV}]}{[\mathbf{ABC}]}$$

Thus, under the conditions of the problem,

$$\mathbf{V} = \frac{[\mathbf{VBC}]}{[\mathbf{ABC}]}\mathbf{A} + \frac{[\mathbf{AVC}]}{[\mathbf{ABC}]}\mathbf{B} + \frac{[\mathbf{ABV}]}{[\mathbf{ABC}]}\mathbf{C} \tag{6}$$

The following special case of this result is often useful. If **V** is any vector parallel to the plane determined by **A** and **B**, then $[\mathbf{ABV}] = 0$ and the last term in the expansion (6) is zero. Hence it follows that *if* **A** *and* **B** *are vectors which are not parallel to the same line and if* **V** *is any vector parallel to the plane determined by* **A** *and* **B**, *then* **V** *can be expressed as a linear combination of* **A** *and* **B**.

Since vector multiplication is not associative [Theorem 3, Sec. 15.2], there are vectors **A**, **B**, and **C** for which $\mathbf{A} \times (\mathbf{B} \times \mathbf{C})$ and $(\mathbf{A} \times \mathbf{B}) \times \mathbf{C}$ denote different vectors. Hence, the position of parentheses as symbols of grouping in a **vector triple product** $\mathbf{A} \times \mathbf{B} \times \mathbf{C}$ is significant. To find a simpler expanded form for a vector triple product interpreted as $\mathbf{A} \times (\mathbf{B} \times \mathbf{C})$, we use the expression (7) and the determinant (8) of Sec. 15.2 to express the cross product of $\mathbf{A}$ and $\mathbf{B} \times \mathbf{C}$ as

$$\mathbf{A} \times (\mathbf{B} \times \mathbf{C}) = \begin{vmatrix} \mathbf{i} & \mathbf{j} & \mathbf{k} \\ a_1 & a_2 & a_3 \\ b_2c_3 - b_3c_2 & b_3c_1 - b_1c_3 & b_1c_2 - b_2c_1 \end{vmatrix} \tag{7}$$

Expanding this symbolic determinant in terms of the (vector) elements of the first row and grouping the positive and the negative terms in each cofactor, we obtain

$$\begin{aligned}\mathbf{i}[b_1(a_2c_2 + a_3c_3) - c_1(a_2b_2 + a_3b_3)]& \\ + \mathbf{j}[b_2(a_1c_1 + a_3c_3) - c_2(a_1b_1 + a_3b_3)]& \\ + \mathbf{k}[b_3(a_1c_1 + a_2c_2) - c_3(a_1b_1 + a_2b_2)]&\end{aligned}$$

If we now add and subtract the products $a_1b_1c_1$, $a_2b_2c_2$, $a_3b_3c_3$ in the respective bracketed expressions and note the dot products we have thus formed, this becomes

$$\mathbf{i}[b_1(\mathbf{A} \cdot \mathbf{C}) - c_1(\mathbf{A} \cdot \mathbf{B})] + \mathbf{j}[b_2(\mathbf{A} \cdot \mathbf{C}) - c_2(\mathbf{A} \cdot \mathbf{B})] + \mathbf{k}[b_3(\mathbf{A} \cdot \mathbf{C}) - c_3(\mathbf{A} \cdot \mathbf{B})]$$

Finally, by regrouping and factoring, then substituting into (7), we get the required formula:

$$\mathbf{A} \times (\mathbf{B} \times \mathbf{C}) = (\mathbf{A} \cdot \mathbf{C})\mathbf{B} - (\mathbf{A} \cdot \mathbf{B})\mathbf{C} \tag{8}$$

By a straightforward application of Eq. (8) we find that

$$(\mathbf{A} \times \mathbf{B}) \times \mathbf{C} = -\mathbf{C} \times (\mathbf{A} \times \mathbf{B}) = -(\mathbf{C} \cdot \mathbf{B})\mathbf{A} + (\mathbf{C} \cdot \mathbf{A})\mathbf{B} \tag{9}$$

which, if $(\mathbf{C} \cdot \mathbf{B})\mathbf{A} \neq (\mathbf{A} \cdot \mathbf{B})\mathbf{C}$, is not equal to $\mathbf{A} \times (\mathbf{B} \times \mathbf{C})$. Thus, we have another proof that vector multiplication is not associative.

With a knowledge of scalar and vector triple products, products involving more than three vectors can be expanded without difficulty. For instance,

$$(\mathbf{A} \times \mathbf{B}) \cdot (\mathbf{C} \times \mathbf{D})$$

can be regarded as the scalar triple product of the vectors $\mathbf{A}$, $\mathbf{B}$, and $\mathbf{C} \times \mathbf{D}$. This, together with Theorem 1 and Eq (8), allows us to write

$$\begin{aligned}(\mathbf{A} \times \mathbf{B}) \cdot (\mathbf{C} \times \mathbf{D}) &= \mathbf{A} \cdot [\mathbf{B} \times (\mathbf{C} \times \mathbf{D})] \\ &= \mathbf{A} \cdot [(\mathbf{B} \cdot \mathbf{D})\mathbf{C} - (\mathbf{B} \cdot \mathbf{C})\mathbf{D}]\end{aligned}$$

or

$$(\mathbf{A} \times \mathbf{B}) \cdot (\mathbf{C} \times \mathbf{D}) = (\mathbf{A} \cdot \mathbf{C})(\mathbf{B} \cdot \mathbf{D}) - (\mathbf{A} \cdot \mathbf{D})(\mathbf{B} \cdot \mathbf{C}) \tag{10}$$

This result is known as **Lagrange's identity.**

We shall consistently regard $\mathbf{A} \times \mathbf{B} \times \mathbf{C} \times \mathbf{D}$ as the vector triple product of $\mathbf{A} \times \mathbf{B}$, $\mathbf{C}$, and $\mathbf{D}$, interpreted as $(\mathbf{A} \times \mathbf{B}) \times (\mathbf{C} \times \mathbf{D})$. Taking this point of view and applying (8), we find

$$\begin{aligned}(\mathbf{A} \times \mathbf{B}) \times (\mathbf{C} \times \mathbf{D}) &= (\mathbf{A} \times \mathbf{B} \cdot \mathbf{D})\mathbf{C} - (\mathbf{A} \times \mathbf{B} \cdot \mathbf{C})\mathbf{D} \\ &= [\mathbf{ABD}]\mathbf{C} - [\mathbf{ABC}]\mathbf{D}\end{aligned} \tag{11}$$

which, assuming $\mathbf{C}$ and $\mathbf{D}$ are linearly independent, is a vector in the plane of $\mathbf{C}$ and $\mathbf{D}$. On the other hand, the anticommutative property of vector multiplication, coupled with (8), gives

$$\begin{aligned}(\mathbf{A} \times \mathbf{B}) \times (\mathbf{C} \times \mathbf{D}) &= -(\mathbf{C} \times \mathbf{D}) \times (\mathbf{A} \times \mathbf{B}) \\ &= -(\mathbf{C} \times \mathbf{D} \cdot \mathbf{B})\mathbf{A} + (\mathbf{C} \times \mathbf{D} \cdot \mathbf{A})\mathbf{B} \\ &= [\mathbf{CDA}]\mathbf{B} - [\mathbf{CDB}]\mathbf{A}\end{aligned} \tag{12}$$

which, if **A** and **B** are linearly independent, is a vector in the plane of **A** and **B**. Eqs. (11) and (12) together show that (**A** × **B**) × (**C** × **D**) is directed along the line of intersection of the plane of **C** and **D** and the plane of **A** and **B**, provided these planes are distinct.

EXERCISES

In Exercises 1–3, find **(a)** [**ABC**], **(b)** **A** × (**B** × **C**), **(c)** (**A** × **B**) × **C**, **(d)** the volume of the parallelepiped having **A** + **C**, **A** − **C**, and **B** as concurrent edges, **(e)** the volume of the parallelepiped having **A** + **C**, **A** − **C**, and **C** as concurrent edges, **(f)** (**A** × **B**) · (**C** × **D**), and **(g)** (**A** × **B**) × (**C** × **D**).

1. $\mathbf{A} = 2\mathbf{i} - 2\mathbf{j} + \mathbf{k}$
$\mathbf{B} = \mathbf{i} + 8\mathbf{j} - 4\mathbf{k}$
$\mathbf{C} = 12\mathbf{i} - 4\mathbf{j} - 3\mathbf{k}$
$\mathbf{D} = \mathbf{i} + 2\mathbf{j} - \mathbf{k}$

2. $\mathbf{A} = 2\mathbf{i} - 3\mathbf{j} + 6\mathbf{k}$
$\mathbf{B} = 10\mathbf{i} + 2\mathbf{j} + 11\mathbf{k}$
$\mathbf{C} = 2\mathbf{i} - 9\mathbf{j} - 6\mathbf{k}$
$\mathbf{D} = 3\mathbf{i} + 2\mathbf{k}$

3. $\mathbf{A} = 10\mathbf{i} + 10\mathbf{j} + 5\mathbf{k}$
$\mathbf{B} = 5\mathbf{i} - 2\mathbf{j} - 14\mathbf{k}$
$\mathbf{C} = 4\mathbf{i} + 7\mathbf{j} - 4\mathbf{k}$
$\mathbf{D} = 2\mathbf{i} - \mathbf{j} + \mathbf{k}$

4. Determine b such that the line through $(5, 0, 3)$ and $(-1, -10, b)$ will be perpendicular to $(\mathbf{i} + \mathbf{k}) \times (\mathbf{i} \times \mathbf{k})$.

5. Given that $\mathbf{u} = 2\mathbf{i} + \mathbf{j} - \mathbf{k}$, $\mathbf{v} = \mathbf{i} - \mathbf{j} + 2\mathbf{k}$, $\mathbf{w} = \mathbf{i} - \mathbf{k}$, and $c = -2$, find the volume of the parallelepiped having $-c\mathbf{u}$, $\mathbf{v}$, and $c\mathbf{w}$ as concurrent edges.

6. Find the volume of the parallelepiped having $\mathbf{U} = -\mathbf{i} + \mathbf{j} + 3\mathbf{k}$, $\mathbf{V} = \mathbf{i} + 2\mathbf{k}$, and $\mathbf{W} = 3\mathbf{i} + \mathbf{j} + \mathbf{k}$ as concurrent edges; also find the scalar c for which $-c\mathbf{i} + 2c\mathbf{j} + \mathbf{k}$ is perpendicular to $\mathbf{U} \times (\mathbf{V} \times \mathbf{W})$.

7. Find all scalars b and c for which the line through $(-2, 1, b)$ and $(4, 1, 2)$ is perpendicular to $\mathbf{i} + 3\mathbf{j} - 2\mathbf{k}$ and parallel to the plane of $\mathbf{i} + c\mathbf{k}$ and $c\mathbf{j} + b\mathbf{k}$.

8. Prove **(a)** that two vectors in 3-space are linearly dependent if and only if they are parallel, and **(b)** that three such vectors are linearly dependent if and only if they are parallel to the same plane.

9. Determine the scalar c for which the vectors $2\mathbf{i} - 5\mathbf{j} + \mathbf{k}$, $-3\mathbf{i} + 2\mathbf{j} + 4\mathbf{k}$, and $\mathbf{j} + c\mathbf{k}$ are linearly dependent.

10. Find an equation of the plane through $(0, 0, 3)$, $(1, 1, 1)$, and $(-1, 1, 2)$ by requiring the scalar triple product of $(x, y, z) - (0, 0, 3)$, $(1, 1, 1) - (0, 0, 3)$, and $(-1, 1, 2) - (0, 0, 3)$ to be zero.

11. Using the method of Exercise 10, find an equation of the plane through $(1, 0, 4)$, $(2, -1, 2)$, and $(4, 4, 1)$.

12. Show that if four vectors **A**, **B**, **C**, and **D** are coplanar, then $(\mathbf{A} \times \mathbf{B}) \times (\mathbf{C} \times \mathbf{D}) = \mathbf{0}$. Is the converse true?

13. Determine a scalar b such that $(\mathbf{B} \times \mathbf{C}) \times (\mathbf{C} \times \mathbf{A}) = b\mathbf{C}$, interpret $|b|$ geometrically, and express $(\mathbf{A} \times \mathbf{B}) \cdot (\mathbf{B} \times \mathbf{C}) \times (\mathbf{C} \times \mathbf{A})$ in terms of b.

14. Prove that four vectors of 3-space are always linearly dependent. *Hint:* Expand $(\mathbf{A} \times \mathbf{B}) \times (\mathbf{C} \times \mathbf{D})$ in two different ways and equate the results.

15. Find the length of the projection of $\mathbf{v} = -7\mathbf{i} + 14\mathbf{j} + 7\mathbf{k}$ along a line perpendicular to $\mathbf{u} = 2\mathbf{i} + 3\mathbf{j} - 6\mathbf{k}$ in the plane of **v** and **u**.

16. Using Lagrange's identity, with $\mathbf{C} = \mathbf{A}$ and $\mathbf{D} = \mathbf{B}$, verify that $|\mathbf{A} \times \mathbf{B}| = |\mathbf{A}|\,|\mathbf{B}|\,|\sin\theta|$.

17. Find all values of c such that the line through $(5, 0, c)$ and $(-1, 10, 25)$ is perpendicular to $\mathbf{i} + 5\mathbf{j} - 2\mathbf{k}$ and also parallel to the plane of $\mathbf{i} + \mathbf{k}$ and $7\mathbf{i} - 5\mathbf{j} - 7\mathbf{k}$.

18. If **A**, **B**, and **C** are any three vectors prove that

$$\mathbf{A} \times (\mathbf{B} \times \mathbf{C}) + \mathbf{B} \times (\mathbf{C} \times \mathbf{A}) + \mathbf{C} \times (\mathbf{A} \times \mathbf{B}) = \mathbf{0}$$

19. Prove that $(\mathbf{A} \times \mathbf{B}) \cdot (\mathbf{C} \times \mathbf{D}) + (\mathbf{B} \times \mathbf{C}) \cdot (\mathbf{A} \times \mathbf{D}) + (\mathbf{C} \times \mathbf{A}) \cdot (\mathbf{B} \times \mathbf{D}) = 0$.

20. If the plane determined by **A** and **B** is perpendicular to the plane determined by **C** and **D**, show that $(\mathbf{A} \times \mathbf{B}) \cdot (\mathbf{C} \times \mathbf{D}) = 0$.

21. Show that the volume of the parallelepiped having **A** + **B**, **B** + **C**, and **C** + **A** as concurrent edges is twice the volume of the parallelepiped having **A**, **B**, and **C** as concurrent edges.

22. Prove that for all real numbers $a_1, a_2, a_3, b_1, b_2, b_3$

$$\left(\sum_{n=1}^{3} a_n b_n\right)^2 \le \left(\sum_{n=1}^{3} a_n^2\right)\left(\sum_{n=1}^{3} b_n^2\right)$$

Hint: Consider $(a_1, a_2, a_3) \cdot (b_1, b_2, b_3)$.

23. Prove that $(\mathbf{A} \times \mathbf{B}) \cdot (\mathbf{B} \times \mathbf{C}) \times (\mathbf{C} \times \mathbf{A}) = [\mathbf{ABC}]^2$.

24. If **A**, **B**, and **C** are any three independent vectors, the vectors

$$\mathbf{U} = \frac{\mathbf{B} \times \mathbf{C}}{[\mathbf{ABC}]} \qquad \mathbf{V} = \frac{\mathbf{C} \times \mathbf{A}}{[\mathbf{ABC}]} \qquad \mathbf{W} = \frac{\mathbf{A} \times \mathbf{B}}{[\mathbf{ABC}]}$$

are said to form a set **reciprocal** to the set **A**, **B**, **C**. Show that

$$\mathbf{A} \cdot \mathbf{U} = \mathbf{B} \cdot \mathbf{V} = \mathbf{C} \cdot \mathbf{W} = 1 \quad \text{and} \quad [\mathbf{UVW}] = \frac{1}{[\mathbf{ABC}]}$$

If $\mathbf{A} = \mathbf{i} + 2\mathbf{j} - 2\mathbf{k}$, $\mathbf{B} = \mathbf{i} + 8\mathbf{j} + 4\mathbf{k}$, and $\mathbf{C} = 12\mathbf{i} - 4\mathbf{j} + 3\mathbf{k}$, express the vector $\mathbf{i} + 2\mathbf{j} + 3\mathbf{k}$ as a linear combination of **A**, **B**, and **C** and also as a linear combination of the vectors **U**, **V**, and **W** of the set reciprocal to **A**, **B**, and **C**.

25. Show that if $\mathbf{A} = a_1\mathbf{i} + a_2\mathbf{j} + a_3\mathbf{k}$, $\mathbf{B} = b_1\mathbf{i} + b_2\mathbf{j} + b_3\mathbf{k}$, $\mathbf{C} = c_1\mathbf{i} + c_2\mathbf{j} + c_3\mathbf{k}$, and $\mathbf{D} = d_1\mathbf{i} + d_2\mathbf{j} + d_3\mathbf{k}$, then the system of equations

$$a_1x + b_1y + c_1z = d_1$$
$$a_2x + b_2y + c_2z = d_2$$
$$a_3x + b_3y + c_3z = d_3$$

is equivalent to the single vector equation $x\mathbf{A} + y\mathbf{B} + z\mathbf{C} = \mathbf{D}$. Assuming that $[\mathbf{ABC}] \neq 0$, solve this vector equation for x, y, and z, and show that the result is equivalent to that obtained from the algebraic form of the system by using Cramer's rule (Theorem 5, Sec. 3.7).

15.4 VECTOR FUNCTIONS OF ONE VARIABLE

The introduction of ideas and methods of calculus is essential to many of the most important theoretical results and practical applications involving vectors and vector functions. If to each member of a nonempty set there corresponds a unique vector, we have a **vector function.** In our study of vector analysis the range of the vector functions we consider will always be some subset of 3-space. Indeed, in our work we shall assume that every vector is expressible as a linear combination of three mutually orthogonal unit vectors **i**, **j**, and **k** or in terms of the components of the vector in the directions of three arbitrary vectors **A**, **B**, **C**, such that $[\mathbf{ABC}] \neq 0$.

DEFINITION 1 A **vector function of one variable** is a vector function whose domain is a set of *real* numbers.

As with functions in general, the value of a vector function **V** corresponding to an arbitrary element t in the domain of **V** is denoted by $\mathbf{V}(t)$, and the function itself is identified interchangeably by either **V** or $\mathbf{V}(t)$.

If **V** is a vector function whose domain is an interval I, then for every t in I the vector $\mathbf{V}(t)$ has three scalar components $V_1(t)$, $V_2(t)$, and $V_3(t)$ in the respective directions of **i**, **j**, and **k**. Hence, **V** is represented on I by

(1) $$\mathbf{V}(t) = V_1(t)\mathbf{i} + V_2(t)\mathbf{j} + V_3(t)\mathbf{k}$$

Each of the real-valued functions $V_1(t)$, $V_2(t)$, and $V_3(t)$ has domain I and is said to be a **component function of V.**

DEFINITION 2 A vector function $\mathbf{V}(t)$ of a scalar variable t is **continuous at** t_0 if and only if all three of its component functions $V_1(t)$, $V_2(t)$, and $V_3(t)$ are continuous at t_0.

A vector function of one variable $\mathbf{V}(t)$ is **continuous on an interval** I if and only if each of its component functions is continuous on I.

The process of taking the limit as t tends to t_0 of a vector function of one variable $\mathbf{V}(t)$ is denoted by $\lim_{t \to t_0} \mathbf{V}(t)$ and is given meaning by the following definition.

DEFINITION 3 For a vector function of one variable $\mathbf{V}(t) = V_1(t)\mathbf{i} + V_2(t)\mathbf{j} + V_3(t)\mathbf{k}$, **the limit as t tends to t_0 of $\mathbf{V}(t)$** is $\mathbf{L} = L_1\mathbf{i} + L_2\mathbf{j} + L_3\mathbf{k}$, that is,

(2) $$\lim_{t \to t_0} \mathbf{V}(t) = \mathbf{L}$$

if and only if $\lim_{t \to t_0} V_1(t) = L_1$, $\lim_{t \to t_0} V_2(t) = L_2$, $\lim_{t \to t_0} V_3(t) = L_3$.

If the independent variable t of a vector function of one variable $\mathbf{V}(t)$ changes an amount Δt, the value of the function will in general change both in magnitude and direction. Specifically, corresponding to the scalar increment Δt we have the vector increment in the value of the function

$$\begin{aligned}\Delta\mathbf{V} &= \mathbf{V}(t + \Delta t) - \mathbf{V}t) \\ &= [V_1(t + \Delta t)\mathbf{i} + V_2(t + \Delta t)\mathbf{j} + V_3(t + \Delta t)\mathbf{k}] - [V_1(t)\mathbf{i} + V_2(t)\mathbf{j} + V_3(t)\mathbf{k}]\end{aligned}$$

or, collecting coefficients on **i**, **j**, and **k**,

(3) $$\Delta\mathbf{V} = \Delta V_1\mathbf{i} + \Delta V_2\mathbf{j} + \Delta V_3\mathbf{k}$$

The limit as $\Delta t \to 0$ of the **difference quotient** $\Delta\mathbf{V}/\Delta t$ serves to define the *derivative* of $\mathbf{V}$, conventionally denoted by $\mathbf{V}'$ or $d\mathbf{V}/dt$.

DEFINITION 4 The **derivative of a vector function of one variable V** is represented by

$$\mathbf{V}'(t) = \lim_{\Delta t \to 0} \frac{\mathbf{V}(t + \Delta t) - \mathbf{V}(t)}{\Delta t} = \lim_{\Delta t \to 0} \frac{\Delta \mathbf{V}}{\Delta t} \tag{4}$$

on the set of all real numbers for which the limit (4) exists.

Using (3) to rewrite Eq. (4) as

$$\frac{d\mathbf{V}}{dt} = \lim_{\Delta t \to 0} \frac{\Delta V_1}{\Delta t}\mathbf{i} + \lim_{\Delta t \to 0} \frac{\Delta V_2}{\Delta t}\mathbf{j} + \lim_{\Delta t \to 0} \frac{\Delta V_3}{\Delta t}\mathbf{k}$$

and replacing limits of difference quotients here by derivatives, we get

$$\frac{d\mathbf{V}}{dt} = \frac{dV_1}{dt}\mathbf{i} + \frac{dV_2}{dt}\mathbf{j} + \frac{dV_3}{dt}\mathbf{k} \tag{5}$$

In words this formula says

THEOREM 1 If all derivatives involved exist, the derivative of a vector function of one variable $\mathbf{V}$ can be found by differentiating the respective components in the component form (1) of $\mathbf{V}$.

Motivated by (5), we denote the differential of $\mathbf{V}$ by $d\mathbf{V}$.

DEFINITION 5 The **differential $d\mathbf{V}$ of a vector function $\mathbf{V}(t)$ of one variable** is

$$d\mathbf{V} = dV_1\mathbf{i} + dV_2\mathbf{j} + dV_3\mathbf{k} \tag{6}$$

For the very important vector

$$\mathbf{R} = x\mathbf{i} + y\mathbf{j} + z\mathbf{k} \tag{7}$$

drawn from the origin to the point (x, y, z), we have

$$d\mathbf{R} = dx\mathbf{i} + dy\mathbf{j} + dz\mathbf{k} \tag{8}$$

A vector function $\mathbf{V}(t)$ of a scalar variable is **differentiable at** $t = t_0$ if and only if $\mathbf{V}'(t_0)$ exists. From the definition of the derivative of a vector function of one variable it follows that sums, differences, and products of such functions can be differentiated by formulas very much like those of ordinary calculus, provided the vector or scalar functions involved are suitably differentiable and the proper order of factors is maintained wherever the order is significant. For example, we have

$$\frac{d(\mathbf{U} \pm \mathbf{V})}{dt} = \frac{d\mathbf{U}}{dt} \pm \frac{d\mathbf{V}}{dt} \tag{9}$$

$$\frac{d(\phi\mathbf{V})}{dt} = \frac{d\phi}{dt}\mathbf{V} + \phi\frac{d\mathbf{V}}{dt} \tag{10}$$

(11) $$\frac{d(\mathbf{U}\cdot\mathbf{V})}{dt} = \frac{d\mathbf{U}}{dt}\cdot\mathbf{V} + \mathbf{U}\cdot\frac{d\mathbf{V}}{dt}$$

(12) $$\frac{d(\mathbf{U}\times\mathbf{V})}{dt} = \frac{d\mathbf{U}}{dt}\times\mathbf{V} + \mathbf{U}\times\frac{d\mathbf{V}}{dt}$$

(13) $$\frac{d[\mathbf{UVW}]}{dt} = \left[\frac{d\mathbf{U}}{dt}\mathbf{VW}\right] + \left[\mathbf{U}\frac{d\mathbf{V}}{dt}\mathbf{W}\right] + \left[\mathbf{UV}\frac{d\mathbf{W}}{dt}\right]$$

(14) $$\frac{d[\mathbf{U}\times(\mathbf{V}\times\mathbf{W})]}{dt} = \frac{d\mathbf{U}}{dt}\times(\mathbf{V}\times\mathbf{W}) + \mathbf{U}\times\left(\frac{d\mathbf{V}}{dt}\times\mathbf{W}\right) + \mathbf{U}\times\left(\mathbf{V}\times\frac{d\mathbf{W}}{dt}\right)$$

Since every vector equation implies three simultaneous scalar equations, it follows from the definition of vector differentiation that the integral of a vector function of one variable is found by integrating each component, provided only that the constant of integration is an arbitrary constant *vector*. Thus, if $\mathbf{V}(t) = V_1(t)\mathbf{i} + V_2(t)\mathbf{j} + V_3(t)\mathbf{k}$, an indefinite integral of $\mathbf{V}(t)$ is

(15) $$\int \mathbf{V}(t)\,dt = \mathbf{i}\int V_1(t)\,dt + \mathbf{j}\int V_2(t)\,dt + \mathbf{k}\int V_3(t)\,dt + \mathbf{C}$$

Modified linear differential equations in which the dependent variable is a vector and the independent variable a scalar can be solved in the usual way provided, again, that the constants of integration are arbitrary constant vectors. Vector methods are especially well suited to the theory of linear ordinary differential and difference equations because they lead to important vector criteria which, among other things, yield well-known results concerning wronskians and Casorati determinants, provide geometric insight into the method of variation of parameters, and motivate a generalization of the cross product (see Exercises 62–72).

The parametric vector equation

$$\mathbf{R}(t) = \mathbf{R}_0 + t\mathbf{A} \qquad -\infty < t < \infty$$

defines the position vector $\mathbf{R}$ of an arbitrary point on a line L in the direction of a vector $\mathbf{A}$ which passes through a point P_0 whose position vector is $\mathbf{R}_0$. The derivative of $\mathbf{R}$ with respect to t is simply the vector $\mathbf{A}$ which determines the direction of L.

Another example of a vector function of one variable is the set of vectors drawn from the origin to the points of a curve C on which the scalar variable t is a parameter. In fact, a general point on C is associated with a unique value of the parameter, say $t = t_1$, and determines with the origin a unique vector $\mathbf{V}(t_1)$ (Fig. 15.7*a*). This correspondence between the values of t and the vectors $\mathbf{V}(t)$ is clearly a vector function of t according to our definition. Conversely, if the values of a continuous vector function $\mathbf{V}(t)$ are drawn from a common origin, their endpoints will define a curve C whose points will be in correspondence with the values of the scalar variable t.

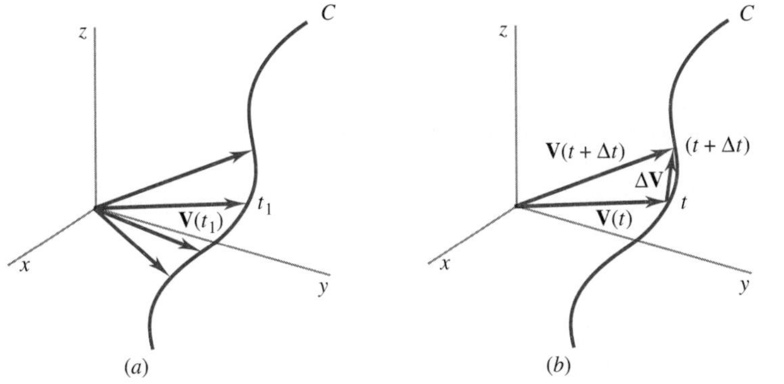

FIGURE 15.7
The geometric interpretation of a vector function of one variable.

This point of view leads to an important geometric interpretation of the derivative $d\mathbf{V}/dt$. For, since Δt is just a scalar, the quotient $\Delta\mathbf{V}/\Delta t$ is a well-defined vector having the same direction as $\Delta\mathbf{V}$ itself. Moreover, as Fig. 15.7*b* shows, the direction of $\Delta\mathbf{V}$ is that of an infinitesimal chord of the curve C. Therefore, as Δt approaches 0, the direction of $\Delta\mathbf{V}$, and hence the direction of $\Delta\mathbf{V}/\Delta t$, approaches the direction of the tangent to C. Thus, we have

THEOREM 2 *Unless $d\mathbf{V}/dt = \mathbf{0}$, as it may at a singular point of the curve C, the derivative $d\mathbf{V}/dt$ is a vector tangent to the curve C which is the locus of the endpoints of the vector $\mathbf{V}(t)$.*

In particular, if the scalar variable t is taken to be the length s of C, measured from some reference point on C, then

$$\left|\frac{d\mathbf{V}}{ds}\right| = \lim_{\Delta s\to 0}\frac{|\Delta\mathbf{V}|}{|\Delta s|} = \lim_{\Delta s\to 0}\frac{\text{infinitesimal chord of } C}{\text{infinitesimal arc of } C} = 1$$

from which there follows

COROLLARY 1 *If s is the arc length of the curve C defined by the terminal points of the vectors $\mathbf{V}(s)$ drawn from the origin to C, then $d\mathbf{V}/ds$ is a unit tangent to C.*

A curve C is **smooth** if and only if it is defined by a vector function $\mathbf{V}(t)$ whose domain is an interval I over which $\mathbf{V}'(t)$ is continuous and never zero. For each value of t in I, the **tangent line** to such a curve C at the point P on C corresponding to t is the line through P that contains $\mathbf{V}'(t)$. Along a smooth curve the direction of the tangent line is said to *vary continuously,* and through each point P of the curve there is a unique plane, the so-called **normal plane** at P, which is perpendicular to the **tangent,** or tangent line, to the curve at P.

EXAMPLE 1

At what point or points is the tangent to the curve $x = t^3$, $y = 5t^2$, $z = 10t$ perpendicular to the tangent at the point where $t = 1$?

From our earlier discussion it is clear that the given curve is equivalent to the vector function

$$\mathbf{V}(t) = t^3\mathbf{i} + 5t^2\mathbf{j} + 10t\mathbf{k}$$

Moreover, a vector tangent to this curve at a point corresponding to a general t is

$$\frac{d\mathbf{V}}{dt} = 3t^2\mathbf{i} + 10t\mathbf{j} + 10\mathbf{k}$$

and, in particular, for $t = 1$, a tangent vector is

$$3\mathbf{i} + 10\mathbf{j} + 10\mathbf{k}$$

From the fact that two nonzero vectors are perpendicular if and only if their dot product is zero, it follows that the tangent at a point on the curve corresponding to a general value t will be perpendicular to the tangent where $t = 1$ if and only if

$$3(3t^2) + 10(10t) + 10(10) = 9t^2 + 100t + 100 = 0$$

This condition holds just for the two values $t = -\frac{10}{9}, -10$. Hence, evaluating the x, y, and z coordinates of the points determined by these values of t, we find that the tangent to the curve at $(-\frac{1000}{729}, \frac{500}{81}, -\frac{100}{9})$ and the tangent at $(-1{,}000, 500, -100)$ are both perpendicular to the tangent to the curve where $t = 1$ and that these are the only points with this property.

EXAMPLE 2

VELOCITY AND ACCELERATION

From the point of view of vector analysis discuss the problem of determining the velocity and acceleration of a particle p moving along a smooth curve C.

To do this, let us suppose that the path C, which is the locus of the instantaneous positions of the moving particle, is defined by the vector function $\mathbf{r}(t)$, where t is the time. In other words $\mathbf{r}(t)$ is the vector extending from the origin to the position of the moving particle at the general time t (Fig. 15.8*a*).

Now let s be the arc length of C. Then, by the chain rule, we can write

$$\frac{d\mathbf{r}}{dt} = \frac{d\mathbf{r}}{ds}\frac{ds}{dt} \tag{16}$$

Since ds/dt is the speed v of p along C, and since $d\mathbf{r}/ds$ is a unit vector tangent to C pointing in the direction of motion, it follows from (16) that at each instant the magnitude and direction of the vector

$$\mathbf{v} = \frac{d\mathbf{r}}{dt} \tag{17}$$

specify the speed and direction, respectively, of the moving particle p. The vector $\mathbf{v}$ defined by (17) as *the first derivative with respect to time of the position vector* $\mathbf{r}(t)$ of p is called the **velocity** of the particle. In terms of $\mathbf{v}$, the **speed** of p is *defined* as $|\mathbf{v}|$.

By definition, the **acceleration a** of p is *the time derivative of the velocity,* or *the second time derivative of* $\mathbf{r}(t)$, i.e.,

$$\mathbf{a} = \frac{d\mathbf{v}}{dt} = \frac{d^2\mathbf{r}}{dt^2} \tag{18}$$

If the general unit tangent vector to C, namely, $d\mathbf{r}/ds$, is denoted by the symbol $\mathbf{T}$, (17) becomes

$$\mathbf{v} = v\mathbf{T}$$

For the acceleration we then have

$$\begin{aligned}\mathbf{a} &= \frac{d\mathbf{v}}{dt} = \frac{d(v\mathbf{T})}{dt} = \frac{dv}{dt}\mathbf{T} + v\frac{d\mathbf{T}}{dt} = \frac{dv}{dt}\mathbf{T} + v\frac{d\mathbf{T}}{ds}\frac{ds}{dt}\\ &= \frac{dv}{dt}\mathbf{T} + v^2\frac{d\mathbf{T}}{ds}\end{aligned} \tag{19}$$

In the first term on the right in (19), the scalar quantity dv/dt is the rate of change of the speed v, and, as we shall soon see, the unit tangent vector $\mathbf{T}$ to C is perpendicular to the vector $d\mathbf{T}/ds$ in the second term. This suggests—and it is easy to verify—that $(dv/dt)\mathbf{T}$ is just the vector projection of the acceleration $\mathbf{a}$ in the direction of $\mathbf{T}$ (Exercise 27), known as the **tangential acceleration** of the moving particle.

To interpret the second term on the right in (19), we observe that since $\mathbf{T}$ is a unit vector, it can vary only in direction. Hence, if the various values of $\mathbf{T}$ are drawn from a common origin, the locus of their end points

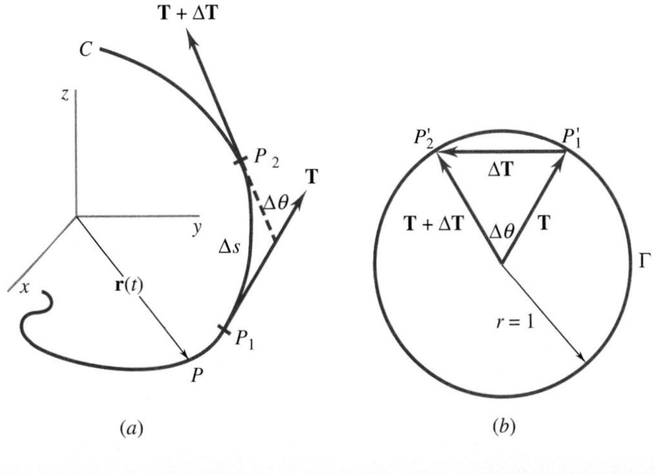

FIGURE 15.8
The unit tangents to a space curve plotted from a common origin.

will be a curve Γ on a sphere of unit radius. Now the length of the increment $\Delta\mathbf{T}$ corresponding to two adjacent points P_1' and P_2' on the curve Γ (Fig. 15.8*b*) is approximately the length of the arc $P_1'P_2'$, which in turn is equal to $\Delta\theta$, where $\Delta\theta$ is the angle between the tangents to C at the points P_1 and P_2, a distance Δs apart (Fig. 15.8*a*). Hence

$$\left|\frac{d\mathbf{T}}{ds}\right| = \lim_{\Delta s\to 0}\left|\frac{\Delta\mathbf{T}}{\Delta s}\right| = \lim_{\Delta s\to 0}\frac{\text{angle between tangents to } C \text{ at } P_1 \text{ and } P_2}{\text{arc length along } C \text{ between } P_1 \text{ and } P_2}$$

$$= \textbf{curvature} \text{ of } C \text{ at } P_1$$

Moreover, from Fig. 15.8*b* it is evident that the limiting position of ΔT as Δs (and hence $\Delta\theta$) approaches 0 is perpendicular to $\mathbf{T}$ in the plane which $\mathbf{T}$ and $\mathbf{T} + \Delta\mathbf{T}$ determine in the limit. If C is a plane curve, this, of course, is the unique plane in which C lies. If C is a twisted curve, this plane, which is known as the **osculating plane,** will vary from point to point along C. Hence, to summarize, *at any point P on the curve C, $d\mathbf{T}/ds$ is a vector which is perpendicular to C in the osculating plane of C at P and whose magnitude is equal to the curvature κ of C at P. If,* finally, we let $\mathbf{N}$ denote a unit normal drawn toward the concave side of C in the osculating plane and define, as usual, the **radius of curvature** of C to be

$$\rho = \frac{1}{\kappa}$$

we can write Eq. (19) in the form

(20)
$$\mathbf{a} = \frac{dv}{dt}\mathbf{T} + \frac{v^2}{\rho}\mathbf{N}$$

which shows that at any point in its path, *the vector acceleration of a moving particle is the sum of a component of magnitude dv/dt along the tangent to the path and a component of magnitude v^2/ρ normal to the path in the osculating plane to the path.*

The so-called **tangential component** of the acceleration (20) is $a_T = dv/dt$, the **normal,** or **centripetal, component** is $a_n = v^2/\rho$, and $|\mathbf{a}|^2 = a_T^2 + a_n^2$. Expressing $\mathbf{a}$ as $\mathbf{a} = a_T\mathbf{T} + a_n\mathbf{N}$ and noting that $\mathbf{v} = v\mathbf{T}$, we see that $\mathbf{v}\cdot\mathbf{a} = a_T v$. If $v \neq 0$, then

(21*a*)
$$a_T = \frac{\mathbf{a}\cdot\mathbf{v}}{|\mathbf{v}|}$$

Hence, a_T is simply the scalar projection of the acceleration in the direction of the velocity. On the other hand, $\mathbf{a}\times\mathbf{v} = |\mathbf{a}|\,|\mathbf{v}|\sin\theta\,\mathbf{B}$, where the unit vector $\mathbf{B}$ in the direction $\mathbf{a}\times\mathbf{v}$ is perpendicular to the plane of $\mathbf{T}$ and $\mathbf{N}$. Taking magnitudes of each member of the last equation, and subsequently substituting a_n for $|\mathbf{a}|\sin\theta$, we get $|\mathbf{a}\times\mathbf{v}| = |\mathbf{v}|a_n$ and then

(21*b*)
$$a_n = \frac{|\mathbf{a}\times\mathbf{v}|}{|\mathbf{v}|}$$

as an alternative to $|a_n|\sin\theta$ and as a formula for the scalar projection of the acceleration in the direction of $\mathbf{N}$, perpendicular to the velocity.

Now let us consider a vector function $\mathbf{r}$ of a single variable t, not necessarily the time, having a given representation,

(22)
$$\mathbf{r}(t) = x(t)\mathbf{i} + y(t)\mathbf{j} + z(t)\mathbf{k}$$

a nonzero derivative, and a third-order derivative on an interval $[a, b]$. The arc length

$$s(t) = \int_a^t |\mathbf{r}'(\tau)|\,d\tau$$

of the curve C traced by the tip of the position vector $\mathbf{r}(t)$ is a strictly monotone increasing function of t. The length of C is

$$l = \int_a^b |\mathbf{r}'(\tau)| \, d\tau$$

and to each value of s such that $0 \leq s \leq l$ there corresponds a unique value of t. Representing by $t(s)$ the function thus defined on $[0, l]$, and substituting into $\mathbf{r}(t)$, we obtain the representation

$$\mathbf{R}(s) = \mathbf{r}[t(s)] \qquad 0 \leq s \leq l \tag{23}$$

of an alternative vector function $\mathbf{R}$ defining the curve C. As in Example 2, $d\mathbf{r}/dt$ is a vector tangent to C, and

$$\mathbf{T} = \frac{d\mathbf{R}}{ds} \tag{24}$$

is a unit vector to C; hence, $\mathbf{T} \cdot \mathbf{T} = 1$. Differentiating this equation, we get

$$\mathbf{T} \cdot \frac{d\mathbf{T}}{ds} + \frac{d\mathbf{T}}{ds} \cdot \mathbf{T} = 2\mathbf{T} \cdot \frac{d\mathbf{T}}{ds} = 0$$

from which it follows that $d\mathbf{T}/ds$ is perpendicular to $\mathbf{T}$. For all values of s such that the *curvature* κ of C has a nonzero value

$$\kappa(s) = |d\mathbf{T}/ds| \tag{25}$$

a vector function $\mathbf{N}$ is defined by

$$\mathbf{N} = \frac{1}{\kappa} \frac{d\mathbf{T}}{ds} \tag{26}$$

whose values $\mathbf{N}(s) = \mathbf{T}'(s)/k(s)$ are all unit vectors. By name, $\mathbf{N}$ is known as the **normal,** or **principal normal,** to the curve C.

Since the **unit tangent vector T** of (24) can change direction only, the geometric significance of the curvature κ is that it gives the magnitude of the rate at which the direction of $\mathbf{T}$ changes with respect to arc length along C. Remember, the reciprocal of κ is the *radius of curvature* $\rho = 1/\kappa$. At a typical point P of C, the vector functions $\mathbf{T}$, $\mathbf{N}$ and the **binormal vector** function $\mathbf{B}$ of C, given by

$$\mathbf{B} = \mathbf{T} \times \mathbf{N} \tag{27}$$

determine three orthonormal vectors which when drawn from P establish a local right-handed coordinate system and, what is more, form a basis for all vectors drawn from P. In other words, any vector drawn from P is a linear combination of these basis vectors. The triple of vector functions $\mathbf{T}$, $\mathbf{N}$, and $\mathbf{B}$ is called the **moving trihedral** of the curve C. The triad of vectors $\mathbf{T}(s)$, $\mathbf{N}(s)$, $\mathbf{B}(s)$, at the point P of C, whose position vector is $\mathbf{R}(s)$, is called the **trihedron** of C at P and the planes through P perpendicular to $\mathbf{T}(s)$, $\mathbf{N}(s)$, and $\mathbf{B}(s)$ are named the **normal, rectifying,** and **osculating** planes of C at P, respectively. These three planes are what is meant by the **fundamental planes** of C at P.

Now, $\mathbf{B}$ like $\mathbf{T}$ is a unit vector function. Hence, $d\mathbf{B}/ds$ is perpendicular to $\mathbf{B}$ and, therefore, lies in the plane of $\mathbf{T}$ and $\mathbf{N}$. Differentiating the equation $\mathbf{T} \cdot \mathbf{B} = 0$, which says that $\mathbf{T}$ is perpendicular to $\mathbf{B}$, and using the relations $d\mathbf{T}/ds = \kappa\mathbf{N}$ and $\mathbf{B} \cdot \kappa\mathbf{N} = 0$, we get

$$\frac{d\mathbf{B}}{ds} \cdot \mathbf{T} + \mathbf{B} \cdot \frac{d\mathbf{T}}{ds} = \frac{d\mathbf{B}}{ds} \cdot \mathbf{T} + \mathbf{B} \cdot \kappa\mathbf{N} = \frac{d\mathbf{B}}{ds} \cdot \mathbf{T} = 0$$

The last equation tells us that $d\mathbf{B}/ds$ is parallel to $\mathbf{N}$. But this implies there is a scalar function $\tau(s)$ such that

$$\frac{d\mathbf{B}}{ds} = -\tau\mathbf{N} \tag{28}$$

The function τ is called the **torsion** of the curve C, and the **radius of torsion** σ is defined by

$$\sigma = \frac{1}{|\tau|} \tag{29}$$

Just as the curvature κ measures the rate at which the tangent $\mathbf{T}$ is changing direction, so the torsion τ measures the rate at which the binormal $\mathbf{B}$ is changing direction, that is, the rate at which the osculating plane is turning, as a point moves along the curve. (See Exercise 41.)

Equations (26) and (28) provide formulas for $d\mathbf{T}/ds$ and $d\mathbf{B}/ds$. To find a formula for $d\mathbf{N}/ds$, we differentiate both sides of $\mathbf{N} = \mathbf{B} \times \mathbf{T}$, getting

$$\frac{d\mathbf{N}}{ds} = \mathbf{B} \times \frac{d\mathbf{T}}{ds} + \frac{d\mathbf{B}}{ds} \times \mathbf{T} = \mathbf{B} \times \kappa\mathbf{N} - \tau\mathbf{N} \times \mathbf{T} = -\kappa\mathbf{T} + \tau\mathbf{B} \tag{30}$$

Equations (26), (28), and (30) yield the so-called **Frenet formulas**† of differential geometry, namely,

$$(a)\ \frac{d\mathbf{T}}{ds} = \kappa\mathbf{N} \qquad (b)\ \frac{d\mathbf{N}}{ds} = \tau\mathbf{B} - \kappa\mathbf{T} \qquad (c)\ \frac{d\mathbf{B}}{ds} = -\tau\mathbf{N} \tag{31}$$

In practice, a space curve C will not be defined by a vector function of the arc length variable s but by a vector function $\mathbf{r}(t)$ of some other independent variable t. Consequently, the use of (25) and (28) for the purpose of finding the curvature κ and the torsion τ of C will usually prove to be rather tedious. So let us develop formulas for κ and τ which involve only derivatives of $\mathbf{r}$ with respect to t, thus making them easy to apply directly. Using a prime to denote differentiation with respect to t, we start with the relations

$$\mathbf{r}' = \mathbf{T}s' \qquad \mathbf{r}'' = \mathbf{T}s'' + \kappa\,\mathbf{N}(s')^2 \qquad \mathbf{r} \times \mathbf{r}'' = \kappa\,\mathbf{B}(s')^3$$

and equate magnitudes of both members in the last equation, noting that $s' = |\mathbf{r}'|$; then we solve for κ, getting

$$\kappa = \frac{|\mathbf{r}' \times \mathbf{r}''|}{|\mathbf{r}'|^3} \tag{32}$$

Keeping the Frenet formulas in mind, we next differentiate both sides of the equation $\mathbf{r}'' = \mathbf{T}s'' + \kappa\mathbf{N}(s')^2$ with respect to t, using the chain rule as we do so, to obtain

$$\mathbf{r}''' = \kappa\,\mathbf{N}s's'' + \mathbf{T}s''' + (\tau\,\mathbf{B} - \kappa\mathbf{T})\kappa(s')^3 + \mathbf{N}\frac{d}{dt}\,[\kappa(s')^2]$$

†Also called the **Frenet-Serret formulas** after the French mathematicians Jean-Frédéric Frenet (1816–1900) and Joseph A. Serret (1819–1885).

Using the preceding expressions for $\mathbf{r}'$, $\mathbf{r}''$, and $\mathbf{r}'''$, and the mutual orthogonality properties of the moving trihedral of C, we find that

$$\mathbf{r}' \times \mathbf{r}'' = \kappa\,\mathbf{B}(s')^3 \qquad |\mathbf{r}' \times \mathbf{r}''|^2 = \kappa^2(s')^6 \qquad [\mathbf{r}'\mathbf{r}''\mathbf{r}'''] = \tau\kappa^2(s')^6$$

Finally, substituting from the second of these equations into the third equation for $\kappa^2(s')^6$, then solving for τ, we have

(33)
$$\tau = \frac{[\mathbf{r}'\mathbf{r}''\mathbf{r}''']}{|\mathbf{r}' \times \mathbf{r}''|^2}$$

A remarkable feature of curves in 3-space definable by vector functions of arc length s is that, under suitable conditions of smoothness, the curvature κ and the torsion τ of such a curve determine that curve uniquely, except for its orientation and position in space. In other words, *all curves which have both the same curvature function $\kappa(s)$ and the same torsion function $\tau(s)$ are congruent* (Exercise 53). For this reason, the equations

(34)
$$\kappa = \kappa(s) \qquad \tau = \tau(s)$$

are called the **natural,** or **intrinsic, equations** of a curve.

EXAMPLE 3

INTRINSIC EQUATIONS OF A HELIX

Find intrinsic equations of the right-handed circular helix H of pitch $2b\pi$ defined by the vector equation

$$\mathbf{r}(t) = a\cos\omega t\,\mathbf{i} + a\sin\omega t\,\mathbf{j} + bt\mathbf{k} \qquad a, b, \omega > 0, \qquad 0 \le t < \infty$$

Since both the curvature κ and torsion τ of H are to depend on arc length, we first use (24), (25), and (28), to write the formulas

(35)
$$(a)\;\; \kappa(s) = |\mathbf{R}''(s)| \qquad (b)\;\; \tau(s) = -\mathbf{N}\cdot\mathbf{B}'(s)$$

Now, the magnitude of $\mathbf{r}'(t) = -a\omega\sin\omega t\,\mathbf{i} + a\omega\cos\omega t\,\mathbf{j} + b\mathbf{k}$ is $\sqrt{a^2\omega^2 + b^2}$, and so the arc length s of H is given by

$$s = \int_0^t |\mathbf{r}'(\tau)|\,d\tau = \sqrt{a^2\omega^2 + b^2}\,t$$

Hence, another vector equation for H is

$$\mathbf{R}(s) = a\cos cs\,\mathbf{i} + a\sin cs\,\mathbf{j} + \frac{bcs}{\omega}\mathbf{k} \qquad \text{where } c = \omega/\sqrt{a^2\omega^2 + b^2}$$

Differentiating $\mathbf{R}(s)$ twice with respect to s, we obtain

$$\mathbf{R}''(s) = -ac^2\cos cs\,\mathbf{i} - ac^2\sin cs\,\mathbf{j} \qquad \text{and} \qquad |\mathbf{R}''(s)| = ac^2$$

Substituting from the last equation into (35a), and eliminating c, we have

$$\kappa(s) = \frac{a\omega^2}{a^2\omega^2 + b^2}$$

With κ thus determined, with $d\mathbf{T}/ds = \mathbf{R}''(s) = -ac^2(\cos cs\,\mathbf{i} + \sin cs\,\mathbf{j})$, and since $ac^2/\kappa = 1$, Eq. (26) reduces to

$$\mathbf{N} = -\cos cs\,\mathbf{i} - \sin cs\,\mathbf{j}$$

Differentiating $\mathbf{R}(s)$ once we get

$$\mathbf{T} = d\mathbf{R}/ds = -ac \sin cs\, \mathbf{i} + ac \cos cs\, \mathbf{j} + \frac{bc}{\omega}\mathbf{k}$$

We next compute and simplify the cross product $\mathbf{T} \times \mathbf{N}$ of (27), which yields

$$\mathbf{B} = \frac{bc}{\omega}(\sin cs\, \mathbf{i} - \cos cs\, \mathbf{j}) + ac\mathbf{k}$$

Differentiating both members here, we have

$$\mathbf{B}'(s) = \frac{bc^2}{\omega}(\cos cs\, \mathbf{i} + \sin cs\, \mathbf{j})$$

From (35b) it follows that

$$\tau = -\mathbf{N} \cdot \mathbf{B}'(s) = \frac{bc^2}{\omega}(\cos cs\, \mathbf{i} + \sin cs\, \mathbf{j}) \cdot (\cos cs\, \mathbf{i} + \sin cs\, \mathbf{j}) = \frac{b\omega}{a^2\omega^2 + b^2}$$

The natural equations of the given helix

$$\kappa(s) = \frac{a\omega^2}{a^2\omega^2 + b^2} \qquad \tau(s) = \frac{b\omega}{a^2\omega^2 + b^2}$$

show that both the curvature and torsion of H are constant.

EXERCISES

1. If $\mathbf{R} = \mathbf{A} \cos \omega t + \mathbf{B} \sin \omega t$, where both $\mathbf{A}$ and $\mathbf{B}$ are arbitrary constant vectors, show that $\mathbf{R} \times d\mathbf{R}/dt$ is a constant and that $d^2\mathbf{R}/dt^2 + \omega^2\mathbf{R} = \mathbf{0}$.
2. If $\mathbf{R}$ is any twice differentiable vector function, show that

$$\frac{d}{dt}\left(\mathbf{R} \times \frac{d\mathbf{R}}{dt}\right) = \mathbf{R} \times \frac{d^2\mathbf{R}}{dt^2}$$

3. Find the derivative of:
   (a) $\mathbf{V} \cdot \dfrac{d\mathbf{V}}{dt} \times \dfrac{d^2\mathbf{V}}{dt^2}$ **(b)** $\mathbf{V} \times \left(\dfrac{d\mathbf{V}}{dt} \times \dfrac{d^2\mathbf{V}}{dt^2}\right)$
4. If $\mathbf{V}$ is an arbitrary differentiable vector function of t, is $|d\mathbf{V}| = d|\mathbf{V}|$?
5. If $\mathbf{V}$ is a differentiable vector function of t, show that $\mathbf{V} \cdot d\mathbf{V}/dt = V(dV/dt)$.
6. A position vector $\mathbf{R} = R\mathbf{u}$ is twice differentiable with respect to t and is never zero. **(a)** Is $\dot{\mathbf{R}} = d\mathbf{R}/dt$ perpendicular to $\mathbf{R}$? **(b)** Is $\dot{\mathbf{u}}$ perpendicular to $\mathbf{u}$? **(c)** Express $\mathbf{R} \times \dot{\mathbf{R}} \cdot \ddot{\mathbf{R}}$ in terms of R, $\mathbf{u}$, $\dot{\mathbf{u}}$, and $\ddot{\mathbf{u}}$, and simplify.
7. Find **(a)** a unit tangent and **(b)** a unit normal to the curve defined by the vector function
   $\mathbf{r}(t) = \cos t\, \mathbf{i} + \sin t\, \mathbf{j} + \mathbf{k}$ at the point where $t = \pi/2$.
8. Work Exercise 7 with
   $\mathbf{r}(t) = \cos t\, \mathbf{i} - \ln \cos t\, \mathbf{j} + \sin t\, \mathbf{k}$ and $t = 0$.
9. What is the angle between the tangents to the curve $x = t$, $y = t^2$, $z = t^3$ and the points where $t = 1$ and $t = -1$?
10. Find the angle between the curves
    $\mathbf{R} = (1 - \theta)\mathbf{i} + 2 \cos \mathbf{j} - (1 - \sin \theta)\mathbf{k}$ and
    $\mathbf{r} = t\, \mathbf{i} + 2t^2\, \mathbf{j} - \dfrac{1}{t}\mathbf{k}$ at their point of intersection.
11. What is the angle between the tangents to the curve $x = t$, $y = t^2$, $z = t^4$ at the points where $t = 1$ and $t = 2$?
12. Find **(a)** parametric equations, **(b)** symmetric equations, and **(c)** a parametric vector equation for the tangent line to the curve $\mathbf{r} = (3t^2 + 1)\mathbf{i} + 4t\mathbf{j} + e^{t-1}\mathbf{k}$ at the point $(4, 4, 1)$.
13. Show that there are no pairs of points on the curve $x = t$, $y = t^2$, $z = t^3$ at which the tangents are parallel. Are there such pairs of points on the curve $x = 3t^4 - 6t^2 + 12t$, $y = 4t^3 - 6t^2$, $z = 12t$? On the curve $x = 15t$, $y = 5t^3$, $z = 15t + 3t^5$?
14. Find an equation of the plane which is perpendicular to the twisted cubic $\mathbf{r} = t\mathbf{i} + 2t^2\mathbf{j} - t^3\mathbf{k}$ at the point $(1, 2, -1)$.
15. If $\mathbf{r} = t^2\mathbf{i} - t^3\mathbf{j} + t^4\mathbf{k}$ is the vector from the origin to a moving particle, find the velocity of the particle when $t = 1$. What is the component of this velocity in the direction of $8\mathbf{i} - \mathbf{j} + 4\mathbf{k}$? What is the acceleration of the particle? What are the tangential and normal components of the acceleration?

The position vector of a particle p at time t is given by

$\mathbf{r}(t) = 6t\,\mathbf{i} + 12t^2\,\mathbf{j} + 8t^3\,\mathbf{k}$. Find all values of t for which the motion of p is

16. **(a)** Parallel to $\mathbf{i} + 4\mathbf{j} + 4\mathbf{k}$
(b) Perpendicular to $\mathbf{i} + 4\mathbf{j} + 16\mathbf{k}$

17. **(a)** Parallel to $\mathbf{i} + 2\mathbf{j} + \mathbf{k}$
(b) Perpendicular to $\mathbf{i} - 5\mathbf{j} + 16\mathbf{k}$

18. **(a)** Parallel to $4\mathbf{i} + 4\mathbf{j} + \mathbf{k}$
(b) Perpendicular to $\mathbf{i} - 16\mathbf{k}$

19. The position vector of a particle p is given by $\mathbf{r}(t) = t^2\mathbf{i} + (2t - 1)\mathbf{j} + (t - t^2)\mathbf{k}$. Find each of the following when $t = 2$: **(a)** the speed of p, **(b)** the acceleration of p, **(c)** a unit tangent to the path, and **(d)** the angle between the velocity and the acceleration of p.

20. A particle p, whose position vector is $\mathbf{r} = t\mathbf{i} - (t + 1)\mathbf{j} + t^2\mathbf{k}$, moves along a curve C. Write **(a)** a parametric vector equation and **(b)** symmetric equations of the tangent line to C at $(1, -2, 1)$. Find all values of t for which p is moving **(c)** perpendicular to $\mathbf{A} = 3\mathbf{i} + \mathbf{j} - \mathbf{k}$ and **(d)** parallel to $\mathbf{A}$. **(e)** What is the component of $\mathbf{A}$ along the direction of motion when $t = 2$?

21. If a particle starts to move from rest at the point $(0, 1, 2)$ with component accelerations $a_x = 1 + t$, $a_y = t^3$, $a_z = 2t - t^2$, find the vector from the origin to the instantaneous position of the particle.

22. A particle starts to move from rest at the point $(1, 2, 3)$ with x, y, and z components of acceleration $1 + 2t$, t^3, and $2t - t^2$, respectively. What is the position vector of the particle?

23. At $t = 0$ a body starts to move from the point $(-1, 0, 1)$ with velocity $\mathbf{v}_0 = \mathbf{i} + 2\mathbf{j} - 3\mathbf{k}$. If the acceleration of the body is $\mathbf{a}(t) = 3t^2\mathbf{i} + \mathbf{j} + (\cos 2t)\mathbf{k}$, find the position of the body at any subsequent time. With what velocity is the body moving at $t = 1$?

24. The position vector of a particle is $\mathbf{r}(t) = (1 + t^3)\mathbf{i} + (1 - t^2)\mathbf{j} + t\mathbf{k}$. Find all values of t for which the direction of motion is **(a)** perpendicular to $\mathbf{A} = 2\mathbf{i} + \frac{5}{2}\mathbf{j} + \mathbf{k}$ and **(b)** parallel to $\mathbf{A}$. **(c)** Find the component of $\mathbf{A}$ in the direction of motion when $t = 1$. **(d)** Where is the particle when its speed is minimal?

25. Solve the differential equation $d\mathbf{V}/dt = \lambda\mathbf{V}$, where λ is a scalar constant.

26. Solve the differential equation $d\mathbf{V}/dt = \lambda(t)\mathbf{V}$.

27. Show that if v is the speed and $\mathbf{T}$ a unit tangent to the path of a moving particle, then $(dv/dt)\mathbf{T}$ is the vector projection of the acceleration $\mathbf{a}$ in the direction of $\mathbf{T}$.
Hint: Substitute from (19) into $\left(\dfrac{\mathbf{a} \cdot \mathbf{T}}{T^2}\right)\mathbf{T}$ for $\mathbf{a}$ and simplify.

In Exercises 28–31, find the tangential and normal components of the acceleration of a particle whose position vector is the given function $\mathbf{r}(t)$.

28. $\mathbf{r}(t) = t\,\mathbf{i} + (1 - t)\mathbf{j} + \frac{1}{2}t^2\mathbf{k}$

29. $\mathbf{r}(t) = (1 + t^2)\mathbf{i} - (t^2 - 1)\mathbf{j} + t^2\mathbf{k}$

30. $\mathbf{r}(t) = a\cos\omega t\,\mathbf{i} + a\sin\omega t\,\mathbf{j} + b\mathbf{k}$

31. $\mathbf{r}(t) = \frac{1}{2}(t\cos t\,\mathbf{i} + t\sin t\,\mathbf{j} + \sqrt{3}t\,\mathbf{k})$

32. Find the length of the helix $\mathbf{r}(t) = 4\cos t\,\mathbf{i} + 4\sin t\,\mathbf{j} + 3t\,\mathbf{k}$ from the point $(4, 0, 0)$ to $(4, 0, 6\pi)$.

33. Find the length of the arc of the curve $\mathbf{r}(t) = e^t(\cos t\,\mathbf{i} + \sin t\,\mathbf{j} + \mathbf{k})$ from where $t = 0$ to where $t = 3$.

34. Show that if $\mathbf{u}(t)$ is differentiable and $|\mathbf{u}| \neq 0$, then

$$\frac{d}{dt}\left(\frac{\mathbf{u}(t)}{|\mathbf{u}(t)|}\right) = \frac{\dot{\mathbf{u}}(\mathbf{u} \cdot \mathbf{u}) - \mathbf{u}(\mathbf{u} \cdot \dot{\mathbf{u}})}{(\mathbf{u} \cdot \mathbf{u})^{3/2}}$$

35. In the case of a plane curve defined by parametric equations $x = f(t)$ and $y = g(t)$, show that Eq. (32) reduces to

$$\kappa = \frac{|f'g'' - g'f''|}{[(f')^2 + (g')^2]^{3/2}}$$

36. In the case of a plane curve which is the graph of a function $y(x)$, show that Eq. (32) reduces to

$$\kappa = \frac{|y''|}{[1 + (y')^2]^{3/2}}$$

37. Find the curvature and radius of curvature of the curve $y = e^x$ at the point $(0, 1)$.

38. Find the curvature of the four-cusped hypocycloid $\mathbf{r}(t) = \cos^3 t\,\mathbf{i} + \sin^3 t\,\mathbf{j}$ at the point where $t = \pi/4$.

39. The position vector of a particle p is $\mathbf{r}(t) = t\,\mathbf{i} + t^2\,\mathbf{j} + \frac{2}{3}t^3\,\mathbf{k}$. Find **(a)** the velocity, acceleration, and speed of p, **(b)** the tangential and centripetal components of the acceleration, **(c)** the curvature of the curve C traversed by p, and **(d)** the minimum radius of curvature of C.

40. Find the curvature and torsion of the conical helix whose vector equation is

$$\mathbf{r}(t) = \tfrac{1}{2}(t\cos t\,\mathbf{i} + t\sin t\,\mathbf{j} + \sqrt{3}t\mathbf{k}) \qquad 0 \le t$$

41. **(a)** Use the relation $d\mathbf{T}/ds = \kappa\mathbf{N}$ to show that if $\mathbf{T}$, $\mathbf{N}$, and $\mathbf{B}$ are the moving trihedral of a curve C, then $d\mathbf{B}/ds = \mathbf{T} \times d\mathbf{N}/ds$.
(b) Deduce that $d\mathbf{B}/ds$ is parallel to $\mathbf{N}$ and, in so doing, derive Eq. (28). In a sense, the torsion τ measures how the moving trihedral twists about C.

42. Using the **Darboux vector** $\mathbf{D} = \tau\mathbf{T} + \kappa\mathbf{B}$, named for the French mathematician Jean Gaston Darboux (1842–1917), write the Frenet formulas as

$$\frac{d\mathbf{T}}{ds} = \mathbf{D} \times \mathbf{T} \qquad \frac{d\mathbf{N}}{ds} = \mathbf{D} \times \mathbf{N} \qquad \frac{d\mathbf{B}}{ds} = \mathbf{D} \times \mathbf{B}$$

43. What is the equation of the osculating plane to the space curve $x = t^4$, $y = t^2$, $z = t^3$ at the point P_1: (x_1, y_1, z_1)? *Hint:* Let P: (x, y, z) be a general point in the osculating plane, and impose the condition that the vector joining P to P_1 be coplanar with the vectors $\mathbf{T}$ and $d\mathbf{T}/ds$ at P_1.

44. What is the equation of the osculating plane to the curve $x = t$, $y = t^2$, $z = t^3$ at the point whose parameter value is $t = 1$? What is the equation of the tangent to this curve at the point $t = 1$? What is the equation of the normal at $t = 1$? What is the equation of the binormal at $t = 1$?

45. Let $\mathbf{T}$ denote the unit tangent at a general point of a curve C. By differentiating the relation $\mathbf{T} \cdot \mathbf{T} = 1$, show that $\mathbf{T}$

is perpendicular to the **curvature vector** $(1/\rho)\mathbf{N}$ at every point where **T** and **N** are defined.

46. The position vector of a point P on a curve C is $\mathbf{r}_0$ and **T**, **N**, **B** is the trihedron of C at P. Write vector equations of the fundamental planes through P, and give the name of the plane defined by each equation.

The unit vectors **T**, **N**, **B**, together with the fundamental planes, are known as the **six elements** of the moving trihedral to a curve. In Exercises 47–50, find these elements at the point of the curve determined by $\mathbf{r}(t)$, where t has the given value.

47. $\mathbf{r}(t) = t\,\mathbf{i} + \frac{1}{2}t^2\,\mathbf{j} + \mathbf{k} \qquad t = 0$

48. $\mathbf{r}(t) = 2t\,\mathbf{i} + t^2\,\mathbf{j} + (1 - t)\mathbf{k} \qquad t = 1$

49. $\mathbf{r}(t) = 3\cos t\,\mathbf{i} + 3\sin t\,\mathbf{j} + 4t\,\mathbf{k} \qquad t = \pi/2$

50. $\mathbf{r}(t) = 3t\,\mathbf{i} + 5\cosh t\,\mathbf{j} + 4t\,\mathbf{k} \qquad t = 0$

51. Two curves C and Γ defined by the respective vector functions $\mathbf{R}(s)$ and $\mathbf{Q}(s)$ have the same curvature and torsion. The moving trihedrals of C and Γ are **T**, **N**, **B** and **t**, **n**, **b** respectively. Show that

$$\mathbf{T}\cdot\mathbf{t} + \mathbf{N}\cdot\mathbf{n} + \mathbf{B}\cdot\mathbf{b} = K \qquad K \text{ a constant}$$

Hint: Use the Frenet formulas and the same curvature and torsion for both curves to simplify the derivative of $\mathbf{T}\cdot\mathbf{t} + \mathbf{N}\cdot\mathbf{n} + \mathbf{B}\cdot\mathbf{b}$.

52. In Exercise 51, evaluate K and show that $\mathbf{T} = \mathbf{t}$, $\mathbf{N} = \mathbf{n}$, and $\mathbf{B} = \mathbf{b}$. *Hint: Translate and rotate* C or Γ, if necessary, so that their initial points and trihedrals at $s = 0$ coincide.

53. Prove that the curves C and Γ of Exercise 51 are congruent. *Hint:* Use the functional identity $\mathbf{T} = \mathbf{t}$ of Exercise 52 and note that $\mathbf{R}(0) = \mathbf{Q}(0)$.

54. Taking $s = 0$ at $x = 0$, find intrinsic equations of the catenary $y = a\cosh x/a$.

55. Taking $s = 0$ at $\theta = \pi$, find intrinsic equations of the cycloid defined parametrically by $x = a(\theta - \sin\theta)$, $y = a(1 - \cos\theta)$.

56. Find intrinsic equations of the conical helix in Exercise 40. *Hint:* Take $s = 0$ at $t = 0$.

57. If $\mathbf{R}_1, \mathbf{R}_2, \ldots, \mathbf{R}_n$ are the vectors from the origin to the respective mass particles $m_1, m_2, \ldots, m_n$, the endpoint of the vector

$$\mathbf{C} = \frac{\sum_{i=1}^{n} m_i\mathbf{R}_i}{\sum_{i=1}^{n} m_i}$$

is called the **center of gravity** of the system of particles. Show that for any vector **R**,

$$\sum_{i=1}^{n} m_i(\mathbf{R} - \mathbf{R}_i)\cdot(\mathbf{R} - \mathbf{R}_i) = m(\mathbf{R} - \mathbf{C})\cdot(\mathbf{R} - \mathbf{C}) + \sum_{i=1}^{n} m_i(\mathbf{C} - \mathbf{R}_i)\cdot(\mathbf{C} - \mathbf{R}_i)$$

where m is the total mass of all the particles.

58. If a particle moves under the influence of a force which is always directed toward the origin, show that $\mathbf{R} \times d^2\mathbf{R}/dt^2 = \mathbf{0}$, where **R** is the vector from the origin to the particle. *Hint:* Newton's law, i.e., mass × acceleration = force, remains correct when the acceleration and the force are interpreted as vector quantities.

59. If $\mathbf{R}(t)$ is the vector from the origin to the instantaneous position of a particle moving along a curve C, show that $\mathbf{R} \times d\mathbf{R}$ is equal to twice the area of the sector defined by the two vectors $\mathbf{R}(t)$ and $\mathbf{R}(t + \Delta t) \equiv \mathbf{R} + d\mathbf{R}$ and the arc of C which they intercept. Hence show that:

(a) If $\mathbf{R} \times d\mathbf{R} = \mathbf{0}$, the vector **R** has a constant direction.

(b) If $\mathbf{R} \times d^2\mathbf{R}/dt^2 = \mathbf{0}$, the particle moves so that the radius vector sweeps out equal area in equal times. This is a generalization of one of the laws of planetary motion discovered by Johannes Kepler (1571–1630).

60. While a disk is rotating in a horizontal plane with constant angular velocity ω, a particle moves radially outward from the center of the disk, its position vector in space being $\mathbf{R}(t) = t\mathbf{b}$, where **b** is a unit vector rotating with the disk. By noting first that **b** can be expressed in the form

$$\mathbf{b} = \cos\omega t\,\mathbf{i} + \sin\omega t\,\mathbf{j}$$

where **i** and **j** are mutually perpendicular vectors which are fixed in space, show that the velocity and acceleration of the particle are, respectively,

$$\dot{\mathbf{R}}(t) = \mathbf{b} + t\dot{\mathbf{b}} \qquad \text{and} \qquad \ddot{\mathbf{R}}(t) = 2\dot{\mathbf{b}} + t\ddot{\mathbf{b}} = 2\dot{\mathbf{b}} - \omega^2 t\mathbf{b}$$

[The component of acceleration corresponding to the term $2\dot{\mathbf{b}}$ is called the **Coriolis acceleration,** after the French physicist Gaspard Gustave de Coriolis (1792–1843).]

61. In Exercise 60, if the position vector of the particle in space is $\mathbf{R}(t) = x(t)\mathbf{i} + y(t)\mathbf{j}$, where **i** and **j** are mutually perpendicular unit vectors fixed with respect to the disk, show that $d\mathbf{i}/dt = \omega\mathbf{j}$ and $d\mathbf{j}/dt = -\omega\mathbf{i}$. Then show that

$$\dot{\mathbf{R}}(t) = (\dot{x}\mathbf{i} + \dot{y}\mathbf{j}) + (-\omega y\mathbf{i} + \omega x\mathbf{j})$$
$$\ddot{\mathbf{R}}(t) = (\ddot{x}\mathbf{i} + \ddot{y}\mathbf{j}) + 2\omega(-\dot{y}\mathbf{i} + \dot{x}\mathbf{j}) - \omega^2(x\mathbf{i} + y\mathbf{j})$$

Verify that the relative velocity vector, $\dot{x}\mathbf{i} + \dot{y}\mathbf{j}$, and the Coriolis acceleration, namely, $2\omega(-\dot{y}\mathbf{i} + \dot{x}\mathbf{j})$, are perpendicular.

The next six exercises relate to the general linear second-order differential equation $Ly = y'' + p_1y' + p_2y = 0$, where p_1 and p_2 are continuous functions of t on some interval I. In these exercises you are asked to formulate various properties of solutions of the equation $Ly = 0$ in terms of the vector functions

$$\mathbf{P} = \mathbf{i} + p_1\mathbf{j} + p_2\mathbf{k} \qquad \text{and} \qquad \mathbf{Y} = y''\mathbf{i} + y'\mathbf{j} + y\mathbf{k}$$

62. Show that y is a solution of the equation $Ly = 0$ if and only if $\mathbf{P}\cdot\mathbf{Y} = 0$.

63. If y_1 and y_2 are two solutions of the equation $Ly = 0$, show that $\mathbf{Y}_1 \times \mathbf{Y}_2 = -W_{12}\mathbf{P}$, where W_{12} is the wronskian of y_1 and y_2. Interpret this result in terms of the linear dependence or independence of y_1 and y_2.

64. From the result of Exercise 63, show that $W_{12}' + p_1 W_{12} = 0$.

65. If y_1, y_2, y_3 are any three solutions of the equation $Ly = 0$, show that $[\mathbf{Y}_1\mathbf{Y}_2\mathbf{Y}_3] = 0$. If $\mathbf{Y}_1 \times \mathbf{Y}_2 \neq \mathbf{0}$, does this imply that there exist constants a and b such that $y_3 = ay_1 + by_2$ at all points of I? Explain.

66. Let y_1, y_2, y_3 be any three solutions of $Ly = 0$, let $\mathbf{U} = y_1\mathbf{i} + y_2\mathbf{j} + y_3\mathbf{k}$, and let $\mathbf{V} = \mathbf{U} \times \mathbf{U}'$. Show that $\mathbf{V}' = -p_1\mathbf{V}$ and $\mathbf{V} = \mathbf{C}e^{-\int p_1\,dt}$.

67. Using the result of Exercise 66, show that $\mathbf{U} \cdot \mathbf{V} = \mathbf{C} \cdot \mathbf{U}e^{-\int p_1\,dt} = 0$. Assuming that y_1 and y_2 are linearly independent, show that the last equation implies that $\mathbf{C} \cdot \mathbf{U} = 0$. Finally, show that this implies the existence of a pair of constants c_1 and c_2 such that $y_3 = c_1y_1 + c_2y_2$ at all points of I.

68. Let $\mathbf{e}_3 = (0, 0, 1)$, $\mathbf{c} = (c_1, c_2, 0)$, and $\mathbf{u} = (y_1, y_2, 0)$ be vectors for which $\mathbf{c} \cdot \mathbf{u}$ is a complementary function of the nonhomogeneous normal differential equation $a_0y'' + a_1y_1' + a_2y = f$ on I. Show that $W_{12} = [\mathbf{e}_3\mathbf{u}\mathbf{u}'] \neq 0$ on I, and use variation of parameters to find a vector function $\mathbf{c}$ such that $y_p = \mathbf{c} \cdot \mathbf{u}$ yields a particular solution of the nonhomogeneous differential equation. *Hint:* Impose the conditions $\mathbf{c}' \cdot \mathbf{e}_3 = 0$, $\mathbf{c}' \cdot \mathbf{u} = 0$, and require that $\mathbf{c}'$ be parallel to $\mathbf{e}_3 \times \mathbf{u}$.

69. By introducing the appropriate definitions, restate the results of Exercises 62 and 63 for the general linear second-order difference equation $(E^2 + p_1E + p_2)y = 0$.

70. Continue Exercise 69 by restating the results of Exercises 64 and 65 for the equation

$$(E^2 + p_1E + p_2)y = 0$$

71. Let $\mathbf{c} = (c_1, c_2, c_3)$ and $\mathbf{u} = (y_1, y_2, y_3)$ be vectors for which $\mathbf{c} \cdot \mathbf{u}$ is a complementary function of the nonhomogeneous third-order difference equation $a_0y_{k+3} + a_1y_{k+2} + a_2y_{k+1} + a_3y_k = f_k$ on the set of positive integers. Show that $C_{123} = C_k = [\mathbf{u}_k\mathbf{u}_{k+1}\mathbf{u}_{k+2}] \neq 0$, identify C_k, and use variation of parameters to find a vector function $\mathbf{c}$ of k such that $y_p = \mathbf{c} \cdot \mathbf{u}$ yields a particular solution of the nonhomogeneous difference equation. *Hint:* Impose the conditions $\Delta\mathbf{c}_k \cdot \mathbf{u}_{k+1} = 0$, $\Delta\mathbf{c}_k \cdot \mathbf{u}_{k+2} = 0$, and $\Delta\mathbf{c}_k \cdot \mathbf{u}_{k+3} = f_0/a_0$, and note that $\Delta\mathbf{c}_k$ is parallel to $\mathbf{u}_{k+1} \times \mathbf{u}_{k+2}$.

72. Let $A = (a_{ij})$ be an $n \times n$ matrix with real row vectors $\mathbf{r}_i = (a_{i1}, a_{i2}, \ldots, a_{in})$ for which the cofactor of a_{ij} in $|A|$ is C_{ij}. Define the **cross product of the** $n - 1$ **vectors** $\mathbf{r}_1, \mathbf{r}_2, \ldots, \mathbf{r}_{n-1}$, written $\mathop{\times}\limits_{i=1}^{n-1} \mathbf{r}_i$, by

(36) $\mathop{\times}\limits_{i=1}^{n-1} \mathbf{r}_i = (C_{n1}, C_{n2}, \ldots, C_{nn}) \qquad n \geq 2$

(a) Show that, for $n = 3$, $\mathop{\times}\limits_{i=1}^{2} \mathbf{r}_i$ particularizes to the cross product $\mathbf{r}_1 \times \mathbf{r}_2$ defined for vectors of 3-space by Eq. (7), Sec. 15.2.

(b) Show that if $\{\mathbf{e}_1, \mathbf{e}_2, \mathbf{e}_3, \mathbf{e}_4\}$ is the standard basis of R^4, then $\mathop{\times}\limits_{i=1}^{3} \mathbf{r}_i$ is given by Laplace's expansion by elements of the last row of

$$\begin{vmatrix} a_{11} & a_{12} & a_{13} & a_{14} \\ a_{21} & a_{22} & a_{23} & a_{24} \\ a_{31} & a_{32} & a_{33} & a_{34} \\ \mathbf{e}_1 & \mathbf{e}_2 & \mathbf{e}_3 & \mathbf{e}_4 \end{vmatrix}$$

(c) Describe a determinantal representation of $\mathop{\times}\limits_{i=1}^{n-1} \mathbf{r}_i$.

(d) Evaluate the **generalized scalar triple,** or **scalar multiple, product** $\mathbf{r}_n \cdot \mathop{\times}\limits_{i=1}^{n-1} \mathbf{r}_i$.

(e) Extend the method of Exercise 68 to find a particular solution of a normal nth-order linear differential equation $Ly = a_0y^{(n)} + a_1y^{(n-1)} + \cdots + a_{n-1}y' + a_ny = f$ on an interval I, where $\mathbf{u} = (y_1, y_2, \ldots, y_n)$ is a basis for $Ly = 0$. *Hint:* Require that $\mathbf{c}' \cdot \mathbf{u}^{(m)} = 0$, $0 \leq m \leq n - 2$, and use generalized cross and scalar triple products.

15.5 THE OPERATOR ∇

A function which assigns a unique scalar to each point in a nonempty subset S of 3-space is called a **scalar point function,** or a **scalar-valued function of position.** In vector analysis, where the scalars are real numbers, such functions are also referred to as **scalar fields.** Vector functions defined on subsets of 3-space are known as **vector fields.** Every scalar or vector field is characterized analytically by a function of a single vector variable or of several real variables.

The rate of change of a differentiable scalar point function $\phi(x, y, z)$ in the direction of a coordinate axis is given by the corresponding partial derivative of ϕ. We now proceed to show that if $\phi(x, y, z)$ possesses continuous partial derivatives with respect to x, y, and z, throughout some region of space, then there is a *vector-valued derivative* of ϕ, denoted by $\nabla\phi$, such that the rate of change of ϕ in an arbitrary direction is given by the scalar projection of $\nabla\phi$ along that direction. The name of this projection, a definition of $\nabla\phi$, and the meaning of the vector operator ∇ will be given as we proceed.

Let the region of continuity of ϕ_x, ϕ_y, ϕ_z be $\mathcal{R}$, let $\mathbf{R} = x\mathbf{i} + y\mathbf{j} + z\mathbf{k}$ be the position vector of an interior point P: (x, y, z) of $\mathcal{R}$, and let the direction of an arbitrary ray, or half-line, l drawn from

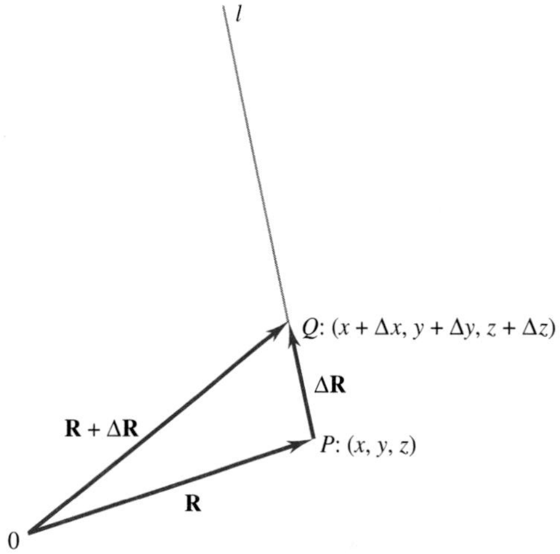

FIGURE 15.9
The position vectors of two neighboring points P and Q.

P be that of a unit vector **u** originating at P. If we move from P along l to a neighboring point Q: $(x + \Delta x, y + \Delta y, z + \Delta z)$ in $\mathcal{R}$ (Fig. 15.9), the function ϕ will change by an amount whose exact value, as derived in calculus for a differentiable function, is

$$\Delta\phi = \frac{\partial\phi}{\partial x}\Delta x + \frac{\partial\phi}{\partial y}\Delta y + \frac{\partial\phi}{\partial z}\Delta z + \epsilon_1\,\Delta x + \epsilon_2\,\Delta y + \epsilon_3\,\Delta z \tag{1}$$

where ϵ_1, ϵ_2, ϵ_3 are quantities which approach zero as Q approaches P, that is, as Δx, Δy, and Δz approach zero. If we divide the change $\Delta\phi$ by the distance $\Delta s = |\Delta\mathbf{R}|$ between P and Q, we obtain a measure of the average rate at which ϕ changes when we move along l from P to Q:

$$\frac{\Delta\phi}{\Delta s} = \frac{\partial\phi}{\partial x}\frac{\Delta x}{\Delta s} + \frac{\partial\phi}{\partial y}\frac{\Delta y}{\Delta s} + \frac{\partial\phi}{\partial z}\frac{\Delta z}{\Delta s} + \epsilon_1\frac{\Delta x}{\Delta s} + \epsilon_2\frac{\Delta y}{\Delta s} + \epsilon_3\frac{\Delta z}{\Delta s} \tag{2}$$

For instance, if $\phi(x, y, z)$ is the temperature at the general point P: (x, y, z), then $\Delta\phi/\Delta s$ is the average rate of change of temperature in the direction in which Δs is measured. In the limit, as $\Delta s \to 0$, the last three terms in (2) become zero, and we have explicitly

$$\frac{d\phi}{ds} = \frac{\partial\phi}{\partial x}\frac{dx}{ds} + \frac{\partial\phi}{\partial y}\frac{dy}{ds} + \frac{\partial\phi}{\partial z}\frac{dz}{ds} \tag{3}$$

The first factor in each product on the right in (3) depends only on ϕ and the coordinates of the point at which the derivatives of ϕ are evaluated. The second factor in each product is independent of ϕ and depends only on the direction in which the derivative is being computed. This observation suggests that $d\phi/ds$ can be thought of as the dot product of two vectors, one depending only on ϕ and the coordinates of P, the other depending only on the direction of ds; and in fact we can write

$$\frac{d\phi}{ds} = \left(\frac{\partial\phi}{\partial x}\mathbf{i} + \frac{\partial\phi}{\partial y}\mathbf{j} + \frac{\partial\phi}{\partial z}\mathbf{k}\right)\cdot\left(\frac{dx}{ds}\mathbf{i} + \frac{dy}{ds}\mathbf{j} + \frac{dz}{ds}\mathbf{k}\right)$$

$$= \left(\frac{\partial\phi}{\partial x}\mathbf{i} + \frac{\partial\phi}{\partial y}\mathbf{j} + \frac{\partial\phi}{\partial z}\mathbf{k}\right)\cdot\frac{d\mathbf{R}}{ds} \tag{4}$$

The vector function

$$\frac{\partial\phi}{\partial x}\mathbf{i} + \frac{\partial\phi}{\partial y}\mathbf{j} + \frac{\partial\phi}{\partial z}\mathbf{k}$$

is known as the **vector-valued derivative** of ϕ. It is also identified in other ways.

DEFINITION 1 Wherever the partial derivatives of ϕ exist, the **gradient of** ϕ, written **grad** ϕ, is the vector field

$$\mathbf{grad}\ \phi = \frac{\partial \phi}{\partial x}\mathbf{i} + \frac{\partial \phi}{\partial y}\mathbf{j} + \frac{\partial \phi}{\partial z}\mathbf{k} \tag{5}$$

The gradient of a function is frequently written in operational form as

$$\mathbf{grad}\ \phi = \left(\mathbf{i}\frac{\partial}{\partial x} + \mathbf{j}\frac{\partial}{\partial y} + \mathbf{k}\frac{\partial}{\partial z}\right)\phi$$

The parenthetic expression operating on ϕ is usually denoted by the symbol $\boldsymbol{\nabla}$ (read ''del''); i.e.,

$$\boldsymbol{\nabla} = \mathbf{i}\frac{\partial}{\partial x} + \mathbf{j}\frac{\partial}{\partial y} + \mathbf{k}\frac{\partial}{\partial z} \tag{6}$$

This gives the relation

$$\mathbf{grad}\ \phi = \boldsymbol{\nabla}\phi \tag{7}$$

Substituting **grad** ϕ into (4) for the vector-valued derivative of ϕ, we get

$$\frac{d\phi}{ds} = (\mathbf{grad}\ \phi)\cdot\frac{d\mathbf{R}}{ds} \tag{8}$$

Since Δs is by definition just the length of $\Delta\mathbf{R}$, it follows that $d\mathbf{R}/ds$ is a unit vector. In particular, if Q approaches P along l, then $\Delta\mathbf{R} = \Delta s\ \mathbf{u}$, $d\mathbf{R}/ds = \mathbf{u}$, and $d\phi/ds$ is denoted by $D_{\mathbf{u}}\phi$. This derivative gives the rate of change of ϕ at P in the direction of **u** and is called the **directional derivative of** ϕ **in the direction of u.** The value of the directional derivative $D_{\mathbf{u}}\phi$ at a fixed point P_0: (x_0, y_0, z_0) is denoted by $D_{\mathbf{u}}\phi(P_0)$, or by $D_{\mathbf{u}}\phi(\mathbf{R}_0)$, where $\mathbf{R}_0 = x_0\mathbf{i} + y_0\mathbf{j} + z_0\mathbf{k}$ is the position vector of P_0. This is stated more precisely in the following definition.

DEFINITION 2 The **directional derivative** $D_{\mathbf{u}}\phi(\mathbf{R}_0)$ **of a function** $\phi(x, y, z)$ **at** $\mathbf{R}_0$ **in the direction of a unit vector u** is given by

$$D_{\mathbf{u}}\phi(\mathbf{R}_0) = \lim_{h\to 0}\frac{\phi(\mathbf{R}_0 + h\mathbf{u}) - \phi(\mathbf{R}_0)}{h} \tag{9}$$

provided the limit exists.

Substituting $D_{\mathbf{u}}\phi$ for $d\phi/ds$ and **u** for $d\mathbf{R}/ds$ in (8), we have

$$D_{\mathbf{u}}\phi = (\mathbf{grad}\ \phi)\cdot\mathbf{u} \tag{10}$$

which shows that the directional derivative of ϕ in the direction of **u** is just the scalar projection of **grad** ϕ, i.e., of $\boldsymbol{\nabla}\phi$, along the direction of **u**. In other words, according to (10),

THEOREM 1 The vector-valued derivative of ϕ, **grad** ϕ, has the property that its component along any direction equals the directional derivative of ϕ in that direction.

This property of **grad** ϕ is illustrated in Fig. 15.10a. Note that the directional derivative of ϕ in the direction of **i**, **j**, or **k** is the corresponding partial derivative of ϕ in that direction (Exercise 16). Since the maximum projection of a vector is the vector itself (Exercise 14),

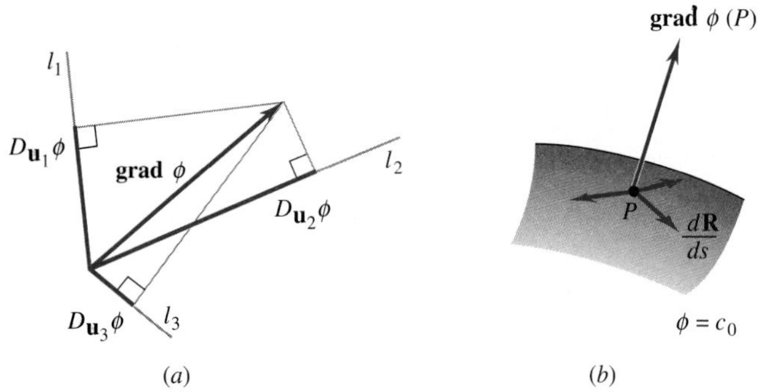

FIGURE 15.10
Geometric properties of the gradient of a function ϕ.

COROLLARY 1 The gradient of ϕ, i.e., $\nabla\phi$, extends in the direction of the greatest rate of change of ϕ and has that rate of change for its length.

Other useful features of **grad** ϕ also deserve our attention. If $\nabla\phi \neq 0$ and we set $\phi(x, y, z) = c$, we obtain, as the parameter c takes on different real values, a family of surfaces known as the **level surfaces**† of ϕ. Since the function ϕ is single-valued, one and only one level surface passes through any given point P. Suppose the equation of this surface is $\phi = c_0$. If we consider the level surface through P and fix our attention on an arbitrary neighboring point Q of the same surface a distance Δs from P, we have $d\phi/ds = 0$, because by definition ϕ has the same value at all points of a level surface. Hence, by Eq. (8),

$$0 = (\mathbf{grad}\ \phi) \cdot \frac{d\mathbf{R}}{ds} \tag{11}$$

for any vector $d\mathbf{R}/ds$ whose direction is the limiting direction of a secant PQ of the level surface. Since such vectors are all tangent to $\phi = c_0$ at the point P, from the vanishing of the dot product in (11) it follows that **grad** ϕ is perpendicular to every tangent to the level surface at P (Fig. 15.10*b*). The set of all these tangents constitutes the tangent plane ρ to the level surface $\phi = c_0$ which passes through P. Consequently, the gradient of ϕ at P, denoted by either **grad** $\phi(P)$ or $\nabla\phi(P)$, is perpendicular to ρ. But **grad** ϕ(P) is perpendicular to $\phi = c_0$ if and only if $\nabla\phi$(P) is perpendicular to ρ. Thus, another important property of **grad** ϕ is that presented in Theorem 2.

THEOREM 2 The gradient of ϕ at any point P is perpendicular to the level surface of ϕ which passes through that point.

Evidently, **grad** ϕ is related to the level surfaces of ϕ in a way which is independent of the particular coordinate system used to describe ϕ. In other words, **grad** ϕ depends only on the intrinsic properties of ϕ. It follows, therefore, that in the expression

$$\mathbf{grad}\ \phi = \frac{\partial\phi}{\partial x}\mathbf{i} + \frac{\partial\phi}{\partial y}\mathbf{j} + \frac{\partial\phi}{\partial z}\mathbf{k}$$

i, **j**, and **k** can be replaced by any other set of mutually perpendicular unit vectors provided that $\partial\phi/\partial x$, $\partial\phi/\partial y$, $\partial\phi/\partial z$ are replaced by the directional derivatives of ϕ along the new axes.

†This name, which is used regardless of the number of independent variables, is suggested by the analogy between the general case and the two-dimensional topographic interpretation in which $\phi(x, y)$ is the elevation at the point (x, y) and the loci $\phi(x, y) = c$ are the contour lines, i.e., curves whose points are all at the same *level* or whose elevation above (or below) the xy plane is a constant.

A vector **N** that is perpendicular to a surface $\phi(x, y, z) = c$ at a point P_0: (x_0, y_0, z_0) of the surface is called a **normal, or normal vector, to the surface at** P_0. The **normal line to** $\phi = c$ **at** P_0 is the line through P_0 that contains a normal **N** at P_0. Since $\nabla\phi(P_0)$ is a normal to $\phi = c$ at P_0, a vector equation of the normal line to the surface at P_0 is [Eq. (9), Sec. 15.2],

$$\mathbf{R} = \mathbf{R}_0 + t\,\nabla\phi(P_0) \qquad t \text{ real} \tag{12}$$

where $\mathbf{R} = x\mathbf{i} + y\mathbf{j} + z\mathbf{k}$ is the position vector of an arbitrary point P: (x, y, z) on the line. A vector equation of the tangent plane to $\phi = c$ at P_0 is [Eq. (12), Sec. 15.2],

$$(\mathbf{R} - \mathbf{R}_0) \cdot \nabla\phi(P_0) = 0 \tag{13}$$

where **R** is now the position vector of an arbitrary point P: (x, y, z) of ρ. A concise form of the differential of $\phi(x, y, z)$, in terms of $\nabla\phi$ and $\mathbf{R} = x\mathbf{i} + y\mathbf{j} + z\mathbf{k}$, is

$$d\phi = \nabla\phi \cdot d\mathbf{R} \tag{14}$$

If ϕ is a differentiable function of a scalar variable ψ, which, in turn, is a differentiable function of x, y, and z then

$$\begin{aligned}
\nabla\phi &= \frac{\partial\phi}{\partial x}\mathbf{i} + \frac{\partial\phi}{\partial y}\mathbf{j} + \frac{\partial\phi}{\partial z}\mathbf{k} \\
&= \frac{d\phi}{d\psi}\frac{\partial\psi}{\partial x}\mathbf{i} + \frac{d\phi}{d\psi}\frac{\partial\psi}{\partial y}\mathbf{j} + \frac{d\phi}{d\psi}\frac{\partial\psi}{\partial z}\mathbf{k} \\
&= \frac{d\phi}{d\psi}\left(\frac{\partial\psi}{\partial x}\mathbf{i} + \frac{\partial\psi}{\partial y}\mathbf{j} + \frac{\partial\psi}{\partial z}\mathbf{k}\right) \\
&= \frac{d\phi}{d\psi}\nabla\psi
\end{aligned} \tag{15}$$

EXAMPLE 1

What is the directional derivative of the function $\phi(x, y, z) = xy^2 + yz^3$ at the point $(2, -1, 1)$ in the direction of the vector $\mathbf{i} + 2\mathbf{j} + 2\mathbf{k}$?

Our first step must be to find the gradient of ϕ at the point $(2, -1, 1)$. It is

$$\begin{aligned}
\nabla\phi &= \frac{\partial(xy^2 + yz^3)}{\partial x}\mathbf{i} + \frac{\partial(xy^2 + yz^3)}{\partial y}\mathbf{j} + \frac{\partial(xy^2 + yz^3)}{\partial z}\mathbf{k}\Bigg|_{2,-1,1} \\
&= y^2\mathbf{i} + (2xy + z^3)\mathbf{j} + 3yz^2\mathbf{k}\Big|_{2,-1,1} \\
&= \mathbf{i} - 3\mathbf{j} - 3\mathbf{k}
\end{aligned}$$

The projection of this in the direction of the given vector will be the required directional derivative. Since this projection can be found at once as the dot product of $\nabla\phi$ and a unit vector in the given direction, we next reduce $\mathbf{i} + 2\mathbf{j} + 2\mathbf{k}$ to a unit vector by dividing it by its magnitude, getting

$$\frac{\mathbf{i} + 2\mathbf{j} + 2\mathbf{k}}{\sqrt{1 + 4 + 4}} = \tfrac{1}{3}\mathbf{i} + \tfrac{2}{3}\mathbf{j} + \tfrac{2}{3}\mathbf{k}$$

The answer to our problem is therefore

$$\nabla\phi\cdot\left(\tfrac{1}{3}\mathbf{i}+\tfrac{2}{3}\mathbf{j}+\tfrac{2}{3}\mathbf{k}\right)=(\mathbf{i}-3\mathbf{j}-3\mathbf{k})\cdot\left(\tfrac{1}{3}\mathbf{i}+\tfrac{2}{3}\mathbf{j}+\tfrac{2}{3}\mathbf{k}\right)=-\tfrac{11}{3}$$

The negative sign, of course, indicates that ϕ decreases in the given direction.

EXAMPLE 2

What is the unit normal to the surface $xy^3z^2 = 4$ at the point $(-1, -1, 2)$?

Let us regard the given surface as a particular level surface of the function $\phi = xy^3z^2$. Then the gradient of this function at the point $(-1, -1, 2)$ will be perpendicular to the level surface through $(-1, -1, 2)$, which is the given surface. When this gradient has been found, the unit normal can be obtained at once by dividing the gradient by its magnitude:

$$\nabla\phi=\left.\frac{\partial(xy^3z^2)}{\partial x}\mathbf{i}+\frac{\partial(xy^3z^2)}{\partial y}\mathbf{j}+\frac{\partial(xy^3z^2)}{\partial z}\mathbf{k}\right|_{-1,-1,2}$$

$$=-4\mathbf{i}-12\mathbf{j}+4\mathbf{k}$$

$$|\nabla\phi|=\sqrt{16+144+16}=4\sqrt{11}$$

$$\frac{\nabla\phi}{|\nabla\phi|}=\frac{-4\mathbf{i}-12\mathbf{j}+4\mathbf{k}}{4\sqrt{11}}=-\frac{1}{\sqrt{11}}\mathbf{i}-\frac{3}{\sqrt{11}}\mathbf{j}+\frac{1}{\sqrt{11}}\mathbf{k}$$

It may be necessary to reverse the direction of this result by multiplying it by -1, depending on which side of the surface we wish the normal to extend.

The vector character of the operator ∇ suggests that we also consider dot and cross products in which it appears as one factor. If $\mathbf{F} = F_1\mathbf{i} + F_2\mathbf{j} + F_3\mathbf{k}$ is a vector field whose components are differentiable functions of x, y, and z, this leads to the combinations

$$\nabla\cdot\mathbf{F}=\left(\mathbf{i}\frac{\partial}{\partial x}+\mathbf{j}\frac{\partial}{\partial y}+\mathbf{k}\frac{\partial}{\partial z}\right)\cdot(F_1\mathbf{i}+F_2\mathbf{j}+F_3\mathbf{k})$$

$$=\frac{\partial F_1}{\partial x}+\frac{\partial F_2}{\partial y}+\frac{\partial F_3}{\partial z} \tag{16}$$

which is known as the **divergence** of the vector function **F**, and

$$\nabla\times\mathbf{F}=\left(\mathbf{i}\frac{\partial}{\partial x}+\mathbf{j}\frac{\partial}{\partial y}+\mathbf{k}\frac{\partial}{\partial z}\right)\times(F_1\mathbf{i}+F_2\mathbf{j}+F_3\mathbf{k})$$

$$=\mathbf{i}\left(\frac{\partial F_3}{\partial y}-\frac{\partial F_2}{\partial z}\right)-\mathbf{j}\left(\frac{\partial F_3}{\partial x}-\frac{\partial F_1}{\partial z}\right)+\mathbf{k}\left(\frac{\partial F_2}{\partial x}-\frac{\partial F_1}{\partial y}\right) \tag{17}$$

which is known as the **curl** of **F**.

The last expression for the curl of **F** can easily be remembered by regarding it as the formal expansion of the determinant

$$\nabla\times\mathbf{F}=\begin{vmatrix}\mathbf{i}&\mathbf{j}&\mathbf{k}\\ \dfrac{\partial}{\partial x}&\dfrac{\partial}{\partial y}&\dfrac{\partial}{\partial z}\\ F_1&F_2&F_3\end{vmatrix} \tag{18}$$

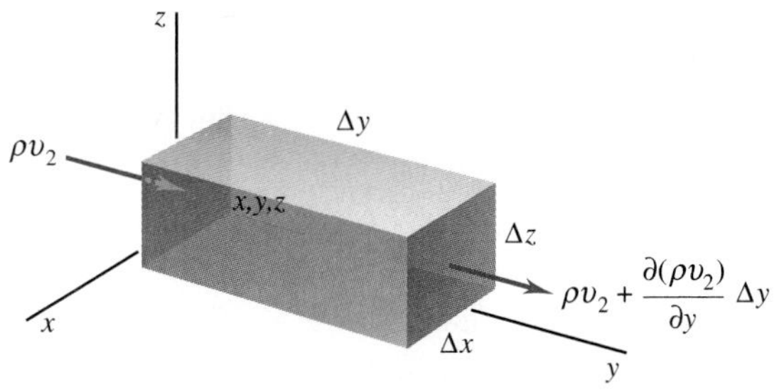

FIGURE 15.11
A typical volume element in a region filled with a moving fluid.

In expanding (18), the order of the factors in each product must be such that the operators in the second row act upon the appropriate functions in the third row. Otherwise, the expansion of the determinant will be another vector operator and not a vector function, as of course it is.

Both the divergence and the curl admit of physical interpretations which justify their names. For instance, to illustrate the significance of the divergence, consider the steady-state motion of a compressible fluid in some region of space and let

$$\mathbf{v} = v_1\mathbf{i} + v_2\mathbf{j} + v_3\mathbf{k}$$

be a vector function representing at each point the instantaneous velocity of the particle of fluid at that point. For any infinitesimal volume element (Fig. 15.11) in the region occupied by the fluid, there will be flow through each of its faces, and as a result the amount of fluid within the element may vary. To measure this variation, let us compute the loss of fluid from the element in the time Δt.

The volume of fluid which passes through one face of the element ΔV in time Δt is approximately equal to the component of the fluid velocity normal to the face times the area of the face times Δt, and the corresponding mass flow is, of course, the product of this volume and the density of the fluid ρ. Hence, computing the loss of fluid through each face in turn (remembering that since the fluid is not assumed to be incompressible, the density as well as the velocity may vary from point to point), we have

Right face: $$\left[\rho v_2 + \frac{\partial(\rho v_2)}{\partial y}\Delta y\right]\Delta x\,\Delta z\,\Delta t$$

Left face: $$-\rho v_2\,\Delta x\,\Delta z\,\Delta t$$

Front face: $$\left[\rho v_1 + \frac{\partial(\rho v_1)}{\partial x}\Delta x\right]\Delta y\,\Delta z\,\Delta t$$

Rear face: $$-\rho v_1\,\Delta y\,\Delta z\,\Delta t$$

Top face: $$\left[\rho v_3 + \frac{\partial(\rho v_3)}{\partial z}\Delta z\right]\Delta x\,\Delta y\,\Delta t$$

Bottom face: $$-\rho v_3\,\Delta x\,\Delta y\,\Delta t$$

If we add these and convert the resulting estimate of the absolute loss of fluid from ΔV in the interval Δt into the loss per unit volume per unit time by dividing by $\Delta V\,\Delta t \equiv \Delta x\,\Delta y\,\Delta z\,\Delta t$, we obtain in the limit

$$\text{Rate of loss per unit volume} = \frac{\partial(\rho v_1)}{\partial x} + \frac{\partial(\rho v_2)}{\partial y} + \frac{\partial(\rho v_3)}{\partial z}$$

which is precisely the divergence of the vector $\rho\mathbf{v}$. Thus fluid mechanics affords one possible interpretation of the divergence as the rate of loss of fluid per unit volume.

If the fluid is incompressible, there can be neither gain nor loss of fluid in a general element. Hence, since the density ρ is constant for an incompressible fluid, we must have

$$\nabla \cdot \rho\mathbf{v} = \rho\nabla \cdot \mathbf{v} = 0 \qquad \text{or} \qquad \nabla \cdot \mathbf{v} = 0 \tag{19}$$

which is known as the **equation of continuity** for incompressible fluids. However, if ΔV encloses a source of fluid, there is a net loss of fluid through the surface of ΔV equal to the amount *diverging* from the source. Similar results, of course, hold for such things as electric and magnetic flux, which exhibit many of the properties of incompressible fluids.

To find a possible interpretation of the curl, let us consider a body rotating with uniform angular speed ω about an axis l. Let us define the **vector angular velocity $\mathbf{\Omega}$** to be a vector of length ω extending along l in the direction in which a right-handed screw would advance if subject to the same rotation as the body. Finally, let $\mathbf{R}$ be the vector drawn from any point O on the axis l to an arbitrary point P in the body.

From Fig. 15.12 it is evident that the radius at which P rotates is $|\mathbf{R}| \cdot |\sin\theta|$. Hence, the linear speed of P is

$$|\mathbf{v}| = \omega|\mathbf{R}| \cdot |\sin\theta| = |\Omega| \cdot |\mathbf{R}| \cdot |\sin\theta| = |\mathbf{\Omega} \times \mathbf{R}|$$

Moreover, the vector velocity $\mathbf{v}$ is directed perpendicular to the plane of $\mathbf{\Omega}$ and $\mathbf{R}$, so that $\mathbf{\Omega}$, $\mathbf{R}$, and $\mathbf{v}$ form a right-handed system. Hence, the cross product $\mathbf{\Omega} \times \mathbf{R}$ gives not only the magnitude of $\mathbf{v}$ but the direction as well.

If we now take the point O as the origin of coordinates, we can write

$$\mathbf{R} = x\mathbf{i} + y\mathbf{j} + z\mathbf{k} \qquad \text{and} \qquad \mathbf{\Omega} = \Omega_1\mathbf{i} + \Omega_2\mathbf{j} + \Omega_3\mathbf{k}$$

Hence, the equation $\mathbf{v} = \mathbf{\Omega} \times \mathbf{R}$ can be written at length in the form

$$\mathbf{v} = (\Omega_2 z - \Omega_3 y)\mathbf{i} - (\Omega_1 z - \Omega_3 x)\mathbf{j} + (\Omega_1 y - \Omega_2 x)\mathbf{k}$$

if we take the curl of $\mathbf{v}$, we therefore have

$$\nabla \times \mathbf{v} = \begin{vmatrix} \mathbf{i} & \mathbf{j} & \mathbf{k} \\ \dfrac{\partial}{\partial x} & \dfrac{\partial}{\partial y} & \dfrac{\partial}{\partial z} \\ \Omega_2 z - \Omega_3 y & -(\Omega_1 z - \Omega_3 x) & \Omega_1 y - \Omega_2 x \end{vmatrix}$$

Expanding this, remembering that $\mathbf{\Omega}$ is a constant vector, we find

$$\nabla \times \mathbf{v} = 2\Omega_1\mathbf{i} + 2\Omega_2\mathbf{j} + 2\Omega_3\mathbf{k} = 2\mathbf{\Omega}$$

$$\mathbf{\Omega} = \tfrac{1}{2}\nabla \times \mathbf{v} \tag{20}$$

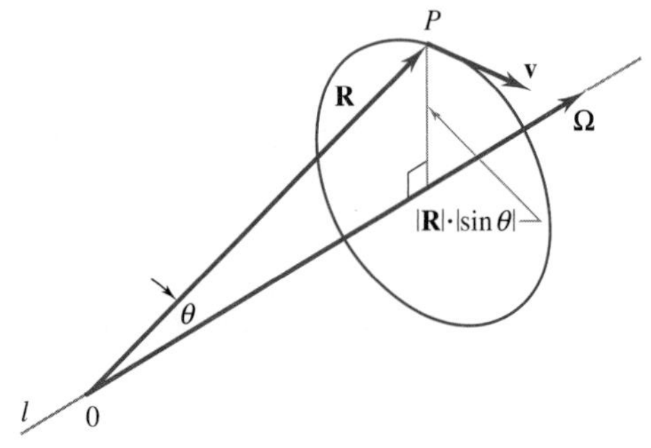

FIGURE 15.12
A physical interpretation of the curl.

The angular velocity of a uniformly rotating body is thus equal to one-half the curl of the linear velocity of any point of the body. The aptness of the name *curl* in this connection is apparent.

The results of applying the operator ∇ to various combinations of scalar and vector functions can be found by the following formulas:†

$$\nabla \cdot \phi\mathbf{v} = \phi\nabla \cdot \mathbf{v} + \mathbf{v} \cdot \nabla\phi \tag{21}$$

$$\nabla \times \phi\mathbf{v} = \phi\nabla \times \mathbf{v} + (\nabla\phi) \times \mathbf{v} \tag{22}$$

$$\nabla \cdot (\mathbf{u} \times \mathbf{v}) = \mathbf{v} \cdot \nabla \times \mathbf{u} - \mathbf{u} \cdot \nabla \times \mathbf{v} \tag{23}$$

$$\nabla \times (\mathbf{u} \times \mathbf{v}) = \mathbf{v} \cdot \nabla\mathbf{u} - \mathbf{u} \cdot \nabla\mathbf{v} + \mathbf{u}\nabla \cdot \mathbf{v} - \mathbf{v}\nabla \cdot \mathbf{u} \tag{24}$$

$$\nabla(\mathbf{u} \cdot \mathbf{v}) = \mathbf{u} \cdot \nabla\mathbf{v} + \mathbf{v} \cdot \nabla\mathbf{u} + \mathbf{u} \times (\nabla \times \mathbf{v}) + \mathbf{v} \times (\nabla \times \mathbf{u}) \tag{25}$$

$$\nabla \times \nabla\phi = \mathbf{0} \tag{26}$$

$$\nabla \cdot \nabla \times \mathbf{v} = 0 \tag{27}$$

$$\nabla \times (\nabla \times \mathbf{v}) = \nabla(\nabla \cdot \mathbf{v}) - \nabla \cdot \nabla\mathbf{v} = \nabla(\nabla \cdot \mathbf{v}) - \nabla^2\mathbf{v} \tag{28}$$

These identities can all be verified by direct expansion. For instance, to prove (21), we have

$$\begin{aligned}\nabla \cdot (\phi\mathbf{v}) &= \nabla \cdot [\phi(v_1\mathbf{i} + v_2\mathbf{j} + v_3\mathbf{k})] \\ &= \frac{\partial(\phi v_1)}{\partial x} + \frac{\partial(\phi v_2)}{\partial y} + \frac{\partial(\phi v_3)}{\partial z} \\ &= \left(\phi\frac{\partial v_1}{\partial x} + v_1\frac{\partial \phi}{\partial x}\right) + \left(\phi\frac{\partial v_2}{\partial y} + v_2\frac{\partial \phi}{\partial y}\right) + \left(\phi\frac{\partial v_3}{\partial z} + v_3\frac{\partial \phi}{\partial z}\right)\end{aligned}$$

which, on regrouping, is simply $\phi\nabla \cdot \mathbf{v} + \mathbf{v} \cdot \nabla\phi$ as asserted.

In general, however, it is easier to establish formulas like those in the above list by treating ∇ as a vector, manipulating the expressions according to the appropriate formulas from vector algebra, and finally giving ∇ its operational meaning. Since ∇ is a linear combination of scalar differential operators which obey the usual product rule of differentiation, i.e., act on the factors in a product one at a time, it is clear that ∇ itself has this property. In other words, we can apply ∇ to products of various sorts by assuming that each of the factors, in turn, is the only one which is variable and adding the partial results so obtained. As a notation to aid us in determining these partial results, it is helpful to attach to ∇, whenever it is followed by more than one factor, a subscript indicating the one factor upon which it is currently allowed to operate.

To prove (22), using the second, more formal, procedure, we suppose first that the scalar function ϕ is a constant; i.e., we let ∇ operate only on the vector $\mathbf{v}$. Then we can write

$$\nabla_v \times (\phi\mathbf{v}) = \phi\nabla \times \mathbf{v}$$

where the subscript has been omitted from the right-hand side, since it is always completely clear what ∇ operates on when it is followed by just one factor. Similarly, if we regard $\mathbf{v}$ as constant and ϕ as variable, we have

$$\nabla_\phi \times (\phi\mathbf{v}) = (\nabla\phi) \times \mathbf{v}$$

where the parentheses now restrict the effect of ∇ to the factor ϕ alone and so make a subscript on ∇ unnecessary. Finally, adding our two partial results, we have

$$\nabla_v \times (\phi\mathbf{v}) + \nabla_\phi \times (\phi\mathbf{v}) \equiv \nabla \times (\phi\mathbf{v}) = \phi\nabla \times \mathbf{v} + (\nabla\phi) \times \mathbf{v}$$

To prove (23), we have, from the cyclic properties of scalar triple products,

$$\nabla_u \cdot (\mathbf{u} \times \mathbf{v}) = \mathbf{v} \cdot \nabla \times \mathbf{u} \qquad \text{and} \qquad \nabla_v \cdot (\mathbf{u} \times \mathbf{v}) = -\mathbf{u} \cdot \nabla \times \mathbf{v}$$

†We must remember, however, that these results are correct only for the cartesian form of the operator ∇ given by Eq. (6). Different formulas arise when ∇ is expressed in terms of more general coordinate systems.

Hence, adding these two partial results, we find

$$\nabla_u \cdot (\mathbf{u} \times \mathbf{v}) + \nabla_v \cdot (\mathbf{u} \times \mathbf{v}) \equiv \nabla \cdot (\mathbf{u} \times \mathbf{v}) = \mathbf{v} \cdot \nabla \times \mathbf{u} - \mathbf{u} \cdot \nabla \times \mathbf{v}$$

To prove (24), we have

$$\begin{aligned} \nabla_u \times (\mathbf{u} \times \mathbf{v}) &= (\nabla_u \cdot \mathbf{v})\mathbf{u} - (\nabla_u \cdot \mathbf{u})\mathbf{v} = \mathbf{v} \cdot \nabla\mathbf{u} - \mathbf{v}\nabla \cdot \mathbf{u} \\ \nabla_v \times (\mathbf{u} \times \mathbf{v}) &= (\nabla_v \cdot \mathbf{v})\mathbf{u} - (\nabla_v \cdot \mathbf{u})\mathbf{v} = \mathbf{u}\nabla \cdot \mathbf{v} - \mathbf{u} \cdot \nabla\mathbf{v} \end{aligned}$$

Adding gives

$$\begin{aligned} \nabla_u \times (\mathbf{u} \times \mathbf{v}) + \nabla_v \times (\mathbf{u} \times \mathbf{v}) &\equiv \nabla \times (\mathbf{u} \times \mathbf{v}) \\ &= \mathbf{v} \cdot \nabla\mathbf{u} - \mathbf{u} \cdot \nabla\mathbf{v} + \mathbf{u}\nabla \cdot \mathbf{v} - \mathbf{v}\nabla \cdot \mathbf{u} \end{aligned}$$

To prove (25), we note that

$$\begin{aligned} \mathbf{u} \times (\nabla \times \mathbf{v}) &\equiv \mathbf{u} \times (\nabla_v \times \mathbf{v}) = (\mathbf{u} \cdot \mathbf{v})\nabla_v - (\mathbf{u} \cdot \nabla)\mathbf{v} \\ &= \nabla_v(\mathbf{u} \cdot \mathbf{v}) - \mathbf{u} \cdot \nabla\mathbf{v} \end{aligned}$$

and

$$\begin{aligned} \mathbf{v} \times (\nabla \times \mathbf{u}) &\equiv \mathbf{v} \times (\nabla_u \times \mathbf{u}) = (\mathbf{v} \cdot \mathbf{u})\nabla_u - (\mathbf{v} \cdot \nabla)\mathbf{u} \\ &= \nabla_u(\mathbf{u} \cdot \mathbf{v}) - \mathbf{v} \cdot \nabla\mathbf{u} \end{aligned}$$

Hence, transposing and adding, we find

$$\begin{aligned} \nabla_u(\mathbf{u} \cdot \mathbf{v}) + \nabla_v(\mathbf{u} \cdot \mathbf{v}) &\equiv \nabla(\mathbf{u} \cdot \mathbf{v}) \\ &= \mathbf{u} \times (\nabla \times \mathbf{v}) + \mathbf{v} \times (\nabla \times \mathbf{u}) + \mathbf{u} \cdot \nabla\mathbf{v} + \mathbf{v} \cdot \nabla\mathbf{u} \end{aligned}$$

The fact that the operational coefficient $\nabla \times \nabla$ in formula (26) appears as the cross product of two identical factors suggests that $\nabla \times \nabla\phi$ is indeed zero, and it is easy to verify that this is the case (Exercise 36). Similarly, the fact that the left member of formula (27) appears as a scalar triple product with two identical factors suggests the truth of the formula, and again it is not difficult to prove that $\nabla \cdot \nabla \times \mathbf{v}$ is always equal to zero (Exercise 37). It should be noted, however, that the structure of these formulas is not sufficient to establish their truth, since the formal scalar triple product $\nabla \times \mathbf{v} \cdot \mathbf{v}$ is not zero in general, even though it has two identical factors (Exercise 39).

Finally, it is easy to establish formula (28) by direct expansion (Exercise 38), or it can be verified by applying the usual rule for expanding a vector triple product:

$$\nabla \times (\nabla \times \mathbf{v}) = (\nabla \cdot \mathbf{v})\nabla - (\nabla \cdot \nabla)\mathbf{v} = \nabla(\nabla \cdot \mathbf{v}) - \nabla^2\mathbf{v}$$

where the conventional symbol ∇^2 has been substituted for the second-order operator

$$\begin{aligned} \nabla \cdot \nabla &= \left(\mathbf{i}\frac{\partial}{\partial x} + \mathbf{j}\frac{\partial}{\partial y} + \mathbf{k}\frac{\partial}{\partial z}\right) \cdot \left(\mathbf{i}\frac{\partial}{\partial x} + \mathbf{j}\frac{\partial}{\partial y} + \mathbf{k}\frac{\partial}{\partial z}\right) \\ &= \frac{\partial^2}{\partial x^2} + \frac{\partial^2}{\partial y^2} + \frac{\partial^2}{\partial z^2} \end{aligned}$$

The scalar operator

(29) $$\nabla^2 = \frac{\partial^2}{\partial x^2} + \frac{\partial^2}{\partial y^2} + \frac{\partial^2}{\partial z^2}$$

(read "del squared") is called the **Laplace operator** in the Cartesian coordinates x, y, and z, the expression $\nabla^2\phi$ is known as the **Laplacian** of ϕ, and $\nabla^2\phi = 0$ is Laplace's equation in rectangular coordinates [Eq. (16), Sec. 11.2]. A function which satisfies this equation and has continuous partial derivatives of order two is called a **harmonic function.**

The divergence $\nabla \cdot \mathbf{F}$ and curl $\nabla \times \mathbf{F}$ of a vector function $\mathbf{F}$ are often denoted by div $\mathbf{F}$ and curl $\mathbf{F}$, respectively. To each differentiable vector field

$$\mathbf{F}(x, y, z) = F_1(x, y, z)\mathbf{i} + F_2(x, y, z)\mathbf{j} + F_3(x, y, z)\mathbf{k}$$

there corresponds a **Jacobi matrix:**†

$$J[\mathbf{F}(x, y, z)] = \begin{bmatrix} \frac{\partial F_1}{\partial x} & \frac{\partial F_1}{\partial y} & \frac{\partial F_1}{\partial z} \\ \frac{\partial F_2}{\partial x} & \frac{\partial F_2}{\partial y} & \frac{\partial F_2}{\partial z} \\ \frac{\partial F_3}{\partial x} & \frac{\partial F_3}{\partial y} & \frac{\partial F_3}{\partial z} \end{bmatrix} \tag{30}$$

The **trace** of this matrix, i.e., the sum of its principal diagonal elements, is

$$\text{div } \mathbf{F} = \frac{\partial F_1}{\partial x} + \frac{\partial F_2}{\partial y} + \frac{\partial F_3}{\partial z}$$

Its determinant is the familiar **jacobian** of calculus $\partial(F_1, F_2, F_3)/\partial(x, y, z)$. The components of

$$\text{curl } \mathbf{F} = \left(\frac{\partial F_3}{\partial y} - \frac{\partial F_2}{\partial z}\right)\mathbf{i} + \left(\frac{\partial F_1}{\partial z} - \frac{\partial F_3}{\partial x}\right)\mathbf{j} + \left(\frac{\partial F_2}{\partial x} - \frac{\partial F_1}{\partial y}\right)\mathbf{k}$$

are the differences, in the orders indicated, of the elements of the matrix located symmetrically with respect to its main diagonal.

EXERCISES

1. Find the gradient of each of the following functions.
(a) $x^2 + 2yz$ **(b)** e^{xyz} **(c)** $x \sin yz$
(d) $x^3 + y^3 - 3xyz$ **(e)** $x^a y^b z^c$

2. Find a unit normal to each of the following surfaces at the specified points.
(a) $x^2 + y^2 + z^2 = 2(x + y + z)$ at $(2, 2, 0)$ and $\left(-\frac{1}{2}, \frac{1}{2}, 1 + \frac{1}{\sqrt{2}}\right)$
(b) $z = x^2 + y^2$ at $(1, -2, 5)$ and $(-1, 1, 2)$
(c) $9x^2 + 9y^2 - 4z^2 = 0$ at $(0, 2, -3)$ and $(\sqrt{2}, \sqrt{2}, 3)$

3. Find the **normal derivative** $d\phi/dn = |\nabla\phi|$ of each of the following functions and its **direction** $\nabla\phi/|\nabla\phi|$.
(a) e^{x+y+z}
(b) $\sqrt{2} \cosh (x + y) + 2z$
(c) $\frac{1}{2}x^2 + 2xy + y^2$ at $(1, 1)$ **(d)** $xy - yz$ at $(1, 1, -1)$

4. **(a)** Find the unit outward normal $\mathbf{n}$ at a general point P: (x, y) on the ellipse

$$b^2x^2 + a^2y^2 = a^2b^2$$

as a vector function of x.
(b) Express the angle θ between $\mathbf{n}$ and the position vector $\mathbf{R} = x\mathbf{i} + y\mathbf{j}$ to P in terms of x.
(c) Discuss the special cases which arise when $a = b$ and when $x = 0$ or $x = \pm a$.

5. What is the angle between the normals to the surface $xy = z^2$ at the points $(1, 4, 2)$ and $(-3, -3, 3)$?

6. Find the directional derivative of each of the following functions at the given point in the direction of the given vector.
(a) $2xy + z^2$, $(1, -1, 3)$, $\mathbf{i} + 2\mathbf{j} + 2\mathbf{k}$
(b) $3x^2 - 2y^2 + z^2 + xy - yz$, $(1, -2, -1)$, $2\mathbf{i} - 2\mathbf{j} - \mathbf{k}$
(c) $(x + y)/(y + z)$, $(1, 1, 1)$, $\mathbf{i} + \mathbf{j} - \mathbf{k}$
(d) $x - 2y + z^2$, $(3, 1, -2)$, $7\mathbf{i} + 6\mathbf{j} + 6\mathbf{k}$

7. Find the directional derivative of the function x^3y^4 in the direction of $\mathbf{i} + \mathbf{j}$.

8. What is the directional derivative of $\phi(x, y) = 2x^2 - 5y^2 + 6x - 2y + 5$ at $(1, -1)$ in the direction of the vector directed from $(1, 5)$ to $(4, 1)$, and what is the maximum rate of change of ϕ at $(1, -1)$ in any direction?

†Karl Gustav Jacob Jacobi (1804–1851) was a German arithmetician and analyst.

9. Find $D_{\mathbf{u}}\phi(\pi/3, 0)$ given that $\phi(x, y) = e^{-y} \sin x + \frac{1}{3}e^{-3y} \sin 3x$ and $\mathbf{u} = (-\mathbf{i} + \sqrt{3}\,\mathbf{j})/2$. What is the maximum rate of change of ϕ at $(\pi/3, 0)$ in any direction?

10. In what direction is the rate of change of a function $\phi(x, y, z)$ at $\mathbf{R}$ minimal? What is this rate of change?

11. The steady-state temperature T of a solid is given by the scalar field $x^2 - (y + z)^2$.
 (a) Find a vector whose magnitude yields the maximum rate of change of T at $(2, 1, 1)$.
 (b) What is the rate of change of T at $(2, 1, 1)$ in the direction of $\mathbf{i} - 2\mathbf{j} + \mathbf{k}$?

12. Show that, in general, $D_{-\mathbf{u}}\phi(\mathbf{R}_0) = -D_{\mathbf{u}}(\mathbf{R}_0)$.

13. **(a)** Find the rate of change of $\phi(x, y, z) = 4x^2 + y^2 - 16z$ at $(2, 4, 2)$ in the direction of a normal to the plane $x + 2y + 2z = 9$.
 (b) What is the maximum rate of change of ϕ at $(2, 4, 2)$ in any direction?

14. Prove Corollary 1 analytically. *Hint:* $|D_{\mathbf{u}}\phi(\mathbf{R}_0)| = |\nabla\phi(\mathbf{R}_0)| \cos\theta$, where θ is the angle between $\nabla\phi(\mathbf{R}_0)$ and $\mathbf{u}$.

15. Find the rate of change of $f(x, y, z) = x^2 + y - z$ at $(2, 2, 1)$ in the direction of an outward normal to the paraboloid of revolution $x^2 + y^2 - z = 1$ at $(1, 1, 1)$.

16. Identify $D_{\mathbf{i}}\phi$, $D_{\mathbf{j}}\phi$, and $D_{\mathbf{k}}\phi$ as partial derivatives of ϕ.

17. What is the rate of change of the function $x^2y^3z^4$ in the direction making equal angles with the positive x, y, and z axes?

18. The steady-state temperature T of a metallic plate in degrees centigrade is given by the scalar field $50 - 5x - 10y + xy$, $|x| \le 2$, $|y| \le 2$. Find $T(1, 1)$ and the direction in which the rate of change of T at $(1, 1)$ is greatest.

19. Write symmetric equations of the normal line defined by Eq. (12).

20. Write a scalar equation of the tangent plane defined by Eq. (13).

21. Write a vector equation and symmetric equations for the line normal to the surface $x^2 - y - z^2 = 0$ at $(5, 9, 4)$.

22. Write an equation of the tangent plane to the surface $4x^2 + y^2 - 16z = 0$ at $(2, 4, 2)$.

23. Determine a such that the tangent plane to the surface $x^3 - 2xy + yz = a + 4$ at the point $(2, 1, a)$ will pass through the origin.

24. Write an equation of the tangent plane to the surface $z = x^2 + y^2$ at $(1, -2, 5)$.

25. Find an equation of the tangent plane and symmetric equations of the normal line to each of the following surfaces at the given point.
 (a) $x^2 + 4y^2 - 4z = 0$, $(2, 2, 5)$
 (b) $x = y^2 + z^2$, $(5, 1, 2)$
 (c) $xy + yz + xz = 11$, $(1, 2, 3)$
 (d) $4x^2 - y^2 + 9z^2 = 36$, $(1, -2, 2)$
 (e) $z = xy^2$, $(2, 1, 2)$
 (f) $4x^2 + y^2 - 16z = 0$, $(2, 4, 2)$

26. Write an equation of the tangent plane and symmetric equations of the normal line to the surface $x^2 - 3y^2 - z^2 - 6xy + y + 2z + 10 = 0$ at $(1, 1, -1)$.

27. Find the distance from the point $(2, 3, 7)$ to the plane that is tangent to the surface $x^2 + y^2 - z = 1$ at $(1, 1, 1)$.

28. A scalar point function ϕ is defined by $\phi(x, y, z) = x^2 - y^2 + 2yz$. Find **(a)** $\nabla\phi$, **(b)** the directional derivative of ϕ at $(1, 1, -1)$ in the direction of $\mathbf{i} - 2\mathbf{k}$, **(c)** a unit normal to the level surface $\phi(x, y, z) = 4$ at $(1, 1, 2)$, and **(d)** the divergence and curl of $\nabla\phi$.

29. Compute the divergence and curl of each of the following vector fields.
 (a) $y^2\mathbf{i} + 2x^2z\mathbf{j} - xyz\mathbf{k}$
 (b) $xyz\mathbf{i} + 3x^2y\mathbf{j} + (xz^2 - y^2z)\mathbf{k}$
 (c) $(z + \sin y)\mathbf{i} - (z - x\cos y)\mathbf{j}$
 (d) $xy^2\mathbf{i} + x^2y\mathbf{j} + z^2\mathbf{k}$
 (e) $(x^2 + yz)\mathbf{i} + (y^2 + xz)\mathbf{j} + (z^2 + xy)\mathbf{k}$
 (f) $\sin y\,\mathbf{i} + \sin z\,\mathbf{j} + \sin x\,\mathbf{k}$

30. Given that $\mathbf{v} = e^z(\sin y\,\mathbf{i} + x\cos y\,\mathbf{j} + x\sin y\,\mathbf{k})$, $\mathbf{w} = x\mathbf{i} - 3y\mathbf{j} + 2z\mathbf{k}$, and $\phi = 5z - e^{-y}\sin x$, find:
 (a) curl $\mathbf{v}$ **(b)** $\nabla^2\phi$ **(c)** div $(\phi\mathbf{w})$

31. Is the function $f(x, y, z) = e^x \cos y - z$ harmonic? Explain.

32. If three functions u, v, and w are connected by a relation $f(u, v, w) = 0$, prove that $[\nabla u\; \nabla v\; \nabla w] = 0$. *Hint:* Consider the dot product of ∇f and $\nabla u \times \nabla v$.

33. Evaluate $[\nabla u\; \nabla v\; \nabla w]$ if u, v, and w are the respective scalar fields:
 (a) $x^2 + 2xy + z^3$, $x + y$, $-2z(x + y)$
 (b) $x + y + z$, $x + y$, $-2xy - 2yz - z^2$

34. Verify Eq. (22) by direct expansion.

35. Verify Eq. (23) by direct expansion.

36. Verify Eq. (26) by direct expansion.

37. Verify Eq. (27) by direct expansion.

38. Verify Eq. (28) by direct expansion.

39. Show by an example that in general $\nabla \times \mathbf{v} \cdot \mathbf{v} \ne 0$.

40. Prove that the curl of any vector whose direction is constant is perpendicular to that direction.

41. What is the generalization of Eq. (15) to the case in which ϕ is a function of u, v, and w, where u, v, and w are scalar functions of x, y, and z?

In Exercises 42–50, $\mathbf{R} = x\mathbf{i} + y\mathbf{j} + z\mathbf{k}$ as usual, and $r = |\mathbf{R}| = \sqrt{x^2 + y^2 + z^2}$.

42. Prove that $\nabla \times \mathbf{R} = \mathbf{0}$. What is $\nabla \cdot \mathbf{R}$?

43. What is $(\mathbf{A} \cdot \nabla)\mathbf{R}$? If $\mathbf{A}$ is an arbitrary constant vector, prove that $\nabla(\mathbf{A} \cdot \mathbf{R}) = \mathbf{A}$.

44. Prove that $\nabla r^n = nr^{n-2}\mathbf{R}$.

45. For what values of n is $\nabla^2 r^n = 0$?

46. Prove that the curl of $f(r)\mathbf{R}$ is identically zero.

47. Determine n so that $\nabla \cdot (r^n\mathbf{R})$ will vanish identically.

48. Prove that $(\mathbf{A} \times \nabla) \times \mathbf{R} = -2\mathbf{A}$. What is $(\mathbf{A} \times \nabla) \cdot \mathbf{R}$?

49. Prove that $\nabla \cdot [(1/r)(\mathbf{A} \times \mathbf{R})] = 0$ for any constant vector $\mathbf{A}$.

50. Prove that

$$\nabla \times \left[\frac{1}{r}(\mathbf{A} \times \mathbf{R})\right] = \frac{1}{r}\mathbf{A} + \frac{\mathbf{A} \cdot \mathbf{R}}{r^3}\mathbf{R}$$

for any constant vector $\mathbf{A}$.

51. If $\mathbf{V}_1$ and $\mathbf{V}_2$ are the vectors which join the fixed points P_1: (x_1, y_1, z_1) and P_2: (x_2, y_2, z_2) to the variable point P: (x, y, z), prove that the gradient of $\mathbf{V}_1 \cdot \mathbf{V}_2$ is $\mathbf{V}_1 + \mathbf{V}_2$. What is $\nabla \cdot (\mathbf{V}_1 \times \mathbf{V}_2)$? What is $\nabla \times (\mathbf{V}_1 \times \mathbf{V}_2)$?

52. Derive the equation of continuity for the non-steady-state flow of a compressible fluid.

53. **(a)** If $u = f(x + y)$ and $v = g(x + y)$, show that $\nabla u \times \nabla v = \mathbf{0}$.
(b) If $u = f(x, y)$, $v = g(x, y)$, $w = h(x, y)$, show that $[\nabla u \; \nabla v \; \nabla w] = 0$.

54. Let $\mathbf{R}(r, \theta)$ be the vector from the origin to the point P whose polar coordinates are (r, θ). **(a)** Show that

$$\mathbf{u} = \frac{\partial \mathbf{R}}{\partial r} \quad \text{and} \quad \mathbf{v} = \frac{1}{r}\frac{\partial \mathbf{R}}{\partial \theta}$$

are unit vectors in the direction of $\mathbf{R}$ and in the direction perpendicular to $\mathbf{R}$. **(b)** Show that $\mathbf{u}$ and $\mathbf{v}$ are functions of θ alone and that $d\mathbf{u}/d\theta = \mathbf{v}$ and $d\mathbf{v}/d\theta = -\mathbf{u}$.

55. Using the results of Exercises 41 and 54, obtain a formula for the gradient of a function ϕ in polar coordinates.

56. Show that the gradient of a harmonic function has zero divergence. *Hint:* div $\nabla\phi = \nabla^2\phi$.

57. Let ϕ be a scalar field whose mixed partial derivatives are all continuous. Find the Jacobi matrix, divergence, and curl of $\nabla\phi = \mathbf{grad}\ \phi = \frac{\partial \phi}{\partial x}\mathbf{i} + \frac{\partial \phi}{\partial y}\mathbf{j} + \frac{\partial \phi}{\partial z}\mathbf{k}$.

58. Use the Jacobi matrix corresponding to the curl of a vector field $\mathbf{v}$ to verify (27) and (28), i.e., to prove that if all mixed partial derivatives are continuous

$$\text{div (curl } \mathbf{v}) = 0 \text{ and curl (curl } \mathbf{v}) = \mathbf{grad}\text{ (div } \mathbf{v}) - \nabla^2\mathbf{v}$$

59. Find the Jacobi matrix, divergence, and curl of each of the following vector fields.
(a) $x\mathbf{i} + y\mathbf{j} + z\mathbf{k}$
(b) $xy^2z^2\,\mathbf{i} + z^2 \sin y\,\mathbf{j} + x^2e^y\,\mathbf{k}$
(c) $\frac{-y}{x^2 + y^2}\mathbf{i} + \frac{x}{x^2 + y^2}\mathbf{j} + \mathbf{k}$

60. If $\mathbf{v} = v_1\,\mathbf{i} + v_2\,\mathbf{j} + v_3\,\mathbf{k}$ is a vector field, what is the meaning of $\nabla^2\mathbf{v}$?

61. Prove that

$$\nabla\phi_1 \times \nabla\phi_2 = \nabla \times (\phi_1\,\nabla\phi_2) = -\nabla \times (\phi_2\,\nabla\phi_1)$$

15.6 LINE, SURFACE, AND VOLUME INTEGRALS

In the rest of our work in vector analysis and in much of the work ahead in the chapters on complex variables, a simple extension of the familiar process of integration known as line integration will be of fundamental importance. Although in vector analysis we are usually concerned with line integrals taken along space curves, it is convenient to begin our discussion with a consideration of line integration along plane curves, since the applications of line integration in our study of complex variables will be exclusively in two dimensions. In both the two- and three-dimensional cases our work will involve only continuous curves which are **sectionally,** or **piecewise, smooth,** i.e., curves which are continuous and consist of a finite number of smooth arcs, or curves, joined end to end, along each of which the direction of the tangent line changes continuously. Every sectionally smooth curve has only a finite number of ''corners'' where the direction of the tangent changes abruptly. Moreover, as we learned in calculus, the length of each such curve between any two of its points is finite. Curves of finite length are said to be **rectifiable.**

Let $F(x, y)$ be a function of x and y defined throughout some region of the xy plane containing a continuous sectionally smooth curve C, and let A and B be two distinct points of C. $F(x, y)$ bears no relation to any equation defining C and is merely a function whose domain contains C. Let the portion or arc of C with endpoints A and B be divided into n adjoining segments Δs_i whose projections on the x and y axes are, respectively, Δx_i and Δy_i, and let (ξ_i, η_i) be the coordinates of an arbitrary point of the segment Δs_i (Fig. 15.13).

If we evaluate the given function $F(x, y)$ at each of the points (ξ_i, η_i) and form the products

$$F(\xi_i, \eta_i)\,\Delta x_i \qquad F(\xi_i, \eta_i)\,\Delta y_i \qquad F(\xi_i, \eta_i)\,\Delta s_i$$

and then sum over all the subdivisions of the arc AB, we have the three sums

$$(1) \qquad \sum_{i=1}^{n} F(\xi_i, \eta_i)\,\Delta x_i \qquad \sum_{i=1}^{n} F(\xi_i, \eta_i)\,\Delta y_i \qquad \sum_{i=1}^{n} F(\xi_i, \eta_i)\,\Delta s_i$$

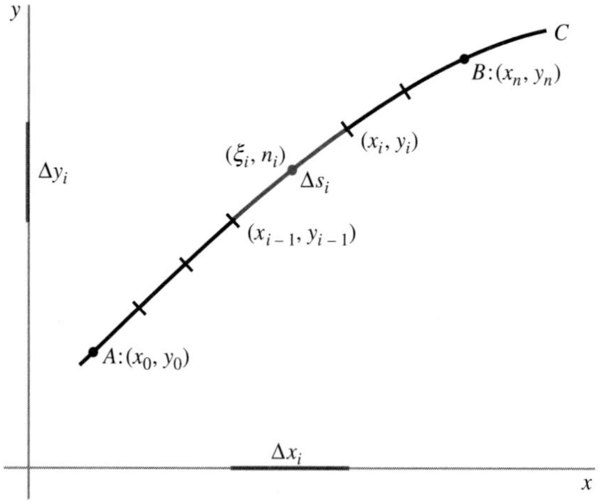

FIGURE 15.13
The subdivision of an arc preparatory to defining a line integral.

The limits of these sums, as n becomes infinite in such a way that the length of each Δs_i approaches zero, are known as **line integrals** and are written, respectively,

$$(2)\qquad (a)\ \int_C F(x, y)\, dx \qquad (b)\ \int_C F(x, y)\, dy \qquad (c)\ \int_C F(x, y)\, ds$$

By name, these integrals are referred to, in order, as the **line,** or **curvilinear, integral of** F **along** C **with respect to** x, **with respect to** y, and **with respect to the arc length** s. The notation for each of these integrals is incomplete because it fails to identify the portion of C, the part from A to B, over which F is to be integrated, nor does it indicate the direction of integration. Since the value of each integral depends on the path of integration and the direction it is traversed, this information must either be described geometrically or specified analytically before a line integral of a known function, symbolized by an expression like one of those in (2), can be evaluated.

It can be shown† that the continuity of F and the sectional smoothness of the path of integration denoted by C are sufficient conditions for the existence of the limits which define the integrals of (2). When these integrals are written, respectively,

$$(3)\qquad (a)\ \int_A^B F(x, y)\, dx \qquad (b)\ \int_A^B F(x, y)\, dy \qquad (c)\ \int_A^B F(x, y)\, ds$$

the limits of integration specify both the portion of C over which F is to be integrated and the direction of integration, namely, from A to B.

In the series (1), whose limits define the three types of integrals in (3), Δx_i and Δy_i are signed quantities, whereas Δs_i is intrinsically positive. Thus the following properties of ordinary definite integrals

a. $\int_A^B c\phi(t)\, dt = c\int_A^B \phi(t)\, dt \quad c$ a constant

b. $\int_A^B [\phi_1(t) \pm \phi_2(t)]\, dt = \int_A^B \phi_1(t)\, dt \pm \int_A^B \phi_2(t)\, dt$

c. $\int_A^B \phi(t)\, dt = -\int_B^A \phi(t)\, dt$

d. $\int_A^P \phi(t)\, dt + \int_P^B \phi(t)\, dt = \int_A^B \phi(t)\, dt$

†See, for instance, David V. Widder, "Advanced Calculus," 2d ed., p. 219, Prentice-Hall, Englewood Cliffs, N.J., 1961.

are equally valid for line integrals of the first two types, provided that throughout each formula the curve joining A and B remains the same. On the other hand, line integrals of the third type, although they do have Properties **a** and **b**, do not have Property **c**, since, in fact,

$$\int_A^B F(x, y)\, ds = \int_B^A F(x, y)\, ds$$

Moreover, Property **d** holds for these integrals if and only if P is between A and B on the path of integration. In general, we shall be more interested in integrals of the first two types than in those of the third.

Much of the initial strangeness of line integrals will disappear if we observe that the ordinary definite integrals of elementary calculus are just line integrals in which the curve C is the x axis and the integrand is a function of x alone. Moreover, the evaluation of line integrals can be reduced to the evaluation of ordinary definite integrals, as the following example shows.

EXAMPLE 1

What is the value of $\int_A^B [1/(x + y)\, dx$ along each of the paths shown in Fig. 15.14?

Before this integral can be evaluated, y must be expressed in terms of x. To do this, we recall from the definition of a line integral that the integrand is always to be evaluated *along the path of integration.* Along the parabolic arc joining A and B, we have $y = x^2$, and making this substitution in the given line integral yields the ordinary definite integral

$$\int_1^2 \frac{dx}{x + x^2} = \int_1^2 \left(\frac{1}{x} - \frac{1}{1 + x}\right) dx = [\ln x - \ln (1 + x)]_1^2 = \ln \tfrac{4}{3}$$

Similarly, along the straight-line path from A to B, we have $y = 3x - 2$, and making this substitution in th integrand of the given integral, we obtain the ordinary definite integral

$$\int_1^2 \frac{dx}{x + (3x - 2)} = \frac{1}{4}[\ln (4x - 2)]_1^2 = \tfrac{1}{4}(\ln 6 - \ln 2) = \tfrac{1}{4} \ln 3$$

To compute the line integral along the path APB, we must perform two integrations, one along AP an one along PB, since the relation expressing y in terms of x is different on these two segments. Along AP the

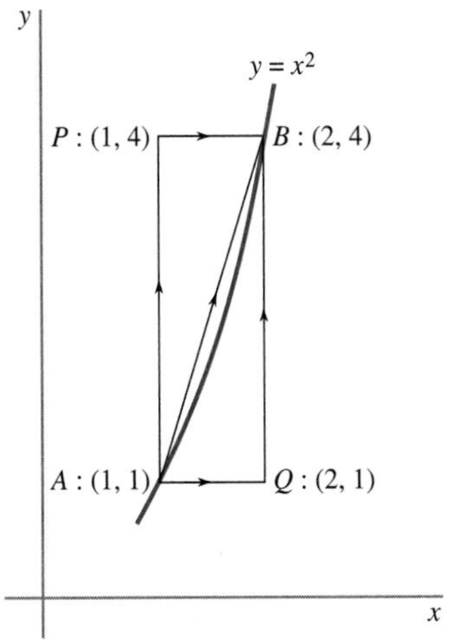

FIGURE 15.14
Possible paths for line integration from A:(1, 1) to B:(2, 4).

integral is obviously zero, since x remains constant and therefore in the sum leading to the integral each Δx_i is zero. Along PB, on which $y = 4$, we have the integral

$$\int_1^2 \frac{dx}{x+4} = [\ln(x+4)]_1^2 = \ln \tfrac{6}{5}$$

which is thus the value of the integral along the entire path APB.

Along the path AQB we again have two integrations to perform. Along AQ, on which $y = 1$, we have the integral

$$\int_1^2 \frac{dx}{x+1} = [\ln(x+1)]_1^2 = \ln \tfrac{3}{2}$$

Along the vertical segment QB the integral is again zero. Hence for the entire path AQB the value of the given integral is $\ln \frac{3}{2}$.

This example not only illustrates the computational details of line integration but also shows that in general a line integral depends not only on the endpoints of the integration but also upon the particular path which joins them.

It is possible, as in ordinary integration, to interpret a line integral as an area. For if we think of the integrand function $F(x, y)$ as defining a surface extending above some region in the xy plane, then the vertical cylindrical surface standing on the arc AB as base, or directrix, will cut the surface $z = F(x, y)$ in some curve, such as the arc PQ shown in Fig. 15.15. This curve is clearly the upper boundary of the portion $ABQP$ of the cylindrical surface which lies above the xy plane, below the surface $z = F(x, y)$, and between the generators AP and BQ. Moreover, the product $F(\xi_i, \eta_i)\,\Delta s_i$ is approximately the area of the vertical strip of this portion of the surface which stands above the infinitesimal base Δs_i. Hence the sum

$$\sum_{i=1}^{n} F(\xi_i, \eta_i)\,\Delta s_i$$

is approximately equal to the curved area $ABQP$, and, in the limit, the integral

$$\int_C F(x, y)\,ds$$

gives this area exactly.

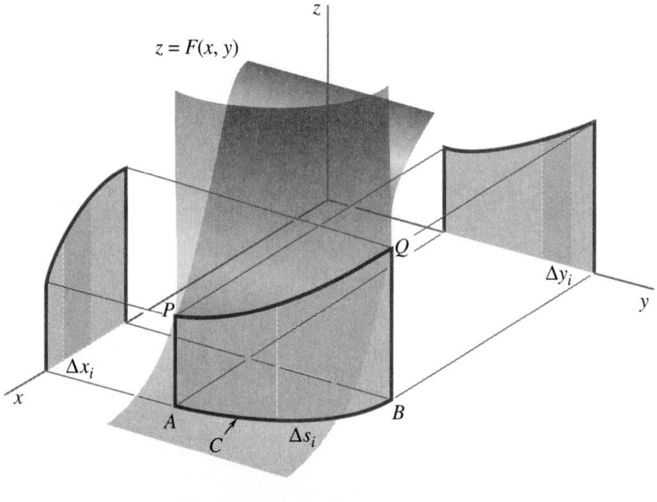

FIGURE 15.15
The interpretation of a line integral as an area.

As portrayed in the figure, with s increasing from A to B, Δx_i is negative. Hence, the product $-F(\xi_i, \eta_i)\,\Delta x_i$ is approximately the area of the projection on the xz plane of the vertical strip standing on Δs_i; the sum

$$-\sum_{i=1}^{n} F(\xi_i, \eta_i)\,\Delta x_i$$

represents approximately the area of the projection on the xz plane of the entire curved area $ABQP$; and in the limit, the integral

$$-\int_C F(x, y)\,dx$$

gives the projected area exactly. In similar fashion, the integral

$$\int_C F(x, y)\,dy$$

represents the area of the projection of $ABQP$ on the yz plane.

Although this geometrical interpretation of line integrals as areas is vivid and easily grasped, it obscures the fact that almost invariably in applications the function $F(x, y)$ describes some physical property of the plane of integration and is actually unrelated to any other region of space.

EXAMPLE 2

WORK REQUIRED TO MOVE A PARTICLE

If a particle is attracted toward the origin by a force whose magnitude is proportional to the distance r of the particle from the origin, how much work is done when the particle is moved from the point (0, 1) to the point (1, 2) along the path $y = 1 + x^2$ assuming a coefficient of friction μ between the particle and the path?

Let θ be the angle which the tangent to the curve at a general point P: (x, y) makes with the x axis; let ϕ be the angle which the radius vector to P makes with the x axis; and let α be the angle between the tangent and the radius vector at P (Fig 15.16). In moving the particle an infinitesimal distance Δs along the path, work must be done against two forces, namely, the tangential component of the central force

$$F_t = F\cos\alpha = kr\cos\alpha$$

and the frictional force

$$F_f = \mu F_n = \mu F\sin\alpha = \mu kr\sin\alpha$$

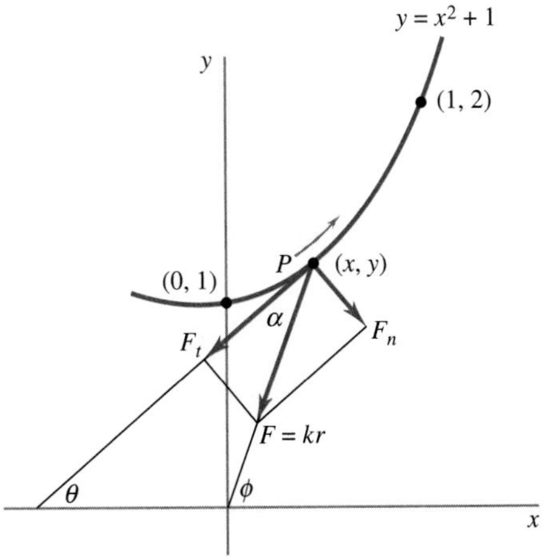

FIGURE 15.16
The resolution of a central force into tangential and normal components along a curve.

arising from the component of the central force which is perpendicular to the path and which acts to press the particle against the path. The infinitesimal amount of work done against these forces in moving a distance Δs along the curve is approximately

$$\Delta W = F_t \, \Delta s + F_f \, \Delta s = (kr \cos \alpha + \mu kr \sin \alpha) \, \Delta s$$

From the exterior-angle theorem of plane geometry, $\alpha = \phi - \theta$. Hence, summing and passing to the limit as $\Delta s \to 0$, we have

$$W = k \int_{0,1}^{1,2} r \cos (\phi - \theta) \, ds + \mu k \int_{0,1}^{1,2} r \sin (\phi - \theta) \, ds$$

$$= k \int_{0,1}^{1,2} r(\cos \phi \cos \theta + \sin \phi \sin \theta) \, ds + \mu k \int_{0,1}^{1,2} r(\sin \phi \cos \theta - \cos \phi \sin \theta) \, ds$$

Now

$$r \cos \phi = x \qquad r \sin \phi = y \qquad \cos \theta \, ds = dx \qquad \sin \theta \, ds = dy$$

Therefore, substituting these into the last expression for W, we have

$$W = k \int_{0,1}^{1,2} (x \, dx + y \, dy) + \mu k \int_{0,1}^{1,2} (y \, dx - x \, dy)$$

The first of these integrals can be written very simply as

$$\frac{k}{2} \int_{0,1}^{1,2} d(x^2 + y^2)$$

which, regardless of the relation between x and y, that is, *independent of the path,* can be integrated at once, giving

$$\frac{k}{2}(x^2 + y^2) \Big|_{0,1}^{1,2} = 2k$$

The second integral in the expression for W is not an exact differential and thus cannot be integrated until y is expressed in terms of x, or vice versa; therefore, as usual, due account must be taken of the path of integration. Now, along the path, we have $y = x^2 + 1$ and $x = \sqrt{y - 1}$. Hence

$$\mu k \int_{0,1}^{1,2} (y \, dx - x \, dy) = \mu k \int_0^1 (x^2 + 1) \, dx - \mu k \int_1^2 \sqrt{y - 1} \, dy$$

$$= \mu k \left[\frac{x^3}{3} + x \right]_0^1 - \mu k \left[\frac{2(y - 1)^{3/2}}{3} \right]_1^2 = \frac{2\mu k}{3}$$

The total amount of work done in the course of the motion is therefore

$$2k + \frac{2\mu k}{3}$$

The first term represents recoverable work stored as potential energy in the system; the second term represents irrecoverable work dissipated as heat through friction.

If $U(x, y)$ and $V(x, y)$ are continuous functions whose domains contain a curve C defined by two parametric equations

$$(4) \qquad x = x(t) \qquad y = y(t) \qquad a \le t \le b$$

or the equivalent position vector

$$\mathbf{R}(t) = x(t)\mathbf{i} + y(t)\mathbf{j} \qquad a \le t \le b \tag{5}$$

the sum of the line integrals along C of U with respect to x, and of V with respect to y, is customarily written

$$\int_C U(x, y)\, dx + V(x, y)\, dy = \int_C U(x, y)\, dx + \int_C V(x, y)\, dy \tag{6}$$

The arc length s of C has the value $s(a) = 0$; consequently, A: $[x(a), y(a)]$ is called the **initial point of** C. The **terminal point of** C is B: $[x(b), y(b)]$. As t increases, so does the arc length of C, and the variable point P: $[x(t), y(t)]$ traces the curve C in the **direction from** A **to** B. If t decreases from b to a, the initial and terminal points of C are interchanged and the direction along C is reversed. The curve C is said to be **closed** if its initial and terminal points coincide, and C is said to **cross itself at a point** if that point is the terminus of $\mathbf{R}(t)$ for two distinct values of t which belong to (a, b). For the purposes of our study, we shall call a closed sectionally smooth curve that does not cross itself a **simple closed curve.**†

In many line integral problems the path of integration C will consist of one or more simple closed curves forming the boundary of a plane region $\mathcal{R}$. Since line integration can be performed in either of two directions around a simple closed curve, it is important that we be able to distinguish between them. This is done by adopting the following criterion.

DEFINITION 1 The **positive direction of integration** around a simple closed curve bounding a plane region $\mathcal{R}$ is the direction in which an observer would move forward if he traversed the curve in such a way that the area of $\mathcal{R}$ was always on his left.

According to this definition, if $\mathcal{R}$ is the interior of a simple closed curve C, then the positive direction around C is the counterclockwise direction. If $\mathcal{R}$ is the region exterior to a simple closed curve C, then the positive direction around C is clockwise. If $\mathcal{R}$ is the region interior to a simple closed curve C_1 and exterior to a second simple closed curve C_2, then the positive direction of traversing the entire boundary of $\mathcal{R}$, namely, $C = C_1 \cup C_2$, is counterclockwise around C_1 and clockwise around C_2. An arrow on, or adjacent to, a curve is often used to indicate the positive direction.

The extension of line integration to paths in three dimensions is easily accomplished. Let $F(x, y, z)$ be a continuous function of x, y, and z, and let C be a continuous, sectionally smooth curve joining the points A and B. Furthermore, let the arc of C between A and B be divided in an arbitrary manner into n subintervals Δs_i whose projections on the coordinate axes are Δx_i, Δy_i, and Δz_i, and let an arbitrary point P_i: (ξ_i, η_i, ζ_i) be chosen in each Δs_i. We now evaluate $F(x, y, z)$ at each of the points P_i and form the sums

$$\sum_{i=1}^{n} F(\xi_i, \eta_i, \zeta_i)\, \Delta x_i \qquad \sum_{i=1}^{n} F(\xi_i, \eta_i, \zeta_i)\, \Delta y_i$$

$$\sum_{i=1}^{n} F(\xi_i, \eta_i, \zeta_i)\, \Delta z_i \qquad \sum_{i=1}^{n} F(\xi_i, \eta_i, \zeta_i)\, \Delta s_i$$

†That this is not the whole story concerning such curves, however, can be inferred from G. T. Whyburn, "What Is a Curve?" *American Mathematics Monthly,* vol. 49, pp. 493–497, October 1942.

The limits of these sums as n becomes infinite in such a way that the length of each Δs_i approaches zero define the respective line integrals

(7) $(a)\ \int_C F(x, y, z)\,dx \qquad (b)\ \int_C F(x, y, z)\,dy \qquad (c)\ \int_C F(x, y, z)\,dz \qquad (d)\ \int_C F(x, y, z)\,ds$

which by name are the **line integrals of F with respect to** x, y, z, and s, in order.

Because of the difficulty of defining a space curve C as the intersection of several surfaces, such curves are ordinarily defined by three parametric equations

(8) $$x = x(t) \qquad y = y(t) \qquad z = z(t) \qquad a \le t \le b$$

or by a parametric position vector

(9) $$\mathbf{R}(t) = x(t)\mathbf{i} + y(t)\mathbf{j} + z(t)\mathbf{k} \qquad a \le t \le b$$

Hence, line integrals in three dimensions, or along a plane curve defined parametrically, are usually evaluated by replacing the position variables in the integrand by their parametric representations. In this way the evaluation of such an integral is reduced to a problem in ordinary Riemann integration involving a single variable.

EXAMPLE 3

What is $\int_C (xy + z^2)\,ds$, where C is the arc of the helix

$$x = \cos t \qquad y = \sin t \qquad z = t$$

which joints the points $(1, 0, 0)$ and $(-1, 0, \pi)$?

Since $(ds)^2 = (dx)^2 + (dy)^2 + (dz)^2$, and since

$$dx = -\sin t\,dt \qquad dy = \cos t\,dt \qquad dz = dt$$

we have at once that

$$ds = \sqrt{\sin^2 t + \cos^2 t + 1}\,|dt| = \sqrt{2}\,|dt|$$

Furthermore, it is clear that the point $(1, 0, 0)$ corresponds to the parametric value $t = 0$ and that the point $(-1, 0, \pi)$ corresponds to the parametric value $t = \pi$. Hence, when the integrand is expressed in terms of the parameter t, the required integral becomes

$$\int_0^\pi (\cos t \sin t + t^2)\sqrt{2}\,dt = \sqrt{2}\left[\frac{\cos^2 t}{2} + \frac{t^3}{3}\right]_0^\pi = \frac{\sqrt{2}\,\pi^3}{3}$$

Both plane and space curves are said to be **regular** if they are sectionally smooth and do not cross themselves (Exercise 18). If a vector field

$$\mathbf{F}(x, y, z) = U(x, y, z)\mathbf{i} + V(x, y, z)\mathbf{j} + W(x, y, z)\mathbf{k}$$

is continuous along a regular curve C parameterized, as in (9), by a position vector $\mathbf{R}$, then the line integral of $\mathbf{F}$ along C exists (Exercise 20) and is denoted by

(10) $$\int_C \mathbf{F} \cdot d\mathbf{R}$$

This integral is commonly written

(11) $$\int_C U(x, y, z)\,dx + V(x, y, z)\,dy + W(x, y, z)\,dz$$

The line integral of a vector field **F** around a regular closed plane curve is symbolized quite often by

(12) $$\oint_C \mathbf{F} \cdot d\mathbf{R} \qquad \text{or} \qquad \oint_C \mathbf{F} \cdot d\mathbf{R}$$

according to whether the direction of integration around C is counterclockwise or clockwise.

Much as the concept of a line integral extends integration over an interval to integration over a curve, surface integrals provide an extension of double integration over a plane region to integration over a region on a curved surface. To see just what a surface integral is, let us start by considering some helpful ideas.

DEFINITION 2 A surface is said to be **smooth** if at each of its points there exists a tangent plane which varies continuously as the point varies continuously on the surface.

DEFINITION 3 A smooth surface is said to be **orientable** if it is **two-sided,** i.e., if it is possible at each point to identify consistently a unique direction normal to the surface.

DEFINITION 4 A surface that can be subdivided by a finite number of sectionally smooth curves into pieces which are all orientable (and therefore smooth) is said to be **regular.**†

With these definitions in mind, let $F(x, y, z)$ be a continuous function of x, y, and z, and let S be a given regular surface or portion of a regular surface in the region where $F(x, y, z)$ is defined. Let S be subdivided in an arbitrary manner into n parts $S_1, \ldots, S_n$, of respective areas $\Delta S_1, \ldots, \Delta S_n$, and in each part let an arbitrary point P_i: (ξ_i, η_i, ζ_i) be chosen. Finally, let $F(x, y, z)$ be evaluated at each of the points P_i. Then the limit of the sum

$$\sum_{i=1}^{n} F(\xi_i, \eta_i, \zeta_i)\, \Delta S_i$$

as n becomes infinite in such a way that not only the area of each S_i but also its maximum chord approaches zero is the **surface integral**

$$\iint_S F(x, y, z)\, dS$$

Similarly, given a function $F(x, y, z)$ and a region of space V, we can subdivide V into arbitrary subregions V_i of respective volumes ΔV_i, then evaluate $F(x, y, z)$ at an arbitrary point P_i: (ξ_i, η_i, ζ_i) in each V_i, and form the sum

$$\sum_{i=1}^{n} F(\xi_i, \eta_i, \zeta_i)\, \Delta V_i$$

†For a discussion of smooth surfaces which are not regular, i.e., smooth one-sided surfaces, see, for instance, Richard Courant and Herbert Robbins, *What Is Mathematics?* pp. 259–264, Oxford, New York, 1969.

The limit of this sum as n becomes infinite in such a way that not only the volume of each V_i but also its maximum chord approaches zero is the **volume integral**

$$\iiint_V F(x, y, z)\, dV$$

We leave a detailed treatment of surface and volume integrals to a standard course in calculus. However, the next example illustrates some computational aspects of such integrals.

EXAMPLE 4

What is the integral of the function x^2z taken over the entire surface of the right circular cylinder of height h which stands on the circle $x^2 + y^2 = a^2$? What is the integral of the given function taken throughout the volume of the cylinder?

To answer the first question, we must perform three integrations; i.e., we must integrate separately over the curved surface, the lower base, and the upper base of the cylinder. In each case, of course, we must employ a subdivision of the appropriate portion of the surface which will lead, if possible, to integrals that can conveniently be evaluated. This is most easily done by using cylindrical coordinates, as shown in Fig. 15.17. Then, on the curved surface, say S_1, we have

$$dS_1 = a\, d\theta\, dz \qquad x = a \cos\theta \qquad z = z$$

and the integral over this portion of the entire surface becomes

$$\iint_{S_1} x^2z\, dS_1 = \int_0^h \int_0^{2\pi} (a\cos\theta)^2 z(a\, d\theta\, dz) = a^3 \int_0^h z \int_0^{2\pi} \cos^2\theta\, d\theta\, dz$$

$$= a^3 \int_0^h z\left[\frac{\theta}{2} + \frac{\sin 2\theta}{4}\right]_0^{2\pi} dz = \pi a^3 \int_0^h z\, dz = \frac{\pi a^3 h^2}{2}$$

On the lower base, say S_2, we have

$$dS_2 = r\, dr\, d\theta \qquad x = r\cos\theta \qquad z = 0$$

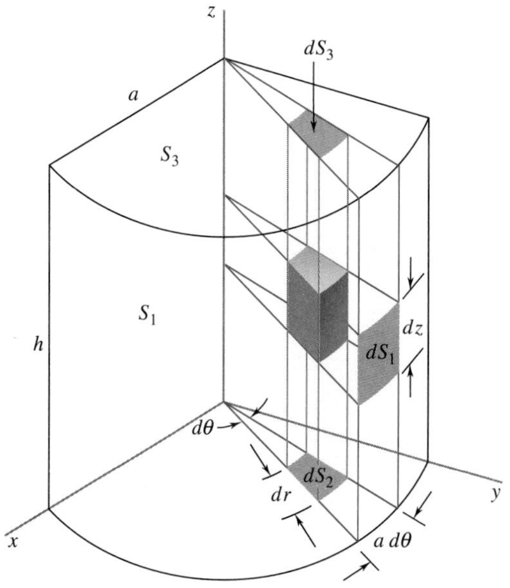

FIGURE 15.17
A typical volume element in cylindrical coordinates.

However, because of the factor z, the integrand vanishes identically on S_2, and without further calculations we have

$$\iint_{S_2} x^2 z\, dS_2 = 0$$

On the upper base, say S_3, we have $dS_3 = r\, dr\, d\theta$, $x = r\cos\theta$, and $z = h$. Hence

$$\iint_{S_3} x^2 z\, dz = \int_0^{2\pi}\int_0^a (r\cos\theta)^2 h(r\, dr\, d\theta) = h\int_0^{2\pi}\cos^2\theta\int_0^a r^3\, dr\, d\theta$$

$$= h\int_0^{2\pi}\cos^2\theta\left[\frac{r^4}{4}\right]_0^a d\theta = \frac{a^4 h}{4}\left[\frac{\theta}{2}+\frac{\sin 2\theta}{4}\right]_0^{2\pi} = \frac{\pi a^4 h}{4}$$

The integral over the entire surface S is, of course, the sum of the integrals over S_1, S_2, and S_3, that is

$$\iint_S x^2 z\, dS = \frac{\pi a^3 h^2}{2} + 0 + \frac{\pi a^4 h}{4} = \frac{\pi a^3 h(2h+a)}{4}$$

In computing the required volume integral, it is also convenient to use cylindrical coordinates. Doing this, we have $dV = r\, dr\, d\theta\, dz$, $x = r\cos\theta$, $z = z$, and the required integral becomes

$$\iiint_V x^2 z\, dV = \int_0^h\int_0^{2\pi}\int_0^a (r\cos\theta)^2 z(r\, dr\, d\theta\, dz) = \frac{a^4}{4}\int_0^h\int_0^{2\pi} z\cos^2\theta\, d\theta\, dz = \frac{\pi a^4}{4}\int_0^h z\, dz = \frac{\pi a^4 h^2}{8}$$

For the most part, our interest in line, surface, and volume integrals will be theoretical rather than computational; i.e., we shall use them far more often in derivations than in numerical calculations. Fundamental among the theorems we shall need for this purpose is **Green's lemma,** which relates the line integral of a function taken around the boundary of a plane region to the surface integral of an associated function taken over the region itself. It reads as follows.

THEOREM 1 If $\mathcal{R}$ is a plane region bounded by a finite number of simple closed curves, and if $U(x, y)$, $V(x, y)$, $\partial U/\partial y$, and $\partial V/\partial x$ are continuous at all points of $\mathcal{R}$ and its boundary C, then

$$\int_C U\, dx + V\, dy = \iint_{\mathcal{R}} \left(\frac{\partial V}{\partial x} - \frac{\partial U}{\partial y}\right) dx\, dy$$

provided the line integral is taken in the positive direction around C.

PROOF Let us first suppose that the boundary of $\mathcal{R}$ is a single simple closed curve C with the property that any line parallel to either of the coordinate axes cuts it in at most two points, and let us draw the horizontal and vertical lines which circumscribe C (Fig. 15.18). Then the arcs $P_4P_1P_2$ and $P_4P_3P_2$ define single-valued functions of x, which we shall call $f_1(x)$ and $f_2(x)$, respectively. Similarly, the arcs $P_1P_4P_3$ and $P_1P_2P_3$ define single-valued functions of y, which we shall call $g_1(y)$ and $g_2(y)$, respectively. Now consider

$$I_1 = \iint_{\mathcal{R}} \frac{\partial V}{\partial x}\, dx\, dy$$

To carry out this integration over $\mathcal{R}$, it is sufficient to integrate with respect to x from the arc $P_1P_4P_3$ to arc $P_1P_2P_3$ and then to integrate with respect to y from c to d. Hence

$$I_1 = \int_c^d\int_{g_1(y)}^{g_2(y)} \frac{\partial V}{\partial x}\, dx\, dy$$

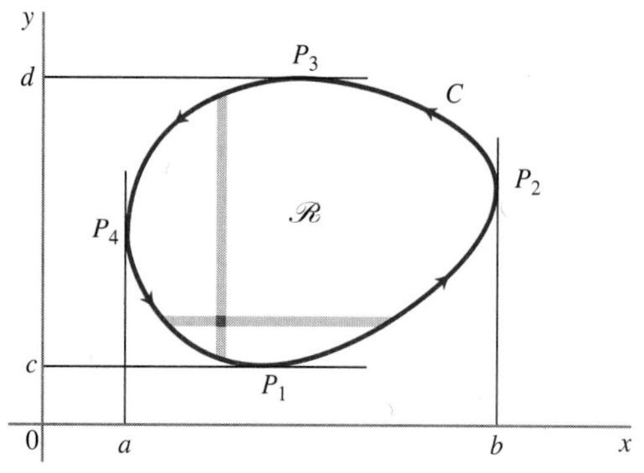

FIGURE 15.18
A plane region and its boundary.

The inner integration can easily be performed, and we find

$$I_1 = \int_c^d V(x, y)\Big|_{g_1(y)}^{g_2(y)} dy = \int_c^d V[g_2(y), y]\, dy - \int_c^d V[g_1(y), y]\, dy$$

$$= \int_c^d V[g_2(y), y]\, dy + \int_d^c V[g_1(y), y]\, dy$$

Obviously, the integrand of the first of these integrals is just $V(x, y)$ evaluated for $x = g_2(y)$. Hence this integral is precisely the line integral

$$\int_c^d V(x, y)\, dy$$

taken along the path $x = g_2(y)$ from P_1 (where $y = c$) to P_3 (where $y = d$). Similarly, the second integral is just the same line integral taken this time along the path $x = g_1(y)$ in the direction from P_3 through P_4 to P_1. Together they constitute the line integral of $V(x, y)$ around the entire closed curve C in the positive direction; hence,

(13) $$\iint_{\mathscr{R}} \frac{\partial V}{\partial x}\, dx\, dy = \int_C V(x, y)\, dy$$

In the same way, if we consider

$$I_2 = \iint_{\mathscr{R}} \frac{\partial U}{\partial y}\, dx\, dy = \iint_{\mathscr{R}} \frac{\partial U}{\partial y}\, dy\, dx$$

we can write more specifically

$$I_2 = \int_a^b \int_{f_1(x)}^{f_2(x)} \frac{\partial U}{\partial y}\, dy\, dx$$

Then, performing the inner integration, we have

$$I_2 = \int_a^b U(x, y)\Big|_{f_1(x)}^{f_2(x)} dx = \int_a^b U[x, f_2(x)]\, dx - \int_a^b U[x, f_1(x)]\, dx$$

$$= -\int_b^a U[x, f_2(x)]\, dx - \int_a^b U[x, f_1(x]\, dx$$

The first of these integrals is just the negative of the line integral of $U(x, y)$ along the path $y = f_2(x)$ in the direction from P_2 through P_3 to P_4. The second is the negative of the integral of $U(x, y)$ along $y = f_1(x)$ from P_4 through P_1 to P_2. Together they constitute the negative of the line integral of $U(x, y)$ entirely around the closed curve C in the same direction (the positive direction) in which we integrated in (13), i.e.,

$$\iint_{\mathcal{R}} \frac{\partial U}{\partial y}\, dx\, dy = -\int_C U(x, y)\, dx \tag{14}$$

Finally, if we subtract (14) from (13) and combine the integrals on each side, we obtain

$$\int_C U\, dx + V\, dy = \iint_{\mathcal{R}} \left(\frac{\partial V}{\partial x} - \frac{\partial U}{\partial y}\right) dx\, dy \tag{15}$$

which establishes Green's lemma for the special regions we have thus far been considering.

It is a simple matter, now, to extend Green's lemma to regions whose boundaries do not satisfy the condition that every line parallel to either of the coordinate axes cuts them in at most two points. For if this is not the case, the region $\mathcal{R}$ can be divided into subregions $\mathcal{R}_i$ whose boundaries C_i do have this property (Fig. 15.19). Then Eq. (15) can be applied to each subregion, following which the addition of these results yields Green's lemma for the general region $\mathcal{R}$ itself. For instance, for the region shown in Fig. 15.19, when Green's lemma is applied to each subregion, we have

$$\int_{C_1} U\, dx + V\, dy = \iint_{\mathcal{R}_1} \left(\frac{\partial V}{\partial x} - \frac{\partial U}{\partial y}\right) dx\, dy$$

$$\int_{C_2} U\, dx + V\, dy = \iint_{\mathcal{R}_2} \left(\frac{\partial V}{\partial x} - \frac{\partial U}{\partial y}\right) dx\, dy$$

$$\int_{C_3} U\, dx + V\, dy = \iint_{\mathcal{R}_3} \left(\frac{\partial V}{\partial x} - \frac{\partial U}{\partial y}\right) dx\, dy$$

$$\int_{C_4} U\, dx + V\, dy = \iint_{\mathcal{R}_4} \left(\frac{\partial V}{\partial x} - \frac{\partial U}{\partial y}\right) dx\, dy$$

When these results are added, the four integrals on the right combine to give exactly

$$\iint_{\mathcal{R}} \left(\frac{\partial V}{\partial x} - \frac{\partial U}{\partial y}\right) dx\, dy$$

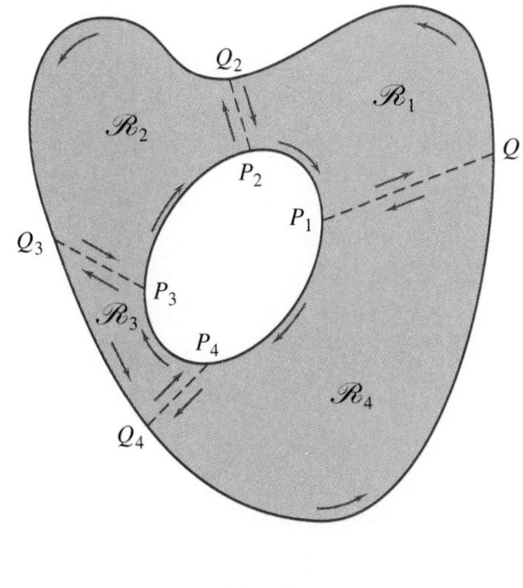

FIGURE 15.19
A plane region $\mathcal{R}$ subdivided into simpler regions $\mathcal{R}_1$, $\mathcal{R}_2$, $\mathcal{R}_3$, $\mathcal{R}_4$.

since $\mathcal{R}_1 \cup \mathcal{R}_2 \cup \mathcal{R}_3 \cup \mathcal{R}_4 = \mathcal{R}$. Moreover, the four line integrals on the left combine to give the line integral around the two curves which form the boundary of $\mathcal{R}$ *plus* a set of line integrals taken along the auxiliary boundary arcs P_iQ_i. Because U and V are continuous throughout $\mathcal{R}$, these integrals cancel in pairs, since each of the arcs P_iQ_i is traversed twice, once in one direction and once in the opposite direction. Hence we are left with

$$\int_C U\,dx + V\,dy = \iint_{\mathcal{R}} \left(\frac{\partial V}{\partial x} - \frac{\partial U}{\partial y}\right) dx\,dy$$

which is the assertion of Green's lemma. ◗

EXERCISES

1. Is the curve $x = t^2$, $y = t^3$, $0 \le t \le 2$, rectifiable? Explain.
2. Work Exercise 1 if the curve is $x = \cos t$, $y = \sin t$, $z = \frac{1}{2}t^2$, $0 \le t \le \pi/2$.
3. Work Exercise 1 if the curve is $y = \ln \cos x$, $0 \le x < \pi/2$.
4. Compute the value of the line integral $\displaystyle\int_C \frac{dy}{x+y}$ along each of the paths shown in Fig. 15.14.
5. Evaluate $\int_{0,1}^{2,3} (2xy - 1)\,dx + (x^2 + 1)\,dy$ along **(a)** $y = x + 1$ and **(b)** $y = \dfrac{x^2}{2} + 1$.
6. Evaluate $\int_{-1,0}^{1,0} y(1 + x)\,dy$ along **(a)** the x axis and **(b)** $y = 1 - x^2$.
7. Evlauate $\int_{0,0}^{1,1} x\,ds$ along **(a)** $y = x$, **(b)** $y = x^2$, and **(c)** $y = x^{3/2}$.
8. Evaluate $\int_C x^2y^2\,ds$ around the circle $x^2 + y^2 = 1$. *Hint:* Use polar coordinates.
9. Evaluate $\int_{0,0}^{1,1} \sqrt{y}\,dx + (x - y)\,dy$ along **(a)** $y = x$ and **(b)** $y^2 = x^3$.
10. Along what curve of the family $y = kx(1 - x)$ does the integral $\int_{0,0}^{1,0} y(x - y)\,dx$ attain its largest value?
11. Along what curve of the family $y = x^n$ does the integral $\int_{0,0}^{1,1} (25xy - 8y^2)\,dx$ attain its largest value?
12. What is the value of the line integral $\int_C 3x^2y^2\,dx + 2x^3y\,dy$ in the positive direction around C, where C is the ellipse $x^2 + 4y^2 = 4$?
13. A particle is attracted toward the origin by a force whose magnitude is proportional to the distance of the particle from the origin. How much work is done if the particle is moved from the point $(0, 1)$ to the point $(1, e)$ along the path $y = e^x$, assuming a coefficient of friction μ between the path and the particle?
14. The force on a particle p is given by the vector field $\mathbf{F}(x, y) = xy\,\mathbf{i} + (x^2 + y^2)\mathbf{j}$. Find the work done by the force as it moves p from $(0, 0)$ to $(4, 2)$ along $x = y^2$.
15. Find the mass of a semicircular wire of radius a whose density is proportional to **(a)** the distance along the wire from one end and **(b)** the straight-line distance to the wire from one end.
16. Show that, under the usual assumptions of continuity on $\mathbf{F}$ and smoothness of C defined parametrically by (4), the value of each integral of (2) is invariant under a strictly increasing differentiable change of the parameter. *Hint:* Express x and y in terms of a parameter u, where $t = \phi(u)$, $c \le u \le d$, $\phi'(u) > 0$ on (c, d), $\phi(c) = a$, and $\phi(d) = b$.
17. Evaluate $\int_C (x^2 - y^2)\,dx - 2xy\,dy$ along the arc of the parabola $y = 2x^2$ from $(0, 0)$ to $(1, 2)$ using the parameterization **(a)** $x = t$, $y = 2t^2$, $0 \le t \le 1$, and then **(b)** $t = \sin u$, $0 \le u \le \pi/2$.
18. A space curve C defined by Eqs. (8) *crosses itself at a point P* if and only if what is true?
19. A curve C is defined by the parametric position vector

$$\mathbf{R}(t) = \begin{cases} \cos t\,\mathbf{i} + \sin t\,\mathbf{j} & -\pi/2 \le t \le \pi/2 \\ \sin t\,\mathbf{j} - \cos t\,\mathbf{k} & \pi/2 \le t \le \pi \\ -\sin t\,\mathbf{i} - \cos t\,\mathbf{k} & \pi \le t \le 2\pi \end{cases}$$

(a) What are the initial and terminal points of C?
(b) Find all points where C crosses itself.

20. Show that the integral (10) exists if $\mathbf{F}$ is continuous on a regular curve C parameterized by (9). *Hint:* Note that (11) is a sum of Riemann integrals.
21. Evaluate $\displaystyle\int_{1,1,1}^{4,0,-2} \left[\frac{x-1}{3}\mathbf{i} + (1-y)^2\mathbf{j} + \frac{z-1}{3}\mathbf{k}\right] \cdot d\mathbf{R}$ along a straight-line path.
22. Evaluate the following integral along a straight path. $\int_{1,-1,0}^{2,1,\pi/2} [3x^2y\,\mathbf{i} + (x^3 + \cos yz)\,\mathbf{j} + y\cos yz\,\mathbf{k}] \cdot d\mathbf{R}$

In Exercises 23–27, for the given vector field $\mathbf{F}$ and curve C, evaluate $\int_C \mathbf{F} \cdot d\mathbf{R}$.

23. $\mathbf{F}(x, y) = xy\,\mathbf{i} + (2x - y)\,\mathbf{j}$;
C: the arc of $y = x^2$ from $(0, 0)$ to $(1, 1)$
24. $\mathbf{F}(x, y) = (x - y)\,\mathbf{i} + (x + y)\,\mathbf{j}$;
C: $x = e^t$, $y = e^{-t}$, $0 \le t \le 1$
25. $\mathbf{F}(x, y, z) = xy\,\mathbf{i} + y^2\,\mathbf{j} - xz\,\mathbf{k}$;
C: $\mathbf{R}(t) = t\,\mathbf{i} - 2t\,\mathbf{j} - \ln t\,\mathbf{k}$, $1 \le t \le 3$
26. $\mathbf{F}(x, y, z) = yz\,\mathbf{i} + xz\,\mathbf{j} + xy\,\mathbf{k}$;
C: $\mathbf{R}(t) = t\,\mathbf{i} + t^2\,\mathbf{j} + t^3\,\mathbf{k}$, $0 \le t \le 1$
27. $\mathbf{F}(x, y, z) = x^2\,\mathbf{i} + y^2\,\mathbf{j} + z^2\,\mathbf{k}$;
C: $x = \cos t$, $y = \sin t$, $z = t$, $0 \le t \le \pi$
28. Evaluate $\int_C (x^2 - y^2)\,ds$ around the circle of radius 3 centered at $(0, 0)$.

29. **(a)** Evaluate $\oint_C (-y\mathbf{i} + x\mathbf{j}) \cdot d\mathbf{R}$, where C is the ellipse $x = 2\cos\theta$, $y = 3\sin\theta$, $0 \le \theta \le 2\pi$.

(b) Evaluate $\oint_C (y\mathbf{i} + x\mathbf{j}) \cdot d\mathbf{R}$, where C is the union of the arc $y = x^2$, $0 \le x \le 2$, and the straight-line segment joining $(0, 0)$ to $(2, 4)$.

30. Evaluate $\iint_S (x + y)z\, dS$, where S is the surface of the cube whose vertices are $(0, 0, 0)$, $(1, 0, 0)$, $(1, 1, 0)$, $(0, 1, 0)$, $(0, 0, 1)$, $(1, 0, 1)$, $(1, 1, 1)$, $(0, 1, 1)$.

31. Evaluate $\iint_S (x + y + z)\, dS$, where S is the part of the surface of the sphere $x^2 + y^2 + z^2 = a^2$ which lies in the first octant. *Hint:* Use spherical coordinates.

32. Evaluate $\iiint_V x^2 z\, dV$, where V is the volume under the surface $x^2 + y^2 + z^2 = a^2$ and above the xy plane.

33. Evaluate $\iiint_V (x^2 + y^2 + z^2)\, dV$ where V is the volume of Exercise 32.

34. Verify Green's lemma for the integral $\int (x^2 + y)\, dx - xy^2\, dy$ taken around the boundary of the square whose vertices are $(0, 0)$, $(1, 0)$, $(1, 1)$, and $(0, 1)$.

35. Verify Green's lemma for the integral $\int (x - y)\, dx + (x + y)\, dy$ taken around the boundary of the finite area in the first quadrant between the curves $y = x^2$ and $y^2 = x$.

36. Verify Green's lemma for $\int (x - 2y)\, dx + x\, dy$ taken around the circle $x^2 + y^2 = a^2$.

37. Use Green's lemma to evaluate $\int_C x^2 y\, dx + y^3\, dy$, where C is the closed path formed by the graphs of $y^3 = x^2$ and $y = x$.

38. If a particle is attracted toward the origin by a force proportional to the nth power of the distance from the origin, show that the work done against this force in moving the particle from the point (x_0, y_0) to the point (x_1, y_1) is independent of the path along which the particle is moved. What is the amount of work done?

39. A particle is attracted toward the origin by a force proportional to the cube of the distance from the origin. How much work is done in moving the particle from the origin to the point $(1, 1)$ if in each case the coefficient of friction between the particle and the path is μ and if motion takes place **(a)** along the path $y = x$, **(b)** along the path $y = x^2$, **(c)** along the x axis to $(1, 0)$ and then vertically to $(1, 1)$, and **(d)** along the y axis to $(0, 1)$ and then horizontally to $(1, 1)$.

40. A sectionally smooth curve C in the xy plane encloses an area A. The position vector to a general point of C is $\mathbf{R}$. Express $\int_C \mathbf{k} \cdot \mathbf{R} \times d\mathbf{R}$ in terms of A.

41. Using Green's lemma, show that the area bounded by any simple closed curve C is given by the formula $A = \frac{1}{2}\int_C x\, dy - y\, dx$. Is this formula correct for regions bounded by more than one simple closed curve?

42. Let $\mathcal{R}$ be the region bounded by a simple closed curve C, let the area of $\mathcal{R}$ be A, and let the coordinates of the center of gravity of $\mathcal{R}$ be $(\bar{x}, \bar{y})$. Use Green's lemma to show that

$$\frac{1}{2}\int_C x^2\, dy = A\bar{x} \qquad \text{and} \qquad \int_C xy\, dy = A\bar{y}$$

43. If I_x and I_y are, respectively, the moments of inertia about the x axis and y axis of the area bounded by a simple closed curve C, show that

$$I_x = \int_C xy^2\, dy \qquad \text{and} \qquad I_y = -\int_C x^2 y\, dx$$

What is the formula for the polar moment of inertia about the origin of the area bounded by C?

44. Discuss the extension of Green's lemma to regions whose boundaries contain segments which are parallel to one or the other of the coordinate axes.

45. If U, V, $\partial U/\partial y$, and $\partial V/\partial x$ are continuous, and if $\partial U/\partial y = \partial V/\partial x$ at all points in the interior of a simple closed curve C, show that $\int_\Gamma U\, dx + V\, dy = 0$ for any simple closed curve Γ which lies entirely within C.

46. Show that Green's lemma fails to hold for the functions

$$U = -\frac{y}{x^2 + y^2} \qquad \text{and} \qquad V = \frac{x}{x^2 + y^2}$$

If $\mathcal{R}$ is the interior of the circle C: $x^2 + y^2 = 1$. Explain.

47. Using Green's lemma, establish the formula

$$\iint_{\mathcal{R}} \left(\frac{\partial^2 F}{\partial x^2} + \frac{\partial^2 F}{\partial y^2}\right) dx\, dy = \int_C \frac{dF}{dn}\, ds$$

where $\mathcal{R}$ is the region bounded by the simple closed curve C and dF/dn is the directional derivative of F in the direction of the outer normal to C.

48. By setting $U = f\dfrac{\partial g}{\partial x}$ and $V = f\dfrac{\partial g}{\partial y}$ in Green's lemma, show that

$$\iint_{\mathcal{R}} \left(\frac{\partial f}{\partial x}\frac{\partial g}{\partial y} - \frac{\partial f}{\partial y}\frac{\partial g}{\partial x}\right) dx\, dy = \int_C f\, dg$$

where $\mathcal{R}$ is the region bounded by the simple closed curve C. What is $\int_C g\, df$?

49. Define $\mathbf{F}(x, y)$ such that Green's lemma reads

$$\int_C \mathbf{F} \cdot d\mathbf{R} = \iint_{\mathcal{R}} \left(\frac{\partial V}{\partial x} - \frac{\partial U}{\partial y}\right) dx\, dy.$$

50. By setting $U = f\dfrac{\partial g}{\partial y}$ and $V = -f\dfrac{\partial g}{\partial x}$ in Green's lemma, show that

$$\iint_{\mathcal{R}} f \cdot \left(\frac{\partial^2 g}{\partial x^2} + \frac{\partial^2 g}{\partial y^2}\right) dx\, dy + \iint_{\mathcal{R}} \left(\frac{\partial f}{\partial x}\frac{\partial g}{\partial x} + \frac{\partial f}{\partial y}\frac{\partial g}{\partial y}\right) dx\, dy = \int_C f\frac{dg}{dn}\, ds$$

where $\mathcal{R}$ is the region bounded by the simple closed curve C and dg/dn is the directional derivative of g in the direction of the outer normal to C.

51. Show that Green's lemma can be written in the form $\int_C \mathbf{F} \cdot \mathbf{N}\, ds = \iint_{\mathcal{R}} \nabla \cdot \mathbf{F}\, dx\, dy$, where $\mathbf{F} = V\mathbf{i} - U\mathbf{j}$ and $\mathbf{N}$ is the outer unit normal to the curve C.

52. Show that Green's lemma can be written in the form $\int_C \mathbf{F} \cdot \mathbf{T}\, ds = \iint_{\mathcal{R}} \nabla \times \mathbf{F} \cdot \mathbf{k}\, dx\, dy$, where $\mathbf{F} = V\mathbf{i} - U\mathbf{j}$ and $\mathbf{T}$ is the unit tangent vector to C.

53. If f satisfies Laplace's equation in a region $\mathcal{R}$, show that

$$\iint_{\mathcal{R}} \left[\left(\frac{\partial f}{\partial x}\right)^2 + \left(\frac{\partial f}{\partial y}\right)^2\right] dx\, dy = \int_C f \frac{df}{dn}\, ds$$

54. Show that

$$\iint_{\mathcal{R}} (f\nabla^2 g - g\nabla^2 f)\, dx\, dy = \int_C \left(f\frac{dg}{dn} - g\frac{df}{dn}\right) ds.$$

55. A particle of mass m moves from A to B along a curve C under the influence of a variable force $\mathbf{F}$. Show that the work done by the force is equal to $\int_A^B \mathbf{F} \cdot \mathbf{dR}$. Show further that $W = \int_{t_a}^{t_b} \mathbf{F} \cdot \mathbf{v}\, dt$, where $\mathbf{v}$ is the vector velocity of the particle and t_a and t_b are the times when the particle is at A and at B. Using Newton's law, show further that $W = \int_{t_a}^{t_b} m(\dot{\mathbf{v}} \cdot \mathbf{v})\, dt$. Finally, by evaluating the last integral, show that the work done is equal to the gain in kinetic energy during the motion.

15.7 INTEGRAL THEOREMS

Most of the integrals we encounter in vector analysis are scalar quantities. For instance, given a vector function $\mathbf{F}(x, y, z)$, we are often interested in the integral of its tangential component along a curve C or in the integral of its normal component over a surface S. In the first case, if $\mathbf{R}$ is the vector from the origin to a general point of C so that $d\mathbf{R}/ds \equiv \mathbf{T}$ is the unit vector tangent to C at a general point, then $\mathbf{F} \cdot \mathbf{T}$ is the tangential component of $\mathbf{F}$ and

$$\int_C \mathbf{F} \cdot \mathbf{T}\, ds = \int_C \mathbf{F} \cdot \frac{d\mathbf{R}}{ds}\, ds = \int_C \mathbf{F} \cdot d\mathbf{R} \tag{1}$$

is the integral of this component along the curve C. The last integral of (1) is of course, reminiscent of Eq. (10), Sec. 15.6. In the second case, if $\mathbf{N}$ is the unit outward normal to S at a general point, then $\mathbf{F} \cdot \mathbf{N}$ is the normal component of $\mathbf{F}$ and

$$\iint_S \mathbf{F} \cdot \mathbf{N}\, dS \tag{2}$$

is the integral of this component over the surface S. Some writers denote the **differential vector** $\mathbf{N}\, dS$ by $d\mathbf{S}$ or $d\mathbf{A}$ and write (2) as

$$\iint_S \mathbf{F} \cdot d\mathbf{S} \quad \text{or} \quad \iint_S \mathbf{F} \cdot d\mathbf{A}$$

Other scalar integrals of frequent occurrence are the surface integral of the normal component of the curl of $\mathbf{F}$

$$\iint_S (\nabla \times \mathbf{F}) \cdot \mathbf{N}\, dS \tag{3}$$

and the volume integral of the divergence of $\mathbf{F}$

$$\iiint_V \nabla \cdot \mathbf{F}\, dV \tag{4}$$

Fundamental in many of the applications of vector analysis is the following so-called **divergence theorem,** which asserts the equality of the integrals (2) and (4) when V is the volume bounded by the closed regular surface S.

THEOREM 1 If $\mathbf{F}(x, y, z)$ and $\nabla \cdot \mathbf{F}$ are continuous over the closed regular surface S and its interior V, and if $\mathbf{N}$ is the unit vector perpendicular to S at a general point and extending outward from S, then

$$\iint_S \mathbf{N} \cdot \mathbf{F}\, dS = \iiint_V \nabla \cdot \mathbf{F}\, dV$$

PROOF To prove this theorem, we shall first suppose that S is a closed surface such that no line parallel to one of the coordinate axes cuts it in more than two points. If $\mathbf{F} = u\mathbf{i} + v\mathbf{j} + w\mathbf{k}$, the assertion of the theorem can be written at length in the form

$$\iint_S \mathbf{N} \cdot (u\mathbf{i} + v\mathbf{j} + w\mathbf{k})\, dS = \iiint_V \left(\frac{\partial u}{\partial x} + \frac{\partial v}{\partial y} + \frac{\partial w}{\partial z}\right) dV$$

or

$$(5) \quad \iint_S \mathbf{N} \cdot \mathbf{i}u\, dS + \iint_S \mathbf{N} \cdot \mathbf{j}v\, dS + \iint_S \mathbf{N} \cdot \mathbf{k}w\, dS$$

$$= \iiint_V \frac{\partial u}{\partial x}\, dV + \iiint_V \frac{\partial v}{\partial y}\, dV + \iiint_V \frac{\partial w}{\partial z}\, dV$$

We shall establish (5) by proving that respective integrals on each side are equal. To do this, let us consider first the integral

$$\iiint_V \frac{\partial w}{\partial z}\, dV$$

Under our assumption that no line parallel to one of the coordinate axes meets S in more than two points, it follows, in particular, that S is a double-valued surface over its projection on the xy plane and hence can be thought of as consisting of a lower half, say S_1, and an upper half, say S_2. Then if we take $dV = dx\, dy\, dz$ and perform the z integration first, we have

$$(6) \qquad \iiint_{z \text{ on } S_1}^{z \text{ on } S_2} \frac{\partial w}{\partial z}\, dz\, dx\, dy = \iint \left(w|_{\text{on } S_2} - w|_{\text{on } S_1}\right) dx\, dy$$

where, of course, x and y range over the area in the xy plane which is the projection of S. Moreover, the elements dS_1 and dS_2 can be defined so that they have $dx\, dy$ as their common projection on the xy plane (Fig. 15.20). Now since $\mathbf{k}$, $\mathbf{N}_1$, and $\mathbf{N}_2$ are all unit vectors, it follows that $\mathbf{k} \cdot \mathbf{N}_1$ and $\mathbf{k} \cdot \mathbf{N}_2$ are, respectively, the cosines of the angles between the normal to the xy plane, namely, $\mathbf{k}$, and the

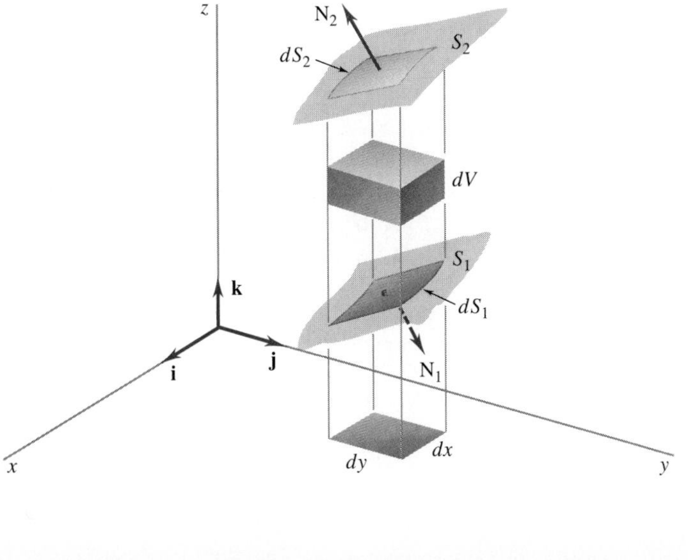

FIGURE 15.20
Integration in the z direction from S_1 to S_2 in the proof of the divergence theorem.

outer normals to dS_1 and dS_2; that is, they are numerically the cosines of the angles through which dS_1 and dS_2 are projected onto the element $dx\,dy$. Hence

$$dx\,dy = -\mathbf{k}\cdot\mathbf{N}_1\,dS_1 = \mathbf{k}\cdot\mathbf{N}_2\,dS_2$$

where the minus sign is necessary in the first equality because the outer normal $\mathbf{N}_1$ to dS_1 makes an angle of more than 90° with the direction of $\mathbf{k}$ and thus $\mathbf{k}\cdot\mathbf{N}_1$ is negative, whereas both $dx\,dy$ and dS_1 are clearly positive. Therefore, substituting for $dx\,dy$ in the right-hand side of (6), i.e., transferring the integration from the common projection of S_1 and S_2 back onto S_1 and S_2 themselves, we have

$$\begin{aligned}\iiint_V \frac{\partial w}{\partial z}\,dV &= \iint w|_{\text{on } S_2}\,dx\,dy - \iint w|_{\text{on } S_1}\,dx\,dy \\ &= \iint w|_{\text{on } S_2}\,\mathbf{k}\cdot\mathbf{N}_2\,dS_2 + \iint w|_{\text{on } S_1}\,\mathbf{k}\cdot\mathbf{N}_1\,dS_1 \\ &= \iint_{S_2} w\mathbf{k}\cdot\mathbf{N}\,dS + \iint_{S_1} w\mathbf{k}\cdot\mathbf{N}\,dS\end{aligned}$$

where the subscripts have been dropped from the integrands as superfluous, since the ranges of integration are now explicitly indicated. Finally, since S_1 and S_2 together make up the entire closed surface S, we can combine the last two integrals, getting

$$\iiint_V \frac{\partial w}{\partial z}\,dV = \iint_S w\mathbf{k}\cdot\mathbf{N}\,dS$$

Similarly we can show that

$$\iiint_V \frac{\partial u}{\partial x}\,dV = \iint_S u\mathbf{i}\cdot\mathbf{N}\,dS \qquad \iiint_V \frac{\partial v}{\partial y}\,dV = \iint_S v\mathbf{j}\cdot\mathbf{N}\,dS$$

Adding the last three equations, we obtain the expanded form (5) of the divergence theorem, under the assumption that S is exactly two-valued over its projections on each of the coordinate planes.

On the other hand, if S does not have this property, then (very much as in our extension of Green's lemma to more general regions in the last section) we can partition its interior V into subregions V_i whose boundaries S_i do have this property. Now, applying our limited result to each of these regions, we obtain a set of equations of the form

$$\iint_{S_i} \mathbf{N}\cdot\mathbf{F}\,dS = \iiint_{V_i} \nabla\cdot\mathbf{F}\,dV$$

If these are added, the sum of the volume integrals is, of course, just the integral of $\nabla\cdot\mathbf{F}$ throughout the entire volume V. The sum of the surface integrals is equal to the integral of $\mathbf{N}\cdot\mathbf{F}$ over the original surface S plus a set of integrals over the auxiliary boundary surfaces which were introduced when V was subdivided. These cancel in pairs, however, since the integration extends twice over each interface, with integrands which are identical except for the oppositely directed unit normals they contain as factors. Thus, our proof can be extended to volumes bounded by general closed regular surfaces, and Theorem 1 is established. ◗

EXAMPLE 1

Prove that

$$\iint_S \mathbf{N}\times\mathbf{F}\,dS = \iiint_V \nabla\times\mathbf{F}\,dV$$

To show this, let us apply the divergence theorem to the vector **F × C**, where **C** is an arbitrary constant vector. Then

$$\iint_S \mathbf{N} \cdot (\mathbf{F} \times \mathbf{C})\, dS = \iiint_V \boldsymbol{\nabla} \cdot (\mathbf{F} \times \mathbf{C})\, dV$$

Now taking advantage of the fact that **C** is a constant vector and that a cyclic permutation of the elements of a scalar triple product leaves the product unchanged, we can write

$$\iint_S \mathbf{C} \cdot \mathbf{N} \times \mathbf{F}\, dS = \iiint_V \mathbf{C} \cdot \boldsymbol{\nabla} \times \mathbf{F}\, dV$$

or, removing the constant vector **C** from each integral,

$$\mathbf{C} \cdot \iint_S \mathbf{N} \times \mathbf{F}\, dS = \mathbf{C} \cdot \iiint_V \boldsymbol{\nabla} \times \mathbf{F}\, dV$$

Since **C** is an arbitrary vector, this equation asserts that the vectors

$$\iint_S \mathbf{N} \times \mathbf{F}\, dS \qquad \text{and} \qquad \iiint_V \boldsymbol{\nabla} \times \mathbf{F}\, dV$$

have equal projections in all directions and hence must be equal, as asserted.

Various important theorems stem from the divergence theorem. For instance, if u and v are two sufficiently differentiable scalar point functions, and if we set

$$\mathbf{F} = u\boldsymbol{\nabla} v$$

then, by Eq. (21), Sec. 15.5,

$$\boldsymbol{\nabla} \cdot \mathbf{F} = \boldsymbol{\nabla} \cdot (u\, \boldsymbol{\nabla} v) = u\boldsymbol{\nabla} \cdot \boldsymbol{\nabla} v + \boldsymbol{\nabla} u \cdot \boldsymbol{\nabla} v = \boldsymbol{\nabla} u \cdot \boldsymbol{\nabla} v + u\, \nabla^2 v$$

Hence, applying the divergence theorem to the vector $\mathbf{F} = u\, \boldsymbol{\nabla} v$, we have

$$\iiint_V (\boldsymbol{\nabla} u \cdot \boldsymbol{\nabla} v + u\, \nabla^2 v)\, dV = \iint_S \mathbf{N} \cdot u\, \boldsymbol{\nabla} v\, dS \tag{7}$$

Similarly, if we interchange the roles of u and v in (7), we obtain

$$\iiint_V (\boldsymbol{\nabla} v \cdot \boldsymbol{\nabla} u + v\, \nabla^2 u)\, dV = \iint_S \mathbf{N} \cdot v\, \boldsymbol{\nabla} u\, dS \tag{8}$$

Finally, if we subtract (8) from (7), we obtain **Green's theorem:**†

THEOREM 2 If V is the volume bounded by a closed regular surface S, and if $u(x, y, z)$ and $v(x, y, z)$ are scalar functions having continuous second partial derivatives, then

$$\iiint_V (u\, \nabla^2 v - v\, \nabla^2 u)\, dV = \iint_S \mathbf{N} \cdot (u\, \boldsymbol{\nabla} v - v\, \boldsymbol{\nabla} u)\, dS$$

where **N** is the outer normal to the surface S which bounds V.

†This theorem should not be confused with *Green's lemma,* Theorem 1, Sec. 15.6.

Another result of some importance can be obtained by applying the divergence theorem to the function $\mathbf{F} = \mathbf{R}/r^3$, where, as usual,

$$\mathbf{R} = x\mathbf{i} + y\mathbf{j} + z\mathbf{k} \qquad \text{and} \qquad r = |\mathbf{R}| = \sqrt{x^2 + y^2 + z^2}$$

Thus, substituting into the divergence theorem, we have

$$\iint_S \mathbf{N} \cdot \frac{\mathbf{R}}{r^3}\, dS = \iiint_V \nabla \cdot \frac{\mathbf{R}}{r^3}\, dV \tag{9}$$

Now, by Eq. (21), Sec. (15.5), and Exercise 44, Sec. (15.5),

$$\nabla \cdot \frac{\mathbf{R}}{r^3} = \frac{1}{r^3}\nabla \cdot \mathbf{R} + \mathbf{R} \cdot \nabla \frac{1}{r^3} = \frac{3}{r^3} + \mathbf{R} \cdot \frac{d(1/r^3)}{dr}\nabla r$$

$$= \frac{3}{r^3} + \mathbf{R} \cdot \left(-\frac{3}{r^4}\frac{\mathbf{R}}{r}\right) = \frac{3}{r^3} - 3\frac{\mathbf{R} \cdot \mathbf{R}}{r^5} = 0$$

Hence, we conclude from (9) that

$$\iint_S \mathbf{N} \cdot \frac{\mathbf{R}}{r^3}\, dS = 0 \tag{10}$$

provided, of course, that r is different from zero at all points on and within S, that is, provided that the origin from which $\mathbf{R}$ is drawn does not lie on S or within the volume enclosed by S.

Since the divergence theorem requires that the function to which it is applied have continuous first partial derivatives throughout the volume of integration, it cannot be applied to $\mathbf{R}/r^3$ if the origin of $\mathbf{R}$ is within S. In this case we therefore modify the region of integration by constructing a sphere S' of radius ϵ having the origin O as center (Fig. 15.21). In the region V' between S and S' the function $\mathbf{R}/r^3$ satisfies the conditions of the divergence theorem, and thus Eq. (10) can properly be applied, giving

$$\iint_{S \cup S'} \mathbf{N} \cdot \frac{\mathbf{R}}{r^3}\, dS = 0 \qquad \text{or} \qquad \iint_S \mathbf{N} \cdot \frac{\mathbf{R}}{r^3}\, dS + \iint_{S'} \mathbf{N} \cdot \frac{\mathbf{R}}{r^3}\, dS = 0 \tag{11}$$

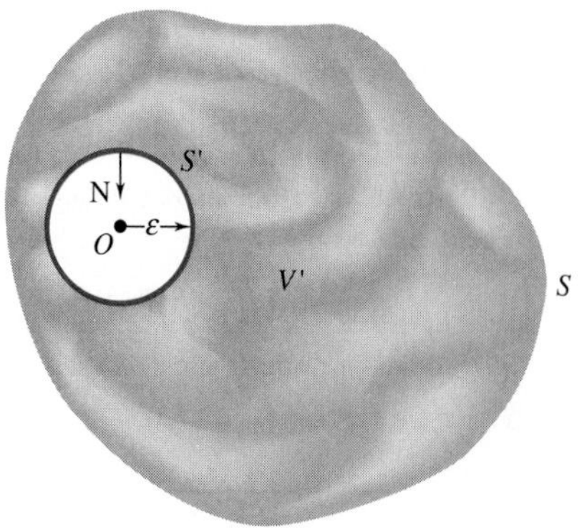

FIGURE 15.21
A singular point excluded from a three-dimensional region by an auxiliary spherical boundary.

At any point of S', the direction of the normal which extends outward from the volume V' is opposite to $\mathbf{R}$. Hence, the unit outer normal to S' is $\mathbf{N} = -\mathbf{R}/\epsilon$, since on S' the length of the radius vector $\mathbf{R}$ is $r = \epsilon$. Therefore, in the last integral,

$$\mathbf{N} \cdot \mathbf{R} = -\frac{\mathbf{R}}{\epsilon} \cdot \mathbf{R} = -\frac{\epsilon^2}{\epsilon} = -\epsilon$$

and Eq. (11) becomes

$$\iint_S \mathbf{N} \cdot \frac{\mathbf{R}}{r^3}\, dS + \iint_{S'} \frac{-\epsilon}{\epsilon^3}\, dS = 0 \qquad \text{or} \qquad \iint_S \mathbf{N} \cdot \frac{\mathbf{R}}{r^3}\, dS = \frac{1}{\epsilon^2} \int_{S'} dS = \frac{4\pi\epsilon^2}{\epsilon^2} = 4\pi$$

This result, coupled with Eq. (10), gives us **Gauss' theorem:**

THEOREM 3 If S is a closed regular surface, then

$$\iint_S \mathbf{N} \cdot \frac{\mathbf{R}}{r^3}\, dS = \begin{cases} 0 & O \text{ outside } S \\ 4\pi & O \text{ inside } S \end{cases}$$

Returning to Eq. (5), we next set $u = \eta f$, $v = \eta g$, and $w = \eta h$ and replace the dot products $\mathbf{N} \cdot \mathbf{i}$, $\mathbf{N} \cdot \mathbf{j}$, $\mathbf{N} \cdot \mathbf{k}$ in that equation by the respective direction cosines $\cos\alpha$, $\cos\beta$, and $\cos\gamma$ of $\mathbf{N}$. This leads to the relation

$$\iint_S \eta(f \cos\alpha + g\cos\beta + h\cos\gamma)\, dS = \iiint_V \left[\frac{\partial(\eta f)}{\partial x} + \frac{\partial(\eta g)}{\partial y} + \frac{\partial(\eta h)}{\partial z}\right] dV$$

Performing the indicated partial differentiations in the integrand of the volume integral, then writing volume integrals in cartesian form and regrouping, we obtain the three-dimensional analog of integration by parts:

$$(5a) \quad \iiint_V \left(f\frac{\partial\eta}{\partial x} + g\frac{\partial\eta}{\partial y} + h\frac{\partial\eta}{\partial z}\right) dx\,dy\,dz = \iint_S \eta(f\cos\alpha + g\cos\beta + h\cos\gamma)\, dS - \iiint_V \eta\left(\frac{\partial f}{\partial x} + \frac{\partial g}{\partial y} + \frac{\partial h}{\partial z}\right) dx\,dy\,dz$$

Still another integral formula of great importance in vector analysis is that of **Stokes' theorem**:†

THEOREM 4 If S is the portion of a regular surface bounded by the closed curve C, and if $\mathbf{F}(x, y, z)$ is a vector function possessing continuous first partial derivatives, then

$$\int_C \mathbf{F} \cdot d\mathbf{R} = \iint_S \mathbf{N} \cdot \nabla \times \mathbf{F}\, dS$$

provided the direction of integration around C is positive with respect to the side of S on which the unit normal $\mathbf{N}$ is drawn.

†Named for the English mathematical physicist G. G. Stokes (1819–1903).

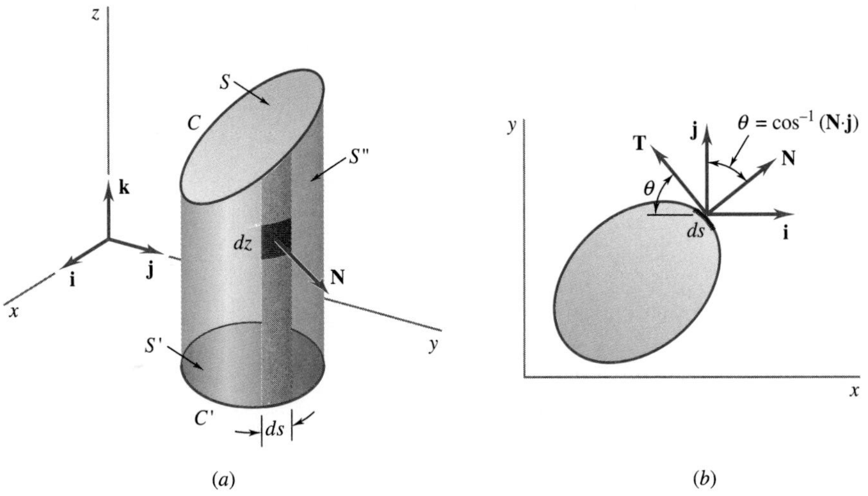

FIGURE 15.22
The closed surface $S \cup S' \cup S''$ employed in the proof of Stokes' theorem.

PROOF To prove this theorem we suppose first that S has the property that it is single-valued relative to its projections on each of the three coordinate planes. If we write $\mathbf{F} = u\mathbf{i} + v\mathbf{j} + w\mathbf{k}$, Stokes' theorem asserts that

$$\int_C u\,dx + v\,dy + w\,dz = \iint_S \mathbf{N}\cdot\nabla\times(u\mathbf{i} + v\mathbf{j} + w\mathbf{k})\,dS$$

$$(12) \qquad = \iint_S \mathbf{N}\cdot\nabla\times u\mathbf{i}\,dS + \iint_S \mathbf{N}\cdot\nabla\times v\mathbf{j}\,dS + \iint_S \mathbf{N}\cdot\nabla\times w\mathbf{k}\,dS$$

and to establish the theorem it is sufficient to show that respective integrals on the two sides of the last equation are equal. We consider first the integral

$$\iint_S \mathbf{N}\cdot\nabla\times u\mathbf{i}\,dS$$

taken over the *closed* surface consisting of S, its projection on the xy plane, say S', and the cylindrical surface, say S'', which projects S onto S' (Fig. 15.22*a*). If we apply the divergence theorem to the vector $\nabla\times u\mathbf{i}$ over this surface and the volume it encloses and note from Eq. (27), Sec. 15.5, that $\nabla\cdot(\nabla\times u\mathbf{i}) = 0$, we obtain

$$\iint_S \mathbf{N}\cdot\nabla\times u\mathbf{i}\,dS + \iint_{S'} \mathbf{N}\cdot\nabla\times u\mathbf{i}\,dS + \iint_{S''} \mathbf{N}\cdot\nabla\times u\mathbf{i}\,dS = \iiint_V \nabla\cdot(\nabla\times u\mathbf{i})\,dV = 0$$

or

$$(13) \qquad \iint_S \mathbf{N}\cdot\nabla\times u\mathbf{i}\,dS = -\iint_{S'} \mathbf{N}\cdot\nabla\times u\mathbf{i}\,dS - \iint_{S''} \mathbf{N}\cdot\nabla\times u\mathbf{i}\,dS$$

Now

$$\nabla\times u\mathbf{i} = \begin{vmatrix} \mathbf{i} & \mathbf{j} & \mathbf{k} \\ \dfrac{\partial}{\partial x} & \dfrac{\partial}{\partial y} & \dfrac{\partial}{\partial z} \\ u & 0 & 0 \end{vmatrix} = \mathbf{j}\,\frac{\partial u}{\partial z} - \mathbf{k}\,\frac{\partial u}{\partial y}$$

Moreover, on S' the outer normal $\mathbf{N}$ is clearly equal to $-\mathbf{k}$. Hence, on S' we have

$$\mathbf{N} \cdot \nabla \times u\mathbf{i} = -\mathbf{k} \cdot \left(\mathbf{j}\frac{\partial u}{\partial z} - \mathbf{k}\frac{\partial u}{\partial y}\right) = \frac{\partial u}{\partial y}$$

and

$$\iint_{S'} \mathbf{N} \cdot \nabla \times u\mathbf{i}\, dS = \iint_{S'} \frac{\partial u}{\partial y}\, dS$$

If we now apply Green's lemma (Theorem 1, Sec. 15.6) to the integral on the right, the last equation becomes

$$\iint_{S'} \mathbf{N} \cdot \nabla \times u\mathbf{i}\, dS = -\int_{C'} u\, dx \tag{14}$$

where C' is the boundary of the region S' which is the projection of S.

To investigate the second integral on the right in Eq. (13), we note first that since S'' is a cylindrical surface whose generators are parallel to the z axis, the normals to S'' are all perpendicular to the vector $\mathbf{k}$. Therefore, on S'' we have

$$\mathbf{N} \cdot \nabla \times u\mathbf{i} = \mathbf{N} \cdot \left(\mathbf{j}\frac{\partial u}{\partial z} - \mathbf{k}\frac{\partial u}{\partial y}\right) = \mathbf{N} \cdot \mathbf{j}\frac{\partial u}{\partial z}$$

Next, taking $dS = dz\, ds$ (Fig. 15.22a), and noting that $\mathbf{N} \cdot \mathbf{j}$ is independent of z, we have

$$\iint_{S''} \mathbf{N} \cdot \nabla \times u\mathbf{i}\, dS = \int_{C'} \mathbf{N} \cdot \mathbf{j} \int_{z \text{ on } S'}^{z \text{ on } S} \frac{\partial u}{\partial z}\, dz\, ds = \int_{C'} (u|_S - u|_{S'})\mathbf{N} \cdot \mathbf{j}\, ds \tag{15}$$

Now $\mathbf{N} \cdot \mathbf{j}$ is equal to the cosine of the angle θ between the normal $\mathbf{N}$ and the positive y axis, and this is numerically equal but opposite in sign to the cosine of the angle between the directed tangent to C' and the positive x axis (Fig 15.22b). Hence $\mathbf{N} \cdot \mathbf{j}\, ds = -dx$, and Eq. (15) becomes

$$\iint_{S''} \mathbf{N} \cdot \nabla \times u\mathbf{i}\, dS = -\int_{C'} u|_S\, dx + \int_{C'} u|_{S'}\, dx \tag{16}$$

In the first integral on the right in (16), the integrand, being evaluated at points of S which are directly above the curve C', is actually evaluated along the curve C. Moreover, because C' is the projection of C in the z direction, the variation of x around C' is exactly the same as the variation of x around C. Hence, in this integral we can properly replace the indicated path of integration C' by the curve C, getting

$$\iint_{S''} \mathbf{N} \cdot \nabla \times u\mathbf{i}\, dS = -\int_{C} u\, dx + \int_{C'} u\, dx \tag{17}$$

Therefore, substituting from (14) and (17) into (13), we have

$$\iint_{S} \mathbf{N} \cdot \nabla \times u\mathbf{i}\, dS = -\left(-\int_{C'} u\, dx\right) - \left(-\int_{C} u\, dx + \int_{C'} u\, dx\right)$$

$$= \int_{C} u\, dx \tag{18}$$

In precisely the same way we can show that

$$\iint_S \mathbf{N} \cdot \nabla \times v\mathbf{j}\, dS = \int_C v\, dy \tag{19}$$

$$\iint_S \mathbf{N} \cdot \nabla \times w\mathbf{k}\, dS = \int_C w\, dz \tag{20}$$

Finally by adding (18), (19), and (20) we obtain Eq. (12).

It is now a simple matter to extend Eq. (12) to surfaces which are not single-valued relative to their projections on the coordinate planes. For if this is not the case, we can always divide S into subregions S_i which do have this property and then apply Eq. (12) to each S_i and its boundary C_i, getting the set of equations

$$\int_{C_1} \mathbf{F} \cdot d\mathbf{R} = \iint_{S_1} \mathbf{N} \cdot \nabla \times \mathbf{F}\, dS$$
$$\cdots\cdots\cdots\cdots$$
$$\int_{C_n} \mathbf{F} \cdot d\mathbf{R} = \iint_{S_n} \mathbf{N} \cdot \nabla \times \mathbf{F}\, dS$$

When these are added, the surface integrals combine to give precisely the surface integral over S itself, since $S_1 \cup \cdots \cup S_n = S$. At the same time, the line integrals combine to give the line integral around the actual boundary of S plus the line integral along all the auxiliary boundary arcs taken twice in opposite directions, just as in Fig. 15.19. Since the latter cancel identically, the line integral around C itself is all that remains and the theorem follows in the general case. ◗

If A and B are two arbitrary points in space, it is often important to know whether the line integral

$$\int_A^B \mathbf{F} \cdot d\mathbf{R} \tag{21}$$

is independent of the path along which the integral is calculated. As a first step in establishing criteria for this, we shall prove

LEMMA 1 If $\int \mathbf{F} \cdot d\mathbf{R}$ is zero around every closed curve in a region, then the integral (21) is independent of the path.

◖ **PROOF** For any two paths APB and AQB joining A and B (Fig. 15.23), by hypothesis,

$$\int_{APB} \mathbf{F} \cdot d\mathbf{R} + \int_{BQA} \mathbf{F} \cdot d\mathbf{R} = 0$$

whence, reversing the direction of integration along BQA and transposing, we have

$$\int_{APB} \mathbf{F} \cdot d\mathbf{R} = \int_{AQB} \mathbf{F} \cdot d\mathbf{R}$$

as asserted. ◗

The converse of Lemma 1 also holds, as we show next.

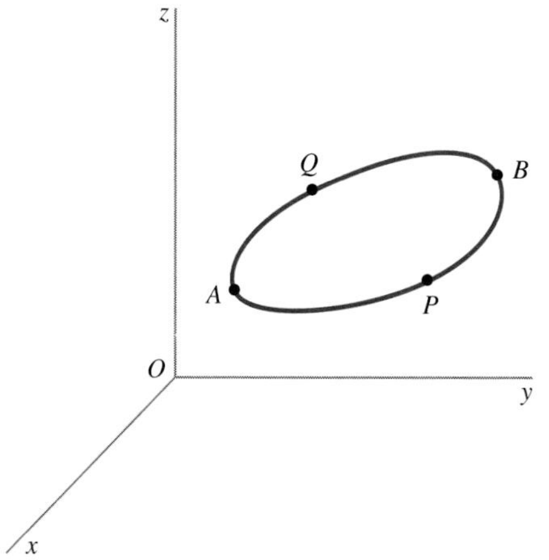

FIGURE 15.23
Two paths from A to B forming a simple closed curve.

LEMMA 2 If the integral (21) is independent of the path, then $\int \mathbf{F} \cdot d\mathbf{R}$ taken around any closed path is zero.

PROOF Let C be any simple closed curve, and let A and B be any two points on C (Fig. 15.23). Then since, by hypothesis, the integral (21) is independent of the path

$$\int_{APB} \mathbf{F} \cdot d\mathbf{R} = \int_{AQB} \mathbf{F} \cdot d\mathbf{R}$$

If we reverse the direction of integration in the integral on the right, we have

$$\int_{APB} \mathbf{F} \cdot d\mathbf{R} = -\int_{BQA} \mathbf{F} \cdot d\mathbf{R}$$

or, transposing,

$$\int_{APB} \mathbf{F} \cdot d\mathbf{R} + \int_{BQA} \mathbf{F} \cdot d\mathbf{R} = \int_{C} \mathbf{F} \cdot d\mathbf{R} = 0$$

as asserted.

Now if the integral (21) is independent of the path, then when we integrate from a fixed point P_0: (x_0, y_0, z_0) to a variable point P: (x, y, z), the result is a function only of the coordinates x, y, z of the variable endpoint. That is, if $\mathbf{F} = u\mathbf{i} + v\mathbf{j} + z\mathbf{k}$, we can appropriately write

$$\int_{P_0}^{P} \mathbf{F} \cdot d\mathbf{R} = \int_{P_0}^{P} u\,dx + v\,dy + w\,dz = \phi(x, y, z)$$

In what follows it will be necessary to know the partial derivatives of the function ϕ defined by the last equation. To obtain these, it is convenient to go back to the fundamental definition of a derivative and write, for the x partial derivative, for instance,

$$\begin{aligned}\frac{\partial \phi}{\partial x} &= \lim_{\Delta x \to 0} \frac{\phi(x + \Delta x, y, z) - \phi(x, y, z)}{\Delta x} \\ &= \lim_{\Delta x \to 0} \frac{1}{\Delta x}\left(\int_{x_0, y_0, z_0}^{x+\Delta x, y, z} u\,dx + v\,dy + w\,dz - \int_{x_0, y_0, z_0}^{x, y, z} u\,dx + v\,dy + w\,dz\right)\end{aligned}$$

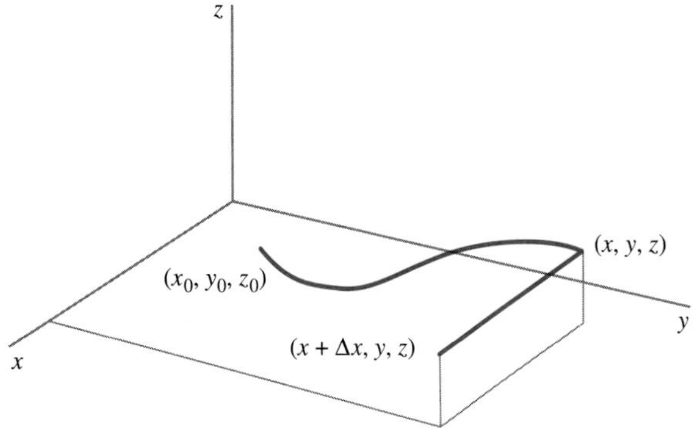

FIGURE 15.24
A convenient path of integration from (x_0, y_0, z_0) through (x, y, z) to $(x + \Delta x, y, z)$.

Since, by hypothesis, these integrals are independent of the path, we can use any paths we find convenient. In particular, in the integral from (x_0, y_0, z_0) to $(x + \Delta x, y, z)$ we shall let the path of integration consist of any smooth curve joining (x_0, y_0, z_0) to (x, y, z) plus the segment of the straight line joining (x, y, z) to $(x + \Delta x, y, z)$ (Fig. 15.24). Then

$$\begin{aligned}\frac{\partial \phi}{\partial x} &= \lim_{\Delta x \to 0} \frac{1}{\Delta x}\left[\left(\int_{x_0,y_0,z_0}^{x,y,z} u\,dx + v\,dy + w\,dz + \int_{x,y,z}^{x+\Delta x,y,z} u\,dx + v\,dy + w\,dz\right)\right. \\ &\qquad \left. - \int_{x_0,y_0,z_0}^{x,y,z} u\,dx + v\,dy + w\,dz\right] \\ &= \lim_{\Delta x \to 0} \frac{1}{\Delta x} \int_{x,y,z}^{x+\Delta x,y,z} u\,dx + v\,dy + w\,dz\end{aligned}$$

Along the path of integration in the last integral we have

$$dy \equiv 0 \qquad \text{and} \qquad dz \equiv 0$$

Hence,

$$\frac{\partial \phi}{\partial x} = \lim_{\Delta x \to 0} \frac{1}{\Delta x} \int_{x}^{x+\Delta x} u\,dx$$

Since u is assumed to be continuous, the law of the mean for integrals can be applied to the last expression and we have

$$\begin{aligned}\frac{\partial \phi}{\partial x} &= \lim_{\Delta x \to 0} \frac{1}{\Delta x}[u(x + \theta\,\Delta x, y, z)\,\Delta x] \qquad 0 \le \theta \le 1 \\ &= u(x, y, z)\end{aligned}$$

In the same way, the partial derivatives with respect to y and z can be determined, and we have the following result.

LEMMA 3 If $\mathbf{F} = u\mathbf{i} + v\mathbf{j} + w\mathbf{k}$ is a continuous vector function of x, y, and z, with the property that

$$\int \mathbf{F} \cdot d\mathbf{R} = \int u\,dx + v\,dy + w\,dz$$

is independent of the path, then the partial derivatives of the function

$$\phi(x, y, z) \equiv \int_{P_0}^{P} \mathbf{F} \cdot d\mathbf{R} = \int_{P_0}^{P} u\,dx + v\,dy + w\,dz$$

are

$$\frac{\partial \phi}{\partial x} = u \qquad \frac{\partial \phi}{\partial y} = v \qquad \frac{\partial \phi}{\partial z} = w$$

We are now in a position to prove

LEMMA 4 If $\mathbf{F} = u\mathbf{i} + v\mathbf{j} + w\mathbf{k}$ is a continuous vector function such that $\int \mathbf{F} \cdot d\mathbf{R}$ is independent of the path, then $\mathbf{F}$ is the gradient of some scalar function ϕ.

PROOF With P_0: (x_0, y_0, z_0) a fixed point and P: (x, y, z) a variable point, and with ϕ defined by

$$\phi(x, y, z) = \int_{P_0}^{P} \mathbf{F} \cdot d\mathbf{R} = \int_{P_0}^{P} u\,dx + v\,dy + w\,dz$$

we have, by Lemma 3,

$$\nabla\phi \equiv \frac{\partial \phi}{\partial x}\mathbf{i} + \frac{\partial \phi}{\partial y}\mathbf{j} + \frac{\partial \phi}{\partial z}\mathbf{k} = u\mathbf{i} + v\mathbf{j} + w\mathbf{k} = \mathbf{F}$$

as asserted.

Whenever convenient, a position vector $\mathbf{R} = x\mathbf{i} + y\mathbf{j} + z\mathbf{k}$ may be denoted by (x, y, z) and thought of as a point, instead of as a directed line segment. Thinking of $\mathbf{R}$ in this way, we say

DEFINITION 1 A vector field $\mathbf{F}$ is a **gradient field** if and only if there is a scalar field $\phi(x, y, z)$ defined on a region $\mathcal{R}$ such that for all $\mathbf{R}$ contained in $\mathcal{R}$

$$\mathbf{F}(\mathbf{R}) = \nabla\phi(\mathbf{R})$$

This definition identifies a gradient field as a vector function which, throughout some region, is the gradient of a scalar point function. Every vector function $\mathbf{F}$ which in a region $\mathcal{R}$ satisfies the hypotheses of Lemma 4 is a gradient field.

Before we can state a correct converse of Lemma 4 we must distinguish between two types of regions in space. On the one hand, a region $\mathcal{R}$ may have the property that every simple closed curve within it can be continuously contracted into a point without having to leave the region at any stage. Regions of this type are said to be **simply connected;** as examples, we have the interior of a sphere, the exterior of a sphere, and the space between two concentric spheres. On the other hand, a region $\mathcal{R}$ may contain simple closed curves which cannot be continuously contracted to a point without having to leave the region at some stage. Such regions are said to be **multiply connected;** as an example we have the space between two infinitely long coaxial cylinders, within which it is clearly impossible to shrink into a single point any closed curve encircling the inner cylindrical boundary. Both the interior and the exterior of a torus are also examples of multiply connected regions. This distinction between simply connected and multiply connected regions applies equally well in the plane, of course, and in our study of functions of complex variables it will often be an important distinction. We are now ready to prove the following converse of Lemma 4.

LEMMA 5 If throughout a simply connected region $\mathcal{R}$ a vector function $\mathbf{F}$ is the gradient of a scalar function ϕ, and if A and B are arbitrary points in $\mathcal{R}$, then $\int_A^B \mathbf{F} \cdot d\mathbf{R}$ is independent of the path joining A and B.

PROOF Let C be an arbitrary closed curve in $\mathcal{R}$, and let S be any regular surface having C as its boundary curve and lying entirely in $\mathcal{R}$. Since $\mathcal{R}$ is simply connected, it is clear that such a surface can always be found. By Stokes' theorem,

$$\int_C \mathbf{F} \cdot d\mathbf{R} = \iint_S \mathbf{N} \cdot \nabla \times \mathbf{F}\, dS = \iint_S \mathbf{N} \cdot \nabla \times \nabla\phi\, dS = 0$$

since, by (26), Sec. 15.5, the curl of any gradient is identically zero. Thus $\int_C \mathbf{F} \cdot d\mathbf{R} = 0$ for every closed curve C in $\mathcal{R}$. Hence, by Lemma 1, it follows that $\int_A^B \mathbf{F} \cdot d\mathbf{R}$ is independent of the path in the *simply connected* region $\mathcal{R}$.

If $\mathcal{R}$ is multiply connected, the preceding proof breaks down because if the curve C encircles an inner boundary of $\mathcal{R}$, then every surface S spanning C will perforce contain points which are not in $\mathcal{R}$ and where, for all we know, ϕ may not exist. If this is the case, Stokes' theorem cannot be applied and the conclusion that $\int \mathbf{F} \cdot d\mathbf{R}$ is independent of the path cannot be inferred. It is easy to show by an example that this can actually happen (Exercise 33).

Finally, we note that if $\mathbf{F}$ is the gradient of a function ϕ, then $\mathbf{F} \cdot d\mathbf{R} = \nabla\phi \cdot d\mathbf{R} = d\phi$; that is, $\mathbf{F} \cdot d\mathbf{R}$ is an exact differential.

To emphasize the results of the preceding lemmas and observations, we summarize them in the following important theorem.

THEOREM 5 If $\mathbf{F} = u\mathbf{i} + v\mathbf{j} + w\mathbf{k}$ is a vector function of x, y, and z possessing continuous first partial derivatives at all points of a simply connected region $\mathcal{R}$, then the following statements are all logically equivalent; i.e., any one of them implies each of the others:

(a) $\int_A^B \mathbf{F} \cdot d\mathbf{R} \equiv \int_A^B u\,dx + v\,dy + w\,dz$ is independent of the path between A and B.

(b) $\int \mathbf{F} \cdot d\mathbf{R} \equiv \int u\,dx + v\,dy + w\,dz$ is zero around every closed curve in $\mathcal{R}$.

(c) $\mathbf{F}$ is the gradient of the scalar point function

$$\phi(x, y, z) = \int_{P_0}^{P} \mathbf{F} \cdot d\mathbf{R} \equiv \int_{P_0}^{P} u\,dx + v\,dy + w\,dz$$

(d) The curl of $\mathbf{F}$ vanishes identically.

(e) $\mathbf{F} \cdot d\mathbf{R} \equiv u\,dx + v\,dy + w\,dz$ is an exact differential.

An especially significant feature of gradient fields is that the respective scalar functions whose gradients they are play the same role in the evaluation of line integrals as do antiderivatives in the evaluation of ordinary definite integrals. Specifically, as an extension of the fundamental theorem of calculus, we have the **fundamental theorem of calculus for line integrals:**

THEOREM 6 If A and B are any two points of a simply connected region $\mathcal{R}$ over which the vector function $\mathbf{F}$ is a gradient field represented by

$$\mathbf{F}(\mathbf{R}) = \nabla\phi(\mathbf{R}) \qquad \mathbf{R} \in \mathcal{R}$$

where ϕ is a scalar field; then, for any piecewise smooth curve C in $\mathcal{R}$ joining A and B,

$$\int_C \mathbf{F} \cdot d\mathbf{R} = \phi(\mathbf{R})\Big|_A^B = \phi(B) - \phi(A)$$

This theorem follows from Properties **a** and **e**, Theorem 5, if C is smooth. If C is piecewise smooth, the same argument applies to the individual smooth segments of C; the values of ϕ at the junction points cancel out (Exercise 39).

By using the expressions for the partial derivatives of ϕ in Lemma 3, and by equating mixed partial derivatives, the following test for gradient fields is easy to derive (Exercise 40).

THEOREM 7 If $\mathbf{F} = u\mathbf{i} + v\mathbf{j} + w\mathbf{k}$ is a vector field with continuous partial derivatives in a simply connected region of 3-space, then $\mathbf{F}$ is a gradient field if and only if

$$\frac{\partial v}{\partial x} = \frac{\partial u}{\partial y} \qquad \frac{\partial w}{\partial y} = \frac{\partial v}{\partial z} \qquad \frac{\partial u}{\partial z} = \frac{\partial w}{\partial x}$$

EXAMPLE 2

USE OF AN ANTIGRADIENT TO INTEGRATE

Evaluate $\int_A^B \mathbf{F} \cdot d\mathbf{R}$ if $A = \left(\frac{\pi}{4}, 1, 1\right)$, $B = (\pi, 0, 2)$, and $\mathbf{F} = u\mathbf{i} + v\mathbf{j} + w\mathbf{k}$, where

$$u = z^2 - y \sin x \qquad v = \cos x - 2z \qquad w = 2xz + z - 2y$$

Computing partial derivatives, we find

$$\frac{\partial v}{\partial x} = -\sin x = \frac{\partial u}{\partial y} \qquad \frac{\partial w}{\partial y} = -2 = \frac{\partial v}{\partial z} \qquad \frac{\partial u}{\partial z} = 2z = \frac{\partial w}{\partial x}$$

Hence, by Theorem 7, $\mathbf{F}$ is a gradient field. To find an **antigradient** of $\mathbf{F}$, i.e., a function ϕ such that $\nabla\phi = \mathbf{F}$, we invoke **a** and **c**, Theorem 5, to choose $(0, 0, 0)$ as a convenient point in the domain of $\mathbf{F}$ and then integrate along the three consecutive straight-line segments from $(0, 0, 0)$ to $(x, 0, 0)$, from $(x, 0, 0)$ to $(x, y, 0)$, and from $(x, y, 0)$ to (x, y, z), obtaining thereby

$$\phi(x, y, z) = \int_{0,0,0}^{x,y,z} \mathbf{F} \cdot d\mathbf{R} = \int_0^x 0\, dt + \int_0^y \cos x\, dt + \int_0^z (2xt - 2y + t)\, dt$$

which, when integrated, reduces to

$$\phi(x, y, z) = y \cos x + xz^2 - 2yz + \frac{z^2}{2}$$

A direct application of Theorem 6, and subsequent simplification, yields

$$\phi(\pi, 0, 2) - \phi\left(\frac{\pi}{4}, 1, 1\right) = (4\pi + 2) - \left(\frac{\sqrt{2}}{2} + \frac{\pi}{4} - 2 + \frac{1}{2}\right) = \frac{15\pi}{4} + \frac{7}{2} - \frac{\sqrt{2}}{2}$$

as the required value of $\int_A^B \mathbf{F} \cdot d\mathbf{R}$.

To signify that every closed curve or surface of a simply connected region $\mathcal{R}$ can be continuously shrunk to a point without passing outside $\mathcal{R}$, such a curve or surface is said to be **reducible.** An **irrotational vector in a region** is a vector function whose integral around every reducible closed curve in the region is zero.

EXERCISES

1. If $\mathbf{F} = 2y\mathbf{i} + x\mathbf{j} + z^2\mathbf{k}$, evaluate $\int_{0,0,0}^{1,1,1} \mathbf{F} \cdot d\mathbf{R}$ along:
   (a) The rectilinear path from $(0, 0, 0)$ to $(1, 0, 0)$ to $(1, 1, 0)$ to $(1, 1, 1)$
   (b) The rectilinear path from $(0, 0, 0)$ to $(1, 1, 0)$ to $(1, 1, 1)$
   (c) The straight line joining $(0, 0, 0)$ to $(1, 1, 1)$
   (d) The curve which is the intersection of the paraboloid $x^2 + y^2 = 2z$ and the plane $x = y$
2. If $\mathbf{F} = x\mathbf{i} + y\mathbf{j} + 2\mathbf{k}$, evaluate $\iint_S \mathbf{F} \cdot \mathbf{N}\, dS$ over:
   (a) The surface of the cube whose vertices are $(0, 0, 0)$, $(1, 0, 0)$, $(1, 1, 0)$, $(0, 1, 0)$, $(0, 0, 1)$, $(1, 0, 1)$, $(1, 1, 1)$, $(0, 1, 1)$

(b) The portion of the plane $x + 2y + 3z = 6$ which lies in the first octant

(c) The entire surface of the sphere $x^2 + y^2 + z^2 = 1$

(d) The portion of the cone $x^2 + y^2 - (1 - z)^2 = 0$ between the plane $z = 0$ and the plane $z = 1$

3. If $\mathbf{F} = y\mathbf{i} + x\mathbf{j} + z^2\mathbf{k}$, evaluate $\iiint_V \nabla \cdot \mathbf{F}\, dV$ throughout:

(a) The volume bounded by the cube whose vertices are $(0, 0, 0)$, $(1, 0, 0)$, $(1, 1, 0)$, $(0, 1, 0)$, $(0, 0, 1)$, $(1, 0, 1)$, $(1, 1, 1)$, $(0, 1, 1)$

(b) The volume cut off from the first octant by the plane $x + 2y + 3z = 6$

(c) The upper half of the volume within the sphere $x^2 + y^2 + z^2 = 1$

(d) The volume under the paraboloid $z = 1 - x^2 - y^2$ and above the plane $z = 0$

4. Write the divergence theorem in cartesian form.

5. Verify the divergence theorem for the function $2xz\mathbf{i} + yz\mathbf{j} + z^2\mathbf{k}$ over the upper half of the sphere $x^2 + y^2 + z^2 = a^2$.

6. Verify the divergence theorem for the function $y\mathbf{i} + x\mathbf{j} + z^2\mathbf{k}$ over the cylindrical region bounded by $x^2 + y^2 = a^2$, $z = 0$, and $z = a$.

7. Verify the divergence theorem for the function $x^2\mathbf{i} + z\mathbf{j} + yz\mathbf{k}$ over the cube whose vertices are $(0, 0, 0)$, $(1, 0, 0)$, $(1, 1, 0)$, $(0, 1, 0)$, $(0, 0, 1)$, $(1, 0, 1)$, $(1, 1, 1)$, and $(0, 1, 1)$.

8. Verify Stokes' theorem for the vector function $xy\mathbf{i} + yz\mathbf{j} + z^2\mathbf{k}$ over the cube described in Exercise 7 if the face of the cube in the xy plane is missing.

9. What is the surface integral of the normal component of the curl of the vector function $(x + y)\mathbf{i} + (y - x)\mathbf{j} + z^3\mathbf{k}$ over the upper half of the sphere $x^2 + y^2 + z^2 = 1$?

10. If at each point of a surface S the vector $\mathbf{F}(x, y, z)$ is perpendicular to S, prove that the curl of $\mathbf{F}$ either vanishes identically or is everywhere tangent to S. *Hint:* Apply Stokes' theorem to $\mathbf{F}$ over the portion of S bounded by an arbitrary closed curve on S.

11. If $\mathbf{F} + y\mathbf{i} = x\mathbf{j}$, and C is the ellipse $b^2x^2 + a^2y^2 = a^2b^2$, use Stokes' theorem to evaluate $\int_C \mathbf{F} \cdot d\mathbf{R}$.

12. Verify Stokes' theorem for $\mathbf{F} = z^3\mathbf{i} + x\mathbf{j} + y^2\mathbf{k}$ over the portion of the paraboloid $z = a^2 - x^2 - y^2$ for which $z \geq 0$.

13. Write Green's theorem in cartesian form.

14. Write Gauss' theorem in cartesian form.

15. Write Stokes' theorem in cartesian form.

16. If S is a closed surface, what is $\iint_S \mathbf{N} \cdot \nabla \times \mathbf{F}\, dS$?

17. If $\mathbf{T}$ is the variable unit tangent to a curve C, what is $\int_C \mathbf{T} \cdot d\mathbf{R}$? Can Stokes' theorem be used to evaluate this integral?

18. If $\mathbf{A}$ is a constant vector and C is a closed curve, show that $\int_C \mathbf{A} \cdot d\mathbf{R} = 0$. What is $\int_C d\mathbf{R}$?

19. If C is a closed curve, show that $\int_C \mathbf{R} \cdot d\mathbf{R} = 0$.

20. If C is a closed curve, show that $\int_C (u\,\nabla v + v\,\nabla u) \cdot d\mathbf{R} = 0$.

21. If S is a closed regular surface, show that $\iint_S \mathbf{N} \cdot \mathbf{R}\, dS = 3V$, where V is the volume enclosed by S.

22. If S is an arbitrary closed surface and $\iint_S \mathbf{N} \cdot \mathbf{F}\, dS = 0$, can we conclude that $\mathbf{F} \equiv \mathbf{0}$? Can we if S is an arbitrary open surface?

23. By applying the divergence theorem to the vector $\phi\mathbf{A}$, where $\mathbf{A}$ is an arbitrary constant vector, show that $\iint_S \phi\mathbf{N}\, dS = \iiint_V \nabla\phi\, dV$. What is $\iint_S \mathbf{N}\, dS$?

24. By applying Stokes' theorem to the vector $\phi\mathbf{A}$, where $\mathbf{A}$ is an arbitrary constant vector, show that $\int_C \phi d\mathbf{R} = \iint_S \mathbf{N} \times \nabla\phi\, dS$.

25. If S is an open surface, what is $\iint_S \mathbf{N} \times \mathbf{R}\, dS$? *Hint:* Use the result of Exercise 24.

26. By applying Stokes' theorem to the vector $\mathbf{F} \times \mathbf{A}$, where $\mathbf{A}$ is an arbitrary constant vector, show that $\int_C d\mathbf{R} \times \mathbf{F} = \iint_S (\mathbf{N} \times \nabla) \times \mathbf{F}\, dS$. What is $\int_C d\mathbf{R} \times \mathbf{R}$?

27. If at each point of a closed surface S the vector $\mathbf{F}(x, y, z)$ is perpendicular to S, prove that $\iiint_V \nabla \times \mathbf{F}\, dV = \mathbf{0}$. *Hint:* Use the result of Example 1.

28. If $\mathbf{A}$ is an arbitrary constant vector, show that $\iint_S \mathbf{N} \times (\mathbf{A} \times \mathbf{R})\, dS = 2V\mathbf{A}$, where V is the volume bounded by the closed surface S. *Hint:* Use the result of Example 1.

29. Extend Gauss' theorem to the case in which O lies *on* the surface S.

30. Show that

$$\iiint_V \psi\nabla^2\phi\, dx\, dy\, dz = \iint_S \psi \frac{d\phi}{dn}\, dS - \iiint_V \left(\frac{\partial\phi}{\partial x}\frac{\partial\psi}{\partial x} + \frac{\partial\phi}{\partial y}\frac{\partial\psi}{\partial y} + \frac{\partial\phi}{\partial z}\frac{\partial\psi}{\partial z}\right) dx\, dy\, dz$$

where $d\phi/dn$ denotes the directional derivative of ϕ in the direction of the outer normal to the closed surface S which bounds the volume V. *Hint:* In Eq. (5a), set $\eta = \psi$, $f = \dfrac{\partial\phi}{\partial x}$, $g = \dfrac{\partial\phi}{\partial y}$, $h = \dfrac{\partial\phi}{\partial z}$, and note that

$$\frac{d\phi}{dn} = \frac{\partial\phi}{\partial x}\cos\alpha + \frac{\partial\phi}{\partial y}\cos\beta + \frac{\partial\phi}{\partial z}\cos\gamma.$$

31. Show that

$$\iiint_V \left(\frac{\partial^2\phi}{\partial x^2} + \frac{\partial^2\phi}{\partial y^2} + \frac{\partial^2\phi}{\partial z^2}\right) dV = \iint_S \frac{d\phi}{dn}\, dS$$

where $d\phi/dn$ is the directional derivative of ϕ in the direction of the outer normal to the closed surface S which bounds the volume V. *Hint:* In Exercise 30, take $\psi = 1$.

32. If $\phi(x, y, z)$ is a solution of Laplace's equation, show that

$$\iiint_V \left[\left(\frac{\partial\phi}{\partial x}\right)^2 + \left(\frac{\partial\phi}{\partial y}\right)^2 + \left(\frac{\partial\phi}{\partial z}\right)^2\right] dV = \iint_S \phi\frac{d\phi}{dn}\, dS$$

where $d\phi/dn$ is the directional derivative of ϕ in the direction of the outer normal to the closed surface S. Hence show also that

$$\iint_S \phi\frac{d\phi}{dn}\, dS > 0$$

if ϕ is a solution of Laplace's equation.

33. Show that although the function

$$\mathbf{F} = \frac{-y}{x^2+y^2}\mathbf{i} + \frac{x}{x^2+y^2}\mathbf{j} + \mathbf{k}$$

is continuous and equal to the gradient of

$$\phi(x, y, z) = \tan^{-1}\frac{y}{x} + z$$

at all points of the region between the two cylinders

$$x^2 + y^2 = \tfrac{1}{4} \quad\text{and}\quad x^2 + y^2 = 4$$

the integral $\int \mathbf{F} \cdot d\mathbf{R}$ is not independent of the path in this region. *Hint:* Take A to be $(-1, 0, 0)$ and B to be $(1, 0, 0)$, and compute $\int_A^B \mathbf{F} \cdot d\mathbf{R}$ along the semicircular arcs of the circle $x^2 + y^2 = 1$, $z = 0$, having A and B as endpoints.

34. Prove directly that Statement **b** of Theorem 5 implies Statement **d**.

35. **(a)** Let S be a closed surface of volume V containing the point P in its interior and let $\mathbf{N}$ be the outer unit normal at a general point of S. Show that the divergence of a vector function $\mathbf{F}$ at P is equal to

$$\nabla \cdot \mathbf{F} = \lim_{V\to 0} \frac{\iint_S \mathbf{F}\cdot\mathbf{N}\, dS}{V}$$

(b) Verify the result of Part **(a)** if $\mathbf{F} = x\mathbf{i} + y\mathbf{j} + 2z\mathbf{k}$, P is the origin, and S is the sphere $x^2 + y^2 + z^2 = r^2$.

36. Show that the volume V bounded by a closed surface S is given by the formula

$$V = \frac{1}{3}\iint_S r\cos\theta\, dS$$

where $r = |\mathbf{R}|$ is the distance from the origin to a general point P of S and θ is the angle between $\mathbf{R}$ and the outer normal to S at P.

37. If $\int_A^B \mathbf{F} \cdot d\mathbf{R} = \int_A^B u\,dx + v\,dy + w\,dz$ is independent of the path from A to B, so that $\mathbf{F}$ is the gradient of some scalar function ϕ, explain how ϕ can be determined.

38. If the curl of $\mathbf{F}(x, y, z)$ vanishes identically, is $\mathbf{F}$ a gradient field? Explain.

39. Prove Theorem 6.

40. Prove Theorem 7.

41. Show that

$$\int_A^B \mathbf{F}\cdot d\mathbf{R} = \int_A^B (2xy^2z + \cos y)\,dx + (2x^2yz - x\sin y + \sin z)\,dy + (x^2y^2 + y\cos z)\,dz$$

is independent of the path, and find a function $\phi(x, y, z)$ whose gradient is $\mathbf{F}$.

42. Show that the vector function $\mathbf{F} = (y+z)\mathbf{i} + (x+z)\mathbf{j} + (x+y)\mathbf{k}$ is a gradient field; then **(a)** find a function $\phi(x, y, z)$ whose gradient is $\mathbf{F}$ and **(b)** evaluate

$$\int_{0,1,1}^{2,2,0} \mathbf{F}\cdot d\mathbf{R}$$

43. For the vector function $\mathbf{F} = (5xy^4 + \sin x)\mathbf{i} + 10x^2y^3\mathbf{j} - z\mathbf{k}$, **(a)** determine $\nabla \times \mathbf{F}$, **(b)** find an antigradient $\phi(x, y, z)$, and **(c)** evaluate $\int_C \mathbf{F}\cdot d\mathbf{R}$, where C is any piecewise smooth curve joining $(0, 0, 0)$ to $(1, 1, 1)$.

In Exercises 44–48, find an antigradient of $\mathbf{F}$ and for the given points A and B evaluate $\int_A^B \mathbf{F}\cdot d\mathbf{R}$.

44. $\mathbf{F} = yz\mathbf{i} + xz\mathbf{j} + xy\mathbf{k}$; $A = (0, 0, 0)$, $B = (1, 1, 1)$

45. $\mathbf{F} = x^2\mathbf{i} + y^2\mathbf{j} + z^2\mathbf{k}$; $A = (1, 0, 0)$, $B(-1, 0, \pi)$

46. $\mathbf{F} = (3x^2 + 2xy + z^2)\mathbf{i} + (3y^2 + 2yz + x^2)\mathbf{j} + (3z^2 + 2xz + y^2)\mathbf{k}$; $A = (-1, -1, -1)$, $B = (1, 2, 3)$

47. $\mathbf{F} = \left(\frac{y}{z} - e^z\right)\mathbf{i} + \left(\frac{x}{z} + 3\right)\mathbf{j} - \left(xe^z + \frac{xy}{z^2}\right)\mathbf{k}$;

$A(-1, -1, 1)$, $B = \left(0, \frac{e+1}{3}, e\right)$

48. $\mathbf{F} = \sinh x\,\mathbf{i} + \cos y\,\mathbf{j} + \frac{1}{z}\mathbf{k}$; $A = (0, 0, 1)$, $B = (10, -2\pi, e^3)$

49. Is $\mathbf{F}(x, y, z)$ irrotational in a region if the curl of $\mathbf{F}$ vanishes identically there? Explain.

50. Is the vector function $\mathbf{F} = 2xyz\,\mathbf{i} + x^2z\,\mathbf{j} + x^2y\,\mathbf{k}$ irrotational? If so, find a scalar function $\phi(x, y, z)$ whose gradient is $\mathbf{F}$.

15.8 FURTHER APPLICATIONS

One of the most important uses of vector analysis is in the concise formulation of physical laws and the derivation of other results from those laws. As a first example of this sort, we shall develop the concept of *potential* and obtain the partial differential equation satisfied by the gravitational potential.

To do this, let us suppose that we have a **field of force** of some kind, or, in other words, let us consider a region of space in which at every point a force vector $\mathbf{F}$ is defined. The field might, for instance, be **gravitational,** in which case $\mathbf{F}(x, y, z)$ would be the force acting on a unit mass at the general point P: (x, y, z) because of the attraction of other masses present in the region. On the other hand, the field might be **electrostatic,** in which case $\mathbf{F}(x, y, z)$ would be the force acting on a unit charge at the general point P: (x, y, z) because of the attraction or repulsion of other charges present in the region. Or the field might be **magnetic,** in which case $\mathbf{F}(x, y, z)$ would be the force acting on a unit magnetic pole situated at the point P: (x, y, z). In any case, the force $\mathbf{F}$ experienced by a unit test body of the appropriate nature is called the **field intensity.**

The amount of work that must be done when a unit test body is moved along an arbitrary curve in the force field defined by a vector function $\mathbf{F}$ is the line integral of the tangential component of $\mathbf{F}$; that is,

$$W = \int_C \mathbf{F} \cdot d\mathbf{R}$$

If there is no dissipation of energy through friction or similar effects, then, according to the law of the conservation of energy, this integral must be zero around every closed path, and hence by **a**, Theorem 5, Sec. 15.7, it must be independent of the path joining any given points A and B, assuming, of course, that the hypotheses of that theorem apply. Fields for which this is the case are said to be **conservative.** Furthermore, according to Theorem 5, Sec. 15.7, it is clear that in a conservative field the force vector $\mathbf{F}$ is the gradient of the scalar function

$$\phi(x, y, z) = \int_{P_0}^{P} \mathbf{F} \cdot d\mathbf{R}$$

The function ϕ is called the **potential function**† of the field. In most problems, the masses or charges which produce $\mathbf{F}$ are given, and it is required to find $\mathbf{F}$ itself. Since $\mathbf{F} = \nabla\phi$, it is clear that knowing ϕ is equivalent to knowing $\mathbf{F}$, and hence the determination of ϕ is of prime importance in most field problems. Such problems and those attending the solution of Laplace's equation [Sec. 15.5] comprise the subject matter of **potential theory.**

Assuming, for definiteness, that we are dealing with a gravitational field, let $\mathbf{F}$ be the field intensity at a general point P: (x, y, z) and let $\Delta\mathbf{F}$ be the contribution to $\mathbf{F}$ due to the infinitesimal mass Δm_1 in an infinitesimal volume $\Delta V_1 = \Delta x_1 \, \Delta y_1 \, \Delta z_1$ enclosing the point P_1: (x_1, y_1, z_1). According to Newton's law of universal gravitation, $\Delta\mathbf{F}$ is a vector whose magnitude is

$$\Delta F = G\frac{1 \cdot \Delta m_1}{r^2}$$

where

$$r^2 = (x - x_1)^2 + (y - y_1)^2 + (z - z_1)^2$$

and whose direction is opposite to that of the vector

$$\mathbf{R} = (x - x_1)\mathbf{i} + (y - y_1)\mathbf{j} + (z - z_1)\mathbf{k}$$

extending from P_1 to P (Fig. 15.25). In other words, if units are so chosen that the constant G in Newton's law is equal to unity, then the field intensity at P due to the infinitesimal mass Δm_1 at P_1 is

$$\Delta\mathbf{F} = -\frac{\Delta m_1}{r^2} \cdot \frac{\mathbf{R}}{r} = -\rho(x_1, y_1, z_1)\,\Delta V_1 \frac{\mathbf{R}}{r^3} \tag{1}$$

where $\rho(x_1, y_1, z_1)$ is the density of the material at the point P_1.

†Many writers define the potential to be $\int_P^{P_0} \mathbf{F} \cdot d\mathbf{R}$, in which case $\mathbf{F} = -\nabla\phi$. In particular, P_0 is often taken to be infinitely distant, so that $\phi = \int_P^{\infty} \mathbf{F} \cdot d\mathbf{R}$.

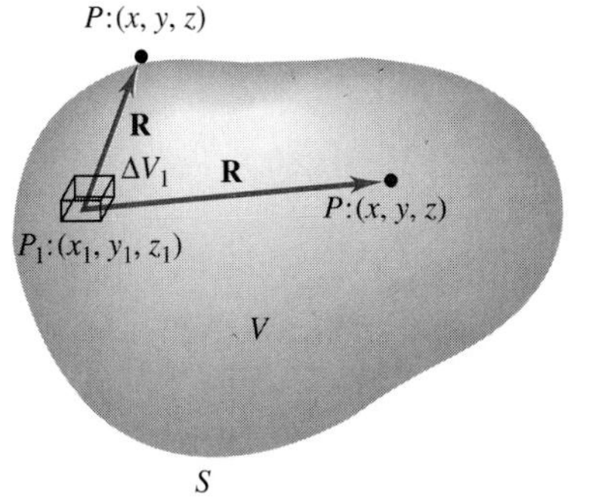

FIGURE 15.25
Calculating the potential at a point P due to the material in a volume element ΔV_1.

Now let S be an arbitrary closed regular surface bounding a volume V, and let I denote the integral over S of the normal component of the force due to all the attracting material in the field. By hypothesis, $\mathbf{F} = \nabla\phi$; hence

$$I = \iint_S \mathbf{N} \cdot \mathbf{F}\, dS = \iint_S \mathbf{N} \cdot \nabla\phi\, dS \tag{2}$$

However, I can also be computed by first determining the part of it, ΔI, due to the material in an arbitrary volume element ΔV_1 and then taking into account all such infinitesimal contributions by integrating over the entire field. From this point of view we have, from (1) and (2).

$$\begin{aligned}\Delta I &= \iint_S \mathbf{N} \cdot \Delta\mathbf{F}\, dS = -\iint_S [\rho(x_1, y_1, z_1)\, \Delta V_1]\mathbf{N} \cdot \frac{\mathbf{R}}{r^3}\, dS \\ &= -\rho(x_1, y_1, z_1)\, \Delta V_1 \iint_S \mathbf{N} \cdot \frac{\mathbf{R}}{r^3}\, dS\end{aligned}$$

since x_1, y_1, z_1 are constant with respect to the x, y, z integration over S. The last integral can, of course, be evaluated by Gauss' theorem (Theorem 3, Sec. 15.7). Specifically, if the origin of $\mathbf{R}$, namely, the point P_1: (x_1, y_1, z_1), is within S, the value of the integral is 4π; otherwise the value of the integral is 0. Hence

$$\Delta I = \begin{cases} -4\pi\rho(x_1, y_1, z_1)\, \Delta V_1 & \Delta V_1 \text{ within } S \\ 0 & \Delta V_1 \text{ outside } S \end{cases}$$

and, therefore, in computing I it is necessary to integrate only over the volume V bounded by S. Doing this, we find

$$I = \int dI = -4\pi \iiint_V \rho(x_1, y_1, z_1)\, dV_1$$

or, since x_1, y_1, z_1 are just dummy variables,

$$I = -4\pi \iiint_V \rho(x, y, z)\, dV \tag{3}$$

Equating the two expressions (2) and (3) which we now have for I, we get

$$\iint_S \mathbf{N} \cdot \nabla\phi\, dS = -4\pi \iiint_V \rho(x, y, z)\, dV$$

If we now apply the divergence theorem to the integral on the left, we have

$$\iiint_V \nabla \cdot (\nabla \phi)\, dV = -4\pi \iiint_V \rho(x, y, z)\, dV$$

or

$$\iiint_V [\nabla^2 \phi + 4\pi\rho(x, y, z)]\, dV = 0$$

Since this holds for an arbitrary volume V, it follows that the integrand must vanish identically,† and therefore that

$$\nabla^2 \phi = -4\pi\rho(x, y, z) \tag{4}$$

This is **Poisson's equation,‡** and we have thus shown that *in regions occupied by matter, the gravitational potential satisfies Poisson's equation.* In empty space $\rho(x, y, z) = 0$, and thus *in empty space the gravitational potential satisfies Laplace's equation*

$$\nabla^2 \phi = 0 \tag{5}$$

Results similar to these hold for the electrostatic and magnetic potentials.

As a second example of the use of vector analysis in formulating physical laws in mathematical terms, we shall now derive *Maxwell's equations* for electric and magnetic fields. To do this we shall have to work with the vector quantities

$\mathbf{E}$ = electric intensity
$\mathbf{H}$ = magnetic intensity
$\mathbf{D} = \epsilon\mathbf{E}$ = electric flux density
$\mathbf{B} = \mu\mathbf{H}$ = magnetic flux density
$\mathbf{J}$ = current density

and the scalars

ϵ = permittivity
μ = permeability
σ = conductivity
Q = charge density
$q = \iiint_V Q\, dV$ = total charge within volume V
$\phi = \iint_S \mathbf{N} \cdot \mathbf{B}\, dS$ = total magnetic flux passing through surface S
$i = \iint_S \mathbf{N} \cdot \mathbf{J}\, dS$ = total current flowing through surface S

These quantities are connected by a number of equations expressing relations discovered experimentally in the early years of the nineteenth century, chiefly by Michael Faraday (1791–1867). In particular, we have:

---

†Suppose that this is not the case, and let P_0 be a point at which the integrand does not vanish. Then. if $\rho(x, y, z)$ and $\nabla^2\phi$ are continuous (as we have implicitly assumed), it follows that throughout some sufficiently small three-dimensional region V_0 enclosing P_0 the integrand has everywhere the same sign it has at P_0. Integrating over V_0, we then obtain an integral which is not equal to zero, contrary to the fact that the integral has been shown to be zero for every volume V.

‡Named for the French mathematical physicist Simeon Denis Poisson (1781–1840).

(a) Faraday's law,

(6) $$\int_C \mathbf{E} \cdot d\mathbf{R} = -\frac{\partial \phi}{\partial t}$$

which asserts that the integral of the tangential component of the electric intensity vector around any closed curve C is equal but opposite in sign to the rate of change of the magnetic flux passing through any surface spanning C.

(b) Ampere's law,

(7) $$\int_C \mathbf{H} \cdot d\mathbf{R} = i$$

which asserts that the integral of the tangential component of the magnetic intensity vector around any closed curve C is equal to the current flowing through any surface spanning C.

(c) Gauss' law for electric fields,

(8) $$\iint_S \mathbf{N} \cdot \mathbf{D}\, dS = q$$

which asserts that the integral of the normal component of the electric flux density over any closed surface S is equal to the total electric charge enclosed by S.

(d) Gauss' law for magnetic fields,

(9) $$\iint_S \mathbf{N} \cdot \mathbf{B}\, dS = 0$$

which asserts that the total magnetic flux passing through any closed surface S is zero.

If we now apply Stokes' theorem to Faraday's law (6), we have

$$\iint_S \mathbf{N} \cdot \nabla \times \mathbf{E}\, dS = -\frac{\partial \phi}{\partial t}$$

and, substituting for ϕ from its definition in terms of $\mathbf{B}$,

$$\iint_S \mathbf{N} \cdot \nabla \times \mathbf{E}\, dS = -\frac{\partial}{\partial t}\left(\iint_S \mathbf{N} \cdot \mathbf{B}\, dS\right) = -\iint_S \mathbf{N} \cdot \frac{\partial \mathbf{B}}{\partial t}\, dS$$

Since S is an arbitrary surface spanning the arbitrary closed curve C, the last equation can hold only if

(10) $$\nabla \times \mathbf{E} = -\frac{\partial \mathbf{B}}{\partial t}$$

Similarly, by applying Stokes' theorem to Ampere's law (7), we obtain

$$\iint_S \mathbf{N} \cdot \nabla \times \mathbf{H}\, dS = i = \iint_S \mathbf{N} \cdot \mathbf{J}\, dS$$

and again, since S is an arbitrary open surface, we conclude that the vectors being integrated over S must be identical; i.e.,

$$\nabla \times \mathbf{H} = \mathbf{J} \tag{11}$$

As Maxwell was the first to realize, the current density $\mathbf{J}$ consists of two parts, namely, a conduction current density

$$\mathbf{J}_c = \sigma \mathbf{E}$$

due to the flow of electric charges, and a displacement current density

$$\mathbf{J}_d = \frac{\partial \mathbf{D}}{\partial t} = \epsilon \frac{\partial \mathbf{E}}{\partial t}$$

due to the time variation of the electric field. Thus,

$$\mathbf{J} = \sigma \mathbf{E} + \epsilon \frac{\partial \mathbf{E}}{\partial t}$$

and (11) becomes

$$\nabla \times \mathbf{H} = \sigma \mathbf{E} + \epsilon \frac{\partial \mathbf{E}}{\partial t} \tag{12}$$

Next we apply the divergence theorem to the first of Gauss' laws, (8), getting

$$\iiint_V \nabla \cdot \mathbf{D}\, dV = q = \iiint_V Q\, dV$$

whence, since V is arbitrary, it follows that

$$\nabla \cdot \mathbf{D} = Q \tag{13}$$

In the same way, by applying the divergence theorem to Gauss' second law, (9), we find that

$$\iiint_V \nabla \cdot \mathbf{B}\, dV = 0$$

and therefore, since V is arbitrary,

$$\nabla \cdot \mathbf{B} = 0 \tag{14}$$

If we take the curl of Eq. (10), we obtain

$$\nabla \times (\nabla \times \mathrm{E}) = -\nabla \times \frac{\partial \mathbf{B}}{\partial t} = -\frac{\partial}{\partial t}(\nabla \times \mathbf{B}) = -\mu \frac{\partial}{\partial t}(\nabla \times \mathbf{H})$$

If we expand the term $\nabla \times (\nabla \times \mathbf{E})$ by means of Eq. (28), Sec. 15.5, the last equation becomes

$$\nabla(\nabla \cdot \mathbf{E}) - \nabla^2 \mathbf{E} = -\mu \frac{\partial}{\partial t}(\nabla \times \mathbf{H})$$

and, substituting for $\nabla \times \mathbf{H}$ from (12),

$$\nabla(\nabla \cdot \mathbf{E}) - \nabla^2 \mathbf{E} = -\mu \frac{\partial}{\partial t}\left(\sigma \mathbf{E} + \epsilon \frac{\partial \mathbf{E}}{\partial t}\right) \tag{15}$$

If the space charge density Q is zero, as it is to a high degree of approximation in both good dielectrics and good conductors, then from (13) and the relation $\mathbf{D} = \epsilon\mathbf{E}$ we see that

$$\nabla \cdot \mathbf{E} = 0$$

Therefore Eq. (15) reduces to

$$\nabla^2\mathbf{E} = \mu\epsilon\frac{\partial^2\mathbf{E}}{\partial t^2} + \mu\sigma\frac{\partial\mathbf{E}}{\partial t}$$

which is **Maxwell's equation for the electric intensity vector E.**

Similarly, if we take the curl of Eq. (12), we obtain

$$\nabla \times (\nabla \times \mathbf{H}) = \nabla \times \left(\sigma\mathbf{E} + \epsilon\frac{\partial\mathbf{E}}{\partial t}\right)$$

and, expanding the left-hand side,

$$\begin{aligned}\nabla(\nabla \cdot \mathbf{H}) - \nabla^2\mathbf{H} &= \sigma\nabla \times \mathbf{E} + \epsilon\nabla \times \frac{\partial\mathbf{E}}{\partial t}\\ &= \sigma\nabla \times \mathbf{E} + \epsilon\frac{\partial}{\partial t}(\nabla \times \mathbf{E})\end{aligned}$$

Now, substituting for $\nabla \times \mathbf{E}$ from (10), we have

$$\nabla(\nabla \cdot \mathbf{H}) - \nabla^2\mathbf{H} = \sigma\left(-\frac{\partial\mathbf{B}}{\partial t}\right) + \epsilon\left(-\frac{\partial^2\mathbf{B}}{\partial t^2}\right)$$

But $\mathbf{B} = \mu\mathbf{H}$, by definition. Hence (14) implies that $\nabla \cdot \mathbf{H} = 0$, and therefore the last equation reduces to

$$\nabla^2\mathbf{H} = \mu\epsilon\frac{\partial^2\mathbf{H}}{\partial t^2} + \mu\sigma\frac{\partial\mathbf{H}}{\partial t}$$

which is **Maxwell's equation for the magnetic intensity vector H.**

For a perfect dielectric, $\sigma = 0$. Hence in this case Maxwell's equations reduce to the three-dimensional wave equations

$$\nabla^2\mathbf{E} = \mu\epsilon\frac{\partial^2\mathbf{E}}{\partial t^2} \qquad \text{and} \qquad \nabla^2\mathbf{H} = \mu\epsilon\frac{\partial^2\mathbf{H}}{\partial t^2}$$

On the other hand, in a good conductor the terms arising from the displacement current, i.e., the terms containing the second time derivatives, are negligible, and Maxwell's equations reduce to

$$\nabla^2\mathbf{E} = \mu\sigma\frac{\partial\mathbf{E}}{\partial t} \qquad \text{and} \qquad \nabla^2\mathbf{H} = \mu\sigma\frac{\partial\mathbf{H}}{\partial t}$$

which are examples of the three-dimensional heat equation.

As a final application of the methods of vector analysis, we shall investigate the question whether a solution of the heat equation satisfying prescribed boundary and initial conditions over a given region is necessarily unique. In our discussion of boundary-value problems in Chap. 11 we proceeded on the assumption that this was the case. Nevertheless, examples have been given† of solutions of the one-dimensional heat equation

$$a^2 \frac{\partial u}{\partial t} = \frac{\partial^2 u}{\partial x^2}$$

which possess derivatives of all orders for all values of x and t, satisfy identical initial conditions everywhere on the entire x axis, and yet are different! Confronted with such a clear-cut failure of intuition, we must regard the uniqueness question as of more than academic interest and any positive result as having important practical significance.

Let us suppose, then, that we are to solve the three-dimensional heat equation

$$a^2 \frac{\partial u}{\partial t} = \nabla^2 u$$

throughout a region V bounded by the closed surface S, subject to the boundary condition

$$u = f(x, y, z, t) \qquad \text{on } S$$

and the initial condition

$$u(x, y, z, 0) = g(x, y, z) \qquad \text{throughout } V$$

Furthermore, let us suppose that we have two solutions of this problem, u_1 and u_2, each of which, with its derivatives through the second, is continuous in V.

If we define a new function

$$w(x, y, z, t) = u_2(x, y, z, t) - u_1(x, y, z, t)$$

it is clear from the linearity of the heat equation that w also satisfies this equation. Moreover, w obviously assumes boundary and initial conditions which are identically zero. Finally, w is continuous and differentiable, since it is the difference of two functions with these properties.

Now consider the volume integral

(16) $$J(t) = \frac{1}{2} \iiint_V w^2(x, y, z, t)\, dV \qquad t \geq 0$$

Clearly, $J(t)$ is a continuous function of t which is always equal to or greater than zero, since its integrand is everywhere nonnegative. Also, since $w = 0$ when $t = 0$, it follows that $J(0) = 0$. Now, differentiating with respect to t inside the integral sign, we have

$$J'(t) = \frac{1}{2} \iiint_V 2w \frac{\partial w}{\partial t}\, dV$$

and thus, since w satisfies the heat equation, on substituting for $\partial w/\partial t$ we have

(17) $$J'(t) = \frac{1}{a^2} \iiint_V w\, \nabla^2 w\, dV$$

†See, for instance, P. C. Rosenbloom and D. V Widder, "A Temperature Function Which Vanishes Identically," *American Mathematics Monthly*, vol. 65, p. 607, October 1958.

According to Eq. (7), Sec. 15.7, with both u and v in that formula taken to be the function w of the present problem, we have

$$\iiint_V (w\,\nabla^2 w + \nabla w \cdot \nabla w)\,dV = \iint_S \mathbf{N} \cdot w\,\nabla w\,dS \tag{18}$$

Since the function w vanishes identically on S, the integral on the right-hand side of (18) is zero, and we have

$$\iiint_V w\,\nabla^2 w\,dV = -\iiint_V \nabla w \cdot \nabla w\,dV$$

Hence, substituting into (17),

$$\begin{aligned} J'(t) &= -\frac{1}{a^2}\iiint_V \nabla w \cdot \nabla w\,dV \\ &= -\frac{1}{a^2}\iiint_V \left[\left(\frac{\partial w}{\partial x}\right)^2 + \left(\frac{\partial w}{\partial y}\right)^2 + \left(\frac{\partial w}{\partial z}\right)^2\right] dV \end{aligned}$$

which shows that

$$J'(t) \le 0 \qquad \text{for} \qquad t \ge 0$$

Now, by the law of the mean,

$$\frac{J(t) - J(0)}{t} = J'(t_1) \qquad 0 < t_1 < t$$

or

$$J(t) = J(0) + tJ'(t_1) \qquad 0 < t_1 < t$$

But we have already verified that $J(0) = 0$. Hence the last equation reduces to

$$J(t) = tJ'(t_1)$$

which shows that

$$J(t) \le 0 \qquad \text{for} \qquad t \ge 0 \tag{19}$$

since we have just proved that $J'(t)$ is nonpositive for $t \ge 0$. However, as we observed earlier, the definition of $J(t)$ shows that

$$J(t) \ge 0 \qquad \text{for} \qquad t \ge 0 \tag{20}$$

The only way in which the inequalities (19) and (20) can simultaneously be fulfilled is for $J(t)$ to be identically zero. But this is possible if and only if the integrand of $J(t)$ vanishes identically. Hence

$$w(x, y, z, t) \equiv u_2(x, y, z, t) - u_1(x, y, z, t) \equiv 0$$

or

$$u_2(x, y, z, t) \equiv u_1(x, y, z, t)$$

Thus, *in bounded regions, twice differentiable solutions of the heat equation satisfying prescribed surface and initial temperature conditions are unique.*

EXERCISES

1. What is the potential function for a central-force field in which the attraction on a particle varies directly as the square of the distance from the origin? Inversely as the distance from the origin?
2. What is the potential function of the force field due to uniform rotation about the z axis?
3. What is the potential function for the gravitational field of a uniform circular disk at any point on the axis of the disk?
4. What is the potential function for the gravitational field of a uniform sphere of radius a and mass M? Show that the attraction of the sphere at a point P a distance r from the center of the sphere is

$$\mathbf{F} = \begin{cases} -\dfrac{M\mathbf{R}}{a^3} & r \le a \\ -\dfrac{M\mathbf{R}}{r^3} & r \ge a \end{cases}$$

5. Show that the electrostatic field intensity at a point P due to a set of charges q_i is equal to

$$\mathbf{E} = -\sum_{i=1}^{n} \frac{q_i}{r_i^3}\mathbf{R}_i$$

where $\mathbf{R}_i$ is the vector from the point P to the point P_i where the charge q_i is located. Verify that $\nabla \cdot \mathbf{E} = 0$ in this case.
6. Show that the work done in bringing a charge of strength q from infinity to a point at a distance of r_0 from a fixed charge q_0 is qq_0/r_0. Using this result, determine the total energy in the electrostatic field defined by the fixed charges $q_1, q_2, \ldots, q_n$ whose mutual distances are $\{r_{ij}\}$.
7. If a **conductor** is defined to be a body in whose interior the electric field is everywhere zero, show that any charge on a conductor must be located entirely on its surface.
8. Let V_1 and V_2 be two regions with respective dielectric constants ϵ_1 and ϵ_2, and let S be the surface of discontinuity which separates them. By applying Gauss' law for electric fields to a closed cylindrical surface of infinitesimal height whose bases are parallel to S in the respective media, show that if there are no charges on S, the normal component of the electric flux density is continuous across S. Similarly, by applying Faraday's law to a rectangle of negligible width whose longer sides are parallel to S in the respective media, prove that if the field is conservative, the tangential component of the electric intensity is continuous across S.
9. What is the electric field in the empty space between the perfectly conducting, infinite planes $y = 0$ and $y = l$ if

$$\mathbf{E}\Big|_{t=0} = \mathbf{i} + \mathbf{k} \quad \text{and} \quad \frac{\partial \mathbf{E}}{\partial t}\Big|_{t=0} = \mathbf{i} - \mathbf{k}?$$

Hint: From the nature of the region of the problem and the initial conditions, it is clear that the field has no component in the y direction and that E_x and E_z are functions only of y.
10. Prove that a solution of the heat equation, possessing continuous second partial derivatives, which takes on prescribed initial values throughout a region V and whose normal derivative takes on prescribed values on the surface S which encloses V is unique.
11. Prove that a solution $u(x, y, z, t)$ of the heat equation possessing continuous second partial derivatives which takes on prescribed initial values throughout a region V and for which the expression

$$u(x, y, z, t) + h^2\mathbf{N} \cdot \nabla u(x, y, z, t)$$

takes on prescribed values on the surface S which encloses V is unique.
12. Prove that a vector function is uniquely determined at all points of a region V if its curl and divergence throughout V and its normal component over S are known. *Hint:* Assume that there are two such functions, verify that their difference must be the gradient of some scalar function w, and then use Eq. (7), Sec. 15.7, with u and v each taken equal to w.
13. Give a vector derivation of the heat equation. *Hint:* Let $u(x, y, z, t)$ be the temperature at any point in the conducting region D. First, compute the rate of heat flow through an arbitrary closed surface S in D by integrating k times the normal component of the gradient of u over S, k being the thermal conductivity of the medium. Then use the divergence theorem to convert this surface integral into a volume integral. Next, find another expression for the rate of change of heat in the volume V bounded by S by integrating $c\rho\, \partial u/\partial t$ throughout V, c and ρ being, respectively, the specific heat and the density of the medium. Finally, equate these two expressions for the rate of change of the heat content of the material in the arbitrary volume V and obtain the heat equation.
14. Let ϕ be a solution of Laplace's equation which possesses continuous second partial derivatives through some region of space D. Show that if ϕ is identically zero at all points of a closed surface S in D, then it vanishes identically throughout the interior of S.
15. Use the result of Exercise 14 to prove the following result: A solution of Laplace's equation which possesses continuous second partial derivatives throughout a region D is uniquely determined throughout any volume V in D by its values on the surface S which bounds V. *Hint:* Assume that there are two such solutions and consider their difference.
16. Consider a continuous fluid having neither sources nor sinks inside a surface S enclosing a volume V. Let

$\mathbf{F} = \rho\mathbf{v}$ where ρ and $\mathbf{v}$ are, respectively, the density and velocity of the fluid. Interpret $\mathbf{F}$ physically.

17. In Exercise 16, interpret the integrals $\iint_S \mathbf{N} \cdot \mathbf{F}\, dS$ and $-\iiint_V \dot{\rho}\, dV$ physically. Are these two integrals equal?

18. Use the divergence theorem to find a volume integral which equals the surface integral of Exercise 17.

19. Derive the **equation of continuity for fluid flow:**

$$\nabla \cdot (\rho\mathbf{v}) = -\dot{\rho} \tag{21}$$

Hint: Equate the volume integrals of Exercises 17 and 18.

20. **(a)** Is Eq. (19), Sec. 15.5, a special case of (21)? Explain.

(b) If there exists a **velocity potential** $\phi(x, y, z, t)$ such that $\nabla\phi = \mathbf{v}$, what equation does ϕ satisfy?

CHAPTER 16

THE CALCULUS OF VARIATIONS

The calculus of variations is concerned with maximizing or minimizing quantities, called *functionals* (Sec. 1.1), which depend not on one or more independent variables, but on one or more functions. These quantities are typically definite integrals involving some function in the integrand which must be chosen to extremize the integral. Necessary conditions for this are obtained in Sec. 16.2. Often, the function to be found is subject to some constraint. For example, one may want the (unknown) equation of the curve in which a chain of given length will hang between two given points so that its potential energy will be a minimum. Necessary conditions for solving problems of this sort are obtained in Sec. 16.3.

Section 16.4 discusses the intimate relation between a Sturm-Liouville problem (Sec. 11.6) and an equivalent problem in the calculus of variations. This is of special interest when a Sturm-Liouville problem must be solved approximately, because, in general, approximate calculations are much more easily, and accurately, carried out with integrals rather than derivatives.

Section 16.5 introduces the concept of *variations,* which are objects related to functionals in roughly the same way that differentials are related to functions. Through their use, in Secs. 16.6 and 16.7, *Hamilton's principle, Lagrange's equations of motion,* and *Hamilton's equations,* the cornerstones of classical dynamics, are established.

Prerequisites for this chapter: multivariable calculus, a calculus-based course in physics, differential equations, Chap. 11.

16.1 INTRODUCTION

One of the fundamental problems in calculus is the determination of the values of x for which a function of one variable, say $f(x)$, assumes its maximum and minimum values. A related problem is the determination of the maxima and minima of a function of more than one variable, say $g(x_1, x_2, \ldots, x_n)$. Sometimes the maxima and minima of g must be found when the variables $x_1, x_2, \ldots, x_n$ are required to satisfy certain conditions of constraint. Significantly more general and more difficult than any of these, however, is a problem of the following kind. Let the domain of a real valued function F be a set of real functions $\{h(x)\}$ where each function $h(x)$ is defined on an interval I. Under what conditions will F take on a maximum or minimum value, and, if such a value exists, for what member of $\{h(x)\}$ does F take on that value? Questions concerned with problems of this sort constitute the branch of mathematics known as the **calculus of variations.** Of necessity our treatment of the subject must be brief; hence we shall assume without proof a number of topics from calculus including, in particular, certain properties of continuous functions and the technique of Lagrange's multipliers.

The features which characterize one important type of variational problem are clearly indicated by the following examples.

EXAMPLE 1

SHORTEST PLANE DISTANCE

Find the shortest smooth plane curve joining two distinct points.

Suppose all the curves and the two given points (x_1, y_1), (x_2, y_2) lie in the xy plane and that $x_1 < x_2$. Each curve is defined by a continuously differentiable function y on $[x_1, x_2]$ which satisfies the conditions $y(x_1) = y_1$ and $y(x_2) = y_2$. Its corresponding length is given by

(1) $$L = \int_{x_1}^{x_2} \sqrt{1 + (y')^2}\, dx$$

Thus, the problem of finding the shortest curve joining (x_1, y_1) and (x_2, y_2) has become that of finding the continuously differentiable function $y = y(x)$ defined on $[x_1, x_2]$ which minimizes the integral (1).

EXAMPLE 2

SMALLEST SURFACE OF REVOLUTION

What plane curve joining the points (x_1, y_1) and (x_2, y_2), and lying above or on the x axis, generates the smallest surface area when revolved about the x axis?

Let us assume that the positive direction of the y axis is upward, that $y_1 > 0$, $y_2 > 0$, that each curve is defined by a continuously differentiable function y which satisfies the conditions $y(x_1) = y_1$, $y(x_2) = y_2$, and that $y(x) \geq 0$ on $[x_1, x_2]$. The surface area S generated by revolving the curve defined by y about the x axis is

(2) $$S = \int_{x_1}^{x_2} 2\pi y\sqrt{1 + (y')^2}\, dx$$

Our problem now is to find, if possible, the continuously differentiable function $y = y(x)$ with domain $[x_1, x_2]$ for which the integral (2) is as small as possible.

EXAMPLE 3

MINIMIZING EXPENSES

A mineshaft is to be driven from (x_1, y_1) to (x_2, y_2) through strata of varying composition (Fig. 16.1). Digging costs per unit length of tunnel are given by a continuous function $c(y)$. All other expenses for the project are given by the cost function $C = 100 + 20\int_{x_1}^{x_2} xy(xy' + y)\, dx$. Along what curve should the shaft run if total expenses are to be minimized?

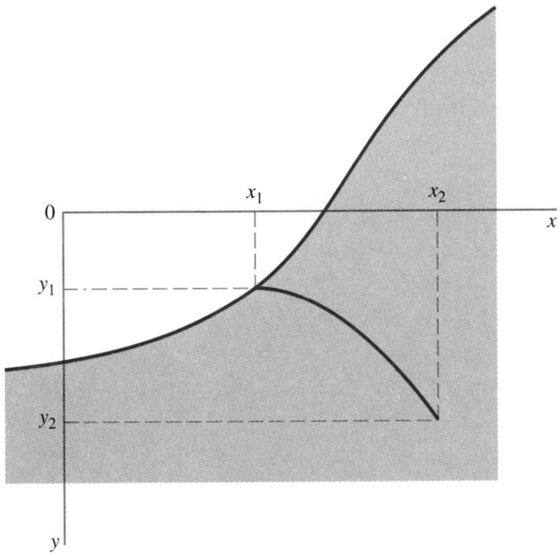

FIGURE 16.1
Illustration of a shaft driven between two points.

Let y be any continuously differentiable function on $[x_1, x_2]$ such that $y(x_1) = y_1$ and $y(x_2) = y_2$. The total cost of digging from (x_1, y_1) to (x_2, y_2) along the curve defined by y is given by $D = \int_{x_1}^{x_2} c(y)\sqrt{1 + (y')^2}\, dx$. Overall expenses corresponding to this curve come to

$$E = C + D = 100 + \int_{x_1}^{x_2} [20xy(xy' + y) + c(y)\sqrt{1 + (y')^2}]\, dx \tag{3}$$

We are to find the function $y = y(x)$ which minimizes E.

Clearly, the integrals (1), (2), and those of Example 3 are all special cases of the more general integral

$$I = \int_{x_1}^{x_2} f(x, y, y')\, dx \tag{4}$$

Now suppose that f is a given function which is twice-differentiable with respect to any combination of its arguments. Also suppose a function $y = y(x)$ exists which is twice-differentiable on $[x_1, x_2]$, which satisfies the **end conditions** $y(x_1) = y_1$ and $y(x_2) = y_2$, and which minimizes or else maximizes I. Under these assumptions, we shall attempt to determine the extremizing function $y(x)$ and, in so doing, try to discover methods of solving problems like the preceding examples.

We shall soon see that the following lemma, or a suitable modification of it, is frequently applicable to variational problems.

LEMMA 1 If g is a continuous function on an interval $[x_1, x_2]$, and if

$$\int_{x_1}^{x_2} \eta(x) g(x)\, dx = 0 \tag{5}$$

for every continuously differentiable function η such that

$$\eta(x_1) = \eta(x_2) = 0 \tag{6}$$

then

$$g(x) \equiv 0 \quad \text{on} \quad [x_1, x_2] \tag{7}$$

PROOF Assume that for some number k in $[x_1, x_2]$, $g(k) \neq 0$. For definiteness, suppose $g(k) > 0$ (otherwise consider $-g$). Since g is continuous, there is a subinterval of $[x_1, x_2]$ throughout which $g(x) > 0$. In fact, numbers a, b, and c exist such that $g(x) \geq c > 0$ on $[a, b]$, where $x_1 < a \leq x \leq b < x_2$. It is readily verified that the function η represented by

$$\eta(x) = \begin{cases} 0 & x_1 \leq x \leq a \\ (x-a)^2(x-b)^2 & a \leq x \leq b \\ 0 & b \leq x \leq x_2 \end{cases}$$

is continuously differentiable on $[x_1, x_2]$ and satisfies (6). For this particular η we have

$$\int_{x_1}^{x_2} \eta(x)g(x)\,dx = \int_a^b (x-a)^2(x-b)^2 g(x)\,dx > 0$$

which contradicts (5). It follows that $g(x) \equiv 0$ on $[x_1, x_2]$ and the lemma is proved.

Sometimes $\eta(x)$ will have to satisfy more stringent conditions than just continuous differentiability. For instance, it may have to be m-times continuously differentiable. Such cases are covered by the following corollary, which is readily proved by constructing an appropriate particular $\eta(x)$.

COROLLARY 1 If a function g is continuous on $[x_1, x_2]$, if m is a nonnegative integer, and if $\int_{x_1}^{x_2} \eta(x)g(x)\,dx = 0$ for all m-times continuously differentiable functions η for which Eqs. (6) hold, then $g(x) \equiv 0$ on $[x_1, x_2]$.

Returning to the problem of finding the twice-differentiable function $y(x)$ which we have assumed extremizes the integral (4), let us consider the set $\{Y\}$ of all twice-differentiable functions Y on $[x_1, x_2]$ which take on the values y_1 and y_2 at x_1 and x_2, respectively. By suitably choosing a real number ϵ, and a twice-differentiable function η which satisfies Eqs. (6), it is possible to represent any member of $\{Y\}$ by an expression of the form

$$Y(x) = y(x) + \epsilon\eta(x) \qquad x_1 \leq x \leq x_2 \tag{8}$$

On the other hand, given any twice-differentiable function η on $[x_1, x_2]$ for which Eqs. (6) hold, the function (8) corresponding to any value of ϵ is clearly a member of $\{Y\}$. Moreover, $y(x)$ is contained in $\{Y\}$. For, no matter what η is chosen, the function (8) reduces to $y(x)$ when $\epsilon = 0$. Members of $\{Y\}$, or functions of the form (8), are referred to as **comparison functions.**

The set of all functions (8), corresponding to some specific η, is a subset of $\{Y\}$ which constitutes a one-parameter family of comparison functions whose derivatives are given by

$$Y'(x) = y'(x) + \epsilon\eta'(x) \qquad x_1 \leq x \leq x_2 \tag{9}$$

Replacing y and y' in (4) by Y and Y', respectively, we obtain the integral

$$I(\epsilon) = \int_{x_1}^{x_2} f(x, Y, Y')\,dx \tag{10}$$

which, for each η, is a function of the parameter ϵ. Since setting $\epsilon = 0$ has the effect of replacing Y and Y' in Eq. (10) by the extremizing function y and its derivative y', $I(\epsilon)$ must take on its extreme value when $\epsilon = 0$. This happens no matter what particular function η is involved in Eqs. (8) and (9). But, if $I(\epsilon)$ has an extremum for $\epsilon = 0$, then

$$I'(0) = 0 \tag{11}$$

Differentiating (10) with respect to ϵ, and taking note of (8) and (9), we have

$$\frac{dI}{d\epsilon} = I'(\epsilon) = \int_{x_1}^{x_2} \left[\frac{\partial f}{\partial Y}\frac{\partial Y}{\partial \epsilon} + \frac{\partial f}{\partial Y'}\frac{\partial Y'}{\partial \epsilon} \right] dx = \int_{x_1}^{x_2} \left[\frac{\partial f}{\partial Y}\eta + \frac{\partial f}{\partial Y'}\eta' \right] dx$$

which, upon setting $\epsilon = 0$, and making use of (11), gives

$$I'(0) = \int_{x_1}^{x_2} \left[\frac{\partial f}{\partial y}\eta + \frac{\partial f}{\partial y'}\eta' \right] dx = 0 \tag{12}$$

Applying integration by parts to the last term in this integral, we obtain

$$I'(0) = \frac{\partial f}{\partial y'}\eta \Big|_{x_1}^{x_2} + \int_{x_1}^{x_2} \left[\frac{\partial f}{\partial y} - \frac{d}{dx}\left(\frac{\partial f}{\partial y'}\right) \right] \eta\, dx = \int_{x_1}^{x_2} \left[\frac{\partial f}{\partial y} - \frac{d}{dx}\left(\frac{\partial f}{\partial y'}\right) \right] \eta\, dx = 0 \tag{13}$$

since, by hypothesis, $\eta(x_1) = \eta(x_2) = 0$. Except for the conditions just noted, and the requirement of being twice-differentiable on $[x_1, x_2]$, η is an arbitrary function. Hence, if f is any function for which $\frac{\partial f}{\partial y} - \frac{d}{dx}\left(\frac{\partial f}{\partial y'}\right)$ is continuous on $[x_1, x_2]$, the last of Eqs. (13) and Corollary 1 imply that *a necessary condition for the integral* (4) *to have an extreme value is that the extremizing function* $y = y(x)$ *satisfy the differential equation*

$$\frac{\partial f}{\partial y} - \frac{d}{dx}\left(\frac{\partial f}{\partial y'}\right) = 0 \qquad x_1 \le x \le x_2 \tag{14}$$

This equation is called the **Euler-Lagrange equation** for the integral (4).

We must be careful to note that Eq. (14) is not a sufficient condition for y to extremize the corresponding integral I of Eq. (4). For a solution of Eq. (14) which satisfies the end conditions $y(x_1) = y_1$, $y(x_2) = y_2$ may fail to maximize or minimize I. The fact that $\epsilon = 0$ is a value which makes $dI/d\epsilon = 0$ implies only that $\epsilon = 0$ is a stationary point of $I(\epsilon)$ at which I may have a maximum or a minimum or *a horizontal inflexion.* A solution of the Euler-Lagrange equation, with prescribed end conditions, thus renders I stationary but not necessarily a maximum or a minimum; and even if an extremum occurs, it may be relative and not absolute. In some problems the requirement that I be rendered stationary by a twice-differentiable function, or even by a single-valued relation, is excessively restrictive. Furthermore, solving the Euler-Lagrange equation may well lead to an implicit rather than an explicit solution, with attendant complications. These observations suggest the need for a more comprehensive theory, but fortunately in elementary applications of the calculus of variations the nature of the problem will usually provide at least informal evidence that some function, such as I, whose value depends on other functions, is indeed maximized or minimized by a certain function in its domain. Let us therefore forgo theory and concentrate on the practical aspects of the subject which, after all, are our primary objectives.

Since $\partial f/\partial y'$ will usually involve x not only explicitly but also implicitly, through y and y', we have

$$\frac{d}{dx}\left(\frac{\partial f}{\partial y'}\right) = \frac{\partial}{\partial x}\left(\frac{\partial f}{\partial y'}\right) + \frac{\partial}{\partial y}\left(\frac{\partial f}{\partial y'}\right)y' + \frac{\partial}{\partial y'}\left(\frac{\partial f}{\partial y'}\right)y''$$

Hence Eq. (14) is equivalent to

$$f_y - f_{y'x} - f_{y'y}y' - f_{y'y'}y'' = 0 \tag{15}$$

In general, this differential equation is of the second order in y.

Various special cases of Eq. (14) are worthy of note:

(a) If the function f in the integrand of (4) does not involve y explicitly, then $\partial f/\partial y \equiv 0$ and Eq. (14) reduces to $\dfrac{d}{dx}\left(\dfrac{\partial f}{\partial y'}\right) = 0$, which leads to the first integral

$$(16) \qquad \frac{\partial f}{\partial y'} = c \qquad c \text{ an arbitrary constant}$$

(b) If f involves neither x nor y explicitly, the partial derivative $\partial f/\partial y'$ is a function of y' alone. Hence Eq. (16) may be regarded as an elementary equation in the unknown y' and every solution must be of the form $y' = a$, where the constant a is a function of c. It follows that, if a function y extremizes an integral (4) in which f is an explicit function of y' alone, then y must be a linear function of x. As an immediate application, we observe that the integrand of (1) involves only y' explicitly. Thus the solution to the problem of Example 1 must be the straight-line segment joining (x_1, y_1) and (x_2, y_2).

(c) It is easy to verify by differentiation that

$$\frac{d}{dx}\left[y'\frac{\partial f}{\partial y'} - f\right] = -y'\left[\frac{\partial f}{\partial y} - \frac{d}{dx}\left(\frac{\partial f}{\partial y'}\right)\right] - \frac{\partial f}{\partial x}$$

If f does not involve x explicitly, then $\partial f/\partial x \equiv 0$ and the Euler-Lagrange equation (14) clearly implies that the first member of the last equation vanishes. A first integral of (14), in this case, is therefore

$$(17) \qquad y'\frac{\partial f}{\partial y'} - f = k \qquad k \text{ an arbitrary constant}$$

(d) If the integrand f of the integral I of Eq. (4) is the total derivative with respect to x of some function $g(x, y)$, then

$$I = \int_{x_1}^{x_2} f(x, y, y')\,dx = \int_{x_1}^{x_2} \frac{d}{dx}g(x, y)\,dx = \int_{(x_1,y_1)}^{(x_2,y_2)} dg(x, y) = g(x_2, y_2) - g(x_1, y_1)$$

This shows that the value of I is independent of the function y relative to which the integration is performed provided y satisfies the end conditions $y(x_1) = y_1$ and $y(x_2) = y_2$. Let us examine the corresponding Euler-Lagrange equation in this case.

By hypothesis, $f = \dfrac{dg}{dx} = \dfrac{\partial g}{\partial x} + \dfrac{\partial g}{\partial y}y'$. From this we find

$$f_y = \frac{\partial^2 g}{\partial y\,\partial x} + \frac{\partial^2 g}{\partial y^2}y' \qquad f_{y'} = \frac{\partial g}{\partial y} \qquad \frac{d}{dx}(f_{y'}) = \frac{\partial^2 g}{\partial x\,\partial y} + \frac{\partial^2 g}{\partial y^2}y'$$

Since $g_{xy} = g_{yx}$, it follows that $f_y - \dfrac{d}{dx}(f_{y'}) \equiv 0$. Thus, in this case, the Euler-Lagrange equation holds for all twice-differentiable functions y on $[x_1, x_2]$.

Now suppose the Euler-Lagrange equation (14) corresponding to I of Eq. (4) is satisfied by all twice-differentiable functions on $[x_1, x_2]$. Then, since the first three terms in the expanded form (15) of Eq. (14) involve no derivatives of y of order higher than the first, the coefficient

$f_{y'y'}$ of y'' in the fourth term must therefore vanish identically [Exercise 53]. But if $f_{y'y'} = 0$, then $f_{y'} = N(x, y)$ and

$$f(x, y, y') = M(x, y) + N(x, y)y' \tag{18}$$

Moreover

$$\frac{\partial f}{\partial y} = \frac{\partial M}{\partial y} + \frac{\partial N}{\partial y}y' \qquad \text{and} \qquad \frac{d}{dx}\left(\frac{\partial f}{\partial y'}\right) = \frac{dN}{dx} = \frac{\partial N}{\partial x} + \frac{\partial N}{\partial y}y'$$

Hence the Euler-Lagrange equation (14) yields the identity

$$\frac{\partial M}{\partial y} + \frac{\partial N}{\partial y}y' - \frac{\partial N}{\partial x} - \frac{\partial N}{\partial y}y' = 0 \qquad \text{or} \qquad \frac{\partial M}{\partial y} = \frac{\partial N}{\partial x}$$

in x and y. The last relation is precisely the condition of Theorem 1, Sec. 1.7, under which f as given by Eq. (18) is exact. We have thus proved that the Euler-Lagrange equation corresponding to the integral (4) is satisfied by all twice-differentiable functions on $[x_1, x_2]$ if and only if a function $g(x, y)$ exists such that $f = dg/dx$. This result has an important corollary, verification of which is asked for in Exercise 38: *A necessary and sufficient condition that the Euler-Lagrange equation for the integral* (4) *be left unaltered by the addition of a term to the integrand f of I is that the added term be the derivative dg/dx of some function $g(x, y)$.*

We are now prepared to discuss Examples 2 and 3 in greater detail.

EXAMPLE 2 **(continued)**

In forming the Euler-Lagrange equation for the integral (2), the factor 2π in the integrand of S can obviously be omitted. Since $f = y\sqrt{1 + (y')^2}$ does not involve x explicitly, a first integral of the Euler-Lagrange equation can be found using Eq. (17). This gives

$$\frac{y(y')^2}{\sqrt{1 + (y')^2}} - y\sqrt{1 + (y')^2} = k \qquad \text{or} \qquad -y = k\sqrt{1 + (y')^2} \qquad \text{or} \qquad y^2 - k^2 = k^2(y')^2$$

where $k < 0$, since $y \geq 0$ and $y \not\equiv 0$. The absence of x from the left-hand member of the last equation suggests that we solve for dx/dy, then separate variables and integrate. In this way, we obtain

$$x = \pm k \int \frac{dy}{\sqrt{y^2 - k^2}} = \pm k \cosh^{-1}\left(-\frac{y}{k}\right) + h \qquad \text{or} \qquad y = -k \cosh\frac{x - h}{k}$$

which for each fixed value of h and k defines a **catenary.** The corresponding surface of revolution about the x axis is called a **catenoid.** If there is a smooth curve joining (x_1, y_1) to (x_2, y_2), and defined by a twice-differentiable function that minimizes S, it must be a catenary which passes through these two points. Accordingly, we next impose the prescribed end conditions on the equation of the family to obtain the two simultaneous equations

$$y_1 = -k \cosh\frac{x_1 - h}{k} \qquad \text{and} \qquad y_2 = -k \cosh\frac{x_2 - h}{k}$$

in the parameters h and k. As it turns out, there are many choices of (x_1, y_1) and (x_2, y_2) for which these equations either have no solution, a unique solution, or two solutions. When a solution does exist, the corresponding catenary may or may not generate a surface of minimum area.†

†For a detailed treatment of this problem, see Gilbert A. Bliss, *Calculus of Variations,* Mathematical Association of America, 1944.

EXAMPLE 3 **(continued)**

To find the function $y = y(x)$ which minimizes E, as given by (3), we first observe that the constant term 100 can be omitted from the minimizing process. The first term in the integrand of Eq. (3) can also be eliminated because $20xy(xy' + y) = (d/dx)(10x^2y^2)$. This signifies, of course, that the cost function C is fixed independent of the route taken by the shaft. The Euler-Lagrange equation corresponding to E is therefore the same as the one for D. As in the preceding example, we use the fact that $f = c(y)\sqrt{1 + (y')^2}$ does not involve x explicitly to obtain a first integral of the Euler-Lagrange equation. Simplifying it, then separating variables and integrating, we have

$$x = \pm k \int \frac{dy}{\sqrt{c^2(y) - k^2}}$$

This is as far as we can go toward solving this problem without being given a specific function $c(y)$.

In some applications an integral $I = \int_{x_1}^{x_2} f(x, y, y')\, dx$, whose limits of integration are fixed, is to be extremized relative to functions whose values at either or both of the end points are unspecified. Assuming that $y(x)$ is the extremizing function, we form comparison functions $Y(x) = y(x) + \epsilon\eta(x)$, just as before, except that now the value of η is left arbitrary at an end where no condition of the form $y(x_i) = y_i$, $i = 1, 2$, is prescribed. The first of Eqs. (13), namely,

$$I'(0) = \frac{\partial f}{\partial y'}\eta \Big|_{x_1}^{x_2} + \int_{x_1}^{x_2} \left[\frac{\partial f}{\partial y} - \frac{d}{dx}\left(\frac{\partial f}{\partial y'}\right)\right] \eta\, dx = 0$$

still holds. In particular, among admissible η's, this equation holds for all η's such that $\eta(x_1) = \eta(x_2) = 0$. For these, as before, we have $\int_{x_1}^{x_2} \left[\frac{\partial f}{\partial y} - \frac{d}{dx}\left(\frac{\partial f}{\partial y'}\right)\right] \eta\, dx = 0$ and, from Lemma 1, we infer that $y(x)$ must be a solution of Eq. (14). But if this is true, then the vanishing of $I'(0)$ requires that $\frac{\partial f}{\partial y'}\eta \Big|_{x_1}^{x_2} = 0$. Finally, at an end where $y(x_i)$ is unspecified and $\eta(x_i)$ therefore remains arbitrary, we may now take $\eta(x_i) = 1$, which implies that

$$\frac{\partial f}{\partial y'} = 0 \qquad \text{at } x = x_i \tag{19}$$

In determining the constants of integration in a solution of a particular Euler-Lagrange equation, Eq. (19), which is called a **natural boundary condition,** takes the place of the simpler condition $y(x_i) = y_i$ at any *fixed* endpoint where the value of y is unspecified.

If, for $i = 1$ or 2, x_i itself is unspecified, and $I = \int_{x_1}^{x_2} f(x, y, y')\, dx$ is to be extremized by a function $y(x)$ which satisfies a relation $g(x, y) = 0$ at x_i, so that $g(x_i, y\{x_i\}) = 0$, it can be shown that $y(x)$ must be a solution of Eq. (14) for which a **natural boundary condition** of the type

$$f_{y'} = \frac{(\partial g/\partial y)f}{(\partial g/\partial x) + (\partial g/\partial y)y'} \tag{20}$$

holds at the *variable* endpoint x_i.† Geometrically, a condition of this sort applies when the curve defined by $y(x)$ has its end at x_i somewhere on the curve specified by $g(x, y) = 0$.

†See Robert Weinstock, *Calculus of Variations,* pp. 39–40, McGraw-Hill, New York, 1952.

EXERCISES

1. **(a)** Prove Corollary 1 for $m = 2$.
   (b) Prove Corollary 1 for a general integer $m \geq 0$.

For each of the following integrals, find the solution of the corresponding Euler-Lagrange equation which satisfies boundary conditions as stated.

2. $\int_1^2 \left[\frac{x^2}{2}(y')^2 + y^2 + x^2 y\right] dx$; $y(1) = y(2) = \frac{5}{4}$
3. $\int_{1/10}^1 y'(1 + x^2 y')\, dx$; $y(\frac{1}{10}) = 19$, $y(1) = 1$
4. $\int_0^1 e^{3x}[(y')^2 - 2y^2]\, dx$; $y(0) = e - 2$, $y(1)$ unspecified
5. $\int_0^1 e^{3x}[(y')^2 - 2y^2]\, dx$; $y(0) = 1$, $y(1)$ unspecified
6. $\int_0^1 e^{-x}[(y')^2 + 2y^2 + 8xy]\, dx$; $y(0)$ unspecified, $y(1) = -(e + 2)/e$
7. $\int_0^\pi [e^{2x}(y')^2 - 5e^{2x}y^2 + 8e^x y]\, dx$; $y(0) = 0$, $y(\pi)$ unspecified
8. $\int_0^\pi [16y^2 - (y')^2 - 136y \sinh x]\, dx$; $y(0)$ unspecified, $y(\pi) = 4 \sinh \pi$
9. $\int_1^3 [x(y')^2 + (y^2/x) + 6x^2 y + 2xy + x^2 y']\, dx$; $y(1)$ unspecified, $y(3) = 3$
10. Find a complete solution of the Euler-Lagrange equation for $\int_{x_1}^{x_2} [y^2 - (y')^2 - 2y \cosh x]\, dx$.
11. **(a)** Find the Euler-Lagrange equation and natural boundary conditions which are satisfied by a function $y(x)$ that renders the integral $\int_0^1 [(y')^2 + k^2 \cos y]\, dx$ stationary.
    (b) Write a first integral of the differential equation.
12. Evaluate the integral $I = \int_1^2 [e^x + (y + xy') \cos xy]\, dx$ when y is a solution of the corresponding Euler-Lagrange equation such that $y(1) = 2\pi$ and $y(2) = \pi/2$.
13. Evaluate the integral $I = \int_0^2 [e^x + (xy' + y) \sin xy]\, dx$ when y is a function which renders I stationary and satisfies the conditions $y(0) = k$ and $y(2) = \pi$.
14. Solve the Euler-Lagrange equation for the integral $I = \int_1^3 [x^{-4}(y')^2 - 6x^{-6}y^2 + 8x^{-5}y]\, dx$ subject to the end conditions $y(1) = 2$, $y(3) = 6$ and find the corresponding stationary value of I.
15. Find all solutions of the Euler-Lagrange equation for $I = \int_0^2 [3x^2 + (y + xy') \cos xy]\, dx$ which satisfy the end conditions $y(0) = b$, b finite, and $y(2) = \pi$. What are the corresponding values of I?
16. Find all solutions of the Euler-Lagrange equation for $\int_{x_1}^{x_2} (1 - \sin xy)(y + xy')\, dx$ which take on prescribed values at x_1 and x_2. Explain the significance of your answer.
17. In Example 2, show that if $(x_1, y_1) = (-a, b)$ and $(x_2, y_2) = (a, b)$ then $h = 0$ so that $b = -k \cosh(a/k)$. Then prove this equation has none, one, or two solutions, depending on the values of a and b.
18. Evaluate the cost function C of Example 3.
19. In Example 3, solve the Euler-Lagrange equation for E if:
    (a) $c(y) = y$ **(b)** $c(y) = \sqrt{y}$
    (c) $c(y)$ is a positive constant
20. Prove that, if y is the solution of the Euler-Lagrange equation for the integral

$$I = \int_{x_1}^{x_2} [p^2(y')^2 + q^2 y^2]\, dx$$

such that $y(x_1) = y_1$ and $y(x_2) = y_2$, the value of I corresponding to y is an absolute minimum relative to twice-differentiable functions which take on the prescribed endpoint values y_1 and y_2. *Hint:* Verify that $I = (p^2 yy')\big|_{x_1}^{x_2}$ and that, if z is an arbitrary twice-differentiable function for which $z(x_1) = z(x_2) = 0$, then

$$\int_{x_1}^{x_2} [p^2 y' z' + q^2 yz]\, dx = 0$$

Finally, replace y by $y + z$ in I.

21. **(a)** Find the twice-differentiable function y for which $y(1) = 4$, $y(2) = 1$, and the value of the integral $I = \int_1^2 [x^2(y')^2 + 2y^2]\, dx$ is a minimum.
    (b) What is the minimum value of I?
    (c) What is the minimum value of I if natural boundary conditions hold at $x = 1$ and $x = 2$?
    (d) What is the value of I if the graph of y is a straight-line segment joining the points $(1, 4)$ and $(2, 1)$?
22. Find the twice-differentiable function y for which $y(1) = 0$, $y(2) = 21$, and the value of the integral $I = \int_1^2 [x^2 y' + \cos x^2)y' + 2y(y - x \sin x^2)]\, dx$ is a minimum. What is the minimum value of I?
23. Find the equation of the curve which joins the points $(0, 1)$ and $(2, 3)$ and along which the integral $\int_0^2 \frac{\sqrt{1 + (y')^2}}{y}\, dx$ is a minimum.
24. Find the natural boundary conditions for the integral

$$I = \int_0^1 [y^2 - yy' + (y')^2]\, dx$$

What is the Euler-Lagrange equation for I? What is the equation of the curve joining P_1: $(0, 1)$ and P_2: $(1, 2)$ on which I is a minimum? What is the function which minimizes I, takes on the value 1 at $x = 0$, and satisfies the natural boundary condition at $x = 1$? What is the function which minimizes I, satisfies the natural boundary condition at $x = 0$, and takes on the value 2 at $x = 1$? What is the function which minimizes I and satisfies the natural boundary condition at $x = 0$ and at $x = 1$?

25. Work Exercise 24 for the integral $\int_1^3 [x^2(y')^2 - yy']\, dx$ and the points P_1: $(1, 1)$ and P_2: $(3, 2)$.
26. If the cost per mile to travel in the first quadrant is equal to $1 + x$, what is the equation of the family of curves along which it is most economical to travel?
27. Determine the plane curve down which a particle will slide without friction from the point (x_1, y_1) to (x_2, y_2) in the shortest time. Such a curve is called a **brachistochrone.** *Hint:* Assume the positive direction of the y

axis is vertically downward and that $x_1 < x_2$. Then, since energy is conserved, the speed v of a particle sliding along any curve is related to the initial speed v_1, and the ordinate y of the particle, by the formula $v = \sqrt{2g(y - y_0)}$, where $y_0 = y_1 - (v_1^2/2g)$. This problem was one of the first calculus of variations problems to be solved.

28. Solve the brachistochrone problem given in Exercise 27 if $(x_1, y_1) = (0, 0)$, $(x_2, y_2) = (1, 1)$, and $v_1 = 0$.

29. **(a)** Show that the brachistochrone from a fixed point (x_1, y_1) to a vertical line $x = x_2$ intersects the vertical line orthogonally. *Hint:* Use Eq. (19).

(b) Extend the result of Part **(a)** to a brachistochrone from (x_1, y_1) to a curve defined by $g(x, y) = 0$. *Hint:* Use Eq. (20).

(c) A brachistochrone is to originate on a curve defined by $h(x, y) = 0$ and end at a fixed point (x_2, y_2). It can be shown that, if the brachistochrone is defined by a function $y(x)$, the condition $\left.\dfrac{\partial h/\partial x}{\partial h/\partial y}\right|_{x_1} = 1/y'(x_2)$ must hold.† Interpret this result geometrically.

(d) A brachistochrone is to be constructed from a given curve $h(x, y) = 0$ to a second curve $g(x, y) = 0$. What relation must these two curves bear to one another at their respective points of intersection with the brachistochrone? *Hint:* See Parts **(b)** and **(c)**.

30. A particle starts from rest at the origin and traverses a brachistochrone to the line $x + 2y - 2 = 0$.

(a) Determine the brachistochrone and its point of intersection with the given line. *Hint:* See Exercise 29**(b)**.

(b) Find the time T of travel along the length of the brachistochrone.

(c) Find the time F it would take for a freely falling particle, starting from rest, to reach the given line.

(d) Prove $T < F$.

(e) Verify that the brachistochrone intersects the given line orthogonally.

Work Exercise 30 with the line defined by each of the following equations.

31. $\sqrt{3}x + 3y = 3$

32. $y = \pi - x$

33. $x + y = 1$

34. $2x + 2y = 3\pi$

35. A particle starts from rest on the line $y = x + 4$ and moves along a brachistochrone to the parabola $y^2 = x$, the positive direction of the y axis being vertically upward.

(a) Find the point of intersection of the brachistochrone and the parabola. *Hint:* See Exercise 29**(d)**.

(b) Find the point on $y = x + 4$ where the brachistochrone originates. *Hint:* See Exercise 29**(c)**.

(c) Find the time of descent T along the brachistochrone.

(d) Assuming the particle starts from rest, find the minimum free-fall time F along a vertical path from $y = x + 4$ to $y^2 = x$. Verify that $T < F$.

36. Work Exercise 35 with the particle starting on the line $y = x + b$, $b > \frac{1}{4}$.

37. A particle starts from rest on the line $y = \frac{3}{4}x + b$, $b > \frac{25}{2}$, and traverses a brachistochrone to the quarter-circle $y = \sqrt{100 - x^2}$, $-10 \le x \le 0$, the positive direction of y being upward. Find the point of intersection of the brachistochrone and the quarter-circle.

38. Prove that the addition of a term to the integrand f of the integral (4) does not change the corresponding Euler-Lagrange equation if and only if the term added is the derivative dg/dx of some function $g(x, y)$.

39. Show that, if $y_1 \ne y_2$, the shortest distance from (x_1, y_1) to the line l defined by $y = y_2$ is along the straight line through (x_1, y_1) which is perpendicular to l.

40. According to **Fermat's principle**‡ *the time required for light to pass from one point to another through any transmitting medium is an extremum relative to possible paths connecting the points.* Apply this principle to determine the integral which must be extremized if we are to find the light path joining two points (x_1, y_1) and (x_2, y_2) of an optical medium in which the speed of light in any direction at an arbitrary point (x, y) is given by a function $u(x, y)$.

41. If, in Exercise 40, u is represented by $\sqrt{2g(y - y_0)}$, where $y_0 = y_1 - (v_1^2/2g)$, how does the light path compare with the brachistochrone of Exercise 27?

42. **(a)** If the first quadrant is filled with a transparent medium in which the speed of light at any point is equal to $1 + x$, apply Fermat's principle to find the equation of the path along which light will travel from the origin to the point $(2, 3)$.

(b) Show that your solution to Part **(a)** is an equation which defines a curve that is intersected more than once by a vertical line over some interval.

43. A **geodesic** on a surface is a curve along which the distance between two points of the surface is a minimum.

(a) Find the geodesics on a right circular cylinder of radius a. *Hint:* In cylindrical coordinates the differential of arc on a cylinder is given by the formula $(ds)^2 = (a\,d\theta)^2 + (dz)^2$.

(b) Find the geodesics on a sphere of radius a. *Hint:* In spherical coordinates, the differential of arc length on a sphere is given by

$$(ds)^2 = a^2[(\sin\phi\, d\theta)^2 + (d\phi)^2]$$

where θ gives the longitude and ϕ the colatitude of a point on the sphere.

(c) Find the geodesics on a right circular cone whose **vertex angle** (the angle between the axis and an element) is α. *Hint:* In spherical coordinates the differ-

†See, for instance, Robert Weinstock, *Calculus of Variations,* pp. 41–43, McGraw-Hill, New York, 1952.
‡Pierre Fermat (1601–1665) was a great French mathematician.

ential of arc on a right circular cone is given by $(ds)^2 = (dr)^2 + (r \sin \alpha \, d\theta)^2$.

Describe, and find parametric equations for, the geodesic on a right circular cylinder of radius a which connects the point $(a, 0, 0)$ to each of the following points.

44. $(a, \pi/2, \pi/6)$ **45.** $(a, \pi/2, -\pi)$

46. $(a, \pi/2, \pi/3)$

47. The points

$$A: (a/\sqrt{2}, a/\sqrt{2}, 0) \text{ and } B: (0, a/\sqrt{2}, -a/\sqrt{2})$$

lie on the surface of a sphere with center $(0, 0, 0)$ and of radius a.

(a) Find an Euler-Lagrange equation for the geodesic connecting A and B whose variables are the spherical (surface) coordinates θ and ϕ, where θ gives the longitude and ϕ the colatitude of a point on the spherical surface.

(b) Write an equation for the geodesic in terms of θ and ϕ.

(c) When regarded as vectors, the given points may be designated by **A** and **B** and an arbitrary point (x, y, z) by **P**. Show that the geodesic which connects **A** and **B** lies in the plane having $\mathbf{A} \times \mathbf{B} \cdot \mathbf{P} = 0$ as an equation.

48. **(a)** Given two distinct points P and Q on the surface S defined by an equation $g(x, y, z) = 0$, find an integral having the form of Eq. (4) which is extremized by the function defining the geodesic on S connecting P and Q. *Hint:* Express the equation $g(x, y, z) = 0$ of the surface in parametric form by introducing surface coordinates u and v, thus obtaining $x = x(u, v)$, $y = y(u, v)$, $z = z(u, v)$; then find $(ds)^2$.

(b) Simplify the integral of Part **(a)** in those cases for which the family of curves $u = c$ is orthogonal to the curves $v = k$ on the surface S.

49. A function $y(x)$, defined on $0 \le x \le x_1$, has the values $y(0) = y_1$ and $y(x_1) = 0$. The area bounded by $x = 0$, $y = 0$, and the graph of $y(x)$ is revolved about the y axis to for a solid of revolution (Fig. 16.2). The solid moves with constant velocity **v**, of magnitude v, in the positive direction of the y axis (with its axis always on the y axis) in a perfect incompressible fluid. Assuming the force of resistance per unit area at any point of the front surface to be proportional to the square of the component of the velocity normal to the surface, and neglecting any forces which act on the trailing circular base of the solid, show that $\int_0^{x_1} \frac{x\,dx}{1 + (y')^2}$ must be minimized to minimize the resistance. Solve the corresponding Euler-Lagrange equation. *Hint:* The force of resistance on an element ΔA of the front surface is given by the product of the resistance per unit area and the projection of ΔA onto the plane of the circular base. This problem dates back to Newton's *Principia* (1687).

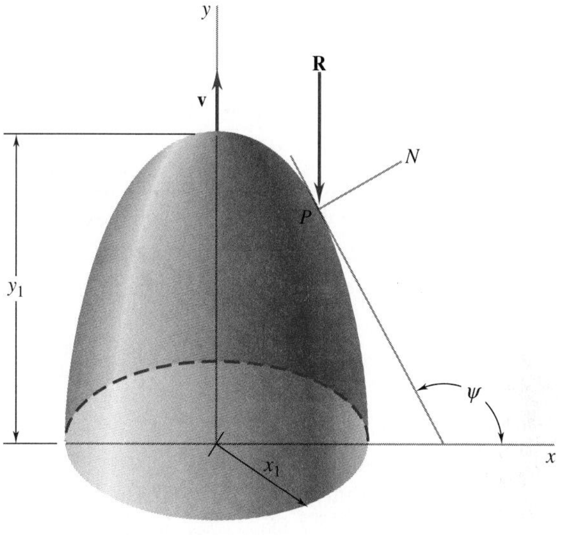

FIGURE 16.2

Newton's solid of minimum resistance, **R**, N, and ψ represent the resistant force per unit area on the front surface, a normal to the surface, and the angle between the direction of the positive x axis and the tangent line to a meridian curve at P, respectively.

50. Show that the Euler-Lagrange equation for the integral $I = \int_{x_1}^{x_2} f(x, y, y', y'')\,dx$ is

$$\frac{\partial f}{\partial y} - \frac{d}{dx}\left(\frac{\partial f}{\partial y'}\right) + \frac{d^2}{dx^2}\left(\frac{\partial f}{\partial y''}\right) = 0$$

51. Find natural boundary conditions at a fixed endpoint for the Euler-Lagrange equation corresponding to the integral given in Exercise 50.

52. Write the Euler-Lagrange equation for the integral $I = \int_{x_1}^{x_2} f(x, y, y', \ldots, y^{(n)})\,dx$.

53. Show that if a differential equation $F(x, y, y') = 0$ has a twice-differentiable solution $y = Y(x)$ on an interval I over which $Y''(x) \ne 0$, if $f(x, y, y')$ is a function for which Eq. (14) holds on I, and if every twice-differentiable function is a solution of $F(x, y, y') - f_{y'y'}y'' = 0$, then there is a function $g(x, y)$ such that $f(x, y, y') = d/dx\, g(x, y)$.

16.2 SYSTEMS OF EULER-LAGRANGE EQUATIONS

In many problems an integral must be extremized whose integrand involves several dependent variables. A simple, but typical, integral of this sort is

$$(1) \qquad I = \int_{t_1}^{t_2} f(t, x, y, \dot{x}, \dot{y})\,dt$$

Let us suppose that t_1 and t_2 are fixed, that $x(t)$ and $y(t)$ are twice-differentiable functions on the interval $[t_1, t_2]$ which take on prescribed values at t_1 and t_2, and that the value of I is an extremum when $x = x(t)$ and $y = y(t)$. To derive equations for the determination of $x(t)$ and $y(t)$, we form one-parameter families of comparison functions

$$X(t) = x(t) + \epsilon\xi(t) \qquad Y(t) = y(t) + \epsilon\eta(t) \tag{2}$$

where ξ and η are arbitrary twice-differentiable functions on $[t_1, t_2]$ for which

$$\xi(t_1) = \xi(t_2) = \eta(t_1) = \eta(t_2) = 0 \tag{3}$$

Conditions (3) guarantee that any two comparison functions $X(t)$ and $Y(t)$ satisfy the same respective end conditions as the extremizing functions $x(t)$ and $y(t)$. Substituting X, Y, $\dot{X}$, and $\dot{Y}$ into Eq. (1), we obtain an integral $I(\epsilon) = \int_{t_1}^{t_2} f(t, X, Y, \dot{X}, \dot{Y})\, dt$ which must be an extremum for $\epsilon = 0$.

Reasoning much as in Sec. 14.1, we find that $I(\epsilon)$ has an extreme value for $\epsilon = 0$ only if

$$I'(0) = \int_{t_1}^{t_2} \left[\frac{\partial f}{\partial x}\xi + \frac{\partial f}{\partial \dot{x}}\dot{\xi} + \frac{\partial f}{\partial y}\eta + \frac{\partial f}{\partial \dot{y}}\dot{\eta}\right] dt = 0 \tag{4}$$

Since this relation must hold for arbitrary functions ξ and η, it must hold, in particular, if ξ is arbitrary and $\eta(t) \equiv 0$. But, if $\eta(t) \equiv 0$, then after an obvious integration by parts, Eq. (4) simplifies to $\int_{t_1}^{t_2} \left[\frac{\partial f}{\partial x} - \frac{d}{dt}\left(\frac{\partial f}{\partial \dot{x}}\right)\right] \xi\, dt = 0$. Similarly, we find that if η is arbitrary but $\xi(t) \equiv 0$, then $\int_{t_1}^{t_2} \left[\frac{\partial f}{\partial y} - \frac{d}{dt}\left(\frac{\partial f}{\partial \dot{y}}\right)\right] \eta\, dt = 0$. From Lemma 1 it follows that the ordered pair $[x(t), y(t)]$ of extremizing functions must be a solution of the system of Euler-Lagrange equations

$$\frac{\partial f}{\partial x} - \frac{d}{dt}\left(\frac{\partial f}{\partial \dot{x}}\right) = 0 \qquad \frac{\partial f}{\partial y} - \frac{d}{dt}\left(\frac{\partial f}{\partial \dot{y}}\right) = 0 \tag{5}$$

As an immediate generalization we have: *A necessary condition for n twice-differentiable functions $x_1(t)$ $x_2(t), \ldots, x_n(t)$ to extremize an integral*

$$I = \int_{t_1}^{t_2} f(t, x_1, x_2, \ldots, x_n, \dot{x}_1, \dot{x}_2, \ldots, \dot{x}_n)\, dt \tag{6}$$

is that the vector function $[x_1(t), x_2(t), \ldots, x_n(t)]$ be a solution of the n simultaneous Euler-Lagrange equations

$$\frac{\partial f}{\partial x_i} - \frac{d}{dt}\left(\frac{\partial f}{\partial \dot{x}_i}\right) = 0 \qquad i = 1, 2, \ldots, n \tag{7}$$

EXAMPLE 1

Write the system of Euler-Lagrange equations corresponding to the integral

$$I = \int_3^7 (8xy + 5x^2 + 5y^2 + \dot{x}^2 + \dot{y}^2)\, dt$$

and find a complete solution of the system.

With $f = 8xy + 5x^2 + 5y^2 + \dot{x}^2 + \dot{y}^2$, Eqs. (7) give the pair of Euler-Lagrange equations

$$f_x - \frac{d}{dt}(f_{\dot{x}}) = 8y + 10x - 2\ddot{x} = 0 \qquad\qquad (D^2 - 5)x - 4y = 0$$

or

$$f_y - \frac{d}{dt}(f_{\dot{y}}) = 8x + 10y - 2\ddot{y} = 0 \qquad\qquad -4x + (D^2 - 5)y = 0$$

Equivalent systems are

$$\begin{aligned}(D^2 - 5)x - 4y &= 0\\ -16x + 4(D^2 - 5)y &= 0\end{aligned} \qquad \text{and} \qquad \begin{aligned}(D^2 - 5)x - 4y &= 0\\ [(D^2 - 5)(D^2 - 5) - 16]x &= 0\end{aligned}$$

Writing the final equation of the last system as $(D^2 - 1)(D^2 - 9)x = 0$ and solving for x, we obtain

$$x = c_1 e^{-t} + c_2 e^{t} + c_3 e^{-3t} + c_4 e^{3t}$$

Using this result in conjunction with the relation $y = \frac{1}{4}(D^2 - 5)x$, which comes from the first equation of the final system, we are led to

$$\begin{bmatrix} x \\ y \end{bmatrix} = c_1 \begin{bmatrix} 1 \\ -1 \end{bmatrix} e^{-t} + c_2 \begin{bmatrix} 1 \\ -1 \end{bmatrix} e^{t} + c_3 \begin{bmatrix} 1 \\ 1 \end{bmatrix} e^{-3t} + c_4 \begin{bmatrix} 1 \\ 1 \end{bmatrix} e^{3t}$$

as a complete solution of the system of Euler-Lagrange equations.

If f does not depend explicitly on some dependent variable x_k, a first integral of the kth of Eqs. (7) is

(8) $$\frac{\partial f}{\partial \dot{x}_k} = c_k \qquad c_k \text{ constant}$$

Now consider the identity (Exercise 7)

(9) $$\frac{d}{dt}\left[\sum_{i=1}^{n} \dot{x}_i \frac{\partial f}{\partial \dot{x}_i} - f\right] = -\sum_{i=1}^{n}\left[\frac{\partial f}{\partial x_i} - \frac{d}{dt}\left(\frac{\partial f}{\partial \dot{x}_i}\right)\right]\dot{x}_i - \frac{\partial f}{\partial t}$$

If f does not involve t explicitly, $\partial f/\partial t = 0$. Hence, the right-hand member of Eq. (9) vanishes in virtue of Eqs. (7), and a first integral of system (7) is

(10) $$\sum_{i=1}^{n} \dot{x}_i \frac{\partial f}{\partial \dot{x}_i} - f = \dot{x}_1 \frac{\partial f}{\partial \dot{x}_1} + \dot{x}_2 \frac{\partial f}{\partial \dot{x}_2} + \cdots + \dot{x}_n \frac{\partial f}{\partial \dot{x}_n} - f = k \qquad k \text{ constant}$$

As in Exercise 42, Sec. 16.1, an implicit solution of the Euler-Lagrange equation for an integral $I = \int_{x_1}^{x_2} f(x, y, y')\, dx$, fulfilling given end conditions $y(x_1) = y_1$, $y(x_2) = y_2$, will often be a multiple-valued relation defining more than one function over intervals other than $[x_1, x_2]$. Thus, it would appear that such a solution must be rejected, since Eq. (14), Sec. 16.1, was derived using only single-valued relations on $[x_1, x_2]$ as comparison functions. However, despite appearances, implicit solutions can generally be accepted. For by introducing parametric representations $x = x(t)$ and $y = y(t)$ of the variables x and y, where $x(t_i) = x_i$ and $y(t_i) = y_i$ for $i = 1, 2$, I can usually be written as $I = \displaystyle\int_{t_1}^{t_2} f\left(x, y, \frac{\dot{y}}{\dot{x}}\right)\dot{x}\, dt$. Functions x and y which extremize this integral need *not* satisfy the requirement that y be a twice-differentiable function of x on $[x_1, x_2]$. With g defined as the function

$$g(x, y, \dot{x}, \dot{y}) = f(x, y, y')\dot{x} = f\left(x, y, \frac{\dot{y}}{\dot{x}}\right)\dot{x} \tag{11}$$

the integral I becomes

$$I = \int_{t_1}^{t_2} g(x, y, \dot{x}, \dot{y})\, dt$$

for which the associated system of Euler-Lagrange equations is

$$\frac{\partial g}{\partial x} - \frac{d}{dt}\left(\frac{\partial g}{\partial \dot{x}}\right) = 0 \qquad \frac{\partial g}{\partial y} - \frac{d}{dt}\left(\frac{\partial g}{\partial \dot{y}}\right) = 0 \tag{12}$$

By using (11), it can be shown that (Exercise 9)

$$\frac{\partial g}{\partial x} - \frac{d}{dt}\left(\frac{\partial g}{\partial \dot{x}}\right) = -\dot{y}\left[\frac{\partial f}{\partial y} - \frac{d}{dx}\left(\frac{\partial f}{\partial y'}\right)\right] \tag{13a}$$

$$\frac{\partial g}{\partial y} - \frac{d}{dt}\left(\frac{\partial g}{\partial \dot{y}}\right) = \dot{x}\left[\frac{\partial f}{\partial y} - \frac{d}{dx}\left(\frac{\partial f}{\partial y'}\right)\right] \tag{13b}$$

Hence, any solution of $\dfrac{\partial f}{\partial y} - \dfrac{d}{dx}\left(\dfrac{\partial f}{\partial y'}\right) = 0$, implicit or not, clearly yields a solution of System (12). We may therefore accept as an extremizing relationship for $\int_{x_1}^{x_2} f(x, y, y')\, dx$ any explicit or implicit solution of the corresponding Euler-Lagrange equation and drop the overly restrictive requirement that I be extremized only with respect to twice-differentiable functions on $[x_1, x_2]$.

Some interesting and important problems involving the minimization of integrals whose integrands contain more than one dependent variable arise in the study of rocket motion. To formulate such problems, we begin with the **general form of Newton's second law,**

$$\frac{d(m\mathbf{v})}{dt} = \sum \mathbf{F}_i \tag{14}$$

which asserts that *the time rate of change of the momentum* $\mathbf{M} = m\mathbf{v}$ *of a body is equal to the vector sum of all external forces acting on the body.*

Now consider a body B comprised of a system of particles S at time t and denote the momentum of S at times t and $t + \Delta t$ by $\mathbf{M}(t)$ and $\mathbf{M}(t + \Delta t)$, respectively. Then, by Newton's second law,

$$\lim_{\Delta t \to 0} \frac{\mathbf{M}(t + \Delta t) - \mathbf{M}(t)}{\Delta t} = \sum \mathbf{F}_i \tag{15}$$

During the time interval Δt, suppose a set of particles s separates from B so that B consists of an aggregate of particles S_1 at time $t + \Delta t$. Denoting the momenta of s and S_1, in turn, by $\Delta\mathbf{M}_s$ and $\mathbf{M}_1(t + \Delta t)$, and noting that at time $t + \Delta t$ the system S consists of s and S_1, we have

$$\mathbf{M}(t + \Delta t) = \mathbf{M}_1(t + \Delta t) + \Delta\mathbf{M}_s \tag{16}$$

Since $\Delta\mathbf{M}_s \to 0$ as $\Delta t \to 0$,

$$\mathbf{M}(t) = \mathbf{M}_1(t) \tag{17}$$

Substituting from (16) and (17) into the left-hand member of (15), then rearranging the numerator and passing to the limit, we obtain

$$\frac{d\mathbf{M}_1}{dt} + \frac{d\mathbf{M}_s}{dt} = \sum \mathbf{F}_i \tag{18}$$

The first term of (18) represents the rate at which the momentum of the body is changing at time t, whereas the second term gives the rate at which momentum is being transferred from the body at the same instant. We thus have the following corollary to Newton's second law: *The time rate of change of the momentum of a body plus the rate at which momentum is being transferred from it due to mass loss is equal to the vector sum of all external forces acting on the body.*

To derive a differential equation which describes the motion of a rocket R, we assume that:

(a) Any decrease in the mass of R is due to the emission of gases from its exhaust nozzles.

(b) With satisfactory accuracy, an average velocity vector $\mathbf{V}(t)$, relative to R, may be attributed to the escaping gases.

The escaping gases, then, have velocity $\mathbf{v} + \mathbf{V}$ relative to a fixed reference frame in space for which $\mathbf{v}$ is the velocity of R. Of course, $\dot{m} = dm/dt$ is the rate at which the mass m of the rocket is changing, and so $-\dot{m}$ is the rate at which R is losing mass. The time rate of change of momentum of the rocket is $d/dt\,(m\mathbf{v})$, and the rate at which momentum is being transferred from R is $-\dot{m}(\mathbf{v} + \mathbf{V})$. Hence, the preceding corollary gives

$$\frac{d}{dt}(m\mathbf{v}) - \dot{m}(\mathbf{v} + \mathbf{V}) = \sum \mathbf{F}_i \tag{19a}$$

or

$$m\dot{\mathbf{v}} - \dot{m}\mathbf{V} = \Sigma\, \mathbf{F}_i \tag{19b}$$

as the **differential equation of rocket motion.**

In a uniform gravitational field, and with drag neglected, $\Sigma\, \mathbf{F}_i = -mg(\mathbf{r}/r)$, where $\mathbf{r}$ is directed upward and g is the gravitational constant. Hence, Eqs. (19) become

$$\dot{\mathbf{v}} + \frac{g}{r}\mathbf{r} = \frac{\dot{m}}{m}\mathbf{V} \tag{20}$$

By confining $\mathbf{V}$ to a single plane, the possible trajectories of a rocket may be limited to plane curves. So let us introduce a rectangular coordinate system into the plane of motion, with the x axis horizontal and the y axis vertical, and then resolve Eq. (20) into rectangular components. Noting that $\mathbf{r}/r$ is a unit vector in the positive direction of y, and letting α denote the smallest nonnegative angle between $\mathbf{V}$ and the positive direction of x, we obtain

$$\ddot{x} = \frac{\dot{m}}{m} V \cos \alpha \tag{21a}$$

$$\ddot{y} + g = \frac{\dot{m}}{m} V \sin \alpha \tag{21b}$$

where V denotes the magnitude of $\mathbf{V}$. For any given rocket, we assume that V is constant.

Squaring Eqs. (21) and adding, then extracting the square root, we find

$$\sqrt{(\ddot{x})^2 + (\ddot{y} + g)^2} = -\frac{\dot{m}}{m} V \tag{22}$$

which, when integrated from t_1 to t_2, gives

$$\int_{t_1}^{t_2} \sqrt{(\ddot{x})^2 + (\ddot{y} + g)^2}\, dt = V \ln\,[m(t_1)/m(t_2)] \qquad t_1 < t_2 \tag{23}$$

Now suppose the rocket R has velocity components

$$\dot{x}(t_1) = \mu_1 \qquad \dot{y}(t_1) = \nu_1 \tag{24}$$

at time t_1, and is required to have velocity components

$$\dot{x}(t_2) = \mu_2 \qquad \dot{y}(t_2) = \nu_2 \tag{25}$$

at time t_2. Of all paths the rocket may pursue in fulfilling Eqs. (24) and (25), we seek one over which a minimum of fuel is expended. A path thus characterized is called an **optimum trajectory** corresponding to the given velocity conditions. Along such a trajectory the **mass ratio** $m(t_1)/m(t_2)$ in Eq. (23) must clearly be a minimum. Thus, to find an optimum trajectory, we must determine functions x and y for which the value of the integral in the left-hand member of Eq. (23) is as small as possible.†

Introducing the substitutions

$$X = \dot{x} \qquad \text{and} \qquad Y = \dot{y} + gt \tag{26}$$

Eq. (23) becomes

$$\int_{t_1}^{t_2} \sqrt{\dot{X}^2 + \dot{Y}^2}\, dt = V \ln\,[m(t_1)/m(t_2)] \tag{27}$$

The problem of finding an optimum trajectory corresponding to Eqs. (24) and (25) is thus reduced to the problem of minimizing the integral

$$I = \int_{t_1}^{t_2} \sqrt{\dot{X}^2 + \dot{Y}^2}\, dt \tag{28}$$

relative to functions X and Y for which

$$X(t_1) = \mu_1 \qquad Y(t_1) = \nu_1 + gt_1 \qquad X(t_2) = \mu_2 \qquad Y(t_2) = \nu_2 + gt_2 \tag{29}$$

Once this has been done, given the initial position of the rocket, we may proceed to determine whatever optimum trajectories R may have.

By replacing $\Sigma\, \mathbf{F}_i$ with appropriate analytic expressions, Eq. (19*b*) can also be adapted to the study of rocket motion in nonuniform force fields, even if drag is present (see Exercises 15 and 16).

---

†Note that we have not, as yet, prescribed initial conditions on the position of the rocket. Neither have we specified a mode of fuel expenditure so as to fix $m(t)$. Thus, recognizing that a minimum of fuel might be expended when burning takes place in any one of a number of different ways, we anticipate that, even with the initial position known, a family of optimum trajectories, and not just one, will correspond to the prescribed velocity conditions.

In each instance, the relevant differential equation can be resolved into scalar components which may be used to obtain an integral whose value must be a minimum if the rocket is to traverse an optimum trajectory corresponding to given auxiliary conditions (see Exercise 17).

EXERCISES

Find a complete solution of the system of Euler-Lagrange equations corresponding to each of the following integrals.

1. $I = \int_{\pi}^{2\pi} (2xy + \dot{x}^2 + \dot{y}^2)\, dt$
2. $I = \int_1^3 (x^2 + y^2 + \dot{x}^2 + \dot{y}^2)\, dt$
3. $I = \int_2^8 (6xy + 5x^2 + 5y^2 + 2\dot{x}^2 + 2\dot{y}^2)\, dt$
4. $I = \int_0^5 (4xy - 6x^2 - 2y^2 + 2\dot{x}^2 + \dot{y}^2)\, dt$
5. Suppose $(x_1, y_1) \neq (x_2, y_2)$ and that $x(t)$ and $y(t)$ are twice-differentiable functions on $[t_1, t_2]$ which minimize the value of $\int_{t_1}^{t_2} \sqrt{\dot{x}^2 + \dot{y}^2}\, dt$ and which satisfy the conditions $x(t_i) = x_i$, $y(t_i) = y_i$, for $i = 1, 2$. What curve is defined by the parametric equations $x = x(t)$, $y = y(t)$?
6. Show that the curve of minimum length connecting two distinct points in 3-space is the straight-line segment connecting the two points.
7. Establish the differential identity (9).
8. Prove that Eq. (10) reduces to an identity, with $k = 0$, if f is homogeneous in $\dot{x}_1, \dot{x}_2, \ldots, \dot{x}_n$ of degree 1. *Hint:* Apply Euler's homogeneous function theorem.
9. Establish Eqs. (13*a*) and (13*b*).
10. Find necessary conditions for twice-differentiable functions $x(t)$ and $y(t)$ to extremize

$$I = \int_{t_1}^{t_2} f(t, x, y, \dot{x}, \dot{y}, \ddot{x}, \ddot{y})\, dt$$

11. Let $P_1{:}\ (\mu_1, \nu_1 + gt_1)$ and $P_2{:}\ (\mu_2, \nu_2 + gt_2)$ be the points in the plane of X and Y determined by Eqs. (29).
   (a) If $\dot{m} \equiv 0$, for $t_1 \le t \le t_2$, how are P_1 and P_2 related?
   (b) For rocket motion along an optimum trajectory, determine $m(t_2)$ in terms of $m(t_1)$ and the coordinates of P_1 and P_2.
   (c) Prove that for optimum flight the **angle of thrust** α must remain constant throughout the motion.
   (d) Verify the results of Table 14.1. *Hint:* Integrate Eqs. (21) and observe that the direction of the thrust is opposite that of **V**.
12. A rocket R has two exhaust jets through which it loses mass at the rates $\dot{m}_1$ and $\dot{m}_2$. The corresponding velocity vectors of the expelled gases, relative to R, are $\mathbf{V}_1$ and $\mathbf{V}_2$.
   (a) Write an equation of motion for R.
   (b) Assuming that $|\mathbf{V}_1| = |\mathbf{V}_2| = V$, prove that during powered flight R expends fuel at a minimum rate when the angle ψ between $\mathbf{V}_1$ and $\mathbf{V}_2$ is zero. *Hint:* Write the equation of motion in the form $\dot{m}_1\mathbf{V}_1 + \dot{m}_2\mathbf{V}_2 = \mathbf{G}$ and take scalar products of each member with itself.

Particularize the equation of rocket motion (19*b*) for each of the following cases.

13. A uniform gravitational field in which there is drag proportional to the square of the velocity.
14. A uniform gravitational field in which the drag varies jointly as the pressure and the square of the velocity, assuming the pressure decreases exponentially with altitude.
15. A drag-free attractive inverse square force field for which k, M, and $\mathbf{r}$ are the gravitational constant, the mass of the attracting center, and the position vector of the rocket, respectively.
16. A general attractive central-force field of intensity $-f(r)\mathbf{r}/r$ in which the drag is proportional to the square of the velocity.
17. A rocket R moves in a central-force field. Its motion is confined to a plane through a fixed point O. Let $\mathbf{r}$ be the position vector of R. The intensity of the field is given by $-f(r)\mathbf{r}/r$ and there exists a potential function $U(r)$ for which $f(r) = dU(r)/dr$. The average velocity vector $\mathbf{V}$ of the escaping gases, relative to R, has constant magnitude V.
   (a) Write an equation of motion for the rocket.
   (b) Resolve the equation of motion of Part **(a)** into components along, and normal to, $\mathbf{r}$. Use ϕ for the polar angle of $\mathbf{r}$ and let α denote the smallest nonnegative angle that can be measured from the positive sense of $\mathbf{r}$ to the positive sense of $\mathbf{V}$.

TABLE 14.1

If, in a constant gravity field, a rocket is to traverse an optimum trajectory corresponding to prescribed velocity conditions for which:

(a) $\mu_2 > \mu_1$, $\nu_2 + gt_2 > \nu_1 + gt_1$	then the constant direction of the thrust must be	**(a)** upward and to the right
(b) $\mu_2 > \mu_1$, $\nu_2 + gt_2 < \nu_1 + gt_1$		**(b)** downward and to the right
(c) $\mu_2 < \mu_1$, $\nu_2 + gt_2 > \nu_1 + gt_1$		**(c)** upward and to the left
(d) $\mu_2 < \mu_1$, $\nu_2 + gt_2 < \nu_1 + gt_1$		**(d)** downward and to the left

(c) During a fixed period of flight $0 \le t \le t_0$, the rocket is to traverse a trajectory over which it expends a minimum of fuel; initial and final velocity conditions must also be satisfied. Formulate the problem of minimizing the fuel consumption as a variational problem.

(d) Show that $C = r^2\dot{\phi}$ and $E = \frac{1}{2}(\dot{r}^2 + r^2\dot{\phi}^2) + U(r)$ are invariant during free flight.†

(e) Letting C and E depart from their role as constants, that is to say, considering powered flight once again, formally transform the variational problem of Part **(c)** into one in which the integral to be minimized has the form $\int_a^{r_0} f(r, C, E, C', E')\,dr$.

(f) State the order of the Euler-Lagrange equations associated with the integrals of Parts **(c)** and **(e).**

18. A rocket is to traverse an ellipse which has one focus at the center of two concentric circles, is tangent internally to the larger circle, and has at least one point in common with the smaller circle. Find the length of the maximum major axis of all such ellipses if the radii of the circles are R and $R + h$.

16.3 THE EXTREMA OF INTEGRALS UNDER CONSTRAINTS

Frequently extreme values of a function of several variables must be found when the variables of the function are subject to certain constraints. Likewise, extreme values of an integral may be required when the functions with respect to which the integral is to be extremized are subject to constraints other than those pertaining to continuity or some kind of end condition. When each new constraint takes the form of a definite integral whose value is fixed, we have what is called an **isoperimetric problem.** This name goes back to the problem of finding a closed curve of given length which encloses maximum area.

Perhaps the simplest kind of isoperimetric problem consists of finding a function $y(x)$ that extremizes an integral

$$I = \int_{x_1}^{x_2} f(x, y, y')\,dx \tag{1}$$

relative to twice-differentiable functions on $[x_1, x_2]$ which satisfy an integral constraint

$$J = \int_{x_1}^{x_2} g(x, y, y')\,dx \qquad J \text{ a fixed constant} \tag{2}$$

as well as a pair of prescribed end conditions $y(x_1) = y_1$, $y(x_2) = y_2$. Both f and g will be assumed to be twice-differentiable with respect to their arguments.

To discover a way of determining $y(x)$, we cannot proceed as before and use only one-parameter families of comparison functions, because any change in value of a *single* parameter would in general alter the value of the integral in Eq. (2), whose value must stay fixed. So, instead, we use two-parameter families of comparison functions

$$Y(x) = y(x) + \epsilon_1\eta_1(x) + \epsilon_2\eta_2(x) \tag{3}$$

where $\eta_1(x)$ and $\eta_2(x)$ are arbitrary twice-differentiable functions on $[x_1, x_2]$ such that

$$\eta_1(x_1) = \eta_1(x_2) = 0 = \eta_2(x_1) = \eta_2(x_2) \tag{4}$$

These conditions ensure that all comparison functions $Y(x)$ satisfy the same end conditions as the extremizing function $y(x)$. Moreover, for every η_1, η_2, $Y(x)$ becomes $y(x)$ if $\epsilon_1 = \epsilon_2 = 0$.

---

†C and E are called **Kepler's constant** and the **total specific energy,** respectively. Kepler was a prominent German astronomer.

Substituting an arbitrary comparison function Y into both Eqs. (1) and (2) for y, we obtain

(5) $$I(\epsilon_1, \epsilon_2) = \int_{x_1}^{x_2} f(x, Y, Y')\,dx \qquad \text{and} \qquad J(\epsilon_1, \epsilon_2) = \int_{x_1}^{x_2} g(x, Y, Y')\,dx$$

Since a functional relation

(6) $$J(\epsilon_1, \epsilon_2) = J \qquad J \text{ constant}$$

exists between ϵ_1 and ϵ_2, these parameters are not independent. Now, by definition, $y(x)$ extremizes I subject to the constraint (2). Therefore, when $\epsilon_1 = \epsilon_2 = 0$, $I(\epsilon_1, \epsilon_2)$ has an extremum relative to values of ϵ_1 and ϵ_2 which satisfy Eq. (6). Our isoperimetric problem has thus become that of finding the condition(s) $y(x)$ must satisfy when it is known that, for $\epsilon_1 = \epsilon_2 = 0$, the ordinary function $I(\epsilon_1, \epsilon_2)$ of two variables ϵ_1, ϵ_2 has an extremum subject to Eq. (6). The method of Lagrange multipliers is well suited to this problem.

To implement the method, we return to Eqs. (5) and set

(7) $$I^*(\epsilon_1, \epsilon_2) = I(\epsilon_1, \epsilon_2) + \lambda J(\epsilon_1, \epsilon_2) = \int_{x_1}^{x_2} f^*(x, Y, Y')\,dx$$

where the **Lagrange multiplier** λ is an arbitrary constant and the function f^* is defined by

(8) $$f^* = f + \lambda g$$

As usual we must have

(9) $$\frac{\partial I^*}{\partial \epsilon_1} = \frac{\partial I^*}{\partial \epsilon_2} = 0 \qquad \text{when } \epsilon_1 = \epsilon_2 = 0$$

Computing these derivatives, with the aid of Eqs. (7) and (3), we find

(10) $$\frac{\partial I^*}{\partial \epsilon_i} = \int_{x_1}^{x_2} \left[\frac{\partial f^*}{\partial Y}\frac{\partial Y}{\partial \epsilon_i} + \frac{\partial f^*}{\partial Y'}\frac{\partial Y'}{\partial \epsilon_i}\right] dx = \int_{x_1}^{x_2} \left[\frac{\partial f^*}{\partial Y}\eta_i + \frac{\partial f^*}{\partial Y'}\eta_i'\right] dx \qquad i = 1, 2$$

Because Y and Y' become the extremizing function y and its derivative y', when $(\epsilon_1, \epsilon_2) = (0, 0)$, we have from Eqs. (9) and (10)

(11) $$\left.\frac{\partial I^*}{\partial \epsilon_i}\right|_{(0,0)} = \int_{x_1}^{x_2} \left[\frac{\partial f^*}{\partial y}\eta_i + \frac{\partial f^*}{\partial y'}\eta_i'\right] dx = 0 \qquad i = 1, 2$$

Integrating the second term in the integrand of (11) by parts and recalling Eqs. (4), we get

(12) $$\int_{x_1}^{x_2} \left[\frac{\partial f^*}{\partial y} - \frac{d}{dx}\left(\frac{\partial f^*}{\partial y'}\right)\right] \eta_i\,dx = 0 \qquad i = 1, 2$$

to which Lemma 1, Sec. 16.1, may be applied, since η_1 and η_2 are arbitrary.

Whether $i = 1$ or 2, we obtain

(13) $$\frac{\partial f^*}{\partial y} - \frac{d}{dx}\left(\frac{\partial f^*}{\partial y'}\right) = 0$$

as the Euler-Lagrange equation which a function y must satisfy if it is to extremize I subject to an integral constraint (2). As a rule, a complete solution to the second-order equation (13) will contain two constants of integration as well as the Lagrange multiplier λ. If an extremizing function exists, definite values for these parameters can be found by imposing the given end conditions and the requirement that the value of the integral in Eq. (2) be as prescribed.

Several direct extensions of the foregoing isoperimetric problem should be mentioned. Most of them can be established by employing suitable families of comparison functions and familiar methods of calculus concerning extrema of functions of several variables. Note the resemblance of these results to certain of our earlier observations in Sec. 16.1.

(a) Let y be a function which renders an integral $I = \int_{x_1}^{x_2} f(x, y, y')\, dx$ stationary relative to twice-differentiable functions on $[x_1, x_2]$ which satisfy prescribed end conditions $y(x_1) = y_1$, $y(x_2) = y_2$, and m integral constraints $J_k = \int_{x_1}^{x_2} g_k(x, y, y')\, dx$, where each J_k is a fixed number, $k = 1, 2, \ldots, m$. Then y is a solution of Eq (13), but with

$$f^* = f + \sum_{k=1}^{m} \lambda_k g_k \tag{14}$$

where the parameters $\lambda_1, \lambda_2, \ldots, \lambda_m$ are Lagrange multipliers.

(b) If in an isoperimetric problem both x_1 and x_2 are fixed but, for $i = 1$ or 2, no value is prescribed for the comparison functions at x_i, then the natural boundary condition $\partial f^*/\partial y' = 0$ must hold at the fixed end x_i. Of course, f^* is defined by an expression of the form given in Eq. (14).

(c) If in an isoperimetric problem, for $i = 1$ or 2, x_i is unspecified and all comparison functions must satisfy a single relation $g(x, y) = 0$ at x_i, then the natural boundary condition $f^*_{y'} = \dfrac{(\partial g/\partial y) f^*}{(\partial g/\partial x) + (\partial g/\partial y) y'}$ must be satisfied by an extremizing function at the variable end-point x_i. Here too f^* has the form of Eq. (14).

(d) If f^* does not involve the dependent variable y explicitly, a first integral of the Euler-Lagrange equation (13) is $\partial f^*/\partial y' = c$, where c is an arbitrary constant.

(e) If f^* involves neither x nor y explicitly, and $y = y(x)$ is an extremizing function, then $y(x)$ must be linear in x.

(f) If f^* does not involve x explicitly, a first integral of Eq. (13) is $y'(\partial f^*/\partial y') - f^* = k$, where k is an arbitrary constant.

EXAMPLE 1

CURVE WITH THE LOWEST CENTER OF GRAVITY

Find the curve of prescribed length $2l$ which joins the points $(-a, b)$ and (a, b) and has its center of gravity as low as possible.

The y coordinate of the center of gravity of the required curve is given by

$$I = \frac{\int_{x=-a}^{x=a} y\, ds}{\int_{x=-a}^{x=a} ds} = \frac{1}{2l}\int_{-a}^{a} y\sqrt{1 + (y')^2}\, dx = \bar{y}$$

since we are given the constraint

$$J = \int_{x=-a}^{x=a} ds = \int_{-a}^{a} \sqrt{1 + (y')^2}\, dx = 2l$$

To minimize I, that is, $\bar{y}$, subject to this constraint, we introduce a Lagrange multiplier of the form $\lambda/2l$, thus obtaining

$$f^* = \frac{1}{2l} y\sqrt{1 + (y')^2} + \frac{\lambda}{2l}\sqrt{1 + (y')^2} = \frac{1}{2l}(y + \lambda)\sqrt{1 + (y')^2}$$

which does not involve x explicitly. Thus, according to list item **(f)**, a first integral of Eq. (13), in this case, is

$$\frac{(y+\lambda)(y')^2}{2l\sqrt{1+(y')^2}} - \frac{(y+\lambda)\sqrt{1+(y')^2}}{2l} = k \qquad \text{or} \qquad \frac{y+\lambda}{\sqrt{1+(y')^2}} = c$$

were $c = -2kl$. Separating variables in the last equation, then integrating, we get

$$x = c\int \frac{dy}{\sqrt{(y+\lambda)^2 - c^2}} \qquad \text{or} \qquad x = c\cosh^{-1}\frac{y+\lambda}{c} + h$$

Solved for y, the last equation becomes

$$y = c\cosh\frac{x-h}{c} - \lambda$$

This is a complete solution of the Euler-Lagrange equation on $[-a, a]$. The end conditions $y(-a) = b$, $y(a) = b$, will be satisfied by this solution if and only if

$$b + \lambda = c\cosh\frac{-a-h}{c} \qquad \text{and} \qquad b + \lambda = c\cosh\frac{a-h}{c}$$

that is to say, if and only if $(a+h)/c = (a-h)/c$. Hence $h = 0$, and so the required curve must be symmetric with respect to the y axis. Furthermore, we must have

$$\lambda = c\cosh\frac{a}{c} - b$$

To satisfy the prescribed length condition requires that

$$\int_{x=-a}^{x=a} ds = \int_{-a}^{a}\sqrt{1+\sinh^2\frac{x}{c}}\,dx = \int_{-a}^{a}\cosh\frac{x}{c}\,dx = 2l$$

whence

$$c\sinh\frac{x}{c}\Big|_{-a}^{a} = 2c\sinh\frac{a}{c} = 2l$$

When c has been found from this equation (by some approximate numerical procedure), the value of λ can be found and the equation of the required curve will be determined. In particular,

$$\lambda = c\cosh\frac{a}{c} - b = c\sqrt{1+\sinh^2\frac{a}{c}} - b = \sqrt{c^2+l^2} - b$$

and, finally,

$$y = c\cosh\frac{x}{c} + b - \sqrt{c^2+l^2}$$

Many applications give rise to isoperimetric problems in which several functions $x_1(t)$, $x_2(t), \ldots, x_n(t)$ are to be found for which the value of an integral

$$I = \int_{t_1}^{t_2} f(t, x_1, x_2, \ldots, x_n, \dot{x}_1, \dot{x}_2, \ldots, \dot{x}_n)\,dt \tag{15}$$

is stationary relative to functions which are twice differentiable on $[t_1, t_2]$ and which also satisfy m integral constraints

$$J_k = \int_{t_1}^{t_2} g_k(t, x_1, x_2, \ldots, x_n, \dot{x}_1, \dot{x}_2, \ldots, \dot{x}_n)\,dt \qquad J_k \text{ fixed} \tag{16}$$

Of course, suitable end conditions must also hold. In any particular case, the required functions, if they exist, must be components of a solution $[x_1(t), x_2(t), \ldots, x_n(t)]$ of the system of Euler-Lagrange equations

$$\frac{\partial f^*}{\partial x_i} - \frac{d}{dt}\left(\frac{\partial f^*}{\partial \dot{x}_i}\right) = 0 \qquad i = 1, 2, \ldots, n \tag{17}$$

where

$$f^* = f + \sum_{k=1}^{m} \lambda_k g_k \tag{18}$$

and $f, g_1, g_2, \ldots, g_m$ denote the respective integrands of the integral being rendered stationary and the integrals of constraint. If f^* is explicitly independent of t, a first integral of system (17) is

$$\sum_{i=1}^{n} \dot{x}_i \frac{\partial f^*}{\partial \dot{x}_i} - f^* = k \qquad k \text{ an arbitrary constant} \tag{19}$$

The constraints placed on an integral (15) need not be integral constraints. Quite often they will occur as differential or *finite* equations

$$g_k(t, x_1, x_2, \ldots, x_n, \dot{x}_1, \dot{x}_2, \ldots, \dot{x}_n) = 0 \qquad k = 1, 2, \ldots, m \qquad m < n \tag{20}$$

containing no integrals at all. (If g_k is represented by an expression which contains none of the derivatives $\dot{x}_1, \dot{x}_2, \ldots, \dot{x}_n$ explicitly, the equation $g_k = 0$ is said to be **finite.**) Variational problems involving an integral (15) and constraints given by (20) can be handled by choosing suitable comparison functions and appropriately modifying the reasoning underlying this and the previous two sections.†

The principal result concerning such problems is this: Let $x_1(t), x_2(t), \ldots, x_n(t)$ be functions for which the integral I of Eq. (15) has a stationary value relative to functions which are twice-differentiable on $[t_1, t_2]$, which take on prescribed values at t_1 and t_2, and which satisfy m consistent and independent differential or finite constraints (20). Then, on $[t_1, t_2]$, the vector function $[x_1(t), x_2(t), \ldots, x_n(t)]$ is a solution of the system of Euler-Lagrange equations (17) except that f^*, instead of being given by Eq. (18) in which the Lagrange multipliers are constants, is given by

$$f^* = f + \sum_{k=1}^{m} \mu_k(t) g_k \tag{21}$$

where, in general, the functions μ_k will not be constant on $[t_1, t_2]$ and the g_k are the functions of Eqs. (20). The restriction $m < n$ is essential here but not to an isoperimetric problem.

EXAMPLE 2

CONSTRAINED MOTION OF A PARTICLE ON A SPHERE

A unit mass particle moves on the surface of the sphere $x^2 + y^2 + z^2 = 1$ from the point $(0, 0, 1)$ to the point $(0, 0, -1)$ in time T. The motion takes place in such a way that the integral of the kinetic energy of the particle $I = \int_0^T \frac{1}{2}[\dot{x}^2 + \dot{y}^2 + \dot{z}^2]\, dt$ is a minimum. Discuss the motion of the particle and find the minimum value of I.

†See Robert Weinstock, *Calculus of Variations,* pp. 57–61, McGraw-Hill, New York, 1952.

Clearly I is to be minimized subject to the finite constraint $g(x, y, z) = x^2 + y^2 + z^2 - 1 = 0$. Writing f^* as $f^* = \frac{1}{2}[\dot{x}^2 + \dot{y}^2 + \dot{z}^2] - [\mu(t)/2][x^2 + y^2 + z^2 - 1]$, we easily find that the related system of Euler-Lagrange equations simplifies to

(22) $$\ddot{x} + \mu x = 0 \qquad \ddot{y} + \mu y = 0 \qquad \ddot{z} + \mu z = 0$$

Observing that

(23) $$\frac{\ddot{x}}{x} = \frac{\ddot{y}}{y} = \frac{\ddot{z}}{z} = -\mu$$

and eliminating μ, we get

(24) $$\ddot{x}y - \ddot{y}x = 0 \qquad \ddot{x}z - \ddot{z}x = 0 \qquad \ddot{y}z - \ddot{z}y = 0$$

Integrating these equations, we have

(25*a*) $$\dot{x}y - \dot{y}x = a$$
(25*b*) $$\dot{x}z - \dot{z}x = b$$
(25*c*) $$\dot{y}z - \dot{z}y = c$$

where a, b, and c are arbitrary constants.

Now let us use the relations

$$x = \sqrt{1 - z^2}\cos\theta \qquad y = \sqrt{1 - z^2}\sin\theta \qquad z = z$$

to introduce cylindrical type coordinates into the first of Eqs. (25). This gives $(z^2 - 1)\dot{\theta} = a$. Since $z(0) = 1$, $a = 0$. Hence, $\dot{\theta} = 0$ and $\theta = \alpha$, α constant. This proves that the particle travels along half of a great circle from $(0, 0, 1)$ to $(0, 0, -1)$.

From the fact that $r = 1$ and $\theta = \alpha$, it is clear that the equations

(26) $$x = \cos\alpha \sin\phi \qquad y = \sin\alpha \sin\phi \qquad z = \cos\phi$$

express x, y, and z in terms of the spherical coordinate ϕ. Recall that $0 \le \phi \le \pi$. Substituting from Eqs. (26) into Eqs. (25*b*) and (25*c*), we obtain $\dot{\phi}\cos\alpha = b$ and $\dot{\phi}\sin\alpha = c$, respectively. Thus we see that $\dot{\phi} = k$ and $\phi = kt + h$, where h and k are constants. Because the particle is initially at $(0, 0, 1)$, $\phi(0) = 0$. Therefore $h = 0$, $\phi(t) = kt$, and Eqs. (26) become

(27) $$x = \cos\alpha \sin kt \qquad y = \sin\alpha \sin kt \qquad z = \cos kt$$

These equations, together with Eqs. (23), imply that $\mu(t) = k^2$ is actually a constant function in this example.

To determine k, we note that $\phi(T) = kT = \pi$, or $k = \pi/T$, because the particle arrives at $(0, 0, -1)$ when $t = T$. The position of the particle, for $0 \le t \le T$, is therefore given by

(28) $$(x, y, z) = \left(\cos\alpha \sin\frac{\pi t}{T}, \sin\alpha \sin\frac{\pi t}{T}, \cos\frac{\pi t}{T}\right)$$

from which we find its kinetic energy to be $\pi^2/2T^2$. Hence, $I_{\min} = \int_0^T (\pi^2/2T^2)\,dt = \pi^2/2T$.

EXERCISES

1. Find the twice-differentiable function y for which $y(0) = 0$, $y(1) = \frac{1}{2}$, the integral $\int_0^1 [(y')^2 + 2y]\,dx$ is rendered stationary, and $\int_0^1 2y\,dx = \frac{1}{6}$.
2. Find the twice-differentiable function y for which $y(0) = 0$, $y(2)$ is unspecified, the integral $\int_0^2 [(y')^2 + 4y]\,dx$ is stationary, and $\int_0^2 y\,dx = 16$.
3. Find an equation for the curve which joins the points $(0, 1)$ and $(1, 2)$, bounds with the vertical line through

$x = 1$ and the coordinate axes a unit area, and has the average value of the square of its slope a minimum.

4. A function y is to render the integral $I = \int_{x_1}^{x_2} [r(x)(y')^2 - q(x)y^2]\, dx$ stationary subject to the constraint $\int_{x_1}^{x_2} p(x)y^2\, dx = 1$. Find a differential equation y must satisfy and a natural boundary condition at x_1.

5. What is the curve of given length that joins the fixed points $(x_1, 0)$ and $(x_2, 0)$ of the positive x axis and cuts off from the first quadrant the maximum area?

6. What is the curve of minimum length that passes through the fixed points $(x_1, 0)$ and $(x_2, 0)$ of the positive x axis and cuts off from the first quadrant a given area?

7. What is the curve of given length that joins the points $(-a, b)$ and (a, b) and generates the minimum surface area when it is revolved about the x axis?

8. Determine the curve that joins the points $(0, y_1)$ and (x_2, y_2), bounds with the vertical line through $x = x_2$ and the coordinate axes a prescribed area A, and has the average value of the square of its slope as small as possible.

9. **(a)** Find all twice-differentiable functions y that satisfy the conditions $y(0) = y(\pi) = 0$ and which render the integral $I = \int_0^{\pi} (y')^2\, dx$ stationary subject to the constraint $1 = \int_0^{\pi} y^2\, dx$.

(b) Rework Part **(a)** with the constraint removed.

10. **(a)** Find all twice-differentiable functions y that satisfy the conditions $y(-\pi) = y(\pi) = 0$ and which render the integral $I = \int_{-\pi}^{\pi} [1 + (y')^2]\, dx$ stationary subject to the constraint $\pi = \int_{-\pi}^{\pi} y^2\, dx$.

(b) Rework Part **(a)** with the constraint removed.

11. For b positive but otherwise unspecified, find the curve of length $l > b$, with endpoints $(0, 0)$ and $(0, b)$, which cuts off from the first quadrant the maximum area A_M, and evaluate A_M.

12. **(a)** Determine the curve of minimum length which originates on the positive y axis, ends at $(1, 0)$, and cuts off from the first quadrant a prescribed area $A > 0$.

(b) Identify the curve of Part **(a)** for which $A = \pi/4$.

13. Let S be the set of all twice-differentiable functions y on $[0, b]$ such that $y(0) = 0$, $y(b) = 0$, and $y(x) \geq 0$ throughout $[0, b]$. The positive number b is unspecified and each curve defined by a member of S is of length l.

(a) Formulate an isoperimetric problem for determining the curve defined by a function of S which when revolved about the x axis generates a surface of minimum area.

(b) Solve the related Euler-Lagrange equation.

(c) Determine $y'(b)$.

(d) Is the isoperimetric problem solvable? Explain.

14. Each curve of a family C is of length l, has $(0, 1)$ and $(1, 2)$ as endpoints, and is defined by a twice-differentiable function y such that $y(x) \geq 0$ on $[0, 1]$. The lines $x = 0$, $y = 0$, $x = 1$, and each curve of C bound a fixed area A.

(a) Formulate an isoperimetric problem for determining the member of C which when revolved about the x axis generates a surface of minimum area.

(b) Write a first integral of the corresponding Euler-Lagrange equation.

(c) Solve the first-order equation of Part **(b)** by expressing x in terms of an integral.

15. A uniform flexible cable of weight w per unit length and of length l hangs in a vertical plane under the influence of gravity with its ends fixed at the points (x_1, y_1) and (x_2, y_2), $x_1 < x_2$. The configuration assumed by the cable is defined by a function $y = y(x)$.

(a) Find the potential energy V of the cable relative to the x axis.

(b) Apply the principle that the value of V is a minimum when the cable is in stable equilibrium to determine the shape of the hanging cable.

16. A perfectly flexible cable of uniform linear density w and length l hangs at rest in stable equilibrium so that its potential energy is minimal. The ends of the cable are secured at $(-1, 0)$ and $(1, 0)$. Find an equation which defines the plane curve in which the cable hangs.

17. Work Exercise 16 with the ends of the cable fixed at $(-1, -y_1)$ and $(1, y_1)$, where $y_1 > 0$.

18. A perfectly flexible cable of length l hangs in stable equilibrium in a vertical plane between two fixed points (x_1, y_1) and (x_2, y_2). The distribution of its mass is *uniform with respect to the horizontal.* Determine the shape of the curve in which the cable hangs.

19. A unit mass particle moves along the circle $x^2 + y^2 = 1$ from the point $(0, 1)$ to the point $(1, 0)$ in time T and in such a way that the integral of the kinetic energy of the particle $I = \int_0^T \frac{1}{2}[\dot{x}^2 + \dot{y}^2]\, dt$ is a minimum. Determine x and y as functions of t and the minimum value of I.

20. During a given time interval $t_1 \leq t \leq t_2$ a rocket is to traverse an optimum trajectory corresponding to prescribed initial and final velocity conditions, while undergoing a specified change of altitude.

(a) If the vector function (x, y) denotes the position of the rocket, formulate an isoperimetric problem for the determination of $X = \dot{x}$ and $Y = \dot{y} + gt$.

(b) Show that the Euler-Lagrange equations of this problem are those which apply when no change of altitude is prescribed, but that the families of optimum trajectories in the two cases are in general different.

21. A particle is attracted toward the origin by a force whose magnitude is proportional to the distance from the origin. Assuming a coefficient of friction μ between the particle and its path, determine the path of a given length which joins the points (x_1, y_1) and (x_2, y_2) and along which the particle will move with the minimum work done against the force of friction.

22. Of all smooth closed non-self-intersecting plane curves having a given length l, determine one which encloses maximum area. *Hint:* Define a typical curve parametrically and use Exercise 41, Sec. 15.6, to obtain the formula $A = \frac{1}{2}\int_{t_1}^{t_2} [x\dot{y} - y\dot{x}]\, dt$ for the area inside a closed curve.

23. If $x = x(t)$, $y = y(t)$, $z = z(t)$ are parametric equations of a geodesic on a surface $G(x, y, z) = 0$, and if $f = \sqrt{\dot{x}^2 + \dot{y}^2 + \dot{z}^2}$, show that

$$\frac{\frac{d}{dt}\left(\frac{\dot{x}}{f}\right)}{\partial G/\partial x} = \frac{\frac{d}{dt}\left(\frac{\dot{y}}{f}\right)}{\partial G/\partial y} = \frac{\frac{d}{dt}\left(\frac{\dot{z}}{f}\right)}{\partial G/\partial z}$$

24. Make use of Exercise 23 to show that a geodesic on a sphere must be an arc of a great circle.

25. A particle moves without friction on the surface $\phi(x, y, z) = 0$ from the point (x_1, y_1, z_1) to the point (x_2, y_2, z_2) in the time T. If the motion takes place in such a way that the average value of the kinetic energy of the particle is a minimum, show that the parametric equations of its path on the surface satisfy the equations

$$\frac{\ddot{x}}{\phi_x} = \frac{\ddot{y}}{\phi_y} = \frac{\ddot{z}}{\phi_z}$$

26. A particle of mass m moves on the ellipsoid of revolution $x^2 + y^2 + a^2z^2 = a^2$ from $(0, 0, 1)$ to $(0, 0, -1)$ in time T, and in such a way that the integral of its kinetic energy over the interval $[0, T]$ is a minimum.

(a) Prove that the motion takes place in a plane.

(b) Find an ordinary differential equation with related auxiliary conditoins for the determination of z.

27. Find equations of the path of a particle moving from $(0, 1, 1)$ to $(1, 0, -1)$ on the cone $x^2 + y^2 = z^2$ if the average kinetic energy of the particle is a minimum. *Hint:* Eliminate the constraint by expressing the kinetic energy in terms of the polar coordinates $r = |z|$ and θ.

28. Work Exercise 27 with the surface being the right circular cylinder $x^2 + y^2 = 1$.

29. Work Exercise 27 with the endpoints of the path now $(0, 2, 2)$ and $(\sqrt{3}, 0, \sqrt{3})$.

30. Under suitable continuity and endpoint conditions, derive the Euler-Lagrange equations for $\int_{x_1}^{x_2} f(x, y, y', y'')\,dx$ by extremizing $\int_{x_1}^{x_2} f(x, y, z, z')\,dx$ subject to the differential constraint $z - y' = 0$.

16.4 STURM-LIOUVILLE PROBLEMS

Not only do variational methods serve to unify the treatment of many applied problems, they also provide an especially effective means of studying differential equations. We have repeatedly observed that the problem of extremizing an integral leads to one or more differential equations. Let us now turn things around and see how a boundary-value problem can be related to a variational problem. Although this cannot always be done, it is possible in some very important cases.

By way of illustration, we start with the general second-order Sturm-Liouville equation

$$[r(x)y']' + [q(x) + \lambda p(x)]y = 0 \tag{1}$$

where the functions r, p, and q are continuous, $r(x) \neq 0$, and $p(x) > 0$ on a fundamental interval $[a, b]$. Multiplying Eq. (1) through by y, then integrating from a to b, and subsequently solving algebraically for λ, we obtain

$$\lambda = \frac{-\int_a^b \{q(x)y^2 + y[r(x)y']'\}\,dx}{\int_a^b p(x)y^2\,dx} = \frac{I}{J} \tag{2}$$

Now, if y is any solution of Eq. (1), cy is a solution also. In fact, since $p(x) > 0$ on $[a, b]$, every solution of Eq. (1) is expressible in terms of solutions which satisfy the normalization condition

$$J = \int_a^b p(x)y^2\,dx = 1 \tag{3}$$

Imposing this constraint, the last equation in (2) becomes

$$-\int_a^b \{q(x)y^2 + y[r(x)y']'\}\,dx = I$$

or, integrating $y[r(x)y']'$ by parts,

$$I = \int_a^b [r(x)(y')^2 - q(x)y^2]\,dx + r(a)y(a)y'(a) - r(b)y(b)y'(b) \tag{4}$$

For a Sturm-Liouville problem consisting of Eq. (1) and a pair of *fixed* endpoint conditions

$$y(a) = 0 \tag{5a}$$
$$y(b) = 0 \tag{5b}$$

the last two terms in Eq. (4) vanish, leaving

$$I = \int_a^b [r(x)(y')^2 - q(x)y^2]\,dx \tag{6}$$

The problem of rendering this integral stationary, subject to the constraint (3), and relative to functions which vanish at $x = a$ and at $x = b$, is clearly an isoperimetric problem. Its corresponding Euler-Lagrange equation is

$$\frac{\partial f^*}{\partial y} - \frac{d}{dx}\left(\frac{\partial f^*}{\partial y'}\right) = 0 \qquad \text{where } f^* = [r(x)(y')^2 - q(x)y^2] - \lambda p(x)y^2$$

It is readily verified that these relations yield Eq. (1). Thus the functions which solve our isoperimetric problem must be normalized solutions of the Sturm-Liouville problem consisting of Eqs. (1) and (5), the normalization being with respect to p. Of course, these solutions are just the normalized characteristic functions $y_1, y_2, \ldots, y_n, \ldots$, which correspond to the characteristic values of the Sturm-Liouville problem.

With the characteristic values $\lambda_1, \lambda_2, \ldots, \lambda_n, \ldots$, arranged in increasing order, it is possible to show that λ_k is the minimum of the integral (6) relative to suitably differentiable functions y which satisfy Eqs. (5), (3), and the $k - 1$ orthogonality relations

$$\int_a^b p(x)y_i y\,dx = 0 \qquad i = 1, 2, \ldots, k-1\dagger$$

That I of Eq. (6) takes on the value of λ_k when y is replaced by y_k can be established as follows. First, replace y by y_k in the integrand of I. Then integrate the term $r(x)(y_k')^2$ by parts, observing that $y_k(a) = y_k(b) = 0$. The result of all this is

$$I = -\int_a^b y_k\{[r(x)y_k']' + q(x)y_k\}\,dx$$

Since $[r(x)y_k']' + q(x)y_k = -\lambda_k p(x)y_k$ and $\int_a^b p(x)y_k^2\,dx = 1$, we have $I = \lambda_k$ as was to be shown.

A Sturm-Liouville problem consisting of Eq. (1) and two *free* endpoint conditions

$$a_1 r(a)y'(a) + a_2 y(a) = 0 \tag{7a}$$
$$b_1 r(b)y'(b) + b_2 y(b) = 0 \tag{7b}$$
$$a_1 b_1 \neq 0$$

†See Robert Weinstock, *Calculus of Variations,* Chaps. 7 and 8, McGraw-Hill, New York, 1952.

can also be related to an isoperimetric problem. In this case, we may set $h = a_2/a_1$ and $k = b_2/b_1$ and utilize Eqs. (7) to write Eq. (4) as

(8) $$I = \int_a^b [r(x)(y')^2 - q(x)y^2]\,dx + ky^2(b) - hy^2(a)$$

The last two terms of this equation can be incorporated into the integral by introducing any continuously differentiable function $g(x)$ on $[a, b]$ for which $g(a) = h$ and $g(b) = k$. Indeed, this enables us to express Eq. (8) as

(9) $$I = \int_a^b \left[r(x)(y')^2 - q(x)y^2 + \frac{d}{dx}(gy^2)\right] dx$$

The isoperimetric problem requiring that this integral be rendered stationary, subject to Eq. (3), and relative to continuously differentiable functions defined on $[a, b]$, also has Eq. (1) as its Euler-Lagrange equation because $(d/dx)[gy^2]$ is an exact derivative. This can, of course, be verified directly. Hence the choice of g has no effect on the stationary values of I. Since no conditions have been imposed on the comparison functions at the endpoints of the specified interval $[a, b]$, a natural boundary condition

$$\frac{\partial f^*}{\partial y'} = 0 \qquad \text{where } f^* = \left[r(x)(y')^2 - q(x)y^2 + \frac{d}{dx}(gy^2)\right] - \lambda p(x)y^2$$

must hold at $x = a$ and at $x = b$. Differentiating f^* partially with respect to y', we find

$$\begin{aligned}\frac{\partial f^*}{\partial y'} &= \frac{\partial}{\partial y'}\{[r(x)(y')^2 - q(x)y^2 + g'y^2 + 2gyy'] - \lambda p(x)y^2\} \\ &= 2r(x)y' + 2gy\end{aligned}$$

and so the condition $r(x)y' + gy = 0$ must hold at each end of $[a, b]$. Substituting $x = a$ and $x = b$, in turn, into this equation we get Eqs. (7). The choice of g is again of no consequence so long as $g(a) = a_2/a_1$ and $g(b) = b_2/b_1$.

In many practical applications a variational formulation of the problem being investigated is easy to derive. To illustrate, we shall find an isoperimetric problem which when solved yields the natural frequencies of a string of weight per unit length $w(x)$ stretched under a tension T between $x = 0$ and $x = l$ and vibrating transversely about a horizontal x axis in the xy plane.

If the change in length of the string during its motion is so small that T can be assumed constant, the potential energy stored in the string by virtue of the work done against T is equal to T times the change of length in the string:

$$T\int_0^l [\sqrt{1 + (y')^2} - 1]\,dx$$

If, further, the vibrations are such that $|y'| \ll 1$, then by expanding $\sqrt{1 + (y')^2}$ by the binomial expansion and retaining only the dominant term in the integrand, we have

$$\frac{T}{2}\int_0^l (y')^2\,dx$$

as the expression for the potential energy stored in the string by virtue of its elongation. Should the string also be subjected to a distributed force of magnitude per unit length in the x direction $|f(x)y|$

acting vertically, for example, an elastic restoring force $-ky$, then the string has additional potential energy given by

$$-\int_0^l \int_0^y f(x)s\,ds\,dx = -\int_0^l \frac{1}{2} f(x)y^2\,dx$$

which is the work required, due to the distributed force, to change the deflection curve of the string from the segment $[0, l]$ of the x axis into the curve determined by y. Thus the total potential energy of the string is

$$\frac{1}{2}\int_0^l [T(y')^2 - f(x)y^2]\,dx$$

Similarly, the total instantaneous kinetic energy of the string is

$$\frac{1}{2}\int_0^l \frac{w(x)}{g}(\dot{y})^2\,dx$$

The displacement y which we have been considering is actually a function of x and t of the form $y = X(x)\cos\omega t$. Hence, substituting into the two energy expressions, and applying the principle that during free vibrations the maximum value of the kinetic energy is equal to the maximum value of the potential energy, we obtain

$$\omega^2 = \frac{\int_0^l [T(X')^2 - f(x)X^2]\,dx}{\int_0^l [w(x)/g]X^2\,dx}$$

This equation is satisfied by a nontrivial function X if and only if it is satisfied by cX, where c is a nonzero parameter. Thus we need consider only functions X which are normal with respect to $w(x)/g$.

The problem of extremizing the integral

$$I = \int_0^l [T(X')^2 - f(x)X^2]\,dx$$

subject to the constraint

$$J = \int_0^l [w(x)/g]X^2\,dx = 1$$

and relative to appropriate comparison functions, is obviously an example of an isoperimetric problem like those just discussed. With

$$f^* = [T(X')^2 - f(x)X^2] - \omega^2[w(x)/g]X^2$$

we find, in the usual way, that the corresponding Euler-Lagrange equation is

$$[TX']' + [f(x) + \omega^2\{w(x)/g\}]X = 0 \tag{10}$$

This is precisely the equation for the space factor X which results when the partial differential equation governing the vibrations of a nonuniform string [Eq. (2), Sec. 11.2] is solved by the method of separation of variables.

If both ends of the string are fixed on the x axis, so that $y(0, t) = y(l, t) = 0$ for all t, then all comparison functions of the isoperimetric problem, and in particular the solutions of Eq. (10), must satisfy the fixed-end conditions $X(0) = 0$ and $X(l) = 0$.

On the other hand, if at both ends of the string the displacement y is unspecified, then all comparison functions, and in particular the solutions of Eq. (10), must satisfy natural boundary conditions of the form

$$\frac{\partial f^*}{\partial X'} = 2TX' = 0 \qquad \text{at } x = 0 \text{ and at } x = l$$

Since $2T \neq 0$, these reduce to the free-end conditions

$$X'(0) = 0 \qquad \text{and} \qquad X'(l) = 0$$

More general free-end conditions of the type

(11) $$TX'(0) + hX(0) = 0 \qquad \text{and} \qquad TX'(l) + kX(l) = 0$$

might also apply. Such conditions arise, for instance, when each end of the string is restrained by a nonzero restoring force ϕ proportional to the displacement, say $\phi(0) = ay_0$ and $\phi(l) = by_l$. The potential energy stored in the system by virtue of the work done against these forces is

$$\tfrac{1}{2}ay_0^2 + \tfrac{1}{2}by_l^2$$

and hence the total potential energy of the system is now

$$\frac{1}{2}\int_0^l [T(y')^2 - f(x)y^2]\,dx + \tfrac{1}{2}ay_0^2 + \tfrac{1}{2}by_l^2$$

To incorporate the last two terms into the integral, let $g(x)$ be an arbitrary differentiable function of x such that $g(0) = -a$ and $g(l) = b$. Then

$$\int_0^l \frac{d}{dx}[g(x)y^2]\,dx = g(x)y^2\Big|_0^l = g(l)y^2(l) - g(0)y^2(0)$$
$$= by_l^2 + ay_0^2$$

Hence the expression for the instantaneous potential energy of the string can be rewritten in the form

$$\frac{1}{2}\int_0^l \left\{T(y')^2 - f(x)y^2 + \frac{d}{dx}[g(x)y^2]\right\} dx$$

Again setting $y = X(x)\cos\omega t$ and equating the maximum values of the potential energy and the kinetic energy, we are led to an isoperimetric problem in which

$$I = \int_0^l \left\{T(X')^2 - f(x)X^2 + \frac{d}{dx}[g(x)X^2]\right\} dx$$

while J is unchanged. With

$$f^* = \{T(X')^2 - f(x)X^2 + g'(x)X^2 + 2g(x)XX'\} - \omega^2[w(x)/g]X^2$$

we find that the natural boundary condition $\partial f^*/\partial X' = 0$ gives

$$TX' + gX = 0$$

which, when evaluated at $x = 0$ and at $x = l$, yields conditions on X of the general form shown in Eq. (11).

Let us finally consider a general nonhomogeneous second-order linear differential equation $a_0(x)y'' + a_1(x)y' + a_2(x)y = \phi(x)$ which is normal on an interval $[a, b]$. From Exercise 29, Sec. 2.12, we know that all such equations can be written in the form

$$[r(x)y']' + q(x)y = w(x) \tag{12}$$

This equation will be the Euler-Lagrange equation for an integral of the type $I = \int_a^b f(x, y, y')\,dx$ if $\partial f/\partial y' = r(x)y'$ and $\partial f/\partial y = w(x) - q(x)y$. From the last two relations we get

$$f = \tfrac{1}{2}r(x)(y')^2 + u(x, y) \qquad \text{and} \qquad f = w(x)y - \tfrac{1}{2}q(x)y^2 + v(x, y')$$

Sufficient conditions for these two representations of f to be identical are that

$$u(x, y) = w(x)y - \tfrac{1}{2}q(x)y^2 \qquad \text{and} \qquad v(x, y') = \tfrac{1}{2}r(x)(y')^2$$

in which case I becomes the **variational integral**

$$I = \int_a^b [\tfrac{1}{2}r(x)(y')^2 - \tfrac{1}{2}q(x)y^2 + w(x)y]\,dx \tag{13}$$

It is a simple matter to verify that Eq. (12) is the Euler-Lagrange equation for this integral. In fact, if an exact derivative $(d/dx)[g(x, y)]$ is added to the integrand of Eq. (13), Eq. (12) is still the corresponding Euler-Lagrange equation.

By defining f^* as usual, with $-\lambda/2$ as the Lagrange multiplier, it is easy to show that the isoperimetric problem involving the integral (13) and an integral constraint

$$J = \int_a^b p(x)y^2\,dx \qquad J \text{ a fixed constant and } p(x) \not\equiv 0 \text{ on } [a, b] \tag{14}$$

has as its Euler-Lagrange equation

$$[ry']' + [q(x) + \lambda p(x)]y = w(x) \tag{15}$$

Of course, Eqs. (12) and (15) will be accompanied by prescribed or natural boundary conditions, depending on how the comparison functions of the related variational problems behave at the endpoints of $[a, b]$.

EXERCISES

Convert each of the following boundary-value problems into a variational problem.

1. $y'' - y = 0$; $y'(-1) = 0$, $y(0) = \cosh 1$
2. $y'' - 2y' + 2y = 0$; $y(0) = 1$, $y(\pi/2) = 0$
3. $y'' + \lambda y = 0$; $y(0) = y(\pi) = 0$
4. $y'' + \lambda y = 0$; $y'(0) = y(\pi) = 0$
5. $xy'' + 2y' + 3y = 1$; $y'(1) = 0$, $y(3) = 1$
6. $xy'' + 2y' + y - 1 = 0$; $y(1) = 2$, $y(2) = 4$
7. $x^3y'' + 3x^2y' + y = x$; $y(1) = 1$, $y'(2) = 0$
8. $x^2y'' + xy' + (1 + \lambda x)y = x^2$; $2y'(2) + 3y(2) = 0$, $3y'(3) + 2y(3) = 0$
9. $x^2y'' + xy' + (\lambda^2x^2 - \nu^2)y = 0$; $x_1y'(x_1) = 0$, $ax_2y'(x_2) - by(x_2) = 0$
10. $y'' - y' + e^x[\cos x + \lambda \cosh x]y = e^x \sin x$; $y'(0) + 2y(0) = 0$, $y(1) = 5$
11. Write the boundary-value problem $[x^2y']' + x(y + 1) = 0$,

$[x^2y']_{x=0} = 0$, $y(1) = 0$ as a variational problem and find the function of the family $y = a(1 - x^2)$ which best approximates the minimizing function.

12. Work Exercise 11 using the family of functions $y = a(1 - x)^2$.

13. Convert the boundary-value problem $2x^2y'' + 6xy' + 2y = 1$, $y(1) = y'(2) = 0$ into a variational problem and find the function of the family $y = a(x^2 - 4x + 3)$ which best approximates the minimizing function.

14. Convert the boundary-value problem $xy'' + 2y' + y + 1 = 0$, $y(0) = 0$, $y'(1) = 0$ into a variational problem and find the quadratic function which best approximates the twice-differentiable function that renders the related variational integral stationary. Is the stationary value a minimum?

15. Convert the boundary-value problem $[(1 + x)y']' + \lambda y = 0$, $y(0) = y(1) = 0$ into a variational problem and approximate the smallest characteristic value by minimizing the related variational integral relative to normalized comparison functions of the form $y = ax(x - 1)$.

16. Work Exercise 15 for the boundary-value problem $y'' + \lambda^2 y = 0$, $y(0) = y(\pi) = 0$ using normalized comparison functions of the form $y = ax(\pi - x)$. How does the approximate value compare to the actual value of λ_1?

17. Work Exercise 16 with $y(0) = y'(\pi) = 0$ as the boundary conditions and using all normalized quadratic functions which satisfy these conditions as comparison functions.

18. **(a)** Find the characteristic values λ_k and the corresponding characteristic functions of the boundary-value problem $x^2y'' + xy' + 9\lambda^2 y = 0$, $y(1) = y(e) = 0$.

(b) Write the boundary-value problem as a Sturm-Liouville system and identify the related weight function $p(x)$.

(c) Normalize the characteristic functions of Part **(a)** with respect to $p(x)$.

(d) Convert the boundary-value problem into a variational problem and directly verify that, for each normalized characteristic function y_k, the value of the variational integral I is $I(y_k) = \lambda_k^2$.

(e) Using only appropriate quadratic functions as comparison functions, obtain an approximate value Λ_1 for λ_1.

(f) If the value of y is unspecified at $x = e$, what boundary condition applies there?

19. In Exercise 16, approximate the first two characteristic values of λ using comparison functions of the form $y = ax(\pi - x) + bx^2(\pi - x)$ in the variational problem.†

20. How can the principle which says that the maximum kinetic energy is equal to the maximum potential energy be established for a freely vibrating string?

21. A function y is to render $\int_a^b [p(x)(y')^2 - q(x)y^2]\, dx$ stationary and at the same time

$$\int_a^b r(x)y^2\, dx = 1$$

is to hold. Find a differential equation which y must satisfy and a natural boundary condition at $x = a$.

22. An integral $I = \int_{x_1}^{x_2} f(x, y, y', y'')\, dx$ is rendered stationary by a function $y = y(x)$ which satisfies a constraint $J = \int_{x_1}^{x_2} g(x, y, y', y'')\, dx$, where J is a fixed constant.

(a) Show that, if the comparison functions, and their derivatives, take on arbitrary values at $x = x_i$, $i = 1$ or 2, then at x_i the function $y(x)$ must satisfy the **natural boundary conditions**

$$\frac{\partial f^*}{\partial y''} = 0 \qquad \text{and} \qquad \frac{\partial f^*}{\partial y'} - \frac{d}{dx}\left(\frac{\partial f^*}{\partial y''}\right) = 0$$

where $f^* = f + \lambda g$

(b) Which of the preceding conditions applies if, at x_i, all comparison functions must have the same value, but their derivatives have different values?

(c) Which of the natural boundary conditions applies if, at x_i, the comparison functions have different values, but their derivatives all have the same value?

23. **(a)** Formulate an isoperimetric problem for which

$$(py'')'' + (qy')' + (r + \lambda s)y = 0 \qquad x_1 \le x \le x_2$$

is the Euler-Lagrange equation.

(b) Describe how the comparison functions of the isoperimetric problem affect the boundary conditions which go with the Euler-Lagrange equation. *Hint:* Use the results of Exercise 22.

24. Is the boundary-value problem $y'' + y' + xe^{-x}y = 0$, $y(-1) = 2$, $y'(1) = -2$ satisfied by a function $y(x)$ that renders $I = \int_{-1}^{1} [e^x(y')^2 - xy^2]\, dx$ stationary relative to all twice-differentiable functions on $[-1, 1]$ which have the value 2 at $x = -1$?

25. Show that the characteristic values of a matrix equation $(\mathrm{A} - \lambda \mathrm{B})\mathbf{x} = \mathbf{0}$ are the stationary values of the quadratic form $\mathbf{x}^T\mathbf{A}\mathbf{x}$ subject to the constraint $\mathbf{x}^T\mathbf{B}\mathbf{x} = 1$.‡

---

†Exercises 15 to 17, 18(**e**), and (19) illustrate the so-called **Rayleigh-Ritz method** of approximating the characteristic values of a boundary-value problem by minimizing a related variational integral relative to a restricted family of normalized functions that satisfy the boundary conditions of the given problem together with all pertinent orthogonality conditions.

‡Several more results concerning matrix characteristic value problems and stationary values of quadratic forms, which closely resemble those of this section concerning, Sturm-Liouville and isoperimetric problems, are treated in C. Ray Wylie, *Advanced Engineering Mathematics,* 4th ed., pp. 600–607, McGraw-Hill, New York, 1975.

16.5 VARIATIONS

A real-valued function F whose domain is a set of real functions $\{y\}$ is sometimes called a **functional** or, more specifically, a **functional of a single independent variable.** Functionals of several independent variables are also of interest. With ordinary functions, the values of the independent variables are numbers. However, with functionals, the values of the independent variables are functions.

In general, the value of a function changes when the values of its independent variables change. The differential provides an estimate of the change experienced by an ordinary function. It is natural to ask how a change in the value of a functional can be estimated. To see how this can be done, let us consider a function $F(x, y, y')$ which, when x is held fixed, becomes a functional defined on a set of functions $\{y\}$, and let us develop an estimate for the change in F corresponding to an assigned change in the value $y(x)$ of a function y in $\{y\}$ *for a fixed value of* x.

If $y(x)$ is changed into

$$Y(x) = y(x) + \epsilon\eta(x) \qquad \epsilon \text{ independent of } x$$

we call the change $\epsilon\eta(x)$ the **variation of** y and denote it by δy, thus

$$\delta y = \epsilon\eta(x) \tag{1}$$

Moreover, from the changed value of y we infer that the changed value of $y'(x)$ is

$$Y'(x) = y'(x) + \epsilon\eta'(x)$$

Hence we have the companion formula

$$\delta y'(x) = \epsilon\eta'(x) \tag{2}$$

for the variation of $y'(x)$. Corresponding to these changes we have the change

$$\Delta F = F(x, y + \epsilon\eta, y' + \epsilon\eta') - F(x, y, y')$$

If we expand the first term on the right in a Maclaurin's expansion in powers of ϵ, we have

$$\Delta F = F(x, y, y') + \left(\frac{\partial F}{\partial y}\eta + \frac{\partial F}{\partial y'}\eta'\right)\epsilon + \left(\frac{\partial^2 F}{\partial y^2}\eta^2 + 2\frac{\partial^2 F}{\partial y\,\partial y'}\eta\eta' + \frac{\partial^2 F}{\partial y'^2}\eta'^2\right)\frac{\epsilon^2}{2!} + \cdots - F(x, y, y')$$

or, neglecting powers of ϵ higher than the first,

$$\Delta F \doteq \frac{\partial F}{\partial y}\eta\epsilon + \frac{\partial F}{\partial y'}\eta'\epsilon$$

$$= \frac{\partial F}{\partial y}\,\delta y + \frac{\partial F}{\partial y'}\,\delta y'$$

By analogy with the differential of a function, we define the last expression to be the **variation of the functional** F and denote it by δF. Accordingly,

$$\delta F = \frac{\partial F}{\partial y}\,\delta y + \frac{\partial F}{\partial y'}\,\delta y' \dagger \tag{3}$$

†By strict analogy with the differential of a function of three variables, we might have expected the definition

$$\delta F = \frac{\partial F}{\partial x}\,\delta x + \frac{\partial F}{\partial y}\,\delta y + \frac{\partial F}{\partial y'}\,\delta y'$$

However, we must remember that the functional is the value of $F(x, y, y')$ at a particular value of x, that is, x is not varied in the calculation of δF, and hence $\delta x = 0$.

In passing, we note that in its simplest form the differential of a function is a first-order approximation to the change in the function as x varies along a particular curve, whereas the variation of a functional is a first-order approximation to the change in the functional at a particular value of x as we vary from curve to curve.

It is interesting and important to note that variations can be calculated by the same rules that apply to differentials. Specifically,

(4) $$\delta(F_1 \pm F_2) = \delta F_1 \pm \delta F_2$$

(5) $$\delta(F_1 F_2) = F_1\,\delta F_2 + F_2\,\delta F_1$$

(6) $$\delta\left(\frac{F_1}{F_2}\right) = \frac{F_2\,\delta F_1 - F_1\,\delta F_2}{F_2^2}$$

(7) $$\delta(F^n) = nF^{n-1}\,\delta F$$

These relations are easy to prove by means of (3). For example, with F replaced by F_1F_2, (3) yields

$$\begin{aligned}\delta(F_1F_2) &= \frac{\partial}{\partial y}(F_1F_2)\,\delta y + \frac{\partial}{\partial y'}(F_1F_2)\,\delta y' \\ &= [F_1(\partial F_2/\partial y) + F_2(\partial F_1/\partial y)]\,\delta y + [F_1(\partial F_2/\partial y') + F_2(\partial F_1/\partial y')]\,\delta y' \\ &= F_1[(\partial F_2/\partial y)\,\delta y + (\partial F_2/\partial y')\,\delta y'] + F_2[(\partial F_1/\partial y)\,\delta y + (\partial F_1/\partial y')\,\delta y'] \\ &= F_1\,\delta F_2 + F_2\,\delta F_1\end{aligned}$$

as asserted by (5). Proofs of (4), (6), and (7) we leave as exercises.

From the definitive relations (1) and (2), and with $D = d/dx$, we have

$$\delta Dy = \delta y' = \epsilon\eta' = \epsilon D\eta = D\epsilon\eta = D\,\delta y$$

hence δ and D commute; i.e., taking the variation of a function $y(x)$ and differentiating it with respect to its *independent variable* are commutative operations.

We can, of course, consider functionals of more than one function, and the variations of such functionals are defined by expressions analogous to Eq. (3). For instance, for the functional $F(x, u, v, u', v')$ we have

(8) $$\delta F = \frac{\partial F}{\partial u}\,\delta u + \frac{\partial F}{\partial v}\,\delta v + \frac{\partial F}{\partial u'}\,\delta u' + \frac{\partial F}{\partial v'}\,\delta v'$$

Equally well, we may consider variations of functionals which depend on functions of more than one variable. For example, for the functional $F(x, y, u, u_x, u_y)$, whose value depends, for fixed x and y, on the function $u(x, y)$, we have

(9) $$\delta F = \frac{\partial F}{\partial u}\,\delta u + \frac{\partial F}{\partial u_x}\,\delta u_x + \frac{\partial F}{\partial u_y}\,\delta u_y$$

For a functional expressed as a definite integral, say the integral

$$I(y) = \int_a^b f(x, y, y')\,dx$$

of the kind we discussed in Sec. 16.1, we have, first of all

$$\Delta I = I(y + \epsilon\eta) - I(y)$$

If the limits of I do not depend on y, we have further

$$\Delta I = \int_a^b f(x, y + \epsilon\eta, y' + \epsilon\eta')\, dx - \int_a^b f(x, y, y')\, dx$$
$$= \int_a^b [f(x, y + \epsilon\eta, y' + \epsilon\eta') - f(x, y, y')]\, dx$$
$$= \int_a^b \Delta f(x, y, y')\, dx$$

The **variation of** I is now defined as the expression resulting when Δf in the last integral is replaced by the first-order approximation δf; that is,

$$\delta I = \int_a^b \delta f(x, y, y')\, dx \tag{10}$$

Writing (10) as

$$\delta \int_a^b f(x, y, y')\, dx = \int_a^b \delta f(x, y, y')\, dx$$

we see that integrating a function $f(x, y, y')$ over an interval $[a, b]$ and taking the variation of $f(x, y, y')$ are commutative operations; i.e., the operators symbolized by $\int_a^b (\quad)\, dx$ and δ commute.

From calculus we recall that a necessary condition for a function to have an extremum is that its *differential* vanish. We can now show, similarly, that a necessary condition for the functional I to have an extremum is that its *variation* vanish. In fact, using the results of the preceding discussion, we can write

$$\delta I = \int_a^b \delta f(x, y, y')\, dx$$
$$= \int_a^b (f_y\, \delta y + f_{y'}\, \delta y')\, dx$$
$$= \int_a^b \left[f_y\, \delta y + f_{y'} \frac{d(\delta y)}{dx}\right] dx$$

Now integrating the last term by parts, with $u = f_{y'}$ and $dv = [d(\delta y)/dx]\, dx$, we have

$$\int_a^b f_{y'} \frac{d(\delta y)}{dx}\, dx = f_{y'}\, \delta y \Big|_a^b - \int_a^b \frac{d(f_{y'})}{dx}\, \delta y\, dx$$

When we assume that at $x = a$ and $x = b$ either the variation $\delta y = \epsilon\eta(x)$ is zero, because $\eta(x)$ is, or a natural boundary condition holds so that $f_{y'}$ vanishes, it follows that the integrated portion of the last equation is equal to zero. Hence we have

$$\delta I = \int_a^b \left[f_y - \frac{d(f_{y'})}{dx}\right] \delta y\, dx$$

Since we have already seen that $f_y - d(f_{y'})/dx = 0$ is a necessary condition for an extremum of I, it follows that δI is also zero at any extremum of I. Conversely, since δy is an arbitrary variation in y, the condition $\delta I = 0$ implies that

$$f_y - \frac{d(f_{y'})}{dx} = 0$$

EXERCISES

With $\delta y = \epsilon\eta(x)$, find δF and ΔF for each of the following functionals.

1. $F(x, y, y') = xyy'$
2. $F(x, y, y') = y^3 + xy'$
3. $F(x, y, y') = x^3y^2y'$
4. $F(x, y, y') = 2e^y - 3x(y')^2$
5. $F(x, y, y') = xy'/y$
6. $F(x, y, y') = e^x + e^y + e^{y'}$
7. Given $F(x, y, y') = (y')^2 + xy$, compute δF and ΔF for $x = x_0$, $y = x^2$, and $\delta y = \epsilon x^n$. Then verifying δF using Eq. (3).
8. With $\delta u = \epsilon\xi(x)$ and $\delta v = \epsilon\eta(x)$, find δF and ΔF for the functional

$$F(x, u, v, u', v') = x(uv - u')$$

9. With $\delta y = 3\epsilon x$, find δI for $I(y) = \int_{-\pi}^{\pi} (2xy + y' \cos x)\, dx$.
10. **(a)** If both y and x are functions of the independent variable t, show that

$$\delta\left(\frac{dy}{dx}\right) \neq \frac{d(\delta y)}{dx}$$

by showing that the correct formula is

$$\delta\left(\frac{dy}{dx}\right) = \frac{d(\delta y)}{dx} - \frac{dy}{dx}\frac{d(\delta x)}{dx}$$

(b) Let $y = 1 + x^2$, where both y and x are functions of t. Calculate $\delta(dy/dx)$ and $d(\delta y)/dx$ when $x = t^3$ and $\delta x = \epsilon t^2$ and verify the formula of Part **(a)**.

11. Derive Formula (4) for $\delta(F_1 \pm F_2)$.
12. Derive Formula (6) for $\delta(F_1/F_2)$.
13. Derive Formula (7) for $\delta(F^n)$.
14. Derive Eq. (8).
15. Assuming that $\delta y = 0$ at $x = a$ and $x = b$, or that y satisfies the natural boundary condition $ry' = 0$ at a and at b, show that the variation of

$$\lambda = \frac{\int_a^b [r(y')^2 - qy^2]\, dx}{\int_a^b py^2\, dx} \quad \text{is given by}$$

$$\delta\lambda = -\frac{2}{J}\int_a^b [(ry')' + qy + \lambda py]\, \delta y\, dx$$

Then show that a function y which renders λ stationary must satisfy the Sturm-Liouville equation $(ry')' + (q + \lambda p)y = 0$ and an appropriate pair of boundary conditions. *Hint:* $J = \int_a^b py^2\, dx$.

16. Show that if $F(x, y, y') = e^{y+y'} f(x)$ then $\delta F = (\delta y + \delta y')F$.

16.6 HAMILTON'S PRINCIPLE AND LAGRANGE EQUATIONS OF MOTION

Although Newton's laws of motion form the fundamental foundation for the study of mechanical phenomena, there are refinements and extensions which often provide more effective methods of investigation. In this section we shall take a brief look at two of these, Hamilton's principle and Lagrange's equation(s).

Let us consider a particle of mass m acted upon by a force $\mathbf{F}$. Let $\mathbf{r}(t)$ be the vector from the origin to the instantaneous position of the particle. Then according to Newton's second law in vector form, the actual path of the particle is that for which

$$m\frac{d^2\mathbf{r}}{dt^2} - \mathbf{F} = \mathbf{0} \tag{1}$$

Now consider any other path joining the points where the particle is located at $t = t_1$ and at $t = t_2$. Such a path is, of course, described by a vector function $\mathbf{r} + \delta\mathbf{r}$, where

$$\delta\mathbf{r}\big|_{t_1} = \delta\mathbf{r}\big|_{t_2} = \mathbf{0}$$

(Fig. 16.3). If we now form the scalar product of (the vector) $\delta\mathbf{r}$ and the terms of Eq. (1), and integrate from t_1 to t_2, we obtain

$$\int_{t_1}^{t_2} (m\ddot{\mathbf{r}} \cdot \delta\mathbf{r} - \mathbf{F} \cdot \delta\mathbf{r})\, dt = 0 \tag{2}$$

Applying integration by parts to the first term in Eq. (2), with $\mathbf{u} = \delta\mathbf{r}$ and $d\mathbf{v} = \ddot{\mathbf{r}}\, dt$, we obtain

$$m\dot{\mathbf{r}} \cdot \delta\mathbf{r}\Big|_{t_1}^{t_2} - m\int_{t_1}^{t_2} \dot{\mathbf{r}} \cdot \delta\dot{\mathbf{r}}\, dt$$

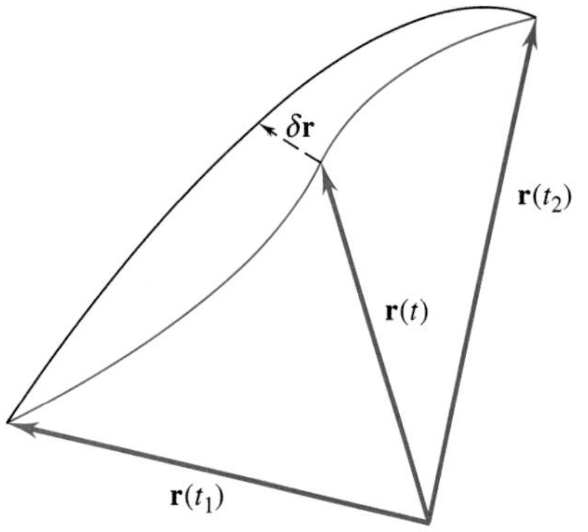

FIGURE 16.3
The actual path of a moving particle and a possible variation of that path.

The integrated term vanishes because of the properties of $\delta\mathbf{r}$. Moreover,

$$m\dot{\mathbf{r}} \cdot \delta\dot{\mathbf{r}} = \frac{m}{2}\,\delta(\dot{\mathbf{r}} \cdot \dot{\mathbf{r}}) = \delta\left(\frac{m}{2}v^2\right) = \delta T$$

where T is the kinetic energy of the moving particle. Hence Eq. (2) can be rewritten

$$\int_{t_1}^{t_2} (\delta T + \mathbf{F} \cdot \delta\mathbf{r})\,dt = 0 \tag{3}$$

This is **Hamilton's principle** in its general form, as applied to the motion of a single mass particle in a force field that can be either conservative or nonconservative.

If $\mathbf{F}$ is conservative, Hamilton's principle assumes an even simpler form. For if $\mathbf{F}$ is conservative, then there exists a scalar function $\phi(x, y, z)$ such that $\mathbf{F} = \nabla\phi$ (see Theorem 5, Sec. 15.7). The function ϕ is called the **potential function,** and $-\phi$ is (to within an additive constant) the potential energy of the particle in the field. Now

$$\mathbf{F} = \nabla\phi = \frac{\partial\phi}{\partial x}\,\mathbf{i} + \frac{\partial\phi}{\partial y}\,\mathbf{j} + \frac{\partial\phi}{\partial z}\,\mathbf{k}$$

and

$$\delta\mathbf{r} = \delta x\,\mathbf{i} + \delta y\,\mathbf{j} + \delta z\,\mathbf{k}$$

Hence

$$\mathbf{F} \cdot \delta\mathbf{r} = \frac{\partial\phi}{\partial x}\,\delta x + \frac{\partial\phi}{\partial y}\,\delta y + \frac{\partial\phi}{\partial z}\,\delta z = \delta\phi$$

and therefore the equation $\int_{t_1}^{t_2} (\delta T + \mathbf{F} \cdot \delta\mathbf{r})\,dt = 0$ can be rewritten

$$\int_{t_1}^{t_2} (\delta T + \delta\phi)\,dt = 0 \qquad \text{or} \qquad \delta\int_{t_1}^{t_2} (T + \phi)\,dt = 0$$

or, finally, since $\phi = -V$, where V is the potential energy of the system,

$$\delta\int_{t_1}^{t_2} (T - V)\,dt = 0 \tag{4}$$

This is *Hamilton's principle for a single mass particle in a conservative field.* The principle can, of course, be extended to a system of discrete particles by summation and to continuous systems by integration.

In many elementary problems, a dynamical system is described in terms of coordinates which are distances. Such a choice of coordinates is not always the best one, however. Sometimes problems involving geometric constraints of one kind or another can be solved more conveniently by choosing other quantities as coordinates, for instance, angles or even areas.

In the particular case of a system of p discrete particles, the effect of geometric constraints, assumed to be constant in time, is to reduce the number of variables required to specify completely the state of the system at any given instant. To see that a set of constraints actually has such an effect, observe that three rectangular coordinates suffice to determine the position vector (x_j, y_j, z_j) of each mass particle m_j, $j = 1, 2, \ldots, p$. If the constraints can be described by $k(<3p)$ consistent and independent equations

$$g_i(x_1, y_1, z_1, \ldots, x_p, y_p, z_p) = 0 \qquad i = 1, 2, \ldots, k \tag{5}$$

these k equations may be used, at least theoretically, to eliminate k of the position variables thus leaving only $3p - k$ independent coordinates.

As has been mentioned, the variables employed in a particular problem need not be rectangular coordinates. They may be any $3p - k = n$ variables $q_1, q_2, \ldots, q_n$ which are independent of one another, which are well suited to a mathematical investigation of the problem at hand, and in terms of which the positions of the p particles can be expressed by means of $3p$ equations

$$x_j = x_j(q_1, \ldots, q_n) \qquad y_j = y_j(q_1, \ldots, q_n) \qquad z_j = z_j(q_1, \ldots, q_n) \qquad j = 1, 2, \ldots, p \tag{6}$$

These $3p$ equations in effect ensure that the constriants (5) are all complied with.

Variables $q_1, q_2, \ldots, q_n$ of the kind just described are called **generalized coordinates.** A set of generalized coordinates for any particular mechanical system is by no means unique; however, each such set must contain the same number of variables.

Let us investigate the behavior of a system of p particles a little further. By differentiating Eqs. (6) with respect to t, and substituting the results

$$\dot{x}_j = \sum_{i=1}^{n} \frac{\partial x_j}{\partial q_i} \dot{q}_i \qquad \dot{y}_j = \sum_{i=1}^{n} \frac{\partial y_j}{\partial q_i} \dot{q}_i \qquad \dot{z}_j = \sum_{i=1}^{n} \frac{\partial z_j}{\partial q_i} \dot{q}_i \tag{7}$$

into the expression

$$T = \frac{1}{2} \sum_{j=1}^{p} m_j(\dot{x}_j^2 + \dot{y}_j^2 + \dot{z}_j^2) \tag{8}$$

we find that the kinetic energy T of the system can be written as a quadratic form

$$T = \dot{\mathbf{q}}^T \mathrm{A}\, \dot{\mathbf{q}} \tag{9}$$

where $\dot{\mathbf{q}}^T = [\dot{q}_1 \dot{q}_2 \cdots \dot{q}_n]$ and the symmetric matrix A is expressed solely in terms of $q_1, q_2, \ldots, q_n$ (Exercise 20). Of course, this implies that T is a homogeneous function of degree 2 in the **generalized velocity components** $\dot{q}_1, \dot{q}_2, \ldots, \dot{q}_n$.

Now in a conservative system, by definition, the potential energy V depends only on the positions of the particles. Hence $V = V(q_1, q_2, \ldots, q_n)$ must be a function of the generalized coordinates alone. The function $L = T - V$ is usually referred to as the **lagrangian** or the **kinetic potential.** Hamilton's principle, when extended to a conservative system of particles, may be stated as follows:

During an arbitrary time interval $[t_1, t_2]$, *the actual motion of a conservative system of particles whose lagrangian is* $T - V = L(q_1, q_2, \ldots, q_n, \dot{q}_1, \dot{q}_2, \ldots, \dot{q}_n)$ *is such that* **Hamilton's integral**

$$\int_{t_1}^{t_2} (T - V)\, dt = \int_{t_1}^{t_2} L\, dt \tag{10}$$

is rendered stationary relative to continuously twice-differentiable functions $q_1, q_2, \ldots, q_n$ *which take on prescribed values at* t_1 *and* t_2.

From this principle and the result embodied in Eqs. (6) and (7), Sec. 16.2, it follows that if $q_1, q_2, \ldots, q_n$ are generalized coordinates of a conservative system of particles they must satisfy the system of Euler-Lagrange equations

$$\frac{\partial L}{\partial q_i} - \frac{d}{dt}\left(\frac{\partial L}{\partial \dot{q}_i}\right) = 0 \qquad i = 1, 2, \ldots, n \tag{11}$$

These equations are known as **Lagrange's equations of motion,** or simply as **Lagrange's equations.**

Since the lagrangian L of a conservative system does not explicitly involve t, a first integral of System (11) is [see Eq. (10), Sec. 16.2]

$$\sum_{i=1}^{n} \dot{q}_i \frac{\partial L}{\partial \dot{q}_i} - L = E \qquad \text{where } E \text{ is a constant} \tag{12}$$

Recalling that V is not a function of the generalized velocity components, we have $\partial L/\partial \dot{q}_i = \partial(T - V)/\partial \dot{q}_i = \partial T/\partial \dot{q}_i$. The series in Eq. (12) thus becomes $\sum_{i=1}^{n} \dot{q}_i \dfrac{\partial T}{\partial \dot{q}_i}$. Using the fact that T is a homogeneous function of degree 2 in $\dot{q}_1, \dot{q}_2, \ldots, \dot{q}_n$, and applying Euler's homogeneous function theorem (Exercise 59, Sec. 1.10), we find that the sum of the preceding series is $2T$. We may therefore write Eq. (12) as

$$2T - (T - V) = T + V = E \tag{13}$$

This result shows that a conservative particle system, when in motion, must move in such a way that the sum of its kinetic and potential energies remains constant. The total energy E of the system is determined when the initial values of all the q_i and $\dot{q}_i$ are assigned.

As our earlier expression $\mathbf{F} \cdot \delta\mathbf{r} = \delta\phi$ for the potential energy of a single particle indicates, in a conservative system of discrete particles, the work done by the various forces when the generalized coordinates $\{q_i\}$ of the system undergo small changes $\{\delta q_i\}$ is

$$\begin{aligned}\delta\phi = -\delta V &= -\left(\frac{\partial V}{\partial q_1}\,\delta q_1 + \frac{\partial V}{\partial q_2}\,\delta q_2 + \cdots + \frac{\delta V}{\delta q_n}\,\delta q_n\right)\\ &= Q_1\,\delta q_1 + Q_2\,\delta q_2 + \cdots + Q_n\,\delta q_n\end{aligned}$$

where we have introduced the conventional symbol Q_i for $-\partial V/\partial q_i$. The typical term in this expression, $Q_i\,\partial q_i$, is the work done in a displacement in which ∂q_i is different from zero but all other δq's are equal to zero. Since q_i is not necessarily a distance, Q_i is not necessarily a force. Nonetheless, the Q's are referred to as **generalized forces.**

Using the relations $L = T - V$, $\partial V/\partial \dot{q}_i = 0$, and $\partial V/\partial q_i = -Q_i$, we find that System (11) can be written

$$\frac{d}{dt}\left(\frac{\partial T}{\partial \dot{q}_i}\right) - \frac{\partial T}{\partial q_i} = Q_i \qquad i = 1, 2, \ldots, n \tag{14}$$

In a nonconservative system, V as well as T may involve the generalized velocities and the generalized coordinates, in which case the relations $\partial V/\partial \dot{q}_i = 0$ no longer apply. Nevertheless, Eqs. (14) are still correct, although we shall not prove the fact, the only difference being that in a nonconservative system the generalized forces cannot be derived from a potential function.

EXAMPLE 1

TORSIONAL VIBRATIONS OF COUPLED DISKS

Using Lagrange's equations, derive the system of differential equations describing the torsional vibrations of the system of elastically coupled disks shown in Fig. 16.4.

Taking as coordinates the angles of twist θ_i of the respective disks, we have at once

$$T = \tfrac{1}{2}(I_1\dot{\theta}_1^2 + I_2\dot{\theta}_2^2 + \cdots + I_i\dot{\theta}_i^2 + \cdots + I_n\dot{\theta}_n^2)$$

Also, since the potential energy stored in a twisted shaft is

$$\tfrac{1}{2} \text{ modulus} \times (\text{angle of twist})^2$$

we have

$$V = \tfrac{1}{2}[k_0\theta_1^2 + k_1(\theta_1 - \theta_2)^2 + \cdots + k_{i-1}(\theta_{i-1} - \theta_i)^2 + k_i(\theta_i - \theta_{i+1})^2 + \cdots + k_{n-1}(\theta_{n-1} - \theta_n)^2 + k_n\theta_n^2]$$

Hence, Lagrange's equations are

$$\begin{aligned} I_1\ddot{\theta}_1 + k_0\theta_1 + k_1(\theta_1 - \theta_2) &= 0 \\ I_2\ddot{\theta}_2 + k_1(\theta_1 - \theta_2)(-1) + k_2(\theta_2 - \theta_3) &= 0 \\ \cdots\cdots\cdots\cdots\cdots\cdots\cdots\cdots\cdots\cdots\cdots\cdots& \\ I_i\ddot{\theta}_i + k_{i-1}(\theta_{i-1} - \theta_i)(-1) + k_i(\theta_i - \theta_{i+1}) &= 0 \\ \cdots\cdots\cdots\cdots\cdots\cdots\cdots\cdots\cdots\cdots\cdots\cdots& \\ I_n\ddot{\theta}_n - k_{n-1}(\theta_{n-1} - \theta_n) + k_n\theta_n &= 0 \end{aligned}$$

These equations are easy to set up by elementary methods based on Newton's law in torsional form, but the use of Lagrange's equations eliminates the need to check the signs of the various torques, which is sometimes a troublesome detail.

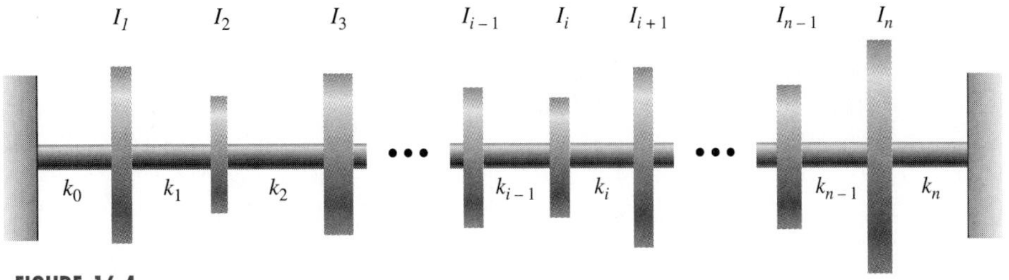

FIGURE 16.4
A system of elastically coupled disks.

EXERCISES

1. Use Lagrange's equation to set up the differential equation of a freely falling body.
2. Use Lagrange's equation to set up the differential equation of a spring-suspended mass.
3. A particle P of mass m starts from rest at $(0, b)$ and slides with negligible friction along the line $y = ax + b$ under the influence of gravity.
   (a) Using x as a generalized coordinate, write the lagrangian function L.
   (b) What is the corresponding Lagrange equation?
   (c) Determine the position of the particle at time t.
   (d) Discuss the motion if $a = 0$ and if $a = \infty$.
4. **(a)** In Exercise 3, use the modified lagrangian
   $$L^* = \frac{m}{2}(\dot{x}^2 + \dot{y}^2) - mgy - \mu(t)[y - ax - b]$$
   to find the x and y components of force acting on P.
   (b) Verify that the resultant force on P is parallel to the line of motion, and find its magnitude.

5. A particle of mass m starts from rest at $(-1, 1)$ and slides with negligible friction along the curve $y = x^2$ under the influence of gravity.

(a) Using x as a generalized coordinate, write the lagrangian L and the corresponding Lagrange equation.

(b) Express the x and y components of velocity of the particle in terms of x, and verify that $\dot{x}$ is maximal at the origin.

6. A particle of mass m starts from rest at $(-a, 0)$ and slides with negligible friction along the lower half of the circle $x^2 + y^2 = a^2$ under the influence of gravity.

(a) Using polar coordinates, and noting that $r \equiv a$, write the Lagrange equation which θ must satisfy.

(b) Express $\dot{\theta}^2$ as a function of θ.

(c) Verify that the particle always has zero energy.

7. (a) Use Lagrange's equation to set up a differential equation for the angular displacement θ from the vertical of a simple pendulum of length l.

(b) Determine the tension $\mu(t)$ in the cord of the pendulum as a function of the magnitude v of the velocity of the suspended mass and θ.

8. (a) Find Lagrange's equations of motion for the mechanical system of Example 2, Sec. 7.5, for general values of k_1, k_2, and l, using m for the mass of the bar and I for its moment of inertia about its midpoint. *Hint:* Assume the ends of the bar move vertically.

(b) Linearize the Lagrange equations by replacing $\sin\theta$ by θ and $\cos\theta$ by 1, and verify that if $k_1 = 24$ lb/in, $k_2 = 15$ lb/in, $l = 48$ in, and W = 64 lb, the linear equations reduce to Eqs. (6) and (7), Sec. 7.5.

9. A mass m hangs from a spring of modulus k. If the point of suspension moves vertically according to the law $s = a \sin \omega t$, use Lagrange's equation to set up the differential equation describing the motion of the weight, and solve the equation if the point of suspension begins its motion when the mass is at rest in its equilibrium position.

10. A pendulum consists of a mass m suspended from a rigid but weightless rod of length l. If the point of suspension of the pendulum moves horizontally according to the law $s = a \sin \omega t$, set up the expressions for T and V in terms of s and the inclination angle θ of the pendulum. Then use Lagrange's equation to obtain the differential equation describing the motion of the pendulum. What appears to be the limiting behavior of the pendulum as $\omega \to \infty$?

11. Masses m_1 and m_2 are suspended, one at each end of a cord of negligible weight and length $l + a$ which passes over a smooth peg of circumference $2a$ as shown in Fig. 16.5a. Initially, the overhangs are of equal length and the masses are released from rest.

(a) Use Hamilton's principle with a constraint to obtain Euler-Lagrange equations in the dependent variables q_1 and q_2.

(b) Derive an equation of motion having q_1 as dependent variable.

(c) Determine q_1 as a function of t.

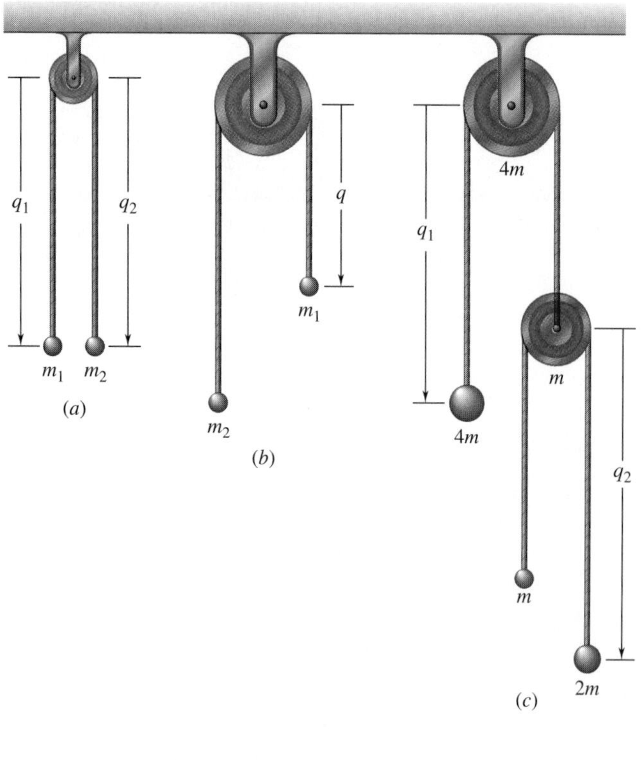

FIGURE 16.5

(d) Find the tension $|\mu|$ in the cord.
(e) If $m_2 > m_1$, when will m_1 reach the peg?

12. Masses m_1 and m_2 are attached to the ends of a cord of negligible weight that passes over a pulley of radius R and weight W as shown in Fig. 16.5*b*. The pulley turns with no appreciable friction. Derive Lagrange's equation of motion with dependent variable q and note its form when $m_1 = m_2$.
13. A mass $4m$ and a movable pulley of mass m are attached to a cord which passes over a fixed pulley of mass $4m$. Another cord with masses m and $2m$ attached to its ends passes over the movable pulley as shown in Fig. 16.5*c*. Assuming that the cords are inextensible and of negligible weight, that the pulleys are frictionless, and that $q_1(0) = 1$ and $\dot{q}_1(0) = 0$, when will $q_1(t) = 2$?
14. A particle moves without friction in the xy plane under such conditions that its potential energy depends only on its distance from the origin. Using Lagrange's equations, set up the differential equations describing its motion, and deduce Kepler's law: *The radius vector from the origin to the particle sweeps out area at a constant rate.*
15. A particle of mass m slides without friction on a straight rod which passes through the origin and rotates with constant velocity about the z axis in such a way that it describes a circular conical surface with vertex at the origin and elements inclined upward at an angle α with the vertical. Use Lagrange's equation to set up the equation of motion of the particle, and find the solution corresponding to the initial conditions $r = r_0$, $\dot{r} = 0$, where r is the distance of the particle from the origin.
16. In Example 1, Sec. 14.7, use Lagrange's equations to obtain the differential equations describing the motion of the system in terms of the normal coordinates y_1, y_2, y_3.
17. The point of suspension of a simple pendulum can slide without friction along a horizontal line against elastic restoring forces, as shown in Fig. 16.6*a*. Letting x be the horizontal displacement of the point of suspension and θ be the angle the pendulum makes with the vertical, determine the kinetic and potential energies of the system. Using Lagrange's equations, obtain the equations of motion of the system. Under the assumption of small vibrations, what are the natural frequencies of the system? *Hint:* Linearize the Lagrange equations by replacing $\sin\theta$, $\cos\theta$ and $\dot{\theta}^2 \sin\theta$ by θ, 1, and 0, respectively.
18. Assuming that the spring remains horizontal, use Lagrange's equations to set up the differential equations describing the motion of the coupled pendulums shown in Fig. 16.6*b*. Assuming small vibrations, what are the natural frequencies of the system?
19. (a) How many generalized coordinates are needed to describe the motion of a particle on a given surface?
(b) Describe a possible choice.
20. Show that the matrix A of Eq. (9) can be written in the form $A = \frac{1}{2}[X^TX + Y^TY + Z^TZ]$ where the elements of X, Y, and Z are given by

$$x_{ji} = \sqrt{m_j}\,\frac{\partial x_j}{\partial q_i} \qquad y_{ji} = \sqrt{m_j}\,\frac{\partial y_j}{\partial q_i} \qquad z_{ji} = \sqrt{m_j}\,\frac{\partial z_j}{\partial q_i}$$
$$1 \le j \le p \qquad 1 \le i \le n$$

and verify that A is symmetric.
21. A mass m under the influence of gravity executes small oscillations about the origin on a frictionless paraboloid $Ax^2 + 2Bxy + Cy^2 = 2z$, where $A > 0$, $B^2 < AC$, and the positive direction of the z axis is upward. Find an equation whose solutions give the natural frequencies of the motion. *Hint:* Use x and y as generalized coordinates and take $\dot{z}^2 = 0$.
22. Assuming that friction is negligible and that the masses are guided along a straight line, find and simplify Lagrange's equations of motion and also find the natural frequencies of the system shown in Fig. 16.7*a* for each combination of parameter values.
(a) $m_1 = 4$, $m_2 = 2$, $k_1 = 8$, $k_2 = 4$
(b) $m_1 = 1$, $m_2 = 2$, $k_1 = 2$, $k_2 = 3$
(c) $m_1 = 2$, $m_2 = 2$, $k_1 = 8$, $k_2 = 5$ and a spring just like the left-hand one is fastened to the right-hand mass so that the configuration becomes symmetrical.

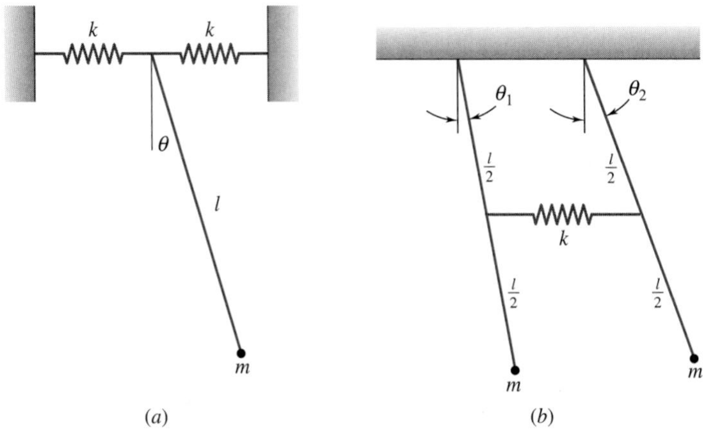

FIGURE 16.6

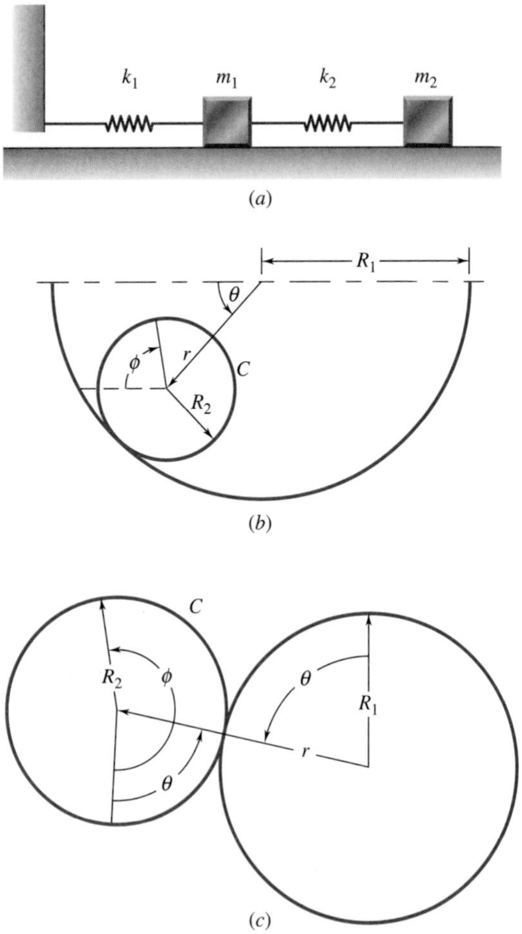

FIGURE 16.7

23. A cylinder C of radius R_2 and mass m rolls over the surface of a stationary cylindrical trough of radius $R_1 > R_2$, the axes of the cylinders remaining parallel during the motion (Fig. 16.7*b*). Friction between the surfaces prevents relative slippage but may otherwise be neglected.
- **(a)** Determine the reactive force **F** of the trough upon the rolling cylinder. *Hint:* Impose an appropriate constraint on the relevant Hamilton integral.
- **(b)** Write an equation of motion for C involving θ and t as its only variables.
- **(c)** Assuming that $\pi/2 - \theta$ remains very small, determine the frequency f of oscillation of C.
- **(d)** Discuss the behavior of f as $R_2 \to R_1$, as $R_2 \to 0$, and

24. A particle P of mass m moves without friction on the lower half of a sphere $r = a$, $a > 0$, under the influence of gravity. Write equations of motion for P having the spherical coordinates θ and ϕ as dependent variables. Assuming $\dot{\theta}(0) = \dot{\phi}(0) = 0$, $\pi - \phi(0) = \alpha$, and that $\pi - \phi$ remains small, discuss the motion of P and find the reactive force of the sphere upon the particle.

25. A cylinder C of radius R_2 and mass m rolls without slippage and with negligible friction over the surface of a stationary cylinder of radius R_1, the axes of the cylinders remaining parallel during the motion (Fig. 16.7*c*).
- **(a)** Write an equation of motion for C with dependent variable θ which holds as long as the cylinders are in contact.
- **(b)** Determine the reactive force $\boldsymbol{\mu}$ of the stationary cylinder upon the rolling cylinder.
- **(c)** Write equations of motion for C which hold after the cylinders separate.

26. Two small metal spheres S_1 and S_2, of equal mass m and electric charge $+e$, are suspended at the ends of insulating threads, of equal length l and negligible weight, which are attached a units apart to a horizontal support (Fig. 16.8). The spheres are set into motion and undergo small displacements (compared to a) in the plane determined by the equilibrium position of the threads. Using q_1 and q_2 to denote the respective displacements of S_1 and S_2 from their equilibrium positions when neither sphere is charged, find:
- **(a)** The related lagrangian function L and Lagrange's equations. *Hint:* When the spheres are $a + q_2 - q_1$ units apart, the electric force of repulsion is $e^2/(a + q_2 - q_1)^2$.
- **(b)** Linear approximations to Lagrange's equations and a complete solution of the linear system. *Hint:*
$$\frac{e^2}{(a + q_2 - q_1)^2} = \frac{e^2}{a^2}\left(1 + \frac{q_2 - q_1}{a}\right)^{-2} \doteq e^2[a - 2(q_2 - q_1)]/a^3.$$
- **(c)** Translations of the origins of q_1 and q_2 which transform the displacements of each sphere into a superposition of two harmonic functions.

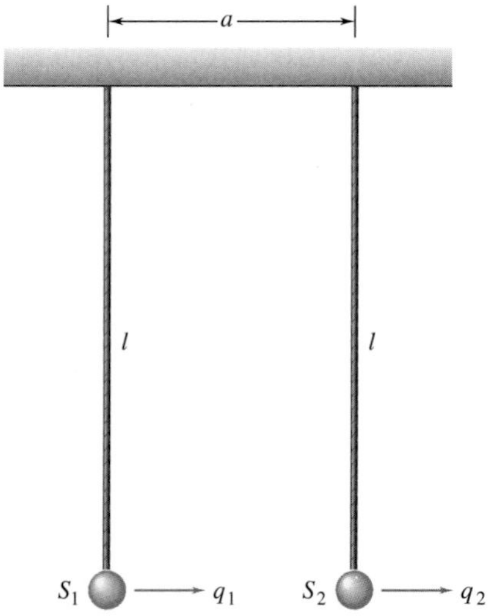

FIGURE 16.8

16.7 HAMILTON'S EQUATIONS

A particle of mass m moving freely with velocity $\mathbf{v}$ has momentum $m\mathbf{v}$ and kinetic energy $T = \frac{1}{2}m\mathbf{v} \cdot \mathbf{v}$. In three dimensions, each rectangular component of the momentum equals the derivative of the kinetic energy $T = \frac{1}{2}m(\dot{x}^2 + \dot{y}^2 + \dot{z}^2)$ with respect to the corresponding velocity component, i.e.,

$$m\dot{x} = \frac{\partial T}{\partial \dot{x}} \qquad m\dot{y} = \frac{\partial T}{\partial \dot{y}} \qquad m\dot{z} = \frac{\partial T}{\partial \dot{z}}$$

For a system of particles having generalized coordinates q_i, $1 \le i \le n$, and kinetic energy $T(q_1, q_2, \ldots, q_n, \dot{q}_1, \dot{q}_2, \ldots, \dot{q}_n)$, the variables

$$(1) \qquad p_i = \frac{\partial T}{\partial \dot{q}_i} \qquad 1 \le i \le n$$

are, by analogy, referred to as **generalized momenta,** or **generalized momentum coordinates,** although these quantities may or may not have the dimensions of momentum. From Eq. (9), Sec. 16.6, the kinetic energy T of the system is a quadratic form

$$(2) \qquad T = \dot{\mathbf{q}}^T \mathrm{A}\, \dot{\mathbf{q}} \qquad \mathrm{A} = \mathrm{A}^T$$

in the generalized velocity components $\dot{q}_i$; thus, Euler's homogeneous function theorem, in conjunction with (1), gives

$$(3) \qquad \sum_{i=1}^{n} \dot{q}_i \frac{\partial T}{\partial \dot{q}_i} = \sum_{i=1}^{n} p_i \dot{q}_i = 2T$$

which shows that each product $p_i \dot{q}_i$ has the dimensions of energy regardless of the nature of the generalized coordinates.

Differentiating (2) with respect to $\dot{q}_i$, and noting that $a_{ij} = a_{ji}$, we get

$$\begin{aligned} p_i &= \frac{\partial}{\partial \dot{q}_i}(\dot{\mathbf{q}}^T)\mathrm{A}\, \dot{\mathbf{q}} + \dot{\mathbf{q}}^T \mathrm{A} \frac{\partial}{\partial \dot{q}_i}(\dot{\mathbf{q}}) \\ &= [a_{i1}\ a_{i2} \cdots a_{in}]\, \dot{\mathbf{q}} + \dot{\mathbf{q}}^T [a_{1i}\ a_{2i} \cdots a_{ni}]^T \\ (4) \qquad &= 2[a_{i1}\dot{q}_1 + a_{i2}\dot{q}_2 + \cdots + a_{in}\dot{q}_n] \qquad 1 \le i \le n \end{aligned}$$

In matrix form these n equations read

$$(5) \qquad \mathbf{p} = 2\mathrm{A}\, \dot{\mathbf{q}}$$

Both (4) and (5) determine each p_i as a linear homogeneous function of $\dot{q}_1, \dot{q}_2, \ldots, \dot{q}_n$. Solving (5) for $\dot{\mathbf{q}}$, we obtain

$$(6) \qquad \dot{\mathbf{q}} = \frac{1}{2}\mathrm{A}^{-1}\mathbf{p}$$

which determines each $\dot{q}_i$ as a linear homogeneous function of $p_1, p_2, \ldots, p_n$.† Let us denote these functions by

$$(7) \qquad \dot{q}_i = f_i(q_1, q_2, \ldots, q_n, p_1, p_2, \ldots, p_n)$$

†This assumes that A is nonsingular, which is a consequence of the positive definite property of T as a quadratic form in $\dot{q}_1, \dot{q}_2, \ldots, \dot{q}_n$ (see Sec. 14.5).

For a conservative system of particles having total energy E, potential energy V, and lagrangian L,

$$E = T + V = 2T - (T - V) = 2T - L \tag{8}$$

or, substituting from (3) for $2T$,

$$E = \sum_{i=1}^{n} p_i \dot{q}_i - L \tag{9}$$

When each $\dot{q}_i$ in the right-hand member of this equation is replaced by the corresponding function f_i of (7), the total energy E is transformed into a function H of the variables $q_1, q_2, \ldots, q_n, p_1, p_2, \ldots, p_n$ called the **hamiltonian** of the particle system. Thus,

$$H(q_1, q_2, \ldots, q_n, p_1, p_2, \ldots, p_n) = \sum_{i=1}^{n} p_i \dot{q}_i - L(q_1, q_2, \ldots, q_n, \dot{q}_1, \dot{q}_2, \ldots, \dot{q}_n) \tag{10}$$

where, in the right-hand member of this equation, each $\dot{q}_i$ stands for the corresponding function $f_i(q_1, q_2, \ldots, q_n, p_1, p_2, \ldots, p_n)$ of Eq. (7). Moreover, from the first of Eqs. (8),

$$H(q_1, q_2, \ldots, q_n, p_1, p_2, \ldots, p_n) = T(q_1, q_2, \ldots, q_n, f_1, f_2, \ldots, f_n) + V(q_1, q_2, \ldots, q_n) \tag{11}$$

that is to say, *the hamiltonian of a conservative system is the sum of the kinetic and potential energies when the kinetic energy is expressed in terms of the q_i and p_i* instead of the q_i and $\dot{q}_i$.

Differentiating the hamiltonian (10) partially with respect to p_i, we get

$$\frac{\partial H}{\partial p_i} = \dot{q}_i + \sum_{j=1}^{n} p_j \frac{\partial \dot{q}_j}{\partial p_i} - \sum_{j=1}^{n} \frac{\partial L}{\partial \dot{q}_j} \frac{\partial \dot{q}_j}{\partial p_i} = \dot{q}_i + \sum_{j=1}^{n} \left(p_j - \frac{\partial L}{\partial \dot{q}_j}\right) \frac{\partial \dot{q}_j}{\partial p_i} \tag{12}$$

Since no $\dot{q}_j$ is an argument of V,

$$\frac{\partial L}{\partial \dot{q}_j} = \frac{\partial T}{\partial \dot{q}_j} = p_j \qquad \text{i.e., } p_j - \frac{\partial L}{\partial \dot{q}_j} = 0 \qquad 1 \le j \le n$$

Hence, (12) reduces to

$$\frac{\partial H}{\partial p_i} = \dot{q}_i \qquad 1 \le i \le n \tag{13}$$

These n equations show that the partial derivative of the hamiltonian with respect to p_i is the ith component f_i of the vector function $\dot{\mathbf{q}}$ set forth in (6) as a solution to Eqs. (4) or (5). However, until the hamiltonian is known, the components of $\dot{\mathbf{q}}$, as given by (13), remain indeterminate. The most convenient way of forming the hamiltonian of a given system is this:

1. Express the potential energy V in terms of the q_i and the kinetic energy T in terms of the q_i and the $\dot{q}_i$.
2. Form and solve the n equations (4) for the $\dot{q}_i$ in terms of the q_i and the p_i.
3. Substitute for the $\dot{q}_i$ in T to obtain $H = T + V$ in terms of $q_1, q_2, \ldots, q_n, p_1, p_2, \ldots, p_n$.

Using (10) to express the lagrangian in terms of the hamiltonian, we get

$$L = \sum_{i=1}^{n} p_i \dot{q}_i - H$$

The related Hamilton integral is

$$\int_{t_1}^{t_2} \left(\sum_{i=1}^{n} p_i \dot{q}_i - H \right) dt \tag{14}$$

According to Hamilton's principle, this integral is rendered stationary by the $2n$ continuously differentiable functions $q_1, q_2, \ldots, q_n, p_1, p_2, \ldots, p_n$ which characterize the actual motion of the system of particles and satisfy the constraints provided by (13), namely,

$$\dot{q}_i - \frac{\partial H}{\partial p_i} = 0 \qquad 1 \le i \le n$$

To derive the differential equations of motion satisfied by the q_i and the p_i, we form the modified lagrangian

$$L^* = \sum_{i=1}^{n} p_i \dot{q}_i - H + \sum_{i=1}^{n} \mu_i(t) \left(\dot{q}_i - \frac{\partial H}{\partial p_i} \right)$$

where $\mu_1, \mu_2, \ldots, \mu_n$ are undetermined functions. For $1 \le i \le n$, we have $\partial L^*/\partial \dot{p}_i = 0$. Thus, the two sets of Lagrange equations

$$(a)\ \frac{\partial L^*}{\partial q_i} - \frac{d}{dt}\left(\frac{\partial L^*}{\partial \dot{q}_i} \right) = 0 \qquad (b)\ \frac{\partial L^*}{\partial p_i} - \frac{d}{dt}\left(\frac{\partial L^*}{\partial \dot{p}_i} \right) = 0 \qquad 1 \le i \le n \tag{15}$$

yield, in turn,

$$-\frac{\partial H}{\partial q_i} - \sum_{j=1}^{n} \mu_j(t) \frac{\partial^2 H}{\partial q_i\, \partial p_j} - \frac{d}{dt}[p_i + \mu_i(t)] = 0 \qquad 1 \le i \le n \tag{16}$$

and

$$\dot{q}_i - \frac{\partial H}{\partial p_i} - \sum_{j=1}^{n} \mu_j(t) \frac{\partial^2 H}{\partial p_i \partial p_j} = 0 \qquad 1 \le i \le n \tag{17}$$

Because of the constraints, Eqs. (17) reduce to

$$\sum_{j=1}^{n} \frac{\partial^2 H}{\partial p_i\, \partial p_j} \mu_j(t) = 0 \qquad 1 \le i \le n \tag{18}$$

The coefficient matrix

$$\mathrm{B} = [\partial^2 H/\partial p_i \partial p_j] \qquad 1 \le i, j \le n \tag{19}$$

of this system of linear equations in $\mu_1, \mu_2, \ldots, \mu_n$ is the matrix $\frac{1}{2}\mathrm{A}^{-1}$ of Eq. (6) and is therefore nonsingular (see Exercise 5). Hence, for $1 \le i \le n$,

$$\mu_i(t) \equiv 0 \qquad \text{and Eqs. (16) become} \qquad -\frac{\partial H}{\partial q_i} - \dot{p}_i = 0$$

These n equations, together with Eqs. (13), form a system of $2n$ first-order differential equations

$$(a)\ \dot{q}_i = \frac{\partial H}{\partial p_i} \qquad (b)\ \dot{p}_i = -\frac{\partial H}{\partial q_i} \qquad 1 \le i \le n \tag{20}$$

known as **Hamilton's equations,** or the **canonical form** of the equations of motion. As we noted earlier, the subset (20a) of Hamilton equations is automatically given by the solution of Eqs. (4) for the $\dot{q}_i$ in terms of the q_i and p_i. Once H is known, these n equations may be checked by computing the partial derivatives of H with respect to the p_i. The other n Hamilton equations (20*b*) require that H be found before their determination. Collectively, Hamilton's equations provide a basis for more sophisticated mathematical techniques applicable to advanced problems of dynamics, celestial mechanics, and atomic structure.

EXERCISES

1. A harmonic oscillator consisting of a mass m attached to the free end of a spring of modulus k is vibrating vertically.
(a) Determine the hamiltonian as a function of the displacement q of the mass from its equilibrium position and the generalized momentum p.
(b) Find Hamilton's equations.
(c) Use Hamilton's equations to derive a harmonic equation of motion.

2. A simple pendulum of length l and mass m has one degree of freedom. Using its angle of swing θ from the vertical as a generalized coordinate find:
(a) The lagrangian as a function of θ and $\dot{\theta}$, and write Lagrange's equation.
(b) The hamiltonian H as a function of θ and the generalized momentum p.
(c) Hamilton's equations, and show that they imply Lagrange's equation.

3. One end of the cord of a spherical pendulum of unit length and mass m is affixed to the origin.
(a) Using the spherical coordinates θ and ϕ of the mass m as generalized coordinates, find the hamiltonian.
(b) Find Hamilton's equations and use them to derive Lagrange's equations.

4. A mass m having potential energy $V(r)$ is moving in the xy plane under the influence of a central force whose line of action is through the origin.
(a) With the polar coordinates of m as generalized coordinates, find the hamiltonian.
(b) Find Hamilton's equations.
(c) Use Hamilton's equations to derive the related Lagrange equations.

5. Show that the matrix B of Eq. (19) is the matrix $\frac{1}{2}\text{A}^{-1}$ of Eq. (6). *Hint:* Note that $\dot{q}_i = \dfrac{\partial H}{\partial p_i} = \sum_{j=1}^{n} b_{ij}p_j$.

6. Use Hamilton's equations to deduce that $dH/dt = 0$ and thus verify that the hamiltonian is of constant value. *Hint:* Multiply (20*a*) by $\dot{p}_i$, multiply (20*b*) by $-\dot{q}_i$, then add and sum over i.

7. According to the **principle of minimum potential energy,** *a system is in stable equilibrium if and only if its potential energy is a minimum.* If neither the kinetic energy T nor the potential energy V depends on the time t, how can Lagrange's general equations of motion $\dfrac{d}{dt}\left(\dfrac{\partial L}{\partial \dot{q}_i}\right) - \dfrac{\partial L}{\partial q_i} = 0,\ 1 \le i \le n$ be used to obtain the equilibrium conditions, and what are they?

8. **(a)** Find the hamiltonian of the pendulum system of Exercise 17, Sec. 16.6. *Hint:* Choose x and θ as generalized coordinates.
(b) Find Hamilton's equations.
(c) Find the values of x and θ for which the system, when static, is in stable equilibrium. *Hint:* See Exercise 7.

9. **(a)** Using the vertical displacements of the masses from their equilibrium positions as generalized coordinates, find the hamiltonian of the oscillatory spring-mass system illustrated in Fig. 16.9*a*.
(b) Find the related Hamilton equations.
(c) Use Hamilton's equations to derive Lagrange's equations.

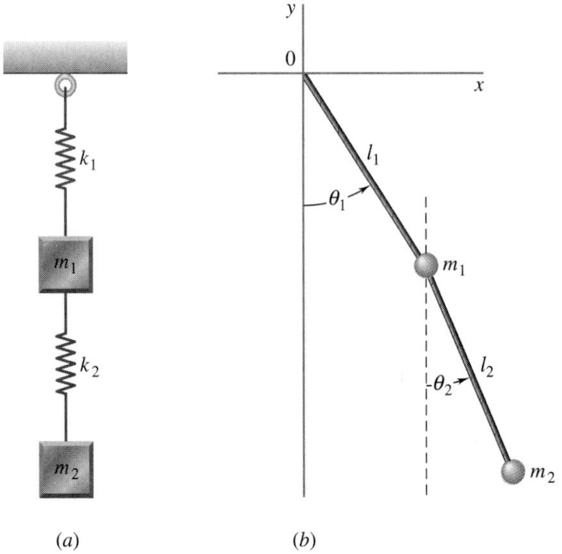

FIGURE 16.9

10. The compound pendulum shown in Fig. 16.9*b* is swinging in the xy plane. Using the indicated angles θ_1 and θ_2 as generalized coordinates, find:
(a) The generalized momenta p_1 and p_2.
(b) The subset (20*a*) of Hamilton's equations.
(c) The hamiltonian of the system.

CHAPTER 17

ANALYTIC FUNCTIONS OF A COMPLEX VARIABLE

In Sec. 3.1, we reviewed the *algebra* of complex numbers. In this chapter, we begin the study of the *calculus* of functions of a complex variable. In Sec. 17.3, after making the rather surprising discovery that most functions of a complex variable, including some very simple ones, do not have a derivative, we develop criteria to identify those that do. These, known as the *Cauchy-Riemann equations,* also provide a formula for the derivative when it exists. Functions which do have derivatives are known as *analytic functions,* a concept we exploited in Secs. 2.9 and 2.10.

With this information available, Sec. 17.4 discusses the definitions and properties of the elementary functions, when their independent variables are allowed to take on complex values. Fortunately, all familiar real identities, including the formulas of differentiation and antidifferentiation, remain valid in the complex plane.

Integration in the complex plane is concerned primarily with line integrals and integrals over areas bounded by closed curves. Using concepts and techniques from Secs. 15.6 and 15.7, including in particular Green's lemma, *Cauchy's theorem* is established (Sec. 17.5) which tells when a contour integral of a complex function is sure to vanish. This theorem has an especially important consequence, known as *Cauchy's integral formula* (Sec. 17.6) which provides a representation for analytic functions in terms of contour integrals. These findings are of fundamental importance in much of complex variable theory and its applications.

The chapter ends with a discussion of the application of analytic functions to two-dimensional fluid flow, using the fact that the real and the imaginary parts of any analytic function define two families of orthogonal trajectories (Sec. 1.14), which can be interpreted as lines of flow and equipotential lines, or similar properties of any other two-dimensional field problem. A number of specific flow patterns are considered in the examples.

Prerequisites for this chapter: multivariable calculus, Sec. 3.1, Secs. 15.6 and 15.7.

Prerequisite for Sec. 17.7: a calculus-based physics course.

17.1 INTRODUCTION

In our work up to this point we have frequently found the use of complex numbers necessary or at least convenient. For instance, we encountered them in Chaps. 2 and 4 in the solution of differential equations and systems of differential equations with constant coefficients. In Chap. 7 they appeared in the complex impedance, which we found useful in the determination of the steady-state behavior of electric circuits. Then in Chaps. 8 and 9 their use led to the important complex exponential forms of both Fourier series and Fourier integrals and ultimately to the inversion formula of Laplace transform theory. Finally, in Chap. 12, we found that certain physical problems required the consideration of Bessel functions of complex arguments.

None of these applications, with the exception of the inversion integral for the Laplace transform, required any knowledge of the properties of complex numbers or functions of a complex variable beyond what is ordinarily acquired in courses in college algebra and calculus. This material we reviewed in Sec. 3.1, which should be consulted, as necessary, for the basic facts about the algebra and geometry of complex numbers. There are, however, large areas of applied mathematics in which familiarity with the theory of functions of a complex variable beyond this minimum is indispensable. In this and the next three chapters we shall develop the major features of this theory and illustrate some of its more striking applications.

17.2 FUNCTIONS OF A COMPLEX VARIABLE

If $z = x + iy$ and $w = u + iv$ are two complex variables and if for each value of z in some portion of the complex plane one or more values of w are defined, then, over that region, w is said to be a **function** of z,† and we write

$$w = f(z)$$

If $w = f(z)$, that is, if

$$w = u + iv = f(x + iy)$$

it follows that the real numbers u and v are themselves determined by the real numbers x and y. Hence the assertion that w is a function of $z = x + iy$ can also be written more explicitly as

$$w = u(x, y) + iv(x, y) \tag{1}$$

where $u(x, y)$ and $v(x, y)$ are real-valued functions of the real variables x and y. Clearly, whenever a value of z is given, values of x and y are thereby provided, and thus one or more values of w are determined by (1).

EXAMPLE 1

If $w = f(z) = (x^2 - y^2) + i(x + y^2)$ and if $z = 1 + 2i$, then $x = 1$ and $y = 2$, and therefore

$$w = f(1 + 2i) = (1^2 - 2^2) + i(1 + 2^2) = -3 + 5i$$

†The best modern usage restricts the word *function* to the case in which to each value of the independent variable there corresponds a single value of the dependent variable. However, it is convenient and not uncommon in complex variable theory to define a function as we have just done and to speak of both *single-valued functions* and *multiple-valued functions*.

If w is defined as a function of z in the form (1), it may be possible by suitable manipulations to rearrange the expression $u(x, y) + iv(x, y)$ so that x and y occur only in the binomial combination $x + iy$. For instance,

$$w = (x^2 - y^2) + 2ixy$$

is immediately recognizable as

$$w = (x + iy)^2 = z^2$$

and

$$w = \frac{x}{x^2 + y^2} - i\frac{y}{x^2 + y^2} = \frac{x - iy}{x^2 + y^2}$$

is nothing but the standard complex form of

$$w = \frac{1}{x + iy} = \frac{1}{z}$$

On the other hand, it may be impossible to express w in a form involving only the explicit combination $x + iy$ without using such "artificial" expressions as $\text{Re}\, z \equiv x$ and $\text{Im}\, z \equiv y$, with which, of course, any formula in x and y can be written in terms of z. For instance, unless we resort to these artificial functions, no rearrangement of the formula

$$w = 7x + 3iy = 4\,\text{Re}\, z + 3z = 7z - 4i\,\text{Im}\, z = 5z + 2\bar{z} = \cdots$$

can reduce w to explicit dependence on z alone. In our work, and in fact in most applications of complex variable theory, the only functions of real interest are those that can be written in terms of z alone without recourse to $\bar{z}$, $\text{Re}\, z$, $\text{Im}\, z$, and similar expressions.

Frequently, our interest in a function will be restricted to its behavior at the points of some specific part of the z plane. However, before we can undertake discussions of this sort, we must define and explain some of the simpler properties of the sets of points we intend to consider.

A **neighborhood** of a point z_0 is a set consisting of all the points satisfying an inequality of the form

$$|z - z_0| < \epsilon \qquad \epsilon > 0$$

Geometrically speaking, a neighborhood of a point z_0 consists of all the points within but not on a circle having z_0 as center. In Fig. 17.1*a*, S_1 is a neighborhood of z_0.

A point z_0 is an **interior point** of a set S if there exists at least one neighborhood of z_0 all of whose points are points of S. In Fig. 17.1*a*, P is an interior point of the set S_1, and in Fig. 17.1*b*, P is an interior point of S_2.

A point z_0 is an **exterior point** of a set S if there exists at least one neighborhood of z_0 none of whose points is a point of S. In Fig. 17.1*a*, R is an exterior point of S_1, and in Fig. 17.1*b*, R is an exterior point of S_2.

A point z_0, which may or may not belong to a set S, is a **boundary point** of S if every neighborhood of z_0 contains both points belonging to S and points not belonging to S. In Fig. 17.1*a*, Q is a boundary point of S_1 although it does not belong to S_1. In Fig. 17.1*b*, Q is a boundary point of S_2 which is also a point of S_2.

A set S is **open** if each of its points is an interior point. Both S_1, in Fig. 17.1*a*, and S_5, in Fig. 17.1*e*, are open sets since, according to their defining inequalities, neither contains any of its boundary points.

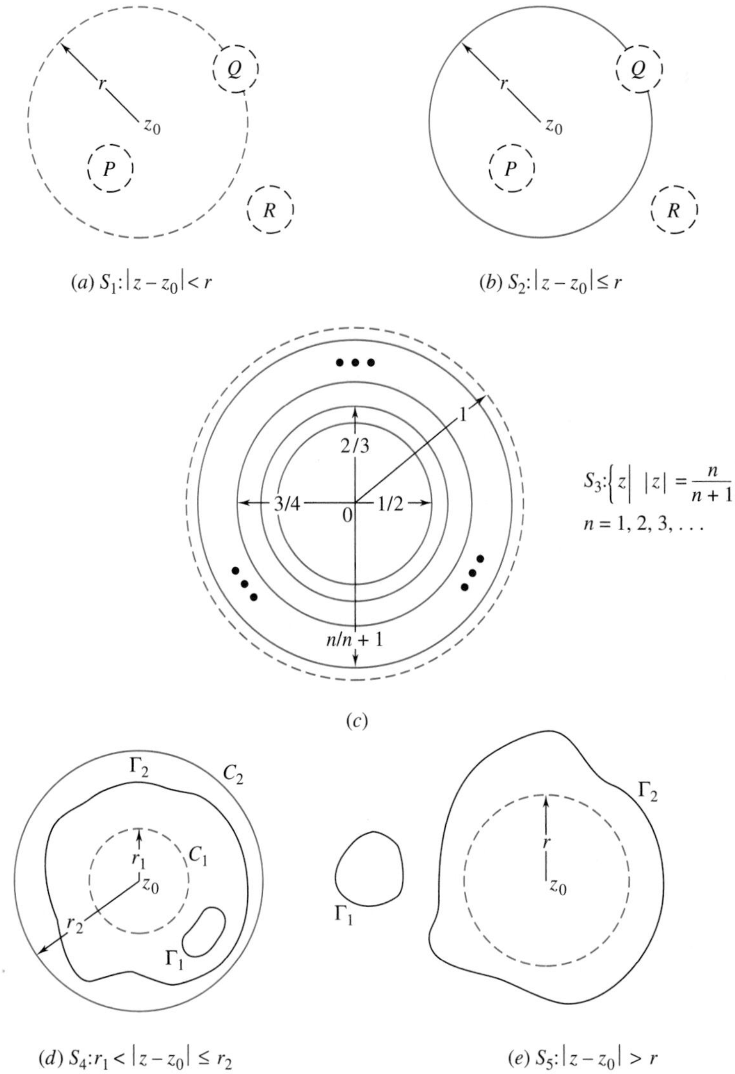

FIGURE 17.1
Typical sets of points in the z plane.

A set S is **closed** if it contains all of its boundary points. In Fig. 17.1*b*, S_2 is closed since all of its boundary points are included in its definition. Of course, there can be sets like S_4 in Fig. 17.1*d* which are neither open nor closed since they contain some but not all of their boundary points.

A point z_0 is a **limit point** of a set S if every neighborhood of z_0 contains points of S distinct from z_0. Every interior point and every boundary point of a set S is also a limit point of S. The use of the word *limit* is motivated by sets such as S_3 in Fig. 17.1*c*. Every point on the circle $|z| = 1$ is a limit point of S_3 because every neighborhood of such a point contains points on infinitely many of the circles which approach $|z| = 1$ as a limit.

A set S is **connected** if every two points in S can be joined, or connected, by a polygonal path all of whose points are points of S. Clearly, the sets S_1, S_2, S_4, and S_5 in Fig. 17.1 are connected.

A set S is **simply connected** if *every* simple closed curve† which can be drawn in the interior of S contains *only* points of S in its interior. The sets S_1 and S_2 are simply connected.

---

†See the footnote, p. 1043.

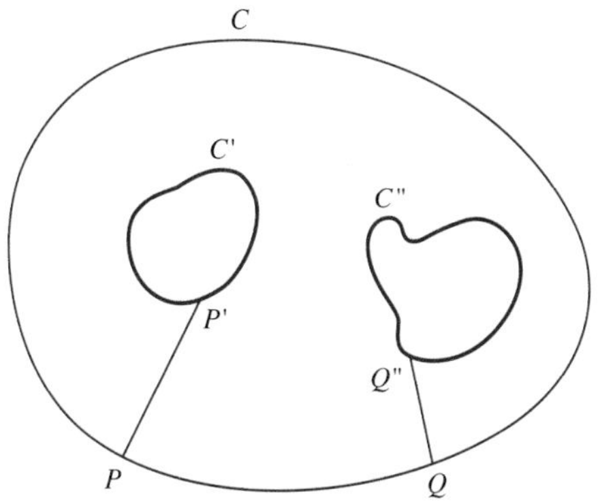

FIGURE 17.2
A multiply connected region made simply connected by the insertion of two cross-cuts.

A set S is **multiply connected** if there exists at least one simple closed curve in the interior of S which encloses one or more points not belonging to S. The set S_4 is multiply connected because, although there are simple closed curves in the interior of S_4 (like Γ_1) which enclose only points belonging to S_4, there are also curves (like Γ_2) which enclose points which do not belong to S_4. The set S_5 is also an example of a multiply connected set.

A **domain** is an open connected set.

A **region** is a domain with all, some, or none of its boundary points.

An **annulus,** or **annular region,** is the region bounded by two concentric circles. The set S_4 in Fig. 17.1*d* is an annulus containing some, but not all of its boundary points.

A set S is **bounded** if there exists a number d such that $|z| < d$ for all z in S. In Fig. 17.1, S_1, S_2, S_3, and S_4 are bounded, and S_5 is unbounded.

Because simply connected regions are in many respects easier to work with than multiply connected regions, it is often desirable to be able to reduce the latter to the former. This can always be done by modifying the given multiply connected region through the introduction of auxiliary boundary arcs, or **cross-cuts,** joining boundary curves that were originally disconnected. The effectiveness of this technique is illustrated in Fig. 17.2, which shows a region originally multiply connected with one outer boundary curve C and two inner boundary curves C' and C''. The introduction of the auxiliary boundary arcs $P'P$ and $Q''Q$ clearly makes it impossible to draw closed curves which lie in the *interior* of the modified region and at the same time enclose either of the inner boundaries C' or C''. The modified region is therefore simply connected, as desired.

It will often be necessary for us to consider the ***limit of a function of*** z as z approaches a limit point z_0 of the domain of the function. The basis for this is the following definition.

DEFINITION 1 If $f(z)$ is a single-valued function of z and w_0 is a complex constant, and if for every $\epsilon > 0$ there exists a positive number $\delta(\epsilon)$ such that $|f(z) - w_0| < \epsilon$ for all z in the domain of f such that $0 < |z - z_0| < \delta(\epsilon)$, then w_0 is said to be the **limit of** $f(z)$ **as** z **approaches** z_0.

In less technical terms, w_0 is the limit of $f(z)$ as z approaches z_0 provided that $f(z)$ can be kept arbitrarily close to w_0 by keeping z sufficiently close to but distinct from z_0.

EXAMPLE 2

If $f(z) = \dfrac{(x+y)^2}{x^2+y^2}$ show that $\lim_{x\to 0}[\lim_{y\to 0} f(z)] = 1$ and $\lim_{y\to 0}[\lim_{x\to 0} f(z)] = 1$ but that $\lim_{z\to 0} f(z)$ does not exist.

Clearly,

$$\lim_{x\to 0}[\lim_{y\to 0} f(z)] = \lim_{x\to 0}\left[\lim_{y\to 0}\frac{(x+y)^2}{x^2+y^2}\right] = \lim_{x\to 0}(1) = 1$$

and

$$\lim_{y\to 0}[\lim_{x\to 0} f(z)] = \lim_{y\to 0}\left[\lim_{x\to 0}\frac{(x+y)^2}{x^2+y^2}\right] = \lim_{y\to 0}(1) = 1$$

as asserted. Thus, along two particular paths, namely, one from $z = z_1$ vertically to the x axis and then along the x axis to $z_0 = 0$ and one from $z = z_1$ horizontally to the y axis and then along the y axis to $z_0 = 0$, the limits of $f(z)$ are the same. However, for $\lim_{z\to 0} f(z)$ to exist, it is necessary that $f(z)$ approach the same limit along *all* paths leading to the origin, and this is not the case. In fact, along the paths $y = mx$ we have

$$\lim_{z\to 0} f(z) = \lim_{z\to 0}\frac{(x+y)^2}{x^2+y^2} = \lim_{x\to 0}\frac{(1+m)^2}{1+m^2} = \frac{(1+m)^2}{1+m^2}$$

The limiting value here clearly depends on m; that is, $f(z)$ approaches different values along different radial lines, and hence no limit exists.

Closely associated with the concept of a limit is the concept of ***continuity.***

DEFINITION 2 The function $f(z)$ is **continuous at the point** z_0 provided that $\lim_{z\to z_0} f(z) = f(z_0)$.

In other words, for a function to be continuous at a point z_0, the function must have both a value at that point and a limit as z approaches that point and the two must be equal. If $f(z)$ is continuous at every point of a region, it is said to be **continuous throughout that region.**

In addition to the fundamental theorems on limits that we encountered in calculus, there are various theorems on continuous functions which we shall need from time to time. For the most part these appear almost self-evident, although their proofs are by no means trivial. We shall merely list them here and refer to standard texts on complex variables for their proofs.†

THEOREM 1 Sums, differences, and products of continuous functions, and quotients of continuous functions, provided the divisor functions are different from zero, are continuous.

THEOREM 2 A continuous function of a continuous function is a continuous function.

In other words, Theorem 2 asserts that *a composition of continuous functions is continuous.*

THEOREM 3 A necessary and sufficient condition that

$$f(z) = u(x, y) + iv(x, y)$$

be continuous is that the real functions $u(x, y)$ and $v(x, y)$ be continuous.

THEOREM 4 If $f(z)$ is continuous at a point z_0 and if $f(z_0) \neq 0$, then there exists a neighborhood of z_0 throughout which $f(z)$ is different from zero.

THEOREM 5 If $f(z)$ is continuous over a bounded, closed region R, then there exists a positive constant M such that $|f(z)| < M$ for all values of z in R.

†See, for instance, R. V. Churchill and J. W. Brown, *Complex Variables and Applications,* 5th ed., pp. 33–42, McGraw-Hill, New York, 1990.

EXERCISES

Evaluate each of the following functions for the indicated value of z.

1. $f(z) = xy + i(x^2 - y^2)$; $z = -1 + 2i$
2. $f(z) = (x^2 - y) + i(x + y^2)$; $z = 2 - 3i$
3. $f(z) = z\bar{z} + z - 2\bar{z} + 1$; $z = 4 + 3i$
4. $f(z) = (z + 1)/(z + i)$; $z = 1 + i$

Express each of the following functions as a polynomial in the binomial argument $z = x + iy$.

5. $f(z) = (x^2 - y^2 - 2y + 1) + 2i(xy + x)$
6. $f(z) = (x^3 - 3xy^2 - y) + i(3x^2y - y^3 + x)$
7. $f(z) = (-x^2 + y^2 - y + 2) + i(x - 2xy)$
8. $f(z) = (2xy + 2x - 1) - i(x^2 - y^2 - 2y)$
9. Express $f(z) = x^2 + iy^2$ in terms of z and $\bar{z}$.
10. Describe each of the following sets of points, telling whether it is open or closed, simply or multiply connected, and bounded or unbounded.
    (a) $\text{Im } z > 0$ (b) $2 \le |z| \le 3$
    (c) $|z - 1| > 4$ (d) $0 \le \text{Re } z \le 1$
    (e) $0 \le \text{Im } z < \text{Re } z$ (f) $\text{Re } z < |z - 1|$
    (g) $\text{Im } z \ge |z - \text{i}|$ (h) $\text{Re } z \ge |\text{Im } z|$
    (i) $|\text{Re } z| \ge |\text{Im } z|$

Using Definition 1, establish each of the following limits.

11. $\lim_{z \to i} (2z - i) = i$
12. $\lim_{z \to (1-i)} (2z + i) = 2 - i$
13. $\lim_{z \to 2i} \dfrac{z^2 + 4}{z - 2i} = 4i$. *Hint:* Factor the numerator.
14. $\lim_{z \to i} \dfrac{z^2 - 3iz - 2}{z - i} = -i$. *Hint:* Factor the numerator.
15. $\lim_{z \to i} \dfrac{z + i}{z} = 2$. *Hint:* At the appropriate point, make a suitable underestimate of the absolute value of the denominator.
16. $\lim_{z \to 1} z^2 = 1$
17. $\lim_{z \to i} z^2 + 4 = 3$

Determine at what points, if any, each of the following functions fails to be continuous and explain why.

18. $\dfrac{z}{z^2 + 1}$
19. $\dfrac{z^2 - 1}{z - 1}$
20. $(x + y^2) + ixy$
21. $\begin{cases} \dfrac{z^2 + 3iz - 2}{z + i} & z \ne -i \\ i & z = -i \end{cases}$
22. $\begin{cases} z^2 + iz + 2 & z \ne i \\ i & z = i \end{cases}$
23. Show that $\lim_{z \to 0} [xy/(x^2 + y^2)]$ does not exist.
24. Show that $\lim_{z \to 0} x^2y/(x^4 + y^2)$ does not exist even though this function approaches the same value along every straight line through the origin.
25. If

$$f(z) = \begin{cases} x \sin \dfrac{1}{y} & y \ne 0 \\ 0 & y = 0 \end{cases}$$

show that $\lim_{y \to 0} [\lim_{x \to 0} f(z)]$ and $\lim_{x \to 0} f(z)$ exist and are equal but that $\lim_{x \to 0} [\lim_{y \to 0} f(z)]$ does not exist.

26. Show that every neighborhood of a limit point of a set S contains infinitely many points of S.
27. If $z_0 = x_0 + iy_0$, show that the condition $0 < |z - z_0| < \delta$ in Definition 1 is equivalent to the conditions

$$0 < |x - x_0| < \delta_1 \qquad \text{and} \qquad 0 < |y - y_0| < \delta_1$$

If δ is known, what is the largest value of δ_1 that will surely serve? If δ_1 is known, what is the largest value of δ that will surely serve?

28. Consider the segment of the real axis for which $0 < x < 1$. Show that as a subset of the real axis this set is open, but that as a subset of the z plane this set is not open.
29. In Fig. 17.1*c*, what is the boundary of S_3? What is the interior of S_3?
30. (a) Identify the set $S = \{z | 3 < |z - 4 + 3i| \le 5\}$ as an annulus.
    (b) Is S a domain? Explain.
    (c) Is S a region? Explain.
    (d) Classify each of the points 0, 1, 4, $-3i$, $4 - 3i$ as a boundary point, interior point, exterior point, or limit point of S; that is, state whichever of these the point may be.

17.3 ANALYTIC FUNCTIONS

We begin with a familiar but crucial definition.

DEFINITION 1 The **derivative of a function of a complex variable** $w = f(z)$ is

$$\frac{dw}{dz} = w' = f'(z) = \lim_{\Delta z \to 0} \frac{f(z + \Delta z) - f(z)}{\Delta z} \tag{1}$$

This definition is formally identical with that for the derivative of a function of a real variable. Moreover, since the general theory of limits is phrased in terms of absolute values, it is valid for complex variables as well as for real variables. Hence it is clear that formulas for the differentiation of combinations of functions of a real variable have identical counterparts in the field of complex variables. In particular, such familiar formulas as

$$\frac{d(w_1 \pm w_2)}{dz} = \frac{dw_1}{dz} \pm \frac{dw_2}{dz}$$

$$\frac{d(w_1 w_2)}{dz} = w_1 \frac{dw_2}{dz} + w_2 \frac{dw_1}{dz}$$

$$\frac{d(w_1/w_2)}{dz} = \frac{w_2(dw_1/dz) - w_1(dw_2/dz)}{w_2^2}$$

$$\frac{d(w^n)}{dz} = nw^{n-1}\frac{dw}{dz}$$

are valid when w_1, w_2, and w are differentiable functions of a complex variable z. Furthermore, after we have defined e^z in the next section, we will find that the familiar formulas for differentiating exponential, trigonometric, and logarithmic functions of a real variable hold equally well for functions of a complex variable. However, $\Delta z = \Delta x + i\,\Delta y$ is itself a complex variable, and the question of just how it is to approach zero involves difficulties which have no counterpart in the differentiation of functions of a real variable.

In Fig. 17.3, it is clear that Δz can approach zero in infinitely many ways; i.e., a point Q: $z + \Delta z$ can approach the point P: z along infinitely many different paths. In particular, Q can approach P along the line AP on which Δx is zero or along the line BP on which Δy is zero. Clearly, *for the derivative of $f(z)$ to exist, it is necessary that the limit of the difference quotient (1) be the same no matter how Δz approaches zero.* How severe a restriction this is can be seen by considering the simple function

$$w = f(z) = \bar{z} = x - iy$$

Giving to z the increment $\Delta z = \Delta x + i\,\Delta y$ means that x changes by the amount Δx and y changes by the amount Δy. Hence, for the given function $x - iy$, we have

$$\frac{f(z + \Delta z) - f(z)}{\Delta z} = \frac{[(x + \Delta x) - i(y + \Delta y)] - (x - iy)}{\Delta x + i\,\Delta y} = \frac{\Delta x - i\,\Delta y}{\Delta x + i\,\Delta y}$$

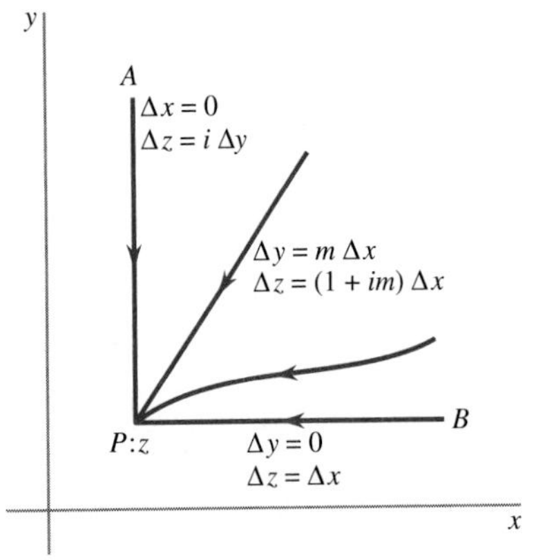

FIGURE 17.3
Various ways in which Δz can approach zero.

Now if Δz is real, so that $\Delta y = 0$, we have

$$\lim_{\Delta z \to 0} \frac{\Delta x - i\,\Delta y}{\Delta x + i\,\Delta y} = \lim_{\Delta x \to 0} \frac{\Delta x}{\Delta x} = 1$$

On the other hand, if Δz is imaginary, so that $\Delta x = 0$, we have

$$\lim_{\Delta z \to 0} \frac{\Delta x - i\,\Delta y}{\Delta x + i\,\Delta y} = \lim_{\Delta y \to 0} \frac{-i\,\Delta y}{i\,\Delta y} = -1$$

More generally, if we let Δz approach zero in such a way that $\Delta y = m\,\Delta x$, we have

$$\lim_{\Delta z \to 0} \frac{\Delta x - i\,\Delta y}{\Delta x + i\,\Delta y} = \lim_{\Delta x \to 0} \frac{\Delta x - im\,\Delta x}{\Delta x + im\,\Delta x} = \frac{1 - im}{1 + im} = \frac{(1 - m^2) - 2im}{1 + m^2}$$

Thus there are infinitely many complex values which the difference quotient for $f(z) = x - iy$ can be made to approach by suitably choosing the manner in which Δz is to approach zero. It is therefore clear that $\bar{z} = x - iy$ has no derivative.

That a function as simple as $f(z) = x - iy$ should have no derivative seems at first glance a discouraging state of affairs. However, there are many functions of z that do have derivatives, and in applications it is these functions which are of importance. Our immediate task is to identify these functions by obtaining conditions for the existence of the derivative of a function of a complex variable.

To do this, consider

$$w = f(z) = u(x, y) + iv(x, y)$$

By definition, if dw/dz exists,

$$\frac{dw}{dz} = \lim_{\Delta z \to 0} \frac{\Delta w}{\Delta z}$$

(2) $$= \lim_{\substack{\Delta x \to 0 \\ \Delta y \to 0}} \frac{[u(x + \Delta x, y + \Delta y) + iv(x + \Delta x, y + \Delta y)] - [u(x, y) + iv(x, y)]}{\Delta x + i\,\Delta y}$$

Now if Δz is real, i.e., if $\Delta y = 0$, we obtain

$$\frac{dw}{dz} = \lim_{\Delta x \to 0} \frac{[u(x + \Delta x, y) + iv(x + \Delta x, y)] - [u(x, y) + iv(x, y)]}{\Delta x}$$

$$= \lim_{\Delta x \to 0} \left[\frac{u(x + \Delta x, y) - u(x, y)}{\Delta x} + i \frac{v(x + \Delta x, y) - v(x, y)}{\Delta x} \right]$$

In the two difference quotients appearing in the last expression, y remains constant. Hence these difference quotients are precisely those whose limits define the partial derivatives of **u** and **v** with respect to x. Hence it appears that

(3) $$\frac{dw}{dz} = \frac{\partial u}{\partial x} + i \frac{\partial v}{\partial x}$$

On the other hand, if Δz is imaginary, i.e., if $\Delta x = 0$, we find from (2) that

$$\frac{dw}{dz} = \lim_{\Delta y \to 0} \frac{[u(x, y + \Delta y) + iv(x, y + \Delta y)] - [u(x, y) + iv(x, y)]}{i\,\Delta y}$$

$$= \lim_{\Delta y \to 0} \left[\frac{u(x, y + \Delta y) - u(x, y)}{i\,\Delta y} + i\frac{v(x, y + \Delta y) - v(x, y)}{i\,\Delta y} \right]$$

In these two difference quotients, x remains constant; hence their limits are the partial derivatives of u and v with respect to y, and we have

$$\frac{dw}{dz} = \frac{1}{i}\frac{\partial u}{\partial y} + \frac{\partial v}{\partial y}$$

or, finally,

$$\frac{dw}{dz} = \frac{\partial v}{\partial y} - i\frac{\partial u}{\partial y} \tag{4}$$

Obviously, if the derivative dw/dz is to exist, it is necessary that the two expressions (3) and (4) which we have just derived for it be the same. Hence

$$\frac{\partial u}{\partial x} + i\frac{\partial v}{\partial x} = \frac{\partial v}{\partial y} - i\frac{\partial u}{\partial y}$$

which requires that

$$\frac{\partial u}{\partial x} = \frac{\partial v}{\partial y} \tag{5a}$$

$$\frac{\partial u}{\partial y} = -\frac{\partial v}{\partial x} \tag{5b}$$

These two extremely important conditions, which are known as the **Cauchy-Riemann equations,**† have arisen here from a consideration of only two of the infinitely many ways in which Δz can approach zero. It is therefore natural to expect that severe additional conditions will be necessary to ensure that along these other paths $\Delta w/\Delta z$ will also approach the same limit dw/dz. This is not the case, however, and it can be proved without great difficulty‡ that if u and v, together with their first partial derivatives u_x, u_y, v_x, v_y are continuous in some neighborhood of the point z_0, then the Cauchy-Riemann equations are not only necessary but also sufficient conditions for the existence of the derivative of $w = f(z) = u(x, y) + iv(x, y)$ throughout that neighborhood.

If $w = f(z)$ possesses a derivative at $z = z_0$ and at every point in some neighborhood of z_0, then $f(z)$ is said to be **analytic** at z_0 and z_0 is called a **regular point** of the function. If $f(z)$ is not analytic at z_0, but if every neighborhood of z_0 contains points at which $f(z)$ is analytic, then z_0 is called a **singular point** of $f(z)$. A function which is analytic at every point of a region R we shall call **analytic in** R. Although most writers use this term, a few substitute such adjectives as **regular** and **holomorphic.** As a summary of our discussion we have the following theorem.

†Named for the French mathematician A. L. Cauchy (1789–1857) and the German mathematician G. F. B. Riemann (1826–1866).

‡See, for instance, R. V. Churchill and J. W. Brown, *Complex Variables and Applications,* 5th ed., pp. 50, 51, McGraw-Hill, New York, 1990.

THEOREM 1 If u and v are real single-valued functions of x and y which with their four first partial derivatives are continuous throughout a region R, then the **Cauchy-Riemann equations**

$$\frac{\partial u}{\partial x} = \frac{\partial v}{\partial y} \quad \text{and} \quad \frac{\partial u}{\partial y} = -\frac{\partial v}{\partial x}$$

are both necessary and sufficient conditions that $f(z) = u(x, y) + iv(x, y)$ be analytic in R. Under these conditions, the derivative of $f(z)$ is given by either

$$f'(z) = \frac{\partial u}{\partial x} + i\frac{\partial v}{\partial x} \quad \text{or} \quad f'(z) = \frac{\partial v}{\partial y} - i\frac{\partial u}{\partial y}$$

EXAMPLE 1

For $f(z) = \bar{z} = x - iy$, we have $u = x$ and $v = -y$. In this case

$$\frac{\partial u}{\partial x} = 1 \qquad \frac{\partial u}{\partial y} = 0 \qquad \frac{\partial v}{\partial x} = 0 \qquad \frac{\partial v}{\partial y} = -1$$

and although the second of the Cauchy-Riemann equations is satisfied everywhere, the first is nowhere satisfied. Hence there is no point in the z plane where $f'(z)$ exists, which of course confirms our earlier investigation of this function.

EXAMPLE 2

For $f(z) = z\bar{z} = x^2 + y^2$, we have $u = x^2 + y^2$ and $v = 0$. In this case the four first partial derivatives

$$\frac{\partial u}{\partial x} = 2x \qquad \frac{\partial u}{\partial y} = 2y \qquad \frac{\partial v}{\partial x} = 0 \qquad \frac{\partial v}{\partial y} = 0$$

are continuous everywhere. However, the Cauchy-Riemann equations, which in this case are, respectively,

$$2x = 0 \quad \text{and} \quad 2y = 0$$

are satisfied only at the origin. Hence $z = 0$ is the only point at which $f'(z)$ exists, and therefore $f(z) = z\bar{z}$ is nowhere analytic.

EXAMPLE 3

For $f(z) = z^2 = (x^2 - y^2) + 2ixy$, we have

$$\frac{\partial u}{\partial x} = 2x \qquad \frac{\partial u}{\partial y} = -2y \qquad \frac{\partial v}{\partial x} = 2y \qquad \frac{\partial v}{\partial y} = 2x$$

and, clearly, the Cauchy-Riemann equations are satisfied everywhere. Moreover, the four first partial derivatives of u and v are everywhere continuous. Hence the derivative $f'(z)$ exists at all points of the z plane, and its value from either (3) or (4) is

$$f'(z) = 2x + 2iy = 2z$$

This of course is exactly what formal differentiation of z^2 by the power rule would give.

Analytic functions have numerous important properties, many of which we shall investigate in later sections. At this point we note only the following.

PROPERTY 1 If $f(z) = u(x, y) + iv(x, y)$ is analytic in a region R and if u and v have continuous second partial derivatives in R, then, in R, u and v satisfy Laplace's equation

$$\frac{\partial^2 \phi}{\partial x^2} + \frac{\partial^2 \phi}{\partial y^2} = 0$$

PROOF Let $f(z) = u(x, y) + iv(x, y)$ be an analytic function of z in a region R. Then in R both u and v must satisfy the Cauchy-Riemann equations, namely,

$$\frac{\partial u}{\partial x} = \frac{\partial v}{\partial y} \qquad \text{and} \qquad \frac{\partial u}{\partial y} = -\frac{\partial v}{\partial x}$$

If we differentiate the first of these with respect to x and the second with respect to y and add the results, we obtain our first assertion:

$$\frac{\partial^2 u}{\partial x^2} = \frac{\partial^2 v}{\partial x\, \partial y}$$

$$\frac{\partial^2 u}{\partial y^2} = -\frac{\partial^2 v}{\partial y\, \partial x}$$

$$\frac{\partial^2 u}{\partial x^2} + \frac{\partial^2 u}{\partial y^2} = 0$$

In exactly the same way we can establish our second assertion, namely, that v also satisfies Laplace's equation.

The existence of the second derivatives and their continuity, which makes the order of differentiation in the cross partial derivatives immaterial, must here be assumed. Later we shall show that every analytic function possesses not only a first derivative but derivatives of *all* orders, which implies the existence and continuity of all the partial derivatives of u and v.

A function possessing continuous second partial derivatives and satisfying Laplace's equation is called a **harmonic function.** Two harmonic functions, u and v, so related that $u + iv$ is an analytic function are called **conjugate harmonic functions.**† This use of the word *conjugate* must not be confused with its use in describing $\bar{z}$, the complex number which is the conjugate of z.

PROPERTY 2 If $w = u(x, y) + iv(x, y)$ is an analytic function of z in a region R, then in R the curves of the family $u(x, y) = c$ are orthogonal trajectories of the curves of the family $v(x, y) = k$, and vice versa.

PROOF To prove this, we compute the slope of the general curve of each family by implicit differentiation, obtaining for the curves $u(x, y) = c$ the expression

$$\frac{dy}{dx} = -\frac{\partial u/\partial x}{\partial u/\partial y} \tag{6}$$

and for the curves $v(x, y) = k$ the expression

$$\frac{dy}{dx} = -\frac{\partial v/\partial x}{\partial v/\partial y} \tag{7}$$

†The order in the pair (u, v) is important, as Exercise 9 makes clear.

By hypothesis, $w = u + iv$ is an analytic function. Hence it follows from Theorem 1 that u and v satisfy the Cauchy-Riemann equations. When these equations are used, Eq. (7) for the slope of the general curve of the family $v(x, y) = k$ can be rewritten

$$\frac{dy}{dx} = \frac{\partial u/\partial y}{\partial u/\partial x}$$

which at any common point is just the negative reciprocal of the slope of the general curve of the family $u(x, y) = c$, as given by Eq. (6). This suffices to prove that the two families of curves are orthogonal trajectories, as asserted. ◗

The importance of Property 1 is apparent when we recall from Chaps. 11 and 12 that Laplace's equation is fundamental in the study of heat flow and potential theory. As we shall see in later sections, there are many problems in steady-state heat flow and field theory which can be solved by finding the proper analytic function and using its real part or its imaginary part as the required solution. Moreover, if and when such a function has been found, Property 2 tells us that the curves $u(x, y) = c$ and $v(x, y) = k$ are the isothermal lines and the lines of heat flow or the equipotential lines and the lines of flux, depending on the nature of the problem.

PROPERTY 3 If in any analytic function $w = u(x, y) + iv(x, y)$ the variables x and y are replaced by their equivalents in terms of z and $\bar{z}$, namely,

$$x = \frac{z + \bar{z}}{2} \qquad \text{and} \qquad y = \frac{z - \bar{z}}{2i}$$

then w will appear as a function of z alone.

◖ **PROOF** Although z and $\bar{z}$ are clearly dependent, since either is determined when the other is given, we can regard w, by virtue of the given substitutions, as formally a function of two new independent variables z and $\bar{z}$. To show that w depends only on z and does not involve $\bar{z}$, it is sufficient to compute $\partial w/\partial \bar{z}$ and verify that it is identically zero. By the chain rule for partial derivatives,

$$\frac{\partial w}{\partial \bar{z}} = \frac{\partial(u + iv)}{\partial \bar{z}} = \frac{\partial u}{\partial \bar{z}} + i\frac{\partial v}{\partial \bar{z}} = \left(\frac{\partial u}{\partial x}\frac{\partial x}{\partial \bar{z}} + \frac{\partial u}{\partial y}\frac{\partial y}{\partial \bar{z}}\right) + i\left(\frac{\partial v}{\partial x}\frac{\partial x}{\partial \bar{z}} + \frac{\partial v}{\partial y}\frac{\partial y}{\partial \bar{z}}\right)$$

Moreover, from the equations expressing x and y in terms of z and $\bar{z}$, we have

$$\frac{\partial x}{\partial \bar{z}} = \frac{1}{2} \qquad \text{and} \qquad \frac{\partial y}{\partial \bar{z}} = -\frac{1}{2i} = \frac{i}{2}$$

Hence we can write

$$\frac{\partial w}{\partial \bar{z}} = \left(\frac{1}{2}\frac{\partial u}{\partial x} + \frac{i}{2}\frac{\partial u}{\partial y}\right) + i\left(\frac{1}{2}\frac{\partial v}{\partial x} + \frac{i}{2}\frac{\partial v}{\partial y}\right) = \frac{1}{2}\left(\frac{\partial u}{\partial x} - \frac{\partial v}{\partial y}\right) + \frac{i}{2}\left(\frac{\partial u}{\partial y} + \frac{\partial v}{\partial x}\right)$$

Since w, by hypothesis, is an analytic function, u and v satisfy the Cauchy-Riemann equations. Hence each of the quantities in parentheses in the last expression is equal to zero. Thus $\partial w/\partial \bar{z} \equiv 0$, and therefore w is independent of $\bar{z}$; that is, w depends on x and y only through the combination $z = x + iy$, as asserted. ◗

EXERCISES

Using Theorem 1, compute the derivative of each of the following functions and verify that the derivative also has the asserted form.

1. $w = -2z^2 + 3z + 4;$ $w' = -4z + 3$

2. $w = 5z^2 - 4z + 2;$ $w' = 10z - 4$

3. $w = z^3;\ w' = 3z^2$

4. $w = 1/z;\ w' = -1/z^2$

5. Where are the Cauchy-Riemann equations satisfied by the function $f(z) = xy^2 + ix^2y$? Where does $f'(z)$ exist? Where is $f(z)$ analytic?

6. At what points in the z plane does $f(z) = (z-2)/(z+1)(z^2+1)$ fail to be analytic?

7. Show that at no point in the z plane does the derivative of $f(z) = \text{Re } z = x$ exist. Does this contradict the fact that according to the rules of calculus, $dx/dx = 1$?

8. If $u + iv$ is an analytic function, under what conditions, if any, will $v + iu$ be analytic?

9. If u and v are conjugate harmonic functions, show that v and $-u$ as well as $-v$ and u are also conjugate harmonic functions, but that v and u are not.

10. Verify by direct substitution that Re z^2 and Im z^2 satisfy Laplace's equation.

11. Work Exercise 10 for Re z^3 and Im z^3.

12. Work Exercise 10 for Re z^4 and Im z^4.

13. Does there exist an analytic function for which $u = x^2 + y$?

14. If u and v are harmonic in a region R, show that

$$\left(\frac{\partial u}{\partial y} - \frac{\partial v}{\partial x}\right) + i\left(\frac{\partial u}{\partial x} + \frac{\partial v}{\partial y}\right)$$

is analytic in R. Show by an example that $u + iv$ need not be analytic in R.

15. Is the converse of Property 2 true, i.e., if $u(x, y) = c$ and $v(x, y) = k$ are orthogonal trajectories, is $u + iv$ necessarily an analytic function? *Hint:* Try to find a counterexample.

16. Show that the various values approached by the difference quotient of $f(z) = \bar{z}$ as $\Delta z \to 0$ along the lines $y = mx$ all lie on a circle. *Hint:* Interpret the result obtained in the text as defining the parametric equations of a curve.

17. Show that $u = e^x \cos y$ is a harmonic function and determine v so that $u + iv$ is an analytic function. *Hint:* Use the first of the Cauchy-Riemann equations to determine $\partial v/\partial y$, then integrate this expression with respect to y, and finally use the second Cauchy-Riemann equation to determine the arbitrary function of x introduced by this integration.

18. Work Exercise 17 if $u = e^x \sin y$.

19. Work Exercise 17 if $u = 2x(3 - y)$. Express $u + iv$ as a function of z.

20. Work Exercise 17 if $v = 2x(3 - y)$. Express $u + iv$ as a function of z.

21. Prove that if $f'(z) \equiv 0$, then $f(z)$ is a constant.

22. If both $f(z)$ and $\overline{f(z)}$ are analytic functions, show that $f(z) = c$.

23. If $f(z)$ is an analytic function for which $u^2 + v^2$ is a constant, show that $f(z)$ is a constant. *Hint:* Differentiate the relation $u^2 + v^2 = c$ with respect to x and then with respect to y and consider the resulting equations as two simultaneous linear equations in u and v.

24. Prove L'Hospital's rule for analytic functions: if $f(z)$ and $g(z)$ are analytic functions in a region containing z_0, if $f(z_0) = g(z_0) = 0$ and if $g'(z_0) \neq 0$, then

$$\lim_{z \to z_0} \frac{f(z)}{g(z)} = \frac{f'(z_0)}{g'(z_0)}$$

25. If $f(x, y)$ is regarded as a function of z and $\bar{z}$ via the substitutions $x = (z + \bar{z})/2$ and $y = (z - \bar{z})/2i$, show that

$$\frac{\partial f}{\partial x} = \frac{\partial f}{\partial z} + \frac{\partial f}{\partial \bar{z}} \quad \text{and} \quad \frac{\partial f}{\partial y} = i\left(\frac{\partial f}{\partial z} - \frac{\partial f}{\partial \bar{z}}\right)$$

Obtain expressions for $\partial f/\partial z$ and $\partial f/\partial \bar{z}$ in terms of the x and y partial derivatives of f.

26. If $f(z)$ is an analytic function, show that

$$\left(\frac{\partial^2}{\partial x^2} + \frac{\partial^2}{\partial y^2}\right)|f(z)|^2 = 4|f'(z)|^2$$

27. If $f(z)$ is an analytic function, show that

$$\left[\frac{\partial}{\partial x}|f(z)|\right]^2 + \left[\frac{\partial}{\partial y}|f(z)|\right]^2 = |f'(z)|^2$$

28. If the analytic function $f(z) = u(x, y) + iv(x, y)$ is expressed in terms of the polar coordinates r and θ, show that

$$\frac{\partial u}{\partial r} = \frac{1}{r}\frac{\partial v}{\partial \theta} \quad \text{and} \quad \frac{\partial v}{\partial r} = -\frac{1}{r}\frac{\partial u}{\partial \theta}$$

29. If the analytic function $f(z) = u(x, y) + iv(x, y)$ is expressed in terms of the polar coordinates r and θ, show that

$$f'(z) = (\cos\theta - i\sin\theta)\frac{\partial f}{\partial r} = -\frac{\sin\theta + i\cos\theta}{r}\frac{\partial f}{\partial \theta}$$

30. Show that at the origin the function

$$f(z) = \begin{cases} \dfrac{x^3 - y^3}{x^2 + y^2} + i\dfrac{x^3 + y^3}{x^2 + y^2} & z \neq 0 \\ 0 & z = 0 \end{cases}$$

satisfies the Cauchy-Riemann equations but does not have a derivative. Explain.

31. Show that $u = 2x(3 - y)$ is harmonic and find a harmonic conjugate v. Express the function $f(z) = u + iv$ in terms of z.

32. Using the results of Exercise 28, show that when an analytic function is expressed in terms of polar coordinates, both its real part and its imaginary part satisfy Laplace's equation in polar coordinates:

$$\frac{\partial^2 \phi}{\partial r^2} + \frac{1}{r}\frac{\partial \phi}{\partial r} + \frac{1}{r^2}\frac{\partial^2 \phi}{\partial \theta^2} = 0$$

33. Work Exercise 31 if $u = x^3 - 3xy^2$.

34. **(a)** Where does the function $f(z) = x^3 - i(y-1)^3$ satisfy the Cauchy-Riemann equations?
(b) Where does $f'(z)$ exist? What is its value there?
(c) Where is $f(z)$ analytic? Explain.

35. Work Exercise 34 if $f(z) = xy^2 + ix^2y$.

36. Show that the function $u = \ln(x^2 + y^2)$ is harmonic and find a corresponding analytic function $f(z) = u + iv$ and its derivative $f'(z)$.

37. **(a)** Does the function $f(z) = \begin{cases} e^{-(1/z^4)} & z \neq 0 \\ 0 & z = 0 \end{cases}$ satisfy the Cauchy-Riemann equations at $z = 0$?
(b) For what values of z is $f(z)$ analytic?
(c) Show that f is continuous at $z = 0$.

17.4 THE ELEMENTARY FUNCTIONS OF z

The exponential function e^z is of fundamental importance not only for its own sake but also as a basis for defining all the other elementary transcendental functions of z. In its definition we seek to preserve as many of the familiar properties of the real exponential function e^x as possible. Specifically, we desire that

a. e^z shall be single-valued and analytic.

b. $de^z/dz = e^z$.

c. e^z should reduce to e^x when Im $z = 0$.

If we let

$$e^z = u + iv \tag{1}$$

and recall from Eq. (3), Sec. 17.3, that the derivative of an analytic function can be written in the form

$$f'(z) = \frac{\partial u}{\partial x} + i\frac{\partial v}{\partial x}$$

then, to satisfy Condition b, we must have

$$\frac{\partial u}{\partial x} + i\frac{\partial v}{\partial x} = u + iv$$

Hence, equating real and imaginary parts, we must have

$$\frac{\partial u}{\partial x} = u \tag{2}$$

$$\frac{\partial v}{\partial x} = v \tag{3}$$

Now Eq. (2) will be satisfied if

$$u = e^x\phi(y) \tag{4}$$

where $\phi(y)$ is any function of y. Furthermore, since e^z is to be analytic (Condition a), u and v must satisfy the Cauchy-Riemann equations. Hence, replacing $\partial v/\partial x$ by $-\partial u/\partial y$, according to the second of these equations, Eq. (3) can be written

$$-\frac{\partial u}{\partial y} = v \tag{5}$$

Differentiating this with respect to y, we obtain

$$\frac{\partial^2 u}{\partial y^2} = -\frac{\partial v}{\partial y}$$

or, replacing $\partial v/\partial y$ with $\partial u/\partial x$ according to the first of the Cauchy-Riemann equations,

$$\frac{\partial^2 u}{\partial y^2} = -\frac{\partial u}{\partial x}$$

Finally, when we use (2), this becomes

$$\frac{\partial^2 u}{\partial y^2} = -u$$

which, on substituting $u = e^x\phi(y)$ from (4), reduces to

$$e^x\phi''(y) = -e^x\phi(y) \qquad \text{or} \qquad \phi''(y) = -\phi(y)$$

This is a simple linear differential equation whose solution can be written down at once:

$$\phi(y) = A\cos y + B\sin y$$

Hence, from (4),

$$u = e^x\phi(y) = e^x(A\cos y + B\sin y)$$

and, from (5),

$$v = -\frac{\partial u}{\partial y} = -e^x(-A\sin y + B\cos y)$$

Therefore, from (1),

$$e^z = u + iv = e^x[(A\cos y + B\sin y) + i(A\sin y - B\cos y)]$$

Finally, if this is to reduce to e^x when $y = 0$, as required by Condition c, we must have

$$e^x = e^x(A - iB)$$

which will be true if and only if $A = 1$ and $B = 0$.

Thus we have been led inevitably to the conclusion that *if* there is a function of z satisfying Conditions a, b, and c, *then* it must be

$$e^z = e^{x+iy} = e^x(\cos y + i\sin y) \tag{6}$$

That this expression does indeed meet our requirements can be checked immediately; hence we adopt it as the definition of e^z.

It is important to note that the right-hand side of (6) is in standard polar form. Hence

$$\text{mod } e^z = |e^z| = e^x \qquad \text{and} \qquad \arg e^z = y$$

The possibility of writing any complex number z in exponential form is now apparent, for, applying (6) with $x = 0$ and $y = \theta$, we have

$$\cos\theta + i\sin\theta = e^{i\theta} \tag{7}$$

and thus

$$z = x + iy = r(\cos\theta + i\sin\theta) = re^{i\theta} \tag{8}$$

The fact that the angle, or argument, of a complex number is actually an exponent explains why the angles of complex numbers are added when the numbers are multiplied and subtracted when the numbers are divided, as we discovered in Sec. 3.1.

Our derivation of Eq. (8) provides the logical justification for the exponential way of writing the roots of a general complex number, which we inferred and used in Sec. 3.1. In fact, from

$$z = re^{i\theta} = re^{i(\theta+2k\pi)}$$

we have

$$z^{1/p} = (re^{i(\theta+2k\pi)})^{1/p} = r^{1/p}e^{i(\theta+2k\pi)/p}$$

In particular, for $r = 1$, $p = n$, and $\theta = 0$ we have the n nth roots of 1 expressed in the form

$$e^{2ki\pi/n} \qquad k = 0, 1, 2, \ldots, n-1$$

Of these, $w_n = e^{2i\pi/n}$ is the primitive nth root of 1. General rational powers of z can of course be handled in the same way.

From the relation

$$e^{i\theta} = \cos\theta + i\sin\theta$$

and its obvious companion

$$e^{-i\theta} = \cos(-\theta) + i\sin(-\theta) = \cos\theta - i\sin\theta$$

we obtain, by addition and subtraction, the so-called **Euler formulas** for the sine and cosine which we inferred in Sec. 3.1:

$$\cos\theta = \frac{e^{i\theta} + e^{-i\theta}}{2} \qquad \text{and} \qquad \sin\theta = \frac{e^{i\theta} - e^{-i\theta}}{2i}$$

On the basis of these equations, we extend the definitions of the sine and cosine into the complex domain by the formulas

(9) $$\cos z = \frac{e^{iz} + e^{-iz}}{2}$$

(10) $$\sin z = \frac{e^{iz} - e^{-iz}}{2i}$$

From these definitions it is easy to establish the validity of such familiar formulas as

$$\cos^2 z + \sin^2 z = 1$$
$$\cos(z_1 \pm z_2) = \cos z_1 \cos z_2 \mp \sin z_1 \sin z_2$$
$$\sin(z_1 \pm z_2) = \sin z_1 \cos z_2 \pm \cos z_1 \sin z_2$$
$$\frac{d(\cos z)}{dz} = -\sin z$$
$$\frac{d(\sin z)}{dz} = \cos z$$

If we expand the exponentials in (9), we find

$$\begin{aligned}\cos z &= \frac{e^{i(x+iy)} + e^{-i(x+iy)}}{2} \\ &= \frac{e^{-y}e^{ix} + e^{y}e^{-ix}}{2} \\ &= \frac{e^{-y}(\cos x + i \sin x) + e^{y}(\cos x - i \sin x)}{2} \\ &= \cos x \frac{e^{y} + e^{-y}}{2} - i \sin x \frac{e^{y} - e^{-y}}{2}\end{aligned}$$

or, using the usual definitions of the hyperbolic functions of real variables,

$$\cos z = \cos (x + iy) = \cos x \cosh y - i \sin x \sinh y \tag{11}$$

Similarly, it is easy to show that

$$\sin z = \sin (x + iy) = \sin x \cosh y + i \cos x \sinh y \tag{12}$$

In particular, taking $x = 0$ in (11) and (12), we find

$$\cos iy = \cosh y \tag{13}$$

$$\sin iy = i \sinh y \tag{14}$$

The remaining trigonometric functions of z are defined in terms of $\cos z$ and $\sin z$ by means of the usual identities.

EXAMPLE 1

What is $\cos (1 + 2i)$?

By direct use of (11), we have

$$\begin{aligned}\cos (1 + 2i) &= \cos 1 \cosh 2 - i \sin 1 \sinh 2 \doteq (0.5403)(3.7622) - i(0.8415)(3.6269) \\ &\doteq 2.033 - 3.052i\end{aligned}$$

EXAMPLE 2

THE ZEROS OF $\sin z$

Prove that the only values of z for which $\sin z = 0$ are the real values $z = 0, \pm\pi, \pm 2\pi, \ldots$

From (12), $\sin z = \sin x \cosh y + i \cos x \sinh y$. Hence if $\sin z$ is to vanish, it is necessary that simultaneously

$$\sin x \cosh y = 0 \qquad \text{and} \qquad \cos x \sinh y = 0$$

Since y is a real number, it follows from the familiar properties of the hyperbolic cosine that $\cosh y \geq 1$. Hence the first of these equations can hold only if $\sin x = 0$, that is, only if

$$x = 0, \pm\pi, \pm 2\pi, \ldots$$

But for these values of x, $\cos x$ is either 1 or -1 and therefore cannot vanish. Thus for the second equation to hold, it is necessary that $\sinh y = 0$. Since y is real, the familiar properties of the hyperbolic sine can be invoked, leading to the conclusion that

$$y = 0$$

Hence the only values of z for which $\sin z = 0$ are of the form

$$z = n\pi + 0i = n\pi \qquad n = 0, \pm 1, \pm 2, \ldots$$

Since a function of a complex variable $w = f(z)$, that is,

$$u + iv = f(x + iy)$$

involves *four* real variables, x, y, u, and v, conventional methods are inadequate for the geometric representation of such functions. Instead, it is customary to regard a function $w = f(z)$ as a transformation between an xy plane, where $z = x + iy$ is plotted, and a uv plane, where $w = u + iv$ is plotted. From this point of view, loci in the z plane are mapped point by point into corresponding loci in the w plane. This is an extremely important idea, and all of Chap. 20 will be devoted to this concept and its applications. Our next example is a preliminary illustration of these ideas, showing how certain lines in the z plane are mapped into curves in the w plane by the particular function $w = \cos z$ that we have just defined.

EXAMPLE 3

THE MAPPING $w = \cos z$

Into what curves in the w plane does the function $w = \cos z$ transform lines of the families $x = c$ and $y = k$ in the z plane?

From Eq. (11), we have $w = \cos z = \cos x \cosh y - i \sin x \sinh y$. Hence

$$u = \cos x \cosh y \qquad \text{and} \qquad v = -\sin x \sinh y$$

If x is given the constant value c, these become the parametric equations of a curve in the w, or uv plane, having y as parameter. To eliminate the parameter y, we solve them for $\cosh y$ and $\sinh y$ and then square and subtract, getting

$$\frac{u^2}{\cos^2 c} - \frac{v^2}{\sin^2 c} = 1 \qquad c \neq \frac{n\pi}{2}$$

which is the equation of a hyperbola.

Similarly, if y is given the constant value k, we have another set of parametric equations, with x as a parameter. To eliminate the parameter this time, we solve for $\cos x$ and $\sin x$ and square and add, obtaining

$$\frac{u^2}{\cosh^2 k} + \frac{v^2}{\sinh^2 k} = 1 \qquad k \neq 0$$

which is the equation of an ellipse. Several members of each of these families of conics are shown in Fig. 17.4.

The curves shown in Fig. 17.4 appear to be orthogonal trajectories, and this is indeed the case. To see why this is true, it is convenient to consider the inverse of the function $w = f(z) = \cos z$, namely,

$$z = f^{-1}(w) = \cos^{-1} w \qquad \text{or} \qquad z = x(u, v) + iy(u, v) = \cos^{-1} w$$

Since the inverse function $\cos^{-1} w$ has a derivative (see Exercise 20), it is analytic and therefore, according to Property 2, Sec. 17.3, its real and imaginary parts, when set equal to constants:

$$x(u, v) = c \qquad \text{and} \qquad y(u, v) = k$$

define two families of orthogonal trajectories, as Fig. 17.4 shows.

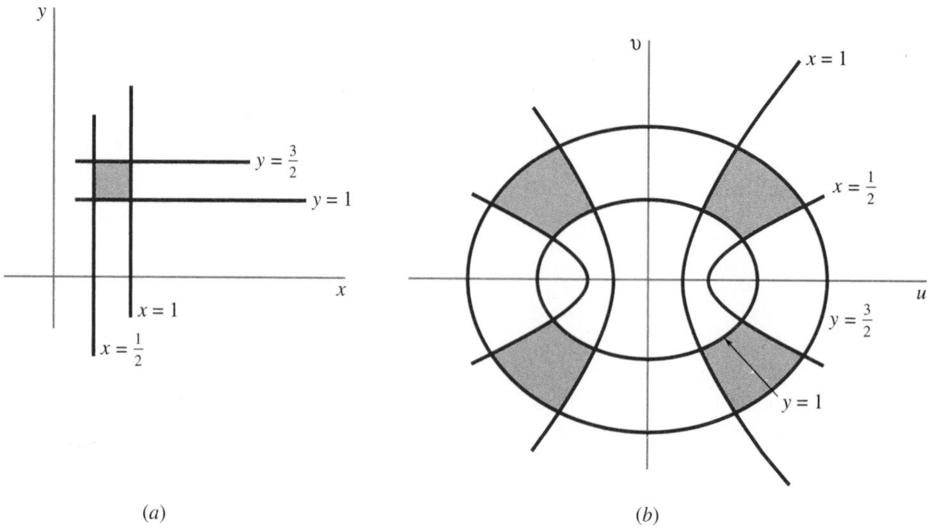

FIGURE 17.4

Typical members of the families $\dfrac{u^2}{\cos^2 x} - \dfrac{v^2}{\sin^2 x} = 1$ and $\dfrac{u^2}{\cosh^2 y} + \dfrac{v^2}{\sinh^2 y} = 1$.

The mapping defined by $w = \cos z$ is not quite as simple as Example 3 made it seem. Additional aspects of it are explored in Exercises 46–51.

The hyperbolic functions of z we define simply by extending the familiar definitions into the complex number field:

(15) $$\cosh z = \frac{e^z + e^{-z}}{2}$$

(16) $$\sinh z = \frac{e^z - e^{-z}}{2}$$

By expanding the exponentials and regrouping, as we did in deriving (11), we obtain without difficulty the formulas

(17) $$\cosh z = \cosh x \cos y + i \sinh x \sin y$$

(18) $$\sinh z = \sinh x \cos y + i \cosh x \sin y$$

In particular, setting $x = 0$, we find

(19) $$\cosh iy = \cos y$$

(20) $$\sinh iy = i \sin y$$

Upon comparing the pair of equations (19) and (20) with the pair (13) and (14) we see a striking resemblance which makes it easier to remember these four identities. The remaining hyperbolic functions are defined in terms of $\cosh z$ and $\sinh z$ via the usual identities.

The logarithm of z we define implicitly as the function $w = \ln z$ which satisfies the equation

(21) $$e^w = z$$

If we let $w = u + iv$ and $z = re^{i\theta}$, Eq. (21) becomes

$$e^{u+iv} = e^u e^{iv} = re^{i\theta}$$

Hence $e^u = r$, or $u = \ln r$, and $v = \theta$. Thus

$$w = u + iv = \ln r + i\theta$$

$$= \ln |z| + i \arg z \tag{22}$$

If we let θ_1 be the **principal argument** of z, that is, the particular argument of z which lies in the interval $-\pi < \theta \leq \pi$, Eq. (22) can be written

$$\ln z = \ln |z| + i(\theta_1 + 2n\pi) \qquad n = 0, \pm 1, \pm 2, \ldots \tag{22a}$$

which shows that the logarithmic function is infinitely many-valued. For any particular value of n a unique branch of the function is determined, and the logarithm becomes effectively single-valued. If $n = 0$, the resulting branch of the logarithmic function is called the **principal value.**

For every n, the corresponding branch of $\ln z$ is obviously discontinuous at $z = 0$. Moreover, for each n the corresponding branch is also discontinuous at every point of the negative real axis. To verify this, we note that if $n = n_0$, the corresponding branch of $\ln z$ is, by Eq. (22*a*) and the restriction that $-\pi < \theta_1 \leq \pi$,

$$\ln |z| + i \arg z \qquad \text{where } (2n_0 - 1)\pi < \arg z \leq (2n_0 + 1)\pi$$

Hence, if P is an arbitrary point on the negative real axis, the limit of $\arg z$ as z approaches P through the second quadrant is $(2n_0 + 1)\pi$, while the limit of $\arg z$ as z approaches P through the third quadrant is $(2n_0 - 1)\pi$. Since these two values are different, it follows that on any particular branch, $\ln z$ does not approach a limit as z approaches an arbitrary point on the negative real axis and therefore is discontinuous at every such point.

At all points except the points on the nonpositive real axis, each branch of $\ln z$ is continuous and analytic. In fact, from the definition

$$\ln z = \ln |z| + i \arg z = \frac{1}{2} \ln (x^2 + y^2) + i \tan^{-1} \frac{y}{x}$$

it is easy to verify that the Cauchy-Riemann equations are satisfied everywhere except at the origin. Moreover, from the preceding discussion, it is clear that

$$u = \frac{1}{2} \ln (x^2 + y^2) \qquad \text{and} \qquad v = \tan^{-1} \frac{y}{x}$$

are continuous except on the nonpositive real axis. Hence, by Theorem 1, Sec. 17.3, it follows that everywhere except on the nonpositive real axis

$$\frac{d(\ln z)}{dz} = \frac{\partial u}{\partial x} + i\frac{\partial v}{\partial x}$$

$$= \frac{x}{x^2 + y^2} - i\frac{y}{x^2 + y^2} = \frac{x - iy}{x^2 + y^2} = \frac{\bar{z}}{z\bar{z}} = \frac{1}{z}$$

as expected.

The familiar laws for the logarithms of real quantities all hold for the logarithms of complex quantities in the following sense. If z_1 and z_2 are given, then among the infinite set of values for $\ln z_1 z_2$ there is *one* for which the relation

$$\ln z_1 z_2 = \ln z_1 + \ln z_2$$

holds. Similarly, among the infinite set of values of $\ln (z_1/z_2)$ there is *one* for which the relation

$$\ln \frac{z_1}{z_2} = \ln z_1 - \ln z_2$$

is correct. Likewise, among the infinite set of values of $\ln z^m$ there is *one* for which the relation

$$\ln z^m = m \ln z$$

is correct. For example, to show that $\ln z_1 z_2 = \ln z_1 + \ln z_2$ in this sense, let $z_1 = r_1 e^{i(\theta_1 + 2n_1\pi)}$ and $z_2 = r_2 e^{i(\theta_2 + 2n_2\pi)}$ where θ_1 and θ_2 are the principal arguments of z_1 and z_2, respectively. Then,

$$\begin{aligned} \ln z_1 + \ln z_2 &= [\ln r_1 + i(\theta_1 + 2n_1\pi)] + [\ln r_2 + i(\theta_2 + 2n_2)\pi] \\ &= (\ln r_1 + \ln r_2) + i[(\theta_1 + \theta_2) + 2(n_1 + n_2)\pi] \\ &= \ln r_1 r_2 + i[(\theta_1 + \theta_2) + 2n_3\pi] \\ &= \ln |z_1 z_2| + i \arg z_1 z_2 \\ &= \ln z_1 z_2 \end{aligned}$$

since $(\theta_1 + \theta_2) + 2(n_1 + n_2)\pi$ is *one* of the arguments of $z_1 z_2$.

However, the familiar laws of logarithms are *not* necessarily true if we insist that all values are to be read from the same branch of the logarithmic function. For instance, in terms of principal values,

$$\ln i = \ln 1 + i\frac{\pi}{2} = i\frac{\pi}{2} \qquad \text{and} \qquad \ln(-1 + i) = \ln\sqrt{2} + i\frac{3\pi}{4}$$

and for the logarithm of the product $i(-1 + i)$ we have by direct calculation

$$\ln i(-1 + i) = \ln(-1 - i) = \ln\sqrt{2} + i\left(-\frac{3\pi}{4}\right)$$

However, adding principal values of the logarithms of $z_1 = i$ and $z_2 = -1 + i$, gives the value

$$i\frac{\pi}{2} + \left(\ln\sqrt{2} + i\frac{3\pi}{4}\right) = \ln\sqrt{2} + i\frac{5\pi}{4}$$

Clearly, the principal value of $\ln i(-1 + i)$ differs from the sum of the principal values of $\ln i$ and $\ln(-1 + i)$ by $2\pi i$, and so the formula $\ln z_1 + \ln z_2 = \ln z_1 z_2$ fails to hold.

For principal values, the proper generalizations of the familiar laws of logarithms are contained in the following theorem whose proof we shall leave as an exercise.

THEOREM 1 The principal value of $\ln z$ satisfies the following relations:

$$\ln z_1 z_2 = \begin{cases} \ln z_1 + \ln z_2 + 2i\pi & -2\pi < \arg z_1 + \arg z_2 \le -\pi \\ \ln z_1 + \ln z_2 & -\pi < \arg z_1 + \arg z_2 \le \pi \\ \ln z_1 + \ln z_2 - 2i\pi & \pi < \arg z_1 + \arg z_2 \le 2\pi \end{cases}$$

$$\ln \frac{z_1}{z_2} = \begin{cases} \ln z_1 - \ln z_2 + 2i\pi & -2\pi < \arg z_1 - \arg z_2 \le -\pi \\ \ln z_1 - \ln z_2 & -\pi < \arg z_1 - \arg z_2 \le \pi \\ \ln z_1 - \ln z_2 - 2i\pi & \pi < \arg z_1 - \arg z_2 \le 2\pi \end{cases}$$

$$\ln z^m = m \ln z - 2ki\pi$$

where k is the unique integer such that $(m/2\pi)\arg z - \frac{1}{2} \le k < (m/2\pi)\arg z + \frac{1}{2}$.

General powers of z are defined by the formula

$$z^\alpha = e^{\alpha \ln z} \tag{23}$$

which generalizes a familiar result for real variables which we frequently found useful in solving linear first-order differential equations. Since $\ln z$ is infinitely many-valued, so too is z^α, in general. Specifically,

$$\begin{aligned} z^\alpha = e^{\alpha \ln z} &= e^{\alpha[\ln |z| + i(\theta_1 + 2n\pi)]} \\ &= e^{\alpha \ln |z|}\, e^{\alpha\theta_1 i} e^{2n\alpha\pi i} \end{aligned}$$

The last factor in this product clearly involves infinitely many different values unless α is a rational number, say p/q, in which case, as we saw in our discussion of Demoivre's theorem in Sec. 3.1, there are only q distinct values.†

EXAMPLE 4

What is the principal value of $(1+i)^i$?

By definition,

$$\begin{aligned} (1+i)^i &= e^{i \ln (1+i)} \\ &= e^{i\{\ln \sqrt{2} + i[(\pi/4) + 2n\pi]\}} \\ &= e^{-[(\pi/4)+2n\pi]}\, e^{i \ln \sqrt{2}} \end{aligned}$$

The principal value of this, obtained by taking $n = 0$, is

$$\begin{aligned} &e^{-(\pi/4)}[\cos(\ln \sqrt{2}) + i \sin(\ln \sqrt{2})] \\ &\doteq e^{-0.7854}(\cos 0.3466 + i \sin 0.3466) \\ &\doteq 0.429 + 0.155i \end{aligned}$$

The inverse trigonometric and hyperbolic functions we define implicitly. For instance,

$$w = \cos^{-1} z$$

is defined as the value or values of w satisfying the equation

$$z = \cos w = \frac{e^{iw} + e^{-iw}}{2}$$

From this, by obvious steps, we obtain successively

$$\begin{aligned} e^{2iw} - 2ze^{iw} + 1 &= 0 \\ e^{iw} &= z \pm \sqrt{z^2 - 1} \end{aligned}$$

and finally, by taking logarithms and solving for w,

$$w = \cos^{-1} z = -i \ln (z \pm \sqrt{z^2 - 1}) \tag{24}$$

†However, in the particular case $z = e$, the expression $z^\alpha = e^\alpha$ is single-valued for all values of α, rational or not, since $e^{(\alpha_r + i\alpha_i)}$ was defined simply as $e^{\alpha_r}(\cos \alpha_i + i \sin \alpha_i)$, which is clearly a unique complex number.

Since the logarithm is infinitely many-valued, so too is $\cos^{-1} z$.
Similarly, we can obtain the formulas

$$\sin^{-1} z = -i \ln (iz \pm \sqrt{1 - z^2}) \tag{25}$$

$$\tan^{-1} z = \frac{i}{2} \ln \frac{i + z}{i - z} \tag{26}$$

$$\cosh^{-1} z = \ln (z \pm \sqrt{z^2 - 1}) \tag{27}$$

$$\sinh^{-1} z = \ln (z \pm \sqrt{z^2 + 1}) \tag{28}$$

$$\tanh^{-1} z = \frac{1}{2} \ln \frac{1 + z}{1 - z} \tag{29}$$

From these, after their principal values have been suitably defined by choosing the plus sign preceding the square root and the principal value of the logarithm in each case, the usual differentiation formulas can be obtained without difficulty.

EXERCISES

1. Express the principal value of each of the following in the form $a + ib$, where a and b are decimal fractions.
(a) $\ln(-10)$ **(b)** $\ln(1 - i\sqrt{3})$ **(c)** $\sin(1 + i)$
(d) $\cosh(1 - i)$ **(e)** i^i **(f)** $\cos^{-1} 2$

2. Express the principal value of each of the following in the form $a + ib$, where a and b are decimal fractions.
(a) $\ln(-3 + 4i)$ **(b)** 2^i **(c)** $\tan i$
(d) $\tanh^{-1} 2$ **(e)** $(1 - i)^{2+i}$ **(f)** $(2 + i)^{1-i}$

3. Show that all the values of $(1 + i)^{1-i}$ have the same direction.

4. What condition(s) on a, b, c, d will ensure that the values of $(a + ib)^{c+id}$ have the same direction?

5. Show that the moduli of the successive values of $(a + ib)^{c+id}$ form a geometric progression.

Show that over the appropriate regions, each of the following is a solution of Laplace's equation.

6. $\text{Re}(\cos z)$, $\text{Im}(\cos z)$ **7.** $\text{Re}(\sin z)$, $\text{Im}(\sin z)$

8. $\text{Re}(\cosh z)$, $\text{Im}(\cosh z)$ **9.** $\text{Re}(\sinh z)$, $\text{Im}(\sinh z)$

10. $\text{Re}(\ln z)$, $\text{Im}(\ln z)$ **11.** $\text{Re}(1/z)$, $\text{Im}(1/z)$

12. Using Eq. (6), prove that $e^{z_1}e^{z_2} = e^{z_1+z_2}$.

13. Prove that $\cos^2 z + \sin^2 z = 1$.

14. Prove that $\cos(z_1 \pm z_2) = \cos z_1 \cos z_2 \mp \sin z_1 \sin z_2$.

15. Prove that $\sin(z_1 \pm z_2) = \sin z_1 \cos z_2 \pm \cos z_1 \sin z_2$.

16. Prove that $d(\cos z)/dz = -\sin z$.

17. Prove that $d(\sin z)/dz = \cos z$.

18. Prove that $d(\tan z)/dz = \sec^2 z$.

19. What is the derivative of **(a)** $\cosh z$? **(b)** $\sinh z$? **(c)** $\tanh z$?

20. Show that: **(a)** $d(\cos^{-1} z)/dz = -1/\sqrt{1 - z^2}$,
(b) $d(\sin^{-1} z)/dz = 1/\sqrt{1 - z^2}$,
and **(c)** $d(\tan^{-1} z)/dz = 1/(1 + z^2)$.

21. What is **(a)** $d(\cosh^{-1} z)/dz$? **(b)** $d(\sinh^{-1} z)/dz$? **(c)** $d(\tanh^{-1} z)/dz$?

22. Prove that there is no value for z for which $e^z = 0$.

23. Find all the values of z for which $\cosh z = 0$.

24. If $g(x, y)$ is a real function of x and y, what is $|e^{ig(x,y)}|$?

25. Find all solutions of the equation $e^z = -2$.

26. Find all solutions of the equation $\sin z = 3$.

27. Find all solutions of the equation $\cosh z = -2$.

28. Prove that $\overline{e^z} = e^{\bar{z}}$. **29.** Prove that $\overline{\cos z} = \cos \bar{z}$

30. Is $\overline{\ln z} = \ln \bar{z}$? **31.** Is $\overline{\sin z} = \sin \bar{z}$?

32. By inspection, $e^0 = 1 > 0$ and $e^{i\pi} = -1 < 0$, yet by Exercise 22 there is no value of z for which $e^z = 0$ even though e^z is everywhere continuous. Explain.

33. Show that Rolle's theorem fails to hold for the function $e^{iz} - 1$ even though the conditions of the theorem appear to be satisfied with respect to the two values $z = 0$ and $z = 2\pi$.

34. **(a)** Show that $|\sin z|^2 = \sin^2 x + \sinh^2 y$. *Hint:* At the appropriate point, express $\cos^2 x$ and $\cosh^2 y$ in terms of $\sin^2 x$ and $\sinh^2 y$, respectively.
(b) Obtain a similar expression for $|\cos z|^2$.

35. **(a)** Show that $|\sinh z|^2 = \sinh^2 x + \sin^2 y$.
(b) Obtain a similar expression for $|\cosh z|^2$.

36. If $z = x + iy$, show that $|\sinh y| \le |\sin z| \le |\cosh y|$.

37. If $|z| \le 1$, show that $|\sin z| \le \frac{6}{5}|z|$.

38. Obtain an upper bound for $|\cos z|$ if $|z| \le 1$.

39. Show that the relation $|f(x + iy)| = |f(x) + f(iy)|$ is satisfied by $f(z) = \sin z$. Is it satisfied by $f(z) = \cos z$?

40. Show that $\tan z = \dfrac{\sin 2x + i \sinh 2y}{\cos 2x + \cosh 2y}$.

41. Show that $\tanh z = \dfrac{\sinh 2x + i \sin 2y}{\cosh 2x + \cos 2y}$.

42. Show that the equation $\tan z = i$ has no solution. Does the equation $\tanh z = i$ have a solution?

43. Show that if $z = x$ is a real root of the equation $\tan z = \tanh z$, then $z = ix$ is also a root.

44. If a is real, show that all roots of the equation $\tan z = az$

are real, with the exception of one pure imaginary root if $0 < a < 2$.

45. If $w = \ln z$, what is the net change in w as z varies continuously in the counterclockwise direction around the curve C and returns to its initial position if **(a)** C is the curve $|z| = 1$? **(b)** C is the curve $|z - 1| = 2$? **(c)** C is the curve $|z - 2| = 1$?

46. In Example 3, show that the ellipses are confocal, with foci at $z = \pm 1$. *Hint:* The foci of the ellipse

$$\frac{x^2}{a^2} + \frac{y^2}{b^2} = 1$$

are the points $\pm c$, where $c^2 = a^2 + b^2$.

47. In Example 3, show that the hyperbolas are all confocal, with foci at $z = \pm 1$.

48. In Example 3, show that the entire w plane is covered in a 1:1 fashion by the images of the points in any strip of the z plane of the form $n\pi \le x \le (n + 1)\pi$. *Hint:* Remember the periodic properties of $\cos x$ and $\sin x$.

49. In Example 3, if $0 < c < \pi/2$, show that the images of the points of the line $x = c$ form only the right half of the hyperbola $u^2/\cos^2 c - v^2/\sin^2 c = 1$. What is the locus of the preimages of the left half of this hyperbola? *Hint:* Work with the parametric equations of the hyperbola.

50. In Example 3, if $k > 0$, show that for $0 < x < \pi$ the images of the points of the line $y = k$ form only the lower half of the ellipse $u^2/\cosh^2 k + v^2/\sinh^2 k = 1$. What is the locus of the preimages of the points of the upper half of this ellipse? *Hint:* Use the parametric equations of the ellipse.

51. In Example 3, what is the locus of the images of the points of the line $x = 0$? What is the locus of the images of the points of the line $y = 0$?

52. If $w = \sin z$, what are the equations of the curves in the uv plane into which the lines $x = c$ and $y = k$ are transformed by w?

53. Work Exercise 52 if $w = \cosh z$.

54. Work Exercise 52 if $w = \sinh z$.

55. Show that for every complex number $z = x + iy$, $|\sin z| \ge |\sin x|$.

56. Prove Theorem 1.

57. For what values of z is $\cos z = 0$?

58. Find all values of $\tan^{-1} 2i$.

17.5 INTEGRATION IN THE COMPLEX PLANE: CAUCHY'S THEOREM

Line integrals in the complex plane are defined as follows. Let $f(z) = u(x, y) + iv(x, y)$ be any continuous function of z, analytic or not, and let C be a sectionally smooth arc joining the points A and B. Divide C into n smaller arcs Δs_k by the points z_k $(k = 1, 2, \ldots, n - 1)$ and let $\Delta z_k = z_k - z_{k-1}$ be the infinitesimal chord determined by Δs_k. Finally in each segment Δs_k on C, choose an arbitrary point $\zeta_k = \xi_k + i\eta_k$ (Fig. 17.5). Then, if it exists, the limit of the sum

$$(1) \qquad \sum_{k=1}^{n} f(\zeta_k)\,\Delta z_k$$

as n becomes infinite in such a way that the length of each chord Δz_k approaches zero is called the **line integral** of $f(z)$ along C:

$$(2) \qquad \int_C f(z)\,dz = \lim_{n\to\infty} \sum_{k=1}^{n} f(\zeta_k)\,\Delta z_k$$

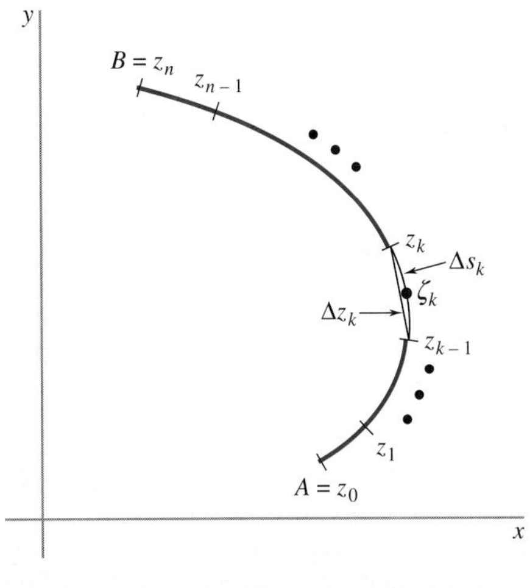

FIGURE 17.5
The subdivision of an arc preparatory to defining a line integral in the complex plane.

In the special case in which A and B coincide and the path of integration forms a closed curve, the integral in (2) is often called a **contour integral** and is sometimes denoted by the symbol

$$\oint f(z)\,dz$$

In working with complex line integrals it is frequently necessary to establish bounds on their absolute values. To do this, let us return to the definitive sum (1) and apply to it the fundamental fact that the absolute value of a sum of complex numbers is less than or equal to the sum of their absolute values [Eq. (11), Sec. 3.1]. Then

$$\left|\sum_{k=1}^{n} f(\zeta_k)\,\Delta z_k\right| \le \sum_{k=1}^{n} |f(\zeta_k)\,\Delta z_k| = \sum_{k=1}^{n} |f(\zeta_k)|\,|\Delta z_k|$$

the last equality following from the fact that the absolute value of a product is equal to the product of the absolute values [Eq. (12f), Sec. 3.1]. As $n \to \infty$, this yields a corresponding inequality for the integrals which are the limits of the respective sums:

(3) $$\left|\int_C f(z)\,dz\right| \le \int_C |f(z)|\,|dz|$$

The integral on the right is the real line integral

$$\int_C \sqrt{u^2+v^2}\sqrt{(dx)^2+(dy)^2} = \int_C \sqrt{u^2+v^2}\,ds$$

where ds is the differential of arc length on C, which of course exists since C is assumed to be sectionally smooth. In particular, if $f(z) \equiv 1$, we have the simple but important result

(4) $$\int_C |dz| = \int_C ds = L$$

where L is the length of the path of integration. Since $f(z)$ is assumed to be continuous on the path of integration, including the endpoints A and B, it follows that $f(z)$ is a bounded function of z on the path of integration; in other words, there exists a constant M such that $|f(z)| \le M$ for all values of z on C. Hence we have, from (3),

$$\left|\int_C f(z)\,dz\right| \le \int_C |f(z)|\,|dz| \le \int_C M|dz| = M\int_C |dz|$$

Therefore, using (4), we obtain the important inequality

(5) $$\left|\int_C f(z)\,dz\right| \le ML$$

where M is any bound for $|f(z)|$ on the path of integration and L is the length of the path of integration.

Complex line integrals can readily be expressed in terms of real line integrals.† For the sum (1) can be written

$$\sum_{k=1}^{n} [u(\xi_k, \eta_k) + iv(\xi_k, \eta_k)](\Delta x_k + i\,\Delta y_k) = \sum_{k=1}^{n} [u(\xi_k, \eta_k)\,\Delta x_k - v(\xi_k, \eta_k)\,\Delta y_k] + i\sum_{k=1}^{n} [v(\xi_k, \eta_k)\,\Delta x_k + u(\xi_k, \eta_k)\,\Delta y_k]$$

and, in the limit, the last expression yields the relation

$$\int_C f(z)\,dz = \int_C u\,dx - v\,dy + i\int_C v\,dx + u\,dy = \int_C (u + iv)(dx + i\,dy) \tag{6}$$

From (6) and the known properties of real line integrals (Sec. 15.6) or directly from the definition (2), it is easy to see that when the same path of integration is used in each integral, we have

$$\int_A^B f(z)\,dz = -\int_B^A f(z)\,dz \tag{7}$$

$$\int_A^B kf(z)\,dz = k\int_A^B f(z)\,dz \tag{8}$$

$$\int_A^B [f(z) \pm g(z)]\,dz = \int_A^B f(z)\,dz \pm \int_A^B g(z)\,dz \tag{9}$$

and if D is a third point on the arc AB,

$$\int_A^B f(z)\,dz = \int_A^D f(z)\,dz + \int_D^B f(z)\,dz \tag{10}$$

In evaluating complex integrals, it is not necessary that they first be reduced to the form (6). Often the path of integration C is a curve with parametric representation

$$z = z(t) = x(t) + iy(t) \qquad t \text{ real}$$

and the integration can be carried out directly in terms of the parameter t. In other words, when $z = z(t)$, we can write

$$\int_A^B f(z)\,dz = \int_{A'}^{B'} f[z(t)]z'(t)\,dt$$

where A' and B' are the values of t that correspond, respectively, to the values $z = A$ and $z = B$, and then integrate with respect to t. Our first example illustrates the important procedure of integration with respect to a parameter.

†A discussion of real line integrals, with examples, is given in Sec. 15.6.

EXAMPLE 1

INTEGRATING $1/(z - z_0)^{n+1}$ AROUND A CIRCLE

If C is a circle of radius r and center z_0 and if n is an integer, determine the value of

$$\int_C \frac{dz}{(z - z_0)^{n+1}}$$

For convenience, let us make the substitution $z - z_0 = re^{i\theta}$ noting that the parameter θ varies from 0 to 2π as z varies around the circle in the counterclockwise direction (Fig. 17.6). Then $dz = rie^{i\theta}\,d\theta$, and the integral becomes

$$\int_0^{2\pi} \frac{rie^{i\theta}\,d\theta}{r^{n+1}e^{i(n+1)\theta}} = \frac{i}{r^n}\int_0^{2\pi} e^{-in\theta}\,d\theta$$

If $n = 0$, this reduces to $i\int_0^{2\pi} d\theta = 2\pi i$. If $n \neq 0$, we have

$$\frac{i}{r^n}\int_0^{2\pi} (\cos n\theta - i \sin n\theta)\,d\theta = 0$$

This is an important result to which we shall have occasion to refer from time to time.

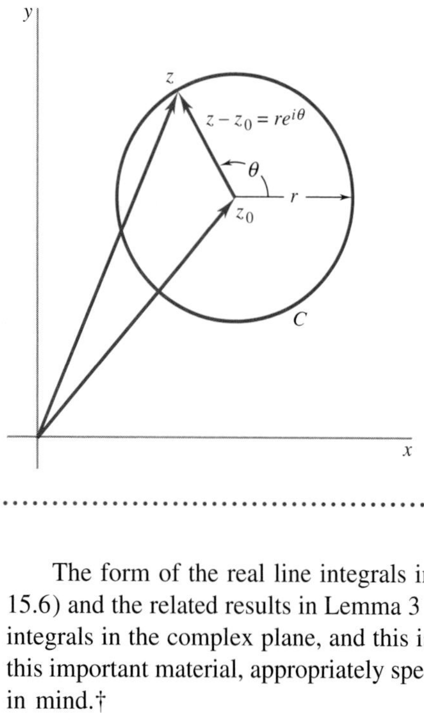

FIGURE 17.6
The circle $z - z_0 = re^{i\theta}$.

The form of the real line integrals in Eq. (6) suggests that Green's lemma (Theorem 1, Sec. 15.6) and the related results in Lemma 3 and Theorem 5, Sec. 15.7, may be useful in studying line integrals in the complex plane, and this is indeed the case. Hence, for ease of reference, we repeat this important material, appropriately specialized to the two-dimensional applications we now have in mind.†

†To avoid confusion with u and v in the standard notation for a function of a complex variable, namely, $f(z) = u + iv$, we here use P and Q in place of the symbols U and V we used in Chap. 15.

THEOREM 1 **(Green's Lemma)** If R is a region, either simply or multiply connected, whose boundary C is sectionally smooth and if $P(x, y)$, $Q(x, y)$, $\partial P/\partial y$, and $\partial Q/\partial x$ are continuous in and on the boundary of R, then

$$\int_C P\,dx + Q\,dy = \int_R\int \left(\frac{\partial Q}{\partial x} - \frac{\partial P}{\partial y}\right) dx\,dy$$

where the integration is taken around C in the positive direction with respect to the interior of R.

THEOREM 2 In any region where $\int P\,dx + Q\,dy$ is independent of the path, the partial derivatives of the function

$$\phi(x, y) = \int_{a,b}^{x,y} P(x, y)\,dx + Q(x, y)\,dy$$

are

$$\frac{\partial \phi}{\partial x} = P(x, y) \qquad \text{and} \qquad \frac{\partial \phi}{\partial y} = Q(x, y)$$

THEOREM 3 If $\partial Q/\partial x = \partial P/\partial y$ at all points of a simply connected region R, then in R the integral

$$\int P\,dx + Q\,dy$$

is independent of the path, and conversely.

As a first application of Green's lemma, we have *Cauchy's theorem,* perhaps the most fundamental and far-reaching result in the theory of analytic functions.

THEOREM 4 **(Cauchy's Theorem)** If R is a region, either simply or multiply connected, whose boundary is sectionally smooth and if $f(z)$ is analytic and $f'(z)$ is continuous in and on the boundary of R, then

$$\int_C f(z)\,dz = 0$$

where C is the entire boundary of R.

PROOF We begin by recalling from Eq. (6) that

$$\int_C f(z)\,dz = \int_C u\,dx - v\,dy + i\int_C v\,dx + u\,dy$$

Now the hypothesis that $f'(z)$ is continuous means that the partial derivatives $\partial u/\partial x$, $\partial u/\partial y$, $\partial v/\partial x$, $\partial v/\partial y$ exist and are continuous throughout R. Hence Green's lemma can be applied to each of the integrals on the right in the last expression, giving

$$\int_C f(z)\,dz = \int_R\int \left(-\frac{\partial v}{\partial x} - \frac{\partial u}{\partial y}\right) dx\,dy + i\int_R\int \left(\frac{\partial u}{\partial x} - \frac{\partial v}{\partial y}\right) dx\,dy$$

However, u and v necessarily satisfy the Cauchy-Riemann equations, since, by hypothesis, $f(z)$ is analytic. Therefore the integrand of each of the double integrals vanishes identically in R, leaving

$$\int_C f(z) = 0$$

as asserted.

Theorem 4 can be proved without assuming that $f'(z)$ is continuous.† The French mathematician Edouard Goursat (1858–1936) was the first to do this, and in his honor the more general form of the theorem is usually referred to as the **Cauchy-Goursat theorem.**

EXAMPLE 2

The function $f(z) = \sin z/(z^2 + 4)$ is analytic everywhere except at the zeros of the denominator, $z = \pm 2i$. Hence, according to Theorem 4, if C is the unit circle, then $\int_C f(z)\,dz = 0$, since neither $z = 2i$ nor $z = -2i$ lies on or within C. In fact, the integral of $f(z)$ is zero around any simple closed curve which does not pass through or contain either of the points $z = \pm 2i$.

If $f(z)$ is analytic in and on the boundary of the multiply connected region R between two simple closed curves, C_1 and C_2, the Cauchy-Goursat theorem tells us that

$$\int_{C_1} f(z)\,dz + \int_{C_2} f(z)\,dz = 0$$

provided that each curve is traversed in the positive direction with respect to the interior of R, as shown in Fig. 17.7*a*. On the other hand, if we reverse the direction of integration around the inner curve C_2 and transpose the resulting integral, we obtain

$$\int_{C_1} f(z)\,dz = \int_{C_2} f(z)\,dz$$

each integration now being performed in the counterclockwise sense, as shown in Fig. 17.7*b*. Since there may be points in the interior of C_2 (which of course is not a part of R) where $f(z)$ is not analytic, we cannot apply the general form of Theorem 4 to the integral of $f(z)$ around either C_1 or C_2 and claim that the integral is zero. However, we have shown that *the integrals have the same value.* The crucial point here is the observation that C_1 can be continuously shrunk or deformed into C_2 without passing through a point where $f(z)$ is not analytic. With this in mind, we can summarize our discussion in the highly important theorem.

THEOREM 5 **(Deformation of Contours)** The line integral of an analytic function around any simple closed curve is equal to the line integral of the same function around any other simple closed curve into which the first can be continuously deformed without passing through a point where $f(z)$ is nonanalytic.

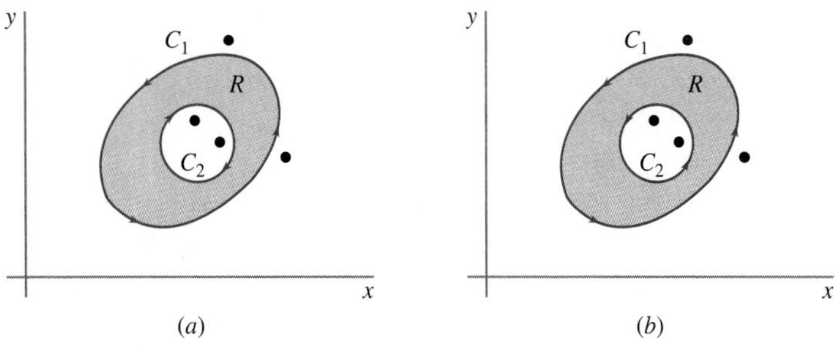

FIGURE 17.7
Contours that can be continuously deformed into each other.

†See, for instance, R. V. Churchill and J. W. Brown, *Complex Variables and Applications,* 5th ed., pp. 108–113, McGraw-Hill, New York, 1990.

EXAMPLE 3

The function $f(z) = 1/(z^2 - 2z)$ is analytic everywhere except at the zeros of its denominator, namely, $z = 0, 2$. The circles C_1: $|z| = 3$ and C_2: $|z - 1| = \frac{3}{2}$ each encircle both of these *singular* points (Fig. 17.8); hence we cannot conclude that either $\int_{C_1} f(z)\,dz$ or $\int_{C_2} f(z)\,dz$ is equal to zero. However, it is clear that C_1 can be continuously deformed into C_2 without passing through a point where $f(z)$ is not analytic. Therefore it follows, by Theorem 5, that the two integrals have the *same* value even though at present we have no convenient way of determining it. On the other hand, if C_3 is the unit circle, we cannot say that $\int_{C_2} f(z)\,dz = \int_{C_3} f(z)\,dz$ since C_2 cannot be continuously deformed into C_3 without passing through the singular point $z = 2$.

Incidentally, since the curve C shown in Fig. 17.8 also encloses the two singular points of $f(z)$, Theorem 5 assures us that the integral of $f(z)$ around C is also equal to the integrals of $f(z)$ around C_1 and C_2. One of the main uses of Theorem 5 in evaluating integrals in the complex plane is to transform the integration from some complicated contour, such as C, into a more convenient one, such as C_1, around which the integral has the same value.

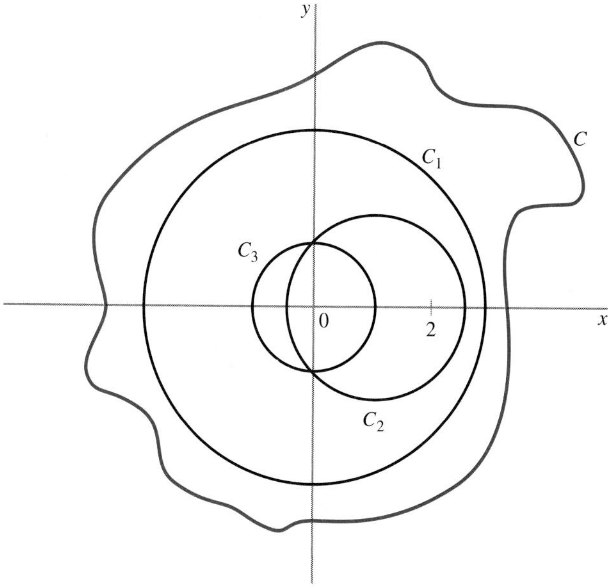

FIGURE 17.8
The contours discussed in Example 3.

Our next theorem is essentially a restatement of Theorem 3 for analytic functions.

THEOREM 6 In any simply connected region R in which $f(z)$ is analytic, the integral $\int_{z_0}^{z} f(z)\,dz$ is independent of the path.

PROOF Let $f(z)$ be analytic in a simply connected region R, let z_0 and z be any two points in R, and let C_1 and C_2 be arbitrary paths from z_0 to z which together form a simple closed curve C in R. By the Cauchy-Goursat theorem, we have

$$\int_C f(z)\,dz = \underset{\text{along } C_1}{\int_{z_0}^{z}} f(z)\,dz + \underset{\text{along } C_2}{\int_{z}^{z_0}} f(z)\,dz = 0$$

or, reversing the direction of integration along C_2 and transposing,

$$\underset{\text{along } C_1}{\int_{z_0}^{z}} f(z)\,dz = \underset{\text{along } C_2}{\int_{z_0}^{z}} f(z)\,dz$$

which is the assertion of the theorem. The possibility that C_1 and C_2 cross each other and do not form a simple closed curve is easy to handle, and we leave it as an exercise (Exercise 23). ◗

By reversing the argument used in the proof of the last theorem, we obtain the following result which is valid in both simply and multiply connected regions.

COROLLARY 1 If $\int_{z_0}^{z} f(z)\,dz$ is independent of the path in a region R, then the integral of $f(z)$ around every simple closed curve in R is zero.

If R is a multiply connected region, the integral of $f(z)$ from z_0 to z is not necessarily independent of the path of integration. However, the values of the integral along different paths are related, as the next theorem tells us.

THEOREM 7 If $f(z)$ is analytic in a multiply connected region R, then the possible values of $\int_{z_0}^{z} f(z)\,dz$ differ by constants independent of the endpoints z_0 and z.

PROOF Let C_1 and C_2 be two paths from z_0 to z in R and, for simplicity, suppose that $C = C_1 \cup C_2$ is a simple closed curve. (The possibility that C is not a *simple* closed curve is considered in Exercise 23.) If C encloses only points of R, as of course it may, then, as in the proof of Theorem 6,

$$\underset{\text{along } C_1}{\int_{z_0}^{z}} f(z)\,dz = \underset{\text{along } C_2}{\int_{z_0}^{z}} f(z)\,dz$$

and the difference between the integrals is zero. On the other hand, if C encloses one or more inner boundaries of R in whose interiors there are singular points of f, the integral around C will in general not be zero. In this case we have

$$\underset{\text{along } C_1}{\int_{z_0}^{z}} f(z)\,dz + \underset{\text{along } C_2}{\int_{z}^{z_0}} f(z)\,dz = \underset{\text{along } C_1}{\int_{z_0}^{z}} f(z)\,dz - \underset{\text{along } C_2}{\int_{z_0}^{z}} f(z)\,dz = K$$

However, by Theorem 5, the value of K depends only on the integrals of $f(z)$ around those inner boundaries of R lying in the interior of C (see Exercise 24), and, clearly, these integrals are constants, independent of z_0 and z.

Using Theorem 6, we can now establish the following pair of theorems which, in simply connected regions, justify the use of antiderivatives in evaluating integrals of analytic functions.

THEOREM 8 If $f(z)$ is analytic throughout a simply connected region R, then $F(z) = \int_{z_0}^{z} f(z)\,dz$ is an analytic function whose derivative at each point of R is $f(z)$.

PROOF Since $f(z)$ is analytic in the simply connected region R, it follows from Theorem 6 that the integral

$$\int_{z_0}^{z} f(z)\,dz$$

is independent of the path and hence is a function only of the upper limit z, say $F(z)$. If we now let $f(z) = u + iv$ and $F(z) = U + iV$, this integral can be written

$$F(z) = U + iV = \int_{x_0,y_0}^{x,y} u\,dx - v\,dy + i\int_{x_0,y_0}^{x,y} v\,dx + u\,dy$$

or, equating real and imaginary parts,

$$U = \int_{x_0,y_0}^{x,y} u\,dx - v\,dy \qquad \text{and} \qquad V = \int_{x_0,y_0}^{x,y} v\,dx + u\,dy$$

By Theorem 2, each of these integrals can be differentiated partially with respect to x and y, and we find

$$\frac{\partial U}{\partial x} = u \qquad \frac{\partial U}{\partial y} = -v \qquad \frac{\partial V}{\partial x} = v \qquad \frac{\partial V}{\partial y} = u$$

From these it is obvious that

$$\frac{\partial U}{\partial x} = \frac{\partial V}{\partial y} \quad \text{and} \quad \frac{\partial U}{\partial y} = -\frac{\partial V}{\partial x}$$

or, in other words, that U and V satisfy the Cauchy-Riemann equations. Moreover, since u and v are continuous because of the hypothesis that $f(z)$ is analytic, it follows that $\partial U/\partial x$, $\partial U/\partial y$, $\partial V/\partial x$, and $\partial V/\partial y$ are continuous. Hence $F(z) = U + iV$ is an analytic function whose derivative is

$$F'(z) = \frac{\partial U}{\partial x} + i\frac{\partial V}{\partial x} = u + iv$$

as asserted. ◗

THEOREM 9 If $F(z)$ is analytic in a simply connected region R, then, provided the path of integration lies entirely in R,

$$\int_{z_0}^{z_1} f(z)\,dz = G(z_1) - G(z_0)$$

where $G(z)$ is any antiderivative of $f(z)$.

◖ **PROOF** Since $f(z)$ is analytic in R, it follows from Theorem 8, that the integral

$$F(z) = \int_{z_0}^{z} f(z)\,dz$$

is an antiderivative of $f(z)$. Therefore, if $G(z)$ is any antiderivative of $f(z)$, then $F'(z) - G'(z) = f(z) - f(z) = 0$. Hence $F(z) = G(z) + c$ (Exercise 21, Sec. 17.3); that is,

$$\int_{z_0}^{z} f(z)\,dz = G(z) + c$$

If we put $z = z_0$ in this identity, we find at once that $c = -G(z_0)$. Subsequently, putting $z = z_1$ yields the assertion of the theorem. ◗

EXAMPLE 4

What is $\int_0^{1+i\pi} (z^2 + \cosh 2z)\,dz$?

The integrand $f(z) = z^2 + \cosh 2z$ is obviously analytic everywhere and has an antiderivative $G(z) = \frac{1}{3}z^3 + \frac{1}{2}\sinh 2z$. Hence, by Theorem 9 and Eq. (18), Sec. 17.4, the value of the integral is

$$\left[\frac{1}{3}z^3 + \frac{1}{2}\sinh 2z\right]_0^{1+i\pi} = \frac{1}{3}(1 + i\pi)^3 + \frac{1}{2}\sinh 2(1 + i\pi) = \frac{1}{3} - \pi^2 + \frac{1}{2}\sinh 2 + \frac{i}{3}(3\pi - \pi^3)$$

In Sec. 1.14 we learned how to find the orthogonal trajectories of the curves of a given family $u(x, y) = c$. Our next theorem provides us with another method of doing this in the important case where u is a harmonic function, that is, a solution of Laplace's equation.

THEOREM 10 If $u(x, y)$ is a solution of Laplace's equation in a region R, then in R there exists an analytic function having u as its real part, namely, $f(z) = u + iv$, where

$$v = \int_{x_0,y_0}^{x,y} -\frac{\partial u}{\partial y}\,dx + \frac{\partial u}{\partial x}\,dy$$

and the path of integration from (x_0, y_0) to (x, y) lies entirely in R.

PROOF Suppose first that R is simply connected. Then in R the integral defining v is independent of the path between the arbitrary fixed point (x_0, y_0) and the variable point (x, y) since the condition for independence provided by Theorem 3 is in this case

$$\frac{\partial(\partial u/\partial x)}{\partial x} = \frac{\partial(-\partial u/\partial y)}{\partial y} \qquad \text{or} \qquad \frac{\partial^2 u}{\partial x^2} = -\frac{\partial^2 u}{\partial y^2}$$

which is true because of the hypothesis that u satisfies Laplace's equation. Theorem 2 can therefore be applied to the integral which defines v, and we have

$$\frac{\partial v}{\partial x} = -\frac{\partial u}{\partial y} \qquad \text{and} \qquad \frac{\partial v}{\partial y} = \frac{\partial u}{\partial x}$$

These are precisely the Cauchy-Riemann equations which, if the derivatives are continuous, are the conditions that $f(z) = u + iv$ be an analytic function. But $\partial u/\partial x$ and $\partial u/\partial y$, hence $\partial v/\partial x$ and $-\partial v/\partial y$, to which these are respectively equal, must be continuous since the partial derivatives $\partial^2 u/\partial x^2$ and $\partial^2 u/\partial y^2$ are known to exist. Hence if R is simply connected, $f(z) = u + iv$ is analytic, as asserted.

On the other hand, if R is multiply connected, then, by Theorem 7, the possible values of v differ at most by constants independent of the endpoints. And, clearly, a constant added to v will not affect the analyticity of $f(z) = u + iv$. This completes the proof of the theorem. That the curves $v(x, y) = k$ are the orthogonal trajectories of the curves $u(x, y) = c$ follows of course from Property 2, Sec. 17.3. ◗

EXAMPLE 5

If $u = y^3 - 3x^2y + y$, determine v so that $u + iv$ is an analytic function.

It is easy to verify that u satisfies Laplace's equation. Hence, taking $x_0 = y_0 = 0$, for convenience, we have by Theorem 10,

$$\begin{aligned} v &= \int_{0,0}^{x,y} -(3y^2 - 3x^2 + 1)\,dx + (-6xy)\,dy \\ &= \int_{0,0}^{x,y} (3x^2 - 1)\,dx - 3(y^2\,dx + 2xy\,dy) = \int_{0,0}^{x,y} (3x^2 - 1)\,dx - 3\int_{0,0}^{x,y} d(xy^2) \\ &= [(x^3 - x) - 3(xy^2)]_{0,0}^{x,y} = x^3 - 3xy^2 - x \end{aligned}$$

Thus

$$\begin{aligned} f(z) = u + iv &= (y^3 - 3x^2y + y) + i(x^3 - 3xy^2 - x) \\ &= i(x^3 + 3ix^2y - 3xy^2 - iy^3) - i(x + iy) \\ &= iz^3 - iz \end{aligned}$$

EXERCISES

1. Evaluate $\int_0^{1+i} (x + y)\, dz$ along each of the following paths.
(**a**) Along the y axis to i and then horizontally to $1 + i$
(**b**) Along the line $y = x$
(**c**) Along the parabola $y = x^2$
(**d**) Along the x axis to 1 and then vertically to $1 + i$

2. Evaluate $\int_0^{1+i} (x + iy)\, dz$ along each of the paths in Exercise 1.

3. Evaluate $\int_0^{1+i} (x^2 + iy)\, dz$ along each of the paths in Exercise 1.

4. Evaluate $\int_1^{i} \bar{z}\, dz$ along each of the following paths.
(**a**) Along the x axis to 0 and then along the y axis to i
(**b**) Along the line $y = 1 - x$
(**c**) Vertically to $1 + i$ and then horizontally to i

5. Verify Theorem 6 by evaluating $\int_0^{3+i} z^2\, dz$ along each of the following paths.
(**a**) Along the real axis to 3 and then vertically to $3 + i$
(**b**) Along the line $y = x/3$
(**c**) Along the y axis to i and then horizontally to $3 + i$

6. Evaluate $\int_0^{3+i} (\bar{z})^2\, dz$ along each of the paths in Exercise 5.

Evaluate each of the following integrals.

7. $\int_0^{1+i} e^{2z}\, dz$ **8.** $\int_0^{i} (z + \sin z)\, dz$

9. $\int_0^{1+i\pi} \cos z\, dz$ **10.** $\int_1^{i} ze^z\, dz$

11. $\int_0^{1+i\pi} z \sin z\, dz$ **12.** $\int_i^{1} \frac{dz}{(z+1)^2}$

13. Around which of the following contours, if any, is $\oint \frac{\sin 2z\, dz}{(z^2+1)(z-1)} = 0$? Around which contours does the integral have the same value?
(**a**) $|z| = \frac{1}{2}$ (**b**) $|z| = 2$
(**c**) $|z - 1| = 2$ (**d**) $|z + 1| = 1$
(**e**) $|z + 1| = \frac{3}{2}$ (**f**) $|z - 1 + i| = \frac{3}{2}$

14. Around which of the following contours, if any, is $\oint \frac{e^z}{z^2 + 2z + 2}\, dz = 0$? Around which contours does the integral have the same value?
(**a**) $|z - i| = \frac{3}{2}$ (**b**) $|z| = 2$
(**c**) $|z + 1 + i| = \frac{3}{2}$ (**d**) $|z - 1| = 3$
(**e**) $|z - 3| = 1$ (**f**) $|z + 1| = \frac{1}{2}$

Whenever possible, determine v so that $u + iv$ is an analytic function if

15. $u = x + y$ **16.** $u = 2x + 3y$
17. $u = e^x \cos y$ **18.** $u = e^x \sin y$
19. $u = x^2 + y^2$ **20.** $u = \cos x \sinh y$
21. $u = \cos x \sin y$ **22.** $u = x^2 + 2xy - y^2$

23. How must the proof of Theorem 6 be modified if C_1 and C_2 form a curve C which crosses itself, as in Fig. 17.9? *Hint:* Apply the argument in the text to each of the loops z_0Pz_0 and PzP.

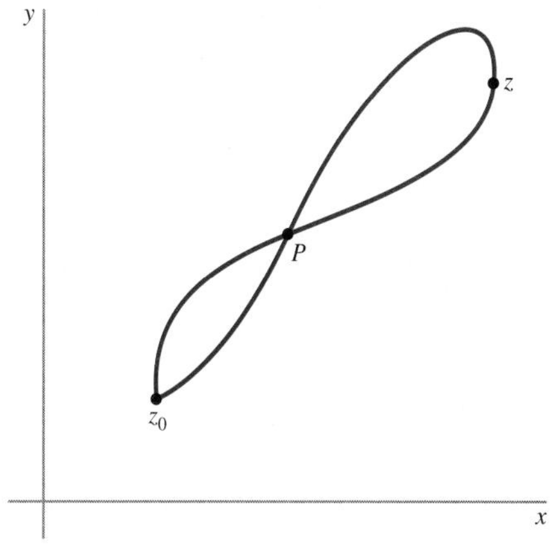

FIGURE 17.9

24. Justify the statement made in the proof of Theorem 7 that according to Theorem 5 the integral of $f(z)$ around $C = C_1 \cup C_2$ is equal to the sum of the integrals of $f(z)$ around the interior boundaries of R that lie within C. *Hint:* Make the interior of C into a simply connected region by introducing the appropriate cross-cuts and then note that the sum of the integrals of $f(z)$ in opposite directions along each cross-cut is zero.

25. Evaluate the following integrals.
(*a*) $\int_{\pi i}^{1} e^z\, dz$
(**b**) $\int_C \frac{z^2+1}{z^2-1}\, dz$, where C is the circle $|z - i| = 1$

26. Evaluate the integral in Example 4 by computing the corresponding real integrals along the x axis from $(0, 0)$ to $(1, 0)$ and then vertically to $(1, \pi)$.

27. Evaluate $\int_{-i}^{i} (x^2 + iy^2)\, dz$ along (**a**) the right-hand half of $|z| = 1$ and (**b**) the line $x = 0$.

28. Obtain an upper bound for the absolute value of the integral $\int_0^{1+i} e^{-z^2}\, dz$ (**a**) along $y = x$, (**b**) along $y = x^2$, and (**c**) along the real axis to 1 and then vertically to $1 + i$.

29. Obtain an upper bound for the absolute value of the integral

$$\frac{1}{2\pi i}\int \frac{e^{2z}}{z^2+1}\,dz$$

taken around the circle $|z| = 3$. What is the value of this integral if the path of integration is the circle $|z| = \frac{1}{2}$?

30. Without evaluating integrals, verify that along a straight-line path joining the limits of integration:

(a) $\left|\int_{i}^{2+i} \frac{dz}{z^2}\right| \le 2$

(b) $\left|\int_{-2}^{2(\sqrt{3}-1)+2i} \frac{dz}{z^2}\right| \le 4$

and show that if C is the first-quadrant arc of the circle $|z| = 2$,

(c) $\left|\int_{C} \frac{dz}{z^2+1}\right| \le \frac{\pi}{3}$

17.6 CAUCHY'S INTEGRAL FORMULA

One of the most important consequences of Cauchy's theorem is *Cauchy's integral formula,* a result which allows us to infer the value of an analytic function at any interior point of a region from a knowledge of the values of the function on the boundary of the region.

THEOREM 1 **(Cauchy's Integral Formula)** If $f(z)$ is analytic within and on the boundary C of a simply connected region R whose boundary C is sectionally smooth and positively oriented, and if z_0 is any point in the interior of R, then the value of f at z_0 is given by the formula

$$f(z_0) = \frac{1}{2\pi i}\int_C \frac{f(z)}{z-z_0}\,dz$$

PROOF Let C_0 be a circle with center at z_0 and arbitrary radius small enough for C_0 to lie entirely in R (Fig. 17.10). Now by hypothesis, $f(z)$ is analytic everywhere within R. Hence the function $f(z)/(z - z_0)$ is analytic everywhere within R except at the one point $z = z_0$. In particular, it is analytic everywhere in the region R' between C and C_0. Hence, by Theorem 5, Sec. 17.5, the integral of this function around C is equal to its integral around C_0. That is,

$$\int_C \frac{f(z)}{z-z_0}\,dz = \int_{C_0} \frac{f(z)}{z-z_0}\,dz = \int_{C_0} \frac{f(z_0) + [f(z) - f(z_0)]}{z-z_0}\,dz$$

(1)
$$= f(z_0)\int_{C_0} \frac{dz}{z-z_0} + \int_{C_0} \frac{f(z)-f(z_0)}{z-z_0}\,dz$$

By Example 1, Sec. 17.5, the first integral on the right is equal to $2\pi i$. Hence the assertion of the theorem will be established if we can show that the last integral vanishes. To do this, we observe that

(2)
$$\left|\int_{C_0} \frac{f(z)-f(z_0)}{z-z_0}\,dz\right| \le \int_{C_0} \frac{|f(z)-f(z_0)|}{|z-z_0|}\,|dz|$$

On C_0 we have $|z - z_0| = \rho$. Moreover, since $f(z)$ is analytic and hence continuous, it follows that, for any $\epsilon > 0$, there exists a δ such that

$$|f(z) - f(z_0)| < \epsilon \qquad \text{provided } |z - z_0| \equiv \rho < \delta$$

Choosing the radius ρ to be less than δ and inserting these estimates in the right-hand member of (2), we therefore have

$$\left|\int_{C_0} \frac{f(z)-f(z_0)}{z-z_0}\,dz\right| < \int_{C_0} \frac{\epsilon}{\rho}\,|dz| = \frac{\epsilon}{\rho}\int_{C_0} |dz| = \frac{\epsilon}{\rho}\,2\pi\rho = 2\pi\epsilon$$

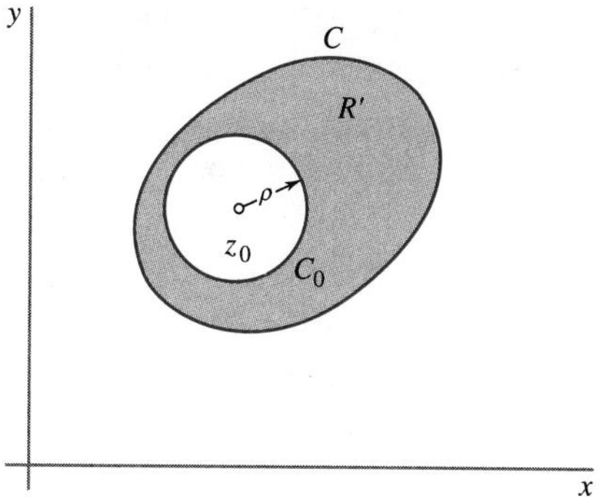

FIGURE 17.10
The circle C_0 used in the proof of Cauchy's integral formula.

Since the integral on the left is independent of ϵ, yet cannot exceed $2\pi\epsilon$, which can be made arbitrarily small, it follows that the absolute value of the integral, and hence the integral itself, is zero. Thus (1) reduces to

$$\int_C \frac{f(z)}{z - z_0}\, dz = f(z_0)2\pi i + 0$$

from which

$$f(z_0) = \frac{1}{2\pi i}\int_C \frac{f(z)}{z - z_0}\, dz$$

as asserted. ◗

Cauchy's integral formula is also true for multiply connected regions, but we shall leave as an exercise the easy modification of our proof required to establish this fact.

EXAMPLE 1

Find the values of

$$\int_C \frac{e^z}{z^2 + 1}\, dz$$

if C is a circle of unit radius with center at **(a)** $z = i$ and **(b)** $z = -i$.

In **(a)** we think of the integral as written in the form

$$\int_C \frac{e^z}{z + i}\frac{dz}{z - i}$$

and identify z_0 as i and $f(z)$ as $e^z/(z + i)$. The function $f(z)$ is analytic everywhere within and on the given circle of unit radius around $z = i$. (In fact, it is analytic everywhere except at $z = -i$.) Therefore we can apply Cauchy's integral formula, getting

$$\int_C \frac{e^z}{z + i}\frac{dz}{z - i} = 2\pi i f(z_0) = 2\pi i f(i) = 2\pi i \frac{e^i}{2i} = \pi(\cos 1 + i \sin 1)$$

In **(b)** we identify z_0 as $-i$ and $f(z)$ as $e^z/(z - i)$. Then Cauchy's integral formula gives immediately

$$\int_C \frac{e^z}{z - i}\frac{dz}{z + i} = 2\pi i f(z_0) = 2\pi i f(-i) = 2\pi i \frac{e^{-i}}{-2i} = -\pi(\cos 1 - i \sin 1)$$

From Cauchy's integral formula, which expresses the value of an analytic function at an interior point of a region R in terms of its values on the boundary of the region, we can readily obtain an expression for the derivative of a function at an interior point of R in terms of the boundary values of the function. In fact, from the definition of a derivative and Cauchy's integral formula, we have

$$
\begin{aligned}
f'(z_0) &= \lim_{\Delta z_0 \to 0} \frac{f(z_0 + \Delta z_0) - f(z_0)}{\Delta z_0} \\
&= \lim_{\Delta z_0 \to 0} \frac{1}{\Delta z_0} \left[\frac{1}{2\pi i} \int_C \frac{f(z)\, dz}{z - (z_0 + \Delta z_0)} - \frac{1}{2\pi i} \int_C \frac{f(z)\, dz}{z - z_0} \right] \\
&= \lim_{\Delta z_0 \to 0} \frac{1}{\Delta z_0} \left\{ \frac{1}{2\pi i} \int_C f(z) \left[\frac{1}{z - (z_0 + \Delta z_0)} - \frac{1}{z - z_0} \right] dz \right\} \\
&= \lim_{\Delta z_0 \to 0} \frac{1}{2\pi i} \int_C \frac{f(z)\, dz}{(z - z_0 - \Delta z_0)(z - z_0)}
\end{aligned}
$$

Assuming that limit of the integral is equal to the integral of the limit in the last expression and letting $\Delta z_0 \to 0$ in the integrand, we obtain the desired result:

$$
f'(z_0) = \frac{1}{2\pi i} \int_C \frac{f(z)\, dz}{(z - z_0)^2}
$$

That the interchange of the operations of integration and taking the limit is legitimate can be established by adding and subtracting the desired final integral, namely,

$$
\frac{1}{2\pi i} \int_C \frac{f(z)\, dz}{(z - z_0)^2}
$$

and then showing that the absolute value of the difference

$$
\frac{1}{2\pi i} \int_C \frac{f(z)\, dz}{(z - z_0 - \Delta z_0)(z - z_0)} - \frac{1}{2\pi i} \int \frac{f(z)\, dz}{(z - z_0)^2} \tag{3}
$$

approaches zero as $\Delta z_0 \to 0$.

Continuing in the same way, we obtain the additional formulas

$$
f''(z_0) = \frac{2!}{2\pi i} \int_C \frac{f(z)\, dz}{(z - z_0)^3}
$$

$$
f'''(z_0) = \frac{3!}{2\pi i} \int_C \frac{f(z)\, dz}{(z - z_0)^4}
$$

$$
\cdots\cdots\cdots\cdots\cdots\cdots
$$

These results could also have been obtained formally by repeated differentiation of Cauchy's integral formula with respect to the parameter z_0.

From the preceding discussion we conclude not only that an analytic function possesses derivatives of all orders but also that each derivative is itself analytic since it too possesses a derivative. This completes the proof of the following theorem.

THEOREM 2 If $f(z)$ is analytic throughout a closed simply connected region R, then at any interior point z_0 of R the derivatives of $f(z)$ of all orders exist and are analytic. Moreover,

$$
f^{(n)}(z_0) = \frac{n!}{2\pi i} \int_C \frac{f(z)\, dz}{(z - z_0)^{n+1}}
$$

where C is the boundary of R.

It is interesting to note that functions of a real variable do not in general possess the derivative properties described by Theorem 2, for at particular points a function of a real variable may possess one or more derivatives without derivatives of all orders existing. For instance, at the origin the function $x^{7/3}$ possesses a first and a second derivative but no derivatives of higher order.

In the proof of Theorem 8, Sec. 17.5, we assumed the analyticity of $f(z) = u + iv$ in R so that we could conclude that $\int_{z_0}^{z} f(z)\,dz$ was independent of the path and that u and v were continuous in R; but nowhere else in the proof did we make any use of the fact that $f(z)$ was analytic. Hence if we assume just these two things, without going as far as assuming that $f(z)$ is analytic, our proof is still valid. In other words, if we assume only that in R, u and v are continuous and $\int_{z_0}^{z} f(z)\,dz$ is independent of the path, or (Corollary 1, Theorem 6, Sec. 17.5) that the integral of $f(z)$ around every simple closed curve in R is zero, it still follows that $F(z) = \int_{z_0}^{z} f(z)\,dz$ is an analytic function whose derivative is $f(z)$. Now, with Theorem 2 available, we can go one step further and assert that $f(z)$, being the derivative of the analytic function $F(z)$, is itself analytic. This in turn establishes the converse of Cauchy's theorem, which is known as **Morera's theorem**†:

THEOREM 3 If $f(z)$ is continuous in a region R and if $\int_C f(z)\,dz = 0$ for every simple closed curve C in R, then $f(z)$ is analytic in R.

Beginning with the formula for $f^{(n)}(z_0)$ provided by Theorem 2, we can now establish **Cauchy's inequality,** introduced by

THEOREM 4 If $f(z)$ is analytic within and on a circle C of radius r with center at z_0, then

$$|f^{(n)}(z_0)| \le \frac{n!M}{r^n} \tag{4}$$

where M is the maximum value of $f(z)$ on C.

PROOF From Theorem 2, we have

$$\begin{aligned} |f^{(n)}(z_0)| &= \left| \frac{n!}{2\pi i} \int_C \frac{f(z)\,dz}{(z - z_0)^{n+1}} \right| \\ &\le \frac{n!}{2\pi} \int_C \frac{|f(z)|\,|dz|}{|z - z_0|^{n+1}} \\ &\le \frac{n!}{2\pi} \frac{M}{r^{n+1}} \int_C |dz| \\ &= \frac{n!}{2\pi} \frac{M}{r^{n+1}} 2\pi r \\ &= \frac{n!M}{r^n} \end{aligned}$$

as asserted.

For the special case $n = 0$, Cauchy's inequality becomes

$$|f(z_0)| \le M$$

which shows that on every circle around z_0, no matter how small, $|f(z)|$ has a maximum value M which is at least as great as $|f(z_0)|$. In other words, we have the following result, usually referred to as the **maximum modulus theorem.**

†Named for the Italian mathematician Giacinto Morera (1856–1909).

THEOREM 5 The absolute value of a nonconstant function $f(z)$ cannot have a maximum at any point where the function is analytic.

If $f(z)$ is analytic at all points of a closed region R, bounded by a simple closed curve C, then the real function $|f(z)|$ must have a maximum at some point of R. By the last theorem, this maximum cannot occur in the interior of R; hence it must occur on the boundary C. Thus we have the following corollary of the maximum modulus theorem.

COROLLARY 1 If $f(z)$ is a nonconstant function which is analytic over a closed region R bounded by a simple closed curve C, then the maximum value of $|f(z)|$ over R occurs on the boundary C.

A similar statement is true for the minimum value of $|f(z)|$ over R, provided that $f(z) \neq 0$ in R, but we shall leave the proof of this fact as an exercise (see Exercise 15).

EXAMPLE 2

EXTREMES OF $|z^2 + 2|$ ON $|z| \leq 1$

If $f(z) = z^2 + 2$, find the extreme values of $|f(z)|$ over the closed region R: $|z| \leq 1$.

Since the only zeros of $f(z)$, namely, $z = \pm\sqrt{2}i$, lie outside the circle C:$|z| = 1$, it follows from Corollary 1, Theorem 5, and the accompanying remarks about minima, that both the maxima and the minima of $|f(z)|$ must occur on C. Hence our problem is actually not a problem involving the two independent variables x and y, ranging over R, but rather a problem involving a single variable, namely, the argument θ of a variable z ranging over C.

Setting $x = \cos\theta$, $y = \sin\theta$, and using familiar trigonometric identities, we find at once that

$$\begin{aligned}|f(z)| = |z^2 + 2| &= |x^2 - y^2 + 2ixy + 2| \\ &= \sqrt{(x^2 - y^2 + 2)^2 + (2xy)^2} \\ &= \sqrt{(\cos 2\theta + 2)^2 + \sin^2 2\theta} \\ &= \sqrt{4\cos 2\theta + 5}\end{aligned}$$

The extreme values of $|f(z)|$ are thus determined by the condition

$$\frac{d|f(z)|}{d\theta} = \frac{-8\sin 2\theta}{2\sqrt{4\cos 2\theta + 5}} = 0$$

Hence, $\theta = 0,\ \pi/2,\ \pi,\ 3\pi/2$. The values 0 and π correspond to maxima of 3; the values $\pi/2$ and $3\pi/2$ correspond to minima of 1.

Had we attempted to solve the problem over R by equating to zero the derivatives $\partial|f(z)|/\partial x$ and $\partial|f(z)|/\partial y$, we would have found after easy simplifications the equations

$$(x^2 + y^2 + 2)x = 0$$
$$(x^2 + y^2 - 2)y = 0$$

The only solution of these equations which corresponds to a point in R, that is, to a point for which $x^2 + y^2 < 1$, is (0, 0). Thus, without a further investigation of points on the boundary C (which would be precisely our first method of solution), the analysis in terms of x and y fails to disclose either the maxima or the minima of $|f(z)|$ over R.

When the boundary curve C in Theorem 1 is a circle, there is an alternative form of Cauchy's integral formula which is of great importance in solving boundary-value problems involving Laplace's equation. To derive it, let $f(z)$ be analytic in and on the circumference of the circle C

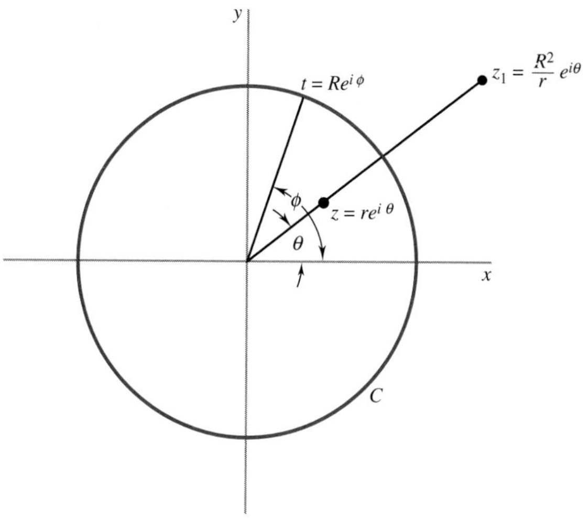

FIGURE 17.11
The circle used in the derivation of Poisson's integral formula for a circle.

whose center is the origin and whose radius is R and let $z = re^{i\theta}$ be an arbitrary point in the interior of C. Then, by Cauchy's integral formula,

$$f(z) = \frac{1}{2\pi i}\int_C \frac{f(t)}{t - z}\,dt \tag{5}$$

where $t = Re^{i\phi}$ is used as a dummy variable of integration.

Now let z_1 be the point whose distance from the origin is R^2/r on the same ray as z (Fig. 17.11). Since $|z| = r$ and $|t| = R$, this implies that

$$z_1 = \frac{R^2}{r}e^{i\theta} = \frac{t\bar{t}}{re^{-i\theta}} = \frac{t\bar{t}}{\bar{z}} \tag{6}$$

Since $R > r$, it is clear that z_1 lies outside the circle C; hence the function $f(z)/(z - z_1)$ is analytic on C and throughout its interior. Therefore, by Cauchy's theorem, we can write

$$0 = \frac{1}{2\pi i}\int_C \frac{f(t)}{t - z_1}\,dt \tag{7}$$

If Eq. (7) is subtracted from Eq. (5) and the integrals combined, we obtain

$$f(z) = \frac{1}{2\pi i}\int_C f(t)\left(\frac{1}{t - z} - \frac{1}{t - z_1}\right)dt$$

or, substituting for z_1 from Eq. (6) and simplifying,

$$\begin{aligned} f(z) &= \frac{1}{2\pi i}\int_C f(t)\left(\frac{1}{t - z} - \frac{\bar{z}}{\bar{z}t - t\bar{t}}\right)dt \\ &= \frac{1}{2\pi i}\int_C f(t)\left[\frac{t\bar{t} - z\bar{z}}{(t - z)(\bar{t} - \bar{z})}\right]\frac{dt}{t} \\ &= \frac{1}{2\pi i}\int_C f(t)\left[\frac{R^2 - r^2}{(t - z)(\bar{t} - \bar{z})}\right]\frac{dt}{t} \end{aligned}$$

It is convenient now to substitute $t = Re^{i\phi}$ and $z = re^{i\theta}$ and perform the integration in terms of the parameter ϕ. Since $dt = Rie^{i\phi}\,d\phi = it\,d\phi$, we thus have

$$f(re^{i\theta}) = \frac{1}{2\pi}\int_0^{2\pi} f(Re^{i\phi})\frac{R^2 - r^2}{(Re^{i\phi} - re^{i\theta})(Re^{-i\phi} - re^{-i\theta})}\,d\phi$$

The denominator of the integrand simplifies at once to

$$R^2 - Rr(e^{i(\phi-\theta)} + e^{-i(\phi-\theta)}) + r^2 = R^2 - 2Rr\cos(\phi - \theta) + r^2$$

Hence for the integral we have

$$f(re^{i\theta}) = \frac{1}{2\pi}\int_0^{2\pi} f(Re^{i\phi})\frac{R^2 - r^2}{R^2 - 2Rr\cos(\phi - \theta) + r^2}\,d\phi$$

Finally, if we write $f(re^{i\theta}) = u(r, \theta) + iv(r, \theta)$ and $f(Re^{i\phi}) = u(R, \phi) + iv(R, \phi)$ and equate the real parts of the last equation, we have

$$(8) \qquad u(r, \theta) = \frac{1}{2\pi}\int_0^{2\pi} \frac{R^2 - r^2}{R^2 - 2Rr\cos(\phi - \theta) + r^2}u(R, \phi)\,d\phi \qquad r < R$$

which is a formula expressing the harmonic function $u(r, \theta)$ over the interior of the circle C in terms of its values on the circumference of C. This important result is known as **Poisson's integral formula for a circle.**† **Poisson's integral formula for a half-plane** is given in Exercise 23. We will gain a clearer understanding of the usefulness of these formulas when we study *conformal mapping* in Chap. 20 and learn that many quite general regions can be mapped onto circles and half-planes.

The hypothesis that $f(z)$ is analytic and therefore continuous on the boundary C appears to be a very serious restriction, for it rules out consideration of problems where a boundary condition such as a temperature or a potential has a finite jump at one or more points, which is often the case. Fortunately, it can be shown that Eq. (8) defines a function which is harmonic in the interior of C provided only that $u(R, \phi)$ is at least piecewise continuous on C.‡

EXERCISES

1. What is the value of $\displaystyle\int_C \frac{3z^2 + 7z + 1}{z + 1}\,dz$

(a) If C is the circle $|z + 1| = 1$?
(b) If C is the circle $|z + i| = 1$?
(c) If C is the ellipse $x^2 + 2y^2 = 8$?

2. What is the value of $\displaystyle\int_C \frac{z + 4}{z^2 + 2z + 5}\,dz$

(a) If C is the circle $|z| = 1$?
(b) If C is the circle $|z + 1 - i| = 2$?
(c) If C is the circle $|z + 1 + i| = 2$?

3. What is the value of $\displaystyle\int_C \frac{\sin 2z}{z^2 - 4z + 5}\,dz$

(a) If C is the circle $|z| = 1$?
(b) If C is the circle $|z - 2i| = 3$?
(c) If C is the circle $|z - 1 + 2i| = 2$?

4. What is the value of $\displaystyle\int_C \frac{e^z}{(z + 1)^2}\,dz$

(a) If C is the circle $|z - 1| = 3$?
(b) If C is the ellipse $x^2 + 2y^2 + 2x = 0$?

5. What is the value of $\displaystyle\int_C \frac{z + 1}{z^3 - 2z^2}\,dz$

(a) If C is the circle $|z| = 1$?
(b) If C is the circle $|z - 2 - i| = 2$?
(c) If C is the circle $|z - 1 - 2i| = 2$?

---

†Named for the French mathematician Siméon Denis Poisson (1781–1840).
‡See, for instance, R. V. Churchill and J. W. Brown, *Complex Variables and Applications,* 5th ed., pp. 307–310, McGraw-Hill, New York, 1990.

Find the extreme values of each of the following functions over the indicated region.

6. $|z + 1|$ over $|z| \le 1$
7. $|z^2 - 3z - 1|$ over $|z| \le 1$
8. $|z^2 + z + 1|$ over $|z| \le 1$
9. $|z^2 + 4z - 2|$ over $|z| \le 2$
10. $|z^2 - z + 2|$ over the unit square and its interior.
11. Complete the proof of Theorem 2 by showing that the absolute value of the difference (3) approaches zero as Δz_0 approaches zero.
12. In the derivation of Poisson's integral formula, show that

$$(t - z)(\bar{t} - \bar{z}) = R^2 - 2Rr\cos(\phi - \theta) + r^2$$

by interpreting the quantity on the left as a distance and then calculating it by the law of cosines.

13. Show that the area bounded by a simple closed curve C is given by the formula

$$A = \frac{1}{2i}\int_C \bar{z}\, dz$$

Hint: Recall Eq. (6), Sec. 17.5, and then use Green's lemma.

14. Prove Morera's theorem under the additional hypothesis that $f'(z)$ is continuous by applying Green's lemma to Eq. (6), Sec. 17.5.
15. Prove that if $f(z)$ is analytic inside and on a simple closed curve C and if $f(z) \ne 0$ inside C, then $|f(z)|$ must assume its minimum value on C. This result is sometimes called the **minimum modulus theorem.** *Hint:* Apply the maximum modulus theorem to $1/f(z)$.
16. Prove that if $f(z)$ is analytic inside and on a circle C with center at z_0, then the average value of $f(z)$ (with respect to θ) on C is $f(z_0)$. This result is sometimes called **Gauss' mean-value theorem.** *Hint:* Make the substitution $z = z_0 + re^{i\theta}$ in Cauchy's integral formula, where r is the radius of C.
17. Show that $\displaystyle\int_0^{2\pi} \cos^{2n}\theta\, d\theta = \int_0^{2\pi} \sin^{2n}\theta\, d\theta = \frac{1}{2^{2n}}\binom{2n}{n}2\pi.$

Hint: Transform the integral by the substitution $z = e^{i\theta}$, then use the binomial theorem and the results of Example 1, Sec. 17.5.

18. If $u(x, y)$ is harmonic, i.e., satisfies Laplace's equation, in the closed region R bounded by a simple closed curve C, prove that the maximum value of $u(x, y)$ in R always occurs on C and not in the interior of C. *Hint:* Apply the maximum modulus theorem to the function $e^{f(z)}$, where $f(z)$ is the analytic function having $u(x, y)$ as its real part.
19. Using Poisson's integral formula (8), show that if $u(x, y)$ is a harmonic function, then its average value (with respect to θ) around any circle is equal to its value at the center of the circle. *Hint:* Imagine the axes translated so that the center of the circle is the origin.
20. Using Theorem 2, show that

$$\frac{x^n}{n!} = \frac{1}{2\pi i}\int_C \frac{e^{xz}}{z^{n+1}}\, dz$$

where C is any simple closed curve encircling the origin.

21. Observing that the result of Exercise 20 can be written

$$\left(\frac{x^n}{n!}\right)^2 = \frac{1}{2\pi i}\int_C \frac{x^n e^{xz}}{n!z^{n+1}}\, dz$$

show that

$$\sum_{n=0}^{\infty}\left(\frac{x^n}{n!}\right)^2 = I_0(2x) = \frac{1}{2\pi}\int_0^{2\pi} e^{2x\cos\theta}\, d\theta$$

where I_0 is the modified Bessel function of order 0 (see Sec. 12.3). *Hint:* Take C to be the circle $|z| = 1$ and then put $z = e^{i\theta}$.

22. Proceeding essentially as in Exercise 21, show that

$$J_0(2x) = \frac{1}{2\pi}\int_0^{2\pi} e^{2ix\sin\theta}\, d\theta$$

where J_0 is the Bessel function of order 0 (see Sec. 12.2).

23. Let $f(z) = u(x, y) + iv(x, y)$ be a function which is analytic in the upper half-plane $\mathrm{Im}(z) \ge 0$ and for which positive constants k and M exist such that

$$|z^k f(z)| < M \qquad \mathrm{Im}(z) \ge 0,\ k > 0$$

Show that for $y > 0$,

$$u(x, y) = \frac{1}{\pi}\int_{-\infty}^{\infty} \frac{yu(t, 0)}{(t - x)^2 + y^2}\, dt$$

This formula is called **Poisson's integral formula for a half-plane.** *Hint:* Let z be a point in the upper half-plane and consider the integrals

$$f(z) = \frac{1}{2\pi i}\int_C \frac{f(t)}{t - z}\, dt \qquad \text{and} \qquad 0 = \frac{1}{2\pi i}\int_C \frac{f(t)}{t - \bar{z}}\, dt$$

where C consists of the upper half of a circle $|z| = R$ with its horizontal diameter. First show that as $R \to \infty$ the contribution to each integral from the semicircular portion of the path approaches zero. Then subtract and combine the remaining integrals along the real axis.

24. Show that although $f(z) = 1/(z^2 + 1)$ has singular points in the interior of the curve C_1: $|z| = 3$, the integral of $f(z)$ around every simple closed curve in the multiply connected region between C_1 and C_2: $|z + 1| = \frac{3}{2}$ is zero.
25. Work Exercise 24 if $f(z) = (iz^2 + 3z)/(z^2 - 3iz - 2)$ and C_1 and C_2 are, respectively, the curves C_1: $|z| = 5$ and C_2: $|z - i| = 3$.
26. According to Rodrigues' formula (Theorem 3, Sec. 12.9), the Legendre polynomial $P_n(x)$ is given by the expression

$$P_n(x) = \frac{1}{2^n n!}\frac{d^n(x^2 - 1)^n}{dx^n}$$

Prove that if C is any simple closed curve enclosing the point $z = x$, then

$$P_n(x) = \frac{1}{2\pi i}\int_C \frac{(z^2 - 1)^n}{(z - x)}\, dz$$

27. Show that Cauchy's integral formula is valid in multiply connected regions.

17.7 ANALYTIC FUNCTIONS AND TWO-DIMENSIONAL FIELD THEORY

In various places in our work up to this point, we have encountered problems whose answers had to be solutions of Laplace's equation. This was the case throughout Chaps. 11 and 12 whenever we dealt with steady-state heat flow. It was also the case in Sec. 12.9 where we first met the idea of the potential of a field of force, and again in Sec. 15.8 where we proved that in empty space the gravitational potential satisfies Laplace's equation. In each of these instances, the problem came first and then, usually by using the method of separation of variables, we had to find the solution by solving Laplace's equation.

Now, with a knowledge of analytic functions available, the situation in two dimensions is effectively reversed. Because we know (Property 1, Sec. 17.3) that both the real and the imaginary parts of any analytic function satisfy Laplace's equation, we now have the solutions. But what are the problems? How do we find them? And are they ones that interest us? Our first example illustrates this curious state of affairs.

EXAMPLE 1

TEMPERATURE AND POTENTIAL PROBLEMS HAVING A KNOWN SOLUTION

Determine the problem(s) whose solution is the real part of the analytic function $f(z) = -iz^2 + 1$.

From the given function, we have

$$-iz^2 + 1 = -i(x^2 - y^2 + 2ixy) + 1 = (2xy + 1) - i(x^2 - y^2)$$

We can now think of $\text{Re}[f(z)] = 2xy + 1$ either as a description of the steady-state temperature in the xy plane or as a description of the potential at any point in the xy plane. Taking the first point of view, the curves of the family $2xy + 1 = c$ become the isothermal curves; and any two of them may be considered the boundaries of a region in which $2xy + 1$ is the temperature distribution corresponding to the boundary conditions $c = T$ existing on those curves. For instance, on the hyperbola $xy = 1$ the temperature is $c = T = 3$ and on the hyperbola $xy = 3$ the temperature is $c = T = 7$. If these curves are maintained at these temperatures, respectively, then the temperature at any point in the region between them is $T(x, y) = 2xy + 1$. Equally well, $T(x, y) = 2xy + 1$ is the temperature between the axes and the hyperbola $xy = 2$ when the axes are maintained at the temperature 1 and the hyperbola is maintained at the temperature 5. In each case, the curves along which heat flow takes place are the portions of the orthogonal trajectories (Property 2, Sec. 17.3),

$$\text{Im}[f(z)] = v(x, y) = -x^2 + y^2 = k$$

which lie in the appropriate region. Typical curves of each family are shown in Fig. 17.12. Clearly, $\text{Re}[f(z)]$ is the solution to infinitely many closely related problems in steady-state heat flow. Moreover, since $\text{Im}[f(z)]$ is also a solution of Laplace's equation, we could equally well have considered

$$\text{Im}[f(z)] = v(x, y) = -x^2 + y^2$$

to be a steady-state temperature distribution in the xy plane with isotherms $-x^2 + y^2 = k$ and lines of heat flow defined by the orthogonal trajectories $u(x, y) = 2xy + 1 = c$.

By suitably modifying $f(z)$, solutions to related problems having the same family of isotherms but different boundary conditions can easily be found. Suppose, for instance, that we want the steady-state temperature distribution in the region between the hyperbolas $xy = 1$ and $xy = 5$ when the curves are maintained at the respective temperatures 100 and 60. If we begin with $f(z) = -iaz^2 + b$, a, b real, we have

$$-iaz^2 + b = -ia(x^2 - y^2 + 2ixy) + b = (2axy + b) - ia(x^2 - y^2)$$

Hence for the temperature function we now have $\text{Re}[f(z)] = 2axy + b$. For this to have the value 100 on the hyperbola $xy = 1$ requires that

$$2a + b = 100$$

and for it to have the value 60 on the hyperbola $xy = 5$ requires that

$$10a + b = 60$$

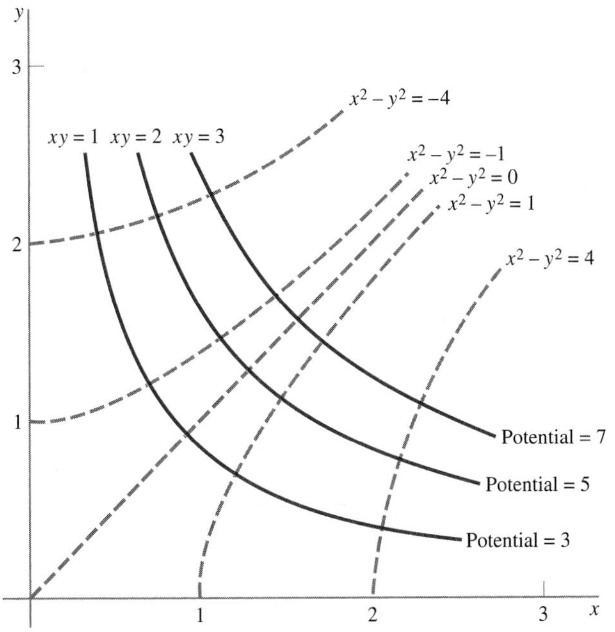

FIGURE 17.12
Equipotential lines and lines of flux defined by the complex potential $f(z) = -iz^2 + 1$.

Solving these equations for a and b, we find that $a = -5$ and $b = 110$. The required temperature function is then $-10xy + 110$ and the related analytic function is $f(z) = 5iz^2 + 110$.

Of course, everything we have said about steady-state temperatures and lines of heat flow can be restated exactly for equipotential lines and lines of flux.

It would be a serious mistake to suppose that the use of analytic functions in applied problems consists of the random checking of particular functions to see if they can be related to problems of practical interest. To emphasize this and to illustrate the systematic application of the theory of functions of a complex variable to an important area of applied mathematics, we now turn our attention from heat flow and potential theory to fluid mechanics. Specifically, we shall consider the two-dimensional steady-state flow of an ideal, i.e., frictionless and incompressible, fluid.

We begin by supposing that the velocity **v** of flow is known at every point in the region of flow. Then by a proper choice of units the rate at which fluid crosses any arc may be determined by integrating the normal component of the fluid velocity along that arc. In particular, if there are no **sources** or **sinks** in the region, i.e., no points at which fluid continuously appears or disappears, then the integral of the normal component of the velocity around any closed curve must be zero since the fluid, being incompressible, cannot accumulate within the curve. This in turn implies that between any two points the line integral of the normal component of the fluid velocity is independent of the path (Theorem 5, Sec. 15.7).

Now consider an arbitrary **streamline** Γ, that is, the path followed by any particle in the fluid during the course of its motion. If O is an arbitrary fixed point in the region of flow and P is an arbitrary point of Γ, then by what we have just said, the rate of flow across all curves joining O and P is a constant. In fact, if P and P' are any two points on Γ, then, since there can be no flow across a streamline and since fluid cannot accumulate in the region OPP', the flow across OP must be the same as the flow across OP'. In other words, the flow across *any* curve joining O to *any* point on Γ is a constant independent of the point. Thus each streamline has associated with it a definite value of what is called the **stream function** Ψ, namely, the common value of the rate of flow across any curve drawn from an arbitrary fixed point O to an arbitrary point of that streamline.

If the stream function is known, the character of the entire flow is known because from the stream function the components of the fluid velocity can be found at once. To see this, we observe

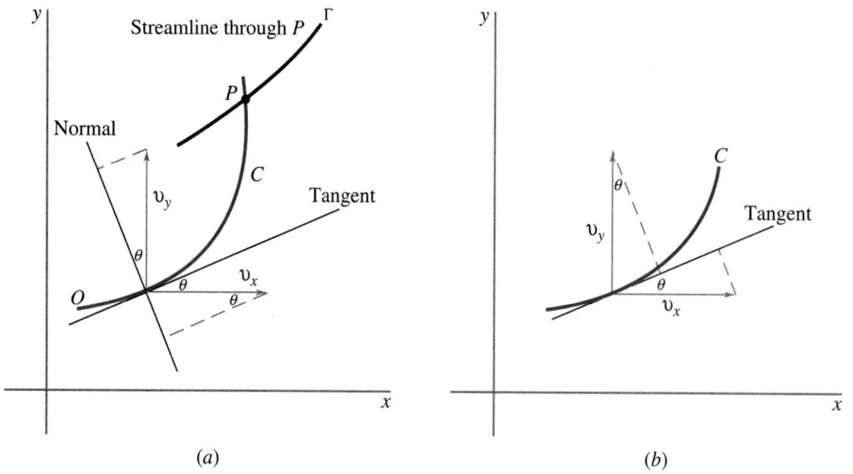

FIGURE 17.13
Fluid flow in the direction of the normal and in the direction of the tangent to a curve C.

from Fig. 17.13*a* that the value of the stream function at an arbitrary point P, that is, the value of the stream function for the (unique) streamline through P, is

$$\begin{aligned} \Psi(x, y) &= \int_O^P (v_x \sin\theta - v_y \cos\theta)\, ds \\ &= \int_O^P v_x\, dy - v_y\, dx \qquad (1) \end{aligned}$$

where v_x and v_y are the components of the fluid velocity $\mathbf{v}$ in the directions of the positive coordinate axes. Since, as we just observed, the integral defining Ψ is independent of the path of integration, it follows from Theorem 2, Sec. 17.5, that

(2) $$v_x = \frac{\partial \Psi}{\partial y} \qquad \text{and} \qquad v_y = -\frac{\partial \Psi}{\partial x}$$

Now consider the line integral of the tangential component of the fluid velocity along any curve C in the region of flow. From Fig. 17.13*b* it is clear that this integral is equal to

(3) $$\int_C (v_x \cos\theta + v_y \sin\theta)\, ds = \int_C v_x\, dx + v_y\, dy$$

This integral is called the **circulation**, and for a closed curve it measure the tendency of the fluid to move, or *circulate*, around the curve. If this integral is independent of the path, then, considering paths originating at the same point, we obtain a function Φ of position only, which is known as the **velocity potential** of the flow

(3*a*) $$\Phi = \int_O^P v_x\, dx + v_y\, dy$$

If we apply Green's lemma to the circulation integral (3) over a closed curve C and its interior R, we have

$$\int_C v_x\,dx + v_y\,dy = \int\int_R \left(\frac{\partial v_y}{\partial x} - \frac{\partial v_x}{\partial y}\right) dx\,dy$$

If $\partial v_y/\partial x - \partial v_x/\partial y \equiv 0$, the circulation around every closed curve is zero and the flow is said to be **irrotational**. In other words, when the flow is irrotational, the circulation integral (3) is independent of the path and the velocity potential (3*a*) exists. Whether or not the flow is irrotational, the quantity

$$\text{(4)} \qquad \Omega = \frac{1}{2}\left(\frac{\partial v_y}{\partial x} - \frac{\partial v_x}{\partial y}\right)$$

is called the **vorticity** of the flow.

In regions where the flow is irrotational and where the velocity potential therefore exists, the velocity components v_x and v_y can be found from the velocity potential (3*a*) as well as from the stream function (1). In fact, applying Theorem 2, Sec. 17.5, to the integral in (3*a*), we have

$$\text{(5)} \qquad v_x = \frac{\partial \Phi}{\partial x} \quad \text{and} \quad v_y = \frac{\partial \Phi}{\partial y}$$

The possibility of applying complex variable theory to fluid mechanics is now apparent. From Eqs. (2) and (5), we have

$$v_x = \frac{\partial \Phi}{\partial x} = \frac{\partial \Psi}{\partial y} \quad \text{and} \quad v_y = \frac{\partial \Phi}{\partial y} = -\frac{\partial \Psi}{\partial x}$$

Hence Φ and Ψ satisfy the Cauchy-Riemann equations. Therefore (assuming the necessary continuity) each satisfies Laplace's equation, and $\Phi + i\Psi$ is an analytic function. Extending an earlier definition, we call the analytic function $f(z) = \Phi(x, y) + i\Psi(x, y)$ the **complex potential** of the flow.

Conversely (that is, in the semiexperimental spirit of Example 1), starting with any analytic function $\Phi + i\Psi$, we may interpret Φ as a velocity potential and Ψ as the corresponding stream function, thus obtaining a description of some particular flow pattern. The nature of the flow can be determined by considering the family of streamlines defined by the equation $\Psi(x, y) = k$, obtained by setting the stream function equal to an arbitrary constant. Since the velocity of the fluid is always in the tangential direction along each streamline, as it is of course along any physical boundary, any streamline can be considered a boundary, and a specific physical flow is thereby determined. Furthermore, by Theorem 1, Sec. 17.3, if $f(z) = \Phi + i\Psi$, then $f'(z) = \partial\Phi/\partial x + i\partial\Psi/\partial x$ and therefore, from Eqs. (2) and (5),

$$\text{(6)} \qquad v_x + iv_y = \frac{\partial \Phi}{\partial x} + i\left(-\frac{\partial \Psi}{\partial x}\right) = \overline{f'(z)}$$

In other words, *the* **velocity field of the flow** *is given by the conjugate of the derivative of the complex potential.*

EXAMPLE 2

UNIFORM FLOW

If $f(z) = \Phi + i\Psi = Uz = U(x + iy)$, then, assuming that U is a real number, $\Phi = Ux$ and $\Psi = Uy$. In this case the streamlines $\Psi = Uy = c$ are lines parallel to the x axis, and the flow pattern is called a **uniform flow of strength** U.

EXAMPLE 3

POINT SOURCES AND SINKS

If $f(z) = \Phi + i\Psi = (A/2\pi)\ln z = (A/2\pi)(\ln r + i\theta)$, where A is a real number, then $\Phi = (A/2\pi)\ln r$ and $\Psi = (A/2\pi)\,\theta$. In this case the streamlines are the curves of the family $\Psi = (A/2\pi)\theta = c_2$, that is, the half-lines extending outward from the origin. The lines on which the velocity potential is constant are the curves of the family $\Phi = (A/2\pi)\ln r = c_1$, that is, circles having the origin as center.

Since we are considering only incompressible fluids, the flow across any simple closed curve encircling the origin is equal to the flow across any circle with center at the origin; that is,

$$\Psi_C = \int_C d\Psi = \int_0^{2\pi} \frac{A}{2\pi}\, d\theta = A$$

If A is positive, fluid is emanating from the origin at a rate equal to A, and the flow pattern is called a **point source of strength** A. If A is negative, fluid is flowing toward the origin and disappearing there at a rate equal to A, and the flow pattern is called a **point sink of strength** A. The circulation around any closed curve, whether A is positive or negative, is

$$\phi_C = \int_C d\Phi = \frac{A}{2\pi}\int d(\ln r) = 0$$

since r, and hence $\ln r$, has the same value at the beginning and end of any closed path.

EXAMPLE 4

A POINT VERTEX

If $f(z) = \Phi + i\Psi = -(iK/2\pi)\ln z = (K/2\pi)(\theta - i\ln r)$, where K is a real number, then $\Phi = (K/2\pi)\theta$ and $\Psi = -(K/2\pi)\ln r$. In this case the streamlines are the curves of the family $\Psi = -(K/2\pi)\ln r = c_2$, that is, the circles having the origin as center. The equipotential lines are the curves of the family $\Phi = (K/2\pi)\theta = c_1$, that is, the half-lines extending outward from the origin.

The flow across any closed curve is

$$\Psi_C = \int_C d\Psi = -\frac{K}{2\pi}\int_0^{2\pi} d(\ln r) = 0$$

The circulation around any simple closed curve encircling the origin is

$$\phi_C = \int_C d\Phi = \frac{K}{2\pi}\int_0^{2\pi} d\theta = K$$

This flow pattern is called a **point vortex of strength** K.

Another flow pattern of considerable interest is the **doublet**, defined to be the limiting configuration of a source and a sink of equal strength which approach each other in such a way that the product of their common strength and the distance between them remains constant. Translating the flow patterns of Example 3 from the origin to the points b and $-b$, we begin with the complex potential

$$f(z) = \frac{A}{2\pi}\ln(z-b) - \frac{A}{2\pi}\ln(z+b)$$

By hypothesis, the product $A \cdot 2b$ is a constant, say μ. Hence $f(z)$ can be written in the form

$$f(z) = \frac{\mu}{4\pi}\,\frac{\ln(z-b) - \ln(z+b)}{b}$$

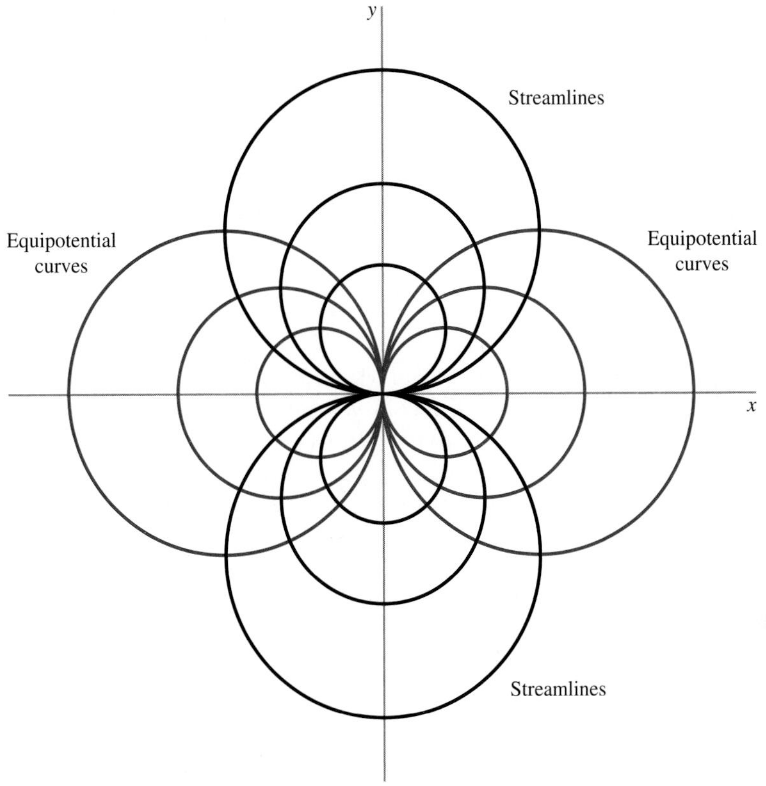

FIGURE 17.14
Typical streamlines and equipotential curves for the doublet defined by $f(z) = -\mu/2\pi z$.

As $b \to 0$ this assumes the indeterminate form 0/0, but evaluating by L'Hospital's rule yields the limiting value

$$-\frac{\mu}{2\pi z}$$

The line along which the source and sink approach each other (the real axis in this case) is called the **axis** of the doublet. Typical streamlines and equipotential lines are shown in Fig. 17.14.

One especially important flow pattern is obtained by superimposing on a doublet a uniform flow in the negative x direction (the direction of the axis of the doublet). In this case we have

$$f(z) = -Uz - \frac{\mu}{2\pi z} = -U\left(z + \frac{a^2}{z}\right) \tag{7}$$

where $a^2 = \mu/2\pi U$ as the complex potential. Hence

$$\Phi + i\Psi = -U\left[x + iy + \frac{a^2(x - iy)}{x^2 + y^2}\right]$$

and therefore the stream function is

$$\Psi = -U\left(y - \frac{a^2 y}{x^2 + y^2}\right) = -Uy\left(1 - \frac{a^2}{x^2 + y^2}\right)$$

From this it is clear that the streamline corresponding to $\Psi = 0$ is composite, consisting of the x axis and the circle $x^2 + y^2 = a^2$. Considering this circle as a rigid object in the fluid, we have thus solved the important problem of finding the flow of an ideal fluid past a smooth circle, or circular cylinder, placed in a uniform stream.

Not unexpectedly, perhaps, the fluid velocity is zero at the points where the streamline $y = 0$ intersects the circle $x^2 + y^2 = a^2$. To confirm this, we note from Eq. (6) that v_x and v_y, and hence the total fluid velocity, will be zero where $f'(z) = 0$. Since, from Eq. (7),

$$f'(z) = -U + \frac{\mu}{2\pi z^2} = -U\left(1 - \frac{a^2}{z^2}\right)$$

it follows that the fluid velocity is zero only when $z = \pm a$, that is, only at the points where $x = \pm a$ and $y = 0$. Points like these where the fluid velocity is zero are called **stagnation points**.

Using **Bernoulli's equation** $p + \frac{1}{2}\rho v^2 = H$, where p is the pressure at any point in the fluid, ρ is the density, $v = |\mathbf{v}|$, and H is a constant, it is now possible to find the resultant force the moving fluid exerts on the circle. From the symmetry of the flow, it is clear that the circle experiences no force in the y direction, i.e., the direction perpendicular to the direction of the stream. Surprisingly, however, it turns out that the circle experiences no force in the x direction either. However, if a point vortex is superimposed on the flow pattern, the velocity of the stream is greater around one side of the circle than it is around the other, and one finds that the stream does indeed exert a nonzero force on the circle. This observation is the starting point of airfoil theory, but we must leave its further development to more specialized texts.†

EXERCISES

1. In Example 1, find the temperature at any point in the portion of the first quadrant between the coordinate axes and the hyperbola $xy = 3$ if the axes are maintained at the temperature 100 and the hyperbola is maintained at the temperature 50.
2. Work Exercise 1 if the axes are maintained at the temperature 50 and the hyperbola is maintained at the temperature 100.
3. In Example 1, find the potential at any point in the region bounded by hyperbolas $x^2 - y^2 = -1$ and $x^2 - y^2 = 1$ if the first hyperbola is maintained at the potential 100 and the second is maintained at the potential 50. *Hint:* Begin with the analytic function $az^2 + b = 0$ with a and b real.
4. What difference, if any, is there between the equipotential lines and the lines of flux defined by $f(z) = -iz^2 + 1$ and by $g(z) = -iaz^2 + b + ci$, a, b, c real?
5. Discuss the flow defined by the complex potential $f(z) = z^2$. What is the velocity of the fluid at the point $1 + 2i$? At the point $-2 + i$? What is the locus of the points where the absolute value of the velocity of the fluid is equal to 1?
6. For a general flow, what is the locus of the points at which the absolute value of the fluid velocity is a constant?
7. What is the potential at any point between the lines $x = -1$ and $x = 2$ if the first line is maintained at the potential 50 and the second line is maintained at the potential 20? *Hint:* Begin with the function $f(z) = az + b$, a, b real.
8. What is the temperature at any point between the lines $y = 2$ and $y = 5$ if the first line is maintained at the temperature 20 and the second line is maintained at the temperature 80?
9. What is the potential at any point between the circles $x^2 + y^2 = 1$ and $x^2 + y^2 = 4$ if the first circle is maintained at the potential 10 and the second is maintained at the potential 50? *Hint:* Assume $f(z) = a \ln z + b$, a, b real.
10. How is the stream function affected if the point O used in its definition is replaced by another point O'?
11. What is the complex potential for a uniform flow in the direction of the line $y = mx$?
12. What is the complex potential for a doublet whose axis is the line $y = mx$?
13. What is the complex potential of a uniform flow in the x direction superimposed on a uniform flow in the y direction? What is the resulting flow pattern?
14. What is the complex potential corresponding to a unit sink at $z = -1$ and a unit source at $z = 1$? Find the ve-

†For a brief introduction to airfoil theory, see C. R. Wylie, Jr., *Advanced Engineering Mathematics,* 1st ed., pp. 416–428, McGraw-Hill, New York, 1951.

locity potential and the stream function and determine the nature of the streamlines and the equipotential curves.

15. What is the complex potential corresponding to unit sources at $z = -1$ and $z = 1$? Find the velocity potential and the stream function and determine the nature of the equipotential curves and the streamlines.

16. If a point vortex is superimposed on the flow whose complex potential is given by Eq. (7), verify that the circle $x^2 + y^2 = a^2$ is still a part of one of the streamlines.

17. If the point vortex whose complex potential is $f(z) = -(iK/2\pi) \ln z$ is superimposed on the flow whose complex potential is given by Eq. (7), determine the stagnation points of the flow. How do the stagnation points vary with K?

18. Find the complex potential for the flow consisting of a vortex of strength 4π superimposed on a uniform flow of strength 1 in the direction of the positive y axis. Determine the velocity potential and the stream function and plot the streamlines $\Psi = -1, 0, 1$.

19. Find the complex potential for the flow consisting of a source of strength 4π superimposed on a uniform flow of strength 1 in the direction of the positive y axis. Determine the velocity potential and the stream function and plot the streamlines $\Psi = -\pi/2, 0, \pi/2$.

20. Find the complex potential of the flow consisting of two vortices of equal strength but of opposite sign placed at $z = -1$ and $z = 1$. Determine the velocity potential and the stream function and describe the streamlines.

21. Find the complex potential, the velocity potential, and the stream function for a source and a sink of strength 2π located, respectively, at $z = -1$ and $z = 1$ in a uniform flow of strength 1 **(a)** in the direction of the positive x axis and **(b)** in the direction of the positive y axis.

22. Discuss the flow pattern consisting of a doublet in a uniform stream if the direction of the stream is perpendicular to the axis of the doublet. Is there a circular streamline?

23. Work Exercise 19 if the source is replaced by a sink of the same strength.

24. Find the velocity potential and the stream function for the flow whose complex potential is $f(z) = \cos^{-1} z$. Show that this can be interpreted as the flow of fluid circulating around a flat plate of length 2. *Hint:* Rewrite the complex potential in the form $z = \cos[f(z)] = \cos(\Phi + i\Psi)$ and proceed very much as in Example 3, Sec. 17.4.

25. Find the velocity potential and the stream function for the flow whose complex potential is $f(z) = \cosh^{-1} z$. Show that this can be interpreted as the flow through an aperture of width 2 in an infinite plate. Note the hint in Exercise 24.

26. Show that the potential throughout the first quadrant is not uniquely determined by the requirement that the potential have a given constant value on the positive x and y axes. Does this contradict your intuition? (In this connection, see the footnote on p. 1074.)

CHAPTER 18

INFINITE SERIES IN THE COMPLEX PLANE

This chapter deals with the properties of infinite series, primarily power series, in the complex plane. In Secs. 18.1–18.2 the concepts of *convergence, absolute convergence,* and *uniform convergence* are discussed, and a number of theorems establishing criteria for uniform convergence and the important properties of uniformly convergent series are proved. In this connection, it should be interesting and perhaps helpful to review the corresponding material on definite integrals in Sec. 10.3.

In Sec. 18.3, these ideas are applied specifically to power series. Then in Sec. 18.4, through the use of Cauchy's integral formula (Sec. 17.6), Taylor's theorem is proved. From it, using the remainder theorem it provides, we establish the important fact that around every point P, where a function f is analytic, f has a unique Taylor expansion which represents it in some neighborhood of P. This is another property of analytic functions, which we exploited in Secs. 2.9 and 2.10.

Around a point where a function is not analytic, it cannot have a Taylor expansion. Nonetheless, in Sec. 18.5 we show that if we are willing to accept negative powers of $z - a$, then around a point P: $z = a$ where a function f is not analytic, it does have one or more power series in $z - a$ which represent it in some deleted neighborhood of P and possibly in one or more annular rings with center P. Such series are called *Laurent expansions* of f. The coefficients in such series are given by definite integrals but cannot be expressed in terms of the derivatives of f as they can in Taylor series. One of them, the coefficient of the term $(z - a)^{-1}$, plays a crucial role in the work of the next chapter.

Prerequisites for this chapter: multivariable calculus, Chap. 17.

18.1 SERIES OF COMPLEX TERMS

Most of the definitions and theorems relating to infinite series of real terms carry over with little or no change to series whose terms are complex. To restate these ideas briefly in the setting of complex analysis, let

$$(1)\qquad \sum_{n=1}^{\infty} f_n(z) = f_1(z) + f_2(z) + \cdots + f_n(z) + \cdots$$

be a series whose terms are complex-valued functions of a complex variable z. Then the **partial sums** of this series are defined to be the finite sums

$$\begin{aligned} S_1(z) &= f_1(z) \\ S_2(z) &= f_1(z) + f_2(z) \\ &\cdots\cdots\cdots \\ S_n(z) &= f_1(z) + f_2(z) + \cdots + f_n(z) \\ &\cdots\cdots\cdots \end{aligned}$$

The series (1) is said to **converge to the sum** $S(z)$ in a region R just in case that for all z in R the limit of the nth partial sum $S_n(z)$ as n becomes infinite is $S(z)$. More precisely, we have the following definition.

DEFINITION 1 The series $\sum_{n=1}^{\infty} f_n(z) = f_1(z) + f_2(z) + \cdots + f_n(z) + \cdots$ **converges** to the **sum** $S(z_0)$ at $z = z_0$ if and only if for every $\epsilon > 0$ there exists an integer N depending in general on z_0 as well as ϵ such that

$$|S(z_0) - S_n(z_0)| < \epsilon \qquad \text{for all } n > N$$

The set of points at which a series converges is called its **region of convergence.** At points where a series does not converge it is said to **diverge** or to be **divergent.**

The difference $S(z_0) - S_n(z_0)$ which appears in Definition 1 is clearly just the **remainder** after the first n terms in $S(z_0)$, say $R_n(z_0)$. Thus *an equivalent requirement that a series converge at* $z = z_0$ *is that the limit of* $|R_n(z_0)|$ *as n becomes infinite is zero.*

If the absolute values of the terms in (1) form a convergent series

$$|f_1(z)| + |f_2(z)| + \cdots + |f_n(z)| + \cdots$$

then (1) is said to be **absolutely convergent.** If the series (1) converges but is not absolutely convergent, it is said to be **conditionally convergent.** Absolute convergence is an important property because it is a sufficient (though not a necessary) condition for ordinary convergence. Moreover the terms of an absolutely convergent series can be rearranged in any manner whatsoever without affecting the sum of the series, whereas rearranging the terms of a conditionally convergent series may alter the sum of the series or even cause the series to diverge.

From the definition of convergence it is easy to prove the following theorem.

THEOREM 1 A necessary and sufficient condition that the series of complex terms $f_1(z) + f_2(z) + \cdots + f_n(z) + \cdots$ should converge is that the series of the real parts and the series of the imaginary parts of these terms should each converge. Moreover, if

$$\sum_{n=1}^{\infty} \text{Re}(f_n) \qquad \text{and} \qquad \sum_{n=1}^{\infty} \text{Im}(f_n)$$

converge to the respective functions $\text{Re}(f)$ and $\text{Im}(f)$, then the given series converges to $\text{Re}(f) + i\,\text{Im}(f) = f(z)$.

Of all the tests for the convergence of infinite series, the most useful is probably the familiar *ratio test,* which applies to series whose terms are complex as well as to series whose terms are real.

THEOREM 2 **(The Ratio Test)** For the series

$$\sum_{n=1}^{\infty} f_n(z) = f_1(z) + f_2(z) + \cdots + f_n(z) + \cdots$$

if

$$\lim_{n\to\infty} \left|\frac{f_{n+1}(z)}{f_n(z)}\right| = |r(z)|$$

then the given series converges absolutely for those values of z for which $0 \le |r(z)| < 1$ and diverges for those values of z for which $|r(z)| > 1$. The values of z for which $|r(z)| = 1$ form the boundary of the region of convergence of the series, and at these points the ratio test provides no information about the convergence or divergence of the series.

EXAMPLE 1

Find the region of convergence of the series

$$1 + \frac{1}{2^2}\left(\frac{z+1}{z-1}\right) + \frac{1}{3^2}\left(\frac{z+1}{z-1}\right)^2 + \cdots + \frac{1}{n^2}\left(\frac{z+1}{z-1}\right)^{n-1} + \cdots$$

Applying the ratio test, we find

$$\left|\frac{f_{n+1}(z)}{f_n(z)}\right| = \left|\frac{\dfrac{1}{(n+1)^2}\left(\dfrac{z+1}{z-1}\right)^n}{\dfrac{1}{n^2}\left(\dfrac{z+1}{z-1}\right)^{n-1}}\right| = \left|\frac{n^2}{(n+1)^2}\,\frac{z+1}{z-1}\right|$$

As n becomes infinite, this ratio approaches $|(z+1)/(z-1)|$. Hence the values of z for which the series surely converges are those in the region defined by the inequality

$$\left|\frac{z+1}{z-1}\right| < 1$$

that is, by

$$|z+1| < |z-1|$$

Since $|z+1|$ is just the distance from z to the point -1 and since $|z-1|$ is just the distance from z to the point 1, z is thus restricted to be nearer the point -1 than to the point 1. In other words, z must lie to the left of the perpendicular bisector of the segment joining -1 and 1; that is, z must lie in the left half of the complex plane. The boundary cases for which the test fails are the points which are equidistant from -1 and 1, that is, the points on the imaginary axis. For these points $|z+1|/|z-1| = 1$ and the related series of absolute values is the convergent real series

$$1 + \frac{1}{2^2} + \frac{1}{3^2} + \cdots + \frac{1}{n^2} + \cdots$$

Hence, for all values of z on the imaginary axis, the given series, being absolutely convergent, is convergent. Thus these points also belong to the region of convergence of the series.

Addition, subtraction, and multiplication of series are covered by the following theorems.

THEOREM 3 For any value of z for which two series converge, their sum or difference can be found by term-by-term addition or subtraction.

THEOREM 4 For any value of z for which two series converge *absolutely,* their product can be found by multiplying them as though they were polynomials, or by means of their Cauchy product (Sec. 2.9).

The important question of when a series can legitimately be differentiated or integrated term by term we shall consider in the next section. But before we can do this, we must introduce the concept of *uniform convergence.*

DEFINITION 2 A series of functions $f_1(z) + f_2(z) + \cdots + f_n(z) + \cdots$, having $S_n(z)$ as its nth partial sum, is said to **converge uniformly** to the function $f(z)$ in a region R, either open or closed, if corresponding to an arbitrary $\epsilon > 0$ there exists a positive integer N, depending on ϵ *but not on* z, such that for every value of z in R,

$$|f(z) - S_n(z)| < \epsilon \qquad \text{for all } n > N$$

In other words, if a series converges uniformly in a region R, then corresponding to any $\epsilon > 0$ there exists an integer N such that *everywhere* in R the sum of the series $f(z)$ can be approximated with an error less than ϵ by using *no more than* N terms of the series. It may well be that fewer than N terms will suffice at most of the points in the region, but *nowhere* will *more* than N be required. This is in sharp contrast to ordinary convergence; for in the neighborhood of certain points in a region of ordinary convergence it may be that no bound can be set on the number of terms required to secure a prescribed degree of accuracy. Our next example explores this matter in detail.

EXAMPLE 2

Discuss the convergence of the series

$$z^2 + \frac{z^2}{1+z^2} + \frac{z^2}{(1+z^2)^2} + \frac{z^2}{(1+z^2)^3} + \cdots + \frac{z^2}{(1+z^2)^{n-1}} + \cdots$$

in the 90° sector bounded by the right halves of the lines $y = \pm x$ (Fig. 18.1*a*).

The given series is a geometric series which will converge for all values of z for which the absolute value of the common ratio, namely,

$$|r| = \left|\frac{1}{1+z^2}\right|$$

is less than 1. Now the angle of z is restricted, by hypothesis, to be between $-\pi/4$ and $\pi/4$. Hence the angle of z^2 must be between $-\pi/2$ and $\pi/2$. Therefore (Fig. 18.1*b*) the vectors representing the numbers 1, z^2, and $1 + z^2$ are the sides of a triangle in which the side $1 + z^2$ is opposite the largest angle. Thus, for every z in the given region D, we have

FIGURE 18.1

A 90° sector before and after modification to exclude its vertex.

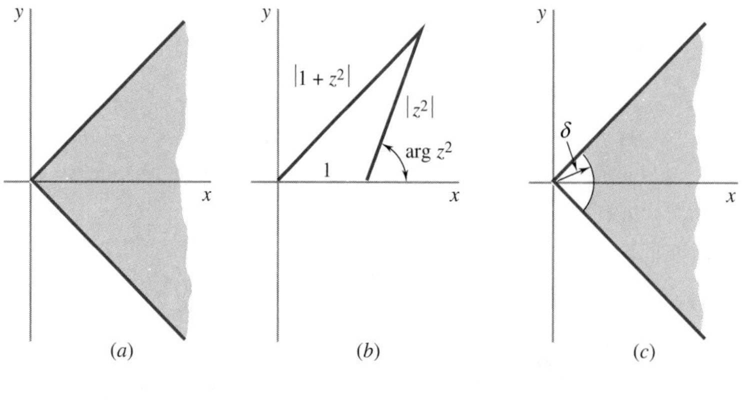

$$|1 + z^2| \geq 1 \qquad \text{and} \qquad \frac{1}{|1 + z^2|} \leq 1$$

and the equality sign holds only for the value $z = 0$. Thus the given series converges for all values of z in D, and its sum (as an infinite geometric progression when $z \neq 0$) is

$$S(z) = \begin{cases} \dfrac{a}{1 - r} = \dfrac{z^2}{1 - 1/(1 + z^2)} = 1 + z^2 & z \neq 0 \\ 0 + 0 + 0 + 0 + \cdots = 0 & z = 0 \end{cases}$$

Now let an arbitrary $\epsilon > 0$ be given, and let us attempt to determine how many terms of the series must be taken in order that

$$|S(z) - S_n(z)| < \epsilon$$

This difference, i.e., the remainder after n terms of the series, is just the geometric series

$$\frac{z^2}{(1 + z^2)^n} + \frac{z^2}{(1 + z^2)^{n+1}} + \frac{z^2}{(1 + z^2)^{n+2}} + \cdots$$

whose sum is

$$R_n(z) = \begin{cases} \dfrac{1}{(1 + z^2)^{n-1}} & z \neq 0 \\ 0 & z = 0 \end{cases}$$

Hence to show that the given series converges *uniformly* in D, we must find, if possible, a value of N such that

$$(2) \qquad |R_n(z)| = \frac{1}{|1 + z^2|^{n-1}} < \epsilon \qquad \text{for all } n > N \text{ and all } z \text{ in } D$$

Now $|1 + z^2| \leq 1 + |z^2| = 1 + |z|^2$. Hence, underestimating $R_n(z)$ by overestimating its denominator, we have

$$|R_n(z)| = \frac{1}{|1 + z^2|^{n-1}} \geq \frac{1}{(1 + |z|^2)^{n-1}}$$

From this inequality we observe that if it should be impossible to find an integer N such that

$$(3) \qquad \frac{1}{(1 + |z|^2)^{n-1}} < \epsilon \qquad \text{for all } n > N \text{ and all } z \text{ in } D$$

then surely it will be impossible to find an integer N which will suffice to keep

$$|R_n(z)| < \epsilon$$

everywhere in D. And this is indeed the case, for if we attempt to solve the inequality (3) for n, we find, by obvious steps,

$$(1 + |z|^2)^{n-1} > \frac{1}{\epsilon}$$

$$(n - 1) \ln (1 + |z|^2) > \ln \frac{1}{\epsilon} = -\ln \epsilon$$

$$n > 1 - \frac{\ln \epsilon}{\ln (1 + |z|^2)}$$

But for values of z within the sector of the problem and sufficiently close to the origin, $\ln (1 + |z|^2)$ can be made arbitrarily close to ln 1, that is, zero. Hence n is unbounded, and there exists no integer N for which (3) holds. Since $|R_n(z)|$ is larger than the fraction in (3), it is clear that the fundamental requirement of uniform convergence (2) cannot be fulfilled. Thus the convergence of the given series in the original region D is nonuniform.

On the other hand, if we restrict z to the infinite region D' bounded by the given rays and a circular arc of small but fixed radius δ, as shown in Fig. 18.1c, the series does converge uniformly. In fact, the law of cosines applied in Fig. 18.1b gives

$$|1 + z^2|^2 = 1 + |z^2|^2 + 2|z^2| \cos(\arg z^2)$$

Reducing the right-hand side by dropping the last term, which is surely nonnegative since $-\pi/2 \le \arg z^2 \le \pi/2$, gives

$$|1 + z^2|^2 \ge 1 + |z^2|^2 = 1 + |z|^4$$

Hence, overestimating $|R_n(z)|$ by underestimating its denominator, we can write

$$|R_n(z)| = \frac{1}{|1 + z^2|^{n-1}} \le \frac{1}{(1 + |z|^4)^{(n-1)/2}}$$

From this it is clear that if we can find an integer N such that

(4) $$\frac{1}{(1 + |z|^4)^{(n-1)/2}} < \epsilon \qquad \begin{array}{l}\text{for all } n > N \\ \text{and all } z \text{ in } D'\end{array}$$

then surely for the same N we shall have

(5) $$|R_n(z)| < \epsilon \qquad \begin{array}{l}\text{for all } n > N \\ \text{and all } z \text{ in } D'\end{array}$$

Hence we attempt to solve the inequality in (4) for n:

$$(1 + |z|^4)^{(n-1)/2} > \frac{1}{\epsilon}$$

$$\frac{n-1}{2} \ln(1 + |z|^4) > \ln \frac{1}{\epsilon} = -\ln \epsilon$$

$$n > 1 - \frac{2 \ln \epsilon}{\ln(1 + |z|^4)}$$

The most unfavorable case, i.e., the largest possible value of the fraction on the right, occurs when $|z|$ is as small as possible. But in the modified region D' which we are now considering, the smallest possible value of $|z|$ is δ, which yields

$$n > 1 - \frac{2 \ln \epsilon}{\ln(1 + \delta^4)}$$

If we choose N to be the first integer equal to or greater than the expression on the right, then (4) will surely hold. But, as we observed above, if (4) is satisfied, so too is (5), and hence in the modified region D' the given series does converge uniformly.

Usually, uniform convergence is established not by a direct application of the definition, as in Example 2, but by the so-called **Weierstrass M test,**† which reads as follows.

†The German mathematician Karl Weierstrass (1815–1897) is often called the "father of modern rigor."

THEOREM 5 If a sequence of positive constants M_n exists such that $|f_n(z)| \le M_n$ for all positive integers n and for all values of z in a given region D, and if the series

$$M_1 + M_2 + M_3 + \cdots + M_n + \cdots$$

is convergent, then the series

$$f_1(z) + f_2(z) + f_3(z) + \cdots + f_n(z) + \cdots$$

converges uniformly in D.

PROOF To prove this, we must show that for any $\epsilon > 0$ there exists an integer N, independent of z, such that for all values of z in D the absolute value of the remainder after n terms of the series of the f's is less than ϵ whenever n exceeds N. To do this, we note that

$$\begin{aligned} |R_n(z)| &= |f_{n+1}(z) + f_{n+2}(z) + \cdots| \\ &\le |f_{n+1}(z)| + |f_{n+2}(z)| + \cdots \\ &\le M_{n+1} + M_{n+2} + \cdots \end{aligned} \tag{6}$$

The last expression is just the remainder after n terms of the series of the M's. Since this series is convergent, by hypothesis, it follows that for every $\epsilon > 0$ there exists an N such that this remainder is less than ϵ for all $n > N$. This value of N, arising as it does from a series of constants, is obviously independent of z. Moreover, from the inequality (6) it is clear that whenever n exceeds this N, $|R_n(z)| < \epsilon$ for all values of z in D. Hence the series of the f's is uniformly convergent, as asserted. Incidentally, this theorem implies a comparison test which proves that the series of the f's is also absolutely convergent. ◗

It is important to bear in mind that the Weierstrass M test is only a sufficient test for uniform convergence, not a necessary one. In other words, there exist uniformly convergent series whose terms cannot be dominated by the respective terms of any convergent series of positive constants.† The M test suffices for almost all applications, however.

EXAMPLE 3

Show that the infinite series $\Sigma_{n=1}^{\infty} (\sin nz/e^{nz})$ converges uniformly in the region R defined by the inequalities $0 \le y \le x - \delta$, where $\delta > 0$

To establish uniform convergence by the Weierstrass M test we must obtain a sequence of positive constants M_n such that

a. $M_n \ge |\sin nz/e^{nz}|$ for all n and all z in R.

b. The series $\Sigma_{n=1}^{\infty} M_n$ converges.

In this example we shall obtain the required constants M_n by suitably overestimating the absolute value of the nth term of the given series.

Using the first result of Exercise 34(**a**), Sec. 17.4, and the formula $e^{nz} = e^{nx}(\cos ny + i \sin ny)$, we have

$$\begin{aligned} \left|\frac{\sin nz}{e^{nz}}\right| &= \frac{\sqrt{\sin^2 nx + \sinh^2 ny}}{e^{nx}} \le \frac{\sqrt{1 + \sinh^2 ny}}{e^{nx}} \\ &= \frac{\cosh ny}{e^{nx}} = \frac{e^{ny} + e^{-ny}}{2e^{nx}} = \frac{e^{n(y-x)} + e^{-n(y+x)}}{2} \end{aligned}$$

†One example of such a series is given in Exercise 11.

The first term in the numerator of the last fraction attains its maximum value when $y - x$ is as large as possible. Similarly, the second attains its maximum value when $y + x$ is as small as possible. From the second of the inequalities which limit y, we have

$$y - x \leq -\delta$$

Also, writing it in the form $x \geq y + \delta$, we have

$$x + y \geq 2y + \delta \geq \delta$$

Using these two estimates, we now have

$$\left|\frac{\sin nz}{e^{nz}}\right| \leq \frac{e^{n(y-x)} + e^{-n(y+x)}}{2} \leq \frac{e^{n(-\delta)} + e^{-n\delta}}{2} = e^{-n\delta}$$

Clearly, for each n, $M_n \equiv e^{-n\delta}$ is a positive constant, independent of z, such that

$$M_n \geq \left|\frac{\sin nz}{e^{nz}}\right| \qquad \text{for each } z \text{ in } R$$

Moreover, since $\delta > 0$, the series $\Sigma_{n=1}^{\infty} M_n \equiv \Sigma_{n=1}^{\infty} e^{-n\delta}$ converges because it is a geometric series whose common ratio $e^{-\delta}$ is between 0 and 1. Thus the given series converges uniformly in R, as asserted.

One useful property of uniformly convergent series is contained in the following theorem.

THEOREM 6 If the terms of a uniformly convergent series are multiplied by any bounded function of z, the resulting series will also converge uniformly.

PROOF Let D be the region of uniform convergence of the series

$$f_1(z) + f_2(z) + \cdots + f_n(z) + \cdots$$

and let g be a function such that

$$|g(z)| < M$$

for all values of z in D. Now since the series of the f's converges uniformly, it follows that corresponding to the infinitesimal ϵ/M there exists an integer N such that

$$|R_n(z)| = |f_{n+1}(z) + f_{n+2}(z) + \cdots| < \frac{\epsilon}{M} \qquad \begin{array}{l}\text{for all } n > N\\ \text{and all } z \text{ in } D\end{array}$$

Hence

$$\begin{aligned}|g(z)f_{n+1}(z) + g(z)f_{n+2}(z) + \cdots| &= |g(z)| \cdot |f_{n+1}(z) + f_{n+2}(z) + \cdots|\\ &< M|f_{n+1}(z) + f_{n+2}(z) + \cdots|\\ &< M\frac{\epsilon}{M}\\ &= \epsilon \qquad \begin{array}{l}\text{for all } n > N\\ \text{and all } z \text{ in } D\end{array}\end{aligned}$$

which is precisely the condition that the product series

$$g(z)f_1(z) + g(z)f_2(z) + \cdots + g(z)f_n(z) + \cdots$$

is uniformly convergent.

EXERCISES

1. Find the region of convergence and the sum of each of the following series.
(a) $1 + (z - i) + (z - i)^2 + (z - i)^3 + \cdots$
(b) $z(1 - z) + z^2(1 - z) + z^3(1 - z) + \cdots$

2. Find the region of convergence and the sum of each of the following series.

(a) $\frac{1}{2}\left(\frac{z+1}{z-1}\right) + \frac{1}{2^2}\left(\frac{z+1}{z-1}\right)^2 + \frac{1}{2^3}\left(\frac{z+1}{z-1}\right)^3 + \cdots$

(b) $\frac{1}{2(z+i)} + \frac{1}{2^2(z+i)^2} + \frac{1}{2^3(z+i)^3} + \cdots$

3. Find the region of convergence of each of the following series.

(a) $1 + \frac{1}{2^2}\left[\frac{\operatorname{Re} z}{z+1}\right] + \frac{1}{3^2}\left[\frac{\operatorname{Re} z}{z+1}\right]^2 + \frac{1}{4^2}\left[\frac{\operatorname{Re} z}{z+1}\right]^3 + \cdots$

(b) $1 + \frac{1}{2^2}\left[\frac{\operatorname{Im} z}{z+1}\right] + \frac{1}{3^2}\left[\frac{\operatorname{Im} z}{z+1}\right]^2 + \frac{1}{4^2}\left[\frac{\operatorname{Im} z}{z+1}\right]^3 + \cdots$

4. Find the region of convergence and the sum of each of the following series.

(a) $\sum_{n=1}^{\infty} e^{-nz}$ **(b)** $\sum_{n=1}^{\infty} e^{-nz^2}$

(c) $\sum_{n=1}^{\infty} e^{-n(z^2+1)}$ **(d)** $\sum_{n=1}^{\infty} e^{n(z^2-z)}$

5. Determine a region of uniform convergence for each of the series in Exercise 4.

6. In Example 2, show that the boundary of the entire region of convergence of the series is the curve $(x^2 - y^2 + 1)^2 + 4x^2y^2 = 1$. Does the series converge in the interior or exterior of this curve? At what points on the boundary does the series converge?

7. Show that the series in Exercise 1(**b**) converges uniformly for $|z| \le \rho < 1$ but does not converge uniformly for $|z| < 1$.

8. **(a)** Where does the series $\sum_{n=1}^{\infty} \frac{z}{(1+z)^{n-1}}$ converge?
(b) What is its sum?
(c) Where does it converge uniformly?

9. **(a)** Where does the series $\sum_{n=1}^{\infty} z(1+z)^{n-1}$ converge?
(b) What is its sum?
(c) Where does it converge uniformly?

10. **(a)** Where does the series $\sum_{n=1}^{\infty} (1-z^2)^{n-1}$ converge?
(b) What is its sum?
(c) Where does it converge uniformly?

11. Show that the series $\frac{1}{1+x^2} - \frac{1}{2+x^2} + \frac{1}{3+x^2} - \frac{1}{4+x^2} + \cdots$ converges uniformly over any interval of the x axis but that this cannot be established by the Weierstrass M test. *Hint:* Recall the properties of alternating real series.

12. **(a)** Where does the series $\frac{1}{1+|z|} - \frac{1}{2+|z|} + \frac{1}{3+|z|} - \frac{1}{4+|z|} + \cdots$ converge?
(b) Where does it converge uniformly?

13. **(a)** Where does the series $\frac{1}{1+|z|} - \frac{1}{1+2|z|} + \frac{1}{1+3|z|} - \frac{1}{1+4|z|} + \cdots$ converge?
(b) Where does it converge uniformly?

14. Determine the region of convergence and the sum of the series

$$\frac{z}{(0z+1)(z+1)} + \frac{z}{(z+1)(2z+1)} + \frac{z}{(2z+1)(3z+1)} + \frac{z}{(3z+1)(4z+1)} + \cdots$$

Show that for $|z| \ge \rho > 1$ the convergence is uniform. *Hint:* Express each term as a sum of partial fractions.

15. Determine the region of convergence and the sum of the series

$$\frac{1}{z(z+1)} + \frac{1}{(z+1)(z+2)} + \frac{1}{(z+2)(z+3)} + \frac{1}{(z+3)(z+4)} + \cdots$$

Where does the series converge uniformly?

16. **(a)** What is the region of convergence of the series $\sum_{n=1}^{\infty} \frac{e^{inz}}{n^{3/2}}$?
(b) Where does the series converge uniformly?

17. Extend the result of Example 3 by showing that the series also converges uniformly in the region defined by the inequalities $0 \le -y \le x - \delta$, where $\delta > 0$.

18. In Example 2, let z be real and plot $S(x)$ and $S_n(x)$ for $n = 1, 2, 4, 8$. How does the behavior suggested by these curves compare with the Gibbs phenomenon discussed in Sec. 9.2?

19. Prove the following theorem: If at all points of a region R its terms are bounded and $\left|\frac{u_{n+1}(z)}{u_n(z)}\right| < \rho < 1$ for $n > N$, then the series $\sum_{n=1}^{\infty} u_n(z)$ converges uniformly in R. *Hint:*

Show first that $|u_{n+k}(z)| < \rho^k |u_n(z)|$; then use the Weierstrass M test.

20. Prove the following theorem: Let the partial sums of the series $u_1(z) + u_2(z) + \cdots + u_n(z) + \cdots$ be **uniformly bounded** in a region R; that is, suppose that there exists a constant K such that $|S_n(z)| < K$ for all $n > 0$ and for all z in R. Let $v_1, v_2, \ldots, v_n, \ldots$, be a monotone sequence of positive constants converging to zero. Then the series

$$v_1u_1(z) + v_2u_2(z) + \cdots + v_nu_n(z) + \cdots$$

converges uniformly in R. *Hint:* In considering the remainder R_n, use the identity

$$\begin{aligned} R_n(z) &= u_{n+1}v_{n+1} + u_{n+2}v_{n+2} + u_{n+3}v_{n+3} + \cdots \\ &= (S_{n+1} - S_n)v_{n+1} + (S_{n+2} - S_{n+1})v_{n+2} \\ &\quad + (S_{n+3} - S_{n+2})v_{n+3} + \cdots \end{aligned}$$

21. Using the result of Exercise 20, show that each of the series $\sum_{n=1}^{\infty} a_n \cos nx$ and $\sum_{n=1}^{\infty} b_n \sin nx$ converges uniformly in $[x_0, x_1]$, where $0 < x_0 < x_1 < 2\pi$, provided that $\{a_n\}$ and $\{b_n\}$ are monotonically decreasing sequences of positive constants which approach zero. *Hint:* Recall the identities

$$\sum_{k=1}^{n} \cos k\theta = \left(\cos \frac{n\theta}{2} \sin \frac{(n+1)\theta}{2}\right) \bigg/ \sin \frac{\theta}{2}$$

$$\sum_{k=1}^{n} \sin k\theta = \left(\sin \frac{n\theta}{2} \sin \frac{(n+1)\theta}{2}\right) \bigg/ \sin \frac{\theta}{2}$$

22. Prove Theorem 1.

23. Find the region of convergence of the series

$$\frac{4 - z^2}{1^2} + \frac{(4 - z^2)^2}{2^2} + \frac{(4 - z^2)^3}{3^2} + \cdots$$

24. What is the region of convergence of the series $\sum_{n=1}^{\infty} \frac{e^{nz}}{n^{3/2}}$? Where does the series converge uniformly?

25. Show that the series $x + x(1-x) + x(1-x)^2 + x(1-x)^3 + \cdots$ converges for $0 \le x < 2$ but that the convergence is not uniform in any subinterval which contains the origin. Is the convergence uniform for $0 < \rho_1 \le x \le \rho_2 < 2$?

18.2 PROPERTIES OF UNIFORMLY CONVERGENT SERIES

Series that converge uniformly have a number of important properties which nonuniformly convergent series may, but usually do not, have. Our first theorem gives us one such result.

THEOREM 1 The sum of a uniformly convergent series of continuous functions is a continuous function.

PROOF Let $f(z) = f_1(z) + f_2(z) + \cdots + f_n(z) + \cdots = S_n(z) + R_n(z)$ be a series of continuous functions which is uniformly convergent in a region R, and let z_0 be an arbitrary point in R. To prove that $f(z)$ is continuous in R we must show, as usual, that for any $\epsilon > 0$ there exists a δ such that if $z_0 + \Delta z$ is also a point of R, then

$$|f(z_0 + \Delta z) - f(z_0)| < \epsilon \qquad \text{for all } |\Delta z| < \delta \tag{1}$$

To do this, we begin by overestimating the difference (1):

$$\begin{aligned} |f(z_0 + \Delta z) - f(z_0)| &= |[S_n(z_0 + \Delta z) + R_n(z_0 + \Delta z)] - [S_n(z_0) + R_n(z_0)]| \\ &\le |S_n(z_0 + \Delta z) - S_n(z_0)| + |R_n(z_0 + \Delta z)| + |R_n(z_0)| \end{aligned} \tag{2}$$

By hypothesis, the given series converges uniformly. Hence for every $\epsilon > 0$ there exists an integer N such that for every z_0 in R,

$$|R_n(z_0 + \Delta z)| < \frac{\epsilon}{3} \tag{3a}$$

$$|R_n(z_0)] < \frac{\epsilon}{3} \tag{3b}$$

for all $n > N$

Furthermore, since the sum of any finite number of continuous functions is continuous, it follows that the nth partial sum $S_n(z)$ is a continuous function in R. In particular, this means that for every z_0 in R and every $\epsilon > 0$, there exists a δ such that

$$|S_n(z_0 + \Delta z) - S_n(z_0)| < \frac{\epsilon}{3} \qquad \text{for all } \Delta z \text{ such that } |\Delta z| < \delta \tag{4}$$

Replacing the terms in the right-hand member of (2) by their respective overestimates (4), (3a) and (3b), we find that

$$|f(z_0 + \Delta z) - f(z_0)| < \frac{\epsilon}{3} + \frac{\epsilon}{3} + \frac{\epsilon}{3} = \epsilon \qquad \text{for all } |\Delta z| < \delta$$

Thus (1) is satisfied, and our proof is complete. ◗

Theorem 1 makes no assertion about the sum of a series of continuous functions if the convergence is nonuniform, and in particular cases the sum may be either continuous or discontinuous. For instance, Example 2, Sec. 18.1, shows that the sum of a series of continuous functions may be discontinuous if the convergence is nonuniform. In fact, in that example we found that the sum of the series

$$z^2 + \frac{z^2}{1+z^2} + \frac{z^2}{(1+z^2)^2} + \cdots + \frac{z^2}{(1+z^2)^{n-1}} + \cdots$$

was

$$S(z) = \begin{cases} 1 + z^2 & z \neq 0 \\ 0 & z = 0 \end{cases}$$

Hence in any neighborhood of $z = 0$, where the convergence is nonuniform, the sum jumps abruptly from $1 + z^2$ to 0, even though each term of the series is continuous for all values of z in the region of convergence. On the other hand, Exercise 6 provides an example of a nonuniformly convergent series of continuous functions whose sum is continuous. Thus uniform convergence is a sufficient but not a necessary condition that a series of continuous functions should have a continuous sum.

One of the most important properties of uniformly convergent series is given by the following theorem.

THEOREM 2 The integral of the sum of a uniformly convergent series of continuous functions along any curve C lying entirely in the region of uniform convergence can be found by term-by-term integration of the series. Moreover, if each term of the series is analytic, so too is the sum.

◖ **PROOF** Let the given series be

$$f(z) = f_1(z) + f_2(z) + \cdots + f_n(z) + \cdots$$

Then, to establish the theorem, we must show that

$$\int_C f(z)\,dz = \int_C f_1(z)\,dz + \int_C f_2(z)\,dz + \cdots + \int_C f_n(z)\,dz + \cdots$$

which, in accordance with the usual definition of convergence, requires that we prove the existence, for every $\epsilon > 0$, of an integer N such that

$$\left| \int_C f(z)\, dz - \sum_{i=1}^{n} \int_C f_i(z)\, dz \right| < \epsilon \qquad \text{for all } n > N$$

Now for any *finite* sum it is true that the integral of a sum is equal to the sum of the integrals. Hence the left-hand member of the last inequality can be written

$$\left| \int_C f(z)\, dz - \int_C \sum_{i=1}^{n} f_i(z)\, dz \right| = \left| \int_C \left[f(z) - \sum_{i=1}^{n} f_i(z) \right] dz \right| = \left| \int_C R_n(z)\, dz \right|$$

Let L be the length of the path of integration. Then, from the uniform convergence of the given series, we know that there exists an integer N such that

$$|R_n(z)| < \frac{\epsilon}{L} \qquad \text{for all } n > N$$

and for all z's in the region of uniform convergence, including in particular all values of z on the path of integration C. If $n > N$, we can therefore write

$$\left| \int_C f(z)\, dz - \sum_{i=1}^{n} \int_C f_i(z)\, dz \right| = \left| \int_C R_n(z)\, dz \right|$$

$$\leq \int_C |R_n(z)| \cdot |dz| < \frac{\epsilon}{L} \int_C |dz| = \frac{\epsilon}{L} L = \epsilon$$

which establishes the first part of the theorem.

To establish the second part, we suppose that the region of uniform convergence D is either simply connected or has been made simply connected by suitable cross-cuts. Then if each term f_i is analytic in D, it follows from Cauchy's theorem that the integral of each term around any simple closed curve in D (or its simply connected modification) is zero. Hence the integral of the sum $f(z)$ around any closed curve is zero, and thus, by Morera's theorem, $f(z)$ is analytic. This completes the proof of the theorem. ◗

As an illustration of the importance of the last theorem, we have the following example which shows that termwise integration of a series which is not uniformly convergent may lead to an incorrect answer.

EXAMPLE 1

Let $f(x) = \sum_{n=1}^{\infty} f_n(x) = \sum_{n=1}^{\infty} [u_n(x) - u_{n-1}(x)]$, where $u_n(x) = nxe^{-nx^2}$. Then the nth partial sum of the series for $f(x)$ is

$$S_n = (u_1 - u_0) + (u_2 - u_1) + \cdots + (u_n - u_{n-1})$$

$$= u_n - u_0 = u_n \qquad \text{since } u_0 = 0$$

Hence

$$f(x) = \lim_{n\to\infty} S_n(x) = \lim_{n\to\infty} u_n(x) = \lim_{n\to\infty} nxe^{-nx^2} = 0$$

for all values of x; and from this it follows trivially that for arbitrary (real) a,

$$\int_0^a f(x)\, dx = \int_0^a 0\, dx = 0$$

To check this against the result obtained by integrating the series for $f(x)$ term by term, we observe first that termwise integration of the series amounts to asserting that

$$\int_0^a f(x)\,dx = \sum_{n=1}^{\infty} \int_0^a f_n(x)\,dx = \lim_{k\to\infty} \sum_{n=1}^{k} \int_0^a f_n(x)\,dx$$

Now a finite sum can always be correctly integrated term by term. Hence in the last expression the operations of integration and summation can be interchanged, and we have

$$\begin{aligned}\int_0^a f(x)\,dx &= \lim_{k\to\infty} \int_0^a \sum_{n=1}^{k} f_n(x)\,dx \\ &= \lim_{k\to\infty} \int_0^a S_k(x)\,dx = \lim_{k\to\infty} \int_0^a kxe^{-kx^2}\,dx \\ &= \lim_{k\to\infty} \left[-\frac{1}{2}e^{-kx^2}\right]_0^a = \lim_{k\to\infty} \left[\frac{1}{2} - \frac{1}{2}e^{-ka^2}\right] = \frac{1}{2}\end{aligned}$$

Thus termwise integration of the series for $f(x)$ between $x = 0$ and $x = a$ does not give the correct answer, namely, zero.

That termwise integration from 0 to a would give an incorrect answer should have been expected on the basis of Theorem 2 because the series for $f(x)$ does not converge uniformly over any interval containing $x = 0$. To see this, we note first that since $f(x) = S_n(x) + R_n(x)$ and since, as we noted earlier, $f(x) \equiv 0$, therefore $R_n(x) = -S_n(x)$. Hence on $[0, a]$ we have $|R_n(x)| = nxe^{-nx^2}$. Thus, if the series for $f(x)$ converges uniformly on $[0, a]$, there must exist an integer N such that $nxe^{-nx^2} < \epsilon$ for x in $[0, a]$ and all $n > N$. However, a simple calculation shows that nxe^{-nx^2} has a maximum value of $\sqrt{n/2e}$ at $x = 1/\sqrt{2n}$. Therefore, no matter how large N may be, there will always be points in the interval $[0, a]$ where, for $n > N$, $|R_n(x)|$ not only fails to be less than ϵ but in fact is arbitrarily large. Hence on $[0, a]$ the convergence is nonuniform, the legitimacy of termwise integration cannot be guaranteed, and in fact integrating term by term gives an incorrect answer.

Theorem 2 and Example 1 should not be misunderstood. Theorem 1 guarantees that *if a series converges uniformly, it can always be legitimately integrated term by term.* Example 1 shows that termwise integration of a series which does not converge uniformly *may* lead to a wrong answer. But neither Theorem 2 nor Example 1 says that integrating a nonuniformly convergent series term by term will *always* yield an incorrect result. In fact, Exercise 11 involves a nonuniformly convergent series for which termwise integration does give the correct answer. In other words, uniform convergence is a sufficient but not a necessary condition for the termwise integrability of an infinite series.

Our next result complements Theorem 2 by giving a sufficient condition for the termwise differentiability of an infinite series.

THEOREM 3 If $f(z)$ is the sum of a uniformly convergent series of analytic functions, then the derivative of $f(z)$ at any interior point of the region of uniform convergence can be found by term-by-term differentiation of the series.

PROOF Let z be a general point of the region of uniform convergence R of the given series and let C be a simple closed curve in R which encircles z and encloses only points of R. If we write the given series in the form

$$f(t) = f_1(t) + f_2(t) + \cdots + f_n(t) + \cdots$$

where t is a variable ranging over the values of z on C, we can multiply it by the bounded function

$$\frac{1}{2\pi i(t - z)^2}$$

and, by Theorem 6, Sec. 18.1, the resulting series

$$\frac{f(t)}{2\pi i(t-z)^2} = \frac{f_1(t)}{2\pi i(t-z)^2} + \frac{f_2(t)}{2\pi i(t-z)^2} + \cdots + \frac{f_n(t)}{2\pi i(t-z)^2} + \cdots$$

will also converge uniformly. Therefore, by Theorem 2, it can be integrated term by term around C, giving

$$\frac{1}{2\pi i}\int_C \frac{f(t)\,dt}{(t-z)^2} = \frac{1}{2\pi i}\int_C \frac{f_1(t)\,dt}{(t-z)^2} + \frac{1}{2\pi i}\int_C \frac{f_2(t)\,dt}{(t-z)^2} + \cdots + \frac{1}{2\pi i}\int_C \frac{f_n(t)\,dt}{(t-z)^2} + \cdots$$

By the first generalization of Cauchy's integral formula (Theorem 2, Sec. 17.6), these integrals are precisely the derivatives of the respective terms of the given series at the point z. Hence

$$f'(z) = f_1'(z) + f_2'(z) + \cdots + f_n'(z) + \cdots$$

which establishes the theorem. ◗

It is interesting and important to note that Theorem 3 does not apply to series of functions of the real variable x. To justify term-by-term differentiation of such series, not uniform convergence of the original series but uniform convergence of the series of derivatives is required. More precisely, we have the following theorem, which is proved in most texts on advanced calculus.†

THEOREM 4 If for all x in the interval $[a, b]$, the series

$$\sum_{n=1}^{\infty} f_n(x) = f_1(x) + f_2(x) + \cdots + f_n(x) + \cdots$$

converges to $f(x)$, if for every value of n, $f_n'(x)$ is continuous on $[a, b]$, and if $\Sigma_{n=1}^{\infty} f_n'(t)$ converges uniformly on $[a, b]$, then $f'(t) = \Sigma_{n=1}^{\infty} f_n'(t)$.

EXERCISES

1. Find the region of convergence of **(a)** the series $S = \sum_{n=1}^{\infty}\left(\frac{z-1}{z+1}\right)^n$ and **(b)** the termwise derivative of S.

2. Work Exercise 1 for the series $S = \sum_{n=1}^{\infty} (1-z^2)^n$.

3. Find the region of convergence of **(a)** the series $S = \sum_{n=1}^{\infty} \frac{1}{n} e^{-nz}$, **(b)** the termwise derivative of S, **(c)** the termwise integral of S.

4. Work Exercise 3 for the series $S = \sum_{n=1}^{\infty} e^{n(z-1)}$.

5. If the region of convergence of the series $S = \sum_{n=1}^{\infty} f_n(z)$ can be established by the ratio test and if the termwise derivative and the termwise integral of S converge, show that except possibly for points on the boundary, the three series have the same region of convergence. *Hint:* Recall L'Hospital's rule.

6. Show by an example that the sum of a nonuniformly convergent series of continuous functions may be continuous. *Hint:* Consider the series $(1-x) + (x-x^2) + (x^2-x^3) + \cdots$ on $[0, 1]$.

7. Show that on $[0, 1]$ the sum of the series

$$\sum_{n=0}^{\infty} \frac{x}{(1+nx)(1+\{n+1\}x)}$$ is a continuous function

even though the series does not converge uniformly. *Hint:* Break up the general term into partial fractions.

8. Find the region of convergence and sum of the series

†See, for example, R. C. Buck, *Advanced Calculus,* 3d ed., p. 270, McGraw-Hill, New York, 1978.

$\sum_{n=1}^{\infty} ne^{-nz}$. *Hint:* Integrate the given series term by term and identify the sum of the resulting series.

9. Work Exercise 8 for the series $\sum_{n=1}^{\infty} nze^{-nz^2}$. *Hint:* Compare the given series with the derivative of the series expansion of the function $1/(e^{z^2} - 1)$.
10. In Sec. 8.3 we showed that if a periodic function has one or more discontinuities in a period, its Fourier coefficients can decrease no faster than c/n. Using Theorems 1 and 5, Sec. 18.1, give another proof of this result.
11. Show that the series $\sum_{n=1}^{\infty} [u_n - u_{n-1}]$, where $u_n = nxe^{-nx}$, can be integrated term by term from $x = 0$ to $x = a$ even though the series does not converge uniformly on this interval.
12. Why doesn't the proof of Theorem 3 justify the termwise differentiation of uniformly convergent series of real terms?
13. In Example 1, show that the series converges uniformly over any interval of the form $[\delta, a]$, where $\delta > 0$.
14. Show that the series $\sum_{n=1}^{\infty} [nxe^{-n^2x^2} - (n-1)xe^{-(n-1)^2x^2}]$ can be integrated term by term from $x = 0$ to $x = a$ even though the series is not uniformly convergent on this interval.
15. If $\sum_{n=1}^{\infty} f_n(x)$ is a series of integrable real functions which is uniformly convergent on $[0, a]$, show that after multiplication by $1/x^{\alpha}$, $0 < \alpha < 1$, the series can still be legitimately integrated term by term from $x = 0$ to $x = a$ even though its terms are now unbounded on the interval of integration.
16. In Example 1, let x be real and plot $S(x)$ and $S_n(x)$ for $n = 1, 2, 4, 8$. How does the behavior suggested by these curves compare with the Gibbs phenomena discussed in Sec. 9.2?

18.3 POWER SERIES

Of all series of complex terms, the most important are those known as *power series.*

DEFINITION 1 A series of the form

$$c_0 + c_1(z-a) + c_2(z-a)^2 + \cdots = \sum_{n=0}^{\infty} c_n(z-a)^n \tag{1}$$

where a and the coefficients $c_0, c_1, c_2, \ldots$ are complex constants, is called a **power series in** $z - a$.

Unlike general complex series, which as we saw in Sec. 18.1 can have regions of convergence of a variety of forms, the region of convergence of a power series is always a circle. More precisely, we have the following theorem.

THEOREM 1 If $L = \lim_{n\to\infty} \left|\frac{c_{n+1}}{c_n}\right|$, then the power series $\sum_{n=0}^{\infty} c_n(z-a)^n$ converges absolutely at all points in the interior of the circle $|z - a| = 1/L$, diverges at all points in the exterior of this circle, and may converge absolutely, converge conditionally, or diverge at points on this circle.

PROOF For the given series, we have for the limit of the absolute value of the ratio of successive terms

$$|r(z)| = \lim_{n\to\infty} \left|\frac{c_{n+1}(z-a)^{n+1}}{c_n(z-a)^n}\right| = \lim_{n\to\infty} \left|\frac{c_{n+1}}{c_n}\right| |z-a| = L|z-a|$$

Thus, according to the ratio test (Theorem 2, Sec. 18.1), the given series converges absolutely for all values of z such that

$$|r(z)| = L|z - a| < 1 \qquad \text{or} \qquad |z - a| < \frac{1}{L}$$

and diverges for all values of z such that

$$|r(z)| = L|z - a| > 1 \qquad \text{or} \qquad |z - a| > \frac{1}{L}$$

Since the ratio test provides no information about the series when $|r(z)| = 1$, the series may converge absolutely, converge conditionally, or diverge in this borderline case. Simple examples exist illustrating each of these possibilities (see Exercises 1–3). This completes the proof. ◗

The nonnegative real number $R = 1/L$ is called the **radius of convergence** of the series (1), and the circle $|z - a| = R = 1/L$ is called the **circle of convergence** of the series.

The important question of where a power series converges *uniformly* is answered by the following theorem.

THEOREM 2 The power series $\sum_{n=0}^{\infty} c_n(z - a)^n$ converges uniformly within and on any circle $|z - a| = r$ whose radius r is less than the radius of convergence of the series, R.

◖ **PROOF** Let z_1 be an arbitrary point in the interior of the circle of convergence of the given series, that is, let z_1 be a point such that $|z_1 - a| = r < R$. By Theorem 1, the power series $\sum_{n=0}^{\infty} c_n(z_1 - a)^n$ converges absolutely, which implies that $\sum_{n=0}^{\infty} |c_n|r^n$ is a convergent series of positive constants. Now let z_0 be any point such that $|z_0 - a| \le |z_1 - a|$, that is, let z_0 be any point within or on the circle $|z - a| = r$. Clearly,

$$|c_n(z_0 - a)^n| \le |c_n|r^n$$

Hence, by the Weierstrass M test (Theorem 5, Sec. 18.1), the series $\sum_{n=0}^{\infty} c_n(z_0 - a)^n$ converges uniformly. Since z_0 is any point in the closed circular region $|z - a| \le r$, the theorem is established. ◗

Theorem 2 should not be misunderstood, for, as Exercise 8 makes clear, the fact that the series $\sum_{n=0}^{\infty} c_n(z - a)^n$ converges uniformly for all z such that $|z - a| \le r < R$ is not equivalent to the assertion that the series converges uniformly for all z such that $|z - a| < R$. In one case, however, Theorem 2 can be strengthened. Suppose that there is at least one point z_2 on the circle of convergence of the series at which the series converges absolutely. Then, if z_0 is any value of z such that

$$|z_0 - a| \le |z_2 - a| = R$$

we have

$$|c_n(z_0 - a)^n| \le |c_n(z_2 - a)^n| = |c_n|R^n$$

Hence since the series $\sum_{n=0}^{\infty} |c_n|R^n$ converges, by hypothesis, it follows by the Weierstrass M test that the series $\sum_{n=0}^{\infty} c_n(z_0 - a)^n$ converges absolutely and uniformly. Thus, summarizing, we have

THEOREM 3 If the series $\sum_{n=0}^{\infty} c_n(z - a)^n$ converges absolutely at one point on the circle of convergence of the series, it converges absolutely and uniformly at all points within and on the circle of convergence.

Because of the uniform convergence guaranteed by Theorem 2, termwise integration and differentiation of the series $\sum_{n=0}^{\infty} c_n(z - a)^n$ is justified by Theorems 2 and 3, Sec. 18.2. Moreover, by applying the ratio test to the series resulting from the integration or differentiation, it is clear that these series will have the same radius of convergence as the original series. This establishes the following theorem.

THEOREM 4 The power series $\Sigma_{n=0}^{\infty} c_n(z-a)^n$ can be legitimately integrated and differentiated within its circle of convergence, and the resulting series will have the same radius of convergence as the original series.

Finally, since each of the series $\Sigma_{n=0}^{\infty} c_n(z-a)^n$ is an analytic function, the following result follows from Theorem 2, p. 1192, and the second part of Theorem 2, Sec. 18.2:

THEOREM 5 Within its circle of convergence, the sum of the series $\Sigma_{n=0}^{\infty} c_n(z-a)^n$ is an analytic function.

EXERCISES

1. Find the radius of convergence of the series $\Sigma_{n=0}^{\infty} z^n$ and show that the series converges at no point on its circle of convergence.
2. Find the radius of convergence of the series $\Sigma_{n=1}^{\infty} z^n/n^2$ and show that the series converges at all points on its circle of convergence.
3. Find the radius of convergence of the series $\Sigma_{n=1}^{\infty} z^n/n$ and show that the series converges conditionally at some points on its circle of convergence and diverges at others.
4. Show by an example that the radius of convergence of a power series may be **(a)** infinite or **(b)** zero.
5. Show that if a power series converges conditionally at one point on its circle of convergence, there is no point on the circle of convergence where it converges absolutely.
6. Find the radius of convergence of each of the following series.

   (a) $\sum_{n=1}^{\infty} 4^n(z-1)^n$ **(b)** $\sum_{n=1}^{\infty} n^4(z-1)^n$

   (c) $\sum_{n=1}^{\infty} nz^{2n}$ **(d)** $\sum_{n=1}^{\infty} \frac{2n}{n^2+1}(z+2)^n$

   (e) $\sum_{n=1}^{\infty} e^{\sqrt{n}}(z+i)^n$
7. Without finding the radii of convergence, show that the series in each of the following pairs have the same radius of convergence.

   (a) $\sum_{n=1}^{\infty} \frac{n+1}{n} z^n, \quad \sum_{n=1}^{\infty} \frac{1}{n} z^{n+1}$

   (b) $\sum_{n=1}^{\infty} \frac{n-1}{n^2} z^{n+1}, \quad \sum_{n=1}^{\infty} \frac{n^2-1}{n} z^{n-1}$
8. Show that the series $\sum_{n=0}^{\infty} \frac{z^n}{2^n}$ does not converge uniformly throughout the interior of its circle of convergence.
9. Show that for the series $\sum_{n=0}^{\infty} c_n(z-a)^n$,

   $R = \lim_{n\to\infty} |c_n/c_{n+1}|$.

18.4 TAYLOR'S EXPANSION

Power series arise most commonly as convenient representations of analytic functions in the neighborhood of some particular point $z = a$. The systematic study of such representations begins with the following result.

THEOREM 1 **(Taylor's Theorem)** If $f(z)$ is analytic throughout the region bounded by a simple closed curve C and if both z and a are interior to C, then

$$f(z) = f(a) + f'(a)(z-a) + f''(a)\frac{(z-a)^2}{2!} + \cdots + f^{(n-1)}(a)\frac{(z-a)^{n-1}}{(n-1)!} + R_n$$

where

$$R_n = \frac{(z-a)^n}{2\pi i}\int_C \frac{f(t)\,dt}{(t-a)^n(t-z)}$$

PROOF We first note that after adding and subtracting a in the denominator of the integrand, Cauchy's integral formula (Theorem 1, Sec. 17.6) can be written in the form

$$f(z) = \frac{1}{2\pi i}\int_C \frac{f(t)\,dt}{t-z} = \frac{1}{2\pi i}\int_C \frac{f(t)\,dt}{(t-a)-(z-a)}$$

$$= \frac{1}{2\pi i}\int_C \frac{f(t)}{t-a}\left[\frac{1}{1-(z-a)/(t-a)}\right] dt$$

From this, by applying the identity

$$\frac{1}{1-u} = 1 + u + u^2 + \cdots + u^{n-1} + \frac{u^n}{1-u}$$

to the factor $1/[1-(z-a)/(t-a)]$ in the last integral, we obtain

$$\begin{aligned} f(z) &= \frac{1}{2\pi i}\int_C \frac{f(t)}{t-a}\left[1 + \frac{z-a}{t-a} + \left(\frac{z-a}{t-a}\right)^2 + \cdots + \left(\frac{z-a}{t-a}\right)^{n-1} + \frac{(z-a)^n/(t-a)^n}{1-(z-a)/(t-a)}\right] dt \\ &= \frac{1}{2\pi i}\int_C \frac{f(t)\,dt}{t-a} + \frac{z-a}{2\pi i}\int_C \frac{f(t)\,dt}{(t-a)^2} + \frac{(z-a)^2}{2\pi i}\int_C \frac{f(t)\,dt}{(t-a)^3} + \cdots \\ &\qquad + \frac{(z-a)^{n-1}}{2\pi i}\int_C \frac{f(t)\,dt}{(t-a)^n} + \frac{(z-a)^n}{2\pi i}\int_C \frac{f(t)\,dt}{(t-a)^n(t-z)} \end{aligned}$$

From the generalizations of Cauchy's integral formula (Theorem 2, Sec. 17.6) it is evident that, except for the necessary factorials, the coefficients of the successive powers of $z-a$ in the first n terms in the last expression are precisely the corresponding derivatives of $f(z)$ evaluated at the point $z = a$. Hence, multiplying and dividing by the required factorials in each term, we have

$$\begin{aligned} f(z) = f(a) + f'(a)(z-a) + f''(a)\frac{(z-a)^2}{2!} + \cdots + f^{(n-1)}(a)\frac{(z-a)^{n-1}}{(n-1)!} \\ + \frac{(z-a)^n}{2\pi i}\int_C \frac{f(t)\,dt}{(t-a)^n(t-z)} \end{aligned}$$

which establishes the theorem. ◗

By **Taylor's series** we mean the infinite expansion corresponding to $f(z)$ suggested by Theorem 1, namely,

$$f(z) \sim f(a) + f'(a)(z-a) + f''(a)\frac{(z-a)^2}{2!} + \cdots + f^{(n-1)}(a)\frac{(z-a)^{n-1}}{(n-1)!} + \cdots$$

To show that this series (usually called **McLaurin's series** if $a = 0$) actually converges to $f(z)$, we must show, as usual, that the absolute value of the difference between $f(z)$ and the sum of the first n terms of the series approaches zero as n becomes infinite. From Taylor's theorem it is evident that this difference is

$$R_n(z) = \frac{(z-a)^n}{2\pi i}\int_C \frac{f(t)\,dt}{(t-a)^n(t-z)}$$

Accordingly, we must determine the values of z for which the absolute value of this remainder approaches zero as n becomes infinite.

To do this, let C_1 and C_2 be two circles of radii r_1 and r_2, with $r_1 < r_2$, having their centers at the point a and lying entirely in the interior of C (Fig. 18.2). Since $f(z)$ is analytic throughout the interior of C, the entire integrand of $R_n(z)$ is analytic in the region between C and C_2 provided that z, like a, lies in the interior of C_2. Under these conditions, the principle of the deformation of contours (Theorem 5, Sec. 17.5) allows us to replace the integral around C by the integral around C_2. If, in addition, z is interior to C_1, then for all values of t on C_2 (the values of t which are now involved in the integration) we have

$$|t-a| = r_2$$

$$|z-a| < r_1$$

$$|t-z| > r_2 - r_1$$

$$|f(t)| \le M$$

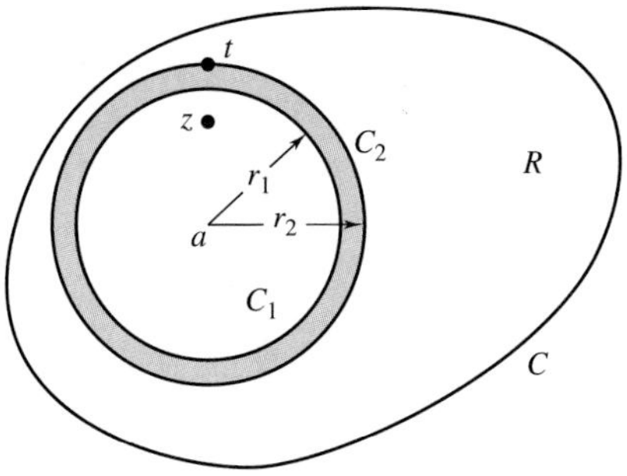

FIGURE 18.2
The circles C_1 and C_2 used in the proof of the convergence of Taylor's series.

where M is the maximum of $|f(z)|$ on C_2. Hence, overestimating factors in the numerator and underestimating factors in the denominator, we have

$$\begin{aligned}
|R_n(z)| &= \left| \frac{(z-a)^n}{2\pi i} \int_{C_2} \frac{f(t)\,dt}{(t-a)^n(t-z)} \right| \\
&\le \frac{|z-a|^n}{|2\pi i|} \int_{C_2} \frac{|f(t)|\,|dt|}{|t-a|^n|t-z|} \\
&< \frac{r_1^n}{2\pi} \int_{C_2} \frac{M\,|dt|}{r_2^n(r_2-r_1)} \\
&= \frac{r_1^n M}{2\pi r_2^n(r_2-r_1)} 2\pi r_2 \\
&= M\left(\frac{r_1}{r_2}\right)^n \frac{r_2}{r_2-r_1}
\end{aligned}$$

Since $0 < r_1 < r_2$, the fraction $(r_1/r_2)^n$ approaches zero as n becomes infinite and therefore the limit of $|R_n(z)|$ is zero. Thus we have established the following important theorem.

THEOREM 2 Taylor's series

$$f(z) = f(a) + f'(a)(z-a) + f''(a)\frac{(z-a)^2}{2!} + \cdots + f^{(n)}(a)\frac{(z-a)^n}{n!} + \cdots$$

is a valid representation of $f(z)$ at all points in the interior of any circle having its center at $z = a$ and within which $f(z)$ is analytic.

If we let Γ be the largest circle with center at $z = a$ and within which $f(z)$ is everywhere analytic, Theorem 2 guarantees that the Taylor's series of $f(z)$ converges to $f(z)$ at all points in the interior of Γ. However, it does not provide any information about the behavior of the series outside Γ. Actually, the Taylor's series of $f(z)$ converges to $f(z)$ only within and possibly on the circle Γ and either diverges or else converges to a limit other than $f(z)$ outside Γ. The next theorem makes these observations more precise.

THEOREM 3 It is impossible for the Taylor's series of a function $f(z)$ to converge to $f(z)$ outside the circle whose center is the point of expansion $z = a$ and whose radius is the distance from a to the nearest singularity of $f(z)$.

PROOF Let α be the singular point of $f(z)$ which is closest to the center of expansion $z = a$ and suppose that the Taylor's series of $f(z)$ converges for some value $z = z_1$ which is farther from a than α is. Since Taylor series are power series, by Theorem 1, Sec. 18.3, z_1 lies within or on the circle of convergence of the series, and so the series must converge at all points within the circle

$|z - a| = |z_1 - a|$. Moreover, by Theorem 5, Sec. 18.3, the sum of the series must be analytic at every such point. Thus the series converges to a function which is analytic at $z = \alpha$, in apparent contradiction of the assumption that $z = \alpha$ is a point where $f(z)$ is not analytic. Hence we must either abandon the supposition that the series converges at points farther from a than the distance $|\alpha - a|$, or else we must accept the fact that the function to which the series converges in the neighborhood of α is different from $f(z)$.

The usual situation is that the series diverges for all values of z such that $|z - a| > |\alpha - a|$, or in other words that the circle in which the series converges to $f(z)$ is actually the circle of convergence of the series. It is possible, however, that the circle in which the series converges to $f(z)$ is smaller than the circle of convergence and between the two circles the series converges to an analytic function which is different from $f(z)$. Exercises 14 and 15 illustrate this possibility. In either case, however, the theorem is established. ◗

Theorem 3 is often useful in determining the interval in which the Taylor's series of a function of a *real* variable converges and represents that function. To illustrate, consider

$$f(z) = \frac{1}{1 + z^2} = a_0 + a_1(z - x_0) + a_2(z - x_0)^2 + a_3(z - x_0)^3 + \cdots$$

According to Theorem 3, this will converge to $f(z)$ throughout the largest circle around $z = x_0$ in which $f(z)$ is analytic. By inspection, $f(z)$ becomes infinite as z approaches $\pm i$, and even though one may be concerned only with real values of z [for which $1/(1 + x^2)$ is everywhere infinitely differentiable], these singularities in the complex plane set an inescapable limit to the interval of convergence on the x axis. We can in fact have convergence to $f(x)$ around $x = x_0$ on the real axis only over the horizontal diameter of the circle of convergence around the point $z = x_0$ in the complex plane.

In many problems, the Taylor's expansion of a function is found not by evaluating the derivatives of the function and then substituting these values into the series provided by Theorem 2 but rather by algebraic manipulations suggested by the nature of the function. In particular, when dealing with **rational fractional functions,** i.e., functions which are quotients of polynomials, it is often advantageous to express them in terms of partial fractions and then expand these various fractions in *binomial series* of the appropriate form through the use of the **generalized binomial theorem,** which we list here for reference.

THEOREM 4 The expansion of $(s + t)^n$ as a **binomial series**

$$(s + t)^n = s^n + ns^{n-1}t + \frac{n(n-1)}{2!}s^{n-2}t^2 + \frac{n(n-1)(n-2)}{3!}s^{n-3}t^3 + \cdots$$

is valid for all values of n if $|s| > |t|$. If $|s| < |t|$, the expansion is valid only if n is a nonnegative integer. If $|s| = |t|$, the expansion is valid if n is a nonnegative integer but may or may not be valid for other values of n.

That such procedures are correct is a consequence of the following simple but important theorem.

THEOREM 5 If $f(z)$ can be represented in the neighborhood of $z = a$ by a series of the form $\sum_{n=0}^{\infty} a_n(z - a)^n$, the representation is unique.

EXAMPLE 1

Find the Taylor expansion of $f(z) = 3/(3z - z^2)$ around $z = 1$.

By familiar partial fraction techniques we find that

$$f(z) = \frac{1}{z} + \frac{1}{3 - z}$$

From this, by obvious rearrangements, we have

$$f(z) = \frac{1}{1+(z-1)} + \frac{1}{2-(z-1)} = [1+(z-1)]^{-1} + [2-(z-1)]^{-1}$$

Applying the binomial expansion to each term in the last expression, we obtain

$$\begin{aligned} f(z) &= 1 - (z-1) + (z-1)^2 - \cdots + (-1)^n(z-1)^n + \cdots \\ &\quad + 2^{-1} + 2^{-2}(z-1) + 2^{-3}(z-1)^2 + \cdots + 2^{-(n+1)}(z-1)^n + \cdots \\ &= \frac{3}{2} - \frac{3}{4}(z-1) + \frac{9}{8}(z-1)^2 - \cdots + \left[\frac{1}{2^{n+1}} + (-1)^n\right](z-1)^n + \cdots \end{aligned}$$

which is the required expansion.

By Theorem 4, the first series is valid in the interior of the circle $|z-1| < 1$, while the second is valid in the interior of the larger circle $|z-1| < 2$. The final series of course is valid only under the more restrictive of these two conditions, i.e., $|z-1| < 1$. This is consistent with Theorem 3, since the distance from the point of expansion $a = 1$ to the nearest point where $f(z)$ is not analytic, namely, $z = 0$, is 1.

Incidentally, when the Taylor's expansion of a function is found by algebraic manipulation of the function rather than by evaluating its derivatives, the numerical values of the derivatives at the point of expansion can be obtained at once from the general coefficient formula

$$a_n = \frac{f^{(n)}(a)}{n!} \qquad \text{or} \qquad f^{(n)}(a) = n!a_n$$

For example, in the present problem,

$$f^{\text{iv}}(1) = 4!\left(\frac{1}{2^5} + 1\right) = \frac{99}{4}$$

As an application of Taylor's expansion, we shall conclude this section by establishing the simple but important result known as the *theorem of Liouville.*†

THEOREM 6 **(Liouville's Theorem)** If $f(z)$ is bounded and analytic for all values of z, then $f(z)$ is a constant.

PROOF We observe first that since $f(z)$ is everywhere analytic, it possesses a power-series expansion about the origin:

$$f(z) = f(0) + f'(0)z + \cdots + f^{(n)}(0)\frac{z^n}{n!} + \cdots$$

which converges and represents it for all values of z. Now if C is an arbitrary circle with the origin as center, it follows from Cauchy's inequality (Theorem 4, Sec. 17.6) that

$$|f^{(n)}(0)| \le \frac{n!M_C}{r^n}$$

where r is the radius of C and M_C is the maximum value of $|f(z)|$ on C. Hence, for the coefficient of z^n in the expansion of $f(z)$, we have

$$\left|\frac{f^{(n)}(0)}{n!}\right| \le \frac{M_C}{r^n} \le \frac{M}{r^n}$$

†Named for the French mathematician Joseph Liouville (1809–1882) but actually due to Cauchy.

where M, the bound on $|f(z)|$ for *all* values of z, is independent of r. Since r can be taken arbitrarily large, it follows, therefore, that the coefficient of z^n is zero for $n = 1, 2, 3, \ldots$. In other words, for all values of z,

$$f(z) = f(0)$$

which proves the theorem. ◗

A function which is analytic for all values of z is called an **entire function** or an **integral function,** and Liouville's theorem thus states that *any entire function which is bounded for all values of z is a constant.*

EXERCISES

1. Expand $f(z) = (z - 1)/(z + 1)$ in a Taylor's series **(a)** about the point $z = 0$ and **(b)** about the point $z = 1$. Determine the radius of convergence of each series.

2. Expand $f(z) = 1/(z^2 + 3z + 2)$ in a Taylor's series **(a)** about the point $z = 0$ and **(b)** about the point $z = 2$. Determine the radius of convergence in each case.

3. Expand $f(z) = 1/z$ in a Taylor's series about the point $z = i$ and determine the radius of convergence of the series.

4. Expand $f(z) = e^z$ in a Taylor's series about each of the following points and tell where each series converges.
(a) $z = i\pi/2$ **(b)** $z = i\pi$ **(c)** $z = 3i\pi/2$

5. Expand $f(z) = \cosh z$ in a Taylor's series about the point $z = i\pi$. What is the radius of convergence of the series?

6. Obtain the first three nonzero terms in the Maclaurin expansion of **(a)** $f(z) = \tan z$ and **(b)** $f(z) = \sec z$. What is the radius of convergence of each series?

7. Without obtaining the series, determine the radius of convergence of each of the following expansions.

(a) $\text{Tan}^{-1} z$ around $z = 1$

(b) $\dfrac{1}{e^z - 1}$ around $z = 4i$

(c) $\dfrac{x}{x^2 + 2x + 10}$ around $x = 0$

(d) $\dfrac{1}{x^2 - 9x + 10}$ around $x = -1$

8. Find the value of the indicated derivative of each of the following functions at the indicated point.
(a) The fifth derivative of $z/(2 + z^2)$ at $z = 0$
(b) The sixth derivative of $1/(z^2 - 4)$ at $z = 1$
(c) The fifth derivative of $1/(z^2 + z^3)$ at $z = 2$

9. Prove that if a function is analytic at a point $z = a$, its Taylor's series in the neighborhood of that point is unique.

10. Show that if the values of an analytic function $f(z)$ are known at the points of an arbitrarily short arc having $z = a$ as one endpoint, then the Taylor's expansion of $f(z)$ around $z = a$ is completely determined.

11. Prove that every polynomial equation $P(z) = 0$ has at least one root. *Hint:* Assume the contrary and apply the theorem of Liouville to the function $f(z) = 1/P(z)$.

12. **(a)** Let $\Sigma_{n=0}^{\infty} a_n(z - a)^n$ be the Taylor's expansion of a function $f(z)$ around $z = a$ and let $z_0 \neq a$ be an arbitrary point in the interior of the circle Γ in which the series converges to $f(z)$. Then, by Theorem 3, Sec. 18.2, all the derivatives of $f(z)$ at $z = z_0$ can be found from the series. Hence, by using these values, the expansion of $f(z)$ around $z = z_0$ can be constructed. Show that in general this new series will converge in the interior of a circle Γ' which lies partly inside and partly outside Γ. When this is the case, the second series is said to be an **analytic continuation** of the first since it provides a series representation of $f(z)$ beyond that provided by the first series.

(b) Show that the series $\displaystyle\sum_{n=0}^{\infty} \frac{z^n}{2^{n+1}}$ and $\displaystyle\sum_{n=0}^{\infty} \frac{(z - i)^n}{(2 - i)^{n+1}}$ are analytic continuations of each other and determine the region common to their respective circles of convergence.

13. Show that no analytic continuation beyond the circle $|z| = 1$ is possible for the function defined by the series

$$f(z) = 1 + z + z^2 + z^4 + z^8 + \cdots = 1 + \sum_{n=0}^{\infty} z^{2^n}$$

Hint: Note first that $z = 1$ is a singular point of the given function. Then note that

$$\begin{aligned} f(z) &= z + f(z^2) \\ &= z + z^2 + f(z^4) \\ &= z + z^2 + z^4 + f(z^8) \\ &\cdots\cdots\cdots \end{aligned}$$

so that not only the value $z = 1$ but also the values of z such that $z^2 = 1$, $z^4 = 1$, $z^8 = 1, \ldots$ are singular points of $f(z)$. Finally, consider the distribution of these singularities around the circumference of the circle of convergence $|z| = 1$. A curve, such as this circle, beyond which a function cannot be extended by analytic continuation is called a **natural boundary** of the function.

14. Find the Maclaurin expansion of

$$f(z) = \int_C \frac{dt}{t - z} \qquad |z| \neq 1$$

if C is the circle $|t| = 1$. Show that this expansion converges for all values of z but represents $f(z)$ only in the interior of C. *Hint:* Recall Cauchy's integral formula and the result of Example 1, Sec. 17.5.

15. Find the Taylor's expansion of the principal value of ln z around the point $z = -1 + i$. Show that the radius of convergence of this series is $\sqrt{2}$ but that the series represents the principal value of ln z only within a circle of radius 1 about the point $z = -1 + i$.

16. **(a)** Find the Maclaurin series representation of the entire function e^z.
(b) If $z = x + 0i$, what does the expansion of Part **(a)** reduce to?

17. **(a)** Find the Maclaurin series representation of the entire function sin z.
(b) Use the series of Part **(a)** to find the Maclaurin series for sinh z. *Hint:* sinh $z = -i \sin iz$ (Sec. 17.4).

18. Find the Taylor series of $1/z$ about the point $z = 1$ **(a)** using Theorem 2 and **(b)** using Theorem 4. *Hint:* $1/z = [1 + (z - 1)]^{-1}$.

19. Show that if $f(z)$ is an entire function which satisfies everywhere the inequality $|f(z)| \le c|z|^k$, where c and k are positive constants, then $f(z)$ is a polynomial of degree not greater than k. *Hint:* Apply Cauchy's inequality to the coefficients in the Maclaurin expansion of $f(z)$.

20. Show that the function $f(z) = \sum_{n=0}^{\infty} z^{n!}$ has a natural boundary. What is it? *Hint:* How are points of the type $z = e^{2\pi ip/q}$, p and q positive integers, distributed on the circle of convergence and how does f behave at each such point?

18.5 LAURENT'S EXPANSION

In many applications it is necessary to expand functions around points at which or in the neighborhood of which the functions are not analytic. The method of Taylor's series is obviously inapplicable in such cases, and a new type of series known as *Laurent's expansion*† is required. This furnishes us with a representation which is valid in the annular ring between two concentric circles, provided that the function being expanded is analytic everywhere on and between the two circles. As with Taylor's series, the function may have singular points outside the larger circle and, as the essentially new feature, it may also have singular points within the inner circle. The price we pay for this is that negative as well as positive powers of $z - a$ now appear in the expansion and the coefficients, even of the positive powers of $z - a$, cannot be expressed in terms of the evaluated derivatives of the function. The precise result is given by the following theorem.

THEOREM 1 **(Laurent's Expansion)** If $f(z)$ is analytic throughout the closed region R bounded by two concentric circles, then at any point in the annular ring bounded by the circles, $f(z)$ can be represented by the series

$$f(z) = \sum_{n=-\infty}^{\infty} a_n(z - a)^n$$

where a is the common center of the circles and

$$a_n = \frac{1}{2\pi i}\int_C \frac{f(t)\,dt}{(t - a)^{n+1}}$$

each integral being taken in the counterclockwise sense around any curve C lying in the annulus and encircling its inner boundary.

PROOF Let z be an arbitrary point in the given annulus. Then, according to Cauchy's integral formula, we can write

$$f(z) = \frac{1}{2\pi i}\int_{C_1 \cup C_2} \frac{f(t)\,dt}{t - z} = \frac{1}{2\pi i}\int_{C_2} \frac{f(t)\,dt}{t - z} + \frac{1}{2\pi i}\int_{C_1} \frac{f(t)\,dt}{t - z}$$

†Named for the French mathematician Hermann Laurent (1841–1908).

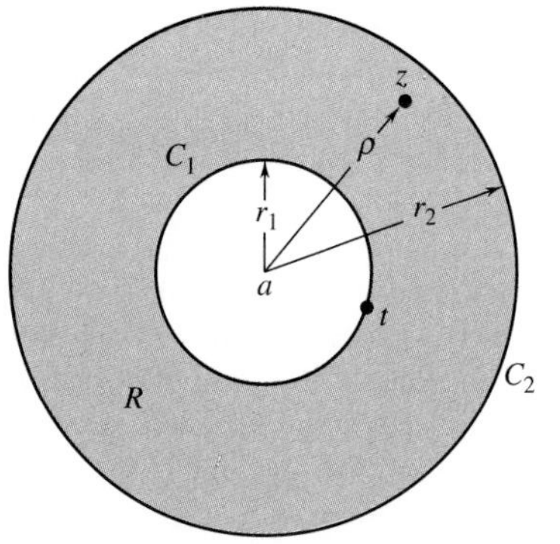

FIGURE 18.3
The circles C_1 and C_2 used in the derivation of Laurent's expansion.

where C_2 is traversed in the counterclockwise direction and C_1 is traversed in the clockwise direction, in order that the entire integration shall be in the positive direction (Fig. 18.3). Reversing the sign of the integral around C_1 and also changing the direction from clockwise to counterclockwise, we can write, very much as in the proof of Taylor's theorem in Sec. 18.4,

$$f(z) = \frac{1}{2\pi i}\int_{C_2}\frac{f(t)\,dt}{t-z} - \frac{1}{2\pi i}\int_{C_1}\frac{f(t)\,dt}{t-z}$$
$$= \frac{1}{2\pi i}\int_{C_2}\frac{f(t)}{t-a}\left[\frac{1}{1-(z-a)/(t-a)}\right]dt + \frac{1}{2\pi i}\int_{C_1}\frac{f(t)}{z-a}\left[\frac{1}{1-(t-a)/(z-a)}\right]dt$$

In each of these modified integrals let us now apply the identity

$$\frac{1}{1-u} = 1 + u + u^2 + \cdots + u^{n-1} + \frac{u^n}{1-u}$$

to the last factor. Then,

$$f(z) = \frac{1}{2\pi i}\int_{C_2}\frac{f(t)}{t-a}\left[1 + \frac{z-a}{t-a} + \cdots + \left(\frac{z-a}{t-a}\right)^{n-1} + \frac{(z-a)^n/(t-a)^n}{1-(z-a)/(t-a)}\right]dt$$
$$+ \frac{1}{2\pi i}\int_{C_1}\frac{f(t)}{z-a}\left[1 + \frac{t-a}{z-a} + \cdots + \left(\frac{t-a}{z-a}\right)^{n-1} + \frac{(t-a)^n/(z-a)^n}{1-(t-a)/(z-a)}\right]dt$$
$$= \frac{1}{2\pi i}\int_{C_2}\frac{f(t)\,dt}{t-a} + \frac{z-a}{2\pi i}\int_{C_2}\frac{f(t)\,dt}{(t-a)^2} + \cdots + \frac{(z-a)^{n-1}}{2\pi i}\int_{C_2}\frac{f(t)\,dt}{(t-a)^n} + R_{n2}$$
$$+ \frac{1}{2\pi i(z-a)}\int_{C_1} f(t)\,dt + \frac{1}{2\pi i(z-a)^2}\int_{C_1}(t-a)f(t)\,dt + \cdots$$
$$+ \frac{1}{2\pi i(z-a)^n}\int_{C_1}(t-a)^{n-1}f(t)\,dt + R_{n1}$$

where

$$R_{n2} = \frac{(z-a)^n}{2\pi i}\int_{C_2}\frac{f(t)\,dt}{(t-a)^n(t-z)}$$
$$R_{n1} = \frac{1}{2\pi i(z-a)^n}\int_{C_1}\frac{(t-a)^n f(t)\,dt}{z-t}$$

The truth of the theorem will be established if we can show that

$$\lim_{n\to\infty} R_{n2} = 0 \qquad \text{and} \qquad \lim_{n\to\infty} R_{n1} = 0$$

The proof of the first of these equations we can pass over without comment because it was given in complete detail in the derivation of Taylor's series in Sec. 18.4. To prove the second, we note that for values of t on C_1 (Fig. 18.3),

$$\begin{aligned} |t-a| &= r_1 \\ |z-a| &= \rho \qquad \text{say, where } \rho > r_1 \\ |z-t| &= |(z-a)-(t-a)| \geq \rho - r_1 \end{aligned}$$

and

$$|f(t)| \leq M$$

where M is the maximum of $|f(z)|$ on C_1. Thus

$$\begin{aligned} |R_{n1}| &= \left| \frac{1}{2\pi i(z-a)^n} \int_{C_1} \frac{(t-a)^n f(t)\,dt}{z-t} \right| \\ &\leq \frac{1}{|2\pi i|\,|z-a|^n} \int_{C_1} \frac{|t-a|^n |f(t)|\,|dt|}{|z-t|} \\ &\leq \frac{r_1^n M}{2\pi\rho^n(\rho-r_1)} \int_{C_1} |dt| \\ &= \frac{M}{2\pi}\left(\frac{r_1}{\rho}\right)^n \frac{2\pi r_1}{\rho - r_1} \\ &= M\left(\frac{r_1}{\rho}\right)^n \frac{r_1}{\rho - r_1} \end{aligned}$$

Since $0 < r_1/\rho < 1$, the last expression approaches zero as n becomes infinite. Hence $\lim_{n\to\infty} R_{n1} = 0$; and thus we have

$$\begin{aligned} f(z) = {} & \frac{1}{2\pi i}\int_{C_2} \frac{f(t)\,dt}{t-a} + \left[\frac{1}{2\pi i}\int_{C_2} \frac{f(t)\,dt}{(t-a)^2}\right](z-a) + \left[\frac{1}{2\pi i}\int_{C_2} \frac{f(t)\,dt}{(t-a)^3}\right](z-a)^2 + \cdots \\ & + \left[\frac{1}{2\pi i}\int_{C_1} f(t)\,dt\right]\frac{1}{z-a} + \left[\frac{1}{2\pi i}\int_{C_1} (t-a)f(t)\,dt\right]\frac{1}{(z-a)^2} + \cdots \end{aligned}$$

Since $f(z)$ is analytic throughout the region between C_1 and C_2, the paths of integration C_1 and C_2 can be replaced by any other curve C within this region and encircling C_1. The resulting integrals are precisely the coefficients a_n described by the theorem; hence our proof is complete. ◗

It should be noted that the coefficients of the positive powers of $z - a$ in Laurent's expansion, although identical in form with the integrals of Theorem 2, Sec. 17.6, *cannot* be replaced by the derivative expressions

$$\frac{f^{(n)}(a)}{n!}$$

as they were in the derivation of Taylor's series, since $f(z)$ is not analytic throughout the entire interior of C_2 (or C) and hence Cauchy's generalized integral formula cannot be applied. Specifically, $f(z)$ may have many points of nonanalyticity within C_1 and therefore within C_2 (or C).

As in the case of Taylor's expansion, the Laurent expansion of a given function in a given annulus is usually not found through the use of the last theorem but rather by algebraic manipulations suggested by the nature of the function. That such procedures are correct follows from the following theorem.

THEOREM 2 The Laurent expansion of a function over a given annulus, if it exists, is unique.

In other words, if an expansion of the Laurent form is found by any process, it must be *the* Laurent expansion of the given function over the given annulus. The following example illustrates the details of such a derivation.

EXAMPLE 1

DISTINCT LAURENT EXPANSIONS AROUND A POINT

Find the Laurent expansion of the function

$$f(z) = \frac{7z - 2}{(z + 1)z(z - 2)}$$

in the annulus $1 < |z + 1| < 3$.

As a preliminary step it is convenient to apply the method of partial fractions to $f(z)$ and express it in the form

$$f(z) = \frac{-3}{z + 1} + \frac{1}{z} + \frac{2}{z - 2}$$

Now, after suitable rearrangement, these fractions can be expanded into infinite series by means of the binomial theorem (Theorem 4, Sec. 18.4) and added to give the required expansion of $f(z)$.

To do this, we observe that since the center of the given annulus is $z = -1$, the series we are seeking must be one involving powers of $z + 1$. Hence we modify the second and third terms in the partial-fraction representation of $f(z)$ so that z will appear in the combination $z + 1$. This gives us the equivalent representation

$$\begin{aligned} f(z) &= \frac{-3}{z + 1} + \frac{1}{(z + 1) - 1} + \frac{2}{(z + 1) - 3} \\ &= -3(z + 1)^{-1} + [(z + 1) - 1]^{-1} + 2[(z + 1) - 3]^{-1} \end{aligned}$$

The first term $-3(z + 1)^{-1}$ is already in suitable form. The second term $[(z + 1) - 1]^{-1}$, if expanded in its present form by the binomial theorem, will give rise to a series of the desired form which will converge for $|z + 1| > 1$, as required. However, the third term $[(z + 1) - 3]^{-1}$ if expanded in its present form will give us a series which converges only for $|z + 1| > 3$, whereas we require an expansion valid for $|z + 1| < 3$. Hence we rewrite this term in the other order $[-3 + (z + 1)]^{-1}$ before expanding it. Now, combining the first term with the binomial series obtained from the other two terms, we have

$$\begin{aligned} f(z) &= -3(z + 1)^{-1} + [(z + 1)^{-1} + (z + 1)^{-2} + (z + 1)^{-3} + \cdots] \\ &\quad + 2\left[-\frac{1}{3} - \frac{z + 1}{9} - \frac{(z + 1)^2}{27} - \frac{(z + 1)^3}{81} - \cdots\right] \\ &= \cdots + (z + 1)^{-3} + (z + 1)^{-2} - 2(z + 1)^{-1} - \tfrac{2}{3} \\ &\quad - \tfrac{2}{9}(z + 1) - \tfrac{2}{27}(z + 1)^2 - \tfrac{2}{81}(z + 1)^3 - \cdots \qquad 1 < |z + 1| < 3 \end{aligned}$$

It is important to note that $f(z)$ has two other Laurent expansions around the point $z = -1$. One is valid in the annular region between a circle of arbitrarily small radius around $z = -1$ and a circle of unit radius around $z = -1$. The other is valid in the exterior of a circle of radius 3 around $z = -1$ (Fig. 18.4). Each of these can

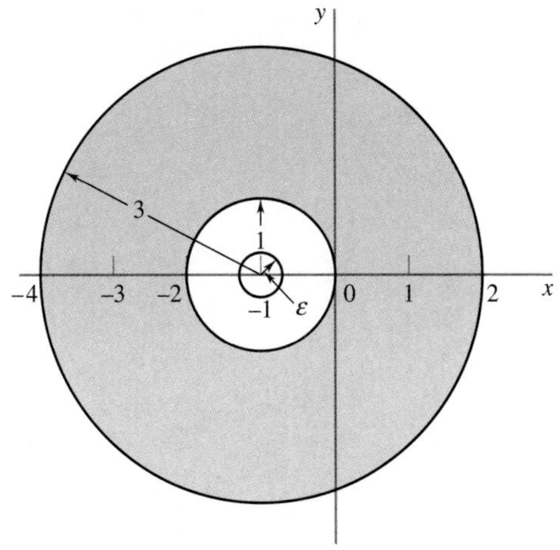

FIGURE 18.4
The regions of validity of the three Laurent expansions of $(7z-2)/(z+1)z(z-2)$ around $z=-1$.

be found, as before, by suitably rearranging the terms in the partial-fraction representation of $f(z)$ and then expanding these terms as binomial series by the binomial theorem. Thus for the innermost region we have

$$\begin{aligned}
f(z) &= -3(z+1)^{-1} + [-1+(z+1)]^{-1} + 2[-3+(z+1)]^{-1} \\
&= -3(z+1)^{-1} + [-1-(z+1)-(z+1)^2-(z+1)^3-\cdots] \\
&\quad + 2\left[-\frac{1}{3}-\frac{z+1}{9}-\frac{(z+1)^2}{27}-\frac{(z+1)^3}{81}-\cdots\right] \\
&= -3(z+1)^{-1} - \tfrac{5}{3} - \tfrac{11}{9}(z+1) - \tfrac{29}{27}(z+1)^2 - \tfrac{83}{81}(z+1)^3 - \cdots \qquad 0<|z+1|<1
\end{aligned}$$

Similarly, in the outermost region we have

$$\begin{aligned}
f(z) &= -3(z+1)^{-1} + [(z+1)-1]^{-1} + 2[(z+1)-3]^{-1} \\
&= -3(z+1)^{-1} + [(z+1)^{-1}+(z+1)^{-2}+(z+1)^{-3}+\cdots] \\
&\quad + 2[(z+1)^{-1}+3(z+1)^{-2}+9(z+1)^{-3}+\cdots] \\
&= \cdots + 19(z+1)^{-3} + 7(z+1)^{-2} \qquad |z+1|>3
\end{aligned}$$

Incidentally, the fact that we have obtained these Laurent expansions without using the general theory means that we can evaluate the integrals in the coefficient formulas by comparing them with the numerical values of the coefficients we have found by independent means. For instance, in the first expansion the coefficient of $(z+1)^{-1}$ is -2. On the other hand, according to the theory of Laurent's expansion, the coefficient of this term is

$$a_{-1} = \frac{1}{2\pi i}\int_C f(z)\,dz = \frac{1}{2\pi i}\int_C \frac{7z-2}{(z+1)z(z-2)}\,dz$$

where C is any closed curve lying in the interior of the circle $|z+1|=3$ and enclosing the circle $|z+1|=1$. Thus, although we have done nothing resembiing an integration, we have nonetheless shown that

$$\frac{1}{2\pi i}\int_C \frac{7z-2}{(z+1)z(z-2)}\,dz = -2 \qquad \text{or} \qquad \int_C \frac{7z-2}{(z+1)z(z-2)}\,dz = -4\pi i$$

a result, incidentally, which could not have been obtained by a direct application of Cauchy's integral formula, as in Example 1, Sec. 17.6.

In the next chapter we shall systematically exploit this possibility as a means of evaluating several important classes of definite integrals.

EXERCISES

1. Expand $f(z) = 1/(z-1)(z-2)$:
(a) For $|z| < 1$ **(b)** For $1 < |z| < 2$
(c) For $2 < |z|$ **(d)** For $0 < |z-1| < 1$
(e) For $|z-1| > 1$ **(f)** For $0 < |z-2| < 1$
(g) For $|z-2| > 1$

2. Obtain two distinct Laurent expansions for $f(z) = (3z+1)/(z^2-1)$ around $z = 1$ and tell where each converges.

3. Expand $f(z) = 1/z^2(z-i)$ in two different Laurent expansions around $z = i$ and tell where each converges.

4. Construct all the Laurent expansions of $f(z) = 1/z(z-1)(z-2)$ around $z = -1$ and tell where each converges.

5. Criticize the following argument: Since (by long division, for instance)

$$\frac{z}{1-z} = z + z^2 + z^3 + z^4 + \cdots$$

and

$$\frac{z}{z-1} = 1 + \frac{1}{z} + \frac{1}{z^2} + \frac{1}{z^3} + \cdots$$

and since $$\frac{z}{1-z} + \frac{z}{z-1} = 0$$

therefore, by adding these two series, we obtain

$$\cdots + \frac{1}{z^3} + \frac{1}{z^2} + \frac{1}{z} + 1 + z + z^2 + z^3 + z^4 + \cdots = 0$$

6. Criticize the following argument: The series

$$\frac{1}{z} + 1 + z + z^2 + z^3 + z^4 + \cdots$$

converges to the sum $S(z) = 1/z(1-z)$ for all values of z such that $|z| < 1$, *including* $z = 0$, since

$$\begin{aligned}|S(z) - S_n(z)| &= \left|\frac{1}{z(1-z)} - \left(\frac{1}{z} + 1 + z + \cdots + z^{n-2}\right)\right| \\ &= \left|\frac{1}{z} + \frac{1}{1-z} - \frac{1}{z} - 1 - z - \cdots - z^{n-2}\right| \\ &= \left|\frac{1}{1-z} - 1 - z - \cdots - z^{n-2}\right| \\ &= \left|\frac{1}{1-z} - \frac{1-z^{n-1}}{1-z}\right| \\ &= \left|\frac{z^{n-1}}{1-z}\right|\end{aligned}$$

and this expression clearly approaches 0 as n becomes infinite for *all* values of z such that $|z| < 1$.

7. By identifying the proper coefficients in the appropriate Laurent expansions, find the value of $\int_C f(z)\,dz$ if C is the circle $|z| = 3$ and $f(z)$ is:

(a) $\dfrac{1}{z(z+2)}$ **(b)** $\dfrac{z+2}{z(z+1)}$ **(c)** $\dfrac{1}{(z+1)^2}$

(d) $\dfrac{1}{z(z+1)^2}$ **(e)** $\dfrac{z}{(z+1)(z+2)}$ **(f)** $\dfrac{1/(z+4)}{z(z+1)}$

Hint: Recall the remarks at the end of Example 1.

8. Prove Theorem 2.

9. If k is a real number such that $k^2 < 1$, prove that

$$\sum_{n=0}^{\infty} k^n \sin(n+1)\theta = \frac{\sin\theta}{1 - 2k\cos\theta + k^2}$$

and

$$\sum_{n=0}^{\infty} k^n \cos(n+1)\theta = \frac{\cos\theta - k}{1 - 2k\cos\theta + k^2}$$

Hint: First expand $1/(z-k)$ in descending powers of z and set $z = e^{i\theta}$. Then set $z = e^{i\theta}$ in $1/(z-k)$ and reduce the denominator to real form. Finally, equate real and imaginary parts of the two expressions for $1/(z-k)$.

10. **(a)** Show that the Laurent expansion of the function $f(z) = \sinh(z + 1/z)$ is $\Sigma_{n=-\infty}^{\infty} a_n z^n$, where

$$a_n = \frac{1}{2\pi}\int_0^{2\pi} \cos n\theta \sinh(2\cos\theta)\,d\theta$$

Hint: In the formula for a_n provided by Theorem 1, take the curve C to be the circle $|z| = 1$. On this circle let the variable of integration t be taken in the form $t = e^{i\theta}$. Finally, verify that the imaginary part of the integral for a_n is equal to zero.

(b) Show that the coefficients in the Laurent expansion of $f(z) = \sin(z + 1/z)$ are given by the formula

$$a_n = \frac{1}{2\pi}\int_0^{2\pi} \cos n\theta \sin(2\cos\theta)\,d\theta$$

(c) What are the coefficients in the Laurent expansion of $\cosh(z + 1/z)$ in powers of z?

(d) What are the coefficients in the Laurent expansion of $\cos(z + 1/z)$ in powers of z?

11. If $F(\theta)$ is the function of θ to which $f(z)$ reduces on the circle $|z| = 1$, show that

$$F(\theta) = \sum_{n=-\infty}^{\infty} a_n e^{in\theta} \quad \text{where} \quad a_n = \frac{1}{2\pi}\int_0^{2\pi} F(\theta)e^{-in\theta}\,d\theta$$

12. Using the result of Exercise 11, show that $F(\theta) =$

$$\frac{1}{2\pi}\int_0^{2\pi} F(\phi)\,d\phi + \frac{1}{\pi}\sum_{n=1}^{\infty}\int_0^{2\pi} F(\phi)\cos n(\theta - \phi)\,d\phi$$

Hint: Substitute a_n, as given in Exercise 11, into the series for $F(\theta)$. Then break up the series into three parts corresponding to $n \le -1$, $n = 0$, $n \ge 1$ and combine the two infinite parts term by term.

13. Show that the Laurent expansion of the function

$$f(z) = \exp\left[\frac{u}{2}\left(z - \frac{1}{z}\right)\right] \qquad |z| > 0$$

is $\sum_{n=-\infty}^{\infty} a_n z^n$, where $a_n = \frac{1}{2\pi}\int_0^{2\pi} \cos(n\theta - u \sin\theta)\, d\theta$,

and verify that a_n is the Bessel function $J_n(u)$. *Hint:* In the integral for a_n given by Theorem 1, let $z = e^{i\theta}$; then recall Eq. (7), Sec. 12.6.

14. Find two different Laurent expansions of the function $f(z) = 1/z^2(1 - z)$ around $z = 0$ and tell where each expansion represents $f(z)$.

15. **(a)** Find the Laurent series representing $f(z) = z/(z - 1)(z - 3)$ for $0 < |z - 1| < 2$.

(b) Find the Laurent series representing the function $f(z) = z/(z^2 + 1)(z + 2)$ in the annulus $0 < |z - i| < 2$.

CHAPTER 19

THE THEORY OF RESIDUES

In Sec. 18.5, we saw that the coefficient a_{-1} of the term $(z - a)^{-1}$ in the Laurent expansion of a function f in the deleted neighborhood of a singular point P: $z = a$ of f is equal to $1/2\pi$ times the integral of f around any circle with center P which does not enclose any other singular points of f. In other words, a_{-1} is all that remains when the series representing f is integrated around such a circle and for this reason is called the *residue* of f at P. The *residue theorem,* which is established in Sec. 19.1, generalizes this result to the case in which the path of integration C is not necessarily a circle and may enclose more than one singularity of f. Not unexpectedly, the result is that the integral of f around such a curve is equal to the sum of the residues of f at its singular points within C. By a judicious choice of the contour C, the residue theorem can be adapted to the evaluation of several important classes of real definite integrals, and this is done in Sec. 19.2.

In Chap. 9 we found that the Fourier integral and its inverse were equally useful in solving problems, but in Chap. 10 we worked exclusively with the Laplace transform and made no use of its inverse because, as we said (Sec. 9.6), to do so would require more complex variable theory than we then had at our disposal. Now, with the theory of residues available, we can make a deeper investigation of the inverse Laplace transform and adapt it to the solution of specific problems, and this is done in Sec. 19.3.

Our final application of the residue theorem in Sec. 19.4 shows us how to determine if a control system is stable or not by analyzing the singular points of its transfer function.

Prerequisites for this chapter: Chaps. 10 and 17.

19.1 THE RESIDUE THEOREM

In Sec. 17.3 we defined a singular point of a function $f(z)$ as a point where $f(z)$ is not analytic but in every neighborhood of which there are points where $f(z)$ is analytic. A particular kind of singular point which we shall study in this chapter is described in our first definition.

DEFINITION 1 If $z = a$ is a singular point of the function $f(z)$ but there exists a neighborhood of a in which there are no other singular points of $f(z)$, then $z = a$ is called an **isolated singular point.**

Clearly, if $z = a$ is an isolated singular point of $f(z)$, then by Theorem 1, Sec. 18.5, $f(z)$ will possess a Laurent expansion around $z = a$ which will represent $f(z)$ in the interior of an annulus $r < |z - a| < R$ whose outer radius R is the distance from a to the nearest of the other singularities of $f(z)$ and whose inner radius r can be taken arbitrarily small. If $r = 0$, the annulus becomes **the deleted neighborhood of** a, $0 < |z - a| < R$.

The essential feature of a Laurent expansion is that it contains at least some negative powers of $z - a$. The next two definitions classify isolated singular points according as a particular one of the associated Laurent expansions contains a finite number or an infinite number of negative powers of $z - a$.

DEFINITION 2 If the Laurent expansion of $f(z)$ in the deleted neighborhood of an isolated singular point contains only a finite number of negative powers of $z - a$, then $z = a$ is called a **pole** of $f(z)$. If $(z - a)^{-m}$ is the (numerically) highest negative power in the expansion, the pole is said to be of **order** m and the sum of all the terms containing negative powers, namely,

$$\frac{a_{-m}}{(z-a)^m} + \cdots + \frac{a_{-2}}{(z-a)^2} + \frac{a_{-1}}{z-a}$$

is called the **principal part** of $f(z)$ at $z = a$.

DEFINITION 3 If the Laurent expansion of $f(z)$ in the deleted neighborhood of an isolated singularity contains infinitely many negative powers of $z - a$, then $z = a$ is said to be an **essential singularity** of $f(z)$.

EXAMPLE 1

Since the function $f(z) = 1/z(z-1)^2$ can be written

$$\frac{1}{z(z-1)^2} = \frac{[1 + (z-1)]^{-1}}{(z-1)^2} = \frac{1 - (z-1) + (z-1)^2 - (z-1)^3 + \cdots}{(z-1)^2}$$

$$= \frac{1}{(z-1)^2} - \frac{1}{z-1} + 1 - (z-1) + \cdots \qquad 0 < |z-1| < 1$$

it follows that $f(z)$ has a pole of order 2 at $z = 1$ and its principal part there is

$$\frac{1}{(z-1)^2} - \frac{1}{z-1}$$

It is important to note that although we can also write

$$\frac{1}{z(z-1)^2} = \frac{[(z-1)+1]^{-1}}{(z-1)^2} = \frac{(z-1)^{-1}-(z-1)^{-2}+(z-1)^{-3}-\cdots}{(z-1)^2}$$

$$= \cdots + \frac{1}{(z-1)^5} - \frac{1}{(z-1)^4} + \frac{1}{(z-1)^3} \qquad |z-1|>1$$

the fact that this expansion contains infinitely many negative powers of $z-a$ does not contradict our observation that $f(z)$ has a pole of order 2 at $z=1$. For this series is valid only *outside* the circle $|z-1|=1$, whereas, by definition, the presence of poles and essential singularities is determined by the structure of the Laurent expansion which is valid in the innermost annulus, or deleted neighborhood of the point in question.

EXAMPLE 2

Since the function $f(z)=e^{1/z}$ is represented for all values of z except $z=0$ by the series

$$e^{1/z} = 1 + \frac{1}{z} + \frac{1}{2!z^2} + \frac{1}{3!z^3} + \frac{1}{4!z^4} + \cdots$$

it has an essential singularity at the origin.

It should be noted that if the terms in the expansion of $f(z)$ around a pole of order m, say $z=a$, are put over a common denominator, $f(z)$ will contain the factor $1/(z-a)^m$. Conversely, if a function $f(z)$ is expressed as a fraction in lowest terms, then, according to the theory of partial fractions, the presence of a factor of the form $(z-a)^m$ in the denominator implies that $f(z)$ has a pole of order m at $z=a$. In most applications this is how the poles of a function are found.

As we suggested in Example 1 at the end of the last chapter, the coefficient a_{-1} of the term containing $(z-a)^{-1}$ in the Laurent expansion of a function $f(z)$ is of great importance because of its connection with the integral of the function through the particular case

$$a_{-1} = \frac{1}{2\pi i}\int_C f(z)\,dz$$

of the formula given by Theorem 1, Sec. 18.5. Specifically, we have the following definition.

DEFINITION 4 The coefficient of $(z-a)^{-1}$ in the Laurent expansion of $f(z)$ in the deleted neighborhood of an isolated singular point $z=a$ is called the **residue** of $f(z)$ at that point.

The usefulness of residues in evaluating contour integrals is made clear by the following theorem.

THEOREM 1 **(The Residue Theorem)** If C is a simple closed curve and if $f(z)$ is analytic on and within C except at a finite number of singular points in the interior of C, then

$$\int_C f(z)\,dz = 2\pi i(r_1 + r_2 + \cdots + r_n)$$

where $r_1, r_2, \ldots, r_n$ are the residues of $f(z)$ at its singular points within C.

PROOF Let C be a simple closed curve enclosing exactly n isolated singular points of a function $f(z)$. If around each of these points we draw a circle so small that it encloses no other singular points (Fig. 19.1), these circles, together with the curve C, form the boundary of a multiply

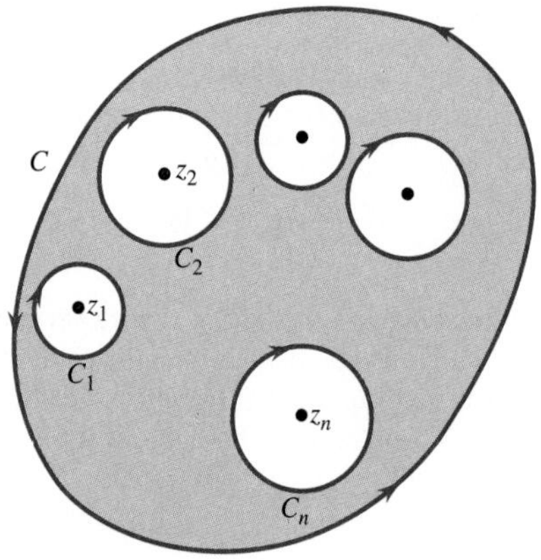

FIGURE 19.1
The circles $C_1, C_2, \ldots, C_n$ enclosing, respectively, the singular points $z_1, z_2, \ldots, z_n$ within a simple closed curve.

connected region in which $f(z)$ is everywhere analytic and to which Cauchy's theorem can therefore be applied. This gives

$$\frac{1}{2\pi i}\int_C f(z)\,dz + \frac{1}{2\pi i}\int_{C_1} f(z)\,dz + \cdots + \frac{1}{2\pi i}\int_{C_n} f(z)\,dz = 0$$

If we reverse the direction of integration around each of the circles and change the sign of each integral to compensate, the last equation can be written

$$\frac{1}{2\pi i}\int_C f(z)\,dz = \frac{1}{2\pi i}\int_{C_1} f(z)\,dz + \cdots + \frac{1}{2\pi i}\int_{C_n} f(z)\,dz$$

where all the integrals are now taken in the counterclockwise sense. Since the integrals on the right are, by definition, just the residues of $f(z)$ at its isolated singularities within C, the theorem is established. ◗

EXAMPLE 3

What is the integral of $f(z) = \dfrac{-3z + 4}{z(z-1)(z-2)}$ around the circle $|z| = \frac{3}{2}$?

In this case, although there are three points where $f(z)$ is singular, namely, the three first-order poles at $z = 0$, $z = 1$, and $z = 2$, only $z = 0$ and $z = 1$ lie within the path of integration. Hence the core of the problem is to find the residues of $f(z)$ at these two points. This requires that we find the Laurent expansions of $f(z)$ around $z = 0$ and $z = 1$ and then identify the coefficients of $1/z$ and $1/(z-1)$ in the respective series. Preparatory to this, it is convenient to construct the partial-fraction representation of $f(z)$, namely,

$$f(z) = \frac{2}{z} - \frac{1}{z-1} - \frac{1}{z-2}$$

Around $z = 0$, both $1/(z-1)$ and $1/(z-2)$ are analytic; hence their expansions are Taylor's series and will contain no negative powers of z. Moreover, $2/z$ is its own Laurent expansion around $z = 0$. Therefore the residue of $f(z)$ at $z = 0$ is simply the coefficient 2. Likewise, around $z = 1$ both $1/z$ and $1/(z-2)$ are analytic; hence their expansions are Taylor expansions and will contain no negative powers of $z - 1$. Moreover, $-1/(z-1)$ is its own Laurent expansion around $z = 1$. Therefore the residue of $f(z)$ at $z = 1$ is simply the coefficient -1. Hence, by the residue theorem,

$$\int_C f(z)\,dz = 2\pi i(2 - 1) = 2\pi i$$

As the last example illustrates, when $f(z)$ is the quotient of two polynomials, its residues are easy to determine once that quotient has been expressed in terms of partial fractions. However, the method of partial fractions can be applied only to rational fractional functions and is often tedious. Hence it is desirable to have a simpler alternative procedure that can be used on almost any function. Such a process is provided by the following considerations. Suppose first that $f(z)$ has a simple, or first-order, pole at $z = a$. It follows then that we can write

$$f(z) = \frac{a_{-1}}{z-a} + a_0 + a_1(z-a) + a_2(z-a)^2 + \cdots$$

If we multiply this identity by $z - a$, we get

$$(z-a)f(z) = a_{-1} + a_0(z-a) + a_1(z-a)^2 + a_2(z-a)^3 + \cdots$$

Now if we let z approach a, every term after the first approaches zero and we obtain for the residue

$$a_{-1} = \lim_{z\to a} (z-a)f(z)$$

If $f(z)$ has a second-order pole at $z = a$, then

$$f(z) = \frac{a_{-2}}{(z-a)^2} + \frac{a_{-1}}{z-a} + a_0 + a_1(z-a) + a_2(z-a)^2 + \cdots$$

To obtain the residue a_{-1} in this case we must first multiply this identity by $(z-a)^2$, getting

$$(z-a)^2 f(z) = a_{-2} + a_{-1}(z-a) + a_0(z-a)^2 + a_1(z-a)^3 + a_2(z-a)^4 + \cdots$$

and then differentiate with respect to z before letting z approach a. The result this time is

$$a_{-1} = \lim_{z\to a} \frac{d}{dz}[(z-a)^2 f(z)]$$

The same procedure can be extended to poles of higher order, leading to the formula contained in the following theorem.

THEOREM 2 If $f(z)$ has a pole of order m at $z = a$, then the residue of $f(z)$ at $z = a$ is

$$a_{-1} = \frac{1}{(m-1)!} \lim_{z\to a} \frac{d^{m-1}}{dz^{m-1}}[(z-a)^m f(z)]$$

In many problems the order of the pole at $z = a$ will not be known in advance. In such cases it is still possible to apply Theorem 2 by taking $m = 1, 2, 3, \ldots,$ in turn, until for the *first* time a finite limit is obtained for a_{-1}. The value of m for which this occurs is the order of the pole, and the value of a_{-1} thus determined is the residue. However, if $f(z)$ has an essential singularity at $z = a$, this process fails and the residue cannot be determined by means of Theorem 2.

EXAMPLE 4

What is the residue of $f(z) = \dfrac{z}{(z-1)^2(z+1)^3}$ at each of its poles?

To find the residue at the second-order pole at $z = 1$, we multiply $f(z)$ by $(z-1)^2$, then differentiate once and take the limit as z approaches 1. The result is

$$a_{-1} = \lim_{z\to 1} \frac{d}{dz}\left[\frac{z}{(z+1)^3}\right] = -\frac{1}{16}$$

To find the residue at the third-order pole at $z = -1$, we multiply $f(z)$ by $(z+1)^3$, then differentiate twice, take the limit as z approaches -1, and multiply by $\frac{1}{2}$. The result is

$$a_{-1} = \lim_{z\to -1} \frac{1}{2}\frac{d^2}{dz^2}\left[\frac{z}{(z-1)^2}\right] = \frac{1}{16}$$

EXAMPLE 5

What is the residue of

$$f(z) = \frac{1+z}{1-\cos z}$$

at the origin?

Here the order of the pole is unknown; so it appears that we may have to proceed tentatively, trying $m = 1, 2, 3, \ldots$, in turn, until for the first time we obtain a finite value for the residue a_{-1}. However, if we replace $\cos z$ by its Maclaurin expansion, we obtain for $f(z)$ the expression

$$\frac{1+z}{1-(1-z^2/2+z^4/24-\cdots)} = \frac{2(1+z)}{z^2(1-z^2/12+\cdots)}$$

and the factor z^2 in the denominator now identifies the pole as of the second order. Hence, applying Theorem 2 with $m = 2$, we have

$$a_{-1} = \lim_{z\to 0}\frac{d}{dz}\left[z^2\frac{2(1+z)}{z^2(1-z^2/12+\cdots)}\right] = \lim_{z\to 0} 2\frac{(1-z^2/12+\cdots)-(1+z)(-z/6+\cdots)}{(1-z^2/12+\cdots)^2} = 2$$

The expansion of a rational fractional function $p(x)/q(x)$ in terms of partial fractions is a well-known procedure we have used in various places throughout this book, most recently in Example 1. Using the residue theorem, we can now establish a result known as the *Mittag-Leffler expansion theorem,*† which generalizes this type of expansion to a much wider class of functions.

THEOREM 3 **(The Mittag-Leffler Theorem)** If $f(z)$ is a function which is analytic at $z = 0$ and whose only singularities are first-order poles at the points $a_1, a_2, a_3, \ldots$, where the residues are, respectively, $b_1, b_2, b_3, \ldots$; if $\{C: |z| = R_N\}$ is a set of circles none of which passes through any singularity of $f(z)$ and whose radii $\{R_N\}$ become infinite as $N \to \infty$; and if on each of these circles $|f(z)|$ is bounded by a constant M which is independent of N, then

$$f(z) = f(0) + \sum_{k=1}^{\infty} b_k\left(\frac{1}{z-a_k} + \frac{1}{a_k}\right)$$

PROOF Suppose that $f(z)$ has first-order poles, with corresponding residues $b_1, b_2, b_3, \ldots$, at $z = a_1, a_2, a_3, \ldots$, and suppose that $z = t$ is not one of these points. Then the function

$$\frac{f(z)}{z-t}$$

has first-order poles at $z = a_1, a_2, a_3, \ldots$ and at $z = t$. At $z = a_k$ the residue of $f(z)/(z-t)$ is, according to Theorem 2,

$$\lim_{z\to a_k}(z-a_k)\frac{f(z)}{z-t} = [\lim_{z\to a_k}(z-a_k)f(z)]\left[\lim_{z\to a_k}\frac{1}{z-t}\right] = \frac{b_k}{a_k-t}$$

†Named for the Swedish mathematician G. M. Mittag-Leffler (1846–1927).

and at $z = t$ the residue is

$$\lim_{z \to t} (z - t)\frac{f(z)}{z - t} = f(t)$$

Hence, applying the residue theorem to the region bounded by C_N, assumed large enough to include the point $z = t$, we have

$$\frac{1}{2\pi i}\int_{C_N} \frac{f(z)}{z - t}\,dz = f(t) + \sum \frac{b_k}{a_k - t} \tag{1}$$

where the summation extends over all the poles of $f(z)$ which lie within C_N. If we set $t = 0$ in this equation, we obtain

$$\frac{1}{2\pi i}\int_{C_N} \frac{f(z)}{z}\,dz = f(0) + \sum \frac{b_k}{a_k}$$

and, subtracting this from (1), we have

$$f(t) - f(0) + \sum b_k\left(\frac{1}{a_k - t} - \frac{1}{a_k}\right) = \frac{1}{2\pi i}\int_{C_N} f(z)\left(\frac{1}{z - t} - \frac{1}{z}\right) dz$$

or

$$f(t) - f(0) - \sum b_k\left(\frac{1}{t - a_k} + \frac{1}{a_k}\right) = \frac{t}{2\pi i}\int_{C_N} f(z)\frac{1}{z(z - t)}\,dz \tag{2}$$

Our next step is to show that the limit of the last integral as $N \to \infty$ is zero. To do this, we note that for values of z on C_N, we have

$$|z - t| \geq |z| - |t| = R_N - |t|$$

Hence, remembering that, by hypothesis, $|f(z)|$ is bounded on C_N by a number M which is independent of N, we find that

$$\left|\frac{t}{2\pi i}\int_{C_N} \frac{f(z)}{z(z - t)}\,dz\right| \leq \frac{|t|}{2\pi}\int_{C_N} \frac{|f(z)|\,|dz|}{|z|\,|z - t|} \leq \frac{|t|}{2\pi}\frac{M}{R_N(R_N - |t|)}\int_{C_N} |dt|$$
$$= \frac{|t|\,M}{R_N - |t|}$$

Since M is a constant, this fraction does approach zero as N, and hence R_N becomes infinite. Therefore, in the limit, Eq. (2) becomes

$$f(t) = f(0) + \sum_{k=1}^{\infty} b_k\left(\frac{1}{t - a_k} + \frac{1}{a_k}\right)$$

Except that t appears in place of z, this is the assertion of the theorem, and our proof is complete. ◗

EXERCISES

1. Find the residue of $f(z) = z/(z^2 + 1)$ **(a)** at $z = i$ and **(b)** at $z = -i$.

2. Find the residue of $f(z) = (z + 1)/[z^2(z - 2)]$ **(a)** at $z = 0$ and **(b)** at $z = 2$.

3. Find the residue of $f(z) = z/(z^2 + 2z + 5)$ at each of its poles.

4. What is the residue of $f(z) = 1/(z + 1)^3$ at $z = -1$?

5. What is the residue of $f(z) = \tan z$ at $z = \pi/2$?

6. What is the residue of $f(z) = z/(\cosh z - \cos z)$ at $z = 0$?

7. What is the residue of $f(z) = 1/(z - \sin z)$ at $z = 0$?

8. What is the residue of $f(z) = 1/(e^z - 1)$ at $z = 0$?

9. If C is the circle $|z| = 4$, evaluate $\int_C f(z)\,dz$ for each of the following functions.

(a) $\dfrac{z}{z^2 - 1}$ **(b)** $\dfrac{z + 1}{z^2(z + 2)}$

(c) $\dfrac{1}{z(z - 2)^3}$ **(d)** $\dfrac{z^2}{(z^2 + 3z + 2)^2}$

(e) $\dfrac{1}{z^2 + z + 1}$ **(f)** $\dfrac{1}{z(z^2 + 6z + 4)}$

10. Work Exercise 9 if C is the circle $|z - 1| = \frac{3}{2}$.

11. If C is the circle $|z| = 2$, evaluate $\int_C f(z)\,dz$ for each of the following functions.

(a) $\tan z$ **(b)** $\dfrac{1}{z \sin z}$ **(c)** $\dfrac{1}{z^2 \sin z}$

(d) $\dfrac{e^{-z}}{z^2}$ **(e)** $ze^{1/z}$ **(f)** $\dfrac{z}{\cos z}$

12. Work Exercise 11 if C is the circle $|z - 1| = \frac{3}{2}$.

13. Apply Theorem 3 to each of the following functions and verify that the result is equivalent to the usual partial fraction expansion.

(a) $\dfrac{3z - 4}{z^2 - 3z + 2}$ **(b)** $\dfrac{1}{z^3 - 5z^2 - 6z}$

(c) $\dfrac{2z + 1}{z^3 + 6z^2 + 11z + 6}$ **(d)** $\dfrac{z^2 - z}{(z + 1)(z^2 + 1)}$

(e) $\dfrac{z + 3}{z^4 - 5z^2 + 4}$

14. Work Exercise 13 for each of the following functions.

(a) $\dfrac{z + 3}{z^4 + 5z^2 + 4}$ **(b)** $\dfrac{z^2}{z^4 - 16}$

(c) $\dfrac{z + 1}{z^3 + 8}$ **(d)** $\dfrac{z^2 - 2}{z^4 + 3z^2 - 4}$

15. Show that the corollary to the first Heaviside expansion theorem (Theorem 1, Sec. 10.9) is equivalent to the following result: If $\Phi(s) = p(s)/q(s)$, where p and q are polynomials such that the degree of q is greater than the degree of p and q is completely factorable into unrepeated real linear factors $(s - a_1)(s - a_2)\cdots(s - a_n)$, then $\mathcal{L}^{-1}[\Phi(s)] = \dfrac{1}{2\pi i}\sum_{i=1}^{n} \text{Res}(a_i)e^{a_i t}$ where $\text{Res}(a_i)$ is the residue of $\Phi(s)$ at $s = a_i$.

16. If $\Phi(t) = p(t)/q(t)$ is a rational fractional function in which the degree of q is at least 2 more than the degree of p, and q is completely factorable into unrepeated real linear factors, show that in the partial fraction expansion of $\Phi(t)$, the sum of the numerators is zero. *Hint:* Consider $\int_C \Phi(z)\,dz$, where C is the circle $|z| = R$, and show that the integral approaches zero as $R \to \infty$.

17. Show that the result given in Exercise 15 is correct whether the unrepeated linear factors of $q(s)$ are real or complex.

18. **(a)** Show that

$$f(z) = \begin{cases} \cot z - 1/z & z \neq 0 \\ 0 & z = 0 \end{cases}$$

is uniformly bounded on the sequence of circles $\{C_N\colon |z| = (2N + 1)\pi/2\}$. *Hint:* Using the results of Exercises 34 and 36, Sec. 17.4, show first that

$$\left|\cot z - \frac{1}{z}\right| = \left|\frac{z\cos z - \sin z}{z \sin z}\right| \le \frac{|z| + 1}{|z|} \times \frac{\cosh y}{(\sin^2 x + \sinh^2 y)^{1/2}}$$

and then consider the portions of C_N for which $|x| < R_N - 1/R_N$ and $|x| \ge R_N - 1/R_N$.

(b) Using Theorem 3 and the result of Part **(a)**, show that

$$\cot z = \frac{1}{z} + \sum_{k=1}^{\infty}\left(\frac{1}{z - n\pi} + \frac{1}{n\pi}\right)$$

19. **(a)** Show that $\cot \pi z$ is uniformly bounded on the squares of the family whose Nth member is shown in Fig. 19.2. *Hint:* Consider separately the portions of

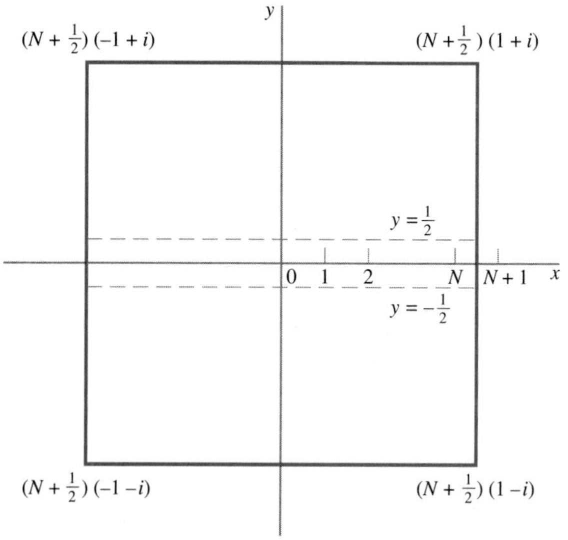

FIGURE 19.2

the square C_N for which $|y| > \frac{1}{2}$ and $|y| \le \frac{1}{2}$, and in the second case show first that if $z = \pm(N + \frac{1}{2}) + iy$, then $|\cot \pi z| = |\tanh \pi y| \le |\tanh \pi/2|$.

(b) Use the result of Part **(a)** to show that if $f(z)$ is a function which has only a finite number of poles, none of which is a point of the set $z = \pm n$, and if on each square of the sequence described in Part **(a)**, $|f(z)| < M/|z|^k$, where M and k are independent of N and $k > 1$, then $\Sigma_{n=-\infty}^{\infty} f(n) = -\Sigma$ residues of $\pi f(z) \cot \pi z$ at the poles of $f(z)$.

20. Show that under the conditions on $f(z)$ described above in Exercise 19**(b)**, $\Sigma_{n=-\infty}^{\infty} (-1)^n f(n) = -\Sigma$ residues of $\pi f(z) \csc \pi z$ at the poles of $f(z)$.

21. Using the result of Exercise 19**(b)** find the sum of each of the following series.

(a) $\displaystyle\sum_{n=-\infty}^{\infty} \frac{1}{n^2 + a^2}$ **(b)** $\displaystyle\sum_{n=-\infty}^{\infty} \frac{1}{(n + a)^2 + b^2}$

(c) $\displaystyle\sum_{n=-\infty}^{\infty} \frac{1}{(n + a)^3}$ **(d)** $\displaystyle\sum_{n=-\infty}^{\infty} \frac{1}{n^4 + 4a^4}$

22. Using the result of Exercise 20, find the sum of each of the following series.

(a) $\displaystyle\sum_{n=-\infty}^{\infty} (-1)^n \frac{1}{n^2 + a^2}$ **(b)** $\displaystyle\sum_{n=-\infty}^{\infty} (-1)^n \frac{1}{(n + a)^2}$

(c) $\displaystyle\sum_{n=-\infty}^{\infty} (-1)^n \frac{n}{n^3 + 8a^3}$

23. Evaluate $\displaystyle\int_C \frac{dz}{z^3(z + 4)}$ if C is the circle **(a)** $|z| = 2$ and **(b)** $|z + 2| = 3$.

24. Evaluate $\displaystyle\int_C \frac{z + 1}{z^3 - 2z^2}\, dz$ if C is the circle **(a)** $|z| = 1$, **(b)** $|z - 2 - i| = 2$, and **(c)** $|z - 1 - 2i| = 2$.

25. Evaluate each of the following integrals.

(a) $\displaystyle\int_C \frac{1 + z}{1 - \cos z}\, dz$, where C is the circle $|z| = 6$. *Hint:* See Example 5.

(b) $\displaystyle\int_0^{2\pi} \frac{d\theta}{5 + 3 \cos \theta}$ *Hint:* Set $z = e^{i\theta}$ and integrate around $|z| = 1$.

19.2 THE EVALUATION OF REAL DEFINITE INTEGRALS

Although we might not expect it, the residue theorem can also be used for the evaluation of several large and important classes of *real* definite integrals. The results in question are contained in the next three theorems.

THEOREM 1 If $R(\cos \theta, \sin \theta)$ is a rational function of $\cos \theta$ and $\sin \theta$ which is finite on the closed interval $0 \le \theta \le 2\pi$, and if $f(z)$ is the function obtained from R by the substitutions

$$\cos \theta = \frac{z + z^{-1}}{2} \qquad \text{and} \qquad \sin \theta = \frac{z - z^{-1}}{2i}$$

then $\int_0^{2\pi} R(\cos \theta, \sin \theta)\, d\theta$ is equal to $2\pi i$ times the sum of the residues of the function $f(z)/iz$ at such of its poles as lie within the unit circle $|z| = 1$.

PROOF As a first step, let us transform the given integral by means of the substitution $z = e^{i\theta}$, according to which

$$\cos \theta = \frac{e^{i\theta} + e^{-i\theta}}{2} = \frac{z + z^{-1}}{2} \qquad \sin \theta = \frac{e^{i\theta} - e^{-i\theta}}{2i} = \frac{z - z^{-1}}{2i} \qquad d\theta = \frac{dz}{iz}$$

Under this transformation the original integrand becomes a rational function of z which we call $f(z)$. Furthermore, as θ ranges from 0 to 2π, the relation $z = e^{i\theta}$ shows that z ranges around the unit circle $|z| = 1$. Hence the transformed integral is

$$\int_C f(z) \frac{dz}{iz}$$

where C is the unit circle. By the residue theorem, the value of this integral is $2\pi i$ times the sum of the residues at those poles of its integrand, namely, $f(z)/iz$, which lie within the unit circle. Since this integral is equal to the original one, the theorem is established.

EXAMPLE 1

Evaluate

$$\int_0^{2\pi} \frac{\cos 2\theta \, d\theta}{1 - 2p \cos \theta + p^2} \qquad -1 < p < 1$$

We note first that by adding and subtracting $2p$, the denominator of the integrand can be written in either of two equivalent forms:

$$\begin{aligned} 1 - 2p \cos \theta + p^2 &= 1 - 2p + p^2 + 2p - 2p \cos \theta = (1 - p)^2 + 2p(1 - \cos \theta) \\ &= 1 + 2p + p^2 - 2p - 2p \cos \theta = (1 + p)^2 - 2p(1 + \cos \theta) \end{aligned}$$

From the first of these it is clear that if $0 \le p < 1$, the denominator is different from zero for all values of θ; and from the second it is clear that if $-1 < p \le 0$, the denominator is also different from zero for all values of θ. Hence if $-1 < p < 1$, the integrand is finite on the closed interval $0 \le \theta \le 2\pi$ and Theorem 1 can be applied. Now

$$\cos 2\theta = \frac{e^{2i\theta} + e^{-2i\theta}}{2} = \frac{z^2 + z^{-2}}{2}$$

and thus the given integral becomes

$$\begin{aligned} \int_C \frac{z^2 + z^{-2}}{2} \frac{1}{1 - 2p(z + z^{-1})/2 + p^2} \frac{dz}{iz} &= \int_C \frac{z^4 + 1}{2z^2} \frac{z}{z - pz^2 - p + p^2 z} \frac{dz}{iz} \\ &= \int_C \frac{(1 + z^4)\, dz}{2iz^2(1 - pz)(z - p)} \end{aligned}$$

Of the three poles of the integrand, only the first-order pole at $z = p$ and the second-order pole at $z = 0$ lie within the unit circle C. For the residue at the former we have

$$\lim_{z \to p} (z - p) \frac{1 + z^4}{2iz^2(1 - pz)(z - p)} = \frac{1 + p^4}{2ip^2(1 - p^2)}$$

For the residue at the second-order pole $z = 0$, we have

$$\begin{aligned} \lim_{z \to 0} \frac{d}{dz} \left[z^2 \frac{1 + z^4}{2iz^2(z - pz^2 - p + p^2 z)} \right] &= \lim_{z \to 0} \frac{(z - pz^2 - p + p^2 z)(4z^3) - (1 + z^4)(1 - 2pz + p^2)}{2i(z - pz^2 - p + p^2 z)^2} \\ &= -\frac{1 + p^2}{2ip^2} \end{aligned}$$

By Theorem 1, the value of the integral is therefore

$$2\pi i \left[\frac{1 + p^4}{2ip^2(1 - p^2)} - \frac{1 + p^2}{2ip^2} \right] = \frac{2\pi p^2}{1 - p^2}$$

THEOREM 2 If $Q(z)$ is a function which is analytic in the upper half of the z plane except at a finite number of poles none of which lies on the real axis, and if $zQ(z)$ converges uniformly to zero when $z \to \infty$ through values for which $0 \le \arg z \le \pi$, then $\int_{-\infty}^{\infty} Q(x)\, dx$ is equal to $2\pi i$ times the sum of the residues at the poles of $Q(z)$ which lie in the upper half-plane.

PROOF Consider a semicircular contour with center at $z = 0$ and with radius R large enough to include all the poles of $Q(z)$ which lie in the upper half z plane (Fig. 19.3). Then, by the residue theorem,

$$\int_{C_1 \cup C_2} Q(z)\, dz = 2\pi i \sum \text{residues of } Q(z) \text{ at all its poles within } C_1 \cup C_2$$

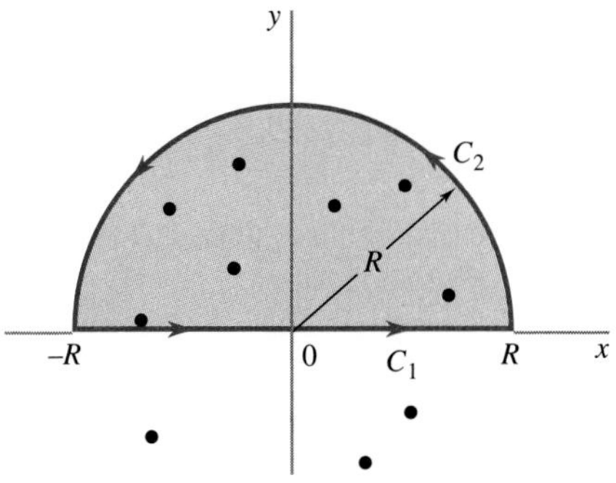

FIGURE 19.3
A semicircular contour enclosing all the poles of a function that lie in the upper half-plane.

or, noting that z is real along C_1,

$$\int_{-R}^{R} Q(x)\,dx + \int_{C_2} Q(z)\,dz = 2\pi i \sum \text{residues}$$

Hence

$$(1) \qquad \left| \int_{-R}^{R} Q(x)\,dx - 2\pi i \sum \text{residues} \right| = \left| -\int_{C_2} Q(z)\,dz \right|$$

and to prove the theorem we must show that the limit of the right-hand side of this equation as $R \to \infty$ is zero. To do this, put $z = Re^{i\theta}$ and $dz = Rie^{i\theta}\,d\theta = iz\,d\theta$ in the integral. Then,

$$\left| -\int_{C_2} Q(z)\,dz \right| = \left| -\int_0^{\pi} Q(z) iz\,d\theta \right| \le \int_0^{\pi} |zQ(z)||d\theta|$$

Now the hypothesis that $zQ(z)$ converges *uniformly* to zero as $z \to \infty$ and $0 \le \arg z \le \pi$ means that given any small positive quantity, say ϵ/π, there exists a radius R_0, depending on ϵ, of course, *but not on* θ, such that

$$|zQ(z)| < \frac{\epsilon}{\pi}$$

whenever $R > R_0$ and $0 \le \theta \le \pi$. Thus for any semicircular arc C_2 whose radius is greater than R_0, we have

$$\int_0^{\pi} |zQ(z)\,d\theta| < \frac{\epsilon}{\pi} \int_0^{\pi} |d\theta| = \epsilon$$

This, coupled with (1), proves that

$$\lim_{R\to\infty} \int_{-R}^{R} Q(x)\,dx = 2\pi i \sum \text{residues}$$

Since the limit on the left is what is meant by $\int_{-\infty}^{\infty} Q(x)\,dx$,† the theorem is established. ◗

---

†Actually, $\lim_{R\to\infty} \int_{-R}^{R} Q(x)\,dx$ is only the **Cauchy principal value** of the integral $\int_{-\infty}^{\infty} Q(x)\,dx$, whose general definition is

$$\lim_{R\to\infty} \int_{-R}^{0} Q(x)\,dx + \lim_{S\to\infty} \int_{0}^{S} Q(x)\,dx$$

where R and S become infinite independently of each other. As the simple function $Q(x) = x$ shows, the principal value of an integral may exist when the integral itself is undefined. However, under the relatively stringent conditions of Theorem 2, the existence of the principal value implies the existence of the integral under its general definition.

In particular, the quotient of two polynomials $p(x)/q(x)$ automatically satisfies all the hypotheses of Theorem 2 whenever the degree of the denominator exceeds the degree of the numerator by at least 2 and $q(x)$ has no real zeros. Hence we have the following important corollary of Theorem 2.

COROLLARY 1 If $p(x)$ and $q(x)$ are real polynomials such that the degree of $q(x)$ is at least 2 more than the degree of $p(x)$ and if $q(x) = 0$ has no real roots, then

$$\int_{-\infty}^{\infty} \frac{p(x)}{q(x)}\,dx = 2\pi i \sum \text{residues of } \frac{p(z)}{q(z)} \text{ at its poles in the upper half-plane}$$

EXAMPLE 2

Evaluate

$$\int_{-\infty}^{\infty} \frac{x^2}{(x^2+a^2)(x^2+b^2)}\,dx \qquad a, b \text{ positive}$$

This is an integral to which Corollary 1 can surely be applied. The only poles of

$$\frac{z^2}{(z^2+a^2)(z^2+b^2)}$$

are at $z = \pm ai, \pm bi$. Of these, only $z = ai$ and $z = bi$ lie in the upper half-plane. At $z = ai$ the residue is

$$\lim_{z\to ai} (z-ai)\frac{z^2}{(z-ai)(z+ai)(z^2+b^2)} = \frac{-a^2}{2ai(-a^2+b^2)} = \frac{a}{2i(a^2-b^2)}$$

From symmetry, the residue at $z = bi$ is obviously $b/2i(b^2 - a^2)$. Hence the value of the integral is

$$2\pi i\left[\frac{a}{2i(a^2-b^2)} + \frac{b}{2i(b^2-a^2)}\right] = \frac{\pi}{a+b}$$

If $Q(z)$ satisfies all the hypotheses of Theorem 2, then so does $e^{imz}Q(z)$, provided $m > 0$. In fact, e^{imz} is analytic everywhere, and under the assumption that $m > 0$ its absolute value is

$$|e^{imz}| = |e^{im(x+iy)}| = |e^{imx}e^{-my}| = e^{-my}$$

which is less than or equal to 1 for all values of y in the upper half-plane. Therefore

$$|e^{imz}zQ(z)| \le |zQ(z)|$$

and thus, if $zQ(z)$ converges uniformly to zero when $z \to \infty$ and $0 \le \arg z \le \pi$, so will $e^{imz}zQ(z)$. Hence the conclusions of Theorem 2 can be applied equally well to $e^{imz}Q(z)$, and we can write

$$(2) \qquad \int_{-\infty}^{\infty} e^{imx}Q(x)\,dx = 2\pi i \sum \text{residues of } e^{imz}Q(z) \text{ at its poles in the upper half-plane}$$

Separating the integral in (2) into its real and its imaginary parts and equating these to the corresponding parts of the expression on the right, we obtain the following useful result.

COROLLARY 2 If $Q(z)$ is analytic in the upper half of the z plane except at a finite number of poles none of which lies on the real axis, and if $zQ(z)$ converges uniformly to zero when z becomes infinite through the upper half-plane, then

$$\int_{-\infty}^{\infty} \cos mx\, Q(x)\,dx = -2\pi \sum \text{imaginary parts of the residues of } e^{imz}Q(z) \text{ at its poles in the upper half-plane}$$

$$\int^{\infty} \sin mx\, Q(x)\,dx = 2\pi \sum \text{real parts of the residues of } e^{imz}Q(z) \text{ at its poles in the upper half-plane}$$

EXAMPLE 3

Evaluate

$$\int_{-\infty}^{\infty} \frac{\cos mx}{1 + x^2}\, dx$$

To do this, we consider the related function $e^{imz}/(1 + z^2)$. The only pole of this function in the upper half-plane is $z = i$, and the residue there is

$$\lim_{z\to i}\,(z - i)\frac{e^{imz}}{(z - i)(z + i)} = \frac{e^{-m}}{2i} = -\frac{ie^{-m}}{2}$$

Hence, by Corollary 2,

$$\int_{-\infty}^{\infty} \frac{\cos mx}{1 + x^2}\, dx = -2\pi\, \mathrm{Im}\left(-\frac{ie^{-m}}{2}\right) = \pi e^{-m}$$

Incidentally, the fact that the residue at $z = i$ is a pure imaginary confirms the observation, obvious from symmetry, that

$$\int_{-\infty}^{\infty} \frac{\sin mx}{1 + x^2}\, dx = 0$$

As a final result on the evaluation of real definite integrals by the method of residues, we have the following theorem, whose proof we omit because of its relative intricacy.†

THEOREM 3 If $Q(z)$ is analytic everywhere in the z plane except at a finite number of poles none of which lies on the positive half of the real axis, and if $z^a Q(z)$ converges uniformly to zero when $z \to 0$ and when $z \to \infty$, then

$$\int_0^{\infty} x^{a-1} Q(x)\, dx = \frac{\pi}{\sin a\pi} \sum \text{residues of } (-z)^{a-1}Q(z) \text{ at all its poles}$$

provided that arg z is taken in the interval $(-\pi, \pi)$.

EXAMPLE 4

Evaluate

$$\int_0^{\infty} \frac{x^{a-1}}{1 + x^2}\, dx \qquad 0 < a < 2$$

For a within the specified range, the conditions of Theorem 3 are fulfilled; hence the given integral is equal to $\pi/(\sin a\pi)$ times the sum of the residues of $(-z)^{a-1}/(1 + z^2)$ at $z = \pm i$. At $z = i$, we have for the residue

$$\lim_{z\to i}\,(z - i)\frac{(-z)^{a-1}}{(z - i)(z + i)} = \frac{(-i)^{a-1}}{2i} = \frac{(e^{-i\pi/2})^{a-1}}{2i} = \frac{e^{-i\pi(a-1)/2}}{2i}$$

At $z = -i$, we find the residue to be

$$\lim_{z\to -i}\,(z + i)\frac{(-z)^{a-1}}{(z + i)(z - i)} = \frac{i^{a-1}}{-2i} = \frac{(e^{i\pi/2})^{a-1}}{-2i} = \frac{e^{i\pi(a-1)/2}}{-2i}$$

†See, for instance, E. T. Whittaker and G. N. Watson, *Modern Analysis,* p. 117, Macmillan, New York, 1943.

The value of the integral is therefore

$$\frac{\pi}{\sin a\pi}\,\frac{e^{i\pi(a-1)/2}-e^{-i\pi(a-1)/2}}{-2i}=-\frac{\pi}{\sin a\pi}\sin\frac{(a-1)\pi}{2}=\frac{\pi}{\sin a\pi}\cos\frac{a\pi}{2}=\frac{\pi}{2\sin(a\pi/2)}$$

For definite integrals not covered by the theorems of this section, evaluation by the method of residues, when possible at all, usually requires considerable ingenuity in selecting the appropriate contour and in eliminating the integrals over all but the desired portion of the contour. Our last example is a problem of this sort. Others, with hints, will be found among the exercises.

EXAMPLE 5

Show that

$$\int_0^\infty \cos x^2\,dx=\int_0^\infty \sin x^2\,dx=\frac{\sqrt{2\pi}}{4}$$

Since $e^{ix^2}=\cos x^2+i\sin x^2$ and since along the x axis the complex variable z reduces to the real variable x, it is reasonable (though hardly obvious) to begin with the function e^{iz^2} and integrate around some contour including the portion of the real axis between 0 and R. If, as $R\to\infty$, the integral of e^{iz^2} along the other portions of the path can be evaluated, combined with the integral along the x axis, or shown to approach zero, we should be able to obtain the required integrals.

The appropriate contour is shown in Fig. 19.4*a*. Since e^{iz^2} is analytic everywhere, we know from Cauchy's theorem that its integral around this contour (or any other contour) is equal to zero. That is,

$$\int_{\overrightarrow{OA}} e^{iz^2}\,dz+\int_{\widehat{AB}} e^{iz^2}\,dz+\int_{\overrightarrow{BO}} e^{iz^2}\,dz=0 \tag{3}$$

Now on $\overrightarrow{OA}$ we have $z=x$ and $dz=dx$, on $\widehat{AB}$ we have $z=Re^{i\theta}$ and $dz=iRe^{i\theta}\,d\theta$, and on $\overrightarrow{BO}$ we have $z=r\left(\cos\frac{\pi}{4}+i\sin\frac{\pi}{4}\right)=re^{i\pi/4}$ and $dz=e^{i\pi/4}\,dr$. Hence (3) can be written

$$\int_0^R e^{ix^2}\,dx+\int_0^{\pi/4} e^{iR^2e^{2i\theta}}iRe^{i\theta}\,d\theta+\int_R^0 e^{ir^2e^{i\pi/2}}e^{i\pi/4}\,dr=0$$

or, simplifying and then transposing the last two terms on the left,

$$\int_0^R(\cos x^2+i\sin x^2)\,dx=e^{i\pi/4}\int_0^R e^{-r^2}\,dr-iR\int_0^{\pi/4} e^{iR^2(\cos 2\theta+i\sin 2\theta)}e^{i\theta}\,d\theta \tag{4}$$

As $R\to\infty$, the first integral on the right becomes $e^{i\pi/4}\int_0^\infty e^{-r^2}\,dr$ and by Exercise 19, Sec. 10.5, this is equal to

$$e^{i\pi/4}\frac{\sqrt{\pi}}{2}=\frac{\sqrt{\pi}}{2}\left(\cos\frac{\pi}{4}+i\sin\frac{\pi}{4}\right)=\frac{\sqrt{2\pi}}{4}(1+i)$$

FIGURE 19.4
The contour required for the evaluation of $\int_0^\infty \cos x^2\,dx$ and $\int_0^\infty \sin x^2\,dx$.

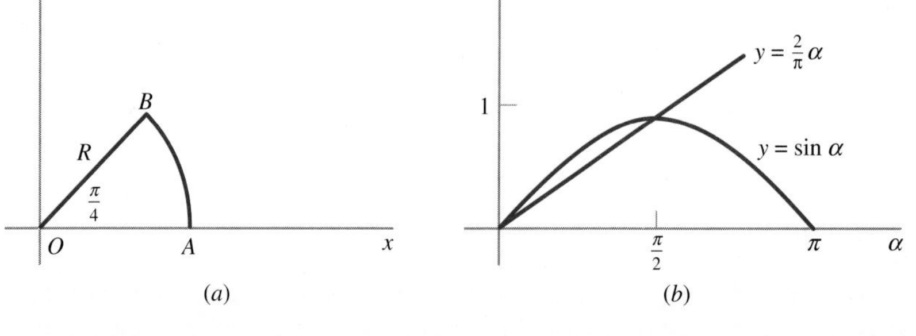

Considering the absolute value of the second integral on the right in (4), we have

$$\left|-iR\int_0^{\pi/4} e^{iR^2(\cos 2\theta+i\sin 2\theta)}e^{i\theta}\,d\theta\right| \le R\int_0^{\pi/4} \left|e^{iR^2\cos 2\theta}\right|\cdot\left|e^{-R^2\sin 2\theta}\right|\cdot\left|e^{i\theta}\right|d\theta = R\int_0^{\pi/4} e^{-R^2\sin 2\theta}\,d\theta$$

If we change the variable of integration from θ to α by the substitution $\alpha = 2\theta$, the last integral becomes

$$\frac{R}{2}\int_0^{\pi/2} e^{-R^2\sin\alpha}\,d\alpha$$

Now from Fig. 19.4*b*, it is clear that

$$\sin\alpha \ge \frac{2\alpha}{\pi} \qquad \text{if } 0 \le \alpha \le \frac{\pi}{2}$$

Hence, overestimating, we have

$$\frac{R}{2}\int_0^{\pi/2} e^{-R^2\sin\alpha}\,d\alpha \le \frac{R}{2}\int_0^{\pi/2} e^{-R^2 2\alpha/\pi}\,d\alpha$$

$$= -\frac{\pi}{4R}\left[e^{-R^2 2\alpha/\pi}\right]_0^{\pi/2} = \frac{\pi}{4R}(1-e^{-R^2})$$

From the last expression, it is clear that as $R \to \infty$ the second integral on the right in (4) approaches zero. Thus (4) becomes

$$\int_0^{\infty} (\cos x^2 + i\sin x^2)\,dx = \frac{\sqrt{2\pi}}{4}(1+i) + 0$$

Hence, equating real and imaginary parts, it follows that

$$\int_0^{\infty} \cos x^2\,dx = \int_0^{\infty} \sin x^2\,dx = \frac{\sqrt{2\pi}}{4}$$

as asserted.

EXERCISES

Evaluate the following integrals by the method of residues.

1. $\displaystyle\int_0^{2\pi} \frac{d\theta}{1-2p\sin\theta+p^2} \qquad -1<p<1$

2. $\displaystyle\int_0^{2\pi} \frac{d\theta}{(a+b\cos\theta)^2} \qquad 0<b<a$

3. $\displaystyle\int_0^{2\pi} \frac{d\theta}{\cos\theta+2\sin\theta+3}$

4. $\displaystyle\int_0^{2\pi} \frac{d\theta}{2\cos\theta+3\sin\theta+7}$

5. $\displaystyle\int_0^{2\pi} \frac{\sin^2\theta\,d\theta}{a+b\cos\theta} \qquad 0<b<a$

6. $\displaystyle\int_0^{\pi} \frac{\cos 2\theta\,d\theta}{5+4\cos\theta}$

7. $\displaystyle\int_{-\infty}^{\infty} \frac{dx}{x^4+a^4}$

8. $\displaystyle\int_{-\infty}^{\infty} \frac{dx}{(1+x^2)^3}$

9. $\displaystyle\int_{-\infty}^{\infty} \frac{x^2\,dx}{1+x^6}$

10. $\displaystyle\int_{-\infty}^{\infty} \frac{x^2\,dx}{(1+x^4)^2}$

11. $\displaystyle\int_0^{\infty} \frac{dx}{(a^2+x^2)^2}$

12. $\displaystyle\int_0^{\infty} \frac{dx}{1+x^6}$

13. $\displaystyle\int_{-\infty}^{\infty} \frac{dx}{(x^2+a^2)(x^2+b^2)(x^2+c^2)}$

14. $\displaystyle\int_{-\infty}^{\infty} \frac{x^2\,dx}{(x^2+a^2)(x^2+b^2)(x^2+c^2)}$

15. Verify the assertion made in the footnote on p. 1217 that the principal value of $\displaystyle\int_{-\infty}^{\infty} x\,dx$ exists, but that under the general definition the integral itself does not.

16. Does the integral $\displaystyle\int_{\infty}^{\infty} \frac{2x+2}{x^2+2x+2}\,dx$ exist? Does its principal value exist?

Evaluate the following integrals by the method of residues.

17. $\displaystyle\int_{-\infty}^{\infty} \frac{\cos mx}{(x-a)^2+b^2}\,dx$ **18.** $\displaystyle\int_{-\infty}^{\infty} \frac{\sin mx}{(x-a)^2+b^2}\,dx$

19. $\displaystyle\int_{0}^{\infty} \frac{\cos mx}{(a^2+x^2)^2}\,dx$ **20.** $\displaystyle\int_{0}^{\infty} \frac{\cos mx}{1+x^4}\,dx$

21. $\displaystyle\int_{-\infty}^{\infty} \frac{\cos mx}{(x^2+a^2)(x^2+b^2)}\,dx$

22. $\displaystyle\int_{-\infty}^{\infty} \frac{x\sin mx}{(x^2+a^2)(x^2+b^2)}\,dx$

23. $\displaystyle\int_{-\infty}^{\infty} \frac{x\sin mx}{1+x^4}\,dx$

24. $\displaystyle\int_{0}^{\infty} \frac{x^{a-1}}{(x+b)(x+c)}\,dx \qquad \begin{matrix} 0<a<2 \\ 0<b,\ c \end{matrix}$

25. $\displaystyle\int_{0}^{\infty} \frac{x^{a-1}}{(x-b)^2+c^2}\,dx \qquad 0<a<2$

26. $\displaystyle\int_{0}^{\infty} \frac{x^{a-1}}{(x+b)(x+c)(x+d)}\,dx \qquad \begin{matrix} 0<a<3 \\ 0<b,\ c,\ d \end{matrix}$

27. $\displaystyle\int_{0}^{\infty} \frac{x^{a-1}}{1+x^3}\,dx \qquad 0<a<3$

28. $\displaystyle\int_{0}^{\infty} \frac{x^{a-1}}{1+x^4}\,dx \qquad 0<a<4$

29. Show that $\Gamma(a)\Gamma(1-a) = \pi/(\sin a\pi)$ if $0<a<1$. *Hint:* Consider the integral $\displaystyle\int_0^\infty \frac{y^{a-1}}{1+y}\,dy$, and evaluate it first by the method of residues and then by making the substitution $y = x/(1-x)$ and expressing it in terms of gamma functions.

30. Show that

$$\int_0^\infty \frac{\sin x}{x}\,dx = \frac{\pi}{2}$$

Hint: Integrate e^{iz}/z around the contour shown in Fig. 19.5 and let $r \to 0$ and $R \to \infty$.

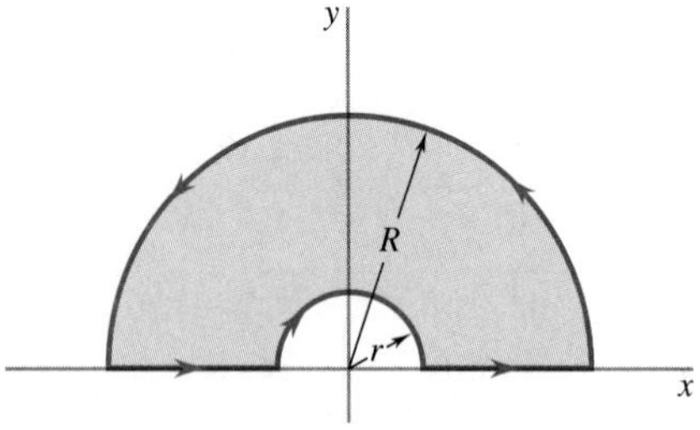

FIGURE 19.5

31. Show that

$$\int_0^\infty \frac{\cos x}{\sqrt{x}}\,dx = \int_0^\infty \frac{\sin x}{\sqrt{x}}\,dx = \sqrt{\frac{\pi}{2}}$$

Hint: Integrate $e^{iz}/\sqrt{z}$ around the contour shown in Fig. 19.6, let $r \to 0$ and $R \to \infty$, and recall (Exercise 19, Sec. 10.5) that

$$\int_0^\infty e^{-x^2}\,dx = \frac{\sqrt{\pi}}{2}$$

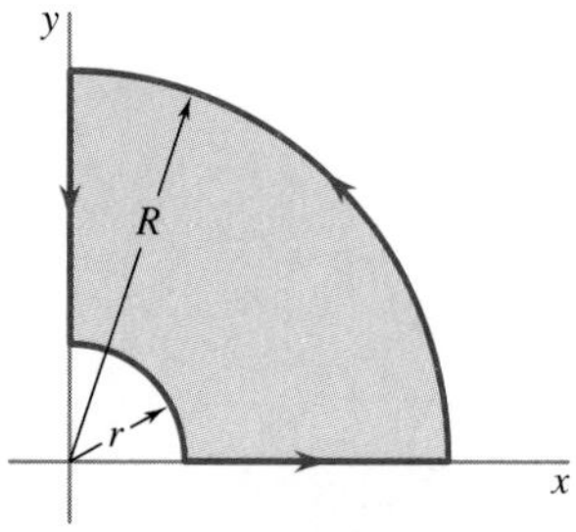

FIGURE 19.6

32. Show that

$$\int_{-\infty}^{\infty} \frac{\cos mx}{e^x + e^{-x}}\,dx = \frac{\pi}{e^{m\pi/2} + e^{-m\pi/2}}$$

Hint: Integrate the function $e^{imz}/(e^z + e^{-z})$ around the contour shown in Fig. 19.7 and let $R \to \infty$.

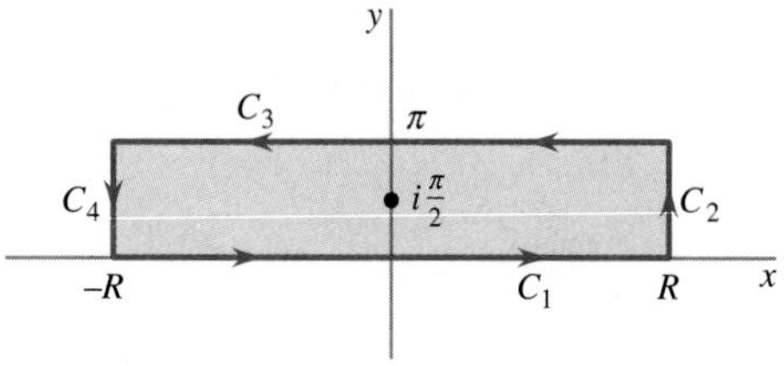

FIGURE 19.7

33. If $f(z)$ has a number of first-order poles on the real axis but otherwise satisfies all the conditions of Theorem 2, show that the principal value of $\displaystyle\int_{-\infty}^{\infty} e^{imx}f(x)\,dx$ is equal to $2\pi i$ times the sum of the residues of $e^{imz}f(z)$ at its poles in the upper half-plane plus $i\pi$ times the sum of the residues of $e^{imz}f(z)$ at its poles on the real axis. *Hint:* Use a contour like that shown in Fig. 19.5, suitably indented around each of the poles of $f(z)$ which lies on the real axis.

34. If n is a positive integer, show that

$$\int_0^{2\pi} \exp(\cos\theta)\cos(n\theta - \sin\theta)\,d\theta = \frac{2\pi}{n!}$$

35. Find the Fourier expansion of the periodic function

$$\frac{1}{a + b\cos\theta} \qquad 0<b<a$$

Discuss from the point of view of Theorem 3, Sec. 8.3, the limiting behavior of the Fourier coefficients of this function as $n \to \infty$.

36. Show that

$$\int_0^\infty \frac{\ln (x^2+1)}{1+x^2}\,dx = \pi \ln 2$$

Hint: Consider the integral of the function $[\ln (z+i)]/(z^2+1)$ around the contour of Fig. 19.8.

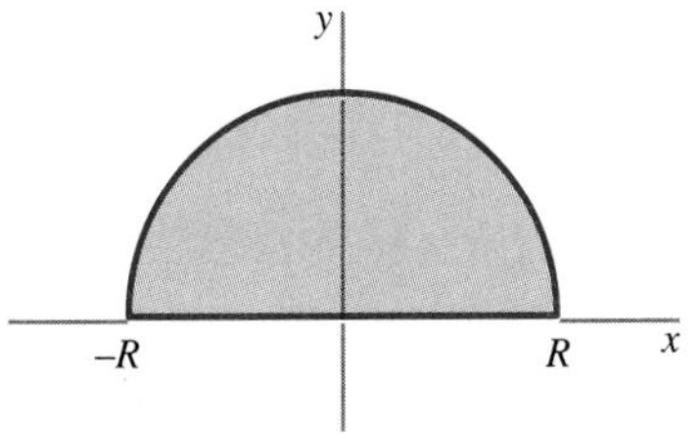

FIGURE 19.8

37. Show that

$$\int_0^\infty \frac{(\ln x)^2}{1+x^2}\,dx = \frac{\pi^3}{8} \qquad \text{and} \qquad \int_0^\infty \frac{\ln x}{1+x^2}\,dx = 0$$

Hint: Integrate the function $(\ln z)^2/(1+z^2)$ around the contour shown in Fig. 19.5.

38. Find the values of the integrals

$$\int_0^\infty \frac{(\ln x)^2}{a^2+x^2}\,dx \qquad \text{and} \qquad \int_0^\infty \frac{\ln x}{a^2+x^2}\,dx$$

39. In the proof of Theorem 2, show that the assumption that $zQ(z)$ converges uniformly to zero through the upper half-plane can be replaced by the condition that in the upper half-plane

$$|Q(z)| \le \frac{M}{|z|^k}$$

where M and k are real constants and $k > 1$.

Evaluate each of the following integrals by applying the method of residues to an appropriate function of z integrated around the contour of Fig. 19.8.

40. $\displaystyle\int_{-\infty}^\infty \frac{x^2-x+2}{x^4+10x^2+9}\,dx$

41. $\displaystyle\int_{-\infty}^\infty \frac{x^2}{(x^2+a^2)(x^2+b^2)}\,dx \qquad a, b > 0$

42. $\displaystyle\int_{-\infty}^\infty \frac{\cos x}{x^2+a^2}\,dx \qquad a > 0$

43. $\displaystyle\int_{-\infty}^\infty \frac{\sin x}{x^2+4x+5}\,dx$

44. Evaluate $\displaystyle\int_0^\infty x^n e^{-x^{1/4}} \sin x^{1/4}\,dx$, $n = 0, 1, 2, \ldots$. *Hint:* Set $x = t^4$, then integrate around the contour given by Fig. 19.6 if $r = 0$.

45. Evaluate $\displaystyle\int_\alpha^\beta \left(\frac{\beta - t}{t-\alpha}\right)^{a-1} \frac{dt}{t}, \quad 0 < \alpha < \beta, \; 0 < a < 2.$

Hint: First set $e^x = (\beta - t)/(t-\alpha)$, then carry out a contour integration around a rectangular contour having vertices at $-S$, R, $R + 2\pi i$, and $-S + 2\pi i$; $R, S > 0$.

19.3 THE LAPLACE INVERSION INTEGRAL

With the residue theorem available, we are now in a position to appreciate more fully the complex inversion integral of Laplace transform theory. In Sec. 9.6 we defined the Laplace transform of a function $f(t)$ to be

$$\mathcal{L}\{f(t)\} = \int_0^\infty f(t)e^{-st}\,dt \tag{1}$$

and we showed, conversely, that f is represented by the complex inversion integral, or **Laplace inversion integral,**

$$f(t) = \frac{1}{2\pi i}\int_{a-i\infty}^{a+i\infty} \mathcal{L}\{f(t)\}e^{st}\,ds \tag{2}$$

s being a complex variable. It is interesting now to reconsider the derivation of (2) in the light of complex-variable theory and to investigate how this formula can be applied to the determination of a function when its transform is known.

In the complex plane let $\phi(z)$ be a function of z which is analytic on the line $x = a$ and in the entire half-plane R to the right of this line. Moreover, let $\phi(z)$ approach zero uniformly as z

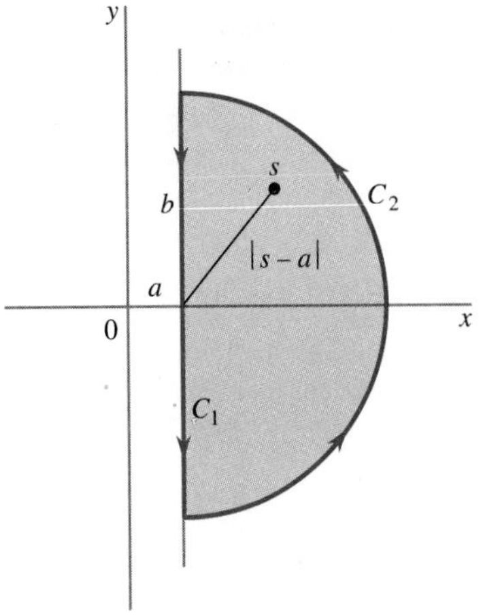

FIGURE 19.9
The contour used to obtain the complex inversion integral.

becomes infinite through this half-plane. Then if s is any point in the half-plane R, we can choose a semicircular contour $C = C_1 \cup C_2$, as shown in Fig. 19.9, and apply Cauchy's integral formula, getting

$$\phi(s) = \frac{1}{2\pi i}\int_C \frac{\phi(z)}{z-s}\,dz = \frac{1}{2\pi i}\int_{a+ib}^{a-ib} \frac{\phi(z)}{z-s}\,dz + \frac{1}{2\pi i}\int_{C_2} \frac{\phi(z)}{z-s}\,dz \tag{3}$$

It is our hope that we can show that the last integral approaches zero as the radius b of C_2 becomes infinite. To do this, we note first that if z is an arbitrary point on C_2, then the minimum value of $|z - s|$ occurs when z is collinear with a and s. Hence for all values of z on C_2 and for b sufficiently large, we have

$$|z - s| \geq b - |s - a| \geq b - |s| - |a| > 0$$

whether a is positive, as shown in Fig. 19.9, or negative. Thus, letting M denote the maximum value of $|\phi(z)|$ on C_2, we have

$$\left|\int_{C_2} \frac{\phi(z)}{z-s}\,dz\right| \leq \int_{C_2} \frac{|\phi(z)|}{|z-s|}\,|dz| \leq \frac{M}{b-|s|-|a|}\int_{C_2} |dz| = \frac{\pi b M}{b-|s|-|a|}$$

As b becomes infinite, the fraction

$$\frac{b}{b-|s|-|a|}$$

approaches 1 and at the same time M approaches zero since, by hypothesis, $|\phi(z)|$ converges uniformly to zero as z becomes infinite through the right half-plane R. Hence

$$\lim_{b\to\infty}\int_{C_2} \frac{\phi(z)}{z-s}\,dz = 0$$

and in the limit we have, from (3),

$$\phi(s) = \lim_{b\to\infty}\frac{1}{2\pi i}\int_{a+ib}^{a-ib} \frac{\phi(z)}{z-s}\,ds = \frac{1}{2\pi i}\int_{a-i\infty}^{a+i\infty} \frac{\phi(z)}{s-z}\,dz$$

Let us now attempt to determine the function of t whose Laplace transform is $\phi(s)$. Taking the inverse of $\phi(s)$ as defined by the last expression, we have

$$\mathcal{L}^{-1}\{\phi(s)\} = f(t) = \mathcal{L}^{-1}\left\{\frac{1}{2\pi i}\int_{a-i\infty}^{a+i\infty}\frac{\phi(z)}{s-z}\,dz\right\}$$

Assuming that the operations of integrating along the vertical line $x = a$ and applying the inverse Laplace transformation can be interchanged, the last equation can be written

$$f(t) = \frac{1}{2\pi i}\int_{a-i\infty}^{a+i\infty}\mathcal{L}^{-1}\left\{\frac{\phi(z)}{s-z}\right\}dz$$

or, since the operator $\mathcal{L}^{-1}$ refers only to the variable s,

$$f(t) = \frac{1}{2\pi i}\int_{a-i\infty}^{a+i\infty}\phi(z)\mathcal{L}^{-1}\left\{\frac{1}{s-z}\right\}dz$$

Now the specific result

$$\mathcal{L}^{-1}\left\{\frac{1}{s-z}\right\} = e^{zt}$$

is known to us through independent reasoning. Hence we have finally

$$f(t) = \frac{1}{2\pi i}\int_{a-i\infty}^{a+i\infty}\phi(z)e^{zt}\,dz = \frac{1}{2\pi i}\int_{a-i\infty}^{a+i\infty}\mathcal{L}\{f(t)\}e^{zt}\,dz$$

Except that the variable of integration is z instead of s, this is exactly Eq. (2). Hence, summarizing, we have the following theorem.

THEOREM 1 The Laplace inversion integral is a line integral in the complex plane taken along a vertical line to the right of all singularities of the transform $\phi(s)$ or along any other path into which this line can legitimately be deformed.

As we have just seen, to obtain the Laplace inversion integral it was necessary to use a contour consisting of a semicircle lying on the right of its diameter on the line $x = a$ (Fig. 19.9). On the other hand, to evaluate the line integral (2) and determine the inverse of a particular transform $\phi(s)$, it is necessary to use a contour consisting of a semicircle lying to the *left* of its diameter on the line $x = a$. This is so because evaluating the integral by the residue theorem requires a contour enclosing all the poles of the transform $\phi(s)$, and these, we assumed, lay to the left of the line $x = a$ (Fig. 19.10). Specifically, we have the following result.

THEOREM 2 If the Laplace transform $\phi(s)$ is an analytic function of s except at a finite number of poles each of which lies to the left of the vertical line Re $s = a$, and if $s\phi(s)$ is bounded as s becomes infinite through the half-plane Re $s \leq a$, then

$$\mathcal{L}^{-1}\{\phi(s)\} = \sum \text{residues of } \phi(s)e^{st} \text{ at each of its poles}$$

PROOF Using the contour shown in Fig. 19.10, we have by the residue theorem,

$$\frac{1}{2\pi i}\int_{a-ib}^{a+ib}\phi(s)e^{st}\,ds + \frac{1}{2\pi i}\int_{C_2}\phi(s)e^{st}\,ds = \sum \text{residues of } \phi(s)e^{st}$$

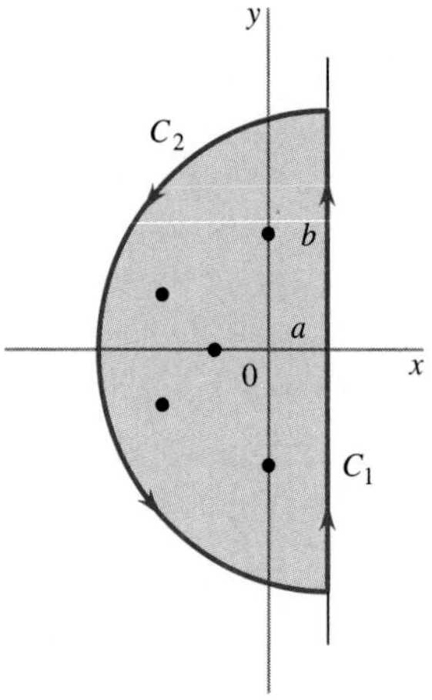

FIGURE 19.10
A typical contour used in evaluation of the complex inversion integral.

Hence

(4) $$\left|\frac{1}{2\pi i}\int_{a-ib}^{a+ib}\phi(s)e^{st}\,ds - \sum \text{residues of } \phi(s)e^{st}\right| = \left|-\frac{1}{2\pi i}\int_{C_2}\phi(s)e^{st}\,ds\right|$$

Again it is our hope that we can show that the absolute value of the integral on the right approaches zero as the radius b of C_2 becomes infinite. To show this, we note first that along C_2 we have

$$s = a + be^{i\theta} = a + b(\cos\theta + i\sin\theta) \qquad \frac{\pi}{2} \le \theta \le \frac{3\pi}{2}$$

and

$$ds = ibe^{i\theta}\,d\theta = i(s-a)\,d\theta$$

and, for sufficiently large s,

$$|s-a| \le |s| + |a| \le 2|s| \qquad \text{and} \qquad |s\phi(s)| \le M$$

Therefore

$$\begin{aligned}
\left|-\frac{1}{2\pi i}\int_{C_2}\phi(s)e^{st}\,ds\right| &\le \frac{1}{2\pi}\int_{C_2}|\phi(s)|\,|e^{st}|\,|ds| \\
&= \frac{1}{2\pi}\int_{\pi/2}^{3\pi/2}|\phi(s)|\,|\exp\{t[a+b(\cos\theta+i\sin\theta)]\}|\,|i(s-a)\,d\theta| \\
&= \frac{1}{2\pi}\int_{\pi/2}^{3\pi/2}|\phi(s)|\,|s-a|\exp[t(a+b\cos\theta)]\,d\theta \\
&\le \frac{1}{2\pi}\int_{\pi/2}^{3\pi/2}2|s\phi(s)|e^{at}\exp(bt\cos\theta)\,d\theta \\
&\le \frac{1}{\pi}Me^{at}\int_{\pi/2}^{3\pi/2}\exp(bt\cos\theta)\,d\theta
\end{aligned}$$

If we now set $\theta = \pi/2 + \alpha$ and then take advantage of the symmetry of the resulting integrand, the last integral becomes

$$\frac{M}{\pi} e^{at} \int_0^{\pi} \exp(-bt \sin \alpha)\, d\alpha = \frac{2M}{\pi} e^{at} \int_0^{\pi/2} \exp(-bt \sin \alpha)\, d\alpha$$

Now, recalling Fig. 19.4*b*, it is evident that

$$\sin \alpha \geq \frac{2\alpha}{\pi} \qquad \text{for } 0 \leq \alpha \leq \frac{\pi}{2}$$

Hence the last integral is overestimated if we replace $\sin \alpha$ in the exponent by the smaller *positive* quantity $2\alpha/\pi$. Doing this and then performing the integration, we have

$$\left| \frac{1}{2\pi i} \int_{C_2} \phi(s) e^{st}\, ds \right| \leq \frac{2M}{\pi} e^{at} \left[\frac{e^{-2bt\alpha/\pi}}{-2bt/\pi} \right]_0^{\pi/2}$$
$$= \frac{2M}{\pi} e^{at} \left[-\frac{\pi}{2bt} (e^{-bt} - 1) \right]$$

For $t > 0$, the last expression clearly approaches zero as b becomes infinite. Hence, returning to Eq. (4), it is clear that

$$\lim_{b\to\infty} \frac{1}{2\pi i} \int_{a-ib}^{a+ib} \phi(s) e^{st}\, ds \equiv \frac{1}{2\pi i} \int_{a-i\infty}^{a+i\infty} \phi(s) e^{st}\, ds$$
$$= \mathcal{L}^{-1}\{\phi(s)\}$$
$$= \sum \text{residues of } \phi(s)e^{st}$$

as asserted. ◗

The proof of Theorem 2 breaks down if $\phi(s)$ has infinitely many poles because then as $b \to \infty$, there will always be semicircles C_2 on which $|s\phi(s)|$ is not bounded. However, to prove the theorem it is not necessary to know that $|s\phi(s)|$ is bounded for *all* values of s to the left of the line Re $s = a$ and outside some sufficiently large semicircle. It is sufficient, in fact, that there should exist a sequence of semicircles $\{(C_2)_N\}$ whose radii $\{b_N\}$ become infinite as N becomes infinite and a number M, independent of N, such that $|s\phi(s)| < M$ on every semicircle in the sequence. If $\phi(s)$ is a function with infinitely many poles which is uniformly bounded on such a sequence of semicircles, it can be shown that Theorem 2 is still valid.†

EXAMPLE 1

What is $\mathcal{L}^{-1}\{1/[(s + a)^2 + b^2]\}$?

Using Theorem 2, we have only to compute the residues of

$$\frac{e^{st}}{(s + a)^2 + b^2}$$

†This is essentially the hypothesis we needed to prove the Mittag-Leffler expansion theorem in Sec. 19.1. For a more detailed discussion of this point, see, for instance, R. V. Churchill, *Operational Mathematics,* 3d ed., pp. 206–216, McGraw-Hill, New York, 1972.

at its two first-order poles $-a \pm ib$. At $s = -a + ib$, we have for the residue

$$\lim_{s\to -a+ib} \frac{[s-(-a+ib)]e^{st}}{[s-(-a+ib)][s-(-a-ib)]} = \frac{e^{(-a+ib)t}}{2ib}$$

and at $s = -a - ib$ we have for the residue

$$\lim_{s\to -a-ib} \frac{[s-(-a-ib)]e^{st}}{[s-(-a+ib)][s-(-a-ib)]} = \frac{e^{(-a-ib)t}}{-2ib}$$

Hence, by Theorem 2,

$$f(t) = \mathcal{L}^{-1}\{\phi(s)\} = \frac{e^{(-a+ib)t}}{2ib} + \frac{e^{(-a-ib)t}}{-2ib}$$

$$= e^{-at}\frac{e^{ibt} - e^{-ibt}}{2ib}$$

$$= \frac{e^{-at}\sin bt}{b}$$

This example of course has been merely a new approach to a result with which we were already familiar. However, in more difficult applications, the use of the complex inversion integral and contour integration is often either the only way or the best way of finding a function when its transform is known.

EXAMPLE 2

Find $\mathcal{L}^{-1}\{1/(s\cosh as)\}$.

Obviously, in this case the function $\phi(s)$ has a first-order pole at $s = 0$. Moreover, since $\cosh as = \cos ias$ [Eq. (13), Sec. 17.4], it follows that $\phi(s)$ has infinitely many other first-order poles, namely, the points where

$$ias = \pm\frac{(2n-1)\pi}{2} \qquad \text{or} \qquad s = \pm\frac{(2n-1)\pi}{2ia} \qquad n = 0, 1, 2, \ldots$$

Therefore, as we observed just before Example 1, $|s\phi(s)|$ is not bounded everywhere in the left half-plane outside some sufficiently large semicircle. However, if we set $s = \sigma + i\omega$, we have, by Eq. (17), Sec. 17.4,

$$|s\phi(s)| = \left|\frac{1}{\cosh as}\right| = \frac{1}{(\cosh^2 a\sigma\cos^2 a\omega + \sinh^2 a\sigma\sin^2 a\omega)^{1/2}}$$

or, using the identities $\cos^2\theta = 1 - \sin^2\theta$ and $\sinh^2\theta = \cosh^2\theta - 1$,

$$|s\phi(s)| = \frac{1}{(\cosh^2 a\sigma - \sin^2 a\omega)^{1/2}}$$

and it is not difficult to show (see Exercise 11) that this expression has a bound independent of N on the semicircles of the set $(C_2)_N\}$ determined by the center $s = a$ and the sequence of points $s_N = N\pi i/a$. Hence the inverse of $\phi(s)$ is simply the sum of the residues of

$$\phi(s)e^{st} = \frac{e^{st}}{s\cosh as}$$

at each of its poles.

At $s = 0$ the residue is

$$\lim_{s\to 0}\frac{se^{st}}{s\cosh as} = 1$$

and at $s = (2n - 1)(\pi/2ia)$ [using L'Hospital's rule and Eqs. (19) and (20). Sec. 17.4, to evaluate the indeterminacy], the residue is

$$\lim_{s\to(2n-1)\pi/2ia} \frac{[s-(2n-1)\pi/2ia]e^{st}}{s\cosh as} = \lim_{s\to(2n-1)\pi/2ia} \frac{[s-(2n-1)\pi/2ia]te^{st}+e^{st}}{\cosh as + as\sinh as}$$

$$= \frac{e^{[(2n-1)\pi/2ia]t}}{[(2n-1)\pi/2ia]a\sinh(2n-1)\pi/2i} = \frac{2(-1)^n e^{(2n-1)\pi t/2ia}}{(2n-1)\pi}$$

Similarly, at $s = -(2n-1)\pi/2ia$, the residue is

$$\frac{2(-1)^n e^{-(2n-1)\pi t/2ia}}{(2n-1)\pi}$$

Hence, pairing the terms corresponding to the same value of n, we have

$$f(t) = \mathcal{L}^{-1}\{\phi(s)\} = 1 + \frac{2}{\pi}\sum_{n=1}^{\infty}\frac{(-1)^n}{2n-1}[e^{(2n-1)\pi t/2ia} + e^{-(2n-1)\pi t/2ia}]$$

$$= 1 + \frac{4}{\pi}\sum_{n=1}^{\infty}\frac{(-1)^n}{2n-1}\cos\frac{(2n-1)\pi t}{2a}$$

EXERCISES

Using the complex inversion integral, find the inverses of the following Laplace transforms. In each case discuss the resemblance of the method of residues to the use of the Heaviside expansion theorems (Sec. 10.9).

1. $\dfrac{1}{(s+1)(s+3)}$ **2.** $\dfrac{1}{(s+2)^2}$ **3.** $\dfrac{1}{s^2+4}$

4. $\dfrac{s}{s^2+4s+13}$ **5.** $\dfrac{1}{s(s^2+1)}$ **6.** $\dfrac{s}{s^3+1}$

7. $\dfrac{s}{(s^2+4)^2}$ **8.** $\dfrac{1}{(s^2+9)(s^2+4)}$

9. $\dfrac{s+1}{(s+2)^2(s+3)}$ **10.** $\dfrac{1}{(s^2+2s+5)^2}$

11. Complete Example 2 by showing that

$$\frac{1}{(\cosh^2 a\sigma - \sin^2 a\omega)^{1/2}}$$

is uniformly bounded on the semicircles $(C_2)_N$ described in the example. *Hint:* Note that it is sufficient to show that $\sin^2 a\omega$ is bounded from 1 when $\sigma = 0$.

12. Complete the solution of Exercise 17, Sec. 11.8, by finding the angular displacement at a general point x.

Find the inverse of each of the following transforms.

13. $\dfrac{1}{s\sinh as}$ **14.** $\dfrac{1}{(s+b)\cosh as}$ **15.** $\dfrac{\sinh x\sqrt{s}}{s\sinh\sqrt{s}}$

16. $[I_0(r\sqrt{s})/sI_0(\sqrt{s})]$, where I_0 is the modified Bessel function of the first kind.

19.4 STABILITY CRITERIA

In the analysis of many physical systems a complete description of the behavior of the system is unnecessary, and all that is required is a knowledge of whether or not the system is **stable,** i.e., whether its response to a bounded excitation remains bounded or becomes infinite as $t \to \infty$. As we shall see in this section, this question can be answered by analyzing the Laplace transform of the response without actually determining the response itself.

Let us consider, then, a system acted upon by an excitation $f(t)$. We begin by supposing that by methods like those described in Chap. 10, we have obtained the Laplace transform of the response of the system $\mathcal{L}\{y(t)\} = \phi(s)$ and that $\phi(s)$ is a rational function of s; i.e.,

$$\phi(s) = \frac{P(s)}{Q(s)}$$

TABLE 19.1

Factor	Term
From unrepeated factors	
1. s	1, that is, $u(t)$
2. $s^2 + b^2$	$\cos bt$, $\sin bt$
3. $s - a$, $a \neq 0$	e^{at}
4. $(s - a)^2 + b^2$, $a \neq 0$	$e^{at} \cos bt$, $e^{at} \sin bt$
From repeated factors	
5. s^n, $n > 1$	t^k, $0 \leq k \leq n - 1$
6. $(s^2 + b^2)^n$, $n > 1$	$t^k \cos bt$, $t^k \sin bt$, $0 \leq k \leq n - 1$
7. $(s - a)^n$, $n > 1$, $a \neq 0$	$t^k e^{at}$, $0 \leq k \leq n - 1$
8. $[(s - a)^2 + b^2]^n$, $n > 1$, $a \neq 0$	$t^k e^{at} \cos bt$, $t^k e^{at} \sin bt$, $0 \leq k \leq n - 1$

where $P(s)$ and $Q(s)$ are real polynomials in the complex variable $s = a + i\omega$. We know from algebra that any real polynomial, such as $Q(s)$, can always be factored into real linear and quadratic factors that may or may not be repeated. Moreover, we know from the Heaviside theorems (Sec. 10.9) that the form of the inverse $y(t) = \mathcal{L}^{-1}\{\phi(s)\}$ is determined completely and solely by the factors of $Q(s)$ and that the only terms that can possibly occur in it are those listed in Table 19.1. Clearly, terms of the forms **1** and **2** are stable in all cases, for although they do not approach zero as $t \to \infty$, they do remain finite. Terms of the forms **3, 4, 7,** and **8** are stable if and only if a is negative, in which case they not only remain finite but in fact approach zero as $t \to \infty$. Terms of the forms **5** and **6** are unstable in all cases since the factor t means that each becomes unbounded as $t \to \infty$. Translating these observations into conditions on the roots of the polynomial equations $Q(s) = 0$, we see that the response $y(t)$ will be stable if and only if the following conditions are met:

a. Every unrepeated real root is nonpositive.

b. Every repeated real root is negative.

c. Every pure imaginary root is unrepeated.

d. Every general complex root has a negative real part.

Geometrically speaking, these conditions are specified by the following theorem.

THEOREM 1 In order for the function

$$y(t) = \mathcal{L}^{-1}\left\{\frac{P(s)}{Q(s)}\right\}$$

to be stable it is necessary and sufficient that the equation $Q(s) = 0$ have no roots to the right of the imaginary axis in the complex s plane and that any roots on the imaginary axis in the s plane be unrepeated.

Various methods are available for determining whether the roots of a polynomial equation all have nonpositive real parts.† In general, however, these are more conveniently formulated as methods for determining whether the roots all have real parts that are strictly negative, and most, though not all, of our results will be of this nature. This is not a serious disadvantage because in practice zero roots and pure imaginary roots, i.e., roots whose real parts are zero, if they occur at all, are usually easily recognizable.

†See, for instance, A. Bronwell, *Advanced Mathematics in Physics and Engineering,* pp. 386–413, McGraw-Hill, New York, 1953, and E. A. Guilleman, *The Mathematics of Circuit Analysis,* pp. 395–409, Wiley, New York, 1953.

A preliminary result of considerable importance is contained in the following theorem.

THEOREM 2 If the real part of each root of the polynomial equation $Q(s) = 0$ is less than or equal to zero, then the coefficients in $Q(s)$ all have the same sign.

PROOF We observe first that it is no specialization to interpret the condition of the theorem as asserting that all coefficients in $Q(s)$ are positive. In fact, the case in which all coefficients are negative can be converted into the case in which all coefficients are positive, and vice versa, simply by multiplying $Q(s) = 0$ by -1, which of course in no way alters the roots of this equation. Now if every root of $Q(s) = 0$ has a nonpositive real part, then the only possible factors of $Q(s)$ are of the form

$$s + a_i \qquad \text{and} \qquad (s + a_j)^2 + b_j^2 \qquad \text{where } a_i,\ a_j \geq 0$$

Since these factors contain only nonnegative terms and since $Q(s)$ is simply the product of a finite number of these factors, no negative terms can arise and every nonzero coefficient in $Q(s)$ must be positive, as asserted.

It is also clear from the preceding argument that if every a is positive, so that every root of $Q(s) = 0$ has negative real part, then there can be no zero coefficients in $Q(s)$; that is, all terms must be present. Hence, restating this observation contrapositively, we have the following corollary.

COROLLARY 1 If one or more terms are missing from $Q(s)$, then the equation $Q(s) = 0$ has at least one root whose real part is nonnegative.

The condition of Theorem 2 is only a necessary and not a sufficient one; i.e., it *cannot* be asserted, conversely, that if the coefficients in $Q(s)$ all have the same sign, then the real part of each root of $Q(s) = 0$ is nonpositive. For instance,

$$s^4 + s^3 + s^2 + 11s + 10 = 0$$

contains only terms with positive coefficients; yet the roots of the equation

$$s^4 + s^3 + s^2 + 11s + 10 = 0$$

are

$$s = -1,\ -2,\ 1 \pm 2i$$

and the two complex roots have positive real parts. On the other hand, it is clear, from the proof of Theorem 2, that we do have the following result.

COROLLARY 2 If $Q(s)$ contains some terms with positive coefficients and some terms with negative coefficients, then the equation $Q(s) = 0$ has at least one root whose real part is positive.

For quadratic equations the necessary condition of Theorem 2 is also sufficient. In fact, if the equation $a_0s^2 + a_1s + a_2 = 0$ contains no negative coefficients, then its roots

$$s = \frac{-a_1 \pm \sqrt{a_1^2 - 4a_0a_2}}{2a_0}$$

are clearly either nonpositive real numbers or conjugate complex numbers with nonpositive real parts.

For cubic equations, a sufficient condition supplementing Theorem 2 is contained in the following result.

THEOREM 3 A necessary and sufficient condition that every root of the cubic equation $a_0s^3 + a_1s^2 + a_2s + a_3 = 0$ have a negative real part is that all coefficients have the same sign and that $a_1a_2 - a_0a_3 > 0$.

PROOF Let us assume for definiteness that the given equation has one real root r and one pair of conjugate complex roots $p \pm iq$. The case in which the equation has three real roots can be handled in exactly the same fashion (Exercise 6). From algebra we recall that the roots, say r_1, r_2, r_3, of any cubic equation are related to the coefficients through the equations

$$\frac{a_1}{a_0} = -(r_1 + r_2 + r_3)$$

$$\frac{a_2}{a_0} = r_1r_2 + r_2r_3 + r_3r_1$$

$$\frac{a_3}{a_0} = -r_1r_2r_3$$

In the present case these become

$$\frac{a_1}{a_0} = -(r + 2p) \tag{1}$$

$$\frac{a_2}{a_0} = p^2 + q^2 + 2pr \tag{2}$$

$$\frac{a_3}{a_0} = -r(p^2 + q^2) \tag{3}$$

From (3) and the assumption that the a's all have the same sign, it follows that $r < 0$. To prove that $p < 0$, we note that the condition $a_1a_2 - a_0a_3 > 0$ can be rewritten, after division by a_0^2, as

$$\frac{a_1}{a_0}\frac{a_2}{a_0} - \frac{a_3}{a_0} > 0$$

When the ratios of the a's are replaced by their equivalents from (1), (2), and (3), this becomes

$$-(r + 2p)(p^2 + q^2 + 2pr) + r(p^2 + q^2) > 0$$

or, simplifying and rearranging,

$$-2p[(p^2 + q^2 + 2pr) + r^2] > 0 \tag{4}$$

Now from (2) and the hypothesis that the a's are all of the same sign, it is evident that $p^2 + q^2 + 2pr > 0$. Hence

$$(p^2 + q^2 + 2pr) + r^2 > 0$$

and it follows from (4) that $p < 0$, as asserted. This proves the sufficiency of the conditions of Theorem 3.

The necessity that all the coefficients in the cubic equation have the same sign follows immediately from (1), (2), and (3), since the right-hand sides of these relations are all positive if $p < 0$ and

$r < 0$. The necessity of the condition $a_1a_2 - a_0a_3 > 0$ follows by reversing the above steps and working backward to this inequality from (4), which is surely true if $p < 0$ and $r < 0$. ◗

The extension of Theorem 3 to polynomial equations of higher degree is contained in the next theorem, which we state without proof.†

THEOREM 4 In the polynomial equation

$$Q(s) = a_0s^n + a_1s^{n-1} + a_2s^{n-2} + \cdots + a_{n-1}s + a_n = 0$$

let every coefficient be positive, and construct the n quantities

$$D_1 = a_1 \qquad D_2 = \begin{vmatrix} a_1 & a_0 \\ a_3 & a_2 \end{vmatrix} \qquad D_3 = \begin{vmatrix} a_1 & a_0 & 0 \\ a_3 & a_2 & a_1 \\ a_5 & a_4 & a_3 \end{vmatrix} \qquad \cdots$$

$$D_n = \begin{vmatrix} a_1 & a_0 & 0 & 0 & 0 & 0 & \cdots & \cdot \\ a_3 & a_2 & a_1 & a_0 & 0 & 0 & \cdots & \cdot \\ a_5 & a_4 & a_3 & a_2 & a_1 & a_0 & \cdots & \cdot \\ \cdots & \cdots & \cdots & \cdots & \cdots & \cdots & \cdots & \cdots \\ a_{2n-1} & a_{2n-2} & a_{2n-3} & a_{2n-4} & a_{2n-5} & a_{2n-6} & \cdots & a_n \end{vmatrix}$$

where, in each determinant, all a's with negative subscripts or with subscripts greater than n are to be replaced by zero. Then a necessary and sufficient condition that each root of $Q(s) = 0$ have negative real part is that each D_i be positive.

The test provided by Theorem 4 is commonly known as the **Routh** or **Routh-Hurwitz stability criterion,** conamed for the English mathematician Edward J. Routh (1831–1907) and the Swiss mathematician H. Hurwitz (1859–1919).

EXAMPLE 1

For the equation $s^5 + s^4 + 2s^3 + s^2 + s + 2 = 0$, we have

$$D_1 = 1 \qquad D_2 = \begin{vmatrix} 1 & 1 \\ 1 & 2 \end{vmatrix} = 1 \qquad D_3 = \begin{vmatrix} 1 & 1 & 0 \\ 1 & 2 & 1 \\ 2 & 1 & 1 \end{vmatrix} = 2$$

$$D_4 = \begin{vmatrix} 1 & 1 & 0 & 0 \\ 1 & 2 & 1 & 1 \\ 2 & 1 & 1 & 2 \\ 0 & 0 & 2 & 1 \end{vmatrix} = -4 \qquad D_5 = \begin{vmatrix} 1 & 1 & 0 & 0 & 0 \\ 1 & 2 & 1 & 1 & 0 \\ 2 & 1 & 1 & 2 & 1 \\ 0 & 0 & 2 & 1 & 1 \\ 0 & 0 & 0 & 0 & 2 \end{vmatrix} = -8$$

Since not all the D's are positive, the given equation has at least one root whose real part is nonnegative. This can be confirmed of course by actually finding the roots of the given equation, which are in fact

$$r_1 = -1 \qquad r_2, r_3 = \frac{1}{2} \pm i\frac{\sqrt{3}}{2} \qquad r_4, r_5 = -\frac{1}{2} \pm i\frac{\sqrt{7}}{2}$$

†See, for instance, J. V. Uspensky, *Theory of Equations,* pp. 304–309, McGraw-Hill, New York, 1948.

A somewhat different method of obtaining information about the location of the roots of an equation $f(z) = 0$, which has the advantage of telling exactly how many roots there are with positive real parts and moreover not being restricted to the case in which $f(z)$ is a polynomial, is based on the following theorem.

THEOREM 5 If $f(z)$ is analytic within and on a closed curve C except at a finite number of poles and if $f(z)$ has neither poles nor zeros on C, then

$$\frac{1}{2\pi i}\int_C \frac{f'(z)}{f(z)}\,dz = N - P$$

where N is the number of zeros of $f(z)$ within C and P is the number of poles of $f(z)$ within C, each counted as many times as its multiplicity.

PROOF Suppose first that at a point $z = a_k$ within $C, f(z)$ has a zero of order n_k. Then $f(z)$ can be written in the form

$$f(z) = (z - a_k)^{n_k}\phi(z)$$

where $\phi(z)$ is nonvanishing and analytic in some neighborhood of $z = a_k$. From this,

$$f'(z) = n_k(z - a_k)^{n_k - 1}\phi(z) + (z - a_k)^{n_k}\phi'(z)$$

and thus

$$\frac{f'(z)}{f(z)} = \frac{n_k(z - a_k)^{n_k-1}\phi(z) + (z - a_k)^{n_k}\phi'(z)}{(z - a_k)^{n_k}\phi(z)} = \frac{n_k}{z - a_k} + \frac{\phi'(z)}{\phi(z)}$$

Since $\phi(z)$, and hence $\phi'(z)$, is analytic at $z = a_k$, and since $\phi(z)$ does not vanish at $z = a_k$, the fraction $\phi'(z)/\phi(z)$ is analytic at $z = a_k$. Hence it is clear from the last expression that $f'(z)/f(z)$ has a simple pole with residue n_k at every point a_k where $f(z)$ has a zero of order n_k. Similarly, if $f(z)$ has a pole of order p_k at the point $z = b_k$, we can write

$$f(z) = \frac{c_{-p_k}}{(z - b_k)^{p_k}} + \frac{c_{-p_k+1}}{(z - b_k)^{p_k-1}} + \cdots + \frac{c_{-1}}{z - b_k} + c_0 + \cdots$$

Hence, putting these fractions over a common denominator, we have, in the neighborhood of $z = b_k$,

$$f(z) = \frac{1}{(z - b_k)^{p_k}}\psi(z) = (z - b_k)^{-p_k}\psi(z)$$

where, since $c_{-p_k} \neq 0$,

$$\psi(z) = c_{-p_k} + c_{-p_k+1}(z - b_k) + c_{-p_k+2}(z - b_k)^2 + \cdots$$

is obviously analytic and nonvanishing at $z = b_k$. Therefore, around b_k,

$$f'(z) = -p_k(z - b_k)^{-p_k-1}\psi(z) + (z - b_k)^{-p_k}\psi'(z)$$

and thus

$$\frac{f'(z)}{f(z)} = \frac{-p_k(z - b_k)^{-p_k-1}\psi(z) + (z - b_k)^{-p_k}\psi'(z)}{(z - b_k)^{-p_k}\psi(z)} = \frac{-p_k}{z - b_k} + \frac{\psi'(z)}{\psi(z)}$$

The last fraction on the right is clearly analytic; hence $f'(z)/f(z)$ has a simple pole with residue $-p_k$ at every point where $f(z)$ has a pole of order p_k. Applying the residue theorem to $f'(z)/f(z)$ over the region bounded by C, we therefore have

$$\int_C \frac{f'(z)}{f(z)}\,dz = 2\pi i \sum \text{residues} = 2\pi i\left(\sum n_k - \sum p_k\right) = 2\pi i(N - P)$$

since $\sum n_k$ is the total multiplicity N of all the zeros of $f(z)$ within C and $\sum p_k$ is the total multiplicity P of all the poles of $f(z)$ within C. Dividing by $2\pi i$, we obtain the assertion of the theorem. ◗

An important alternative form of the last theorem can be derived by noting that

$$\frac{1}{2\pi i}\int_C \frac{f'(z)}{f(z)}\,dz = \frac{1}{2\pi i}\int_C d[\ln f(z)]$$

Hence, performing the integration,

$$N - P = \frac{1}{2\pi i}[\text{variation of } \ln f(z) \equiv \ln|f(z)| + i \arg f(z) \text{ in going completely around } C]$$

Clearly, $\ln|f(z)|$ is the same at the beginning and at the end of one full circuit around any closed curve. Therefore

$$N - P = \frac{1}{2\pi i}[\text{variation of } i \arg f(z) \text{ around } C]$$

$$= \frac{\text{variation of } \arg f(z) \text{ around } C}{2\pi}$$

In particular, if $f(z)$ is analytic everywhere within C, so that $P = 0$, we have the following important result commonly known as the **principle of the argument.**

COROLLARY 1 If $f(z)$ is analytic within and on a closed curve C and does not vanish on C, then the number of zeros of $f(z)$ within C is equal to $1/2\pi$ times the net variation in the argument of $f(z)$ as z traverses the curve C in the counterclockwise direction.

In geometric terms, this means that if the locus of $w = u + iv = f(z)$ is plotted for values of z ranging around the given contour C, then the number of times this locus encircles the origin in the w plane is the number of zeros of $f(z)$ within C. Moreover, although the possibility that $f(z) = 0$ on C was excluded in the statement of Corollary 1, it is clear that *if* $f(z)$ has a zero on C, then the image curve passes *through* the origin in the w plane.

To use the last theorem and its corollary to determine whether or not each of the roots of a polynomial equation $Q(z) = 0$ has a negative real part, we proceed as follows. In the z plane let the contour C consist of the segment of the imaginary axis between $-R$ and R together with the semicircle lying in the right half-plane and having this segment as diameter (Fig. 19.11). Since a polynomial equation has only a finite number of roots, it is clear that if R is taken sufficiently large, any roots of $Q(z) = 0$ which lie in the right half-plane, i.e., any roots which have positive real parts, will lie within C.

Now let z range over the contour C, and in an auxiliary $w = u + iv$ plane let the locus of the corresponding values of $w = Q(z)$ be plotted. If this curve does not enclose the origin in the w plane, then according to the corollary of Theorem 5, $Q(z) = 0$ has no roots in the right half-plane. If, further, this curve does not pass through the origin in the w plane, then $Q(z) = 0$ has no roots on the imaginary axis either; i.e., all roots of $Q(z) = 0$ have negative real parts. On the other hand, if

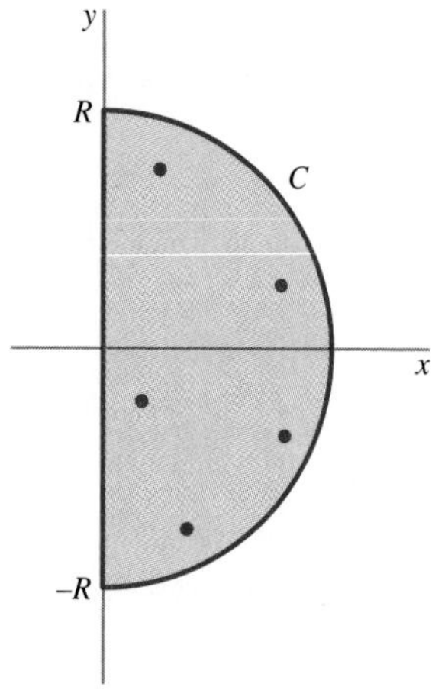

FIGURE 19.11
A semicircular contour enclosing all zeros of a function that lie in the right half-plane.

the image curve encircles the origin in the w plane a net number of times k, then $Q(z) = 0$ has k roots in the right half-plane, i.e., has k roots with positive real parts. Moreover, for every time this curve passes through the origin in the w plane there is a root of $Q(z) = 0$ lying on the imaginary axis in the z plane. Distinct pure imaginary roots of $Q(z) = 0$ thus give rise to a multiple point at the origin in the w plane, the tangents at the multiple point being distinct. A repeated pure imaginary root in the z plane similarly gives rise, in general, to a cusp at the origin in the w plane.

The labor of plotting the image curve in the w plane can be reduced considerably by letting $R \to \infty$. The image of the semicircular portion of C then recedes to infinity in the w plane, and without any plotting its contribution to possible encirclements of the origin can be determined in the following way. On the semicircle we have

$$z = Re^{i\theta} \qquad -\frac{\pi}{2} \le \theta \le \frac{\pi}{2}$$

For the images of these values of z and Q of degree n, we have

$$\begin{aligned} w = Q(z) = Q(Re^{i\theta}) &= a_0(Re^{i\theta})^n + a_1(Re^{i\theta})^{n-1} + a_2(Re^{i\theta})^{n-2} + \cdots + a_n \\ &= R^n\left(a_0 e^{in\theta} + \frac{1}{R}a_1 e^{i(n-1)\theta} + \frac{1}{R^2}a_2 e^{i(n-2)\theta} + \cdots + \frac{1}{R^n}a_n\right) \end{aligned}$$

Hence, for arbitrarily large values of R, the lengths of the terms in $Q(z)$, after the first, are vanishingly small in comparison with the length of the first term. Thus, regardless of the angles of these terms, $Q(z)$ is arbitrarily close to its leading term

$$R^n a_0 e^{in\theta}$$

Therefore, as z traverses the semicircular portion of C in the positive direction, with $\theta = \arg z$ varying from $-\pi/2$ to $\pi/2$, the argument of its image

$$w \doteq a_0 R^n e^{in\theta}$$

varies from $-n\pi/2$ to $n\pi/2$, which represents a net variation in arg w, that is, arg $Q(z)$, of $n\pi$. Hence if $w = Q(z)$ is plotted only for z varying from $i\infty$ to $-i\infty$ along the imaginary axis and the net change in the argument of w is noted, with its proper sign, of course, this change plus $n\pi$ will give the net change as the entire contour C is traversed. This change divided by 2π gives the net number of times the image curve encircles the origin in the w plane, and this number is equal to the number of roots of $Q(z) = 0$ in the right half of the z plane. The labor of plotting can be reduced still further by noting that for polynomials with real coefficients, such as we encounter as Laplace transforms, we have

$$Q(\bar{z}) = \overline{Q(z)}$$

and hence the plot of $Q(z)$ for values of z on the lower half of the imaginary axis is just the reflection in the real axis of the plot of $Q(z)$ for values of z on the upper half of the imaginary axis.

EXAMPLE 2

A STABLE INVERSE LAPLACE TRANSFORM

Discuss the stability of $y(t)$ if

$$\mathcal{L}\{y(t)\} = \frac{s^2 + 1}{s^3 + s^2 + 4s + 1}$$

As we pointed out before, the stability of $y(t)$ is determined solely by the location of the zeros of the denominator of $\mathcal{L}\{y(t)\}$. Hence we begin by plotting

$$w = Q(s) = s^3 + s^2 + 4s + 1$$

for values of s on the imaginary s axis, i.e., for $s = i\omega$ and ω ranging from ∞ to $-\infty$. Parametric equations for the image curve are easily obtained, for we have

$$Q(i\omega) = -i\omega^3 - \omega^2 + 4i\omega + 1$$

and so the real and imaginary parts of $w = u + iv$ are

$$u = 1 - \omega^2 \qquad \text{and} \qquad v = 4\omega - \omega^3$$

Figure 19.12 shows a plot of this curve, together with a plot of the variation of arg w, obtained by inspection

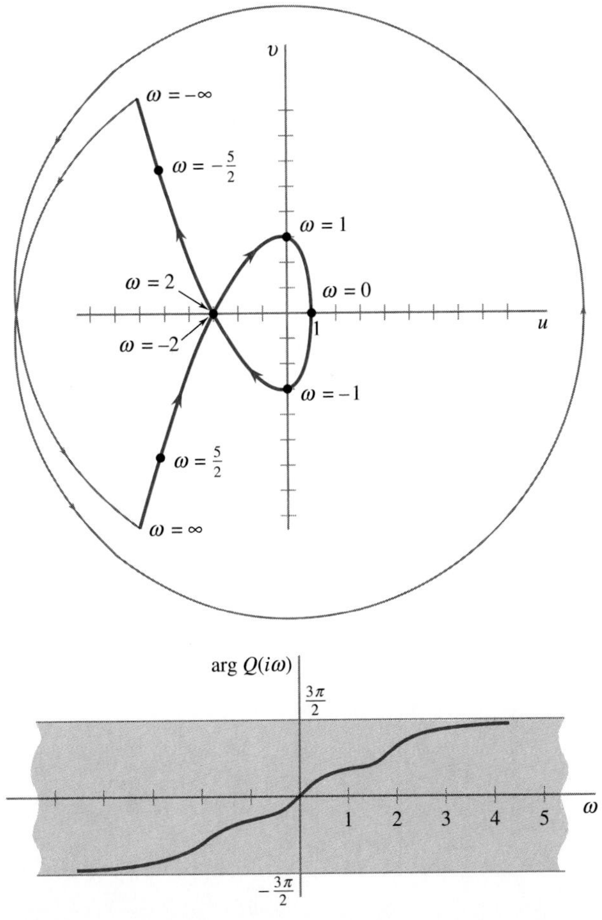

FIGURE 19.12
Plots of $Q(s) = s^3 + s^2 + 4s + 1$ and arg $Q(s)$ for $s = i\omega$.

from the graph of $w = Q(i\omega)$. Evidently, as s traverses the imaginary axis from $i\infty$ to $-i\infty$, arg w varies from $3\pi/2$ (when $\omega = \infty$) to $-3\pi/2$ (when $\omega = -\infty$), which is a net variation of -3π. This value, added to the value $n\pi \equiv 3\pi$ contributed, according to our preceding discussion, by the semicircular portion of the contour (Fig. 19.11) gives a net variation of zero as the entire contour is traversed. Hence $Q(s)$ has no roots in the right half of the s plane. Moreover, since the image curve does not pass through the origin in the w plane, $Q(s)$ has no zeros on the imaginary axis. Therefore, by our earlier discussion, the inverse $y(t)$ is stable.

EXAMPLE 3

AN UNSTABLE INVERSE LAPLACE TRANSFORM

Discuss the stability of $y(t)$ if $\mathscr{L}\{y(t)\} = \dfrac{s-2}{s^3+s^2+s+4}$.

Proceeding exactly as in Example 2, we obtain from

$$Q(i\omega) = -i\omega^3 - \omega^2 + i\omega + 4$$

the parametric equations

$$u = 4 - \omega^2 \qquad \text{and} \qquad v = \omega - \omega^3$$

and the image curve shown in Fig. 19.13. In this case, as s traverses the imaginary axis from $i\infty$ to $-i\infty$, arg w varies from $3\pi/2$ (when $\omega = \infty$) to $5\pi/2$ (when $\omega = -\infty$), which is a net variation of $5\pi/2 - 3\pi/2 = \pi$. Hence, adding the variation $n\pi \equiv 3\pi$ contributed by the semicircular portion of the contour C (Fig. 19.11), we obtain 4π for the net variation in arg w as the entire contour is traversed. Dividing this by 2π, we obtain 2 as the number of zeros of $Q(s)$ in the right half-plane. Therefore the inverse in this case is unstable.

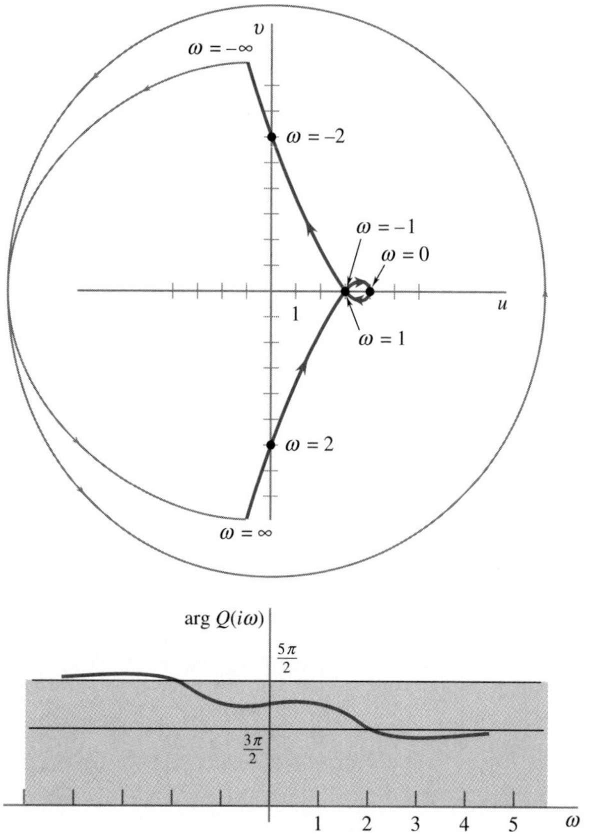

FIGURE 19.13
Plots of $Q(s) = s^3 + s^2 + s + 4$ and arg $Q(s)$ for $s = i\omega$.

Theorem 5 finds its best known application in the so-called **Nyquist stability criterion,** which is a modification of the preceding process especially well adapted to the stability analysis of closed-loop control systems. One common problem in engineering is to make the output $x_o(t)$ of a system follow quickly and accurately changes made in the input $x_i(t)$ to the system. In an **open-loop system** (Fig. 19.14*a*) this is often difficult to accomplish; specifically, prolonged oscillation of $x_o(t)$ about its desired value may well follow an abrupt change in the input $x_i(t)$ to some desired new value. One possible way to remedy this situation is to construct a **feedback loop,** like the one shown in Fig. 19.14*b*, to sample the output and feed it back to a differential device which will in turn transmit the **error signal** $x_i(t) - x_o(t)$ as a modified or corrected input to the original system. More generally, the output $x_o(t)$ may be (and usually is) modified by some additional device in the feedback loop to produce the **feedback signal** $x_f(t)$ before it is fed to the differential device (Fig. 19.14*c*).

In Fig. 19.14, let $G_1(s)$ and $G_2(s)$ be the transfer functions of the original system and the feedback loop, respectively. Then from the definition of a transfer function as the ratio of the transformed output to the transformed input (Sec. 10.11), we can write

$$\mathcal{L}\{x_o(t)\} = G_1(s)[\mathcal{L}\{x_i(t)\} - \mathcal{L}\{x_f(t)\}]$$

$$\mathcal{L}\{x_f(t)\} = G_2(t)\mathcal{L}\{x_o(t)\}$$

If we eliminate $\mathcal{L}\{x_f(t)\}$ between these two equations, we obtain at once

$$\mathcal{L}\{x_o(t)\} = \frac{G_1(s)}{1 + G_1(s)G_2(s)}\mathcal{L}\{x_i(t)\}$$

Evidently, $G_1(s)/[1 + G_1(s)G_2(s)]$ is the overall transfer function of the entire closed-loop system.

The question of the stability of a feedback system is of great importance and, as discussed earlier, can be answered by an examination of the Laplace transform of the output, namely,

$$\frac{G_1(s)}{1 + G_1(s)G_2(s)}\mathcal{L}\{x_i(t)\}$$

If the original system without the feedback loop is stable for the input $x_i(t)$, as we shall suppose, then the product $G_1(s)\mathcal{L}\{x_i(t)\}$ can have no poles in the right half of the s plane and the stability of the overall system depends solely on the location of the zeros of the denominator

$$1 + G_1(s)G_2(s)$$

Hence, as before, we plot the locus of the function

$$w(s) = 1 + G_1(s)G_2(s)$$

as s ranges over the contour of Fig. 19.11.

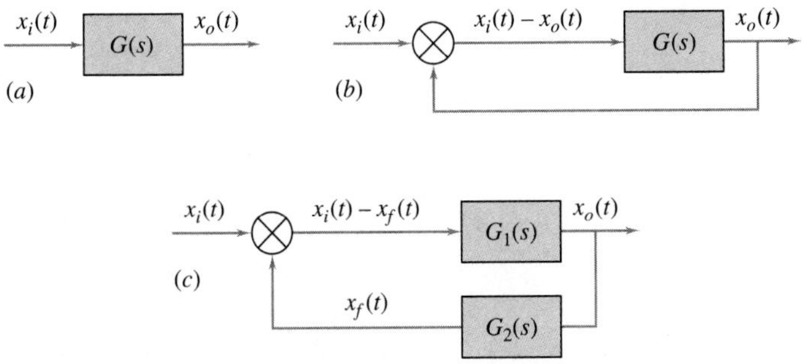

FIGURE 19.14
Systems with feedback loops.

In this case, since $G_1(s)$ and $G_2(s)$ are themselves Laplace transforms, each approaches zero as R becomes infinite (Corollary 3, Theorem 2, Sec. 10.2). Hence the image of the semicircular portion of the contour C shrinks to the single point $w = 1$ as $R \to \infty$ and therefore contributes nothing to the total variation of arg w as s traverses C. Thus to determine stability, it is necessary only to plot $w(s) = 1 + G_1(s)G_2(s)$ for values of s on the imaginary axis and determine whether or not the resulting curve encloses the origin. Moreover, as we pointed out before, this curve can be constructed simply by plotting $1 + G_1(i\omega)G_2(i\omega)$ for positive values of ω and then reflecting the resulting arc in the real axis. In practice, instead of plotting $w = 1 + G_1(i\omega)G_2(i\omega)$ and observing whether the image curve encircles the origin, it is customary to plot $w = G_1(i\omega)G_2(i\omega)$ and observe whether it encircles the point $w = -1$. The equivalence of the two procedures is obvious.

It would take us too far afield and involve us in too many details of a purely engineering nature to discuss the applications of the Nyquist stability criterion to specific, nontrivial closed-loop systems. Such applications appear in large numbers in books on servomechanisms, to which we must turn for illustrations and further information.†

EXERCISES

1. Verify that $f(z) = z^2 + 1$ has one zero inside the circle $|z - i| = 1$ and two zeros inside the circle $|z| = 2$ by plotting $f(z)$ as z varies around the respective circles.

Using the geometric approach based on the corollary of Theorem 5, determine whether the following equations have any roots with nonnegative real parts. Check by using Theorem 4.

2. $s^3 + s + 9 = 0$

3. $s^3 + 6s^2 + 10s + 6 = 0$

4. $s^4 + 2s^3 + 7s^2 + 4s + 10 = 0$

5. $s^4 + s^3 + s^2 + 10s + 10 = 0$

6. Prove Theorem 3 under the assumption that the cubic has three real roots.

7. Using the principle of the argument, prove the following result, known as **Rouché's theorem.** If $f(z)$ and $g(z)$ are analytic within and on a simple closed curve C and if $|g(z)| < |f(z)|$ on C, then $f(z) + g(z)$ and $f(z)$ have the same number of zeros in the interior of C. *Hint:* Verify that arg $(f + g) = \arg f + \arg\,[1 + (g/f)]$ and then show that the variation in arg $[1 + (g/f)]$ as z traverses C is zero.

8. Using the result of Exercise 7, show that if $g(z)$ is analytic within and on the circle C: $|z| = 1$ and if $|g(z)| < 1$ for z on C, then there is exactly one point z_0 in the interior of C such that $g(z_0) = z_0$.

9. Using the result of Exercise 7, show that if $|a| > e$, then the equation $az^n - e^z = 0$ has exactly n roots in the interior of the circle C: $|z| = 1$.

10. Show that all the roots of the equation $z^7 - z^3 + 12 = 0$ are located in the annulus bounded by the circle $|z| = 1$ and the circle $|z| = 2$.

11. Verify the conclusion of Example 2 by finding the roots of the equation $Q(s) = s^3 + s^2 + 4s + 1 = 0$.

12. Verify the conclusion of Example 3 by finding the roots of the equation $Q(s) = s^3 + s^2 + s + 4 = 0$.

13. Use the principle of the argument to show that the equation $z^2 - 2z + 2 = 0$ has exactly one root in the first quadrant. *Hint:* Consider the contour of Fig. 19.6, with $r = 0$, and let $R \to \infty$.

14. In what quadrants do the roots of the equation $z^4 + z^3 + 4z^2 + 2z + 3 = 0$ lie? *Hint:* First show that there are no real or pure imaginary roots, then use the principle of the argument as in Exercise 13.

†See, for instance, G. J. Thaler and R. G. Brown, *Analysis and Design of Feedback Control Systems,* 2d ed., McGraw-Hill, New York, 1960, or H. Chesnut and R. W. Mayer, *Servo-mechanisms and Regulating System Design,* Wiley, New York, 1951.

CHAPTER 20

CONFORMAL MAPPING

Functions of a complex variable are represented geometrically not by drawing curves and surfaces, as in ordinary analytic geometry, but by regarding them as point transformations between corresponding regions in the xy plane of the independent variable and the uv plane of the function. If such a transformation has the property that infinitesimal configurations and their images are arbitrarily close to being similar, the transformation is said to be *conformal*. In Sec. 20.2, we show that in the neighborhood of a regular point, the mapping effected by an analytic function is always conformal. Moreover, around a singular point, the departure from similarity is uniquely determined by the nature of the singular point.

In Sec. 20.3, the transformation effected by a general linear fractional function is studied in detail, and a number of useful mappings involving such regions as circles and their exteriors, half-planes, quadrants, and angles in general are determined. These are often useful in solving two-dimensional boundary-value problems because, as we show in Sec. 20.2, solutions of Laplace's equation remain solutions when transformed by an analytic function. Hence if the complicated region in which a problem is to be solved can be mapped conformally onto a simpler region where it can be solved by the methods of Chap. 11, for instance, that solution, transformed back into the original region, becomes the required solution. Example 5, Sec. 20.3, illustrates this possibility.

Although it is theoretically possible to set up a conformal transformation between any two simply connected regions, the only mappings of any practical value besides those discussed in Sec. 20.3 are those between general polygonal regions and a half-plane. These can be accomplished by the *Schwarz-Christoffel* transformation, which is discussed in Sec. 20.4. Its usefulness is often limited by the complexity of the integrals that are involved.

Prerequisities for this chapter: Chaps. 17 and 19.

20.1 THE GEOMETRIC REPRESENTATION OF FUNCTIONS OF z

Although in Sec. 19.4, in our discussion of stability criteria, we plotted the values of a function $w = f(z)$ for certain values of z, namely, those on a particular semicircular contour, we have not yet attempted to provide a geometric representation for $w = f(z)$ when z ranges over the entire complex plane. To do so now requires a decided departure from the conventional methods of cartesian plotting which associate a curve with a real function of one variable $y = g(x)$ and a surface with a function of two real variables, $z = h(x, y)$. In the complex domain, a relation $w = f(z)$, defined for all z, that is,

$$u + iv = f(x + iy)$$

involves *four* real variables, namely, the two independent variables x and y and the two dependent variables u and v. Hence a space of *four* dimensions is required if we are to plot $w = f(z)$ in the cartesian fashion. The difficulties in this approach are obvious, and in its place the following procedure (see Example 3, Sec. 17.4) is universally used for the geometric representation of functions of a complex variable z.

Let there be given two planes, one the z plane (in which the point $z = x + iy$ is to be plotted) and the other the w plane (in which the point $w = u + iv$ is to be plotted). A function $w = f(z)$ is now represented not by a locus of points in a space of four dimensions but by a correspondence between the points of these two cartesian planes. Whenever a point is given in the z plane, the function $w = f(z)$ determines one or more values of $u + iv$ and hence one or more points in the w plane. As z ranges over any configuration in the z plane, the corresponding point $w = u + iv$ describes some configuration in the w plane. The function $w = f(z)$ thus defines a *mapping* or a *transformation* of the z plane into the w plane and in turn is represented by this mapping.

EXAMPLE 1

THE MAPPING EFFECTED BY $w = z^2$

Discuss the way in which the z plane is mapped onto the w plane by the function $w = z^2$.

In this case we have

$$w = u + iv = z^2 = (x + iy)^2 = x^2 - y^2 + 2ixy$$

and thus

$$u = x^2 - y^2 \qquad \text{and} \qquad v = 2xy \tag{1}$$

These are the equations of the transformation from the z plane to the w plane. From them, many features of the mapping can easily be inferred.

For instance, lines parallel to the y axis, i.e., lines with equations $x = c_1$, map into curves in the w plane whose parametric equations are, from (1),

$$u = c_1^2 - y^2 \qquad \text{and} \qquad v = 2c_1 y$$

Eliminating the parameter y, we obtain the equation

$$u = c_1^2 - \frac{v^2}{4c_1^2}$$

This defines a family of parabolas having the origin in the w plane as focus and the line $v = 0$ as axis, all opening to the left (Fig. 20.1). Similarly, lines parallel to the x axis, i.e., lines with equations $y = c_2$, map into curves in the w plane whose parametric equations are

$$u = x^2 - c_2^2 \qquad \text{and} \qquad v = 2c_2 x$$

Eliminating x we obtain

$$u = \frac{v^2}{4c_2^2} - c_2^2$$

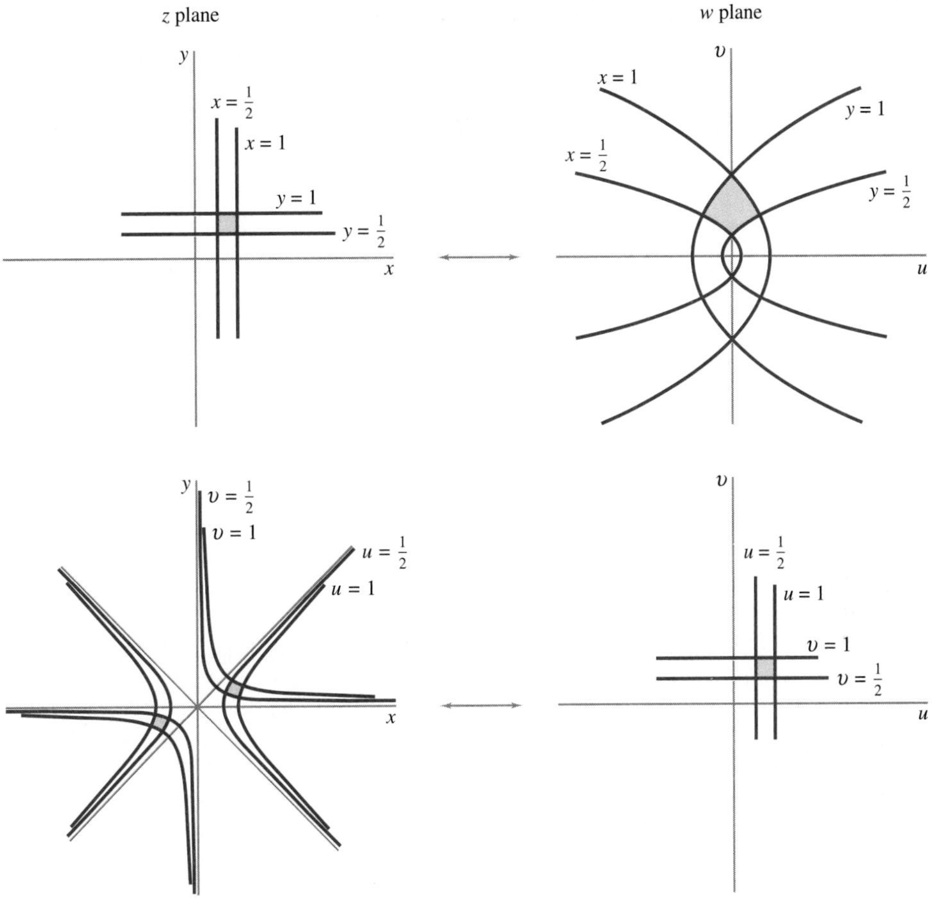

FIGURE 20.1
The mapping of certain lines by the function $w = z^2$.

which is also the equation of a family of parabolas having the origin as focus and the line $v = 0$ as axis, but this time all opening to the right.

Mapping from the w plane back onto the z plane is even more immediate. From (1), it is clear that the lines $u = k_1$ ($k_1 \neq 0$) correspond to the rectangular hyperbolas

$$x^2 - y^2 = k_1$$

and the lines $v = k_2$ ($k_2 \neq 0$) correspond to the rectangular hyperbolas

$$xy = \tfrac{1}{2}k_2$$

The images of other curves, or regions, can be found in the same fashion, with varying degrees of difficulty. For instance, to find the curve into which the line

$$y = 2x + 1$$

is transformed, we must eliminate x and y between this equation and the equations of the transformation (1). To do this, we first substitute for y in Eqs. (1), getting

$$u = x^2 - (2x + 1)^2 = -3x^2 - 4x - 1$$

$$v = 2x(2x + 1) = 4x^2 + 2x$$

Now if we regard these as two simultaneous equations in the quantities x and x^2, we can solve, getting

$$x = \frac{4u + 3v + 4}{-10} \qquad \text{and} \qquad x^2 = \frac{u + 2v + 1}{5}$$

Hence

$$\frac{u + 2v + 1}{5} = \left(\frac{4u + 3v + 4}{-10}\right)^2$$

or, collecting terms,

$$16u^2 + 24uv + 9v^2 + 12u - 16v = 4$$

which is the equation of a parabola.

Although w is a single-valued function of z, the converse is not true. In fact, when w is given, z may be either of the two square roots of w. Because of this, the mapping from the z plane to the w plane covers the latter twice, as Fig. 20.2 suggests. This of course is nothing but a graphic representation of the now familiar fact that the angles of complex numbers are doubled when the numbers are squared.

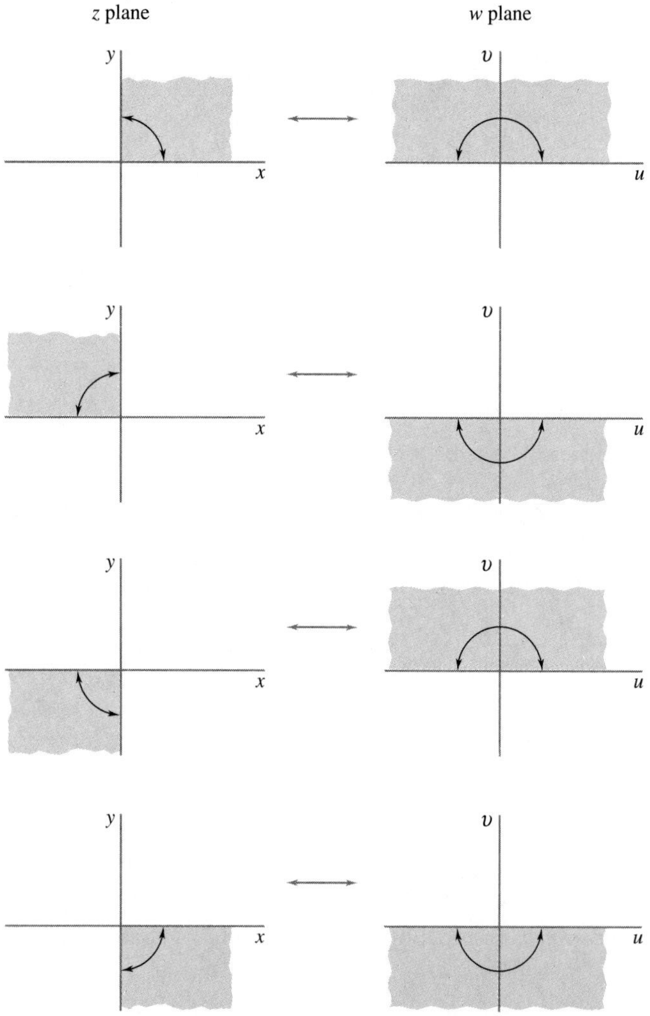

FIGURE 20.2
The two-valued character of the mapping defined by $z = w^{1/2}$.

EXERCISES

1. What is the image of the circle $x^2 + y^2 = a^2$ under the mapping $w = z^2$? What is the image of the upper half of the circle? *Hint:* Note that on the given circle $z = ae^{i\theta}$.
2. Work Exercise 1 if the mapping is defined by $w = z^3$.
3. Discuss the mapping between the z plane and the w plane defined by the function $w = \bar{z}$.
4. Discuss the mapping between the z plane and the w plane defined by the function $w = \bar{z}^2$.
5. What relation, if any, exists between the mappings defined by $w = f(z)$ and $w = f(\bar{z})$?
6. Discuss the mapping defined by the function $w = 2iz + 1$. Show that under this mapping the image of every circle is another circle.
7. Discuss the mapping defined by the function $w = 1/z$. What are the images of the lines of the families $x = c$ and $y = k$? What are the preimages of the lines of the families $u = a$ and $v = b$?
8. Show that in general the mapping $w = z^2$ transforms a straight line into a parabola. Are there any exceptions?
9. **(a)** What are the equations of the mapping defined by the function $w = (2 + i)z - (1 + 3i)$? What are the equations of the mapping from the w plane onto the z plane?
   (b) Show that every circle in the z plane is mapped into a circle in the w plane.
10. Find the equations of the mapping defined by the function $w = (z - i)/z$ and show that every circle through the origin in the z plane is transformed into a straight line. *Hint:* Recall the complex form of the equation of a circle from Exercise 23, Sec. 3.1.
11. If $w = 1/(z + 1)$, what are the equations of the mapping from the z plane onto the w plane? What are the equations of the mapping from the w plane onto the z plane?
12. What are the equations of the mapping defined by the function $w = z^3$? What is the image of the circle $x^2 + y^2 = a^2$? What is the image of the line $x = 1$? Plot the preimage of the line $u = 1$.
13. Show that in the mapping defined by the function $w = (x^2 - y^2) + kixy$ the preimages of the lines $u = a$ and $v = b$ are always perpendicular. On the other hand, show that the images of the lines $x = c$ and $y = d$ are perpendicular if and only if $k = 2$. Is there any significance in the value $k = 2$?
14. For what values of a, b, and c, if any, does the function

$$w = ax + iby + c \qquad a, b, c \text{ real}$$

transform perpendicular lines in the z plane into perpendicular lines in the w plane?

15. If $w = z^2$, what are the equations of the mapping from the w plane onto the z plane?
16. If $x = f(t)$ and $y = g(t)$ are the parametric equations of a smooth curve in the z plane, show that $z = f(w) + ig(w)$ maps the curve onto the real axis in the w plane.
17. **(a)** If $w = e^z$, what are the equations of the mapping of the z plane onto the w plane? What are the equations of the mapping of the w plane onto the z plane? How are these related to the inverse of the function $w = e^z$?
    (b) What are the images of the lines $x = c$? What are the images of the lines $y = k$?
    (c) Show that the strip $-\pi < y \leq \pi$ is mapped onto the entire w plane. What is the preimage of the line $v = 0$? What are the images of the positive halves of the strips $-\pi < y \leq -\pi/2$, $-\pi/2 < y \leq 0$, $0 < y \leq \pi/2$, $\pi/2 < y \leq \pi$? What are the images of the left halves of these strips? What other regions in the z plane will map onto the entire w plane?
    (d) What is the image in the fundamental strip $-\pi < y \leq \pi$ of the line $u = c$? Of the line $v = k$?
18. Discuss the mapping defined by the function $w = \sin z$.
19. If $w = i(1 - z)/(1 + z)$, determine whether the imaginary part of $w = u + iv$ is negative, positive, or zero when $|z| < 1$.
20. What is the image of the infinite strip $x > 0$, $0 < y < 1$ under the mapping $w = i/z$?

20.2 CONFORMAL MAPPING

In the last section we saw that various functions of a complex variable map the xy plane onto the uv plane. We now propose to investigate in more general terms the character of transformations for which the mapping function $w = f(z) = u(x, y) + iv(x, y)$ is analytic.

At the outset it is important to know when the equations of the transformation can be solved (at least theoretically) for x and y as single-valued functions of u and v, that is, when the transformation has a single-valued inverse. The condition for this, as established in most texts on advanced calculus,† is simply that the **jacobian determinant** of the transformation

$$J\left(\frac{u, v}{x, y}\right) = \begin{vmatrix} \partial u/\partial x & \partial u/\partial y \\ \partial v/\partial x & \partial v/\partial y \end{vmatrix}$$

†See, for instance, R. C. Buck, *Advanced Calculus*, 3d ed., pp. 362–366, McGraw-Hill, New York, 1978.

be different from zero. Since $w = f(z)$ is here assumed to be analytic, u and v must satisfy the Cauchy-Riemann equations (Eqs. $5a$ and $5b$, Sec. 17.3). Hence, substituting into the jacobian determinant, we have

$$J\left(\frac{u, v}{x, y}\right) = \begin{vmatrix} \partial u/\partial x & -\partial v/\partial x \\ \partial v/\partial x & \partial u/\partial x \end{vmatrix} = \left(\frac{\partial u}{\partial x}\right)^2 + \left(\frac{\partial v}{\partial x}\right)^2 = \left|\frac{\partial u}{\partial x} + i\frac{\partial v}{\partial x}\right|^2 = |f'(z)|^2$$

which establishes the following result.

THEOREM 1 If $f(z)$ is analytic, the function $w = f(z)$ will have a single-valued inverse in the neighborhood of any point where the derivative of $f(z)$ is different from zero.

To describe the exceptional case when $f'(z) = 0$, we have the following definition.

DEFINTION 1 A point where $f'(z) = 0$ is called a **critical point** of the transformation $w = f(z)$.

To investigate the difference between the behavior of the transformation $w = f(z)$ around a critical point and around an **ordinary point** where $f'(z) \neq 0$, consider a value of z and its image $w = f(z)$, where $f(z)$ is analytic, and let

$$\Delta z = |\Delta z| e^{i\theta} \qquad \text{and} \qquad \Delta w = |\Delta w| e^{i\phi}$$

be corresponding increments of these quantities (Fig. 20.3). Then,

$$f'(z) = \lim_{\Delta z \to 0} \frac{\Delta w}{\Delta z} = \lim_{\Delta z \to 0} \frac{|\Delta w| e^{i\phi}}{|\Delta z| e^{i\theta}} = \lim_{\Delta z \to 0} \left|\frac{\Delta w}{\Delta z}\right| e^{i(\phi - \theta)}$$

From this it is apparent that

$$\lim_{\Delta z \to 0} \left|\frac{\Delta w}{\Delta z}\right| = |f'(z)| \qquad \text{and} \qquad \lim_{\Delta z \to 0} (\phi - \theta) = \arg f'(z)$$

or, to an arbitrary degree of approximation,

(1) $$|\Delta w| = |f'(z)| \cdot |\Delta z|$$

and

$$\phi = \theta + \arg f'(z)$$

or

(2) $$\arg \Delta w = \arg \Delta z + \arg f'(z)$$

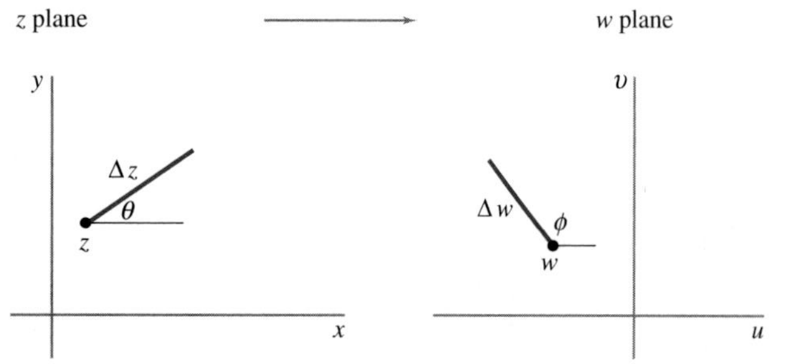

FIGURE 20.3
Δz and its image Δw under a mapping $w = f(z)$.

The fact that $f'(z)$ exists [which of course it does since $f(z)$ is assumed to be analytic] means that both $|f'(z)|$ and $\arg f'(z)$ are independent of the manner in which $\Delta z \to 0$. In other words, they depend solely on z and not on the limiting orientation of the increment Δz. Hence from (1) we draw the following conclusion.

THEOREM 2 In the mapping defined by an analytic function $w = f(z)$, the lengths of infinitesimal segments, regardless of their direction, are altered by a factor $|f'(z)|$ which depends only on the point from which they are drawn.

From Theorem 2 and the intrinsic relation between length and area, we have the following corollary.

COROLLARY 1 Around any point z, infinitesimal areas are altered by the factor $|f'(z)|^2$, that is, by $J(u, v/x, y)$.

Similarly, we conclude from (2) that in general the difference between the angles of an infinitesimal segment and its image is independent of the direction of the segment and depends only on the point from which the segment is drawn. In particular, two infinitesimal segments forming an angle will both be rotated in the same direction by the same amount, namely, $\arg f'(z)$. Hence the measure of the angle between them will be left invariant by the transformation.

However, at a critical point, where $f'(z) = 0$, $\arg f'(z)$ is undefined and we cannot assert that angles are preserved. To investigate this case, suppose that $f'(z)$ has an n-fold zero at the critical point $z = z_0$. Then $f'(z)$ must contain the factor $(z - z_0)^n$, and hence we can write

$$f'(z) = (n + 1)a(z - z_0)^n + (n + 2)b(z - z_0)^{n+1} + \cdots \qquad a \neq 0$$

where $a, b, \ldots,$ are complex coefficients of no concern to us and the factors $n + 1, n + 2, \ldots,$ have been inserted for convenience in integrating $f'(z)$ to obtain $f(z)$:

$$f(z) = f(z_0) + a(z - z_0)^{n+1} + b(z - z_0)^{n+2} + \cdots$$

If in this expression we transpose $f(z_0)$, set

$$z - z_0 = \Delta z \qquad \text{and} \qquad f(z) - f(z_0) = \Delta w$$

and divide by $a(\Delta z)^{n+1}$, we obtain

$$\frac{\Delta w}{a(\Delta z)^{n+1}} = 1 + \frac{b}{a}\Delta z + \cdots$$

As $\Delta z \to 0$, the right-hand side approaches 1. Therefore

$$\lim_{\Delta z \to 0} (\arg \Delta w) - \lim_{\Delta z \to 0} [\arg a(\Delta z)^{n+1}] = \arg 1 = 0$$

or, to an arbitrary degree of approximation,

$$\arg \Delta w = \arg a + (n + 1) \arg \Delta z$$

Now let Δz_1 and Δz_2 be two infinitesimal segments which make an angle θ with each other and let Δw_1 and Δw_2 be their images. From the last equation we have

$$\arg \Delta w_1 = \arg a + (n + 1) \arg \Delta z_1$$
$$\arg \Delta w_2 = \arg a + (n + 1) \arg \Delta z_2$$

Hence, subtracting, we have

$$\arg \Delta w_2 - \arg \Delta w_1 = (n + 1)(\arg \Delta z_2 - \arg \Delta z_1) = (n + 1)\theta$$

Thus we have established the following theorem.

THEOREM 3 In the mapping defined by an analytic function $w = f(z)$, angles are in general preserved in magnitude and in sense. The only exception to this occurs when the vertex of the angle is a critical point where $f'(z)$ has an n-fold zero, in which case the measure of the angle is altered by the factor $n + 1$.

Example 1 of Sec. 20.1 is an excellent illustration of the behavior described by Theorem 3. The mapping function $w = f(z) = z^2$ is everywhere analytic and, as Fig. 20.1 suggests, angle measures are in general preserved. However, the derivative $f'(z) = 2z$ has a simple zero at $z = 0$ and, as Fig. 20.2 indicates, angles with vertex at the origin in the z plane are not preserved by the given transformation but are doubled.

A transformation which preserves the magnitudes of angles is said to be **isogonal.** A transformation which preserves the sense as well as the magnitude of angles is said to be **conformal.** If $f(z)$ is an analytic function, it follows from Theorem 3 that in the neighborhood of any point where $f'(z) \neq 0$, the transformation defined by $w = f(z)$ is conformal. Conversely, it can be shown† that if the mapping

$$u = u(x, y) \qquad \text{and} \qquad v = v(x, y)$$

is conformal, and if the first partial derivatives of u and v are continuous, then $w = u + iv = f(z)$ is an analytic function. Because of the properties guaranteed by Theorems 2 and 3, it is clear that under a conformal transformation an infinitesimal configuration and its image *conform,* in the sense of being approximately similar. This is not true, however, for large configurations, which may bear little or no resemblance to their images.

EXAMPLE 1

A CHANGE OF AREA UNDER THE MAPPING $w = z^2$

What is the area of the region into which the unit square whose vertices are the points $z = 0, 1, 1 + i, i$ is transformed by the transformation $w = z^2$?

Let S denote the unit square in the z plane and let R denote the region in the w plane into which S is mapped by the transformation $w = z^2$. Then, as usual, the required area is simply

$$\iint_R dA$$

To evaluate this integral, we observe from Corollary 1, Theorem 2, that if the element of area in the z plane is chosen to be $dx\,dy$, then the corresponding element of area in the w plane is

$$dA = |f'(z)|^2\,dx\,dy = |2z|^2\,dx\,dy = 4(x^2 + y^2)\,dx\,dy$$

and we have

$$\iint_R dA = \iint_S |f'(z)|^2\,dx\,dy = 4\int_0^1\int_0^1 (x^2 + y^2)\,dx\,dy = \frac{8}{3}$$

†See, for instance, E. G. Phillips, *Functions of a Complex Variable,* pp. 35–36, Interscience, New York, 1945.

One important reason for studying conformal transformations is that solutions of Laplace's equation remain solutions of Laplace's equation when subjected to a conformal transformation. More precisely, we have the following theorem.

THEOREM 4 If $\phi(x, y)$ is a solution of the equation

$$\frac{\partial^2 \phi}{\partial x^2} + \frac{\partial^2 \phi}{\partial y^2} = 0$$

then when $\phi(x, y)$ is transformed into a function of u and v by a conformal transformation, it will satisfy the equation

$$\frac{\partial^2 \phi}{\partial u^2} + \frac{\partial^2 \phi}{\partial v^2} = 0$$

everywhere except possibly at the images of the points where the derivative of the mapping function is equal to zero.

PROOF Let $w = u(x, y) + iv(x, y)$ define a conformal transformation by means of which $\phi(x, y)$ is transformed into a function of u and v. Then,

$$\frac{\partial \phi}{\partial x} = \frac{\partial \phi}{\partial u}\frac{\partial u}{\partial x} + \frac{\partial \phi}{\partial v}\frac{\partial v}{\partial x} \quad \text{and} \quad \frac{\partial \phi}{\partial y} = \frac{\partial \phi}{\partial u}\frac{\partial u}{\partial y} + \frac{\partial \phi}{\partial v}\frac{\partial v}{\partial y}$$

A second differentiation of each of these yields the results

$$\frac{\partial^2 \phi}{\partial x^2} = \frac{\partial \phi}{\partial u}\frac{\partial^2 u}{\partial x^2} + \left(\frac{\partial^2 \phi}{\partial u^2}\frac{\partial u}{\partial x} + \frac{\partial^2 \phi}{\partial v\,\partial u}\frac{\partial v}{\partial x}\right)\frac{\partial u}{\partial x} + \frac{\partial \phi}{\partial v}\frac{\partial^2 v}{\partial x^2} + \left(\frac{\partial^2 \phi}{\partial u\,\partial v}\frac{\partial u}{\partial x} + \frac{\partial^2 \phi}{\partial v^2}\frac{\partial v}{\partial x}\right)\frac{\partial v}{\partial x}$$

$$\frac{\partial^2 \phi}{\partial y^2} = \frac{\partial \phi}{\partial u}\frac{\partial^2 u}{\partial y^2} + \left(\frac{\partial^2 \phi}{\partial u^2}\frac{\partial u}{\partial y} + \frac{\partial^2 \phi}{\partial v\partial u}\frac{\partial v}{\partial y}\right)\frac{\partial u}{\partial y} + \frac{\partial \phi}{\partial v}\frac{\partial^2 v}{\partial y^2} + \left(\frac{\partial^2 \phi}{\partial u\,\partial v}\frac{\partial u}{\partial y} + \frac{\partial^2 \phi}{\partial v^2}\frac{\partial v}{\partial y}\right)\frac{\partial v}{\partial y}$$

When these are added, we obtain

$$\begin{aligned}\frac{\partial^2 \phi}{\partial x^2} + \frac{\partial^2 \phi}{\partial y^2} &= \frac{\partial \phi}{\partial u}\left(\frac{\partial^2 u}{\partial x^2} + \frac{\partial^2 u}{\partial y^2}\right) + \frac{\partial^2 \phi}{\partial u^2}\left[\left(\frac{\partial u}{\partial x}\right)^2 + \left(\frac{\partial u}{\partial y}\right)^2\right] \\ &\quad + 2\frac{\partial^2 \phi}{\partial u\,\partial v}\left(\frac{\partial u}{\partial x}\frac{\partial v}{\partial x} + \frac{\partial u}{\partial y}\frac{\partial v}{\partial y}\right) + \frac{\partial \phi}{\partial v}\left(\frac{\partial^2 v}{\partial x^2} + \frac{\partial^2 v}{\partial y^2}\right) + \frac{\partial^2 \phi}{\partial v^2}\left[\left(\frac{\partial v}{\partial x}\right)^2 + \left(\frac{\partial v}{\partial y}\right)^2\right]\end{aligned}$$

Since $w = u + iv$ is analytic, by hypothesis, u and v themselves satisfy Laplace's equation. Hence the first and fourth groups of terms on the right vanish identically. Moreover, u and v also satisfy the Cauchy-Riemann equations; hence the third group of terms also vanishes identically. By using the Cauchy-Riemann equations again, what remains can be written

$$\begin{aligned}\frac{\partial^2 \phi}{\partial x^2} + \frac{\partial^2 \phi}{\partial y^2} &= \frac{\partial^2 \phi}{\partial u^2}\left[\left(\frac{\partial u}{\partial x}\right)^2 + \left(-\frac{\partial v}{\partial x}\right)^2\right] + \frac{\partial^2 \phi}{\partial v^2}\left[\left(\frac{\partial v}{\partial x}\right)^2 + \left(\frac{\partial u}{\partial x}\right)^2\right] \\ &= \left[\left(\frac{\partial u}{\partial x}\right)^2 + \left(\frac{\partial v}{\partial x}\right)^2\right]\left(\frac{\partial^2 \phi}{\partial u^2} + \frac{\partial^2 \phi}{\partial v^2}\right) \\ &= |f'(z)|^2\left(\frac{\partial^2 \phi}{\partial u^2} + \frac{\partial^2 \phi}{\partial v^2}\right)\end{aligned}$$

Thus, at any point where the transformation is conformal, that is, where $f'(z) \neq 0$,

$$\frac{\partial^2 \phi}{\partial x^2} + \frac{\partial^2 \phi}{\partial y^2} = 0 \qquad \text{implies} \qquad \frac{\partial^2 \phi}{\partial u^2} + \frac{\partial^2 \phi}{\partial v^2} = 0$$

as asserted. ◗

Suppose now that it is required to solve Laplace's equation, subject to certain boundary conditions, within a region R. Unless R is of a very simple shape, a direct attack upon the problem will usually be exceedingly difficult. However, it may be possible to find a conformal transformation which will convert R into some simpler region R' such as a circle or a half-plane, in which Laplace's equation can be solved, subject of course to the transformed boundary conditions. If this is the case, the resulting solution, when carried back to R by the inverse transformation, will be the required solution of the original problem.

EXERCISES

1. In Example 1, sketch the region into which the unit square in the z plane is transformed and find its area by direct integration in the w plane.

2. Work Example 1 if the transformation is **(a)** $w = (a + ib)z + (c + id)$ and **(b)** $w = z^3$.

3. **(a)** What are the critical points of the transformation $w = 3z - z^3$?
(b) What is the locus of points at which the magnification of lengths is equal to 1?
(c) Wht is the locus of points at which infinitesimal segments are rotated through 45°?
(d) What is the locus of points at which infinitesimal segments are rotated through 90°?

4. Are there any points at which infinitesimal segments are left unchanged in length and direction **(a)** by the transformation $w = 2iz + z^2$? **(b)** By the transformation $w = z^2 + z^3$?

5. **(a)** What is the length of the curve into which the upper half of the circle $|z| = a$ is transformed by the function $w = 1/z$?
(b) What is the length of the arc into which $w = 1/z$ transforms the segment of $y = 1 - x$ which lies in the first quadrant?

6. If $u = 2x^2 + y^2$ and $v = y^2/x$, show that the curves $u =$ constant and $v =$ constant are orthogonal trajectories but that the transformation defined by $w = u + iv$ is not conformal. Give a specific illustration of the latter fact.

7. Using the equations obtained by differentiating the relations $u = u(x, y)$ and $v = v(x, y)$ with respect to u and then with respect to v, show that

$$x_u(u, v) = \frac{1}{J\left(\frac{u, v}{x, y}\right)} v_y(x, y), \quad x_v(u, v) = \frac{-1}{J\left(\frac{u, v}{x, y}\right)} u_y(x, y)$$

$$y_u(u, v) = \frac{-1}{J\left(\frac{u, v}{x, y}\right)} v_x(x, y), \quad y_v(u, v) = \frac{1}{J\left(\frac{u, v}{x, y}\right)} u_x(x, y)$$

8. Verify the relations given in Exercise 7 for the function $w = 2z + 3$.

9. Verify the relations given in Exercise 7 for the function $w = z^2$.

10. If $w = f(z)$ is analytic at $z = z_0$ and $f'(z_0) \neq 0$, then f has an inverse in some neighborhood of z_0. Show that this inverse is an analytic function. *Hint:* Let $z = g(w) = x(u, v) + iy(u, v)$ be the inverse and use the relations of Exercise 7 to show that $x(u, v)$ and $y(u, v)$ satisfy the Cauchy-Riemann equations.

11. In Exercise 10, show that the derivative of the inverse function is different from zero at z_0; that is, show that the inverse mapping is also conformal. *Hint:* Recall the relation $g[f(z)] = z$ and then apply the chain rule of differentiation.

12. Show that $J\left(\frac{u, v}{x, y}\right) = 1/J\left(\frac{x, y}{u, v}\right)$. *Hint:* Recall the rule for multiplying two determinants.

13. What is the most general transformation which will map an arbitrary square in the z plane onto a congruent square in the w plane? *Hint:* Note that the transformation must involve a linear function of z.

14. A region S of the z plane is mapped onto a region R of the w plane of area A by an analytic function on S: $w = f(z) = u + iv$. Use the formula of calculus for changing variables in double integrals

$$A = \iint_S \left| J\left(\frac{u, v}{x, y}\right) \right| dx\, dy$$

to verify that

$$A = \iint_S |f'(z)|^2\, dx\, dy$$

15. The function $w = f(z)$ maps $|z| < R_0$ conformally onto a domain of area A. Prove that $A \geq \pi|f'(0)|^2 R_0^2$. *Hint:* Start with the final formula of Exercise 14, change variables of integration to polar coordinates, expand $f(z)$ as a series $\sum_{n=0}^{\infty} a_n z^n$, and show that if $r < R_0$,

$$\frac{1}{2\pi}\int_0^{2\pi} |f(re^{i\theta})|^2\, d\theta = \sum_{n=0}^{\infty} |a_n|^2 r^{2n}$$

20.3 THE BILINEAR TRANSFORMATION

The simplest class of conformal transformations, and one of the most important, is the class of **bilinear** or **linear fractional** or **Möbius† transformations,** defined by the family of functions

$$(1) \qquad w = \frac{az + b}{cz + d} \qquad ad - bc \neq 0$$

The restriction $ad - bc \neq 0$ is necessary because, if $ad = bc$, then $a/c = b/d$ and the numerator and denominator of w are proportional. As a consequence, w is a constant, independent of z, and thus the entire z plane is mapped into the same point in the w plane, a trivial case in which we have no interest. (See also Exercise 8.)

Our next observation about bilinear transformations is this:

THEOREM 1 Every bilinear transformation can be obtained as a composition of the three special transformations

$$\textbf{(a)}\ w = z + \lambda \qquad \textbf{(b)}\ w = \mu z \qquad \textbf{(c)}\ w = \frac{1}{z}$$

PROOF If $c \neq 0$, the general bilinear transformation is equivalent to the following chain of special transformations:

$$w_1 = z + \frac{d}{c}$$

$$w_2 = cw_1 = cz + d$$

$$w_3 = \frac{1}{w_2} = \frac{1}{cz + d}$$

$$w_4 = \frac{bc - ad}{c} w_3 = \frac{bc - ad}{c(cz + d)}$$

$$w = w_4 + \frac{a}{c} = \frac{bc - ad}{c(cz + d)} + \frac{a}{c} = \frac{az + b}{cz + d}$$

If $c = 0$, it is clear from the restriction $ad - bc \neq 0$ that neither a nor d can be zero. Hence we can write

$$w_1 = z + \frac{b}{a}$$

$$w = \frac{a}{d} w_1 = \frac{a}{d}\left(z + \frac{b}{a}\right) = \frac{az + b}{d}$$

Thus we have shown that whether $c = 0$ or $c \neq 0$, any bilinear transformation can be expressed as a chain of transformations of the special forms (**a**), (**b**), and (**c**).

Each of the transformations (**a**), (**b**), (**c**) appearing in Theorem 1 has an interesting geometric interpretation. Moreover, if we agree to extend the idea of a circle to include straight lines, then each of the transformations (**a**), (**b**), (**c**) has the property that it transforms a circle into a circle. As

†Named for the German mathematician A. F. Möbius (1790–1868).

we shall soon see, it is this property that makes the bilinear transformation useful in solving two-dimensional boundary-value problems.

For the transformation (**a**), the image w of a general point z is found by adding a constant vector λ to z. In other words, the transformation is just a **translation** through a distance $|\lambda|$ in the direction of arg λ. Since a translation is a rigid motion, it is clear that any translation carries a circle into a circle.

The transformation (**b**) can be thought of as the result of rotating each z through an angle equal to arg μ and then multiplying its length by the factor $|\mu|$. Such a transformation is usually called a **rotation and expansion** if $|\mu| > 1$, and a **rotation and contraction** if $|\mu| < 1$. If $|\mu| = 1$, the transformation is of course a **pure rotation.** To verify that $w = \mu z$ also transforms circles into circles, it is convenient to express the cartesian equation of a general circle

$$a(x^2 + y^2) + bx + cy + d = 0 \begin{cases} a, b, c, d \text{ real} \\ b^2 + c^2 \geq 4ad \end{cases}$$

in terms of z and $\bar{z}$ by means of the relations

$$x = \frac{z + \bar{z}}{2} \qquad y = \frac{z - \bar{z}}{2i} \qquad x^2 + y^2 = z\bar{z}$$

The result is

$$az\bar{z} + \frac{b - ic}{2} z + \frac{b + ic}{2} \bar{z} + d = 0$$

or, renaming the coefficients,

$$(A + \bar{A})z\bar{z} + Bz + \overline{Bz} + (D + \bar{D}) = 0 \tag{2}$$

where now A, B, C, and D are arbitrary complex numbers, subject to the condition $B\bar{B} \geq (A + \bar{A})(D + \bar{D})$, derived from the original condition $b^2 + c^2 \geq 4ad$, which ensures that the radius of the circle is real. If the substitution

$$z = \frac{w}{\mu} \qquad \mu \neq 0$$

is made in (2), we obtain the equation of the transformed curve

$$(A + \bar{A})\frac{w}{\mu}\frac{\bar{w}}{\bar{\mu}} + B\frac{w}{\mu} + \bar{B}\frac{\bar{w}}{\bar{\mu}} + (D + \bar{D}) = 0$$

or

$$(A + \bar{A})w\bar{w} + (B\bar{\mu})w + (\bar{B}\mu)\bar{w} + (D + \bar{D})\mu\bar{\mu} = 0 \tag{3}$$

Since the coefficients of the first and last terms in (3) are real and since the coefficients of the linear terms w and $\bar{w}$ are conjugates, this equation has the same structure as (2) and hence will also represent a circle provided its coefficients satisfy the condition necessary for the radius to be real. For the locus described by Eq. (3), this condition is

$$(B\bar{\mu})(\bar{B}\mu) \geq (A + \bar{A})(D + \bar{D})\mu\bar{\mu}$$

or, dividing by $\mu\bar{\mu}$, which is necessarily positive,

$$B\bar{B} \geq (A + \bar{A})(D + \bar{D})$$

which is true by hypothesis. If $a = 0$, so that $A + \overline{A} = 0$, both the given circle and its image are straight lines.

For the transformation (**c**) we can write

$$w = \frac{1}{z} = \frac{\overline{z}}{z\overline{z}} \tag{4}$$

which shows that w is of length $1/|z|$ and has the direction of $\overline{z}$. To describe the geometric process by which a point with these characteristics can be obtained from a given point z, we must first define the general process known as **inversion:** Let C be a circle with center O and radius r and let P be any point in the plane of C. Then the **inverse** of P with respect to C is the point P' on the ray OP for which

$$OP \cdot OP' = r^2 \tag{5}$$

From the symmetry of this relation it is clear that P is also the inverse of P'. Geometrically, a point and its inverse are related as follows. From any point P outside a circle C with center O let the two tangents to C be drawn, and let the points of contact of these tangents be joined (Fig. 20.4). The intersection of this chord with the line OP is the inverse P' of P. Conversely, let P' be any point in the interior of C. At P' erect a perpendicular to OP' and at either of the points at which this perpendicular meets C let the tangent to C be drawn. The intersection of this tangent and the line OP' is the inverse P of P'. The consistency of these constructions with the definitive property (5) is evident, since in Fig. 20.4 the triangles $OP'T_1$ and OT_1P are similar, i.e.,

$$\Delta OP'T_1 \sim \Delta OT_1P$$

and therefore

$$\frac{OP'}{OT_1} = \frac{OT_1}{OP}$$

or

$$OP \cdot OP' = (OT_1)^2 = r^2$$

It is evident now that for the transformation (**c**), the construction of w from z requires that the inverse of z in the *unit circle* be found and then reflected in the real axis; for the first of these steps gives a complex number whose length is $1/|z|$, and the second achieves the direction of $\overline{z}$, as required by (4).

To show that circles are also transformed into circles by the transformation (**c**), let the substitution $z = 1/w$ be made in the **self-conjugate form** of the equation of a circle (2). This gives

$$(A + \overline{A})\frac{1}{w}\frac{1}{\overline{w}} + \frac{B}{w} + \frac{\overline{B}}{\overline{w}} + (D + \overline{D}) = 0$$

or

$$(D + \overline{D})w\overline{w} + \overline{B}w + B\overline{w} + (A + \overline{A}) = 0$$

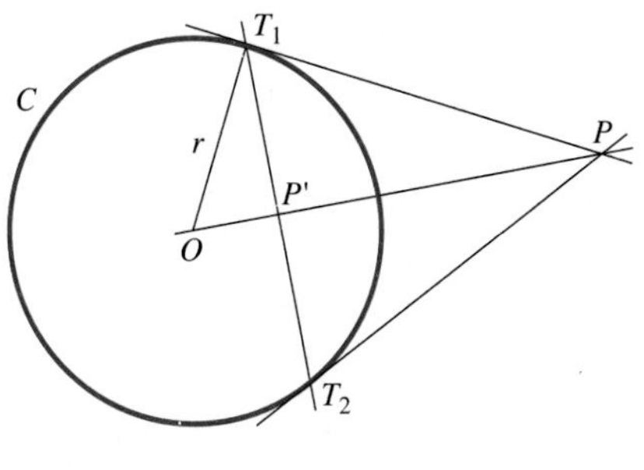

FIGURE 20.4
The geometric relation between a point and its inverse.

which is also the equation of a circle with real radius. If $A + \bar{A} = 0$, the original circle reduces to a straight line whose image is a circle passing through the origin since its equation contains no constant term. Conversely, any circle passing through the origin is transformed into a straight line.

Since, as we have just seen, each of the transformations (**a**), (**b**), and (**c**) transforms a circle into a circle and since, according to Theorem 1, every bilinear transformation is equivalent to a chain of these special transformations, it is clear that the assertion of the following theorem has now been established.

THEOREM 2 Under the general bilinear transformation, circles are transformed into circles.

The general bilinear transformation

$$w = \frac{az + b}{cz + d} \tag{1}$$

depends on three essential constants, namely, the ratios of any three of the coefficients a, b, c, d to the fourth. Hence it is evident that three conditions are necessary to determine a bilinear transformation. In particular, the requirement that three distinct values of z, say z_1, z_2, z_3, have specified distinct images w_1, w_2, w_3 leads to a unique transformation.

Although the transformation which sends three given points into three specified image points can be found by imposing these conditions on the general equation (1) and then solving the resulting three equations for the ratios of the coefficients, it is generally simpler to make use of the fact that if w_1, w_2, w_3, w_4 are, respectively, the images of z_1, z_2, z_3, z_4, then

$$\frac{(w_1 - w_2)(w_3 - w_4)}{(w_1 - w_4)(w_3 - w_2)} = \frac{(z_1 - z_2)(z_3 - z_4)}{(z_1 - z_4)(z_3 - z_2)}$$

To establish this relation, we observe that

$$w_i - w_j = \frac{az_i + b}{cz_i + d} - \frac{az_j + b}{cz_j + d} = \frac{(ad - bc)(z_i - z_j)}{(cz_i + d)(cz_j + d)}$$

Hence

$$\frac{(w_1 - w_2)(w_3 - w_4)}{(w_1 - w_4)(w_3 - w_2)} = \frac{\dfrac{(ad - bc)(z_1 - z_2)}{(cz_1 + d)(cz_2 + d)} \dfrac{(ad - bc)(z_3 - z_4)}{(cz_3 + d)(cz_4 + d)}}{\dfrac{(ad - bc)(z_1 - z_4)}{(cz_1 + d)(cz_4 + d)} \dfrac{(ad - bc)(z_3 - z_2)}{(cz_3 + d)(cz_2 + d)}}$$

$$= \frac{(z_1 - z_2)(z_3 - z_4)}{(z_1 - z_4)(z_3 - z_2)} \qquad ad - bc \neq 0$$

The last fraction is called the **cross ratio** or **anharmonic ratio** of the four numbers z_1, z_2, z_3, z_4. Hence the result we have just established can be formulated as the following theorem.

THEOREM 3 The cross ratio of four points is invariant under a bilinear transformation.

Suppose now that it is required to find the transformation which sends z_1, z_2, z_3 into w_1, w_2, w_3, respectively. If w is the image of a general point z under this transformation, then, according to Theorem 3, the cross ratio of w_1, w_2, w_3, and w must equal the cross ratio of z_1, z_2, z_3, and z. That is,

$$\frac{(w_1 - w_2)(w_3 - w)}{(w_1 - w)(w_3 - w_2)} = \frac{(z_1 - z_2)(z_3 - z)}{(z_1 - z)(z_3 - z_2)}$$

This equation is clearly bilinear in w and z and is satisfied by the three pairs of corresponding values (z_1, w_1), (z_2, w_2), (z_3, w_3). Moreover, everything in it is known except the variables w and z themselves. Hence it is necessary only to solve for w in terms of z to obtain the required transformation in standard form.

EXAMPLE 1

What is the bilinear transformation which sends the points $z = -1, 0, 1$ into the points $w = 0, i, 3i$, respectively?

Setting up the appropriate cross ratios, we have

$$\frac{(0 - i)(3i - w)}{(0 - w)(3i - i)} = \frac{(-1 - 0)(1 - z)}{(-1 - z)(1 - 0)}$$

or

$$\frac{3 + iw}{-2iw} = \frac{-1 + z}{-1 - z}$$

Finally, solving this equation for w, we obtain without difficulty

$$w = -3i\frac{z + 1}{z - 3}$$

EXAMPLE 2

What is the most general bilinear transformation which maps the upper half of the z plane onto the interior of the unit circle in the w plane (Fig. 20.5)?

Let the required transformation be

$$w = \frac{az + b}{cz + d}$$

Since the boundaries of corresponding regions must correspond under any transformation, the unit circle in the w plane must be the image of the real axis in the z plane. Therefore, for all *real* values of z, we must have

$$|w| = \frac{|az + b|}{|cz + d|} = \frac{|a|}{|c|}\frac{|z + b/a|}{|z + d/c|} = 1 \qquad a, c \neq 0$$

In particular, from the limiting case $|z| \to \infty$, we find

$$\frac{|a|}{|c|} = 1$$

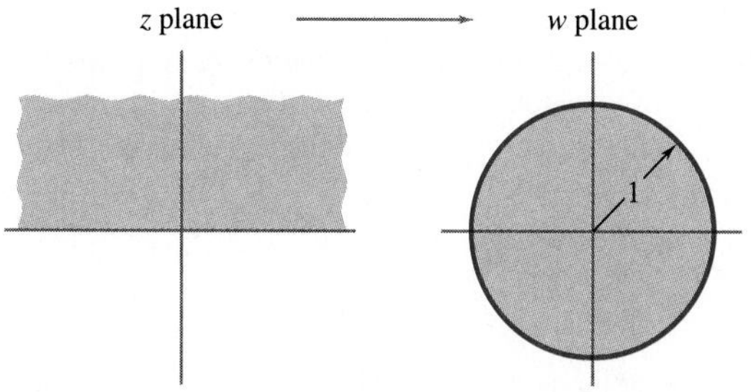

FIGURE 20.5
The upper half of the z plane to be mapped onto the interior of the unit circle in the w plane.

In other words, a/c is a complex number of absolute value 1, say $e^{i\theta}$. From this we conclude that for all real values of z,

$$\left|z + \frac{b}{a}\right| = \left|z + \frac{d}{c}\right| \quad \text{or} \quad \left|z - \left(-\frac{b}{a}\right)\right| = \left|z - \left(-\frac{d}{c}\right)\right|$$

The last equation expresses the fact that the complex numbers $-b/a$ and $-d/c$ are equally far from an arbitrary point z on the real axis, which is possible if and only if the real axis in the z plane is the perpendicular bisector of the segment joining the points $-b/a$ and $-d/c$. Thus $-b/a$ and $-d/c$ must be conjugate complex numbers, say λ and $\bar{\lambda}$, and hence we can write

$$w = \frac{az + b}{cz + d}$$

$$= \frac{a}{c} \cdot \frac{z + b/a}{z + d/c}$$

$$= \frac{a}{c} \cdot \frac{z - \lambda}{z - \bar{\lambda}}$$

$$= e^{i\theta} \frac{z - \lambda}{z - \bar{\lambda}} \tag{6}$$

So far we have enforced only the condition that the boundaries of the two regions correspond. It is now necessary to make sure that the regions themselves correspond as required and that the upper half of the z plane has not been mapped onto the *outside* of the unit circle in the w plane. This is most easily verified by determining the image of some convenient point, say $z = \lambda$. Clearly, the point $z = \lambda$ maps into the point $w = 0$, which is certainly inside the circle $|w| = 1$. Thus if λ is restricted to lie in the *upper* half of the z plane, the solution is complete.

As a special case of some interest, let $e^{i\theta} = -1$, and let λ be a pure imaginary, say i. Then

$$w = -\frac{z - i}{z + i} \tag{7}$$

Now

$$\text{Im } w = \frac{w - \bar{w}}{2i} = -\frac{1}{2i}\left(\frac{z - i}{z + i} - \frac{\bar{z} + i}{\bar{z} - i}\right)$$

or, reducing to a common denominator and simplifying,

$$\text{Im } w = \frac{z + \bar{z}}{(z + i)(\bar{z} - i)}$$

The denominator of the last fraction is the product of $z + i$ and its conjugate $\bar{z} - i$ and hence is a positive quantity. Thus the imaginary part of w will be positive if and only if $z + \bar{z}$ is positive. Since $z + \bar{z}$ is equal to twice the real part of z, this shows that the transformation (7) not only maps the upper half of the z plane onto the interior of the unit circle $|w| = 1$ but does it in such a way that the first quadrant of the z plane (where Re $z > 0$) corresponds to the upper half of the circle (where Im $z > 0$) and the second quadrant of the z plane corresponds to the lower half of the circle. In the opposite direction, the inverse transformation

$$z = -i\frac{w - 1}{w + 1} \tag{8}$$

maps the interior of the circle $|w| = 1$ onto the upper half of the z plane in such a way that the upper half of the circle maps onto the first quadrant of the z plane.

EXAMPLE 3

Find a transformation which will map an infinite sector of angle $\pi/4$ onto the interior of the unit circle.

Since the boundary of the sector consists of portions of *two* straight lines while its image is to be a *single* circle, it is apparent that the mapping cannot be accomplished by a bilinear transformation alone. However, a simple combination of a power function and a linear fractional transformation will define a suitable transformation. Specifically, the transformation

$$t = z^4$$

will open out the given sector in the z plane into the upper half of the auxiliary t plane (Fig. 20.6). Following this, the upper half of the t plane can be mapped onto the unit circle in the w plane by any transformation of the family (6) which we obtained in the last example, say

$$w = \frac{t - i}{t + i}$$

Taking the composition of these two mappings, we obtain for the required transformation

$$w = \frac{z^4 - i}{z^4 + i}$$

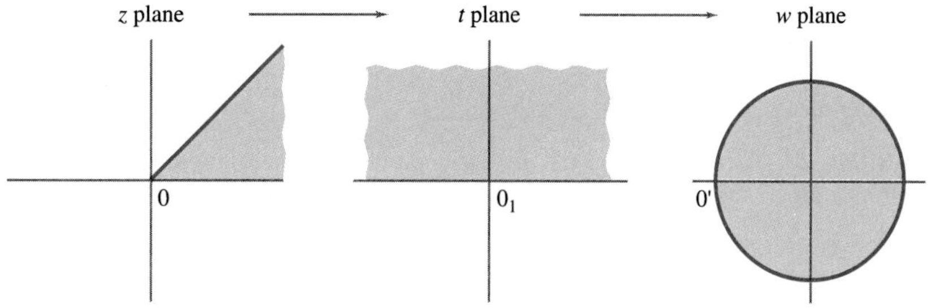

FIGURE 20.6
The two transformations needed to map an infinite sector onto the interior of the unit circle.

EXAMPLE 4

Find a transformation which will map a 60° sector of the unit circle in the z plane onto the upper half of the w plane.

At first glance it would seem that this problem could be solved simply by opening the given sector into a full circle by the transformation

$$t = z^6$$

and then mapping the circle from the t plane onto the upper half of the w plane by means of the inverse of one of the transformations of the family (6) which we obtained in Example 2, for instance the transformation (8). This method fails, however, because the circular region obtained in the t plane in this case is *not* of the type considered in Example 2. The latter consisted of a simple circular boundary plus its interior, whereas the former consists of the interior of a circle cut along a radius, since the radius $O'A' \equiv O'B'$ is actually the image of the two boundary radii OA and OB of the given sector (Fig. 20.7).

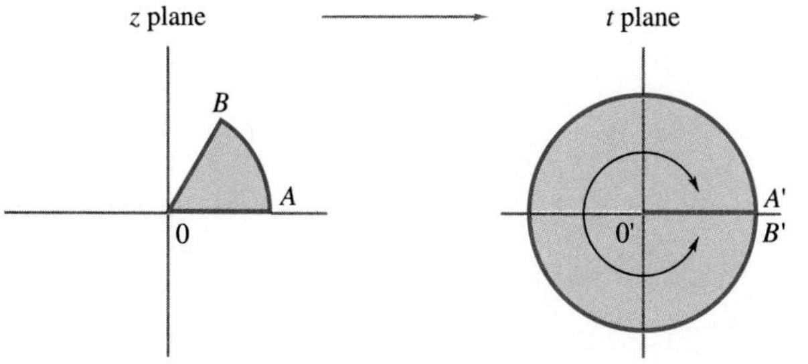

FIGURE 20.7
A circular sector opened out into a circular region cut along a radius.

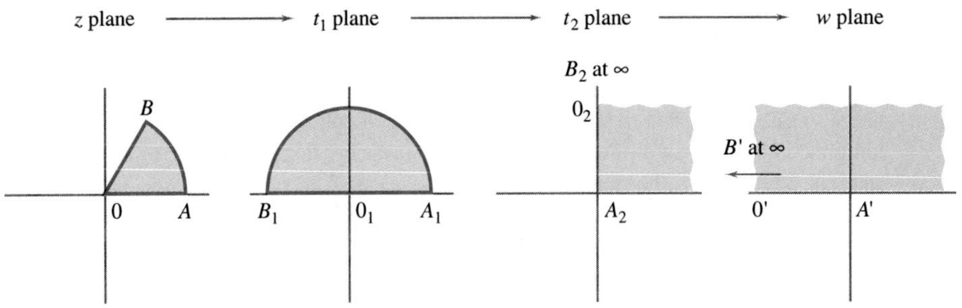

FIGURE 20.8
The sequence of transformations necessary to map a circular sector onto a half-plane.

To avoid this difficulty, let us first map the sector onto a semicircle by the transformation

$$t_1 = z^3$$

Then let us map the semicircle from the t_1 plane onto the first quadrant of the t_2 plane by means of the transformation (8)

$$t_2 = -i\frac{t_1 - 1}{t_1 + 1}$$

Finally (Fig. 20.8), let us open out the first quadrant of the t_2 plane into the upper half of the w plane by the transformation

$$w = t_2^2$$

Taking the composition of these three transformations, we find

$$w = -\left(\frac{z^3 - 1}{z^3 + 1}\right)^2$$

as the required mapping.

EXAMPLE 5

TEMPERATURES IN A QUARTER-PLANE METALLIC SHEET

A thin sheet of metal coincides with the first quadrant of the z plane. The upper and lower faces of the sheet are insulated so that heat flow in the sheet is strictly two-dimensional. Find the steady-state temperature at any point of the sheet if the boundary conditions are those shown in Fig. 20.9a.

FIGURE 20.9
An infinite 90° sector mapped, with its boundary conditions, onto a half-plane.

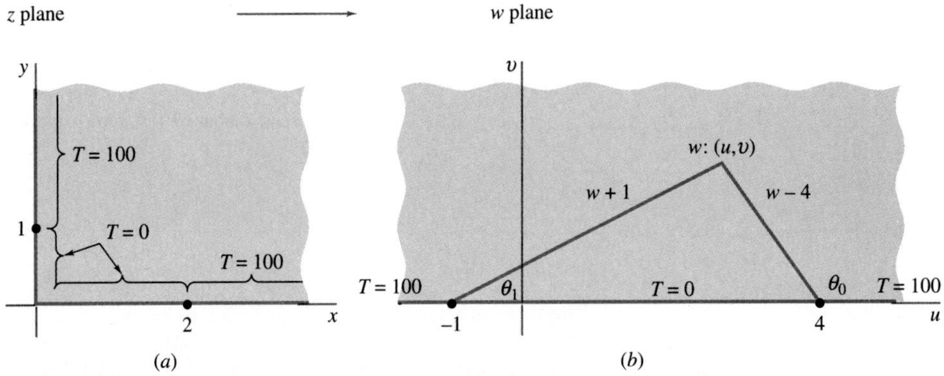

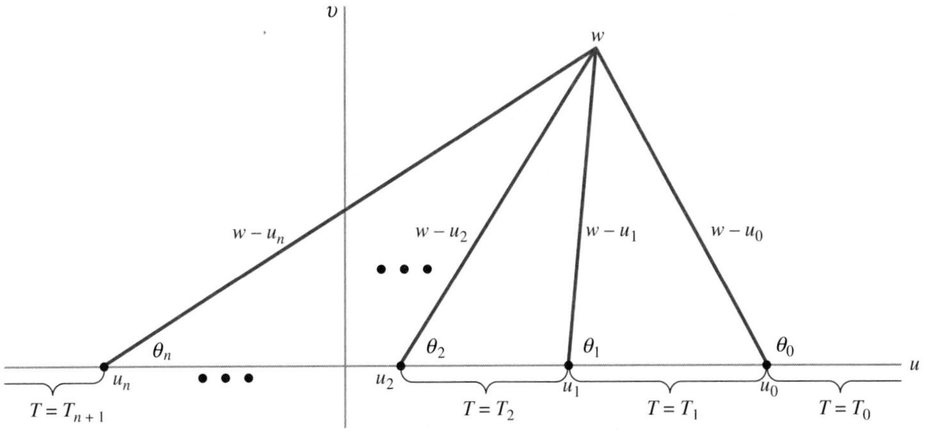

FIGURE 20.10
The behavior of arg $(w - u_i)$ as w varies along the real axis.

Under the assumptions of the problem, the flow of heat in the sheet is two-dimensional, and we must accordingly solve Laplace's equation, i.e., the two-dimensional steady-state heat equation derived in Sec. 11.2.

$$\frac{\partial^2 T}{\partial x^2} + \frac{\partial^2 T}{\partial y^2} = 0$$

subject to the given conditions along the boundaries of the first quadrant. To do this, it is convenient to map the first quadrant of the z plane onto the upper half of the w plane by the transformation

$$w = z^2 = (x^2 - y^2) + 2ixy$$

This reduces the problem to that of finding a solution of Laplace's equation in the upper half of the w plane which assumes along the real axis the transformed boundary conditions shown in Fig. 20.9*b*.

We have long since discovered that both the real part and the imaginary part of any analytic function satisfy Laplace's equation. Therefore if we can find a function of w which is analytic in the upper half-plane and whose real part or imaginary part takes on the relevant boundary values when w is real, we shall have the required solution. To obtain such a function to use in our problem, we observe first that, if $u_0 > u_1 > \cdots > u_n$,

$$(9)\quad f(w) = iT_0 + \frac{1}{\pi}[(T_1 - T_0)\ln(w - u_0) + (T_2 - T_1)\ln(w - u_1) + \cdots + (T_{n+1} - T_n)\ln(w - u_n)]$$

is analytic everywhere in the upper half-plane. Hence its imaginary part

$$(10)\quad \text{Im}[f(w)] = T_0 + \frac{1}{\pi}[(T_1 - T_0)\arg(w - u_0) + (T_2 - T_1)\arg(w - u_1) + \cdots + (T_{n+1} - T_n)\arg(w - u_n)]$$

will be a solution of Laplace's equation everywhere in the upper half-plane. Moreover, along the real axis, this solution takes on the boundary values shown in Fig. 20.10. To verify this, we observe from Fig. 20.10 that the complex number $w - u_i$ is represented by the vector joining the fixed point u_i to the variable point w, and thus arg $(w - u_i)$ is simply the inclination angle θ_i of this vector. Hence the function (10) can be rewritten

$$(11)\qquad \text{Im}[f(w)] = T(w) = T_0 + \frac{1}{\pi}[(T_1 - T_0)\theta_0 + (T_2 - T_1)\theta_1 + \cdots + (T_{n+1} - T_n)\theta_n]$$

Again referring to Fig. 20.10, it is clear that for all values of w on the real axis to the right of u_0, each of the θ's is zero. Hence from (11) we see that T reduces to the constant value T_0 along this portion of the real axis.

Furthermore, when w lies between u_1 and u_0, the angle θ_0 is equal to π while all the other θ's are still zero. Hence along this segment (11) reduces to

$$T = T_0 + \frac{1}{\pi}[(T_1 - T_0)\pi] = T_1$$

Similarly, for values of w between u_2 and u_1, the angles θ_1 and θ_0 are each equal to π, but all other θ's are equal to zero. Hence along this segment of the real axis, we have

$$T = T_0 + \frac{1}{\pi}[(T_1 - T_0)\pi + (T_2 - T_1)\pi] = T_2$$

Continuing in this fashion, we can verify that T, as defined by either (10) or (11), not only is a solution of Laplace's equation, being the imaginary part of the analytic function (9), but also assumes along the real axis the boundary values shown in Fig. 20.10.

When we specialize these observations to our problem, it appears that the solution we require for the temperature T in the w plane is

$$\begin{aligned} T &= 100 + \frac{1}{\pi}[(0 - 100)\theta_0 + (100 - 0)\theta_1] \\ &= 100 + \frac{100}{\pi}(\theta_1 - \theta_0) = \frac{100}{\pi}[\pi + (\theta_1 - \theta_0)] \end{aligned}$$

Multiplying by $\pi/100$ and then taking the tangent of both sides of the last equation, we have

$$\begin{aligned} \tan\frac{\pi T}{100} &= \tan\,[\pi + (\theta_1 - \theta_0)] = \tan\,(\theta_1 - \theta_0) \\ &= \frac{\tan\theta_1 - \tan\theta_0}{1 + \tan\theta_0 \tan\theta_1} \end{aligned}$$

Substituting for $\tan\theta_0$ and $\tan\theta_1$ their values, as read from Fig. 20.9b, we obtain from the last expression

$$\begin{aligned} \tan\frac{\pi T}{100} &= \frac{v/(u+1) - v/(u-4)}{1 + v^2/(u+1)(u-4)} \\ &= \frac{-5v}{u^2 + v^2 - 3u - 4} \qquad (12) \end{aligned}$$

which is the solution of the transformed problem in the w plane. Returning to the z plane by means of the mapping equations

$$u = x^2 - y^2 \qquad \text{and} \qquad v = 2xy$$

we thus find, from (12), that

$$T = \frac{100}{\pi}\tan^{-1}\frac{-10xy}{(x^2 + y^2)^2 - 3x^2 + 3y^2 - 4}$$

is the solution to the original problem given in the z plane.

EXERCISES

1. What is the cross ratio of the four points $P_1 = 0$, $P_2 = 1$, $P_3 = 1 + i$, $P_4 = i$? What is the cross ratio if the points are taken in the order P_2, P_1, P_3, P_4? What is the cross ratio if the points are taken in the order P_4, P_2, P_3, P_1? Do other orderings of the points yield other values of the cross ratio?

2. What is the cross ratio of the four fourth roots of -1 taken in the order of increasing arguments?
3. What is the cross ratio of the four complex sixth roots of 1 taken in the order of increasing arguments?
4. What is the inverse of the bilinear transformation $w = (az + b)/(cz + d)$?
5. What is the bilinear transformation which sends the points $z = 0, 1, i$ into the points $w = -\frac{1}{2}, 0, -1 + i$, respectively?
6. What is the bilinear transformation which sends the points $z = 0, i, -i$ into the points $w = i, 1, i/2$, respectively?
7. What is the bilinear transformation which sends the points $z = 0, 1, i$ into the points $w = i, (1 + i)/2, \infty$, respectively?
8. Show that if a transformation of the form $w = (az + b)/(cz + d)$ maps z_1 into w_1 and z_2 into w_1, then either $z_1 = z_2$ or $ad - bc = 0$.
9. Show that if $ad - bc \neq 0$, then the transformation $w = (az + b)/(cz + d)$ has no critical points and is therefore everywhere conformal.
10. Show that in general there are two points which are left invariant by a bilinear transformation, thought of as a mapping of the z plane onto itself. Are there any bilinear transformations which leave only one point invariant? No points invariant?
11. What are the invariant points of the transformation $z' = (z + i)/(-iz + 1)$?
12. What are the invariant points of the transformation $z' = (2z + i)/(iz + 4)$?
13. Find the invariant points of the transformation $z' = (5z - 2)/(z + 2)$ and prove that these two points, together with an arbitrary point z and its image z' form a set of four point whose cross ratio is independent of z.
14. What is the bilinear transformation which sends the points $z = 0, -1, \infty$ into the points $w = -1, -2 - i, i$, respectively? What is the image of the circle $|z| = 1$ under this transformation?
15. Under what conditions, if any, will the transformation $z' = (az + b)/(cz + d)$ have the property that if z' is the image of z, then z is the image of z'?
16. Find the equations of the transformation of inversion in the circle $x^2 + y^2 = 1$ and show that under this transformation a circle is mapped onto itself if and only if it is perpendicular to the circle defining the inversion.
17. What is the most general bilinear transformation which maps the upper half of the z plane onto the lower half of the w plane?
18. Prove that $w = z/(1 - z)$ maps the upper half of the z plane onto the upper half of the w plane. What is the image of the circle $|z| = 1$ under this transformation?
19. Find a transformation which will map an infinite sector of angle $\pi/3$ onto the interior of the unit circle.
20. Show that along the circle $|cz + d| = \sqrt{|ad - bc|}$ the transformation $w = (az + b)/(cz + d)$ does not alter the lengths of infinitesimal segments. What happens to segments inside this circle? Outside this circle? What is the locus of points where infinitesimal segments are not rotated?
21. Find a transformation which will map a 45° sector of the unit circle in the z plane onto the upper half of the w plane.
22. Find a transformation which will map the upper half of the unit circle onto the entire unit circle.
23. Show that if $|c| = |d|$, then the transformation $w = (az + b)/(cz + d)$ maps the unit circle in the z plane onto a straight line in the w plane.
24. Prove that four points z_1, z_2, z_3, z_4 lie on a circle if and only if their cross ratio is real. *Hint:* Take z_n in the form $z_n = z_0 + re^{i\theta_n}$ in proving the necessity of the condition.
25. Find the steady-state temperature distribution in a sheet of metal coinciding with the first quadrant of the z plane if $T = 100°C$ along the positive x axis and $T = 0°C$ along the positive y axis.
26. Find the steady-state temperature distribution in the upper half of the z plane if the portion of the x axis between $x = -1$ and $x = 1$ is maintained at the temperature $T = 100°C$ while the rest of the x axis is maintained at the temperature $T = 0°C$.
27. Work Exercise 26 if the portion of the x axis between $x = -1$ and $x = 1$ is maintained at the temperature $T = 50°C$ while the rest of the x axis is maintained at the temperature $T = 100°C$.
28. Find the steady-state temperature distribution in a sheet of metal coinciding with the interior of a 60° angle in standard position in the z plane if $T = 0°C$ along the horizontal side of the angle and 100°C along the other side.
29. Find the steady-state temperature distribution in a sheet of metal coinciding with the first quadrant of the z plane if $T = 100°C$ along the positive y axis, $T = 50°C$ between $x = 0$ and $x = 3$, and $T = 0°C$ to the right of $x = 3$ on the x axis.
30. Find the steady-state temperature distribution in a sheet of metal coinciding with the interior of the unit circle in the z plane if the upper half of the circumference of the circle is kept at the temperature $T = 100°C$ and the lower half of the circumference is kept at the temperature $T = 0°C$. *Hint:* Recall Formula (8).
31. Find the steady-state temperature distribution in a sheet of metal coinciding with the upper half of the unit circle in the z plane if the curved portion of the boundary is kept at the temperature $T = 100°C$ and the bounding diameter is kept at the temperature $T = 0°C$. *Hint:* Recall the results of Example 4.
32. Show that $w = z + 1/z$ maps the portion of the upper half of the z plane exterior to the circle $|z| = 1$ onto the entire upper half of the w plane. Use this result to find the steady-state temperature distribution in the upper half of the z plane exterior to the unit circle if $T = 100°C$ along the linear portion of the boundary and $T = 0°C$ along the circular portion of the boundary.
33. Find the steady-state temperature distribution in the portion of the unit circle which lies in the first quadrant of the z plane if the radial boundaries are maintained at the

temperature $T = 100°C$ while the curved boundary is maintained at the temperature $T = 0°C$.

34. Find a transformation which will map the interior of the unit circle in the z plane, with the radius from $z = 0$ to $z = 1$ regarded as part of the boundary, onto the upper half of the w plane. Explain how this result can be used to find the steady-state temperature distribution in the interior of a circle whose circumference is maintained at the temperature $T = 0°C$ and in which one radius is maintained at the temperature $T = 50°C$.

35. Find the steady-state temperature distribution in the portion of the z plane exterior to the parabola $y^2 = 4(1 - x)$ if the arc of the parabola above the x axis is maintained at the temperature $T = 100°C$ while the arc below the x axis is maintained at the temperature $T = 0°C$. Plot the 25°, 50°, and 75° isotherms. *Hint:* Represent the parabola parametrically by the equations $x = 1 - t^2$, $y = 2t$ and use the result of Exercise 16, Sec. 20.1.

36. **(a)** Find a bilinear transformation which will map the exterior of the unit circle in the z plane onto the upper half of the w plane.
(b) Use this result to find the steady-state temperature distribution in the portion of the z plane exterior to the unit circle if the upper half of the circumference is maintained at the temperature $T = 100°C$ while the lower half is maintained at the temperature $T = 0°C$. Plot the 25°, 50°, and 75° isotherms.

37. Work Exercise 36(**b**) if the arc of the circle in the third quadrant is maintained at the temperature $T = 0°C$ while the rest of the circumference is maintained at the temperature $T = 100°C$.

38. If z_1 and z_2 are invariant points of the transformation $t = (4z + 2i)/(iz + 1)$, where $|z_1| < |z_2|$, what is the bilinear transformation which maps $t(0)$, z_1, and z_2 into 0, 1, and 2, respectively?

39. Verify that under the transformation

$$w = k\frac{z + a}{z - a} \qquad a \text{ real}$$

every circle which passes through the two points $z = -a$, a is transformed into a straight line through the origin in the w plane. If C_1 is the circle of this family whose y intercept is p, and if k is real, show that the image of C_1 is the straight line through the origin whose inclination angle is $\phi = -2 \cot^{-1}(p/a)$. Finally, show that if C_2 is a circle intersecting C_1 at an angle α at $z = -a$ and at $z = a$, then the transformation

$$w = \exp\left(2i \cot^{-1}\frac{p}{a}\right)\left(\frac{z + a}{z - a}\right)$$

maps the crescent-shaped region between C_1 and C_2 onto the interior of an angle of measure α in standard position in the w plane.

20.4 THE SCHWARZ-CHRISTOFFEL TRANSFORMATION

In general, the conformal transformation of one given region onto another is exceedingly difficult. However, the *existence* of such a transformation is assured by the following theorem, due to Riemann.

THEOREM 1 Either of two bounded simply connected regions can be mapped conformably onto the other.

The determination of the specific function which accomplishes a required mapping, however, is usually out of the question. In fact, in addition to the simple regions which we found could be mapped by means of the elementary functions, the only class of regions for which conformal transformations of practical interest exist are those bounded by polygons having a finite number of vertices (one or more of which may lie at infinity). These can always be mapped onto a half-plane (and hence onto any region into which a half-plane can be transformed) by means of a transformation which we shall now discuss.

To see how this can be done, we first recall the mapping properties of the power function

$$w = z^m$$

Since this transformation has the property (Theorem 3, Sec. 20.2) that it alters by the factor m any angle with vertex at the origin, it follows that the transformation

$$(1) \qquad w - w_1 = (z - x_1)^{\alpha_1/\pi} \qquad \frac{dw}{dz} = \frac{\alpha_1}{\pi}(z - x_1)^{(\alpha_1/\pi)-1}$$

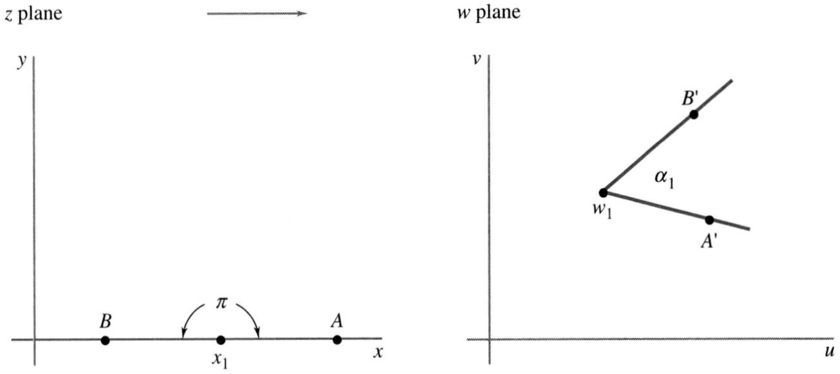

FIGURE 20.11
The effect of the transformation $w - w_1 = (z - z_1)^{\alpha_1/\pi}$.

will take the x axis, viewed as a straight angle with vertex at the point $z = x_1$, bend it at the point x_1, and fold it into an angle of measure

$$\left(\frac{\alpha_1}{\pi}\right)\pi = \alpha_1$$

with vertex at the point $w = w_1$ (Fig. 20.11). Clearly, if this could be done simultaneously for a number of points $x_1, x_2, \ldots, x_n$ on the x axis, the x axis would be mapped into a polygon whose angles were, respectively, $\alpha_1, \alpha_2, \ldots, \alpha_n$, and conversely; and the biggest step in the solution of our problem would be taken. This is actually possible, and the transformation which accomplishes it, suggested by the form of the derivative of the function in (1), is defined by

$$(2) \qquad \frac{dw}{dz} = K(z - x_1)^{(\alpha_1/\pi)-1}(z - x_2)^{(\alpha_2/\pi)-1} \cdots (z - x_n)^{(\alpha_n/\pi)-1}$$

To verify this, we begin with a point z on the x axis to the left of all the points $x_1, x_2, \ldots, x_n$ and investigate the locus of its image as it moves to the right along the x axis (Fig. 20.12*a*).† By taking arguments in (2), we obtain at once the relation

$$(3) \quad \arg dw = \arg K + \left(\frac{\alpha_1}{\pi} - 1\right)\arg(z - x_1) + \left(\frac{\alpha_2}{\pi} - 1\right)\arg(z - x_2) + \cdots$$
$$+ \left(\frac{\alpha_n}{\pi} - 1\right)\arg(z - x_n) + \arg dz$$

From this it is apparent that until z reaches x_1, every term on the right remains constant since $z - x_1, z - x_2, \ldots, z - x_n$ are all negative real numbers and hence have π for their respective arguments, and since dz is positive and therefore has 0 for its argument. Thus the image point w traces a straight line since the argument of the increment dw remains constant. However, as z passes

†By reasoning in this fashion, we are actually considering the inverse problem of mapping the upper half of the z plane onto a polygonal region in the w plane. As a consequence, in our applications, we will be mapping a polygonal region *given in the w plane* onto the upper half of the z plane. This interchange of the usual roles of the z and w planes should cause no confusion, however.

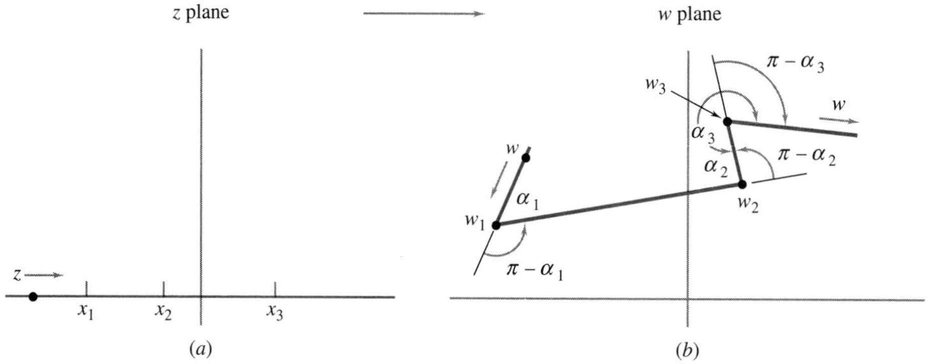

FIGURE 20.12
The mapping of the real axis in the z plane into a polygon with prescribed angles in the w plane.

through x_1, the difference $z - x_1$ changes abruptly from negative to positive, and therefore arg $(z - x_1)$ changes abruptly from π to 0. Hence, arg dw changes by the amount

$$\left(\frac{\alpha_1}{\pi} - 1\right)(-\pi) = \pi - \alpha_1$$

But, from Fig. 20.12*b* it is evident that this is the precise amount through which it is necessary to turn if w is to begin to move in the direction of the next side of the polygon. As z moves from x_1 to x_2, the same situation exists. The argument of dw remains constant, and thus w moves in a straight line until z reaches x_2. Here $z - x_2$ changes abruptly from negative to positive, its argument jumps immediately from π to 0, and, as a consequence, arg dw increases by the amount $\pi - \alpha_2$, which is just the amount of rotation required to give the direction of the next side of the polygon.

Thus as z traverses the x axis, it is clear that w moves along the boundary of a polygon whose interior angles are precisely the given angles $\alpha_1, \alpha_2, \ldots, \alpha_n$. Moreover, it is evident that the region which is mapped onto the half-plane is the region which contains these angles. The required transformation will be obtained if we can ensure that the lengths of the sides of the polygon, as well as its angles, have the correct values.

Now the mapping function w, obtained by integrating (2), is

(4) $$w = K \int [(z - x_1)^{(\alpha_1/\pi)-1}(z - x_2)^{(\alpha_2/\pi)-1} \cdots (z - x_n)^{(\alpha_n/\pi)-1}]\, dz + C$$

and this can be thought of as the composition of the two transformations

(5) $$t = \int [(z - x_1)^{(\alpha_1/\pi)-1}(z - x_2)^{(\alpha_2/\pi)-1} \cdots (z - x_n)^{(\alpha_n/\pi)-1}]\, dz$$

(6) $$w = Kt + C$$

The first of these transforms the x axis into some polygon which the second then translates, rotates, and either stretches or shrinks, as the case may be. If, then, the polygon determined by (5) is similar to the given polygon, the constants in (6) can always be chosen so as to make the two polygons coincide.

For two polygons to be similar, not only must corresponding angles be congruent but corresponding sides must also be proportional. For triangles this is automatically the case. For quadrilaterals one further condition is required, namely, that two pairs of corresponding sides have the same ratio. For pentagons, two such conditions are required, and, in general, for polygons of n sides, $n-3$ conditions, over and above the congruence of corresponding angles, are necessary to ensure similarity. Hence, in mapping a polygon of n sides onto a half-plane, three of the points $x_1, x_2, \ldots, x_n$, which are the images of the vertices of the polygon, can be assigned arbitrarily, following which the remaining $n-3$ image points are fixed by conditions of similarity. In many important problems, one vertex of the polygon, often an infinite one, will correspond to $z=\infty$. This of course accounts for one of the three assignments of image points which can be made arbitrarily. As a consequence, there will be one less finite image point in the expression (2) for dw/dz, and only *two* of the finite image points $x_1, x_2, \ldots, x_{n-1}$ can be specified arbitrarily. In either case, the resulting transformation is known as the **Schwarz-Christoffel transformation.**† Obviously, since w is analytic everywhere except possibly at the points $x_1, x_2, \ldots, x_n$, the transformation is conformal over the interior of the two regions. In practice, the usefulness of the Schwarz-Christoffel transformation is often limited by the complexity of the integral which defines the mapping function.

EXAMPLE 1

Find the transformation which maps the semi-infinite strip in the w plane shown in Fig. 20.13a onto the upper half of the z plane, as indicated in Fig. 20.13b.

Clearly, $\alpha_1 = \alpha_2 = \pi/2$, and therefore the required transformation is defined by

$$\frac{dw}{dz} = K(z+1)^{(\pi/2)/\pi-1}(z-1)^{(\pi/2)/\pi-1} = K(z+1)^{-1/2}(z-1)^{-1/2}$$

Hence

$$w = K\int \frac{dz}{(z^2-1)^{1/2}} = K\cosh^{-1} z + C$$

Since $w=0$ is to correspond to $z=1$ (that is, $O \to O'$, in Fig. 20.13), we have

$$0 = K\cosh^{-1} 1 + C \qquad \text{or} \qquad C = 0$$

Also, $w = i\pi$ (A in Fig. 20.13a) is to correspond to $z=-1$ (A' in Fig. 20.13b), and therefore

$$i\pi = K\cosh^{-1}(-1) = K(i\pi) \qquad \text{or} \qquad K = 1$$

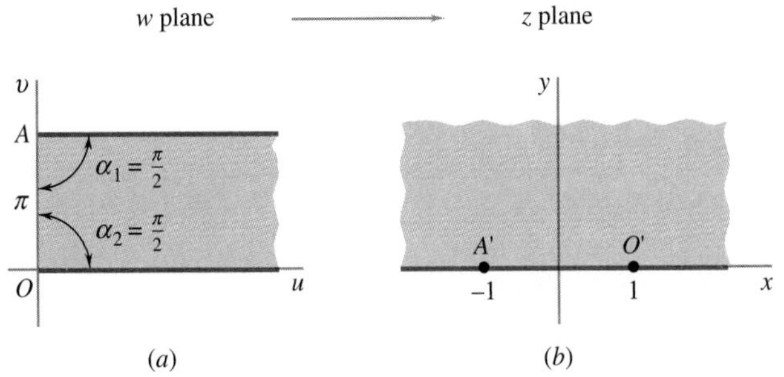

FIGURE 20.13
A semi-infinite strip mapped onto a half-plane.

†Named for the German mathematician H. A. Schwarz (1843–1921) and the Swiss mathematician E. B. Christoffel (1829–1900) who discovered it independently about 1865.

The required transformation is thus $w = \cosh^{-1} z$, or

$$z = \cosh w$$

Broken down into its real and imaginary parts, the equation of the transformation becomes [Eq. (17), Sec. 17.4]

$$x + iy = \cosh u \cos v + i \sinh u \sin v$$

or

$$x = \cosh u \cos v \qquad \text{and} \qquad y = \sinh u \sin v$$

Eliminating v and u in turn, we have further

$$\frac{x^2}{\cosh^2 u} + \frac{y^2}{\sinh^2 u} = 1 \qquad \text{and} \qquad \frac{x^2}{\cos^2 v} - \frac{y^2}{\sin^2 v} = 1$$

which, if necessary, can be solved for u and v in terms of x and y.

EXAMPLE 2

Find the transformation which maps the infinite region in the w plane shown in Fig. 20.14*a* onto the upper half of the z plane, as indicated.

With images assigned as shown and with the angle at the finite vertex A identified as $\alpha_1 = 2\pi$ and the angle at the infinite vertex B identified as $\alpha_2 = 0$, we have

$$\frac{dw}{dz} = K(z+1)^{(2\pi/\pi)-1} z^{(0/\pi)-1} = K\left(1 + \frac{1}{z}\right)$$

and

$$w = K(z + \ln z) + C \tag{7}$$

To determine the constants K and C, we write (7) in the form

$$u + iv = (K_1 + iK_2)(x + iy + \ln |z| + i \arg z) + C_1 + iC_2$$

from which, by equating imaginary parts, we obtain

$$v = K_1 y + K_2 x + K_2 \ln |z| + K_1 \arg z + C_2 \tag{8}$$

When w becomes infinite along AB, on which $v = \pi$, the image point z approaches zero along the negative real axis, on which $y = 0$ and $\arg z = \pi$. Hence, from (8),

$$\pi = \lim_{z \to 0^-} (K_1 \cdot 0 + K_2 x + K_2 \ln |z| + K_1 \pi + C_2)$$

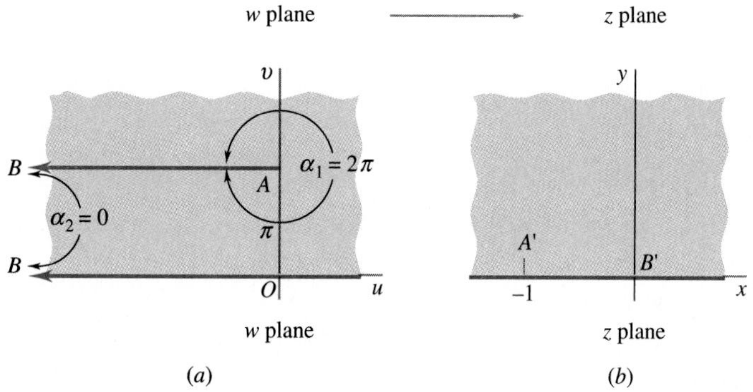

FIGURE 20.14
A semi-infinite channel to be mapped onto a half-plane.

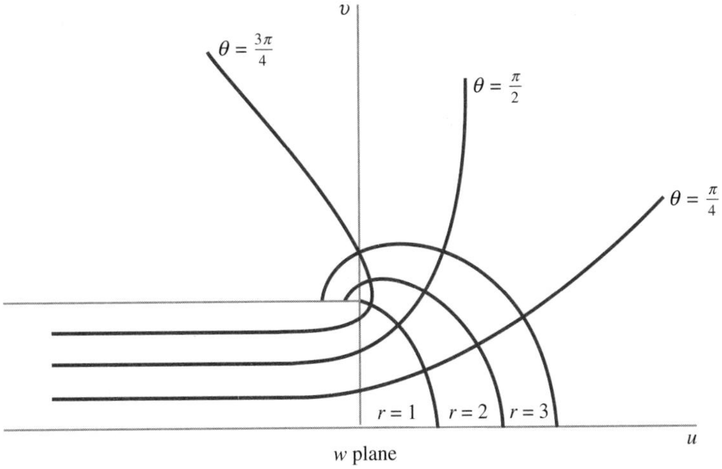

FIGURE 20.15
Typical streamlines for fluid flow from a long, straight channel.

Obviously K_2 must be zero to keep $\ln |z|$ from making the right-hand member infinite. Hence

$$\pi = K_1\pi + C_2 \tag{9}$$

Also, as w becomes infinite along OB, on which $v = 0$, the image point z approaches zero along the positive real axis, on which $y = 0$ and $\arg z = 0$. Hence, using (8) again, we have

$$0 = \lim_{z\to 0^+} (K_1 \cdot 0 + C_2) = C_2$$

Therefore, $C_2 = 0$, and so from (9) we find that $K_1 = 1$. Thus (7) reduces to

$$w = z + \ln z + C_1$$

Finally, the point A, that is, $w = i\pi$, must map into the point A', that is, $z = -1$. Hence from the last equation,

$$\begin{aligned} i\pi &= -1 + \ln(-1) + C_1 \\ &= -1 + i\pi + C_1 \end{aligned}$$

and so C_1 must equal 1. The required mapping function is therefore

$$w = z + \ln z + 1$$

Figure 20.15 shows the curves in the w plane which correspond to the semicircles $r = 1, 2, 3$ and the rays $\theta = \pi/4$, $\pi/2$, $3\pi/4$ in the z plane. This figure can be shown to represent either the lines of equal velocity potential and the streamlines for the flow of an ideal incompressible fluid from an infinite straight channel or the lines of flux and the equipotential lines for a parallel-plate capacitor.

Our final example is one which illustrates how the Schwarz-Christoffel transformation often leads to unfamiliar integrals which limit its usefulness until new functions are investigated and their properties are determined.

EXAMPLE 3

Find the transformation which maps the interior of the triangle in the w plane shown in Fig. 20.16a onto the upper half of the z plane as indicated.

With the given assignment of images, remembering that the vertex R is mapped into the point at infinity, we have

$$\frac{dw}{dz} = K(z-0)^{(\alpha/\pi)-1}(z-1)^{(\beta/\pi)-1}$$

or, factoring -1 from $(z-1)^{(\beta/\pi)-1}$ and setting $K_1 = K(-1)^{(\beta/\pi)-1}$

$$\frac{dw}{dz} = K_1 z^{(\alpha/\pi)-1}(1-z)^{(\beta/\pi)-1}$$

It is convenient in this case to introduce the dummy variable t and integrate between the limits of 0 and z. This gives us

$$w = K_1 \int_0^z t^{(\alpha/\pi)-1}(1-t)^{(\beta/\pi)-1}\,dt + C$$

Since $w = 0$ corresponds to $z = 0$, it follows that $C = 0$. Likewise, since $w = 1$ corresponds to $z = 1$, we have

$$1 = K_1 \int_0^1 t^{(\alpha/\pi)-1}(1-t)^{(\beta/\pi)-1}\,dt$$

The last integral is the beta function $B(\alpha/\pi, \beta/\pi)$. We encountered this integral in Exercise 18, Sec. 10.5, where we found its values to be

$$\frac{\Gamma(\alpha/\pi)\Gamma(\beta/\pi)}{\Gamma\left(\dfrac{\alpha+\beta}{\pi}\right)}$$

Thus $K_1 = 1/B(\alpha/\pi, \beta/\pi)$, and

$$w = \frac{1}{B(\alpha/\pi, \beta/\pi)} \int_0^z t^{(\alpha/\pi)-1}(1-t)^{(\beta/\pi)-1}\,dt$$

In the last expression z of course is complex and the integral is very difficult to evaluate. When z is a real number between 0 and 1, the integral is the so-called **incomplete beta function,** which is of considerable interest in statistics.

FIGURE 20.16
A triangle mapped onto a half-plane.

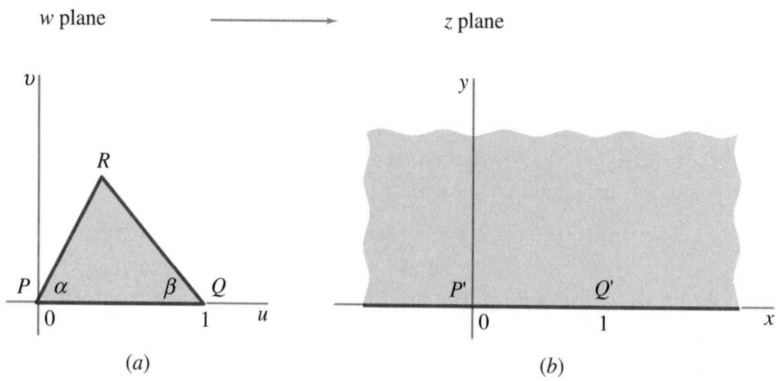

EXERCISES

1. Find the transformation which will map the region in the w plane shown in Fig. 20.17a onto the upper half of the z plane, as indicated.
2. Using the results of Exercise 1, find the steady-state temperature distribution in the w plane if the upper side of the negative u axis is kept at the temperature $T = 100°C$ and the lower side of the negative u axis is kept at the temperature $T = 0°C$.
3. Using the results of Exercise 1, find the steady-state temperature distribution in the w plane if the portion of the negative u axis between -1 and 0 is maintained at the temperature $T = 100°C$ while the rest of the negative u axis is maintained at the temperature $T = 0°C$.
4. Using the results of Example 1, find the steady-state temperature distribution in the semi-infinite strip in the w plane shown in Fig. 20.13a if the portions of the upper and lower boundaries to the right of the line $\text{Re}\, w = \cosh^{-1} 2$ are kept at the temperature $T = 0°C$ and the rest of the boundary is kept at the temperature $T = 100°C$.
5. Rework Example 1 using the mapping indicated in Fig. 20.18.
6. Rework Exercise 4 using the mapping obtained in Exercise 5.
7. Find the transformation which will map the doubly infinite strip in the w plane shown in Fig. 20.19a onto the upper half of the z plane, as indicated.
8. Using the results of Exercise 7, find the steady-state temperature distribution in the infinite strip shown in Fig. 20.19 if the lower boundary is kept at the temperature $T = 100°C$ and the upper boundary is kept at the temperature $T = 0°C$.
9. Work Exercise 8 if the negative half of the lower boundary and the positive half of the upper boundary are maintained at the temperature $T = 100°C$ while the rest of the boundary is maintained at the temperature $T = 0°C$.
10. Find the transformation which will map the exterior of the first quadrant in the w plane shown in Fig. 20.20a onto the upper half of the z plane, as indicated.
11. Using the results of Exercise 10, find the equations of the isotherms in the exterior of the first quadrant of the w plane if the positive half of the u axis is kept at the temperature $T = 100°C$ and the positive half of the v axis is kept at the temperature $T = 0°C$.
12. Find the transformation which will map the region in the w plane shown in Fig. 20.21a onto the upper half of the z plane, as indicated.

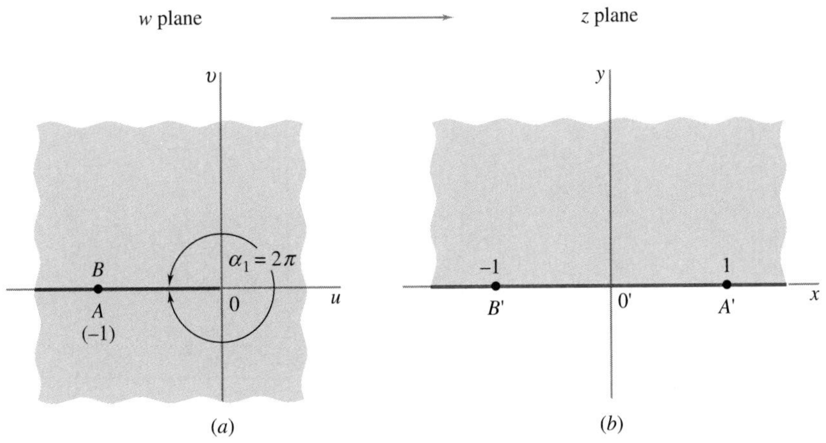

FIGURE 20.17

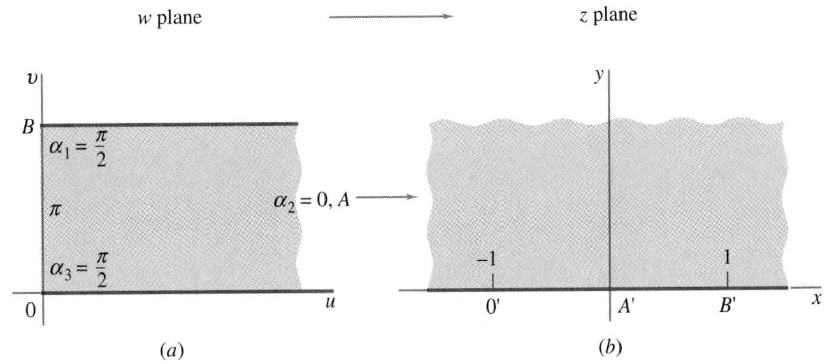

FIGURE 20.18

13. Using the results of Exercise 12, find the equations of the curves in the w plane which correspond to the lines $y = c$ in the z plane. (These are the streamlines of a perfect incompressible fluid flowing past an idealized vertical obstacle.)
14. Find the transformation which will map the region in the w plane shown in Fig. 20.22a onto the upper half of the z plane, as indicated.
15. Find the transformation which will map the region in the w plane shown in Fig. 20.23a onto the upper half of the z plane, as indicated.
16. Find the transformation which will map the region in the w plane shown in Fig. 20.24a onto the upper half of the z plane, as indicated.
17. Find the integral which defines the transformation which maps the rectangle in the w plane shown in Fig. 20.25a onto the upper half of the z plane, as indicated. Discuss, as far as possible, the determination of the constants appearing in the answer. (This integral is known as an *elliptic integral* of the first kind; see Exercise 48, Sec. 2.11).

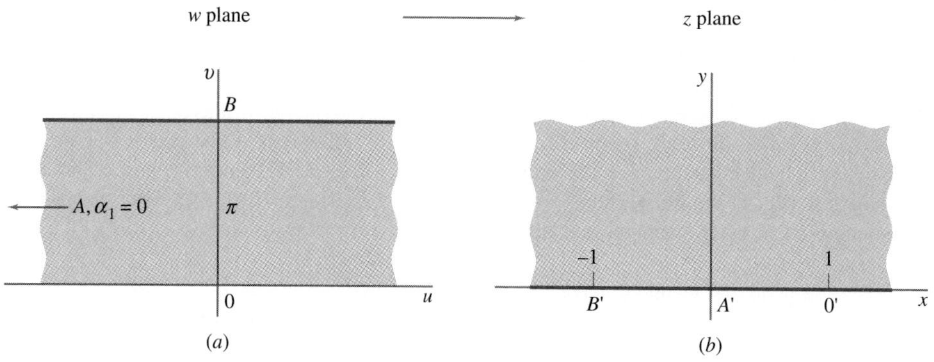

FIGURE 20.19

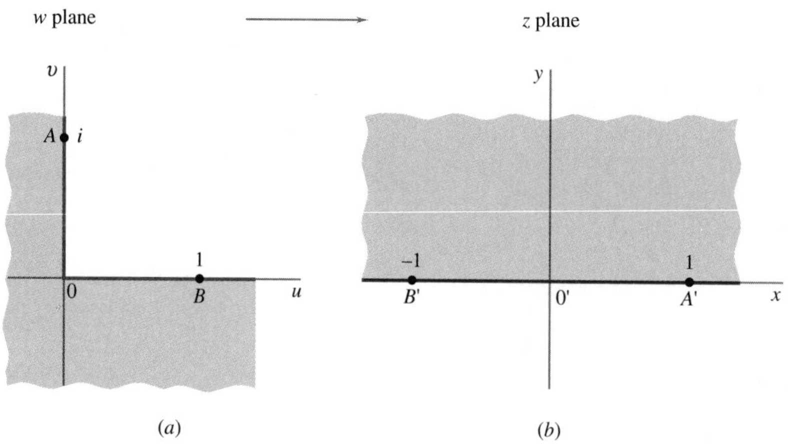

FIGURE 20.20

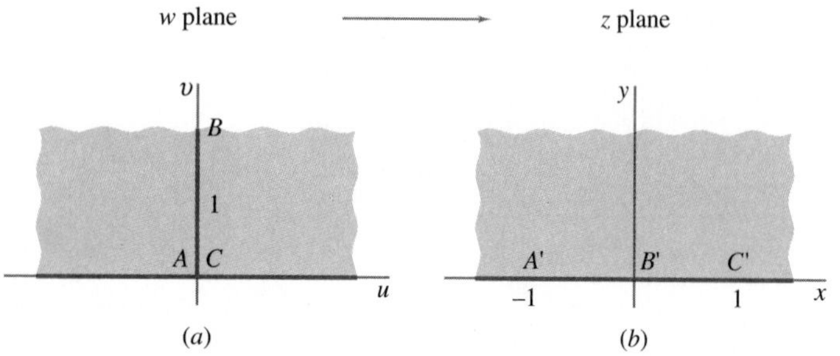

FIGURE 20.21

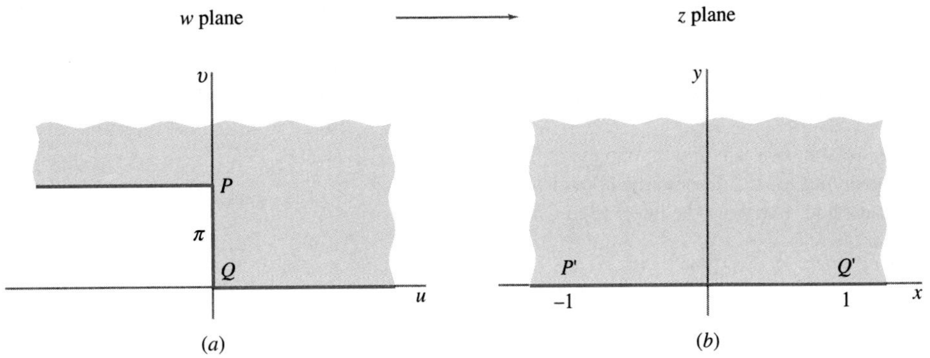

FIGURE 20.22

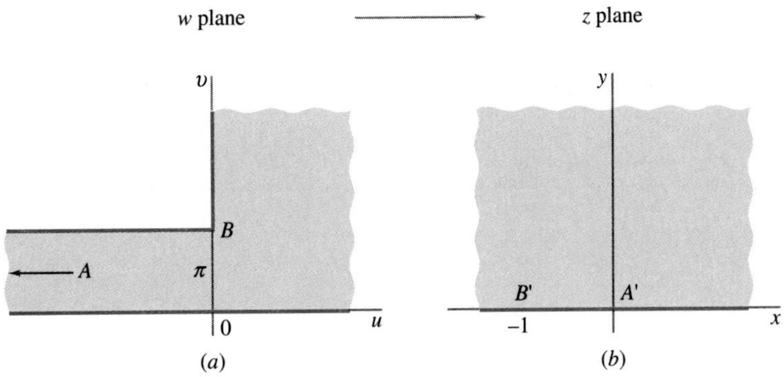

FIGURE 20.23

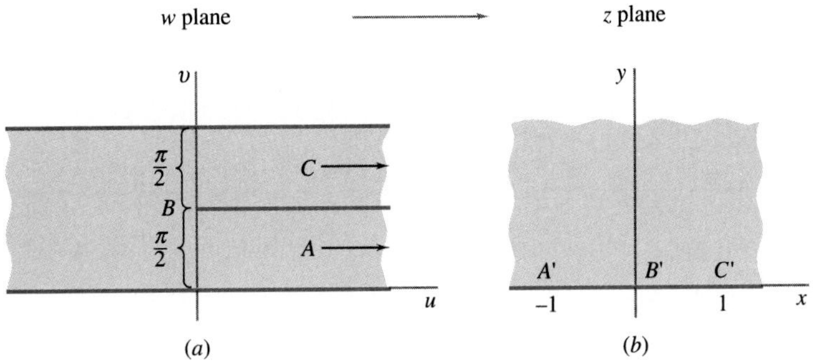

FIGURE 20.24

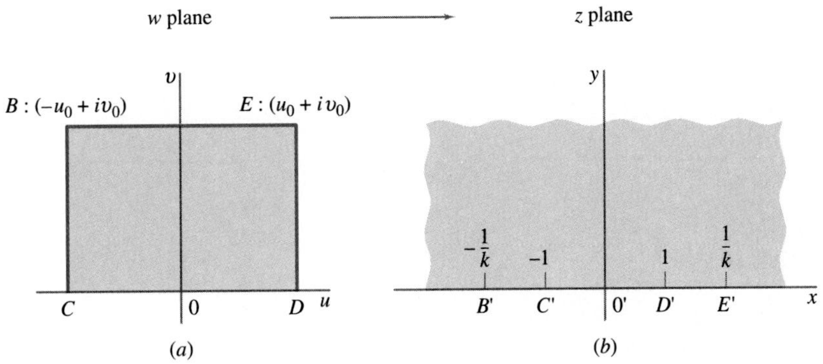

FIGURE 20.25

18. Find the transformation which will map the region in the w plane shown in Fig. 20.26a onto the upper half of the z plane, as indicated. Discuss the case where b approaches zero and verify that the transformation reduces to the one obtained in Exercise 12.

19. Find the transformation which will map the region in the w plane shown in Fig. 20.27a onto the upper half of the z plane, as indicated. Carry out the integration when $\alpha = 0$ and $\alpha = \pi/2$.

20. Find the transformation which will map the region in the w plane shown in Fig. 20.28a onto the upper half of the z plane, as indicated.

21. In the derivation of the Schwarz-Christoffel transformation, what is the direction of the path of w until z reaches x_1?

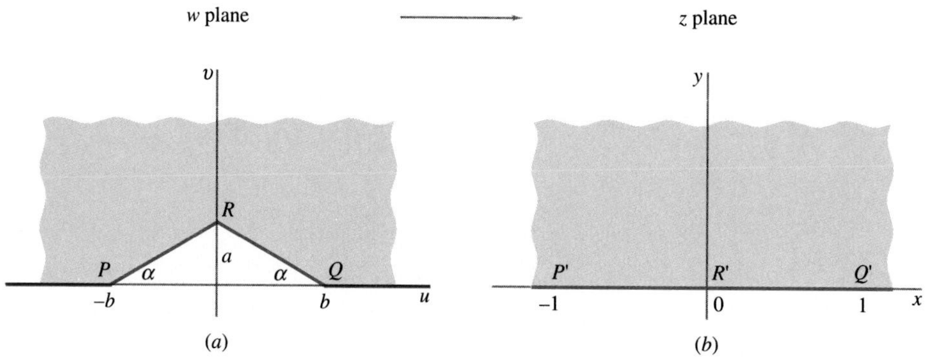

FIGURE 20.26

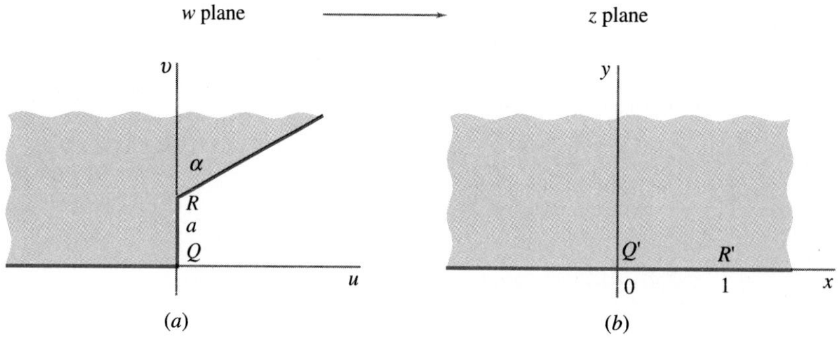

FIGURE 20.27

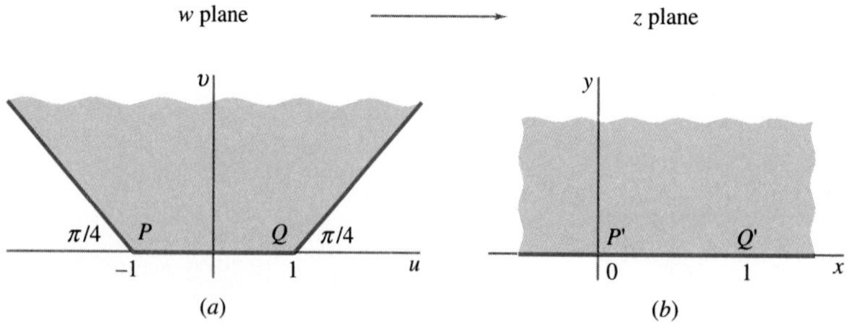

FIGURE 20.28

ANSWERS TO ODD-NUMBERED EXERCISES

CHAPTER 1

Sec. 1.1, page 6

1. **(a)** 0 **(b)** 1 **(c)** $2 + \ln 5$
3. **(a)** 0 **(b)** $1 + \pi/4$ **(c)** $1 - \pi/4$
5. **(a)** 2 **(b)** 0 **(c)** $\frac{5}{2}$
7. **(a)** 0 **(b)** 0 **(c)** $-\frac{4}{3}$
9. $2(1 - \cos^2 t) + t$, $2 \sin^2 t + t$, $1 - \cos 2t + t$
11. **(a)** The eleventh **(b)** 11 **(c)** 75 weeks
17. **(a)** π **(b)** 0 **(c)** 2 **(d)** 100
19. y dependent, x independent

Sec. 1.2, page 9

1. Second order, ordinary, linear
3. Third order, ordinary, linear
5. Second order, ordinary, linear
7. First order, ordinary, linear in x, linear in y, linear in the product xy; the equation is nonlinear.
9. Fourth order, partial, linear
11. Second order, partial, linear
13. Second order, partial, linear in u, nonlinear in v; the equation is nonlinear.
15. Nonlinear in x and in y
17. Linear in x and in y

Sec. 1.3, page 14

11. **(a)** $m = -\frac{1}{3}, \frac{1}{2}$ **(b)** $m = -2, 2, 3$
13. **(a)** $m = 1, 3$ **(b)** $m = 1, 2$; $y = x^{1/3}$ is a solution on every interval not containing $x = 0$.
15. **(a)** None; the function does not satisfy the equation. The function is a solution of **(b)**, **(c)**, and **(d)** on all subintervals of $(-\infty, \infty)$.
17. $y = c + x + \cos 2x$
19. $y = c_1 + c_2 x - \ln |\cos x|$
21. $y = x + \frac{1}{2}x^2 + \frac{3}{2}e^{-x^2} + c$
23. $y = c_1 + c_2 x + c_3 x^2 + \ln |x(\sec x + \tan x)|$
25. $y = (3xe^{3x} - e^{3x} + 1)/9$
27. $y = 2x \operatorname{Tan}^{-1} x - \ln (1 + x^2)$
29. $y = 2(\sin^3 x - 3 \sin x + 1)$
31. Be^x, where $B = Ae^k$
33. $C \ln |x|$, where $C = ab$
35. $P \sin x + Q \cos x$, where $P = A \cos b + C \cos d$ and $Q = A \sin b + C \sin d$
37. $P \cosh^2 \theta + Q \sinh^2 \theta$, where $P = a + c$, $Q = b + c$
39. $\dfrac{P}{x + 1} + \dfrac{Q}{x + 2}$, where $P = A + C$, $Q = B - C$

Sec. 1.4, page 17

1. Complete solution: $y = 3x^{4/3} + c$; $y = 3x^{4/3}$
3. General integral: $4x^2 + 9y^2 = c$, $c > 0$; $4x^2 + 9y^2 = 36$; $y = -\frac{2}{3}\sqrt{9 - x^2}$, $-3 < x < 3$
5. **(a)** $x^2 - y^2 = c$ **(b)** Standard equilateral hyperbolas and their asymptotes
(c)

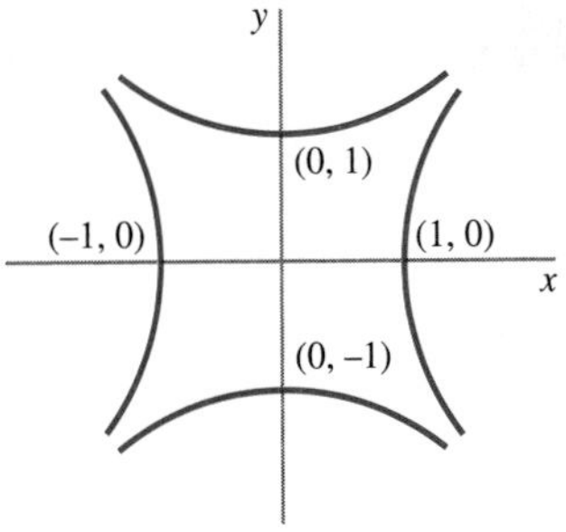

(c)

(d)

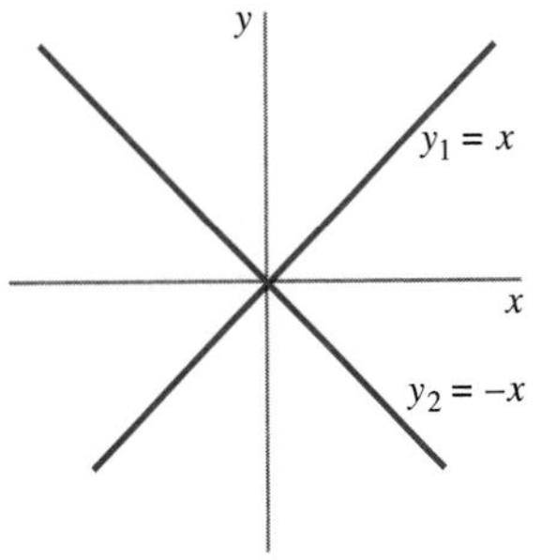

(d)

7. **(a)** $x = \sin y + c$
(b) $y = \operatorname{Sin}^{-1} x$

(c) $y = 2\pi + \text{Sin}^{-1} x$

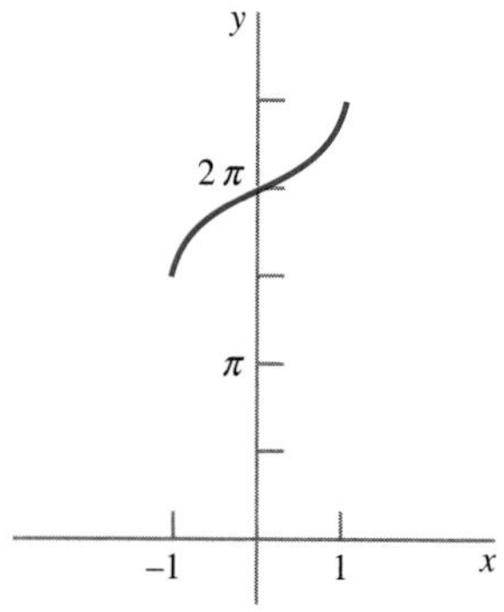

9. $y = \begin{cases} c_1x^2 - 1 & x \le 0 \\ c_2x^2 - 1 & x \ge 0 \end{cases}$

11. $y = \begin{cases} c_1(x^2 - 1)^2 & x \le -1 \\ c_2(x^2 - 1)^2 & -1 \le x \le 1 \\ c_3(x^2 - 1)^2 & x \ge 1 \end{cases}$

Sec. 1.5, page 19

1. $y'' = 0$
3. **(a)** $i\,dE = E\,di$ **(b)** $A\,dI = 2A\,dA$
5. $(\cos x + \sin x)y'' + (2 \sin x)y' - (\cos x - \sin x)y = 0$
7. $y'' + 4y' + 4y = 0$ **9.** $x^2y'' - 2xy' + 2y = 0$
11. $(y^2 - 1)y'' - y(y')^2 = 0$
13. $(x^2 - y^2)y'' - x(y')^3 + y(y')^2 + xy' - y = 0$
15. $(x^2 + 2x)y^{\text{iv}} - (x^2 - 6)y''' - (2x + 6)y'' = 0$
17. $xy''' + 3y'' = 0$ **19.** $3xy''' + 5y'' = 0$
21. $yy' - x = 0$ **23.** $2yy'' = (y')^2$
25. $[(2 + x) \cos x + (1 - x) \sin x]y''' + 2(x \cos x + 2 \sin x)y'' + [(3 + x) \sin x - (2 - x) \cos x]y' = 0$

Sec. 1.6, page 22

1. $y = 1 + e^{\sin x}$ on $(-\infty, \infty)$; unique
3. $y = 1 - \cos(\sin x)$ on $(-\infty, \infty)$; unique
5. $y = x \,\text{Sin}^{-1} x + (1 - x^2)^{1/2} + 4$ on $-1 \le x \le 1$; unique
7. $y = \tan x$ on $-\pi/2 < x < \pi/2$; unique
9. $y = x$, $y = -x$ on $(-\infty, \infty)$; not unique
11. $y = (x + 1)^2$ on $(-1, \infty)$; unique
13. No solution
15. $y = -1$ on $(-\infty, \infty)$; unique
17. Complete solution on $(-\infty, \infty)$: $y = \ln \cosh x + ax + b$
Particular solution on $(-\infty, \infty)$: $y = \ln \cosh x$
19. Complete solution on $|x| < 1$: $y = \sin^{-1} x + c_1x + c_2$
The initial-value problem has no solution.
21. Complete solution on $(-\infty, \infty)$: $y = x \sin x + c_1x + c_2$
Initial-value problem solution on $(-\infty, \infty)$: $y = x \sin x$
23. Complete solution on either of the disjoint intervals $(-\infty, 0)$ or $(0, \infty)$: $y = x^4 + x + c_1 \ln |x| + c_2$
Initial-value problem solution on $(-\infty, \infty)$: $y = x^4 + x$
25. $-2 < x < 2$ **27.** $\frac{3}{4} < x < \frac{5}{4}$
31. No, it is only a verification.

Sec. 1.7, page 27

1. $x^3 + xy^2 = c$ **3.** $x^2y + \frac{1}{4}x^4 + \frac{1}{3}y^3 = c$
5. $x + \sin x - y = c$ **7.** $xy^2 - y^3 - 2y = c$
9. $x^2y + xe^y = c$
11. $(x^2 + y^2)^{3/2} + 3xy = c$ **13.** $x^2 + x \,\text{Tan}^{-1} y = c$
15. $x \sin y + \cos xy = c$ **17.** $x^3 + y^2 + y \ln |x| = c$
19. $y = \frac{1}{2}(x + \sqrt{4 - x^2})$, $|x| \le 2$
21. $x^2y + xe^y = 2 + \ln 2$
23. $3x + 3x^2y + y^3 + 26 = 0$
25. $y = -x^3/\ln x^3$, $x > 0$
27. $x \tan y + y \,\text{Tan}^{-1} x - 2y \,\text{Tan}^{-1} y + \ln(1 + y^2) + \pi/2 = \ln 2$

Sec. 1.8, page 30

1. $\sqrt{x^2 + y^2} = e^y + c$; $u = 1/\sqrt{x^2 + y^2}$
3. $2xy + \ln(x^2 + y^2) = c$; $u = 2/(x^2 + y^2)$
5. $x^3y^4 + x^3 = c$; $u = x^2$
7. $y = 2 \,\text{Tan}^{-1}(y/x) + c$; $u = 1/(x^2 + y^2)$
9. $3 \ln(x^2 + y^2) + 2 \,\text{Tan}^{-1}(y/x) = c$; $u = 2/(x^2 + y^2)$
11. $2(x/y) + 3 \ln |y| = c$; $u = 1/y^2$
13. $x^2y^2 = x^2 - y^2 + c$; $u = 2xy$
15. $ye^{x/y} + x \sin xy = c$; $u = 1/y$
17. $xy + \ln |\cos(y/x)| = c$; $u = 1/x^2$
19. $e^x \sin^2 y + y \cos y - \sin y = c$; $u = \sin y$
21. $x^y = cy^x$; exact
23. $x \ln |xy| = c$; $u = 1/y$
25. $x^5y^4 - x^4y^3 \sin y = c$
31. **(a)** $\phi = c \exp\left[\int \left(\frac{\partial M}{\partial y} - \frac{\partial N}{\partial x}\right) \frac{dx}{N}\right]$
(b) $\phi = k \exp\left[\int \left(\frac{\partial N}{\partial x} - \frac{\partial M}{\partial y}\right) \frac{dy}{M}\right]$

Sec. 1.9, page 34

1. $\sin x \cos y = c$ **3.** $\sin^2 x = cy$
5. $y = cx^3$ **7.** $(x - 1) = c(x + 1)e^{y^2}$
9. $e^{x^2} + 2(y + 1)e^{-y} = c$ **11.** $y = c/(1 + 3c^3 \sin x)^{1/3}$
13. $x(y - 1) = c(y - 2)$ **15.** $x^2e^{\tan y} \cos y = c$
17. $\cos x = cy$, $x \ne n\pi$ **19.** $y^2 = cxe^x$
21. $y = \dfrac{(x + 3)^3 + 2c(x + 1)^3}{(x + 3)^3 - c(x + 1)^3}$ **23.** $y = x + c$
25. $x^2e^{\tanh y} \,\text{sech}\, y = c$
27. $\text{Tan}^{-1}(x + 2) + \ln |y| - \ln |y + 1| = c$
29. $y = \ln |\cos(x - c)| + k$ **31.** $y = ke^{cx}$
33. $x^2 + \ln |y| + y = 5$ **35.** $y = 2/\sqrt{x}$
37. $y = (x^2 - 4)/5$ **39.** $y = (5x^2 + 1)^2 - \frac{1}{2}$
41. $x = e^y y^{1/3}$
43. $r = (-\cos 2\theta)^{1/2}$, $\pi/4 \le \theta \le 3\pi/4$
45. **(a)** $x = 0$ **(b)** $x = 1$ **(c)** $y = -2\pi$ **(d)** $y = 3\pi$
(e) $x = 0$; $x = 1$; $y = n\pi$, n an integer
(f) $x = c(x - 1)e^{\cot y}$ **(g)** $x = 2(x - 1)e^{\cot y}$

47. No; yes, $y = \begin{cases} 1 - 2x^3 & x \le 0 \\ 1 + x^3 & 0 \le x \end{cases}$. Theorem 1, Sec. 1.6, applies to $y' = 3(y-1)/x$ on $0 < x$ but not on an interval containing 0.
53. $y = 1 - 2x - \ln|c - x|$
55. $y = 1 + x + \dfrac{1 + 2ce^{-3x}}{1 - ce^{-3x}}$

27. $y = xe^{-x} + 3e^{-x}$
29. $y = (1 - e^{1-x})/x$
31. $y = e^{-x^2}(1 + 2e^{-x})$
33. $y = 1/x + 1/x^2$
35. $y = x^3 - 4x$
37. $y = -1 - \cos 2x$
39. $y = x$
41. $y = 1 + 5/(\tan x + \sec x)$
43. $x = y^{5/3} + 4y^{1/3}$
45. $y = e^{-x}(\cosh x - \operatorname{sech} x)$
47. $y = \begin{cases} x^2 & x \le 0 \\ x^2 - 2x^3 & 0 \le x \end{cases}$

Sec. 1.10, page 38

1. Nonhomogeneous
3. Nonhomogeneous
5. The expression as a whole is nonhomogeneous, although the first major term is homogeneous of degree 1 and the second is homogeneous of degree 2.
13. $(x + y)^2 = cx$
15. $x = ce^{x/y}$
17. $y = \dfrac{x}{2}\left[1 + \sqrt{3}\tan\left(\dfrac{\sqrt{3}}{2}\ln|cx|\right)\right]$
19. $c(x^2 + y^2)^3 = x^2$; $c \ge 0$
21. $y + xe^{x/y} = c$
23. $y \sinh(x/y) = c$
25. $y^2 = c(x^2 + 2xy - y^2 \ln y^2)$
27. $x = ce^{2/[1+\tan(y/2x)]}$
29. $\ln y^2 + e^{(x/y)^2} = c$
31. $y = x + c$; parallel lines of slope 1
33. $x^2 - 2xy - 3y^2 = c$. If $c \ne 0$, the curves are hyperbolas. If $c = 0$, the two lines $y = x/3$ and $y = -x$ are obtained.
35. $2\,\mathrm{Tan}^{-1}(y/x) = \ln[|c|(x^2 + y^2)]$. The curves of this family are complicated transcendental curves.
37. $y = (1 - x^{5/4})^{4/5}$; $0 \le x \le 1$
39. $y = \frac{1}{4}x^2 - 1$
41. $y^2 = (x^2 \ln|x|)/(1 + \ln|x|)$
43. $y^3 = x^3 - x^{3/2}$
45. $xe^{\tan(y/2x)} = 2e$
49. $(y - x + 3)^4 = c(y + 2x - \frac{3}{4})$; $y = \frac{3}{4} - 2x$
51. $4x + 2y + 1 = ce^{2y}$
53. $4x^2 - 4xy + y^2 - 10x = c$
55. The equation $y' = 2x/(3x - y)$ is a counterexample since its solution $(y - x)^2 = c(y - 2x)$ defines a family of conics although $b + c = 3 \ne 0$.
57. $\theta = \ln|cr|$ or $\mathrm{Tan}^{-1}(y/x) = \ln(|c|\sqrt{x^2 + y^2})$ (Compare the solution processes here and in Exercise 35.)
59. If $f(x_1, x_2, \ldots, x_m)$ is a homogeneous function of degree n in all of the variables, then $\sum_{i=1}^{m} x_i \dfrac{\partial f}{\partial x_i} = nf$
63. $xe^{y/x} = c$
65. $2(y/x)^{1/2} - \ln|x| = c$

Sec. 1.11, page 41

1. $y = c/x^2 + x^3/5$
3. $y = ce^{3x} - e^{2x}$
5. $y = \frac{2}{7}x^3 + c/\sqrt{x}$
7. $y = e^{-x^2}(x + c) - \frac{1}{2}$
9. $y = \frac{2}{3}\sin^2 x + c/\sin x$
11. $y = 2 + c\sqrt{1 - x^2}$
13. $y = \frac{1}{2}(x + 1)^4 + c(x + 1)^2$
15. $y = (1/x^2)(ce^{x^2} - \frac{1}{2})$
17. $y = (c + \tan x)\sec x$
19. $x = cy^{-3} + 4y^2/5$
21. $x = 1 + c \operatorname{sech} y$
23. $y = c_1e^x + c_2 - x$
25. $y = c\ln|x - 1| + k + (x - 1)^3/9$

Sec. 1.12, page 44

1. $y^2 = (x + c)/x^4$
3. $y^3 = 1 + ce^{-x^3}$
5. $y^3 = cx^3 - 3x^2$
7. $y = 1/(1 + x + ce^x)$; $y = 0$
9. $y = (x - 3 + ce^{-x/3})^3$; $y = 0$
11. $y^3 = 1/(cx + x^2)$; $y = 0$
13. $y = (\cos x)/(c + \sin^2 x)$; $y = 0$
15. $x = 1/(1 + ce^y)$; $x = 0$
17. $y^2 = c(x^2 + y^2)^3$
19. $y = k + (x - c)^2/4k$
21. $y = 1/(1 + e^x)$
23. $y = -1$
25. $x^2y^2 + y^4 + 2y^2 = 4$
27. $y^2 = 2/(1 + 2x + e^{2x})$
29. $y^2 = 1/[xe^{2x}(3x - 1)]$
31. $y^{2/3} = x(1 - 1/\ln|x|)$
33. $y = e^{2x}x^{2(1-x)}$
37. $y = 1 + 1/(1 - x + ce^{-x})$
39. $y = x + 5x/(c - x^5)$
41. $y = e^x + 2/(ce^{-3x} - e^{-x})$
45. $y = mx - 4m^3$; $y^2 = x^3/27$
47. $y = mx + 1/(1 + m)$; $y = 2\sqrt{x} - x$
49. $y = mx - m^2/4$; $y = x^2$

Sec. 1.13, page 46

1. $y = ce^{kx}$
3. $y = \frac{1}{2}e^x - c_1e^{-x} + c_2$
5. $y = -x + c_1 \sin x + c_2$
7. $y^{1/2} = ke^{ax} + c$
9. A first integration yields $y' = -1 - 1/(c - x^2)$. If $c = -c_1^2$, then $y = (1/c_1)\,\mathrm{tan}^{-1}(x/c_1) - x - c_2$. If $c = 0$, then $y = -1/x - x + c_2$. If $c = c_1^2$, then $y = (1/2c_1)\ln|(c_1 - x)/(c_1 + x)| - x - c_2$.
11. $y = 2\tan(2x + \mathrm{Tan}^{-1}\frac{1}{2})$
13. $y = 1 + (1 - x/2)^2$
15. $y = 2 - x + x\ln|x|$
17. $y = 5 - \ln x^2$
19. $y = e^{-x} + 1$

Sec. 1.14, page 50

1. **(a)** $x = k$ **(b)** 10 **(c)** 6
3. **(a)** $b(x^2 + y^2 - 1) = y$
(b) No; $y = 0$ corresponds to $b = 0$. **(c)** Yes
5. $xy = k$
7. $16y^3 = 9(x + k)^2$
9. $y(3x^2 + y^2) = k$
11. $\sin x \sinh y = k$
13. **(a)** $y^2 + 2x = k$
(b) $y = e^x$ and $y^2 = 1 - 2x$

(c)

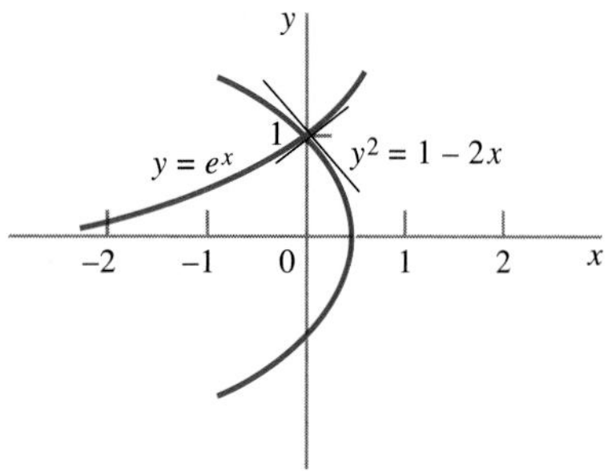

(c)

15. **(a)** The coordinate axes and all parabolas having the origin as vertex and the x axis as axis

(b) $x^2 + \frac{1}{2}y^2 = k^2$, $k > 0$

(c)

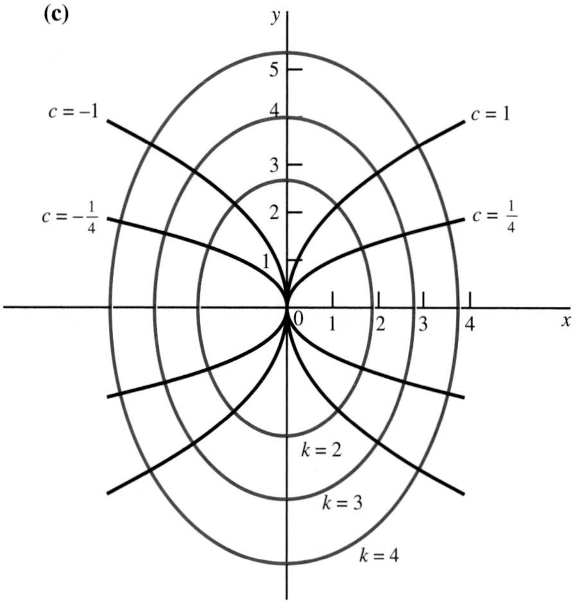

(c)

(d) Yes

(e) All members of the given family pass through (0, 0), but none of the ellipses do. Exactly one curve of each family passes through $(x, y) \neq (0, 0)$.

17. **(a)** Since $x^2 + y^2 \geq 1$ and $|y| \leq 1$, the orthogonal trajectories are defined by $y^2 = ke^{x^2+y^2}$, with $0 \leq k < 1/e$; and by $x = 0$, with $|y| \leq 1$.

(b) Equation of the contour: $x^2 + 4y^2 = 4$; equation of the trajectory: $2e^{5/2}y^2 = e^{x^2+y^2}$

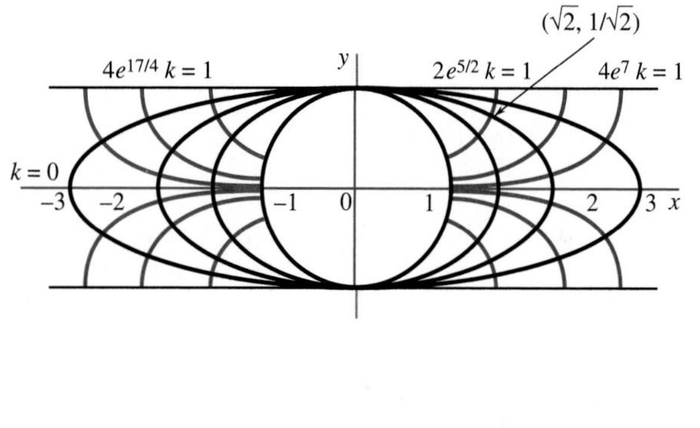

19. The equations of the two families of curves are $x^2 + y^2 = cx$ and $x^2 + y^2 = ky$, as in Exercise 10.

Sec. 1.15, page 63

1. $Q = 100e^{-t/20}$; $Q \leq 100$ for all $t \geq 0$

3. $Q = 2(100 + t) - 150[100/(100 + t)]^2$; 44 min; 171.2 lb

5. 50 lb; $Q > 20$ for all $t > 0$, but of course $Q = 20$ initially.

7. 3 h, 50 min, 18 s

9. Approximately 305 days; approximately 1675 acre-ft

11. $y^{5/2} = h^{5/2} - (5r^2h^2\sqrt{2g}\,t)/2R^2$; $t = (R^2/5r^2)\sqrt{2h/g}$

13. $y^{1/2} = -\dfrac{r^2}{R^2}\sqrt{\dfrac{g}{2}}t + h$; $t = \dfrac{R^2}{r^2}\sqrt{\dfrac{2h}{g}}$

15. $y^{3/2} = 8 - 3r^2\sqrt{\dfrac{g}{2}}t$; $t = \dfrac{8}{3r^2}\sqrt{\dfrac{2}{g}}$

17. $y = kx^4$

19. $y^{-1/2} = \dfrac{h\sqrt{2g}}{2R^2}t + h^{-1/2}$; the tank will never be completely empty.

21. $\alpha \ln \dfrac{\alpha - \sqrt{h}}{\alpha - \sqrt{y}} + \sqrt{h} - \sqrt{y} = \dfrac{a}{A}\sqrt{\dfrac{g}{2}}t$, where $\alpha = \dfrac{Q}{a\sqrt{2g}}$; the limiting depth is $y = \alpha^2$.

23. $Q = 40 - 4320/(t + 108)$ **25.** $Q = 20t/(t + 48)$

27. $9 \ln(1 - Q/60) - 8 \ln(1 - Q/40) = 0.0192t$

29. $\dfrac{4Q}{40 - Q} + \ln\left(1 - \dfrac{Q}{40}\right) = 0.0294t$

31. $Q = 30 - 1500/(t + 50)$; $t = 50$ min

33. $Q = 90\dfrac{1 - e^{-0.00645t}}{3 - 2e^{-0.00645t}}$; $t = 44.6$ min

35. $B = \dfrac{A_0B_0}{B_0 + (A_0 - B_0)e^{-A_0kt}}$ **37.** $T = 50[1 + 3(\frac{2}{3})^{t/2}]$°F

39. $Q = [k(T_1 - T_0)]/h$; $T = T_0 - (T_0 - T_1)(x/h)$

41. $T = 120 - 65e^{-t/25}$

43. $T = T_a - (T_a - T_0)\left(1 + \dfrac{a - b}{P}t\right)^{-a/(a-b)}$

45. $T = 60 + 40e^{-0.0462t}$; $t = 21.2$ min

53. Verify that $P^2 = (m\pi^2/2k)L^3$

55. $N = N_0e^{(k_b - k_d)t}$, where k_b and k_d are, respectively, the proportionality constants in the birth rates and death rates.

57. If a is the proportionality constant in the birthrate and $bN + c$ is the proportionality constant in the deathrate, then

(a) If $a > b$, $N_0 = c/(a - b)$, then $N = c/(a - b)$.

(b) If $a > b$, $N_0 > \dfrac{c}{a-b}$, then for
$$0 < t < \frac{1}{c}\ln\frac{N_0}{N_0 - c/(a-b)},$$
$$N = \frac{c}{a-b}\,\frac{N_0}{N_0 - [N_0 - c/(a-b)]e^{ct}}.$$
(c) If $a > b$, $N_0 < \dfrac{c}{a-b}$, then
$$N = \frac{c}{a-b}\,\frac{N_0}{N_0 + [c/(a-b) - N_0]e^{ct}}.$$
(d) If $a = b$, then $N = N_0 e^{-ct}$.
(e) If $a < b$, then
$$N = \frac{c}{b-a}\,\frac{N_0}{[N_0 + c/(b-a)]e^{ct} - N_0}.$$

59. $N = N_0 \exp[(b_1 - d_1)t - (b_2 - d_2)t^2/2]$. The population has an extremum when $t = (b_1 - d_1)/(b_2 - d_2)$. It is a maximum if $b_2 - d_2 > 0$, and a minimum if $b_2 - d_2 < 0$. If $(b_1 - d_1)/(b_2 - d_2) \geq 0$, the extremum occurs in "real time," i.e., for $t \geq 0$. As $t \to \infty$,

$N \to 0$	if $b_2 - d_2 > 0$
$N \to \infty$	if $b_2 - d_2 < 0$
$N \to 0$	if $b_2 - d_2 = 0$ and $b_1 - d_1 < 0$
$N \to \infty$	if $b_2 - d_2 = 0$ and $b_1 - d_1 > 0$
$N = N_0$	if $b_1 - d_1 = b_2 - d_2 = 0$

61. $\ln\dfrac{N}{N_0} = \dfrac{b_1}{2b_2}\ln\left|\dfrac{b_2+t}{b_2-t}\right| - \dfrac{d_1}{2d_2}\ln\left|\dfrac{d_2+t}{d_2-t}\right|$; as $t \to \infty$, $\ln(N/N_0) \to 0$, which is a meaningful limit. This is not relevant, however, since the terms on the right in the expression for $\ln(N/N_0)$ are undefined, and the process comes to an end with N being 0, N_0, or ∞ at whichever of the times $t = |b_2|$ or $t = |d_2|$ is the smaller.

63. $R = R_0 e^{-0.000436t}$; 660 yr; 4.3% of R_0; 3.9% of R_0; 2.9% of R_0

65. **(a)** $x = x_0 e^{-0.000126t}$ **(b)** 13,200 B.C. **(c)** 4400 B.C.

67. **(b)** $p = 14.7e^{-0.0000385h}$; the height of the atmosphere is infinite. **(c)** 36,000 ft **(d)** $\frac{7}{2}p^{2/7} = -0.0000753h + 7.544$. The predicted height of the atmosphere is 100,000 ft.

69. $P = P_0 e^{pt/100}$

71. $y = (480 - t)/40$ ft, $0 \leq t \leq 480$

73. $y = h - (k_1/k_2\rho)(1 - e^{-k_2 t})$; $h \leq k_1/k_2\rho$, where k_1 and k_2 are, respectively, the proportionality constants in the rates of evaporation and condensation; ρ is the density of the substance.

75. $x_0 + v_0/k$

77. **(a)** $v = 4(1 + e^{-t})$ ft/s
(b) Yes, $x(1) = 4(2 - 1/e)$ ft > 6 ft.

79. 2.4 revolutions

81. $\omega =$
$$\begin{cases} 100\sqrt{3}\tan[1.40 - (6\sqrt{3}/5)t] & 0 \leq t \leq 0.42 \\ (20.1 - 24t)^2 & 0.42 \leq t \leq 0.84 \end{cases}$$
$\omega = 0$ when $t = 0.84$ min

83. $v = \dfrac{w}{k}(1 - e^{-kgt/w})$; $s = w\left(\dfrac{kgt - w + we^{-kgt/w}}{k^2 g}\right)$

85. $v = v_\infty \tanh(gt/v_\infty)$; $s = (v_\infty^2/g)\ln\cosh(gt/v_\infty)$

87. $x = x_0 \cos\sqrt{k/m}\,t$; $v = -x_0\sqrt{k/m}\sin\sqrt{k/m}\,t$

89. $x = x_0 \cosh\sqrt{k/m}\,t$; $v = x_0\sqrt{k/m}\sinh\sqrt{k/m}\,t$

91. $v = -\sqrt{\dfrac{2gr^2}{y_0}}\sqrt{\dfrac{y_0 - y}{y}}$; $t = \sqrt{\dfrac{y_0}{2gr^2}}\left(\sqrt{y_0 y - y^2} + \dfrac{y_0}{2}\cos^{-1}\dfrac{2y - y_0}{y_0}\right)$, where r is the radius of the earth

93. $s = \dfrac{g\sin\alpha}{3}t^2$

95. $Q = EC(1 - e^{-t/RC})$, $t_{1/2} = 0.693RC$

97. $Q = Q_0 e^{-t/RC}$

99. $i = \dfrac{E_0}{R^2 + \omega^2 L^2}[R\cos\omega t + \omega L\sin\omega t - Re^{-Rt/L}]$

101. $y = \ln|\sec x| + a$

103. $y = \dfrac{w}{2H}x^2$, where H is the tension in the cable at the lowest point and w is the load per unit horizontal distance.

105. $y = \dfrac{x^2\omega^2}{2g} + h - \dfrac{\omega^2 r^2}{4g}$

107. $r = r_0 e^{\rho\pi r_0^2 x/2w}$, where x is the distance from the upper base to a general cross section.

109. $x = -\sqrt{l^2 - y^2} + l\ln\dfrac{l + \sqrt{l^2 - y^2}}{y}$

111. 11:23 A.M.

113. $Q = \dfrac{A}{k} - \left(\dfrac{A}{k} - Q_0\right)e^{-kt/V}$ **115.** 24.1 min; 48.2 min

117. 66.2 min

119. $x = \frac{1}{4}[(3x_0 + y_0) + (x_0 - y_0)e^{-4kAt/3V}]$

CHAPTER 2

Sec. 2.1, page 75

1. All subintervals of $(-\infty, \infty)$
3. All subintervals of $(10, \infty)$
5. $(-\infty, \infty)$
7. There is no such interval.
9. $2m < x < 2m + 1$, m an integer
11. $(-\infty, -e)$, $(-e, 0)$, $(0, e)$, (e, ∞)
17. $y(x) = x$; no, because the differential equation is not normal on any interval containing $x = 0$.
19. $y = 0$ **21.** $y = x^3$
23. $y = \ln x$, $0 < x$
25. No, because the leading coefficient has the value 0 at the point $x = 0$ of the interval $-2 \leq x \leq 2$.

Sec. 2.2, page 84

1. $y_1 = x, y_2 = x^2 + 9$, and $c_1y_1 + c_2y_2$ are all solutions on $(-\infty, \infty)$; the differential equation is normal on any interval which does not contain $x = -3$ or $x = 3$.

3. $y_1 = x$ is a solution on $(-\infty, \infty)$; $y_2 = x \ln|x|$ on $(-\infty, 0)$ and $(0, \infty)$; $c_1y_1 + c_2y_2$ on $(-\infty, 0)$ and $(0, \infty)$; the differential equation is normal on $(-\infty, 0)$ and $(0, \infty)$.

5. $y_1 = e^{-x}$ is a solution on $(-\infty, \infty)$; $y_2 = 1/x$ on $(-\infty, 0)$ and $(0, \infty)$; $c_1y_1 + c_2y_2$ on $(-\infty, 0)$ and $(0, \infty)$; the differential equation is normal on $(-\infty, 0)$, $(0, 1)$, and $(1, \infty)$.

7. $y_1 = |x|^{1/2}, y_2 = |x|^{-1/2}$, and $c_1y_1 + c_2y_2$ are all solutions on $(-\infty, 0)$ and $(0, \infty)$; the differential equation is normal on $(-\infty, 0)$ and $(0, \infty)$.

9. $y_1 = \sin x$ is a solution on $(-\infty, \infty)$; $y_2 = \tan x$ and $c_1y_1 + c_2y_2$ on $\left(\frac{2m-1}{2}\pi, \frac{2m+1}{2}\pi\right)$; the differential equation is normal on any interval which contains none of the values $m\pi/2$, m an integer.

11. The first equation is linear; the second is nonlinear.

13. The first equation is linear; the second is nonlinear.

15. **(b)** Yes; $c_1(x^2 + 2) + c_2(3)$, with $c_1 = 1 - \frac{3}{2}c_2$ or $c_1 = 0$.

19. $y'' = 0$

21. The functions are linearly independent.

23. The functions are linearly independent.

25. The functions are linearly dependent; they satisfy the relation $2 \cosh x - e^x - e^{-x} \equiv 0$.

27. The functions are linearly dependent; they satisfy the relation $2(x^2 - 1) - 3(x^2 + x + 1) + (x^2 + 3x + 5) \equiv 0$.

29. The functions are linearly independent.

31. The functions are linearly dependent; they satisfy the relation $6 \cos x \sin 2x + 3(2 \sin x \cos 2x) - 2(3 \sin 3x) \equiv 0$.

33. **(b)** $e^{-2x} = \cosh 2x - \sinh 2x$

39. **(a)** P is never at the origin because $W(t) = r^2(t)\dot{\theta}(t) \neq 0$, so that $r(t) \neq 0$.
(b) The direction of rotation never changes since $\dot{\theta}(t)$ is never zero.

43. $-6e^{2x}$

47. Let y_1 be any nontrivial solution of $a_0(x)y'' + a_1(x)y' + a_2(x)y = 0$. Then in general any other solution satisfies Abel's formula $y_1y' - y_1'y = k \exp[-\int a_1(x)/a_0(x)\, dx]$, which is a nonhomogeneous first-order linear differential equation. It can be solved by the method given in Sec. 1.11 to find a second independent solution, y_2.

53. $xy'' - 2y' = 0$

55. $3x(3x - 7)y'' - (9x^2 - 28)y' + (21x - 28)y = 0$

57. $y''(2 \cosh x + \sinh x) + 3y' \cosh x - 2y(\cosh x + 2 \sinh x) = 0$

61. $y = c_1e^x + c_2e^{-x}$; $y = k_1 \cosh x + k_2 \sinh x$

63. $y = c_1 \sin(x + \pi/4) + c_2 \sin(x - \pi/4)$; $y = k_1 \cos x + k_2 \sin x$

Sec. 2.3, page 90

1. $1/\pi, 13$ **3.** $4/\pi, 5$

5. $\frac{1}{10}, 13$

7. $y = c_1 \cos 2x + c_2 \sin 2x + 4x + \frac{8}{5}e^x$

9. $y = c_1 \cos(x/2) + c_2 \sin(x/2) + 2e^x - 5 \cos x$

11. $y = c_1 \cos \pi x + c_2 \sin \pi x - \sin 2\pi x$

13. $f(\theta) = a \cos \theta$; $g(\theta) = b \sin \theta$; the ellipse $x^2/a^2 + y^2/b^2 = 1$

15. $c_1 \cosh 3x + c_2 \sinh 3x$

17. $c_1 \cosh 6x + c_2 \sinh 6x$

19. $y = \cos x + \sin x + 2e^{2x}$

21. $y = \sinh x + \sin x - 2x$

23. $y = \cos x + \sin x + e^x$

25. $y = c_1 \cosh \omega x + c_2 \sinh \omega x$

Sec. 2.4, page 95

1. $y = c_1e^x + c_2e^{3x} + e^{-x}/8$ **3.** $y = c_1 + c_2e^x + e^{-x}$

5. $y = c_1e^{-x} + c_2e^{2x} - \frac{1}{3}xe^{-x}$

7. $y = c_1 \sin x + c_2 \cos x - (x \cos x)/2$

9. $y = c_1 \sin x + c_2 \cos x - (\cos x) \ln|\sec x + \tan x|$

11. $y = c_1 + c_2x^{1/2} + 8x^{5/4}$

13. $y = c_1x + c_2(1/x) - (\ln|x|)/2x$

15. $y = c_1x^3 + c_2(1/x^2) + (x^3/50)(5 \ln^2|x| - 2 \ln|x|)$

17. $y = c_1x + c_2x \ln|x| + 1/4x$

19. $y = c_1x + c_2x^2 - x \ln|x| - (x/2) \ln^2|x|$

21. $y = c_1e^x + c_2e^{2x} + e^x \ln(1 + e^x) + e^{2x} \ln(1 + e^{-x})$.

23. $y = c_1 + c_2e^{x^2} + \frac{1}{4}(1 - x^2 + x^4/2)e^{x^2}$

25. $y = c_1e^x + c_2e^{x^2} + (2x - 3)e^{x^2+x}$

29. $Y = \frac{1}{2}x^2e^x \ln|x|$

31. $Y = \ln|\sec x + \tan x| - x \cos x + (\sin x) \ln|\cos x|$

33. $Y = x \cosh x - (\sinh x) \ln \cosh x - \tan^{-1}(\sinh x)$

35. $y = \frac{1}{2}e^x + xe^{-x}$ **37.** $y = x^{3/2}(c_1 \ln|x| + c_2)$

39. $y = c_1e^x + c_2e^{-x} + c_3xe^{-x}$ **41.** $y = (c_1 + c_2x)e^{x^2}$

43. $y = c_1x + c_2x(1 + x)$ **45.** $y = c_1e^{2x} + c_2(x + 1)$

47. $y = c_1x^2 + c_2/x^2$

49. $y = c_1e^{2x} + c_2(x + 1) - \frac{1}{2} - (x + 1) \ln|x + 1|$

53. $y = c_1e^x + c_2e^{2x} - xe^x$

55. $y = c_1e^x + c_2e^{2x} - xe^x$

57. $y = c_1 \cos x + c_2 \sin x + (x \sin x)/2$

59. $y = c_1x + c_2\left(1 + \frac{1}{x}\right) + \frac{x}{2}\left[\ln|x(x + 2)| - \frac{1}{x(x + 2)}\right] + \left(1 + \frac{1}{x}\right)\left(2 \ln|x + 2| + \frac{1}{x + 2} - x\right)$

Sec. 2.5, page 105

1. $y' - y = 0$ **3.** $y'' + 36y = 0$

5. $y'' - y = 0$ **7.** $y^{(n+1)} = 0$

9. $y'' + 2y' + y = 0$ **13.** $1/(1 + x^2)$

15. $\dfrac{8x^3(2+x^2)}{1+x^2}$ **17.** $\dfrac{x[1-\ln(1+x^2)^2]}{(1+x^2)^3}$

19. 2 **21.** 0

23. $Df(x)$ is the same as $f'(x)$, whereas $f(x)D$ is an operator.

25. For L_1L_2 to exist, f must be in the domain of L_2 and L_2f must be in the domain of L_1; for L_2L_1 to exist, f must be in the domain of L_1 and L_1f must be in the domain of L_2.

29. The common value of the expressions is $-21\cos 3x - 7\sin 3x$

31. If and only if $r_2(x) = r_1(x) + c$

33. $r(x) = -1$

35. $y = c_1x + c_2x\int(e^x/x)\,dx$

37. $y = c_1e^x + c_2e^{2x}$ **39.** $y = c_1e^{-2x} + c_2e^x$

41. $y = c_1e^{-x} + c_2e^{-4x}$ **43.** $y = c_1 + c_2e^{-5x}$

45. $y = c_1\cosh\sqrt{2}x + c_2\sinh\sqrt{2}x$

47. $y = c_1e^{\sqrt{3}x} + c_2xe^{\sqrt{3}x}$ **49.** $y = c_1e^{2x/3} + c_2xe^{2x/3}$

51. $y = e^{-5x}(c_1\cos x + c_2\sin x)$

53. $y = e^{-3x}(c_1\cos\sqrt{2}\pi x + c_2\sin\sqrt{2}\pi x)$

55. $y = \frac{1}{4}(e^{2x} + 3e^{-2x})$

57. $y = 2e^{x/2}\sinh(\sqrt{5}/2)x$ **59.** $y = 5(1 + e^{3x/5})$

61. $y = -e^{-x}(4\cos 2x + \sin 2x)$

63. There is no such solution.

65. $y = e^{-2x}\cos(x/2)$ **67.** $-2 < b < 6$

69. $\lambda = n$; $y_n = B_n\sin nx$, B_n arbitrary

71. $\lambda = (2n+1)/2$; $y_n = B_n\sin[(2n+1)x/2]$, B_n arbitrary

73. The only solution is the trivial solution $y = 0$.

75. **(c)** The functions $x(t)$ and $y(t)$ need not be solutions of a differential equation.

81. **(c)** A homogeneous second-order equation whose characteristic equation has equal roots can be written in the form $(D-a)^2y = 0$. Part **b** now provides a solution immediately since we know that if $y = (c_1 + c_2x)e^{ax}$, then $(D-a)^2y = 0$, as required.

83. D^0 is the identity operator since, by definition, $D^0f = f$. If, for suitable functions f we define D^{-1} such that $D^{-1}f(x) = \int_a^x f(s)\,ds$, then $DD^{-1}f = f$, i.e., $DD^{-1} = D^0$. However, $D^{-1}Df(x) = f(x) - f(a)$, so that $D^{-1}D \neq D^0$ unless $f(a) = 0$. This difficulty prevents D^{-1}, as defined, from being both a left and a right inverse, or *the* inverse of D except for a very special class of functions.

85. $y = \dfrac{A\sin x - B\cos x}{x(A\cos x + B\sin x)}$

87. $y = \dfrac{(A-2B)\sin x - (2A+B)\cos x}{(A\cos x + B\sin x)\sin x}$

89. Yes, theoretically. However, the equation obtained by substituting $y = u(x)xe^{px}$ into Eq. (21) is $xu'' + 2u' + q^2xu = 0$, and we are unable to solve this at the present time. [This equation can be solved in terms of Bessel functions (see Chap. 12), and its solutions turn out to be $(\cos qx)/x$ and $(\sin qx)/x$, as expected.]

Sec. 2.6, page 110

1. $y = c_1 + c_2e^x + c_2e^{2x}$

3. $y = c_1e^{-2x} + c_2\cos x + c_3\sin x$

5. $y = c_1e^x + c_2\cos x + c_3\sin x$

7. $y = c_1e^{-2x} + (c_2 + c_3x)e^{3x}$

9. $y = c_1e^{x/5} + c_2e^{7x} + c_3e^{-3x}$

11. $y = e^{-x}(c_1 + c_2x + c_3\cos x + c_4\sin x)$

13. $y = c_1e^{-x} + c_2e^x + (c_3 + c_4x)e^{x/2}$

15. $y = e^{x/2}[(c_1 + c_2x)\cos(\sqrt{3}x/2) + (c_3 + c_4x)\sin(\sqrt{3}x/2)]$

17. $y = c_1e^{-x} + (c_2 + c_3x + c_4x^2)e^{2x}$

19. $y = (c_1 + c_2x)e^{-2x} + (c_3 + c_4x + c_5x^2)e^{2x}$

21. $y = c_1 + c_2x + c_3x^2 + c_4x^3 + (c_5 + c_6x)e^{-x} + (c_7 + c_8x)e^x$

23. $y = c_1 + c_2e^{-x} + (c_3 + c_4x + c_5x^2)e^{2x} + (c_6 + c_7x)\cos x + (c_8 + c_9x)\sin x$

25. $y''' - 2y'' - 3y' = 0$

27. $y''' + 2y'' - 4y' - 8y = 0$

29. $(D-5)(D-2)^3y = 0$

31. $(D^4 - 65D^2 - 1296)y = 0$

33. $y = e^x - 2e^{2x} + e^{3x}$ **35.** $y = xe^{-x}$

37. $y = \frac{1}{5}(2e^{2x} + 3\cos x + \sin x)$

39. $y = \cos\pi x$ **41.** $y = e^{x/2}\sin x$

43. $y = e^{5x}(\frac{1}{12}\sin 12x - x\cos 12x)$

45. $y = e^{-x} - e^{-3x}$ **47.** $y = \sum_{n=0}^{\infty} x^n/n! = e^x$

49. Only for values of λ which satisfy the equation $\tanh\lambda = \tan\lambda$. $y_n = A_n(\sin\lambda_n\sinh\lambda_nx - \sinh\lambda_n\sin\lambda_nx)$, where λ_n is the nth positive root of the equation $\tanh\lambda = \tan\lambda$.

57. No, because y_1, y_2, y_3 will not be linearly independent.

Sec. 2.7, page 120

1. $y = c_1e^{-x} + c_2e^{-3x} + x/3 - \frac{7}{9}$

3. $y = c_1e^{-2x} + c_2e^{x/2} + e^{2x} - 5$

5. $y = c_1e^{x/3} + c_2e^{-x/7} - 74 + 8x - x^2$

7. $y = c_1e^x + c_2e^{2x} + 1 - xe^x$

9. $y = c_1e^x + c_2e^{-x} + \frac{1}{2}xe^x + \frac{2}{3}e^{2x}$

11. $y = c_1e^{-2x} + c_2e^{-3x} + 3xe^{-2x} + \frac{1}{30}e^{3x}$

13. $y = e^{2x}(c_1\cos x + c_2\sin x + 1) + 3(\sin x + \cos x)$

15. $y = e^{-2x}(c_1\cos 3x + c_2\sin 3x) + (2\cos 3x + \sin 3x)/20$

17. $y = e^{-x}(c_1\cos 3x + c_2\sin 3x) + \frac{5}{2}x^2 - x - \frac{3}{10} - \frac{1}{3}e^{-x}$

19. $y = c_1\cos x + c_2\sin x + [e^x(\sin x - 2\cos x)]/5$

21. $y = e^x[c_1\cos 2x + c_2\sin 2x - \frac{1}{4}(\cos 2x)\ln|\sec 2x + \tan 2x|]$

23. $y = e^{-2x}(c_1 + c_2x - \ln|x|)$

25. This exercise involves duplication between the nonhomogeneous term and the complementary function since $\cosh x$ contains the term e^x which is a part of the complementary function. One method of solution is to replace $\cosh x$ by $(e^x + e^{-x})/2$ and proceed as usual. If $\cosh x$ is retained, the particular solution has the form $Y = Ax\cosh x + Bx\sinh x + C\cosh x + E\sinh x$ rather than just $Y = Ax\cosh x + Bx\sinh x$. A complete solution is

$$y = c_1e^x + c_2e^{4x} - \frac{x}{6}\cosh x - \frac{x}{6}\sinh x - \frac{1}{10}\sinh x$$
$$= c_1e^x + c_2e^{4x} - \tfrac{1}{6}xe^x + \tfrac{1}{20}e^{-x}$$

27. $y = c_1 \cos \frac{x}{2} + c_2 \sin \frac{x}{2} + (632 \cos x + 376 \sin x - 520x \cos x + 65x \sin x)e^x$

29. $y = c_1 e^{-x} + c_2 x e^{-x} - \frac{3 \cos 2x}{50} + \frac{2 \sin 2x}{25} + \frac{1}{2}$

31. $y = 2xe^{-x}$ **33.** $y = \frac{1}{2}x(1 + x)e^{2x}$

35. $y = -2e^{-2x}(\cos x + 3 \sin x) + 2e^x$

37. $y = e^{-x}(1 - x + x^3/6)$

39. As $\lambda \to k$, the particular integral Y approaches $-(t \cos kt)/2k$ which is a particular integral of the limiting differential equation $y'' + k^2 y = \sin kt$.

41. If and only if $c_1 + c_2 = 1$

43. $y = c_1 e^{-x} + c_2 e^{-2x} + c_3 e^{-3x} + x - 3$

45. $y = c_1 \cos x + c_2 \sin x + c_3 \cos 3x + c_4 \sin 3x - \frac{1}{15} \cos 2x$

47. $y = c_1 e^{-x} + (c_2 \cos x + c_3 \sin x)e^{-2x} + 1 + 13xe^{-x}$

49. $y = c_1 e^x + c_2 x e^x - \frac{1}{2}x^2 e^x + \frac{1}{6}x^3 e^x$

51. $y = c_1 e^{-x} + c_2 x e^{-x} + (\frac{1}{2}x^2 \ln |x| - \frac{3}{4}x^2)e^{-x}$

53. $y = c_1 e^x + (c_2 \cos 2x + c_3 \sin 2x)e^{-x} + \frac{1}{8}xe^x$

55. $y = c_1 \cos x + c_2 \sin x - 1 - x \cos x + (\sin x) \ln (1 + \sin x)$

57. $y = c_1 e^{-2x} + c_2 x e^{-2x} + c_3 e^{3x} + 2 - 3x + xe^{3x}$

59. $y = c_1 + c_2 \cosh x + c_3 \sinh x + x \cosh x - (\sinh x) \ln \cosh x - \text{Tan}^{-1} \sinh x$

61. $y = c_1 + c_2 e^x + c_3 e^{2x} + \frac{1}{4}(x^2 + 3x) - xe^x$

63. $y = c_1 + e^{-x}(c_2 + c_3 x + c_4 x^2) + 2x + x^2$

65. $y = c_1 + c_2 \cos x + c_3 \sin x - 2x \cos x$

67. $y = c_1 + c_2 x + c_3 \cos x + c_4 \sin x + \frac{1}{4}x^4 - 3x^2 + x \sin x + 2x \cos x$

69. $y = (c_1 + c_2 x + c_3 x^2)e^{-x} - \frac{1}{2}x^3 e^{-x}$

71. $y = c_1 e^x + c_2 e^{-x} + c_3 \cos 3x + c_4 \sin 3x - x^2 - \frac{16}{9} - \frac{1}{5} \sin 2x$

73. $y = c_1 \cos 2x + c_2 \sin 2x + c_3 \cosh 2x + c_4 \sinh 2x - \frac{1}{15}e^x$

75. $y = c_1 + c_2 x + c_3 \cosh \sqrt{2}x + c_4 \sinh \sqrt{2}x + c_5 \cos 2x + c_6 \sin 2x + c_7 x \cos 2x + c_8 x \sin 2x + e^{2x}$

77. $Y \sim -\left[\frac{1}{x} + \frac{2!}{x^3} + \frac{4!}{x^5} + \frac{6!}{x^7} + \cdots\right]$ There is no value of x for which this series converges.

79. **(a)** Given the equation $p(D)y = e^{rx}$, where r is a root of the characteristic equation, i.e., where r is a value such that $p(r) = 0$, assume a particular integral of the form $Y = Axe^{rx}$. Then, using the formula of Exercise 82 **(b)**, Sec. 2.5, we have $p(D)(Axe^{rx}) = Ap(r)xe^{rx} + Ap'(r)e^{rx} = 0 + Ap'(r)e^{rx}$. Finally, this will yield e^{rx}, as required, if A has the value $1/p'(r)$. Furthermore, since r is a simple root of $p(D) = 0$ by hypothesis, it follows that $p'(r) \neq 0$. Hence A is well defined. **(c)** Given $p(D)y = e^{rx}$, where r is a double root of the characteristic equation, i.e., where $p(r) = p'(r) = 0$, assume that $Y = ax^2 e^{rx}$. Then by the result of Part **b**, $p(D)(Ax^2 e^{rx}) = Ap(r)x^2 e^{rx} + 2Ap'(r)xe^{rx} + Ap''(r)e^{rx} = 0 + 0 + Ap''(r)e^{rx}$, and this will equal e^{rx}, as required, if A is chosen so that $A = 1/p''(r) = 1/2a$.

81. If the nonhomogeneous term $f(x)$ is a solution of a constant-coefficient homogeneous linear differential equation, or if $f(x)$ possesses only a finite number of linearly independent derivatives

83. $Y = e^x(A \cos 3x + B \sin 3x)$

85. $Y = (c_1 + c_2 x)x^3 e^{2x} + (c_3 + c_4 x + c_5 x^2)e^{3x}$

87. $Y = A \cos 2x$

89. $Y = c_1 e^x + (c_2 x^3 + c_3 x^4 + c_4 x^5) \cos 2x + (c_5 x^3 + c_6 x^4 + c_7 x^5) \sin 2x$

91. $Y = x^3(c_1 + c_2 x + c_3 x^2)e^x + (c_4 + c_5 x) \cos x + (c_6 + c_7 x) \sin x$

93. $Y = x^4(c_1 + c_2 x + c_3 x^2 + c_4 x^3) + (c_5 + c_6 x + c_7 x^2)e^{-x} + (c_8 \cos 2x + c_9 \sin 2x)xe^{-x}$

95. $y = \frac{1}{3}(x - 1)e^x$ **97.** $y = 1 + \frac{3}{7}xe^x$

99. $y = (1 - x)^3$

101. $y = \cos x + \sin x + e^x$

103. $y = \sin x \tan x \quad |x| < \pi/2$

107. $Y = \frac{1}{2} \int_{x_0}^x f(s)[e^{x-s} - 2e^{2(x-s)} + e^{3(x-s)}]\, ds$

109. $Y = \frac{1}{2} \int_{x_0}^x f(s)[(x - s)e^{x-s} - \sinh (x - s)]\, ds$

111. $y = c_1 e^{(-a+b)x} + c_2 e^{(-a-b)x} + \frac{1}{2b} \int_{x_0}^x f(s)[e^{(-a+b)(x-s)} - e^{(-a-b)(x-s)}]\, ds$

113. $y = e^{-ax}(c_1 \cos bx + c_2 \sin bx) + \frac{1}{b} \int_{x_0}^x f(s)e^{-a(x-s)} \sin b(x - s)\, ds$

115. $y = c_1 + c_2 \cos x + c_3 \sin x - x \sin x + x + \ln |\tan (x/2)| - \cos x \ln |\sin x|$

117. $Y = \int_0^x \sqrt{s} \sin (x - s)\, ds$. Unlike the formal series obtained in Exercise 78, this expression is meaningful for all x such that $x > 0$.

Sec. 2.8, page 125

1. $y = c_1(1/x) + c_2 x$

3. $y = c_1 x + c_2 x \ln |x| + x^5/16$

5. $y = x(c_1 \cos \ln |x| + c_2 \sin \ln |x|) + 1 + \ln |x| + \frac{1}{2} \ln^2 |x|$

7. $y = c_1/\sqrt{x} + c_2/x + 2 + x/2$

9. $y = c_1 x^{1/3} + c_2 x + c_3 x^2$

11. $y = c_1 \cos \ln |x| + c_2 \sin \ln |x| - (\cos \ln |x|) \ln| \sec \ln |x| + \tan \ln |x||$

13. $y = c_1 x + c_2 x^2 + c_3 x^3 + 10x \ln |x|$

15. $y = c_1 x^{-1/3} + c_2 x^3 + c_3 x^3 \ln |x| + 3x^2$

17. $y = c_1 + c_2(1/x) + c_3 x^3 - \frac{1}{2}x^2 - \frac{1}{6} \ln^2 |x| - \frac{4}{9} \ln |x|$

19. $y = c_1 x^{-2} + |x|^{1/2}(c_2 + c_3 \cos \ln x^2 + c_4 \sin \ln x^2) + 1 + \ln |x|$

21. $y = 1/4x + x \ln x,\ x > 0$ **23.** $y = x^2,\ x > 0$

25. $y = 1/x^2 + x \ln x^2,\ x > 0$

27. $y = \frac{1}{9}(5x^3 + 4 + \ln x^3),\ x > 0$

29. $y = x(1 - \ln x - \frac{1}{2} \ln^2 x),\ x > 0$

31. $y = x + \cos \ln x^2 - 2 \sin \ln x^2,\ x > 0$

33. $y = x - \frac{x}{5} \cos \ln x^2 - \frac{2x}{5} \sin \ln x^2,\ x > 0$

35. $y = 13 + x^2 \ln x,\ x > 0$

37. $y = c_1 x^{-\sqrt{3}} + c_2 x^{\sqrt{3}} + c_3 x^{-3} + c_4 x^3$

39. Yes

Sec. 2.9, page 140

1. $(n+1)e^{nx}$

3. $\dfrac{n(n^2-1)}{6}$

7. 0

21. $W(0) = 1$

27. $\text{Tan}^{-1}\, x = \sum_{n=0}^{\infty} \dfrac{(-1)^n}{2n+1} x^{2n+1}, \; |x| < 1$

29. $R = 1$

31. $R = e$

33. $y = c_0 \sum_{n=0}^{\infty} \dfrac{(\lambda x)^n}{n!} = c_0 e^{\lambda x}$

35. $y = c_0 \sum_{k=0}^{\infty} \dfrac{(\lambda x)^{2k}}{(2k)!} = c_1 \dfrac{1}{\lambda} \sum_{k=0}^{\infty} \dfrac{(\lambda x)^{2k+1}}{(2k+1)!} = c_0 \cosh \lambda x + c_1 \dfrac{1}{\lambda} \sinh \lambda x$

37. **(a)** $y = 1 - \dfrac{1}{2!}x^2 - \dfrac{2}{3!}x^3 - \dfrac{3}{4!}x^4 - \cdots = \sum_{n=0}^{\infty} \dfrac{1-n}{n!} x^n = e^x - xe^x$

(b) $y = \dfrac{1}{1!}x + \dfrac{2}{2!}x^2 + \dfrac{3}{3!}x^3 + \dfrac{4}{4!}x^4 + \cdots = \sum_{n=1}^{\infty} \dfrac{n}{n!} x^n = x \sum_{n=0}^{\infty} \dfrac{x^n}{n!} = xe^x$

39. $y = c_0 \sum_{k=0}^{\infty} \dfrac{(-1)^k 2^k k!}{(2k)!} x^{2k} + c_1 \sum_{k=0}^{\infty} \dfrac{(-1)^k}{2^k k!} x^{2k+1}, \; |x| < \infty$

41. $y = c_0 \left[1 - \dfrac{x^2}{25}\right] + c_1 \left[x - \sum_{k=1}^{\infty} \dfrac{x^{2k+1}}{5^{2k}(4k^2-1)}\right], \; |x| < 5$

43. $y = c_0 \sum_{k=0}^{\infty} \dfrac{(2k+1)!}{(2^k k!)^2} x^{2k} + c_1 \sum_{k=0}^{\infty} \dfrac{2^{2k} k!(k+1)!}{(2k+1)!} x^{2k+1}, \; |x| < 1$

45. $y = c_0 \left(1 - \dfrac{1}{16}x^4\right) - 3c_1 \sum_{k=0}^{\infty} \dfrac{x^{4k+1}}{16^k (4k-3)(4k+1)}, \; |x| < 2$

47. $y = c_0 [1 + x^2] + c_1 \left[x - \sum_{k=1}^{\infty} \dfrac{(-1)^k x^{2k+1}}{(4k^2-1)}\right], \; |x| < 1$

49. $y = c_0 \left[1 - \sum_{k=1}^{\infty} \dfrac{(-1)^k (2k)! x^{2k}}{(2k-1)(2^k k!)^2}\right] + c_1 x, \; |x| < 1$

51. $y = c_0 \left[1 + \sum_{k=1}^{\infty} \dfrac{(-1)^k x^{3k}}{2 \cdot 5 \cdot 8 \cdots (3k-1)}\right] + c_1 \sum_{k=0}^{\infty} \dfrac{(-1)^k x^{3k+1}}{3^k k!} + c_2 \left[x^2 + \sum_{k=1}^{\infty} \dfrac{(-1)^k x^{3k+2}}{4 \cdot 7 \cdot 10 \cdots (3k+1)}\right], \; |x| < \infty$

53. $y = 1 + 6x - 3x^2 - 2x^3$

55. $y = 1 - x + \sum_{k=1}^{\infty} \dfrac{(2k)! x^{2k}}{[2^k (2k-1) k!]^2}, \; |x| < 1$

57. $\sum_{n=0}^{\infty} \left[\dfrac{(-1)^n}{2^n n!} \sum_{k=0}^{n} \dfrac{(-1)^k}{2k+1} \binom{n}{k}\right] x^{2n}$

59. $y_1 \doteq 1 - x^2 + x^3 - \frac{13}{12}x^4; \; y_2 \doteq x - \frac{1}{2}x^2 + \frac{1}{6}x^3 - \frac{1}{8}x^4$

61. $y_1 \doteq 1 + \frac{1}{6}x^3 + \frac{1}{8}x^5 + \frac{1}{180}x^6; \; y_2 \doteq x + \frac{1}{2}x^3 + \frac{1}{12}x^4 + \frac{3}{8}x^5$

63. $y_1 \doteq 1 - \frac{1}{2}x^2 - \frac{1}{6}x^3 - \frac{1}{6}x^4; \; y_2 \doteq x - \frac{1}{3}x^3 - \frac{1}{12}x^4 + \frac{1}{24}x^5$

65. $y = c_0 \left[1 + 3 \sum_{k=0}^{\infty} \dfrac{(x-2)^{2k}}{(2k-3)(2k-1)}\right] + c_1[(x-2) - (x-2)^3], \; 1 < x < 3$

67. $y = c_0 \sum_{k=0}^{\infty} \dfrac{(2k+1)}{2^k k!}(x+3)^{2k} + c_1 \sum_{k=0}^{\infty} \dfrac{2^k (k+1)!}{(2k+1)!}(x+3)^{2k+1}, \; |x+3| < \infty$

69. A complex number x_0 is a **singular point** of Eq. (1) if and only if $a_0(x_0) = 0$.

Sec. 2.10, page 156

1. **(a)** There are no singular points **(b)** $x = 0$, regular **(c)** $x = 1$, $x = -1$, both regular **(d)** $x = 0$, irregular **(e)** $x = 0$, irregular; $x = -\frac{1}{2}$, regular **(f)** $x = 0$, $x = -1$, both regular; $x = 2$, irregular

3. If either $p(x)$ or $q(x)$ has a removable discontinuity at the singular point.

5. $2r^2 - r = 0, \; x = 0$

7. $3r^2 - 2r - 1 = 0, \; x = 0$

9. $3r^2 + 5r = 0, \; x = \frac{1}{2}; \; 3r^2 - 4r = 0, \; x = -1$

11. **(a)** $(2n+3)c_n - c_{n-1} = 0$ **(b)** $2nd_n - d_{n-1} = 0$

13. Power series suffice to solve Eq. (1) around an ordinary point.

15. $y_1 = \sum_{n=0}^{\infty} \dfrac{(-1)^n (n+1)! 2^{2n} x^{n+1}}{(2n+3)!}, \; |x| < \infty;$

$y_2 = |x|^{-1/2} \sum_{n=0}^{\infty} \dfrac{(-1)^n x^n}{n!} = |x|^{-1/2} e^{-x}, \; 0 < |x| < \infty$

17. $y_1 = x, \; |x| < \infty; \; y_2 = |x|^{-1/2}, \; 0 < |x| < \infty$

19. $y_1 = |x|^{1/2} \sum_{n=0}^{\infty} \dfrac{(-1)^n x^n}{(n!)^2}, \; 0 < |x| < \infty;$

$y_2 = y_1 \ln|x| - 2|x|^{1/2} \sum_{n=1}^{\infty} \dfrac{(-1)^n H_n x^n}{(n!)^2}, \; 0 < |x| < \infty$

21. $y_1 = \sum_{n=0}^{\infty} x^n = \dfrac{1}{1-x}, \; |x| < 1; \; y_2 = y_1 \ln|x|, \; 0 < |x| < 1$

23. $y_1 = \sum_{n=0}^{\infty} \dfrac{x^{n+1}}{(n+2)!}, \; |x| < \infty; \; y_2 = x^{-1} + 1, \; 0 < |x| < \infty$

25. $y_1 = \sum_{n=0}^{\infty} \frac{x^n}{4n^2 - 1}, |x| < 1;$
$y_2 = x^{-1/2}(1 - x), 0 < |x| < \infty$

27. $y_1 = |x|^{1/2} \sum_{n=0}^{\infty} \frac{x^n}{n!} = |x|^{1/2} e^x, 0 < |x| < \infty;$
$y_2 = y_1 \ln|x|, 0 < |x| < \infty$

29. $y_1 = \sum_{n=0}^{\infty} \frac{(-1)^n x^{n+2}}{(n + 2)!}, |x| < \infty; y_2 = 1 - x, |x| < \infty$

31. $y_1 = x - \frac{2}{3}x^2, |x| < \infty;$
$y_2 = |x|^{1/2} \sum_{n=0}^{\infty} \frac{x^n}{n!(2n - 3)(2n - 1)}, 0 < |x| < \infty$

33. $y_1 = \sum_{n=0}^{\infty} \frac{(-1)^n x^n}{n!} = e^{-x}, |x| < \infty;$
$y_2 = y_1 \ln|x| - \sum_{n=1}^{\infty} \frac{(-1)^n H_n x^n}{n!}, 0 < |x| < \infty$

35. $y_1 = x^2 \sum_{n=0}^{\infty} \frac{(-1)^n x^n}{n!} = x^2 e^{-x}, |x| < \infty; y_2 =$
$|x|^{1/2}\left[1 + 2x - \sum_{n=0}^{\infty} \frac{(-1)^n 2^{2n-2}(n - 2)! x^n}{(2n - 3)!}\right], 0 < |x| < \infty$

37. $y_1 = |x|^{-1} \sum_{n=0}^{\infty} \frac{(-1)^n x^n}{(n!)^2}, 0 < |x| < \infty;$
$y_2 = y_1 \ln|x| - |x|^{-1} \sum_{n=1}^{\infty} \frac{2(-1)^n H_n x^n}{(n!)^2}, 0 < |x| < \infty$

39. $y_1 = \sum_{n=0}^{\infty} \frac{(-1)^n x^{n+2}}{n!(n + 2)!}, |x| < \infty; y_2 = 1 + x +$
$\sum_{n=3}^{\infty} \frac{(-1)^n (H_{n-2} + H_n - H_2) x^n}{(n - 2)! n!} - y_1 \ln|x|, 0 < |x| < \infty$

41. $y_1 = |x|^{1/2} \sum_{n=0}^{\infty} \frac{x^n}{2^n n!}, 0 < |x| < \infty; y_2 =$
$y_1 \ln|x| + |x|^{-1/2}\left[2 - \sum_{n=2}^{\infty} \frac{H_{n-1} x^n}{2^{n-1}(n - 1)!}\right], 0 < |x| < \infty$

43. $y_1 = |x|^{1/2} \sum_{n=0}^{\infty} \frac{(-1)^n 2^{2n}(n + 1)!(n + 1) x^n}{(2n + 3)!}, 0 < |x| < \infty;$
$y_2 = \sum_{n=0}^{\infty} \frac{(-1)^{n+1}(2n - 1) x^{n-1}}{n!}, 0 < |x| < \infty$

45. $y_1 = |x|^{1/2}, 0 < |x| < \infty; y_2 = x^{-1}, 0 < |x| < \infty$
47. $y_1 = x^3, |x| < \infty; y_2 = 1 - x, |x| < \infty$
49. $y_1 = x, |x| < \infty; y_2 = 1 + x \ln|x|, 0 < |x| < \infty$
51. $y_1 = x^{-1} e^x, y_2 = x^{-1} + 1$
57. $y_1 = |x|^a \cosh x, y_2 = |x|^a \sinh x$

59. $y_1 = x - 1, y_2 = \sum_{n=1}^{\infty} \frac{(-1)^n (x - 1)^{n+1}}{n! n} +$
$(x - 1) \ln|x - 1|, 0 < |x - 1| < \infty$

61. $y_1 = x - 1;$ $y_2 = 1 - \sum_{n=2}^{\infty} \frac{(-1)^n (n + 1)(x - 1)^n}{n - 1} +$
$2(x - 1) \ln(x - 1), 1 < x < 2$

63. None exists. **67.** $y = \frac{a \cos x + b \sin x}{1 + x}$

69. $x = a$ is an ordinary point if $P(x)$, $Q(x)$, and $R(x)$ are analytic at $x = a$. $x = a$ is a regular singular point if at least one of the functions $P(x)$, $Q(x)$, $R(x)$ is not analytic at $x = a$ but each of the functions $(x - a)P(x)$, $(x - a)^2 Q(x)$, $(x - a)^3 R(x)$ is analytic at $x = a$. $x = a$ is an irregular singular point if at least one of the functions $(x - a)P(x)$, $(x - a)^2 Q(x)$, $(x - a)^3 R(x)$ is not analytic at $x = a$. The indicial equation comes from substituting $y = \sum_{n=0}^{\infty} c_n (x - a)^{n+r}$, $c_0 \neq 0$, into the differential equation and equating the coefficient of x^r is the result to zero.

71. $y = a \sum_{n=0}^{\infty} \frac{(-1)^n x^{2n+2}}{2^n (n + 1)!(2n + 1)!} + b \sum_{n=0}^{\infty} \frac{(-1)^n 2^n n! x^{2n+1}}{(2n)!(2n + 1)!}$

73. $y = a\left[x^2 + \sum_{n=1}^{\infty} \frac{(-1)^n x^{3n+2}}{5 \cdot 8 \cdot 11 \cdots (3n + 2)}\right] +$
$b\left[x + \sum_{n=1}^{\infty} \frac{(-1)^n x^{3n+1}}{4 \cdot 7 \cdot 10 \cdots (3n + 1)}\right] + c \sum_{n=0}^{\infty} \frac{(-1)^n x^{3n}}{3^n n!}$

75. $y = a \sum_{n=0}^{\infty} \frac{(-1)^n x^{n+2}}{(n - 1)(n!)^2}$

Sec. 2.11, page 168

1. $k = 1$ kg/cm
3. $y = -\cos 14t + \frac{1}{2} \sin 14t$
5. $y = 10 \cos 8t$
7. **(a)** $\frac{d^2y}{dt^2} + \frac{kg}{W} y = 0$ **(b)** $y = -2 \cos 12t$
(c) $y = \frac{1}{4} \sin 12t$ **(d)** $y = 4 \cos 12t + \frac{2}{3} \sin 12t$
9. 6.75 in **11.** $f(t) = 2ma\omega^2 \sinh \omega t$

13. $y = \begin{cases} \frac{-wx^2}{6EI}(3L - 2x) & 0 \le x \le L/2 \\ \frac{wL^2}{24EI}(L - 6x) & L/2 \le x \le L \end{cases}$ $y(L) = -\frac{5wL^3}{24EI}$

15. $y_n = B_n \sin \frac{(2n + 1)\pi}{2L} x$

17. $y = \tan \theta \left(x - \sqrt{\frac{EI}{F \cos \theta}} \frac{\sinh \sqrt{F \cos \theta / EI}\, x}{\cosh \sqrt{F \cos \theta / EI}\, L}\right)$
$y_{\text{tip}} = L \tan \theta \left(1 - \frac{1}{L} \sqrt{\frac{EI}{F \cos \theta}} \tanh \sqrt{\frac{F \cos \theta}{EI}} L\right)$
where the origin is taken at the free end of the beam.

19. $y(L) = -L^5/30EI$

21. $y = -\dfrac{ax(x-1)^2(2x+1)}{48EI} - \dfrac{b(2\sin\pi x + \pi x^3 - \pi^2 x)}{2EI\pi^4}$

23. $y = \dfrac{P}{F}\left(x - \sqrt{\dfrac{EI}{F}}\,\dfrac{\sinh\sqrt{F/EI}\,x}{\cosh\sqrt{F/EI}\,L}\right)$. The deflection of the free end relative to the fixed end is simply the value of y at $x = L$, namely, $\dfrac{P}{F}\left(L - \sqrt{\dfrac{EI}{F}}\tanh\sqrt{\dfrac{F}{EI}}\,L\right)$

25. $(1/4\pi)\sqrt{kg/W}$ cycles per unit time

27. $(3/4\pi)\sqrt{kg/W}$ cycles per unit time

29. $\dfrac{1}{2\pi}\sqrt{\dfrac{kgr^2}{WR^2 + Ig}}$ cycles per unit time

31. $(1/8\pi)\sqrt{kg/W}$ cycles per unit time

33. $y = \dfrac{a}{2}\left(\cosh\sqrt{\dfrac{2wgR^2}{wLR^2 + Ig}}\,t - 1\right)$

35. $y = a\cosh\sqrt{g/L}\,t$

37. $y = (5\cosh 2t - 4)$ ft; $0 \le t \le \frac{1}{2}\cosh^{-1}\frac{8}{5}$ s

39. $y = \left(8\cosh\dfrac{2}{\sqrt{3}}t - 5\right)$ ft

41. $y = 10[\cosh(t/2) - 1]$ ft; valid until $y = b$

43. $\omega_n = (z_n^2/L^2)\sqrt{EIg/A\rho}$, where z_n is the nth root of the equation $\cos z\cosh z = 1$. The deflection corresponding to ω_n is

$$y_n = (\cos z_n - \cosh z_n)\left(\cos z_n\frac{x}{L} - \cosh z_n\frac{x}{L}\right) + (\sin z_n + \sinh z_n)\left(\sin z_n\frac{x}{L} - \sinh z_n\frac{x}{L}\right)$$

45. $\mathrm{f} = \dfrac{1}{2\pi}\sqrt{\dfrac{3g[4kl^2 - (2W + w)L]}{2L^2(w + 3W)}}$ cycles per unit time provided $4kl^2 > (2W + w)L$. If this is not the case, the differential equation has hyperbolic functions in its solution, and the system is unstable.

47. $\mathrm{f} = (1/2\pi)\sqrt{g/l}$, $\mathrm{p} = 2\pi\sqrt{l/g}$

49. True period $\doteq$ 1.18 times the approximate period.

Sec. 2.12, page 185

1. $g(x, s) = \begin{cases} (l - s)x/Tl & 0 \le x \le s \\ (l - x)s/Tl & s \le x \le l \end{cases}$

3. $y(x) = \begin{cases} 5Px/4T & 0 \le x \le \frac{1}{4}l \\ P(l + x)/4T & \frac{1}{4}l \le x \le \frac{1}{2}l \\ 3P(l - x)/4T & \frac{1}{2}l \le x \le l \end{cases}$

5. See Fig. A2.1.

7. $y(x) = \begin{cases} -Px/T & 0 \le x \le \frac{1}{4}l \\ P(2x - l)/2T & \frac{1}{4}l \le x \le \frac{1}{2}l \\ 0 & \frac{1}{2}l \le x \le l \end{cases}$

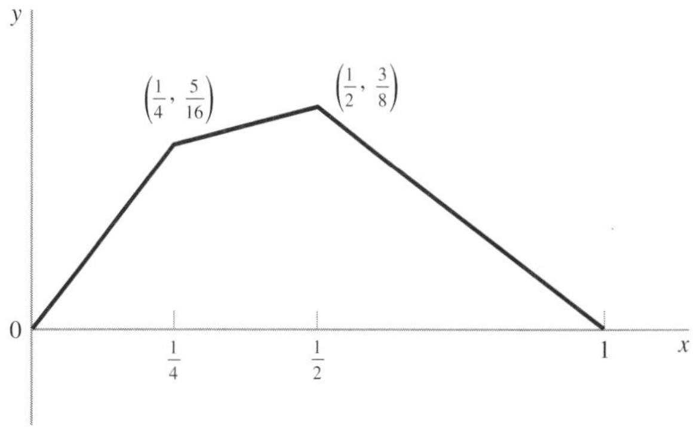

FIGURE A2.1

9. $y(x) = \dfrac{x^2(x^2 - l^2)}{6T}$

11. $y(x) = \begin{cases} \dfrac{1}{T}\left[\dfrac{x^3}{6} - x\left(\dfrac{l^2}{6} - \dfrac{3P}{4}\right)\right] & 0 \le x \le \dfrac{1}{4}l \\ \dfrac{1}{T}\left[\dfrac{x^3}{6} - x\left(\dfrac{l^2}{6} + \dfrac{P}{4}\right) + \dfrac{Pl}{4}\right] & \dfrac{1}{4}l \le x \le l \end{cases}$

13. $y(x) = [x(1 - x)(x^2 - x - 1]/12$

15. **(a)** $y(x) = \begin{cases} \dfrac{l^2}{\pi^2 T}\sin\dfrac{\pi x}{l} + \dfrac{2x}{3} & 0 \le x \le \dfrac{1}{3}l \\ \dfrac{l^2}{\pi^2 T}\sin\dfrac{\pi x}{l} + \dfrac{l - x}{3} & \dfrac{1}{3}l \le x \le l \end{cases}$

(b) See Fig. A2.2.

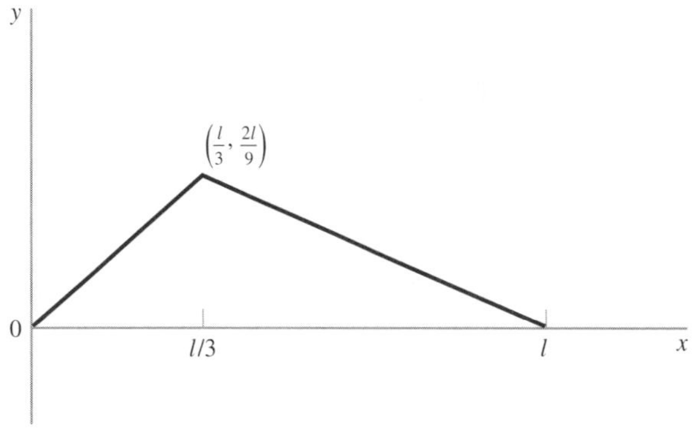

FIGURE A2.2

17. $g(x, s) = \begin{cases} e^s(1 - e^{-x}) & 0 \le x \le s \\ e^s - 1 & s \le x \le \pi \end{cases}$

19. $g(x, s) = \begin{cases} \cos s\sin x & 0 \le x \le s \\ \cos x\sin s & s \le x \le \pi \end{cases}$

21. $g(x, s) = \begin{cases} x & 0 \le x \le s \\ s & s \le x \le l \end{cases}$

23.

$$g(x, s) = \begin{cases} \dfrac{1}{2k}\left[-\cot\dfrac{k}{2}\cos k(x - s) + \sin k(x - s)\right] & 0 \le x \le s \\ \dfrac{1}{2k}\left[-\cot\dfrac{k}{2}\cos k(x - s) - \sin k(x - s)\right] & s \le x \le 1 \end{cases}$$

25. **(a)** $g(x,s)=\begin{cases}\dfrac{\cosh k(b-s)\sinh kx}{k\cosh kb} & 0\le x\le s\\ \dfrac{\cosh k(b-x)\sinh ks}{k\cosh kb} & s\le x\le b\end{cases}$

(b) $g(x,s)=\begin{cases}\dfrac{\sinh k(b-s)\cosh kx}{k\cosh kb} & 0\le x\le s\\ \dfrac{\sinh k(b-x)\cosh ks}{k\cosh kb} & s\le x\le b\end{cases}$

(c)

$$g(x,s)=\begin{cases}\dfrac{\cosh k(b-s)\cosh k(x-a)}{k\sinh k(b-a)} & a\le x\le s\\ \dfrac{\cosh k(b-x)\cosh k(s-a)}{k\sinh k(b-a)} & s\le x\le b\end{cases}$$

(d)

$$g(x,s)=\begin{cases}\dfrac{\sinh k(b-s)[\sinh k(x-a)+k\cosh k(x-a)]}{k\sinh k(b-a)+k^2\cosh k(b-a)}\\ a\le x\le s\\ \dfrac{\sinh k(b-x)[\sinh k(s-a)+k\cosh k(s-a)]}{k\sinh k(b-a)+k^2\cosh k(b-a)}\\ s\le x\le b\end{cases}$$

27.

$$g(x,s)=\begin{cases}\dfrac{(s^3-2e^3e^{3s})(x^3-10e^6e^{3x})}{6e^3(1-5e^3)s^3(s-1)^2e^{3s}} & -2\le x\le s\\ \dfrac{(x^3-2e^3e^{3x})(s^3-10e^6e^{3s})}{6e^3(1-5e^3)s^3(s-1)^2e^{3s}} & s\le x\le -1\end{cases}$$

35. $g(x,s)=\begin{cases}\frac{1}{2}e^{s/2}\sinh(x/2) & 0\le x\le s\\ \frac{1}{2}e^{x/2}\sinh(s/2) & s\le x\le 2\end{cases}$

$y(x)=xe^{x/2}-2e^2\sinh\dfrac{x}{2}$

37. $g(x,s)=\begin{cases}-e^{2s}\sin s\, e^{-2x}\cos x & 0\le x\le s\\ -e^{2s}\cos s\, e^{-2x}\sin x & s\le x\le \pi\end{cases}$

$y(x)=e^{-2x}(2\sin x-\sin 2x)$

39. $g(x,s)=\begin{cases}x^6/121s^7 & 1\le x\le s\\ x^{7/3}/121s^{10/3} & s\le x\le 2\end{cases}$

$y(x)=3x^{7/3}-x^6[3+11\ln(2/x)]$

43. $\phi(x)=\dfrac{\omega^2\rho}{E_sg}\displaystyle\int_0^x s\phi(s)\,ds+\dfrac{\omega^2\rho x}{E_sg}\int_x^l \phi(s)\,ds$

$\omega_n=\dfrac{(2n+1)\pi}{2l}\sqrt{\dfrac{E_sg}{\rho}}$ rad per unit time

45. $g(x,s)=\begin{cases}x^2(x-3s)/6EI & 0\le x\le s\\ s^2(s-3x)/6EI & s\le x\le l\end{cases}$
and is symmetric in x and s.

47. **(a)** $y(l)=-5Pl^3/48EI$ **(b)** At $x=l/3$

49. $y(x)=-(wx^2/24EI)[(x-2l)^2+2l^2]$

51. $y(x)=(x^2/120EI)(x^3-5lx^2+10l^2x-10l^3)$;
$y(l)=-l^5/30EI$; $EIy''(0)=-l^3/6$

53. $y(l)=-19l^5/120EI$

55. $g(x,s)=\begin{cases}\dfrac{x(l-s)(x^2-2sl+s^2)}{6EIl} & 0\le x\le s\\ \dfrac{s(l-x)(s^2-2xl+x^2)}{6EIl} & s\le x\le l\end{cases}$

57.

$$g(x,s)=\begin{cases}\dfrac{x^2[2l^3(x-3s)+s^2(3l-s)(3l-x)]}{12EIl^3} & 0\le x\le s\\ \dfrac{s^2[2l^3(s-3x)+x^2(3l-x)(3l-s)]}{12EIl^3} & s\le x\le l\end{cases}$$

59. **(a)** $\kappa(x,s)=\begin{cases}[\cosh ks\sinh kx]/k & 0\le x\le s\\ [\cosh kx\sinh ks]/k & s\le x\le 1\end{cases}$

(b) $g(x,s)=\begin{cases}\dfrac{\sinh kx\sinh k(1-s)}{k\sinh k} & 0\le x\le s\\ \dfrac{\sinh ks\sinh k(1-x)}{k\sinh k} & s\le x\le 1\end{cases}$

CHAPTER 3

Sec. 3.1, page 200

1. A number and its conjugate are reflections in the x axis. A number and its negative are reflections in the origin.

3. **(a)** $\frac{1}{50}+\frac{2}{50}i$ **(b)** $-115-236i$ **(c)** $\frac{5}{41}+\frac{4}{41}i$
(d) $-30-140i$ **(e)** $-124-32i$ **(f)** $-19+8i$

5. **(a)** $\frac{1}{5}$ **(b)** $\frac{90}{41}$ **(c)** $\frac{8}{5}$ **(d)** $\sqrt{41}$ **(e)** $10\sqrt{205}$
(f) $-30+140i$ **(g)** $-94+108i$

11. $r_1=\sqrt{2}\left(\cos\dfrac{\pi}{12}+i\sin\dfrac{\pi}{12}\right)=\sqrt{2}e^{i\pi/12}$

$r_2=\sqrt{2}\left(\cos\dfrac{5\pi}{12}+i\sin\dfrac{5\pi}{12}\right)=\sqrt{2}e^{5i\pi/12}$

$r_3=\sqrt{2}\left(\cos\dfrac{9\pi}{12}+i\sin\dfrac{9\pi}{12}\right)=\sqrt{2}e^{9i\pi/12}$

$r_4=\sqrt{2}\left(\cos\dfrac{13\pi}{12}+i\sin\dfrac{13\pi}{12}\right)=\sqrt{2}e^{13i\pi/12}$

$r_5=\sqrt{2}\left(\cos\dfrac{17\pi}{12}+i\sin\dfrac{17\pi}{12}\right)=\sqrt{2}e^{17i\pi/12}$

$r_6=\sqrt{2}\left(\cos\dfrac{21\pi}{12}+i\sin\dfrac{21\pi}{12}\right)=\sqrt{2}e^{21i\pi/12}$

15. $e^{-\pi}$

23. Center $=\left(-\dfrac{B+\bar{B}}{2(A+\bar{A})},\dfrac{B-\bar{B}}{2i(A+\bar{A})}\right)$;

radius $=\dfrac{B\bar{B}-(A+\bar{A})(C+\bar{C})}{A+\bar{A}}$

25. **(a)** $4\sqrt{2}e^{7i\pi/12}$ **(b)** $24\sqrt{2}e^{-i\pi/12}$

(c) $120e^{(\phi+\pi/6)i}$, $\phi=\tan^{-1}\left(\dfrac{-3}{4}\right)$

(d) $120\sqrt{2}e^{(\phi-\pi/12)i}$, $\phi=\tan^{-1}\left(\dfrac{-3}{4}\right)$

41. (1, 0), (0, 1)
45. It says that $x + iy = (x, y)$.

Sec. 3.2, page 207

1. $(x, y, z) = (-3, 6, -3i)$ **3.** $(x, y) = (15, -12)$
5. $z = -1, 1, -i, i$ **7.** $x = 0, -2, 2$
9. $x = n\pi$, n an integer **11.** $\mathbf{u} \cdot \mathbf{v} = 7$, $(\frac{2}{3}, \frac{1}{3}, \frac{2}{3})$
13. $\mathbf{u} \cdot \mathbf{v} = -210$, $(\frac{1}{3}, \frac{2}{5}, \frac{8}{15}, \frac{2}{3})$
15. $\mathbf{u} \cdot \mathbf{v} = 28$, $(\frac{7}{11}, -\frac{4}{11}, \frac{2}{11}, -\frac{4}{11}, \frac{6}{11})$
17. Orthonormal
19. The vectors are mutually orthogonal, but the set $\{\mathbf{u}, \mathbf{v}, \mathbf{w}\}$ is neither orthonormal nor orthogonal since it contains the zero vector.
21. Orthogonal but not orthonormal
25. $t = -5, 1, 4$
27. $\mathbf{u} \cdot \mathbf{v} = 4 - 8i$; $\bar{\mathbf{u}} \cdot \mathbf{v} = 8i$; $\mathbf{u} \cdot \bar{\mathbf{v}} = -8i$; $l_h(\mathbf{u}) = 2$; $l_h(\mathbf{v}) = 6$; $\mathbf{u}$ and $\mathbf{v}$ are not orthogonal.
29. $\mathbf{u} \cdot \mathbf{v} = -6 + 6i$; $\bar{\mathbf{u}} \cdot \mathbf{v} = \mathbf{u} \cdot \bar{\mathbf{v}} = 0$; $l_h(\mathbf{u}) = 2$; $l_h(\mathbf{v}) = 6$; $\mathbf{u}$ and $\mathbf{v}$ are orthogonal in the hermitian sense.
31. There is a unique vector $(4, -2, 1)$, corresponding to $x = -2$, which is perpendicular to $(1, 3, 2)$ and $(1, 1, -2)$.
33. The vector $\left(\frac{-7}{19}, \frac{-1}{19}, \frac{5}{19}\right)$, corresponding to $k = \frac{13}{19}$.
35. **(a)** No **(b)** Yes, any scalar multiple of $(1, -2, 3)$
37. **(a)** $\left(\frac{2}{9}, \frac{-4}{9}, \frac{5}{9}, \frac{-2}{9}\right)$ **(b)** $(3, 6, 6, 2)$, $(24, 24, 12, 2)$
39. **(a)** $(2 - 6i, 4i, 2 - 5i, 6i)$, $(2 + 6i, 4i, 2 - 5i, -6i)$
(b) $(2 - 12\sqrt{2}i, 4i, 2 - 5i, 12\sqrt{2}i)$,
$(2 + 12\sqrt{2}i, 4i, 2 - 5i, -12\sqrt{2}i)$
(c) $(2, 4i, 2 - 5i, 0)$

Sec. 3.3, page 214

1. **(a)** $(5, 2, i, -1)$; $\begin{bmatrix} i \\ 0 \\ 2 \end{bmatrix}$; $1 + i$

(b) $(-16, -9)$; $\begin{bmatrix} 4 \\ 16 \\ -9 \\ -1 \end{bmatrix}$; 17

(c) $(-i, 1, i)$; $\begin{bmatrix} i \\ 0 \\ 1 \end{bmatrix}$; $1 + i$ **(d)** $(1, 0, i, 0)$; $\begin{bmatrix} 0 \\ i \\ 0 \\ 1 \end{bmatrix}$; $4i$

3. $z = 0, 2$ **5.** $z = i, -i$

7. **(a)** $\begin{bmatrix} 6 & 13 \\ 19 & 47 \end{bmatrix}$ **(b)** $\begin{bmatrix} 182 & -211 \\ -205 & -273 \end{bmatrix}$

(c) $\begin{bmatrix} 30 & -35 \\ -29 & -49 \end{bmatrix}$

9. **(a)** $\begin{bmatrix} 5 & -19 & -6 \\ 1 & 4 & 17 \end{bmatrix}$ **(b)** $\begin{bmatrix} -5 \\ -21 \end{bmatrix}$

11. **(a)** [0] **(b)** $\begin{bmatrix} 1 & -2 \\ 5 & -4 \\ -1 & 2 \end{bmatrix}$

13. If and only if the matrix is square
15. **(a)** A, B square, of the same order, and commutative

(b) $A = \begin{bmatrix} 1 & 1 \\ 1 & 1 \end{bmatrix}$ $B = \begin{bmatrix} 0 & 1 \\ -1 & 0 \end{bmatrix}$

17. $\begin{bmatrix} ab & -a^2 \\ b^2 & -ab \end{bmatrix}$; a, b arbitrary complex numbers

19. $x = 1$, $y = 3$ **21.** $x = \frac{3}{2}$, $y = 2$
23. **(a)** Rows 2 and 3 of A are interchanged in BA.
(b) Columns 2 and 3 of A are interchanged in AB.

25. **(a)** $\begin{bmatrix} 1 \\ 1 \\ 1 \end{bmatrix}$ **(b)** [0]

27. **(a)** $x = -2$, $y = 3$, $z = 0$, $w = -1$
(b) $x = 1$, $y = 2$, $z = -2i$

29. **(a)** $X = \begin{bmatrix} 11 & 14 \\ -4 & -5 \end{bmatrix}$ **(b)** $X = \begin{bmatrix} 7 & -2 \\ 4 & -1 \end{bmatrix}$

(c) $X = \begin{bmatrix} -5 & -14 \\ 4 & 11 \end{bmatrix}$ **(d)** $X = \begin{bmatrix} -1 & 2 \\ -4 & 7 \end{bmatrix}$

Sec. 3.4, page 224

1. **(a)** $\begin{bmatrix} 6 \\ -5 \\ 2 \\ 4 \\ 0 \end{bmatrix}$; {6} **(b)** $\begin{bmatrix} 9 & -5 \\ -1 & 3 \\ 6 & 0 \\ 7 & 4 \end{bmatrix}$; {9, 3}

(c) $\begin{bmatrix} 2 & -6 & 7 \\ 8 & 0 & -1 \end{bmatrix}$; {2, 0}

(d) $\begin{bmatrix} 1 & -3 & 4 \\ -3 & 0 & 2 \\ 4 & 2 & 5 \end{bmatrix}$; {1, 0, 5}

3. **(a)** $S_1 = \begin{bmatrix} 2 & 8 \\ -5 & 6 \end{bmatrix}$

$S_2 = \begin{bmatrix} -3 & 7 & 0 \\ 1 & -9 & 5 \end{bmatrix}$ $S_3 = \begin{bmatrix} 10 & 0 \\ 0 & 1 \\ -1 & -2 \end{bmatrix}$

(b) $S_1 S_2 S_3 = \begin{bmatrix} -20 & -138 \\ 180 & -149 \end{bmatrix}$

$(S_1 S_2 S_3)^T = S_3^T S_2^T S_1^T = \begin{bmatrix} -20 & 180 \\ -138 & -149 \end{bmatrix}$

(c) No
7. **(a)** $(AB)^* = B^*A^* = [2 \quad 1 - i]$

(b) $(AB)^* = B^*A^* = \begin{bmatrix} 4+7i & 6+9i \\ -1-5i & 3-5i \end{bmatrix}$

9. (a) **11.** (a), (b), and (c)

13. None (not a square matrix)

15. **(a)** $\begin{bmatrix} 8 & 0 & 0 \\ 0 & -6 & 0 \\ 0 & 0 & 10 \end{bmatrix}$ **(b)** $\begin{bmatrix} 1 & 0 & 0 \\ 0 & 0 & 0 \\ 0 & 0 & 0 \end{bmatrix}$

(c) $\begin{bmatrix} i & 0 & 0 & 0 \\ 0 & 1-i & 0 & 0 \\ 0 & 0 & 1 & 0 \\ 0 & 0 & 0 & 1+i \end{bmatrix}$ **(d)** $\begin{bmatrix} 1 & 0 & 0 & 0 \\ 0 & 1 & 0 & 0 \\ 0 & 0 & 1 & 0 \\ 0 & 0 & 0 & 1 \end{bmatrix}$

17. $DA = \begin{bmatrix} 3 & -6 \\ 0 & 0 \\ 6 & -2 \\ 2 & -7 \end{bmatrix}$ AD undefined

19. $DA = \begin{bmatrix} 6i & 0 & 3 & 9i \\ 0 & 0 & 0 & 0 \\ -2i & 2 & 0 & 2i \\ -1 & -1 & -1 & -1 \end{bmatrix}$

$AD = \begin{bmatrix} 6i & 0 & 2 & -3i \\ 0 & 0 & 16 & 0 \\ -3i & 0 & 0 & -i \\ 3 & 0 & 2 & -1 \end{bmatrix}$

23. $X = \begin{bmatrix} -2a & -2b \\ a & b \end{bmatrix}$ for arbitrary a and b, $a^2 + b^2 \neq 0$; $XA = O$ if and only if $a + 2b = 0$; no.

25. **(a)** $X = \begin{bmatrix} -3 & -3 \\ \frac{5}{2} & 2 \end{bmatrix}$ **(b)** $X = \begin{bmatrix} \frac{4}{3} & 2 \\ -\frac{5}{3} & -2 \end{bmatrix}$

(c) $X = \begin{bmatrix} 0 & 0 \\ \frac{1}{2} & 1 \end{bmatrix}$ **(d)** $X = \begin{bmatrix} 1 & 2 \\ -1 & -2 \end{bmatrix}$

(e) No solution

31. If and only if $AB = BA$

33. **(a)** (e), $\{2i, 0, 3i\}$, $\{-13, 0, 13\}$
(b) (a), (c), (e), $\{0, 0, 0\}$, $\{-2, 0, 2\}$
(c) (d), $\{1, -2, 3\}$, $\{i, -2, -i\}$
(d) (a), (b), (d), $\{7, -2, 5\}$, $\{4, -2, 4\}$

39. $(ABC)' = A'BC + AB'C + ABC'$. No, unless A and A′ commute.

41. $$\frac{dA}{dt} = D\begin{bmatrix} \cos\lambda_1 t & 0 & \cdots & 0 \\ 0 & \cos\lambda_2 t & \cdots & 0 \\ \cdots & \cdots & \cdots & \cdots \\ 0 & 0 & \cdots & \cos\lambda_n t \end{bmatrix}$$

$$\frac{d^2A}{dt^2} = -D^2A$$

45. **(a)** $\begin{bmatrix} 1 & 0 & 0 & 0 \\ 0 & 0 & 1 & 0 \\ 0 & 1 & 0 & 0 \\ 0 & 0 & 0 & 1 \end{bmatrix}$ **(b)** $\begin{bmatrix} 0 & 0 & 0 & 1 \\ 0 & 0 & 1 & 0 \\ 0 & 1 & 0 & 0 \\ 1 & 0 & 0 & 0 \end{bmatrix}$

(c) $\begin{bmatrix} 0 & a & 0 & 0 \\ 0 & 0 & 0 & b \\ 0 & 0 & c & 0 \\ d & 0 & 0 & 0 \end{bmatrix}$

47. $\begin{bmatrix} 0 & 0 & 1 & 0 & 0 \\ 1 & 0 & 0 & 0 & 0 \\ 0 & 0 & 0 & 1 & 0 \\ 0 & 0 & 0 & 0 & 1 \\ 0 & 1 & 0 & 0 & 0 \end{bmatrix}$

Sec. 3.5, page 238

1. **(a)** 80 **(b)** 0 **(c)** 4 **(d)** −100 **(e)** −40 **(f)** 0

5. The values of the determinants in (a), (b), and (c) are, respectively, 3, −13, −13.

7. If n is odd, $|A| = 0$.

9. **(a)** $k = 3, -6$ **(b)** $k = -1, \pm 3i$ **(c)** $k = 0, \pm 2$

13. $D_n = aD_{n-1} - bcD_{n-2}$; $D_n = 2^{n+1} - 1$

15. Yes. In this case the equation reduces to the equation of the line on which the three points lie.

Sec. 3.6, page 253

1. **(a)** $(x, y) = (-3, 6)$; the solution is the point of intersection of the lines defined by the given equations.
(b) An inconsistent system; the lines defined by the given equations are parallel.
(c) $(x, y) = c(\frac{6}{5}, 1) + (\frac{7}{5}, 0)$; the equations define coincident lines.
(d) $(x, y, z) = c(1, 2, 1) + (2, 3, 0)$; the planes defined by the given equations have a line in common.
(e) An inconsistent system; the planes defined by the given equations have no point in common.
(f) $(x, y, z) = (-1, -1, 1)$; the planes defined by the given equations have but one point in common.

3. $\begin{bmatrix} x \\ y \\ z \\ w \end{bmatrix} = c\begin{bmatrix} 5 \\ -6 \\ 2 \\ 1 \end{bmatrix} + \begin{bmatrix} 1 \\ 1 \\ 0 \\ 0 \end{bmatrix}$

5. $\begin{bmatrix} x \\ y \\ z \\ w \end{bmatrix} = c_1\begin{bmatrix} -7 \\ 1 \\ 0 \\ 5 \end{bmatrix} + c_2\begin{bmatrix} -11 \\ 0 \\ 1 \\ 7 \end{bmatrix}$

7. $\begin{bmatrix} x \\ y \\ z \\ w \end{bmatrix} = \begin{bmatrix} -66 \\ 27 \\ 6 \\ 4 \end{bmatrix}$

9. $\begin{bmatrix} x \\ y \\ z \\ w \end{bmatrix} = c_1\begin{bmatrix} -1 \\ 1 \\ 1 \\ 0 \end{bmatrix} + c_2\begin{bmatrix} -2 \\ 1 \\ 0 \\ 4 \end{bmatrix} + \begin{bmatrix} 2 \\ -\frac{1}{2} \\ 0 \\ 0 \end{bmatrix}$

11. There is no solution.

13. $\begin{bmatrix} x_1 \\ x_2 \\ x_3 \\ x_4 \\ x_5 \end{bmatrix} = c_1 \begin{bmatrix} 1 \\ 1 \\ 0 \\ 0 \\ 0 \end{bmatrix} + c_2 \begin{bmatrix} -16 \\ 6 \\ 2 \\ 4 \\ 1 \end{bmatrix} + \begin{bmatrix} 1 \\ 0 \\ 2 \\ 3 \\ 0 \end{bmatrix}$

15. $\begin{bmatrix} x_1 \\ x_2 \\ x_3 \\ x_4 \\ x_5 \end{bmatrix} = c_1 \begin{bmatrix} 0 \\ 3 \\ -2 \\ 0 \\ 1 \end{bmatrix} + c_2 \begin{bmatrix} 5 \\ -6 \\ 2 \\ 1 \\ 0 \end{bmatrix} + \begin{bmatrix} 1 \\ 1 \\ 0 \\ 0 \\ 0 \end{bmatrix}$

17. **(a)** $k = \pm 2$ **(b)** $k = \frac{15}{7}$ **(c)** $k = \frac{5}{2}$

19. A simple example is $\begin{bmatrix} 1 & -1 \\ 2 & -2 \\ 3 & -3 \end{bmatrix} \begin{bmatrix} x \\ y \end{bmatrix} = \begin{bmatrix} 0 \\ 0 \\ 0 \end{bmatrix}$ which has $\begin{bmatrix} x \\ y \end{bmatrix} = \begin{bmatrix} 1 \\ 1 \end{bmatrix}$ as one of its solutions.

25. $2\mathbf{v}_1 - 3\mathbf{v}_2 + 7\mathbf{v}_3 = \mathbf{0}$; hence the vectors are linearly dependent.

27. The vectors are linearly dependent since $2\mathbf{v}_1 = \mathbf{v}_2 + \mathbf{v}_3$.

29. The k $m \times n$ matrices $M_1, M_2, \ldots, M_k$ are linearly dependent if and only if $\Sigma_{i=1}^{k} c_i M_i = O$ with at least one of the c's $\neq 0$?

31. M_1, M_2, and M_3 are linearly independent.

33. $c = a + b$ **35.** $m + n + 1$

Sec. 3.7, page 266

1. **(a)** Singular **(b)** Nonsingular **(c)** Nonsingular **(d)** Neither, the matrix is not square **(e)** Singular

3. **(a)** $k = -1, 2$ **(b)** All nonintegral values of k **(c)** All values of k **(d)** All values of k

7. **(a)** $\text{adj A} = \begin{bmatrix} 4 & -2 \\ -3 & 1 \end{bmatrix}$ $\quad A^{-1} = -\frac{1}{2}\begin{bmatrix} 4 & -2 \\ -3 & 1 \end{bmatrix}$

(b) $\text{adj A} = \begin{bmatrix} 3 & 11 & 1 \\ -11 & -5 & 14 \\ 12 & -9 & 4 \end{bmatrix}$

$A^{-1} = \frac{1}{53}\begin{bmatrix} 3 & 11 & 1 \\ -11 & -5 & 14 \\ 12 & -9 & 4 \end{bmatrix}$

(c) $\text{adj A} = \begin{bmatrix} -4 & 1 & 1 \\ 8 & -2 & -2 \\ -4 & 1 & 1 \end{bmatrix}$

Since $|A| = 0$, A^{-1} does not exist.

(d) $\text{adj A} = \begin{bmatrix} -14 & 21 & 7 \\ 6 & -9 & -3 \\ 10 & -15 & -5 \end{bmatrix}$

Since $|A| = 0$, A^{-1} does not exist.

9. **(a)** $(x_1, x_2, x_3) = (\frac{5}{4}, \frac{5}{4}, \frac{1}{2})$
(b) $(x_1, x_2, x_3) = (-159, -205, -78)$
(c) $(w, x, y, z) = (7, -1, 3, 2)$
(d) $(w, x, y, z) = (5, 2, 0, -1)$

15. **(a)** $|A| = 1 \qquad A^{-1} = \begin{bmatrix} 3 & -1 \\ -2 & 1 \end{bmatrix}$

(b) $|A| = -2 \qquad A^{-1} = \begin{bmatrix} -\frac{5}{2} & \frac{3}{2} \\ 2 & -1 \end{bmatrix}$

(c) $|A| = -1 \qquad A^{-1} = \begin{bmatrix} -4 & 3 & -1 \\ 5 & -3 & 1 \\ -2 & 1 & 0 \end{bmatrix}$

17. **(a)** Only if A is nonsingular

27. $A^{-1} = -\frac{1}{18}\begin{bmatrix} 5 & -1 & -7 \\ -1 & -7 & 5 \\ -7 & 5 & -1 \end{bmatrix}$

$A^2 = \begin{bmatrix} 14 & 11 & 11 \\ 11 & 14 & 11 \\ 11 & 11 & 14 \end{bmatrix}$

$(A^2)^{-1} = \frac{1}{324}\begin{bmatrix} 75 & -33 & -33 \\ -33 & 75 & -33 \\ -33 & -33 & 75 \end{bmatrix}$

31. $x_1 = c(k + 5p - 4n) \geq 0$
$x_2 = c(-3k + 3p) \geq 0 \qquad k \leq 30,\ p \leq 20,\ n \leq 30$
$x_3 = c(k - 7p + 2n) \geq 0$
$x_4 = c(k - p + 2n - 60) \geq 0$

33. k, p, and n must satisfy the relation $-16k + p + 5n = 0$. When they do,
$x_1 = 6k - p \geq 0$
$x_2 = -2k + 2p \geq 0 \qquad k \leq 10,\ p \leq 30,\ n \leq 30$
$x_3 = 50 - 4k - p \geq 0$

35. No, since the forms need not be diagonal matrices.

37. Because the cofactors of the elements in the third row, namely, $10(40 - k - p)$, $10(k - p)$, and $20(-10 + p)$, cannot all be zero.

39. $\begin{bmatrix} x_1 \\ x_2 \\ x_3 \end{bmatrix} = \begin{bmatrix} 1 \\ -1 \\ 1 \end{bmatrix}$

Sec. 3.8, page 274

1. **(a)** $\lambda_1 = 1$, $\mathbf{x} = c_1 \begin{bmatrix} 4 \\ 1 \\ -3 \end{bmatrix}$, $c_1 \neq 0$;

$\lambda_2 = 2$, $\mathbf{x} = c_2 \begin{bmatrix} 3 \\ 1 \\ -2 \end{bmatrix}$, $c_2 \neq 0$

The characteristic vectors are not orthogonal.

(b) $\lambda_1 = 1$, $\mathbf{x} = c_1 \begin{bmatrix} 1 \\ 3 \\ 0 \end{bmatrix}$, $c_1 \neq 0$; $\lambda_2 = 2$, $\mathbf{x} = c_2 \begin{bmatrix} 2 \\ 1 \\ 2 \end{bmatrix}$, $c_2 \neq 0$

The characteristic vectors are not orthogonal.

(c) $\lambda_1 = -2$, $\mathbf{x} = c_1\begin{bmatrix}1\\-1\\0\end{bmatrix}$, $c_1 \neq 0$;

$\lambda_2 = 9$, $\mathbf{x} = c_2\begin{bmatrix}2\\2\\-1\end{bmatrix}$, $c_2 \neq 0$;

$\lambda_3 = -18$, $\mathbf{x} = c_3\begin{bmatrix}1\\1\\4\end{bmatrix}$, $c_3 \neq 0$

The characteristic vectors are orthogonal.

(d) $\lambda_1 = 0$, $\mathbf{x} = c_1\begin{bmatrix}1\\1\\1\end{bmatrix}$, $c_1 \neq 0$; $\lambda_2 = 1$,

$\mathbf{x} = c_2\begin{bmatrix}1\\-1\\2\end{bmatrix}$, $c_2 \neq 0$; $\lambda_3 = 2$, $\mathbf{x} = c_3\begin{bmatrix}2\\1\\2\end{bmatrix}$, $c_3 \neq 0$

The characteristic vectors are not orthogonal.

(e) $\lambda_1 = -1$, $\mathbf{x} = c_1\begin{bmatrix}6\\2\\-7\end{bmatrix}$, $c_1 \neq 0$; $\lambda_2 = 1$,

$\mathbf{x} = c_2\begin{bmatrix}0\\1\\-1\end{bmatrix}$, $c_2 \neq 0$; $\lambda_3 = 4$, $\mathbf{x} = c_3\begin{bmatrix}3\\1\\-1\end{bmatrix}$,

$c_3 \neq 0$

The characteristic vectors are not orthogonal.

(f) $\lambda_1 = 1$, $\mathbf{x} = c_1\begin{bmatrix}1\\1\\0\end{bmatrix} + c_2\begin{bmatrix}1\\0\\1\end{bmatrix}$, $c_1^2 + c_2^2 \neq 0$;

$\lambda_2 = 6$, $\mathbf{x} = c_3\begin{bmatrix}1\\1\\1\end{bmatrix}$, $c_3 \neq 0$

The characteristic vectors are not orthogonal.

3. **(a)** $-1/\sqrt{2} \le k \le 1/\sqrt{2}$

(b) $\lambda_1 = 0$, $\mathbf{x} = c_1(k^2 + 6, 2k, -k^2 - 2)$, $c_1 \neq 0$

$$\lambda_2 = \begin{cases} 3 - \sqrt{1 - 2k^2} & k \neq 0 \\ 2 & k = 0 \end{cases}$$

$$\mathbf{x} = \begin{cases} c_2(1 - k^2 - \sqrt{1 - 2k^2}, k\sqrt{1 - 2k^2} - k, 1 - k^2 - \sqrt{1 - 2k^2}) & k \neq 0 \\ c_2(0, 1, 0) \qquad c_2 \neq 0 & k = 0 \end{cases}$$

$\lambda_3 = 3 + \sqrt{1 - 2k^2}$,

$\mathbf{x} = c_3(k^2 - 1 - \sqrt{1 - 2k^2}, k + k\sqrt{1 - 2k^2}, k^2 - 1 - \sqrt{1 - 2k^2})$, $c_3 \neq 0$

(c) $\lambda = 4$, $\mathbf{x} = -2c_3(1, 0, 1)$, $c_3 \neq 0$ (evaluate λ_3 and the corresponding $\mathbf{x}$ in Part **b** when $k = 0$)

(d) $k = \pm 1/\sqrt{2}$

$$\lambda_1 = 0, \mathbf{x} = \begin{cases} c_1\left(\frac{13}{2}, \sqrt{2}, \frac{-5}{2}\right), c_1 \neq 0, k = \frac{1}{\sqrt{2}} \\ c_1\left(\frac{13}{2}, -\sqrt{2}, \frac{-5}{2}\right), c_1 \neq 0, k = \frac{-1}{\sqrt{2}} \end{cases}$$

$$\lambda_2 = 3, \mathbf{x} = \begin{cases} c_2(1, -\sqrt{2}, 1), c_2 \neq 0, k = \frac{1}{\sqrt{2}} \\ c_2(1, \sqrt{2}, 1), c_2 \neq 0, k = \frac{-1}{\sqrt{2}} \end{cases}$$

9. $T = T_1T_2$: $\begin{array}{l} x'' = -x\cos\alpha - y\sin\alpha \\ y'' = -x\sin\alpha + y\cos\alpha \end{array}$

$\begin{vmatrix} -\cos\alpha - \lambda & -\sin\alpha \\ -\sin\alpha & \cos\alpha - \lambda \end{vmatrix} = 0$; $\lambda = 1$, $\dfrac{y}{x} = -\cot\dfrac{\alpha}{2}$;

$\lambda = -1$, $\dfrac{y}{x} = \tan\dfrac{\alpha}{2}$

11. **(a)** $T = T_2T_1$: $\begin{array}{l} x'' = x\sin\alpha + y\cos\alpha \\ y'' = x\cos\alpha - y\sin\alpha \end{array}$

$\begin{vmatrix} \sin\alpha - \lambda & \cos\alpha \\ \cos\alpha & -\sin\alpha - \lambda \end{vmatrix} = 0$; $\lambda = 1$, $\dfrac{y}{x} = \dfrac{1 - \sin\alpha}{\cos\alpha}$;

$\lambda = -1$, $\dfrac{y}{x} = -\dfrac{1 + \sin\alpha}{\cos\alpha}$

(b) $T = T_1T_2$: $\begin{array}{l} x'' = -x\sin\alpha + y\cos\alpha \\ y'' = x\cos\alpha + y\sin\alpha \end{array}$

$\begin{vmatrix} -\sin\alpha - \lambda & \cos\alpha \\ \cos\alpha & \sin\alpha - \lambda \end{vmatrix} = 0$; $\lambda = 1$, $\dfrac{y}{x} = \dfrac{1 + \sin\alpha}{\cos\alpha}$;

$\lambda = -1$, $\dfrac{y}{x} = -\dfrac{\cos\alpha}{1 - \sin\alpha}$

15. If B is nonsingular, one can multiply through by B^{-1}, getting $(\lambda B^{-1}B - B^{-1}A)\mathbf{x} = (\lambda I - C)\mathbf{x} = \mathbf{0}$, where $C = B^{-1}A$.

CHAPTER 4

Sec. 4.1, page 279

1. **(a)** Linear, nonhomogeneous **(b)** Nonlinear

3. **(a)** $x_1'' + x_1' - x_2'' + 3x_3' = 0$
$x_1'' - x_1 + 2x_2' + x_2 - 3x_3' = 0$
$x_1'' + 2x_1' + x_1 - x_2' - 3x_3 = 0$

(b) $x_1'' + 2x_1' + x_1 - x_2' - x_2 + 2x_3'' + 4x_3' + 2x_3 = 1$
$x_1'' - x_1' + x_2'' + x_3' = e^{-t}$
$x_1' - x_1 - 3x_2'' + 6x_2' - 3x_2 + x_3' = 0$

5. **(a)** $\begin{bmatrix}0 & 0\\0 & 0\end{bmatrix}$ **(b)** $\begin{bmatrix}0 & 0\\0 & 0\end{bmatrix}$ **(c)** $4\begin{bmatrix}-t & -b\\-b & -t\end{bmatrix}$

7. $\mathbf{x}$ is a solution.

9. If the third equation is subtracted from 3 times the first equation, the result is $0 = t$.

Sec. 4.2, page 286

1. $x = 4$
$y = -5 + \frac{1}{3}e^{-t}$

3. $x = -\frac{2}{3}c_1e^{t} - \frac{1}{3}c_2e^{-2t} - \frac{9}{2}$;
$y = c_1e^{t} + c_2e^{-2t} + \frac{15}{2} + \frac{1}{2}e^{-t}$

5. $x = 3c_2\cos 3t - 3c_1\sin 3t + t - \frac{34}{9}$
$y = c_1\cos 3t + c_2\sin 3t + \dfrac{2t}{9} - \dfrac{5}{9}$

7. $x = e^{-t}[(-c_1 + c_2)\cos t - (c_1 + c_2)\sin t] - 3\sin t + 2\cos t$
$y = e^{-t}(c_1 \cos t + c_2 \sin t) - \cos t + 3\sin t$

9. $x = (-2c_1 + c_2)e^t - 2c_2te^t - \frac{6}{5}c_3e^{3t}$
$y = c_1e^t + c_2te^t + c_3e^{3t}$

11. $x = -\frac{1}{2}c_1e^t + \frac{1}{2}c_2e^{-t} + \frac{1}{2}c_3e^{3t} - 2e^{2t}$
$y = -c_2e^{-t} + c_3e^{3t} - 2e^{2t}$
$z = c_1e^t + c_2e^{-t} + c_3e^{3t} + 2e^{2t}$

13. $x = e^t;\ y = e^t;\ z = e^t$

15. $x = c_1e^{-t} + c_2e^t + \int_0^t \sinh(t-s)[-z''(s) - z'(s) + z(s)]\,ds$
$y = c_1e^{-t} - c_2e^t + \int_0^t \sinh(t-s)[-z''(s) + 2z(s)]\,ds$
$z =$ any twice-differentiable function of t

17. $x = c_1e^{-t} + c_2e^{-2t} + c_3\cos 2t + c_4\sin 2t - 3$
$y = -c_1e^{-t} - \frac{8}{3}c_2e^{-2t} - 2(c_3 + c_4)\cos 2t + 2(c_3 - c_4)\sin 2t + 2$

19. **(a)** $x = 4\sin 2t - 2;\ y = -3\sin 2t - \cos 2t + 1$
(b) $x = 3e^t + 2\sin 2t + 2\cos 2t - 2;$
$y = -2e^t - \sin 2t - 2\cos 2t + 1$

21. $3(D-2)x + 2(D-2)y = 0$
$2(D+3)x - (D+3)y = 0$

23. **(a)** For all values of a and b, the differential equation $(aD + b)x - (bD - a)y = 0$ is satisfied by the expressions given for x and y. Any two nonproportional members of this family will form a system meeting the required condition.
(b) The expressions $x = c_1\cos t + c_2\sin t$ and $y = c_1\sin t - 2c_2\cos t$ will satisfy the equation $(aD + b)x + (cD + d)y = 0$ for all values of c_1 and c_2 if and only if $a = b = c = d = 0$.

25. $(5D - 4)x - (4D - 5)y = 0$
$(5D^2 - 8)x - 3Dy = 0$

27. If D times the second equation in Exercise 17 is subtracted from $D - 1$ times the first equation, the result is $3y = (D^3 + D^2 + D - 2)x$, which yields y without introducing any additional arbitrary constants. In Exercise 19, if $D + 5$ times the second equation is subtracted from the first equation, the result is $24y = -4 + (D^2 - 3D - 14)x$, which yields y without introducing any additional constants. This method can be used in general.

29. $Q_1 = 100(1 - e^{-t/10});\ Q_2 = 100(1 + e^{-t/10})$

31. **(a)** $cp_1\dfrac{du_1}{dt} + (k_1 + qc)u_1 - qcu_2 = h + k_1T_1$

$-qcu_1 + cp_2\dfrac{du_2}{dt} + (k_2 + qc)u_2 = k_2T_2$

(b) $u_1 = -33.9e^{-t/100} + 0.5e^{-27t/100} + 83.4$
$u_2 = -20.3e^{-t/100} - 7.6e^{-27t/100} + 78.0$
(c) $u_1 = 1.4e^{-31t/100} - 29.8e^{-t/100} + 88.4$
$u_2 = -7.1e^{-31t/100} - 6.0e^{-t/100} + 73.1$

39. **(a)** $a - c + 1 \neq 0$
(b) $a - c + 1 = 0,\ 2a + b - c - 1 \neq 0$
(c) $a - c + 1 = 0,\ 2a + b - c - 1 = 0,\ c - 2b \neq 0$
(d) $a = b = 1, c = 2$; that is, the two equations are identical.

41. **(a)** $\begin{bmatrix} x \\ y \end{bmatrix} = c_1\begin{bmatrix} 1 \\ 2 \end{bmatrix}e^{2t} + c_2\begin{bmatrix} 1 \\ 3 \end{bmatrix}e^{-3t}$

(b) $\begin{bmatrix} x \\ y \end{bmatrix} = c_1\begin{bmatrix} 1 \\ 1 \end{bmatrix}e^{-t} + c_2\begin{bmatrix} 3 \\ 4 \end{bmatrix}e^{2t}$

43. **(a)** $\begin{bmatrix} x \\ y \end{bmatrix} = c\begin{bmatrix} 0 \\ 1 \end{bmatrix}e^{-2t} + \begin{bmatrix} -4 \\ 1 \end{bmatrix} + \begin{bmatrix} 1 \\ -3t \end{bmatrix}e^{-2t}$

(b) $\begin{bmatrix} x \\ y \end{bmatrix} = c_1\begin{bmatrix} -5 \\ 1 \end{bmatrix}e^{-t} + c_2\begin{bmatrix} -2 \\ 1 \end{bmatrix}e^{2t} + \begin{bmatrix} -11 \\ 5 \end{bmatrix}e^t + \begin{bmatrix} -3 \\ 1 \end{bmatrix}$

45. **(a)** $\begin{bmatrix} x \\ y \end{bmatrix} = c_1\begin{bmatrix} 1 \\ 1 \end{bmatrix}e^t + c_2\begin{bmatrix} 1 \\ 2 \end{bmatrix}e^{2t} - \dfrac{1}{4}\begin{bmatrix} 9 + 6t \\ 6 + 4t \end{bmatrix}$

(b) $\begin{bmatrix} x \\ y \end{bmatrix} = c_1\begin{bmatrix} 1 \\ 1 \end{bmatrix}e^t + c_2\begin{bmatrix} t \\ t + 1 \end{bmatrix}e^t - \begin{bmatrix} 9 + 4t \\ 4 + 2t \end{bmatrix}$

47. **(a)** $\begin{bmatrix} x \\ y \end{bmatrix} = c\begin{bmatrix} -3 \\ 1 \end{bmatrix}e^t + \dfrac{1}{2}\begin{bmatrix} -1 \\ 1 \end{bmatrix}\sin t$

(b) $\begin{bmatrix} x \\ y \end{bmatrix} = c_1e^t\begin{bmatrix} \cos t \\ \sin t \end{bmatrix} + c_2e^t\begin{bmatrix} \sin t \\ -\cos t \end{bmatrix} + \dfrac{1}{5}\begin{bmatrix} -2\cos t - \sin t \\ 4\cos t + 3\sin t \end{bmatrix}$

49. **(a)** $\begin{bmatrix} x \\ y \end{bmatrix} = c_1\begin{bmatrix} -3\sin 3t \\ \cos 3t \end{bmatrix} + c_2\begin{bmatrix} 3\cos 3t \\ \sin 3t \end{bmatrix} + \dfrac{1}{3}\begin{bmatrix} 0 \\ -1 \end{bmatrix} + \dfrac{1}{3}\begin{bmatrix} 3\cos 3t \\ \sin 3t \end{bmatrix}\ln|\sec 3t + \tan 3t|$

(b) $\begin{bmatrix} x \\ y \end{bmatrix} = c_1\begin{bmatrix} \cos t \\ -\sin t \end{bmatrix} + c_2\begin{bmatrix} \sin t \\ \cos t \end{bmatrix} - \begin{bmatrix} 0 \\ 1 \end{bmatrix} + \begin{bmatrix} -\cos t \\ \sin t \end{bmatrix}\ln|\sec t + \tan t|$

51. **(a)** $\begin{bmatrix} x \\ y \\ z \\ w \end{bmatrix} = c_1\begin{bmatrix} 0 \\ 1 \\ 2 \\ 2 \end{bmatrix} + c_2\begin{bmatrix} 0 \\ t + 1 \\ 2t + \frac{1}{2} \\ 2t \end{bmatrix} + c_3\begin{bmatrix} 1 \\ 0 \\ 0 \\ 0 \end{bmatrix}e^{2t} + c_4\begin{bmatrix} 0 \\ 1 \\ 0 \\ 0 \end{bmatrix}e^{2t}$

(b) $\begin{bmatrix} x \\ y \\ z \end{bmatrix} = c_1\begin{bmatrix} 1 \\ 1 \\ 1 \end{bmatrix}e^t + c_2\begin{bmatrix} 0 \\ 2 \\ 1 \end{bmatrix}e^{2t} + c_3\begin{bmatrix} -1 \\ 5 \\ 3 \end{bmatrix}e^{3t} + \begin{bmatrix} 1 \\ -2 \\ 0 \end{bmatrix}e^t$

Sec. 4.3, page 293

1. **(a)** $(0, 1]$ **(b)** $(-\infty, -2), (-2, 0), (0, \infty)$

3. **(a)** $x_1' = 4x_1 - x_2 + 5x_3 + t$
$x_2' = 3x_1 + 2x_3 + 2e^{-t}$
$x_3' = x_2 + 6x_3$

(b) $x_1' = (1+t)x_2 - 2tx_3 + 7te^t$
$x_2' = (t-2)x_1 + t^2x_3 - 7t$
$x_3' = 3tx_1 + (t-3)x_2 + 7t^2$

5. (a) $\begin{bmatrix} x_1 \\ x_2 \end{bmatrix}' = \begin{bmatrix} 0 & 1 \\ 4 & 3 \end{bmatrix}\begin{bmatrix} x_1 \\ x_2 \end{bmatrix} + \begin{bmatrix} 0 \\ 1 \end{bmatrix}\frac{1}{2}\ln|t|,\ \mathbf{x}(1) = \begin{bmatrix} 1 \\ -1 \end{bmatrix}$

(b) $\begin{bmatrix} x_1 \\ x_2 \\ x_3 \end{bmatrix}' = \begin{bmatrix} 0 & 1 & 0 \\ 0 & 0 & 1 \\ 0 & -13e^t & 17t \end{bmatrix}\begin{bmatrix} x_1 \\ x_2 \\ x_3 \end{bmatrix} + \begin{bmatrix} 0 \\ 0 \\ 1 \end{bmatrix}(11+7e^{-t}),$

$\mathbf{x}(0) = \begin{bmatrix} 1 \\ 2 \\ 3 \end{bmatrix}$

7. $x_1(t_0) = k_1,\ x_2(t_0) = k_2,\ \ldots,\ x_n(t_0) = k_n$

9. (a) $y = c_1 + c_2e^t - t - t^2/2$ **(b)** $x_1' = x_2$
$x_2' = x_2 + t$

(c) $\begin{bmatrix} x_1 \\ x_2 \end{bmatrix} = c_1\begin{bmatrix} 1 \\ 0 \end{bmatrix} + c_2\begin{bmatrix} 1 \\ 1 \end{bmatrix}e^t - \begin{bmatrix} t + \dfrac{t^2}{2} \\ 1+t \end{bmatrix}$

(d) Use x_1 of Part c.

13. (a) Independent **(b)** Independent **(c)** Dependent **(d)** Dependent **(e)** Independent

Sec. 4.4, page 304

1. (a) $\mathbf{x}_1, \mathbf{x}_2, \mathbf{x}_3$ are fundamental solutions on any interval that does not contain $t = 0$.

(b) $w(t) = t$ is zero when $t = 0$; this does not contradict Theorem 3 since the differential system is not normal on an interval that contains $t = 0$.

(c) $\mathbf{x}_3(t) = \begin{bmatrix} 1 \\ 0 \\ 0 \end{bmatrix}\mathbf{w}(t) + \begin{bmatrix} 0 \\ 1 \\ 0 \end{bmatrix}\mathbf{w}'(t)$

(d) Only **(i)** and **(iii)** are solutions.

9. (b) $\mathbf{v} = \frac{1}{2}t^2\mathbf{x}_1$

11. (a) $x_1 = c_1e^{-t} + c_2te^{-t},$
$x_2 = (-3c_1 + c_2)e^{-t} - 3c_2te^{-t}$

(b) $\begin{bmatrix} 1 \\ -3 \end{bmatrix}$ **(c)** $\dfrac{1}{3}e^{2t}\begin{bmatrix} 1 \\ -3 \end{bmatrix}$ **(d)** $\dfrac{1}{2}te^{-t}\begin{bmatrix} 2+5t \\ 4-15t \end{bmatrix}$

(e) $(t-1)\begin{bmatrix} 1 \\ -3 \end{bmatrix}$

13. $\mathbf{v} = \begin{bmatrix} 1 + \frac{1}{2}\sin 2t \\ -1 + \frac{1}{2}\sin 2t \end{bmatrix}$ **15.** $\mathbf{v} = \begin{bmatrix} t/2 - \ln|t| \\ 2t\ln|t| - \frac{1}{2}t^2 \end{bmatrix}$

17. $\mathbf{x} = e^{-t}\begin{bmatrix} 2\sin 2t \\ -\cos 2t - \sin 2t \end{bmatrix}$

19. $\mathbf{x} = \begin{bmatrix} 2t - (1+2t)\ln t \\ 2t(1 - \ln t) \end{bmatrix}$

21. $\mathbf{x} = \begin{bmatrix} 2 \\ 4 \end{bmatrix}e^{5t} + \begin{bmatrix} \cos t \\ \sin t - 3\cos t \end{bmatrix}$

23. $\mathbf{x} = \begin{bmatrix} 3 \\ 4 \end{bmatrix}e^{2t}(1 + \ln t)$

25. $\mathbf{x} = \begin{bmatrix} 1 \\ 2 \end{bmatrix}e^t(1 + \tan t)$

27. $\mathbf{x} = \begin{bmatrix} 1 \\ 1 \end{bmatrix}e^{4t} + \begin{bmatrix} 2 \\ -2 \end{bmatrix}t^{3/2}e^{-2t}$

29. $\mathbf{x} = \begin{bmatrix} 2 \\ 0 \\ 0 \end{bmatrix}e^{2t} - \begin{bmatrix} 1 \\ 1 \\ 0 \end{bmatrix}e^{-t}$ **31.** $\mathbf{x} = \begin{bmatrix} -1 \\ -2 \\ 1 \end{bmatrix}\operatorname{Tan}^{-1} t$

Sec. 4.5, page 318

1. $\mathbf{x} = c_1\begin{bmatrix} -3\cos 2t - 2\sin 2t \\ \cos 2t \end{bmatrix}e^t + c_2\begin{bmatrix} 2\cos 2t - 3\sin 2t \\ \sin 2t \end{bmatrix}e^t$

3. $\mathbf{x} = c_1\begin{bmatrix} 1 \\ -1 \end{bmatrix}e^{t/2} + c_2\begin{bmatrix} 1+t \\ -\frac{3}{2} - t \end{bmatrix}e^{t/2}$

5. $\mathbf{x} = c_1\begin{bmatrix} 1 \\ -1 \end{bmatrix}e^{-t} + c_2\begin{bmatrix} 1-2t \\ 2t \end{bmatrix}e^{-t}$

7. $\mathbf{x} = c_1\begin{bmatrix} 3\cos t - \sin t \\ 5\cos t \end{bmatrix} + c_2\begin{bmatrix} \cos t + 3\sin t \\ 5\sin t \end{bmatrix}$

9. $\mathbf{x} = c_1\begin{bmatrix} 2\cos t \\ 7\cos t + \sin t \end{bmatrix}e^{-3t} + c_2\begin{bmatrix} 2\sin t \\ 7\sin t - \cos t \end{bmatrix}e^{-3t};$

$\mathbf{x} = \begin{bmatrix} \cos t - \sin t \\ 4\cos t - 3\sin t \end{bmatrix}10e^{-3t}$

11. $\mathbf{x} = c_1\begin{bmatrix} 1 \\ 3 \end{bmatrix}e^{-t} + c_2\begin{bmatrix} 1 \\ 1 \end{bmatrix}e^t + \begin{bmatrix} -1 \\ -2+t \end{bmatrix}$

13. $\mathbf{x} = c_1\begin{bmatrix} 3 \\ -1 \\ 6 \end{bmatrix}e^{-t} + c_2\begin{bmatrix} 1 \\ 0 \\ 1 \end{bmatrix}e^t + c_3\begin{bmatrix} 0 \\ 1 \\ 0 \end{bmatrix}e^{2t};$

$\mathbf{x} = 3\begin{bmatrix} 1 \\ 0 \\ 1 \end{bmatrix}e^t + \begin{bmatrix} 0 \\ 1 \\ 0 \end{bmatrix}e^{2t}$

15. $\mathbf{x} = c_1\begin{bmatrix} 3 \\ -2 \\ 12 \end{bmatrix}e^{-t} + c_2\begin{bmatrix} 1 \\ 0 \\ 2 \end{bmatrix}e^t + c_3\begin{bmatrix} 0 \\ 1 \\ 0 \end{bmatrix}e^{2t};$

$\mathbf{x} = \begin{bmatrix} -3 \\ 2 \\ -12 \end{bmatrix}e^{-t} + \begin{bmatrix} 2 \\ 0 \\ 4 \end{bmatrix}e^t$

17. $\mathbf{x} = c_1\begin{bmatrix} 1 \\ -1 \\ 2 \end{bmatrix}e^t + c_2\begin{bmatrix} 2t-1 \\ -2t-1 \\ 4t \end{bmatrix}e^t + c_3\begin{bmatrix} -\frac{3}{2} - t + t^2 \\ \frac{1}{2} - t - t^2 \\ 2t^2 \end{bmatrix}e^t + \begin{bmatrix} 2 - \frac{5}{8}e^{-t} \\ -1 - \frac{1}{8}e^{-t} \\ 4 - e^{-t} \end{bmatrix}$

19. $x = ce^{-3t} - \frac{1}{2};\ y = ce^{-3t} + \frac{1}{2}$

21. $x = -b_1e^t - 2b_2e^{2t} + t;\ y = b_1e^t + b_2e^{2t} + 1$

23. $x = -2b_1 + b_2e^{-5t} + 5e^t + e^{-t}$
$y = b_1 - 3b_2e^{-5t} - 3e^t + e^{-t}$

25. $x = a_1e^t + 4a_2e^{2t} + 5a_3e^{3t} - 5e^{-t} + 2$
$y = -a_1e^t - 5a_2e^{2t} - 7a_3e^{3t} + 3e^{-t} - 1$

27. $x = 2a_1 \cos t + 2a_2 \sin t + b_1 \cos 2t + b_2 \sin 2t + 2 \sin 3t$
$y = -3a_1 \cos t - 3a_2 \sin t - 3b_1 \cos 2t - 3b_2 \sin 2t - 3 \sin 3t$

29. $x = 4ae^{-t} + 14be^{-2t} + ce^{-3t} + e^t$
$y = -9ae^{-t} - 30be^{-2t} - 2ce^{-3t}$
$z = 6ae^{-t} + 17be^{-2t} + ce^{-3t} - e^{-t}$

31. $\begin{bmatrix} x \\ y \end{bmatrix} = c_1\begin{bmatrix} 2 \\ -1 \end{bmatrix}e^t + c_2\begin{bmatrix} 1 \\ 1 \end{bmatrix}e^{-2t} + \frac{1}{2}\begin{bmatrix} 2 \\ 1 \end{bmatrix}e^{-t}$

33. $\begin{bmatrix} x \\ y \end{bmatrix} = c_1\begin{bmatrix} 4 \\ -3 \end{bmatrix}e^t + c_2\begin{bmatrix} 5 \\ -3 \end{bmatrix}e^{-2t} + \begin{bmatrix} 1 \\ 0 \end{bmatrix}e^t$

35. $\begin{bmatrix} x \\ y \end{bmatrix} = c_1e^{-2t}\left(\begin{bmatrix} 0 \\ 1 \end{bmatrix}\cos t - \begin{bmatrix} 1 \\ -1 \end{bmatrix}\sin t\right) + c_2e^{-2t}\left(\begin{bmatrix} 1 \\ -1 \end{bmatrix}\cos t + \begin{bmatrix} 0 \\ 1 \end{bmatrix}\sin t\right) + \begin{bmatrix} 1 \\ -1 \end{bmatrix}e^t$

37. $\begin{bmatrix} x \\ y \end{bmatrix} = c_1e^{-t}\left(\begin{bmatrix} 1 \\ 1 \end{bmatrix}\cos 2t - \begin{bmatrix} 2 \\ -4 \end{bmatrix}\sin 2t\right) + c_2e^{-t}\left(\begin{bmatrix} 2 \\ -4 \end{bmatrix}\cos 2t + \begin{bmatrix} 1 \\ 1 \end{bmatrix}\sin 2t\right) + \frac{1}{6}\begin{bmatrix} -1 \\ 5 \end{bmatrix}e^{-t}$

39. **(a)** $\frac{wl}{3gT}\ddot{y}_1 + 2y_1 - y_2 = 0;\ -y_1 + \frac{wl}{3gT}\ddot{y}_2 + 2y_2 = 0$
(b) $\omega_1 = \sqrt{3gT/wl};\ \omega_2 = 3\sqrt{gT/wl}$
(c) $y_1 = A_1 \cos \lambda t + B_1 \sin \lambda t + C_1 \cos \sqrt{3}\lambda t + E_1 \sin \sqrt{3}\lambda t$
$y_2 = A_1 \cos \lambda t + B_1 \sin \lambda t - C_1 \cos \sqrt{3}\lambda t - E_1 \sin \sqrt{3}\lambda t$
where $\lambda = \sqrt{3gT/wl}$. At the frequency ω_1 the particles vibrate in phase with each other through equal amplitudes. At the frequency ω_2 the particles vibrate through equal amplitudes but in opposite phase.

41. $x = 2c_1e^t + 4c_2te^t;\ y = -(3c_1 + c_2)e^t - 6c_2te^t$

53. $w(t) = -e^{2t}$

55. $x = c_1\begin{bmatrix} 2 \\ -1 \\ 0 \end{bmatrix}e^t + c_2\begin{bmatrix} -\frac{1}{2} + 2t \\ -t \\ \frac{1}{2} \end{bmatrix}e^t + c_3\begin{bmatrix} -1 + t + 2t^2 \\ -t - t^2 \\ t \end{bmatrix}e^t$

57. **(a)** $\frac{1}{2}\begin{bmatrix} 1 \\ -5 \end{bmatrix}e^t$ **(b)** $\begin{bmatrix} -t - 2 \\ 2t + 2 \end{bmatrix}e^{2t}$ **(c)** $\frac{1}{2}\begin{bmatrix} 3t - 1 \\ -3t + 2 \end{bmatrix}e^{3t}$
(d) $\frac{1}{4}\begin{bmatrix} -t^2 - 2t - 2 \\ 2t^2 + 2t + 2 \end{bmatrix}e^{2t}$

59. **(a)** $\begin{bmatrix} D & 0 \\ 0 & D \end{bmatrix}$ **(b)** $\begin{bmatrix} 2D & -1 \\ 0 & D-1 \end{bmatrix}$ **(c)** $\begin{bmatrix} 1 & 0 \\ 0 & D-2 \end{bmatrix}$
(d) $\begin{bmatrix} 0 & 1 \\ D+1 & 0 \end{bmatrix}$ **(e)** $\begin{bmatrix} 0 & D-2 \\ D+1 & 0 \end{bmatrix}$
(f) $\begin{bmatrix} D & 0 \\ 0 & D-2 \end{bmatrix}$ **(g)** $\begin{bmatrix} 2D & -D \\ 2 & D^2 \end{bmatrix}$ **(h)** $\begin{bmatrix} D & -2 \\ 2 & D \end{bmatrix}$

61. **(a)** $\begin{bmatrix} x \\ y \\ z \end{bmatrix} = \begin{bmatrix} c_1 + \frac{t}{2} \\ 2c_1 + t + \frac{3}{2} \\ -1 \end{bmatrix}e^{-t} + \begin{bmatrix} c_2 + c_3t \\ c_3 \\ -2c_3 \end{bmatrix}e^t + \begin{bmatrix} 1 \\ 1 \\ 0 \end{bmatrix}$
(b) $\begin{bmatrix} x \\ y \\ z \end{bmatrix} = \begin{bmatrix} -c_2 \sin t + c_3 \cos t \\ 0 \\ c_2 \cos t + c_3 \sin t + 1 \end{bmatrix}e^t + \begin{bmatrix} 0 \\ c_1 - \cos t \\ 0 \end{bmatrix}e^{2t}$

CHAPTER 5

Sec. 5.2, page 327

3.

x	$f(x)$				
0.0	1.0000				
		5.1250			
0.5	3.5625		−0.7500		
		3.6250		5.5000	
2.0	9.0000		15.7500		1.0000
		43.0000		9.0000	
3.0	52.0000		42.7500		1.0000
		107.1250		14.5000	
3.5	105.5625		100.7500		
		409.3750			
6.0	1129.0000				

5. **(a)** $\Delta f = 2hx + h^2 + 3h;\ \Delta^2 f = 2h^2$
(b) $\Delta f = 3hx^2 + 3h^2x + h^3 - 2hx - h^2 + 2h$
$\Delta^2 f = 6h^2x + 6h^3 - 2h^2$
$\Delta^3 f = 6h^3$

7. $\Delta^n a^x = a^x(a^h - 1)^n$ **9.** See Theorem 1, Sec. 2.2.

15. $\Delta^k x_h^{(n)} = h^k n(n-1)\cdots(n-k+1)x_h^{(n-k)}$

17. In MacLaurin's expansion, the coefficients are found by successive differentiation and evaluation. In this procedure, the coefficients are found by successive differencing and evaluating.

19. Use the method of synthetic division, divide $p(x)$ and its successive quotients by x, $x - 1$, $x - 2$, . . . , respectively. The successive remainders, beginning with the last, are the coefficients of $x^{(n)}$, $x^{(n-1)}$, $x^{(n-2)}$, ...

21. $\Delta^k x^{(-n)} = (-1)^k n(n+1)(n+2)\cdots(n+k-1)x^{(-n-k)}$

23. $x^{(-n)} = \frac{\Delta x^{(-n+1)}}{-n+1},\ n \neq 1;\ x^{(-n)} = \frac{\Delta^2 x^{(-n+2)}}{(-n+1)(-n+2)}$

25. $x^{(-2)} - 2x^{(-3)} + 2x^{(-4)}$

29. In the process of definite integration, the values of the antiderivative at the ends of the range of integration are subtracted. In the process of summation, the values of the antidifference at the ends of the range of summation are subtracted.

31. **(a)** $\frac{\sin \frac{1}{2}a(n+1) \sin \frac{1}{2}an}{\sin \frac{1}{2}a}$ **(b)** $\frac{\cos \frac{1}{2}a(n+1) \sin \frac{1}{2}an}{\sin \frac{1}{2}a}$

Sec. 5.3, page 337

3. **(a)** 7.08520 **(b)** 7.47663 **5.** $x^3 - 3x^2 + 17$

11. $f_0 - \dfrac{(f_{01} - f_{12} + f_{02})^2}{4f_{012}}$

15. $e^x = 1 + r(1 - e^{-1}) + \dfrac{r(r+1)}{2!}(1 - e^{-1})^2 + \dfrac{r(r+1)(r+2)}{24}(1 - e^{-1})^3 + \cdots$

$x = x_0 + rh = 0 + r \cdot 1 = r$

21. **(a)** $S(x) = \begin{cases} 5.5 - x & 3.0 \le x \le 4.5 \\ -1.7 + 0.6x & 4.5 \le x \le 7.0 \\ 9.0 - x & 7.0 \le x \le 9.0 \end{cases}$

(b) 1.3

23. **(a)** $S(x) = \begin{cases} 1 + x^3 & 0 \le x \le 2 \\ 25 - 36x + 18x^2 - 2x^3 & 2 \le x \le 4 \\ -103 + 60x - 6x^2 & 4 \le x \le 6 \end{cases}$

(b) No

25. **(a)** $S(x) = \begin{cases} 2x - kx^2 - kx^3 & -1 \le x \le 0 \\ 2x - kx^2 + kx^3 & 0 \le x \le 1 \end{cases}$

(b) No, because a natural spline must be a cubic, not a linear, spline.

Sec. 5.4, page 343

1. $f'(200) \doteq 0.00500000$, correct value $\doteq 0.00500000$
$f''(200) \doteq -0.00002499$, correct value $\doteq -0.00002500$
$f'''(200) \doteq 0.00000024$, correct value $\doteq 0.00000025$
$f'(205) \doteq 0.00487806$, correct value $\doteq 0.00487805$
$f''(205) \doteq -0.00002377$, correct value $\doteq -0.00002380$
$f'''(205) \doteq 0.00000025$, correct value $\doteq 0.00000023$

3. $f'(0) \doteq 0.2005$, $f'(10) \doteq 1.170$

5.

x	$\int_0^x e^{-t^2}\,dt$	x	$\int_0^x e^{-t^2}\,dt$
0.0	0.00000	0.6	0.53446
0.1	0.09950	0.7	0.59996
0.2	0.19704	0.8	0.65696
0.3	0.29078	0.9	0.70556
0.4	0.37908	1.0	0.74620
0.5	0.46063		

7. $I = \dfrac{hk}{9}\sum_{i,j} w_{ij}f_{ij}$, where

$w_{ij} = \begin{cases} 1 & i = j = 0 \\ 2 & i = 0,\ j \text{ even};\ j = 0,\ i \text{ even} \\ 4 & i = 0,\ j \text{ odd};\ j = 0,\ i \text{ odd}; \\ & i,\ j \text{ both even and } >0 \\ 8 & i \text{ odd},\ j \text{ even and } >0;\ j \text{ odd},\ i \text{ even and } >0 \\ 16 & i,\ j \text{ both odd} \end{cases}$

9. Yes; Eq. (1) yields identical results.

Sec. 5.5, page 349

3. **(a)** $y = \frac{721}{245}x - \frac{260}{245}x^2$ **(b)** No; it equals $-\frac{157}{49}$
(c) $\frac{1}{5}E^2 \doteq 0.693$
(d) It seems improbable, at first glance. However, the fact that the assumed curve must always pass through (0, 0) seriously limits its ability to fit the data.

5. One of the f's must be a constant.

7. **(a)** $x \doteq 1.683$, $y \doteq -1.847$ **(b)** $x \doteq 1.739$, $y \doteq -1.811$ The factor 4 acts as a weight which attaches relatively more significance to the third equation.

9. $y = 0..889x^{3.496}$

11. $y = 0.873 + 1.690x$

13. $y = 1.545 - 0.664x$. No; on $[0, \pi]$ the least-square approximation is $y = 12/\pi^2 - (24/\pi^3)x = 1.216 - 0.774x$.

15. $y = x - \frac{77}{509}x^3$

17. After y_i' has been replaced by $(1/h)(y_{i+1} - y_i)$, it is a straightforward matter to find a by the method of least squares. After a has been determined, A can be found by applying the method of least squares to the equations

$$y_1 = Ae^{ax_1} \qquad y_2 = Ae^{ax_2} \qquad \ldots \qquad y_n = Ae^{ax_n}$$

in which A is the only unknown.

19. The procedure suggested in Exercises 17 and 18 is preferable to linearizing by taking logarithms because it does not introduce any unwarranted weighting of the data. It is clearly simpler than linearizing through the use of Taylor's series, although the repeated use of Taylor's series ultimately leads to a smaller average squared error.

Sec. 5.6, page 360

1. $y(0.4) \doteq -1.4682$, $y(0.6) \doteq -1.7379$

3. $y(0.1) \doteq -1.100$, $y(0.2) \doteq -1.209$, $y(0.3) \doteq -1.326$

5. $y(1.1) \doteq 1.0048$, $y(1.2) \doteq 1.0187$, $y(1.3) \doteq 1.0408$. These values are correct to four decimal places.

7. Solving the initial-value problem $y' = -2xy$, $y = 1$ when $x = 0$, yields the following values:

x	e^{-x^2}
0.0	1.00000
0.1	0.99005
0.2	0.96079
0.3	0.91393
0.4	0.85214
0.5	0.77880

These values are correct to five decimal places.

9.

x	y	z
0.0	0.00000	1.00000
0.1	0.10500	0.99983
0.2	0.21994	0.99866
0.3	0.34467	0.99550

11. $y(0.4) \doteq 0.7407$, $y(0.5) \doteq 0.7131$, $y(0.6) \doteq 0.6977$ These values are correct to four decimal places.

13.

x	y
0.0	0.0000
0.1	0.0050
0.2	0.0200
0.3	0.0451
0.4	0.0805
0.5	0.1256

15.

x	y
0.4	0.7406
0.5	0.7130
0.6	0.6975

17. $y(0.7) = -1.8763$
$y(0.8) = -2.0145$

21. $y = y_0 + hy_0'\left(\frac{x}{h}\right) + (-3y_0 + 3y_1 - 2hy_0' - hy_1')\left(\frac{x}{h}\right)^2 + (2y_0 - 2y_1 + hy_0' + hy_1')\left(\frac{x}{h}\right)^3$

$y_2 = 5y_0 - 4y_1 + 2hy_0' + 4hy_1'$.
One way to obtain a closed formula is to read y_2 from the polynomial of minimum degree which takes on the values y_0, y_0', y_1', and y_2' at $x = 0, h, 2h$, respectively. The result is $y_2 = y_0 + (h/3)(y_0' + 4y_1' + y_2')$.

The first formula for y_2 can be used as an open formula in the same way that Milne's formula is used: With x_0 and y_0 given, y_0' can be found from the given differential equation. Then after y_1 has been found, by means of Taylor's series, for instance, y_1' can be found from the differential equation. With these data known, the open formula can be used to determine y_2. With y_2 available, y_2' can be found from the differential equation, and the closed formula can be used to check and correct y_2. Once a satisfactory value of y_2 has been determined, y_3 can be found similarly, and so on as far as desired.

23. $y = y_0 + \left(-y_0 + y_1 - \frac{1}{3}h^2y_0'' - \frac{1}{6}h^2y_1''\right)\left(\frac{x}{h}\right) + \frac{1}{2}y_0''\left(\frac{x}{h}\right)^2 - \frac{1}{6}h^2(y_0'' - y_1'')\left(\frac{x}{3}\right)^3$

$y_2 = -y_0 + 2y_1 + h^2y_1''$. An accompanying closed formula can be obtained by fitting a cubic polynomial to y_2'' (which can be obtained from the given differential equation as soon as a first approximation for y_2 is obtained) and any three of the data y_0, y_0'', y_1, y_1'', provided only two second-derivative values are used. This formula can be used exactly like the formula in Exercise 21 is used, except that here it is the value of y'', rather than y', that is determined from the given differential equation.

25. **(a)** $\frac{3}{8}h^4y_{n-2}^{\text{iv}}$ **(b)** $-\frac{1}{24}h^4y_{n-2}^{\text{iv}}$

Sec. 5.7, page 374

1. $(E^2 - 5E + 6)y = 0$ **3.** $(E^2 + 1)y = 3^x$

5. Each of the roots of $F(E) = 0$ is 1 more than the corresponding root of $f(\Delta) = 0$.

9.		11.		13.	
$y_0 =$	1	$y_0 =$	1	$y_0 =$	1
$y_1 =$	-1	$y_1 =$	4	$y_1 =$	1
$y_2 =$	-4	$y_2 =$	2	$y_2 =$	-2
$y_3 =$	-6	$y_3 =$	-6	$y_3 =$	-4
$y_4 =$	-4	$y_4 =$	-10	$y_4 =$	-4
$y_5 =$	4	$y_5 =$	2	$y_5 =$	-4

15.

$y_{-4} = \frac{23}{8}$	$y_1 = -1$
$y_{-3} = \frac{11}{4}$	$y_2 = -5$
$y_{-2} = \frac{5}{2}$	$y_3 = -13$
$y_{-1} = 2$	$y_4 = -29$
$y_0 = 1$	

17. $y = c_15^n + c_2(-2)^n - 2^{n-2}$

19. $y = c_14^n + c_25^n + \frac{1}{2}\,3^n$

21. $y = c_12^n + c_2(-1)^n + \frac{1}{10}4^n$

23. $y = c_13^n + c_2(-2)^n - n^2 - 3n - 1$

25. $y = c_1(-1)^n + c_2n(-1)^n + \frac{1}{4}$

27. $y = 2^n\left(c_1 \cos\frac{\pi n}{2} + c_2 \sin\frac{\pi n}{2}\right) + \frac{\cos(n-2) + 4\cos n}{17 + 8\cos 2}$

29. $y = c_1 + c_2(-1)^n + 3^n\left(c_3 \cos\frac{\pi n}{2} + c_4 \sin\frac{\pi n}{2}\right) + n$

31. $y = 2^{n/2}\left(\cos\frac{\pi n}{4} - 2\sin\frac{\pi n}{4}\right)$

$y = \frac{3}{2} - \frac{1}{2}\cos\frac{\pi n}{2} - \frac{3}{2}\sin\frac{\pi n}{2}$

$y = 2^{n/2}(\cos n\theta + \sqrt{7}\sin n\theta);\ \theta = \tan^{-1}\sqrt{7}$

37. **(a)** $[n(n+1)]/2$ **(b)** $[4n^4 - n]/3$ **(c)** $[n^2(n+1)^2]/4$

39.

$1 + n$	$a = 2$
$\frac{\sin(n+1)\mu}{\sin\mu}$, $\cos\mu = \frac{a}{2}$	$-2 < a < 2$
$(-1)^n(1+n)$	$a = -2$
$(-1)^n\frac{\sinh(n+1)\mu}{\sinh\mu}$, $\cosh\mu = -\frac{a}{2}$	$a < -2$

41. $V_n = \frac{V_0}{2^N + 1}\left(\frac{2^N}{2^n} + 2^n\right)$

43. $I_n = \frac{\pi\sin n\lambda}{\sin\lambda}$

45. $x_n = \frac{-\sinh n\mu}{2k\cosh\frac{2n+1}{2}\mu\sinh\frac{\mu}{2}}F_0$,

$\mu = \cosh^{-1}\left(1 + \frac{\lambda}{2}\right)$, $\lambda > 0$

47. $V = V_0\frac{\sinh(N-n)\mu}{\sinh N\mu}$. After noting that $e^\mu = 2$, the two answers are readily shown to be equivalent.

51. $y_n = -\frac{11}{5}(-1)^n - \frac{14}{5}(-6)^n + 2$
$z_n = -\frac{22}{5}(-1)^n + \frac{42}{5}(-6)^n - 2$

53. **(b)** Δ^{-1} has the property that $\Delta^{-1}\phi(n) \equiv \theta(n)$ is a function such that $\Delta\theta(n) \equiv \Delta\Delta^{-1}\phi(n) \equiv \phi(n)$.

55. $A_k = -\sum_{i=j}^{k-1}\frac{f_iv_{i+1}}{C_{i+1}} + c_1$

$B_k = \sum_{i=j}^{k-1}\frac{f_iu_{i+1}}{C_{i+1}} + c_2$, where C_{i+1} denotes the **casoratian,** or Casorati determinant, of u and v evaluated at $k + 1$.

57. $Y_k = \frac{1}{50}(3k - 5k^2)$

59. $c_1\begin{bmatrix} 5 \\ -3 \end{bmatrix} + c_2\begin{bmatrix} 4 \\ -3 \end{bmatrix}(-2)^k$

Sec. 5.8, page 379

1. The solution obtained by Euler's method is $y = c(1 + Ah)^{(x_n - x_0)/h}$ which approaches the true solution $y = ce^{Ax}$ for all values of A as $h \to 0$.

3. The formula is numerically stable for the equation $y' = Ay$ for all values of A as $h \to 0$.
5. The formula is numerically stable for the equation $y' = Ay$ only if $A > 0$.

Sec. 5.9, page 389

1. $y_n = \frac{5}{2}2^n - \frac{3}{4}4^n$
3. $y_n = n3^{n-1}$
5. $y_n = -\frac{5}{2} + 2^{n+2} - \frac{1}{2}\,3^{n+1}$
7. $y_n = \frac{1}{3}[(-1)^n - (-4)^n]$; $z_n = \frac{1}{3}[2(-1)^n + (-4)^n]$
9. $y_n = 2^{n+1}\dfrac{\sin (n+1)\phi}{\sin \phi} + 2^{n+2}\dfrac{\sin n\phi}{\sin \phi}$

   $z_n = -2^n \dfrac{\sin (n+1)\phi + \sin n\phi}{\sin \phi}$ $\qquad \phi = \cos^{-1}(-\frac{3}{4})$
11. **(a)** $f_n = 0, 1, 1, 1, 1, \ldots$ **(b)** $f_n = 0, 0, 1, 1, 1, \ldots$
19. $G(\sin na) = \dfrac{s \sin a}{1 - 2s\cos a + s^2}$
21. $G[\sin (n+1)a] = \dfrac{\sin a}{1 - 2s \cos a + s^2}$
23. $G[\sin (na + b)] = \dfrac{s \sin a \cos b}{1 - 2s\cos a + s^2} + \dfrac{\sin b}{1 + a^2n^2s^2}$
25. $\frac{1}{2}(n+1)(n+2)$
27. $y_n = \displaystyle\int_0^t \frac{(u-t)^n}{n!} f'(u)\,du + \sum_{j=0}^{n} \frac{(-t)^j}{j!} h_{n-j}$
29. $y_n = \displaystyle\int_0^t \frac{(t^2-u^2)^n}{n!} f'(u)\,du + \sum_{j=0}^{n} \frac{t^{2j}}{j!} h_{n-j}$
31. $\dfrac{1 - e^a s \cos b}{1 - 2se^a \cos b + e^{2a}s^2}$
33. $\dfrac{d}{ds}[sG(f_n)]$

CHAPTER 6

Sec. 6.2, page 396

1. $(-1, 2)$
3. $(1, 1), (1, -1), (-1, 1), (-1, -1)$
5. $x = c_1e^t + c_2e^{2t}$, $y = c_1e^t + c_2e^{2t}$; $x - y = k(2x - y)^2$
7. $x = c_1 \cos t + c_2 \sin t$, $y = -c_1 \sin t + c_2 \cos t$; $x^2 + y^2 = k^2$
9. $x = c_1e^t + c_2e^{-t}$, $y = c_1e^t - c_2e^{-t}$; $x^2 - y^2 = k$
11. $x = c_1e^{-t} + c_2e^t$, $y = c_2e^t$; $xy - y^2 = k$
13. $x^4 + x^2 + y^2 = k$. The trajectories are closed curves in the xy plane. We cannot find parametric equations for the trajectories since the equation for x requires elliptic functions for its solution.
15. **(a)** $x = t^3 - 2t + 2$, $y = 3t^2 - 2$
    (b) $x = t^3 - 11t + 15$, $y = 3t^2 - 11$
    Infinitely many. More than one trajectory of a nonautonomous system may pass through a particular point. Only one trajectory of an autonomous system can pass through a particular point.

Sec. 6.3, page 408

1. Saddle point, necessarily stable
3. Stable node
5. Unstable node
7. No, because the property of stability is not included

Sec. 6.4, page 417

1. Saddle point, necessarily unstable
3. Saddle point, necessarily unstable
5. Unstable spiral point
7. Unstable node
9. $x^2y + 3xy - 4x^2 + 2y^2 = k$
11. The critical point $(0, 0)$ is an asymptotically stable node. The critical point $(1, -\frac{1}{2})$ is a saddle point, necessarily unstable.
13. The critical point $(0, 0)$ is an unstable node. The critical point $(-1, 1)$ is a saddle point, necessarily unstable.
15. The critical point $(1, 1)$ is an asymptotically stable spiral point. The critical point $(-1, 1)$ is a saddle point, necessarily unstable.
17. In Example 2, the formula for the slope of a trajectory is

$$m_t = -\frac{r\cos\theta + r^2 \sin^2\theta}{r \sin\theta} = -\frac{\cos\theta}{\sin\theta} - r\sin\theta$$
$$= m_c - r\sin\theta$$

    Since $r \sin \theta$ can be either positive or negative, it cannot be concluded that $m_t > m_c$, as in Example 3.
21. The parameter h is the total energy, kinetic plus potential, in the system.
25. **(a)** The point $(x_0, 0)$
    (b) A pair of curves intersecting at $(x_0, 0)$ and serving as asymptotes for the two families of hyperbola-like trajectories.
27. The critical point $(0, 0)$ is a center and necessarily stable.
29. The system has critical points at $(0, 0)$, $(1/\sqrt{2}, 0)$, $(-1/\sqrt{2}, 0)$. V has a minimum at $x = 0$, hence $(0, 0)$ is a center and necessarily stable. V has maxima at $(1/\sqrt{2}, 0)$ and $(-1/\sqrt{2}, 0)$, hence these are saddle points and necessarily unstable. If $h = \frac{1}{4}$, the corresponding path curve consists of the two parabolas $2y = 1 - 2x^2$ and $2y = 2x^2 - 1$. The closed trajectories lie in the finite region bounded by the two parabolas.
31. **(a)** The critical points are $(0, 0)$, which is a saddle point, and $(\frac{2}{3}, 0)$, which is a center.
    (b) The critical points are $(0, 0)$, which is a center, and $(\frac{2}{3}, 0)$, which is a saddle point.

Sec. 6.5, page 424

1. Use $4x^2 + y^2$.
3. Use $-x^2 + y^2$.
5. Asymptotically stable
7. Unstable
9. Unstable
11. $\dot{V}$ is negative semidefinite for $V = x^2 + y^2$ and is negative definite for $V = 3x^2 + y^2$.

Sec. 6.6, page 431

1. $r = 2$, $\theta = t + t_0$, stable
3. $r = 1$, $\theta = t + t_0$, stable; $r = 2$, $\theta = t + t_0$, unstable
5. $r = 1$, $\theta = -t + t_0$, stable
   $r = 2$, $\theta = -t + t_0$, semistable
   $r = 3$, $\theta = -t + t_0$, unstable

7. (a) $r = \dfrac{2}{(1 + ke^{-8t})^{1/2}}$, $\theta = t + t_0$

   (b) $\ln\left|\dfrac{r}{r-1}\right| - \dfrac{1}{r-1} = t + k$, $\theta = -t + t_0$

   (c) $\dfrac{r(r-2)}{(r-1)^2} = ke^{2t}$, $\theta = t + t_0$

21. If Γ is a closed trajectory defined by the particular periodic solution $x = f(t)$, $y = g(t)$, then $x = f(t + t_0)$, $y = g(t + t_0)$ are other periodic solutions defining the same closed trajectory.

CHAPTER 7

Sec. 7.2, page 441

1. $k = k_1k_2/(k_1 + k_2)$ 3. $\frac{1}{3}M_s$
5. In equilibrium, the upper spring is stretched an amount equal to

$$\frac{w + k_2(l - l_1 - l_2)}{k_1 + k_2}$$

7. $CL\dfrac{d^2i_i}{dt^2} + \dfrac{L}{R}\dfrac{di_i}{dt} + i_i = I_0 \cos \omega t$

   $CL\dfrac{d^2i_r}{dt^2} + \dfrac{L}{R}\dfrac{di_r}{dt} + i_r = -\dfrac{\omega L}{R}I_0 \sin \omega t$

   $CL\dfrac{d^2i_c}{dt^2} + \dfrac{L}{R}\dfrac{di_c}{dt} + i_c = -\omega^2 LCI_0 \cos \omega t$

9. $\ddot{h} + \dfrac{2g}{l}h = 0$ 11. $\ddot{\theta} + \dfrac{2g}{3(R-r)}\theta = 0$

13. $\dfrac{d^2(\theta_1 - \theta_2)}{dt^2} + k\dfrac{I_1 + I_2}{I_1I_2}(\theta_1 - \theta_2) = 0$

Sec. 7.3, page 456

1. $y = (2 \cos 12t + \frac{1}{6} \sin 12t)$ in
3. $\ddot{x} - 2\dot{x} + x = e^t/t$; $x = c_1e^t + c_2te^t + te^t \ln t$, $t > 0$
5. $y = 6(4t - 1)e^{-24t}$ in; when $t = \frac{1}{4}$ s
7. $y = -4 \cos 2t + \sin 2t$; $y(\pi/2) = 4$ ft
9. $y = \sqrt{2} \cos (24t - 3\pi/4)$ in; $\omega_r = \frac{4}{3}\sqrt{94}$ rad/s
11. $y = \frac{4}{7} \sin (21t - \pi/4)$ in; $\omega_r = 7\sqrt{\frac{23}{2}} \doteq 23.74$ rad/s
13. $y = \frac{5}{2} \cos \frac{48}{5}t - \frac{5}{2}$
15. $y = F_0\dfrac{e^{-32t} - e^{-8t}}{120} + F_0\dfrac{\sin 16t}{120}$
17. (a) $y = (36/\sqrt{3361}) \cos (5t - \alpha)$, $\alpha = \tan^{-1} \frac{15}{56}$
    (b) $y = (36/\sqrt{1261}) \cos (10t - \alpha)$, $\alpha = \tan^{-1} (-\frac{30}{19})$
    (c) $y = (4/\sqrt{281}) \cos (15t - \alpha)$, $\alpha = \tan^{-1} (-\frac{5}{16})$
19. $c = c_c = 120\sqrt{15}$; $y = -5t \exp [-(4\sqrt{15}/5)t]$
21. $M = \dfrac{1}{1 + \omega^2/\omega_n^2}$
23. 12.52 Hz
25. $y \doteq 0.91 \sin (15t - 48°34')$
27. $y = -e^{-6t}(2 \cos 8t + \frac{3}{2} \sin 8t) + 2$
29. $y = \dfrac{F_0}{241}[e^{-4.2t}(-3 \cos 14.4t + \frac{29}{24} \sin 14.4t) + 3e^{-10t}]$
31. The first integer greater than $\dfrac{\ln 2}{2\pi}\dfrac{\sqrt{1 - (c/c_c)^2}}{c/c_c}$
35. Yes; the particular solutions $y = e^{-rt}(\cosh st \pm \sinh st)$ are not included in either of the given forms.
37. (a) $y = (A + Bt) \exp (-\omega_n t) + \delta_{st}\dfrac{1}{1 + (\omega/\omega_n)^2} \cos (\omega t - \alpha)$

    (b) $y = A \exp [-(c/c_c) - \sqrt{(c/c_c)^2 - 1}]\omega_n t$
    $+ B \exp [-(c/c_c) + \sqrt{(c/c_c)^2 - 1}]\omega_n t + \delta_{st} M \cos (\omega t - \alpha)$
45. $\dfrac{1}{2(c/c_c)[1 - (c/c_c)^2]^{1/2}}$
49. $\omega_n = \sqrt{2g/l}$. The natural frequency ω_n does not depend on either the mass M_0 or the polar moment of inertia I_0 of the disk or on the length a, since these parameters are eliminated from the differential equation by the relation $I_0 = \frac{1}{2}M_0a^2$. However, if a body of known mass M but unknown moment of inertia I is placed on the disk so that the line about which the polar moment of inertia of the body is to be determined coincides with the polar axis of the disk, a repetition of the analysis leads to the frequency $\omega_n = \sqrt{2\dfrac{I_0 + Ma^2}{I_0 + I}\dfrac{g}{l}}$. With ω_n determined experimentally, I is the only unknown in this equation.
51. $2(w/g)v_0^2 + cv_0y_0 > 0$
55. $p_d = 4\pi/\sqrt{626} \doteq 0.5$ s; $\omega_r = 12$ rad/s; $Y_{ss} = (34/\sqrt{313}) \cos (12t - \alpha)$ in, $\alpha = \tan^{-1} (12\sqrt{2}/5) \doteq 73.58°$; $\alpha = 135°$ when $\omega = (5 + 11\sqrt{3})/\sqrt{2} \doteq 17.01$ rad/s
57. $Y_{ss} \doteq (F_0/50)[2.28 \cos (15t - 2°10') + 2.77 \cos (16t - 2°48')]$. Because of the near equality of the frequencies and the amplitudes of the impressed forces, the phenomenon of beats is to be expected.
59. $\alpha = \tan^{-1} \sqrt{(c_c/c)^2 - 2}$
61. The energy loss per cycle is $c\pi A^2\omega$, where A is the amplitude of the driving force.

Sec. 7.4, page 467

1. Across the resistor: $E = iR = 40(e^{-200t} - e^{-800t})$

   Across the inductor: $E = L\dfrac{di}{dt} = -8e^{-200t} + 32e^{-800t}$

   Across the capacitor: $E = (1/C) \int_0^t i\, dt = -32e^{-200t} + 8e^{-800t} + 24$

3. $i = \frac{1}{15}e^{-20,000t} - \frac{1}{15}e^{-5000t}$

5. $i = -\frac{1}{8}e^{-6000t}\sin 8000t$

7. $i = \frac{625}{4}te^{-2500t}$

9. $i = -50\pi Ae^{-500t} + 500(50\pi A + 500B)te^{-500t} + 50\pi A\cos 50\pi t - 50\pi B\sin 50\pi t$

where $$A = \frac{110[\frac{1}{4}\times 10^6 - (50\pi)^2]}{[\frac{1}{4}\times 10^6 - (50\pi)^2]^2 + (50{,}000\pi)^2}$$

$$B = \frac{110(50{,}000\pi)}{[\frac{1}{4}\times 10^6 - (50\pi)^2]^2 + (50{,}000\pi)^2}$$

11. $t = 0.00039$

13. $E_{ss} = \frac{150}{109}\sin(150t + \delta)$, where $\delta = \tan^{-1}\frac{91}{60}$

17. If an impedance Z consists of n impedances $Z_1, Z_2, \ldots, Z_n$ connected in parallel, then $1/Z = 1/Z_1 + 1/Z_2 + \cdots + 1/Z_n$.

21. $|Z|$ is a minimum for the natural frequency $\Omega_n = 1/\sqrt{LC}$. For the magnification ratio the maximum always occurs at a frequency below the natural frequency. This involves no contradiction since the magnification ratio M relates F and y, whose electrical analogs are E and Q, whereas the impedance relates E and $i = dQ/dt$.

25. $i = \frac{3}{50}\cos(60\pi t + \pi/6)$ A; $\Omega_n = 60\pi\sqrt{2}$ rad/s; $\omega_r = 30\pi\sqrt{2}$ rad/s

Sec. 7.5, page 474

1. $\omega_1 = 1$, $a_1{:}a_2 = 1{:}1$; $\omega_2 = 2$, $a_1{:}a_2 = -2{:}1$

3. **(a)** $\omega_1 = \sqrt{3}$, $a_1{:}a_2 = 3{:}1$; $\omega_2 = \sqrt{13}$, $a_1{:}a_2 = -1{:}3$
(b) $\omega_1 = 2$, $a_1{:}a_2 = 1{:}1$; $\omega_2 = 3$, $a_1{:}a_2 = 1{:}-1$
(c) $\omega_1 = \sqrt{\frac{3}{2}}$, $a_1{:}a_2 = 1{:}2$; $\omega_2 = \sqrt{3}$, $a_1{:}a_2 = 1{:}-1$.
(d) $\omega_1 = 1/\sqrt{2}$, $a_1{:}a_2 = 1{:}\frac{2}{3}$; $\omega_2 = \sqrt{\frac{3}{2}}$, $a_1{:}a_2 = 1{:}-\frac{2}{3}$
(e) $\omega_1 = 1$, $a_1{:}a_2 = 1{:}2$; $\omega_2 = 2$, $a_1{:}a_2 = 1{:}-1$.

5. **(a)** $\omega_1 = \frac{1}{2}$, $a_1{:}a_2{:}a_3 = 3{:}3{:}2$; $\omega_2 = 1$, $a_1{:}a_2{:}a_3 = 3{:}0{:}-10$; $\omega_3 = 2$; $a_1{:}a_2{:}a_3 = 3{:}-12{:}2$
(b) $\omega_1 = 1/\sqrt{2}$, $a_1{:}a_2 = 1{:}\frac{3}{2}$; $\omega_2 = \sqrt{6}$, $a_1{:}a_2 = 1{:}-\frac{1}{3}$

7. $$X = \begin{bmatrix} x_1 \\ x_2 \end{bmatrix} = -\frac{1}{2}\begin{bmatrix} 1 \\ 1 \end{bmatrix}\cos t + \frac{1}{2}\begin{bmatrix} 3 \\ -1 \end{bmatrix}\cos\sqrt{5}t$$

9. $$Y = \begin{bmatrix} y_1 \\ y_2 \end{bmatrix} = \frac{16}{3}\begin{bmatrix} 1 \\ 2 \end{bmatrix}\cos t - \frac{4}{3}\begin{bmatrix} 1 \\ -1 \end{bmatrix}\cos 2t + \begin{bmatrix} 4 \\ 12 \end{bmatrix}$$

11. $$x_1 = \frac{(\omega^6 + 6\omega^4 - 11\omega^2 + 8) - j(\omega^5 - 3\omega^3 + 6\omega)}{(\omega^4 - 6\omega^2 + 4)^2 + (-2\omega^3 + 5\omega)^2}\frac{F_0}{4}e^{j\omega t}$$

$$x_2 = \frac{(\omega^4 - 6\omega^2 + 4) - j(-2\omega^3 + 5\omega)}{(\omega^4 - 6\omega^2 + 4)^2 + (-2\omega^3 + 5\omega)^2}\frac{F_0}{2}e^{j\omega t}$$

x_1 lags the driving force by θ_1/ω units of time, where

$$\theta_1 = \tan^{-1}\frac{\omega^5 - 3\omega^3 + 6\omega}{-\omega^6 + 6\omega^4 - 11\omega^2 + 8}$$

x_2 lags the driving force by θ_2/ω units of time, where

$$\theta_2 = \tan^{-1}\frac{-2\omega^3 + 5\omega}{\omega^4 - 6\omega^2 + 4}$$

15. Not necessarily

17. $\omega_1 = \sqrt{k/M}$, $a_1{:}a_2 = 1{:}1$; $\omega_2 = \sqrt{(k + 2k_2)/M}$, $a_1{:}a_2 = 1{:}-1$; where $k = k_1 = k_2$ and $M = M_1 = M_2$
$x_1 + x_2 = A\cos\omega_1 t + B\sin\omega_1 t$
$x_1 - x_2 = C\cos\omega_2 t + D\sin\omega_2 t$
Beating would be anticipated if $\omega_1 \doteq \omega_2$, that is, if the coupling spring is very weak.

19. Because of symmetry, the normal modes of the system are vertical motion without rotation, with frequency $\omega_1 = 12\sqrt{2}$, and pure rotation without translation, with frequency $\omega_2 = 12\sqrt{6}$.

21. **(a)** The system is being driven at resonance and infinite amplitudes build up since a particular integral is $x_1 = \frac{1}{8}t\sin t + \frac{1}{4}\cos t$, $x_2 = \frac{1}{8}\cos t$.
(b) The system is being driven at resonance and infinite amplitudes build up since a particular integral is $x_1 = -\frac{1}{8}t\sin t + \frac{1}{12}\cos t$, $x_2 = -\frac{1}{8}\cos t$.
(c) The system is being driven at resonance, but infinite amplitudes do not build up since a particular integral is $x_1 = \frac{1}{4}\cos t$, $x_2 = -\frac{1}{12}\cos t$. If the forces applied to M_1 and M_2 are $\cos\omega t$ and $-\cos\omega t$, respectively, a particular integral is

$$x_1 = \frac{1 - \omega^2}{(\omega^2 - 1)(\omega^2 - 5)}\cos\omega t$$

$$x_2 = \frac{\omega^2 - 1}{(\omega^2 - 1)(\omega^2 - 5)}\cos\omega t$$

As $\omega \to 1$, these become $x_1 = \frac{1}{4}\cos t$ and $x_2 = -\frac{1}{12}\cos\omega t$, as in Part **(c)**. No, infinite amplitudes do not always occur.

23. $$i_1 = \frac{Q_0}{11RC}\left[-\frac{1}{6}e^{-t/12RC} - 9e^{-t/RC}\right]$$

$$i_2 = \frac{Q_0}{11RC}\left[\frac{1}{2}e^{-t/12RC} - 6e^{-t/RC}\right]$$

25. $$i_1 = \frac{E}{R}\left[-\frac{9}{17}e^{-3Rt/5L} - \frac{8}{17}e^{-4Rt/L} + 1\right]$$

$$i_2 = \frac{E}{R}\left[\frac{6}{17}e^{-3Rt/5L} - \frac{6}{17}e^{-4Rt/L}\right]$$

27. $$i_1 = \frac{E_0e^{j\omega t}(Z_2 + Z_3)}{Z_1Z_2 + Z_2Z_3 + Z_3Z_1};\ i_2 = \frac{E_0e^{j\omega t}Z_3}{Z_1Z_2 + Z_2Z_3 + Z_3Z_1};$$

$$i_3 = \frac{E_0e^{j\omega t}Z_2}{Z_1Z_2 + Z_2Z_3 + Z_3Z_1}$$

29. $$i_1 = E_0e^{j\omega t}\left[Z_1 + \frac{Z_2Z_3}{Z_2 + Z_3} + \frac{Z_4Z_5}{Z_4 + Z_5}\right];$$

$$i_2 = i_1\frac{Z_3}{Z_2 + Z_3};\ i_3 = i_1\frac{Z_2}{Z_2 + Z_3};\ i_4 = i_1\frac{Z_5}{Z_4 + Z_5};$$

$$i_5 = i_1\frac{Z_4}{Z_4 + Z_5}$$

31. $\omega_1 = 1/\sqrt{2LC}$; $\omega_2 = 2/\sqrt{LC}$

33. $\omega_1 = 1/2\sqrt{LC}$; $\omega_2 = 2/\sqrt{LC}$

Sec. 7.6, page 487

1. If N is even, deflections are symmetric about the midpoint of the system; if N is odd, they are skew-symmetric.
3. No
5. $\theta_n : \theta_0 = 0.062439, 0.000195, 0.000002$
7. $\omega_N = 2\sqrt{\frac{k}{M}} \sin \frac{N\pi}{2(n+1)}$; $(A_j)_N = \sin \frac{N\pi j}{n+1}$, $N = 1, 2, \ldots, n$
9. $\omega_N = 2\sqrt{\frac{c}{I}} \sin \frac{N\pi}{2n}$; $(A_j)_N = \cos \frac{jN\pi}{n}$, $N, j = 1, 2, \ldots, n$
11. $\omega_N = 2\sqrt{\frac{c}{I}} \sin \frac{(2N+1)\pi}{4n}$, $N = 0, 1, \ldots, n-1$

    $(A_j)_N = \cos \frac{(2N+1)j\pi}{4n}$, $j = 1, 2, \ldots, n$
13. Velocity $= l\omega/\mu$; length of the wave $= 2\pi l/\mu$; no, since $\theta_x = \cos(\omega t + \mu x)$ represents a wave traveling to the left from $x = \infty$.
15. $\theta_x = (-1)^x e^{-\mu x} \cos \omega t$.
17. It is not hard to verify that $i_k = A \sin(\omega t - k\mu)$, where $\cos \mu = 1 + L/2C$.

Sec. 7.7, page 495

3. $\frac{d^2z}{dT^2} + \frac{1}{\nu RC}\frac{dZ}{dT} + \frac{1}{\nu^2 LC} Z = -\frac{\omega I_0}{\nu^2 \epsilon C} \sin \frac{\omega T}{\nu}$, where $Z = \frac{e}{\epsilon}$ and $T = \nu t$, ϵ is an arbitrary voltage, and ν is an arbitrary frequency

5. $$\frac{d^2x_1}{dT^2} + \frac{c}{M_1\nu}\frac{dX_1}{dT} + \frac{k_1+k_2}{M_1\nu^2}X_1 - \frac{k_2 s_2}{M_1\nu^2 s_1}X_2 = 0$$
$$-\frac{k_2 s_1}{M_2\nu^2 s_2}X_1 + \frac{d^2X_2}{dT^2} + \frac{k_2}{M_2\nu^2}X_2 = \frac{F_0}{M_2 s_2 \nu^2}\cos\frac{\omega}{\nu}T$$
where ν is an arbitrary frequency, s_1 and s_2 are arbitrary distances, $X_i = s_i y_i$, and $T = \nu t$.

7. $$\frac{d^2X_1}{dT^2} + \frac{R}{L_1\nu}\frac{dX_1}{dT} + \frac{C_1+C_2}{C_1C_2L_1\nu^2}X_1 - \frac{q_2}{L_1C_2\nu^2 q_1}X_2 = 0$$
$$-\frac{q_1}{C_2L_2\nu^2 q_2}X_1 + \frac{d^2X_2}{dT^2} + \frac{1}{C_2L_2\nu^2}X_2 = \frac{E_0}{L_2 q_2 \nu^2}\cos\frac{\omega}{\nu}T$$
where ν is an arbitrary frequency, q_1 and q_2 are arbitrary charges, $X_i = q_i Q_i$, and $T = \nu t$. (*Note:* Q_2 is not the charge on the capacitor in the second loop but rather a fictitious charge defined by the relation $i_2 = dQ_2/dt$, where i_2 is the current flowing through the inductor in the second loop.)

9. **(a)** See Fig. A7.1*a*. **(b)** See Fig. A7.1*b*.

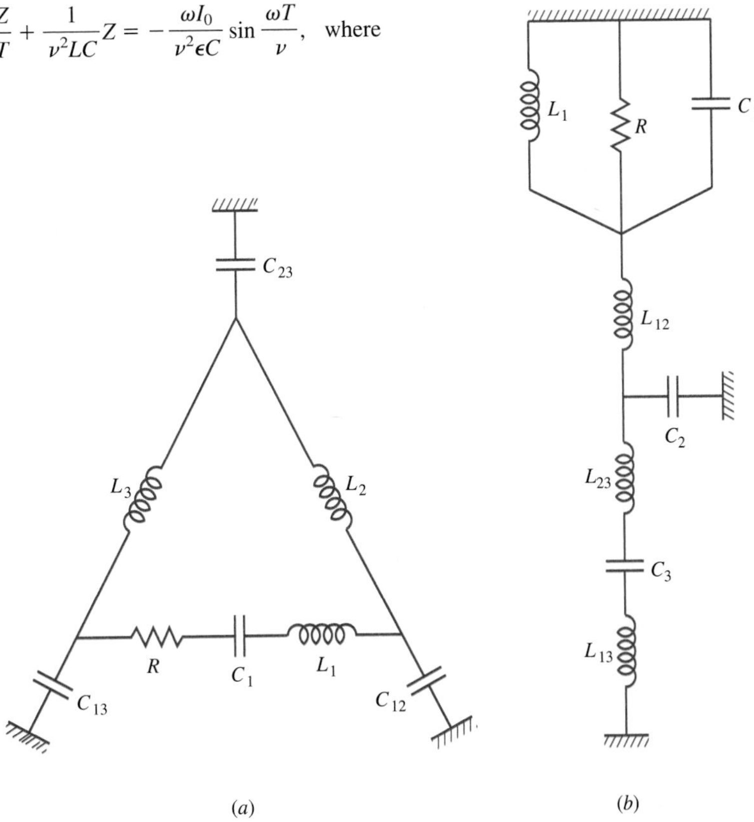

FIGURE A7.1

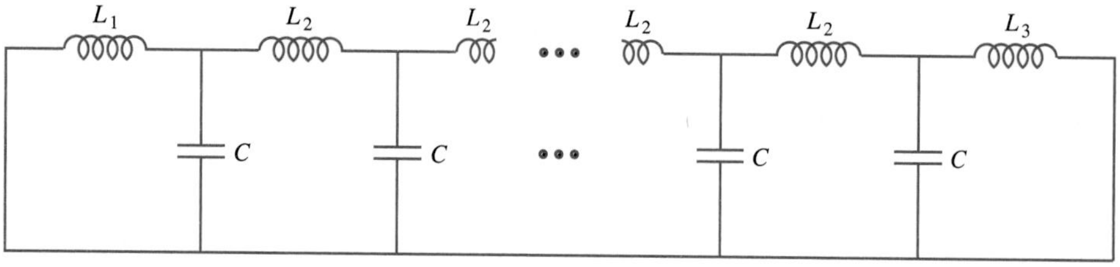

FIGURE A7.2

11. The series analog is shown in Fig. A7.2.

$$\frac{d^2X_1}{dT^2}+\frac{c}{M_1\nu^2}X_1-\frac{cs_2}{M_1\nu^2s_1}X_2=0$$

$$\cdots\cdots\cdots\cdots\cdots\cdots\cdots\cdots$$

$$\frac{d^2X_k}{dT^2}+\frac{2c}{M_2\nu^2}X_2-\frac{cs_{k-1}}{M_2\nu^2s_k}X_{k-1}-\frac{cs_{k+1}}{M_2\nu^2s_k}X_{k+1}=0$$

$$\cdots\cdots\cdots\cdots\cdots\cdots\cdots\cdots$$

$$\frac{d^2X_n}{dT^2}+\frac{c}{M_3\nu^2}X_n-\frac{cs_{n-1}}{M_3\nu^2s_n}X_{n-1}=0$$

where $T=\nu t$, $X_k=x_k/s_k$, $k=1,2,\ldots,n$, ν is an arbitrary frequency, and $s_1, s_2, \ldots, s_n$ are arbitrary distances. The dimensionless equations for the series analog are obtained immediately from the above equations by replacing M_i by L_i, $i=1, 2, 3$, replacing c by $1/C$, and replacing s_k by q_k, $k=1,2,\ldots,n$, where the q_k's are arbitrary charges.

13. The equivalent mechanical and electric circuits are shown in Fig. A7.3. If $k_1=k_2$, or if $C_2=0$, the system is uncoupled, i.e., consists of two independent systems with one degree of freedom.

CHAPTER 8

Sec. 8.1, page 500

1. **(a)** 2π **(b)** 2 **(c)** $2\pi/5$ **(d)** $4\pi/3$ **(e)** π **(f)** π **(g)** $2p$ **(h)** 1 **(i)** 2π **(j)** 4
3. All values of d
5. Yes
7. **(a)** $2p$ **(b)** $2p$ **9.** $2p/3$ **11.** 2π
13. **(a)** $P(38)=\overline{P}(2)=9336$ during February
(b) $P(153)-P(74)=\overline{P}(9)-\overline{P}(2)=1141$ is the September average minus the February average.
(c) 6664
(d) Rainbows in April when $\overline{P}(4)=10{,}352$; browns in October when $\overline{P}(10)=11{,}000$
15. The set of all rational numbers.

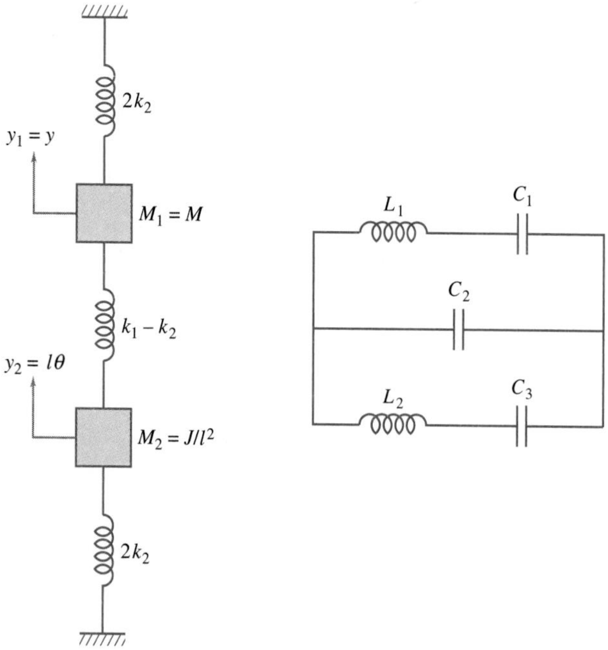

FIGURE A7.3

Sec. 8.2, page 507

1. $\dfrac{1}{2} + \dfrac{2}{\pi}\left(\sin \pi t + \dfrac{\sin 3\pi t}{3} + \dfrac{\sin 5\pi t}{5} + \cdots\right)$

3. $a_0 = \dfrac{1}{2}$, $a_n = \dfrac{1}{n\pi}\sin\dfrac{n\pi}{2}$; $b_n = \dfrac{1}{n\pi}\left(1 - \cos\dfrac{n\pi}{2}\right)$

5. $b_n = \begin{cases} 2/\pi n & n = 1, 3, 5, \ldots \\ 4/\pi n & n = 2, 6, 10, \ldots; \\ 0 & n = 4, 8, 12, \ldots \end{cases}$ $a_n = 0$, $n = 0, 1, 2, \ldots$

7. $\dfrac{2}{\pi}\left(\sin \pi t - \dfrac{\sin 2\pi t}{2} + \dfrac{\sin 3\pi t}{3} - \dfrac{\sin 4\pi t}{4} + \cdots\right)$

9. $\dfrac{1}{4} - \dfrac{2}{\pi^2}\left(\cos \pi t + \dfrac{1}{3^2}\cos 3\pi t + \dfrac{1}{5^2}\cos 5\pi t + \cdots\right)$ $+ \dfrac{1}{\pi}\left(\sin \pi t - \dfrac{1}{2}\sin 2\pi t + \dfrac{1}{3}\sin 3\pi t - \dfrac{1}{4}\sin 4\pi t + \cdots\right)$

11. $a_0 = \dfrac{4}{\pi}$, $a_n = -\dfrac{4}{(4n^2 - 1)\pi}$; $b_n \equiv 0$

13. $a_0 = 2(1 - e^{-1})$, $a_n = \dfrac{2(1 - e^{-1})}{1 + 4n^2\pi^2}$; $b_n = \dfrac{4n\pi(1 - e^{-1})}{1 + 4n^2\pi^2}$

15. $\dfrac{2a^2}{3} + \dfrac{4a^2}{\pi^2}\sum_{n=1}^{\infty}(-1)^{n+1}\dfrac{1}{n^2}\cos\dfrac{n\pi t}{a}$

17. $\dfrac{1}{\pi} + \dfrac{1}{2}\cos\dfrac{\pi t}{2} + \dfrac{2}{\pi}\sum_{n=1}^{\infty}(-1)^{n+1}\dfrac{\cos n\pi t}{4n^2 - 1}$

19. $\dfrac{1}{4}(3\sin t - \sin 3t)$

23. **(a)** $1 + \dfrac{2}{\pi}\sum_{n=1}^{\infty}(-1)^{n+1}\dfrac{\sin n\pi t}{n}$ **(b)** $\dfrac{\pi}{4}$ **(c)** -1

25. $f(t) = t^{1/3}$, $-1 < t < 1$

Sec. 8.3, page 517

1. $\dfrac{3}{2} - \dfrac{2}{\pi}\sum_{n=1}^{\infty}\dfrac{\sin(2n - 1)t}{2n - 1}$

3. $a_0 = \dfrac{1}{2}$, $a_{2n} = 0$, $a_{2n-1} = \dfrac{2}{(2n - 1)^2\pi^2}$; $b_n = -\dfrac{1}{n\pi}$

5. $a_0 = 2$, $a_n = 0$, $n \neq 0$; $b_n = \dfrac{2}{n\pi}$

7. $a_0 = p$, $a_{2n} = 0$, $a_{2n-1} = \dfrac{-4p}{(2n - 1)^2\pi^2}$; $b_n = 0$

9. $\dfrac{17}{12} - \dfrac{4}{\pi^2}\sum_{n=0}^{\infty}\dfrac{\cos(2n + 1)\pi t}{(2n + 1)^2} + \dfrac{1}{2\pi^2}\sum_{n=1}^{\infty}\dfrac{\cos 2n\pi t}{n^2} - \dfrac{4}{\pi^3}\sum_{n=0}^{\infty}\dfrac{\sin(2n + 1)\pi t}{(2n + 1)^3}$

11. $a_0 = \dfrac{7}{12}$, $a_{2n} = \dfrac{5}{4n^2\pi^2}$, $a_{2n-1} = \dfrac{12 - 5(2n - 1)^2\pi^2}{(2n - 1)^4\pi^4}$; $b_{2n} = \dfrac{3}{4n^3\pi^3}$, $b_{2n-1} = \dfrac{-2}{(2n - 1)^3\pi^3}$

13. $a_0 = \dfrac{e^{2kp} - 1}{kp}$, $a_n = -\dfrac{kp(1 - e^{2kp})}{n^2\pi^2 + k^2p^2}$; $b_n = \dfrac{n\pi(1 - e^{2kp})}{n^2\pi^2 + k^2p^2}$

15. $\dfrac{8}{\pi}\sum_{n=1}^{\infty}\dfrac{n}{4n^2 - 1}\sin 2nt$

17. $\dfrac{1}{a}\sinh a + 2a\sinh a\sum_{n=1}^{\infty}\dfrac{(-1)^n}{a^2 + n^2\pi^2}\cos\dfrac{n\pi t}{a}$

19. $a_0 = e - 1 + \sinh 1$, $a_n = \dfrac{(-1)^n(e + \sinh 1) - 1}{1 + n^2\pi^2}$; $b_n = \dfrac{(-1)^{n+1}n\pi\sinh 1}{1 + n^2\pi^2}$

21. $a_0 = 3$, $a_2 = \dfrac{1}{2}$, $a_n = 0$, $n \neq 0, 2$; $b_{2n} = 0$, $b_{2n-1} = \dfrac{8}{\pi(2n - 1)(2n + 1)(2n - 3)}$

23. c/n^3 **25.** c/n **27.** c/n^3

29. $a = -1$. Yes; try putting $t = \frac{1}{2}$ and $t = \frac{1}{3}$.

31. The function is discontinuous.

33. $\displaystyle\int_{-1}^{t} f(s)\,ds = \frac{1}{4} + \frac{t}{2} - \frac{2}{\pi^2}\sum_{n=1}^{\infty}\frac{\cos n\pi t}{n^2}$

$\displaystyle\int_{-1}^{t}\int_{-1}^{r} f(s)\,ds\,dr = \frac{t}{4} + \frac{1}{4}t^2 - \frac{2}{\pi^3}\sum_{n=1}^{\infty}\frac{\sin n\pi t}{n^3}$

These of course are not Fourier series since they contain, respectively, the algebraic terms $t/2$ and $t/4 + \frac{1}{4}t^2$. If these are replaced by their Fourier expansions over $(-1, 1)$, the required Fourier series will be obtained

$$\frac{1}{4} + \frac{1}{\pi}\sum_{n=1}^{\infty}(-1)^{n+1}\frac{\sin n\pi t}{n} - \frac{2}{\pi^2}\sum_{n=1}^{\infty}\frac{\cos n\pi t}{n^2}$$

and

$$\frac{1}{2\pi}\sum_{n=1}^{\infty}(-1)^{n+1}\frac{\sin n\pi t}{n} + \frac{1}{4}\left[\frac{1}{3} + \frac{4}{\pi^2}\sum_{n=1}^{\infty}(-1)^n\frac{\cos n\pi t}{n^2}\right] - \frac{2}{\pi^3}\sum_{n=1}^{\infty}\frac{\sin n\pi t}{n^3}$$

35. Pure resonance will occur since, in particular,

$a_{15} = 4/225\pi^2 \neq 0$, and the natural frequency of the system is $\omega_n = 15$.

37. Pure resonance cannot occur since the Fourier expansion of $f(t)$ does not involve a term of either of the forms $\cos 15t$ or $\sin 15t$.

Sec. 8.4, page 522

1. $a_0 = 1,\ a_n = \dfrac{2}{n\pi}\sin\dfrac{n\pi}{2};\ b_n = \dfrac{2}{n\pi}\left(1 - \cos\dfrac{n\pi}{2}\right)$

3. $a_0 = \dfrac{p}{2},\ a_n = \dfrac{4p}{n^2\pi^2}\cos\dfrac{n\pi}{2};$

$b_n = \dfrac{2p}{n\pi}\left(1 - \dfrac{2}{n\pi}\sin\dfrac{n\pi}{2}\right)$

5. $a + \dfrac{4a}{\pi}\dfrac{\sin(2n-1)t}{2n-1}$

9. $a_0 = a + b,\ a_{2n} = 0,\ a_{2n-1} = \dfrac{2(a-b)(-1)^{n-1}}{\pi(2n-1)};\ b_n = 0$

11. $8\sum_{n=1}^{\infty}\dfrac{n}{4n^2-1}\sin 2n\pi t$

13. $a_0 = 1,\ a_{2n-1} = 0,\ a_{2n} = -\dfrac{1}{n^2\pi^2};$

$b_n = \dfrac{2(-1)^{n+1}n\pi\sinh 1}{n^2\pi^2+1}$

15. $\dfrac{1}{3} + \dfrac{4}{\pi^2}\sum_{n=1}^{\infty}\dfrac{(-1)^n}{n^2}\cos n\pi t + \dfrac{3}{4}\sin\pi t - \dfrac{1}{4}\sin 3\pi t$

17. $\dfrac{2l}{\pi}\sum_{n=1}^{\infty}(-1)^{n+1}\dfrac{1}{n}\sin\dfrac{n\pi t}{l};$

$\dfrac{l}{2} - \dfrac{4l}{\pi^2}\sum_{n=0}^{\infty}\dfrac{1}{(2n+1)^2}\cos\dfrac{(2n+1)\pi t}{l}$

19. $\dfrac{8}{l}\sum_{n=0}^{\infty}\dfrac{1}{(2n+1)^3}\sin\dfrac{(2n+1)\pi t}{l};$

$\dfrac{\pi^2}{6} - \sum_{n=1}^{\infty}\dfrac{1}{n^2}\cos\dfrac{2n\pi t}{l}$

21. $\dfrac{al}{4} - \dfrac{2al}{\pi^2}\sum_{n=1}^{\infty}\dfrac{1}{(2n-1)^2}\cos\dfrac{2(2n-1)\pi t}{l};$

$\dfrac{4al}{\pi^2}\sum_{n=0}^{\infty}\dfrac{(-1)^n}{(2n+1)^2}\sin\dfrac{(2n+1)\pi t}{l}$

23. $\dfrac{8}{\pi}\sum_{n=0}^{\infty}\dfrac{1}{4-(2n+1)^2}\cos\dfrac{(2n+1)t}{2};\ \sin t$

25. $a_n \equiv 0,\ b_n = \dfrac{(-1)^{n+1}2n\sin a\pi}{\pi(n^2-a^2)};$

$a_n = \dfrac{2a[1-(-1)^n\cos a\pi]}{\pi(a^2-n^2)},\ b_n \equiv 0$

29. $b_{2n} \equiv 0,\ b_{2n+1} = \dfrac{4p}{(2n+1)\pi} - \dfrac{(-1)^n 8p}{(2n+1)^2\pi^2}$

31. The coefficients in the half-range sine expansion decrease as c/n^3. The coefficients in the half-range cosine expansion decrease as c/n^4.

33. The coefficients in the half-range sine expansion decrease as c/n. The coefficients in the half-range cosine expansion decrease faster than any fixed negative power of n since the derivatives of all orders are continuous. (See Exercise 35, Sec. 19.2.)

35. Extend $f(t)$ from p to $2p$ so that $f(2p - t) = -f(t)$.

$$a_n = \begin{cases} 0 & n \text{ even} \\ \dfrac{2}{p}\displaystyle\int_0^p f(t)\cos\dfrac{n\pi t}{p}\,dt & n \text{ odd}\end{cases}$$

39. $a_0 = 100,\ a_n = \dfrac{200}{n\pi}\left(-\sin\dfrac{n\pi}{4} + \dfrac{\sin n\pi}{2} - \sin\dfrac{3n\pi}{4}\right),$

$b_n \equiv 0;$

$b_n = \dfrac{200}{n\pi}\left(\cos\dfrac{n\pi}{4} - \cos\dfrac{n\pi}{2} + \cos\dfrac{3n\pi}{4} - \cos n\pi\right),$

$a_n \equiv 0$

Sec. 8.5, page 527

1. $A_n = \dfrac{1}{n\pi}\sqrt{2\left(1 - \cos\dfrac{n\pi}{2}\right)};\ \gamma_n = \dfrac{n\pi}{4};\ \delta_n = \dfrac{\pi}{2} - \gamma_n$

3. $A_n = \dfrac{1 - e^{-p}\cos n\pi}{(n^2\pi^2 + p^2)^{1/2}};\ \gamma_n = -\tan^{-1}\dfrac{n\pi}{p};\ \delta_n = \dfrac{\pi}{2} - \gamma_n$

5. $A_n = \dfrac{2\sinh p}{(n^2\pi^2 + p^2)^{1/2}};\ \gamma_n = -\tan^{-1}\dfrac{n\pi}{p};\ \delta_n = \dfrac{\pi}{2} - \gamma_n$

9. $c_0 = 0,\ c_n = \dfrac{1}{2ni\pi}(1 - e^{-ni\pi})^2$

11. $c_n = \dfrac{(-1)^n\sinh p}{p + ni\pi}$

13. $c_n = \dfrac{2}{(1-4n^2)\pi}$

15. $c_n = \dfrac{(-1)\sinh 1}{1 + n^2\pi^2}$

17. $\dfrac{3}{2} + \dfrac{2}{\pi}\sum_{n=1}^{\infty}\dfrac{1}{2n-1}\sin\dfrac{(2n-1)\pi t}{2}$

19. $\dfrac{\pi}{4} - \dfrac{2}{\pi}\sum_{n=1}^{\infty}\dfrac{1}{(2n-1)^2}\cos(2n-1)t -$

$3\sum_{n=1}^{\infty}\dfrac{(-1)^n}{n}\sin nt$

21. $c_{2n} = \dfrac{3}{4n}e^{i\pi/2},\ c_{-2n} = \dfrac{3}{4n}e^{-i\pi/2}$; for c_{2n-1} and $c_{-(2n-1)}$,

we have $A_{2n-1} = \dfrac{\sqrt{4 + 9(2n-1)^2\pi^2}}{\pi(2n-1)^2}$ and

$\gamma_{2n-1} = \tan^{-1}\dfrac{3\pi(1-2n)}{2}.$

23. $c_n = \dfrac{1}{n\pi} \sin \dfrac{n\pi a}{p}$

Sec. 8.6, page 540

1. $y_{ss} = \dfrac{80}{\pi} \sum_{n=1}^{\infty} \dfrac{\sin 2n\pi t}{(100 - n^2\pi^2)n}$

3. $y_{ss} \doteq 0.190 \sin (2\pi t - 0.056) + 0.142 \sin (6\pi t - 0.389) + 0.047 \sin (10\pi t - 2.790) + 0.011 \sin (14\pi t - 2.987) + \cdots$

5. $y_{ss} \doteq 0.174 - 0.151 \cos (2\pi t - 0.028) - 0.040 \cos (6\pi t - 0.202) - 0.008 \cos (10\pi t - 2.960) + \cdots$

7. $y_{ss} \doteq F_0[\frac{1}{120} + 0.011 \sin (\pi t - 0.022) + 0.009 \sin (3\pi t - 0.163) + 0.003 \sin (5\pi t - 2.996) + \cdots]$

9. $y_{ss} \doteq F_0[0.011 \sin (2\pi t - 0.027) - 0.081 \sin (4\pi t - 2.175) + 0.002 \sin (6\pi t - 3.095) - \cdots]$

11. **(a)** $(y_1)_{ss} = -\dfrac{40}{\pi} \sum_{n=1}^{\infty} \dfrac{\lambda_n^2 - 2}{(\lambda_n^2 - 1)(\lambda_n^2 - 4)\lambda_n} \sin \lambda_n t,$

$\lambda_n = (4n - 2)\pi$

$(y_2)_{ss} = \dfrac{80}{\pi} \sum_{n=1}^{\infty} \dfrac{1}{(\lambda_n^2 - 1)(\lambda_n^2 - 4)\lambda_n} \sin \lambda_n t$

(b) $(y_1)_{ss} = \dfrac{80}{\pi} \sum_{n=1}^{\infty} \dfrac{1}{(\lambda_n^2 - 1)(\lambda_n^2 - 4)\lambda_n} \sin \lambda_n t,$

$\lambda_n = (4n - 2)\pi$

$(y_2)_{ss} = -\dfrac{80}{\pi} \sum_{n=1}^{\infty} \dfrac{\lambda_n^2 - 3}{(\lambda_n^2 - 1)(\lambda_n^2 - 4)\lambda_n} \sin \lambda_n t$

13. $i_{ss} = -E_0 \sum_{n=-\infty}^{\infty} \dfrac{1}{4(2n + 1)^2\pi^2 - 2500} e^{200ni\pi t}$

15. **(a)** $i_{ss} = E_0 \sum_{n=-\infty}^{\infty} \dfrac{ie^{200ni\pi t}}{200[600n\pi + i(200n^2\pi^2 - 1250)]},$

$n \neq 0$

(b) $i_{ss} = 4 \sum_{n=-\infty}^{\infty} \dfrac{n(1 - 4ni)e^{200ni\pi t}}{(1 - 16n^2)[12n\pi + i(4n^2\pi^2 - 25)]},$

$n \neq 0$

17. **(a)** $y = \dfrac{1}{2} \cosh t + \left[\dfrac{2}{\pi} \sum_{n=1}^{\infty} \dfrac{1}{1 + (2n - 1)^2}\right] \sinh t$

$- \dfrac{1}{2} - \dfrac{2}{\pi} \sum_{n=1}^{\infty} \dfrac{1}{(2n - 1) + (2n - 1)^3} \sin (2n - 1)t$

(b) $y = -\dfrac{1}{2} \cos t +$

$\dfrac{1}{\pi}\left[1 - 2 \sum_{n=2}^{\infty} \dfrac{1}{1 - (2n - 1)^2}\right] \sin t$

$+ \dfrac{1}{2} - \dfrac{1}{\pi} t \cos t +$

$\dfrac{2}{\pi} \sum_{n=2}^{\infty} \dfrac{1}{(2n - 1) - (2n - 1)^3} \sin (2n - 1)t$

(c) $y = \dfrac{3}{2} \cosh t + \left[\dfrac{2}{\pi} \sum_{n=1}^{\infty} \dfrac{1}{1 + (2n - 1)^2}\right] \sinh t$

$- \dfrac{1}{2} - \dfrac{2}{\pi} \sum_{n=1}^{\infty} \dfrac{1}{(2n - 1) + (2n - 1)^3} \sin (2n - 1)t$

(d) $y = \dfrac{1}{2} \sin t + \dfrac{1}{\pi}\left[1 - 2 \sum_{n=2}^{\infty} \dfrac{1}{1 - (2n - 1)^2}\right] \cos t$

$+ \dfrac{1}{2} - \dfrac{1}{\pi} t \cos t + \dfrac{2}{\pi} \sum_{n=2}^{\infty} \dfrac{\sin (2n - 1)t}{(2n - 1) - (2n - 1)^3}$

(e) $y = \left[1 - 2 \sum_{n=1}^{\infty} B_n + \sum_{n=1}^{\infty} (2n - 1)A_n\right] e^t$

$+ \left[-\dfrac{1}{2} + \sum_{n=1}^{\infty} B_n - \sum_{n=1}^{\infty} (2n - 1)A_n\right] e^{2t}$

$- \dfrac{1}{2} + \sum_{n=1}^{\infty} A_n \sin (2n - 1)t + \sum_{n=1}^{\infty} B_n \cos (2n - 1)t$

where

$A_n = \dfrac{2}{(2n - 1)\pi} \dfrac{2 - (2n - 1)^2}{[2 - (2n - 1)^2]^2 + [3(2n - 1)]^2}$

$B_n = \dfrac{2}{(2n - 1)\pi} \dfrac{3(2n - 1)}{[2 - (2n - 1)^2]^2 + [3(2n - 1)]^2}$

19. In each case, the error is the integral of the square of f over one period minus the sum of the squares of the first three nonzero Fourier coefficients of f.

21. **(a)** $a = 0.098$, $\cos x \sim 0.098 - 0.393x^2$
(b) $a = 0.431$, $\cos x \sim 1.000 - 0.431x^2$
(c) $a = 0.562$, $\cos x \sim 0.842 - 0.250x^2$
Each function is an even quadratic polynomial, similar to the first two terms in the MacLaurin expansion of $\cos x$.

25. $y = \dfrac{2kl^5}{\pi^5 EI} \sum_{n=1}^{\infty} \dfrac{(-1)^n}{n^5} \sin \dfrac{n\pi x}{l}$

27. $y = -\dfrac{P_0 l^4}{EI\pi^4}\left(\sin \dfrac{\pi x}{l} + \dfrac{1}{81} \sin \dfrac{3\pi x}{l}\right)$

29. $b_n = \dfrac{2P_0 l^4}{EIan^5\pi^5} \sin \dfrac{n\pi a}{l} (-1)^{(n+1)/2}, \quad n$ odd; $\quad b_n = 0,$

n even

31. The exact deflection curve is

$y(x) = -\dfrac{1}{360EI}(3x^5 - 10l^2x^3 + 7l^4x)$

At $x = l/2$ this gives the deflection $-0.006510l^5/EI$. The first term in the Fourier series gives the value $-0.006536l^5/EI$. The first two terms give the value $-0.006509l^5/EI$.

33. Yes. By noting the point in the frequency range where the behavior of the coefficients changes from c/n to c/n^2, a rough estimate of a can be obtained, and this in turn can be used in Fig. 8.15 to obtain a rough estimate of how fast the switch opens and closes.

35. **(a)** $r(t) = a + \dfrac{c(b-a)}{\pi} + \dfrac{2(b-a)}{\pi}\sum_{n=1}^{\infty}\dfrac{\sin cn \cos nt}{n}$

(b) If $a = b$, $r(t) = a$; if $c = 0$, $r(t) = a$; if $c = \pi$, $r(t) = b$.

(c) $\pi - c = 2\sum_{n=1}^{\infty}\dfrac{\sin nc}{n}$

(d) $c = 2\sum_{n=1}^{\infty}(-1)^{n+1}\dfrac{\sin nc}{n}$

(e) $r_{\text{rms}} =$
$$\sqrt{\left[a + \frac{c(b-a)}{\pi}\right]^2 + 2\left[\frac{b-a}{\pi}\right]^2\sum_{n=1}^{\infty}\frac{\sin^2 nc}{n^2}}$$

(f) If $a = b$, $r_{\text{rms}} = a$; if $c = 0$, $r_{\text{rms}} = a$; if $c = \pi$, $r_{\text{rms}} = b$.

37. $y(x) = -\dfrac{8l^4}{\pi^5 T}\sum_{n=1}^{\infty}\dfrac{1}{(2n-1)^5}\sin\dfrac{(2n-1)\pi x}{l}$

39. $w(x)$ must be expressed as a cosine series; and in addition to a cosine series with coefficients to be determined, the assumed expression for the deflection curve must include a binomial of the form $b_0(l^2 - 4x^2)$ to take account of the term $\frac{1}{2}B_0$ in the expansion of $w(x)$.

CHAPTER 9

Sec. 9.1, page 554

1. **(a)** $B(\omega) = \dfrac{2}{\pi}\dfrac{1-\cos\omega}{\omega}$;

$$f(t) = \frac{2}{\pi}\int_0^\infty \frac{1-\cos\omega}{\omega}\sin\omega t\, d\omega$$

(b) $B(\omega) = \dfrac{2}{\pi}\dfrac{\sin\omega - \omega\cos\omega}{\omega^2}$;

$$f(t) = \frac{2}{\pi}\int_0^\infty \frac{\sin\omega - \omega\cos\omega}{\omega^2}\sin\omega t\, d\omega$$

(c) $A(\omega) = \begin{cases} 1/\pi & \omega = 0 \\ \dfrac{2}{\pi}\dfrac{1-\cos\omega}{\omega^2} & \omega > 0 \end{cases}$;

$$f(t) = \frac{2}{\pi}\int_0^\infty \frac{1-\cos\omega}{\omega^2}\cos\omega t\, d\omega$$

(d) $B(\omega) = \dfrac{2}{\pi}\dfrac{\omega - \sin\omega}{\omega^2}$;

$$f(t) = \frac{2}{\pi}\int_0^\infty \frac{\omega - \sin\omega}{\omega^2}\sin\omega t\, d\omega$$

5. See Fig. A9.1 **7.** See Fig. A9.2.

9. $$f(t) = \frac{1}{2\pi}\int_{-\infty}^{\infty}\frac{e^{i\omega t}}{\ln 2 + i\omega}d\omega = \frac{1}{\pi}\int_0^\infty \frac{(\ln 2)\cos\omega t + \omega\sin\omega t}{\ln^2 2 + \omega^2}d\omega$$

11. $f(t) =$
$$\frac{1}{2\pi i}\int_{-\infty}^{\infty}\frac{2(1-\omega^2)(1-i\omega)e^{i\omega} + \omega^2 - 2 + i\omega^3 e^{-i\pi\omega/2}}{\omega^3(1-\omega^2)}e^{i\omega t}\,d\omega$$
$$= \frac{1}{\pi}\int_0^\infty \frac{a(\omega)\cos\omega t + b(\omega)\sin\omega t}{\omega^3(\omega^2-1)}d\omega$$
$$a(\omega) = 2(\omega^2-1)(\sin\omega - \omega\cos\omega) - \omega^3\cos(\pi\omega/2)$$
$$b(\omega) = 2(\omega^2-1)(\cos\omega + \omega\sin\omega) + 2 - \omega^2 - \omega^3\sin(\pi\omega/2)$$

13. $f(t) =$
$$\frac{1}{2\pi}\int_{-\infty}^{\infty}\frac{(1+i\omega)\sinh(1-i\omega)p - (1-i\omega)\sinh(1+i\omega)}{1+\omega^2}e^{i\omega t}\,d\omega$$
$$= \frac{2}{\pi}\int_0^\infty \frac{\cosh p\sin\omega p - \omega\sinh p\cos\omega p}{1+\omega^2}\sin\omega t\, d\omega$$

15. $$f(t) = \frac{1}{\pi}\left[(a+b)\int_0^\infty \frac{\sin\omega}{\omega}\cos\omega t\, d\omega + (a-b)\int_0^\infty \frac{1-\cos\omega}{\omega}\sin\omega t\, d\omega\right]$$

17. $$f(t) = \frac{2}{\pi}\int_0^\infty\left[\frac{\sin\omega}{\omega}\cos\omega t + \frac{\sin\omega - \omega\cos\omega}{\omega^2}\sin\omega t\right]d\omega$$

19. $$f(t) = \frac{2}{\pi}\int_0^\infty \frac{\cos(\omega\pi/2)\cos\omega t}{1-\omega^2}d\omega$$

21. $$f(t) = \frac{1}{\pi}\left[\int_0^\infty \frac{3\omega\sin\omega + 2(\cos\omega - 1)}{\omega^2}\cos\omega t\, d\omega + \int_0^\infty \frac{\cos\omega - 1}{\omega}\sin\omega t\, d\omega\right]$$

23. $$f(t) = \frac{4}{\pi}\int_0^\infty \frac{\sin\omega - \omega\cos\omega}{\omega^3}\cos\omega t\, d\omega$$

25. $f(t) =$
$$\frac{1}{\pi}\int_0^\infty \frac{\omega[2 + (b-a)]\sin\omega + (b-a)(\cos\omega - 1)}{\omega^2}\cos\omega t\, d\omega + \frac{a+b}{\pi}\int_0^\infty \frac{\sin\omega - \omega\cos\omega}{\omega^2}\sin\omega t\, d\omega$$

27. $$f(t) = \frac{1}{\pi}\int_0^\infty \frac{a\cos\omega(t-b) - \omega\sin\omega(t-b)}{a^2+\omega^2}d\omega$$

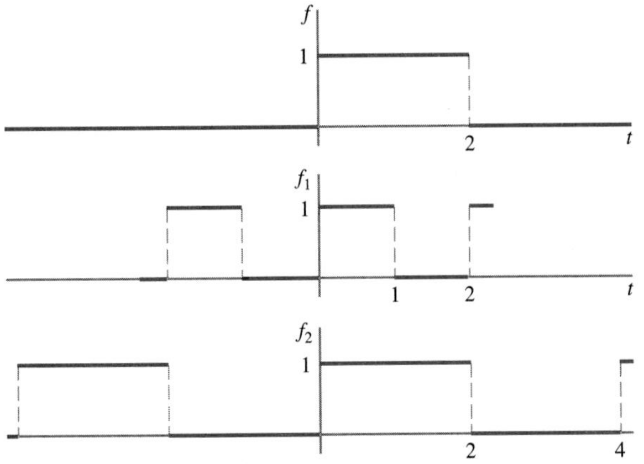

FIGURE A9.1

29. $f(t) =$

$$\frac{2}{\pi}\int_0^\infty \frac{k\sin c\omega + (d-k)\sin b\omega - d\sin a\omega}{\omega}\cos\omega t\, d\omega$$

$f(t) =$

$$\frac{2}{\pi}\int_0^\infty \frac{d\cos a\omega + (k-d)\cos b\omega - k\cos c\omega}{\omega}\sin\omega t\, d\omega$$

31. $$f(t) = \frac{2}{\pi e}\int_0^\infty \frac{1+\omega^2+\omega\sin\omega-\cos\omega}{\omega^2(1+\omega^2)}\cos\omega t\, d\omega$$

$$f(t) = \frac{2}{\pi e}\int_0^\infty \frac{\omega-\sin\omega+\omega^3(1+\cos\omega)}{\omega^2(1+\omega^2)}\sin\omega t\, d\omega$$

33. $$f(t) = \frac{2}{\pi}\int_0^\infty \frac{\sin\pi\omega}{\omega(1-\omega^2)}\cos\omega t\, d\omega;$$

$$f(t) = \frac{2}{\pi}\int_0^\infty \frac{1-2\omega^2-\cos\pi\omega}{\omega(1-\omega^2)}\sin\omega t\, d\omega$$

35. f is continuous; f' is discontinuous; $C(\omega)\to 0$ as c/ω^2.

37. f is continuous; f' is discontinuous; $C(\omega)\to 0$ as c/ω^2.

49. $A(\omega) = \sqrt{\frac{2}{\pi}}e^{-\omega^2/2};$

$$f(t) = \sqrt{\frac{2}{\pi}}\int_0^\infty e^{-\omega^2/2}\cos\omega t\, d\omega;\quad \int_0^\infty e^{-\omega^2/2}\, d\omega = \sqrt{\frac{\pi}{2}}$$

51. $A(\omega) = e^{-\omega}$; $f(t = \int_0^\infty e^{-\omega}\cos\omega t\, d\omega$; $\pi/2$

53. $r(\omega) = 1/2\pi(a^2+\omega^2)^{1/2}$; $\theta(\omega) = \omega b - \tan^{-1}(\omega/a)$

55. $r(\omega) = |\sin\omega/2|/\pi\omega$; $\theta(\omega) = -3\omega/2$

FIGURE A9.2

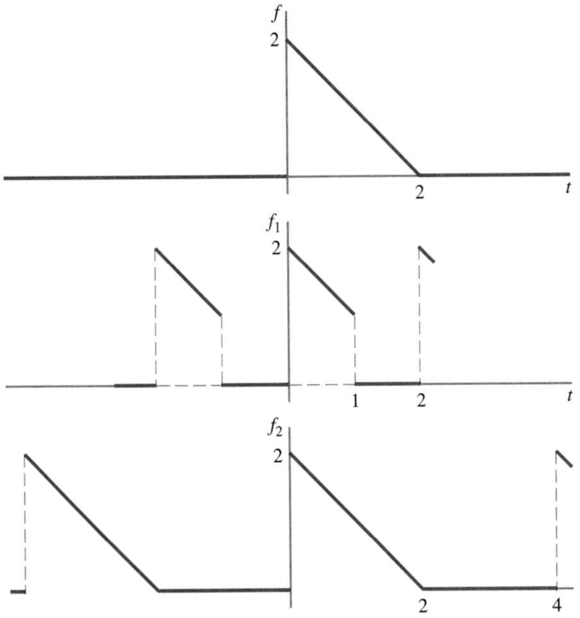

59. $A(\omega) =$

$$-\frac{1}{\pi\omega}\sum_k J_k \sin \omega t_k - \frac{1}{\pi\omega^2}\sum_k J_k' \cos \omega t_k' - \frac{1}{\omega^2}A''(\omega)$$

$B(\omega) =$

$$\frac{1}{\pi\omega}\sum_k J_k \cos \omega t_k - \frac{1}{\pi\omega^2}\sum_k J_k' \sin \omega t_k' - \frac{1}{\omega^2}B''(\omega)$$

61. $f(t) = \dfrac{2}{\pi}\displaystyle\int_0^\infty \frac{1 - \cos \pi\omega}{\omega^2} \cos \omega t \, d\omega$

63. $f(t) = 2\displaystyle\int_0^\infty \frac{\sin \pi\omega - \pi\omega \cos \pi\omega}{\omega^2} \sin \omega t \, d\omega$

65. $f(t) = 2\displaystyle\int_0^\infty \frac{\sin \omega}{\pi^2 - \omega^2} \sin \omega t \, d\omega$

67. $f(t) = \dfrac{1}{\pi}\displaystyle\int_0^\infty \frac{\omega(1 + \cos \omega) - 2 \sin \omega}{\omega^3} \cos \omega t \, d\omega +$

$$\frac{1}{\pi}\int_0^\infty \frac{2(1 - \omega^2) \cos \omega - 2 + 3\omega \sin \omega}{\omega^3} \sin \omega t \, d\omega$$

69. $f(t) = \dfrac{4}{\pi}\displaystyle\int_0^\infty \frac{\cos 2\omega - 1}{\omega^2} \cos \omega t \, d\omega$

$f(t) = \dfrac{4}{\pi}\displaystyle\int_0^\infty \frac{\sin 2\omega - 2\omega}{\omega^2} \sin \omega t \, d\omega$

71. $f(t) = \dfrac{4}{\pi}\displaystyle\int_0^\infty \frac{\pi \cos 2\omega + 2\omega \sin \omega}{\pi^2 - 4\omega^2} \cos \omega t \, d\omega$

$f(t) = \dfrac{4}{\pi}\displaystyle\int_0^\infty \frac{\pi \sin 2\omega - 2\omega \cos \omega}{\pi^2 - 4\omega^2} \sin \omega t \, d\omega$

73. $(1/n)[\text{Si}(b^n) - \text{Si}(a^n)]$ **75.** $\frac{1}{2}[\text{Si}(2b) - \text{Si}(2a)]$

77. $\dfrac{\cos a}{a} - \dfrac{\cos b}{b} - \text{Si}\, b + \text{Si}\, a$

79. $\frac{1}{4}[\text{Si}\, 2 + \text{Si}\, 4 - \text{Si}\, 6]$

81. $(1/\lambda)[\text{Ci}(e^{\lambda b}) - \text{Ci}(e^{\lambda a})]$

83. $\frac{1}{2}[\text{Ci}(b) - \text{Ci}(a)] - \frac{1}{2}[\text{Ci}(3b) - \text{Ci}(3a)]$

85. $e\text{Ei}(2)$ **87.** $\text{Ei}(1) - e\text{Ei}(2)$

Sec. 9.2, page 561

1. $P(\omega_0, t) = \dfrac{1}{\pi}[2\, \text{Si}(\omega_0 t) - \text{Si}\, \omega_0(t - 1) - \text{Si}\, \omega_0(t + 1)]$

3. $P(\omega_0, t) = \dfrac{1}{\pi}[\text{Si}\, \omega_0(1 + t) + \text{Si}\, \omega_0(2 - t)]$

5. $P(\omega_0, t) = \dfrac{1}{\pi}\left[\dfrac{\cos \omega_0(1 + t) - \cos \omega_0(1 - t)]}{\omega_0}\right] +$

$$\frac{t}{\pi}[\text{Si}\, \omega_0(1 - t) + \text{Si}\, \omega_0(1 + t)]$$

7. $P(\omega_0, t) = \dfrac{1}{\pi}[\text{Si}\, \omega_0(t - p) - 2\, \text{Si}\, \omega_0 t + \text{Si}\, \omega_0(t + p)]$

9. Same as Exercise 5

11. $P(\omega_0, t) = (1/\pi)[(1 - \cos \omega_0) \cos \omega_0 t +$
$(1 + t)\, \text{Si}\, \omega_0(1 + t) + (1 - t)\, \text{Si}\, \omega_0(t - 1)]$

Sec. 9.3, page 576

1. **(a)** $\dfrac{1 - \cos 2\omega}{\omega^2}$ **(b)** $\dfrac{2(1 - \cos \omega)e^{-i\omega}}{\omega^2}$

(c) $\dfrac{4(1 - \cos \omega)}{\omega^2}$ **(d)** $\dfrac{2(1 - \cos 2\omega)}{\omega^2}$

(e) $\dfrac{2(\cos \omega - \cos 2\omega)}{\omega^2}$ **(f)** $\dfrac{-2(1 - \cos \omega)e^{-i\omega}}{i\omega}$

3. **(a)** $f(t) = \begin{cases} 0 & t < 0 \\ e^{-t} - e^{-2t} & t > 0 \end{cases}$

(b) $f(t) = \begin{cases} 0 & t < 0 \\ -e^{-t} + 2e^{-2t} & t > 0 \end{cases}$

(c) $f(t) = \begin{cases} 0 & t < 0 \\ e^{-(1-i)t} & t > 0 \end{cases}$

(d) $f(t) = \begin{cases} 0 & t < 0 \\ e^{-2(t-1)} & t > 0 \end{cases}$

(e) $f(t) = \begin{cases} 0 & t < 0 \\ (t - 2)e^{-3(t-2)} & t > 0 \end{cases}$

5. **(b)** If the inverse of $1/(a + i\omega)$ is $\phi(t) + i\theta(t)$, then

$\phi(t) = \begin{cases} \frac{1}{2}e^{at} & t < 0 \\ \frac{1}{2}e^{-at} & t > 0 \end{cases}$ and $\theta(t) = \begin{cases} -\frac{1}{2}e^{at} & t < 0 \\ \frac{1}{2}e^{-at} & t > 0 \end{cases}$

13. $f(t) = \begin{cases} 0 & t < 0 \\ e^{-(t-3)} - e^{-2(t-3)} & t > 3 \end{cases}$

15. **(b)** $\mathcal{F}[f''(t)] = -\omega^2\mathcal{F}[f(t)] - i\omega f(0) - f'(0)$

19. By hypothesis, the arbitrary pulse is given by the series $f(t)p(t) = (a_0 + a_1 t + a_2 t^2 + \cdots + a_n t^n + \cdots)p(t)$. Hence $\mathcal{F}[f(t)p(t)] = a_0\mathcal{F}[p(t)] + a_1\mathcal{F}[tp(t)] + \cdots + a_n\mathcal{F}[t^n p(t)] + \cdots$, and these transforms are known.

23. $\mathcal{F}[f(t)] = -i\dfrac{\sin(\omega - \omega_0)T}{\omega - \omega_0} + i\dfrac{\sin(\omega - \omega_0)T}{\omega - \omega_0}$

25. **(a)** $F(\omega) = \dfrac{-i}{\sqrt{2\pi}}F_s(\omega)$

(b) $f(t) = \begin{cases} \dfrac{e^{-ak} - 1}{2\pi a}e^{at} & -\infty < t \le 0 \\ \dfrac{e^{-a(k-t)} - e^{-at}}{2\pi a} & 0 \le t \le k \\ \dfrac{e^{ak} - 1}{2\pi a}e^{-at} & k \le t < \infty \end{cases}$

Sec. 9.4, page 586

1. $Y = \dfrac{2}{\pi}\displaystyle\int_0^\infty \frac{(b - \omega^2) \cos \omega t + a\omega \sin \omega t}{(b - \omega^2)^2 + (a\omega)^2} \frac{\sin \omega}{\omega} d\omega$

3. $Y =$

$$\frac{2}{\pi}\int_0^\infty \frac{-a\omega \cos \omega t + (b - \omega^2) \sin \omega t}{(b - \omega^2)^2 + (a\omega)^2} \frac{\sin \omega - \omega \cos \omega}{\omega^2} d\omega$$

5. $y(x) = \dfrac{-2P_0}{\pi l^2}\displaystyle\int_0^\infty \frac{1 - \cos \omega l}{\omega^2(EI\omega^4 + k)} \cos \omega x \, d\omega,$

$$\lim_{l\to 0} y(x) = \frac{-P_0}{\pi}\int_0^\infty \frac{\cos \omega x}{EI\omega^4 + k}\,d\omega$$

7. $y(x) = \dfrac{-2P_0}{\pi l}\displaystyle\int_0^\infty \frac{(1-\cos\omega l)\sin\omega x}{\omega(EI\omega^4 + k)}\,d\omega$, $\lim\limits_{l\to 0} y(x) = 0$

9. $$y(x) = \begin{cases} -b[-2e^{ax}u(x) + e^{a(x-l)}u(x-l) + e^{a(x+l)}u(x+l)] & -\infty < x \le -l \\ -b[4a(x+l) - 2e^{ax}u(x) + e^{a(x-l)}u(x-l) + e^{-a(x+l)}v(x+l)] & -l \le x \le 0 \\ -b[-4a(x-l) - 2e^{-ax}v(x) + e^{a(x-l)}u(x-l) + e^{-a(x+l)}v(x+l)] & 0 \le x \le l \\ -b[-2e^{-ax}v(x) + e^{-a(x-l)}v(x-l) + e^{-a(x+l)}v(x+l) & l \le x \le \infty \end{cases}$$

where $a = (k/4EI)^{1/4}$, $b = P_0/(4kal^2)$, and u and v are the functions given by $u(x) = \cos ax + \sin ax$ and $v(x) = \cos ax - \sin ax$

11. $$y(x) = \begin{cases} -\dfrac{P_0}{2kl}[2e^{ax}\cos ax - e^{a(x-l)}\cos a(x-l) - e^{a(x+l)}\cos a(x+l)] & -\infty < x \le -l \\ -\dfrac{P_0}{2kl}[-2 + 2e^{ax}\cos ax - e^{a(x-l)}\cos a(x-l) + e^{-a(x+l)}\cos a(x+l)] & -l \le x \le 0 \\ -\dfrac{P_0}{2kl}[2 - 2e^{-ax}\cos ax - e^{a(x-l)}\cos a(x-l) + e^{-a(x+l)}\cos a(x+l)] & 0 \le x \le l \\ -\dfrac{P_0}{2kl}[-2e^{-ax}\cos ax + e^{-a(x-l)}\cos a(x-l) + e^{-a(x+l)}\cos a(x+l)] & l \le x < \infty \end{cases}$$

13. The deflection of the semi-infinite beam may be found by suitably extending the beam into an infinite beam and finding the deflection of the infinite beam on $(-\infty, \infty)$

15. **(a)** $$Y(t) = \frac{1}{\pi}\int_{-\infty}^\infty \frac{(2-\omega^2) - 3i\omega}{(1+\omega^2)(4+\omega^2)}\frac{\sin\omega}{\omega}e^{i\omega t}\,d\omega$$

$$\mathscr{F}[Y(t)] = \frac{2[(2-\omega^2) - 3i\omega]\sin\omega}{\omega(1+\omega^2)(4+\omega^2)}$$

(b) $F(\omega) = \dfrac{2\sin\omega}{\omega}$; $\mathscr{F}[y(t)] = \dfrac{2\sin\omega}{\omega[(i\omega)^2 + 3i\omega + 2]}$

(c) Both $Y(t)$ and $y(t)$ represent the same function; thus $y(t)$ provides an evaluation of $Y(t)$, and $Y(t)$ is a Fourier integral representation of $y(t)$.

(d) $\mathscr{F}[y(t)] = \dfrac{2\sin\omega}{\omega(1+i\omega)(2+i\omega)}$;

$$Y(t) = \frac{1}{\pi}\int_{-\infty}^\infty \frac{\sin\omega}{\omega(1+i\omega)(2+i\omega)}e^{i\omega t}\,d\omega$$

19. $Y = \frac{1}{2}t^2e^{-t}$

21. $$Y(t) = \begin{cases} 0 & -\infty < t \le 0 \\ \frac{1}{2}(e^t - e^{-t}) & 0 \le t \le 1 \\ \frac{1}{2}(e^2 - 1)e^{-t} & 1 \le t < \infty \end{cases}$$

provides a particular solution on $(-\infty, 0)$, $(0, 1)$, and $(1, \infty)$.

23. $Y = 4e^{2t}$

25. $Y = 4\cos 3t + 3\sin 3t$ **27.** $Y = -e^{t/3}$

29. $$Y = \begin{cases} \dfrac{2a}{15}(6e^{t/2}\sinh\dfrac{b}{2} - e^{3t}\sinh 3b) & -\infty < t \le -b \\ \dfrac{a}{15}(5 - 6e^{-b/2}e^{t/2} + e^{-3b}e^{3t}) & -b \le t \le b \\ 0 & b \le t < \infty \end{cases}$$

gives nontrivial solutions on the normal intervals $(-\infty, -b)$ and $(-b, b)$, and a trivial solution on (b, ∞).

31. $y = te^{-t}$

33. $y = 3e^{-t/3} - 2(1+4t)e^{-3t}$

35. $y = ke^{-at}$, $0 \le t$. Yes. Since the differential equation is normal on $(-\infty, \infty)$, on every interval containing $t = 0$, the solution of the initial-value problem is unique.

37. **(a)** $W(s) = \dfrac{1}{(s-a)(s-b)}$;

$$W(i\omega) = \frac{1}{b-a}\left[\frac{-1}{-a+i\omega} - \frac{-1}{-b+i\omega}\right];$$

$$w(t) = \begin{cases} \dfrac{1}{b-a}(e^{at} - e^{-bt}) & t < 0 \\ 0 & 0 < t \end{cases}$$

(b) $m < a$

(c) $Y = \dfrac{e^{mt}}{(m-a)(m-b)}$; $m \ne a, b$

Sec. 9.5, page 593

5. **(a)** $F(\omega) = \dfrac{1 - e^{-3ai\omega}}{3ai\omega}$ **(b)** $\lim\limits_{a\to 0} F(\omega) = 1$

7. **(a)** $I = \dfrac{\sin 3a}{3a}$ **(b)** $\lim\limits_{a\to 0} I = 1 = \cos 3t|_{t=0}$

9. **(b)** $I = \dfrac{a\tan^{-1}(a/2) + \ln 4 - \ln(a^2+4)}{a^2}$

(c) $\lim\limits_{a\to 0} I = \frac{1}{4}$

(d) Yes, since $\frac{1}{4} = 1/(4+t^2)|_{t=0}$

11. **(b)** $F(\omega) = e^{-i\omega}\dfrac{1 - 2e^{i\omega a} + e^{-2ia\omega}}{a^2}$

(c) $\lim\limits_{a\to 0} F(\omega) = i\omega e^{-i\omega}$

(d) $I = \dfrac{\cos 2 - 2\cos 2(1+a) + \cos 2(1+2a)}{2a^2}$

(e) $\lim\limits_{a\to 0} I = -2\cos 2$

(f) The unit doublet appears to have the same effect as $-\delta'(t)$.

15. **(a)** $(\pi/2)[\delta(\omega - \omega_0) + \delta(\omega + \omega_0)] + i\omega/(\omega_0^2 - \omega^2)$
(b) $(\pi/2i)[\delta(\omega - \omega_0) - \delta(\omega + \omega_0)] + \omega_0/(\omega_0^2 - \omega^2)$

19. $$Y(t) = \begin{cases} -1 & t \le 0 \\ 1 - 10e^{-4t} + 8e^{-5t} & 0 \le t \end{cases}$$

gives nontrivial solutions on the normal intervals $(-\infty, 0)$ and $(0, \infty)$.

Sec. 9.6, page 597

1. The transform of te^{-at} is $G(\omega) = 1/(a + i\omega)^2$.

3. No; yes; $G(\omega) = \dfrac{a + i\omega}{(a + i\omega)^2 + k^2}$

CHAPTER 10

Sec. 10.1, page 601

1. $y = c_1e^{2t} + c_2e^{3t} + \frac{1}{2}e^t$

3. $y = c_1e^{-t} + c_2e^{-3t} + \frac{1}{3}t - \frac{4}{9}$

5. $y = c_1te^t + c_2e^t + e^{2t}$

7. $y = c_1e^t + c_2e^{2t} + c_3e^{3t} - 1$

Sec. 10.2, page 606

1. **(a)** Yes **(b)** No **(c)** No **(d)** Yes **(e)** Yes **(f)** Yes

3. **(a)** 0 **(b)** 0 **(c)** 0 **(d)** k **(e)** k **(f)** 0

9. No; for instance, the abscissa of convergence for e^{-t} is $\alpha_0 = -1$, while for $\int_0^t e^{-t}\,dt$ it is $\alpha_1 = 0$.

Sec. 10.3, page 610

7. No; the integral in $G'(s)$ does not converge uniformly.

9. Yes; the integral in $G'(s)$ converges uniformly.

Sec. 10.4, page 615

3. Yes

7. **(a)** $\mathscr{L}\{\sin at\} = a/(s^2 + a^2)$
(b) $\mathscr{L}\{\cos at\} = s/(s^2 + a^2)$

11. $S(f') = -nC(f)$; $C(f') = (-1)^n f(\pi) - f(0) + nS(f)$

13. If $T(f')$ and $T(f'')$ are not to involve the evaluation of f or any of its derivatives, it is necessary that $K(s, a) = K(s, b) = 0$ and that $\left.\dfrac{\partial K(s, t)}{\partial t}\right|_{t=a} = \left.\dfrac{\partial K(s, t)}{\partial t}\right|_{t=b} = 0$. If $\phi(s, t)$ is an arbitrary differentiable function which is bounded at $t = a$ and $t = b$, these conditions are met by any kernel of the form $K(s, t) = (t - a)^2(t - b)^2\phi(s, t)$

15. By including the voltage Q_0/C across the capacitor when $t = 0$ as an additional term in the integrodifferential equation.

Sec. 10.5, page 621

1. See Fig. A10.1.

3. **(a)** $\dfrac{s}{s^2 - k^2}$ **(b)** $\dfrac{2}{s^3} + \dfrac{2}{s^2} + \dfrac{1}{s}$
(c) $\dfrac{1}{s} - \dfrac{6}{s - 2} + \dfrac{9}{s - 4}$ **(d)** $\dfrac{s + 2}{s^2 + 4}$

5. $\dfrac{1}{2}\left(\dfrac{1}{s} + \dfrac{s}{s^2 + 4b^2}\right)$

7. **(a)** $y = \frac{1}{6}e^{-5t} + \frac{5}{6}e^t$ **(b)** $y = -\frac{1}{4}e^{2t} - \frac{3}{4}e^{-2t}$
(c) $y = 2\cos(t/2) + 2\sin(t/2)$ **(d)** $y = \frac{1}{4} - \frac{1}{4}\cos 2t$

9. $z = 4e^{-4t} + 2e^t$

11. $y = -\frac{2}{3}e^{-t} + \frac{8}{3}e^{4t}$; $z = -\frac{1}{3}e^{-t} - \frac{2}{3}e^{-4t}$

13. $y = u(t) - 2e^{-2t}$; $z = -\frac{1}{2}u(t) + \frac{3}{2}e^{-2t}$

15. **(a)** $\dfrac{1}{(\ln c)^{c+1}}\Gamma(c + 1)$ **(b)** $\Gamma(\frac{1}{2}) = \sqrt{\pi}$
(c) $\dfrac{1}{(m + 1)^{n+1}}\Gamma(n + 1)$

17. $\dfrac{\sqrt{\pi}}{2}\dfrac{\Gamma[(k + 1)/2]}{\Gamma[(k/2) + 1]}$, $k > -1$

21. **(a)** $\dfrac{\sqrt{\pi}}{2}\dfrac{\Gamma(3/4)}{\Gamma(5/4)}$ **(b)** $\frac{1}{2}\Gamma(\frac{1}{4})\Gamma(\frac{3}{4})$ **(c)** $\dfrac{\Gamma(\frac{5}{6})\Gamma(\frac{7}{8})}{2\Gamma(\frac{41}{24})}$

25. Simply interpret $y_0, y_0', z_0, z_0', \ldots,$ as arbitrary constants when transforming the equations.

Sec. 10.6, page 628

1. **(a)** $\dfrac{s - 2}{s^2 - 4s + 29}$ **(b)** $\dfrac{15\sqrt{\pi}}{8(s + 1)^{7/2}}$ **(c)** $\dfrac{e^{-as}}{s}$

3. **(a)** $\dfrac{3\sqrt{\pi}}{(s - 2)^{5/2}}$ **(b)** $\dfrac{1}{s^2 - 4}$ **(c)** $\dfrac{1}{s}$

5. **(a)** $\dfrac{1}{s^2(s + 1)}$ **(b)** $\dfrac{2 - s^2}{2s^3}$ **(c)** $\dfrac{2e^{-s}(s + 1)}{s^3}$

7. **(a)** $\dfrac{abs}{[(s - b)^2 + a^2][(s + b)^2 + a^2]}$
(b) $\dfrac{2}{s^2(s^2 + 4)} - \dfrac{\pi}{4s}$ **(c)** $\left(\dfrac{s\cos 1 - \sin 1}{s^2 + \perp}\right)e^{-s}$

9. **(a)** $\dfrac{2}{s(s^2 + 4)} + \dfrac{e^{-s}}{s^3}(s^2 + 2s + 2)$ **(b)** $\dfrac{2(1 - e^{-\pi s})}{s^2 + 4}$
(c) $\dfrac{(s + 1)\cos 1 + \sin 1}{(s + 1)^2 + 1}$

11. $\left(\dfrac{1}{s^2} - \dfrac{1}{s}\right)(e^{-s} - e^{-3s})$

13. $\dfrac{1}{s^2}(e^{-s} - 2e^{-2s} + e^{-3s})$

15. **(a)** $\frac{1}{2}t^2e^{-5t}(3 - 5t)$ **(b)** $\frac{1}{5}(3e^{2t} + 2e^{-3t})$ **(c)** $e^{-2t}\cos t$

17. **(a)** $\frac{1}{2}t^2e^{5t}(1 + t)$ **(b)** $\frac{1}{2}(3e^{3t} - e^{-t})$ **(c)** $te^{-2t} - t^2e^{-2t}$

19. **(a)** $[1 - u(t - \pi)]\cos t$
(b) $u(t - 1)[e^{2(t-1)} - e^{t-1}] + u(t - 2)[e^{2(t-2)} - e^{t-2}]$
(c) $\frac{1}{3}\sinh(t - 2)\,u(t - 2)$

21. **(a)** $y = -\dfrac{1}{2} + \dfrac{e^{2t}}{10} + \dfrac{1}{5}(2\cos t - \sin t)$
(b) $y = e^{2t}(-3 + 7t + t^3/6) + 4\cos t + 3\sin t$
(c) $y = te^{-t} + \frac{1}{2}(t - 1)^2e^{-(t-1)}u(t - 1)$

23. **(a)** $y = t(\cos t + \sin t)$
(b) $y = \dfrac{e^{-t} - e^{-2t}(\cos 3t + 2\sin 3t)}{5}$
(c) $y = e^{-2t}(\cos 3t + \sin 3t) + \frac{1}{13}\{u(t - 3) - e^{-2(t-3)}[\cos 3(t - 3) + \frac{2}{3}\sin 3(t - 3)]u(t - 3)\}$

25. **(a)** $y = \frac{2}{15}e^{-t}t^{3/2}(5 - 2t)$
(b) $y = \frac{1}{4}[(7 + 2t)e^{-t} - 3e^{-3t}]$
(c) $y = \cos t + u(t - 2)[2\sin(t - 2) + 1 - \cos(t - 2)]$

27. **(a)** $y = (c_1 + c_2t)e^t + 1$ **(b)** $y = (c_1 + c_2t)e^{-kt}$

29. **(a)** $(x, y) = (-3 - 11e^t, 1 + 5e^t)$
(b) $(x, y) = [e^{-t}\sin t, 2e^{-3t} - 2e^{-t}(\cos 2t + \sin 2t)]$
(c) $(x, y) = (-2 + \frac{3}{2}e^{-t} + \frac{1}{2}e^{3t}, 1 - \frac{1}{2}e^{-t} + \frac{1}{2}e^{3t} - e^{2t})$
(d) $\begin{bmatrix} x \\ y \end{bmatrix} = \begin{bmatrix} 4 + 2\cos\sqrt{3}t - 6\cos t \\ 16 - 4\cos\sqrt{3}t - 12\cos t \end{bmatrix}$

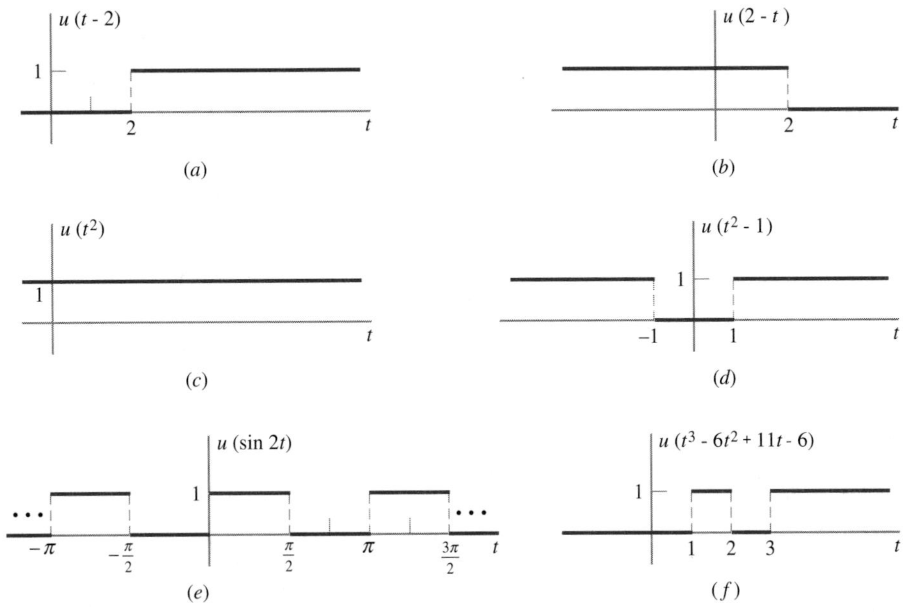

FIGURE A10.1

31. $(\sqrt{\pi}+1)/2$ **33.** $\mathcal{L}\{t^{1/2}\} = \sqrt{\pi}/2s^{3/2}$

35. s

37. $y = (t^4 + 9t + 3)e^{-3t}$ in

39. $y = -aw(2c^4 - \frac{4}{3}c^3x + x^4/24)$

Sec. 10.7, page 633

1. $\dfrac{s^2-9}{(s^2+9)^2}$ **3.** $\dfrac{2}{(s+1)^3}$

5. $\dfrac{2s(s^2-12)}{(s^2+4)^3}$ **7.** $\dfrac{2s^3+15s^2+36s+39}{(s^3+6s^2+13s)^2}$

9. $\dfrac{4}{[(s+3)^2+4]^2}$ **11.** $\dfrac{6}{s(s+2)^4}$

13. $\ln\dfrac{\sqrt{s^2+9}}{s}$ **15.** $\cot^{-1}\left(\dfrac{s+3}{2}\right)$

17. $\dfrac{1}{s+3}\cot^{-1}\left(\dfrac{s+3}{2}\right)$ **19.** $\dfrac{(-2t^3+3t^2)e^{-2t}}{6}$

21. $t - 1 + e^{-t}$ **23.** $\dfrac{2(1-\cosh t)}{t}$

25. $\dfrac{1+e^{-t}-2\cos t}{t}$ **27.** $\displaystyle\int_0^t \frac{\sin u}{u}\,du$ or $\mathrm{Si}(t)$

29. $2(\sinh t - t\cosh t)/t^2$. The number 2 in the given transform, though it appears to play no role in the determination of the inverse, is necessary in order that $\lim_{s\to\infty}\phi(s) = 0$, as required by Corollary 3, Theorem 2, Sec. 10.2.

31. $2te^{-3t/2}\sinh\dfrac{t}{2}$

33. $y = \cos 2t + \frac{1}{2}\sin 2t + \dfrac{t}{2}\sin 2t$

35. $y = e^{-2t} + te^{-2t} + \frac{1}{2}(t-1)^2e^{-2(t-1)}u(t-1)$

37. $y = (t\sin t)/2$ **39.** $y = y_0e^{-t} + \dfrac{c}{6}t^3e^{-t}$

41. $y = c_1(\cos t + t\sin t) + c_2(\sin t - t\cos t)$

43. $y = e^{-t} - 1 + \displaystyle\int_0^t \frac{1-e^{-\tau}}{\tau}\,d\tau$

Sec. 10.8, page 638

1. $f(0^+) = 1$, $\lim_{t\to\infty} f(t) = 0$

3. $f(0^+) = 1$, $\lim_{t\to\infty} f(t) = 0$

7. $\ln(\sqrt{5}/2)$

9. $\tan^{-1}\frac{3}{2} - \tan^{-1} 3$

11. $\tan^{-1}(p/q) - \tan^{-1}(a/b)$

13. $\ln(\sqrt{5}/3)$

15. $f'(t)$, $f''(t)$, and $f'''(t)$ must be piecewise regular and of exponential order. Yes.

19. In using Theorem 1, Sec. 10.3, to establish the necessary region of uniform convergence. In other words, if α_0 is not negative, then the values of s in the neighborhood of $s = 0$ are not in the region of uniform convergence and we have no assurance that the limit can be taken inside the integral sign.

Sec. 10.9, page 644

1. $\frac{5}{2}e^{-t} - 9e^{-2t} + \frac{15}{2}e^{-3t}$ **3.** $-e^{-t} + e^{-2t} + 2te^{-2t}$

5. $\frac{1}{5}(e^{-t} - \cos 2t + 3\sin 2t)$

7. $\frac{1}{25}(3\cos t + 4\sin t - 3e^{-2t} - 10te^{-2t})$

9. $(-1 + \cos t + 2\sin t)e^{-2t}$

11. $\frac{1}{5}[-e^{-2t}(14\cos t + 3\sin t) + e^{-3t}(14\cos t + 22\sin t)]$

13. $y = \frac{1}{3}e^{2t} - \frac{1}{2}e^{t} + \frac{1}{6}e^{-t} + \frac{1}{2}u(t-2) + \frac{1}{6}e^{2(t-2)}u(t-2) - \frac{1}{2}e^{t-2}u(t-2) - \frac{1}{6}e^{-(t-2)}u(t-2)$

15. $y = \frac{1}{4}[e^{-t}(t^2 + 2t + 1) - \cos t - \sin t]$

17. $y = \frac{1}{50}(9\cos t - 37\sin t - 9e^{-3t} + 10te^{-3t})$
$z = \frac{1}{50}(-11\cos t + 48\sin t + 11e^{-3t} - 15te^{-3t})$

21. The inverse, $(t/4) \sin 2t$, is of course the same by either method.

Sec. 10.10, page 654

3. $\dfrac{\coth(\pi s/2)}{s^2+1}$ **5.** $\dfrac{2}{(s^2+4)(1+e^{-\pi s})}$

7. $\dfrac{1-(1+as)e^{-as}}{s^2(1-e^{-2as})}$

9. $(-1)^n\phi_6(\tau, 2, 2) + \phi_6(t, 2, 2) + (-1)^n\phi_6(\tau, 3, 2) + \phi_6(t, 3, 2)$

11. $(-1)^n\phi_6(\tau, 1, 1) + \phi_6(t, 1, 1) - [(-1)^n\phi_6(\tau, 2, 1) + \phi_6(t, 2, 1)] - [(-1)^n\phi_{12}(\tau, 2, 1) + \phi_{12}(t, 2, 1)]$

13. $y = \phi_2(t, 2) - [(-1)^n\phi_8(\tau, 0, 1, 1) + \phi_8(t, 0, 1, 1)]$

15. $y = -e^{-t} + 2e^{-3t} + [(-1)^n\phi_6(\tau, 1, 2) + \phi_6(t, 1, 2)] - [(-1)^n\phi_6(\tau, 3, 2) + \phi_6(t, 3, 2)]$

17. $f(t) = \phi_2(t, 1) - \phi_2[(t-1), 2]u(t-1)$. Figure A10.2 shows a plot of f.

19. $f(t) = \begin{cases} \sin 2t & 0 < t < \pi \\ 0 & \pi < t < 2\pi \end{cases}$; $f(t) = \begin{cases} \frac{1}{2}\sin 4t & 0 < t < \pi \\ 0 & \pi < t < 2\pi \end{cases}$

The graph of $f(t)$ when $a = 0$, $b = 2$, and $k = \pi$ is a sine wave of period π from which every second (full) cycle is missing.

21. The converse is true only if $f(t)$ is identically zero between k and $2k$.

23. Because the corresponding ϕ's are undefined for these sets of values, which in turn happens because in the finite geometric series which are summed in the derivation of the inverses the common ratio in either 1 or -1.

25. **(a)** $\dfrac{e^s-1}{s(e^s-2)}$ **(b)** $\dfrac{e^s-1}{s(e^s+1)}$

27. $e^{-(n+1)}$, $n < t < n+1$

31. $f(t) = \phi_2(t, 2)$, shifted one unit to the right and cut off to the left of $t = 1$.

33. $\mathscr{L}\{f_n(t)\} = e^{-nks}(1 - e^{-ks})\mathscr{L}\{f(t)\}$

35. **(a)** $\phi(s) = \phi_0(s)/(1 - e^{-ks})$ **(b)** With $\phi(s) = \mathscr{L}\{f(t)\}$ and $\phi_0(s) = \int_0^k f(t)e^{-st}\,dt$, the two results are identical.

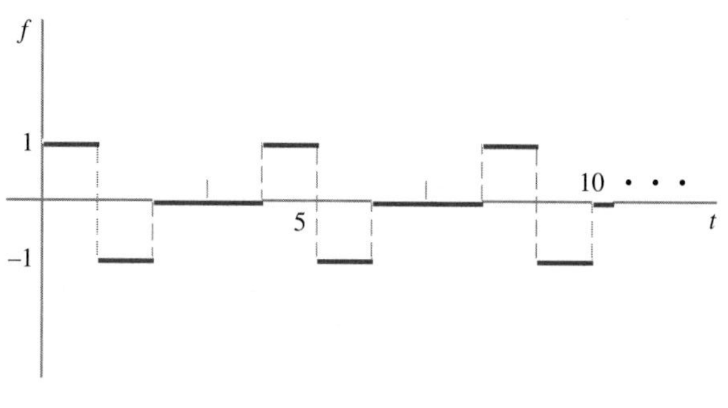

FIGURE A10.2

37. $\phi(s) \equiv \mathscr{L}\{f(t)\} = \dfrac{s}{(s^2+1)(1-e^{-\pi s})}$

39. $\phi(s) \equiv \mathscr{L}\{f(t)\} = \dfrac{1}{s(1+e^{-s})}$

Sec. 10.11, page 665

1. $\int_0^t e^{(t-\lambda)}e^{2\lambda}\,d\lambda = e^{2t} - e^t$. The two solutions are the same.

3. $\frac{1}{16}(\sin 2t - 2t\cos 2t)$

5. $\frac{1}{216}(t \sin 3t - 3t^2 \cos 3t)$

7. $e^{-t} - \frac{1}{2}(\sin t - t\cos t)$

9. $D(t) - 2\cos 2t + \frac{3}{2}\sin 2t$

11. $D(t-1) - \delta(t-1) + e^{-(t-1)}u(t-1)$

13. $Y(t) = \dfrac{1}{b-a}\displaystyle\int_0^t (e^{-a\lambda} - e^{-b\lambda})f(t-\lambda)\,d\lambda = \dfrac{1}{b-a}\displaystyle\int_0^t f(\lambda)[e^{-a(t-\lambda)} - e^{-b(t-\lambda)}]\,d\lambda$

17. $[(1 - e^{-as})/as]^2$; 1

19. **(a)** $y = u(t) - e^{-t}$ **(b)** $y = e^{-t} - e^{-2t}$ **(c)** $y = \frac{1}{4}(e^{3t} - e^{-t})$ **(d)** $y = \frac{1}{4}(e^{-t} - e^{-5t})$

21. **(a)** $y = e^{-t}\cosh 2t$ **(b)** $y = e^{-t}\cos 2t$ **(c)** $y = e^{-4t}\cosh t$ **(d)** $y = e^{-4t}\cos 2t$

23. **(a)** $y = -2u(t) + e^{-t} + e^t$ **(b)** $y = 2u(t) - \cos t$

25. $y = te^{-t} + (t - t_0)e^{-(t-t_0)}u(t - t_0)$. No subsequent impulse can bring the system into a state of rest. However, the simultaneous application of an impulse of strength $-t_0e^{-t_0}$ and a doublet of strength $-(t_0+1)e^{-t_0}$ will bring the system to rest at a displacement of $y = t_0e^{-t_0}$.

29. $A(t) = \frac{1}{2} - e^{-t} + \frac{1}{2}e^{-2t}$; $h(t) = e^{-t} - e^{-2t}$

31. $A(t) = \dfrac{3\times 10^{-3}}{11}(2e^{-2000t} - e^{-1000t/6})$

$h(t) = \dfrac{1}{2\times 10^3}\left[I(t) - \dfrac{12}{11}e^{-2000t} + \dfrac{1}{22}e^{-1000t/6}\right]$

Note: The term $I(t)$ arises because there is no inductor in the circuit and hence the current builds up *instantaneously* to a nonzero value.

33. **(b)** $\dfrac{2}{\sqrt{\pi}}\displaystyle\int_0^t e^{-\tau^2}\,d\tau$

35. **(a)** $\dfrac{2\sin t}{\sqrt{\pi}}\displaystyle\int_0^{\sqrt{t}} \cos u^2\,du - \dfrac{2\cos t}{\sqrt{\pi}}\displaystyle\int_0^{\sqrt{t}} \sin u^2\,du$

(b) $\dfrac{2\cos t}{\sqrt{\pi}}\displaystyle\int_0^{\sqrt{t}} \cos u^2\,du + \dfrac{2\sin t}{\sqrt{\pi}}\displaystyle\int_0^{\sqrt{t}} \sin u^2\,du$

41. **(a)** $\begin{cases} 0 & t < a \\ f(t-a) & 0 \le a \le t \end{cases}$

(b) $\begin{cases} 0 & t < a \\ \int_a^t f(t-\lambda)\,d\lambda & 0 \le a \le t \end{cases}$

43. It depends on what kinds of functions are acceptable solutions. Since $f * x = g \rightarrow \mathscr{L}(x) = \mathscr{L}\{g\}/\mathscr{L}\{f\}$, it is possible that $\mathscr{L}\{x\}$ will not be the transform of any "respectable" function. This is the case, for example, if $f = g$.

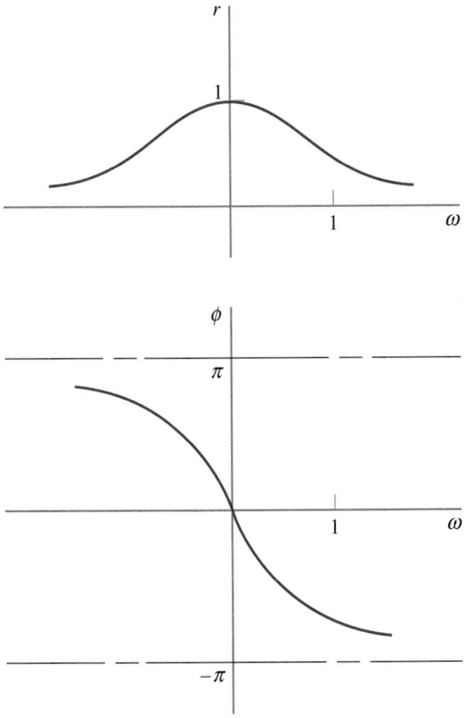

FIGURE A10.3

45. $y = 3t$

47. $y = y(0) + \dfrac{t^{b-1} * f(t)}{\Gamma(b)\Gamma(1-b)}$

49. $y = t^2 - 4t^{3/2}$

Sec. 10.12, page 670

1. **(a)** $Z(z) = z/(z - e^{-aT})$ **(b)** $Z(z) = z/(ze^{aT} - 1)$

3. $Z(z) = \dfrac{z^2 - z\cos \omega t}{z^2 - 2z\cos\omega t + 1}$ 5. $Z(z) = \dfrac{z(z+1)T^2}{(z-1)^2}$

7. 0, 0, 1, 0, 4, 0, 4^2, 0, 4^3, 0, ⋯

9. Spectrum: $\dfrac{1}{1+\omega^2}e^{-2i\tan^{-1}\omega}$. See Fig. A10.3.

CHAPTER 11

Sec. 11.2, page 682

5. **(a)** Assume $\Phi(x, y) = Ae^{mx+ny}$ as a particular integral.
   (b) Assume $\Phi(x, y) = A\sin(mx + ny)$ as a particular integral.
   (c) Assume $\Phi(x, y) = A\cos(mx + ny)$ as a particular integral.
   (d) Assume a particular integral of any one of the forms

$$\Phi(x, y) = x^2(Ax^2 + Bxy + Cy^2)$$
$$\Phi(x, y) = xy(Ax^2 + Bxy + Cy^2)$$
$$\Phi(x, y) = y^2(Ax^2 + Bxy + Cy^2)$$

   (e) Assume $\Phi(x, y)$ equal to a general homogeneous polynomial of degree n in x and y multiplied by x^2, xy, or y^2.

7. Yes

9. **(a)** $A = 0$; $\lambda = n\pi a/l$; B, C, D arbitrary
   (b) $\lambda = 0$; A, B, C, D arbitrary

13. **(a)** $u(x) = 100 - 50\dfrac{\ln(1+x)}{\ln 2}$
   (b) $u(x) = \frac{1}{3}(300 - 100x - 50x^2)$

17. $\dfrac{\partial^2 u}{\partial x^2} = \dfrac{1}{k}\dfrac{\partial u}{\partial t}$

19. $u(r) = 100\dfrac{\ln r}{\ln 2}$ 21. $u(r) = (60/r) + 40$

23. $\dfrac{\partial^2 u}{\partial x^2} = a^2\dfrac{\partial u}{\partial t} + b^2(u - u_0)$, where $a^2 = \dfrac{c\rho}{kg}$, $b^2 = \dfrac{2h}{kr}$, k is the thermal conductivity, and h is the proportionality constant in Newton's law of cooling. If the cross sections are not circular, $b^2 = hp/kA$, where p is the perimeter of the cross-sectional area A.

27. $\dfrac{\partial^2 u}{\partial x^2} + \dfrac{\partial^2 u}{\partial y^2} = a^2\dfrac{\partial u}{\partial t} + 2\dfrac{h}{k}(u - u_0)$, where $a^2 = \dfrac{c\rho}{kg}$, h = surface conductivity, and u_0 = (constant) temperature of surrounding medium.

29. $\dfrac{\partial}{\partial r}\left(r\dfrac{\partial u}{\partial r}\right) = a^2 r\dfrac{\partial u}{\partial t} + 2r\dfrac{h}{k}(u - u_0)$

31. $\dfrac{\partial^2 z}{\partial t^2} = \dfrac{Tg}{w(x, y)}\left(\dfrac{\partial^2 z}{\partial x^2} + \dfrac{\partial^2 z}{\partial y^2}\right) + \dfrac{q}{w(x, y)}f(x, y, z, \dot{z}, t)$

33. $a^2\dfrac{\partial u}{\partial t} = \dfrac{\partial^2 u}{\partial r^2} + \dfrac{1}{r}\dfrac{\partial u}{\partial r} + \dfrac{1}{r^2}\dfrac{\partial^2 u}{\partial \theta^2}$, where $a^2 = \dfrac{c\rho}{kg}$

37. **(a)** $a^2\left(r^2\dfrac{\partial^2 w}{\partial r^2} - r\dfrac{\partial w}{\partial r} + w\right) = r^2\dfrac{\partial^2 w}{\partial t^2}$
   (b) $a^2\dfrac{\partial^2 w}{\partial r^2} = \dfrac{\partial^2 w}{\partial t^2}$

39. Methods **(a)**, **(b)**, and **(d)** all yield

$$y(x) = \begin{cases} \dfrac{-kx}{48EI}(2x^3 - 6lx^2 + 9l^3) & 0 \le x \le l \\ \dfrac{-kl}{48EI}(x - 2l)(2x^2 - 8lx + l^2) & l \le x \le 2l \end{cases}$$

Method **(c)** yields

$$y(x) = \frac{-2kl^4}{EI\pi^4}\left[16\sum_{n=1}^{\infty}\frac{1}{(2n-1)^5}\sin\frac{(2n-1)\pi x}{2l} + \sum_{n=1}^{\infty}\frac{1}{(2n-1)^5}\sin\frac{(2n-1)\pi x}{l}\right]$$

Sec. 11.3, page 693

1. **(a)** $y(x, t) = \sin x\cos at$ **(b)** $y(x, t) = \sin 2x\cos 2at$

3. $y(x, t) = \sin mx\cos mat + (1/na)\sin nx\sin nat$

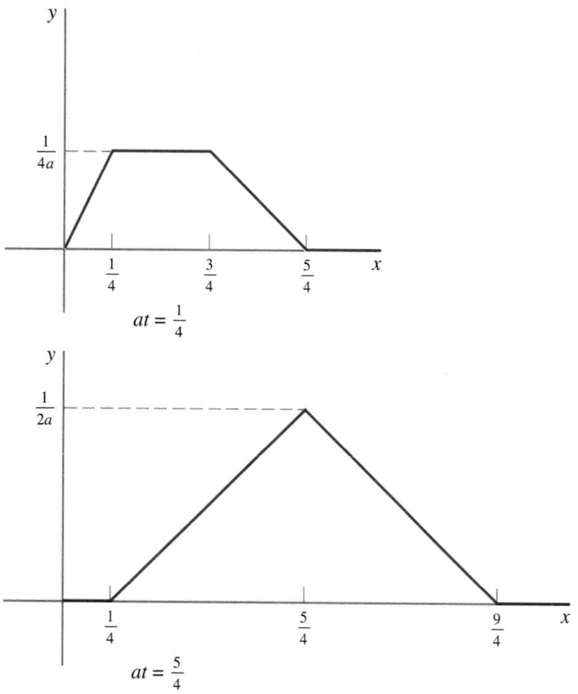

FIGURE A11.1

5. See Fig. A11.1.

7.

$0 < at < 1$	
$x < -at - 1;$	$y = 0$
$-at - 1 < x < at - 1;$	$y = \dfrac{x + at + 1}{2a}$
$at - 1 < x < -at + 1;$	$y = t$
$-at + 1 < x < at + 1;$	$y = \dfrac{-x + at + 1}{2a}$
$at + 1 < x;$	$y = 0$
$1 < at$	
$x < -at - 1;$	$y = 0$
$-at - 1 < x < -at + 1;$	$y = \dfrac{x + at + 1}{2a}$
$-at + 1 < x < at - 1;$	$y = 1/a$
$at - 1 < x < at + 1;$	$y = \dfrac{-x + at + 1}{2a}$
$at + 1 < x;$	$y = 0$

9. $\text{KE} = \begin{cases} (1 - at/2)\rho & 0 < at < 1 \\ \frac{1}{2}\rho & 1 < at \end{cases}$

ρ = mass/unit length

11. $x < -at;$ $\quad y = 0$

$-at < x < at;$ $\quad y = \dfrac{1}{2a} \ln (x + at + 1)$

$at < x;$ $\quad y = \dfrac{1}{2a} \ln \dfrac{x + at + 1}{x - at + 1}$

13. $x < -at;$ $\quad y = 0$

$-at < x < at;$ $\quad y = (1/2a)(1 - e^{-(x+at)})$

$at < x;$ $\quad y = (1/a)e^{-x} \sinh at$

15. $x < -at;$ $\quad y = (1/a)e^{x} \sinh at$

$-at < x < at;$ $\quad y = (1/a)[1 - e^{-at} \cosh x]$

$at < x;$ $\quad y = (1/a)e^{-x} \sinh at$

19. $\theta(x) = a\phi'(x)$

21. For no left-moving wave:

$\phi(x) = (1/a)(1 + \cos x)[u(x + \pi) - u(x - \pi)]$

For no right-moving wave:

$\phi(x) = -(1/a)(1 + \cos x)[u(x + \pi) - u(x - \pi)]$

25. $x \le -at;$ $\quad y = e^{x}(x \cosh at + at \sinh at)$

$-at \le x \le at;$ $\quad y = e^{-at}(x \cosh x - at \sinh x)$

$at \le x;$ $\quad y = e^{-x}(x \cosh at - at \sinh at)$

27. $$y(10, t) = \begin{cases} 0 & 0 \le at \le 6 \\ (at - 6)/2a & 6 \le at \le 9 \\ 3/2a & 9 \le at \le 11 \\ (14 - at)/2a & 11 \le at \le 14 \\ 0 & 14 \le at \end{cases}$$

29. $y(x, t) = \frac{1}{2}(1 - |x - at|)[u(x - at + 1) - u(x - at - 1)]$
$+ \frac{1}{2}(1 - |x + at|)[u(x + at + 1) - u(x + at - 1)]$
$v(0, t) = -a,\ 0 \le at < 1;\ v(0, t) = 0,\ 1 < at$

31. $y(x, t) = \frac{1}{2} \cos (x - at)[u(x - at + \pi/2) -$
$u(x - at - \pi/2)] + \frac{1}{2} \cos (x + at)[u(x + at + \pi/2) -$
$u(x + at - \pi/2)]$

$$v(0, t) = \begin{cases} -a \sin at & 0 \le at < \pi/2 \\ 0 & \pi/2 < at \end{cases}$$

33. $$y(x, 0) = \phi(x) = \begin{cases} x^2 - 1 + k \sin \pi x & |x| \le 1 \\ 0 & |x| \ge 1 \end{cases}$$

35. $y(\pi, at) = \frac{1}{2}(\pi - at)^2 \sin at;$ $\quad \dot{y}(\pi, at) =$
$(a/2)(\pi - at)[(\pi - at) \cos at - 2 \sin at],\ 0 < at < 2\pi$
$y(\pi, at) = 0;$ $\quad \dot{y}(\pi, at) = 0,$ $\quad 2\pi < at$

37. $y(\pi, at) = 1 + \cos at;\ \dot{y}(\pi, at) = -a \sin at,\ 0 \le at \le \pi$
$y(\pi, at) = 0;$ $\quad \dot{y}(\pi, at) = 0,$ $\quad \pi \le at$

39. $y(x, t) =$

$$\begin{aligned} &\tfrac{1}{2} \cos (x - at)[u(x - at - \pi/2) - u(x - at - 3\pi/2)] \\ &- \tfrac{1}{2} \cos (x - at)[u(x - at + 3\pi/2) - u(x - at + \pi/2)] \\ &+ \tfrac{1}{2} \cos (x + at)[u(x + at - \pi/2) - u(x + at - 3\pi/2)] \\ &- \tfrac{1}{2} \cos (x + at)[u(x + at + 3\pi/2) - u(x + at + \pi/2)] \end{aligned}$$

$$v(\pi, t) = \begin{cases} a \sin at & 0 < at < \pi/2 \\ 0 & \pi/2 < at < 3\pi/2 \\ (a/2) \sin at & 3\pi/2 < at < 5\pi/2 \\ 0 & 5\pi/2 < at \end{cases}$$

Sec. 11.4, page 699

3. **(a)** Parabolic at points on the parabola $x^2 - y - 1 = 0$; elliptic at points in the interior of this parabola; hyperbolic at points outside this parabola
(b) Parabolic at points on the hyperbola $x^2 - xy - y = 0$, elliptic at points in the interior of this hyperbola, hyperbolic at points outside this hyperbola
(c) Parabolic at points on the circle $(x - 1)^2 + y^2 = 1$, elliptic at points in the interior of this circle, hyperbolic at points outside this circle

9. **(a)** $0 \le t \le l$, $y(x, t) = \begin{cases} \sin 2\pi(t - x) & x \le t \\ 0 & x \ge t \end{cases}$

(b) $l \le t \le 2l$, $y(x, t) =$
$\begin{cases} \sin 2\pi(t - x) & x \le 2l - t \\ \sin 2\pi(t - x) - \sin 2\pi(t + x - 2l) & x \ge 2l - t \end{cases}$

(c) $2l \le t \le 3l$, $y(x, t) =$
$\begin{cases} \sin 2\pi(t - x) + \sin 2\pi(t - x - 2l) & x \le t - 2l \\ \sin 2\pi(t - x) & x \ge t - 2l \end{cases}$

Sec. 11.5, page 714

1. **(a)** A shaft connecting two objects of high polar moment of inertia, for example, the shaft connecting a rotary engine and an airplane propeller
(b) The hull of a ship twisting under the influence of waves hitting the bow and stern at different times
(c) A tall chimney or a radio antenna tower

3. **(a)** $\sin 2\lambda l/a = 0$; $\lambda_n = n\pi a/2l$, $y_n = \cos n\pi(l - x)/2l$
(b) $\sin 2\lambda l/a = 0$, $\lambda_n = n\pi a/2l$, $y_n = \sin n\pi(x + l)/2l$
(c) $\cos 2\lambda l/a = 0$, $\lambda_n = (2n + 1)\pi a/4l$, $y_n = \sin (2n + 1)\pi(x + l)/4$

5. Doubling the tension multiplies the frequency by $\sqrt{2}$. Because it is easier to change the length quickly and accurately than it is to change the tension.

7. Equations **(a)**–**(d)** and **(f)**–**(k)** can be separated. Equation **(e)** cannot be separated.

9. **(a)** $\theta(x, t) =$

$$\frac{8l^2}{\pi^3} \sum_{m=1}^{\infty} \frac{1}{(2m - 1)^3} \sin \frac{(2m - 1)\pi x}{l} \cos \frac{(2m - 1)\pi at}{l}$$

11. **(a)** $u(x, y) = \sum_{n=1}^{\infty} B_n \sin n\pi x \sinh n\pi y$,

$$B_n = \frac{2}{\sinh n\pi} \int_0^1 f(x) \sin n\pi x \, dx$$

(b) $u(x, y) =$

$$\frac{400}{\pi} \sum_{m=1}^{\infty} \frac{\sin (2m - 1)\, \pi x \cosh (2m - 1)\pi y}{(2m - 1) \cosh [(2m - 1)\pi]}$$

(c) $u(x, y) = \sum_{m=0}^{\infty} A_n \cos n\pi x \cosh n\pi(1 - y)$,

$$A_n = \frac{2}{\cosh n\pi} \int_0^1 f(x) \cos n\pi x \, dx$$

(d) $u(x, y) =$

$$\frac{400}{\pi} \sum_{m=0}^{\infty} (-1)^m \frac{\cos [\frac{1}{2}(2m + 1)\pi x] \cosh [\frac{1}{2}(2m + 1)\pi y]}{(2m + 1) \cosh [\frac{1}{2}(2m + 1)\pi]}$$

(e) $u(x, y) = \dfrac{\cos (\pi x/2) \sinh (\pi y/2)}{\sinh (\pi/2)}$

13. **(a)** $u(x, y) = 100$ **(b)** $u(x, y) = 100x$

(c) $u(x, y) =$

$$\frac{400}{\pi} \sum_{n=1}^{\infty} \frac{1}{2n - 1} \sin \frac{(2n - 1)\pi x}{2} \exp \left[\frac{-(2n - 1)\pi y}{2}\right]$$

(d) $u(x, y) = 100x + \dfrac{200}{\pi} \sum_{n=1}^{\infty} \dfrac{(-1)^n}{n} e^{-n\pi y} \sin n\pi x$

(e) $u(x, y) = \dfrac{2}{\pi} - \dfrac{4}{\pi} \sum_{n=1}^{\infty} \dfrac{1}{4n^2 - 1} e^{-2n\pi y} \cos 2n\pi x$

15. $y(x, t) = \dfrac{4l}{\pi^2} \sum_{n=1}^{\infty} \dfrac{1}{n^2} \sin \dfrac{n\pi}{2} \sin \dfrac{n\pi x}{l} \cos \dfrac{n\pi at}{l}$

17. $y(x, t) = \dfrac{4l}{\pi^3} \sum_{n=1}^{\infty} \dfrac{1}{n^3} \sin \dfrac{n\pi}{2} \sin \dfrac{n\pi x}{l} \sin \dfrac{n\pi at}{l}$

19. $y(x, t) = \sum_{n=1}^{\infty} B_n \sin \dfrac{n\pi x}{l} \sin \dfrac{n\pi at}{l}$, where

$$B_n = \frac{2}{n\pi a} \sin \frac{n\pi}{2} \left[\sin \frac{n\pi k}{2l} \middle/ \frac{n\pi k}{2l}\right]$$

As $k \to 0$, $B_n \to (2/n\pi a) \sin (n\pi/2)$, which we infer to be the coefficient formula when the string is set in motion by a unit impulse applied at its center.

21. $C(x, t) =$

$$\tfrac{1}{6}C_0 l^2 - C_0 \frac{l^2}{\pi^2} \sum_{m=1}^{\infty} \frac{1}{m^2} \cos \frac{2m\pi x}{l} \exp \frac{-4m^2\pi^2 t}{a^2 l^2}$$

23. The restrictions are $\int_0^l f(x)\, dx = 0$ and $\int_0^l g(x)\, dx = 0$. The first restriction implies that the integrated initial angular displacement is zero, which will always be the case if the origin of θ is suitably chosen. Since the shaft is of uniform cross section, the second condition implies that $\int_0^l I\dot{\theta}(x, 0)\, dx = 0$, which is precisely the condition that the total angular momentum of the shaft is initially (and hence permanently) zero. In other words, the restriction on $g(x)$ implies that the vibration being studied is not superposed on a uniform rotation.

25. $u(x, t) = \dfrac{400}{\pi} \sum_{\substack{m=1 \\ m \text{ odd}}}^{\infty} \dfrac{1}{m} \sin \dfrac{m\pi x}{l} \exp \dfrac{-m^2\pi^2 t}{a^2 l^2}$

27. **(a)** $u(x, t) =$

$$\frac{100x}{l} + \frac{200}{\pi} \sum_{n=1}^{\infty} \frac{(-1)^n}{n} \sin \frac{n\pi x}{l} \exp \frac{-n^2\pi^2 t}{a^2 l^2}$$

(b) $u(x, t) = \dfrac{50x}{l} + 50 + \sum_{n=1}^{\infty} B_n \sin \dfrac{n\pi x}{l} \exp \dfrac{-n^2\pi^2 t}{a^2 l^2}$,

where $B_n = \begin{cases} 100/n\pi & n \text{ even} \\ -300/n\pi & n \text{ odd} \end{cases}$

29. $u\left(\dfrac{l}{2}, t\right) = \dfrac{400}{\pi}\left(z - \dfrac{1}{3}z^9 + \dfrac{1}{5}z^{25} - \cdots\right)$, where $z = e^{-\pi^2 t/a^2 l^2}$

From this example it appears that the temperature at $x = l/2$ begins to decrease immediately; that is, the cooling effect of the zero end conditions is felt instantly at $x = l/2$. In other words, according to the one-dimensional heat equation, thermal disturbances propagate with infinite velocity.

31. **(a)** $u(x, t) = e^{-4\alpha t} \sin 2x$

(b) $u(x, y) = 2y + \dfrac{3 \sinh y \cos x}{\sinh 2} - \dfrac{2 \sinh 2y \cos 2x}{\sinh 4}$

33. **(a)** $u(x, y) = 100(1 - y);\ 0 \le x,\ 0 \le y \le 1$

(b) $u(x, y) = e^{-x} \cos y + e^{-3x} \cos 3y +$

$$b\left[1 + \frac{4}{\pi} \sum_{n=1}^{\infty} \frac{(-1)^n}{2n-1} e^{-(2n-1)x} \cos (2n-1)y\right]$$

$0 \le x,\ 0 \le y \le \pi/2$

35. **(a)** $u(x, t) = 10 + e^{-4\alpha t} \sin 2x;\ 0 \le x \le \pi,\ 0 \le t$

(b) $u(x, t) = [100(h - z)]/h$

Sec. 11.6, page 742

1. $a = 0,\ b = -3,\ h(x) = 1 - 3x^2$

3. **(a)** $a = b$ **(b)** $p(x) = \begin{cases} -x & -c < x < 0 \\ \dfrac{a}{b}x & 0 < x < c \end{cases}$

5. $\lambda_n = n/4,\ y_n = \sin n\pi x$

7. $\lambda_n = 3n\pi/2,\ y_n = \sin (n\pi x/2)$

9. $u(x, t) = \sum_{n=1}^{\infty} B_n \sin \lambda_n(l - x) \exp \dfrac{-\lambda_n^2 t}{a^2}$, where

$$B_n = \int_0^l f(x) \sin \lambda_n(l - x)\, dx \Big/ \int_0^l \sin^2 \lambda_n(l - x)\, dx$$

and $\tan \lambda l = \dfrac{-\lambda}{h}$, as before.

11. $u(x, t) = \dfrac{70xh}{1 + hl} + \sum_{n=1}^{\infty} B_n \sin \lambda_n x \exp \dfrac{-\lambda_n^2 t}{a^2}$, where

$$B_n = \int_0^l \left[f(x) - \frac{70xh}{1 + hl}\right] \sin \lambda_n x\, dx \Big/ \int_0^l \sin^2 \lambda_n x\, dx$$

and $\tan \lambda l = \dfrac{-\lambda}{h}$, as before.

13. $u(x, t) = \sum_{n=1}^{\infty} B_n \sin \lambda_n x \exp \dfrac{-\lambda_n^2 t}{a^2}$, where

$B_n = \int_0^l f(x) \sin \lambda_n x\, dx \Big/ \int_0^l \sin^2 \lambda_n x\, dx$ and

$\tan z = -\dfrac{2\beta z}{1 - (\beta z)^2}$ or $\tan \dfrac{z}{2} = \begin{cases} -\beta z \\ 1/\beta z \end{cases}$, where

$z = \lambda l,\ \beta = 1/hl$

15. **(a)** $u(x, y) = \sum_{n=1}^{\infty} A_n \cos \lambda_n x \sinh \lambda_n y$, where

$$A_n = \int_0^a f(x) \cos \lambda_n x\, dx \Big/ \sinh \lambda_n b \int_0^a \cos^2 \lambda_n x\, dx$$

and $\tan \lambda a = h/\lambda$

(b) $u(x, y) = \sum_{n=1}^{\infty} A_n \cosh \lambda_n x \sin \lambda_n y$, where

$$A_n = \int_0^b f(y) \sin \lambda_n y\, dy \Big/ \cosh \lambda_n a \int_0^b \sin^2 \lambda_n y\, dy$$

and $\tan \lambda b = -\lambda/h$

17. **(a)** $u(x, t) = \dfrac{400}{\pi} \sum_{\substack{n=1 \\ n \text{ odd}}}^{\infty} \dfrac{1}{n} \sin \dfrac{n\pi x}{l} \exp \dfrac{-(n^2\pi^2 + b^2l^2)t}{a^2l^2}$

(b) $u(x, t) = \dfrac{T_0 \sinh bx}{\sinh bl} +$

$$\sum_{n=1}^{\infty} \frac{2T_0 n\pi(-1)^n}{n^2\pi^2 + b^2l^2} \sin \frac{n\pi x}{l} \exp \frac{-(n^2\pi^2 + b^2l^2)t}{a^2l^2}$$

19. $u(x, y) = \dfrac{400}{\pi} \sum_{\substack{n=1 \\ n \text{ odd}}}^{\infty} \dfrac{\sin n\pi x \sinh (n^2\pi^2 + b^2)^{1/2} y}{n \sinh (n^2\pi^2 + b^2)^{1/2}}$,

$b^2 = \dfrac{2h}{k}$

21. $u(x) = \begin{cases} \dfrac{100k_2 x}{a_1k_2 + a_2k_1} & 0 \le x \le a_1 \\ \dfrac{100k_1(x - a_1 - a_2)}{a_1k_2 + a_2k_1} + 100 & a_1 \le x \le a_1 + a_2 \end{cases}$

23. The characteristic or frequency equation is $\tan z = -\alpha z$, where $z = \lambda l/a$ and $\alpha = E_s J/kl$, k being the proportionality constant in the elastic end constraint.

25. The frequency equation is $\tan z = 2\alpha z/[(\alpha z)^2 - 1]$, where $z = \lambda l/a$ and $\alpha = E_s J/kl$.

27. **(a)** 340 Hz **(b)** 15.8 Hz **(c)** 22.3 Hz

29. The frequency equation is $\tan z = \pm r_1^2/r_2^2$, where $z = \lambda l/a$.

33. From the frequency equations obtained in Example 3 and in Exercise 32**(b)** and **(c)**, it follows that except for the first two or three frequencies, the natural frequencies of identical cantilever beams, fixed-fixed beams, and free-free beams are almost identical.

35. The frequency equation is $\sin z[2 \cos z \cosh z - rz(\cosh z \sin z - \cos z \sinh z)] = 0$, where $z = l\sqrt{\lambda}/a$ and $r = M/M_b$, M_b being the mass of the beam. When $M = 0$, the system reduces to a simple hinged-hinged beam of length $2l$. When $M = \infty$, the system reduces to two identical hinged-fixed beams of length l. In each case, the general frequency equation reduces to the appropriate special case.

39. $\int_a^b f^2(x)\, dx - \sum_{i=1}^{n} c_i^2$

41. **(a)** $y_n = \sin (n\pi \ln x)$ **(b)** $y_n = \cos (n\pi \ln x]$

(c) $y_n = \cos [(2n + 1)\pi \ln x]$

(d) $y_n = \sin (\lambda_n \ln x)$, where λ_n is the nth root of the equation $\lambda \cos \lambda = e \sin \lambda$

47. **(a)** $y(x) = \sum_{n=1}^{\infty} B_n X_n(x)$, where $X_n(x)$ is the nth characteristic function of a beam of length l which is fixed at both ends (see Exercise 32) and

$$B_n = -\frac{4}{EIl}\frac{1}{[X_n''(l)]^2}\int_0^l w(x)X_n(x)\,dx.$$

(b) $y(x) = \sum_{n=1}^{\infty} B_n X_n(x)$, where

$$B_n = \frac{2}{EIlX_n'(l)X_n'''(l)}\int_0^l w(x)X_n(x)\,dx$$

(c) In appearance, the formula for B_n is exactly what it is in Part **b.** The only difference is in the definition of $X_n(x)$ in the two problems.

49. (a) $X(x) = -\frac{\lambda^2\rho}{E_s g}\int_0^x \frac{1}{J(s)}\int_0^s J(r)X(r)\,dr\,ds +$

$$\frac{\lambda^2\rho}{E_s g}\int_0^l \frac{1}{J(s)}\int_0^s J(r)X(r)\,dr\,ds\left(\int_0^x \frac{ds}{J(s)}\Big/\int_0^l \frac{ds}{J(s)}\right)$$

(b) $X(x) = -\frac{\lambda^2\rho}{E_s g}\int_0^x \frac{1}{J(s)}\int_0^s J(r)X(r)\,dr\,ds +$

$$\frac{\lambda^2\rho}{E_s g}\frac{\int_0^l J(r)I(r)\,dr}{\int_0^l J(r)\,dr},$$

where $I(x) = \int_0^x \frac{1}{J(s)}\int_0^s J(r)X(r)\,dr\,ds$

51. After ϕ_1 has been determined, a second arbitrary starting function can be selected. This will have an expansion in terms of the ϕ's, say $X_0 = a_1\phi_1 + a_2\phi_2 + a_3\phi_3 + \cdots$, in which a_1 can now be determined since ϕ_1 is known. Then the function $X_0 - a_1\phi_1 = a_2\phi_2 + a_3\phi_3 + \cdots$ is one in which the coefficient of ϕ_1 is obviously zero. Thus the iterative process described in Exercise 50 must converge to the second characteristic value and the corresponding characteristic function.

However, since ϕ_1 is only approximately determined, a_1 is not exactly zero. Hence as the iteration proceeds, the process will very slowly tend to converge both to the smallest characteristic value and to the corresponding characteristic function unless steps are taken to prevent this. This requires that after every iteration, or at least after every few iterations, the current approximation to ϕ_2 be "purified" by determining the coefficient of ϕ_1 in its expansion and subtracting the corresponding term from the current approximation. Clearly, this process can be extended to the determination of $\phi_3, \phi_4, \ldots$.

53. $X(x) = \frac{\omega^2}{Tg}\left[\frac{x}{l}\int_0^l\int_0^s w(r)X(r)\,dr\,ds - \int_0^x\int_0^s w(r)X(r)\,dr\,ds\right]$

55. The problem reduces to the solution of the corresponding problem for a uniform sheet of length $a_1 + a_2$.

57. $u(x, y) = \sum_{n=1}^{\infty} K_n X_n(x)\sinh\lambda_n y$. The characteristic functions and the characteristic equation are the same as in Example 5. The expression for K_n is exactly the same as in Example 5 except that $\sinh\lambda_n b$ takes the place of $\cosh\lambda_n b$ in the denominator.

63. Those of Exercise 62 since realizable physical conditions are almost always of this form.

65. Depending on the numerical value of the modulus of the spring, one of the characteristic functions may be algebraic or hyperbolic.

69. $u(x, t) = \sum_{n=1}^{\infty} A_n X_n(x)\exp(-\lambda_n^2 t)$, where

$$X_n(x) = \begin{cases}\sin\lambda_n a_2 \sin\lambda_n a_1(1 + x) & -1 < x < 0\\ \sin\lambda_n a_1 \sin\lambda_n a_2(1 - x) & 0 < x < 1\end{cases}$$

$$A_n = \frac{200(a_1k_1 \sin\lambda_n a_2 + a_2k_2\sin\lambda_n a_1)}{[a_1^2k_1\sin^2\lambda_n a_2 + a_2^2k_2\sin^2\lambda_n a_1]\lambda_n}$$

where the λ's are the roots of the characteristic equation

$$k_1a_1\tan\lambda a_2 + k_2a_2\tan\lambda a_1 = 0$$

71. (a) Yes **(b)** Yes **73.** $abhk > 0$

75. $F(z) \equiv f_1(z)f_4(z) - f_2(z)f_3(z) = 0$, where $f_1(z) = \sin z$, $f_4(z) = -16\sin z$, $f_2(z) = \cos z$, $f_3(z) = -\cos z$; $W_{12} = -1$, $W_{34} = 16$, so that $W_{12}W_{34} < 0$ in accordance with Theorem 7, $z = \lambda/a$.

77. (a)

$$\begin{cases}\text{I.} & (\theta_1)_{tt} = a_1^2(\theta_1)_{xx},\ -1 < x < 0;\\ & (\theta_2)_{tt} = a_2^2(\theta_2)_{xx},\ 0 < x < 1;\ 0 < t\\ \text{II.} & \theta_1(-1, t) = 0; \qquad a_i^2 = (E_s)_i g/\rho_i,\ i = 1, 2\\ \text{III.} & \theta_1(0^-, t) - \theta_2(0^+, t) = 0\\ \text{IV.} & (E_s)_1J_1(\theta_1)_x(0^-, t) - (E_s)_2J_2(\theta_2)_x(0^+, t) = 0\\ \text{V.} & (\theta_2)_x(1, t) = 0\end{cases}$$

(b)

$$\begin{cases}\text{1.} & X_1'' + \frac{\lambda^2}{a_1^2}X_1 = 0,\ -1 < x < 0;\\ & X_2'' + \frac{\lambda^2}{a_2^2}X_2 = 0,\ 0 < x < 1\\ \text{0;1.} & X_1(-1) = 0\\ \text{1,1.} & X_1(0^-) - X_2(0^+) = 0\\ \text{1,2.} & b_1X_1'(0^-) - b_2X_2'(0^+) = 0;\ b_i = (E_s)_iJ_i;\ i = 1, 2\\ \text{1;2.} & X_2'(1) = 0\end{cases}$$

(c) $P(x) = \begin{cases}\rho_1/(E_s)_1 g & -1 < x < 0\\ J_2\rho_2/[(E_s)_1J_1g] & 0 < x < 1\end{cases}$

$\lambda_1 \in (0, \frac{1}{8})$, $\lambda_2 \in (\frac{1}{4}, \frac{3}{8})$, $\lambda_3 = \frac{1}{2}$

Sec. 11.7, page 763

1. In each case, $u_{ss} = 0$ and $u(x, y, t) =$

$$\sum_{m=1}^{\infty}\sum_{n=0}^{\infty} E_{mn}\sin m\pi x\cos n\pi y\exp\frac{-(m^2 + n^2)\pi^2 t}{a^2}$$

In **(a)**, $E_{mn} = 0$, $n \neq 0$; $E_{m0} = \begin{cases}0 & m \text{ even}\\ 400/m\pi & m \text{ odd}\end{cases}$

In **(b)**, $E_{mn} = \begin{cases} 0 & n \text{ even}, n \neq 0 \\ (-1)^m 8/mn^2\pi^3 & n \text{ odd} \end{cases}$

$E_{m0} = (-1)^{m+1}/m\pi$

In **(c)**, $E_{mn} = \begin{cases} 0 & m \text{ or } n \text{ even}, n \neq 0 \\ -16/mn^2\pi^3 & m \text{ and } n \text{ odd} \end{cases}$

$E_{m0} = \begin{cases} -2/m\pi & m \text{ even} \\ 4/m\pi & m \text{ odd} \end{cases}$

3. In each case, $u_{ss} = 0$ and $u(x, y, t) =$

$$\sum_{m=1}^{\infty}\sum_{n=1}^{\infty} E_{mn} \sin m\pi x \sin n\pi x \exp \frac{-(m^2+n^2)\pi^2 t}{a^2}$$

In **(a)**, $E_{mn} = \begin{cases} 0 & m \text{ or } n \text{ even} \\ 1600/mn\pi^2 & m \text{ and } n \text{ odd} \end{cases}$

In **(b)**, $E_{mn} = 4(-1)^{m+n}/mn\pi^2$

In **(c)**, $E_{mn} = \dfrac{4}{mn\pi^2}[2(-1)^{m+n} - (-1)^m - (-1)^n]$

5. $u(x, y, t) =$

$$\sum_{m=0}^{\infty}\sum_{n=0}^{\infty} E_{mn} \cos m\pi x \cos n\pi x \exp \frac{-(m^2+n^2)\pi^2 t}{a^2}$$

$$E_{mn} = 4\int_0^1\int_0^1 g(x, y) \cos m\pi x \cos n\pi y \, dx\, dy, \quad mn \neq 0$$

If one of the pair (m, n) is zero, the coefficient 4 in the formula for E_{mn} is to be replaced by 2. If both m and n are zero, the coefficient 4 in the formula for E_{mn} is to be replaced by 1.

7. **(a)** $y(x, t) = \dfrac{l^2}{3\pi^2 a^2} \sin \dfrac{\pi x}{l}\left(2 \sin \dfrac{\pi a t}{l} - \sin \dfrac{2\pi a t}{l}\right)$

(b) $y(x, t) = \dfrac{l^2}{6\pi^2 a^2} \sin \dfrac{2\pi x}{l}\left(2 \sin \dfrac{\pi a t}{l} - \sin \dfrac{2\pi a t}{l}\right)$

(c) $y(x, t) = \dfrac{l^2}{3\pi^2 a^2} \sin \dfrac{2\pi x}{l}\left(\cos \dfrac{\pi a t}{l} - \cos \dfrac{2\pi a t}{l}\right)$

In each case, although the string is being driven at one of its natural frequencies, resonance does not occur.

9. If $\omega = m\pi a/l$, then when particular integrals are found for the terms in the assumed solution

$Y(x, t) =$

$$\left(\sum_{n=1}^{\infty} D_n \sin \frac{n\pi x}{l}\right) \sin \omega t = \left(\sum_{n=1}^{\infty} D_n \sin \frac{n\pi x}{l}\right) \sin \frac{m\pi a t}{l}$$

the term $D_m \sin \dfrac{m\pi x}{l} \sin \dfrac{m\pi a t}{l}$ must be replaced by the term $D_m \sin \dfrac{m\pi x}{l}\left(t \cos \dfrac{m\pi a t}{l}\right)$ and a particular integral found for this modified term. Unless $D_m = 0$, the presence of the factor t implies an infinite buildup of amplitudes.

11. If $\omega = n\pi a/l$, the boundary conditions on Φ cannot be successfully imposed.

13. The first method goes through without change provided that $\Phi(x)$ is expanded in a series of the appropriate characteristic functions

$\sin \dfrac{n\pi x}{l}$	for a fixed-fixed shaft
$\cos \dfrac{n\pi x}{l}$	for a free-free shaft
$\sin \dfrac{2n+1}{2}\dfrac{\pi x}{l}$	for a fixed-free shaft

The second method goes through without change provided the appropriate boundary conditions are imposed on Φ:

$\Phi(0) = \Phi(l) = 0$	for a fixed-fixed shaft
$\Phi'(0) = \Phi'(l) = 0$	for a free-free shaft
$\Phi(0) = \Phi'(l) = 0$	for a fixed-free shaft

15. **(a)** If the impressed force in Example 3 is of the form $f(x, t) = \phi(x)\theta(t)$, where $\theta(t)$ is periodic, $\theta(t)$ can be expanded in a Fourier series and the first method of Example 3 can be applied to each of the terms $\phi(x)[A_n \cos(n\pi t/p) + B_n \sin(n\pi t/p)]$ where $2p$ is the period of θ.

(b) If $\theta(t)$ is not periodic, it can be represented as a Fourier integral and the methods of Example 3 can be applied to each of the infinitesimal components of the integral and the total response can be found by integration.

17. **(a)** $y(x, t) = \sin \dfrac{\pi x}{l} \cos \sqrt{\dfrac{\pi^2 a^2}{l^2} + c^2}\, t$

(b) $y(x, t) =$

$$\frac{2}{(\pi^2 a^2 + l^2 c^2)^{1/2}} \sin \frac{\pi x}{l} \sin \sqrt{\frac{\pi^2 a^2}{l^2} + c^2}\, t$$

(c) $y(x, t) = \displaystyle\sum_{n=1}^{\infty} D_n \sin \frac{n\pi x}{l} \sin \sqrt{\frac{n^2\pi^2 a^2}{l^2} + c^2}\, t$

where $D_n = \dfrac{4}{n\pi(n^2\pi^2 a^2 + c^2 l^2)^{1/2}}\left(\sin \dfrac{n\pi}{2} \sin \dfrac{n\pi}{4}\right)$

19. A solution of the form $Y_{ss} = A(x) \sin \omega t + B(x) \cos \omega t$ can be found. Both the magnification ratio and the phase shift appear to depend on x as well as on the parameters of the system.

21. $u(x, y, t) =$

$$\sum_{m=0}^{\infty}\sum_{n=0}^{\infty} E_{mn} \cos m\pi x \cos n\pi y \exp \frac{-(b^2 + m^2\pi^2 + n^2\pi^2)t}{a^2}$$

where $\quad E_{mn} = \alpha \displaystyle\int_0^1\int_0^1 g(x, y) \cos m\pi x \cos n\pi y \, dx\, dy$

and $\alpha = 4, 2,$ or 1, according as neither m nor n is zero, exactly one of the pair (m, n) is zero, or both m and n are zero.

25. $u(x, t) = \int_0^\infty B(\lambda) \cos \lambda x e^{-\lambda^2 t/a^2}\, d\lambda$, where

$B(\lambda) = \frac{2}{\pi}\int_0^\infty f(x) \cos \lambda x\, dx$

27. $u(x, t) = \int_0^\infty B(\lambda) \sin \lambda x e^{-(\lambda^2+b^2)t/a^2}\, d\lambda$, where

$B(\lambda) = \frac{2}{\pi}\int_0^\infty f(x) \sin \lambda x\, dx$, $b = hp/kA$, h being the surface conductivity, A the cross-sectional area, and p the perimeter of the cross section.

29. $u(x, t) =$

$$\frac{u_0}{\cosh 2\lambda - \cos 2\lambda}\Big[\sin \omega t \cos \frac{\lambda x}{l} \cosh\left(2 - \frac{x}{l}\right)\lambda - \cos \omega t \sin \frac{\lambda x}{l} \sinh\left(2 - \frac{x}{l}\right)\lambda - \sin \omega t \cos\left(2 - \frac{x}{l}\right)\lambda \cosh \frac{\lambda x}{l} + \cos \omega t \sin\left(2 - \frac{x}{l}\right)\lambda \sinh \frac{\lambda x}{l}\Big]$$

where $\lambda = al\sqrt{\omega/2}$. By applying this formula to each term in the Fourier expansion of an arbitrary periodic end condition, the steady-state temperature distribution produced by such an end condition can be determined.

31. $u(x, t) =$

$$\sum_{n=1}^{\infty} C_n \cos \frac{(2n-1)\pi x}{2l} \exp \frac{-(2n-1)^2\pi^2 t}{a^2 l^2} + \Phi(x)$$

where $\Phi(x) = \frac{1}{k}\int_x^l \int_0^s \phi(r)\, dr\, ds$ and

$C_n = \frac{2}{l}\int_0^l [g(x) - \Phi(x)] \cos \frac{(2n-1)\pi x}{2l}\, dx$

33. $u(x, y, t) =$

$$\sum_{n=1}^{\infty}\int_0^\infty E(n, \lambda) \sin n\pi x \cos \lambda y \exp\left[\frac{-(n^2\pi^2 + \lambda^2)t}{a^2}\right] d\lambda$$

where $E(n, \lambda) = \frac{4}{\pi}\int_0^\infty \int_0^1 g(x, y) \sin n\pi x \cos \lambda y\, dx\, dy$

35. $u(x, y, t) =$

$$\sum_{n=1}^{\infty}\int_0^\infty E(n, \lambda) \sin n\pi x \sin \lambda y \exp\left[\frac{-(b^2 + n^2\pi^2 + \lambda^2)t}{a^2}\right] d\lambda$$

where $E(n, \lambda) = \frac{4}{\pi}\int_0^\infty \int_0^1 g(x, y) \sin n\pi x \sin \lambda y\, dx\, dy$

and $b^2 = \frac{2h}{k}$

37. (a) $u(x, y, z) =$

$$\sum_{m=1}^{\infty}\sum_{n=1}^{\infty} E_{mn} \sin m\pi x \sin n\pi y \cosh \sqrt{m^2 + n^2}\pi(1 - z)$$

where $E_{mn} =$

$$\frac{4}{\cosh (m^2 + n^2)^{1/2}\pi}\int_0^1 \int_0^1 f(x, y) \sin m\pi x \sin n\pi y\, dx\, dy$$

(b) $u(x, y, z) =$

$$\sum_{m=0}^{\infty}\sum_{n=0}^{\infty} E_{mn} \cos m\pi x \cos n\pi y \sinh \sqrt{m^2 + n^2}\pi(1 - z)$$

where $E_{mn} =$

$$\frac{\alpha}{\sinh (m^2 + n^2)^{1/2}\pi}\int_0^1 \int_0^1 f(x, y) \cos m\pi x \cos n\pi y\, dx\, d$$

and $\alpha = 4$, 2, or 1, according as neither m nor n is zero, exactly one of the pair (m, n) is zero, or both m and n are zero.

(c) $u(x, y, z) =$

$$\sum_{m=0}^{\infty}\sum_{n=1}^{\infty} E_{mn} \cos m\pi x \sin n\pi y \sinh \sqrt{m^2 + n^2}\pi(1 - z$$

where $E_{mn} =$

$$\frac{\alpha}{\sinh (m^2 + n^2)^{1/2}\pi}\int_0^1 \int_0^1 f(x, y) \cos m\pi x \sin n\pi y\, dx\, d$$

and $\alpha = 4$ or 2, according as m is different from zero or equal to zero.

39. (a) $u(x, y) = \int_0^\infty B(\lambda) \sinh \lambda x \sin \lambda y\, d\lambda$

where $B(\lambda) = \frac{2}{\pi \sinh \lambda}\int_0^\infty f(y) \sin \lambda y\, dy$

(b) $u(x, y) = \int_0^\infty A(\lambda) \cosh \lambda x \sin \lambda y\, d\lambda$

where $A(\lambda) = \frac{2}{\pi \cosh \lambda}\int_0^\infty f(y) \sin \lambda y\, dy$

41. (a) $u(x, y) = \int_0^\infty B(\lambda) e^{-\lambda x} \sin \lambda y\, d\lambda$

where $B(\lambda) = \frac{200(1 - \cos \lambda a)}{\lambda\pi}$

(b) $u(x, y) = \int_0^\infty A(\lambda) e^{-\lambda x} \cos \lambda y\, d\lambda$

where $A(\lambda) = \frac{200 \sin a\lambda}{\lambda\pi}$

43. $y(x, t) = -\frac{a^2}{EI}\left(\sum_{n=1}^{\infty}\frac{B_n}{\lambda_n^2 - \omega^2} \sin z_n \frac{x}{l}\right) \sin \omega t$

where B_n is the coefficient of $\sin z_n(x/l)$ in the expansion of $\phi(x)$ in terms of the characteristic functions of the beam and $\lambda = n^2\pi^2 a/l^2$, $z_n = l\sqrt{\lambda_n/a}$, and $a^2 = EIg/A\rho$.

Substitute $Y(x, t) = \Phi(x) \sin \omega t$ into the equation

$$a^2 \frac{\partial^4 y}{\partial x^4} = -\frac{\partial^2 y}{\partial t^2} - b^2\Phi(x) \sin \omega t$$

Solve for Φ and then determine the four constants appearing in Φ so that the end conditions for the beam are satisfied.

45. Simply express $\Phi(x)$ in terms of the characteristic functions of the beam in question and then proceed as in Exercise 43.

47. $y(x, t) = \dfrac{u(z) f(zx/l) - v(z)h(zx/l)}{\cos z \sinh z - \sin z \cosh z} A \sin \omega t$
where $u(y) = \sin y + \sinh y$, $f(y) = \cos y - \cosh y$, $v(y) + \cos y + \cosh y$, $h(y) = \sin y - \sinh y$, and $z = \sqrt{\omega/a}\, l$.

49. $(\cos z \cosh z + 1) + \alpha(\sin z \cosh z - \cos z \sinh z) = 0$, where $\alpha = l^3k/EIz^3$ and $\lambda = az^2/l^2$

51. $u(x, t) = \dfrac{2}{\pi} \displaystyle\int_0^\infty \int_0^\infty \exp\left(\frac{-\lambda^2 t}{a^2}\right) f(s) \cos \lambda s \cos \lambda x \, ds \, d\lambda$

53. $u(x, y) = \dfrac{2q}{\pi k} \displaystyle\int_0^\infty \frac{1 - \cos \lambda a}{\lambda^2} e^{-\lambda x} \sin \lambda y \, d\lambda$

55. $A(\lambda) = \dfrac{1}{\pi} \displaystyle\int_0^\infty - E_0 e^{-as} \cos bs \sin \lambda s \, ds$

$B(\lambda) = \dfrac{1}{\pi q} \displaystyle\int_0^\infty E_0 e^{-as} (\omega \sin bs - p \cos bs) \sin \lambda s \, ds$

Sec. 11.8, page 771

1. $y(x, t) = \sin (m\pi x/l) \cos (m\pi at/l)$
3. Yes; expand the arbitrary initial conditions in Fourier series and apply the results of Exercises 1 and 2 to each term.
5. Yes; expand $g(x)$ in a Fourier series and apply the result of Exercise 4 to each term. Yes; expand $h(t)$ in a Fourier series and apply the result of Exercise 4 to each term.
7. $\theta(x, t) = \sin \dfrac{(2n + 1)\pi x}{2l} \cos \dfrac{(2n + 1)\pi at}{2l}$
9. $\theta(x, t) = \cos \dfrac{n\pi x}{l} \cos \dfrac{n\pi at}{l}$
11. $\theta(x, t) = \dfrac{ax}{2\sqrt{\pi} t^{3/2}} \exp \dfrac{-a^2x^2}{4t}$
13. $y(x, t) = \begin{cases} (g/2a^2)(x^2 - 2axt) & x < at \\ -\frac{1}{2}gt^2 & x \geq at \end{cases}$
15. $y(x, t) = \frac{1}{2}F_0\left[(t - \dfrac{x}{a}) u\left(t - \dfrac{x}{a}\right) - t^2\right]$

$$-\tfrac{1}{2}F_0 \begin{cases} \dfrac{v^2}{a_2 - v^2}\left[\left(t - \dfrac{x}{v}\right)^2 u\left(t - \dfrac{x}{v}\right) - \left(t - \dfrac{x}{a}\right)^2 u\left(t - \dfrac{x}{a}\right)^2\right] & v \neq a \\ \dfrac{1}{a}\left[x\left(t - \dfrac{x}{a}\right) u\left(t - \dfrac{x}{a}\right) u\left(t - \dfrac{x}{a}\right)\right] & v = a \end{cases}$$

17. $\mathscr{L}\{\theta(x, t)\} = \dfrac{aT_0 \sinh (xs/a)}{E_s J s^2 \cosh (ls/a)}$. When $x = l$, this becomes $(aT_0/E_s Js^2) \tanh (ls/a)$, and applying Theorem 4, Sec. 10.4, to the result of Example 1, Sec. 10.10, it follows that $\theta(l, t)$ is the integral of the rectangular wave shown in Fig. 10.10. In other words, $\theta(l, t)$ is a triangular wave of period $4l/a$.
19. $\theta(l, t) = D(t) - D(t - b)u(t - b)$, where $D(t)$ is the deflection of the end of the shaft in Exercise 17. If $b = 2l/a$, $\theta(l, t)$ is an alternating triangular wave after $t = b$. If $b = 4l/a$, $\theta(l, t)$ is identically zero after $t = b$.

Sec. 11.9, page 781

1. After five iterations taking advantage of symmetry, the Liebmann procedure yields the following values

0.0871	0.1200	0.0871
0.0396	0.0553	0.0396
0.0154	0.0216	0.0154

3. The new starting values can all be found by means of Eq. (3).
5. **(a)** 5 **(b)** 7 **(c)** 6 **(d)** 6
9. The solution over the entire grid can be obtained by symmetry from the following values at the mesh points in the lower left-hand quadrant of the figure:

29.2	61.8		
25.8	56.2		
18.0	37.1	56.2	61.8
9.8	18.0	25.8	29.2

11.

94.4	79.8	75.4	79.8	94.4
63.6	46.1	41.9	46.1	63.6

13.

0.344	0.438	0.344
0.438	0.562	0.438
0.344	0.438	0.344

15. The approximations to the values of u at the mesh points on the left half of the rod for the first 10 values of t are (reading upward from the initial temperature distribution)

33.34	33.34	32.34	29.84	28.81	27.34
34.33	34.33	32.34	30.34	27.34	27.28
35.25	35.25	33.41	29.44	27.28	25.25
36.88	36.88	33.62	29.94	25.25	24.62
38.50	38.50	35.25	28.75	24.62	21.75
41.25	41.25	35.75	29.25	21.75	20.00
44.00	44.00	38.50	27.50	20.00	16.00
49.00	49.00	39.00	28.00	16.00	12.00
54.00	54.00	44.00	24.00	12.00	8.00
68.00	68.00	40.00	20.00	8.00	4.00
100.00	64.00	36.00	16.00	4.00	0.00

17. The approximations to the values of y at the mesh points on the left half of the string for the first 10 nonzero values of t are (reading upward from the initial displacement)

0.000	−0.207	−0.400	−0.566	−0.693	−0.773	−0.800
0.000	−0.159	−0.307	−0.434	−0.532	−0.593	−0.614
0.000	−0.100	−0.193	−0.273	−0.334	−0.373	−0.386
0.000	−0.034	−0.066	−0.093	−0.114	−0.127	−0.132
0.000	0.034	0.066	0.093	0.114	0.127	0.132
0.000	0.100	0.193	0.273	0.334	0.373	0.386
0.000	0.159	0.307	0.434	0.532	0.593	0.614
0.000	0.207	0.400	0.566	0.659	0.773	0.800
0.000	0.241	0.466	0.659	0.807	0.900	0.932
0.000	0.259	0.500	0.707	0.866	0.966	1.000
0.000	0.259	0.500	0.707	0.866	0.966	1.000

19. Yes, if the value of m is adjusted from point to point to take account of the varying parameters that are contained in a

CHAPTER 12

Sec. 12.1, page 784

1. (a) $\dfrac{a^2T'}{T} = \dfrac{rR'' + R'}{R} = \mu$

(b) $\dfrac{a^2T'}{T} = \dfrac{r^2R'' + 2rR'}{R} = \mu$

Sec. 12.2, page 790

1. To produce a linear combination of solutions which assumes the indeterminate from 0/0 when $\nu \to n$

3. The characteristic equation is $J_0(\lambda a) = 0$. The characteristic functions are $\{J_0(\lambda_n r)\}$, $n = 1, 2, 3, \ldots$. The required expansion is

$$-100 = \sum_{n=1}^{\infty} A_n J_0(\lambda_n r)$$

Before the A's can be determined, we must know that the characteristic functions are orthogonal, possibly with respect to a weight function, on $[0, a]$.

7. $t_1 = 3\pi/4 \doteq 2.356$, $t_2 = 7\pi/4 \doteq 5.498$, $t_3 = 11\pi/4 \doteq 8.693$, $t_4 = 15\pi/4 \doteq 11.781$. The first four zeros of $J_0(z)$, correct to three decimal places, are $z_1 = 2.405$, $z_2 = 5.520$, $z_3 = 8.654$, $z_4 = 11.792$.

11. Yes, since J_n and J_{-n} are dependent and therefore have a vanishing wronskian

13. $-\pi J_\nu(x)/2Y_\nu(x) + c$

15. $Y = \dfrac{2}{\pi}\left[-J_n(x)\displaystyle\int_a^x Y_n(s)f(s)\,ds + Y_n(x)\int_a^x J_n(s)f(s)\,ds\right]$

19. $a_{n+3} = -\dfrac{(n+2)^2a_{n+2} + a_{n+1} + a_n}{(n+2)(n+3)}$, $n \geq 0$

$y_0 = 1 - \frac{1}{2}(t-1)^2 + \frac{1}{6}(t-1)^3 - \frac{1}{12}(t-1)^4 + \cdots$

$y_1 = (t-1) - \frac{1}{2}(t-1)^2 + \frac{1}{6}(t-1)^3 - \frac{1}{6}(t-1)^4 + \cdots$

21. $y = 13J_0(x)$ **23.** $y = cJ_{3/2}(x)$

25. (a) $y_n(x) = A_n J_1(z_n x/10)$, $A_n \neq 0$ an arbitrary constant, $n = 1, 2, 3, \ldots$, where z_n is the nth positive zero of $J_1(z)$ in order of increasing magnitude

(b) $y_1(x) = J_1(0.3832x)$, $y_2(x) = J_1(0.7016x)$, $y_3(x) = J_1(1.0174x)$

Sec. 12.3, page 797

1. To produce a linear combination of solutions which assumes the indeterminate form 0/0 when $\nu \to 0$

11. $u(r) = AI_0(br) + BK_0(br)$, where

$$A = \frac{100K_0(b)}{I_0(2b)K_0(b) - I_0(b)K_0(2b)}$$

$$B = \frac{-100I_0(b)}{I_0(2b)K_0(b) - I_0(b)K_0(2b)}$$

where $b = 2h/k$ and h and k are, respectively, the surface and thermal conductivities.

Sec. 12.4, page 800

3. $y = x[c_1J_2(2\sqrt{x}) + c_2Y_2(2\sqrt{x})]$

5. $y = \sqrt{x}[c_1J_0(2\sqrt{x}) + c_2Y_0(2\sqrt{x})]$

7. $y = x[c_1J_0(2\sqrt{x}) + c_2Y_0(2\sqrt{x})]$

9. $y = (1/x)[c_1J_0(2\sqrt{x}) + c_2Y_0(2\sqrt{x})]$

11. $y = [c_1J_0(\frac{2}{3}x^{3/2}) + c_2Y_0(\frac{2}{3}x^{3/2})]$

13. $y = x[c_1J_{1/3}(2x^3/3) + c_2J_{-1/3}(2x^3/3)]$

15. $y = x^{5/2}e^{-x/2}[c_1I_{1/2}(x/2) + c_2I_{-1/2}(x/2)]$

17. $y = x^{3/2}[c_1I_{1/3}(\frac{2}{3}x^{3/2}) + c_2I_{-1/3}(\frac{2}{3}x^{3/2})]$

19. $y = \sqrt{x}e^{-x}[c_1J_{3/4}(x^2/2) + c_2J_{-3/4}(x^2/2)]$

21. No

23. $y = c_1I_0\left(\dfrac{2}{|b|}t^{1/2}\right) + c_2K_0\left(\dfrac{2}{|b|}t^{1/2}\right)$, $t = a + bx$

25. $y = c_1J_0\left(\dfrac{2\sqrt{a}}{|m|}e^{mx/2}\right) + c_2Y_0\left(\dfrac{2\sqrt{a}}{|m|}e^{mx/2}\right)$

27. $b_0b_1 = 2c_1$, $b_1^2 = 4c_2$

29. $y = \sqrt{x}[\exp(1+x)^2/4][c_1I_{1/3}(\frac{2}{3}x^{3/2}) + c_2I_{-1/3}(\frac{2}{3}x^{3/2})]$

31. $y = \sqrt{x}e^{-x^3/3}\left[c_1I_1\left(\dfrac{2\sqrt{3}}{3}x^{3/2}\right) + c_2K_1\left(\dfrac{2\sqrt{3}}{3}x^{3/2}\right)\right]$

35. $y = x^{3/2}[c_1J_2(2\sqrt{2x}) + c_2Y_2(2\sqrt{2x}) + c_3I_2(2\sqrt{2x}) + c_4K_2(2\sqrt{2x})]$

39. (a) $y = x^{-2}[c_1I_{4/3}(-\frac{2}{3}x^{-3/2}) + c_2I_{-4/3}(-\frac{2}{3}x^{-3/2})]$

(b) $y = x^{13/2}e^{-2x^3}[c_1I_{5/2}(x) + c_2I_{-5/2}(x)]$

41. $y = 5t\sin t$

43. $f_n = (1/2\pi l)\sqrt{(Eg/\rho)}z_n$, where z_n is the nth positive zero of $J_{1/2}(z) = n\pi$, $n = 1, 2, 3, \ldots$

Sec. 12.5, page 806

1. $\lambda x^\nu J_\nu(\lambda x)$, $-\lambda x^\nu J_{\nu+1}(\lambda x)$

3. $2xJ_0(\lambda x) - \lambda x^2J_1(\lambda x)$

5. $J_{3/2}(x) = \sqrt{\dfrac{2}{\pi x}}\left(\dfrac{\sin x}{x} - \cos x\right)$,

$J_{-3/2}(x) = -\sqrt{\dfrac{2}{\pi x}}\left(\sin x + \dfrac{\cos x}{x}\right)$

9. (a) $2\lambda x^3J_0(\lambda x^2)$ **(b)** $2\lambda^2x^2J_0(\lambda x)J_1(\lambda x)$

(c) $J_0(\lambda x^2) - 2\lambda x^2J_1(\lambda x^2)$

13. (a) $x^2J_1^2(x) + c$

(b) $(15x^4 - 45x^2)J_1(x) - (x^5 - 15x^3 + 45x)J_0(x) + 45\int J_0(x)\,dx$

17. (a) $(1 + x)J_1(1 - x) - 2J_0(1 - x) + \int J_0(1 - x)\,dx$
(b) $(1/\lambda)\,x \ln x\, J_1(\lambda x) + (1/\lambda^2)J_0(\lambda x) + c$

19. (a) $xI_1(x) + c$ **(b)** $x^2 I_1(x) - xI_0(x) + \int I_0(x)\,dx$
(c) $xI_0(x) + \int I_0(x)\,dx$ **(d)** $x^2 I_2(x) + c$

25. $-(2/\pi x)\sin \nu\pi$

31. Yes

33. (b) $xJ_1(x)\sin x + J_0(x)(x\cos x - \sin x) + c$

35. (a) $\dfrac{x^2}{3}[J_0(x)\sin x - J_1(x)\cos x] + \dfrac{x}{3}J_1(x)\sin x + c$

(b) $-\dfrac{x^2}{3}[J_0(x)\sin x - J_1(x)\cos x] + \dfrac{2x}{3}J_1(x)\sin x + c$

43. $y = c_1 I_0(\lambda x) + c_2 K_0(\lambda x) + c_3 x I_1(\lambda x) + c_4 x K_1(\lambda x)$

45. The equation should read

$$\left[\frac{d^2}{dx^2} + \frac{1}{x}\frac{d}{dx} + \left(\lambda^2 - \frac{n^2}{x^2}\right)\right]^2 y = 0$$

47. $a/(a^2 + \lambda^2)^{3/2}$

49. $\dfrac{\lambda}{(a^2 + \lambda^2)^{1/2}[a + (a^2 + \lambda^2)^{1/2}]}$

51. $\dfrac{a^2 - 2\lambda^2}{(a^2 + \lambda^2)^{3/2}}$

53. (a) $2\sqrt{x}\,J_1(\sqrt{x}) + c$ **(b)** $-4J_0(\sqrt{x}) - 2\sqrt{x}\,J_1(\sqrt{x}) + c$

55. (a) $J_4(x) = \left(\dfrac{48}{x^3} - \dfrac{8}{x}\right)J_1(x) + \left(1 - \dfrac{24}{x^2}\right)J_0(x)$

(b) $J_5(x) = \left(\dfrac{384}{x^4} - \dfrac{72}{x^2} + 1\right)J_1(x) - \left(\dfrac{192}{x^3} - \dfrac{12}{x}\right)J_0(x)$

(c) $\int J_3(x)\,dx = J_0(x) - \dfrac{4}{x}J_1(x) + c$

(d) $\int_0^x \lambda^5 J_0(\lambda)\,d\lambda = x(x^2 - 8)[4x\,J_0(x) + (x^2 - 8)J_1(x)]$

57. (a) $y(x) = \pi J_{5/2}(x) \equiv$

$$\sqrt{\frac{2\pi}{x}}\left[\left(\frac{3}{x^2} - 1\right)\sin x - \frac{3}{x}\cos x\right]$$

(b) 0 **(c)** 0 **(d)** 0

Sec. 12.6, page 811

1. $1 = J_0(x) + 2\sum_{k=1}^{\infty} J_{2k}(x)$, exactly as in Example 1.

3. $\cos(x\cos\phi) = J_0(x) + 2\sum_{k=1}^{\infty}(-1)^k J_{2k}(x)\cos 2k\phi$

$\sin(x\cos\phi) = 2\sum_{k=1}^{\infty}(-1)^{k+1}J_{2k-1}(x)\cos(2k-1)\phi$

5. (b) $\displaystyle\int_0^{2\pi} e^{ix\cos\phi}\cos n\phi\,d\phi =$

$$\begin{cases} 2i\pi(-1)^{(n-1)/2}J_n(x) & n \text{ odd} \\ 2\pi(-1)^{n/2}J_n(x) & n \text{ even} \end{cases}$$

c. $\displaystyle\int_0^{2\pi} e^{ix\cos\phi}\sin n\phi\,d\phi = 0$

9. Equation (4), Sec. 12.5, is obtained.

13. $e^{x\cos\phi} = \sum_{n=-\infty}^{\infty} I_n(x)\cos n\phi \qquad \sum_{n=-\infty}^{\infty} I_n(x)\sin n\phi = 0$

17. $J_n(x + y) = \sum_{k=-\infty}^{\infty} J_k(x)J_{n-k}(y)$

Sec. 12.7, page 818

1. $b_1 = b_2 = 0$: annular ring with $u = 0$ on each circular boundary
$b_1 \neq 0$, $b_2 = 0$: annular ring with $u = 0$ on outer boundary and either insulated or free-escape condition on inner boundary
$b_1 = 0$, $b_2 \neq 0$: annular ring with $u = 0$ on inner boundary and either insulated or free-escape condition on outer boundary
$b_1 \neq 0$, $b_2 \neq 0$: annular ring with either insulated or free-escape condition on each boundary
$b_2 = 0$: solid disk with $u = 0$ on outer boundary
$b_2 \neq 0$: solid disk with either insulated or free-escape condition on outer boundary

7. (b) Yes

13. $\dfrac{x}{\lambda_1^2 - \lambda_2^2}[I_\nu(\lambda_2 x)I_\nu'(\lambda_1 x) - I_\nu(\lambda_1 x)I_\nu'(\lambda_2 x)]$

15. $\lambda \doteq 0.60$. Yes, $\lambda \doteq 0.73$. No.

17. $1 = \sum_{n=1}^{\infty} A_n J_0(\lambda_n x)$, $A_n = \dfrac{2}{3\lambda_n J_1(3\lambda_n)}$, $J_0(3\lambda_n) = 0$

19. $1 = A_0 I_0(\lambda_0 x) + \sum_{n=1}^{\infty} A_n J_0(\lambda_n x)$, $I_0(3\lambda_0) - \lambda_0 I_1(3\lambda_0) = 0$,
$\lambda_0 \doteq 1.19$, $J_0(3\lambda_n) + \lambda_n J_1(3\lambda_n) = 0$,
$A_0 = \dfrac{2}{3(\lambda_0^2 - 1)\lambda_0 I_1(3\lambda_0)}$, $A_n = \dfrac{2}{3(\lambda_n^2 + 1)\lambda_n J_1(3\lambda_n n)}$,
$n = 1, 2, 3, \ldots$

21. $x = \sum_{n=1}^{\infty} A_n J_1(\lambda_n x)$, $A_n = -\dfrac{3}{\lambda_n J_0(3\lambda_n)}$, $J_1(3\lambda_n) = 0$

23. $x = \sum_{n=1}^{\infty} A_n J_1(\lambda_n x)$, $A_n = \dfrac{24\lambda_n}{(144\lambda_n^2 - 7)J_1(3\lambda_n)}$,
$-J_1(3\lambda_n) + 12\lambda_n J_2(3\lambda_n) = 0$

25. $x = A_0 I_1(\lambda_0 x) + \sum_{n=1}^{\infty} J_1(\lambda_n x), 5I_1(3\lambda_0) - 6\lambda_0 I_0(3\lambda_0) = 0$,
$\lambda_0 \doteq 0.60$, $5J_1(3\lambda_n) - 6\lambda_n J_0(3\lambda_n) = 0$,
$A_0 = \dfrac{12}{(36\lambda_0^2 - 5)I_1(3\lambda_0)}$, $A_n = \dfrac{12}{(36\lambda_n^2 + 5)J_1(3\lambda_n)}$

27. $4x - x^3 = \sum_{n=1}^{\infty} A_n J_1(\lambda_n x), \quad A_n = \dfrac{32(\lambda_n^2 - 3)}{\lambda_n^2(4\lambda_n^2 - 3)J_1(2\lambda_n)},$
$3J_1(2\lambda_n) - 2\lambda_n J_0(2\lambda_n) = 0$

29. $x^2 = \sum_{n=1}^{\infty} A_n J_n(\lambda_n x),\ A_n = \dfrac{4}{(\lambda_n^2 - 1)J_2(2\lambda_n)},$
$\lambda_n J_1(2\lambda_n) - J_2(2\lambda_n) = 0$

31. $\sqrt{x} = \sqrt{2\pi} \sum_{n=1}^{\infty} (-1)^{n+1} \dfrac{1}{\sqrt{n}} J_{1/2}(nx)$. When $J_{1/2}(nx)$ is replaced by $\sqrt{\dfrac{2}{n\pi x}} \sin nx$, this series reduces at once to $\dfrac{2}{\sqrt{x}} \sum_{n=1}^{\infty} (-1)^{n+1} \dfrac{\sin nx}{n}$, which is just $\dfrac{1}{\sqrt{x}}$ times the Fourier sine series for x.

33. $J_\nu(\lambda)J_{-\nu}(2\lambda) - J_\nu(2\lambda)J_{-\nu}(\lambda) = 0$. If ν is an integer, $J_{-\nu}$ must be replaced by Y_ν.

35. $J_\nu'(\lambda)J_{-\nu}'(2\lambda) - J_\nu'(2\lambda)J_{-\nu}'(\lambda) = 0$. If ν is an integer, $J_{-\nu}$ must be replaced by Y_ν.

39. $\frac{1}{32}$

41. $1 = \sum_{n=1}^{\infty} a_n J_1(\lambda_n x)$, where
$$a_n = \frac{-2}{\lambda_n J_0^2(\lambda_n)}\left[J_0(\lambda_n) - \int_0^1 J_0(\lambda_n x)\,dx\right]$$
and $J_1(\lambda_n) = 0$

Sec. 12.8, page 836

1. $J_k[2\omega\sqrt{(k+1)l/g}] = 0$

3. $y = \sum_{n=1}^{\infty} A_n J_0(2\lambda_n\sqrt{x}) \cos(\lambda_n\sqrt{g}t),$
$$A_n = \frac{\alpha}{2\lambda_n^3\sqrt{l}J_1(2\lambda_n\sqrt{l})}$$
$J_0(2\lambda_n\sqrt{l}) = 0$

5. $z = \sum_{n=1}^{\infty} A_n J_0(\lambda_n r) \cos \lambda_n at,\ J_0(\lambda_n B) = 0$
$$A_n = \frac{2vbJ_1(\lambda_n b)}{a\lambda_n^2 B^2 J_1^2(\lambda_n B)}$$

7. The natural frequencies are λ_{mn}/a, where λ_{mn} is the mth root of the equation $J_n(\lambda b) = 0$ and b is the radius of the drumhead. The nodal lines are the concentric circles on which $J_n(\lambda_{mn} r) = 0$ and the radial lines on which $A_n \cos n\theta + B_n \sin n\theta = 0$, where A_n and B_n depend on the initial conditions.

9. λ_{mn}/b, where λ_{mn} is the mth root of the equation $J_n(\lambda b) = 0$, $n = 1, 2, 3, \ldots$, and b is the radius of the semicircle

11. $u(x) = u_0 + (u_w - u_0)\dfrac{\cosh \alpha x}{\cosh \alpha a}$, where $\alpha^2 = \dfrac{2h}{kw}$

13. $u(r, \theta, z) = \sum_{n=1}^{\infty} \sin n\theta \sum_{m=1}^{\infty} A_{nm} J_n(\lambda_{nm} r) \sinh \lambda_{nm} z$

where $A_{nm} = \dfrac{2\lambda_{nm}^2 \int_0^b rG_n(r)J_n(\lambda_{nm} r)\,dr}{\sinh(\lambda_{nm} h)J_n^2(\lambda_{nm} b)[(\lambda_{nm} b)^2 - n^2]}$

$G_n(r) = \dfrac{2}{\pi}\displaystyle\int_0^\pi f(r, \theta) \sin n\theta\, d\theta$, and the λ's are determined from the equation $J_n'(\lambda b) = 0$.

15. The required solution is the sum of the solutions of the following two problems: **(a)** $u(r, 0, z) = u(r, \pi, z) = 0$, $u(b, \theta, z) = 0$, $u(r, \theta, 0) = 0$, $u(r, \theta, h) = f(r, \theta)$ and **(b)** $u(r, 0, z) = u(r, \pi, z) = 0$, $u(r, \theta, 0) = u(r, \theta, h) = 0$, $u(b, \theta, z) = g(\theta, z)$

17. $u(r, z) = \dfrac{1}{2}B_0 + \sum_{n=1}^{\infty} B_n I_0\left(\dfrac{n\pi r}{h}\right) \cos \dfrac{n\pi z}{h}$, where
$$B_n = \frac{2}{hI_0(n\pi b/h)} \int_0^h f(z) \cos \frac{n\pi z}{h}\,dz$$

19. $u(r, z) = \dfrac{100(h - z)}{h} - \dfrac{100}{\pi} \sum_{n=1}^{\infty} \dfrac{1}{n} I_0\left(\dfrac{2n\pi r}{h}\right) \sin \dfrac{2n\pi z}{h}$

21. $u(r, z) = \sum_{n=1}^{\infty} B_n J_0(\lambda_n r) \sin \lambda_n z$, where $J_0(\lambda_n b) = 0$ and
$$B_n = \frac{8}{\lambda_n^3 b \sinh \lambda_n h J_1(\lambda_n b)}$$

23. $u(r, \theta, t) = \sum_{n=1}^{\infty}\left[\sum_{m=1}^{\infty} A_{mn} J_n(\lambda_{mn} r) \exp \dfrac{-\lambda_{mn}^2 t}{a^2}\right] \sin n\theta,$
where $J_n(\lambda_{mn} b) = 0$ and
$$A_{mn} = \begin{cases} \dfrac{\dfrac{400}{n\pi}\displaystyle\int_0^b rJ_n(\lambda_{mn} r)\,dr}{\displaystyle\int_0^b rJ_n^2(\lambda_{mn} r)\,dr} & n \text{ odd} \\ 0 & n \text{ even} \end{cases}$$

25. $u(r, t) = \sum_{n=1}^{\infty} A_n Y_0(\lambda_n r) \exp \dfrac{-\lambda_n^2 t}{a^2}$, where
$J_0(\lambda_n B)Y_0(\lambda_n b) - J_0(\lambda_n b)Y_0(\lambda_n B) = 0,$
$A_n = \dfrac{200}{\lambda_n[BY_1(\lambda_n B) + bY_1(\lambda_n b)]}$, and
$Y_i(\lambda r) = Y_i(\lambda B)J_i(\lambda r) - J_i(\lambda B)Y_i(\lambda r),\ i = 1, 2$

27. $u(r, z, t) =$
$\sum_{m=1}^{\infty}\sum_{n=1}^{\infty} A_{mn} J_0(\nu_m r) \sin \lambda_n z \exp \dfrac{-(\nu_m^2 + \lambda_n^2)t}{a^2}$, where
$\lambda_n = n\pi/h,\ J_0(\nu_m b) = 0$, and
$$A_{mn} = \begin{cases} \dfrac{800}{\nu_m b\lambda_n h J_1(\nu_m b)} & n \text{ odd} \\ 0 & n \text{ even} \end{cases}$$

41. $e^{-2t}J_0(3t)$

43. $e^{2t}I_0(2t)$

45. $\displaystyle\int_0^t I_0(a\lambda)J_0[a(t - \lambda)]\,d\lambda$

53. $2J_1(t) - \sin t$

59. $e_n(t) = \dfrac{1}{(LC)^n} \dfrac{\Gamma[(2n+1)/2]}{\Gamma(2n)} \left(\dfrac{2t}{\beta}\right)^{n-1/2} e^{-\alpha t} J_{n-1/2}(\beta t)$
where $\alpha = R/2L$ and $\beta^2 = (1/LC) - (R^2/4L^2)$

63. **(a)** $\omega_1 \doteq 3180$ Hz, $\omega_2 \doteq 6360$ Hz
(b) $\omega_1 \doteq 3880$ Hz, $\omega_2 \doteq 7100$ Hz
(c) $\omega_1 \doteq 4550$ Hz, $\omega_2 \doteq 7820$ Hz

65. $J_{1/3}(\lambda)J_{-1/3}(\lambda s^{3/2}) - J_{1/3}(\lambda s^{3/2})J_{-1/3}(\lambda) = 0$, where $s = 1 + \alpha l$ and $\lambda = (2\omega/3\alpha)\sqrt{(w_0/Tg)}$

67. $x(t) = e^{\alpha t/2}[c_1 J_1(be^{\alpha t/2}) + c_2 Y_1(be^{\alpha t/2})]$, where $b = 2k/|\alpha|\sqrt{m_0}$ and c_1 and c_2 are determined from the equations

$$c_1 J_1(b) + c_2 Y_1(b) = x_0$$
$$c_1 J_0(b) + c_2 Y_0(b) = 0$$

69. $x(t) = (1 + \alpha t)\{c_1 J_{2/3}[\lambda(1 + \alpha t)^{3/2}] + c_2 J_{-2/3}[\lambda(1 + \alpha t)^{3/2}]\}$ where c_1 and c_2 are determined from the equations

$$c_1 J_{2/3}(\lambda) + c_2 J_{-2/3}(\lambda) = x_0$$
$$c_1 J_{-1/3}(\lambda) - c_2 J_{1/3}(\lambda) = 0$$

and $\lambda = 2k/3\alpha\sqrt{m_0}$.

71. $Y(x) = \tan\theta\left[\dfrac{\sqrt{x}I_1(2a/\sqrt{x})}{aI_0(2a\sqrt{l})} - x\right]$, where

$$a^2 = \frac{12F\cos\theta}{Ebk^3}$$

75. $u(r, z) = \displaystyle\int_0^\infty \lambda e^{-\lambda z}\left[\frac{4a}{\lambda^3}J_1(\lambda a) - \frac{2a^2}{\lambda^2}J_0(\lambda a)\right]J_0(\lambda r)\,d\lambda$

77. $u(r, t) = aT_0 \displaystyle\int_0^\infty e^{-\lambda^2 kt/c\rho} J_1(\lambda a) J_0(\lambda r)\,d\lambda$

79. $u(r, z) = \dfrac{2}{\pi}\displaystyle\sum_{n=1}^\infty \frac{\int_0^\pi f(z)\sin nz\,dz}{I_0(n)} I_0(nr)\sin nz$

81. **(a)** $u(r, z) = 2\displaystyle\sum_{n=1}^\infty \frac{J_0(\lambda_n r)\cosh\lambda_n z}{\lambda_n J_1(\lambda_n)\cosh\lambda_n}$; $J_0(\lambda_n) = 0$, $\lambda_n > 0$

(b) $u(r, z) = 1 - 8\displaystyle\sum_{n=1}^\infty \frac{J_0(\lambda_n r)\cosh\lambda_n z}{\lambda_n^3 J_1(\lambda_n)\cosh\lambda_n}$; $J_0(\lambda_n) = 0$, $\lambda_n > 0$

83. $u = u_0 + c_1 I_0(2\alpha\sqrt{x}) + K_0(2\alpha\sqrt{x})$, where c_1 and c_2 are determined by the equations

$$u_w - u_0 = c_1 I_0(2\alpha\sqrt{a}) + c_2 K_0(2\alpha\sqrt{a})$$
$$u_c - u_0 = c_1 I_0(2\alpha\sqrt{c}) + c_2 K_0(2\alpha\sqrt{c})$$

and u_c is determined by the equation

$$u_c = u_0 + \frac{a}{c}(u_w - u_0)$$
$$+ \frac{c_1}{c}[cI_2(2\alpha\sqrt{c}) - aI_2(2\alpha\sqrt{a}) - \frac{\sqrt{a}}{\alpha}I_1(2\alpha\sqrt{a})]$$
$$+ \frac{c_2}{c}[cK_2(2\alpha\sqrt{c}) - aK_2(2\alpha\sqrt{a}) + \frac{\sqrt{a}}{\alpha}K_1(2\alpha\sqrt{a})]$$

85. Letting $a^2 = A\rho/EI$, the critical values of the parameters are determined by the values of a which satisfy the equation $J_{-1/3}(\frac{2}{3}al^{3/2}) = 0$.

87. $J_1(2\mu\sqrt{l})I_2(2\mu\sqrt{l}) - J_2(2\mu\sqrt{l})I_1(2\mu\sqrt{l}) = 0$, where $\mu^4 = 4\rho\omega^2/Egk^2$

89. $\sin(\omega l/a) = 0$

91. $u(r, \theta, t) = 100\left(\dfrac{r}{b}\right)^n +$

$$\sum_{n=1}^\infty \sum_{m=1}^\infty A_{nm} J_n(\lambda_{nm} r)\sin n\theta \exp\left(\frac{-\lambda_{nm}^2 t}{a^2}\right),$$

where n ranges over the odd positive integers, m ranges over all the positive integers, λ_{nm} is the mth one of the roots of the equation $J_n(\lambda b) = 0$ and

$$A_{nm} = \frac{800\int_0^b [1 - (r/b)^n] r J_n(\lambda_{nm} r)\,dr}{\pi b^2 J_{n+1}^2(\lambda_{nm} b)}$$

93. $u(r, t) = 100\dfrac{\ln r - \ln r_2}{\ln r_1 - \ln r_2} +$

$$\sum_{n=1}^\infty A_n[Y_0(\lambda_n r_2)J_0(\lambda_n r) - J_0(\lambda_n r_2)Y_0(\lambda_n r)]e^{-\lambda_n^2 t/a^2}$$

where λ_n is the nth one of the roots of the equation

$$Y_0(\lambda r_1)J_0(\lambda r_2) - J_0(\lambda r_1)Y_0(\lambda r_2) = 0$$

and $A_n =$

$$\frac{-100\int_{r_1}^{r_2} \frac{\ln r/r_2}{\ln r_1/r_2}[Y_0(\lambda_n r_2)J_0(\lambda_n r) - J_0(\lambda_n r_2)Y_0(\lambda_n r)]r\,dr}{(r_2^2/2)[Y_0(\lambda_n r_2)J_1(\lambda_n r_2) - J_0(\lambda_n r_2)Y_1(\lambda_n r_2)]^2 - (r_1^2/2)[Y_0(\lambda_n r_2)J_1(\lambda_n r_1) - J_0(\lambda_n r_2)Y_1(\lambda_n r_1)]^2}$$

95. The frequency equation is
$F(\lambda) \equiv f_1(\lambda)f_4(\lambda) - f_2(\lambda)f_3(\lambda) = 0$, where
$f_1(\lambda) = \beta^{-3/2}J_{3/2}(\beta\lambda)$
$\quad = \sqrt{2/\pi}\beta^{-3}\lambda^{-3/2}[\sin\beta\lambda - \beta\lambda\cos\beta\lambda]$,
$f_2(\lambda) = \beta^{-3/2}\lambda J'_{3/2}(\beta\lambda) - \frac{3}{2}\beta^{-5/2}J_{3/2}(\beta\lambda)$
$\quad = -\sqrt{2/\pi}\beta^{-4}\lambda^{-3/2}[(3 - \beta^2\lambda^2)\sin\beta\lambda - 3\beta\lambda\cos\beta\lambda]$,
$f_3(\lambda) = \sin(1 - \beta)c\lambda$,
$f_4(\lambda) = -\mu c\lambda\cos(1 - \beta)c\lambda$
and where $c = a_1/a_2$, $a_1^2 = G_1 g/\rho_1$, $a_2^2 = G_2 g/\rho_2$, $\mu = \eta/\beta^4$, $\eta = G_2 r_0^4/G_1 b^4$, g denotes the acceleration of gravity, and the nth circular frequency ω_n equals $a_1\lambda_n$.

Sec. 12.9, page 853

5. $\dfrac{(-1)^n(2n+2)!}{2^{2n+1}n!(n+1)!}$

9. $x^2 = \frac{1}{3}[P_0(x) + 2P_2(x)]$, $x^3 = \frac{1}{5}[3P_1(x) + 2P_3(x)]$, $x^4 = \frac{1}{35}[7P_0(x) + 20P_2(x) + 8P_4(x)]$

19. $f(x) = \sum_{n=0}^{\infty} A_n P_n(x)$, where

$$A_n = \begin{cases} \frac{1}{2} & n = 0 \\ \frac{1}{2}[P_{n-1}(0) - P_{n+1}(0)] & n \neq 0 \end{cases}$$

21. $f(x) = \sum_{n=0}^{\infty} A_n P_n(x)$, $A_0 = \frac{1}{4}$, $A_1 = \frac{1}{2}$, $A_n = 0$ n odd and $\neq 1$,

$$A_n = -\frac{2n+1}{2}\left[\frac{n+1}{(2n+1)(2n+3)}P_{n+2}(0) + \frac{1}{(2n-1)(2n+3)}P_n(0) - \frac{n}{(2n-1)(2n+1)}P_{n-2}(0)\right],$$

n even and $\neq 0$

25. $$u(r, \theta) = 50 + 50\sum_{n=1}^{\infty} \frac{1}{b^{2n-1}}[P_{2n}(0) - P_{2n-2}(0)]r^{2n-1}P_{2n-1}(\cos\theta)$$

27. $u(r, \theta) = 25 + 25\sum_{n=1}^{\infty} A_{2n}r^{2n}P_{2n}(\cos\theta)$, where $A_{2n} =$

$$\frac{4n+1}{b^{2n}}\left[\frac{2n+1}{(4n+1)(4n+3)}P_{2n+2}(0) + \frac{1}{(4n-1)(4n+3)}P_{2n}(0) - \frac{2n}{(4n-1)(4n+1)}P_{2n-2}(0)\right]$$

29. $$u(r, \theta, t) = u_{ss} + \sum_{n=0}^{\infty}\sum_{m=1}^{\infty} A_{mn}\frac{1}{\sqrt{r}}J_{n+1/2}(\lambda_{mn}r)P_n(\cos\theta)\exp\frac{-\lambda_{mn}^2 t}{a^2}$$

$$A_{mn} = \frac{2\int_0^b B_n(r)\sqrt{r}J_{n+1/2}(\lambda_{mn}r)\,dr}{b^2J^2_{(2n+3)/2}(\lambda_{mn}b)}$$

where $B_n(r) = \dfrac{2n+1}{2}\int_0^{\pi}(-u_{ss})P_n(\cos\theta)\,d\theta$, u_{ss} is the expression for the steady-state temperature distribution obtained in Example 3, and $J_{n+1/2}(\lambda_{mn}b) = 0$

31. If $r > a$, $$V = \frac{2M}{a}\left[\frac{1}{2}\frac{a}{r} - \frac{1\cdot 1}{2\cdot 4}\left(\frac{a}{r}\right)^3 P_2(\cos\theta) + \frac{1\cdot 1\cdot 3}{2\cdot 4\cdot 6}\left(\frac{a}{r}\right)^5 P_4(\cos\theta) - \frac{1\cdot 1\cdot 3\cdot 5}{2\cdot 4\cdot 6\cdot 8}\left(\frac{a}{r}\right)^7 P_6(\cos\theta) - \cdots\right]$$

If $r < a$, $$V = \frac{2M}{a}\left[1 - \frac{r}{a}P_1(\cos\theta) + \frac{1}{2}\left(\frac{r}{a}\right)^2 P_2(\cos\theta) - \frac{1\cdot 1}{2\cdot 4}\left(\frac{r}{a}\right)^4 P_4(\cos\theta) + \frac{1\cdot 1\cdot 3}{2\cdot 4\cdot 6}\left(\frac{r}{a}\right)^6 P_6(\cos\theta) - \cdots\right]$$

43. $u(r) = \sum_{n=0}^{\infty}\left(A_n r^n + \dfrac{B_n}{r^{n+1}}\right)P_n(\cos\theta)$, where A_n and B_n are determined by the equations

$$A_n b_1^n + \frac{B_n}{b_1^{n+1}} = \frac{2n+1}{2}\int_0^{\pi} T_0 \sin\theta\, P_n(\cos\theta)\,d\theta$$

$$A_n b_2^n + \frac{B_n}{b_2^{n+1}} = \frac{2n+1}{2}\int_0^{\pi} T_1 \sin\theta\, P_n(\cos\theta)\,d\theta$$

When the coefficients A_n and B_n are evaluated, the result is identical to the answer obtained in Exercise 40, Sec. 1.15.

CHAPTER 13

Sec. 13.1, page 868

1. The trivial function with domain $\{1, 2, 3, 4\}$, that is, $\mathbf{0}_4 = (0, 0, 0, 0)$
3. The pair comprised of **s** and **w**.
5. **(a)** No **(b)** Yes
7. **(b)** The lengths of $\mathbf{b}_1$, $\mathbf{b}_2$, and $\mathbf{b}_3$ are $\sqrt{3}$, $\sqrt{2}$, and $\sqrt{6}$, respectively. **(c)** $(1, 0, 5) = 2\mathbf{b}_1 - 2\mathbf{b}_2 + \mathbf{b}_3$
9. **(b)** $\mathbf{v}_1 = \frac{1}{4}\mathbf{u}_1 + \frac{1}{20}\mathbf{u}_2 - \frac{1}{3}\mathbf{u}_3 + \frac{1}{30}\mathbf{u}_4$, $\mathbf{v}_2 = \frac{1}{2}\mathbf{u}_1 + \frac{4}{5}\mathbf{u}_2 + \frac{1}{6}\mathbf{u}_3 + \frac{1}{30}\mathbf{u}_4$, $\mathbf{v}_3 = \frac{1}{2}\mathbf{u}_1 + \frac{3}{10}\mathbf{u}_2 + \frac{1}{5}\mathbf{u}_4$, $\mathbf{v}_4 = \mathbf{u}_1 + \mathbf{u}_2 + \mathbf{u}_3$
11. I $(f, g) = (g, f)$, II $(cf, g) = (f, cg) = c(f, g)$, III $(g, f_1 + f_2) = (g, f_1) + (g, f_2)$, and IV $(g_1 + g_2, f) = (g_1, f) + (g_2, f)$ may be verified by using familiar properties of integrals; this justifies the substitution process.
13. The vector $\mathbf{u}_4 = (1, -3, 2, 1)$ is orthogonal to $\mathbf{u}_1$, $\mathbf{u}_2$, and $\mathbf{u}_3$.
15. $f(x) = x^2 = \frac{1}{3}P_0(x) + \frac{2}{3}P_2(x)$
17. The trivial expansion of $g(x)$ does prove that in the space of cubic functions the set of P's is incomplete.
19. The two complex-valued functions f and g are **orthogonal in the hermitian sense on an interval** $[a, b]$ if and only if $\int_a^b \overline{f(x)}\, g(x)\,dx = 0$.
21. $(w_1, w_2) = 1$. No.
23. **(a)** $\|f\| = \sqrt{\pi}$ **(b)** $\|g\| = \sqrt{e-2}$ **(c)** $\|h\| = \sqrt{\pi}$ **(d)** $\|q\| = \sqrt{2}$
25. A is defined by $A(i, j) = (-1)^{i+j}(i + j)$ for $1 \le i \le 3$ and $1 \le j \le 4$.
27. **(a)** No; the submatrices must also be transposed. **(b)** $A\mathbf{y}_j$, where $\mathbf{y}_j$ is the jth column vector of B **(c)** $B(i, j) = 1 + \cos ij\pi$ for $1 \le i \le 2$ and $1 \le j \le 3$
29. R^3

31.

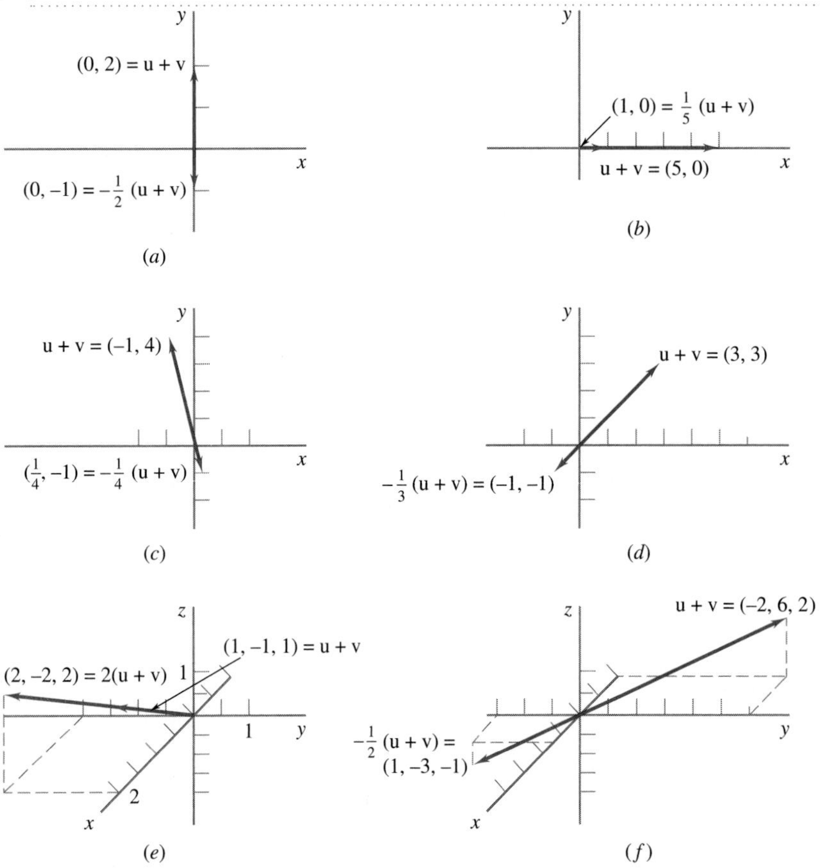

33. **(a)** $D = \{-2, 2\}$ **(b)** $D = \{x|0 \leq x \leq 1\}$ **(c)** $D = \{x|1 \leq |x| \leq 2\}$ **(d)** Impossible
35. **(a)** False **(b)** False† **(c)** False **(d)** False **(e)** True
37. Yes **39.** No **41.** No
43. No **45.** Yes **47.** No
49. Yes **51.** Yes **53.** Yes
55. **(a)** $\mathbf{v} = \mathbf{v}_1 + 3\mathbf{v}_2 - \frac{1}{2}\mathbf{v}_3$ **(b)** Impossible **(c)** $\mathbf{v} = k(\mathbf{v}_1 + 3\mathbf{v}_2 + \mathbf{v}_3)$, k arbitrary
57. **(b)** The lengths of B_1, B_2, B_3, and B_4 are $\sqrt{3}$, $\sqrt{3}$, $\sqrt{3}$, and $2\sqrt{3}$, respectively.
(c) $\begin{bmatrix} 1 & 0 \\ 0 & 1 \end{bmatrix} = \frac{1}{3}B_1 + \frac{2}{3}B_2 + \frac{1}{6}B_4$

$\begin{bmatrix} 1 & 2 \\ 3 & 4 \end{bmatrix} = 2B_1 + \frac{2}{3}B_2 + \frac{5}{3}B_3 + \frac{5}{6}B_4$

(d) $A = E_1 + 2E_2 + 3E_3 + 4E_4$

Sec 13.2, page 876

1. **(a)** Yes **(b)** Yes **(c)** Yes **(d)** A line defined by $x + y = 0$
3. **(a)** No; not closed under the operations of R^3 **(b)** Yes **(c)** Yes **(d)** No; not closed under addition **(e)** Yes **(f)** No; not closed under scalar multiplication **(g)** Yes **(h)** No; the set is empty. **(i)** Yes **(j)** No; not closed under multiplication by complex numbers
5. **(a)** Yes **(b)** Yes **(c)** No; no closure **(d)** No; not closed under scalar multiplication. **(e)** Yes **(f)** Yes
9. $t = 1$; $(2, 2, 0)$, $(-2, 2, 1)$; these vectors span the plane $x - y + 4z = 0$. $t = -1$; $(0, -2, 0)$, $(-2, 0, -1)$; these vectors span the plane $x - 2z = 0$. $t = -2$; $(-1, -4, 3)$, $(-2, -1, -2)$; these vectors span the plane $11x - 8y - 7z = 0$.
13. **(a)** Yes **(b)** No **(c)** Yes **(d)** No **(e)** Yes **(f)** Yes
15. **(a)** All p in P such that $p(x) = a + bx + cx^2$; a, b, c parameters **(b)** All p in P such that $p(x) = x(a + bx^2)$; a, b parameters **(c)** All p in P such that $p(x) = (x - 1)(a + bx + cx^2)$; a, b, c parameters
17. **(a), (b), (c),** and **(e)**
33. $(c_1, c_2, c_3, c_4) = (1, -1, -2, 1)$
39. $a + c = 0$ and $b + 2c + d = 0$ **41.** No
49. **(a)** Linearly independent **(b)** Linearly independent **(c)** Linearly dependent **(d)** Linearly independent

†Multiplication by real numbers is not the same operation as multiplication by complex numbers.

Sec. 13.3, page 884

1. $k = -9, 2$
3. Yes; $(1, 0, 1, 0) = -3\mathbf{u}_1 + 2\mathbf{u}_2$;
$(0, 1, 0, -1) = 2\mathbf{u}_1 - \mathbf{u}_2$ and $\begin{vmatrix} -3 & 2 \\ 2 & -1 \end{vmatrix} = -1 \neq 0$
5. Basis: $\mathbf{e}_1, \mathbf{e}_2, \ldots, \mathbf{e}_n$; dim $C^n = n$
7. **(a)** Basis: $\{e_1, e_2, e_3\}$; dim $V = 3$
(b) Basis $\{(1, -1, 2), (0, 3, -1)\}$; dim $V = 2$
(c) Basis: $\{(1, 3, 0, 0), (0, 0, 1, 0), (0, 0, 0, 1)\}$; dim $V = 3$
(d) Basis: $\{(1, -1, -2, 0), (0, 0, 0, 1)$; dim $V = 2$
(e) Basis: $\{1, x, x^2\}$; dim $V = 3$
(f) Basis: $\{2 - x^2, 3x + 5x^2\}$; dim $V = 2$
(g) Basis: $\{1, x, x^2\}$; dim $V = 2$
(h) Basis: $\{1, x, x^2, x^3\}$; dim $V = 4$
(i) Basis: $\begin{bmatrix} 0 & 0 \\ 5 & -1 \end{bmatrix}, \begin{bmatrix} 0 & 1 \\ -1 & 1 \end{bmatrix}, \begin{bmatrix} 1 & 1 \\ 1 & 1 \end{bmatrix}$; dim $V = 3$
(j) Basis: the given matrices; dim $V = 4$
(k) Basis: $\{1, \cos 2x\}$; dim $V = 2$
(l) Basis: $\{1, e^{2x}, e^{-2x}\}$; dim $V = 3$
13. No
17. Nine 3×3 matrices each having a 1 in a different row and column and the other eight entries 0; dimension 9
19. **(a)** $x^2 = \frac{1}{2}(2 - x^2) - \frac{1}{2}(2 - 3x^2)$
(b) $1 + x = 3(2 - x^2) - (x^3 - x) - (2 - 3x^2) - (3 - x^3)$
(c) $x^3 = \frac{9}{4}(2 - x^2) - \frac{3}{4}(2 - 3x^2) - (3 - x^3)$
(d) $x + x^2 = \frac{11}{4}(2 - x^2) - (x^3 - x) - \frac{5}{4}(2 - 3x^2) - (3 - x^3)$
21. $\langle 2, -3, 4, -5 \rangle$
23. **(a)** $(5, -2, -1)$ **(b)** $(4, -2, 2)$ **(c)** $(5, -4, 5)$ **(d)** $(0, 0, 0)$ **(e)** $(8, -6, 10)$ **(f)** $(3, 0, -1)$
27. $\{(2, 0, -6, 1), (3, -1, 0, 6), (0, 0, 1, 0), (0, 0, 0, 1)\}$
31. $r = 2$; dim $U = 2$; dim $V = 2$
33. The vectors $\mathbf{u}_k$ where, for each positive integer k, $\mathbf{u}_k$ has an infinite sequence of real components only one of which, the kth, is not zero
35. Only $\mathbf{u}$ belongs to the Hilbert space.

Sec. 13.4, page 893

1. **(a)** $2/\sqrt{5}$ **(b)** -1 **(c)** $4/\sqrt{65}$ **(d)** $-\frac{13}{21}$ **(e)** $-\frac{4}{9}$ **(f)** 0 **(g)** $\sqrt{6}/4$ **(h)** $\frac{3}{16}$
3. **(a)** $\pi/4$ **(b)** π **(c)** $2\pi/3$ **(d)** $3\pi/4$ **(e)** $\pi/3$ **(f)** $\text{Cos}^{-1}(-\frac{2}{15}) \doteq 98°$ **(g)** $\pi/3$ **(h)** $2\pi/3$
5. $k = 2 - \sqrt{3}$
7. **(a)** $\pi - \alpha$ **(b)** $\pi - \alpha$ **(c)** α
13. **(a)** $(6, 8)$ **(b)** $\left(\frac{-3}{25}, \frac{21}{25}\right)$ **(c)** $(\pi^3 - \pi e^2, e^3 - e\pi^2)$ **(d)** $(1, -2, 4)$ **(e)** $(-2, 3, -2)$ **(f)** $(-4, 2, -4)$ **(g)** $(-1, 2, 7, 2, -1)$ **(h)** $(2, 0, -4, 0, 4\pi, 0, 2\pi)$
19. **(a)** $\frac{1}{5}(3, 4), \frac{1}{5}(-4, 3)$
(b) $\frac{1}{7}(2, 3, 6), \frac{1}{\sqrt{61}}(3, 6, -4)$
(c) $\frac{1}{15}(2, -14, 5), \frac{1}{3}(2, 1, 2)$
(d) $\frac{1}{\sqrt{2}}(1, 1, 0), \frac{1}{\sqrt{6}}(1, -1, 2), \frac{1}{\sqrt{3}}(-1, 1, 1)$
(e) $\frac{1}{3}(1, 2, 2), \frac{1}{\sqrt{2}}(0, 1, -1), \frac{1}{3\sqrt{2}}(-4, 1, 1)$
(f) $\frac{1}{2}(1, 1, 1, 1), \frac{1}{2\sqrt{11}}(5, 1, -3, -3), \frac{1}{\sqrt{22}}(2, -4, 1, 1)$
(g) $\frac{1}{2}(1, -1, 1, -1), \frac{1}{2\sqrt{5}}(1, 1, 3, 3), \frac{1}{\sqrt{170}}(13, 8, -6, -1)$
(h) $\frac{1}{9}(2, -8, 2, 3), \frac{1}{\sqrt{82}}(5, 4, 5, 4), \frac{1}{\sqrt{2}}(1, 0, -1, 0)$
21. $u_1 = 1$, $u_2 = x - \frac{1}{2}$, $u_3 = x^2 - x + \frac{1}{6}$, $u_4 = x^3 - \frac{3}{2}x^2 + \frac{3}{5}x - \frac{1}{20}$
25. The constant term a_0 is $\pi/2$ and, for $n \geq 1$, $a_n = 0$, $b_n = -1/n$, $y(t) = \frac{\pi}{2} - \sum_{n=1}^{\infty} \sin 2nt/n$
27. The constant term a_0 is $\pi/2$ and, for $n \geq 1$, $a_n = 0$,
$$b_n = \frac{(-1)^n - 1}{n} = \begin{cases} 0 & n \text{ even} \\ -2/n & n \text{ odd} \end{cases}$$
$$y_1(t) = \frac{\pi}{2} + \sum_{n=1}^{\infty} \frac{(-1)^n - 1}{n} \sin nt = \frac{\pi}{2} - 2\sum_{k=1}^{\infty} \frac{\sin(2k-1)t}{2k-1}$$
29. The scalar projections of Y_1 are $a_0 = \pi/2$ and, for $n \geq 1$, $a_n = 0$, $b_n = -1/n$; those of Y are $a_0 = \pi/2$ and, for $n \geq 1$, $a_n = 0$, $b_n = [(-1)^n - 1]/n$.

Sec. 13.5, page 899

1. **(a)** Linear **(b)** Linear **(c)** Nonlinear **(d)** Nonlinear **(e)** Linear **(f)** Linear **(g)** Linear **(h)** Nonlinear **(i)** Linear **(j)** Nonlinear
3. **(a)** $\begin{bmatrix} 3 \\ -1 \end{bmatrix}$ **(b)** $K_A = \{\mathbf{0}\}$ **(c)** Yes, since $K_A = \{\mathbf{0}\}$ **(d)** Yes **(e)** None is in K_A; all four are in R_A.
5. **(a)** $\begin{bmatrix} 2 \\ 3 \end{bmatrix}$ **(b)** $K_A = \{\mathbf{0}\}$ **(c)** Yes, since $K_A = \{\mathbf{0}\}$ **(d)** Yes **(e)** None is in K_A; all four are in R_A.
7. **(a)** $(-5, -3, 2)$ **(b)** All vectors $(0, c, 0)$ with c real, or the y axis. **(c)** No, $K_T \neq \{\mathbf{0}\}$
9. $(-1, 2, 1, 4)$
11. **(b), (c),** and **(e)**
13. **(a), (b),** and **(c)**
15. **(a, g), (b, h), (c, e), (d, f)**

17. No; C^n and R^n are not vector spaces over the same set of scalars.

21. **(a)** A: $R^3 \to R^4$; $A = \begin{bmatrix} 1 & 1 & 0 \\ 1 & 0 & -1 \\ 0 & 1 & 1 \\ 1 & -1 & 1 \end{bmatrix}$

(b) The columns of A
(c) $\dim R_A = 3$

23. **(a)** A: $R^4 \to R^3$; $A = \begin{bmatrix} 1 & 2 & 1 & 3 \\ 2 & -1 & 1 & 1 \\ 1 & 0 & 1 & 1 \end{bmatrix}$

(b) The first three columns of A
(c) $\dim R_A = 3$

27. $N = n - r$ **29.** No

Sec. 13.6, page 906

1. **(b)** Yes; since A and B are in the space of all linear transformations from R^3 into R^2, so are all linear combinations of these two matrices.
(c) The range of A − B is comprised of all vectors $\begin{bmatrix} c \\ -2c \end{bmatrix}$ with c real, or the line $y = -2x$.
(d) Geometrically, it is obvious that the line $y = -2x$ is a linear space.
(e) All vectors $(2k + 1, -5k + 1, -5k)$ with k real, or the line $(x - 1)/2 = (y + 1)/-5 = z/-5$

3. **(a)** The intersection of the operator domains of T_1, T_2, and T_3 is $\mathscr{C}^{(2)}[-2, 2]$. $\mathscr{C}^{(2)}[-2, 2]$ is a subspace $\mathscr{C}^{(1)}[-2, 2]$ which is a subspace of $\mathscr{C}[-2, 2]$.
(b) Yes; $(T_1 + T_2 + T_3) \cos x = 5 \cos x - (x - 1) \sin x$
(c) Yes; $(T_1 + T_2 + T_3) \operatorname{Tan}^{-1} x =$
$(x + 6) \operatorname{Tan}^{-1} x - \frac{1}{2} \ln (1 + x^2) + \dfrac{x(x^2 - 1)}{(x^2 + 1)^2}$
(d) $\tan x$ is *not* continuous over $-2 \le x \le 2$.
(e) Yes; $(T_1 + T_2 + T_3) \sinh x =$
$7 \sinh x + (x + 1) \cosh x$
(f) $\ln |x|$ is *not* continuous over $-2 \le x \le 2$

5. No. In general $V_1 \cup V_2$ will not be a linear space.

7. $T_1T_2 = T_2T_1 = O$; $T_1T_2(x, y) = \mathbf{0}$

9. $T_1T_2 = \begin{bmatrix} -1 & 1 \\ 4 & -4 \end{bmatrix}$, $T_2T_1 = \begin{bmatrix} 0 & -10 \\ 0 & -5 \end{bmatrix}$; $T_1T_2 \ne T_2T_1$
$T_1T_2(x, y) = (-x + y, 4x - 4y)$;
$T_2T_1(x, y) = (-10y, -5y)$

11. $T_2T_1(w, x, y, z) = (w + y, -x + 2y + z)$;
$T_2T_1 = \begin{bmatrix} 1 & 0 & 1 & 0 \\ 0 & -1 & 2 & 1 \end{bmatrix}$; T_1T_2 is undefined.

13. $T_1T_2 = T_2T_1 = -4x^2D^2$; $T_1T_2(9 + 6x - x^2) = 8x^2$

15. $T_1T_2 = xD + 1$; $T_2T_1 = xD$; $T_1T_2(1 + x^3) = 4x^3 + 1$; $T_2T_1(1 + x^3) = 3x^3$

17. $T_1T_2 = T_2T_1 = abx^2D^2 + (ab + ak + bh)xD + hk$;
$T_1T_2x^{10} = (100ab + 10\{ak + bh\} + hk)x^{10}$

19. **(a)** $T_1p(x) = \int_0^x (x - s)p(s)\, ds$; $T_2p(x) = \int_0^x sp(s)\, ds$
(b) $T_1T_2p(x) = \dfrac{1}{2} \int_0^x t(x - t)^2 p(t)\, dt$;
$T_2T_1p(x) = \dfrac{1}{6} \int_0^x (2x^3 - 3tx^2 + t^3)p(t)\, dt$
(c) No
(d) $T_1T_2(30x) = x^5/2$; $T_2T_1(30x) = x^5$
(e) $(T_1 + T_2)p(x) = \int_0^x \int_0^x p(s)\, ds\, dt = x \int_0^x p(s)\, ds$
(f) $(T_1 + T_2)(30x) = 15x^3$
(g) $T_2(T_1 + T_2)p(x) = \dfrac{1}{3} \int_0^x (x^3 - t^3)p(t)\, dt$
(h) $T_2(T_1 + T_2)(30x) = 3x^5$

21. **(a)** Neither **(b)** Unitary **(c)** Unitary and orthogonal **(d)** Unitary and orthogonal

25. $T^2(x, y) =$
$([a^2 + c^2]x + c[a + b]y, c[a + b]x + [b^2 + c^2]y)$

27. **(a)** $A = \begin{bmatrix} a & c \\ c & b \end{bmatrix}$ **(b)** Yes, Yes **29.** No

Sec. 13.7, page 912

1. **(a)** $[-1, 1]$ **(b)** $[-1, 1]$ **(c)** The inverse $\operatorname{Cos}^{-1} x$ of $\operatorname{Cos} x$ is the **principal arc,** or **inverse, cosine function.** Its domain is $[-1, 1]$; its range is $[0, \pi]$.

3. **(a)** Linear **(b)** Linear **(c)** Linear **(d)** Nonlinear **(e)** Nonlinear **(f)** Linear **(g)** Linear **(h)** Linear **(i)** Linear **(j)** Nonlinear **(k)** Nonlinear **(l)** Linear

5. **(a)** $k = 1$; a complete solution of Eq. (13**i**) is $y = c_1x + c_2 \cos x + c_3 \sin x$.
(b) $k = 1$; a fundamental matrix of the first-order system is

$$\begin{bmatrix} \cos x & \sin x & x \\ -\sin x & \cos x & 1 \\ -\cos x & -\sin x & 0 \end{bmatrix}$$

(c) $y = c(\cos x - \sin x)$

7. $k = 1$ **9.** $k = 1$

CHAPTER 14

Sec. 14.1, page 921

1. **(a)** $P = \begin{bmatrix} 0 & 1 & 0 & 0 & 0 \\ 0 & 0 & 0 & 0 & 1 \\ 0 & 1 & 0 & 1 & 0 \\ 1 & 0 & 1 & 0 & 1 \\ 1 & 0 & 1 & 0 & 0 \end{bmatrix}$ $P^2 = \begin{bmatrix} 1 & 0 & 1 & 0 & 1 \\ 1 & 0 & 1 & 0 & 0 \\ 1 & 0 & 1 & 0 & 2 \\ 1 & 2 & 1 & 1 & 1 \\ 0 & 2 & 0 & 1 & 1 \end{bmatrix}$

(b) $P = \begin{bmatrix} 0 & 1 & 0 & 0 & 1 \\ 0 & 0 & 0 & 1 & 1 \\ 0 & 1 & 0 & 0 & 0 \\ 0 & 0 & 1 & 0 & 1 \\ 1 & 0 & 0 & 1 & 0 \end{bmatrix}$ $\quad P^2 = \begin{bmatrix} 1 & 0 & 0 & 2 & 1 \\ 1 & 0 & 1 & 1 & 1 \\ 0 & 0 & 0 & 1 & 1 \\ 1 & 1 & 0 & 1 & 0 \\ 0 & 1 & 1 & 0 & 2 \end{bmatrix}$

(c) $P = \begin{bmatrix} 0 & 0 & 0 & 1 & 0 \\ 1 & 0 & 0 & 1 & 0 \\ 0 & 1 & 0 & 0 & 0 \\ 1 & 1 & 1 & 0 & 1 \\ 1 & 0 & 0 & 1 & 0 \end{bmatrix}$ $\quad P^2 = \begin{bmatrix} 1 & 1 & 1 & 0 & 1 \\ 1 & 1 & 1 & 1 & 1 \\ 1 & 0 & 0 & 1 & 0 \\ 2 & 2 & 0 & 3 & 0 \\ 1 & 1 & 1 & 1 & 1 \end{bmatrix}$

(d) $P = \begin{bmatrix} 0 & 1 & 0 & 0 & 0 \\ 0 & 0 & 0 & 1 & 0 \\ 0 & 0 & 0 & 1 & 0 \\ 1 & 0 & 1 & 0 & 1 \\ 0 & 0 & 0 & 1 & 0 \end{bmatrix}$ $\quad P^2 = \begin{bmatrix} 0 & 0 & 0 & 1 & 0 \\ 1 & 0 & 1 & 0 & 1 \\ 1 & 0 & 1 & 0 & 1 \\ 0 & 1 & 0 & 2 & 0 \\ 1 & 0 & 1 & 0 & 1 \end{bmatrix}$

5. $P = \begin{bmatrix} 1 & 0 & 0 & 0 \\ \frac{1}{3} & \frac{1}{3} & \frac{1}{3} & 0 \\ 0 & \frac{1}{3} & \frac{1}{3} & \frac{1}{3} \\ 0 & 0 & 0 & 1 \end{bmatrix}$ $\quad P^2 = \begin{bmatrix} 1 & 0 & 0 & 0 \\ \frac{4}{9} & \frac{2}{9} & \frac{2}{9} & \frac{1}{9} \\ \frac{1}{9} & \frac{2}{9} & \frac{2}{9} & \frac{4}{9} \\ 0 & 0 & 0 & 1 \end{bmatrix}$

$P^3 = \begin{bmatrix} 1 & 0 & 0 & 0 \\ \frac{14}{27} & \frac{4}{27} & \frac{4}{27} & \frac{5}{27} \\ \frac{5}{27} & \frac{4}{27} & \frac{4}{27} & \frac{14}{27} \\ 0 & 0 & 0 & 1 \end{bmatrix}$ $\quad P^4 = \begin{bmatrix} 1 & 0 & 0 & 0 \\ \frac{46}{81} & \frac{8}{81} & \frac{8}{81} & \frac{19}{81} \\ \frac{19}{81} & \frac{8}{81} & \frac{8}{81} & \frac{46}{81} \\ 0 & 0 & 0 & 1 \end{bmatrix}$

$P_{21}^{(4)} - P_{21}^{(3)} = \frac{4}{81},\ 1 - P_{21}^{(4)} = \frac{35}{81}$

7. $P = P^2 = P^3 = \cdots = \frac{1}{6}\begin{bmatrix} 1 & 1 & 1 & 3 \\ 1 & 1 & 1 & 3 \\ 1 & 1 & 1 & 3 \\ 1 & 1 & 1 & 3 \end{bmatrix}$

11. $K = -\dfrac{162EI}{26L^3}\begin{bmatrix} 80 & -46 & 12 \\ -46 & 44 & -16 \\ 12 & -16 & 7 \end{bmatrix}$;

$K^{-1} = -\dfrac{L^3}{162EI}\begin{bmatrix} 2 & 5 & 8 \\ 5 & 16 & 28 \\ 8 & 28 & 54 \end{bmatrix}$

15. (a) $\begin{bmatrix} 1 & 0 & 0 & 0 \\ 0 & 0 & 1 & 0 \\ 0 & 1 & 0 & 0 \\ 0 & 0 & 0 & 1 \end{bmatrix}$ **(b)** $\begin{bmatrix} 0 & 0 & 0 & 1 \\ 0 & 0 & 1 & 0 \\ 0 & 1 & 0 & 0 \\ 1 & 0 & 0 & 0 \end{bmatrix}$

(c) $\begin{bmatrix} 0 & a & 0 & 0 \\ 0 & 0 & 0 & b \\ 0 & 0 & c & 0 \\ d & 0 & 0 & 0 \end{bmatrix}$

17. $P_5 = \begin{bmatrix} 0 & 0 & 1 & 0 & 2 \\ 1 & 0 & 0 & 0 & -1 \\ 0 & 2 & 0 & 3 & 0 \\ 1 & 0 & 0 & 0 & 1 \\ 0 & 1 & 1 & 1 & 0 \end{bmatrix}$

19. The adjoint of M_j is $\bar{z}$. The inverse of M_j is $1/z$. The characteristic values of M_j are z and $\bar{z}$.

Sec. 14.2, page 934

1. (a) $-16i$ **(b)** $\begin{bmatrix} 4 & 0 & 0 & 0 \\ 0 & 0 & 0 & 4 \\ 0 & 0 & 4 & 0 \\ 0 & 4 & 0 & 0 \end{bmatrix}$

3. $\begin{bmatrix} 1 & 1 & 1 & 1 & 1 & 1 & 1 & 1 \\ 1 & w^7 & w^6 & w^5 & w^4 & w^3 & w^2 & w \\ 1 & w^6 & w^4 & w^2 & 1 & w^6 & w^4 & w^2 \\ 1 & w^5 & w^2 & w^7 & w^4 & w & w^6 & w^3 \\ 1 & w^4 & 1 & w^4 & 1 & w^4 & 1 & w^4 \\ 1 & w^3 & w^6 & w & w^4 & w^7 & w^2 & w^5 \\ 1 & w^2 & w^4 & w^6 & 1 & w^2 & w^4 & w^6 \\ 1 & w & w^2 & w^3 & w^4 & w^5 & w^6 & w^7 \end{bmatrix}$

5. $\begin{bmatrix} 1 & 0 & 1 & 0 \\ 0 & 1 & 0 & i \\ 1 & 0 & -1 & 0 \\ 0 & 1 & 0 & -i \end{bmatrix}\begin{bmatrix} 1 & 1 & 0 & 0 \\ 1 & -1 & 0 & 0 \\ 0 & 0 & 1 & 1 \\ 0 & 0 & 1 & -1 \end{bmatrix}\begin{bmatrix} 1 & 0 & 0 & 0 \\ 0 & 0 & 1 & 0 \\ 0 & 1 & 0 & 0 \\ 0 & 0 & 0 & 1 \end{bmatrix}$

$= \begin{bmatrix} 1 & 1 & 1 & 1 \\ 1 & i & -1 & -i \\ 1 & -1 & 1 & -1 \\ 1 & -i & -1 & i \end{bmatrix}$

7. $F = [3 \quad 1 \quad -1 \quad 1]^T$; $f(t) = 1 + \cos t + \cos 3t$

9. (a) $a_1 = a_3$ **(b)** $a_0 = a_2 = 0,\ a_1 = -a_3$

11. $a_0 = a_4 = 0,\ a_1 = -a_7,\ a_2 = -a_6,\ a_3 = -a_5$

13. $A = \frac{1}{4}[3 \quad (-1+i) \quad -1 \quad (-1-i)]^T$

$f(t) = \frac{1}{2}(3 - \cos t - \cos 2t - \cos 3t - \sin t + \sin 3t)$

15. (a) $F = [4 \quad 0 \quad 0 \quad 0 \quad 4 \quad 0 \quad 0 \quad 0]^T$,

$f(t) = 1 + \cos 2t + \cos 4t + \cos 6t$

(b) $f(t) = 1 + \cos t + \cos 2t + \cos 3t$

Sec. 14.3, page 941

5. (a) $r = 2$ **(b)** $r = 3$ **(c)** $r = 3$ **(d)** $r = 2$

7. If $A = \begin{bmatrix} 6 & 4 & 2 \\ 9 & 6 & 3 \\ -3 & -2 & -1 \end{bmatrix}$ and $B = \begin{bmatrix} 0 & 1 & -2 \\ -1 & 0 & 3 \\ 2 & -3 & 0 \end{bmatrix}$,

the rank of AB is 0 and the rank of BA is 1.

15. $P_6P_5P_4P_3P_2P_1B = I$, where

$P_1 = \begin{bmatrix} 1 & 0 & -1 \\ 0 & 1 & 0 \\ 0 & 0 & 1 \end{bmatrix}$ $\quad P_2 = \begin{bmatrix} 1 & 0 & 0 \\ 0 & 1 & 0 \\ -1 & 0 & 1 \end{bmatrix}$

$P_3 = \begin{bmatrix} 1 & 0 & 0 \\ 0 & 1 & 0 \\ 0 & -1 & 1 \end{bmatrix}$ $\quad P_4 = \begin{bmatrix} 1 & 0 & 1 \\ 0 & 1 & 0 \\ 0 & 0 & 1 \end{bmatrix}$

$P_5 = \begin{bmatrix} 1 & 0 & 0 \\ 0 & 1 & 1 \\ 0 & 0 & 1 \end{bmatrix}$ $\quad P_6 = \begin{bmatrix} 1 & 0 & 0 \\ 0 & 1 & 0 \\ 0 & 0 & -1 \end{bmatrix}$

$$B^{-1} = \begin{bmatrix} 0 & -1 & 1 \\ -1 & 0 & 2 \\ 1 & 1 & -2 \end{bmatrix}$$

17. $B = PAQ$, where $P = \begin{bmatrix} -2 & 1 \\ 1 & 0 \end{bmatrix}$ and

$$Q = \begin{bmatrix} 1 & 0 & -1 \\ 0 & 1 & 1 \\ 0 & 0 & 1 \end{bmatrix}$$

19. $B = PAQ$, where $P = \begin{bmatrix} 1 & 0 & 0 \\ 0 & 1 & 0 \\ -2 & 0 & 1 \end{bmatrix}$ and

$$Q = \begin{bmatrix} 1 & 0 & 0 \\ 0 & 1 & 0 \\ 0 & -1 & 2 \end{bmatrix}$$

31. **(a)** $\begin{bmatrix} 1 & 0 & 0 & 0 \\ 0 & 0 & 1 & 0 \\ 0 & 1 & 0 & 0 \\ 0 & 0 & 0 & 1 \end{bmatrix}$ **(b)** $\begin{bmatrix} 0 & 0 & 0 & 1 \\ 0 & 0 & 1 & 0 \\ 0 & 1 & 0 & 0 \\ 1 & 0 & 0 & 0 \end{bmatrix}$

(c) $\begin{bmatrix} 0 & 0 & a & 0 \\ 0 & 0 & 0 & b \\ 0 & c & 0 & 0 \\ d & 0 & 0 & 0 \end{bmatrix}$

These changes cannot be accomplished by postmultiplications since postmultiplication can accomplish only column changes.

33. Premultiply by $\begin{bmatrix} 0 & 0 & 1 & 0 \\ 0 & 0 & 0 & 1 \\ 1 & 0 & 0 & 0 \\ 0 & 1 & 0 & 0 \end{bmatrix}$, postmultiply by

$$\begin{bmatrix} 0 & 0 & 0 & 1 \\ 0 & 1 & 0 & 0 \\ 0 & 0 & 1 & 0 \\ 1 & 0 & 0 & 0 \end{bmatrix}.$$

35. **(a)** $A = \begin{bmatrix} 1 & 2 \\ 3 & 4 \end{bmatrix}$ is of rank 2; $B = \begin{bmatrix} 2 & 1 \\ 4 & 2 \end{bmatrix}$ is of rank 1; $AB = \begin{bmatrix} 10 & 5 \\ 22 & 11 \end{bmatrix}$ is of rank 1.

(b) $A = \begin{bmatrix} 2 & -1 \\ 4 & -2 \end{bmatrix}$ is of rank 1; $B = \begin{bmatrix} 2 & 1 \\ 4 & 2 \end{bmatrix}$ is of rank 1; $AB = \begin{bmatrix} 0 & 0 \\ 0 & 0 \end{bmatrix}$ is of rank 0.

Sec. 14.4, page 949

1. **(a)** $g(x, s) = \begin{cases} \cos s \sin x & 0 \le x \le s \\ \sin s \cos x & s \le x \le \pi \end{cases}$

(b) $u(x) = \sin 2x + 2 \sin x$ **(c)** $u(x) = \sin 2x$

3. $u(x) = x^3$

5. $u(x) = \dfrac{11}{4} + \left(\dfrac{\pi}{8} + \dfrac{1}{4} - \dfrac{\sqrt{2}}{2}\right)x - \dfrac{x^2}{2} - 2\cos x + \dfrac{1}{4}\cos 2x$

7. $u(x) = 7 - \frac{7}{2}x - 5x^3 + \frac{5}{2}x^4$

9. $\begin{bmatrix} y_1 \\ y_2 \end{bmatrix} = \begin{bmatrix} x - x\ln(1+x) \\ x - (x+1)\ln(1+x) \end{bmatrix} e^x$

11. A change of sign in the right-hand member of (5) is required.

Sec. 14.5, page 956

1. **(a)** Positive-definite **(b)** Negative-definite **(c)** Indefinite **(d)** Positive-semidefinite **(e)** Indefinite

3.
$$\begin{aligned} y_1 &= -\tfrac{1}{2}x_1 + 2x_2 + 3x_3 + 2x_4 \\ y_2 &= \tfrac{3}{2}x_1 + x_3 + 2x_4 \\ y_3 &= x_1 \\ y_4 &= x_3 \end{aligned}$$
and $Q = y_1^2 + y_2^2 - \frac{3}{2}y_3^2 + y_4^2$ or, equally well,
$$\begin{aligned} z_1 &= \tfrac{1}{2}x_1 + x_2 + 2x_3 + 2x_4 \\ z_2 &= -\tfrac{1}{2}x_1 + 4x_2 + 4x_3 \\ z_3 &= x_2 + x_3 \\ z_4 &= x_3 \end{aligned}$$
and $Q = 2z_1^2 + \frac{1}{2}z_2^2 - 6z_3^2 + z_4^2$

In each case, after transformation, Q contains three terms with positive coefficients and one term whose coefficient is negative. (This is an example of **Sylvester's law of inertia:** if a quadratic form is reduced to a sum of squares by each of two nonsingular linear transformations, the number of terms with positive coefficients will be the same in each case.)

5. **(a)** Positive-semidefinite **(b)** Indefinite **(c)** Positive-semidefinite **(d)** Indefinite **(e)** Indefinite **(f)** Indefinite

13. $PE = \frac{1}{2}k_1x_1^2 + \frac{1}{2}k_{12}(x_2 - x_1)^2 + \frac{1}{2}k_{13}(x_3 - x_1)^2 + \frac{1}{2}k_{23}(x_3 - x_2)^2 + \frac{1}{2}k_3x_3^2$ The matrix of this quadratic form is just $-\frac{1}{2}$ the matrix K.

15. $\mathbf{u}_1 = \begin{bmatrix} 1 \\ 0 \\ 0 \end{bmatrix}$ $\mathbf{u}_2 = \dfrac{1}{\sqrt{2}}\begin{bmatrix} 0 \\ 1 \\ 0 \end{bmatrix}$ $\mathbf{u}_3 = \dfrac{1}{\sqrt{3}}\begin{bmatrix} 0 \\ 0 \\ 1 \end{bmatrix}$

17. If $\mathbf{P}$ and $\mathbf{Q}$ are the coordinate vectors of two points on a line l, then l has two, one, or no (real) intersections with S according as $(\mathbf{P}^T\mathrm{A}\mathbf{Q})^2 - (\mathbf{P}^T\mathrm{A}\mathbf{P})(\mathbf{Q}^T\mathrm{A}\mathbf{Q})$ is greater than, equal to, or less than zero.

Sec. 14.6, page 975

3. **(a)** $\lambda_1 = 1$, $(\mathbf{x}_1)_1 = \dfrac{1}{2\sqrt{2}}\begin{bmatrix} 1 \\ 0 \\ 1 \end{bmatrix}$, $(\mathbf{x}_1)_2 = \dfrac{1}{6\sqrt{2}}\begin{bmatrix} -1 \\ 4 \\ 3 \end{bmatrix}$;

$\lambda_2 = 2$, $\mathbf{x}_2 = \dfrac{1}{6}\begin{bmatrix} 1 \\ 2 \\ -3 \end{bmatrix}$

(b) $\lambda_1 = 1$, $(\mathbf{x}_1)_1 = \frac{1}{3}\begin{bmatrix} 1 \\ 1 \\ 0 \end{bmatrix}$, $(\mathbf{x}_1)_2 = \frac{1}{6}\begin{bmatrix} 1 \\ -2 \\ 3 \end{bmatrix}$;

$\lambda_2 = 2$, $\mathbf{x}_2 = \frac{1}{6}\begin{bmatrix} 1 \\ 2 \\ 3 \end{bmatrix}$

5. Unless B is positive-definite, the n (orthogonal) characteristic vectors cannot be normalized with respect to B.

9. If $A = \begin{bmatrix} 1 & 2 \\ 2 & -1 \end{bmatrix}$ and $B = \begin{bmatrix} 1 & 0 \\ 0 & -1 \end{bmatrix}$, the roots of the equation $|A - \lambda B| = 0$ are $1 \pm 2i$.

13. No. For example, $A = \begin{bmatrix} 1 & 1 \\ -1 & -1 \end{bmatrix}$ is a nonnull *unsymmetric* matrix whose characteristic values are all zero.

19. $\lambda_1 = \frac{7}{4}$, $\mathbf{x}_1 = \begin{bmatrix} 1 \\ -4 \\ 2 \end{bmatrix}$

21. (a) $\lambda_2 = 1.56$, $\mathbf{x}_1 = \begin{bmatrix} 4.96 \\ -2.60 \\ 1.00 \end{bmatrix}$; $\lambda_3 = 0.033$,

$\mathbf{x}_3 = \begin{bmatrix} 1.27 \\ 1.72 \\ 1.00 \end{bmatrix}$

(b) $\lambda_2 = 1$, $\mathbf{x}_2 = \begin{bmatrix} -1 \\ 1 \\ 1 \end{bmatrix}$; $\lambda_3 = \frac{1}{4}$, $\mathbf{x}_3 = \begin{bmatrix} 2 \\ 4 \\ 1 \end{bmatrix}$

23. $[1 \quad r_1 \quad r_1^2 \quad r_1^3 \quad \cdots \quad r_1^{n-1}]$

27. If and only if $\lambda_1 = \lambda_2$

29. $\mathbf{x} = \frac{5}{1-\omega^2}\begin{bmatrix} 1 \\ 0 \\ -1 \end{bmatrix} \sin \omega t$, $\omega \neq 1$

31. $\mathbf{x} = \frac{3}{8}\begin{bmatrix} 2 \\ 3 \\ 3 \end{bmatrix} \cos\frac{t}{2} + \frac{1}{5}\begin{bmatrix} 1 \\ 0 \\ -1 \end{bmatrix} \cos t +$

$\frac{1}{40}\begin{bmatrix} 2 \\ -5 \\ 3 \end{bmatrix} \cos\frac{3t}{2} - \frac{1}{4}\begin{bmatrix} 2 \\ 3 \\ 3 \end{bmatrix} \sin\frac{t}{2} -$

$\frac{11}{5}\begin{bmatrix} 1 \\ 0 \\ -1 \end{bmatrix} \sin t - \frac{11}{60}\begin{bmatrix} 2 \\ -5 \\ 3 \end{bmatrix} \sin\frac{3t}{2}$

33. $\mathbf{x} =$

$\frac{1}{3}\begin{bmatrix} 1 \\ 2 \\ 2 \end{bmatrix} \cos\frac{t}{\sqrt{2}} + \frac{3}{4}\begin{bmatrix} 1 \\ 0 \\ -2 \end{bmatrix} \cos t - \frac{1}{12}\begin{bmatrix} 1 \\ -4 \\ 2 \end{bmatrix} \cos\sqrt{2}t$

$+ \frac{2\sqrt{2}}{3}\begin{bmatrix} 1 \\ 2 \\ 2 \end{bmatrix} \sin\frac{t}{\sqrt{2}} + \frac{\sqrt{2}}{6}\begin{bmatrix} 1 \\ -4 \\ 2 \end{bmatrix} \sin\sqrt{2}t$

35. $\omega_1 = 1$, $a_1 = \begin{bmatrix} 1 \\ 1 \\ 1 \end{bmatrix}$; $\omega_2 = 2$, $(a_2)_1 = \begin{bmatrix} 1 \\ -1 \\ 0 \end{bmatrix}$,

$(a_2)_2 = \begin{bmatrix} 1 \\ 0 \\ -1 \end{bmatrix}$

37. $\omega_1 = \frac{3}{4}$, $a_1 = \begin{bmatrix} 4 \\ 2 \\ 1 \end{bmatrix}$; $\omega_2 = 1$, $(a_2)_1 = \begin{bmatrix} 1 \\ -1 \\ 0 \end{bmatrix}$,

$(a_2)_2 = \begin{bmatrix} 1 \\ 0 \\ -1 \end{bmatrix}$

47. $\begin{bmatrix} 40 \\ 20 \\ 40 \end{bmatrix} \to \begin{bmatrix} 38.0 \\ 20.0 \\ 42.0 \end{bmatrix} \to \begin{bmatrix} 36.8 \\ 20.0 \\ 43.2 \end{bmatrix} \to \begin{bmatrix} 36.1 \\ 20.0 \\ 43.9 \end{bmatrix} \to \begin{bmatrix} 35.6 \\ 20.0 \\ 44.4 \end{bmatrix} \to$

$\begin{bmatrix} 35.4 \\ 20.0 \\ 44.6 \end{bmatrix}$

Sec. 14.7, page 983

1. (a) $P = \begin{bmatrix} 1 & 0 \\ -3 & 1 \end{bmatrix}$ $Q = \begin{bmatrix} 1 & -2 \\ 0 & 1 \end{bmatrix}$

(b) $P = \begin{bmatrix} 1 & 0 \\ 3 & 1 \end{bmatrix}$ $Q = \begin{bmatrix} 0 & 1 \\ 1 & 1 \end{bmatrix}$

(c) $P = \begin{bmatrix} 3 & 0 & 0 \\ -6 & 3 & 0 \\ 2 & -1 & 3 \end{bmatrix}$ $Q = \begin{bmatrix} 1 & 1 & -1 \\ 0 & 1 & 0 \\ 0 & 0 & 1 \end{bmatrix}$

(d) $P = \begin{bmatrix} 1 & 0 & 0 \\ -1 & 1 & 0 \\ 2 & 3 & 1 \end{bmatrix}$ $Q = \begin{bmatrix} 1 & 0 & -3 \\ 0 & 1 & -2 \\ 0 & 0 & 1 \end{bmatrix}$

3. (a) $\begin{bmatrix} x_1 \\ x_2 \end{bmatrix} = \begin{bmatrix} 1/\sqrt{3} & 2/\sqrt{6} \\ 1/\sqrt{3} & -1/\sqrt{6} \end{bmatrix}\begin{bmatrix} y_1 \\ y_2 \end{bmatrix}$

(b) $\begin{bmatrix} x_1 \\ x_2 \end{bmatrix} = \begin{bmatrix} 1/\sqrt{6} & 1/\sqrt{3} \\ -2/\sqrt{6} & 1/\sqrt{3} \end{bmatrix}\begin{bmatrix} y_1 \\ y_2 \end{bmatrix}$

(c) $\begin{bmatrix} x_1 \\ x_2 \end{bmatrix} = \begin{bmatrix} \frac{1}{2} & 3/\sqrt{12} \\ -\frac{1}{2} & 1/\sqrt{12} \end{bmatrix}\begin{bmatrix} y_1 \\ y_2 \end{bmatrix}$

(d) $\begin{bmatrix} x_1 \\ x_2 \end{bmatrix} = \begin{bmatrix} 1 & 1 \\ -1 & 0 \end{bmatrix} \begin{bmatrix} y_1 \\ y_2 \end{bmatrix}$

(e) $\begin{bmatrix} x_1 \\ x_2 \\ x_3 \end{bmatrix} = \begin{bmatrix} 1/2\sqrt{6} & \frac{1}{4} & 1/4\sqrt{3} \\ 1/\sqrt{6} & 0 & -1/\sqrt{3} \\ 1/\sqrt{6} & -\frac{1}{2} & 1/2\sqrt{3} \end{bmatrix} \begin{bmatrix} y_1 \\ y_2 \\ y_3 \end{bmatrix}$

(f) $\begin{bmatrix} x_1 \\ x_2 \\ x_3 \end{bmatrix} = \begin{bmatrix} \frac{1}{3} & -\frac{1}{3} & \frac{1}{6} \\ \frac{2}{3} & \frac{1}{3} & -\frac{2}{3} \\ \frac{1}{6} & \frac{1}{3} & \frac{1}{3} \end{bmatrix} \begin{bmatrix} y_1 \\ y_2 \\ y_3 \end{bmatrix}$

(g) $\begin{bmatrix} x_1 \\ x_2 \\ x_3 \end{bmatrix} = \frac{1}{2} \begin{bmatrix} 0 & 1 & 1 \\ 1 & 0 & -1 \\ 1 & -1 & 0 \end{bmatrix} \begin{bmatrix} y_1 \\ y_2 \\ y_3 \end{bmatrix}$

(h) $\begin{bmatrix} x_1 \\ x_1 \\ x_3 \end{bmatrix} = \begin{bmatrix} 1/2\sqrt{5} & 4/9\sqrt{5} & \frac{1}{3} \\ -1/2\sqrt{5} & 1/9\sqrt{5} & \frac{1}{3} \\ 0 & -5/9\sqrt{5} & \frac{1}{3} \end{bmatrix} \begin{bmatrix} y_1 \\ y_2 \\ y_3 \end{bmatrix}$

5. **(a)** $S = \begin{bmatrix} 1 & 2 \\ 2 & 5 \end{bmatrix} \qquad S^{-1} = \begin{bmatrix} 5 & -2 \\ -2 & 1 \end{bmatrix}$

(b) $S = \begin{bmatrix} 1 & 1 \\ 1 & -1 \end{bmatrix} \qquad S^{-1} = \frac{1}{2}\begin{bmatrix} 1 & 1 \\ 1 & -1 \end{bmatrix}$

(c) $S = \begin{bmatrix} 1 & 1 \\ -2 & 1 \end{bmatrix} \qquad S^{-1} = \frac{1}{3}\begin{bmatrix} 1 & -1 \\ 2 & 1 \end{bmatrix}$

(d) $S = \begin{bmatrix} 1 & 1 & 1 \\ 1 & 1 & -1 \\ 1 & -1 & 1 \end{bmatrix} \qquad S^{-1} = \frac{1}{2}\begin{bmatrix} 0 & 1 & 1 \\ 1 & 0 & -1 \\ 1 & -1 & 0 \end{bmatrix}$

(e) $S = \begin{bmatrix} 1 & 0 & 3 \\ 0 & 1 & -1 \\ 0 & 1 & 1 \end{bmatrix} \qquad S^{-1} = \frac{1}{2}\begin{bmatrix} 2 & 3 & -3 \\ 0 & 1 & 1 \\ 0 & -1 & 1 \end{bmatrix}$

(f) $S = \begin{bmatrix} 1 & 0 & 1 \\ 1 & 1 & -1 \\ 0 & -1 & 1 \end{bmatrix} \qquad S^{-1} = \begin{bmatrix} 0 & 1 & 1 \\ 1 & -1 & -2 \\ 1 & -1 & -1 \end{bmatrix}$

7. $y_1 = \frac{\sqrt{6}}{3}(2x_1 + x_2 + x_3)$

$y_2 = 2x_1 - x_3$

$y_3 = \frac{\sqrt{3}}{3}(2x_1 - 2x_2 + x_3)$

Sec. 14.8, page 992

5. Yes

9. For all values of a and for all nonzero values of b, the given equations are satisfied, respectively, by the following matrices:

(a) $X = \begin{bmatrix} a & b \\ \frac{-a^2 + 2a + 3}{b} & 2 - a \end{bmatrix}$

(b) $X = \begin{bmatrix} a & b \\ \frac{-4a^2 + 4a - 3}{b} & 4 - a \end{bmatrix}$

(c) $X = \begin{bmatrix} a & b \\ \frac{-a^2 + 4a + 5}{b} & 4 - a \end{bmatrix}$

(d) $X = \frac{1}{b}\begin{bmatrix} b & ab & 0 \\ 0 & 2b & 0 \\ -2 & -2a & 3b \end{bmatrix}$

15. **(a)** $X_1 = \begin{bmatrix} 0 & 2 \\ -1 & 3 \end{bmatrix}, X_2 = \begin{bmatrix} -1 & 4 \\ -2 & 5 \end{bmatrix},$

$X_3 = \begin{bmatrix} 6 & -4 \\ 2 & 0 \end{bmatrix}, X_4 = \begin{bmatrix} 5 & -2 \\ 1 & 2 \end{bmatrix}$

(b) $X_1 = \begin{bmatrix} -4 & 3 \\ -2 & 1 \end{bmatrix}, X_2 = \begin{bmatrix} 4 & -9 \\ 6 & -11 \end{bmatrix},$

$X_3 = \begin{bmatrix} -10 & 9 \\ -6 & 5 \end{bmatrix}, X_4 = \begin{bmatrix} -2 & -3 \\ 2 & -7 \end{bmatrix}$

(c) $X_1 = \begin{bmatrix} 0 & 2 \\ -1 & 3 \end{bmatrix}$ **(d)** $X_1 = \begin{bmatrix} 2 & -1 & -1 \\ -3 & 4 & 5 \\ 3 & -3 & -4 \end{bmatrix}$

17. **(a)** $\begin{bmatrix} 60 & 19 \\ -57 & -16 \end{bmatrix}$ **(b)** $\begin{bmatrix} -3 & -6 \\ 9 & 12 \end{bmatrix}$ **(c)** $\begin{bmatrix} 0 & -6 \\ 3 & 9 \end{bmatrix}$

(d) $\begin{bmatrix} -12 & -27 & -9 \\ 3 & 12 & 3 \\ 9 & 9 & 6 \end{bmatrix}$ **(e)** $\begin{bmatrix} 6 & 35 & 35 \\ 3 & 76 & 73 \\ -3 & -35 & -32 \end{bmatrix}$

25. $67A^2 - 178A + 116I$

27. $A^n = (-2^n + 3^n)A + (3 \cdot 2^n - 2 \cdot 3^n)I$

29. $\lambda_1 = 2,\ x_1 = \begin{bmatrix} 3 \\ 1 \end{bmatrix};\ \lambda_2 = 4,\ x_2 = \begin{bmatrix} 1 \\ 1 \end{bmatrix}$

31. $\sin A = (\frac{1}{3} \sin 3)A$

CHAPTER 15

Sec. 15.1, page 1001

1. **(a)** 3, 9, 13 **(b)** $\frac{1}{3}(2, -2, 1)$, $\frac{1}{13}(12, -4, -3)$ **(c)** -18 **(d)** $-\frac{8}{9}$ **(e)** $(-4, 4, -2)$ **(f)** $\text{Cos}^{-1}(-\frac{2}{3})$

3. **(a)** 15, 15, 9 **(b)** $\frac{1}{3}(2, 2, 1)$, $\frac{1}{9}(4, 7, -4)$ **(c)** -40 **(d)** $\frac{62}{15}$ **(e)** $-\frac{8}{9}(2, 2, 1)$ **(f)** $\text{Cos}^{-1}(-\frac{8}{45})$

5. Not necessarily; not necessarily; not necessarily; yes

7. **0, 0** **9.** $(1/\sqrt{83})(3\mathbf{i} + 5\mathbf{j} + 7\mathbf{k})$ and $-(1/\sqrt{83})(3\mathbf{i} + 5\mathbf{j} + 7\mathbf{k})$

11. **(a)** $\mathbf{P} = 2\mathbf{i}$, $\mathbf{Q} = -3\mathbf{i} + 3\mathbf{j}$, $\mathbf{R} = -4\mathbf{k}$, $\mathbf{Q} - \mathbf{R} = -3\mathbf{i} + 3\mathbf{j} + 4\mathbf{k}$
(b) $-3\mathbf{i} + 9\mathbf{j} + 8\mathbf{k} = 3\mathbf{P} + 3\mathbf{Q} - 2\mathbf{R}$

13. $\mathbf{R} = -\mathbf{i} + 2\mathbf{j} + 5\mathbf{k}$

15. $\text{Cos}^{-1}(1/\sqrt{3}) \doteq 0.9553$ rad

17. -2, $-\frac{1}{7}(4\mathbf{i} + 6\mathbf{j} - 12\mathbf{k})$

33. The magnitude of the force must exceed $2000\sqrt{3}$ lb.

Sec. 15.2, page 1006

1. $-2\mathbf{i} + 3\mathbf{j} + 13\mathbf{k}$, $\sqrt{182}$, $\theta = \pi/2$

3. $-4\mathbf{i} + 3\mathbf{j} + \mathbf{k}$, $\sqrt{26}$, $\theta = \text{Sin}^{-1}(\sqrt{\frac{13}{21}})$

5. $5\mathbf{i} - 5\mathbf{j} + 5\mathbf{k}$, $5\sqrt{3}$, $\theta = \text{Sin}^{-1}(5/2\sqrt{7})$

7. $10\mathbf{i} + 18\mathbf{j} + 16\mathbf{k}$ **9.** $-75\mathbf{i} + 60\mathbf{j} + 30\mathbf{k}$

11. $\mathbf{i} - 5\mathbf{j} - 4\mathbf{k}$ **13.** $(-\mathbf{j} - 2\mathbf{k})/\sqrt{5}$

15. $5\sqrt{2}$ **17.** $12\sqrt{2}$

27. (a) $\dfrac{x-1}{-10} = \dfrac{y-4}{16} = \dfrac{z-3}{1}$

(b) $(-9, 20, 4)$, $(11, -12, 2)$

29. $x = 2t \quad y = 1 + t \quad z = -1 - 3t$

31. (a) $\dfrac{3\pi}{4}$ (b) $\dfrac{x-1}{2} = \dfrac{y-2}{-2} = z + 1$

33. (a) $x = x_0 \quad \dfrac{y - y_0}{a_2} = \dfrac{z - z_0}{a_3}$

(b) $x = x_0 + a_1 t \quad y = y_0 \quad z = z_0$

35. (a) $\sqrt{293}$ (b) $(5, -5, 10)$

(c) $\dfrac{x-1}{4} = \dfrac{y-1}{-6} = \dfrac{z-1}{9}$

37. $x + 2y + 3z = 14$ **39.** $5x + 3y - 9z + 10 = 0$

41. (b) With $\mathbf{n} = (a, b, c)$,

$$d = \frac{|a(x_1 - x_0) + b(y_1 - y_0) + c(z_1 - z_0)|}{(a^2 + b^2 + c_2)^{1/2}}$$

43. $d = \sqrt{14}$ **45.** $\begin{aligned} 2x - y + 2z &= 2 \\ 2x - y + 2z &= 8 \end{aligned}$

49. $6/\sqrt{179}$

57. No, because $\mathbf{F} \times \mathbf{R}$ is opposite the direction in which $\mathbf{F}$ would cause a right-handed screw to advance

Sec. 15.3, page 1012

1. (a) -90 (b) $245\mathbf{i} + 160\mathbf{j} - 170\mathbf{k}$ (c) $45\mathbf{i} + 216\mathbf{j} - 108\mathbf{k}$ (d) -180 (e) 0 (f) 585 (g) $90(\mathbf{i} + 2\mathbf{j} - \mathbf{k})$

3. (a) 915 (b) $610\mathbf{i} + 100\mathbf{j} - 1420\mathbf{k}$ (c) $-170\mathbf{i} - 800\mathbf{j} - 1570\mathbf{k}$ (d) 1830 (e) 0 (f) -1110 (g) $-3810\mathbf{i} - 2250\mathbf{j} + 1065\mathbf{k}$

5. 16 **7.** $b = -1$, $c = 0, \frac{1}{2}$

9. $c = -1$ **11.** $11x - 3y + 7z = 39$

13. $b = [\mathbf{ABC}]$; $|b|$ is the volume of the parallelepiped having $\mathbf{A}$, $\mathbf{B}$, and $\mathbf{C}$ as concurrent edges; b^2.

15. $\sqrt{290}$ **17.** $c = 3$

25. $x = [\mathbf{DBC}]/[\mathbf{ABC}]$, $y = [\mathbf{ADC}]/[\mathbf{ABC}]$, $z = [\mathbf{ABD}]/[\mathbf{ABC}]$

Sec. 15.4, page 1022

3. (a) $\mathbf{V} \cdot \dfrac{d\mathbf{V}}{dt} \times \dfrac{d^3\mathbf{V}}{dt^3}$

(b) $\dfrac{d\mathbf{V}}{dt} \times \left(\dfrac{d\mathbf{V}}{dt} \times \dfrac{d^2\mathbf{V}}{dt^2}\right) + \mathbf{V} \times \left(\dfrac{d\mathbf{V}}{dt} \times \dfrac{d^3\mathbf{V}}{dt^3}\right)$

7. (a) $-\mathbf{i}$ (b) $-\mathbf{j}$ **9.** $\text{Cos}^{-1}(\frac{3}{7})$

11. $\text{Cos}^{-1}(137/\sqrt{21}\sqrt{1041})$

13. Yes; the tangents at $t = 0$ and $t = 1$ are parallel. Yes; for all values of t the tangents at t and $-t$ are parallel.

15. $\mathbf{v}(1) = 2\mathbf{i} - 3\mathbf{j} + 4\mathbf{k}$; $\frac{35}{9}$; $\mathbf{a}(t) = 2\mathbf{i} - 6t\mathbf{j} + 12t^2\mathbf{k}$; $a_T = \dfrac{2(2 + 9t^2 + 24t^4)|t|}{t(4 + 9t^2 + 16t^4)^{1/2}}$; $a_n = 2|t|\sqrt{\dfrac{9 + 64t^2 + 36t^4}{4 + 9t^2 + 16t^4}}$

17. (a) $t = \frac{1}{2}$ (b) $t = \frac{1}{4}, \frac{1}{16}$

19. (a) $\sqrt{29}$ (b) $2\mathbf{i} - 2\mathbf{k}$ (c) $(4\mathbf{i} + 2\mathbf{j} - 3\mathbf{k})/\sqrt{29}$ (d) $\text{Cos}^{-1}(7/\sqrt{58})$

21. $\mathbf{r} = \left(\dfrac{t^2}{2} + \dfrac{t^3}{6}\right)\mathbf{i} + \left(\dfrac{t^5}{20} + 1\right)\mathbf{j} + \left(\dfrac{t^3}{3} - \dfrac{t^4}{12} + 2\right)\mathbf{k}$

23. $\mathbf{r} = \left(\dfrac{t^4}{4} + t - 1\right)\mathbf{i} + \left(\dfrac{t^2}{2} + 2t\right)\mathbf{j} - \left(\dfrac{1}{4}\cos 2t + 3t - \dfrac{5}{4}\right)\mathbf{k}$; $\mathbf{v}(1) = 2\mathbf{i} + 3\mathbf{j} + \left(\dfrac{\sin 2}{2} - 3\right)\mathbf{k}$

25. $\mathbf{V} = \mathbf{C}e^{\lambda t}$

29. $a_T = \begin{cases} 2\sqrt{3} & t > 0 \\ -2\sqrt{3} & t < 0 \end{cases}$; $\quad a_n = 0,\ t \neq 0$

31. $a_T = \dfrac{t}{2(4 + t^2)^{1/2}}$, $\quad a_n = \dfrac{(t^4 + 7t^2 + 16)^{1/2}}{2(4 + t^2)^{1/2}}$

33. $\sqrt{3}(e^3 - 1)$ **37.** $\kappa = 1/2\sqrt{2}$, $\rho = 2\sqrt{2}$

39. (a) $\mathbf{v} = \mathbf{i} + 2t\mathbf{j} + 2t^2\mathbf{k}$, $\mathbf{a} = 2\mathbf{j} + 4t\mathbf{k}$, $v = 1 + 2t^2$ (b) $a_T = 4t$, $a_n = 2$ (c) $\kappa = 2/(1 + 2t^2)^2$ (d) $\rho_{\min} = \frac{1}{2}$

43. $\begin{vmatrix} x - t^4 & y - t^2 & z - t^3 \\ 4t^3 & 2t & 3t^2 \\ 12t^2 & 2 & 6t \end{vmatrix} = 0$

47. $\mathbf{T}, \mathbf{N}, \mathbf{B} = \mathbf{i}, \mathbf{j}, \mathbf{k}$; normal, rectifying, osculating planes: $x = 0$, $y = 0$, $z = 1$, respectively

49. $\mathbf{T} = (-3\mathbf{i} + 4\mathbf{k})/5$, $\mathbf{N} = -\mathbf{j}$, $\mathbf{B} = (4\mathbf{i} + 3\mathbf{k})/5$; normal plane: $3x - 4z + 8\pi = 0$, rectifying plane: $y = 3$, osculating plane: $4x + 3z - 6\pi = 0$

55. $\kappa(s) = 1/\sqrt{16a^2 - s^2}$, $\tau(s) \equiv 0$

63. $\mathbf{Y}_1 \times \mathbf{Y}_2 = \mathbf{0}$ if and only if y_1 and y_2 are linearly dependent.

65. Yes, because $\mathbf{Y}_1 \times \mathbf{Y}_2 \neq \mathbf{0}$ implies y_1 and y_2 are linearly independent

69. Briefly, y is a solution of the difference equation if and only if $\mathbf{P} \cdot \mathbf{Y} = 0$, where $\mathbf{P} = (1, p_1, p_2)$, $\mathbf{Y} = (y_{k+2}, y_{k+1}, y_k)$; and $\mathbf{Y}_1 \times \mathbf{Y}_2 = -C_{12}\mathbf{P}$, where $C_{12} = \begin{vmatrix} y_{1,k} & y_{1,k+1} \\ y_{2,k} & y_{2,k+1} \end{vmatrix}$ is the Casorati determinant of y_1 and y_2.

71. C_k is the Casorati determinant of y_1, y_2, and y_3;

$$\mathbf{c}_k = \sum (\mathbf{u}_{k+1} \times \mathbf{u}_{k+2})\left(\frac{f_k}{a_{0,k}C_{k+1}}\right)$$

Sec. 15.5, page 1035

1. (a) $2(x\mathbf{i} + z\mathbf{j} + y\mathbf{k})$

(b) $e^{xyz}(yz\mathbf{i} + xz\mathbf{j} + xy\mathbf{k})$

(c) $\sin yz\,\mathbf{i} + xz\cos yz\,\mathbf{j} + xy\cos yz\,\mathbf{k}$

(d) $(3x^2 - 3yz)\mathbf{i} + (3y^2 - 3xz)\mathbf{j} - 3xy\mathbf{k}$

(e) $x^a y^b z^c\left(\dfrac{a}{x}\mathbf{i} + \dfrac{b}{y}\mathbf{j} + \dfrac{c}{z}\mathbf{k}\right)$

3. (a) $\sqrt{3}e^{x+y+z}$, $(\mathbf{i} + \mathbf{j} + \mathbf{k})/\sqrt{3}$

(b) $2\cosh(x + y)$, $[(\mathbf{i} + \mathbf{j})\sqrt{2}\sinh(x + y) + 2k]/2\cosh(x + y)$

(c) 5, $(3\mathbf{i} + 4\mathbf{j})/5$ (d) $\sqrt{6}$, $(\mathbf{i} + 2\mathbf{j} - \mathbf{k})/\sqrt{6}$

5. $\text{Cos}^{-1}(-1/\sqrt{22})$ **7.** $x^2y^3(3y+4x)/\sqrt{2}$ **9.** $-\frac{1}{2}$, 1
11. **(a)** $4\mathbf{i} - 4\mathbf{j} - 4\mathbf{k}$ **(b)** $8/\sqrt{6}$
13. **(a)** 0 **(b)** 24 **15.** $\frac{11}{3}$
17. $(2xy^3z^4 + 3x^2y^2z^4 + 4x^2y^3z^3)/\sqrt{3}$
19. $\dfrac{x-x_0}{\phi_x(x_0,y_0,z_0)} = \dfrac{y-y_0}{\phi_y(x_0,y_0,z_0)} = \dfrac{z-z_0}{\phi_z(x_0,y_0,z_0)}$
21. $\mathbf{R} = (5\mathbf{i}+9\mathbf{j}+4\mathbf{k}) + t(10\mathbf{i}-\mathbf{j}-8\mathbf{k})$,
$\dfrac{x-5}{10} = \dfrac{y-9}{-1} = \dfrac{z-4}{-8}$
23. $a = -8$
25. **(a)** $x+4y-z=5$, $\dfrac{x-2}{1} = \dfrac{y-2}{4} = \dfrac{z-5}{-1}$
(b) $x-2y-4z=-5$, $\dfrac{x-5}{-1} = \dfrac{y-1}{2} = \dfrac{z-2}{4}$
(c) $5x+4y+3z=22$, $\dfrac{x-1}{5} = \dfrac{y-2}{4} = \dfrac{z-3}{3}$
(d) $2x+y+9z=18$, $\dfrac{x-1}{2} = y+2 = \dfrac{z-2}{9}$
(e) $x+4y-z=4$, $\dfrac{x-2}{1} = \dfrac{y-1}{4} = \dfrac{z-2}{-1}$
(f) $2x+y-2z=4$, $\dfrac{x-2}{2} = \dfrac{y-4}{1} = \dfrac{z-2}{-2}$
27. 0
29. **(a)** $-xy$, $-x(2x+z)\mathbf{i} + yz\mathbf{j} + 2(2xz-y)\mathbf{k}$
(b) $yz + 3x^2 + 2xz - y^2$, $-2yz\mathbf{i} + (xy - z^2)\mathbf{j} + x(6y-z)\mathbf{k}$
(c) $-x\sin y$, $\mathbf{i}+\mathbf{j}$ **(d)** y^2+x^2+2z, $\mathbf{0}$
(e) $2(x+y+z)$, $\mathbf{0}$ **(f)** 0, $-\cos z\mathbf{i} - \cos x\mathbf{j} - \cos y\mathbf{k}$
31. Yes; the second-order partial derivatives of f are continuous and $\nabla^2 f = 0$.
33. **(a)** $-4y(x+y)$ **(b)** 0
39. If $\mathbf{v} = y\mathbf{i} + z\mathbf{j} + x\mathbf{k}$, then $\boldsymbol{\Delta} \times \mathbf{v} \cdot \mathbf{v} = -y - z - x \neq 0$.
41. $\nabla\phi = \dfrac{\partial\phi}{\partial u}\nabla u + \dfrac{\partial\phi}{\partial v}\nabla v + \dfrac{\partial\phi}{\partial w}\nabla w$
43. **A** **45.** $n = -1, 0$
47. $n = -3$ **51.** 0, $2(\mathbf{V}_1 - \mathbf{V}_2)$
55. $\dfrac{\partial\phi}{\partial r}\left(\dfrac{\partial\mathbf{R}}{\partial r}\right) + \dfrac{1}{r}\dfrac{\partial\phi}{\partial\theta}\left(\dfrac{1}{r}\dfrac{\partial\mathbf{R}}{\partial\theta}\right)$
57.
$$\begin{bmatrix} \dfrac{\partial^2\phi}{\partial x^2} & \dfrac{\partial^2\phi}{\partial y\,\partial x} & \dfrac{\partial^2\phi}{\partial z\,\partial x} \\ \dfrac{\partial^2\phi}{\partial x\,\partial y} & \dfrac{\partial^2\phi}{\partial y^2} & \dfrac{\partial^2\phi}{\partial z\,\partial y} \\ \dfrac{\partial^2\phi}{\partial x\,\partial z} & \dfrac{\partial^2\phi}{\partial y\,\partial z} & \dfrac{\partial^2\phi}{\partial z^2} \end{bmatrix}, \ \operatorname{div}(\mathbf{grad}\,\phi) = \nabla^2\phi, \ \operatorname{curl}(\mathbf{grad}\,\phi) = \mathbf{0}$$
59. **(a)** $\begin{bmatrix} 1 & 0 & 0 \\ 0 & 1 & 0 \\ 0 & 0 & 1 \end{bmatrix}$, 3, 0
(b) $\begin{bmatrix} y^2z^2 & 2xyz^2 & 2xy^2z \\ 0 & z^2\cos y & 2z\sin y \\ 2xe^y & x^2e^y & 0 \end{bmatrix}$, $y^2z^2 + z^2\cos y$,
$(x^2y - 2z\sin y)\mathbf{i} + 2x(y^2z - e^y)\mathbf{j} - 2xy^2z\mathbf{k}$
(c) $\begin{bmatrix} \dfrac{2xy}{(x^2+y^2)^2} & \dfrac{y^2-x^2}{(x^2+y^2)^2} & 0 \\ \dfrac{y^2-x^2}{(x^2+y^2)^2} & \dfrac{-2xy}{(x^2+y^2)^2} & 0 \\ 0 & 0 & 0 \end{bmatrix}$, 0, $\mathbf{0}$

Sec. 15.6, page 1050

1. Yes, its length $\frac{8}{27}(10\sqrt{10} - 1)$ is finite.
3. No, for $0 < b < \dfrac{\pi}{2}$, the arc length is given by
$s(b) = \ln\dfrac{\tan(b/2 + \pi/4)}{\tan(\pi/4)}$ and $\lim_{b\to\pi/2} s(b) = \infty$.
5. **(a)** 12 **(b)** 12
7. **(a)** $\sqrt{2}/2$ **(b)** $(5\sqrt{5}-1)/12$ **(c)** $\dfrac{64 + 247\sqrt{13}}{1215}$
9. **(a)** $\frac{2}{3}$ **(b)** $\frac{47}{70}$
11. $n = \frac{1}{2}$
13. $W = ke^2/2 + k\mu(e-2)$
15. **(a)** $ka^2\pi^2/4$ **(b)** $4ka^2$
17. **(a)** $-\frac{11}{3}$ **(b)** $-\frac{11}{3}$
19. **(a)** Initial point: $(0,-1,0)$; terminal point: $(0,0,-1)$ **(b)** $(1,0,0)$
21. $\frac{8}{3}$ **23.** $\frac{13}{12}$
25. $-(\frac{254}{3} + \ln 27)$ **27.** $(\pi^3 - 2)/3$
29. **(a)** 12π **(b)** 0
31. $\frac{3}{4}\pi a^3$ **33.** $\frac{2}{5}\pi a^5$
35. The common value of the integrals is $\frac{2}{3}$.
37. $-\frac{1}{44}$
39. **(a)** k **(b)** $k + 12\mu k/35$ **(c)** $k + 4\mu k/3$ **(d)** $k + 4\mu k/3$
41. Yes
43. $I_p = \int_C -x^2y\,dx + xy^2\,dy$
49. $\mathbf{F}(x,y) = U(x,y)\mathbf{i} + V(x,y)\mathbf{j}$

Sec. 15.7, page 1065

1. **(a)** $\frac{4}{3}$ **(b)** $\frac{11}{6}$ **(c)** $\frac{11}{6}$ **(d)** $\frac{11}{6}$
3. **(a)** 1 **(b)** 6 **(c)** $\pi/2$ **(d)** $\pi/3$
5. The common value of the integrals is $5\pi a^4/4$.
7. The common value of the integrals is $\frac{3}{2}$.
9. -2π **11.** $2\pi ab$
13. $$\iiint_V \left[u\left(\frac{\partial^2 v}{\partial x^2} + \frac{\partial^2 v}{\partial y^2} + \frac{\partial^2 v}{\partial z^2}\right) - v\left(\frac{\partial^2 u}{\partial x^2} + \frac{\partial^2 u}{\partial y^2} + \frac{\partial^2 u}{\partial z^2}\right)\right] dV = \iint_S \left(u\frac{dv}{dn} - v\frac{du}{dn}\right) dS$$
15. $$\int_C u\,dx + v\,dy + w\,dz = \iint_S \left[\left(\frac{\partial w}{\partial y} - \frac{\partial v}{\partial z}\right)\cos\alpha + \left(\frac{\partial u}{\partial z} - \frac{\partial w}{\partial x}\right)\cos\beta + \left(\frac{\partial v}{\partial x} - \frac{\partial u}{\partial y}\right)\cos\gamma\right] dS$$
17. The length of the curve. No, because the vector function $\mathbf{T}$ is defined *only* on the curve C.

23. 0 **25.** $\frac{1}{2}\int_C \mathbf{R}\cdot\mathbf{R}\,d\mathbf{R}$

29. If 0 is a point at which the surface has a tangent plane, the integral in Gauss' theorem is equal to 2π. If 0 is a singular point on S, the integral may have any value between 0 and 4π.

33. The integrals along the upper and lower semicircles are, respectively, $-\pi$ and π.

37. As in Example 2; or another way is to integrate u with respect to x and to this add the integral with respect to y of the terms in v which do not contain x; finally to this sum add the integral with respect to z of the terms in w which are independent of x and y.

41. $\phi = x\cos y + x^2y^2z + y\sin z$

43. **(a)** 0 **(b)** $\phi = \dfrac{5x^2y^4}{2} - \cos x - \dfrac{z^2}{2}$ **(c)** $3 - \cos 1$

45. $\phi = \frac{1}{3}(x^3 + y^3 + z^3)$, $(\pi^3 - 2)/3$

47. $\phi = xy/z - xe^z + 3y$, 3

49. Yes; because, in Theorem 5, **d** implies **b**.

Sec. 15.8, page 1076

1. $-G(r^3/3)$, $-G\ln r$

3. $(2M/a^2)(\sqrt{a^2+z^2} - z)$

9. $$\mathbf{E} = \mathbf{i}\sum_{n=1}^{\infty}\left(\frac{4}{n\pi}\cos\frac{n\pi at}{l} + \frac{4}{n^2\pi^2 a}\sin\frac{n\pi at}{l}\right)\sin\frac{n\pi y}{l}$$
$$+\,\mathbf{k}\sum_{n=1}^{\infty}\left(\frac{4}{n\pi}\cos\frac{n\pi at}{l} - \frac{4l}{n^2\pi^2 a}\sin\frac{n\pi at}{l}\right)\sin\frac{n\pi y}{l}$$
$$a^2 = \frac{1}{\mu\epsilon},\ n \text{ odd}$$

17. The surface integral represents the excess of the rate of efflux of fluid mass out of, over the influx into, the volume V bounded by the surface S; the volume integral represents the rate at which fluid mass in V is decreasing. Yes.

CHAPTER 16

Sec. 16.1, page 1087

3. $y = 2/x - 1$ **5.** $y = \dfrac{e^{1-2x} - 2e^{-x}}{e - 2}$

7. $y = e^{-x}(1 - \cos 2x)$

9. $y = -\dfrac{27}{40x} - \dfrac{23x}{10} + \dfrac{3x^3}{8}$

11. **(a)** $2y'' + k^2\sin y = 0$; $y'(0) = 0$, $y'(1) = 0$
(b) $(y')^2 - k^2\cos y = c$

13. $I = e^2 - 1 = [e^x - \cos xy]_{(0,\,k)}^{(2,\,\pi)}$

15. All twice-differentiable functions on [0, 2] which satisfy the end conditions. For all such functions, $I = 8$.

19. **(a)** $y = k_1\cosh[(x - k_2)/k_1]$, $k_1 > 0$
(b) $(x - a)^2 = 4b(y - b)$
(c) $y = ax + b$

21. **(a)** $y = 4/x^2$ **(b)** $I_{\min} = 28$ **(c)** 0 **(d)** 35

23. $y = \sqrt{10 - (x-3)^2}$; $0 \le x \le 2$

25. $2x^2y' - y = 0$ at $x = 1$ and $x = 3$; $xy'' + 2y' = 0$; $y = -\dfrac{3}{2x} + \dfrac{5}{2}$; $y = -\dfrac{3}{4x} + \dfrac{7}{4}$; $y = -\dfrac{3}{4x} + \dfrac{9}{4}$; $y = 0$

27. The inverted cycloid with parametric equations $x = x_0 + a(\theta - \sin\theta)$; $y = y_0 + a(1 - \cos\theta)$, where the constants x_0 and a are to be determined so that the curve will pass through the given points (x_1, y_1) and (x_2, y_2).

29. **(b)** The condition $(-g_x/g_y)y' = -1$ is found to hold at $x = x_2$. Since $y'(x_2)$ denotes the slope of the brachistochrone at the point (x_2, y_2) where it intersects the curve $g(x, y) = 0$, and since the slope of the curve defined by $g(x, y) = 0$ at (x_2, y_2) is given by $-g_x/g_y$ evaluated at that point, the two curves are orthogonal at (x_2, y_2)
(c) The tangent to the brachistochrone at the right-hand endpoint is perpendicular to the tangent to the given curve $h(x, y) = 0$ at the left-hand endpoint.
(d) The tangent lines to the given curves where they intersect the brachistrochrone must be parallel.

31. **(a)** Parametric equations of the brachistrochrone are $x = (3\sqrt{3}/\pi)(\theta - \sin\theta)$, $y = (3\sqrt{3}/\pi)(1 - \cos\theta)$; $0 \le \theta \le \pi/3$; $(x_2, y_2) = (\sqrt{3} - 9/2\pi, 3\sqrt{3}/2\pi)$
(b) $T = \sqrt{\pi/\sqrt{3}g}$ **(c)** $F = \sqrt{2/g}$

33. **(a)** Parametric equations of the brachistrochrone are $x = (2/\pi)(\theta - \sin\theta)$, $y = (2/\pi)(1 - \cos\theta)$; $0 \le \theta \le \pi/2$; $(x_2, y_2) = [(\pi - 2)/\pi, 2/\pi]$
(b) $T = \sqrt{\pi/2g}$ **(c)** $F = \sqrt{2/g}$

35. **(a)** $(\frac{1}{4}, \frac{1}{2})$ **(b)** $(x_1, y_1) = (15/2\pi - \frac{7}{4}, 15/2\pi + \frac{1}{2})$
(c) $T = \frac{1}{2}\sqrt{15\pi/2g}$ **(d)** $F = \sqrt{15/2g}$

37. $(-6, 8)$

41. The light path is the arc of the cycloid of Exercise 27.

43. **(a)** The spirals of the family $\theta = mz + b$, where θ and z are the usual cylindrical coordinates
(b) The great-circle arcs $\theta = \pm\sin^{-1}\dfrac{\cot\phi}{[(a/c)^2 - 1]^{1/2}} + k$, where a is the radius of the sphere
(c) The spirals of the family $\theta = \dfrac{1}{\sin\alpha}\sec^{-1}\left(\dfrac{r}{k}\right) + b$, where r and θ are spherical coordinates

45. The arc of the helix with parametric equations $x = a\cos\theta$, $y = a\sin\theta$, $z = -2\theta$, $0 \le \theta \le \pi/2$

47. **(a)** $\dfrac{d}{d\phi}\left\{\dfrac{(\sin^2\phi)\theta'}{[1 + (\sin^2\phi)(\theta')^2]^{1/2}}\right\} = 0$
(b) $\sqrt{2}\sin\left(\dfrac{\pi}{4} - \theta\right) = \cot\phi$, where θ increases monotonically from $\pi/4$ to $\pi/2$ as ϕ increases from $\pi/2$ to $3\pi/4$
(c) The equation of Part **b** yields $x - y = z$ as the rectangular equation of the plane in which the geodesic must lie, and $\mathbf{A}\times\mathbf{B}\cdot\mathbf{P} = 0$ is equivalent to this equation.

49. $\dfrac{xy'}{[1 + (y')^2]^2} = c$ is a first integral; with $v = y'$,

$x = (c/v)(1 + v^2)^2$ and $y = k + c(v^2 + \frac{3}{4}v^4 - \ln|v|)$ give parametric equations of solutions.

51. $\dfrac{\partial f}{\partial y'} - \dfrac{d}{dx}\left(\dfrac{\partial f}{\partial y''}\right) = 0, \quad \dfrac{\partial f}{\partial y''} = 0$

Sec. 16.2, page 1095

1. $\begin{bmatrix} x \\ y \end{bmatrix} = c_1\begin{bmatrix} 1 \\ 1 \end{bmatrix}e^{-t} + c_2\begin{bmatrix} 1 \\ 1 \end{bmatrix}e^{t} + c_3\begin{bmatrix} 1 \\ -1 \end{bmatrix}\cos t + c_4\begin{bmatrix} 1 \\ -1 \end{bmatrix}\sin t$

3. $\begin{bmatrix} x \\ y \end{bmatrix} = c_1\begin{bmatrix} 1 \\ -1 \end{bmatrix}e^{-t} + c_2\begin{bmatrix} 1 \\ -1 \end{bmatrix}e^{t} + c_3\begin{bmatrix} 1 \\ 1 \end{bmatrix}e^{-2t} + c_4\begin{bmatrix} 1 \\ 1 \end{bmatrix}e^{2t}$

5. The straight-line segment connecting (x_1, y_1) and (x_2, y_2).

11. **(a)** $P_1 = P_2$ **(b)** $m(t_2) = m(t_1)e^{-L/V}$, where $L = \sqrt{(\mu_2 - \mu_1)^2 + [(\nu_2 - \nu_1) + g(t_2 - t_1)]^2}$

13. $\dot{\mathbf{v}} + g\dfrac{\mathbf{r}}{r} + \dfrac{\kappa v}{m}\mathbf{v} = \dfrac{\dot{m}}{m}\mathbf{V}$; $\kappa > 0$ a proportionality constant.

15. $\dot{\mathbf{v}} + \dfrac{kM}{r^3}\mathbf{r} = \dfrac{\dot{m}}{m}\mathbf{V}$

17. **(a)** $\dot{\mathbf{v}} + f(r)\dfrac{\mathbf{r}}{r} = \dfrac{\dot{m}}{m}\mathbf{V}$

(b) $\ddot{r} - r\dot{\phi}^2 + f(r) = V\dfrac{\dot{m}}{m}\cos\alpha, r\ddot{\phi} + 2\dot{r}\dot{\phi} = V\dfrac{\dot{m}}{m}\sin\alpha$

(c) Since

$$\int_0^{t_0} \sqrt{(r\ddot{\phi} + 2\dot{r}\dot{\phi})^2 + [\ddot{r} - r\dot{\phi}^2 + f(r)]^2}\, dt =$$

$V\ln\dfrac{m(0)}{m(t_0)}$, we are to extremize the integral in the left-hand member subject to the given velocity conditions.

(e) The integral of Part **c** becomes

$$\int_a^{r_0} \sqrt{\left(\frac{C'}{r}\right)^2 + \frac{(E' - CC'/r^2)^2}{2E - C^2/r^2 - 2U(r)}}\, dr$$

where $a = r(0)$ and $r_0 = r(t_0)$.

(f) Part **c**, order 4; Part **e**, order 2.

Sec. 16.3, page 1101

1. $y = x^2 - \frac{1}{2}x$ **3.** $y = 3x^2 - 2x + 1$

5. $(x - h)^2 + (y + k)^2 = \lambda^2$, where $h = \frac{1}{2}(x_1 + x_2)$, $k^2 = \lambda^2 - a^2, a = \frac{1}{2}(x_2 - x_1), x_2 - x_1 = \pm 2\lambda\sin(l/2\lambda)$, and l is the given length.

7. $y = k\cosh(x/k) - \lambda$, where $\sinh(a/k) = l/2k$ and $\lambda = -b \pm \sqrt{k^2 + l^2/4}$

9. **(a)** $y_n(x) = \pm\sqrt{2/\pi}\sin nx, n = 1, 2, 3, \ldots$

(b) $y(x) \equiv 0$

11. $y = \sqrt{x(2l/\pi - x)}$, $0 \le x \le 2l/\pi$, defines a semicircle which is the curve for which $A = A_M$, $A_M = l^2/2\pi$.

13. **(a)** Minimize $\int_0^b 2\pi y\sqrt{1 + (y')^2}\, dx$ subject to $l = \int_0^b \sqrt{1 + (y')^2}\, dx$

(b) $y = a\cosh[(x - b)/a] - \lambda$

(c) $y'(b)$ infinite

(d) No, because the conditions $y(0) = y(b) = 0$ imply that $b = 0$

15. **(a)** $V = w\displaystyle\int_{x_1}^{x_2} y\sqrt{1 + (y')^2}\, dx$

(b) $y = -\dfrac{\lambda}{w} + \dfrac{b}{w}\cosh\dfrac{w(x - a)}{b}$

17. $y = b\cosh\dfrac{x + a}{b} - \sqrt{b^2\cosh^2\dfrac{a}{b} + \dfrac{l^2}{4}}, \quad -1 \le x \le 1,$

where $\dfrac{l}{2} = b\sinh\dfrac{1}{b}\cosh\dfrac{a}{b}$ and $\sqrt{\dfrac{l^2}{4} - y_1^2} = b\sinh\dfrac{1}{b}$

19. $x = \cos\dfrac{\pi}{2}\left(1 - \dfrac{t}{T}\right), \quad y = \sin\dfrac{\pi}{2}\left(1 - \dfrac{t}{T}\right), \quad 0 \le t \le T;$

$I_{\min} = \dfrac{\pi^2}{8T}$

21. The path is the circular arc of the prescribed length which joins the given points and is concave upward

27. $x = \begin{cases} 0 & 0 \le t \le T/2 \\ (2t - T)/T & T/2 \le t \le T \end{cases}$

$y = \begin{cases} (T - 2t)/T & 0 \le t \le T/2 \\ 0 & T/2 \le t \le T \end{cases}$

$z = (T - 2t)/T, \; 0 \le t \le T$

29. $x = r\cos\theta$, $y = r\sin\theta$, $z = r$, $0 \le t \le T$, where

$r = \sqrt{4 + \dfrac{t(t - 2\beta)}{T(2\beta - T)}},$

$\theta = \dfrac{\pi}{2} - \sqrt{2}\left(\operatorname{Tan}^{-1}\dfrac{t - \beta}{k} + \operatorname{Tan}^{-1}\dfrac{\beta}{k}\right)$

$k = \dfrac{1}{[4T(2\beta - T) - \beta^2]^{1/2}}$

$\beta = \dfrac{2[2(1 + 7\delta) + \sqrt{3(1 + \delta)}]T}{1 + 49\delta} \doteq 0.647T$ since

$\delta = \tan^2\dfrac{\pi}{2\sqrt{2}} \doteq 4.072$

Sec. 16.4, page 1108

1. $I = \int_{-1}^0 [(y')^2 + y^2]\, dx$; all comparison functions must satisfy $\partial f/\partial y' = 0$ at $x = -1$ and have the value $\cosh 1$ at $x = 0$.

3. $I = \int_0^\pi (y')^2\, dx$, $J = \int_0^\pi y^2\, dx$; all comparison functions must vanish at $x = 0$ and at $x = \pi$.

5. $I = \int_1^3 [\frac{1}{2}x^2(y')^2 - \frac{3}{2}xy^2 + xy]\, dx$; all comparison functions must have the value 1 at $x = 3$; at $x = 1$ they satisfy $\partial f/\partial y' = 0$, that is, $y'(1) = 0$.

7. $I = \int_1^2 [\frac{1}{2}x^3(y')^2 - \frac{1}{2}y^2 + xy]\,dx$; all comparison functions must have the value 1 at $x = 1$; at $x = 2$ they satisfy $\partial f/\partial y' = 0$, that is, $y'(2) = 0$.

9. $I = \int_{x_1}^{x_2} \left\{x(y')^2 + \frac{v^2}{x}y^2 + \frac{d}{dx}[g(x)y^2]\right\} dx,$

$J = \int_{x_1}^{x_2} xy^2\,dx$, $g(x_1) = 0$, $g(x_2) = \frac{-b}{a}$; all comparison functions must satisfy natural boundary conditions at $x = x_1$ and $x = x_2$.

11. $I = \int_0^1 [x^2(y')^2 - xy^2 - 2xy]\,dx$, $y = \frac{15}{38}(1 - x^2)$

13. $I = \int_1^2 [x^3(y')^2 - xy^2 + xy]\,dx$, $y = \frac{65}{228}(x^2 - 4x + 3)$

15. $I = \int_0^1 (1 + x)(y')^2\,dx$, $1 = \int_0^1 y^2\,dx$, $\lambda_1 \doteq 15$

17. $I = \int_0^{\pi} (y')^2\,dx$, $1 = \int_0^1 y^2\,dx$, $\lambda_1 \doteq 0.503$. The exact value is $\lambda_1 = \frac{1}{2}$.

19. $\lambda_1 \doteq \sqrt{10/\pi} \doteq 1.007$, $\lambda_2 \doteq \sqrt{42/\pi} \doteq 2.063$; the exact values are $\lambda_1 = 1$ and $\lambda_2 = 2$.

21. $(py')' + (q + \lambda r)y = 0$, $p(a)y'(a) = 0$

23. **(a)** $I = \int_{x_1}^{x_2} [p(y'')^2 - q(y')^2 + ry^2]\,dx$,
$J = \int_{x_1}^{x_2} sy^2\,dx$ = fixed constant
(b) At an end x_i, $i = 1$ or 2, where the comparison functions are not required to have a prescribed value, a natural boundary condition of the form $(py'')' + qy' = 0$ applies. At an end x_i, where the derivatives of the comparison functions may have different values, a natural boundary condition of the form $py'' = 0$ applies. If all comparison functions must have the same value y_i at x_i, the boundary condition $y(x_i) = y_i$ applies. If the derivatives of the comparison functions must all have the same value y_i' at x_i, the boundary condition $y'(x_i) = y_i'$ applies.

Sec. 16.5, page 1113

1. $\delta F = [xy'\eta(x) + xy\eta'(x)]\epsilon$, $\Delta F = \delta F + x\eta(x)\eta'(x)\epsilon^2$

3. $\delta F = x^3y[2y'\eta(x) + y\eta'(x)]\epsilon$
$\Delta F = \delta F + x^3\eta(x)[y'\eta(x) + 2y\eta'(x)]\epsilon^2 + x^3\eta^2(x)\eta'(x)\epsilon^3$

5. $\delta F = x\left[\frac{\eta(x)}{y}\right]'\epsilon,$

$$\Delta F = \delta F + x\left[\frac{\eta(x)}{y}\right]' \sum_{k=1}^{\infty} (-1)^k \left[\frac{\eta(x)}{y}\right]^k \epsilon^{k+1}$$

7. $\delta F = (4nx_0^n + x_0^{n+1})\epsilon$, $\Delta F = \delta F + n^2x_0^{2(n-1)}\epsilon^2$

9. $\delta I = 4\pi^3\epsilon$

Sec. 16.6, page 1117

1. $\ddot{y} = -g$

3. **(a)** $L = \frac{m}{2}(1 + a^2)\dot{x}^2 - mg(ax + b)$
(b) $(1 + a^2)\ddot{x} + ag = 0$
(c) $(x, y) = (0, b) - (1, a)\frac{agt^2}{2(1 + a^2)}$
(d) If $a = 0$, $(x, y) \equiv (0, b)$; the particle remains motionless. If $a = \infty$, $(x, y) = (0, b - gt^2/2)$; the particle falls freely.

5. **(a)** $L = \frac{m}{2}(1 + 4x^2)\dot{x}^2 - mgx^2$,
$(1 + 4x^2)\ddot{x} + 4x\dot{x}^2 + 2gx = 0$
(b) $\dot{x} = \sqrt{\frac{2g(1 - x^2)}{1 + 4x^2}}$; $\dot{y} = 2x\sqrt{\frac{2g(1 - x^2)}{1 + 4x^2}}$

7. **(a)** $\ddot{\theta} + \frac{g}{l}\sin\theta = 0$ **(b)** $\mu(t) = \frac{mv^2}{l} - mg\cos\theta$

9. $\ddot{y} + \frac{k}{m}y = ka\sin\omega t$;

$$y = \frac{a}{1 - \omega^2 m/k}\left(\sin\omega t - \omega\sqrt{\frac{m}{k}}\sin\sqrt{\frac{k}{m}}t\right)$$

11. **(a)** $m_1\ddot{q}_1 = m_1 g + \mu(t)$, $m_2\ddot{q}_2 = m_2 g + \mu(t)$
(b) $(m_1 + m_2)\ddot{q}_1 = (m_1 - m_2)g$
(c) $q_1 = \frac{1}{2}\left(\frac{m_1 - m_2}{m_1 + m_2}gt^2 + l\right)$ **(d)** $|\mu(t)| \equiv \frac{2m_1m_2g}{m_1 + m_2}$
(e) When $t = \sqrt{\frac{(m_2 + m_1)l}{(m_2 - m_1)g}}$

13. When $t = \sqrt{34}/4$

15. $\ddot{r} - r\omega^2\sin^2\alpha = -g\cos\alpha$

$$r = \left[r_0 - \frac{g\cos\alpha}{\omega^2\sin^2\alpha}\right]\cosh[(\omega\sin\alpha)t] + \frac{g\cos\alpha}{\omega^2\sin^2\alpha}$$

17. PE $= mgl(1 - \cos\theta) + kx^2$; KE $= (m/2)(\dot{x}^2 + 2\dot{x}\dot{\theta}l\cos\theta + l^2\dot{\theta}^2)$; $\ddot{x}\cos\theta + l\ddot{\theta} + g\sin\theta = 0$; $m\ddot{x} + ml\ddot{\theta}\cos\theta - m(\dot{\theta})^2 l\sin\theta + 2kx = 0$. There is a single natural frequency $\omega^2 = 2kg/(mg + 2kl)$.

19. **(a)** 2 **(b)** Use surface coordinates

21. $\omega^4 - g(A + C)\omega^2 + g^2(AC - B^2) = 0$; there are two values of ω^2 unless $A = C$ and $B = 0$.

23. **(a)** Component of **F** along **r**: $-m(g\sin\theta + r\dot{\theta}^2)$ which acts in the direction of $-\mathbf{r}$. Component of **F** transverse to **r**: $-\frac{1}{3}mg\cos\theta$ which acts in the direction of decreasing θ when $\cos\theta > 0$ and in the direction of increasing θ when $\cos\theta < 0$.
(b) $\ddot{\theta} - (2g/3r)\cos\theta = 0$, where $r = R_1 - R_2$
(c) $f = \frac{1}{2\pi}\sqrt{\frac{2g}{3(R_1 - R_2)}}$
(d) As $R_2 \to R_1, f \to \infty$; as $R_2 \to 0, f \to \frac{1}{2}\pi\sqrt{2g/3R_1}$; as $R_1 \to \infty, f \to 0$.

25. **(a)** $\frac{3}{2}r\ddot{\theta} - g\sin\theta = 0$
(b) $\boldsymbol{\mu}(t) =$

$$\begin{cases} (mg\cos\theta - mr\dot{\theta}^2, -\frac{1}{3}mg\sin\theta) & mg\cos\theta - mr\dot{\theta}^2 \ge 0 \\ (0, 0) & mg\cos\theta - mr\dot{\theta}^2 < 0 \end{cases}$$

vector components of $\boldsymbol{\mu}(t)$ being along r and normal thereto, positive in the increasing directions of r and θ.
(c) $\ddot{r} - r\dot{\theta}^2 + g\cos\theta = 0$, $(d/dt)(r^2\dot{\theta}) - rg\sin\theta = 0$, $\dot{\phi} = 0$

Sec. 16.7, page 1124

1. **(a)** $H = p^2/2m + mgq + (k/2)(q^2 - 2eq)$, where e is the elongation of the spring at equilibrium
(b) $\dot{q} = p/m$, $\dot{p} = -kq$ **(c)** $\ddot{q} + (k/m)q = 0$

3. **(a)** $H = (1/2m)[(p_\theta/\sin\phi)^2 + p_\phi^2] + mg\cos\phi$

(b) $\dot\theta = p_\theta/(m\sin^2\phi)$, $\dot\phi = p_\phi/m$, $\dot p_\theta = 0$,
$\dot p_\phi = \dfrac{p_\theta^2\cos\phi}{m\sin^3\phi} + mg\sin\phi$; $(d/dt)(\dot\theta\sin^2\phi) = 0$,
$\ddot\phi - \frac{1}{2}\dot\theta^2\sin 2\phi - g\sin\phi = 0$

7. Write Lagrange's equations as $\dfrac{d}{dt}\left(\dfrac{\partial T}{\partial \dot q_i}\right) - \dfrac{\partial T}{\partial q_i} + \dfrac{\partial V}{\partial q_i} = 0$ and delete all derivatives of T to obtain the required conditions: $\partial V/\partial q_i = 0$, $1 \le i \le n$

9. **(a)** $H = p_1^2/2m_1 + p_2^2/2m_2 + \frac{1}{2}k_1(x_1+e_1)^2 + \frac{1}{2}k_2[(x_2-x_1)+e_2]^2 - m_1gx_1 - m_2gx_2$, where e_1 and e_2 denote the elongations of the springs at static equilibrium.
(b) $\dot x = p_1/m_1$, $\dot x_2 = p_2/m_2$, $\dot p_1 = -k_1x_1 + k_2(x_2-x_1)$, $\dot p_2 = k_2(x_1-x_2)$
(c) $m_1\ddot x_1 = -k_1x_1 + k_2(x_2-x_1)$, $m\ddot x_2 = k_2(x_1-x_2)$

CHAPTER 17

Sec. 17.2, page 1131

1. $-2-3i$ **3.** $22+9i$
5. $z^2+2iz+1$ **7.** $-z^2+iz+2$
9. $(z^2+\bar z^2)\dfrac{1-i}{4} + z\bar z\dfrac{1+i}{2}$
19. $z=1$; $f(1)$ is not defined.
21. Continuous everywhere since $f(-i) = \lim_{z\to -i} f(z)$.
23. Along $y = mx$, $\lim_{z\to 0}\dfrac{xy}{x^2+y^2} = \lim_{x\to 0}\dfrac{mx^2}{x^2+(mx)^2}$, which depends on m
27. If δ is known, $\delta_1 \le \delta/\sqrt{2}$. If δ_1 is known, $\delta = \delta_1$.
29. Every point of the set is a boundary point. The set has no interior points.

Sec. 17.3, page 1138

5. Only at the origin; only at the origin; $f(z)$ is nowhere analytic.
7. No
13. No, since x^2+y does not satisfy Laplace's equation.
15. No. For instance, $ye^{2x} = c$ and $x - y^2 = k$ are orthogonal trajectories, but clearly $ye^{2x} + i(x-y^2)$ is not an analytic function.
17. $v = e^x\sin y$, $f(z) = e^z$ **19.** $f(z) = iz^2 + 6z$
25. $\dfrac{\partial f}{\partial z} = \dfrac{1}{2}\left(\dfrac{\partial f}{\partial x} - i\dfrac{\partial f}{\partial y}\right)$, $\dfrac{\partial f}{\partial \bar z} = \dfrac{1}{2}\left(\dfrac{\partial f}{\partial x} + i\dfrac{\partial f}{\partial y}\right)$
31. $v = x^2 - y^2 + 6y$, $f(z) = 6z + iz^2$
33. $v = 3x^2y - y^3 + c$, $f(z) = z^3 + ci$
35. **(a)** Only at $z = 0$
(b) Only at $z=0$; $f'(0) = 0$
(c) Nowhere, because there is no neighborhood of $z=0$ over which $f'(z)$ exists
37. **(a)** Yes **(b)** For all bounded values of z except $z=0$

Sec. 17.4, page 1148

1. **(a)** $1 + i\pi \doteq 1.000 + 3.142i$
(b) $\ln 2 - i(\pi/3) \doteq 0.693 - 1.047i$
(c) $1.299 + 0.635i$
(d) $0.834 - 2.164i$
(e) $e^{-\pi/2} \doteq 0.208$
(f) $-1.317i$
3. $(1+i)^{1-i} = \exp[(\ln\sqrt{2} + \pi/4 + 2n\pi) + i(-\ln\sqrt{2} + \pi/4 + 2n\pi)$, and the arguments of the different values differ only by integral multiples of 2π.
5. The common ratio is $e^{-2\pi d}$.
19. **(a)** $\dfrac{d\cosh z}{dz} = \sinh z$ **(b)** $\dfrac{d\sinh z}{dz} = \cosh z$
(c) $\dfrac{d\tanh z}{dz} = \operatorname{sech}^2 z$
21. **(a)** $\dfrac{d\cosh^{-1} z}{dz} = \dfrac{1}{(z^2-1)^{1/2}}$
(b) $\dfrac{d\sinh^{-1} z}{dz} = \dfrac{1}{(z^2+1)^{1/2}}$
(c) $\dfrac{d\tanh^{-1} z}{dz} = \dfrac{1}{1-z^2}$
23. $i(\pi/2 + n\pi)$ **25.** $\ln 2 + i(2n+1)\pi$
27. $\cosh^{-1} 2 + i(2n+1)\pi$ **31.** Yes
35. **(b)** $|\cosh z|^2 = \sinh^2 x + \cos^2 y$
39. No
45. **(a)** $2\pi i$ **(b)** $2\pi i$ **(c)** 0 **49.** The line $x = \pi - c$
51. The portion of the u axis on which $u \ge 1$
The closed interval $[-1, 1]$ on the u axis
53. $\dfrac{u^2}{\cosh^2 c} + \dfrac{v^2}{\sinh^2 c} = 1$ if $c \ne 0$; the interval $[-1, 1]$ on the u axis if $c = 0$
$\dfrac{u^2}{\cos^2 k} - \dfrac{v^2}{\sin^2 k} = 1$ if $k \ne 0$; the half-ray $[1, \infty)$ on the u axis if $k = 0$
57. $\pm\pi/2, \pm 3\pi/2, \pm 5\pi/2, \ldots$

Sec. 17.5, page 1159

1. **(a)** $(3+i)/2$ **(b)** $1+i$ **(c)** $(8+9i)/6$ **(d)** $(1+3i)/2$
3. **(a)** $(-1+6i)/6$ **(b)** $(-1+5i)/6$ **(c)** $(-1+5i)/6$
(d) $(-1+6i)/6$
5. The common value is $(18+26i)/3$.
7. $\frac{1}{2}[e^2(\cos 2 + i\sin 2) - 1]$
9. $\sin 1\cosh\pi + i\cos 1\sinh\pi$
11. $-(1+i\pi)\cos(1+i\pi) + \sin(1+i\pi)$
13. **(a)** $I = 0$ **(b)** $I \ne 0$ **(c)** $I \ne 0$ **(d)** $I = 0$ **(e)** $I \ne 0$
(f) $I \ne 0$; $I_a = I_d = 0$, $I_b = I_c \ne 0$
15. $v = (y - x)$ **17.** $v = e^x\sin y$
19. v does not exist since u does not satisfy Laplace's equation.
21. v does not exist since u does not satisfy Laplace's equation.
25. **(a)** $1+e$ **(b)** 0 **27.** **(a)** $-\frac{2}{3} + \frac{4}{3}i$ **(b)** $-\frac{2}{3}$

29. $3e^6/8$, $2e/3$. (Each of these is a very crude estimate because when the methods of Chap. 19 are used, the value of the integral around $|z| = 3$ is sin 2 and by Cauchy's theorem the value of the integral around $|z| = \frac{1}{2}$ is 0.)

Sec. 17.6, page 1166

1. **(a)** $-6\pi i$ **(b)** 0 **(c)** $-6\pi i$
3. **(a)** 0 **(b)** 0 **(c)** $-\pi \sin(4 - 2i)$
5. **(a)** $-3\pi i/2$ **(b)** $3\pi i/2$ **(c)** 0
7. Maxima: $\sqrt{13}$ at $z = \pm i$; minima: 3 at $z = \pm 1$
9. Maxima: $6\sqrt{3}$ at $z = 1 \pm i\sqrt{3}$; minima: 10 at $z = 2$, 6 at $z = -2$

Sec. 17.7, page 1174

1. $T = -\frac{50}{3}xy + 100$; $f(z) = \frac{25}{3}iz + 100$
3. $V = -25(x^2 - y^2) + 75$; $f(z) = -25z^2 + 100$
5. The streamlines are the rectangular hyperbolas $2xy = k$, $v(1 + 2i) = 2 - 4i$, $v(-2 + i) = -4 - 2i$. Along the circle $r = \frac{1}{2}$, the speed is equal to 1.
7. $V = -10x + 40$; $f(z) = -10z + 40$
9. The real potential is $u(r) = \dfrac{40 \ln r}{\ln 4}$; the complex potential is $f(z) = \dfrac{40}{\ln 4}z + 10$.
11. $f(z) = a(1 - im)z$, a real
13. If Uz and $-iVz$ are uniform flows parallel to the positive x and y axes, respectively, the superposition of these flows is a uniform flow in the direction of the line $Uy = Vx$.
15. $f(z) = \dfrac{1}{2\pi}\ln(z^2 - 1)$; $\Phi = \dfrac{1}{2\pi}\ln|z^2 - 1|$;
$\Psi = \dfrac{1}{2\pi}[\arg(z + 1) + \arg(z - 1)]$
The equipotential curves are the loci of points which vary so that the product of their distances from $z = -1$ and $z = 1$ is a constant. These curves are the so-called **ovals of Cassini,** and the streamlines are the orthogonal trajectories of these curves.
17. The stagnation points occur where $\theta = -\sin^{-1}(K/4\pi aU)$. If $K = 0$, the stagnation points are at the ends of the horizontal diameter of the circle $r = a$. As $k \to 4\pi aU$, the stagnation points move symmetrically around the lower half of the circle and coalesce at $\theta = -\pi/2$ when $K = 4\pi aU$. If $K > 4\pi aU$, the stagnation points leave the circle and occur in the body of the fluid.
19. $f(z) = -iz + 2\ln z$; $\Phi = y + 2\ln r$; $\Psi = -x + 2\tan^{-1}(y/x)$ or, equally well, $r = (2\theta - \Psi)/\cos\theta$, where (r, θ) are polar coordinates in the xy plane. The streamlines are shown in Fig. A17.1.
21. **(a)** $f(z) = \ln\dfrac{z+1}{z-1} + z$; $\Phi = \ln\left|\dfrac{z+1}{z-1}\right| + x$,
$\Psi = \arg(z + 1) - \arg(z - 1) + y$
(b) $f(z) = \ln\dfrac{z+1}{z-1} - iz$; $\Phi = \ln\left|\dfrac{z+1}{z-1}\right| + y$,
$\Psi = \arg(z + 1) - \arg(z - 1) - x$

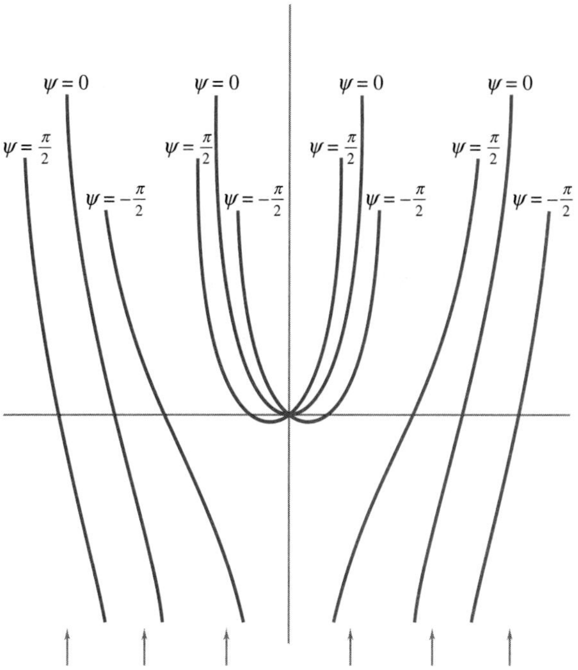

FIGURE A17.1 Streamlines for a source in a uniform stream

23. $f(z) = -iz - 2\ln z$; $\Phi = -x - 2\ln r$,
$\Psi = -x - 2\tan^{-1}(y/x)$
For each value of Ψ, the streamline $\Psi = c$ in this exercise is the reflection in the x axis of the streamline $\Psi = c$ in Exercise 19, with the direction of flow reversed.
25. $\Phi = \ln|z + \sqrt{z^2 + 1}|$, $\Psi = \arg[z + \sqrt{z^2 + 1}]$. The equipotential curves are the ellipses $x^2/\cosh^2\Phi + y^2/\sinh^2\Phi = 1$. The streamlines are the hyperbolas $x^2/\cos^2\Psi - y^2/\sin^2\Psi = 1$. These can be obtained most easily by considering the inverse function $z = \cosh w$ and proceeding as in Example 3, Sec. 17.4. Note that every streamline must intersect the x axis between $x = -1$ and $x = 1$.

CHAPTER 18

Sec. 18.1, page 1185

1. **(a)** The series converges to $1/(1 + i - z)$ in the interior of the circle $|z - i| = 1$.
(b) The series converges to the sum
$$S = \begin{cases} z & |z| < 1 \\ 0 & z = 1 \end{cases}$$
3. **(a)** The parabola $y^2 + 2x + 1 = 0$ and its exterior
(b) All values of z except $z = -1$
5. **(a)** On and to the right of any vertical line $x = \delta > 0$
(b) On and in the interior of any hyperbola of the family $x^2 - y^2 = \delta > 0$
(c) On and in the exterior of any hyperbola of the family $x^2 - y^2 + 1 = \delta > 0$

(d) On and in the exterior of any hyperbola of the family $x^2 - y^2 - x = \delta > 0$

9. (a) The series converges for $z = 0$ and for $|z + 1| < 1$.

(b) $S = \begin{cases} 0 & z = 0 \\ -1 & |z+1| < 1 \end{cases}$

(c) The series converges uniformly for $|z + 1| < \delta < 1$.

13. (a) The series converges for all z except $z = 0$.

(b) The series converges uniformly for all z such that $|z| > \delta > 0$.

15. The series converges to the sum $1/z$ for all values of z except $z = 0, -1, -2, \ldots$. The convergence is uniform over the entire region of convergence.

23. The multiply connected region consisting of the curve $(x^2 - y^2 - 4)^2 + 4x^2y^2 = 0$ and its interior

25. Yes

Sec. 18.2, page 1190

1. Both series converge in the open half-plane to the right of the imaginary axis.

3. All three series converge where $|e^{-z}| < 1$, that is, in the open half-plane to the right of the imaginary axis.

9. The series converges to the sum $S = ze^{z^2}/(e^{z^2} - 1)^2$ for all z such that $|e^{-z^2}| < 1$, that is, where $x^2 - y^2 > 0$.

Sec. 18.3, page 1193

1. $r = 1$. When $|z| = 1$, $u_n = \cos n\theta + i \sin n\theta$. Unless $\theta = 0, \pi/2, \pi, 3\pi/4$, neither $\cos n\theta$ nor $\sin \theta$ approaches 0 as $n \to \infty$. If $\theta = 0, \pi/2, \pi, 3\pi/2$, one of the pair $\cos \theta$, $\sin n\theta$ is zero, but the other is either 1 or -1 and does not approach zero as $n \to \infty$.

3. $r = 1$. The series converges conditionally at $z = -1$ and diverges at $z = 1$.

7. (a) The first series is the derivative of the second.

(b) The second series is the second derivative of the first.

Sec. 18.4, page 1198

1. (a) $f(z) = -1 + 2z - 2z^2 + 2z^3 - \cdots, |z| < 1$

(b) $f(z) = \dfrac{z-1}{2} - \dfrac{(z-1)^2}{4} + \dfrac{(z-1)^3}{8} - \dfrac{(z-1)^4}{16} + \cdots, |z - 1| < 2$

3. $1/z = -i - (z - i) + i(z - i)^2 + (z - i)^3 - i(z - i)^4 - \cdots, |z - i| < 1$

5. $\cosh z = -1 - \dfrac{(z - i\pi)^2}{2!} - \dfrac{(z - i\pi)^4}{4!} - \dfrac{(z - i\pi)^6}{6!} - \cdots, |z| < \infty$

7. (a) $\sqrt{2}$ **(b)** $2\pi - 4$ **(c)** $\sqrt{10}$ **(d)** $\sqrt{17}$

15. $\ln z = \dfrac{1}{2} \ln 2 + \dfrac{3i\pi}{4} - \sum_{n=1}^{\infty} \dfrac{1}{n}\left(\dfrac{1+i}{2}\right)^n [z - (-1 + i)]^n$

The series converges to the principal value of $\ln z$ only within a circle of radius 1 about $z = -1 + i$ because there is a singular point (at $z = -1$) at a distance of 1 from $z = -1 + i$.

17. (a) $\sin z = \sum_{n=0}^{\infty} (-1)^n \dfrac{z^{2n+1}}{(2n+1)!}, |z| < \infty$

(b) $\sinh z = \sum_{n=0}^{\infty} \dfrac{z^{2n+1}}{(2n+1)!}, |z| < \infty$

Sec. 18.5, page 1204

1. (a) $f(z) = \frac{1}{2} + \frac{3}{4}z + \frac{7}{8}z^2 + \frac{15}{16}z^3 + \cdots$

(b) $f(z) = \cdots - \dfrac{1}{z^3} - \dfrac{1}{z^2} - \dfrac{1}{z} - \dfrac{1}{2} - \dfrac{z}{4} - \dfrac{z^2}{8} - \dfrac{z^3}{16} - \cdots$

(c) $f(z) = \cdots + \dfrac{15}{z^5} + \dfrac{7}{z^4} + \dfrac{3}{z^3} + \dfrac{1}{z^2}$

(d) $f(z) = -\dfrac{1}{z-1} - 1 - (z - 1) - (z - 1)^2 - \cdots$

(e) $f(z) = \cdots + \dfrac{1}{(z-1)^4} + \dfrac{1}{(z-1)^3} + \dfrac{1}{(z-1)^2}$

(f) $f(z) = \dfrac{1}{z-2} - 1 + (z - 2) - (z - 2)^2 + (z - 2)^3 - \cdots$

(g) $f(z) = \cdots + \dfrac{1}{(z-2)^4} - \dfrac{1}{(z-2)^3} + \dfrac{1}{(z-2)^2}$

3. $f(z) = -\dfrac{1}{z-i} - 2i + 3(z - i) - 4i(z - i)^2 + 5(z - i)^3 - \cdots, 0 < |z - i| < 1$

$f(z) = \cdots + \dfrac{4i}{(z-i)^6} - \dfrac{3}{(z-i)^5} - \dfrac{2i}{(z-i)^4} + \dfrac{1}{(z-i)^3}, |z + i| > 1$

5. The first series converges only for $|z| < 1$, and the second converges only for $|z| > 1$. Hence there is no value of z for which the two series are simultaneously valid.

7. (a) 0 **(b)** $2i\pi$ **(c)** 0 **(d)** 0 **(e)** $2i\pi$ **(f)** $-i\pi/6$

15. (a) $-\dfrac{1}{2}(z - 1)^{-1} - 3 \sum_{n=0}^{\infty} \dfrac{(z-1)^n}{2^{n+2}}$

(b) $\dfrac{1}{10}(2 - i)(z - i)^{-1} + \dfrac{1}{10} \sum_{n=0}^{\infty} (-1)^n \left[\dfrac{2+i}{(2i)^{n+1}} - \dfrac{4}{(2+i)^{n+1}}\right](z - i)^n$

CHAPTER 19

Sec. 19.1, page 1214

1. (a) $\frac{1}{2}$ (b) $\frac{1}{2}$
3. (a) At $z = -1 + 2i$ the residue is $(2 + i)/4$.
   (b) At $z = -1 - 2i$ the residue is $(2 - i)/4$
5. -1
7. $3/10$
9. (a) $2\pi i$ (b) 0 (c) 0 (d) 0 (e) 0 (f) $(5 - 3\sqrt{5})\pi i/20$
11. (a) $-4\pi i$ (b) 0 (c) $i\pi/3$ (d) $-2\pi i$ (e) πi (f) $-2\pi^2 i$
21. (a) $\dfrac{\pi \coth \pi a}{a}$ (b) $\dfrac{\pi \sinh 2\pi b}{2b(\cosh^2 \pi b - \cos^2 \pi a)}$
   (c) $\dfrac{\pi^3 \cos \pi a}{\sin^3 \pi a}$ (d) $\dfrac{\pi(\sinh 2\pi a + \sin 2\pi a)}{4a^3(\cosh 2\pi a - \cos 2\pi a)}$
23. (a) $\pi i/32$ (b) 0
25. (a) $4\pi i$ (b) $\pi/2$

Sec. 19.2, page 1221

1. $\dfrac{2\pi}{1 - p^2}$
3. π
5. $\dfrac{2\pi}{b^2}(a - \sqrt{a^2 - b^2})$
7. $\dfrac{\pi}{\sqrt{2}a^3}$
9. $\dfrac{\pi}{3}$
11. $\dfrac{\pi}{4a^3}$
13. $\pi\dfrac{a^3(b - c) + b^3(c - a) + c^3(a - b)}{abc(a^2 - b^2)(b^2 - c^2)(c^2 - a^2)}$
17. $\dfrac{\pi e^{-mb} \cos ma}{b}$
19. $\dfrac{\pi(1 + am)e^{-am}}{4a^3}$
21. $\dfrac{\pi}{a^2 - b^2}\left(\dfrac{e^{-bm}}{b} - \dfrac{e^{-am}}{a}\right)$
23. $\pi \exp\left(-\dfrac{m}{\sqrt{2}}\right) \sin \dfrac{m}{\sqrt{2}}$
25. $\dfrac{\pi}{\sin a\pi} \dfrac{r^{a-1} \sin (a - 1)\theta}{c}$, where $r = \sqrt{a^2 + b^2}$ and $\theta = \tan^{-1}\left(\dfrac{-c}{-b}\right)$
27. $\dfrac{\pi}{3 \sin a\pi}\left(1 + 2 \cos \dfrac{2\pi a}{3}\right)$
35. $a_n = \dfrac{2(-1)^n}{(a^2 - b^2)^{1/2}}\left[\tan\left(\dfrac{1}{2} \csc^{-1} \dfrac{a}{b}\right)\right]^n$. As $n \to \infty$, a_n approaches zero more rapidly than the reciprocal of any fixed power of n. This of course is implied by Theorem 3, Sec. 8.3, since all derivatives of the given function are everywhere continuous.
41. $\dfrac{\pi}{a + b}$
43. $-\dfrac{\pi \sin 2}{e}$
45. $\dfrac{\pi}{\sin a\pi}\left[1 - \left(\dfrac{\beta}{\alpha}\right)^{a-1}\right]$

Sec. 19.3, page 1229

1. $\frac{1}{2}(e^{-t} - e^{-3t})$
3. $(\sin 2t)/2$
5. $1 - \cos t$
7. $(t \sin 2t)/4$
9. $(2 - t)e^{-2t} - 2e^{-3t}$
13. $\dfrac{t}{a} + \dfrac{2}{\pi}\sum_{n=1}^{\infty} \dfrac{(-1)^n}{n} \sin \dfrac{n\pi t}{a}$
15. $x + \dfrac{2}{\pi}\sum_{n=1}^{\infty} \dfrac{(-1)^n}{n} \exp(-n^2\pi^2 t) \sin n\pi x$

Sec. 19.4, page 1240

3. $D_1 = 6, D_2 = 4, D_3 = 324$. Hence each root has negative real part. Actually, the roots are approximately -3.77, $-1.12 \pm 0.59i$.
5. $D_1 = 1$, $D_2 = -9$, $D_3 = -100$, $D_4 = -1000$. Hence there is at least one root with nonnegative real part.
11. $s = -0.262, -0.369 \pm 1.916i$

CHAPTER 20

Sec. 20.1, page 1245

1. The upper half of the circle $x^2 + y^2 = a^2$ is mapped once onto the entire circle $u^2 + v^2 = a^4$. The entire circle $x^2 + y^2 = a^2$ is mapped twice onto $u^2 + v^2 = a^4$.
3. The transformation $w = \bar{z}$ is equivalent to reflection in the real axis.
5. The transformation $w = f(\bar{z})$ is equivalent to the transformation $w = f(z)$ followed by reflection in the real axis.
7. The equations of the transformation are $u = x/(x^2 + y^2)$ and $v = -y/(x^2 + y^2)$. The lines $x = c$ and $y = k$ are transformed, respectively, into the circles $u = c(u^2 + v^2)$ and $-v = k(u^2 + v^2)$. The preimages of the lines $u = a$ and $v = b$ are, respectively, the circles $x = a(x^2 + y^2)$ and $-y = b(x^2 + y^2)$.
9. (a) $u = 2x - y - 1$, $v = x + 2y - 3$; $x = (2u + v + 5)/5$, $y = (-u + 2v + 5)/5$
11. $u = \dfrac{x + 1}{x^2 + y^2 + 2x + 1}$, $v = \dfrac{-y}{x^2 + y^2 + 2x + 1}$
   $x = \dfrac{-u^2 - v^2 + u}{u^2 + v^2}$, $y = \dfrac{-v}{u^2 + v^2}$
13. $k = 2$ is the only value for which the mapping function is analytic.
15. $x^2 = \dfrac{\sqrt{u^2 + v^2} + u}{2}$, $y^2 = \dfrac{\sqrt{u^2 + v^2} - u}{2}$
17. (a) $u = e^x \cos y$, $v = e^x \sin y$; $x = \frac{1}{2} \ln (x^2 + y^2)$, $y = \tan^{-1}(v/u)$; x and y are, respectively, the real and the imaginary parts of the inverse function $z = \ln w$.
   (b) The circles $|w| = c$; the rays $\arg w = k$
   (c) $y = 0$. The required mappings are indicated in Fig.

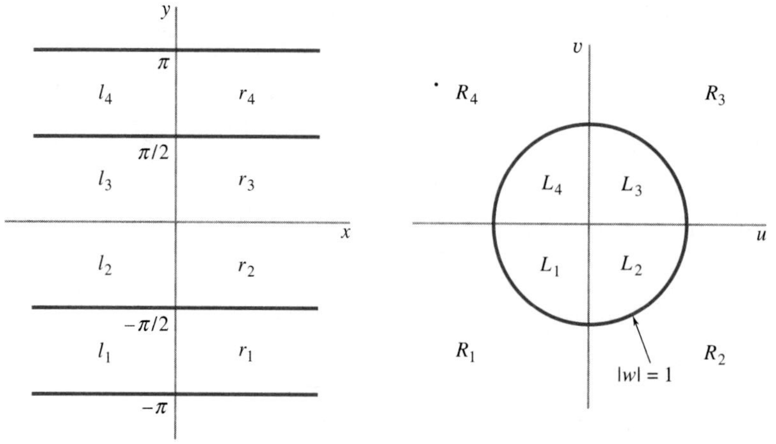

FIGURE A20.1

A20.1. Any strip of the form $(2n-1)\pi < y \leq (2n+1)\pi$ will map into the entire w plane.

(d) $y = \sec^{-1}(e^x/c),\ -\pi < y \leq \pi$

$y = \csc^{-1}(e^x/k),\ -\pi < y \leq \pi$

19. Positive

Sec. 20.2, page 1250

1. See Fig. A20.2, $\frac{8}{3}$.

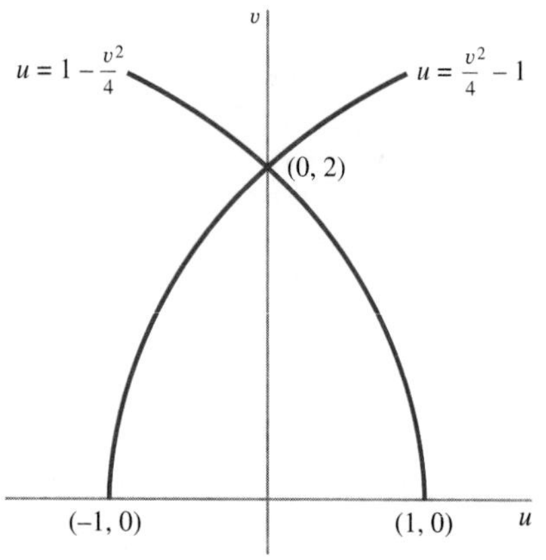

FIGURE A20.2

3. **(a)** $z = \pm 1$ **(b)** $z^2 - 1 = \frac{1}{3}$ or $(x^2 - y^2 - 1)^2 + 4x^2y^2 = \frac{1}{9}$ **(c)** $x^2 - 2xy - y^2 = 1$ **(d)** $x^2 - y^2 = 1$

5. **(a)** π/a **(b)** $\pi/\sqrt{2}$

13. $w = az + b$, where a and b are complex numbers and $|a| = 1$

Sec. 20.3, page 1260

1. $-1, \frac{1}{2}, 2$; yes

3. $-\frac{1}{3}$

5. $w = \dfrac{z-1}{iz+2}$

7. $w = \dfrac{i}{iz+1}$

11. $z = \pm i$

13. $z = 1, 2$. The value of the cross ratio is $\frac{1}{4}$.

15. If and only if $a = d$

17. $w = \dfrac{az+b}{cz+d}$, where a, b, c, d are all real and $ad - bc < 0$

19. $w = \dfrac{z^3 - 1}{z^3 + i}$

21. $w = -\left(\dfrac{z^4-1}{z^4+1}\right)^2$

25. $T = \dfrac{100}{\pi}\tan^{-1}\dfrac{2xy}{y^2 - x^2}$

27. $T = \dfrac{50}{\pi}\tan^{-1}\dfrac{-2y}{x^2 + y^2 - 1}$

29. $T = \dfrac{50}{\pi}\tan^{-1}\dfrac{4xy(x^2 - y^2) - 18xy}{(x^2 - y^2)^2 - 9(x^2 - y^2) - 4x^2y^2}$

31. $T = \dfrac{100}{\pi}\tan^{-1}\dfrac{4y(x^2 + y^2 - 1)}{4y^2 - (x^2 + y^2 - 1)^2}$

33. $T = \dfrac{100v}{\pi u}$, where $w = u + iv = -\left(\dfrac{z^2-1}{z^2+1}\right)^2$

35. $T = -\dfrac{100}{\pi}\tan^{-1}\dfrac{v}{u}$, where $x = -u^2 + v^2 - 2v + 1$ and $y = 2u - 2uv$

37. $T = \dfrac{100}{\pi}\tan^{-1}\dfrac{v}{u-1}$, where $x = \dfrac{1 - u^2 - v^2}{u^2 + v^2 - 2v + 1}$ and $y = \dfrac{-2u}{u^2 + v^2 - 2v + 1}$

Sec. 20.4, page 1269

1. $w = -z^2$

3. $\cos\dfrac{\theta}{2} = \dfrac{r-1}{2\sqrt{r}}\tan\dfrac{\pi T}{2}$, where (r, θ) are polar coordinates in the w plane.

7. $z = e^w$

9. $\tan(\pi T/100) = \tan v/\tanh u$

11. $T = (100/\pi)(\pi + \frac{2}{3}\theta)$, where θ is the polar angle in the w plane

13. $v^2 = c^2 + \dfrac{c^2}{u^2 + c^2}$

15. $w = 2\sqrt{z+1} + \ln \dfrac{\sqrt{z+1} - 1}{\sqrt{z+1} + 1}$

17. $w = A \displaystyle\int_0^z \frac{dt}{[(1-t^2)(1-k^2t^2)]^{1/2}}$.

Since $w = u_0$ corresponds to $z = 1$,

$$u_0 = A \int_0^1 \frac{dt}{[(1-t^2)(1-k^2t^2)]^{1/2}} = AK(k)$$

where $K(k)$ denotes the complete elliptic integral of the first kind of modulus k. Therefore $A = u_0/K(k)$.

19. $w = K \int_0^z t^{-1/2}(t-1)^{\alpha/\pi}\,dt$, where
$K = ai/\int_0^1 t^{-1/2}(t-1)^{\alpha/\pi}\,dt$
If $\alpha = 0$, $w = ai\sqrt{z}$.
If $\alpha = \pi/2$, $w = (2ai/\pi)[\sin^{-1}\sqrt{z} + \sqrt{z(1-z)}]$.

21. $\text{Arg}\, dw = n\pi - \sum_{j=1}^{n} \alpha_j + \arg K$

INDEX*

Page numbers followed by an *e* indicate exercise.
Page numbers followed by an *n* indicate footnote.

E

G

J

K

L

M

N

O

Q

R

U

V

W

Y

Z